BIOLOGY

fifth edition

Eldra Pearl Solomon
University of South Florida

Linda R. Berg
St. Petersburg Junior College

Diana W. Martin
Rutgers University

®

SAUNDERS COLLEGE PUBLISHING
HARCOURT BRACE COLLEGE PUBLISHERS

Fort Worth   Philadelphia   San Diego   New York   Orlando   Austin   San Antonio   Toronto   Montreal   London   Sydney   Tokyo

To our families, friends, and colleagues
who gave freely of their love, support, knowledge, and time
as we labored over this revision of BIOLOGY . . .
Especially, to . . .
Rabbi Theodore and Freda Brod
Kathleen M. Heide, Ph.D.
Alan Berg
Jennifer Berg
Margaret Martin

ABOUT THE COVER

An elusive cougar *(Felis concolor)* turns abruptly and pauses when she hears a twig snap during a winter storm in Montana. Her ears are directed to catch the slightest sound, and her eyes search the surroundings for any sign of movement. Her long, heavy tail provides balance when she climbs rocks, stalks prey, or reverses direction. These and other adaptations are the products of evolution and enable the cougar to survive in a harsh, unforgiving environment (© Charles Sleicher/Tony Stone Images).

Publisher: Emily Barrosse
Executive Editor: Edith Beard Brady
Product Manager: Erik Fahlgren
Developmental Editors: Gabrielle Goodman, Cathleen Petree
Project Editors: Nancy Lubars, Ted Lewis
Production Manager: Joanne Cassetti
Art Director: Caroline McGowan
Text and Cover Designer: Ruth Hoover

Cover Credit: © Charles Sleicher/Tony Stone Images

BIOLOGY, Fifth Edition
ISBN: 0-03-022299-0
Library of Congress Catalog Card Number: 98-85056

Address for domestic orders:
Saunders College Publishing, 6277 Sea Harbor Drive, Orlando, FL 32887-6777.
1-800-782-4479

Address for international orders:
International Customer Service, Harcourt Brace & Company
6277 Sea Harbor Drive, Orlando, FL 32887-6777.
(407) 345-3800
Fax (407) 345-4060
email hbintl@harcourtbrace.com

Address for editorial correspondence:
Saunders College Publishing, 150 S. Independence Mall West, Philadelphia, PA 19106-3412

Web Site Address
http://www.hbcollege.com

Printed in the United States of America

9012345678 032 10 98765432

As we enter the 21st century, we reflect on the important contributions of the biological sciences to modern society. As biologists have studied life, we have gained greater understanding of human life processes and have become more aware of our interdependence with the vast diversity of organisms with which we share our planet. With new advances in biological research, our lives have become healthier, safer, more comfortable, and also more challenging.

One of our principal goals in developing *Biology*, fifth edition, has been to share with beginning biology students our sense of excitement about the biological sciences. We want students to understand and appreciate the diverse organisms on Earth, their remarkable adaptations to the environment, and their evolutionary and ecological relationships. Special emphasis is placed on the basic unity of life and the fundamental similarities of the challenges that have been faced and solved by all living organisms. We are very aware of our responsibility to impress upon our readers our interdependence with the many life forms with which we share planet Earth.

IMPORTANT FEATURES OF THE FIFTH EDITION

The evolution of *Biology* through its editions reflects the advances in the biological sciences and in biological education. Every effort has been made to update its content and pedagogy so that the book accurately presents modern biology.

As in the fourth edition, a significant effort was made in the preparation of this edition to speak directly to students and determine how they use the text and what might be done to enhance its efficacy in teaching biology. To that end, focus groups of students from many colleges and universities were conducted to gauge how students felt we could improve the utility of the text, artwork, and pedagogy. Students met with our editors and told them how they study, which pedagogical elements are most helpful, and which pieces of art are effective (or ineffective) in helping them learn the subject matter.

Themes

Throughout the book, we emphasize three basic themes of biology: transmission of information, evolution of life, and flow of energy through living systems. As we introduce the concepts of modern biology, we explain how these themes are connected and how life depends on them. In this new edition, we also emphasize the process of science and the role of the many scientists who have contributed to our current understanding of biology.

The author team

Rapid advances in the biological sciences require a multi-author team who specialize in particular areas of biology. The author team of the fifth edition includes Dr. Eldra Solomon, zoologist and physiologist; botanist Dr. Linda Berg; and cell biologist/geneticist Dr. Diana Martin. We also acknowledge the legacy of Dr. Claude A. Villee, Professor Emeritus, Harvard University, who contributed to previous editions. All of the authors are experienced college biology teachers.

Tools for learning

Learning the principles of biology is a challenging endeavor. A variety of learning aids are included within the textbook to help the student achieve mastery of the concepts presented.

1. Each **chapter introduction** provides an overview to the chapter contents and ties it to material in previous chapters.
2. **Learning Objectives** that follow the chapter introduction indicate in behavioral terms what the student must be able to do in order to demonstrate mastery of the material in the chapter.
3. **Concept-statement heads** introduce each section, previewing and summarizing the key idea(s) that will be discussed in that section. The subheadings provide a concise outline of the chapter's contents.
4. **Making the Connection** boxes and **Focus On** boxes facilitate integration of concepts, spark interest, and reinforce the scientific process. For example, *Making the Connection: Terrestrial-Aquatic Linkages and the Detritus Connection* in Chapter 54 describes recent field evidence that supports a long-held assumption that a forest provides a stream flowing through it with an important energy input in the form of detritus. *Focus On Mammalian Cloning* in Chapter 16 emphasizes the scientific process as

...undbreaking research of Ian Wilmut and ... cloning a sheep from a cultured mammary

...**Cutting Edge** boxes present exciting research ar... ...t are currently being explored. In Chapter 19, for ...ple, *On the Cutting Edge: Explaining the Rapid Loss* ... *Cichlid Diversity in Lake Victoria* discusses how hybridization caused by human-induced changes in habitat is destroying cichlid species in Lake Victoria.

6. Numerous **tables,** many of them illustrated, summarize and organize material presented in the text. *Table 52–1: Ecological Interactions among Species* in Chapter 52 is an example of a table that is new to the fifth edition.

7. Carefully rendered **illustrations,** many of them new in this edition, support concepts covered in the text. Many of the illustrations are sequential, with close-ups "exploded" to reveal greater detail. Composite pieces of line art and photographs help students to interpret electron micrographs. **Scale bars** accompany micrographs to provide information regarding magnification. As requested by student focus groups, **figure legends** in the fifth edition identify each micrograph as a LM, TEM, or SEM. Figures 17–15 (phylogenetic tree of selected primates) and 22–4 (monophyletic, paraphyletic, and polyphyletic taxa) are examples of line art that are new to the fifth edition.

8. **Sequence Summaries** highlight with a beige background reviews of sequential material discussed in the text. For example, Chapter 33 contains two new sequence summaries, phloem loading at the source and phloem unloading at the sink.

9. **Boldface terms** facilitate identification of key terms and their definitions, and provide emphasis. Chapter 55, for example, introduces and defines the term *biotic pollution* and then discusses its impact on threatened and endangered species.

10. A **Summary With Key Terms** in outline form at the end of each chapter provides a review of the material presented. New to this edition is the integration of boldfaced selected key terms with the summary, enabling students to study vocabulary words in context with related concepts.

11. A **Post-Test** provides students with the opportunity to evaluate their mastery of the material in the chapter. In the fifth edition, these questions are multiple choice, as requested by students and professors. For example, examine the Post-Test at the end of Chapter 3; some questions are based on the recall of important terms, whereas other questions challenge students to integrate their knowledge. Answers are provided in Appendix A.

12. **Review Questions** test knowledge of important concepts and applications. They are designed to help students test their mastery of the chapter learning objectives. New to the fifth edition are quantitative problems on population genetics (see Chapter 18 Review Questions). Answers to all Review Questions in chapters 10 and 18 are provided in Appendix A.

13. **You Make the Connection** questions (1) challenge the student to relate information in the chapter to principles in other chapters or (2) require the student to apply concepts to new situations. These questions can be used for class discussions or essay assignments. Examples of You Make the Connection questions that are new to the fifth edition include numbers 1, 2, and 4 in Chapter 7.

14. A list of **Recommended Readings** at the end of each chapter provides references for further learning. All lists have been updated from the last edition; examine the Recommended Readings for Chapter 52, all of which are new to this edition. Each Recommended Readings list ends with the address of our website, where students can find links to additional resources on the World Wide Web.

15. **Career Visions** present a variety of professional possibilities in the biological sciences for students to explore. An interview with a professional who majored in biology is presented at the end of each of the book's eight parts. Interviewees discuss how they decided on and prepared for their career and what they do professionally. Two examples of careers new to the fifth edition are renal dietitian and aquarist.

16. **To the Student,** which appears immediately before the Contents, is newly conceived in this edition to offer students specific, tried-and-true advice on how to prepare, study, and succeed in introductory biology courses.

17. A separate **Glossary,** expanded for the fifth edition, facilitates rapid definition of terms. As in the fourth edition, vertical color bars along the margin facilitate rapid access to the Glossary.

18. **Appendices** provide post-test answers, the complete periodic table of the elements (new to this edition), an overview of biological classification, career information, understanding biological terms, and abbreviations used in the text. Scientific measurement is located inside the back cover for easy access.

THE ORGANIZATION OF *BIOLOGY,* FIFTH EDITION

Educators present the major topics of an introductory biology course in a variety of orders. A lack of consensus regarding sequence of topics is understandable because reasonable arguments can be advanced for each of the many possible combinations and permutations. All aspects of biology are intimately related, and each could be grasped much more readily if all other topics had been mastered previously. Because this feat cannot be accomplished, each instructor must select the topic sequence that seems most reasonable. For this reason, we have carefully designed each of the eight parts so that they do not depend heavily on preceding chapters and parts. The eight parts and their chapters can be presented in any number of sequences with pedagogical success.

The content updates and changes to the fifth edition are far too numerous to list. In this brief survey of the eight parts of *Biology*, we include representative examples of what is new.

PART 1: THE ORGANIZATION OF LIFE

Chapter 1, A View of Life: Basic Concepts of Biology, introduces several major concepts of biology, including the fundamental similarities of all living things; the organization of life on individual and ecological levels; the transfer of information; the evolution of life; the diversity of life and how biologists classify organisms; energy transfer among organisms; and how science works. Chapters 2 and 3 focus on the molecular level of organization and lay the foundations in chemistry needed for an understanding of biological processes. Chapters 4 and 5 focus on the cellular level of organization, with emphasis on recent advances in cell biology. New and expanded discussions of cell signaling appear in the information transfer section of Chapter 1 and in Chapter 5.

PART 2: ENERGY TRANSFER THROUGH LIVING SYSTEMS

Part 2, which focuses on the metabolism and energy transactions involved in life processes, has been extensively revised for this edition. Chapter 6 introduces energy in cells and organisms. Chapters 7 and 8, which can be taught in either order, discuss the metabolic adaptations by which organisms obtain and use energy through cellular respiration and photosynthesis. Examples of new line art in Chapter 6 include Figure 6–2, which distinguishes between closed and open systems, and Figure 6–16, which illustrates feedback inhibition.

PART 3: THE CONTINUITY OF LIFE: GENETICS

Because knowledge is advancing so rapidly in the field of genetics, Part 3 has been extensively updated. This unit begins with a discussion of mitosis and meiosis in Chapter 9. Chapter 10 describes Mendelian genetics and related patterns of inheritance. Chapter 11 discusses the structure and replication of DNA, and Chapter 12 presents RNA and protein synthesis. Gene regulation is discussed in Chapter 13. In Chapter 14, we focus on genetic engineering; and in Chapter 15, we focus on human genetics. In Chapter 16, we introduce the role of genes in development, including the latest findings in this exciting and rapidly changing area of biology. The fifth edition places greater emphasis on the significance of genetics in our lives; Chapter 15, for example, contains a new section on the issue of genetic discrimination.

PART 4: THE CONTINUITY OF LIFE: EVOLUTION

The unit on evolution has been revised for this edition. Chapter 17 introduces Darwinian evolution and presents scientific evidence for evolution. In Chapter 18, we examine evolution at the population level. Chapter 19 describes the evolution of new species and discusses aspects of macroevolution. Chapter 20 summarizes the evolutionary history of life on Earth. In Chapter 21, we recount the evolution of the primates, including humans. New to the fifth edition are a section on the testing of evolutionary hypotheses experimentally in Chapter 17 and a discussion of recently discovered fossil finds in early hominid evolution in Chapter 21.

PART 5: THE DIVERSITY OF LIFE

An evolutionary framework is used in our survey of the kingdoms of organisms. In Chapter 22, we discuss why and how organisms are classified. Chapter 23 is devoted to the viruses and bacteria. Chapter 24 describes the protists, and Chapter 25 describes the fungi. Chapters 26 and 27 present the members of the plant kingdom. Chapters 28 through 30 focus on the diversity of animals. The discussion of each group of organisms focuses on their evolutionary relationships and on their structural and functional adaptations. In this edition, discussion of both the kingdom/domain debate and cladistics has been expanded in Chapter 22.

PART 6: STRUCTURE AND LIFE PROCESSES IN PLANTS

This part integrates plant structure and function, beginning in Chapter 31 with a discussion of plant structure, growth, and differentiation. Chapters 32 through 34 discuss the structural and physiological adaptations of leaves, stems, and roots. Chapter 35 describes reproduction in flowering plants, including asexual reproduction, flowers, fruits, and seeds. Chapter 36 focuses on growth responses and regulation of growth. New topics include *On the Cutting Edge: Mycorrhizae and Carbon Exchange between Plant Species in a Forest Ecosystem* in Chapter 34, *Making the Connection: Flowers and Evolution* in Chapter 35, and new material on the basic mode of action of all plant hormones in Chapter 36.

PART 7: STRUCTURE AND LIFE PROCESSES IN ANIMALS

This part emphasizes the structural, functional, and behavioral adaptations that help animals meet environmental challenges.

As each system of the animal body is discussed, a comparative approach is used to examine how various animal groups have solved similar and diverse problems. Chapter 37 is devoted to the architecture of the animal body, emphasizing the various tissues and organ systems. Chapters 38 through 49 present animal life processes. After a comparison of how different animal groups carry on a particular process, such as digestion, gas exchange, and internal transport, each chapter considers the human adaptations for that process. The unit ends with a discussion of behavioral adaptations in Chapter 50. The recent explosion of research on receptors and signal transduction has been integrated throughout Part 7. Also new to the fifth edition is a discussion of the mechanisms of learning in Chapter 40.

PART 8: THE INTERACTIONS OF LIFE: ECOLOGY

The ecology unit has been reorganized and revised extensively. Chapters 51 through 54 provide the foundations of the ecology of populations, communities, ecosystems, and the biosphere, with the final chapter (55) focusing on environmental problems caused by humans. New topics include the relationship of human population growth and resource consumption to environmental degradation in Chapter 51, how ancient and modern communities from a biogeographical context in Chapter 52, the complex relationship between productivity and biological diversity in Chapter 53, and expanded coverage of conservation biology in Chapter 55.

SUPPLEMENTS

To further facilitate learning and teaching, a supplement package has been carefully designed for the student and instructor.

The **Study Guide** by Ronald S. Daniel of California State Polytechnic University, Pomona, Sharon C. Daniel of Orange Coast College, and Ronald L. Taylor includes extensive study outlines, vocabulary exercises, thought questions, multiple choice self-tests, and coloring book exercises.

A Problem-Based Guide to Basic Genetics by Donald Cronkite of Hope College and *Chemistry for Biologists* by I. Edward Alcamo of State University of New York, Farmingdale provide students with extra support in two difficult content areas.

The **Instructor's Resource Manual** by Jane Aloi of Saddleback College features lecture outlines, research and discussion topics, teaching suggestions, and further suggested readings, as well as a new section in each chapter called "Connections and Resources," to help instructors integrate all of the supplements offered with *Biology* into their teaching programs. The **Test Bank** by Jane Aloi, Oriana Kim of Saddleback College and Santa Monica College, and Adel Rajab of Saddleback College and Cypress College comprises approximately 2,200 questions of three types; multiple choice, true/false, and essay/short answer. The same material is also

found in the **Computerized Test Bank,** which is available for the IBM PC, Windows™, and Macintosh platforms.

Other important components of *Biology's* supplement package are a set of 250 full-color **Overhead Transparencies** based on diagrams in the book; a set of 150 **Electron Micrograph Overhead Transparencies;** and 50 **General Sequence Overhead Transparencies,** which contain topics displayed in a series of layers.

A **Laboratory Manual** written by Russell V. Skavaril, Mary M. Finnen, and Steven M. Lawton, all of Ohio State University, and an accompanying **Laboratory Instructor's Manual** are available. Also available is a **Laboratory Manual** written by Carolyn Eberhard of Cornell University. Additionally, a custom publication service, **Saunders Select,** is available: a wide variety of individual laboratory exercises may be selected and combined in a single volume.

The following supplementary texts are available:

Handbook on Teaching Undergraduate Science Courses: A Survival Training Manual by Gordon E. Uno of the University of Oklahoma is a complete guide for new instructors or those who are looking for new ideas, and is the ideal, everything-you-need-to-know companion for teaching assistants.

Student-Active Science: Models of Innovation in College Science Teaching edited by Ann P. McNeal and Charlene D'Avanzo of Hampshire College is a collection of articles by college faculty and administrators from across the country who came together to attend an NSF-sponsored conference on inquiry-based science teaching. At the forefront of innovative biology teaching, these attendees offer insights on how to encourage student activities that are investigative, collaborative, and foster higher order thinking skills.

Thinking Toward Solutions: Problem-Based Learning Activities in General Biology by Deborah Allen and Barbara Duch of the University of Delaware presents complex, open-ended problems that cover all aspects of biology from cells to ecology. Addressing real-world applications of biology with questions of ethics, economics, and daily living, the problems foster critical thinking, cooperative learning, and problem-solving skills. An accompanying **Instructor's Manual** provides detailed, practical suggestions on how to use the problems in any size course and with a variety of teaching styles. *Thinking Toward Solutions* also has an Internet component, which offers links to resources that assist students in solving each problem.

Plain Talk About Using the Web in Teaching by Arthur L. Buikema, Jr. and Brian Ward of Virginia Polytechnic Institute and State University is a basic handbook on intelligently designing a course that fully integrates computer technology and the World Wide Web.

Multimedia offerings

The Process of Science: Discovering Biology™ interactive CD-ROM allows students to explore the discoveries of some of the most important concepts in biology. In the *Interactive Investigations* mode, students retrace the steps of scientists' experiments and discoveries using the scientific process as a road map. An *Investigator's Notepad* allows students to track their progress through each investigation, pose new questions to pursue, or initiate a discussion with the instructor or other students via an Internet connection. *Concept Tutorials* provide students with essential background information in general biology for the course and the investigations.

Instructor's Resource CD-ROM (IRC) for Biology 1999 contains all of the illustrations from the fifth edition of *Biology* as well as those from *The Science of Genetics* by Atherly, Girton, and McDonald. This CD-ROM is available for Windows™ and Macintosh platforms as a presentation tool to be used in conjunction with commercial presentation packages, such as PowerPoint™ and Persuasion™. The IRC imagery can be used as masters to custom-make additional overhead transparencies.

The **Biology Survival Kit CD-ROM** is available for both IBM and Macintosh formats. It includes **BioXL+**™, a series of interactive animations that guide students through essential biological concept and provide real-world applications, and **Biology A₂Z**™ **The Dynamic Glossary,** which provides text definitions, audio pronunciations, graphics and animations for approximately 800 key terms and concepts commonly presented in introductory biology.

LifeWeb, our new Life Sciences Online Magazine, is updated monthly and features articles, reviews of recent research findings, stories focusing on applications and public policy interactive online experiments, a calendar of events, a searchable glossary, and an extensive quizzing and testing program. The fifth edition of *Biology* also has its own book-specific section of the magazine, which includes links to chapter-related resources on the World Wide Web, instructor presentation and teaching tools, and student study tools. Visit LifeWeb at http://www.saunderscollege.com/lifesci/; if you want to go directly to the *Biology* site, go to http://www.saunderscollege.com/lifesci/html and click on Solomon/Berg/Martin Biology.

WebCT is a world wide web course management tool that allows professors to create course-specific websites. Using a simple point-and-click interface, anyone familiar with today's Internet browsers can create a fully functional website that can include timed online quizzes, image archives, bulletin boards, chat rooms, online syllabus and lecture notes, and more.

The **Saunders General Biology Videodisc** has been prepared to enhance lecture or laboratory presentation of material that is difficult to visualize. The 60-minute Videodisc contains more than 1500 still images and a collection of video clips.

Saunders College Publishing may provide complementary instructional aids and supplements or supplement packages to those adopters qualified under our adoption policy. Please contact your sales representative for more information. If as an adopter or potential user you receive supplements you do not need, please return them to your sales representative or send them to

Attn: Returns Department
Troy Warehouse
465 South Lincoln Drive
Troy, MO 63379

ACKNOWLEDGMENTS

The development and production of the fifth edition of *Biology* required extensive interaction and cooperation among the authors and many individuals in our home and professional environments. We appreciate the valuable input and support from editors, colleagues, students, family, and friends.

We are grateful to the editorial and production staffs at Saunders College Publishing for their help and support throughout this project. We thank our publisher Emily Barrosse and Executive Editor Edith Beard Brady for their support, enthusiasm, and ideas. Our Developmental Editors Gabrielle Goodman and Cathleen Petree efficiently guided the project through the revision process, providing us with many thoughtful reviews and useful suggestions.

We are grateful to our talented Art Editor Carlyn Iverson for her new and exciting ideas for reconceptualizing and improving the art. We thank Photo Editor Amy Dunleavy for helping us find outstanding photographs that enhance the text.

We appreciate the help of our Project Editors Nancy Lubars and Ted Lewis who expertly guided the project through the complexities of production. We thank Senior Art Director Caroline McGowan for coordinating the art program and design. We appreciate the efforts of Erik Fahlgren, Marketing Manager, in marketing our book. All of these dedicated professionals and many others at Saunders provided the skill and attention needed to produce *Biology,* fifth edition. We thank them for their help and support throughout this project.

We thank our families and friends for their understanding, support, and encouragement as we struggled through many revisions and deadlines. We especially thank Mical Solomon, Dr. Amy Solomon, Graeme McMurray, Dr. Kathleen M. Heide, Alan and Jennifer Berg, and Dr. Charles Martin and Margaret Martin for their input and support.

Our colleagues and students who have used our book have provided valuable input by sharing their responses to the fourth edition of *Biology.* We thank them and ask again for their comments and suggestions as they use this new edition. We can be reached through our editors at Saunders College Publishing. We express our thanks to the many biologists who have read the manuscript during various stages of its development and provided us with valuable suggestions for improving it. The authors particularly acknowledge the contributions of James Adams of Dalton College. Fifth edition reviewers include:

Dawn Adams, *Baylor University*
James Adams, *Dalton College*
Surinder Aggarwal, *Michigan State University*
Julie Aires, *Florida Community College, Jacksonville*
Venita Allison, *Southern Methodist University*
David Asch, *Youngstown State University*
Edward Ashworth, *Purdue University*
Susan Bandoni, *State University of New York, Geneseo*
Chris Beard, *Carnegie Museum of Natural History*
David Begun, *University of Texas, Austin*
Vincent Bellis, *East Carolina University*

Del Blackburn, *Clark College*

Gary Booth, *Brigham Young University*

Robert Boyd, *Auburn University*

Dean Bratis, *Delaware County Community College*

Anne Britt, *University of California, Davis*

William Brooks, *Florida Atlantic University*

George Brown, *Iowa State University*

Gary Brusca, *Humboldt State University*

Arthur Buikema, *Virginia Tech*

David Carlberg, *California State University, Long Beach*

Barry Chess, *Pasadena City College*

W. Dennis Clark, *Arizona State University*

William Cohen, *University of Kentucky*

Mark Condon, *Dutchess Community College*

Kenneth Curry, *University of Southern Mississippi*

Anne Cusic, *University of Alabama, Birmingham*

Henry Daniell, *Auburn University*

Thomas Davis, *University of New Hampshire*

Jonathan Day, *Pennsylvania State University*

Patricia DeLeon, *University of Delaware*

Jean DeSaix, *University of North Carolina at Chapel Hill*

Laura DiCaprio, *Ohio University*

Linda Dion, *University of Delaware*

Penny Dobbins, *Syracuse University*

Inge Eley, *Hudson Valley Community College*

Robert Evans, *Rutgers University, Camden*

David Firmage, *Colby College*

Steven K. Fisher, *University of California, Santa Barbara*

Malinda Fitzgerald, *Christian Brothers University*

Jim Florini, *Syracuse University*

Bruce Fowles, *Colby College*

Douglas Gaffin, *University of Oklahoma*

Gary Galbreath, *Northwestern University*

Dan Gebo, *Northern Illinois University*

Robert P. George, *University of Wyoming*

Michael Ghedotti, *University of Kansas*

Florence Gleason, *University of Minnesota*

Elizabeth Godrick, *Boston University*

Wayne Goodey, *University of British Columbia*

Thomas Goodwin, *Andrews University*

Katharine B. Gregg, *West Virginia Wesleyan College*

Thaddeus Grudzien, *Oakland University*

David Hale, *Texas A&M University*

Paul K. Hayes, *University of Britsol*

Fritz Hertel, *University of California, Los Angeles*

Martinez Hewlett, *University of Arizona*

Linden Higgins, *University of Texas, Austin*

Betsy Hirsch, *University of Minnesota*

Helmut Hirsch, *State University of New York, Albany*

Robert Holmberg, *Athabasca University*

Jill VanWort Hood, *University of Texas, Arlington*

Linda Hsu, *Seton Hall University*

Joan Hudson, *Sam Houston State University*

Stephen Hudson, *Furman University*

Gerard Iwantsch, *Fordham University*

Alice Jacklet, *State University of New York, Albany*

Charles Janson, *State University of New York, Stony Brook*

Claudia Jolls, *East Carolina University*

Richard Karp, *University of Cincinnati*

Glenn Kasparian, *Brookhaven Community College*

Phil Keeting, *West Virginia University*

M.A.Q. Khan, *University of Illinois, Chicago*

William Kimbel, *Institute of Human Origins*

Robert Kitchin, *University of Wyoming*

Paul Kugrens, *Colorado State University*

Vaughn A. Langman, *Louisiana State University*

Ralph Larson, *San Francisco State University*

Brenda Leicht, *University of Iowa*

Joe Leverich, *St. Louis University*

Harvey Lillywhite, *University of Florida*

Roger M. Lloyd, *Florida Community College, Jacksonville*

Heather Lorimer, *Youngstown State University*

Sharook Madon, *Pace University, Westchester*

Dennis Matthews, *University of New Hampshire*

Henry M. McHenry, *University of California, Davis*

Michael Meighan, *University of California, Berkeley*

Charles Mims, *University of Georgia*

Darrell Moore, *East Tennessee State University*

Michael Morgan, *University of Wisconsin, Green Bay*

Alison Mostrom, *Philadelphia College of Pharmacy and Science*

James Murray, *University of Virginia*

Richard Myers, *Southwest Missouri State University*

William H. Nelson, *Morgan State University*

Glenn R. Parsons, *University of Mississippi*

Greg Paulson, *Shippensburg University*

Jerome Perry, *North Carolina State University*

Greg Phillips, *Blinn Bryan College*

Ruth Pitkin, *Shippensburg University*

Thomas Pitzer, *Florida International University*

Susan Pross, *University of South Florida*

Jerry Purcell, *San Antonio College*

James Pushnik, *Chico State University*

Richard Racusen, *University of Maryland*

Robert Roberson, *Arizona State University*

Laurel Roberts, *University of Pittsburgh*

Jean Saillant, *University of Michigan, Dearborn*

Mimi A. Sayed, *Michigan State University*

Edward Schneider, *University of California, Santa Barbara*

Janet L. Schottel, *University of Minnesota*

Brian Schwartz, *Columbus State University*

William A. Searcy, *University of Miami*

Duane W. Sears, *University of California, Santa Barbara*

David Seigler, *University of Illinois*

Mark Sheridan, *North Dakota State University*

David Shomay, *University of Illinois, Chicago*

Dennis M. Smith, *Wellesley College*

Bruce Stallsmith, *University of Massachusetts, Boston*

Karen Steudel, *University of Wisconsin, Madison*

Chris Tarp, *Contra Costa College*

Walter Taylor, *University of Central Florida*

Robert M. Timm, *University of Kansas*

Gordon Uno, *University of Oklahoma*
Frederick Utech, *Carnegie Museum of Natural History*
Joseph Vanable, *Purdue University*
Charles Walcott, *Cornell University*
Fred Wasserman, *Boston University*
Matthew White, *Ohio University*
John Whitmarsh, *University of Illinois, Urbana*
Donald Whitmore, *University of Texas, Arlington*
Leslie Williams, *Mountain View College*
Steven Woeste, *Scholl College*
David Woodruff, *University of California, San Diego*

We also thank the students who participated in our focus groups for their insightful comments and observations about our book and biology textbooks and courses in general. These students are:

Felisha Avery, *Mountain View College, Texas Women's University*
Leslie Baker, *Eastfield College, Texas Tech University*
Chris Churchman, *Furman University, North Lake Community College*
Alan Cohen, *University of Michigan–Ann Arbor*
Jana A. Damphousse, *Oakland University*
Karen Davis, *Tarrant County Jr. College*
Anjali Cherise D'Souza, *North Lake Community College, University of Texas–Austin*
Estelle S. D'Souza, *University of Toledo*
Katharine Edmund, *University of Michigan–Ann Arbor*
Cory Fajardo, *East Carolina University*
Michael A. Fox, *North Lake Community College, Parker Chiropractic College*

Hannah A. M. Gilkenson, *University of Michigan–Ann Arbor*
Shuaib A. Gill, *Wayne State University*
Lindsay Goodman, *Eastfield College, University of Texas–Austin*
Kelly B. Hall, *North Lake Community College, Texas A&M University*
Shelli Hornberger, *Mountain View College, Stephen F. Austin University*
Rebekah L. Hunter, *Eastern Michigan University*
William DeVaughn Hunt, *East Carolina University*
James Patrick Jarvis, *East Carolina University*
Jennifer Ray Jones, *North Lake Community College, Texas Tech University*
Jenny Kerekles, *University of Michigan–Ann Arbor*
Evelyn Knox, *East Carolina University*
Charles W. Luce, *East Carolina University*
Beth L. Measamer, *Brookhaven College*
Kimberly S. Miller, *Eastfield Community College, Abilene Christian University*
Michael Jason Miller, *East Carolina University*
Ngoc Quang Nguyen, *University of North Texas, Texas Women's University*
Mark Nolan, *University of Texas–Arlington*
Anthony Orlando, *University of Michigan–Ann Arbor*
Stacy Peebles, *Tarrant County Jr. College*
Mary A. Radlick, *Oakland University*
Cynthia Rainey, *Texas A&M University*
Gus Reese, *University of Texas–Arlington*
Renee Sandora, *Eastern Michigan University*
Jennifer Schklair, *Mountain View Community College*
Nicholas John Urbanczyk, *University of Michigan–Dearborn*
Jill Wauldron, *Oakland University*

To the Student

Biology is one of the most varied and challenging subjects one can study. The thousands of students we have taught have differed in their life goals and learning styles. Some have had excellent backgrounds in science, others poor ones. Regardless of their backgrounds, it is common for students taking their first college biology course to find that they must work harder than they expected. You can make the task easier by using approaches to learning that have been successful for a broad range of our students over the years.

MAKE A STUDY SCHEDULE

Many college professors suggest that students study three hours for every hour spent in class. This major investment in study time is one of the main differences between high school and college. To succeed academically, college students must learn how to manage their time effectively. The actual number of hours you spend studying biology will vary depending on your course load and personal responsibilities, such as work schedules and family commitments.

The most successful students are often those who are best organized. It is helpful to make a detailed daily calendar at the beginning of the semester. Mark off the hours that you are in each class, along with transportation time to and from class if you are a commuter. After you have received course syllabi, add to your calendar the dates of all exams, quizzes, papers, and reports. It also helps to add an entry one week before each major exam or assignment so they do not catch you unaware. Now add your work schedule and other personal commitments to your calendar. Making a calendar helps you find convenient study times.

Many of our successful biology students set aside two hours a day to study biology, as opposed to a weekly marathon session for eight or ten hours during the weekend (it never happens). Put your study hours into your daily calendar, and make every effort to adhere to your schedule.

TEXT MATERIAL VERSUS LECTURE NOTES

Some professors test almost exclusively on material covered in lecture, whereas others rely on their students' learning most, or even all, of the content in assigned chapters. Try to ascertain from your professor what his or her requirements are, be-cause the way you study will vary according to how and on what you are being tested.

How to study when professors test lecture material

If lectures are the main source of examination questions, your lecture notes must be as complete and organized as possible. Before going to class, skim over the chapter, identifying key terms and examining the main figures, so that you will be able to take effective lecture notes. You should spend no more than one hour on this.

You should rewrite your notes within 24 hours after class. Before rewriting, however, read the notes and make marginal notes about anything that is not clear. Then read the corresponding material in your text. Highlight or underline any sections that clarify questions you had in your notes. Read the *entire* chapter, including parts that are not covered in lecture. This extra information will give you breadth and help you form key concepts.

After reading the text, you are ready to rewrite your notes, incorporating relevant material from the text. It also helps to use the Glossary to help define unfamiliar terms.

Many students think that developing a set of flash cards of key terms and concepts is a good way to study, but in our experience, flash cards produce mixed results. They are not effective when the student tries to second-guess the professor. ("She won't ask this, so I won't make a flash card of it.") Flash cards are also a hindrance when they are relied on exclusively. Studying flash cards instead of reading the text is a bit like reading the first page of each chapter in a mystery novel. It's hard to fill in the missing parts because you are learning the facts in an unconnected way. Flash cards are a useful tool, however, to help you learn scientific terminology. They are portable and can be used at times when other studying is not possible, for example, when riding a bus.

How to study when professors test material in the book

If the assigned readings in the text are going to be tested, you must use your text intensively. After reading the chapter introduction, read through the list of Learning Objectives for the chapter. These objectives are written in behavioral terms, that is, you must "do" something in order to demonstrate mastery. The objectives provide you with a concrete set of goals for the chapter. You will revisit the Learning Objectives later in the study process.

Read the chapter *actively*. Many students read and study passively. An active learner always has questions in mind and is constantly making connections. For example, there are many processes that must be understood in biology. Do not try to blindly memorize these; instead think about causes and effects, so that every process becomes a story. Eventually you will see that many processes are connected by common elements.

You will probably have to read each chapter two or three times before mastering the material. The second and third times through will be much easier than the first because you will be reinforcing concepts that you have already partially learned.

After reading the chapter, you should write a four- to six-page chapter outline by using the subheads as the body of the outline (first-level heads are boldface and all caps; second-level heads are boldface but not all caps). Flesh out your outline by adding important concepts and boldface terms with definitions. You will use this outline when preparing for the exam.

Now it is time to test yourself. Answer the Post-Test questions, and check your answers in the Appendix. Write answers to each of the Review Questions and You Make the Connection questions. Finally, go back to the Learning Objectives and try to answer them. If your professor has told you that some or all of the exam will be short-answer or essay format, you should answer each Learning Objective in writing. Remember that this is a self-test. If you do not know an answer to a question, find it in the text. If you can't find the answer, use the index.

LEARN THE VOCABULARY

One stumbling block for many students is the necessity of learning a great deal of terminology. In fact, it would be much more difficult to learn and communicate if we did not have this terminology, for words are really tools for thinking. Learning terminology generally becomes easier if you realize that most biological terms are modular. They are composed of mostly Latin and Greek roots; once you learn many of these, you will find you have a good idea of the meaning of a new word even before it is defined. For this reason we have in-

cluded an Appendix on Understanding Biological Terms. To be sure you understand the precise definition of a term, you will want to use the Index and Glossary. The more you use biological terms in speech and writing, the more comfortable you will be.

FORM A STUDY GROUP

Active learning is facilitated if you do some of your studying in a small group. In a study group, the roles of teacher and learner should be interchangeable: the best way to be sure you understand is to teach. A study group allows you to be challenged in a nonthreatening environment and can provide some emotional support. Study groups are effective learning tools when combined with individual study of text and lecture notes. If, however, you and other members of your study group have not prepared for your meetings by studying individually in advance, the study session can be a waste of time.

HOW TO PREPARE FOR AN EXAM

Your calendar tells you it is now one week before your first biology exam. If you have been following these suggestions, you are well prepared and will only need some last-minute reviewing. No all-nighters will be required.

During the week prior to the exam, spend two hours each day actively studying your lecture notes or chapter outlines. It helps many students to read these notes out loud (most people listen to what they say!). Begin with the first lecture/chapter covered on the exam, and continue in the order on the lecture syllabus. Stop when you have reached the end of your two-hour study period. The following day, begin where you stopped the previous day. When you reach the end of your notes, start at the beginning and study them a second time. The material should be *very* familiar to you by the second or third time around. At this stage, use your textbook only to answer questions or clarify important points.

The night before the exam, do a little light studying, eat a nutritious dinner, and get a full night's sleep. You will arrive in class on exam day with a well-rested body (and brain) and the self-confidence that goes with being well prepared.

Contents Overview

Contents

PART 2

Energy Transfer Through Living Systems 133

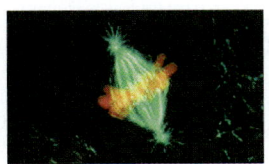

PART 3

The Continuity of Life: Genetics *197*

PART 4

The Continuity of Life: Evolution *369*

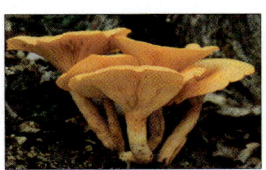

PART 5

The Diversity of Life 465

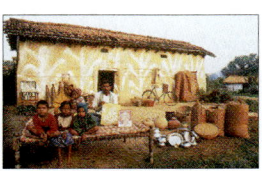

The Organization of Life

Biologists study the details of living organisms and all of their interactions. The Louisiana heron (*Hydranassa tricolor*) and the turtle (*Chrysemys nelsoni*) on which it has perched are complex organisms that make their home in the Florida cypress swamp.   *(Mark J. Thomas/Dembinsky Photo Associates)*

CHAPTER 1

A View of Life

In the 1950s developmental biologists Robert Briggs and Thomas King, then at the Institute for Cancer Research in Philadelphia, developed techniques for cloning frogs. They removed the nucleus (the structure in the cell that contains the genetic material) of a frog egg cell and replaced it with a nucleus from the cell of a developing frog embryo. Briggs and King found that nuclei from very early embryos were able to direct development to the tadpole stage, but nuclei from more developed embryos appeared to lose that ability. In the 1960s embryologist John Gurdon, now at Cambridge University, and his colleagues, using different techniques, showed that at least some nuclei retained the ability to direct development. Gurdon removed the nucleus of a frog egg cell and replaced it with the nucleus from a tadpole cell. The egg developed into a tadpole that was a clone, a genetic duplicate, of the animal that contributed the nucleus.

The research of Briggs and King and of Gurdon challenged the long-held belief that during development, the DNA (genetic material) of each cell becomes committed to coding for a specific type of cell and can no longer direct the development of an entire organism. However, attempts to clone other species were unsuccessful and many biologists concluded that frogs were an exception. Certainly, it was thought, cloning would not be possible in humans and other mammals. With the development of more sophisticated techniques, biologists became more optimistic about cloning mammals but speculated it would be well into the 21st century before this could be accomplished.

In 1997 biologist Ian Wilmut of the Roslin Institute in Edinburgh, Scotland, and his colleagues stunned the scientific community with their report in the journal *Nature* that they had successfully cloned a sheep. These researchers transferred nuclei from cultured mammary gland cells from a six-year-old ewe into specially prepared eggs from another sheep. One of these eggs developed into the clone that was named Dolly. Dolly represented the first time that DNA from an *adult mammal* had been successfully used to create a new, genetically identical animal. Shown here is Dolly with *her* new lamb, Bonnie. Bonnie was conceived naturally, further evidence that Dolly is a fully-functioning animal. (More about cloning and the techniques used to create Dolly are discussed in Chapter 16, *Genes and Development*.) Since Dolly, fetal cells have been used to clone cows, sheep, pigs and other mammals.

The recent successes in cloning have generated a great deal of controversy about possible ethical implications. A major concern centers on the future possibility of cloning humans.

(Roslin Institute/PA News Photo Library)

Wilmut has stated that his goal in cloning Dolly was not to make copies of animals, but to develop techniques for making "precise genetic changes in cells." Already, cloning techniques have been combined with genetic engineering procedures to produce animals like Polly, a lamb that carries a human gene, and Genie, a sow that produces a human clotting protein in her milk. Such *transgenic* animals may become a routine source of relatively inexpensive human proteins needed by individuals suffering from medical conditions from hemophilia to strokes and heart attacks.

Among other applications for cloning technology are the preservation of endangered species and the development of animal strains that will be more resistant to disease or produce low-fat milk or eggs. Genetically engineered, cloned cow or pig organs might be transplanted to ailing human patients. Human tissues could be engineered to treat spinal cord injuries or human diseases like leukemia or diabetes. Society will need to balance the potential benefits to health, medicine, and conservation with the concerns about misuse of cloning techniques.

The ethical questions posed by new technology in genetics and developmental biology represent one group of issues that demand a biologically literate society. As we prepare to enter the 21st century, we face many challenges, such as the expanding human population, diminishing natural resources, decreasing biodiversity, and curing diseases such as cancer and AIDS. Meeting these challenges will require the combined efforts of biologists and other scientists, politicians, and biologically informed citizens.

LEARNING OBJECTIVES

AFTER YOU HAVE STUDIED THIS CHAPTER YOU SHOULD BE ABLE TO

1. Define biology and discuss its applications to human life and society.
2. Distinguish between living and nonliving things by describing the features that characterize living organisms.
3. Relate metabolism to homeostasis and give specific examples of these life processes.
4. Summarize the importance of information transfer to living systems, giving specific examples.
5. Give a brief overview of the theory of evolution and explain why it is the principal unifying concept in biology.
6. Apply the theory of natural selection to any given adaptation, suggesting a logical explanation of how the adaptation may have evolved.
7. Construct a hierarchy of biological organization, including individual and ecological levels.
8. Demonstrate the binomial system of nomenclature using several specific examples and classify an organism (a human, for example) according to kingdom, phylum, class, order, family, genus, and species.
9. Contrast the six kingdoms of living organisms and cite examples of each group.
10. Contrast the roles of producers, consumers, and decomposers and cite examples of their interdependence.
11. Design an experiment to test a given hypothesis, using the procedure and terminology of the scientific method.

WHAT IS BIOLOGY?

Biology is the science of life. More than any other discipline, biology helps us understand ourselves and the millions of other organisms with which we share our planet. As biologists continue to study interrelationships of organisms, they enhance our awareness of our own impact on our environment. Whatever your college major or career goals, a knowledge of biological concepts is a vital tool for understanding our world and for meeting many of the personal and global challenges that confront us. Applications of basic research in biology have provided us with the technology to transplant hearts, manipulate genes, and increase world food production. For example, research in molecular biology and genetics has led to new insights into disease processes, leading to the new science of gene therapy.

In this first chapter we will introduce three basic themes of biology: **evolution of life, transmission of information, and flow of energy through living systems.** Scientists have accumulated a wealth of evidence showing that the diverse life forms on our planet are related and that organisms have evolved through time from earlier life forms. The process of evolution is the framework for the science of biology and a major theme of this book.

The process of evolution, as well as the survival and function of every organism, depends on the orderly transmission of information. At the molecular level, instructions for producing and maintaining each living organism and each new generation are encoded in the DNA molecules that make up the genes. At higher levels, the activities of organisms are coordinated by many forms of chemical signaling. Animals also use chemical, as well as behavioral, signals to communicate with one another. For example, many female animals release chemical substances that attract males.

Energy is required to maintain the precise order that characterizes living systems. Maintaining the chemical transactions and cellular organization essential to life requires a continuous input of energy. We begin our study of biology by developing a more precise understanding of what life is.

LIFE CAN BE DEFINED IN TERMS OF THE CHARACTERISTICS OF ORGANISMS

We can easily recognize that an oak tree, a butterfly, and a lamb are living, whereas a rock is not. Despite their diversity, the organisms that inhabit our planet share a common set of characteristics that distinguish them from nonliving things. These features include a precise kind of organization; growth and development; self-regulated metabolism; movement; the ability to respond to stimuli; reproduction; and adaptation to environmental change. We consider each of these characteristics in the following sections.

(a)

25 μm

(b)

Figure 1–1 **Unicellular and multicellular life forms.** *(a)* Unicellular organisms are generally smaller and consist of one intricate cell that performs all of the functions essential to life. Ciliates, such as this *Paramecium*, move about by beating their hairlike cilia. *(b)* Multicellular organisms, like this African buffalo *(Syncerus caffer)* and the plants on which it grazes, may consist of millions of cells specialized to perform specific functions.   *(a, Roland Birke/Peter Arnold, Inc.; b, McMurray Photography)*

Organisms are composed of cells

Living organisms are highly organized, and, as discussed later in this chapter, we can identify a hierarchy of biological organization. As expressed in the **cell theory,** one of the fundamental unifying concepts of biology, all living organisms are composed of basic units called **cells.** Two German scientists are credited with the cell theory. Matthias Schleiden (in 1838) and Theodor Schwann (in 1839) were the first to report that plants and animals consist of groups of cells. Although they vary greatly in size and appearance, all organisms are composed of these small building blocks. Some of the simplest life forms, such as bacteria, are *unicellular,* meaning that each consists of a single cell. In contrast, the body of a lamb or a maple tree is made of billions of cells (Fig. 1–1). In such complex *multicellular* organisms, life processes depend on the coordinated functions of component cells that may be organized to form tissues, organs, and organ systems.

Organisms grow and develop

Some nonliving things appear to grow. For example, salt crystals form and enlarge in a supersaturated salt solution. However, this is not growth in the biological sense. Biological **growth** is an increase in the size of individual cells of an organism or in the number of cells, or both (Fig. 1–2). Growth may be uniform in the various parts of an organism, or it may

be greater in some parts than in others, causing the body proportions to change as growth occurs.

Some organisms, most trees, for example, continue to grow throughout their lives. Many animals have a defined growth period that terminates when a characteristic adult size is reached.

Figure 1–2 **Biological growth.** The young African elephant *(Loxodonta africana)* eats and grows until it reaches the adult size of its parents, the largest living land animals. These elephants were photographed in Kenya.   *(McMurray Photography)*

One of the remarkable aspects of the growth process is that each part of the organism continues to function as it grows.

Living organisms develop as well as grow. **Development** includes all the changes that take place during the life of an organism. Each human, like many other organisms, begins life as a fertilized egg which then grows and develops. The structures and body form that develop are exquisitely adapted to the functions the organism must perform.

Organisms regulate their metabolic processes

Within all organisms, chemical reactions and energy transformations occur that are essential to nutrition, growth and repair of cells, and conversion of energy into usable forms. The sum of all the chemical activities of the organism is its **metabolism.** Metabolic processes occur continuously in every living organism, and they must be carefully regulated to maintain **homeostasis,** a balanced internal state. When enough of some cellular product has been made, its manufacture must be decreased or turned off. When a particular substance is needed, cellular processes that produce it must be turned on. These **homeostatic mechanisms** are self-regulating control systems that are remarkably sensitive and efficient.

The regulation of glucose (a simple sugar) concentration in the blood of complex animals is a good example of a homeostatic mechanism. The circulatory system delivers glucose and other nutrients to all the cells. Most cells require a constant supply of glucose, which they break down to obtain energy. When the concentration of glucose in the blood rises above normal limits, it is stored in the liver and in muscle cells. When the concentration begins to fall (between meals), stored nutrients are converted to glucose so that the concentration in the blood returns to normal levels. When glucose becomes depleted, we also feel hungry and restore nutrients by eating.

Movement is a basic property of cells

Organisms move as they interact with the environment, and in fact the living material within their cells is in continuous motion. In some organisms, locomotion results from the slow oozing of the cell (a process called *amoeboid motion*); in others, from the beating of tiny hairlike extensions of the cell called **cilia** or longer structures called **flagella** (Fig. 1–3). Most animals move very obviously; they wiggle, crawl, swim, run, or fly by contracting muscles. A few animals, such as sponges, corals, and oysters, have free-swimming larval stages but do not move from place to place as adults. Even though these adults, described as **sessile,** remain firmly attached to some surface, they may have cilia or flagella. These structures beat rhythmically, moving the surrounding water that brings food and other necessities to the organism.

Although plants move more slowly than most animals, they do move. For example, plants orient their leaves to the sun and grow toward light. In some plants, for example the Venus flytrap, movement is obvious, even dramatic (described in the next section).

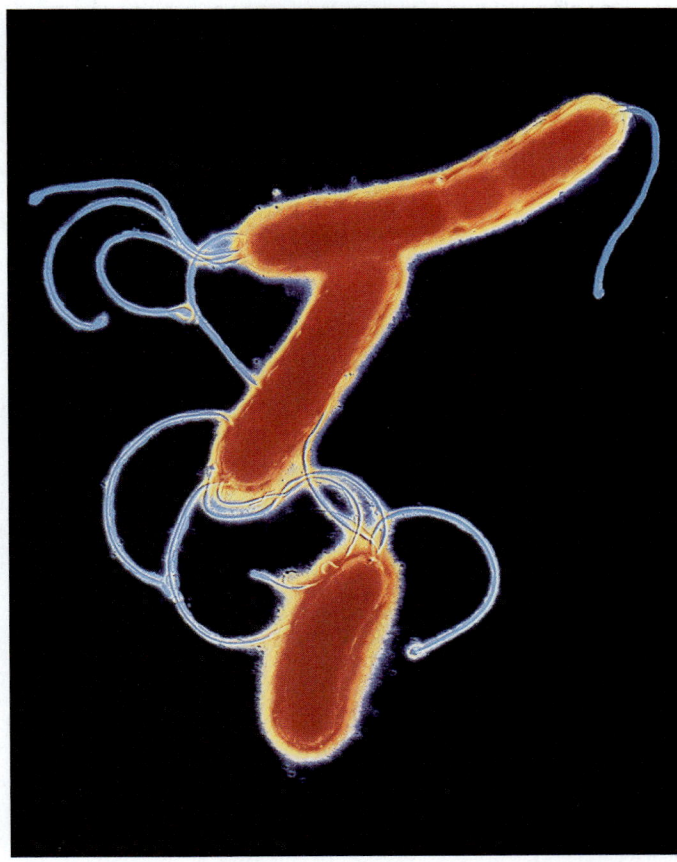

1 μm

Figure 1–3 Biological movement. These bacteria *(Helicobacter pylori),* equipped with flagella for locomotion, have been linked to stomach ulcers. *(A. B. Dowsett/Science Photo Library/Photo Researchers, Inc.)*

Organisms respond to stimuli

All forms of life respond to **stimuli,** physical or chemical changes in their internal or external environment. Stimuli that evoke a response in most organisms are changes in the color, intensity, or direction of light; changes in temperature, pressure, or sound; and changes in the chemical composition of the surrounding soil, air, or water. In simple organisms, the entire organism may be sensitive to stimuli. Certain unicellular organisms, for example, respond to bright light by retreating. In complex animals such as polar bears or humans, certain cells of the body are highly specialized to respond to specific types of stimuli. For example, cells in the retina of the eye respond to light.

Although their responses may not be as obvious as those of animals, plants do respond to light, gravity, water, touch, and other stimuli. Many plant responses involve different rates of growth of various parts of the plant body. A few plants, such as the Venus flytrap of the Carolina swamps (Fig. 1–4), are remarkably sensitive to touch and can catch insects. Their leaves are hinged along the midrib and they have a scent that attracts insects. Trigger hairs on the leaf surface detect the arrival of an

(a)

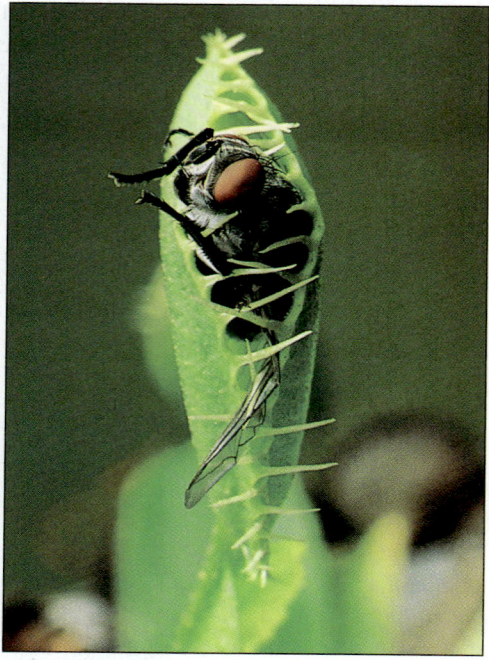

(b)

insect and stimulate the leaf to fold. When the edges come together, the hairs interlock, preventing escape of the prey. The leaf then secretes enzymes that kill and digest the insect. The Venus flytrap is usually found in soil that is deficient in nitrogen. The plant obtains part of the nitrogen required for its growth from the insect prey it "eats."

Organisms reproduce

At one time worms were thought to arise spontaneously from horsehairs in a water trough, maggots from decaying meat, and frogs from the mud of the Nile. We now know that each can come only from previously existing organisms. If any one characteristic can be said to be the very essence of life, it is the ability of an organism to reproduce its kind.

In simple organisms such as the amoeba, reproduction may be **asexual,** that is, without the fusion of egg and sperm to form a fertilized egg (Fig. 1–5). When an amoeba has grown to a certain size, it reproduces by splitting in half to form two new amoebas. Before an amoeba divides, its hereditary material (genes) duplicate and one complete set is distributed to each new cell. Except for size, each new amoeba is similar to the parent cell. The only way that variation occurs among asex-

(a) Asexual reproduction

100 μm

(b) Sexual reproduction

▶ **Figure 1–5** **Asexual and sexual reproduction.** *(a)* Asexual reproduction in *Difflugia,* a unicellular amoeba. In asexual reproduction, one individual gives rise to two or more offspring that are similar to the parent. *(b)* A pair of tropical flies mating. In sexual reproduction, typically two parents each contribute a gamete (sperm or egg). Gametes join to produce the offspring, which is a combination of the traits of both parents. *(a, Visuals Unlimited/Cabisco; b, L.E. Gilbert, University of Texas at Austin/Biological Photo Service)*

Figure 1–6 **Adaptations.** These Burchell's zebras (*Equus burchelli*), photographed at Ngorongoro Crater in Tanzania, are behaviorally adapted to position themselves to watch for lions and other predators. Stripes are thought to be an adaptation for visually protecting themselves against predators. They serve as camouflage or to break up form when spotted from a distance. The zebra stomach is adapted for feeding on coarse grass passed over by other grazers, an adaptation that helps them survive when food is scarce. *(McMurray Photography)*

ually reproducing organisms is by genetic mutation, a permanent change in the genes.

In most plants and animals, **sexual reproduction** is carried out by the production of specialized egg and sperm cells that fuse to form a fertilized egg. The new organism develops from the fertilized egg. Offspring produced by sexual reproduction are the product of the interaction of various genes contributed by both the mother and the father. Such genetic variation provides raw material for the vital processes of evolution and adaptation.

Populations evolve and become adapted to the environment

The ability of a population to evolve (change over time) and adapt to its environment enables it to survive in a changing world. **Adaptations** are traits that enhance an organism's ability to survive in a particular environment. They may be structural, physiological, behavioral, or a combination of all three (Fig. 1–6). The long, flexible tongue of the frog is an adaptation for catching insects, and the thick fur coat of the polar bear is an adaptation for surviving frigid temperatures. Every biologically successful organism is a complex collection of coordinated adaptations produced through evolutionary processes.

INFORMATION MUST BE TRANSMITTED WITHIN AND BETWEEN INDIVIDUALS

In order for an organism to grow, develop, carry on self-regulated metabolism, move, respond, and reproduce, it must have precise instructions. The information an organism needs to carry on these life processes is coded and delivered in the form of chemical substances and electrical impulses.

DNA transmits information from one generation to the next

Humans give birth only to human babies, not to giraffes or rose bushes. In organisms that reproduce sexually, each offspring is a combination of the traits of its parents. In 1953, James Watson and Francis Crick worked out the structure of deoxyribonucleic acid, more simply known as **DNA.** This chemical substance makes up the **genes,** the units of hereditary material. The work of Watson and Crick led to the understanding of the genetic code that transmits information from generation to generation. This code works somewhat like our alphabet; it can spell an amazing variety of instructions for making organisms as diverse as bacteria, frogs, and redwood trees. The genetic code is a dramatic example of the unity of life because it is used to specify instructions for making every living organism (Fig. 1–7).

Figure 1–7 **DNA.** An organism's ability to transmit information from one generation to the next is essential to the continuity of life. In all organisms, the hereditary material is DNA. This computer-generated image shows the double-helix configuration of DNA. *(Will and Deni McIntyre/Photo Researchers, Inc.)*

Information is transmitted by chemical and electrical signals

Genes control the development and functioning of every organism. DNA contains the "recipes" for making all of the **proteins** needed by the organism. Proteins are very large molecules that are important in determining structure and function of cells and tissues. Brain cells are different from muscle cells, in large part because they have different types of proteins. Some proteins are important in communication within and among cells. Certain proteins on the surface of a cell serve as markers so that other cells "recognize" them. Other cell surface proteins serve as receptors that combine with chemical messengers.

Cells use proteins and many other types of ions and molecules to communicate with one another. In a multicellular organism chemical compounds secreted by cells help regulate growth, development, and metabolic processes in other cells. The mechanisms involved in **cell signaling** are complex, often involving multistep biochemical sequences, and cell signaling is currently an area of intense research. A major focus has been the transfer of information among cells of the immune system. A better understanding of how cells communicate promises new insights into how the body protects itself against disease organisms. Learning to manipulate cell signaling may lead to new methods of delivering drugs into cells and new treatments for cancer and other diseases. Examples of cell signaling will be discussed throughout this book.

Hormones are molecules that function as chemical messengers that transmit information from one part of an organism to another. A hormone can signal cells to produce or secrete a certain protein or other substance.

Many organisms use electrical signals to transmit information. Most animals have nervous systems that transmit information by way of both electrical impulses and chemical compounds known as **neurotransmitters.** Information transmitted from one part of the body to another is important in regulating life processes. In complex animals, the nervous system transmits signals from sensory receptors like the eyes and ears to the brain, giving the animal information about its outside environment.

Information must also be transmitted from one organism to another. Mechanisms for this type of communication include release of chemicals, visual displays, and sounds.

EVOLUTION IS THE PRIMARY UNIFYING CONCEPT OF BIOLOGY

The theory of **evolution,** which explains how populations of organisms have changed over time, has become the greatest unifying concept of biology. Some element of an evolutionary perspective is present in every specialized field within biology. Biologists try to understand the structure, function, and behavior of organisms and their interactions with one another by considering them in light of the long, continuing process of evolution. Although evolution is discussed in depth in

Chapters 17 through 21, we present a brief overview here to give you the background necessary to understand other aspects of biology.

Species adapt in response to changes in their environment

Every organism is the product of complex interactions between environmental conditions and the genes of its ancestors. If every organism of a species[1] were exactly like every other, any change in the environment might be disastrous to all, and the species would become extinct. Adaptation to changes in the environment involves changes in populations rather than in individual organisms. Such adaptations are the result of evolutionary processes that occur over time and involve many generations.

Natural selection is an important mechanism by which evolution proceeds

Although the concept of evolution had been discussed by philosophers and naturalists through the ages, Charles Darwin and Alfred Wallace first brought the theory of evolution to general attention and suggested a plausible mechanism, **natural selection,** to explain it. In his book *On the Origin of Species by Means of Natural Selection,* published in 1859, Darwin synthesized many new findings in geology and biology. He presented a wealth of evidence that the present forms of life on Earth descended, with modifications, from previously existing forms. Darwin's book raised a storm of controversy in both religion and science, some of which still lingers.

Darwin's theory of evolution has helped shape the biological sciences to the present day. His work generated a great wave of scientific observation and research that has provided much additional evidence that evolution is responsible for the great diversity of organisms present on our planet. Even today, the details of the process of evolution are a major focus of investigation and debate.

Darwin based his theory of natural selection on the following four observations: (1) Individual members of a species show some variation from one another. (2) Organisms produce many more offspring than will survive to reproduce (Fig. 1–8). (3) Organisms compete for necessary resources like food, sunlight, and space. Individuals who happen to have characteristics that give them some advantage are more likely to survive. (4) The survivors live to reproduce and pass their adaptations for survival on to their offspring. Thus, the best adapted individuals of a population leave, on average, more offspring than do other individuals. Because of this differential reproduction, a greater proportion of the population becomes adapted to the prevailing environmental conditions. The environment *selects* the best adapted organisms for survival.

[1]A species is a group of organisms with similar structure, function, and behavior; in nature they breed only with each other. Members of a species have a common gene pool and share a common ancestry.

Figure 1-8 **Egg masses of the wood frog (Rana sylvatica).**
Many more eggs are produced than can possibly develop into adult frogs. Random events are largely responsible for determining which of these developing frogs will hatch, reach adulthood, and reproduce. However, certain traits possessed by each organism will also contribute to its probability for success in its environment. Not all organisms are as prolific as the frog, but the generalization that more organisms are produced than survive is true throughout the living world. (*J. Serrao/Photo Researchers, Inc.*)

Darwin did not know about DNA or understand the mechanisms of inheritance. We now understand that the variations among organisms are a result of different varieties of genes that code for each characteristic. The ultimate source of these variations is random **mutations,** chemical changes in DNA that persist and can be inherited. Mutations modify genes; by this process they provide the raw material for evolution.

Populations evolve as a result of selective pressures from changes in the environment

All the genes present in a population make up its gene pool. By virtue of its gene pool, a population is a reservoir of variation. Natural selection acts on individuals within a population. Selection favors individuals with genes that specify traits that enable them to cope effectively with pressures exerted by the environment. These organisms are most likely to survive and produce offspring. As these successful organisms pass on their genetic recipe for survival, their traits become more widely distributed in the population. Over time, as organisms continue to change (and as the environment itself changes, bringing different selective pressures), the members of the population become better adapted to their environment and less like their ancestors.

An interesting case of evolution in action has been documented in England since 1850. The tree trunks in a certain region of England were once white because of a type of lichen that grew on them. (A lichen is a compound organism usually consisting of an alga and fungus.) The common peppered moth was beautifully adapted for resting on these white tree trunks because its light color blended with the trunks and protected it from predatory birds (Fig. 1–9). At that time black peppered moths were rare.

Figure 1-9 **Evolution in action.** Both a dark and a light peppered moth can be seen on the tree trunk. Which is most likely to become food for the bird? Note that the tree trunk is light in color because it is covered by lichens. (*John D. Cunningham/Visuals Unlimited*)

Then humans changed the environment. They built industries that polluted the air with soot, killing the lichens and coloring the tree trunks black. The light-colored moths became easy prey to the birds. The black moths blended with the dark trunks and escaped the sharp eyes of predators. In these new surroundings, the dark moths were better adapted and were selected for survival. Eventually, more than 90% of the peppered moths in the industrial areas of England were dark. This adaptation is known as *industrial melanism.* Interestingly, with efforts to control air pollution, there has been an increase in the population of the light-colored moths.

Adaptation of the peppered moth was studied in the 1950s by H. B. D. Kettlewell of Oxford, who marked hundreds of male moths with a spot of paint under their wings and then released them in both rural and industrial areas. Observers reported that birds preyed on the moths that were more visible. After a period of time, surviving moths were recaptured by attracting them with light or females. Based on observation and on the percentage of each type of moth recaptured, these studies confirmed that significantly more dark moths survived in industrial areas and more light moths survived in rural areas.

BIOLOGICAL ORGANIZATION IS HIERARCHICAL

Whether we study a single complex organism or the world of life as a whole, we can identify a hierarchy of biological organization (Fig. 1–10). At every level, structure and function are precisely coordinated. One way to study a particular level is by looking at its components. For example, biologists can learn about cells by studying atoms and molecules. Learning about a structure by studying its parts is known as **reductionism.** However, the whole is more than the sum of its parts. Each level has **emergent properties,** characteristics not found at lower levels. For example, populations have emergent properties such as density, age structure, and birth and death rates. The individuals that make up a population lack these characteristics.

Organisms have several levels of organization

The **chemical level,** the most basic level of organization, includes atoms and molecules. An **atom** is the smallest unit of a chemical element (fundamental substance) that retains the characteristic properties of that element. For example, an atom of iron is the smallest possible amount of iron. Atoms combine chemically to form **molecules.** Two atoms of hydrogen combine with one atom of oxygen to form a single molecule of water. Although composed of two types of atoms that are gases, water is a liquid with very different properties, an example of emergent properties.

At the **cellular level** many different types of atoms and molecules associate with one another to form cells. However, a cell is much more than a heap of atoms and molecules. Its emergent properties make it the basic structural and functional unit of life, the simplest component of living matter that can carry on all of the activities necessary for life. Every cell is surrounded by a **plasma membrane** that regulates the passage of materials between the cell and its surrounding environment. All cells have specialized molecules that contain genetic instructions. Cells typically have internal structures called **organelles** that are specialized to perform specific functions.

Two fundamentally different types of cells are known. Bacteria have **prokaryotic cells.** All other organisms are characterized by their **eukaryotic cells.** Unlike the structurally simpler prokaryotic cells, eukaryotic cells typically contain a variety of membrane-bounded organelles, including a **nucleus** which houses DNA.

During the evolution of multicellular organisms, cells associated to form **tissues.** For example most animals have muscle tissue and nervous tissue, and plants have epidermis (a tissue that serves as a protective covering). In most complex organisms, tissues organize into functional structures, called **organs,** such as the heart and stomach in animals and roots and leaves in plants. In animals, each major group of biological functions is performed by a coordinated group of tissues and organs called an **organ system.** The circulatory and digestive systems are examples of organ systems. Functioning together with great precision, organ systems make up a complex multicellular **organism.** Again, emergent properties are evident. An organism is much more than its component organ systems.

Several levels of ecological organization can be identified

Organisms interact to form still more complex levels of biological organization. All of the members of one species that live in the same geographic area at the same time make up a **population.** The populations of organisms that inhabit a particular area and interact with one another form a **community.** A community can consist of hundreds of different types of organisms. As populations within a community evolve, the community changes.

A community together with its nonliving environment is referred to as an **ecosystem.** An ecosystem can be as small as a pond (or even a puddle) or as vast as the Great Plains of North America or the Arctic tundra. All of planet Earth's ecosystems together are known as the **biosphere.** The biosphere includes all of Earth that is inhabited by living organisms—the atmosphere, the hydrosphere (water in any form), and the lithosphere (Earth's crust). The study of how organisms relate to one another and to their physical environment is called **ecology** (derived from the Greek *oikos,* meaning "house").

Figure 1-10 Biological organization. Atoms join to form molecules of varying size, including very large molecules, or macromolecules. Atoms and molecules form organelles such as the nucleus or mitochondrion (the site of many energy transformations) of the cell. Cells associate to form tissues, such as bone tissue. Tissues form organs, such as bones, which, in turn, make up organ systems. The skeletal system and other organ systems work together to make up the functioning organism. A population is composed of organisms of the same species. The populations of different species that inhabit a particular area make up a community, which together with the nonliving environment, form an ecosystem. Planet Earth and all of its communities make up the biosphere.

TABLE 1-1   Classification of Domestic Cat, Human, and White Oak

Category	Classification of Cat	Classification of Human	Classification of White Oak
Kingdom	Animalia	Animalia	Plantae
Phylum	Chordata	Chordata	Anthophyta
Subphylum	Vertebrata	Vertebrata	None
Class	Mammalia	Mammalia	Dicotyledones
Order	Carnivora	Primates	Fagales
Family	Felidae	Hominidae	Fagaceae
Genus and specific epithet	*Felis catus*	*Homo sapiens*	*Quercus alba*

MILLIONS OF SPECIES HAVE EVOLVED

About 1.75 million species of extant (currently living) organisms have been scientifically identified, and biologists estimate that millions more remain to be discovered. In order to study life, we need a system for organizing, naming, and classifying its myriad forms. That system is part of **taxonomy,** the science of naming and classifying organisms. Biologists who specialize in classification are **taxonomists.**

Biologists use a binomial system for naming organisms

In the 18th century Carolus Linnaeus, a Swedish botanist, developed a hierarchical system of naming and classifying organisms that, with some modification, is still used today. The basic unit of classification is the **species.** Closely related species are grouped together in the next higher level of classification, the **genus** (pl. *genera*).

The Linnaean system of naming species is referred to as the **binomial system of nomenclature** because each species is assigned a two-part name. The first part of the name is the genus, and the second part, the **specific epithet,** designates a particular species belonging to that genus. The specific epithet is often a descriptive word expressing some quality of the organism. It is always used together with the full or abbreviated generic name preceding it. The generic name is always capitalized; the specific epithet is generally not capitalized. Both names are always italicized or underlined. For example, the dog, *Canis familiaris* (abbreviated *C. familiaris*), and the timber wolf, *Canis lupus (C. lupus),* belong to the same genus. The cat, *Felis catus,* belongs to a different genus. The scientific name of the American white oak is *Quercus alba,* whereas the name of the European white oak is *Quercus robur.* Another tree, the white willow, *Salix alba,* belongs to a different genus. The scientific name for our own species is *Homo sapiens* ("wise man").

Taxonomic classification is hierarchical

Just as species may be grouped together in a common genus, a number of related genera can be grouped in a more inclu-sive group, a **family** (Table 1–1). In turn, families may be grouped into **orders,** orders into **classes,** and classes into **phyla.** Note that each group is more inclusive. The family Canidae, which includes all doglike carnivores (animals that eat mainly meat), consists of 12 genera and about 34 living species. Family Canidae, along with family Ursidae (bears), family Felidae (catlike animals), and several other families that eat mainly meat, is placed in order Carnivora. Order Carnivora, order Primates (the order to which humans belong), and several other orders belong to class Mammalia (mammals). Class Mammalia, class Aves (birds), class Reptilia (reptiles), and four other classes are grouped together as subphylum Vertebrata. The vertebrates belong to phylum Chordata, which is part of kingdom Animalia.

Organisms can be assigned to six kingdoms

In the system of classification used in this book, every organism is assigned to one of six **kingdoms:** Archaebacteria, Eubacteria, Protista, Fungi, Plantae, or Animalia (Fig. 1–11). Bacteria are unicellular, and they differ from all other organisms in that they are prokaryotes; their cells lack a discrete nucleus and other membrane-bounded organelles. Among the prokaryotes two very distinct groups have been recognized, so in the

▶ **Figure 1–11   A survey of the kingdoms of life.** *(a)* These bacteria (*Methanosarcina mazei*) members of kingdom Archaebacteria, produce methane. *(b)* The large rod-shaped bacterium *Bacillus anthracis,* a member of kingdom Eubacteria, is the causative agent of anthrax, a disease of cattle and sheep that can infect humans. *(c)* Unicellar protozoa (*Tetrahymena* sp.) are classified in kingdom Protista. *(d)* Mushrooms, such as these fly agaric mushrooms (*Amanita muscaria*), belong to kingdom Fungi. *(e)* The plant kingdom claims many beautiful and diverse forms such as the lady's slipper (*Phragmipedium caricinum*). *(f)* Among the fiercest members of the animal kingdom, lions (*Panthero leo*) are also among the most sociable. The largest of the big cats, lions live in prides (groups).   *(a, R. Robinson/Visuals Unlimited, b, CNRI/Science Photo Library/Photo Researchers, Inc.; c, David M. Phillips/Visuals Unlimited; d, Ulf Sjostedt/FPG International; e, John Arnaldi; f, McMurray Photography.)*

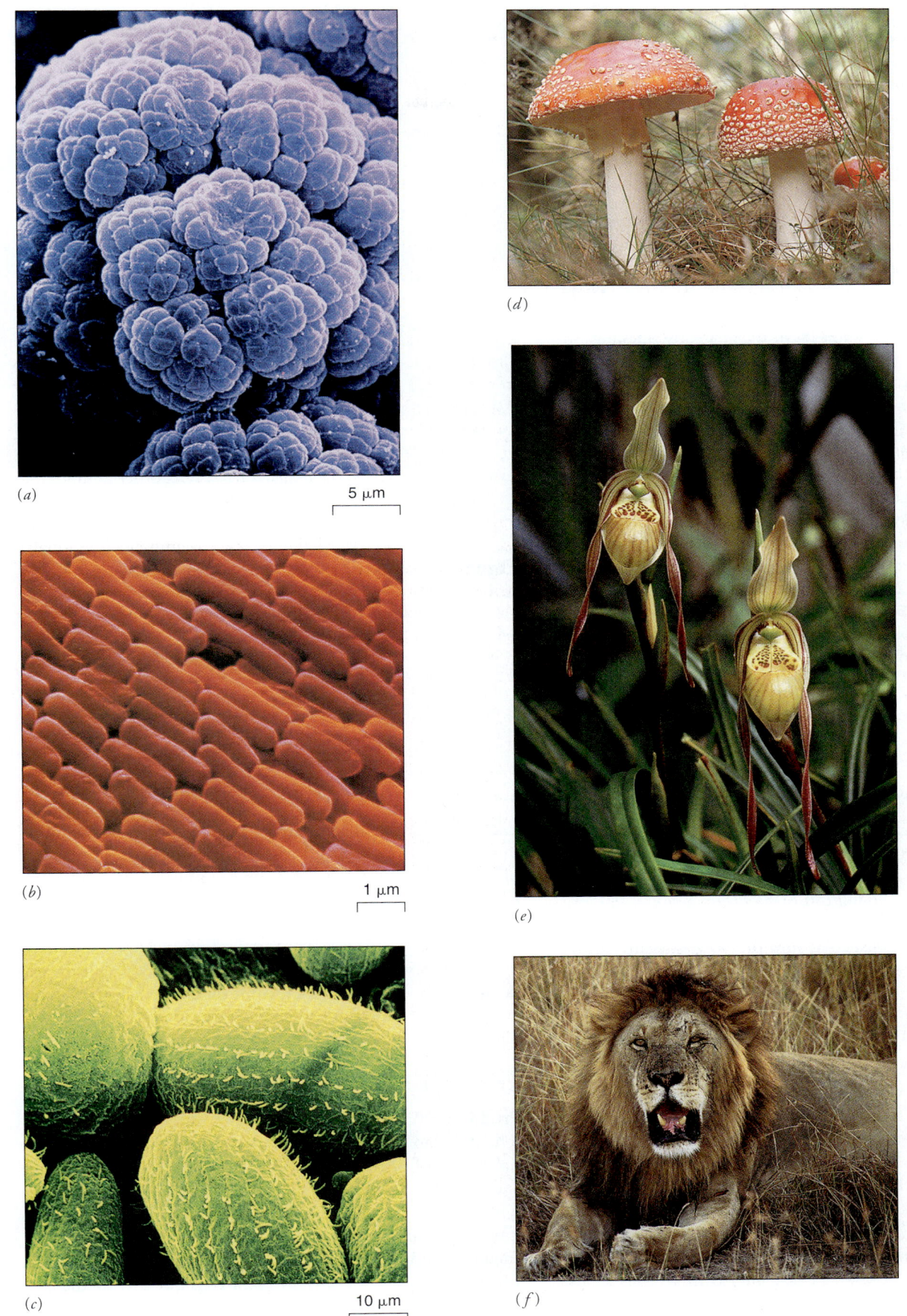

(a)

5 μm

(b)

1 μm

(c)

10 μm

(d)

(e)

(f)

six-kingdom system, bacteria are assigned to two kingdoms, **Archaebacteria** and **Eubacteria.** Kingdom **Protista** consists of protozoa, algae, water molds, and slime molds. These are single-celled or simple multicellular organisms. Some protists are adapted to carry out **photosynthesis,** the process in which light energy is converted to the chemical energy of food molecules.

Kingdom **Fungi** is composed of the yeasts, mildews, molds, and mushrooms. These organisms do not photosynthesize. They obtain their nutrients by secreting digestive enzymes into food and then absorbing the predigested food.

Members of kingdom **Plantae** are complex multicellular organisms adapted to carry out photosynthesis. Among characteristic plant features are the *cuticle* (a waxy covering over aerial parts that reduces water loss), *stomata* (tiny openings in stems and leaves for gas exchange), and multicellular *gametangia* (organs that protect developing reproductive cells). Kingdom Plantae includes both nonvascular plants (mosses) and vascular plants (ferns, conifers, and flowering plants).

Kingdom **Animalia** is made up of multicellular organisms that must eat other organisms for nourishment. Complex animals have high degrees of tissue specialization and body organization; these have evolved along with motility, complex sense organs, nervous systems, and muscular systems.

A more detailed presentation of the kingdoms can be found in Chapters 22 through 30, and classification is summarized in Appendix C. We refer to these groups repeatedly throughout this book as we consider the many kinds of challenges faced by living organisms and the various adaptations that have evolved in response to these challenges.

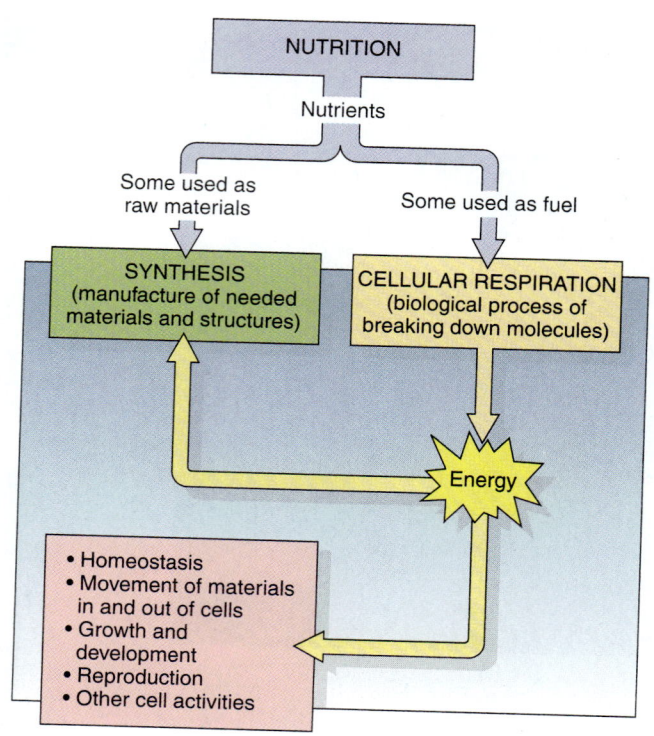

Figure 1–12 **Relationships of metabolic processes.** These processes occur continuously in every living organism. Some of the nutrients in food are used to synthesize needed materials and cell parts. Other nutrients are used as fuel for cellular respiration, a process that releases energy stored in food. This energy is needed for synthesis and for other forms of cellular work.

LIFE DEPENDS ON CONTINUOUS INPUT OF ENERGY

A continuous input of energy from the Sun enables life to exist. Every activity of a living cell or organism requires energy. Whenever energy is used to perform biological work, some is converted to heat and dispersed into the environment.

Energy flows through cells and organisms

Recall that all of the energy transformations and chemical processes that occur within an organism are referred to as metabolism. Energy is necessary in order to carry on the metabolic activities essential for growth, repair, and maintenance. Each cell of an organism requires nutrients. Some nutrients are used as fuel for **cellular respiration,** a process during which some of the energy stored in the nutrient molecules is released for use by the cells (Fig. 1–12). This energy can be used for cellular work or for synthesis of needed materials such as new cellular components.

Energy flows through ecosystems

Like individual organisms, ecosystems depend on a continuous input of energy. A self-sufficient ecosystem contains three types of organisms—producers, consumers, and decomposers—and has a physical environment appropriate for their survival. These organisms depend on each other and on the environment for nutrients, energy, oxygen, and carbon dioxide. However, there is a one-way flow of energy through ecosystems. Organisms can neither create energy nor use it with complete efficiency. During every energy transaction, some energy is lost to biological systems as it is dispersed into the environment as heat (Fig. 1–13).

Producers manufacture their own food

Producers, or **autotrophs,** are plants, algae, and certain bacteria that can produce their own food from simple raw materials. Most of these organisms use sunlight as an energy source and carry out photosynthesis, in which complex molecules are synthesized from carbon dioxide and water. The light energy is transformed into chemical energy, which is stored within the chemical bonds of the food molecules produced. Oxygen, which is required not only by plant cells but also by the cells

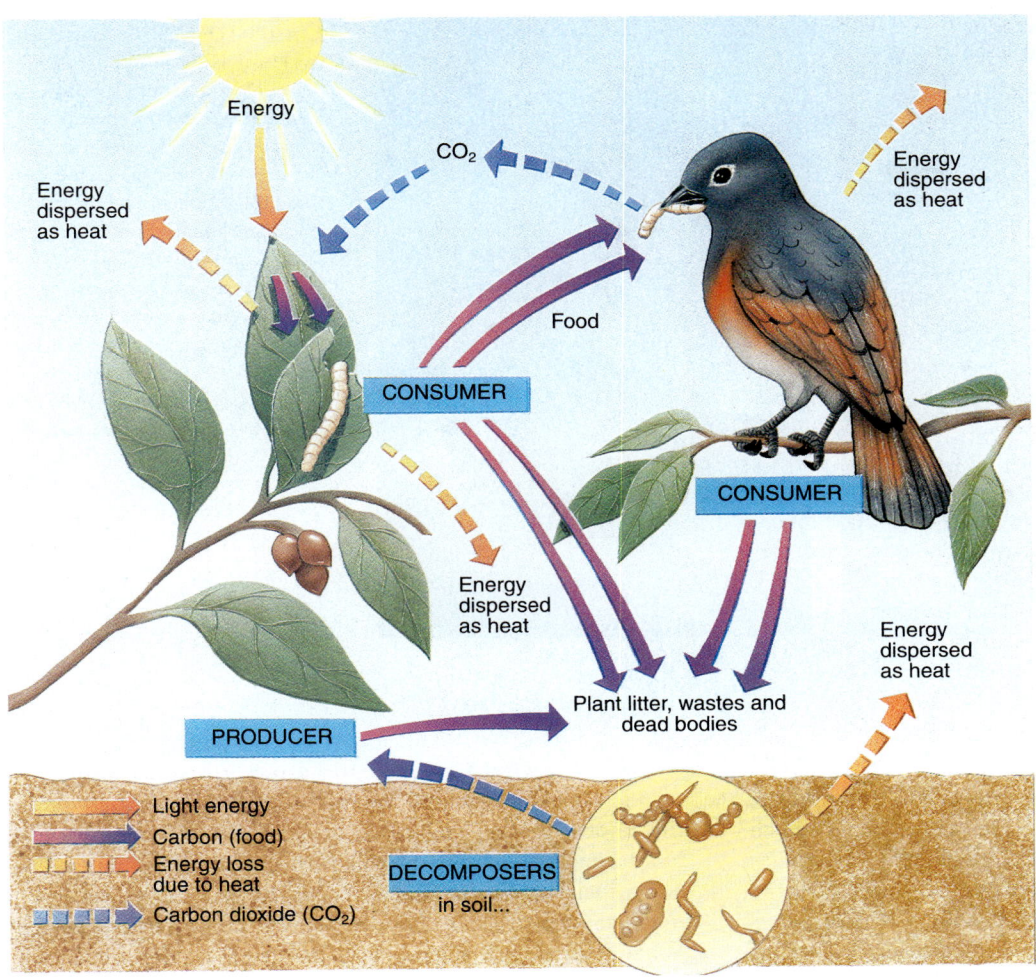

Figure 1–13 **Energy flow.** Continuous energy input from the Sun operates the biosphere. During photosynthesis, producers use the energy from sunlight to make complex molecules from carbon dioxide and water. Consumers obtain energy, carbon, and other needed materials when they eat producers. Wastes and dead organic material supply decomposers with energy and carbon. During every energy transaction some energy is lost to biological systems as it is dispersed as heat.

of most other organisms, is produced as a byproduct of photosynthesis:

Carbon dioxide + Water + Energy $\longrightarrow$ Sugars (food) + Oxygen

Consumers obtain energy by eating producers

Animals are **consumers,** or **heterotrophs,** organisms that depend on producers for food, energy, and oxygen. Consumers obtain energy by breaking down sugars and other food molecules originally produced during photosynthesis. Recall that the biological process of breaking down sugars and other fuel molecules is known as cellular respiration. When chemical bonds are broken during cellular respiration, their stored energy is made available for life processes.

Sugars (and other fuel molecules) + Oxygen $\longrightarrow$ Carbon dioxide + Water + Energy

Consumers contribute to the balance of the ecosystem. For example, consumers produce carbon dioxide needed by produc-

ers. The metabolism of consumers and producers helps maintain the life-sustaining mixture of gases in the atmosphere.

Decomposers obtain energy from wastes and dead organisms

Bacteria and fungi are **decomposers,** heterotrophs that obtain nutrients by breaking down wastes, dead leaves, and the bodies of dead organisms. In their process of obtaining energy, decomposers make the components of wastes and dead organisms available for reuse. If decomposers did not exist, nutrients would remain locked up in dead bodies, and the supply of elements required by living systems would soon be exhausted.

BIOLOGY IS STUDIED USING THE SCIENTIFIC METHOD

This book is a starting point for your exploration of biology. It will provide you with tools to become a part of this fascinating science and a more informed inhabitant of our planet. Biologists work both in laboratories and out in the field (Fig. 1–14). Their investigations range from the study of molecu-

(a)

(b)

Figure 1-14 **Biologists at work.** *(a)* This biologist studying the rainforest canopy in Costa Rica is part of an international effort to study and preserve tropical rain forests. Researchers study the interactions of organisms and the effects of human activities on the rain forests. *(b)* These researchers, working on developing an AIDS vaccine, are examining a DNA sequencing autoradiogram over a light box. *(a, Mark Moffett/Minden Pictures; b, Hank Morgan/Photo Researchers, Inc.)*

lar biology and viruses to the interactions of the communities of our biosphere. Perhaps you will decide to become a research biologist and help unravel the complexities of the human brain; discover new hormones that cause plants to flower; identify new species of animals or bacteria; or develop new treatments for diseases such as cancer, AIDS, or heart disease. Or perhaps you will choose to enter an applied field of biology such as forestry, dentistry, medicine, or veterinary medicine.

Biology is a **science.** The word *science* comes from a Latin word meaning "to know." Science is a way of thinking and a method of investigating the world around us in a systematic manner. Science enables us to uncover ever more about the world we live in and leads us to an expanded appreciation of our universe.

The **process of science** is investigative, dynamic, and often controversial. It changes over time as it is influenced by cultural, social, and historical contexts as well as by the personalities of scientists themselves. The observations made, the range of questions posed, and the design of experiments depend on the creativity of the individual scientist. In contrast, the **scientific method** involves a series of ordered steps and is a framework used by most scientists.

Using the scientific method, scientists make careful observations, ask critical questions, develop **hypotheses** (testable statements), make predictions that can be tested, and perform experiments to test their predictions (Fig. 1-15). They interpret the results of their experiments and draw conclusions from them. Even results that do not support the hypothesis may be valuable and may lead to new hypotheses. If the results support a hypothesis, a scientist may use them to generate related hypotheses.

Science is systematic. Scientists organize, and often quantify, knowledge, making it readily accessible to all who wish to build on its foundation. In this way science is both a personal and a social endeavor. Science is not mysterious. Anyone who understands its rules and procedures can take on its challenges. What distinguishes science is its insistence on rigorous methods to examine a problem. Science seeks to give us precise knowledge about those aspects of the world that are accessible to its methods of inquiry. It is not a replacement for philosophy, religion, or art. Being a scientist does not prevent one from participating in other fields of human endeavor, just as being an artist does not prevent one from practicing science.

Science requires systematic thought processes

Two types of systematic thought processes used by scientists are deduction and induction. With **deductive reasoning,** we begin with supplied information, called *premises,* and draw conclusions on the basis of that information. Deduction proceeds from general principles to specific conclusions. For example, if we accept the premise that all birds have wings, and the second premise that sparrows are birds, we can conclude deductively that sparrows have wings (Fig. 1-16). Deduction helps us discover relationships among known facts. The **hypothetico-deductive approach** emphasizes the use of deductive reasoning to test hypotheses.

Scientists also use a **hypothetico-inductive approach** that focuses on discovering new general principles. **Inductive reasoning** is the opposite of deduction. We begin with specific observations and draw a conclusion or discover a general prin-

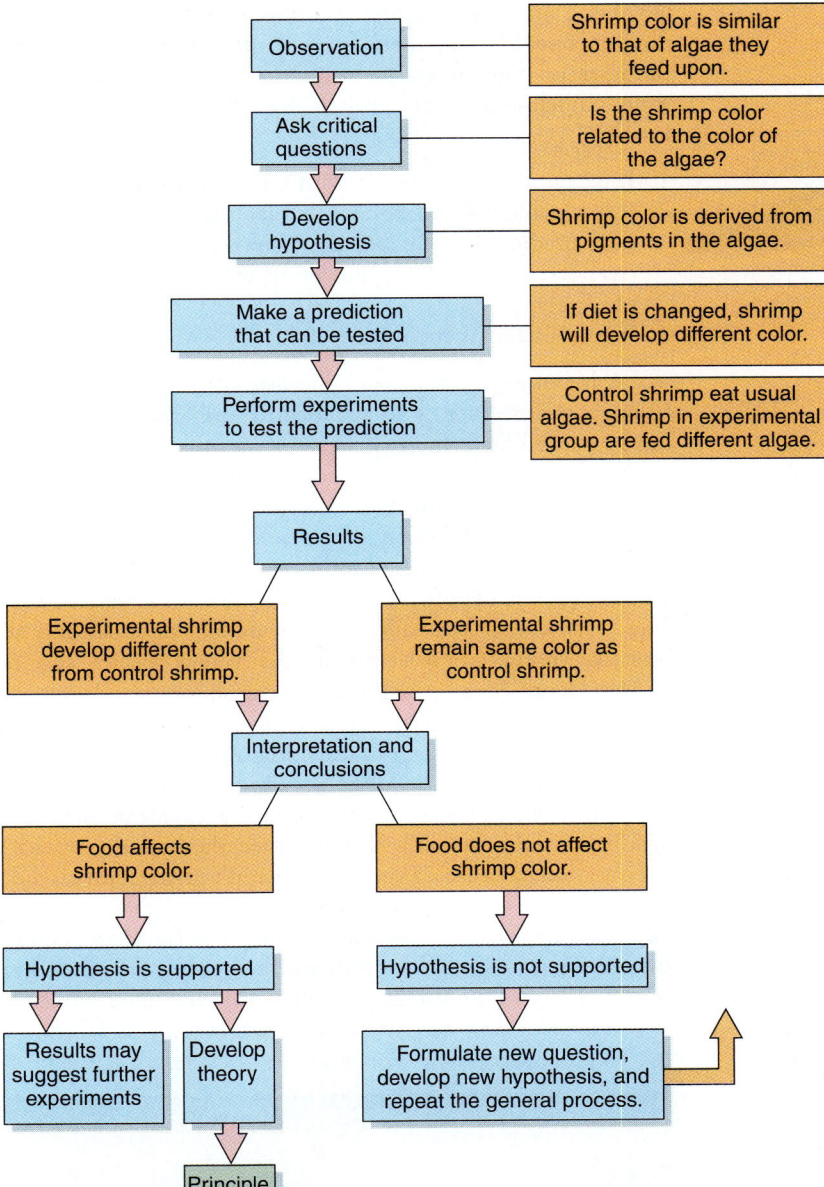

Figure 1–15 The scientific method. Scientists use the scientific method as a framework for their research.

ciple. For example, if we know that sparrows have wings and are birds, and we know that robins, eagles, pigeons, and hawks have wings and are birds, we might induce that all birds have wings. In this way, the inductive method can be used to organize raw data into manageable categories by answering the question, What do all these facts have in common?

A weakness of inductive reasoning is that conclusions generalize the facts to all possible examples. We go from many observed examples to all possible examples when we formulate the general principle. This is known as the **inductive leap.** Without it, we could not arrive at generalizations. However, we must be sensitive to exceptions and to the possibility that the conclusion is not valid. For example, the kiwi bird of New Zealand does not have functional wings! The generalizations that inductive conclusions contain come from the creative insight of the human mind, and creativity, however admirable, is not infallible.

Figure 1–16 **All birds have wings.** A diagrammatic example of a syllogism, the classic form of deductive reasoning.

Scientists make careful observations and ask critical questions

Significant discoveries are usually made by those who are in the habit of looking critically at nature. Chance and luck are often involved in recognizing a phenomenon or problem. In 1928 the British bacteriologist Alexander Fleming observed that one of his bacterial cultures had become invaded by a blue mold. He almost discarded it, but before he did, he noticed that the area contaminated by the mold was surrounded by a zone where bacterial colonies did not grow well.

The bacteria were disease organisms of the genus *Staphylococcus,* which can cause boils and skin infections. Anything that could kill them was interesting! Fleming saved the mold, a variety of *Penicillium* (blue bread mold). It was subsequently discovered that the mold produced a substance that slowed reproduction of the bacterial population but was usually harmless to laboratory animals and humans. The substance was penicillin, one of the first antibiotics.

We may wonder how many times the same type of mold grew on the cultures of other bacteriologists who failed to make the connection and simply threw away their contaminated cultures. Fleming benefited from chance, but his mind was prepared to make observations and formulate critical questions, and his pen was prepared to publish them. It was left to others, however, to develop the practical applications. Though Fleming recognized the potential practical benefit of penicillin, he did not vigorously promote it, and more than ten years passed before the drug was put to significant use.

A hypothesis is a testable statement

In the early stages of an investigation, a scientist typically thinks of many possible hypotheses and hopes that the right one is among them. He or she then decides which, if any, could and should be subjected to experimental test. Why not test them all? Time and money are important considerations in conducting research. We must establish priority among the hypotheses in order to decide which to test first. Fortunately, some guidelines do exist. A good hypothesis exhibits the following:

1. It is reasonably consistent with well established facts.
2. It is capable of being tested; that is, it should generate definite predictions, whether the results are positive or negative. Test results should also be repeatable by independent observers.
3. It is falsifiable, which means it can be proved false.

A hypothesis cannot really be proved true, but in theory (though not necessarily in practice) a well stated hypothesis can be proved false. If one believes in an unfalsifiable hypothesis (e.g., the existence of invisible and undetectable angels), it must be on grounds other than scientific ones.

Consider the following hypothesis: All female mammals (animals that have hair and produce milk for their young) bear live young. The hypothesis was based on the observations that dogs, cats, cows, lions, and humans are all mammals and all bear live young. Consider further that a new species, species X, was identified as a mammal. Biologists predicted that females of species X would bear live young. When a female of the new species gave birth to offspring, this supported the hypothesis. Yet it did not really *prove* the hypothesis.

Before the Southern Hemisphere was explored, most individuals would probably have accepted the hypothesis without question, because all known furry, milk-giving animals did, in fact, bear live young. But it was discovered that two Australian animals (the duck-billed platypus and the spiny anteater) had fur, produced milk for their young, but laid eggs (Fig. 1–17). The hypothesis, as stated, was false no matter how

Figure 1–17 **Is this animal a mammal?** The duck-billed platypus is classified as a mammal because it has fur and produces milk for its young. However, unlike most mammals, it lays eggs. *(Tom McHugh/Photo Researchers, Inc.)*

many times it had previously been supported. As a result, biologists either had to consider the platypus and the spiny anteater as nonmammals, or to broaden their definition of mammals to include them (they chose the latter).

A hypothesis is not true just because some of its predictions (the ones we happen to have thought of or have thus far been able to test) have been shown to be true. After all, they could be true by coincidence. Failure to observe a predicted outcome does not make a hypothesis false, but it does not show that the hypothesis is true, either.

A prediction is a logical consequence of a hypothesis

A hypothesis is an abstract idea, so there is no way to test it directly. But hypotheses suggest certain logical consequences, that is, observable things that cannot be false if the hypothesis is true. On the other hand, if the hypothesis is, in fact, false, other definite predictions should disclose that. As used here, then, a **prediction** is a deductive, logical consequence of a hypothesis. It does not have to be a future event.

Predictions can be tested by experiment

A prediction can be tested by controlled experiments. Early biologists observed that the nucleus was the most prominent part of the cell, and they hypothesized that it might be essential for the well-being of the cell. They predicted that if the nucleus were removed from the cell, the cell would die. Experiments were performed in which the nucleus of a single-celled amoeba was removed surgically with a micro-loop. After this surgery, the amoeba continued to live and move but it did not grow, and after a few days it died. These results suggested that the nucleus is necessary for the metabolic processes that provide for growth and cell reproduction (Fig. 1–18).

But, the investigators asked, what if the operation itself and not the loss of the nucleus caused the amoeba to die? They performed a *controlled* experiment in which two groups of amoebas were subjected to the same operative trauma. However, in the **experimental group** the nucleus was removed, whereas in the **control group** it was not. An experimental group ideally differs from a control group only with respect to the variable being studied. In the control group, a micro-loop was inserted into each amoeba and pushed around inside the cell to simulate the removal of the nucleus; then the needle was withdrawn, leaving the nucleus inside. Amoebas treated with such a sham operation recovered and subsequently grew and divided, but the amoebas without nuclei died. This experiment provided data that it was the removal of the nucleus and not simply the operation that caused the death of the amoebas. The data supported the hypothesis that the nucleus is essential for the well-being of the cell.

Let us consider another example of a scientific study. Steven N. Blair and his colleagues reported in 1996 in the *Journal of the American Medical Association* the results of their study of the relationship between cardiac fitness and cardio-

(a) Experimental

(b) Control

Figure 1–18   **Testing a prediction.** A controlled experiment tested the prediction that if the nucleus is removed from a cell, the cell would die. The data gathered from this and similar experiments provided support for the hypothesis that the nucleus is essential for the well-being of the cell. *(a)* When its nucleus is surgically removed with a micro-loop, the amoeba dies. *(b)* Control amoebas subjected to similar surgical procedures (including insertion of a micro-loop), but without actual removal of the nucleus, do not die.

vascular disease. The subjects were more than 25,000 men and 7,000 women, ages 20 to 88 years, who had completed a preventive medical examination. The investigators used treadmill time to group participants into fitness categories and then followed these subjects over several years. Outcome measures were cardiovascular disease and mortality from all causes. The investigators reported that low fitness and cigarette smoking were significant predictors of mortality in men and women. Moderately fit men and women had a significantly lower death rate compared to those in the low-fitness group.

In scientific studies, care must be taken to avoid **bias.** For example, to prevent bias, most medical experiments today are carried out in **double-blind** fashion. When a drug is being tested, one group of patients is given the new medication, while a second similar group of patients is given a placebo, a harmless starch pill similar in size, shape, color, and taste to the pill being tested. This is a double-blind study because neither the patient nor the physician knows who is getting the experimental drug and who is getting the placebo. The pills or treatments are coded in some way, and only after the experiment is over and the results are recorded is the code broken. Not all experiments can be so neatly designed; for one thing, it is often difficult to establish appropriate controls.

Scientists interpret the results of experiments and make conclusions

Scientists gather data in an experiment, interpret their results, and then formulate conclusions. For example, in the amoeba experiment described above, investigators concluded that the nucleus was essential for the well-being of the cell. One reason for inaccurate conclusions is **sampling error.** Since *all* cases of what is being studied cannot be observed or tested (scientists cannot study every amoeba), we must be content with a sample, or subset, of them. Yet how can we know whether that sample is truly representative of whatever we are studying? In the first place, if the sample is too small, it may be different due to random factors. A study with only two, or even ten, amoebas might not yield reliable data that could be generalized to other amoebas. This problem can usually be solved by using large numbers of subjects and applying the mathematics of statistical analysis (Fig. 1–19).

We must also ensure that the sample is typical of the group that we intend to study. Scientists use statistical techniques to ensure that there is no consistent bias in the way that experimental samples are chosen.

Even if a conclusion is based on results from a carefully designed experiment, it is still possible that new observations or results from other experiments can challenge the conclusion. If we test a large number of cases, we are more likely to draw accurate scientific conclusions. The scientist seeks to state with confidence that any specific conclusion has a certain statistical probability of being correct.

Experiments must also be replicated. Approximately one year after Wilmut (see Chapter Opener) reported that he had cloned a lamb from adult DNA, some biologists challenged his work. Wilmut's critics were concerned that he had not repeated his experiment and that his results had not yet been replicated by other biologists. They pointed out that Wilmut himself reported that his cloning efforts had been successful only once out of about 400 attempts. Before scientists fully accept the experimental results, the study must be repeated in at least one other laboratory.

A well-supported hypothesis may lead to a theory

Nonscientists often use the word *theory* incorrectly to refer to a hypothesis. A **theory** can be developed only when a hypothesis has been supported by consistent results from many observations or experiments. A good theory relates facts that previously appeared to be unrelated. A good theory grows, building on additional facts as they become known. It predicts new facts and suggests new relationships among phenomena. It may even suggest practical applications.

Figure 1–19 **Statistical probability.** Taking a single sample can result in sampling error. The greater the number of samples we take of an unknown, the more likely we can make valid assumptions about it.

A good theory, by showing the relationships among classes of facts, simplifies and clarifies our understanding of natural phenomena. As Einstein wrote, "In the whole history of science from Greek philosophy to modern physics, there have been constant attempts to reduce the apparent complexity of natural phenomena to simple, fundamental ideas and relations."

A theory that, over a long period of time, has yielded true predictions and is thus almost universally accepted is referred to as a scientific **principle.** The term **law** is sometimes used for a principle judged to be of great basic importance, such as the law of gravity.

Science has ethical dimensions

Researchers who publish their work in scientific journals describe their experiments in sufficient detail to be independently performed by others. This permits objective observers to detect errors or bias in the original study and helps to guard against the occasional odd result caused by random or uncontrolled factors, as well as those tainted by dishonesty on the part of the original researcher.

Scientific investigation depends on commitment to such practical ideals as truthfulness and the obligation to communicate results. Honesty is particularly important in science. Consider the great (though temporary) damage done whenever an unprincipled or even desperate researcher, whose career might depend on publication of a research study, knowingly disseminates false data. Until the deception is uncovered, researchers might devote many thousands of dollars or hours of precious professional labor to futile lines of research inspired by erroneous reports.

Fortunately, science tends to be self-correcting through the consistent use of the scientific process itself. Sooner or later, someone's experimental results are bound to cast doubt on false data.

Scientists face many important ethical issues surrounding such issues as cloning, research on human embryos, human and animal experimentation, and applications of genetic engineering. Many of these concerns will be discussed in this book.

S U M M A R Y   W I T H   K E Y   T E R M S

I. **Biology** is the study of life. Basic themes of biology include the evolution of life, the transmission of information, and the flow of energy through organisms.

II. A living organism is able to grow and develop, carry on self-regulated metabolism, move, respond to stimuli, and reproduce. In addition, species evolve and adapt to their environment.
   A. All living organisms are composed of one or more **cells.**
   B. Organisms grow by increasing the size and number of their cells.
   C. **Metabolism** refers to all the chemical activities that take place in the organism, including the chemical reactions essential to nutrition, growth and repair, and the conversion of energy to usable forms. **Homeostasis** is the tendency of organisms to maintain a constant internal environment.
   D. Movement, although not necessarily locomotion, is characteristic of living organisms. Some organisms use tiny extensions of the cell called **cilia,** or longer **flagella,** to move from place to place. Other organisms are **sessile** and remain rooted to some surface.
   E. Organisms respond to **stimuli,** physical or chemical changes in their external or internal environment.
   F. In **asexual reproduction** offspring are typically identical to the single parent; in **sexual reproduction** offspring are typically the product of genes contributed by two parents.
   G. Populations evolve and become adapted to their environment. **Adaptations** are traits that increase an organism's ability to survive in its environment.

III. Organisms transmit information chemically, electrically, and behaviorally.
   A. DNA, which makes up the **genes,** contains the instructions for the development of an organism and for carrying out life processes.
      1. Information encoded in **DNA** is transmitted from one generation to the next.
      2. DNA codes for **proteins,** which are important in determining the structure and function of cells and tissues.
   B. **Hormones** are chemical messengers that transmit messages from one part of an organism to another.
   C. Many organisms use electrical signals to transmit information; most animals have nervous systems that transmit electrical impulses and release **neurotransmitters.**

IV. **Evolution** is the process by which populations of organisms change over time in response to changes in the environment.
   A. **Natural selection,** the mechanism by which evolution proceeds, favors organisms with traits that enable them to cope with environmental changes. These organisms are most likely to survive and to produce offspring.
   B. Charles Darwin based his theory of natural selection on his observations that individuals of a species vary; organisms produce more offspring than survive to reproduce; individuals that are best adapted to their environment are more likely to survive and reproduce; as successful organisms pass on their genes, their traits become more widely distributed in the population.
   C. The source of variation in a population is random **mutation.**

V. Biological organization is hierarchical.
   A. A complex **organism** is organized at the **chemical, cellular, tissue, organ,** and **organ system levels.**
   B. The basic unit of ecological organization is the **population.** Various populations form **communities;** a community and its physical environment are referred to as an **ecosystem;** all of our planet's communities and ecosystems together make up the **biosphere.**

VI. Millions of species have evolved on our planet.
   A. Taxonomic classification is hierarchical; it includes **species, genus, family, order, class, phylum,** and **kingdom.**
   B. Biologists use a **binomial system of nomenclature** in which the name of each species includes a genus name and a **specific epithet.**
   C. Bacteria have **prokaryotic cells;** all other organisms have **eukaryotic cells.**
   D. Organisms can be classified into six kingdoms: **Archaebacteria, Eubacteria, Protista** (protozoa, algae, water molds, and slime molds), **Fungi** (molds and yeasts), **Plantae,** and **Animalia.**

VII. Life depends on continuous energy input from the Sun. Activities of living cells require energy.
   A. During **photosynthesis** plants and many other types of producers use the energy of sunlight to synthesize complex molecules from carbon dioxide and water.
   B. During **cellular respiration,** cells capture the energy stored in nutrients by producers. Some of that energy is then used to synthesize needed materials or to carry on other cell activities.
   C. A self-sufficient ecosystem includes **producers,** or **autotrophs,** that make their own food, **consumers** that eat producers or organisms that have eaten producers, and **decomposers** that obtain energy by breaking down wastes and dead organisms. Consumers and decomposers are **heterotrophs,** organisms that depend on producers as an energy source and for food and oxygen.

VIII. The **process of science** is a dynamic approach to investigation. The **scientific method** is a framework that scientists use in their work; it includes observing, recognizing a problem or stating a critical question, developing a hypothesis, making a prediction that can be tested, performing experiments, interpreting results, and drawing conclusions that support or falsify the hypothesis.

A. **Deductive reasoning** and **inductive reasoning** are two categories of systematic thought processes that can be used in the scientific method.
B. A **hypothesis** is a testable statement about the nature of an observation or relationship.
C. A properly designed scientific experiment includes both a **control group** and an **experimental group,** and must be as free as possible from **bias.** The experimental group differs from a control group only with respect to the variable being studied.
D. When a hypothesis has been supported by conclusions from many experiments, scientists may develop a **theory** based on it. A well established and tested theory may be referred to as a scientific **principle.**
E. Science has important ethical dimensions.

POST-TEST

1. Metabolism (a) is the sum of all the chemical activities of an organism (b) results from an increase in the number of cells (c) is characteristic of plant and animal kingdoms only (d) refers to chemical changes in an organism's environment (e) does not take place in producers
2. Homeostasis (a) is the tendency of organisms to maintain a constant internal environment (b) generally depends on the action of cilia (c) is the long-term response of organisms to changes in their environment (d) occurs at the ecosystem level, not in cells or organisms (e) may be sexual or asexual
3. Structures used by some organisms for locomotion are (a) cilia (b) flagella (c) nuclei (d) answers a, b, and c are correct (e) answers a and b only are correct
4. The splitting of an amoeba into two is best described as an example of (a) locomotion (b) neurotransmission (c) asexual reproduction (d) sexual reproduction (e) metabolism
5. Cells (a) are the building blocks of living organisms (b) always have nuclei (c) are not found among the bacteria (d) answers a, b, and c are correct (e) answers a and b only are correct
6. An increase in size or number of cells best describes (a) homeostasis (b) biological growth (c) the chemical level of organization (d) asexual reproduction (e) adaptation
7. DNA (a) makes up the genes (b) transmits information from one species to another (c) cannot be changed (d) is a neurotransmitter (e) is produced during cellular respiration
8. Cellular respiration (a) is a process whereby sunlight is used to synthesize cellular components with the release of energy (b) occurs in heterotrophs only (c) is carried on by both autotrophs and heterotrophs (d) causes chemical changes in DNA (e) occurs in response to environmental changes
9. Which of the following is a correct sequence of levels of biological organization? (a) cellular, organ, tissue, organ system (b) chemical, cellular, organ, tissue (c) chemical, cellular, tissue, organ (d) tissue, organ, cellular, organ system (e) chemical, cellular, population, species
10. Which of the following is a correct sequence of levels of biological organization? (a) organism, population, ecosystem, community (b) organism, population, community, ecosystem (c) population, biosphere, ecosystem, community (d) species, population, ecosystem, community (e) population, species, community, biosphere
11. Darwin suggested that evolution takes place by (a) mutation (b) changes in the individuals of a species (c) natural selection (d) interaction of hormones (e) homeostatic responses to each change in the environment
12. Protozoa are assigned to kingdom (a) Protista (b) Fungi (c) Archaebacteria (d) Animalia (e) Plantae
13. Yeasts and molds are assigned to kingdom (a) Protista (b) Fungi (c) Archaebacteria (d) Animalia (e) Plantae
14. In the binomial system of nomenclature, the first part of an organism's name designates the (a) specific epithet (b) genus (c) class (d) kingdom (e) phylum
15. Which of the following is a correct sequence of levels of classification? (a) genus, species, family, order, class, phylum, kingdom (b) genus, species, order, phylum, class, kingdom (c) genus, species, order, family, class, phylum, kingdom (d) species, genus, family, order, class, phylum, kingdom (e) species, genus, order, family, class, kingdom, phylum
16. A testable statement is a(an) (a) theory (b) hypothesis (c) principle (d) inductive leap (e) critical question
17. Ideally, an experimental group differs from a control group (a) only with respect to the hypothesis being tested (b) only with respect to the variable being studied (c) by being less subject to bias (d) because it is less vulnerable to sampling error (e) because its subjects are more reliable.

REVIEW QUESTIONS

1. Contrast a living organism with a nonliving object.
2. In what ways might the metabolisms of an oak tree and a tiger be similar? Relate these similarities to the biological themes of transmission of information, energy, and evolution.
3. What would be the consequences if an organism's homeostatic mechanisms failed? Explain your answer.
4. What components do you think might be present in a balanced forest ecosystem? In what ways are consumers dependent on producers? On decomposers? Include energy considerations in your answer.
5. Why do you suppose that the binomial system of nomenclature has survived for more than 200 years and is still used by biologists?
6. How might you explain the sharp claws and teeth of tigers in terms of natural selection?
7. What is meant by a "controlled" experiment?
8. Make a prediction and devise a suitably controlled experiment to test each of the following hypotheses: (a) A type of mold found in your garden does not produce an effective antibiotic. (b) The rate of growth of a bean seedling is affected by temperature. (c) Estrogen alleviates the symptoms of Alzheimer's disease in elderly women.

YOU MAKE THE CONNECTION

1. How might a firm understanding of evolutionary processes be helpful to a biologist who is doing research in (a) animal behavior, (b) ecology, and (c) the development of a vaccine against HIV, the virus that causes AIDS?
2. If you could influence U. S. policy on continuing cloning research, what position would you take? Explain. What would be your position on the use of gene-modified animals for producing drugs like insulin or blood-clotting factors?

RECOMMENDED READINGS

Cohen, J. "Can Cloning Help Save Beleaguered Species?" *Science,* Vol. 276, 30 May, 1997. A brief discussion of the benefits and concerns of cloning endangered species.

Mirsky, S., and J. Rennie. "What Cloning Means for Gene Therapy," *Scientific American,* Vol. 276, Jun. 1997. A special report about the benefits of combining cloning technology with other biotechnologies.

Moore, J.A. *Science as a Way of Knowing: The Foundations of Modern Biology.* Harvard University Press, Cambridge, 1993. An account of scientific thought as related to the history of modern biology.

Science, Vol. 277, 25 Jul. 1997. Special Issue: Human-Dominated Ecosystems. The authors examine the global consequences of human activity on several ecosystems.

Scientific American, Vol. 273, No. 3, Sep. 1995. Special Issue: Key Technologies for the 21st Century. This issue includes several interesting articles on medical technology and on "Energy and the Environment."

Velander. W.H., Lubon, H., and W.N. Drohan. "Transgenic Livestock as Drug Factories," *Scientific American,* Vol. 276, No. 1, Jan. 1997. A discussion of some exciting new medical applications of genetic techniques.

● Visit our website at **http://www.saunderscollege.com/lifesci/titles.html** and click on Solomon/Berg/Martin Biology for links to chapter-related resources on the World Wide Web.

CHAPTER 2

Atoms and Molecules: The Chemical Basis of Life

This jaguar and the plants of the rain forest, as well as an abundance of insects and microorganisms, share fundamental similarities in their chemical composition and basic metabolic processes. These chemical similarities provide strong evidence for the evolution of all organisms from a common ancestor and explain why much of what biologists learn from studying bacteria or rats in laboratories can be applied to other organisms, including humans. Furthermore, the basic physical principles governing organisms are not unique to living things, for they apply to nonliving systems as well.

Today much attention is given to **molecular biology**—the chemistry and physics of the molecules that constitute living things. A molecular biologist might study how proteins interact with DNA in ways that control the expression of certain genes, or might investigate the precise interactions among a cell's atoms and molecules that maintain the energy flow essential to life. However, an understanding of chemistry is essential to *all* biologists. An evolutionary biologist might study evolutionary relationships by comparing proteins produced by different types of organisms. An ecologist might study the biological effects of changes in the salinity of the water in an estuary. A botanist might be a "chemical prospector," seeking new sources of medicines from plants.

In this chapter we lay a foundation for understanding how the structure of atoms determines the way they form chemical bonds to produce complex compounds. Most of our discussion will center around small, simple substances known as **inorganic compounds.** Among the biologically important groups of inorganic compounds are water, simple acids and bases, and simple salts. We pay particular attention to water, the most abundant substance on Earth's surface and in organisms, and we examine how its unique properties affect living things as well as their nonliving environment. In Chapter 3 we extend our discussion to **organic compounds,** which are generally large and complex and which always contain carbon atoms joined together to form the backbone, or skeleton, of the molecule.

(Frans Lanting/Minden Pictures)

1. Name the principal chemical elements in living things and give an important function of each.
2. Compare the physical properties (mass and charge) and the locations of electrons, protons, and neutrons.
3. Distinguish between the atomic number and the mass number of an element.
4. Define the term *electron orbital,* and relate orbitals to energy levels.
5. Explain how the number of valence electrons of an atom is related to its chemical properties.
6. Distinguish among covalent bonds, hydrogen bonds, and ionic bonds. Compare them in terms of the mechanisms by which they form and their relative bond strengths.
7. Explain how cations and anions form and how they interact.

8. Distinguish between the terms *oxidation* and *reduction* and relate these processes to the transfer of energy.
9. Draw a simple ball-and-stick model of a water molecule, indicating the regions of partial positive and partial negative charge. Show how hydrogen bonds form between adjacent water molecules and explain how these are responsible for many of the properties of water.
10. Contrast acids and bases and discuss their properties.
11. Convert the hydrogen ion concentration (moles per liter) of a solution to a pH value. Describe how buffers help minimize changes in pH.
12. Describe the composition of a salt and explain why salts are important in organisms.

ELEMENTS ARE NOT CHANGED IN NORMAL CHEMICAL REACTIONS

Elements are substances that cannot be broken down into simpler substances by ordinary chemical reactions. Scientists have assigned each element a **chemical symbol:** usually the first letter or first and second letters of the English or Latin name of the element. For example, O is the symbol for oxygen, C for carbon, H for hydrogen, N for nitrogen, and Na for sodium (Latin *natrium*).

Just four elements—oxygen, carbon, hydrogen, and nitrogen—are responsible for over 96% of the mass of most organisms. Others, such as calcium, phosphorus, potassium, and magnesium, are also consistently present but in smaller quantities. Some elements, such as iodine and copper, are known as *trace elements* because they are present only in minute amounts. Table 2–1 lists the elements that make up two representative organisms, a human and a typical nonwoody plant, and briefly explains why each is important.

ATOMS ARE THE FUNDAMENTAL PARTICLES OF ELEMENTS

An **atom** is the smallest portion of an element that retains its chemical properties. Atoms are much smaller than the tiniest particle visible under a light microscope. By special scanning tunneling electron microscopy, with magnification as high as × 5 million, researchers have been able to photograph the positions of some large atoms in molecules.

Physicists have discovered a number of subatomic particles, but for our purposes we need consider only three: electrons, protons, and neutrons. An **electron** is a particle that carries a unit of negative electrical charge; a **proton** carries a unit of positive charge; and a **neutron** is an uncharged particle. In an electrically neutral atom, the number of electrons is equal to the number of protons.

Clustered together, protons and neutrons compose the **atomic nucleus.** Electrons, however, have no fixed locations and move rapidly through space outside the atomic nucleus.

An atom is uniquely identified by its number of protons

Each kind of element has a fixed number of protons in the atomic nucleus. This number, called the **atomic number,** is written as a subscript to the left of the chemical symbol. Thus $_1$H indicates that the hydrogen nucleus contains one proton, and $_8$O that the oxygen nucleus contains eight protons. It is the atomic number, the number of protons in its nucleus, that determines an atom's identity.

The **periodic table** (Fig. 2–1 and Appendix B) is a chart in which elements are arranged in order by atomic number. As will become evident in subsequent discussions, the periodic table is an extremely useful device because it allows us to simultaneously correlate a great many of the relationships among the various elements.

Figure 2–1 includes representations of the **electron configurations** of several elements important in organisms. These *Bohr models,* which show the electrons arranged in a series of concentric circles around the nucleus, are convenient to use, but inaccurate. As we will see, electrons do not actually circle the nucleus in fixed concentric pathways.

Protons plus neutrons determine atomic mass

The mass of a subatomic particle is exceedingly small, much too small to be conveniently expressed in grams or even micrograms. Such masses are expressed in terms of the **atomic**

TABLE 2-1 Elements That Make Up Some Representative Organisms

Element and Chemical Symbol	Approximate % of Total Mass of Human Body	Approximate % of Total Mass of Nonwoody Plant	Importance or Functions
Oxygen (O)	65	78	Required for cellular respiration; present in most organic compounds; component of water
Carbon (C)	18	11	Forms backbone of organic molecules; each carbon atom can form four bonds with other atoms
Hydrogen (H)	10	9	Present in most organic compounds; component of water; hydrogen ion (H^+) is involved in some energy transfers
Nitrogen (N)	3	*	Component of proteins and nucleic acids; component of chlorophyll in plants
Calcium (Ca)	1.5	*	Structural component of bones and teeth; calcium ion (Ca^{2+}) is important in muscle contraction, conduction of nerve impulses, and blood clotting; associated with plant cell wall
Phosphorus (P)	1	*	Component of nucleic acids and of phospholipids in membranes; important in energy transfer reactions; structural component of bone
Potassium (K)	*	*	Potassium ion (K^+) is principal positive ion (cation) in interstitial (tissue) fluid of animals; important in nerve function; affects muscle contraction; controls opening of stomata in plants
Sulfur (S)	*	*	Component of most proteins
Sodium (Na)	*	*	Sodium ion (Na^+) is principal positive ion (cation) in interstitial (tissue) fluid of animals; important in fluid balance; essential for conduction of nerve impulses; not essential in most plants
Magnesium (Mg)	*	*	Needed in blood and other tissues of animals; activates many enzymes; component of chlorophyll in plants
Chlorine (Cl)	*	*	Chloride ion (Cl^-) is principal negative ion (anion) in interstitial (tissue) fluid of animals; important in water balance; essential for photosynthesis
Iron (Fe)	*	*	Component of hemoglobin in animals; activates certain enzymes

*The asterisk indicates that these elements represent less than 1% of the total mass. Other elements found in very small (trace) amounts in animals, plants, or both include iodine (I), manganese (Mn), copper (Cu), zinc (Zn), cobalt (Co), fluorine (F), molybdenum (Mo), selenium (Se), boron (B), silicon (Si), and a few others.

mass unit (amu), also called the **dalton** in honor of John Dalton, who formulated an atomic theory in the early 1800s. One amu is equal to the approximate mass of a proton or neutron. Protons and neutrons make up almost all of the mass of an atom. Each electron has only about 1/1800 of the mass of a proton or neutron.

The **atomic mass** of an atom is a number that indicates how massive it is compared with another atom. This value is determined by adding the number of protons to the number of neutrons and expressing the result in atomic mass units or daltons.[1] The mass of the electrons is ignored because it is so small. The atomic mass number is indicated by a superscript to the left of the chemical symbol. The common form of the oxygen atom, with eight protons and eight neutrons in its nu-

[1]Unlike weight, mass is independent of the force of gravity. For convenience, however, we will consider mass and weight to be equivalent. Atomic weight has the same numerical value as atomic mass, but has no units.

Figure 2–1 **The periodic table.** Note the Bohr models depicting the electron configuration of some biologically important atoms. Although the Bohr model does not depict electron configurations accurately, it is commonly used because of its simplicity and convenience.

cleus, has an atomic number of 8 and a mass of 16 atomic mass units. It is indicated by the symbol $^{16}_{8}O$.

The characteristics of protons, electrons, and neutrons are summarized below:

Particle	Charge	Approximate Mass	Location
Proton	positive	1 amu	nucleus
Neutron	neutral	1 amu	nucleus
Electron	negative	approx. 1/1800 amu	outside nucleus

Isotopes differ in number of neutrons

Most elements consist of a mixture of atoms with different numbers of neutrons and thus different masses. Such atoms are called **isotopes.** Isotopes of the same element have the same number of protons and electrons; only the number of neutrons varies. The three isotopes of hydrogen, $^{1}_{1}H$ (ordinary hydrogen), $^{2}_{1}H$ (deuterium), and $^{3}_{1}H$ (tritium), contain zero, one, and two neutrons, respectively. Bohr models of two isotopes of carbon, $^{12}_{6}C$ and $^{14}_{6}C$, are illustrated in Figure 2–2. The mass of an element is expressed as an average of the masses of its isotopes (weighted by their relative abundance in nature). For example,

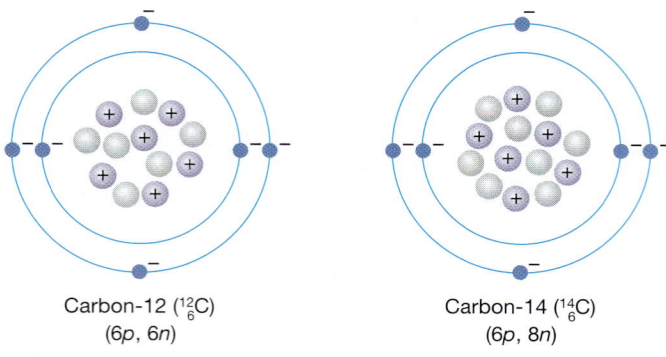

Carbon-12 ($^{12}_6$C)
(6p, 6n)

Carbon-14 ($^{14}_6$C)
(6p, 8n)

Figure 2–2   **Isotopes differ in atomic mass.** Carbon-12 ($^{12}_6$C) is the most common isotope of carbon. Its nucleus contains six protons and six neutrons, so its atomic mass is 12. Carbon-14 ($^{14}_6$C) is a rare radioactive carbon isotope. Because it contains eight neutrons, its atomic mass is 14.

Concentrated silver grains

50 μm

Figure 2–3   **Autoradiography.** The chromosomes of the fruit fly, *Drosophila melanogaster*, shown in this light micrograph have been covered with a special type of x-ray film in which silver grains (*dark spots*) are produced when tritium (^{3}H) undergoes radioactive decay. The concentrations of silver grains (*arrows*) mark the locations of specific DNA molecules.   ©*(Peter J. Bryant/Biological Photo Service)*

the atomic mass of hydrogen is not 1.0 amu, but 1.0079 amu, reflecting the natural occurrence of small amounts of deuterium and tritium.

All isotopes of a given element have essentially the same chemical characteristics. However, some isotopes are unstable and tend to break down, or decay, to a more stable isotope (usually becoming a different element). For example, the radioactive decay of $^{14}_6$C yields the common form of nitrogen, $^{14}_7$N. Such isotopes are termed **radioisotopes** because they emit radiation when they decay. (The radioactive decay of $^{14}_6$C occurs as a neutron decomposes to form a proton and a fast-moving electron, which is emitted from the atom as a form of radiation known as a β particle.) Sophisticated instruments allow scientists to detect and measure this and other types of radiation. Radioactive decay can also be detected by a method known as **autoradiography,** in which radiation causes the appearance of dark silver grains in special x-ray film (Fig. 2–3).

Despite the difference in the number of neutrons, the different isotopes of a given element are usually metabolized and/or localized in the organism in a similar way. For this reason, radioisotopes such as ^{3}H (tritium), ^{14}C, and ^{32}P have been extremely valuable research tools in studies ranging from determining the age of fossils, to DNA synthesis, to sugar transport in plants.

In medicine, radioisotopes are used for both diagnosis and treatment. The location and/or metabolism of a sugar, hormone, or drug can be followed in the body by labeling the substance with a radioisotope such as carbon-14 or tritium. For example, the active component in marijuana (tetrahydrocannabinol, or THC) can be labeled and administered intravenously. Then the amount of radioactivity in the blood and urine can be measured at successive intervals. Results of such measurements have determined that for several weeks this compound remains in the blood, and products of its metabolism can be detected in the urine. Radioisotopes are also used to test thyroid gland function, to measure the rate of red blood cell production, and to study many other aspects of body func-

tion and chemistry. Because radiation can interfere with cell division, radioisotopes have been used in the treatment of cancer (a disease often characterized by rapidly dividing cells).

Electrons occupy orbitals corresponding to energy levels

Electrons move through characteristic regions of three-dimensional space, termed **orbitals.** Each orbital contains a maximum of two electrons. Because it is impossible to know an electron's position at any given time, orbitals are most accurately depicted as "electron clouds," shaded areas whose density is proportional to the probability that an electron is present there at any given instant. The energy of an electron depends on the orbital it occupies. Electrons in orbitals with similar energies, said to be at the same **principal energy level,** make up an **electron shell.** These are illustrated in Figure 2–4.

In general, electrons in a shell distant from the nucleus have greater energy than those in a shell close to the nucleus. This is because energy is required to move a negatively charged electron farther away from the positively charged nucleus. The most energetic electrons, known as **valence electrons,** are said to occupy the **valence shell.** The valence shell is represented as the outermost concentric ring in a Bohr model.

An electron can move to an orbital farther from the nucleus by receiving more energy, or it can give up energy and sink to a lower energy level in an orbital nearer the nucleus. Changes in electron energy levels are important in energy conversions in organisms. For example, in photosynthesis (see Chapter 8) light energy absorbed by chlorophyll molecules causes electrons to move to a higher energy level.

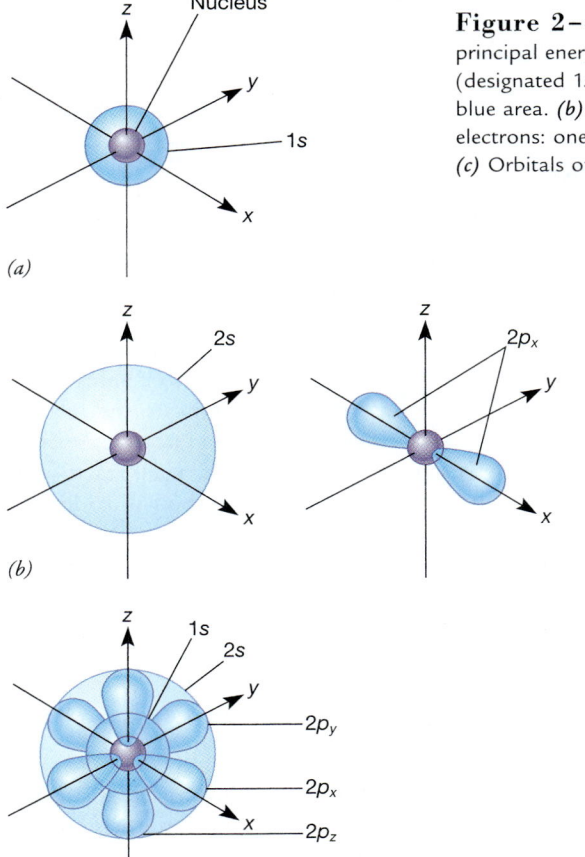

Figure 2-4 **Atomic orbitals.** Each orbital is represented as an "electron cloud." *(a)* The first principal energy level contains a maximum of two electrons, occupying a single spherical orbital (designated 1s). The electrons depicted in the diagram could be present anywhere within the deep blue area. *(b)* The second principal energy level includes four orbitals, each with a maximum of two electrons: one spherical (2s), and three dumbbell-shaped (2p) orbitals at right angles to each other. *(c)* Orbitals of the first and second principal energy levels are shown superimposed.

ATOMS UNDERGO CHEMICAL REACTIONS

The chemical behavior of an atom is determined primarily by the number and arrangement of the valence electrons. The valence shell of hydrogen or helium is full (i.e., stable) when it contains two electrons. The valence shell of any other atom is full when it contains eight electrons. When the valence shell is not full, the atom tends to lose, gain, or share electrons to achieve a full outer shell. (The valence shells of all isotopes of an element are the same; this is why they have similar chemical properties and can substitute for each other in chemical reactions.)

Elements that fall into the same vertical column (said to belong to the same *group*) of the periodic table have similar chemical properties because their valence shells have similar tendencies to lose, gain, or share electrons. For example, chlorine and bromine, included in a group commonly known as the halogens, are highly reactive. Because their valence shells have seven electrons, they tend to gain an electron in chemical reactions. By contrast, hydrogen, sodium, and potassium each have a single valence electron, which they tend to give up or share with another atom. Helium (He) and neon (Ne) belong to a group referred to as the "noble gases." They are quite un-

reactive because their valence shells are full. Note the valence shells of some the elements important in organisms, including carbon, hydrogen, oxygen, and nitrogen, in Figure 2–1.

Atoms form molecules and compounds

Two or more atoms may combine chemically to form units called **molecules.** For example, when two atoms of oxygen combine chemically, a molecule of oxygen is formed. Atoms of *different* elements can combine to form chemical compounds. A **chemical compound** consists of two or more different elements combined in a fixed ratio. For example, water is a chemical compound consisting of hydrogen and oxygen in a ratio of 2:1. Water happens to be a molecular compound, with each molecule consisting of two atoms of hydrogen and one of oxygen. However, as we shall see, not all compounds are made up of molecules.

The properties of a chemical compound can be quite different from those of its component elements: at room temperature, water is usually a liquid, whereas hydrogen and oxygen are gases.

A substance can be described by a chemical formula

A **chemical formula** is a shorthand expression that describes the chemical composition of a substance. Chemical symbols indicate the types of atoms present, and subscript numbers indicate the ratios among the atoms. There are several types of chemical formulas, each providing specific kinds of information.

In a **molecular formula,** the subscripts indicate the actual numbers of each type of atom in a molecule. The formula for molecular oxygen, O_2, tells us that this molecule consists of two atoms of oxygen. The molecular formula for water, H_2O, indicates that each molecule consists of two atoms of hydrogen and one atom of oxygen. (Note that when a single atom of one type is present, the subscript number 1 is never written.)

Another type of formula is the **structural formula,** which shows not only the types and numbers of atoms in a molecule, but also their arrangement. From the molecular formula for water, H_2O, you would not know whether the atoms were arranged H—H—O or H—O—H. The structural formula, H—O—H, settles the matter, indicating that the two hydrogen atoms are attached to the oxygen atom.

One mole of any substance contains the same number of units

The **molecular mass** of a compound is the sum of the atomic masses of the component atoms of a single molecule; thus, the molecular mass of water, H_2O, is $(2 \times 1 \text{ amu}) + (16 \text{ amu})$, or 18 amu. (Owing to the presence of isotopes, atomic mass values are not whole numbers. However, for easy calculation each atomic mass value has been rounded off to a whole number.) The molecular mass of glucose ($C_6H_{12}O_6$), a simple sugar that is a key compound in cellular metabolism, is $(6 \times 12 \text{ amu}) + (12 \times 1 \text{ amu}) + (6 \times 16 \text{ amu})$, or 180 amu.

The amount of an element or compound whose mass in grams is equivalent to its atomic or molecular mass is termed 1 **mole.** Thus 1 mole of glucose has a mass of 180 grams. The mole is a useful concept because it allows us to make meaningful comparisons between atoms and molecules of very different mass. This is because *one mole of any substance always has exactly the same number of units,* whether they be small atoms or large molecules. The very large number of units in a mole, 6.02×10^{23}, is known as **Avogadro's number,** named for the Italian physicist, Amadeo Avogadro, who first calculated it. Thus 1 mole (180 grams) of glucose contains 6.02×10^{23} molecules, as does 1 mole (2 grams) of molecular hydrogen (H_2). Although it is impossible to count atoms and molecules individually, this fact allows a scientist to count them simply by weighing a sample. Molecular biologists usually deal with smaller values, either millimoles (a mmole is one thousandth of a mole) or micromoles (a μmole is one millionth of a mole).

The mole concept allows us to make useful comparisons among solutions. A 1-molar solution, represented by 1 M, contains 1 mole of that substance dissolved in 1 liter of solution. For example, we can compare 1 liter of a 1 M solution of glucose with 1 liter of a 1 M solution of sucrose (table sugar). They differ in the mass of the dissolved sugar (180 g and 340 g, respectively), but they each contain 6.02×10^{23} sugar molecules.

Chemical equations describe chemical reactions

During any moment in the life of an organism, be it a mushroom or a butterfly, many complex chemical reactions are taking place. Chemical reactions—for example, the reaction be-

tween glucose and oxygen—can be described by means of chemical equations:

$$C_6H_{12}O_6 + 6\,O_2 \longrightarrow 6\,CO_2 + 6\,H_2O + \text{Energy}$$

Glucose   Oxygen   Carbon dioxide   Water

In a chemical equation, the **reactants** (the substances that participate in the reaction) are generally written on the left side, and the **products** (the substances formed by the reaction) are written on the right side. The arrow means "yields" and indicates the direction in which the reaction tends to proceed.

Chemical compounds react with each other in quantitatively precise ways. The numbers preceding the chemical symbols or formulas (known as *coefficients*) indicate the relative number of atoms or molecules reacting. For example, 1 mole of glucose burned in a fire or metabolized in a cell reacts with 6 moles of oxygen to form 6 moles of carbon dioxide and 6 moles of water. Many reactions can proceed in the reverse direction (to the left) as well as in the forward direction (to the right); at **equilibrium** the rates of the forward and reverse reactions are equal (see Chapter 6). Reversible reactions are indicated by double arrows:

$$CO_2 + H_2O \rightleftharpoons H_2CO_3$$

Carbon dioxide   Water   Carbonic acid

In this example, the arrows are drawn different lengths to indicate that when the reaction reaches equilibrium there will be more reactants (CO_2 and H_2O) than product (H_2CO_3).

ATOMS ARE JOINED BY CHEMICAL BONDS

The atoms of a compound are held together by forces of attraction called **chemical bonds.** Each bond represents a certain amount of chemical energy. **Bond energy** is the energy necessary to break a bond. The valence electrons dictate how many bonds an atom can participate in. The two principal types of strong chemical bonds are covalent bonds and ionic bonds.

In covalent bonds electrons are shared

Covalent bonds involve the sharing of electrons between atoms in a way that results in each having a filled valence shell. A compound consisting mainly of covalent bonds is called a **covalent compound.** A simple example of a covalent bond is the joining of two hydrogen atoms in a molecule of hydrogen gas, H_2. Each atom of hydrogen has one electron, but two electrons are required to complete its valence shell. The hydrogen atoms have equal capacities to attract electrons, so neither donates an electron to the other. Instead, the two hydrogen atoms share their single electrons so that each of the two electrons is attracted simultaneously to the two protons in the two hydrogen nuclei. The two electrons thus whirl around *both* atomic nuclei, joining the two atoms.

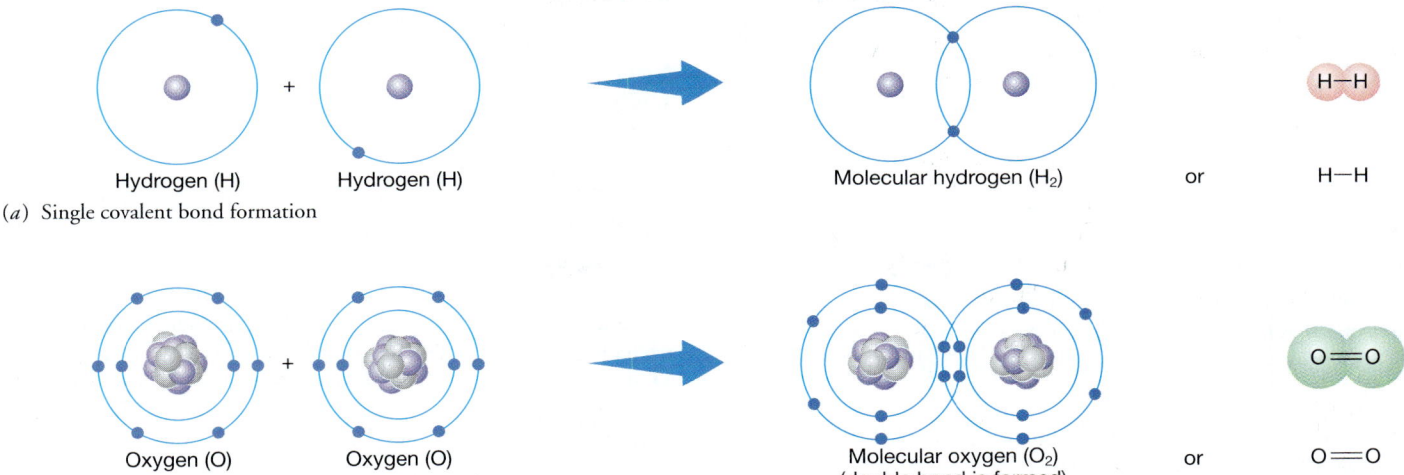

(a) Single covalent bond formation

(b) Double covalent bond formation

Figure 2–5 **Electron sharing in covalent compounds.** *(a)* Two hydrogen atoms achieve stability by sharing electrons, thereby forming a molecule of hydrogen. In the structural formula shown on the right, the straight line between the hydrogen atoms represents a single covalent bond. *(b)* In molecular oxygen, two oxygen atoms share two pairs of electrons, forming a double covalent bond.

A simple way of representing the electrons in the valence shell of an atom is to use dots placed around the chemical symbol of the element. Such a representation is called the *Lewis structure* of the atom. In a water molecule, two hydrogen atoms are covalently bonded to an oxygen atom:

$$\text{H} \cdot + \text{H} \cdot + \cdot \ddot{\text{O}} \cdot \longrightarrow \text{H} \colon \ddot{\text{O}} \colon \text{H}$$

Oxygen has six valence electrons; by sharing electrons with two hydrogen atoms, it completes its valence shell of eight. At the same time each hydrogen atom obtains a complete valence shell of two. (Note that in the structural formula H—O—H, each pair of shared electrons constitutes a covalent bond, represented by a solid line. Unshared electrons are usually omitted in a structural formula.)

The carbon atom has four electrons in its valence shell. These four electrons are available for covalent bonding:

$$\cdot \dot{\text{C}} \cdot$$

When one carbon and four hydrogen atoms share electrons, a molecule of methane, CH_4, is formed:

$$\begin{array}{ccc} & & \text{H} \\ & & | \\ \text{H} \colon \ddot{\text{C}} \colon \text{H} & \text{or} & \text{H}—\text{C}—\text{H} \\ \text{H} & & | \\ & & \text{H} \end{array}$$

Lewis structure   Structural formula

The nitrogen atom has five electrons in its valence shell. Recall that each orbital can hold a maximum of two electrons. Usually two electrons occupy one orbital, leaving three available for sharing with other atoms:

$$\cdot \ddot{\text{N}} \cdot$$

When a nitrogen atom shares electrons with three hydrogen atoms, a molecule of ammonia, NH_3, is formed:

$$\begin{array}{ccc} \text{H} \colon \ddot{\text{N}} \colon \text{H} & \text{or} & \text{H}—\text{N}—\text{H} \\ \text{H} & & | \\ & & \text{H} \end{array}$$

Lewis structure   Structural formula

When one pair of electrons is shared between two atoms, the covalent bond is referred to as a **single covalent bond** (Fig. 2–5*a*). Two oxygen atoms may achieve stability by forming covalent bonds with one another. Each oxygen atom has six electrons in its outer shell. To become stable, the two atoms share two pairs of electrons, forming molecular oxygen (Fig. 2–5*b*). When two pairs of electrons are shared in this way, the covalent bond is called a **double covalent bond,** which is represented by two parallel solid lines. Similarly, a **triple covalent bond** (represented by three parallel solid lines) is formed when three pairs of electrons are shared between two atoms.

The number of covalent bonds usually formed by the atoms commonly present in biologically important molecules is summarized as follows:

Atom	Symbol	Covalent Bonds
Hydrogen	H	1
Oxygen	O	2
Carbon	C	4
Nitrogen	N	3
Phosphorus	P	5
Sulfur	S	2

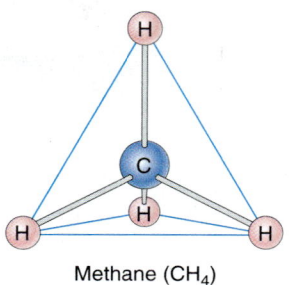

Figure 2–6 Methane. The four hydrogens are located at the corners of a tetrahedron.

The function of a molecule is related to its shape

Each kind of molecule has a characteristic size and a general overall shape. Although the shape of a molecule may change (within certain limits), the functions of molecules in living cells are largely dictated by their geometric shapes. A molecule that consists of two atoms, for example, is linear. Molecules composed of more than two atoms may have more complicated shapes. The geometric shape of a molecule provides the optimal distance between the atoms to counteract the repulsion of electron pairs.

When an atom forms covalent bonds with other atoms, the orbitals in the valence shell may become rearranged, thereby affecting the shape of the resulting molecule. For example, when four hydrogen atoms combine with a carbon atom to form a molecule of methane (CH_4), the valence shell orbitals of the carbon become rearranged such that a geometric structure known as a *tetrahedron* is formed, with one hydrogen atom present at each of its four corners (Fig. 2–6).

Covalent bonds can be nonpolar or polar

The atoms of each element have a characteristic affinity for electrons. **Electronegativity** is a measure of an atom's attraction for electrons in chemical bonds. Very electronegative atoms are sometimes called "electron-greedy." When covalently

bound atoms have similar electronegativities, the electrons are shared equally, and the covalent bond is described as **nonpolar.** The covalent bond of the hydrogen molecule is nonpolar, as are the covalent bonds of molecular oxygen and methane.

In a covalent bond between two different elements, such as oxygen and hydrogen, the electronegativities of the atoms may be different. If so, electrons are pulled closer to the atomic nucleus of the element with the greater electron affinity (in this case, oxygen). A covalent bond between atoms that differ in electronegativity is called a **polar covalent bond.** Such a bond has two dissimilar ends (or poles), one with a partial positive charge and the other with a partial negative charge. Each of the two covalent bonds in water is polar because there is a partial positive charge at the hydrogen end of the bond and a partial negative charge at the oxygen end, where the "shared" electrons are more likely to be found (Fig. 2–7).

Covalent bonds differ in their degree of polarity, ranging from those in which the electrons are exactly shared (as in the nonpolar hydrogen molecule) to those in which the electrons are much closer to one atom than to the other (as in water). Oxygen is quite electronegative and forms polar covalent bonds with carbon, hydrogen, and many other atoms. Nitrogen is also relatively electronegative, although less so than oxygen.

A molecule with one or more polar covalent bonds can be polar even though it is electrically neutral as a whole. A **polar molecule** has one end with a partial positive charge and another end with a partial negative charge. One example is water (Fig. 2–7). The polar bonds between the hydrogens and the oxygens are arranged in a "V" shape, rather than linearly. The oxygen end therefore constitutes the negative pole, and the end with the two hydrogens is the positive pole.

Ionic bonds form between cations and anions

Some atoms or groups of atoms are not electrically neutral. A particle with one or more units of electrical charge is called an **ion.** An atom becomes an ion if it gains or loses one or more electrons. An atom with one, two, or three electrons in its va-

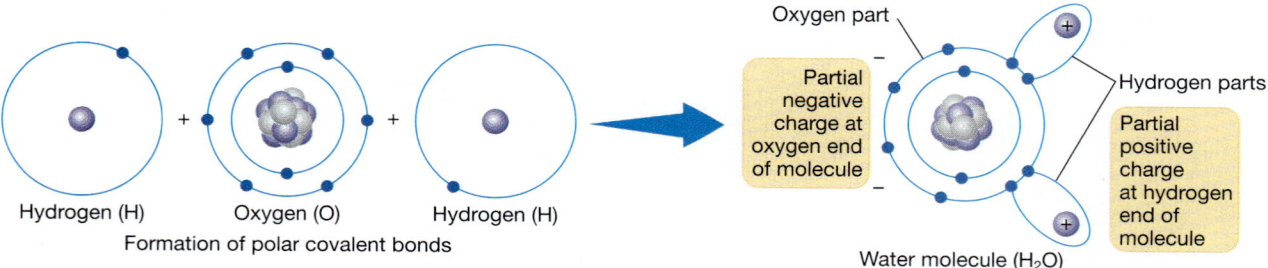

Figure 2–7 Water, a polar molecule. Note that the electrons tend to stay closer to the nucleus of the oxygen atom than to the hydrogen nuclei. This results in a partial negative charge on the oxygen portion of the molecule and a partial positive charge at the hydrogen end. Although the water molecule as a whole is electrically neutral, it is a polar covalent compound.

Figure 2–8 Ions and biological processes. Sodium, potassium, and chloride ions are essential for this nerve cell to stimulate these muscle fibers. Calcium ions in the muscle cell are required for muscle contraction. *(D.W. Fawcett)*

lence shell tends to lose electrons to other atoms. Such an atom then becomes positively charged because its nucleus contains more protons than the number of electrons orbiting around the nucleus. These positively charged ions are termed **cations.** Atoms with five, six, or seven valence electrons tend to gain electrons from other atoms and become negatively charged **anions.**

The properties of ions are very different from those of the electrically neutral atoms from which they were derived. For example, although chlorine gas is a poison, chloride ions (Cl⁻) are essential to life. Because their electrical charges provide a basis for many interactions, cations and anions are involved in energy transformations within the cell, the transmission of nerve impulses, muscle contraction, and many other life processes (Fig. 2–8).

A group of atoms can also become an ion (polyatomic ion). Unlike a single atom, a group of atoms can lose or gain protons (derived from hydrogen atoms) as well as electrons. Therefore, a group of atoms can become a cation if it loses one or more electrons or gains one or more protons. A group of atoms becomes an anion if it gains one or more electrons or loses one or more protons.

An **ionic bond** forms as a consequence of the attraction between the positive charge of a cation and the negative charge of an anion. An **ionic compound** is a substance consisting of anions and cations bonded together by their opposite charges.

A good example of how ionic bonds are formed is the attraction between sodium ions and chloride ions. A sodium atom has one electron in its valence shell. It cannot fill its valence shell by obtaining seven electrons from other atoms, for it would then have a very large unbalanced negative charge. Instead, it gives up its single valence electron to a very electronegative atom, such as chlorine, which acts as an electron acceptor (Fig. 2–9). Chlorine cannot give up the seven electrons in its valence shell, because it would then have a vast positive charge. Instead it strips an electron from an electron donor (sodium in this example) to complete its valence shell.

When sodium reacts with chlorine, its valence electron is transferred completely to chlorine. Sodium is now a cation, with one unit of positive charge (Na⁺). Chlorine is now an anion, a chloride ion with one unit of negative charge (Cl⁻).

11 protons
12 neutrons

17 protons
18 neutrons

Sodium (Na)   and   Chlorine (Cl)   Sodium chloride (NaCl)

Salt crystal

Cl⁻ Na⁺ Cl⁻
Na⁺ Cl⁻ Na⁺
Na⁺

Arrangement of atoms in a crystal of salt

Figure 2–9 Ionic bonding. Sodium becomes a positively charged ion when it donates its single valence electron to chlorine, which has seven valence electrons. With this additional electron, chlorine completes its valence shell and becomes a negatively charged chloride ion. These ions are attracted to one another by their unlike electrical charges, forming the ionic compound sodium chloride.

Figure 2–10 **Hydration of an ionic compound.** When the crystal of NaCl is added to water, the sodium and chloride ions are pulled apart as the partial negative ends of the water molecules are attracted to the positive sodium ions, and the partial positive ends of the water molecules are attracted to the negative chloride ions. When the NaCl is dissolved, each Na^+ and Cl^- is surrounded by water molecules electrically attracted to it.

These ions attract each other as a result of their opposite charges. They are held together by this electrical attraction in ionic bonds to form NaCl, sodium chloride,[2] or common table salt.

The term *molecule* does not adequately explain the properties of ionic compounds such as NaCl. When NaCl is in its solid crystal state, each ion is actually surrounded by six ions of opposite charge. The molecular formula NaCl indicates that sodium ions and chloride ions are present in a one-to-one ratio, but in the actual crystal, no discrete molecules composed of one Na^+ and one Cl^- ion are present.

Compounds joined by ionic bonds, such as sodium chloride, have a tendency to *dissociate* (separate) into their individual ions when placed in water:

$$NaCl \xrightarrow{\text{in } H_2O} Na^+ + Cl^-$$

Sodium chloride   Sodium ion Chloride ion

[2]In both covalent and ionic binary compounds (*binary* denotes compounds consisting of two elements), the element having the greater attraction for electrons is named second, and an -*ide* ending is added to the stem name, e.g., sodium chloride, hydrogen fluoride. The -*ide* ending is also used to indicate an anion, as in chloride (Cl^-) and hydroxide (OH^-).

In the solid form of an ionic compound, the ionic bonds are very strong. Water, however, is an excellent **solvent;** as a liquid it is capable of dissolving many substances, particularly those that are polar or ionic. This is because of the polarity of water molecules. The localized partial positive charges (on the hydrogen atoms) and partial negative charges (on the oxygen atom) on each water molecule attract the anions and cations on the surface of an ionic solid. As a result, the solid dissolves. A dissolved substance is referred to as a **solute.** In solution, each cation and anion of the ionic compound is surrounded by oppositely charged ends of the water molecules (Fig. 2–10). This process is known as **hydration.** Hydrated ions still interact with each other to some extent, but the transient ionic bonds formed are much weaker than those in a solid crystal.

Hydrogen bonds are weak attractions

Another type of bond important in organisms is the **hydrogen bond.** When hydrogen combines with oxygen (or with another relatively electronegative atom such as nitrogen), it has a partial positive charge because its electron spends more time closer to the electronegative atom. Hydrogen bonds tend to form between an atom with a partial negative charge and a hydrogen atom that is covalently bonded to oxygen or nitrogen (Fig. 2–11). The atoms involved may be in two parts of the same large molecule or in two different molecules. Water molecules interact with each other extensively through hydrogen bond formation.

Hydrogen bonds are readily formed and broken. Although individually relatively weak, hydrogen bonds are collectively strong when present in large numbers. Furthermore, they have a specific length and orientation. As we will see in Chapter 3, these features are very important in helping determine the three-dimensional structure of large molecules such as DNA and proteins.

Figure 2–11 **Hydrogen bonding.** A hydrogen atom in a water molecule has a partial positive charge because of its polar covalent bond with oxygen. Nitrogen is relatively electronegative and, in molecules like ammonia (NH_3), has a partial negative charge because of its polar covalent bonds with hydrogen. Here, the nitrogen atom of the ammonia molecule is joined by a hydrogen bond to a hydrogen atom of a water molecule. A hydrogen bond is generally indicated by a dotted line.

ELECTRONS AND THEIR ENERGY ARE TRANSFERRED IN REDOX REACTIONS

Many of the energy conversions that go on in a cell involve reactions in which an electron is transferred from one substance to another. This is because the transfer of an electron also involves the transfer of the energy of that electron. Such an electron transfer is called an oxidation-reduction, or **redox reaction.** Both cellular respiration (Chapter 7) and photosynthesis (Chapter 8) are essentially redox processes.

Rusting, which is the combination of iron (symbol Fe) with oxygen, is a simple illustration of oxidation and reduction:

$$4\,Fe + 3\,O_2 \longrightarrow 2\,Fe_2O_3$$
$$\text{Iron (III) oxide}$$

Oxidation and reduction always occur together, but initially we will discuss them separately. **Oxidation** is a chemical process in which an atom, ion, or molecule loses electrons. In rusting, each iron atom becomes oxidized as it loses three electrons.

$$4\,Fe \longrightarrow 4\,Fe^{3+} + 12\,e^-$$

The e^- is a symbol for an electron; the $+$ superscript represents an electron deficit. (When an atom loses an electron, it acquires one unit of positive charge from the excess of one proton. In our example, each iron atom loses three electrons and acquires three units of positive charge.)

You will recall that the oxygen atom is very electronegative, able to remove electrons from other atoms. In this reaction, oxygen gains electrons from iron.

$$3\,O_2 + 12\,e^- \longrightarrow 6\,O^{2-}$$

Oxygen becomes reduced when it accepts electrons from the iron. **Reduction** is a chemical process in which an atom, ion, or molecule *gains* electrons. (The term reduction refers to the fact that the gain of an electron results in the *reduction* of any positive charge that might be present.)

Redox reactions occur simultaneously because one substance must accept the electrons that are removed from the other. In a redox reaction, one component, the *oxidizing agent,* accepts one or more electrons and becomes reduced. Oxidizing agents other than oxygen are known, but oxygen is such a common one that its name was given to the process. Another reaction component, the *reducing agent,* gives up one or more electrons and becomes oxidized.

In our example there was a complete transfer of electrons from iron to oxygen. Some redox reactions are not so obvious, however. In these, electrons simply move farther from the reducing agent and closer to the oxidizing agent.

Electrons are not easily removed from covalent compounds unless an entire atom is removed. In cells, oxidation often involves the removal of a hydrogen *atom* (an electron plus a proton that "goes along for the ride") from a compound; reduction often involves the addition of hydrogen (see Chapter 6).

WATER IS ESSENTIAL TO LIFE

A large part of the mass of most organisms is water. In human tissues the percentage of water ranges from 20% in bones to 85% in brain cells. About 70% of our total body weight is water; as much as 95% of a jellyfish and certain plants is water. Water is the source, through photosynthesis (see Chapter 8), of the oxygen in the air we breathe, and its hydrogen atoms become incorporated into many organic compounds. Water is also the solvent for most biological reactions and a reactant or product in many chemical reactions.

Water is not only important inside organisms but also is one of the principal environmental factors affecting them. Many organisms live within the ocean or in freshwater rivers, lakes, or puddles. Water's unique combination of physical and chemical properties has permitted living things to originate, survive, and evolve on Earth (Fig. 2–12).

Water molecules are polar

As discussed previously, water molecules are polar, that is, one end of each molecule bears a partial positive charge and the other a partial negative charge (see Fig. 2–7). The water molecules in liquid water and in ice associate by hydrogen bonds. The hydrogen atom of one water molecule, with its partial positive charge, is attracted to the oxygen atom of a neighboring water molecule, with its partial negative charge, forming a hydrogen bond. An oxygen in a water molecule has two regions of partial negative charge, and each of the two hydrogens has a partial positive charge. Each water molecule can therefore form hydrogen bonds with a maximum of four neighboring water molecules (Fig. 2–13).

(a)   100 µm   *(b)*   10 µm

Figure 2–12   **Tardigrade.** *(a)* Commonly known as "water bears," tardigrades such as these members of the genus *Echiniscus* are small (less than 1.2 mm long) animals that normally live in moist habitats, such as thin films of water on mosses. *(b)* When subjected to dessication tardigrades assume a barrel-shaped form known as a *tun,* remaining in this state, motionless but alive, for as long as 100 years. When rehydrated they assume their normal appearance and activities. *(a, Diane R. Nelson; b, Robert O. Schuster, courtesy of Diane R. Nelson)*

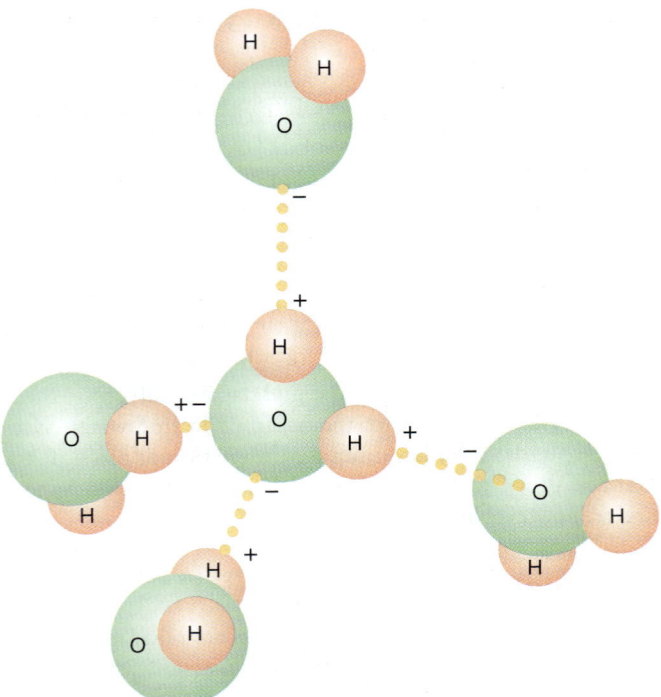

Figure 2–13 **Hydrogen bonding of water molecules.** Each water molecule can form hydrogen bonds (dotted lines) with as many as four neighboring water molecules.

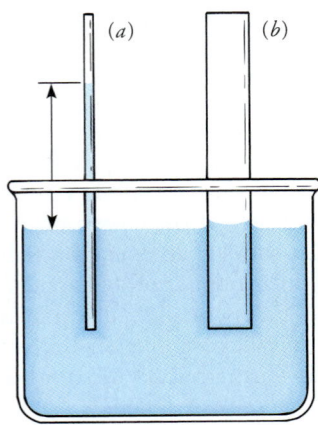

Figure 2–14 **Capillary action.** *(a)* In a narrow tube, adhesive forces attract water molecules to the glass wall of the tube. Other water molecules inside the tube are then "pulled along" by cohesive forces, which are actually due to hydrogen bonds between the water molecules. *(b)* In the wider tube, a smaller percentage of the water molecules line the glass wall. Because of this, the adhesive forces are not strong enough to overcome the cohesive forces of the water beneath the surface level of the container, and water in the tube rises only slightly.

Water is the principal solvent in organisms

Because its molecules are polar, water is an excellent solvent, a liquid capable of dissolving many different kinds of substances, especially polar and ionic compounds. Previously in this chapter, we discussed how polar water molecules pull the ions of ionic compounds apart so that they dissociate (see Fig. 2–10). Because of its solvent properties and the tendency of the atoms in certain compounds to form ions when in solution, water plays an important role in facilitating chemical reactions. Substances that interact readily with water are said to be **hydrophilic** ("water-loving"). Examples include table sugar (a polar compound) and NaCl (an ionic compound), which dissolve readily in water. Not all substances in organisms are hydrophilic, however. Many **hydrophobic** ("water-fearing") substances found in living things are especially important because of their ability to form structures that are not dissolved by water. Examples, to be discussed more fully in Chapter 3, include fats and other nonpolar substances.

Hydrogen bonding makes water cohesive and adhesive

Water molecules have a very strong tendency to stick to each other; that is, they are **cohesive.** This is due to the hydrogen bonds among the molecules. Water molecules also stick to many other kinds of substances, most notably those with charged groups of atoms or molecules on their surfaces. These **adhesive** forces explain how water makes things wet.

A combination of adhesive and cohesive forces accounts for the tendency, termed **capillary action,** of water to move in narrow tubes, even against the force of gravity (Fig. 2–14). For example, water moves through the microscopic spaces between soil particles to the roots of plants by capillary action. Because of the cohesive nature of water molecules, any force exerted on part of a column of water will be transmitted to the column as a whole. The major mechanism of water movement in plants (see Chapter 33) depends on this fact.

Water has a high degree of **surface tension** because of the cohesiveness of its molecules, which have a much greater attraction for each other than for molecules in the air. Thus, water molecules at the surface crowd together, producing a strong layer as they are pulled downward by the attraction of other water molecules beneath them (Fig. 2–15).

Water helps maintain a stable temperature

Raising the temperature of a substance involves adding heat energy to make its molecules move faster, that is, to increase the **kinetic energy** (energy of motion) of the molecules (see Chapter 6). The term **heat** refers to the *total* amount of kinetic energy in a sample of a substance; **temperature** refers to the *average* kinetic energy of the particles. Water has a high **specific heat;** that is, the amount of energy required to raise the temperature of water is quite large. A **calorie** is a unit of heat energy (equivalent to 4.184 joules) that equals the amount

Figure 2–15 **Surface tension of water.** Hydrogen bonding between water molecules is responsible for the surface tension of water, which is strong enough to support these water striders, and causes the dimpled appearance of the surface. Although they are denser than water, water striders can walk on the surface of a pond because fine hairs at the ends of their legs spread their weight over a large area. *(Dennis Drenner)*

of heat required to raise the temperature of 1 gram of water 1 degree Celsius. The specific heat of water is therefore 1 calorie per gram of water per degree Celsius. Most other common substances have much lower specific heat values.

The high specific heat of water results from the hydrogen bonding of its molecules. Some of the hydrogen bonds holding the water molecules together must first be broken to permit the molecules to move more freely. Much of the energy added to the system is used up in breaking the hydrogen bonds, and only a portion of the heat energy is available to speed the movement of the water molecules (thereby increasing the temperature of the water). Conversely, when liquid water changes

to ice, additional hydrogen bonds must be formed, liberating a great deal of heat into the environment.

Because so much heat input is required to raise the temperature of water (and so much heat is lost when the temperature is lowered), the ocean and other large bodies of water have relatively constant temperatures. Thus, many organisms living in the ocean are provided with a relatively constant environmental temperature. The properties of water are crucial in stabilizing temperatures on the surface of Earth. Although surface water is only a thin film relative to Earth's volume, the quantity is enormous compared to the exposed land mass. This relatively large mass of water resists both the warming effect of heat and the cooling effect of low temperatures. In addition, hydrogen bonding gives ice unique properties that have important environmental consequences (see *Making the Connection: Hydrogen Bonding and the Environment*).

The high water content of organisms helps them maintain relatively constant internal temperatures. Such minimizing of temperature fluctuations is important because biological reactions can take place only within a relatively narrow temperature range.

Because its molecules are held together by hydrogen bonds, water has a high **heat of vaporization.** To change 1 gram of liquid water into 1 gram of water vapor, 540 calories of heat are required. The heat of vaporization of most other common liquid substances is much less. As a sample of water is heated, some molecules are moving much faster than others (that is, they have more heat energy). These faster moving molecules are more likely to escape the liquid phase and enter the vapor phase (Fig. 2–16). When they do, they take their heat energy with them (thus lowering the temperature of the sample, in a process called **evaporative cooling**). For this reason the human body can dissipate excess heat as sweat evaporates from the skin, and a leaf can keep cool in the bright sunlight as water evaporates from its surface.

MAKING THE CONNECTION

HYDROGEN BONDING AND THE ENVIRONMENT

Why does ice float? This is because hydrogen bonds contribute to another important property of water. Liquid water expands as it freezes because the hydrogen bonds joining the water molecules in the crystalline lattice keep the molecules far enough apart to give ice a density about 10% less than the density of liquid water (see Fig. 2–16). When ice has been heated enough to raise its temperature above 0° C (32° F), these hydrogen bonds among the water molecules are broken, freeing the molecules to slip closer together. The density of water is greatest at 4° C, above which water begins to expand again as the speed of its molecules increases. As a result, ice floats on the denser cold water.

This unusual property of water has been important in enabling life as we know it to appear, survive, and evolve on Earth. If ice had a greater density than water, it would sink; eventually all ponds, lakes, and even the ocean would freeze solid from the bottom to the surface, making life impossible. When a body of deep water cools, it becomes covered with floating ice. The ice insulates the liquid water below it, preventing it from freezing and permitting a variety of animals and plants to survive below the icy surface.

Figure 2–16 Three forms of water. *(a)* When water boils, as in this hot spring at Yellowstone National Park, many hydrogen bonds are broken, causing steam, consisting of minuscule water droplets, to form. If most of the remaining hydrogen bonds are subsequently broken, the molecules begin to move freely and rapidly as water vapor (a gas). *(b)* Water molecules in a liquid state continually form, break, and re-form hydrogen bonds with each other. *(c)* In ice, each water molecule participates in four hydrogen bonds with adjacent molecules, resulting in a regular, evenly distanced crystalline lattice structure. Because the water molecules move apart slightly as the hydrogen bonds form, water expands as it freezes; thus ice floats on water. *(a, Woodbridge Wilson/National Park Service; b, Gary R. Bonner; c, Barbara O'Donnell/Biological Photo Service)*

(a) Steam becoming water vapor (gas)

(b) Water (liquid)

(c) Ice (solid)

100° C

50° C

0°C

ACIDS ARE PROTON DONORS; BASES ARE PROTON ACCEPTORS

Water molecules have a slight tendency to **ionize,** that is, to dissociate into hydrogen ions (H^+) and hydroxide ions (OH^-).[3] In pure water, a very small number of water molecules ionize. This slight tendency of water to dissociate is reversible as hydrogen ions and hydroxide ions reunite to form water:

$$HOH \rightleftharpoons H^+ + OH^-$$

[3]The H^+ immediately combines with a negatively charged region of a water molecule, forming a hydronium ion (H_3O^+). However, by convention H^+, rather than the more accurate H_3O^+, is used.

Because each water molecule splits into one hydrogen ion and one hydroxide ion, the concentrations of hydrogen ions and hydroxide ions in pure water are exactly equal (0.0000001 or 10^{-7} moles per liter for each ion). Such a solution is said to be **neutral,** neither acidic nor basic (alkaline).

An **acid** is a substance that dissociates in solution to yield hydrogen ions (H^+) and an anion.

$$\text{Acid} \longrightarrow H^+ + \text{Anion}$$

An acid is a proton *donor*. (Recall that a hydrogen ion, or H^+, is nothing more than a proton.) An acidic solution has a hydrogen ion concentration that is higher than its hydroxide ion concentration. Acidic solutions turn blue litmus paper red and have a sour taste. Hydrochloric acid (HCl) and sulfuric acid (H_2SO_4) are examples of inorganic acids. Lactic acid ($CH_3CHOHCOOH$) from sour milk and acetic acid (CH_3COOH) from vinegar are two common organic acids.

A **base** is defined as a proton *acceptor*. Most bases are substances that dissociate to yield a hydroxide ion (OH^-) and a cation when dissolved in water. A hydroxide ion can act as a base by accepting a proton (H^+) to form water. Sodium hydroxide (NaOH) is a common inorganic base.

$$NaOH \longrightarrow Na^+ + OH^-$$
$$OH^- + H^+ \longrightarrow H_2O$$

Some bases do not dissociate to yield hydroxide ions directly. For example, ammonia (NH_3) acts as a base by accepting a proton from water, producing an ammonium ion (NH_4^+) and releasing a hydroxide ion.

$$NH_3 + H_2O \longrightarrow NH_4^+ + OH^-$$

A basic solution is one in which the hydrogen ion concentration is lower than the hydroxide ion concentration. Basic solutions turn red litmus paper blue and feel slippery to the touch. In later chapters we will encounter a number of organic bases, such as the purine and pyrimidine bases that are components of nucleic acids.

pH is a convenient measure of acidity

The degree of a solution's acidity is generally expressed in terms of **pH,** defined as the *negative logarithm (base 10) of the hydrogen ion concentration (expressed in moles per liter).*

$$pH = -\log_{10}[H^+]$$

The brackets refer to concentration; therefore the term $[H^+]$ means "the concentration of hydrogen ions," which is expressed in moles per liter because we are interested in the *number* of hydrogen ions per liter. Because the range of possible pH values is very broad, a logarithmic scale (with a tenfold difference between successive units) is more convenient than a linear scale.

Hydrogen ion concentrations are nearly always less than 1 mole per liter. One gram of hydrogen ions dissolved in 1 liter of water (a 1-molar solution) may not sound very impressive, but such a solution would be extremely acidic. The logarithm

TABLE 2-2	The Relationship Between pH and Hydrogen Ion Concentration		
Substance	$[H^+]$*	log $[H^+]$	pH
Gastric juice	0.01, 10^{-2}	−2	2
Pure water, neutral solution	0.0000001, 10^{-7}	−7	7
Household ammonia	0.00000000001, 10^{-11}	−11	11

*$[H^+]$ = hydrogen ion concentration (moles/L)

of a number less than one is a negative number; thus the *negative* logarithm corresponds to a *positive* pH value.

Whole number pH values are easy to calculate (Table 2–2). For instance, consider our example of pure water, which has a hydrogen ion concentration of 0.0000001 (10^{-7}) moles per liter. The logarithm is −7. The negative logarithm is 7; therefore the pH is 7.

If the hydrogen ion concentration of a solution is known, the hydroxide ion concentration can be easily calculated. The product of the hydrogen ion concentration and the hydroxide ion concentration is 1×10^{-14}:

$$[H^+][OH^-] = 1 \times 10^{-14}$$

In pure (freshly distilled) water, the hydrogen ion concentration is 10^{-7}; therefore the hydroxide concentration is also 10^{-7}. Such a solution, in which the concentrations are equal, is said to be neutral. Acidic solutions (those with an excess of hydrogen ions) have pH values smaller than 7. For example, the hydrogen ion concentration of a solution with pH 1 is ten times that of a solution with pH 2. Basic solutions (those with an excess of hydroxide ions) have pH values greater than 7.

The pH values of some common substances are shown in Figure 2–17. Although some very acidic compartments exist within cells (Chapter 4), most of the interior of an animal or plant cell is neither strongly acidic nor strongly basic but instead an essentially neutral mixture of acidic and basic substances. Although certain bacteria are adapted to life in extremely acidic environments (Chapter 23), a substantial change in pH is incompatible with life for most cells (Fig. 2–18). The pH of most types of plant and animal cells (and their environment) ordinarily ranges from around 7.2 to 7.4.

Buffers minimize pH change

Many homeostatic mechanisms operate to maintain appropriate pH values. For example, the pH of human blood is about 7.4 and must be maintained within very narrow limits. Should the blood become too acidic (for example, as a result of res-

Distilled
Water
Milk 6.6 ┐ ┌ Blood 7.4
Black coffee 5.0 ┐ ┌ Egg white 8.0
Tomato 4.6 ┐ ┌ Baking soda 9.0
Vinegar 3.0 ┐ ┌ Milk of magnesia 10.5
Stomach gastric juice 2.0 ┐ ┌ Household ammonia 11.0
┌ Bleach 13.0
Hydrochloric acid 0.8 ┐ ┌ Oven cleaner 13.8

0.0 1.0 2.0 3.0 4.0 5.0 6.0 7.0 8.0 9.0 10.0 11.0 12.0 13.0 14.0

Neutral

Strong
acid

Strong
base

Figure 2–17   **pH values.** The pH meter is an electronic device used to measure the acidity of a solution. A neutral solution (pH 7) has equal concentrations of H^+ and OH^-. Acidic solutions have pH values lower than 7, while basic solutions have pH values higher than 7.

Figure 2–18   **Acid rain damage.** Oxides of sulfur emitted from fossil fuel plants and other industries, and oxides of nitrogen from automobile exhaust, are converted in the moist atmosphere into acids that become dispersed over wide areas. Unlike unpolluted rain (average pH 5.6), the pH of acid rain has been measured at 4.2 and even lower. Plants, such as these trees photographed in the Black Forest, Germany, may be damaged when the resulting increase in soil acidity causes certain minerals, particularly calcium ions, to leach out of the soil. *(Hans Reinhard/Bruce Coleman)*

piratory disease), coma and death may result. Excessive alkalinity can result in overexcitability of the nervous system and even convulsions. Organisms contain many natural buffers. A **buffer** is a substance or combination of substances that resists changes in pH when an acid or base is added. A buffering system includes a weak acid or a weak base (Fig. 2–19). A weak acid or weak base does not ionize completely. That is, at any given instant only a fraction of the molecules are ionized; most are undissociated.

One of the most common buffering systems is found in the blood of vertebrates (see Chapter 44). Carbon dioxide, produced as a waste product of cellular metabolism, enters the blood, the main constituent of which is water. The carbon dioxide reacts with the water to form carbonic acid, a weak acid that dissociates to yield a hydrogen ion and a bicarbonate ion. The buffering system is described by the following expression:

$$CO_2 \;+\; H_2O \;\rightleftharpoons\; H_2CO_3 \;\rightleftharpoons\; H^+ \;+\; HCO_3^-$$

Carbon dioxide   Water   Carbonic acid   Bicarbonate ion

As indicated by the arrows, all the reactions are reversible. Because carbonic acid is a weak acid, undissociated molecules are always present, as are all the other components of the system. The expression describes the system when it is at equilibrium,

Figure 2-19 **Buffering.** Sodium bicarbonate, which ionizes to form bicarbonate ions (HCO_3^-), is sometimes used as a remedy for excess stomach acid. The bubbles are CO_2 produced by the reaction between a weak acid (citric acid) and the sodium bicarbonate. *(Charles D. Winters)*

TABLE 2-3	Some Biologically Important Ions	
Name	**Formula**	**Charge**
Sodium	Na^+	1+
Potassium	K^+	1+
Hydrogen	H^+	1+
Magnesium	Mg^{2+}	2+
Calcium	Ca^{2+}	2+
Iron	Fe^{2+} or Fe^{3+}	2+ [iron(II)] or 3+ [iron(III)]
Ammonium	NH_4^+	1+
Chloride	Cl^-	1−
Iodide	I^-	1−
Carbonate	C_3O^{2-}	2−
Bicarbonate	HCO_3^-	1−
Phosphate	P_4O^{3-}	3−
Acetate	CH_3COO^-	1−
Sulfate	S_4O^{2-}	2−
Hydroxide	OH^-	1−
Nitrate	NO_3^-	1−
Nitrite	NO_2^-	1−

when the rates of the forward and reverse reactions are equal and the relative concentrations of the components are not changing. If a system is at equilibrium, it can be "shifted to the right" by adding reactants or removing products. Conversely, it can be "shifted to the left" by adding products or removing reactants. Hydrogen ions are the important products to consider in this system.

The addition of excess hydrogen ions temporarily shifts the system to the left, as they combine with the bicarbonate ions to form carbonic acid. Eventually a new equilibrium is established; at this point the hydrogen ion concentration is similar to the original concentration.

If hydroxide ions are added, they combine with the hydrogen ions to form water, effectively removing a product and thus shifting the system to the right. More carbonic acid then ionizes, replacing the hydrogen ions that were removed.

Organisms contain many weak acids and weak bases, thus maintaining an essential reserve of buffering capacity and avoiding pH extremes.

An acid and a base react to form a salt

When an acid and a base are mixed together, the H^+ of the acid unites with the OH^- of the base to form a molecule of water. The remainder of the acid (an anion) combines with the remainder of the base (a cation) to form a salt. For example, hydrochloric acid reacts with sodium hydroxide to form water and sodium chloride:

$$HCl + NaOH \longrightarrow H_2O + NaCl$$

A **salt** is a compound in which the hydrogen ion of an acid is replaced by some other cation. Sodium chloride, NaCl, is a compound in which the hydrogen ion of HCl has been replaced by the cation Na^+.

When a salt, an acid, or a base is dissolved in water, its dissociated ions can conduct an electrical current; these substances are called **electrolytes.** Sugars, alcohols, and many other substances do not form ions when dissolved in water; they do not conduct an electrical current and are referred to as **nonelectrolytes.**

Cells and extracellular fluids (such as blood) of animals and plants contain a variety of dissolved salts that are the source of the many important mineral ions essential for fluid balance and acid-base balance. The concentrations and relative amounts of the various cations and anions are kept remarkably constant. Any marked change results in impaired cellular functions and may lead to death. Nitrates and ammonium from the soil are the important nitrogen sources for plants. In animals, nerve and muscle function, blood clotting, bone formation, and many other aspects of body function depend on ions. Sodium, potassium, calcium, and magnesium are the chief cations present; chloride, bicarbonate, phosphate, and sulfate are important anions (Table 2–3).

I. The chemical composition and metabolic processes of all organisms are very similar. The physical and chemical principles that govern nonliving things also govern organisms.

II. Organisms are made up of small, simple, **inorganic compounds** as well as large, complex, carbon-containing **organic compounds.**

III. An **element** is a substance that cannot be decomposed into simpler substances by normal chemical reactions. Four elements—carbon, hydrogen, oxygen, and nitrogen—make up 96% or more of an organism's mass.

IV. Each **atom** is composed of a nucleus containing **protons** and **neutrons.** In the space outside the nucleus, **electrons** move rapidly in **orbitals** that correspond to energy levels.
   A. An atom is identified by its number of protons (**atomic number**).
   B. Atoms of the same element with different numbers of neutrons (different **atomic masses**) are **isotopes.** Some isotopes are radioactive (**radioisotopes**).
   C. In an electrically neutral atom, the number of protons equals the number of electrons.
   D. The chemical properties of an atom are determined chiefly by the number and arrangement of its most energetic electrons, known as **valence electrons.**

V. Different atoms are joined by chemical bonds to form **compounds.** A **chemical formula** gives the types and relative numbers of atoms in a substance.
   A. One **mole** (the atomic or molecular mass in grams) of any substance contains 6.02×10^{23} atoms, molecules, or ions.
   B. **Covalent bonds** are strong, stable bonds formed when atoms share **valence electrons,** forming **molecules.** A **molecular formula** gives the actual numbers of each type of atom in a molecule; a **structural formula** shows their arrangement.
      1. Covalent bonds are **nonpolar** if the electrons are shared equally between the two atoms.
      2. Covalent bonds are **polar** if one atom is more electronegative (has a greater affinity for electrons) than the other.
   C. An **ionic bond** is formed between a positively charged **cation** and a negatively charged **anion.**
   D. **Hydrogen bonds** are relatively weak bonds formed when a hydrogen atom with a partial positive charge is attracted to an atom (usually oxygen or nitrogen) with a partial negative charge already

bonded to another molecule or in another part of the same molecule.

VI. **Oxidation** and **reduction (redox)** reactions are chemical processes in which electrons (and their energy) are transferred from a reducing agent to an oxidizing agent.

VII. Water accounts for a large part of the mass of most organisms. It is important in many chemical reactions within living things and has unique properties that also affect the environment.
   A. Water is a **polar molecule** because one end has a partial positive charge and the other has a partial negative charge.
   B. Because its molecules are polar, water is an excellent **solvent** for ionic or polar **solutes.**
   C. Water molecules are **cohesive** because they form hydrogen bonds with each other; they are also **adhesive** through hydrogen bonding to substances with ionic or polar regions.
   D. Water has a high **specific heat,** which helps organisms maintain a relatively constant internal temperature; this property also helps keep the ocean and other large bodies of water at a constant temperature.
   E. Water has a high **heat of vaporization.** Molecules entering the vapor phase carry a great deal of heat, which accounts for **evaporative cooling.**
   F. The fact that ice is less dense than liquid water makes the environment less extreme.
   G. Water has a slight tendency to **ionize,** that is, to dissociate to form hydrogen ions (protons, H^+) and hydroxide ions (OH^-).

VIII. **Acids** are proton (H^+) donors; **bases** are proton acceptors. Many bases dissociate in solution to yield hydroxide ions, which then accept protons.
   A. The **pH scale** is a logarithmic expression of the hydrogen ion concentration of a solution. As a solution becomes more acidic, its pH falls below 7 (neutrality). As a solution becomes more basic (alkaline), its pH rises above 7.
   B. A buffering system is based on a weak acid or a weak base. A **buffer** resists changes in the pH of a solution when acids or bases are added.
   C. A **salt** is a compound in which the hydrogen atom of an acid is replaced by some other cation. Salts provide the many mineral ions essential for life functions.

1. Which of the following elements is mismatched with its properties or function? (a) carbon—forms the backbone of organic compounds (b) nitrogen—component of proteins (c) hydrogen—very electronegative (d) oxygen—can participate in hydrogen bonding (e) all of the above are correctly matched

2. Which of the following applies to a neutron? (a) positive charge and located in an orbital (b) negligible mass and located in the nucleus (c) positive charge and located in the nucleus (d) uncharged and located in the nucleus (e) uncharged and located in an orbital

3. $^{32}_{15}P$, a radioactive form of phosphorus, has (a) an atomic number of 32 (b) an atomic mass of 15 (c) an atomic mass of 47 (d) 32 electrons (e) 17 neutrons

4. Which of the following facts allows you to determine that atom A and atom B are isotopes of the same element? (a) they each have six protons (b) they each have four neutrons (c) in each, the sum of their electrons and neutrons is 14 (d) they each have four valence electrons (e) they each have a mass number of 14

5. Sodium and potassium behave similarly in chemical reactions. This is because (a) they have the same number of neutrons (b) each has a single valence electron (c) they have the same atomic mass (d) they have the same number of electrons (e) they have the same number of protons

6. The orbitals comprising an atom's valence electron shell (a) are arranged as concentric spheres (b) contain the atom's least energetic electrons (c) may change shape when covalent bonds are formed (d) never contain more than one electron each (e) more than one of the above is correct

7. Which of the following bonds and properties are correctly matched? (a) ionic bonds—strong only if the participating ions are hydrated (b) hydrogen bonds—responsible for bonding oxygen and hydrogen to form a single water molecule (c) polar covalent bonds—can occur between two atoms of the same element (d) covalent bonds—may be single, double, or triple (e) hydrogen bonds—stronger than covalent bonds

8. In a redox reaction a) energy is transferred from a reducing agent to an oxidizing agent (b) a reducing agent becomes oxidized as it accepts an electron (c) an oxidizing agent accepts a proton (d) a reducing agent donates a proton (e) the electrons in an atom move from its valence shell to a shell closer to its nucleus

9. A solution with a pH of 2 has a hydrogen ion concentration that is _____ the hydrogen ion concentration of a solution with a pH of 4. (a) 1/2 (b) 1/100 (c) two times (d) ten times (e) one hundred times

10. The high heat of vaporization of water accounts for (a) evaporative cooling (b) the fact that ice floats (c) the fact that heat is liberated when ice forms (d) the cohesive properties of water (e) capillary action

11. NaOH and HCl react to form Na^+, Cl^-, and water. Which of the following statements is true? (a) Na^+ is an anion, and Cl^- is a cation (b) Na^+ and Cl^- are both anions (c) a hydrogen bond can form between Na^+ and Cl^- (d) Na^+ and Cl^- are electrolytes (e) Na^+ is an acid, and Cl^- is a base

REVIEW QUESTIONS

1. What is the relationship between molecules and compounds? Are all compounds composed of molecules?
2. What are the ways an atom or molecule can become an anion or a cation?
3. What is a radioisotope? Why is it able to substitute for an ordinary (non-radioactive) atom of the same element in a molecule? What are some of the ways radioisotopes are used in biological research?
4. How do ionic and covalent bonds differ?
5. Why does water form hydrogen bonds? List some of the properties of water that result from hydrogen bonding. How do these properties contribute to the role of water as an essential component of organisms?
6. How can weak forces, such as hydrogen bonds, have significant effects in organisms?

7. A solution has a hydrogen ion concentration of 0.01 moles/liter. What is its pH? What is its hydroxide ion concentration? How would this solution differ from one with a pH of 1?
8. Why are buffers important in organisms? Give a specific example of how a buffering system works.
9. Differentiate clearly among acids, bases, and salts.
10. Why must oxidation and reduction occur simultaneously? Why are redox reactions important in some energy transfers?
11. Describe a reversible reaction that is at equilibrium. What would be the consequences of adding or removing a reactant or a product?

YOU MAKE THE CONNECTION

1. Element A has two electrons in its valence shell (which is complete when it contains eight electrons). Would you expect element A to share, donate, or accept electrons? What would you expect of element B, which has four valence electrons, and element C, which has seven?
2. A hydrogen bond formed between two water molecules is only about one-twentieth as strong as a covalent bond between hydrogen and oxygen. In what ways would the physical properties of water be different if these hydrogen bonds were stronger (e.g., one-tenth the strength of covalent bonds)?
3. Consider the following reaction (in water).

$$HCl \longrightarrow H^+ + Cl^-$$

Name the reactant(s) and product(s). Does the expression indicate that the reaction is reversible? Could HCl be used as a buffer?

RECOMMENDED READINGS

Atkins, P.W. *Periodic Kingdom*. Basic Books, Harper Collins Publishers, New York, 1995. In this imaginative work the periodic table is described as a landscape inhabited by elements whose properties are determined by the region in which they reside.

Hedin, L.O. and G.E. Likens. "Atmosphere Dust and Acid Rain." *Scientific American*, Vol. 275, No. 6 Dec. 1996. This article examines the idea that recent reductions in basic compounds attached to dust particles in the atmosphere may be increasing the environmental damage done by acid rain.

Joesten, M.D. and J.L. Wood. *World of Chemistry*. Saunders College Publishing, Philadelphia, 1996. A very readable introduction to chemistry.

Zimmer, C. "Wet, Wild and Weird." *Discover*, Dec. 1992. Computer simulations illustrate the many ways water molecules can interact through hydrogen bonding.

● Visit our website at **http://www.saunderscollege.com/lifesci/titles.html** and click on Solomon/Berg/Martin Biology for links to chapter-related resources on the World Wide Web.

CHAPTER 3

The Chemistry of Life: Organic Compounds

Both inorganic and organic forms of carbon occur widely in nature. The human, plant, paper, and the wood of the desk in the photograph all contain **organic compounds,** formed from backbones of covalently bonded carbon. Inorganic elemental carbon is represented by the diamond and the pencil "lead," which is called graphite. Some very simple carbon compounds are also considered inorganic, especially if the carbon is not bonded to hydrogen. Organic compounds are so named because at one time they were thought to be produced only by living (organic) organisms. In 1928 the German chemist Friedrich Wühler synthesized urea, a metabolic waste product. Since that time, scientists have learned to synthesize many organic molecules and have discovered organic compounds not found in organisms.

In this chapter we focus on some of the major groups of organic compounds important in organisms, including carbohydrates, lipids, proteins, and nucleic acids (DNA and RNA). Organic compounds are the main structural components of cells and tissues. They participate in and regulate metabolic reactions, transmit information, and provide energy for life processes. Evolution involves chemical changes in the organic compounds produced by organisms.

More than five million organic compounds have been identified. Perhaps because it can form a greater variety of molecules than any other element, carbon is the central component of all organic compounds. There are many reasons that organic molecules are so diverse. **Hydrocarbons,** organic compounds consisting only of carbon and hydrogen, can be produced in a wide variety of three-dimensional shapes. Furthermore, the carbon atom can form bonds with a greater number of different elements than any other type of atom. The addi-

(Kenneth Knott/Fine Light Photography)

tion of chemical groups containing atoms of other elements, especially oxygen, nitrogen, phosphorus, and sulfur, can profoundly change the properties of an organic molecule.

Further diversity is provided by the fact that a great many organic compounds found in organisms are extremely large **macromolecules.** Cells construct these from simpler modular subunits. For example, protein molecules are built from smaller compounds called amino acids.

AFTER YOU HAVE STUDIED THIS CHAPTER YOU SHOULD BE ABLE TO

1. Distinguish between organic and inorganic compounds.
2. Describe the properties of carbon that make it the central component of organic compounds.
3. Distinguish among the three principal types of isomers.
4. Identify the major functional groups present in organic compounds and describe their properties.
5. Compare the functions and chemical compositions of the major groups of organic compounds: carbohydrates, lipids, proteins, and nucleic acids.
6. Distinguish among monosaccharides, disaccharides, and polysaccharides. Compare storage polysaccharides with structural polysaccharides.
7. Distinguish among neutral fats, phospholipids, and steroids; describe the compositions, characteristics, and biological functions of each.
8. Sketch the structure of an amino acid. Explain how amino acids are grouped into classes based on the characteristics of their side chains.
9. Distinguish among the levels of organization of protein molecules.
10. Describe the components of a nucleotide. Name some nucleic acids and discuss the importance of these compounds in living organisms.

CARBON ATOMS FORM AN ENORMOUS VARIETY OF STRUCTURES

Carbon has unique properties that permit formation of the carbon backbones of the large, complex molecules essential to life. Because a carbon atom has four valence electrons, it can complete its valence shell by forming a total of four covalent bonds. Each bond can link it to another carbon atom or to an atom of a different element. Carbon is particularly well suited to serve as the backbone of a large molecule because carbon-to-carbon bonds are strong and not easily broken. However, they are not so strong that it would be impossible for cells to break them. Carbon-to-carbon bonds are not limited to single bonds (based on sharing of one electron pair). Two carbon atoms can share two electron pairs with each other, forming double bonds:

In some compounds, triple carbon-to-carbon bonds are formed:

$$-C\equiv C-$$

Carbon chains can be unbranched or branched, and carbon atoms can also be joined into rings (Fig. 3–1). Rings and chains are joined in some compounds.

The molecules in the cell are analogous to the components of a machine. Each component has a shape that permits it to fill certain roles and to interact with other components (often with a complementary shape). Similarly, the shape of a molecule is important in determining its biological properties and function. Carbon atoms are able to link to each other and to other atoms to produce a wide variety of three-dimensional molecular shapes. This is because the four covalent bonds of carbon do not form in a single plane. Instead, as discussed in Chapter 2, the valence electron orbitals become elongated and project from the carbon atom toward the corners of a tetrahedron (Fig. 3–2). The structure is highly symmetrical, with an angle of about 109.5 degrees between any two of these

bonds. Keep in mind that, for simplicity, many of the figures in this book are drawn as two-dimensional graphic representations of three-dimensional molecules. For example, hydrocarbon chains, such as those seen in Figure 3–1, are not actually straight but have a three-dimensional zigzag structure.

Generally, there is freedom of rotation around each carbon-to-carbon single bond. This property permits organic molecules to be flexible and to assume a variety of shapes, depending on the extent to which each single bond is rotated. Double and triple bonds do not permit rotation, so regions of a molecule with such bonds tend to be inflexible.

One reason for the very great number of possible carbon-containing compounds is the fact that the same components usually can link together in more than one pattern, generating an even wider variety of molecular shapes.

ISOMERS HAVE THE SAME MOLECULAR FORMULA, BUT DIFFERENT STRUCTURES

Compounds with the same molecular formulas but different structures and thus different properties are called **isomers.** Isomers do not have identical physical or chemical properties and may have different common names. Cells can distinguish between isomers. Usually, one isomer is biologically active and the other is not. Three types of isomers are structural isomers, geometric isomers, and enantiomers.

Structural isomers are compounds that differ in the covalent arrangements of their atoms. For example, Figure 3–3a illustrates two structural isomers with the molecular formula C_2H_6O. Similarly, there are two structural isomers of the four-carbon hydrocarbon butane (C_4H_{10}), one with a straight chain and the other with a branched chain (isobutane). Large compounds have more possible structural isomers. There are only two structural isomers of butane, but there can be up to 366,319 isomers of $C_{20}H_{42}$.

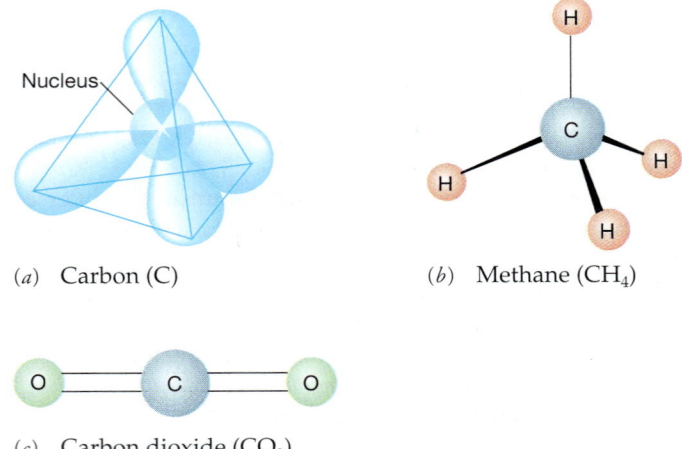

Ethane

(a) Carbon atoms can form chains of varying length

Propane

1-Butene   **2-Butene**

(b) Carbon atoms may form double bonds with one another

Cyclopentane   **Benzene**

(d) Carbon atoms can form rings

Isobutane   **Isopentane**

(c) Carbon atoms can form branched chains

Histidine (an amino acid)

(e) Rings and chains may be joined

Figure 3–1   Organic molecules. Note that each carbon atom forms four covalent bonds, producing a wide variety of shapes.

(a) Carbon (C)

(b) Methane (CH_4)

(c) Carbon dioxide (CO_2)

Nucleus

Figure 3–2   Carbon bonding. The three-dimensional arrangement of the bonds of a carbon atom (a) is responsible for the tetrahedral architecture of methane (b). (c) In carbon dioxide oxygen atoms are joined to a central carbon by polar double bonds. Because the bonds are arranged linearly, carbon dioxide does not have a positively charged end and a negatively charged end, and is therefore a nonpolar molecule.

Geometric isomers are compounds that are identical in the arrangement of their covalent bonds, but different in the spatial arrangement of groups of atoms. Geometric isomers are present in some compounds with carbon-to-carbon double bonds. Because double bonds are not flexible like single bonds, atoms joined to the carbons of a double bond cannot rotate freely about the axis of the bonds. The *cis-trans* isomers may be drawn as shown in Figure 3–3*b*. The designation *cis* (Latin, "on this side") indicates that the two larger components are on the same side of the double bond. If they are on opposite sides of the double bond, the compound is designated a *trans* (Latin, "across") isomer.

Enantiomers are molecules that are mirror images of one another (Fig. 3–3*c*). Recall that the four groups bonded to a single carbon atom are arranged at the vertices of a tetrahedron. If the four bonded groups are all different, the central carbon is described as asymmetric. Figure 3–3*d* illustrates that the four groups can be arranged about the asymmetric carbon in two different ways that are mirror images of each other. The two molecules are enantiomers if they cannot be superimposed on one another no matter how they are rotated in space. Although enantiomers have similar chemical properties and iden-

Ethanol (C₂H₆O) Dimethyl ether (C₂H₆O) *trans*-2-butene *cis*-2-butene

(a) Structural isomers   (b) Geometric isomers

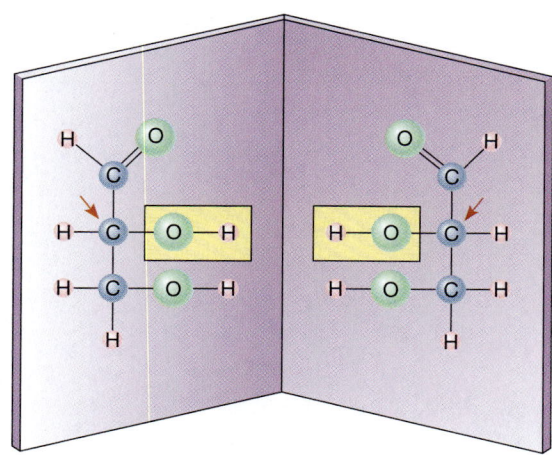

(c) Enantiomers are mirror images   (d) Enantiomers of glyceraldehyde

Figure 3–3 Isomers. Isomers have the same molecular formula, but their atoms are arranged differently. *(a)* Structural isomers differ in the covalent arrangement of their atoms. *(b)* Geometric, or *cis-trans,* isomers have identical covalent bonds but differ in the order in which groups of atoms are arranged in space. *(c)* and *(d)* Enantiomers are isomers that are mirror images of one another. The arrows indicate the asymmetric carbons.   *(c, Dennis Drenner)*

tical physical properties, cells recognize the difference in shape and usually only one form is found in organisms.

The existence of isomers is not the only source of variety among organic molecules. The addition of various combinations of atoms, known as functional groups, can generate a vast array of molecules with differing properties.

FUNCTIONAL GROUPS CHANGE THE PROPERTIES OF ORGANIC MOLECULES

Because covalent bonds between hydrogen and carbon are nonpolar, hydrocarbons lack distinct charged regions For this reason, hydrocarbons are insoluble in water and tend to cluster together, through **hydrophobic** ("water-fearing") **interactions.** This is not because bonds form among the hydrocarbons, but rather because the hydrogen-bonded water molecules exclude them and, in a sense, drive them together. Because of their hydrophobic nature, hydrocarbons do not interact readily with most other compounds.

However, the characteristics of an organic molecule can be changed dramatically by replacing one of the hydrogens with a group of atoms known as a **functional group.** Functional groups help determine the types of chemical reactions in which the compound participates. Most functional groups readily form associations, such as ionic and hydrogen bonds, with other molecules. Polar and ionic functional groups are hydrophilic because they associate strongly with polar water molecules.

The properties of the major classes of organic compounds —carbohydrates, lipids, proteins, and nucleic acids—are largely a consequence of the types and arrangement of functional groups they contain. When we know what kinds of functional groups are present in an organic compound, we can predict its chemical behavior. As you read the rest of this section please refer to Table 3–1 for the complete structural formulas of these groups, as well as additional information. Note that the symbol R is used to represent the *remainder* of the molecule of which each functional group is a part.

The **hydroxyl group** (abbreviated R—OH) must not be confused with the hydroxide ion (OH⁻) discussed in Chapter 2. The hydroxyl group is polar due to the presence of a

TABLE 3–1 Some Biologically Important Functional Groups

Functional group	Structural formula	Class of compounds characterized by group	Description
Hydroxyl	R—OH	Alcohols $H-\overset{\displaystyle H}{\underset{\displaystyle H}{C}}-\overset{\displaystyle H}{\underset{\displaystyle H}{C}}-OH$ Ethanol	Polar because electronegative oxygen attracts covalent electrons
Carbonyl	$R-\overset{\displaystyle O}{\overset{\|}{C}}-H$	Aldehydes $H-\overset{\displaystyle O}{\overset{\|}{C}}-H$ Formaldehyde	Carbonyl group carbon is bonded to at least one H atom; polar because electronegative oxygen attracts covalent electrons
	$R-\overset{\displaystyle O}{\overset{\|}{C}}-R$	Ketones $H-\overset{\displaystyle H}{\underset{\displaystyle H}{C}}-\overset{\displaystyle O}{\overset{\|}{C}}-\overset{\displaystyle H}{\underset{\displaystyle H}{C}}-H$ Acetone	Carbonyl group carbon is bonded to two other carbons; polar because electronegative oxygen attracts covalent electrons
Carboxyl	$R-\overset{\displaystyle O}{\overset{\|}{C}}-OH$ Nonionized $R-\overset{\displaystyle O}{\overset{\|}{C}}-O^-$ + H$^+$ Ionized	Carboxylic acids (organic acids) $R-\overset{\displaystyle NH_2}{\underset{\displaystyle H}{C}}-\overset{\displaystyle O}{\overset{\|}{C}}-OH$ Amino acid	Weakly acidic; can release an H$^+$ ion
Amino	$R-N\overset{\displaystyle H}{\underset{\displaystyle H}{}}$ Nonionized $R-N^+\overset{\displaystyle H}{\underset{\displaystyle H}{-H}}$ Ionized	Amines $R-\overset{\displaystyle NH_2}{\underset{\displaystyle H}{C}}-\overset{\displaystyle O}{\overset{\|}{C}}-OH$ Amino acid	Weakly basic; can accept an H$^+$ ion
Methyl	R—CH$_3$	Component of many organic compounds $H-\overset{\displaystyle H}{\underset{\displaystyle H}{C}}-\overset{\displaystyle H}{\underset{\displaystyle H}{C}}-H$ Ethane	Hydrocarbon; nonpolar

TABLE 3–1 continued

Functional group	Structural formula	Class of compounds characterized by group	Description
Phosphate	$$R{-}O{-}\overset{\overset{\displaystyle O}{\|}}{P}{-}OH$$ $$\|$$ $$OH$$ Nonionized $$R{-}O{-}\overset{\overset{\displaystyle O}{\|}}{P}{-}O^- + 2\,H^+$$ $$\|$$ $$O^-$$ Ionized	Organic phosphates $$HO{-}\overset{\overset{\displaystyle O}{\|}}{P}{-}O{-}R$$ $$\|$$ $$OH$$ Phosphate ester (as found in ATP)	Weakly acidic; one or two H^+ ions can be released
Sulfhydryl	$R{-}SH$	Thiols $$H{-}\overset{\overset{\displaystyle H}{\|}}{C}{-}\overset{\overset{\displaystyle H}{\|}}{C}{-}\overset{\overset{\displaystyle O}{\|}}{C}{-}OH$$ $$\quad\;\; SH \;\; NH_2$$ Cysteine	Helps stabilize internal structure of proteins

strongly electronegative oxygen atom. If a hydroxyl group replaces one of the hydrogens of a hydrocarbon, the resulting molecule can have significantly altered properties. For example, ethane (Fig. 3–1a) is a hydrocarbon that is a gas at room temperature. If a hydrogen is replaced by a hydroxyl group, the resulting molecule is ethyl alcohol, or ethanol, which is found in alcoholic beverages (Fig. 3–3a). Ethanol is somewhat cohesive because the polar hydroxyl groups of adjacent molecules interact; it is therefore liquid at room temperature. Unlike ethane, ethyl alcohol can dissolve in water because the polar hydroxyl groups interact with the polar water molecules.

The **carbonyl group** consists of a carbon atom that has a double covalent bond with an oxygen atom. This double bond is polar because of the electronegativity of the oxygen; thus the carbonyl group is hydrophilic. The position of the carbonyl group in the molecule determines the class to which the molecule belongs. An **aldehyde** has a carbonyl group positioned at the end of the carbon skeleton (abbreviated R—CHO); a **ketone** has an internal carbonyl group (abbreviated R—CO—R).

The **carboxyl group** (abbreviated R—COOH) consists of a carbon atom joined by a double covalent bond to an oxygen atom, and by a single covalent bond to another oxygen, which is in turn bonded to a hydrogen atom. Two electronegative oxygen atoms in such close proximity establish an extremely polarized condition, which can cause the hydrogen

atom to be stripped of its electron and released as a hydrogen ion (H^+). The carboxyl group then has one unit of negative charge (R—COO$^-$):

$$R{-}C{\overset{\displaystyle O}{\underset{\underset{\displaystyle H}{\displaystyle O}}{\big\langle}}} \longrightarrow R{-}C{\overset{\displaystyle O}{\underset{\displaystyle O^-}{\big\langle}}} + H^+$$

Carboxyl groups are weakly acidic; only a fraction of the molecules ionize in this way. This group can therefore exist in one of two hydrophilic states: ionic or polar. Carboxyl groups are essential constituents of amino acids.

An **amino group** (abbreviated R—NH$_2$) includes a nitrogen atom covalently bonded to two hydrogen atoms. Amino groups are weakly basic because they are able to accept a hydrogen ion (proton), thus acquiring a unit of positive charge. Amino groups are components of amino acids and of nucleic acids.

A **phosphate group** (abbreviated R—PO$_4$H$_2$) is weakly acidic. The attraction of electrons by the oxygens can result in the release of one or two hydrogen ions, producing ionized forms with one or two units of negative charge. Phosphates are constituents of nucleic acids and certain lipids.

The **sulfhydryl group** (abbreviated R—SH), consisting of an atom of sulfur covalently bonded to a hydrogen atom, is found in molecules called *thiols*. As we will see, amino acids that contain a sulfhydryl group can make important contributions to the structure of proteins.

The **methyl group** (abbreviated R—CH_3), a common hydrocarbon group, is nonpolar.

MANY BIOLOGICAL MOLECULES ARE POLYMERS

Many biological molecules such as proteins and nucleic acids are very large, consisting of thousands of atoms. Such giant molecules are known as **macromolecules.** Most macromolecules are **polymers,** produced by linking small organic compounds called **monomers** (Fig. 3–4). Just as all the words in this book have been written by arranging the 26 letters of the alphabet in various combinations, monomers can be grouped together to form an almost infinite variety of larger molecules. Just as we use different words to convey information, cells use different molecules to convey information. The thousands of different complex organic compounds present in organisms are constructed from about 40 small, simple monomers. For example, the 20 monomers called amino acids can be linked end-to-end in countless ways to form the polymers we know as proteins.

Each organism is unique because of differences in the monomer sequence within its DNA, the polymer that constitutes the information in the genes. Cells and tissues within the same organism are also different, due to variations in their component polymers. Muscle tissue and brain tissue differ in large part because of differences in the types and sequences of amino acids in proteins. Ultimately this protein structure is dictated by the sequence of monomers within the DNA of the organism, which is expressed somewhat differently in each cell type.

Polymers can be degraded to their component monomers by **hydrolysis** (which means "to break with water"). In a reaction regulated by a specific enzyme,[1] a hydrogen from a water molecule attaches to one monomer, and a hydroxyl from water attaches to the adjacent monomer.

The synthetic process by which monomers are covalently linked is called **condensation.** Because the *equivalent* of a molecule of water is removed during the reactions that combine monomers, the term *dehydration synthesis* is sometimes used to describe the process. However, in biological systems the synthesis of a polymer is not simply the reverse of hydrolysis, even though the net effect is the opposite of hydrolysis. Synthetic processes require energy and are regulated by different enzymes.

In the following sections we will examine carbohydrates, lipids, proteins, and nucleic acids. Our discussion will begin

<hr/>

[1]An enzyme (see Chapter 6) is a protein catalyst that accelerates a specific chemical reaction.

Figure 3–4   **A simple polymer.** This small polymer of polyethylene is formed by linking two-carbon ethylene (C_2H_4) monomers. One such monomer is outlined.

with the smaller, simpler forms of these compounds and will extend to the linking of these monomers to form macromolecules.

CARBOHYDRATES INCLUDE SUGARS, STARCHES, AND CELLULOSE

Sugars, starches, and cellulose are **carbohydrates.** Sugars and **starches** serve as energy sources for cells; **cellulose** is the main structural component of the walls that surround plant cells. Carbohydrates contain carbon, hydrogen, and oxygen atoms in a ratio of approximately one carbon to two hydrogens to one oxygen $(CH_2O)_n$. The term *carbohydrate*, meaning "hydrate (water) of carbon," reflects the 2:1 ratio of hydrogen to oxygen, the same ratio found in water (H_2O). Carbohydrates contain one sugar unit (*mono*saccharides), two sugar units (*di*saccharides), or many sugar units (*poly*saccharides).

Monosaccharides are simple sugars

Monosaccharides typically contain from three to seven carbon atoms. In a monosaccharide, a hydroxyl group is bonded to each carbon except one; that carbon is double-bonded to an oxygen atom, forming a carbonyl group. If the carbonyl group is at the end of the chain, the monosaccharide is an aldehyde; if the carbonyl group is at any other position, the monosaccharide is a ketone. (By convention, the numbering of the carbon skeleton of a sugar begins with the carbon at or nearest the carbonyl end of the open chain.) The large number of polar hydroxyl groups, plus the carbonyl group, gives a monosaccharide hydrophilic properties.

Figure 3–5 shows simplified, two-dimensional representations of some common monosaccharides. The simplest carbohydrates are the three-carbon sugars (trioses): glyceraldehyde and dihydroxyacetone. Ribose and deoxyribose are common pentoses, sugars that contain five carbons; they are components of nucleic acids (DNA, RNA, and related compounds). Glucose, fructose, galactose, and other six-carbon sugars are called **hexoses.** (Note that the names of carbohydrates typically end in *-ose*.)

Glucose ($C_6H_{12}O_6$), the most abundant monosaccharide, is used as an energy source in most organisms. During cellular

Figure 3–5 **Monosaccharides.** Shown are two-dimensional chain structures of *(a)* 3-carbon trioses, *(b)* 5-carbon pentoses, and *(c)* 6-carbon hexoses. Although it is convenient to show monosaccharides in this form, the pentoses and hexoses are more accurately depicted as ring structures, as in Figure 3–6.

respiration (Chapter 7), cells oxidize glucose molecules, converting the stored energy to a form that can be readily used for cellular work. Glucose is also used as a component in the synthesis of other types of compounds such as amino acids and fatty acids. So important is glucose in metabolism that its concentration is carefully kept at a homeostatic (relatively constant) level in the blood of humans and other complex animals.

Glucose and fructose are structural isomers: they have identical molecular formulas, but their atoms are arranged differently. In fructose (a ketone) the double-bonded oxygen is linked to a carbon within the chain rather than to a terminal carbon as in glucose (which is an aldehyde). Because of these differences, the two sugars have different properties. For example, fructose tastes sweeter than glucose.

Glucose and galactose are both hexoses and aldehydes.

However, they are mirror images (enantiomers) because they differ in the arrangement of the atoms attached to asymmetric carbon atom 4.

The "stick" formulas in Figure 3–5 give a clear but somewhat unrealistic picture of the structures of some common monosaccharides. As has been discussed, molecules are not two-dimensional; in fact, the properties of each compound depend largely on its three-dimensional structure. Thus, three-dimensional formulas are helpful in understanding the relationship between molecular structure and biological function. Molecules of glucose and other pentoses and hexoses in solution are actually rings, rather than extended straight carbon chains.

Glucose in solution (as in the cell) typically exists as a ring of five carbons and one oxygen. It assumes this configuration

Alpha-Glucose
(ring form)

Formation of glucose ring

Beta-Glucose
(ring form)

(a) Forms of glucose

CH₂OH

HO

OH

OH

OH

Alpha-Glucose

CH₂OH

HO

OH

OH

OH

Beta-Glucose

(b) Simplified ring-structure

Figure 3–6 **α and β forms of glucose.** *(a)* When dissolved in water, glucose undergoes a rearrangement of its atoms, forming one of two possible ring structures, α-glucose or β-glucose. Although the drawing does not attempt to show the complete three-dimensional structure, the thick, tapered bonds in the lower portion of each ring represent the part of the molecule that would project out of the page toward you. *(b)* The essential differences between α-glucose and β-glucose are more readily apparent in these simplified structures. By convention, a carbon atom is assumed to be present at each angle in the ring unless another atom is shown. Most hydrogen atoms have been omitted.

when its atoms undergo a rearrangement, permitting a covalent bond to connect carbon 1 to the oxygen attached to carbon 5 (Fig. 3–6). When glucose forms a ring, two isomeric forms are possible, differing only in the orientation of the hydroxyl (—OH) group attached to carbon 1. When this hydroxyl group is on the same side of the plane of the ring as the —CH₂OH side group, the glucose is designated β-glucose. When it is on the side (with respect to the plane of the ring) opposite the —CH₂OH side group, the compound is designated α-glucose.

Disaccharides consist of two monosaccharide units

A disaccharide (two sugars) contains two monosaccharide rings joined by a **glycosidic linkage,** consisting of a central oxygen

covalently bonded to two carbons, one in each ring (Figure 3–7). The glycosidic linkage of a disaccharide generally forms between carbon 1 of one molecule and carbon 4 of the other molecule. The disaccharide maltose (malt sugar) consists of two covalently linked α-glucose units. Sucrose, common table sugar, consists of a glucose unit combined with a fructose unit. Lactose (the sugar present in milk) is composed of one molecule of glucose and one of galactose.

A disaccharide can be hydrolyzed, that is, split by the addition of water, into two monosaccharide units. During digestion, maltose is hydrolyzed to form two molecules of glucose:

Maltose + Water $\longrightarrow$ Glucose + Glucose

Similarly, sucrose is hydrolyzed to form glucose and fructose:

Sucrose + Water $\longrightarrow$ Glucose + Fructose

Figure 3–7 **Hydrolysis of disaccharides.** *(a)* Maltose may be broken down (as it is during digestion) to form two molecules of glucose. The highlighted glycosidic linkage is broken in a hydrolysis reaction, which requires the addition of water. *(b)* Sucrose can be hydrolyzed to yield a molecule of glucose and a molecule of fructose. Note that an enzyme (a protein catalyst) is needed to promote these reactions.

Polysaccharides can store energy or provide structure

The most abundant carbohydrates are the **polysaccharides,** a group that includes starches, glycogen, and cellulose. A polysaccharide is a macromolecule consisting of repeating units of simple sugars, usually glucose. Although the precise number of sugar units varies, thousands of units are typically present in a single molecule. The polysaccharide may be a single long chain or a branched chain. Because they are composed of different isomers and because the units may be arranged differently, polysaccharides vary in their properties. Those that can be easily broken down to their subunits are well suited for energy storage, whereas the macromolecular three-dimensional architecture of others makes them particularly well suited to form stable structures.

Starch is the main storage carbohydrate of plants

Starch, the typical form of carbohydrate used for energy storage in plants, is a polymer consisting of α-glucose subunits. These monomers are joined by α 1—4 linkages, which means that carbon 1 of one glucose is linked to carbon 4 of the next glucose in the chain (Fig. 3–8). Starch occurs in two forms, amylose and amylopectin. Amylose, the simpler form, is unbranched. Amylopectin, the more common form, usually consists of about 1000 units in a branched chain.

Plant cells store starch mainly as granules within specialized organelles called **amyloplasts** (Fig. 3–8*a*). When energy is needed for cellular work, the plant can hydrolyze the starch, releasing the glucose subunits. Humans and other animals that eat plant foods have enzymes to hydrolyze starch.

Glycogen is the main storage carbohydrate of animals

Glycogen (sometimes referred to as *animal starch*) is the form in which glucose is stored as an energy source in animal tissues. It is similar in structure to plant starch, but more extensively branched and more water-soluble. Glycogen is stored mainly in liver and muscle cells.

Cellulose is a structural carbohydrate

Carbohydrates are the most abundant group of organic compounds on Earth, and **cellulose** is the most abundant carbohydrate; it accounts for 50% or more of all the carbon in plants (Fig. 3–9). Cellulose is a structural carbohydrate. Wood is about half cellulose, and cotton is at least 90% cellulose. Plant

100 µm

(a)

(b)

CH₂OH

Starch

(c)

Figure 3—8 **Starch, a storage polysaccharide.** *(a)* Starch (stained purple) is stored in special-ized organelles, called *amyloplasts,* in these cells of a buttercup root. *(b)* Starch is made up of highly branched chains; the arrows indicate the branch points. Each chain is actually in the form of a coil or helix, stabilized by hydrogen bonds between the glucose subunits. *(c)* Starch is composed of α-glucose molecules joined by glycosidic bonds. At the branch points are bonds between carbon 6 of the glucose in the straight chain and carbon 1 of the glucose in the branching chain.   *(a, Ed Reschke)*

cells are surrounded by strong supporting cell walls consisting mainly of cellulose.

Cellulose is an insoluble polysaccharide composed of many glucose molecules joined together. The bonds joining these sugar units are different from those in starch. Recall that starch is composed of α-glucose subunits, joined by α 1—4 glycosidic linkages. Cellulose contains β-glucose monomers joined by β 1—4 linkages. These bonds are not split by the enzymes that hydrolyze the alpha linkages in starch. Humans, like most organisms, do not have enzymes that can digest cel-lulose and therefore cannot use it as a nutrient. Because cel-lulose remains fibrous, as discussed in Chapter 45, it helps keep the digestive tract functioning properly.

Some microorganisms can digest cellulose to glucose. In fact, cellulose-digesting bacteria live in the digestive systems of cows and sheep, enabling these grass-eating animals to obtain nourishment from cellulose. Similarly, the digestive systems of termites contain microorganisms that digest cellulose (see Fig. 24—5*a*).

Cellulose molecules have characteristics that make them well suited for a structural role. The β-glucose subunits are joined in a way that allows extensive hydrogen bonding among

Figure 3–9 **Cellulose, a structural polysaccharide.** *(a)* An electron micrograph of cellulose fibers from a cell wall. The fibers visible in the photograph consist of bundles of cellulose molecules, interacting through hydrogen bonds. *(b and c)* The cellulose molecule is an unbranched polysaccharide composed of approximately 10,000 β-glucose units joined by glycosidic bonds. *(a, Omikron/Photo Researchers, Inc.)*

different cellulose molecules. Thus, cellulose molecules aggregate in long bundles of fibers, as shown in Figure 3–9a.

Some modified and complex carbohydrates have special roles

Many derivatives of monosaccharides are important biological molecules. Some form important structural components. The amino sugars galactosamine and glucosamine are compounds in which a hydroxyl group (—OH) is replaced by an amino group (—NH₂). Galactosamine is present in cartilage, a constituent of the skeletal system of vertebrates. *N*-acetyl glucosamine (NAG) subunits, joined by glycosidic bonds, compose **chitin**, a main component of the external skeletons of insects, crayfish, and other arthropods (Fig. 3–10), and of the cell walls of fungi. Chitin forms very tough structures because, as in cellulose, its molecules interact through multiple hydrogen bonds.

Carbohydrates may also be combined with proteins to form **glycoproteins**, compounds present on the outer surface of cells other than bacteria. Some of these carbohydrate chains allow cells to adhere to one another, while others provide protection. Most proteins secreted by cells are glycoproteins. Carbohydrates can combine with lipids to form **glycolipids,** compounds present on the surfaces of animal cells that are thought to allow cells to recognize and interact with one another (see Chapters 4 and 5).

LIPIDS ARE FATS OR FATLIKE SUBSTANCES

Lipids are a heterogeneous group of compounds that are defined, not by their structure, but rather by the fact that they are soluble in nonpolar solvents (ether, chloroform, etc.) and are relatively insoluble in water. Lipid molecules have these properties because they consist mainly of carbon and hydrogen, with few oxygen-containing functional groups. Oxygen

Figure 3–10 **Chitin, a structural polysaccharide.** *(a)* Chitin is a polymer composed of *N*-acetyl glucosamine (NAG) subunits. *(b)* Chitin is an important component of the exoskeleton (outer covering) that this dragonfly is shedding. *(b, Dwight R. Kuhn)*

atoms are commonly found in hydrophilic functional groups; therefore lipids, with little oxygen, tend to be hydrophobic. Among the biologically important groups of lipids are neutral fats, phospholipids, carotenoids (orange and yellow plant pigments), steroids, and waxes. Some lipids are used for energy storage, others serve as structural components of cellular membranes, and some are important hormones.

Neutral fats contain glycerol and fatty acids

The most abundant lipids in living organisms are the **neutral fats.** These compounds are an economical form of reserve fuel storage because they yield more than twice as much energy per gram as do carbohydrates. Carbohydrates and proteins can be transformed by enzymes into fats and stored within the cells of adipose (fat) tissue of animals and in some seeds and fruits of plants.

A neutral fat consists of glycerol joined to one, two, or three fatty acids. **Glycerol** is a three-carbon alcohol that contains three hydroxyl (—OH) groups (Fig. 3–11). A **fatty acid** is a long, unbranched hydrocarbon chain with a carboxyl group (—COOH) at one end. About 30 different fatty acids are commonly found in lipids, and they typically have an even number of carbon atoms. For example, butyric acid, present in rancid butter, has four carbon atoms. Oleic acid, with 18 carbons, is the most widely distributed fatty acid in nature and is found in most animal and plant fats.

Saturated fatty acids contain the maximum possible number of hydrogen atoms. Fats high in saturated fatty acids, such as animal fat and solid vegetable shortening, tend to be solid at room temperature. This is because even electrically neutral, nonpolar molecules can develop transient regions of weak positive charge and weak negative charge, as the constant motion of their electrons causes some regions to have a temporary excess of electrons, while others have a temporary electron deficit. These slight opposite charges result in attractions, known as **van der Waals forces,** between adjacent molecules. Although van der Waals interactions are individually weak, they can be strong when many occur among long hydrocarbon chains.

Unsaturated fatty acids include one or more adjacent pairs of carbon atoms joined by a double bond. Therefore they are not fully saturated with hydrogen. Fatty acids with one double bond are called **monounsaturated fatty acids,** while those with more than one double bond are **polyunsaturated fatty acids.** Fats containing a high proportion of monounsaturated or polyunsaturated fatty acids tend to be liquid at room temperature. This is because each double bond produces a bend in the hydrocarbon chain that prevents it from aligning closely

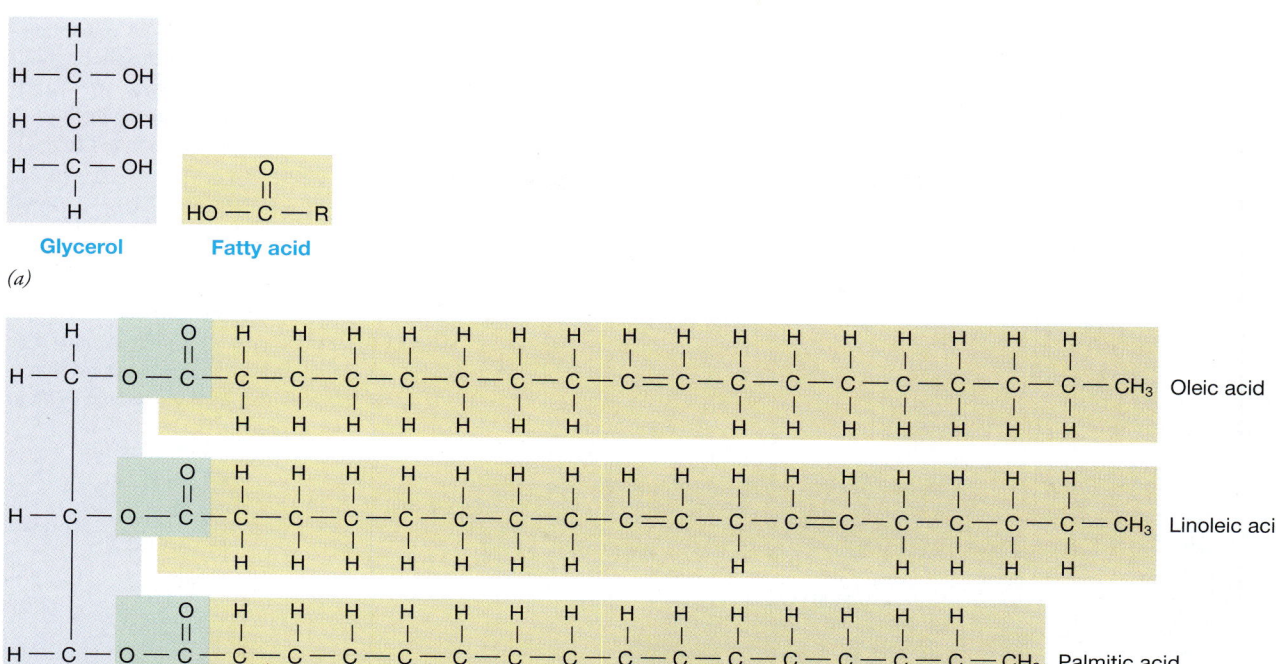

(a)

Glycerol   Fatty acid

(b)

Oleic acid

Linoleic acid

Palmitic acid

A triacylglycerol

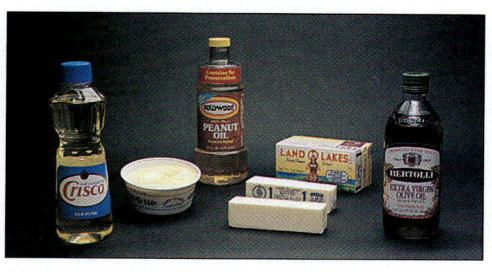

(c)

Figure 3–11   Neutral fats. *(a)* Glycerol and fatty acids are the components of neutral fats. Every fatty acid contains a carboxyl (—COOH) group, plus a long hydrocarbon (represented by "R") that is specific for that particular fatty acid. *(b)* Glycerol is attached to fatty acids by ester linkages, shown in green. Note that oleic and linoleic acid are unsaturated fatty acids. They are drawn as straight chains, but the molecules are actually bent or kinked wherever a carbon-to-carbon double bond appears. *(c)* Commonly used cooking fats contain triacylglycerols.   *(c, Kenneth Knott/Fine Light Photography)*

with an adjacent chain, thereby limiting van der Waals interactions.

At least two unsaturated fatty acids (linoleic acid and arachidonic acid) are essential nutrients that must be obtained from food because the human body cannot synthesize them. However, the amounts required are small, and deficiencies are rarely seen. There is no dietary requirement for saturated fatty acids.

When a glycerol molecule combines chemically with one fatty acid, a **monoacylglycerol** (sometimes called *monoglyceride*) is formed. **Diacylglycerols** (or *diglycerides*) and **triacylglycerols** (or *triglycerides*) contain two or three fatty acids, respectively. In each condensation reaction, the equivalent of a water molecule is removed as one of the glycerol's hydroxyl groups reacts with the carboxyl group of a fatty acid. Such a reaction between a hydroxyl group and a carboxyl group results in the formation of a covalent linkage known as an **ester linkage** (Fig. 3–11*b*). During digestion, neutral fats are hydrolyzed to produce fatty acids and glycerol (see Chapter 45).

Phospholipids are components of cellular membranes

Phospholipids belong to a group of lipids, called **amphipathic lipids,** in which one end of each molecule is hydrophilic and the other end is hydrophobic. The two ends of a phospholipid differ both physically and chemically. A phospholipid consists of a glycerol molecule attached at one end to two fatty acids, and at the other end to a phosphate group linked to an organic compound such as choline. The organic compound usually contains nitrogen (Fig. 3–12). (Note that phosphorus and nitrogen are absent in the neutral fats.) The fatty acid portion of the molecule (containing the two hydrocarbon "tails") is hydrophobic and not soluble in water. However, the portion composed of glycerol, phosphate, and the organic base (the "head" of the molecule) is ionized and readily water-soluble. The amphipathic properties of these lipid molecules make them uniquely suited to form cellular membranes, structures that will be discussed in Chapters 4 and 5.

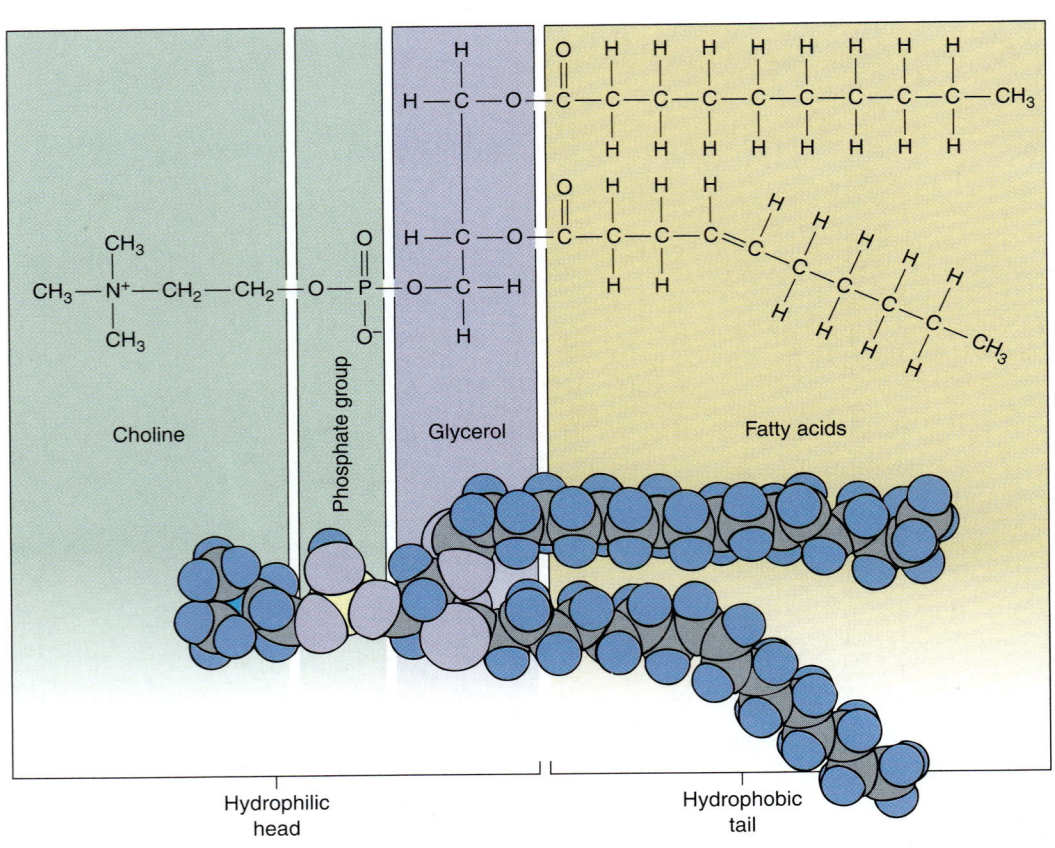

Figure 3–12 A phospholipid. Note the hydrophobic tail, made up of two fatty acids, and the hydrophilic head, which includes a glycerol bonded to a phosphate group, which is in turn bonded to an organic group that can vary. Choline is the organic group in the molecule shown, lecithin (or phosphatidylcholine). The lower fatty acid in the figure is monounsaturated; it contains one double bond that produces a characteristic bend in the chain.

Choline

Phosphate group

Glycerol

Fatty acids

Hydrophilic head

Hydrophobic tail

LECITHIN

Carotenoid plant pigments are derived from isoprene units

The orange and yellow plant pigments called **carotenoids** are classified with the lipids because they are insoluble in water and have an oily consistency. These pigments, found in the cells of all plants, play a role in photosynthesis. As discussed in *Making the Connection: Molecules that Absorb Light,* carotenoid molecules and many other important pigments consist of five-carbon hydrocarbon monomers known as *isoprene units.*

Steroids contain four rings of carbon atoms

A **steroid** consists of carbon atoms arranged in four attached rings; three of the rings contain six carbon atoms, and the fourth contains five (Fig. 3–13). The length and structure of the side chains that extend from these rings distinguish one

Figure 3–13 Steroids. Four attached rings—three six-carbon rings and one with five carbons—make up a steroid. Note that some carbons are shared by two rings. In these simplified structures, a carbon atom is present at each angle of a ring; the hydrogen atoms attached directly to the ring have not been drawn. *(a)* Cholesterol is an essential component of animal cellular membranes. *(b)* Cortisol is a steroid hormone secreted by the adrenal glands. Notice that cortisol differs from cholesterol in its attached functional groups.

Indicates double bond

Cholesterol

(a)

Cortisol

(b)

MOLECULES THAT ABSORB LIGHT

What do some biologically important pigments have in common? Many are formed from 5-carbon isoprene monomers (*a*). One such molecule is β-carotene (*b*), a yellow plant pigment that plays a role in photosynthesis. (The dashed lines indicate the boundaries of the individual isoprene units within β-carotene.) Most animals can break carotenoids at the point of cleavage marked in the diagram and convert them to vitamin A (*c*). Vitamin A can then be converted to the visual pigment retinal (*d*). Retinal can exist in more than one, three-dimensional form (not shown).

Interestingly, eyes are found in three different lines of animals: the mollusks, insects, and vertebrates. There is no evidence that they share a common ancestor equipped with eyes, although they may have had a common ancestor with primitive photoreceptors. They all have the same compound involved in the process of light reception, retinal, a molecule that is apparently uniquely fitted for this role.

Notice that all these molecules have a pattern of double bonds alternating with single bonds. The electrons that make up these bonds can move about relatively easily when light strikes the molecule. Such molecules tend to be highly colored pigments because they strongly absorb light of certain wavelengths and reflect other wavelengths.

β carotene, vitamin A, and retinal are composed of isoprene subunits.

Isoprene

(*a*)

β-Carotene

(*b*)

Point of cleavage

Vitamin A₁

(*c*)

Retinal

(*d*)

steroid from another. Steroids are synthesized from isoprene units.

Among the steroids of biological importance are cholesterol, bile salts, reproductive hormones, and cortisol and other hormones secreted by the adrenal cortex. Cholesterol is a structural component of animal cell membranes; plant cell membranes contain molecules similar to cholesterol. Bile salts emulsify fats in the intestine so that they can be enzymatically hydrolyzed. Steroid hormones regulate certain aspects of metabolism in a variety of animals, including vertebrates, insects, and crabs.

Some chemical mediators are derived from fatty acids

Animal cells secrete chemicals that permit them to communicate with each other or to regulate their own activities. Some chemical mediators are produced by the modification of fatty acids that have been removed from membrane phospholipids. These include prostaglandins, which have varied roles, including promoting inflammation and smooth muscle contraction. Certain hormones, such as the juvenile hormone of insects, are also fatty acid derivatives (Chapter 47).

PROTEINS ARE MACROMOLECULES FORMED FROM AMINO ACIDS

Because **proteins** are extraordinarily versatile macromolecules, they are of central importance in the chemistry of life. Proteins can be assembled into a variety of shapes, allowing them to serve as major structural components of cells and tissues. For this reason, growth and repair, as well as maintenance of the organism, depend on an adequate supply of these compounds. Most **enzymes** (molecules that speed up the thousands of different chemical reactions that take place in an organism) are proteins. Proteins serve in a great many other specialized capacities, including cellular motors, hormones and other chemical messengers, regulators of cellular activities, and defenders against foreign invaders.

The protein constituents of a cell are the clues to its lifestyle. Each cell type contains characteristic forms, distributions, and amounts of protein that largely determine what the cell looks like and how it functions. A muscle cell contains large amounts of the proteins myosin and actin, which are responsible for its appearance as well as for its ability to contract. The protein hemoglobin, found in red blood cells, is responsible for the specialized function of oxygen transport.

Although carbohydrates and lipids tend to have the same structures among different species, most proteins are species-specific; that is, their structures vary from species to species. The specific proteins present (determined by the instructions in the genes) are largely responsible for differences among species. Thus, the proteins in the cells of a dog vary somewhat from those of a fox and even more from those of an oak tree. The degree of difference in the proteins of two species is thought to depend on evolutionary relationships. Distantly related organisms have proteins that differ more markedly than those of closely related forms.

Some proteins differ slightly even among individuals of the same species. Although it is common for individuals to have many identical proteins, most individuals are biochemically unique in the sense that no two are likely to have *all* identical proteins unless they are also genetically identical (i.e., identical twins or members of closely inbred strains).

Amino acids are the subunits of proteins

Amino acids, the constituents of proteins, have an amino group ($-NH_2$) and a carboxyl group ($-COOH$) bonded to the same asymmetric carbon atom, known as the **alpha carbon.** There are about 20 amino acids commonly found in proteins, each uniquely identified by the variable group (R group) bonded to the alpha carbon (Fig. 3–14). Glycine, the simplest amino acid, has a hydrogen atom as its R group; alanine has a methyl ($-CH_3$) group.

Amino acids in solution at neutral pH are mainly dipolar ions. This is generally how amino acids exist at cellular pH. Each carboxyl group ($-COOH$) donates a proton and becomes dissociated ($-COO^-$), while each amino group ($-NH_2$) accepts a proton and becomes $-NH_3^+$ (Fig. 3–15). Because of their amino and carboxyl groups, amino acids in solution resist changes in acidity and alkalinity and so are important biological buffers.

The amino acids are grouped in Figure 3–14 by the properties of their side chains. These broad groupings actually include amino acids with a fairly wide range of properties. Amino acids classified as having *nonpolar* side chains tend to have hydrophobic properties, whereas those classified as *polar* are more hydrophilic. An acidic amino acid has a side chain that contains a carboxyl group. At cellular pH the carboxyl group is dissociated, giving the R group a negative charge. A basic amino acid becomes positively charged when the amino group in its side chain accepts a hydrogen ion. Acidic and basic side chains are ionic at cellular pH and therefore hydrophilic.

In addition to the 20 common amino acids, some proteins have unusual ones. These rare amino acids are produced by the modification of common ones after they have become part of a protein. For example, lysine and proline may be converted to hydroxylysine and hydroxyproline after they have been incorporated into collagen. These amino acids can form cross links between the peptide chains that make up collagen. Such cross links are responsible for the firmness and great strength of the collagen molecule, which is a major component of cartilage, bone, and other connective tissues.

With some exceptions, bacteria and plants can synthesize all their needed amino acids from simpler substances. If the proper raw materials are available, the cells of humans and animals can manufacture some, but not all, of the biologically significant amino acids. Those that animals cannot synthesize and so must obtain from the diet are known as **essential amino acids.** Animals differ in their biosynthetic capacities; what is

POLAR

Asparagine
Asn

Glutamine
Gln

Tyrosine
Tyr

Serine
Ser

Threonine
Thr

ELECTRICALLY CHARGED

ACIDIC

BASIC

Aspartic Acid
Asp

Glutamic Acid
Glu

Arginine
Arg

Lysine
Lys

Histidine
His

NON-POLAR

Glycine
Gly

Alanine
Ala

Valine
Val

Leucine
Leu

Isoleucine
Ile

Tryptophan
Trp

Proline
Pro

Cysteine
Cys

Methionine
Met

Phenylalanine
Phe

Figure 3–14 **The twenty common amino acids.** Amino acids designated *polar* have relatively hydrophilic side chains, while the side chains of those referred to as *nonpolar* are relatively hydrophobic. Carboxyl groups and amino groups are electrically charged at cellular pH; therefore acidic and basic amino acids are hydrophilic. The three-letter symbols are the conventional abbreviations for the amino acids.

Figure 3–15 **An amino acid at pH 7.** In living cells, amino acids exist mainly as dipolar ions.

an essential amino acid for one species may not be for another. The essential amino acids for humans include isoleucine, leucine, lysine, methionine, phenylalanine, threonine, tryptophan, valine, histidine, and (in children) arginine.

Peptide bonds join amino acids

Amino acids combine chemically with one another by a condensation reaction that bonds the carboxyl carbon of one molecule to the amino nitrogen of another (Fig. 3–16). The covalent carbon-to-nitrogen bond linking two amino acids together is called a **peptide bond.** When two amino acids combine, a **dipeptide** is formed; a longer chain of amino acids is a **polypeptide.** A protein consists of one or more polypeptide chains. Each polypeptide has a free amino group at one end and a free carboxyl group (belonging to the last amino acid added to the chain) at the opposite end. The other amino and carboxyl groups of the amino acid monomers (except those in side chains) are part of the peptide bonds. The complex process by which polypeptides are synthesized is discussed in Chapter 12.

A polypeptide may contain hundreds of amino acids joined in a specific linear order. The backbone of the chain includes the repeating sequence

$$N-C-C-N-C-C-N-C-C$$

plus all other atoms *except those in the R groups.* The R groups of the amino acids extend from this backbone.

An almost infinite variety of protein molecules is possible, differing from one another in the number, types, and sequences of amino acids they contain. The 20 types of amino acids found in proteins may be thought of as letters of a protein alphabet; each protein is a very long sentence made up of amino acid letters.

Proteins have four levels of organization

The polypeptide chains making up a protein are twisted or folded to form a macromolecule with a specific *conformation,* or three-dimensional shape. Some polypeptide chains form long fibers. **Globular** proteins are tightly folded into compact, roughly spherical shapes. There is a close relationship between a protein's conformation and its function. For example, a typical enzyme is a globular protein with a unique shape that permits it to catalyze a specific chemical reaction. Similarly, the shape of a protein hormone enables it to combine with receptors on its target cell (the cell the hormone acts upon).

Four main levels of protein organization can be recognized: primary, secondary, tertiary, and quaternary. An analogy for secondary and tertiary structure is depicted in Figure 3–17.

(a)

Glycine   Alanine   Glycylalanine (a dipeptide)

(b)

Glycylalanine   Cysteine   Free amino group   Glycylalanylcysteine (a tripeptide)   Free carboxyl group

Figure 3–16 **Peptide bonds.** *(a)* A dipeptide is formed by the removal of the equivalent of a water molecule from the carboxyl group of one amino acid and the amino group of another amino acid. The resulting peptide bond is a covalent, carbon-to-nitrogen bond. *(b)* The carboxyl group of the dipeptide reacts with the amino group of a third amino acid to form a chain of three amino acids (a tripeptide, or small polypeptide). Additional amino acids can be added to form a long polypeptide chain with a free amino group at one end and a free carboxyl group at the other.

Figure 3–17 **Levels of protein structure.** A telephone cord provides a familiar analogy for the levels of protein structure. *(a)* Secondary structure, in this case analogous to an α-helix. *(b)* Tertiary structure, in which interactions among side chains cause the molecule to fold back on itself.

Primary structure is the amino acid sequence

The sequence of amino acids, joined by peptide bonds, is the **primary structure** of a polypeptide chain. As discussed in Chapter 12, this sequence is specified by the instructions in a gene. Using analytical methods developed in the early 1950s, investigators can determine the exact sequence of amino acids in a protein molecule. Insulin, a hormone secreted by the pancreas and used in the treatment of diabetes, was the first protein for which the exact sequence of amino acids in the polypeptide chains was identified. Insulin is a very small protein, consisting of 51 amino acid units in two linked chains, each with its own primary sequence (Fig. 3–18).

Primary structure is always represented in a simple, linear, "beads-on-a-string" form. However, the overall conformation of a protein is far more complex, involving interactions among the various amino acids that comprise the primary structure of the molecule. Therefore, the higher orders of structure—secondary, tertiary, and quaternary—ultimately derive from the specific amino acid sequence (i.e., the primary structure).

Secondary structure results from hydrogen bonding involving the backbone

Some regions of a polypeptide exhibit **secondary structure,** which is highly regular because it is maintained by hydrogen

bonds between certain atoms of the polypeptide chain's uniform backbone.

A common secondary structure in protein molecules is the **α-helix,** a region where a polypeptide chain forms a uniform spiral coil (Fig. 3–19*a*). The helical structure is determined and maintained by the formation of hydrogen bonds between the backbones of the amino acids in successive turns of the spiral coil. Each hydrogen bond forms between an oxygen with a partial negative charge and a hydrogen with a partial positive charge. The oxygen is part of the remnant of the carboxyl group of one amino acid; the hydrogen is part of the remnant of the amino group of the fourth amino acid down the chain. Thus 3.6 amino acids are included in each complete turn of the helix. Every amino acid in an α-helix is hydrogen-bonded in this way.

The α-helix is the basic structural unit of some fibrous proteins that make up wool, hair, skin, and nails. The elasticity of these fibers is due to a combination of physical factors (the helical shape) and chemical factors (hydrogen bonding). Although hydrogen bonds maintain the helical structure, these bonds can be broken, allowing the fibers to stretch under tension (like a telephone cord). When the tension is released, the fibers recoil and hydrogen bonds reform. This is why human hairs can stretch to some extent and then snap back to their original length.

Another type of secondary structure is the **β-pleated sheet**[2] (Fig. 3–19*b*). The hydrogen bonding in a β-pleated sheet takes place between different polypeptide chains, or different regions of a polypeptide chain that has turned back on itself. Each chain is fully extended, but because each has a zigzag structure, the resulting "sheet" has an overall pleated conformation (much like a sheet of paper that has been folded to make a fan). Although the pleated sheet is strong and flexible, it is not elastic. This is because the distance between the pleats is fixed, determined by the strong covalent bonds of the polypeptide backbones. Fibroin, the protein of silk, is characterized by a β-pleated sheet structure, as are the cores of many globular proteins.

It is not uncommon for a single polypeptide chain to include both α-helical regions and regions with β-pleated sheet

[2]Note that the designations α and β refer simply to the order in which these two types of secondary structures were discovered.

Figure 3–18 **Primary structure of a polypeptide.** Insulin is a very small protein made up of two polypeptides, each with its own primary structure. The linear sequence of amino acids is indicated by ovals containing their abbreviated names (see Fig. 3–14).

KEY: ● Carbon atom ● Oxygen atom ● Nitrogen atom ● Hydrogen atom ○ R group

(a)

(b)

Figure 3–19 **Secondary structure of a protein.** *(a)* Note that the R groups project out from the sides of the α-helix. (The R groups have been omitted in the simplified diagram at left.) *(b)* In a β-pleated sheet, half the R groups project above the sheet and the other half project below it. *(c)* The strength and elasticity of a spider's web result from combining proteins with β-pleated sheet conformations and those with α-helical regions. *(c, Skip Moody/Dembinsky Photo Associates)*

(c)

conformations. In addition, the properties of some complex biological materials result from such combinations. A spider's web is composed of a material that is extremely strong, flexible, and elastic. Once again we see function and structure working together, as these properties derive from the fact that it is a composite of proteins with α-helical conformations (providing elasticity) and others with β-pleated sheet conformations (providing strength) (Fig. 3–19c).

Tertiary structure depends on interactions among side chains

The **tertiary structure** of a protein molecule is the overall shape assumed by each individual polypeptide chain (Fig. 3–20). This three-dimensional structure is determined by four main factors that involve *interactions among R groups (side chains) belonging to the same polypeptide chain.*

(a)   (b)

Figure 3-20 **Tertiary structure of a protein.** *(a)* Hydrogen bonds, hydrophobic interactions, and ionic attractions between R groups hold the parts of the molecule in the designated shape. Disulfide bonds are covalent bonds between the sulfur atoms of two cysteines. *(b)* Schematic drawing of the tertiary structure of a polypeptide. α-helical regions are represented as blue tubes lettered A through F; β-pleated sheets are the gray arrows numbered 1 through 12. Green lines represent connecting regions. Although the molecule seems very complicated, it is a single polypeptide chain, starting at the amino end (bottom left) and terminating at the carboxyl end. Most of the bends and foldbacks that give the molecule its overall conformation (tertiary structure) are stabilized by R-group interactions. This polypeptide is a subunit of a DNA-binding protein (CAP) from the bacterium *Escherichia coli*.

1. Hydrogen bonds form between R groups of certain amino acid subunits.
2. Ionic attraction can occur between an R group with a unit of positive charge and one with a unit of negative charge.
3. Hydrophobic interactions result from the tendency of nonpolar R groups to be excluded by the surrounding water and therefore to associate in the interior of the globular structure.
4. Covalent bonds known as disulfide bonds or disulfide bridges (—S—S—) may link the sulfur atoms of two cysteine subunits belonging to the same chain. A disulfide bridge forms when the sulfhydryl groups of two cysteines react; the two hydrogens are removed and the two sulfur atoms that remain become covalently linked.

Quaternary structure results from interactions among polypeptides

Many functional proteins are composed of two or more polypeptide chains, interacting in specific ways to form the biologically active molecule. **Quaternary structure** is the resulting three-dimensional architecture of these polypeptide chains

(each with its own primary, secondary, and tertiary structure). The same types of interactions that produce secondary and tertiary structure can also contribute to quaternary structure; these include hydrogen bonding, ionic bonding, hydrophobic interactions, and disulfide bridges.

For example, Figure 3-18 illustrates not only primary structure, but also an aspect of quaternary structure; i.e., the two disulfide bridges joining the separate polypeptide chains that make up a functional insulin molecule. (Note that the disulfide bridges of insulin would contribute to tertiary structure if they were between cysteines belonging to the *same* polypeptide chain.) A functional antibody molecule consists of four polypeptide chains joined by disulfide bridges (Chapter 43). Disulfide bridges are a common feature of proteins, such as antibodies and insulin, that are secreted from cells; these strong bonds stabilize the molecules in the extracellular environment.

Hemoglobin, the protein in red blood cells that is responsible for oxygen transport, is an example of a globular protein with quaternary structure (Fig. 3-21). Hemoglobin consists of 574 amino acids arranged in four polypeptide chains: two identical chains called alpha chains and two identical chains called beta chains.

Beta chains
(β-globin)

Heme

Alpha chains
(α-globin)

Hemoglobin

Figure 3-21 **Quaternary structure of hemoglobin.** Hemoglobin has quaternary structure because it consists of more than one polypeptide. Each polypeptide is joined to an iron-containing molecule, a heme.

The amino acid sequence of a protein determines its conformation

Under defined experimental conditions *in vitro* (that is, outside a living cell) a polypeptide can be demonstrated to spontaneously undergo folding processes that allow it to attain its normal, functional conformation. For example, researchers at the University of Illinois at Urbana-Champaign recently conducted an experiment in which they completely unfolded myoglobin, a polypeptide that stores oxygen in muscle cells, and then used sophisticated technology to track the refolding process. They found that within a few fractions of a microsecond the molecule had coiled up to form α-helices, and formation of the tertiary structure was completed within four microseconds.

This and other types of evidence support the widely held conclusion that amino acid sequence is the ultimate determinant of protein conformation. However, because conditions *in vivo* (in the cell) are quite different from defined laboratory conditions, proteins do not necessarily always fold spontaneously. On the contrary, in recent years it has been learned that proteins known as **molecular chaperones** mediate the folding of certain proteins. Chaperones are thought to make the folding process more orderly and efficient, and to prevent partially folded proteins from becoming inappropriately aggregated. However, there is no evidence that chaperones actually dictate the folding pattern. For this reason, the existence of chaperones is not an argument against the idea that amino acid sequence determines conformation.

Protein conformation is studied through a variety of methods

The conformation of a protein is ascertained directly through sophisticated types of analysis, such as the x-ray diffraction studies discussed in Chapter 11. Because these studies are tedious and costly, efforts have been made to develop alternative approaches. Today protein amino acid sequences can be determined rapidly through the application of genetic engineering techniques (Chapter 14). A variety of efforts are being made to effectively use these data to predict a protein's architecture. As we have seen, side chains can interact in relatively predictable ways, through ionic bonds, hydrogen bonds, etc. In addition, regions with certain types of side chains appear more likely to form α-helices or β-pleated sheets. Very complex computer programs are used to make such predictions, but these are imprecise because of the many possible combinations of folding patterns.

Computers are an essential part of yet another strategy. Once the amino acid sequence of a polypeptide has been determined, researchers use computers to search large databases, many accessible through the Internet, to find polypeptides with similar sequences. If the conformations of any of those polypeptides or portions of them are already known, this information can be extrapolated to make similar correlations between amino acid sequence and three-dimensional structure for the protein under investigation. These predictions are becoming increasingly reliable as more information is added to the databases on a daily basis.

Protein conformation determines function

The overall structure of a protein helps determine its biological activity. A single protein may have more than one distinct structural region, each with its own function. Many proteins are modular, consisting of two or more globular regions, called *domains,* connected by less compact regions of the polypeptide chain. Each domain may have a different function. For example, a protein might have one domain that attaches it to a membrane and another that allows it to act as an enzyme.

The biological activity of a protein can be disrupted by a change in amino acid sequence that results in a change in conformation. For example, the genetic disease known as *sickle cell anemia* (see Chapter 15) is due to a mutation that causes the substitution of the amino acid valine for glutamic acid at position 6 (the sixth amino acid from the amino end) in the beta chain of hemoglobin. The substitution of valine (which has a nonpolar side chain) for glutamic acid (which has a charged side chain) makes the hemoglobin less soluble and more likely to form crystal-like structures. This alteration of the hemoglobin affects the red blood cells, changing them to the crescent or sickle shapes that characterize this disease.

Changes in the three-dimensional structure of a protein also disrupt its biological activity. When a protein is heated, subjected to significant pH changes, or treated with any of a number of chemicals, its structure can become disordered and the coiled peptide chains can unfold to give a more random conformation. This unfolding, which is mainly due to the disruption of hydrogen bonds and ionic bonds, is typically accompanied by a loss of normal function. Such changes in shape and the accompanying loss of biological activity are termed **denaturation** of the protein. For example, a denatured enzyme would lose its ability to catalyze a chemical reaction. An everyday example of denaturation occurs when we fry an egg. The consistency of the egg white protein, known as albumin, changes to a solid. Denaturation generally cannot be reversed (you can't "unfry" an egg). However, under certain conditions, some proteins have been denatured and have returned to their original shape and biological activity when normal environmental conditions have been restored.

DNA AND RNA ARE NUCLEIC ACIDS

Nucleic acids transmit hereditary information and determine what proteins a cell manufactures. There are two classes of nucleic acids found in cells: **ribonucleic acids (RNA)** and **deoxyribonucleic acids (DNA).** DNA comprises the genes, the hereditary material of the cell, and contains instructions for making all the proteins, as well as all the RNA, needed by the organism. RNA is required as a direct participant in the complex process in which amino acids are linked to form polypeptides. Like proteins, nucleic acids are large, complex molecules. The name *nucleic acid* reflects the fact that they are acidic and were first identified, by Friederich Miescher in 1870, in the nuclei of pus cells.

Nucleic acids consist of nucleotide subunits

Nucleic acids are polymers of **nucleotides,** molecular units that consist of (1) a five-carbon sugar, either ribose or deoxyribose, (2) one or more phosphate groups, which make the molecule acidic, and (3) a nitrogenous base, a ring compound that contains nitrogen. The nitrogenous base may be either a double-ringed purine or a single-ringed pyrimidine (Fig. 3–22).

DNA commonly contains the purines adenine (A) and guanine (G), the pyrimidines cytosine (C) and thymine (T), the sugar deoxyribose, and phosphate. RNA contains the purines adenine and guanine, and the pyrimidines cytosine and uracil (U), together with the sugar ribose, and phosphate.

The molecules of nucleic acids are made of linear chains of nucleotides, which are joined by **phosphodiester linkages,** each consisting of a phosphate group and the covalent bonds that attach it to the sugars of adjacent nucleotides (Fig. 3–23). Note that each nucleotide is defined by its particular base and that nucleotides can be joined in any sequence. A nucleic acid molecule is uniquely defined by its specific sequence of nucleotides, which constitutes a kind of code (see Chapter 12). While RNA is usually composed of one nucleotide chain, DNA consists of two nucleotide chains held together by hydrogen bonds and entwined around each other in a double helix (see Fig. 1–7).

(a)

(b)

(c)

Figure 3–22 **Nucleotides.** *(a)* The three major pyrimidine bases found in nucleotides are cytosine, thymine (in DNA only), and uracil (in RNA only). *(b)* The two major purine bases found in nucleotides are adenine and guanine. The hydrogens indicated by the boxes are removed when the base is attached to a sugar. *(c)* A nucleotide, adenosine monophosphate (AMP).

Some nucleotides are important in energy transfers and other cellular functions

In addition to their importance as subunits of DNA and RNA, nucleotides serve other vital functions in living cells. **Adenosine triphosphate (ATP),** composed of adenine, ribose, and three phosphates (Fig. 3–24), is of major importance as the primary energy currency of all cells (see Chapter 6). The two terminal phosphate groups are joined to the nucleotide by un-

A nucleotide

A phosphodiester linkage

Figure 3–23 **RNA.** Nucleotides, each with a specific base, are joined by phosphodiester linkages. P, phosphate; S, the sugar ribose; G, guanine; C, cytosine; A, adenine; U, uracil.

Cyclic AMP

Figure 3–25 **Cyclic AMP.** The single phosphate is part of a ring connecting two regions of the ribose.

Phosphate groups

Adenine

Ribose

Figure 3–24 **ATP, a nucleotide.** The two terminal phosphate groups are joined by unstable bonds (indicated by wavy lines). These bonds permit the phosphates to be transferred to other molecules, making them more reactive.

Nicotinamide

Adenine

Figure 3–26 **NAD^+, an important hydrogen (electron) acceptor.** The nicotinamide portion of the molecule accepts hydrogen and becomes reduced in the process. The resulting reduced molecule, known as NADH, is an electron donor.

stable covalent bonds, traditionally indicated by wavy lines. ATP can transfer a phosphate group to another molecule, making that molecule more reactive. In this way ATP is able to donate some of its chemical energy. Most of the readily available chemical energy of the cell is associated with the phosphate groups of ATP.

A nucleotide may be converted to an alternative form with specific cellular functions. ATP, for example, is converted to **cyclic AMP** (c-AMP or cyclic adenosine monophosphate) by the enzyme adenylyl cyclase (Fig. 3–25). Cyclic AMP regulates certain cellular functions and is important in the mechanism by which some hormones act (see Chapters 13, 39, and 47).

Cells contain several dinucleotides, which are of great importance in metabolic processes. For example, as discussed in Chapter 6, **nicotinamide adenine dinucleotide** (Fig. 3–26)

has a primary role in biological oxidations and reductions within cells. It can exist in an oxidized form (**NAD$^+$**) that is converted to a reduced form (**NADH**) when it accepts electrons (in association with hydrogen). These electrons, along with their energy, can be transferred to other molecules.

BIOLOGICAL MOLECULES CAN BE RECOGNIZED BY THEIR KEY FEATURES

Although the fundamental classes of biological molecules may seem to form a bewildering array, one can learn to distinguish them readily by understanding their chief attributes. These are summarized in Table 3–2.

TABLE 3–2 Some of the Groups of Biologically Important Organic Compounds

Class of compounds	Component elements	Description	How to recognize	Principal function in living systems
Carbohydrates	C, H, O	Contain approximately 1 C:2 H:1 O (but make allowance for loss of oxygen and hydrogen when sugar units are linked)	Count the carbons, hydrogens, and oxygens.	Cellular fuel; energy storage; structural component of plant cell walls; component of other compounds such as nucleic acids and glycoproteins
		1. Monosaccharides (simple sugars) Mainly five-carbon (pentose) molecules such as ribose or six-carbon (hexose) molecules such as glucose and fructose	Look for the ring shapes:	Cellular fuel; components of other compounds
		2. Disaccharides. Two sugar units linked by a glycosidic bond, e.g., maltose, sucrose	Count sugar units	Components of other compounds; form of sugar transported in plants
		3. Polysaccharides. Many sugar units linked by glycosidic bonds, e.g., glycogen, cellulose	Count sugar units	Energy storage; structural components of plant cell walls
Lipids	C, H, O	Contain much less oxygen relative to carbon and hydrogen than do carbohydrates		Energy storage; cellular fuel, structural components of cells; thermal insulation
		1. Neutral fats. Combination of glycerol with one to three fatty acids. Monoacylglycerol contains one fatty acid; diacylglycerol contains two fatty acids; triacylglycerol contains three fatty acids. If fatty acids contain double carbon-to-carbon linkages (C=C), they are unsaturated; otherwise they are saturated	Look for glycerol at one end of molecule:	Cellular fuel; energy storage

continued

TABLE 3-2 continued

Class of compounds	Component elements	Description	How to recognize	Principal function in living systems
		2. Phospholipids. Composed of glycerol attached to one or two fatty acids and to an organic base containing phosphorus	Look for glycerol and side chain containing phosphorus and nitrogen.	Components of cell membranes
		3. Steroids. Complex molecules containing carbon atoms arranged in four attached rings. (Three rings contain six carbon atoms each, and the fourth ring contains five.)	Look for four attached rings:	Some are hormones, others include cholesterol, bile salts, vitamin D, components of cell membranes
		4. Carotenoids. Orange and yellow pigments; consist of isoprene units	Look for isoprene units. $H_2C=C-C=CH_2$ with H and CH_3	Retinal (important in photoreception) and vitamin A are formed from carotenoids.
Proteins	C, H, O, N. usually S	One or more polypeptides (chains of amino acids) coiled or folded in characteristic shapes	Look for amino acid units joined by C—N bonds.	Serve as enzymes; structural components; muscle proteins; hemoglobin
Nucleic acids	C, H, O. N, P	Backbone composed of alternating pentose and phosphate groups, from which nitrogenous bases project. DNA contains the sugar deoxyribose and the bases guanine, cytosine, adenine, and thymine. RNA contains the sugar ribose and the bases guanine, cytosine, adenine, and uracil. Each molecular subunit, called a *nucleotide,* consists of a pentose, a phosphate, and a nitrogenous base.	Look for a pentose-phosphate backbone. DNA forms a double helix.	Storage, transmission, and expression of genetic information

SUMMARY WITH KEY TERMS

I. The major groups of biologically important organic compounds are carbohydrates, lipids, proteins, and nucleic acids.

II. The properties of carbon atoms make them extraordinarily versatile, able to form the backbones of the large variety of organic compounds essential to life.

A. Each carbon atom can form four covalent bonds with four other atoms; these can be single, double, or triple bonds.

B. Carbon forms covalent bonds with a greater number of different elements than does any other type of atom. Carbon atoms can form straight or branched chains or can join into rings.

III. **Isomers** are compounds with the same molecular formula but different structures.

A. **Structural isomers** differ in the covalent arrangements of their atoms.

B. **Geometric isomers,** or *cis-trans* isomers, differ in the spatial arrangements of their atoms.

C. **Enantiomers** are isomers that are mirror images of each other. Cells can distinguish between these configurations.

IV. Organic compounds are made up of specific functional groups with characteristic properties.

A. **Hydrocarbons** are nonpolar and hydrophobic.

B. Polar and ionic functional groups interact with each other and dissolve in water.

C. Partial charges on atoms at opposite ends of a bond are responsible for the polar property of a functional group. **Hydroxyl** and **carbonyl** groups are polar.

D. **Carboxyl** and **phosphate** groups are acidic, becoming negatively charged when they release hydrogen ions. The **amino** group is basic, becoming positively charged when it accepts a hydrogen ion.

V. Long chains of similar organic compounds linked together are called **polymers.** Large polymers such as polysaccharides, proteins, and DNA are referred to as **macromolecules.**

VI. **Carbohydrates** contain carbon, hydrogen, and oxygen in a ratio of approximately one carbon to two hydrogens to one oxygen.

A. **Monosaccharides** are simple sugars such as glucose, fructose, and ribose.

B. Two monosaccharides can be joined by a glycosidic linkage, forming a **disaccharide** such as maltose or sucrose.

C. Most carbohydrates are **polysaccharides,** long chains of repeating units of a simple sugar.

1. Carbohydrates are typically stored in plants as **starch** and in animals as **glycogen.**

2. The cell walls of plants are composed mainly of the polysaccharide **cellulose.**

VII. **Lipid** molecules are composed mainly of hydrocarbon-containing regions, with few oxygen-containing (polar or ionic) functional groups. Lipids have a greasy or oily consistency and are relatively insoluble in water.

A. **Neutral fats** are used for fuel storage. A fat consists of a molecule of **glycerol** combined with one to three **fatty acids.**

1. **Monoacylglycerols, diacylglycerols,** and **triacylglycerols** are neutral fats containing one, two, and three fatty acids, respectively.

2. A fatty acid can be either **saturated** with hydrogen, or **unsaturated.**

B. **Phospholipids** are structural components of cellular membranes.

C. **Steroid** molecules contain carbon atoms arranged in four attached rings. Cholesterol, bile salts, and certain hormones are important steroids.

VIII. **Proteins** are large, complex molecules made of simpler subunits, called **amino acids,** joined by **peptide bonds.**

A. Proteins are the most versatile class of biological molecules, serving as **enzymes,** structural components, cellular regulators, etc.

B. Proteins are composed of various linear sequences of 20 different amino acids. Two amino acids combine to form a **dipeptide.** A longer chain of amino acids is a **polypeptide.**

1. All amino acids contain an amino group and a carboxyl group, but vary in their side chains. The side chains of amino acids dictate their chemical properties.

2. Amino acids generally exist as dipolar ions at cellular pH and serve as important biological buffers.

C. Four levels of organization can be distinguished in protein molecules.

1. **Primary structure** is the linear sequence of amino acids in the polypeptide chain.

2. **Secondary structure** is a regular conformation, such as an **α-helix** or a **β-pleated sheet,** due to hydrogen bonding between elements of the uniform backbone of the polypeptide.

3. **Tertiary structure** is the overall shape of the polypeptide chains, as dictated by chemical properties and interactions of the side chains of specific amino acids. Hydrogen bonding, ionic bonds, hydrophobic interactions, and disulfide bridges contribute to tertiary structure.

4. **Quaternary structure** is determined by the association of two or more polypeptide chains.

IX. The **nucleic acids** DNA and RNA store and transfer information that governs the sequence of amino acids in proteins and ultimately the structure and function of the organism.

A. Nucleic acids are composed of long chains of **nucleotide** subunits, each composed of a purine or pyrimidine nitrogenous base, a five-carbon sugar (ribose or deoxyribose), and a phosphate group.

B. **ATP** (adenosine triphosphate) is a nucleotide of special significance in energy metabolism. **NAD$^+$** is also involved in energy metabolism through its role as an electron (hydrogen) acceptor in biological oxidations.

POST-TEST

1. Which of the following would be considered to be an inorganic form of carbon? (a) H_2CO_3 (b) C_2H_4 (c) CH_3COOH (d) b and c (e) all of the above are inorganic

2. Carbon is particularly well suited to be the backbone of organic molecules because (a) it can form both covalent bonds and ionic bonds (b) its covalent bonds are very irregularly arranged in three-dimensional space (c) its covalent bonds are the strongest chemical bonds known (d) it can bond to a large number of other types of atoms (e) all of the bonds it forms are polar

3. The structures depicted below are

$$\begin{array}{cc} \text{CH}_3\ \text{CH}_3 & \text{H}\quad\text{CH}_3 \\ |\quad | & |\quad | \\ \text{H—C—C—H} & \text{H—C—C—H} \\ |\quad | & |\quad | \\ \text{H}\quad\text{H} & \text{CH}_3\ \text{H} \end{array}$$

(a) enantiomers (b) different views of the same molecule (c) geometric *(cis-trans)* isomers (d) both geometric isomers and enantiomers (e) structural isomers

4. Which of the following are generally hydrophobic? (a) polar molecules and hydrocarbons (b) ions and hydrocarbons (c) nonpolar molecules and ions (d) polar molecules and ions (e) none of the above

5. Which of the following is a nonpolar molecule? (a) water (H_2O) (b) ammonia (NH_3) (c) methane (CH_4) (d) ethane (C_2H_6) (e) more than one of the above

6. Which of the following functional groups normally acts as an acid? (a) hydroxyl (b) carbonyl (c) sulfhydryl (d) phosphate (e) amino

7. A monosaccharide designated as an aldehyde sugar contains (a) a terminal carboxyl group (b) an internal carboxyl group (c) a terminal carbonyl group (d) an internal carbonyl group (e) a terminal carboxyl group and an internal carbonyl group

8. Structural polysaccharides typically (a) have extensive hydrogen bonding between adjacent molecules (b) are much more hydrophilic than storage polysaccharides (c) have much stronger covalent bonds than do storage polysaccharides (d) consist of alternating α-glucose and β-glucose subunits (e) form helical structures in the cell

9. Fatty acids are components of (a) phospholipids and carotenoids (b) carotenoids and triacylglycerol (c) steroids and triacylglycerol (d) phospholipids and triacylglycerol (e) carotenoids and steroids

10. Saturated fatty acids are so named because they are saturated with (a) hydrogen (b) water (c) hydroxyl groups (d) glycerol (e) double bonds

11. Which pair of amino acid side groups would be most likely to associate with each other by an ionic bond?
    1. —CH$_3$
    2. —CH$_2$—COO$^-$
    3. —CH$_2$—CH$_2$—NH$_3^+$
    4. —CH$_2$—CH$_2$—COO$^-$
    5. —CH$_2$—OH
    (a) 1 and 2 (b) 2 and 4 (c) 1 and 5 (d) 2 and 5 (e) 3 and 4

12. Which of the following levels of protein structure may be affected by hydrogen bonding? (a) primary and secondary (b) primary and tertiary (c) secondary, tertiary, and quaternary (d) primary, secondary, and tertiary (e) primary, secondary, tertiary, and quaternary

13. Each phosphodiester linkage in DNA or RNA includes a phosphate joined by covalent bonds to (a) two bases (b) two sugars (c) two additional phosphates (d) a sugar, a base, and a phosphate (e) a sugar and a base

REVIEW QUESTIONS

1. Marble is composed of the carbon-containing compound calcium carbonate ($CaCO_3$). Should calcium carbonate be considered an organic molecule? Why or why not?
2. What are some of the ways that the features of carbon-to-carbon bonds influence the stability and three-dimensional structure of organic molecules?
3. Draw pairs of simple sketches comparing two (a) structural isomers, (b) geometric isomers, and (c) enantiomers.
4. Sketch the following functional groups: methyl, amino, carbonyl, hydroxyl, carboxyl, and phosphate. Classify each as either nonpolar, polar, acidic, or basic.
5. What features related to hydrogen bonding give storage polysaccharides, such as starch and glycogen, very different properties from structural polysaccharides, such as cellulose and chitin?
6. Draw a structural formula of a simple amino acid and identify the carboxyl group, amino group, and R group.
7. How does the primary structure of a polypeptide influence its secondary and tertiary structures? How can the conformation of a protein be disrupted?
8. Compare the functions of proteins and nucleic acids.

YOU MAKE THE CONNECTION

1. Like oxygen, sulfur forms two covalent bonds. However, it is far less electronegative. In fact, it is approximately as electronegative as carbon. How would the properties of the various classes of biological molecules be altered if you were to replace all the oxygen atoms with sulfur atoms?
2. How would you reconcile the following statements? (1) All organisms share a common ancestry. (2) Individual organisms are generally biochemically unique.

RECOMMENDED READINGS

Atkins, P.W. *Molecules.* Scientific American Library, W.H. Freeman and Co., New York, 1987. A fascinating account of the relationship between the properties and the molecular structures of various substances.

Bettelheim, F.A. and J. March. *Introduction to General, Organic and Biochemistry,* 4th ed. Saunders College Publishing, Philadelphia, 1995. A very readable reference text for those who would like to know more about the chemistry basic to life.

Garrett, R.H. and C.M. Grisham. *Biochemistry.* Saunders College Publishing, Philadelphia, 1995. A comprehensive, advanced biochemistry text.

Richards, F.M. "The Protein Folding Problem." *Scientific American,* Vol. 264, No. 1, Jan. 1991. A discussion of the mechanisms involved when a protein folds into its biologically active shape.

● Visit our website at **http://www.saunderscollege.com/lifesci/titles.html** and click on Solomon/Berg/Martin Biology for links to chapter-related resources on the World Wide Web.

CHAPTER 4

Organization of the Cell

Cells are a dramatic example of the underlying unity of all living things. When we examine a wide range of seemingly very diverse organisms, ranging from simple bacteria to the most complex plants and animals, we find striking similarities at the cell level. This is a reflection of their evolution from a common ancestor, as well as the fact that living things have many common needs. Careful studies of shared cell features help us trace the evolutionary history of various groups of organisms and furnish powerful evidence that all organisms alive today had a common origin.

Each cell is a virtual microcosm of life, for it is the smallest unit that can carry out all life activities. When provided with essential nutrients and an appropriate environment, some cells can be kept alive and growing in the laboratory for many years. By contrast, no isolated cell part is capable of sustained survival. Composed of a vast array of inorganic and organic ions and molecules including water, salts, carbohydrates, lipids, proteins, and nucleic acids, most cells have all the physical and chemical components needed for their own maintenance, growth, and division. Genetic information is stored in DNA molecules and is faithfully replicated and passed on to each new generation of cells during cell division. Information in DNA codes for specific proteins that in turn determine cell structure and function. In this chapter and those that follow we will discuss how cells use many of the chemical materials introduced in Chapters 2 and 3.

Cells exchange materials and energy with the environment. All living cells need one or more sources of energy, but a cell rarely obtains energy in a form that is immediately usable. Cells convert energy from one form to another, and that energy is used to carry out various activities, ranging from mechanical work to chemical synthesis. Cells convert energy to a convenient form, usually chemical energy stored in ATP (adenosine triphosphate; see Chapter 3). Although the specifics vary, the basic strategies cells use for energy conversion are very similar. The chemical reactions that convert energy from one form to another are essentially the same in all cells, from bacteria to those of complex plants and animals.

Cells are the building blocks of complex multicellular organisms. Although cells are basically similar, they are also ex-

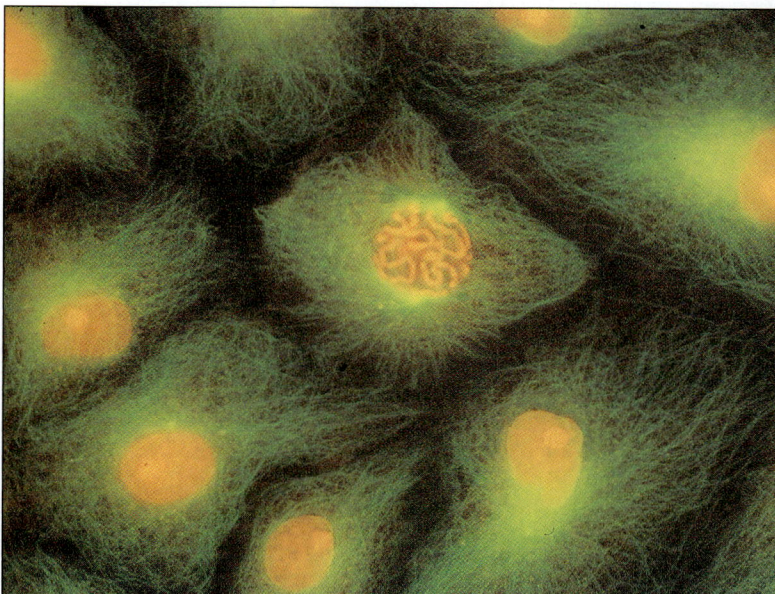

(Courtesy of Dr. John M. Murray, Department of Cell and Developmental Biology, University of Pennsylvania)

traordinarily versatile. They can be modified in a variety of ways to carry out specialized functions.

Biologists have long labored to understand the mechanisms of cell structure and function. The cells shown here were stained with fluorescent antibodies (specific proteins) that bound to proteins associated with DNA *(orange)* and to a protein (tubulin) in one of the cells' skeletal elements, microtubules *(green)*. This type of microscopy, known as *confocal fluorescence microscopy,* shows the extensive distribution of microtubules in these cells. The cell skeleton (cytoskeleton) functions in maintaining cell shape, in cell movement, and in transport of materials within the cell. Investigation of the cytoskeleton is currently a very active and exciting area of research.

1. Explain why the cell is considered the basic unit of life and discuss some of the implications of the cell theory.
2. Compare and contrast the general characteristics of prokaryotic and eukaryotic cells.
3. Compare size relationships among different cells and cell structures.
4. Explain why the relationship between surface area and volume of a cell is important in determining cell size limits.
5. Describe the structure of the nucleus and relate this information to its function in eukaryotic cells.
6. Distinguish between smooth and rough endoplasmic reticulum in terms of both structure and function and discuss the relationship between the endoplasmic reticulum and other internal membranes in the cell.
7. Trace the path of certain proteins synthesized in the rough endoplasmic reticulum as they are subsequently processed, modified, and sorted by the Golgi complex.
8. Describe the functions of lysosomes. Explain what can happen when they fail to carry out their functions or when they leak.
9. Compare the functions of chloroplasts and mitochondria and discuss ATP synthesis by each of these organelles.
10. Explain the importance of the cytoskeleton to the cell and describe the structures of the major types of fibers that make up the cytoskeleton.
11. Relate the structural features of cilia and flagella to the way in which these organelles are able to move.

THE CELL IS THE BASIC UNIT OF LIFE

Two German scientists, botanist Matthias Schleiden in 1838 and zoologist Theodor Schwann in 1839, were the first to point out that all plants and animals are composed of cells. Later, Rudolph Virchow, a physician, observed cells dividing and giving rise to daughter cells. In 1855, Virchow proposed that new cells are formed only by the division of previously existing cells. In other words, cells do not arise by spontaneous generation from nonliving matter, a belief that had prevailed for centuries.

The work of Schleiden, Schwann, and Virchow gave rise to the **cell theory,** the unifying concept that cells are the basic living units of organization and function in all organisms and that all cells come from other cells. About 1880 another biologist, August Weismann, added an important corollary to Virchow's concept by pointing out that the ancestry of all the cells alive today can be traced back to ancient times. Evidence that all presently living cells have a common origin is provided by the basic similarities in their structures and in the molecules of which they are made.

CELL ORGANIZATION AND SIZE PERMIT HOMEOSTASIS

Recall from Chapter 1 that **homeostasis** is the process of maintaining an internal environment that is appropriate and supportive to life. Cells experience constant changes in their environments such as deviations in salt concentration, pH, and temperature. In order for its biochemical mechanisms to function, the cell must work continuously to restore appropriate conditions.

Organization is basically similar in all cells

Every cell must be able to keep its contents together and also keep them separated from the external environment. For this reason all cells, from bacteria to human cells, are enclosed by a structurally distinctive surface membrane commonly known as the **plasma membrane.** Cells must also be able to accumulate materials and energy stores and to exchange materials with the environment, usually in a highly regulated fashion. Therefore the plasma membrane must serve as an extremely selective barrier, making the interior of the cell an enclosed compartment with a chemical composition quite different from that outside.

Typically, cells have internal structures, called **organelles,** specialized to carry on life activities such as converting energy to usable forms, synthesizing needed compounds, and manufacturing structures essential to function and reproduction. Each cell has genetic instructions in the form of DNA, which is concentrated in a limited region of the cell.

Cell size is limited

Although their sizes vary over a wide range (Fig. 4–1), most cells are microscopic, and very small units are required to measure them and their internal structures. The basic unit of linear measurement in the metric system (see inside back cover) is the meter (m), which is just a little longer than a yard. A millimeter (mm) is 1/1000 of a meter and is about as long as the bar enclosed in parentheses (-). The micrometer (μm) is the most convenient unit for measuring cells. A bar 1 μm long is far too short to be seen with the unaided eye, for it is 1/1,000,000 (one-millionth) of a meter, or 1/1000 of a millimeter, long. Most of us have difficulty thinking about units

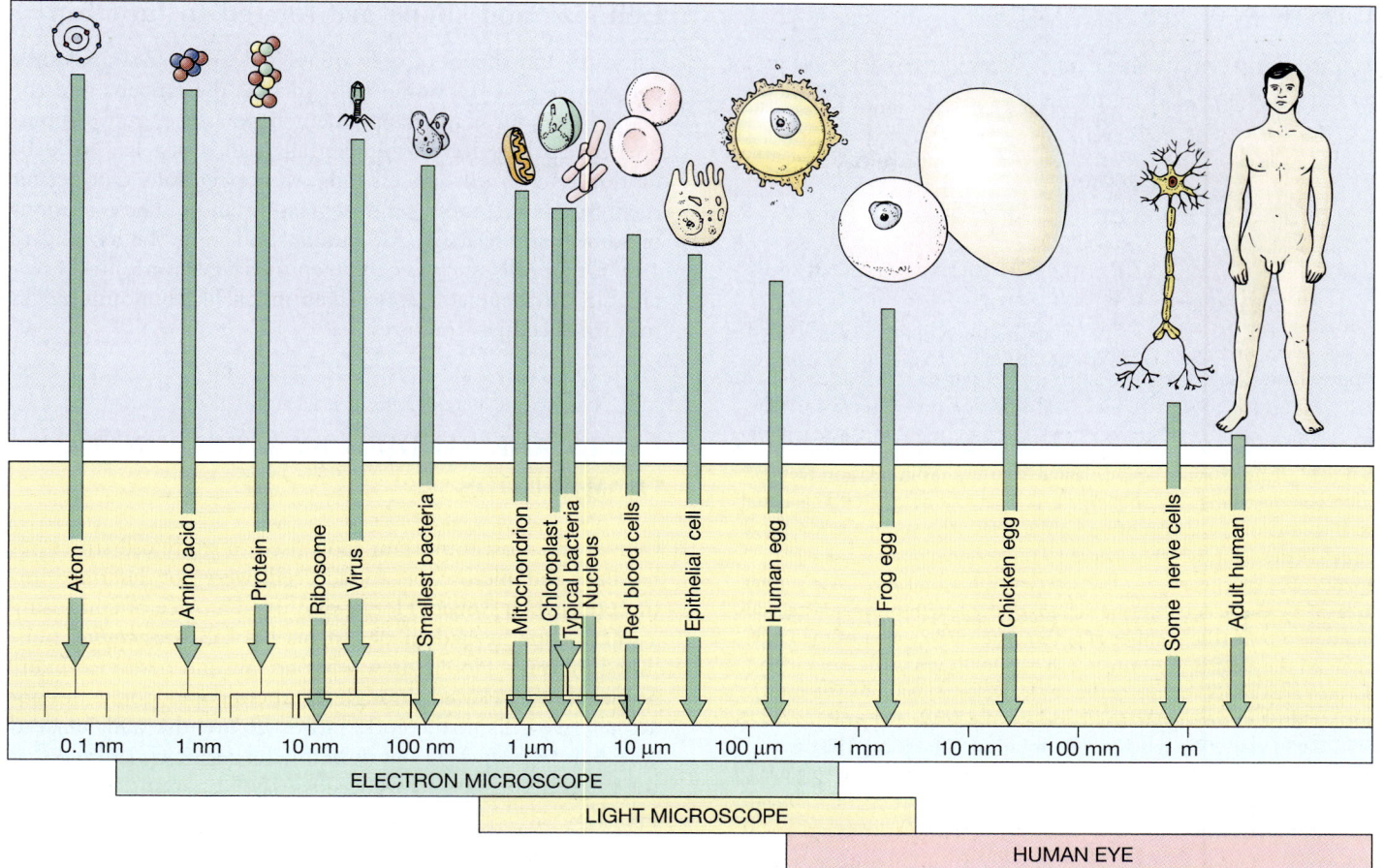

Figure 4-1 **Biological size and cell diversity.** Relative size from chemical to organismic levels is most conveniently compared using a logarithmic scale (multiples of ten). The prokaryotic cells of bacteria typically range in size from less than 1 to 10 μm long; their small size enables them to grow and divide rapidly. Eukaryotic cells (cells of all other organisms) are typically 10 to 100 μm in diameter; the majority are between 10 and 30 μm. The nuclei of animal and plant cells range from about 3 to 10 μm in diameter. Mitochondria are about the size of small bacteria, whereas chloroplasts are usually larger, about 5 μm long. Ova (egg cells) are among the largest cells. Although microscopic, some nerve cells are very long, specialized to transmit messages from one part of the body to another. The cells shown here are not drawn to scale.

that are too small to see, but it is very helpful to remember that a micrometer has the same relationship to a millimeter that a millimeter has to a meter (1/1000).

As small as it is, the micrometer is actually too large to measure most cellular components. For these purposes we use the nanometer (nm), which is 1/1,000,000,000 (one-billionth) of a meter, or 1/1000 of a micrometer. To mentally move down to the world of the nanometer, recall that a millimeter is one thousandth of a meter, a micrometer is one thousandth of a millimeter, and a nanometer is one thousandth of a micrometer.

A good light microscope allows us to see most types of bacterial cells, and some specialized algae and animal cells are large enough to be seen with the naked eye. A human egg cell, for example, is about 130 μm in diameter, or approximately the size of the period at the end of this sentence. The largest cells are birds' eggs, but they are atypical because both the yolk and the egg white consist of food reserves. The functioning part of the cell is a small mass on the surface of the yolk.

Why are most cells so small? If you consider what a cell must do to grow and survive, it may be easier to understand the reasons for its small size. A cell must take in food and other materials and must rid itself of waste products generated by metabolic reactions. Everything that enters or leaves a cell must pass through its plasma membrane. The plasma membrane contains specialized "pumps" and "gates" that selectively regu-

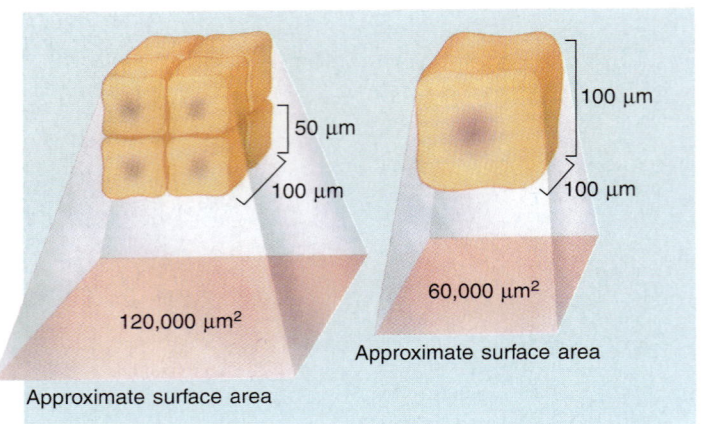

Figure 4–2 **Surface area to volume ratio.** Eight small cells have a much greater surface area (plasma membrane) in relation to their total volume than does one large cell. The volume of a cube is equal to the cube of the length of one of its sides. The volume of the large cube on the right, which is equal to the total volume of the eight small cubes on the left, is $(100 \ \mu m)^3 = 1,000,000 \ \mu m^3$. The surface area of a cube is equal to the square of the length of one of its sides, multiplied by 6 (the number of sides). The surface area of the large cube on the right is $(100 \ \mu m)^2 \times 6 = 60,000 \ \mu m^2$. The ratio of its surface area to its volume (surface area:volume) is 0.06. The total surface area of the eight small cubes on the left is $(50 \ \mu m)^2 \times 6 \times 8 = 120,000 \ \mu m^2$. The ratio of their total surface area to their total volume is 0.12—double the surface-to-volume ratio of the single large cube.

late the passage of materials into and out of the cell. The plasma membrane must be large enough relative to the cell volume to keep up with the demands of regulating the passage of materials. This means that a critical factor in determining cell size is the ratio of its surface area to its volume (Fig. 4–2).

As a cell becomes larger its volume increases at a greater rate than its surface area (i.e., its plasma membrane), which effectively places an upper limit on cell size. Above some critical size, the number of molecules required by the cell could not be transported into the cell fast enough to sustain its needs. In addition, the cell would not be able to regulate its concentration of various ions or efficiently export its wastes.

Of course, not all cells are spherical or cuboidal. Some very large cells have relatively favorable ratios of surface area to volume because of their shapes. In fact, some variations in cell shape represent a strategy for increasing the ratio of surface area to volume. For example, many large plant cells are long and thin, which increases their surface-to-volume ratio. Some cells, such as epithelial cells, have finger-like projections of the plasma membrane, called *microvilli,* that significantly increase the surface area used to absorb nutrients and other materials.

Another reason that cells are small is that, once inside, molecules must be transported to the locations where they are converted into other forms. Because cells are small, the distances molecules travel within them are relatively short, which speeds up many cellular activities.

Cell size and shape are related to function

The sizes and shapes of cells are related to the functions they perform (Fig. 4–1). Some cells, such as the amoeba and the white blood cell, can change their shape as they move about. Sperm cells have long, whiplike tails, called flagella, for locomotion. Nerve cells possess long, thin extensions that permit them to transmit messages over great distances. The extensions of some nerve cells in the human body may be as long as 1 m. Other cells, such as certain epithelial cells, are almost rectangular in shape and are stacked much like building blocks to form sheetlike structures.

CELLS ARE STUDIED BY A COMBINATION OF METHODS

One of the most important tools used to study cell structures has been the microscope. In fact, cells were not described until 1665, when Robert Hooke examined a piece of cork using a microscope he had made. In his book *Micrographia,* published in 1665, Hooke drew what he saw and described many objects that he viewed through his microscope. Hooke did not actually see cells in the cork; rather, he saw the walls of dead cork cells (Fig. 4–3a). Not until much later was it realized that the interior enclosed by the walls is the important part of living cells.

Light microscopes are used to study stained or living cells

A few years after Hooke described dead cork cells, the Dutch naturalist Anton van Leeuwenhoek viewed living cells with small lenses that he made. However, he did not share his lens-making techniques, and more than a century passed before biologists realized the importance of microscopes and what they could reveal. It was not until the early 19th century that microscopes were sufficiently developed for biologists to begin their study of cells.

The **light microscope (LM),** the type used by most students, consists of a tube with glass lenses at each end. (Because it contains several lenses, the light microscope is sometimes called a compound microscope.) Visible light passes through the specimen being observed and through the lenses. Light is refracted (bent) by the lenses, magnifying the image.

Two features of a microscope determine how clearly a small object can be viewed: magnification and resolving power. *Magnification* is the ratio of the size of the image seen with the microscope to the actual size of the object. The best light microscopes usually magnify an object no more than 1000 times. **Resolution,** or **resolving power,** is the capacity to distinguish fine detail in an image. This is defined as the minimum distance between two points at which they can both be seen separately rather than as a single, blurred point. Resolving power depends on the quality of the lenses and the *wave-*

Figure 4–3 **Viewing cells with various types of microscopes.** *(a)* Using a crude microscope that he constructed, Robert Hooke looked at a thin slice of cork and drew what he saw. Hooke used the term *cell* because the tissue reminded him of the small rooms that monks lived in during that period. More sophisticated microscopes and techniques permit biologists to view cells in more detail. Unstained epithelial cells from the skin of human cheek are compared using *(b)* bright field (transmitted light), *(c)* dark field, *(d)* phase contrast, and *(e)* Nomarski differential interference microscopy. The phase contrast and differential interference microscopes enhance detail by increasing the differences in optical density in different regions of the cells. *(a, from Hooke's Micrographica, 1665; b–e, Jim Solliday/Biological Photo Service)*

length of the illuminating light. As the wavelength decreases, the resolution increases. The visible light used by light microscopes has wavelengths ranging from about 400 nm (violet) to 700 nm (red); this limits the resolution of the light microscope to details no smaller than the diameter of a small bacterial cell (about 1 micrometer).

By the early 20th century, refined versions of the light microscope, as well as certain organic compounds that specifically stain different cell structures, became available. These enabled biologists to discover that cells contain a number of different internal structures, the organelles. The contribution of organic chemists in the development of biological stains was essential to this understanding, because the interior of many cells is transparent. Most of the methods used to prepare and stain cells for observation, however, also kill them in the process.

More recently, sophisticated types of light microscopes have been developed that use interfering waves of light to enhance the internal structures of cells. With *phase contrast* and *Nomarski differential interference microscopes,* some internal structures can be seen in unstained living cells (Fig. 4–3 *d* and

e). One of the most striking things that can be observed with these microscopes is that living cells contain numerous internal structures that are constantly changing shape and location.

Fluorescence microscopes are used to detect the locations of specific molecules in cells. Fluorescent stains (like paints that glow under black light) are molecules that absorb light energy of one wavelength and then release some of that energy as light of a longer wavelength. One such stain binds specifically to DNA molecules and emits green light after absorbing ultraviolet light. Cells can be stained, and the location of the DNA can be determined, by observing the source of the green fluorescent light within the cell. Some fluorescent stains can be chemically bonded to *antibodies,* protein molecules important in internal defense. The antibody can then bind to a highly specific region of a molecule in the cell. A single type of antibody molecule can bind to only one type of structure, such as a part of a specific protein or some of the sugars in a specific polysaccharide. Purified fluorescent antibodies known to bind to a specific protein isolated from a cell can be used to determine where that protein is located. Powerful computer imaging methods have allowed the development of the *confocal flu-*

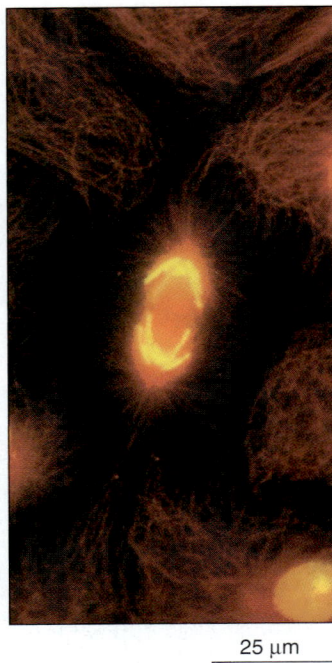

25 μm

Figure 4–4 **Confocal fluorescence micrograph of cultured animal cells.** The cell in the center is dividing. The DNA of the chromosomes is yellow; the microtubules (skeletal elements of the cells) are red. *(Courtesy of Dr. John M. Murray, Department of Cell and Developmental Biology, University of Pennsylvania)*

orescence microscope, which greatly improves the resolution of structures labeled by fluorescent dyes (Fig. 4–4 and LM in chapter introduction).

Cell biologists are developing new techniques for viewing cells, using computers, lasers, and photodetectors. Computer-based image processing synthesizes multiple images to produce three-dimensional views.

Electron microscopes provide a high resolution image that can be greatly magnified

Even with improved microscopes and techniques for staining cells, ordinary light microscopes can distinguish only the gross details of many cell parts. In most cases all that can be seen clearly is the outline of a structure and its ability to be stained by some dyes and not by others. Not until the development of the **electron microscope (EM),** which came into wide use in the 1950s, were researchers able to study the fine details, or **ultrastructure,** of cells.

Whereas light microscopes magnify an object no more than about 1000 times, the electron microscope can magnify it 250,000 times or more. And while the best light microscopes have about 500 times more resolution than the human eye, the electron microscope multiplies the resolving power by more than 10,000 (Fig. 4–5). This is because electrons have very short wavelengths, on the order of about 0.1 to 0.2 nm. Al-

though such resolution is difficult to achieve with biological material, it can be approached when isolated molecules such as proteins or DNA are examined.

The image formed by the electron microscope cannot be seen directly. The electron beam itself consists of energized electrons, which, because of their negative charge, can be focused by electromagnets just as images are focused by glass lenses in a light microscope (Fig. 4–5). For **transmission electron microscopy (TEM),** the specimen is embedded in plastic and then cut into extraordinarily thin sections (50 to 100 nm thick) with a glass or diamond knife. A section is then placed on a small metal grid. The electron beam passes through the specimen and then falls onto a photographic plate or a fluorescent screen that works much like a television screen. When you look at transmission electron micrographs in this chapter (and elsewhere), keep in mind that each represents only a thin cross section of a cell.

To reconstruct a three-dimensional view of the cell, it is necessary to study many consecutive sectional views (called *serial sections*) through the object. To understand the enormity of such a task, try imagining what it would be like to reconstruct the contents of your home from a set of hundreds of consecutive 5-cm sections.

Special methods using antibody molecules that have very tiny gold particles bound to them allow the detection of specific molecules in electron microscope images. The dense gold particles block the electron beam and identify the location of the proteins recognized by the antibodies as precise black spots on the electron micrograph.

In another type of electron microscope, the **scanning electron microscope (SEM),** the electron beam does not pass through the specimen. Instead, the specimen is coated with a thin film of gold or some other metal. When the electron beam strikes various points on the surface of the specimen, secondary electrons are emitted whose intensity varies with the contour of the surface. The recorded emission patterns of the secondary electrons give a three-dimensional picture of the surface (Fig. 4–5). The SEM provides information about the shape and external features of the specimen that cannot be obtained with the TEM.

Note that the light microscope, TEM, and SEM are focused by similar principles. A beam of light or an electron beam is directed by the condenser lens onto the specimen and is magnified by the objective lens and the eyepiece in the light microscope or by the objective lens and the projector lens in the TEM. The TEM image is focused onto a fluorescent screen, and the SEM image is viewed on a type of television screen. Lenses in the electron microscopes are actually magnets that bend the beam of electrons.

Cell fractionation procedures permit study of cell components

The EM is a powerful tool for studying cell structure, but it has limitations. The methods used to prepare cells for electron microscopy kill them and may even alter their structure. Fur-

Light microscope

- Light beam
- Ocular lens
- Objective lens
- Specimen
- Condenser lens
- Light source

Transmission electron microscope

- Electron gun
- Electron beam
- First condenser lens (magnet)
- Specimen
- Projector lens (magnet)
- Film or screen

Scanning electron microscope

- Second condenser lens
- Scanning coil
- Final (objective) lens
- Cathode ray tube synchronized with scanning coil
- Secondary electrons
- Specimen
- Electron detector

(a) 100 μm

(b) 1 μm

(c) 100 μm

Figure 4–5 Comparison of light and electron microscopy. Distinctive images of cells, such as the protist *Paramecium* shown in the photomicrographs, are provided by three types of microscopes. *(a)* A phase-contrast light microscope can be used to view stained or living cells, but at relatively low resolution. *(b)* The transmission electron microscope (TEM) produces a high resolution image that can be greatly magnified. Because of the high magnification, only a small part of the *Paramecium* is shown in the photograph. *(c)* The scanning electron microscope (SEM) is used to provide a clear view of surface features. *(Photos courtesy of T.K. Maugel/University of Maryland)*

thermore, electron microscopy provides few clues about the functions of organelles and other cell components. To determine what organelles actually do, researchers have to be able to purify different parts of cells so that they can be studied by physical and chemical methods.

Cell fractionation procedures are methods for purifying organelles. Generally, cells are broken apart as gently as possible, and the mixture, referred to as the cell extract, is subjected to centrifugal force by spinning in a device called a **centrifuge.** An ultracentrifuge, a very powerful centrifuge, can spin at

Figure 4–6 Cell fractionation. Differential centrifugation permits cell biologists to separate cell structures into various fractions by spinning the suspension at increasing revolutions per minute. Membranes and organelles from the resuspended pellets can then be further purified by density gradient centrifugation, shown as the last step in the figure.

speeds exceeding 100,000 rpm (revolutions per minute), generating a centrifugal force of 500,000 Gs (a G is equal to the force of gravity). Centrifugal force separates the extract into two fractions: a pellet and a supernatant. The *pellet,* containing heavier materials, such as nuclei, packed together, forms at the bottom of the tube. The *supernatant,* the liquid above the pellet, contains lighter particles, dissolved molecules, and ions (Fig. 4–6).

The supernatant can be centrifuged again at a higher speed to obtain a pellet that contains the next heaviest cell components, for example mitochondria and chloroplasts. In **differential centrifugation,** the supernatant is spun at successively higher speeds, permitting various cell components to be separated on the basis of their different sizes and densities.

Cell components in the resuspended pellets can be further purified by **density gradient centrifugation.** In this procedure, the resuspended pellet is placed in a layer on top of a density gradient, usually made up of a solution of sucrose and water. The concentration of sucrose (table sugar) is highest at the bottom of the tube and decreases gradually so that it is lowest at the top. Because the densities of organelles differ, each will migrate during centrifugation and form a band at the position in the gradient where its own density equals that of the sucrose solution.

Purified organelles can be examined to determine what kinds of proteins and other molecules they might contain, as well as the nature of the chemical reactions that take place within them. Today, cell biologists often use a combination of experimental approaches to study the functions of cellular structures.

PROKARYOTIC CELLS ARE STRUCTURALLY SIMPLER THAN EUKARYOTIC CELLS

Recall from Chapter 1 that two basic types of cells are known: **prokaryotic** and **eukaryotic cells.** Only bacteria have prokaryotic cells. A major difference between prokaryotic and eukaryotic cells is that the DNA of prokaryotic cells is not enclosed in a nucleus. In fact the term prokaryotic means "before the nucleus." In the prokaryotic cell, the DNA is located in a limited region of the cell called a **nuclear area,** or **nucleoid** that is not enclosed by a membrane (Fig. 4–7). Other types of internal membrane-bounded organelles are also absent in prokaryotic cells. These cells are typically smaller than eukaryotic cells. In fact, the average prokaryotic cell is only about one-tenth the diameter of the average eukaryotic cell.

Figure 4–7 **TEM of a prokaryotic cell.** This bacterium, *Bacillus subtilis,* which is dividing, has a prominent cell wall surrounding the plasma membrane. The nuclear areas are clearly visible. *(Courtesy of A. Ryter)*

Like eukaryotic cells, prokaryotic cells have a plasma membrane that confines the contents of the cell to an internal compartment. In some prokaryotic cells the plasma membrane may be folded inward to form a complex of membranes along which the cell's energy-transforming reactions take place. Most prokaryotic cells have **cell walls**—structures that enclose the entire cell, including the plasma membrane. Many prokaryotes have **flagella** (sing., *flagellum*), long fibers that project from the surface of the cell. Flagella operate like propellers and are important in locomotion.

The dense internal material of the bacterial cell contains **ribosomes,** small complexes of RNA and protein that synthesize polypeptides (discussed later in this chapter), as well as storage granules that hold glycogen, lipid, or phosphate compounds. The ribosomes of prokaryotic cells are smaller than those found in eukaryotic cells. Prokaryotic cells are discussed in more detail in Chapter 23.

EUKARYOTIC CELLS ARE CHARACTERIZED BY MEMBRANE-BOUNDED ORGANELLES

Eukaryotic cells are characterized by highly organized membrane-bounded organelles. The most prominent of these is the *nucleus,* which contains the hereditary material, DNA. In fact, the name eukaryote means "true nucleus." Table 4–1 summarizes the types of organelles typically found in eukaryotic cells. Some organelles may be found only in specific cells. For example, chloroplasts, structures that trap sunlight for energy conversion, are found only in cells that carry on photosynthesis, such as certain plant cells. The many specialized organelles of eukaryotic cells allow them to overcome some of the problems associated with large size, permitting eukaryotic cells to be considerably larger than prokaryotic cells.

Early biologists thought that the cell consisted of a homogeneous jelly, which they called *protoplasm.* With the elec-

tron microscope and other modern research tools, perception of the environment within the cell has been greatly expanded. We now know that the cell is highly organized and complex (Figs. 4–8 to 4–11). It has its own control center, internal transportation system, power plants, factories for making needed materials, packaging plants, and even a "self-destruct" system. Biologists refer to the part of the cell outside the nucleus as **cytoplasm** and the part of the cell within the nucleus as **nucleoplasm.** Various organelles are suspended within the fluid component of the cytoplasm, which is referred to as the **cytosol.** Therefore, the term *cytoplasm* includes both the cytosol and all the organelles other than the nucleus.

Membranes divide the cell into compartments

Membranes have unique properties that enable membranous organelles to carry out a wide variety of functions. For example, cell membranes never have free ends; therefore a membranous organelle always contains at least one enclosed internal space or compartment. These membrane-bounded compartments allow certain cell activities to be localized within specific enclosed regions of the cell. Reactants that are located in only a small part of the total cell volume are far more likely to come in contact, and the rate of the reaction can be dramatically increased. Membrane-bounded compartments keep certain reactive compounds away from other parts of the cell that might be adversely affected by them. Compartmentalizing also permits many different activities to go on simultaneously.

Membranes also allow the storage of energy. The membrane provides a barrier that is analogous to a dam on a river. A difference in the concentration of some substance on the two sides of a membrane is a form of stored energy or *potential energy* (see Chapter 6). As particles of the substance move across the membrane from the side of higher concentration to the side of lower concentration, the cell can convert some of this potential energy to the chemical energy of ATP molecules. This process of energy conversion (discussed in Chapters 7 and 8) is a basic mechanism that cells use to capture and convert energy to sustain life.

Membranes also serve as important work surfaces. For example, a number of chemical reactions in cells are carried out by enzymes that are bound to membranes. Because the enzymes that carry out successive steps of a series of reactions are organized close together on a membrane surface, certain series of chemical reactions can occur more rapidly.

In a eukaryotic cell a number of membranes are generally considered to be part of the internal membrane system, or **endomembrane system.** Look at Figures 4–8 and 4–9 and notice how membranes divide the cell into many compartments, including the cell itself (bounded by the plasma membrane), the nucleus, endoplasmic reticulum, Golgi complexes, lysosomes, and vesicles and vacuoles. (Although it is not internal,

TABLE 4-1 Eukaryotic Cell Structures and Their Functions

Structure	Description	Function
Cell Nucleus		
Nucleus	Large structure surrounded by double membrane; contains nucleolus and chromosomes	Information in DNA is transcribed in RNA synthesis; specifies cellular proteins
Nucleolus	Granular body wthin nucleus; consists of RNA and protein	Site of ribosomal RNA synthesis; ribosome subunit assembly
Chromosomes	Composed of a complex of DNA and protein known as chromatin; condense during cell division, becoming visible as rodlike chromosomes	Contain genes (units of hereditary information) that govern structure and activity of cell
Cytoplasmic Organelles		
Plasma membrane	Membrane boundary of cell	Encloses cellular contents; regulates movement of materials in and out of cell; helps maintain cell shape; communicates with other cells (also present in prokaryotes)
Endoplasmic reticulum (ER)	Network of internal membranes extending through cytoplasm	Synthetic site of lipids and many proteins; origin of intracellular transport vesicles carrying proteins
Smooth	Lacks ribosomes on outer surface	Lipid biosynthesis; drug detoxification
Rough	Ribosomes stud outer surface	Manufacture of many proteins destined for secretion or for incorporation into membranes
Ribosomes	Granules composed of RNA and protein; some attached to ER, some free in cytosol	Synthesize polypeptides in both prokaryotes and eukaryotes
Golgi complex	Stacks of flattened membrane sacs	Modifies proteins; packages secreted proteins; sorts other proteins to vacuoles and other organelles
Lysosomes	Membranous sacs (in animals)	Contain enzymes to break down ingested materials, secretions, wastes
Vacuoles	Membranous sacs (mostly in plants, fungi, algae)	Transport and store materials, wastes, water

the plasma membrane is also included because of its participation in the activities of the endomembrane system.)

Some organelles have direct connections between their membranes and compartments. Others transport materials in **vesicles,** small membrane-bounded sacs. Vesicles also carry materials from one organelle to another. Through a complex series of steps, a vesicle can form as a "bud" from one membrane and then be transported to another membrane to which it fuses, thus delivering its contents into another compartment. Mitochondria and chloroplasts are also separate compartments but are not generally considered part of the endomembrane system because they function independently of other membranous organelles.

The cell nucleus contains DNA

Typically, the **nucleus** is the most prominent organelle in the cell. It is usually spherical or oval and averages 5 μm in di-

ameter. Owing to its size and the fact that it often occupies a relatively fixed position near the center of the cell, some early investigators guessed long before experimental evidence was available that the nucleus served as the control center of the cell (see *Focus On: Acetabularia*). Most cells have one nucleus, although there are exceptions.

The **nuclear envelope** consists of two concentric membranes that separate the nuclear contents from the surrounding cytoplasm (Fig. 4–12). These membranes are separated by about 20–40 nm. At intervals the membranes come together to form **nuclear pores** that regulate the passage of certain materials between nucleoplasm and cytoplasm.

Most of the cell's DNA is located inside the nucleus. Recall that the DNA molecules make up the **genes,** which contain the chemically coded instructions for producing most of the proteins needed by the cell. The nucleus controls protein synthesis by sending messenger RNA molecules to the cytoplasm (where proteins are manufactured).

TABLE 4–1   continued

Structure	Description	Function
Peroxisomes	Membranous sacs containing a variety of enzymes	Sites of many diverse metabolic reactions
Mitochondria	Sacs consisting of two membranes; inner membrane is folded to form cristae and encloses matrix	Site of most reactions of cellular respiration; transformation of energy originating from glucose or lipids into ATP energy
Plastids (e.g., chloroplasts)	Double membrane structure enclosing internal thylakoid membranes; chloroplasts contain chlorophyll in thylakoid membranes	Chlorophyll captures light energy; ATP and other energy-rich compounds are formed and then used to convert CO_2 to glucose
Cytoskeleton		
Microtubules	Hollow tubes made of subunits of tubulin protein	Provide structural support; have role in cell and organelle movement and cell division; components of cilia, flagella, centrioles, basal bodies
Microfilaments	Solid, rodlike structures consisting of actin protein	Provide structural support; play role in cell and organelle movement and cell division
Intermediate filaments	Stable, tough fibers made of polypeptides	Help strengthen cytoskeleton; stabilize cell shape
Centrioles	Pair of hollow cylinders located near nucleus; each centriole consists of nine microtubule triplets (9 × 3 structure)	Mitotic spindle forms between centrioles during animal cell division; may anchor and organize microtubule formation in animal cells; absent in most plants
Cilia	Relatively short projections extending from surface of cell; covered by plasma membrane; made of two central and nine peripheral microtubules (9 + 2 structure)	Movement of some single-celled organisms; used to move materials on surface of some tissues
Flagella	Long projections made of two central and nine peripheral microtubules (9 + 2 structure); extend from surface of cell; covered by plasma membrane	Cellular locomotion by sperm cells and some single-celled eukaryotes

DNA is associated with proteins, forming a complex known as **chromatin,** which appears as a network of granules and strands in cells that are not dividing. Although chromatin appears disorganized, it is not. Because DNA molecules are extremely long and thin, they must be packed inside the nucleus in a very regular fashion. In dividing cells, the chromatin condenses and becomes visible as distinct threadlike structures called **chromosomes.** If the DNA in the 46 chromosomes of one human cell could be stretched end to end, it would extend for two meters.

Most nuclei have one or more compact structures called **nucleoli** (sing., *nucleolus*). A nucleolus is *not* membrane-bounded and usually stains differently from the surrounding chromatin. Each nucleolus contains a nucleolar organizer, made up of chromosomal regions containing instructions for making the type of RNA in ribosomes. This ribosomal RNA is synthesized in the nucleolus. The proteins needed to make ribosomes are synthesized in the cytoplasm and imported into the nucleolus. Ribosomal RNA and proteins are then assembled into ribosomal subunits that leave the nucleus through the nuclear pores.

Ribosomes manufacture proteins

Ribosomes are small granules consisting of RNA and protein. Free ribosomes are suspended in the cytosol. Other ribosomes are associated with certain internal membranes within the cell. Each ribosome has two main components: a large subunit and a small subunit. Each subunit contains ribosomal RNA and several ribosomal proteins. Ribosomes, which contain the enzyme necessary to form peptide bonds, are tiny manufacturing plants that assemble proteins. Protein synthesis is discussed in detail in Chapter 12. (See *Making the Connection: Biological Molecules and Cell Control* on p. 91 for a brief overview of DNA, RNA, and protein synthesis)

(Text continues on p.87)

Cristae

Membranous sacs

Golgi complex

Mitochondrion

Cell wall

Plasma membrane

Vacuole

Granum

Stroma

Smooth ER

Nuclear envelope

Nucleolus

Nuclear pores

Rough ER

Chromatin

Ribosomes

Chloroplast

Rough and smooth endoplasmic reticulum (ER)

Nucleus

Figure 4–8 Composite diagram of a plant cell. The TEMs show certain structures or areas of the cell. Some plant cells do not have all the organelles shown. For example, cells that carry on photosynthesis contain chloroplasts, whereas root cells do not. Chloroplasts, a cell wall, and prominent vacuoles are characteristic of plant cells. Many of the other organelles, such as the nucleus, mitochondria, and endoplasmic reticulum, are also found in protist, fungal, and animal cells. *(Clockwise from top left: D.W. Fawcett; D.W. Fawcett and R. Bolender; D.W. Fawcett/Visuals Unlimited; R. Bolender and D.W. Fawcett; E.H. Newcomb and W.P. Wergin, Biological Photo Service)*

Chromatin

Nuclear
envelope

Nuclear
pores

Nucleolus

Nucleus

Membranous
sacs of
Golgi

Golgi complex

Plasma
membrane

Nuclear
envelope

Lysosome

Cristae

Ribosomes

Rough
ER

Smooth ER

Rough and smooth
endoplastic reticulum (ER)

Centrioles

Mitochondrion

Figure 4-9   Composite diagram of an animal cell. This generalized animal cell is shown in realistic context surrounded by adjacent cells that cause it to be slightly compressed. TEMs show the structure of various organelles. Depending on the cell type, certain organelles may be more or less prominent.   *(Clockwise from top left: D.W. Fawcett; D.W. Fawcett and R. Bolender; D.W. Fawcett; B.F. King, Biological Photo Service; R. Bolender and D.W. Fawcett/Visuals Unlimited)*

(a)

(a)

5 μm

Nucleus

Starch grain

Rough endoplasmic reticulum

Ribosomes

Vacuole

Plasma membrane

Prolamellar body

Golgi complex

Intercellular space

Chloroplasts

Cell wall

(b)

Figure 4–10 **TEM of a plant cell paired with interpretive drawing.** Most of this cross section of a cell from the leaf of a young bean plant *(Phaseolus vulgaris)* is dominated by a vacuole. Prolamellar bodies are membranous regions typically seen in developing chloroplasts. *(a, courtesy of Dr. Kenneth Miller, Brown University)*

5 μm

Plasma membrane

Chromatin   Nucleolus   Nucleus

Golgi complex

Ribosomes

Zymogen granules

Rough endoplasmic reticulum

Mitochondria

Smooth endoplasmic reticulum

(b)

Figure 4–11 **TEM of a human pancreas cell paired with interpretive drawing.** Most of the structures of a typical animal cell are present. However, like most cells, this one has certain structures associated with its specialized functions. Pancreas cells such as the one shown here secrete large amounts of digestive enzymes. The large, dark, circular bodies in the TEM in *(a)* and the corresponding tan structures in *(b)* are zymogen granules containing inactive enzymes. When released from the cell, they catalyze chemical reactions such as the breakdown of peptide bonds of ingested proteins in the intestine. Most of the membranes visible in this section are part of the rough endoplasmic reticulum, an organelle specialized to manufacture protein. *(a, Dr. Susumu Ito, Harvard Medical School)*

Figure 4–12 **Structure of the cell nucleus.** *(a)* The TEM and interpretive drawing show that the nuclear envelope, composed of two concentric membranes, is perforated by nuclear pores. Each pore is surrounded by a complex of proteins. The outer membrane of the nuclear envelope is continuous with the membrane of the endoplasmic reticulum. The nucleolus is not bounded by a membrane. *(b)* TEM of nuclear pores. A technique known as freeze-fracture was used to split the membrane. *(c)* The nuclear pores, which are made up of proteins, form channels between the nucleoplasm and cytoplasm. *(a, D.W. Fawcett; b, R. Kessel-G. Shih/Visuals Unlimited)*

Labels in figure: Rough ER, Nucleolus, Nuclear pore, Chromatin, Nuclear pores, Nuclear envelope, 2 μm, *(a)*, *(b)*, 0.25 μm, Outer nuclear membrane, Nuclear pore, *(c)*, Nuclear pore proteins, Inner nuclear membrane

The endoplasmic reticulum is a major manufacturing center

One of the most prominent features in the electron micrograph in Figure 4–11 is a maze of parallel internal membranes that encircle the nucleus and extend into many regions of the cytoplasm. This complex of membranes is the **endoplasmic reticulum (ER),** which forms a significant part of the total volume of the cytoplasm in many cells. A higher-magnification TEM of the ER is shown in Figure 4–13. Remember that a TEM represents only a thin cross section of the cell, so there is a tendency to interpret the ER as a series of tubes. In fact, many ER membranes consist of a series of tightly packed and flattened saclike structures that form interconnected compartments within the cytoplasm.

The internal space enclosed by the membranes is called the ER **lumen.** In most cells the ER lumen forms a single internal compartment that is continuous with the outer membrane of the nuclear envelope (Fig. 4–12). The compartment formed between the two nuclear membranes is thus connected to the ER lumen. The membranes of other organelles are not directly connected to the ER and appear to form distinct and separate compartments within the cytoplasm.

The ER membranes and lumen contain a large variety of enzymes that catalyze many different types of chemical reactions. In some cases the membranes serve as a framework for systems of enzymes that carry out sequential biochemical reactions. The two surfaces of the membrane contain different sets of enzymes and represent regions of the cell with different synthetic capabilities, just as different regions of a factory are used to make different parts of a particular product. Still other enzymes are located within the ER lumen.

Two distinct regions of the ER can be distinguished in TEMs: rough ER and smooth ER. Although these regions have different functions, their membranes are connected and their internal spaces are continuous. **Rough ER** has ribosomes attached to it and consequently appears rough in electron micrographs. Notice in Figure 4–13 that one membrane face (the lumen side) appears to be bare, while the other membrane face (the cytosolic side) is studded with ribosomes that appear as dark particles.

ACETABULARIA: THE MERMAID'S WINEGLASS AND THE CONTROL OF CELL ACTIVITIES

As discussed in Chapter 1, we can learn something about the role of the nucleus by removing it from a cell and examining the consequences. When the nucleus of a single-celled amoeba is removed with a microneedle, the amoeba continues to live and move, but it ceases to grow and dies after a few days. A control amoeba, subjected to similar trauma but without removal of the nucleus, does not die. We conclude that the nucleus is necessary for the metabolic processes that provide for growth and cell reproduction.

Acetabularia Is Easy to Manipulate Because It Is a Single Giant Cell

To the romantically inclined, the little seaweed *Acetabularia* resembles a mermaid's wineglass, although the literal translation of its name, "vinegar cup," is somewhat less elegant (Fig. a).

In the 19th century, biologists discovered that this marine eukaryotic alga consists of a single cell. At about 5 cm in length, *Acetabularia* is small for a seaweed but gigantic for a cell. It consists of a root-like **holdfast**, a long cylindrical **stalk**, and a cuplike **cap**. The nucleus is found in the holdfast, about as far away from the cap as it can be.

Regeneration Experiments Demonstrated That the Cap Shape Is under the Control of Something in the Stalk or the Holdfast

If the cap of *Acetabularia* is removed experimentally, another one grows after a few weeks. Such behavior, common among simple organisms, is called **regeneration**. This fact attracted the attention of investigators, especially J. Hämmerling and J. Brachet, who became interested in whether a relationship exists between the nucleus and the physical characteristics of the alga. Because of its great size, *Acetabularia* could be subjected to surgery that would be impossible with smaller cells. During the 1930s and 1940s these researchers performed brilliant experiments that in many ways laid the foundation for much of our modern knowledge of the nucleus. Two species were used for most experiments, *A. mediterranea*, which has a smooth cap, and *A. crenulata*, which has a cap divided into a series of finger-like projections.

(a) **Acetabularia.**   *(a, L. Sims/Visuals Unlimited)*

The kind of cap that is regenerated depends on the species of *Acetabularia* used in the experiment. As you might expect, *A. crenulata* regenerates a "cren" cap, and *A. mediterranea* regenerates a "med" cap. But it is possible to graft together two capless algae of different species. Through this union, they regenerate a common cap that has characteristics intermediate between those of the two species involved (Fig. b). Thus, it is clear that something about the lower part of the cell controls cap shape.

Stalk Exchange Experiments Indicated That Short-Term Control Can Be Exerted by the Stalk, but Long-Term Control Is in the Holdfast

It is possible to attach a section of *Acetabularia* to a holdfast that is not its own by telescoping the cell walls of the two into one another. In this way the stalks and holdfasts of different species may be intermixed.

First, we take *A. mediterranea* and *A. crenulata* and remove their caps. Then we sever the stalks from the holdfasts. Finally, we exchange the parts (Fig. c). What happens? Not, perhaps, what you would expect! The caps that regenerate are characteristic not of the species donating the holdfasts but of those donating the stalks!

However, if the caps are removed once again, this time the caps that regenerate are characteristic of the species that donated the holdfasts. This continues to be the case no matter how many more times the regenerated caps are removed.

From all these results Hämmerling and Brachet deduced that the ultimate control of the *Acetabularia* cell is associated with the holdfast. Because there is a time lag be-

Cap
Stalk
Nucleus
Holdfast

A. mediterranea   A. crenulata

(b)

Stalks and holdfasts exchanged

First regenerated caps   Second regenerated caps

(c)

fore the holdfast appears to take over, they hypothesized that it produces some temporary cytoplasmic messenger substance whereby it exerts its control. They further hypothesized that initially the grafted stalks still contain enough of the substance from their former holdfasts to regenerate a cap of the former shape. But this still leaves us with the question of what it is about the holdfast that accounts for its apparent control. An obvious suspect is the nucleus.

Nuclear Exchange Experiments Demonstrated That the Nucleus Is the Ultimate Source of Information for the Control of Cellular Activities

If the nucleus is removed and the cap cut off, a new cap regenerates (Fig. d). *Acetabularia,* however, can usually regenerate only once without a nucleus. If the nucleus of another species is now inserted and the cap is cut off once again, a new cap regenerates that is characteristic of the species of the nucleus (Fig. e)! If two kinds of nuclei are inserted, the regenerated cap is intermediate in shape between those of the species that donated the nuclei.

As a result of these and other experiments, biologists began to develop certain basic ideas. The control of the cell exerted by the holdfast is attributable to the nucleus that is located there. Further, the nucleus is the apparent source of some "messenger substance" that can temporarily exert control but is limited in quantity and cannot be produced without the nucleus (Fig. f). This information helped provide a starting point for research on the role of the nucleic acids in the control of all cells.

These ideas have been extended in our modern view of information flow and control in the cell. We now know that the nucleus of eukaryotes controls the cell's activities because it contains DNA (deoxyribonucleic acid), the ultimate source of biological information. DNA can pass on its information to successive generations because it is able to precisely replicate itself. The information in the DNA is used to specify the sequence of amino acids in all the proteins of the cell. To carry out its mission, DNA uses ribonucleic acid (RNA) as the cytoplasmic messenger substance.

Eventually

(d)

Eventually

(e)

The characteristics of the cell are governed by the messenger substance, and therefore ultimately by the nucleus.

Messenger substance

The nucleus produces the messenger substance

(f)

Figure 4–13 **The endoplasmic reticulum (ER).** The TEM shows both rough and smooth ER in a liver cell. *(R. Bolender and D.W. Fawcett/Visuals Unlimited)*

Ribosomes

Mitochondrion

Rough endoplasmic reticulum

1 μm

Smooth endoplasmic reticulum

The ER plays a central role in the synthesis and assembly of proteins. Many proteins that are exported from the cell (such as digestive enzymes) and those destined for other organelles are synthesized on ribosomes attached to the ER membrane. Recent studies suggest that the ribosome forms a tight seal with the ER membrane. The ribosome contains a tunnel that connects to an ER pore, or *translocon.* Proteins are transported through this extended tunnel (through the ER membrane) into the ER lumen. In the ER lumen, proteins may be modified by enzymes that add complex carbohydrates or lipids to them. Other enzymes, called **chaperones,** in the ER lumen catalyze the efficient folding of proteins into proper conformations. The proteins are then transferred to other compartments by small **transport vesicles,** which bud off of the ER membrane and are then inserted into the membrane of some target compartment.

Smooth ER is more tubular and does not have ribosomes bound to it, so its outer membrane surfaces appear smooth. The smooth ER is the primary site of phospholipid, steroid, and fatty acid metabolism. While the smooth ER may be a minor membrane component in some cells, extensive amounts of smooth ER are present in others. For example, extensive smooth ER is visible in human liver cells, which synthesize

and process cholesterol and other lipids and serve as a major detoxification site. Enzymes located along the smooth ER of liver cells break down toxic chemicals such as carcinogens (cancer-causing agents). These compounds are then converted to water-soluble products that can be excreted.

The Golgi complex processes and packages proteins

The **Golgi complex** (also known as the *Golgi body* or *Golgi apparatus*) was first described in 1898 by the Italian microscopist Camillo Golgi, who found a way to specifically stain that organelle. In many cells the Golgi complex consists of stacks of flattened membranous sacs, which may be distended in certain regions because they are filled with cellular products (Fig. 4–14). Each of the flattened sacs has an internal space, or lumen. However, unlike the endoplasmic reticulum, most of these internal spaces of the Golgi complex and the membranes that form them are not continuous. Hence a Golgi complex contains a number of separate compartments, as well as some that are interconnected.

Each Golgi stack has three areas referred to as *cis* and *trans faces,* with a *medial* region between. Typically, the *cis* face is lo-

BIOLOGICAL MOLECULES AND CELL CONTROL

How does a cell store and use genetic information? The cell stores information in the form of deoxyribonucleic acid (DNA). The DNA molecule contains a linear sequence of components called nucleotides. In all cells this sequence of nucleotides serves as a code that specifies the amino acid sequence (primary structure) in proteins. Proteins function as structural components of cells and as **enzymes,** molecules that regulate virtually every chemical reaction that takes place in the cell. By specifying the structure of enzymes and other proteins, DNA directs the metabolism of the cell.

The cell uses a second nucleic acid, ribonucleic acid (RNA), as an intermediary. The sequence of bases in DNA that codes for a specific protein is copied as a sequence of bases in a type of RNA, called **messenger RNA.** This process is known as **transcription.** In eukaryotic cells, DNA is stored in the nucleus and transcription occurs there.

Messenger RNA leaves the nucleus and attaches to ribosomes along the rough ER (or to free ribosomes in the cytosol); two other types of RNA are also involved in protein synthesis. The complex process of assembling a chain of specific amino acids using the code in messenger RNA is called **translation.** After proteins are manufactured they may be transported to the Golgi complex where they are modified and sent on to other locations in the cell.

When a cell divides, the information stored in DNA must be reproduced and passed intact to the two daughter cells. DNA has the unique ability to make an exact duplicate of itself through a process called **replication** (see Chapter 11).

① Immediately after synthesis on ribosomes, glycoproteins are found in the ER.

Ribosomes

Rough ER

Glycoprotein

Golgi complex

cis face

② Minutes later some of the labeled glycoproteins have migrated to the inner layers of the Golgi complex.

③ A short time later, the labeled glycoproteins can be seen at the outer face of the Golgi complex. Many are inside vesicles.

trans face

Plasma membrane

0.5 μm

④ In the final stages of secretion, labeled glycoproteins can be seen in vesicles between the Golgi complex and the plasma membrane. Some of the vesicles fuse with the plasma membrane and release their contents outside the cell.

Figure 4–14 **TEM and interpretive drawing of the Golgi complex.** Glycoproteins are transported from the rough ER to the Golgi where they are modified. This diagram shows the passage of glycoproteins through the Golgi complex during the secretory cycle of a mucous-secreting goblet cell that lines the intestine. Mucus is a complex mixture of covalently linked proteins and carbohydrates. *(D.W. Fawcett and R. Bolender)* cated nearest the nucleus and functions to receive materials from transport vesicles from the ER. The *trans* face, nearest to the plasma membrane, packages molecules in vesicles and transports them out of the Golgi.

In a cross-sectional view like that in the TEM in Figure 4–14, many of the ends of the sheetlike layers of Golgi membranes are distended, an arrangement that is characteristic of well developed Golgi complexes in many types of cells. In some animal cells the Golgi complex is often located at one side of the nucleus; in other animal and in plant cells there are many Golgi complexes, usually consisting of separate stacks of membranes dispersed throughout the cell. Cells that secrete large amounts of glycoproteins have large numbers of Golgi stacks. Golgi complexes of plant cells produce extracellular polysaccharides that are used as components of the cell wall.

The Golgi complex functions principally as an apparatus for processing, sorting, and modifying proteins. Cell biologists have studied the function of the Golgi complex by radioactively labeling newly manufactured amino acids or carbohydrates and observing their movement. Glycoproteins are first located in the rough ER and only later in the Golgi complex. The proteins are transported from the rough ER to the *cis* face of the Golgi complex in small transport vesicles formed from the ER membrane. After the glycoprotein molecules are released into the Golgi complex, they are enclosed in new vesicles that shuttle them from one compartment to another within the Golgi. While moving through the Golgi complex, proteins are modified in different ways, resulting in the formation of complex biological molecules. For example, the carbohydrate part of the glycoprotein (first added to proteins in the rough ER) may be modified.

In some cases the carbohydrate added to protein within the Golgi complex may be used as a "sorting signal" that routes the protein to a specific organelle. The glycoproteins are then packaged in secretory vesicles in the *trans* region. These vesicles pinch off from the Golgi membrane and transport their contents to a specific destination. Some vesicles fuse with the plasma membrane; the vesicle membrane becomes part of the plasma membrane and the glycoproteins are secreted from the cell. Other vesicles may store glycoproteins for secretion at a later time, while still others are routed to various organelles of the endomembrane system. In animal cells, the Golgi complex also manufactures lysosomes.

Lysosomes are compartments for digestion

Small sacs of digestive enzymes called **lysosomes** are dispersed in the cytoplasm of most eukaryotic cells (Fig. 4–15). The enzymes in these organelles break down complex molecules—lipids, proteins, carbohydrates, and nucleic acids—that originate inside or outside the cell. About 40 different digestive enzymes have been identified in lysosomes; most are active under rather acidic conditions (pH 5). The powerful enzymes and low pH maintained by the lysosome is a good example of the importance of separating functions into different compartments. Under most normal conditions, the lysosome mem-

5 μm

Figure 4–15 Lysosomes. The dark vesicles in this TEM are lysosomes. These compartments separate powerful digestive enzymes from the rest of the cell. Primary lysosomes bud off from the Golgi complex. After a lysosome encounters material to be digested, it is known as a secondary lysosome. The large vesicles shown here are secondary lysosomes containing various material being digested. *(Don Fawcett/ Photo Researchers, Inc.)*

brane confines its enzymes and their actions. Some forms of tissue damage have been related to "leaky" lysosomes.

Primary lysosomes are formed by budding from the Golgi complex. Their hydrolytic enzymes are synthesized in the rough ER. As these enzymes pass through the lumen of the ER, sugars are attached to each molecule, identifying it. This signal permits the Golgi complex to appropriately sort the enzyme to the lysosomes rather than export it from the cell.

Lysosomes degrade bacteria or debris ingested by scavenger cells. The ingested matter is enclosed in a vesicle formed from part of the plasma membrane. One or more primary lysosomes fuse with the vesicle containing the ingested material, forming a larger vesicle called a *secondary lysosome.* In the secondary lysosome the powerful enzymes come in contact with the ingested molecules and degrade them into their components.

Lysosomal enzymes are also released into the cell in some normal processes. For example, under some conditions lysosomes break down organelles so that their components can be recycled or used as an energy source. Programmed cell death, or **apoptosis,** is a normal part of development and maintenance. For example, during the metamorphosis of a tadpole to a frog, lysosomes break down the cells of the tadpole tail. And during human development, the hand is webbed until, through apoptosis, lysosomes destroy the tissue between the fingers. Cells in the upper layer of human skin and in the intestinal wall are continuously destroyed by apoptosis and replaced by new cells. Inappropriate initiation or inhibition of apoptosis may contribute to a variety of diseases including cancer, AIDS, and Alzheimer's disease.

In certain genetic diseases of humans, known as lysosomal storage diseases, one of the normally present digestive enzymes is lacking. Its substrate (substance that the enzyme would normally break down) accumulates in the lysosomes, ultimately interfering with cellular activities. An example is Tay-Sachs disease (see Chapter 15), in which a normal lipid cannot be broken down in brain cells. The lipid accumulates in the cells, resulting in mental retardation and death.

Peroxisomes metabolize small organic compounds

Peroxisomes are membrane-bounded organelles containing enzymes that catalyze an assortment of metabolic reactions in which hydrogen is transferred from various compounds to oxygen (Fig. 4–16). During these reactions, hydrogen peroxide (H_2O_2), a substance toxic to the cell, is produced as a byproduct. Peroxisomes contain catalase, an enzyme that splits hydrogen peroxide, rendering it harmless.

Peroxisomes are found in large numbers in cells that synthesize, store, or degrade lipids. In plant seeds, specialized peroxisomes, called **glyoxysomes,** contain enzymes that convert stored fats to sugars. The sugars are used by the young plant as an energy source and as a component needed to synthesize other compounds. Animal cells lack glyoxysomes and cannot convert fatty acids into sugars.

Mitochondria
Chloroplasts
Peroxisomes

1 μm

Figure 4–16 **Peroxisomes.** In this TEM of a tobacco leaf cell, peroxisomes are seen in close association with chloroplasts and mitochondria. These organelles may cooperate in carrying out some metabolic processes. *(E.H. Newcomb and S.E. Frederick/Biological Photo Service)*

When yeast cells grow in an alcohol-rich medium, they manufacture a large number of peroxisomes. These peroxisomes contain an enzyme that degrades the alcohol. Peroxisomes in human liver and kidney cells detoxify ethanol, the alcohol in alcoholic beverages.

Vacuoles are large, fluid-filled sacs with a variety of functions

Although lysosomes have been identified in almost all kinds of eukaryotic cells, their occurrence in plant and fungal cells is open to debate. Many of the functions carried out in animal cells by lysosomes are performed in plant and fungal cells by a large, single, membrane-bounded sac referred to as the **vacuole.** The vacuolar membrane, part of the endomembrane system, is referred to as the **tonoplast.** The term *vacuole,* which means "empty," refers to the fact that these organelles have no internal structure. Although the terms *vacuole* and *vesicle* are sometimes used interchangeably, vacuoles are usually larger structures, sometimes produced by the merging of many vesicles.

Vacuoles play a significant role in plant growth and development. Immature plant cells are generally small and contain numerous small vacuoles. As water accumulates in these vacuoles, they tend to coalesce, forming a large central vacuole. A plant cell increases in size mainly by adding water to this central vacuole.

As much as 90% of the volume of a plant cell may be occupied by a large central vacuole containing water, as well as stored food, salts, pigments, and wastes (Figs. 4–8 and 4–10). The vacuole may serve as a storage compartment for inorganic compounds and for molecules such as proteins in seeds. Plants lack organ systems for disposing of toxic metabolic waste products. Wastes may be recycled in the vacuole or they may aggregate and form small crystals inside the vacuole.

Compounds that are noxious to herbivores (animals that eat plants) may also be stored in some plant vacuoles as a means of defense. Plant vacuoles are lysosome-like in their ability to break down unneeded organelles and other cellular components.

Vacuoles have numerous other functions and are also present in many types of animal cells and in single-celled protists. Most protozoa have **food vacuoles** which fuse with lysosomes so that the food they contain can be digested (Fig. 4–17). Many protozoa also have **contractile vacuoles** which remove excess water from the cell (see Chapter 24).

Mitochondria and chloroplasts are energy-converting organelles

When a cell obtains energy from its environment, it is usually in the form of chemical energy in food molecules (such as glucose) or in the form of light energy. These types of energy must be converted to forms that can be used more conveniently by cells. Some energy conversions go on in the cytosol, but other

15 μm

Figure 4–17 **LM of food vacuoles.** This protozoon, *Chilodonella,* has ingested many small, photosynthetic protists called diatoms *(dark areas)* that have been enclosed in vacuoles. From the number of diatoms scattered about its cell, one might judge that *Chilodonella* has a rather voracious appetite. *(M.I. Walker/Photo Researchers, Inc.)*

types take place in mitochondria and chloroplasts, organelles that are specialized to facilitate conversion of energy from one form to another. Chemical energy is most commonly stored in ATP. Recall from Chapter 3 that the chemical energy of ATP can be used to drive a variety of chemical reactions in the cell.

Figure 4–18 summarizes the main activities that take place in mitochondria, found in almost all eukaryotic cells (including algae and plants), and in chloroplasts, found only in algae and certain plant cells. In addition to their central roles in energy metabolism, mitochondria and chloroplasts contain small amounts of DNA, and they provide important clues about the evolution of eukaryotic cells (see *Making the Connection: Mitochondria, Chloroplasts, and Cell Evolution*).

Mitochondria make ATP through cellular respiration

Virtually all eukaryotic cells (plant, animal, fungal, and protist) contain complex organelles called **mitochondria** (sing., *mitochondrion*). These organelles are the site of **aerobic respiration,** a process that includes most of the reactions that convert the chemical energy present in certain foods to ATP (see Chapter 7). Aerobic respiration requires oxygen and results in the release of carbon atoms from food molecules as carbon dioxide (a waste product).

Mitochondria are most numerous in cells that are very active and therefore have high energy requirements. More than 1000 mitochondria have been counted in a single liver cell!

Figure 4–18 **Cellular respiration and photosynthesis.** In cellular respiration, which takes place in the mitochondria of virtually all eukaryotic cells, the chemical energy of glucose is converted to chemical energy in the form of ATP. Photosynthesis, which is carried out in chloroplasts in some plant and algal cells, converts light energy to ATP and other forms of chemical energy. This energy is used to synthesize glucose from carbon dioxide and water.

Outer membrane

Inner membrane

Matrix

Cristae

0.25 μm

Figure 4–19 **Mitochondria.** Aerobic respiration takes place within mitochondria. Cristae are evident in the TEM as well as in the drawing; the drawing reveals the relationship between the inner and outer mitochondrial membranes. *(D.W. Fawcett)*

trix, the space enclosed by the inner mitochondrial membrane. Cristae greatly increase the surface area of the membrane, providing a surface for the chemical reactions that transform the chemical energy in food molecules into the energy of ATP. The membrane contains the complex series of enzymes and other proteins needed for these reactions.

In a mammalian cell, each mitochondrion has 5–10 identical, circular molecules of DNA, accounting for up to 1% of the total DNA in the cell. In 1988 two research teams independently linked mutations in mitochondrial DNA with certain genetic diseases. Douglas C. Wallace at Emory University reported that an inherited mutation in a mitochondrial gene was associated with a form of young-adult blindness. Ian J. Holt and his colleagues at the Institute of Neurology in London linked mitochondrial mutations to progressive muscle disorders. Investigators have found evidence for links between mitochondrial mutations and several other diseases, including some cases of Alzheimer's disease and type 2 diabetes.

Chloroplasts convert light energy to chemical energy through photosynthesis

Certain plant and algal cells carry out a complex set of energy conversion reactions known as *photosynthesis* (see Chapters 1 and 8). **Chloroplasts** are organelles that contain **chlorophyll,** a green pigment that traps light energy for photosynthesis. Chloroplasts also contain a variety of yellow and orange, light-absorbing pigments known as **carotenoids** (see Chapter 3). A unicellular alga may have only a single large chloroplast, whereas a leaf cell may have 20 to 100. Chloroplasts tend to be somewhat larger than mitochondria, with lengths typically ranging from about 5 to 10 μm or longer.

Chloroplasts are typically complex disc-shaped structures and, like mitochondria, have a complex system of folded membranes (Fig. 4–20; see Chapter 8 for more detailed descriptions of structure). Two membranes, separated by a small space, separate the chloroplast from the cytosol. The inner membrane encloses a fluid-filled space called the **stroma,** which contains enzymes responsible for producing carbohydrates from carbon dioxide and water, using energy trapped from sunlight. The inner chloroplast membrane may fold inward to form a system of internal membranes, consisting of an interconnected set of flat, disclike sacs called **thylakoids.** The thylakoids are arranged in stacks called **grana** (sing., *granum*).

The thylakoid membranes enclose a third, innermost compartment within the chloroplast, called the **thylakoid interior space.** The thylakoid membranes, in which chlorophyll is found, are similar to the inner mitochondrial membranes in that they are involved in the formation of ATP. Energy trapped from sunlight by the chlorophyll molecules is used to excite electrons; the energy in these excited electrons is then used to form ATP and other molecules that can transfer chemical energy. This chemical energy is used to produce carbohydrates from carbon dioxide and water in the stroma.

Chloroplasts belong to a group of organelles known as **plastids** that produce and store food materials in cells of plants

Mitochondria vary in size, ranging from 2 to 8 μm in length and are capable of changing size and shape rapidly. Mitochondria usually give rise to other mitochondria by growth and subsequent division.

Each mitochondrion is bounded by a double membrane, which forms two *different* compartments within the organelle: the intermembrane space and the matrix (Fig. 4–19; see Chapter 7 for more detailed descriptions of mitochondrial structure). The **intermembrane space** is the compartment formed between the outer and inner mitochondrial membranes. The **matrix,** the compartment enclosed by the inner mitochondrial membrane, contains enzymes that break down food molecules and convert their energy to other forms of chemical energy.

The **outer mitochondrial membrane** is smooth and allows many small molecules to pass through it. By contrast, the **inner mitochondrial membrane** has numerous folds and strictly regulates the types of molecules that can move across it. The folds, called **cristae** (sing., *crista*), extend into the **ma-**

MITOCHONDRIA, CHLOROPLASTS, AND CELL EVOLUTION

What is the evolutionary relationship between prokaryotic cells and the more complex cells of eukaryotes? Mitochondria and chloroplasts have provided valuable insights because these organelles have been shown to have many prokaryote features. For example, although most of the DNA in eukaryotic cells resides in the nucleus, both mitochondria and chloroplasts (as well as other plastids) have DNA molecules in their inner compartments. These DNA molecules code for a small number of the proteins found in these organelles. These proteins are synthesized on mitochondrial or chloroplast ribosomes, which are similar to the ribosomes of prokaryotes. The majority of the mitochondrial and chloroplast proteins, however, are controlled by nuclear genes, manufactured on free ribosomes outside the organelles, and then transported to their appropriate locations within.

The existence of a separate set of ribosomes and DNA molecules in mitochondria and chloroplasts and their similarity in size

to many bacteria, along with other prokaryote-like characteristics, provide support for the **endosymbiont theory** (see Chapter 20). According to this theory mitochondria and chloroplasts evolved from prokaryotic organisms that took up residence inside larger cells, and eventually lost the ability to function as autonomous organisms.

Evolutionary biologists are currently sequencing mitochondrial genes from protists in order to determine the mitochondrion's evolution from bacterium to highly specialized organelle. Evidence suggests that during this process, some genes moved from the early mitochondrion to the nucleus. One such gene codes for a protein known as chaperonin 60 that helps other proteins fold appropriately. Protists that have retained this gene in their mitochondria are thought to have evolved earlier than those in which the chaperonin 60 gene is found in the nucleus.

and algae. All plastids develop from **proplastids,** precursor organelles found in less specialized plant cells, particularly in growing, undeveloped tissues. Depending on the special functions a cell will eventually have, its proplastids can mature into a variety of specialized mature plastids. These are extremely

versatile organelles; in fact, under certain conditions even mature plastids can convert from one form to another.

Chloroplasts are produced when proplastids are stimulated by exposure to light. **Chromoplasts** contain pigments that give certain flowers and fruits their characteristic colors; these at-

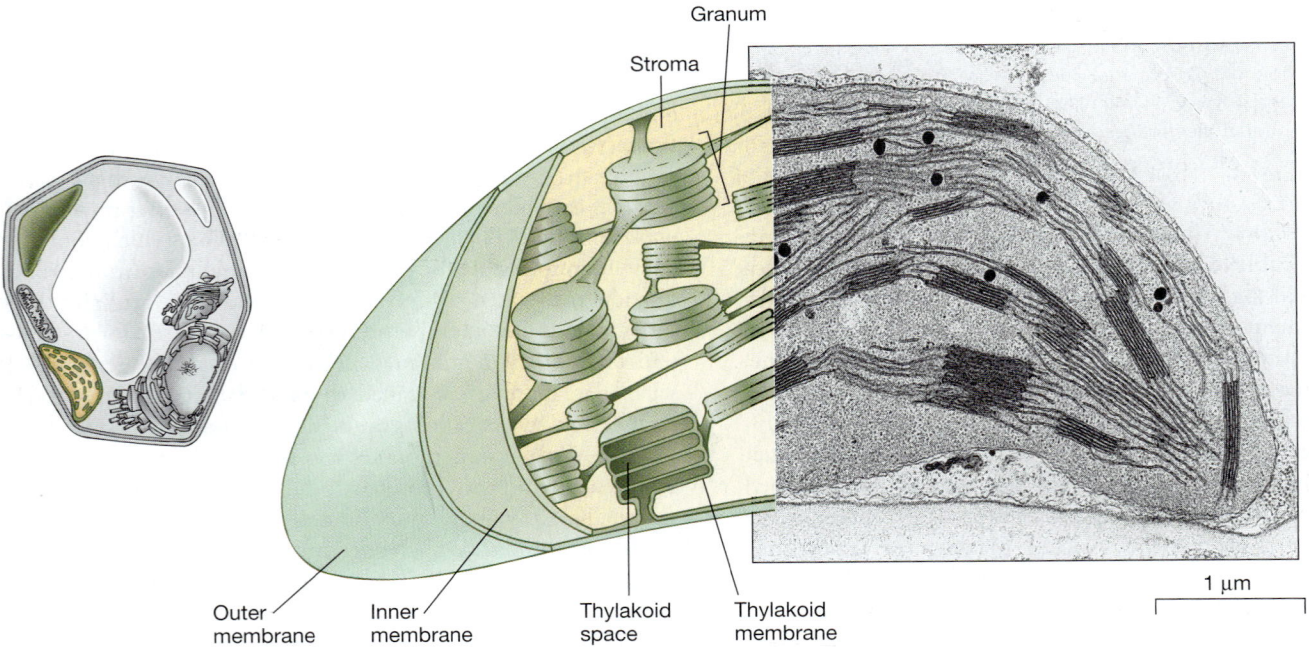

1 µm

Figure 4–20 The chloroplast, organelle of photosynthesis. The TEM shows part of a chloroplast from a corn leaf cell. Chlorophyll and other photosynthetic pigments are found in the thylakoid membranes. One granum has been cut open to show the thylakoid space. The inner chloroplast membrane may or may not be continuous with the thylakoid membrane (as shown). *(E.H. Newcomb and W.P. Wergin/Biological Photo Service)* tract animals as pollinators or as seed dispersers. **Leukoplasts** are unpigmented plastids; they include **amyloplasts** (see Fig. 3–8), which store starch in the cells of many seeds, roots, and tubers (e.g., potatoes).

EUKARYOTIC CELLS CONTAIN A CYTOSKELETON

We have seen that there are many different sizes and shapes of cells. When we watch cells growing in the laboratory, we see that they may change shape and in many cases move about. The shapes of cells and their ability to move are determined in large part by the **cytoskeleton,** a dense network of protein fibers (Fig. 4–21). In addition to providing mechanical support, the cytoskeleton functions in transport of materials within the cell and in cell division.

The cytoskeleton is highly dynamic and constantly changing. Its framework is made of three types of protein filaments: **microtubules, microfilaments** (also known as actin filaments), and **intermediate filaments.** Both microfilaments and microtubules are fibers formed from beadlike, globular protein subunits, which can be rapidly assembled and disassembled. Intermediate filaments are made from fibrous protein subunits and are more stable than microtubules and microfilaments.

Microtubules are hollow cylinders

Microtubules, the thickest filaments of the cytoskeleton, are about 25 nm in diameter and up to several micrometers in length. In addition to playing a structural role in the formation of the cytoskeleton, these extremely adaptable structures are involved in the movement of chromosomes during cell division. They serve as tracks for several other kinds of intracellular movement and are the major structural components of cilia and flagella—specialized structures used in some cell movements.

Microtubules consist of two very similar proteins: α and β **tubulin.** These proteins combine to form a dimer (recall from Chapter 3 that a dimer forms from the association of two similar, simpler units, referred to as monomers). In 1998, biophysicists Kenneth Downing and Eva Nogales reported in the journal *Nature* that, using electron microscopy, they were able to determine the three-dimensional structure of tubulin. A microtubule elongates by the addition of tubulin dimers (Fig. 4–22). The energy for microtubule elongation is provided by GTP, a molecule similar to ATP. Downing and Nogales found that both α and β tubulin have GTP-binding sites. However, only the GTP bound to β tubulin breaks down to GDP. Each microtubule has polarity, and its two ends are referred to as *plus* and *minus.* The plus end elongates more rapidly. Microtubules can be disassembled by the removal of dimers, which can then be recycled to form microtubules in other parts of the cell.

For microtubules to act as a structural framework or participate in cell movement, they must be anchored to other parts of the cell. In nondividing cells, the minus ends of microtubules appear to extend from regions called **microtubule-organizing centers (MTOC).** In most eukaryotic cells, the main MTOC is in the **centrosome,** a structure that is important in cell division. In almost all animal cells, the centrosome contains two structures called **centrioles** (Fig. 4–23). These structures, which are at right angles to each other, are known as *9 × 3 structures;* they consist of nine sets of three microtubules arranged to form a hollow cylinder. The centrioles are duplicated before cell division and may play a role in some types of microtubule assembly. Most plant cells have a microtubule-organizing center but lack centrioles. This suggests either that centrioles are not essential to most microtubule assembly processes or that alternative assembly mechanisms are possible.

The ability of microtubules to assemble and disassemble rapidly can be seen during cell division, when much of the cytoskeleton appears to break down (see Chapter 9). Many of the tubulin subunits organize into a structure called the **spindle,** which serves as a framework for the orderly distribution of chromosomes when the cell divides.

MTOC

Plasma membrane

Microtubule   Intermediate filament   Microfilament

Figure 4–21   The cytoskeleton. Eukaryotic cells contain a cytoskeleton consisting of networks of several types of fibers, including microtubules, microfilaments, and intermediate filaments. The cytoskeleton contributes to the shape of the cell, anchors organelles, and sometimes rapidly changes shape during cellular locomotion.

(a)

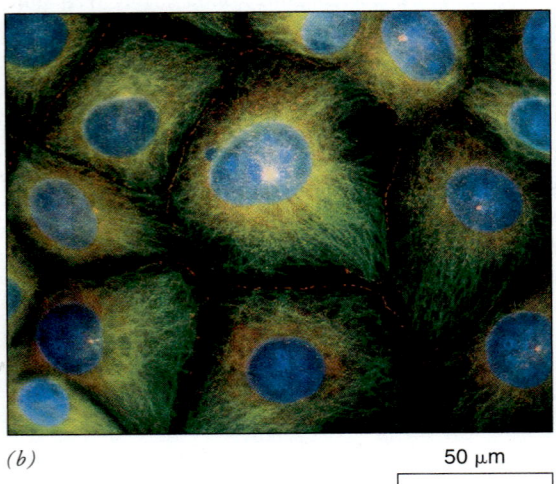

(b)

50 μm

Figure 4–22 **Organization of microtubules.** *(a)* Microtubules are manufactured in the cell by adding dimers of α-tubulin and β-tubulin to an end of the hollow cylinder. Notice that the cylinder has polarity. The end shown at the top of the figure is the fast-growing or plus end; the opposite end is the minus end. Each turn of the spiral requires 13 dimers. *(b)* Confocal fluorescence LM showing microtubules in green. A microtubule-organizing center *(pink dot)* is visible near most of the cell nuclei *(blue)*. *(b, Nancy Kedersha)*

Several **microtubule-associated proteins (MAPs)** have been identified and classified into two groups: fibrous MAPs and motors. *Fibrous MAPs* cross-link microtubules so that they form bundles. Such bundles help form the shapes of many types of cells. *Motors* use ATP energy to produce movement. George Langford of Dartmouth College and other investigators are studying the mechanisms by which the cell moves organelles and other materials within the cell. Nerve cells typically have long extensions called axons that transmit signals to

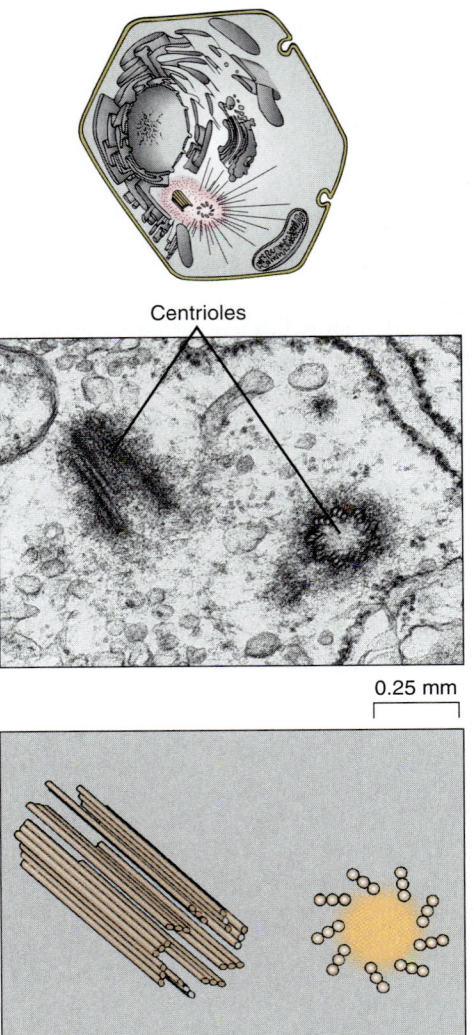

Centrioles

0.25 mm

Figure 4–23 **Centrioles.** The TEM is paired with an interpretive drawing. The centrioles are positioned at right angles to each other, near the nucleus of a nondividing animal cell. Note the 9 × 3 arrangement of microtubules. The centriole on the left has been cut longitudinally and the one on the right transversely. *(B.F. King/Biological Photo Service)*

other nerve cells, muscle cells, or cells that produce hormones. Because of its length and availability and because other cells have similar mechanisms, investigators have used the axon as a model system to study transport of organelles within the cell. Mitochondria, transport and secretory vesicles, and other organelles attach to microtubules that serve as tracks along which organelles can be moved to different cellular locations.

One motor protein, named *kinesin,* moves organelles toward the plus end of a microtubule (Fig. 4–24). *Dynein,* another motor protein, transports organelles in the opposite direction, toward the minus end. This is referred to as retrograde transport. In 1997, Waterman Storer of the University of Pennsylvania and his colleagues reported that dynein is essential but not sufficient for retrograde transport. These investigators determined that a protein complex called dynactin is required.

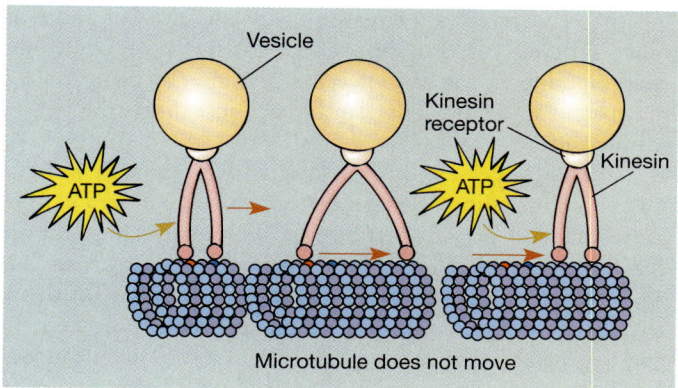

Figure 4-24 Hypothetical model of a kinesin motor. A kinesin molecule attaches to a specific receptor on the vesicle. Energy from ATP allows the kinesin molecule to change its conformation and "walk" along the microtubule, carrying the vesicle along.

Dynactin binds to both microtubules and dynein and appears to function in transport, linking the organelle, microtubule, and dynein.

Cilia and flagella are composed of microtubules

Projecting from surfaces of many cells are thin, movable structures important in cell movement. If a cell has one, or only a few, of these appendages and if they are relatively long (up to 1 mm or longer) relative to the size of the cell, they are called flagella (sing., *flagellum*). If the cell has many short (typically 2–10 μm long) appendages, they are called **cilia** (sing., *cilium*). Both cilia and flagella are used by cells to move through a watery environment or to pass liquids and particles across the cell surface. These structures are commonly found on unicellular and small multicellular organisms. In animals and certain plants, flagella serve as the tails of sperm cells. In animals cilia commonly occur on the surfaces of cells that line internal ducts of the body (e.g., respiratory passageways).

Eukaryotic cilia and flagella are structurally alike. Each consists of a slender, cylindrical stalk covered by an extension of the plasma membrane. The core of the stalk contains a group of microtubules arranged so that there are nine attached pairs of tubules around the circumference and two unpaired microtubules in the center (Fig. 4–25). This *9 + 2* arrangement of microtubules is characteristic of virtually all eukaryotic cilia and flagella, another example of the unity of organisms that reflects their common origin.

The microtubules move by sliding in pairs past each other. The sliding force is generated by dynein proteins, which are attached to the microtubules like small arms. These proteins use the energy from ATP to power the cilia or flagella. The dynein proteins (arms) on one pair of tubules change their shape and "walk" along the adjacent pair. Thus, the microtubules on one side of a cilium or a flagellum extend farther toward the tip than those on the other side. This sliding of microtubules translates into a bending motion (Fig. 4–25*b*). Cilia typically move like oars, alternating power and recovery strokes and exerting a force that is parallel to the cell surface. A flagellum moves like a whip, exerting a force perpendicular to the cell surface.

(b)

(c)   0.5 µm

(d)   0.5 µm

Figure 4-25 Structure of cilia. A cilium (or flagellum) contains microtubules in a 9 + 2 arrangement. *(a)* This three-dimensional representation shows nine attached microtubule pairs (doublets) arranged in a cylinder, with two unattached microtubules in the center. The "arms" are made of dynein, a motor protein that uses energy from ATP to bend the cilia by "walking" up and down the neighboring pair of microtubules. The dynein arms, shown widely spaced for clarity, are actually much closer together along the longitudinal axis. *(b)* The dynein arms move the microtubules by forming and breaking cross bridges on the adjacent microtubules, so that one tubule "walks" along its neighbor. *(c)* TEM of cross sections through cilia showing the 9 + 2 arrangement of microtubules. *(d)* TEM of a longitudinal section of three cilia of the protist *Tetrahymena*, an organism often used in genetic research. Some of the interior microtubules are visible. *(c, d, W.L. Dentler/Biological Photo Service)*

Each cilium or flagellum is anchored in the cell by a **basal body,** which has nine sets of three microtubules in a cylindrical array (a *9 × 3 structure*). The basal body appears to be the organizing structure for the cilium or flagellum when it first begins to form. However, experiments have shown that as growth proceeds, the tubulin subunits are added much faster to the tips of the microtubules than to the base. Basal bodies and centrioles appear to be functionally related as well as structurally similar. In fact, centrioles are typically found in the cells of organisms that are capable of producing flagellated or ciliated cells; these include animals, certain protists, and a few plants. Both basal bodies and centrioles give rise to new basal bodies and centrioles.

Microfilaments consist of intertwined strings of actin

Microfilaments, also called actin filaments, are flexible, solid fibers about 7 nm in diameter. Each microfilament consists of two intertwined strings of beadlike **actin** molecules (Fig.

(a)

100 µm

(b)

Figure 4–26 **Microfilaments.** *(a)* Many bundles of aggregated microfilaments *(green)* are evident in this confocal fluorescence LM of fibroblasts (cells found in connective tissue). *(b)* An individual microfilament consists of two intertwined strings of beadlike actin molecules. *(a, Nancy Kedersha/ImmunoGen, Inc.)*

4–26). Actin molecules cross-link with one another and with other proteins to form bundles of fibers that provide mechanical support for various cell structures. In many cells, a network of microfilaments can be seen just inside the plasma membrane.

In muscle cells, actin is associated with another protein, myosin, to form fibers that generate the forces involved in muscle contraction (see Chapter 38). In nonmuscle cells, actin can also associate with myosin, forming contractile structures that are involved in various cell movements. Actin filaments themselves cannot contract, but they can generate movement by rapidly assembling and disassembling. Actin filaments associated with myosin are involved in transient functions, such as cell division in animals, in which contraction of a ring of actin associated with myosin causes the constriction of the cell to form two daughter cells (see Chapter 9). Certain organelles in the giant axons of the squid move along microfilaments. A type of myosin appears to be the motor for this transport.

Many types of cells have **microvilli,** projections of the plasma membrane that increase the surface area of the cell for transporting materials across the plasma membrane. Microvilli contain bundles of microfilaments, and they extend and retract as a result of the building and breaking down of these microfilaments.

Intermediate filaments help stabilize cell shape

Intermediate filaments are very stable, tough fibers made of polypeptides that can vary widely in size among different cell types and different species of animals. Typically, they are 8 to 10 nm in diameter. These fibers are thought to help strengthen the cytoskeleton, stabilizing cell shape. They are abundant in parts of a cell that may be subject to mechanical stress. It is not clear whether intermediate filaments are involved in cellular functions beyond their structural role.

The assembly of intermediate filaments is probably irreversible; unpolymerized subunits are not abundant in cells. Cells may be able to regulate the length of intermediate filaments, however, by use of enzymes that break down their polypeptides into smaller fragments.

Intermediate filaments form a sheath called the **nuclear lamina,** just inside the nuclear envelope. These filaments are important in the disorganization and reorganization of the nucleus during cell division.

AN EXTRACELLULAR MATRIX SURROUNDS MOST CELLS

Most eukaryotic cells are surrounded by a **glycocalyx,** or **cell coat,** formed by polysaccharide side chains of proteins and lipids that are part of the plasma membrane. Certain mole-

Figure 4–27 **Extracellular matrix (ECM).** Fibronectins, glycoproteins of the ECM, bind to integrins and other receptors in the plasma membrane.

cules of the cell coat permit cells to recognize one another, make contact, and in some cases to form associations. Other molecules of the cell coat contribute to the mechanical strength of multicellular tissues.

Many animal cells are also surrounded by an **extracellular matrix (ECM),** consisting of a gel of carbohydrates and fibrous proteins (Fig. 4–27). The main structural protein in the ECM is **collagen,** which forms very tough fibers. Certain glycoproteins of the ECM, called **fibronectins,** bind to protein receptors that extend from the plasma membrane. The main membrane receptors for the ECM are **integrins.** These proteins have a variety of functions, including anchoring the external ECM to the microfilaments of the internal cytoskeleton. Binding of proteins to integrins also appears to be important in cell movement and in organizing the cytoskeleton so that cells assume a definite structure.

Most bacteria, fungi, and plant cells are surrounded by a **cell wall** and proteins. Plant cells are surrounded by thick cell walls that contain multiple layers of the polysaccharide **cellulose** (see Fig. 3–9). Other polysaccharides in the plant cell wall form cross links between the bundles of cellulose fibers. Each cellulose fiber layer runs in a different direction from the adjacent layer, giving the cell wall great mechanical strength.

A growing plant cell secretes a thin, flexible *primary cell wall,* which can stretch and expand as the cell increases its size (Fig. 4–28). After the cell stops growing, either new wall material is secreted that thickens and solidifies the primary wall or multiple layers of a *secondary cell wall* with a different chemical composition are formed between the primary wall and the plasma membrane. Wood is made mainly of secondary cell walls. Between the primary cell walls of adjacent cells lies the

middle lamella, a layer of gluelike polysaccharides called *pectins.* The middle lamella causes the cells to adhere tightly to one another.

Figure 4–28 **Plant cell walls.** The cell walls of two adjacent plant cells are labeled in this TEM. The cells are cemented together by the middle lamella, a layer of gluelike polysaccharides called pectins. A growing plant cell first secretes a thin primary wall that is flexible and can stretch as the cell grows. The thicker layers of the secondary wall are secreted inside the primary wall after the cell stops growing. *(Biophoto Associates)*

I. The **cell** is considered the basic unit of life because it is the smallest self-sufficient unit of living material.

II. The **cell theory** states that organisms are composed of cells and that all cells arise by division of preexisting cells.

III. Every cell is surrounded by a **plasma membrane** that forms a cytoplasmic compartment, inside of which are found the contents of the cell.

   A. All cells have genetic material in the form of **DNA.**

   B. Cells have internal structures called **organelles.**

   C. A critical factor in determining cell size is the ratio of the plasma membrane (surface area) to the cell's volume; the plasma membrane must be large enough to regulate the passage of materials into and out of the cell.

   D. Cell size and shape are related to function and are limited by the need to maintain homeostasis.

IV. Biologists have learned much about cellular structure by studying cells with **light microscopes** and **electron microscopes** and with a variety of chemical methods.

   A. The electron microscope has superior **resolving power,** enabling investigators to see details of cell structures not observable with conventional microscopes.

   B. Cell biologists use **cell fractionation** techniques and other biochemical methods to gain information about the function of cellular structures.

V. **Prokaryotic cells** are bounded by a plasma membrane and have a **nuclear area,** but a membrane-bounded nucleus is absent; these cells have little or no internal membrane organization. Prokaryotes typically have a **cell wall** and **ribosomes** and may have **flagella** (sing., **flagellum**).

VI. **Eukaryotic cells** have a membrane-bounded **nucleus** and **cytoplasm,** which is organized into organelles; the fluid component of the cytoplasm is the **cytosol.**

   A. Plant cells differ from animal cells in that they possess rigid cell walls, **plastids,** and large **vacuoles;** cells of most plants lack centrioles.

   B. Membranes divide the cell into membrane-bounded compartments that allow cells to conduct specialized activities within small areas of the cytoplasm, concentrate molecules, and organize metabolic reactions. A network of membranes forms the **endomembrane system.** Small membrane-bounded sacs, called **vesicles,** transport materials between compartments.

   C. The nucleus, the control center of the cell, contains genetic information in the form of DNA.

     1. The nucleus is bounded by a **nuclear envelope** consisting of a double membrane perforated with **nuclear pores** that communicate with the cytoplasm.

     2. DNA in the nucleus associates with protein to form **chromatin.** During cell division, chromatin condenses and the chromosomes become visible.

     3. The **nucleolus** is a region in the nucleus that is the site of ribosomal RNA synthesis and ribosome assembly.

   D. The **endoplasmic reticulum (ER)** is a network of folded internal membranes with many functions.

     1. **Rough ER** is studded along its outer walls with ribosomes, which manufacture proteins. Proteins synthesized on rough ER can be transferred to other cell membranes or secreted from the cells by **transport vesicles,** which are formed by membrane budding.

     2. **Smooth ER** is the site of lipid synthesis and detoxifying enzymes.

   E. The **Golgi complex** consists of stacks of flattened membranous sacs that process, sort, and modify proteins synthesized on the ER. It adds carbohydrates and lipids to proteins and can route proteins, by way of **secretory vesicles,** to the plasma membrane for export from the cell, or to other destinations. The Golgi complex also manufactures lysosomes.

   F. **Lysosomes** function in intracellular digestion; they contain enzymes that break down both worn-out cell structures and substances taken into cells.

   G. **Peroxisomes** are membrane-bounded sacs that contain enzymes that catalyze a variety of reactions in which hydrogen peroxide is formed as a byproduct. Peroxisomes contain catalase, an enzyme that splits hydrogen peroxide.

   H. Vacuoles are important in plant growth and development. Many protists have **food vacuoles** and **contractile vacuoles.** Vacuoles may be formed by the merging of many vesicles.

   I. **Mitochondria,** the sites of aerobic respiration, are double-membrane organelles in which the inner membrane is folded, forming **cristae** that increase the surface area of the membrane. The cristae and the compartment enclosed by the inner membrane, the **matrix,** contain enzymes for the reactions of cellular respiration.

   J. Cells of algae and plants contain plastids; **chloroplasts,** the sites of photosynthesis, are double-membraned plastids. Typically, the inner membrane encloses a fluid-filled space, the **stroma,** and stacks, called **grana,** of disclike sacs, the **thylakoids. Chlorophyll,** the green pigment that traps light energy during **photosynthesis,** is found in the thylakoid membranes.

VII. The **cytoskeleton** is a dynamic internal framework made of at least three types of fibers: microtubules, microfilaments, and intermediate filaments. The cytoskeleton provides structural support and functions in various types of cell movement, including transport of materials in the cell.

   A. **Microtubules** are hollow cylinders assembled from subunits of the protein **tubulin.**

     1. In cells that are not dividing, the minus ends of microtubules appear to be anchored in **microtubule-organizing centers (MTOCs).** The main MTOC is the **centrosome.** In most animal cells the centrosome contains two **centrioles.** Each centriole has a 9×3 arrangement of microtubules.

     2. **Microtubule-associated proteins (MAPs)** include fibrous MAPs and motors. Two motor proteins are kinesin and dynein.

     3. **Cilia** and flagella are important in cell movement. Each consists of a $9 + 2$ arrangement of microtubules, and each is anchored in the cell by a **basal body** that has a 9×3 organization of microtubules.

   B. **Microfilaments,** or **actin filaments,** formed from subunits of the protein actin, are important in cell movement.

   C. **Intermediate filaments** are stable structures formed from several different types of protein; they are thought to stabilize cell shape.

VIII. Most cells are surrounded by a **glycocalyx,** or **cell coat,** formed by polysaccharides extending from the plasma membrane.

   A. Many animal cells are also surrounded by an **extracellular matrix (ECM)** consisting of carbohydrates and protein. **Fibronectins** are glycoproteins of the ECM that bind to **integrins**—receptor proteins in the plasma membrane.

   B. Most bacteria, fungi, and plant cells are surrounded by a cell wall made of carbohydrates and protein. Plant cells secrete cellulose and other polysaccharides to form rigid cell walls.

1. The ability of a microscope to reveal fine detail is known as (a) magnification (b) resolving power (c) cell fractionation (d) scanning electron microscopy (e) phase contrast
2. A plasma membrane is characteristic of (a) all cells (b) prokaryotic cells only (c) eukaryotic cells only (d) animal cells only (e) eukaryotic cells except for plant cells
3. Detailed information about the shape and external features of a specimen can best be obtained by using a (a) differential centrifuge (b) fluorescence microscope (c) transmission electron microscope (d) scanning electron microscope (e) light microscope
4. In eukaryotic cells DNA may be found in (a) chromosomes (b) chromatin (c) mitochondria (d) answers a, b, and c are correct (e) answers a and b only are correct
5. Which of the following structures would NOT be found in prokaryotic cells? (a) cell wall (b) ribosomes (c) nuclear area (d) nucleus (e) flagellum
6. Which of the following is most closely associated with protein synthesis? (a) ribosomes (b) smooth ER (c) mitochondria (d) microfilaments (e) lysosomes
7. Which of the following is most closely associated with the breakdown of ingested material? (a) ribosomes (b) smooth ER (c) mitochondria (d) microfilaments (e) lysosomes
8. Which of the following is most closely associated with photosynthesis?
   (a) basal bodies (b) smooth ER (c) cristae (d) thylakoids (e) MTOCs
9. A 9 + 2 arrangement of microtubules best describes (a) cilia (b) centrosomes (c) basal bodies (d) microfilaments (e) microvilli
10. Which sequence most accurately describes information flow in the eukaryotic cell? (a) DNA in nucleus → messenger RNA → ribosomes → protein synthesis (b) DNA in nucleus → ribosomal RNA → mitochondria → protein synthesis (c) RNA in nucleus → messenger DNA → ribosomes → protein synthesis (d) DNA in nucleus → messenger RNA → Golgi complex → protein synthesis (e) DNA in nucleus → messenger RNA → smooth ER → protein synthesis
11. Which sequence most accurately describes glycoprotein processing in the eukaryotic cell? (a) smooth ER → transport vesicle → *cis* region of Golgi → *trans* region of Golgi → plasma membrane or other organelle (b) rough ER → transport vesicle → *cis* region of Golgi → *trans* region of Golgi → plasma membrane or other organelle (c) rough ER → transport vesicle → *trans* region of Golgi → *cis* region of Golgi → plasma membrane or other organelle (d) rough ER → nucleus → *cis* region of Golgi → *trans* region of Golgi → plasma membrane or other organelle (e) smooth ER → transport vesicle → *cis* region of Golgi → chloroplast
12. Which of the following are part of the cytoskeleton? (a) microfilaments (b) lysosomes (c) peroxisomes (d) ribosomes (e) endoplasmic reticulum
13. Which of the following function in cell movement? (a) microtubules (b) cristae (c) grana (d) smooth ER (e) rough ER

REVIEW QUESTIONS

1. Describe the basic needs of all living things and explain how a cell is able to meet these needs. Why is the cell theory important to an understanding of how living things function?
2. Compare and contrast prokaryotic and eukaryotic cells.
3. Draw a chloroplast and a mitochondrion. Label the membranes and their compartments.
4. Describe the functions of each of the following
   (a) ribosomes
   (b) smooth endoplasmic reticulum
   (c) Golgi complex
   (d) microtubules
5. Trace the path of a protein from its site of synthesis to its final destination for the following
   (a) a secreted protein
   (b) a protein found inside a lysosome
   (c) a protein associated with the plasma membrane
6. Contrast microfilaments and microtubules. Compare their structures and the different roles they play in cell structure and function.
7. Why are lysosomes sometimes referred to as the "self-destruct system" of the cell?
8. Describe plant cell walls. How are they formed?
9. Label the diagrams of the animal and plant cell. How is the structure of each organelle related to its function? Use Figures 4–8 and 4–9 to check your answers.

YOU MAKE THE CONNECTION

1. Why does a eukaryotic cell need both membranous organelles and fibrous cytoskeletal components?
2. Describe three examples illustrating the correlation between cell structure and function. (*Hint:* Think of mitochondrial structure.)
3. The *Acetabularia* experiments described in this chapter suggest that DNA is much more stable in the cell than messenger RNA. Is this advantageous or disadvantageous to the cell? Why? How can *Acetabularia* live for a few days after its nucleus is removed?

RECOMMENDED READINGS

Alberts, B., D. Bray, J. Lewis, M. Raff, K. Roberts, and J.D. Watson. *Molecular Biology of the Cell,* 3rd ed. Garland, New York, 1994. A comprehensive, well written, and well illustrated text on cell biology.

Duke, R.C., D.M. Oljcius, and J. Ding-E Young. "Cell Suicide in Health and Disease." *Scientific American,* Vol. 275, No. 6, Dec. 1996. A discussion of the role of apoptosis in regulation and disease.

Glover, D.M., C. Gonzalez, and J.W. Raff. "The Centrosome." *Scientific American,* Vol. 268, No. 6, Jun. 1993. The centrosome helps direct the assembly of the cytoskeleton.

Horwitz, A.F. "Integrins and Health." *Scientific American,* Vol. 276, No. 5, May 1997. Integrins are proteins in the plasma membrane that link microtubules of the cytoskeleton to the extracellular matrix.

Science, Vol. 279, No. 5350, Jan. 1998. Special Issue: Frontiers in Cell Biology: The Cytoskeleton. Several articles and an editorial discuss the latest findings on cytoskeletal structure and function.

Stossel, T.P. "The Machinery of Cell Crawling." *Scientific American,* Vol. 271, No. 3, Sept. 1994. The cytoskeleton functions in cell locomotion.

Tobin, A.J. and R.E. Morel. *Asking About Cells.* Saunders College Publishing, Philadelphia, 1997. A clearly written, excellent introduction to cell biology.

Wallace, D.C. "Mitochondrial DNA in Aging and Disease." *Scientific American,* Vol. 277, No. 2, Aug. 1997. Mutations in mitochondrial DNA have been linked to several human diseases.

● Visit our website at **http://www.saunderscollege.com/lifesci/titles.html** and click on Solomon/Berg/Martin Biology for links to chapter-related resources on the World Wide Web.

CHAPTER 5

Biological Membranes

To carry out the many chemical reactions necessary to sustain life, a cell must maintain an appropriate internal environment. The plasma membrane that surrounds every cell physically separates the cell from the outside world and defines the cell as a distinct entity. The plasma membrane helps maintain a life-supporting internal environment by regulating the passage of materials into and out of the cell. Many biologists view the evolution of biological membranes as an essential step in the origin of life. One can argue further that membranes made the evolution of complex cells possible because the extensive internal membranes of eukaryotes form multiple compartments with unique environments for highly specialized activities.

Cell biologists have found that biological membranes are not inanimate walls; they are complex and dynamic structures made from lipid and protein molecules that are in constant motion. The unusual properties of membranes allow them to perform many functions in addition to defining the cell as a compartment and regulating passage of materials. These functions include participating in many chemical reactions, transmitting signals and information between the environment and the interior of the cell, and acting as an essential part of energy transfer and storage systems (see Chapters 7 and 8).

One exciting area of cell membrane research focuses on membrane proteins. Many proteins associated with the plasma membrane are enzymes. Others function in transport of materials or in transfer of information. Still others are important in connecting cells to one another to form tissues. Investigators are studying how membrane proteins function in health and disease. The human skin cells shown in the photograph were grown in culture and stained with fluorescent antibodies. Membrane proteins known as cadherins are seen as green belts around each cell in this sheet of cells. These proteins are responsible for calcium-dependent adhesion between cells that form sheets. An absence of these proteins has been observed when cells become malignant. (In the cells shown in the photograph, the nuclei appear as blue spheres; myosin in the cell appears red.)

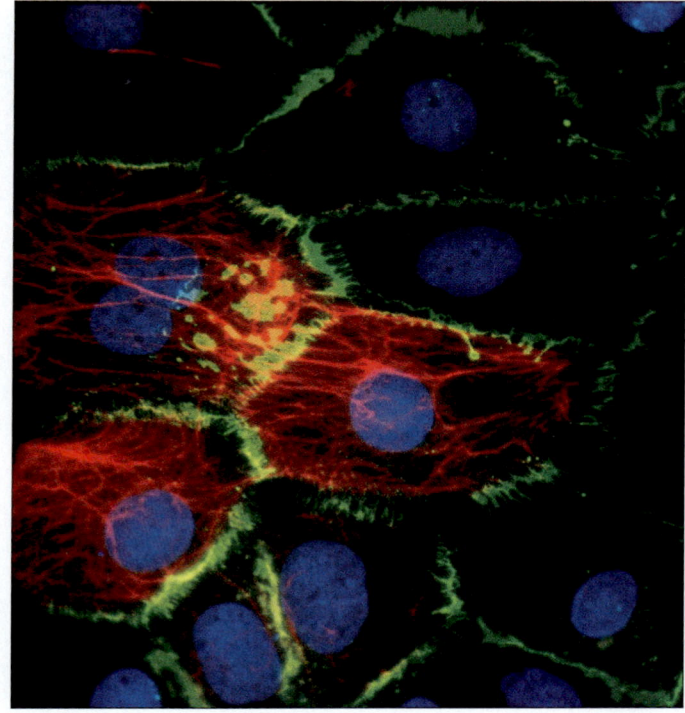

(Nancy Kedersha)

In this chapter, we first consider what is known about the composition and structure of biological membranes. We also consider how information can cross the plasma membrane through a signal relay system. We then survey how various materials, ranging from simple to complex molecules, and even particles, move across membranes. Finally, specialized structures that permit interactions between membranes of different cells are examined. Although most of our discussion centers on the structure and functions of plasma membranes, most of the concepts are also applicable to other membranes of the cell.

AFTER YOU HAVE STUDIED THIS CHAPTER YOU SHOULD BE ABLE TO

1. Evaluate the importance of membranes to the cell, emphasizing their various functions.
2. Make a detailed sketch of the fluid mosaic model of cell membrane structure.
3. Explain how the properties of the lipid bilayer are responsible for many of the physical properties of a cell membrane.
4. Explain how the various classes of membrane proteins associate with the lipid bilayer and discuss the different roles of membrane proteins.
5. Contrast the physical processes of simple diffusion and osmosis with the carrier-mediated physiological processes by which materials are transported across cell membranes.

6. Solve simple problems involving osmosis; for example, predict whether cells will swell or shrink under various osmotic conditions.
7. Summarize the main ways that small hydrophilic molecules can move across membranes.
8. Differentiate between the processes of facilitated diffusion and active transport and discuss the ways in which energy is supplied to active transport systems.
9. Compare endocytotic and exocytotic transport mechanisms.
10. Describe the structures and compare the functions of desmosomes, tight junctions, gap junctions, and plasmodesmata.

BIOLOGICAL MEMBRANES ARE LIPID BILAYERS WITH ASSOCIATED PROTEINS

Long before the development of the electron microscope, it was known that membranes are composed of both lipids and proteins. Work by researchers in the 1920s and 1930s had provided clues that the core of cell membranes is composed of lipids, mostly phospholipids (see Chapter 3).

Phospholipids form bilayers in water

Phospholipids are primarily responsible for the physical properties of biological membranes. This is because certain phospholipids have unique attributes, including features that allow them to form bilayered structures. A phospholipid contains two fatty acid chains linked to two of the three carbons of a glycerol molecule. The fatty acid chains make up the nonpolar, hydrophobic ("water-fearing") portion of the phospholipid. Bonded to the third carbon of the glycerol is a negatively charged, hydrophilic ("water-loving") phosphate group, which in turn is linked to a polar, hydrophilic organic group. Molecules of this type, which have distinct hydrophobic and hydrophilic regions, are called **amphipathic** molecules. All lipids that make up the core of biological membranes have amphipathic characteristics.

Because one end of each phospholipid associates freely with water and the opposite end does not, the most stable orientation for them to assume in water results in the formation of a bilayer structure (Fig. 5–1). This arrangement allows the hydrophilic heads of the phospholipids to be in contact with the aqueous medium, while their oily tails, the hydrophobic fatty acid chains, are buried in the interior of the structure away from the water molecules.

Amphipathic properties alone do not predict the ability of lipids to associate as a bilayer. Shape is also important. Phospholipids tend to have uniform widths; their roughly cylindrical shapes, together with their amphipathic properties, are responsible for bilayer formation. In summary, phospholipids form bilayers because the molecules have (1) two distinct re-

gions, one strongly hydrophobic and the other strongly hydrophilic (making them strongly amphipathic), and (2) cylindrical shapes that allow them to associate with water most easily as a bilayer structure.

Many common detergents are amphipathic molecules, each containing a single hydrocarbon chain (like a fatty acid) at one end and a hydrophilic region at the other. These molecules are roughly cone-shaped, with the hydrophilic end forming the broad base and the hydrocarbon tail leading to the point. Because of their shapes, these molecules do not associate as bilayers but instead tend to form spherical structures in water (Fig. 5–1). Detergents are able to "solubilize" oil because the oil molecules associate with the hydrophobic interiors of the spheres.

Current data support a fluid mosaic model of membrane structure

By examining the plasma membrane of the mammalian red blood cell and comparing the surface area of the membrane

Detergent in water	Phospholipids in water

Hydrophilic heads

Hydrophobic tails

(a)   *(b)*

Figure 5–1 **Lipid membranes.** The ability of lipids to associate in water depends on their amphipathic properties and their shapes. *(a)* Detergent molecules are roughly cone-shaped amphipathic molecules that associate in water as spherical structures. *(b)* Phospholipids associate as bilayers in water because they are roughly cylindrical amphipathic molecules. The hydrophobic fatty acid chains associate with each other and are not exposed to the water. The hydrophilic phospholipid heads are in contact with the water.

Membrane proteins

Phospholipid bilayer

Membrane proteins

(a) The Davson-Danielli "Sandwich" model

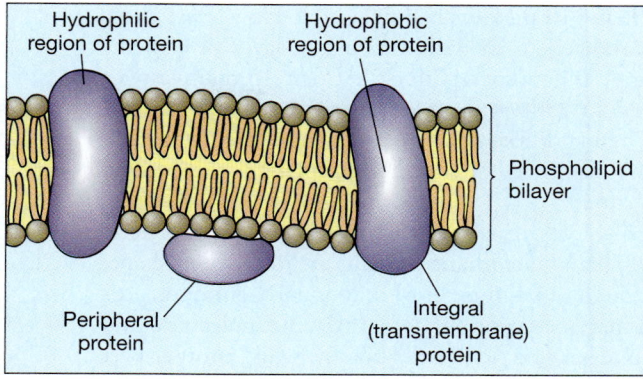

Hydrophilic region of protein

Hydrophobic region of protein

Phospholipid bilayer

Peripheral protein

Integral (transmembrane) protein

(b) Fluid mosaic model

Figure 5-2 **Two models of membrane structure.** *(a)* According to the Davson-Danielli model, the membrane is a sandwich of phospholipids spread between two layers of protein. Although accepted for more than 20 years, this model was shown to be incorrect. *(b)* According to the fluid mosaic model, a cell membrane is made up of a fluid lipid bilayer with a constantly changing "mosaic pattern" of associated proteins.

with the total number of lipid molecules per cell, early investigators were able to calculate that the phospholipids are probably arranged so that the membrane is no more than two phospholipid molecules thick. These findings, together with other data, led H. Davson and J.F. Danielli in 1935 to propose a model in which they envisioned a membrane as a kind of "sandwich" consisting of a *lipid bilayer* (a double layer of lipid) between two protein layers (Fig. 5–2*a*). This very useful model had a great influence on the direction of membrane research for over 20 years. Models are very important in the scientific process; good ones not only explain the available data, but are testable. That is, scientists can use the model to help them develop hypotheses that can be tested experimentally (see Chapter 1).

With the development of the electron microscope in the

1950s, cell biologists were able to see the plasma membrane for the first time. One of their most striking observations was how uniform and thin the membranes are. The plasma membrane is no more than 10 nanometers thick. The electron microscope revealed a three-layered structure, something like a railroad track, with two dark layers separated by a lighter layer (Fig. 5–3). Their findings seemed to support the protein-lipid-protein sandwich model.

During the 1960s, a paradox emerged regarding arrangement of the proteins. It was widely assumed that membrane proteins were very uniform and had shapes that would allow them to lie like thin sheets on the membrane surface. When proteins from membranes were purified by cell fractionation, however, they were found to be far from uniform; in fact, they varied widely in composition and size. Some proteins were

Cell interior

Plasma membrane

Outside cell

0.1 µm

Figure 5-3 **TEM of the plasma membrane of a mammalian red blood cell.** The plasma membrane separates the cytoplasm (darker region) from the external environment (lighter region). The hydrophilic heads of the phospholipids are seen as the parallel dark lines, while the hydrophobic tails are visible as the light zone between them. *(Omikron/Photo Researchers, Inc.)*

much larger than investigators had imagined. How could these proteins be arranged to fit within a surface layer of a membrane less than 10 nanometers thick? At first, some investigators attempted to address these objections by modifying the model with the hypothesis that the proteins on the membrane surfaces were a flattened, extended form, perhaps a β-pleated sheet (see Chapter 3).

Other cell biologists found that instead of having sheet-like structures, many membrane proteins are rounded in shape, or globular. Studies of a number of individual membrane proteins showed that one region (or domain) of the molecule could always be found on one side of the bilayer, while another part of the protein might be located on the opposite side. It appeared that, rather than forming a thin surface layer, many membrane proteins extend completely through the lipid bilayer. Thus, membranes appear to contain many different types of proteins of different shapes and sizes that are associated with the bilayer in a mosaic pattern.

In 1972, S.J. Singer and G.L. Nicolson proposed a model of membrane structure that represented a synthesis of the known properties of biological membranes. According to their **fluid mosaic model,** a cell membrane consists of a fluid bilayer of phospholipid molecules in which the proteins are embedded or otherwise associated, much like the tiles in a mosaic picture. This mosaic pattern is not static, however, because the positions of the proteins are constantly changing as they move about like icebergs in a fluid sea of phospholipids. This model has provided great impetus to research; it has been repeatedly tested and has been shown to accurately predict the properties of many kinds of cell membranes. Figure 5–2*b* depicts the plasma membrane of a eukaryotic cell; prokaryotic plasma membranes are discussed in Chapter 23.

Biological membranes are two-dimensional fluids

An important physical property of phospholipid bilayers is that they behave as *liquid crystals*. The bilayers are crystal-like in that the lipid molecules form an ordered array with the heads on the outside and fatty acid chains on the inside; they are liq-

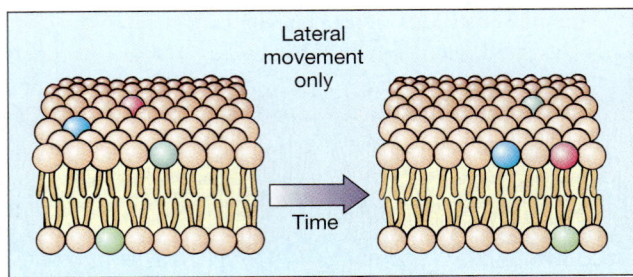

Figure 5–4 Membrane fluidity. The ordered arrangement of phospholipid molecules makes the cell membrane a fluid crystal. However, the hydrocarbon chains are in constant motion, allowing each molecule to move laterally on the same side of the bilayer.

uid-like in that, despite the orderly arrangement of the molecules, their hydrocarbon chains are in constant motion. Thus molecules are free to rotate and can move laterally within their single layer (Fig. 5–4). Such movement gives the bilayer the property of a *two-dimensional fluid*. Under normal conditions this means that a single phospholipid molecule can travel laterally across the surface of a eukaryotic cell in seconds.

The fluid-like qualities of lipid bilayers also allow molecules embedded in them to move along the plane of the membrane (as long as they are not anchored in some way). This was elegantly demonstrated by David Frye and Michael Ediden in 1970. They conducted experiments in which they followed the movement of membrane proteins on the surface of two cells that had been joined together (Fig. 5–5). When the plasma membranes of a mouse cell and a human cell are fused, within minutes at least some of the membrane proteins from each cell migrate and become randomly distributed over the single continuous plasma membrane that surrounds the joined cells. The experiments of Frye and Ediden demonstrated that the fluidity of the lipids in the membrane allows many of the proteins to move, producing an ever-changing configuration.

If a membrane is to function properly, its lipids must be in a state of optimal fluidity. The structure of a membrane is weakened if its lipids are too fluid. On the other hand, it has been shown that many membrane functions, such as the trans-

Human cell   Mouse cell
Labeled membrane proteins

Human-mouse hybrid cell forming

Proteins randomly distributed

Figure 5–5 Mobility of membrane proteins. An elegant series of experiments by Frye and Ediden demonstrated that at least some membrane proteins are highly mobile entities in a two-dimensional fluid. Membrane proteins of mouse cells and human cells were labeled with fluorescent dye markers in two different colors. When the plasma membranes of a mouse cell and a human cell were fused, mouse proteins were observed migrating to the human side and human proteins to the mouse side. After a short time mouse and human proteins became randomly distributed on the cell surface.

port of certain substances, are inhibited or cease if the lipid bilayer is too rigid. At normal temperatures, cell membranes are fluid. However, the motion of the fatty acid chains is slowed at low temperatures. If the temperature decreases to a critical point, the membrane is converted to a more solid gel state.

Certain properties of membrane lipids have significant effects on the fluidity of the bilayer. Recall from Chapter 3 that molecules are free to rotate around single carbon-to-carbon covalent bonds. Because most of the bonds in hydrocarbon chains are single bonds, the chains themselves can undergo very rapid twisting motions that increase as the temperature increases.

The fluid state of the membrane depends on its component lipids. You have probably noticed that when melted butter is left at room temperature, it solidifies. Vegetable oils, however, remain liquid at room temperature. Recall from our discussion of fats in Chapter 3 that butter is a saturated fat, so it has no double bonds in its fatty acids. In contrast, a vegetable oil may be polyunsaturated with two or more double bonds in its fatty acid chains. Double bonds produce "bends" in the molecules that prevent the hydrocarbon chains from coming close together. In this way unsaturated fats lower the temperature at which oil or membrane lipids solidify.

Many organisms have regulatory mechanisms that allow them to maintain their membranes in an optimally fluid state. For example, some organisms that are unable to maintain a constant internal temperature can compensate for temperature changes by altering the fatty acid content of their membrane lipids. When grown at colder temperatures, such organisms are found to have relatively high proportions of unsaturated fatty acids in their membrane lipids.

Some membrane lipids have the ability to help stabilize membrane fluidity within certain limits. One such "fluidity buffer" is cholesterol, a steroid found in animal cell membranes. A cholesterol molecule is largely hydrophobic, but is slightly amphipathic due to the presence of a single hydroxyl group (see Fig. 3–13a). This hydroxyl group associates with the hydrophilic heads of the phospholipids; the hydrophobic remainder of the cholesterol molecule fits between the fatty acid hydrocarbon chains (Fig. 5–6).

Figure 5–6   Structure of the cell membrane according to the fluid mosaic model. Although the lipid bilayer consists mainly of phospholipids, other lipids such as cholesterol are present. Peripheral proteins are loosely associated with the bilayer, while integral proteins are tightly bound. The integral proteins shown here are transmembrane proteins that extend through the bilayer. They have hydrophilic regions on both sides of the bilayer, connected by a membrane-spanning α-helix with hydrophobic amino acid side chains (not shown). Glycolipids (carbohydrates attached to lipids) and glycoproteins (carbohydrates attached to proteins) are exposed on the extracellular surface; they play roles in cell recognition and adhesion.

At low temperatures the cholesterol molecules act as "spacers" between the hydrocarbon chains, restricting molecular interactions that would promote solidifying. Cholesterol also helps prevent the membrane from becoming weakened or unstable at higher temperatures. This is because the cholesterol molecules interact strongly with the portions of the hydrocarbon chains closest to the phospholipid head. This interaction restricts motion in these regions. Plant cells have steroids other than cholesterol that carry out similar functions.

Biological membranes fuse and form closed vesicles

Lipid bilayers, particularly those in the liquid-crystalline state, have additional important physical properties. Bilayers tend to resist forming free ends; as a result, they are self-sealing and under most conditions spontaneously round up to form closed vesicles. Fluid bilayers are also flexible, allowing cell membranes to change shape without breaking. Finally, under appropriate conditions lipid bilayers have the ability to fuse with other bilayers.

Membrane fusion is an important cellular process. Recall from Chapter 4 that when a vesicle fuses with another membrane, both membrane bilayers and their compartments become continuous. Various transport and secretory vesicles form from and also merge with membranes of the ER and Golgi complex, allowing materials to be transferred from one compartment to another. A secretory vesicle can fuse with the plasma membrane when a product is secreted from the cell. This process, known as exocytosis, will be discussed later in this chapter. In endocytosis, large molecules are brought into the cell from the outside by the formation of vesicles from the plasma membrane.

Membrane proteins include integral and peripheral proteins

The two major classes of membrane proteins, integral proteins and peripheral proteins, are defined by how tightly they are associated with the lipid bilayer (Fig. 5–6). **Integral membrane proteins** are firmly bound to the membrane. Cell biologists usually can release them only by disrupting the bilayer with detergents. These proteins are amphipathic. Their hydrophilic regions extend out of the cell or into the cytoplasm, while their hydrophobic regions interact with the tails of the membrane phospholipids.

Some integral proteins do not extend all the way through the membrane. Many others, called **transmembrane proteins,** extend completely through the membrane. Some span the membrane only once, while others wind back and forth as many as 24 times. The most common kind of transmembrane protein is an α-helix (see Chapter 3) with hydrophobic amino acid side chains projecting out from the helix into the hydrophobic region of the lipid bilayer (Fig. 5–6).

Peripheral membrane proteins are not embedded in the lipid bilayer. They are located on the inner or outer surfaces of the plasma membrane, usually bound to exposed regions of integral proteins by noncovalent interactions. Peripheral proteins can be easily removed from the membrane without disrupting the structure of the bilayer.

Proteins are oriented asymmetrically across the bilayer

One of the most remarkable demonstrations that proteins are actually embedded in the lipid bilayer comes from freeze-fracture electron microscopy (Fig. 5–7), which enables investiga-

0.1 μm

Figure 5–7 Asymmetry of the plasma membrane. *(a)* In the freeze-fracture method, the path of membrane cleavage is along the hydrophobic interior of the lipid bilayer, resulting in two complementary fracture faces: (1) an inner half-membrane presenting the P-face (or protoplasmic face), from which project the majority of the membrane proteins, and (2) a relatively smooth, outer half-membrane presenting the E-face (or external face), which shows fewer protein particles. In a good fracture, particles are visible on both of the inside faces of the fractured membrane, as shown here. These particles are transmembrane proteins inserted into the lipid bilayer. Freeze-fractured bilayers of lipids alone do not have particles on the fracture planes. *(b)* A freeze-fracture TEM. Notice the greater number of proteins on the P-face of the membrane. *(b, D.W. Fawcett)* tors to literally see the membrane from "inside out." When cellular membranes are examined in this way, numerous particles are observed on the fracture faces. These particles are clearly integral membrane proteins because they are never seen in freeze-fractured artificial lipid bilayers. These findings profoundly influenced Singer and Nicolson in their development of the fluid mosaic model.

When the two sides of a membrane are compared (as in Fig. 5–7), large numbers of particles are found on one side and very few on the other. This does not necessarily mean that there are more proteins on one side of the membrane than on the other but rather that most are more firmly attached to a given side. Thus, the protein molecules are *asymmetrically oriented*. Each side of a membrane has different characteristics because each type of protein is oriented in the bilayer in only one way. Proteins are not randomly placed into membranes; asymmetry is produced by the highly specific way in which each protein is inserted into the bilayer.

Membrane proteins that will become part of the inner surface of the plasma membrane are manufactured by free ribo-

somes and move to the membrane through the cytoplasm. Membrane proteins that will be associated with the cell's outer surface are manufactured like proteins destined to be exported from the cell. As discussed in Chapter 4, these proteins are initially formed by ribosomes on the rough endoplasmic reticulum (ER). They pass through the ER membrane into the lumen, where sugars are added, making them glycoproteins. Only a part of each protein passes through the ER membrane, so each completed protein has some regions that are located in the ER lumen and other regions that remain in the cytosol. Enzymes that attach the sugars to certain amino acids on the protein are found only in the lumen of the ER. Thus, carbohydrates can be added only to the parts of proteins that are located in that compartment.

Follow the vesicle budding and membrane fusion events that are part of the transport process from top to bottom in Figure 5–8. You can see that the same region of the protein that protruded into the ER lumen is also transferred to the lumen of the Golgi complex. There additional enzymes further modify the carbohydrate chains. Within the Golgi complex,

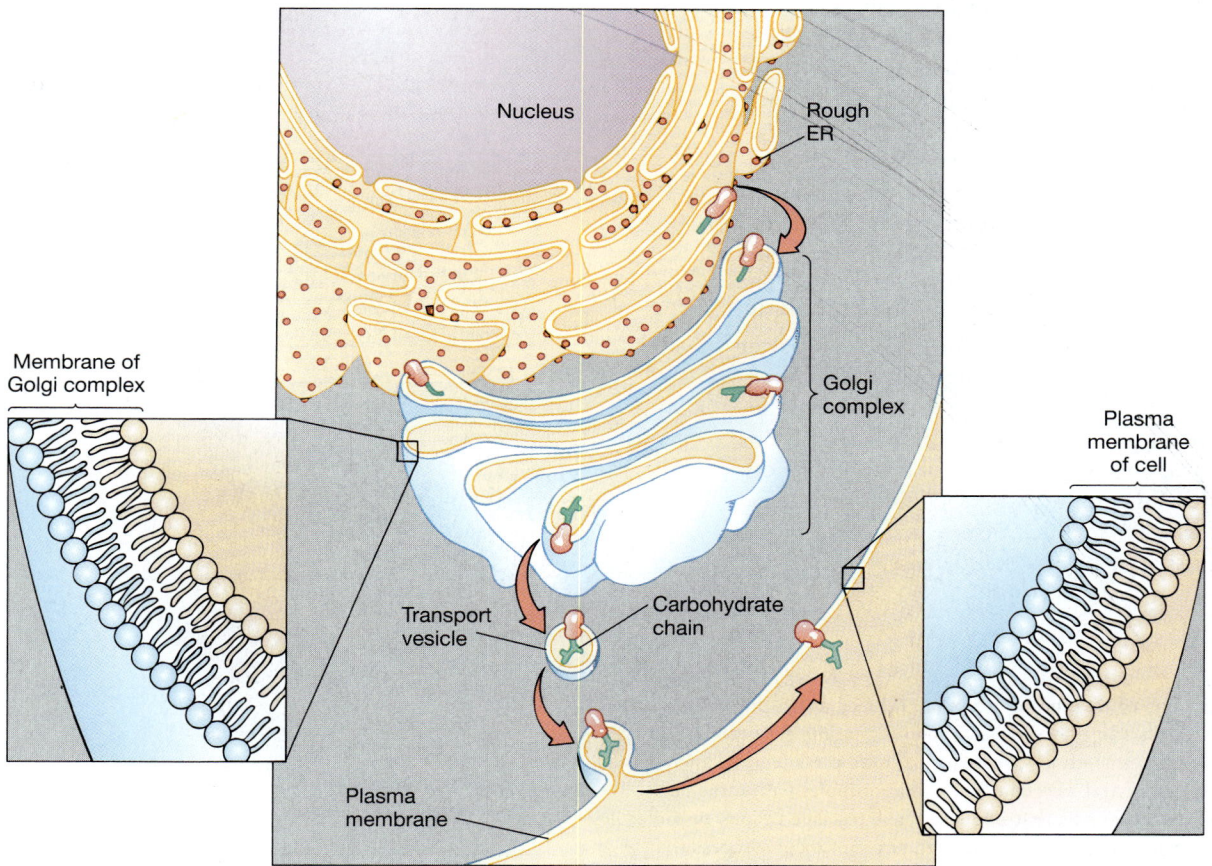

Figure 5–8 Formation of plasma membrane. The surface of the rough ER membrane that faces the lumen of the rough ER also faces the lumen of the Golgi complex and vesicles. When a vesicle fuses with the plasma membrane its inner surface becomes the extracellular surface of the plasma membrane. Note that the orientation of a protein in the plasma membrane is also a consequence of the pathway of its synthesis and transport in the cell. Carbohydrates added to proteins in the ER and then modified in the Golgi complex are associated with the extracellular surface of the plasma membrane.

the glycoprotein is sorted and directed to the plasma membrane. The modified region of the protein remains inside a membrane compartment of a secretory vesicle as it buds from the Golgi complex. When the secretory vesicle fuses with the plasma membrane, the carbohydrate chain becomes the part of the membrane protein that extends to the exterior of the cell surface.

ER lumen $\longrightarrow$ transport vesicle $\longrightarrow$ vesicles in Golgi (transport to successive compartments) $\longrightarrow$ secretory vesicle $\longrightarrow$ plasma membrane

Membrane proteins function in transport, information transfer, and as enzymes

Why should a membrane such as the plasma membrane illustrated in Figure 5–6 require so many different proteins? This diversity reflects the number of activities that take place in or on the membrane. Generally, plasma membrane proteins fall into several broad functional groups (Fig. 5–9). Some membrane proteins form junctions between adjacent cells (discussed later in this chapter). Others attach to cytoskeletal elements. For example, the integrins described in Chapter 4 serve as receptors, or docking sites, for proteins of the extracellular matrix and also bind to microfilaments inside the cell.

A number of membrane proteins are involved in the transport of molecules across the membrane. Some form channels that selectively allow the passage of certain ions or molecules. Other proteins form pumps that use ATP to actively transport solutes across the membrane. Some membrane proteins are enzymes that catalyze reactions that take place near the cell surface. In some membranes, for example the mitochondrial or chloroplast membranes, a sequence of enzymes may be organized to regulate a series of reactions such as occurs in cellular respiration or photosynthesis. Some membrane proteins transmit signals to neighboring cells. Membrane proteins can serve as receptors that function in cell recognition, permitting cells to identify each other for such diverse activities as joining cells to form a tissue or for internal defense.

Figure 5–9 **Functions of membrane proteins.** *(a)* Cell adhesion proteins attach membranes of adjacent cells and may serve as anchoring points for networks of cytoskeletal elements. *(b)* Some membrane proteins, for example integrins, anchor the cell to the extracellular matrix and also connect to microtubules within the cell. *(c)* Transport proteins form channels that allow selective passage of ions or molecules. *(d)* Some transport proteins pump solutes across the membrane, a process that requires a direct input of energy. *(e)* Membrane-bound enzymes catalyze reactions that take place within or along the surface of the membrane. *(f)* Some receptor proteins function in cell recognition. For example, the cells of every human have distinctive MHC receptors that identify them as part of the body; cells of bacteria have different surface proteins that are recognized as foreign and stimulate immune defenses that destroy the bacteria. *(g)* Some receptors bind with signal molecules such as hormones and transmit information into the cell by signal transduction.

Many receptor proteins receive information from the environment, for example from hormones, and transmit it to the cell interior (see *Making the Connection: Information Transfer across the Plasma Membrane*). Signal molecules may transmit information by **signal transduction.** In this process, the receptor converts an extracellular signal into an intracellular signal that affects some function of the cell.

Each component of a signal transduction system acts as a relay "switch," which can be in an activated ("on") state or an inactive ("off") state. The first component may be a receptor, a transmembrane protein with a domain exposed on the extracellular surface. In a typical sequence of events, the binding of the external signal activates the receptor by changing its shape. The activated receptor then changes the conformation of a second protein, which then becomes activated. Ultimately these interactions result in the activation of a specific enzyme bound to the membrane, which may activate intracellular enzymes or catalyze the production of large numbers of intracellular signal molecules. In this way the original signal received by the receptor protein is amplified many times, and the metabolism of the cell may be profoundly altered.

CELL MEMBRANES ARE SELECTIVELY PERMEABLE

Whether a membrane permits a substance to pass through it depends on the size and charge of the substance and on the composition of the membrane. A membrane is said to be *permeable* to a given substance if it permits that substance to pass through, and impermeable if it does not. A **selectively permeable membrane** allows some but not other substances to pass through it readily. In general, biological membranes are most permeable to small molecules and to lipid-soluble substances able to pass through the hydrophobic interior of the bilayer.

Although they are polar (and therefore not lipid-soluble), water molecules can rapidly cross a fluid lipid bilayer. They can pass through a membrane because they are small enough to pass through gaps that occur as a fatty acid chain momentarily moves out of the way. Gases such as oxygen, carbon dioxide, and nitrogen; small polar molecules like glycerol; plus larger, nonpolar (hydrophobic) substances such as hydrocarbons also move through the lipid bilayer rapidly. Slightly larger polar molecules, such as glucose, and charged ions of any size pass through the bilayer much more slowly.

Although the bilayer is relatively impermeable to ions, cells must be able to move ions, as well as large and small polar molecules such as amino acids and sugars, across membranes. The permeability of membranes to those substances is due primarily to the activities of specialized membrane proteins. Biological membranes surrounding cells, nuclei, vacuoles, mitochondria, chloroplasts, and other organelles are selectively permeable to different types of molecules.

In response to varying environmental conditions or cellular needs, a plasma membrane may be a barrier to a particular substance at one time and actively promote its passage at another time. By regulating chemical traffic in this way, a cell can exert some control over its own internal ionic and molecular composition, which can be very different from that on the outside. In the nonliving world, materials move passively by physical processes such as diffusion. In living organisms, some particles can diffuse across the bilayer, while other materials can be moved rapidly by physiological processes such as active transport, exocytosis, and endocytosis (discussed later in this chapter). Physiological transport mechanisms require a direct expenditure of metabolic energy by the cell.

Random motion of particles leads to diffusion

Some substances pass into or out of cells and move about within cells by simple **diffusion,** a physical process based on random motion. All atoms and molecules possess kinetic energy, or energy of motion, at temperatures above absolute zero (0° Kelvin, −273° Celsius, or −459.4°F). Matter may exist as a solid, liquid, or gas, depending on the freedom of movement of its constituent particles. The particles of a solid are closely packed, and the forces of attraction between them allow them to vibrate but not to move around. In a liquid the particles are farther apart; the attractions are weaker, and the particles move about with considerable freedom. In a gas the particles are so far apart that intermolecular forces are negligible; molecular movement is restricted only by the walls of the container that encloses the gas. This means that atoms and molecules in liquids and gases move in a kind of "random walk," changing directions as they collide.

Although the movement of the individual particles is undirected and unpredictable, we can nevertheless make predictions about the behavior of groups of particles. If the particles (atoms, ions, or molecules) are not evenly distributed, then at least two regions exist, one with a higher concentration of particles and the other with a lower concentration. Such a difference in the concentration of a substance from one place to another is a **concentration gradient.**

In the phenomenon of diffusion, the random motion of particles results in their net movement "down" their own concentration gradient (from the region of higher concentration to the one of lower concentration). This does not mean that individual particles are prohibited from moving "against" the gradient. However, because there are initially more particles in the region of high concentration, it logically follows that more particles move randomly from there into the low-concentration region than vice versa.

Diffusion can occur rapidly over very short distances. The rate of diffusion is determined by the movement of the particles, which in turn is a function of their size and shape, their electrical charges, and the temperature. As the temperature rises, particles move faster and the rate of diffusion increases.

Particles of different substances in a mixture diffuse independently of each other. If particles are not added to or re-

INFORMATION TRANSFER ACROSS THE PLASMA MEMBRANE

How can an extracellular hormone or other regulatory molecule transmit information to the cell interior without physically crossing the plasma membrane? Most signal molecules rely on systems of interacting integral membrane proteins to transmit the information by signal transduction. After the **ligand,** a signaling molecule such as a hormone or other regulatory molecule, binds to the membrane receptor, the signal is relayed through a sequence of proteins (see figure). The signaling molecule is sometimes referred to as the first messenger. Often the pathway involves **protein kinases,** enzymes that transfer phosphate groups from ATP to certain proteins. The phosphorylated protein is activated and affects some metabolic activity or brings about some structural change.

Some ligand-receptor complexes bind to and activate certain integral membrane proteins, referred to as **G proteins.** In 1994 Dr. Alfred G. Gilman and Dr. Martin Rodbell were awarded a Nobel Prize for their research on G proteins. These proteins were so named because the active form is bound to GTP, or guanosine triphosphate, a molecule similar to ATP but containing the base guanine instead of adenine. G proteins catalyze the hydrolysis of GTP to GDP, a process that releases energy.

In a complex sequence of events, a G protein relays the message from the receptor to an enzyme such as adenylyl cyclase that catalyzes the production of a **second messenger,** often cyclic AMP. Typically, the second messenger activates protein kinases, enzymes that activate certain proteins by phosphorylating them. This sequence of reactions, beginning with the binding of the signaling molecule to the receptor, leads to a change in some cell function.

G proteins are involved in a number of important signal transductions, including the action of many hormones (see Chapter 47). Some G proteins regulate channels that allow ions to cross the plasma membrane, and still others play important roles in the senses of sight, smell, and taste (see Chapter 41).

Ras proteins, a group of GTP-binding proteins that function somewhat like G proteins, are thought to be important in signal transduction necessary for many cell activities. Fibroblasts (a type of connective tissue cell) require the presence of two growth factors (epidermal growth factor and platelet-derived growth factor) for DNA synthesis. However, when investigators inactivated Ras proteins by injecting antibodies to them into the fibroblasts, the growth factors were no longer effective. Data from this and similar experiments led to the conclusion that Ras is important in signal transduction involving growth factors. Ras links growth factor receptors with a class of enzymes known as serine kinases.

Cell biologists have demonstrated that when certain ligands bind to integrins (transmembrane proteins that connect the cell to the extracellular matrix) in the plasma membrane, certain signal transduction pathways are activated. **Growth factors** also turn on signaling pathways, and it is thought that growth factors and certain molecules of the extracellular matrix may modulate each other's messages. Integrins also respond to information received from inside the cell. This inside-out signaling affects how selective integrins are about the molecules to which they bind and how strongly they bind to them.

---

moved from the system, a state of **equilibrium,** a condition of no net change in the system, is ultimately reached. At equilibrium the particles are uniformly distributed.

More commonly in organisms, equilibrium is never attained. For example, carbon dioxide continually forms within a human cell as sugars and other molecules are metabolized during the process of aerobic respiration. Carbon dioxide readily diffuses across the plasma membrane but then is rapidly removed by the blood. This limits the opportunity for the molecules to reenter the cell, so a sharp concentration gradient of carbon dioxide molecules always exists across the membrane. Two special cases of diffusion are dialysis and osmosis.

Dialysis is the diffusion of a solute across a selectively permeable membrane

To demonstrate **dialysis,** one can fill a cellophane bag[1] with a sugar solution and immerse it in a beaker of pure water (Fig. 5–10). If the cellophane membrane is permeable to sugar as well as to water, both sugar and water molecules will pass through it and the concentrations of sugar molecules in the water on the two sides of the membrane will eventually become equal. Subsequently, both solute and water molecules will continue to cross the membrane, but there will be no net change in their concentrations.

The principle of dialysis has many practical applications. For example, a dialysis machine can be used to cleanse the blood of wastes when the kidneys do not function properly. Waste products in the form of small molecules diffuse readily across the artificial membrane in the dialysis apparatus. Wastes can thus be removed from the blood, while blood cells, blood proteins, and other large molecules are retained.

---

[1]Cellophane is often used as an "artificial membrane." It is made from cellulose and can be formed into a thin sheet that allows the passage of water molecules. Such membranes can be constructed with varying permeability to different solutes, and can be very different from biological membranes in their permeability. (When cellophane is used to package foods, it is coated to make it impermeable to air and water.)

Transfer of information across the plasma membrane. *(a)* A signal molecule binds with a receptor in the plasma membrane. *(b)* The signal molecule-receptor complex activates a G protein. *(c)* G protein activates an enzyme that catalyzes the production of a second messenger such as cyclic AMP (cAMP). *(d)* cAMP then activates one or more enzymes such as protein kinases. The enzymes may phosphorylate proteins, which then alter the activity of the cell in some way.

Figure 5-10 **Dialysis.** *(a)* A cellophane bag, filled with a mixture of sugar, water, and large molecules such as proteins, is immersed in a beaker of pure water. The cellophane acts as a selectively permeable membrane, permitting passage of the sugar and water molecules *(arrows),* but preventing passage of larger molecules. *(b)* The arrows indicate net movement of sugar molecules through the membrane into the water of the beaker. *(c)* Eventually the sugar becomes distributed equally between the two compartments. Although sugar and water molecules continue to diffuse back and forth *(arrows),* net movement is zero.

Osmosis is the diffusion of water (solvent) across a selectively permeable membrane

The selective permeability of cell membranes results in another special kind of diffusion called **osmosis.** This process involves the movement of *solvent* (in this case, water) molecules through a selectively permeable membrane. The water molecules pass freely in both directions, but, as in all types of diffusion, *net* movement is from the region where the water molecules are more concentrated to the region where they are less concentrated. Most solute molecules cannot diffuse freely through selectively permeable membranes of the cell.

The principles involved in osmosis can be illustrated using an apparatus called a U-tube (Fig. 5–11). The U-tube is divided into two sections by a selectively permeable membrane that allows solvent (water) molecules to pass freely but excludes solute molecules (e.g., sugar, salt). A water/solute solution is placed on one side, and pure water is placed on the other. The side containing the solute dissolved in the water has a lower effective concentration of water than the pure water side. This is because the solute particles, which are charged (ionic) or polar, interact with the partial electrical charges on the polar water molecules. Many of the water molecules are thus "bound up" and no longer free to diffuse across the membrane.

Because of the difference in effective water concentration, there is net movement of water molecules from the pure water side (with a high effective concentration of water) to the water/solute side (with a lower effective concentration of water). As a result, the fluid level drops on the pure water side and rises on the water/solute side. Because the solute molecules do not diffuse across the membrane, equilibrium is never attained. Net movement of water continues, and the fluid level continues to rise on the side containing the solute. The weight of the rising column of fluid eventually exerts enough pressure to stop further changes in fluid levels, although water molecules continue to pass through the selectively permeable membrane in both directions.

We define the **osmotic pressure** of a solution as the tendency of water to move into that solution by osmosis. In our U-tube example, we could measure the osmotic pressure by inserting a piston on the water/solute side of the tube and measuring how much pressure must be exerted by the piston to prevent the rise of fluid on that side of the tube. A solution with a high solute concentration has a low effective water concentration and a high osmotic pressure; conversely, a solution with a low solute concentration has a high effective concentration of water and a low osmotic pressure.

Two solutions may be isotonic to each other, or one may be relatively hypertonic and the other relatively hypotonic

Dissolved in the fluid compartment of every living cell are salts, sugars, and other substances that give that fluid a certain osmotic pressure. When a cell is placed in a fluid with exactly the same osmotic pressure, no net movement of water mole-

Figure 5–11 **Osmosis.** The U-tube contains pure water on the right and water plus a solute on the left, separated by a selectively permeable membrane. Water molecules are able to cross the membrane in both directions *(red arrows)*. Solute molecules are unable to cross *(green arrows)*. The fluid level rises on the left and falls on the right because net movement of water *(black arrow)* is to the left. The force that must be exerted by the piston in order to prevent the rise in fluid level is equal to the osmotic pressure of the solution.

cules occurs, either into or out of the cell. The cell neither swells nor shrinks. Such a fluid is said to be **isotonic,** of equal solute concentration, to the fluid within the cell (Table 5–1). Normally, our blood plasma (the fluid component of blood) and all our other body fluids are isotonic to our cells; they contain a concentration of water equal to that in the cells. A solution of 0.9% sodium chloride (sometimes called *physiological saline*) is isotonic to the cells of humans and other mammals. Human red blood cells placed in 0.9% sodium chloride neither shrink nor swell (Fig. 5–12*a*).

If the surrounding fluid has a concentration of dissolved substances greater than the concentration within the cell, it has a higher osmotic pressure than the cell and is said to be **hypertonic** to the cell. Because the hypertonic solution has a

TABLE 5-1   Osmotic Terminology

Solute Concentration in Solution A	Solute Concentration in Solution B	Tonicity	Direction of Net Movement of Water
Greater	Less	A hypertonic to B B hypotonic to A	B to A
Less	Greater	B hypertonic to A A hypotonic to B	A to B
Equal	Equal	A and B are isotonic to each other	No net movement

lower effective water concentration, a cell placed in such a solution shrinks as it loses water by osmosis. Human red blood cells placed in a solution of 1.3% sodium chloride shrink and are said to be *crenated* (Fig. 5–12b). If a cell that has a cell wall is placed in a hypertonic medium, it loses water to its surroundings. Its contents shrink, and the plasma membrane separates from the cell wall, a process known as **plasmolysis** (Fig. 5–13). Plasmolysis occurs in plants when the soil or water around them contains high concentrations of salts or fertilizers.

If the surrounding fluid contains a lower concentration of dissolved materials than does the cell, it has a lower osmotic pressure and is said to be **hypotonic** to the cell; water then enters the cell and causes it to swell. Red blood cells placed in a solution of 0.6% sodium chloride gain water, swell (Fig. 5–12c), and may eventually burst. Many cells that normally live in hypotonic environments have adaptations to prevent excessive water accumulation. For example, certain protists such as *Paramecium* have a contractile vacuole that they use to expel excess water.

Figure 5–12   How cells respond to osmotic pressure differences. *(a)* When a cell is placed in an isotonic solution, water molecules pass in and out of the cell, but the net movement is zero. *(b)* When a cell is placed in a hypertonic solution, there is a net movement of water out of the cell *(arrow)*, and the cell becomes dehydrated, crenated (shrunken), and may die. *(c)* When a cell is placed in a hypotonic solution, there is a net movement of water molecules into the cell *(arrow)*, causing the cell to swell or even burst.   *(SEMs of human red blood cells courtesy of Dr. R.F. Baker, University of Southern California Medical School)*

(a) Isotonic solution   10 μm

(b) Hypertonic solution

(c) Hypotonic solution

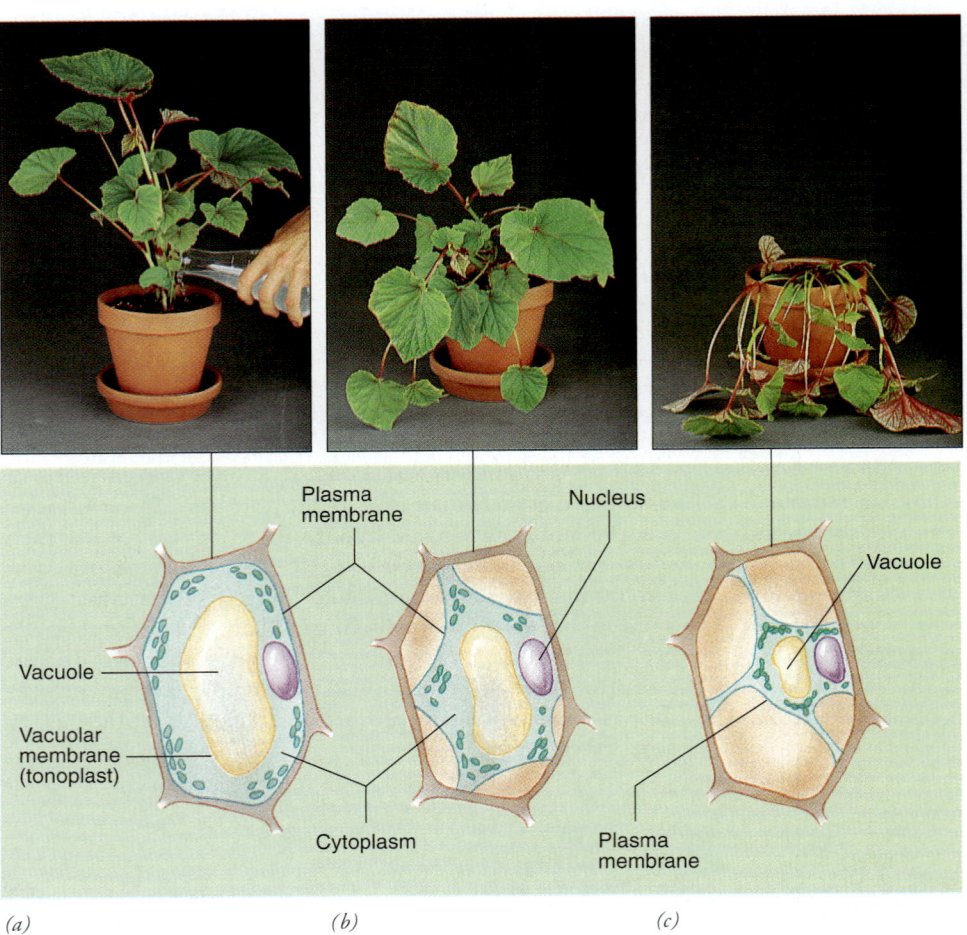

Figure 5–13 **Plasmolysis.** *(a)* In hypotonic surroundings, the vacuole of a plant cell fills, but the rigid cell walls prevent the cell from expanding. The cells of this healthy begonia plant are turgid. *(b)*, *(c)* When the begonia plant is exposed to a hypertonic solution, its cells become plasmolyzed as they lose water. The plant wilts and eventually dies. *(Dennis Drenner)*

Plasma membrane

Nucleus

Vacuole

Vacuole

Vacuolar membrane (tonoplast)

Cytoplasm

Plasma membrane

(a)   *(b)*   *(c)*

Turgor pressure is the internal hydrostatic pressure usually present in walled cells

The relatively rigid cell walls of plant cells, algae, bacteria, and fungi enable these cells to withstand, without bursting, an external medium that is very dilute, containing only a very low concentration of solutes. Because of the substances dissolved in the cytoplasm, the cells are hypertonic to the outside medium (conversely, the outside medium is hypotonic to the cytoplasm). Water moves into the cells by osmosis, filling their central vacuoles and distending the cells. The cells swell, building up a pressure, termed **turgor pressure,** against the rigid cell walls (Fig. 5–13*a*). The cell walls can be stretched only very slightly, and a steady state is reached when their resistance to stretching prevents any further increase in cell size and thereby halts the net movement of water molecules into the cells (although, of course, molecules continue to move back and forth across the plasma membrane). Turgor pressure in the cells is an important factor in providing support for the body of nonwoody plants. Thus, lettuce becomes limp in a salty salad dressing and a picked flower wilts from lack of water.

Carrier-mediated transport of solutes requires special integral membrane proteins

A lipid bilayer is relatively impermeable to most large polar molecules. This is advantageous to cells for a number of reasons. Most of the compounds required in metabolism are polar, and the impermeability of the plasma membrane prevents their loss by diffusion. A lipid bilayer is also impermeable to ions, which play important roles in many physiological processes. Some ions, such as calcium ions, are used as intracellular signals, and changes in their cytoplasmic concentration trigger changes in a number of cellular processes (such as muscle contraction; see Chapter 38). As a cell controls the influx and efflux of ions, it is able to directly or indirectly control many metabolic activities (see *Focus On: How the Patch Clamp Technique Has Revolutionized the Study of Ion Channels*).

Cells also must continually acquire essential polar nutrient molecules such as glucose and amino acids. To transport ions and nutrients through membranes, systems of carrier proteins apparently evolved very early in the origin of cells. This transfer of solutes by proteins located within the membrane is

HOW THE PATCH CLAMP TECHNIQUE HAS REVOLUTIONIZED THE STUDY OF ION CHANNELS

Because of their electrical charges, which provide a basis for various interactions, ions play an essential role in most cellular processes. Their electrical charges, however, prevent them from crossing a lipid bilayer by simple diffusion. For this reason, every membrane of every cell contains numerous ion carriers, or *ion channels*. Some ion channels, referred to as "ion pumps," require a direct input of metabolic energy, while others provide for facilitated diffusion.

Movement of ions across a membrane can result in a charge difference (electrochemical gradient). If the cell is large enough, this charge difference (usually expressed in millivolts, mV) can be measured by using two microelectrodes connected to an extremely sensitive oscilloscope or voltmeter (see figure). One of the microelectrodes is inserted into the cell and the other is placed just outside the plasma membrane. Although valuable, these techniques have serious limitations, for they cannot be used on smaller cells and do not provide information on the function of individual ion channels.

In the mid-1970s Erwin Neher and Bert Sakmann developed a method, known as the **patch clamp technique,** that allows researchers to study single ion channels of very small cells, In this technique, a micropipette is tightly sealed to a patch of membrane so small that it generally contains only a single ion channel. The flow of ions through the channel can be measured using an extremely sensitive recording device. This basic technique has been modified in many ways and has been applied to studies of the roles of ion channels in a wide range of cellular processes in both plants and animals. For example, studies of single ion channels enabled researchers to demonstrate that the genetic disease cystic fibrosis (see Chapter 15) is caused by a defect in a specific type of chloride ion channel. Because of the far-reaching implications of their work, Neher and Sakmann were awarded a Nobel Prize in 1991.

Patch clamp technique. *(a)* The difference in electrical charge across a membrane can be measured using microelectrodes and a voltmeter. *(b)* A micropipette can be used to form a tight seal with a patch of plasma membrane. The membrane can be pulled away from the rest of the cell, and the flow of ions through a single ion channel can be studied.

termed **carrier-mediated transport.** The two forms of carrier-mediated transport—facilitated diffusion and carrier-mediated active transport—differ in their capabilities and energy sources.

Facilitated diffusion occurs down a concentration gradient

In all processes in which substances move across membranes by passive diffusion, the net transfer of those molecules from one side to the other occurs as a result of a concentration gradient. If the membrane is permeable to a substance, there is net movement from the side of the membrane where it is more highly concentrated to the side where it is less concentrated. Such a gradient across the membrane is actually a form of stored energy. A concentration gradient can be established as a result of certain processes taking place in the cell. The stored energy of the concentration gradient is released when molecules move from a region of high concentration to one of low concentration; movement down a concentration gradient is therefore spontaneous. (These types of energy and spontaneous processes are discussed in greater detail in Chapter 6.)

In the type of transport known as **facilitated diffusion,** the membrane may be made permeable to a solute, such as an ion or a polar molecule, by a specific *carrier* or *transport protein* (Fig. 5–14). An important example of a carrier that works by facilitated diffusion is glucose permease, a transmembrane protein that transports glucose into red blood cells. These cells keep the internal concentration of glucose low by immediately adding a phosphate group to entering glucose molecules, converting them to highly charged glucose phosphates that cannot pass back through the membrane. Because glucose phosphate is a different molecule, it does not contribute to the glucose concentration gradient. Thus a steep concentration gradient for glucose is continually maintained, and glucose

rapidly diffuses into the cell, only to be immediately changed to the phosphorylated form.

The mechanism of facilitated diffusion for glucose has been studied by using **liposomes,** artificial vesicles surrounded by phospholipid bilayers. The phospholipid membrane of a liposome does not permit the passage of glucose unless researchers introduce the transport protein glucose permease to the membrane. Glucose permease and similar transport proteins temporarily bind to the molecules they transport. This mechanism appears to be similar to the way an enzyme binds with its substrate, the molecule it regulates (see Chapter 6). And as in enzyme action, binding apparently changes the shape of the carrier. This allows the glucose molecule to be released on the inside of the cell. According to this model, when the glucose is released into the cytoplasm, the transport protein reverts to its original structure and is available to bind another glucose molecule on the outside of the cell.

Another similarity to enzyme action is that transport proteins become saturated when the transported molecule is at high concentration. This may be because there are a finite number of transport molecules available; when the concentrations of solute molecules to be transported reaches a certain level, all of the transport molecules are working.

Some carrier-mediated active transport systems "pump" substances against their concentration gradients

Although adequate amounts of some substances can be transported across cell membranes by diffusion, a cell often needs to move solutes against a concentration gradient. Many substances are required by the cell in concentrations higher than those outside the cell. These molecules are moved across cellular membranes by **carrier-mediated active transport** mechanisms. Because this active transport requires that particles be "pumped" from a region of low concentration to a region of high concentration (i.e., *against a concentration gradient*), transport must be coupled to an energy source such as ATP.

One of the most striking examples of an active transport mechanism is the **sodium-potassium pump** found in virtually all animal cells (Fig. 5–15). The pump is a group of specific proteins in the plasma membrane that uses energy in the form of ATP to exchange sodium ions on the inside of the cell for potassium ions on the outside of the cell. The exchange is unequal, so that usually only two potassium ions are imported inside for every three sodium ions exported. Because these particular concentration gradients involve ions, an electrical potential (separation of electrical charges) is generated across the membrane, and we say that the membrane is polarized.

Both sodium and potassium ions are positively charged, but because there are fewer potassium ions inside relative to the sodium ions outside, the inside of the cell is negatively charged relative to the outside. The action of sodium-potassium pumps helps maintain a charge separation across the plasma membrane. Because there is both a concentration difference and an electrical charge difference on the two sides of

Figure 5–14 **Facilitated diffusion.** A transport protein in the membrane binds a solute particle. The transport protein changes its shape, opening a channel through the membrane. A specific solute can be transported from the inside of the cell to the outside or from the outside to the inside, but net movement is always from a region of higher solute concentration to a region of lower concentration. The potential energy of a concentration gradient is needed.

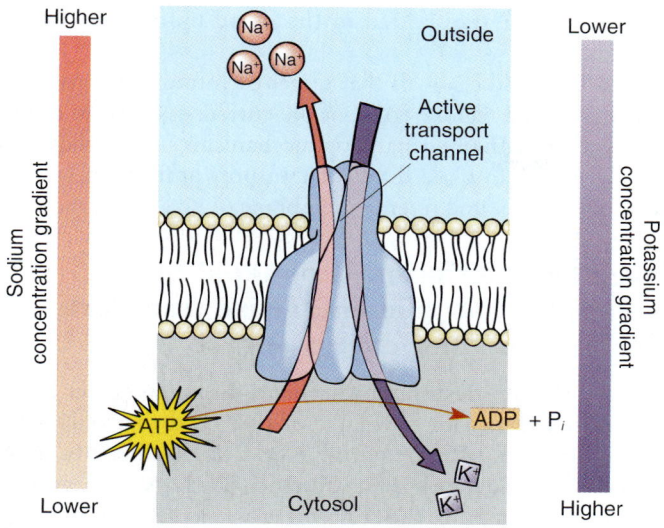

(a)

Figure 5–15 **The sodium-potassium pump.** *(a)* The sodium-potassium pump is an active transport system that requires energy from ATP. Each complete pumping cycle uses one molecule of ATP; three sodium ions are exported, and two potassium ions are imported. *(b)* A model illustrating the sodium-potassium pumping cycle. (1) Three sodium ions bind to the transport protein. (2) A phosphate group is transferred from ATP to the transport protein. (3) The transport protein undergoes a conformational change, releasing three sodium ions outside the cell. (4) Two potassium ions bind to the transport protein. (5) The phosphate is released. (6) The transport protein returns to its original shape: two potassium ions are released inside the cell.

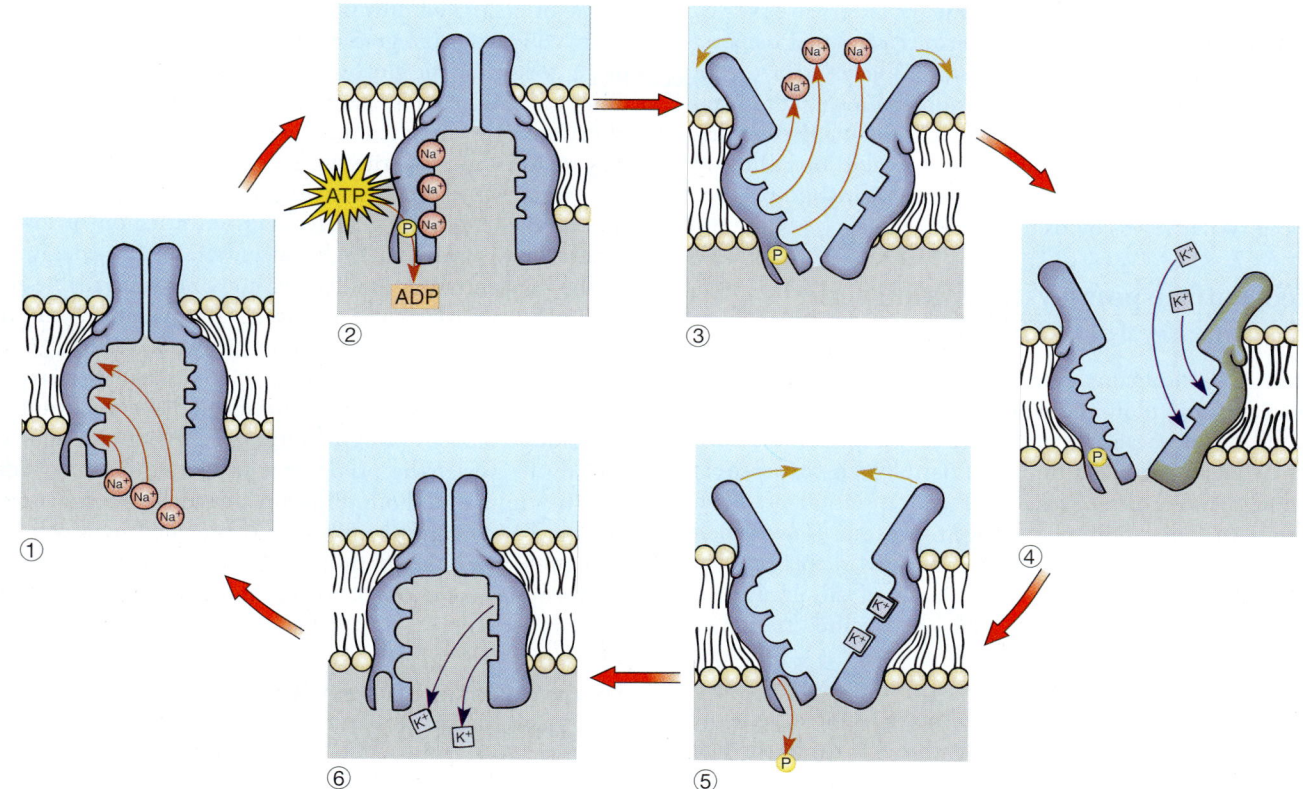

(b)

the membrane, this gradient is referred to as an *electrochemical* gradient. Such gradients also store energy (like water stored behind a dam) that can be used to drive other transport systems. So important is the electrochemical gradient produced by these pumps that some cells (e.g., nerve cells) expend 70% of their total energy just to power this one transport system.

Sodium-potassium pumps (as well as all other ATP-driven pumps) are transmembrane proteins that extend entirely through the membrane. By undergoing a series of con-

formational changes (that is, changes in shape), the pumps are able to exchange sodium for potassium across the plasma membrane. Unlike facilitated diffusion, at least one of the conformational changes in the pump cycle requires energy, which is provided by ATP. The shape of the pump protein changes as a phosphate group (from ATP) first binds to it and is subsequently removed later in the pump cycle.

The use of electrochemical potentials for energy storage is not confined to the plasma membrane of animal cells. Plant and fungal cells use ATP-driven plasma membrane pumps to transfer protons from the cytoplasm of their cells to the outside. Removal of positively charged protons from the cytoplasm of these cells results in a large difference in the concentration of protons, such that the outside of the cells is relatively positively charged and the inside of the plasma membrane is relatively negatively charged. The energy stored in these electrochemical gradients can be made available to do certain kinds of cell work.

Other proton pumps can be used in "reverse" to synthesize ATP. As we will discuss in Chapters 7 and 8, bacteria, mitochondria, and chloroplasts use energy from food or from light to establish proton concentration gradients. When the protons diffuse through the proton carriers from a region of high proton concentration to one of low concentration, ATP is synthesized. These electrochemical gradients form the basis for the major energy-conversion system in virtually all cells.

Ion pumps have other important roles. For example, they are instrumental in the ability of an animal cell to equalize the osmotic pressures of its cytoplasm and its external environment. If an animal cell does not control its internal osmotic pressure, its contents will become hypertonic relative to the exterior. Water will enter the cell by osmosis, causing it to swell and possibly burst (see Fig. 5–12c). By controlling the ion distribution across the membrane, the cell is able to indirectly control the movement of water, for when ions are pumped out of the cell, water leaves by osmosis.

Linked cotransport systems indirectly provide energy for active transport

The electrochemical concentration gradients generated by the sodium-potassium pump (and other pumps) provide sufficient energy to power the active transport of other essential substances. In these systems a transport protein can **cotransport** the required molecules *against* their concentration gradient, while sodium, potassium, or hydrogen ions move *down* their gradient. Energy from ATP is used indirectly in this process, for it produces the ion gradient; the energy of this gradient is then used to drive the active transport of a required substance against its gradient.

In some cells more than one system may work to transport a given substance. For example, the transport of glucose from the intestine to the blood occurs through a thin sheet of epithelial cells that line the intestine. The surface that is exposed to the intestine has many **microvilli** (sing., *microvillus*), finger-like extensions that effectively increase the surface area of the membrane available for absorption. The glucose transport protein on that region of the cell surface is part of an active transport system for glucose that is "driven" by the cotransport of sodium. The sodium concentration inside the cell is kept low by an ATP-requiring sodium-potassium pump that transports sodium out of the cell and into the blood. Because of its high concentration inside the cell (relative to the blood),

glucose can be transported to the blood by facilitated diffusion.

What are the signals that target each transport protein to its appropriate region? Some of the current research in cell biology focuses on understanding mechanisms such as those that allow the cell to place different transport proteins in separate regions of the same plasma membrane.

Facilitated diffusion is powered by a concentration gradient; active transport requires another energy source

It is a common misconception that diffusion, whether simple or facilitated, is somehow "free of cost" and that only active transport mechanisms require energy. Because diffusion always involves net movement of a substance down its concentration gradient, we say that the concentration gradient "powers" the process. However, energy is required to do the work of establishing and maintaining the gradient. Think back to the example of facilitated diffusion of glucose. The cell maintains a steep concentration gradient (high outside, low inside) by phosphorylating the glucose molecules once they enter the cell. One ATP molecule is spent for every glucose molecule phosphorylated (not to mention such additional costs as the energy required to make the enzymes that carry out the reaction).

An active transport system can work *against* a concentration gradient, pumping materials from a region of low concentration to a region of high concentration. The energy stored in the concentration gradient is not only unavailable to the system, but actually works against it. For this reason some other source of energy must be provided. As we have seen, in many cases ATP energy is used directly. In a cotransport system, energy is provided by a concentration gradient for some other substance (e.g., an ion). Of course, ATP energy is required indirectly to power the pump that produces the ion gradient.

To summarize, both diffusion and active transport require energy. The energy for diffusion is provided by a concentration gradient for the substance being transported. Active transport requires some other, usually more direct, expenditure of metabolic energy.

In exocytosis and endocytosis large particles are transported by vesicles or vacuoles

In both simple and facilitated diffusion, and in carrier-mediated active transport, individual molecules and ions pass through the plasma membrane. Larger quantities of material, such as particles of food or even whole cells, must also be moved into or out of cells. Such work requires that cells expend energy directly, making it a form of active transport.

In **exocytosis,** a cell ejects waste products or specific secretion products such as hormones by the fusion of a vesicle with the plasma membrane (Fig. 5–16). Exocytosis results in the incorporation of the membrane of the secretory vesicle into *(a)*   *(b)*   0.25 μm

Figure 5–16   **Exocytosis.** *(a)* A vesicle approaches the plasma membrane, fuses with it, and releases its contents outside the cell. *(b)* TEM showing exocytosis of the protein components of milk by a mammary gland cell.   *(b, A. Ichikawa/from D.W. Fawcett)*

the plasma membrane, as well as the release of the contents of the vesicle from the cell. This is also the primary mechanism by which plasma membranes grow larger.

In **endocytosis,** materials are taken into the cell. Several types of endocytotic mechanisms operate in biological systems including phagocytosis, pinocytosis, and receptor-mediated endocytosis. In **phagocytosis** (literally, "cell-eating"), the cell ingests large solid particles such as bacteria or food (Fig. 5–17). Phagocytosis is a mechanism used by certain protists and by several types of vertebrate white blood cells to ingest particles, some of which are as large as an entire bacterium. During ingestion, folds of the plasma membrane enclose the particle, which has bound to the surface of the cell, forming a large membranous sac, or vacuole. When the membrane has encircled the particle, it fuses at the point of contact. The vacuole then fuses with lysosomes, and the ingested material is degraded.

In the form of endocytosis known as **pinocytosis** ("cell-drinking"), the cell takes in dissolved materials. Tiny droplets of fluid are trapped by folds in the plasma membrane (Fig. 5–18), which pinch off into the cytosol as tiny vesicles. The liquid contents of these vesicles are then slowly transferred into the cytosol; the vesicles may become progressively smaller, to the point that they appear to vanish.

In a third type of endocytosis, called **receptor-mediated endocytosis,** specific molecules combine with *receptor proteins* embedded in the plasma membrane. A molecule that binds specifically to a receptor is called a **ligand.** The receptors are concentrated in *coated pits,* depressed regions on the cytoplasmic surface of the plasma membrane. Each pit is coated by a layer of a protein, called *clathrin,* found just below the plasma membrane. After a ligand binds with a receptor, the coated pit forms a *coated vesicle* by endocytosis.

Cholesterol in the blood is taken up by animal cells by receptor-mediated endocytosis. Much of the receptor-mediated endocytosis pathway was detailed through studies by M. Brown and J. Goldstein of the receptor for low-density lipoprotein (LDL), a primary cholesterol carrier in blood. In 1986 these investigators were awarded the Nobel Prize for their pioneering work. Their findings have important medical implications because cholesterol that remains in the blood instead of entering the cells can become deposited in the artery walls, increasing the risk of heart attack.

In Figure 5–19 the uptake of a low-density lipoprotein (LDL) particle is shown. Seconds after the vesicle moves into the cytoplasm, the coating dissociates from it, leaving an uncoated vesicle, called an *endosome,* free in the cytoplasm. The endosome typically forms two vesicles. One contains the

Figure 5–17 **Phagocytosis.** In this type of endocytosis, a cell ingests relatively large solid particles. *(a)* Steps of endocytosis: (1) Folds of the plasma membrane surround the particle to be ingested, forming a small vacuole around it. (2) The vacuole then pinches off inside the cell. (3) Lysosomes may fuse with the vacuole and pour their potent enzymes onto the ingested material. *(b)* The white blood cell (known as a neutrophil) shown in this TEM is phagocytizing bacteria. The vacuoles contain bacteria that have already been phagocytized, while other bacteria are still outside the cell. The granules in the cytosol contain digestive enzymes. *(D.W. Fawcett)*

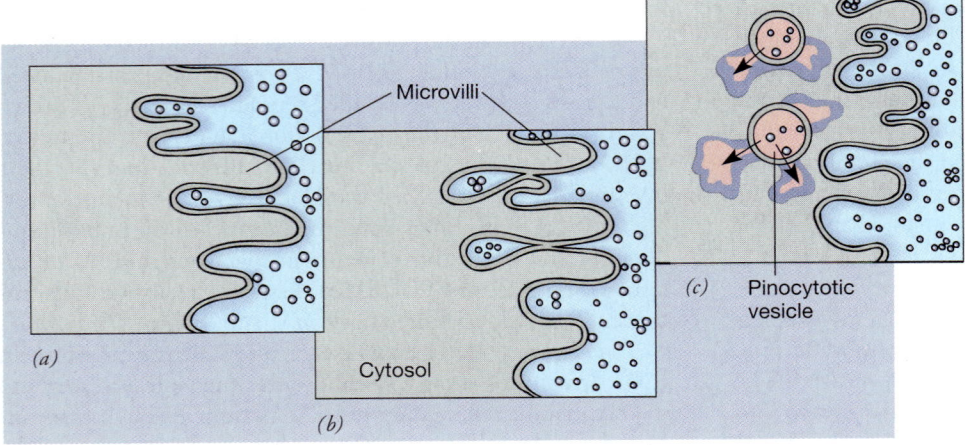

Figure 5–18 **Pinocytosis or "cell-drinking."** *(a)* Tiny droplets of fluid are trapped by folds of the plasma membrane. These pinch off *(b)* into the cytosol as small fluid-filled vesicles. The contents of these vesicles are then slowly transferred to the cytosol.

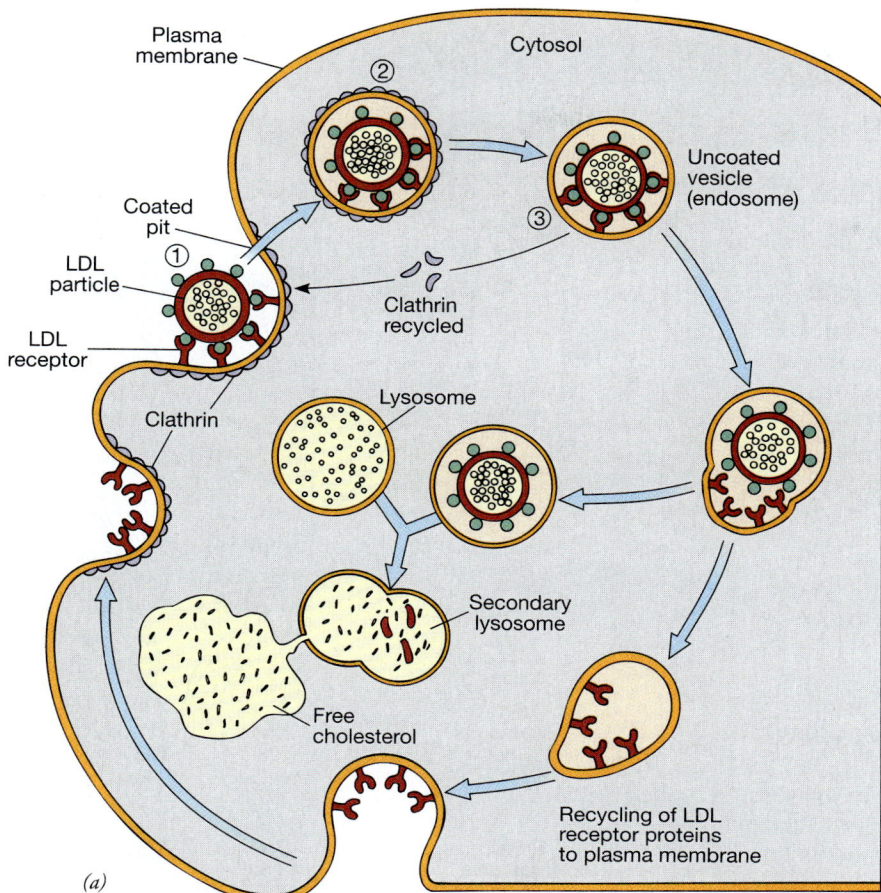

(a)

Figure 5–19 **Receptor-mediated endocytosis.** *(a)* Uptake of low-density lipoprotein (LDL) particles which transport cholesterol in the blood: (1) LDL attaches to specific receptors in coated pits on the plasma membrane. (2) Endocytosis results in the formation of a coated vesicle in the cytosol. (3) Seconds later the coat is removed. The vesicle is now called an endosome. The receptors are returned to the plasma membrane and recycled. The vesicle containing LDL particles fuses with lysosomes to form a secondary lysosome. Hydrolytic enzymes then release the cholesterol from the particles for use by the cell. *(b–e)* Series of TEMs showing the formation of a coated vesicle from a coated pit. *(b–e, from Perry, M.M., and A.B. Gilbert, J. Cell. Sci. 39:257–272, 1979)*

(b)   0.25 μm

(c)   0.25 μm

(d)   0.25 μm

(e)   0.25 μm

receptors, the other the LDL particle. The receptors are returned to the plasma membrane where they are recycled. The remaining vesicle fuses with a lysosome and its contents are then digested and released into the cytosol.

Ligand binds to receptors in coated pits of plasma membrane ⟶ coated vesicle forms by endocytosis ⟶ coating detaches from vesicle ⟶ endosome divides:
    ⟶ One vesicle returns receptors to plasma membrane where they are recycled
    ⟶ Other portion of endosome fuses with lysosome ⟶ contents are digested and released into the cytosol

The recycling of LDL receptors to the plasma membrane through vesicles illustrates a problem common to all cells that employ endocytotic and exocytotic mechanisms. In cells that are constantly involved in secretion, an equivalent amount of membrane must be returned to the interior of the cell for each vesicle that fuses with the plasma membrane; if it is not, the cell surface will keep expanding even though the growth of the cell itself may be arrested. A similar situation exists for cells that use endocytosis. A macrophage, for example, ingests the equivalent of its entire surface membrane in about 30 minutes, requiring an equivalent amount of recycling or new membrane synthesis for the cell to maintain its surface area.

JUNCTIONS ARE SPECIALIZED CONTACTS BETWEEN CELLS

Cells in close contact with each other typically develop specialized intercellular junctions that involve their plasma membranes as well as other components. These structures may allow neighboring cells to form strong connections with each other, prevent passage of materials, or establish rapid communication between adjacent cells. In animals there are three common types of intercellular contacts: desmosomes, tight junctions, and gap junctions. Plant cells are connected by plasmodesmata.

Desmosomes are points of attachment between some animal cells

Adjacent epithelial cells, such as those found in the outer layer of the skin, are so tightly bound to each other that strong mechanical forces are required to separate them. They are held together by anchoring junctions called **desmosomes** (Fig. 5–20). Each desmosome includes components of two adjacent cells. It is made up of regions of dense material associated with the cytosolic sides of the two plasma membranes, plus protein filaments that cross the narrow intercellular space between them.

Desmosomes are anchored to systems of intermediate filaments inside the cells. Thus the intermediate filament networks of adjacent cells are connected so that mechanical stresses are distributed throughout the tissue. The function of the desmosomes appears to be purely mechanical; they hold cells together at one point like a rivet or a spot weld. As a result, cells can form strong sheets, and substances can still pass freely through the spaces between the plasma membranes.

Tight junctions seal off intercellular spaces between some animal cells

Tight junctions are literally areas of tight connections between the membranes of adjacent cells. These connections are so tight that no space remains between the cells; substances cannot leak between the cells. TEMs of tight junctions show that in the region of the junction the membranes of the two cells are in actual contact with each other, held together by proteins linking the two cells (Fig. 5–21).

Cells connected by tight junctions seal off body cavities. For example, tight junctions between cells lining the intestine prevent substances in the intestine from entering the body or the blood by passing around the cells. The sheet of cells thus

Figure 5–20 Desmosomes. The dense structure in the TEM is a desmosome. Each desmosome consists of a pair of button-like discs associated with the plasma membranes of adjacent cells, plus the intercellular protein filaments that connect them. Intermediate filaments in the cells are attached to the discs and are connected to other desmosomes. *(D.W. Fawcett)*

Microvillus

Lumen of
the intestine

Tight
junction

Intercellular
space

Cell #1　　Cell #2

(a)　　0.1 µm

(b)

Rows of
tight junction
proteins

Plasma
membranes

Intercellular
space

Figure 5–21　Tight junctions. These junctions prevent passage of materials through spaces between cells. Tight junctions occur at the points of contact between two cells and extend completely around the cells. *(a)* This TEM shows points of fusion between the plasma membranes of adjacent cells lining the intestine. One tight junction is marked by the box. *(b)* The diagram shows that a tight junction is formed by linkages between rows of proteins of adjacent cells. These proteins are tightly packed in rows that seal off the intercellular space.　　*(a, G.E. Palade)*

acts as a selective barrier. Food substances must be transported across the plasma membranes and through the intestinal cells before they enter the blood. This arrangement helps prevent toxins and other unwanted materials from entering the blood.

Gap junctions permit transfer of small molecules and ions

A third type of intercellular connection in animal cells, the **gap junction,** is like the desmosome in that it bridges the space between cells; however, the space it spans is somewhat narrower (Fig. 5–22). Gap junctions also differ in that they not only connect the membranes, but also contain channels connecting the cytoplasm of adjacent cells. A gap junction consists of a hexagonal array of proteins (connexin) forming a cluster of channels, each of which is about 1 to 2 nanometers in diameter. Small inorganic molecules (e.g., ions) and some biological molecules (e.g., derivatives of ATP) can pass through the pores, but larger molecules are excluded. When appropriate marker substances are injected into one of a group of cells connected by gap junctions, the marker passes rapidly into the adjacent cells but does not enter the space between the cells.

Cells are able to control the passage of materials through gap junctions by opening and closing the pores (Fig. 5–22*d*). There is evidence that the open and closed states are regulated mainly by the intracellular concentrations of certain ions.

Gap junctions provide for rapid chemical and electrical communications between cells. Cells in the pancreas, for example, are linked together by gap junctions in such a way that if one of a group of cells is stimulated to secrete insulin, the

(a)

(b)

(c)

0.25 µm

Closed   Open

(d)

0.1 µm

Figure 5–22   **Gap junctions.** These connections permit transfer of small molecules and ions between adjacent cells. *(a)* A TEM of a gap junction *(between the arrows)*. *(b)* Model of a gap junction based on electron-microscopic and x-ray diffraction data. The two membranes contain cylinders composed of six protein subunits arranged to form a pore. Two cylinders from opposite membranes are joined to form a pore about 1.5 to 2.0 nanometers in diameter connecting the cytoplasmic compartments of the two cells. *(c)* Freeze-fracture replica of the P-face of a gap junction between two ovarian cells of a mouse, showing the numerous protein particles present. *(d)* Model illustrating how a gap junction pore might open and close.   *(a, D.W. Fawcett; c, E. Anderson, J. Morph. 156:339–366, 1978)*

signal is passed through the junctions to the other cells in the cluster, ensuring a coordinated response to the initial signal. Gap junctions allow some nerve cells to be electrically coupled. Heart muscle cells are linked by gap junctions that permit the flow of ions necessary to synchronize contractions.

Plasmodesmata allow movement of certain molecules and ions between plant cells

Plant cells do not need desmosomes for strength because they have cell walls. However, these same walls would isolate the cells, preventing them from communicating. For this reason, plant cells require connections that are functionally equivalent

to the gap junctions of some animal cells. **Plasmodesmata** (sing., *plasmodesma*) are 20- to 40-nanometer-wide channels through adjacent cell walls connecting the cytoplasm of neighboring cells (Fig. 5–23). The plasma membranes of adjacent cells are therefore continuous with each other through the plasmodesmata. Most plasmodesmata contain a cylindrical membranous structure, called the *desmotubule,* which also runs through the opening and connects the endoplasmic reticulum of the two adjacent cells.

Plasmodesmata generally allow molecules and ions, but not organelles, to pass through the openings from cell to cell. The movement of ions through the plasmodesmata allows for a very slow type of electrical signaling in plants.

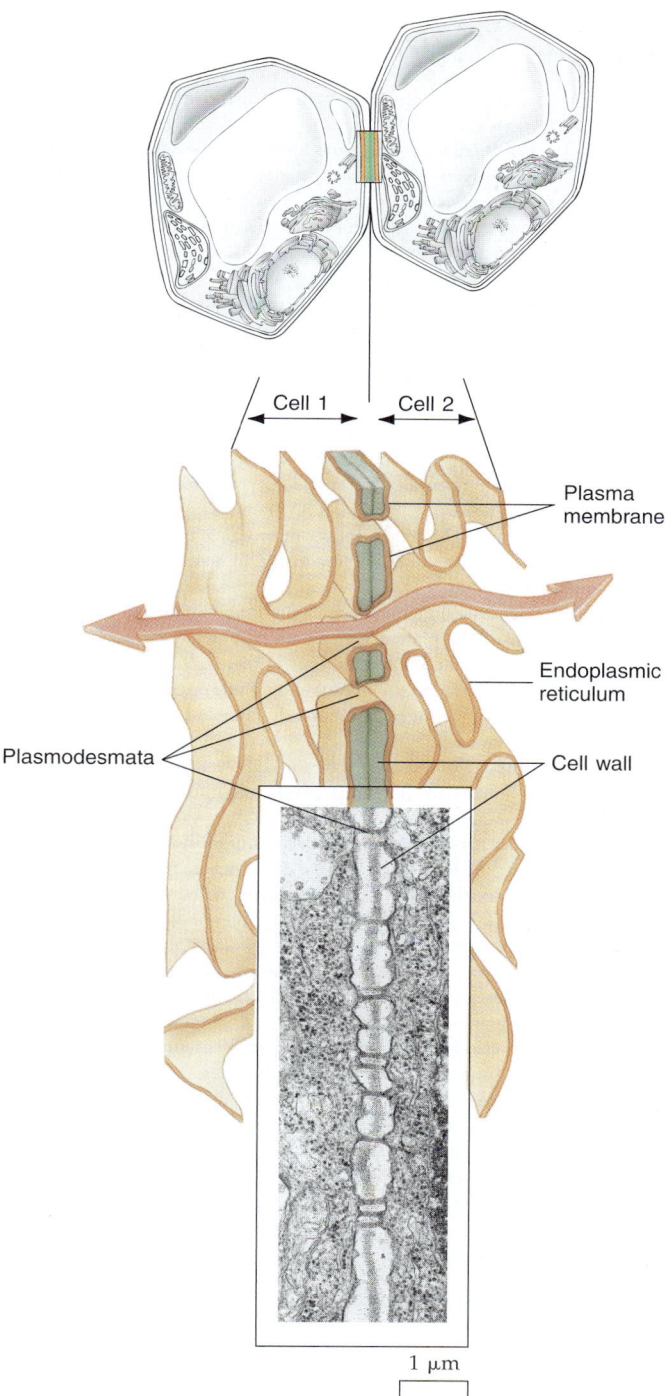

Figure 5–23 **Plasmodesmata.** TEM and line art of cytoplasmic channels through the cell walls of adjacent plant cells that allow passage of water, ions, and small molecules *(wide arrows)*. The channels are lined with the fused plasma membranes of the two adjacent cells. *(E.H. Newcomb, Biological Photo Service)*

Cell 1   Cell 2

Plasma membrane

Endoplasmic reticulum

Plasmodesmata

Cell wall

1 μm

SUMMARY WITH KEY TERMS

I. Cell membranes are complex structures that (1) physically separate the interior of the cell from the extracellular environment and (2) form compartments within the cells of eukaryotes that allow them to perform complex functions. Membranes have many different structural and functional roles.

  A. They regulate the passage of materials.

  B. They receive information that permits the cell to sense changes in its environment and respond to them.

  C. They contain specialized structures that allow specific contacts and communications with other cells.

  D. They participate in and serve as surfaces for biochemical reactions.

II. According to the **fluid mosaic model,** membranes consist of a fluid phospholipid bilayer in which a variety of proteins are embedded.

  A. The phospholipid molecules are **amphipathic:** they have hydrophobic and hydrophilic regions.

  B. The lipid bilayer is arranged in such a way that the hydrophilic heads of the phospholipids are at the two surfaces of the bilayer and their hydrophobic fatty acid chains are in the interior.

C. In almost all biological membranes, the lipids of the bilayer are in a fluid or liquid-crystalline state, which allows the molecules to move rapidly in the plane of the membrane.

D. **Integral membrane proteins** are embedded in the bilayer with their hydrophilic surfaces exposed to the aqueous environment and their hydrophobic surfaces in contact with the hydrophobic interior of the bilayer. **Transmembrane proteins** are integral proteins that extend completely through the membrane.

E. **Peripheral membrane proteins** are associated with the surface of the bilayer, usually bound to integral proteins, and are easily removed without disrupting the structure of the membrane.

F. Membrane proteins, lipids, and carbohydrates are asymmetrically positioned with respect to the bilayer so that one side of the membrane has a different composition and structure than the other.

G. Many materials are transported from one part of the cell to another in vesicles that bud from various cell membranes and then fuse with some other membrane.

H. Membrane proteins have various functions including transport of materials, acting as enzymes or receptors, and structurally linking cells together.

I. In **signal transduction,** a receptor converts an extracellular signal into an intracellular signal that causes some change in the cell.
   1. Signal transduction typically involves a sequence of molecules that relay information from one to another.
   2. The pathway often involves protein kinases, enzymes that phosphorylate specific proteins.

III. Biological membranes are **selectively permeable membranes;** that is, they allow the passage of some substances but not others.

A. Some molecules pass through membranes by simple **diffusion.**
   1. Diffusion is the net movement of a substance down its concentration gradient (from a region of greater concentration to one of lower concentration).
   2. **Dialysis** is the diffusion of a solute across a membrane.
   3. **Osmosis** is a kind of diffusion in which molecules of water pass through a selectively permeable membrane from a region where water has a higher effective concentration to a region where its effective concentration is lower.
   4. The **osmotic pressure** of a solution is determined by its concentration of dissolved substances (solutes). Cells regulate their internal osmotic pressures to prevent shrinking or bursting.
   5. An **isotonic** solution has an equal solute concentration to another fluid, e.g., the fluid within the cell. When placed in a **hypertonic** solution, one that has a greater solute concentration than the cell, cells lose water to the surroundings and undergo **plasmolysis,** a process in which the plasma membrane separates from the cell wall. When cells are placed in a **hypotonic** solution, one with a lower concentration of dissolved materials relative to the cell, water enters the cells and causes them to swell.

6. Plant cells can withstand high internal hydrostatic pressure because their cell walls prevent them from expanding and bursting. When water moves into cells by osmosis, it fills the central vacuoles. The cells swell, building up **turgor pressure** against the rigid cell walls.

B. In **carrier-mediated transport** special membrane proteins move ions or molecules across a membrane.
   1. Some substances pass through membranes by **facilitated diffusion,** a form of carrier-mediated transport that uses the energy of a concentration gradient for the substance being transported and cannot work against the gradient.
   2. In **carrier-mediated active transport** the cell expends metabolic energy to move ions or molecules across a membrane *against* a concentration gradient. The **sodium-potassium pump** uses ATP to pump sodium ions out of the cell and potassium ions into the cell.

C. In **cotransport** an ATP-powered pump such as the sodium-potassium pump transports ions or some other solute and indirectly powers the active transport of other solutes by maintaining a concentration gradient.

D. In **exocytosis,** the cell ejects waste products or secretes substances such as hormones or mucus by fusion of a vesicle with the plasma membrane.

E. In **endocytosis** materials such as food may be moved into the cell; a portion of the plasma membrane envelops the material, enclosing it in a vesicle or vacuole that is then released inside the cell.
   1. In **phagocytosis,** the plasma membrane encloses a particle such as a bacterium or protist, forms a vacuole around it, and moves it into the cell.
   2. In **pinocytosis,** the cell takes in dissolved materials by forming tiny vesicles around droplets of fluid trapped by folds of the plasma membrane.
   3. In **receptor-mediated endocytosis, ligands** bind to specific receptors in coated pits along the plasma membrane. These pits, coated by the protein clathrin, form coated vesicles by endocytosis.

IV. Plasma membranes of many eukaryotic cells contain specialized structures that permit contact and communication with other cells.

A. **Desmosomes** weld adjacent animal cells together to form strong tissues.

B. **Tight junctions** seal membranes of adjacent animal cells together, preventing substances from passing between cells.

C. **Gap junctions** are protein complexes that form channels in membranes, allowing communication between the cytoplasm of adjacent animal cells.

D. **Plasmodesmata** are channels connecting adjacent plant cells. Openings in the cell walls allow the plasma membranes and cytoplasm to be continuous, thus permitting certain molecules and ions to pass from cell to cell.

POST-TEST

1. Lipids that make up the core of biological membranes (a) are amphipathic (b) have hydrophobic and hydrophilic regions (c) are typically in a fluid state (d) answers a, b, and c are correct (e) answers a and c only

2. According to the fluid mosaic model, membranes consist of (a) a lipid-protein sandwich (b) mainly phospholipids with scattered nucleic acids (c) a fluid phospholipid bilayer in which proteins are embedded (d) a fluid phospholipid bilayer in which carbohydrates are embedded (e) none of the preceding is correct

3. Transmembrane proteins (a) are peripheral proteins (b) are integral proteins (c) extend completely through the membrane (d) answers a, b, and c are correct (e) answers b and c only

4. Which of the following is *not* a function of the plasma membrane? (a) transports materials (b) helps to structurally link cells together (c) manufactures proteins (d) anchors the cell to the extracellular matrix (e) has receptors that relay signals

5. Which of the following processes require(s) the cell to expend metabolic energy directly (e.g., from ATP)? (a) active transport (b) facilitated diffusion (c) dialysis (d) osmosis (e) diffusion

6. Which of the following are examples of carrier-mediated transport? (a) active transport (b) facilitated diffusion (c) dialysis (d) answers a, b, and c are correct (e) answers a and b only

7. Transport of sodium by sodium-potassium pumps is an example of (a) active transport (b) pinocytosis (c) dialysis (d) exocytosis (e) facilitated diffusion

8. The cell takes in dissolved materials by forming tiny vesicles around fluid droplets trapped by folds of the plasma membrane. This process is (a) ac-

tive transport (b) pinocytosis (c) receptor-mediated endocytosis (d) exocytosis (e) facilitated diffusion

9. When plant cells are in a hypotonic medium, they (a) undergo plasmolysis (b) build up turgor pressure (c) wilt (d) answers a, b, and c are correct (e) answers a and c only are correct

10. Which sequence most accurately describes receptor-mediated endocytosis? (a) ligand binds to receptors in coated vesicle → vesicle enters cytosol by cotransport mechanisms → clathrin accumulates around vesicle (b) ligand binds to receptors in coated pit → pit forms coated vesicle by endocytosis → clathrin coating detaches from vesicle (c) ATP binds to receptors in coated vesicle → vesicle enters cytosol by facilitated diffusion → protein coat dissolves (d) ligand binds to receptors in coated pit → pit forms coated vesicle by phagocytosis → coating detaches from

vesicle (e) clathrin binds to receptors in coated pit → pit forms coated vesicle by endocytosis → protein coating forms around vesicle

11. In signal transduction (a) an extracellular signal is converted to an intracellular signal (b) a signal is relayed through a series of molecules in the membrane (c) the metabolism of the cell is always inhibited (d) answers a, b, and c are correct (e) answers a and b only

12. Anchoring junctions that hold cells together at one point like a spot weld are (a) tight junctions (b) microfilaments (c) desmosomes (d) gap junctions (e) plasmodesmata

13. Junctions that permit the transfer of water, ions, and molecules between adjacent plant cells are (a) tight junctions (b) microfilaments (c) desmosomes (d) gap junctions (e) plasmodesmata

REVIEW QUESTIONS

1. What molecules are responsible for the physical properties of a cell membrane?

2. Illustrate how a transmembrane protein might be positioned in a lipid bilayer. How do the hydrophilic and hydrophobic regions of the protein affect its orientation?

3. Describe the pathway used by cells to place carbohydrates on plasma membrane proteins. Explain why this pathway results in the carbohydrate groups being on only one side of the lipid bilayer.

4. What is the source of energy for diffusion? State a rule for predicting the movement of particles along their concentration gradient. Is the rule different for facilitated diffusion compared with simple diffusion?

5. Distinguish between osmosis and dialysis.

6. Predict the consequences if a plant cell were to be placed in a relatively (a) isotonic, (b) hypertonic, or (c) hypotonic environment. Would you have to modify any of your predictions for an animal cell?

7. What are some of the functions of the plasma membrane? Discuss the nature of the proteins that carry out those functions and explain how their properties make them especially adapted for their functions.

8. Identify a common energy source for active transport. In what ways are facilitated diffusion and carrier-mediated active transport similar? In what ways do they differ?

9. Draw a diagram illustrating how membrane lipid bilayers fuse during the processes of exocytosis and endocytosis. Is one the exact reverse of the other? Why or why not?

10. Discriminate between the processes of phagocytosis and pinocytosis.

11. How are desmosomes and tight junctions functionally similar? How do they differ? Do they share any structural similarities?

12. What is the justification for considering gap junctions and plasmodesmata to be functionally similar? How do they differ structurally?

13. Label the diagram of a typical plasma membrane. Use Figure 5–6 to check your answer.

YOU MAKE THE CONNECTION

1. Why can't larger polar molecules and ions cross a lipid bilayer? Would it be advantageous to the cell if they could?

2. Most cells do not actively transport water, yet water is essential to life. How, then, are cells able to control their water content?

3. You prepare a salad with dressing in the morning but find that it is limp and unappetizing by lunch time. Why?

4. Most adjacent living cells in a plant are connected by plasmodesmata. On the other hand, only certain adjacent animal cells are associated through gap junctions. Why?

RECOMMENDED READINGS

Bayley, H. "Building Doors into Cells." *Scientific American,* Vol. 277, No. 3, Sept. 1997. Investigators are using recombinant DNA technology to create artificial pores in cell membranes. The technique has many clinical applications, including delivery of drugs.

Lasic, D.D. "Liposomes," *Science & Medicine,* Vol. 3, No. 3, May/Jun. 1996. Liposomes, artificial vesicles that can be produced commercially, are being investigated as vehicles for delivering drugs to specific cell types in the body.

Linder, M.E., and A. Gilman. "G Proteins." *Scientific American,* Vol. 267, No. 1, Jul. 1992. Pioneers in cell-signaling research discuss the many roles of G proteins.

Nehr, E., and B. Sakmann. "The Patch Clamp Technique." *Scientific American,* Vol. 266, No.3, Mar. 1992. The developers of the patch clamp technique discuss its varied applications.

Rothman, J.E., and L. Orci. "Budding Vesicles in Living Cells." *Scientific American,* Vol. 274, No. 3, Mar. 1996. The authors discuss the exciting process of discovering how cells form transport vesicles.

Vogel, S. "Dealing Honestly with Diffusion." *The American Biology Teacher,* Oct., 1994. An explanation of why most macroscopic phenomena attributed to diffusion actually have other explanations. This article emphasizes the fact that diffusion is rapid only over extremely short distances.

● Visit our website at **http://www.saunderscollege.com/lifesci/titles.html** and click on Solomon/Berg/Martin Biology for links to chapter-related resources on the World Wide Web.

Pharmacist

NABIL ELKAREH

Nabil Elkareh is a retail pharmacist in Orlando, Florida. Born in Beirut, Lebanon, Nabil came to Dallas, Texas in 1986 to visit a brother. He decided to stay because of the war back home and switched his tourist visa to a student visa. He enrolled at North Texas State University as a biology major. After two years, Nabil applied his biology credits to the pharmacy program. He completed his B.S. in Pharmacy in his fifth year. Nabil has always been fascinated by the way chemicals interact with the human body, and he now enjoys counseling patients about the interactions of their prescriptions. Nabil takes seriously his role as the responsible link between doctor and patient.

Did you know what you wanted to major in at the time you enrolled?

Yes, I had already made up my mind about pharmacy, but I was also interested in the medical field. So, I started out as a biology major until I could decide on my professional track.

Did you go through a five-year pharmacy program then? Did this include some clinical experience?

Yes. I applied my two years of biology toward a pharmacy degree, and then spent the next three years in pharmacy school. The requirements of the biology degree are the same as the requirements of the pharmacy degree until the last two years.

In my fifth year, I worked for five weeks in a hospital pharmacy. For a comparison, I then worked five weeks in a retail pharmacy setting. After that, the program required ten weeks in a clinical

rotation at a hospital. This means that I rotated with doctors and interns. Finally, I was sent on rounds for one week with a nutritionist. I did my rounds alone following that, as a pharmacist intern on an assigned floor of the hospital.

Have there been any changes in the requirements of a pharmacy degree since you graduated?

In the future, a pharmacy doctorate or Pharm.D. could become a job requirement. This means a sixth year of school, which emphasizes clinical and product-related issues and more time in a variety of practice settings. The course work emphasizes disease states and therapeutics. Hospital pharmacy jobs today prefer Pharm.D. graduates. In fact, some schools now only offer Pharm.D. programs.

How does your background in biology help you in your career?

It gives me personal satisfaction and a deeper understanding of disease biology. For example, we study cancer in pharmacy, but we stress the drugs more than the disease state. With a biology degree you have some knowledge of oncogenes (malfunctioning genes thought to cause cancer). You understand what causes cancer, how you can prevent it, and how the drugs work. Biology gives more of the human body background—from heredity to environment.

How did you get your retail pharmacy position?

I took my pharmacy board exam in Florida. Various pharmacies had employment booths outside the test center. I interviewed on the spot and was hired soon afterwards. There are many openings for pharmacists today.

What is the difference between retail pharmacy and hospital pharmacy?

In retail, counseling is reduced. Instead of counseling with every prescription, you ask the patient if they have questions. If I'm not dealing with many prescriptions, I can

counsel each patient. But if I did that regularly, I would have a line-up of patients.

At hospital sites, on the other hand, counseling is frequent, as is interaction with the doctors. Pharmacists are more involved because they witness patient care inside the hospital. They walk with the doctors during clinical rotations and discuss drug effectiveness. Hospital pharmacists do not have to handle 400 people and their prescriptions daily. That makes a big difference.

What are your daily activities?

I dispense pills with the aid of technicians who enter prescriptions from the doctor in the computer. I verify each prescription to make sure the medication is correct, checking to see if there are any allergies or drug interactions. The computer will flag these. If anything is flagged, I go through the patient's profile, then double-check the medication. I either call the doctor or I tell the patient to observe themselves for a few days to see how they feel. I try to counsel the patients whenever I can. I enjoy counseling.

How do you keep up with the changes in the field?

Florida state law requires that pharmacists take fifteen hours per year of continuing education. Other states may have different requirements. You can fulfill this by attending some of the many seminars offered by the pharmaceutical companies, whether it's about antidepressants, hormones, or cancer medications. As a refresher, a talk is given about the drugs that have been available for a long time. Then they present the new drugs.

Where do you see yourself five years from now?

Pharmacy supervisor and intern recruiter for the company. I hope to manage operations, supervise 20 to 30 pharmacies, making sure that they follow company policies, as well as state and federal laws. Every career step counts. I could also work in research for drug companies or for the FDA. Another option is to get onto hospital committees that advise pharmaceutical houses on formulary drugs, which are the drugs the hospital carries and dispenses. The formulary is a regularly revised list of drugs. It is the current clinical judgment of the hospital staff that these drugs are the best in their class. For that reason, it is a shorter list than what a retail pharmacy carries.

Energy Transfer Through Living Systems

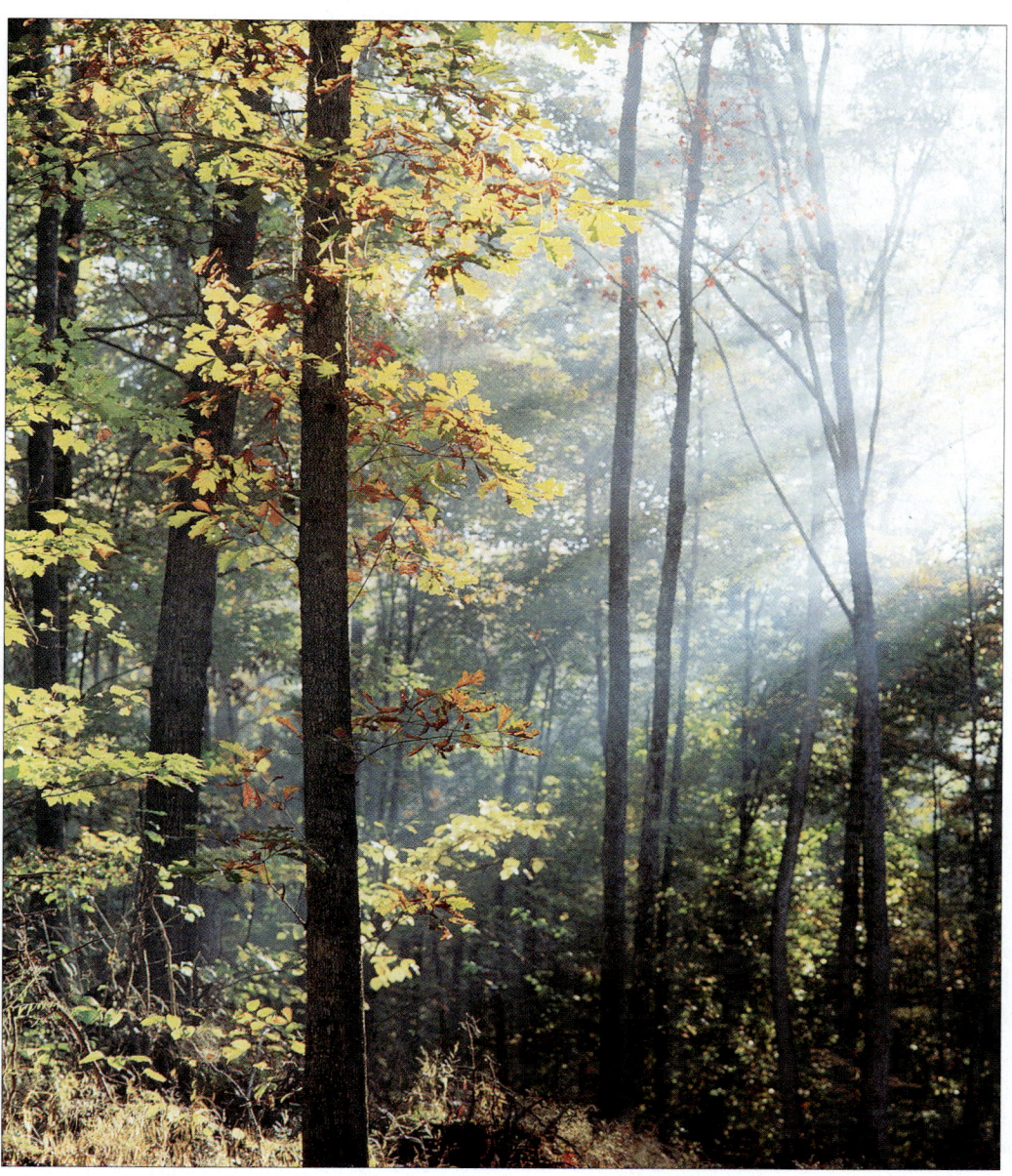

Sunlight filters through a forest clearing in West Virginia. Almost all life ultimately depends on solar energy. *(Michael Hubrich/Photo Researchers, Inc.)*

CHAPTER 6

Energy and Metabolism

All living things require energy because life processes involve work. It may seem obvious that cells need energy to grow and reproduce, but even nongrowing cells need energy simply to maintain themselves. The sun is the ultimate source of almost all the energy that powers life. Plants and other photosynthetic organisms capture a tiny portion of the sun's energy and, in the process of photosynthesis, convert it to chemical energy in organic molecules. The chemical energy captured by photosynthesis and stored in seeds and leaves is transferred to animals, such as this black-tailed prairie dog, when they eat. Plants, animals, or other organisms need the energy stored in these organic molecules, and they commonly use the process of cellular respiration to break them apart and convert their energy to more immediately usable forms.

Because energy cannot be created or destroyed, cells have no way to produce new energy. Energy is captured from the environment, temporarily stored, and then used to perform biological work. However, not all of the captured energy can be used for work; at every step some inevitably becomes converted to heat and is dispersed back into the environment.

Cells obtain energy in many forms, but seldom can that energy be used directly to power cellular processes. For this reason cells have mechanisms that convert energy from one form to another. Because most of the components of these energy conversion systems evolved very early in the history of life, many aspects of energy metabolism tend to be very similar in a wide range of organisms.

This chapter focuses on some of the basic principles that govern how cells capture, transfer, store, and use energy. We discuss the functions of ATP and other molecules used in energy conversions, including those that transfer electrons in redox reactions. We also pay particular attention to the essential role of enzymes in cellular energy dynamics. In Chapter 7 we will explore some of the main metabolic pathways used in cellular respiration, and in Chapter 8 we will discuss the energy transformations of photosynthesis. The flow of energy in ecosystems is discussed in Chapter 53.

(Barbara Gerlach/Visuals Unlimited)

AFTER YOU HAVE STUDIED THIS CHAPTER YOU SHOULD BE ABLE TO

1. Define energy, emphasizing how it is related to work and to heat.
2. Use examples to contrast potential energy and kinetic energy.
3. State the first and second laws of thermodynamics and discuss the implications of these laws as they relate to organisms.
4. Discuss how changes in free energy in a reaction are related to changes in entropy and enthalpy.
5. Compare the energy dynamics of a reaction at equilibrium with the dynamics of a reaction not at equilibrium.
6. Distinguish between exergonic and endergonic reactions and give examples of how they may be coupled.
7. Explain how the chemical structure of ATP allows it to transfer a phosphate group. Discuss the central role of ATP in the overall energy metabolism of the cell.
8. Relate the transfer of electrons (or hydrogen atoms) to the transfer of energy.
9. Explain how an enzyme lowers the required energy of activation for a reaction.
10. Describe some of the ways enzymes are regulated.

BIOLOGICAL WORK REQUIRES ENERGY

Energy, one of the most important concepts in biology, can be understood in the context of **matter,** which is anything that has mass and takes up space. **Energy** can be defined as the capacity to do **work,** which is any change in the state or motion of matter.

Biologists generally express energy in units of work **(kilojoules, kJ)** or units of heat energy **(kilocalories, kcal).** One kilocalorie equals 4.184 kilojoules. Because heat energy cannot do cellular work, the kilojoule is the unit preferred by most biologists today (see *Making the Connection: Energy, Work, and Heat*). However, we will use both units because references to the kilocalorie are common in the scientific literature.

Many of the activities performed by an organism are mechanical. At this very moment you are expending considerable energy to carry out such activities as breathing and circulating your blood. In these processes, the state or position of matter is changed in some way. However, these forms of mechanical work are the consequence of cellular activities. For example, the cells of the heart muscle use a great deal of energy to contract, thereby pumping the blood through your body. As we will see, however, not all of the work of cells is mechanical. A great deal of it is chemical. For example, heart muscle cells expend energy to synthesize the proteins required for contraction. Energy can be converted to many different forms, including not only mechanical and chemical energy, but also heat energy, electrical energy, and radiant energy.

MAKING THE CONNECTION

ENERGY, WORK, AND HEAT

Why can we express energy both in units of work (kilojoules) and in units of heat energy (kilocalories)? We can because of a conceptual breakthrough that came about in the 1800s, after the invention of the steam engine. Scientists studying the connections among the heat energy that powered the engine, the mechanical work that the engine was able to perform, and the heat that was transferred to the environment were able to demonstrate that all these forms of energy are interconvertible.

Today we know that not only mechanical work but all forms of energy can be converted to heat. In fact, the study of energy and its transformations has been named thermodynamics, that is, heat dynamics. (Recall from Chapter 2 that *heat* refers to the total amount of kinetic energy in a sample of a substance, whereas *temperature* refers to the average kinetic energy of the particles.) **Heat energy** is energy that can flow from an object with a higher temperature (known as the heat source) to an object with a lower temperature (the heat sink).

Cells cannot work as heat engines because they are isothermal; they are too small to have regions that differ in temperature. Therefore, heat cannot be used to do biological work. Nevertheless, the fact that all forms of energy can be converted to heat is useful to scientists because heat energy is particularly convenient to measure.

Nutritionists use the kilocalorie to express the potential energy of foods and usually refer to it as a Calorie (with a capital C; see Chapter 45). For example, the energy content of 10 grams, about 2 teaspoons, of table sugar (sucrose) is about 36 Calories (36 kcal or 151 kJ), whereas the energy content of 20 potato chips is about 150 Calories (150 kcal or 628 kJ). A person weighing 58 kilograms (130 pounds) uses about 1 kcal (4.184 kJ) per minute to maintain the body while sleeping and up to 10 kcal (41.84 kJ) per minute when engaged in strenuous activity.

Organisms carry out conversions between potential energy and kinetic energy

When an archer draws a bow, **kinetic energy,** which is energy of motion, is used and work is done (Fig. 6–1). The resulting tension in the bow and string represents stored energy, or **potential energy.** Potential energy is the capacity to do work owing to position or state. When the string is released, this potential energy is converted to kinetic energy in the motion of the bow, which propels the arrow.

Most of the actions of an organism involve a complex series of energy transformations that occur as kinetic energy is converted to potential energy or as potential energy is converted to kinetic energy. For example, potential energy derived from chemical energy of food molecules is converted to kinetic energy in the muscles of the archer.

TWO LAWS OF THERMODYNAMICS GOVERN ENERGY TRANSFORMATIONS

All the activities of our universe, from the life and death of cells to the life and death of stars, are governed by **thermodynamics,** which is the study of energy and its transformations. When considering thermodynamics, scientists use the term *system* to refer to an object that is being studied, whether it is a cell, an organism, or planet Earth. The rest of the universe other than the system being studied is known as the *surroundings*. A **closed system** is one that does not exchange energy or matter with its surroundings, whereas an **open system** is one that can exchange matter and energy with its surroundings (Fig. 6–2). There are two laws about energy that apply to all things in the universe. These are known as the first and second laws of thermodynamics.

The total energy in the universe does not change

According to the **first law of thermodynamics,** energy cannot be created or destroyed, although it can be transferred or changed from one form to another. As far as we know, the energy present in the universe at its formation, approximately 15 to 20 billion years ago, equals the amount of energy present in the universe today.[1] This is all the energy that can ever be present in the universe. Similarly, the energy of any system and its surroundings is constant. A system may absorb energy from its surroundings, or it may give up some energy into its surroundings, but the total energy content of that system and its surroundings is always the same.

---

[1]Technically, mass is a form of energy, and so we should say that the total mass-energy of the universe is a constant. Energy can be produced from mass (recall Einstein's famous equation $E = mc^2$). This is the basis behind the energy generated by the sun and stars. More than 4 billion kilograms of matter per second are converted to energy in our sun.

Figure 6–1 Potential versus kinetic energy. The potential chemical energy released by cellular respiration is converted to kinetic energy in the muscles, which do the work of drawing the bow. The potential energy stored in the drawn bow is transformed into kinetic energy as the bowstring pushes the arrow toward its target.

As specified by the first law of thermodynamics, then, organisms cannot create the energy that they require to live. Instead, they must capture energy from the environment to use for biological work, a process involving the transformation of energy from one form to another. In photosynthesis, for example, plants absorb the radiant energy of the sun and convert it into the chemical energy contained in the bonds of carbohydrate molecules. Some of that chemical energy may later be transformed by the plant to do various types of cellular work or by some animal that eats the plant and might convert it to the mechanical energy of muscle contraction or some other needed form.

The entropy of the universe is increasing

As each energy transformation occurs, some of the energy is converted to heat energy that is then given off into the cooler surroundings. This energy can never again be used by any organism for biological work; it is lost from the biological point of view. However, it is not really gone from a thermodynamic

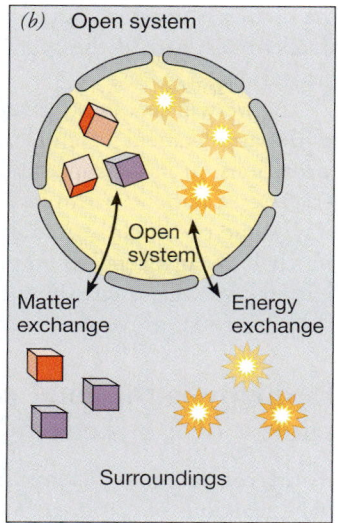

Figure 6–2 Closed and open systems. *(a)* Matter and energy are not exchanged between a closed system and its surroundings. *(b)* Matter and energy are exchanged between an open system and its surroundings. *(Adapted from Tobin and Morel)*

point of view because it still exists in the surroundings. For example, the use of food to enable us to walk or run does not destroy the chemical energy that was once present in the food molecules. After we have performed the task of walking or running, the energy still exists in the surroundings as heat.

The **second law of thermodynamics** can be stated most simply as follows: when energy is converted from one form to another, some usable energy, that is, energy available to do work, is degraded into a less usable form, heat, that disperses into the surroundings. As a result, the amount of usable energy available to do work in the universe decreases over time.

It is important to understand that the second law of thermodynamics is consistent with the first law; that is, the total amount of energy in the universe is *not* decreasing with time. However, the energy available to do work is being degraded to less-usable energy with time.

Less-usable energy is more diffuse, or disorganized. **Entropy (S)** is a measure of this disorder or randomness; organized, usable energy has a low entropy, whereas disorganized energy such as heat has a high entropy (Fig. 6–3). The total entropy of the universe is continuously increasing in all natural processes. It may be that at some time, billions of years from now, all energy will exist as heat uniformly distributed throughout the universe. If that happens, the universe will cease to operate because no work will be possible. Everything will be at the same temperature, so there will be no way to convert the thermal energy of the universe into usable mechanical energy.

Another way to explain the second law of thermodynamics, then, is that entropy, or disorder, in a closed system tends to increase spontaneously over time. (The word *spontaneously* in this context means that entropy occurs naturally rather than being caused by some external influence.)

(a)

(b)

Figure 6–3 Entropy. *(a)* Entropy as demonstrated by frogs. The frogs are placed on one side of a box with partitions to represent a highly ordered system. As the frogs hop into the other side and back again so that they are randomly distributed throughout both sides of the box, they represent a system of greater entropy. *(b)* Similarly, the beaker on the left, in which all marbles of the same color are located together, represents a highly organized system with low entropy. The beaker on the right, in which the marbles are randomly arranged, regardless of color, represents a more disorganized system with greater entropy. *(a, Adapted from Tobin and Morel; b, Dennis Drenner)*

As a result of the second law of thermodynamics, no process requiring an energy conversion is ever 100% efficient, because much of the energy is dispersed as heat, resulting in an increase in entropy. For example, an automobile engine, which converts the chemical energy of gasoline to mechanical energy, is between 20% and 30% efficient. That is, only 20% to 30% of the original energy stored in the chemical bonds of

Figure 6–4 **LM of club moss stem cross section.** The highly organized cellular structure of this *Lycopodium clavatum* stem is developed and maintained only by the constant input of energy from the sun. Thus, the high degree of organization of living things does not refute the second law of thermodynamics. *(John D. Cunningham/Visuals Unlimited)*

the gasoline molecules is actually transformed into mechanical energy; the other 70% to 80% is dissipated as waste heat. Energy utilization in our cells is about 40% efficient, with the remaining energy given to the surroundings as heat.

Organisms have a high degree of organization, and at first glance they appear to refute the second law of thermodynamics (Fig. 6–4). As organisms grow and develop, they maintain a high level of order and do not appear to become more disorganized. However, organisms are able to maintain their degree of order over time only with the constant input of energy from their surroundings. That is why plants must photosynthesize and animals must eat. Although the order of organisms might tend to increase temporarily, the total entropy of the universe (organisms plus surroundings) will increase over time.

METABOLIC REACTIONS INVOLVE ENERGY TRANSFORMATIONS

The chemical reactions that enable an organism to carry on its activities—to grow, move, maintain and repair itself, reproduce, and respond to stimuli—together make up its **metabolism.** Metabolism was defined in Chapter 1 as the sum of all the chemical activities that take place in an organism. An organism's metabolism consists of many intersecting series of chemical reactions, or pathways, which are of two main types. **Anabolism** refers to the various pathways in which complex molecules are synthesized from simpler substances, such as the linking of amino acids to form proteins. **Catabolism** includes

the pathways in which larger molecules are broken down into smaller ones, such as the degradation of starch to form monosaccharides.

As we will see, these changes not only involve alterations in the arrangement of atoms, but also various energy transformations. Catabolism and anabolism are complementary processes; catabolic pathways involve an overall release of energy, some of which is used to power the anabolic pathways, which have an overall energy requirement. In the following sections we will discuss how to predict whether a particular chemical reaction requires energy or releases it.

Enthalpy is the total potential energy of a system

In the course of any chemical reaction, including the metabolic reactions of a cell, chemical bonds break, and new and different bonds may form. Every specific type of chemical bond has a certain amount of **bond energy,** defined as the energy required to break that bond. The total bond energy is essentially equivalent to the total potential energy of the system, a quantity known as **enthalpy (H).** Because energy can be conveniently measured as heat, enthalpy is often referred to as the heat content of the system.

Free energy is energy that is available to do cellular work

Entropy and enthalpy are related by a third dimension of energy, termed **free energy (G),** which is the amount of energy available to do work under the conditions of a biochemical reaction. Free energy, the only kind of energy that can do cellular work, is the aspect of thermodynamics of greatest interest to a biologist.

Entropy (S) and free energy (G) are related inversely; as entropy increases, the amount of free energy decreases. The two are related by the following equation:

$$G = H - TS$$

in which G is the free energy, H is the enthalpy of the system, T is the absolute temperature expressed in degrees Kelvin, and S is entropy. If we assume that entropy is zero, the free energy is simply equal to the total potential energy (enthalpy); an increase in entropy reduces the amount of free energy.

What is the significance of the temperature (T)? Remember that as the temperature increases, there is an increase in random molecular motion that contributes to disorder and multiplies the effect of the entropy term.

Chemical reactions involve changes in free energy

Biologists need ways to analyze the role of energy in the many reactions that comprise metabolism. Although the total free energy of a system (G) cannot be effectively measured, the equation $G = H - TS$ can be extended to predict whether any particular chemical reaction will release energy or require an input of energy. This is because *changes* in free energy can be measured. We use the Greek letter delta (Δ) to denote any change that occurs in the system between its initial state before the reaction and its final state after the reaction. To express what happens with respect to energy in a chemical reaction, the equation becomes:

$$\Delta G = \Delta H - T\Delta S$$

Notice that the temperature does not change; it is held constant during the reaction. Thus the change in free energy (ΔG) during the reaction is equal to the change in enthalpy (ΔH) minus the product of the absolute temperature (T) multiplied by the change in entropy (ΔS). ΔG and ΔH are expressed in kilojoules or kilocalories per mole; ΔS is expressed in kilojoules per degree or in kilocalories per degree.

Free energy decreases during an exergonic reaction

In accordance with the second law of thermodynamics, no chemical reaction is 100% efficient. No reaction can take place without a decrease in enthalpy, an increase in entropy, or both (see *Making the Connection: Energy and Diffusion*). For this reason, the total free energy of the system in its final state is always less than the total free energy of the system in its initial state. When calculated in this way, ΔG is a negative number. Such a reaction, with a $-\Delta G$, is referred to as an **exergonic reaction** (Fig. 6–5*a*).

An exergonic reaction releases energy and is said to be a spontaneous or a "downhill" reaction. The term *spontaneous* may give the false impression that such reactions are always instantaneous. In fact, spontaneous reactions do not necessarily occur readily; some are extremely slow. This is because energy, known as activation energy, is required to initiate every reaction, even a spontaneous one. Activation energy will be discussed later in the chapter.

Free energy increases during an endergonic reaction

An **endergonic reaction** is a reaction in which there is a gain of free energy (Fig. 6–5*b*). Because the free energy of the products is greater than the free energy of the reactants, ΔG has a positive value. Such a reaction cannot take place in isolation. Instead, it must occur in such a way that energy can be supplied from the surroundings. Of course, many energy-requiring reactions take place in cells, and, as we will see, metabolic mechanisms have evolved that supply the energy needed to "drive" these nonspontaneous cellular reactions in a particular direction.

MAKING THE CONNECTION

ENERGY AND DIFFUSION

What is the source of energy for diffusion? In Chapter 5 we saw that randomly moving particles can diffuse down their own **concentration gradient** (see figure). That is, although the movements of the individual particles are random, net movement of the group of particles seems to be directional. What provides the energy for this seemingly directed process? A concentration gradient, with a region of higher concentration and another region of lower concentration, is an orderly state. A cell must expend energy to produce a concentration gradient. Because work must be done to produce this order, the concentration gradient is a form of potential energy. As the particles move about randomly, disorder increases. Although there is no change in enthalpy, entropy increases. The process is spontaneous because there is an overall decrease in free energy ($-\Delta G$); diffusion is paid for by an increase in entropy.

In cellular respiration and photosynthesis, the potential energy stored in a concentration gradient of hydrogen ions (H^+) can be transformed into chemical energy in ATP as the hydrogen ions pass through a membrane down the concentration gradient. This important concept, known as **chemiosmosis,** will be discussed further in Chapters 7 and 8.

Energy, entropy, and diffusion. The tendency of entropy to increase can be used to produce work, in this case, diffusion. *(Left)* A concentration gradient is a form of potential energy. *(Right)* When molecules are evenly distributed, they have a high entropy.

(a) **Exergonic reaction**
(spontaneous; energy-releasing)

(b) **Endergonic reaction**
(not spontaneous; energy-requiring)

Figure 6–5 **Exergonic and endergonic reactions.** *(a)* In an exergonic reaction there is a net loss of free energy. The products have less free energy than was present in the reactants, and the reaction proceeds spontaneously. *(b)* In an endergonic reaction there is a net gain in free energy. The products have more free energy than was present in the reactants. An endergonic reaction occurs only if energy lost from some other system is fed into the reaction. *(Adapted from Tobin and Morel)*

Free energy changes depend on the concentrations of reactants and products

According to the second law of thermodynamics, any process that increases entropy (*S*) can do work. Differences in concentration of a substance, for example, between two different parts of a cell, represent a more orderly state than when the substance is diffused homogeneously throughout the cell. We have seen that free energy changes in any chemical reaction depend mainly on the difference in bond energies (enthalpy, *H*) between reactants and products. Free energy also depends on *concentrations* of both reactants and products. The change in molecules from a more concentrated to a less concentrated state increases entropy because it is movement from a more orderly to a less orderly state.

In most biochemical reactions there is little intrinsic free energy difference between reactants and products. Such reactions are reversible, a fact that is indicated by drawing double arrows ($\rightleftharpoons$) between the reactants and the products.

$$A \rightleftharpoons B$$

At the beginning of a reaction, only the reactant molecules (A) may be present. As the reaction proceeds, the concentration of the reactant molecules decreases, and the concentration of the product molecules (B) increases. As the concentration of the product molecules increases, they may have enough free energy to initiate the reverse reaction. The reaction thus proceeds in both directions simultaneously; if undisturbed it could eventually reach a state of **dynamic equilibrium,** in which the rate of the reverse reaction is equal to the rate of the forward reaction. At equilibrium there is no net change in the system; every forward reaction is balanced by a reverse reaction.

Knowledge that a system is at equilibrium tells us nothing about the relative concentrations of reactants and products at equilibrium. If the reactants have much greater intrinsic free energy than the products, the reaction goes almost to completion; that is, it reaches equilibrium at a point at which most of the reactants have been converted to products. Reactions in which the reactants have much less intrinsic free energy than

the products reach equilibrium at a point where very few of the reactant molecules have been converted to products.

If we increase the initial concentration of A, then the equilibrium will "shift to the right," and more A will be converted to B; a similar effect can be obtained if B is removed from the reaction mixture. The opposite effect occurs if the concentration of B is increased, or if A is removed; here the equilibrium "shifts to the left." The actual free energy change that occurs during a reaction is defined mathematically to include these effects, which are a consequence of the relative initial concentrations of reactants and products.

Cells manipulate the relative concentrations of reactants and products of almost every reaction. Cellular reactions are virtually never at equilibrium. By displacing their reactions far from equilibrium, cells are able to supply energy to endergonic reactions and direct their metabolism in accordance with their needs.

Cells drive endergonic reactions by coupling them to exergonic reactions

Many metabolic reactions such as protein synthesis are anabolic and endergonic. Because an endergonic reaction cannot take place without an input of energy, endergonic reactions are coupled to exergonic reactions. In **coupled reactions,** the thermodynamically favorable exergonic reaction provides the energy required to drive the thermodynamically unfavorable endergonic reaction. The endergonic reaction can proceed only if it absorbs free energy released by the exergonic reaction to which it is coupled.

Consider the free energy change, ΔG, in the following reaction:

(1) $A \longrightarrow B \qquad \Delta G = +20.9$ kJ/mole ($+5$ kcal/mole)

Because ΔG has a positive value, we know that the product of this reaction has more free energy than the reactant. This is an endergonic reaction. It is not spontaneous and does not take place without an energy source.

By contrast, consider the following reaction:

(2) C $\longrightarrow$ D $\Delta G = -33.5$ kJ/mole (-8 kcal/mole)

The negative value of ΔG tells us that the free energy of the reactant is greater than the free energy of the product. This exergonic reaction can proceed spontaneously.

We can sum up Reactions 1 and 2 as follows:

(1) A $\longrightarrow$ B $\Delta G = +20.9$ kJ/mole ($+5$ kcal/mole)
(2) C $\longrightarrow$ D $\Delta G = -33.5$ kJ/mole (-8 kcal/mole)
        Overall $\Delta G = -12.6$ kJ/mole (-3 kcal/mole)

Because thermodynamics considers the overall changes in these two reactions, which show a net negative value of ΔG, the two reactions taken together are exergonic.

The fact that we can write reactions this way is a useful bookkeeping device, but it does not mean that an exergonic reaction can mysteriously transfer energy to an endergonic "bystander" reaction. However, these reactions can be coupled if their pathways are altered such that they are linked by a common intermediate. Reactions 1 and 2 might be coupled by an intermediate (I) in the following way:

(3) A + C $\longrightarrow$ I $\Delta G = -8.4$ kJ/mole (-2 kcal/mole)
(4) I $\longrightarrow$ B + D $\Delta G = -4.2$ kJ/mole (-1 kcal/mole)
        Overall $\Delta G = -12.6$ kJ/mole (-3 kcal/mole)

Note that Reactions 3 and 4 are sequential. Thus the reaction pathways have changed, but overall the reactants and products are the same, and the free energy change is the same.

Generally, for each endergonic reaction occurring in a living cell, there is a coupled exergonic reaction to drive it. Often, the exergonic reaction involves the breakdown of adenosine triphosphate (ATP). We now examine specific examples of the role of ATP in energy coupling.

ATP IS THE ENERGY CURRENCY OF THE CELL

In all living cells, energy is temporarily packaged within a remarkable chemical compound called **adenosine triphosphate (ATP),** which holds readily available energy for very short periods of time. We may think of ATP as the energy currency of the cell. When you work to earn money, you might say that your energy is symbolically stored in the money you earn. The energy the cell requires for immediate use is temporarily stored in ATP, which is like cash. When you earn extra money, you might deposit some in the bank; similarly, a cell might deposit energy in the chemical bonds of lipids, starch, or glycogen. Moreover, just as you dare not make less money than you spend, so too the cell must avoid energy bankruptcy, which would mean its death. Finally, just as you (alas) do not keep what you make very long, so too the cell continuously spends its ATP, which must be replaced immediately.

ATP is a nucleotide consisting of three main parts: adenine, a nitrogen-containing organic base; ribose, a five-carbon sugar; and three phosphate groups, identifiable as phosphorus atoms surrounded by oxygen atoms (Fig. 6–6a). Notice that the phosphate groups are bonded to the end of the molecule in a series, rather like three cars behind a locomotive, and, like the cars of a train, they can be attached and detached.

ATP donates energy through the transfer of a phosphate group

When the terminal phosphate is removed from ATP, the remaining molecule is **adenosine diphosphate (ADP)** (Fig. 6–6b). If the phosphate group is not transferred to another molecule, it is released as inorganic phosphate (P_i). This is an exergonic reaction. ATP is sometimes called a "high-energy" compound because the hydrolysis reaction that releases a phosphate has a relatively large $-\Delta G$.[2]

(5) ATP + H_2O $\longrightarrow$ ADP + P_i
        $\Delta G = -32$ kJ/mole (or -7.6 kcal/mole)

Reaction 5 can be coupled to endergonic reactions in cells. Consider the following endergonic reaction in which the disaccharide sucrose is formed from two monosaccharides, glucose and fructose.

(6) Glucose + Fructose $\longrightarrow$ Sucrose + H_2O
        $\Delta G = +27$ kJ/mole (or $+6.5$ kcal/mole)

With a free energy change of -32 kJ/mole (-7.6 kcal/ mole), the hydrolysis of ATP in Reaction 5 can drive Reaction 6, but only if the reactions can be coupled through a common intermediate. The following series of reactions is a simplified version of an alternative pathway used by some bacteria.

(7) Glucose + ATP $\longrightarrow$ Glucose-P + ADP

(8) Glucose-P + Fructose $\longrightarrow$ Sucrose + P_i

Reaction 7 is a **phosphorylation reaction,** one in which a phosphate group is transferred to some other compound. Glucose is phosphorylated to form glucose phosphate (glucose-P), the intermediate that links the two reactions. Glucose-P, which corresponds to "I" in Reactions 3 and 4, reacts exergonically with fructose to form sucrose. For energy coupling to work in this way, Reactions 7 and 8 must occur in sequence. It is convenient to summarize the reactions in the following way:

(9) Glucose + Fructose + ATP $\longrightarrow$ Sucrose + ADP + P_i
        $\Delta G = -5$ kJ/mole (-1.2 kcal/mole)

When encountering an equation written in this way, remember that it is actually a summary of a series of reactions and that transitory intermediate products are sometimes not shown.

[2]Calculations of the free energy of ATP hydrolysis vary somewhat, but range between about -28 and -37 kJ/mole (-6.8 to -8.7 kcal/mole).

Figure 6–6 ATP. *(a)* The energy currency of all living things, ATP is composed of adenine, ribose, and three phosphate groups. *(b)* The hydrolysis of ATP, an exergonic reaction, yields ADP and inorganic phosphate.

Adenine

NH$_2$

Phosphate groups

Ribose

OH   OH

(a) Adenosine triphosphate (ATP)

Phosphate groups

H$_2$O

ATP

ADP

Inorganic phosphate

(b)

ATP links exergonic and endergonic reactions

We have just discussed how the transfer of a phosphate group from ATP to some other compound can be coupled to endergonic reactions in the cell. Conversely, adding a phosphate group to AMP (forming ADP) or to ADP (forming ATP) requires coupling to exergonic reactions in the cell.

$$\text{AMP} + \text{P}_i + \text{Energy} \longrightarrow \text{ADP}$$
$$\text{ADP} + \text{P}_i + \text{Energy} \longrightarrow \text{ATP}$$

Thus ATP occupies an intermediate position in the metabolism of the cell and is an important link between exergonic reactions, which are generally components of catabolic pathways, and endergonic reactions, which are generally part of anabolic pathways (Fig. 6–7).

The cell maintains a very high ratio of ATP to ADP

The cell maintains the ratio of ATP to ADP far from the equilibrium point. ATP is constantly formed from ADP and inorganic phosphate as nutrients are oxidized in cellular respiration or as the radiant energy of sunlight is trapped in photosynthesis (see Chapters 7 and 8). At any point in time,

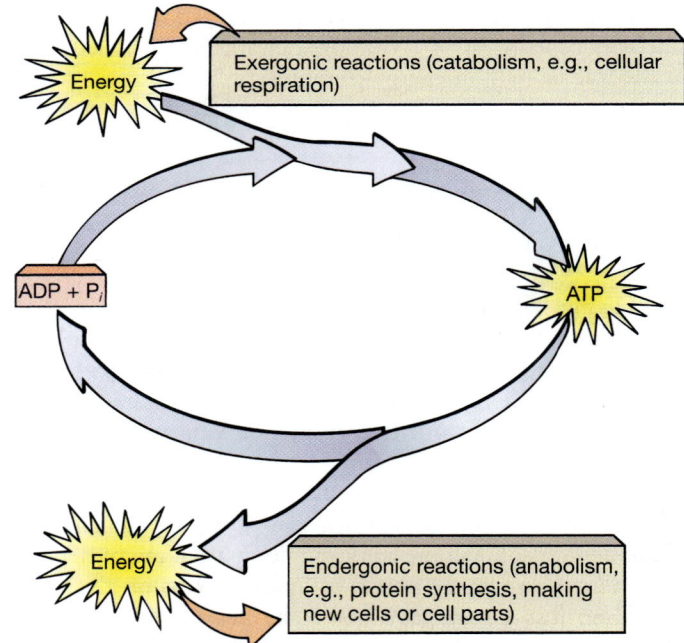

Energy

Exergonic reactions (catabolism, e.g., cellular respiration)

ADP + P$_i$

ATP

Energy

Endergonic reactions (anabolism, e.g., protein synthesis, making new cells or cell parts)

Figure 6–7 ATP links exergonic and endergonic reactions. Because ATP is responsible for coupling many exergonic and endergonic reactions, it is an important link between catabolism and anabolism in cells.

a typical cell contains more than ten ATP molecules for every ADP molecule. The fact that the cell maintains the ATP concentration at such a high level (relative to the concentration of ADP) makes its hydrolysis reaction even more strongly exergonic and more able to drive the endergonic reactions to which it is coupled.

Although the cell maintains a high ratio of ATP to ADP, large quantities of ATP cannot be stored in the cell. The concentration of ATP is always very low, less than 1 millimole per liter. In fact, studies suggest that a bacterial cell has no more than a one-second supply of ATP. Thus, ATP molecules are used almost as quickly as they are produced. A human at rest uses about 45 kilograms (99 pounds) of ATP each day, but the amount present in the body at any given moment is less than 1 gram (0.035 ounce). Every second in every cell, an estimated 10 million molecules of ATP are made from ADP and phosphate, and an equal number of ATPs transfer their phosphate groups along with their energy to whatever chemical reactions may require them.

CELLS TRANSFER ENERGY BY REDOX REACTIONS

We have seen that cells can transfer energy through the transfer of a phosphate group from ATP. Energy can also be transferred through the transfer of electrons. As discussed in Chapter 2, **oxidation** is the chemical process in which a substance loses electrons, whereas **reduction** is the complementary process in which a substance gains electrons. Because electrons released during an oxidation reaction cannot exist in the free state in living cells, every oxidation reaction must be accompanied by a reduction reaction, in which the electrons are accepted by another atom, ion, or molecule. Oxidation and reduction reactions are often called **redox reactions** because they occur simultaneously. The substance that becomes oxidized gives up energy as it releases electrons, and the substance that becomes reduced receives energy as it gains electrons.

Redox reactions often occur in a series as electrons are transferred from one molecule to another. These electron transfers, which are equivalent to energy transfers, are an essential part of cellular respiration, photosynthesis, and many other chemical reactions. Redox reactions, for example, release the energy stored in food molecules so that ATP can be synthesized using that energy.

Most electron carriers carry hydrogen atoms

Generally it is not easy to remove one or more electrons from a covalent compound; it is much easier to remove a whole atom. For this reason, redox reactions in cells usually involve the transfer of a hydrogen atom rather than just an electron. A hydrogen atom contains an electron and a proton that does not participate in the oxidation/reduction.

When an electron, either singly or as part of a hydrogen atom, is removed from an organic compound, it takes with it

Figure 6–8 NAD⁺. The acceptor molecule NAD⁺ is composed of two nucleotides joined together. The oxidized form, NAD⁺, becomes reduced (NADH) by the transfer of two electrons and one proton from another organic compound, which becomes oxidized in the process.

some of the energy stored in the chemical bond of which it was a part. That electron, along with its energy, is transferred to an acceptor molecule. An electron progressively loses free energy as it is transferred from one acceptor to another.

One of the most frequently encountered acceptor molecules is **nicotinamide adenine dinucleotide (NAD⁺).** When NAD⁺ becomes reduced, it temporarily stores large amounts of free energy. Here is a generalized equation showing the transfer of hydrogen from a compound we call X to NAD⁺:

$$\underset{\text{Oxidized}}{XH_2 + NAD^+} \longrightarrow \underset{\text{Reduced}}{X + NADH + H^+}$$

Note that the NAD⁺ becomes reduced when it combines with hydrogen. NAD⁺ is an ion with a net charge of +1. When two electrons and one proton are added, the charge is neutralized and the reduced form of the compound, **NADH,** is produced (Fig. 6–8).[3] Some of the energy stored in the bonds holding the hydrogen atoms to molecule X has been transferred by this redox reaction and is temporarily held by NADH. When NADH transfers the electrons to some other molecule, some of their energy is transferred. This energy is usually then transferred through a complex series of reactions that result in the formation of ATP (see Chapter 7).

Nicotine adenine dinucleotide phosphate (NADP⁺) is a hydrogen acceptor that is chemically similar to NAD⁺ but

---

[3]Although the correct way to write the reduced form of NAD⁺ is NADH + H⁺, for simplicity we will present the reduced form as NADH in this and succeeding chapters.

with an extra phosphate group. Unlike NADH, the reduced form of $NADP^+$ (abbreviated **NADPH**) is not involved in ATP synthesis. Instead, the electrons of NADPH are used more directly to provide energy for certain reactions, including certain essential reactions of photosynthesis (see Chapter 8).

Other important hydrogen acceptors or electron acceptors include **flavin adenine dinucleotide (FAD)** and the **cytochromes.** FAD is a nucleotide that accepts hydrogen atoms and their electrons; its reduced form is **FADH₂.** The cytochromes are proteins that contain iron; the iron component accepts electrons from hydrogen atoms and then transfers these electrons to some other compound. Like NAD^+ and $NADP^+$, FAD and the cytochromes are electron transfer agents. Each can exist in a reduced state in which it has more free energy or in an oxidized state in which it has less. Each is an essential component of many redox reaction sequences in cells.

ENZYMES ARE CHEMICAL REGULATORS

The principles of thermodynamics help us predict whether a reaction can occur, but they tell us nothing about the speed of the reaction. The breakdown of glucose, for example, is an exergonic reaction, yet a glucose solution keeps virtually indefinitely in a bottle if kept free of bacteria and molds and not subjected to high temperature or strong acids or bases. Cells cannot wait for centuries for glucose to break down, nor can they use extreme conditions to cleave glucose molecules. Cells regulate the rates of chemical reactions with **enzymes,** which are protein **catalysts** that affect the speed of a chemical reaction without being consumed by the reaction.[4]

Cells require a steady release of energy, and they must be able to regulate that release to meet metabolic energy requirements. Metabolism generally proceeds by a series of steps so that a molecule may go through as many as 20 or 30 chemical transformations before it reaches some final state. Even then, the seemingly completed molecule may enter yet another chemical pathway and become totally transformed or consumed to produce energy. The changing needs of the cell require a system of flexible metabolic control. The key directors of this control system are enzymes.

The catalytic ability of some enzymes is truly remarkable. For example, hydrogen peroxide (H_2O_2) breaks down extremely slowly if the reaction is uncatalyzed, but a single molecule of the enzyme catalase brings about the decomposition of 5 million molecules of hydrogen peroxide per minute at 0°C! Catalase protects cells because hydrogen peroxide is a poisonous substance produced as a byproduct of some cellular reactions. The bombardier beetle uses the enzyme catalase as a defense mechanism (Fig. 6–9).

[4]In recent years scientists have learned that protein enzymes are not the only cellular catalysts; some types of RNA molecules have catalytic activity as well (see Chapter 12).

Figure 6–9 Catalase. When threatened, a bombardier beetle uses the enzyme catalase to decompose hydrogen peroxide. The oxygen gas formed in the decomposition ejects water and other chemicals with explosive force. Because the reaction releases a great deal of heat, the water comes out as steam. (The beetle is immobilized by a wire attached to its back by a drop of adhesive. His leg was just prodded with the dissecting needle on the left to trigger the ejection.) *(Thomas Eisner and Daniel Aneshansley/Cornell University)*

All reactions have a required energy of activation

All reactions have an energy barrier known as the **energy of activation (E_A)** or **activation energy.** The energy barrier is the energy required to break the existing bonds and begin the reaction. In a population of molecules of any kind, some have a relatively high energy content, while others have a lower energy content. Only molecules with a relatively high energy content are likely to react to form the product (Fig. 6–10*a*).

Even a strongly exergonic reaction, one that releases a substantial quantity of energy as it proceeds, may be prevented from proceeding by the activation energy required to begin the reaction. For example, molecular hydrogen and molecular oxygen can react explosively to form water:

$$2 H_2 + O_2 \longrightarrow 2 H_2O$$

This reaction is spontaneous (exergonic), yet hydrogen and oxygen can be safely mixed as long as all sparks are kept away. This is because the required energy of activation for this particular reaction is relatively high. A tiny spark provides the activation energy that allows a few molecules to react. Their reaction liberates so much heat that the rest react, producing an explosion. Such an explosion occurred on the Hindenburg, an airship that used hydrogen gas, which is lighter than air, for buoyancy (Fig. 6–11). At the end of a transatlantic voyage in 1937, it exploded on landing, probably because a small spark supplied the activation energy for the reaction of hydrogen with oxygen from the air.

(a) Uncatalyzed reaction, high activation energy

(b) Catalyzed reaction, low activation energy

Figure 6–10 **Kinetic energy in molecules and activation energy, as demonstrated by frogs.** *(a)* If a reaction is uncatalyzed, only a small fraction of reactant molecules (frogs) have sufficient energy to overcome the barrier of activation energy and undergo a chemical reaction to form product molecules (jump into the adjacent compartment). *(b)* An enzyme lowers the activation energy barrier and increases the fraction of molecules (frogs) that can react (jump). *(Adapted from Tobin and Morel)*

An enzyme lowers a reaction's activation energy

As do all catalysts, enzymes affect the rate of a reaction by lowering the energy needed to initiate the reaction. An enzyme greatly reduces the activation energy necessary to initiate a chemical reaction (Figs. 6–10*b* and 6–12). If molecules need less energy to react because the activation barrier is lowered, a larger fraction of the reactant molecules reacts at any one time.

As a result, the reaction proceeds more quickly.

Although an enzyme lowers the activation energy for a reaction, it has no effect on the overall free energy change. That is, an enzyme can only promote a chemical reaction that could proceed without it. No catalyst can cause a reaction to proceed in a thermodynamically unfavorable direction or can influence the final concentrations of reactants and products if the reaction goes to equilibrium. Enzymes simply speed up reaction rates.

Figure 6–11 **The Hindenburg explosion.** This disaster resulted when a spark triggered an explosive, exergonic reaction between hydrogen in the airship and oxygen in the atmosphere. *(Archive Photos)*

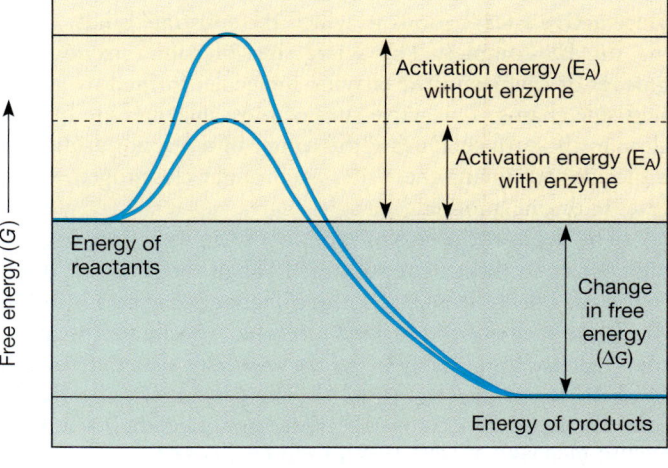

Figure 6–12 **Activation energy and enzymes.** An enzyme speeds up a reaction by lowering its activation energy (E_A). In the presence of an enzyme, it takes less activation energy for reacting molecules to complete a reaction.

(a)

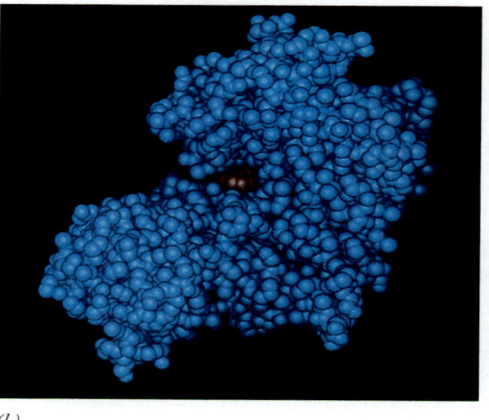

(b)

Figure 6-13 Active site and induced fit. *(a)* Computer graphic model of the enzyme hexokinase *(blue)* and its substrate, glucose *(red)*, before forming an ES complex. The active site of the enzyme is the furrow where glucose will bind. *(b)* The binding of glucose to the active site of hexokinase changes the shape of the enzyme, a phenomenon known as induced fit. Hexokinase, which is involved in cellular respiration, catalyzes the transfer of a phosphate group from ATP to glucose. *(Courtesy of Thomas A. Steitz)*

An enzyme works by forming an enzyme-substrate complex

An uncatalyzed reaction depends on random collisions among reactants. Because of its ordered structure, an enzyme is able to reduce this reliance on random events and thereby control the reaction. The enzyme is thought to accomplish this by forming an unstable intermediate complex with the **substrate,** the substance on which it acts. When the **enzyme-substrate complex,** or **ES complex,** breaks up, the product is released; the original enzyme molecule is regenerated and is free to form a new ES complex.

$$\text{Enzyme + Substrate(s)} \longrightarrow \text{ES complex}$$
$$\text{ES complex} \longrightarrow \text{Enzyme + Product(s)}$$

The enzyme itself is not permanently altered or consumed by the reaction and can be reused.

As shown in Figure 6–13*a*, every enzyme contains one or more **active sites,** regions to which the substrate binds, forming the ES complex. The active sites of some enzymes are grooves or cavities in the enzyme molecule, formed by amino acid side chains. The active sites of most enzymes are located close to the surface. During the course of a reaction, substrate molecules occupying these sites are brought close together and react with one another.

The shape of the enzyme does not seem to be exactly complementary to that of the substrate. When the substrate binds to the enzyme molecule, it causes a change, known as **induced fit,** in the shape of the enzyme molecule. Usually the shape of the substrate also changes slightly, in a way that may distort its chemical bonds (Fig. 6–13*b*). The proximity and orientation of the reactants, together with strains in their chemical bonds, facilitate the breakage of old bonds and the formation of new ones. Thus the substrate is changed into product, which moves away from the enzyme. The enzyme is then free to catalyze the reaction of more substrate molecules to form more product molecules.

Most enzyme names end in *-ase*

Enzymes are usually named by the addition of the suffix *-ase* to the name of the substrate. The enzyme sucrase, for example, splits sucrose into glucose and fructose. A few enzymes retain traditional names that do not end in *-ase;* some of these end in *-zyme.* For example, lysozyme (from the Greek *lysis,* "to dissolve") is an enzyme found in tears and saliva; this enzyme breaks down bacterial cell walls. Other examples of enzymes with traditional names include pepsin and trypsin, which break internal peptide bonds in proteins.

Enzymes are specific

Virtually every chemical reaction that takes place in an organism is catalyzed by an enzyme. Because there is a close relationship between the shape of the active site and the shape of the substrate, the majority of enzymes are highly specific. Most are capable of catalyzing only a few closely related chemical reactions or, in many cases, only one particular reaction. For example, the enzyme urease, which decomposes urea to ammonia and carbon dioxide, attacks no other substrate. The enzyme sucrase splits only sucrose; it does not act on other disaccharides such as maltose or lactose.

A few enzymes are specific only to the extent that they require the substrate to have a certain kind of chemical bond. For example, lipase secreted by the pancreas splits the ester linkages connecting the glycerol and fatty acids of a wide variety of fats.

Enzymes that catalyze similar reactions are classified into groups, although each particular enzyme in the group may catalyze only one specific reaction. Some of the important classes of enzymes and their roles are listed in Table 6–1. Each class is divided into many subclasses. For example, sucrase, mentioned above, is referred to as a glycosidase because it cleaves a glycosidic linkage. Glycosidases are a subclass of the hydrolases.

TABLE 6-1 Some Important Classes of Enzymes

Enzyme Class	Function
Oxidoreductases	Catalyze oxidation-reduction reactions
Transferases	Catalyze the transfer of a functional group from a donor molecule to an acceptor molecule
Hydrolases	Catalyze hydrolysis reactions
Isomerases	Catalyze conversion of a molecule from one isomeric form to another
Ligases	Catalyze certain reactions in which two molecules are joined
Lyases	Catalyze certain reactions in which double bonds are formed or broken.

centration. Any departure from optimal conditions adversely affects enzyme activity.

Each enzyme has an optimal temperature

Most enzymes have an optimal temperature at which the rate of reaction is fastest. For human enzymes, the temperature optima are near body temperature (35° to 40°C). Enzymatic reactions occur slowly or not at all at low temperatures. As the temperature increases, molecular motion increases, resulting in more molecular collisions. The rates of most enzyme-controlled reactions therefore increase as the temperature increases, within limits (Fig. 6–14a). High temperatures rapidly denature most enzymes; the molecular conformation of the protein

Many enzymes require cofactors

Some enzymes, for example, pepsin, which is secreted by the stomach, consist only of protein. Others have two components: a protein referred to as the **apoenzyme** and an additional chemical component called a **cofactor.** Neither the apoenzyme nor the cofactor alone has catalytic activity; only when the two are combined does the enzyme function. A cofactor may be inorganic, or it may be an organic molecule.

Some enzymes require a specific metal ion as a cofactor. Two very common inorganic cofactors are magnesium ions and calcium ions. Most of the trace elements, such as iron, copper, zinc, and manganese, all of which are required in very small amounts, function as cofactors.

An organic, nonpolypeptide compound that binds to the apoenzyme and serves as a cofactor is called a **coenzyme.** Most coenzymes are carrier molecules that transfer electrons or part of a substrate from one molecule to another. Some examples of coenzymes have already been introduced in this chapter. NADH, NADPH, and $FADH_2$ are coenzymes; they transfer electrons. ATP functions as a coenzyme; it is responsible for transferring phosphate groups. Yet another coenzyme, **coenzyme A,** is involved in the transfer of groups derived from organic acids. Most vitamins, which are organic compounds that an organism requires in small amounts but cannot synthesize itself, are coenzymes or components of coenzymes (see Table 45–4).

Enzymes are most effective at optimal conditions

Enzymes generally work best under certain narrowly defined conditions, such as appropriate temperature, pH, and ion con-

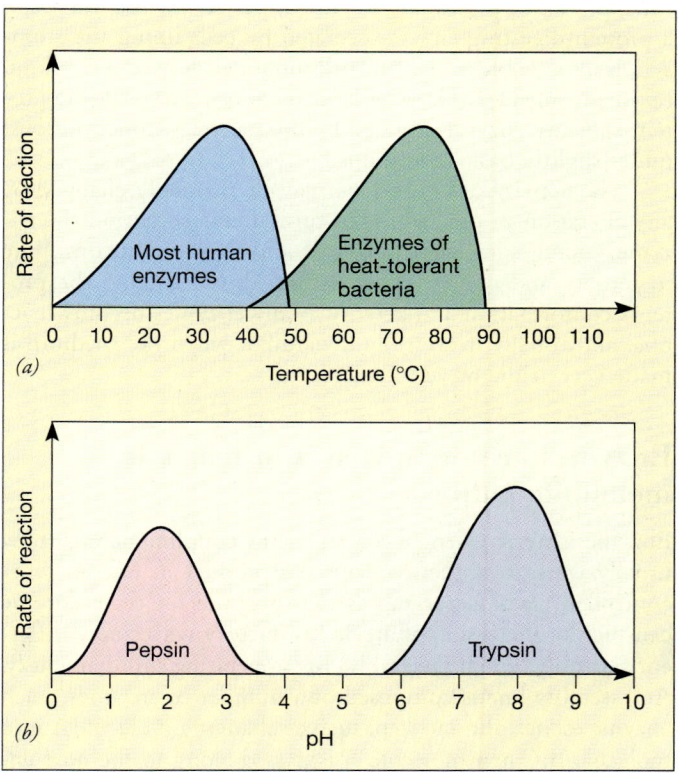

Figure 6-14 Effect of temperature and pH on enzyme activity. Substrate and enzyme concentrations are held constant in the reactions illustrated. *(a)* Generalized curves for the effect of temperature on enzyme activity. As temperature increases, enzyme activity increases until it reaches an optimal temperature. Enzyme activity abruptly falls after it exceeds the optimal temperature because the enzyme, being a protein, denatures. *(b)* Enzyme activity is very sensitive to pH. Pepsin is a protein-digesting enzyme in the very acidic stomach juice. Trypsin, secreted by the pancreas into the slightly alkaline small intestine, digests polypeptides.

becomes altered as the hydrogen bonds responsible for its secondary, tertiary, and quaternary structures are broken. Because this inactivation is usually not reversible, activity is not regained when the enzyme is cooled.

Most organisms are killed by even a short exposure to high temperature; their enzymes are denatured, and they are unable to continue metabolism. A few remarkable exceptions to this rule exist: certain species of bacteria can survive in the waters of hot springs, such as those in Yellowstone Park, where the temperature is almost 100°C; these organisms are responsible for the brilliant colors in the terraces of the hot springs. Still other bacteria live at temperatures much above that of boiling water, near deep-sea vents, where the extreme pressure keeps water in its liquid state (see Chapter 23 and *Focus On: Life Without the Sun* in Chapter 52).

Each enzyme has an optimal pH

Most enzymes are active only over a narrow pH range and have an optimal pH at which the rate of reaction is fastest. The optimal pH for most human enzymes is between 6 and 8. Pepsin, a protein-digesting enzyme secreted by cells lining the stomach, is remarkable in that it works only in a very acid medium, optimally at pH 2 (Fig. 6–14b). In contrast, trypsin, the protein-splitting enzyme secreted by the pancreas, functions best under slightly basic conditions.

The activity of an enzyme may be markedly changed by any alteration in pH, which in turn alters charges on the enzyme. Changes in charge affect the ionic bonds that contribute to tertiary and quaternary structure, thus changing the protein's conformation and activity. Many enzymes become inactive, and usually irreversibly denatured, when the medium is made very acidic or very basic.

Enzymes are organized into teams in metabolic pathways

Enzymes play an essential role in energy coupling because they usually work in sequence, with the product of one enzyme-controlled reaction serving as the substrate for the next. We can picture the inside of a cell as a factory with many different assembly (and disassembly) lines operating simultaneously. An assembly line is composed of a number of enzymes. Each enzyme carries out one step, such as changing molecule A into molecule B. Then molecule B is passed along to the next enzyme, which converts it into molecule C, and so on. Such a series of reactions is referred to as a **metabolic pathway.**

$$\text{A} \xrightarrow{\text{Enzyme 1}} \text{B} \xrightarrow{\text{Enzyme 2}} \text{C}$$

Each of these reactions is theoretically reversible, and the fact that it is catalyzed by an enzyme does not change that fact. An enzyme does not itself determine the direction of the reaction it catalyzes. However, the overall reaction sequence is portrayed as proceeding from left to right. You will recall that if there is little intrinsic free energy difference between the re-

actants and products for a particular reaction, its direction will be determined mainly by the relative concentrations of reactants and products.

In biological pathways, both intermediate and final products are often removed and converted to other chemical compounds. Such removal drives the sequence of reactions in a particular direction. Let us assume that Reactant A is being constantly supplied and that its concentration remains constant. Enzyme 1 converts Reactant A to Product B. The concentration of B is always lower than the concentration of A because B is removed as it is converted to C in the reaction catalyzed by Enzyme 2. If C is removed as quickly as it is formed (perhaps by leaving the cell), the entire pathway is "pulled" toward C.

The cell regulates enzymatic activity

Enzymes regulate the chemistry of the cell, but what controls the enzymes? One mechanism depends simply on controlling the amount of enzyme produced. The synthesis of each type of enzyme is directed by a specific gene. The gene, in turn, may be switched on by a signal from a hormone or by some other type of cellular product. When the gene is switched on, the enzyme is synthesized. The amount of enzyme present then influences the rate of the reaction.

If the pH and temperature are kept constant, the rate of the reaction can be affected by the substrate concentration or by the enzyme concentration. If an excess of substrate is present, the enzyme concentration is the rate-limiting factor. The initial rate of the reaction is then directly proportional to the concentration of enzyme present (Fig. 6–15a).

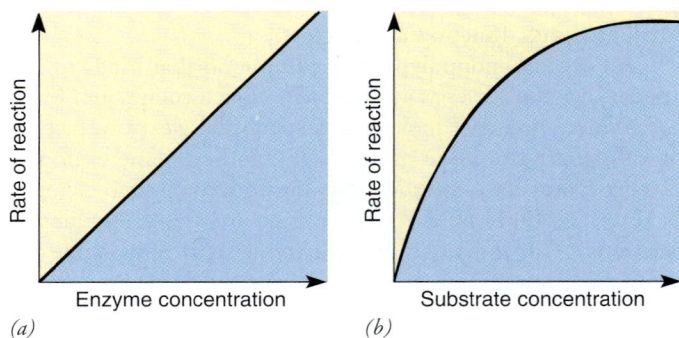

(a)   (b)

Figure 6–15 **Effect of enzyme concentration and substrate concentration on the rate of a reaction.** *(a)* In this example, the rate of reaction is measured at different enzyme concentrations, and an excess of substrate is present at all times. The rate of the reaction is therefore directly proportional to the enzyme concentration. *(b)* In this example, the rate of reaction is measured at different substrate concentrations, and enzyme concentration is constant. If the substrate concentration is relatively low, then the reaction rate is directly proportional to substrate concentration. However, higher substrate concentrations do not increase the reaction rate because the enzyme molecules become saturated with substrate.

If the enzyme concentration is kept constant, the initial rate of an enzymatic reaction is proportional to the concentration of substrate present. Substrate concentration is the rate-limiting factor at lower concentrations; the rate of the reaction is therefore directly proportional to the substrate concentration. However, at higher substrate concentrations the enzyme molecules become saturated with substrate, and increasing the substrate concentration does not increase the reaction rate (Fig. 6–15b).

The product of one enzymatic reaction may control the activity of another enzyme, especially in a complex sequence of enzymatic reactions. For example, in the following metabolic pathway,

$$A \xrightarrow{\text{Enzyme 1}} B \xrightarrow{\text{Enzyme 2}} C \xrightarrow{\text{Enzyme 3}} D \xrightarrow{\text{Enzyme 4}} E$$

each step is catalyzed by a different enzyme, and the final product E may inhibit the activity of Enzyme 1. When the concentration of E is low, the sequence of reactions proceeds rapidly. However, an increasing concentration of E serves as a signal for Enzyme 1 to slow down and eventually to stop functioning. Inhibition of Enzyme 1 stops the entire reaction sequence. This type of enzyme regulation, in which the formation of a product inhibits an earlier reaction in the sequence, is called **feedback inhibition** (Fig. 6–16).

Another important method of enzymatic control depends on the activation of enzyme molecules. In their inactive form the active sites of the enzyme are inappropriately shaped, so that the substrates do not fit. Among the factors that influence the shape of the enzyme are pH, the concentration of certain ions, and the addition of phosphate groups to certain amino acids in the enzyme.

Some enzymes possess a receptor site, called an **allosteric site,** on some region of the enzyme molecule other than the active site. (The word *allosteric* means "another space.") Substances that affect enzyme activity by binding to allosteric sites are called **allosteric regulators.** Some allosteric regulators are inhibitors that keep the enzyme in its inactive shape. Other allosteric regulators are activators that result in an enzyme with a functional active site.

The enzyme cyclic AMP-dependent protein kinase is an

Figure 6–16 Feedback inhibition. Bacteria synthesize the amino acid isoleucine from the amino acid threonine. The isoleucine synthetic pathway involves five steps, each catalyzed by a different enzyme. When enough isoleucine accumulates in the cell, the isoleucine inhibits the enzyme that catalyzes the first step in this pathway.

allosteric enzyme with a regulator that is a protein that binds reversibly to the allosteric site and inactivates the enzyme. Protein kinase is in this inactive form most of the time (Fig. 6–17). When protein kinase activity is needed, the compound cyclic AMP (cAMP; see Fig. 3–25) contacts the enzyme-inhibitor complex and removes the inhibitory protein, thereby activating the protein kinase. Activation of protein kinases by cAMP is an important aspect of the mechanism of action of certain hormones (see Chapter 47).

Enzymes can be inhibited by certain chemical agents

Most enzymes may be inhibited or even destroyed by certain chemical agents. Enzyme inhibition may be reversible or irreversible. **Reversible inhibition** occurs when an inhibitor forms

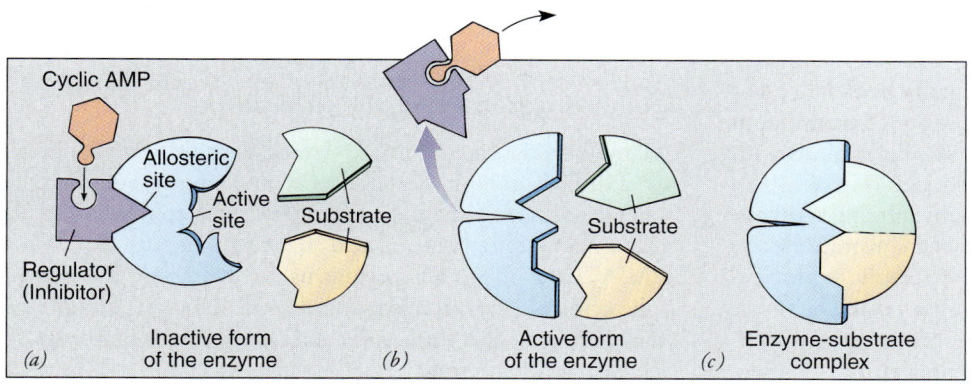

Figure 6–17 Allosteric enzyme.
(a) The enzyme protein kinase is inhibited by a regulatory protein that binds reversibly to its allosteric site. When the enzyme is in this inactive form, the shape of the active site is modified so that the substrate cannot combine with it. *(b)* Cyclic AMP removes the allosteric inhibitor and activates the enzyme. The substrate can then combine with the active site *(c).*

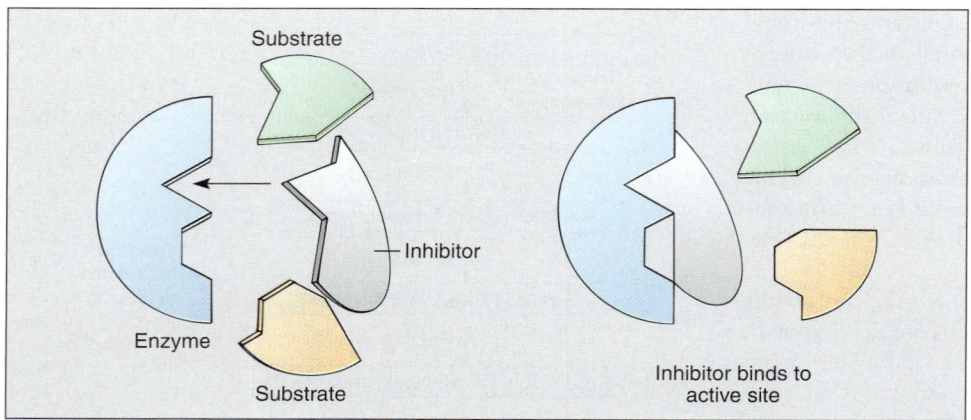

Figure 6–18 Competitive and noncompetitive inhibition. *(a)* In competitive inhibition, the inhibitor competes with the normal substrate for the active site of the enzyme. A competitive inhibitor occupies the active site only temporarily. *(b)* In noncompetitive inhibition, the inhibitor binds with the enzyme at a site other than the active site, altering the shape of the enzyme and thereby inactivating it. Noncompetitive inhibition may be reversible.

(a)   Competitive inhibition

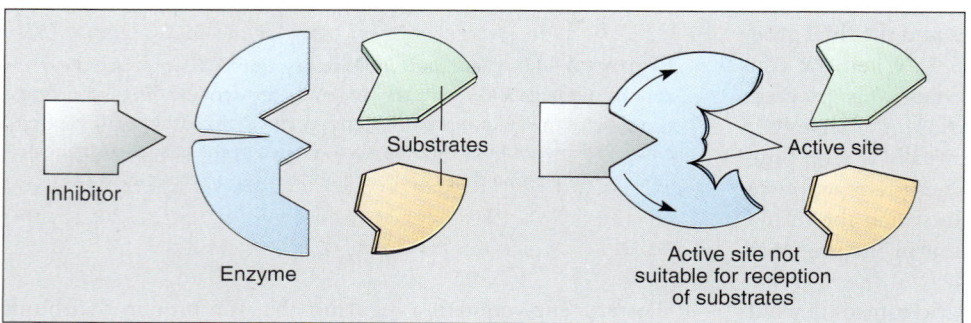

(b)   Noncompetitive inhibition

weak chemical bonds with the enzyme. Reversible inhibition can be competitive or noncompetitive.

In **competitive inhibition,** the inhibitor competes with the normal substrate for binding to the active site of the enzyme (Fig. 6–18*a*). Usually a competitive inhibitor is structurally similar to the normal substrate and so fits into the active site and combines with the enzyme. However, it is not similar enough to substitute fully for the normal substrate in the chemical reaction, and the enzyme cannot attack it to form product molecules. A competitive inhibitor occupies the active site only temporarily and does not permanently damage the enzyme. In competitive inhibition, an active site is occupied by the inhibitor part of the time and by the normal substrate part of the time. If the concentration of the substrate is increased relative to the concentration of the inhibitor, the active site will usually be occupied by the substrate. Competitive inhibition is demonstrated experimentally by the fact that it can be reversed by increasing the substrate concentration.

In **noncompetitive inhibition,** the inhibitor binds with the enzyme at a site other than the active site (Fig. 6–18*b*). Such an inhibitor inactivates the enzyme by altering its shape so that the active site cannot bind with the substrate. Many important noncompetitive inhibitors are metabolic substances that regulate enzyme activity by combining reversibly with the enzyme. Noncompetitive inhibition has some features in common with allosteric inhibition discussed previously.

In **irreversible inhibition,** an inhibitor permanently inactivates or destroys an enzyme when it combines with one of its functional groups. Many poisons are irreversible enzyme inhibitors. For example, heavy metals such as mercury and lead bind irreversibly to and denature many proteins, including enzymes. Certain nerve gases poison the enzyme acetylcholinesterase, which is important to the function of nerves and muscles. Cytochrome oxidase, one of the enzymes that transports electrons in cellular respiration, is especially sensitive to cyanide. Death results from cyanide poisoning because cytochrome oxidase is irreversibly inhibited and can no longer transfer electrons from substrate to oxygen. A number of insecticides and drugs are irreversible enzyme inhibitors. Irreversible inhibition may also occur if a protein is denatured by heat or organic solvents.

Some drugs are enzyme inhibitors

Many bacterial infections are treated with drugs that directly or indirectly inhibit bacterial enzyme activity. For example, sulfa drugs have a chemical structure similar to that of the nutrient para-aminobenzoic acid (PABA) (Fig. 6–19). When PABA is available, microorganisms can synthesize the vitamin folic acid, which is necessary for growth. Humans do not synthesize folic acid from PABA, and that is why sulfa drugs selectively affect bacteria.

Para-aminobenzoic acid (PABA)

Generic sulfonamide (Sulfa drug)

Figure 6–19 **Para-aminobenzoic acid and sulfonamides.** Sulfa drugs owe their antibiotic properties to their similarity in structure to para-aminobenzoic acid, a precursor in the synthesis of folic acid. Sulfa drugs block the synthesis of folic acid, an important vitamin necessary for growth. Animals, including humans, obtain folic acid in their diets, but many bacteria synthesize it.

When a sulfa drug is present, competitive inhibition occurs within the bacterium: the drug competes with PABA for the active site of the bacterial enzyme. When bacteria use the sulfa drug instead of PABA, they synthesize a compound that cannot be used to make folic acid. Therefore, the bacterial cells are unable to grow.

Penicillin and related antibiotics irreversibly inhibit a bacterial enzyme called transpeptidase. This enzyme is responsible for establishing some of the chemical linkages in the bacterial cell wall. Susceptible bacteria cannot produce properly constructed cell walls and are prevented from multiplying effectively. Human cells do not have cell walls and do not employ this enzyme. Thus, except for individuals allergic to it, penicillin is harmless to humans. Unfortunately, during the years since it was introduced, resistance to penicillin has evolved in many bacterial strains. The resistant bacteria fight back with an enzyme of their own, penicillinase, which breaks down the penicillin and renders it ineffective. Because bacteria evolve at such a rapid rate, drug resistance is a growing problem in medical practice (see *Making the Connection: Tuberculosis, Bacterial Resistance to Antibiotics, and Evolution* in Chapter 17). Although new antibacterial drugs are constantly under development, certain serious infections, such as tuberculosis, are becoming increasingly difficult to treat.

SUMMARY WITH KEY TERMS

I. **Energy** can be defined as the capacity to do work (expressed in **kilojoules, kJ**).
   A. All life depends on a continuous input of energy. Most producers capture energy during photosynthesis and incorporate some of it into the chemical bonds of organic compounds. Some of this chemical energy then becomes available to consumers and decomposers.
   B. All forms of energy are interconvertible.
      1. **Potential energy** is stored energy; **kinetic energy** is energy of motion.
      2. Energy can be conveniently measured as **heat energy;** the unit of heat energy is the **kilocalorie (kcal),** which is equal to 4.184 kilojoules. Heat energy cannot do cellular work.

II. The **first law of thermodynamics** states that energy cannot be created or destroyed but can be transferred and changed in form. The **second law of thermodynamics** states that disorder (**entropy**) in the universe is continuously increasing.
   A. The first law explains why organisms cannot produce energy but must continuously capture it from the surroundings.
   B. The second law explains why no process requiring energy is ever 100% efficient. In every energy transaction, some energy is dissipated as heat, which contributes to entropy.

III. When a chemical reaction is in a state of **dynamic equilibrium,** the rate of change in one direction is exactly the same as the rate of change in the opposite direction; the system can do no work because the **free energy** difference between the reactants and products is zero.
   A. As entropy (S) increases, the amount of free energy (G) decreases, as shown in the equation $G = H - TS$, in which G is the free energy, H is the **enthalpy** (total potential energy of the system), T is the absolute temperature (expressed in degrees Kelvin), and S is entropy.
   B. The equation $\Delta G = \Delta H - T\Delta S$ indicates that the change in free energy (ΔG) during a chemical reaction is equal to the change in enthalpy (ΔH) minus the product of the absolute temperature (T) multiplied by the change in entropy (ΔS).

IV. A **spontaneous reaction** releases free energy and can perform work.
   A. Free energy decreases in an **exergonic reaction.** Exergonic reactions are spontaneous.
   B. Free energy increases in an **endergonic reaction.** The input of free energy required to drive an endergonic reaction may be supplied by **coupling** it to an exergonic reaction.

V. **Adenosine triphosphate (ATP)** is the immediate energy currency of the cell; it generally transfers energy through the transfer of its terminal phosphate group to acceptor molecules.
   A. ATP is formed by the **phosphorylation** of **ADP,** an endergonic process that requires an input of energy.
   B. ATP is the common cellular link between exergonic and endergonic reactions and between **catabolism** and **anabolism.**

VI. Energy can be transferred in **oxidation-reduction (redox) reactions.**
   A. A substance that becomes oxidized gives up one or more electrons (and energy) to a substance that becomes reduced. Electrons are typically transferred as part of hydrogen atoms.
   B. **NAD$^+$** and **NADP$^+$** accept electrons as part of hydrogen atoms and become reduced to form NADH and NADPH, respectively. These electrons (along with some of their energy) can be transferred to other acceptors.

VII. An **enzyme** is a biological **catalyst;** it greatly increases the speed of a chemical reaction without being consumed.
   A. An enzyme lowers the **activation energy** necessary to get a reaction going.
   B. An **active site** of an enzyme is a three-dimensional region where **substrates** come into close contact and thereby react more readily. A substrate binds to an active site, causing an **induced fit** in which the shape of the enzyme changes slightly.
   C. Some enzymes consist of an **apoenzyme** and a **cofactor.**
      1. Most inorganic cofactors are metal ions.
      2. A **coenzyme** is an organic cofactor; many coenzymes transfer electrons or part of a substrate from one molecule to another.
   D. Enzymes work best at specific temperature and pH conditions.

E. A cell can regulate enzymatic activity by controlling the amount of enzyme produced and by regulating metabolic conditions that influence the shape of the enzyme.
   1. Some enzymes have **allosteric sites,** noncatalytic sites to which a substance can bind, changing the enzyme's activity.
   2. Allosteric enzymes are subject to **feedback inhibition,** in which the formation of an end product inhibits an earlier reaction in the sequence.
F. Most enzymes can be inhibited by certain chemical substances. Inhibition may be reversible or irreversible.
   1. **Reversible inhibition** occurs when an inhibitor forms weak chemical bonds with the enzyme. Reversible inhibition may be **competitive,** in which the inhibitor competes with the substrate for the active site, or **noncompetitive,** in which the inhibitor binds with the enzyme at a site other than the active site.
   2. **Irreversible inhibition** occurs when an inhibitor combines with an enzyme and permanently inactivates it.

POST-TEST

1. According to the first law of thermodynamics (a) energy may be changed from one form to another but is neither created nor destroyed (b) much of the work an organism does is mechanical work (c) the disorder of the universe is increasing (d) free energy is available to do cellular work (e) a cell is in a state of dynamic equilibrium
2. According to the second law of thermodynamics (a) energy may be changed from one form to another but is neither created nor destroyed (b) much of the work an organism does is mechanical work (c) the disorder of the universe is increasing (d) free energy is available to do cellular work (e) a cell is in a state of dynamic equilibrium
3. In thermodynamics, _____ is a measure of the amount of disorder in the system. (a) bond energy (b) catabolism (c) entropy (d) enthalpy (e) work
4. The _____ of a system is that part of the total energy available to do cellular work. (a) activation energy (b) bond energy (c) kinetic energy (d) free energy (e) heat energy
5. A reaction that requires a net input of free energy is described as (a) exergonic (b) endergonic (c) spontaneous (d) both a and c (e) both b and c
6. A reaction that releases energy is described as (a) exergonic (b) endergonic (c) spontaneous (d) both a and c (e) both b and c
7. A spontaneous reaction is one in which the change in free energy (ΔG) has a _____ value. (a) positive (b) negative (c) positive or negative (d) none of these (ΔG has no value)
8. To drive a reaction that requires an input of energy (a) an enzyme-substrate complex must form (b) the concentration of ATP must be decreased (c) the activation energy must be increased (d) some reaction that yields energy must be coupled to it (e) some reaction that requires energy must be coupled to it
9. The energy required to initiate a reaction is called (a) activation energy (b) bond energy (c) kinetic energy (d) free energy (e) heat energy
10. A biological catalyst that affects the rate of a chemical reaction without being consumed by the reaction is a(an) (a) product (b) cofactor (c) coenzyme (d) substrate (e) enzyme
11. The region of an enzyme molecule that combines with the substrate is the (a) allosteric site (b) reactant (c) active site (d) coenzyme (e) product
12. This inhibitor binds to the active site of an enzyme. (a) noncompetitive inhibitor (b) competitive inhibitor (c) irreversible inhibitor (d) allosteric regulator
13. Which of the following reactions could be coupled to an endergonic reaction with $\Delta G = +3.56$ kJ/mole? (a) A $\rightarrow$ B, $\Delta G = +6.08$ kJ/mole (b) C $\rightarrow$ D, $\Delta G = +3.56$ kJ/mole (c) E $\rightarrow$ F, $\Delta G = -1.22$ kJ/mole (d) G $\rightarrow$ H, $\Delta G = -5.91$ kJ/mole
14. In the following reaction series, which enzyme is most likely to have an allosteric site to which the end product E binds?

Enzyme 1   Enzyme 2   Enzyme 3   Enzyme 4

A $\longrightarrow$ B $\longrightarrow$ C $\longrightarrow$ D $\longrightarrow$ E

(a) enzyme 1 (b) enzyme 2 (c) enzyme 3 (d) enzyme 4

REVIEW QUESTIONS

1. Why can we express energy either in kilojoules or in kilocalories? Which is more convenient to measure? Which has more meaning in biology?
2. You exert tension on a spring and then release it. Explain how these actions relate to work, potential energy, and kinetic energy.
3. Life is sometimes described as a constant struggle against the second law of thermodynamics. How do organisms succeed in this struggle without violating the second law?
4. Consider the free energy change in a reaction in which enthalpy decreases and entropy increases. Is ΔG zero, or does it have a positive value or a negative value? Is the reaction endergonic or exergonic?
5. Why do coupled reactions typically have common intermediates? Give a generalized example involving ATP. Why is ATP able to serve as an important link between exergonic and endergonic reactions?
6. What is activation energy? What effect does an enzyme have on activation energy?
7. Give the function of each of the following (a) active site of an enzyme (b) coenzyme (c) allosteric site
8. Describe three factors that influence how an enzyme functions.
9. Label the diagram. Use Figure 6–12 to check your answers.

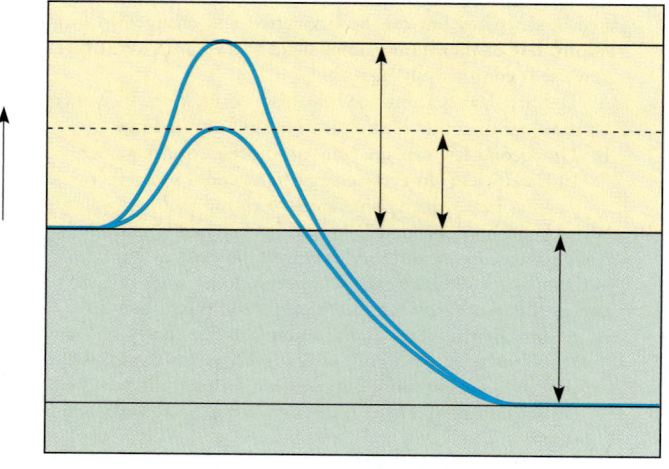

YOU MAKE THE CONNECTION

1. Reaction 1 and Reaction 2 happen to have the same free energy change: $\Delta G = -41.8$ kJ/mole (-10 kcal/mole). Reaction 1 is at equilibrium, but Reaction 2 is far from equilibrium. Is either reaction capable of performing work? If so, which one?
2. You are doing an experiment in which you are measuring the rate at which succinic acid is converted to fumaric acid by the enzyme succinic dehydrogenase. You decide to add a little malonic acid to make things interesting. You observe that the reaction rate slows markedly and conclude that malonic acid must be acting as an inhibitor. Design an experiment that will help you decide if malonic acid is acting as a competitive inhibitor or a noncompetitive inhibitor.
3. Based on what you have learned in this chapter, explain why an extremely high fever (above 105°F or 40°C) is often lethal.

RECOMMENDED READINGS

Adams, S. "No Way Back." *New Scientist,* 22 Oct. 1994. Examines the second law of thermodynamics.

Atkins, P.W. *The Second Law.* W.H. Freeman, San Francisco, 1984. A basic, understandable introduction to thermodynamics with an extensive section devoted to its biological implications.

Hinrichs, R.A. *Energy: Its Use and the Environment,* 2nd ed. Saunders College Publishing, Philadelphia, 1996. The focus of this introductory text is the physical principles behind energy use and its effects on the environment.

Tobin, A.J., and R.E. Morel. *Asking About Cells.* Saunders College Publishing, Philadelphia, 1997. A readable cell biology text with excellent coverage of cellular energetics.

• Visit our website at **http://www.saunderscollege.com/lifesci/titles.html** and click on Solomon/Berg/Martin Biology for links to chapter-related resources on the World Wide Web.

CHAPTER 7

How Cells Make ATP: Energy-Releasing Pathways

Cells are tiny factories that process materials on the molecular level, through thousands of metabolic reactions. Cells exist in a dynamic state and are continuously building up and breaking down the many different cellular constituents. As you learned in Chapter 6, metabolism has two complementary components: **catabolism,** which is the splitting of molecules into smaller components, and **anabolism,** the synthesis of complex molecules from simpler building blocks. Anabolic reactions in cells produce proteins, nucleic acids, lipids, polysaccharides, and other complex molecules that help to maintain the cell or the organism of which it is a part. Most anabolic reactions are endergonic and require ATP or some other energy source to drive them.

Every organism must extract energy from the organic food molecules that it either manufactures by photosynthesis or captures from the environment. The gerenuks in the photograph, for example, eat the leaves of thorny shrubs and trees to obtain energy. In gerenuks and other complex animals, food is first broken down by the digestive system. During digestion, proteins are split into their component amino acids, carbohydrates are digested to simple sugars, and fats are split into glycerol and fatty acids. These nutrients are then absorbed into the blood and transported to all the cells. Each cell converts the energy in the chemical bonds of nutrients to ATP energy in a process known as **cellular respiration.** (The term *cellular respiration* is used to distinguish these cellular processes from *organismic respiration,* the exchange of oxygen and carbon dioxide with the environment by animals that have special modifications, such as lungs or gills, for gas exchange.)

Cellular respiration may be either aerobic or anaerobic. **Aerobic** respiration requires molecular oxygen (O_2), whereas **anaerobic** pathways, which include anaerobic respiration and fermentation, do not require oxygen. Most cells use aerobic respiration, which is by far the most common pathway and the main subject of this chapter. All three pathways—aerobic respiration, anaerobic respiration, and fermentation—are exergonic and release free energy.

(Renee Lynn/Photo Researchers, Inc.)

AFTER YOU HAVE STUDIED THIS CHAPTER YOU SHOULD BE ABLE TO

1. Write a summary reaction for aerobic respiration, showing which reactant becomes oxidized and which becomes reduced.
2. List and give a brief overview of the four stages of aerobic respiration, indicate where each stage takes place in a eukaryotic cell, and add up the energy captured (as ATP, NADH, and FADH$_2$) in each stage.
3. Draw a diagram illustrating chemiosmosis and explain (1) how a gradient of protons is established across the inner mitochondrial membrane and (2) the process by which the proton gradient drives ATP synthesis.
4. Summarize how the products of protein and lipid catabolism enter the same metabolic pathway that oxidizes glucose.
5. Compare and contrast aerobic and anaerobic pathways used by cells to extract free energy from nutrients; include the mechanism of ATP formation, the final electron acceptor, and the end products.
6. Summarize the basic similarities of alcohol and lactate fermentation.

AEROBIC RESPIRATION IS A REDOX PROCESS

Most eukaryotes and prokaryotes use a form of cellular respiration requiring oxygen and hence carry out **aerobic respiration.** During aerobic respiration, nutrients are catabolized to carbon dioxide and water. Most cells of plants, animals, protists, fungi, and bacteria use aerobic respiration to obtain energy from glucose. The overall reaction pathway for the aerobic respiration of glucose is summarized as follows:

$$C_6H_{12}O_6 + 6\ O_2 + 6\ H_2O \longrightarrow$$
$$6\ CO_2 + 12\ H_2O + \text{Energy (as ATP)}$$

Note that water is shown on both sides of the equation; this is because it is a reactant in some reactions and a product in others.

For purposes of discussion, the equation for aerobic respiration can be simplified to indicate that there is a net yield of water:

$$\overbrace{C_6H_{12}O_6 + 6\ O_2 \longrightarrow 6\ \underbrace{CO_2 + 6\ H_2O}}^{\text{Oxidation}} + \text{Energy (as ATP)}$$

If we analyze this summary reaction, it appears that carbon dioxide is produced by the removal of hydrogen atoms from glucose. Conversely, water appears to be formed as the hydrogen atoms are accepted by oxygen. Because the transfer of hydrogen atoms is equivalent to the transfer of electrons, this is a redox process in which glucose is oxidized and oxygen is reduced (see Chapter 6).

The products of the reaction would be the same if the glucose were simply placed in a test tube and burned in the presence of oxygen. However, if cells were to burn glucose, its energy would be released all at once as heat, which would not only be unavailable to the cell but would actually destroy it. For this reason, cells do not transfer hydrogen atoms directly from glucose to oxygen. Aerobic respiration is a redox process in which electrons associated with the hydrogen atoms in glucose are transferred to oxygen in a series of about 30 steps (Fig. 7–1). During this process, the free energy of the electrons is used for ATP synthesis.

AEROBIC RESPIRATION HAS FOUR STAGES

The chemical reactions of the aerobic respiration of glucose can be grouped into four stages (Fig. 7–2 and Table 7–1). In eukaryotes, the first stage (glycolysis) takes place in the cytosol, and the rest take place inside the mitochondria. Most bacteria also carry out these processes, but because these cells lack mitochondria, the reactions of aerobic respiration occur in the cytosol and in association with the plasma membrane.

1. **Glycolysis.** A six-carbon glucose molecule is converted to two, three-carbon molecules of pyruvate,[1] and ATP and NADH are formed.[2]
2. **Formation of acetyl coenzyme A.** Each pyruvate enters a mitochondrion and is oxidized to a two-carbon group (acetate) that combines with coenzyme A, forming acetyl coenzyme A. NADH is produced and carbon dioxide is released as a waste product.
3. **The citric acid cycle.** The acetate group of acetyl coenzyme A combines with a four-carbon molecule (oxaloacetate) to form a six-carbon molecule (citrate). In the course of the cycle, citrate is recycled to oxaloacetate and carbon dioxide is released as a waste product. Energy is captured as ATP and the reduced, high-energy compounds NADH and FADH$_2$.
4. **The electron transport chain and chemiosmosis.** The electrons removed from glucose during the preceding stages are transferred from NADH and FADH$_2$ to a chain of electron acceptor compounds. As the electrons are passed from one electron acceptor to another, some of their energy is used to pump hydrogen ions (protons) across the inner mitochondrial membrane, forming a proton gradient. In a process known as chemiosmosis, to be described later, the energy of this proton gradient is used to produce ATP.

---

[1]Pyruvate and many other compounds in cellular respiration exist as anions at the pH found in the cell. They sometimes associate with H$^+$ to form acids. For example, pyruvate forms pyruvic acid. In some textbooks these compounds are presented in the acid form.

[2]Although the correct way to write the reduced form of NAD$^+$ is NADH + H$^+$, for simplicity we will present the reduced form as NADH throughout the chapter.

Figure 7–1 **Changes in free energy.** The release of energy from a glucose molecule is analogous to the liberation of energy by a falling object. The total energy released is the same whether it occurs all at once or in a series of steps.

Most reactions involved in aerobic respiration are one of three types: dehydrogenations, decarboxylations, and those that we will informally categorize as preparation reactions. **Dehydrogenations** are reactions in which two hydrogen atoms (actually, two electrons plus two protons) are removed from the substrate and transferred to NAD^+ or FAD. **Decarboxylations** are reactions in which part of a carboxyl group (—COOH) is removed from the substrate as a molecule of CO_2. The carbon dioxide we exhale with each breath is derived from decarboxylations that occur in our cells. The rest of the reactions are preparation reactions in which molecules undergo rearrangements and other changes so that they can subsequently undergo further dehydrogenations or decarboxylations. As we examine the individual reactions of aerobic respiration, we will encounter many examples of these three basic types.

In following the reactions of aerobic respiration, it helps to do some bookkeeping as you go along. Because glucose is the starting material, it is useful to express changes on a per

glucose basis. We will be paying particular attention to changes in the number of carbon atoms per molecule and to steps where some type of energy transfer takes place.

In glycolysis, glucose yields two pyruvates

Glycolysis (from Greek words meaning "sugar-splitting") does not require oxygen and can proceed under aerobic or anaerobic conditions. Figure 7–3 shows a simple summary of glycolysis, in which a glucose molecule comprising six carbons is converted to two molecules of pyruvate, a three-carbon molecule. Some of the energy in the glucose is captured; there is a net yield of two ATP molecules and two NADH molecules. The reactions of glycolysis take place in the cytosol, where the necessary reactants, such as ADP, NAD^+, and inorganic phosphates, float freely and are used as needed.

The glycolysis pathway consists of a series of reactions, each of which is catalyzed by a specific enzyme (Fig. 7–4).

(Text continues on p. 160)

Figure 7–2 **The four stages of aerobic respiration.** The first stage, glycolysis, occurs in the cytosol. Pyruvate, the product of glycolysis, enters a mitochondrion, where cellular respiration continues with the formation of acetyl coenzyme A, the citric acid cycle, and electron transport/chemiosmosis. Most ATP is synthesized by chemiosmosis.

TABLE 7–1 Summary of Aerobic Respiration

Phase	Summary	Some Starting Materials	Some End Products
1. Glycolysis (in cytosol)	Series of reactions in which glucose is degraded to pyruvate; net profit of 2 ATPs; hydrogen atoms are transferred to carriers; can proceed anaerobically	Glucose, ATP, NAD^+ ADP, P_i	Pyruvate, ATP, NADH
2. Formation of acetyl CoA (in mitochondria)	Pyruvate is degraded and combined with coenzyme A to form acetyl CoA; hydrogen atoms are transferred to carriers; CO_2 is released	Pyruvate, coenzyme A, NAD^+	Acetyl CoA, CO_2, NADH
3. Citric acid cycle (in mitochondria)	Series of reactions in which the acetyl portion of acetyl CoA is degraded to CO_2; hydrogen atoms are transferred to carriers; ATP is synthesized	Acetyl CoA, H_2O, NAD^+, FAD, ADP, P_i	CO_2, NADH, $FADH_2$, ATP
4. Electron transport and chemiosmosis (in mitochondria)	Chain of several electron transport molecules; electrons are passed along chain; energy released is used to form a proton gradient; ATP is synthesized as protons diffuse down the gradient; oxygen is final electron acceptor	NADH, $FADH_2$, oxygen, ADP, P_i	ATP, H_2O, NAD^+, FAD

Figure 7–3 Overview of glycolysis. The energy investment phase of glycolysis *(left column)* leads to the splitting of sugar; ATP and NADH are produced during the energy capture phase *(right column)*. During glycolysis, each glucose molecule is converted to two pyruvates, with a net yield of two ATP molecules and two NADH molecules.

GLYCOLYSIS

Energy investment phase and splitting of glucose

Glucose

3 steps → Two ATP / Two ADP

Fructose-1,6-bisphosphate

Glyceraldehyde phosphate (G3P)   Glyceraldehyde phosphate (G3P)

Energy capture phase

G3P   G3P

NAD^+ → NADH
Two ADP → Two ATP
5 steps

NAD^+ → NADH
Two ADP → Two ATP

Pyruvate   Pyruvate

Subtotal (per glucose)	− Two ATPs (invested)	Four ATPs Two NADH

| Net yield (per glucose) | Two ATPs and Two NADH ||

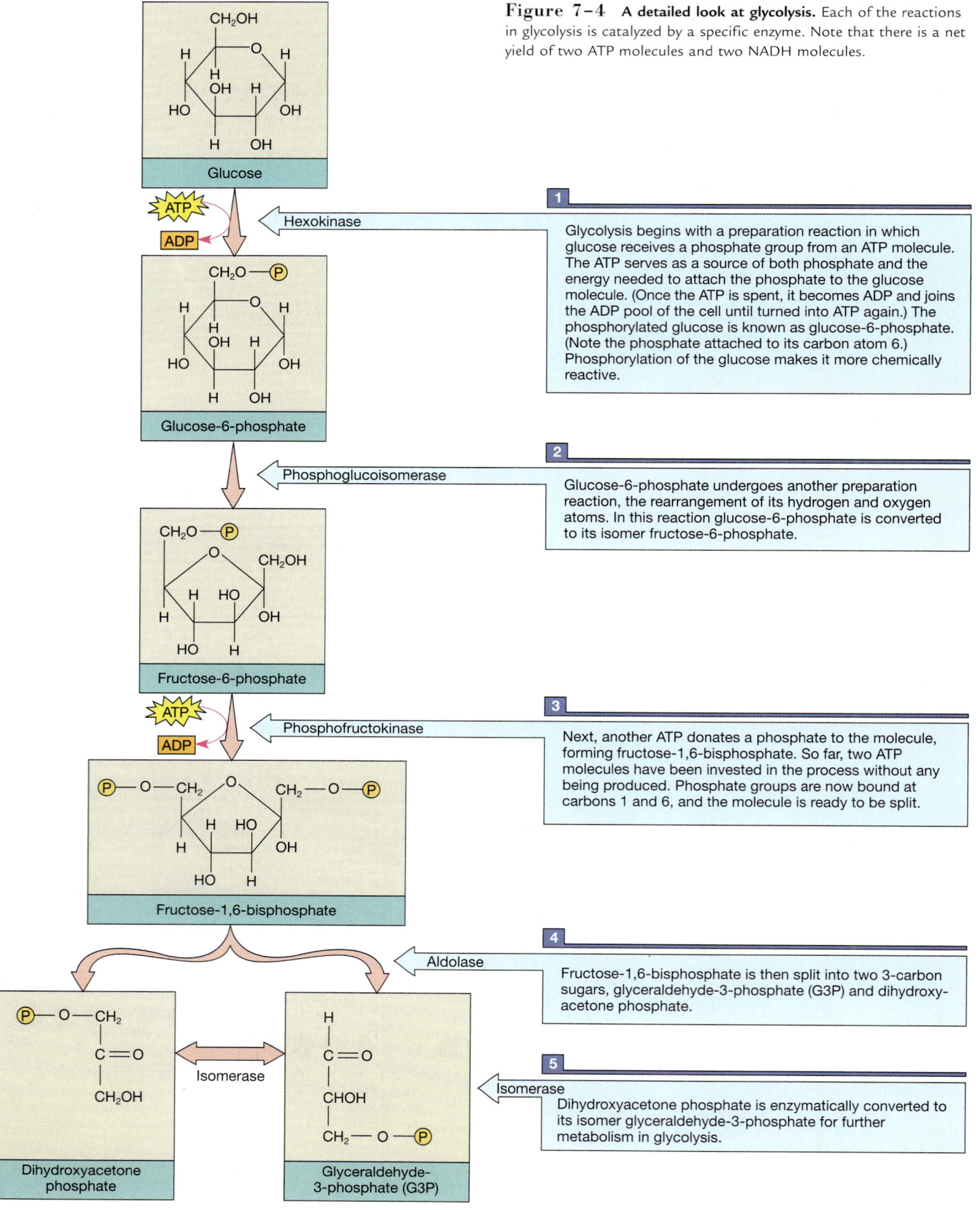

Figure 7–4 A detailed look at glycolysis. Each of the reactions in glycolysis is catalyzed by a specific enzyme. Note that there is a net yield of two ATP molecules and two NADH molecules.

1

Glycolysis begins with a preparation reaction in which glucose receives a phosphate group from an ATP molecule. The ATP serves as a source of both phosphate and the energy needed to attach the phosphate to the glucose molecule. (Once the ATP is spent, it becomes ADP and joins the ADP pool of the cell until turned into ATP again.) The phosphorylated glucose is known as glucose-6-phosphate. (Note the phosphate attached to its carbon atom 6.) Phosphorylation of the glucose makes it more chemically reactive.

2

Glucose-6-phosphate undergoes another preparation reaction, the rearrangement of its hydrogen and oxygen atoms. In this reaction glucose-6-phosphate is converted to its isomer fructose-6-phosphate.

3

Next, another ATP donates a phosphate to the molecule, forming fructose-1,6-bisphosphate. So far, two ATP molecules have been invested in the process without any being produced. Phosphate groups are now bound at carbons 1 and 6, and the molecule is ready to be split.

4

Fructose-1,6-bisphosphate is then split into two 3-carbon sugars, glyceraldehyde-3-phosphate (G3P) and dihydroxy-acetone phosphate.

5

Dihydroxyacetone phosphate is enzymatically converted to its isomer glyceraldehyde-3-phosphate for further metabolism in glycolysis.

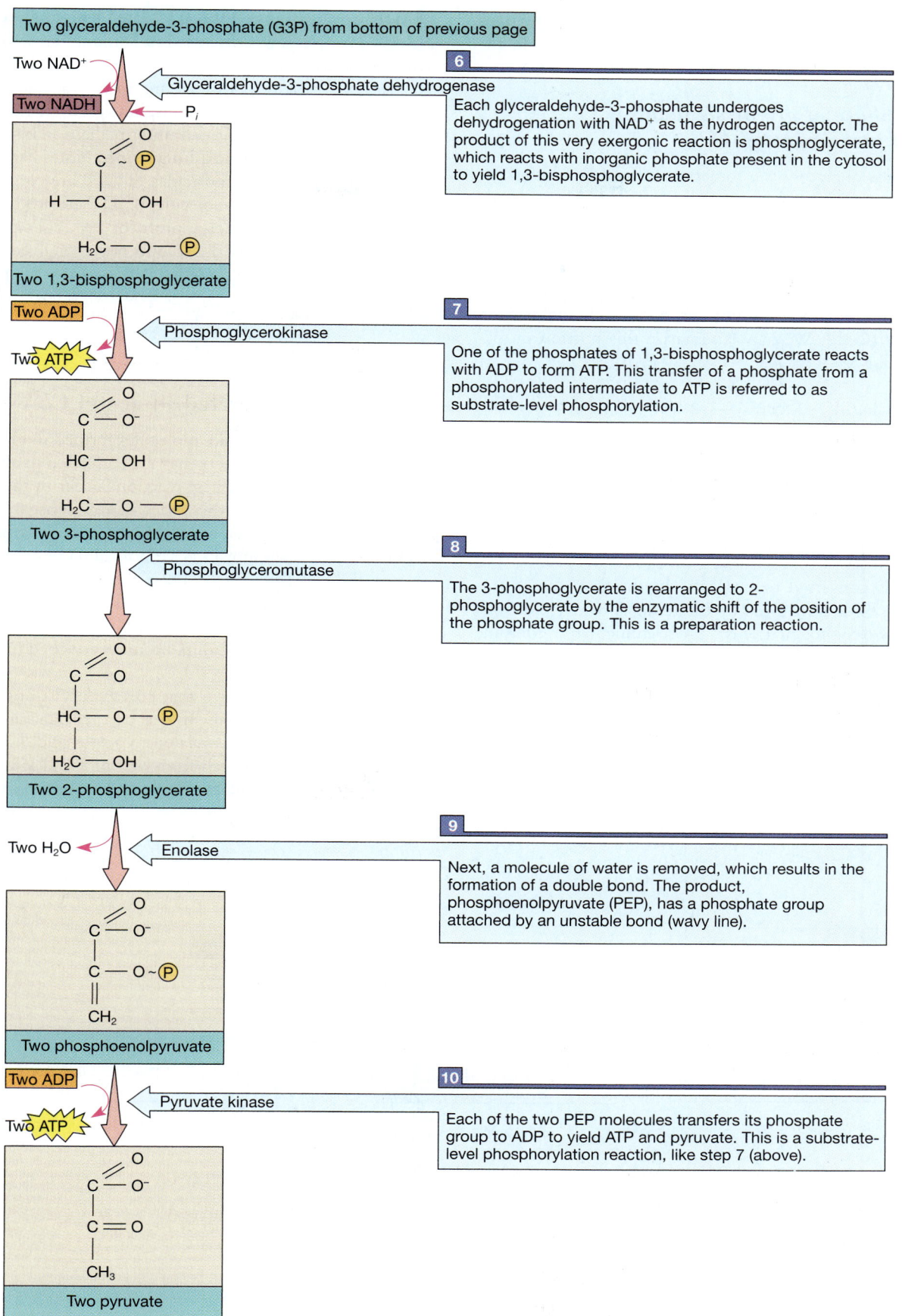

Two glyceraldehyde-3-phosphate (G3P) from bottom of previous page

Two NAD+

Two NADH

P_i

6

Glyceraldehyde-3-phosphate dehydrogenase

Each glyceraldehyde-3-phosphate undergoes dehydrogenation with NAD+ as the hydrogen acceptor. The product of this very exergonic reaction is phosphoglycerate, which reacts with inorganic phosphate present in the cytosol to yield 1,3-bisphosphoglycerate.

Two 1,3-bisphosphoglycerate

Two ADP

Two ATP

7

Phosphoglycerokinase

One of the phosphates of 1,3-bisphosphoglycerate reacts with ADP to form ATP. This transfer of a phosphate from a phosphorylated intermediate to ATP is referred to as substrate-level phosphorylation.

Two 3-phosphoglycerate

8

Phosphoglyceromutase

The 3-phosphoglycerate is rearranged to 2-phosphoglycerate by the enzymatic shift of the position of the phosphate group. This is a preparation reaction.

Two 2-phosphoglycerate

Two H_2O

9

Enolase

Next, a molecule of water is removed, which results in the formation of a double bond. The product, phosphoenolpyruvate (PEP), has a phosphate group attached by an unstable bond (wavy line).

Two phosphoenolpyruvate

Two ADP

Two ATP

10

Pyruvate kinase

Each of the two PEP molecules transfers its phosphate group to ADP to yield ATP and pyruvate. This is a substrate-level phosphorylation reaction, like step 7 (above).

Two pyruvate

Glycolysis is divided into two major phases: the first includes endergonic reactions that require ATP, while the second includes exergonic reactions that yield ATP and NADH.

The first phase of glycolysis requires an initial investment of ATP

The first phase of glycolysis is sometimes referred to as the "energy investment" phase. Glucose is a relatively stable molecule and is not easily broken down. In two separate **phosphorylation reactions,** a phosphate group is transferred from ATP to the sugar. The resulting phosphorylated sugar (fructose-1,6-bisphosphate) is less stable and is broken enzymatically into two molecules of a three-carbon compound: glyceraldehyde-3-phosphate (G3P). We may summarize this portion of glycolysis as follows:

$$\text{Glucose} + 2 \text{ ATP} \longrightarrow 2 \text{ G3P} + 2 \text{ ADP}$$

Six-carbon compound   Three-carbon compound

The second phase of glycolysis yields NADH and ATP

Each G3P is converted to pyruvate. In the first step of this process each G3P is oxidized by the removal of two electrons (as part of two hydrogen atoms). These immediately combine with the hydrogen carrier molecule, NAD^+:

$$NAD^+ + 2 \text{ H} \longrightarrow NADH + H^+$$

Oxidized   (From G3P)   Reduced

Because there are two G3P molecules for every glucose, two NADH are formed. The energy of the electrons carried by NADH can be used to form ATP later. The process by which this is accomplished is discussed in conjunction with the electron transport chain.

In two of the reactions leading to the formation of pyruvate, ATP is formed when a phosphate group is transferred to ADP from a phosphorylated intermediate. This process is called **substrate-level phosphorylation.** Note that in the first phase of glycolysis two molecules of ATP are consumed, but in the second phase four molecules of ATP are produced. Thus, glycolysis yields a net energy profit of *two* ATPs per glucose.

We may summarize the second phase of glycolysis as follows:

$$2 \text{ G3P} + 2 \text{ NAD}^+ + 4 \text{ ADP} \longrightarrow$$
$$2 \text{ Pyruvate} + 2 \text{ NADH} + 4 \text{ ATP}$$

Pyruvate is converted to acetyl CoA

In eukaryotes, the pyruvate molecules formed in glycolysis enter the mitochondria, where they are converted to **acetyl coenzyme A (acetyl CoA).** These reactions occur in the cytosol of aerobic prokaryotes. In this series of reactions, pyruvate undergoes a process known as oxidative decarboxylation. First, a carboxyl group is removed as carbon dioxide, which diffuses out of the cell (Fig. 7–5). Then the remaining two-carbon fragment is oxidized, and the electrons removed during the oxidation are accepted by NAD^+. Finally, the oxidized two-carbon fragment, an acetyl group, becomes attached to **coenzyme A,** yielding acetyl CoA.

Recall from Chapter 6 that coenzyme A transfers groups derived from organic acids. In this case, coenzyme A transfers an acetyl group, which is related to acetic acid. Coenzyme A is manufactured in the cell from one of the B vitamins, pantothenic acid. The attachment of an acetyl group to coenzyme

Figure 7–5 **Formation of acetyl coenzyme A.** Pyruvate, a three-carbon molecule that is the end product of glycolysis, enters the mitochondrion and undergoes oxidative decarboxylation. First, the carboxyl group is split off as carbon dioxide. Then the remaining two-carbon fragment is oxidized, and its electrons are transferred to NAD^+. Finally, the oxidized two-carbon group, an acetyl group, is attached to coenzyme A. Coenzyme A has a sulfur atom that forms a very unstable bond, shown as a wavy line, with the acetyl group.

A is catalyzed by a multienzyme complex that contains several copies of each of three different enzymes. The overall reaction for the formation of acetyl coenzyme A is the following:

$$2 \text{ Pyruvate} + 2 \text{ NAD}^+ + 2 \text{ CoA} \longrightarrow$$
$$2 \text{ Acetyl CoA} + 2 \text{ NADH} + 2 \text{ CO}_2$$

Note that the original glucose molecule has now been partially oxidized, yielding two acetyl groups and two CO_2 molecules. The electrons removed have reduced NAD^+ to NADH. At this point in aerobic respiration, four NADH molecules have been formed as a result of the catabolism of a single glucose molecule: two during glycolysis and two during the formation of acetyl CoA from pyruvate.

The citric acid cycle oxidizes acetyl CoA

The **citric acid cycle** is also known as the **tricarboxylic acid (TCA) cycle** and as the **Krebs cycle** after Hans Krebs, who assembled the accumulated contributions of many scientists and worked out the details of the cycle in the 1930s. A simple summary of the citric acid cycle, which takes place in the mitochondria, is given in Figure 7–6. The eight steps of the citric acid cycle are shown in Figure 7–7. Each reaction is catalyzed by a specific enzyme.

The first reaction of the cycle occurs when acetyl CoA transfers its two-carbon acetyl group to the four-carbon acceptor compound **oxaloacetate,** forming **citrate,** a six-carbon compound:

$$\underset{\substack{\text{Four-carbon} \\ \text{compound}}}{\text{Oxaloacetate}} + \underset{\substack{\text{Two-carbon} \\ \text{compound}}}{\text{Acetyl CoA}} \longrightarrow \underset{\substack{\text{Six-carbon} \\ \text{compound}}}{\text{Citrate}} + \text{CoA}$$

The citrate then goes through a series of chemical transformations, losing first one and then a second carboxyl group as CO_2. Most of the energy made available by the oxidative steps of the cycle is transferred as energy-rich electrons to NAD^+, forming NADH. For each acetyl group that enters the citric acid cycle, three molecules of NADH are produced. Electrons

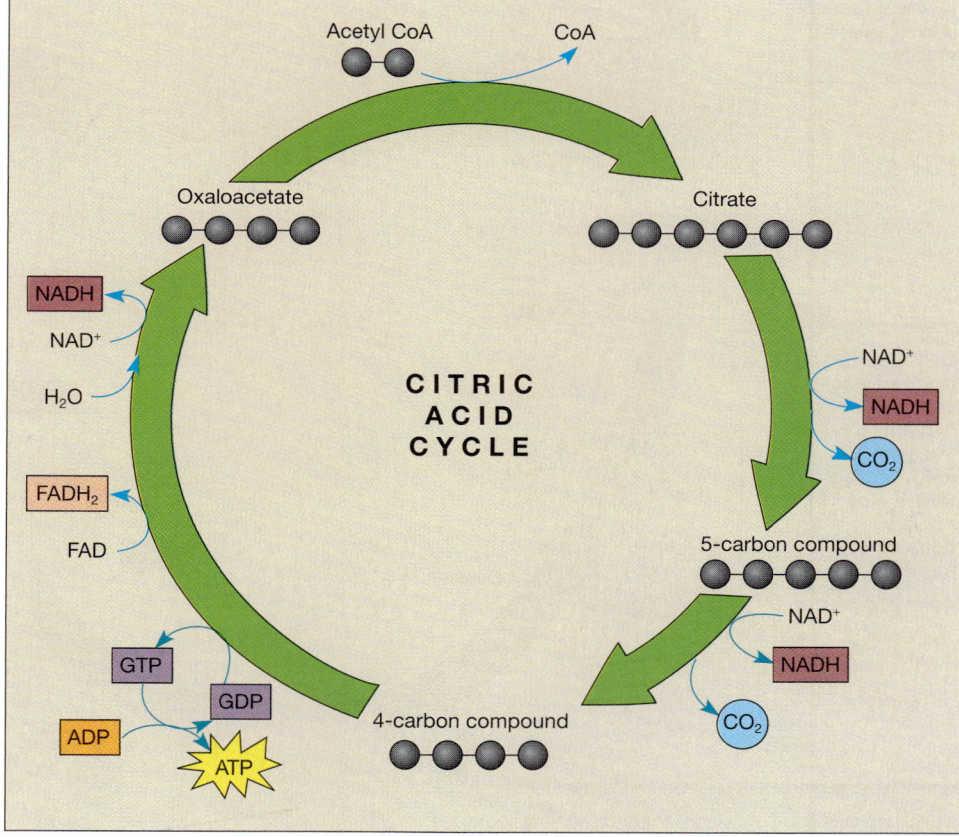

Figure 7–6 Overview of the citric acid cycle. Two acetyl groups enter the citric acid cycle for every glucose. Each two-carbon acetyl group combines with the four-carbon compound oxaloacetate to form the six-carbon compound citrate. Two CO_2 molecules are removed to regenerate oxaloacetate, and in the process, energy is captured as one ATP, three NADH, and one $FADH_2$ per acetyl group (or two ATPs, six NADH, and two $FADH_2$ per glucose).

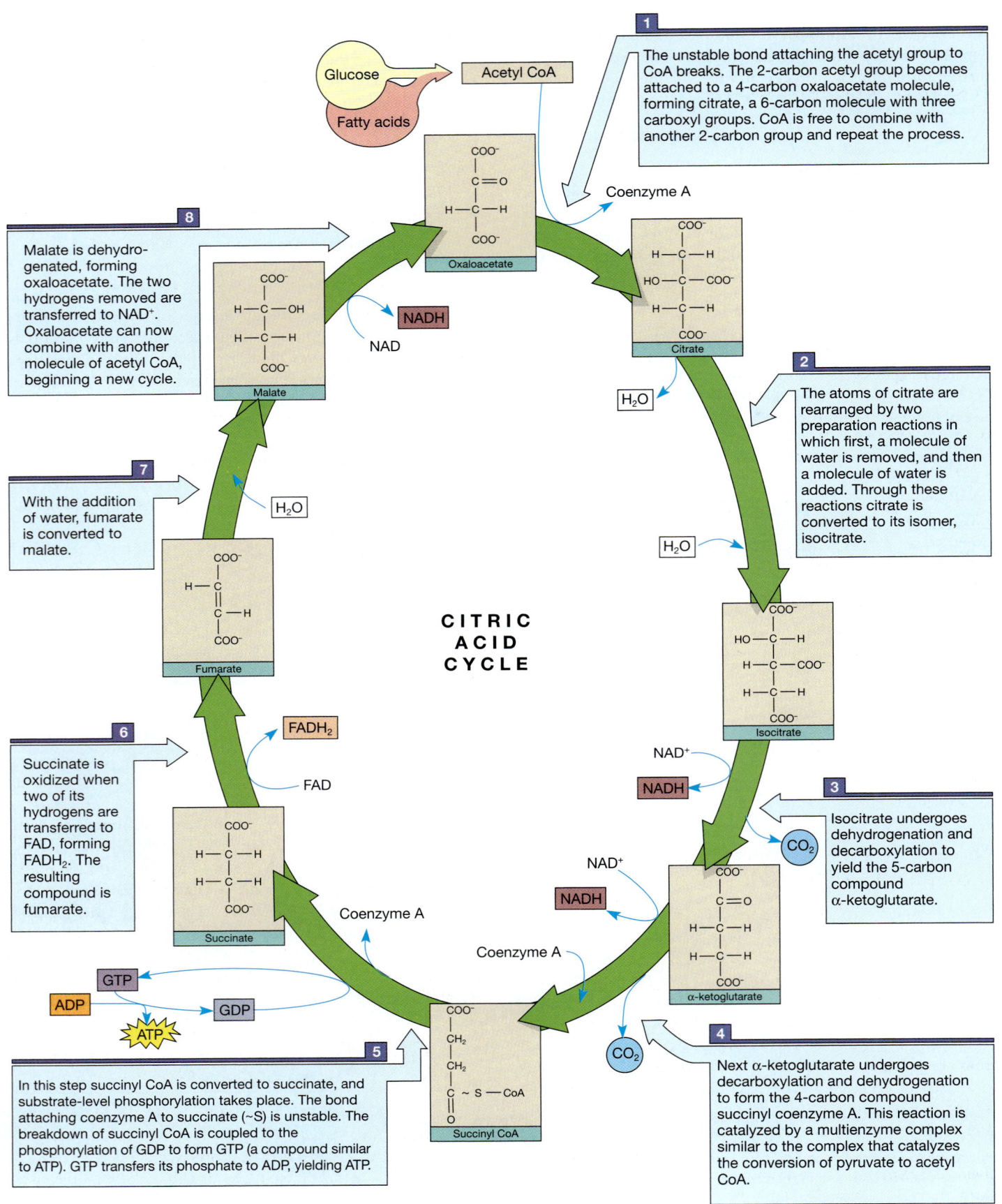

Glucose

Fatty acids

Acetyl CoA

1

The unstable bond attaching the acetyl group to CoA breaks. The 2-carbon acetyl group becomes attached to a 4-carbon oxaloacetate molecule, forming citrate, a 6-carbon molecule with three carboxyl groups. CoA is free to combine with another 2-carbon group and repeat the process.

Coenzyme A

Oxaloacetate

COO⁻
C=O
H—C—H
COO⁻

Citrate

COO⁻
H—C—H
HO—C—COO⁻
H—C—H
COO⁻

H_2O

2

The atoms of citrate are rearranged by two preparation reactions in which first, a molecule of water is removed, and then a molecule of water is added. Through these reactions citrate is converted to its isomer, isocitrate.

H_2O

Isocitrate

COO⁻
HO—C—H
H—C—COO⁻
H—C—H
COO⁻

NAD⁺

NADH

3

Isocitrate undergoes dehydrogenation and decarboxylation to yield the 5-carbon compound α-ketoglutarate.

CO_2

8

Malate is dehydrogenated, forming oxaloacetate. The two hydrogens removed are transferred to NAD⁺. Oxaloacetate can now combine with another molecule of acetyl CoA, beginning a new cycle.

Malate

COO⁻
H—C—OH
H—C—H
COO⁻

NADH

NAD

CITRIC ACID CYCLE

NAD⁺

NADH

α-ketoglutarate

COO⁻
C=O
H—C—H
H—C—H
COO⁻

Coenzyme A

CO_2

7

With the addition of water, fumarate is converted to malate.

H_2O

Fumarate

COO⁻
H—C
C—H
COO⁻

6

Succinate is oxidized when two of its hydrogens are transferred to FAD, forming FADH₂. The resulting compound is fumarate.

FADH₂

FAD

Succinate

COO⁻
H—C—H
H—C—H
COO⁻

Coenzyme A

4

Next α-ketoglutarate undergoes decarboxylation and dehydrogenation to form the 4-carbon compound succinyl coenzyme A. This reaction is catalyzed by a multienzyme complex similar to the complex that catalyzes the conversion of pyruvate to acetyl CoA.

GTP

ADP

GDP

ATP

Succinyl CoA

COO⁻
CH₂
CH₂
C ~ S —CoA
O

5

In this step succinyl CoA is converted to succinate, and substrate-level phosphorylation takes place. The bond attaching coenzyme A to succinate (~S) is unstable. The breakdown of succinyl CoA is coupled to the phosphorylation of GDP to form GTP (a compound similar to ATP). GTP transfers its phosphate to ADP, yielding ATP.

are also transferred to the electron acceptor FAD, forming $FADH_2$.

In the course of the citric acid cycle, two molecules of CO_2 and the equivalent of eight hydrogen atoms (eight protons and eight electrons) are removed, forming three NADH and one $FADH_2$. You may wonder why more hydrogen is generated by these reactions than entered the cycle with the acetyl CoA molecule. These hydrogen atoms come from water molecules that are added during the reactions of the cycle. The CO_2 produced accounts for the two carbon atoms of the acetyl group that entered the citric acid cycle. At the end of each cycle, the four-carbon oxaloacetate has been regenerated, and the cycle can continue.

Because two acetyl CoA molecules are produced from each glucose molecule, two cycles are required per glucose. After two turns of the cycle, the original glucose has lost all of its carbons and may be regarded as having been completely consumed. To summarize, the citric acid cycle yields 4 CO_2, 6 NADH, 2 $FADH_2$, and 2 ATP per glucose molecule.

At the end of the citric acid cycle, glucose has been completely catabolized. Only four molecules of ATP have been formed by substrate-level phosphorylation: two during glycolysis and two during the citric acid cycle. At this time, most of the energy of the glucose is in the form of high-energy electrons in NADH and $FADH_2$. Their energy will be used to synthesize additional ATP through the electron transport chain and chemiosmosis.

The electron transport chain is coupled to ATP synthesis

Let us consider the fate of all the electrons removed from a molecule of glucose during glycolysis, acetyl CoA formation, and the citric acid cycle. Recall that these electrons were transferred as part of hydrogen atoms to the acceptors NAD^+ and FAD, forming NADH and $FADH_2$. These reduced compounds now enter the **electron transport chain,** where the high-energy electrons of their hydrogen atoms are shuttled from one acceptor to another. As the electrons are passed along in a series of exergonic redox reactions, some of their energy is used to drive the synthesis of ATP, which is an endergonic process. Because ATP synthesis (by phosphorylation of ADP) is coupled to the redox reactions in the electron transport chain, the entire process is known as **oxidative phosphorylation.**

The electron transport chain transfers electrons from NADH and $FADH_2$ to oxygen

The electron transport chain is a series of electron carriers embedded in the inner membrane of the mitochondrion of eukaryotes and in the plasma membrane of aerobic prokaryotes. Like NADH and $FADH_2$, each carrier can exist in an oxidized form or a reduced form. Electrons pass down the electron transport chain in a series of redox reactions: each acceptor molecule is alternately reduced as it accepts electrons and oxidized as it gives up electrons. The electrons entering the electron transport chain have a relatively high energy content. They lose some of their energy at each step as they pass along the chain of electron carriers.

The electron transport chain consists of four complexes, or groups, of acceptors (Fig. 7–8). Each complex is a multiprotein aggregate. Members of the electron transport chain include flavin mononucleotide (FMN), ubiquinone (CoQ), and a group of closely related proteins called **cytochromes.** The last cytochrome in the chain, cytochrome a_3, passes two electrons to oxygen. The electrons simultaneously unite with protons from the surrounding medium to form hydrogen, and the chemical reaction between hydrogen and oxygen produces water.

Because oxygen is the final electron acceptor in the electron transport chain, organisms that respire aerobically require oxygen. What happens when cells that are strict aerobes are deprived of oxygen? When no oxygen is available to accept them, the last cytochrome in the chain is stuck with its electrons. When that occurs, each acceptor molecule in the chain remains stuck with electrons (i.e., is reduced), and the entire chain is blocked all the way back to NADH. Because oxidative phosphorylation is coupled to electron transport, no further ATPs are produced by way of the electron transport chain. Most cells of complex organisms cannot live long without oxygen because the amount of ATP they produce in its absence is insufficient to sustain life processes.

Lack of oxygen is not the only factor that interferes with the electron transport chain. Some poisons, including cyanide, inhibit the normal activity of the cytochromes. Cyanide binds tightly to the iron in cytochrome a_3, making it unable to transport electrons on to oxygen. This blocks the further passage of electrons through the chain, halting ATP production.

Although the flow of electrons in electron transport is usually tightly coupled to the production of ATP, some organisms are able to uncouple the two processes to produce heat (see *Making the Connection: Electron Transport and Heat*).

The chemiosmotic model explains the coupling of ATP synthesis to electron transport

For decades scientists were aware that oxidative phosphorylation occurs in mitochondria, and many experiments had shown that the transfer of two electrons from each NADH to oxygen (via the electron transport chain) usually results in the production of up to three ATP molecules. However, for a long

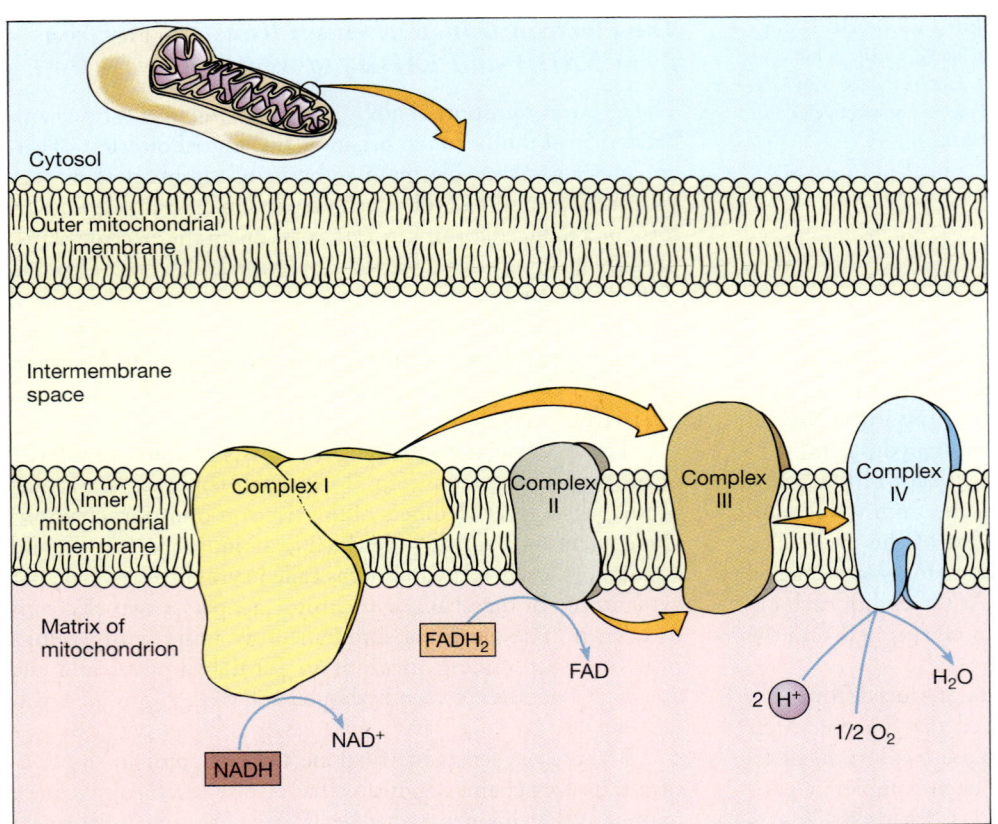

Figure 7–8 Overview of the electron transport chain. Electrons fall to successively lower energy levels as they are passed along the four complexes of the electron transport chain, which are located in the mitochondrial inner membrane. (The orange arrows indicate the pathway of electrons.) The carriers within each complex are alternately reduced and oxidized as they accept and donate electrons. The terminal acceptor is oxygen; one of the two atoms of an oxygen molecule (written $\frac{1}{2}$ O_2) accepts two electrons, which are added to two protons from the surrounding medium to produce water.

MAKING THE CONNECTION

ELECTRON TRANSPORT AND HEAT

What is the source of our body heat? Essentially, it is a byproduct of various exergonic reactions, especially those involving the electron transport chains in our mitochondria. Some organisms are able to produce unusually large amounts of heat by uncoupling electron transport from ATP production. As a consequence, most of the energy of glucose is converted to heat rather than to ATP energy.

Even some plants, which are not generally considered "warm" organisms, have this capability. Skunk cabbage *(Symplocarpus foetidus)*, for example, lives in North American swamps and wet woodlands and generally flowers during February and March when the ground is still covered with snow *(see figure)*. Its uncoupled mitochondria generate large amounts of heat, enabling the plant to melt the snow and presumably attract pollinators. The flower temperature of skunk cabbage is 15° to 22°C (59° to 72°F) when the air surrounding it is −15° to 10°C (5° to 50°F). Skunk cabbage flowers maintain this temperature for two weeks or more. Other plants, such as splitleaf philodendron *(Philodendron selloum)* and sacred lotus *(Nelumbo nucifera)*, also generate heat when they bloom and maintain their temperatures within precise limits.

Some plants generate as much or more heat per gram of tissue than animals in flight, which have long been considered the greatest heat producers in the living world. The European plant lords-and-ladies *(Arum maculatum)*, for example, produces 0.4 joules (0.1 cal) of heat per second per gram of tissue, whereas a hummingbird in flight produces 0.24 joules (0.06 cal) per second per gram of tissue.

Skunk cabbage *(Symplocarpus foetidus)*. This plant not only produces a significant amount of heat when it flowers, but regulates its temperature within a specific range. *(Earth Scenes © 1999 Leonard Lee Rue III)*

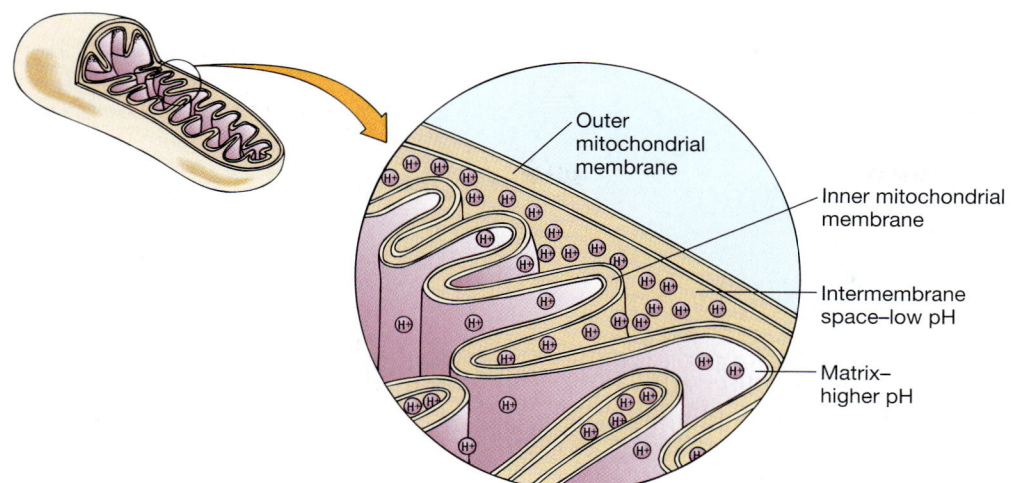

Figure 7–9 **Accumulation of protons (H$^+$) within the intermembrane space.** As electrons move down the electron transport chain, the electron transport complexes move protons (H$^+$) from the matrix to the intermembrane space, creating a proton gradient. The higher concentration of H$^+$ in the intermembrane space lowers the pH.

time, the connection between ATP synthesis and electron transport remained a mystery. Then, in 1961 Peter Mitchell proposed the **chemiosmotic model,** based on his experiments with bacteria. His model was so radical that it was not accepted immediately, but by 1978 so much supporting evidence had accumulated in support of the chemiosmotic model that Peter Mitchell was awarded a Nobel Prize.

Because the respiratory electron transport chain is located in the plasma membrane of an aerobic bacterial cell, the bacterial plasma membrane can be considered comparable to the inner mitochondrial membrane. Mitchell was able to show that if bacterial cells were placed in an acidic environment (that is, an environment with a high hydrogen ion, or proton, concentration), the cells would synthesize ATP even if no electron transport were taking place. On the basis of these and other experiments, Mitchell proposed that electron transport and ATP synthesis are coupled by means of a proton gradient across the inner mitochondrial membrane in eukaryotes (or across the plasma membrane in bacteria).

The proton gradient is established by the electron transport chain; some of the energy released as electrons pass down the electron transport chain is used to pump protons (H$^+$) across a membrane. In eukaryotes the protons are pumped across the inner mitochondrial membrane into the intermembrane space, that is, the space between the inner and outer mitochondrial membranes (Fig. 7–9). Hence the inner mitochondrial membrane separates a space with a higher concentration of protons (the intermembrane space) from a space with a lower concentration of protons (the mitochondrial matrix).

Protons are pumped across the inner mitochondrial membrane by three of the four electron transport complexes (Fig. 7–10). The result is a proton gradient across the inner mitochondrial membrane. This gradient is a form of potential energy, like water behind a dam; it can be harnessed to provide the energy for ATP synthesis.

Diffusion of protons through the inner mitochondrial membrane is limited to specific channels formed by the enzyme complex **ATP synthase,** a transmembrane protein. Portions of these complexes project from the inner surface of the membrane (the surface that faces the matrix) and are visible by electron microscopy. Diffusion of the protons down their gradient, through the ATP synthase complex, is exergonic because the entropy of the system increases. This exergonic process provides the energy for ATP production, although the exact mechanism by which ATP synthase catalyzes the phosphorylation of ADP is still incompletely understood.

Chemiosmosis is a fundamental mechanism of energy coupling in cells; it allows exergonic redox processes to drive the endergonic reaction in which ATP is produced by phosphorylating ADP. In photosynthesis (see Chapter 8), ATP is produced by a comparable process.

AEROBIC RESPIRATION OF ONE GLUCOSE YIELDS A MAXIMUM OF 36 TO 38 ATPS

Let us now review where biologically useful energy is captured in aerobic respiration and calculate the total energy yield from the complete oxidation of glucose. Table 7–2 summarizes the arithmetic involved.

1. In glycolysis, glucose is activated by the addition of phosphates from 2 ATP molecules and converted ultimately to 2 pyruvates + 2 NADH + 4 ATPs, yielding a net profit of 2 ATPs.
2. The 2 pyruvates are metabolized to 2 acetyl CoA + 2 CO_2 + 2 NADH.
3. In the citric acid cycle the 2 acetyl CoA molecules are metabolized to 4 CO_2 + 6 NADH + 2 FADH$_2$ + 2 ATPs.

Because the oxidation of NADH in the electron transport chain yields up to 3 ATPs per molecule, the 10 NADH molecules can yield up to 30 ATPs. The 2 NADH molecules from glycolysis, however, yield either 2 or 3 ATPs each. This is because certain types of eukaryotic cells must expend energy to shuttle the NADH produced by glycol-

Figure 7–10 A detailed look at electron transport and chemiosmosis. The electron transport chain in the inner mitochondrial membrane includes three proton pumps that are located in three of the four electron transport complexes. The energy released during electron transport is used to transport protons (H^+) from the mitochondrial matrix to the intermembrane space, where a high concentration of protons accumulates. The protons are prevented from diffusing back into the matrix except through special channels in ATP synthase in the inner membrane. The flow of the protons through ATP synthase generates ATP.

ysis across the mitochondrial membrane (see *Focus On: Shuttles across the Mitochondrial Membrane*). Prokaryotic cells lack mitochondria; hence they have no need to shuttle NADH molecules. For this reason, bacteria are able to generate three ATPs for every NADH, even those produced during glycolysis.

Thus, the maximum number of ATPs formed using the energy from NADH is 28 to 30. The oxidation of $FADH_2$ yields 2 ATPs per molecule, so the 2 $FADH_2$ molecules produced in the citric acid cycle yield 4 ATPs.

4. Summing all the ATPs (2 from glycolysis, 2 from the citric acid cycle, and 32 to 34 from electron transport and

TABLE 7-2 Energy Yield from the Complete Oxidation of Glucose by Aerobic Respiration

1. Net ATP profit from glycolysis		2 ATP (substrate-level phosphorylation)
Also from glycolysis:	2 NADH $\longrightarrow$	4–6 ATP (oxidative phosphorylation)
2. 2 pyruvate to 2 acetyl CoA	2 NADH $\longrightarrow$	6 ATP (oxidative phosphorylation)
3. 2 acetyl CoA through citric acid cycle		2 ATP (substrate-level phosphorylation)
and electron transport system	6 NADH $\longrightarrow$	18 ATP (oxidative phosphorylation)
	2 FADH$_2$ $\longrightarrow$	4 ATP (oxidative phosphorylation)
Total ATP Profit		36–38 ATP

chemiosmosis), we see that the complete aerobic metabolism of 1 molecule of glucose yields a maximum of 36 to 38 ATPs. Note that most of the ATPs are generated by oxidative phosphorylation, which involves the electron transport chain and chemiosmosis. Only 4 ATPs are formed by substrate-level phosphorylation in glycolysis and the citric acid cycle.

We can analyze the efficiency of the overall process of aerobic respiration by comparing the free energy captured as ATP to the total free energy in a glucose molecule. You will recall from Chapter 6 that, although heat energy cannot power cellular reactions, it is convenient to measure energy as heat. This can be done through the use of a calorimeter, an instrument that measures the heat of a reaction. A sample is placed in a compartment surrounded by a chamber of water. As the sample burns (becomes oxidized), the temperature of the water rises, providing a measure of the heat released during the reaction.

When one mole of glucose is burned in a calorimeter, some 686 kcal (2879.2 kJ) are released as heat. The free energy temporarily held in the phosphate bonds of ATP is about 7.6 kcal (31.8 kJ) per mole; when 36 to 38 ATPs are generated during the aerobic respiration of glucose, the free energy trapped in ATP amounts to 7.6 kcal per mole × 36, or about 274 kcal (1146.4 kJ) per mole. Thus, the efficiency of aerobic respiration is 274/686, or about 40%. (By comparison, a steam power plant has an efficiency of 35% to 36% in converting its fuel energy into electricity.) The remainder of the energy in the glucose is released as heat.

NUTRIENTS OTHER THAN GLUCOSE ALSO PROVIDE ENERGY

Many organisms depend on nutrients other than glucose as a source of energy. Humans and many other animals usually obtain more of their energy by oxidizing fatty acids than by oxidizing glucose. Amino acids derived from protein digestion are also used as fuel molecules. Such nutrients are transformed

FOCUS ON

SHUTTLES ACROSS THE MITOCHONDRIAL MEMBRANE

The inner mitochondrial membrane is not permeable to NADH, which is a large molecule. Therefore the NADH molecules produced in the cytosol during glycolysis cannot diffuse into the mitochondria to transfer their electrons to the electron transport chain. Unlike ATP and ADP, NADH does not have a carrier protein to transport it across the membrane. Instead, several systems have evolved to transfer just the *electrons* of NADH, not the NADH molecules themselves, into the mitochondria.

In liver, kidney, and heart cells, a special shuttle system transfers the electrons from NADH through the inner mitochondrial membrane to an NAD$^+$ molecule in the matrix. These electrons are transferred to the electron transport chain in the inner mitochondrial membrane, and up to three molecules of ATP are produced per pair of electrons.

In skeletal muscle, brain, and some other types of cells, another type of shuttle operates. Because this shuttle requires more energy than the shuttle in liver, kidney, and heart cells, the electrons are at a lower energy level when they enter the electron transport chain. They are accepted by coenzyme Q rather than by NAD$^+$ and so generate a maximum of 2 ATP molecules per pair of electrons. This is why the number of ATPs produced by aerobic respiration of 1 molecule of glucose in skeletal muscle cells is 36 rather than 38.

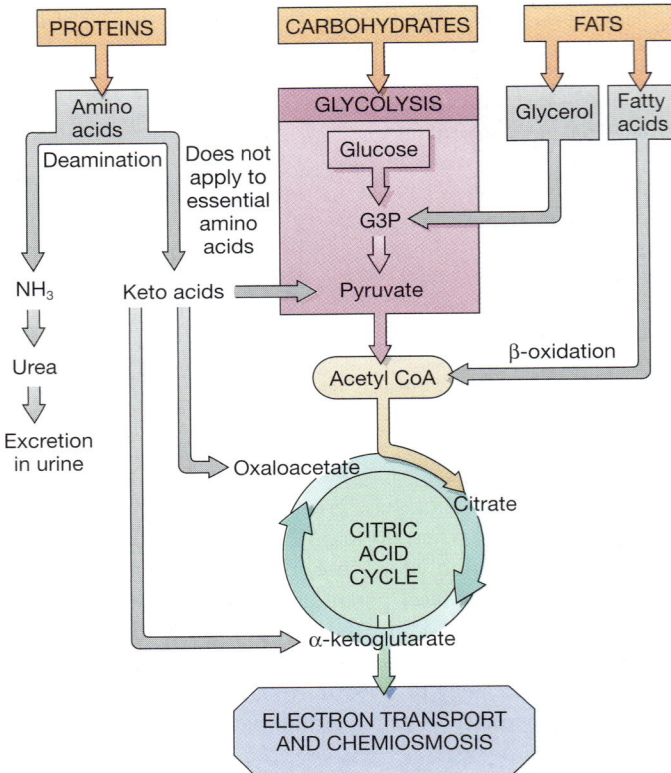

Figure 7–11 **Energy from carbohydrates, proteins, and fats.** Products of the catabolism of carbohydrates, proteins, and fats enter glycolysis or the citric acid cycle at various points. This diagram is greatly simplified and illustrates only a few of the principal catabolic pathways.

two-carbon acetyl groups that are bound to coenzyme A; that is, fatty acids are converted to acetyl CoA. This process, which occurs in the mitochondrial matrix, is called β-**oxidation.** Acetyl CoA molecules formed by β-oxidation enter the citric acid cycle.

CELLS REGULATE AEROBIC RESPIRATION

Aerobic respiration requires a steady input of fuel molecules and oxygen. Under normal conditions these materials are adequately provided and do not affect the rate of respiration. Instead, the rate of aerobic respiration is regulated by how much ADP and phosphate are available. In a resting muscle cell, for example, ATP synthesis continues until most of the ADP has been converted to ATP. At this point oxidative phosphorylation slows considerably. Because electron flow is tightly coupled to oxidative phosphorylation, the flow of electrons also slows, which in turn slows down the citric acid cycle.

When ATP transfers energy to power an energy-requiring process like muscle contraction, many molecules of ATP are hydrolyzed. The ADP molecules produced can then accept phosphate to become ATP once again; aerobic respiration speeds up until most of the ADP has again been converted to ATP.

The control of most metabolic pathways is exerted on a particular enzyme that catalyzes a reaction early in the pathway (see Fig. 6–16 and discussion of feedback inhibition in Chapter 6). The regulated enzyme is usually inhibited by the presence of the end product of the pathway. One of the important control points in aerobic respiration in mammals is phosphofructokinase, an enzyme that catalyzes an early reaction of glycolysis (Fig. 7–12). The active site of phosphofructokinase binds ATP and fructose-6-phosphate. However, the enzyme is inhibited by the presence of very high levels of ATP and activated by the presence of ADP and AMP (adenosine monophosphate, a molecule formed when two phosphates are removed from ATP). Therefore, this enzyme is inactivated when ATP levels are high and activated when they are low.[3] Phosphofructokinase possesses different allosteric sites for both enzyme inhibitors (in this case, ATP) and enzyme activators (in this case, ADP and AMP).

When respiration produces more ATP than the cell currently needs, some of the excess ATP binds to the allosteric inhibitor site of phosphofructokinase, changing its conformation so that it is no longer active. Thus, glycolysis and aerobic respiration slow down and less ATP is produced. As excess ATP is used by the cell, ADP and AMP are produced. Now the allosteric inhibitor site of phosphofructokinase is no longer occupied by ATP. Instead, the allosteric activator sites are filled with ADP and AMP. Thus the enzyme becomes activated and binds ATP to its active site; respiration proceeds, generating more ATP.

into one of the metabolic intermediates that are fed into glycolysis or the citric acid cycle (Fig. 7–11).

Amino acids are metabolized by reactions in which the amino group is first removed, a process called **deamination.** In mammals and some other animals, the amino group is converted to urea and excreted, but the carbon chain is metabolized and eventually is used as a reactant in one of the steps of the citric acid cycle. The amino acid alanine, for example, undergoes deamination to become pyruvate, the amino acid glutamate is converted to α-ketoglutarate, and the amino acid aspartate yields oxaloacetate. Ultimately, the carbon chains of all the amino acids are metabolized in this way.

Each gram of lipid in the diet contains 9 kcal (38 kJ), more than twice as much energy as one gram of glucose or amino acids, which have about 4 kcal (17 kJ) per gram. Lipids are rich in energy because they are highly reduced; that is, they have many hydrogen atoms and few oxygen atoms. When completely oxidized in aerobic respiration, a molecule of a six-carbon fatty acid generates up to 44 ATPs (compared with 36 to 38 ATPs for a molecule of glucose, which also has six carbons).

Both the glycerol and fatty acid components of a neutral fat (see Chapter 3) are used as fuel; phosphate is added to glycerol, converting it to G3P or another compound that enters glycolysis. Fatty acids are oxidized and split enzymatically into

---

[3]Other materials, including citrate, also affect the activity of phosphofructokinase.

GLYCOLYSIS

Glucose

High levels of ATP inhibit phosphofructokinase

Fructose phosphate

Fructose bisphosphate

Pyruvate

ATP

Acetyl CoA

CITRIC ACID CYCLE

ELECTRON TRANSPORT AND CHEMIOSMOSIS

Figure 7–12 **Regulation of aerobic respiration.** When enough ATP accumulates in the cell, the ATP inhibits phosphofructokinase, an enzyme that catalyzes an early step in glycolysis. This feedback inhibition controls the rate of aerobic respiration, matching it to the cell's demands for energy.

an inorganic substance such as nitrate (NO_3^-) or sulfate (SO_4^{2-}) replaces molecular oxygen as the terminal electron acceptor. The end products of this type of anaerobic respiration are carbon dioxide, one or more reduced inorganic substances, and ATP. One representative type of anaerobic respiration, summarized below, is part of the biogeochemical cycle known as the nitrogen cycle (see Chapter 53).

$$C_6H_{12}O_6 + 12\ KNO_3 \longrightarrow$$

Potassium
nitrate

$$6\ CO_2 + 6\ H_2O + 12\ KNO_2 + Energy\ (as\ ATP)$$

Potassium
nitrite

Certain other bacteria, as well as some fungi, regularly use **fermentation,** an anaerobic pathway that does not involve an electron transport chain. During fermentation only two ATPs are formed per glucose (by phosphorylation at the substrate level during glycolysis). One might expect that a cell that obtains energy from glycolysis would produce pyruvate, the end product of glycolysis. However, this cannot happen because every cell has a limited supply of NAD^+. If virtually all NAD^+ becomes reduced to NADH during glycolysis, then glycolysis stops, and no more ATP can be produced. In fermentation, NADH molecules transfer their hydrogen atoms to organic molecules, thus regenerating the NAD^+ needed to keep glycolysis going. The resulting, relatively reduced, organic molecules (commonly alcohol or lactate) tend to be toxic to the cells and are essentially waste products. Table 7–3 summarizes features of aerobic respiration, anaerobic respiration, and fermentation.

ANAEROBIC RESPIRATION AND FERMENTATION DO NOT REQUIRE OXYGEN

Anaerobic respiration, which does not use oxygen as the final electron acceptor, is performed by some types of bacteria that live in such anaerobic environments as waterlogged soil, stagnant ponds, or animal intestines. As in aerobic respiration, electrons are transferred in anaerobic respiration from glucose to NADH; they then pass down an electron transport chain that is coupled to ATP synthesis by chemiosmosis. However,

TABLE 7–3 **A Comparison of Aerobic Respiration, Anaerobic Respiration, and Fermentation**

	Aerobic Respiration	Anaerobic Respiration	Fermentation
Immediate Fate of Electrons in NADH	Transferred to an electron transport chain	Transferred to an electron transport chain	Transferred to an organic molecule
Terminal Electron Acceptor of Electron Transport Chain	O_2	Inorganic substances such as NO_3^- or SO_4^{2-}	(No electron transport chain)
Reduced Product(s) Formed	Water	Relatively reduced inorganic substances	Relatively reduced organic compounds (commonly alcohol or lactate)
Mechanism of ATP Synthesis	Chemiosmosis; also substrate-level phosphorylation	Chemiosmosis; also substrate-level phosphorylation	Substrate-level phosphorylation only (during glycolysis)

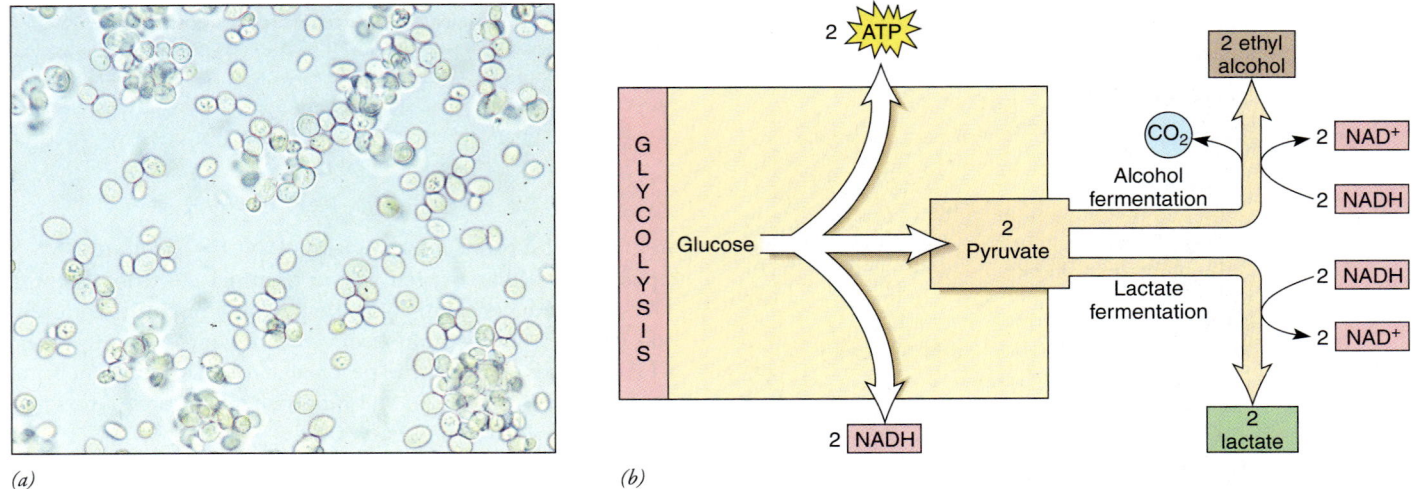

(a)

(b)

Figure 7-13 **Fermentation.** *(a)* LM of live brewer's yeast *(Saccharomyces cerevisiae)*. Yeast cells possess mitochondria and carry on aerobic respiration when O_2 is present. In the absence of O_2, yeasts carry on alcohol fermentation. *(b)* Glycolysis is the first part of fermentation pathways. In alcohol fermentation, CO_2 is split off, and the two-carbon compound ethyl alcohol is the end product. In lactate fermentation, the final product is the three-carbon compound lactate. In both alcohol and lactate fermentation, there is a net gain of only two ATPs per molecule of glucose. *(Dwight R. Kuhn)*

Alcohol fermentation and lactate fermentation are inefficient

Yeasts are **facultative anaerobes.** These eukaryotic, unicellular fungi have mitochondria and carry out aerobic respiration when oxygen is available but switch to **alcohol fermentation** when deprived of oxygen (Fig. 7–13*a*). They have enzymes that decarboxylate pyruvate, releasing carbon dioxide and forming a two-carbon compound called acetaldehyde. NADH produced during glycolysis transfers hydrogen atoms to acetaldehyde, reducing it to form **ethyl alcohol** (Fig. 7–13*b*). Alcohol fermentation is the basis for the production of beer, wine, and other alcoholic beverages. Yeast cells are also used in baking to produce the carbon dioxide that causes dough to rise; the alcohol evaporates during baking.

Certain fungi and bacteria perform **lactate (lactic acid) fermentation.** In this alternative pathway, NADH produced during glycolysis transfers hydrogen atoms to pyruvate, reducing it to form **lactate** (see Fig. 7–13*b*). The ability of some bacteria to produce lactate is exploited by humans, who use them to make yogurt and to ferment cabbage for sauerkraut.

Lactate is also produced during strenuous activity in the muscle cells of humans and other complex animals. If the amount of oxygen delivered to muscle cells is insufficient to support aerobic respiration, the cells shift briefly to lactate fermentation. The shift is only temporary, however, and oxygen is required for sustained work. As lactate accumulates in mus-

cle cells, it contributes to fatigue and muscle cramps.[4] About 80% of the lactate is eventually exported to the liver, where it is used to regenerate more glucose for the muscle cells. The remaining 20% of the lactate is metabolized in muscle cells in the presence of oxygen. This explains why you continue to breathe heavily after you have stopped exercising; the additional oxygen is needed to oxidize lactate, thereby restoring the muscle cells to their normal state.

Although humans can only use lactate fermentation to produce ATP for a few minutes, a few animals can live without oxygen for much longer periods. The red-eared slider can remain underwater for as long as two weeks (Fig. 7–14). It relies on lactate fermentation for ATP production.

Both alcohol fermentation and lactate fermentation are very inefficient because the fuel is only partially oxidized. Alcohol, the end product of fermentation by yeast cells, can be burned and can even be used as automobile fuel; obviously, it contains a great deal of energy that the yeast cells are unable to extract using anaerobic methods. Lactate, a three-carbon compound, contains even more energy than the two-carbon alcohol. In contrast, all available energy is removed during aerobic respiration because the fuel molecules become completely

---

[4]Why exercise causes fatigue and muscle cramps is incompletely understood at this time, but these conditions may be related to lactate buildup, insufficient oxygen, and/or the depletion of fuel molecules.

Figure 7–14 **Long-term use of lactate fermentation for ATP production.** The red-eared slider *(Trachemys scripta)* can stay submerged for as long as two weeks. *(Cleveland P. Hickman, Jr./Visuals Unlimited)*

oxidized to CO_2. A net profit of only 2 ATPs is produced by the fermentation of one molecule of glucose, compared with up to 36 to 38 ATPs when oxygen is available.

The inefficiency of fermentation necessitates a large supply of fuel. For example, skeletal muscle cells, which often metabolize anaerobically for short periods, store large quantities of glucose in the form of glycogen. To perform the same amount of work, a cell engaged in fermentation must consume up to 20 times more glucose or other carbohydrate per second than a cell using aerobic respiration.

SUMMARY WITH KEY TERMS

I. Metabolism, the total of all the chemical reactions that occur in cells, has two complementary components.
   A. **Catabolism** releases energy by breaking down molecules into simpler molecules. Catabolic pathways include aerobic respiration, anaerobic respiration, and fermentation.
   B. **Anabolism** is the synthesis of complex molecules from simpler building blocks. Protein synthesis is an example of anabolism.
II. During **aerobic respiration,** a fuel molecule such as glucose is oxidized to form carbon dioxide and water.
   A. Aerobic respiration is a redox process in which electrons are transferred from glucose (which becomes **oxidized**) to oxygen (which becomes **reduced**).
   B. Up to 36 to 38 **ATP**s are produced per molecule of glucose.
III. The chemical reactions of aerobic respiration occur in four stages: glycolysis, formation of acetyl CoA, the citric acid cycle, and the electron transport chain/chemiosmosis.
   A. During **glycolysis,** which occurs in the cytosol, a molecule of glucose is degraded to two molecules of **pyruvate.**
      1. Two ATP molecules (net) are produced by **substrate-level phosphorylation** during glycolysis.
      2. Four hydrogen atoms are removed from the fuel molecule and used to produce two **NADH.**
   B. The two pyruvate molecules each lose a molecule of carbon dioxide, and the remaining **acetyl groups** each combine with **coenzyme A,** producing two molecules of **acetyl CoA.** One NADH is produced as each pyruvate is converted to acetyl CoA.
   C. Each acetyl CoA enters the **citric acid cycle** by combining with a four-carbon compound, oxaloacetate, to form **citrate,** a six-carbon compound. Two acetyl CoA molecules enter the cycle for every glucose molecule.
      1. For every two carbons that enter the cycle as part of an acetyl

CoA molecule, two leave as carbon dioxide.
      2. For every acetyl CoA, hydrogen atoms are transferred to three NAD^+ and one **FAD;** only one ATP is produced by substrate-level phosphorylation.
   D. Hydrogen atoms (or their electrons) removed from fuel molecules are transferred from one electron acceptor to another down an **electron transport chain** located in the mitochondrial inner membrane.
      1. Water is formed when oxygen combines with H^+ and with electrons from the electron transport chain.
      2. According to the **chemiosmotic model,** some of the energy of the electrons in the electron transport chain is used to establish a proton gradient across the inner mitochondrial membrane.
      3. The diffusion of protons through the membrane from the intermembrane space to the mitochondrial matrix (through channels formed by the enzyme **ATP synthase**) provides the energy needed to synthesize ATP.
IV. Organic nutrients other than glucose are converted to appropriate compounds and fed into glycolysis or the citric acid cycle.
   A. Amino acids are **deaminated** and their carbon skeletons are converted to metabolic intermediates of glycolysis and the citric acid cycle.
   B. Both the glycerol and fatty acid components of lipids are oxidized as fuel. Fatty acids are converted to acetyl CoA molecules by the process of **β-oxidation.**
V. In **anaerobic respiration,** electrons are transferred from fuel molecules to an electron transport chain; the final electron acceptor is an inorganic substance such as nitrate or sulfate, not molecular oxygen.
VI. **Fermentation** is an anaerobic process that does not use an electron transport chain. There is a net gain of only two ATPs per glucose; these are produced during glycolysis. To maintain the supply of NAD^+ es-

sential for glycolysis, hydrogen atoms are transferred from NADH to an organic compound derived from the initial nutrient.

A. Yeast cells carry out **alcohol fermentation,** in which **ethyl alcohol** and carbon dioxide are the final waste products.

B. Certain fungi, bacteria, and animal cells carry out **lactate fermentation,** in which hydrogen atoms are added to pyruvate to form **lactate,** a waste product.

Summary Reactions for Aerobic Respiration

Summary reaction for the complete oxidation of glucose:

$$C_6H_{12}O_6 + 6\ O_2 + 6\ H_2O \longrightarrow$$
$$6\ CO_2 + 12\ H_2O + \text{Energy (36 to 38 ATP)}$$

Summary reaction for glycolysis:

$$C_6H_{12}O_6 + 2\ ATP + 2\ ADP + 2\ P_i + 2\ NAD^+ \longrightarrow$$
$$2\ \text{Pyruvate} + 4\ ATP + 2\ NADH + H_2O$$

Summary reaction for the conversion of pyruvate to acetyl CoA:

$$2\ \text{Pyruvate} + 2\ \text{Coenzyme A} + 2\ NAD^+ \longrightarrow$$
$$2\ \text{Acetyl CoA} + 2\ CO_2 + 2\ NADH$$

Summary reaction for the citric acid cycle:

$$2\ \text{Acetyl CoA} + 6\ NAD^+ + 2\ FAD + 2\ ADP + 2\ P_i + 2\ H_2O$$
$$\longrightarrow 4\ CO_2 + 6\ NADH + 2\ FADH_2 + 2\ ATP + 2\ CoA$$

Summary reactions for the processing of the hydrogen atoms of NADH and $FADH_2$ in the electron transport chain:

$$NADH + 3\ ADP + 3\ P_i + \frac{1}{2}\ O_2 \longrightarrow NAD^+ + 3\ ATP + H_2O$$

$$FADH_2 + 2\ ADP + 2\ P_i + \frac{1}{2}\ O_2 \longrightarrow FAD + 2\ ATP + H_2O$$

Summary Reactions for Fermentation

Summary reaction for lactate fermentation:

$$C_6H_{12}O_6 \longrightarrow 2\ \text{Lactate} + \text{Energy (2 ATP)}$$

Summary reactions for alcohol fermentation:

$$C_6H_{12}O_6 \longrightarrow 2\ CO_2 + 2\ \text{Ethyl alcohol} + \text{Energy (2 ATP)}$$

POST-TEST

1. The process of splitting larger molecules into smaller ones is an aspect of metabolism called (a) anabolism (b) fermentation (c) catabolism (d) oxidative phosphorylation (e) chemiosmosis
2. The synthetic aspect of metabolism is referred to as (a) anabolism (b) fermentation (c) catabolism (d) oxidative phosphorylation (e) chemiosmosis
3. A chemical process during which a substance gains electrons is called (a) oxidation (b) oxidative phosphorylation (c) deamination (d) reduction (e) dehydrogenation
4. The pathway through which glucose is degraded to pyruvate is referred to as (a) aerobic respiration (b) the citric acid cycle (c) the oxidation of pyruvate (d) alcohol fermentation (e) glycolysis
5. The reactions of _____ take place within the cytosol of eukaryotic cells. (a) glycolysis (b) oxidation of pyruvate (c) the citric acid cycle (d) chemiosmosis (e) the electron transport chain
6. Before pyruvate enters the citric acid cycle, it is decarboxylated, oxidized, and combined with coenzyme A, forming acetyl CoA, carbon dioxide, and one molecule of (a) NADH (b) $FADH_2$ (c) ATP (d) ADP (e) $C_6H_{12}O_6$
7. In the first step of the citric acid cycle, acetyl CoA reacts with oxaloacetate to form (a) pyruvate (b) citrate (c) NADH (d) ATP (e) CO_2
8. Dehydrogenase enzymes remove hydrogen atoms from fuel molecules and transfer them to acceptors such as (a) O_2 and H_2O (b) ATP and FAD (c) NAD^+ and FAD (d) CO_2 and H_2O (e) CoA and pyruvate
9. In the process of _____, electron transport and ATP synthesis are coupled by a proton gradient across the inner mitochondrial membrane. (a) chemiosmosis (b) deamination (c) anaerobic respiration (d) glycolysis (e) decarboxylation
10. Which of the following molecules provides the greatest yield of ATP in a cell carrying out aerobic respiration? (a) glucose (b) pyruvate (c) acetyl CoA (d) NADH (e) oxaloacetate
11. A net profit of only 2 ATPs can be produced anaerobically from the _____ of 1 molecule of glucose, compared with a maximum of 38 ATPs produced in _____. (a) fermentation; anaerobic respiration (b) aerobic respiration; fermentation (c) aerobic respiration; anaerobic respiration (d) dehydrogenation; decarboxylation (e) fermentation; aerobic respiration
12. When deprived of oxygen, yeast cells obtain energy by fermentation, producing carbon dioxide, ATP, and (a) acetyl CoA (b) ethyl alcohol (c) lactate (d) pyruvate (e) citrate
13. During strenuous muscle activity, the pyruvate in muscle cells may accept hydrogen from NADH to become _____. (a) acetyl CoA (b) ethyl alcohol (c) lactate (d) pyruvate (e) citrate

REVIEW QUESTIONS

1. What is the specific role of oxygen in most cells? What happens when cells that can only respire aerobically are deprived of oxygen?
2. Mitochondria are often referred to as the "power plants" of the cell. Justify this with a specific explanation.
3. Refer to Figure 7–7, the diagram of the steps in the citric acid cycle. Look at each reaction and, without reading the description, determine if it is a dehydrogenation, decarboxylation, or preparation reaction.
4. Sketch a mitochondrion and indicate the locations of the electron transport chain and the proton gradient that drives ATP production.
5. Why is each of the following essential to chemiosmotic ATP synthesis? (a) electron transport chain (b) proton gradient (c) ATP synthase complex
6. Explain the roles of the following in aerobic respiration: (a) NAD^+ and FAD (b) oxygen
7. Sum up how much energy (as ATP) is made available to the cell from a single glucose molecule by the operation of glycolysis, the formation of acetyl CoA, the citric acid cycle, and the electron transport chain.
8. Trace the fate of hydrogen atoms removed from glucose during glycolysis when oxygen is present in muscle cells; compare this to the fate of hydrogen atoms removed from glucose when the amount of available oxygen is insufficient to support aerobic respiration.
9. Compare the ATP yields of aerobic respiration and fermentation.
10. Label the figure on p. 173. Use Figure 7–2 to check your answers.

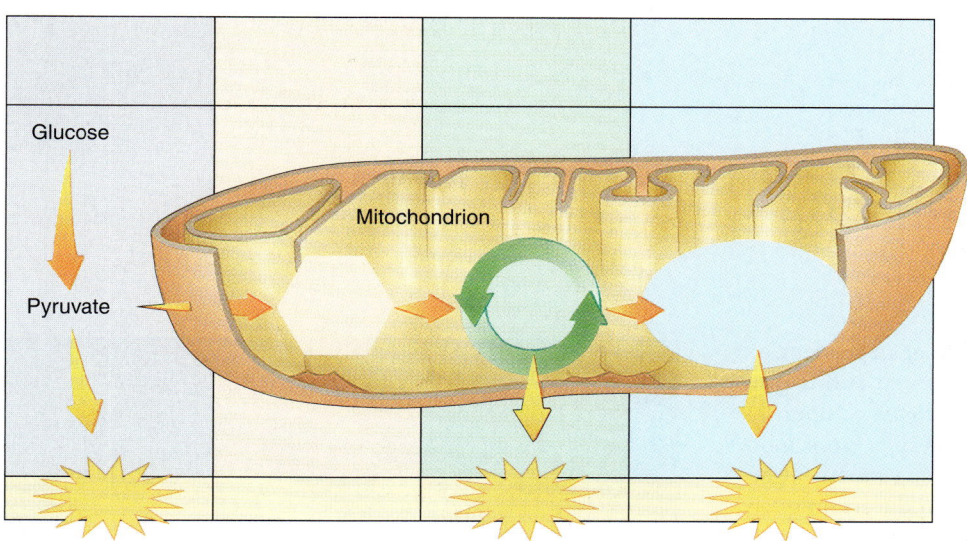

Glucose

Pyruvate

Mitochondrion

YOU MAKE THE CONNECTION

1. The reactions of glycolysis are identical in *all* organisms—bacteria, protists, fungi, plants, and animals—that obtain energy from glucose catabolism. What does this universality suggest about the evolution of glycolysis?
2. How are the endergonic reactions of the first phase of glycolysis coupled to the hydrolysis of ATP, which is exergonic? How are the exergonic reactions of the second phase of glycolysis coupled to the endergonic synthesis of ATP and NADH?

3. What is the role of the mitochondrial inner membrane in the coupling of electron transport and ATP synthesis?
4. Based on what you have learned in this chapter, explain why a school child can run 17 miles per hour in a 100-meter dash, but a trained athlete can run only about 11.5 miles per hour in a 26-mile marathon.

RECOMMENDED READINGS

Fackelmann, K. "Power Failure." *Science News,* Vol. 152, 27 Sept. 1997. Defects in mitochondria can lead to several different human diseases.

Garrett, R.H., and C.M. Grisham. *Biochemistry,* 2nd ed. Saunders College Publishing, Philadelphia, 1999. A comprehensive biochemistry text with good coverage of cellular respiration and related aspects of metabolism.

Seymour, R.S. "Plants That Warm Themselves." *Scientific American,* Mar.

1997. Some plants produce a significant amount of heat when they flower, and a few precisely regulate their temperature.

Tobin, A.J., and R.E. Morel. *Asking About Cells.* Saunders College Publishing, Philadelphia, 1997. Contains a readable and detailed account of cellular energy metabolism.

● Visit our website at **http://www.saunderscollege.com/lifesci/titles.html** and click on Solomon/Berg/Martin Biology for links to chapter-related resources on the World Wide Web.

CHAPTER 8

Photosynthesis: Capturing Energy

All living things are either producers or dependent on the activities of producers. Producers are **autotrophs** (from the Greek *auto*, "self," and *trophos*, "nourishing"), organisms that can make organic molecules from inorganic raw materials. Most producers, including plants, algae, and certain bacteria, are **photoautotrophs,** producers uniquely capable of absorbing and converting light energy into stored chemical energy of organic molecules by the process of **photosynthesis.** These blue lupines (*Lupinus hirsutus*) and the trees behind them, for example, are photoautotrophs. Photoautotrophs use light energy to make ATP and other molecules that temporarily hold chemical energy but are unstable and cannot be stockpiled in the cell. Their energy drives the anabolic pathway by which a photosynthetic cell synthesizes stable organic molecules from the simple inorganic compounds, carbon dioxide and water. These organic compounds are used not only as starting materials to synthesize all the other organic compounds the photosynthetic organism needs (such as complex carbohydrates, amino acids, and lipids) but also for energy storage. They include relatively reduced compounds, such as glucose and other carbohydrates, that can be subsequently oxidized by aerobic respiration or by some other catabolic pathway (see Chapter 7).

Consumers and decomposers are **heterotrophs** (from the Greek *heter*, "other," and *trophos*, "nourishing"), organisms that cannot make their own organic compounds and so must obtain them from other organisms. Consumers live by feeding on producers or on organisms that have eaten producers, whereas decomposers live by breaking down dead organic material. Photosynthesis sustains not only plants and most other producers but also indirectly supports almost all animals and other organisms in the biosphere. Each year photosynthetic organisms convert carbon dioxide into billions of tons of organic molecules. The chemical energy stored in these molecules fuels the metabolic reactions that sustain almost all life.

(Skip Moody/Dembinsky Photo Associates)

AFTER YOU HAVE STUDIED THIS CHAPTER YOU SHOULD BE ABLE TO

1. Explain the relationship between a wavelength of light and its energy and describe the physical properties of light.
2. Describe what can happen to an electron in a biological molecule such as chlorophyll when a photon of light energy is absorbed.
3. Diagram the internal structure of a chloroplast and explain how its components interact and facilitate the process of photosynthesis.
4. Write a summary reaction for photosynthesis, showing the origin and fate of each substance involved.
5. Describe photosynthesis as a redox process.

6. Distinguish between the light-dependent reactions and carbon fixation reactions of photosynthesis.
7. Contrast cyclic and noncyclic photophosphorylation.
8. Explain how a proton (H^+) gradient is established across the thylakoid membrane and how this gradient functions in ATP synthesis.
9. Summarize the chemical reactions involved in the conversion of CO_2 to carbohydrate in the Calvin cycle and indicate the roles of ATP and NADPH in the process.
10. Discuss how the C_4 pathway increases the effectiveness of the Calvin cycle in certain types of plants.

LIGHT IS COMPOSED OF PARTICLES THAT TRAVEL AS WAVES

Because most life on our planet depends on light, either directly or indirectly, it is important to understand the nature of light and how it permits photosynthesis to occur. Light is a very small portion of a vast, continuous spectrum of radiation called the electromagnetic spectrum (Fig. 8–1). All radiations in this spectrum travel in waves. A **wavelength** is the distance from one wave peak to the next. At one end of the electro-

magnetic spectrum are gamma rays, which have very short wavelengths (measured in nanometers). At the other end of the spectrum are radio waves with wavelengths so long they can be measured in kilometers. The portion of the electromagnetic spectrum from 380 nanometers to 760 nanometers is called the visible spectrum because humans can see it. The visible spectrum includes all the colors of the rainbow; violet has the shortest wavelength, and red has the longest (Fig. 8–2).

Light behaves not only as waves do but also as particles. Light is composed of small particles, or packets, of energy called **photons.** The energy of a photon is inversely proportional to its wavelength; shorter wavelength light has more energy per photon than does longer wavelength light.

Why does photosynthesis depend on light detectable by the human eye (visible light) rather than on some other wavelength of radiation? We can only speculate on the answer. One reason may be that much of the radiation reaching our planet from the sun is within this portion of the electromagnetic spectrum; thus, organisms may have evolved the ability to use visible light because it was the most abundant form available. Perhaps more important, radiation within the visible light portion of the spectrum excites certain types of biological molecules, moving electrons into higher energy levels. Radiation with wavelengths longer than those of visible light does not possess

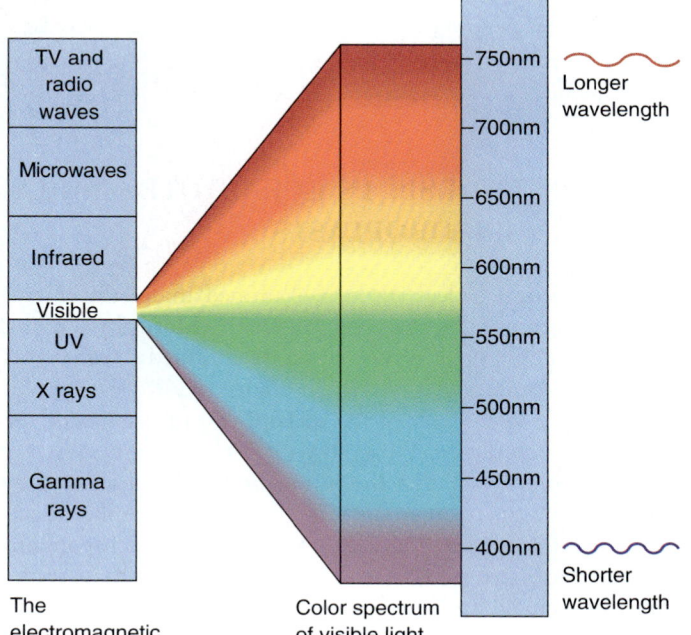

Figure 8–1 The electromagnetic spectrum. Visible light represents a small fraction of the electromagnetic spectrum and consists of a mixture of wavelengths ranging from approximately 380 nm to 760 nm. The energy from visible light is used in photosynthesis.

Figure 8–2 Radiation emitted from the sun. Electromagnetic radiation from the sun contains ultraviolet radiation and visible light of varying colors and wavelengths.

Photon is absorbed by an excitable electron that moves into a higher energy level.

Photon

Nucleus

Ground state e⁻ level

Electron

The electron *may* return to ground level emitting a less energetic photon.

Former location of electron at a lower energy level

Less energetic long wavelength photon is emitted

Electron acceptor molecule

The electron *may* be accepted by an electron acceptor molecule.

Figure 8–3 Interactions between light and atoms or molecules. *(Upper left)* When a photon of light energy strikes an atom, or a molecule of which the atom is a part, the energy of the photon may push an electron to an orbital farther from the nucleus (i.e., into a higher energy level). *(Upper right)* If the electron "falls" back to the next lower energy level, a less energetic (longer wavelength) photon is emitted. *(Lower left)* If the appropriate electron acceptors are available, the electron may leave the atom. During photosynthesis an electron acceptor captures the energetic electron and passes it to a chain of acceptors.

enough energy to excite these biological molecules. Radiation with wavelengths shorter than those of visible light is so energetic that it disrupts the bonds of many biological molecules.

When a molecule absorbs a photon of light energy, one of its electrons is energized. One of two things then happens, depending on the atom and its surroundings (Fig. 8–3). The electron may return to its ground state.[1] If this happens, its energy is dissipated as heat or as light of a longer wavelength than the wavelength of the absorbed light; this emission of light is called **fluorescence.** Alternatively, the energized electron may leave the atom and be accepted by an electron acceptor molecule; this is what occurs in photosynthesis.

Now that we have an understanding of some of the properties of light, we will consider the cellular location where light is used for photosynthesis.

[1]The lowest energy state an atom possesses is called the ground state, but energy can be added to an electron so that it attains a higher energy level. When an electron is raised to a higher energy level than its ground state, the atom is said to be *excited,* or *energized.*

PHOTOSYNTHESIS IN EUKARYOTES TAKES PLACE IN CHLOROPLASTS

When a section of leaf tissue is examined under the microscope, we can see that the green pigment, **chlorophyll,** is not uniformly distributed in the cell but is confined to organelles called **chloroplasts** (Fig. 8–4). In plants, chloroplasts are located mainly in the cells of the **mesophyll,** a tissue inside the leaf. Each mesophyll cell has 20 to 100 chloroplasts.

The chloroplast, like the mitochondrion, is bounded by an outer and an inner membrane. The inner membrane encloses a fluid-filled region called the **stroma,** which contains most of the enzymes required to produce carbohydrate molecules. Suspended in the stroma is a third system of membranes that forms an interconnected set of flat, disclike sacs called **thylakoids.** The thylakoid membrane encloses a fluid-filled interior space, the **thylakoid interior space.** In some regions, thylakoid sacs are arranged in stacks called **grana** (sing., *granum*). Each granum looks something like a stack of coins, with each "coin" being a thylakoid disc. Some thylakoid mem-

Cuticle
Upper epidermis
Palisade mesophyll

Spongy mesophyll

Stoma
Xylem
Phloem } Vein
Bundle sheath

(a) Leaf cross section

Thylakoids Outer membrane Intermembrane space
Inner membrane

Granum (stack of thylakoids) Stroma Thylakoid membrane Thylakoid interior space

(c)

(b) LM of plant cells with chloroplasts

10 μm

Figure 8–4 Where photosynthesis occurs in the plant.
(a) A leaf cross section reveals that the mesophyll is the photosynthetic tissue. CO_2 enters the leaf through tiny pores called stomata (sing., stoma), and H_2O is carried to the mesophyll in veins. *(b)* LM of plant cells with numerous chloroplasts. *(c)* Chloroplast structure. The pigments necessary for the light-capturing reactions of photosynthesis are part of thylakoid membranes, whereas the enzymes for the synthesis of carbohydrate molecules are in the stroma. It is not known if the inner chloroplast membrane is continuous with the thylakoid system *(as shown)* or not. *(b, M. Eichelberger/Visuals Unlimited)*

branes extend from one granum to another. Chlorophyll and other photosynthetic pigments are part of the thylakoid membranes. These membranes, like the inner mitochondrial membrane (see Chapter 7), are involved in ATP synthesis. (Photosynthetic prokaryotes have no chloroplasts, but thylakoid membranes are often arranged around the periphery of the cell as infoldings of the plasma membrane.)

Chlorophyll is found in the thylakoid membrane

Thylakoid membranes contain several kinds of pigments, which are substances that absorb visible light. Different pigments absorb light of different wavelengths. Chlorophyll, the main pigment of photosynthesis, absorbs light primarily in the

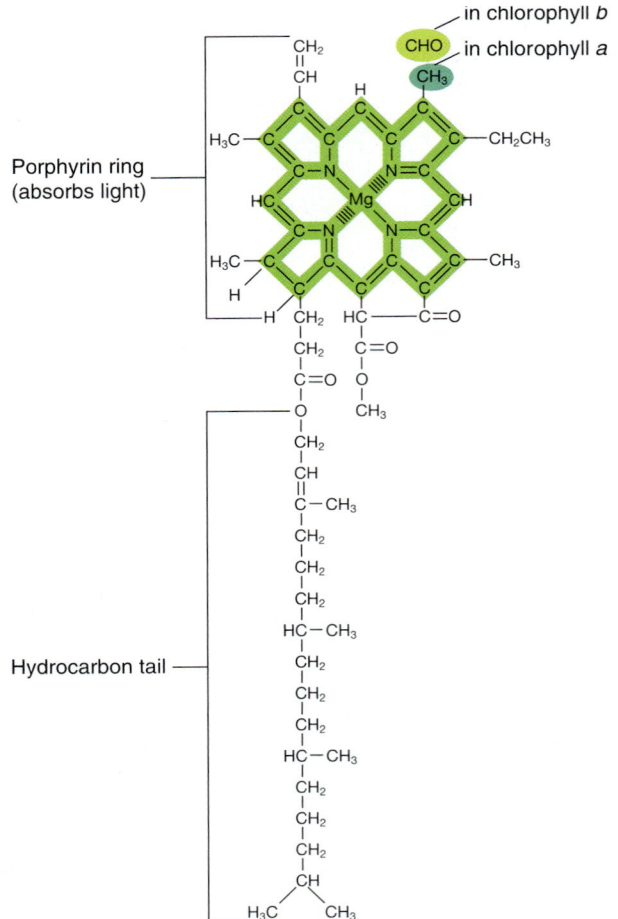

Figure 8–5 Chlorophyll structure. Chlorophyll consists of a porphyrin ring and a hydrocarbon tail. The porphyrin ring, with a magnesium atom in its center, contains a system of alternating double and single bonds; these are commonly found in molecules that strongly absorb visible light. At the top right corner of the diagram, the methyl group (—CH_3) distinguishes chlorophyll *a* from chlorophyll *b*, which has a carbonyl group (—CHO) in this position.

blue and red regions of the visible spectrum. Green light is not appreciably absorbed by chlorophyll. Plants usually appear green because most of the green light that strikes them is scattered or reflected.

A chlorophyll molecule (Fig. 8–5) has two main parts: one captures energy and the other holds the molecule in place in the thylakoid membrane. Light energy is absorbed by a complex ring, called a *porphyrin ring*, made up of joined smaller rings composed of carbon and nitrogen atoms. The porphyrin ring of chlorophyll is strikingly similar to the heme portion of the red pigment hemoglobin in red blood cells. However, unlike heme, which contains an atom of iron in the center of the ring, chlorophyll contains an atom of magnesium in that position. The chlorophyll molecule is embedded in the thylakoid membrane by a long hydrophobic hydrocarbon tail. Because of their shape, many chlorophyll molecules can be grouped together like a stack of saucers. Each thylakoid membrane is filled

with precisely oriented chlorophyll molecules, an arrangement that permits transfer of energy from one molecule to another.

There are several kinds of chlorophyll. The most important is **chlorophyll *a*,** the pigment that initiates the light-dependent reactions. **Chlorophyll *b*** is an accessory pigment that also participates in photosynthesis. It differs from chlorophyll *a* only in a functional group on the porphyrin ring: the methyl group (—CH_3) in chlorophyll *a* is replaced in chlorophyll *b* by a terminal carbonyl group (—CHO). This difference shifts the wavelengths of light absorbed and reflected by chlorophyll *b*, making it yellow-green, whereas chlorophyll *a* is bright green.

Chloroplasts also have other accessory photosynthetic pigments, such as **carotenoids,** which are yellow and orange (for a brief discussion of the carotenoid β-carotene, see *Making the Connection: Molecules That Absorb Light* in Chapter 3). Carotenoids absorb different wavelengths of light than chlorophyll does and so broaden the spectrum of light that provides energy for photosynthesis. Chlorophyll may be excited by light directly or indirectly by energy passed to it from accessory pigments that have become excited by light. When a carotenoid molecule is excited, its energy can be transferred to chlorophyll *a*.

CHLOROPHYLL IS THE MAIN PHOTOSYNTHETIC PIGMENT

As we have seen, the thylakoid membrane contains more than one kind of pigment. It is possible to determine which of these pigments is mainly responsible for photosynthesis by comparing the wavelengths of light absorbed by each pigment with the wavelengths of light that are most effective in promoting the reactions of photosynthesis. An instrument called a spectrophotometer is used to measure the relative abilities of different pigments to absorb different wavelengths of light. The **absorption spectrum** of a pigment is a plot of its absorption of light of different wavelengths. Figure 8–6*a* shows the absorption spectra for chlorophylls *a* and *b*.

The relative effectiveness of these different wavelengths of light in photosynthesis is given by an **action spectrum** of photosynthesis (Fig. 8–6*b*). The first action spectrum was obtained in one of the classic experiments in biology. In 1883 the German biologist T. W. Engelmann carried out an experiment that took advantage of the shape of the chloroplast in a species of *Spirogyra,* a green alga that occurs as long, filamentous strands in freshwater habitats, especially slow-moving or still waters (Fig. 8–7*a*). The individual cells of *Spirogyra* are exquisite, each containing a long, spiral-shaped, emerald-green chloroplast embedded in cytoplasm. Engelmann exposed these cells to a color spectrum produced by passing light through a prism. He reasoned that if chlorophyll were indeed responsible for photosynthesis, then it would take place most rapidly in the areas where the chloroplast was illuminated by the colors most readily absorbed by chlorophyll.

(a)   (b)

Figure 8–6 The absorption spectra for chlorophylls *a* and *b* and the action spectrum for photosynthesis. *(a)* Chlorophylls *a* and *b* absorb light mainly in the blue and red regions of visible light. *(b)* The action spectrum of photosynthesis shows how effective various wavelengths of light are in powering photosynthesis.

Yet how could photosynthesis be measured in those technologically unsophisticated days? Engelmann knew that photosynthesis produces oxygen and that certain motile bacteria are attracted to areas of high oxygen concentration (Fig. 8–7*b*). He determined the action spectrum of photosynthesis by observing that the bacteria swam toward the portions of

Spirogyra located in the red and blue regions of the spectrum. The fact that the bacteria did not move toward red and blue areas when *Spirogyra* was absent showed that bacteria are not merely attracted to any region where red or blue light is present. Because the action spectrum of photosynthesis closely matched the absorption spectrum of chlorophyll, Engelmann

(a)   (b)

Figure 8–7 The first action spectrum of photosynthesis. *(a)* LM of filaments of *Spirogyra* sp., the green alga that Engelmann used in his classic experiment. *(b)* Engelmann illuminated a filament of *Spirogyra* with light that had been passed through a prism. In this way, different parts of the filament were exposed to different wavelengths of light. He estimated the rate of photosynthesis indirectly, by observing the movement of aerobic bacteria toward the portions of the algal filament emitting the most oxygen. Watching through a microscope, Engelmann observed that the bacteria aggregated most densely along the cells in the blue and red portions of the spectrum. This indicated that blue and red light works most effectively for photosynthesis. *(a, T.E. Adams/Visuals Unlimited)*

PHOTOSYNTHESIS AND AEROBIC RESPIRATION

In what ways are photosynthesis and aerobic respiration alike, and how do they differ? Both are redox processes that are intimately connected with the energy requirements of organisms. Although the series of steps by which photosynthesis and aerobic respiration occur are quite different, their overall equations are almost exactly opposite.

In photosynthesis:

$$6\ CO_2 + 12\ H_2O \xrightarrow{\text{Light energy}} C_6H_{12}O_6 + 6\ O_2 + 6\ H_2O$$

In aerobic respiration:

$$C_6H_{12}O_6 + 6\ O_2 + 6\ H_2O \longrightarrow 6\ CO_2 + 12\ H_2O + \text{ATP energy}$$

The following table compares other aspects of these processes.

	Photosynthesis	Aerobic Respiration
Raw materials	CO_2, H_2O	$C_6H_{12}O_6$, O_2
End products	$C_6H_{12}O_6$, O_2	CO_2, H_2O
Which cells have these processes?	Cells that contain chlorophyll (certain cells of plants, algae, and some bacteria)	Every actively metabolizing cell has aerobic respiration or some other energy-releasing pathway.
Sites involved (in eukaryotic cells)	Chloroplasts	Cytosol (glycolysis); mitochondria
ATP production	By photophosphorylation (a chemiosmotic process)	By substrate-level phosphorylation and by oxidative phosphorylation (a chemiosmotic process)
Principal electron transfer compound	$NADP^+$ is reduced to form $NADPH$*	NAD^+ is reduced to form $NADH$*
Location of electron transport chain	Thylakoid membrane	Mitochondrial inner membrane (cristae)
Source of electrons for electron transport chain	In noncyclic phosphorylation: H_2O (undergoes photolysis to yield electrons, protons, and oxygen)	Immediate source: NADH, $FADH_2$ Ultimate source: glucose or other carbohydrate
Terminal electron acceptor for electron transport chain	In noncyclic phosphorylation: $NADP^+$ (becomes reduced to form NADPH)	O_2 (becomes reduced to form H_2O)

*NADPH and NADH are very similar hydrogen (i.e., electron) carriers, differing only in a single phosphate group. However, NADPH generally works with enzymes in anabolic pathways, such as photosynthesis. NADH is associated with catabolic pathways, such as cellular respiration.

concluded that chlorophyll in the chloroplasts (and not another compound in another organelle) is responsible for photosynthesis. Numerous studies using sophisticated instruments have since confirmed Engelmann's conclusions.

The action spectrum of photosynthesis does not parallel the absorption spectrum of chlorophyll exactly (Fig. 8–6). This difference occurs because accessory pigments, such as carotenoids, transfer some of the energy of excitation produced by green light to chlorophyll molecules. The presence of these accessory photosynthetic pigments can be demonstrated by chemical analysis of almost any leaf, although it is obvious in temperate climates when leaves change color in the fall. Toward the end of the growing season, chlorophyll breaks down (and its magnesium is stored in the permanent tissues of the tree), leaving accessory pigments in the leaves.

PHOTOSYNTHESIS IS THE CONVERSION OF LIGHT ENERGY TO CHEMICAL BOND ENERGY

During photosynthesis, a cell uses light energy captured by chlorophyll to power the synthesis of carbohydrates. The overall reactions of photosynthesis can be summarized as follows:

$$6\ CO_2 + 12\ H_2O \xrightarrow[\text{Chlorophyll}]{\text{Light}} C_6H_{12}O_6 + 6\ O_2 + 6\ H_2O$$

Carbon dioxide   Water   Glucose   Oxygen   Water

The equation is typically written in the form given above, with H_2O on both sides, because water is a reactant in some reactions and a product in others. However, because there is no

Chloroplast — Granum — Stroma — Thylakoid

Calvin cycle

ADP

ATP

NADP⁺

NADPH

Light-dependent reactions

Carbon fixation reactions

Carbohydrates

H₂O

O₂

CO₂

Figure 8–8 A summary of photosynthesis. Photosynthesis consists of light-dependent reactions, which occur in association with the thylakoids, and carbon fixation reactions, which occur in the stroma.

net yield of H_2O, we can simplify the summary equation of photosynthesis for purposes of discussion:

$$6\,CO_2 + 6\,H_2O \xrightarrow[\text{Chlorophyll}]{\text{Light}} C_6H_{12}O_6 + 6\,O_2$$

Carbon dioxide   Water   Glucose   Oxygen

When we analyze this process, it appears that hydrogen atoms are transferred from H_2O to CO_2 to form carbohydrate, and so we recognize it as an oxidation-reduction (redox) reaction. As you learned in Chapter 6, in a redox reaction one or more electrons, usually as part of one or more hydrogen atoms, are transferred from an electron donor (a reducing agent) to an electron acceptor (an oxidizing agent).

$$6\,CO_2 + 6\,H_2O \xrightarrow[\text{Chlorophyll}]{\text{Light}} C_6H_{12}O_6 + 6\,O_2$$

—————— Reduction ——————

—————— Oxidation ——————

Carbon dioxide   Water   Glucose   Oxygen

When the electrons are transferred, some of their energy is transferred as well. However, the summary equation of photosynthesis is somewhat misleading because no direct transfer of hydrogen atoms actually occurs. The summary equation describes what happens but not how it happens. The "how" is much more complex and involves many steps, many of which are redox reactions. (See *Making the Connection: Photosynthesis and Aerobic Respiration* on page 180 for a comparison of these two biologically important redox processes.)

The reactions of photosynthesis are divided into two parts: the light-dependent reactions (the *photo* part of photosynthe-

sis) and the carbon fixation reactions (the *synthesis* part of photosynthesis). Each set of reactions occurs in a different part of the chloroplast: the light-dependent reactions in association with the thylakoids, and the carbon fixation reactions in the stroma (Fig. 8–8).

ATP and NADPH are the products of the light-dependent reactions

Light energy is converted to chemical energy in the **light-dependent reactions,** which are associated with the thylakoids. The light-dependent reactions begin as chlorophyll captures light energy, which causes one of its electrons to move to a higher energy state. The energized electron is transferred to an acceptor molecule and is replaced by an electron from H_2O. When this happens, H_2O is split and molecular oxygen is released (Fig. 8–9). Some of the energy of the energized electrons is used to phosphorylate ADP, forming ATP. In addition, the coenzyme **nicotinamide adenine dinucleotide phosphate (NADP⁺)** is reduced, forming NADPH.[2] The products of the light-dependent reactions, ATP and NADPH, are both needed in the endergonic carbon fixation reactions.

Carbohydrates are produced during the carbon fixation reactions

The ATP and NADPH molecules produced during the light-dependent phase are suited for transferring chemical energy

[2]Although the correct way to write the reduced form of NADP⁺ is NADPH + H⁺, for simplicity's sake we present the reduced form as NADPH throughout the chapter.

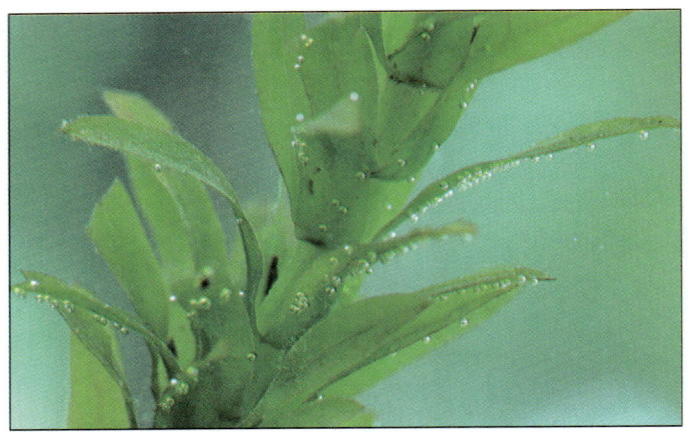

Figure 8–9 Oxygen produced by photosynthesis. On sunny days the oxygen released by aquatic plants may sometimes be visible as bubbles in the water. This plant (*Elodea* sp.) is actively carrying on photosynthesis. *(Bernd Wittich/Visuals Unlimited)*

but not for long-term energy storage. For this reason, some of their energy is transferred to chemical bonds in carbohydrates, which can be produced in large quantities and stored for future use. Known as **carbon fixation** or **CO₂ fixation,** these reactions "fix" carbon atoms from CO_2 to existing skeletons of organic molecules. Because the carbon fixation reactions have no direct requirement for light, they are sometimes referred to as the light-independent reactions or the "dark" reactions. However, they certainly do not require darkness; in fact, many of the enzymes involved are much more active in

the light than in the dark. Furthermore, carbon fixation reactions depend on the products of the light-dependent reactions. Carbon fixation reactions take place in the stroma.

Now that we have presented an overview of photosynthesis, let's examine the entire process more closely. (Table 8–1 provides a detailed summary of photosynthesis.)

THE LIGHT-DEPENDENT REACTIONS CONVERT LIGHT ENERGY TO CHEMICAL ENERGY

In the light-dependent reactions, the radiant energy from sunlight is used to make ATP and to reduce $NADP^+$, forming NADPH. The light energy captured by chlorophyll is temporarily stored in these two compounds. The light-dependent reactions are summarized as follows:

$$12 \text{ H}_2\text{O} + 12 \text{ NADP}^+ + 18 \text{ ADP} + 18 \text{ P}_i \xrightarrow[\text{Chlorophyll}]{\text{Light}}$$
$$6 \text{ O}_2 + 12 \text{ NADPH} + 18 \text{ ATP}$$

Photosystems I and II include antenna complexes that trap light

The light-dependent reactions of photosynthesis begin when chlorophyll *a* and/or accessory pigments absorb light. According to the currently accepted model, chlorophyll molecules and accessory pigments are organized in the thylakoid membrane

---

TABLE 8–1 Summary of Photosynthesis

Reaction Series	Summary of Process	Needed Materials	End Products
A. Light-dependent reactions (take place in thylakoid membranes)	Energy from sunlight used to split water, manufacture ATP, and reduce $NADP^+$		
1. Photochemical reactions	Chlorophyll activated; reaction center gives up photoexcited electron to electron acceptor	Light energy; pigments (chlorophyll)	Electrons
2. Electron transport	Electrons are transported along chain of electron acceptors in thylakoid membranes; electrons reduce $NADP^+$; splitting of water provides some of H^+ that accumulates inside thylakoid space	Electrons, $NADP^+$, H_2O, electron acceptors	NADPH, O_2
3. Chemiosmosis	H^+ are permitted to move across the thylakoid membrane down their gradient; they cross the membrane through special channels in ATP synthase complex; energy released is used to produce ATP	Proton gradient, ADP + P_i	ATP
B. Carbon fixation reactions (take place in stroma)	Carbon fixation: carbon dioxide is used to make carbohydrate	Ribulose bisphosphate, CO_2, ATP, NADPH, necessary enzymes	Carbohydrates, ADP + P_i, $NADP^+$

Figure 8–10 An antenna complex. Chlorophyll molecules and accessory pigments are arranged in arrays that act as light-harvesting antennae. When a molecule in the antenna complex absorbs a photon, the energy of that photon is funneled into the reaction center. When this energy reaches a chlorophyll in the reaction center, an electron becomes energized and is accepted by a primary electron acceptor.

into units called **antenna complexes.** The pigments are arranged as highly ordered groups of 200 to 300 molecules associated with specific enzymes and other proteins. Each antenna complex traps light and transfers the energy to a **reaction center,** a complex of chlorophyll molecules and proteins that participates directly in photosynthesis (Fig. 8–10). Light energy is converted to chemical energy in the reaction centers by a series of electron transfer reactions.

Two types of photosynthetic units, designated Photosystem I and Photosystem II, are involved in photosynthesis. Their reaction centers are distinguishable because they are associated with proteins in a way that causes a slight shift in their absorption spectra. Ordinary chlorophyll *a* has a strong absorption peak at about 660 nanometers. In contrast, the chlorophyll *a* molecules that make up the reaction center associated with **Photosystem I** have an absorption peak at 700 nanometers and are referred to as **P700.** (The *P* in P700 stands for "pigment.") The reaction center of **Photosystem II (P680)** is made up of chlorophyll *a* molecules with an absorption peak of about 680 nanometers.

When a pigment molecule absorbs light energy, that energy is passed from one pigment molecule to another until it reaches the reaction center. When the energy reaches a molecule of P700 or P680 at the reaction center, an electron is then raised to a higher energy level. This energized electron can be donated to an electron acceptor that becomes reduced in the process.

Noncyclic photophosphorylation produces ATP and NADPH

Both Photosystems I and II are used in **noncyclic photophosphorylation,** which is the more common light-dependent reaction. In noncyclic photophosphorylation, light ener-

gizes electrons, which pass down an **electron transport chain** from the ultimate electron source, H_2O, to the terminal electron acceptor, $NADP^+$. The zigzag pathway of electrons shown in Figure 8–11 is sometimes called the **Z scheme.** For every two electrons that enter this pathway, there is an energy yield of as much as two ATP molecules and one NADPH molecule.

We begin our discussion of noncyclic photophosphorylation with the events associated with Photosystem I. A pigment molecule in an antenna complex associated with Photosystem I absorbs a photon of light. The absorbed energy is transferred to the reaction center, where it excites an electron in a molecule of P700. This energized electron is transferred to a primary electron acceptor, which transfers it to ferredoxin, a membrane-bound, iron-containing protein. Ferredoxin then transfers the electron to $NADP^+$. The electron transport chain must furnish two electrons to reduce $NADP^+$ to NADPH. When $NADP^+$ accepts the two electrons, they unite with a proton (H^+); hence the reduced form of $NADP^+$ is NADPH, which is released into the stroma. P700 becomes positively charged when it gives up an electron to the primary electron acceptor; the missing electron is replaced by one donated by Photosystem II.

Like Photosystem I, Photosystem II is activated when a pigment molecule in an antenna complex absorbs a photon of light energy. The energy is transferred to the reaction center, where it causes an electron in a molecule of P680 to move to a higher energy level. This energized electron is accepted by a primary electron acceptor and then passes through a chain of acceptor molecules until it is donated to P700 in Photosystem I.

A molecule of P680 that has given up an energized electron to the primary electron acceptor is positively charged. This P680 molecule is an oxidizing agent so strong that it is capable of pulling electrons away from an oxygen atom that is part of a H_2O molecule. In a reaction catalyzed by a unique enzyme, the process of **photolysis** (literally "light-splitting") breaks water into its components: two electrons, two protons, and oxygen. Each electron is donated to a P680 molecule, and the protons are released into the thylakoid interior space. Because oxygen does not exist in atomic form, the oxygen produced by splitting one H_2O molecule is written $\frac{1}{2} O_2$. Two H_2O molecules must be split to yield one molecule of oxygen (O_2), which is ultimately released into the atmosphere. The photolysis of H_2O is a remarkable reaction, but its name is somewhat misleading because it implies that H_2O is broken by light. Actually, light breaks H_2O indirectly, by oxidizing P680 molecules.

Noncyclic photophosphorylation is a continuous linear process

In the presence of light, there is a continuous, one-way flow of electrons from the ultimate electron source, H_2O, to the terminal electron acceptor, $NADP^+$. Water undergoes enzymatically catalyzed photolysis to replace energized electrons donated to the electron transport chain by molecules of P680 in Photosystem II. These electrons travel down the electron trans-

Figure 8–11 Noncyclic photophosphorylation. In noncyclic photophosphorylation, also called the Z scheme, the formation of ATP is coupled to the one-way flow of energized electrons from H_2O to $NADP^+$. Single electrons actually pass down the electron transport chain; two are shown in the figure because two are required to form one NADPH. Electrons are supplied to the system from the splitting of H_2O by Photosystem II, with the release of O_2 as a byproduct. When Photosystem II is activated by absorbing photons, electrons are passed along an electron transport chain and are eventually donated to Photosystem I. Electrons in Photosystem I are "re-energized" by the absorption of additional light energy and are passed to $NADP^+$.

port chain that connects Photosystem II with Photosystem I and replace the energized electrons donated by P700 that ultimately reduce $NADP^+$.

As electrons are transferred along the electron transport chain that connects Photosystem II with Photosystem I, they lose energy. Some of the energy released is used to pump protons across the thylakoid membrane, from the stroma to the thylakoid interior space, producing a proton gradient. The energy of this proton gradient is harnessed to produce ATP from ADP by chemiosmosis, which is discussed shortly. ATP and NADPH, the products of the light-dependent reactions, are released into the stroma, where both are required in the carbon fixation reactions.

Cyclic photophosphorylation produces ATP but no NADPH

Only Photosystem I is involved in **cyclic photophosphorylation,** the simplest light-dependent reaction. The pathway is cyclic because energized electrons that originate from P700 at the reaction center eventually return to P700. In the presence of light, there is a continuous flow of electrons through an electron transport chain within the thylakoid membrane. As they are passed from one acceptor to another, the electrons lose energy, some of which is used to pump protons across the thylakoid membrane. An enzyme (ATP synthase, discussed shortly) in the thylakoid membrane uses the energy of the pro-

TABLE 8–2   A Comparison of Noncyclic and Cyclic Photophosphorylation

	Noncyclic photophosphorylation	Cyclic photophosphorylation
Electron source	H_2O	None—electrons cycle through the system
Oxygen released?	Yes (from H_2O)	No
Terminal electron acceptor	$NADP^+$	None—electrons cycle through the system
Form in which energy is temporarily captured	ATP (by chemiosmosis); NADPH	ATP (by chemiosmosis)
Photosystem(s) required	PSI (P700) and PSII (P680)	PSI (P700) only

THE EVOLUTION OF PHOTOSYSTEMS I AND II

Photosynthesis is an extremely ancient biological process that has apparently changed a great deal since it first appeared more than 3 billion years ago. The use of light energy to manufacture organic molecules first evolved in ancient bacteria that were similar to the green sulfur bacteria existing today.

Initially there was only one photosystem, Photosystem I, which used a green pigment called *bacteriochlorophyll* to gather light energy. Photosystem I operated alone, generating ATP from light energy by cyclic photophosphorylation. However, cyclic photophosphorylation does not provide the reducing power of NADPH, which is needed to manufacture carbohydrate molecules from CO_2. (Recall that in cyclic photophosphorylation, the electrons from

chlorophyll are not passed to $NADP^+$ but are instead returned to chlorophyll.) Ancient photosynthetic bacteria, like some of their modern counterparts, used electron donors such as hydrogen sulfide (H_2S) rather than H_2O to generate the reducing power needed to manufacture carbohydrates in photosynthesis:

$$H_2S \longrightarrow S + 2\,H^+ + 2\,e^-$$

$$2\,H^+ + 2\,e^- + NADP^+ \longrightarrow$$
$$NADPH + H^+$$

This process is not very efficient, however; bacteriochlorophyll has enough oxidative potential to extract electrons from H_2S but not sufficient oxidative potential to extract electrons from H_2O.

Around 3.1 to 3.5 billion years ago, a new group of prokaryotes called **cyanobacteria** evolved (see Chapter 20). These ancient cyanobacteria were probably similar to modern cyanobacteria, which have light-requiring reactions that are similar to those of photosynthetic eukaryotes, including plants. They possess chlorophyll *a* instead of bacteriochlorophyll and can carry out noncyclic photophosphorylation because they have Photosystem II in addition to Photosystem I. Water provides the electrons required to generate NADPH, which in turn provides the reducing power required to manufacture carbohydrate molecules from CO_2.

ton gradient to manufacture ATP. NADPH is not produced, H_2O is not split, and oxygen is not generated. By itself, cyclic photophosphorylation could not serve as the basis of photosynthesis because, as we explain in the next section, NADPH is required to reduce CO_2 to carbohydrate.

The significance of cyclic photophosphorylation to photosynthesis in plants is not yet certain. Cyclic photophosphorylation may occur in plant cells when there is too little $NADP^+$ to accept electrons from ferredoxin. Biologists generally agree that this process was used by ancient bacteria to produce ATP from light energy (see *Focus On: The Evolution of Photosystems I and II*). A reaction pathway analogous to cyclic photophosphorylation in plants is present in some modern photosynthetic bacteria. Noncyclic and cyclic photophosphorylation are compared in Table 8–2 on page 184.

ATP synthesis occurs by chemiosmosis

Each member of the electron transport chain, embedded in the thylakoid membrane, can exist in an oxidized (lower energy) form and a reduced (higher energy) form. The electron accepted from P680 by the primary electron acceptor is highly energized; it is passed from one carrier to the next in a series of exergonic redox reactions, losing some of its energy at each step. Some of the energy given up by the electron is not lost by the system, however; it is used to drive the synthesis of ATP (an endergonic reaction). Because the synthesis of ATP (that is, the phosphorylation of ADP) is coupled to the transport of electrons that have been energized by photons of light, the process is called **photophosphorylation.**

The chemiosmotic model explains the coupling of ATP synthesis and electron transport

As discussed earlier, the pigments and electron acceptors of the light-dependent reactions are embedded in the thylakoid membrane. Energy released from electrons traveling through the chain of acceptors is used to pump protons from the stroma, across the thylakoid membrane, and into the thylakoid interior space (Fig. 8–12). Thus, the pumping of protons results

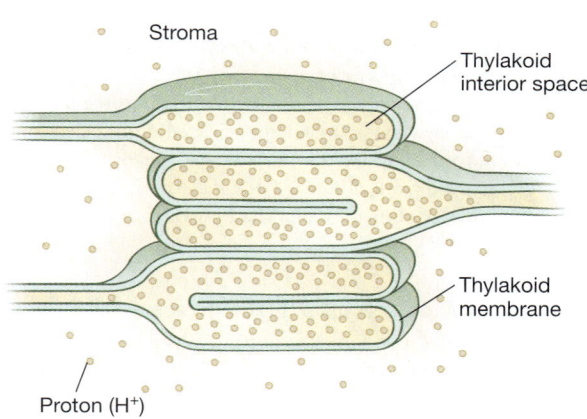

Figure 8–12 Accumulation of protons (H^+) within the thylakoid interior space. As electrons move down the electron transport chain, protons move from the stroma to the thylakoid interior space, creating a proton gradient. The greater concentration of H^+ in the thylakoid interior space lowers the pH.

Figure 8–13 A detailed look at electron transport and chemiosmosis. The orange arrows indicate the pathway of electrons along the electron transport chain in the thylakoid membrane. The electron carriers within the membrane become alternately reduced and oxidized as they accept and donate electrons. The energy released during electron transport is used to transport H^+ from the stroma to the thylakoid interior space, where a high concentration of H^+ accumulates. The H^+ are prevented from diffusing back into the stroma except through special channels in ATP synthase in the thylakoid membrane. The flow of the H^+ through ATP synthase generates ATP.

in the formation of a proton gradient across the thylakoid membrane. Because protons are actually hydrogen ions (H^+), the accumulation of protons causes the pH of the thylakoid interior to fall to a pH of about 5 in the thylakoid interior space, as compared to a pH of about 8 in the stroma. This difference of about 3 pH units across the thylakoid membrane means there is an approximately 1000-fold difference in hydrogen ion concentration.

The proton gradient has a great deal of free energy be-

cause of its state of low entropy. How does the chloroplast convert that energy to a more useful form? According to the general principles of diffusion, the concentrated protons inside the thylakoid might be expected to diffuse out readily. However, they are prevented from doing so because the thylakoid membrane is impermeable to H^+ except through certain channels formed by an enzyme called **ATP synthase.** ATP synthase, a transmembrane protein, forms complexes so large they can be seen in electron micrographs; these project into the stroma. As the protons diffuse through an ATP synthase complex, free energy decreases as a consequence of an increase in entropy. Each ATP synthase complex couples this exergonic process of diffusion down a concentration gradient to the endergonic process of phosphorylation of ADP to form ATP, which is released into the stroma (Fig. 8–13).

The mechanism by which the phosphorylation of ADP is coupled to diffusion down a proton gradient is called **chemiosmosis.** As the essential connection between the electron transport chain and the phosphorylation of ADP, chemiosmosis is a basic mechanism of energy coupling in cells. You may recall from Chapter 7 that chemiosmosis also occurs in cellular respiration.

THE CARBON FIXATION REACTIONS REQUIRE ATP AND NADPH

In carbon fixation, the energy of ATP and NADPH is used in the formation of organic molecules from CO_2. The carbon fixation reactions may be summarized as follows:

$$12 \text{ NADPH} + 18 \text{ ATP} + 6 \text{ CO}_2 \longrightarrow$$
$$C_6H_{12}O_6 + 12 \text{ NADP}^+ + 18 \text{ ADP} + 18 \text{ P}_i + 6 \text{ H}_2O$$

Most plants use the Calvin (C₃) cycle to fix carbon

Carbon fixation occurs in the stroma through a sequence of reactions known as the **Calvin cycle.** M. Calvin, A. Benson,

and others at the University of California were able to elucidate the details of this cycle; for his work, Calvin was awarded a Nobel Prize in 1961 (Fig. 8–14).

The Calvin cycle begins when a molecule of CO_2 reacts with a highly reactive phosphorylated five-carbon compound, **ribulose bisphosphate (RuBP).** This reaction is catalyzed by the enzyme **ribulose bisphosphate carboxylase,** also known as **Rubisco.** The product of this reaction is an unstable, six-carbon intermediate, which immediately breaks down into two molecules of **phosphoglycerate (PGA)** with three carbons each. The carbon that was originally part of a CO_2 molecule is now part of a carbon skeleton; the carbon has been "fixed." Because the product of the initial carbon fixation reaction is a three-carbon compound, the Calvin cycle is also known as the **C₃ pathway.** A total of six carbons must be fixed in this way to produce the equivalent of one molecule of glucose or some other hexose (six-carbon sugar), such as fructose. At the end of each cycle, the starting material, RuBP, is re-formed.

With the energy and reducing power from ATP and NADPH (both produced in the light-dependent reactions), the PGA molecules are converted to **glyceraldehyde-3-phosphate (G3P).** As shown in Figure 8–15, for every six carbons that enter the cycle as CO_2, six carbons can leave the system as two molecules of G3P, to be used in carbohydrate synthesis. Each of these three carbon molecules of G3P is essentially half a hexose molecule. (In fact, you may recall that G3P is a key intermediate in the splitting of sugar in glycolysis; see Figs. 7–3 and 7–4.) The reaction of two molecules of G3P is exergonic and can lead to the formation of glucose or fructose. In some plants, glucose and fructose are then joined to produce sucrose (common table sugar). Sucrose can be harvested from

Figure 8–14 Experimental apparatus of Calvin and Benson. The classic experiments carried out by Calvin, Benson, and others in the 1950s elucidated the steps in carbon fixation of photosynthesis. Calvin and his colleagues grew algae in the green "lollipop." CO_2 labeled with ^{14}C was bubbled through the algae, and they were periodically killed by dumping the "lollipop" contents into a beaker of boiling alcohol. By identifying which compounds contained the ^{14}C at different times, Calvin was able to determine the steps of carbon fixation in photosynthesis. *(Courtesy of Melvin Calvin, University of California, Berkeley)*

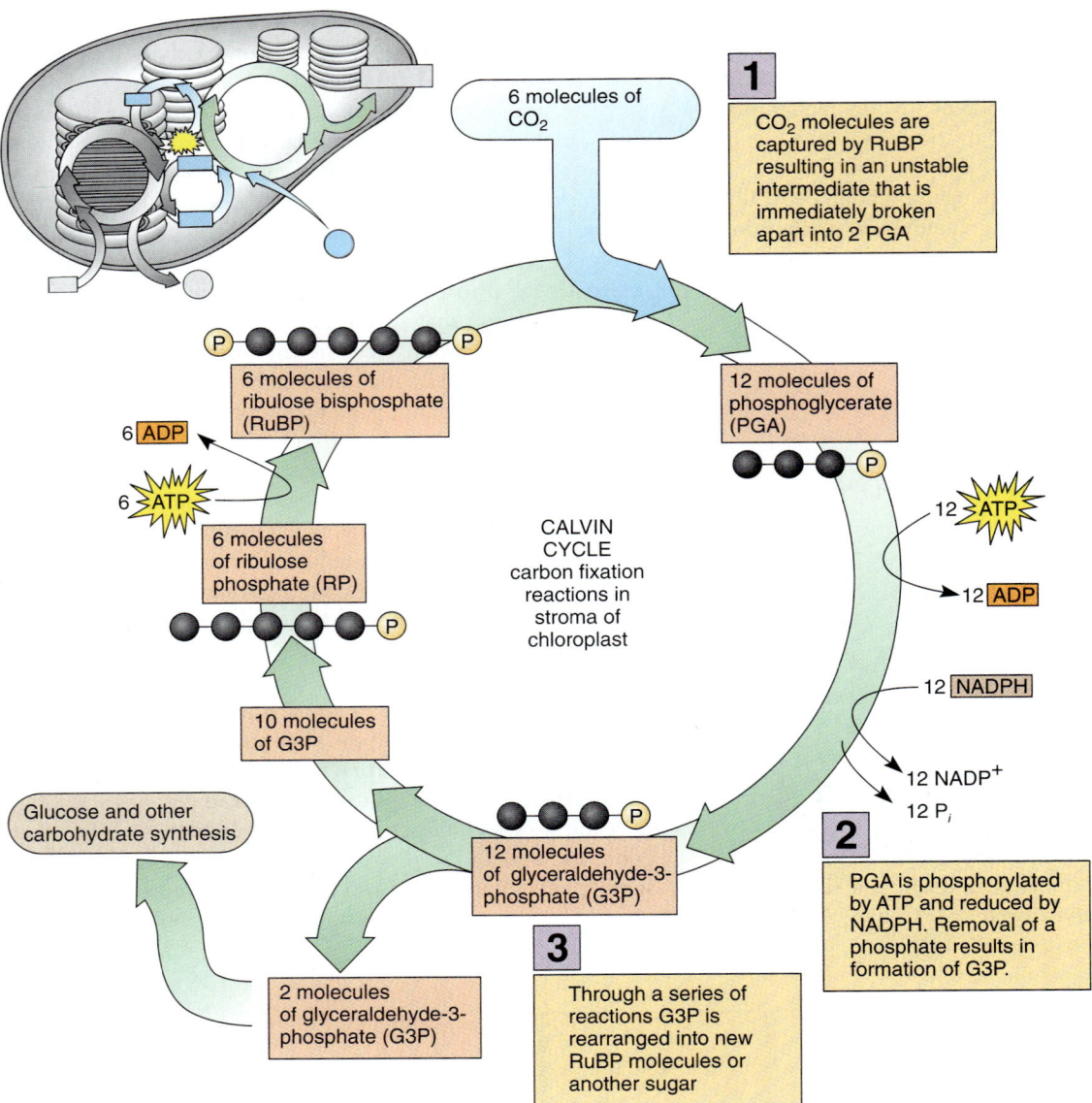

Figure 8–15 A detailed look at the Calvin cycle. This diagram, which shows carbon atoms as black balls, demonstrates that six molecules of CO_2 must be "fixed" (incorporated into preexisting carbon skeletons) to produce one molecule of a six-carbon sugar such as glucose. Two glyceraldehyde-3-phosphate (G3P) molecules "leave" the cycle for every glucose formed. Although these reactions do not require light directly, the energy that drives the Calvin cycle comes from ATP and NADPH, which are the products of the light-dependent reactions.

sugar cane, sugar beets, and maple sap. The plant cell also uses glucose to produce starch or cellulose.

Notice that, although two G3P molecules are removed from the cycle, ten G3P molecules remain; this represents 30 carbon atoms in all. Through a complex series of reactions, these 30 carbons and their associated atoms are rearranged into six molecules of ribulose phosphate, each of which becomes phosphorylated to produce RuBP, the very five-carbon compound with which the cycle started. These RuBP molecules

can begin the process of CO_2 fixation and eventual G3P production once again.

In summary, the inputs required for the carbon fixation reactions are six molecules of CO_2, phosphates transferred from ATP, and electrons (as hydrogen) from NADPH. In the end, the six carbons from the CO_2 can be accounted for by the harvest of a hexose molecule. The remaining G3P molecules are used to synthesize the RuBP molecules with which more CO_2 molecules may combine.

NUTRITION AND METABOLIC DIVERSITY

How can we categorize organisms on the basis of their nutritional needs? It is helpful to understand that nutrition has two main components: (1) how the organism obtains the carbon atoms required to make up the carbon skeletons of its organic molecules, and (2) how the organism obtains energy. Organisms obtain carbon in two ways. **Autotrophs** are able to carry out carbon fixation; they use CO_2 as a carbon source. **Heterotrophs** cannot fix carbon; they use preformed organic molecules (produced by other organisms) as a carbon source.

Energy can come from chemical nutrients or from light. **Photoautotrophs** use light as their primary energy source. Plants, algae, and some bacteria are photoautotrophs (see figures *a*, *b*, and *c*); they use light as the source of the energy required to carry out

carbon fixation. **Chemoautotrophs** are bacteria that obtain their energy from the oxidation of reduced inorganic molecules such as hydrogen sulfide (H_2S), nitrite (NO_2^-), or ammonia (NH_3). Some of this energy is then used to carry out carbon fixation. (See *Focus On: Life without the Sun* in Chapter 52 for a discussion of one group of chemoautotrophs.)

All animals, fungi, and most bacteria are **chemoheterotrophs** and use preformed organic molecules as a source of both energy and carbon. A few plants are chemoheterotrophs as well (see figure *d*). A few bacteria are **photoheterotrophs,** organisms able to use light energy but unable to carry out carbon fixation. Photoheterotrophs must obtain carbon from organic sources (as "food").

(a)

Photoautotrophs. *(a)* The prayer plant (*Maranta* sp.). Plants are the main photoautotrophs on land. *(b)* A kelp (*Macrocystis integrifolia*), a type of alga. Algae range in size from those that are microscopic to large seaweeds. They are important photoautotrophs in aquatic environments. *(c)* A photoautotrophic cyanobacterium (*Nostoc* sp.) that photosynthesizes like plants and algae. *(d)* Although the vast majority of plants are photoautotrophs, there are exceptions. This plant, commonly known as Indian pipes (*Monotropa uniflora*), lives underground (except for its flowers, which are shown) and has lost the capacity to perform photosynthesis. It is a chemoheterotroph that obtains energy and carbon from organic molecules produced by other plants and supplied by soil-dwelling fungi. *(a, Jerome Wexler/PhotoResearchers, Inc; b, Flip Nicklin/Minden Pictures; c, R. Calentine/Visuals Unlimited; d, Carlyn Iverson)*

(b)

(c)

5 μm

(d)

The initial carbon fixation step differs in C₄ plants and in CAM plants

Because CO_2 is not a very abundant gas (comprising only about 0.03% of the atmosphere), it is not easy for plants to obtain the CO_2 they need. This problem is complicated by the fact that gas exchange can occur only across a moist surface. The surfaces of leaves and other exposed plant parts are covered with a waterproof layer that helps prevent excess loss of water vapor. Entry and exit of gases is therefore limited to tiny pores, called **stomata** (sing., *stoma*), usually concentrated on

the undersides of the leaves. These openings lead to the interior of the leaf, which is made up of chloroplast-containing cells, known as the **mesophyll,** with many air spaces and a very high concentration of water vapor. The stomata open and close in response to such environmental factors as water content and light intensity. When conditions are hot and dry, the stomata close to reduce the loss of water vapor. As a result, the supply of CO_2 is greatly diminished. Ironically, CO_2 is potentially less available at the very times when maximum sunlight is available to power the light-dependent reactions.

Many plant species living in hot, dry environments have

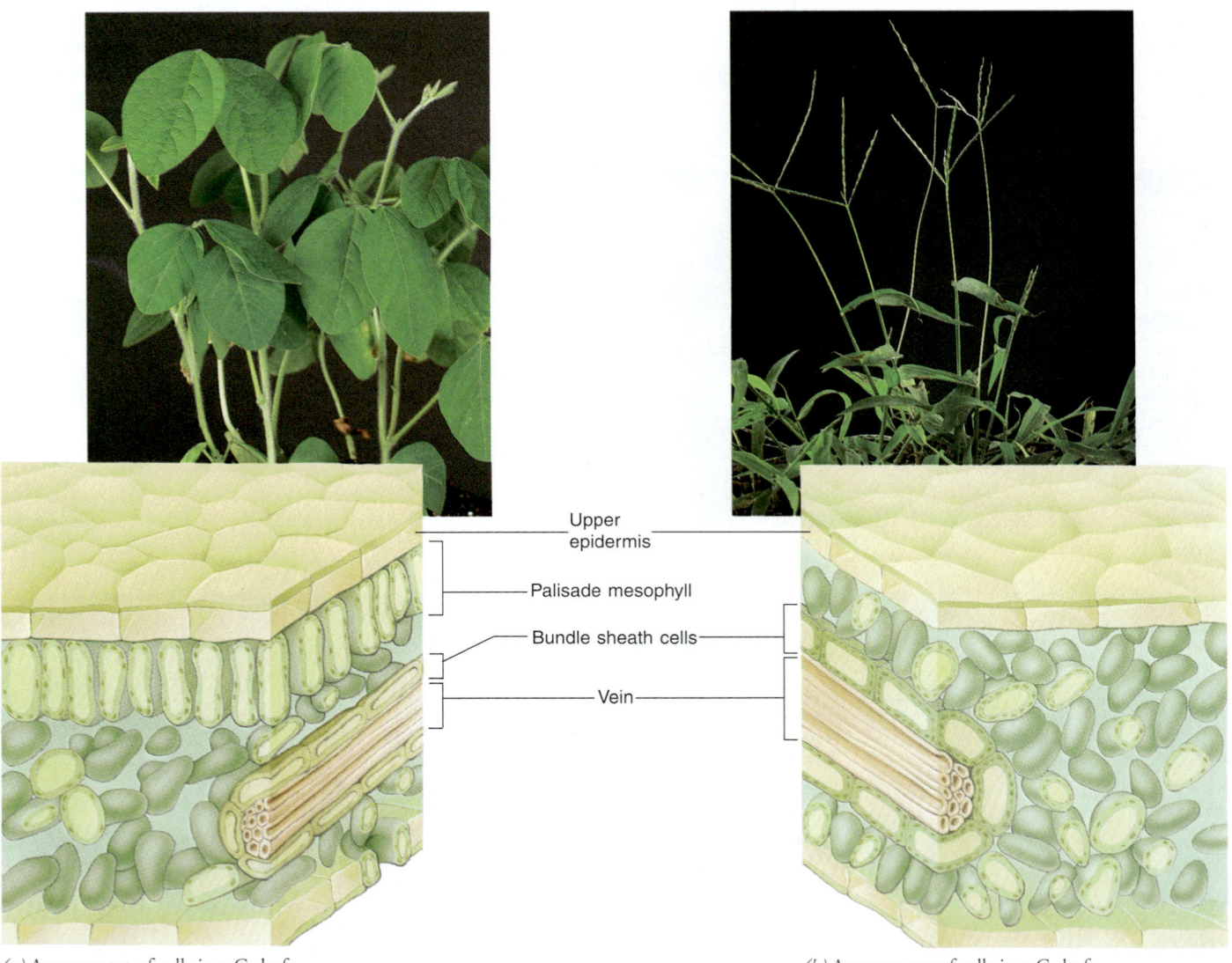

(a) Arrangement of cells in a C₃ leaf

(b) Arrangement of cells in a C₄ leaf

Figure 8–16 Comparison of C₃ and C₄ anatomy. *(a)* In C₃ plants, such as soybeans, the Calvin cycle takes place in the mesophyll cells. *(b)* In C₄ plants, such as crabgrass, reactions that fix CO_2 into four-carbon compounds take place in the mesophyll cells. The four-carbon compounds are transferred from the mesophyll cells to the bundle sheath cells, where the Calvin cycle takes place. *(a and b, Dennis Drenner)* adaptations that allow them initially to fix CO_2 through one of two pathways that help minimize water loss. These pathways, known as the C_4 pathway and the CAM pathway, take place in the cytosol. Both the C_4 and CAM pathways merely *precede* the Calvin cycle (C_3 pathway); they do not replace it.

The C_4 pathway efficiently fixes CO_2 at low concentrations

Some plants, known as **C_4 plants,** first fix CO_2 into a four-carbon compound, **oxaloacetate,** prior to the C_3 pathway. The C_4 pathway not only occurs before the C_3 pathway, it also occurs in different cells.

Leaf anatomy is usually distinctive in C_4 plants. In addition to having mesophyll cells, C_4 leaves have prominent chloroplast-containing **bundle sheath cells** (Fig. 8–16). These cells tightly encircle the veins of the leaf. The mesophyll cells in C_4 plants are closely associated with the bundle sheath cells. The **C_4 pathway** (also called the **Hatch-Slack pathway,** after M. D. Hatch and C. R. Slack, who worked out many of its steps) occurs in the mesophyll cells, whereas the Calvin cycle takes place within the bundle sheath cells.

The key component of the C_4 pathway is a remarkable enzyme that has an extremely high affinity for CO_2, binding it effectively even at unusually low concentrations. This enzyme, **PEP carboxylase,** catalyzes the reaction by which CO_2 reacts with the three-carbon compound **phosphoenolpyruvate (PEP),** forming oxaloacetate (Fig. 8–17).

In a step that requires NADPH, oxaloacetate is converted to some other four-carbon compound, usually malate. The malate then passes to chloroplasts within bundle sheath cells, where a different enzyme catalyzes the decarboxylation of malate to yield pyruvate (which has three carbons) and CO_2. NADPH is formed, replacing the one used earlier.

$$\text{Malate} + \text{NADP}^+ \longrightarrow \text{Pyruvate} + CO_2 + \text{NADPH}$$

The CO_2 released in the bundle sheath cell combines with ribulose bisphosphate and goes through the Calvin cycle in the usual manner. The pyruvate formed in the decarboxylation reaction returns to the mesophyll cell, where it reacts with ATP to regenerate phosphoenolpyruvate.

The role of the C_4 pathway is to efficiently capture CO_2 and ultimately increase its concentration within the bundle sheath cells. The concentration of CO_2 within the bundle sheath cells is about 10 to 60 times greater than the concentration in the mesophyll cells of plants having only the C_3 pathway.

The combined C_3–C_4 pathway involves the expenditure of 30 ATPs per hexose, rather than the 18 ATPs used by the C_3 pathway alone. The extra energy expense is worthwhile at high light intensity because it ensures a high concentration of CO_2 in the bundle sheath cells and permits them to carry on photosynthesis at a rapid rate. The C_4 pathway is present in addition to the C_3 pathway in many plant species and apparently has evolved independently several times. Because PEP

Figure 8–17 Summary of the C_4 pathway. CO_2 combines with phosphoenolpyruvate (PEP) in the chloroplasts of mesophyll cells, forming a four-carbon compound that is converted to malate. Malate goes to the chloroplasts of bundle sheath cells, where it is decarboxylated. The CO_2 thus released in the bundle sheath cell is used to make carbohydrate by way of the Calvin cycle.

carboxylase fixes CO_2 so efficiently, C_4 plants do not need to have their stomata open as much; they therefore tolerate higher temperatures and higher light intensities, lose less water by transpiration (evaporation), and have higher rates of photosynthesis and growth than plants that use only the Calvin cycle. Among the many quick-growing and aggressive plants that use the C_4 pathway are sugar cane, corn, and crabgrass. If sunlight is abundant, the yields of C_4 crop plants can be two to three times greater than those of C_3 plants. If this pathway could be incorporated into more of our crop plants by genetic manipulation, we might be able to greatly increase food production in some parts of the world.

When light is abundant, the rate of photosynthesis is limited by the concentration of CO_2, so C_4 plants, with their higher levels of CO_2 in bundle sheath cells, have the advantage. At lower light intensities and temperatures, C_3 plants are favored. For example, winter rye, a C_3 plant, grows lavishly in cool weather when crabgrass cannot because it requires more energy to fix CO_2.

CAM plants fix CO₂ at night

Plants living in very dry, or *xeric,* conditions have a number of structural adaptations that enable them to survive. Many xeric plants have physiological adaptations as well. For example, their stomata may open during the cooler night and close during the hot day to reduce water loss from transpiration. This is in contrast to most plants, which have stomata that are open during the day and closed at night. But xeric plants that have their stomata closed during the day cannot exchange gases for photosynthesis. (Recall that other plants typically fix CO_2 during the day, when sunlight is available.)

Many xeric plants evolved a special carbon fixation pathway called **crassulacean acid metabolism (CAM)** that in effect solves this dilemma. The name comes from the stonecrop plant family (the Crassulaceae), which possesses the CAM pathway, although it has evolved independently in some members of more than 25 other plant families, including the cactus family (Cactaceae), the lily family (Liliaceae), and the orchid family (Orchidaceae) (Fig. 8–18).

CAM plants use the enzyme PEP carboxylase to fix CO_2 during the night when stomata are open, forming oxaloacetate, which is converted to malate and stored in cell vacuoles. During the day, when stomata are closed and gas exchange cannot occur between the plant and the atmosphere, CO_2 is removed from malate by a decarboxylation reaction. Now the CO_2 is available *within the leaf tissue* to be fixed into sugar by the Calvin cycle (C₃ pathway).

The CAM pathway is very similar to the C₄ pathway but with important differences. C₄ plants initially fix CO_2 into four-carbon organic acids in mesophyll cells. The acids are later decarboxylated to produce CO_2, which is fixed by the C₃ pathway in the bundle sheath cells. In other words, the C₄ and C₃ pathways occur in *different locations* within the leaf of a C₄ plant. In CAM plants, the initial fixation of CO_2 occurs at night. Decarboxylation of malate and subsequent production of sugar from CO_2 by the normal C₃ photosynthetic pathway occur during the day. In other words, the CAM and C₃ pathways occur at *different times* within the same cell of a CAM plant.

Although it does not promote rapid growth the way that the C₄ pathway does, the CAM pathway is a very successful adaptation to xeric conditions. CAM plants are able to exchange gases for photosynthesis and to reduce water loss significantly. Plants with CAM photosynthesis survive in deserts where neither C₃ nor C₄ plants can.

Photorespiration reduces photosynthetic efficiency

Many C₃ plants, including certain agriculturally important crops such as soybeans, wheat, and potatoes, do not yield as much carbohydrate from photosynthesis as might be expected. This reduction in yield is especially significant during very hot spells in summer. On hot, dry days plants close their stomata to conserve water. Once the stomata close, photosynthesis rapidly uses up the CO_2 remaining in the leaf and produces O_2, which accumulates in the chloroplasts. Recall that the enzyme RuBP carboxylase (Rubisco) is responsible for CO_2 fixation in the Calvin cycle by attaching CO_2 to RuBP. O_2 competes with CO_2 for binding to the active site of Rubisco. Therefore, when chloroplast oxygen levels are high and CO_2 levels are low, Rubisco is more likely to catalyze the reaction of RuBP with O_2 instead of with CO_2. When this occurs, some of the intermediates involved in the Calvin cycle are degraded to CO_2 and H_2O. This process is called **photorespi-**

Figure 8–18 A typical CAM plant. Painted lady *(Echeveria derenbergii),* a member of the family Crassulaceae, is a CAM plant that originated in xeric habitats in Mexico. *(Michel Viard/Peter Arnold, Inc.)*

ration because (1) it occurs in the presence of light; (2) it requires oxygen, like aerobic respiration; and (3) it produces CO_2 and H_2O, like aerobic respiration. Unlike aerobic respiration, however, ATP is not produced during photorespiration. Photorespiration reduces photosynthetic efficiency because it removes some of the intermediates used in the Calvin cycle.

The reasons for photorespiration are incompletely understood, although it is thought to possibly reflect the origin of Rubisco at an ancient time when CO_2 levels were high and molecular oxygen levels were low.

Photorespiration is negligible in C_4 plants because the concentration of CO_2 in bundle sheath cells (where Rubisco is present) is always high. However, many important crop plants are C_3 plants that carry out photorespiration. This is yet another reason that some scientists are attempting to transfer genes for the C_4 pathway to C_3 crops such as soybeans and wheat. If this genetic transfer is accomplished, these plants should be able to produce much more carbohydrate during hot weather.

SUMMARY WITH KEY TERMS

I. Most producers are **photoautotrophs** and use light as an energy source for manufacturing organic compounds from CO_2 and H_2O.
II. Light behaves as both a wave and a particle. Particles of light energy, called **photons,** can excite biological molecules. The resulting energized electrons may be accepted by electron acceptor compounds.
III. In plants, **photosynthesis** occurs in **chloroplasts,** which are located mainly within **mesophyll** cells inside the leaf. During photosynthesis, light energy is captured and converted to the chemical energy of carbohydrates; oxygen is released as a byproduct.
   A. Chloroplasts are organelles bounded by a double membrane; the inner membrane encloses the **stroma** in which membranous, saclike **thylakoids** are suspended. Thylakoids arranged in stacks are called **grana.**
   B. **Chlorophyll** and other photosynthetic pigments are components of the thylakoid membranes of chloroplasts. Each thylakoid encloses a **thylakoid interior space.**
   C. The **absorption spectra** of chlorophylls *a* and *b* are very similar to the **action spectrum** for photosynthesis.
   D. **Photosystems I** and **II** are the two types of photosynthetic units involved in photosynthesis. Each photosynthetic unit includes an antenna complex and a reaction center.
      1. Chlorophyll molecules and accessory pigments are organized into **antenna complexes.**
      2. Only a special chlorophyll *a* in the **reaction center** actually gives up its energized electrons to a nearby electron acceptor.
IV. During the noncyclic **light-dependent reactions,** known as **noncyclic photophosphorylation,** ATP and **NADPH** are formed.
   A. The electrons in Photosystem I are energized by the absorption of light and passed through a chain of electron acceptors to $NADP^+$.
   B. Electrons given up by P700 in Photosystem I are replaced by electrons from P680 in Photosystem II.
   C. Electrons given up by P680 in Photosystem II are replaced by electrons made available by the **photolysis** of H_2O; oxygen is released in the process.
   D. A series of redox reactions takes place as energized electrons are passed along a chain of electron acceptors from Photosystem II to Photosystem I. Some of the energy is used to pump protons across the thylakoid membrane, providing the energy to generate ATP by **chemiosmosis.**
V. During the cyclic light-dependent reactions, known as **cyclic photophosphorylation,** electrons from Photosystem I are eventually returned to Photosystem I.
   A. ATP is produced by chemiosmosis.
   B. No NADPH or oxygen is generated.
VI. During the **carbon fixation reactions,** the energy of ATP and NADPH is used to manufacture carbohydrate molecules from CO_2.
   A. The carbon fixation reactions proceed by way of the **Calvin cycle,** also known as the **C_3 pathway.**
      1. CO_2 is combined with **ribulose bisphosphate (RuBP),** a five-carbon sugar, by the enzyme **ribulose bisphosphate carboxylase (Rubisco).**
      2. For every six CO_2 molecules fixed, two molecules of **glyceraldehyde-3-phosphate (G3P)** can leave the cycle.
      3. Two G3P molecules are required to produce the equivalent of one molecule of glucose; the remaining G3P molecules are modified to regenerate RuBP.
   B. In the **C_4 pathway,** the enzyme **PEP carboxylase** binds CO_2 effectively, even when CO_2 is at a low concentration.
      1. C_4 reactions take place within mesophyll cells. The CO_2 is fixed in **oxaloacetate,** which is then converted to **malate.**
      2. The malate moves into a **bundle sheath cell,** and CO_2 is removed from it. The released CO_2 then enters the Calvin cycle.
   C. The **crassulacean acid metabolism (CAM)** pathway is similar to the C_4 pathway. PEP carboxylase fixes carbon at night in the mesophyll cells, and the Calvin cycle occurs during the day in the same cells.
VII. In **photorespiration,** C_3 plants consume oxygen and generate CO_2 but do not produce ATP. Photorespiration, which decreases photosynthetic efficiency, occurs on bright, hot, dry days when plants close their stomata, conserving water but preventing the passage of CO_2 into the leaf.

Summary Equations for Photosynthesis

The light-dependent reactions (noncyclic photophosphorylation):

$$12\,H_2O + 12\,NADP^+ + 18\,ADP + 18\,P_i \xrightarrow[\text{Chlorophyll}]{\text{Light}} 6\,O_2 + 12\,NADPH + 18\,ATP$$

The carbon fixation reactions (Calvin cycle):

$$12\,NADPH + 18\,ATP + 6\,CO_2 \longrightarrow C_6H_{12}O_6 + 12\,NADP^+ + 18\,ADP + 18\,P_i + 6\,H_2O$$

By canceling out the common items on opposite sides of the arrows in these two coupled equations, we obtain the simplified overall equation for photosynthesis:

$$\underset{\substack{\text{Carbon} \\ \text{dioxide}}}{6\,CO_2} + \underset{\text{Water}}{12\,H_2O} \xrightarrow[\text{Chlorophyll}]{\text{Light}} \underset{\text{Glucose}}{C_6H_{12}O_6} + \underset{\text{Oxygen}}{6\,O_2} + \underset{\text{Water}}{6\,H_2O}$$

1. Where is chlorophyll located in the chloroplast? (a) thylakoid membranes (b) stroma (c) mitochondrial matrix (d) thylakoid interior space (e) between the inner and outer membranes
2. In photolysis, some of the energy captured by chlorophyll is used to split (a) CO_2 (b) ATP (c) NADPH (d) H_2O (e) both b and c
3. Light is composed of particles of energy called (a) carotenoids (b) reaction centers (c) photons (d) antenna complexes (e) photosystems
4. The relative effectiveness of different wavelengths of light in photosynthesis is demonstrated by (a) an action spectrum (b) photolysis (c) carbon fixation reactions (d) a photoheterotroph (e) an absorption spectrum
5. In plants, the final electron acceptor in the light-dependent reactions is: (a) $NADP^+$ (b) CO_2 (c) H_2O (d) $\frac{1}{2} O_2$ (e) G3P
6. In addition to chlorophyll, most plants contain accessory photosynthetic pigments such as (a) PEP (b) G3P (c) carotenoids (d) PGA (e) $NADP^+$
7. In _____, electrons that have been energized by light contribute their energy to add phosphate to ADP, producing ATP. (a) crassulacean acid metabolism (b) the Calvin cycle (c) photorespiration (d) C_4 pathways (e) photophosphorylation
8. In _____, there is a one-way flow of electrons to $NADP^+$, forming NADPH. (a) crassulacean acid metabolism (b) the Calvin cycle (c) photorespiration (d) cyclic photophosphorylation (e) noncyclic photophosphorylation
9. The mechanism by which electron transport is coupled to ATP production is called (a) chemiosmosis (b) crassulacean acid metabolism (c) fluorescence (d) the C_3 pathway (e) the C_4 pathway
10. In photosynthesis in eukaryotes, the transfer of electrons through a sequence of electron acceptors provides energy to pump protons across the (a) chloroplast outer membrane (b) chloroplast inner membrane (c) thylakoid membrane (d) inner mitochondrial membrane (e) plasma membrane
11. The inputs for _____ are CO_2, NADPH, and ATP. (a) cyclic photophosphorylation (b) the carbon fixation reactions (c) noncyclic photophosphorylation (d) Photosystems I and II (e) chemiosmosis
12. The Calvin cycle begins when CO_2 reacts with (a) phosphoenolpyruvate (b) glyceraldehyde-3-phosphate (c) ribulose bisphosphate (d) oxaloacetate (e) phosphoglycerate
13. The enzyme directly responsible for almost all carbon fixation on Earth is (a) Rubisco (b) PEP carboxylase (c) ATP synthase (d) phosphofructokinase (e) ligase
14. In C_4 plants, C_4 and C_3 pathways occur at different _____, whereas in CAM plants, CAM and C_3 pathways occur at different _____. (a) times of day; locations within the leaf (b) seasons; locations within the leaf (c) locations within the leaf; times of day (d) locations within the leaf; seasons (e) times of day; seasons

R E V I E W   Q U E S T I O N S

1. Why does photosynthesis require light energy?
2. What is the role of chlorophyll in photosynthesis?
3. What is the significance of the fact that the combined absorption spectra of chlorophyll *a* and *b* roughly parallel the action spectrum of photosynthesis? Why do they not coincide exactly?
4. How is oxygen produced during photosynthesis?
5. In noncyclic photophosphorylation, what molecule becomes phosphorylated? Why is the phosphorylation process referred to as *photo*phosphorylation? Why is it said to be noncyclic?
6. How are ATP and NADPH produced and used in the process of photosynthesis?
7. How are carbohydrates produced from CO_2 in photosynthesis?
8. How do the C_4 and CAM pathways improve photosynthesis in hot, dry environments?
9. Label the following figure. Use Figure 8–8 to check your answers.

YOU MAKE THE CONNECTION

1. Only some plant cells have chloroplasts, but *all* actively metabolizing plant cells have mitochondria. Why?
2. Explain why the proton gradient formed during chemiosmosis represents a state of low entropy. (You may have to refer to the discussion of entropy in Chapter 6.)
3. The electrons in glucose have relatively high free energies. How did they become so energetic?
4. Rubisco has been described as "the most abundant protein in the world." Explain how an enzyme, which is used repeatedly and is therefore present in very small amounts in cells, could be the world's most abundant protein.
5. What strategies might be employed in the future to increase world food supply? Base your answer on your knowledge of photosynthesis and related processes.

RECOMMENDED READINGS

Balzani, V. "Greener Way to Solar Power." *New Scientist,* Vol. 12, Nov. 1994. Chemists are imitating the light-gathering photosystems of plants in an attempt to convert solar energy to electricity.

Govindjee, and W.J. Coleman. "How Plants Make Oxygen." *Scientific American,* Feb. 1990. Probes the photosynthetic process of using solar energy to split water molecules into oxygen, protons, and electrons.

Hendry, G. "Making, Breaking, and Remaking Chlorophyll." *Natural History,* May 1990. Examines the endless process by which plants make chlorophyll in the spring and break it down in the fall.

Taiz, L., and E. Zeiger. *Plant Physiology.* Benjamin Cummings, Redwood City, CA, 1991. An in-depth examination of the photochemistry of photosynthesis.

● Visit our website at **http://www.saunderscollege.com/lifesci/titles.html** and click on Solomon/Berg/Martin Biology for links to chapter-related resources on the World Wide Web.

Renal Dietitian

SUE KARR

Sue Karr is a renal dietitian who works in an outpatient dialysis clinic in Boulder, Colorado. She manages the nutritional care of kidney disease patients who come to the clinic for their dialysis treatments. She attended the University of Colorado at Boulder and received her B.A. in Environmental, Population, and Organismic Biology in 1993. Because she had always wanted to work with people in an area of health care that had a strong scientific basis, she went on to complete a Master's Degree in Nutrition from the University of Minnesota, where she also completed her registered dietitian credential.

Did you know what you wanted to major in when you entered college?

I had no idea. I took enough prerequisites for any degree my first year. Then I decided I wanted to work in health care but I was not quite sure what area. That's when I thought I would major in biology because no matter what health care career route I eventually took, it was a way to get the basic coursework.

At what point did you choose nutrition?

What sparked me most was a nutrition course. Although the University of Colorado did not offer a nutrition program, Colorado did offer one very good introductory nutrition course, which brought together areas that I had been interested in like wellness, fitness, and eating healthy foods. I saw it also as an area where I could work with people within a strong scientific framework.

I spent some time speaking with the professor of that nutrition course, and with people who work in the community in different areas of nutrition. We had a career center that I used as well. In addition, I did some volunteer work related to health care and nutrition: I was a driver for Meals on Wheels; I volunteered in an assisted living facility helping with some of the nutritional considerations of the elderly; I also had a part-time job at a nursing home.

What is the difference between a dietitian and a nutritionist?

A dietitian is someone who has met the requirements of the American Dietetic Association to be a registered dietitian (R.D.). To become a registered dietitian, one must have at least a Bachelor's Degree and have taken required courses such as human anatomy and physiology, chemistry, clinical nutrition, food science, and food service management, among others. A registered dietitian also must complete a clinical internship, usually six months to a year, under the supervision of registered dietitians in the community. After completing the internship, one is then eligible to take the national credentialing exam for registered dietitians. The term "nutritionist" does not necessarily specify any particular training. I could call myself a nutritionist, but the title "dietitian" is more specific for me.

Do you believe that your biology background helped you get into graduate school?

Absolutely. I came to my graduate program already having a good background in the basic sciences.

How does understanding basic biology concepts such as energy transfer and metabolism help you on a daily basis?

In my practice, one of the most important things I do is to make sure that my clients are at a weight that is appropriate. I have overweight and underweight clients. Understanding what factors influence their energy needs helps me assess what their nutritional needs may be. Also, knowing how the body uses the different energy components of proteins, carbohydrates, and fats helps me develop an individualized plan for them that has the right nutrient distribution.

How do you calculate the energy needs of your patients?

I calculate the energy needs for my patients based on their activity level, height, and weight. For practical reasons, BMR (basal metabolic rate) calculations are rarely available. I have all my new clients complete a three-day diet record so I can estimate how closely they are meeting their energy and protein needs. I may modify energy intake recommendations to help my clients reach goals for weight loss or weight gain.

What exactly is your job and how long have you been doing it?

Since August 1997 I have been working at an outpatient dialysis clinic, which is for people who have kidney failure. We work with about sixty clients who are on hemodialysis. They come to the clinic three days a week to get a dialysis treatment, which functions as an artificial kidney that removes metabolic wastes, excess ions, and other substances from their blood.

Is there some kind of nutritional problem common to kidney disease?

There are many issues that come up when the kidney is not working because the kidney is an important filter in our body. The kidney rids the body of excess minerals and metabolic waste products, such as the breakdown products of proteins, and controls fluid balance. I look at certain minerals like potassium and phosphorus, which can be problematic if they build up in the body.

What advice would you give to someone who is considering your career?

Talk to as many professionals as possible who are actually in the field. I did talk to a few people, but I had no idea how many different areas of nutrition are available. There are nutritionists who work in private practice, in the community, in industry, and apart from health care. I recommend volunteering, too. Volunteer work is valuable not only for experience in an area to see if it interests you, but also to strengthen any application to a graduate program.

The Continuity of Life: Genetics

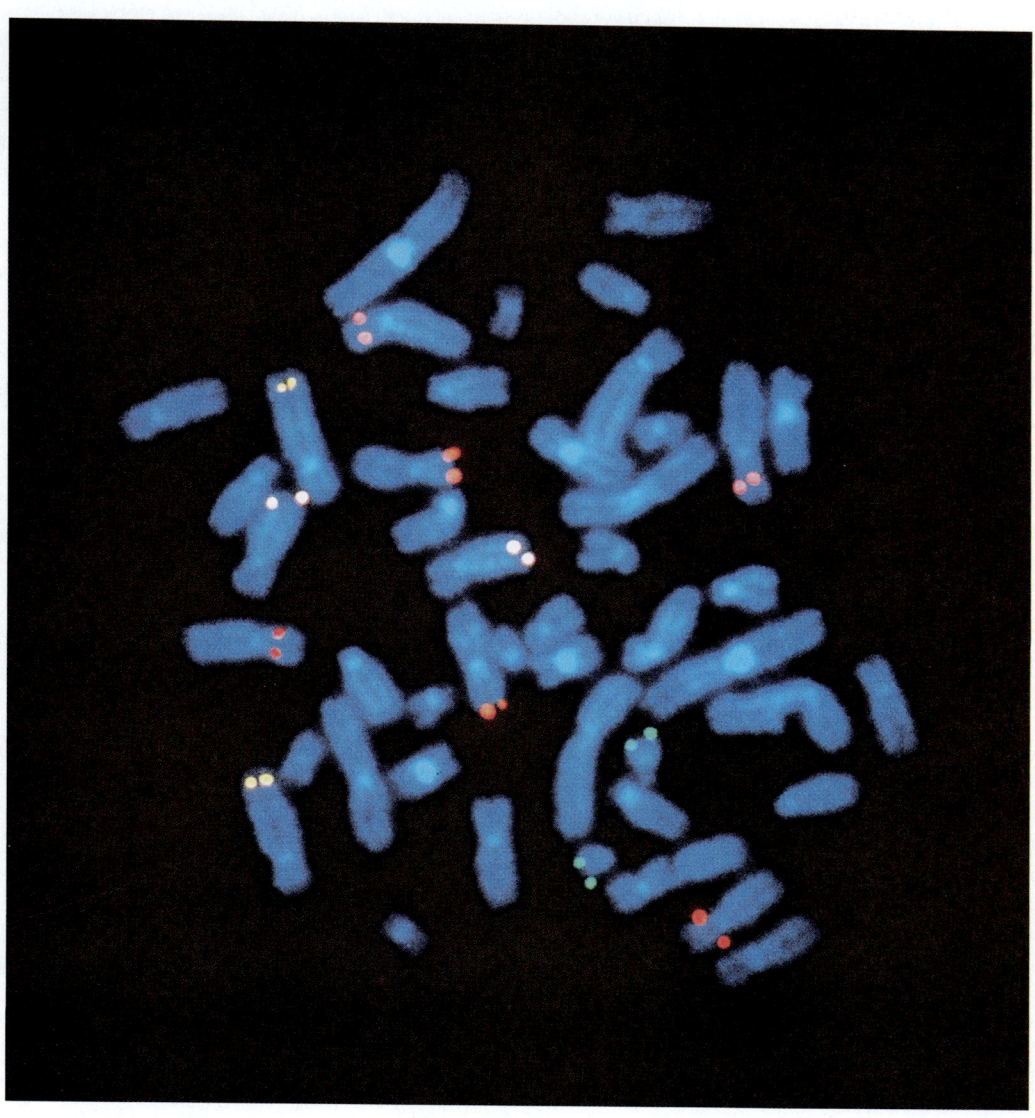

Researchers in human genetics have identified the genes responsible for many hereditary diseases. The colored dots in this fluorescence micrograph indicate the positions of specific genes on these human chromosomes.   (Drs. T. Ried and D. Ward/Peter Arnold, Inc.)

CHAPTER 9

Chromosomes, Mitosis, and Meiosis

Every organism, even the simplest, contains a massive amount of information in the form of DNA. The DNA is organized into informational units called **genes** that ultimately control all aspects of the life of the organism.

A eukaryotic nucleus contains multiple DNA molecules, each of which is packaged with proteins and assembled into a structure called a **chromosome.** This organization, which prevents the tangling of the long, thin molecules, is essential to the highly organized processes by which DNA is distributed during cell division.

All cells are formed by the division of preexisting cells. When a cell divides, the information contained in the DNA first must be precisely duplicated and the copies then transmitted to each daughter cell through a complex series of processes. Most cell divisions in the body cells of eukaryotes involve a process called **mitosis,** which ensures that each daughter cell, such as the human cells in this false-color SEM, receives one copy of every chromosome (and therefore one copy of every gene) from the parent cell.

Prokaryotic cells contain much less DNA, which is usually circular and packaged with very few associated proteins. Although the distribution of genetic material in dividing prokaryotic cells is a simpler process, it nevertheless must be very precise if the daughter cells are to be genetically identical to the parent cell. Bacterial reproduction is described in Chapter 23.

In eukaryotes, sexual reproduction occurs when two sex cells, or **gametes,** fuse to form a single cell called a **zygote.** In higher plants and animals, the gametes are the eggs and sperm. Each gamete contains only half the number of parental chromosomes, thereby preventing the zygotes from having twice as many chromosomes as the parents. For this reason, sexual life cycles include a special type of cell division, called **meiosis,** which reduces the chromosome number by half.

(CNRI/Science Photo Library/Photo Researchers, Inc.)

AFTER YOU HAVE STUDIED THIS CHAPTER YOU SHOULD BE ABLE TO

1. Discuss the significance of chromosomes in terms of their information content.
2. Identify the stages in the eukaryotic cell cycle, describe the principal events characteristic of each, and point out some ways in which the cycle is controlled.
3. Illustrate the structure of a duplicated chromosome, labeling the sister chromatids, sister centromeres, and sister kinetochores.
4. Explain the significance of mitosis and diagram the process.

5. Discriminate between asexual and sexual reproduction.
6. Distinguish between haploid and diploid cells and define homologous chromosomes.
7. Explain the significance of meiosis and diagram the process.
8. Contrast mitosis and meiosis, emphasizing how differences in events lead to different outcomes.
9. Compare the roles of mitosis and meiosis and of haploidy and diploidy in various generalized life cycles.

EUKARYOTIC CHROMOSOMES CONTAIN DNA AND PROTEIN

The major carriers of genetic information in eukaryotes are the **chromosomes** contained within the cell nucleus. Although the term *chromosome* means "colored body," chromosomes are virtually colorless; the name refers to their ability to be stained by certain dyes.

Chromosomes are made up of **chromatin,** a complex material that consists of fibers containing protein and deoxyribonucleic acid (DNA). When a cell is not dividing, the chromosomes are present but in an extended, partially unraveled form. The chromatin consists of long, thin threads that are somewhat aggregated, which gives them a granular appearance when viewed with the electron microscope (see Fig. 4–12). At the time of cell division, the chromatin fibers condense and the chromosomes become visible as distinct structures (Fig. 9–1). The structure of chromatin is described in more detail in Chapter 11.

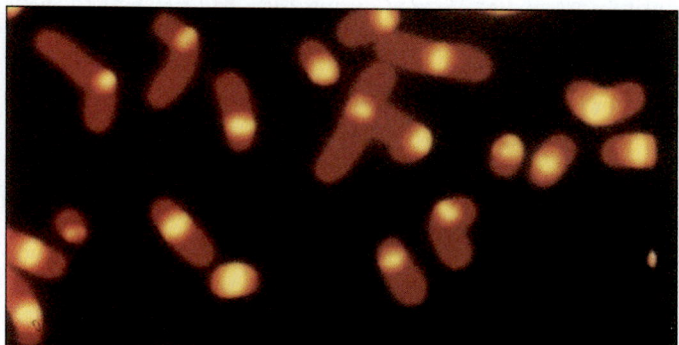

10 μm

Figure 9–1 Chromosomes. The human chromosomes shown in this confocal fluorescence LM have been stained with a fluorescent antibody that binds to the centromere regions *(yellow)*. *(Courtesy of Oncor, Inc.)*

DNA is organized into informational units called genes

Each chromosome may contain hundreds or even thousands of **genes.** For example, humans are thought to have 70,000 to 100,000 genes. The precise number is not known, although scientists are engaged in a massive coordinated effort to make this determination as part of the Human Genome Project (see Chapter 15). As will be evident in succeeding chapters, our concept of the gene has changed considerably since the beginnings of the science of genetics, but our definitions have always centered on the gene as an informational unit. By providing information needed to carry out one or more specific cellular functions, a gene ultimately affects some characteristic of the organism. For example, we speak of genes controlling eye color in humans, wing length in flies, seed color in peas, and so on.

Chromosomes of different species differ in number and informational content

Every individual of a given species has a characteristic number of chromosomes in most nuclei of its body cells. For example, most human body cells have exactly 46 chromosomes.

Humans are not unique in having 46 chromosomes; some other species of animals and plants also have 46, whereas others have different chromosome numbers. A certain species of roundworm has only two chromosomes in each cell, while some crabs have as many as 200 and some ferns have more than 1000. Most animal and plant species have between 10 and 50 chromosomes. Numbers above and below this are uncommon.

Humans are not humans merely because they have 46 chromosomes; in fact, some humans have abnormal karyotypes (chromosome assortments) with more or fewer than 46 (see Chapter 15). The *number* of chromosomes is not what makes each species unique but rather the *information* specified by the genes in the chromosomes.

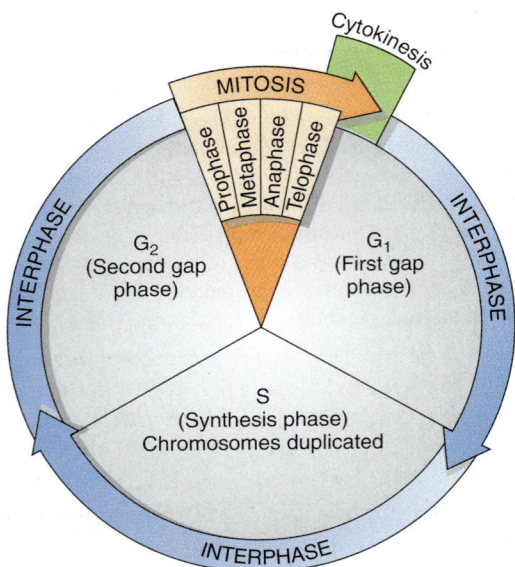

Figure 9–2 **The cell cycle.** The generation time, the time required to complete one cycle, includes cell division (mitosis and cytokinesis) and interphase. Actual times vary with the species, cell type, and growth conditions.

5 μm

Figure 9–3 **The S phase.** The silver grains (black dots) lying on top of the bean chromosomes in this autoradiogram mark locations where the chromosomes incorporated ³H-labeled thymidine, a DNA precursor, during the S phase of the interphase that preceded mitosis. *(Professor J. H. Taylor, Proceedings of the National Academy of Science, 43:122,1957, courtesy of New York Academy of Medicine)*

THE CELL CYCLE IS A SEQUENCE OF CELL GROWTH AND DIVISION

Usually when cells reach a certain size, they must either stop growing or divide. Some cells, such as nerve, skeletal muscle, and red blood cells, do not normally divide once they are mature. The activities of growing and dividing cells can be described in terms of the life cycle of the cell, or the **cell cycle.**

In cells capable of dividing, the cell cycle is the period from the beginning of one division to the beginning of the next and is represented in diagrams as a circle (Fig. 9–2). The time it takes to complete one cell cycle is the **generation time.** The generation time can vary widely, but in actively growing plant and animal cells it is often about 8 to 20 hours.

Cell division involves two main processes, mitosis and cytokinesis. **Mitosis,** a complex process involving the nucleus, ensures that each new nucleus receives the same number and types of chromosomes as were present in the original nucleus. **Cytokinesis,** which generally begins before mitosis is complete, is the division of the cytoplasm of the cell to form two cells. Multinucleate cells are formed if mitosis is not followed by cytokinesis; this is a normal condition for some kinds of cells.

Chromosomes become duplicated during interphase

Most of the life of the cell is spent in **interphase,** the stage between successive cell divisions. The cell is very active during this time, synthesizing needed materials and growing. Most proteins and other materials are synthesized throughout interphase. In the early 1950s, researchers demonstrated that the chromosomes are synthesized at a relatively restricted interval during interphase and not during early mitosis, as had been previously thought. They used isotopes such as ³H, to synthesize radioactive thymidine, a nucleotide (see Chapter 3) that is incorporated specifically into DNA as it is synthesized. After the radioactive thymidine had been supplied for a brief period to actively growing cells, a fraction of the cells could be shown by autoradiography (see Chapter 2) to have silver grains over their nuclei (Fig. 9–3). These labeled cells were those that had been engaged in DNA replication (known as the **synthesis phase,** or **S phase**) during the experiment. The S phase provides the major landmark that is the basis for subdividing interphase, (see Fig. 9–2). Other chromosomal constituents, such as the chromosomal proteins, are also synthesized during the S phase. Chromosome duplication is a complex process discussed in Chapter 11.

The time between mitosis and the beginning of the S phase is termed the **G₁ phase** (G stands for *gap,* an interval during which no DNA synthesis occurs). Growth takes place during the G₁ phase, which is usually the most variable in length and also the longest. Toward the end of G_1 there is increased activity of enzymes required for DNA synthesis; these enzymes, along with many other factors, make it possible for the cell to enter the S phase. Cells that are not dividing usually become arrested prior to the onset of the S phase. They are said to be in a state called G_0, which is not part of the cell cycle per se.

After it completes the S phase, the cell enters a second gap phase, the **G_2 phase.** At this time, increased protein synthesis occurs as the final steps in the cell's preparation for division take place. The completion of the G_2 phase is marked by the beginning of mitosis. The sequence of the substages of interphase is therefore:

$$G_1 \text{ phase} \rightarrow S \text{ phase} \rightarrow G_2 \text{ phase}$$

Mitosis ensures orderly distribution of chromosomes

Each mitotic division is a continuous process. However, for descriptive purposes mitosis has been divided into stages. Refer to Figure 9–4 as you read the descriptions of these stages as they would occur in a typical plant or animal cell:

$$\text{Prophase} \rightarrow \text{Metaphase} \rightarrow \text{Anaphase} \rightarrow \text{Telophase}$$

During prophase duplicated chromosomes become visible with the microscope

The first stage of mitosis, **prophase,** begins when the long chromatin threads begin to condense and appear as mitotic chromosomes. This condensation is accomplished mainly by a coiling process in which chromosomes become simultaneously shorter and thicker. The chromatin can then be distributed to the daughter cells without tangling.

When stained with certain dyes and viewed through the light microscope, chromosomes are visible as darkly staining bodies during prophase. Each chromosome has been duplicated during the preceding S phase and consists of a pair of identical units, termed **sister chromatids.** Each chromatid includes a constricted region called the **centromere.** Sister chromatids are tightly associated in the vicinity of their centromeres (Fig. 9–5). Although the chemical basis for this close association is not completely understood, evidence suggests that special DNA sequences and special proteins that bind to those DNA sequences are involved. Attached to each centromere is a **kinetochore,** a structure formed from proteins to which microtubules can bind.

A dividing cell is usually described as a globe, with an equator that determines the midplane, or equatorial plane, and two opposite poles. This terminology is used for all cells regardless of their actual shape.

Microtubules radiate from each pole and some of these protein fibers elongate toward the chromosomes, forming a complex structure known as the **mitotic spindle** (Fig. 9–6).

The mitotic spindle is responsible for the separation of the chromosomes during anaphase.

Animal cells differ from the cells of complex plants in the details of mitotic spindle formation. In both types of dividing cells, each pole contains a region, referred to as a **microtubule-organizing center,** from which the microtubules grow outward. When viewed with the electron microscope, microtubule-organizing centers in the cells of higher plants consist of rather dense fibrillar matter with little or no discernible structure.

In contrast, animal cells have a pair of **centrioles** (see Chapter 4) in the middle of each microtubule-organizing center. The centrioles are surrounded by fibrils that make up the **pericentriolar material.** The ends of the spindle microtubules are found in the pericentriolar material, but they do not actually touch the centrioles themselves. Evidence that the pericentriolar material may be functionally similar to the material in the microtubule organizing centers of plants derives from the fact that they are similar in appearance and in the specific types of proteins present.

Each of the two centrioles becomes duplicated during interphase, yielding two pairs of centrioles. Late in prophase, microtubules radiate from the pericentriolar material surrounding the centrioles, and one pair of centrioles migrates to each pole. The migration of the centrioles to the poles essentially marks the migration of the microtubule organizing centers. Additional microtubules form clusters extending outward in many directions from the microtubule organizing centers at the poles; these structures are called **asters** (Fig. 9–6). The function of the asters is not well understood, although there is evidence that these structures play a role in cytokinesis.

It is likely that both plant and animal spindles are organized by similar microtubule organizing centers. Although centrioles were long thought to be required for spindle formation in animal cells, their apparent involvement is probably coincidental. Because centrioles usually arise in association with preexisting centrioles, the localization of the centrioles in the microtubule organizing center may have evolved to provide for the orderly distribution of these organelles to the daughter cells. Centrioles are important to animal cells because they are involved in the formation of the basal bodies of cilia and flagella (see Chapter 4). (New basal bodies can also form if preexisting basal bodies are present, as in the cells of some protists.) Centrioles are not found in the cells of flowering plants and more advanced gymnosperms. Both of these groups lack flagellated sperm and other flagellated or ciliated cells (see Chapter 27). On the other hand, the sperm cells of mosses and ferns (see Chapter 26) are flagellated and possess centrioles.

During prophase, the nucleolus (see Chapter 4) diminishes in size and usually disappears. Toward the end of prophase, the nuclear envelope breaks down, and each chromatid becomes attached to some of the spindle microtubules at its kinetochore. The chromosomes then move back and forth from pole to pole and finally become aligned along the equatorial plane of the cell, midway between the two poles.

INTERPHASE	EARLY PROPHASE	LATE PROPHASE	METAPHASE
The cell is carrying out its normal life activities. The chromosomes become duplicated.	Nuclear envelope begins to disappear. Nucleolus disappears. Long thread-like bodies of chromatin become evident and begin to condense as visible chromosomes.	Chromosomes continue to shorten and thicken. Spindle forms between the centrioles, which have moved to the poles of the cell.	Spindle fibers attach to the kinetochores of the chromosomes. Chromosomes line up along the equatorial plane of the cell.

(a)

25 μm

INTERPHASE   EARLY PROPHASE   LATE PROPHASE   METAPHASE

PLANT

ANIMAL

25 μm

(b)

At metaphase duplicated chromosomes line up on the midplane

The period during which the chromosomes are lined up along the equatorial plane of the cell constitutes **metaphase.** The mitotic spindle is complete. It is composed of two types of microtubules: **polar microtubules** extend from each pole to the equatorial region, where they generally overlap. **Kinetochore microtubules** extend from each pole and attach to the kinetochores of the chromosomes (see Fig. 9–6). At mitotic metaphase the individual sister kinetochores are attached by kinetochore microtubules to *opposite* poles of the cell.

During metaphase each chromatid is completely condensed and appears quite thick and distinct. Because individ-

ANAPHASE

Chromatids separate at their centromeres, and one group of chromosomes moves toward each pole.

TELOPHASE

The events of prophase are reversed with the nuclei reforming, and cell division is completed as cytokinesis produces two daughter cells.

INTERPHASE

The daughter cells formed are genetically (and usually physically) identical to the parent cell except for size.

ANAPHASE

TELOPHASE

INTERPHASE

Figure 9–4 **Cell division.** Interphase and the stages of mitosis are similar in plant and animal cells. Although both types have microtubule organizing centers, the plant cells lack centrioles. (**a**) The drawings depict generalized animal cells with a diploid chromosome number of four. The sizes of the nuclei and chromosomes are exaggerated to show the structures more clearly. (**b**) The upper row of LMs depicts stained chromosomes in sectioned cells of the onion, *Allium cepa.* Cells of an animal (the whitefish) are shown in the lower row. The chromosomes have been stained and the cells flattened on microscope slides. (Plant cells, 1st interphase through telophase, *Ed Reschke;* 2nd interphase, *Carolina Biological Supply Company/Phototake.* Animal cells, *Michael Abbey/Science Source/Photo Researchers, Inc.*)

ual chromosomes can be seen more distinctly at metaphase than at any other time, they are usually photographed at this stage to be studied for certain chromosome abnormalities (see Chapter 15).

During anaphase chromosomes move toward the poles

Anaphase begins as the protein tethers holding the sister chromatids together in the vicinity of their centromeres are released.

Each chromatid is now referred to as an independent chromosome. The now separate chromosomes slowly move to opposite poles. The kinetochores of the chromosomes, still attached to kinetochore microtubules, lead the way, with the chromosome arms trailing behind. Anaphase ends when all the chromosomes have reached the poles.

The overall mechanism of chromosome movement in anaphase is still poorly understood, although significant progress is being made in this area. Microtubules lack elastic or contractile properties. So how do the chromosomes move

Figure 9–5   A metaphase chromosome. An SEM *(bottom)* is paired with an interpretive drawing. The sister chromatids, each consisting of tightly coiled chromatin fibers, are tightly associated at their centromere regions, indicated by the brace. Associated with each centromere is a structure known as a kinetochore, which serves as a microtubule attachment site. The kinetochores and microtubules are not evident in the SEM.   *(E.J. DuPraw)*

apart? Are they pushed or pulled, or do other forces operate?

Chromosome movements are studied in a variety of ways. Numbers of microtubules present at a particular stage or after certain treatments can be determined through careful analysis of electron micrographs. It is possible to physically perturb living cells that are dividing, using laser beams or mechanical devices known as micromanipulators. Skilled researchers can move chromosomes, break their connections to microtubules, and even remove them from the cell entirely.

There is a considerable body of evidence that kinetochore microtubules shorten during anaphase. Because microtubules are constantly changing structures, with tubulin subunits being constantly removed and others being added, many current hypotheses to explain anaphase movement are based on suggestions that chromosomes move poleward as tubulin subunits disassemble from the ends of the microtubules. One suggestion has been that kinetochore microtubules are disassembled

Figure 9–6   The mitotic spindle. Kinetochore microtubules and polar microtubules are found in both plant and animal cells, whereas most plant cells lack centrioles and astral microtubules. (*a*) One end of each microtubule of this animal cell is associated with one of the poles. Astral microtubules *(green)* radiate in all directions, forming the aster. Kinetochore microtubules *(red)* connect the kinetochores to the poles, and polar microtubules *(blue)* overlap at the midplane. (*b*) Fluorescent-stained LM of animal cell at metaphase with well defined spindle and asters (chromosomes *orange*, microtubules *green*). (*b, CNRI/Phototake, NYC*).

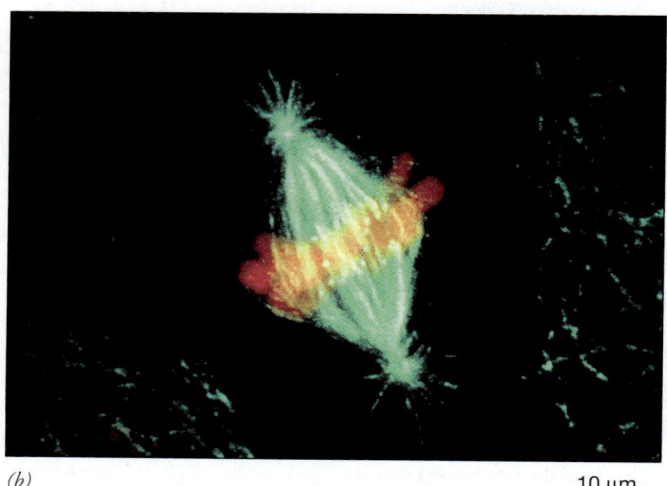

at the kinetochores themselves, implying that chromosomes play an active role in their own movements. However, this idea is not supported by recent findings. It has been shown that if all the chromosomes are removed from a living cell after the spindle has formed, the kinetochore microtubules shorten just as they would if the chromosomes were present. One possible interpretation is that microtubule shortening takes place at the poles and that the chromosomes are transported passively.

A second phenomenon also plays a role in chromosome separation. During anaphase the spindle as a whole elongates, at least partly because polar microtubules originating at opposite poles are associated with motors that enable them to slide past one another at the equator, decreasing the degree to which they are overlapped and thereby "pushing" the poles apart. This mechanism indirectly causes the chromosomes to move apart because they are attached to the poles.

During telophase two separate nuclei are formed

The final stage of mitosis, telophase, is characterized by a return to interphase-like conditions. The chromosomes decondense by uncoiling. A new nuclear envelope forms around each set of chromosomes, made at least in part from small vesicles and other components derived from the old nuclear envelope. The spindle microtubules disappear, and the nucleoli reorganize.

Cytokinesis is the formation of two separate daughter cells

Cytokinesis, the division of the cytoplasm to yield two daughter cells, usually overlaps mitosis, generally beginning during telophase. Cytokinesis of an animal cell begins as a ring of actin microfilaments forms at right angles to the spindle, encircling the cell in the equatorial region (Fig. 9–7a). The ring contracts, producing a furrow that gradually deepens and separates the cytoplasm into two daughter cells, each with a complete nucleus.

In plant cells, cytokinesis occurs by the formation of a **cell plate** (Fig. 9–7b), a partition constructed in the equatorial region of the spindle and growing laterally to the cell wall. The cell plate forms as a line of vesicles that originate in the Golgi complex (see Chapter 4). The vesicles contain materials to construct both a primary cell wall for each daughter cell and a middle lamella that will cement the primary cell walls together. The vesicle membranes fuse to become the plasma membrane of each daughter cell.

Mitosis typically produces two cells genetically identical to the parent cell

The remarkable regularity of the process of cell division ensures that each of the daughter nuclei receives exactly the same number and kinds of chromosomes that the parent cell had. Thus, with a few exceptions, every cell of a multicellular organism has exactly the same genetic makeup. If a cell receives

more or fewer than the characteristic number of chromosomes through some malfunction of the cell division process, the resulting cell may show marked abnormalities and be unable to survive.

Most cytoplasmic organelles are distributed randomly to the daughter cells

Mitosis provides for the orderly distribution of chromosomes (and of centrioles, if present), but what about the various cytoplasmic organelles? For example, all eukaryotic cells, including plant cells, require mitochondria. Likewise, photosynthetic plant cells cannot carry out photosynthesis without chloroplasts. These organelles contain their own DNA and appear to form by the division of previously existing mitochondria or plastids or their precursors. However, they generally divide during interphase, not when the cell divides. Because many copies of each organelle are present in each cell, organelles are apportioned more or less equally between the daughter cells at cytokinesis.

The cell cycle is controlled by an internal genetic program interacting with external signals

When conditions are optimal, some prokaryotic cells (which divide nonmitotically; see Chapter 23) can divide every 20 minutes. The generation times of eukaryotic cells are generally much longer, although the frequency of cell division varies widely among different species and among different tissues of the same species. Some cells in the central nervous system usually cease dividing after the first few months of life, whereas blood-forming cells, digestive tract cells, and skin cells divide frequently throughout the life of the organism. Under optimal conditions of nutrition, temperature, and pH, the eukaryotic cell cycle length is constant for any cell type. Under less favorable conditions, however, the generation time may be longer.

According to a considerable body of evidence that has accumulated in recent years, certain fundamental mechanisms of genetic control of the cell cycle are common to all eukaryotes. Among the key components of the regulatory system are **protein kinases,** enzymes that activate or inactivate other proteins by adding phosphate groups (phosphorylation). The particular protein kinases involved in controlling the cell cycle are called cyclin-dependent protein kinases because they are only active when complexed with regulatory proteins called **cyclins.** The cyclins are so-named because their levels fluctuate predictably during the cell cycle (i.e., they "cycle").

When a specific cyclin-dependent protein kinase (Cdk) forms a complex with a specific cyclin, it actively phosphorylates certain cellular proteins. Some of these proteins, including certain enzymes, become activated when they are phosphorylated. Conversely, phosphorylation inactivates some other enzymes and proteins. As some enzymes are activated and others are inactivated by phosphorylation, the activities of

Figure 9-7 Cytokinesis in plant and animal cells. The nuclei in both figures are at the telophase stage. (**a**) This TEM shows a cleavage furrow forming in the equatorial region of a cultured animal cell undergoing cytokinesis. (**b**) Cytokinesis is occurring by cell plate formation in this TEM of a maple leaf cell, *Acer saccharinum*. (a, T.E. Schroeder, University of Washington/Biological Photo Service; b, E.H. Newcomb and B.A. Palevitz, University of Wisconsin/Biological Photo Service)

Cleavage furrow

Ring of contractile microfilaments

(a)

10 μm

Nucleus

Small vesicles fuse, forming larger vesicles

Vesicles gather on equatorial plane

Cell plate forming

(b)

5 μm

Eventually, one large vesicle exists

New cell walls (from vesicle contents)

New plasma membranes (from vesicle membranes)

the cell (as they relate to the steps of the cell cycle) change (Fig. 9–8). Although not all of the details are understood, these systems of regulating the cell cycle have been highly conserved during the evolution of eukaryotes; they are found in organ-isms as diverse as yeast (a unicellular fungus), clams, frogs, and plants.

Certain drugs can stop the cell cycle. Some of these prevent DNA synthesis, whereas others inhibit the synthesis of

Figure 9-8 **Genetic control of the cell cycle.** In this example, *mitosis-promoting factor (MPF)* is required for the cell to make the transition from G_2 to mitosis. Active MPF is produced when an inactive cyclin-dependent protein kinase enzyme (Cdk) becomes complexed with cyclin B. MPF then phosphorylates many proteins, thereby activating those needed for mitosis and inactivating those that would impede mitosis. Later in mitosis, MPF becomes inactive as cyclin B is degraded. These events are repeated when cyclin B levels rise again in G_2 of the next cell cycle.

proteins that control the cycle, as well as the synthesis of structural proteins that contribute to the mitotic spindle. Because cancer cells often divide much more rapidly than most normal body cells, they can be most affected by these drugs. Many of the side effects of certain anticancer drugs (e.g., nausea, hair loss) are due to the drugs' effects on rapidly dividing cells in the digestive system, hair follicles, and so forth.

Colchicine, a drug used to block cell division in eukaryotic cells, binds with unpolymerized tubulin subunits, preventing them from being added to the spindle microtubules. Under these conditions, the rate of microtubule breakdown far exceeds the rate of microtubule assembly, resulting in the disappearance of the spindle. Although the sister chromatids eventually become detached from one another and each becomes a chromosome, they cannot move to the poles in the absence of the spindle. As a result, a cell may end up with extra sets of chromosomes (a condition known as **polyploidy,** which is discussed in the next section).

In plant cells, certain hormones are known to stimulate mitosis. These include the **cytokinins,** a group of plant hormones that promote mitosis both in normal growth and in wound healing (see Chapter 36).

Protein **growth factors,** which are active at extremely low concentrations, stimulate mitosis in certain animal cells. Of the approximately 50 protein growth factors known, some act only on specific cell classes, while others work over a broader range. For example, the effects of erythropoetin are limited to cells that will develop into red blood cells, whereas epidermal growth factor stimulates many cell types to divide. In addition, hormones such as certain steroid hormones can act as growth factors (see Chapter 47).

SEXUAL LIFE CYCLES REQUIRE A MECHANISM TO REDUCE THE CHROMOSOME NUMBER

Although the details of the reproductive process vary greatly among different kinds of eukaryotes, we can distinguish two basic types of reproduction: asexual and sexual. In **asexual reproduction** a single parent usually splits, buds, or fragments to produce two or more individuals (Fig. 9–9). In most forms of asexual reproduction, all the cells are produced by mitosis, so their genes and inherited traits are identical to those of the parent. Such a group of genetically identical organisms is termed a **clone.** Asexual reproduction is usually a rapid process; it permits organisms well adapted to their environment to produce new generations of similarly adapted organisms.

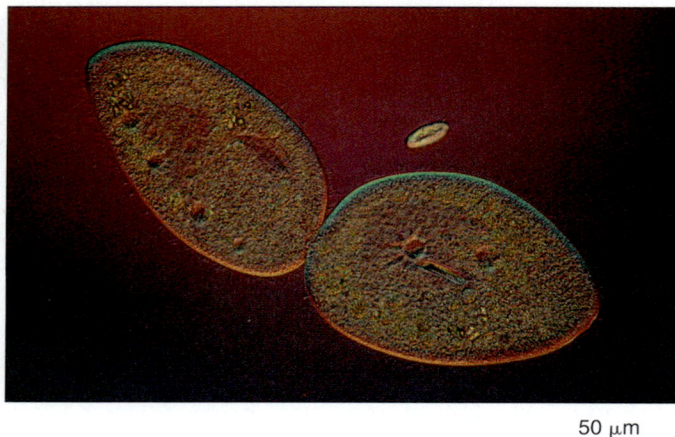

50 μm

Figure 9–9 **Binary fission.** This cell of *Paramecium caudatum,* a ciliate protozoon, is reproducing asexually by binary fission. Binary fission in *Paramecium* includes mitosis. *(M. Abbey/Photo Researchers, Inc.)*

Figure 9–10 **A human karyotype.** Although these are duplicated chromosomes (late prophase), the sister chromatids are not clearly evident because they are closely aligned throughout their lengths. As discussed in Chapter 10, the X and Y chromosomes of this normal male are not strictly homologous; normal females have two X chromosomes and no Y chromosome. The process of karyotyping is discussed in Chapter 15. *(Courtesy of Dr. Leonard Sciorra)*

In contrast, **sexual reproduction** involves the union of two specialized sex cells, or **gametes,** to form a single cell called a **zygote.** Usually the gametes are contributed by two different parents, but in some cases a single parent furnishes both gametes. In the case of animals and plants, the egg and the sperm cells are the gametes, and the fertilized egg is the zygote. Offspring produced sexually are not genetically identical to their parents, so some may be able to survive environmental changes or other stresses better than either parent, whereas others, with a different combination of traits, may be less likely to survive.

Because you now understand the roles of chromosomes in inheritance, you may recognize a problem in eukaryotic sexual reproduction: if each gamete has the same number of chromosomes as did the parental cell that produced it, then the zygote would be expected to have twice as many chromosomes. This doubling would occur generation after generation. How do organisms avoid producing zygotes with ever-increasing chromosome numbers? To answer this question, we need more information about the types of chromosomes found in cells.

Each chromosome found in a somatic (body) cell of a higher plant or animal normally has a partner chromosome. The two partners, known as **homologous chromosomes,** are similar in size, shape, and the position of their centromeres. When stained by special techniques, the members of a pair generally share a characteristic pattern of bands. In most species, chromosomes vary enough in their morphological features that cytologists can distinguish the different homologous pairs and match up the partners. The 46 chromosomes in human cells constitute 23 different homologous pairs (Fig. 9–10). The most important feature of homologous chromosomes is that they carry similar, but not necessarily identical, genetic information. For example, members of a pair of homologous chromosomes might each carry a gene that specifies hemoglo-

bin structure; however, one member might have the information for normal hemoglobin, whereas the other might specify the abnormal form of hemoglobin associated with sickle cell anemia (see Chapter 15). Homologous chromosomes can therefore be contrasted with the two members of a pair of sister chromatids, which are precisely identical to each other.

A *set* of chromosomes has one of each kind of chromosome; in other words, it contains one member of each homologous pair. If a cell or nucleus contains two sets of chromosomes, it is said to have a **diploid** chromosome number. If it has only a single set of chromosomes, it has the **haploid** number. In humans the diploid chromosome number is 46 and the haploid number is 23. When a sperm and egg fuse at fertilization, each gamete is haploid, contributing one set of chromosomes; the diploid number is thereby restored in the fertilized egg (zygote). When the zygote divides by mitosis to form the first two cells of the embryo, each daughter cell receives the diploid number of chromosomes, and this is repeated in subsequent mitotic divisions. Thus, most human body cells are diploid.

If a cell or an individual has three or more sets of chromosomes, we say that it is **polyploid.** Polyploidy is relatively rare among animals but quite common among plants (see

Figure 9–11 **Polyploidy.** Modern bread wheat is a hexaploid plant. *(Sharon Cummings/Dembinsky Photo Associates)*

DIPLOID CELLS UNDERGO MEIOSIS TO FORM HAPLOID CELLS

We have examined the process of mitosis, which ensures that each daughter cell receives exactly the same number and kinds of chromosomes that the parent cell had. A diploid cell that undergoes mitosis produces two diploid cells; similarly a mitotic haploid cell produces two haploid cells. A division resulting in a reduction in chromosome number is called **meiosis.** The term *meiosis* means "to make smaller," referring to the fact that the chromosome number is reduced by one-half. In meiosis a diploid cell undergoes two cell divisions, potentially yielding four haploid cells.

The position of meiosis in the life cycle varies among groups

Because sexual reproduction is characterized by the fusion of two haploid sex cells to form a diploid zygote, it follows that, in a sexual life cycle, meiosis must occur before gametes can be produced.

In animals and a few other organisms meiosis leads directly to gamete formation (Figure 9–12*a*). The body (somatic) cells of an individual organism multiply by mitosis and are diploid; the only haploid cells produced are the gametes. These are formed when certain **germ line** cells undergo meiosis. The formation of gametes is known as **gametogenesis.** Male gametogenesis, termed **spermatogenesis,** results in the formation of four haploid sperm cells for each cell that enters meiosis.

In contrast, female gametogenesis, termed **oogenesis,** results in the formation of a single egg cell, or **ovum,** for every cell that enters meiosis. This is accomplished by a process that apportions virtually all of the cytoplasm to only one of the two nuclei at each of the meiotic divisions. At the end of the first meiotic division, one nucleus is retained and the other, called the first **polar body,** is excluded from the cell and ultimately degenerates. Similarly, at the end of the second division, one nucleus becomes the second polar body and the other nucleus survives. In this way, one haploid nucleus becomes the recipient of most of the accumulated cytoplasm and nutrients from the original meiotic cell. (See Figure 48-13 for a more detailed description.)

However, although meiosis occurs at some point in a sexual life cycle, it does not always *immediately* precede gamete formation. Many simple eukaryotes (including some fungi and algae) remain haploid (their cells dividing mitotically) throughout most of their lives, with individuals being unicellular or multicellular. Two haploid gametes (produced by mitosis) fuse to produce a diploid zygote that undergoes meiosis to restore the haploid state (Figure 9–12*b*). Examples of these types of life cycles can be found in Chapters 24 and 25.

The most complex life cycles are displayed by plants and some algae (Figure 9–12*c*). These life cycles, characterized by an **alternation of generations,** consist of a multicellular diploid stage, termed the **sporophyte generation,** and a mul-

Chapter 19). In fact, polyploidy has been an important factor in plant evolution. As many as 80% of all flowering plants are polyploid. Polyploid plants are often larger and hardier than diploid members of the same group. Many commercially important plants are polyploid. Modern bread wheat, *Triticum aestivum* (Fig. 9–11), is a hexaploid with 42 chromosomes, derived from three different diploid species with 14 chromosomes each.

The abbreviation for the chromosome number found in the gametes of a particular species is *n,* and the zygotic chromosome number is given as *2n.* If the organism is not polyploid, the haploid chromosome number is equal to n and the diploid number is equal to $2n$. For example, in humans $n = 23$ and $2n = 46$. For simplicity, in the rest of this chapter we assume that the organisms used as examples are not polyploid. We therefore use the designations diploid and $2n$, and haploid and n, interchangeably, although these terms are not strictly synonymous.

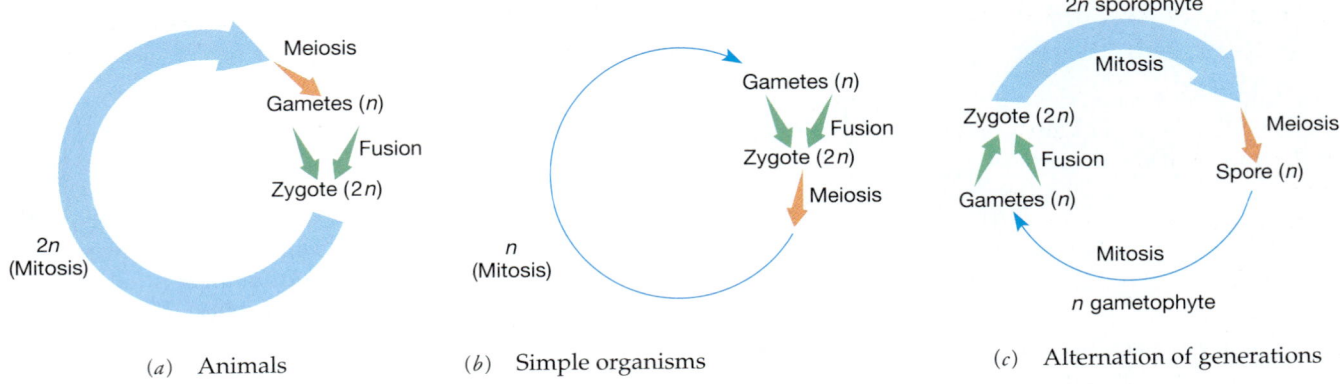

(a) Animals   (b) Simple organisms   (c) Alternation of generations

Figure 9–12   Representative life cycles.

ticellular haploid stage, termed the **gametophyte generation.** Diploid sporophyte cells undergo meiosis to form haploid spores, each of which then divides mitotically to produce a multicellular haploid gametophyte. Gametophytes produce gametes by mitosis. The female and male gametes (eggs and sperm cells) then fuse to form a diploid zygote that divides mitotically to produce a multicellular, diploid sporophyte.

In higher plants, including flowering plants, the diploid sporophyte—which includes the roots, stems, and leaves of the plant body—is the dominant form. The gametophytes are small and inconspicuous. For example, a pollen grain contains a haploid male gametophyte that forms haploid sperm by mitosis. More detailed descriptions of alternation of generations can be found in Chapters 26 and 27.

Figure 9–13   **Meiosis.** Two nuclear divisions, meiosis I and meiosis II, are required. In this illustration, the process begins with a cell that has a diploid chromosome number of four and ends with the formation of four haploid cells with two chromosomes each. The maternal chromosomes are shown in pink; the paternal chromosomes are purple.

PREMEIOTIC INTERPHASE
Interphase preceding meiosis; DNA replicates.

PROPHASE I
Homologous chromosomes synapse, forming tetrads; crossing-over occurs. Nuclear envelope breaks down.

METAPHASE I
Tetrads line up on equatorial plane of cell. Tetrads held together at chiasmata (sites of prior crossing-over).

ANAPHASE I
Homologous chromosomes separate and move to opposite poles. Note that sister chromatids remain attached at their centromeres.

TELOPHASE I
One of each pair of homologous chromosomes is at each pole. Cytokinesis occurs.

INTERKINESIS
DNA does not replicate. Note that the chromatids are still joined. Chromosomes do not completely elongate.

Meiosis produces haploid cells with unique gene combinations

The events of meiosis are similar to the events of mitosis, with four important differences: (1) Meiosis involves two successive nuclear and cytoplasmic divisions, producing up to four cells. (2) Despite two successive nuclear divisions, the DNA and other chromosomal components are duplicated only once, during the interphase preceding the first meiotic division. (3) Each of the four cells produced by meiosis contains the haploid chromosome number, that is, only one set with only one representative of each homologous pair. (4) During meiosis, the genetic information from both parents is shuffled, so each resulting haploid cell has a virtually unique combination of genes.

In meiosis homologous chromosomes become distributed into different daughter cells

Meiosis typically consists of two nuclear and cytoplasmic divisions, designated the *first* and *second meiotic divisions,* or simply **meiosis I** and **meiosis II.** Each includes prophase, metaphase, anaphase, and telophase stages. During meiosis I, the members of each homologous pair of chromosomes first join together and then separate and move into different nuclei. In meiosis II the sister chromatids that make up each chromosome separate and are distributed to the nuclei of the daughter cells. The following discussion describes meiosis in an animal with a diploid chromosome number of four. Refer to Figures 9–13 and 9–14 as you read.

Prophase I includes synapsis and crossing-over

As in mitosis, the chromosomes are duplicated during the S phase of interphase, before meiosis actually begins. Each duplicated chromosome consists of two chromatids. During *prophase I,* while the chromatids are still elongated and thin, the homologous chromosomes come to lie lengthwise side by side. This process is called **synapsis,** which means "fastening together." In our example, because the diploid number is four, there are two homologous pairs.

It is customary when discussing higher organisms to refer to one member of each homologous pair as a **maternal homologue** because it was originally inherited from the female parent, and to the other as a **paternal homologue** because it

PROPHASE II	METAPHASE II	ANAPHASE II	TELOPHASE II	
Chromosomes condense again.	Chromosomes line up along an equatorial plane of the cell.	Sister chromatids separate, chromosomes move to opposite poles.	Nuclei form at opposite poles of each cell and cytokinesis occurs.	**HAPLOID CELLS** Four gametes (animals) or four spores (plants) are produced.

(a)

(b)

(c)

(d)

(e)

(f)

(g)

(h)

25 μm

Figure 9–14 **Meiosis in the trumpet lily, *Lilium longiflorum*.** The chromosomes shown in these LMs have been stained and the cells flattened on microscope slides. (***a***) Mid-prophase I. (***b***) Late prophase I. (***c***) Metaphase I. (***d***) Anaphase I. (***e***) Prophase II. (***f***) Metaphase II. (***g***) Anaphase II. (***h***) Four daughter cells. *(Clare Hasenkampf, University of Toronto/Biological Photo Service)*

was contributed by the male parent during the formation of the zygote. Because each chromosome was duplicated during the premeiotic interphase and now consists of two chromatids, synapsis results in the association of *four* chromatids. The resulting complex is known as a **bivalent** or a **tetrad.** The term bivalent, in which the prefix *bi-* refers to the two homologous chromosomes, is commonly used by cytogeneticists (scientists who study inheritance at the cellular level, particularly through the analysis of chromosomes). The term tetrad (*tetra* means "four") is preferred by some geneticists interested in following the fates of the four chromatids. We will use tetrad in further discussions.

The number of tetrads per prophase I cell is equal to the haploid chromosome number. In our example there are two tetrads; in a human cell at prophase I there are 23 tetrads (and a total of 92 chromatids).

Homologous chromosomes become closely associated during synapsis. Electron microscopic observations reveal that a characteristic structure, known as the **synaptonemal complex,** forms between the synapsed homologues (Fig. 9–15). This structure holds the synapsed homologues together and is

thought to play a role in **crossing-over,** a process in which genetic material is exchanged between homologous (nonsister) chromatids. In crossing-over, enzymes break homologous chromatids and then join them to produce new combinations of genes. The resulting **genetic recombination** greatly enhances the amount of genetic variation among sexually produced offspring. This process is discussed in more detail in Chapter 10.

In many species, prophase I is a lengthy phase during which the cell grows and synthesizes nutrients. This is especially true during the formation of some egg cells because materials need to be made for the benefit of the future embryo. In many types of meiotic cells, the chromosomes assume unusual shapes during this phase. For example, lampbrush chromosomes, found in the female meiotic cells (oocytes) of some amphibians, are composed of hundreds of pairs of loops of chromatin projecting from the chromatid axis. They owe their name to their resemblance to the brushes used to clean old-fashioned oil lamps (Fig. 9–16). The loops are sites of intense synthesis of RNA, which is used to direct the synthesis of specific proteins.

In addition to the unique processes of synapsis and cross-

Chromosome

Synaptonemal
complex

Chromosome

Paternal
sister
chromatids

(a)

0.5 μm

Protein

Chromatin

(b)

Maternal
sister chromatids

Figure 9–15 A synaptonemal complex. These synapsing homologous chromosomes in meiotic prophase I are held together by a synaptonemal complex, composed mainly of protein. Synaptonemal complexes are thought to be involved in crossing-over and may have other functions. (*a*) A TEM of a synaptonemal complex. (*b*) A three-dimensional model of a tetrad with a complete synaptonemal complex. *(a, D. Von Wettstein, 1971 Proceedings of the National Academy of Science 68, pp. 851–855)*

50μm

Figure 9–16 Lampbrush chromosomes. This LM shows parts of several tetrads from a female meiotic cell (oocyte) of the newt *Triturus viridescens*. The loops, composed of chromatin, are sites of intense RNA synthesis. *(Dennis Gould)*

Figure 9–17 A meiotic tetrad. (*a*) This LM is of a tetrad during late prophase I of a male meiotic cell (spermatocyte) from a salamander. The four chromatids are labelled C, the kinetochores are visible at K, and chiasmata (resulting from crossing-over) are indicated at each X. (*b*) Interpretive drawing indicating the structure of the tetrad. The paternal chromatids are purple, and the maternal chromatids are pink. (*a, Courtesy of J. Kezer*)

(a)

1 μm

(b)

ing-over, events similar to those seen during mitotic prophase also take place. A spindle composed of microtubules and other components forms. If centrioles are present (as in animal cells), one pair moves to each pole, and astral microtubules are formed. The nuclear envelope disappears in late prophase I, and in favorable material the structure of the tetrads can be seen clearly with the microscope (Fig. 9–17). The sister chromatids continue to be closely aligned along their lengths, but the homologous chromosomes are no longer closely associated, and their centromeres (and kinetochores) are separated from one another. In late prophase I, the homologous chromosomes are held together only at specialized regions, termed **chiasmata** (sing., *chiasma*). Each chiasma originates at a site of crossing-over, that is, a site at which homologous chromatids previously broke, exchanged genetic material, and rejoined, producing an X-shaped configuration.

During meiosis I homologous chromosomes separate

Prophase I ends when the tetrads become aligned on the equatorial plane; the cell is now said to be at **metaphase I.** Both sister kinetochores of one chromosome are attached by spindle fibers to the same pole, and both kinetochores of the homologous chromosome are attached to the opposite pole. (By contrast, in mitosis, sister kinetochores are attached to opposite poles.) During **anaphase I,** the homologous chromosomes of each pair separate, or disjoin, and move toward opposite poles. Each pole receives a random mixture of maternal and paternal chromosomes, but only one member of each homologous pair is present at each pole. The sister chromatids are

united at their centromere regions. Again, this differs from mitotic anaphase, in which the sister chromatids pass to opposite poles.

During **telophase I,** the chromatids generally decondense somewhat, the nuclear envelope may reorganize, and cytokinesis may take place. Each telophase I nucleus contains the haploid number of chromosomes, but each chromosome is a duplicated chromosome. In our example, there are two duplicated chromosomes at each pole, for a total of four chromatids; in humans there are 23 duplicated chromosomes (46 chromatids) at each pole.

An interphase-like stage usually follows. Because it is not a true interphase (i.e., there is no S phase) it is given the name **interkinesis.** Interkinesis is very brief in most organisms and absent in some.

Chromatids separate in meiosis II

Because the chromosomes usually remain partially condensed between divisions, the prophase of the second meiotic division is brief. **Prophase II** is similar to mitotic prophase in many respects. There is no pairing of homologous chromosomes (indeed, only one member of each pair is present in each nucleus) and no crossing-over.

During **metaphase II,** the chromosomes line up on the equatorial planes of their cells. The first and second metaphases can be easily distinguished in diagrams; at metaphase I the chromatids are arranged in bundles of four (tetrads), and at metaphase II they are in groups of two (as in mitotic metaphase). This is not always so obvious in living cells.

During **anaphase II** the chromatids, attached to spindle

MITOSIS	MEIOSIS

PROPHASE / PROPHASE I

Nucleus

No synapsis of homologous chromosomes

Synapsis of homologous chromosomes to form tetrads

ANAPHASE / ANAPHASE I

Sister centromeres move to opposite poles

Homologous centromeres move to opposite poles

DAUGHTER CELLS / PROPHASE II

Two 2n cells with unduplicated chromosomes

Two n cells with duplicated chromosomes

ANAPHASE II

Sister centromeres move to opposite poles

HAPLOID CELLS

Four n cells with unduplicated chromosomes

Figure 9–18   Mitosis and meiosis. The diagram compares the events and the outcomes of mitosis and meiosis, in each case beginning with a diploid cell with four chromosomes (i.e., two pairs of homologous chromosomes). Because the chromosomes were duplicated in the previous interphase, each chromosome consists of two sister chromatids. The chromosomes derived from one parent are shown in purple, and those from the other parent are pink. The homologous partner chromosomes can be recognized because they are similar in size and shape. Some of the stages have been omitted for simplicity.

fibers at their kinetochores, separate and move to opposite poles, just as they would at mitotic anaphase. As in mitosis, each former chromatid is now referred to as a chromosome. Thus, at **telophase II** there is one representative for each homologous pair at each pole. Each is an unduplicated chromosome. Nuclear envelopes then re-form, the chromosomes gradually elongate to form chromatin threads, and cytokinesis occurs.

The two successive divisions yield four haploid nuclei, each containing *one* of each kind of chromosome. Each resulting haploid cell has a different combination of genes. This genetic variation has two sources: (1) During meiosis the maternal and paternal chromosomes are "shuffled" so that one member of each pair becomes randomly distributed to the poles at anaphase I. (2) DNA segments are exchanged between maternal and paternal homologues during crossing-over. The genetic consequences of these events are discussed in more detail in Chapter 10.

THE EVENTS OF MITOSIS AND MEIOSIS LEAD TO CONTRASTING OUTCOMES

Although mitosis and meiosis share many similar features, specific distinctions between these processes result in the formation of different types of cells (Fig. 9–18).

Mitosis is a single division in which *sister chromatids* separate from each other. These are distributed to the two daughter cells, which are genetically identical to each other and to the original cell. Homologous chromosomes do not associate physically at any time in mitosis.

In meiosis, a diploid cell undergoes two successive divisions, meiosis I and meiosis II. In prophase I of meiosis the homologous chromosomes undergo synapsis to form tetrads. If we ignore crossing-over, we can say that *homologous chromo-* *somes* separate during meiosis I, and *sister chromatids* separate during meiosis II. It is also correct to say that homologous centromeres (or kinetochores) separate during meiosis I, and sister centromeres (or kinetochores) disjoin during meiosis II. Meiosis ends with the formation of four, genetically different, haploid daughter cells. The fates of these cells depend on the type of life cycle; in animals they differentiate as gametes, while in plants they become spores.

SUMMARY WITH KEY TERMS

I. In the production of a new generation, cells transfer genetic information from parent to offspring.
II. **Genes** are made of DNA, which is complexed with protein to form the **chromatin** fibers that make up **chromosomes.**
  A. A **diploid** organism of a given species has a characteristic number of chromosome pairs per cell.
  B. The two members of each chromosome pair, called **homologous chromosomes,** are similar in length, shape, and other structural features, and carry genes affecting the same kinds of attributes of the organism.
III. The eukaryotic **cell cycle** is the period from the beginning of one division to the beginning of the next; the time required to complete one cycle is the **generation time.**
  A. **Interphase** can be divided into the first gap phase (G_1), the chromosomal synthesis phase (S), and the second gap phase (G_2).
    1. During the **G_1 phase,** the cell grows and prepares for the S phase.
    2. During the **S phase,** DNA and the chromosomal proteins are synthesized.
    3. During the **G_2 phase,** protein synthesis increases in preparation for cell division.
  B. During **mitosis,** identical chromosomes are distributed to each pole of the cell, and a nuclear envelope forms around each set.
    1. During **prophase,** duplicated chromosomes, each composed of a pair of **sister chromatids** associated with each other in the vicinity of their **centromeres,** become visible with the microscope. The nucleolus disappears, the nuclear envelope breaks down, and the **mitotic spindle** begins to form.
    2. During **metaphase** the chromosomes are aligned on the equatorial plane of the cell; the mitotic spindle is complete and the **kinetochores** of the sister chromatids are attached by microtubules to opposite poles of the cell.
    3. During **anaphase** the sister chromatids become separated and move to opposite poles. Each former chromatid is now referred to as a chromosome.
    4. During **telophase** a nuclear envelope re-forms around each set of chromosomes, nucleoli become apparent, the chromosomes uncoil, and the spindle disappears.
  C. During **cytokinesis,** which generally begins in telophase and therefore overlaps mitosis, the cytoplasm divides to form two individual cells.

1. In animal cells a ring of microfilaments contracts, producing a furrow that divides the cytoplasm.
2. In plant cells the **cell plate** provides materials for new plasma membranes and cell walls.
IV. There are two major forms of reproduction: asexual and sexual.
  A. Offspring produced by **asexual reproduction** usually have hereditary traits identical to those of the single parent. These offspring constitute a **clone.** Usually all the cells involved are produced by mitosis.
  B. In **sexual reproduction** two haploid sex cells, or **gametes,** fuse to form a single diploid **zygote.** This process is balanced by **meiosis** at some point in the life cycle.
    1. The somatic cells of animals are diploid; the only haploid cells are the gametes (produced by **gametogenesis,** which in animals includes meiosis.)
    2. Simple eukaryotes may be regularly haploid; the only diploid stage is the zygote, which undergoes meiosis to restore the haploid state.
    3. Plants and some algae have **alternation of generations.** A multicellular diploid **sporophyte** forms haploid spores by meiosis. Each spore divides mitotically to form a multicellular haploid **gametophyte,** which produces gametes by mitosis. Two haploid gametes then fuse to form a diploid zygote, which divides mitotically to produce a new diploid sporophyte.
V. A diploid cell undergoing meiosis completes two successive cell divisions to give rise to four haploid cells.
  A. **Meiosis I** begins with **prophase I;** the members of a homologous pair of chromosomes become physically joined by a process known as **synapsis** and undergo **crossing-over,** a process of **genetic recombination** during which segments of DNA strands are exchanged between homologous (nonsister) chromatids.
  B. At metaphase I **tetrads,** each composed of a pair of homologous chromosomes held together by one or more **chiasmata,** line up on the equatorial plane.
  C. The members of each pair of homologous chromosomes separate during meiotic **anaphase I** and are distributed to different nuclei. Each nucleus contains the haploid number of chromosomes; each chromosome consists of two chromatids.
  D. During **meiosis II** the two chromatids of each chromosome separate and one is distributed to each daughter cell. Each former chromatid is now referred to as a chromosome.

POST-TEST

1. Chromatin fibers include (a) DNA and structural polysaccharides (b) RNA and phospholipids (c) protein and carbohydrate (d) DNA and protein (e) triacylglycerol and steroids
2. The term *S phase* refers to (a) DNA synthesis during interphase (b) synthesis of chromosomal proteins during prophase (c) active RNA synthesis in lampbrush chromosomes (d) synapsis of homologous chromosomes (e) fusion of gametes in sexual reproduction
3. At which of the following stages do human skin cell nuclei have the same DNA content? (a) early mitotic prophase; mitotic telophase (b) G_1; G_2 (c) G_1; early mitotic prophase (d) G_1; mitotic telophase (e) G_2; mitotic telophase
4. In a cell at _____ each chromosome consists of a pair of attached chromatids. (a) mitotic prophase (b) meiotic prophase II (c) meiotic prophase I (d) meiotic anaphase I (e) all of the above
5. In an animal cell at mitotic metaphase, you would expect to find (a) two pairs of centrioles located on the metaphase plate (b) a pair of

centrioles inside the nucleus (c) a pair of centrioles within each microtubule organizing center (d) a centriole within each centromere (e) no centrioles

6. Cell plate formation usually begins during (a) telophase in a plant cell (b) telophase in an animal cell (c) G_2 in a plant cell (d) G_2 in an animal cell (e) a and b are correct

7. The life cycle of a sexually reproducing organism includes (a) mitosis (b) meiosis (c) fusion of sex cells (d) b and c (e) a, b, and c

8. Which of the following are genetically identical? (a) two cells resulting from meiosis I (b) two cells resulting from meiosis II (c) four cells resulting from meiosis I followed by meiosis II (d) two cells resulting from a mitotic division (e) all of the above

9. You would expect to find a synaptonemal complex in a cell at (a) mitotic prophase (b) meiotic prophase I (c) meiotic prophase II (d) meiotic anaphase I (e) meiotic anaphase II

10. A particular plant species has a diploid chromosome number of 20. A haploid cell of that species at mitotic prophase contains a total of _____ chromosomes and _____ chromatids. (a) 20; 20 (b) 20; 40 (c) 10; 10 (d) 10; 20 (e) none of the above because haploid cells cannot undergo mitosis

11. A diploid nucleus at early mitotic prophase has _____ set(s) of chromosomes; a diploid nucleus at mitotic telophase has _____ chromosome set(s). (a) 1; 1 (b) 1; 2 (c) 2; 2 (d) 2; 1 (e) not enough information has been given.

12. A chiasma links a pair of (a) homologous chromosomes at meiotic metaphase II (b) homologous chromosomes at meiotic metaphase I (c) sister chromatids at meiotic metaphase II (d) sister chromatids at mitotic metaphase (e) sister chromatids at meiotic metaphase I

REVIEW QUESTIONS

1. Two species may have the same chromosome number and yet have very different attributes. Explain.

2. Sketch a duplicated chromosome and label the sister chromatids, the centromeres, and the kinetochores. What are the functions of centromeres and of kinetochores?

3. How does the DNA content of the cell change from the beginning of interphase to the end of interphase? Does the number of chromatids change? Does the number of chromosomes change?

4. Are homologous chromosomes present in a diploid cell? Are they present in a haploid cell?

5. How does meiosis differ from mitosis? Are there any points of similarity

between these two processes? Explain.

6. What kinds of life cycles include a multicellular haploid stage? Can haploid cells divide by mitosis? By meiosis?

7. Assume that an animal has a diploid chromosome number of ten. (a) How many chromosomes would it have in a typical body cell, such as a skin cell? (b) How many chromosomes would be present in a cell at mitotic prophase? How many chromatids? (c) How many chromosomes would be present in each daughter cell produced by mitosis? Are these duplicated chromosomes? (d) How many tetrads would form in prophase I of meiosis? (e) How many chromosomes would be present in each gamete? Are these duplicated chromosomes?

YOU MAKE THE CONNECTION

1. Decide whether each of the following is an example of sexual or asexual reproduction and state why. (a) A diploid queen honeybee produces haploid eggs by meiosis. Some of these eggs are never fertilized and develop into haploid male honeybees (drones). (b) Haploid male honeybees produce haploid sperm by mitosis. These sperm fertilize haploid eggs produced by the queen, resulting in the development of diploid female worker bees. (c) Seeds develop after a flower has been pollinated with pollen from a different plant of the same species. (d) Seeds develop after a flower has been pollinated with pollen from the same plant. (e) A cutting from a plant develops roots after it has been placed in water. The plant survives and grows after it is transplanted to soil.

RECOMMENDED READINGS

Alberts, B., D. Bray, J. Lewis, M. Raff, K. Roberts, and J.D. Watson. *Molecular Biology of the Cell,* 3rd ed. Garland, New York, 1994. An extensive, detailed, and well written discussion of cell growth and division, covering the control of cell division, the cell cycle, and the events of mitosis and meiosis.

Haber, J.E. "Searching for a Partner." *Science,* Vol. 279, 6 Feb. 1998. A discussion of experimental evidence that the order of events in meiotic prophase is not the same in all organisms.

McKim, K.S. and R.S. Hawley. "Chromosomal Control of Meiotic Cell Division." *Science,* Vol. 270, 8 Dec. 1995. This article discusses the many ways that chromosomes actively participate in meiosis.

Nasmyth, K. "Viewpoint: Putting the Cell Cycle in Order." *Science,* Vol. 274, 6 Dec. 1996. This overview introduces a series of articles in the same issue that explore various aspects of the cell cycle.

Nicklas, R. Bruce. "How Cells Get the Right Chromosomes." *Science,* Vol. 275, 31 Jan. 1997. A pioneer in the micromanipulation of chromosomes in living cells reviews the mechanisms that ensure proper distribution of chromosomes in mitosis and meiosis.

Pluta, A.F., A.M. Mackay, A.M. Ainsztein, I.G. Goldberg and W.C. Earnshaw. "The Centromere: Hub of Chromosomal Activities." *Science,* Vol. 270, 8 Dec. 1995. This article explores the importance of centromeres throughout the cell cycle.

● Visit our website at **http://www.saunderscollege.com/lifesci/titles.html** and click on Solomon/Berg/Martin Biology for links to chapter-related resources on the World Wide Web.

CHAPTER 10

The Basic Principles of Heredity

Heredity, the transmission of genetic information from parent to offspring, is generally a very regular process, following predictable patterns. The basic rules of inheritance in eukaryotes were first discovered by Gregor Mendel (1822–1884), an abbot who bred pea plants, depicted here in his monastery garden at Brünn, Austria (now Brno, Czech Republic). Mendel was the first scientist to effectively apply quantitative methods to the study of inheritance. He did not merely describe his observations; he planned his experiments carefully, recorded the data, and subjected the results to mathematical analysis. Although his work was unappreciated in his lifetime, it was rediscovered in 1900. His major findings, including those now known as Mendel's principles of segregation and independent assortment, became the foundation of the science of **genetics.**

During the decades following the rediscovery of Mendel's findings, geneticists initially extended Mendel's principles by correlating the transmission of genetic information from generation to generation with the behavior of chromosomes during meiosis. They also refined his methods and, through their studies on a variety of organisms, both verified Mendel's findings and added to a growing list of so-called exceptions to his principles. These include such phenomena as linkage, sex linkage, and polygenic inheritance.

Some geneticists were very active in the development of the emerging science of statistical analysis (which had been in its infancy in Mendel's time), thereby providing scientists with increasingly sophisticated ways to analyze and interpret experimental data. These statistical methods were also essential to the study of the genetic makeup of natural populations of organisms. The results of these investigations on the genetics of populations were combined with Charles Darwin's theory of evolution by natural selection to develop a unified modern theory of evolution, firmly based on genetic principles (see Chapters 17 and 18).

Geneticists study not only the transmission of genes, but also the expression of genetic information. As you will see in this and succeeding chapters, our understanding of the relationship between an organism's genes and its characteristics has become increasingly sophisticated as we have learned more about the flow of information in cells.

(The Bettmann Archive)

AFTER YOU HAVE STUDIED THIS CHAPTER YOU SHOULD BE ABLE TO

1. Define and use correctly the terms *allele, locus, genotype, phenotype, dominant, recessive, homozygous, heterozygous,* and *test cross.*
2. Apply Mendel's principles to solve genetics problems involving monohybrid and dihybrid crosses.
3. Apply the product rule and sum rule appropriately when predicting the outcomes of genetic crosses.
4. Solve genetics problems involving incomplete dominance, epistasis, polygenes, multiple alleles, and X-linked inheritance.
5. Explain some of the ways in which genes may interact to affect the phenotype; discuss how it is possible for a single gene to affect many features of the organism simultaneously.
6. Analyze data from a test cross involving alleles of two loci.

Show how such data can be used to distinguish between independent assortment and linkage. Relate independent assortment and linkage to specific events in meiosis.
7. Discuss the genetic determination of sex and the role of the Y chromosome in determining male sex in humans; contrast the mechanism of sex determination in humans and other mammals with that in various other animals and some plants; compare dosage compensation of X-linked genes in mammals and *Drosophila.*
8. Assess the effects of inbreeding versus outbreeding on a population; discuss the genetic basis of hybrid vigor.

MENDEL FIRST DEMONSTRATED THE PRINCIPLES OF INHERITANCE

Gregor Mendel was not the first plant breeder; at the time he began his work, **hybrid** plants and animals (offspring of two genetically dissimilar parents) had been known for a long time. When Mendel began his breeding experiments in 1856, two main facts about inheritance were widely recognized: (1) All hybrid plants that are offspring of the same kinds of parents are similar in appearance. (2) When these hybrids are mated to each other they do not breed true; their offspring show a mixture of traits. Some look like their parents, and some have features like their grandparents. Mendel's genius lay in his ability to recognize a pattern in the way the parental traits reappear in the offspring of hybrids. No one before had categorized and counted the offspring and analyzed these regular patterns over several generations.

Just as geneticists do today, Mendel chose the organism for his experiments very carefully. The garden pea, *Pisum sativum,* had a number of advantages. Pea plants are easy to grow, and many varieties were available through commercial sources. It is impossible to study inheritance without such genetic **variation.** (If every person in the population had blue eyes, it would be impossible to study the inheritance of eye color.) Another advantage of pea plants is that controlled pollinations are relatively easy to conduct. Pea flowers (Fig. 10–1) have both male and female parts, and are naturally self-pollinated. However, the male anthers (pollen-producing parts of the flower) can be removed to prevent self-fertilization. Pollen from a different source can then be applied to the stigma (receptive surface of the female part). Pea flowers are easily protected from other sources of pollen because the reproductive structures are completely enclosed by the petals. Although Mendel did not mention having done this, plant hybridizers usually cover the flowers with small bags to provide additional protection from pollinating insects.

Although his original pea seeds were obtained from commercial sources, Mendel did some important preliminary work

before he started his actual experiments. For several years he worked to develop genetically pure, or **true-breeding,** lines for a number of inherited features. Today we use the term **phenotype** to refer to the physical appearance of an organism. A true-breeding line produces only offspring expressing the same phenotype (e.g., round seeds or tall plants), generation after generation. During this time he apparently chose those characteristics of his pea strains that could be studied most easily, and probably discarded or ignored a number of others. He probably made the initial observations that would later form the basis of his theories.

Mendel eventually chose strains representing seven clearly contrasting pairs of phenotypes: yellow versus green seeds, round versus wrinkled seeds, green versus yellow pods, tall versus short plants, inflated versus constricted pods, white seed coats versus gray seed coats, and flowers borne on the ends of the stems versus flowers appearing all along the stems. Other plant breeders typically studied hybrids between parents that differed in many, often not clearly defined, ways. Mendel's re-

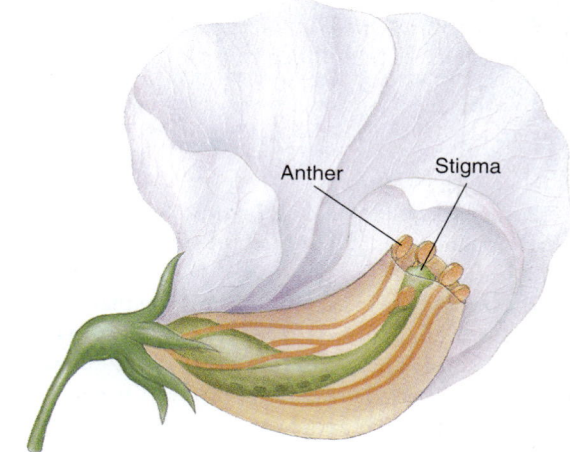

Figure 10–1 **Reproductive structures of a pea flower.** This cut-away view shows the pollen-producing anthers and the stigma, that portion of the female part of the flower that receives the pollen.

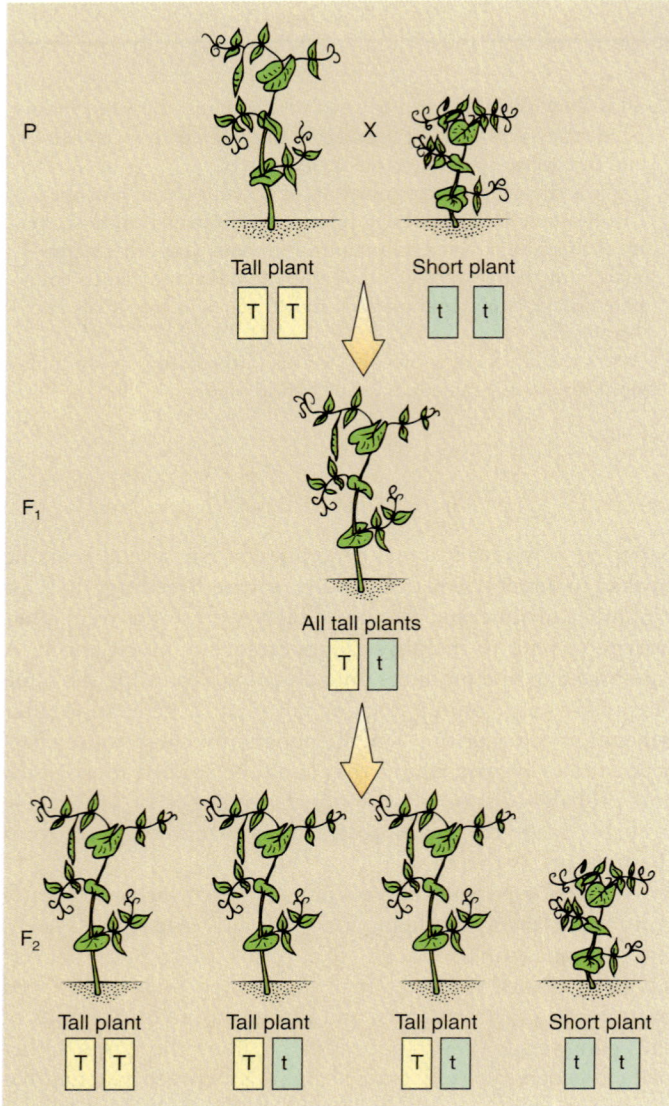

Figure 10–2 **One of Gregor Mendel's many pea plant crosses.** Crossing a tall pea plant with a short pea plant yielded only tall offspring in the F₁ generation. However, when these F₁ individuals self-pollinated, or when two F₁ individuals were crossed, the resulting F₂ generation included tall and short plants in a ratio of about 3:1.

F₂ generation, was produced by a cross between F₁ individuals, or by self-pollination of F₁ individuals. Mendel's F₂ generation in this experiment included 787 tall plants and 277 short plants.

Most breeders of Mendel's time thought that inheritance was controlled by fluids that blended together when hybrids were formed. One implication of this idea is that a hybrid should be intermediate between the two parents. However, in Mendel's experiments, hereditary factors from one of the parents apparently masked expression of the hereditary factors from the other parent in an F₁ hybrid. The factor expressed in the F₁ generation (tallness in our example) is said to be **dominant;** the one hidden (shortness) is said to be **recessive.** Although we know today that dominance is not always observed (exceptions are considered later in this chapter), the fact that dominance can occur conflicted with the notion of blending inheritance.

Mendel's results argued against blending inheritance in yet another way. Once two fluids have blended, it is very difficult to imagine how they can be separated. However, in the example just discussed, the hereditary factor(s) that controlled shortness clearly had not been lost or blended inseparably with the hereditary factor(s) that controlled tallness in the F₁ generation, because shortness reappeared in the F₂ generation. Mendel therefore proposed that each kind of inherited feature of an organism is controlled by two factors that behave like particles and are present in every individual. To Mendel these "hereditary factors" were abstractions, because he knew nothing of chromosomes and DNA. They are essentially what we call **genes** today, so we will use that term in our discussion. Today we know that genes are not particles, but treating them as such allowed Mendel to develop precise mathematical models, testable by experiment, to predict the patterns by which genes are transmitted from generation to generation.

Mendel's experiments led to his discovery and explanation of the major principles of heredity, which we now know as the principles of segregation and independent assortment. We consider the first now and the second later in the chapter.

The principle of segregation states that alleles separate before gametes are formed

Today we use the term **alleles** to refer to the alternative forms of a gene. In the example in Figure 10–2, each F₁-generation tall plant had two different alleles that control plant height: one for tallness (which we designate T) and one for shortness (which we designate t), but because the tall gene was dominant, these plants were tall. To explain his experimental results, Mendel proposed an idea that we now refer to as the principle of segregation. Using modern terminology, the **principle of segregation** states that, in order for sexual reproduction to occur, the two alleles carried by an individual parent must become separated (segregated). As a result, each sex cell (egg or sperm) formed contains only one allele of each pair. An essential feature of the process is that the alleles remain intact (one does not "contaminate" or eliminate the other); thus, re-

sults were much easier to analyze because he chose easily distinguishable phenotypes and limited the genetic variation studied in each experiment.

Mendel began his experiments by crossing plants from two different true-breeding lines with contrasting phenotypes; these genetically pure individuals constituted the **parental,** or **P, generation.** In every case, the members of the first generation of offspring all looked alike and resembled one of the two parents. For example, when he crossed tall plants with short plants, all the progeny were tall (Fig. 10–2). These offspring were the **first filial** (*filial* comes from Latin for "sons and daughters") generation, or **F₁ generation.** The second filial generation, or

Metaphase I

Metaphase II

T

t

Figure 10–3 **The chromosomal basis for segregation.** The separation of homologous chromosomes during meiosis results in the segregation of alleles. Note that one-half of the gametes will carry *T* and one-half will carry *t*.

meiosis (Fig. 10–3). (Recall from Chapter 9 that in all sexual life cycles, meiosis must occur at some point prior to gamete formation.) Later, at the time of fertilization, each haploid gamete contributes one chromosome from each homologous pair and therefore one gene for each gene pair (either *T* or *t* in our example). Although gametes and fertilization were known at the time Mendel carried out his research, mitosis and meiosis had not yet been discovered. It is truly remarkable that Mendel was able to formulate his ideas mainly on the basis of mathematical abstractions. Today his principles are much easier to understand because we are able to think about them in concrete terms by relating the transmission of genes to the behavior of chromosomes.

Mendel reported these and other findings (discussed later in this chapter) at a meeting of the Brünn Society for the Study of Natural Science; he published his results in the transactions of that society in 1866. At that time biology was largely a descriptive science, and biologists had little interest in applying quantitative and experimental methods such as Mendel had used. The importance of his results and his interpretations of those results was not appreciated by other biologists of the time, and his findings were neglected for nearly 35 years.

In 1900 Hugo DeVries in Holland, Karl Correns in Germany, and Erich von Tschermak in Austria each rediscovered Mendel's paper and found that it provided explanations for their own research findings. They gave credit to Mendel by naming the basic laws of inheritance after him. By this time biologists had a much greater appreciation of the value of quantitative experimental methods. The details of mitosis, meiosis, and fertilization had been described, and in 1903 W. S. Sutton pointed out the connection between Mendel's segregation of genes and the separation of homologous chromosomes during meiosis. The time was right for wider acceptance and extension of these ideas and their implications.

Alleles occupy corresponding loci on homologous chromosomes

Today we know that each chromatid is made up of one long DNA molecule and that each gene is actually a segment of that DNA molecule. We also know that homologous chromosomes usually have similar genes located in corresponding positions. The term **locus**[1] (pl., *loci*) was originally coined to designate the location of a particular gene on the chromosome (Fig. 10–4). Of course we are actually referring to a segment of the DNA that has the information required to control some aspect of the organism. One locus may be involved in determining seed color, another seed shape, still another the shape of the pods, and so on. The existence of a particular locus can be inferred (by traditional genetic methods at least) only if at least two allelic variants of that locus, producing contrasting phenotypes (e.g., yellow peas versus green peas) are available

---

[1]In mathematics a locus is a dimensionless point; a genetic locus, being a segment of DNA, is obviously not dimensionless!

cessive alleles are not lost and so can reappear in the F_2 generation.

In our example, before the F_1 plants formed gametes, the allele for tallness separated (segregated) from the allele for shortness, so that half of the gametes contained a *T* allele and the other half a *t* allele. The random process of fertilization led to three possible combinations of alleles in the F_2 offspring: one-fourth with two tallness alleles *(TT)*, one-fourth with two shortness alleles *(tt)*, and one-half with one allele for tallness and one for shortness *(Tt)*. Because both *TT* and *Tt* plants are tall, on average Mendel expected approximately three-fourths (787/1064) to express the phenotype of the dominant allele (tall) and about one-fourth (277/1064) the phenotype of the recessive allele (short). (The mathematical reasoning behind these predictions will be explained shortly.)

Today we know that segregation of alleles is a direct result of the separation of homologous chromosomes during

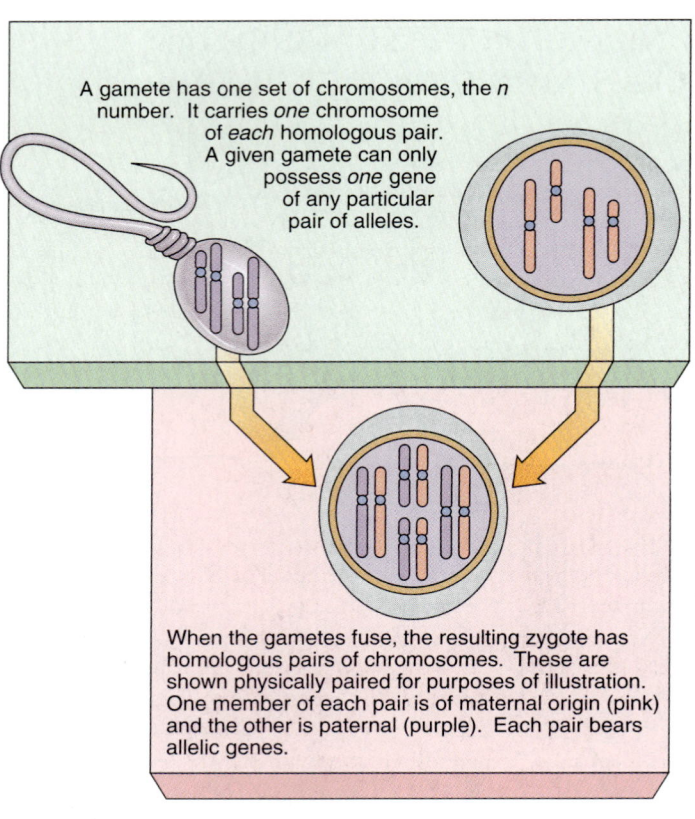

Figure 10-4 Loci and their alleles.

for study. In the simplest cases an individual can express one (yellow) or the other (green) but not both.

Thus alleles are genes that govern variations of the same feature (yellow versus green seed color) and occupy corresponding loci on homologous chromosomes. Each allele (variant) of a locus is assigned a single letter (or group of letters) as its symbol. Although more complicated forms of notation are often used by geneticists, it is customary when working simple genetics problems to indicate a dominant allele with a capital letter and a recessive allele with the same letter in lowercase. The choice of the letter is generally determined by the first allelic variant found for that locus. For example, the dominant allele that governs the yellow color of the seed might be

designated Y, and the recessive allele responsible for the green color would then be designated y. Because discovery of the yellow allele made identification of this locus possible, we refer to the locus as the *yellow* locus, although pea seeds are most commonly green.

Remember that the term *locus* designates not only a position on a chromosome but also a type of gene controlling a particular kind of characteristic; thus, Y (yellow) and y (green) represent a specific pair of alleles of a locus involved in determining seed color in peas. *Although you may initially be uncomfortable with the fact that geneticists sometimes use the term gene to specify a locus and at other times to specify one of the alleles of that locus, the meaning is usually clear from the context.*

A MONOHYBRID CROSS INVOLVES INDIVIDUALS WITH DIFFERENT ALLELES OF A GIVEN LOCUS

The basic principles of genetics and the use of genetic terms are best illustrated by examples. In the simplest case, a **monohybrid cross,** the inheritance of two alleles of a single locus is studied. Our first example in this section deals with the expected ratios in the F_2 generation, as did our previous example of Mendel's work on tall and short pea plants.

Heterozygotes carry two different alleles of a locus; homozygotes carry identical alleles

Figure 10–5 illustrates a monohybrid cross featuring a locus that governs coat color in guinea pigs. The female comes from a true breeding line of black guinea pigs. We say that she is **homozygous** for black because the two alleles she carries for this locus are identical. The brown male is also from a true breeding line and is homozygous for brown. What color would you expect the F_1 offspring to be? Dark brown? Spotted? It is impossible to make such a prediction without more information.

In this particular case, the F_1 offspring are black, but they are **heterozygous,** meaning that they carry two different alleles for this locus. The allele for brown coat color can be expressed only in a homozygous brown individual; it is referred to as a recessive allele. The allele for black coat color can be expressed in both homozygous black and heterozygous individuals; it is a dominant allele. On the basis of this information, we can use standard notation to designate the dominant black allele as *B* and the recessive brown allele as *b*.

During meiosis in the female parent *(BB)*, the two *B* alleles separate according to Mendel's principle of segregation so that each egg has only one *B* allele. In the male *(bb)*, the two *b* alleles separate so that each sperm has only one *b* allele. The fertilization of each *B* egg by a *b* sperm results in heterozygous F_1 offspring, each with the alleles *Bb*. That is, each individual has one allele for brown coat and one for black coat. Because this is the only possible combination of alleles present in the eggs and sperm, all the F_1 offspring are *Bb*.

A Punnett square predicts the ratios of genotypes and phenotypes of the offspring of a cross

During meiosis in heterozygous black guinea pigs *(Bb)*, the chromosome containing the *B* allele becomes separated from its homologue (the chromosome containing the *b* allele), so each normal sperm or egg contains *B* or *b*, but never both. Gametes containing *B* alleles and those containing *b* alleles are formed in equal numbers by heterozygous *Bb* individuals. Because no special attraction or repulsion occurs between an egg

Figure 10–5 A monohybrid cross. In this example, a homozygous black guinea pig is mated with a homozygous brown guinea pig. The F_1 generation includes only black individuals. However, the mating of two of these offspring yields F_2 generation offspring in the expected ratio of 3 black : 1 brown, indicating that the F_1 individuals are heterozygous. The corresponding F_2 genotypic ratio is 1*BB* : 2*Bb* : 1*bb*.

and a sperm containing the same allele, fertilization is a random process.

As illustrated in Figure 10–5, the possible combinations of eggs and sperm at fertilization can be represented in the form of a "checkerboard" known as a **Punnett square,** devised by an early geneticist, Sir Reginald Punnett. The types of gametes (and their expected frequencies) from one parent are represented across the top, and those from the other parent are indicated along the left side. The squares are then filled in with the resulting F_2 zygote combinations. Three-fourths of all F_2 offspring have the genetic constitution *BB* or *Bb* and are phenotypically black; one-fourth have the genetic constitution *bb* and are phenotypically brown. The genetic mechanism responsible for the approximate 3:1 F$_2$ ratios (called *monohybrid F$_2$ phenotypic ratios*) obtained by Mendel in his pea-breeding experiments is again evident.

The phenotype of an individual does not always reveal its genotype

An organism's phenotype is its appearance (in a given environment) with respect to a certain inherited feature. However, because some alleles may be dominant and others recessive, we cannot always determine which alleles are carried by an organism simply by looking at it. The *genetic constitution* of that organism, most often expressed in symbols, is its **genotype.** In the cross we have been considering, the genotype of the female parent is homozygous dominant, *BB,* and her phenotype is black. The genotype of the male parent is homozygous recessive, *bb,* and his phenotype is brown. The genotype of all the F$_1$ offspring is heterozygous, *Bb,* and their phenotype is black. To prevent confusion we always indicate the genotype of a heterozygous individual by writing the symbol for the dominant allele first and the recessive allele second (always *Bb,* never *bB*).

The phenomenon of dominance partly explains why an individual may resemble one parent more than the other, even if the two parents make equal contributions to their offspring's genetic constitution. Dominance is not predictable and can be determined only by experiment. In one species of animal, black coat may be dominant to brown; in another species, brown may be dominant to black.

(a)

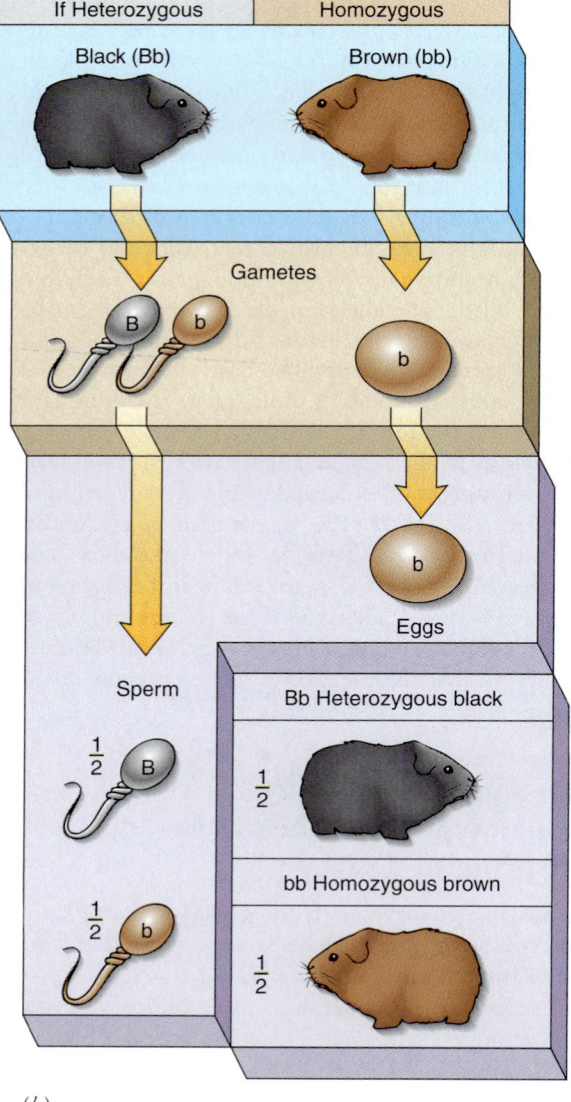

(b)

Figure 10–6 **A test cross.** In this illustration, a test cross is used to determine the genotype of a black guinea pig. *(a)* If a black guinea pig is mated with a brown guinea pig and all the offspring are black, the black parent probably has a homozygous genotype. *(b)* If any of the offspring are brown, the black guinea pig must be heterozygous. The expected ratio is 1 black:1 brown.

A test cross can detect heterozygosity

Guinea pigs with the genotypes *BB* and *Bb* are alike phenotypically; they both have black coats. How, then, can we know the genotype of a black guinea pig? Geneticists can accomplish this by performing a **test cross,** in which an individual of unknown genotype is crossed with a homozygous recessive individual (Fig. 10–6). In a test cross, the alleles carried by the gametes from the parent of unknown genotype are never "hidden" in the offspring by dominant alleles contributed by the other parent. Therefore, one can deduce the genotypes of all the classes of offspring directly from their phenotypes. If all the offspring were black, what inference would you make about the genotype of the black parent? If any of the offspring were brown, what conclusion would you draw regarding the genotype of the black parent? Would you be more certain about one of these inferences than the other?

Mendel did just these sorts of experiments, breeding F₁ (tall) pea plants with homozygous recessive *(tt)* short ones. If his ideas were correct, the F₁ individuals were expected to be heterozygous, producing equal numbers of *T* and *t* gametes. Because the homozygous short parents were expected to produce only *t* gametes, Mendel predicted that he would obtain equal numbers of tall *(Tt)* and short *(tt)* offspring. His results agreed with his predictions, thereby providing additional evidence supporting the hypothesis that there is 1:1 segregation of the alleles of a heterozygous parent. Thus, Mendel's principle of segregation not only explained the known facts, such as the 3:1 monohybrid F₂ phenotypic ratio, but also enabled him to successfully anticipate the results of other experiments, in this case the 1:1 test cross phenotypic ratio.

THE RULES OF PROBABILITY PREDICT THE LIKELIHOOD OF GENETIC EVENTS

All genetic ratios are properly expressed in terms of probabilities. In the examples just discussed, among the offspring of two individuals heterozygous for the same gene pair, the ratio of the phenotypes of the dominant and recessive alleles is 3:1. The probability of an event is its expected frequency; therefore we can say that there are 3 chances in 4 (3/4) that any particular individual offspring of two heterozygous individuals will express the dominant allele phenotype and 1 chance in 4 (1/4) that it will express the recessive allele phenotype. Although we sometimes speak in terms of percentages, probabilities must always be calculated as fractions (e.g., 3/4) or decimal fractions (e.g., 0.75). If an event is certain to occur, its probability is 1; if it is certain not to occur, its probability is 0. A probability can be 0, 1, or some number between 0 and 1.

Often we wish to *combine* two or more probabilities. The Punnett square, which we use to predict the results of genetic

Figure 10–7 **The rules of probability.** For each coin toss the probability of heads is 1/2 and the probability of tails is also 1/2. Because the outcome of the first toss is independent of the outcome of the second, the combined probabilities of the outcomes of successive tosses are calculated by multiplying their individual probabilities (according to the product rule: 1/2 × 1/2 = 1/4). These same rules of probability are used to predict genetic events.

crosses, is a device that allows us to combine probabilities. When we use a Punnett square we are intuitively following two important rules known as the **product rule** and the **sum rule.**

The product rule predicts the combined probabilities of independent events

Events are independent if the occurrence of one does not affect the probability that the other will occur. For example, the probability of obtaining heads on the first toss of a coin is 1/2; the probability of obtaining heads on the second toss (an independent event) is also 1/2. If two or more events are *independent* of each other, the probability of their both occurring is the *product* of their individual probabilities. If this seems strange to you, keep in mind that when we multiply two numbers that are less than 1, the product is a smaller number. Therefore, the probability of obtaining heads two times in a row is 1/2 × 1/2 = 1/4, or 1 chance in 4 (Fig. 10–7).

Similarly, we can apply the product rule to genetic events. If both parents are *Bb*, what is the probability that they will produce a child who is *bb*? For the child to be *bb*, he or she must receive a *b* gamete from each parent. The probability of a *b* egg is 1/2 and the probability of a *b* sperm is also 1/2. Like the outcomes of the coin tosses, these probabilities are independent, so we combine them by the product rule (1/2 × 1/2 = 1/4). You may wish to check this result using a Punnett square.

The sum rule predicts the combined probabilities of mutually exclusive events

In some cases there is more than one way to obtain a specific outcome. These different ways are called *mutually exclusive* events because no more than one of them can happen; i.e., if one of them occurs, the other(s) cannot. For example, if both parents are *Bb*, what is the probability that their first child will also have the *Bb* genotype? There are two different ways these parents can have a *Bb* child: either a *B* egg combines with a *b* sperm (probability 1/4), or a *b* egg combines with a *B* sperm (probability 1/4).

Naturally, if there is more than one way to obtain a result, the chances of its being obtained are improved; we therefore combine the probabilities of mutually exclusive events by *summing* (adding) their individual probabilities. The probability of obtaining a *Bb* child in our example is therefore 1/4 + 1/4 = 1/2. (Because there is only one way these heterozygous parents can produce a homozygous recessive child, *bb*, that probability is only 1/4. The probability of a homozygous dominant child, *BB*, is likewise 1/4.)

The rules of probability can be applied to a variety of calculations

The rules of probability have wide applications. For example, what are the probabilities that a family with two (and only two) children will have two girls, two boys, or one girl and one boy? For purposes of discussion we will assume that male and female births are equally probable. The probability of having a girl first is 1/2, and the probability of having a girl second is also 1/2. These are independent events, so we combine their probabilities by multiplying: 1/2 × 1/2 = 1/4. Similarly, the probability of having two boys is also 1/4.

In families with both a girl and a boy, the girl can be born first or the boy can be born first. The probability that a girl will be born first is 1/2, and the probability that a boy will be born second is also 1/2. We use the product rule to combine the probabilities of these two independent events: 1/2 × 1/2 = 1/4. Similarly, the probability that a boy will be born first and a girl second is also 1/4. These two kinds of families represent mutually exclusive outcomes, that is, two different ways of obtaining a family with one boy and one girl. Having two different ways of obtaining the desired result improves our chances, so we use the sum rule to combine the probabilities: 1/4 + 1/4 = 1/2. Notice that the probabilities of the three types of families—both boys (1/4), both girls (1/4), and one girl, one boy (1/2)—add up to 1. This serves as a useful check that the calculations have been done correctly. You may also wish to confirm these results by making a Punnett square.

In working with probabilities, it is important to keep in mind a point that many gamblers forget: chance has no memory. This means that if events are truly random, past events have no influence on the probability of the occurrence of independent future events. For example, if two brown-eyed people have a child, what is the probability that it will have blue eyes? If their first child has blue eyes, what is the probability that their second child will also have blue eyes? The color of the iris of the human eye is controlled by alleles at several loci, but alleles at one locus are primarily responsible. The allele for brown eye color, *B*, is usually dominant to the allele for blue, *b*. If the two brown-eyed parents are heterozygous, there is 1 chance in 4 that any child of theirs will have blue eyes. Each fertilization is a separate, independent event; its result is not affected by the results of any previous fertilizations. If these two, heterozygous, brown-eyed parents already have three brown-eyed children and are expecting their fourth child, what is the probability that the child will have blue eyes? The uninformed might guess that this one *must* have blue eyes, but in fact there is still only 1 chance in 4 that the child will have blue eyes and 3 chances in 4 that the child will have brown eyes.

If two heterozygous people marry and *plan* to have four children, what is the probability that all four will have brown eyes? The probability of brown eyes for each child is 3/4, so we combine these independent events by the product rule: 3/4 × 3/4 × 3/4 × 3/4 = 81/256 or 0.32. Why is the answer to this question so different from the answer to the previous question? Remember that for the brown-eyed children that were already born, chance (3/4) is replaced by certainty (1), so the calculation becomes 1 × 1 × 1 × 3/4 = 3/4. The chance that the fourth (as yet unborn) child will have brown eyes is therefore 3/4 (and the chance of blue eyes is 1/4).

When working probability problems, common sense is more important than blindly memorizing rules. Examine your results to see if they appear reasonable; if they do not, you should reevaluate your assumptions.

A DIHYBRID CROSS INVOLVES INDIVIDUALS THAT HAVE DIFFERENT ALLELES OF TWO LOCI

Simple monohybrid crosses each involve a pair of alleles of a single locus. Mendel also analyzed crosses involving alleles of two or more loci. A mating between individuals with different alleles at two loci is called a **dihybrid cross.** Consider the case when two pairs of alleles are located in nonhomologous chromosomes (i.e., one pair of alleles is located in one pair of homologous chromosomes, and the other pair of alleles is located in a *different* pair of homologous chromosomes). Each pair of alleles is inherited independently; that is, each pair segregates during meiosis independently of the other.

An example of a dihybrid cross carried through the F_2 generation is illustrated in Figure 10–8. When a homozygous, black, short-haired guinea pig (*BBSS*, because black is dominant to brown and short hair is dominant to long hair) and a homozygous, brown, long-haired guinea pig (*bbss*) are mated, the *BBSS* animal produces gametes that are all *BS*, and the *bbss* individual produces gametes that are all *bs*. Each gamete contains one and only one allele for each of the two loci. The

Figure 10–8 A dihybrid cross. When a black, short-haired guinea pig is crossed with a brown, long-haired one, all the offspring are black and have short hair. However, when two members of the F₁ generation are crossed, the ratio of phenotypes is 9:3:3:1. Note that the two pairs of alleles considered here assort independently.

union of the *BS* and *bs* gametes yields only individuals with the genotype *BbSs*. All these F₁ offspring are heterozygous for hair color and for hair length, and all are phenotypically black and short-haired.

The principle of independent assortment states that the alleles of different loci on nonhomologous chromosomes are randomly distributed into gametes

Each F₁ guinea pig produces four kinds of gametes with equal probability: *BS, Bs, bS,* and *bs.* Hence, the Punnett square has 16 (4²) squares representing the zygotes, some of which are genotypically or phenotypically alike. There are 9 chances in 16 of obtaining a black, short-haired individual; 3 chances in 16 of obtaining a black, long-haired individual; 3 chances in 16 of obtaining a brown, short-haired individual; and 1 chance in 16 of obtaining a brown, long-haired individual. This

9:3:3:1 phenotypic ratio is expected in a dihybrid F₂ if the hair color and hair length loci are on nonhomologous chromosomes.

On the basis of similar results, Mendel formulated the principle of inheritance now called Mendel's **principle of independent assortment,** which states that members of any gene pair segregate from one another independently of the members of the other gene pairs. This occurs in a regular way that ensures that each gamete contains one allele for each locus, but the alleles of different loci are assorted at random with respect to each other in the gametes. (Mendel only reported on the results of crosses in which the genes assorted independently. As you will soon see, independent assortment does not always occur.)

Procedures used in solving genetics problems that illustrate Mendel's principles of segregation and independent assortment are summarized at the end of the chapter in *Focus On: Solving Genetics Problems* on page 240 and *Focus On: Deducing Genotypes* on page 241.

METAPHASE
I

METAPHASE
II

bs   BS   bS   Bs

Figure 10−9 Meiosis and independent assortment. There are two equally likely ways that two different pairs of homologous chromosomes can line up at metaphase I and be subsequently distributed. A cell with the orientation shown at the left produces half *BS* and half *bs* gametes. Conversely, the cell at the right produces half *Bs* and half *bS* gametes. Because approximately half of the meiotic cells are of each type, the ratio of the four possible types of gametes is 1:1:1:1.

The mechanics of meiosis are the basis for independent assortment

Today we recognize that independent assortment is related to the events of meiosis. It occurs because there are two different ways in which two pairs of homologous chromosomes can be arranged at metaphase I of meiosis. These occur randomly, with approximately half the meiotic cells having one orientation, and the other half having the opposite orientation. The orientation of the homologous chromosomes on the metaphase plate then determines the way they subsequently separate and are distributed into the haploid cells (Fig. 10−9).

Linked genes do not assort independently

Independent assortment does not apply if the two loci are located in the same pair of homologous chromosomes. In fruit flies there is a locus controlling wing shape (the dominant allele *V* for normal wings and the recessive allele *v* for vestigial wings) and another locus controlling body color (the dominant allele *B* for gray and the recessive allele *b* for black). If a homozygous *BBVV* fly is crossed with a homozygous *bbvv* fly, the F₁ flies all have gray bodies and normal wings, and their genotype is *BbVv* (Fig. 10−10).

Because these loci happen to be located in the *same pair*

Figure 10–10 A two-point test cross to detect linkage. In fruit flies, loci for wing length (vestigial versus normal wings) and for body color (black versus gray body) are linked; they are located on a homologous chromosome pair. Linkage can be recognized when an excess of parental type offspring and a deficiency of recombinant type offspring are produced.

of homologous chromosomes, their alleles do not assort independently; instead they tend to be inherited together and are said to be **linked.** Linkage is most readily observed by analyzing the results of a test cross in which heterozygous F₁ flies *(BbVv)* are mated with homozygous recessive *(bbvv)* flies (see bottom of Fig. 10–10). Because heterozygous individuals are mated to homozygous recessive individuals, this test cross is similar to the test cross described previously. However, it is called a **two-point test cross** because alleles of two loci are involved.

If the loci governing these characteristics were on different chromosomes (unlinked), the heterozygous parent in a test cross would produce four kinds of gametes (*BV, Bv, bV,* and *bv*) in equal numbers. As a result of this independent assortment, offspring with new gene combinations not present in the parental generation would be produced. Any process that leads to new gene combinations is called **recombination.** In our example, *Bv* and *bV* are both **recombinant** gametes. The other two kinds of gametes, *BV* and *bv,* are called **parental gametes** because they are identical to the gametes produced by the P generation. Of course, the homozygous recessive parent produces only one kind of gamete, *bv.* Thus, if independent assortment were to occur in the F₁ flies, approximately 1/4 of

the test-cross offspring would be gray-bodied and normal-winged *(BbVv),* 1/4 black-bodied and normal-winged *(bbVv),* 1/4 gray-bodied and vestigial-winged *(Bbvv),* and 1/4 black-bodied and vestigial-winged *(bbvv).* Notice that the two-point test cross allows us to determine the genotypes of the offspring directly from their phenotypes.

By contrast, the loci in our example behave differently because they are linked. Alleles at different loci on a given chromosome tend to be inherited together because chromosomes pair and separate during meiosis as units and therefore tend to be inherited as units. If linkage were complete, only parental type flies, with gray bodies and normal wings *(BbVv),* or black bodies and vestigial wings *(bbvv),* would be produced. However, in our example, the progeny also include some gray-bodied, vestigial-winged flies and some black-bodied, normal-winged flies. These are recombinant type flies, having received a recombinant gamete from the heterozygous F₁ parent. Each recombinant gamete arose by crossing-over between these loci in a meiotic cell of a heterozygous female[2] fly. Recall from

[2]Fruit flies are unusual in that crossing-over occurs only in females and not in males. It is far more common for crossing-over to occur in both sexes of a species.

Chapter 9 that when chromosomes pair and undergo synapsis, **crossing-over** occurs as homologous (nonsister) chromatids exchange segments of chromosomal material by a process of breakage and rejoining (Fig. 10–11).

The linear order of linked genes on a chromosome is determined by calculating the frequency of crossing-over

In our example, about 20% of the offspring are recombinant types: gray flies with vestigial wings, *Bbvv* (approximately 10% of the total); and black flies with normal wings, *bbVv* (also about 10% of the total). The remaining 80% are parental types. These data can be used to calculate the percentage of crossing-over between the loci. This is done by adding the number of individuals in the two recombinant classes of offspring (10 + 10), dividing by the *total number of offspring* (40 + 40 + 10 + 10), and multiplying by 100. Thus, the *V* locus and the *B* locus have 20% recombination between them.

During a single meiotic division, several exchanges may occur at different points along the length of each homologous chromosome pair. In general, a crossover is more likely to occur between two loci if they are far apart on the chromosome and less likely to occur if they are close together. Because this rough correlation exists between the frequency of recombination between two loci and the linear distance between them, a genetic map of the chromosome can be generated by converting the percentage of recombination to **map units.** By convention, 1% recombination between two loci equals a distance of 1 map unit, so the loci in our example are 20 map units apart.

The frequencies of recombination between specific linked loci have been measured in a number of species. All of the experimental results are consistent with the hypothesis that

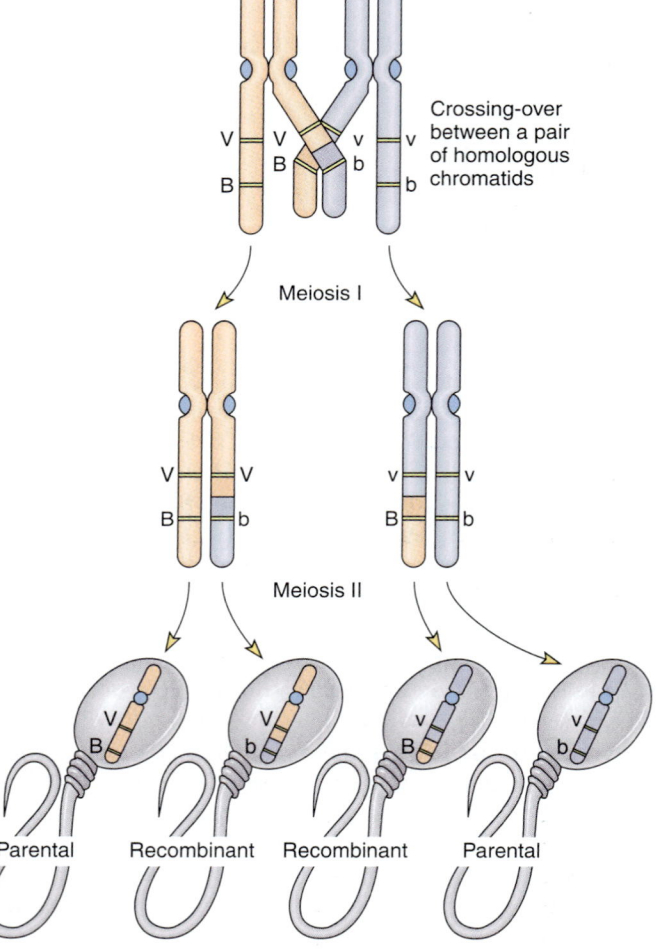

Figure 10–11

Figure 10–11 **Crossing-over.** The exchange of segments between chromatids of homologous chromosomes permits the recombination of linked genes. Genes located far apart on a chromosome have a greater probability of being separated by an exchange of segments than do genes that are closer together.

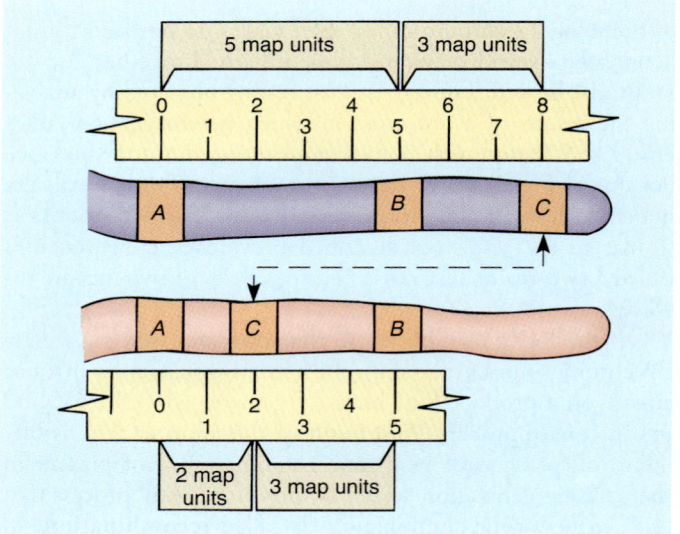

Figure 10–12 **Gene mapping.** Gene order (i.e., which locus lies between the other two) is determined by the percentage of recombination between each of the possible pairs. In this hypothetical example, the percentage of recombination between locus *A* and locus *B* is 5% (corresponding to 5 map units) and that between *B* and *C* is 3% (3 map units). If the recombination between *A* and *C* is 8% (8 map units), *B* must be in the middle. However, if the recombination between *A* and *C* is 2%, then *C* must be in the middle.

genes are present in a linear order in the chromosomes. Figure 10–12 illustrates the traditional method for determining the order of genes in a chromosome.

More than one crossover between two loci in a single tetrad can occur in a given cell undergoing meiosis. We can observe only the frequency of offspring receiving recombinant gametes from the heterozygous parent, not the actual number of crossovers. In fact, the actual frequency of crossing-over is slightly more than the observed frequency of recombinant gametes. This is because the simultaneous occurrence of two crossovers involving the same two homologous chromatids reconstitutes the original combination of genes (Fig. 10–13). When two loci are relatively close together, this effect is minimized.

The genes in a particular chromosome tend to be inherited together and, therefore, are said to constitute a **linkage group.** The number of linkage groups determined by genetic tests is equal to the number of pairs of chromosomes. By putting together the results of many crosses, scientists painstakingly developed detailed linkage maps for a number of eukaryotes, including the fruit fly (which has four pairs of chromosomes), the mouse, yeast, and *Neurospora* (a fungus). In addition, special genetic methods have made possible the development of a detailed map for *Escherichia coli,* a bacterium with a single, circular DNA molecule, and a number of other prokaryotes and viruses. Much more sophisticated maps of chromosomes have been made by means of recombinant DNA technology (see Chapter 14). These methods have been particularly useful in producing maps of human chromosomes through the Human Genome Project (see Chapter 15).

SEX IS COMMONLY DETERMINED BY SPECIAL SEX CHROMOSOMES

Mechanisms of sex determination vary considerably (see *Making the Connection: Mechanisms of Sex Determination*). Most animals have special sex chromosomes. Typically, members of one sex (the **homogametic** sex) have a pair of similar sex chromosomes and produce gametes that are all identical in sex chromosome constitution. The members of the other sex (the **heterogametic** sex) have two different sex chromosomes and produce two kinds of gametes, each bearing a single kind of sex chromosome.

The females of many animal species (including humans) are homogametic; their cells contain two **X chromosomes.** In contrast, the males are **heterogametic,** having a single X chromosome and a smaller **Y chromosome.** Human males have 22 pairs of **autosomes,** which are chromosomes other than the sex chromosomes, plus one X chromosome and one Y chromosome; females have 22 pairs of autosomes plus two X chromosomes.

The Y chromosome determines male sex in most species of mammals

Do human males have a male phenotype because they have only one X chromosome or because they have a Y chromosome? Much of the evidence bearing on this question comes from studies of persons with abnormal sex chromosome constitutions (see Chapter 15). A person with an XXY constitution is a nearly normal male in external appearance but has underdeveloped testes (Klinefelter syndrome). A person with one X but no Y chromosome has the appearance of an immature female (Turner syndrome). An embryo with a Y but no X does not survive. Hence all individuals require at least one X, and the Y is the male-determining chromosome. In fact, several genes on the Y that are involved in male determination have been identified.

The X and Y chromosomes are not truly homologous, that is, they are not similar in size, shape, and genetic constitution. However, they have a short homologous "pairing region" that allows them to synapse and separate from one another during meiosis. Half the sperm contain an X chromosome and half contain a Y chromosome. All normal eggs bear a single X chromosome (Fig. 10–14). Fertilization of an X-bearing egg by an X-bearing sperm results in an XX (female) zygote; fertilization by a Y-bearing sperm results in an XY (male) zygote.

Figure 10–13 Double crossing-over. If the same homologous chromatids undergo double crossing-over between the genes of interest, recombinant gametes are not formed.

MECHANISMS OF SEX DETERMINATION

What determines the sex of an organism? Genes are the most important sex determinants in most organisms, although in some species sex is controlled mainly by the environment. The major sex-determining genes of most animals are carried by sex chromosomes.

An XX/XY sex chromosome mechanism, similar to that of humans, operates in many species of animals. However, it is not universal, and many of the details may vary. For example, the fruit fly, *Drosophila*, has homogametic (XX) females and heterogametic (XY) males, but the Y is not male-determining; a fruit fly with an X chromosome and no Y chromosome has a male phenotype. In birds and butterflies the mechanism is reversed, with homogametic males (the equivalent of XX) and heterogametic females (the equivalent of XY).

In **hermaphroditic** organisms, organs of both sexes are found in the same individual. Hermaphroditic animals (see Chapters 28, 29, and 48) do not have sex chromosomes. Most flowering plants are hermaphrodites. The male and female sexual organs may be in the same flowers. If they are in separate flowers on the same plant, the plants are said to be **monoecious;** corn, walnuts, and oaks are examples. Far fewer flowering plants are **dioecious,** having male and female floral organs on separate plants. A few dioecious plants, such as asparagus, apparently have sex chromosomes, although they are not necessarily comparable to those of animals.

We would expect to have equal numbers of X- and Y-bearing sperm and a 1 : 1 ratio of females to males. In fact, however, more males are conceived than females and more males die before birth. Even at birth the ratio is not 1 : 1; about 106 boys are born for every 100 girls. It is not known why this occurs, but the Y-bearing sperm is assumed to have some competitive advantage.

X-linked genes have unusual inheritance patterns

The human X chromosome contains many loci that are required in both sexes, whereas the Y chromosome contains only a few genes, including one or more genes for maleness. Genes located in the X chromosome, such as colorblindness and hemophilia, are sometimes called **sex-linked** genes. It is more appropriate, however, to refer to them as **X-linked** genes because they follow the transmission pattern of the X chromosome and, strictly speaking, are not linked to the sex of the organism per se.

A female receives one X from her mother and one X from her father. A male receives his Y chromosome, which makes him male, from his father. From his mother he inherits a single X chromosome and therefore all of his X-linked genes. In the male, every X chromosome allele present is expressed, whether that allele was dominant or recessive in the female parent. A male is neither homozygous nor heterozygous for his X-linked loci; instead he is always **hemizygous** for every X-linked locus (*hemi* means "half").

We will use a simple system of notation for problems involving X linkage, indicating the X and incorporating specific alleles as superscripts. For example, the symbol X^c signifies a recessive X-linked allele for colorblindness and X^C the dominant X-linked allele for normal color vision. The Y chromosome is written without superscripts because it does not carry the locus of interest. Two recessive X-linked alleles must be present in a female for the abnormal phenotype to be expressed, whereas in the hemizygous male a single abnormal allele is expressed. As a practical consequence, these abnormal alleles are usually expressed only in male offspring, although they may be carried by a female.

To be expressed in a female, a recessive X-linked allele must be inherited from both parents. A colorblind female, for

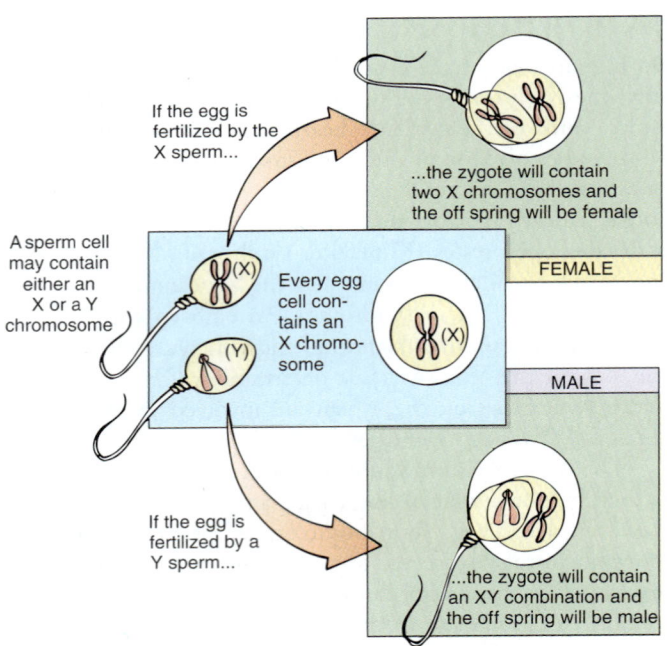

Figure 10–14 **Sex is determined by the sperm.** An X-bearing sperm produces a female; a Y-bearing sperm produces a male.

Figure 10–15 **X-linked recessive colorblindness.** Note that the Y chromosome does not carry a gene for color vision. *(a)* To be colorblind, a female must inherit alleles for colorblindness from both parents. *(b)* If a normal male mates with a carrier female, one-half of their sons would be expected to be colorblind and one-half of their daughters would be expected to be carriers.

example, must have a colorblind father and a mother who is heterozygous or homozygous for colorblindness (Fig. 10–15). Such a combination is unusual. In contrast, a colorblind male need only have a mother who is heterozygous for colorblindness; his father can be normal. Hence, X-linked recessive traits are generally much more common in males than in females, a fact that may partially explain why human male embryos are more likely to die.

Dosage compensation equalizes the expression of X-linked genes in males and females

The X chromosome contains numerous genes required by both sexes, yet a normal female has two copies ("doses") for each locus, whereas a normal male has only one. **Dosage compensation** is a mechanism that makes the two doses in the female and the single dose in the male equivalent. Male fruit flies accomplish this by making their single X chromosome more active. In most tissues the metabolic activity of a single male X chromosome is equal to the combined metabolic activity of the two X chromosomes present in the female.

Dosage compensation in mammals generally involves inactivation of one of the two X chromosomes in the female. During interphase a dark spot of chromatin, called a **Barr**

body, is visible at the edge of the nucleus of each female mammalian cell (Fig. 10–16). The Barr body has been found to represent a dense, dark-staining, and metabolically inactive X chromosome. The other X chromosome resembles the metabolically active autosomes; during interphase it is a greatly extended thread that is not evident by light microscopy. From this and other evidence, the British geneticist Mary Lyon has suggested that in any one cell of a female mammal, only one of the two X chromosomes is active; the other is inactive and

10 μm

Figure 10–16 **A Barr body.** The darkly-stained Barr body *(arrow)* at the edge of the nucleus in this LM is an inactivated X chromosome. The entire cell is not shown. *(Omikron/Photo Researchers, Inc.)*

Figure 10–17 A calico cat. This cat has X-linked genes for both black and yellow (or orange) pigmentation of the fur, but because of random X chromosome inactivation, black is expressed in some clones of cells and yellow (or orange) is expressed in others. Because other genes affecting fur color are also present, white patches are usually evident as well. *(Larime Photographic/Dembinsky Photo Associates)*

Sex-influenced genes are autosomal, but their expression is affected by the individual's sex

Not all characteristics that differ in the two sexes are X-linked. Certain **sex-influenced** traits are inherited through autosomal genes, but the *expression* of alleles at these loci can be altered or influenced by the sex of the animal. Therefore, males and females with the same genotype with respect to these loci may have different phenotypes.

Pattern baldness in humans, characterized by premature loss of hair on the front and top of the head, but not on the sides, is far more common among males than among females. It has been proposed that a single pair of alleles is involved, with the allele responsible for pattern baldness being dominant in males and recessive in females. Because of this unusual situation, we modify our notation, designating the pattern baldness allele as B_1 and the allele for normal hair growth as B_2. Individuals with the genotype B_1B_1 show pattern baldness, regardless of sex. Persons with a B_1B_2 genotype are bald if they are male but not bald if they are female. Individuals with the genotype B_2B_2 are not bald, regardless of sex.

Evidence suggests that the expression of most sex-influenced traits is strongly modified by sex hormones. For example, male hormones (see Chapter 48) are strongly implicated in the expression of pattern baldness.

is visible as a Barr body. (Actually, X chromosome inactivation is never complete; a small fraction of the genes are expressed.)

Because only one X chromosome is active in any one cell and because X chromosome inactivation is a random event, a female mammal that is heterozygous at an X-linked locus expresses one of the alleles in about half her cells and the other allele in the other half. This is sometimes (but not always) evident in the phenotype. Mice and cats have several X-linked genes for certain coat colors. Females that are heterozygous for such genes may show patches of one coat color in the midst of areas of the other coat color. This phenomenon, termed **variegation,** is evident in calico (Fig. 10–17) and tortoiseshell cats. Early in development, when relatively few cells are present, X chromosome inactivation occurs randomly in each cell. When any one of these cells divides by mitosis, the cells of the resulting clone (group of genetically identical cells) all have the same active X chromosome, and, therefore, a patch of cells that all express the same color develops.

Why, you might ask, is variegation not always apparent in females heterozygous at X-linked loci? The answer is that, although variegation usually occurs, we may need to use special techniques to observe it. For example, colorblindness is due to abnormal pigments in the cone cells in the retina of the eye (Chapter 41). In at least one type of red-green colorblindness, the retina of a heterozygous female actually contains patches of abnormal cones, but the patches of normal cones are sufficient to provide normal color vision.

THE RELATIONSHIP BETWEEN GENOTYPE AND PHENOTYPE IS OFTEN COMPLEX

The relationship between a given locus and the characteristic it controls may be simple: a single pair of alleles of a locus may regulate the appearance of a single characteristic of the organism (e.g., tall versus short in garden peas). Alternatively, the relationship may be more complex: a pair of alleles may participate in the control of several characteristics, or alleles of many loci may cooperate to regulate the appearance of a single characteristic. Not surprisingly, these more complex relationships are quite common.

As you will learn in Chapters 11 and 12, each locus is a segment of DNA in which genetic information is stored. Most loci code for specific types of proteins, and it is the presence of a protein, such as an enzyme, that usually provides the chemical basis for the genetic trait. Because most biologically important molecules are synthesized by complex metabolic pathways requiring a number of enzymes, it is not difficult to appreciate why relationships between genes and the characteristics of the organism are complex.

We may assess the phenotype on one or many levels. It may be a morphological characteristic such as shape, size, or color. It may be a physiological characteristic or even a biochemical trait, such as the presence or absence of a specific enzyme required for the metabolism of some specific molecule.

Figure 10–18 **Incomplete dominance in Japanese four o'clocks.** If a pair of alleles is incompletely dominant to each other, a heterozygote has a phenotype intermediate between its parents. Two incompletely dominant alleles, R^1 and R^2, are responsible for red, white, and pink flower colors. Red-flowered plants are R^1R^1; white-flowered plants are R^2R^2, and heterozygotes (R^1R^2) are pink. Note that uppercase notation is used for both alleles, because neither is recessive to the other.

In addition, the phenotypic expression of genes may be altered by changes in the environmental conditions under which the organism develops.

Dominance is not always complete

Studies of the inheritance of many traits in a wide variety of organisms have clearly shown that one member of a pair of alleles may not be completely dominant to the other. In such instances it is improper to use the terms *dominant* and *recessive*.

For example, the plants commonly known as Japanese four o'clocks may have red or white flowers. Each color breeds true when these plants are self-pollinated. What flower color might we expect in the offspring of a cross between a red-flowering plant and a white-flowering one? Without knowing which is dominant, we might predict that all would have red flowers or all would have white flowers. This cross was first made by the German botanist Karl Correns (one of the rediscoverers of Mendel's work), who found that all F_1 offspring have pink flowers! Does this result in any way prove that Mendel's assumptions about inheritance are wrong? Did the parental characteristics blend inseparably in the offspring? Quite the contrary, for when two of these pink-flowered plants are crossed, red-flowered, pink-flowered, and white-flowered offspring appear in a ratio of 1:2:1 (Fig. 10–18).

In this instance, as in all other aspects of the scientific process, results that differ from those predicted prompt scientists to reexamine and modify their assumptions to account for the exceptional results. The pink-flowered plants are clearly the heterozygous individuals, and neither the red allele nor the white allele is completely dominant. When the heterozygote has a phenotype that is intermediate between those of its two parents, the genes are said to show **incomplete dominance.** In these crosses the genotypic and phenotypic ratios are identical.

Incomplete dominance is not unique to Japanese four o'clocks. Red- and white-flowered sweet pea plants also produce pink-flowered plants when crossed, and numerous additional examples of incomplete dominance are known in both plants and animals.

In both cattle and horses, reddish coat color is not completely dominant to white coat color. Heterozygous individuals have a mixture of reddish colored hairs and white hairs, which is called roan. If you saw a white mare nursing a roan foal, what would you guess was the coat color of the foal's father? Because the reddish and white colors are expressed independently (hair by hair) in the roan heterozygote, we sometimes refer to this as a case of **codominance.** Strictly speaking, incomplete dominance refers to instances in which the heterozygote is intermediate in phenotype, and codominance refers to instances in which the heterozygote simultaneously expresses the phenotypes of both types of homozygotes. The human ABO blood group (see Chapter 15) provides a classic example of codominant alleles, as well as an example of a locus with multiple alleles.

Multiple alleles for a locus may exist in a population

The examples given so far have dealt with situations in which each locus was represented by a maximum of two allelic variants, and in most of these examples one of the alleles has been dominant and one recessive. It is true that a single diploid individual has a maximum of two different alleles for a particular locus and that a haploid gamete has only one allele for each locus. However, if we survey a population, we may find more than two alleles for a particular locus. If three or more alleles for a given locus exist within the population, we say that locus has **multiple alleles.** A great many loci can be shown to have multiple alleles if the population is surveyed carefully. Some alleles can be identified by the activity of a certain enzyme or by some other biochemical feature but do not produce an obvious phenotype. Others produce a readily recognizable phenotype, and certain patterns of dominance can be discerned when the alleles are combined in various ways.

In rabbits, for example, a C allele causes a fully colored coat. The homozygous recessive genotype, cc, causes albino coat color. There are two additional allelic variants of the same locus, c^h and c^{ch}. The genotype $c^h c^h$ causes the "Himalayan" pattern, in which the body is white but the tips of the ears, nose, tail, and legs are colored (similar to the color pattern of a Siamese cat). An individual with the genotype $c^{ch} c^{ch}$ has the "chinchilla" pattern, in which the entire body has a light gray color. On the basis of the results of genetic crosses, these alleles can be arranged in the following series: $C > c^h > c^{ch} > c$. Each allele is dominant to those following it and recessive to those preceding it. For example, a $c^h c^{ch}$ rabbit has the "Himalayan" pattern, whereas a $c^{ch} c$ rabbit has the "chinchilla" pattern. In some other series of multiple alleles, certain alleles may be codominant and others incompletely dominant; hence the heterozygotes commonly have phenotypes intermediate between those of their parents.

A single gene may affect multiple aspects of the phenotype

In the examples presented so far, the relationship between a gene and its phenotype has been direct, precise, and exact, and the loci considered have controlled the appearance of single traits. However, the relationship of gene to characteristic may be quite complex. Most genes probably have many different effects, a quality referred to as **pleiotropy.** This is dramatically evident in many genetic diseases, such as cystic fibrosis and sickle cell anemia (see Chapter 15), in which multiple symptoms are caused by a single pair of alleles. Most cases of pleiotropy can be traced to a single fundamental cause. For example, a defective enzyme may affect the functioning of many types of cells.

Alleles of different loci may interact to produce a phenotype

Several pairs of alleles may interact to affect a single phenotype, or one pair may inhibit or reverse the effect of another pair. More than 12 pairs of alleles interact in various ways to produce coat color in rabbits, and more than 100 pairs are concerned with eye color and shape in fruit flies.

One type of gene interaction is illustrated by the inheritance of combs in poultry, where two genes may interact to produce a novel phenotype (Fig. 10–19). The allele for a rose comb, R, is dominant to that for a single comb, r. An unlinked gene pair governs the inheritance of a pea comb, P, versus a single comb, p. A single-combed fowl must therefore be homozygous for the recessive allele at both loci *(pprr)*; a pea-combed fowl is either *PPrr* or *Pprr*; and a rose-combed fowl is either *ppRR* or *ppRr*. When an R and P occur together, the phenotype is neither a pea nor a rose comb but a completely different type, called a *walnut comb*. The walnut comb phenotype is produced whenever a fowl has one or two R alleles, plus one or two P alleles (that is, *PPRR, PpRR, PPRr,* or *PpRr*). What would you predict about the types of combs among the offspring of two heterozygous walnut-combed fowl, *PpRr*? How does this form of gene interaction affect the ratio of phenotypes in the F_2 generation? Is it the typical Mendelian $9:3:3:1$ ratio?

| Single comb
pprr | Pea comb
PPrr or Pprr | Walnut comb
P_R_ | Rose comb
ppRR or ppRr |

Figure 10–19 Gene interaction.
Two gene pairs govern these types of genetically determined combs in roosters. Note that four genotypes—*PPRR, PpRR, PPRr,* and *PpRr*—all have walnut combs.

Epistasis is a common type of gene interaction in which the presence of certain alleles of one locus can prevent the expression of alleles of a different locus. (Literally, epistasis means "standing upon.") We have already seen that coat color in guinea pigs can be determined by the *B* and *b* allelic pair, with the *B* allele for black coat dominant to the *b* allele for brown coat. The expression of either phenotype, however, depends on the presence of a dominant allele at yet another locus. This allele, *C*, codes for the enzyme tyrosinase, which converts a colorless precursor to the pigment melanin and hence is required for the production of any kind of pigment. The recessive allele (*c*) codes for an inactive form of the enzyme. Thus, an animal that is homozygous recessive for this allele lacks the enzyme and produces no melanin. It is therefore a white-coated, pink-eyed albino, regardless of the combination of *B* and *b* alleles. Albinism, or lack of melanin pigment, is not restricted to guinea pigs, but is found in humans (see Chapter 15 introduction) and a variety of other animals (Fig. 10–20).

When an albino guinea pig with the genotype *ccBB* is mated to a brown guinea pig with the genotype *CCbb*, the F_1 generation is black coated, *CcBb*. When two such animals are mated, their offspring appear black-coated, brown-coated, and albino in a ratio of 9:3:4. (Make a Punnett square to verify this.)

You might wonder why heterozygous *Cc* individuals do not show at least some lightening of the coat color, given the fact that they produce only about half the normal amount of tyrosinase enzyme. It turns out that half the normal amount of enzyme is usually adequate to produce normal amounts of pigment. This type of situation applies to many enzymes, and explains many (although certainly not all) cases of dominance.

Polygenes act additively to produce a phenotype

The inherited components of many human characteristics, such as height, body form, and skin color, are not inherited through alleles at a single locus. The same holds true for many commercially important characteristics in domestic plants and animals, such as milk and egg production. Alleles at several, perhaps many, different loci affect each characteristic. The term **polygenic inheritance** is applied when multiple independent pairs of genes have similar and additive effects on the same characteristic.

Polygenes are responsible for the inheritance of skin color in humans. It is now thought that alleles representing four or more different loci are involved in determining skin color, but for simplicity the principle of polygenic inheritance can be illustrated with pairs of alleles at only three unlinked loci (Fig 10–21). These can be designated *A* and *a*, *B* and *b*, and *C* and *c*. The capital letters represent incompletely dominant alleles producing dark skin. The more capital letters, the darker the skin, because the alleles affect skin color in an additive fashion. A person with the darkest skin would have the genotype *AABBCC,* and a person with the lightest skin would have the genotype *aabbcc.* The F_1 offspring of an *aabbcc* person and an

Figure 10–20 **Epistasis.** Homozygous alleles for albinism exhibit epistasis, masking the expression of alleles of other loci that govern production of melanin pigment. Albino individuals, such as this albino koala, occur occasionally in nature. *(Tom McHugh/Photo Researchers, Inc.)*

AABBCC person are all *AaBbCc* and have an intermediate skin color. The F_2 offspring of two such triple heterozygotes would have skin colors ranging from very dark to very light.

Polygenic inheritance is therefore characterized by an F_1 generation that is intermediate between the two completely homozygous parents and by an F_2 generation that shows wide variation between the two parental types. Most of the F_2-generation individuals have one of the intermediate phenotypes; only a few show the extreme phenotypes of the grandparents (P generation). On average, only 1 of 64 is as dark as the very dark grandparent, and only 1 of 64 is as light as the very light grandparent (Fig. 10–22). The alleles *A, B,* and *C* each produce about the same amount of darkening of the skin; hence, the genotypes *AaBbCc, AABbcc, AAbbCc, AaBBcc, aaBBCc, AabbCC,* and *aaBbCC* all produce similar intermediate phenotypes.

The model used here for the inheritance of skin color in humans is a rather simple example of polygenic inheritance because only three major allelic pairs are used. The inheritance of height in humans involves alleles representing ten or more

Figure 10-21 **Polygenic inheritance.** This simplified model assumes that skin color in humans is governed by alleles of three unlinked loci. The alleles producing dark skin (*A*, *B*, and *C*) are represented by capital letters, but they are not dominant. Instead they have additive effects. If one parent is very dark and the other very light, their children (F₁) are intermediate in skin color. A wide range of skin colors are expected in the F₂. The number of dark dots (each signifying an allele producing dark skin) is counted to determine the phenotype. The results are summarized in Figure 10-22.

loci. Because many allelic pairs are involved and because height is modified by a variety of environmental conditions, the height of adults ranges from perhaps 125 to 215 cm. If we were to measure the heights of 1000 adult American men selected at random, we would find that only a few are as tall as 215 cm or as short as 125 cm. The heights of most would cluster around the mean, about 170 cm. When the number of men at each height is plotted against height (in centimeters) and the points are connected, the result is a bell-shaped curve, called a **normal distribution curve** (Fig. 10-23).

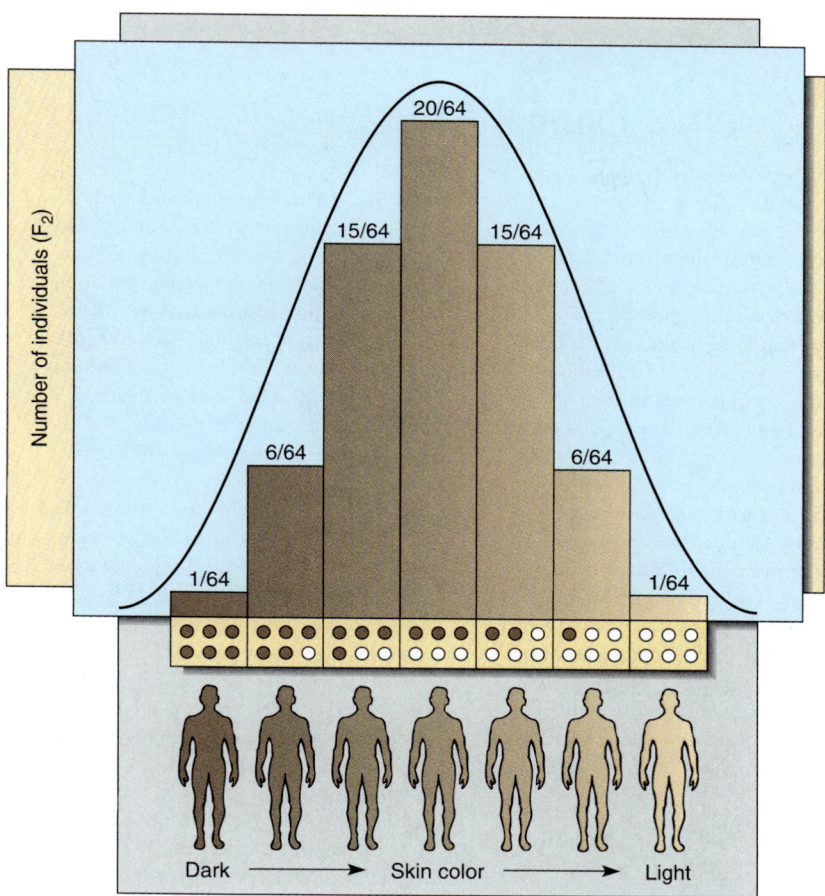

Figure 10-22 **Distribution of phenotypes in polygenic inheritance.** The bars indicate the expected phenotypic ratios in the F$_2$ generation shown in Figure 10-21. This expected distribution of phenotypes is consistent with the superimposed normal curve.

Figure 10-23 **Polygenic inheritance in a population.** The distribution of heights of 1083 adult males approximates a normal curve, which is consistent with continuous phenotypic variation in the population. The bars indicate the actual number of men whose heights were within the unit range. For example, there were 163 men whose heights were between 170 and 173 cm.

SELECTION, INBREEDING, AND OUTBREEDING ARE USED TO DEVELOP IMPROVED STRAINS

How do geneticists go about establishing a breed of cow that will give more milk, a strain of hens that will lay bigger eggs, or a variety of corn with more kernels per ear? By selecting organisms that manifest the desired phenotype and using these organisms in further matings, a true-breeding strain with the commercially advantageous phenotype is gradually developed. Such a strain is expected to be homozygous for all of the genes involved, whether they be dominant, recessive, or additive in their effects.

There is a limit to the effectiveness of breeding by selection. When a strain becomes homozygous for all of the genes involved, further selective breeding cannot increase the desired quality. Moreover, because of **inbreeding**—the mating of two closely related individuals—the strain may become homozygous for multiple undesirable traits as well. Evidence suggests that human inbreeding increases the frequency of genetic dis-

ease in the population, although the individual risk is relatively small (Chapter 15).

The mating of individuals of totally unrelated strains, termed **outbreeding**, frequently leads to offspring much better adapted for survival than either parent. Such improvement reflects a phenomenon called **hybrid vigor.** A large proportion of the corn, wheat, and other crops grown in the United States consists of hybrid strains. Each year the seed to grow these crops must be obtained by mating the original strains. The hybrids are heterozygous at a great many loci and give rise, even when self-fertilized, to a wide variety of forms, few of which are as good as the original hybrid. (The seeds produced by F$_1$ hybrid corn plants are not normally planted, but eaten instead!)

The reason for hybrid vigor has long been a matter of debate, and, in fact, there may be multiple causes. One explanation may be that each of the parental strains is homozygous for certain undesirable recessive genes, but any two strains are homozygous for different undesirable genes. Each strain contains dominant genes to make up for the recessive undesirable genes of the other strain. The hybrid offspring would express

SOLVING GENETICS PROBLEMS

Simple Mendelian genetics problems are like puzzles. They can be fun and easy to work if you follow certain conventions and are methodical in your approach.

1. Always use standard designations for the generations. The generation with which a particular genetic experiment is begun is called the P, or *parental, generation*. Offspring of this generation (the "children") are called the F_1, or *first filial, generation*. The offspring resulting when two F_1 individuals are bred constitute the F_2, or second filial, generation (the "grandchildren").

2. Write down a key for the symbols you are using for the allelic variants of each locus. Use uppercase to designate a dominant allele and lowercase to designate a recessive allele. Use the same letter of the alphabet to designate both alleles of a particular locus. If you are not told which is dominant and which is recessive, the phenotype of the F_1 generation is a good clue.

3. Determine the genotypes of the parents of each cross by making use of the following types of evidence:
   a. Are they from true-breeding lines? If so, they should be homozygous.
   b. Can their genotypes be reliably deduced from their phenotypes? This is usually true if they express the recessive phenotype.
   c. Do the phenotypes of their offspring provide any information? See *Focus On: Deducing Genotypes* for an exam-

ple of how these determinations can be made.

4. Indicate the possible kinds of gametes formed by each of the parents. It is helpful to draw a circle around the symbols for each kind of gamete.
   a. If it is a monohybrid cross, we must apply the principle of segregation; i.e., a heterozygote *Aa* forms two kinds of gametes: *A* and *a*. Of course a homozygote, such as *aa*, forms only one kind of gamete: *a*.
   b. If it is a dihybrid cross, we must apply both the principle of segregation *and* the principle of independent assortment. For example, an individual heterozygous for two loci would have the genotype *AaBb*. *A* segregates from *a*, and *B* segregates from *b*. The assortment of *A* and *a* into gametes is independent of the assortment of *B* and *b*. Therefore *A* is equally likely to end up in a gamete with *B* or *b*. The same is true for *a*.

	Segregation of the alleles of each locus

	Independent assortment of alleles of different loci

5. Set up a Punnett square, placing the possible types of gametes from one parent down the left side and the possible types from the other parent across the

top.

6. Fill in the Punnett square and read off (and sum up) the genotypic and phenotypic ratios of the offspring. Avoid confusion by consistently placing the dominant allele first and the recessive allele second in heterozygotes (*Aa*, never *aA*). If it is a dihybrid cross, it is very important to always write the two alleles of one locus first and the two alleles of the other locus second. It does not matter which locus you choose to write first, but once you have decided on the order, it is critical that you maintain it consistently. This means that if the individual is heterozygous for both loci you will always use the form *AaBb*. Writing this particular genotype as *aBbA*, for example, would cause confusion.

7. If you do not need to know the frequencies of all of the expected genotypes and phenotypes, you may use the rules of probability as a shortcut. For example, if both parents are *AaBb*, what is the probability of an *AABB* offspring? To be *AA*, the offspring must receive an *A* gamete from each parent. The probability that a given gamete is *A* is 1/2 and each gamete represents an independent event, so we combine their probabilities by multiplying ($1/2 \times 1/2 = 1/4$). The probability of *BB* is calculated similarly and is also 1/4. The probability of *AA* is independent of the probability of *BB*, so again we use the product rule to obtain their combined probabilities ($1/4 \times 1/4 = 1/16$).

---

all the desirable (dominant) traits and none of the undesirable (recessive) traits of the two parental strains.

Alternatively, hybrid vigor may be due to **heterozygote advantage,** i.e., the superiority of the heterozygous genotype to either homozygous genotype. The key may be that a particular allele may have advantages under one set of conditions, but that a different allele may be favored when conditions change. Therefore an individual with two different alleles for a locus may be able to function over a wider range of conditions. This view is supported by experimental evidence that the metabolism of an individual heterozygous at many loci tends to be more stable and less affected by environmental changes.

A case of heterozygote advantage with a rather specific explanation is seen in humans who are heterozygous for the recessive sickle cell anemia allele *(s)* and the normal dominant allele *(S)*. These *Ss* individuals appear to have increased resistance to the parasite that lives inside red blood cells and causes malaria. Such resistance is a significant advantage in areas of the world where malaria is still uncontrolled. Homozygous normal individuals *(SS)* appear to be less resistant to malaria; homozygous sickle cell individuals *(ss)* are at a distinct disadvantage due to severe anemia and other serious effects of the sickle cell allele (see Chapters 15 and 18).

DEDUCING GENOTYPES

The science of genetics resembles mathematics in that it consists of a few basic principles, which, once grasped, enable the student to solve a wide variety of problems. Very often the genotypes of the parents can be deduced from the phenotypes of their offspring. In chickens, for example, the allele for rose comb (*R*) is dominant to the allele for single comb (*r*). Suppose that a cock is mated to three different hens, as shown in the figure. The cock and hens A and C have rose combs; hen B has a single comb. Breeding the cock with hen A produces a rose-combed chick, with hen B a single-combed chick, and with hen C a single-combed chick. What types of offspring can be expected from further matings of the cock with these hens?

Because the allele for single comb, *r*, is recessive, all of the hens and chicks that are phenotypically single-combed must be *rr*. We can deduce that hen B and the offspring of hens B and C are genotypically *rr*.

All individuals that are phenotypically rose-combed must have at least one *R* allele. The fact that the offspring of the cock and hen B was single-combed proves that the cock is heterozygous *Rr*, because, although the single-combed chick received one *r* allele from its mother, it must have received the second one from its father.

The fact that the offspring of the cock and hen C had a single comb proves that hen C is heterozygous, *Rr*. It is impossible

to decide from the data given whether hen A is homozygous *RR* or heterozygous *Rr*; further breeding would be necessary to determine this. (Can you suggest an appropriate mating?)

Additional matings of the cock with hen B should result in half rose-combed and half single-combed individuals; additional matings of the cock with hen C should produce three-fourths rose-combed and one-fourth single-combed chicks.

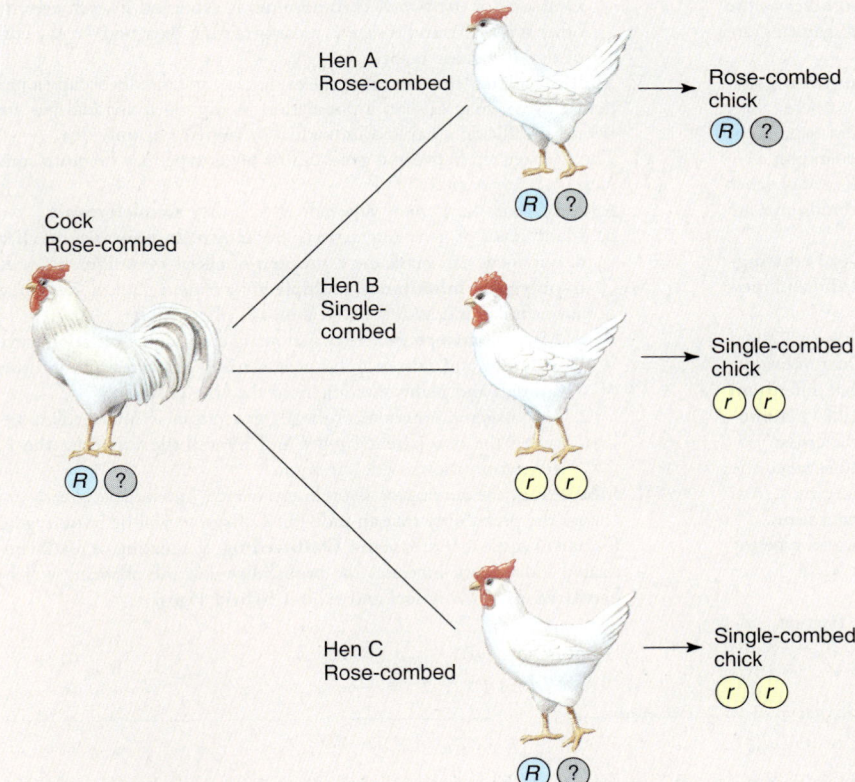

Using phenotypes to deduce genotypes. In many cases the phenotypes of the offspring can provide information on the genotypes of the parents.

I. **Genetics** is the study of the structure, expression, and transmission of genetic information. The transfer of genetic information from parent to offspring is called **heredity.**

II. Mendel's inferences about inheritance obtained from his garden pea-breeding experiments have been tested repeatedly in all kinds of diploid organisms and found to be generally true. These principles have been extended and now can be stated in a more modern form.

A. Today we know that the genes are in chromosomes; the site a gene occupies in the chromosome is its **locus.**

B. Different forms of a particular gene are **alleles;** they occupy corresponding loci on homologous chromosomes. Genes therefore exist as pairs of alleles in diploid individuals.

C. An individual that carries two identical alleles is said to be **homozygous** for that locus. If the two alleles are different, that individual is said to be **heterozygous** for that locus.

D. One allele (the **dominant allele**) may mask the expression of the other allele (the **recessive allele**) in a heterozygous individual. For this reason two individuals with the same appearance (**phenotype**) may differ from each other in genetic constitution (**genotype**).

E. Dominance does not always apply, and alleles can be **incompletely dominant,** in which the heterozygote is intermediate in phenotype, or **codominant,** in which the heterozygote simultaneously expresses the phenotypes of both homozygotes.

F. According to Mendel's **principle of segregation,** during meiosis the alleles for each locus separate, or segregate, from each other as the homologous chromosomes separate. When haploid gametes are formed, each contains only one allele for each locus.

G. According to Mendel's **principle of independent assortment,** alleles of different loci are distributed randomly into the gametes. This can result in recombination, that is, production of new gene combinations that were not present in the parental (P) generation.

H. Each chromosome behaves genetically as if it were composed of genes arranged in a linear order. Genes in the same chromosome are **linked.**

   1. Recombination of linked genes can occur as a result of **crossing-over** (breaking and rejoining of homologous chromatids) in meiotic prophase I.

   2. By measuring the frequency of recombination between linked genes, it is possible to construct a genetic map of a chromosome.

I. A cross between homozygous parents (**P generation**) that differ from each other with respect to their alleles at one locus is called a **monohybrid cross;** if they differ at two loci, it is a **dihybrid cross.**

   1. The first generation of offspring is heterozygous and is called the first filial, or **F$_1$, generation;** the generation produced by a cross of two F$_1$ individuals is the second filial, or **F$_2$, generation.**

   2. A **test cross** is between an individual of unknown genotype (or an F$_1$ individual) and a homozygous recessive individual.

III. Genetic ratios can be expressed in terms of probabilities.

A. Any probability is expressed as a fraction or decimal fraction, calculated as the number of favorable events divided by the total number of events. This can range from 0 (an impossible event) to 1 (a certain event).

B. According to the **product rule,** the probability of two independent events occurring together can be obtained by multiplying the probabilities of each occurring separately.

C. According to the **sum rule,** the probability of an outcome that can be obtained in more than one way can be calculated by adding the separate probabilities.

IV. The sex of humans and many other animals is determined by the **X** and **Y sex chromosomes** or their equivalents. **Autosomes** are chromosomes other than sex chromosomes.

A. Normal female mammals have two X chromosomes; normal males have one X and one Y.

B. The fertilization of an X-bearing egg by an X-bearing sperm results in a female (XX) zygote. The fertilization of an X-bearing egg by a Y-bearing sperm results in a male (XY) zygote.

C. The Y chromosome is responsible for determining male sex in mammals.

D. The X chromosome contains many important genes unrelated to sex determination that are required by both males and females. A male receives all his **X-linked** genes from his mother. A female receives X-linked genes from both parents.

E. A female mammal shows **dosage compensation** of X-linked genes. Only one of the two X chromosomes is expressed in each cell; the other is inactive and is seen as a dark-staining **Barr body** at the edge of the interphase nucleus.

V. **Multiple alleles** (three or more alleles that can potentially occupy a particular locus) may exist in a population. A diploid individual has any two of the alleles; a haploid individual or gamete has only one.

VI. The relationship between a gene and its phenotype may be quite complex.

A. Most genes have many different effects; they are **pleiotropic.**

B. Many types of gene interactions are known. In **epistasis,** an allele of one locus can mask the expression of alleles of a different locus.

C. In **polygenic inheritance,** multiple independent pairs of genes may have similar and additive effects on the phenotype.

   1. Many human characteristics showing continuous variation, such as height and skin color, as well as many characteristics in other animals and plants, are inherited through polygenes.

   2. In polygenic inheritance, the F$_1$ generation is intermediate between the two parental types and shows little variation; the F$_2$ generation shows wide variation.

VII. **Inbreeding,** the mating of two closely related individuals, greatly increases the probability that an individual offspring will be homozygous for one or more recessive genes. **Outbreeding,** the mating of totally unrelated individuals, increases the probability that the offspring will be heterozygous at many loci and exhibit **hybrid vigor.**

P O S T - T E S T

1. One reason why Mendel was able to discover the basic principles of inheritance is that he (a) understood the behavior of chromosomes in mitosis and meiosis (b) studied a wide variety of experimental organisms (c) began by establishing true-breeding lines (d) studied various types of linkage (e) all of the above

2.–5. Use the following information to answer questions 2 through 5:
In peas, the allele for round seeds (*R*) is dominant to that for wrinkled seeds (*r*); the allele for yellow seeds (*Y*) is dominant to that for green seeds (*y*). These loci are unlinked. Plants from a true-breeding line with round, green seeds are crossed with plants from a true-breeding line with wrinkled, yellow seeds. These parents constitute the P generation.

2. The genotypes of the P generation are: (a) *RRrr* and *YYyy* (b) *RrYy* (c) *RRYY* and *rryy* (d) *RRyy* and *rrYY* (e) *RR* and *YY*

3. What are the expected genotypes of the F$_1$ hybrids produced by the described cross? (a) *RRrr* and *YYyy* (b) *RrYy* (c) *RRYY* and *rryy* (d) *RRyy* and *rrYY* (e) *RR* and *YY*

4. What kinds of gametes can the F$_1$ individuals produce? (a) *RR* and *YY* (b) *Rr* and *Yy* (c) *RR*, *rr*, *YY*, and *yy* (d) *R*, *r*, *Y*, and *y* (e) *RY*, *Ry*, *rY*, and *ry*

5. What is the expected proportion of F$_2$ wrinkled, yellow seeds? (a) 9/16 (b) 1/16 (c) 3/16 (d) 1/4 (e) zero

6. One of the autosomal loci controlling eye color in fruit flies has two alleles, one for brown eyes and the other for red eyes. Fruit flies from a true-breeding line with brown eyes were crossed with flies from a true-breeding line with red eyes. The F$_1$ flies had red eyes. What conclusion can be drawn from this experiment? (a) these alleles underwent independent assortment (b) these alleles underwent segregation (c) these genes are X-linked (d) the allele for red eyes is dominant to the allele for brown eyes (e) all of the above are true

7. The F$_1$ flies described in question 6 were mated with red-eyed flies from a true-breeding line. What phenotypes would you expect the offspring to have? (a) all red eyes (b) all brown eyes (c) half red eyes and half brown eyes (d) red-eyed females and brown-eyed males (e) brown-eyed females and red-eyed males

8. The type of cross described in question 7 is a(n) (a) F$_2$ cross (b) dihybrid cross (c) test cross (d) two-point test cross (e) none of the above

9. Individuals of genotype *AaBb* were crossed with *aabb* individuals. Approximately equal numbers of the following classes of offspring were produced: *AaBb*, *Aabb*; *aaBb*, and *aabb*. These results illustrate Mendel's principle(s) of (a) linkage (b) independent assortment (c) segregation (d) a and c (e) b and c

10. Assume that the ratio of females to males is 1:1. A couple already has two daughters and no sons. If they plan to have a total of six children, what is the probability that they will have a family of all girls? (a) 1/4 (b) 1/8 (c) 1/16 (d) 1/32 (e) 1/64

11. Red-green colorblindness is an X-linked recessive disorder in humans. Your friend is the daughter of a colorblind father. Her mother had normal color vision, but her maternal grandfather was colorblind. What is the probability that your friend is colorblind? (a) 1 (b) 1/2 (c) 1/4 (d) 3/4 (e) zero

12. When homozygous, a particular allele of a locus in rats causes abnormalities of the cartilage throughout the body, an enlarged heart, slow development, and death. This is an example of (a) pleiotropy (b) polygenic inheritance (c) epistasis (d) codominance (e) dosage compensation

REVIEW QUESTIONS

1. In peas, yellow seed color is dominant to green. Predict the phenotypes (and their proportions) of the offspring of the following crosses: (a) homozygous yellow × green; (b) heterozygous yellow × green; (c) heterozygous yellow × homozygous yellow; (d) heterozygous yellow × heterozygous yellow.

2. If two animals heterozygous for a single pair of alleles are mated and have 200 offspring, about how many would be expected to have the phenotype of the dominant allele (i.e., to look like the parents)?

3. When two long-winged flies were mated, the offspring included 77 with long wings and 24 with short wings. Is the short-winged condition dominant or recessive? What are the genotypes of the parents?

4. A blue-eyed man, both of whose parents were brown-eyed, married a brown-eyed woman whose father was blue-eyed and whose mother was brown-eyed. If brown is dominant to blue, what are the genotypes of the individuals involved?

5. Outline a breeding procedure whereby a true-breeding strain of red cattle could be established from a roan bull and a white cow.

6. What is the probability of rolling a seven with a pair of dice? Which is a more likely outcome, rolling a six with a pair of dice or rolling an eight?

7. In rabbits, spotted coat (*S*) is dominant to solid color (*s*), and black (*B*) is dominant to brown (*b*). These loci are unlinked. A brown, spotted rabbit from a pure line is mated to a solid black one, also from a pure line. What are the genotypes of the parents? What would be the genotype and phenotype of an F$_1$ rabbit? What would be the expected genotypes and phenotypes of the F$_2$ generation?

8. The long hair of Persian cats is recessive to the short hair of Siamese cats, but the black coat color of Persians is dominant to the brown-and-tan coat color of Siamese. Make up appropriate symbols for the alleles of these two unlinked loci. If a pure black, long-haired Persian is mated to a pure brown-and-tan, short-haired Siamese, what will be the appearance of the F$_1$ offspring? If two of these F$_1$ cats are mated, what is the chance that a long-haired, brown-and-tan cat will be produced in the F$_2$ generation? (Use the shortcut probability method to obtain your answer; then check it with a Punnett square.)

9. The expression of an allele called *frizzle* in fowl causes abnormalities of the feathers. As a consequence, the animal's body temperature is lowered, adversely affecting the functions of many internal organs. When one gene affects many characteristics of the organism in this way, we say that gene is _____.

10. A walnut-combed rooster is mated to three hens. Hen A, which is walnut-combed, has offspring in the ratio of 3 walnut:1 rose. Hen B, which is pea-combed, has offspring in the ratio of 3 walnut:3 pea:1 rose:1 single. Hen C, which is walnut-combed, has only walnut-combed offspring. What are the genotypes of the rooster and the three hens?

11. What kinds of matings result in the following phenotypic ratios? (a) 3:1 (b) 1:1 (c) 9:3:3:1 (d) 1:1:1:1.

12. The weight of the fruit in a certain variety of squash is determined by two pairs of genes: *AABB* produces fruits weighing 4 pounds each, and *aabb* produces fruits weighing 2 pounds each. Each allele represented by a capital letter adds 0.5 pound to the weight. When a plant that produces 4-pound fruits is crossed with a plant that produces 2-pound fruits, all of the offspring produce fruits that weigh 3 pounds each. What would be the weights of the fruits produced by the F$_2$ plants if two of these F$_1$ plants were crossed?

13. The X-linked *barred* locus in chickens controls the pattern of the feathers, with the alleles *B* for barred pattern and *b* for no bars. If a barred female (X^{B}Y) is mated to a nonbarred male (X^{b}X^b), what will be the appearance of the male and female progeny? (Recall that in birds males are homogametic and females are heterogametic.) Do you see any commercial usefulness for this result? (*Hint:* It is notoriously difficult to determine the sex of newly hatched chicks.)

14. Individuals of genotype *AaBb* were mated to individuals of genotype *aabb*. One thousand offspring were counted, with the following results: 474 *Aabb*; 480 *aaBb*; 20 *AaBb*; 26 *aabb*. What is this type of cross known as? Are these loci linked? What are the two parental classes and the two recombinant classes of offspring? What is the percentage of recombination between these two loci? How many map units apart are they?

15. Genes *A* and *B* are 6 map units apart, and *A* and *C* are 4 map units apart. Which gene is in the middle if *B* and *C* are 10 map units apart? Which is in the middle if *B* and *C* are 2 map units apart?

Answers to these Review Questions are included in Appendix A with the Post-Test answers.

1. Would the development of the science of genetics in the 20th century have been any different if Gregor Mendel had never lived?
2. Sketch a series of diagrams showing each of the following, making sure to end each series with haploid gametes: (a) how a pair of alleles for a single locus segregates in meiosis, (b) how the alleles of two unlinked loci assort independently in meiosis, (c) how the alleles of two linked loci undergo genetic recombination.
3. Can you always ascertain an organism's genotype for a particular locus if you know its phenotype? Conversely, if you are given an organism's genotype for a locus, can you always reliably predict its phenotype? Explain.

RECOMMENDED READINGS

Corcos, A., and F. Monaghan. *Mendel's Experiments on Plant Hybrids: A Guided Study.* Rutgers University Press, New Brunswick, 1993. This interpretive study of Mendel's paper includes information on his life as a monk and a scientist.

Heim, W.G. "What Is A Recessive Allele?" *The American Biology Teacher,* Feb. 1991. A discussion of the molecular basis for the traits Mendel studied in garden peas.

Mendel, G. "Experiments in plant hybridization." Reprinted in *Genetics: Readings from Scientific American.* W.H. Freeman and Co., San Francisco, 1990. Try reading this translation of Mendel's classic paper from the perspective of other scientists of his time who lacked knowledge of chromosomes, mitosis, and meiosis.

There are a number of well written college level genetics texts that cover the principles of genetics in eukaryotes. The following are two representative examples:

Griffiths, A.J.F., J.H. Miller, D.T. Suzuki, R.C. Lewontin, and W.M. Gelbart. *An Introduction to Genetic Analysis,* 6th ed. W.H. Freeman, New York, 1996.

Russell, P. *Genetics,* 5th ed. Benjamin/Cummings, Menlo Park, CA, 1998.

● Visit our website at **http://www.saunderscollege.com/lifesci/titles.html** and click on Solomon/Berg/Martin Biology for links to chapter-related resources on the World Wide Web.

CHAPTER 11

DNA: The Carrier of Genetic Information

Following the rediscovery of Mendel's principles, geneticists conducted elegant experiments to learn how genes are arranged in chromosomes and how they are transmitted from generation to generation. However, two very basic questions remained unanswered: What are genes made of? How do genes work? Although studies of inheritance patterns did not answer these questions, they provided a foundation that allowed scientists to make predictions about the chemical nature of genes and how they might function.

It was obvious that the chemical of which genes are made should have the ability to store information in a form that could be retrieved and used by the cell. But genes had other properties that had to be accounted for as well. Countless genetic experiments on a wide variety of organisms had demonstrated that genes are usually quite stable, being passed unchanged from generation to generation. However, occasionally a gene was observed to convert to a different form; such genetic changes, called **mutations,** were then transmitted unchanged to future generations.

As the science of genetics was developing, the science of biochemistry was flourishing as well. Not surprisingly, there was an increased effort to correlate the known properties of genes with the nature of the various biological molecules. What kind of molecule could store information? How could that information be retrieved and used to direct cellular functions? What kind of molecule could be relatively stable, but have the capacity to change under some circumstances?

Some scientists thought that the problem could never be solved. They thought the information required by a cell to be so complex that no one type of molecule could function as the genetic material. However, as more was learned about the central role that proteins play in virtually every aspect of cellular structure and metabolism, other scientists considered them the prime candidates for the genetic material. In this chapter we discuss how the nucleic acid shown in the photograph, **deoxyribonucleic acid (DNA),** not protein, was found to be the molecule responsible for inheritance, and we examine the unique features of DNA that allow it to carry out this role.

(Prof. K. Seddon & Dr. T. Evans, Queen Univ., Belfast/Science Photo Library/ Photo Researchers, Inc.)

AFTER YOU HAVE STUDIED THIS CHAPTER YOU SHOULD BE ABLE TO

1. Summarize the evidence that accumulated during the 1940s and early 1950s demonstrating that DNA is the genetic material.
2. Relate the chemical and physical features of DNA to the structure proposed by Watson and Crick.
3. Sketch how nucleotide subunits are linked together to form a single DNA strand.
4. Illustrate how the two strands of DNA are oriented with respect to each other.
5. State the base-pairing rules for DNA and describe how complementary bases bind to each other.

6. Cite experimental evidence that allowed scientists to differentiate between semiconservative replication of DNA and alternative models (conservative and dispersive replication).
7. Summarize how DNA replicates and identify some of the unique features of the process.
8. Explain the special constraints on DNA replication that cause it to be (1) bidirectional and (2) discontinuous in one strand and continuous in the other.
9. Compare the organization of DNA in prokaryotic and eukaryotic cells.

MOST GENES CARRY INFORMATION FOR MAKING PROTEINS

The idea that enzymes (which we now know are proteins) and genes are related in some way was first clearly stated by an English physician, Archibald Garrod, who proposed that certain inherited human diseases are caused by a block in a sequence of chemical reactions within the body.

In the first edition of his book, *Inborn Errors of Metabolism* (1908), Garrod discussed a genetic disease called **alkaptonuria,** which has a simple autosomal recessive inheritance pattern. The condition involves the metabolic pathway that breaks down the amino acids phenylalanine and tyrosine, ultimately converting them to carbon dioxide and water. The urine of affected individuals contains an intermediate in this pathway, homogentisic acid, which turns black when exposed to air (Fig. 11–1).

Garrod hypothesized that persons with alkaptonuria lack the enzyme that normally oxidizes homogentisic acid and that this metabolic block causes homogentisic acid to accumulate in their tissues and blood, and to be excreted in their urine. Before the second edition of his book had been published in 1923, it was found that affected persons do indeed lack the enzyme that oxidizes homogentisic acid. Garrod's hypothesis was correct: a mutation in a specific gene could be associated with the absence of a specific enzyme. Shortly thereafter, in 1926, James Sumner purified a different enzyme, urease, and showed it to be a protein; this finding was the first clear evidence that enzymes are proteins.

Despite the implications of these findings, little work was done in this area, primarily because genetically transmitted errors in metabolism appeared to be rare. The lack of experimental subjects made genetic testing and statistical analysis very difficult.

A major advance in understanding the relationship between genes and enzymes came in the early 1940s, when George Beadle and Edward Tatum developed a new approach to the problem. Most efforts until that time had centered on

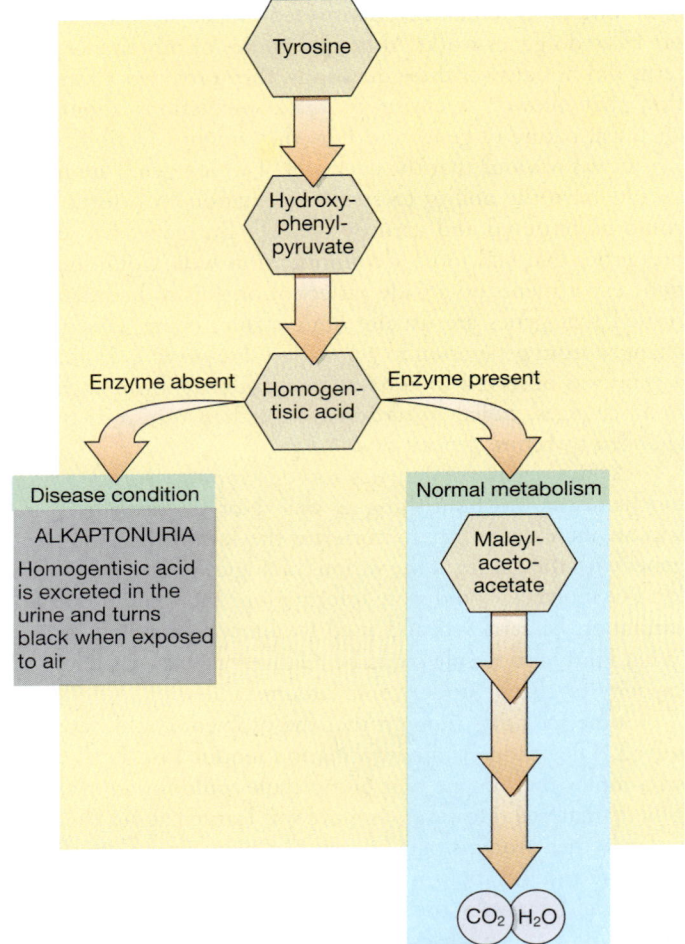

Figure 11–1 An "inborn error of metabolism." Garrod proposed that the alkaptonuria allele causes the absence of a specific enzyme, one that is part of the pathway by which the amino acid tyrosine is catabolized. That enzyme normally converts homogentisic acid to maleylacetoacetate. Homogentisic acid thus accumulates in the blood and is excreted through the urine. When the homogentisic acid in the urine comes in contact with air, it oxidizes and turns black.

studying known loci and attempting to determine what biochemical reactions they affected. Experimenters examined previously identified loci, such as those controlling eye color in *Drosophila* or pigments in plants. They found that specific phenotypes are controlled by a series of biosynthetic reactions, but it was not clear to the investigators whether the genes themselves were acting as enzymes or if they determined the workings of the enzymes in more complex ways.

Beadle and Tatum decided to take the opposite approach. Rather than try to identify the enzymes affected by single genes, they decided to look for mutations interfering with the known metabolic reactions that produce essential molecules such as amino acids and vitamins. They studied the bread mold *Neurospora* (a fungus) for several reasons. First, wild-type[1] *Neurospora* is easy to grow in culture. It can make all of its essential biological molecules when it is grown on a simple minimal medium containing only sugar, salts, and the vitamin biotin. A mutant that cannot make a substance such as an amino acid can still grow if it is simply added to the growth medium.

Second, *Neurospora* grows primarily as a haploid organism. Thus, a recessive mutant allele can be immediately identified because there is no homologous chromosome that could carry a dominant allele that would mask its expression.

Third, the life cycle of *Neurospora* facilitates certain types of manipulations and genetic analysis. *Neurospora* produces large numbers of asexual haploid spores; these can grow and divide mitotically to produce more haploid cells. Two haploid cells can fuse to produce a zygote, which undergoes meiosis to form haploid sexual spores. Thus, researchers can use sexual crosses to perform genetic analyses of isolated mutants. (For an illustration of the generalized life cycles of simple organisms, see Figure 9–12; a more detailed life cycle of organisms similar to *Neurospora* is given in Figure 25–6.)

Beadle and Tatum began by exposing large numbers of wild-type *Neurospora* asexual spores to x rays or ultraviolet radiation to induce mutations. Because each spore contains multiple haploid nuclei, the irradiated cells were mated with another strain, forming zygotes that underwent meiosis to produce haploid sexual spores, each with a single haploid nucleus. (Why did the experimenters not simply irradiate uninucleate sexual spores? The answer is that it is easy to obtain large numbers of asexual spores and they are far more sensitive to the effects of radiation than are sexual spores.) The isolated sexual spores were allowed to grow on a complete medium containing all the amino acids and vitamins normally made by *Neurospora*. Each strain was also tested on a minimal medium. If an isolated strain grew on the complete medium, but failed to grow after transfer to the minimal medium, Beadle and Tatum reasoned that it carried a mutation that made it unable to produce one of the compounds essential for growth. Fur-

ther testing of the mutant on media containing different combinations of amino acids, purines, vitamins, and so on enabled the investigators to determine the exact compound that was required (Fig. 11–2).

Their findings can be illustrated with a class of mutants that require the amino acid arginine. Beadle and Tatum found that some of the arginine-requiring mutants could grow on ornithine or citrulline, as well as arginine; others could grow on citrulline or arginine; and still others could grow only on arginine (Fig. 11–3*a*). This information was then used to deduce the order of these intermediates in the biochemical pathway leading to arginine (Fig. 11–3*b*).

Using this approach, Beadle and Tatum analyzed large numbers of mutants affecting several metabolic pathways. Each mutant strain was verified by special genetic crossing experiments to have a mutation in only one gene locus. They found that for each individual gene locus identified, only one enzyme was affected. This one-to-one correspondence between genes and enzymes was succinctly stated as the *one gene, one enzyme hypothesis*.

Through the discoveries of Beadle and Tatum and others, the sciences of genetics and biochemistry became ever more closely allied, leading to an evolution of the definition of the gene and additional predictions regarding its chemical nature. The idea that a gene encodes the information required to produce a single enzyme held for almost a decade, until additional findings required a modification of this definition.

It became evident that genes control not only enzymes, but other proteins as well. Linus Pauling and his coworkers were able to demonstrate that the structure of hemoglobin can be altered by a mutation of a single locus. This particular mutant form of hemoglobin is associated with the genetic disease sickle cell anemia (Chapter 15). In addition, various studies showed that many proteins are constructed from two or more polypeptide chains, each of which may be controlled by a different locus. For example, hemoglobin was shown to contain two types of polypeptide chains, the α and β subunits (see Chapter 3). Sickle cell anemia results from a mutation affecting the β subunits.

The definition of a gene was therefore extended to state that one gene is responsible for one polypeptide chain. Even this definition has proved to be only partially correct, although we still define a gene in terms of its product (Chapter 12).

EVIDENCE THAT DNA IS THE HEREDITARY MATERIAL WAS FIRST FOUND IN MICROORGANISMS

During the 1940s most geneticists and biochemists were convinced that the genetic material must be protein. Proteins were known to contain more than 20 different kinds of amino acids in many different combinations, allowing each type of protein

---

[1] *Wild type* refers to the genotypes and phenotypes most commonly found in natural populations of a particular species. Wild-type alleles are generally thought of as "normal" or nonmutant alleles.

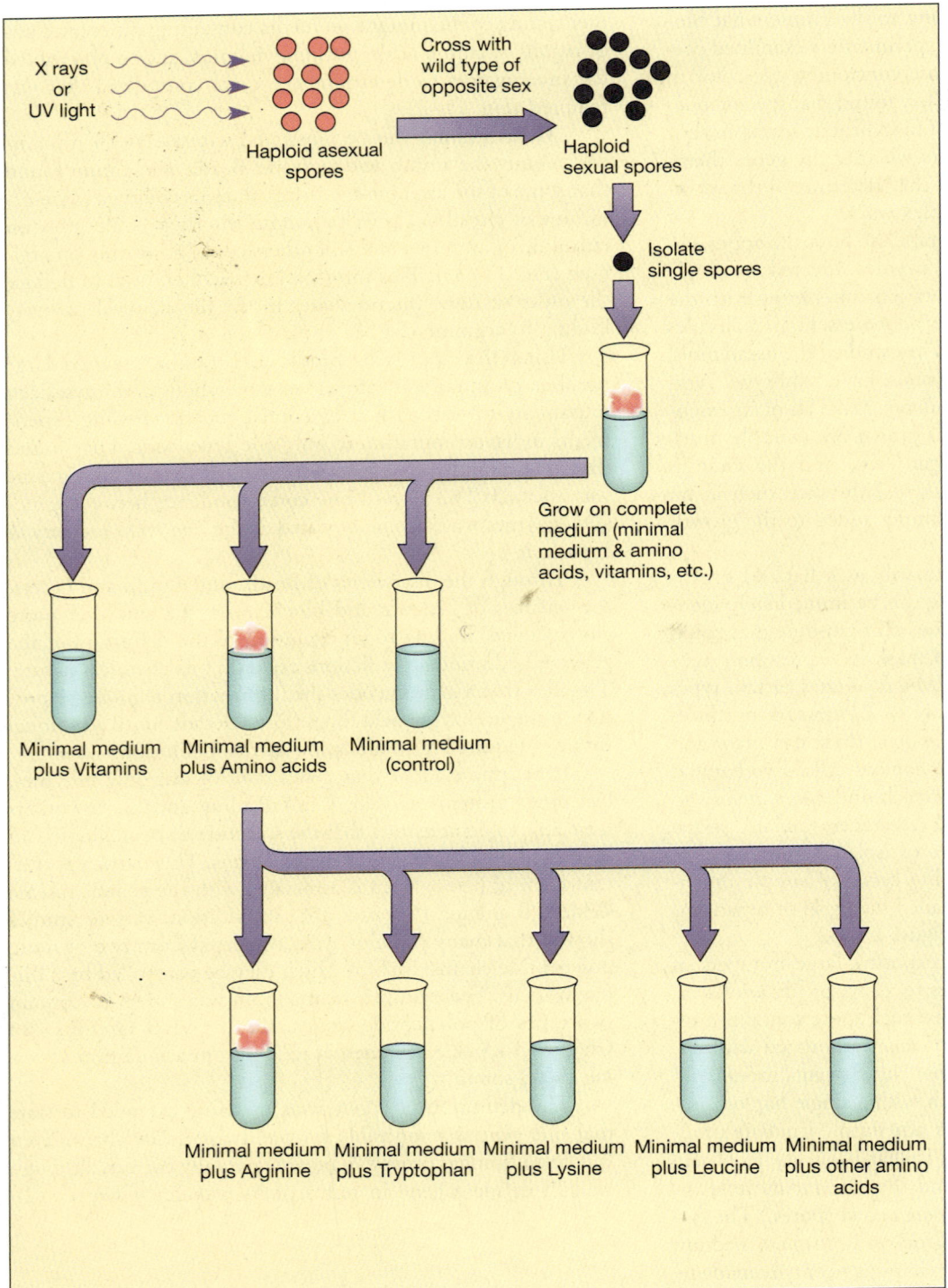

Figure 11–2 Mutations affecting biochemical pathways. Beadle and Tatum irradiated multi-nucleate haploid asexual spores of *Neurospora* to produce random mutations; cultures derived from these spores were then mated with another strain to produce uninucleate haploid sexual spores. Cultures were first established on complete media; subsequent failure to grow on a minimal medium indicated a blocked step in a biochemical pathway. The specific nutritional requirement was determined by testing for growth on minimal media supplemented with individual vitamins or amino acids. In this example, the medium containing the amino acid arginine supports growth, indicating that the mutation affects some part of the arginine biosynthetic pathway.

Figure 11–3 Genes and enzymes. *(a)* In this example, Beadle and Tatum tested a number of different mutant strains that require the amino acid arginine. These were grouped (I, II or III) by their response to various intermediates in the metabolic pathway leading to arginine, which were provided as supplements to the minimal growth medium. *(b)* Analysis of the experimental results led to this model for a portion of the arginine biosynthetic pathway (which is actually a cycle). Because Group I strains grew on minimal medium supplemented with ornithine, citrulline, or arginine, they were thought to be missing enzyme A, required for formation of all three compounds. Group II strains were thought to be missing enzyme B because they were unable to grow on minimal medium supplemented with ornithine but allowed the conversion of citrulline to arginine. Group III strains were thought to be missing enzyme C because they grew on minimal medium to which only arginine had been added. A one-to-one correspondence between each specific enzyme and a specific gene locus was verified by special genetic crosses (not shown).

to have unique properties. It had been shown that genes control the production of proteins. Given their obvious complexity and diversity compared with other molecules, proteins seemed to be the stuff of which genes are made.

In contrast, DNA and other nucleic acids were known to be made of only four nucleotides, and what was known about their arrangement made them seem relatively uninteresting to most scientists. For this reason, several early clues to the role of DNA were not widely recognized.

In 1928 Frederick Griffith made a curious observation concerning two strains of pneumococcus bacteria. A smooth (S) strain, named for its formation of smooth colonies on a solid growth medium, was known to be **virulent,** or lethal. When it was injected into mice, the animals contracted pneumonia and died. A related rough (R) strain, which forms colonies with a rough surface, was known to be **avirulent,** or nonlethal. Griffith found that when a mixture of *heat-killed,* virulent S-strain cells and live avirulent R-strain cells was in-

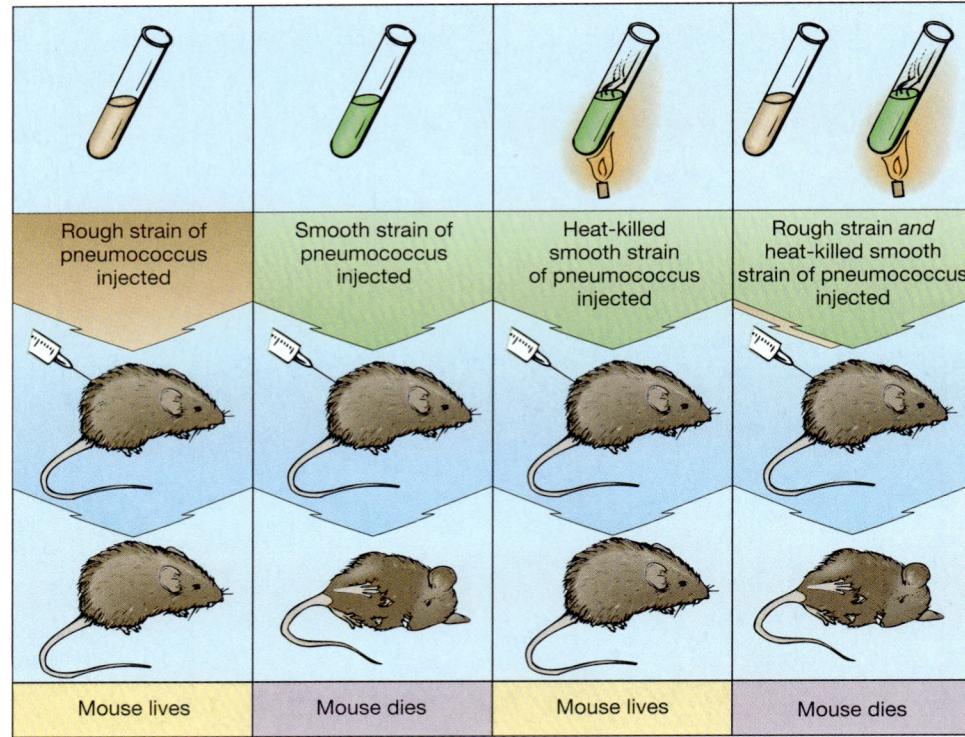

Figure 11–4 **Griffith's transformation experiments.** Although neither the rough (R) strain nor the heat-killed smooth (S) strain could kill a mouse, a combination of the two did. Autopsy of the dead mouse showed the presence of living, S strain pneumococci. These results indicated that some substance in the heat-killed S strain was responsible for the transformation of the living R strain to a virulent form. Avery and his colleagues later showed that purified DNA isolated from the S-strain confers virulence on the R-strain bacteria, establishing that the DNA carries the necessary information for bacterial transformation.

Rough strain of pneumococcus injected

Smooth strain of pneumococcus injected

Heat-killed smooth strain of pneumococcus injected

Rough strain *and* heat-killed smooth strain of pneumococcus injected

Mouse lives

Mouse dies

Mouse lives

Mouse dies

jected into mice, a high proportion of the mice died. Griffith was then able to isolate living S-strain cells from the dead mice.

Because neither the heat-killed S strain nor the living R strain could be converted to the living virulent form when injected by itself, something in the heat-killed cells appeared to convert the avirulent cells to the lethal form. This phenomenon, called **transformation,** was thought to be caused by some chemical substance (called the "transforming principle") in the dead bacteria that "transformed" a related strain to a genetically stable new form (Fig. 11–4).

In 1944, O. T. Avery, C. M. MacLeod, and M. McCarty of the Rockefeller Institute chemically identified Griffith's transforming principle as DNA. Although today we consider their findings to be the first demonstration that DNA is the genetic material, not all scientists of the time were convinced. One argument given was that the DNA preparations used might have been contaminated with a tiny amount of protein, which might have been responsible for the results. This was not a trivial objection, because it was well known that a very small amount of an enzyme could have significant biological effects.

During the next few years, new evidence accumulated that the haploid nuclei of pollen grains and gametes such as sperm contain only half the amount of DNA found in diploid somatic cells of the same species. Because the idea that genes are on chromosomes was generally accepted, these findings correlating DNA content with chromosome number provided strong circumstantial evidence of DNA's importance in inheritance.

In 1952 Alfred Hershey and Martha Chase performed a series of elegant experiments on the reproduction of **bacte-**riophages (viruses that infect bacteria; see Chapter 23) (Fig. 11–5). Their demonstration that bacteriophages inject their DNA into bacterial cells, leaving most of their protein on the outside, emphasized the significance of DNA in viral reproduction and was seen by many as another important indication of the role of DNA as the hereditary material.

THE STRUCTURE OF DNA ALLOWS IT TO CARRY INFORMATION AND TO BE FAITHFULLY DUPLICATED

DNA was not widely accepted as the genetic material until James Watson and Francis Crick proposed a model for its structure that had extraordinary explanatory power. The story of how the structure of DNA came to be determined is one of the most remarkable chapters in the history of modern biology.

A great deal was known about the physical and chemical properties of DNA when Watson and Crick became interested in the problem. Their all-important contribution was to integrate all this information into a model that demonstrated how the molecule can both carry information and serve as its own template (pattern) for duplication.

Nucleotides can be covalently linked in any order to form long polymers

As discussed in Chapter 3, each DNA building block is a **nucleotide** consisting of a pentose sugar **(deoxyribose),** a phosphate, and one of four nitrogenous bases (Fig. 11–6*a*). The

Figure 11–5 **The Hershey-Chase experiments.** Although bacteriophage protein coats labeled with the radioactive isotope ^{35}S *(left)* could be separated from infected bacterial cells without interfering with viral reproduction, viral DNA labeled with the radioactive isotope ^{32}P *(right)* could not, thus demonstrating that the DNA enters the cells and is required for synthesis of new protein coats and DNA.

Figure labels:
- Bacterial viruses grown in ^{35}S to label protein coat *or* ^{32}P to label DNA
- Viruses infect bacteria
- Agitate cells in blender
- Separate by centrifugation
- ^{35}S labeled protein in the supernatant
- Bacteria in the pellet contain ^{32}P labeled DNA
- Viral reproduction inside bacterial cells from the pellet
- Cell lysis

nitrogenous base is attached to the 1′ carbon of the sugar, and the phosphate is attached to the 5′ carbon.[2] The bases include two **purines—adenine (A)** and **guanine (G)**—and two **pyrimidines—thymine (T)** and **cytosine (C).**

The nucleotides are linked by covalent bonds to form an alternating sugar-phosphate backbone. The 3′ carbon of one sugar is bonded to the 5′ phosphate of the adjacent sugar to form a 3′, 5′ **phosphodiester linkage** (Fig. 11–6*b*). It is therefore possible to form a polymer of indefinite length, with the nucleotides linked in any order. We now know that most DNA molecules found in cells are millions of bases long. Figure 11–6*a* also illustrates that a single polynucleotide chain is directional. No matter how long the chain may be, one end (the **5′ end**) has a 5′ carbon and the other (the **3′ end**) has a 3′ carbon that is not linked to another nucleotide.

DNA is made of two polynucleotide chains intertwined to form a double helix

Important information about the structure of DNA came from **x-ray diffraction** studies on crystals of purified DNA, carried out by Rosalind Franklin in the laboratory of M. H. F. Wilkins. X-ray diffraction is a powerful method for determining distances between atoms of molecules arranged in a regular, repeating crystalline structure (Fig. 11–7). X rays have such extremely short wavelengths that they can be scattered by the electrons surrounding the atoms in a molecule. Atoms with dense electron clouds (e.g., phosphorus, oxygen) tend to deflect electrons more strongly than do atoms with lower atomic numbers.

When a crystal is exposed to an intense beam of x rays, the regular arrangement of the atoms in the crystal causes the x rays to be diffracted, or bent, in specific ways. The pattern of diffracted x rays is seen on film as dark spots. Mathematical analysis of the arrangement and distances between the spots can then be used to determine precise distances between atoms and their orientation within the molecules.

Franklin had already produced x-ray crystallographic films of DNA patterns when Watson and Crick began to pursue the problem of DNA structure. The pictures clearly showed that DNA has a type of helical structure, and three major types of regular, repeating patterns in the molecule with the dimensions 0.34 nanometer, 3.4 nanometers, and 2.0 nanometers were evident. Franklin and Wilkins had inferred from these patterns that the nucleotide bases (which are flat molecules) are stacked like rungs of a ladder. Using this information, Watson and Crick began to build scale models of the DNA components and then fit them together to agree with the experimental data.

After a number of trials, the two worked out a model that fit the existing data (Fig. 11–8). The nucleotide chains conformed to the dimensions of the x-ray data only if each DNA molecule consisted of *two* polynucleotide chains arranged in a

---

[2]It is conventional to number the atoms in a molecule, using a system devised by organic chemists. In nucleic acid chemistry the "prime" designations, such as 2′, designate individual carbon atoms in the sugar ring, to distinguish them from carbon atoms in the base.

Figure 11–6 The nucleotide subunits of DNA. *(a)* A single strand of DNA consists of a backbone made of phosphate groups alternating with the sugar deoxyribose *(tan)*. Linked to the 1′ carbon of each sugar is one of four nitrogenous bases. The purine bases, adenine and guanine, have two-ring structures; the pyrimidine bases, thymine and cytosine, have one-ring structures. Note the polarity of the polynucleotide chain, with the 5′ end at the top of the figure and the 3′ end at the bottom. *(b)* Sugars of adjacent nucleotides are joined by phosphodiester linkages.

Figure 11–7 X-ray diffraction image of DNA. Important clues about DNA structure are provided by detailed mathematical analysis of measurements of x-ray diffraction images of the lithium salt of DNA. The diagonal pattern of spots stretching from 11 o'clock to 5 o'clock and from 1 o'clock to 7 o'clock provides evidence for the helical structure of DNA. The elongated horizontal patterns at the top and bottom indicate that the purine and pyrimidine bases are stacked 0.34 nanometers apart and are perpendicular to the axis of the DNA molecule. *(Dr. S.D. Dover, Division of Biomolecular Sciences, Kings College, London)*

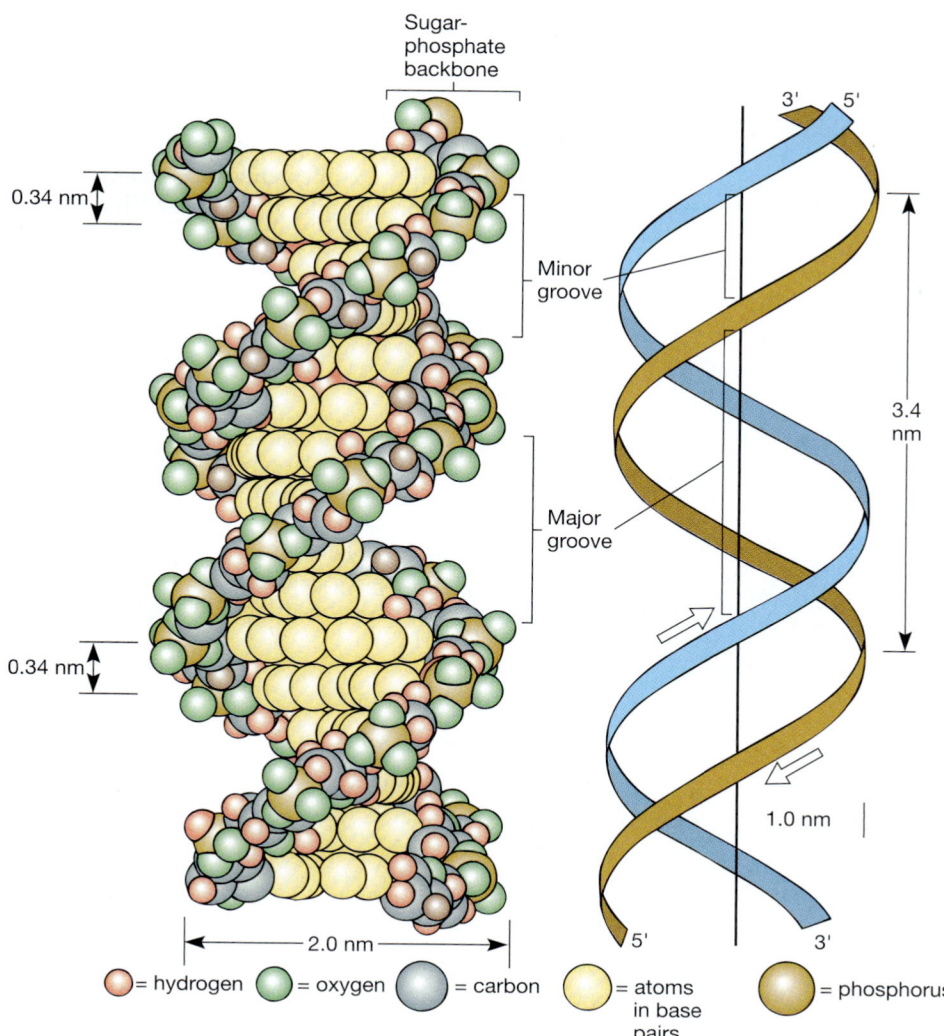

Sugar-phosphate backbone

0.34 nm

Minor groove

Major groove

0.34 nm

2.0 nm

3' 5'

3.4 nm

1.0 nm

5' 3'

● = hydrogen ● = oxygen ● = carbon ● = atoms in base pairs ● = phosphorus

Figure 11–8 Structure of the DNA molecule. On the left is a space-filling model of the DNA double helix. The measurements on the diagrammatic model on the right match those derived from x-ray diffraction images. The ribbons represent the sugar-phosphate backbone of each strand; the thick arrows indicate that the two strands extend in opposite directions.

coiled **double helix.** In their model, the sugar-phosphate backbones of the two chains form the outside of the helix. The bases belonging to the two chains associate as pairs in the center. The reasons for the repeating patterns of 0.34-nanometer and 3.4-nanometer measurements are readily apparent from the model: each pair of bases is exactly 0.34 nanometer from the adjacent pairs above and below. Because exactly ten base pairs are present in each full turn of the helix, each turn is 3.4 nanometers high. To fit the data, the two chains must run in opposite directions; therefore, each end of the double helix must have an exposed 5′ phosphate on one strand and an exposed 3′ hydroxyl group on the other. Because the two strands run in opposite directions, they are said to be **antiparallel** to each other.

In double-stranded DNA, hydrogen bonds form between adenine and thymine and between guanine and cytosine

Other features of the model integrated important information about the chemical composition of DNA with the x-ray dif-

fraction data. By 1950, the base composition of DNA from a number of organisms and tissues had been determined by Erwin Chargaff and his coworkers at Columbia University. They found a simple relationship among the bases that turned out to be an important clue to the structure of DNA. Regardless of the source of the DNA, in Chargaff's words, the "ratios of purines to pyrimidines and also of adenine to thymine and of guanine to cytosine were not far from 1." In other words, in DNA molecules, A=T and G=C.

The x-ray diffraction studies indicated that the double helix has a precise and constant width, as shown by the 2.0-nanometer reflections. This finding is actually connected to Chargaff's rules. Notice in Figure 11–6 that the pyrimidines (cytosine and thymine) contain only one ring of atoms, whereas the larger purines (guanine and adenine), contain two rings. Study of the models made it clear to Watson and Crick that if each cross-rung of the ladder were to contain one purine and one pyrimidine, the width of the helix at that point would be exactly 2.0 nanometers; the combination of two purines (each of which is 1.2 nanometers wide) would be wider and that of two pyrimidines would be narrower.

Further examination of the model showed that adenine can pair with thymine (and guanine with cytosine) in such a way that hydrogen bonds form between them; the opposite combinations, cytosine with adenine and guanine with thymine, do not lead to favorable hydrogen bonding.

The nature of the hydrogen bonding between adenine and thymine and between guanine and cytosine is shown in Figure 11–9. Two hydrogen bonds can form between adenine and thymine, and three between guanine and cytosine. This concept of *specific base-pairing* neatly explains Chargaff's rules. The amount of cytosine has to equal the amount of guanine, be-

cause every cytosine in one chain must have a paired guanine in the other chain. Similarly, every adenine in the first chain must have a thymine in the second chain. Thus, the sequences of bases in the two chains are **complementary,** but not identical, to each other. In other words, the sequence of nucleotides in one chain dictates the complementary sequence of nucleotides in the other. For example, if one strand has the sequence:

3'—AGCTAC—5'

Figure 11–9 **Hydrogen bonding between bases.** The two strands of a DNA double helix are associated by hydrogen bonding between the bases. *(a)* The two sugar-phosphate chains run in opposite directions. This orientation permits the complementary bases to pair. *(b)* Diagram of the hydrogen bonding between base pairs adenine (A) and thymine (T) *(top)* and guanine (G) and cytosine (C) *(bottom)*. The AT pair has two hydrogen bonds; the GC pair has three.

(a)

(b)

then the other strand has the complementary sequence:

5′—TCGATG—3′

The double-helix model strongly suggested that the sequence of bases in DNA can provide for the storage of genetic information. Although there are restrictions on how the bases pair with each other, the number of possible sequences of bases in a strand is virtually unlimited. Because a DNA molecule in a cell can be millions of nucleotides long, it can store enormous amounts of information, usually comprising a large number of genes.

DNA REPLICATION IS SEMICONSERVATIVE

Two immediately apparent and distinctive features of the Watson-Crick model made it seem more plausible that DNA is the genetic material. We have already mentioned that DNA can carry coded information in its sequence of bases. The model also suggested a way in which information in DNA could be precisely copied, a process known as **DNA replication.** The importance of the replication mechanism was known

to Watson and Crick, who noted in a classic and now famous understatement at the end of their first brief paper, "It has not escaped our notice that the specific pairing we have postulated immediately suggests a possible copying mechanism for the genetic material."

The model suggested that, because the nucleotides pair with each other in a complementary fashion, each strand of the DNA molecule could serve as a template, or pattern, for the synthesis of the opposite strand. It would simply be necessary for the hydrogen bonds between the two strands to break and the two chains to separate. Each half-helix could then pair with complementary nucleotides to replace its missing partner. The result would be two DNA double helices, each identical to the original one and consisting of one original strand from the parent molecule and one newly synthesized complementary strand. This type of information copying is known as a **semiconservative replication** mechanism. The recognition that DNA could be copied in this way suggested how DNA could provide a third essential characteristic of genetic material—the ability to mutate (see *Making the Connection: Mutations and the Structure of DNA*).

Although the semiconservative replication mechanism suggested by Watson and Crick was (and is) a simple and compelling model, experimental proof was needed to establish that DNA in fact duplicates in that manner. First it was necessary

---

MAKING THE CONNECTION

MUTATIONS AND THE STRUCTURE OF DNA

How do new genetic variants arise? This question was of great interest to geneticists, who had long known that mutations, or genetic changes, could arise in genes and then be transmitted faithfully to succeeding generations. When the double-helix model was proposed, it seemed plausible that mutations could represent a change in the sequence of bases in the DNA. One could predict that if DNA is copied by a mechanism involving complementary base pairing, any change in the sequence of bases on one strand would result in a new sequence of complementary bases during the next replication cycle. The new base sequence would then be passed on to daughter molecules by the same mechanism used to copy the original genetic material, as if no change had occurred.

In the example shown in the figure, an adenine base in one of the DNA strands has been changed to guanine. This could occur by a rare error in DNA replication or by one of several other known mechanisms. (There are systems of enzymes that repair errors when they occur, but not all mutations are corrected properly.) When the DNA molecule is replicated again, one of the strands gives rise to a molecule exactly like the parent strand; the other (mutated) strand gives rise to a molecule with a new combination of bases that will be stably transmitted generation after generation.

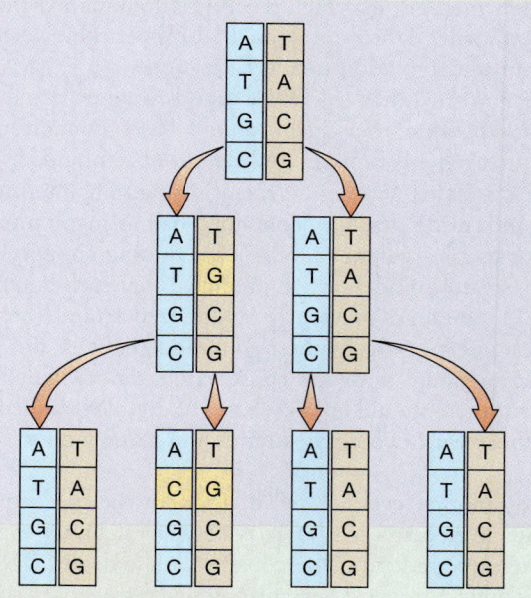

Perpetuation of a mutation. The process of DNA replication can stabilize a mutation so that it will be passed to future generations.

to rule out several other possibilities. For example, with a *conservative replication* mechanism, both parent (or old) strands would remain together, and the two newly synthesized strands would form a second double helix. As a third alternative, the parental and newly synthesized strands might become randomly mixed during the replication process; this possibility was known as *dispersive replication.* To discriminate among the semiconservative replication mechanism and the other possibilities, it was necessary to distinguish between old and newly synthesized strands of DNA.

One way to accomplish this is to use a heavy-nitrogen isotope, nitrogen-15 (ordinary nitrogen is nitrogen-14), to label DNA strands by making them more dense. Large molecules such as DNA can be separated on the basis of differences in their density, using a technique known as **density gradient centrifugation.** When DNA is mixed with a solution containing cesium chloride (CsCl) and centrifuged at high speed, the solution forms a density gradient in the centrifuge tube, ranging from a low density region at the top to a region of highest density at the bottom. During centrifugation the DNA molecules migrate to the region of the gradient identical to their own density.

In 1957, Matthew Meselson and Franklin Stahl grew cells of the bacterium *Escherichia coli* on a medium that contained nitrogen-15 in the form of ammonium chloride (NH_4Cl). The cells used the nitrogen-15 to synthesize bases, which then became incorporated into DNA (Fig. 11–10). The resulting heavy nitrogen-containing DNA molecules were extracted from some of the cells. When they were subjected to density gradient centrifugation, they accumulated in the high-density region of the gradient. The rest of the bacteria (which also contained nitrogen-15-labeled DNA) were transferred to a different growth medium in which the NH_4Cl contained the naturally abundant, lighter nitrogen-14 isotope; they were then allowed to undergo additional cell divisions.

The newly synthesized DNA strands were expected to be less dense because they incorporated bases containing the lighter nitrogen-14 isotope. Indeed, double-stranded DNA from cells isolated after one generation had an intermediate density, indicating that they contained half as many nitrogen-15 atoms as the "parent" DNA. This finding supported the semiconservative model, which predicted that each double helix should contain a previously synthesized strand (heavy in this case) and a newly synthesized strand (light in this case). It was inconsistent with the conservative model, which predicted that there should be two classes of double stranded molecules, those with two heavy strands and those with two light strands.

After another cycle of cell division in the medium with the lighter nitrogen-14 isotope, two types of DNA appeared in the density gradient. One consisted of "hybrid" DNA helices (with one nitrogen-15 strand and one nitrogen-14 strand), whereas the other contained only DNA with the naturally occurring light isotope. This finding refuted the dispersive model, which predicted that all strands should have intermediate density. Instead, each strand of the parental double-helix molecule was conserved, but in a *different* daughter molecule, exactly as predicted by the semiconservative replication model.

DNA replication is complex and has a number of unique features

Although the general principles of DNA replication are simple and straightforward predictions from the Watson-Crick model, the process actually requires a complex "replication machine" containing a large number of proteins and enzymes. Many of the essential features of DNA replication are universal, although some differences exist between prokaryotes and eukaryotes because their DNA is organized differently. In bacterial cells such as *E. coli,* most or all of the DNA is in the form of a single, *circular,* double-stranded molecule. Each unreplicated eukaryotic chromosome contains a single, *linear,* double-stranded molecule associated with a great deal of protein.

DNA strands must be unwound during replication

Watson and Crick recognized that in their double-helix model the two DNA strands are wrapped around one another like the strands of a rope. If we try to pull the strands apart, the rope must either rotate or twist into tighter coils. We would expect similar things to happen when complementary DNA strands are separated for replication. Separating the two strands of DNA is accomplished by **DNA helicase enzymes** that travel along the helix, opening the double helix as they move. Once the strands are separated, **helix-destabilizing proteins** bind to single DNA strands, preventing re-formation of the double helix until the strands are copied. Enzymes called **topoisomerases** produce breaks in the DNA molecules and then rejoin the strands, relieving strain and effectively preventing the formation of knots during replication.

DNA synthesis always proceeds in a 5′ → 3′ direction

The enzymes that catalyze the linking together of the nucleotide subunits are called **DNA polymerases.** They have several limitations that contribute to the complexity of the replication process. They are able to add nucleotides only to the 3′ end of a growing polynucleotide strand, and this strand must be paired with the strand being copied (Fig. 11–11). Nucleotides with three phosphate groups are used as substrates for the polymerization reaction. As the nucleotides are linked together, two of the phosphates are removed. Like the hydrolysis of ATP, these reactions are strongly exergonic (see Chapter 6) and do not require additional energy. Because the new polynucleotide chain is elongated by the linkage of the 5′

Figure 11–10 Meselson and Stahl's experiments. The bacterium *Escherichia coli* was grown in heavy-nitrogen (^{15}N) growth medium for many generations. Some of the cells were then transferred to light-nitrogen (^{14}N) medium. DNA was isolated from cells after growth on ^{15}N medium one generation after transfer to ^{14}N medium, and two generations after transfer to ^{14}N medium. The density of the molecules in each group *(bottom)* matches the labeling pattern expected if DNA is replicated semi-conservatively *(top)*.

Within the figure:

Hypothesis: Semiconservative replication
Parental DNA — First generation — Second generation

Hypothesis testing
Bacteria grown in ^{15}N. All DNA is heavy. — Some cells transferred to ^{14}N (light) medium — First generation — Some cells continue to grow in ^{14}N medium — Second generation — DNA isolated from cells

Cesium chloride CsCl — DNA mixed with CsCl solution and placed in an ultracentrifuge and centrifuged at very high speed for ~48 hours. — CsCl + DNA

High density — Low density — Centrifuge tube — CsCl — Greater concentration of CsCl at the bottom of the tube is due to sedimentation under centrifugal force.

Light DNA — ^{14}N–^{15}N hybrid DNA — Heavy DNA — DNA molecules move to positions where their density equals that of the CsCl solution.

Results
The location of DNA molecules within the centrifuge tube can be determined by UV optics. DNA solutions absorb strongly at 260 nm.
Heavy DNA — Before transfer to ^{14}N
^{14}N–^{15}N hybrid DNA — One cell generation after transfer to ^{14}N
Light DNA — ^{14}N–^{15}N hybrid DNA — Two cell generations after transfer to ^{14}N

phosphate group of the next nucleotide subunit to the 3′ hydroxyl group of the sugar at the end of the preexisting strand, the new strand of DNA always grows in the 5′ → 3′ direction.

DNA synthesis requires an RNA primer

A second limitation of the DNA polymerases is that they can add nucleotides only to the 3′ end of an *existing* polynucleotide strand. So how can DNA synthesis be initiated once the two strands are separated? The answer is that a short piece (usually about five nucleotides) of an **RNA primer** is first synthesized at the point of initiation of replication (Fig. 11–12).

RNA, or **ribonucleic acid** (see Chapters 3 and 12), is a nucleic acid polymer consisting of nucleotide subunits that can associate by complementary base-pairing with the single-stranded DNA template. The RNA primer is synthesized by a protein complex known as a **primosome,** which includes an enzyme that is able to start a new strand of RNA opposite a

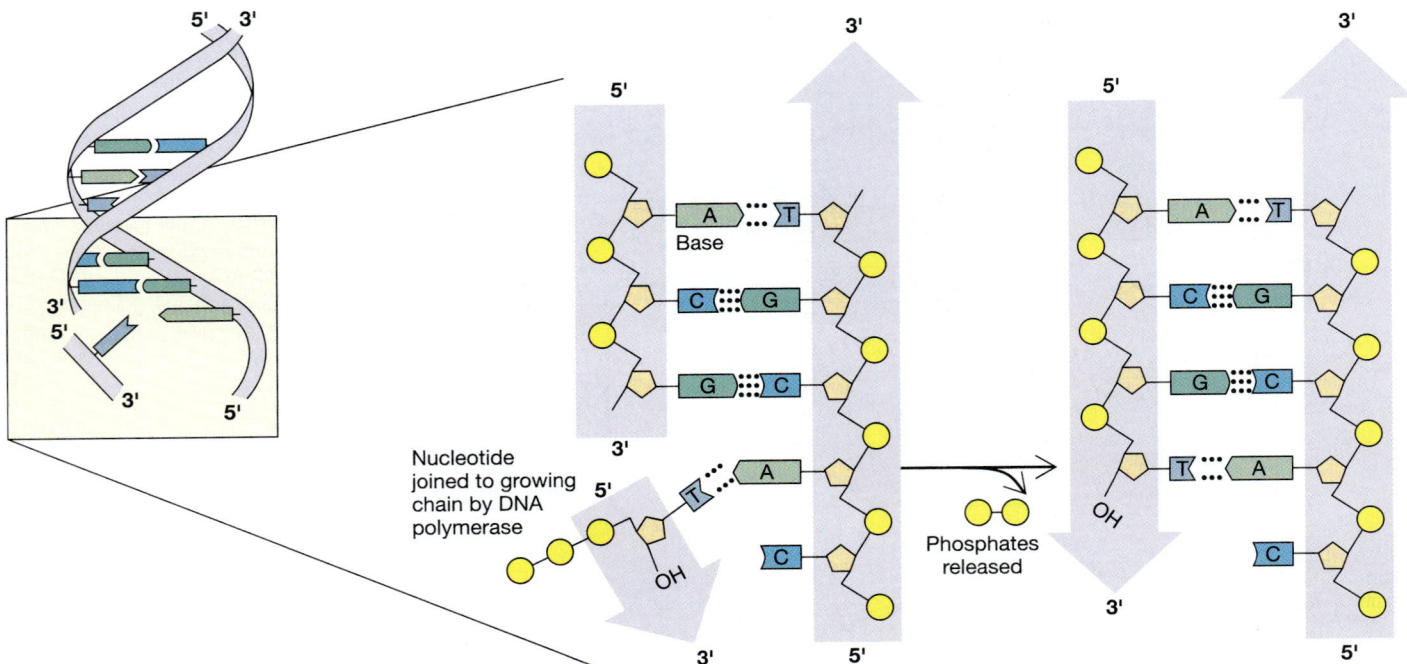

Figure 11–11   DNA replication. The polymerase enzymes that catalyze the polymerization reactions add one nucleotide at a time to the 3′ end of a growing chain.

DNA strand. After a few nucleotides have been added, the primosome is displaced by DNA polymerase, which can then add subunits to the 3′ end of the short RNA primer. The primer is later degraded by specific enzymes, and the space is filled in with DNA.

DNA replication is discontinuous in one strand and continuous in the other

A major obstacle in understanding DNA replication was the fact that the complementary DNA strands are antiparallel. Because DNA synthesis proceeds only in the direction of 5′ → 3′ (which means that the strand being copied is being read in a 3′ → 5′ direction), it would seem necessary to copy one of the strands starting at one end of the double helix and the other strand starting at the opposite end. We know, however, that this is not the case.

DNA replication begins at specific sites on the DNA molecule, termed **origins of replication,** and both strands are replicated at the same time at a Y-shaped structure called the **replication fork** (Fig. 11–13). The position of the replication fork is constantly moving as replication proceeds. Two identical DNA polymerase molecules are responsible for replication. One of these adds nucleotides to the 3′ end of the new strand that is always growing *toward* the replication fork. Because this strand can be formed smoothly and continuously, it is called the **leading strand.**

A separate (but identical) DNA polymerase molecule adds nucleotides to the 3′ end of the other new strand, termed the **lagging strand.** The lagging strand is always growing *away* from the replication fork. Only short pieces can be synthesized because the DNA polymerase enzyme would need to move far away from the fork if it were to add continuously to the 3′ end of that strand. These 100- to 1000-nucleotide pieces are called **Okazaki fragments** after their discoverer, Reijii Okazaki.

Each Okazaki fragment is initiated by a separate RNA primer and is then extended toward the 5′ end of the previously synthesized fragment by DNA polymerase. When the RNA primer of the previously synthesized fragment is reached, the primer is degraded and the resulting gap is filled in with DNA. The fragments are then joined together by **DNA ligase,** an enzyme that links the 3′ end of one DNA fragment to the 5′ end of another.

It has been suggested that simultaneous synthesis of both the leading and lagging strands is possible because the lagging strand (and the DNA strand from which it is copied) forms a loop. This loop allows DNA polymerase to synthesize the lagging strand while remaining close the replication fork as it works.

Most DNA synthesis is bidirectional

When double-stranded DNA is separated, two replication forks are formed, allowing the molecule to be replicated in

DNA synthesis begins at a specific base sequence, termed the *origin of replication*.

Strands are separated at the origin of replication and unwound by DNA helicase, which "walks" along the DNA molecule preceding the DNA-synthesizing enzymes. Single-stranded regions are prevented from re-forming into double strands by helix-destabilizing proteins, which bind to single-stranded DNA. The region of active DNA synthesis is associated with the "replication fork," formed at the junction of the single strands and the double-stranded region. Both strands are synthesized in the vicinity of the fork (each in a 5' → 3' direction).

As the new strands continue to grow in the first direction, unwinding and replication initiate on the other side of the origin of replication, forming a second replication fork. Thus replication proceeds in opposite directions.

Completion of replication results in the formation of two daughter molecules, each containing one old and one newly synthesized strand. Each double helix is a chromatid of a duplicated eukaryotic chromosome.

Figure 11–12 **Overview of DNA replication.** This process requires a number of steps involving several enzymes and RNA primers.

both directions from the origin of replication. Prokaryotic cells usually have only one origin of replication on each circular DNA molecule (Fig. 11–14*a*), so the two replication forks proceed around the circle and eventually meet at the other side to complete the formation of two new DNA molecules.

A eukaryotic chromosome is composed of one, extremely long, linear DNA molecule, so the process is speeded up by having multiple origins of replication (Fig. 11–14*b–d*). Synthesis continues at each replication fork until it meets one coming from the opposite direction, resulting in the formation of a chromosome containing two DNA double helices (each of which corresponds to a chromatid). The ends of eukaryotic chromosomes; known as **telomeres,** present special problems in replication (see *On the Cutting Edge: Telomeres, Cellular Aging, and Cancer on page 263*).

DNA IN CHROMOSOMES IS PACKAGED IN A HIGHLY ORGANIZED WAY

Prokaryotic and eukaryotic cells differ markedly in their DNA content as well as in the organization of DNA molecules. An *E. coli* cell normally contains about 4×10^6 base pairs (almost 1.35 millimeters) of DNA in its single circular DNA molecule. In fact, the total length of the DNA is about 1000 times greater than the length of the cell itself. Therefore the DNA molecule must, with the help of special proteins, be twisted and folded compactly to fit inside the bacterial cell.

A typical eukaryotic cell contains much more DNA than a bacterium does, and it is organized in the nucleus as multi-

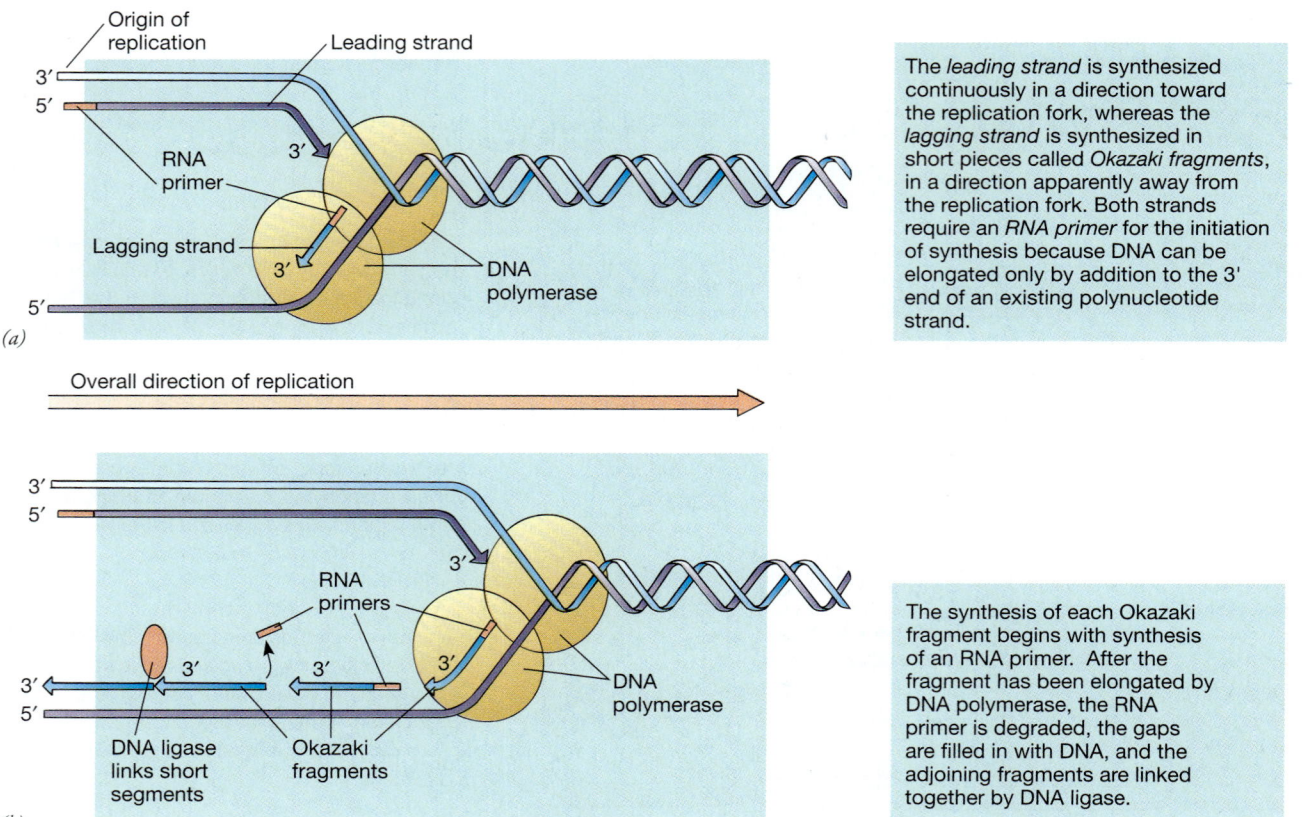

The *leading strand* is synthesized continuously in a direction toward the replication fork, whereas the *lagging strand* is synthesized in short pieces called *Okazaki fragments*, in a direction apparently away from the replication fork. Both strands require an *RNA primer* for the initiation of synthesis because DNA can be elongated only by addition to the 3' end of an existing polynucleotide strand.

Overall direction of replication

The synthesis of each Okazaki fragment begins with synthesis of an RNA primer. After the fragment has been elongated by DNA polymerase, the RNA primer is degraded, the gaps are filled in with DNA, and the adjoining fragments are linked together by DNA ligase.

Figure 11–13 **Leading and lagging DNA strands.** Because elongation can proceed only in a 5' → 3' direction, the two strands at the replication fork are copied in different ways, each by a separate DNA polymerase molecule.

ple chromosomes; these vary widely in size and number among different species. Although a human cell nucleus is about the size of a large bacterial cell, it contains almost 1000 times the amount of DNA found in *E. coli*. The haploid DNA content of a human cell is about 3×10^9 base pairs; if stretched end to end, it would be almost 1 meter long.

In eukaryotes, DNA, which is negatively charged, is associated with positively charged basic proteins known as **histones**[3] to form structures called **nucleosomes.** The fundamental unit of the nucleosome complex consists of a beadlike structure with 146 base pairs of DNA wrapped around a disc-shaped core of eight histone molecules (two each of four different histone types) (Fig. 11–15). Although the nucleosome was originally defined as a bead plus a DNA segment that links it to an adjacent bead, today the term more commonly refers

only to the bead itself (i.e., the eight histones and the DNA wrapped around them).

The nucleosomes are part of the **chromatin,** the complex of nucleic acids and protein that makes up the chromosomes. The higher order structures of chromatin are illustrated in Figure 11–16. For example, a fifth type of histone, known as histone H1, is associated with the linker DNA and is responsible for packing adjacent nucleosomes (each of which is 11 nm in diameter) together to form a 30-nm-diameter thread. These 30-nm-diameter threads form large coiled loops held together by a set of nonhistone **scaffolding proteins.** The loops then interact in complex ways to form the chromatin found in a condensed metaphase chromosome.

Nucleosomes function like tiny spools, thereby preventing DNA strands from becoming tangled. The importance of this role is underscored by Figure 11–17, which illustrates the dense packing of the DNA fibers of a mouse chromosome after the histones have been removed. However, their role is more than structural, for their arrangement also affects the activity of the DNA with which they are associated (Chapter 13).

---

[3]A few types of eukaryotic cells lack histones. Conversely, histones do occur in one group of prokaryotes, the Archaebacteria (Chapter 23).

Figure 11-14 **Bidirectional DNA replication.** The leading strands and lagging strands are not represented in the illustrations. *(a)* The circular DNA in *E. coli* has only one origin of replication. DNA synthesis proceeds from that point in both directions until the two replication forks meet. *(b)* This TEM shows two replication forks *(arrows)* in a segment of a eukaryotic chromosome that has been partly replicated. *(c)* Eukaryotic chromosomal DNA contains multiple origins of replication. DNA synthesis proceeds in both directions from each origin until adjacent "replication bubbles" eventually merge *(d)*. *(b, Courtesy of D.S. Hogness and H.J. Kriegstein)*

(a)

(b)   340 nm

Replication fork

(c)

(d)

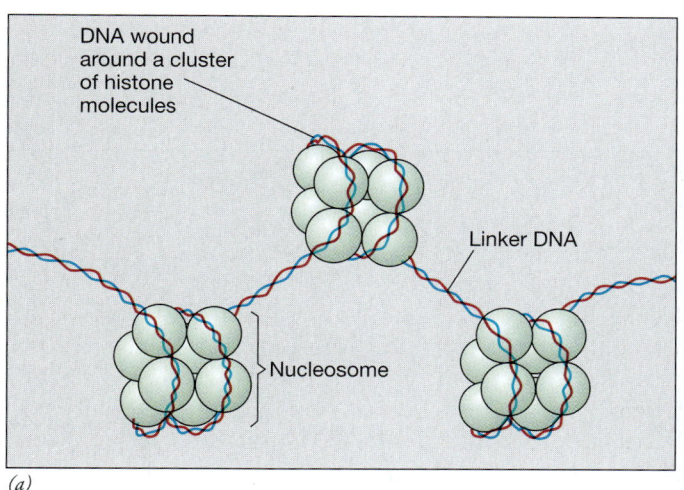

DNA wound around a cluster of histone molecules

Linker DNA

Nucleosome

(a)

(b)   100 nm

Figure 11-15 **Nucleosomes.** *(a)* A model for the structure of a nucleosome. Each nucleosome bead contains a set of eight histone molecules; these form a protein core around which the double-stranded DNA is wound. The DNA surrounding the histone consists of 146 nucleotide pairs; another segment of DNA, about 60 nucleotide pairs long, links nucleosome beads. *(b)* TEM of nucleosomes from the nucleus of a chicken red blood cell. Normally nucleosomes are packed more closely together, but the preparation procedure has spread them apart, revealing the DNA linkers. *(b, courtesy of D.E. Olins and A.L. Olins)*

← 1400 nm →

← 700 nm →

300 nm

30 nm fiber

Condensed
chromosome

Scaffolding
protein

Extended chromatin

Condensed chromatin

30 nm

DNA wound around
a cluster of histone
molecules

Packed nucleosomes

11 nm

Nucleosomes

2 nm

DNA double helix

Figure 11–16 **Organization of a eukaryotic chromosome.**
(Visuals Unlimited/K.G. Murti)

Figure 11–17 **A chromosome depleted of histones.** Note how densely packed the DNA fibrils are in this TEM of a mouse chromosome, even though they have been released from the histone proteins that organize them into tightly coiled structures. The dark structure extending from left to right across the bottom of the photograph is composed of scaffolding proteins. *(Courtesy of U. Laemmli, from Cell 12:817, 1988. Copyright by Cell Press)*

2 μm

Telomerase, Cellular Aging, and Cancer

HYPOTHESIS: The number of cell divisions that normal human somatic cells can undergo when grown in culture is limited by the length of their telomeric DNA.

METHOD: To test the prediction that cells will be able to undergo more divisions if their telomeric DNA is lengthened, researchers used a virus to introduce the gene coding for the catalytic subunit of telomerase into cultured normal human cells, which normally lack telomerase activity.

RESULTS: The cells produced active telomerase, which elongated the telomeres. The cells continued to actively proliferate for many cell cycles beyond the point at which cell division would normally cease.

CONCLUSION: The life spans of cells can be extended by lengthening the telomeric DNA of their chromosomes.

Unlike prokaryotic DNA, which is circular, eukaryotic chromosomes have free ends. This simple fact has far-reaching consequences. Because of the discontinuous way in which DNA polymerases work, they are unable to complete replication neatly when they reach the end of the DNA, and must therefore leave a small portion unreplicated. This situation is less dangerous than it sounds because chromosomes have end caps (telomeres) that do not contain protein-coding genes, but instead consist of short, simple, noncoding DNA sequences that are repeated many times. Therefore, although a small amount of telomeric DNA fails to replicate each time the cell divides, no essential genetic information is lost.

Telomeric DNA can be lengthened by a special DNA replication enzyme known as **telomerase.** This enzyme, which was discovered in 1984 by Carol W. Greider and Elizabeth H. Blackburn, is typically present in cells that can divide an unlimited number of times, including unicellular organisms and many types of cancer cells.

In animals such as humans, active telomerase is usually present in germ line cells (the cells that give rise to eggs and sperm) but not in the normal cells of the body, known as somatic cells. There has been much speculation that this lack of telomerase activity could contribute to the phenomenon known as **cellular aging.** Cellular aging has been analyzed since the 1960s, following the pioneering studies of Leonard Hayflick, who showed that normal human somatic cells grown in culture lose their ability to divide after a limited number of cell divisions. Furthermore, the number of cell divisions is determined by the age of the individual from whom the cells were taken. Cells from a 70 year old can divide only 20 to 30 times, as compared with those from an infant, which can divide 80 to 90 times.

Although correlations between the ability of cells to undergo unlimited divisions and the presence of telomerase activity have been repeatedly noted, this connection remained controversial. Proof of a causal relationship was lacking until Andrea G. Bodnar and her colleagues at the Geron Corporation teamed up with researchers from the University of Texas Southwestern Medical Center to conduct a direct test.* Using the techniques of recombinant DNA technology (see Chapter 14), they infected cultured normal human cells with a virus that carried the genetic information coding for the catalytic subunit of telomerase. Not only did the cells produce active telomerase, which elongated the telomeres significantly, but the cells continued to divide long past the point at which cell divisions would normally cease. These findings have revived interest in telomeres, particularly with respect to their apparent dual roles in aging and cancer, and are leading to more hypotheses that can be experimentally tested.

*Bodnar, A.G., M. Ouellette, M. Frolkis, S.E. Holt, C.-P. Chiu, G.B. Morin, C.B. Harley, J.W. Shay, S. Lichsteiner, and W.E. Wright. "Extension of Life-Span by Introduction of Telomerase into Normal Human Cells." *Science,* Vol. 279, 16 Jan. 1998.

SUMMARY WITH KEY TERMS

I. Many early geneticists thought that genes were made of proteins. Proteins were known to be complex and variable, whereas nucleic acids were thought of as rather simple molecules with a limited ability to store information.

   A. Garrod's work on inborn errors of metabolism and that of Beadle and Tatum with *Neurospora* mutants suggested that each protein is specified by a single gene.

   B. Several lines of evidence supported the idea that DNA is the genetic material.

      1. In **transformation** experiments, the DNA of one strain of bacteria can endow related bacteria with new genetic characteristics.

      2. When a bacterial cell becomes infected with a virus, only the DNA from the virus enters the cell; this DNA is sufficient for the virus to reproduce and form new virus particles.

   C. Watson and Crick's studies on the structure of DNA demonstrated how information can be stored in the molecule's structure and how DNA molecules can serve as templates for their own duplication.

II. DNA is a very regular polymer of **nucleotides.**

   A. Each nucleotide subunit contains a nitrogenous base, which may be one of the **purines (adenine** or **guanine)** or one of the **pyrimidines (thymine** or **cytosine).** Each base is covalently linked to a five-carbon sugar, **deoxyribose,** which is covalently bonded to a phosphate group.

B. The backbone of each single DNA chain is formed by alternating sugar and phosphate groups, joined by covalent **phosphodiester linkages.** Each phosphate group is attached to the 5′ carbon of one deoxyribose and to the 3′ carbon of the neighboring deoxyribose.

C. Each DNA molecule is composed of two polynucleotide chains that associate as a **double helix.** The two chains are **antiparallel** (meaning they run in opposite directions); at each end of the DNA molecule one chain has an exposed 5′ deoxyribose carbon (the **5′ end**) and the other has an exposed 3′ deoxyribose carbon (the **3′ end**).

D. The two chains of the helix are held together by hydrogen bonding between specific base pairs. Adenine (A) forms two hydrogen bonds with thymine (T); guanine (G) forms three hydrogen bonds with cytosine (C).

    1. **Complementary base-pairing** between A and T and between G and C is the basis of Chargaff's rules, which state that A=T and G=C.

    2. Because the two strands of DNA are held together by complementary base-pairing, it is possible to predict the base sequence of one strand if one knows the base sequence of the other strand.

III. During **DNA replication,** the two strands of the double helix unwind. Each strand serves as a template for the formation of a new complementary strand.

A. DNA replication is **semiconservative;** that is, each daughter double helix contains one strand from the parent molecule and one newly synthesized strand.

B. DNA replication is a complex process requiring a number of different enzymes.

    1. The enzyme that adds new deoxyribonucleotides to a growing DNA strand is a **DNA polymerase.**

    2. Additional enzymes and other proteins are required to unwind and stabilize the separated DNA helix and to form **RNA primers.** **Topoisomerases** prevent tangling and knotting, and **DNA ligase** links together fragments of newly synthesized DNA.

C. DNA synthesis always proceeds in a 5′ → 3′ direction. This requires that one DNA strand (the **lagging strand**) be synthesized discontinuously, as short **Okazaki fragments.** The opposite strand (the **leading strand**) is synthesized continuously.

D. DNA replication is bidirectional, starting at the **origin of replication** and proceeding in both directions from that point. A eukaryotic chromosome may have multiple origins of replication and may be replicating at many points along its length at any one time.

E. Eukaryotic chromosome ends, known as **telomeres,** shorten slightly with each cell cycle, but can be extended by the **telomerase** enzyme. The absence of telomerase activity in certain cells may be a cause of **cellular aging.**

IV. DNA is organized in a cell.

A. Prokaryotic cells usually have circular DNA molecules.

B. Eukaryotic chromosomes have several levels of organization.

    1. The DNA is associated with **histones** (basic proteins) to form **nucleosomes,** each of which consists of a histone bead with DNA wrapped around it.

    2. The nucleosomes are organized into large coiled loops held together by nonhistone **scaffolding proteins.**

    3. DNA molecules are much longer than the nuclei or the cells that contain them. The organization of DNA into chromosomes allows the DNA to be accurately replicated and segregated into daughter cells without tangling.

POST-TEST

1. Which of the following inspired Avery and his coworkers to do the experiments that demonstrated that the transforming principle in bacteria is DNA? (a) the fact that A=T and G=C (b) Beadle and Tatum's work on biochemical pathways in *Neurospora* (c) Meselson and Stahl's studies on DNA replication in *E. coli* (d) Griffith's experiments on smooth and rough strains of pneumococcus (e) Hershey and Chase's experiments on the reproduction of bacteriophages

2. Which of the following was commonly thought to be true (by scientists) at the time that Beadle and Tatum began their work on *Neurospora?* (a) enzymes are proteins (b) DNA specifies the structure of proteins (c) DNA replication is semiconservative (d) DNA is a double helix (e) the number of purines in DNA is equal to the number of pyrimidines

3. The statement "DNA replicates by a semiconservative mechanism" means that (a) only one DNA strand is copied (b) first one DNA strand is copied, and then the other strand is copied (c) the two strands of a double helix have identical base sequences (d) some portions of a single DNA strand are old, and other portions are newly synthesized (e) each double helix consists of one old and one newly synthesized strand

4. Multiple origins of replication (a) speed up replication of eukaryotic chromosomes (b) allow the lagging strands and leading strands to be synthesized at different replication forks (c) help to relieve strain as the double helix is unwound (d) prevent mutations (e) are necessary for the replication of a circular DNA molecule in bacteria

5. Topoisomerases (a) synthesize DNA (b) synthesize RNA primers (c) join Okazaki fragments (d) break and rejoin DNA to resolve knots that have formed (e) prevent single DNA strands from joining to form a double helix

6. A phosphate in DNA is (a) hydrogen bonded to a base (b) covalently linked to two bases (c) covalently linked to two deoxyriboses (d) hydrogen-bonded to two additional phosphates (e) covalently linked to a base, a deoxyribose, and another phosphate

7. Which of the following depicts the relative arrangement of the complementary strands of a DNA double helix?
(a) 5′—5′ (b) 3′—5′ (c) 3′—3′ (d) 5′—5′ (e) 3′—5′
    3′—3′   3′—5′   3′—3′   5′—5′   5′—3′

8. A lagging strand is formed by (a) joining primers (b) joining Okazaki fragments (c) joining leading strands (d) breaking up a leading strand (e) joining primers, Okazaki fragments, and leading strands

9. The immediate source of energy for DNA replication is (a) the hydrolysis of nucleoside triphosphates (b) the oxidation of NADPH (c) the hydrolysis of ATP (d) electron transport (e) the breaking of hydrogen bonds

10. A nucleosome consists of (a) DNA and scaffolding proteins (b) scaffolding proteins and histones (c) DNA and histones (d) DNA, histones, and scaffolding proteins (e) histones only

REVIEW QUESTIONS

1. How did the experiments of Avery and coworkers point to DNA as the essential genetic material? Did the Hershey-Chase experiment establish that DNA is the genetic material in all organisms? Did either of these experiments demonstrate how DNA could function as the chemical basis of genes?
2. Sketch the structure of a single strand of DNA. What types of subunits make up the chain? How are they linked?
3. Describe the structure of double-stranded DNA as determined by Watson and Crick.
4. Does a single strand of DNA obey Chargaff's rules? How do Chargaff's rules relate to the structure of DNA?
5. What are some of the mechanical problems encountered in DNA replication? How are they dealt with by the cell?
6. Why is DNA replication continuous for one strand but discontinuous for the other?
7. Compare the structures of a bacterial DNA molecule and a eukaryotic chromosome. What effects do these differences have on replication?
8. Describe how both prokaryotic and eukaryotic cells cope with the large discrepancy between the length of their DNA molecules and the size of the cell or nucleus.

YOU MAKE THE CONNECTION

1. What characteristics must a molecule have if it is to serve as genetic material? What important features of the structure of DNA are consistent with its role as the chemical basis of heredity?
2. In Chapter 10 we discussed the fact that heritable variation is essential for the study of inheritance. What role did mutant strains play in Beadle and Tatum's development of the one gene, one enzyme hypothesis?

RECOMMENDED READINGS

Greider, C.W. and E.H. Blackburn. "Telomeres, Telomerase, and Cancer." *Scientific American,* Vol. 274, No. 2, Feb. 1996. The discoverers of telomerase discuss the possible roles of telomere shortening and lengthening in cancer and aging.

Judson, H.F. *The Eighth Day of Creation: Makers of the Revolution in Biology.* Simon & Schuster, New York, 1979. A beautifully written and fascinating account of the early history of molecular biology.

Pool, R. "Dr. Tinkertoy." *Discover,* Feb. 1997. Chemist Ned Seeman uses short DNA molecules as construction components that join together by complementary base-pairing.

Rennie, J. "DNA's New Twists." *Scientific American,* Mar. 1993. A fascinating summary of novel studies in genetics that challenge previous ideas and open the way for a new understanding of the role of DNA in evolution and in genetic diseases.

Rebek, J. Jr. "Synthetic Self-Replicating Molecules." *Scientific American,* Jul. 1994. Experiments on certain molecules fabricated in the laboratory demonstrate that self-duplication is not unique to DNA.

Watson, J.D. *The Double Helix.* Atheneum, New York, 1968. Watson's view of the discovery of the structure of DNA. Somewhat controversial, but entertaining and insightful reading.

Watson, J.D., and F.H.C. Crick. "Molecular Structure of Nucleic Acids: A Structure for Deoxyribose Nucleic Acid." *Nature* Vol. 171, 1953. Watson and Crick's original report—a simple, clearly written two-page paper that shook the scientific world.

- Visit our website at **http://www.saunderscollege.com/lifesci/titles.html** and click on Solomon/Berg/Martin Biology for links to chapter-related resources on the World Wide Web.

CHAPTER 12

RNA and Protein Synthesis: The Expression of Genetic Information

In Chapter 11 we saw that a base sequence in DNA is replicated by a cell so that it can be passed accurately to its descendents. The basic features of the DNA double helix originally described by Watson and Crick are now known to be the same in all cells studied to date, from bacteria to humans.

By the mid-1950s it became evident that the genetic information in DNA contains the code for all the proteins needed by the cell. However, more than a decade of intense investigation by many scientists was required before a fundamental understanding of how cells are able to convert DNA information into amino acid sequences of proteins could be developed. Much of that understanding came from studying the functions of bacterial genes. After the discovery of the structure of DNA, prokaryotic cells quickly became the organisms of choice for these investigations because they could be grown quickly and easily and because they seemed to contain only the minimal amount of DNA needed for growth and reproduction. The validity, as well as the utility, of this approach has been repeatedly confirmed, as researchers have learned that all organisms share fundamental genetic similarities, which include sharing the same genetic code.

In this chapter we examine at the molecular level how DNA is able to affect the phenotype of the organism through a process known as **gene expression.** Gene expression is a complex series of events by which the information contained in the sequence of bases in DNA is decoded and used to specify the makeup of the proteins in the cell. The proteins produced then affect the phenotype in some way; these effects can range from obvious visible physical traits to subtle changes observable only at the biochemical level. The TEM illustrates **transcription,** the first major step of gene expression. In transcription, RNA molecules *(lateral strands)* are synthesized as complementary copies of a DNA template *(central axis)*. A second major step of gene expression is **translation,** or protein synthesis. In Chapter 13 we consider some of the ways the entire process is controlled.

We first focus our attention on gene expression in prokaryotic cells because these cells are best understood. We then extend our discussion to include eukaryotic cells. Our understanding of these cells is improving rapidly as a result of groundwork laid by study of the simpler bacterial systems.

(Professor Oscar Miller/Science Photo Library/Photo Researchers, Inc.)

AFTER YOU HAVE STUDIED THIS CHAPTER YOU SHOULD BE ABLE TO

1. Outline the flow of genetic information in cells, from DNA to protein.
2. Compare the structures of DNA and RNA and explain how the structure of each is related to its role in the cell.
3. Compare the processes of transcription and replication, identifying both similarities and differences.
4. Identify the features of tRNA that are important in decoding genetic information and converting it into "protein language."
5. Explain how the ribosome functions in protein synthesis.
6. Diagram the processes of initiation, chain elongation, and chain termination in protein synthesis.

7. Compare eukaryotic and prokaryotic mRNAs and explain the functional significance of their structural differences.
8. Analyze the differences in translation in prokaryotic and eukaryotic cells.
9. Explain why the genetic code is said to be redundant and virtually universal. Discuss how these features may reflect the evolutionary history of the code.
10. Give examples of the different classes of mutations that affect the base sequence of DNA and demonstrate the effects that each has on the protein produced.

DNA IS TRANSCRIBED TO FORM RNA; RNA IS TRANSLATED TO FORM PROTEIN

Although the sequence of bases in DNA determines the sequence of amino acids in proteins, the information in DNA is not used directly. Instead, a related nucleic acid, **RNA,** or **ribonucleic acid,** serves as an intermediary between DNA and protein (Fig. 12–1). Like DNA, RNA is a polymer of nucleotides, but it has some important differences.

RNA is usually single-stranded, although internal regions of some RNAs may have complementary sequences that allow the strand to fold back and pair to form short, double-stranded segments. As shown in Figure 12–1, the sugar in RNA is **ribose,** which is similar to deoxyribose of DNA, but with an extra hydroxyl group. The base **uracil** substitutes for thymine and, like thymine, is a pyrimidine that can form two hydrogen bonds with adenine. Hence, uracil and adenine are a complementary pair.

When a protein-coding gene is expressed, an RNA copy is made of the information in the DNA (Fig. 12–2). This process resembles DNA replication in that the sequence of bases in the RNA strand is determined by complementary base-pairing with one of the DNA strands. Because RNA synthesis involves making a copy of information in one kind of nucleic acid (DNA) in the form of another nucleic acid (RNA), we refer to this process as **transcription** ("copying"). The RNA that carries the specific information for making a protein is called **messenger RNA,** or **mRNA.**

In the second stage of gene expression, the transcribed information in the mRNA is used to specify the amino acid sequence of a protein. This process is called **translation** because it involves conversion of the "nucleic acid language" in the mRNA molecule into the "amino acid language" of the protein.

The protein-coding information in mRNA is contained in its sequence of bases. A sequence of three consecutive bases,

Figure 12–1 **Nucleotide structure of RNA.** The nucleotide subunits of RNA are joined by 5′ → 3′ phosphodiester linkages, like those found in DNA. Adenine, guanine, and cytosine are present, as in DNA, but the base uracil replaces thymine. All four nucleotide types contain the five-carbon sugar ribose, which has a hydroxyl group on its 2′ carbon atom. (Compare with the deoxyribose of DNA, which has only a hydrogen at the 2′ position.)

Figure 12-2 Overview of transcription and translation. In transcription, messenger RNA is synthesized as a complementary copy of one of the DNA strands. Messenger RNA carries genetic information in the form of sets of three bases called codons, each of which specifies one amino acid. Messenger RNA codons are translated consecutively, thus specifying the linear sequence of amino acids in the polypeptide chain.

called a **codon,** specifies one amino acid. For example, one codon that corresponds to the amino acid threonine is 5′—ACG—3′ (Table 12–1). Because each codon requires three nucleotides, the code is referred to as a **triplet code.**

Translation requires cellular machinery that can recognize and decode the mRNA codons. **Transfer RNAs (tRNAs)** are critical parts of the decoding machinery because they act as "adapters" that provide a connection between amino acids and nucleic acids. This is possible because each tRNA can (1) link with a specific amino acid and (2) recognize the appropriate mRNA codon for that particular amino acid. A particular tRNA can recognize a specific codon because it has a sequence of three bases, called the **anticodon,** that associates with the mRNA codon by complementary base-pairing. The exact anticodon that is complementary to the codon for threonine in our example is 3′—UGC—5′.

Translation requires that (1) tRNA anticodons be bonded

to the mRNA and (2) the amino acids carried by the tRNAs be linked together in the order specified by the mRNA. This is accomplished by **ribosomes** (see Chapter 4), complex organelles composed of two different subunits, each containing a number of proteins and **ribosomal RNA (rRNA).** Ribosomes attach to one end of the mRNA and travel along it, thereby allowing the tRNAs to attach sequentially to the mRNA. In this way the amino acids become properly positioned to be joined by peptide bonds in the correct sequence to form a polypeptide.

The Watson and Crick model of DNA showed it to be a linear sequence of four nucleotides. If each nucleotide were to code for a single amino acid, it would be possible to specify only 4 amino acids, not the 20 found in the vast variety of cellular proteins. Scientists saw that the DNA bases could serve as a four-letter "alphabet" and hypothesized that three-letter combinations of the four bases (4^3) would make it possible to

TABLE 12-1

TABLE 12-1 — The Genetic Code: Codons of mRNA that Specify a Given Amino Acid

First Position (5′ end)	Second Position	Third Position (3′ end)			
		U	C	A	G
U	U	UUU UUC — Phenylalanine		UUA — Leucine	UUG
	C	UCU UCC — Serine		UCA	UCG
	A	UAU UAC — Tyrosine		UAA (Stop)	UAG (Stop)
	G	UGU UGC — Cysteine		UGA (Stop)	UGG — Tryptophan
C	U	CUU CUC — Leucine		CUA	CUG
	C	CCU CCC — Proline		CCA	CCG
	A	CAU CAC — Histidine		CAA — Glutamine	CAG
	G	CGU CGC — Arginine		CGA	CGG
A	U	AUU AUC AUA — Isoleucine			AUG (start) — Methionine
	C	ACU ACC — Threonine		ACA	ACG
	A	AAU AAC — Asparagine		AAA — Lysine	AAG
	G	AGU AGC — Serine		AGA — Arginine	AGG
G	U	GUU GUC — Valine		GUA	GUG
	C	GCU GCC — Alanine		GCA	GCG
	A	GAU GAC — Aspartic acid		GAA — Glutamic acid	GAG
	G	GGU GGC — Glycine		GGA	GGG

form a total of 64 "words," more than sufficient to specify all the naturally occurring amino acids. By 1967 the "cracking" of the genetic code was completed, verifying the existence of the three-base triplet code that is common to all organisms.

TRANSCRIPTION IS THE SYNTHESIS OF RNA FROM A DNA TEMPLATE

Three main kinds of RNA are transcribed from DNA: ribosomal RNA (rRNA) and transfer RNA (tRNA), as well as messenger RNA (mRNA). Most RNA is synthesized by **DNA-dependent RNA polymerases,** enzymes that are present in all cells. These enzymes require DNA as a template and have many similarities to the DNA polymerases. Like DNA polymerases, they synthesize nucleic acids in a 5′ to 3′ direction. They use nucleotides with three phosphate groups as substrates, removing two of the phosphates as the subunits are covalently linked to the 3′ end of the RNA (Fig. 12–3). Like DNA replication and the hydrolysis of ATP, these reactions are strongly exergonic (see Chapter 6).

Whenever nucleic acid molecules associate by complementary base-pairing, the two strands are antiparallel. Just as the two paired strands of DNA are antiparallel (see Chapter 11), the transcribed strand of the DNA and the complementary RNA strand are also antiparallel.

Messenger RNA contains base sequences that code for protein

An RNA polymerase begins transcription after it binds to a DNA sequence known as a **promoter;** the promoter is not transcribed. Unlike DNA synthesis, RNA synthesis does not require a primer. However, transcription requires a number of proteins in addition to RNA polymerase; these are discussed in Chapter 13. As illustrated in Figure 12–3, the first nucleotide at the 5′ end of a new mRNA chain retains its triphosphate group, but as each additional nucleotide is incorporated at the 3′ end of the growing RNA molecule, two of its phosphates are removed in an exergonic reaction that leaves the remaining phosphate to become part of the sugar-phosphate backbone (as in DNA). The last nucleotide to be incorporated has an exposed 3′ hydroxyl group. The termination of transcription, like its initiation, is controlled by a set of specific base sequences. These sequences at the end of the gene act as "stop" signals for the RNA polymerase.

It is conventional to refer to a sequence of bases in a gene or the mRNA sequence transcribed from it as *upstream* or *downstream* of some reference point. **Upstream** means toward the 5′ end of the mRNA sequence or the 3′ end of the transcribed DNA strand. **Downstream** means toward the 3′ end of the RNA or the 5′ end of the transcribed DNA strand.

Upstream

5′—A—T—G—A—C—T—3′ (nontranscribed DNA strand)

3′—T—A—C—T—G—A—5′ (transcribed DNA strand)

Downstream

Direction of transcription →

Triphosphate 5′—A—U—G—A—C—U—3′ OH (RNA)

In the bacterium *Escherichia coli,* transcription of a gene is initiated when RNA polymerase (with the help of another protein) recognizes a specific promoter sequence of bases upstream from the protein-coding sequence. Different genes may have slightly different promoter sequences, so the cell can direct which genes are transcribed at any one time (Chapter 13).

Figure 12–3 **Transcription.** The unwinding of the DNA double helix allows bases of incoming nucleotides with three phosphates to pair with complementary bases on the DNA template strand *(right)*. RNA polymerase cleaves two phosphates from each nucleotide *(not shown)* and covalently links the remaining phosphate to the 3′ end of the growing RNA chain. Thus, RNA, like DNA, is synthesized in a 5′ → 3′ direction.

Bacterial promoters are usually about 40 bases long and are positioned in the DNA just upstream of the point at which transcription will begin. Once the polymerase has recognized the correct promoter, it unwinds the helix and begins transcription.

Usually only one of the strands in a protein-coding region of DNA is transcribed. This strand, referred to as the *transcribed strand* or the *template strand,* is complementary to the mRNA (Fig. 12–4*a*). For example, consider a segment of DNA that contains the following DNA base sequence in the template strand:

3′—TAACGGTCT—5′

If the complementary DNA strand

5′—ATTGCCAGA—3′

were to be transcribed, a message specifying an entirely different (and generally nonfunctional) protein would be produced. However, the fact that only one strand is transcribed does not mean that the same strand is always the template throughout the length of a chromosome-sized DNA molecule. Instead, a particular strand may serve as the transcribed strand for some genes and the nontranscribed strand for others (Fig. 12–4*b*).

Figure 12–4 **Synthesis of mRNA.** *(a)* The mRNA is synthesized in a 5′ → 3′ direction from the template strand of the DNA molecule. Transcription starts downstream from a DNA promoter sequence, to which the RNA polymerase initially binds. Termination sequences, found downstream from the protein-coding sequences, signal the RNA polymerase to stop transcription and be released from the DNA. *(b)* Usually only one of the two strands is transcribed for a given gene, but the opposite strand may be transcribed for a neighboring gene. Each transcript starts at its own promoter.

Messenger RNA contains additional base sequences that do not directly code for protein

A completed bacterial RNA contains more than the nucleotide sequence that codes for a protein. RNA polymerase starts transcription of a gene well upstream of the protein-coding sequences. As a result, the mRNA has a noncoding **leader sequence** at its 5′ end. The leader contains recognition signals for ribosome binding, which allow the ribosomes to be properly positioned to translate the message. The leader sequence is followed by the **coding sequence,** which contains the actual messages for the proteins. In bacterial cells, one or more proteins may be encoded by a single mRNA molecule (see Chapter 13). At the end of each coding sequence is a special **termination,** or **stop, codon.** The stop codons—UAA, UGA, or UAG (see Table 12–1)—do not code for amino acids, but

instead specify the end of the protein. These are followed by noncoding 3′ trailing sequences, which can vary in length (Fig. 12–5).

DURING TRANSLATION, THE NUCLEIC ACID MESSAGE IS DECODED

Translation, or protein synthesis, adds another level of complexity to the process of information transfer because it involves the conversion of the four-base nucleic acid code to the 20-amino acid alphabet of proteins. The structural differences between a polynucleotide chain and a polypeptide chain are so great that no simple way exists for amino acids to interact directly with an mRNA molecule to make a protein. Translation therefore requires the coordinated functioning of more

Figure 12–5   Bacterial mRNA. This figure compares a bacterial mRNA with the region of DNA from which it was transcribed. RNA polymerase recognizes, but does not transcribe, promoter sequences in the DNA. Initiation of RNA synthesis occurs five to eight bases downstream from the promoter. Ribosome-recognition sites are located in the 5′ mRNA leader sequences. Protein-coding sequences begin at a start codon, which follows the leader sequences, and end at a downstream termination codon near the 3′ end of the molecule. Noncoding trailing sequences, which can vary in length, follow the protein-coding sequences.

than 100 kinds of macromolecules, including the protein and RNA components of the ribosomes, mRNA, and amino acids linked to tRNAs.

An amino acid is attached to transfer RNA before becoming incorporated into a polypeptide

Amino acids are joined together by peptide bonds to form proteins (see Chapter 3). This joining involves linking the amino and carboxyl groups of adjacent amino acids. Peptide bond formation is only one aspect of the translation process, however, because the amino acids must be linked in the correct sequence specified by the codons in the mRNA.

Francis Crick recognized this problem and proposed that a molecule was needed to serve as an "adapter" in protein synthesis and bridge the gap between mRNA and proteins. Crick's adapters turned out to be transfer RNA (tRNA) molecules. DNA contains special tRNA genes that are transcribed to form the tRNAs. Amino acids are covalently linked to their respective tRNA molecules by specific enzymes called **aminoacyl-tRNA synthetases,** which use ATP as an energy source (Fig. 12–6). The resulting complexes, called **aminoacyl-tRNAs,** are able to bind to the mRNA coding sequence so as to align the amino acids in the correct order to form the polypeptide chain.

Transfer RNA molecules have specialized regions with specific functions

Although tRNA molecules are considerably smaller than mRNA or rRNA molecules, they have a complex structure. A

tRNA molecule must have several properties:

1. It must be recognized by a specific aminoacyl-tRNA synthetase that adds the correct amino acid.
2. It must have a region that serves as the attachment site for the amino acid.
3. It must be recognized by ribosomes.
4. It must have an **anticodon,** a specific complementary binding sequence for the correct mRNA codon.

The tRNAs are polynucleotide chains about 70 nucleotides long, each with a number of unique base sequences, as well as some sequences that are common to all (Fig. 12–7). Complementary base-pairing within each tRNA molecule causes it to be doubled back and folded. Three or more loops of unpaired nucleotides are formed, one of which contains the anticodon triplet. The amino acid binding site is at the 3′ end of the molecule. The carboxyl group of the amino acid is bound to the exposed 3′ hydroxyl group of the terminal nucleotide, leaving the amino group free to participate in peptide bond formation. The pattern of folding results in a constant distance between the anticodon and amino acid in all tRNAs examined, allowing for precise positioning of the amino acids during translation.

The components of the translational machinery come together at the ribosomes

The importance of ribosomes and of protein synthesis in cellular metabolism is exemplified by a rapidly growing *E. coli* cell, which contains some 15,000 ribosomes—nearly one-third of the total mass of the cell. Although prokaryotic and eu-

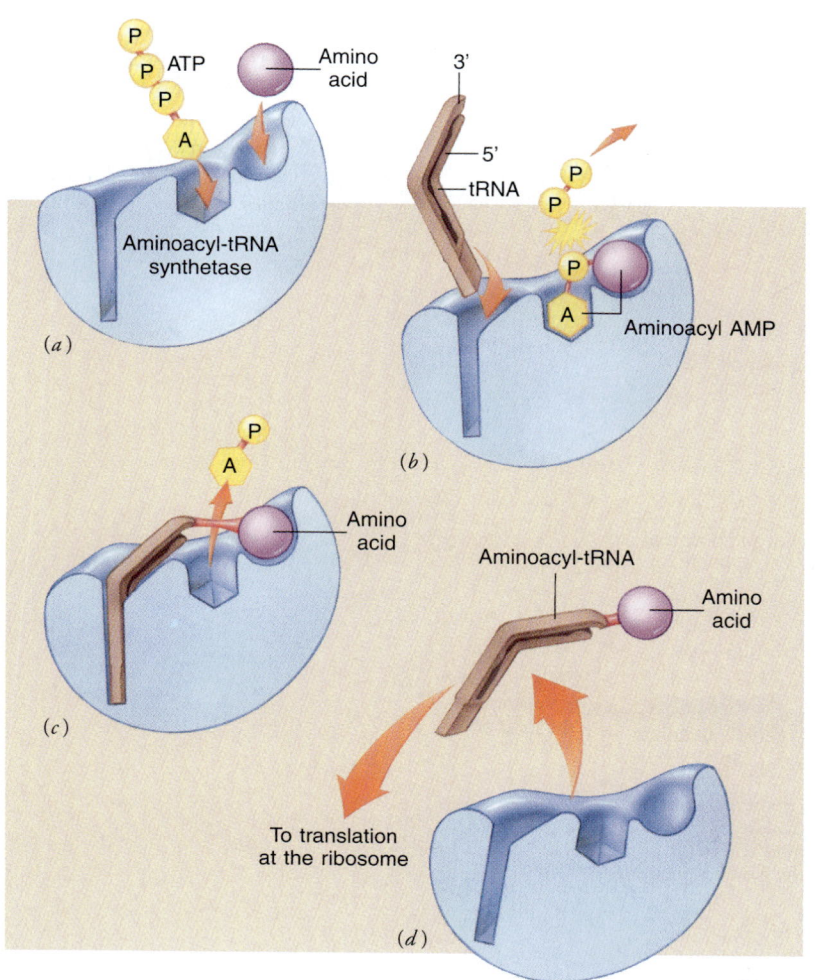

▶ **Figure 12–6** **Formation of aminoacyl-tRNA.** *(a)* Aminoacyl-tRNA synthetase catalyzes an ATP-requiring reaction *(b)* in which the carboxyl group of the amino acid becomes attached to the 3' end of the tRNA, *(c)*, and the resulting aminoacyl-tRNA complex is released *(d)*.

▼ **Figure 12–7** **Three representations of a tRNA molecule.** The genetic code is "read" by tRNA molecules, which have characteristic structures. *(a)* The three-dimensional shape of a tRNA molecule is determined by hydrogen bonds that form between complementary bases. *(b)* One loop contains the triplet anticodon; these unpaired bases can pair with a complementary mRNA triplet codon. The amino acid is attached to the terminal ribose at the 3'OH end. *(c)* This schematic diagram of an aminoacyl tRNA shows an amino acid attached to its tRNA by its carboxyl group, leaving its amino group exposed for peptide bond formation.

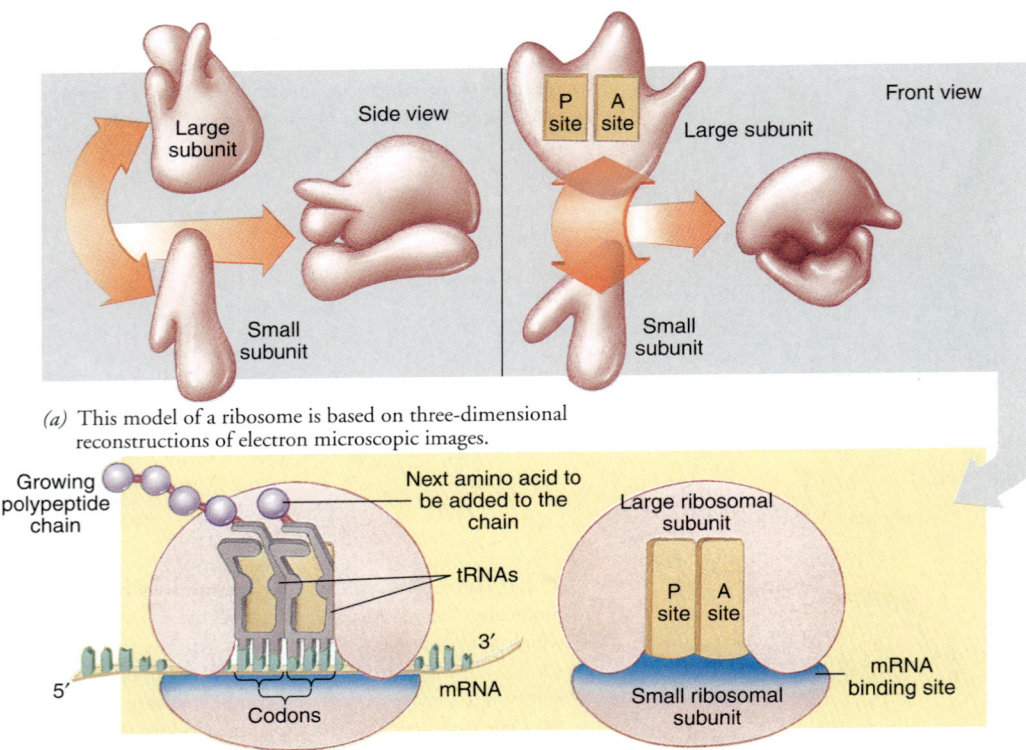

Figure 12-8 **Ribosomal structure.** A ribosome *(a)* consists of two subunits, one larger and one smaller, and *(b)* contains two binding sites for aminoacyl tRNA molecules.

(a) This model of a ribosome is based on three-dimensional reconstructions of electron microscopic images.

(b) The mRNA passes through a groove formed between the two ribosomal subunits. A ribosome contains two binding sites for tRNAs that recognize adjacent codons. The A site (aminoacyl-tRNA site) binds an aminoacyl tRNA that will be used to add an amino acid to the growing chain. The P site (peptidyl-tRNA site) binds the tRNA that is linked to the growing polypeptide chain.

karyotic ribosomes are not identical, ribosomes from all organisms are composed of two subunits made up of protein and ribosomal RNA, which is transcribed from DNA. Ribosomal RNA does not transfer specific information, but instead has structural and catalytic functions.

Each ribosomal subunit can be isolated intact in the laboratory and then separated into each of its RNA and protein constituents. For example, it has been found that in bacteria the smaller of these subunits contains 21 proteins and one ribosomal RNA molecule, and the larger contains 35 proteins and two ribosomal RNA molecules. Under certain conditions it is possible to reassemble each subunit into a functional form by adding each component in its correct order. Through this approach, together with sophisticated electron microscopic studies, it has been possible to determine the three-dimensional structure of the ribosome (Fig. 12–8*a*), as well as how it is assembled in the living cell. The large subunit contains a depression on one surface into which the small subunit fits. The mRNA fits in a groove formed between the contact surfaces of the two subunits.

Within each ribosome are two depressions, the **A** and **P** binding sites for tRNA molecules (Fig. 12–8*b*). The tRNA holding the **p**olypeptide chain occupies the P site. The A site is so named because the **a**minoacyl-tRNA delivering the next amino acid in the sequence binds at this location. Following peptide bond formation between the amino acid at the A site

and the end of the growing polypeptide chain, the tRNA (now with the entire polypeptide chain attached) moves to the P site of the ribosome, leaving the A site available for the next aminoacyl-tRNA molecule.

One of the roles of the ribosome is to hold the mRNA template, the aminoacyl-tRNA, and the growing peptide chain in the correct orientation so that the genetic code can be read and the peptide bond formed.

Translation includes initiation, elongation, and termination

For purposes of discussion, the process of protein synthesis is generally divided into three distinct stages: **initiation,** repeating cycles of **elongation,** and **termination.**

The initiation process consists of several steps and requires a number of proteins, called **initiation factors.** Initiation begins with the loading of a special **initiation tRNA** onto the small ribosomal subunit. In all organisms the codon for the initiation of protein synthesis is AUG, which codes for the amino acid methionine (Fig. 12–9).

Once the initiator tRNA is loaded on the small subunit, the initiation complex binds to the special **ribosome-recognition sequences** near the 5′ end of the mRNA; these are upstream of the coding sequences. Binding results in alignment of the anticodon of the initiator tRNA with the AUG initia-

(a)	Before translation begins, the ribosomal subunits are dissociated into small and large subunits.	Large ribosomal subunit — P site — A site — Small ribosomal subunit — mRNA binding site
(b)	An initiation complex forms, consisting of the small ribosomal subunit, mRNA, and formylated methionine tRNA.	MET — tRNA — 5′ — mRNA — 3′ — mRNA binding site
(c)	The large ribosomal subunit binds to the initiation complex. The process of peptide elongation begins with the addition of the second tRNA with its amino acid.	Amino acid — tRNA — MET — tRNA — P site — A site — 5′ — 3′

Figure 12-9 **Initiation of protein synthesis.** The small ribosomal subunit participates in the formation of an initiation complex, which then associates with the large subunit. The initiation complex shown is found in *E. coli.*

tion codon of the mRNA. The large ribosomal subunit then binds to the complex, forming the completed ribosome.

The addition of other amino acids to the growing polypeptide is called **elongation.** The initiator tRNA is bound to the P site of the ribosome, leaving the A site unoccupied so that it can be filled by the aminoacyl-tRNA specified by the next codon. Figure 12–10 outlines the events involved in elongation. The appropriate aminoacyl-tRNA binds to the A site by specific base-pairing of its anticodon with the complementary mRNA codon. This binding step requires energy, in this case supplied by GTP (guanosine triphosphate, an energy transfer molecule similar to ATP).

The amino group of the amino acid at the A site is now aligned with the carboxyl group of the preceding amino acid at the P site. Peptide bond formation then takes place between the amino group of the new amino acid and the carboxyl group of the preceding amino acid. In this process, the amino acid attached at the P site is released from its tRNA and becomes

attached to the aminoacyl-tRNA at the A site. This reaction is spontaneous (i.e., it does not require additional energy) because energy was transferred from ATP during the formation of the aminoacyl-tRNA. It does, however, require an enzyme (known as *peptidyl transferase*). Remarkably, there is evidence that this enzyme is not a protein, but an rRNA component of the large ribosomal subunit. Such an RNA catalyst is known as a **ribozyme.**

Recall from Chapter 3 that polypeptide chains have direction, or polarity. The amino acid on one end has a free amino group (the amino end), and the amino acid at the other end has a free carboxyl group (the carboxyl end). Protein synthesis always proceeds from the amino end to the carboxyl end of the growing peptide chain.

After the peptide bond is formed, the tRNA molecule in the P site is released. The growing peptide chain, which is now attached to the tRNA in the A site, is then translocated to the P site, leaving the A site open for the next aminoacyl tRNA.

Amino
acids

tRNA with an
amino acid

P A

① The polypeptide
chain is attached to
the tRNA that carries
the amino acid most
recently added to the
chain. This tRNA is in
the P site of the
ribosome.

5'   3'

P A

② A tRNA with its
specific amino acid
attached has bound
to the A site. Base
pairs have formed
between the
anticodon of tRNA
and the codon of
mRNA.

5'   3'

P A

5'   3'

③ The growing
polypeptide chain is
detached from the
tRNA molecule in the
P site and joined by
a peptide bond to the
amino acid linked to
the tRNA at the A
site.

Cytoplasm

P A

5'   3'

④ The released tRNA
joins the cytoplasmic
pool of tRNA and can
bind with another
amino acid.

P A

5'   3'

⑤ In the translocation
step the mRNA and
tRNA move, thereby
transferring the
growing polypeptide
chain to the P site.

Figure 12–10 **Elongation.** Each repetition of the elongation
process adds one amino acid to the growing polypeptide chain.

This **translocation** process requires energy, which is again supplied by GTP.

The ribosome and the message move in relation to each
other so that the mRNA codon specifying the next amino acid
in the polypeptide chain becomes positioned in the unoccupied A site. This process involves movement of the ribosome
in the 3' direction along the mRNA molecule; thus, translation of the mRNA always proceeds in a 5' to 3' direction. The
end of the mRNA molecule that is synthesized first during
transcription is also the first to be translated to form a polypeptide. Formation of each peptide bond requires only about 1/20
of a second; by repeating the elongation cycle an average-sized
protein of about 360 amino acids can be assembled in about
18 seconds.

The synthesis of the polypeptide chain is terminated by
"release factors" that recognize the termination, or stop, codon
at the end of the coding sequence. Recognition of a termination codon by the release factors causes the ribosome to dis-

(a)

0.5 μm

(b)

Inactive DNA segment

Direction of RNA synthesis

RNA polymerase

Active DNA segment

Direction of protein synthesis

Polyribosome

Growing polypeptide chains

Ribosome

Messenger RNA

Figure 12–11 **Coupled transcription and translation in bacteria.** *(a)* TEM of two strands of *E. coli* DNA, one inactive and the other actively producing mRNA. Protein synthesis begins while the mRNA is being completed, as multiple ribosomes attach to the mRNA to form a polyribosome. *(b)* Diagrammatic representation of the coupled transcription and translation processes. *(Courtesy of Dr. Barbara Hamkalo, University of California, Irvine)*

sociate into its two subunits, which can then be used to form a new initiation complex with another mRNA molecule.

A polyribosome is a complex of one mRNA and many ribosomes

In *E. coli* and other prokaryotes, transcription and translation are *coupled* (Fig. 12–11). Ribosomes can bind to the 5′ end of the growing mRNA and initiate translation long before the message is completed. As many as 15 ribosomes may be bound to a single mRNA molecule. Messenger RNA molecules bound to clusters of ribosomes are referred to as **polyribosomes,** or sometimes **polysomes.** (Polyribosomes also occur in eukaryotic cells.)

Although a number of polypeptide chains can be actively synthesized on a single messenger RNA at any one time, the half-life (the time it takes for half of the molecules to be degraded) of mRNA molecules in bacterial cells is only about 2 minutes. Usually, degradation of the 5′ end of the mRNA begins even before the first polypeptide has been synthesized. Once the ribosome recognition sequences are degraded, no more ribosomes can attach to the mRNA and initiate protein synthesis.

TRANSCRIPTION AND TRANSLATION ARE MORE COMPLEX IN EUKARYOTES THAN IN PROKARYOTES

Although the basic mechanisms of transcription and translation are quite similar in all organisms, some significant differences exist between eukaryotes and prokaryotes, particularly with regard to the characteristics of their mRNAs. Whereas bacterial mRNAs are used immediately after transcription without further processing, eukaryotic mRNA molecules undergo specific **posttranscriptional modification and processing.**

Although bacterial mRNA is translated as it is being transcribed from the DNA, eukaryotic RNA is not. Eukaryotic chromosomes are confined to the nucleus of the cell, and protein synthesis takes place in the cytoplasm. The mRNA must be transported through the nuclear envelope and into the cytoplasm before it can be translated. In addition, the original transcript must be modified in several ways (while it is still in the nucleus) before it becomes competent for transport and translation.

Modification of the eukaryotic message begins when the

"SPLIT GENES" AND EVOLUTION

Why do introns occur in most eukaryotic nuclear genes but not in the genes of most prokaryotes (or of mitochondria and chloroplasts)? How did this remarkable genetic system involving interrupted coding sequences ("split genes") evolve and why has it survived? It seems incredible that as much as 75% of the original transcript of a eukaryotic nuclear gene has to be removed to make a working message.

Scientific debate has centered around two major questions: How did "split genes" first originate? What role have they played in the evolution of organisms that have them? One idea that is central to both questions has been advanced. This is the proposal that, although proteins are synthesized as continuous linear amino acid sequences, they actually are *modular* in that they are made up of various functional regions called **domains.** For example, the active site of an enzyme might comprise one domain. A different domain might enable that enzyme to bind to a particular cellular structure, and yet another might be a site involved in allosteric regulation (see Chapter 6).

In the early 1980s, Walter Gilbert of Harvard University proposed that exons are nucleotide sequences that code for different structural and functional protein domains. This has turned out to be only partially true. Analyses of the DNA and amino acid sequences of a number of eukaryotic genes have shown that most exons are too small to code for an entire protein domain. However, a block of several exons can code for a domain.

Gilbert further postulated that new proteins with new functions can emerge rapidly when novel combinations of exons are produced by genetic recombination within intron regions of genes that code for different proteins. This hypothesis has become known as *evolution by "exon shuffling."* For example, the low-density lipoprotein (LDL) receptor protein (a protein found on the surface of human cells that binds to cholesterol transport molecules; see

Chapter 5) has a number of domains that are related to parts of several other proteins with totally different functions.

Evidence supporting evolution by exon shuffling has led some scientists, including Gilbert, to speculate that the number of basic "exon families" (and their corresponding protein domains) is actually relatively small, and that all of the diverse array of proteins found today evolved from just a few thousand domain prototypes. They argue that these were coded for by "mini-genes" (corresponding to exons) separated by "spacers" (corresponding to introns) in organisms that were ancestral to both prokaryotes and eukaryotes. According to their view, known as the *exon theory of genes,* prokaryotes subsequently lost their introns and retained only their exons.

Many scientists disagree; they contend that while exon shuffling probably has been important in the evolution of recently evolved proteins, such as the LDL receptor, these findings do not necessarily support the exon theory of genes. They consider studies of proteins thought to be of "ancient" origin (that is, present in the common ancestor of modern prokaryotes and eukaryotes) to be a better test. Sophisticated biochemical and statistical analyses of such ancient genes have so far failed to reveal any correspondence between exons and protein structure. Thus the exon theory of genes has become less attractive.

Another view is that introns first evolved in the nucleus of an early eukaryote and were propagated as mobile genetic elements, known as *transposons* (discussed later in the chapter). Regardless of how split genes originated, intron excision provides one of the many ways in which present-day eukaryotes regulate the expression of their genes (see Chapter 13). This opportunity for control, together with the fact that eukaryotic RNAs are far more stable than those of prokaryotes, may balance the energy cost of maintaining a large load of noncoding DNA.

---

growing RNA transcript is about 20 to 30 nucleotides long. At that point enzymes add a **cap** to the 5′ end of the mRNA chain. The cap is in the form of an unusual nucleotide, 7-methylguanylate, which is guanosine monophosphate with a methyl group added to one of the nitrogens in the base. Eukaryotic ribosomes cannot bind to an uncapped message.

Capping may also protect the RNA from certain types of degradation and may therefore be partially responsible for the fact that eukaryotic mRNAs are much more stable than prokaryotic mRNAs. Eukaryotic mRNAs have half-lives ranging from 30 minutes to as long as 24 hours; the average half-life of an mRNA molecule in a mammalian cell is about 10 hours (compared with 2 minutes in a bacterial cell).

A second modification of eukaryotic mRNA occurs at the 3′ end of the molecule. Near the 3′ end of a completed message there is usually a sequence of bases that serves as a signal for the addition of a "tail" with many adenines, known as a **polyadenylated** (or **poly-A**) **tail.** Within about one minute of

completion of the transcript, enzymes in the nucleus recognize this **polyadenylation signal** and cut the mRNA molecule at that site. This is followed by the enzymatic addition of a string of 100 to 250 adenine nucleotides to the 3′ end. The function of polyadenylation is not completely understood, although there is evidence that it helps in the export of the mRNA from the nucleus and may stabilize some mRNAs against degradation in the cytoplasm.

Both noncoding and coding sequences are transcribed from eukaryotic genes

One of the greatest surprises in the history of molecular biology was the finding that most eukaryotic genes have **interrupted coding sequences;** that is, there are long sequences of bases within the protein-coding sequences of the gene that do not code for amino acids in the final protein product! The noncoding regions within the gene are called **introns** (**inter-**

vening sequences), as opposed to **exons** (**ex**pressed sequences) which are parts of the protein-coding sequence. The reason for this complex structure of eukaryotic genes is a matter of on-going debate among molecular biologists (see *Making the Connection: "Split Genes" and Evolution*).

A typical eukaryotic gene may have multiple exons and introns, although the number is quite variable. For example, the β-globin gene, which produces one component of hemoglobin, contains 2 introns; the ovalbumin gene of egg white contains 7; and the gene specifying another egg-white protein, conalbumin, contains 16. In many cases the combined lengths of the introns are much greater than those of the exon se-

quences. For instance, the ovalbumin gene contains about 7700 base pairs, whereas the total of all the exon sequences is only 1859 base pairs.

When a gene that contains introns is transcribed, the entire gene is copied as a large RNA transcript called **precursor mRNA,** or **pre-mRNA.** This molecule contains both exon and intron sequences. (Note that the terms *intron* and *exon* refer to corresponding nucleotide sequences in both DNA and RNA.) For the pre-mRNA to be made into a functional message, not only must it be capped and have a poly-A tail added, but the introns must be removed and the exons spliced together to form a continuous protein-coding message (Fig. 12–12). The

Figure 12–12 **Posttranscriptional modification of eukaryotic RNA.** (*a*) A DNA sequence containing both exons and introns is transcribed by RNA polymerase to make the primary transcript, or mRNA precursor. As it is synthesized, the pre-mRNA is "capped" by the addition of a modified base to its 5' end. (*b*) A poly-A tail (50 to 200 nucleotides long) is added to the 3' end; introns are removed, and the exons are spliced together. (*c*) The mature mRNA is transported through the nuclear envelope and into the cytosol (*d*) to be translated.

splicing reactions are mediated by special base sequences within and to either side of the introns. Splicing itself can occur by several different mechanisms. In many instances it involves the association of **small nuclear ribonucleoprotein complexes (snRNPs)** which bind to the introns and catalyze the excision and splicing reactions. In some cases the RNA within the intron acts as a ribozyme, splicing itself without the use of protein enzymes.

THE GENETIC CODE IS READ AS A SERIES OF CODONS

Before the genetic code was deciphered, a number of scientists had become interested in how a genetic code might work. In 1961, Francis Crick and his coworkers concluded from a mathematical analysis that the code was based on nonoverlapping triplets of bases. They predicted that the code is read, one triplet at a time, from a fixed starting point that establishes the **reading frame.** Because there are no "commas" separating the triplets, an alteration in the reading frame would result in the incorporation of incorrect amino acids.

Experimental evidence allowing the assignment of specific triplets to specific amino acids was first obtained by Marshall Nirenberg and Heinrich Matthaei. By constructing artificial mRNA molecules with known base sequences, they were able to determine which amino acids would be incorporated into protein in purified in vitro protein synthetic systems. For example, when the synthetic mRNA polyuridylic acid (UUUU-UUUUU . . .) was added to a mixture of purified ribosomes, aminoacyl tRNAs, and essential cofactors needed to synthesize protein, only phenylalanine was incorporated into the resulting polypeptide chain. The inference that UUU is the triplet that codes for phenylalanine was inescapable. Similar experiments showed that polyadenylic acid (AAAAAAAAA . . .) codes for a polypeptide of lysine, and polycytidylic acid (CCCCC-CCCC . . .) codes for a polypeptide of proline.

By using mixed nucleotide polymers (such as a random polymer of A and C) as artificial messengers, it became possible to assign the other nucleotide triplet codons to specific amino acids. However, three of the codons, UAA, UGA, and UAG, were not found to specify any amino acid. These codons (the stop, or termination, codons mentioned earlier) are now known to be the signals that specify the end of the coding sequence for a polypeptide chain.

Taken together, these experiments led to the coding assignments of all 64 possible codons, listed in Table 12–1, which are essentially universal in living organisms (see *Making the Connection: The Genetic Code and Evolution*). Investigators were also able to demonstrate conclusively that the code is a nonoverlapping triplet code.

Remember that the genetic code we define and use is an mRNA code. The tRNA anticodon sequences as well as the DNA sequence from which the message is transcribed are complementary to the sequences shown in Table 12–1. For example, the mRNA codon for the amino acid methionine is 5'—AUG—3'. It is transcribed from the DNA base sequence 3'—TAC—5', and the corresponding tRNA anticodon is 3'—UAC—5'.

The genetic code is redundant

Given the fact that there are 64 possible codons, it is not surprising that certain amino acids are specified by more than one codon. This redundancy in the code has certain characteristic patterns. The codons CCU, CCC, CCA, and CCG are synonymous in that they all code for the amino acid proline. The only difference among the four codons involves the nucleotide at the 3' end of the triplet. Although the code may be read three nucleotides at a time, only the first two nucleotides appear to contain specific information for proline. A similar pattern can be seen for many other amino acids. Only methionine and tryptophan have single-triplet codes. All other amino acids are specified by two to six different codons.

There are 61 codons that specify amino acids. Although most cells contain only about 40 different tRNA molecules, some of these tRNAs can pair with more than one codon, so all the codons can still be used. This apparent breach of the base-pairing rules was first proposed by Francis Crick as the **wobble hypothesis.** Crick reasoned that the third nucleotide of a tRNA anticodon (which is the 5′ base of that sequence) may sometimes be capable of forming hydrogen bonds with more than one kind of third nucleotide (the 3′ base) of an mRNA codon. Investigators later established this experimentally by determining the anticodon sequences of tRNA molecules and testing their specificities in artificial systems. Some tRNA molecules can recognize as many as three separate codons specifying the same amino acid.

A GENE IS DEFINED AS A FUNCTIONAL UNIT

In Chapter 11 we traced the development of ideas regarding the nature of the gene. For a time it was useful to define a gene as a sequence of nucleotides that codes for one polypeptide chain. As we have continued to learn more about how genes work, we have revised our definition. We now know that some genes are transcribed to produce RNA molecules such as rRNA and tRNA, whereas others specify the RNA component of the small nuclear ribonucleoprotein complexes used to modify complex mRNA molecules. Studies have also shown that in eukaryotic cells a single gene may be capable of producing more than one polypeptide chain by modifications in the way the mRNA is processed (see Chapter 13).

It is perhaps most useful to define a gene in terms of its product. A **gene** can therefore be thought of as a nucleotide sequence that carries the information needed to produce a specific RNA or protein product. Thus a gene includes noncoding regulatory sequences, as well as coding sequences.

MUTATIONS ARE CHANGES IN DNA

One of the first major discoveries about genes was that they can undergo changes, called **mutations.** We now know that mutations are caused by changes in the nucleotide sequence of the DNA. However, the overall observed mutation rate is much lower than the frequency of damage to DNA, because all organisms have special systems of enzymes that can repair certain kinds of alterations in the DNA.

As explained in Chapter 11, once the DNA sequence has been changed, DNA replication copies the altered sequence just as it would copy a normal sequence, making the mutation stable over an indefinite number of generations. In most cases the mutant gene has no greater tendency than the original gene to mutate again. Mutations provide the diversity of genetic material that makes it possible to study inheritance and the molecular nature of genes. As we will see in later chapters, mutations also provide the variation necessary for evolution to occur within a given species.

Genes can be altered by mutation in a number of ways (Fig. 12–13 on p. 284). The simplest type of mutation, called a **base substitution mutation,** involves a change in only one pair of nucleotides. It is now possible to determine where a specific mutation occurs in a gene by using recombinant DNA methods to isolate the gene and determine its sequence of bases (see Chapter 14). Often these mutations result from errors in base pairing that occurred during the replication process. For example, an AT base pair might be replaced by a GC, CG, or TA pair. Such a mutation may cause the altered DNA to be transcribed as an altered mRNA. The altered mRNA may then be translated into a peptide chain with only one amino acid different from the normal sequence.

Base substitutions that result in the replacement of one amino acid by another are sometimes referred to as **missense mutations.** Missense mutations can have a wide range of effects. If the amino acid substitution occurs at or near the active site of an enzyme, the activity of the altered protein may be decreased or even destroyed. Some missense mutations involve a change in an amino acid that is not part of the active site. Others may result in the substitution of a closely related amino acid (one with very similar chemical characteristics). Such mutations may be *silent* (undetectable), at least if one simply examines its effects on the whole organism. Because silent mutations occur relatively frequently, the true number of mutations in an organism or a species is much greater than what is actually observed.

Nonsense mutations are base substitutions that convert an amino acid-specifying codon to a termination codon. A nonsense mutation usually destroys the function of the gene product; in the case of a protein-specifying gene, the part of the polypeptide chain that follows the termination codon is missing.

In **frameshift mutations,** one or two nucleotide pairs are inserted into or deleted from the molecule, causing an alteration of the *reading frame.* As a result of this shift, codons downstream of the insertion site specify an entirely new sequence of amino acids. Depending on where the insertion or deletion occurs in the gene, a number of different effects can be generated. In addition to producing an entirely new polypeptide sequence immediately after the change, frameshift mutations usually produce a stop or termination codon within a short distance of the mutation. This codon terminates the already altered polypeptide chain. A frame shift in a gene specifying an enzyme usually results in a loss of enzyme activity. If the enzyme is an essential one, the effect on the organism can be disastrous.

Other types of mutations may be due to a change in chromosome structure (see Chapters 13, 15, and 16). These changes usually have a wide range of effects because they involve large numbers of genes.

One type of mutation is caused by DNA sequences that "jump" into the middle of a gene. These movable sequences

REVERSE TRANSCRIPTION, JUMPING GENES, AND PSEUDOGENES

For several decades, one of the central premises of molecular biology was that genetic information always flows from DNA to RNA to protein. An important exception to this rule was discovered by Howard Temin in 1964 through his studies of certain viruses. Although viruses are noncellular, they contain a single type of nucleic acid and are capable of reproducing in a host cell. Temin was studying certain unusual, cancer-causing tumor viruses that have RNA, rather than DNA, as their genetic material. He found that infection of a host cell by one of these particular viruses is blocked by inhibitors of DNA synthesis and also by inhibitors of transcription. These findings suggested that DNA synthesis and transcription are required for the multiplication of RNA tumor viruses and that there must be a way for information to flow in the "reverse" direction (that is, from RNA to DNA).

Temin proposed that a **DNA provirus** is formed as an intermediary in the replication of RNA tumor viruses. This hypothesis required a new kind of enzyme that would synthesize DNA using RNA as a template. In 1970 Temin and David Baltimore dis-

covered just such an enzyme, and in 1975 they shared the Nobel Prize for their discovery. This RNA-directed DNA polymerase, also known as **reverse transcriptase,** was found in all RNA tumor viruses. (Some RNA viruses that do not produce tumors, however, are replicated directly without using a DNA intermediate.)

After an RNA tumor virus enters the host cell, the viral reverse transcriptase synthesizes a DNA strand that is complementary to the viral RNA. Next, a complementary DNA strand is synthesized, thus completing the double-stranded DNA provirus, which is then integrated into the host cell's DNA. The provirus DNA is transcribed, and the resulting viral mRNA is translated to form specific viral proteins. Additional viral RNA molecules are produced and then incorporated into mature virus particles enclosed by protein coats. Because of their reversal of the usual direction of information flow, such viruses have become known as **retroviruses** (Fig. A). The AIDS virus (HIV-1) is the most widely known retrovirus.

Until a few years ago, reverse transcription was thought to be associated only

with retroviruses. Evidence now suggests that reverse transcription may be quite common, which may partially explain such curious phenomena as "jumping" genes and pseudogenes. Jumping genes, now known as **mobile genetic elements, transposable elements,** or **transposons,** were discovered in maize (corn) by Barbara McClintock in the 1950s. She observed that certain genes appeared to be "turned off" and "turned on" spontaneously. She deduced that the mechanism involved a gene that moved from one region of a chromosome to another, where it would either activate or inactivate genes in that vicinity. It was not until the development of recombinant DNA methods (see Chapter 14) and the discovery of jumping genes in a wide variety of organisms that this phenomenon began to be understood. In recognition of her insightful findings, McClintock was awarded a Nobel Prize in 1983.

Transposons are segments of DNA that range from a few hundred to several thousand bases. The elements themselves seem to require a special **transposase** enzyme in order to be incorporated into a new location within the chromosome. The longer elements may contain other genes that "go along for the ride."

Many transposable elements have been found to have similarities to retroviruses. Their DNA has unusual base sequences at each end, and their genes are remarkably similar, especially those that code for the proteins required for reverse transcription and integration into the chromosome.

Experiments in Gerald Fink's laboratory have provided evidence that reverse transcriptase is involved in the mechanism by which some transposons move. In these cases the DNA sequence itself does not jump from one location to another; instead, the *information moves through an RNA intermediate* (Fig. B). Fink and his colleagues used recombinant DNA methods (see Chapter 14) to insert an intron into the

Figure A   Infection cycle of an RNA tumor virus.

of DNA, called **transposons,** not only disrupt the functions of some genes but under some conditions also activate previously inactive genes (see *Focus On: Reverse Transcription, Jumping Genes, and Pseudogenes*).

All the mutations discussed so far can occur infrequently but spontaneously, as a consequence either of mistakes in DNA replication or defects in the mitotic or meiotic separation of chromosomes. Some regions of DNA, known as mutational **hot spots,** are much more likely than others to undergo mutation. An example would be a short stretch of repeated nu-

DNA sequence of a yeast transposon as a way of identifying it. They then set up conditions that allowed them to recover and analyze the transposed sequence once it had "jumped." When the transposed sequence appeared at a new location, the intron had been removed, just as introns are removed during the processing of normal mRNA molecules.

Because RNA processing enzymes are known only to splice RNA, it appeared that the transposed DNA sequence had been produced from a processed RNA copy of the original DNA, rather than from the DNA itself. This would have required the RNA sequence to be converted back to DNA by reverse transcriptase activity within the yeast cells.

Other evidence of nonviral reverse transcriptase activities in cells comes from analyses of **pseudogenes,** which closely resemble certain types of normal genes in mammalian cells. Pseudogenes are DNA sequences that are almost identical to those of normal genes, except that they are riddled with mutations that prevent them from functioning in normal protein synthesis. Many pseudogenes resemble DNA copies of mRNA, for where a normal gene would contain one or more introns, these pseudogenes do not. In addition, many pseudogene DNA sequences end with long poly-A tails. One hypothesis concerning the origin of pseudogenes is that some of them may be derived from the processed mRNAs of normal genes. Reverse transcriptase may have synthesized DNA copies, which were reinserted into the chromosome. Because they lack promoter sequences, they are not expressed and simply act as excess baggage, silently accumulating mutations. It is estimated that there may be hundreds or thousands of such sequences in normal human DNA.

Figure B Transposon movement through an RNA intermediate.

cleotides, which can cause DNA polymerase to "slip."

Mutations in certain genes can increase the overall mutation rate. For example, a mutation in a gene coding for DNA polymerase might make DNA replication less precise, or a mutation in a gene coding for a repair enzyme might cause more mutations to persist rather than to be repaired.

Not all mutations occur spontaneously; many of the types of mutations discussed above can also be caused by agents known as **mutagens.** Among these are various types of radiation, including x rays, gamma rays, cosmic rays, and ultravio-

Figure 12–13 Mutations. Missense and nonsense mutations are types of base-substitutions. A missense mutation results in a protein of normal length, but with an amino acid substitution. A nonsense mutation, caused by conversion of an amino acid-specifying codon to a termination codon, results in the production of a truncated protein, which is usually not functional. A frameshift mutation, which results from the insertion or deletion of one or two bases, causes the base sequence following the mutation to shift to a new reading frame. A frame shift may produce a termination codon downstream from the mutation (which would have the same effect as a non-sense mutation caused by base substitution), or it may produce an entirely new amino acid sequence.

let rays. Some chemical mutagens react with and modify specific bases in the DNA, leading to mistakes in complementary base pairing when the DNA molecule is replicated. Other mutagens are inserted into the DNA molecule and change the normal reading frame during replication.

Nevertheless, some new mutations do persist. In fact, each of us is very likely to have some mutant gene that was not present in either of our parents. Although some of these mutations can produce an altered phenotype, most are not noticeable because they are recessive.

Mutations that occur in the cells of the body (somatic cells) are not passed on to the offspring. However, these mutations are of concern because there is a close relationship between somatic mutations and cancer. Many mutagens are also **carcinogens,** agents that produce cancer.

SUMMARY WITH KEY TERMS

I. The mechanism by which information encoded in DNA is used to specify the sequences of amino acids in proteins involves two processes: transcription and translation.
  A. During **transcription,** an RNA molecule that is complementary to the transcribed or template DNA strand is synthesized. **Messenger RNA (mRNA)** molecules contain information that specifies the amino acid sequences of polypeptide chains.
  B. During **translation,** a polypeptide chain specified by the mRNA is synthesized.
    1. Each triplet (three-base sequence) in the mRNA constitutes a **codon,** which specifies one amino acid in the polypeptide chain.
    2. Translation requires tRNAs and complex machinery, including ribosomes.

II. Messenger RNA is synthesized by **DNA-dependent RNA polymerase** enzymes.
  A. RNA is formed from nucleotide subunits, each of which contains the sugar ribose, a base (uracil, adenine, guanine, or cytosine), and three phosphates.
  B. RNA polymerase initially binds to a special DNA sequence called the **promoter** region.
  C. Like DNA, RNA subunits are covalently joined by a 5′—3′ linkage to form an alternating sugar-phosphate backbone. The same base-pairing rules are followed as in DNA replication, except that uracil is substituted for thymine.
  D. RNA synthesis proceeds in a 5′ → 3′ direction, which means that the template DNA strand is "read" in a 3′ → 5′ direction.

III. **Transfer RNAs (tRNAs)** are the "decoding" molecules in the translation process.
  A. Each tRNA molecule is specific for only one amino acid.
    1. One part of the molecule contains a three-base **anticodon,** which is complementary to a codon of the mRNA.
    2. Attached to one end of the tRNA molecule is the amino acid specified by the complementary mRNA codon.
  B. Amino acids are covalently bound to tRNA by **aminoacyl-tRNA synthetase** enzymes.

IV. **Ribosomes** bring together all of the mechanical machinery necessary for translation. They couple the tRNAs to their proper codons on the mRNA, facilitate the formation of peptide bonds between amino acids, and translocate the mRNA so that the next codon can be read.
  A. Each ribosome is made of a large and a small subunit; each subunit contains **ribosomal RNA (rRNA)** and a large number of proteins.
  B. **Initiation,** the first stage of translation, includes the binding of the small ribosomal subunit protein, plus initiation factors and the initiation tRNA, to the 5′ region of the mRNA, followed by binding of the large ribosomal subunit.
  C. **Elongation** is a cyclic process in which amino acids are added one by one to the growing polypeptide chain.
    1. Elongation proceeds in a 5′ → 3′ direction along the mRNA.
    2. The polypeptide chain grows from its amino end to its carboxyl end.
  D. **Termination,** the final stage of translation, occurs when the ribosome reaches one of three special **termination,** or **stop, codons,** which triggers release of the completed polypeptide chain.

V. In bacterial cells, transcription and translation are coupled. Translation of the mRNA molecule usually begins before the 3′ end of the transcript is completed. A single mRNA molecule can be translated by groups of ribosomes called **polyribosomes.**

VI. The basic features of transcription and translation are the same in prokaryotic and eukaryotic cells, but eukaryotic genes and their mRNA molecules are more complex than those of bacteria.
  A. After transcription, eukaryotic mRNA molecules are **capped** at the 5′ end with a modified guanosine triphosphate. Many also have a **tail** of poly-A nucleotides added at the 3′ end. These modifications appear to protect the molecules from degradation, giving them longer lifetimes than bacterial mRNA.
  B. In many eukaryotic genes the coding regions, called **exons,** are interrupted by noncoding regions, called **introns.** Both introns and exons are transcribed, but the introns are later removed from the mRNA precursor, and the exons are spliced together to produce a continuous protein-coding sequence.

VII. The genetic code is defined at the mRNA level. There are 61 codons that code for amino acids, plus three codons that serve as stop signals.
  A. The start signal for all proteins is the codon AUG, which also specifies the amino acid methionine.
  B. The genetic code is virtually universal, strongly suggesting that all organisms are descended from a common ancestor. The only exceptions to the standard code are a few minor variations.
  C. The code is said to be redundant because some amino acids are specified by more than one codon.
  D. The genetic code is read from mRNA as a series of nonoverlapping triplets that specify a single sequence of amino acids.

VIII. A **gene** can be defined as a sequence of nucleotides (plus closely associated regulatory sequences) that can be transcribed to yield a specific protein or RNA product.

IX. Types of **mutations** range from disruption of the structure of a chromosome to a change in only a single pair of nucleotide bases.
  A. A **base substitution** can destroy the function of a protein if a codon is altered such that it specifies a different amino acid (**missense mutation**) or becomes a termination codon (**nonsense mutation**). A base substitution has minimal effects if the amino acid is not altered or if the codon is changed to specify a similar amino acid.
  B. Insertion or deletion of one or two base pairs in a gene invariably destroys the function of that protein because it results in a **frameshift mutation** that changes the codon sequences downstream from the mutation.
  C. Mutations can be produced by errors in DNA replication, by physical agents such as x rays or ultraviolet rays, or by chemical **mutagens.** Mutations can also occur through **transposons** or **mobile genetic elements,** also known as "jumping genes," which move from one part of a chromosome to another, disrupting the function of a part of the DNA. Some damage to DNA can be repaired by special systems of enzymes.

1. The genetic code is defined as a series of _____ in _____
(a) anticodons; tRNA (b) codons; DNA (c) anticodons; mRNA (d) codons; mRNA (e) codons and anticodons; rRNA

2. Transcription is the process by which _____ is/are synthesized. (a) mRNA (b) mRNA and tRNA (c) mRNA, tRNA, and rRNA (d) protein (e) mRNA, tRNA, rRNA, and protein

3. RNA differs from DNA in that the base _____ is substituted for _____. (a) adenine; uracil (b) uracil; thymine (c) guanine; uracil (d) cytosine; guanine (e) guanine; adenine

4. RNA grows in the _____ direction, as DNA-dependent RNA polymerase moves along the template DNA strand in the _____ direction. (a) $5' \rightarrow 3'$; $3' \rightarrow 5'$ (b) $3' \rightarrow 5'$; $3' \rightarrow 5'$ (c) $5' \rightarrow 3'$; $5' \rightarrow 3'$ (d) $3' \rightarrow 3'$; $5' \rightarrow 5'$ (e) $5' \rightarrow 5'$; $3' \rightarrow 3'$

5. Which of the following is/are NOT found in a prokaryotic mRNA molecule? (a) protein termination codon (b) upstream leader sequences (c) downstream trailing sequences (d) start codon (e) promoter sequences

6. Which of the following is/are typically removed from pre-mRNA during nuclear processing in eukaryotes? (a) upstream leader sequences (b) poly-A tail (c) introns (d) exons (e) all of the above are removed

7. Which of the following is a spontaneous process, with no direct requirement for ATP or GTP? (a) formation of a peptide bond (b) translocation of the ribosome (c) formation of aminoacyl-tRNA (d) a and b

(e) all of the above are spontaneous processes

8. Select the events of the elongation cycle of protein synthesis from the following list and place them in the proper sequence.
1. peptide bond formation
2. binding of the small ribosomal subunit to the 5' end of the mRNA
3. binding of aminoacyl-tRNA to the A site
4. translocation of the ribosome
(a) $1 \rightarrow 3 \rightarrow 2 \rightarrow 4$ (b) $3 \rightarrow 1 \rightarrow 4$ (c) $3 \rightarrow 1 \rightarrow 3 \rightarrow 2$ (d) $1 \rightarrow 3 \rightarrow 4$ (e) $4 \rightarrow 2 \rightarrow 1 \rightarrow 3$

9. The statement "the genetic code is redundant" means that (a) some codons specify punctuation (stop and start signals) rather than amino acids (b) some codons specify more than one amino acid (c) certain amino acids can be specified by more than one codon (d) in some cases, the third nucleotide of an anticodon may be able to pair with more than one kind of base in a codon (e) all organisms have essentially the same genetic code

10. A nonsense mutation (a) causes one amino acid to be substituted for another in a polypeptide chain (b) results from the deletion of one or more bases, leading to a shift in the reading frame (c) results from the insertion of one or more bases, leading to a shift in the reading frame (d) results from the insertion of a transposon (e) usually results in the formation of an abnormally short polypeptide chain

R E V I E W   Q U E S T I O N S

1. A certain transcribed DNA strand has the following nucleotide sequence:

$3'$ — TACTGCATAATGATT — $5'$

What would be the sequence of codons in the mRNA transcribed from this strand? What would be the nucleotide sequence of the complementary nontranscribed DNA strand? What would be the exact anticodon for each codon? Use Table 12–1 to determine the amino acid sequence of the polypeptide. Be sure to label the 5' and 3' ends of the nucleic acids, and the carboxyl and amino ends of the polypeptide.

2. What are ribosomes made of? Do ribosomes themselves carry information to specify the amino acid sequence of proteins?

3. In what ways are DNA polymerase and RNA polymerase similar? How do they differ?

4. Describe initiation, elongation, and termination of protein synthesis.

5. Explain how the genetic code was deciphered. What experimental procedures needed to be developed before this could be accomplished?

6. What are the main types of mutations? What effects does each have on the protein product produced?

7. Why can't amino acids become incorporated into polypeptides without the aid of tRNA?

Y O U   M A K E   T H E   C O N N E C T I O N

1. How many amino acids could be specified if two bases coded for one amino acid? Why is redundancy important to the idea of wobble? If you could "reinvent" the genetic code, would you make any changes? Why or why not?

2. Compare and contrast the formation of mRNA in prokaryotic and eu-

karyotic cells. How do the differences affect the way in which each type of mRNA is translated? Does one system have any obvious advantage in terms of energy cost? Which system offers greater opportunities for control of gene expression?

R E C O M M E N D E D   R E A D I N G S

Craig, P.P. "Jumping Genes: Barbara McClintock's Scientific Legacy." *Carnegie Institution of Washington Perspectives in Science,* No. 6, 1994. Copies may be obtained for $1.00 each by calling (202) 939–1121 or by writing The Carnegie Institution of Washington, 1530 P Street N.W., Washington, DC, 20005-1910. A fascinating illustrated essay on the life of Barbara McClintock and the far-reaching implications of her work.

Doolittle, R.F., and P. Bork. "Evolutionarily Mobile Modules in Proteins." *Scientific American,* Vol. 269, No. 4, Oct. 1993. A discussion of the possible roles of exons and introns in protein evolution.

Hayes, B. "The Invention of the Genetic Code." *American Scientist,* Vol. 86, No. 1, Jan.–Feb. 1998. This article relates the history of early conjectures

regarding the nature of the genetic code and how the genetic code was finally solved.

Landick, R. and J.W. Roberts. "The Shrewd Grasp of RNA Polymerase." *Science,* Vol. 273, 12 Jul. 1996. A summary of what is known about the physical association between RNA polymerase and DNA during transcription.

Voytas, D.F. "Retroelements in Genome Organization." *Science,* Vol. 274, 1 Nov. 1996. Transposable elements that replicate through an RNA intermediate are surprisingly common in maize (corn) and many other organisms.

● Visit our website at **http://www.saunderscollege.com/lifesci/titles.html** and click on Solomon/Berg/Martin Biology for links to chapter-related resources on the World Wide Web.

CHAPTER 13

Gene Regulation: The Control of Gene Expression

E ach type of cell in a multicellular organism has a characteristic shape, carries out very specific activities, and makes a distinct set of proteins. Yet, with few exceptions, they all contain the same genetic information. Why, then, are they not identical in structure and molecular composition? This is because genes are regulated, and only certain subsets of the total genetic information are expressed in any given cell. For example, the chromosomes in the LM are from a cell in the salivary gland of the larva of the fruit fly, *Drosophila*. The expanded, "puffed" chromosomal regions (*arrows*) contain genes that are actively transcribed in that cell, while the more compact banded regions include genes that are transcribed at very low rates, or not at all.

What are the mechanisms controlling the expression of a gene? Let us consider a gene that codes for a protein that is an enzyme. Expression of that gene involves not only transcription to form mRNA; the mRNA must be translated into protein, and the protein must actively catalyze a specific reaction. Gene expression, then, is the result of a series of processes, each of which can be controlled in many different ways. The control mechanisms require information in the form of various signals (some originating within the cell and others coming from the environment) that interact with DNA, RNA, or protein.

Some of the main strategies used to regulate gene expression include controlling (1) the amount of mRNA that is available, (2) the rate of translation of the mRNA, and (3) the activity of the protein product.

In bacterial cells, energy efficiency and economical use of resources are usually of primary importance. As a result, most gene regulation in prokaryotes involves transcriptional level control of genes whose products are involved in resource utilization. In eukaryotes there is a much greater emphasis on fine-tuning the control systems, which is consistent with the greater complexity of these cells and the need to control development in multicellular organisms (see Chapter 16). Consequently, eukaryotic gene regulation occurs at many levels.

(Peter J. Bryant/Biological Photo Service)

1. Explain why the organization of genes into operons is advantageous to bacteria. Explain why some genes, such as those of the lactose operon, are inducible, while others, such as those of the tryptophan operon, are repressible.
2. Sketch the main elements of an inducible operon, such as the lactose operon, and explain the functions of the operator and promoter regions.
3. Diagram the major components of an mRNA molecule produced by the lactose operon and relate that structure to the organization of the operon.
4. Differentiate between positive and negative control and show how both types of control operate in the regulation of the lactose operon.
5. Sketch the structure of a typical eukaryotic gene and the DNA sequences involved in the regulation of that gene.
6. Give examples of some of the ways eukaryotic DNA-binding proteins bind to DNA.
7. Illustrate how a change in chromosome structure might affect the activity of a gene.
8. List two ways that a gene in a multicellular organism might be able to produce different products in different types of cells.
9. Identify some of the types of regulatory controls that can be exerted in eukaryotes after mature mRNA is formed.

GENE REGULATION IN PROKARYOTES IS ECONOMICAL

A cell of *Escherichia coli,* a bacterium that is a common inhabitant of the intestine, has 4288 genes that code for proteins, approximately 62% of which have known functions. Some of these genes encode proteins that are always needed (e.g., enzymes involved in glycolysis). These genes, which are constantly transcribed, are called **constitutive genes.** Other proteins are needed only when the bacterium is growing under special conditions.

For instance, the bacteria living in the colon of an adult cow are not normally exposed to the milk sugar lactose. If those cells were to end up in the colon of a calf, however, they would have lactose available as a source of energy. This poses a dilemma. Should a bacterial cell invest energy and materials to produce lactose-metabolizing enzymes just in case it ends up in the digestive system of a calf? Given that the average lifetime of an actively growing *E. coli* cell is about 30 minutes, such a strategy appears wasteful. Yet if *E. coli* cells do not have the capacity to produce those enzymes, they might starve in the midst of an abundant food supply. *E. coli* handles this problem by regulating many of its enzymes in order to efficiently use available organic molecules.

Cells have two basic ways of controlling their metabolic activity. They can regulate the *activity* of certain enzymes (how effectively an enzyme molecule works), and/or they can control the *number* of enzyme molecules present in each cell. Some enzymes may be regulated in both ways in the same type of cell. An *E. coli* cell growing on glucose is estimated to need about 800 different enzymes. Some of these must be present in large numbers, whereas only a few molecules of others are required. In order for the cell to function properly, each enzyme must be efficiently controlled.

Operons in prokaryotes permit coordinated control of functionally related genes

The French researchers François Jacob and Jacques Monod are credited with the first demonstration, in 1961, of how some genes are regulated at the biochemical level. They studied the genes that code for the enzymes that metabolize the disaccharide lactose (see Chapter 3). For *E. coli* to use lactose as an energy source, the sugar must be cleaved into the monosaccharides glucose and galactose by the enzyme β-galactosidase. (β-galactosidase also converts some lactose molecules to a structural isomer, *allolactose.*) Galactose is then converted to glucose by another enzyme, and the resulting two glucose molecules are further broken down by the glycolysis pathway (see Chapter 7).

E. coli cells growing on glucose contain very little of the β-galactosidase enzyme, perhaps no more than one to three molecules per cell. However, cells grown on lactose have as many as 3000 β-galactosidase molecules per cell, accounting for about 3% of the total cellular protein. Levels of two other enzymes, galactose permease and galactoside transacetylase, also increase when the cells are grown on lactose. The permease is needed to transport lactose efficiently across the bacterial plasma membrane; without it, only small amounts of lactose can enter the cell. The transacetylase may be involved in a minor aspect of lactose metabolism, although its function is not clear.

Jacob and Monod were able to identify mutant strains of *E. coli* in which a single genetic defect resulted in the loss of all three enzymes. This finding, along with other information, led them to the conclusion that the coding DNA sequences for all three enzymes are linked together as a unit on the bacterial DNA and are subject to a common control mechanism. Each protein coding sequence is known as a **structural gene.** Jacob and Monod coined the term **operon** to refer to a gene

Figure 13–1 **The lactose operon.** The structural genes coding for the three enzymes used by *E. coli* to metabolize the disaccharide lactose are transcribed as part of a single mRNA molecule. *(a)* In the absence of lactose, a repressor protein, encoded by an adjacent gene, binds a region known as the operator. By preventing RNA polymerase from binding to the promoter, the bound repressor protein blocks transcription of the structural genes. *(b)* When lactose is present, it is converted to allolactose, which binds to the repressor at an allosteric site, altering the structure of the protein so that it can no longer bind to the operator. This allows RNA polymerase to bind to the promoter and transcribe the structural genes.

complex consisting of a group of structural genes with related functions, plus the closely linked DNA sequences responsible for controlling them. The structural genes of the lactose operon —lacZ, lacY, and lacA—code for β-galactosidase, galactose permease, and galactoside transacetylase, respectively. (Fig. 13–1).

Transcription of the lactose operon begins as RNA polymerase binds to a single promoter site upstream from the coding sequences. It then proceeds to transcribe the DNA, forming a single mRNA molecule that contains the coding information for all three enzymes. This mRNA contains translation termination and initiation codons between the enzyme-coding sequences; hence it is translated to form three separate protein molecules. Because all three enzymes are translated from the same mRNA molecule, their synthesis can be coordinated by turning a single molecular "switch" off or on.

The switch that controls mRNA synthesis is called the **operator;** it is a sequence of bases that overlaps part of the promoter region and is upstream from the first structural gene in the operon. When lactose is absent, a **repressor protein** called the **lactose repressor** binds tightly to the operator region. Because the repressor protein is large enough to cover part of the promoter sequence, RNA polymerase is unable to bind to the

lactose promoter site, and transcription of the lactose operon is effectively blocked.

The lactose repressor protein is encoded by a regulatory gene, which in this case is an adjacent structural gene located upstream from the promoter site. Unlike the lactose operon genes, the repressor gene is constitutive; that is, it is always "on," so small amounts of the repressor protein are produced continuously. This protein binds specifically to the lactose operator sequence. When cells are grown in the absence of lactose, the operator site is nearly always occupied by a repressor molecule. Only on rare occasions, when the operator site is briefly free of the repressor, can RNA polymerase bind and initiate transcription of the structural genes. Because *E. coli* mRNA is rapidly degraded (having a half-life of about 2 to 4 minutes), very few proteins are translated from that mRNA.

Lactose is able to "turn on," or *induce,* the transcription of the lactose operon because the lactose repressor protein contains a second functional region separate from its DNA binding site; this is an allosteric binding site for allolactose. (Recall from Chapter 6 that an allosteric regulator, such as allolactose in this case, binds to a site in a protein other than its active site, changing its conformation and thereby altering its function.) If lactose is present in the growth medium, a few molecules are able to enter the cell and are converted to allolactose by the few β-galactosidase molecules present. When a molecule of allolactose binds to the repressor at the allosteric site, it alters the conformation of the protein so that its DNA binding site can no longer recognize the operator. When all the repressor molecules have allolactose bound to them and are therefore inactivated, RNA polymerase binds to the unblocked promoter, and the operon is actively transcribed.

E. coli cell continues to produce β-galactosidase and the other lactose operon proteins until virtually all of the lactose is used up. When intracellular levels of lactose drop, the repressor proteins no longer have the allolactose sugar bound to them. They then assume a conformation that allows them to bind to the operator region and shut down transcription of the operon.

How were Jacob and Monod able to dissect the lactose operon? Their approach centered around the use of mutant strains, which even today play an essential role in allowing researchers to unravel the components of a regulatory system. Mutant strains permit investigators to map the positions of the genes on the DNA and to infer normal gene functions by studying what happens when they are missing or altered. This information is usually combined with results of direct biochemical studies.

For example, Jacob and Monod studied various constitutive mutants; these transcribed the structural genes of the lactose operon at a significant rate even in the absence of lactose. One group of constitutive mutants had abnormal genes with map positions just outside the lactose operon itself. Using special genetic strains, it was possible to show that these particular mutations always caused constitutive expression, regardless of their location in the genome.

On the basis of these findings, Jacob and Monod hypothesized the existence of a repressor gene that is responsible for producing a repressor molecule (later found to be a protein). Although the specific defect may vary, the members of this group of constitutive mutants do not produce active repressor proteins; hence no binding to the lactose operator and promoter takes place, and the lactose operon is transcribed constitutively.

The genes responsible for the behavior of a second group of constitutive mutants had map positions within the lactose operon but did not directly involve any of the three structural genes. The members of this group were found to produce normal repressor molecules but to have abnormal operator sequences incapable of binding the repressor.

In contrast to the constitutive mutants, other mutants failed to transcribe the lactose operon even when lactose was present. Some of these abnormal genes had the same map position as the hypothesized regulatory gene. They were eventually found to have an altered binding site on the repressor protein that prevented allolactose from binding, although the ability of the repressor to bind to the operator was unaffected. Once bound to the operator, such a mutant repressor remains bound, keeping the operon "turned off."

An inducible gene is not transcribed unless a specific inducer inactivates its repressor

The lactose operon is called an **inducible system.** An inducible gene or operon is usually controlled by a repressor that keeps it in the "off" state. The presence of an **inducer molecule** (in this case allolactose) renders the repressor inactive so that the gene or operon is turned "on." Inducible genes or operons usually code for enzymes that are part of catabolic pathways; these break down molecules to provide both energy and components for anabolic reactions. This type of regulatory system enables the cell to save the energy costs of making enzymes when no substrates are available for them to act on.

A repressible gene is transcribed unless a specific repressor/corepressor complex is bound to the DNA

Another type of gene regulation system in bacteria is associated mainly with anabolic pathways in which amino acids, nucleotides, and other essential molecules are synthesized from simpler precursors. Regulation of these pathways normally involves **repressible** enzymes, which are coded for by repressible genes.

Repressible genes and operons are usually "on"; they are turned off only under special conditions. In most cases the molecular signal used to regulate these genes is the end product of the metabolic pathway. When the supply of the end product (e.g., an amino acid) is low, all enzymes in the pathway are actively synthesized. When intracellular levels of the end product are high, enzyme synthesis is repressed. Because compounds such as amino acids are continuously needed by the growing cell, the most effective strategy is to keep the genes that control their production "on" except when a large supply of the amino acid is available. The ability to turn the genes off allows cells to avoid overproduction of amino acids and other molecules that are essential but energetically expensive to make.

The tryptophan operon is an example of a repressible system. In both *E. coli* and a related bacterium, *Salmonella,* the operon consists of five structural genes that code for the enzymes required for synthesis of the amino acid tryptophan; these are clustered together as a transcriptional unit with a single promoter and a single operator (Fig. 13–2). A distant repressor gene codes for a diffusible repressor protein, which differs from the lactose repressor in that it is synthesized in an inactive form that is unable to bind to the operator region of the tryptophan operon.

The DNA-binding site of the repressor becomes effective only when tryptophan, its **corepressor,** binds to an allosteric site on the repressor. When intracellular tryptophan levels are low, the repressor protein is inactive and unable to bind to the operator region of the DNA. As the concentration of intracellular tryptophan rises, some tryptophan binds to the allosteric site of the repressor, altering its conformation so that it binds tightly to the operator. This has the effect of switching the operon off, thereby blocking transcription.

(a)

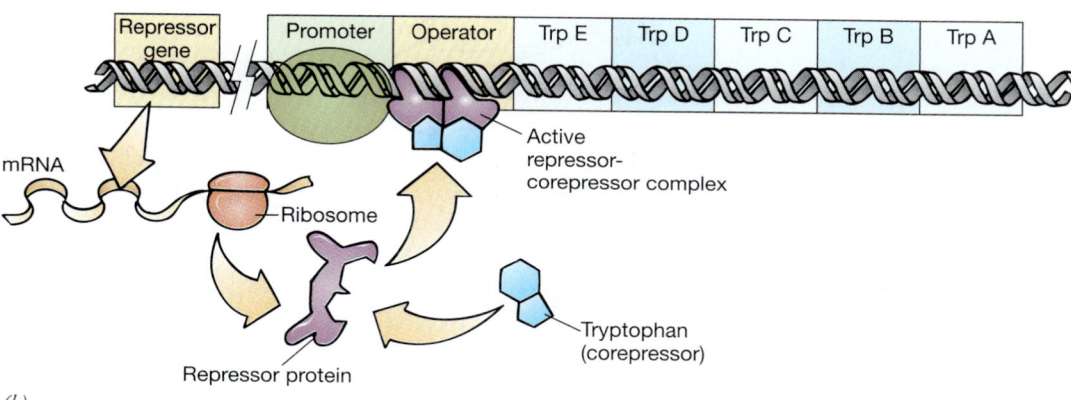

(b)

Figure 13–2 **The tryptophan operon.** Genes coding for enzymes that synthesize the amino acid tryptophan are organized in a repressible operon. *(a)* A regulatory gene encodes a repressor protein that is initially inactive; it is unable to prevent transcription because it cannot bind to the operator. *(b)* When intracellular tryptophan levels are high, the amino acid binds to an allosteric site on the repressor protein, changing its conformation. The resulting active form of the repressor binds to the operator region, blocking transcription of the operon until tryptophan is again required by the cell.

Negative regulators repress transcription; positive regulators activate transcription

The features of the lactose and tryptophan operons described so far are examples of **negative control.** Systems under negative control are those in which the DNA-binding regulatory protein is a repressor that turns off transcription of the gene. Some regulatory systems involve **positive control,** that is, regulation by **activator proteins** that bind to DNA and thereby stimulate transcription of a gene. The lactose operon contains both negative and positive controlling elements (Fig. 13–3).

Positive control of the lactose operon requires that the cell be able to sense the absence of the sugar glucose, which is the initial substrate in the glycolysis pathway (see Chapter 7). Lactose, like glucose, is a catabolite and can undergo stepwise breakdown to yield energy. However, because lactose must first be converted to glucose, it is most efficient for *E. coli* cells to use the available supply of glucose first. By using glucose as the preferred substrate, the cell is spared the considerable energy cost of making additional enzymes such as β-galactosidase.

The lactose operon has a very inefficient promoter sequence; that is, it has a low affinity for RNA polymerase even when the repressor protein is inactivated. However, a DNA se-

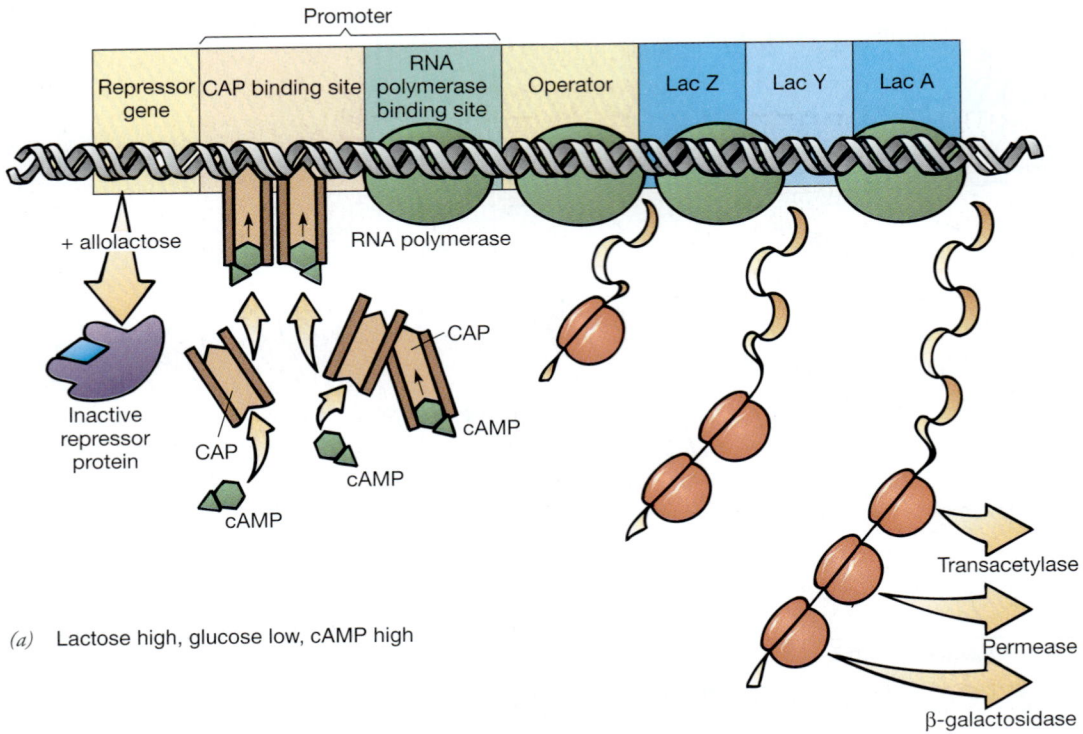

(a) Lactose high, glucose low, cAMP high

(b) Lactose high, glucose high, cAMP low

Figure 13–3 **Positive control of the lactose operon.** The lactose promoter by itself is weak and binds RNA polymerase inefficiently even when the lactose repressor is inactive. The CAP is an allosteric regulator that can bind to a sequence of bases in the promoter, allowing RNA polymerase to bind efficiently, thereby stimulating transcription of the operon. The CAP molecule contains two polypeptides. When each has cAMP bound to its allosteric site, the protein can bind to the DNA sequence. The cell's cAMP concentration is inversely proportional to the glucose concentration. *(a)* When glucose levels are low, cAMP binds to CAP, which then activates transcription of the operon by binding to the DNA. *(b)* When glucose levels are high, cAMP is low. CAP is in an inactive form and cannot stimulate transcription. Therefore, only low-level transcription occurs.

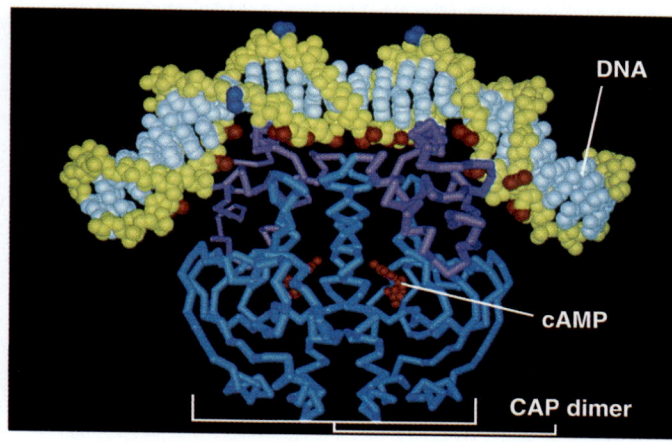

Figure 13–4 **The binding of CAP to DNA.** This computer-generated picture illustrates the bend formed in the DNA double helix when it binds to CAP. CAP is a dimer consisting of two identical polypeptide chains, each of which binds one molecule of cAMP. *(Courtesy of S.C. Schultz, G.C. Shields, and T.A. Steitz, Yale University)*

quence adjacent to the promoter site is a binding site for another protein, called the **catabolite gene activator protein (CAP).**

CAP increases the affinity of the promoter region for RNA polymerase, allowing the enzyme to recognize the promoter efficiently and to bind tightly to the DNA. In its active form CAP has **cyclic AMP,** or **cAMP,** an alternative form of adenosine monophosphate (see Fig. 3–25), bound to an allosteric

site.[1] As the cells become depleted of glucose, cAMP levels increase. The cAMP molecules bind to CAP, and the resulting complex then binds to the CAP-binding site near the lactose operon promoter and stimulates the transcription of the operon in the presence of lactose (Fig. 13–4). Thus, the operon is fully active only if lactose is present and intracellular glucose levels are low. The properties of negative and positive control systems are summarized in Table 13–1.

A regulon is a group of functionally related operons controlled by a common regulator

CAP differs from the lactose and tryptophan repressors in that it can control transcription of operons involved in the metabolism of a number of other catabolites, such as the sugars galactose, arabinose, and maltose, as well as of lactose. A group of operons controlled by one regulator of this type is generally referred to as a **regulon** (Fig. 13–5).

Other multigene systems in bacteria are also controlled in this manner. For example, genes involved in nitrogen and phosphate metabolism are organized into regulons that consist of multiple sets of operons controlled by one or more combinations of regulatory genes. Other complex multigene systems respond to changes in environmental conditions, such as rapid shifts in temperature, exposure to radiation, changes in osmotic pressure, and changes in oxygen levels. Specific mutants often provide clues to the existence of a regulon system. A single

[1]For this reason, CAP is also known as CRP, which stands for **c**AMP **r**eceptor **p**rotein.

TABLE 13–1 Types of Transcriptional Control in Prokaryotes*

NEGATIVE CONTROL

Inducible genes

Repressor protein alone	→	**Active repressor "turns off" regulated gene(s)**
Lactose repressor alone	→	Lactose operon not transcribed
Repressor protein + inducer	→	**Inactive repressor/inducer complex fails to "turn off" regulated gene(s)**
Lactose repressor + allolactose	→	Lactose operon transcribed

Repressible genes

Repressor protein alone	→	**Inactive repressor fails to "turn off" regulated gene(s)**
Tryptophan repressor alone	→	Typtophan operon transcribed
Repressor protein + corepressor	→	**Active repressor-corepressor complex "turns off" regulated gene(s)**
Tryptophan repressor + tryptophan	→	Tryptophan operon not transcribed

POSITIVE CONTROL

Activator protein alone	→	**Activator alone cannot stimulate transcription of regulated gene(s)**
CAP alone	→	Transcription of lactose operon not stimulated
Activator protein + coactivator	→	**Functional activator-coactivator complex stimulates transcription of regulated gene(s)**
CAP + cAMP	→	Transcription of lactose operon stimulated

*A general description of each type is followed by a specific example. A negative regulator is a repressor that "turns off" transcription of the regulated gene(s). Conversely, a positive regulator is an activator that stimulates transcription.

When glucose levels are low, cAMP levels increase, activating CAPs, which bind to their recognition sites in the promoter regions of all operons. If the inducer for an operon is available, its repressor is inactivated and transcription of the message takes place at a rapid rate.

High cAMP

Operons

Arabinose operon   Lactose operon   Maltose operon   Galactose operon

+Arabinose   +Lactose   +Maltose   +Galactose

Proteins

Figure 13–5 **A regulon.** Operons that convert several different sugars to glucose in *E. coli* make up a regulon that is under positive control by CAP.

mutation that destroys the activity of CAP, for example, prevents the cell from efficiently metabolizing not only lactose but many other sugars also regulated by CAP.

Not all constitutive genes are transcribed at the same rate

Many of the gene products encoded by the *E. coli* DNA are needed only under certain environmental or nutritional conditions. As we have seen, these genes are generally regulated at the level of transcription. They can be turned on and off as metabolic and environmental conditions change. By contrast, constitutive genes are continuously transcribed, but they are not necessarily transcribed (or their mRNAs translated) at the same rate. Some enzymes work more effectively or are more stable than others and consequently need to be present in smaller amounts. Constitutive genes that encode proteins required in large amounts are generally transcribed more rapidly than genes coding for proteins required at lower levels. The transcription rate of these genes is controlled by their promoter sequences. Constitutive genes with efficient ("strong") promoters bind RNA polymerase more frequently and consequently transcribe more mRNA molecules than those with inefficient ("weak") promoters.

Genes coding for repressor or activator proteins that regulate metabolic enzymes are usually constitutive and produce their protein products constantly. Because each cell usually needs relatively few molecules of any specific repressor or activator protein, promoters for those genes tend to be relatively weak.

Some posttranscriptional regulation occurs in prokaryotes

As we have seen, much of the variability in protein levels in *E. coli* is determined by **transcriptional level control.** However, regulatory mechanisms that operate after transcription, known as **posttranscriptional controls,** also occur, operating at various levels of gene expression.

Translational controls are posttranscriptional controls that regulate the rate at which a particular mRNA molecule is translated. Because the lifetime of an mRNA molecule in a bacterial cell is very short, a molecule that is translated rapidly can produce more proteins than one that is translated slowly. Some mRNA molecules in *E. coli* are translated as much as 1000 times faster than others. Most of the differences appear to be due to the rate at which ribosomes can attach to the mRNA and begin translation.

Posttranslational controls generally act as switches that activate or inactivate one or more existing enzymes. These systems allow the cell to respond to changes in the intracellular concentrations of essential molecules, such as amino acids, by rapidly adjusting the activities of its enzymes. A common posttranslational control adjusts the rate of synthesis in a metabolic pathway through **feedback inhibition** (see Chapter 6). The end product binds to the first enzyme in the pathway at an allosteric site, temporarily inactivating the enzyme. When the first enzyme in the pathway does not function, all of the succeeding enzymes are deprived of substrates. Notice that this differs from the end-product repression of the tryptophan operon discussed previously. In that case, the end product of

REGULATION IN PROKARYOTES AND EUKARYOTES

Why do prokaryotic and eukaryotic cells have distinctly different strategies for regulating the activity of their genes? In large part these differences reflect the ways in which the organisms make their living. Bacterial cells exist independently, and each cell must be able to perform all of its own essential functions. Because they grow rapidly and have relatively short lifetimes, bacterial cells carry little excess baggage.

The dominant theme of prokaryotic gene regulation is *economy*, and controlling transcription is usually the most cost-effective way to regulate gene expression. The organization of related genes into operons and regulons that can be rapidly turned on and off as units allows these cells to synthesize only the gene products needed at any particular time. This type of regulation requires rapid turnover of mRNA molecules to prevent messages from accumulating and continuing to be translated when they are not needed. Bacteria rarely regulate enzyme levels by degrading proteins. Once the synthesis of a protein ends, the previously synthesized protein molecules are diluted so rapidly in subsequent cell divisions that breaking them down is usually not necessary. Only when cells are starved or deprived of essential amino acids are protein-digesting

enzymes used to recycle their amino acids by breaking down proteins no longer needed for survival.

Eukaryotic cells have different regulatory requirements. In multicellular organisms, groups of cells cooperate with each other in a division of labor. Because a single gene may need to be regulated in different ways in different types of cells, eukaryotic gene regulation is complex, occurring not only at the level of transcription but also at other levels of gene expression. Eukaryotic cells also usually have long lifetimes during which they may need to respond repeatedly to many different stimuli. Rather than synthesize new enzymes each time they respond to a stimulus, these cells make extensive use of preformed enzymes and other proteins that can be rapidly converted from an inactive to an active state.

Much of the emphasis of gene regulation in multicellular organisms is on *specificity* in the form and function of the cells in each tissue. Each type of cell has certain genes that are active and others that may never be used. Apparently the adaptive advantages of cellular cooperation in eukaryotes far outweigh the detrimental effects of carrying a load of inactive genes through many cell divisions.

---

the pathway prevented the formation of new enzymes. Feedback inhibition acts as a fine-tuning mechanism that regulates the activity of the existing enzymes in a metabolic pathway.

GENE REGULATION IN EUKARYOTES IS MULTIFACETED

Like bacteria, eukaryotic cells must respond to changes in their environment by turning appropriate sets of genes on and off. Multicellular eukaryotes require additional modes of regulation that permit individual cells to become committed to specialized roles and groups of cells to organize into tissues and organs (see *Making the Connection: Regulation in Prokaryotes and Eukaryotes*). In previous chapters we observed that all aspects of information transfer, including replication, transcription, and translation, are far more complicated in eukaryotes. Not surprisingly, this complexity provides additional opportunities for control of gene expression.

Unlike many of the prokaryotic genes, most eukaryotic genes are not found in operon-like clusters. However, each eukaryotic gene has specific regulatory sequences that are essential in the control of transcription.

Many of the "housekeeping" enzymes (those needed by all cells) appear to be encoded by constitutive genes, which are expressed in all cells at all times. Some inducible genes have also been found; these respond to environmental threats or stimuli such as heavy metal ingestion, viral infection, and heat

shock. For example, when a cell is exposed to high temperature, many proteins fail to fold properly. A system that appears to be sensitive to these unfolded proteins causes genes known as heat-shock genes to be transcribed, resulting in the production of heat-shock proteins. Although the functions of most of the heat-shock proteins are not known, some are molecular chaperones (see Chapter 3), which are responsible for helping other proteins attain their proper conformation.

Some genes appear to be inducible only during certain periods in the life of the organism; they are thought to be controlled by **temporal regulation** mechanisms. Finally, a number of genes are under the control of **tissue-specific regulation.** For example, a gene involved in the production of a particular enzyme may be regulated by one stimulus (e.g., a hormone) in muscle tissue, by an entirely different stimulus in pancreatic cells, and by a third stimulus in liver cells. These types of regulation are explored in more detail in Chapter 16.

Eukaryotic transcription is controlled at many sites and by many different regulatory molecules

Various DNA segments, including upstream promoter elements, enhancers, and transcription initiation sites, are important in transcriptional control. In addition, the rate of transcription is affected by regulatory proteins known as transcription factors and by the way the DNA is organized in the chromosome.

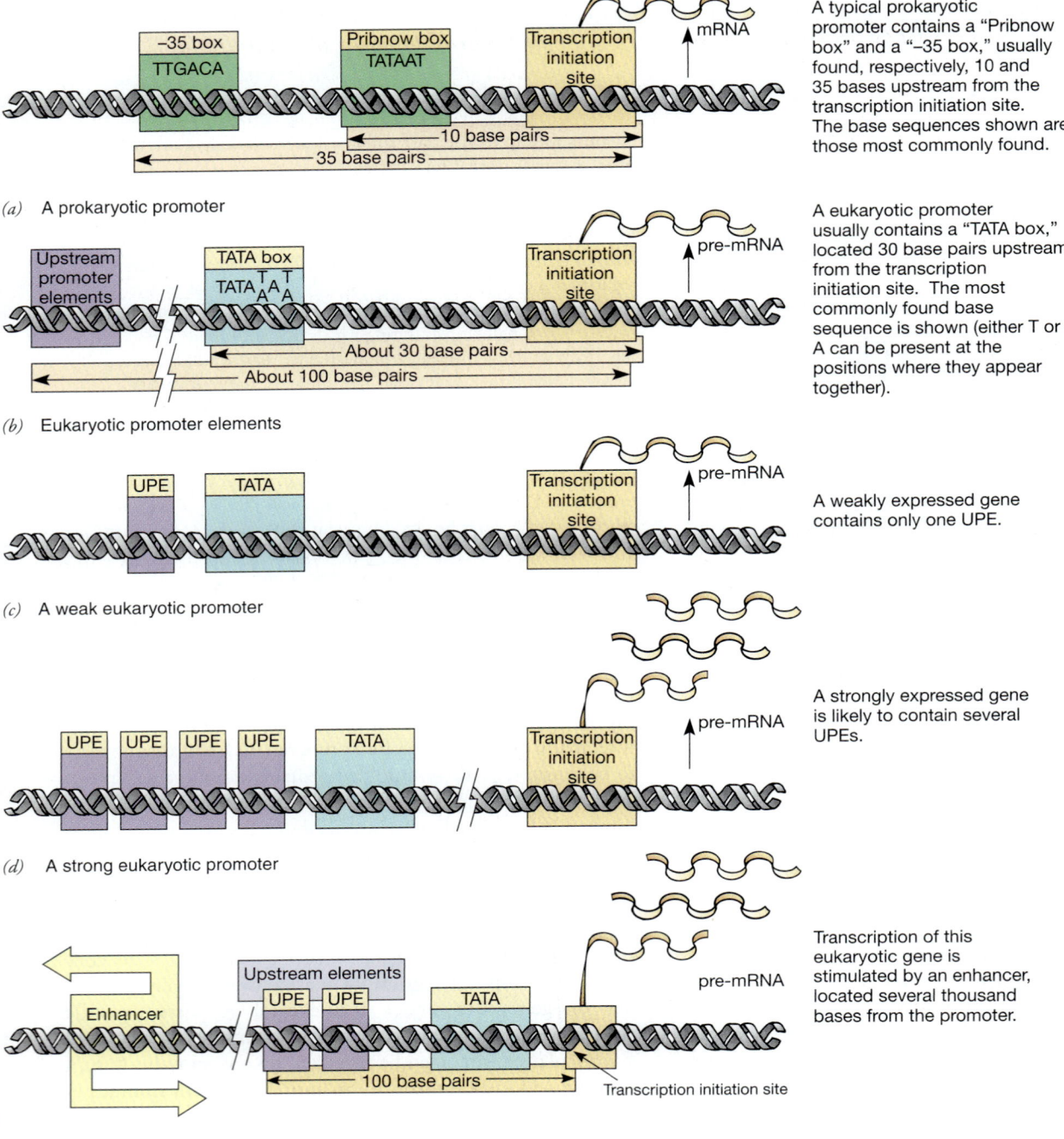

(a) A prokaryotic promoter

−35 box TTGACA

Pribnow box TATAAT

Transcription initiation site

mRNA

10 base pairs

35 base pairs

A typical prokaryotic promoter contains a "Pribnow box" and a "−35 box," usually found, respectively, 10 and 35 bases upstream from the transcription initiation site. The base sequences shown are those most commonly found.

(b) Eukaryotic promoter elements

Upstream promoter elements

TATA box TATA T A T A A

Transcription initiation site

pre-mRNA

About 30 base pairs

About 100 base pairs

A eukaryotic promoter usually contains a "TATA box," located 30 base pairs upstream from the transcription initiation site. The most commonly found base sequence is shown (either T or A can be present at the positions where they appear together).

(c) A weak eukaryotic promoter

UPE

TATA

Transcription initiation site

pre-mRNA

A weakly expressed gene contains only one UPE.

(d) A strong eukaryotic promoter

UPE UPE UPE UPE

TATA

Transcription initiation site

pre-mRNA

A strongly expressed gene is likely to contain several UPEs.

(e) A strong eukaryotic promoter plus an enhancer

Enhancer

Upstream elements

UPE UPE

TATA

pre-mRNA

100 base pairs

Transcription initiation site

Transcription of this eukaryotic gene is stimulated by an enhancer, located several thousand bases from the promoter.

Eukaryotic promoters vary in efficiency, depending on their upstream promoter elements

In eukaryotic as well as prokaryotic cells, the transcription of any gene requires a transcription initiation site, plus a promoter to which RNA polymerase binds. A prokaryotic promoter (Fig. 13–6a) includes certain characteristic base sequences, known as a *Pribnow box* and a *−35 box* (35 base pairs upstream from the transcription initiation site). In multicellular eukaryotes RNA polymerase binds to a sequence of bases, known as a **TATA box,** about 30 base pairs upstream from the transcription initiation site (Fig. 13–6b). A eukaryotic promoter generally contains one or more sequences of 8 to 12 bases within a short distance upstream of the RNA polymerase-binding site. These have been given various names; we will refer to them as **upstream promoter elements (UPEs).** The efficiency of the promoter seems to depend on the number and type of UPEs. Thus, a constitutive gene containing only one UPE is generally weakly expressed, whereas one containing five or six UPEs is usually actively transcribed (Fig. 13–6c and d).

Enhancers are DNA sequences that increase the rate of transcription

Regulated eukaryotic genes commonly require not only the upstream promoter elements but also DNA sequences called **enhancers.** Whereas the promoter elements are required for accurate and efficient initiation of mRNA synthesis, enhancers increase the *rate* of RNA synthesis after initiation, often by several orders of magnitude (Fig. 13–6e).

Enhancer sequences are remarkable in many ways. Although present in all cells, a particular enhancer is functional only in certain types of cells. An enhancer can regulate a gene on the same DNA molecule from very long distances (up to thousands of base pairs away from the promoter) and can be either upstream or downstream of the promoters it controls. Furthermore, if an enhancer sequence is experimentally cut out of the DNA and inverted, it still regulates the gene it normally controls. As we shall see, evidence suggests that at least some enhancers work by interacting with proteins that regulate transcription.

Transcription factors are regulatory proteins that have several functional domains and may work in various combinations

We previously discussed some DNA-binding proteins that regulate transcription in prokaryotes. These **transcription factors** include the lactose repressor, the tryptophan repressor, and the catabolite gene activator protein (CAP). Although transcription factors were first studied in prokaryotes, many have been identified in eukaryotes.

It is useful to compare prokaryotic and eukaryotic transcription factors. Many transcription factors are modular molecules; that is, they have more than one **domain** (region with its own tertiary structure), and each domain has a different function. For example, every prokaryotic regulator has a DNA-binding domain, plus at least one other domain. Examples include domains that can activate RNA polymerase, or bind inducers (such as allolactose) or corepressors (such as tryptophan).

The DNA-binding regions of the lactose repressor and CAP contain two α-helical segments. Each, known as a *recognition helix,* is inserted into a major groove of the DNA without unwinding the double helix (Fig. 13–7a). Recall from Chapter 3 that an α-helix of a protein is arranged such that

Figure 13–7 Transcription factors. *(a)* Generalized view of a recognition α-helix inserted into the major groove of the DNA. The rest of the protein is not shown. Specific amino acid side chains *(not shown)* are thought to form hydrogen bonds with the edges of specific base pairs. *(b)* Regions of certain transcription factors form projections known as "zinc fingers," which can insert into the grooves of the DNA and bind to specific base sequences. The colored circles represent amino acids that bind to the zinc atoms and form the loop. *(c)* Some zinc-finger proteins have multiple projections, each of which is thought to fit into a separate groove in the DNA. *(d)* This leucine zipper protein is a dimer, held together by hydrophobic interactions involving side chains of leucine and other amino acids.

the peptide bonds are in the interior of the helix and the functional groups of the amino acids are on the surface of the helix. This structure allows certain functional groups of specific amino acids to form hydrogen bonds with specific base pairs in the DNA. These hydrogen bonds differ from those involved in complementary base-pairing because they form between specific amino acids and the *edges* of the base pairs.

Eukaryotic regulators, like those of prokaryotes, may be either activators or repressors; we discuss only eukaryotic activators here because they seem to be more common and have been studied more.

Each activator has several functional domains, including a DNA-binding region. In some cases, the binding region consists of one or more recognition α-helices that can be inserted into specific regions of the DNA in a manner similar to that described previously for some prokaryotic regulators. Some other activators have multiple "zinc fingers," loops of amino acids held together by zinc ions (Fig. 13–7*b* and *c*). Certain amino acid functional groups exposed in each finger have been shown to recognize specific DNA sequences.

Many activators are functional only as pairs or *dimers*, and these have special domains required for dimer formation. Many of these transcription factors are known as **leucine zipper proteins** because they are held together by the side chains of leucine and other hydrophobic amino acids (Fig. 13–7*d*). In some cases the two polypeptides that make up the dimer may be identical and form a *homodimer*. In other instances they are different, and the resulting *heterodimer* may have very different regulatory properties. For a very simple and speculative example, let us assume that three regulatory proteins—A, B, and C—are involved in controlling a particular set of genes. These three proteins might associate as dimers in six different ways: three kinds of homodimers (AA, BB, and CC) and three kinds of heterodimers (AB, AC, and BC). Such multiple combinations of regulatory proteins have the potential to greatly increase the number of possible ways that transcription can be controlled.

Transcription factors interact with the general transcription machinery

Transcription in eukaryotes requires multiple regulatory proteins that are bound to different parts of the promoter. The "general transcription machinery" is a protein complex that binds to the TATA region of the promoter near the transcription initiation site. That complex is required for RNA polymerase to bind and initiate transcription. Other combinations of regulatory proteins are bound to the more distant enhancer or UPE regions of the promoter. Those proteins then make contact with the general machinery and control the activity of the RNA polymerase.

Each activator must have at least two functional domains: a DNA recognition site that usually binds to an enhancer or UPE, and a "gene activation site" that contacts the target in the general transcriptional complex. Both enhancers and UPEs apparently become functional when specific regulatory proteins are bound to them. The DNA between the enhancer and promoter sequences is thought to form a loop that allows an activator bound to an enhancer to come in contact with one or more target proteins associated with the general transcriptional complex. When this occurs, transcription is stimulated (Fig. 13–8).

(a) Little or no transcription

(b) High rate of transcription

Figure 13–8 Stimulation of transcription by an enhancer. *(a)* This gene is transcribed at a very low rate or not at all, even though the general transcriptional machinery, including RNA polymerase, is bound to the promoter. *(b)* A regulatory protein that functions as a transcriptional activator becomes bound to an enhancer. The intervening DNA forms a loop, allowing the activator to contact one or more target proteins in the general transcriptional machinery, thereby increasing the rate of transcription.

Figure 13-9 **Transcription initiation at two different sites in the same gene.** The yeast invertase gene encodes an mRNA with two start codons (AUG) in its coding region. If transcription is initiated at a downstream site, a short mRNA is made that yields, when translated, an intracellular form of the enzyme. If transcription starts at a point upstream from both AUG codons, a longer message is made. Translation begins at the first start codon, producing an enzyme with a "signal sequence" at the amino-terminal end of the polypeptide chain that targets the protein to the endoplasmic reticulum and then to the Golgi complex for secretion from the cell.

Transcriptional units may overlap in eukaryotes

A given gene may sometimes be transcribed in more than one way, and somewhat different forms of its protein product may be produced as a result. For example, the enzyme invertase, which cleaves the disaccharide sucrose, exists in two forms in yeast: an intracellular form and a form secreted from the cell into the growth medium. Both forms are encoded by the same gene (Fig. 13–9), but their mRNAs are transcribed from two different **transcription initiation sites.** The longer mRNA encodes the extracellular form of the enzyme, which is a longer polypeptide with extra amino acids at its amino-terminal end. Those amino acids serve as a "signal sequence" indicating that the protein is to be processed through the Golgi complex and secreted from the cell. The smaller mRNA encodes a protein that lacks the signal sequence and remains in the cytosol.

Other genes have tissue-specific overlapping transcriptional units. The primary pre-mRNA transcript of the gene for amylase (an enzyme that breaks down starch) in the mouse salivary gland is several thousand bases longer than its counterpart in the mouse liver. This is because in the salivary gland, transcription starts at a site on the DNA that is farther upstream. Although the coding portions of the mRNAs in both cell types are identical after splicing and processing, transcription occurs about 100 times more frequently in the salivary gland than in the liver, resulting in the production of higher levels of the salivary amylase enzyme.

The organization of the chromosome may affect the expression of some genes

A chromosome is not simply a bearer of genes. Various arrangements of its ordered components can result in increased or decreased expression of the genes it contains.

Multiple copies of genes A single gene cannot always provide enough copies of its mRNA to meet the cell's needs. The requirement for high levels of certain products may be met if multiple copies of the genes that encode them are present in the chromosome. Genes of this type, whose products are essential for all cells, may occur as tandemly repeated gene sequences in all cells. Other genes, which may be required by only a small group of cells, may be selectively replicated in those cells in a process called **gene amplification** (see Chapter 16).

Within an array of repeated genes, each copy is almost identical to the others. Histone genes, which code for the proteins that associate with DNA to form nucleosomes (see Chapter 11), are usually found as multiple copies of 50 to 500 genes in cells of multicellular organisms.

Genes for rRNA and tRNA also occur in multiple copies in all cells. To ensure that the rRNA molecules are made in equal amounts, the RNA genes are arranged as multiple transcription units, each containing one copy of each of the three rRNA genes (Fig. 13–10). Most eukaryotic species contain 150 to 450 such transcription units per cell. The demand for rRNA is so great in actively growing mammalian cells that, although hundreds of copies of the genes are present, each gene is usually copied simultaneously by approximately 100 RNA polymerase enzymes.

Gene inactivation by changes in chromatin structure In multicellular eukaryotes, only a subset of the genes present in a cell are active at any one time. The inactivated genes differ among cell types and in many cases seem to be irreversibly dormant.

Some of the inactive genes appear to be associated with highly compacted chromatin, which can be seen as densely

Transcription unit

Transcription unit

Transcription unit

Spacer region

Spacer region

rRNA coding sequences

(a)

Transcription unit (enlarged)

RNA polymerase

(b)

Precursor ribosomal RNAs

Figure 13–10 **Repeated gene sequences.** Multiple copies of genes may be required when large amounts of their products are needed by the cell. *(a)* Human rRNA genes are organized in transcription units, each of which codes for three different rRNAs. The transcription units are arranged as 200 to 300 tandemly repeated copies (of which only three are shown here), separated by nontranscribed spacer units. *(b)* As each transcription unit is transcribed, it must be maximally loaded with RNA polymerase in order to meet the great need for rRNA in actively growing cells. Each precursor rRNA will be processed to yield three different rRNA molecules. *(Not to scale.)*

staining regions of chromosomes during cell division. These regions of chromatin remain tightly coiled throughout the cell cycle, and even during interphase are visible as darkly staining fibers called **heterochromatin.** Evidence suggests that the DNA of heterochromatin is not transcribed. When one of the two X chromosomes is inactivated in female mammals, most of the inactive X chromosome becomes heterochromatic and is seen as the Barr body (see Chapter 10). Active genes are associated with a more loosely packed chromatin structure called **euchromatin** (Fig. 13–11).

Gene inactivation by DNA methylation Inactive genes of vertebrates and some other organisms typically exhibit a pattern of **DNA methylation** in which the DNA has been chemically altered by enzymes that add methyl groups to certain cytosines. (The resulting 5-methylcytosine is still able to base pair with guanine in the usual way.) There is evidence that certain proteins selectively bind to methylated DNA and make it inaccessible to the transcription machinery.

DNA methylation is thought to reinforce gene inactivation, rather than to serve as the initial mechanism. It appears

Heterochromatin

(a)

Chromatin decondensation

Nucleosome

Histones

DNA

(b)

Euchromatin

Transcribed region

Figure 13–11 **The effect of chromatin structure on transcription.** *(a)* An inactive region of DNA (heterochromatin) is organized into tightly associated nucleosomes. *(b)* Active genes are associated with decondensed chromatin (euchromatin). Chromatin decondensation is often a response to specific inducing signals. The loosely packed chromatin increases the accessibility to RNA polymerase required for transcription of the region.

that once a gene has been turned off by some other means, DNA methylation ensures that it will remain inactive. For example, the DNA of the inactive X of a female mammal becomes methylated after the chromosome has become a condensed Barr body. Each time the DNA replicates, the methylation enzymes perpetuate the preexisting methylation pattern; hence the DNA continues to be transcriptionally inactive in both daughter cells.

The long-lived, highly processed mRNAs of eukaryotes provide many opportunities for posttranscriptional control

The half-life of prokaryotic mRNA is usually measured in minutes; eukaryotic mRNA, even when it turns over rapidly, is far more stable. Prokaryotic mRNA is transcribed in a form that can be translated immediately. In contrast, eukaryotic mRNA molecules require further modification and processing before they can be used in protein synthesis (see Chapter 12). The message is capped, polyadenylated, spliced, and then transported from the nucleus to the cytoplasm to initiate translation. These events represent potential control points at which translation of the message and production of its encoded protein can be regulated.

Some pre-mRNAs can be processed in more than one way

Several forms of regulation involving mRNA processing have been discovered. In some instances, the same gene is used to produce one type of protein in one tissue and a related but somewhat different type of protein in another tissue. This is possible because some genes produce pre-mRNA molecules that have multiple splicing patterns; that is, they can be spliced in more than one way depending on the tissue. Typically, such a gene includes at least one segment that can be either an intron or an exon. Through **differential mRNA processing,** the cells in each tissue produce their own version of mRNA corresponding to the particular gene (Fig. 13–12).

The stability of mRNA molecules can vary

Controlling the lifetime of a particular kind of mRNA molecule makes it possible to control the number of protein molecules translated from it. In some cases messenger RNA stability is under hormonal control. This is true for mRNA that codes for vitellogenin, a protein synthesized in the livers of certain female animals such as frogs and chickens and then transported to the oviduct, where it is used in the formation of egg yolk proteins.

Vitellogenin synthesis is regulated by the hormone estradiol. When estradiol levels are high, the half-life of vitellogenin mRNA in frog liver is about 500 hours. When cells are deprived of estradiol, the half-life of the mRNA drops rapidly to less than 165 hours. This leads to a rapid decrease in cellular vitellogenin mRNA levels and decreased synthesis of the vitellogenin protein. In addition to affecting the stability of the mRNA, the hormone seems to control the rate at which the mRNA is synthesized.

The activity of eukaryotic proteins may be altered by posttranslational chemical modifications

The ultimate phenotypic expression of a gene may also be controlled by regulation of the activity of the gene product. As in bacteria, many metabolic pathways in eukaryotes contain allosteric enzymes that are regulated through feedback inhibition. In addition, many eukaryotic proteins are extensively modified after they are synthesized.

In *proteolytic processing* the proteins are synthesized as inactive precursors, which are converted to an active form by removal of a portion of the polypeptide chain. Other proteins may be regulated in part by a process of *selective degradation,* which keeps their numbers constant within the cell.

Chemical modification, through the addition or removal of functional groups, can reversibly alter the activity of an enzyme. One very common way of modifying the activity of an enzyme or other protein is the addition or removal of phosphate groups. These alterations allow the cell to respond rapidly to certain hormones (Chapter 47), or to fast-changing environmental or nutritional conditions.

Figure 13–12 Differential mRNA processing. In some cases a complex transcriptional unit can be processed in more than one way to yield two or more mRNAs, each of which encodes a related, but different, protein. In this generalized example the gene contains a segment that can be an exon in tissue A *(left),* but an intron in tissue B *(right).*

I. Most regulated genes in bacteria are organized into **operons,** each of which may encode several proteins.

A. Each operon has a single **promoter** region upstream from the protein-coding regions.

B. The **operator** is a sequence of bases that overlaps the promoter and serves as the regulatory switch responsible for **transcriptional level control** of the operon.

1. A **repressor protein** binds specifically to the operator sequence and blocks transcription by preventing RNA polymerase from binding to the promoter.

2. When the repressor is not bound to the operator, RNA polymerase can bind to the promoter and transcription can proceed.

C. An **inducible** operon such as the lactose operon is normally turned off. The repressor protein is synthesized in an active form that binds to the operator. If lactose is present, it is converted to allolactose (the **inducer**), which binds an allosteric site on the repressor protein, causing it to change shape. The altered repressor cannot bind to the operator, and the operon is transcribed.

D. A **repressible** operon such as the tryptophan operon is normally turned on. The repressor protein is synthesized in an inactive form that cannot bind to the operator. A metabolite (usually the end product of a metabolic pathway) acts as a **corepressor.** When intracellular corepressor levels are high, one of the molecules binds to an allosteric site on the repressor, changing its shape so that it can bind to the operator and thereby turn off transcription of the operon.

E. Repressible and inducible operons are under **negative control.** When the repressor protein binds to the operator, transcription of the operon is turned off.

F. Some inducible operons are also under **positive control.** A separate protein can bind to the DNA and stimulate transcription of the gene.

1. The lactose operon is activated by **CAP (catabolite gene activator protein),** which binds to the promoter region, stimulating transcription by binding RNA polymerase tightly.

2. To bind to the lactose operon, CAP requires cAMP (cyclic adenosine monophosphate). Levels of cAMP increase as levels of glucose decrease.

G. A group of operons can be organized into a multigene system, known as a **regulon,** which is controlled by a single regulatory protein. CAP activates a regulon associated with the metabolism of carbohydrates.

II. **Constitutive genes** are neither inducible nor repressible; they are active at all times. Regulatory proteins such as CAP and the repressor proteins are produced constitutively. The activity of these genes is controlled by how efficiently RNA polymerase binds to their promoter regions.

III. **Transcription factors** are DNA-binding regulatory proteins that work by recognizing and binding to specific base sequences in the DNA.

IV. Some **posttranscriptional controls** operate in prokaryotes.

A. A **translational level control** regulates the rate of translation of a particular mRNA.

B. **Posttranslational controls** include **feedback inhibition** of key enzymes in some metabolic pathways.

V. Eukaryotic genes are generally not organized into operons. Regulation of eukaryotic genes can occur at the levels of transcription, mRNA processing, translation, and the protein product.

A. The promoter of a regulated eukaryotic gene consists of an RNA polymerase-binding site and short DNA sequences known as **upstream promoter elements (UPEs).** The efficiency of the promoter is determined by the number and types of UPEs within the promoter region.

B. Inducible eukaryotic genes are controlled by **enhancer** elements, which can operate thousands of bases away from the promoter. Proteins that bind to enhancers appear to facilitate the binding of RNA polymerase to the promoter.

C. As in prokaryotes, eukaryotic genes are controlled by transcription factors. Many of these are transcriptional activators; others are transcriptional repressors.

D. The activity of eukaryotic genes is affected by chromosome structure.

1. Some genes whose products are required in large amounts exist as multiple copies in the chromosome. Other genes may be selectively amplified by DNA replication in some cells.

2. Genes can be inactivated by changes in chromosome structure. Densely packed regions of chromosomes called **heterochromatin** contain inactive genes. Active genes are associated with a loosely packed chromatin structure called **euchromatin.**

3. **DNA methylation** is a mechanism that perpetuates gene inactivation.

E. Many eukaryotic genes are regulated after the RNA transcript is made.

1. Gene regulation can occur as a consequence of mRNA processing. In some cases a single gene can produce different forms of a protein, depending on how the pre-mRNA is polyadenylated or spliced.

2. Certain regulatory mechanisms increase the stability of mRNA, allowing more proteins to be formed per mRNA molecule prior to degradation.

3. Posttranslational control of eukaryotic genes can occur by feedback inhibition or by modification of the protein structure.

---

1. At a time when the lactose operon is actively transcribed
   (a) the operator is bound to the inducer
   (b) the lactose repressor is bound to the promoter
   (c) the operator is not bound to the promoter
   (d) the gene coding for the repressor is not expressed constitutively
   (e) the lactose repressor is bound to the inducer

2. A repressible operon codes for the enzymes of the following pathway. Which component of the pathway is most likely to be the corepressor for that operon?

A ⟶ B ⟶ C ⟶ D
   Enzyme 1   Enzyme 2   Enzyme 3

   (a) substance A   (d) enzyme 1
   (b) substance B or C   (e) enzyme 3
   (c) substance D

3. An mRNA molecule transcribed from the lactose operon contains nucleotide sequences complementary to
   (a) structural genes coding for enzymes
   (b) the operator region
   (c) the promoter region
   (d) the repressor gene
   (e) introns

4. Feedback inhibition is an example of control at the _____ level.
   (a) transcriptional
   (b) translational
   (c) posttranslational
   (d) replicational
   (e) all of the above

5. Which of the following control mechanisms is generally the most economical in terms of conserving energy and resources?

(a) control by means of operons and regulons
(b) feedback inhibition
(c) selective degradation of mRNA
(d) selective degradation of enzymes
(e) gene amplification

6. A repressible operon, such as the tryptophan operon, is "off" when
(a) the gene that codes for the repressor is expressed constitutively
(b) the repressor-corepressor complex binds to the operator
(c) the repressor binds to the structural genes
(d) the corepressor binds to RNA polymerase
(e) CAP binds to the promoter

7. Which of the following is an example of positive control?
(a) transcription can occur when a repressor binds to an inducer
(b) transcription cannot occur when a repressor binds to a corepressor
(c) transcription is stimulated when a transcription activator binds to DNA
(d) a and b

(e) a and c

8. Which of the following are typically absent in prokaryotes?
(a) enhancers   (d) promoters
(b) transcription factors   (e) operators
(c) repressors

9. The "zipper" of a leucine zipper protein attaches
(a) specific amino acids to specific DNA base pairs
(b) two polypeptide chains to each other
(c) one DNA region to another DNA region
(d) amino acids to zinc atoms
(e) RNA polymerase to the operator

10. Inactive genes tend to be associated with
(a) highly condensed chromatin, known as euchromatin
(b) decondensed chromatin, known as euchromatin
(c) highly condensed chromatin, known as heterochromatin
(d) decondensed chromatin, known as heterochromatin
(e) chromatin that is not organized as nucleosomes

REVIEW QUESTIONS

1. Make a sketch of the lactose operon and briefly describe its function. Be sure to include the following elements: (a) structural genes, (b) promoter, (c) operator, (d) CAP-binding site.
2. What structural features does the tryptophan operon have in common with the lactose operon? What features are different?
3. Why do we define the tryptophan operon as repressible and the lactose operon as inducible?
4. How is glucose involved in the positive control of the lactose operon? How is CAP similar to the lactose repressor protein? How is it different?

5. Compare the structure of a prokaryotic promoter region with known eukaryotic promoter regions. How does the regulation of inducible eukaryotic genes differ from the regulation of inducible prokaryotic genes?
6. Explain why it is necessary for certain genes in eukaryotic cells to be present in multiple copies.
7. How can the activity of some eukaryotic genes be affected by the structure of the chromosome?
8. Make a sketch showing how differential mRNA processing can give rise to different forms of a eukaryotic protein.

YOU MAKE THE CONNECTION

1. Develop a simple hypothesis that would explain the behavior of each of the following types of mutants in *E. coli*:
(a) *Mutant a:* The map position of this mutation is in the tryptophan operon. The mutant cells are constitutive; that is, they produce all of the enzymes coded for by the tryptophan operon, even if large amounts of tryptophan are present in the growth medium.
(b) *Mutant b:* The map position of this mutation is in the tryptophan operon. The mutant cells do not produce any of the enzymes coded for by the tryptophan operon under any conditions.
(c) *Mutant c:* The map position of this mutation is some distance from the tryptophan operon. The mutant cells are constitutive; that is, they produce all of the enzymes coded for by the tryptophan operon, even if the growth medium contains large amounts of tryptophan.

(d) *Mutant d:* The map position of this mutation is some distance from the tryptophan operon. The mutant cells do not produce any of the enzymes coded for by the tryptophan operon under any conditions.
2. Compare the types of bacterial genes associated with inducible operons, those associated with repressible operons, and those that are constitutive. Predict the category into which each of the following would most likely fit: (a) a gene that codes for RNA polymerase; (b) a gene that codes for an enzyme required to break down maltose; (c) a gene that codes for an enzyme used in the synthesis of adenine.
3. The regulatory gene that codes for the tryptophan repressor is not tightly linked to the tryptophan operon. Would it be advantageous if it were? Explain your answer.

RECOMMENDED READINGS

Buratowski, S. "Mechanisms of Gene Activation." *Science*, Vol. 270, 15 Dec. 1995. An overview of how regulators are able to affect gene transcription.
Pennisi, E. "Chemical Shackles for Genes?" *Science*, Vol. 273, 2 Aug. 1996. DNA methylation is involved in the regulation of gene activity in plants as well as vertebrates.
Stein, G.S., J.L. Stein, A.J. van Wijnen, and J.B. Lian. "The Maturation of

a Cell." *American Scientist*, Vol. 84, Jan.-Feb. 1996. The authors describe the activation of genes in the maturation of a bone cell.
Tjian, R. "Molecular Machines That Control Genes." *Scientific American*, Vol. 272, No. 2, Feb. 1995. A discussion of the transcriptional machinery, with a perspective on how this information might be used in the future to treat certain diseases.

● Visit our website at **http://www.saunderscollege.com/lifesci/titles.html** and click on Solomon/Berg/Martin Biology for links to chapter-related resources on the World Wide Web.

CHAPTER 14

Genetic Engineering

Beginning in the mid-1970s, a revolution in the field of biology occurred as the development of **recombinant DNA technology** led to radically new research approaches. This technology has had a major impact not only in genetic studies but also in areas ranging from cell biology to evolution.

Recombinant DNA techniques were initially developed to permit scientists to obtain a great many copies of any specific DNA segment in order to study it biochemically. This was first done by introducing foreign DNA into the cells of microorganisms. Under the right conditions, this DNA is replicated and transmitted to the daughter cells when a cell divides. In this way a particular DNA sequence can be amplified, or **cloned,** to provide millions of identical copies that can be isolated in pure form. Today methods for cloning DNA in vitro (outside of a living organism) have become increasingly important.

Studies of cloned **DNA sequences** have been of immense value in allowing scientists to understand the organization of genes and the relationship between genes and their products. In fact, most of our knowledge of the complex structure and control of eukaryotic genes (see Chapters 12 and 13) is derived from the application of these methods.

Recombinant DNA technology also has many practical applications. One of the rapidly advancing areas of study today is **genetic engineering**—the modification of the DNA of an organism to produce new genes with new characteristics. Genetic engineering can take many forms, ranging from the production of strains of bacteria that manufacture useful protein products to the development of plants and animals that express foreign genes. Discoveries in molecular genetics have led to technologies that today are applied in such diverse areas as medicine and the pharmaceutical industries, foods and agriculture, and others. For example, the scientists in the photograph are using genetic engineering techniques to introduce desirable characteristics (perhaps disease resistance or increased yield) into tomatoes.

(Robert Holmgren/Peter Arnold, Inc.)

AFTER YOU HAVE STUDIED THIS CHAPTER YOU SHOULD BE ABLE TO

1. Draw a sketch that demonstrates how a typical restriction enzyme cuts DNA molecules and give examples of the ways in which these enzymes are used in recombinant DNA technology.
2. Summarize the properties of plasmids that allow them to be used as DNA cloning vectors.
3. Distinguish between a genomic DNA library and a cDNA library.
4. Explain why one would clone the same eukaryotic gene from both a genomic library and a cDNA library.
5. Identify some of the uses of DNA hybridization probes.

6. Describe how specific primers can be used to amplify specific genes from a mixture of genomic DNA or cDNA.
7. Draw a diagram that illustrates the most widely used DNA sequencing technique.
8. List some important proteins and other products that can be produced by genetic engineering techniques.
9. List some of the difficulties encountered in using *Escherichia coli* to produce proteins coded by eukaryotic genes and explain the rationale behind using transgenic plants and animals to solve some of those problems.

RECOMBINANT DNA METHODS GREW OUT OF RESEARCH IN MICROBIAL GENETICS

Recombinant DNA technology was not developed quickly. It actually began with the first genetic studies of bacteria and the viruses that infect them, the **bacteriophages** (literally "bacteria eaters"; see Chapters 11 and 23). Only after decades of basic research and the accumulation of extensive knowledge did the current technology become feasible and available to the many scientists who now use these methods.

Bacteria have provided researchers with special enzymes, known as restriction enzymes, that cut DNA molecules only in specific places. Bacteria, viruses, and other microorganisms also provide a means by which many identical copies of DNA molecules can be made. This is accomplished in vitro through the use of certain bacterial DNA polymerases and in vivo by introducing recombinant DNA molecules into bacteria or other microorganisms, such as yeast. Recombinant DNA technology also depends on a fundamental property of nucleic acid molecules: the ability of complementary nucleotide sequences to pair by means of specific hydrogen bonding. Because new methods of genetic engineering are continually emerging, we will not attempt to explore them all. (Several Internet sites, such as "Access Excellence" and the "National Center for Biotechnology Information," are good sources of updated information.) Instead, we will discuss some of the major approaches that have provided a foundation for these developments.

Restriction enzymes are "molecular scissors"

A major breakthrough in the development of recombinant DNA technology was the discovery of bacterial enzymes called **restriction enzymes,** which are able to cut DNA molecules only at specific base sequences. For example, a restriction enzyme known as Hind III recognizes and cuts a DNA molecule at the base sequence 5′—AAGCTT—3′, whereas the sequence

5′—GAATTC—3′ is cut by another, known as EcoRI.[1] Recall from Chapter 11 that during infection a bacteriophage injects its DNA into a bacterial cell. Such a cell can defend itself if it possesses restriction enzymes capable of attacking the bacteriophage DNA. The bacteria protect their own DNA from breakdown by modifying it after replication. An enzyme adds a methyl group to one or more bases in each restriction site so that the restriction enzyme is unable to recognize and cut the DNA. Purification of restriction enzymes has enabled scientists to cut DNA from chromosomes into shorter fragments in a controlled way.

Many of the restriction enzymes used for recombinant DNA studies cut **palindromic** sequences, which means that the base sequence of one strand reads the same as its complement, but in the opposite direction. (Thus, the complement of our example, 5′—AAGCTT—3′, reads 3′—TTCGAA—5′.) By cutting both strands of the DNA, but in a staggered fashion, these enzymes produce fragments with identical, complementary, single-stranded ends. These ends are called "sticky ends" because they can pair (by hydrogen bonding) with the complementary, single-stranded ends of other DNA molecules that have been cut with the same enzyme (Fig. 14–1). Once two molecules have been joined together in this way, they can be treated with **DNA ligase** (see Chapter 11), an enzyme that covalently links the two DNA fragments to form a stable recombinant DNA molecule.

Restriction enzymes vary widely in the number of DNA bases that they recognize, ranging from as few as 4 to as many as 23 bases. If the restriction sites are randomly distributed in the DNA, we expect the restriction sequence of a "four-base cutter" to occur on the average of every 4^4, or 256, bases. A four-base cutter would therefore produce fragments with an average length of 256 bases, whereas a six-base cutter would produce fragments averaging 4^6, or 4096, bases.

---

[1] The names of restriction enzymes are generally derived from the names of the bacteria from which they originally isolated. Hence Hind III and EcoRI are derived from *Hemophilus influenzae* and *Escherichia coli* respectively.

Figure 14–1 Cutting DNA with a restriction enzyme. Many restriction enzymes, like Hind III, cut DNA at sequences that are palindromic, thereby producing complementary sticky ends.

Recombinant DNA is formed when DNA is spliced into a vector

Most recombinant DNA molecules are isolated and amplified by introducing them into cells of the bacterium *E. coli*. To isolate a specific piece of DNA after it has been cut by a restriction enzyme, that fragment must first be incorporated into a suitable carrier, or **vector molecule.**

Bacteriophages or special DNA molecules called plasmids are commonly used as vectors (Fig. 14–2). Recall that bacterial DNA is in the form of a circle; a **plasmid** is a separate, much smaller, circular DNA molecule that may be present and able to replicate inside a bacterial cell. These plasmids can be isolated from bacterial cells in pure form and then be introduced into other cells by a method called **transformation** (see Chapter 11), which involves altering the bacterial cell wall to make it permeable to the plasmid DNA molecules. Once a plasmid enters a cell, it is replicated and distributed to the daughter cells during cell division. Plasmids do not carry genes that are essential to the *E. coli* cells under normal conditions, but they often carry genes that are useful under some environmental conditions, such as those that confer resistance to particular antibiotics.

The plasmids now used in recombinant DNA work have been extensively "engineered" to include a number of features helpful in the isolation and analysis of cloned DNA. Among these are: (1) an origin of replication, (2) one or more sites that can be cut by restriction enzymes, and (3) genes that allow researchers to select cells that have been transformed by recombinant plasmids. These are genes that permit transformed cells to grow under specified conditions that do not allow growth of untransformed cells. Genes that confer resistance to certain antibiotics, and genes that allow the cells to use a particular nutrient are examples (Fig. 14–3).

A limiting property of any vector, however, is the size of the DNA fragment that it can effectively carry. The size of a DNA segment is often given in kilobases, with 1 **kilobase (kb)** being equal to 1000 bases. Fragments smaller than 10 kb can usually be inserted into plasmids for use in *E. coli*. However, larger fragments require the use of bacteriophage vectors, which can handle up to 15 kb of DNA.

Recombinant DNA can also be introduced into cells of more complex organisms. For example, engineered viruses are used as vectors in mammalian cells. These viruses have been disabled in such a way that they do not kill the cells they infect; instead their DNA, and any foreign DNA they carry, becomes incorporated into the chromosomes of the cell following infection. As discussed later, other methods have been developed that do not require a biological vector.

DNA can be cloned inside cells

Because a single gene is only a small part of the total DNA in an organism, isolating the piece of DNA containing that particular gene is like finding a needle in a haystack. A powerful detector is needed. Today there are many methods that permit the isolation of a gene from an organism. We will start with methods that begin with cloning inside bacterial cells to construct a library, or gene bank, using human DNA as an example, although the procedure can be applied to any organism.

A genomic library contains fragments of all the DNA in the genome

The total DNA per cell is referred to as a **genome.** For example, if DNA is extracted from human cells, we refer to it as human genomic DNA. A **genomic library** is a collection of DNA fragments that are more or less representative of all the DNA in the genome. Each fragment is spliced into a plasmid, which is usually contained inside a bacterial cell.

The first step in producing a genomic library is to cut the DNA with a restriction enzyme, generating a population of DNA fragments. These fragments vary in size and in the genetic information they carry, but they all have identical sticky

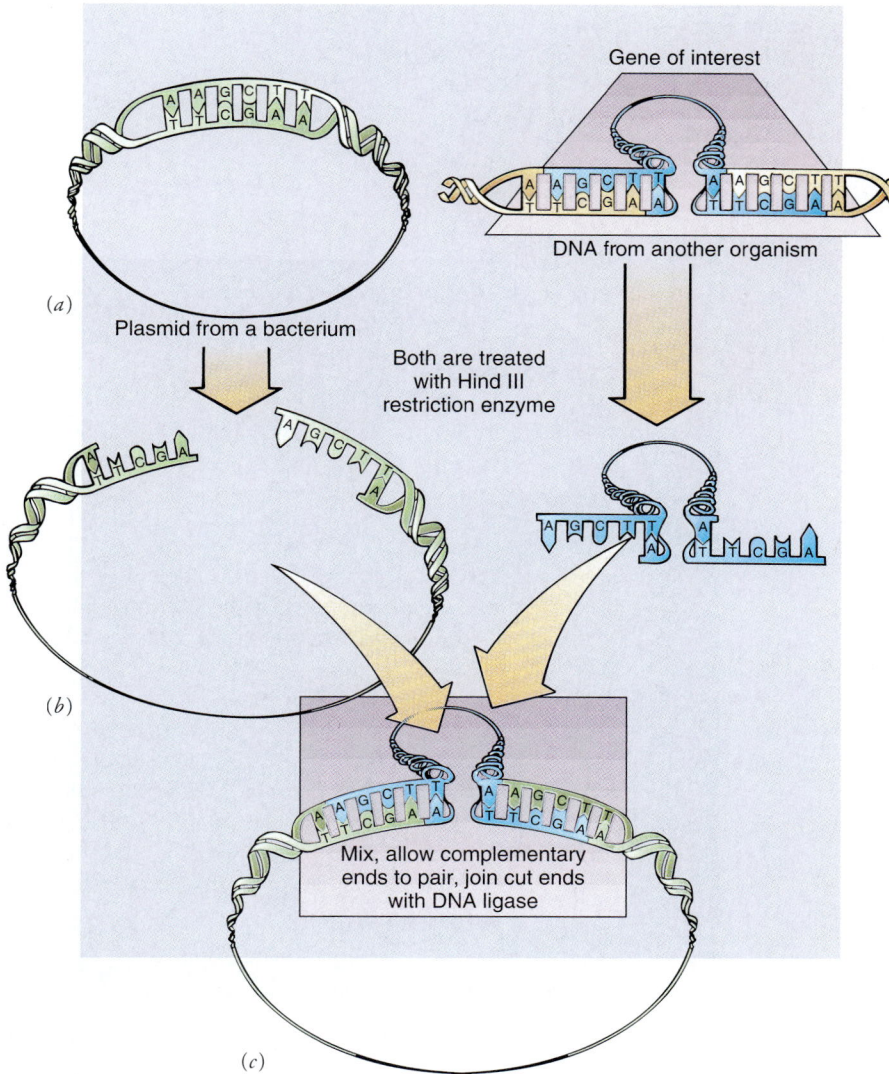

Figure 14–2 **Splicing DNA into a vector.** In this example, a circular plasmid from a bacterium is the vector. (*a*) The plasmids and the DNA from another organism are cut by the same restriction enzyme to produce (*b*) linear molecules with complementary single-stranded ends. (*c*) The recombinant DNA is constructed by mixing the two types of molecules so that their sticky ends pair. DNA ligase then forms covalent 5′ → 3′ phosphodiester linkages at the junctions.

Gene of interest

DNA from another organism

Plasmid from a bacterium

Both are treated with Hind III restriction enzyme

Mix, allow complementary ends to pair, join cut ends with DNA ligase

ends. Plasmid DNA to be used as a vector is treated with the same restriction enzyme, which converts the circular plasmids into linear molecules with sticky ends complementary to those of the human DNA fragments. Recombinant plasmids are produced by first mixing the two kinds of DNA (human and plasmid) together under conditions that promote hydrogen bonding of complementary bases. Then DNA ligase is used to covalently bond the paired ends of the plasmid and human DNA.

The recombinant plasmids are inserted into antibiotic-sensitive *E. coli* cells by transformation. Because the ratio of plasmids to cells is kept very low, it is rare for a cell to receive more than one plasmid molecule, and not all cells receive a plasmid. The normally antibiotic-sensitive cells are incubated on a nutrient medium that includes antibiotics, so only cells that have incorporated a plasmid (which contains a gene for

antibiotic resistance) are able to grow (Fig. 14–4). In addition, the plasmid has usually been engineered in ways that permit researchers to select cells containing *recombinant* plasmids.

A genomic library contains redundancies; that is, many human DNA sequences have been duplicated many times, purely by chance. However, each individual recombinant plasmid (analogous to a book in the library) contains only a single fragment of the total human genome. Each of these fragments is usually smaller than a gene; therefore several fragments must be isolated to study the complete gene.

To allow identification of a plasmid containing a sequence of interest, each plasmid must be amplified, or cloned, until there are millions of copies to work with. This process occurs as the *E. coli* cells grow and divide. A dilute sample of the bacterial culture is spread on solid growth medium, so that the cells will be widely separated. When each cell reproduces it

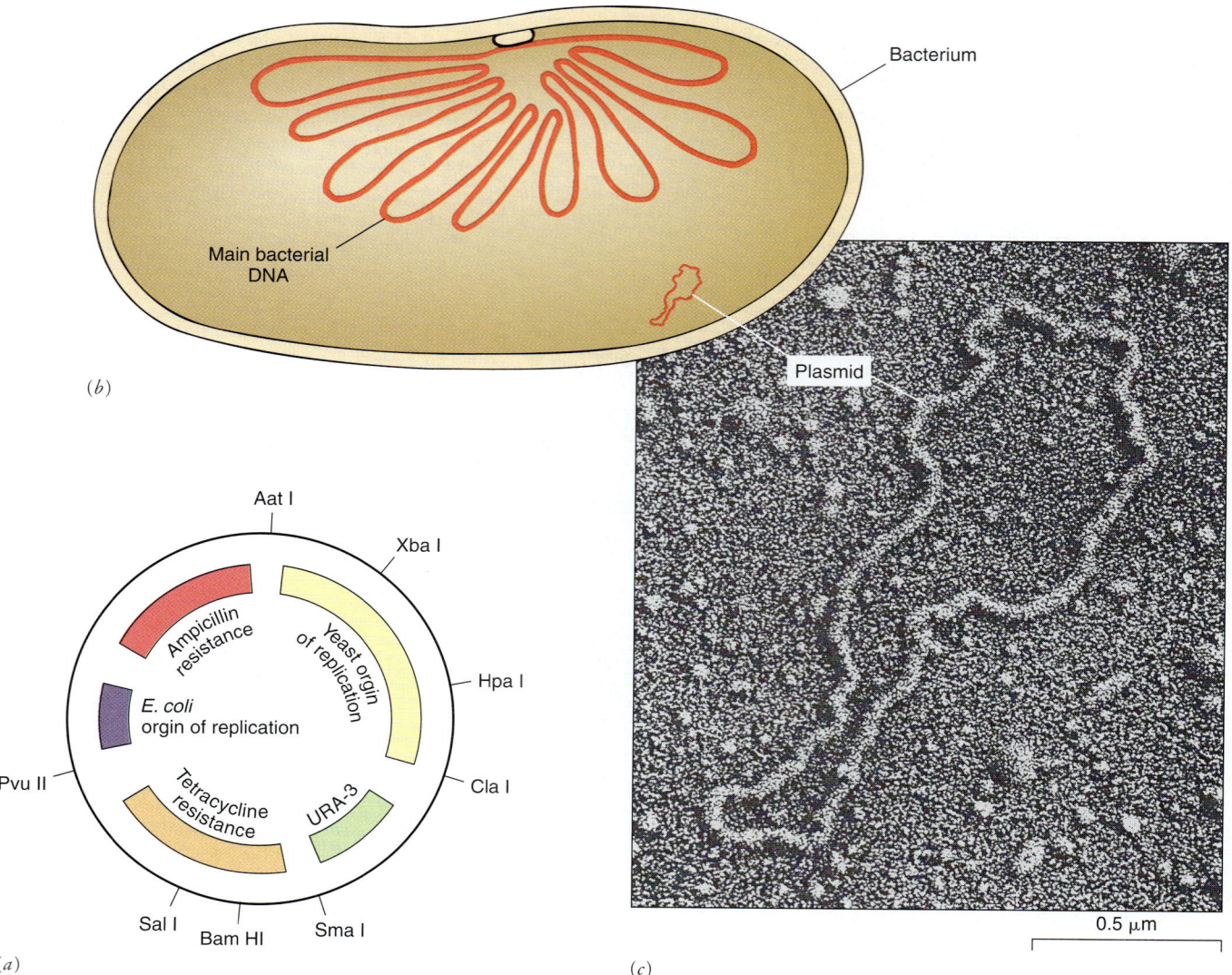

Figure 14–3 **Plasmids.** (*a*) This genetically engineered plasmid vector has many useful features. It was constructed from DNA fragments isolated from plasmids, *E. coli* genes, and yeast genes. The two origins of replication, one for *E. coli* and one for yeast, *Saccharomyces cerevisiae*, allow it to replicate independently in either type of cell. Letters on the outer circle designate sites for restriction enzymes that cut the plasmid only at that position. Resistance genes for the antibiotics ampicillin and tetracycline, and the yeast URA-3 gene are also shown. The URA-3 gene is useful when transforming yeast cells lacking an enzyme required for uracil synthesis. Cells that take up the plasmid are able to grow on a uracil-deficient medium. (*b*) The relative sizes of a plasmid and the main DNA of a bacterium. (*c*) TEM of a plasmid from *E. coli*. (*c, Dr. Stanley Cohen/Science Photo Library/Photo Researchers, Inc.*)

gives rise to a **colony,** which is a clone of genetically identical cells. All the cells of a particular colony contain the same recombinant plasmid, so during this process a specific sequence of human DNA has also been cloned. The major task is to determine which colony (out of thousands) contains a cloned fragment of interest. There are a number of ways in which specific DNA sequences can be identified.

A specific DNA sequence can be detected by a complementary genetic probe

A common approach to the problem of detecting the DNA of interest involves the use of a **genetic probe,** which is usually a radioactively labeled segment of RNA or single-stranded DNA that can **hybridize** (become attached by base pairing)

(a) A specific restriction enzyme is used to cut human DNA into fragments; these are then spliced into a complementary restriction site in a plasmid that contains an antibiotic resistance gene (R).

Sites of cleavage

Human DNA

Fragment 1   Fragment 2   Fragment 3   Fragment 4

(a)

Cut with a restriction enzyme

Fragment of interest

(b)

(b) The recombinant plasmids are used to transform an antibiotic-sensitive *E. coli* strain under conditions that ensure that each bacterial cell receives no more than one plasmid molecule.

Gene for resistance to antibiotic

Bacteria with plasmid live and multiply

Plate with antibiotic-containing medium

Bacteria without plasmid fail to grow

(c) The cells are then grown on antibiotic-containing solid nutrient medium. Only those that receive the plasmid survive.

(c)

Figure 14-4   Cloning genomic DNA.

---

to complementary DNA sequences in the target gene. Suppose that a researcher wishes to identify a gene that codes for a specific protein. If at least part of the amino acid sequence of that protein is known, it is possible to synthesize a radioactive, single-stranded DNA fragment that could code for that sequence. This is not as simple as it may sound: because the genetic code is redundant, a specific amino acid sequence could potentially be coded for by a number of different base sequences (see Chapter 12). One approach to this problem has been to synthesize a mixture of probes, each of which could code for the desired amino acid sequence.

Genetic probes can be used in a variety of ways. For example, cells from *E. coli* colonies containing recombinant plasmids can be transferred to a nitrocellulose filter, which then becomes a *replica* of the colonies. The cells on the filter are treated chemically to lyse them and to cause the DNA to become denatured (single-stranded). The filter is then incubated with the radioactive probe mixture to allow the probes to hybridize with any complementary DNA that may be present. Each spot on the filter containing DNA complementary to

that particular probe becomes radioactive and can be detected by autoradiography (see Chapter 2), using a special x-ray film. Each spot on the film therefore identifies a colony containing a plasmid that includes the DNA of interest (Fig. 14–5).

A cDNA library is complementary to mRNA and does not contain introns

For reasons that will be discussed below, researchers frequently wish to avoid cloning introns and other parts of genes that do not directly code for proteins. In such cases they construct libraries consisting of DNA copies of eukaryotic mRNA. The copies, known as **complementary DNA (cDNA)** because they are complementary to RNA, lack introns. **Reverse transcriptase** (see Chapter 12) is used to synthesize single-stranded cDNA, which is then separated from the RNA and made double stranded by DNA polymerase (Fig. 14–6). A **cDNA library** is formed when the double-stranded cDNA molecules are inserted into plasmid or virus vectors, which then multiply in *E. coli* cells.

Figure 14–5 **Use of a genetic probe.** A radioactive nucleic acid probe (which can be either RNA or single-stranded DNA) reveals the presence of complementary DNA sequences.

Figure 14–6 **Formation of cDNA.** Enzymes, including reverse transcriptase, are used to produce a mixture of cDNA molecules similar to the one illustrated.

Cloning a gene from both a cDNA library and a genomic library has several advantages. Analysis of the genomic DNA clones gives useful information about the structure of the gene in the chromosome and the structure of the primary pre-mRNA transcript, as well as nontranscribed regulatory regions.

Analysis of cDNA clones allows investigators to determine certain characteristics of the protein encoded by the gene, including its exact amino acid sequence. The structure of the processed mRNA can also be studied. Furthermore, because the cDNA copy of the mRNA does not contain intron sequences, comparison of the cDNA and genomic DNA base sequences reveals the locations of intron and exon coding sequences in the gene.

Cloned cDNA sequences are also useful when it is desirable to produce a eukaryotic protein in *E. coli.* When an intron-containing human gene such as the gene for human growth hormone is introduced into *E. coli,* the bacterium is unable to remove the introns from the transcribed RNA to make a functional mRNA for the production of its protein product. If a cDNA clone of the gene is inserted into the bacterium, however, its transcript contains an uninterrupted coding region. A functional protein can be synthesized if the gene is inserted downstream of an appropriate bacterial promoter.

The polymerase chain reaction is a technique for amplifying DNA in vitro

The methods to amplify a specific DNA sequence described above all involve cloning DNA in cells, usually those of bacteria. These processes are time-consuming and require an adequate DNA sample as starting material. The **polymerase chain reaction (PCR)** technique allows researchers to amplify a tiny sample of DNA millions of times in a few hours (Fig. 14–7).

DNA polymerase uses nucleotides and primers to replicate a DNA sequence in vitro, thereby producing two DNA molecules. The two strands of each molecule are then separated by heating and replicated again, so then there are four, double-stranded molecules. After the next cycle of heating and replication there are eight molecules, and so on, with the number of DNA molecules doubling in each cycle. After only 20 heating and cooling cycles (which are carried out using automated equipment) this exponential process yields 2^{20}, or over 1 million, copies of the target sequence!

Because the reaction can only be carried out efficiently if the DNA polymerase can remain stable through many heating cycles, a special heat-resistant DNA polymerase, known as *Taq* polymerase is used. The name of this enzyme reflects its source, *Thermus aquaticus,* a bacterium that lives in hot springs

Figure 14–7 **The polymerase chain reaction.** The number of double-stranded DNA molecules doubles each time the cycle of heating and cooling is repeated. The initial reaction mixture (*not shown*) includes a very small amount of double-stranded DNA, DNA precursors (deoxyribonucleotides), specific nucleic acid primers, and *Taq* polymerase.

1. Double-stranded DNA

2. Single-stranded DNA
Heat
Cool

3. Primers attach to DNA

4. DNA polymerase extends primers

Repeat from step 2

in Yellowstone Park. (Similar enzymes can be found in bacteria living in deep-sea thermal vents; see *Focus On: Life without the Sun,* Chapter 52.)

The PCR technique is particularly valuable because only specific *target sequences* are replicated. Recall from Chapter 11 that DNA polymerase can add nucleotides only to a preexisting polynucleotide strand. In the cell, DNA synthesis begins with the formation of a short primer, which is then extended by DNA polymerase. In the PCR technique, chemically synthesized primers with a specified nucleotide sequence are included in the reaction mixture. The primers attach to complementary target sequences of the single-stranded DNA, thereby designating the starting point for replication by DNA polymerase. In this way, a specific sequence can be cloned from an unpurified mixture of DNA sequences.

The PCR technique has virtually limitless applications. It allows the amplification and analysis of tiny DNA samples from seemingly unlikely sources, ranging from crime scenes to archaeological remains; for example, in 1997 the first analysis of DNA obtained from the bones of Neandertals was reported (Chapter 21).

If the PCR technique has a flaw, it is the fact that it is almost too sensitive. Even a tiny amount of contaminant DNA in a sample could potentially become amplified and lead to an erroneous conclusion. Researchers are constantly improving their methods to avoid this and other technical pitfalls.

Gel electrophoresis is the most widely used technique to separate macromolecules

Mixtures of certain macromolecules such as polypeptides, DNA fragments, or RNA can be separated by **gel electrophoresis,** a method that exploits the fact that these molecules carry charged groups that cause them to migrate in an electrical field. Figure 14–8 illustrates gel electrophoresis of DNA molecules. Because nucleic acids are negatively charged (see Chapters 11 and 12), both DNA and RNA migrate through the gel toward the positive pole of the electrical field. Because the gel retards the movement of the large molecules more than the small molecules, the rate at which they travel is inversely proportional to their length (molecular weight). Including DNA fragments of known size as standards allows accurate measurement of the molecular weights of the unknown fragments.

It is sometimes easier to work with the DNA fragments separated by gel electrophoresis if they are transferred to a nitrocellulose filter, which picks up the DNA much as a blotter picks up ink. The DNA on the filter is then denatured and incubated with a radioactive genetic probe, which hybridizes with any complementary fragments. When the "blot," which is essentially a replica of the gel, is used for autoradiography (Chapter 2), the resulting spots on the x-ray film correspond to the locations of the fragments in the gel that are complementary to the probe. This type of **blot hybridization** (called a **Southern blot** after its inventor, E.M. Southern) has widespread applications. It is often used to diagnose certain types of genetic disorders.

Similar blotting techniques are used to study RNA and proteins. When RNA molecules separated by electrophoresis are transferred to a membrane, the result is, rather in jest, called a **Northern blot.** In the same spirit, the term **Western blot** is applied to a blot consisting of polypeptides previously separated by gel electrophoresis. (So far, no one has invented a type of blot that could be called an "Eastern blot.") In the case of Western blotting, the polypeptides of interest are recognized by radioactive antibody molecules that bind to them specifically.

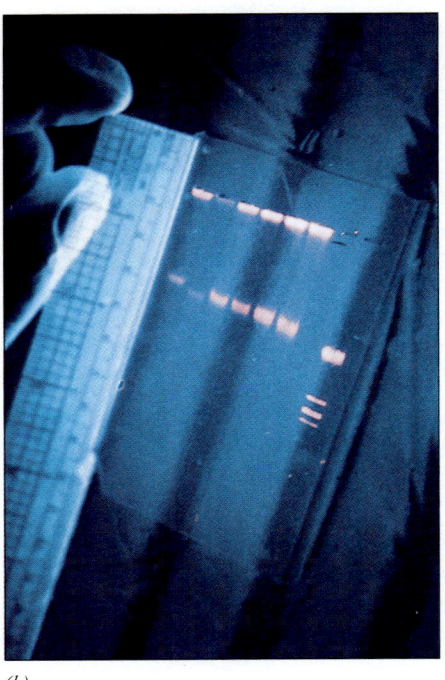

Figure 14–8 Gel electrophoresis. Charged molecules, such as DNA, RNA, or protein, can be separated based on the rate at which they migrate in an electrical field. *(a)* An electrical field is set up in a gel material, consisting of agarose or poly-acrylamide, which is poured as a thin slab on a glass or Plexiglas holder. After the gel has solidified, samples containing a mixture of macromolecules of different sizes are loaded in wells formed at one end of the gel, and then an electrical current is applied. The smallest DNA fragments *(green)* travel the longest distance. *(b)* A gel containing separated DNA fragments. The gel is stained with ethidium bromide, a dye that binds to DNA and is fluorescent under UV light. *(b, Michael Gabridge/Visuals Unlimited)*

A great deal of information can be inferred from a DNA nucleotide sequence

A cloned piece of DNA can be used as a research tool for a wide variety of applications. Even if the purpose of cloning the gene is to obtain the encoded protein for some industrial or pharmaceutical process, a great deal must be known about the gene and how it functions before it can be "engineered" for a particular application. The usual first step is to determine the sequence of nucleotides.

The most commonly used method of DNA sequencing is based on the fact that a replicating DNA strand that has incorporated a particular type of modified nucleotide, known as a dideoxynucleotide, cannot elongate beyond that point. Unlike a "normal" deoxynucleotide (which lacks a hydroxyl group on its 5′ carbon), a dideoxynucleotide also lacks a hydroxyl group on its 3′ carbon (Figure 14–9*a*). (Recall from Chapter 11 that a 3′ hydroxyl group is needed to react each time a phosphodiester linkage is formed.)

Four different reaction mixtures are prepared. Each contains DNA polymerase, appropriate radioactively labeled primers, and all four deoxynucleotides needed to synthesize DNA: dATP, dCTP, dGTP, and dTTP. Each also includes a small amount of only one of the four dideoxynucleotides: ddATP, ddCTP, ddGTP, or ddTTP (Figure 14–9*b*).

For example, consider how the reaction proceeds in the mixture that includes ddATP. At each site where adenine is specified, occasionally a growing strand will incorporate a ddATP and will be unable to elongate further. Consequently, a mixture of DNA fragments of varying lengths is formed in the reaction mixture. Each fragment that contains a ddATP marks a specific location where adenine is normally found. Similarly, in the reaction mixture that includes ddCTP, each fragment that contains ddCTP marks the position of a cytosine, and so on (Figure 14–9*c*).

The radioactive fragments from each reaction are denatured and then separated by gel electrophoresis, with each reaction mixture (corresponding to A, T, G, or C) occupying its own lane in the gel. The positions of the fragments in the gel can then be determined by autoradiography (Fig. 14–9*d* and *e*). Because the high resolution of the gel makes it possible to distinguish between fragments that differ in length by only a single nucleotide, one can read off the sequence one base at a time, beginning with the shortest fragment. To follow the example in Figure 14–9*d*, if the shortest fragment is in the "G" lane, then the first base is G; similarly, if the next shortest frag-

(a) Dideoxynucleotides are modified nucleotides that lack a 3' hydroxyl group and thus block further elongation of a new DNA chain.

(b) Four different reaction mixtures are used to sequence a DNA fragment; each contains a small amount of a single dideoxynucleotide, such as dideoxy ATP (adenine), and larger amounts of the four normal deoxynucleotides.

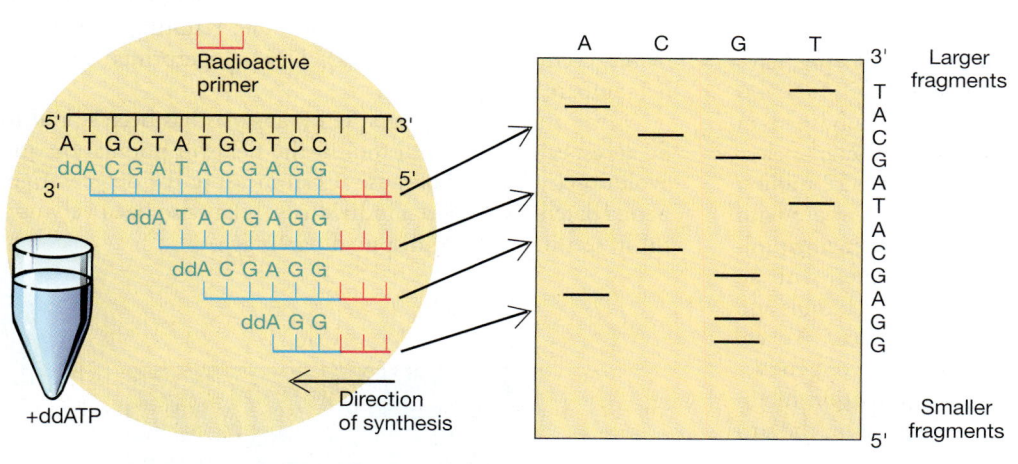

(c) The random incorporation of dideoxy ATP into the growing chain generates a series of smaller DNA fragments ending at all the possible positions where adenine is found in the original fragment.

(d) The radioactive products of each reaction mixture are separated by gel electrophoresis and located by exposing the gel to x-ray film. The nucleotide sequence of the newly synthesized DNA is read directly from the film (5' → 3').

(e) An exposed x-ray film of a DNA sequencing gel. The four lanes represent A, C, G, and T dideoxy reaction mixes, respectively.

Figure 14–9   DNA sequencing.   (e, Courtesy of B. Slatko, New England Biolabs) ment is also in the "G" lane, then the next base is also G; if the third shortest fragment is in the "A" lane, then the third base is A, and so on.

Knowing the DNA sequence of a cloned gene allows investigators to identify which parts of the DNA molecule contain the actual protein-coding sequences, as well as which parts may be regulatory regions involved in gene expression (see Chapter 13). Signals involved in mRNA processing and modification can be recognized, and the amino acid sequence of the encoded protein can be inferred directly from the base sequence. Prior to the development of DNA-sequencing methods, protein sequences were determined by laborious methods from highly purified protein samples. Although protein microsequencing technology has also advanced rapidly, in most cases cloning and sequencing a gene is easier than purifying and sequencing the encoded protein.

Because many parts of the process can now be automated, DNA sequencing can be done very rapidly. These advances in sequencing technology have made it possible for researchers to study the nucleotide sequences of a wide variety of organisms, both prokaryotic and eukaryotic. Much of this research received its initial impetus in conjunction with the Human Genome Project, to be discussed in Chapter 15. The genomes of a number of prokaryotes, both parasitic and free-living, have been completely sequenced. The complete sequence of the 4.6 million base pairs (4288 genes) of *E. coli* was published in 1997. This was preceded by another major landmark, electronic publication of the complete sequence of the 12 million base pairs (6223 genes) of yeast, a unicellular eukaryote with an unusually small genome, in 1996. Ongoing sequencing projects encompass a variety of multicellular eukaryotes, with an emphasis on those that have been important research tools and on others of agricultural and medical importance. We are actually on the verge of an extraordinary explosion of gene sequence data because much more advanced automation methods, in which thousands of bases can be sequenced on the surface of a single microchip, will soon be widely used.

DNA sequence information is now kept in large computer databases, many of which can be accessed through the Internet. Examples include databases maintained by the National Center for Biotechnology Information (a service of the U. S. National Library of Medicine and the National Institutes of Health) and by the Human Genome Organization (HUGO). These allow investigators to compare newly discovered sequences with those already known and to utilize many other kinds of information. By searching for DNA (and amino acid) sequences in a database, researchers can gain a great deal of insight into the function and structure of the gene product, as well as the evolutionary relationships among genes.

Restriction fragment length polymorphisms (RFLPs) are a measure of genetic relationships

The variability of genes within a population can be studied in a number of different ways. A direct approach is to determine

DNA sequences. The previously mentioned advances in DNA sequencing technology will make this the most widely used method in the future.

A more traditional procedure uses restriction enzymes. It is based on the fact that random DNA mutations and recombination may result in individuals differing in the number and location of sites where a particular restriction enzyme cuts the DNA. Therefore, each individual differs in the lengths of the fragments produced by that enzyme. Such **restriction fragment length polymorphisms** (commonly known as **RFLPs,** or "Riflips") can be used to determine how closely related different members of the population are. (The term *polymorphism* literally means "many forms." A genetic polymorphism is said to exist if individuals of two or more discrete genetic types, or "morphs," are found in a population or species; see Chapter 18.)

Restriction enzymes are used to cut the DNA from two or more individuals, and the fragments are separated by gel electrophoresis (with the DNA from each individual in a separate lane). A Southern blot is made of the DNA on the gel, which is then denatured and allowed to hybridize with a genetic probe representing a sequence that is repeated and interspersed throughout the genome. The resulting patterns of bands, commonly referred to as "DNA fingerprints" can then be compared (Fig. 14–10).

Over the past few years RFLP analysis has been found to be an especially powerful tool in the fields of population and evolutionary biology because it can measure the degree of genetic relatedness between individuals. It has also been very useful in settling cases of disputed parentage.

Figure 14–10 A restriction fragment length polymorphism (RFLP). The lanes marked "M" and "F" contain DNA from a mother and father, respectively, and the two marked "C" contain DNA from their children. Note that every band present in one of the children is also found in at least one of the parents. *(David Parker/Science Photo Library/Photo Researchers, Inc.)*

The most controversial use of this technology is in the field of forensics. If even small amounts of blood, semen, or other DNA-containing tissue are left at the scene of a crime, one or more target DNA sequences can be amplified by the PCR technique, cut with appropriate restriction enzymes, and subjected to electrophoresis.

If applied properly, DNA fingerprinting has the power to identify the guilty with a high degree of certainty. Conversely, it can exonerate the innocent. In fact, a number of convicted persons have won new trials and have been subsequently released from incarceration based on the application of DNA fingerprinting to physical evidence from the crime scene. Such evidence has been ruled admissible in many court cases, including certain trials that have received a great deal of attention in recent years. One limitation arises from the fact that the DNA samples are usually small and may have been degraded. Obviously, great care must be taken to prevent contamination of the samples. This is especially crucial if the PCR technique is to be used to amplify the DNA.

The development of DNA fingerprinting sparked a lively debate over how "unique" each individual pattern might be. For example, some scientists argued that a pattern that is quite rare in the general population might be more common in a particular ethnic group, and this might significantly affect the calculated probability of a match. As data on the frequency of particular patterns in various populations have accumulated, it has been learned that these concerns are of less practical importance than once thought. For example, the odds that two persons taken at random from the general population would have identical DNA fingerprints may be as low as one in several billion. If two persons are members of the same ethnic group the odds of a match may increase, but are usually still extremely low (perhaps one in several million).

GENETIC ENGINEERING HAS MANY APPLICATIONS

Recombinant DNA technology has provided not only a new and unique set of tools for examining fundamental questions about how living cells work but also new approaches to problems of applied technology in many other fields. In some cases the production of genetically engineered proteins and organisms has begun to have considerable impact on our lives. The most striking of these have been in the fields of pharmacology and medicine.

Human insulin produced by *E. coli* was one of the first genetically engineered proteins to be commercially produced. Prior to the development of the altered bacterium to produce the human hormone, insulin was derived exclusively from other animals. Many diabetic persons become allergic to the insulin from animal sources because its amino acid sequence differs slightly from human insulin. The ability to produce the human hormone by recombinant DNA methods has resulted in significant medical benefits to diabetics.

Genetically engineered human growth hormone (see Chapter 47) is required by some children to overcome growth deficiencies. Human growth hormone could previously be obtained only from cadavers. Only small amounts were available, and evidence suggested that some of the preparations were contaminated with infectious agents. The list of products that can be produced by genetic engineering is ever growing. These include treatments for multiple sclerosis, certain cancers, heart attacks, and certain forms of anemia. Recombinant DNA technology is also increasingly used to produce vaccines that provide safe and effective immunity against infectious diseases such as hepatitis B.

Additional engineering is required for a recombinant eukaryotic gene to be expressed in bacteria

Even if a gene has been isolated and successfully introduced into *E. coli,* the bacterium does not necessarily make the encoded protein in large quantities. Several obstacles stand in the way of producing gene products of eukaryotes in bacteria. One is that the gene has to be correctly associated with an appropriate set of regulatory and promoter sequences that the bacterial RNA polymerase can recognize. Recall from Chapters 12 and 13 that the regulatory regions of prokaryotic and eukaryotic genes are quite different. A usual approach to this problem is to combine the amino acid coding portion of a eukaryotic gene with a bacterial promoter sequence that can be strongly expressed. Some eukaryotic genes, for example, are fused to the lactose operon regulatory region (see Chapter 13); the protein product of the eukaryotic gene is synthesized when the bacterium is fed lactose in the growth medium.

We have already discussed the fact that bacterial cells cannot process RNA molecules containing eukaryotic intron sequences and that one solution to this problem is to introduce a cDNA copy of the gene. Other problems may arise in the expression of a recombinant protein in *E. coli* because of differences in the ways the proteins are expressed in prokaryotic and eukaryotic cells.

Insulin, for example, is made in human cells from a large polypeptide that is folded in a specific way by the formation of three disulfide bonds (see Fig. 3–18), each joining two cysteines (sulfur-containing amino acids). After the polypeptide is folded, parts of the polypeptide are removed by proteolytic (protein-digesting) enzymes, leaving the insulin as two separate polypeptide chains held together by the disulfide bonds. *E. coli* lacks the specific enzymes necessary to cut the larger protein and is not able to fold the molecule properly. To overcome these problems, the gene was engineered to produce the two polypeptides separately. The recombinant proteins are then purified from the cells and allowed to associate in vitro. This procedure results in a relatively low yield of the active hormone, because the insulin can fold in several ways, only one of which results in a functional hormone. It has been possible to circumvent some of these problems by introducing the gene into eukaryotic cells such as yeast or other fungi, or cultured mammalian cells, that contain the protein-processing machinery required to produce fully functional proteins. As we will see in the next section, some types of genetically engineered plants or animals can also produce foreign proteins.

Transgenic organisms have incorporated foreign DNA into their cells

Plants and animals that have incorporated foreign genes are referred to as **transgenic** organisms. A number of approaches are being used to insert foreign genes into plant or animal cells.

Viruses are often used as vectors, although other methods, such as direct injection of DNA into cells, have also been used.

Transgenic animals are valuable in research and may have commercial uses

One approach to genetic engineering of animal proteins is to use live animals that have incorporated a foreign gene to make the recombinant protein. These transgenic animals are usually produced by microinjecting the DNA of a particular gene into the nucleus of a recipient fertilized egg cell or embryo cells (Fig. 14–11; also see Fig. 16–16*a*). The eggs are then im-

Figure 14–11 **Transgenic mice.** (*a*) How to make a giant mouse. (*b*) The mouse on the right is normal, while the mouse on the left is a transgenic animal that expresses rat growth hormone. *(Photo by R.L. Brinster, University of Pennsylvania Medical School)*

Rat growth hormone genes are cloned

Mouse metallothionein genes are cloned

Rat growth hormone gene and metallothionein genes are combined

Recombined DNA is injected into mouse embryo cells

Rat growth hormone gene

Metallothionein gene promoter

Embryo is implanted in host mother and...

...develops normally

Baby mouse is treated with a small amount of zinc

Zinc stimulates release of large quantities of rat growth hormone,

...which causes development of giant adult mouse. (shown next to normal sized mouse)

(*a*)

(*b*)

THE GENETICS OF MICE AND HUMANS

How can the study of the genetics of other organisms further our understanding of human genetics? In this chapter we have seen how recombinant DNA technology was developed using knowledge of the genetics of bacteria, yeast, and other microorganisms. As the available techniques have become more sophisticated, it has become possible to genetically engineer complex organisms, producing transgenic plants and animals.

The laboratory mouse, *Mus,* has become a particularly useful model for studying human genes. One extremely powerful research tool is **gene targeting,** a procedure in which a single gene is chosen and "knocked out" (inactivated) in a mouse. The roles of the inactivated gene can be determined by observing the phenotype of the mice bearing the knockout gene. Because at least 99% of the loci of mice have human counterparts (although the specific alleles are usually different), information about knockout genes in mice provides details about human genes as well.

Gene targeting, pioneered by Mario Capecchi, a molecular geneticist at the University of Utah School of Medicine, is a rather complex and lengthy procedure; it takes about a year to develop a new strain of knockout mice. First a nonfunctional (knockout) gene is introduced into special mouse cells, known as embryo-derived stem (ES) cells. ES cells are particularly easy to handle because, like cancer cells, they can be grown in culture indefinitely. Most important, if they are placed into a mouse embryo, they are capable of dividing and producing all of the cell types normally found in the mouse. In a tiny fraction of these ES cells, the introduced gene will become physically associated with the corresponding gene in a chromosome. If this occurs, the two will tend to exchange DNA segments in a poorly understood process known as *homologous recombination.* In this way, the normal allele in the mouse chromosome is replaced by the knockout allele.

Researchers inject ES cells they hope are carrying a knockout gene into early mouse embryos and allow the mice to develop to maturity. The mice are then bred for several generations, allowing researchers to eventually select any offspring that might be homozygous for the knockout gene.

If the gene is not lethal when inactivated, the researchers generally study homozygous animals that carry the knockout gene in every cell. However, because many genes are essential to life, researchers have modified the knockout technique to develop strains in which a specific gene is selectively inactivated in only one cell type. Today more than 250 different strains of knockout mice, each displaying its own characteristic phenotype, have been developed in various research labs, and the number continues to grow.

Gene targeting is providing answers to basic biological questions, including the development of embryos, the development of the nervous system, and the normal functioning of the immune system.

Gene targeting has great potential for revealing more about various human diseases, more than 5000 of which have a genetic component. Gene targeting is being used to study cancer, heart disease, respiratory diseases such as cystic fibrosis (see *On the Cutting Edge: Using a Mouse Model to Study a Human Genetic Disease,* Chapter 15), sickle cell anemia, and other health problems.

---

planted into the uterus of a female and allowed to develop.

In a classic pioneering study of this type, reported by the laboratory of R.L. Brinster in 1983, transgenic mice carrying a gene for rat growth hormone were produced. First the gene for growth hormone was isolated from a library of genomic rat DNA. It was then combined with the promoter region of a mouse gene that normally produces metallothionein, a protein whose synthesis is stimulated by the presence of heavy metals such as zinc. The metallothionein regulatory sequences were used as a switch to turn the production of rat growth hormone on and off at will. After the engineered gene was injected into mouse embryo cells, the embryos were implanted into the uterus of a mouse and allowed to develop. Embryos in which the gene transplant had been successful grew rapidly when exposed to small amounts of zinc. One mouse, which developed from an embryo that had received two copies of the growth hormone gene, grew to more than double the normal size. As might be expected, such mice are often able to transmit their increased growth capability to their offspring.

Transgenic offspring have already been shown to have valuable research applications in a wide range of studies. These include regulation of gene expression, immune system function, genetic diseases, viral diseases, and genes responsible for the development of cancer (see *Making the Connection: The Genetics of Mice and Humans*).

Transgenic animals have been used to develop strains that secrete important proteins in milk. For example, the gene for a human blood clotting factor has been introduced into sheep. These recombinant genes have been fused to the regulatory sequences of the milk protein genes and are therefore activated only in mammary tissues involved in milk production.

The advantage of obtaining the protein from milk is that potentially it can be produced in large quantities and can be harvested simply by milking the animal. The protein is then purified from the milk. The animals are not harmed by the introduction of the gene, and, because usually the progeny of the transgenic animal also produce the recombinant protein, transgenic strains can be established simply by breeding the animals.

Sometimes viruses are used as recombinant DNA vectors. RNA viruses called **retroviruses** make DNA copies of themselves by reverse transcription (see *Focus On: Reverse Transcription, Jumping Genes, and Pseudogenes* in Chapter 12). Sometimes the DNA copies become integrated into the host chromosomes, where they are replicated along with host DNA. For example, genetically altered mouse leukemia viruses are retroviruses that can be used as vectors to incorporate recombinant genes into cultured cells. Under certain conditions genes carried by the engineered virus can be expressed in the animal cells to produce genetically engineered proteins.

A major disadvantage of introducing genes into cultured animal cells is that the yields of the proteins encoded by the foreign DNA carried by the viruses are generally low. However, these types of vectors show some promise as a means of **gene therapy** for human genetic disorders.

Transgenic plants are increasingly important in agriculture

Plants have been selectively bred for thousands of years. The success of such efforts depends on the presence of desirable traits in the variety of plant being selected, or in closely related wild or domesticated plants whose traits can be transferred by cross-breeding. Primitive varieties or closely related species of cultivated plants often have traits, such as disease resistance, that could be advantageously introduced into varieties more suited to modern needs. If genes are introduced into plants from strains or species with which they do not ordinarily interbreed, the possibilities for improvement are greatly increased. Unfortunately, a suitable vector for the introduction of recombinant genes into many types of plant cells has proved very difficult to find. The most widely used vector system employs the crown gall bacterium, *Agrobacterium tumefaciens.* This bacterium normally produces plant tumors by introducing a special plasmid, called the *Ti* (for *tumor-inducing*) *plasmid,* into the cells of its host (Fig. 14–12). The Ti plasmid induces abnormal growth by forcing the plant cells to produce elevated levels of a plant growth hormone called cytokinin (see Chapter 36).

It is possible to "disarm" the Ti plasmid so that it does not induce tumor formation and then to use it as a vector to insert genes into plant cells. The cells into which the altered plasmid is introduced are essentially normal except for the genes that have been inserted. Genes placed in the plant genome in this fashion may be transmitted sexually, via seeds, to the next generation, but they can also be propagated asexually if desired.

Unfortunately, not all plants take up DNA readily, and this is particularly true of the grain plants that are the major food source for humans. One very useful approach has been the development of a genetic "shotgun." Microscopic gold fragments are coated with DNA and then shot into plant cells, penetrating the cell walls. Some of the cells retain the DNA and are transformed by it. Those cells can then be cultured and used to regenerate an entire plant (see Chapter 16). For example, such an approach has been successfully used to transfer a gene for resistance to a bacterial disease from a wild rice variety into cultivated rice.

An additional complication of plant genetic engineering is that a number of important plant genes are located in the DNA of the chloroplasts (see Chapter 4). Chloroplasts are essential in photosynthesis, which is the basis for plant productivity. Obviously, it is useful to develop methods for changing

Figure 14–12 **Transformation of plant cells.** This SEM shows *Agrobacterium tumefaciens* infecting cultured plant cells. The close contact permits the transfer of plasmid DNA from the bacteria to the larger plant cells. *(Courtesy of Ann G. Matthysse)*

the portion of the plant's DNA that resides within the chloroplast. Methods of chloroplast engineering are currently the focus of intense research.

The applications of transgenic plants are not limited to disease resistance and increased production of crops. Like some transgenic animals, certain transgenic plants can potentially be used to produce large quantities of medically important proteins, such as antibodies. The developers of this technology are conducting ongoing field trials to demonstrate the large-scale feasibility of these production methods.

SAFETY GUIDELINES HAVE BEEN DEVELOPED FOR RECOMBINANT DNA TECHNOLOGY

People who have experienced the direct applications of recombinant DNA technology today would undoubtedly agree that those developments have been important and beneficial. When the new technology was introduced in the early 1970s, however, many scientists considered the potential misuses to be at least equally significant. The possibility that an organism with undesirable environmental effects might be accidentally produced was a concern. Totally new strains of bacteria or other organisms, with which the world has no previous experience, might be difficult to control. This possibility was recognized by the scientists who developed the recombinant DNA methods and led them to insist on stringent guidelines for making the new technology safe.

Recent history has failed to bear out these worries. Experiments over the past years in thousands of university and industrial laboratories have seen recombinant DNA manipulations carried out safely. One of the initial concerns—the accidental release of laboratory bacterial strains containing dangerous genes into the environment—has turned out to be groundless. Laboratory strains of *E. coli* are poor competition for the wild strains in the outside world and quickly perish. Experiments thought to entail unusual risks are carried out in special facilities designed to contain dangerous disease-causing organisms and allow researchers to work with them safely. The fears of accidentally cloning a dangerous gene or releasing a dangerous organism into the environment seem to be laid to rest. This does not mean, however, that *intentional* manipulations of dangerous genes are not a possibility.

Most scientists today recognize the importance of recombinant DNA technology and agree that the perceived threat to humans and the environment was overestimated. Many of the restrictive guidelines for using recombinant DNA have been relaxed as the safety of the experiments has been established. Stringent restrictions still exist, however, in certain areas of recombinant DNA research where there are known dangers, or where questions about possible effects on the environment are still unanswered.

These restrictions are most evident in research that pro- poses to introduce recombinant organisms into the wild, such as agricultural strains of plants whose seeds or pollen might spread in an uncontrolled manner. A great deal of research activity is now concentrated on determining the effects of introducing recombinant organisms into a natural environment. Carefully conducted tests have shown that recombinant organisms are not dangerous to the environment simply because they are recombinant. However, it is important to assess the biology of each new recombinant organism. In this way scientists will be able to determine if it has characteristics that might cause it to present an environmental hazard under certain conditions. For example, if a transgenic crop plant has been engineered to resist an herbicide, might that gene be transferred, via pollen or by some other route, to that plant's weedy relatives, generating herbicide-resistant "superweeds?" Other concerns relate to plants that have been engineered to produce pesticides, such as insecticides. For example, a transgenic plant may produce a relatively small amount of an insecticide, which may be sufficient to provide protection if the targeted insect population is not very numerous. However, the presence of low levels of the insecticide could potentially provide ideal conditions for selection for resistant individuals in the insect population.

SUMMARY WITH KEY TERMS

I. **Recombinant DNA** technology is concerned with isolating and amplifying specific sequences of DNA by incorporating them into **vector** DNA molecules. The resulting recombinant DNA can then be propagated and amplified in organisms such as *E. coli*.
   A. **Restriction enzymes** are used to cut DNA into specific fragments.
      1. Each type of restriction enzyme recognizes and cuts DNA at a highly specific base sequence.
      2. Many restriction enzymes cleave DNA sequences to produce complementary, single-stranded cut ends (sticky ends).
   B. The most common recombinant DNA vectors are constructed from naturally occurring circular DNA molecules called **plasmids,** or from bacterial viruses called **bacteriophages;** both of these are found in some bacteria.
   C. Recombinant DNA molecules are often constructed by allowing the ends of a DNA fragment and a plasmid (which have both been cut with the same restriction enzyme) to associate by complementary base pairing. The DNA strands are then covalently linked by **DNA ligase** to form the recombinant molecule.
   D. Parts of genes are isolated from recombinant DNA libraries, which are mixtures of DNA fragments inserted into appropriate vectors.
      1. **Genomic libraries** are formed from the total DNA of an organism. Genes present in recombinant DNA genomic libraries from eukaryotes contain introns. Those genes can be amplified in *E. coli*, but the protein is not properly expressed.
      2. When a **cDNA library** is produced, **reverse transcriptase** is used to make DNA copies of mRNA isolated from eukaryotic cells; these are then incorporated into recombinant DNA vectors. Because the introns have been removed from mRNA molecules, eukaryotic genes in cDNA libraries can sometimes be expressed in *E. coli* to make their protein products.
   E. The **polymerase chain reaction (PCR)** is a widely used, usually automated, technique in which a particular DNA sequence can be targeted by specific primers and then cloned in vitro by a special heat-resistant DNA polymerase.
   F. Analysis of a cloned sequence can yield useful information about the gene and its protein and can enable investigators to identify and sub-clone DNA fragments for use as molecular probes.
      1. A **DNA sequence** gives information about the structure of the gene and the probable amino acid sequence of the encoded proteins. It can be compared to other sequences stored in massive databases.
      2. A radioactive DNA or RNA sequence can be used as a **genetic probe** to identify complementary nucleic acid sequences. In the **Southern blot technique,** DNA fragments are separated by **gel electrophoresis** and then blotted onto a nitrocellulose filter. The radioactive probe is then hybridized by complementary base-pairing to the DNA bound to the filter, and the radioactive band or bands of DNA can be identified by autoradiography.
   G. The degree of genetic relationship among the individuals in a population can be determined by comparing nucleotide sequences, or it can be estimated by studying **restriction fragment length polymorphisms (RFLPs).**

II. **Genetic engineering** is a technology that uses genetic and recombinant DNA methods to devise new combinations of genes to produce improved pharmaceutical and agricultural products.
   A. Genes isolated from one organism can be modified and expressed in other organisms ranging from *E. coli* to **transgenic** plants and animals.
      1. Expression of eukaryotic proteins in bacteria such as *E. coli* requires that the gene be linked to regulatory elements that the bacterium can recognize. In addition, bacterial cells do not contain many of the enzymes needed for the posttranslational processing of eukaryotic proteins.
      2. Expression of eukaryotic genes in eukaryotic host organisms shows great promise, because the processing and modification machinery for eukaryotic proteins is already present in these cells.
         a. Production of important pharmaceutical products can be engineered in transgenic animals and possibly in plants.
         b. Genetic engineering of plants and domestic animals holds the promise of increasing the availability of food.
   B. Recombinant DNA technology is carried out under certain safety guidelines.

1. A plasmid (a) can be used as a DNA vector (b) is a type of bacteriophage (c) is a type of cDNA (d) is a retrovirus (e) b and c
2. DNA molecules with complementary "sticky ends" associate by (a) covalent bonds (b) hydrogen bonds (c) ionic bonds (d) disulfide bonds (e) phosphodiester linkages
3. Human DNA and a particular plasmid both have sites that can be cut by the restriction enzymes Hind III and EcoRI. To make recombinant DNA, one should (a) cut the plasmid with EcoRI and the human DNA with Hind III (b) use EcoRI to cut both the plasmid and the human DNA (c) use Hind III to cut both the plasmid and the human DNA (d) a or b (e) b or c
4. Which of the following sequences is *not* palindromic?
   (a) 5'—AAGCTT—3'   (b) 5'—GATC—3'
       3'—TTCGAA—5'   3'—CTAG—5'
   (c) 5'—GAATTC—3'   (d) 5'—CTAA—3'
       3'—CTTAAG—5'   3'—GATT—5'
   (e) b and d
5. The PCR technique uses (a) heat-resistant DNA polymerase (b) reverse transcriptase (c) DNA ligase (d) a and b (e) b and c
6. A cDNA clone contains (a) introns (b) exons (c) anticodons (d) a and b (e) b and c
7. The dideoxynucleotides ddATP, ddTTP, ddGTP, and ddCTP are important in DNA sequencing because they (a) cause premature termination of a growing DNA strand (b) are used as primers (c) cause the DNA fragments that contain them to migrate more slowly through a sequencing gel (d) are not affected by high temperatures (e) have more energy than deoxynucleotides
8. In restriction fragment length polymorphism (RFLP) analysis to determine parentage, (a) every band present in a child would be expected to be present in both of the true parents (b) every band present in a child would be expected to be present in at least one of the true parents (c) every band present in a true parent would be expected to be present in all of the children (d) a and b (e) b and c
9. In the Southern blot technique, _____ is/are transferred from a gel to a special nitrocellulose filter. (a) protein (b) RNA (c) DNA (d) bacterial colonies (e) reverse transcriptase
10. Gel electrophoresis separates nucleic acids on the basis of differences in (a) length (b) charge (c) nucleotide sequence (d) relative proportions of adenine and guanine (e) relative proportions of thymine and cytosine
11. The Ti plasmid, carried by *Agrobacterium tumefaciens,* is especially useful for introducing genes into (a) *E. coli* (b) plants (c) animals (d) yeast (e) all eukaryotes

R E V I E W   Q U E S T I O N S

1. What is meant by the term *genetic engineering?*
2. What are restriction enzymes? How are they used in recombinant DNA research?
3. What characteristics should be engineered into a plasmid to make it a useful cloning vector?
4. Diagram the process by which recombinant DNA molecules are usually constructed.
5. How is a gene library constructed? What are the relative merits of genomic libraries and cDNA libraries?
6. Sketch an example illustrating how a restriction map of a gene is made.
7. Why is the PCR technique valuable?

Y O U   M A K E   T H E   C O N N E C T I O N

1. What are some of the problems that might arise if you were trying to produce a eukaryotic protein in a bacterium? How might some of these problems be solved by using transgenic plants or animals?
2. Would genetic engineering be possible if we did not know a great deal about the genetics of bacteria? Explain.
3. What are some of the ecological concerns regarding transgenic organisms? What kinds of information are needed to allay these fears?

R E C O M M E N D E D   R E A D I N G S

Beardsley, T. "The Food Genome Project." *Scientific American,* Vol. 277, No. 5, Nov. 1997. A number of current and prospective genome sequencing projects involve plants and animals of agricultural importance.

Capecchi, M.R. "Targeted Gene Replacement." *Scientific American,* Vol. 270, No. 3, Mar. 1994. Biologists can now produce mice that contain a mutation in any gene desired, thereby enabling them to determine the function of that gene.

Gibbs, W.W. "Plantibodies." *Scientific American,* Vol. 277, No. 5, Nov. 1997. A look at ongoing field trials of plants such as corn and soybeans that have been genetically engineered to produce antibodies and other proteins of medical importance.

Miller, R.V. "Bacterial Gene Swapping in Nature." *Scientific American,* Vol. 278, No. 1, Jan. 1998. This article assesses the potential mechanisms and extent of transfer of genes to other microbes by genetically engineered bacteria released into natural environments.

Mullis, K.B. "The Unusual Origin of the Polymerase Chain Reaction." *Scientific American,* Vol. 262, No. 4, Apr. 1990. A highly personal, first-hand account of the development of the PCR technique and an excellent illustration of the nature of scientific insight.

Ronald, P.C. "Making Rice Disease-Resistant." *Scientific American,* Vol. 277, No. 5, Nov. 1997. The applications of genetic engineering to rice breeding should increase the food supply for the two billion people worldwide for whom rice is the mainstay of the diet.

● Visit our website at **http://www.saunderscollege.com/lifesci/titles.html** and click on Solomon/Berg/Martin Biology for links to chapter-related resources on the World Wide Web.

CHAPTER 15

Human Genetics

The principles of genetics that apply to other diploid eukaryotic organisms also apply to humans. There are, however, some important differences between human genetic research and genetic research on other organisms. To study aspects of inheritance in other species, geneticists ideally (1) have standard stocks of genetically identical individuals, that is, **isogenic strains,** that are homozygous at virtually all of their loci; (2) conduct **controlled matings** between members of different isogenic strains; and (3) raise the offspring under carefully controlled conditions. Of course, the human population is very diverse, and individuals are heterozygous for many genes. In addition, human families are small, and 20 to 30 years or more elapse between generations. It is therefore virtually impossible, as well as unethical, to conduct research in this way with humans.

Despite the inherent difficulties, knowledge in **human genetics,** the science of inherited variation in humans, is progressing very rapidly. Traditionally, human genetics was examined using such approaches as population studies of large extended families. For example, **albinism**—the lack of pigmentation in the skin, hair, and eyes—is a rare human disorder that is common among the Cuna people that live on the San Blas Islands of Panama. The photograph shows a young albino child from this population. From studies of such human groups, we have inferred that albinism is usually inherited as an autosomal recessive trait. Recessiveness is inferred because the parents of afflicted individuals usually are not affected.

More recently, biochemical and molecular studies have indicated that there are several kinds of albinism. Some albinos lack one of the enzymes needed to synthesize the brown pigment melanin, whereas other albinos produce the necessary enzyme, but it cannot enter the pigment cells where the precursors of melanin are produced.

The field of human genetics has been greatly facilitated by the medical attention given to genetic diseases in humans and by the use of genetic engineering technology. The extensive medical records of diseases serve as a very useful data pool on which hypotheses may be based and against which they may be tested.

Genetic studies of other organisms also have provided invaluable insights. Indeed, many phenomena in human inheritance that were initially puzzling have been explained by solving analogous problems in the inheritance of bacteria, yeasts,

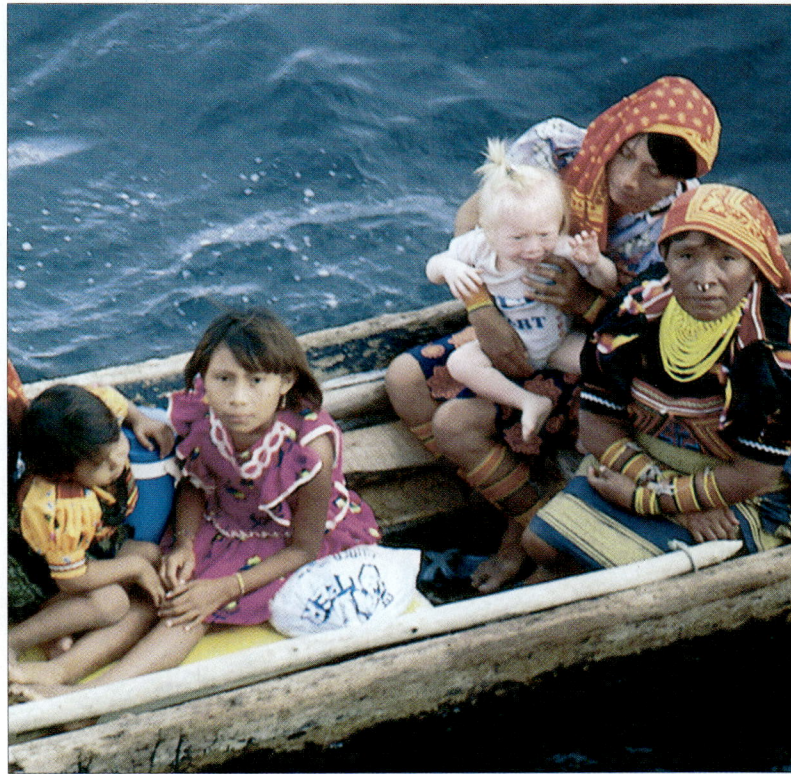

(Anna Zuckerman/Photo Edit)

fruit flies, or mice. Studies with mice, for example, indicated that mice contain several genes (polygenes) for hair color. Mice that are homozygous for the recessive albino allele, however, are albino even though they may have genes for other coat colors. Thus, the gene for albinism is epistatic in mice. (Recall from Chapter 10 that *epistasis* is a gene interaction in which the presence of a particular allele of one gene pair determines whether certain alleles of another gene pair are expressed.) Extending this knowledge of mouse inheritance to humans, it has now been demonstrated that human hair color is also polygenic, and albinism in humans exhibits epistasis.

AFTER YOU HAVE STUDIED THIS CHAPTER YOU SHOULD BE ABLE TO

1. Given a simple pedigree for a particular trait, determine the probable genotypes for all individuals in the pedigree.
2. Distinguish between environmentally induced and inherited abnormalities and between chromosome abnormalities and single gene defects.
3. Make a sketch illustrating how nondisjunction can occur in meiosis. Show how nondisjunction can be responsible for specific chromosome abnormalities such as Down syndrome, Klinefelter syndrome, and Turner syndrome.
4. Describe how amniocentesis is used in the prenatal diagnosis of human genetic abnormalities; state the relative advantages and disadvantages of amniocentesis and chorionic villus sampling.

5. State whether each of the following genetic defects is inherited as an autosomal recessive, autosomal dominant, or X-linked recessive: phenylketonuria, sickle cell anemia, cystic fibrosis, Tay-Sachs disease, Huntington disease, and hemophilia A.
6. Discuss the scope and implications of genetic counseling.
7. Relate each ABO blood type to the appropriate genotype(s). Explain the genetic and physiological basis for Rh incompatibility between a mother and a fetus.
8. Explain why quantitative traits in humans are thought to be under the control of polygenes.
9. Discuss the implications of the Human Genome Project, including the costs and possible benefits.
10. List some of the ways that genetics affects human society.

THE STUDY OF HUMAN GENETICS REQUIRES ALTERNATIVE METHODS

Early studies of human heredity usually dealt with readily identified pairs of contrasting traits and their distribution among members of a family, as illustrated by the **pedigree** in Figure 15–1. This method is still useful, but because human families tend to be small and information on certain family members may be lacking, it has serious limitations.

Human geneticists therefore also use methods that allow them to make inferences about a trait's mode of inheritance based on studies of its distribution in an entire population. By

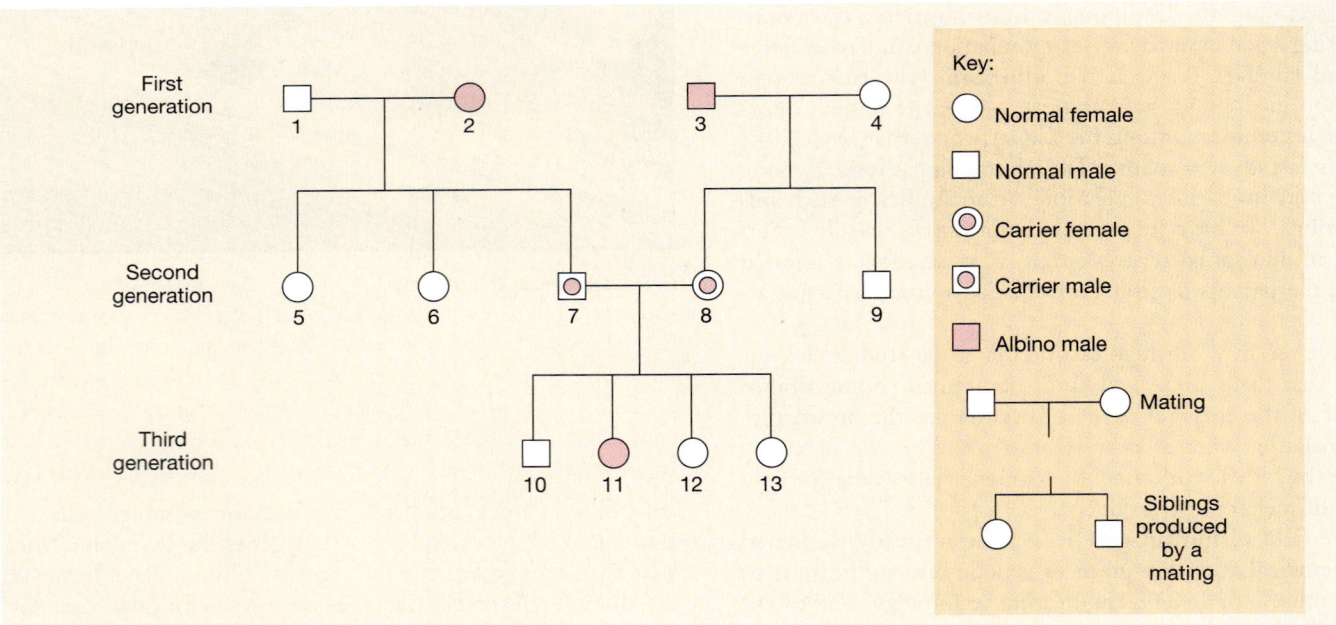

Figure 15–1 Pedigree analysis of albinism. By studying family histories, it is often possible to determine the genetic nature of the trait being studied. Consider 11, an albino girl with two phenotypically normal parents, 7 and 8. Albinism cannot be a dominant allele because if it were, at least one of 11's parents would have to be an albino. Also, albinism cannot be an X-linked recessive allele because if it were, her father would have to be an albino (and her mother would have to be a heterozygous carrier). This pedigree is easily explained if albinism is inherited as an autosomal (not carried on a sex chromosome) recessive allele; with an autosomal recessive allele, two phenotypically normal parents could produce an albino offspring.

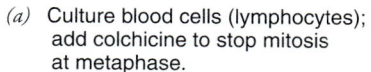

(a) Culture blood cells (lymphocytes); add colchicine to stop mitosis at metaphase.

(b) Spin down cells in a centrifuge; discard supernatant.

Add hypotonic salt solution and resuspend cells.

(c) Add fixative (preservative) to cell suspension; add one drop to a slide, dry, and stain.

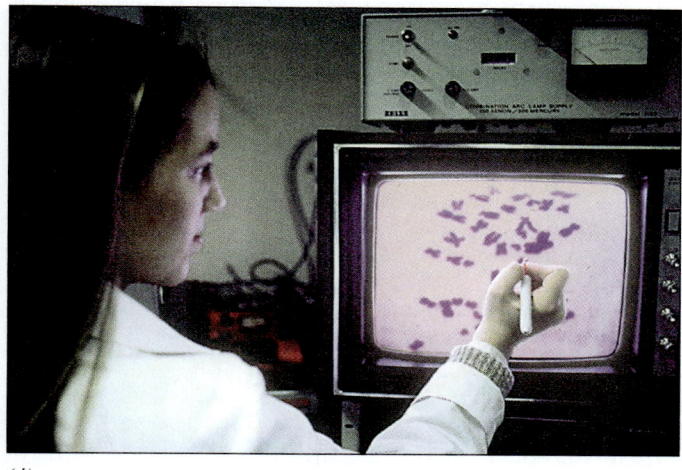

(d)

Figure 15–2 Preparation of a karyotype. Karyotypes are used to help identify some chromosome abnormalities. *(d, SIU Peter Arnold, Inc.)*

applying the laws of probability to data obtained from a relatively large sample of individuals who are representative of the population, it is often possible to determine if the mode of inheritance is simple or complex and if more than one locus is involved. Some of the methods used by population geneticists are discussed in Chapter 18. As we will see later in this chapter, human inheritance can be studied most effectively by combining these and other approaches with the methods of molecular biology and recombinant DNA technology.

Karyotyping is the analysis of chromosomes

Cytogenetics is the study of chromosomes and their role in inheritance. Many of the basic principles of genetics were dis-

covered by experiments with simpler organisms, in which it was possible to relate genetic data to the number and structure of specific chromosomes. Some of the organisms used in genetics, such as the fruit fly *Drosophila,* have very few chromosomes (only four pairs in *Drosophila*). In *Drosophila* salivary glands and certain other tissues, the chromosomes are large enough that their structural details are readily evident. This organism, therefore, has provided unique opportunities for correlating certain kinds of genetic changes with certain kinds of alterations in chromosome structure.

Recall from Chapter 10 that the number of chromosomes for the human species is 46: 44 autosomes (22 pairs) and 2 sex chromosomes (1 pair). The term **karyotype** refers both to the chromosome composition of an individual and to a photomicrograph showing that composition. In karyotyping, cells from the bone marrow, skin, or blood (white blood cells) are cultured and then treated with the drug **colchicine,** which arrests them at mitotic metaphase by preventing the assembly of spindle microtubules (Fig. 15–2). Next they are placed into a hypotonic solution that causes them to swell; this spreads out the chromosomes so they can be readily observed. The cells are then flattened on microscope slides, and the chromosomes are stained to reveal the patterns of bands, which are unique for each homologous pair.

After the microscopic view has been entered into a computer, the homologous pairs are electronically matched and placed together. Chromosomes are identified by length, position of the centromere, banding patterns, and other features such as *satellites,* which are darkly staining regions. The largest chromosome (chromosome 1) is about five times as long as the smallest one (chromosome 22), but there are only slight size differences among some of the intermediate-sized chromosomes.

ABNORMALITIES IN CHROMOSOME NUMBER CAUSE CERTAIN HUMAN DISORDERS

Polyploidy, the presence of multiple chromosome sets, is common in plants but rare in animals. It may arise from failure of chromosomes to separate during meiosis, or from fertilization of an egg by more than one sperm. Polyploidy is lethal in humans and many other animals when it occurs in all the cells of the body. Triploidy ($3n$) is sometimes found in embryos that have been spontaneously aborted in early pregnancy. The few triploid or tetraploid ($4n$) individuals that have been born alive and survived for a few days have been found to contain a mixture of diploid and polyploid cells.

Abnormalities involving the presence or absence of a single extra chromosome are much more common in humans. These conditions are called **aneuploidies.** Recall that ordinarily there are two of each kind of chromosome. An individual with an extra chromosome, that is, with three of one kind, is said to be **trisomic** for that kind of chromosome. An individual lacking one member of a pair of chromosomes is said to be **monosomic.** Table 15–1 summarizes some disorders produced by aneuploidies.

Aneuploidies generally arise as a result of an abnormal meiotic (or mitotic) division in which chromosomes fail to separate at anaphase. This phenomenon is called **nondisjunction.** In meiosis, chromosome nondisjunction may occur during the first or second meiotic division (or both). For example, two X chromosomes that fail to separate at either the first or the second meiotic division might both enter the egg nucleus. Alternatively, the two joined X chromosomes might go into a polar body, leaving the egg with no X chromosome. (Recall from Chapter 9 that a polar body is a nonfunctional haploid cell produced during oogenesis.)

Nondisjunction of the XY pair in the male might lead to the formation of sperm with both an X and a Y chromosome, or sperm with neither an X nor a Y chromosome. Similarly, nondisjunction at the second meiotic division can produce sperm with two X's or two Y's. Some of these examples of meiotic nondisjunction are illustrated in Figure 15–3. When an abnormal gamete unites with a normal one, the resulting zygote has a chromosome abnormality that is present in every cell of the body.

Nondisjunction during a mitotic division leads to the establishment of a clone of abnormal cells in an otherwise normal individual. Such an individual therefore contains a mixture of normal and abnormal cells.

Persons with Down syndrome are usually trisomic for chromosome 21

Down syndrome[1] is one of the most common chromosome abnormalities in humans. It was named after John Down, the British physician who first described the condition in 1866. Affected individuals have abnormalities of the face, eyelids, tongue, hands, and other parts of the body, and are mentally and physically retarded (Fig. 15–4a). They are also unusually susceptible to certain diseases, such as leukemia and Alzheimer's disease.

---

[1] The term *syndrome* refers to a set of symptoms that usually occur together in a particular disorder.

TABLE 15–1 Chromosome Abnormalities: Some Disorders Produced by Aneuploidies

Karyotype	Common Name	Clinical Description
Trisomy 13	Patau syndrome	Multiple defects, with death by age 1 to 3 months
Trisomy 18	Edwards syndrome	Ear deformities, heart defects, spasticity, and other damage; death by age 1 year
Trisomy 21	Down syndrome	Overall frequency is about 1 in 700 live births. True trisomy is most often found among children of older (age 40+) mothers, but translocation resulting in the equivalent of trisomy is not age-related. A similar, though less marked, influence is exerted by the age of the father. Trisomy 21 is characterized by a fold of skin above the eye, varying degrees of mental retardation, short stature, protruding furrowed tongue, transverse palmar crease, cardiac deformities, and increased risk of leukemia and Alzheimer's disease.
Trisomy 22	—	Similar to Down syndrome but with more skeletal deformities
XO	Turner syndrome	Short stature, webbed neck, sometimes slight mental retardation; ovaries degenerate in late embryonic life, leading to rudimentary sexual characteristics; gender is female; no Barr bodies
XXY	Klinefelter syndrome	Male with slowly degenerating testes, enlarged breasts; one Barr body per cell
XYY	XYY karyotype	Unusually tall male with heavy acne, some tendency to mild mental retardation
XXX	Triplo-X	Despite three X chromosomes, usually fertile females with normal intelligence; two Barr bodies per cell

(a)

(b)

Figure 15–3 **Meiotic nondisjunction.** In these examples of nondisjunction of the sex chromosomes in the human male, only the X *(red)* and Y *(brown)* chromosomes are shown. *(a)* Nondisjunction in the first meiotic division results in two XY sperm and two sperm with neither an X nor a Y. *(b)* Second-division nondisjunction of the X chromosome results in one sperm with two X chromosomes, two with one Y each, and one with no sex chromosomes. Similarly, nondisjunction of the Y results in one sperm with two Y chromosomes, two with one X each, and one with no sex chromosome.

(a)

Figure 15–4 **Down syndrome.** *(a)* This male child with Down syndrome is working on a science experiment in his kindergarten class. *(b)* Note the presence of an extra chromosome 21 in this karyotype of a male with Down syndrome. *(a, Richard Hutchings/Photo Researchers, Inc.; b, Courtesy of Dr. Leonard Sciorra)*

(b)

Cytogenetic studies have revealed that most people with Down syndrome have 47 chromosomes because they are trisomic for all or part of chromosome 21, one of the smaller chromosomes (Fig. 15–4b). Nondisjunction during meiosis is thought to be responsible for the presence of the extra chromosome. Although no genetic information is missing in these individuals, the extra copies of chromosome 21 genes bring about some type of genetic imbalance that is responsible for abnormal physical and mental development. Down syndrome is quite variable in expression, with some individuals far more severely affected than others. Researchers are using genetic engineering methods to attempt to pinpoint genes on chromosome 21 that affect mental development, as well as possible oncogenes (see *Focus On: Oncogenes and Cancer* in Chapter 16) and genes that may be involved in Alzheimer's disease.

Down syndrome occurs in only about 0.15% of all births, but its incidence increases markedly with increasing maternal age (Fig. 15–5). It is 100 times more likely in the offspring of mothers who are 45 years of age or older than it is in the offspring of mothers who are under 19 years of age. The occurrence of Down syndrome is affected much less by the age of the father. The reason for the relationship between increased

incidence in Down syndrome and maternal age is not known. Several hypotheses have been proposed to explain the maternal age effect, for example, one hypothesis is that an aging womb is less likely to reject an abnormal fetus, but none are supported unequivocally.

In general, chromosome aneuploidies involving the autosomes are devastating in their consequences. Other than Down syndrome, very few autosomal trisomies are known (see Table 15–1). The condition known as autosomal **monosomy,** in which only one member of a pair is present, is apparently incompatible with life because it is not seen in live births.

Most sex chromosome aneuploidies are less severe than autosome aneuploidies

Sex chromosome aneuploidies appear to be relatively well tolerated (see Table 15–1), apparently at least in part because of the phenomenon of dosage compensation discussed in Chapter 10. According to the *single active X hypothesis,* mammals compensate for extra X chromosome material by rendering all but one X chromosome inactive. The inactive X is seen as a Barr body, a region of darkly staining, condensed chromatin next to the nuclear envelope of an interphase nucleus (see Fig. 10–16).

The presence of the Barr body in the cells of normal females (but not of normal males) has been used as an initial screen to determine whether an individual is genetically female or male. As we will see in our discussion of sex chromosome aneuploidies, however, the Barr body test has serious limitations.

Persons with **Klinefelter syndrome** have 47 chromosomes, including two X's and one Y. They have small testes, produce few or no sperm, and are therefore sterile. Evidence that the Y chromosome is the major determinant of the male phenotype has been substantiated by the fact that there is at least one gene on the Y chromosome that appears to act as a genetic switch, directing male development. People with Klinefelter syndrome tend to be unusually tall and to have female-like breasts. About half show some degree of mental retardation, but many live relatively normal lives. However, when their cells are examined they are found to have one Barr body per cell. On the basis of such a test, they would be erroneously classified as females.

We designate the sex chromosome constitution for **Turner syndrome,** in which an individual has only one X chromosome and no Y chromosome, as XO; the O refers to the absence of a second sex chromosome. Because of the absence of the strong male-determining effect of the Y chromosome, these persons develop as females. However, both their internal and external genital structures are undeveloped, and they are sterile. Apparently a second X chromosome is necessary for the normal development of the ovaries in a female embryo. Examination of their cells reveals no Barr bodies because there is no extra X chromosome to be inactivated. Using the standards of the Barr body test, such an individual would be classified erroneously as a male.

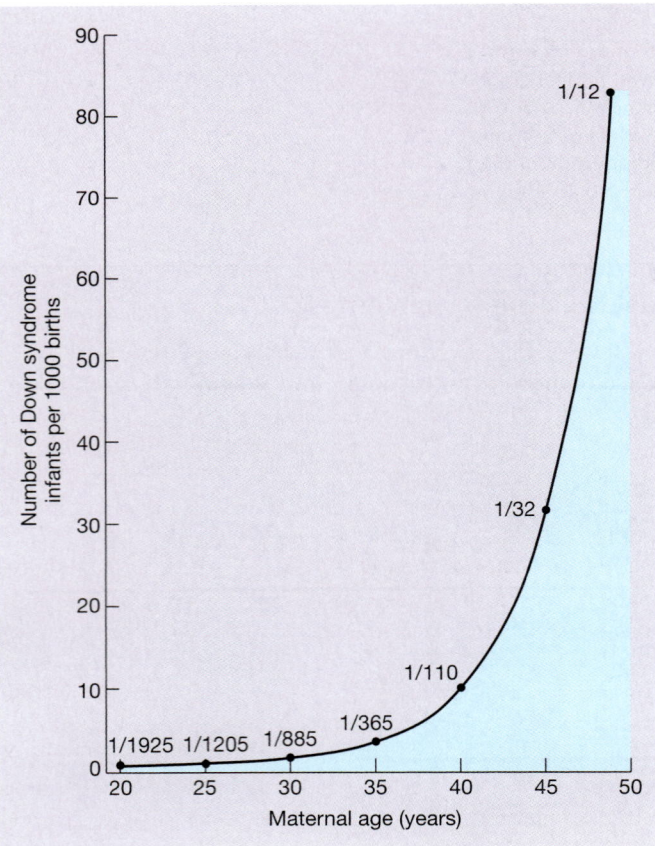

Figure 15–5 Down syndrome and maternal age. The probability of the birth of an infant with Down syndrome rises as maternal age increases. (The fractions indicate estimated probabilities of Down syndrome.)

People with an X chromosome plus two Y chromosomes are phenotypically males, and they are fertile. Other characteristics of these individuals (unusually tall, with severe acne) hardly merit the term *syndrome;* hence the designation **XYY karyotype.** Some years ago there was a widely publicized suggestion that persons with this condition are more likely to display criminal tendencies and to be imprisoned, but further studies have shown this to be incorrect.

Aneuploidies usually result in prenatal death

Recognizable chromosome abnormalities are seen in less than 1% of all live births, but substantial evidence suggests that the rate at conception is much higher. At least 17% to 20% of pregnancies recognized at 8 weeks will end in spontaneous abortion (miscarriage). Approximately half of these spontaneously aborted embryos have major chromosome abnormalities, including autosomal trisomies (e.g., trisomy 21), triploidy and tetraploidy, and Turner syndrome (XO). Autosomal monosomies are exceedingly rare. It is unlikely that they never occur. It is far more probable that autosomal monosomy is so incompatible with life that a spontaneous abortion occurs very early, before the woman is even aware that she is pregnant. Some investigators place surprisingly high estimates (50% or more) on the rate of loss of very early embryos. It is widely assumed that chromosome abnormalities are responsible for a substantial fraction of these.

ABNORMALITIES IN CHROMOSOME STRUCTURE CAUSE CERTAIN DISORDERS

Chromosome abnormalities are not only caused by changes in chromosome *number* but also by distinct changes in the *structure* of one or more chromosomes. Here we consider three simple examples of structural abnormalities: translocations, deletions, and fragile sites.

Translocation is attachment of part of a chromosome to a nonhomologous chromosome

In some cases part of one chromosome may break off and attach to a nonhomologous chromosome (a **translocation**), or two nonhomologous chromosomes may exchange parts (a **reciprocal translocation**). The consequences of translocations vary considerably but include situations in which some genes are missing (**deletions**) and extra copies of other genes are present (**duplications**).

In about 4% of individuals with Down syndrome, only 46 chromosomes are present, but one is abnormal. Extra genetic material from chromosome 21 has been translocated onto one of the larger chromosomes, such as chromosome 14. We refer to the abnormal translocation chromosome as a *14/21*

chromosome. Affected persons have one chromosome 14, one 14/21 chromosome, and two normal copies of chromosome 21. All or part of the genetic material from chromosome 21 is thus present in triplicate. When the karyotypes of such an individual's parents are studied, either the mother or the father is usually found to have only 45 chromosomes, although he or she is generally phenotypically normal. Such a person has one chromosome 14, one 14/21 chromosome, and one chromosome 21. Although the karyotype is abnormal, there is no extra genetic material. In contrast to trisomy 21, this translocation form of Down syndrome can run in families, and its incidence is not related to maternal age.

A deletion is loss of part of a chromosome

Sometimes chromosomes break but fail to rejoin. Such breaks result in deletions of as little as a few base pairs to as much as an entire chromosome arm. As you might expect, large deletions are lethal, whereas small deletions cause several recognizable human disorders.

The most frequently encountered deletion disorder in human infants is **cri-du-chat syndrome,** in which part of the short arm of chromosome 5 is deleted. As in most deletions, the exact point of breakage in chromosome 5 varies from one person to another; some cases of cri-du-chat involve a small loss, whereas others involve a more substantial loss. Infants born with cri-du-chat syndrome typically have a small head with altered features described as a "moon face," severe mental retardation, and a distinctive cry that sounds like a kitten mewing. (The name *cri-du-chat* literally means "cry of the cat.") Affected individuals usually die in infancy or childhood.

Fragile sites are weak points at specific sites in chromatids

A **fragile site,** in which part of a chromatid appears to be attached to the rest of the chromosome by a thin thread of DNA, may occur at a specific location on both chromatids of a particular chromosome; such sites have been identified on the X chromosome as well as on certain autosomes. The location of a fragile site is exactly the same in all of an individual's cells as well as in cells of other family members.

In **fragile X syndrome** the fragile site occurs near the tip on the X chromosome (Fig. 15–6). At the tip of an X chromosome, a nucleotide triplet (CGG, or cytosine, guanine, guanine) is normally repeated up to 50 times. The fragile X chromosome, however, repeats CGG from 200 to more than 1000 times. Fragile X syndrome occurs in 1 out of every 2000 males and in 1 out of every 4000 females. The effects of fragile X syndrome range from mild learning and attention disabilities to severe mental retardation and hyperactivity in affected males. Females with fragile X syndrome are usually heterozygous (because their other X chromosome is normal) and generally have normal intelligence, although occasionally a heterozygous female has borderline intelligence. The discovery of the fragile X gene in 1991 and the development of the first

Figure 15–6 Fragile X syndrome. *(a)* Region on the X chromosome associated with fragile X syndrome. *(b)* The normal allele at the tip of the X chromosome repeats the nucleotide triplet CGG up to 50 times. *(c)* The defective allele repeats CGG more than 200 times.

fragile X mouse model in 1994 have provided researchers with ways to test potential treatments, including gene replacement therapy.

MOST GENETIC DISEASES ARE INHERITED AS AUTOSOMAL RECESSIVE TRAITS

We have seen that several human disorders involve chromosome abnormalities. Hundreds of human disorders, however, involve enzyme defects caused by mutations of single genes. These disorders, sometimes referred to as **inborn errors of metabolism,** include phenylketonuria (PKU) and alkaptonuria (see Chapter 11). Both of these genetic disorders involve blocks in the metabolism of the amino acid phenyalanine. Not all human genetic diseases have a simple inheritance pattern, but most of those that do are transmitted as autosomal recessive traits and so are expressed only in the homozygous state.

Phenylketonuria (PKU) results from an enzyme deficiency

Phenylketonuria (PKU), which is most common in persons of western European descent, is an autosomal recessive disease. Homozygous recessive individuals lack an enzyme that converts the amino acid phenylalanine to another amino acid, tyrosine. These persons instead convert phenylalanine into toxic phenylketones that accumulate and damage the central nervous system. The ultimate result is severe mental retardation. A homozygous PKU infant is usually healthy at birth because its mother, who is heterozygous, produces enough enzyme to prevent phenylalanine accumulation before birth. However, during infancy and early childhood, the toxic products eventually cause irreversible damage to the central nervous system.

In the 1950s it was found that if PKU infants are identified and placed on a special low-phenylalanine diet early enough, the symptoms can be dramatically alleviated. Biochemical tests for PKU have been developed, and screening of newborns through a simple blood test is required in the United States. More than 90% of all infants born in the United States are tested. Because of these screening programs and the availability of effective treatment, thousands of children have not developed severe mental retardation. Most such children are able to discontinue the diet by adolescence. Although they still produce phenylketones, the sensitive period is past.

Ironically, the success of PKU treatment in childhood presents a new problem today. If a homozygous female who was saved from mental retardation becomes pregnant, the high phenylalanine levels in her blood can damage the brain of the fetus she is carrying, even though that fetus is heterozygous. Therefore, she must resume the diet, preferably before becoming pregnant. This procedure is usually (although not always) successful in preventing the effects of **maternal PKU.** Therefore it is especially important that females with PKU be aware of their condition so that they may obtain appropriate counseling and medical treatment during pregnancy.

Sickle cell anemia results from a hemoglobin defect

Sickle cell anemia is inherited as an autosomal recessive trait. The disease is most common in persons of African descent, and about 1 in 12 African-Americans is heterozygous for it. The blood cells of a person with sickle cell anemia are shaped like sickles, or half-moons, whereas normal red blood cells are biconcave discs.

The mutation that causes sickle cell anemia was first identified more than 50 years ago. The sickle cell contains abnormal hemoglobin molecules, which have the amino acid valine instead of glutamic acid at position 6 (the sixth amino acid from the amino terminal end) in the beta chain (see Chapter 3). The substitution of an amino acid with a hydrophobic, uncharged side chain (valine) for one with a hydrophilic, charged side chain (glutamic acid) makes the hemoglobin less soluble. As a result, it tends to form crystal-like structures that change the shape of the red blood cells (Fig. 15–7). This sickling occurs in the veins after the oxygen has been released from the hemoglobin. The blood cells' abnormal sickled shape slows blood flow and blocks small blood vessels, with resulting tissue damage (from lack of oxygen and essential nutrients) and painful episodes. Sickled red blood cells also have short life spans, leading to severe anemia in affected persons. Individuals with sickle cell anemia generally lead shortened, painful lives.

Available treatments for sickle cell anemia include pain-relief measures, transfusions, and some forms of drug therapy, but these are of limited effectiveness. Ongoing research is directed toward eventually providing gene replacement therapy for these individuals. Bone marrow transplants are also a promising future treatment for seriously ill individuals. The development of a mouse model of sickle cell anemia, which

Figure 15–7 **Sickled red blood cells.** Some of the red blood cells from this individual with sickle cell anemia show an abnormal sickle shape. *(G.W. Willis/Biological Photo Service)*

was reported in 1997, will allow researchers to test potential drug and genetic therapies.

The reason that the sickle cell allele (*s*) occurs at a higher frequency in parts of Africa is well known. Individuals who are heterozygous (*Ss*) and carry alleles for both normal hemoglobin (*S*) and sickle cell hemoglobin are more resistant to falciparum malaria, a severe form of malaria that is often fatal. Areas in Africa where falciparum malaria occurs correlate well with areas in which the frequency of the sickle cell allele is more common in the human population. Thus, individuals possessing one copy of the mutant sickle cell allele have a selective advantage over homozygous individuals, both *SS* and *ss*. This phenomenon, known as **heterozygote advantage,** is discussed further in Chapter 18.

Cystic fibrosis results from defective ion transport

Cystic fibrosis is the most common autosomal recessive disorder in children of European descent. About 1 of every 20 persons in the United States is a heterozygous carrier of the cystic fibrosis gene. This disorder is characterized by abnormal secretions in the body; its most severe effect is on the respiratory system, which produces abnormally viscous mucus. The cilia that line the bronchi (see Chapter 44) cannot easily remove the mucus, and it thus becomes a culture medium for dangerous bacteria. These bacteria or their toxins attack the surrounding tissues, leading to recurring pneumonia and other complications. The heavy mucus also occurs elsewhere in the body (e.g., in the ducts of the pancreas and liver and in the intestines), causing digestive difficulties and other effects.

The gene responsible for cystic fibrosis codes for a protein that controls the transport of chloride ions across cell membranes. The mutant protein is responsible for the production of the unusually thick mucus that eventually leads to tissue damage in the respiratory and digestive systems. Although many mutant forms exist and these vary somewhat in severity of symptoms, the disease is usually very serious.

Antibiotics are used to control bacterial infections, and daily physical therapy is required to clear mucus from the respiratory system (Fig. 15–8). Treatment with an enzyme (produced by recombinant DNA technology) that breaks down the mucus is also helpful. Without treatment, death would occur in infancy. With treatment, about 50% of affected persons live into their 20s, only to die in what should be the prime of life after having spent the equivalent of about four years in the hospital. Because of the serious limitations of available treatments, gene replacement therapy for cystic fibrosis is under development.

The most severe mutant allele for cystic fibrosis predominates in northern Europe, and another, somewhat less serious, mutant allele is more prevalent in southern Europe. Presumably these mutant alleles are independent mutations that have been maintained by natural selection. Some experimental evidence supports the hypothesis that heterozygous individuals

Figure 15–8 **Treatment of cystic fibrosis.** The traditional treatment for cystic fibrosis has been chest percussion, or gentle pounding on the chest, to clear mucus from clogged airways in the lungs. *(© 1999 Abraham Menashe)*

Using a Mouse Model to Study a Human Genetic Disease

HYPOTHESIS: The alleles responsible for cystic fibrosis confer a heterozygote advantage because heterozygous individuals are less likely to die from certain types of life-threatening diarrhea.

METHOD: The effects of cholera toxin, which is produced by a bacterial infection and causes severe diarrhea, were studied in mice. The responses of three groups of mice were evaluated (1) mice homozygous for an allele that causes cystic fibrosis, (2) heterozygous mice, and (3) normal mice.

RESULTS: Mice homozygous for a cystic fibrosis allele did not respond to cholera toxin. Cholera toxin caused heterozygotes to lose only half as much fluid from the cells of the intestinal lining as did homozygous "normal" animals.

CONCLUSION: These results support the hypothesis that individuals heterozygous for cystic fibrosis are less likely to die from certain kinds of diarrhea.

Human geneticists have long been puzzled by the fact that heterozygosity for cystic fibrosis alleles is so common among Caucasians (1/20), particularly among Northern Europeans and those of Northern European descent. Recent advances in our understanding of the molecular biology of this genetic disease have led to a new explanation of why these alleles are maintained in the population. This new hypothesis is that heterozygous individuals are less likely to die from certain types of potentially fatal diarrhea.

In severe diarrhea, large amounts of water and electrolytes are lost from the intestine. If unchecked, particularly in infants and young children, this condition can result in death. It has been found that the allele that causes cystic fibrosis is a mutant form of a gene involved in controlling the body's water and electrolyte balance. This gene has been cloned and found to code for a plasma membrane chloride ion channel protein known as the *CFTR protein*. (CFTR stands for *cystic fibrosis transmembrane conductance regulator*.) This ion channel is responsible for transporting chloride ions out of the cells lining the digestive tract and the respiratory system. When the chloride ions leave the cells, water follows by osmosis. Thus the normal secretions of these cells are relatively watery.

Because the cells of individuals with cystic fibrosis lack normal ion channels, their secretions have a very low water content. Cells of heterozygous individuals have only half the usual number of functional CFTR ion channels, but these are sufficient to maintain adequate fluidity of their secretions. However, this ion channel deficiency might be an advantage if an individual is infected by a pathogen that produces a toxin that causes the ion channels to remain constantly open (precipitating diarrhea). With only half as many ion channels, a heterozygote might lose only half as much chloride and water.

Research on any disease is greatly facilitated if an animal model can be used for experimentation. Such a model became available when strains of mice homozygous for cystic fibrosis, as well as those that were heterozygous, were produced by targeted gene replacement (see *Making the Connection: The Genetics of Mice and Humans* in Chapter 14).

Sherif E. Gabriel and his colleagues at the University of North Carolina used the cystic fibrosis mouse model to test the heterozygote advantage hypothesis.* They treated mice with cholera toxin produced by *Vibrio cholerae,* the bacterium that causes cholera. Cholera toxin is known to affect the functioning of the CFTR ion channels, causing the uncontrolled loss of chloride ions and water.

Their results neatly fit the predictions of the heterozygote advantage hypothesis. Animals homozygous for cystic fibrosis, that is, those with no CFTR channels, did not lose any fluid through their intestinal cells when exposed to cholera toxin. The toxin caused heterozygotes (with only half the normal number of CFTR channels) to lose only half as much fluid as mice with the normal number of channels.

Despite the success of this demonstration, it is not thought that cholera itself is the selective force responsible for the high incidence of cystic fibrosis alleles today. Cystic fibrosis is thought to have arisen over 50,000 years ago, whereas European cholera epidemics were first recognized in the early 1800s. Rather, other diarrhea-causing infections are the probable culprits.

Researchers have continued to test candidates likely to be responsible for a possible heterozygote advantage of cystic fibrosis alleles. In 1998 Gerald Pier and his colleagues at the Harvard Medical School, University of Bristol, and University of Cambridge announced that the cystic fibrosis allele protects against typhoid fever, a diarrheal disease caused by the bacterium *Salmonella typhi*[†]. This bacterium enters human gastrointestinal cells using the CFTR channel protein, but does not easily enter cells expressing a cystic fibrosis allele. These data are supported by studies using mice homozygous and heterozygous for a cystic fibrosis allele: no *S. typhi* entered cells of mice homozygous for the cystic fibrosis allele, and 86% fewer *S. typhi* entered heterozygous mouse cells compared with homozygous normal mouse cells.

*Gabriel, S.E., K.N. Brigman, B.H. Koller, R.C. Boucher, and M.J. Stutts. "Cystic Fibrosis Heterozygote Resistance to Cholera Toxin in the Cystic Fibrosis Mouse Model." *Science,* Vol. 266, 7 Oct. 1994.

†Pier, G.B., M. Grout, T. Zaidi, G. Meluleni, S.S. Mueschenborn, G. Banting, R. Ratcliff, M.J. Evans, and W.H. Colledge. "*Salmonella typhi* Uses CFTR to Enter Intestinal Epithelial Cells." *Nature,* Vol. 393, 7 May 1998.

Tay-Sachs disease results from abnormal lipid metabolism in the brain

Tay-Sachs disease is an autosomal recessive disease that affects the central nervous system and results in blindness and severe mental retardation. The symptoms begin within the first year of life and result in death before the age of five. Because of the absence of an enzyme, a normal membrane lipid in the brain cells fails to break down properly and accumulates in the lysosomes. Although research is ongoing, no effective treatment for Tay-Sachs disease is available at this time. However, an effective strategy for the treatment of Tay-Sachs in a mouse model was reported in 1997. This treatment, which involves oral administration of an inhibitor that reduces the synthesis of the lipid that accumulates in the lysosomes, offers the hope of an effective way to deal with Tay-Sachs in humans in the future.

The abnormal allele is especially common in the United States among Jews whose ancestors came from Eastern Europe (Ashkenazi Jews). By contrast, Jews whose ancestors came from the Mediterranean region (Sephardic Jews) have a very low frequency of the allele.

History provides one hypothesis for the high frequency of the Tay-Sachs allele in Ashkenazi Jews. During the Middle Ages, Eastern European Jews were subjected to widespread persecution. They were forced into ghettos, and the population decreased (a phenomenon known to population geneticists as a **population bottleneck;** see Chapter 18). A coincidental high frequency of the Tay-Sachs allele in the small surviving population, which remained isolated and married within their group, might explain the high frequency (1/28) of this allele in the Ashkenazi Jewish population today. However, molecular analysis has revealed that there are at least two mutant alleles for Tay-Sachs disease in that population. The simplest hypothesis is that these mutations arose independently and were maintained because they conferred some kind of heterozygote advantage. People living in ghettos were vulnerable to a variety of infections, such as tuberculosis; resistance to one of these diseases could have been the basis of a heterozygote advantage.

A FEW GENETIC DISEASES ARE INHERITED AS AUTOSOMAL DOMINANT TRAITS

Huntington disease, named after George Huntington, the American physician who first described it in 1872, is due to a rare autosomal dominant allele that affects the nervous system. The disease causes severe mental and physical deterioration, uncontrollable muscle spasms, personality changes, and ultimately death. No effective treatment has been found. Every

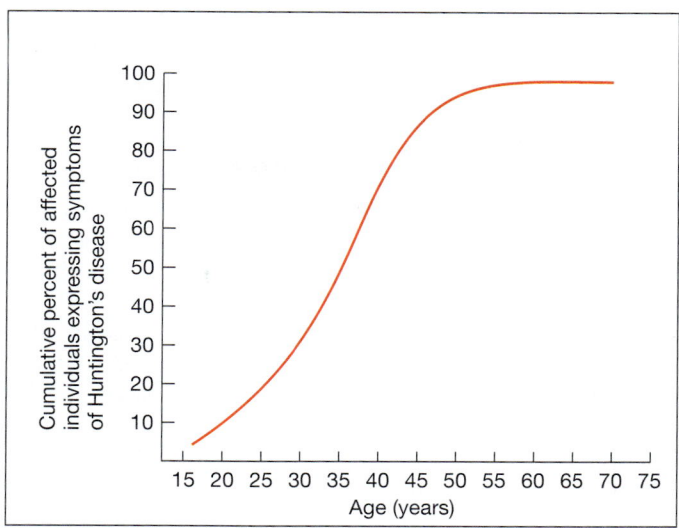

Figure 15-9 **Cumulative percent of persons with Huntington disease who express symptoms.** The onset of Huntington disease occurs relatively late in life (usually between the ages of 30 and 45 for most affected individuals). *(Adapted from Reed, T.E. and J.H. Chandler, 1958,* Am. J. Hum. Genet., *Vol. 10)*

child of an affected individual has a 50% chance of also being affected (and, if a carrier, of passing the abnormal allele to his or her offspring). Ordinarily we would expect a dominant allele with such devastating effects to occur only as a new mutation and not to be transmitted to future generations. This disease is characterized by onset relatively late in life (almost all heterozygotes have developed the disease by age 50, however), so an individual may have children before the disease develops (Fig. 15–9).

More than a decade of basic research was required to pin down the gene after its chromosome location was first identified by recombinant DNA technology. The complexity of the task required extensive international collaboration among researchers in different disciplines. One gene isolation strategy used was to isolate DNA from persons who carry the abnormal allele and to compare it to DNA from close relatives who do not have the allele. This approach is generally done by restriction fragment mapping (see Chapter 14).

One advantage of studying close relatives is that one can minimize other genetic differences and concentrate on the gene of interest. The optimal approach is to study large families, carefully construct pedigrees, and obtain DNA samples from affected and unaffected individuals across generations. A large extended family with a very high incidence of Huntington disease was discovered in Venezuela and has been the subject of exhaustive pedigree and DNA analysis. This made possible the identification of a **DNA marker** for Huntington disease. A DNA marker is not the DNA of the gene of interest, but it is the next best thing: a piece of DNA that is closely associated with the abnormal allele on the chromosome and inherited along with it.

The gene responsible for Huntington disease was finally cloned in 1993. The mutation is a nucleotide triplet (CAG, or cytosine, adenine, guanine) that is repeated many times; the normal allele repeats CAG from 6 to 35 times, whereas the mutant allele repeats CAG from 40 to more than 100 times. Because the triplet CAG codes for the amino acid glutamine, the resulting proteins have long strands of glutamine. Interestingly, the number of nucleotide triplet repeats seems to be important in determining the age of onset and the severity of the disease.

Much research is now directed at studying how the mutation is linked to neurodegeneration in the brain. A mouse model of Huntington is providing valuable clues about the development of the disease.

The Huntington disease DNA marker became the basis for tests that allowed those at risk to learn if they carry the allele. The decision to be tested for any genetic disease is understandably a highly personal one. Certainly, the information can be very useful for those who must make decisions such as whether or not to have children. However, someone who tests positive for the Huntington disease allele must then live with the virtual certainty of eventually developing this devastating and incurable disease. It is hoped that affected persons who choose to be identified before the onset of symptoms may ultimately contribute to the development of effective treatments.

SOME GENETIC DISEASES ARE INHERITED AS X-LINKED RECESSIVE TRAITS

Hemophilia A was once referred to as a disease of royalty because of its high incidence among male descendants of Queen Victoria, but it is also found in many nonroyal pedigrees. Characterized by the lack of a blood-clotting factor, Factor VIII, it causes severe internal bleeding in the head, joints, and other areas from even a slight wound. Because the mode of inheritance is X-linked recessive, affected persons are almost exclusively male, having inherited the abnormal allele on the X chromosome from their heterozygous carrier mothers.

Treatments for hemophilia A consist of blood transfusions and administration of Factor VIII by injection. Unfortunately, these treatments are costly, and, because a lot of blood and blood products come from paid human donors, many Factor VIII preparations were contaminated with HIV-1, the virus that causes AIDS, during the 1980s. Factor VIII produced using recombinant DNA technology (see Chapter 14) provides a safer source of the clotting factor because human donors are not used.

BOTH GENETIC AND ENVIRONMENTAL FACTORS CAUSE BIRTH DEFECTS

A **birth defect,** or congenital defect, is one that is present at birth; it may or may not be inherited. We have seen that some genetic birth defects are caused by chromosome abnormalities, whereas others are caused by a change in a single gene, that is, by a mutation involving a single locus.

Some birth defects are not genetic but are caused by environmental factors that affect prenatal development. For example, if a woman contracts the viral disease rubella (commonly known as German measles) during the first three months of pregnancy, a substantial risk exists that her offspring will develop blindness, deafness, or damage to the heart or brain. Also, when a woman drinks even small amounts of alcohol during pregnancy, the baby may be born with **fetal alcohol syndrome,** which includes a variety of mental and physical defects. Low birth weight and certain structural abnormalities in body organs have been associated with as few as two drinks a day by the mother. Other environmental factors that have been linked to birth defects are discussed in Chapter 49.

Some birth defects can be detected before birth

Genetic abnormalities may become apparent during early intrauterine life or not until late in adult life. Given that early detection increases the possibilities for prevention or alleviation of the effects of genetic abnormalities, efforts have been made over the years to detect such abnormalities before birth. In the past 20 years physicians have become increasingly successful at prenatal diagnosis and treatment, including transfusion and surgical correction of malformations. Intrauterine diagnosis of a number of genetic abnormalities is now possible.

In one diagnostic technique known as **amniocentesis,** a sample of the fluid surrounding the fetus (the *amniotic fluid*) is obtained. A needle is inserted through the walls of the pregnant woman's abdomen and into the uterine cavity, where some of the amniotic fluid is withdrawn from the amniotic cavity into a syringe. This procedure is safe, partly because the positions of the fetus, placenta, and the needle can be determined through **ultrasound imaging** (Fig. 15–10).

The amniotic fluid contains living cells sloughed off the body of the fetus and hence genetically identical to the cells of the fetus. These amniotic fluid cells can be cultured in the laboratory. After two to three weeks, dividing cells from the culture can be karyotyped to detect chromosome abnormalities (Fig. 15-11).

Amniocentesis is performed on pregnant women whose fetuses have a higher than normal risk of Down syndrome. Many other tests have been developed to detect a number of simply inherited genetic disorders, but these disorders are rare enough that the tests are usually done only if a particular problem is suspected. Enzyme deficiencies can often be detected through incubation of cells recovered from amniotic fluid with the appropriate substrate and measurement of the product; this technique has been useful in the prenatal diagnosis of disorders such as Tay-Sachs disease. The tests for a number of other diseases, including sickle cell anemia, Huntington disease, and cystic fibrosis, are less direct, requiring the use of genetic engineering methods. Methods for detecting many more genetic diseases are now being actively sought by researchers.

Figure 15–10 **Ultrasound image of a 12 ½-week-old fetus.** A fetus of this age is typically about 8.5 centimeters long; the head is to the left, and the fetus appears to be sucking its thumb. *(Courtesy of F. R. Batzer, M.D., Philadelphia Fertility Institute)*

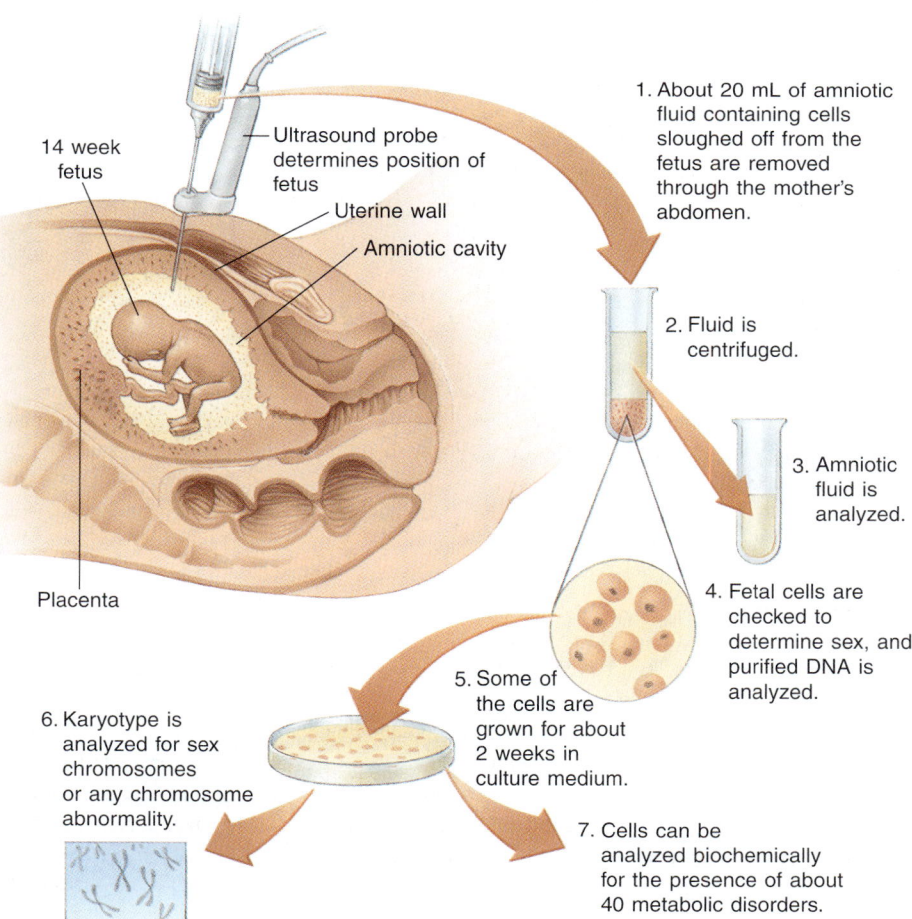

14 week fetus

Ultrasound probe determines position of fetus

Uterine wall

Amniotic cavity

Placenta

1. About 20 mL of amniotic fluid containing cells sloughed off from the fetus are removed through the mother's abdomen.

2. Fluid is centrifuged.

3. Amniotic fluid is analyzed.

4. Fetal cells are checked to determine sex, and purified DNA is analyzed.

5. Some of the cells are grown for about 2 weeks in culture medium.

6. Karyotype is analyzed for sex chromosomes or any chromosome abnormality.

7. Cells can be analyzed biochemically for the presence of about 40 metabolic disorders.

Figure 15–11 **Amniocentesis.** Certain genetic diseases and other abnormal conditions can be diagnosed prenatally by amniocentesis.

Amniocentesis is also useful in detecting a condition known as *spina bifida,* in which the spinal cord does not close properly during development. A relatively common (about 1 in 300 births) nongenetic malformation, this birth defect is associated with abnormally high levels of a normally occurring protein, alpha-fetoprotein, in the amniotic fluid. Abnormally low levels are associated with Down syndrome. Alpha-fetoprotein levels can also be measured by testing the mother's blood, but the results are less reliable, and any unusual findings should be confirmed by amniocentesis or other tests.

One problem with amniocentesis is that most of the conditions it detects are incurable, and the results are generally not obtained until well into the second trimester when abortion is both psychologically and medically more difficult than earlier. Therefore, efforts have been made to develop tests that yield results earlier in the pregnancy. One such test, **chorionic villus sampling (CVS)**, involves removing and studying cells that will form the fetal contribution to the placenta (Fig. 15–12). CVS may be associated with a slightly higher risk of infection or miscarriage than is amniocentesis, but its advantage is that results can usually be obtained within the first trimester.

Although both amniocentesis and CVS can diagnose certain genetic disorders with a high degree of accuracy, they are not foolproof, and many disorders cannot be diagnosed. Therefore, the lack of an abnormal finding is no guarantee of a normal pregnancy. The frequency of errors in amniocentesis and CVS diagnosis has not been determined, but it is quite low.

GENETIC COUNSELORS EDUCATE PEOPLE ABOUT GENETIC DISEASES

Couples who are concerned about the risk of abnormality in their children, either because they have had an abnormal child or have a relative affected by a hereditary disease, may seek **genetic counseling.** Genetics clinics are available in most metropolitan centers.

Genetic counselors, who have specialized training in human genetics, give people information needed to make reproductive decisions. Advice, however, can be given only in terms of the *probability* that any given offspring will have a particular condition. The geneticist needs complete family histories of both the man and the woman and may use tests for the detection of heterozygous carriers of certain conditions. When a disease involves only a single gene locus, probabilities can usually be easily calculated. For example, if one prospective parent is affected with a trait that is inherited as an autosomal dominant disorder, such as Huntington disease, the probability that any given child will have the disease is 0.5.

The birth to phenotypically normal parents of a child affected with an autosomal recessive trait, such as albinism or PKU, establishes that both parents are heterozygous carriers. The probability that any subsequent child will be affected is therefore 0.25. (In this context, the term *carrier* is used specifically to refer to an individual who is heterozygous for a re-

1. Transabdominal sampling technique

Withdrawn chorionic villi cells

Ultrasound probe

Catheter

2. Cervical sampling technique

Syringe

Withdrawn chorionic villi cells

or

Chorionic villi   Catheter

Cells are cultured; biochemical tests and karyotyping are performed

Figure 15–12   Chorionic villus sampling (CVS). This test allows the early diagnosis of some genetic abnormalities. Samples may be obtained by inserting a needle (1) through the uterine wall or through (2) the cervical opening.

cessive allele that causes a genetic disease. Homozygous recessive persons are referred to as *affected* individuals; they are not called *carriers* even though they also carry alleles for the disease.)

For a disease inherited through a recessive allele on the X chromosome, such as hemophilia A, the probability depends on the genotypes of the parents. A normal woman and an affected man will have daughters who are carriers and sons who are normal. The probability that the son of a carrier mother and a normal father will be affected is 0.5; the probability that their daughter will be a carrier is also 0.5.

It is now possible to detect carriers of several genetic diseases. In these cases counseling can be provided when both prospective parents are heterozygous. For diseases that involve an enzyme defect, carriers often show only half the level of enzyme activity characteristic of normal homozygotes. Voluntary screening programs have been set up by synagogues and other organizations to detect couples who are carriers of Tay-Sachs disease among Jews of Eastern European descent.

Persons heterozygous for sickle cell anemia can be readily identified with a simple blood test. They have a mixture of normal and abnormal hemoglobin in their red blood cells, with about 45% of their total hemoglobin being abnormal. Such persons, said to have **sickle cell trait,** are not ill, and their blood cells do not usually undergo sickling (although they can be made to do so by reducing the amount of oxygen). Carrier testing for some other diseases, such as cystic fibrosis and hemophilia A, is much more complicated and is usually done only if family history suggests that a person may be a carrier.

It is very important that identified carriers receive appropriate genetic counseling. A genetic counselor is trained not only to give information needed for reproductive decisions but also to help individuals understand their situation and avoid feeling stigmatized.

Inquiries are commonly made about mental retardation, epilepsy, deafness, congenital heart disease, and other conditions. It is possible that some environmental factor may have played a role in producing the abnormality in the affected child. Did the mother have an infectious disease during pregnancy (e.g., rubella)? Was she receiving some kind of drug therapy, or was she subjected to ionizing radiation? Had the father been exposed to any potentially hazardous agents? By dissecting the environmental contributions, the geneticist can more accurately estimate the probability of the trait's recurrence in subsequent offspring.

GENE REPLACEMENT THERAPY IS BEING USED FOR SEVERAL GENETIC DISEASES

Because many difficulties are inherent in treating most serious genetic diseases, scientists have dreamed of developing actual cures. Today, genetic engineering is bringing these dreams closer to reality. Such therapy could take two main forms.

One approach would be to introduce copies of a normal gene into a fertilized egg, using modifications of the technology already used to produce transgenic animals (see Chapter 14). In some transgenic animals the introduced gene can remain stable from generation to generation, constituting a true "genetic cure." However, this approach raises such complex ethical problems when applied to humans that it is not being actively pursued at this time.

A second strategy receiving increased attention today is to introduce the normal gene into only some body cells (somatic cell gene therapy). The rationale is that, although a particular gene may be present in all cells, it is expressed only in some (see Chapter 16). Expression of the normal allele in only the cells that require it may be sufficient to give a normal phenotype. Needless to say, this approach presents a number of technical obstacles.

The solutions to these problems must be tailored to the nature of the gene itself, as well as to its product and the types of cells in which it must be expressed. First the gene must be cloned and the DNA introduced into the appropriate cells. One of the most successful techniques is to use a virus as the vector. Ideally the virus should infect a high percentage of the cells and facilitate the integration of the introduced gene into a chromosome. Most important, the virus should do no harm, especially over the long term. This is a large order, and a great deal of attention is being paid to the development of viral strains with just such desirable characteristics.

The overall process can be illustrated by the first gene replacement therapy approved for clinical trials. Individuals with a genetic disorder called *severe combined immune deficiency* (SCID) have an immune system that is essentially nonfunctional. Because an affected baby has no defenses, it will die unless isolated from possible sources of infection. Because the cells that constitute the immune system are certain white blood cells that originate in the bone marrow, SCID has been treated successfully in some cases by transplanting healthy bone marrow cells from a normal donor.

About one-quarter of all affected individuals are known to lack a specific enzyme, adenosine deaminase (ADA); it is the absence of ADA in the immune system cells that causes affected individuals to die. These individuals have been treated successfully by providing them with ADA, linked to a molecule that makes the enzyme more stable in the bloodstream. This enzyme-replacement therapy is not a cure, for the ADA treatments must continue throughout life.

Gene replacement therapy for this condition was thought to be feasible because the normal ADA gene had been cloned and successfully introduced into cells through a viral vector. In the first clinical trials, in 1990, a virus was used to introduce the ADA gene into white blood cells from SCID individuals; these cells were then transfused back into the individuals. The results have been encouraging, but because white blood cells have limited lifetimes, such treatments must be repeated.

Although many obstacles must be overcome, gene therapies for a number of other genetic diseases (including cystic

Figure 15–13 **Gene replacement therapy for cystic fibrosis.** A vector carrying the normal allele is introduced into the affected individual's respiratory system through a fiberoptic bronchoscope. The physician (*left front*) is positioning a catheter through the bronchoscope, enabling the vector to be administered. *(Courtesy of Dr. Ronald Crystal, The New York Hospital—Cornell Medical Center)*

fibrosis) are undergoing development or are being tested on individuals in clinical trials (Fig. 15–13). Scientists are currently addressing some of the unique problems presented by each disease. Not all types of cells can be removed from the body, infected with a virus, and then replaced. For example, successful gene replacement therapy for cystic fibrosis will require introduction of normal genes directly into lung cells. A method that has shown some promise is to use as the vector a weakened form of the virus that causes the common cold. The field of gene replacement therapy is expected to develop considerably in the future, although such treatments may not become routinely available for some years.

MUCH NATURAL VARIATION EXISTS IN THE HUMAN POPULATION

One does not need to be a geneticist to recognize that humans are very diverse, and it is widely acknowledged that much human variation has a genetic basis. However, it is difficult to determine the genetic contribution to characteristics that are hard to assess, such as intelligence and behavior.

Humans have several genetically determined blood groups

It is not difficult to understand why so much knowledge of human variation is based on studies of blood. Blood samples are relatively easy to obtain, and blood is a complex tissue consisting of a number of cell types and extracellular molecules that can be studied.

The ABO alleles control the expression of certain red blood cell antigens

The human blood cell types O, A, B, and AB are inherited through multiple alleles representing a single locus. Allele I^A provides the code for the synthesis of a specific glycoprotein, antigen A, which is expressed on the surface of the red blood cells. (Immunity is discussed in Chapter 43; for now we define antigens simply as substances capable of stimulating an immune response.) Allele I^B leads to the production of a different (but related) glycoprotein, antigen B. The allele i^O does not code for an antigen, although it is allelic to I^A and I^B. Allele i^O is recessive to the other two. Neither allele I^A nor allele I^B is dominant to the other; they are both expressed phenotypically in the heterozygote and are therefore **codominant.**

Persons with the genotype $I^A I^A$ or $I^A i^O$ have **blood type A** (Table 15–2); those with genotype $I^B I^B$ or $I^B i^O$ have **blood type B**; and those with genotype $i^O i^O$ have **blood type O.** When both the I^A and I^B alleles are present, both antigen A

TABLE 15–2	ABO Blood Types*			Frequency in U.S. Population (%)	
Phenotype (blood type)	Genotypes	Antigen on RBC	Antibodies in Plasma	Western European Descent	African Descent
A	$I^A I^A$, $I^A i^O$	A	Anti-B	45	29
B	$I^B I^B$, $I^B i^O$	B	Anti-A	8	17
AB	$I^A I^B$	A, B	None	4	4
O	$i^O i^O$	None	Anti-A, anti-B	43	50

*This table and the discussion of the ABO system have been simplified somewhat. Note that persons produce antibodies against the antigens *lacking* on their own red blood cells (RBCs).

First pregnancy

Subsequent pregnancy

Blood vessel of the mother

Placenta

(a) A few Rh⁺ (antigen D-bearing) RBCs leak across the placenta from the fetus into the mother's blood.

(b) The mother produces anti-D antibodies in response to D antigens on Rh⁺ RBCs.

(c) Anti-D antibodies cross the placenta and enter the blood of the fetus. Hemolysis of Rh⁺ blood occurs. The fetus may develop erythroblastosis fetalis.

Rh⁻ RBC of mother

Rh⁺ RBC of fetus with Rh antigen on surface

Anti-D antibody made against Rh⁺ RBC

Hemolysis of Rh⁺ RBC

Figure 15–14 **Rh incompatibility.** Serious problems can occur when an Rh-negative woman and an Rh-positive man produce Rh-positive offspring.

and antigen B are produced in the red blood cells; persons with this $I^A I^B$ genotype have **blood type AB.**

Antibodies anti-A and anti-B are proteins that appear in the plasma (the fluid component of blood) of persons lacking the corresponding antigens on their red blood cells. **Antibodies** are proteins produced by the immune system that combine with specific antigens; hence, anti-A combines with antigen A. Because of their specificity for the corresponding antigens, these antibodies are used in standard tests to determine blood type.

Determining blood types was one of the traditional ways of settling cases of disputed parentage. However, blood type tests can never prove that a certain person *is* the parent of a particular child; they can determine only whether he or she *could* be. Could a man with blood type AB be the father of a child with blood type O? Could a woman with blood type O be the mother of a child with blood type AB? Could a type-B child with a type-A mother have a type-A father or a type-O father?[2]

More than a dozen other sets of blood types, including the Rh group (discussed in the next section), are inherited through other loci, independently of the ABO blood types. Determining some of these types in a given person may be useful in establishing relationships that could not be made with certainty by ABO blood typing alone.

Today more sophisticated types of genetic tests are used to determine parentage. These include **DNA fingerprinting** (see Chapter 14) and **tissue typing,** which is an examination of inherited antigens found on the surfaces of the body's cells

(see Chapter 43). Theoretically, only identical twins would be expected to have the same DNA fingerprint and the same tissue type. If properly performed, these tests have greater than 99% certainty and can come close to proving parentage.

Rh alleles control the expression of other red blood cell antigens

Named for the rhesus monkeys in whose blood it was first found, the Rh system consists of at least eight different kinds of Rh antigens, each referred to as an **Rh factor.** By far the most important of these factors is **antigen D.** About 85% of those United States residents who are of Western European descent are Rh-positive. This means that they have antigen D on the surfaces of their red blood cells (in addition to the antigens of the ABO system and other blood groups). The 15% or so of this population who are Rh-negative have no antigen D. Unlike the situation discussed for the ABO blood group, Rh-negative persons do not naturally produce antibodies against antigen D (anti-D). However, they produce anti-D antibodies if they are exposed to Rh-positive blood. The allele coding for antigen D is dominant to the allele for the absence of antigen D. Hence, Rh-negative persons are homozygous recessive, and Rh-positive persons are heterozygous or homozygous dominant.

Although several kinds of maternal-fetal blood type incompatibilities are known, **Rh incompatibility** is probably the most important (Fig. 15–14). If a woman is Rh-negative and the father of the fetus she is carrying is Rh-positive, the fetus may also be Rh-positive, having inherited the D allele from the father. Ordinarily no mixing of maternal and fetal blood

[2]The answer to all these questions is no.

occurs; nutrients, oxygen, and other substances are exchanged between these two circulatory systems across the placenta. However, late in pregnancy or during the birth process, a small quantity of blood from the fetus may pass through some defect in the placenta.

The fetus's red blood cells, which bear antigen D, activate the mother's white blood cells, inducing them to form antibodies to antigen D. If the woman becomes pregnant again, her sensitized white blood cells produce anti-D antibodies that can cross an intact placenta and enter the fetal blood. There they combine with the antigen D molecules on the surface of the fetal red blood cells, causing the cells to rupture. Breakdown products of the hemoglobin released into the circulation damage many organs, including the brain. In extreme cases of this disease, known as **erythroblastosis fetalis,** so many fetal red blood cells are destroyed that the fetus may die.

When Rh-incompatibility problems are suspected, fetal blood can be exchanged by transfusion before birth, but this is a risky procedure. Rh-negative women are treated just after childbirth (or at termination of pregnancy by miscarriage or abortion) with a preparation of anti-D antibodies known as RhoGAM. These antibodies apparently clear the Rh-positive fetal red blood cells from the mother's blood very quickly, minimizing the chance for her own white blood cells to be sensitized. The antibodies are also soon eliminated from her body. As a result, if she becomes pregnant again, her blood does not contain the anti-D that could harm her baby.

Quantitative traits are controlled by polygenes

Many human characteristics are **quantitative traits;** that is, they represent some measurable quantity such as height. Such characteristics usually show continuous phenotypic variation in the population because of the number of loci involved, which can range from a few to a great many (polygenic inheritance), and also because of environmental factors, whose role is both significant and difficult to quantify. In humans, examples of polygenic inheritance include height and skin color (see Figs. 10–21 and 10–23).

Many common physical traits are inherited

We are often curious about the inheritance of certain physical characteristics. Some of these traits have relatively complex modes of inheritance. For example, dark hair color is due to heavy deposits of the pigment melanin in the hair shaft. There are probably multiple alleles of the locus governing the deposition of melanin, the "darker" alleles being somewhat, but incompletely, dominant to the "lighter" alleles. Red hair color is governed by another unlinked locus with at least two incompletely dominant alleles, one coding for reddish pigment, the other for lack of such pigment. The appearance of red hair is determined by the presence of one or two "red" alleles, but the expression of these alleles can be partially or completely blocked by other genes that code for heavy melanin deposits.

Eye color is determined by the pattern of melanin distribution in the iris. Blue pigment does not exist, only brown or yellowish melanin. If the melanin is deposited in such a way that much of the light is reflected back from the eye, the iris appears blue. Although inheritance of eye color is not simple, the inheritance patterns seen in most families can be explained if we assume that one locus is involved and that the alleles that govern the darker colors are dominant to the alleles that govern the lighter colors (e.g., blue, gray, hazel). It is not uncommon for two dark-eyed parents to have a light-eyed child; it is relatively rare for two light-eyed parents to have a dark-eyed child, but this does occasionally occur.

Several other human characteristics show relatively simple inheritance patterns. For example, the allele for dimples is generally dominant to the allele for no dimples, and the allele for freckling is usually dominant to that for no freckles.

THE HUMAN GENOME PROJECT IS STUDYING ALL HUMAN GENES

Along with advances in technology to determine the base sequences of DNA came the realization that it is possible to study the human genome in great depth. Some scientists envisioned a coordinated effort to determine the total informational content of the human genome. The **Human Genome Project,** initiated in the late 1980s, is an international undertaking that will occur over a 15-year period; researchers hope it will be completed by the year 2003, but information as we go to press indicates that the completion date will probably be a few years later. In its most direct sense the information content is the sequence of 3 billion base pairs in a haploid human genome. However, only a tiny fraction of human DNA is known to code for protein or RNA; the rest (95% or more) is either noncoding or has some function that has not yet been identified.

The project includes not only sequencing but also a broad-based, multifaceted strategy that involves scientists from many disciplines. This approach has concentrated on various types of mapping studies that will allow us to understand the physical and functional relationships among genes and groups of genes as revealed by their order on the chromosomes.

Comparative mapping studies are being carried out simultaneously on several other organisms, especially the laboratory mouse, which has long been the favorite model organism for the study of mammalian genetics. Comparisons of the DNA sequences and chromosome organization of related genes and clusters of genes from different species are powerful tools for identifying the elements essential for their functions. Large-scale analyses of base sequences are also underway. Automation is making the task of sequencing less laborious, and powerful computer programs manage and analyze the data.

The Human Genome Project has been controversial for several reasons. Many scientists are concerned that the Human Genome Project has been funded at the expense of other meritorious science that is equally or more likely to improve the well-being of humankind. Some scientists have also argued that the conventional approach of first identifying an important gene, then cloning and studying it, is scientifically more interesting and more cost-effective over the long run. Supporters of the project, however, argue that many important genes that might be very difficult to identify will be uncovered in the course of the investigation. Even apparently nonfunctional DNA (sometimes called "junk DNA") may turn out to be important.

Our knowledge of human genetics clearly will continue to expand at a great rate over the coming years. The information gained from the comparative mapping studies of other organisms will add greatly to our understanding of evolutionary relationships. Many genes known to be responsible for genetic diseases are being studied, and many others appear to be associated with predispositions to diseases such as heart disease and cancer. Understanding the relationships between genes and disease promises more effective treatments of causes, not just symptoms, in the future.

BOTH HUMAN GENETICS AND BELIEFS ABOUT GENETICS AFFECT SOCIETY

Many misconceptions exist about genetic diseases and their effects on society. Some people erroneously think of certain individuals or populations as genetically unfit and responsible for many of society's ills. They argue, for example, that medical treatment of persons affected with genetic diseases, especially those who are able to reproduce, increases the frequency of abnormal alleles in the population. This is true for autosomal dominant and X-linked diseases, but most genetic diseases that are simply inherited show an autosomal recessive inheritance pattern. Only homozygous persons actually have the disease; heterozygous carriers, present in far greater numbers in the population, are phenotypically normal.

For example, if 1 in 20 persons in the United States is heterozygous for cystic fibrosis, the chance that two parents will both be heterozygous is $1/20 \times 1/20 = 1/400$. On average, one-fourth of the children of such a couple would have cystic fibrosis, so the frequency of affected individuals in the population is about $1/400 \times 1/4 = 1/1600$. Because their numbers are usually very small compared with heterozygotes, reproduction by homozygotes contributes very little to overall frequencies of abnormal alleles.

Abnormal alleles are present in *all* individuals and *all* ethnic groups; no one is exempt. According to one estimate, each of us is heterozygous for several (3 to 15) very harmful alleles, any of which could cause debilitating illness or death in the homozygous state. Why, then, aren't genetic diseases more common? Each of us has many thousands of essential genes, any of which can be mutated. It is very unlikely that the abnormal alleles carried by one person are also carried by that person's mate. Of course, this possibility is more likely if the harmful allele is a relatively common one, such as the one responsible for cystic fibrosis.

Relatives are more likely than nonrelatives to carry the same harmful alleles, having inherited them from a common ancestor. In fact, a greater than normal frequency of a particular genetic disease among offspring of **consanguineous matings** (matings of close relatives) is often the first clue that the mode of inheritance is autosomal recessive. The offspring of consanguineous matings have a small but significantly increased risk of genetic disease. In fact, they can account for a disproportionately high percentage of those individuals in the population with autosomal recessive disorders. Because of this perceived social cost, first-cousin marriages are prohibited in many states in the United States. However, consanguineous marriages are still relatively common in many developing countries, where other factors may outweigh the genetic costs. A high incidence of infectious disease, for example, may result in so many deaths that the health effects of relatively uncommon genetic diseases are largely ignored. The ability of a society to concern itself with genetic diseases is a luxury that comes with affluence.

Genetic discrimination has provoked heated debate

One of the fastest growing areas of medical diagnostics is genetic testing, and the number of new genetic tests that screen for diseases such as cystic fibrosis, sickle cell anemia, Huntington disease, colon cancer, and breast cancer increases each year. However, genetic testing raises many social, ethical, and legal issues that society must address.

One of the most difficult issues is whether genetic information should be made available to health and life insurance companies. Many people think genetic information should not be given to insurance companies, but others, including employers, insurers, and many organizations representing people affected by genetic disorders, say such a view is unrealistic. If people use genetic tests to help them decide when to buy insurance and how much to buy, then insurers insist they should also have access to this information. Insurers say they need access to genetic data to help calculate equitable premiums (the whole idea of insurance is to average risk over a large population). However, there are concerns that insurers might use the results of genetic tests to discriminate against people with genetic diseases or to deny them coverage.

Doctors are concerned that people at risk for a particular genetic disease might delay being tested because they fear genetic discrimination from insurers and employers. The perception of genetic discrimination already exists in society. In a 1996 study, 25% of 332 people with family histories of one or more genetic disorders thought they had been refused life insurance, 22% thought they had been refused health insurance, and 13% thought they had been denied employment because of genetic discrimination.

Making the issue even more complicated is the fact that genetic tests are sometimes difficult to interpret, in part because there are many complex interactions between genes and the environment. If a woman tests positive for a gene that has been linked to breast cancer, for example, she is at significant risk, but testing positive does not necessarily mean that she will get breast cancer. Moreover, if she gets breast cancer, the age of onset and severity of the disease are not predicted by genetic tests. These uncertainties also make it difficult to decide what form of medical intervention, from frequent mammograms to surgical removal of healthy breasts, is appropriate.

The Ethical, Legal, and Social Implications (ELSI) Research Program of the National Human Genome Research Institute has developed principles designed to protect people against genetic discrimination. As we go to press, several bills are up for consideration by Congress that prohibit workplace discrimination, health discrimination, and invasion of privacy.

SUMMARY WITH KEY TERMS

I. **Human genetics** is the science of inherited variation in humans.
   A. Geneticists investigating human inheritance cannot make specific crosses of pure genetic strains; instead, they must rely on studies of populations, analyses of family **pedigrees,** and molecular studies.
   B. Studies of the **karyotype** (the number and kinds of chromosomes present in the nucleus) permit detection of individuals with various chromosome abnormalities.
II. **Aneuploidy,** in which there are either missing or extra copies of certain chromosomes, causes certain human disorders. Aneuploidies include **trisomy,** in which an individual possesses an extra chromosome, and **monosomy,** in which one member of a pair of chromosomes is lacking.
   A. **Trisomy 21,** the most common form of **Down syndrome,** and **Kleinfelter syndrome** (XXY) are examples of trisomy.
   B. **Turner syndrome** (XO) is an example of monosomy.
   C. Trisomy and monosomy are caused by **nondisjunction,** in which sister chromatids or homologous chromosomes fail to disjoin (move apart) properly during meiosis or mitosis.
III. Structural abnormalities in chromosomes cause certain human disorders.
   A. In a **translocation,** part of one chromosome becomes attached to another. About 4% of individuals with Down syndrome have a translocation in which part of chromosome 21 is attached onto one of the larger chromosomes, such as chromosome 14.
   B. In a **deletion,** chromosomes break but fail to rejoin. The deletion may range in size from a few base pairs to an entire chromosome arm. The most common deletion disorder in human infants is **cri-du-chat syndrome,** in which part of the short arm of chromosome 5 is deleted.
   C. **Fragile sites** may occur at specific locations on both chromatids of a particular chromosome. In **fragile X syndrome** the fragile site occurs near the tip on the X chromosome; at that site the nucleotide triplet CGG is repeated many more times than is normal.
IV. Most human genetic diseases that show a simple inheritance pattern are transmitted as autosomal recessive traits.
   A. **Phenylketonuria (PKU)** is an autosomal recessive disorder in which toxic phenylketones damage the developing nervous system.
   B. **Sickle cell anemia** is an autosomal recessive disorder in which abnormal hemoglobin (the protein needed to carry oxygen in the blood) is produced.
   C. **Cystic fibrosis** is an autosomal recessive disorder in which abnormal secretions are produced in the respiratory and digestive systems.
   D. **Tay-Sachs disease** is an autosomal recessive disorder caused by abnormal lipid metabolism in the brain.
   E. **Huntington disease** has an autosomal dominant inheritance pattern. It results in mental and physical deterioration, usually beginning in middle age.
   F. **Hemophilia A** is an X-linked recessive disorder. It results in a defect in one of the components of blood required for clotting.
V. Some genetic diseases and chromosome abnormalities can be diagnosed before birth.
   A. In **amniocentesis,** the amniotic fluid surrounding the fetus is sampled and the fetal cells suspended in the fluid are cultured and screened for genetic defects. Amniocentesis provides results in the second trimester of pregnancy.
   B. In **chorionic villus sampling (CVS),** some chorion cells are removed and studied. CVS provides results in the first trimester of pregnancy.
VI. **Genetic counselors** can advise prospective parents with a family history of genetic disease about the probabilities of having offspring with **birth defects.**
VII. Variation in the human population is exemplified by human blood types, among which are the **ABO blood group** and the **Rh system.**
VIII. The effect of human genetics on society is complex.
   A. The fact that a particular abnormal allele is especially common in a certain population does not mean that group has a higher frequency of abnormal alleles in general.
   B. Some alleles that cause a genetic disease when they are homozygous may confer a **heterozygote advantage,** at least in a particular environment. For example, the allele responsible for sickle cell anemia in homozygous individuals appears to confer resistance to malaria in heterozygotes.
   C. Because most abnormal alleles are recessive, they are manifested phenotypically only in homozygotes, who constitute a tiny fraction of the individuals with the allele. Virtually every individual in the population is a heterozygous carrier of several abnormal alleles.

POST-TEST

1. The array of chromosomes present in a given cell is called a(an) (a) karyotype (b) ABO blood group (c) aneuploidy (d) inborn error of metabolism (e) translocation
2. An abnormality that is present and evident at birth is called a(an) (a) karyotype (b) birth defect (c) quantitative trait (d) ultrasound image (e) DNA marker
3. An abnormality in which there is one more or one fewer than the normal number of chromosomes is called a(an) (a) karyotype (b) inborn error of metabolism (c) aneuploidy (d) DNA marker (e) translocation
4. A person with one extra chromosome (three of one kind) is said to be (a) monosomic (b) triploid (c) trisomic (d) consanguineous (e) isogenic
5. A person who is missing one chromosome, having only one member of a pair, is said to be (a) monosomic (b) haploid (c) trisomic (d) consanguineous (e) isogenic

6. The failure of chromosomes to separate normally during cell division is called (a) amniocentesis (b) an inborn error of metabolism (c) nuclear sexing (d) translocation (e) nondisjunction

7. The transfer of a part of one chromosome to a nonhomologous chromosome is called (a) a karyotype (b) an inborn error of metabolism (c) nuclear sexing (d) a translocation (e) nondisjunction

8. Individuals with trisomy 21, or _____ , are mentally and physically retarded and have abnormalities of the face, tongue, and eyelids. (a) Down syndrome (b) Kleinfelter syndrome (c) Turner syndrome (d) Huntington disease (e) Rh incompatibility

9. An XXY individual has the disorder known as _____ ; an XO individual has _____ . (a) Down syndrome; hemophilia A (b) Kleinfelter syndrome; Down syndrome (c) Turner syndrome; Tay-Sachs disease (d) Rh incompatibility; Turner syndrome (e) Kleinfelter syndrome; Turner syndrome

10. An inherited disorder caused by a defective or absent enzyme is called a(an) (a) karyotype (b) quantitative trait (c) ultrasound image (d) inborn error of metabolism (e) DNA marker

11. In _____ , a genetic mutation codes for an abnormal hemoglobin molecule that is less soluble than usual and more likely than normal to crystallize and deform the shape of the red blood cell. (a) Down syndrome (b) Tay-Sachs disease (c) sickle cell anemia (d) PKU (e) hemophilia A

12. In a person with _____ , the mucus is abnormally viscous and tends to plug the ducts of the pancreas and liver and to accumulate in the lungs. (a) Down syndrome (b) Tay-Sachs disease (c) sickle cell anemia (d) PKU (e) cystic fibrosis

13. During this procedure, a sample of the fluid that surrounds the fetus is obtained by insertion of a needle through the walls of the abdomen and uterus. (a) DNA marker (b) chorionic villus sampling (c) ultrasound imaging (d) consanguineous mating (e) amniocentesis

REVIEW QUESTIONS

1. What means have been devised for overcoming some of the difficulties in studying human inheritance?
2. What is meant by nondisjunction? What are some human abnormalities that appear to be the result of nondisjunction?
3. What is meant by inborn errors of metabolism? Give an example.
4. To be expressed, an autosomal recessive genetic disease must be homozygous. What relationship does this fact have to consanguineous matings?
5. Are all birth defects hereditary? Explain.
6. What are some of the ways that carriers of certain genetic diseases can be identified?
7. What are the relative advantages and disadvantages of amniocentesis and chorionic villus sampling?

8. Examine the following pedigrees and decide whether each disorder is most likely inherited by an autosomal recessive, an autosomal dominant, or an X-linked recessive allele. Determine the probable genotypes for all persons shown in the following three pedigrees.

(a)　　　(b)　　　(c)

9. Complete the following table by checking the correct box for each genetic disorder.

Disease	Chromosome Abnormality	Autosomal Recessive	Autosomal Dominant	X-Linked Recessive
Down syndrome				
Tay-Sachs disease				
Phenylketonuria				
Hemophilia A				
Sickle cell anemia				
Turner syndrome				
Huntington disease				
Kleinfelter syndrome				
Cri-du-chat syndrome				
Fragile X syndrome				
Cystic fibrosis				

YOU MAKE THE CONNECTION

1. Mrs. Doe and Mrs. Smith had babies in the same hospital at the same time. Mrs. Doe took home a girl and named her Nancy. Mrs. Smith took home a boy and named him Richard. However, because she was sure that she had given birth to a girl, she filed suit against the hospital. Blood tests showed that Mr. Smith was blood type A, Mrs. Smith was type B, and Mr. and Mrs. Doe were both type AB. Nancy was type A, and Richard was type B. Had an exchange occurred? What other kinds of genetic information would help resolve this question?

2. Imagine that you are a genetic counselor. What advice or suggestions might you give in the following situations?

   a. A couple has come for advice because the woman had a sister who died of Tay-Sachs disease.

   b. A pregnant woman has learned that as a newborn she suffered a mild case of erythroblastosis fetalis; she is concerned that she might have a similarly affected child.

   c. A young man and woman who are not related are engaged to be married. However, they have learned that the man's parents are first cousins. They are worried about the possibility of increased risk of genetic defects in their own children.

   d. A young woman's paternal uncle (her father's brother) has hemophilia A. Her father is free of the disease, and there has never been a case of hemophilia A in her mother's family. Should she be concerned about the possibility of hemophilia A in her own children?

   e. A 20-year-old man is seeking counseling because his father has just been diagnosed with Huntington disease.

3. A deletion of part of an X chromosome may be lethal in a male but cause few problems in a female. Explain.

4. A common misconception about human genetics is that a person's genes alone determine his or her destiny. Explain why this myth is incorrect. How is the perpetuation of this myth harmful to society?

RECOMMENDED READINGS

Friedmann, T. "Overcoming the Obstacles to Gene Therapy." *Scientific American,* Jun. 1997. What remains to be done before gene replacement therapy can live up to its promise.

Glausiusz, J. "Hidden Benefits." *Discover,* Mar. 1995. Discusses the hypothesis that heterozygosity for cystic fibrosis may confer resistance to certain types of life-threatening diarrhea.

Lewis, R. "The Evolution of a Classic Genetic Tool." *BioScience,* Vol. 44, No. 11, Dec. 1994. Discusses the history and significance of pedigrees, including their use as a clinical tool to diagnose genetic disorders.

McKusick, V.A. *Mendelian Inheritance in Man: A Catalog of Human Genes and Genetic Disorders,* 12th ed. Johns Hopkins University Press, Baltimore, 1998. A comprehensive survey of human genetics.

Rothenberg, K., et al. "Genetic Information and the Workplace: Legislative Approaches and Policy Changes." *Science,* Vol. 275, 21 Mar. 1997. Discusses state and federal initiatives that deal with genetic discrimination in the workplace.

Welsh, M.J., and A.E. Smith. "Cystic Fibrosis." *Scientific American,* Dec. 1995. Includes all aspects of this genetic disorder, from isolation of the responsible gene to strategies for treatment.

● Visit our website at **http://www.saunderscollege.com/lifesci/titles.html** and click on Solomon/Berg/Martin Biology for links to chapter-related resources on the World Wide Web.

CHAPTER 16

Genes and Development

The study of **development,** the process by which cells specialize and organize into a complex organism, encompasses some of the most fascinating and difficult problems in biology today. During the many cell divisions required for a single cell to develop into a multicellular organism, groups of cells become gradually committed to specific patterns of gene activity through a process called **determination.** The final step leading to cell specialization is cell **differentiation.** A differentiated cell can be recognized by its characteristic appearance and activities.

Morphogenesis, the development of form, is an even more intriguing piece of the developmental puzzle. It involves a multistep process known as **pattern formation** by which cells in specific locations become progressively organized into recognizable structures.

Until the late 1970s, little was known about how certain genes interact with various signals from other genes and from the environment to control development. Although certain genes affecting developmental pathways had been identified, their specific functions in the organism were not well understood. Because these networks are too complex to unravel using only traditional methods, it had been thought that it might be impossible to understand development. However, rapid progress in recombinant DNA technology led scientists to renew their search for developmental mutants and to apply the most sophisticated techniques to study them.

Today scientists interested in development study a variety of carefully chosen mutant organisms with altered developmental patterns. They use the tools of genetic engineering combined with more conventional descriptive and experimental approaches to derive fresh insights into the role of genetic information in the control of development. The organism in the photograph is *Caenorhabditis,* a nematode worm that has a number of attributes that make it unusually attractive for developmental studies. For example, its small transparent body allows researchers to locate cells in which a specific developmentally important protein is active; these cells can be recognized because they have been genetically engineered to produce a green fluorescent protein known as GFP. Work with

(Courtesy of Dr. Martin Chalfie, Columbia University)

such organisms has profound implications for our understanding of both normal human development and the kinds of malfunctions that can lead to birth defects and even "normal" aging. Although these worms may seem to have little in common with humans, scientists are learning that many of the genes important in development are quite similar in a wide range of organisms. These similarities have led to new ways to unravel evolutionary relationships, through the study of developmentally important genetic mechanisms that appear to be deeply rooted in the evolutionary history of multicellular organisms.

AFTER YOU HAVE STUDIED THIS CHAPTER YOU SHOULD BE ABLE TO

1. Distinguish between cell determination and cell differentiation.
2. Relate the process of pattern formation to morphogenesis.
3. Describe the kinds of experiments that indicate the totipotency of some differentiated plant cells and some animal nuclei. Discuss how these findings support the idea of nuclear equivalence.
4. Identify the attributes of an organism that would make it especially useful in studies on the genetic control of development. Discuss the value of transgenic organisms in research on the genetic control of development.
5. Indicate the features of the development and genetics of *Drosophila, Caenorhabditis,* and the mouse *(Mus)* that have made these organisms so valuable to researchers.
6. Distinguish among maternal effect genes, zygotic genes, and homeotic genes in *Drosophila.*
7. Explain the relationship between transcription factors and genes that control development. Provide some examples of genes that are known to function as genetic switches in development.
8. Define the phenomena of induction and programmed cell death and give examples of the roles they play in development.
9. Describe the functions of some homeotic-like genes in plants.
10. Point out some of the known exceptions to the general phenomenon of nuclear equivalence.

CELL DIFFERENTIATION USUALLY DOES NOT INVOLVE CHANGES IN DNA

The human body, like those of other vertebrates, contains more than 200 recognizably different types of cells (Fig. 16–1). Combinations of those cells are organized into remarkably diverse and complex structures such as the eye, the hand, and the brain, each capable of carrying out many sophisticated activities. Most remarkable of all, however, is the fact that all the structures of the body and the different cells within them are descended from a single fertilized egg.

All multicellular organisms undergo complex patterns of development. The root cells of plants, for example, have structures and functions very different from those of the various types of cells located in leaves. Remarkable diversity can also be found at the molecular level; most strikingly, each type of plant or animal cell makes a highly specific set of proteins (Fig. 16–2). In some cases, such as the protein hemoglobin in red blood cells, one cell-specific protein may make up more than 90% of the total mass of protein in the cell. Other cells may have a complement of cell-specific proteins that are each present in small amounts but still play an essential role. However, because certain proteins are required in every type of cell (all cells, for example, require certain enzymes for glycolysis), cell-specific proteins usually make up only a fraction of the total number of different kinds of proteins.

One explanation for the fact that each type of differentiated cell makes a unique set of proteins might be that during development each group of cells loses the genes it does not need and retains only those that are required. With just a few exceptions, however, this does not seem to be true. According to the concept of **nuclear equivalence,** the nuclei of essentially all differentiated adult cells of an individual are genetically (though not necessarily metabolically) identical to each other and to the nucleus of the fertilized egg cell from which

they descended. This means that virtually all the **somatic**[1] cells in an adult have the same genes, but different cells express different subsets of these genes.

The evidence for nuclear equivalence comes from cases in which differentiated cells or their nuclei have been found to be capable of supporting normal development. Such cells or nuclei are said to be **totipotent.**

A totipotent nucleus contains all the information required to direct normal development

In plants it is possible to demonstrate that at least some differentiated cells can be induced to become the equivalent of embryonic cells (Fig. 16–3). **Tissue culture** techniques are used to isolate individual cells from certain plants and to allow them to grow in a nutrient medium.

In some of the first experiments on cell totipotency in plants, conducted at Cornell University by F. C. Steward and others, root cells from a carrot were induced to divide in a liquid nutrient medium and to form groups of cells called "embryoid" (embryo-like) bodies. These clumps of dividing cells could then be transferred to an agar medium, which provides nutrients plus a solid supporting structure for the developing plant cells. After transfer to the agar, some of the embryoid cells gave rise to roots, stems, and leaves. The resulting small plants, called plantlets to distinguish them from true seedlings,

---

[1]Somatic cells are cells of the body and are distinguished from *germ-line cells,* which ultimately give rise to a new generation. In animals, germ-line cells, whose descendants ultimately undergo meiosis and differentiate as gametes, are generally set aside early in development. In plants, the distinction between somatic cells and germ-line cells is not clear-cut, and the determination that certain cells will undergo meiosis is made much later in development.

Figure 16–1 **Vertebrate cell lineages.** Repeated divisions of the fertilized egg *(bottom of the figure)* result in the establishment of tissues containing groups of specialized cells (see Chapters 37 and 49 for a discussion of how these are formed). Germ-line cells (cells that produce the gametes) are set aside early in development. Somatic cells progress along the developmental pathways, undergoing a series of commitments that progressively determine their fates.

were then transplanted to soil, where they ultimately developed into adult plants capable of producing flowers and viable seeds. If these plants are all derived from the same parent plant, they are genetically alike and therefore constitute a **clone.** The methods of plant tissue culture are now extensively used to produce genetically engineered plants, for they allow the regeneration of whole plants from individual cells that have incorporated recombinant DNA molecules (see Chapter 36, *Focus On: Cell and Tissue Culture*).

Similar experiments have been attempted with animal cells, but so far it has not been possible to induce a fully differentiated somatic cell to behave like a zygote. Instead, it has been possible to test whether steps in the process of determination are reversible by transplanting the *nucleus* of a cell in a relatively late stage of development into an egg cell that has

been enucleated (i.e., its own nucleus has been destroyed) (Fig. 16–4).

In the 1950s R. Briggs and T. J. King conducted experiments in which nuclei from amphibian cells at different stages of development were transplanted into egg cells. Some of the transplants proceeded normally through a number of developmental stages, and a few even developed into normal tadpoles. As a rule, the nuclei transplanted from cells at earlier stages were most likely to support development to the tadpole stage. As the fate of the cells became more and more determined, the probability that a transplanted nucleus could control normal development diminished rapidly.

In experiments carried out in England in the 1960s by J. Gurdon, in a few cases nuclei isolated from the specialized intestinal cells of a tadpole were able to direct development up

(a)   (b)

Figure 16–2   Cell-specific proteins. The spots in the photographs are proteins from *(a)* muscle and *(b)* liver cells of a mouse. The proteins were separated by two-dimensional gel electrophoresis, a method that separates the proteins in the horizontal direction by their electric charge, followed by a second separation in the vertical direction by molecular weight. Several hundred proteins can be distinguished in each panel. The spots that are labeled with numbers are present in all tissues, but notice that many of them are present in different amounts from one tissue to another. The proteins that are labeled with letters are found only in that specific tissue.   *(Patrick O'Farrell, from Darnell, Lodish, and Baltimore,* Molecular Cell Biology, *Fig. 12–1, Scientific American Books)*

to the tadpole stage. This occurred infrequently, but in such experiments success counts more than failure, and we can safely conclude that at least some nuclei of differentiated animal cells are in fact totipotent.

For many years these successes with amphibia could not be repeated with mammalian embryos, leading many developmental biologists to conclude that some fundamental feature of mammalian reproductive biology might be an impen-

Figure 16–3   Cell totipotency. A complete carrot plant can develop from differentiated somatic cells. Discs of phloem cells, which are specialized for nutrient transport, were isolated from carrot root tissues. When the cells were cultured in a liquid nutrient medium, individual phloem cells divided to form clumps of undifferentiated cells. These clumps (embryoids), which closely resembled plant embryos in their early stages of development, then progressed to form embryonic shoots and roots. Transferring the embryonic tissue to a solid nutrient medium stimulated the tissues to form small plants, called plantlets, which then developed into mature plants.

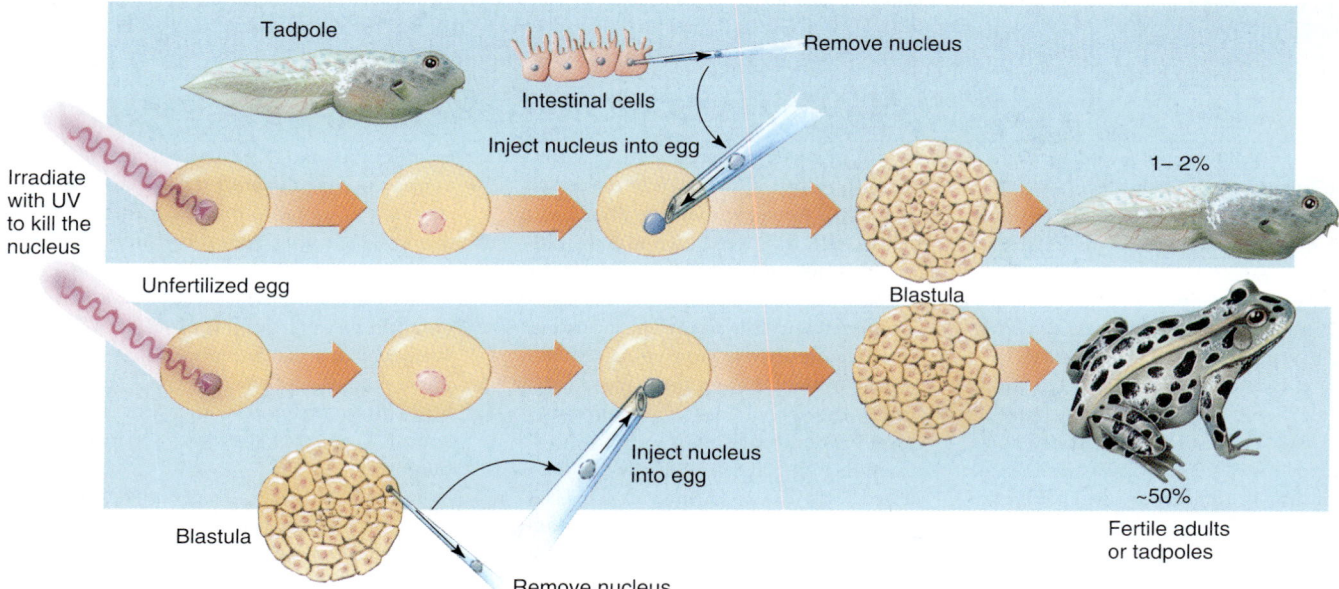

Figure 16–4 **Nuclear totipotency.** In nuclear transplantation experiments in amphibians, nuclei of differentiated cells at different stages of development were injected into eggs whose own nuclei had been destroyed by ultraviolet radiation. As shown in the lower panel, if a nucleus was taken from a cell at the blastula stage of development (when cell division has produced about 1000 cells formed in the shape of a ball), about half the time it would program normal development that would result in a tadpole or fertile adult. However, as seen in the upper panel, most trials using nuclei from tadpole intestinal cells (a much later developmental stage) resulted in no growth, probably as a result of damage to the egg or the nucleus by the procedure. In a small number of trials (about 1–2% of the total) normal development proceeded to the tadpole stage, indicating that the genes necessary to program development to that point were still present.

etrable barrier to mammalian cloning. This perception changed markedly in 1996 and 1997 with the first reports of the birth of cloned mammals (see *Focus On: Mammalian Cloning*).

It may not be particularly surprising that nuclei do not usually lose genetic material during development; after all there should be no loss of chromosomes in the course of normal mitotic cell divisions. These demonstrations of nuclear totipotency also imply that apparently inactive genes are capable of being reactivated when cells or nuclei are placed in a suitable environment. Nevertheless, it is important to recognize that the nuclei of embryonic cells progressively undergo metabolic changes that make it more difficult to remain in a totipotent state. This is especially true of animal nuclei, although various kinds of animals may differ considerably in this regard.

Most differences among cells are due to differential gene expression

Because genes do not appear to be lost regularly during development, the differences in the molecular composition of cells must occur by regulating the activities of different genes. This process of developmental gene regulation is often referred to as **differential gene expression.** As discussed in Chapter 13, the expression of eukaryotic genes can be regulated in many different ways and at many levels. For example, a particular enzyme may be produced in an inactive form and then be activated later. However, much of the regulation that is important in development occurs at the transcriptional level. The transcription of certain sets of genes is repressed, whereas other sets are activated. Even expression of genes that are constitutive (i.e. constantly transcribed; see Chapter 13) and active in all cells can be regulated during development so that the *quantity* of each product varies from one tissue type to another.

We can think of differentiation as a series of pathways leading from a single cell to cells in each of the different specialized tissues, arranged in an appropriate pattern. There are times when a cell makes genetic "commitments" to the developmental path its descendants will follow. These commitments gradually restrict the development of the descendants to a limited set of final tissue types. Determination, then, is a progressive fixation of the fate of a cell's descendants.

As the development of a cell becomes determined along a particular differentiation pathway, the physical appearance of the cell may not change significantly. Nevertheless, when a particular stage of determination is complete, the changes in the cell usually become self-perpetuating and are not easily reversed. Differentiation, then, is usually the last stage in the developmental process. At this stage, a precursor cell becomes structurally and functionally recognizable as a bone cell, for example, and its pattern of gene activity is different from that of a nerve cell.

MAMMALIAN CLONING

In 1996 Ian Wilmut and coworkers at the Roslin Institute in Scotland reported that they had succeeded in cloning sheep, using nuclei from early sheep embryos (blastocyst stage; see Chapter 49). These scientists subsequently received worldwide attention in early 1997 when they announced the birth of a lamb, nicknamed "Dolly," derived from a cultured mammary gland cell (from an adult sheep) fused with an enucleated sheep's egg. The resulting cell divided and developed into an embryo that was then cultured in vitro until it reached a stage at which it could be transferred to a host mother. As might be expected, the overall success rate was low: out of 277 fused cells, only 29 developed into embryos that could be transferred, and Dolly was the only live lamb produced (see Chapter 1). These researchers have also produced cloned transgenic lambs derived from fetal cells.

Why was Wilmut's team apparently successful when so many other researchers had failed? Applying the basic principles of cell biology, they recognized that the cell cycles of the egg cytoplasm and the donor nucleus were incompatible; i.e., the egg cell is a noncycling cell, arrested at metaphase II, while the actively growing donor cell is usually in the DNA synthesis phase (S), or in G_2. By withholding certain nutrients from the donor cells, they were able to cause them to enter a noncycling state referred to as G_0 (see Chapter 9). This had the effect of synchronizing the donor nucleus to the cell cycle of the egg. They then used an electrical shock to cause the donor nucleus to fuse readily with the egg and initiate the development of the embryo.

Although an extremely high level of technical expertise is required, these and other researchers have been able to use modifications and extensions of these techniques to produce cloned transgenic calves, and there is reason to think that the list of mammals that have been successfully cloned will continue to grow. The production of transgenic animals continues to be the main focus of this line of research (see Chapter 14), and it is expected that when this technology is perfected it will be possible to efficiently produce transgenic farm animals that can be used for a variety of purposes.

Results such as these continue to fuel an ongoing debate regarding the potential for human cloning and its ethical implications. In the United States, the National Bioethics Advisory Commission, whose recommendations are posted on the Internet, has been established to study this and other questions:
http://www.nih.gov/nbac.htm.

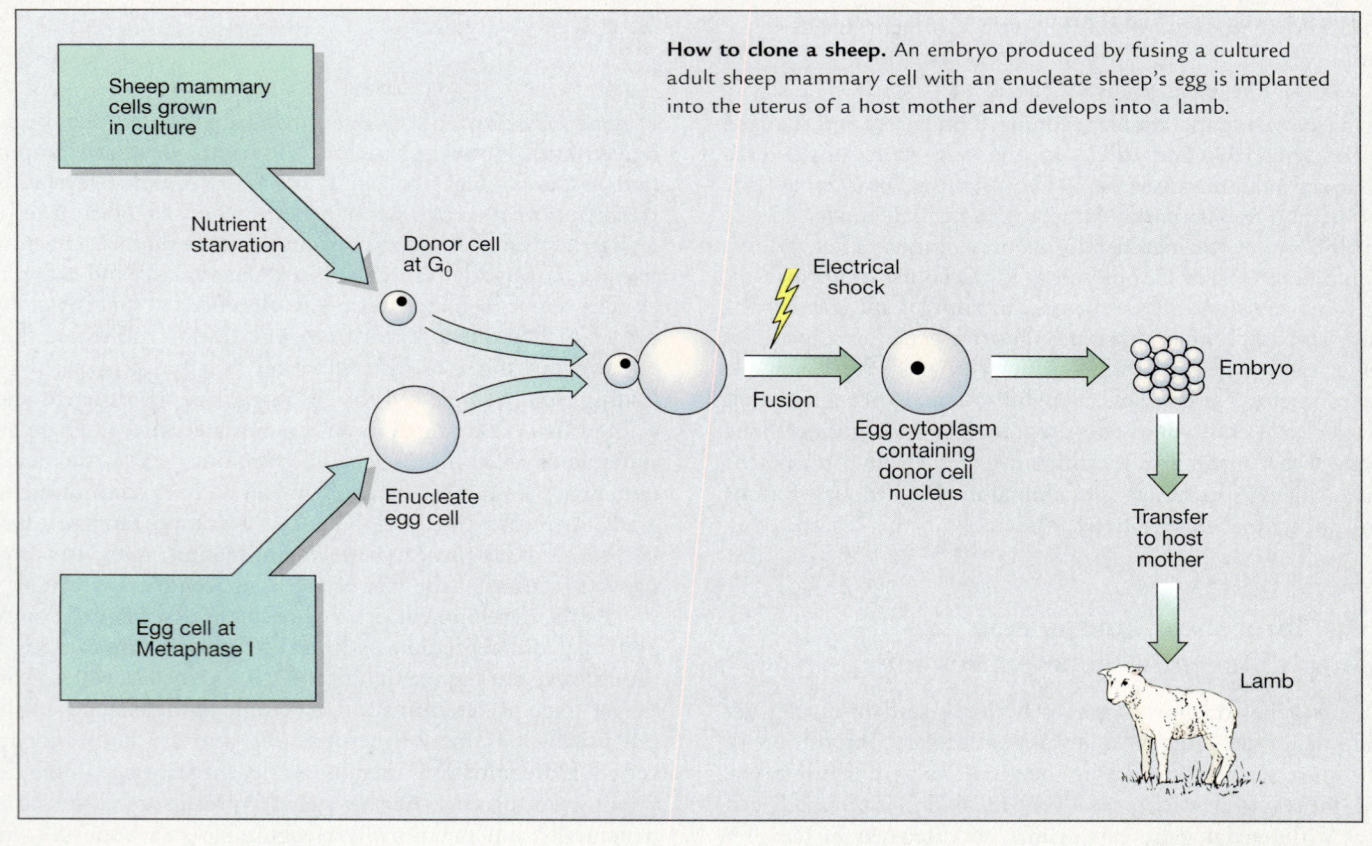

How to clone a sheep. An embryo produced by fusing a cultured adult sheep mammary cell with an enucleate sheep's egg is implanted into the uterus of a host mother and develops into a lamb.

MOLECULAR GENETICS IS REVOLUTIONIZING THE STUDY OF DEVELOPMENT

Development has been an important area of research for many years, and considerable effort has been expended on studying the development of invertebrate and vertebrate animals. By identifying patterns of tissue development in different animals, researchers have been able to identify similarities, as well as differences, in the basic plan of development from a fertilized egg to an adult in organisms ranging from the sea urchin to mammals (see Chapter 49).

In addition to descriptive studies, a number of classic experiments have established important evidence concerning how groups of cells become determined along particular developmental pathways. Researchers have developed elaborate screening programs to detect mutations that allow them to identify large numbers of developmental genes in both plants and animals. They then exploit a wide variety of molecular genetic techniques and other sophisticated methodologies to determine how those genes work and how they interact to coordinate developmental processes.

Certain organisms are well suited for studies on the genetic control of development

In studies of the genetic control of development, the choice of an organism to use as an experimental system has become increasingly important. One of the most powerful approaches involves the isolation of mutants with arrested or abnormal development at a particular stage. Not all organisms have useful characteristics that allow developmental mutants to be isolated and maintained for future study. The genetics of the fruit fly, *Drosophila melanogaster,* is so thoroughly understood that this organism has become one of the most important systems for such studies. Other organisms such as the nematode worm, *Caenorhabditis elegans,* and the laboratory mouse, *Mus musculus,* as well as various plants and some simple eukaryotes, have also become important in developmental genetics. Each of these organisms has attributes that make it particularly useful for examining certain aspects of development.

DROSOPHILA MELANOGASTER PROVIDES RESEARCHERS WITH A WEALTH OF DEVELOPMENTAL MUTANTS

Undoubtedly the most extensive (and spectacular) examples of genes that control development have been identified in the fruit fly, *Drosophila melanogaster.* One of the main advantages of using *Drosophila* as a research organism is the abundance of mutants (including developmental mutants) available for study and the relative ease with which a new mutation can be directly mapped on the chromosomes. The genetic analysis is greatly facilitated by special chromosomes found in certain tissues with large, metabolically active cells, including the salivary glands of the larvae. These **polytene** ("many-stranded") chromosomes (Fig. 16–5) are unusual interphase chromo-

(a)   10 μm

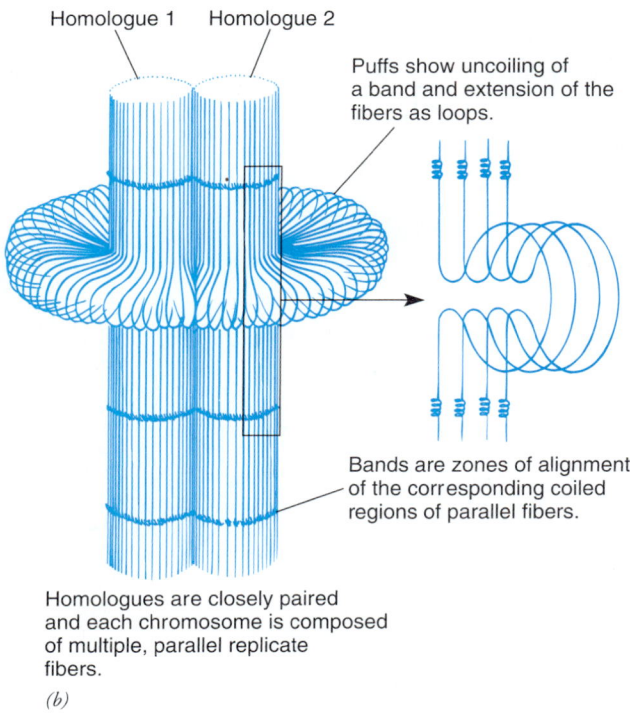

Homologue 1   Homologue 2

Puffs show uncoiling of a band and extension of the fibers as loops.

Bands are zones of alignment of the corresponding coiled regions of parallel fibers.

Homologues are closely paired and each chromosome is composed of multiple, parallel replicate fibers.

(b)

Figure 16–5 *Drosophila* **polytene chromosomes.** These large chromosomes, found in cells of the salivary gland and some other tissues, aid in locating genes. *(a)* A region of a polytene chromosome showing the pattern of stained bands of condensed chromatin (B) and decondensed puffed bands (P), which are the sites of intense gene activity. Although the chromosome banding patterns vary from tissue to tissue, those in a particular tissue are constant and can be associated with the locations of mutant genes by genetic mapping and DNA hybridization methods. *(b)* In contrast to the chromosomes of most somatic cells, homologous polytene chromosomes are paired and each consists of more than 1000 parallel longitudinal DNA fibers. *(a, Courtesy of U. Clever)* somes formed when the DNA replicates many times but without mitosis and cytokinesis.

A typical polytene chromosome may consist of more than 1000 DNA double helices (along with associated histones and other proteins) aligned side by side. Polytene chromosomes are therefore quite large and show a pattern of bands that is very useful in assigning a particular gene to a particular location on the chromosome. When a gene is active, the chromosome band in which it resides uncoils and forms a **puff**, which is a site of intense RNA synthesis. This evidence of gene activity is similar to that observed in lampbrush chromosomes of certain female meiotic cells (see Chapter 9).

Once the chromosomal position of the mutant gene is determined, the gene can be cloned, using a technique called **chromosome walking,** from a nearby gene that has been previously cloned. Studies of *Drosophila* are also facilitated by the fact that foreign DNA can be injected into eggs and become incorporated into the fly's DNA in a process called **transformation** (by analogy with transformation in prokaryotes).

The *Drosophila* life cycle includes egg, larval, pupal, and adult stages

The life cycle of *Drosophila* consists of several distinct stages (Fig. 16–6). After the egg is fertilized, a period of embryogenesis occurs during which the zygote develops into a sexually immature form known as a **larva** (pl., *larvae*). After hatching from the egg, each larva undergoes several molts (shedding of the external covering or cuticle). Each molt allows an increase in size until the larva is ready to pupate. **Pupation** involves a molt and the hardening of the new external cuticle, so that the pupa is completely encased. The insect then undergoes a complete **metamorphosis** (change in form). During that time, most of the larval tissues degenerate and other tissues differentiate to form the body parts of the sexually mature adult fly.

The larvae are wormlike in appearance and look nothing like the adult flies. However, very early in embryogenesis of the developing larvae, precursor cells of many of the adult structures are organized as relatively undifferentiated paired structures called **imaginal discs.** This term comes from **imago,** the name given to the adult form of the insect. Each imaginal disc occupies a definite position in the larva and will form a specific structure, such as a wing or a leg, in the adult body (Fig. 16–7). The discs are formed by the time embryogenesis is complete and the larva is ready to begin feeding. In some respects the larva can be thought of as a complex developmental stage that is simply used to feed and nurture the precursor cells that give rise to the adult fly (which is the only form that can reproduce).

The organization of the precursors of the adult structures, including the imaginal discs, is under complex genetic control. So far more than 50 genes have been identified that specify the formation of the discs, their positions within the larva, and their ultimate functions within the adult fly. Those genes have

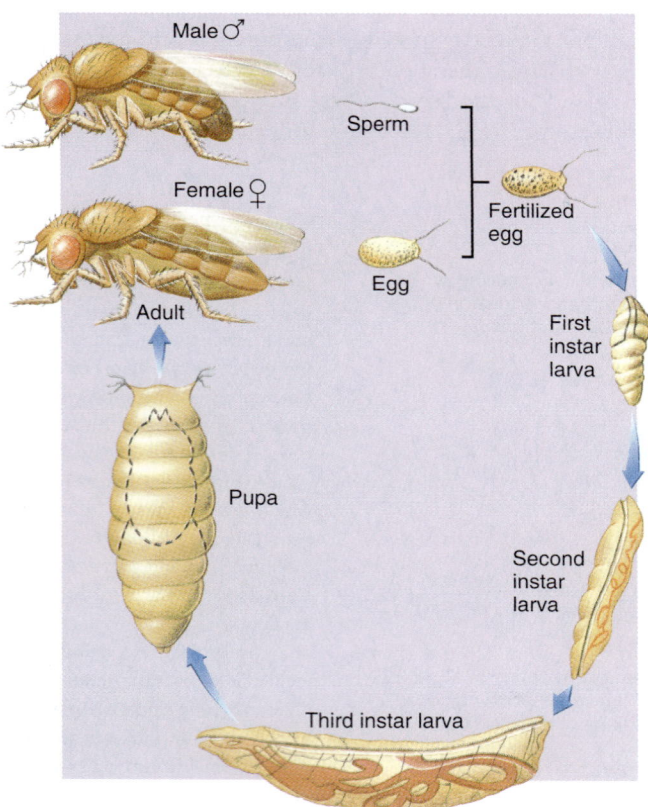

Figure 16–6 ***Drosophila* development.** A *Drosophila* passes through a number of stages as it develops from the egg to the adult fly. (The dotted lines within the pupa represent the animal undergoing metamorphosis.)

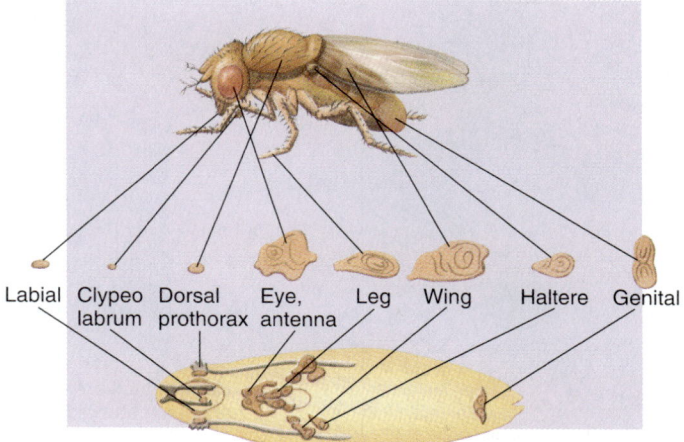

Figure 16–7 **Imaginal discs.** Each pair of discs in a *Drosophila* larva develops into a specific pair of structures in the adult fly.

been identified through mutations that either prevent certain discs from forming or alter their structure or ultimate fate.

Many *Drosophila* developmental mutants affect the body plan

Many developmental mutants of *Drosophila* have been identified. Their effects on development in various combinations have been examined and studied extensively at the molecular level. In our discussion we pay particular attention to those that affect the segmented body plan of the organism, both in the larva and in the adult.

Maternal effect genes organize the egg cytoplasm

The earliest stages of *Drosophila* development are controlled by maternal genes that act to organize the structure of the egg

cell. As the egg develops in the ovary of the female, stores of messenger RNA (mRNA), along with yolk proteins and other cytoplasmic molecules, are passed into it from the surrounding maternal cells. Therefore, all these mRNA molecules are transcribed exclusively from genes found in the mother. The genes that code for these mRNA molecules are referred to as **maternal effect genes.** Analysis of mutants defective in these genes has revealed that many are involved in establishing the polarity of the embryo by designating which parts of the egg are dorsal or ventral and which are anterior or posterior (see Chapter 28).

Figure 16–8*a* illustrates concentration gradients for particular types of maternal mRNA in the very early embryo. These mRNA transcripts of some of the maternal effect genes can be identified by their ability to hybridize with radioactive DNA probes derived from cloned genes. Alternatively, their protein products can be identified by antibodies that specifically bind to them. The protein produced by translation of the

(a) Nuclear migration (1.25 h, ~ 128 nuclei)

(b) Syncytial blastoderm (2 h, ~ 1500 nuclei)

(c) Cellular blastoderm (2.5 h, ~ 5000 cells)

Figure 16–8 **Early development of *Drosophila*.** Longitudinal sections illustrating the morphology of the embryo at different times after fertilization *(left)* are matched with greatly simplified representations of the patterns of gene activity at each stage *(right)*. *(a)* At 1.25 hours (about 128 nuclei). Between the seventh and eighth nuclear divisions, the nuclei start to migrate to the periphery of the egg. The products of several maternal genes can be located in different regions of the egg. The crosses mark the location of maternal mRNA transcribed from a gene that defines the anterior (head) end of the egg. The pink shading represents a concentration gradient of a maternal mRNA extending from the anterior to the posterior end. *(b)* At 2

hours (about 1500 nuclei). Most of the nuclei have reached the perimeter of the egg and have started to make their own mRNA. The maternal mRNA shown in pink in the previous panel is now being transcribed from the corresponding zygotic gene by the nuclei in the anterior part of the embryo. The mRNA from a zygotic gap gene is transcribed from cells in only one segment in the middle of the embryo. *(c)* At 2.5 hours (about 5000 cells). Membranes start to form around nuclei located at the perimeter of the egg. Messenger RNAs can be detected as a series of stripes around the embryo. *(After Akam, Development 101:1–22.)*

Is There a Master Control Gene for Eye Development?

HYPOTHESIS: The *eyeless* locus of *Drosophila* and homologous loci found in other animals are very similar versions of a "master control gene" for eye development.

METHOD: The wild type allele of the *eyeless* locus of *Drosophila* was activated in imaginal discs other than the eye discs in which it is normally expressed. The normal allele of a related locus in mice (*Pax-6*, also known as *Small eye*) was expressed in *Drosophila* in similar experiments.

RESULTS: Normal compound eyes were produced ectopically, that is, in unusual locations, in the structures that developed from *Drosophila* imaginal discs in which either the normal fly allele or the normal mouse allele was activated. Eyes were formed on wings, legs, and antennae.

CONCLUSION: These results support the hypothesis that the *eyeless* locus of *Drosophila* and the *Pax-6* locus of the mouse function similarly as developmental switches in eye formation.

Is it possible that there is a master control gene for the development of eyes? If we survey the animal kingdom, we find several types of apparently very dissimilar types of eyes. For example, if we compare the eyes of vertebrates with the compound eyes of insects (see Chapter 41), they appear so different morphologically that it would seem unlikely that they could develop under the control of similar genes.

The first clues that there could be a link between the genes responsible for eye formation in *Drosophila* and in vertebrates came from the work of R. Quiring and other researchers working in the laboratory of Walter Gehring in Basel, Switzerland.* They discovered a *Drosophila* transcription factor that was coded by *eyeless,* a well known *Drosophila* locus. Mutant alleles of the *eyeless* locus cause the compound eyes to be small or absent. Subsequently, these researchers found that the nucleotide sequence of the *eyeless* locus is very similar to that of the *Pax-6* locus of mice (also known as *Small eye*). Mutant *Pax-6* alleles cause the mice to develop small eyes when heterozygous, and no eyes when homozygous. There is

even a human counterpart to these genes; the eyes of a person heterozygous for a mutant allele of the locus known as *Aniridia* have defects of the iris, cornea, lens, and retina.

G. Halder and P. Callaerts, two other scientists in Gehring's laboratory, tested the hypothesis that these genes are master switches in eye development.[†] They knew that the *eyeless* locus is normally active in the paired eye imaginal discs. They predicted that if the *eyeless* locus were really a master switch, it would be capable of causing eye development wherever it was active. Using a technique known as *targeted gene expression,* they activated the normal *eyeless* allele in various imaginal discs. The experiment worked; compound eyes formed on legs, wings, and antennae! Even more surprisingly, similar results were obtained when normal *Pax-6* alleles (from mice) were expressed in *Drosophila.*

These findings are consistent with the hypothesis that the transcription factors encoded by *eyeless* and by *Pax-6* act as developmental switches, turning on the genes necessary for the formation of the compound eyes of *Drosophila* and vertebrate eyes, respectively.

One interpretation of these findings is that a single "master control gene" for the formation of a "prototypic eye" evolved only once and that both vertebrate and invertebrate eyes evolved from that prototype. Others contend that "master control genes" originated long before the evolution of the structures they currently control. According to this view, such genes provide for general types of control mechanisms, which were subsequently recruited for specific purposes. Studies of recently discovered genes that also play important roles in vertebrate and/or invertebrate eye development may help resolve these issues and will almost certainly raise new questions to be answered.[‡]

Compound eye on a *Drosophila* leg. *(BIOZENTRUM/University of Basel, Switzerland/Courtesy of Professor W.H. Gehring)*

*Quiring, R., U. Walldorf, U. Klotter, and W.J. Gehring. "Homology of the *eyeless* Gene of *Drosophila* to the *Small eye* Gene in Mice and Aniridia in Humans." *Science,* Vol. 165, 5 Aug. 1994.
[†]Halder, G., P. Callaerts, and W.J. Gehring. "Introduction of Ectopic Eyes by Targeted Expression of the *eyeless* Gene in *Drosophila*." *Science,* Vol. 267, 24 March 1995.
[‡]Roush, W. "A 'Master Control' Gene for Fly Eyes Shares its Power." *Science,* Vol. 275, 31 Jan. 1997.

mRNA appears to be part of a system of determinants that organize the early pattern of development in the embryo. A combination of these protein gradients may provide positional information that specifies the fate of each nucleus or cell within the embryo. That information may then be interpreted by a cell as signals specifying the developmental path it should follow. For example, due to the absence of specific signals in the egg, maternal effect mutations can produce an embryo with two heads or two posterior ends.

In many cases, the phenotype associated with a maternal effect mutation can be reversed by injecting normal maternal mRNA into the mutant embryo. When this is done, the fly develops normally, indicating that the gene product is needed only for a short time at the earliest stages of development.

Zygotic segmentation genes continue and extend the developmental program

Immediately after fertilization, the zygote nucleus in the *Drosophila* egg divides, beginning a remarkable series of 13 mitotic divisions. Each of these divisions takes only 5 or 10 minutes, which means that the DNA in the nuclei is replicated constantly at a very rapid rate. During that time the nuclei do not synthesize RNA. Cytokinesis does not take place, and the several thousand nuclei produced by the first seven divisions remain at the center of the embryo until the eighth division occurs.

At that time, most of the nuclei start to migrate out from the center to the periphery of the embryo (Fig. 16–8*b*). Embryonic mRNA production begins, and some of the **zygotic genes** begin to be expressed. (It is customary to refer to the genes of the embryo itself as zygotic genes, even though the embryo is no longer a zygote.) Certain zygotic genes begin to extend the developmental program beyond the pattern established by the maternal genome.

So far geneticists have identified at least 24 **zygotic segmentation genes** that are responsible for generating a repeating pattern of segments within the embryo. The segmentation genes, illustrated in Figure 16–9, fall into three classes—gap genes, pair-rule genes, and segment polarity genes—representing a rough hierarchy of gene action.

The **gap genes** are apparently the first sets of zygotic genes to act. These genes seem to interpret the maternal anterior-posterior information in the egg and begin organization of the segments. A mutation in one of the gap genes usually causes one or more missing segments in an embryo.

The other two classes of segmentation genes do not act on small groups of segments but rather affect all of the segments. For example, mutations in the **pair-rule** genes delete every other segment, whereas mutations in genes belonging to the **segment polarity** class produce segments in which one part is missing and the remaining part is duplicated as a mirror image. The effects of the different classes of mutants are summarized in Table 16–1.

Each gene can be shown to have distinctive times and places in the embryo in which it is active (Fig. 16–10). The

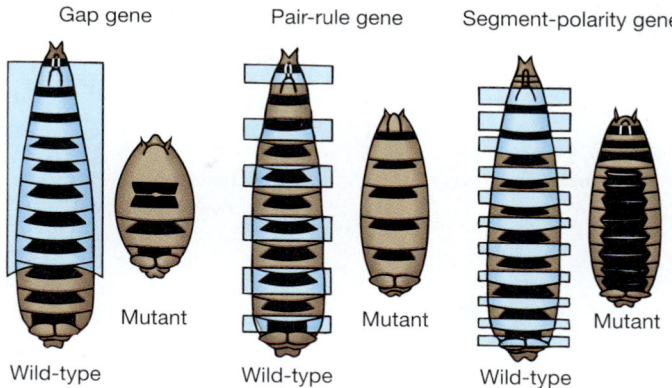

Figure 16–9 **Three classes of zygotic segmentation genes in *Drosophila*.** Gap genes, pair-rule genes, and segment-polarity genes control the pattern of segments in a *Drosophila* embryo. The blue bands mark the regions in which the protein products of these genes are normally expressed in wild type embryos. When mutated, each gene produces a phenotype that is characteristic of the class to which it belongs. *(After Nüsslein-Volhard and Wieschaus, 1980)*

observed pattern of expression of the maternal and zygotic genes controlling segmentation indicates that cells destined to form adult structures are determined by a progressive series of developmental decisions. First, the anterior-posterior (head-to-tail) axis and the dorsal and ventral regions of the embryo are determined by maternal segmentation genes thought to form gradients of **morphogens** in the egg. (A morphogen is a chemical agent that affects the differentiation of cells and the development of form.)

Zygotic segmentation genes then respond to the amounts of various morphogens at each location to control the production of a series of segments from the head to the posterior region. Then, within each segment, other genes are activated that read the position of the segment and interpret that information to specify which body part that segment should become. Within every segment, each cell's position is further specified so that it now has a specific "address" that is designated by combinations of the activities of the various regulatory genes.

It is thought that the zygotic segmentation genes act in sequence, with the gap genes acting first, then the pair-rule genes, and finally the segment polarity genes. In addition, members of each group can interact with each other. Each time a new group of genes acts, cells of a particular group become more finely restricted in the way that they will develop. As the embryo develops, it is progressively subdivided into smaller specified regions.

Most, if not all, of the segmentation genes (maternal and zygotic) code for **transcription factors** (see Chapter 13). For example, some of the segmentation genes code for a "zinc-finger" type of DNA-binding regulatory protein (see Fig. 13–7*b*,*c*). Others code for other types of transcription factors; these are discussed in the next section. The fact that many of the

TABLE 16–1 Classes of Genes Involved in Pattern Formation of Embryonic Segments in *Drosophila*

Type of Gene	Site of Gene Activity	Effects of Mutant Alleles and Proposed Function(s) of Genes
Maternal effect genes	Maternal tissues (ovary)	Many maternal effect mutations alter the polarity of the embryo; initiate pattern formation by activating regulatory genes in nuclei in certain locations in embryo
Zygotic segmentation genes		
Gap genes	Embryo	Mutant alleles cause one or more segments to be missing; some may influence activity of pair-rule genes, segment polarity genes, and homeotic genes
Pair-rule genes	Embryo	When mutated, cause alternate segments to be missing; some may influence activity of segment polarity genes and homeotic genes
Segment polarity genes	Embryo	Mutant alleles delete part of every segment; replace with mirror image of remaining structure; may influence activity of homeotic genes
Homeotic genes	Embryo	Homeotic mutations cause parts of fly to form structures normally formed in other segments; control the identities of the segments

genes involved in the control of development code for transcription factors indicates that those proteins indeed act as genetic "switches" regulating the expression of other genes. Once proteins that function as transcription factors have been identified, it is possible to use the purified proteins to identify the DNA target sequences to which they bind. This approach has been increasingly useful in identifying additional parts of the regulatory pathway involved in different stages of development. Transcription factors also play a role in cancer (see *Focus on: Oncogenes and Cancer*).

Homeotic genes specify the identity of each segment

One function of the zygotic segmentation genes is to regulate the expression of a separate set of genes that actually designates the final adult structure formed by each of the imaginal discs. These genes are called **homeotic genes.** Because of their involvement in segment identity, mutations in homeotic genes cause one body part to be substituted for another and therefore produce some very peculiar changes in the adult. Among the most striking examples are the *Antennapedia* mutants, which have legs that grow from the head at a position where the antennae would normally be found (Fig. 16–11).

Homeotic genes in *Drosophila* were originally identified by the altered phenotypes produced by mutant alleles. When geneticists analyzed the DNA sequences of a number of homeotic genes, they discovered a short DNA sequence of approximately 180 base pairs that is characteristic of many homeotic genes as well as some other genes that play a role in development. This sequence has been termed the **homeobox.** Using the homeobox as a molecular probe made it possible to clone new homeotic genes in *Drosophila* that had not been previously identified. Surprisingly, the homeobox probe has detected homologous DNA sequences in a wide range of other organisms, including humans. This finding generated consid-

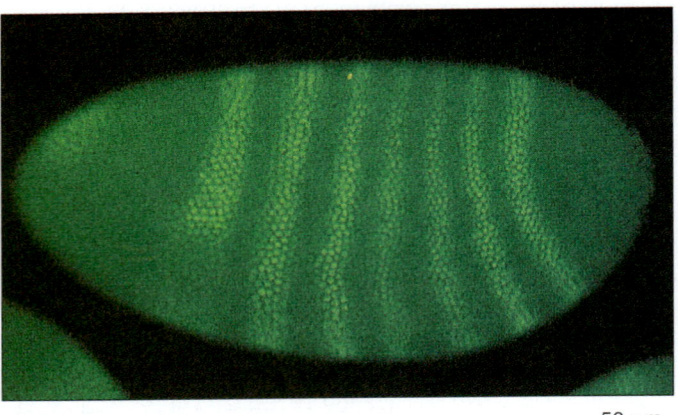

50 µm

Figure 16–10 Activity of a developmental gene. The bright bands in this fluorescence LM reveal the presence of mRNA transcribed from one of the zygotic pair-rule loci known as *fushi tarazu* (Japanese for "not enough segments"). The segments of the larva that are normally derived from these bands are absent when this locus is mutated. *(Courtesy of Steve Paddock, Jim Langeland, Sean Carroll, Howard Hughes Medical Institute, University of Wisconsin)*

Figure 16–11 The **Antennapedia locus.** *Antennapedia* mutations of *Drosophila* cause homeotic transformations in which the antennae are replaced by legs or parts of legs. *(a)* Head of a normal fly and a fly with an *Antennapedia* mutation. *(b)* SEM of the head of a fly with a mutant allele of the *Antennapedia* locus that produces an extreme phenotype. Most of the mutant alleles of this locus produce only incomplete legs in place of the structures of the antennae. *(b, Dr. Thomas Kaufman)*

erable excitement because developmental mutants can be difficult to obtain in many organisms, especially vertebrates. The homeobox has allowed researchers to identify and clone a number of genes thought to control development in complex organisms (see *Making the Connection: Evolution of Gene Complexes That Control the Body Plan*).

The homeobox sequences of a large number of genes have been determined. Comparisons have shown that the DNA sequence itself has been highly conserved during evolution and shows remarkable similarities among organisms as diverse as sea urchins, yeasts, and humans. Each homeobox codes for a protein functional region called a **homeodomain,** consisting of 60 amino acids that form four alpha helices. One of these serves as a recognition helix, which can bind to specific DNA sequences and affect transcription. Thus the products of the homeotic genes, like those of the earlier acting segmentation genes, are transcription factors. In fact, some of the segmentation genes also contain homeoboxes.

CAENORHABDITIS ELEGANS HAS A VERY RIGID EARLY DEVELOPMENTAL PATTERN

Caenorhabditis elegans, a roundworm or nematode (see Chapter 28), has one of the simplest systems of genetic control of development. The study of this animal was begun in the 1960s by Sydney Brenner, a molecular biologist. Today it is an important tool for answering basic questions about the development of individual cells within a multicellular organism.

Even as an adult, *Caenorhabditis* is only 1.5 millimeters long and contains only about 1000 somatic cells (the exact number depends on the sex) and about 2000 germ-line cells. Individuals can be either **hermaphrodites** (organisms with both sexes in the same individual) or males. Hermaphroditic individuals are self-fertilizing, which makes it easy to obtain offspring homozygous for newly induced recessive mutations. The availability of males that can mate with the hermaphrodites makes it possible to do genetic crosses as well.

Because the worm's body is transparent, researchers can follow the development of literally every one of its somatic cells (Fig. 16–12 on p.358) using a Nomarski differential interference microscope (see Chapter 4). As a result of efforts by several laboratories, the lineage of each somatic cell in the adult has now been determined. Those studies have shown that the nematode has a very rigid developmental pattern. After fertilization, the egg undergoes repeated divisions to produce about 550 cells that make up the small, sexually immature larva. After the larva hatches from the egg case, further cell divisions give rise to the adult worm.

The lineage of each somatic cell in the adult can be traced to a single cell in a small group of **stem cells,** or **founder cells,** that are formed early in development (Fig. 16–13 on p.359). If a particular founder cell is destroyed or removed, the structures that would normally develop from that cell are missing. Such a rigid developmental pattern, in which the fates of the cells are largely predetermined, is referred to as **mosaic development.** Each cell has a specific fate in the embryo, just as in art a specific tile forms a particular part of the pattern of a mosaic picture.

ONCOGENES AND CANCER

A cancer cell lacks normal biological inhibitions. Normal cells are tightly regulated by control mechanisms that cause them to divide when necessary and prevent them from growing and dividing at inappropriate times. Cells of many tissues in the adult are normally prevented from dividing; they reproduce only to replace a neighboring cell that has died or become damaged. Cancer cells have escaped such controls and can divide continuously.

As a consequence of their abnormal growth pattern, some cancer cells eventually form a mass of tissue called a **tumor.** If the tumor remains at the spot where it originated, it can usually be removed by surgery. One of the major problems with certain forms of cancer is that the cells can escape from the controls that maintain them in their proper location. These cells can **metastasize,** or spread, to different parts of the body, invading other tissues and forming multiple tumors. Lung cancer, for example, is particularly deadly because its cells are highly metastatic and can enter the blood and spread to form tumors in other parts of the lungs, or in other organs such as the liver and the brain. Tumors with cells that can metastasize are referred to as **malignant tumors.**

We now know that cancer is a disease caused by altered gene expression. Using recombinant DNA methods, researchers identified many of the genes that transform normal cells into cancer cells when they function abnormally. Each kind of cancer cell apparently owes its traits to at least one, and possibly several, of a relatively small set of genes known as **oncogenes** (cancer-causing genes). Oncogenes arise from changes in the expression of certain genes called **proto-oncogenes,** which are *normal* genes found in all cells and involved in the control of growth and development.

Oncogenes were first discovered in viruses that can infect mammalian cells and transform them into cancer cells *(malignant transformation).* Such viruses can incorporate DNA sequences of the proto-oncogenes into their own nucleic acid. In some cases, the viruses alter the expression of the proto-oncogenes, converting them into oncogenes. This may happen if the DNA sequences come under the control of viral regulatory elements, which cause the gene to be transcribed at much higher than normal levels, or if the captured gene mutates so that its protein product is more active than the product of the normal proto-oncogene.

A proto-oncogene in a cell that has not been infected by a virus can also mutate and become an oncogene. One of the first oncogenes identified was isolated from a bladder tumor. In the cell that gave rise to the tumor, a proto-oncogene had undergone a single base-pair mutation; the result was that the amino acid glycine was replaced by a valine in the protein product of the gene. This subtle change was apparently a critical factor in the conversion of the normal cell into a cancer cell.

By means of recombinant DNA technology and other techniques of molecular biology, it has been possible for researchers to identify more than 60 oncogenes and their corresponding proto-oncogenes. Because the fundamental controls of normal cell division and differentiation probably evolved very early in the evolutionary history of eukaryotes, it is not surprising that very similar proto-oncogenes have been found in a diverse array of organisms, ranging from yeasts to humans. For example, the proto-oncogene counterpart of the oncogene found in some bladder tumors (mentioned previously) has also been found in yeast cells. Some of these controls are illustrated in greatly simplified form in the accompanying figure. The growth and division of cells can be triggered by one or more external signal molecules (see *Making the Connection: Information Transfer Across the Plasma Membrane,* Chapter 5). These substances, known as **growth factors,** bind to specific **growth factor receptors** associated with the cell surface, initiating a cascade of events inside the cell. Often the growth factor receptor complex acts as a **protein kinase** (an enzyme that phosphorylates proteins), which then phosphorylates specific amino acids of a number of cytoplasmic proteins. This posttranslational modification usually results in the activation of previously inactive enzymes. These activated enzymes are then able to catalyze the activation of certain nuclear proteins, many of which are **transcription factors.**

Activated transcription factors bind to their DNA targets and stimulate transcription of specific sets of genes that initiate growth and cell division. Even in the simplified scenario presented in the figure, it is evident that multiple steps are required to control cell proliferation. Remarkably, the proto-oncogenes that encode the products responsible for many of these steps have been identified. The current list of known proto-oncogenes includes genes that code for various growth factors or growth factor

---

It was originally thought that each organ in *Caenorhabditis* is derived from only one founder cell. Detailed analysis of cell lineages, however, reveals that many of the structures found in the adult, such as the nervous system and the musculature, are in fact derived from more than one founder cell (see Figure 16–13*h*). Conversely, a few lineages have been identified in which a nerve cell and a muscle cell are derived from the division of a single cell. A number of mutations affecting cell lineages have been isolated, and many of these appear to have properties that would be expected of genes involved in control of developmental decisions.

By using microscopic laser beams small enough to destroy individual cells, it is possible to determine what influence one cell may have on the development of a neighbor. Consistent with the rigid pattern of cell lineages, destruction of an individual cell in *Caenorhabditis* results, in most cases, in the absence of all of the structures derived from that cell, but with the normal differentiation of all of the neighboring somatic

Simplified view of part of a growth control cascade. In this example, a growth factor stimulates cell growth. The growth factor receptor, as well as some of the other components of the system, are coded for by proto-oncogenes. When a proto-oncogene mutates, becoming an oncogene, the cell grows and divides even in the absence of the growth factor. Conversely, other growth control cascades are regulated by growth inhibiting factors. The receptor for the growth inhibiting factor, as well as other parts of the system, are coded for by tumor suppressor genes. When a tumor suppressor gene mutates, the cell grows and divides even if the growth inhibiting factor is present.

Labels in figure: Plasma membrane; Cytosol; Growth factor; Growth factor receptor; Phosphorylated proteins (active); Proteins (inactive); Inactive nuclear proteins; Active nuclear proteins (e.g. activated transcription factors); Nuclear envelope; Stimulation of transcription of specific genes; Cell growth and division

receptors and genes that respond to stimulation by growth factors (including a number of transcription factors). When one of these proto-oncogenes is expressed inappropriately, the cell may misinterpret the signal and respond by growing and dividing.

Not all genes that cause cancer when mutated are proto-oncogenes. About one half of all cancers are caused by a mutation in a **tumor suppressor gene.** These genes, also known as **anti-oncogenes,** normally interact with growth inhibiting factors to block cell division. When mutated they lose their ability to "put on the brakes," and uncontrolled growth ensues.

Certain oncogenes appear to be particularly common and are found in a variety of tumors. However, a change in a single proto-oncogene is usually insufficient to cause a cell to become malignant. The development of cancer is usually a multistep process involving both oncogenes and mutated tumor suppressor genes. Additional factors, such as the inappropriate activation of the enzyme responsible for the maintenance of telomeres, also play a role (see Chapter 11, *On the Cutting Edge: Telomerase, Cellular Aging, and Cancer*). As more of these genes are discovered and their complex interactions are unraveled, we will gain a fuller understanding of the control of growth and development. This understanding is leading to improved diagnosis and treatment of various cancers.

cells. This suggests that development in each cell is regulated through its own internal program.

However, there are cases in which differentiation of a cell can be influenced by interactions with particular neighboring cells, a phenomenon known as **induction.** One example is the formation of the vulva (pl., *vulvae*), the structure through which the eggs are laid. A single nondividing cell, called the *anchor cell,* is a part of the ovary (the structure in which the germ-line cells undergo meiosis to produce the eggs). The anchor cell attaches to the ovary and to a point on the outer surface of the animal, triggering the formation of a passage through which the eggs pass to the outside. When the anchor cell is present, cells on the surface organize to form the vulva and its opening. If the anchor cell is destroyed by a laser beam, however, the vulva does not form and the cells that would normally form the vulva remain as surface cells (Fig. 16–14 on p.360). The anchor cell therefore induces the surface cells to form a vulva.

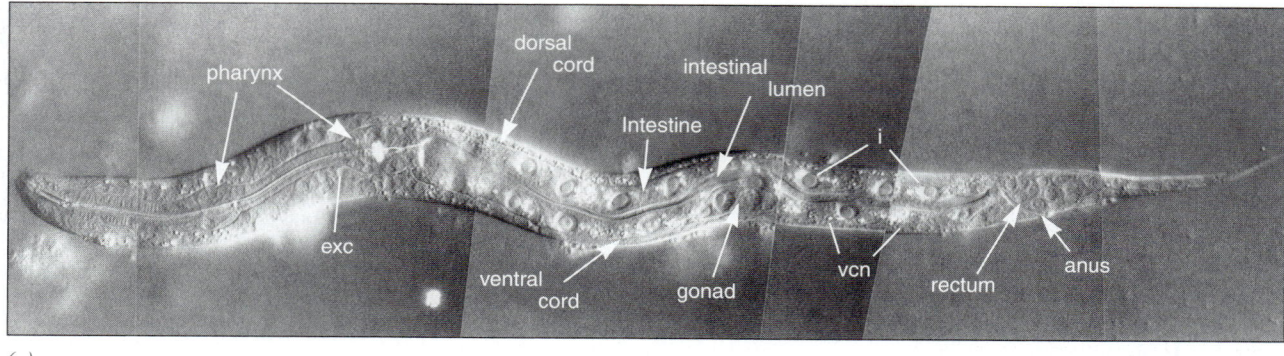

(a)

250 μm

Figure 16–12 *Caenorhabditis elegans.* This transparent organism has a fixed number of somatic cells. *(a)* Nomarski interference LM of the adult hermaphrodite nematode. (The abbreviated labels "exc," "i," and "vcn" refer to certain cells of the excretory, digestive and nervous system respectively.) *(b)* Diagram illustrating structures in the adult hermaphrodite. The sperm-producing structures are not shown. (*a, Courtesy of Dr. John Sulston, Medical Research Council; b, from Walbot and Holder,* Developmental Biology, *p. 607, Figure 22.6a, Random House*)

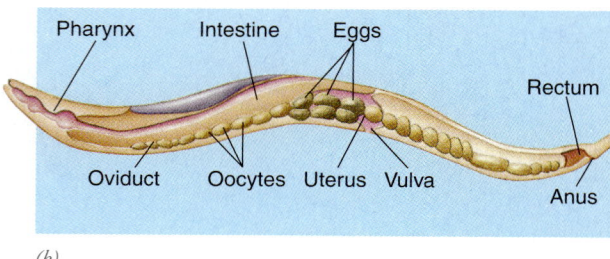

(b)

Analysis of certain cell lineage mutations has been useful in understanding such inductive interactions. For example, several types of mutations cause more than one vulva to form. In such mutant animals, multiple vulvae form even if the anchor cell is destroyed. Thus, the mutant cells do not require an inductive signal from an anchor cell to form a vulva. Evidently in these mutants the gene or genes responsible for vulva formation are constitutive. Conversely, mutants lacking a vulva are also known. In some of these, the cells that would normally form the vulva appear unable to respond to the inducing signal from the anchor cell.

During development in *Caenorhabditis,* a number of instances occur in which cells die shortly after they are produced. This phenomenon, known as **apoptosis** or **programmed cell death** (see Chapter 4) has been observed in a wide variety of organisms, both plant and animal. For example, the human hand is formed as a webbed structure, but the fingers become individualized when the cells between them die. In *Caenorhabditis,* these programmed cell deaths are under genetic control, and several mutants have been isolated that alter the pattern of these deaths. The loci identified by these mutations, in addition to mutations with similar effects in other organisms, are being analyzed at the molecular level and should shed considerable light on the general phenomena of cellular aging and programmed cell death.

Mutations are also known that appear to identify genes involved in developmental timing, known as **chronogenes.** One such locus has recessive alleles that cause certain cells to adopt fates that would ordinarily be seen later in development. Dominant alleles of the same locus cause certain cells to adopt fates that would usually be expressed earlier. Such genes appear to be good candidates for "switches" that control developmental timing.

Genes that contain homeobox-like sequences have been discovered in *Caenorhabditis.* They are sufficiently different from the *Drosophila* homeobox genes that they were not identified by molecular probes from *Drosophila.* Now that the *Caenorhabditis* sequences can be used as probes, researchers are identifying additional homeobox genes in this worm and other organisms.

THE MOUSE IS A MODEL FOR MAMMALIAN DEVELOPMENT

Mammalian embryos develop in markedly different ways from the embryos of *Drosophila* and *Caenorhabditis.* The laboratory mouse, *Mus,* is the best studied example of early mammalian development.

Cells of very early mouse embryos are totipotent

The early development of the mouse and other mammals is similar in many ways to human development, which is de-

(a)   (b)   (c)

(d)   (e)   (f)   25 µm

(g)

(h)

Figure 16–13 **Cell lineages of C. elegans.** All somatic cells of *C. elegans* are derived from five somatic founder cells produced during the early cell divisions of the embryo. *(a–f)* Nomarski interference LMs showing the early cell divisions of the embryo. *(g)* A lineage map showing the origins of the five somatic founder cells *(blue).* The cell shown in white will give rise to the germ cells. *(h)* This lineage map traces the development of the cells that form the intestine. *(a–f, E. Schierenberg, from G. von Ehrenstein and E. Schierenberg, in* Nematodes as Biological Models, *Vol 1, B. Zuckerman, ed., New York: Academic Press, 1980)*

scribed in detail in Chapter 49. During the early developmental period, the embryo lives free in the reproductive tract of the female. It then implants in the wall of the uterus, after which its needs are met by the mother. Consequently, mammalian eggs are very small and contain little in the way of food reserves. Almost all research on mouse development has concentrated on the stages leading to implantation because during those stages the embryo is free-living and can be

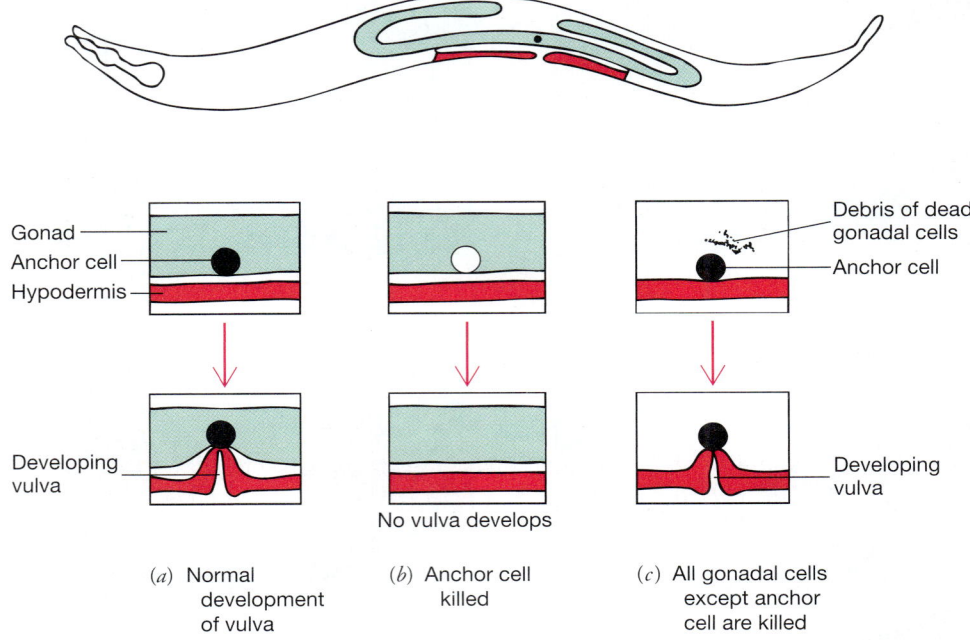

Figure 16–14 **Induction.** A single anchor cell induces neighboring cells to form the vulva in *C. elegans*. This schematic diagram shows how laser destruction of single cells or a group of cells can be used to demonstrate the influence of a cell on its neighbors.

Gonad
Anchor cell
Hypodermis

Debris of dead gonadal cells
Anchor cell

Developing vulva

No vulva develops

Developing vulva

(*a*) Normal development of vulva

(*b*) Anchor cell killed

(*c*) All gonadal cells except anchor cell are killed

experimentally manipulated. During that period, critical developmental commitments take place that have a significant effect on the future organization of the embryo.

Following fertilization, a series of cell divisions gives rise to a loosely packed group of cells. It has been possible to show that all the cells in the very early mouse embryo are equivalent. For example, at the two-cell stage of mouse embryogenesis, one of the two cells can be destroyed by pricking it with a fine needle. Implanting the remaining cell into the uterus of

a surrogate mother in most cases leads to the development of a normal mouse.

Conversely, two embryos at the eight-cell stage of development can be fused together and implanted into a surrogate mother, resulting in the development of a normal-sized mouse (Fig. 16–15). By using two embryos with different genetic markers (such as coat color), it can be demonstrated that the resulting mouse has four genetic parents. These mice have fur with patches of different colors derived from clusters of ge-

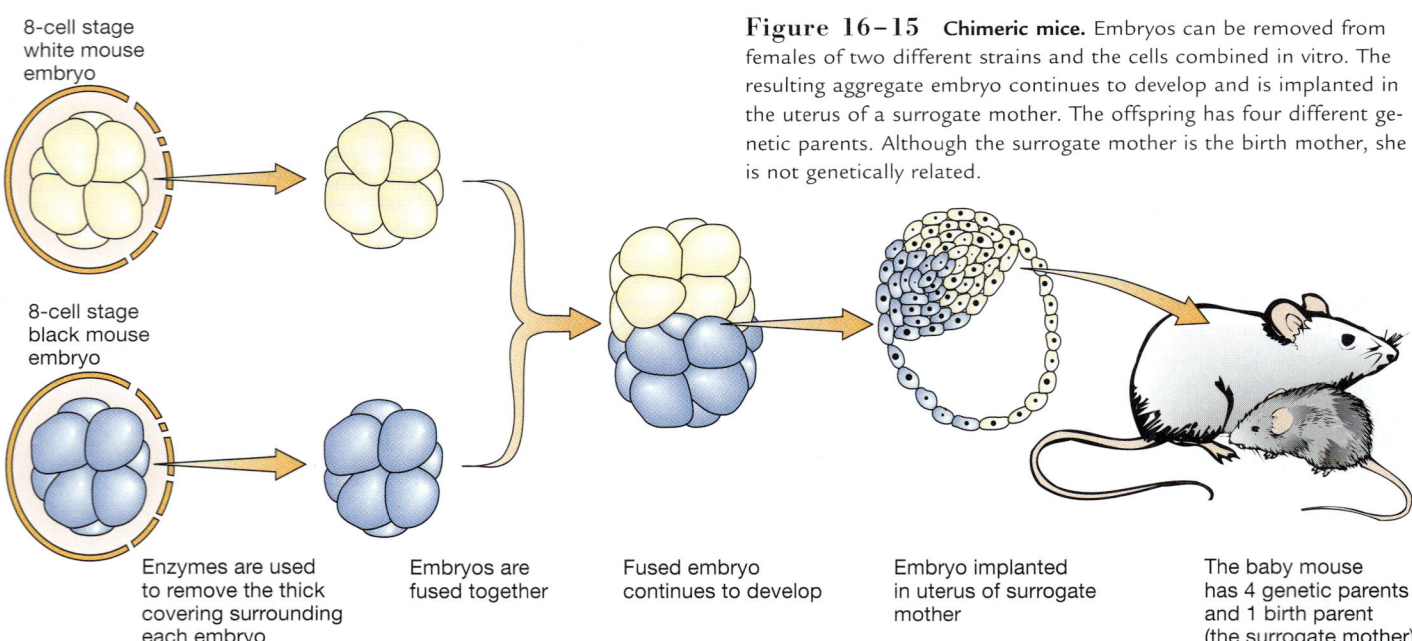

8-cell stage white mouse embryo

8-cell stage black mouse embryo

Figure 16–15 **Chimeric mice.** Embryos can be removed from females of two different strains and the cells combined in vitro. The resulting aggregate embryo continues to develop and is implanted in the uterus of a surrogate mother. The offspring has four different genetic parents. Although the surrogate mother is the birth mother, she is not genetically related.

Enzymes are used to remove the thick covering surrounding each embryo

Embryos are fused together

Fused embryo continues to develop

Embryo implanted in uterus of surrogate mother

The baby mouse has 4 genetic parents and 1 birth parent (the surrogate mother)

EVOLUTION OF GENE COMPLEXES THAT CONTROL THE BODY PLAN

Why do scientists use a wide variety of organisms to study genes that control development? Not only do certain organisms have particular advantages for certain types of investigations, but comparisons among organisms are providing important (and often surprising) insights into evolutionary relationships. This approach is exemplified by studies on clusters of homeobox-containing genes, now generally referred to as Hox genes, which were initially discovered in *Drosophila*.

The Hox genes of *Drosophila* form two adjacent groups on the chromosome: the *Antennapedia* complex and the *bithorax* complex. As homeobox-containing genes have been identified in other animals, including other arthropods and annelids (segmented worms), it has been found that these genes are also clustered and that their organization is remarkably similar to that seen in *Drosophila*. The figure compares the organization of the Hox gene clusters of *Drosophila, Caenorhabditis,* and the mouse. These images are matched with the regions where they are expressed in the animals. Remarkably, the *Drosophila* and mouse Hox genes are located in the same order along the chromosome, although the correlation is less clear for *Caenorhabditis*. Furthermore, the order of the genes on the chromosome reflects the order of the corresponding segments they control (from anterior to posterior) in the animal. This organization apparently reflects the need for these genes to be transcribed in a specific sequence.

Drosophila has only the *Antennapedia/bithorax* complex. However, vertebrates have four similar Hox complexes, each located in a different chromosome. These complexes probably arose through gene duplication. The fact that extra copies of these genes are present helps explain why mutations causing homeotic-like transformations are seldom seen in vertebrate animals. However, one particular type of Hox mutation that has been described in both mice and humans causes abnormalities in the limbs and genitalia. The involvement of the genitalia provides a further explanation for the rarity of these mutant alleles, because affected individuals are unlikely to reproduce.

It is now thought that the homeobox genes are generally responsible for specifying position in the developing animal embryo, particularly the position along the anterior-posterior axis. The fact that very similar developmental controls are seen in organisms as diverse as insects, unsegmented roundworms, and vertebrates (including humans) indicates that the basic mechanism evolved early and has been highly conserved in all animals that have an anterior-posterior axis, even those that are not segmented. The system has apparently been modified in segmented animals such as insects and vertebrates to provide for control of segmentation and specification of segment identity. The idea that homeobox genes are involved in specifying position in the embryo has been strengthened by findings that they not only control the formation of the body axis but also have a role in vertebrate limb development. It is becoming clear that once a successful way of controlling groups of genes and integrating their activities evolved, it was retained, although it has apparently been modified in various ways to provide for alterations of the body plan.

The finding of homeobox-like genes in plants suggests that these genes may be of ancient origin and may in fact be the genes that made multicellularity possible. Further investigations may allow researchers to develop an overall model of how the rudiments of morphogenesis are controlled in both plants and animals. These systems of master genes that control development are proving to be a rich source of "molecular fossils" that are illuminating evolutionary history in new and exciting ways.

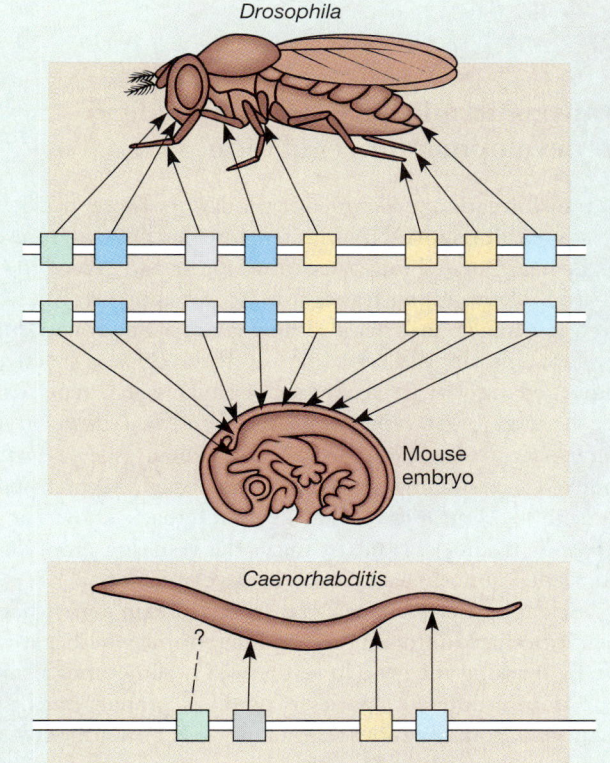

Hox gene clusters. The Hox gene clusters of *Drosophila* and the laboratory mouse are correlated with the parts of the body in which each gene is expressed. (Only one of the four mouse clusters is shown. Although not evident in the figure, some of these regions of expression overlap.) Note that in each organism the order of the genes on the chromosome reflects their spatial order of expression in the embryo. The most anteriorly expressed genes are shown to the left, while those expressed most posteriorly are at the right. *Caenorhabditis* also has similar clustered genes, although their relationships are less well understood. (*After Lenyon et al., Science 253: 516*)

(a)   Nucleus   25 μm

(b)

Figure 16-16   Producing a transgenic mouse. *(a)* Cloned DNA fragments are injected into the nucleus of a fertilized mouse egg by a glass needle, shown entering from the right, which is about 1 μm in diameter at the tip. The egg is held in place by suction on a holding pipet *(left)*. *(b)* The injected eggs are then surgically transferred to a surrogate mother. The presence of the foreign gene can be examined in the transgenic animal, or the animal can be bred to establish a transgenic line of mice.   *(a, R. L. Brinster, University of Pennsylvania School of Veterinary Medicine)*

netically different cells. Animals formed in this way are called **chimeras.** (The term *chimera,* derived from the name of a mythical beast that had the head of a lion, the body of a goat, and the tail of a snake, is used today to refer to any organism that contains two or more kinds of genetically dissimilar cells arising from different zygotes.) Chimeras have been important in allowing the use of genetically marked cells to trace the fates of certain cells during development.

The responses of mouse embryos to these kinds of manipulations are in marked contrast to the mosaic or predetermined nature of early *Caenorhabditis* development, in which the destruction of one of the founder cells results in loss of a significant portion of the embryo. For this reason, we say that the mouse (and presumably other mammals) has highly **regulative development.** This means that the early embryo acts as a self-regulating whole that can accommodate missing or extra parts. On the other hand, it has not been possible so far to demonstrate totipotency of either cells or nuclei from slightly later stages of mouse development. (This may change soon, given the success with nuclear transplantation experiments on sheep, discussed earlier in the chapter.)

Transgenic mice are used in studies on developmental regulation

In transformation experiments similar to those done with *Drosophila,* foreign DNA injected into fertilized mouse eggs can be incorporated into the chromosomes and expressed (Fig. 16–16). The resulting **transgenic** (see Chapter 14) mice have given researchers insights into how genes are activated during development. In addition, mouse genes can be inactivated ("knocked out") by the technique of targeted gene replacement (see Chapter 14, *Making the Connection: The Genetics of Mice and Humans*). For example, a locus known as *engrailed* is a homeobox-containing gene that functions as a segment polarity gene in *Drosophila.* When its counterpart, known as *Engrailed-1,* is knocked out in mice, the resulting embryos exhibit lethal abnormalities in brain development.

Scientists can identify a transgene (foreign gene) that has been introduced into a mouse and determine whether it is active by marking the gene in several ways. Sometimes a similar gene from a different species is used; its protein can be distinguished from the mouse protein by specific antibodies. It is also possible to construct a "hybrid gene" that contains the regulatory elements of a mouse gene of interest together with part of another gene that codes for a "reporter" protein, such as an enzyme not normally found in the mouse. Such studies have been important in showing which DNA sequences of a mouse homeobox gene determine where the gene is expressed in the embryo.

Many developmentally controlled genes have been introduced into mice and have yielded important information about gene regulation. Most important, when developmentally controlled genes from other species such as humans or rats have been introduced into mice, they are regulated in the same way that they normally are in the donor animal. For example, when introduced into the mouse, human genes encoding insulin, globin, and crystallin—which are normally expressed in cells of the pancreas, blood, and eye lens, respectively—are expressed only in those same tissues in the mouse. The fact that these genes are correctly expressed in their appropriate tissues indicates that the signals for tissue-specific gene expression are highly conserved through evolution. This is an exciting finding because it means that information on the regulation of genes controlling development in one organism can have valuable applications to other organisms, such as humans.

HOMEOTIC-LIKE MUTATIONS OCCUR IN PLANTS

Certain well characterized plants are also being used in the study of the genetic control of development. Many of these are economically important crop plants such as corn, *Zea mays.* A number of genes with developmental effects are known in corn, including some that can be thought of as analogous to the homeotic genes of *Drosophila.*

Another plant being used increasingly to study genetics and development in plants is a member of the mustard family, *Arabidopsis.* Although *Arabidopsis* itself is of no economic importance, it has several advantages for research. The plant is quite small, so thousands of individuals can be grown in limited space. Chemical mutagens can be used to produce mutant strains, and a number of developmental mutants, including some that have homeotic-like characteristics, have been isolated. For example, the genes that control flower development have been particularly well characterized (Fig. 16–17). The plant has a very small and simple genome, which greatly facilitates cloning of genes. In addition, cloned foreign genes can be inserted into *Arabidopsis* cells, and these can be integrated into the chromosomes and expressed. These transformed cells can be induced to differentiate into transgenic plants.

Several homeotic-like genes in plants have been shown to code for transcription factors. For example, homeobox-containing genes are involved in the development of the shoot (above-ground portion) of the corn plant. Similarly, the genes

(a)

(b)

Figure 16–17 **Homeotic transformations in *Arabidopsis.* (a)** A normal flower of *A. thaliana* has four outer leafy green sepals (hidden by the petals), four white petals, six stamens (the male reproductive structures), and a central pistil (the female reproductive structure). **(b)** This homeotic mutant has only sepals and petals because the gene normally responsible for the development of the stamens and pistil is missing. *(Dr. Elliot Meyerowitz, California Institute of Technology)*

that specify the identities of the parts of the *Arabidopsis* flower code for transcription factors, although they do not contain homeoboxes. Interestingly, these genes appear to be controlled by regulatory genes that are very similar to genes that are known to regulate the expression of homeotic genes in *Drosophila.* Now that suitable molecular probes are available from plants, many more such genes are being identified in a wide range of organisms. This information will lead to a deeper understanding of the functions and evolutionary history of these genes.

SOME EXCEPTIONS TO THE PRINCIPLE OF NUCLEAR EQUIVALENCE HAVE BEEN FOUND

Although the concept of nuclear equivalence appears to apply to most cells in higher organisms, certain types of developmental regulation can involve physical changes in the DNA. Such changes in the structure of the genome are not common.

Genomic rearrangements involve structural changes in DNA

The activity of some genes may be modified during development by different types of **genomic rearrangements** that lead to actual physical changes in the structure of the gene. In some cases, parts of genes are rearranged to make new coding sequences. This is an important mechanism for the development of the immune system (see Chapter 43).

Another type of rearrangement involves the replacement of an active gene with a copy of a "silent" gene located on a different part of the same chromosome. The baker's yeast *Saccharomyces cerevisiae* is a simple eukaryote that has two sexes or mating types called *a* and α. The mating type of a cell is determined by an active gene located close to the middle of one of the yeast chromosomes; this is called the mating type locus. At some distance on either side of the active gene are two silent genes called *MAT a* and *MAT α*. If a copy of the *MAT a* gene occupies the mating type locus, the mating type is *a*; if a copy of the *MAT α* gene occupies that site, the mating type of the cell is α. These yeast strains can switch their mating type from one form to the other as frequently as every generation (Fig. 16–18). Each time this occurs, the gene located at the mating type locus is removed and replaced by a DNA sequence copied from the silent gene that corresponds to the opposite mating type.

A somewhat similar system of gene replacement takes place in the unicellular parasite *Trypanosoma brucei*, which is a protozoon (see Chapter 24). Trypanosomes carried by the tsetse fly in Africa cause sleeping sickness in humans and related diseases in other animals (Fig. 16–19). When the parasite infects humans, it is able to defeat the immune system by repeatedly changing the glycoprotein molecules that are exposed on the surface of its cell.

Unlike yeast, which has only two basic copies of the mating type gene per cell, the trypanosome cell contains as many as 1000 different genes for cell surface molecules. The differences among their amino acid sequences are so great that an antibody that recognizes one of them would not recognize another. Only one or a few of those copies are expressed at any one time, depending on which copy is present at an **expression site,** which is usually located near the end of a chromosome. The genes in the expression site are exchanged in about

Figure 16–18   Mating type switching in yeast. The active form of the mating type gene resides at the *MAT* locus near the center of the chromosome. Silent copies of the *a* and α genes are located near either end of the chromosome. During cell division, a copy of the opposite mating type is transferred to the MAT locus, resulting in the reversal of the mating type of the cell by the new resident gene.

one out of every 10^4 to 10^6 cells, causing trypanosomes with novel antigens to appear every 7 to 10 days. Thus the infection is maintained by constant supply of new cells that cannot be recognized by the immune system. Although gene replacement clearly offers a mechanism that could serve as a regulatory "genetic switch," it is not known at present whether these mechanisms are relevant to development in multicellular eukaryotes.

Gene amplification increases the number of copies of specific genes

Some gene products are required in such large quantities during certain stages of development that a single copy of a gene cannot be transcribed, nor can its mRNA be translated, rapidly enough to fill the needs of the developing cells. In certain cases, the number of gene copies may be increased, through a process known as **gene amplification,** to meet the demand. For example, the *Drosophila* chorion (eggshell) gene product is a protein made specifically in cells of the insect oviduct. These cells make massive amounts of the particular protein that envelops and protects the fertilized egg. The demand for chorion mRNA in those cells is met by specifically amplifying the gene by DNA replication so that the DNA in that small region of the chromosome is copied many times (Fig. 16–20). In other cells of the insect body, however, the gene appears to exist as a single copy in the chromosome.

(a)

2.5 μm

Figure 16−19 **Gene rearrangement in trypanosomes.** *(a)* LM of blood infected with *Trypanosoma brucei.* *(b)* Trypanosomes change the molecules coating their surfaces frequently and are thus able to "outrun" the immune system of their human host. Each cell contains as many as a thousand silent genes, each coding for a different surface coat glycoprotein. Only one of those genes, which is located at a position near the end of a chromosome, called the expression site, is active at any one time. As the trypanosomes multiply in the blood, occasionally a copy of one of the silent genes replaces the gene currently in the expression site, leading to the production of organisms with a new surface glycoprotein that is not recognized by the immune system. *(a, Ed Reschke)*

Expression site

A   B   C   D   E   F

mRNA

Translation

Surface glycoprotein A

Trypanosome with glycoprotein A surface

Surface membrane containing glycoprotein A molecules

A

D   B   C   D   E   F

mRNA

Translation

Surface glycoprotein D

Trypanosome with glycoprotein D molecules

(b)

Drosophila chorion gene

Gene amplification by repeated DNA replication of chorion gene region

Chorion gene in ovarian cell

Figure 16−20 **Gene amplification.** In *Drosophila,* multiple replications of a small region of the chromosome result in the amplification of the chorion (egg shell) protein genes. Replication is initiated at a discrete chromosome origin of replication for each copy of the gene that is produced, and is randomly terminated, resulting in a series of forked structures in the chromosome.

THE STUDY OF DEVELOPMENTAL BIOLOGY PRESENTS MANY FUTURE CHALLENGES

New model organisms, each with characteristics that uniquely suit it for developmental studies, are now being added to the list of well characterized experimental systems. In a herculean effort spearheaded by Christiane Nüsslein-Volhard of the University of Tubingen, Germany, researchers have identified over 2000 developmental mutants of the zebrafish (*Danio rerio*) and are now embarking on molecular genetic studies. Previously best known to tropical fish enthusiasts, the zebrafish was chosen as a research subject mainly for its transparent embryos (see chapter opener 49).

Scientists are now learning how genes are activated, inactivated, and modified and how batteries of master regulatory genes interact to control development. Eventually researchers expect to understand not only how differentiation and morphogenesis are controlled but also how the basic control systems have evolved. The identification of certain features common to many organisms, such as homeobox genes, will make the task easier, but the work has just begun. Many complex interactions remain to be explored, and many revelations await us.

SUMMARY WITH KEY TERMS

I. **Development** is the process by which the descendants of a single cell specialize and organize into a complex organism.
  A. An organism contains many types of cells that are specialized both structurally and chemically to carry out specific functions. These cells are the product of a process of gradual commitment, called cell **determination,** which ultimately leads to the final step in cell specialization, called cell **differentiation.**
  B. **Morphogenesis,** the development of form, occurs through stages, referred to as **pattern formation,** in which the various specialized cells become organized into structures.
  C. There is no evidence that genes are normally lost during most developmental processes.
    1. At least some nuclei from differentiated plant and animal cells are **totipotent** and contain all the genetic material that would be present in the nucleus of a zygote.
    2. **Nuclear equivalence** is the concept that, with a few exceptions, all of the nuclei of the differentiated somatic cells of an organism are identical to each other and to the nucleus of the single cell from which they descended.
    3. Differences among various cell types are apparently due to **differential gene expression.**
II. Several organisms have characteristics that make them especially useful in studies of the genetic control of development.
  A. Many types of developmental mutants have been identified in the fruit fly, *Drosophila melanogaster.* Many of these mutations affect the segmented body plan of the organism.
    1. The earliest developmental program to operate in the egg is established by **maternal effect genes;** these are active prior to fertilization, and some produce gradients of **morphogens,** thereby affecting the segmentation pattern that is progressively established in the embryo.
    2. **Zygotic segmentation genes** do not become active until much later, when the embryo is no longer a zygote. They continue and extend the developmental program initiated by the maternal effect genes.
    3. The zygotic segmentation genes and their products interact with each other and with the products of the maternal effect genes according to a hierarchical pattern, with certain earlier-acting genes controlling particular later-acting genes.
    4. The later-acting **homeotic genes** are responsible for specifying the identity of each segment.
    5. Many of the segmentation genes are known to code for **transcription factors.** Some of these contain a DNA sequence called a **homeobox,** which codes for a protein with a DNA-binding region called a **homeodomain.**
  B. *Caenorhabditis elegans* is a roundworm with **mosaic development,** an extremely rigid developmental pattern in which the fates of cells are largely predetermined.
    1. The lineage of every somatic cell in the adult is known, and each can be traced to a single **founder cell** in the early embryo.
    2. A number of mutations affecting cell lineages have been identified, and many of these appear to identify genes that control developmental processes such as **induction** (developmental interactions with neighboring cells), **programmed cell death (apoptosis),** and developmental timing (i.e. **chronogenes).**
  C. The laboratory mouse, *Mus,* is extensively used in studies of mammalian development.
    1. In contrast to *Caenorhabditis,* the mouse shows **regulative development,** which means that the very early embryo is a self-regulating whole and can develop normally even if it has extra or missing cells.
    2. Transgenic mice have been extremely useful in determining how genes are activated and regulated during development.
  D. Genes affecting the developmental pattern have also been identified in certain plants, including *Zea mays* (corn) and *Arabidopsis.* Some of these code for transcription factors.
  E. Some homeobox genes are organized into complexes that appear to be systems of master genes specifying an organism's body plan. Remarkable parallels exist between the homeobox complex of *Drosophila* and those of other animals, including the laboratory mouse and *Caenorhabditis.*
III. A few exceptions to the general rule of nuclear equivalence are known. Among these are **genomic rearrangement** (physical rearrangements of the DNA) and **gene amplification,** which provides more copies of certain genes for transcription.

POST-TEST

1. Morphogenesis occurs through a series of stages known as (a) differentiation (b) determination (c) pattern formation (d) totipotency (e) selection
2. The "cloning" experiments carried out on amphibians demonstrated that (a) all differentiated amphibian cells are totipotent (b) some differentiated amphibian cells are totipotent (c) all nuclei from differentiated amphibian cells are totipotent (d) some nuclei from differentiated amphibian cells are totipotent (e) the mechanism of cellular differentiation always requires the loss of certain genes
3. *Drosophila* is a particularly good model for developmental studies because (a) a large number of developmental mutants are available (b) it has a fixed number of somatic cells in the adult (c) its embryos are trans-

parent (d) it is a vertebrate (e) all of the above

4. The anterior-posterior axis of a *Drosophila* embryo is first established by certain (a) homeotic genes (b) maternal effect genes (c) zygotic segmentation genes (d) chronogenes (e) pair-rule genes

5. You discover a new *Drosophila* mutant in which mouthparts appear where the antennae are normally found. You would predict that the mutated gene is most likely a (a) homeotic gene (b) gap gene (c) pair-rule gene (d) maternal effect gene (e) segment polarity gene

6. Most zygotic segmentation genes code for (a) special transfer RNAs (b) enzymes (c) transcription factors (d) histones (e) transport proteins

7. The developmental pattern of *Caenorhabditis* is said to be highly mosaic because (a) development is controlled by gradients of morphogens (b) part of the embryo fails to develop if a founder cell is destroyed (c) some individuals are self-fertilizing hermaphrodites (d) all development is controlled by maternal effect genes (e) programmed cell deaths never occur

8. Which of the following illustrates the regulative nature of early mouse development? (a) the mouse embryo is free-living prior to implantation in the uterus (b) it is possible to produce a transgenic mouse (c) it is possible to produce a mouse in which a specific gene has been "knocked out" (d) genes related to *Drosophila* homeotic genes have been identified in mice (e) a chimeric mouse can be produced by fusing two mouse embryos

9. *Arabidopsis* is useful as a model organism for the study of plant development because (a) it is of great economic importance (b) it has large polytene chromosomes (c) many developmental mutants have been isolated (d) it contains a large amount of DNA per cell (e) it has a rigid developmental pattern

10. The mating type switching system of yeast is an example of (a) nuclear equivalence (b) genomic rearrangement (c) gene amplification (d) cell totipotency (e) pattern formation

REVIEW QUESTIONS

1. Development consists of four main processes: determination, differentiation of cells, pattern formation, and morphogenesis. Define each process and describe how they relate to one another.

2. What lines of evidence support the concept of nuclear equivalence?

3. What are the relative merits of *Drosophila, Caenorhabditis, Mus,* and *Arabidopsis* as model organisms for the study of development?

4. What is the value of homeotic genes in developmental studies?

5. Describe how transgenic organisms are useful in the study of gene regulation in development.

6. Give some examples of genomic rearrangements that are known to occur as a part of some developmental processes.

7. Under what conditions are examples of gene amplification seen?

8. What are oncogenes, and what is their relationship to cellular genes involved in the control of normal growth and development?

YOU MAKE THE CONNECTION

1. Why is an understanding of gene regulation in eukaryotes crucial to an understanding of developmental processes?

2. Why do developmental biologists need to understand biological diversity?

Why is it necessary for scientists to study development in more than one type of organism?

RECOMMENDED READINGS

Ameisen, J.C. "The Origin of Programmed Cell Death." *Science,* Vol. 272, 31 May 1996. The occurrence of programmed cell deaths in unicellular organisms provides clues about the evolution of this process.

Barinaga, M. "Looking to Development's Future." *Science,* Vol. 266, 28 Oct. 1994. This article introduces a special section entitled "Frontiers in Biology: Development," which includes a collection of articles and perspectives by well known developmental biologists. A wide range of organisms, including *Drosophila, C. elegans,* the mouse, and plants, are discussed.

Cavenee W.K., and R.L. White. "The Genetic Basis of Cancer." *Scientific American,* Vol. 272, No. 3, Mar. 1995. Discusses the roles of both oncogenes and mutated tumor suppressor genes in malignancy.

Duke, R.C., D.M. Ojcius, and J.D-E. Young. "Cell Suicide in Health and Disease." *Scientific American,* Vol. 275, No. 6, Dec. 1996. The abnormal regulation of apoptosis can result in various types of disease.

Fletcher, C. "A Garden of Mutants." *Discover,* Aug. 1995. A simplified discussion of the genetic control of flower development in *Arabidopsis.*

McGinnis, W., and M. Kuziora. "Molecular Architects of Body Design." *Scientific American,* Vol. 270, No. 2, Feb. 1994. A comparison of the homeotic genes of *Drosophila* and of vertebrates.

Meyerowitz, E.M. "The Genetics of Flower Development." *Scientific American,* Vol. 271, No. 5, Nov. 1994. Presents a model to explain the genetic control of flower development in *Arabidopsis.*

Nüsslein-Volhard, Christiane. "Gradients That Organize Embryo Development." *Scientific American,* Vol. 275, No. 2, Aug. 1996. The author, who shared the 1995 Nobel Prize for Physiology or Medicine with Eric Wieschaus and Edward B. Lewis, describes the gene interactions that control *Drosophila* development.

● Visit our website at **http://www.saunderscollege.com/lifesci/titles.html** and click on Solomon/Berg/Martin Biology for links to chapter-related resources on the World Wide Web.

Biotechnology Patent Lawyer

KIT CHAN

K it Chan came to the United States with a degree in biology and biochemistry from The Chinese University of Hong Kong. He completed his undergraduate studies during a time when there was much excitement about the prospects of genetic engineering. Kit was particularly fascinated by molecular biology. He went on to obtain a Ph.D. in molecular virology at Baylor College of Medicine in Houston and proceeded to become a laboratory scientist. After working in research science for several years, he attended Columbia University School of Law in New York City and now applies both his scientific and legal knowledge to patent law. In addition to his biotechnology patent practice of eight years, he teaches patent law, intellectual property law, and technology transfer at the City University of New York Law School.

How did you decide to become a lawyer?

After I finished my graduate studies, I began my postdoctoral work at Cold Spring Harbor Laboratory on Long Island. I was working in molecular oncology—how an oncogene is turned on and how that affects the cell, particularly the cell cycle. The oncogene I was working on, called *myc,* encodes a nuclear protein that regulates transcription. My project was to try to see how the protein is made and how it interacts with different cellular components. At that time there were constant discussions at Cold Spring Harbor Laboratory about the Human Genome Project and the legal ramifications of biotechnology. I did some

research and discovered that I could apply my biology background to a career in law as a biotechnology patent lawyer.

How does your biology background help you in your law practice?

A biology background is useful in evaluating the prospects for obtaining a biotechnology patent. For example, if someone discovers a gene that is responsible for Alzheimer's disease, the patent application could include the DNA and RNA probes for this gene. The application might also include the protein that is encoded by the Alzheimer's disease gene and antibodies directed against the protein.

My biology background also enables me to determine the further ramifications of an invention. For instance, if someone wishes to patent a drug for the herpes virus, I know that the same drug might block the replication of other viruses as well. This helps me devise a patent application that covers other potential uses of the drug.

What does an inventor obtain by filing a patent?

The patent is a contract between the government and the applicant. For a definite period of time, an applicant obtains exclusive rights to his or her invention, and in exchange, the inventor provides full disclosure of the invention in the public domain. Currently, the patent expiration period is twenty years. This means that while the patent holder retains rights to the invention, other scientists can have access to the invention and may be able to build upon the original work.

What are some of the legal issues arising from the Human Genome Project and any new genetic technologies on the horizon?

In my law classes, I teach that there are often conflicting perspectives surrounding the potential uses of genetic information. These include the issues of individual privacy and possible discrimination, and the interests of the insurance industry. I have been active in the science and law reading

of legal position papers concerning these issues for the New York City Bar Association. The ethical position is that, for each individual, the genetic profile should be confidential. That sounds easy: Most of us feel that our medical records should be confidential; therefore, so should our genome information. On the other hand, the insurance companies' position is that if you carry, say, an alcoholism or a tobacco preference gene, why shouldn't you have to pay higher premiums than people who are not genetically predisposed to alcohol or smoking high risk factors?

Another debate is about the Employment Discrimination Act. If a company knows that statistically, you are more likely to die at the age of 48 because you are at high risk for cardiac disease based on your genetic profile, whereas I am more likely to live until age 90 based on mine, why shouldn't they hire me instead of hiring you?

The same kind of genome information can be an advantage. Right now, in terms of technology, we are working toward highly individualized medicine. The medication that you take and the medication that I take should be different. The reason is that we have distinct genetic make-ups, which could affect the effective dosage. When someone needs a combination therapy, we should be able to look at his or her genome map and say, "Oh, you have the alcohol dehydrogenase gene isoform 3 or you have LDL isoform 1," and then combine a beta blocker with the effective dosage. Then, you would not be likely to overdose.

Have you noticed a great many changes in the biotechnology patent laws recently?

The recent laws have made it easier and better for biotechnology patents. Five years ago it was a lot harder to get a patent issued. For example, if you were claiming an AIDS therapeutic composition, the Patent Office would require official FDA approval, involving a complete, double-blind study with a placebo and statistically significant results. This is an extremely costly and time-consuming process. Now the Patent Office is more receptive and understands the technology involved in the research and development phase, and appreciates that one purpose of the patent is to encourage the inventor to explore further the commercial significance of the invention. All in all, in my opinion, we are going in the right direction letting more patents be issued so as to encourage the industry. I feel it's very exciting these days.

The Continuity of Life: Evolution

These fossilized remains of a prehistoric fish *(Priscacara serata)* were discovered in Wyoming. *(Alfred Pasieka/Science Photo Library/Photo Researchers, Inc.)*

CHAPTER 17

Introduction to Darwinian Evolution

Life is thought to have originated only once. The biological diversity represented by the millions of species currently living on our planet has evolved from a single ancestor during Earth's long history. Thus, organisms that are radically different, such as slime molds and crocodiles, are in fact distantly related to one another and are linked through numerous intermediate ancestors to a single, common ancestor. The British naturalist Charles Darwin (1809–1882), shown here at age 31, developed a remarkably simple, scientifically testable mechanism to explain this. He argued persuasively that all the species that exist today, as well as the countless extinct species that existed in the past, arose from earlier ones by a process of gradual divergence (separation), or evolution.

Evolution can be defined as the accumulation of heritable changes within populations over time. (**Heritable** traits are those that are genetically based and therefore are able to be passed to offspring. A **population** is a group of individuals of one species that live in the same geographical area at the same time.) Evolution does not refer to changes that occur in an individual within its lifetime, but to changes in the characteristics of populations over many generations. These changes may be so small that they are difficult to detect or so great that the population differs markedly from its ancestral population. Eventually, two populations may diverge to such a degree that we refer to them as different species. (The concept of species is developed extensively in Chapter 19. For now, a simple working definition is that a species comprises one or more populations of similar organisms that are capable of interbreeding with one another.)

The concept of evolution is the cornerstone of biology because it links all fields of the life sciences into a unified body of knowledge. Biologists attempt to understand both the remarkable variety as well as the fundamental similarities of organisms within the context of evolution. Thus, evolution allows biologists to compare common threads among organisms as seemingly different as bacteria, whales, lilies, and tapeworms. Evolution also has important practical applications. Agriculture, for example, must deal with the evolution of pesticide resistance in insects and other pests. Likewise, medicine must respond to the rapid evolutionary potential of disease-causing bacteria.

This chapter discusses Darwin and the scientific development of the theory of evolution by natural selection. It also presents several kinds of evidence that support evolution, including fossils, comparative anatomy, developmental biology, biogeography, molecular biology, and experimental evidence.

(The Granger Collection, New York)

1. Discuss the historical development of the theory of evolution.
2. Define evolution and explain the four premises of evolution by natural selection as proposed by Charles Darwin.
3. Compare the synthetic theory of evolution with Darwin's original theory of evolution.
4. Summarize the evidence for evolution obtained from the fossil record.
5. Summarize the evidence for evolution derived from comparative anatomy.
6. Define and give examples of homologous features, homoplastic features, and vestigial structures.
7. Explain how developmental biology provides insights into the evolutionary process.
8. Define biogeography and summarize how the distribution of organisms supports evolution.
9. Explain how scientists make inferences about evolutionary relationships from the sequence of amino acids in proteins and the sequence of nucleotides in DNA.
10. Give an example of how evolutionary hypotheses can be tested experimentally.

IDEAS ABOUT EVOLUTION ORIGINATED BEFORE DARWIN

Although Darwin is universally associated with evolution, ideas of evolution predate Darwin by centuries. Aristotle (384–322 B.C.) saw much evidence of natural affinities among organisms. This led him to arrange all of the organisms he knew in a "Scale of Nature" that extended from the exceedingly simple to the most complex. He visualized organisms as being imperfect but "moving toward a more perfect state." This idea has been interpreted by some scientific historians as the forerunner of evolutionary theory, but Aristotle was vague on the nature of this "movement toward perfection" and certainly did not propose that the process of evolution was driven by natural selection. Furthermore, modern evolutionary theory now recognizes that evolution does not move toward more "perfect" states, nor even necessarily toward greater complexity.

Long before Darwin, fossils had been discovered embedded in rocks. Some of these corresponded to parts of familiar species, but others were strangely unlike any known species. Fossils were often found in unexpected contexts. Marine invertebrates (sea animals without backbones), for example, were sometimes discovered in rocks high on mountains. Leonardo da Vinci was among the first to correctly interpret these unusual finds as the remains of animals that had existed in previous ages but had become extinct.

The French naturalist Jean Baptiste de Lamarck (1744–1829) was the first scientist to propose that organisms undergo change over time as a result of some natural phenomenon rather than divine intervention. In his *Philosophie Zoologique,* published in 1809, Lamarck presented a possible explanation for how organisms evolved. Lamarck thought that all organisms were endowed with a vital force that drove them to change toward greater complexity over time. He also thought that organisms could pass traits acquired during their lifetimes to their offspring. For example, Lamarck suggested that the long neck of the giraffe developed when a short-necked ancestor stretched its neck to browse on the leaves of trees. Its offspring inherited the longer neck, which stretched still further as they ate.

This process, repeated over many generations, supposedly resulted in the long necks of modern giraffes.

The proposed mechanism for Lamarckian evolution was discredited when the basis of heredity was later discovered. It remained for Darwin to discover the mechanism of evolution by natural selection.

DARWIN'S VOYAGE WAS THE BASIS FOR HIS THEORY OF EVOLUTION

Darwin, the son of a prominent physician, was sent at the age of 15 to study medicine at the University of Edinburgh. Finding himself unsuited for medicine, he transferred to Cambridge University to study theology. During that time, he became the protégé of the Reverend John Henslow, who was a professor of botany. Henslow encouraged Darwin's interest in the natural world. Shortly after receiving his degree, Darwin embarked as a naturalist on the H.M.S. *Beagle,* which was taking a five-year exploratory cruise around the world to prepare navigation charts for the British Navy.

The *Beagle* left Plymouth, England, in 1831 and cruised along the east and west coasts of South America (Fig. 17–1). While other members of the crew mapped the coasts and harbors, Darwin spent many weeks ashore studying the animals, plants, fossils, and geological formations of both coastal and inland regions, areas that had not been extensively explored. He collected and catalogued thousands of plant and animal specimens and kept notes of his observations—information that would become essential in the development of his theory.

The *Beagle* spent almost two months at the Galapagos Islands, 965 kilometers (600 miles) west of Ecuador, where Darwin continued his observations and collections. He compared the animals and plants of the Galapagos with those of the South American mainland. He was particularly impressed by their similarities and wondered why the organisms of the Galapagos should resemble those from South America more than those from other islands in different parts of the world. More-

Figure 17-1 **The voyage of H.M.S. Beagle.** The five-year voyage began in Plymouth, England *(gold star),* in 1831. Darwin's observations of the Galapagos Islands were the basis for his theory of evolution by natural selection.

over, although there were similarities between Galapagos and South American species, there were also distinct differences. There were even recognizable differences in the reptiles and birds from one island to the next (Fig. 17–2). After he returned home, Darwin pondered these observations and attempted to develop a satisfactory explanation for the distribution of species among the islands.

Darwin drew on several lines of evidence when considering how species might have originated. Despite the work of Lamarck, the general notion in the mid-1800s was that the Earth was too young for organisms to have changed significantly since they had first appeared. During the early 19th century, however, geologists advanced the idea that mountains, valleys, and other physical features of Earth's surface did not originate in their present forms, but developed slowly over long periods of time by the geological processes of volcanic activity, uplift, erosion, and glaciation. Darwin took *Principles of Geology,* published by geologist Charles Lyell in 1830, with

(a)   (b)   (c)

Figure 17-2 **Three species of Galapagos finches.** Darwin inferred that these birds are derived from a common ancestral population of seed-eating birds from South America. Variation in their beaks is the result of adaptation to different kinds of food. *(a)* The cactus finch *(Geospiza scandens)* feeds on the fleshy parts of cacti. *(b)* The large ground finch *(Geospiza magnirostris)* has an extremely heavy, nutcracker-type bill adapted for eating thick, hard-walled seeds. *(c)* The woodpecker finch *(Camarhyncus pallidus)* has insectivorous habits similar to those of woodpeckers but lacks the complex beak and tongue adaptations that permit woodpeckers to reach their prey. The adaptations of the woodpecker finch to this lifestyle are almost entirely behavioral—it digs insects out of bark and crevices using cactus spines, twigs, or even dead leaves. *(a and b, Frans Lanting/Minden Pictures; c, Miguel Castro/Photo Researchers, Inc.)*

Figure 17–3 **Artificial selection in chickens.** Shown is a chicken that was deliberately bred to resemble Big Bird on "Sesame Street." Many show breeds exist that exhibit a great deal of variation. The chickens we eat are not a recognizable breed but are hybrids bred for their meat or egg production. *(© Eric Sander)*

Figure 17–4 **Artificial selection in *Brassica oleracea*.** An enlarged terminal bud (the "head") was selected in cabbage, flower clusters in broccoli and cauliflower, axillary buds in brussels sprouts, leaves in collards and kale, and stems in kohlrabi. *(John Arnaldi)*

him on his voyage and studied it carefully. Lyell provided an important concept for Darwin—that the slow pace of geological processes, which still occur today, indicated that the Earth was extremely old.

Other important evidence that influenced Darwin was the fact that breeders and farmers could develop many varieties of domesticated animals in just a few generations (Fig. 17–3). This was accomplished by choosing certain traits and breeding only individuals that exhibited the desired traits, a procedure known as **artificial selection.** Breeders, for example, had produced numerous dog varieties—bloodhounds, Dalmatians, Airedales, border collies, and Pekinese, to name a few—by artificial selection.

Many plant varieties were also produced by artificial selection. For example, cabbage, broccoli, brussels sprouts, cauliflower, collard greens, kale, and kohlrabi are distinct vegetable crops that are all members of the same species, *Brassica oleracea* (Fig. 17–4). All seven were produced by selective breed-

ing of the colewort, or wild cabbage, a leafy plant native to Europe and Asia. Beginning more than 4000 years ago, some farmers artificially selected wild cabbage plants that formed overlapping leaves. Over time, these leaves became so prominant that the plants, which resembled modern-day cabbages, became recognized as separate and distinct from their wild cabbage ancestor. Other farmers selected different features of the wild cabbage, giving rise to the other modifications. For example, kohlrabi was produced by selection for an enlarged storage stem, and Brussels sprouts by selection for enlarged axillary buds. Thus, humans are responsible for the evolution of *Brassica oleracea* into seven distinct vegetable crops. Darwin was aware of artificial selection and thought that a similar selective process occurred in nature.

The ideas of Thomas Malthus (1766–1834), a British clergyman and economist, were another important influence on Darwin. Malthus noted in *An Essay on the Principle of Population as It Affects the Future Improvement of Society,* published in 1798, that population growth is not always desirable—a view contrary to the beliefs of his day. He observed that populations have the capacity to increase geometrically ($1 \rightarrow 2 \rightarrow 4 \rightarrow 8 \rightarrow 16$) and thus outstrip the food supply, which has the capacity to increase only arithmetically ($1 \rightarrow 2 \rightarrow 3 \rightarrow 4 \rightarrow 5$). In the case of humans, Malthus suggested that the conflict between population and food supply generates famine, disease, and war, which serve as inevitable brakes on population growth.

DARWIN PROPOSED THAT EVOLUTION OCCURS BY NATURAL SELECTION

Darwin's years of observing the habits of animals and plants had introduced him to the struggle for existence described by Malthus. It occurred to Darwin that in this struggle inherited variations favorable to survival would tend to be preserved, while unfavorable ones would be eliminated. The result would be **adaptation**—evolutionary modification that improves the chances of survival and reproductive success of the population—to the environment. Eventually, the accumulation of modifications might result in a new species. Time was the only thing required for new species to originate, and the geologists of the era, including Lyell, had supplied evidence that Earth was indeed old enough to provide an adequate period of time.

Darwin had at last obtained a workable mechanism of evolution, that of **natural selection,** in which better adapted organisms are more likely to survive and become the parents of the next generation. As a result of natural selection, the population changes over time; the frequency of favorable traits increases in successive generations while less favorable traits become scarce or disappear. Darwin spent the next 20 years formulating his arguments for natural selection, accumulating an immense body of evidence to support his theory, and corresponding with other scientists.

As Darwin was pondering his ideas, Alfred Russell Wallace (1823–1913), a British naturalist who was studying the plants and animals of the Malay Archipelago, was similarly struck by the diversity of species and the peculiarities of their distribution. He wrote a brief essay on this subject and sent it to Darwin, by then a world-renowned biologist, asking his opinion. Darwin recognized his own theory and realized that Wallace had independently arrived at the same conclusion—that evolution occurs by natural selection. Darwin's colleagues persuaded him to present Wallace's manuscript along with an abstract of his own work, which he had prepared and circulated to a few friends several years earlier. Both papers were presented in July 1858 at a London meeting of the Linnaean Society. Darwin's monumental book, *The Origin of Species by Natural Selection; or, The Preservation of Favored Races in the Struggle for Life,* was published in 1859. Wallace's book, *Contributions to the Theory of Natural Selection,* was published in 1870, eight years after he returned from the Malay Archipelago.

Darwin's mechanism of evolution by natural selection consists of four observations about the natural world: overproduction, variation, limits on population growth, and differential reproductive success.

1. **Overproduction.** Each species has the capacity to produce more offspring than will survive to maturity. Through reproduction, natural populations may geometrically increase in number over time. For example, if each breeding pair of elephants produces six offspring during its 90-year life span, in 750 years a single pair of elephants will give rise to a population of 19 million! Yet elephants have not overrun the planet.

2. **Variation.** The individuals in a population exhibit variation. Each individual has a unique combination of traits, such as size, color, and ability to tolerate harsh environmental conditions. Some traits improve an individual's chances of survival and reproductive success, whereas others do not. It is important to remember that the variation necessary for evolution by natural selection must be heritable (Fig. 17–5). (Although Darwin recognized the importance to evolution of inherited variation, he did not know the mechanism of inheritance.)

3. **Limits on population growth,** or a struggle for existence. There is only so much food, water, light, growing space, and other resources available to a population, and, consequently, organisms compete with one another for these lim-

Figure 17–5 **Genetic variation in emerald tree boas *(Corallus caninus).*** These snakes were caught in a small section of forest in French Guiana. Many snake species exhibit considerable variation in their coloration and patterns. *(J. Sauvanet/Peter Arnold, Inc.)*

ited resources. Because there are more individuals than the environment can support, not all will survive to reproduce. Other limits on population growth include predators, disease organisms, and unfavorable weather conditions.

4. **Differential reproductive success.** Those individuals that possess the most favorable combination of characteristics (those that make individuals better adapted to their environment) are more likely to survive and reproduce. Because offspring tend to resemble their parents, the next generation obtains the parents' heritable traits. Successful reproduction is the key to natural selection: the best adapted individuals are those that reproduce most successfully, whereas less fit individuals die prematurely or produce fewer or inferior offspring.

Over time, enough changes may accumulate in geographically separated populations (often with slightly different environments) to produce new species. Darwin noted that the Galapagos finches appear to have evolved in this way. The different islands of the Galapagos kept the finches isolated from one another, thereby allowing them to diverge, or take separate evolutionary pathways, into separate species. The evolution of new species is considered in greater detail in Chapter 19.

THE SYNTHETIC THEORY OF EVOLUTION COMBINES DARWIN'S THEORY WITH GENETICS

One of the premises on which Darwin based his theory of evolution by natural selection is that individuals transmit traits to the next generation. However, Darwin was unable to explain *how* this occurs or *why* individuals vary within a population. Although he was a contemporary of Gregor Mendel (see Chapter 10), who elucidated the basic patterns of inheritance, Darwin was apparently not acquainted with Mendel's work, which was not recognized by the scientific community until the early part of the 20th century.

During the 1930s and 1940s, biologists combined the principles of genetics with Darwin's theory of natural selection to develop a unified explanation of evolution known as **neo-Darwinism** or, more commonly, the **synthetic theory of evolution.**[1] (*Synthesis* in this context refers to combining parts of several previous theories to form a unified whole.)

The synthetic theory of evolution is a conceptual breakthrough that explains Darwin's observation of variation among offspring in terms of **mutation,** or changes in DNA, such as nucleotide substitutions. That is, mutation provides the genetic variability on which natural selection acts during evolu-

tion. The synthetic theory of evolution, which emphasizes the genetics of *populations* (rather than individuals) as the central focus of evolution, has held up well since it was developed (see Chapter 18). It has dominated the thinking and research of biologists working in many areas, and has resulted in an enormous accumulation of scientific evidence to support evolution by natural selection. (See *Making the Connection: Tuberculosis, Bacterial Resistance to Antibiotics, and Evolution* for a discussion of a serious health issue that is related to natural selection in bacteria.)

Most biologists accept the basic principles of the synthetic theory of evolution but have recently scrutinized some of its aspects. For example, what is the role of chance in evolution? How rapidly do new species evolve? These questions have arisen in part from a reevaluation of the fossil record and in part from discoveries in molecular aspects of inheritance. Such critical analyses are an integral part of the scientific process because they stimulate additional observation and experimentation, along with reexamination of previous evidence. Science is an ongoing process, and information obtained in the future may require modifications to certain parts of the synthetic theory of evolution.

MANY TYPES OF SCIENTIFIC EVIDENCE SUPPORT EVOLUTION

A vast body of experimental evidence supports evolution, including observations from the fossil record, comparative anatomy, developmental biology, biogeography, and molecular biology. In addition, evolutionary hypotheses are increasingly being tested experimentally.

The fossil record provides strong evidence for evolution

Perhaps the most direct evidence for evolution comes from the discovery, identification, and interpretation of **fossils,** which are the remains or traces typically left in sedimentary rock by previous organisms. (The term *fossil* comes from the Latin word *fossilis,* meaning "something dug up.") The fossil record shows a progression from the earliest, single-celled organisms to the many single-celled and multicellular organisms living today. To date, paleontologists (scientists who study fossils) have described and named about 300,000 fossil species, and more are being discovered all the time.

Although most fossils are preserved in sedimentary rock, some more recent remains have been exceptionally well preserved in bogs, tar, amber (ancient tree resin), or ice (Fig. 17–6). For example, the remains of a woolly mammoth deep-frozen in Siberian ice for more than 25,000 years were so well preserved that part of its DNA could be analyzed.

Fossils provide a record of ancient organisms and some understanding of where and when they lived. Using fossils of organisms from different geological ages, the lines of descent

---

[1]Some of the founders of the synthetic theory of evolution were biologists Theodosius Dobzhansky, Ronald Fisher, Julian Huxley, Ernst Mayr, George Gaylord Simpson, and G. Ledyard Stebbins, Jr.

(a)

(b)

(c)

(d)

(e)

Figure 17–6 Fossils develop in different ways. *(a)* Although some fossils contain traces of organic matter, all that remains in this fossil of a seed fern leaf is an impression, or imprint, in the rock. *(b)* Petrified wood from the Petrified Forest National Park in Arizona consists of trees that were buried and infiltrated with minerals. *(c)* A two-million-year-old insect fossil (a midge) was embedded in amber. *(d)* A cast fossil of ancient echinoderms called crinoids formed when the crinoids decomposed, leaving a mold that later filled with dissolved minerals that hardened. *(e)* Dinosaur footprints, each 75 to 90 centimeters (2.5 to 3 feet) in length, provide clues about the locomotion, behavior, and ecology of these extinct animals. *(a, Carolina Biological Supply Company/Phototake, NYC; b, Kenneth Murray/Photo Researchers, Inc.; c, Alfred Pasieka/Science Photo Library/Photo Researchers, Inc.; d, A.J. Copley/Visuals Unlimited; e, Scott Berner/Visuals Unlimited)*

TUBERCULOSIS, BACTERIAL RESISTANCE TO ANTIBIOTICS, AND EVOLUTION

What is antibiotic-resistant tuberculosis, and how is it related to evolution? Beginning in the late 1980s, an alarming increase in the incidence of tuberculosis (TB) has been documented by the U. S. Centers for Disease Control and Prevention in Atlanta. In the thirty or so years before that time, the number of cases of TB had declined in the United States, largely as a result of treating TB with antibiotics, which are drugs intended to harm or kill bacteria and other microorganisms.

Many people are exposed to the bacteria that cause TB, but only people who are very young, very old, or weakened from some other disease usually exhibit symptoms. One of the reasons for the current outbreak of TB is the HIV virus (the virus that causes AIDS), which incapacitates the immune system. Patients who are infected with HIV and who formerly had symptom-free TB are at greater risk for deadly infections.

Tuberculosis exhibits a disturbing trend in which drug-resistant strains of the bacteria that cause TB have developed in the past decade. These strains are resistant to one or more antibiotics that traditionally were used to treat TB. Drug-resistant TB is deadly: as many as 80% of the people infected with multi-drug-resistant TB (MDR-TB) die within two months of diagnosis—even with medical care.

The development of bacterial resistance to antibiotics is an ex- ample of evolution. Bacteria are continually evolving, even inside the bodies of human and animal hosts. When an antibiotic is used to treat a bacterial infection, a few bacteria may survive because they are genetically resistant to the antibiotic*, and they pass these genes to future generations. As a result, the bacterial population contains a larger percentage of antibiotic-resistant bacteria than before.

Drug resistance is usually found in patients who were previously treated for TB, and quite often human behavior is a factor in the development of drug resistance. A person infected with TB must take three to ten pills of antibiotics each day for at least six months. After the first week or two of treatment, the person usually feels better; many patients decide to quit taking their medication at this point. When this happens, the TB bacteria still lurking in their bodies—those with a resistance to the antibiotic formerly used—rally. The evolution of a strain of bacteria resistant to several drugs is a worst-case scenario. MDR-TB is extremely difficult to treat effectively and, as mentioned previously, is often fatal.

*Bacteria develop genetic resistance through mutations (see Chapters 11 and 12) and through acquiring new genes from plasmids (see Chapters 11 and 14, discussion of transformation) or viruses (see Chapter 23, discussion of transduction).

(evolutionary relationships) that gave rise to modern-day organisms can sometimes be inferred. In many instances, fossils provide direct evidence of the origin of new species from pre-existing species, including transitional forms.

For example, during the 1980s and 1990s, paleontologists have discovered several fossil intermediates in whale evolution (Fig. 17–7). Scientific evidence indicates that whales descended from a now-extinct group of four-legged terrestrial mammals called mesonychids. (The closest living relatives of whales and porpoises are hoofed mammals such as antelopes, deer, and giraffes.) About 50 to 60 million years ago, some descendants of mesonychids had adapted to swimming in shallow seas. Fossils of *Ambulocetus natans,* a 50-million-year-old whale discovered in Pakistan, had many features of modern whales but also possessed hind limbs and feet. (Modern whales do not have hind limbs, although vestigial pelvic and hindlimb bones persist. Vestigial structures are discussed later in this chapter.) In addition to swimming, this ancient whale moved about on land, perhaps as sea lions do today. By 40 million years ago, the whale transition from land to ocean was almost complete. Egyptian fossils of *Dorudon atrox,* a 6-meter (18-foot) whale that lived at that time, indicate a streamlined body (like modern-day whales) along with a pair of reduced, 15-centimeter (6-inch) hind limbs too small to be used in locomotion.

The formation and preservation of a fossil require that an organism be buried under conditions that slow or prevent the decay process. This is most likely to occur if an organism's remains are covered quickly by a sediment of fine soil particles suspended in water. In this way remains of aquatic organisms may be trapped in bogs, mud flats, sand bars, or deltas. Remains of terrestrial organisms that lived on a flood plain may also be covered by water-borne sediments or, if the organism lived in an arid region, by wind-blown sand. Over time, the sediments harden to form sedimentary rock, and the organism's remains are usually replaced by minerals so that many details of its structure, even cellular details, remain.

The fossil record is not a random sample of past life, but instead is biased toward aquatic organisms and those living in the few terrestrial habitats conducive to fossil formation. Relatively few fossils of tropical rainforest organisms have been found, for example, because their remains decay extremely rapidly on the forest floor, before fossils can develop. Another reason for bias in the fossil record is that organisms with hard body parts such as bones and shells are more likely to form fossils than those with soft body parts.

Because of the nature of the scientific process, each fossil discovery represents a separate "test" of the theory of evolution. If any of the tests fail, the theory would have to be modified to fit the existing evidence. The discovery, for example,

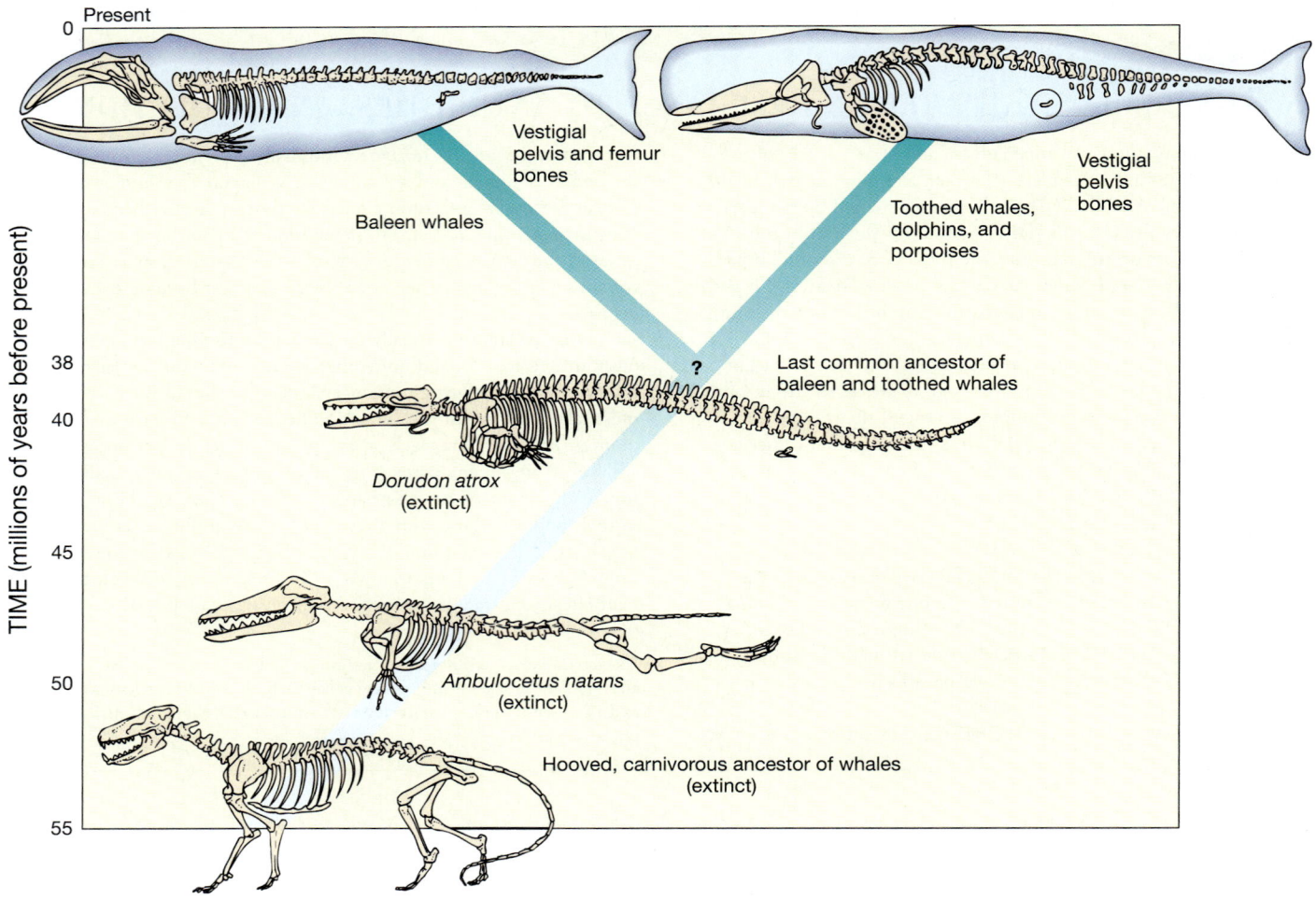

Figure 17–7 Fossil intermediates in whale evolution. The now-extinct *Ambulocetus natans* was a transitional form between modern whale descendants and their terrestrial ancestors. *Ambulocetus* possessed a number of recognizable whale features and retained the hind limbs of its four-legged ancestors. The later *Dorudon atrox* was more streamlined and possessed tiny hind limbs that were nonfunctional. The exact relationships of *Ambulocetus* and *Dorudon* to other whales are uncertain; they may or may not have been direct ancestors of modern whales, as shown in the diagram. *(Adapted from Gingerich and Uhen, University of Michigan Museum of Paleontology, and from Thewissen, Madar, and Hussain. See Recommended Readings for complete citations.)*

of fossil remains of humans *(Homo sapiens)* in Precambrian rocks, which are more than 570 million years old, would falsify the theory of evolution as currently proposed. However, Precambrian rocks examined to date contain fossils of simple organisms, such as cyanobacteria, that are thought to have evolved early in the history of life. The earliest fossils of *Homo sapiens* with anatomically modern features do not appear in the fossil record until approximately 100,000 years ago (see Chapter 21).

Various methods can be used to determine the age of fossils

Because layers of sedimentary rock occur naturally in the sequence of their deposition, with the more recent layers on top

of the older, earlier ones (Fig. 17–8), most fossils are dated by their relative position in sedimentary rock. However, geological events occurring after the rocks were initially formed have occasionally changed the relationships of some rock layers. Geologists identify specific sedimentary rocks not only by their positions in layers but also by features such as mineral content and by the fossilized remains of certain organisms, known as **index fossils,** that characterize a specific layer over large geographical areas. Index fossils are fossils of organisms that existed for a relatively short geological time but were preserved as fossils in large numbers. With this information, geologists can arrange rock layers and the fossils they contain in chronological order and identify comparable layers in widely separated locations.

Radioactive isotopes, also called **radioisotopes,** present in

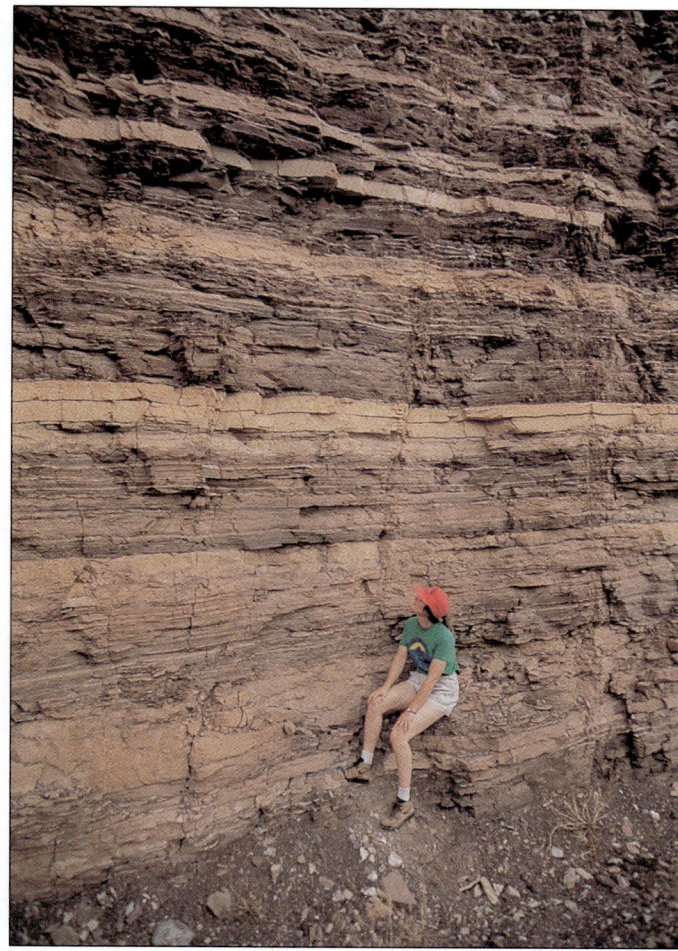

Figure 17–8 **Layers characterize sedimentary rock.** The younger layers overlie the older layers. The layers shown here were exposed when a road cut was made for a highway in Utah. *(Tom Bean)*

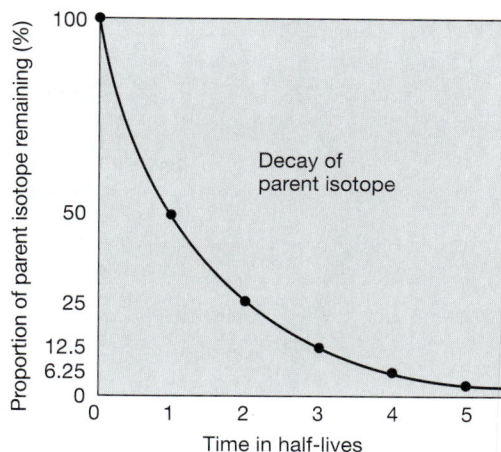

Figure 17–9 **Radioisotope decay.** At time zero, the radioactive clock begins ticking. At this point a sample is composed entirely of the radioisotope. After one half-life, only 50% of the original radioisotope remains. During each succeeding half-life, one half of the remaining radioisotope is converted to decay product(s).

a rock provide a means to accurately measure its age (see Chapter 2). Radioisotopes emit invisible radiations. As a radioisotope emits radiation, its nucleus changes into the nucleus of a different element in a process known as **radioactive decay.** For example, the radioactive nucleus of uranium-235 decays over time into lead-207.

Each radioisotope has its own characteristic rate of decay. The period of time required for one half of the atoms of a radioisotope to change into a different atom is known as its **half-life** (Fig. 17–9). Radioisotopes differ significantly in their half-lives. For example, the half-life of iodine-132 is only 2.4 hours, whereas the half-life of uranium-235 is 704 million years. The half-life of a particular radioisotope is constant and does not vary with temperature, pressure, or any other environmental factor.

The age of a fossil is estimated by measuring the relative proportions of the original radioisotope and its decay product. For example, the half-life of potassium-40 is 1.3 billion years, meaning that in 1.3 billion years half of the radioactive potas-

sium will have decayed into its decay product, argon-40. The radioactive clock begins ticking when the rock solidifies. The rock initially contains some potassium, but no argon. Because argon is a gas, it escapes from hot rock as soon as it forms, but when potassium decays in rock that has cooled and solidified, the argon accumulates in the crystalline structure of the rock. If the ratio of potassium-40 to argon-40 in the rock being tested is 1:1, the rock is 1.3 billion years old.

Several radioisotopes are commonly used to date fossils. These include potassium-40 (half-life 1.3 billion years), uranium-235 (half-life 704 million years), and carbon-14 (half-life 5730 years). Potassium-40, with its long half-life, can be used to date fossils that are many hundreds of millions of years old. Radioisotopes other than carbon-14 are used to date the *rock* in which fossils are found, whereas carbon-14 is used to date the *carbon remains* of anything that was once living, such as wood, bones, or shells. Whenever possible, the age of a fossil is independently verified using two or more different radioisotopes.

Carbon-14, which is continuously produced in the atmosphere from nitrogen-14 (by cosmic radiation), subsequently decays back to nitrogen-14. Because the formation and the decay of carbon-14 occur at constant rates, the ratio of carbon-14 to carbon-12 (the more abundant, stable isotope of carbon) is constant in the atmosphere. Since each organism absorbs carbon from the atmosphere,[2] its ratio of carbon-14 to carbon-12 is the same as the atmosphere. When an organism dies, however, it no longer absorbs carbon, and the proportion of carbon-14 in its remains declines as carbon-14 de-

[2]Organisms absorb carbon from the atmosphere either directly (by photosynthesis) or indirectly (by consuming photosynthetic organisms).

Figure 17–10 Homology in animals. The human arm, cat forelimb, whale flipper, and bat wing have a basic underlying similarity of structure because they are derived from a common ancestor.

cays to nitrogen-14. Because of its relatively short half-life, carbon-14 is useful for dating fossils that are 50,000 years old or less.

Comparative anatomy of related species demonstrates similarities in their structures

Comparing the structural details of features found in different but related organisms reveals a basic similarity. Such features that are derived from the same structure in a common ancestor are termed **homologous features.** For example, consider the limb bones of mammals. A human arm, a cat forelimb, a whale front flipper, and a bat wing, although quite different in appearance, have strikingly similar arrangements of bones, muscles, and nerves. Figure 17–10 shows a comparison of their skeletal structures. Each has a single bone (the humerus) in the part of the limb nearest the trunk of the body, followed by the two bones (radius and ulna) of the forearm, a group of bones (carpals) in the wrist, and a variable number of digits (metacarpals and phalanges). This similarity is particularly striking because arms, forelimbs, flippers, and wings are used in different ways for different functions, and there is no overriding mechanical reason for them to be so similar structurally. Similar arrangements of parts of the forelimb are evident in ancestral reptiles and amphibians and even in the first fishes that came out of water onto land hundreds of millions of years ago.

Leaves are one example of a homologous feature in plants. In many plant species, leaves have been modified for functions other than photosynthesis. A cactus spine and a pea tendril, although quite different in appearance, are homologous because both are modified leaves (Fig. 17–11). The spine protects the succulent stem tissue of the cactus, whereas the ten-

dril, which winds around a small object once it makes contact, helps support the climbing stem of the pea plant. Such basic structural similarities in organs used in different ways are precisely the expected outcome of a common evolutionary origin. The basic structure present in a common ancestor was modified in different ways for different functions as various descendants subsequently evolved.

Not all species with "similar" features have descended from a more recent common ancestor, however. Features that are not homologous but simply have similar functions in distantly related organisms are termed **homoplastic features.** For example, the wings of various distantly related flying animals, such as insects and birds, are homoplastic features that have evolved over time to meet the common function of flight, though they are different in more fundamental aspects. Bird wings are modified forelimbs supported by bones, whereas insect wings may have evolved from gill-like appendages present in the aquatic ancestors of insects. Spines, which are modified leaves, and thorns, which are modified stems, are an example of homoplasy in plants. Spines and thorns resemble one another superficially but are homoplastic features that evolved to solve the common need for protection from herbivores (Fig. 17–12).

Like homology, homoplasy offers crucial evidence of evolution. Homoplastic features are of evolutionary interest because they demonstrate that organisms with separate ancestries may adapt in similar ways to similar environmental demands. Such independent evolution of similar structures in distantly related organisms is known as **convergent evolution.** Aardvarks, anteaters, and pangolins are an excellent example of convergent evolution (Fig. 17–13). They resemble one another in lifestyle and appearance—all have strong, sharp claws to dig open ant and termite mounds and elongated snouts with long, sticky tongues to catch these insects. Yet aardvarks, anteaters, and pangolins evolved from three distantly related orders of

Spine

(a)

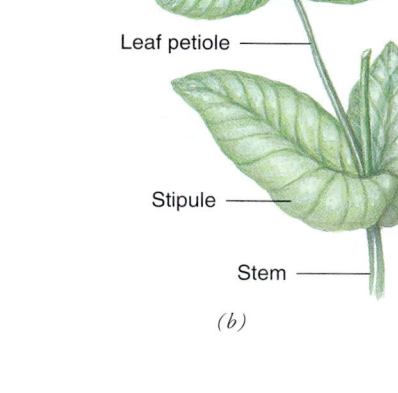

Tendril

Leaflet

Leaf petiole

Stipule

Stem

(b)

Figure 17–11 **Homology in plants.** *(a)* The spines of the fish-hook cactus *(Ferocactus wislizenii)* are modified leaves, as are *(b)* the tendrils of the garden pea *(Pisum sativum)*. Leaves of the garden pea are compound, and the terminal leaflets are modified into tendrils that are frequently branched.

mammals. (See Chapter 22 for further discussion of homology and homoplasy.)

Comparative anatomy reveals the existence of **vestigial structures.** Many organisms contain organs or parts of organs that are seemingly nonfunctional and degenerate, often undersized or lacking some essential part. Vestigial structures are remnants of more developed structures that were present and functional in ancestral organisms. In the human body, more than 100 structures are considered vestigial, including the appendix, coccyx (fused tail bones), body hair, wisdom teeth, and the muscles that move our ears. Whales (see Fig. 17–7) and pythons have vestigial hindlimb bones; pigs have vestigial toes

Shoot (develops from axillary bud)

Spine (midrib of leaf)

Leaf scar

(a)

Thorn (develops from axillary bud)

(b)

Figure 17–12 **Homoplasy.** *(a)* A spine of Japanese barberry *(Berberis thunbergii)* is a modified leaf. (In this example, the spine is actually the midrib of the original leaf, which has been shed.) *(b)* Thorns of downy hawthorn *(Crataegus mollis)* are modified stems that develop from axillary buds.

(a)

(b)

(c)

Figure 17–13 **Convergent evolution.** Three distantly related mammals exhibit structural similarities as a result of convergent evolution. These species adapted independently to eat ants and termites in similar grassland/forest environments in different parts of the world. *(a)* The aardvark *(Orycteropus afer)* is native to central, southern, and eastern Africa. *(b)* A giant anteater *(Myrmecophaga tridactyla)* at a termite mound. The anteater is native to Latin America, from southern Mexico to northern Argentina. *(c)* The pangolin *(Manis crassicaudata)* is native to Africa and southern and southeastern Asia. *(a, Kjell B. Sandved/Visuals Unlimited; b, Gunter Ziesler/Peter Arnold, Inc.; c, Mandal Ranjit/Photo Researchers, Inc.)*

that do not touch the ground; wingless birds have vestigial wing bones; and many blind, burrowing or cave-dwelling animals have nonfunctioning, vestigial eyes.

The occasional presence of a vestigial structure is to be expected as a species adapts to a changing mode of life. Some structures become much less important for survival and may end up as vestiges. When a structure no longer confers a selective advantage, it usually becomes smaller and loses much or all of its function with the passage of time. Since the presence of the vestigial structure is usually not harmful to the organism, however, selective pressure for completely eliminating it is weak, and the vestigial structure can be found in many subsequent generations.

Vertebrates have retained some developmental features of their ancestors

All vertebrates have similar patterns of embryological development that indicate they share a common ancestor. The resemblance among embryos of different vertebrates is closer than the resemblance among adults. In fact, it is difficult to distinguish among the early embryos of a reptile, a bird, a pig, or a human (Fig. 17–14). Segmented muscles, gill pouches, a tubular heart without left and right sides, a system of arteries known as aortic arches in the gill region, and many other features are found in all vertebrate embryos. All of these structures are necessary and functional in the developing fish. The small segmented muscles of the fish embryo give rise to the segmented muscles used by the adult fish in swimming. The gill pouches break through to the surface as gill slits. The adult fish heart remains undivided because it pumps blood forward to the gills that develop in association with the aortic arches.

Since none of these embryonic features persists in the adults of reptiles, birds, or mammals, why are these fishlike structures present in their embryos? Evolution is a conservative process, and natural selection builds on what has come before rather than starting from scratch. Terrestrial vertebrates are thought to have evolved from fishlike ancestors; therefore, they share some of the early stages of development still found in fish today. The accumulation of genetic changes over time in these vertebrates has modified the basic body plan laid out in fish development.

The distribution of plants and animals supports evolution

The study of the past and present geographical distribution of organisms is called **biogeography.** Darwin was interested in biogeography and considered why the species found on ocean islands tend to resemble species of the nearest mainland, even if the environment is different, but not to resemble species on islands with similar environments in other parts of the world. Darwin studied the plants and animals of two sets of arid islands—the Cape Verde Islands, some 650 kilometers (about 400 miles) west of Dakar, Africa, and the Galapagos Islands, a comparable distance west of Ecuador, South America. On each group of islands, the plants and terrestrial animals were indigenous (native), but those of the Cape Verdes resembled African species and those of the Galapagos resembled South American species. The similarities of Galapagos species to South American species were particularly striking considering that the Galapagos Islands are dry and rocky and the nearest part of South America is humid and has a lush tropical rain forest. Darwin concluded that species from the neighboring continent migrated or were carried to the islands, where they subsequently adapted to the new environment and, in the process, evolved into new species.

If evolution was not a factor in the distribution of species, we would expect to find a given species everywhere that it could survive. However, the geographical distribution of organisms that actually exists makes sense in the context of evo-

Figure 17–14 **Similarities in vertebrate embryological development.** The early stages of embryonic development are almost identical in different vertebrate species. Numerous structural similarities are shared by the early stages, including the presence of gill pouches and a tail.

lution. For example, Australia, which has been a separate land mass for millions of years, has distinctive organisms. Australia has populations of egg-laying mammals (monotremes) and pouched mammals (marsupials) not found anywhere else. Two hundred million years ago, Australia and the other continents were joined together in a major land mass (see Chapter 19). Over the course of millions of years, the Australian continent gradually separated from the others. The monotremes and marsupials in Australia continued to thrive and diversify. The isolation of Australia also prevented placental mammals, which arose elsewhere at a later time, from competing with its monotremes and marsupials. In other areas of the world where placental mammals occurred, most monotremes and marsupials became extinct. (Biogeography is discussed further in Chapter 52.)

Molecular comparisons among organisms provide evidence for evolution

Evidence for evolutionary relationships is provided by similarities and differences in the biochemistry and molecular biology of various organisms. Indeed, lines of descent based solely on biochemical and molecular characters closely resemble lines of descent based on structural and fossil evidence. Molecular evidence for evolution includes the universal genetic code and the conserved sequences of amino acids in proteins and of nucleotides in DNA.

The genetic code is universal

Organisms owe their characteristics to the types of proteins that they possess, which in turn are determined by the sequence of nucleotides in their mRNA (as specified by the order of nucleotides in their DNA). Evidence that all life is related comes from the fact that all organisms use a genetic code that is virtually identical. Recall from Chapter 12 that the genetic code specifies a triplet (a sequence of three nucleotides in DNA) that codes for a particular codon (a sequence of three nucleotides in mRNA) that codes for a particular amino acid in a polypeptide chain. For example, "AAA" in DNA codes for "UUU" in mRNA, which codes for the amino acid phenylalanine in organisms as diverse as shrimp, humans, bacteria, and tulips. In fact, "AAA" codes for phenylalanine in *all* organisms examined to date.

The universality of the genetic code—no other code has been found in any organism—is compelling evidence that all organisms arose from a common ancestor. The genetic code has been maintained and transmitted through all branches of the evolutionary tree since its origin in some extremely early (and successful) organism.

Proteins contain a record of evolutionary change

Darwin's theory that all species are related through descent with modification from the earliest organisms has been further

TABLE 17-1	Differences in Nucleotide Sequences in DNA as Evidence of Phylogenetic Relationships*	
Species Pairs	**Percent Divergence in a Selected DNA Sequence†**	
Human–chimpanzee	1.7	
Human–gorilla	1.8	
Human–orangutan	3.3	
Human–gibbon	4.3	
Human–rhesus monkey (Old World monkey)	7.0	
Human–spider monkey (New World monkey)	10.8	
Human–tarsier	24.6	

*From Goodman, M. et al. "Primate Evolution at the DNA Level and a Classification of Hominoids." *Journal of Molecular Evolution*, Vol. 30, 1990.

†Noncoding sequences of β-globin genes.

the nature and number of underlying DNA base-pair changes that must have occurred during evolution. Such molecular information is determined directly by **DNA sequencing,** in which the order of nucleotide bases in DNA is determined. Generally, the more closely species are thought to be related on the basis of other scientific evidence, the greater the percentage of nucleotide sequences that their DNA molecules have in common. By using the DNA sequence data in Table 17–1, for example, you can conclude that the closest living relative of humans is the chimpanzee (because its DNA has the lowest percentage differences in the sequence examined). Which of the primates in Table 17–1 is the most distantly related to humans?

A **phylogenetic tree,** a diagram showing lines of descent, can be derived from differences in a given DNA nucleotide sequence. Such a phylogenetic tree for selected primates is depicted in Figure 17–15. (See *Making the Connection: Molecular Clocks, Evolution, and Genetics* for a discussion of how DNA sequencing is used to estimate the time of divergence between two closely related species or taxonomic groups.)

supported by evidence from molecular biology. Investigations of the sequence of amino acids in proteins that play the same roles in many species have revealed both great similarities and certain specific differences.

Even organisms that are remotely related, such as humans, fruit flies, sunflowers, and yeasts, share some proteins, such as cytochrome *c*, which is part of the electron transport chain in aerobic respiration. To survive, all aerobic organisms need a respiratory protein with the same basic structure and function as the cytochrome *c* of their common ancestor. Consequently, not all of the amino acids that confer the structural and functional features of cytochrome *c* are free to change. Any mutations that changed the amino acid sequence at structurally important sites of the cytochrome *c* molecule would have been harmful, and natural selection would have prevented such mutations from being passed to future generations. However, in the course of the long, independent evolution of different organisms, mutations have resulted in the substitution of many amino acids at less important locations in the cytochrome *c* molecule. The greater the differences in the amino acid sequences of their cytochrome *c* molecules, the longer it is thought to have been since two organisms diverged.

DNA contains a record of evolutionary change

Because a protein's amino acid sequences are coded in DNA,[3] the differences in amino acid sequences indirectly demonstrate

[3]Of course, not all DNA codes for proteins (witness introns and tRNA genes). However, DNA sequencing of non-protein-coding DNA is also useful in determining evolutionary relationships.

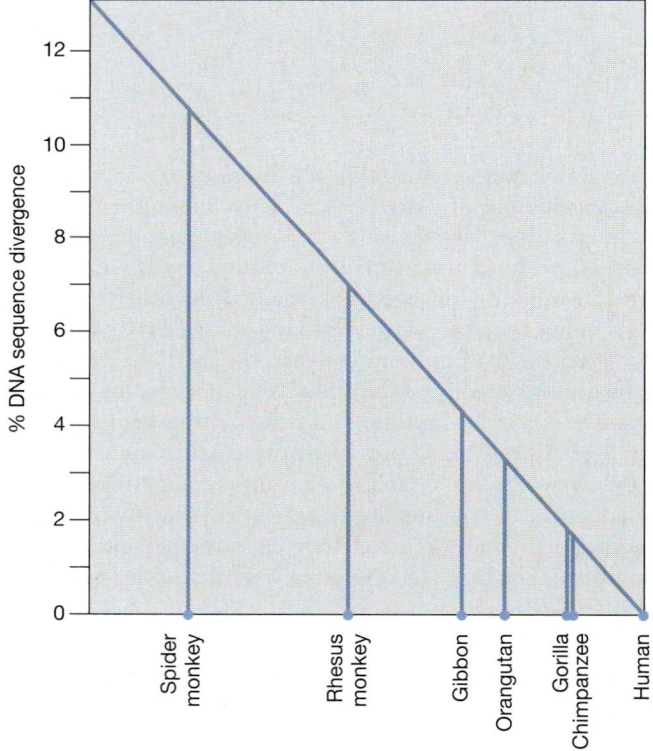

Figure 17-15 Phylogenetic tree of selected primates. This diagram, which is based on the DNA sequence percentage differences given in Table 17-1, is compatible with fossil and structural evidence. Closely related organisms, such as chimpanzees and humans, show fewer differences in their nucleotide sequences. More distantly related organisms, such as spider monkeys and humans, show greater differences in their nucleotide sequences.

MOLECULAR CLOCKS, EVOLUTION, AND GENETICS

Can biologists interested in studying evolutionary relationships make use of the fact that mutations in DNA nucleotide sequences have been occurring throughout the course of evolution? Within a given taxonomic group, mutations are assumed to have occurred at a fairly steady rate over millions of years. Thus, the more differences you find in the same sequences of DNA of one species compared with another, the more time must have elapsed since the two diverged from a common ancestor. (The *same* DNA sequence refers to the nucleotide codes for homologous sections of DNA in the species being compared.)

From the number of alterations in the same DNA sequence taken from different organisms, we can develop a **molecular clock** to estimate the time of divergence between two closely related species or higher taxonomic groups. The clock is calibrated by comparing the number of nucleotide changes with the dates of evolutionary branch points that are known from the fossil record.

Such molecular clocks can be used to complement geological estimates of the divergence of species or to assign tentative dates to evolutionary events that lack fossil evidence. Molecular clocks are also used, along with fossil evidence and structural data, to help reconstruct **phylogeny,** which is the evolutionary history of a group of related species (see Chapter 22). By assigning tentative dates to the divergence of species, molecular clocks show the relative order of branch points in phylogeny.

Molecular clocks must be developed and interpreted with care. The rates of mutation appear to vary among different genes and among distantly related taxonomic groups. Therefore, although some mutations occur at a fairly uniform rate for species that are closely related, a single molecular clock for all genes in all species cannot be established.

Evolutionary hypotheses can be tested experimentally

Increasingly, biologists are designing imaginative experiments, often in natural settings, to test evolutionary hypotheses. David Reznick from the University of California at Riverside and John Endler from James Cook University in Australia, for example, have studied evolution in guppy populations in Venezuela and Trinidad, a small island in the southern Caribbean.

Reznick and Endler observed that different streams have different kinds and numbers of fishes that prey on guppies. Predatory fish that prey on larger guppies are present at lower elevations in all streams; these areas of intense predation pressure are known as *high predation habitats.* Predators are often excluded from tributaries or upstream areas by rapids and waterfalls. The areas above such barriers are known as *low predation habitats* because they contain only one species of small predatory fish that occasionally eat smaller guppies.

Differences in predation are correlated with many differences in the guppies, including male coloration, behavior, and attributes known as *life history traits,* including age and size at sexual maturity, the number of offspring per litter, the size of the offspring, and the frequency of reproduction. For example, guppy adults are larger in streams at higher elevations and smaller in streams at lower elevations. Reznick and colleagues have studied these life history traits and considered the role that predators might have played in the evolution of differences in life histories.

Reznick and his colleagues paid particular attention to the differences in guppy size, hypothesizing that these differences are related to predator preferences. However, they first had to

rule out the possibility that some unknown environmental factor was responsible for the size differences. To determine this, they captured adult female guppies from high and low predation habitats in Trinidad and sent them to their laboratory. They exploited a convenient property of guppy reproductive biology, which is that adult females are virtually always pregnant. Because these females store sperm, each female was able to produce a series of litters when she was isolated in the laboratory. The offspring of each female were then mated to produce a second generation. All the guppies were reared in identical predator free laboratory environments.

The second generation of guppies from low predation environments grew into larger, later maturing adults. Conversely, the second generation from high predation environments grew into smaller adults that matured at an earlier age. It is assumed that any differences among guppy populations that persist after two generations in identical laboratory environments have a genetic basis. This experiment therefore demonstrated that differences in size between the two populations were inherited; that is, they were not caused by some nongenetic factor, such as the availability of food.

Genetic differences between the two guppy populations are of particular interest because ecologists had predicted before this study that higher mortality rates found in high predation habitats would select for earlier maturity and smaller size at maturity. These predicted life history differences are consistent with the genetic differences observed among these populations of guppies.

Do predators actually cause these differences to evolve? Reznick and colleagues tested this evolutionary hypothesis by conducting field experiments in Trinidad. Taking advantage of waterfalls that prevent upstream movement of guppies

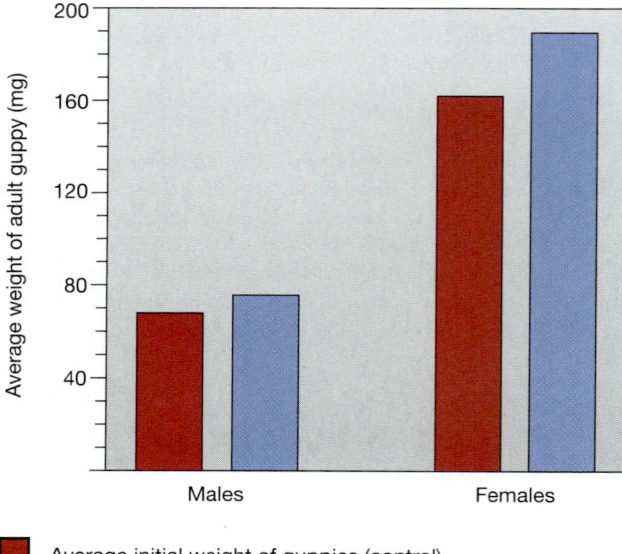

Average weight of adult guppy (mg)

Males
Females

■ Average initial weight of guppies (control)

■ Average weight of guppies 18 generations after being transferred to stream with different selective predation (experimental)

Figure 17–16 **Experimental evidence of natural selection in guppies.** Male and female guppies from a stream in which the predators preferred large adult guppies as prey *(red bars)* were transferred to a stream in which the predators preferred juveniles and small adults. After 11 years, the descendants of these guppies *(blue bars)* were measurably larger in size, compared to their ancestors. *(Data from Reznick et al, Science, Vol. 275, 28 March 1997.)*

and/or guppy predators, they moved either guppies or guppy predators over such barriers. For example, guppies from a high predation habitat were introduced into a low predation habitat by moving them over a barrier waterfall, into a section of stream that was free of guppies and large predators. The only fish species that lived in this section of stream prior to the introduction was the predator that occasionally preyed on small guppies.

Eleven years later, they captured adult females from the introduction site (low predation habitat) and the control site below the barrier waterfall (high predation habitat). They bred these females in their laboratory and compared the life histories of succeeding generations, as in the earlier study. The descendants of guppies introduced into the low predation habitat matured at an older age and larger size than the descendants of guppies collected from the control site below the waterfall (Fig. 17–16). They also produced fewer, but larger, offspring. The life histories of the introduced fish had therefore evolved to be similar to those of fish that are typically found in such low predation habitats. These and other studies have also demonstrated that the predators have played an active role in the evolution of other traits, such as average number of offspring produced during the lifetime of an individual female (fecundity), male coloration, and behavior.

Such experiments demonstrate that evolution is not only real, but that it is occurring now, driven by selective forces, such as predation, that can be experimentally manipulated. Darwin incorrectly assumed evolution is so gradual that humans cannot observe it. As Jonathan Weiner, author of *The Beak of the Finch: A Story of Evolution in Our Time,* puts it, "Darwin did not know the strength of his own theory. He vastly underestimated the power of natural selection. Its action is neither rare nor slow. It leads to evolution daily and hourly, all around us, and we can watch."

SUMMARY WITH KEY TERMS

I. **Evolution,** the genetic change in a **population** of organisms over time, is the unifying concept of biology.

II. Charles Darwin and Alfred Wallace independently proposed the theory of evolution by **natural selection,** which is based on four observations:

A. *Overproduction:* Each species produces more offspring than will survive to maturity.

B. *Variation:* Genetic variation exists among the individuals in a population.

C. *Limits on population growth:* Organisms compete with one another for the resources needed for life, such as food, living space, water, light, and so on.

D. *Differential reproductive success:* The offspring with the most favorable combination of characteristics are most likely to survive and reproduce, passing those genetic characteristics to the next generation. Thus, natural selection results in **adaptations,** which are evolutionary modifications that improve the chances of survival and repro-

ductive success in a particular environment.

E. Over time, enough changes may accumulate in geographically separated populations (often with slightly different environments) to produce new species.

III. The **synthetic theory of evolution,** or **neo-Darwinism,** combines Darwin's theory of evolution by natural selection with modern genetics to explain how species adapt to their environment.

A. **Mutation** provides the genetic variability that natural selection acts on during evolution.

B. The synthetic theory of evolution emphasizes the genetics of populations rather than of individuals.

IV. The concept that evolution has occurred and is occurring is now well documented.

A. Direct evidence of evolution comes from **fossils,** the remains or traces of ancient organisms.

B. Evidence supporting evolution is derived from comparative anatomy.

1. **Homologous** features have basic structural similarities, even though the structures may be used in different ways, because they derive from the same structure in a common ancestor. Homologous features indicate evolutionary affinities among the organisms possessing them.
2. **Homoplastic** features have similar functions in quite different, distantly related organisms. Homoplastic features demonstrate **convergent evolution,** in which organisms with separate ancestries adapt in similar ways to comparable environmental demands.
3. **Vestigial structures** are nonfunctional or degenerate remnants of structures that were present and functional in ancestral organisms. Structures occasionally become vestigial as species adapt to different modes of life.

C. Developmental biology provides evidence of evolution.
1. The embryos of different vertebrates are more similar than the adults.
2. The accumulation of genetic changes since organisms **diverged,**

or took separate evolutionary pathways, has modified the pattern of development in more complex vertebrate embryos.

D. **Biogeography,** the distribution of plants and animals, supports evolution. Areas that have been separated from the rest of the world for a long time have organisms that are unique to those areas.

E. Molecular biology provides evidence of evolution.
1. The universality of the genetic code is compelling evidence that all life is related.
2. The sequence of amino acids in common proteins reveals greater similarities in closely related species.
3. A greater proportion of the nucleotide sequence in DNA is identical in closely related organisms.

F. Experimental data provide evidence of evolution.
1. Reznick has studied the effects of predation intensity on the evolution of guppy populations in both the laboratory and nature.
2. Such experiments are a powerful way for investigators to test the underlying processes of natural selection.

POST-TEST

1. Evolution is based on which of the following concepts? (a) organisms share a common origin (b) over time, organisms have diverged from a common ancestor (c) an animal's body parts can change over its lifetime, and these acquired changes can be passed to the next generation (d) a and b are correct (e) a, b, and c are correct
2. Jewish and Muslim men have been circumcised for many generations, yet this practice has had no effect on the penile foreskin of their offspring. This observation disproves evolution as envisioned by (a) Lamarck (b) Darwin (c) Wallace (d) a and c are correct (e) b and c are correct
3. Which of the following is *least* likely to have occurred after the first small population of finches reached the Galapagos Islands from the South American mainland? (a) after many generations, the finches became increasingly different from the original population (b) over time, the finches adapted to their new environment (c) after many generations, the finches were unchanged and unmodified in any way (d) the finches were unable to survive in their new home and died out
4. Which of the following is *not* part of Darwin's mechanism of evolution? (a) differential reproductive success (b) variation in a population (c) inheritance of acquired (nongenetic) traits (d) overproduction of offspring (e) struggle for existence
5. The fossil record (a) usually occurs in sedimentary rock (b) sometimes appears fragmentary (c) is relatively complete for tropical rainforest organisms but incomplete for aquatic organisms (d) a and b are correct (e) a, b, and c are correct

6. The molecular record found inside cells suggests that evolutionary changes are caused by an accumulation of (a) traits acquired through need (b) alterations in the order of nucleotides in DNA (c) characters acquired during an individual's lifetime (d) hormones (e) environmental changes
7. Darwin proposed _____ as the mechanism by which evolutionary change takes place. (a) natural selection (b) homoplasy (c) convergent evolution (d) mutation (e) homology
8. In _____ , the selecting agent is the environment, whereas in _____ , the selecting agent is humans. (a) natural selection; convergent evolution (b) mutation; artificial selection (c) homoplasy; homology (d) artificial selection; natural selection (e) natural selection; artificial selection
9. Structures that are similar in underlying form in different species due to a common evolutionary origin are called (a) homoplastic (b) homologous (c) vestigial (d) convergent (e) synthetic
10. Aardvarks, anteaters, and pangolins are similar in structure and form as a result of (a) homology (b) convergent evolution (c) biogeography (d) vestigial structures (e) artificial selection
11. The species of the Galapagos Islands (a) are similar to those on the Cape Verde Islands (b) are similar to those on the South American mainland (c) are identical to those on the Cape Verde Islands (d) are identical to those on the South American mainland (e) are similar to those on both the African and South American mainlands

REVIEW QUESTIONS

1. Explain briefly the concept of evolution by natural selection.
2. Why are only inherited variations important in the evolutionary process?
3. What part of Darwin's theory was he unable to explain? How does the synthetic theory of evolution explain this?
4. How do scientists date fossils? How do fossils provide evidence of evolution?
5. Distinguish among homologous features, homoplastic features, and vestigial structures. How does each provide evidence of evolution?

6. How does developmental biology provide evidence of a common ancestry for vertebrates as diverse as reptiles, birds, pigs, and humans?
7. Explain why there are many species of marsupials in Australia and only a few elsewhere.
8. What is indicated if the DNA from two species is found to be almost identical?
9. Explain how predator preference drives the evolution of guppies.

YOU MAKE THE CONNECTION

1. Although most salamanders have four legs, a few species that live in shallow water lack hind limbs and have extremely tiny forelimbs *(see photograph)*. Develop a hypothesis to explain how limbless salamanders came about according to Darwin's mechanism of evolution by natural selection. How could you test your hypothesis?

2. What adaptations must an animal possess in order to swim in the ocean? Why are such genetically different organisms as porpoises, which are mammals, and sharks, which are fish, so similar in form?

3. The human fetus grows a coat of fine hair, the lanugo, that is shed before or shortly after birth. Chimpanzee and other primate fetuses also grow coats of hair, but they are not shed. Explain these observations based on what you have learned in this chapter.

4. Write a short paragraph explaining each of the following statements:
   a. Natural selection chooses from among the individuals in a population those most suited to *current* environmental conditions. It does not guarantee survival under future conditions.
   b. Individuals do not evolve, but populations do.
   c. The organisms that exist today do so because their ancestors possessed traits that allowed them and their offspring to thrive.
   d. At the molecular level, evolution takes place by the replacement of one nucleotide by another.
   e. Evolution is said to have occurred within a population when measurable genetic changes are detected.

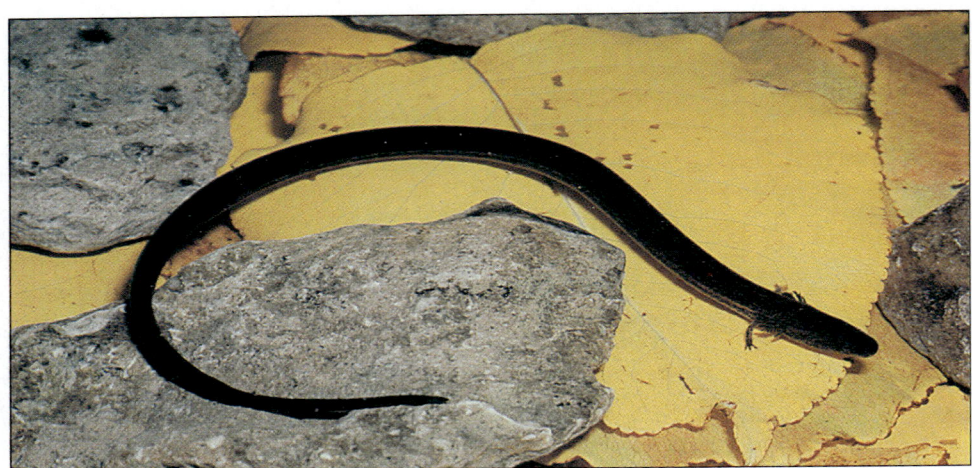

The narrow-striped dwarf siren (*Pseudobranchus striatus axanthus*). This aquatic salamander, which is native to Florida, resembles an eel. *(Suzanne L. Collins and Joseph T. Collins/Photo Researchers, Inc.)*

RECOMMENDED READINGS

Gingerich, P.D., and M.D. Uhen. "*Ancalecetus simonsi,* a New Dorudontine Archaeocete (Mammalia, Cetacea) from the Early Late Eocene of Wadi Hitan, Egypt." *Museum of Paleontology,* University of Michigan, Vol. 29, No. 13, 29 Nov. 1996. This paper on fossil whale ancestors is the source of part of Figure 17–7.

Gould, S.J. "The Paradox of the Visibly Relevant." *Natural History,* Dec. 1997/Jan. 1998. This eminent paleontologist argues that short-term evolutionary studies, while important, cannot explain long-term evolutionary patterns observed in fossils.

McComas, W.F. "The Discovery and Nature of Evolution by Natural Selection: Misconceptions and Lessons from the History of Science." *The American Biology Teacher,* Vol. 59, No. 8, Oct. 1997. The author provides some fascinating insights into Darwin's development of his theory of evolution by natural selection.

Moore, John A. *Science as a Way of Knowing: The Foundations of Modern Biology.* Harvard University Press, Cambridge, 1993. Contains an excellent introduction to the history of evolutionary thought.

Pennisi, E., and W. Roush. "Developing a New View of Evolution." *Science,* Vol. 277, 4 Jul. 1997. How developmental biologists are elucidating some of the mysteries of evolution by studying the genes that control embryonic development.

Porter, D.M., and P.W. Graham. *The Portable Darwin.* Penguin Books, New York, 1993. This collection of Darwin's writings reveals the diverse interests of the man who profoundly changed the intellectual climate of the 19th and 20th centuries.

Smith, J.M. "Bacteria Break the Antibiotic Bank." *Natural History,* Jun. 1994. Bacterial evolution is causing the spread of antibiotic resistance among many disease-causing bacteria.

Thewissen, J.G.M., S.I. Madar, and S.T. Hussain. *Ambulocetus natans,* an Eocene Cetacean (Mammalia) from Pakistan." *CFS,* Vol. 191, 28 Jun. 1996. This paper describes a recently discovered fossil intermediate in whale evolution. It is the source of part of Figure 17–7.

Weiner, J. *The Beak of the Finch: A Story of Evolution in Our Time.* Knopf, New York, 1994. This Pulitzer Prize-winning book focuses on the research of Peter and Rosemary Grant on evolution in the Galapagos finches.

Wiley, J.P., Jr. "Feathered Flights of Fancy." *Smithsonian,* Jan. 1997. This photo essay shows some of the remarkable varieties of chickens that have been produced by artificial selection.

● Visit our website at **http://www.saunderscollege.com/lifesci/titles.html** and click on Solomon/Berg/Martin Biology for links to chapter-related resources on the World Wide Web.

CHAPTER 18

Evolutionary Change in Populations

As we saw in Chapter 17, evolution occurs in populations, not individuals. Although natural selection results from differential survival and reproduction of individuals, individuals do not evolve during their lifetimes. Evolutionary change, which includes modifications in structure, physiology, ecology, and behavior, is inherited from one generation to the next. Although Darwin recognized that evolution occurs in populations, he did not understand how traits are passed to successive generations. One of the most significant advances in biology since Darwin's time has been the demonstration of the genetic basis of evolution.

Recall from Chapter 17 that a **population** consists of all the individuals of the same species that live in a particular place at the same time. Individuals within a population vary in many recognizable traits. A population of snails, for example, may vary in shell size, weight, or color. Some of this variation is due to the environment, and some is due to heredity. Evolution is concerned with heritable traits, that is, with genetic variation as represented by the number, frequency, and kinds of alleles in a population. For example, the photograph shows genetic variation in the shell patterns and colors in a snail species *(Cepaea nemoralis),* native to Scotland. Variation in shell color may have adaptive value in these snails, as some colors predominate in cooler environments, whereas other colors are more common in warmer habitats.

Each population possesses a **gene pool,** which is the total genetic material of all the individuals that make up that population and includes all the alleles for all the genes present in the population. Because diploid *(2n)* organisms possess a maximum of two different alleles at each genetic locus, a single individual typically has only a small fraction of the alleles present in a population's gene pool. The genetic variation that is evident among individuals in a given population indicates that each individual has a different combination of the alleles in the gene pool.

(G.I. Bernard/Animals Animals)

The evolution of populations is best understood in terms of allele frequencies. An **allele frequency** is the percentage of a specific allele of a given gene locus in the population. If the allele frequencies remain constant from generation to generation, the population is not undergoing evolutionary change, and it is said to be in **genetic equilibrium.** Changes in allele frequencies over successive generations, however, indicate that evolution has occurred.

In this chapter we contrast genetic equilibrium with evolutionary change and discuss the five factors responsible for evolutionary change: nonrandom mating, mutation, genetic drift, gene flow, and natural selection. We then consider genetic variation as the raw material for evolution.

AFTER YOU HAVE STUDIED THIS CHAPTER YOU SHOULD BE ABLE TO

1. Define population, allele frequency, genetic equilibrium, and microevolution.
2. Discuss the significance of the Hardy-Weinberg principle as it relates to evolution and list the five conditions required for genetic equilibrium.
3. Use the Hardy-Weinberg principle to solve problems involving populations at genetic equilibrium.
4. Discuss how each of the following alters allele frequencies in populations: nonrandom mating, mutation, genetic drift, gene flow, and natural selection.

5. Distinguish among stabilizing selection, directional selection, and disruptive selection, and give an example of each.
6. Describe the nature and extent of genetic variation, including genetic polymorphism, balanced polymorphism, neutral variation, and geographical variation.
7. Explain how the sickle cell allele illustrates heterozygote advantage.
8. Relate how frequency-dependent selection affects genetic variation.

THE HARDY-WEINBERG PRINCIPLE DESCRIBES GENETIC EQUILIBRIUM

Consider a trait determined by a single gene locus with two alleles, *A* and *a*, that follow simple Mendelian dominance. When you examine a hypothetical natural population of 1000 individuals, you observe that only 90 exhibit the recessive phenotype characteristic of the genotype *aa*. The remaining 910 individuals exhibit the dominant phenotype and are either *AA* or *Aa*. You might assume that, after many generations, genetic recombination during sexual reproduction would cause the dominant allele to become more common in the population. You might even assume that the recessive allele would eventually disappear altogether. However, your assumptions would be incorrect, as the frequencies of alleles and genotypes would not change from generation to generation unless influenced by outside factors that will be discussed shortly.

A population whose allele and genotype frequencies do not change from generation to generation is said to be at genetic equilibrium. The explanation for the stability of successive generations in populations at genetic equilibrium was provided independently by Godfrey Hardy, an English mathematician, and Wilhelm Weinberg, a German physician, in 1908. They pointed out that the expected frequencies of various genotypes in a population can be described mathematically. The resulting **Hardy-Weinberg principle** shows that in large populations the process of inheritance does not by itself cause changes in allele frequencies. It also explains why dominant alleles are not necessarily more common than recessive ones. The Hardy-Weinberg principle represents an ideal situation that probably never occurs in the natural world. However, it is useful because it provides a model to help us understand the real world. Knowledge of the Hardy-Weinberg principle is essential to understanding the mechanisms of evolutionary change in sexually reproducing populations.

We now expand our original example to illustrate the Hardy-Weinberg principle. The frequency of either allele, *A* or *a*, is described by a number that ranges from zero to one. An allele that is totally absent from the population has a frequency of zero. If all the alleles of a given locus are the same in the population, then the frequency of that allele is one.

Because only two alleles, *A* and *a*, exist at the locus in our example, the sum of their frequencies must equal one. If we let *p* represent the frequency of the dominant (*A*) allele in the population, and *q* the frequency of the recessive (*a*) allele, then we can summarize their relationship with a simple binomial equation, $p + q = 1$. When we know the value of either *p* or *q*, we can calculate the value of the other: $p = 1 - q$ and $q = 1 - p$.

Because $p + q = 1$, then $(p + q)^2 = 1$. This binomial equation can be expanded to describe the relationship of the allele frequencies to the genotypes in the population. When it is expanded, we obtain the frequency of the offspring genotypes:

$$p^2 \quad + \quad 2pq \quad + \quad q^2 \quad = \quad 1$$

Frequency of *AA*   Frequency of *Aa*   Frequency of *aa*   All the individuals in a population

From the fact that we had 90 homozygous recessive individuals in our population of 1000, we infer that the frequency of the *aa* genotype, q^2, is 90/1000, or 0.09. Since q^2 equals 0.09, *q* is equal to the square root of 0.09, or 0.3. From the relationship between *p* and *q*, we conclude that $p = 1 - q = 1 - 0.3 = 0.7$.

Based on this information, we can calculate the frequency of homozygous dominant individuals (*AA*): $p^2 = 0.7 \times 0.7 = 0.49$ (Fig. 18–1). The frequency of heterozygous individuals, *Aa*, would be: $2pq = 2 \times 0.7 \times 0.3 = 0.42$. Thus, approximately 490 individuals are expected to be homozygous dominant, and 420 heterozygous. Note that the sum of homozygous dominant and heterozygous individuals equals 910, the number with which we began.

Any population in which the distribution of genotypes conforms to the relation $p^2 + 2pq + q^2 = 1$, whatever the absolute values for *p* and *q* may be, is at genetic equilibrium. *The Hardy-Weinberg principle of genetic equilibrium tells us what to expect when a sexually reproducing population is not evolving.* The proportion of alleles in successive generations will always be the same, provided the following five conditions are met:

Genotypes	AA	Aa	aa
Frequency of genotypes in population	0.49	0.42	0.09
Frequency of alleles in gametes	$A = 0.49 +$ $0.21 = 0.7$		$a = 0.21 +$ $0.9 = 0.3$

(a)

Segregation of alleles
and random fertilization:

(b)

Figure 18–1 **The Hardy-Weinberg principle.** (*a*) How to calculate the frequencies of the alleles A and a in the gametes. (*b*) When eggs and sperm containing A or a alleles unite randomly, the frequency of each of the possible genotypes (*AA, Aa, aa*) among the offspring is calculated by multiplying the frequencies of the alleles A and a in eggs and sperm.

1. **Random mating.** In unrestricted random mating, each individual in a population has an equal chance of mating with any individual of the opposite sex. In our example, the individuals represented by the genotypes *AA, Aa,* and *aa* must mate with one another at random and must not select their mates on the basis of genotype.
2. **No net mutations.** There must be no mutations that convert *A* into *a* or vice versa. That is, the frequencies of *A* and *a* in the population must not change due to mutations.
3. **Large population size.** Allele frequencies in a small population are more likely to be changed by random fluctuations (that is, by genetic drift, discussed shortly) than are allele frequencies in a large population.
4. **No migration.** There can be no exchange of genes with other populations that might have different allele frequencies. In other words, there can be no migration of individuals into or out of a population.

5. **No natural selection.** If natural selection is occurring, certain phenotypes (and their corresponding genotypes) are favored over others. Consequently, the allele frequencies will change, and the population will evolve.

The Hardy-Weinberg principle allows biologists to calculate allele frequencies in a given population if we know the genotype frequencies, and vice versa. These values can be used as a basis of comparison with a population's allele or genotype frequencies in succeeding generations. During that time, if the allele or genotype frequencies deviate from the values predicted by the Hardy-Weinberg principle, then the population is evolving. (You can test your understanding of the Hardy-Weinberg principle by solving the problems in the Review Questions at the end of this chapter.)

MICROEVOLUTION OCCURS WHEN A POPULATION'S ALLELE OR GENOTYPE FREQUENCIES CHANGE

Evolution represents a departure from the Hardy-Weinberg principle of genetic equilibrium, and the degree of departure between the observed allele or genotype frequencies and those expected by the Hardy-Weinberg principle indicates the amount of evolutionary change. This type of evolution—generation-to-generation changes in allele or genotype frequencies *within* a population—is sometimes referred to as **microevolution** because it involves relatively small or minor changes that occur, usually over a few generations. Changes in the allele frequencies of a population result from five microevolutionary processes: nonrandom mating, mutation, genetic drift, gene flow, and natural selection. When one or more of these processes is acting on a population, allele or genotype frequencies will change from one generation to the next.

NONRANDOM MATING CHANGES GENOTYPE FREQUENCIES

When individuals select mates on the basis of phenotype (thereby selecting the corresponding genotype), they can bring about evolutionary change in the population. Two examples of nonrandom mating are inbreeding and assortative mating.

In many populations, individuals mate more often with close neighbors than with more distant members of the population. As a result, neighbors tend to be more closely related—that is, genetically similar—to one another. The mating of genetically similar individuals that are more closely related than if they had been chosen at random from the entire population is known as **inbreeding.** Homozygosity increases with each successive generation of inbreeding. The most extreme example of inbreeding is self-fertilization, which is particularly common in plants.

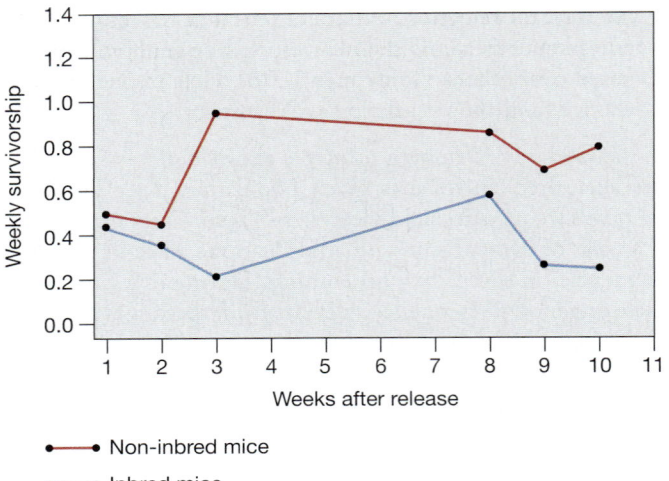

Figure 18-2 **Survivorship of inbred and noninbred mice.** The mouse population was sampled six times during a 10-week period. Noninbred mice *(red)* had a higher survivorship than inbred mice *(blue)*. *(Adapted from Jiménez et al.,* Science, *Vol. 266, 14 Oct. 1994.)*

Inbreeding does not appear to be detrimental in some populations, but in others it causes **inbreeding depression,** in which inbred individuals have lower fitness than noninbred individuals. Inbreeding depression—as evidenced by fertility declines and high juvenile mortality—is thought to be caused by the expression of harmful recessive alleles as homozygosity increases with inbreeding.

Several studies in the 1990s have provided direct evidence of inbreeding depression in nature. For example, white-footed mice *(Peromyscus leucopus)* were taken from a field and used to develop both inbred and noninbred populations in the labo-

ratory. When these laboratory-bred populations were returned to nature, their survivorship was estimated from release-recapture data. The noninbred mice had a statistically significant higher rate of survivorship (Fig. 18–2).

Assortative mating, in which individuals select mates by their phenotypes, is another example of nonrandom mating. For example, biologists selected two phenotypes—high bristle number and low bristle number—in a fruit fly *(Drosophila melanogaster)* population. Although they made no effort to control mating, they observed that the flies preferentially mated with those of similar phenotypes, that is, females with high bristle number tended to mate with males with high bristle number, and females with low bristle number tended to mate with males with low bristle number. Assortative mating is also practiced in many human societies, in which men and women tend to marry individuals like themselves in such characters as height, race, or intelligence. Like inbreeding, assortative mating increases homozygosity at the expense of heterozygosity in the population, but it does not change the overall allele frequencies in the population. Because nonrandom mating changes genotype frequencies, however, it does cause populations to evolve.

MUTATION INCREASES VARIATION WITHIN A POPULATION

Variation is introduced into a population through **mutation,** which is an unpredictable change in DNA (Fig. 18–3). Mutations, which are the source of all new alleles, can result from (1) a change in the nucleotide base pairs of a gene, (2) a rearrangement of genes within chromosomes so that their interactions produce different effects, or (3) a change in the chro-

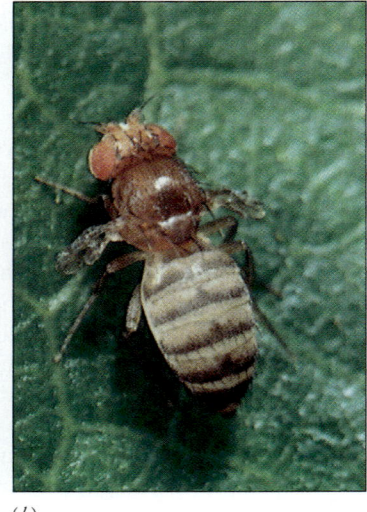

(a)   *(b)*

Figure 18-3 **Fruit fly *(Drosophila melanogaster)* mutation.** *(a)* A normal fly. *(b)* A mutant with vestigial wings. Because mutations are random changes in genetic material, most mutations are neutral or harmful to the organism. Yet for island-dwelling insects, fully developed wings might be more of a disadvantage than an advantage, permitting the insect to be too easily blown away from land. Perhaps for this reason, flies and other insects that live on small islands frequently have reduced wings or are entirely wingless. *(a, b, Peter J. Bryant/Biological Photo Service)* mosomes. Mutations occur unpredictably and spontaneously. The rate of mutation appears to be relatively constant for a particular gene but may vary by several orders of magnitude among genes within a single species and among different species.

Not all mutations pass from one generation to the next. Those occurring in somatic (body) cells are not heritable. When an individual with such a mutation dies, the mutation is lost. Some mutations, however, occur in reproductive cells. These mutations may or may not overtly affect the offspring, because most of the DNA in a cell is "silent" and does not code for specific polypeptides or proteins that are responsible for physical characteristics. Even if a mutation occurs in the DNA that codes for a polypeptide, it may still have little effect on the structure or function of that polypeptide (we discuss such *neutral variation* later in the chapter). However, when the polypeptide is sufficiently altered to change its function, the mutation is usually harmful. By acting against seriously abnormal phenotypes, natural selection eliminates or reduces to low frequencies the most harmful mutations. Small mutations, even those with slightly harmful phenotypic effects, have a better chance of being incorporated into the population, where at some later time, under different environmental conditions, they may produce traits that are helpful or adaptive for the population.

Mutations do not determine the *direction* of evolutionary change. Consider a population living in an increasingly dry environment. A mutation producing a new allele that helps an individual adapt to dry conditions is no more likely to occur than one for adapting to wet conditions or one with no relationship to the changing environment. The production of new mutations simply increases the genetic variability that can be acted on by natural selection and, therefore, the potential for new adaptations.

Mutation by itself causes small deviations in allele frequencies from those predicted by the Hardy-Weinberg principle. Although allele frequencies may be changed by mutation, these changes are typically several orders of magnitude smaller than changes caused by other evolutionary forces, such as genetic drift. As an evolutionary force, mutation is usually negligible, but it is important as the ultimate source of variation.

IN GENETIC DRIFT RANDOM EVENTS CHANGE ALLELE FREQUENCIES

The size of a population has important effects on allele frequencies because random events, or chance, will tend to cause changes of relatively greater magnitude in a small population. If a population consists of only a few individuals, an allele present at a low frequency in the population could be completely lost purely by chance. Such an event would be most unlikely in a large population. For example, consider two populations, one with 10,000 individuals and one with 10 individuals. If

an uncommon recessive allele occurs at a frequency of 10%, or 0.1, then 1900 individuals in the large population possess the allele.[1] That same frequency, 0.1, in the smaller population means that only about 2 individuals possess the allele.[2] From this exercise, it is easy to see that there is a greater likelihood of losing the rare allele from the smaller population than from the larger one. Predators, for example, might happen to kill one or two individuals possessing the uncommon allele in the smaller population purely by chance.

The production of random evolutionary changes in small breeding populations is known as **genetic drift.** Genetic drift results in allele frequency changes in a population from one generation to another. One allele may be eliminated from the population by chance, regardless of whether that allele is beneficial, harmful, or of no particular advantage or disadvantage. Thus, genetic drift can decrease genetic variation *within* a population, although it tends to increase the genetic differences *among* different populations.

When genetic bottlenecks occur, genetic drift becomes a major evolutionary force

Because of fluctuations in the environment, such as depletion in food supply or an outbreak of disease, a population may periodically experience a rapid and marked decrease in the number of individuals. The population is said to go through a **genetic bottleneck** during which genetic drift can occur in the small population of survivors. As the population again increases in size, many allele frequencies may be quite different from those in the population preceding the decline.

Scientists hypothesize that genetic variation in the cheetah was considerably reduced by a genetic bottleneck that occurred at the end of the last Ice Age, some 10,000 years ago. At that time, cheetahs nearly became extinct, perhaps due to overhunting by humans. The few surviving cheetahs possessed greatly reduced genetic variability, and as a result, the cheetah population today is nearly genetically uniform or homogeneous (see *Making the Connection: Cheetahs, Genetics, and Ecology*).

The founder effect occurs when a few "founders" establish a new colony

When one or a few individuals from a large population establish, or found, a colony (as when a few birds separate from the rest of the flock and fly to a new area), they bring with them only a small fraction of the genetic variation present in the original population. As a result, the only alleles represented

[1] $2pq + q^2 = 2(0.9)(0.1) + (0.1)^2 = 0.18 + 0.01 = 0.19; 0.19 \times 10,000 = 1900$

[2] $0.19 \times 10 = 1.9$

CHEETAHS, GENETICS, AND ECOLOGY

Are cheetahs *(Acinonyx jubatus)* endangered because of their extreme genetic uniformity or because of loss of habitat? The outlook for the cheetah's long-term survival is far from certain. There are currently fewer than 10,000 cheetahs in sub-Saharan Africa and northern Iran. For many years, lack of genetic diversity and inbreeding depression were considered the primary factors responsible for the cheetah's plight.

Evidence of the extreme genetic uniformity in cheetahs was demonstrated in the 1980s when biologists discovered that unrelated cheetahs accepted skin grafts from one another. (Normally, only identical twins accept skin grafts so readily.) Geneticists hypothesized that inbreeding depression was responsible for problems with cheetahs in captive breeding programs. Compared to other large cats, cheetah males exhibit low sperm counts, many of their sperm have abnormalities, and cheetah females do not bear as many offspring. Also, many cheetah offspring have health problems and are more susceptible to disease.

Notwithstanding the cheetah's genetic uniformity, ecologists who study cheetahs in nature say that inbreeding depression is not impairing the cheetah's chance of breeding success. They have observed that the main cause of juvenile mortality is predation—most young cheetahs are killed by lions or hyenas, not by the consequences of defective genes. Moreover, ecologists say that male cheetahs have no difficulty siring offspring in nature, despite their low sperm counts and high rates of sperm abnormalities.

Ecologists would like the focus on saving the cheetah to switch from genetic to environmental factors, that is, from captive breeding in zoos to what ecologists perceive as the real threat to the cheetah's survival—loss of habitat. While geneticists agree that protection of habitat should be the main focus in attempts to save the

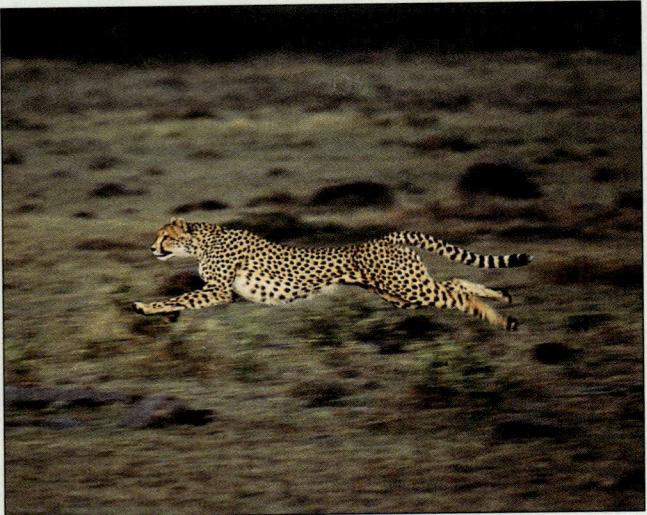

The endangered cheetah. The most specialized member of the cat family, the cheetah is the world's fastest animal and has been clocked at 110 kilometers per hour (70 mph) for short distances. Despite its speed, this fascinating animal is a somewhat timid predator that often gives up its prey to more aggressive animals such as lions, vultures, and hyenas. *(Fritz Pölking/Visuals Unlimited)*

cheetah (and *all* endangered species) from extinction, they also emphasize that lack of genetic diversity should remain an important consideration in conservation strategies to save the cheetah.

---

among their descendants will be those few that the colonizers happened to possess. Typically, the allele frequencies in the newly founded population are very different from those of the parent population. The genetic drift that results when a small number of individuals from a large population colonize a new area is called the **founder effect.**

The founder effect has been observed in populations of wild plants on islands off the Pacific coast of Canada. The Canadian mainland has several wild species of small, weedy annuals in the daisy family. These plants produce wind-dispersed seeds with fluffy parachutes similar to dandelion fruits. The seeds of mainland populations range in size from small to large, as do the fluffy parachutes. Sampling of these plants on 240 islands off the Canadian coast over a 10-year period revealed that the youngest island populations produced signifi-

cantly smaller seeds than mainland populations. (Ages of island populations were easy to estimate because new colonizations occurred frequently.) This observation illustrates both the founder effect (only small seeds with large parachutes remain aloft to be blown by the wind to the nearby islands) and rapid evolutionary change by natural selection.

The Finnish people may also illustrate the founder effect (Fig. 18–4). Geneticists who sampled DNA from Finns and from the European population at large found that Finns exhibit considerably less genetic variation than other Europeans. This evidence supports the hypothesis that Finns are descended from a small group of people who settled in what is now Finland some 4000 years ago and, because of the geography and climate, remained separate from other European societies for centuries.

Figure 18-4 Finns and the founder effect. The Finnish people are thought to have descended from a small founding population that remained separate from the rest of Europe for centuries. *(Courtesy of Embassy of Finland/MattiTirri)*

GENE FLOW GENERALLY INCREASES VARIATION WITHIN A POPULATION

Members of a species tend to be distributed in local populations that are genetically isolated to some degree from other populations. For example, the bullfrogs of one pond form a population separated from those in an adjacent pond. Some exchanges occur by migration between ponds, but the frogs in one pond are much more likely to mate with those in the same pond. Members of most species tend to be distributed in such local populations. Because each population is isolated to some extent from other populations of the species, they can have distinct genetic traits and gene pools.

The migration of breeding individuals between populations causes a corresponding movement of alleles, or **gene flow,** that can have significant evolutionary consequences. As alleles flow from one population to another, they usually increase the amount of genetic variability within the recipient population. If sufficient gene flow occurs between two populations, they become more similar genetically. Because gene flow has a tendency to reduce the amount of variation between two populations, it tends to counteract the effects of natural selection and genetic drift, both of which often cause populations to become increasingly distinct.

If migration by members of a population is considerable, and if populations differ in their allele frequencies, then significant genetic changes can result. For example, by 10,000 years ago modern humans occupied all of Earth's major land areas except a few islands. Because the population density was low, the small, isolated human populations underwent random genetic drift and natural selection. More recently (during the past 300 years or so), major migrations have caused an increase in gene flow, significantly altering allele frequencies within previously isolated human populations.

NATURAL SELECTION CHANGES ALLELE FREQUENCIES IN A WAY THAT INCREASES ADAPTATION

Natural selection is the mechanism of evolution first proposed by Darwin in which members of a population that possess more successful adaptations to the environment are more likely to survive and reproduce (see Chapter 17). Over successive generations, the proportion of favorable alleles increases in the population. In contrast to other microevolutionary processes (nonrandom mating, mutation, genetic drift, and gene flow), natural selection leads to adaptive evolutionary change. As a result, a population becomes better adapted to its environment. Natural selection not only explains why organisms are well adapted to the environments in which they live, but also helps to account for the remarkable diversity of life. Natural selection enables populations to change, thereby adapting to different environments and different ways of life.

Natural selection is the differential reproduction of individuals with different traits, or phenotypes (and therefore different genotypes), in response to the environment. Natural selection preserves individuals with favorable phenotypes and eliminates those with unfavorable phenotypes. Individuals that are able to survive and produce fertile offspring have a selective advantage.

The mechanism of natural selection does not cause the development of a "perfect" organism. Rather, it weeds out those individuals whose phenotypes are less adapted to environmental challenges, while allowing better adapted individuals to survive and pass their alleles to their offspring. By reducing the frequency of alleles that result in the expression of less favorable traits, the probability that favorable alleles responsible for an adaptation will come together in the offspring is increased.

Natural selection operates on an organism's phenotype

Natural selection does not act directly on an organism's genotype. Instead, it acts on the phenotype, which is, at least in part, an expression of the genotype. Natural selection acts on the heritable component of the phenotype, which represents an interaction of all the alleles in the organism's genotype. It is rare that a single gene has complete control over a single phenotypic trait, such as Mendel originally observed in garden peas. Much more common is the interaction of several genes

Stabilizing selection   Directional selection   Disruptive selection

Number of individuals

(a) Phenotypic variation

(b) In a stable environment stresses select against extreme phenotypes, making the population more uniform.

(c) Environmental changes favor phenotypes at one extreme, causing the normal distribution to shift.

(d) Environmental changes favor phenotypes at both extremes of the normal distribution, causing a split.

Figure 18–5   **Modes of selection.** (*a*) A trait that is under polygenic control (in this example, wing colors in a hypothetical population of beetles) exhibits a normal distribution of phenotypes in the absence of selection. (*b*) As a result of stabilizing selection, variation about the mean is reduced. (*c*) Directional selection shifts the curve in one direction. (*d*) Disruptive selection results in two or more peaks.

for the expression of a single trait (see Chapter 10). Many plant and animal characteristics are under this type of polygenic control.

When traits (human height, for example) are under polygenic control, a range of phenotypes occurs, with most of the population located in the median range and fewer at either extreme. This is a normal distribution or standard bell curve (Fig. 18–5*a*; see also Fig. 10–22). Three kinds of selection occur that cause changes in the normal distribution of phenotypes in a population: stabilizing, directional, and disruptive selection. Although we consider each process separately, their influences generally overlap in nature.

Stabilizing selection favors intermediate phenotypes

The process of natural selection associated with a population that is well adapted to its environment is known as **stabilizing selection.** Most populations are probably under the influence of stabilizing forces most of the time. Stabilizing selection selects against phenotypic extremes. In other words, individuals with an average, or intermediate, phenotype are favored.

One of the most widely studied cases of stabilizing selection involves human birth weight, which is under polygenic control and is also influenced by environmental factors. Extensive data from hospitals have shown that infants born with intermediate weights are more likely to survive (Fig. 18–6). Infants at either extreme (too small or too large) have higher rates of mortality. When newborn infants are too small, their body systems are immature, and when they are too large, they have difficult deliveries because they cannot pass as easily

through the cervix and vagina. Stabilizing selection operates to reduce the variability in birth weight so that it is close to the weight with the minimum mortality rate.

Because stabilizing selection tends to decrease variation by favoring individuals near the mean of the normal distribution

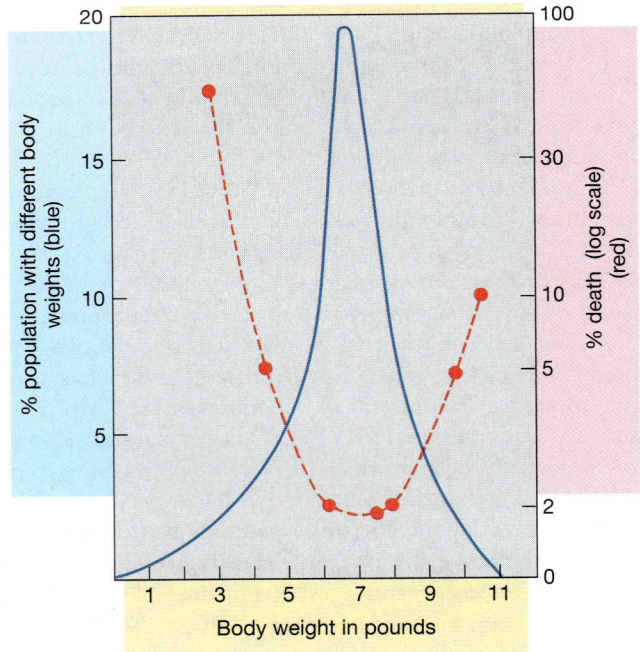

Figure 18–6   **Stabilizing selection.** Infants with very low or very high birth weights have higher death rates. The blue line indicates the percentage of infants born at each weight; the red line indicates percentage of deaths at each birth weight.

at the expense of those at either extreme, the bell curve narrows (Fig. 18–5*b*). Although stabilizing selection decreases the amount of variation in a population, variation is rarely eliminated by this process because other microevolutionary processes act against a decrease in variation. For example, mutation is slowly but continually adding to the genetic variation within a population.

Directional selection favors one phenotype over another

If an environment changes over time, **directional selection** may favor phenotypes at one of the extremes of the normal distribution (Fig. 18–5*c*). Over successive generations, one phenotype gradually replaces another. So, for example, if greater size is advantageous in a new environment, larger individuals will become increasingly common in the population. Directional selection can only occur, however, if alleles favored under the new circumstances are already present in the population.

Darwin's Galapagos finches provide an excellent example of directional selection. Since the 1970s, Peter and Rosemary Grant of Princeton University have studied the Galapagos finches, including a meticulous analysis of their eating habits and beak sizes during an extended drought that was followed by a flood on one of the islands (Fig. 18–7). During the drought, the number of insects and small seeds declined, and large, heavy seeds became the finches' primary food source. Many finches died during this time, and most of the survivors were larger birds whose beaks were larger and deeper. In a few generations, these larger birds became more common in the population. Following the flood, however, smaller seeds became the primary food source, and smaller finches with average-sized beaks were favored. In this example, natural selection is directional: during the drought, natural selection operated in favor of the larger phenotype, whereas following the flood, selection occurred in the opposite direction, favoring the smaller phenotype. The peppered moth population studied in England (see Chapter 1) and the guppy populations studied in Venezuela and Trinidad (see Chapter 17) are other examples of directional selection.

Disruptive selection favors phenotypic extremes

Sometimes extreme changes in the environment may favor two or more different phenotypes at the expense of the mean. That is, more than one phenotype may be favored in the new environment. **Disruptive selection** is a special type of directional selection in which there is a trend in several directions rather than just one (Fig. 18–5*d*). It results in a divergence, or splitting apart, of distinct groups of individuals within a population. Disruptive selection selects against the average, or intermediate, phenotype.

Beaks favored during drought years

Beaks favored after the rains return

(a)

(b)

Figure 18–7   Directional selection. The beaks of Galapagos finches are highly variable, and biologists observed that different phenotypes were favored in different climate conditions. (*a*) Larger birds with large, deep beaks were favored during a period of extended drought when the main available food was large, heavy seeds. (*b*) Smaller finches with average-sized beaks were favored when the rains returned and the birds' primary food source was smaller seeds.

Limited food supply during a severe drought caused a population of finches on another island in the Galapagos to experience disruptive selection. The finch population initially exhibited a variety of beak sizes and shapes. Because the only food available during the drought was wood-boring insects and seeds from cactus fruits, natural selection favored birds with beaks suitable for obtaining these types of food. Finches with longer beaks survived because they could open cactus fruits, and those with wider beaks survived because they could strip off tree bark to expose insects. However, finches with intermediate beaks were unable to use either food source efficiently and consequently had a low survival rate.

Natural selection can induce change in the types and frequencies of alleles in populations only if there is preexisting inherited variation. Genetic variation is the raw material for evolutionary change, as it provides the diversity on which natural selection can act. Without genetic variation, evolution cannot occur. We now examine the genetic basis for variation that is acted on by natural selection.

GENETIC VARIATION IS NECESSARY FOR NATURAL SELECTION

Populations contain abundant genetic variation that was originally introduced by mutation. Sexual reproduction, with its associated crossing-over, independent assortment of chromosomes during meiosis, and random union of gametes, also contributes to genetic variation. The sexual process allows the variability introduced by mutation to be combined in new ways, which may be expressed as new phenotypes.

The effect of recombination can be surprisingly great. Recall from Chapter 10 that nine different genotypes are generated in a dihybrid cross ($AaBb \times AaBb$) involving only two loci, each with only two alleles.[3] If we were dealing with five different unlinked loci, each with six alleles, the number of different genotypes possible would be 4,084,101![4] Because most organisms possess thousands of loci, the number of possible allele combinations is staggering. Some of the combinations may be adaptively superior and favored by natural selection.

Genetic polymorphism exists in genes and the proteins for which they code

One way of evaluating genetic variation in a population is to examine **genetic polymorphism,** which is the presence in a population of two or more alleles for a given locus. Genetic polymorphism is extensive in populations, although many of the alleles are present at low frequencies. Much of genetic polymorphism is not evident because it does not produce distinct phenotypes.

One way that biologists estimate the total amount of genetic polymorphism in populations is by comparing the different forms of a particular protein. Each form consists of a slightly different amino acid sequence that is coded for by a different allele. For example, tissue extracts containing a particular enzyme may be analyzed by gel electrophoresis (see Chapter 14) for different individuals. In gel electrophoresis, the enzymes are placed in slots on a gel, and an electric current is applied that causes each enzyme to migrate across the gel. Slight variations in amino acid sequences in the different forms of a particular enzyme cause each to migrate at a different rate, which can be detected using special stains or radioactive labels. Table 18–1 shows the degree of polymorphism in selected plant and animal groups based on gel elec-

trophoresis of several enzymes. Note that genetic polymorphism tends to be greater in plants than in animals.

Determining the sequence of nucleotides in DNA from individuals in a population provides a *direct* estimate of genetic polymorphism. One method of DNA sequencing is described in Figure 14–9. DNA sequencing of specific genes in an increasing number of organisms, including humans, indicates that genetic polymorphism in most populations is extensive.

Balanced polymorphism can exist for long periods of time

Balanced polymorphism is a special type of genetic polymorphism in which two or more alleles persist in a population over many generations as a result of natural selection. Heterozygote advantage and frequency-dependent selection are mechanisms that preserve balanced polymorphism.

TABLE 18–1 Genetic Polymorphism of Selected Enzymes within Plant and Animal Species

Organism	Number of Species Examined	Percentage of Enzymes Studied that are Polymorphic
Plants		
Gymnosperms	55	70.9
Flowering plants (monocots)	111	59.2
Flowering plants (dicots)	329	44.8
Invertebrates		
Marine snails	5	17.5
Land snails	5	43.7
Insects	23	32.9
Vertebrates		
Fishes	51	15.2
Amphibians	13	26.9
Reptiles	17	21.9
Birds	7	15.0
Mammals	46	14.7

Plant data adapted from Hamrick and Godt, 1990. Animal data adapted from Hartl, 1980, and Hedrick, 1983.

[3]The nine genotypes that result from $AaBb \times AaBb$ are AABB, AABb, AAbb, AaBB, AaBb, Aabb, aaBB, aaBb, and aabb.

[4]There are 21 different ways that six alleles can combine in diploid individuals. Because there are five different genes, each with six alleles, the total number of different genotypes possible is 21^5, or 4,084,101.

Genetic variation may be maintained by heterozygote advantage

We have seen that natural selection often causes unfavorable alleles to be eliminated from a population while favorable alleles are retained. However, natural selection sometimes helps to maintain genetic diversity, including alleles that are unfavorable in the homozygous state, in a population. This happens, for example, when the heterozygote, *Aa*, has a higher degree of **fitness**—that is, is better able to make a genetic contribution to subsequent generations—than either homozygote, *AA* or *aa*. This phenomenon, known as **heterozygote advantage,** is demonstrated in humans by the selective advantage of heterozygous carriers of the sickle cell allele.

The mutant allele (*s*) for sickle cell anemia produces an altered hemoglobin that deforms or sickles the red blood cells, making them more likely to form dangerous blockages in capillaries and to be destroyed in the liver, spleen, or bone marrow (see Chapter 15). Individuals who are homozygous for the sickle cell allele (*ss*) usually die at an early age if medical treatment is not available.

Heterozygous individuals carry alleles for both normal (*S*) and sickle cell hemoglobin. The heterozygous condition (*Ss*) causes an individual to be more resistant to a type of severe malaria (falciparum malaria) than those individuals who are homozygous for the normal hemoglobin allele (*SS*). In a heterozygous individual, each allele produces its own specific kind of hemoglobin, and the red blood cells contain the two kinds in roughly equivalent amounts. Such cells do not ordinarily sickle as readily as cells containing only the *s* allele, and the red blood cells containing the abnormal hemoglobin are more resistant to infection by the malaria-causing parasite, which lives in red blood cells, than are the red blood cells containing only normal hemoglobin.

Each of the two types of homozygous individuals is at a disadvantage. Those homozygous for the sickle cell allele are likely to die of sickle cell anemia, whereas those homozygous for the normal allele may suffer or die of malaria. The heterozygote is therefore more fit than either homozygote. In parts of Africa, the Middle East, and southern Asia where falciparum malaria is prevalent, heterozygous individuals survive in greater numbers than either homozygote (Fig. 18–8). The *s* allele is maintained at a high frequency in the population even though the homozygous recessive condition is almost always lethal.

What happens to the frequency of *s* alleles in Africans and others who possess it when they migrate to the United States and other nonmalarial countries? As might be expected, the frequency of the *s* allele gradually declines in such populations because it confers a selective disadvantage by causing sickle cell anemia in homozygous individuals but no longer confers a selective advantage by preventing malaria in heterozygous individuals. The *s* allele never disappears from the population, however, because it is "hidden" from selection in heterozygous individuals and because it is frequently reintroduced into the

Figure 18–8 **Heterozygote advantage.** The distribution of sickle cell anemia *(bars)* is compared with the distribution of falciparum malaria *(dark tan region)*. The correlation strongly suggests that the resistance of heterozygous individuals to malaria has served to balance the harmful effects of sickle cell anemia.

African American population by gene flow from the African population.

Genetic variation may be maintained by frequency-dependent selection

Thus far in our discussion of natural selection, we have assumed that the fitness of particular phenotypes (and their corresponding genotypes) is independent of their frequency in the population. There are, however, cases of **frequency-dependent selection,** in which the fitness of a particular phenotype depends on how frequently it appears in the population. Often, a phenotype has a greater selective value when it is rare than when it is common in the population. Such phenotypes lose their selective advantage as they become more common.

Frequency-dependent selection often acts to maintain genetic variation in populations of prey species. In this case, the predator catches and consumes the more common phenotype, but may ignore the rarer phenotypes. Consequently, the less common phenotype makes a greater relative contribution to the next generation. Frequency-dependent selection has been demonstrated with aquatic insects called water boatmen (Fig. 18–9), which have three distinct color phenotypes. When all three phenotypes are present at equal frequencies, fish are more likely to consume the most obvious (i.e., least camouflaged) form. However, in populations where one phenotype is present in greater numbers than the other two, the most abundant form is preferentially eaten by fish, regardless of its color. Thus, frequency-dependent selection acts to decrease the frequency of the more common phenotypes (and their genotypes) and increase the frequency of the less common types.

Figure 18–9 **Frequency-dependent selection.** Water boatmen are aquatic insects that swim in ponds and streams by using their middle and hind legs as oars. These insects occur in three color forms (only one is shown). Frequency-dependent selection operates to maintain all three phenotypes within a given population. *(Stephen Dalton/ Photo Researchers, Inc.)*

The extent of neutral variation in organisms is difficult to determine. It is relatively easy to demonstrate that an allele is beneficial or harmful, provided that its effect is observable. But the variation in alleles that involves only slight differences in the proteins they code for may or may not be neutral. These alleles may be influencing the organism in subtle ways that are difficult to measure or assess. Also, an allele that is neutral in one environment may be beneficial or harmful in another.

Populations in different geographical areas often exhibit genetic variation

In addition to the genetic variation among individuals within a population, genetic differences often exist among different populations within the same species, a phenomenon known as *geographical variation.* One type of geographical variation is a **cline,** which is a gradual geographical change in a species' phenotype and genotype frequencies. A cline exhibits variation in the expression of such traits as color, size, shape, physiology, or behavior through a series of geographically separate populations. Clines are common among species with continuous ranges over large geographical areas. For example, the body sizes of many widely distributed birds and mammals increase gradually as the latitude increases, presumably because larger animals are better able to withstand the colder temperatures of winter.

The common yarrow *(Achillea millefolium),* a wildflower that grows in a variety of North American habitats from lowlands to mountain highlands, exhibits clinal variation in height in response to different climates at different altitudes. Although substantial variation exists among individuals within each population, individuals in populations at higher altitudes are, on average, shorter than those at lower altitudes. The genetic basis of these clinal differences can be experimentally demonstrated by growing a series of populations from different geographical areas in the same environment (Fig. 18–10). Despite being exposed to identical environmental conditions, each experimental population exhibits the traits characteristic of the altitude from which it was collected.

Neutral variation may give no selective advantage or disadvantage

Some of the genetic variation observed in a population may confer no apparent selective advantage or disadvantage in a particular environment. For example, random changes in DNA that do not alter protein structure usually do not affect the phenotype. Variation that does not alter the ability of an individual to survive and reproduce and is, therefore, not adaptive is called **neutral variation.**

Figure 18-10 **Clinal variation in yarrow** *(Achillea millefolium).* Seeds from widely dispersed populations in the Sierra Nevada of California and Nevada were collected and grown in a uniform environment, revealing genetic differences in height that were related to the altitude where the plants were collected.

SUMMARY WITH KEY TERMS

I. Each population possesses a **gene pool,** which includes all the alleles for all of the genes present in the population. An **allele frequency** is the percentage of a specific allele of a given gene locus in the population.
   A. If the allele frequencies remain constant from generation to generation, the population is not undergoing evolutionary change and is said to be at **genetic equilibrium.**
   B. If the allele frequencies change over successive generations, the population is undergoing **microevolution.**
II. The **Hardy-Weinberg principle** states that in a population at genetic equilibrium, allele and genotype frequencies do not change from generation to generation.
III. Allele frequencies may be changed by nonrandom mating, mutation, genetic drift, gene flow, and natural selection.
   A. In nonrandom mating, individuals select mates on the basis of genotype (or phenotype).
      1. **Inbreeding** is the mating of genetically similar individuals that are more closely related than if they had been chosen at random from the entire population.
      2. In **assortative mating** individuals select mates by their phenotypes.
   B. New alleles originate as **mutations.**
   C. **Genetic drift** is a random change in the allele frequencies of a small population. The changes caused by genetic drift are usually not adaptive.
   D. The migration of individuals between populations causes a corresponding movement of alleles, or **gene flow,** that can cause changes in allele frequencies.
   E. Changes in allele frequencies that lead to adaptation are caused by **natural selection.**
      1. Natural selection operates on an organism's phenotype.
      2. Natural selection can change the genetic composition of a population in a favorable direction for a particular environment.
         a. **Stabilizing selection** favors the mean at the expense of phenotypic extremes.
         b. **Directional selection** favors one phenotypic extreme over another, causing a shift in the phenotypic mean.
         c. **Disruptive selection** favors two or more phenotypic extremes.
IV. Most populations have abundant genetic variability.
   A. **Genetic polymorphism** is the presence in a population of two or more alleles for a given locus.
   B. **Heterozygote advantage** occurs when the heterozygote exhibits greater fitness than either homozygote. Both dominant and recessive alleles are maintained in the population.
   C. In **frequency-dependent selection,** a genotype's selective value varies with its frequency of occurrence.
   D. Genetic variation that confers no detectable selective advantage is called **neutral variation.**
   E. Geographical variation is genetic variation that exists among different populations within the same species. A **cline** is a gradual geographical change in a species' phenotype and genotype frequencies.

1. The genetic description of an individual is its genotype, whereas the genetic description of a population is its (a) phenotype (b) gene pool (c) genetic drift (d) founder effect (e) changes in allele frequencies

2. If a population's allele frequencies remain constant from generation to generation (a) the population is undergoing evolutionary change (b) the population is said to be at genetic equilibrium (c) microevolution has taken place (d) directional selection is occurring, but only for a few generations (e) genetic drift is a significant evolutionary force

3. Comparing the different forms of a particular protein in a population provides biologists with an estimate of (a) genetic drift (b) genetic polymorphism (c) gene flow (d) heterozygote advantage (e) frequency-dependent selection

4. The continued presence of the allele that causes sickle cell anemia in areas where malaria is prevalent demonstrates which of the following phenomena? (a) genetic polymorphism (b) frequency-dependent selection (c) heterozygote advantage (d) genetic drift (e) population bottleneck

5. Frequency-dependent selection often acts to maintain _____ in a population. (a) assortative mating (b) genetic drift (c) gene flow (d) genetic variation (e) stabilizing selection

6. According to the Hardy-Weinberg principle (a) allele frequencies are not dependent on dominance or recessiveness but remain essentially unchanged from generation to generation (b) the sum of allele frequencies for a given locus is always greater than 1 (c) if a locus has a single allele, its frequency must be 0 (d) allele frequencies change from generation to generation (e) the process of inheritance, by itself, causes changes in allele frequencies

7. What is the correct equation for the Hardy-Weinberg principle? (a) $p^2 + pq + 2q^2 = 1$ (b) $p^2 + 2pq + 2q^2 = 1$ (c) $2p^2 + 2pq + 2q^2 = 1$ (d) $p^2 + pq + q^2 = 1$ (e) $p^2 + 2pq + q^2 = 1$

8. The Hardy–Weinberg principle is applicable provided (a) population size is small (b) migration only occurs at the beginning of the breeding season (c) mutations occur at a constant rate (d) matings occur exclusively between individuals of the same genotype (e) natural selection does not occur

9. Which of the following is *not* an evolutionary agent that causes change in allele frequencies? (a) mutation (b) natural selection (c) genetic drift (d) random mating (e) gene flow from migration

10. Mutation (a) leads to adaptive evolutionary change (b) adds to the genetic variation of a population (c) is the result of genetic drift (d) almost always benefits the organism (e) a and b are correct

11. Which of the following is *not* true of natural selection? (a) Natural selection acts to preserve favorable traits and eliminate unfavorable traits. (b) The offspring of individuals that are better adapted to the environment will make up a larger proportion of the next generation. (c) Natural selection directs the course of evolution by preserving the traits acquired during an individual's lifetime. (d) Natural selection acts on a population's genetic variability, which arises through mutation. (e) Natural selection may result in changes in allele frequencies in a population.

12. In _____ , individuals with a phenotype near the phenotypic mean of the population are favored over those with phenotypic extremes. (a) microevolution (b) stabilizing selection (c) directional selection (d) disruptive selection (e) genetic equilibrium

R E V I E W   Q U E S T I O N S

1. In a human population of 1000, 840 are tongue rollers (*TT* or *Tt*), and 160 are non–tongue rollers (*tt*). What is the frequency of the dominant allele (*T*) in the population?

2. In a population at genetic equilibrium, the frequency of allele *A* is 0.5. What is the frequency of the homozygous dominant genotype (*AA*)? What is the frequency of the heterozygous genotype (*Aa*)?

3. If 96% of the garden peas in a population at genetic equilibrium are tall (*TT* or *Tt*), what is the frequency of the dominant allele (*T*)?

4. If 16% of the individuals in a population at genetic equilibrium are recessive (*aa*), what is the frequency of the recessive allele in the population? What is the frequency of the dominant allele?

5. A biologist catches 1600 moths using special insect traps. Of this population, 64 exhibit the recessive phenotype (*bb*), and 1536 the dominant phenotype (*BB* or *Bb*). Assuming that this is a random sample of the wild population, is the population at genetic equilibrium?

6. The genotype frequencies of a population are determined to be 0.6 *AA*, 0.0 *Aa* (there are no heterozygotes), and 0.4 *aa*. Its allele frequencies are $p = 0.6$ and $q = 0.4$. Is this population at genetic equilibrium? Why or why not?

7. The chemical phenylthiocarbamide (PTC) tastes bitter to most people but is tasteless to others. About 30% of Americans are PTC nontasters (*tt*), and 70% are PTC tasters (*TT* or *Tt*). Assuming the population is at genetic equilibrium, estimate the frequency of the nontaster (*t*) and taster (*T*) alleles.

8. If a population of 2000 is at genetic equilibrium but contains only 180 individuals with the recessive phenotype (*rr*), what is the expected frequency of the recessive allele (*r*) nine generations later?

9. If the genotype frequencies in a population at genetic equilibrium are 0.36 *TT*, 0.48 *Tt*, and 0.16 *tt*, what are the allele frequencies of *T* and *t*?

10. If the genotype frequencies in a population at genetic equilibrium are 0.64 *TT*, 0.32 *Tt*, and 0.04 *tt*, what are the allele frequencies of *T* and *t*?

11. The frequency of the allele *A* in a population at genetic equilibrium is 0.7. What is the expected frequency of the *Aa* genotype?

Answers to these Review Questions are included in the Appendix with the Post-Test answers.

Y O U   M A K E   T H E   C O N N E C T I O N

1. Why are mutations almost always neutral or harmful?
2. Explain this apparent paradox: We discuss evolution in terms of *genotype*

fitness (the selective advantage that a particular genotype confers on an individual), yet natural selection acts on an organism's *phenotype*.

RECOMMENDED READINGS

Case, T.J. "Natural Selection out on a Limb." *Nature*, Vol. 387, 1 May 1997. This "News and Views" article discusses the significance of a long-term experiment (also in this issue) that documents microevolution in lizards introduced on several small Caribbean islands.

Cohn, J.P. "Genetics for Wildlife Conservation." *BioScience*, Vol. 40, No. 3, Mar. 1990. Genetic analysis of endangered species offers hope for their survival.

Dawkins, R. "God's Utility Function." *Scientific American*, Vol. 273, No. 5, Nov. 1995. According to the author, there is no grand purpose in nature, and life can be explained by the contest among genes for survival.

Grant, P.R. "Natural Selection and Darwin's Finches." *Scientific American*, Vol. 265, No. 4, Oct. 1991. A study of the finches of the Galapagos Islands reveals natural selection in action during a drought.

King, R.B.,. and R. Lawson. "Microevolution in Island Water Snakes." *BioScience*, Vol. 47, No. 5, May 1997. The authors demonstrated the interaction between natural selection and gene flow in determining color pattern differences in populations of Lake Erie water snakes.

Mayr, E. *Population, Species, and Evolution*. Harvard University Press, Cambridge, 1970. A classic discussion of evolution.

Weiner, Jonathan. *The Beak of the Finch: A Story of Evolution in Our Time*. Alfred A. Knopf, New York, 1994. This Pulitzer Prize-winning book focuses on the research of Peter and Rosemary Grant on evolution in the Galapagos finches.

● Visit our website at **http://www.saunderscollege.com/lifesci/titles.html** and click on Solomon/Berg/Martin Biology for links to chapter-related resources on the World Wide Web.

CHAPTER 19

Speciation and Macroevolution

Thus far in our study of evolution, we have examined natural selection and how populations evolve (Chapters 17 and 18). We now focus our attention on how species and higher taxa (for example, new classes)[1] evolve. We do not know exactly how many species exist, but biologists estimate there may be something on the order of 13 to 14 million different species. About 1.75 million species have been scientifically named and described. These include 250,000 plant species, 42,000 vertebrate animals, and some 750,000 insects. The concept of distinct kinds of organisms, known as **species** (from Latin, meaning "kind") is not new. However, every definition of species has some sort of limitation. Linnaeus, the 18th century biologist who is considered the founder of modern taxonomy, classified plants into separate species based on structural differences (see Chapter 22). This method, known as the *morphological species concept,* is still used to help characterize species, but structure alone is not adequate to explain what constitutes a species. For example, dogs come in a wide variety of sizes and shapes, yet all dogs are clearly the same kind of organism and are classified as members of the same species.

Population genetics did much to clarify the concept of species. According to the **biological species concept,** first expressed by Ernst Mayr in 1940, a species consists of groups of populations whose members are capable of interbreeding in nature to produce fertile offspring and do not interbreed with, that is, are reproductively isolated from, members of different species. In other words, each species has a gene pool that is isolated from that of other species, and each is restricted by reproductive barriers from interbreeding with other species.

One of the problems with the biological species concept is that it applies only to sexually reproducing organisms. Organisms that reproduce asexually do not interbreed, so we cannot think of them in terms of reproductive isolation. These organisms and extinct organisms are classified on the basis of structural and biochemical characteristics. Another potential problem with the biological species concept is that organisms assigned to different species may successfully interbreed if

(Gary Retherford/Photo Researchers, Inc.)

brought into the artificial environment and caging of a wildlife ranch, circus, zoo, greenhouse, aquarium, or laboratory. The photo shows a "zebrass," a sterile hybrid between a zebra and a donkey that retains features of both parental species. Although such matings may occur under artificial conditions (such as the wildlife ranch in Texas where this cross took place), zebras and donkeys do not interbreed in the wild.

This chapter discusses reproductive barriers that isolate species from one another, the possible evolutionary mechanisms that explain how the millions of species that live today or lived in the past originated from ancestral species, and the rates of evolutionary change. We then examine macroevolution, which is large-scale phenotypic changes (such as the appearance of wings with feathers during the evolution of birds from reptiles) that permit the evolution of taxonomic groups higher than the species level—new genera, families, orders, classes, and even phyla.

---

[1]The taxonomic categories (taxa) above the level of species are artificial constructs used by humans to indicate degrees of relatedness among organisms. Thus, closely related species are grouped into the same genus (pl., *genera*), similar genera into the same family, similar families into the same order, similar orders into the same class, and similar classes into the same phylum.

AFTER YOU HAVE STUDIED THIS CHAPTER YOU SHOULD BE ABLE TO

1. Define a species and explain the limitations of the biological species concept.
2. Explain the significance of reproductive isolating mechanisms and distinguish among the different prezygotic and postzygotic barriers.
3. Explain the mechanism of allopatric speciation and give an example.
4. Explain the mechanisms of sympatric speciation and give both plant and animal examples.
5. Take either side in a debate on the pace of evolution by representing the opposing views of gradualism and punctuated equilibrium.
6. Define macroevolution and distinguish among microevolution, speciation, and macroevolution.
7. Discuss macroevolution in the context of novel features, including preadaptations, allometric growth, and paedomorphosis.
8. Discuss the macroevolutionary significance of adaptive radiation and extinction.
9. Explain how the course of evolution was affected by continental drift.

SPECIES ARE REPRODUCTIVELY ISOLATED IN VARIOUS WAYS

A number of **reproductive isolating mechanisms** prevent interbreeding between two different species whose ranges overlap. These mechanisms preserve the genetic integrity of each species because gene flow between species is prevented. Most species have two or more mechanisms that block a chance occurrence of individuals from two different species overcoming a single reproductive isolating mechanism. Most work before fertilization occurs (prezygotic), whereas others work after fertilization has taken place (postzygotic) (Table 19–1).

Prezygotic barriers interfere with fertilization

Prezygotic barriers are reproductive isolating mechanisms that prevent fertilization from taking place. Because male and female gametes never come into contact, an interspecific zygote (fertilized egg formed by the union of an egg from one species and a sperm from another species) is never produced. Prezygotic barriers include temporal isolation, behavioral isolation, mechanical isolation, and gametic isolation.

Sometimes genetic exchange between two groups is prevented because they reproduce at different times of the day, season, or year. Such examples demonstrate **temporal isolation.** For example, two very similar species of fruit flies, *Drosophila pseudoobscura* and *D. persimilis,* have ranges that overlap to a great extent, but they do not interbreed. *Drosophila pseudoobscura* is sexually active only in the afternoon and *D. persimilis* only in the morning. Similarly, two frog species have overlapping ranges in eastern Canada and the United States. The wood frog *(Rana sylvatica)* usually mates in late March or early April, when the water temperature is about 7.2°C (45°F), whereas the northern leopard frog *(R. pipiens)* usually mates in mid-April, when the water temperature is 12.8°C (55°F) (Fig. 19–1).

Many animal species exchange a distinctive series of signals before mating. Such courtship behaviors illustrate **behavioral isolation** (also known as **sexual isolation**). Bowerbirds, for example, exhibit species-specific courtship patterns. The male satin bowerbird of Australia constructs an elaborate bower of twigs, adding decorative blue parrot feathers and white flowers at the entrance (Fig. 19–2). When a female approaches the bower, the male dances about her, holding a particularly eye-catching decoration in his beak. While dancing, he sings a

TABLE 19–1	Reproductive Isolating Mechanisms
Mechanism	**How It Works**
Prezygotic Barriers	Prevent fertilization
Temporal isolation	Similar species reproduce at different times
Behavioral isolation	Similar species have distinctive courtship behaviors
Mechanical isolation	Similar species have structural differences in their reproductive organs
Gametic isolation	Gametes of similar species are chemically incompatible
Postzygotic barriers	Reduce viability or fertility of hybrid
Hybrid inviability	Interspecific hybrid dies at early stage of embryonic development
Hybrid sterility	Interspecific hybrid survives to adulthood but is unable to reproduce successfully
Hybrid breakdown	Offspring of interspecific hybrid are unable to reproduce successfully

(a)

(b)

(c)

Figure 19–1 **Temporal isolation in wood and leopard frogs.** (*a*) The wood frog (*Rana sylvatica*) mates in early spring, often before the ice has completely melted in the ponds. (*b*) The leopard frog (*R. pipiens*) typically mates a few weeks later. (*c*) Graph of peak mating activity in wood and leopard frogs. In nature, wood and leopard frogs do not interbreed, although they have done so in the laboratory. (*a, L. & D. Klein/Photo Researchers, Inc.; b, Rod Planck/Photo Researchers, Inc.*)

courtship song that consists of a variety of sounds, including buzzes and laughlike hoots. These specific courtship behaviors keep similar bird species reproductively isolated from the satin bowerbird. If a male and female of two different species begin courtship, it stops when one member does not recognize or respond to the signals of the other. Another example of behavioral isolation involves the wood frogs and northern leopard frogs (just discussed as an example of temporal isolation). Males of these two species have very specific vocalizations to attract females of their species for breeding. These vocalizations reinforce the reproductive isolation of these species.

Sometimes members of different species court and even attempt copulation, but the incompatible structures of their genital organs prevent successful mating. Structural differences that inhibit mating between species produce **mechanical isolation.** For example, many flowering plant species have physical differences in their flower parts that help them maintain their reproductive isolation from one another. In such plants, the flower parts are adapted for specific insect pollinators. Two species of sage, for example, have overlapping ranges in southern California. Black sage (*Salvia mellifera*), which is pollinated by small bees, has a floral structure different from that of white sage (*S. apiana*), which is pollinated by large carpenter bees (Fig. 19–3). Interestingly, black and white sage are also prevented from mating by a temporal barrier; black sage flowers in early spring, and white sage flowers in late spring and early summer. Presumably, mechanical isolation prevents insects from cross-pollinating the two species should they happen to flower at the same time.

If mating has taken place between two species, their gametes may still not combine. Molecular and chemical differences between species cause **gametic isolation,** in which the egg and sperm of different species are incompatible. In aquatic animals that release their eggs and sperm into the surrounding water simultaneously, interspecific fertilization is extremely rare (Fig. 19–4). The surface of the egg contains specific proteins that bind only to complementary molecules on the surface of sperm cells of the same species (see Chapter 49). A similar type of molecular recognition often occurs between pollen grains and the stigma (receptive surface of the female part of the flower) so that pollen does not germinate on the stigma of a different plant species.

Postzygotic barriers prevent gene flow when fertilization occurs

Fertilization sometimes occurs between gametes of two different species despite the existence of prezygotic barriers. When this happens, **postzygotic barriers** that increase the likelihood of reproductive failure come into play. Generally, the embryo of an interspecific hybrid spontaneously aborts. Embryonic development is a complex process requiring the precise inter-

Figure 19–2 **Behavioral isolation in bowerbirds.** Each bowerbird species has highly specialized courtship patterns that prevent its mating with another species. The male satin bowerbird (*shown*) constructs an enclosed place, or bower, of twigs to attract a female. Note the flowers and blue decorations that he has arranged at the entrance to his bower. (*Patti Murray*)

(a)

(b)

Figure 19–3 **Mechanical isolation in black and white sage.** Differences in floral structures between black and white sage allow them to be pollinated by different insects. Because the two species exploit different pollinators, they cannot interbreed. (*a*) The petal of the black sage functions as a landing platform for small bees. Larger bees cannot fit on this platform. (*b*) The larger landing platform and longer stamens of white sage allow pollination by larger California carpenter bees (a different species). If smaller bees land on white sage, their bodies do not brush against the stamens. (The upper part of the white sage flower has been removed.)

Figure 19–4 **Gametic isolation in sponges.** A basket sponge (*Xestospongia nutz*) releases a cloud of gametes into the water. If one of these gametes encounters a gamete of a different species biochemical differences will prevent their union. *(Marty Snyderman/Visuals Unlimited)*

action and coordination of many genes. Apparently the genes from parents belonging to different species do not interact properly in regulating the mechanisms for normal development. In this case, reproductive isolation occurs by **hybrid inviability.** For example, nearly all the hybrids die in the embryonic stage when the eggs of a bullfrog are fertilized artificially with sperm from a leopard frog. Similarly, in crosses between different species of irises, the embryos die before reaching maturity.

If an interspecific hybrid does live, it may not be able to reproduce. There are several reasons for this. Hybrid animals may exhibit courtship behaviors incompatible with those of either parental species and, as a result, they will not mate. More often, **hybrid sterility** occurs when problems during meiosis cause the gametes of an interspecific hybrid to be abnormal. Hybrid sterility is particularly common if the two parental species have different chromosome numbers. For example, a mule is the offspring of a female horse ($2n = 64$) and a male donkey ($2n = 62$) (Fig. 19–5). This type of union almost always results in sterile offspring ($2n = 63$) because chromosomal synapsis (the pairing of homologous chromosomes during meiosis) and segregation cannot occur properly.

Occasionally, a fertile interspecific F_1 hybrid develops that produces a second generation (F_2) from a cross between two

F_1 hybrids or a backcross between an F_1 hybrid and one of the parental species. The F_2 hybrid may exhibit **hybrid breakdown,** the inability of a hybrid to reproduce due to some defect. For example, hybrid breakdown in the F_2 generation of a cross between two sunflower species was 80%. In other words, 80% of the F_2 generation were defective in some way and could not reproduce successfully. Hybrid breakdown can also occur in the F_3 and later generations.

Figure 19–5 **Hybrid sterility in mules.** Mules are interspecific hybrids formed by mating a female horse with a male donkey. Although the mule *(center)* exhibits valuable characteristics of each of its parents, it is sterile. *(John Eastcott/Yva Momatiuk/Animals Animals)*

REPRODUCTIVE ISOLATION IS THE KEY TO SPECIATION

We are now ready to consider how entirely new species may arise from previously existing ones. The evolution of a new species, known as **speciation,** occurs when a population becomes reproductively isolated from other members of the species. Over time the gene pools of the two separated populations begin to diverge in genetic composition. When a population is sufficiently different from its ancestral species that no genetic exchange can occur between them, we say that speciation has occurred. Such a situation is thought to arise in two ways, through allopatric and sympatric speciation.

(a) Allopatric speciation

(b) Sympatric speciation (in plants)

Range of a single species

Geographical isolation of a small population

Natural selection, mutation, and genetic drift cause increased genetic divergence between the two populations

Later, if the two populations are reunited, they may be so different that they cannot interbreed

Species 1 range

Species 2 range

Area where the ranges of the two species overlap

Two species in the same area

Individuals from two species interbreed

Although sterile, a small population is maintained by asexual reproduction

Chromosome doubling occurs ⟶ Sexual reproduction now possible

New species

The new species may expand its range

Three species (the original species plus the allopolyploid)

Figure 19–6 Comparison of allopatric and sympatric speciation. (*a*) Allopatric speciation. (*b*) Sympatric speciation in plants. (The role of sympatric speciation in animal evolution is unclear.)

Long physical isolation and different selective pressures result in allopatric speciation

Speciation that occurs when one population becomes geographically separated from the rest of the species and subsequently evolves by natural selection and/or genetic drift is known as **allopatric speciation** (from the Greek *allo,* "different," and *patri,* "native land"). Allopatric speciation is thought to be the most common method of speciation, and the evolution of new animal species has been almost exclusively by allopatric speciation (Fig. 19–6*a*).

The geographical isolation required for allopatric speciation may occur in several ways. Earth's surface is in a constant state of change. Such change includes rivers shifting their courses; glaciers migrating; mountain ranges forming; land bridges developing that separate previously united aquatic populations; and large lakes diminishing into several smaller, geographically separated pools.

What might be an imposing geographical barrier to one species may be of no consequence to another. Birds and cattails, for example, do not become isolated when a lake subsides into smaller pools; birds can easily fly from one pool to another, and cattails disperse their pollen and fruits by air currents. Fish, on the other hand, are usually unable to cross the land barriers between the pools and so become reproductively isolated. In the Death Valley region of California and Nevada, large interconnected lakes formed during wetter climates of the last Ice Age. These lakes were populated by one or several species of pupfish. Over time, the climate became drier, and the large lakes dried up, leaving isolated pools. Presumably, each pool contained a small population of pupfish that gradually diverged from the common ancestral species by genetic drift and natural selection. Today, there are more than 20 species of pupfish, and many, such as the Devil's Hole pupfish *(Cyprinodon diabolis)* and the Owens pupfish *(C. radiosus),* are restricted to one or two isolated springs (Fig. 19–7).

Allopatric speciation also occurs when a small population migrates and colonizes a new area away from the range of the original species. This colony is geographically isolated from its parental species, and the small microevolutionary changes that accumulate in the isolated gene pool over many generations may eventually be sufficient to form a new species. Because islands provide the geographical isolation required for allopatric speciation, they offer excellent opportunities to study this mechanism. The Galapagos Islands and the Hawaiian Islands, for example, probably were originally colonized by a few individuals of a few species. The hundreds of unique species presently found on each island presumably descended from these original colonizers (Fig. 19–8).

Speciation is more likely to occur if the original isolated population is small. Recall that genetic drift, including the

Figure 19–7 **Allopatric speciation of pupfish *(Cyprinodon).*** Shown is one of the more than 20 pupfish species that apparently evolved when larger lakes in southern Nevada dried up about 10,000 years ago, leaving behind small, isolated desert pools fed by springs. The pupfish's short, stubby body is characteristic of fish that live in springs; fish that live in larger bodies of water are more streamlined. *(Steinhart Aquarium, Tom McHugh/Photo Researchers, Inc.)*

Figure 19–8 **Allopatric speciation of the Hawaiian goose (the nene).** Nene (pronounced "nay–nay," *Branta sandvicensis*) are geese found only on volcanic mountains on the geographically isolated islands of Hawaii and Maui, which are some 4200 kilometers (2600 miles) from the nearest continent. Compared to other geese, the feet of Hawaiian geese are not completely webbed, their toenails are longer and stronger, and their foot pads are thicker; these adaptations enable Hawaiian geese to walk easily on lava flows. Nene are thought to have evolved from a small population of geese that originated in North America. *(M.J. Rauzon/VIREO)*

(a)

(b)

Figure 19-9 Allopatric speciation in progress. Some scientists consider the Kaibab squirrel (*a*) and the Abert squirrel (*b*) as distinct populations of the same species (*Sciurus aberti*). Because the Kaibab and Abert squirrels are reproductively isolated from each other, however, some scientists have classified the Kaibab squirrel as a different species (*S. kaibabensis*). (*a, Tom and Pat Leeson; b, Kent and Donna Dannen*)

founder effect, is more consequential in small populations (see Chapter 18). Genetic drift tends to result in rapid changes in allele frequencies in the small, isolated population. The divergence caused by genetic drift is further accentuated by the different selective pressures of the new environment to which the population is exposed.

The Kaibab squirrel is an example of allopatric speciation in progress

About ten thousand years ago, when the American Southwest was less arid, the forests in the area supported a tree squirrel with conspicuous tufts of hair sprouting from its ears. A small tree squirrel population living on the Kaibab Plateau of the Grand Canyon became geographically isolated when the climate changed, causing areas to the north, west, and east to become desert. Just a few miles to the south lived the rest of the squirrels, known as Abert squirrels, but the two groups were separated by the Grand Canyon. With changes over time in both its appearance and its ecology, the Kaibab squirrel is on its way to becoming a new species.

During its many years of geographical isolation, the small population of Kaibab squirrels has diverged from the widely distributed Abert squirrels in a number of ways. Perhaps most evident are changes in fur color. The Kaibab squirrel now has a white tail and a black belly, in contrast to the gray tail and white belly of the Abert squirrel (Fig. 19–9). It is not clear why these striking changes arose in Kaibab squirrels.

Porto Santo rabbits may be an example of extremely rapid allopatric speciation

Allopatric speciation has the potential to occur quite rapidly. Early in the 15th century, a small population of rabbits was released on Porto Santo, a small island off the coast of Portugal. Because there were no other rabbits or competitors and no predators on the island, the rabbits thrived. By the 19th century, these rabbits were markedly different from their European ancestors. They were only half as large (weighing slightly more than 500 grams, or 1.1 pounds), with a different color pattern and a more nocturnal lifestyle. Most significantly, attempts to mate Porto Santo rabbits with mainland European rabbits failed. Many biologists concluded that, within 400 years, an extremely short time in evolutionary history, a new species of rabbit had evolved.

Not all biologists agree that the Porto Santo rabbit is a new species. The objection stems from a more recent breeding experiment and is based on biologists' lack of a consensus about the definition of a species. In the experiment, newborn Porto Santo rabbits were raised by foster mothers of the wild Mediterranean rabbit. When they reached adulthood, these Porto Santo rabbits mated successfully with Mediterranean rabbits to produce healthy, fertile offspring. To some biologists, this experiment clearly demonstrated that Porto Santo rabbits are not a separate species but instead are an example of speciation in progress, much like the Kaibab squirrels just discussed. Other biologists think the Porto Santo rabbit is a separate species because it does not interbreed with other rabbits under natural conditions. They point out that the breeding experiment was successful only after the baby Porto Santo rabbits were raised under artificial conditions that probably modified their natural behavior.

Two populations diverge in the same physical location by sympatric speciation

Although geographical isolation is an important factor in many cases of evolution, it may not be an absolute requirement. In **sympatric speciation** (from the Greek *sym,* "together," and *patri,* "native land"), a new species develops within the same geographical region as the parental species (Fig. 19–6*b*). The divergence of two gene pools in the same geographical range is especially common in plants. The role of sympatric speciation in animal evolution, however, is unclear and, until recently, has been difficult to demonstrate in nature.

How does sympatric speciation occur in plants? As discussed earlier, the union of two gametes from different species rarely forms viable offspring; if offspring are produced, they are usually sterile. Before gametes form, meiosis occurs to reduce the chromosome number (see Chapter 9). For the chromosomes to be parceled correctly into the gametes, homologous chromosome pairs must come together (a process called *synapsis*) during prophase I. This cannot usually occur in interspecific hybrid offspring because the chromosomes are not homologous. However, if the chromosome number doubles *before* meiosis, then pairing of homologous chromosomes can take place. While not a common occurrence, this spontaneous doubling of chromosomes has been documented in a variety of plants and a few animals. It produces nuclei with multiple sets of chromosomes.

Polyploidy, the possession of more than two sets of chromosomes, is a major factor in plant evolution. When polyploidy occurs in conjunction with **hybridization** (sexual reproduction between individuals from different species), it is known as **allopolyploidy.** Allopolyploidy can produce a fertile interspecific hybrid because the polyploid condition provides the homologous chromosome pairs necessary for synapsis during meiosis. As a result, gametes may be viable (Fig. 19–10). An allopolyploid, that is, an interspecific hybrid produced by allopolyploidy, can reproduce with itself (self-fertilize) or with a similar individual. However, allopolyploids are reproductively isolated from both parents because their gametes have a different number of chromosomes than those of either parent.

If a population of allopolyploids (that is, a new species) becomes established, selective pressures cause one of three outcomes. One, the new species may be unable to compete successfully against species that are already established, and so it becomes extinct. Two, the allopolyploid individuals may assume a new role in the environment and so coexist with both parental species. Three, the new species may successfully compete with either or both of its parental species. If it has a combination of traits that confers greater fitness than one or both parental species for all or part of the original range of the parent(s), the hybrid species may replace the parent(s).

Although allopolyploidy is extremely rare in animals, it is considered a significant factor in the evolution of flowering plant species. As many as 80% of all flowering plant species are thought to be polyploids, and most of these are allopolyploids (see *Making the Connection: Polyploidy, Guard Cells, and Angiosperm Evolution* in Chapter 27). Moreover, allopolyploidy provides a mechanism for extremely rapid speciation. A single generation is all that is needed to form a new, reproductively isolated species. Allopolyploidy may explain the rapid appearance of many flowering plant species in the fossil record and their remarkable diversity (about 235,000 species) today.

The kew primrose *(Primula kewensis)* is an example of sympatric speciation that was documented at the Royal Botanic Gardens at Kew, England, in 1898 (Fig. 19–11). The interspecific hybrid of two primrose species, *P. floribunda* (2n =

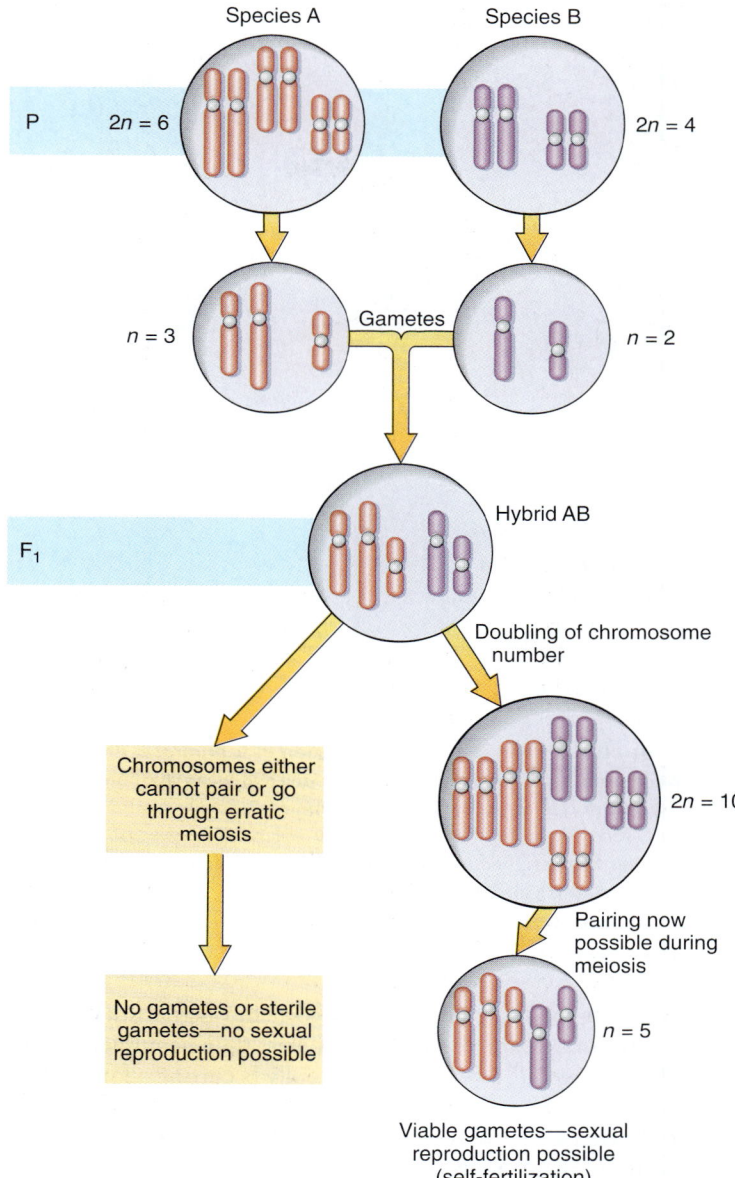

Figure 19–10 **Sympatric speciation by allopolyploidy in plants.** When two species (designated the P generation) successfully interbreed, the interspecific hybrid offspring (the F_1 generation) are almost always sterile *(bottom left)*. If the chromosomes double prior to the onset of meiosis, proper synapsis and segregation of the chromosomes can occur, and viable gametes can form *(bottom right)*. (Unduplicated chromosomes are shown for clarity.)

18) and *P. verticillata* (2n = 18), *P. kewensis* had a chromosome number of 18 but was sterile. Then, at three different times, it was reported to have spontaneously formed a fertile branch, which was an allopolyploid (2n = 36) that produced viable seeds of *P. kewensis*.

The mechanism of sympatric speciation has been experimentally verified for many plant species. One example is a

Figure 19–11 **Sympatric speciation of a primrose.** An allopolyploid primrose, *Primula kewensis,* arose in 1898 as an allopolyploid derived from the interspecific hybridization of *P. floribunda* and *P. verticillata.* Today *P. kewensis* is a popular houseplant. (The specific epithet *kewensis* was assigned in recognition of the species' place of origin, the Royal Botanic Gardens at Kew, England.)

Primula floribunda	Primula kewensis	Primula verticillata

group of species, collectively called hemp nettles, that occurs in temperate parts of Europe and Asia. One hemp nettle, *Galeopsis tetrahit* ($2n = 32$), is a naturally occurring allopolyploid thought to have formed by the hybridization of two species, *G. pubescens* ($2n = 16$) and *G. speciosa* ($2n = 16$). This process occurred in nature but was experimentally reproduced. *Galeopsis pubescens* and *G. speciosa* were crossed to produce F_1 hybrids, most of which were sterile. Nevertheless, both F_2 and F_3 generations were produced. The F_3 generation included a polyploid plant with $2n = 32$ that self-fertilized to yield fertile F_4 offspring that could not mate with either of the parental species. These allopolyploid plants had the same appearance and chromosome number as the naturally occurring *G. tetrahit.* When the experimentally produced plants were crossed with the naturally occurring *G. tetrahit,* a fertile F_1 generation was formed. Thus, the experiment duplicated the speciation process that occurred in nature.

Changing food preferences can cause sympatric speciation in animals

Biologists have observed the occurrence of sympatric speciation in animals, but its significance—how often it occurs and under what conditions—is still actively debated. Many examples involve parasitic insects and rely on genetic mechanisms other than polyploidy. For example, in the 1860s in the Hudson River Valley of New York, a population of fruit flies *(Rhagoletis pomonella)* parasitic on the small, red fruits of native hawthorn trees was documented to have switched to a new host, domestic apples, which had been introduced from Europe. Although the sister populations (hawthorn maggots and apple maggots) continue to occupy the same geographical area, no gene flow occurs between them because they eat, mate, and lay their eggs on different hosts. In other words, because the hawthorn and apple maggots are reproductively isolated from each other, they have effectively become separate species. Most entomologists, however, still recognize hawthorn and apple maggots as a single species.

In situations like this, it is thought that a mutation arises in an individual and spreads through a small group of insects by sexual reproduction. The mutation reproductively isolates the insects from the rest of the population by allowing them a different ecological opportunity—in this case, to parasitize a different host species. Additional mutations may occur that cause the sister populations to diverge even further. In 1988 biologists reported that hawthorn and apple maggots have genetic differences in three chromosomal regions and identified genetic markers associated with changes in the timing of fly development. Both hawthorn and apple larvae (maggots) tunnel out of the fruit before winter, drop to the ground, and burrow into the soil. However, hawthorn maggots emerge from the ground as adults later in the summer than apple maggots, a difference that further contributes to their reproductive isolation.

Biologists have studied the speciation of colorful fishes known as cichlids (pronounced "sik-lids") in several East African lakes. The different species of cichlids in a given lake have remarkably different eating habits. Some graze on algae; some consume dead organic material at the bottom of the lake; and others are predatory and eat plankton (microscopic aquatic organisms), insect larvae, and other cichlid species. Their food preferences are related to size (smaller cichlids consume plankton, for example), which in turn is related to mating preference (small, plankton-eating cichlids mate only with other small, plankton-eating cichlids). Recent DNA sequence data indicate that the cichlid species within each lake are more closely related to one another than they are to fishes in nearby lakes or rivers. These molecular data suggest that cichlid species evolved sympatrically rather than by repeated colonizations by fish populations in nearby rivers.

How rapidly did sympatric speciation occur in cichlids? In 1996 scientists published seismic and drill core data that suggest that the 500 endemic (that is, found nowhere else) cichlid species in Lake Victoria evolved in a remarkably short time—less than 12,400 years. This inference is based on evidence that Lake Victoria dried up completely during the late Pleistocene (about 12,400 years ago) when much of north and equatorial Africa was arid. It appears that the cichlids evolved

Explaining the Rapid Loss of Cichlid Diversity in Lake Victoria

HYPOTHESIS: Cichlids are disappearing from Lake Victoria because of changes in their habitat.

METHOD: Take field measurements in 22 different locations in Lake Victoria. Verify these field observations with laboratory experiments.

RESULTS: Increasing turbidity of Lake Victoria's water is preventing females of cichlid species from choosing mates of their own species, which they do based on color variations among the males of different species. As a result, females are interbreeding with males of other species, and the number of cichlid species declines as the distinctive gene pools of different species are merging.

CONCLUSION: Cichlid diversity will continue to decline unless effective measures are taken to reduce pollution in Lake Victoria immediately.

The hundreds of small, colorful cichlid species that diversified in Lake Victoria have always intrigued evolutionary biologists. However, many of these species are rapidly disappearing, and about half of the 500 species that biologists estimate lived in Lake Victoria in 1978 are now extinct. The Nile perch, a voracious predator that was deliberately introduced into the lake in 1960 to stimulate the local fishing economy and whose population exploded during the 1980s, has been blamed for the extinction of most of the cichlid species that have disappeared.

Biologists from the University of Leiden in the Netherlands observed, however, that species were also disappearing in habitats of the lake where the Nile perch is known to have little or no impact. Ole Seehausen and his colleagues measured environmental features at 22 localities with such habitats and found strong correlations between bright and diverse male colors, the number of species, and clear, well lit water, that is, the better the light in a given part of the lake, the more colorful the males and the greater the number of species present at a specific site.* Each species has its own distinct male coloration, which is used by females of that species to choose mates—an example of sexual isolation (discussed earlier in the chapter). In clear water, members of each species do not interbreed with members of other species. (**Sexual selection,** which is choosing a mate based on its color or some other characteristic, is discussed in Chapter 50.)

*Seehausen, O., J.J.M. van Alphen, and F. Witte, "Cichlid Fish Diversity Threatened by Eutrophication That Curbs Sexual Selection." *Science,* Vol. 277, 19 Sept. 1997.

As nearby forests have been cut, soil erosion has made the water of Lake Victoria more turbid (cloudy). Agricultural practices in the area have also contributed fertilizer as well as sediment pollution. Water transparency in deep lake water has decreased from an average of 6.8 meters in the 1920s to 2.2 meters in the 1990s, whereas water transparency in shallow areas along the shoreline has decreased from 3 meters to 1.5 meters in the past decade.

Seehausen demonstrated that as the water became more turbid from pollution, light could not penetrate as effectively, and the spectrum of visible colors became narrower. He hypothesized that, as a result, females could not distinguish males of their own species from males of closely related species. As discussed in the chapter, cichlids are young species, having evolved during the past 12,000 years. Their relative youth means that these species may not have yet evolved additional reproductive isolating mechanisms other than mate preference based on color. It also means that closely related species can be expected to interbreed without loss of fertility. With increasing turbidity, males lost their bright colors because they were no longer favored by sexual selection, and females have mated with males of other species, producing fertile, healthy offspring. Thus, males are losing their bright colors, and many species are being replaced by one or a few hybrids.

This hypothesis was verified in laboratory experiments. When aquaria were well lit, females of two closely related cichlid species (one blue, the other one red) consistently chose mates of their own species over those of the other species. When biologists blocked the light to simulate the turbid conditions in parts of Lake Victoria, however, females often chose males of the other species.

after the climate became wetter and Lake Victoria refilled. If future scientific experiments substantiate this conclusion, it means that the evolution of Lake Victoria's cichlids is the fastest known for such a large number of vertebrate species. (See *On the Cutting Edge: Explaining the Rapid Loss of Cichlid Diversity in Lake Victoria* for a discussion of how human activities have altered the evolutionary mechanism that maintains cichlid diversity.)

Reproductive isolation breaks down in hybrid zones

When two populations have significantly diverged as a result of geographical separation, there is no easy way to determine if the speciation process is complete (recall the controversy about whether Porto Santo rabbits are a separate species or a geographical race). If such populations, subspecies, or species come into contact, they may hybridize where they meet, forming a **hybrid zone,** or area of overlap in which interbreeding occurs. Hybrid zones are typically narrow, presumably because the hybrids are not well adapted for either parental environment, and the hybrid population is typically very small compared to the parental populations.

On the Great Plains of North America, red-shafted and yellow-shafted flickers (a type of woodpecker) meet and interbreed. The red-shafted flicker, named for its red underwings and tail, is found in the western part of North America from the Great Plains to the Pacific Ocean. The yellow-shafted flicker, which has yellow underwings and tail, ranges east of the Rockies. Hybrid flickers, which form a stable hybrid zone from Texas to southern Alaska, are varied in appearance, although many have orange-colored underwings and tails.

Biologists are not in agreement about whether the red-shafted and yellow-shafted flickers are separate species or geographical races (that is, subspecies; see Chapter 22). According to the biological species concept, if red-shafted and yellow-shafted flickers are two species, they should maintain their reproductive isolation. On the other hand, the flicker hybrid zone has not expanded, that is, the two types of flickers have maintained their distinctiveness and have not rejoined into a single, freely interbreeding population.

The study of hybrid zones has made important contributions to what is known about speciation. As in other fields of science, disagreements and differences of opinion are an important part of the scientific process because they stimulate new ideas, hypotheses, and experimental tests that expand our base of scientific knowledge.

EVOLUTIONARY CHANGE CAN OCCUR RAPIDLY OR GRADUALLY

Does the fossil record provide clues about how rapidly new species arise? Biologists have long recognized that the fossil record lacks many transitional forms; the starting points (ancestral species) and the end points (new species) are present, but the intermediate stages in the evolution from one species to another are often "missing." This observation has traditionally been explained by the incompleteness of the fossil record. Biologists have attempted to fill in the missing parts, much as a writer might fill in the middle of a novel when the beginning and end are already there.

Two different models, punctuated equilibrium and gradualism, have been developed to explain evolution as observed in the fossil record (Fig. 19–12). The **punctuated equilibrium** model was proposed by paleontologists who question whether the fossil record really is as incomplete as it initially appeared. First advanced by Stephen Jay Gould and Niles Eldredge in 1972, the punctuated equilibrium model suggests that the fossil record accurately reflects evolution as it actually occurs in the history of a species, with long periods of **stasis**

(no evolutionary change) punctuated, or interrupted, by short periods of rapid speciation that are perhaps triggered by changes in the environment. Thus, speciation normally proceeds in "spurts." These relatively short periods of active evolution (for example, one hundred thousand years) are followed by long periods (for example, two million years) of stability.

With punctuated equilibrium, speciation can occur in a relatively short period of time. Keep in mind, however, that a "short" amount of time for speciation may be thousands of years. Such a span is short when compared with the several million years of a species' existence. Biologists who support the idea of punctuated equilibrium emphasize that sympatric speciation and even allopatric speciation can occur in such relatively short periods. Punctuated equilibrium accounts for the abrupt appearance of a new species in the fossil record, with little or no evidence of intermediate forms. That is, proponents think that few transitional forms appear in the fossil record because few transitional forms occurred during speciation.

In contrast, the traditional view of evolution espouses the **gradualism** model, in which evolution proceeds continuously over long periods of time. Gradualism is rarely observed in the fossil record because the record is incomplete. (Recall from Chapter 17 that the conditions required for fossil formation are quite precise. The vast majority of organisms decompose when they die, leaving no trace of their existence.) Occasionally, a complete fossil record of transitional forms is discovered and cited as a strong case for gradualism. The gradualism model maintains that populations slowly diverge from one another by the gradual accumulation of adaptive characteristics within each population. These adaptive characteristics accumulate as a result of different selective pressures encountered in different environments.

The abundant fossil evidence of long periods with no change in a species has been used to argue against the gradualism model of evolution. Gradualists, however, maintain that any periods of stasis evident in the fossil record are the result of stabilizing selection (see Chapter 18). They also emphasize that stasis in fossils is deceptive because fossils do not reveal all aspects of evolution. While fossils display changes in external structure and skeletal structure, genetic changes in physiology, internal structure, and behavior—all of which also represent evolution—are not evident. Gradualists recognize rapid evolution only when strong directional selection occurs.

Thus, while biologists agree that natural selection is the primary mechanism for evolution, they may disagree as to the timing, or pace, of evolution during a given species' existence. Many biologists embrace both models to explain the fossil record and contend that the pace of evolution may be abrupt in certain instances and gradual in others, and that neither punctuated equilibrium nor gradualism exclusively characterize the complex changes associated with evolution. Other biologists do not view the distinction between punctuated equilibrium and gradualism as real. They suggest that genetic changes occur gradually and at a roughly constant pace and

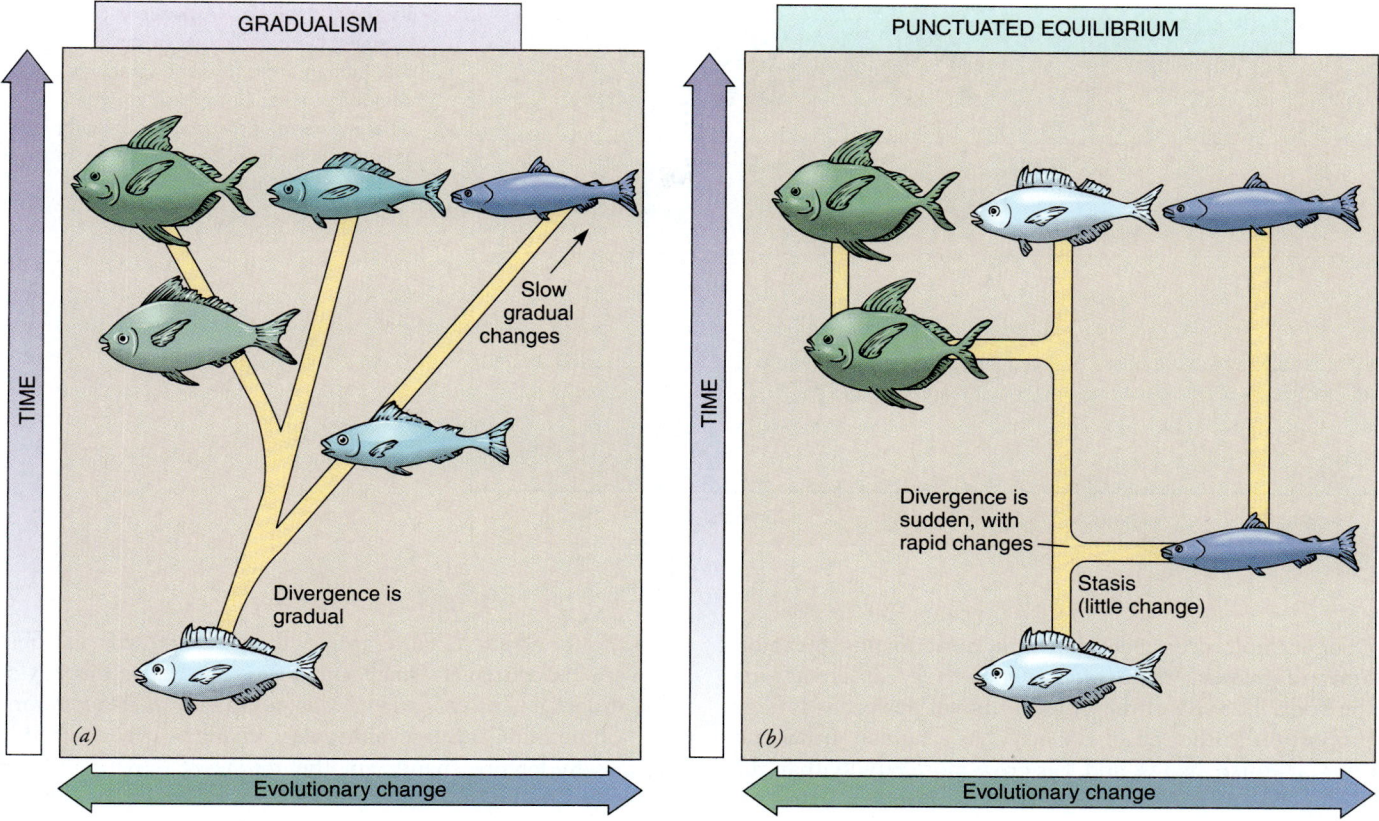

Figure 19–12 **Gradualism and punctuated equilibrium.** (*a*) In gradualism a slow, steady change in species occurs over time. (*b*) In punctuated equilibrium, long periods of stasis are interrupted by short periods of rapid speciation.

that the majority of these mutations do not affect speciation. When the mutations that *do* affect speciation occur, they are dramatic and produce a pattern consistent with the punctuated equilibrium model.

MACROEVOLUTION INVOLVES MAJOR EVOLUTIONARY EVENTS

Macroevolution refers to dramatic changes that occur over long time spans in evolution. One concern of macroevolution is to explain evolutionary novelties, which are large phenotypic changes such as the appearance of wings with feathers during the evolution of birds from reptiles. These phenotypic changes are so great that the new species possessing them are assigned to different genera or higher taxonomic categories. Studies of macroevolution also attempt to discover and explain major changes in species diversity through time, such as occur during adaptive radiation, when many species appear, and mass extinction, when many species disappear. Thus, evolutionary novelties, adaptive radiation, and mass extinction are important aspects of macroevolution.

Evolutionary novelties originate through modifications of preexisting structures

New designs arise from structures already in existence. A change in the basic pattern of an organism can produce something unique, such as wings on insects, flowers on plants, and feathers on birds. Usually these evolutionary novelties are variations of some preexisting structures, called **preadaptations,** that originally fulfilled one role, but were subsequently modified in a way that was adaptive for a different role. Feathers, which evolved from reptilian scales and may have originally provided thermal insulation in primitive birds, represent a preadaptation for flight. That is, with gradual modification, feathers evolved to function in flight as well as to fulfill their original thermoregulatory role. Similarly, the mammalian middle ear bones originated from modified jaw bones of reptiles.

How do such evolutionary novelties originate? Many are probably due to changes during development, which is the orderly sequence of events that occur as an organism grows and matures. Regulatory genes may exert control over hundreds of other genes (see Chapter 16), and very slight genetic changes in regulatory genes could ultimately cause major structural changes in the organism.

| Newborn | 2 years | 5 years | 15 years | Adult |

Figure 19–13 Allometric growth in humans. Different stages in the growth of a human are drawn the same size to demonstrate varied rates of growth for different parts of the body. For example, as humans develop, their legs grow more rapidly than their heads.

For example, during development, most organisms exhibit **allometric growth,** varied rates of growth for different parts of the body. The size of the head in human newborns is large in proportion to the rest of the body. As a human grows and matures, its torso, hands, and legs grow more rapidly than the head (Fig. 19–13). Allometric growth is found in many organisms, including the male fiddler crab with its single, oversized claw, and the ocean sunfish with its enlarged tail (Fig. 19–14). If growth rates are altered even slightly, drastic changes in the shape of an organism may result, changes that may or may not be adaptive. For example, allometric growth may help explain the extremely small and relatively useless forelegs of the dinosaur *Tyrannosaurus rex* as compared to its ancestors.

Sometimes novel evolutionary changes occur when a species undergoes changes in the *timing* of development in comparison to its ancestor. Consider, for example, the changes that would occur if juvenile characteristics were retained in the adult stage, a phenomenon known as **paedomorphosis.** Adults of some salamander species have gills, a feature found only in the larval (immature) stages of other salamanders. Retention

Tail

approx. 1 mm

Newly hatched ocean sunfish

(a)

Adult ocean sunfish

(b)

approx. 3.4 meters

(c)

Figure 19–14 Allometric growth in the ocean sunfish. The tail end of an ocean sunfish grows faster than the head end, resulting in the unique shape of the adult. (*a*) A newly hatched ocean sunfish, only one mm long, has an extremely small tail. (*b*) This allometric transformation can be visualized by drawing rectangular coordinate lines through a picture of the juvenile fish and then changing the coordinate lines mathematically. (*c*) An ocean sunfish, also known as a mola, swims off the coast of southern California. The adult ocean sunfish is 3.4 meters (11 feet) long and weighs one metric ton (2206 pounds). *(Richard Herrmann)*

Figure 19–15 **Paedomorphosis in a salamander.** An adult axolotl salamander *(Ambystoma mexicanum)* retains the juvenile characteristic of external gills. This evolutionary novelty allows the axolotl to remain permanently aquatic and to reproduce without developing typical adult characteristics. *(Jane Burton/Bruce Coleman, Inc.)*

of gills throughout life obviously alters the salamander's behavioral and ecological characteristics (Fig. 19–15). Perhaps such salamanders succeeded because they had a selective advantage over "normal" adult salamanders, that is, by remaining aquatic, they did not have to compete for food with the terrestrial adult forms of related species. The gilled forms also escaped the typical predators of terrestrial salamanders (although they had other predators in their aquatic environment).

Adaptive radiation is the diversification of an ancestral species into many species

Adaptive radiation is the evolutionary diversification of many related species from one or a few ancestral species in a relatively short period of time. The concept of adaptive zones was developed to help explain why adaptive radiations take place. **Adaptive zones** are new ecological opportunities that were not exploited by an ancestral organism. At the species level, an adaptive zone is essentially identical to one or more similar *ecological niches* (the functional roles of species within a community; see Chapter 52). Examples of adaptive zones include nocturnal flying to catch small insects, grazing on grass while migrating across a savanna, and swimming at the ocean's surface to filter out plankton. When many adaptive zones are empty, as was the case of Lake Victoria when it refilled some 12,000 years ago (discussed earlier in the chapter), available

species such as the cichlids are able to rapidly diversify and exploit them.

Because islands have fewer species than do mainland areas of similar size, latitude, and topography, vacant adaptive zones are more common on islands than on continents. Consider the Hawaiian honeycreepers, a group of related birds found on the Hawaiian Islands. When the honeycreeper ancestors reached Hawaii, few other birds were present. The succeeding generations of honeycreepers quickly diversified into many new species and, in the process, occupied the many available adaptive zones that are normally occupied by finches, honey-eaters, creepers, and woodpeckers on the mainland. The diversity of their bills, like those of Galapagos finches (see Chapter 17), is a particularly good illustration of adaptive radiation (Fig. 19–16). Some honeycreeper bills are curved to extract nectar out of tubular flowers, for example, whereas others are short and thickened for ripping away bark in search of insects.

Another example of vacant adaptive zones involves the Hawaiian silverswords, 28 species of closely related plants found only on the Hawaiian Islands. When the silversword ancestor, a California plant related to daisies, reached the Hawaiian Islands, many diverse environments, such as exposed lava flows, dry woodlands, moist forests, and bogs, were present. The succeeding generations of silverswords quickly diversified, occupying the many adaptive zones available to them. The diversity in their leaves, which changed during the course of natural selection to enable different populations to adapt to various levels of light and moisture, is a particularly good illustration of adaptive radiation (Fig. 19–17). Leaves of silverswords that are adapted to shady moist forests are large, for example, whereas those of silverswords living in exposed dry areas are small. The leaves of silverswords living on exposed volcanic slopes are covered with dense silvery hairs that may reflect some of the intense ultraviolet radiation off the plant.

Adaptive radiation appears to be more common during periods of major environmental change, but it is difficult to determine if these changes actually induce adaptive radiation. It is possible that major environmental change indirectly affects adaptive radiation by increasing the rate of extinction. Extinction produces empty adaptive zones, which provide new opportunities for those species that remain. Mammals, for example, had existed as small nocturnal insectivores (insect eaters) for millions of years before undergoing adaptive radiation leading to the modern mammalian orders. This radiation was presumably triggered by the extinction of the dinosaurs. Mammals diversified and exploited a variety of adaptive zones relatively soon after the dinosaurs' demise. Flying bats, running gazelles, burrowing moles, and swimming whales all originated from the small, insect-eating, ancestral mammals.

The appearance of novel features is usually associated with major periods of adaptive radiation. For example, shells and skeletons may have been the evolutionary novelties responsible for a period of adaptive radiation at the beginning of the Paleozoic era (see Chapter 20), in which most animal phyla, living and extinct, appeared. Care must be taken in interpret-

Figure 19–16 Adaptive radiation in the Hawaiian honeycreepers. Compare the various beak shapes and methods of obtaining food. Many honeycreeper species are now extinct as a result of human activities, including the destruction of habitat and the introduction of predators such as rats, dogs, and pigs.

Rips away bark to find insects
Maui parrot bill

Kauai

Sips flower nectar
'I'iwi

Oahu

Forages among leaves and branches
Maui creeper

Maui

Chisels holes in bark to get insects
Akiapolaau

Extinct
Sipped flower nectar
Black mamo

Hawaii

Extinct
Habits unknown
Ula-ai-hawane

Picks food from cracks in the bark
Akialoa

Feeds on snails and invertebrates
Poo-uli

Sips flower nectar
Crested honeycreeper

Sips flower nectar
Apapane

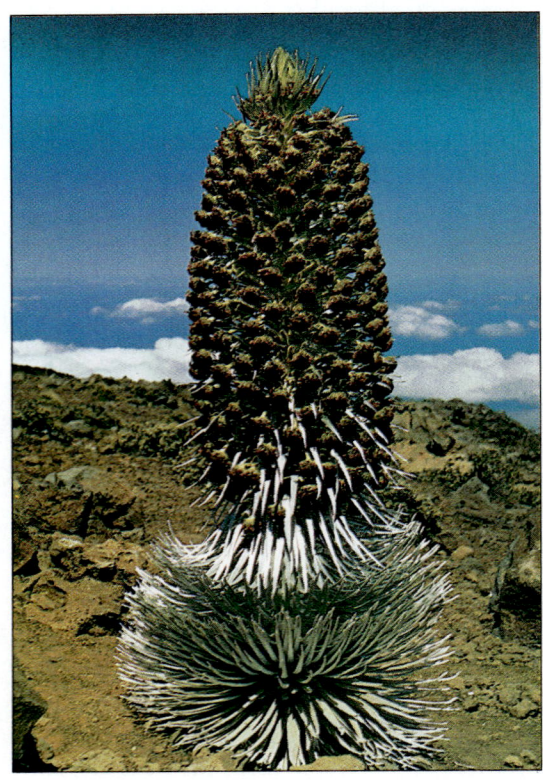

(a)

(b)

Figure 19–17 Adaptive radiation in the Hawaiian silverswords. The silversword ancestor was a California plant similar to the daisy. (*a*) A silversword species (*Argyroxyphium sandwicense*) found only in Haleakala National Park on the island of Maui. (*b*) Leaf shapes of the 28 Hawaiian silverswords, all at a scale of × 0.25. Leaf color, which varies considerably among the species, is not shown. (*a*, H. Gordon Morris, University of Tennessee; *b*, "Plant Speciation and the Founder Principle" by G.L. Stebbins, from Genetics, Speciation, and the Founder Principle, *edited by Luther V. Giddings, Anderson et al. Copyright 1989 by Oxford University Press, Inc. Used by permission of Oxford University Press, Inc.)*

ing a cause-and-effect relationship between the appearance of a novel feature and adaptive radiation, however. It is tempting to take a simplistic approach and state, for example, that the evolution of the flower facilitated the adaptive radiation of thousands of species of flowering plants. It is true that the flowering plants diversified after the evolution of the flower, which may have presented a more competitive method of sexual reproduction because it permitted pollination by insects and other animals. However, adaptive radiation in the flowering plants may instead be a consequence of other adaptations that evolved. (Chapter 27 discusses other flowering plant adaptations as well as their highly successful mode of reproduction.)

Extinction is an important aspect of evolution

Extinction, the end of a lineage, occurs when the last individual of a species dies. The loss is permanent, for once a species is extinct it can never reappear. Extinctions have occurred continually since the origin of life; by one estimate, only one species is alive today for every 2000 that have become extinct. Extinction is the eventual fate of all species, in the same way that death is the eventual fate of all individual organisms.

Although extinction has a negative short-term impact on biological diversity, it can facilitate evolution over a period of thousands to millions of years. As mentioned previously, when species become extinct, their adaptive zones become vacant. Consequently, those organisms still living are presented with new opportunities for speciation and can diverge to fill the unoccupied niches. In other words, the extinct species are eventually replaced by new species.

During the long history of life, extinction appears to have occurred at two different rates. The continuous, low-level extinction of species is sometimes called **background extinction.** In contrast, five or possibly six times during Earth's history, **mass extinctions** of numerous species and higher taxa have taken place in both terrestrial and marine environments. The most recent mass extinction, which occurred about 65 million years ago, killed off the dinosaurs (Fig. 19–18). The time span over which a mass extinction occurred may have lasted several million years, but that is relatively short compared with the 3.5 billion years or so of Earth's history of life. Each period of mass extinction has been followed by a period of adaptive radiation of some of the surviving groups.

The causes of past episodes of mass extinction are not well understood. Both environmental and biological factors seem to have been involved. Major changes in climate could have adversely affected those plants and animals that lacked the genetic flexibility to adapt. Marine organisms, in particular, are adapted to a steady, unchanging climate. If Earth's tempera-

Figure 19–18 **Mass extinction of the archosaurs.** At the end of the Cretaceous period, approximately 65 million years ago, a mass extinction of many organisms, including the dinosaurs, occurred. Most of the archosaurs (one of five main groups of reptiles) became extinct. The only lines to survive were crocodiles and birds, both of which are archosauran descendants.

ture were to increase or decrease by just a few degrees overall, many marine species would probably perish.

It is also possible that past mass extinctions were due to changes in the environment induced by catastrophes. If a large comet or small asteroid collided with Earth, for example, the dust ejected into the atmosphere on impact could have blocked much of the sunlight. In addition to disrupting the food chain by killing many plants (and therefore terrestrial animals), this event would have lowered Earth's temperature, leading to the death of many marine organisms. Evidence that the extinction of dinosaurs was caused by an extraterrestrial object's collision with Earth continues to accumulate (see Chapter 20).

Biological factors also trigger extinction. Competition among species may lead to the extinction of those species that cannot compete effectively. The human species, in particular, has had a profound impact on the rate of extinction. The habitats of many animal and plant species have been altered or destroyed by humans, and habitat destruction can result in a species' extinction. The U. N. Global Biodiversity Assessment, released in late 1995 and based on the work of about 1500 scientists from around the world, estimates that more than 31,000 plant and animal species are currently threatened with extinction. As many as one-fifth of all species may become extinct within the next 30 years. The Center for Plant Conser-

Figure 19–19 **Continential drift.** (a) The supercontinent Pangaea, about 240 million years BP (before present). (b) Breakup of Pangaea into Laurasia (Northern Hemisphere) and Gondwana (Southern Hemisphere), 120 million years BP. (c) Further separation of land masses, 60 million years BP. Note that Europe and North America were still joined and that India was a separate land mass. (d) The continents today. (e) Projected positions of the continents in 50 million years.

(a)   240 million years ago (Triassic period)

(b)   120 million years ago (Cretaceous period)

(c)   60 million years ago (early Tertiary period)

(d)   Today

(e)   50 million years from now

vation, for example, estimates that about 4000 native plant species are of conservation concern in the United States; this number represents about 20% of U. S. plant species. Some biologists think that we have entered the greatest mass extinction episode in Earth's history. (Extinction is discussed further in Chapter 55.)

Earth's geological history is related to macroevolution and biogeography

In 1915 the German scientist Alfred Wegener, who had noted a correspondence between the geographical shapes of South America and Africa, proposed that all the land masses had at one time been joined into one huge supercontinent, which he called Pangaea (Fig. 19–19*a*). He further suggested that Pangaea had subsequently broken apart and that the various land masses had separated in a process known as **continental drift.** Wegener did not know of any mechanism that could have caused continental drift, and so his idea, although debated initially, was largely ignored.

In the 1960s, scientific evidence accumulated that provided the explanation for continental drift. Earth's crust is composed of seven large plates (plus a few smaller ones) that float on the mantle, which is the mostly solid[2] layer of Earth lying beneath the crust and above the core. The land masses are situated on some of these plates. As the plates move, the continents change their relative positions (Fig. 19–19*b*, *c*, *d*, and *e*). The movement of the crustal plates is termed **plate tectonics.**

Any area where two plates meet is a site of intense geological activity. Earthquakes and volcanoes are common in such a region. Both San Francisco, noted for its earthquakes, and the Mount Saint Helens volcano are situated where two plates meet. If land masses lie on the edges of two adjacent plates, mountains may form. The Himalayas formed when the plate carrying India rammed into the plate carrying Asia. When two plates grind together, one of them is sometimes buried under the other in a process known as subduction. When two plates move apart, a ridge of lava forms between them. The Atlantic Ocean is getting larger because of the expanding zone of lava along the mid-Atlantic ridge, where two plates are separating.

Knowledge that the continents were at one time connected and have since drifted apart is useful in explaining the geographical distribution of plants and animals, or **biogeography** (Fig. 19–20). Likewise, continental drift has played a major role in the evolution of different organisms. When Pangaea originally formed during the late Permian period, it brought together terrestrial species that had evolved separately from one another, leading to competition and some extinctions. Marine life was adversely affected, in part because, with the continents

[2]Most of the rock in the upper portion of the mantle is solid, although one or two percent is melted. Because of its higher temperature, the solid rock of the mantle is more plastic than the solid rock of the lithosphere above it.

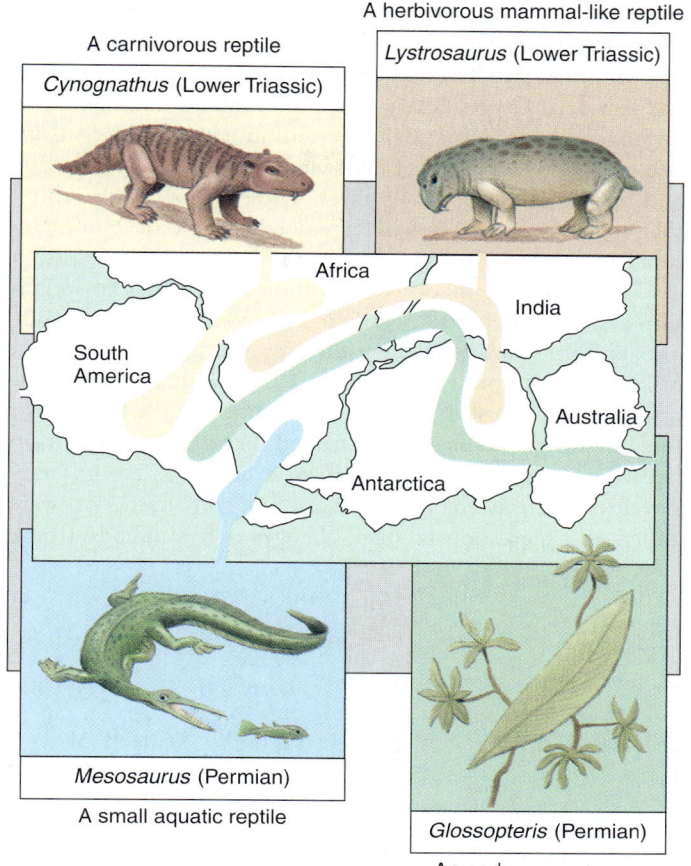

A carnivorous reptile
Cynognathus (Lower Triassic)

A herbivorous mammal-like reptile
Lystrosaurus (Lower Triassic)

Mesosaurus (Permian)
A small aquatic reptile

Glossopteris (Permian)
An early gymnosperm

Figure 19–20 **The distribution of four fossils on five continents.** These biogeographical data suggest that the continents were once joined. *(Adapted from Colbert)*

joined as one large mass, less coastline existed. (Because coastal areas are shallower, they contain high concentrations of marine species.) Pangaea separated into several land masses approximately 180 million years ago. As the continents began to drift apart, populations became geographically isolated in different environmental conditions and began to diverge along separate evolutionary pathways. As a result, the plants, animals, and other organisms of previously connected continents, South America and Africa for example, differ. Continental drift also caused gradual changes in ocean and atmospheric currents that have profoundly influenced the biogeography and evolution of organisms.

IS MICROEVOLUTION RELATED TO SPECIATION AND MACROEVOLUTION?

The concepts presented in Chapters 17 and 18 represent the **synthetic theory of evolution,** in which mutation provides the genetic variation on which natural selection acts. The synthetic theory of evolution combines Darwin's theory with important aspects of genetics. All aspects of the synthetic theory of evolution have been tested and verified at the population and subspecies levels. Many biologists contend that microevolutionary processes (natural selection, mutation, genetic drift, and gene flow) account for the genetic variation within species and for the origin of new species. These biologists also think that macroevolution can be explained by microevolutionary processes.

The synthetic theory of evolution as it relates to speciation and macroevolution is supported by a considerable body of data from many fields. Consider, for example, the evolution of amphibians from fish, which was a major macroevolutionary event in the history of vertebrates. Study of the few known fossil intermediates has demonstrated that the transition from aquatic fish to terrestrial amphibian occurred as evolutionary novelties, such as changes in the limbs and skull roof, were added as a succession of small changes over a period of nine

to fourteen million years. This time scale is sufficient to have allowed natural selection and other microevolutionary processes to have produced the novel characters.

Although few biologists doubt the role of natural selection and microevolution in generating specific adaptations, some question the *extent* of microevolution's role in the overall pattern of life's history. These biologists ask whether speciation and macroevolution have been dominated by microevolutionary processes or by external, chance events (an impact by an asteroid, for example). Chance events do not "care" about adaptive superiority but instead lead to the random extinction or survival of species. In the case of an asteroid impact, for example, those species that survive may do so because they were "lucky" enough to be in a protected environment at the time of impact. If chance events have been the overriding factor during life's history, then microevolution cannot be the exclusive explanation for the biological diversity we have today.

SUMMARY WITH KEY TERMS

I. According to the **biological species concept,** a **species** consists of groups of populations whose members freely interbreed in nature to produce fertile offspring, and do not freely interbreed with members of different species.

II. **Reproductive isolating mechanisms** restrict the gene flow between species.

   A. **Prezygotic barriers** are reproductive isolating mechanisms that prevent fertilization from taking place.

      1. **Temporal isolation** occurs when two species reproduce at different times of the day, season, or year.

      2. In **behavioral isolation,** distinctive courtship behaviors prevent mating between species.

      3. **Mechanical isolation** is due to incompatible structural differences in the reproductive organs of similar species.

      4. In **gametic isolation,** gametes from different species are incompatible due to molecular and chemical differences.

   B. **Postzygotic barriers** are reproductive isolating mechanisms that prevent gene flow after fertilization has taken place.

      1. **Hybrid inviability** is the death of interspecific embryos during development.

      2. **Hybrid sterility** prevents interspecific hybrids that survive to adulthood from reproducing successfully.

      3. **Hybrid breakdown** prevents the offspring of hybrids that survive to adulthood and successfully reproduce from reproducing beyond one or a few generations.

III. **Speciation** is the evolution of a new species from an ancestral population.

   A. **Allopatric speciation** occurs when one population becomes geographically isolated from the rest of the species and subsequently diverges.

   B. **Sympatric speciation** does not require geographical isolation.

      1. Sympatric speciation in plants results almost exclusively from **allopolyploidy,** in which a **polyploid** individual (one with more than two sets of chromosomes) is derived from two species.

      2. Sympatric speciation occurs in animals, but how often it occurs and under what conditions remain to be determined. Sympatric

speciation in animals relies on genetic mechanisms other than polyploidy.

   C. A **hybrid zone** is an area in which two related populations, subspecies, or species meet and interbreed. Hybrid zones are typically narrow, presumably because the hybrids are not well adapted for either parental environment.

IV. The pace of evolution is currently being debated.

   A. According to the **punctuated equilibrium** model, evolution of species proceeds in spurts. Short periods of active speciation intersperse long periods of stasis.

   B. According to the **gradualism** model, populations slowly diverge from one another by the accumulation of adaptive characteristics within a population.

V. **Macroevolution** refers to dramatic evolutionary changes that occur over long time spans.

   A. Macroevolution includes the appearance of evolutionary novelties, which are phenotypic changes so great that the new species possessing them are assigned to different genera or higher taxonomic categories.

      1. Evolutionary novelties may be due to changes during development. Slight genetic changes in regulatory genes, for example, could ultimately cause major structural changes in the organism.

      2. Evolutionary novelties may originate from **preadaptations,** structures that originally fulfilled one role but changed in a way that was adaptive for a different role.

      3. Changes in **allometric growth,** varied rates of growth for different parts of the body, result in overall changes in the shape of an organism.

      4. **Paedomorphosis,** the retention of juvenile characteristics in the adult, can occur due to changes in the timing of development.

   B. **Adaptive radiation** is the process of diversification of an ancestral species into many new species.

   C. **Extinction** is the death of a species. When species become extinct, the **adaptive zones** that they occupied become vacant, allowing other species to evolve and fill them.

1. Two populations belong to the same species if (a) their members freely interbreed in nature (b) individuals from the two populations produce fertile offspring (c) their members do not interbreed with individuals of different species (d) a and c are correct (e) a, b, and c are correct
2. A prezygotic barrier prevents (a) the union of egg and sperm (b) reproductive success by an interspecific hybrid (c) the development of the zygote into an embryo (d) allopolyploidy from occurring (e) changes in allometric growth
3. The reproductive isolating mechanism in which two closely related species live in the same geographical area but reproduce at different times is (a) temporal isolation (b) behavioral isolation (c) mechanical isolation (d) gametic isolation (e) hybrid inviability
4. Interspecific hybrids, if they live, are (a) always sterile (b) always fertile (c) usually sterile (d) usually fertile (e) never sterile
5. The first step leading to allopatric speciation is (a) hybrid inviability (b) hybrid breakdown (c) adaptive radiation (d) geographical isolation (e) paedomorphosis
6. Sympatric speciation (a) is most common in animals (b) does not require geographical isolation (c) accounts for the evolution of the Hawaiian nene (d) involves the accumulation of gradual genetic changes (e) always takes thousands of years
7. Which of the following evolutionary processes is associated with allopolyploidy? (a) gradualism (b) allometric growth (c) sympatric speciation (d) mass extinction (e) preadaptation
8. According to punctuated equilibrium model (a) populations slowly diverge from one another (b) evolution of species occurs in spurts interspersed with long periods of stasis (c) evolutionary novelties originate from preadaptations (d) reproductive isolating mechanisms restrict the gene flow between species (e) the fossil record, being incomplete, does not accurately reflect evolution as it actually occurred
9. The evolutionary conversion of reptilian scales into a bird's feathers is an example of (a) allometric growth (b) paedomorphosis (c) gradualism (d) hybrid breakdown (e) preadaptation
10. Adaptive radiation is common following a period of mass extinction, probably because (a) the survivors of a mass extinction are remarkably well adapted to their environment (b) the unchanging environment following a mass extinction drives the evolutionary process (c) many adaptive zones are empty (d) many ecological niches are filled (e) the environment induces changes in the timing of development for many species
11. Adaptive radiations do not appear to have ever occurred (a) on isolated islands (b) in birds such as honeycreepers (c) in fishes such as cichlids (d) in environments with many existing species

R E V I E W   Q U E S T I O N S

1. Compare the biological species concept with the morphological species concept.
2. Give an example of each of the following: (a) temporal isolation; (b) behavioral isolation; (c) mechanical isolation; (d) gametic isolation.
3. Describe the three types of postzygotic barriers and give an example of each.
4. Identify at least five geographical barriers that might lead to allopatric speciation.
5. Explain how hybridization and polyploidy can cause a new plant species to form in as little as one generation.
6. If you were in a debate and had to support the gradualism model, what would you say? How would you support the punctuated equilibrium model? Are these two ideas mutually exclusive?
7. Why are evolutionary novelties a concern of scientists studying macroevolution?
8. Give an example of each of the following: (a) preadaptation; (b) allometric growth; (c) paedomorphosis.
9. What roles do extinction and adaptive radiation play in macroevolution?
10. How does continental drift occur?
11. Explain why fossils of *Mesosaurus*, an extinct reptile that could not swim across open water, are found in the southern parts of both Africa and South America.

Y O U   M A K E   T H E   C O N N E C T I O N

1. Why is allopatric speciation more likely to occur if the original isolated population is small?
2. Using the definition of the biological species concept given in the chapter introduction, is the Porto Santo rabbit an example of a speciation event that occurred in historical times, or is it an example of speciation in progress? Explain your answer.
3. Based on what you have learned in the chapter, hypothesize what the common ancestor of the more than 20 species of desert pupfish looked like. (*Hint:* the ancestral species lived in one or more large lakes.) How could you test your hypothesis?
4. Could hawthorn and apple maggots be considered an example of assortative mating (discussed in Chapter 18)? Explain your answer.
5. Since extinction is a natural process, should humans be concerned about the current mass extinctions that we are causing? Why or why not?

R E C O M M E N D E D   R E A D I N G S

Arnold, M.L. "Natural Hybridization and Louisiana Irises: Defining a Major Factor in Plant Evolution." *BioScience,* Vol. 43, No. 3, Mar. 1994. Interspecific hybridization appears to be a significant factor in the evolution of the Louisiana irises.

Campbell, D.G. "Splendid Isolation in Thingvallavatn." *Natural History,* Jun. 1996. Examines evolution in action in this 10,000-year-old lake in southwestern Iceland.

Dalziel, I.W.D. "Earth before Pangaea." *Scientific American,* Vol. 272, No. 1,

Jan. 1995. This article discusses continental drift that occurred *before* Pangaea formed some 260 million years ago.

Eldredge, N. *Reinventing Darwin: The Great Debate at the High Table of Evolutionary Theory.* Wiley, New York, 1995. A lively, if partisan, survey of the intense debate in evolutionary theory stimulated by punctuated equilibrium and other ideas. (Recall that Niles Eldredge was one of the first to advance the punctuated equilibrium model.)

Jackson, J. and A. Cheetham. "On the Importance of Doing Nothing." *Natural History,* Jun. 1994. A study of fossil bryozoa supports the punctuated equilibrium model.

Mayr, E. *Animal Species and Evolution.* Harvard University Press, Cambridge, 1963. A classic on animal evolution by the biologist who introduced the biological species concept and is a strong proponent of the allopatric speciation model.

Milner, R. "Ernst Mayr at 93." *Natural History,* May 1997. An interview with one of the 20th century's greatest biologists.

Morell, V. "The Life Machine." *Earth,* Aug. 1997. Earth scientists think that the Amazon's dynamic geological history may be responsible for the evolution of the region's remarkably diverse array of species.

Turner, G. "Small Fry Go Big Time." *New Scientist,* 2 Aug. 1997. An overview of the evolution of cichlid species in Lake Victoria and other African great lakes.

● Visit our website at **http://www.saunderscollege.com/lifesci/titles.html** and click on Solomon/Berg/Martin Biology for links to chapter-related resources on the World Wide Web.

CHAPTER 20

The Origin and Evolutionary History of Life

The preceding three chapters were concerned with the evolution of organisms, but we have not yet dealt with what many regard as the fundamental question of biological evolution: how did life begin? Biologists generally accept the hypothesis that life developed from nonliving matter, but exactly how this process, called **chemical evolution,** occurred is not certain. Chemical evolution probably involved several stages. Current models suggest that small organic molecules first formed spontaneously and accumulated over time. These molecules may have been able to accumulate rather than being broken down (as occurs today) because conditions were different. The two factors presently responsible for breaking down organic molecules—free oxygen and living organisms— were absent from early Earth.

Large organic macromolecules such as proteins and nucleic acids could have then assembled from the smaller molecules. The macromolecules interacted with one another, combining into more complicated structures that could eventually metabolize and replicate. Natural selection favored macromolecular assemblages with cell-like structures. Their descendants eventually became the first true cells. After the first cells originated, they diverged over several billion years into the rich biological diversity that characterizes our planet today. Photosynthesis, aerobic respiration, and eukaryotic cell structure represent several major advances that developed during the evolutionary history of life.

Geological evidence, in particular the fossil record, provides us with much of what is known about the history of life, such as what kinds of organisms existed and where and when they lived. Certain organisms appear in the fossil record, then disappear and are replaced by others. Initially, unicellular prokaryotes predominated, followed by unicellular eukaryotes. The first multicellular eukaryotes—soft-bodied animals that did not leave many fossils—appeared in the ocean approximately 630 million years ago. Shelled animals and many other marine invertebrates (animals without backbones) appeared next, as exemplified by the trilobites in the photograph. Trilobites were members of a large group of primitive aquatic arthropods that lived during the Paleozoic era. Marine invertebrates were followed by the first vertebrates. The first fishes

(William E. Ferguson)

with jaws appeared and diversified; some of these gave rise to amphibians, which also spread and diversified. About 300 million years ago, amphibians gave rise to reptiles, which diversified and populated the land. Reptiles in turn gave rise independently to birds and to mammals. Plants underwent a comparable evolutionary history and diversification.

In this chapter we survey life over a vast span of time, starting some 3.8 billion years ago when our planet was relatively young. We examine proposed models about how life began and trace life's long evolutionary history from its beginnings to the present.

AFTER YOU HAVE STUDIED THIS CHAPTER YOU SHOULD BE ABLE TO

1. Describe the conditions that are thought to have existed on early Earth.
2. Contrast various hypotheses about life's origins.
3. Outline the major steps hypothesized to have occurred in the origin of cells.
4. Explain how the evolution of photosynthetic autotrophs affected both the atmosphere and other organisms.
5. Describe the endosymbiont theory and summarize the evidence supporting it.
6. List the geological eras in chronological order and give approximate dates for each.
7. Briefly describe the distinguishing organisms and major biological events of Precambrian time and of the Paleozoic, Mesozoic, and Cenozoic eras.

EARLY EARTH PROVIDED THE CONDITIONS FOR CHEMICAL EVOLUTION

Many biologists speculate that life originated only once and that life's beginnings occurred under environmental conditions quite different from those of today. We must therefore examine the conditions of early Earth to understand the origin of life. Although we will never be certain about the exact conditions that existed when life arose, scientific evidence from a number of sources provides us with valuable clues that help us formulate plausible scenarios.

Astrophysicists and geologists have determined that Earth is approximately 4.6 billion years old. The atmosphere of early Earth apparently included carbon dioxide (CO_2), water vapor (H_2O), carbon monoxide (CO), hydrogen (H_2), and nitrogen (N_2). It may also have contained some ammonia (NH_3), hydrogen sulfide (H_2S), and methane (CH_4), although these reduced molecules may have been rapidly broken down by ul-

traviolet radiation from the sun. The early atmosphere probably contained little or no free oxygen (O_2).

Four requirements must have existed for the chemical evolution of life: little or no free oxygen, a source of energy, the availability of chemical building blocks, and time. First, life could have begun only in the absence of free oxygen. Oxygen is quite reactive and would have oxidized the organic molecules that are necessary building blocks in the origin of life. Earth's early atmosphere was probably strongly reducing, which means that any free oxygen would have reacted with other elements to form oxides. Thus, oxygen would have been tied up in compounds.

The origin of life would also have required energy to do the work of building biological molecules from simple inorganic chemicals. Early Earth was a place of high energy with violent thunderstorms, widespread volcanic activity, bombardment from meteorites and other extraterrestrial objects, and intense radiation, including ultraviolet radiation from the sun (Fig. 20–1). The young sun probably produced more ul-

Figure 20–1 **Conditions on early Earth.** The strongly reducing atmosphere lacked oxygen; volcanoes erupted, spewing gases that contributed to the atmosphere; and violent thunderstorms produced torrential rainfall that eroded the land. Meteorites and other extraterrestrial objects continually bombarded Earth, causing cataclysmic changes in the crust, ocean, and atmosphere. *(Courtesy of Reader's Digest Books. Drawing by H.K. Wimmer)*

traviolet radiation than it does today, and the ancient Earth had no protective ozone layer to filter it.

A third requirement would have been the presence of the chemical building blocks needed for chemical evolution. These included water, dissolved inorganic minerals (present as ions), and the gases present in the early atmosphere. A final requirement for the origin of life was time for molecules to accumulate and react with one another. Earth is approximately 4.6 billion years old, and the earliest traces of life are approximately 3.8 billion years old; therefore, life had a maximum of 800 million years to get started.

Organic molecules formed on primitive Earth

Because organic molecules are the building materials for organisms, it is reasonable to first consider how they might have originated. The concept that simple organic molecules such as sugars, nucleotide bases, and amino acids could form spontaneously from simpler raw materials was first advanced in the 1920s by two scientists working independently: A. I. Oparin, a Russian biochemist, and J. B. S. Haldane, a Scottish physiologist and geneticist.

Their hypothesis was tested in the 1950s by American biochemists Stanley Miller and Harold Urey, who designed a closed apparatus that simulated conditions that presumably existed on early Earth (Fig. 20–2). They exposed an atmosphere rich in H_2, CH_4, H_2O, and NH_3 to an electrical discharge that simulated lightning. Their analysis of the chemicals produced in a week revealed that amino acids and other organic molecules had formed. Although more recent data suggest that Earth's early atmosphere was not rich in methane or ammonia, similar experiments using different combinations of gases have produced a wide variety of organic molecules that are important in contemporary organisms. These include all 20 amino acids, several sugars, lipids, the nucleotide bases of RNA and DNA, and ATP (when phosphate is present). Thus, before life began, its chemical building blocks may have been accumulating as a necessary step in chemical evolution.

Oparin envisioned that the organic molecules would, over vast spans of time, accumulate in the shallow seas to form a "sea of organic soup." Under such conditions, he envisioned smaller organic molecules (monomers) combining to form larger ones (polymers). Evidence gathered since Oparin's time indicates that organic polymers may have formed and accumulated on rock or clay surfaces rather than in the primordial seas. Clay, which consists of microscopic particles of weathered rock, is particularly intriguing as a possible site for early polymerizations because it binds organic monomers and contains zinc and iron ions that might have served as catalysts. Laboratory experiments have confirmed that organic polymers form spontaneously from monomers on hot rock or clay surfaces.

In a different scenario of chemical evolution, early polymerizations leading to the origin of life may have occurred in

Figure 20–2 **Testing a hypothesis about chemical evolution.** Diagram of the apparatus that Miller and Urey used to simulate the reducing atmosphere of early Earth. An electrical spark was produced in the upper right flask to simulate lightning. The gases present in the flask reacted together, forming a number of simple organic compounds that accumulated in the trap at the bottom.

cracks in the deep-ocean floor where hot water and minerals such as metal sulfides spew forth. Such **hydrothermal vents** would have been better protected than Earth's surface from the catastrophic effects of meteorite bombardment. Today these hot springs produce precursors of biological molecules and of energy-rich "food," including hydrogen sulfide and methane.

After the first polymers formed, could they have assembled spontaneously into more complex structures? Scientists have synthesized several different **protobionts,** which are assemblages of abiotically produced (that is, not produced by organisms) organic polymers. They have been able to recover protobionts that resemble living cells in several ways, thus providing clues as to how aggregations of complex nonliving molecules took that "giant leap" and became living cells. These protobionts exhibit many functional and structural attributes of living cells. They often divide in half (binary fission) after they have sufficiently "grown." Protobionts maintain an internal chemical environment that is different from the external environment (homeostasis), and some of them show the

2 μm

Figure 20–3 **Microspheres.** These tiny protobionts exhibit some of the properties of life. *(Steven Brooke and Richard LeDuc)*

beginnings of metabolism (catalytic activity). They are highly organized, considering their relatively simple composition.

Microspheres are a type of protobiont formed by adding water to abiotically formed polypeptides (Fig. 20–3). Some microspheres demonstrate excitability: they produce an electrical potential across their surfaces, reminiscent of electrochemical gradients in cells. Microspheres can also absorb materials from their surroundings (selective permeability) and respond to changes in osmotic pressure as though they were enveloped by membranes, even though they contain no lipid.

THE FIRST CELLS PROBABLY ASSEMBLED FROM ORGANIC MOLECULES

The study of protobionts allows us to appreciate that relatively simple "pre-cells" can exhibit some of the properties of contemporary life. However, it is a major step (or several steps) to go from simple molecular aggregates such as protobionts to living cells. Although much has been learned about how organic molecules may have formed on primitive Earth, the problem of how pre-cells evolved into living cells remains to be solved.

It is not known exactly when life first appeared on Earth. **Microfossils** (ancient remains of microscopic life) indicate that cells were thriving 3.5 billion years ago, and nonfossil evidence published in 1996 indicates that life existed even earlier, at least 3.8 billion years ago. (The significance of this older date is discussed in *Making the Connection: The Earliest Date for Life and Speculations about How Life Arose.*)

The earliest cells were prokaryotic. Australian and South African rocks have yielded microscopic fossils of prokaryotic cells 3.1 to 3.5 billion years old. **Stromatolites,** another type of fossil evidence of the earliest cells, are rocklike columns composed of many minute layers of prokaryotic cells, usually

(a)

(b)

Figure 20–4 **Stromatolites.** *(a)* Living stromatolites at Hamlin Pool in Western Australia are composed of mats of cyanobacteria and minerals like calcium carbonate and are several thousand years old. *(b)* Cutaway view of a fossil stromatolite showing the layers of cyanobacteria and sediments that accumulated over time. This stromatolite, also from Western Australia, is about 3.5 billion years old. *(a, Fred Bavendam/Peter Arnold, Inc.; b, Biological Photo Service)*

cyanobacteria (Fig. 20–4). Over time, sediment collects around the cells and mineralizes. Meanwhile, a new layer of living cells grows over the older, dead cells. Fossil stromatolite reefs are found in a number of places in the world, including Great Slave Lake in Canada and the Gunflint Iron Formations along Lake Superior in the United States. Some fossil stromatolites are extremely ancient. One group in Western Australia, for example, is several billion years old. Living stromatolite reefs are still found, for example, in hot springs and in warm, shallow pools of fresh and salt water.

We have said that the origin of cells from macromolecular assemblages was a major step in the origin of life. Actually,

THE EARLIEST DATE FOR LIFE AND SPECULATIONS ABOUT HOW LIFE AROSE

Why is it significant that Earth's earliest life may have existed 3.8 billion years ago? When American, Australian, and British scientists reported in 1996 that they had evidence of life some 300 million years earlier than the earliest established date of 3.5 billion years, the announcement generated a great deal of attention in the popular press as well as in scientific circles.

A new instrument called an ion microprobe was employed to determine carbon isotope ratios in some of Earth's oldest rocks, which were collected from Akilia Island off the southwestern coast of Greenland. Analysis of these rocks revealed carbon traces of the remains of ancient organisms. The lack of actual fossils in the rocks is the result of metamorphosis; the intense heat and pressure to which the rock was subjected would have destroyed all fossil traces. The new method of analysis appears to be reliable, and assuming the earlier date for life withstands future scientific scrutiny, it has significance as to when life arose and possibly where it arose.

To understand the significance of the new date, we must consider the early history of the solar system, when Earth was pelted by a hail of comets, asteroids, and meteorites—many as large as 100 kilometers (161 miles) in diameter. This barrage began soon after Earth formed and gradually tapered off about 3.9 billion years ago. Scientists have long assumed that it would have been impossible for life to have begun during the bombardment. Each impact raised Earth's temperature, which did not begin to cool until the bombardment diminished. If the new date is correct and life was already thriving 3.8 billion years ago, then it had only about 100 million years in which to begin, a vanishingly short time in evolutionary terms.

Scientists are trying to understand the significance of the 100-million-year window for chemical evolution. Some speculate that life may have begun during the bombardment but somehow survived—perhaps in hydrothermal ocean vents (mentioned in the text). Other scientists have revived an old idea known as **panspermia**, which suggests that life did not originate on Earth, but began elsewhere in the galaxy—and drifted through space to Earth. If life originated elsewhere and was then introduced to Earth, more time is gained for the origin of life. However, if panspermia accounts for the origin of life on Earth, we still must elucidate a satisfactory explanation for the origin of life; the problem is simply transferred to another planet!

---

the evolution of cells probably occurred in a series of small steps. One of the most significant parts of that process would have been the evolution of molecular reproduction.

A crucial step in the origin of cells was molecular reproduction

In living cells, genetic information is stored in the nucleic acid DNA, which is transcribed into the message in RNA, which in turn is translated into the proper amino acid sequence in proteins. All three macromolecules in the DNA → RNA → protein sequence contain precise information, but only DNA and RNA are capable of self-replication, although only in the presence of the proper enzymes. Because both RNA and DNA can form spontaneously on clay in much the same way that other organic polymers do, the question becomes which molecule, DNA or RNA, first appeared in the prebiotic world.

Some scientists have suggested that RNA was the first informational molecule to evolve in the progression toward a self-sustaining, self-reproducing cell and that proteins and DNA came along later. According to the proposed model of the **RNA world,** the chemistry of prebiotic Earth gave rise to self-replicating RNA molecules that functioned as both enzyme and substrates for their own replication. We represent the replication as a circular arrow:

One of the surprising features of RNA is that it often has catalytic properties; such enzymatic RNAs are called **ribozymes.** In contemporary cells, ribozymes help catalyze the synthesis of RNA and process precursors into rRNA, tRNA, and mRNA (see Chapter 12). Before the evolution of true cells, ribozymes may have catalyzed their own replication in the clays, shallow rock pools, or hydrothermal vents where life originated. When RNA strands are added to a test tube containing RNA nucleotides but no enzymes, the nucleotides combine to form short RNA molecules. The rate of this reaction is increased if zinc is added as a catalyst. (Recall that zinc is bound to clay.)

In the RNA world, ribozymes (catalytic RNAs) initially catalyzed protein synthesis; only later did protein enzymes catalyze RNA synthesis.

$$RNA \longrightarrow protein$$

Interestingly, RNA can direct protein synthesis by catalyzing peptide bond formation. Some single-stranded RNA molecules fold back on themselves as a result of interactions among the nucleotides composing the RNA strand. Sometimes the conformation (shape) of the folded RNA molecule is such that it weakly binds to an amino acid. If amino acids are held close together by RNA molecules, they may bond together, forming a polypeptide.

We have considered how the evolution of informational molecules may have given rise to DNA and later to proteins. If a self-replicating RNA capable of coding for proteins appeared before DNA, how did DNA, the universal molecule of heredity, become involved in translation? Perhaps RNA made double-stranded copies of itself that eventually evolved into DNA.

$$DNA \leftarrow RNA \rightarrow protein$$

The incorporation of DNA into the information transfer system would have been advantageous because the double helix conformation of DNA is more stable (that is, less reactive) than the single-stranded conformation of RNA. Such stability in a molecule that stores genetic information would have provided a decided advantage in the prebiotic world (as it does today).

In the DNA/RNA/protein world, then, DNA became the information storage molecule, RNA remained involved in protein synthesis, and protein enzymes catalyzed most cellular reactions, including DNA replication, RNA synthesis, and protein synthesis.

$$\overset{\curvearrowright}{DNA} \rightarrow RNA \rightarrow protein$$

RNA is still a necessary component of the information transfer system because DNA is not catalytic. Thus natural selection at the molecular level favored the DNA → RNA → protein information sequence. Once DNA was incorporated into this sequence, RNA molecules assumed their present role as an intermediary in the transfer of genetic information.

Several additional steps had to occur before a true living cell could evolve from macromolecular aggregations. For example, the genetic code must have arisen extremely early in the prebiotic world because all organisms possess it, but how did it originate? Also, how did a plasma membrane of lipid and protein come to envelop the pre-cell assemblages, thereby permitting the accumulation of some molecules and the exclusion of others?

The first cells were probably heterotrophs, not autotrophs

The earliest cells probably obtained the organic molecules they needed from the environment, rather than synthesizing them. These primitive **heterotrophs** probably consumed many types of organic molecules that had spontaneously formed—sugars, nucleotides, and amino acids, to name a few. By fermenting these organic compounds, they obtained the energy needed to support life. Fermentation is, of course, an anaerobic process (that is, performed in the absence of oxygen), and the first cells were almost certainly **anaerobes.**

When the supply of spontaneously generated organic molecules was gradually depleted, only certain organisms could survive. Mutations had probably already occurred that permitted some cells to obtain energy directly from sunlight, perhaps by using sunlight to make ATP. These cells, which did not require the energy-rich organic compounds that were now in short supply in the environment, had a distinct selective advantage.

Photosynthesis requires not only light energy but also a source of electrons, which are used to reduce CO_2 when organic molecules such as glucose are synthesized (see Chapter 8). Most likely, the first photosynthetic **autotrophs** (organisms that produce their own food from simple raw materials) used the energy of sunlight to split hydrogen-rich molecules such as H_2S, releasing elemental sulfur (not oxygen) in the process. Indeed, the green sulfur bacteria and the purple sulfur bacteria still use H_2S as a hydrogen source for photosynthesis.[1]

The first photosynthetic autotrophs to obtain hydrogen by splitting water were the cyanobacteria. Water was quite abundant on early Earth, as it is today, and the selective advantage that splitting water bestowed on them allowed the cyanobacteria to thrive. In the process of splitting water, oxygen was released as a gas (O_2). Initially, the oxygen released from photosynthesis oxidized minerals in the ocean and in Earth's crust, and, as a result, oxygen did not begin to accumulate in the atmosphere for a long period of time. Eventually, however, oxygen began to accumulate in the ocean and the atmosphere.

The timing of the events just described can be estimated on the basis of geological and fossil evidence. Fossils from that period, which include rocks that contain traces of chlorophyll, as well as the fossil stromatolites discussed previously, indicate that the first photosynthetic organisms appeared approximately 3.1 to 3.5 billion years ago. This evidence suggests that heterotrophic forms existed even earlier.

Aerobes appeared after oxygen increased in the atmosphere

By two billion years ago, the cyanobacteria had produced sufficient oxygen to begin significantly changing the composition of the atmosphere. The increase in atmospheric oxygen had a profound effect on life. Obligate anaerobes (those organisms that cannot use oxygen for cellular respiration) were poisoned by the oxygen, and many species undoubtedly perished. Some anaerobes, however, survived in environments where oxygen does not penetrate; others developed ways to neutralize the oxygen so that it could not harm them. Some organisms, called **aerobes,** developed a respiratory pathway that *used* the oxygen to extract more energy from food and convert it to ATP energy. Aerobic respiration was joined to the existing anaerobic process of glycolysis.

The evolution of organisms that could use oxygen in their metabolism had several consequences. Organisms that respire aerobically gain much more energy from a single molecule of glucose than anaerobes gain by fermentation. As a result, the newly evolved aerobic organisms were more efficient and more competitive than anaerobes. Coupled with the poisonous na-

[1]A third group of bacteria, the purple nonsulfur bacteria, uses other organic molecules or hydrogen gas as a hydrogen source.

ture of oxygen to anaerobes, the efficiency of aerobes forced anaerobes into relatively minor roles. Today the vast majority of organisms, including plants, animals, and most fungi, protists, and prokaryotes, use aerobic respiration, while only a few bacteria and even fewer protists and fungi are anaerobic.

The evolution of aerobic respiration had a stabilizing effect on both oxygen and carbon dioxide in the biosphere. Photosynthetic organisms used carbon dioxide as a source of carbon for the synthesis of organic compounds. This raw material would have been depleted from the atmosphere in a relatively short time without the advent of aerobic respiration, which releases carbon dioxide as a waste product from the complete breakdown of organic molecules. Carbon thus started cycling in the biosphere, moving from the nonliving physical environment, to photosynthetic organisms, to heterotrophs that ate the plants (see Chapter 53). Carbon was released back into the physical environment as carbon dioxide by aerobic respiration, and the carbon cycle continued. In a similar manner, oxygen was produced by photosynthesis and used during aerobic respiration.

Another significant consequence of photosynthesis occurred in the upper atmosphere, where molecular oxygen reacted to form **ozone**, O_3 (Fig. 20–5). A layer of ozone eventually blanketed Earth, preventing much of the sun's ultraviolet radiation from penetrating to the surface. With the ozone layer's protection from the mutagenic effect of ultraviolet radiation, organisms were able to live closer to the surface in aquatic environments and eventually to move onto land. Because the energy in ultraviolet radiation was necessary to form organic molecules, however, their abiotic synthesis decreased.

Eukaryotic cells descended from prokaryotic cells

Eukaryotes appeared in the fossil record 1.9 to 2.1 billion years ago. They arose from prokaryotes. Recall from Chapter 4 that prokaryotic cells lack nuclear envelopes as well as other membranous organelles such as mitochondria and chloroplasts. How did eukaryotic cells arise from prokaryotes?

The **endosymbiont theory,** advanced by Lynn Margulis, declares that organelles such as mitochondria and chloroplasts may each have originated from mutually advantageous symbiotic relationships between two prokaryotic organisms (Fig. 20–6). Chloroplasts apparently evolved from photosynthetic bacteria (perhaps cyanobacteria) that lived inside larger heterotrophic cells, while mitochondria presumably evolved from aerobic bacteria (perhaps purple bacteria) that lived inside larger anaerobic cells. Thus, early eukaryotic cells were composed of formerly free-living prokaryotes.

How did these bacteria come to be **endosymbionts,** which are organisms that live symbiotically inside a host cell? They may have originally been ingested, but not digested, by a host cell. Once so incorporated, they could have survived and reproduced along with the host cell so that future generations of the host also contained endosymbionts. The two organisms developed a mutualistic relationship in which each contributed

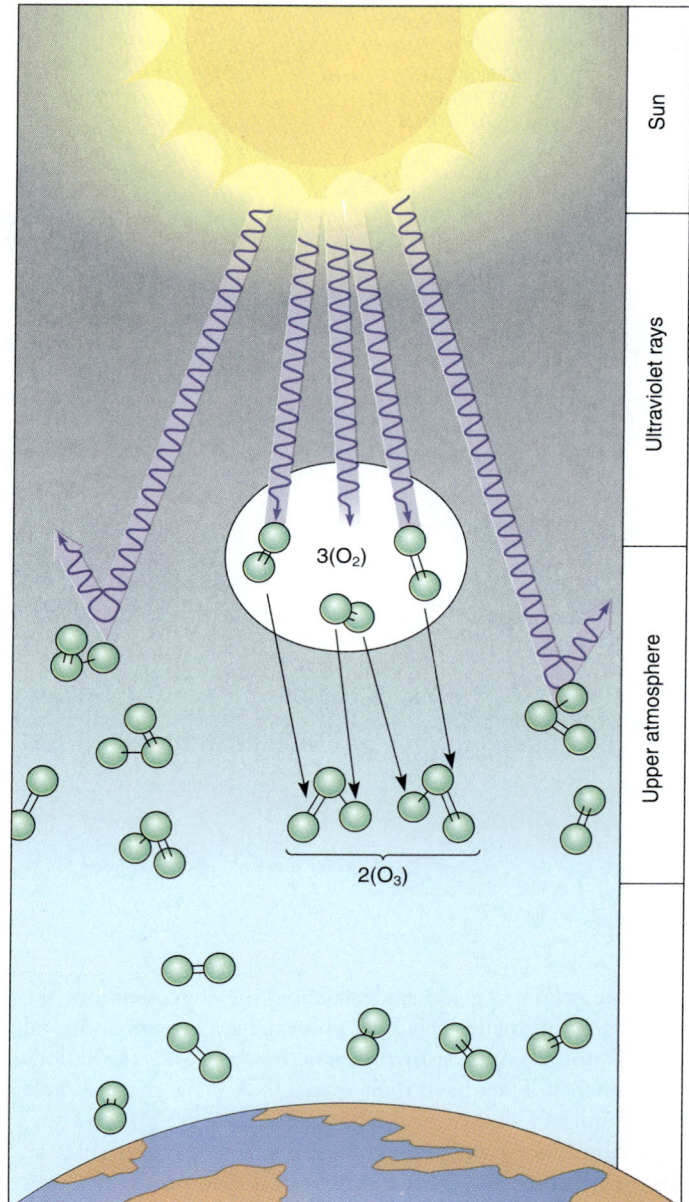

Figure 20–5 **How ozone forms.** Ozone (O_3) forms in the upper atmosphere when ultraviolet radiation from the sun breaks the double bonds of oxygen molecules.

something to the other. Eventually the endosymbiont lost the ability to exist outside its host, and the host cell lost the ability to survive without its endosymbionts. This theory stipulates that each of these partners brought to the relationship something the other lacked. For example, mitochondria provided the ability to carry out the aerobic respiration lacking in the original anaerobic host cell; chloroplasts provided the ability to use a simple carbon source (carbon dioxide) to produce needed organic molecules. The host cell provided a safe habitat and raw materials or nutrients.

ORIGINAL PROKARYOTIC HOST CELL

DNA

Aerobic bacteria

Multiple invaginations of the plasma membrane

The bacteria become mitochondria

Endoplasmic reticulum and nuclear envelope form from the plasma membrane invaginations (not part of endosymbiont theory)

EUKARYOTIC CELLS: ANIMALS, FUNGI, SOME PROTISTS

Photosynthetic bacteria...

...become chloroplasts

EUKARYOTIC CELLS: PLANTS, SOME PROTISTS

Figure 20–6 **The endosymbiont theory.** Chloroplasts and mitochondria of eukaryotic cells are thought to have originated from various bacteria that lived as endosymbionts inside other cells.

The principal evidence in favor of the endosymbiont theory is that mitochondria and chloroplasts possess some (although not all) of their own genetic material and translational components. They have their own DNA (as a circular molecule similar to that of prokaryotes; see Chapter 23) and their own ribosomes (which resemble prokaryotic rather than eukaryotic ribosomes). Mitochondria and chloroplasts also possess some of the machinery for protein synthesis, including tRNA molecules, and are able to conduct protein synthesis on a limited scale independent of the nucleus. Furthermore, it is possible to poison mitochondria and chloroplasts with an antibiotic that affects prokaryotic but not eukaryotic cells. As discussed in Chapter 4, mitochondria and chloroplasts are enveloped by double membranes. The outer membrane apparently developed from the invagination of the host cell's plasma membrane, while the inner membrane is derived from the endosymbiont's plasma membrane.

A number of endosymbiotic relationships exist today. Many corals have algae living as endosymbionts within their cells (see *Making the Connection: Coral Reefs, Interdependence of Organisms, and Environmental Issues* in Chapter 28). In the gut of the termite lives a protozoon *(Myxotricha paradoxa)* that in turn has several different endosymbionts, including spiro-

chete bacteria that are attached to the protozoon and function as whiplike flagella, allowing it to move.

The endosymbiont theory does not completely explain the evolution of eukaryotic cells from prokaryotes. It does not explain, for example, how the genetic material in the nucleus came to be surrounded by a membranous envelope. A few biologists reject the endosymbiont theory and subscribe to the **autogenous model,** in which eukaryotes arose from prokaryotes by the proliferation of internal membranes to form cellular compartments; these internal membranes were derived from the prokaryotic plasma membrane. Regardless of how eukaryotic cells evolved, their advent set the stage for further evolutionary developments.

THE FOSSIL RECORD PROVIDES US WITH CLUES TO THE HISTORY OF LIFE

The sequence of biological, climate, and geological events that make up the history of life is recorded in rocks and fossils. The sediments of Earth's crust consist of five major rock strata (layers), each subdivided into minor strata, lying one on top of the other. Very few places on Earth are covered by all layers, but the strata that are present typically occur in the correct order, with younger rocks on top of older ones. These sheets of rock were formed by the accumulation of mud and sand at the bottoms of the ocean, seas, and lakes. Each layer contains certain characteristic fossils that serve to identify deposits made at approximately the same time in different parts of the world.

Geologists divide Earth's 4.6-billion-year history into units of time based on major geological, climate, and biological events. Relatively little is known about Earth from its beginnings approximately 4.6 billion years ago up to 570 million years ago, a period known informally as **Precambrian time.** The fossil record of ancient organisms is abundant beginning about 570 million years ago. This most recent time, from 570 million years ago to the present, is divided into three **eras** based primarily on organisms that were characteristic of each era (Table 20–1). Eras are subdivided into **periods,** which in turn are composed of **epochs.**

Evidence of living cells is found in Precambrian deposits

Signs of Precambrian life date back to about 3.8 billion years ago. Not much physical evidence is available to us because the rocks of Precambrian time, being extremely ancient, are deeply buried in most parts of the world. Precambrian rocks are exposed in a few places, including the bottom of the Grand Canyon and along the shores of Lake Superior. More than 400 Precambrian rock formations have revealed microfossils.

During Precambrian time, widespread volcanic activity and giant upheavals raised mountains, and the heat, pressure, and churning associated with these movements probably destroyed most of whatever fossils may have been formed. Some evidence of life still remains as traces of graphite or pure carbon, which may be the transformed remains of primitive life. These remains are especially abundant in what were the ocean and seas of that time. Additionally, fossils resembling cyanobacteria have been recovered from several Precambrian formations. The fossils found in later (more recent) Precambrian rocks show unambiguous examples of some major groups of bacteria, fungi, protists (including multicellular algae), and animals.

One rich source of Precambrian fossil deposits is the Ediacaran Hills (pronounced "ee-dee-ack´a-ran") in South Australia. **Ediacaran fossils,** the oldest known fossils of multicellular animals, are from very late in Precambrian time—from 600 to 570 million years ago. Biologists have not yet resolved the phylogenetic affinities of the simple, soft-bodied animals found there and at other Precambrian sites around the world. Some of these animals appear to be early examples of jellyfish, soft corals, segmented worms, mollusks, and soft-bodied arthropods, whereas others show no resemblance to any other known fossil or living organism (Fig. 20–7). If this interpretation is correct, then at least some of the Ediacaran animals were ancestral to animals that followed. Other biologists who have studied the fossils, however, think that the Ediacaran an-

imals have a body plan that is different from all known animal phyla. If this interpretation is correct, then these animals probably went extinct by the end of the Precambrian and, therefore, would not be directly related to modern animals.

A considerable diversity of organisms evolved during the Paleozoic era

The **Paleozoic era** began approximately 570 million years ago and lasted approximately 222 million years. It is divided into six periods: Cambrian, Ordovician, Silurian, Devonian, Carboniferous, and Permian.

The oldest subdivision of the Paleozoic era, the Cambrian period, is represented by rocks rich in fossils. Evolution was in such high gear, with the sudden appearance of many new animal groups, that this period has been nicknamed the **Cambrian explosion.** Fossils of all of the contemporary animal phyla are present, along with many bizarre, extinct phyla, in marine sediments. The sea floor was covered with sponges, corals, sea lilies, sea stars, snails, clamlike bivalves, primitive squidlike cephalopods, lamp shells (brachiopods), trilobites (see chapter opening photograph; also see Fig. 29–12), at least one primitive chordate, and other marine animals. Burrowing, grazing, and predatory animals all became established in the marine environment. Scientists have not determined the factor or factors responsible for the Cambrian explosion, which has been unmatched in the evolutionary history of life. There is, however, some evidence that oxygen concentrations, which had continued to gradually increase in the atmosphere, passed some critical environmental threshold (10% of present-day oxygen, or higher) late in Precambrian time. Scientists who

Figure 20–7   A Precambrian fossil. The organism, from the Ediacaran Hills of South Australia, lived in shallow marine waters. It is unlike any known modern organism.   *(William E. Ferguson)*

TABLE 20-1 Some Important Biological Events in Geological Time

Time	Era	Period	Epoch	Geological/Climatic Conditions
10,000 years ago to present	Cenozoic	Quaternary	Holocene	End of last Ice Age; warmer climate; higher sea levels as glaciers melt
2	Cenozoic	Quaternary	Pleistocene	Multiple ice ages; glaciers in Northern Hemisphere
5	Cenozoic	Tertiary	Pliocene	Uplift and mountain-building; volcanoes; climate much cooler; North and South America join at Isthmus of Panama
25	Cenozoic	Tertiary	Miocene	Mountains form; climate drier and cooler
38	Cenozoic	Tertiary	Oligocene	Rise of Alps and Himalayas; most land low; volcanic activity in Rockies; climate cool and dry
55	Cenozoic	Tertiary	Eocene	Climate warmer
65	Cenozoic	Tertiary	Paleocene	Continental seas disappear; climate mild to cool and wet
144	Mesozoic	Cretaceous		Continents separate; most continents low; large inland seas and swamps; climate warm
213	Mesozoic	Jurassic		Continents low; inland seas; mountains form; continental drift begins; climate mild
248	Mesozoic	Triassic		Many mountains form; widespread deserts; climate warm and dry
286	Paleozoic	Permian		Glaciers; continents rise and merge as Pangaea; climate variable
360	Paleozoic	Carboniferous		Lands low and swampy; climate warm and humid, becoming cooler later
408	Paleozoic	Devonian		Glaciers; inland seas
438	Paleozoic	Silurian		Most continents remain covered by seas; climate warm
505	Paleozoic	Ordovician		Sea covers most continents
570	Paleozoic	Cambrian		Oldest rocks with abundant fossils; lands low; climate mild and wet

Million Years before Present

advocate the *oxygen enrichment hypothesis* note that until late in Precambrian time, Earth possessed insufficient oxygen to support larger animals. The most important fossil site that documents the Cambrian explosion is the **Burgess Shale** in British Columbia (Fig. 20–8).

The major animal body plans, which are discussed in Chapter 28, were established so early in the history of the eukaryotes that little change of a basic nature has occurred since.

This probably indicates that, early in the Cambrian period, each animal phylum had reached a degree of adaptation that allowed it to exploit its environment and accommodate changes in its surroundings with relatively limited modifications in its body plan.

According to geologists, the continents were gradually flooded during the Cambrian period. In the Ordovician period, much of what is now land was covered by shallow seas

TABLE 20-1 Continued

Plants and Microorganisms	Animals
Decline of some woody plants; rise of herbaceous plants	Age of *Homo sapiens*
Extinction of some plant species	Extinction of many large mammals at end
Expansion of extensive grasslands and deserts; decline of forests	Many grazing mammals; large carnivorous mammals; first known human-like primates
Flowering plants continue to diversify	Great diversity of grazing mammals and songbirds
Spread of forests; flowering plant communities expand	Apes appear; present mammalian families are represented
Flowering plants dominant	Modern mammalian orders appear and diversify; modern bird orders appear
Semitropical vegetation (flowering plants and conifers) widespread	Primitive mammals diversify rapidly
Rise of flowering plants	Dinosaurs reach peak, then become extinct at end; toothed birds become extinct; primitive mammals
Gymnosperms common	Large, specialized dinosaurs; first toothed birds; primitive insectivorous mammals diversify
Gymnosperms dominant; ferns common	First dinosaurs; first mammals
Conifers diversify; cycads appear	Modern insects appear; mammal-like reptiles; extinction of many Paleozoic invertebrates and vertebrates at end of Permian
Forests of ferns, club mosses, horsetails, and gymnosperms; mosses and liverworts	First reptiles; spread of ancient amphibians; many insect forms; ancient sharks abundant
Vascular plants diversify and become well established; first forests; gymnosperms appear; bryophytes appear	Many trilobites; fishes with jaws appear and diversify; amphibians appear; wingless insects appear
Algae dominant in aquatic environments; vascular plants appear	Jawless fishes diversify; coral reefs common; terrestrial arthropods
Marine algae dominant; fossil spores of terrestrial plants	Invertebrates dominant; coral reefs appear; first fishes appear
Algae; bacteria and cyanobacteria; fungi	Age of marine invertebrates; modern and extinct animal phyla represented; first chordates

in which there was another burst of evolutionary diversification, although not as dramatic as the Cambrian explosion. The Ordovician seas were inhabited by giant cephalopods, squid-like animals with straight shells 5 to 7 meters (16 to 23 feet) long and 30 centimeters (12 inches) in diameter. Coral reefs first appeared during this period, as did the earliest vertebrates, which were small, jawless, bony-armored fishes called *ostraco-derms* (Fig. 20–9). Lacking jaws, these fishes typically had round or slitlike mouth openings that may have sucked in small food particles from the water or scooped up bottom organic debris. Ordovician deposits also contain fossil spores of ter-restrial (land-dwelling) plants, which suggests that the colonization of land had begun.

During the Silurian period, jawless fishes diversified considerably, and jawed fishes first appeared. Definitive evidence of two life forms of great biological significance appeared in

(a)   *(b)*   *(c)*

Figure 20–8   **Fossils from the Cambrian explosion.** *(a) Marrella splendens* was a small arthropod. *(b)* In spite of its unusual appearance, *Wiwaxia* was actually a worm that was distantly related to earthworms. It had scaly armor and needle-like spines for protection. *(c) Waptia fieldensis* was a crustacean that may have been an ancestor of modern crustaceans such as shrimp. All three of these fossils were obtained from the Burgess Shale in the Canadian Rockies.   *(a, b, c, Chip Clark)*

the Silurian period: terrestrial plants and air-breathing animals. The first known plants resembled ferns in that they possessed vascular (conducting) tissue and reproduced by spores. The evolution of plants allowed animals to colonize the land because plants provided the first terrestrial animals with food and shelter. All air-breathing land animals discovered in Silurian rocks were arthropods—millipedes, spider-like arthropods, and possibly centipedes. Interestingly, from an ecological perspective, the energy flow from plants to animals probably occurred via detritus, which is organic debris from decomposing organisms, rather than directly from living plant material. Millipedes eat plant detritus today, and spiders and centipedes prey on other animals.

A great variety of fishes appeared in the Devonian period. In fact, the Devonian period is frequently called the Age of Fishes. Jawless ostracoderms persisted into the Devonian, but this period also witnessed the explosive radiation of fishes with jaws, an adaptation that enables a vertebrate to chew and bite. Armored *placoderms,* an extinct group of jawed fishes, diversified to exploit varied lifestyles, from bottom-dwelling filter-feeders to the most voracious predators of the time (see Fig.

30–10*b*). Appearing in Devonian deposits are sharks and the two predominant types of bony fish: lobe-finned fishes (including the coelacanths and lungfishes) and ray-finned fishes, which gave rise to the major orders of modern fishes. Coelacanths, primitive bony fish with lobed fins, were originally thought to have become extinct; in 1938, however, the first living coelacanth was discovered in the deep waters off the coast of Madagascar (see Fig. 30–15). This discovery was of great scientific significance because it gave paleontologists an opportunity to test their hypotheses about fossil coelacanths by comparing them to the living species. Lungfishes, an ancient group of air-breathing fish, were most common during the Devonian period; only about six species persist today, in South America, Africa, and Australia.

Upper (more recent) Devonian sediments contain fossil remains of salamander-like amphibians that were often quite large, with short necks and heavy, muscular tails (see Fig. 30–17). These animals, whose skulls were encased in bony armor, were in many respects quite similar to the lobe-finned fishes, one of which may have been their immediate ancestor; for example, early amphibians possessed fishlike tail fins and

(b)

(a)

(c)

Figure 20–9   **Representative ostracoderms.**
(a) Thelodus, *(b) Pterapsis,* and *(c) Jamoytius* are fossil ostracoderms, primitive jawless fishes that appeared in the Devonian period.

Figure 20–12 **Saurischian and ornithischian dinosaurs.** The two orders of dinosaurs are distinguished primarily by differences in their pelvic bones. *(a)* The saurischian pelvis. Note the opening in the hip socket, a trait possessed by no quadrupedal vertebrates other than dinosaurs. *(b)* The ornithischian pelvis also has the hole in the hip socket but differs from the saurischian pelvis in that it has a backward-directed extension of the pubis.

Saurischians

(a)

Ilium
Pubis
Ischium
Coelophysis

Ornithischians

(b)

Ilium
Pubis
Ischium
Stegosaurus

Figure 20–13 **Reconstruction of a skull of *Carcharodontosaurus*.** The fossil remains of this fearsome predator were discovered in North Africa. *(Paul Sereno ©1999. Reprinted with permission of* Discover *Magazine.)*

The Cenozoic era is known as the Age of Mammals

With equal justice the **Cenozoic era** could be called the Age of Mammals, the Age of Birds, the Age of Insects, or the Age of Flowering Plants. This era is marked by the appearance of all these forms in great variety and numbers of species. The Cenozoic era extends from 65 million years ago to the present. It is subdivided into two periods: the Tertiary period, encompassing some 63 million years, and the Quaternary period, which covers the last 2 million years. The Tertiary period is subdivided into five epochs, named from earliest to latest: Paleocene, Eocene, Oligocene, Miocene, and Pliocene. The Quaternary period is subdivided into the Pleistocene and Holocene epochs.

Flowering plants, which arose during the Cretaceous period, continued to diversify during the Cenozoic era. During the Paleocene and Eocene epochs, fossils indicate that tropical to semitropical plant communities extended to relatively high

MAMMALIAN DIVERSITY AND THE CARRYING CAPACITY OF THE ENVIRONMENT

Has mammalian diversity peaked or is it declining? The fossil record indicates that during the 10 million years or so following the extinction of the dinosaurs, mammals underwent considerable adaptive radiation. According to conventional wisdom, mammals have continued to diversify, and the present is the time of maximum mammalian diversity.

John Alroy, a paleontologist at the Smithsonian Institution, disagrees. He assembled an extensive database of all mammalian genera* and species that occurred in North America (excluding bats and marine mammals), from the demise of the dinosaurs, about 65 million years ago, to the present. His data indicate that for the past 55 million years, North America has been home to a constant number (that is, a diversity equilibrium) of mammalian species.[†] (The actual *number* of species at the equilibrium point is not known, but Alroy's method provides diversity *estimates* at different time intervals that are accurate relative to one another.) What is remarkable about Alroy's data is that, despite the repeated turnover of players during the past 55 million years as existing mammalian species became extinct and others evolved, the number of mammalian species remains constant.

Why has North America been home to a constant number of mammalian species for the past 55 million years? Why not more or less? Alroy's research has sparked a connection between paleontology and ecology. Some scientists suggest that the North American continent has a finite amount of food and other resources to support mammals. In ecological terms, an environment's ability to support a group of organisms is known as its *carrying capacity* (see

Chapter 51). These scientists suggest that North America's carrying capacity for mammalian species is fixed.

Although either extinction or speciation could be the mechanism that causes diversity equilibrium, Alroy's data indicate that only speciation is a factor. When the number of species rises above the carrying capacity, greater competition among species decreases the rate of speciation so that mammalian diversity ceases to grow. When the species number falls below the carrying capacity, mammalian diversification increases because competition among species decreases and, as a result, the rate of speciation increases.

Alroy's data also indicate that mammals as a group are robust in an evolutionary sense. During the past 65 million years, mammals have not experienced a large mass extinction event of the magnitude of the one at the end of the Cretaceous period. The mammalian line has, however, undergone important changes in rates of extinction. The largest mass extinction event of the Cenozoic era occurred at the very end of the Pleistocene epoch. In addition to its extraordinary rapidity (less than 1000 years), the end-Pleistocene event is marked by numerous other features that implicate humans as the main cause.

*Recall from Chapter 1 that a genus (pl., *genera*) is a group of closely related species.

[†]Alroy, J. "Constant Extinction, Constrained Diversification, and Uncoordinated Stasis in North American Mammals." *Paleogeography, Paleoclimatography, Paleoecology,* Vol. 127, 1996.

latitudes. Palms, for example, are found in Eocene deposits in Wyoming. Later in the Cenozoic era, there is evidence of more open habitats. Grasslands and savannas occurred throughout much of North America during the Miocene epoch, with deserts developing later in the Pliocene and Pleistocene epochs. During the Pleistocene epoch, plant communities changed dynamically in response to fluctuating climates associated with the multiple advances and retreats of continental glaciers.

During the Paleocene epoch, an explosive radiation of primitive mammals occurred (see *Making the Connection: Mammalian Diversity and the Carrying Capacity of the Environment*). Most of these were small forest dwellers that are not closely related to modern mammals. During the Eocene epoch, there was an explosive radiation of mammals; all of the modern orders first appeared and diversified. Again, many of the mammals were small, but there were also some larger herbivores—the *titanotheres,* for example, which got progressively larger during the Eocene epoch (Fig. 20–14*a*).

During the Oligocene epoch, many modern families of mammals appeared, including the first fossil apes in Africa. A number of lineages showed specializations that suggest a more open type of habitat, such as grassland or savanna. For example, many of the animals were larger and had longer legs for running, specialized teeth for chewing coarse vegetation or for preying on animals, and increases in their relative brain sizes. These specializations continued in the Miocene and Pliocene epochs. Especially diverse during these epochs were hoofed mammals such as horses, which underwent adaptive radiation to include both browsing and grazing lifestyles. Carnivores specialized for long-distance running after prey also appeared in the Pliocene epoch. (Before that, carnivores were primarily ambush predators.) Human ancestors are found during the Pliocene epoch, about 4.4 million years ago, in Africa; the genus *Homo* appeared approximately 2.3 million years ago. (Primate evolution, including human evolution, is discussed in Chapter 21.)

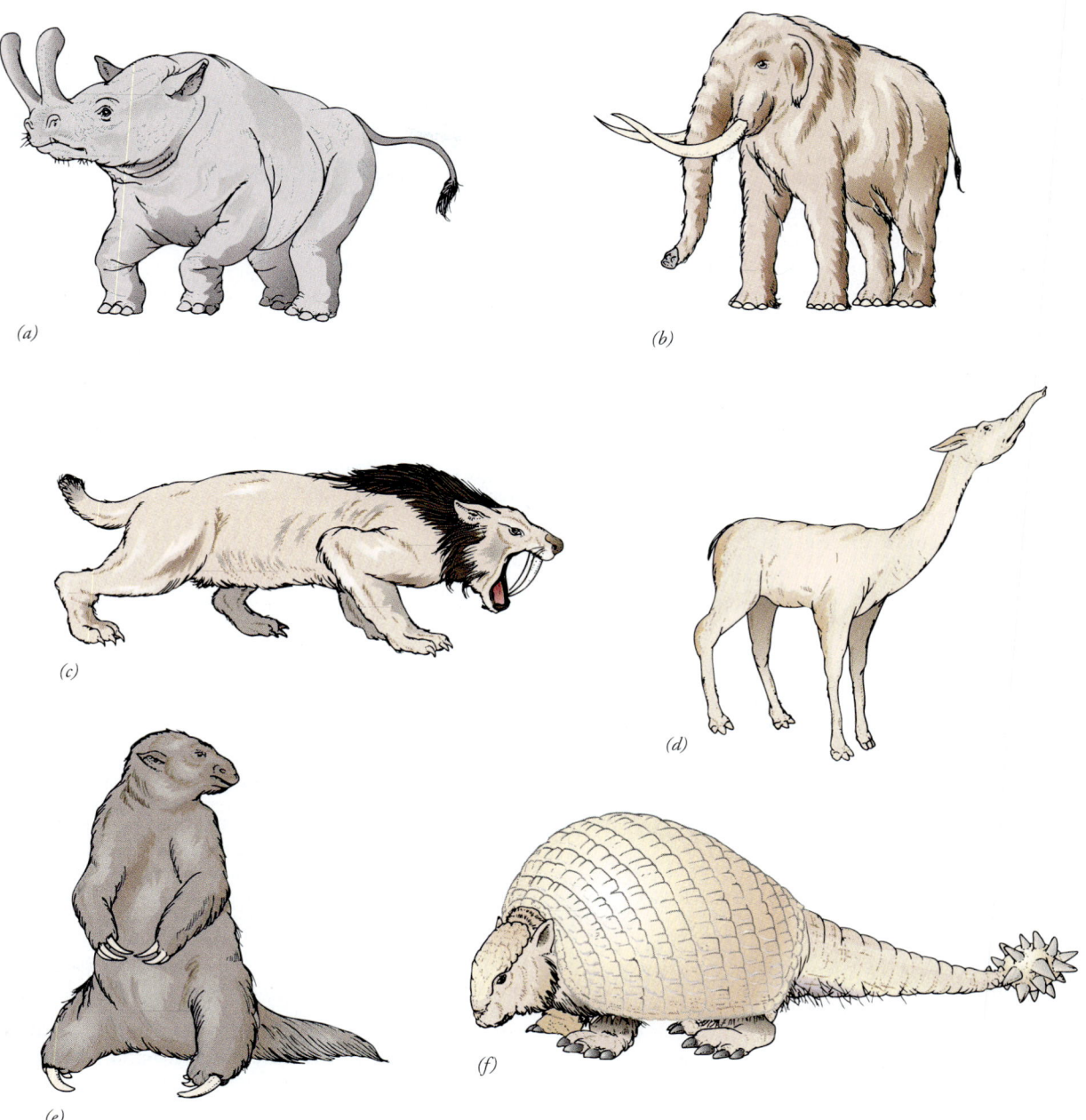

Figure 20–14 **Representative North and South American mammals of the Cenozoic era.**
(a) Brontotherium was a titanothere from the late Eocene epoch. *(b –f)* Mammals of the Pliocene and
Pleistocene epochs. *(b)* The elephant-like mastodon (*Mammut*) was at home in forests, lakes, and
rivers. *(c)* The saber-toothed cat (*Smilodon*) was found in both North and South America. Its enlarged
canine teeth were approximately 20 centimeters (8 inches) long and were curved like sabers. *(d)* This
camel-like mammal (*Macrauchenia*) from South America probably browsed in forest clearings.
Macrauchenia is usually depicted with a short trunk because, like elephants, its nasal openings are on
the skull roof rather than in the front of the skull. *(e) Megatherium* was a South American giant ground
sloth. As long as 6 meters (20 feet), *Megatherium* had 18-centimeter (7-inch) claws that may have been
used to strip bark from trees. Other paleontologists think *Megatherium* used its claws to stab prey.
(f) The giant armadillo (*Glyptodon*) lived on the pampas of South America. Encased in bony armor,
Glyptodon was about 2 meters (6.5 feet) long. Later glyptodont species reached 4 meters (13 feet).

The Pliocene and Pleistocene epochs witnessed a spectacular North and South American large-mammal fauna, including mastodons, saber-toothed cats, camels, giant ground sloths, giant armadillos, and numerous other species (Fig. 20–14*b*–*f*). However, many of the large mammals became extinct at the end of the Pleistocene, perhaps due to climate change—the Pleistocene epoch was marked by several ice ages —and/or the influence of humans, who had spread from Africa to Europe and Asia, and later to North and South America by crossing a land bridge between Siberia and Alaska. Strong archaeological evidence exists that this mass extinction event is concurrent with the appearance of human hunters who possessed Clovis spear-point technology.

During the Eocene epoch, there was an explosive radiation of birds, which acquired specializations for many different habitats. The jaws and beak of the flightless giant bird *Diatryma,* for example, were adapted primarily for crushing and slicing vegetation in Eocene forests, marshes, and grasslands (Fig. 20–15). The songbirds diversified extensively during the Miocene epoch to become the most diverse order of living birds.

Figure 20–15   **A representative bird from the Eocene epoch.** *Diatryma,* a giant, flightless bird, stood 2.1 meters (7 feet) tall and weighed about 175 kilograms (385 pounds).

SUMMARY WITH KEY TERMS

I. Biologists generally agree that life originated from nonliving matter by **chemical evolution.** Although chemical evolution is very difficult to test experimentally, a number of hypotheses about the origin of life are testable.
   A. Four requirements for chemical evolution are:
      1. The absence of oxygen, which would have reacted with and oxidized abiotically-produced organic molecules.
      2. Energy to form organic molecules.
      3. Chemical building blocks, including water, minerals, and gases present in the atmosphere, to form organic molecules.
      4. Sufficient time for molecules to accumulate and react.
   B. Four steps are hypothesized in chemical evolution.
      1. Small organic molecules formed spontaneously and accumulated.
      2. Macromolecules assembled from the small organic molecules.
      3. Macromolecular assemblages called **protobionts** formed from macromolecules.
         a. RNA molecules that could replicate themselves and carry out necessary chemical reactions may have been the first informational molecules to evolve.
         b. Natural selection at the molecular level resulted in the DNA → RNA → protein information sequence.
      4. Cells arose from the macromolecular assemblages.
II. The oldest cells in the fossil record are 3.1 to 3.5 billion years old. Nonfossil evidence places earliest life at 3.8 billion years old.
   A. The first cells were prokaryotic **heterotrophs** that obtained organic molecules from the environment. They were almost certainly **anaerobes.**
   B. Later, **autotrophs,** organisms that produce their own organic molecules by photosynthesis, arose.
   C. The evolution of photosynthesis ultimately changed early life because it generated oxygen, which accumulated in the atmosphere, and per-

mitted the evolution of **aerobes,** organisms that could use oxygen for a more efficient type of cellular respiration.
   D. Eukaryotic cells arose from prokaryotic cells. According to the **endosymbiont theory,** certain eukaryotic organelles (mitochondria and chloroplasts) evolved from prokaryotic **endosymbionts** within larger prokaryotic hosts.
III. Earth's history is divided into **eras, periods,** and **epochs.**
   A. During **Precambrian time,** which extended from approximately 4.6 billion years ago up to 570 million years ago, life began and diverged into different groups of bacteria, protists (including algae), fungi, and animals.
      1. Signs of Precambrian life date back to about 3.8 billion years ago.
      2. One rich source of Precambrian fossil deposits is the Ediacaran Hills in South Australia. **Ediacaran fossils** of multicellular animals are from very late in Precambrian time—from 590 to 570 million years ago.
   B. During the **Paleozoic era,** which began approximately 570 million years ago and lasted approximately 222 million years, all major groups of plants, except for flowering plants, and all animal phyla appeared.
      1. During the Cambrian period, the pace of evolution was so rapid that this period has been nicknamed the **Cambrian explosion.** The most important fossil site that documents the Cambrian explosion is the **Burgess Shale** in British Columbia.
      2. Fish and amphibians flourished, and reptiles appeared and diversified during the Paleozoic era.
      3. The greatest mass extinction of all time occurred at the end of the Paleozoic era, some 250 million years ago. More than 90% of all existing marine species became extinct at this time, as well as many terrestrial species.
   C. The **Mesozoic era** began about 248 million years ago and lasted some 183 million years.

1. The Mesozoic era was characterized by the appearance of flowering plants and the evolutionary diversification of reptiles. Dinosaurs, which descended from early reptiles, dominated Earth during the Mesozoic era. Insects flourished, and birds and early mammals evolved.
2. At the end of the Cretaceous period, 65 million years ago, a great many animals abruptly became extinct. A collision of Earth with a large extraterrestrial body may have resulted in dramatic climate changes that played a role in this mass extinction episode.

D. In the **Cenozoic era,** which extends from 65 million years ago to the present time, flowering plants, birds, insects, and mammals diversified greatly.
   1. All the modern orders of mammals appeared and diversified during the Eocene epoch.
   2. Birds diversified during the Eocene epoch, adapting to various lifestyles and habitats. The songbirds diversified extensively during the Miocene epoch.

POST-TEST

1. Energy, the absence of oxygen, chemical building blocks, and time were the requirements for (a) chemical evolution (b) biological evolution (c) the Cambrian explosion (d) the mass extinction episode at the end of the Cretaceous period (e) continental drift
2. Protobionts (a) form spontaneously in hydrothermal vents in the ocean floor (b) obtain the organic molecules they need from the environment (c) are assemblages of abiotically produced organic polymers that resemble living cells in several ways (d) use sunlight to split hydrogen sulfide (e) are fossilized mats of cyanobacteria
3. Many scientists think that _____ was the first information molecule to evolve. (a) DNA (b) RNA (c) a protein (d) an amino acid (e) a lipid
4. According to the endosymbiont theory (a) life originated from nonliving matter (b) the pace of evolution quickened at the start of the Cambrian period (c) chloroplasts, mitochondria, and possibly other organelles originated from symbiotic relationships among prokaryotic organisms (d) banded iron formations reflect the buildup of sufficient oxygen in the atmosphere to oxidize iron at Earth's surface (e) the first photosynthetic organisms appeared 3.1 to 3.5 billion years ago

5. All geological time prior to the beginning of the Paleozoic era some 570 million years ago is informally known as (a) the Cenozoic era (b) the Paleozoic era (c) the Mesozoic era (d) Precambrian time (e) the Cambrian period
6. The correct chronological order of geological eras, starting with the oldest, is (a) Paleozoic, Cenozoic, and Mesozoic (b) Mesozoic, Cenozoic, and Paleozoic (c) Mesozoic, Paleozoic, and Cenozoic (d) Paleozoic, Mesozoic, and Cenozoic (e) Cenozoic, Paleozoic, and Mesozoic
7. The greatest time of evolutionary diversification in the history of life occurred during the (a) Cambrian period (b) Ordovician period (c) Silurian period (d) Carboniferous period (e) Permian period
8. The greatest mass extinction episode in the history of life occurred at what boundary? (a) Pliocene-Pleistocene (b) Paleozoic-Mesozoic (c) Mesozoic-Cenozoic (d) Cambrian-Ordovician (e) Triassic-Jurassic
9. The Age of Reptiles corresponds to the (a) Paleozoic era (b) Mesozoic era (c) Cenozoic era (d) Pleistocene epoch (e) Permian period
10. Flowering plants and mammals diversified and became dominant during the (a) Paleozoic era (b) Mesozoic era (c) Cenozoic era (d) Devonian period (e) Cambrian period

REVIEW QUESTIONS

1. What are the four requirements for chemical evolution, and why is each essential?
2. How did the presence of molecular oxygen in the atmosphere affect early life?
3. Give at least two types of evidence that support the endosymbiont theory.

4. Put the following organisms in order of appearance in the fossil record, starting with the earliest (a) eukaryotic cells, multicellular organisms, prokaryotic cells; (b) reptiles, mammals, amphibians, fish; (c) flowering plants, ferns, gymnosperms.

YOU MAKE THE CONNECTION

1. If you were experimenting on how protobionts evolved into cells and you developed a protobiont that was capable of self-replication, would you consider it a living cell? Why or why not?
2. Why did the evolution of complex multicellular organisms such as plants

and animals have to be preceded by the evolution of oxygen-producing photosynthesis?
3. How might studying outer space help us reconstruct the evolutionary history of life on Earth?

RECOMMENDED READINGS

Agenbroad, L.D. "This Land/South Dakota: Mammoth Site." *Natural History,* Oct. 1997. Hot Springs Mammoth Site in South Dakota has the world's largest known accumulation of mammoth fossils.

Currier, P.J. "The Great Dinosaur Egg Hunt." *National Geographic,* Vol. 189, No. 5. May 1996. Fossil dinosaur eggs, embryos, and nests provide fascinating information about how dinosaurs lived.

DiSilvestro, R.L. "In Quest of the Origin of Birds." *BioScience,* Vol. 47, No. 8, Sept. 1997. A balanced examination of the two opposing ideas about the immediate ancestors of birds (dinosaurs and thecodonts).

Erwin, D.H. "The Mother of Mass Extinctions." *Scientific American,* Vol. 275, No. 1, Jul. 1996. The worst mass extinction episode in Earth's history occurred at the end of the Permian period.

Levin, H.L. *The Earth Through Time,* 5th ed. Saunders College Publishing, Philadelphia, 1996. Contains detailed discussions of the biological events of each geological era.

Morell, V. "A Cold, Hard Look at Dinosaurs." *Discover,* Dec. 1996. A comparison of nasal structure in skulls of living animals and extinct dinosaur skulls suggests that dinosaurs were ectothermic (cold-blooded).

Nash, J.M. "When Life Exploded." *Time,* 4 Dec. 1995. Examines the wealth of Cambrian fossils from such famous locations as the Burgess Shale in the Canadian Rockies.

Thomas, D.A. and J.O. Farlow. "Tracking a Dinosaur Attack." *Scientific American,* Vol. 277, No. 6, Dec. 1997. Fossil dinosaur imprints provide a dramatic story of a quadrupedal herbivorous dinosaur under attack by a bipedal carnivorous dinosaur.

Wright, K. "When Life Was Odd." *Discover,* Mar. 1997. Examines the bizarre Ediacaran fauna that existed during Precambrian times.

Zimmer, C. "First Cell." *Discover,* Nov. 1995. This article highlights some exciting research on chemical evolution.

● Visit our website at **http://www.saunderscollege.com/lifesci/titles.html** and click on Solomon/Berg/Martin Biology for links to chapter-related resources on the World Wide Web.

CHAPTER 21

The Evolution of Primates

Twelve years after Darwin wrote *The Origin of Species by Natural Selection,* he published another controversial book, *The Descent of Man,* which addressed human evolution. In it, Darwin hypothesized that humans and apes share a common ancestry. For nearly a century after Darwin, fossil evidence of human ancestry remained fairly incomplete. However, research over the last few decades, especially in Africa, has yielded fossils that provide an increasingly clear answer to the question, "Where did we come from?" Fossil evidence has allowed **paleoanthropologists,** scientists who study human evolution, to infer not only the structure but also the habits of early humans.

Humans and other primates, such as lemurs, tarsiers, monkeys, and apes, are mammals, members of the class Mammalia. Mammals are **endothermic** (they use metabolic energy to maintain a constant body temperature); produce body hair for such functions as insulation, protective coloration, and waterproofing; and feed their young with milk from mammary glands. Most mammals are **viviparous,** which means that their eggs develop into young offspring within the female body.

It is currently thought that the three groups of living mammals—the monotremes, marsupials, and placental mammals—are all derived from the same lineage. The **monotremes** are mammals, such as the duck-billed platypus, that lay eggs. The **marsupials,** such as kangaroos and opossums, carry their young in an abdominal pouch after giving birth to them in a very underdeveloped condition. **Placental mammals,** the largest and most successful group, possess a **placenta,** an organ that exchanges materials between the mother and the embryo/fetus developing in the uterus; placental mammals give birth to their young in a more developed condition than marsupials.

Mammals arose from mammal-like reptiles known as therapsids more than 200 million years ago, during the Mesozoic era (see Chapter 20). These early mammals remained a minor component of life on Earth for almost 150 million years be-

(Warren and Genny Garst/Tom Stack & Associates)

fore rapidly diversifying during the Cenozoic era (the last 65 million years). The first primates appeared at this time, apparently descendants of small shrewlike placental mammals that lived in trees and ate insects, much like the modern tree shrew pictured here. Many traits of humans and the 200 other living primate species are related to their **arboreal** (tree-dwelling) past.

1. Describe the structural adaptations that primates possess for life in trees and explain why even primates that live on the ground have these adaptations.
2. List the two semiorders of primates and give representative examples of each.
3. Distinguish among mammals, primates, anthropoids, hominoids, and hominids (hominines).
4. Describe skeletal and skull differences between apes and hominids.
5. Compare the following early hominids: *Ardipithecus ramidus*, *Australopithecus anamensis*, *Australopithecus afarensis*, and *Australopithecus africanus*.
6. Distinguish among the following members of genus *Homo*: *H. habilis*, *H. ergaster*, *H. erectus*, *H. heidelbergensis*, *H. neanderthalensis*, and *H. sapiens*.
7. Discuss the current debate over the origin of modern humans and briefly describe the opposing "out of Africa" and "multiregional" hypotheses.
8. Describe cultural evolution and its impact on the biosphere.

EARLY PRIMATE EVOLUTION REFLECTED AN ARBOREAL EXISTENCE

Fossil evidence indicates that the first true primates appeared by the early Eocene about 55 million years ago. These early primates had digits with nails, and their eyes were directed somewhat forward. The climate was milder then, and early primates were widely distributed over much of North America, Europe, and Asia. (Recall from Chapter 19 that North America was still attached to Europe at that time.) As the climate became cooler and drier toward the end of the Eocene epoch, many of these early primates became extinct.

TABLE 21–1	A Classification of Living Groups in the Order Primates

Semiorder Strepsirhini
　Suborder Lemuriformes
　　Family Cheirogaleidae (dwarf lemurs)
　　Family Lemuridae (lemurs)
　　Family Indriidae (indris)
　　Family Daubentoniidae (aye-ayes)
　　Family Lorisidae (lorises)
　　Family Galagidae (galagos)

Semiorder Haplorhini
　Suborder Tarsiformes
　　Family Tarsiidae (tarsiers)
　Suborder Anthropoidea
　　Family Callitrichidae (marmosets)
　　Family Cebidae (New World monkeys)
　　Family Cercopithecidae (Old World monkeys)
　　Family Hylobatidae (gibbons)
　　Family Pongidae (orangutans)
　　Family Hominidae (gorillas, chimpanzees, and humans)

Several novel adaptations evolved in early primates that allowed them to live in trees. One of the most significant features of primates is that they have five grasping digits: four fingers plus an opposable thumb or toe. The partially or fully opposable first digit enables primates to grasp objects such as branches. Nails (instead of claws) provide a protective covering for the tips of the digits, and the fleshy pads at the ends of the digits are sensitive to touch. Another arboreal feature is long, slender limbs that rotate freely at hips and shoulders, giving primates full mobility to climb and search for food in the tree tops. The location of eyes in front of the head provides stereoscopic, or three-dimensional, vision. Stereoscopic vision is essential for arboreal animals, especially those that leap from branch to branch, because an error in depth perception might cause a fatal fall. In addition to sharp sight, hearing is acute in primates.

Primates share several other characteristics, including a relatively large brain size. It has been suggested that increased sensory input associated with their sharp vision and greater agility favored the evolution of larger brains. Primates also have complex social behaviors. Females usually bear one offspring at a time; the baby is helpless and requires a long period of nurturing and protection.

LIVING PRIMATES ARE CLASSIFIED INTO TWO SEMIORDERS

Most biologists currently divide the order Primates into two semiorders (Table 21–1 and Fig. 21–1). The semiorder Strepsirhini includes lemurs, galagos, and lorises, whereas the semiorder Haplorhini includes tarsiers and **anthropoids** (monkeys, apes, and humans). This classification scheme indicates that tarsiers share a more recent common ancestor with anthropoids than do lemurs.

All lemurs are restricted to the island of Madagascar off the coast of Africa (Fig. 21–2). Because of extensive habitat destruction and hunting, they are highly endangered. Lorises, which are found in tropical areas of Southeast Asia and Africa,

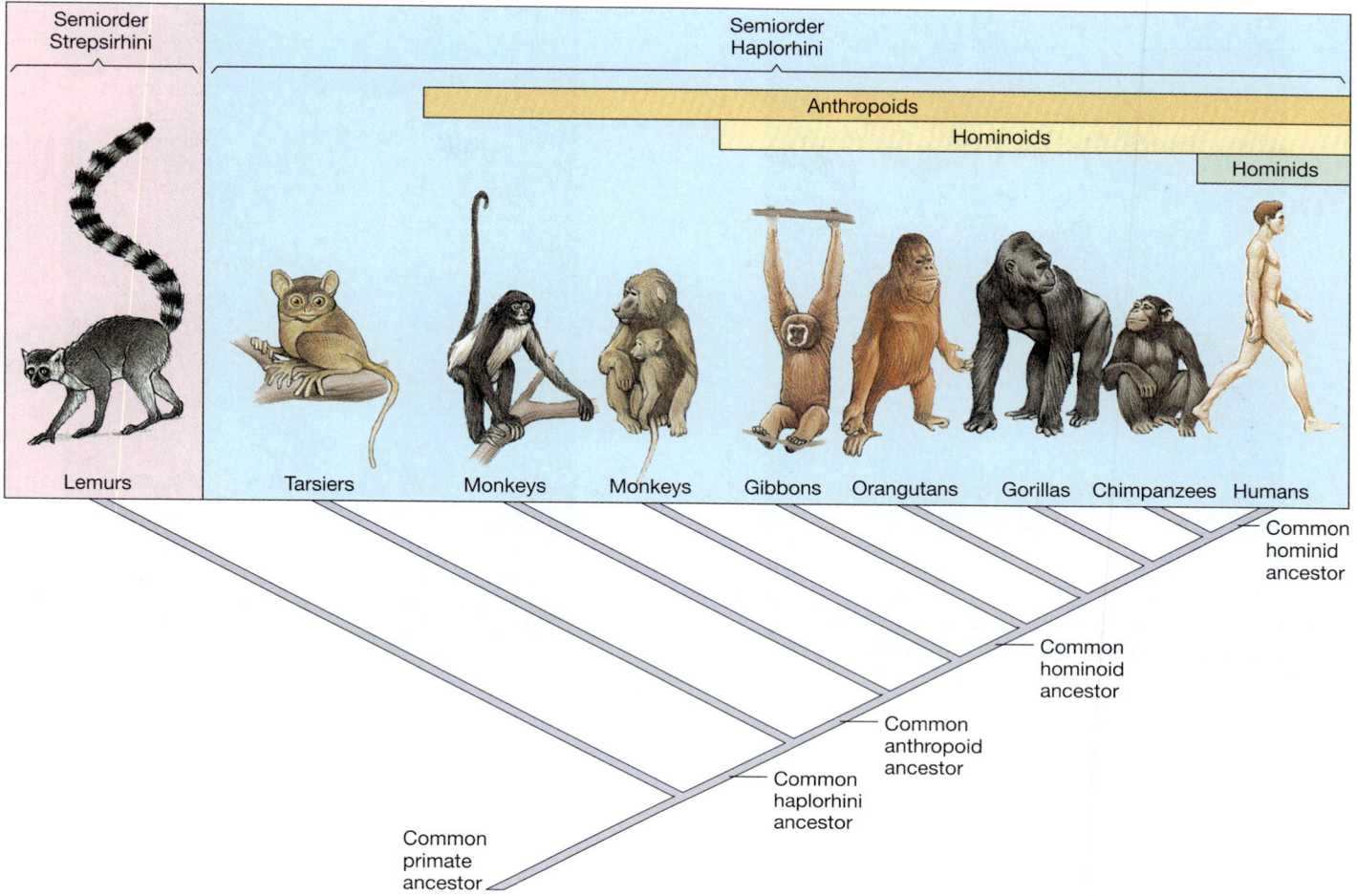

Figure 21-1 **Primate evolution.** This diagram shows hypothetical phylogenetic relationships among living primates, based on current scientific evidence. (Figures are not drawn to scale.)

resemble lemurs in many respects, as do galagos, which live in sub-Saharan Africa. Lemurs, lorises, and galagos have retained some early mammalian features, such as elongated, pointed faces and more lateral-facing eyes.

Tarsiers are found in rain forests of Indonesia and the Philippines. They are small primates the size of squirrels and are very adept climbers and leapers (Fig. 21–3). These nocturnal primates resemble anthropoids in a number of ways, including their shortened snouts and forward-pointing eyes.

The suborder Anthropoidea includes monkeys, apes, and humans

Anthropoid primates arose during the middle Eocene epoch, at least 45 million years ago. Several different fossil anthropoids have been identified from Asia and North Africa. However, there is currently no scientific consensus about the relationships of these fossil groups to one another or to living anthropoids. Evidence indicates that anthropoids originated in

Africa or Asia, quickly spread throughout Europe, Asia, and Africa, and arrived in South America later.

Probably the most significant difference between anthropoids and other primates is in the size of their brains. The cerebrum, in particular, is more developed in monkeys, apes, and humans, where it functions as the center for learning, voluntary movement, and interpretation of sensation.

Monkeys are generally diurnal (active during the day) tree dwellers. They tend to eat fruit and leaves, with nuts, seeds, buds, insects, spiders, birds' eggs, and even small vertebrates playing a smaller part in their diets. The two main groups of monkeys, New World monkeys and Old World monkeys, are named for the hemispheres where they diversified; monkeys in South and Central America are called New World monkeys, whereas monkeys in Africa, Asia, and Europe are called Old World monkeys. New and Old World monkeys have been evolving separately for millions of years. Fossil evidence indicates that monkeys lived in South America as long as 25 million years ago, but one of the most important unanswered

Figure 21-2 Lemurs. A mother ring-tailed lemur (*Lemus catta*) and her baby share a piece of fruit. Lemurs are native to Madagascar. (*Frans Lanting/Minden Pictures*)

Figure 21-3 Tarsiers. The huge eyes of the tarsier (*Tarsius bancanus*) help it find insects, lizards, and other prey when it hunts at night. Tarsiers live in the rain forests of Indonesia and the Philippines. (*Doug Wechsler*)

questions in anthropoid evolution concerns *how* monkeys got to South America. Africa and South America had already drifted apart (see Chapter 19), so the ancestors of New World monkeys may have rafted on logs from Africa to South America or dispersed from Asia to North America to South America.

New World monkeys are restricted to Central and South America and include marmosets, capuchins, howler monkeys, squirrel monkeys, and spider monkeys. New World monkeys are arboreal and some possess long, slender limbs that permit easy movement in the trees (Fig. 21-4*a*). Some have **prehensile** tails capable of wrapping around branches and serving as

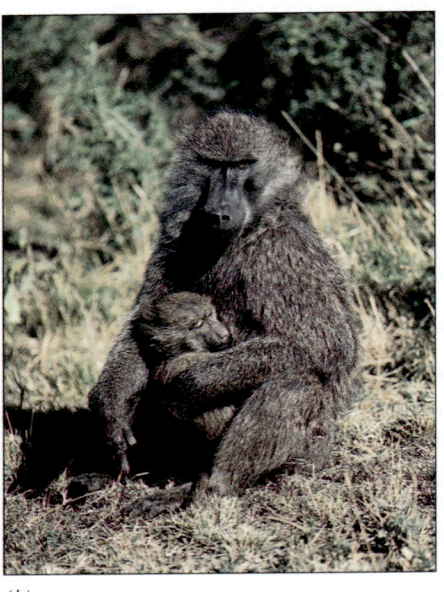

Figure 21-4 New and Old World monkeys. (*a*) The white-faced monkey (*Cebus capucinus*) has a prehensile tail and a flattened nose with nostrils directed to the side. (*b*) The Anubis baboon (*Papio anubis*) is an Old World monkey native to Africa. Note how its nostrils are directed downward (*a, C.C. Lockwood/DRK Photo; b, S. Meyers/Okapia/Photo Researchers, Inc.*)

(*a*) (*b*)

fifth limbs. Some New World monkeys have shorter thumbs, and in certain cases the thumbs are totally absent. Their facial anatomy is different from that of the Old World monkeys; they have flattened noses with the nostrils opening to the side. They live in groups and exhibit complex social behavior.

Old World monkeys are distributed in tropical parts of Africa and Asia. In addition to baboons and macaques (pronounced muh-kacks'), the Old World monkeys include guenons, mangabeys, langurs, and colobus monkeys. Most Old World monkeys are arboreal, although some, such as baboons and macaques, spend much of their time on the ground (Fig. 21–4*b*). The ground dwellers, which are **quadrupedal** ("four-footed"; they walk on all fours), arose from arboreal monkeys. None of the Old World monkeys has a prehensile tail, and some have extremely short tails. They have a fully opposable thumb, and unlike the New World monkeys, their nostrils are closer together and directed downward. Old World monkeys are very social animals.

Modern classification places apes and humans in three families

Old World monkeys shared a common ancestor with the **hominoids,** a group composed of apes and **hominids** (humans and their ancestors; also called **hominines**). One of the earliest anthropoids was discovered in Egypt and named *Aegyptopithecus* (Fig. 21–5*a*). *Aegyptopithecus,* a cat-sized, forest-dwelling arboreal monkey with a few apelike characteristics, lived during the Oligocene, approximately 35 million years ago. During the Miocene epoch, which began approximately 25 million years ago, apes and Old World monkeys diverged. At least 30 different early hominoids lived during the Miocene epoch, but most of them became extinct and were not the common ancestor of modern apes and humans. Fossils of an early, forest-dwelling, chimpanzee-sized ape, *Dryopithecus,* are of special interest because this hominoid may have given rise to modern apes as well as to the human line (Fig. 21–5*b*). Dryopithecines were arboreal, arm-swinging apes. The dryopithecines were distributed widely across Europe, Africa, and Asia. As the climate gradually cooled and became drier, their range became more limited.

Dryopithecus is not the only candidate for the lineage that gave rise to modern apes and humans. Fossil finds of two other early apes (*Kenyapithecus* and *Morotopithecus*) were each reported in 1996 to be the earliest fossil evidence discovered so far of an ape with a body plan similar to those of modern apes and humans. The new fossils have generated many questions about the relationships among the various early apes, and as these and future fossil finds are evaluated, they may lead to a rearrangement of ancestors in the hominoid family tree.

Many biologists classify the five genera of hominoids alive today into three families (Fig. 21–6): Gibbons (*Hylobates*) are known as lesser apes and are placed in the family Hylobatidae. Orangutans (*Pongo*) are placed in the family Pongidae. The family Homininae includes gorillas (*Gorilla*), chimpanzees (*Pan*), and humans (*Homo*).

(*a*) Early anthropoid, *Aegyptopithecus*

(*b*) Early ape, *Dryopithecus*

Figure 21–5 *Aegyptopithecus* and *Dryopithecus.* (*a*) Fossils of *Aegyptopithecus,* an early anthropoid, were discovered in Egypt. (*b*) *Dryopithecus,* a more advanced ape, may have given rise to modern hominoids. (Figures are not drawn to scale.)

Gibbons are natural acrobats that can **brachiate,** or arm-swing, with their weight supported by one arm at a time. Orangutans are also brachiating tree dwellers, but chimpanzees, bonobos, and especially gorillas have adapted to life on the ground. They have retained long arms typical of brachiating primates but use these to assist in quadrupedal walking, sometimes known as **knuckle-walking** because of the way they fold their digits when moving. Apes, like humans, lack tails. They are generally much larger than monkeys; gibbons are a notable exception.

Evidence of the close relatedness of orangutans, gorillas, chimps, and humans is abundant at the molecular level. The amino acid sequence of the chimpanzee's hemoglobin is identical to that of the human; those of the gorilla and rhesus monkey differ from the human's in 2 and 15 amino acids, respectively. DNA sequence analyses indicate that chimpanzees are likely to be our nearest living relatives among the apes (see Table 17–1 and Fig. 17–15). Molecular and fossil evidence demonstrates that gorillas may have diverged from the chimpanzee and hominid lines some 8 to 10 million years ago, whereas chimpanzees and hominids probably separated about 6 million years ago.

Figure 21–6 **The five hominoid genera.**
(a) White-handed gibbons (*Hylobates lar*) are extremely acrobatic and often move through the trees by brachiation. *(b)* An orangutan (*Pongo pygmaeus*) mother and baby. *(c)* A young lowland gorilla (*Gorilla gorilla*) in knuckle-walking stance. *(d)* A bonobo chimpanzee (*Pan paniscus*) grooms another member of the group. Bonobos are endemic to a single country, Zaire. *(e)* A group of adolescent humans (*Homo sapiens*). *(a, Joe McDonald/Visuals Unlimited; b, A. Compost/Peter Arnold, Inc.; c, Nancy Adams/Tom Stack & Associates; d, K. & K. Ammann/Bruce Coleman, Inc.; e, Tim Davis/Photo Researchers, Inc.)*

THE FOSSIL RECORD PROVIDES CLUES TO HOMINID EVOLUTION

Scientists have a growing storehouse of hundreds of fossil hominids, which provide much useful information about general trends in the body design, appearance, and behavior of ancestral humans. It is evident, for example, that early hominids adopted a **bipedal** (two-footed) posture before their brains enlarged. Despite the wealth of fossil evidence, however, interpretations of hominid characteristics, taxonomy, and phylogeny continue to be vigorously debated, and every discovery raises new questions. Furthermore, hominid evolution, like other scientific fields, is influenced by the different perspectives of the various workers studying it. The lack of a scien-

Foramen magnum at the center base of the skull

Complex curvature of human spine

Simply curved spine

Figure 21–7 Gorilla and human skeletons. When gorilla and human skeletons are compared, the skeletal adaptations for bipedalism in humans become apparent.

Foramen magnum at the center rear of skull

Shorter, broader pelvis (front view)

Tall, narrow pelvis (front view)

First toe not opposable, and all toes aligned

First toe opposable and not aligned with others

Human skeleton

Gorilla skeleton

tific consensus regarding certain aspects of hominid evolution is, therefore, an expected part of the scientific process.

Evolutionary changes from the earliest hominids to modern humans are evident in some of the characteristics of the skeleton and skull. Compared with the ape skeleton, the human skeleton possesses distinct differences that reflect our ability to stand erect and walk on two feet (Fig. 21–7). These differences also reflect the habitat change for early hominids, from an arboreal existence in the forest to a life spent at least partly on the ground. The curvature of the human spine provides better balance and weight distribution for bipedal locomotion. The human pelvis is shorter and broader than the ape pelvis, providing a better attachment of muscles used for upright walking. The hole in the base of the skull for the spinal cord, called the **foramen magnum,** is located in the middle of the rear of the skull in apes. In contrast, the human foramen magnum is centered in the skull base, positioning the head for erect walking. An increase in the length of the legs relative to the arms, and alignment of the big toe with the rest of the toes, further adapted the early hominids for bipedalism.

Another major trend in hominid evolution was an increase in the size of the brain relative to the size of the body (Fig. 21–8). In addition, the ape skull possesses prominent bony ridges above the eye sockets, whereas these **supraorbital ridges** are lacking in modern human skulls. Human faces are flatter than those of apes, and the jaws are different. The arrangement of teeth in the ape jaw is somewhat rectangular, compared with a rounded, or U-shaped, arrangement in humans. Apes have larger front teeth (canines and incisors) than do humans, and their canines are especially large. Gorillas and orangutans also have larger back teeth (premolars and molars) than humans.

The earliest hominids belong to the genus *Ardipithecus*

Hominid evolution began in Africa. The earliest hominid, which belongs to the genus *Ardipithecus,* appeared about 4.4 million years ago (Fig. 21–9). *Ardipithecus* gave rise to *Australopithecus,* a genus that includes several species that lived between 4 and 1.25 million years ago. These two genera of early

Figure 21–8   Gorilla and human heads. *(a)* The ape skull has a pronounced supraorbital ridge. *(b)* The human skull is flatter in the front and has a more pronounced chin. The human brain, particularly the cerebrum *(purple)*, is larger than that of an ape, and the human jaw is structured so that the teeth are arranged in a U shape. Human canines and incisors are also smaller than those of apes.

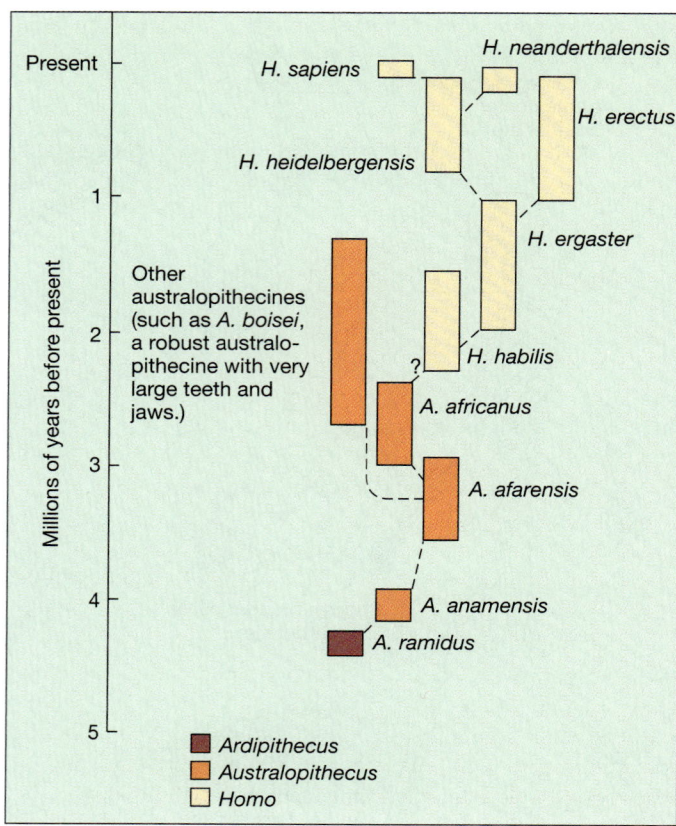

Figure 21–9   One interpretation of hominid evolution. Paleoanthropologists are not in complete agreement about certain specific details of the human family tree, and there are many possible interpretations of the human lineage.

hominids, often referred to as **australopithecines** or "southern man apes," had longer arms, shorter legs, and smaller brains relative to modern humans. The actual number of australopithecine species for which fossil evidence has been found is under debate. Differences in the relatively few skeletal fragments could indicate either variation among individuals within a species or evidence of separate species. Most paleoanthropologists recognize at least six species of australopithecines.

The first fossils of the earliest hominids were discovered in 1992 and assigned to ***Ardipithecus ramidus*** in 1995. The specific epithet *ramidus* is derived from a word meaning "root" in the Afar language, spoken in the region of Ethiopia where the fossils were found. This hominid, which is more primitive than any other known hominid, is quite close to the "root" of the human family tree, that is, to the last common ancestor of bipedal hominids and quadrupedal African great apes. Because no leg bones were found in the initial discovery, it has not yet been determined if *A. ramidus* was bipedal. Future discoveries may clarify this important point.

The genus *Australopithecus* contains the immediate ancestors of the genus *Homo*

Hominids that existed between 3.9 and 4.2 million years ago are assigned to the species ***Australopithecus anamensis,*** first named in 1995 by Maeve Leakey and her coworkers from fossils discovered in East Africa. This hominid species presumably arose from *Ardipithecus ramidus*. The jaws of *A. anamensis* are like those of modern apes, whereas the teeth are more like those of later hominids. Males were about 1.55 meters (5 feet 1 inch) in height, and females were about 1.30 meters (4 feet 3 inches) tall.[1] A tibia (leg bone) indicates that *A. anamensis* had an upright posture and was bipedal, although it also may have foraged in the trees. Thus, bipedalism occurred early in human evolution and may have been the first human adaptation.

---

[1]All heights given in this text represent average approximations based on available fossil data. It is reasonable to assume that individuals within a hominid species varied considerably in height, much like the variation observed in present-day humans.

Australopithecus afarensis, another primitive hominid, may have arisen from *A. anamensis.* Many fossils of *A. afarensis* skeletal remains have been discovered in Africa, including a remarkably complete skeleton nicknamed "Lucy" found in Ethiopia in 1974 by a team led by Donald Johanson. Lucy, a small hominid approximately 1.04 meters (3 feet 5 inches)—males were about 1.50 meters, or 4 feet 11 inches—is thought to be about 3.2 million years old. In 1976, beautifully preserved fossil footprints of three *A. afarensis* individuals who walked more than 3.6 million years ago were discovered by Mary Leakey and coworkers. In 1994 the first adult skull of *A. afarensis* was discovered by William Kimbel and other paleoanthropologists. The skull, characterized by a relatively small brain, pronounced supraorbital ridges, a jutting jaw, and large canine teeth, is an estimated 3 million years old. It is probable that *A. afarensis* did not construct tools or make fires, as no evidence of tools or fire has been found at fossil sites.

Many paleoanthropologists think *A. afarensis* gave rise to several australopithecine species, including ***Australopithecus africanus,*** which appeared approximately 3 million years ago. The first *A. africanus* fossil was discovered in South Africa in 1924, and since then hundreds have been found. Males were about 1.37 meters (4 feet 6 inches), and females were about 1.14 meters (3 feet 9 inches). This hominid walked erect and possessed hands and teeth that were distinctly human-like. Based on characteristics of the teeth, it is thought that *A. africanus* ate both plants and animals. Like *A. afarensis,* it had a small brain, more like that of its primate ancestors than of present-day humans.

Homo habilis is the oldest member of the genus *Homo*

The first hominid to have enough uniquely human features to be placed in the same genus as modern humans is ***Homo habilis.*** *Homo habilis* was a small hominid (males were about 1.32 meters, or 4 feet 4 inches, and females were about 1.17 meters, or 3 feet 10 inches) with a larger brain and smaller teeth than the australopithecines. This early human appeared approximately 2.3 million years ago and persisted for more than 0.75 million years. Fossils of *H. habilis* have been found in numerous areas in Africa. These sites contain primitive tools, stones that had been chipped, cracked, or hammered to make sharp edges for cutting or scraping.[2] Oldowan pebble choppers, for example, were probably used to cut through animal hides to obtain meat.

The relationship between the australopithecines and *H. habilis* is not clear. Using physical characteristics of their fossilized skeletons as evidence, many paleoanthropologists have inferred that the australopithecines were ancestors of *H. habilis.* Discoveries of additional fossils may help clarify these relationships.

Homo erectus apparently evolved from *Homo habilis*

Numerous fossils of ***Homo erectus*** have been found throughout Africa and Asia. *Homo erectus* is thought to have originated in Africa about 2 million years ago and then to have spread quickly to Europe and Asia. The oldest fossils of *H. erectus* that have been found in Southeast Asia, for example, may be as old as 1.8 million years, although the most widely accepted date is about 1 million years. Peking man and Java man, discovered in Asia, were later examples of *H. erectus,* which existed until at least 200,000 years ago; some populations of *H. erectus* may have persisted until as recently as 27,000 to 53,000 years ago.

Homo erectus was taller than *H. habilis;* males were about 1.78 meters (5 feet 10 inches), and females were about 1.60 meters (5 feet 3 inches). Its brain, which was larger than that of *H. habilis,* got progressively larger during the course of its existence. Its skull, although larger, did not possess totally modern features, retaining the heavy supraorbital ridge and projecting face that are more characteristic of its ape ancestors (Fig. 21–10).

The increased mental faculties associated with an increased brain size enabled these early humans to make more advanced stone tools, known as Acheulian tools, including hand axes and other implements that have been interpreted as choppers, borers, and scrapers. Their intelligence also allowed them to survive in cold areas. *Homo erectus* obtained food by hunting or scavenging and may have worn clothing, built fires, and lived in caves or shelters. To date, no evidence of weapons has been unearthed at *Homo erectus* sites.

Ideas regarding *Homo erectus,* like many other aspects of human evolution, are changing with each new fossil discovery. Many scientists now hypothesize that the fossils classified as *Homo erectus* really represent two species: ***Homo ergaster,*** an earlier African species, and *Homo erectus,* a later eastern Asian offshoot. Researchers who support this split speculate that *Homo ergaster* may be the direct ancestor of later humans, whereas *Homo erectus* may be an evolutionary dead end. It is hoped that future fossil discoveries will help clarify the status of *Homo erectus.*

Archaic *Homo sapiens* appeared about 800,000 years ago

Archaic *Homo sapiens* are regionally diverse descendants of *Homo erectus* or *Homo ergaster* that lived in Africa, Asia, and Europe from about 800,000 to 100,000 years ago. They thus overlapped both *Homo erectus* populations and the later-appearing Neandertals (discussed shortly). Males were about 1.75 meters (5 feet 9 inches), and females were about 1.57 meters (5 feet 2 inches). Some researchers classify archaic *Homo sapiens* as a separate species, ***Homo heidelbergensis.***

---

[2]The oldest known stone tools, discovered in the mid-1990s by Sileshi Semaw and colleagues, were found in Gona, Ethiopia. These ancient tools were made some 2.5 million years ago, but because no hominid remains have been found at the site yet, it is not known who made them. Evidence exists that certain australopithecines (for example, *Australopithecus robustus*) fashioned crude tools.

Figure 21–10 A replica of a *Homo erectus* skull. Note the receding forehead, pronounced supraorbital ridge, and projecting face and jaws. *(Dennis Drenner)*

Receding
forehead

Pronounced
supraorbital
ridge

Projecting
face/jaws

Neandertals appeared approximately 230,000 years ago

Neandertals, formerly spelled Neanderthals,[3] were first discovered in the Neander Valley in Germany. They lived throughout Europe and western Asia from about 230,000 to 30,000 years ago. These early humans had short, sturdy builds; males were about 1.65 meters (5 feet 5 inches) and weighed an estimated 83.9 kilograms (185 pounds), and females were about 1.55 meters (5 feet 1 inch) and weighed an estimated 79.4 kilograms (175 pounds). Their faces projected slightly, their chins and foreheads receded, they had heavy brow ridges and jaw bones, and their brains and front teeth were larger than those of modern humans. Their nasal cavities contained triangular bony projections not found in modern humans; scientists have suggested that these projections provided larger surface areas in Neandertal sinuses, enabling them to better warm the cold air of Ice Age Eurasia as air traveled through the head to the lungs.

Scientists have not reached a consensus about whether the Neandertals are a separate species from modern humans. Many think that anatomical differences between Neandertals and modern humans mean that they were separate species, *Homo neanderthalensis* and *Homo sapiens.* Other scientists disagree and think that Neandertals were a race of *Homo sapiens.*

Neandertal tools include the oldest known spear points; known as Mousterian tools, Neandertal tools were more sophisticated than those of *H. erectus.* Studies of Neandertal sites indicate that they hunted large animals. The existence of skeletons of elderly Neandertals and of Neandertals with healed fractures demonstrates that they cared for the aged and the sick, an indication of advanced social cooperation. They apparently had rituals, possibly of religious significance, and buried their dead. The presence of food, weapons, and flowers in some of their graves suggests that they possessed the abstract concept of an afterlife.

The disappearance of the Neandertals some 30,000 years ago is a mystery that has sparked debate among paleoanthropologists. Other groups of *H. sapiens* with more modern features coexisted for tens of thousands of years with the Neandertals. It is possible that the Neandertals interbred with these humans, diluting their features beyond recognition, although a 1997 analysis of mitochondrial DNA (mtDNA) extracted from a Neandertal bone[4] suggests that Neandertals are an evolutionary dead end that did not interbreed with more modern humans. Alternatively, perhaps the other humans out-competed or exterminated the Neandertals, leading to their extinction.

The origin of modern *Homo sapiens* is hotly debated

Homo sapiens with anatomically modern features existed in Africa at least 100,000 years ago. Males were about 1.75 meters (5 feet 9 inches), and females were about 1.60 meters (5

---

[3]The silent "h" has been dropped in modern German.

[4]The Neandertal mtDNA sequence extracted by Svante Pääbo and his colleagues differs significantly from all modern human mtDNA sequences; the Neandertal mtDNA is, however, more similar to human than to chimpanzee mtDNA.

Figure 21–11 Cro-Magnon cave paintings. These are some of the earliest known examples of human art. Discovered in Lascaux, France, these images of reindeer have been interpreted as having religious significance, possibly to guarantee a successful hunt. *(Photo by J. Beckett/D. Stipkovich, courtesy Department of Library Services, American Museum of Natural History)*

feet 3 inches). The *H. sapiens* skull lacked a heavy brow ridge and possessed a distinct chin. The **Cro-Magnon** culture that existed in France and Spain about 30,000 years ago exemplifies these humans. Their weapons and tools were complex and often made of materials other than stone, including bone, ivory, and wood. They made stone blades that were extremely sharp. Cro-Magnons developed art, including cave paintings, engravings, and sculpture, possibly for ritualistic purposes (Fig. 21–11). Their sophisticated tools and art indicate that they may have possessed language, which would have been used to transmit their culture to younger generations.

Two opposing hypotheses currently exist about the origin of these modern humans: the "out of Africa" hypothesis and the "multiregional" hypothesis. The "out of Africa" hypothesis holds that modern *H. sapiens* arose in Africa and then migrated to Europe and Asia, displacing the more primitive humans living there. According to the "multiregional" hypothesis, modern humans originated from *H. erectus* as separately evolving populations living in several parts of Africa, Asia, and Europe. Each of these populations evolved in its own distinctive way but occasionally met and interbred with other populations, thereby preventing complete reproductive isolation; the variation found today in different geographical populations therefore represents a continuation of this multiregional process.

Data from *Homo* fossils, as well as molecular biology and population genetics studies of modern humans, have been cited in support of both hypotheses, and both have vigorous defenders and strong detractors. Such disagreement is an important part of the scientific process because it stimulates research that may ultimately resolve the issue. (See *Making the Con-*

nection: DNA and Human Evolution for a discussion of some of the current molecular data being used to help unravel details of human evolution.)

HUMANS UNDERGO CULTURAL EVOLUTION

Genetically speaking, humans are not very different from other primates. At the level of our DNA sequences, we are roughly 98% identical to gorillas and 99% identical to chimpanzees. Our relatively few genetic differences, however, give rise to several important distinguishing features, such as greater intelligence and the ability to capitalize on it through **cultural evolution,** which is the transmission of knowledge from one generation to the next. Human culture is dynamic; it is modified as we obtain new knowledge. Human cultural evolution is generally divided into three stages: (1) the development of hunter-gatherer societies; (2) the development of agriculture; and (3) the Industrial Revolution.

Early humans were hunters and gatherers who relied on what was available in their immediate environment. They were nomadic, and as the resources in a given area were exhausted or as the population increased, they migrated to a different area. These societies required a division of labor and the ability to make tools and weapons, which were needed not only to kill game but also to scrape hides, dig up roots and tubers, and cook food. Although we are not certain when hunting was incorporated into human society, we do know that it declined in importance approximately 15,000 years ago. This may have

DNA AND HUMAN EVOLUTION

Can molecular biology provide clues about the origin of modern humans? The "out of Africa" hypothesis was originally supported by studies in the late 1980s of mitochondrial DNA from various human populations. In 1992 the statistical assumptions used in one analysis of mitochondrial DNA were found to be erroneous, leading to questions about the validity of this purported demonstration of the "out of Africa" hypothesis. Several other comparative molecular studies, however, have all produced essentially the same answer, that modern humans are descended from an early human population that lived in southern Africa.

One of the largest studies to date, reported in 1996 by Sarah Tishkoff,[*] supports an African origin for humans. Tishkoff examined the genetic variation in two stretches of DNA on human chromosome 12 from 1600 people living in 42 different populations around the world (13 African, 2 Middle Eastern, 7 European, 9 Asian, 3 Pacific, and 8 Amerindian). The DNA segments varied depending on where the populations lived. Based on her research, Tishkoff divided the present-day human population into three groups: sub-Saharan Africans,[†] northeastern Africans, and non-Africans. The sub-Saharan Africans exhibited the greatest genetic diversity, whereas the non-African populations were the least diverse. These findings are consistent with the predictions of the "out of Africa" hypothesis for two reasons. First, according to the hypothesis, the sub-Saharan populations are expected to be more diverse because they are older and have had a longer time to accumulate that diversity. Second, the small populations that emigrated

from Africa could not have been representative of the total diversity present in the larger African population (recall the discussion in Chapter 18 on the effect of genetic drift on allele frequencies).

Tishkoff's work is significant because it provided scientists with an important way to evaluate human origins. Her research did not disprove the "multiregional" hypothesis, but pointed the direction for additional research on other segments of human DNA. A series of recent genetic studies of both mitochondrial and nuclear DNA has strengthened the case for Africa as the birthplace of modern humans. Some of this research has shown, however, that the "out of Africa" hypothesis may not be as simple as originally envisioned, that is, modern humans from Africa may not have completely replaced the humans on other continents but may have interbred with them. While many of the human genes that were analyzed have demonstrated an African ancestry, for example, a few appear to have arisen in Asia and to have been introduced at a later time into African populations, probably by migration from Asia to Africa. Thus, ancestral human populations in both Africa and Asia may have contributed to the gene pool of modern humans.

[*]Tishkoff, S. A. et al. "Global Patterns of Linkage Disequilibrium at the CD4 Locus and Modern Human Origins." *Science,* Vol. 271, 8 Mar., 1996.

[†]Sub-Saharan Africa refers to all countries located south of the Sahara Desert.

---

been due to a decrease in the abundance of large animals, triggered in part by overhunting. A few isolated groups of hunter-gatherer societies, including the Inuit of northern polar regions and the Mbuti of Africa, have survived into the 20th century.

Development of agriculture resulted in a more dependable food supply

Evidence that humans had begun to cultivate crops approximately 10,000 years ago includes the presence of agricultural tools and plant material at archaeological sites. Agriculture, which involves keeping animals as well as cultivating plants, resulted in a more dependable food supply. Recent archaeological evidence suggests that agriculture arose in several steps. Although there is much variation from one site to another, plant cultivation, in combination with hunting, usually occurred first. Animal domestication followed later. Agriculture, in turn, often led to more permanent dwellings because considerable time was invested in growing crops in one area. Villages and cities often grew up around the farmlands, but the connection between agriculture and the establishment of vil-

lages and towns is complicated by recent discoveries. For example, Abu Hureyra in Syria was a village founded *before* agriculture arose. The villagers subsisted on the rich plant life of the area and the migrating herds of gazelle. Once people turned to agriculture, however, they seldom went back to hunting and gathering to obtain food.

Archaeological evidence indicates that agriculture developed independently in several different regions. There were three main centers of agriculture and several minor ones. Each of the main centers was associated with cultivation of a cereal crop, although other foods were grown as well. Cereals are grasses, which are members of the monocot group of flowering plants (see Chapter 27). The cereals associated with the three main centers of agriculture are wheat, corn, and rice.

Wheat was cultivated in the semiarid regions along the eastern edge of the Mediterranean. Other crops that originated there include peas, lentils, grapes, and olives. Central and South America were the sites of the maize culture. Squash, chili peppers, beans, and potatoes were also cultivated there. In southern China, evidence exists of the early cultivation of rice and other crops such as soybeans.

Corn, wheat, and rice are all propagated by seed, which requires fairly sophisticated agricultural practices. Growing plants that could be propagated vegetatively may have occurred earlier. Plants raised in this manner, such as bananas, yams, potatoes, and manioc, do not preserve as well as grains because of their high water content. For that reason, we may have no archaeological evidence of their cultivation.

Other advances in agriculture include the domestication of animals, which were kept to supply food, milk, and hides. In the Old World, animals were also used to prepare fields for planting. Another major advance in agriculture was irrigation, which began more than 5000 years ago in Egypt.

Producing food agriculturally was more time-consuming than hunting and gathering, but it was also more productive. In hunter-gatherer societies, everyone shares the responsibility for obtaining food. In agricultural societies, fewer people are needed to provide food for everyone. Thus agriculture freed some people to pursue other endeavors, including religion, art, and various crafts.

Cultural evolution has had a profound impact on the biosphere

Cultural evolution has had far-reaching effects on both human society and on other organisms. The Industrial Revolution, which began in the 18th century, caused populations to concentrate in urban areas near centers of manufacturing. Advances in agriculture encouraged urbanization, as fewer and fewer people were needed in rural areas to produce food for everyone. The spread of industrialization increased the demand for natural resources to supply the raw materials for industry.

Cultural evolution has permitted the human population to expand so dramatically that there are serious questions about Earth's ability to support so many people indefinitely (see Chapter 51). In 1996 about 840 million people lacked access to the food needed to be healthy and lead productive lives. To further compound the problem, the United Nations projects that three billion *additional* people will be added to the world population, which reached 6 billion in 1999, in the next three to four decades.

Cultural evolution has resulted in large-scale disruption and degradation of the environment. Tropical rain forests and other natural environments are rapidly being eliminated. Soil, water, and air pollution occur in many places. Since World War II, soil degradation due to poor agricultural practices, overgrazing, and deforestation has occurred in an area equal to 17% of the Earth's total vegetated surface area. Many species cannot adapt to the rapid environmental changes caused by humans and thus are becoming extinct. The decrease in biological diversity due to extinction is alarming.

On a positive note, we are aware of the damage we are causing, and we have the intelligence to modify our behavior to improve these conditions. Education, including the study of biology, may help future generations develop environmental sensitivity, making cultural evolution our salvation rather than our destruction.

SUMMARY WITH KEY TERMS

I. Primates arose from small, **arboreal** (tree-dwelling), shrewlike mammals.
  A. Primates are adapted for an arboreal existence by the presence of five grasping digits, including an opposable thumb or toe; long, slender limbs that move freely at the hips and shoulders; and eyes located in front of the head.
  B. Primates are divided into two semiorders, Strepsirhini and Haplorhini.
    1. The semiorder Strepsirhini includes lemurs.
    2. The semiorder Haplorhini includes tarsiers and **anthropoids** (monkeys, apes, and humans).
II. Anthropoids arose from early primate ancestors.
  A. The early anthropoids branched into two groups, the New and Old World monkeys.
  B. **Hominoids** (apes and humans) arose from the Old World monkey lineage.
  C. There are four modern genera of apes: gibbons, orangutans, gorillas, and chimpanzees.
III. The **hominid** (or **hominine**) line consists of humans and their ancestors.
  A. Hominid evolution began in Africa.
    1. The earliest known hominids belong to *Ardipithecus ramidus,* which appeared about 4.4 million years ago.
    2. *Ardipithecus ramidus* presumably gave rise to *Australopithecus anamensis.* The genus *Australopithecus,* which includes at least six species that lived between about 4 and 1.25 million years ago,

contains the immediate ancestors of the genus *Homo.*
    3. *Ardipithecus* and *Australopithecus* species are often referred to as **australopithecines.** Australopithecines were **bipedal** (walked on two feet), a hominid feature. (It is not yet known if *Ardipithecus* was bipedal.)
  B. *Homo habilis* was the earliest known hominid with some of the human features lacking in the australopithecines, including a slightly larger brain. *H. habilis* fashioned crude tools from stone.
  C. *Homo erectus* had a larger brain than *H. habilis,* made more sophisticated tools, and may have worn clothing, built fires, and lived in caves or shelters. Some scientists now think that fossils identified as *Homo erectus* represent two different species, *Homo ergaster,* an earlier African species that gave rise to archaic *Homo sapiens,* and *Homo erectus,* a later Asian offshoot that may be an evolutionary dead end.
  D. **Archaic *Homo sapiens*** lived in Africa, Asia, and Europe from about 800,000 to 100,000 years ago. Some researchers classify archaic *Homo sapiens* as a separate species, *Homo heidelbergensis.*
  E. **Neandertals** existed from about 230,000 to 30,000 years ago.
    1. Neandertals had short, sturdy builds, receding chins and foreheads, heavy brow ridges and jaw bones, larger front teeth, and nasal cavities with unusual triangular bony projections.
    2. Many scientists think that Neandertals were a separate species, *Homo neanderthalensis,* whereas some scientists think Neandertals were a race of modern humans.

3. The disappearance of Neandertals is a mystery.
F. Anatomically modern humans (***Homo sapiens***) existed 100,000 years ago.
  1. The **Cro-Magnon** culture, which existed in France and Spain about 35,000 years ago, exemplifies early modern humans.
  2. The origin of modern humans is controversial. Two different hypotheses, the "out of Africa" and the "multiregional" hypotheses, purport to explain the origin of modern humans. Comparative molecular data favor the African origin of modern humans.
IV. **Cultural evolution** is the transmission of knowledge from one generation to the next.
  A. Large human brain size makes cultural evolution possible.
  B. Two significant advances in cultural evolution were the development of agriculture and the Industrial Revolution.

POST-TEST

1. The anthropoids are more closely related to _____ than to _____. (a) tarsiers; lemurs (b) lemurs; monkeys (c) tree shrews; tarsiers (d) lemurs; tarsiers (e) tree shrews; monkeys
2. Unlike Old World monkeys, some New World monkeys possess (a) body hair (b) five grasping digits (c) a well developed cerebrum (d) a bipedal walk (e) a prehensile tail
3. Apes and humans are collectively called (a) mammals (b) primates (c) anthropoids (d) hominoids (e) hominids
4. The _____ in humans is centered at the base of the skull, positioning the head for erect walking. (a) supraorbital ridge (b) foramen magnum (c) pelvis (d) sagittal crest (e) femur
5. The oldest evidence of bipedalism is found in the tibia of (a) *Australopithecus afarensis* (b) *Australopithecus anamensis* (c) *Australopithecus africanus* (d) *Homo erectus* (e) *Homo habilis*
6. Humans and their immediate ancestors are collectively called (a) mammals (b) primates (c) anthropoids (d) hominoids (e) hominids
7. The earliest hominids belong to the genus (a) *Aegyptopithecus* (b) *Dryopithecus* (c) *Ardipithecus* (d) *Australopithecus* (e) *Homo*
8. The earliest hominid to be placed in the genus *Homo* is (a) *H. habilis* (b) *H. ergaster* (c) *H. erectus* (d) *H. heildelbergensis* (e) *H. neanderthalensis*
9. Some scientists now think that fossils identified as *Homo erectus* represent which two different species? (a) *H. habilis* and *H. erectus* (b) *H. ergaster* and *H. erectus* (c) *H. heildelbergensis* and *H. ergaster* (d) *H. neanderthalensis* and *H. erectus* (e) *H. neanderthalensis* and *H. sapiens*
10. Archaic *Homo sapiens* appeared about _____ years ago. (a) 5 million (b) 800,000 (c) 230,000 (d) 100,000 (e) 5000
11. _____ were an early group of humans with short, sturdy builds and heavy brow ridges that lived throughout Europe and western Asia from about 230,000 to 30,000 years ago. (a) Australopithecines (b) Dryopithecines (c) Archaic *Homo sapiens* (d) Neandertals (e) Cro-Magnons

REVIEW QUESTIONS

1. Distinguish between each of the following pairs: (a) mammals and primates (b) anthropoids and hominoids (c) hominoids and hominids (d) australopithecines and the genus *Homo*
2. Describe three different ways primates are adapted to an arboreal existence.
3. Identify at least three differences between the skulls of apes and humans.
4. Explain at least three ways in which an ape skeleton differs from a human skeleton.
5. Distinguish between each of the following pairs (a) *Homo habilis* and *Homo erectus* (b) *Homo ergaster* and *Homo erectus* (c) *Homo erectus* and *Homo heidelbergensis* (d) *Homo neanderthalensis* and *Homo sapiens*
6. Describe the two currently proposed hypotheses that explain where modern humans originated.
7. What is cultural evolution, and how has it affected Earth?

YOU MAKE THE CONNECTION

1. What types of as-yet-undiscovered scientific evidence would help explain how monkeys got to South America from the Old World?
2. Why are classifying fossil hominid species and determining their evolutionary relationships to one another such controversial endeavors?
3. Which hypothesis of the origin of modern humans ("out of Africa" or "multiregional") seems to more closely match our understanding of evolutionary processes in general? Explain your answer.
4. If you were evaluating whether other early humans exterminated the Neandertals, what kinds of archaeological evidence might you look for?
5. The remains of Cro-Magnons have been found in southern Europe alongside reindeer bones, but reindeer currently exist only in northern Europe and Asia. Can you explain the apparent discrepancy?

RECOMMENDED READINGS

Fullick A., and P. Fullick. "The Human Story." *New Scientist,* Vol. 154, 7 Jun. 1997. A short summary of human evolution.

Gore, R. "Expanding Worlds." *National Geographic,* Vol. 191, May 1997. *Homo erectus* is thought to have been a meat-eater that roamed throughout Africa, Asia, and Europe, beginning almost 2 million years ago.

Gore, R. "The First Humans." *National Geographic,* Vol. 192, Jul. 1997. Considers fossil hominid discoveries in Europe that date from about 1 million years ago to the near-present.

Gore, R. "The First Steps." *National Geographic,* Vol. 191, Feb. 1997. This article reviews what is currently known about australopithecines and early members of the genus *Homo.*

Gore, R. "Neandertals." *National Geographic,* Vol. 189, Jan. 1996. The mysteries surrounding early humans known as Neandertals are explored.

Gore, R. "Tracking the First of Our Kind." *National Geographic,* Vol. 192, Sept. 1997. Fossilized footprints of anatomically modern humans that were discovered in South Africa date to 117,000 years ago.

Johanson, D.C. "Face-to-Face with Lucy's Family." *National Geographic,* Vol. 189, Mar. 1996. Discusses the anatomical features and behaviors of *Australopithecus afarensis.*

Kahn, P., and A. Gibbons. "DNA from an Extinct Human." *Science,* Vol. 277, 11 Jul. 1997. This research news article highlights the successful attempt to extract and analyze DNA from a fossil Neandertal bone.

Leakey, M. "The Farthest Horizon." *National Geographic,* Vol. 188, Sept. 1995. Highlights the remarkable discovery of an early hominid that is transitional between *Ardipithecus ramidus* and *Australopithecus afarensis.* (This early hominid was later named *Australopithecus anamensis.*)

Leakey, M. and A. Walker. "Early Hominid Fossils from Africa." *Scientific American,* Vol. 276, No. 6, Jun. 1997. An overview of recent exciting discoveries of early hominid fossils (*Ardipithecus* and *Australopithecus*) and their significance to human evolution.

Lewin, R. *Patterns in Evolution: The New Molecular View.* Scientific American Library, New York, 1996. Examines how genetic information provides insights into evolution, including the evolution of primates.

Shreeve, J. "The Neandertal Peace." *Discover,* Sept. 1995. This article is adapted from a book by the author entitled *The Neandertal Enigma: Solving the Mystery of Modern Human Origins,* published by Viking in 1996.

● Visit our website at **http://www.saunderscollege.com/lifesci/titles.html** and click on Solomon/Berg/Martin Biology for links to chapter-related resources on the World Wide Web.

Middle School Science Teacher

HENRY ORTIZ

Henry Ortiz graduated with a B.S. in Biology from California State University, Northridge, in 1989. He teaches science in the bilingual program of a Los Angles middle school. Henry also trains teachers for the Globe program headquartered in Washington, DC. In this program, students around the world participate in scientific research by collecting field data and feeding it via the Internet to the scientists who are conducting the research.

Henry emigrated from Cuba to Los Angeles when he was only four years old, along with his mother and five brothers. His father, a physician, did not receive permission to leave Cuba for five more years. Henry's immersion into English classes as a Spanish-speaking child in Los Angeles gives him special sensitivity to the learning difficulties of non-native students. His biology background offers a strong foundation in the basic principles of physical sciences as well as life sciences. It is the interconnectedness of all the sciences that fascinates him.

How did you find yourself at California State Northridge?

I went to junior high and high school in the Burbank-Glendale area in Southern California. I thought about going to UCLA but I wanted to live at home, and Northridge is close. At UCLA most people live in the dorms. I thought Northridge was a good choice. I was not sure what I wanted to do in life so I appreciated the personal attention I got from professors at Northridge. I liked biology, so I majored in it.

What aspect of biology most intrigued you?

What really interested me was the way that knowledge changes, especially in the sciences. It is fascinating how the body of scientific knowledge continually builds on everything we know. Many other aspects are interesting too, such as views on how the Earth's systems evolve. Thinking logically, looking at evidence, and coming up with logical explanations have always been important to me. My family encouraged this way of thinking.

Did your father want you to pursue a medical career?

Yes, and I considered it. For a while, my future was medicine or a laboratory profession. But, I did not want to be in school that long. I wanted to have more free time for traveling and outdoor sports because they are important to me. So I decided to give teaching a try.

When did you get your teaching credential?

I graduated in 1989 with a biology degree. After that, it took me a year to earn my teaching credential. The credential program is like a fifth year of college. I took a methods-in-teaching-science class, a course in secondary education, and fulfilled the student teaching requirement. I taught for a year, then took a leave of absence. I went to Spain.

Why Spain?

I had always wanted to go to Europe, but I also wanted to perfect my Spanish. So, I spent a year at the University of Madrid. I took courses in Spanish like phonetics, grammar, art, history of Spain, and history of Madrid. These classes were interesting and sharpened my Spanish communication skills.

How did you decide on teaching at the middle school level?

I did my student teaching in chemistry at the high school level. It was a fun subject to teach. The students liked the class. I even have some letters that were forwarded to me from students who wanted me to come back.

After that, I was ready to handle a class by myself. So, I went to a middle school that needed bilingual teachers. Since I was bilingual, I applied and got the job. It was a good decision. I did well, although it was a pretty tough school with a difficult group of students. I received an award that year for excellence in student teaching.

Was it your bilingual ability or your biology degree that helped you get the job?

I think it was both. In our district, having a science degree is very advantageous. Having a bilingual degree is also a plus, so both were important. I probably could have gotten a job as a science teacher at any school that needed a science teacher, but also at any school that needed a bilingual teacher. Both were important.

How much of your time do you spend speaking Spanish in the classroom?

We have a bilingual department at our school. I instruct in Spanish 50% of the time. The students who have come here recently from another country go into the bilingual program. They take core classes like history, science, and math in their native language. Sometimes they will have two periods of ESL (English as a Second Language) so that they learn English while learning the core courses in their own language. The reasoning is to keep them from falling behind other students in those primary courses. As they learn English, they will then be able to transfer that knowledge from their language.

What kinds of printed material do you use?

There are major publishers in this country who have published exact translations from the English text. You can have a text that has English on one side of the page and Spanish on the other side of the page. I have some of those books. That way a teacher doesn't need different books for different classes. In addition, the teacher can keep the English and the Spanish students at the same level and pace. It makes it easier to prepare lessons.

Do you only teach bilingual classes?

No. Right now I have two bilingual science classes, one bilingual math class, and an honors physical science eighth grade class in English. I have mostly gifted students in the honors class. It's like teaching tenth graders. They come up with very clever ideas. It's fun to watch them learn, especially when they learn from each other on certain projects. It's a very good class.

What are the obstacles you face in teaching?

The biggest challenge for me are the students who do not value their education. I keep trying to reach these students. However, it's important for me to realize when to dedicate less time to them, so that the other students are not neglected. I like to teach to the top of the class. Some people don't agree with that. I give the students a lot of information. Some of it might be over their heads, but the better students are going to get it. Many of the C-average students will grasp some of it. Even a few of the underachievers learn something.

Do your colleagues have similar backgrounds to yours?

We have a few teachers that went to the same school that I did. Two of them also have science backgrounds. Many of the other teachers do not have science backgrounds. Some of them are even credentialed to teach science through an emergency credential but they don't have any background in science.

I'm unclear about the emergency credential.

It's a credential that is given to them either by the state or the district, which allows them to teach in the district while they work toward their credential. It's like on-the-job training while they are taking courses in college.

Do you feel like your biology degree gave you any advantages over some of these other teachers?

It is possible to teach high school science without a science degree, but because schools prefer teachers who have a background in science, my biology degree was an advantage. If you get the principles down in science, you will be a better science teacher. Some of the teachers who do not have a science background might understand the subject superficially because they've read it, but they might not grasp the scientific thinking behind it.

Can you describe your teaching style?

Like many teachers, I take a constructivist approach, connecting certain scientific ideas to build up the students' knowledge base. My biology degree adequately prepared me to teach students chemistry and physics concepts even though I wasn't a chemistry or physics major. Most of these concepts are very basic.

I don't know what makes me unique, unless it's just the relationships I develop with the students because I have been in their situation. Most of the students I teach have come from another country. If not, their parents did. I remind them that I came from another country, too.

A good teacher gets to know the students, develops a relationship with them, and shows them the type of person he or she is. It is important to show students that you understand them. Then, it becomes easier to help them understand their lessons. If you don't take a personal approach to teaching, half the class will be totally isolated. They won't care because you don't seem to care.

Do you have much interaction with the parents?

I try to maintain contact. For example, tomorrow I have a meeting with the bilingual parents. I'm going to talk to them about what I do in class. Then, they will perform an activity with some of the students. The parents don't have much time because they're working. Usually the parents who come to school are the parents who are very interested, but also have the time to meet.

What activity do you have planned for tomorrow?

We're going to be doing an ocean depth sounding to show how scientists measure depth in oceanography. We are also going to demonstrate how scientists can tell what the ocean bottom looks like by doing a geologic probe activity.

Do you take field trips?

Occasionally we take fields trips to Yosemite National Park with the students. The Yosemite Institute has a program where we spend a week hiking and learning about the relationship of humans to nature and our responsibility as stewards of the Earth.

We've also taken trips to Temescal Canyon in Pacific Palisades, California. It's a three-day program in the Santa Monica Mountains doing field work. The students do a "quadrat study," taking population counts of different organisms in a measured area, learning how to sample, and to work as a team.

We have another trip to the beach at San Pedro. We do similar activities, but out on the ocean. We take marine science measurements from a boat, work in the lab, and at the Cabrillo Museum. We also do a tide pool activity where we classify the different organisms that are present in various parts of the tidal zone.

What is the hardest concept for students at the eighth grade level to grasp about science?

The students have learned what science is, but they haven't learned how to do science. They still think of scientists as guys wearing white coats in laboratories. Some teachers don't emphasize that science is a process, that everyone can do science. It's not this magical activity that only chosen people can do. These misconceptions may continue even after the students are out of junior high school.

Also, students don't know how to apply a scientific concept to different situations. They might learn a specific principle, such as how to calculate the density of a particular type of matter. However, they don't know how to apply that principle to biology, chemistry, and physics. Students need the whole picture. That's the emphasis in science teaching now. I would like to teach students how to think, how to take a body of knowledge, and derive some sort of conclusion from it.

Do you participate in any ongoing teacher education programs?

Yes, I do teacher training for some agencies. I am a member of NSTA (National Science Teachers Association). I also train for the Globe Program out of Washington, DC.

Globe is a program where students, guided by their teachers, do scientific protocols that scientists have designed. We train teachers to do the protocols as the scientists would like to have them done, including how to go on-line and how to enter the data. Students send their data through the Internet to a scientific database. The scientists use this database as part of their research. For example, one scientist is doing a study on atmosphere, how clouds influence the temperature on Earth. Another scientist is classifying the different soils that are found in different parts of the world. Still another scientist is doing validation of land satellite images. Globe is a good program because the students are doing the science by collecting data from the field and supplying it electronically to the researchers. They are contributing information to a world-wide database, which involves over 60 countries.

Who are you training for Globe?

Globe offers workshops for teachers. The organization usually trains between 15 to 45 teachers at a time in different cities. There are usually one or two scientists on the team, a few teachers who have been doing the program, plus a facilitator and a technology person. We meet the day before training to get everything ready. Over the next four days, we train teachers to do the protocols as the scientists would like to have them done, how to go on-line, and how to enter the data. I have been training in the soils protocols mostly and have done some of the hydrology, atmosphere and G.P.S. protocols.

What advice would you give to someone considering your career?

If you are interested in continually learning while you teach, go for it. You have to stay current when you teach science. Do not go into science teaching if you don't want to stay current. That can be boring to you and your students. That will burn you out.

How do you stay current?

I read. I run workshops. I go to workshops. For example, I just attended a six-

week workshop at UCLA called LIMS— Leadership in Marine Science. We learned how to integrate marine biology as a theme in our class subjects. I learned how to use brain biology to teach earth science, physical science, chemistry and biology to secondary level students. That was a very positive experience.

What do you read to keep up to date?

I like *Scientific American* and some of the NSTA journals like *The Journal of College Teaching*. And I read other magazines like *National Geographic*.

Do you see yourself getting an advanced degree?

Yes, I do. I have been thinking about it for the last two years. I would like to get a Master's Degree in science—maybe biology or one of the earth sciences.

Does the school district require that for advancement?

No, they don't. If you take a certain number of units you get more pay. I have taken so many units since my undergraduate degree, that I'm maxed out. Even though I don't need any more units, I would still like to get an advanced degree for my own personal education.

Where do you see yourself in five years? Still teaching?

I might be helping other teachers increase their effectiveness as teachers, helping others develop their materials, new curricula, methods, or activities. Or, perhaps, I will be running programs like the one we have in Temescal Canyon. Right now I'm part-time teaching in the classroom and part-time coordinating some of our district's environmental education programs. I like the classroom, but perhaps in the future I'll get more involved in teacher training.

The Diversity of Life

A tortoise *(Geochelone elephantopus)* from Isla Española in the Galapagos Islands, Ecuador.   (Joe McDonald/Tom Stack & Associates)

CHAPTER 22

Understanding Diversity: Systematics

About two million species[1] of living organisms have been described, and new species are being discovered almost daily. Biologists speculate that at least several million additional species remain to be discovered. The dwarf monkey (about 10 centimeters long and weighing about 190 grams) shown here represents a new primate species identified in 1997 in the Amazon jungle of Brazil. To study the diverse life forms that share our planet and to effectively communicate our findings, we need to organize our knowledge of them. The scientific study of the diversity of organisms and their evolutionary relationships is called **systematics.** An important aspect of systematics is **taxonomy,** the science of naming, describing, and classifying organisms. The term **classification** means ordering organisms into groups based on their similarities or relationships. Classifying organisms is a complex and often controversial endeavor.

Imagine that you were going to develop a classification system. How would you use what you already know about living things to assign them to categories? Would you place insects, bats, and birds in one category because they all have wings and fly? And would you, perhaps, place squid, whales, fish, penguins, and Olympic backstroke champions in another category because they all swim? Or would you classify organisms according to a culinary scheme, placing lobsters and tuna in the same part of the menu, perhaps identifying them as "seafood"? Any of these schemes might be valid, depending on your purpose. Similar methods have been used throughout history. Animals, for example, were classified by St. Augustine in the fourth century as useful, harmful, or superfluous—to humans. During the Renaissance, scholars began to develop categories based more on the characteristics of the organisms themselves. These categories were originally arranged roughly in order from simple to complex organisms.

Of the many classification systems that were developed, the one designed by Carolus Linnaeus in the mid-18th century (described briefly in Chapter 1) has survived with some modification to the present day. He devised a system for assigning species to a hierarchy of increasingly general groups. Linnaeus probably had no theory of evolution in mind when he set up his system. Neither did he have any concept of the vast number of extant (living) and extinct organisms that would later be discovered. Yet his system has proved to be re-

(AP/Wide World Photos)

markably flexible and adaptable to new biological knowledge and theory. Very few other 18th-century inventions survive today in a form that would still be recognizable to their originators. Most biologists currently use a hierarchical system that includes domain and/or kingdom, phylum, class, order, family, genus, and species, as well as many intermediate categories such as subphylum or suborder.

---

[1]Recall that a **species** is a group of similar organisms defined by their common gene pool and ability to interbreed in their natural environment; members of a species are reproductively isolated from other organisms.

1. Offer at least two justifications for the use of scientific names and classifications of organisms.
2. Arrange the Linnaean categories in hierarchical fashion, from most inclusive to least inclusive.
3. List the three domains and six kingdoms of organisms introduced in this chapter, give the rationales for and against these systems of classification, and describe the distinguishing characters of the organisms assigned to each.
4. Critically summarize the difficulties encountered in choosing taxonomic criteria.
5. Contrast monophyletic, paraphyletic, and polyphyletic taxa.
6. Apply the concept of shared derived characteristics to the classification of organisms.
7. Describe the methods of molecular biology now used by taxonomists, and summarize their advantages.
8. Contrast two approaches to systematics: phylogenetic systematics (cladistics) and classical evolutionary taxonomy.

ORGANISMS ARE NAMED USING THE BINOMIAL SYSTEM OF NOMENCLATURE

Before the mid-18th century, each species had a lengthy descriptive name, sometimes composed of ten or more Latin words! Then Linnaeus simplified scientific classification, developing a **binomial system of nomenclature** in which each species is assigned a unique two-part name. The first part designates the **genus** (pl., *genera*), and the second part is called the **specific epithet.** Note that the specific epithet is *not* the species. In fact, the same specific epithet can be used as the second name of species in different genera. For example, *Quercus alba* is the scientific name for the white oak, and *Salix alba* is the name for the white willow. (*Alba* comes from a Latin word meaning "white.") Thus, both parts of the name must be used to accurately identify the species.

The genus name is always capitalized, whereas the specific epithet is usually not. Both names must be underlined or italicized. The genus name can be used alone to designate all species in the genus (e.g., *Quercus* is the genus that includes all oak species). However, the specific epithet is never used alone; it must always be preceded by the full or abbreviated genus name (e.g., *Quercus alba* or *Q. alba*).

Scientific names are generally derived from Greek or Latin roots or from Latin versions of the names of persons, places, or characteristics. For example, the generic (genus) name for the bacterium *Escherichia coli* is based on the name of the scientist Theodor Escherich, who first described it. The specific epithet *coli* reminds us that *E. coli* lives in the colon (large intestine).

Scientific names permit taxonomy to be a truly international study. Even though the common names of an organism may vary in different locations and languages, an organism can be universally identified by its scientific name. A researcher in Puerto Rico can know exactly which organisms were used in a study published by a Russian scientist and therefore can repeat or extend the Russian's experiments using the same species.

Subspecies may become species

The species is the basic unit of classification, but not the smallest grouping in use. Geographically different populations within a species often display certain consistent characteristics that distinguish them from other populations of the same species. If they interbreed in nature, however, they are not truly separate species, but **subspecies.** For some kinds of microorganisms, such as bacteria, the term **strain** is used and for plants, subspecies or **variety.** The full scientific name of a subspecies consists of three names: the generic name; the specific epithet; and the subspecies, strain, or variety name. For example, peaches are *Prunus persica* var. *persica,* and nectarines, a smooth-skinned variety of peaches, are *Prunus persica* var. *nucipersica.*

Experts are usually able to distinguish among the subspecies of a particular species. However, subspecies may grade imperceptibly into one another at the borders of their geographical ranges, where there is opportunity for interbreeding. Some of these subspecies may be in the process of becoming reproductively isolated and may, in the course of time, become separate species. Thus, subspecies often provide an opportunity for field studies of gene pools and of the speciation process.

EACH TAXONOMIC LEVEL IS MORE GENERAL THAN THE ONE BELOW IT

A **taxon** (pl., *taxa*) is a formal grouping of organisms at any given level, such as the species, genus, or phylum. For example, class Mammalia is a taxon that includes many different orders. Similarly, subphylum Vertebrata is a taxon that contains seven classes, including Amphibia, Reptilia, and Mammalia.

The range of taxonomic categories from species to kingdom forms a hierarchy (Fig. 22–1; Table 22–1). Closely related species are assigned to the same genus, and closely re-

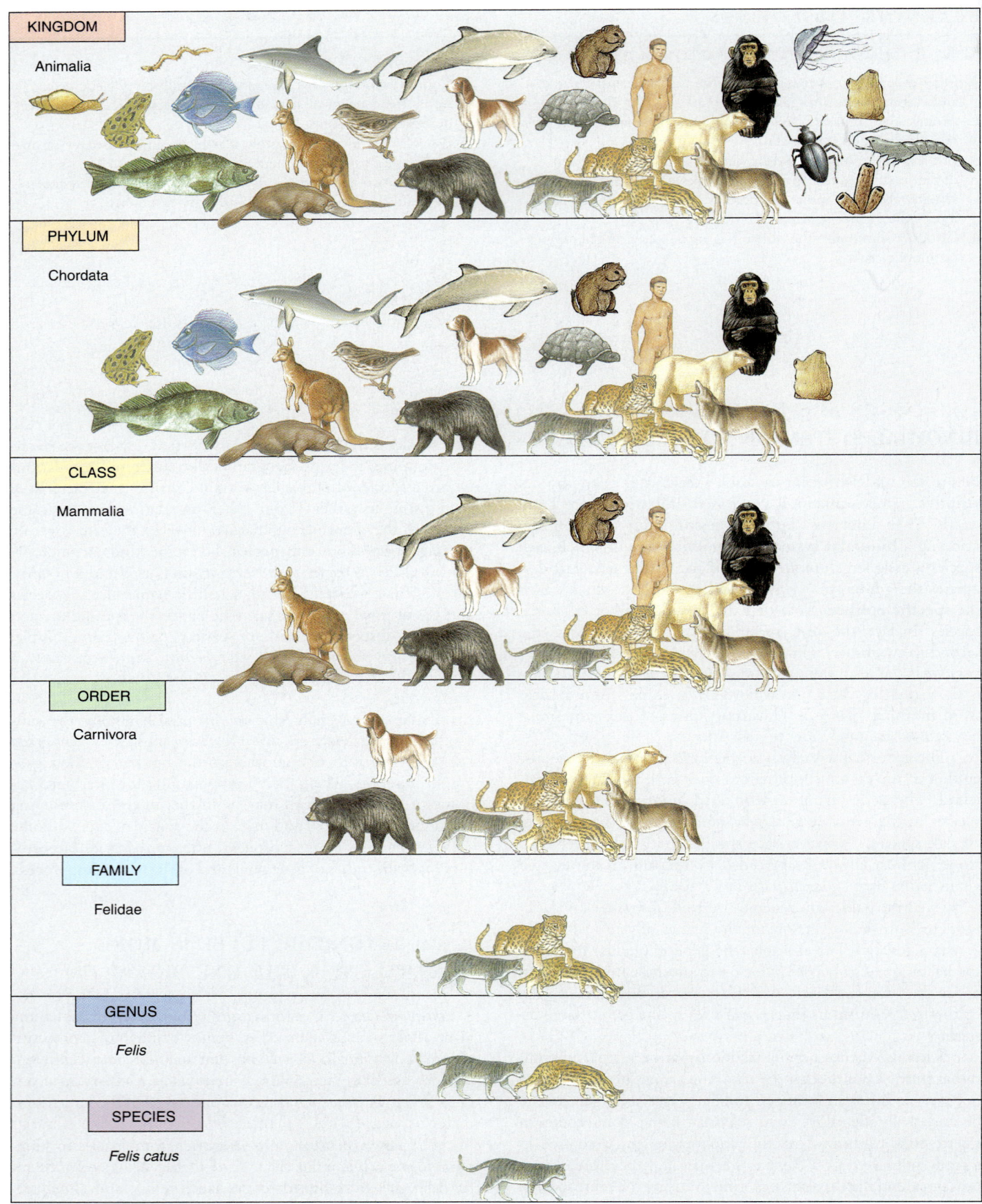

Figure 22-1 **The principal categories used in classification.** The domestic cat *(Felis catus)* is classified here to illustrate the hierarchical organization of our taxonomic system. Each level is more inclusive than the one below it.

KINGDOM
Animalia

PHYLUM
Chordata

CLASS
Mammalia

ORDER
Carnivora

FAMILY
Felidae

GENUS
Felis

SPECIES
Felis catus

TABLE 22-1 Classification of Corn

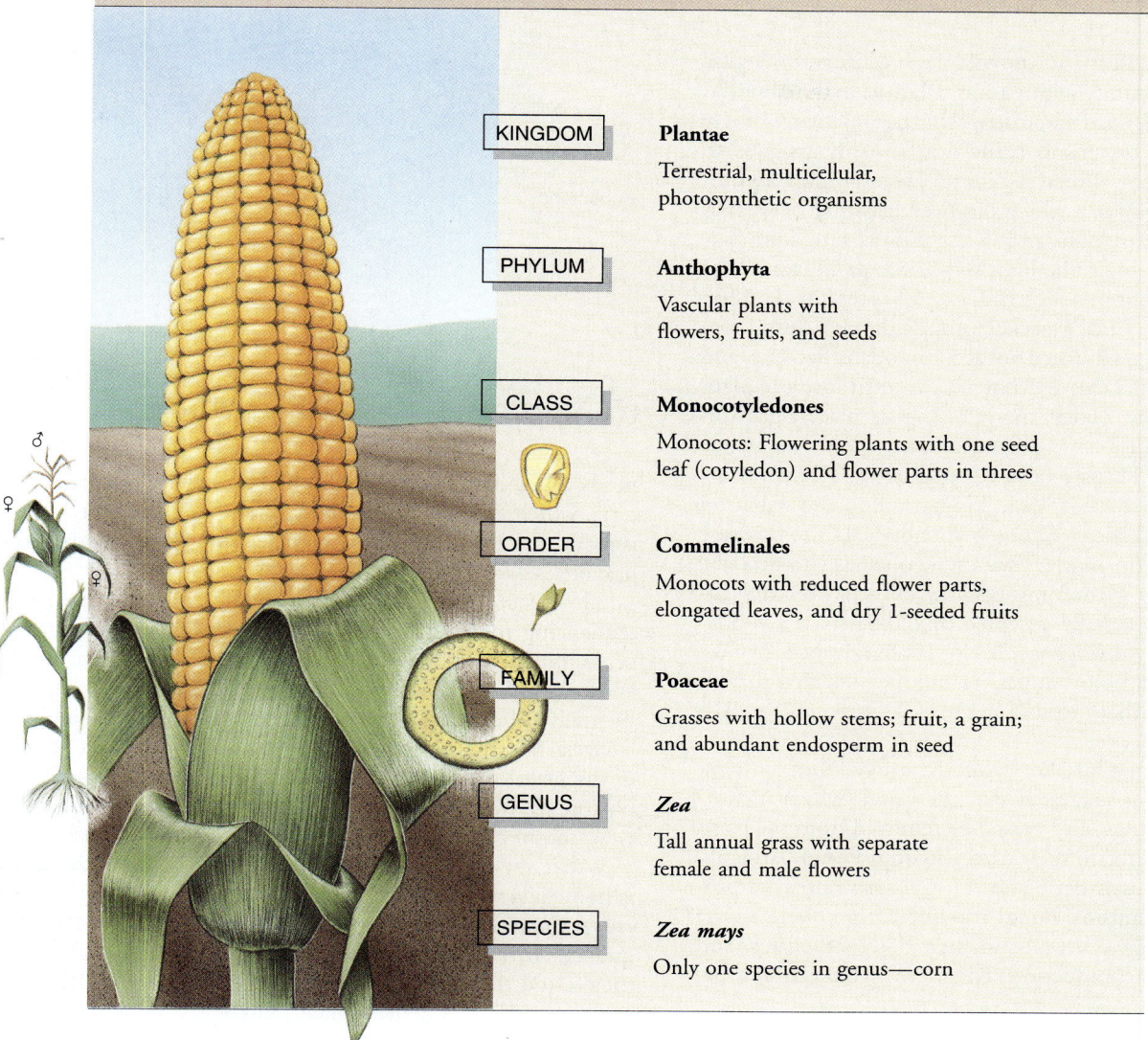

KINGDOM	**Plantae**
	Terrestrial, multicellular, photosynthetic organisms
PHYLUM	**Anthophyta**
	Vascular plants with flowers, fruits, and seeds
CLASS	**Monocotyledones**
	Monocots: Flowering plants with one seed leaf (cotyledon) and flower parts in threes
ORDER	**Commelinales**
	Monocots with reduced flower parts, elongated leaves, and dry 1-seeded fruits
FAMILY	**Poaceae**
	Grasses with hollow stems; fruit, a grain; and abundant endosperm in seed
GENUS	***Zea***
	Tall annual grass with separate female and male flowers
SPECIES	***Zea mays***
	Only one species in genus—corn

lated genera may be grouped together in a single **family.** Families are grouped into orders, orders into **classes,** classes into **phyla,**[2] and phyla into **kingdoms** and/or **domains.** These groupings can also be separated into subgroupings, for example, subphyla or superclasses.

[2]Plants and fungi have been traditionally classified in divisions rather than phyla. However, the International Botanical Congress has approved the phylum designation for plants and fungi.

Just how organisms are grouped can vary according to the judgment of taxonomists and their criteria and decisions for classification. Some taxonomists ignore minor variations and group organisms into already existing taxa. This practice is referred to as "lumping." Other taxonomists subdivide taxa on the basis of minor differences, establishing separate categories for forms that do not fall naturally into one of the existing classifications. This practice is called "splitting." Lumpers acknowledge very few animal and plant phyla, whereas "splitters" may recognize numerous phyla.

NEW DATA INFLUENCE CLASSIFICATION IN KINGDOMS

From the time of Aristotle to the mid-19th century, biologists divided organisms into two kingdoms, **Plantae** and **Animalia.** After the development of microscopes, it became increasingly obvious that many organisms could not be easily assigned to either the plant or the animal kingdom. For example, the unicellular organism *Euglena,* which has been classified at various times in the plant kingdom and in the animal kingdom, carries on photosynthesis in the light but in the dark uses its flagellum to move about in search of food (Chapter 24). In 1866 a German biologist, Ernst Haeckel, proposed that a third kingdom, **Protista,** be established to accommodate bacteria and other microorganisms that did not appear to fit into the plant or animal kingdoms. Today, many biologists place algae (including multicellular forms), protozoa, water molds, and slime molds in kingdom Protista (also referred to as *Protoctista*).

In 1937 the French marine biologist Edouard Chatton suggested the term *procariotique* ("before nucleus") to describe bacteria, and the term *eucariotique* ("true nucleus") to describe all other cells. This dichotomy between prokaryotes and eukaryotes is now universally accepted by biologists as a fundamental evolutionary divergence. In the 1960s advances in electron microscopy and biochemical techniques revealed further cellular differences that inspired many new proposals for classifying organisms.

In 1969 R. H. Whittaker proposed a five-kingdom classification. Whittaker suggested that the fungi (which include the mushrooms, molds, and yeasts) be removed from the plant kingdom and classified in their own kingdom **Fungi.** After all, fungi are not photosynthetic and must absorb nutrients produced by other organisms. Fungi also differ from plants in the composition of their cell walls, in their body structures, and in their modes of reproduction. Kingdom **Prokaryotae** (for-

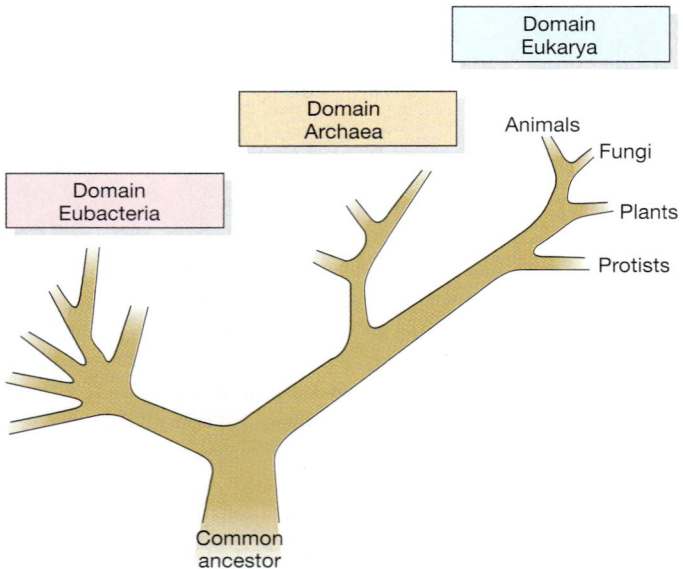

Figure 22–2 The three domains.

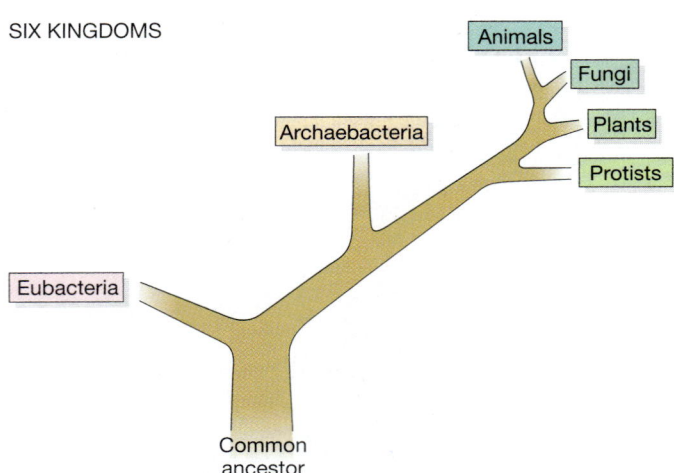

Figure 22–3 The six-kingdom system of classification.

merly called *Monera*) was established to accommodate the bacteria, which are fundamentally different from all other organisms in that they lack distinct nuclei and other membranous organelles.

In the late 1970s, Carl Woese and his colleagues used gene sequencing methods to challenge the long-held view that all of the prokaryotes were closely related and very similar to one another. These researchers proposed that there are two fundamentally different groups of bacteria, Archaebacteria and Eubacteria, and that the prokaryotes account for two out of three of the major branches of organisms. This concept gained support in 1996 when Carol J. Bult of the Institute for Genomic Research in Rockville, Maryland reported in the journal *Science* that she and her colleagues had sequenced the complete genome of a methane-producing archaebacterium *Methanococcus jannaschii.* When these researchers compared gene sequences with those of two previously sequenced eubacteria, they found that less than half of the genes matched. Based on this molecular evidence, many biologists now divide the prokaryotes into kingdom **Eubacteria** and kingdom **Archaebacteria,** resulting in a six-kingdom system.

Based on fundamental molecular differences among the Eubacteria, Archaebacteria, and eukaryotes, many systematists now use a level of classification above the kingdom, called a **domain.** They classify organisms in three domains: **Archaea** (which corresponds to kingdom Archaebacteria), **Eubacteria** (also called *Bacteria*), and **Eukarya** (eukaryotes; Fig. 22–2). Gene sequencing indicates that the Archaea have a combination of bacteria-like and eukaryote-like genes, and appear to be more closely related to the Eukarya than to the Eubacteria.

The evolution of systematics reflects the creative and dynamic process of science. Systematists have been very responsive to new data, and consequently classification of organisms is a challenging and continuously changing process. In this edition of *Biology,* we use the six-kingdom approach: Eubacteria, Archaebacteria, Protista, Fungi, Plantae, and Animalia (Fig. 22–3 and Table 22–2). Because many microbiologists prefer the three-domain approach, we use a domain approach in our discussion of bacteria in Chapter 23.

TABLE 22-2 Six Kingdoms: Eubacteria, Archaebacteria, Protista, Fungi, Plantae, and Animalia

Kingdom	Characteristics	Ecological Role and Comments
Eubacteria	Prokaryotes (lack distinct nuclei and other membranous organelles); unicellular; microscopic; cell walls generally composed of peptidoglycan.	Most are decomposers; some parasitic (and pathogenic); some chemosynthetic autotrophs; some photosynthetic; important in recycling nitrogen and other elements; some used in industrial processes.
Archaebacteria	Prokaryotes; unicellular; microscopic; cell walls do not have peptidoglycan; differ biochemically from eubacteria.	Methanogens are anaerobes that inhabit sewage, swamps, and animal digestive tracts; extreme halophiles inhabit salty environments; extreme thermophiles inhabit hot, sometimes acidic environments.
Protista	Eukaryotes; mainly unicellular or simple multicellular. Three informal groups (not taxa) include protozoa; algae; and slime molds and water molds.	
Protozoa	Microscopic; heterotrophic; most move by means of flagella, cilia, or pseudopodia.	Important part of zooplankton; important in many food webs; some are parasitic (and pathogenic).
Algae	Photosynthetic; sometimes hard to differentiate from protozoa; some have brown or red pigments in addition to chlorophyll; some have alternation of generations.	Very important producers, especially in marine and freshwater ecosystems; produce oxygen.
Slime molds and water molds	Heterotrophic; reproduce by forming spores.	Aquatic or terrestrial; varied modes of nutrition.
Fungi	Eukaryotes; heterotrophic; absorb nutrients; do not photosynthesize; body composed of threadlike hyphae that form tangled masses that infiltrate food or habitat; cell walls of chitin.	Decomposers; some parasitic (and pathogenic); some used as food; yeast used in making bread and alcoholic beverages; some used to make industrial chemicals or antibiotics; responsible for much spoilage and crop loss.
Plantae	Eukaryotes; multicellular; photosynthetic; possess multicellular reproductive organs; alternation of generations; cell walls of cellulose.	Terrestrial biosphere depends on plants in their role as primary producers; important source of oxygen in Earth's atmosphere.
Animalia	Eukaryotes; multicellular heterotrophs; many exhibit tissue differentiation and complex organ systems; most able to move about by muscular contraction; specialized nervous tissue coordinates responses to stimuli.	Consumers; some specialized as herbivores, carnivores, or detritus feeders.

SYSTEMATICS IS CONCERNED WITH RECONSTRUCTING PHYLOGENY

Systematists use systems of classification to reconstruct the evolutionary relationships, or **phylogeny** (literally, "production of phyla"), of organisms. Once these relationships are established, the classification of organisms can be based on common ancestry. A population has a dimension in space—its geographical range—and also a dimension in time. Each population extends backward in time. Somewhat like branches of a tree, a population may diverge from other populations sufficiently to become a new species. Species have various degrees of evolutionary relationship with one another, depending on the degree of genetic divergence since their populations branched from a common ancestor.

Taxa should reflect evolutionary relationships

Systematists recognize three kinds of evolutionary groupings: monophyletic, paraphyletic, and polyphyletic. A **monophyletic group** contains all of the descendants of the most recent common ancestor (Fig. 22–4a). Mammals, for example, form a monophyletic taxon because all mammals are thought to have evolved from a common ancestral mammal, and all descendants of this ancestor are mammals. Monophyletic taxa are, therefore, natural groupings because they represent true evolutionary relationships and include all close relatives. A **paraphyletic group** is a group that contains a common ancestor and some, but not all, of its descendants.

Some currently recognized groups are probably **polyphyletic,** consisting of several evolutionary lines and not including a common ancestor (Fig. 22–4b). The members of such a taxon might have been mistakenly grouped together because they share similar features arising from convergent evolution (see Chapter 17). Thus, by definition, polyphyletic taxa misrepresent evolutionary relationships. For this reason, taxonomists attempt to avoid constructing polyphyletic taxa.

Homologous structures are important criteria for classification

Just how to group species into higher taxa—genera, families, orders, classes, or phyla—can be a difficult decision. Traditionally, many biologists have based their judgments about the degree of relationship on the extent of similarity among living species, and, when available, on the fossil record.

When they evaluate similarities, systematists consider structural, physiological, behavioral, and molecular traits. When comparing structural similarities, they look for homology in different organisms. Recall from Chapter 17 that **homology** refers to the presence in two or more species of a structure derived from a common ancestor. In contrast, the possession by two or more species of similar structures not derived from a shared ancestry sometimes occurs when unrelated or distantly related organisms become adapted to similar environmental conditions. For example, the shark and the dolphin have similar, but independently derived, body forms because they have become adapted to similar environments (aquatic) and lifestyles (predatory). Such similarity of attributes that appear to be homologous, but are actually independently acquired, is referred to as **homoplasy.**

Shared derived characters provide clues about phylogeny

Organisms sharing many homologous structures are considered to be closely related, while organisms sharing few homologous characters are presumed less closely related. However, distinguishing between homology and homoplasy is not always straightforward. Therefore, the appropriate choice of which similarities to use to show evolutionary relationships is extremely important.

How does a systematist interpret the significance of these similarities? In making decisions about taxonomic relationships, the systematist first examines the characteristics found in the largest group of organisms being studied and interprets them as indicating the most remote common ancestry. These **ancestral characters,** or **pleisomorphic characters,** are traits that were present in an ancestral species and that remain essentially unchanged in its descendants. For example, the vertebral column, present in all vertebrates, is an ancestral character for study of classes within the vertebrate subphylum. Studying the presence or absence of the vertebral column would not help us distinguish between various classes of vertebrates, e.g., between amphibians and mammals.

Derived characters, or **synapomorphic characters,** are traits found in two or more taxa that are present in their closest common ancestor. They are homologous traits not present in earlier ancestors because they evolved more recently. Birds

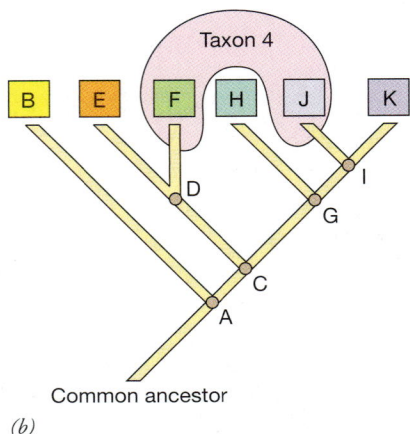

Figure 22–4 **Evolutionary relationships.** *(a)* Monophyletic and paraphyletic groups. Taxon 1 and taxon 3 are monophyletic. Each includes a common ancestor and all of its descendants. Taxon 2 is paraphyletic; it does not include all of the descendants of a common ancestor. *(b)* Taxon 4 is polyphyletic; members of this group do not share a close common ancestor.

and reptiles, for example, have a common ancestor, and both share the ancestral character of laying eggs. The feathers and beaks of birds, however, are derived characters that are not shared by reptiles.

A feature viewed as a *derived character* in a more inclusive (broader) taxon may also be considered an *ancestral character* in a less inclusive (narrower) taxon. More recent common ancestry is indicated by classification into less and less inclusive taxonomic groups with more and more specific shared derived characters. For example, the three small bones in the middle ear are useful in identifying a branch point between reptiles and mammals. The evolution of this derived character was a unique event, and only mammals have these bones. However, if we compare mammals with one another, the three ear bones are an ancestral character because all mammals have them. Consequently, they have no value in distinguishing among mammalian taxa. Other characters must be used to establish branch points among the mammals.

If we compare the mammals, dogs, goats, and dolphins, we find that dogs and goats have hair whereas dolphins do not. However, hair is an ancient trait in mammals and therefore cannot be used as evidence that dogs and goats share a more recent common ancestor. In contrast, the virtual absence of hair in dolphins is a derived character within mammals. When we compare dogs, dolphins, and whales, we find that dolphins and whales share this derived character, providing evidence that these animals branched from a common ancestor not shared by dogs.

Biologists carefully choose taxonomic criteria

Both fishes and dolphins have streamlined body forms, but this characteristic is homoplastic and does not indicate close evolutionary relationships. The dolphin shares important homologous derived characters with mammals such as humans: the ability to breathe air, nurse young, and use metabolic energy to maintain a constant body temperature (endothermy). Thus, the dolphin is classified as a mammal and is thought to have descended from a terrestrial mammalian ancestor.

Although dolphins have more shared derived characters in common with humans than with fishes, some characters are shared by all three of these animals. Among these ancestral characters are a dorsal tubular nerve cord and, during embryonic development, a notochord (skeletal rod) and rudimentary gill slits. These shared ancestral characters indicate a common ancestry and serve as a basis for classification. The ancestry is more remote between the dolphin and the fish than between the dolphin and the human. Therefore, while fishes, humans, and dolphins are grouped together in a more inclusive taxon, the phylum Chordata, humans and dolphins are also classified together in class Mammalia, a less inclusive taxon within phylum Chordata, indicating their closer relationship.

Deciding which traits best illustrate evolutionary relationships can be challenging. What, for example, are the most important taxonomic characteristics of a bird? We might list

Figure 22–5 Is this animal a bird? A few mammals share important characteristics with birds. The duck-billed platypus, a monotreme, lays eggs, has a beak, and lacks teeth. However, it does not have feathers, and it nourishes its young with milk secreted from mammary glands. *(Jean Philippe Varin/Jacana/Photo Researchers, Inc.)*

feathers, beak, wings, absence of teeth, egg-laying, and endothermy. Some mammals, for example, monotremes such as the duck-billed platypus, have many of these same characteristics: beaks, endothermy, absence of teeth, and egg-laying. Yet we do not classify them as birds (Fig. 22–5). No mammal, however, has feathers. Is this trait absolutely diagnostic of birds? According to the conventional taxonomic wisdom, the presence or absence of feathers determines what is and is not a bird. This applies only to modern birds, however. Some extinct reptiles may have had feathers.

Usually, organisms are classified on the basis of a combination of traits rather than on any single trait. The significance of these combinations is determined inductively, that is, by an integration and interpretation of the data. Such induction is a necessary part of the process of science. Taxonomists may hypothesize, for example, that birds should all have beaks, feathers, no teeth, and so on. Then, they reexamine the living world and observe whether there are organisms that might reasonably be called birds that do not fit the current definition of "birdness." If not, the definition is permitted to stand. If too many exceptions emerge, the definition may be modified or abandoned. Sometimes, the taxonomist persuades the world that an apparent exception, the bat, for instance, resembles a bird only superficially and should not be considered one. The bat possesses all of the basic characteristics of a mammal, such as hair and mammary glands that produce milk for the young.

Systematics is a dynamic science that proceeds by the constant reevaluation of data, hypotheses, and theoretical constructs. As new data are discovered and old data are subjected to reinterpretation, the ideas of systematists change. In 1995,

MOLECULAR BIOLOGY, EVOLUTION, AND TAXONOMY

How do molecular biology and evolution come together in the laboratory to provide data that can be used by systematists? Advances in molecular biology have provided the tools for biologists to compare the macromolecules of various organisms. Amino acid sequencing techniques, immunological methods, and DNA and RNA sequencing are among the procedures now used to compare macromolecules.

In Chapter 17 we discussed the respiratory protein cytochrome *c* present in all aerobic organisms. Although the structure and function of cytochrome *c* are similar in all aerobic organisms, some differences in amino acid sequences exist among species. Comparisons of these differences among about 100 species provide a good example of how data gained through amino acid sequencing contribute to taxonomic decisions. Interestingly, chimpanzee cytochrome *c* has the same amino acid sequences as human cytochrome *c,* but in a more distantly related primate, the rhesus monkey, 1 of the 104 amino acids in the sequence is different. In the dog, a nonprimate, 13 amino acids are different. Taxonomists use this type of information to help make decisions about classifying organisms. Such decisions are generally based on data derived from studies of different proteins.

We have learned that among related species the DNA sequences for the same structural genes are very similar. Detailed restriction maps within large homologous regions of chromosomes of related organisms are also very similar (see Chapter 14). For example, the DNA region that codes for the equivalent of the human hemoglobin beta chain has been mapped in several primates. Even though the gorilla diverged from the human line 8 to 10 million years ago, 65 of the 70 restriction sites are identical.

Researchers have determined the nucleotide sequence of a portion of DNA from each of three species of primates (humans, gorillas, and chimpanzees). From their analysis of this 7000 nucleotide sequence, the investigators inferred a common ancestral gene. The simplest branching pattern that would account for the results suggests that the gorilla split off first from the common ancestor it shared with the chimpanzee and human (see Fig. 17–15). The chimpanzee and human lines diverged later (probably about 6 million years ago).

---

for example, a new species of animal *(Symbion pandora)* was discovered inhabiting lobster mouthparts. Its combination of traits did not fit those of any existing phylum, so phylum Cycliophora was established to accommodate this animal.

Molecular biology provides taxonomic tools

When a new species evolves, it does not always exhibit obvious phenotypic differences relative to closely related species. For example, two distinct species of fruit flies may appear identical. Some of their DNA, proteins, and other molecules, however, are different. Variations in the structure of specific macromolecules among species, just like differences in anatomical structure, result from mutations. Macromolecules that are functionally similar in two different types of organisms are considered homologous if their subunit structure is similar.

The emerging science of **molecular taxonomy** focuses on molecular structure (see *Making the Connection: Molecular Biology, Evolution, and Taxonomy*). Differences in subunit sequences of macromolecules provide new data useful in determining evolutionary relationships. Methods that permit taxonomists to compare the nucleotide sequences of nucleic acids and the amino acid sequences of proteins have become important taxonomic tools. Such comparisons provide systematists with valuable information about the degree of relatedness of two organisms. The greater the correspondence in subunit sequences, the more closely organisms are considered to be related. The number of differences in DNA or RNA nucleotide sequences or in amino acid sequences in two groups of organisms may reflect how much time has passed since the groups branched from a common ancestor. (This can be true only if the rates of change have remained constant.) Thus, specific genes and specific proteins can be used as **molecular clocks** (see Chapter 17). Biologists can use such clocks to estimate the time of divergence of two groups from a common ancestor.

Many systematists are currently comparing ribosomal RNA structure to help determine phylogenies. The division of organisms into three domains was based, in large part, on the comparison of ribosomal RNA by Carl Woese and his research team at the University of Illinois. All known organisms have ribosomes that function in protein synthesis, and certain ribosomal RNA nucleotide sequences have been highly conserved in evolution. Prokaryotic ribosomes all contain three types of RNA named in order of increasing size: 5S, 16S, and 23S. The 5S and 16S RNAs have been extensively used to determine evolutionary relationships among bacteria. A common technique used to clone single-species ribosomal RNA genes is the PCR (polymerase chain reaction) discussed in Chapter 14.

Comparison of ribosomal RNA sequences was used to challenge the once widely accepted idea that fungi are more closely related to plants than to animals. In a study reported in *Science* in 1993, investigators suggested that, based on ribosomal RNA analysis, fungi are more closely related to animals than to plants. These biologists hypothesize that animals and fungi share a more recent common ancestor, perhaps a flagellated protist.

In 1997 molecular taxonomist Robert Wayne of the University of California, Los Angeles and his international team of researchers reported in the journal *Science* that mitochon-

Figure 22–6 **Molecular taxonomy.** Diagram of the relationships of 8 wolves and 15 dogs based on differences in mitochondrial DNA. 1030 base pairs of a stretch of mitochondrial DNA were compared. Wolf and dog sequences were very similar, differing by at most 12 substitutions. Dog and coyote sequences differed by at least 20 substitutions. Four clades were identified. (D, dog; W, wolf) *(From Carles Vila et al, published in* Science, *Vol. 276 (5319) 13 June 1997)*

drial DNA sequences were analyzed from 162 wolves at 27 different geographical locations and compared to those from domestic dogs of 67 different breeds (Fig. 22–6). Their results supported the hypothesis that dogs evolved more than 100,000 years ago from a common ancestor that was a wolf. Dog and wolf nucleotide sequences were similar, differing by no more than 12 substitutions. In contrast, dog sequences differed by at least 20 substitutions from jackal and coyote DNA. Their data suggest that repeated genetic exchange has taken place between dog and wolf populations, providing variation for selection. Using DNA sequences as a molecular clock, these investigators proposed that dogs evolved from wolves about 135,000 years ago, much earlier than the 14,000 year date formerly estimated.

SYSTEMATISTS USE TWO MAIN APPROACHES

In determining the relationships among organisms, systematists use a variety of methods and data. Which types of data are used more depends on the systematist's approach. **Phenetics,** or **numerical taxonomy,** was an early approach to the quantitative analysis of characters. First gaining attention of systematists in the 1960s, phenetics is based on as many phenotypic similarities (that is, similarities in *appearance*) as possible. In the phenetic approach, computer programs are used to analyze data and group organisms according to the number of shared characteristics. No attempt is made to determine whether their similarities arose from a common ancestor or from convergent evolution. Pheneticists argue that it is not important to try to differentiate between homology and homoplasy because many more similarities are due to homology than to homoplasy. Thus, the number of similarities that two organisms have in common reflects the degree of homology.

A taxonomist who follows the phenetic system might explain that dolphins and porpoises are classified as mammals rather than as fish because they share more similarities with mammals. Phenetics is not widely used by taxonomists today because the use of homoplastic similarities can lead to inaccurate conclusions about evolutionary relationships. However, the use of computers for making quantitative comparisons of characters has been an important contribution to systematics.

Phenetic techniques are currently used in molecular taxonomy. For example, if each amino acid in a protein is considered as a trait, amino acid sequences of various animals can be determined in the laboratory and compared by computer. The information about differences in amino acid sequences can be used to construct phylogenetic diagrams. Species are placed at relative distances from each other, reflecting the extent of difference in amino acid sequence.

Two major approaches to the classification of organisms that are widely used today are **phylogenetic systematics,** known as **cladistics,** and **classical evolutionary taxonomy,** or simply, evolutionary taxonomy. Although we discuss them separately here, many modern systematists use aspects of both approaches in their work. In contrast to phenetics, cladism and classical evolutionary taxonomy intentionally include the use of evolutionary data.

Phylogenetic systematics (cladistics) emphasizes phylogeny

The cladistic approach to taxonomy emphasizes common ancestry, rather than phenotypic similarity, as the basis for classification. Cladistics uses shared derived characters to reconstruct phylogenies. According to cladistics, dolphins are classified with mammals rather than with fishes because dolphins and mammals share derived characters and thus have a more recent common ancestor than dolphins and fishes.

To determine which attributes are shared derived characters, cladism employs **outgroup analysis.** Three or more related groups are compared. If two groups have a characteristic that the third lacks, the attribute may be a shared derived character for the two groups. If all three groups share the characteristic, it is an ancestral character. Recall that these terms are relative. An attribute that is an ancestral character in one comparison may be a shared derived character in another.

Cladism determines the evolutionary relationships of organisms and expresses them in branching diagrams called

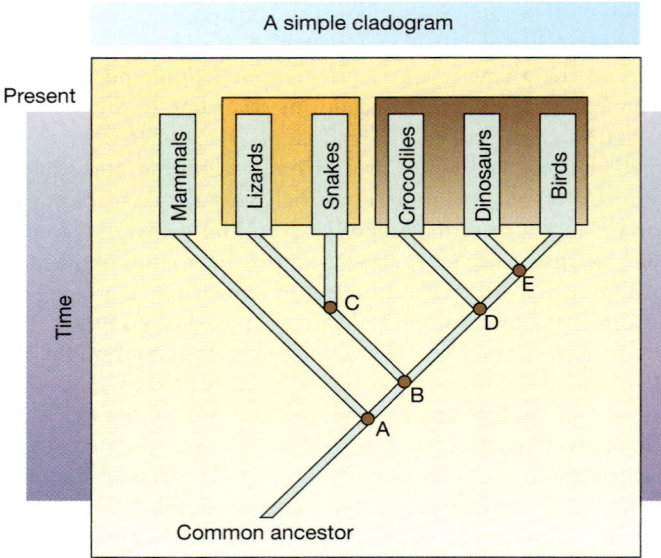

A simple cladogram

Present

Time

Mammals · Lizards · Snakes · Crocodiles · Dinosaurs · Birds

C · E · D · B · A

Common ancestor

Figure 22–7 **The cladistic approach.** Cladists classify birds and some reptiles together because they have a common ancestor. Node D represents the common ancestor of crocodiles, dinosaurs, and birds.

TABLE 22–3	A Comparison of Eight Chordates								
		CHARACTERS						
		Vertebrae (backbones)	Jaws	Tetrapod (4 limbs)	Amniotic egg	Mammary glands	Opposable thumb	Upright posture
A = Absent								
P = Present								
TAXA								
Amphioxus (outgroup)		A	A	A	A	A	A	A
Hagfish		P	A	A	A	A	A	A
Sunfish		P	P	A	A	A	A	A
Newt		P	P	P	A	A	A	A
Lizard		P	P	P	P	A	A	A
Bear		P	P	P	P	P	A	A
Chimpanzee		P	P	P	P	P	P	A
Human		P	P	P	P	P	P	P

cladograms. Each branch on a cladogram represents the divergence, or splitting, of two or more new groups from a common ancestor. Consider the evolutionary grouping of mammals, lizards, snakes, crocodiles, dinosaurs, and birds (Fig. 22–7). Birds, along with dinosaurs, are thought to share a common ancestor with modern crocodiles and alligators (point D on Fig. 22–7). Crocodiles, dinosaurs, and birds, then, comprise a monophyletic group, and cladists would classify them in the same **clade,** which is a monophyletic group in a cladogram. Similarly, snakes and lizards form a clade that is the closest group to birds, dinosaurs, and crocodiles. Finally, mammals form an additional clade. Cladists base their assessment on shared *derived* characters that can be structural, behavioral, physiological, or molecular. The characters must be homologous.

Phylogenies can be constructed by building and interpreting cladograms[3]

The first step in constructing a cladogram is to select the taxa, which may consist of individuals, species, genera, or other taxonomic levels. Here we use a representative group of eight chordates (Table 22–3). The next step is to select the homologous characters to be analyzed. In our example, we use seven characters. For each character, we must define all of the different conditions, or states, as they exist in our taxa. For simplicity, we will consider our characters to have only two dif-

ferent states: present or absent. Keep in mind that many characters used in cladistics have more than two states. For example, black, brown, yellow, and red may be only a few of the many possible states for the character of hair color.

The last, and often the most difficult, step in preparing the data is to organize the character states into their correct evolutionary order. The most common method of accomplishing this task is by outgroup analysis. An *outgroup* is a taxon that is considered to have diverged earlier than any of the other taxa under investigation and thus to represent an approximation of the ancestral condition. In our example, amphioxus is the chosen outgroup. Therefore, the character state "absent" is the ancestral condition, and the character state "present" is the derived condition for all seven characters in Table 22–3.

Constructing a cladogram from the data

Our objective is to construct a cladogram that requires the fewest number of evolutionary changes in the characters. Recall that, in cladistics, taxa are grouped by the presence of shared derived characters. In order to form a valid monophyletic group, all members must share at least one derived character. Membership in a clade cannot be established by shared ancestral characters.

In our example, notice that all taxa except the outgroup possess vertebrae. We may therefore conclude that these seven vertebrate taxa form a valid clade. Next, among the seven vertebrate taxa, notice that jaws are present in all groups except for hagfish. Using these data, we may construct a preliminary cladogram (Figure 22–8a). The base of the cladogram represents the common ancestor for all taxa being analyzed. Each branch point (referred to as a *node*) represents the immediate

[3]This discussion of building and interpreting cladograms is based on an essay contributed by Dr. John Beneski, Department of Biology, West Chester University, West Chester, PA.

Figure 22-8 **Building a cladogram.** In this example, amphioxus is the chosen outgroup that represents an approximation of the ancestral condition.

(a)

(b)

(c)

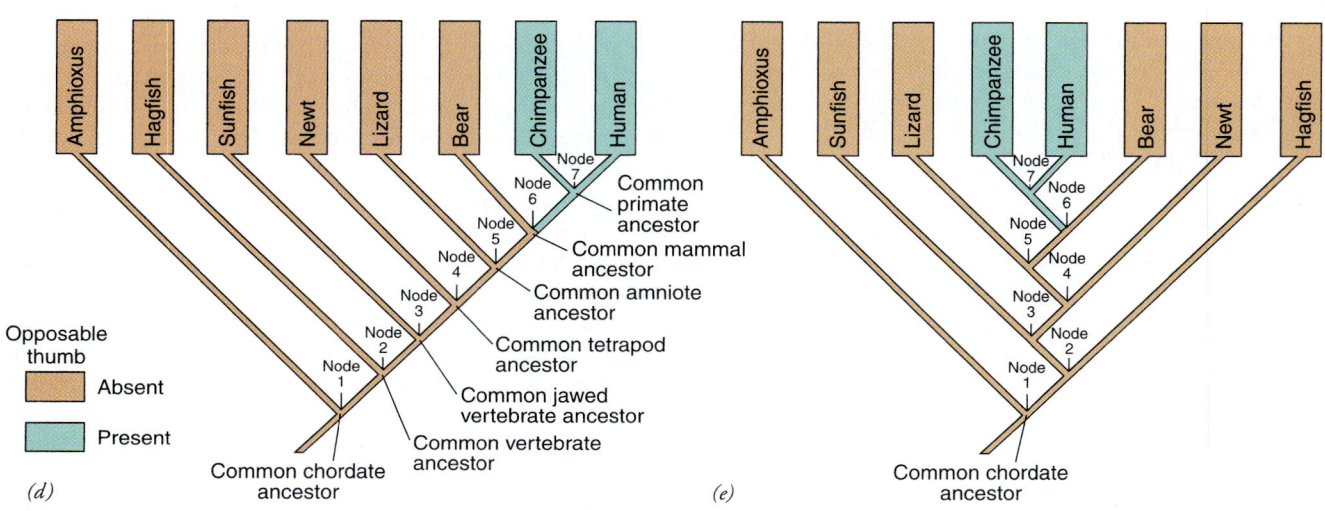

(d)

(e)

common ancestor of the next monophyletic group depicted in the cladogram. In Fig. 22–8*a*, node 1 represents the common chordate ancestor from which the outgroup (amphioxus) and the seven vertebrate taxa evolved. Similarly, node 2 represents the common ancestor of the vertebrates and node 3, the common ancestor of the jawed vertebrates. Continuing with this procedure, notice that among the six jawed taxa, all but sunfish are tetrapods (Fig. 22–8*b*). Among the five tetrapods, all but newts have amniotic eggs (Fig. 22–8*c*). The branching process is continued using Table 22–3 data until all clades are established (Fig. 22–8*d*).

How to interpret the cladogram

In Figure 22–8*d*, notice that humans and chimpanzees are more similar to each other than to any other clade. This relationship is indicated by the presence of a common ancestor at node 7. In the same way, bears are more similar to the human-chimpanzee clade than to any other clade, as indicated by the common ancestor at node 6. In comparing the nodes, the order of divergence is indicated by distance from the base of the diagram. The further a node is located up the cladogram, the more recently the group diverged. In our example, node 7 represents the most recent divergence, and node 1 represents the most ancient divergence. Thus, in our example, humans are closely related to chimpanzees (through node 7) but more distantly related to bears (through node 6). Therefore, humans and chimpanzees would be assigned to a less inclusive taxon (order Primates) whereas humans, chimpanzees, and bears would be assigned to a broader, more inclusive taxon (class Mammalia). In addition, the cladogram reveals that lizards are more closely related to the mammal clade than to newts, sunfish, or any other clade. Can you explain why?

In interpreting cladograms, two points must be kept in mind. First, the relationships among taxa can only be determined by tracing along the branches back to the most recent common ancestor (i.e., node) and not by the relative placement of the branches along the horizontal axis. It is possible to represent the same relationships with many different branching diagrams. For example, the cladogram in Figure 22–8*e* is equivalent to the one in Figure 22–8*d*. (You should verify this by comparing the numbered nodes and by checking the relationships described earlier.) Second, the cladogram does not establish ancestor-descendant relationships among taxa. In other words, a cladogram does not suggest that a taxon gave rise to any other taxon; rather it tells us which taxa shared a common ancestor and how recently they shared a common ancestor. The ancestor itself remains unspecified.

Classical evolutionary taxonomy allows paraphyletic groups

Classical evolutionary taxonomy, or simply evolutionary taxonomy, uses a system of phylogenetic classification and presents evolutionary relationships in phylogenetic trees. Evolutionary taxonomists consider both evolutionary branching (as

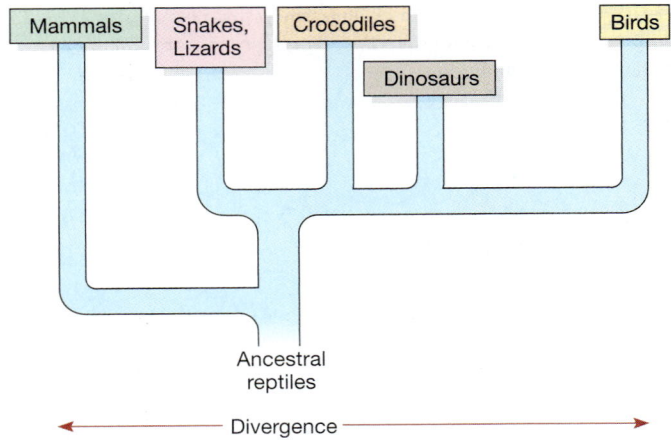

Figure 22–9 **The classical evolutionary taxonomy approach.** These taxonomists consider both common ancestry and extent of divergence that has occurred since two taxa split. The branching points and degrees of difference in the evolution of the major groups of reptiles are shown. Snakes, lizards, and crocodiles are most similar, but birds, dinosaurs, and crocodiles are most closely related because they branched most recently from a common ancestor.

do cladists) and the extent of divergence (structural and other changes) that has occurred in a lineage since it branched from a stem group (Fig. 22–9). They use a combination of shared ancestral characters and shared derived characters to evaluate both similarities and differences among groups.

A systematist using the evolutionary approach might explain that dolphins are mammals rather than fish because they share many characteristics with other mammals and because these characteristics can be traced to a common ancestor. Like cladists, evolutionary taxonomists classify organisms in the same taxon according to their shared characteristics only if those traits are derived from a demonstrable common ancestor. However, evolutionary taxonomists also consider the significance of the adaptations possessed by related organisms. If, for example, egg-laying mammals could be shown to have significantly different adaptations than other mammals, the evolutionary taxonomist might erect a separate taxon to accommodate them. On the other hand, common ancestry, although necessary for inclusion in the same category, would not by itself be sufficient grounds for inclusion (as it would be if a cladistic approach were used).

Many, perhaps most, of the taxa currently recognized by evolutionary taxonomists are monophyletic, based on the possession of shared derived characters. These taxa are also recognized by cladists. For example, proponents of both approaches agree that birds are a monophyletic taxon based on shared derived characters such as feathers. However, evolutionary taxonomy also recognizes paraphyletic taxa, groups that include some, but not all subgroups of organisms that share the same most recent common ancestor. Paraphyletic groups are based on shared ancestral characters. Cladists, for example, do not recognize reptiles as a natural grouping containing snakes, lizards, crocodiles, dinosaurs, and turtles because they do not form a monophyletic clade. They recognize a class Reptilia that includes birds. In contrast, evolutionary taxonomists do recognize class Reptilia as a valid group that does not include birds, even though it is paraphyletic. It does not include all of the subgroups, such as birds, that evolved from the ancestral reptile. Evolutionary taxonomists base their decision on different data; they consider that reptiles share numerous ancestral traits, such as ectothermy. They assign birds to a separate class because they have diverged markedly from the reptiles.

SUMMARY WITH KEY TERMS

I. **Systematics** is the scientific study of the diversity of organisms and their evolutionary relationships.
   A. **Taxonomy** is the branch of systematics devoted to naming, describing, and classifying animals.
   B. The process of assigning organisms into groups based on their similarities or relationships is **classification.**
II. Biologists use a system of classification based on the **binomial system** developed by Linnaeus in the mid-18th century.
   A. In this system the basic unit of classification is the **species.** Different populations within the same species that have consistent distinguishing characteristics are called **subspecies.** In plants subspecies may be referred to as **varieties;** in bacteria they are known as **strains.**
   B. The name of each species has two parts: the **genus** name and the **specific epithet.** For example, the scientific name of the human is *Homo sapiens,* and that of the white oak is *Quercus alba.*
III. The hierarchical system of classification currently used includes **kingdom, phylum, class, order, family, genus,** and **species.**
IV. The three-**domain** classification system assigns organisms to domain **Archaea,** domain **Eubacteria,** or domain **Eukarya.** The six-kingdom classification recognizes the kingdoms **Eubacteria, Archaebacteria, Protista, Fungi, Plantae,** and **Animalia.**
V. Modern systematics is based on similarity, as determined by shared characteristics, and on evolutionary relationships, or **phylogeny.**
   A. A **monophyletic group** includes all of the descendants of the most recent common ancestor. A **paraphyletic group** consists of a common ancestor and some, but not all, of its descendants. The organisms in a **polyphyletic group** evolved from different ancestors.

   B. **Homology** implies evolution from a common ancestor. **Homoplasy** refers to the evolution of similar, but independently evolved, characters that are not homologous.
   C. **Shared ancestral characters,** or **pleisomorphic characters,** suggest a distant common ancestor. **Shared derived characters,** or **synapomorphic characters,** indicate a more recent common ancestor.
   D. **Molecular taxonomy** provides methods for comparing macromolecules such as nucleic acids and proteins as a tool for confirming evolutionary relationships. Comparison of nucleotide sequences in ribosomal RNA has led to important taxonomic decisions regarding domains, kingdoms, and species.
VI. Two main approaches to taxonomy are **phylogenetic systematics,** also known as **cladistics,** and **classical evolutionary taxonomy.**
   A. **Phenetics,** or numerical taxonomy, is based on similarities of many characters; this was an early approach to the quantitative analysis of characters. Organisms were classified according to the number of characteristics they shared without trying to determine whether their similarities were homologous or homoplastic.
   B. Cladistics insists that taxa be monophyletic. Each monophyletic taxon, or **clade,** consists of a common ancestor and all of its descendants and is based on shared derived characters.
      1. Cladists use **outgroup analysis** to determine which characters are ancestral and which are derived.
      2. Shared derived characters are used to determine relationships that can be illustrated in diagrams called **cladograms.**
   C. Classical evolutionary taxonomy considers both evolutionary branching and the extent of divergence. Evolutionary taxonomy is based on shared ancestral characters as well as shared derived characters.

POST-TEST

1. The science of describing, naming, and classifying organisms is (a) systematics (b) taxonomy (c) cladistics (phylogenetic systematics) (d) phenetics (e) evolutionary taxonomy
2. Using the binomial system of nomenclature, the scientific name of each species consists of two parts (a) class, specific epithet (b) family, genus (c) genus, specific epithet (d) family, species (e) genus, species
3. The mold that produces penicillin is *Penicillium notatum. Penicillium* is the name of its (a) genus (b) order (c) family (d) species (e) specific epithet
4. Closely related genera may be grouped together in a single (a) phylum (b) domain (c) species (d) family (e) kingdom
5. Related classes are grouped together in the same (a) genus (b) phylum (c) order (d) paraphyletic taxon (e) family
6. In the six-kingdom system, the kingdom that includes the protozoa is (a) Plantae (b) Protista (c) Archaea (d) Eukarya (e) Fungi
7. Decomposers such as molds and mushrooms belong to kingdom (a) Plantae (b) Protista (c) Archaebacteria (d) Eukarya (e) Fungi
8. A taxon that contains a recent common ancestor and all of its descendants is (a) polyphyletic (b) paraphyletic (c) monophyletic (d) phyletic (e) pleisomorphic
9. The presence of homologous structures in different organisms suggests

that (a) they evolved from a common ancestor (b) convergent evolution has occurred (c) they belong to a polyphyletic group (d) answers a,b, and c are correct (e) answers a and c only are correct
10. The dolphin and the human both have the ability to nurse their young, whereas the less closely related fish does not. The ability to nurse their young is a (a) shared derived character of mammals (b) shared ancestral character of all vertebrates (c) pleisomorphy (d) homologous behavior (e) two of the preceding answers are correct
11. Relative constancy in the rates of DNA and protein evolution permits biologists to use these macromolecules as (a) molecular clocks (b) polymerase chains (c) clades (d) paraphyletic clues (e) two of the preceding answers are correct
12. Phenetics is (a) a numerical taxonomy based on phenotypic similarities (b) an approach to taxonomy that emphasizes common ancestry (c) an approach that emphasizes polyphyletic groups (d) an approach used in molecular taxonomy (e) two of the preceding answers are correct
13. Systematists who classify crocodiles and birds in the same taxon because they are monophyletic follow a(n) (a) phyletic approach (b) cladistic approach (c) evolutionary taxonomy approach (d) polyphyletic approach (e) two of the preceding answers are correct

REVIEW QUESTIONS

1. Briefly describe the binomial system of nomenclature.
2. Define the terms: (a) species (b) class (c) phylum.
3. What are the advantages of a "six-kingdom" system over a "two-kingdom" one? What types of organisms are especially difficult to assign a place in the taxonomic hierarchy?
4. In which kingdom would you classify each of the following? (a) oak tree (b) amoeba (c) *Escherichia coli* (a bacterium) (d) tapeworm (e) black bread mold
5. Compare the cladistic and classical evolutionary approaches to taxonomy.
6. Of what use to a taxonomist is knowledge of the amino acid sequences of the proteins of various organisms?
7. What kind of grouping is represented by the shaded area?

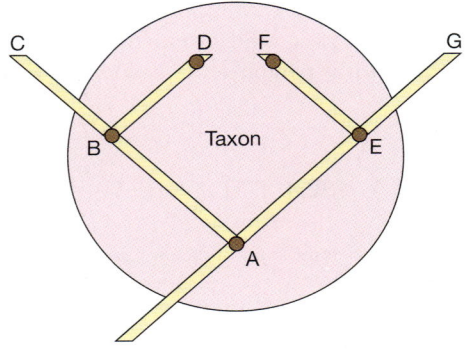

YOU MAKE THE CONNECTION

1. What difficulties do we encounter when we attempt to use the concept of a species? Are members of a genus similar because they share a common ancestor, or do they belong to the same genus because they are similar? How might your answer vary depending on which approach to taxonomy you are following?
2. After many years of being considered old-fashioned, the science of systematics has reemerged on the cutting edge of biological research. Why do you think this shift has occurred?
3. The TATA-binding protein (TBP) is thought to be necessary for transcription in all eukaryotic cell nuclei. Studies show that archaebacteria, but not eubacteria, have a protein structurally and functionally similar to TBP. What does this similarity suggest regarding the evolution of archaebacteria and eukaryotes? How might knowledge of this similarity affect how systematists classify these organisms?
4. Based on the information in question 3 and on studies that indicate that archaebacterial gene sequences are more similar to those in eukaryotes than to those in eubacteria, do you think systematists should abandon the six kingdoms and classify organisms in three domains? Why or why not?

RECOMMENDED READINGS

Davis, J.I. "Phylogenetics, Molecular Variation, and Species Concepts." *BioScience,* Vol., 46 No.7, Jul./Aug. 1996. The author discusses definitions of species and how they relate to phylogenetics.

Funch, P., and R.M. Kristensen. "Cycliophora Is a New Phylum with Affinities to Entoprocta and Ectoprocta." *Nature,* Vol. 378, 14 Dec. 1995. Description of a new species that resulted in the recognition of a new phylum.

Gaffney, E.S., L. Dingus, and M.K. Smith. "Why Cladistics?" *Natural History,* Jun. 1995. An interesting introduction to building cladograms using the dinosaurs as an example.

May, Robert M. "How Many Species Inhabit the Earth?" *Scientific American,* Vol. 267, No. 4, Oct. 1992. An argument for the importance of identifying and classifying the organisms that inhabit our planet; this information affects environmental issues.

Olsen, G. J., and C.R. Woese. "Archaeal Genomics: An Overview." *Cell,* Vol. 89, 27 Jun. 1977. A discussion of the genome of archaebacterium *M. jannaschii* and its evolutionary relationships to eubacteria and eukarya.

Pennisi, E. "Genome Data Shake Tree of Life." Research News. *Science,* Vol. 280, 1 May 1998. Researchers are finding unexpected connections between domains.

Vila, C., Savolainen, P., Maldonado, J.E., Amorim, I.R., Rice, J.E., Honeycutt, R.L., Crandall, K.A., Lundeberg, J., and Wayne, R.K. "Multiple and Ancient Origins of the Domestic Dog." *Science,* Vol. 276, 13 Jun. 1997. Mitochondrial DNA sequences were analyzed and compared in wolves, dogs, and jackals.

Wainright, P.O., G. Hinkle, M.L. Sogin, and S.K. Stickel. "Monophyletic Origins of the Metazoa: An Evolutionary Link with Fungi." *Science,* Vol. 260, 16 Apr. 1993. The authors use comparisons of ribosomal RNA sequences to hypothesize evolutionary relationships of animals and fungi.

Whittaker, R.H. "New Concepts of Kingdoms of Organisms." *Science,* Vol. 163, 1969. Whittaker's original proposal for classifying organisms according to a five-kingdom system.

Williams, P.A. "Evolutionary Taxonomy versus Cladism." *Encyclopedia of Environmental Biology,* Vol. 1. Academic Press, San Diego 1995. An excellent discussion of current approaches to systematics.

● Visit our website at **http://www.saunderscollege.com/lifesci/titles.html** and click on Solomon/Berg/Martin Biology for links to chapter-related resources on the World Wide Web.

CHAPTER 23

Viruses and Bacteria

Bacteria have inhabited our planet for more than 3.5 billion years, whereas eukaryotes evolved about 2 billion years ago. Approximately one-half of the biomass, the mass of living material, on Earth consists of microorganisms, mostly bacteria and fungi. (In contrast, plants account for about 35% and animals for about 15% of the biomass of our planet.) Anton van Leeuwenhoek discovered bacteria and other microorganisms in 1674 when he looked at a drop of lake water through a glass lens. During the late 1800s, many microorganisms, including bacteria, fungi, and protozoa, were identified as **pathogens,** agents that cause disease. The bacterium *(Streptococcus pyogenes)* shown dividing in the TEM is a human pathogen that inhabits the human nose and throat. This bacterium can cause scarlet fever and inflammation of the heart tissue. The strain shown here is resistant to antibiotics, and infection can be fatal.

Although bacteria cause many diseases, including respiratory infections and food poisoning in humans, only a small minority of bacterial species are pathogens. In fact, bacteria play an essential role in the biosphere as decomposers, breaking down organic molecules into their components. Along with fungi, bacteria are nature's recyclers. Without these microorganisms, the available carbon, nitrogen, phosphorus, and sulfur would eventually be tied up in the wastes and dead bodies of plants and animals. Life as we know it would cease to exist because of the lack of raw materials for the synthesis of new cellular components.

Some bacteria are producers that carry on photosynthesis. Other bacteria perform a key role in agriculture by converting atmospheric nitrogen to ammonia and then to nitrates, a form that can be used by plants (see Figure 53–10). The conversion of nitrogen enables plants and animals (because they eat plants) to manufacture essential nitrogen-containing compounds such as proteins and nucleic acids.

In contrast to bacteria, viruses do not consist of cells. Biologists consider them to be nonliving particles. Viruses have no common ancestor and thus are not assigned to a kingdom. Why, then, include them here in a unit on the diversity of life? As we will discuss in this chapter, viruses evolved from various prokaryotic and eukaryotic cells. They contain the nucleic acid necessary to make copies of themselves, and they reproduce by invading living cells and commandeering their metabolic machinery. Viruses are intimately involved with living organisms, and it seems appropriate to discuss the diversity

(Dr. Kari Lounatmaa/Science Photo Library/Photo Researchers, Inc.)

and importance of these tiny, but potent, particles. Viruses are responsible for a wide variety of diseases in plants and animals. Among the viral diseases that affect humans are common colds, influenza, chickenpox, rabies, and AIDS.

This chapter examines the diversity and characteristics of viruses, bacteria, and the smaller viroids and prions. They are *not* a natural group of closely related organisms, and we discuss them in a single chapter only for convenience.

1. Describe the structure of a virus and compare a virus with a free-living cell.
2. Speculate about the evolutionary origin of viruses.
3. Characterize bacteriophages and contrast a lytic cycle with a lysogenic cycle.
4. Explain how viruses infect animals and plants.
5. Describe the reproductive cycle of a retrovirus such as HIV.
6. Characterize viroids and prions.
7. Compare the three domains: Eubacteria, Archaea, and Eukarya.
8. Describe asexual reproduction in bacteria and summarize three

mechanisms (transformation, conjugation, and transduction) that may lead to genetic recombination.
9. Characterize the metabolic diversity of autotrophic and heterotrophic bacteria, including aerobes, facultative anaerobes, and obligate anaerobes.
10. Distinguish among the three main groups of archaea and among several groups of eubacteria as described in Table 23–3. Give examples of each group.
11. Discuss the ecological roles of bacteria, their importance as pathogens, and their commercial importance.

VIRUSES ARE INFECTIOUS AGENTS THAT ARE NOT ASSIGNED TO ANY OF THE SIX KINGDOMS

During the late 1800s, botanists searched for the cause of tobacco mosaic disease, which stunts the growth of tobacco plants and gives the infected tobacco leaves a spotted, mosaic appearance. They found that the disease could be transmitted to healthy plants by daubing their leaves with the sap of diseased plants. In 1892 Dmitrii Ivanowsky, a Russian botanist, showed that the sap was infective even after it had been passed through filters fine enough to remove particles the size of all known bacteria. A few years later his work was expanded by Martinus Beijerinck who provided evidence that the agent that caused tobacco mosaic disease had many characterstics of a living organism. He hypothesized that the agent could only reproduce within a living cell.

Early in the 20th century, scientists discovered infectious agents that could cause disease in animals or kill bacteria. Like the agents that cause tobacco mosaic disease, these pathogens passed through filters that removed known bacteria and were so small that they could not be seen with the light microscope. Curiously, they could not be grown in laboratory cultures unless living cells were present. These pathogens that infected plants or animals came to be known as viruses. Those that killed bacteria were called **bacteriophages** ("bacteria eaters"), or **phages.**

A virus particle consists of nucleic acid surrounded by a protein coat

A **virus,** or **virion,** is a tiny, infectious particle consisting of a nucleic acid core (its genetic material) surrounded by a protein coat called a **capsid.** Some viruses are also surrounded by an outer membranous envelope containing proteins, lipids, carbohydrates, and traces of metals. A typical small virus, like

the poliovirus, is about 20 nanometers in diameter (about the size of a ribosome), whereas a larger virus, such as the poxvirus that causes smallpox, might be 400 nm in length and 200 nm in width.

Viruses are not cellular and cannot independently perform metabolic activities. They do not have the components necessary to carry on cellular respiration or to synthesize proteins and other molecules. All living organisms contain both DNA and RNA, but a virus contains either DNA or RNA, not both. Viruses can reproduce, but only within the complex environment of the living cells they infect. In a sense, viruses come "alive" only when they infect a cell. Viruses have genetic information that can force the host cell to replicate the viral nucleic acid and to synthesize the capsid and envelope components. The genetic information in a virus can take over the translational or transcriptional mechanisms of the host cell.

The shape of a virus is determined by the organization of protein subunits that make up the capsid. Viral capsids are generally either helical, polyhedral, or a complex combination of both shapes (Fig. 23–1). Helical viruses, such as the tobacco mosaic virus, appear as long rods or threads. The capsid is a hollow cylinder. The RNA fits into a groove in the proteins. Polyhedral viruses, such as the adenovirus (which causes a number of human illnesses, including some respiratory infections), appear to be somewhat spherical. The T4 phage that infects *Escherichia coli* consists of a polyhedral "head" attached to a helical "tail."

Viruses may have "escaped" from cells

Where did viruses come from? The hypothesis currently considered most likely is that viruses are bits of nucleic acid that "escaped" from cellular organisms. According to this view, some viruses may trace their origin to animal cells, others to plant cells, and still others to bacterial cells. Their multiple origins might explain why viruses are species-specific; perhaps they infect only those species that are closely related to the or-

Figure 23–1 **Virus structure.** Viruses are typically either helical or polyhedral in shape, or a combination of both. *(a)* TEM of tobacco mosaic virus, which is a helical virus. *(b)* TEM of adenovirus, which has a polyhedral capsid with projecting protein spikes. *(c)* TEM of the bacteriophage known as T₄, which has a polyhedral head and a helical tail. *(a, Omikron/Science/Photo Researchers, Inc. b, Oliver Meckes/E.O.S./Gelderblum/ Photo Researchers, Inc.; c, Lee D. Simon/Science Source/Photo Researchers, Inc.)*

ganisms from which they originated. This hypothesis is supported by the genetic similarity between a virus and its host cell—a closer similarity than exists between one virus and another.

Phages are viruses that attack bacteria

Much of our knowledge of viruses has come from studying phages because they can be cultured easily within living bacteria in the laboratory. As mentioned in Chapter 14, biologists use phages in genetic engineering research. Phages are among the most complex viruses (Fig. 23–1*c*). Their most common structure consists of a long nucleic acid molecule (usually DNA) coiled within a polyhedral head. Many phages have a tail attached to the head. The phage may use fibers extending from the tail to attach to a bacterium.

A lytic reproductive cycle destroys the host cell

A viral reproductive cycle can be lytic or temperate. In a **lytic** cycle, the virus lyses (destroys) the host cell. When the virus infects a susceptible host cell, it forces the host to use its metabolic machinery to replicate viral particles. Viruses that have a lytic cycle are described as **virulent** (lethal).

Five steps are typical in viral reproduction (see Fig. 23–2):

1. **Attachment (or absorption).** The virus attaches to receptors on the host cell wall.
2. **Penetration.** The nucleic acid of the virus moves through the plasma membrane and into the cytoplasm of the host cell. The capsid of a phage remains on the outside. In contrast, many viruses that infect animal cells enter the host cell intact.
3. **Replication.** The viral genome contains all the information necessary to produce new viruses. Once inside, the virus induces the host cell to synthesize the necessary components for its replication.
4. **Assembly.** The newly synthesized viral components are assembled into new viruses.
5. **Release.** Assembled viruses are released from the cell. Generally, lytic enzymes destroy the host cell.

The new viruses infect other cells, and the process begins anew. The time required for viral reproduction from attachment to the bacterium to the release of new viruses is approximately 30 to 35 minutes.

Temperate viruses can integrate their DNA into the host DNA

Temperate viruses do not always destroy their hosts. In a **lysogenic cycle** the viral genome becomes integrated with and is replicated along with the host DNA. In the case of some bacterial viruses, the phage DNA becomes integrated into the host bacterial DNA and is then referred to as a **prophage.** When the bacterial DNA replicates, the prophage also replicates (Fig. 23–3). The viral genes that code for viral structural proteins may be repressed indefinitely. Bacterial cells carrying prophages are called **lysogenic cells.** Certain external conditions (such as ultraviolet light and x rays) can cause temperate viruses to revert to a lytic cycle and then destroy their host. Sometimes temperate viruses become lytic spontaneously.

Bacterial cells containing certain temperate viruses may exhibit new properties. This is called **lysogenic conversion.** An interesting example involves the bacterium *(Corynebacterium diphtheriae)* that causes diphtheria. Two strains of this species exist, one that produces a toxin (and causes diphtheria) and one that does not. The only difference between these two strains is that the toxin-producing bacteria contain a specific temperate phage. The phage DNA codes for the powerful toxin that causes the symptoms of diphtheria. In the same way, the bacterium *(Clostridium botulinum)* that causes botu-

Labels in figure:
- RNA inside capsid
- Capsid
- *(a)* 0.1 μm
- Capsid with antenna-like fibers
- DNA inside capsid
- *(b)* 0.05 μm
- DNA inside capsid
- Capsid
- Tail
- Tail fibers
- Emerging DNA
- *(c)* 0.25 μm

Step 1. Attachment
Phage attaches to the cell surface of a bacterium.

Phages

Bacterium

Step 2. Penetration
Phage DNA enters the bacterial cell.

Bacterial DNA

Step 3. Replication
Phage DNA is replicated. Phage proteins are synthesized.

Phage DNA
Phage Protein

Step 4. Assembly
Phage components are assembled into mature viruses.

Step 5. Release
The bacterial cell lyses and releases many phages that can then infect other cells.

(a)

(b)

0.25 μm

Figure 23–2 **Lytic cycle.** The host cell is destroyed in a lytic infection. *(a)* The sequence of events in a lytic infection are (1) attachment, (2) penetration, (3) replication, (4) assembly, and (5) release. *(b)* TEM of phages infecting a bacterium, *Escherichia coli.* (b, Lee D. Simon/Science Source/Photo Researchers, Inc.)

Step 1. Attachment
Phage attaches to the cell surface of a bacterium.

Step 2. Penetration
Phage DNA enters the bacterial cell.

Prophage

Step 3. Integration
Phage DNA integrates into bacterial DNA.

Step 4. Replication
Integrated prophage replicates when bacterial DNA replicates.

These cells may exhibit new properties

Figure 23–3 **Lysogenic cycle.** Temperate phages can integrate their nucleic acid into the host cell DNA, making it a lysogenic cell. The sequence of events includes (1) attachment, (2) penetration, (3) integration into bacterial DNA, and (4) replication when the bacterial DNA replicates.

lism (a serious form of food poisoning) is harmless unless it contains certain prophage DNA that induces synthesis of toxin.

Some viruses infect animal cells

Hundreds of different viruses infect humans and other animals. Most viruses cannot survive very long outside a living host cell, so their survival depends on their being transmitted from animal to animal.

The type of attachment proteins on the surface of a virus determines what type of cell it can infect. Some viruses, such as the adenoviruses, have fibers that project from the capsid and are thought to help the virus adhere to complementary receptor sites on the host cell. Other viruses, such as those that cause herpes, influenza, and rabies, are surrounded by a lipoprotein envelope with projecting glycoprotein spikes that aid in attachment to a host cell.

Receptor sites vary with each species and sometimes with each type of tissue. Thus, most human viruses can infect only humans, because their attachment proteins combine only with receptor sites found on human cell surfaces. The measles virus and pox viruses can infect many types of human tissue because their attachment proteins combine with receptor sites on a variety of cells. In contrast, polioviruses attach to specific types of cells, such as those that line the digestive tract.

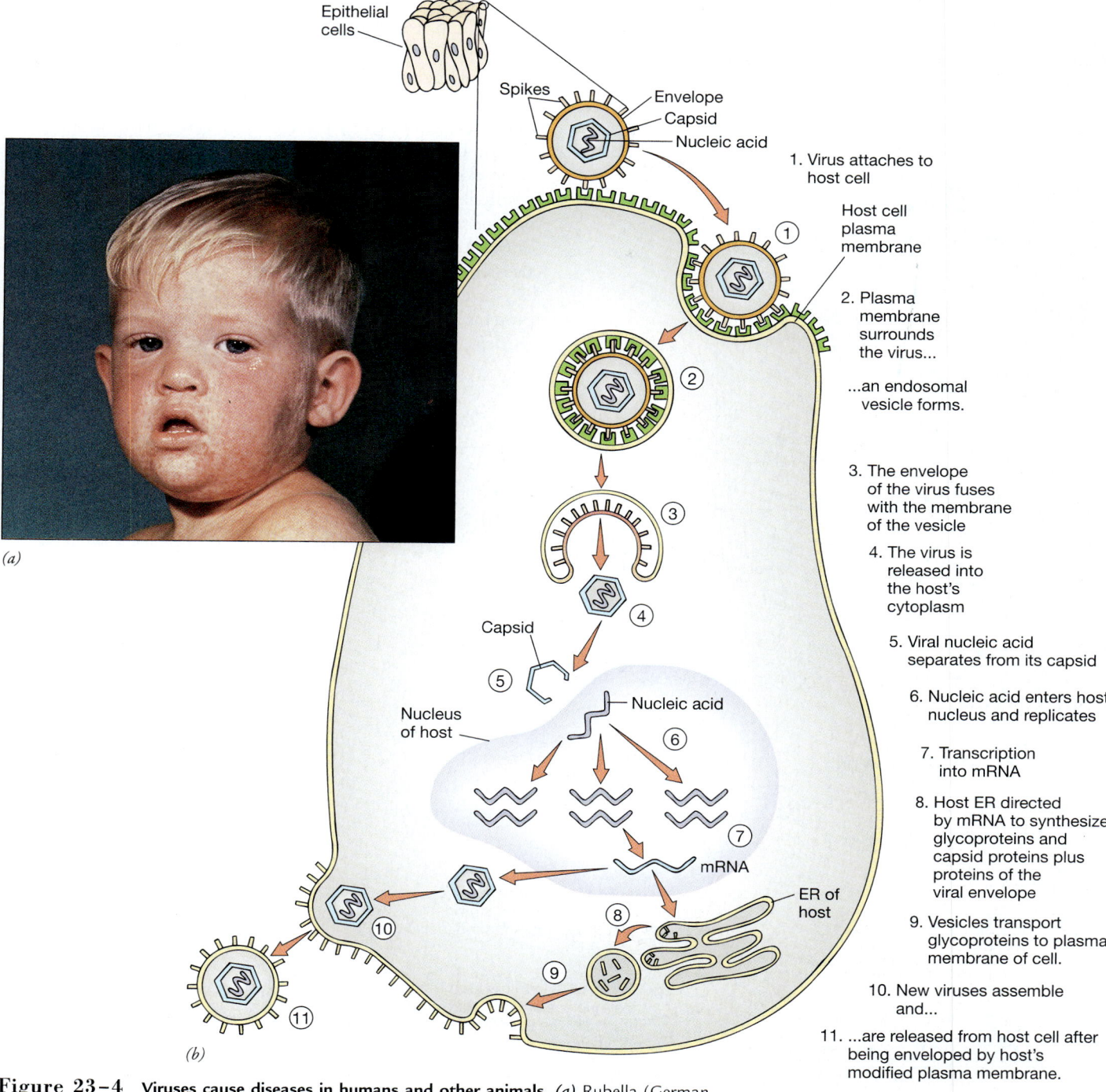

Figure 23–4 **Viruses cause diseases in humans and other animals.** *(a)* Rubella (German measles) is caused by an RNA virus spread by close contact. When contracted during pregnancy, it can cause birth defects. Vaccination has greatly decreased the incidence of this disease. *(b)* The steps by which some viruses enter animal cells by endocytosis. *(a, Centers for Disease Control and Prevention, U.S. Dept. of Health and Human Services)*

Viruses have several ways to penetrate animal cells. After attachment to a host-cell receptor, some enveloped viruses fuse with the animal cell's plasma membrane. The viral capsid and nucleic acid are both released into the animal cell. Other viruses enter the host cell by endocytosis. In this process, the plasma membrane of the animal cell invaginates to form a membrane-bounded vesicle that contains the virus (Fig. 23–4).

Like other viruses, those that infect animal cells multiply and produce new virus particles. During the time that viral nucleic acid is being replicated and viral proteins are being synthesized, the synthesis of host DNA, RNA, and protein may be inhibited.

Animal viruses can contain either DNA or RNA. In DNA viruses, the synthesis of viral DNA and protein is similar to the processes by which the host cell would normally carry out its own DNA and protein synthesis. In most RNA viruses, transcription takes place with the help of an RNA-dependent RNA polymerase. However, **retroviruses** are RNA viruses that have a DNA polymerase called **reverse transcriptase** used to transcribe the RNA genome into a DNA intermediate (Fig. 23–5). This DNA is integrated into the host DNA by an enzyme also carried by the virus. Copies of the viral RNA are synthesized as the incorporated DNA is transcribed by host polymerases. The human immunodeficiency virus (HIV) that causes AIDS is a retrovirus. Certain cancer-causing viruses are also retroviruses.

After the viral genes are transcribed, the viral structural proteins are synthesized. The capsid is produced, and new virus particles are assembled. Viruses that do not have an outer en-

1. The HIV virus attaches to a host cell plasma membrane.

2. The virus enters the host cytoplasm and the capsid is removed.

3. Reverse transcriptase catalyzes the synthesis of single-stranded (ss) DNA that is complementary to the viral RNA.

4. The DNA strand then serves as a template for the synthesis of a complementary DNA strand resulting in double-stranded (ds) DNA.

5. The ds DNA is transferred to the host nucleus and the enzyme integrase integrates the DNA into the host chromosome.

6. Host enzymes transcribe viral RNA.

7. Viral RNA leaves the nucleus, viral proteins are synthesized at host ribosomes and the virus is assembled.

8. The virus buds from the host cell, picking up the viral envelope in the process.

Figure 23–5 Life cycle of HIV, a retrovirus that attacks cells of the immune system known as helper T cells.

EMERGING VIRUSES

0.5 μm

TEM of Ebola virus. *(Courtesy of Frederick A. Murphy)*

The 1995 Ebola virus outbreak serves as a grim reminder that pathogens can strike quickly and fatally. According to the U. S. Centers for Disease Control and Prevention, more than 200 new, continual, or reemerging pathogens have the potential to strike globally. Historically, new viral strains have claimed many human lives. For example, in 1918 an influenza epidemic killed more than 20 million people. Even at the level of our current knowledge about viruses and epidemiology, just how well could we contain a particularly virulent virus? How well are we containing HIV (the virus that causes AIDS), which claims more than two million lives each year?

Let us consider the deadly Ebola virus as an example of an emerging virus. In 1995 an outbreak of the Ebola virus occurred in Zaire, a country in Central Africa. Named after a river in Zaire, Ebola is an elongated, single-stranded RNA virus that causes a fatal disease. Within about three days of infection, victims develop a fever and weakness, followed by a rash and vomiting. The victim hemorrhages internally, and bleeds from the mouth, eyes, ears, and other body openings. Internal organs shut down, and 50% to 90% of victims die within about two weeks after infection.

Because early symptoms of Ebola resemble those of influenza or dysentery, health care workers did not immediately recognize the danger in the 1995 outbreak. When they became suspicious, blood samples from 16 patients were flown to the U. S. Centers for Disease Control and Prevention in Atlanta. Within 36 hours the Ebola virus was identified in blood from 14 of the 16 patients.

As soon as Ebola was identified in the 1995 outbreak, global health care teams, including epidemiologists, were sent into Central Africa to prevent its spread. Health officials confirmed that, like HIV (the virus that causes AIDS), Ebola is not spread by casual contact, but by direct contact with infected body fluids. Because victims of Ebola hemorrhage, infected blood appears to be the principal method of transmission. Infection-control methods, such as using gloves, gowns, and masks, were implemented. Patients were quarantined, and travel out of the infected area was monitored. Global cooperation and rapid response successfully contained the 1995 Ebola outbreak.

An important aspect of understanding an infectious agent is identifying *patient zero,* the first patient who contracted the virus. If the origin of the epidemic can be found, investigators might be able to trace where the virus came from and how it infects humans. Ebola is known to infect chimpanzees, but because they die quickly they are not thought to be the reservoir species. However, some other primate host might be the reservoir that would allow the virus to maintain itself indefinitely. Researchers continue to search for clues to the origin of Ebola and are working on the development of effective antiviral drugs and vaccines.

Ebola outbreaks have occurred several times during the past few decades. The virus was first identified in 1976 in Zaire and Sudan (outbreaks that killed more than 400 people). Between outbreaks the virus remains hidden. Like many other RNA viruses, Ebola makes frequent mistakes when it duplicates its RNA. This leads to a high mutation rate and thus to the rapid evolution of new strains.

Human activity, including social factors, such as urbanization and global travel, contributes to epidemics of infectious disease. For example, as human populations concentrate in cities, large numbers of people are in close contact, permitting rapid spread of viruses. Living conditions, including sanitation, nutrition, physical stress, level of health care, and sexual practices, are important factors in the spread of disease. In the United States and other developed countries, infectious disease accounts for about 4% to 8% of deaths compared to death rates of 30% to 50% in developing regions.

Changing natural habitats can create the conditions necessary for new pathogens to emerge. For example, cutting down forests can bring disease-carrying insects into contact with humans. Sometimes, even naturally occurring ecological changes can spawn outbreaks of disease. The 1993 outbreak of hantavirus in the Southwestern U.S. killed more than 50 persons. A mild winter coupled with heavy rainfall resulted in a large crop of seeds, which supported a population explosion of field mice. The mice carry the virus. Lessons learned from dealing with emerging viruses and the resurgence of old ones will help us contain future epidemics, but much more research is needed.

---

velope exit by cell lysis. The plasma membrane ruptures, releasing many new viral particles. Enveloped viruses receive their lipoprotein envelopes as they leave the infected cell.

Viral proteins damage the host cell in a variety of ways. These proteins may alter the permeability of the plasma membrane or may inhibit synthesis of host nucleic acids or proteins. Viruses sometimes damage or kill their host cells by their sheer numbers. A poliovirus may produce 100,000 new viruses within a single host cell!

Most of us suffer from two to six viral infections each year, including common colds. Other human diseases caused by viruses include chickenpox, herpes simplex (one type causes genital herpes), mumps, rubella (German measles), rubeola (measles), rabies, warts, infectious mononucleosis, influenza, hepatitis, and AIDS (Table 23–1 and *Focus On: Emerging Viruses*). Animal viruses cause hog cholera, foot-and-mouth disease, canine distemper, swine influenza, and certain types of cancer (for example, feline leukemia).

TABLE 23-1 Animal Viruses

Group	Diseases Caused	Characteristics
DNA Viruses		
Poxviruses	Smallpox, cowpox, and economically important diseases of domestic fowl	Large, complex double-stranded DNA; replicate in the cytoplasm of the host cell. The vaccinia (cowpox) virus is used to produce genetically engineered vaccines
Herpesviruses	Herpes simplex type 1 (cold sores); herpes simplex type 2 (genital herpes, a sexually transmitted disease); varicella-zoster (chickenpox and shingles); the Epstein–Barr virus causes infectious mononucleosis and Burkitt's lymphoma	Medium to large, enveloped viruses; double-stranded DNA; replicate in host nucleus; frequently cause latent infections; some cause tumors
Adenoviruses	About 40 types known to infect human respiratory and intestinal tracts; common cause of sore throat, tonsillitis, and conjunctivitis; other varieties infect other animals	Double-stranded DNA; replicates in host nucleus
Papovaviruses	Human warts and some degenerative brain diseases; some cancers	Double-stranded DNA; the virus SV40 has been used as a vector to transport genes into cells
Parvoviruses	Infections in dogs, swine, arthropods, rodents; cause gastroenteritis in humans after eating infected shellfish	Single-stranded DNA; some require a helper virus in order to multiply
RNA Viruses		
Picornaviruses	About 70 types infect humans, including polioviruses; hepatitis A virus; enteroviruses infect intestine; rhinoviruses infect respiratory tract and are main cause of human colds; coxsackievirus and echovirus cause aseptic meningitis	Diverse group of small viruses; single-stranded RNA that can serve as mRNA
Togaviruses	Rubella	Large, diverse group of medium-sized, enveloped viruses; single-stranded RNA that can serve as mRNA; many transmitted by arthropods
Orthomyxoviruses	Influenza in humans and other animals	Medium-sized viruses that often exhibit projecting spikes; single-stranded RNA that serves as template for mRNA synthesis
Paramyxoviruses	Rubeola (measles); mumps; distemper in dogs	Resemble orthomyxoviruses but somewhat larger
Rhabdoviruses	Rabies	Single-stranded RNA
Reoviruses	Vomiting and diarrhea; encephalitis	Double-stranded RNA; no envelope
Retroviruses	AIDS; some types of cancer	RNA viruses that contain reverse transcriptase for transcribing the RNA genome into DNA; two identical molecules of single-stranded RNA
Flavivirus	Yellow fever	Single-stranded RNA
Filoviruses	Cause hemorrhagic fever; Ebola virus in this group	Single-stranded RNA
Bunyaviruses	St. Louis encephalitis	Single-stranded RNA

Viruses can infect plant cells

Viral diseases can be spread among plants by insects such as aphids and leafhoppers as they feed on plant tissues. Plant viruses can also be inherited by way of infected seeds or by asexual propagation. Once a plant is infected, the virus spreads through the plant body by passing through plasmodesmata (cytoplasmic connections) that penetrate the walls between adjacent cells (see Fig. 5–23). The genome of most plant viruses consists of RNA.

Symptoms of viral infection include reduced plant size, and spots, streaks, or mottled patterns on leaves, flowers, or fruits (Fig. 23–6). Infected crops almost always produce lower yields. Cures are not known for most viral diseases of plants, and so it is common to burn plants that have been infected.

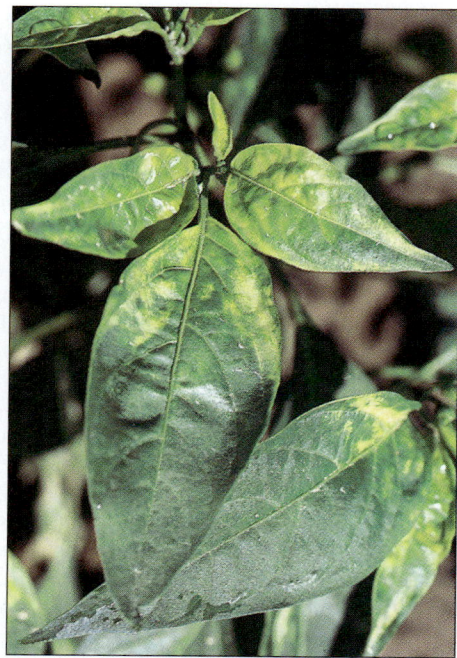

Figure 23–6 **Plant viruses.** *(a)* Virus-streaked tulips. The virus that causes this relatively harmless disease affects pigment formation in the petals. *(b)* Pepper leaves infected with tobacco mosaic virus. The leaf is characteristically mottled with light green areas. *(a, Kenneth M. Corbett; b, Jack M. Bostrack/Visuals Unlimited)*

(a)   *(b)*

Some agricultural scientists are focusing their efforts on prevention of viral disease by developing virus-resistant strains of important crop plants.

VIROIDS AND PRIONS ARE SMALLER THAN VIRUSES

The discovery of infective agents even smaller than viruses has challenged old ideas. Until **viroids** were studied, most biologists assumed that protein was necessary for an infectious agent to duplicate itself. However, no proteins are associated with viroids, and evidence suggests that the viroid genome does not code for any proteins. Each viroid consists of a very short strand of RNA (only 250–400 nucleotides) with no protective coat. Viroids are copied by host RNA polymerases; the viroid is used as a template. Viroids cause a variety of plant diseases and may also infect animals. These infective agents are generally found within the host cell nucleus and may interfere with gene regulation.

Even more heretical is the idea that a pathogen could exist and transfer information without nucleic acids. The **prion,** a protein-like infectious particle, has been linked to a group of fatal degenerative brain diseases. These diseases are called transmissible spongiform encephalopathies (TSEs) because the infected brain appears to develop holes, becoming somewhat spongelike. One of the most studied prion diseases is scrapie in sheep and goats. When infected, animals lose coordination, become irritable, and itch so severely that they scrape off their wool or hair.

A disease similar to scrapie became epidemic in cows in the United Kingdom in the 1990s. More than 160,000 cases of bovine spongiform encephalopathy (referred to as "mad cow disease") were documented, and almost 1 million cattle were estimated to be infected. This outbreak led to a ban on the export and consumption of British cattle. Scientists hypothesized that the cattle became infected when they ate feed that had been mixed with sheep offal—brains and other organs—containing prions. In the spring of 1996, at least ten people died of what appeared to be a human variety of the cow disease, providing evidence that the disease is transmissible from cow to human. This disease, referred to as variant Creutzfeldt-Jakob disease (vCJD), appears to be related to a more common genetic type of CJD.

In 1997 Stanley Prusiner, professor of neurology and biochemistry at the University of California School of Medicine, San Francisco, won the Nobel Prize in Physiology or Medicine "for his discovery of prions—a new biological principle of infection." Prusiner began his studies of prions in the early 1970s, motivated by the death of a patient from CJD. He found that the infective agent was not affected by radiation (which typically mutates nucleic acids), and he could not find DNA or RNA in the particles. In 1982 he named the infective agent *prion* for "proteinaceous infectious particle." Researchers have determined the structure of the prion protein, which consists of 208 amino acids. Despite extensive searching by many investigators, no nucleic acid component has been found.

Because nucleic acids are the molecules replicated during cell division and reproduction, how prions reproduce has been of great biological interest. Prusiner and other researchers have shown that mammals have a gene that codes the prion protein. Although its function is not known, this protein is normally harmless. The prion protein sometimes converts to a different shape, an insoluble variant that accumulates in the brains of patients with TSE. Genetically engineered mice that lack the prion protein gene are immune to TSE infection.

According to Prusiner, mutations in the prion protein gene increase the risk of the protein changing shape. The prion protein changes shape by a posttranslational process that involves refolding of the molecule. Twenty different mutations in the prion gene have been identified that have been linked with inherited prion diseases. Prusiner has genetically engineered mice that produce the mutant prion protein found in humans with inherited TSE. These mice develop TSE.

Many biologists remain skeptical about prions. Some think that eventually a viral component will still be discovered. Investigators agree that proving the prion theory will require production of insoluble prion protein in a nonbiological system in order to ensure no viral contamination. Researchers would then need to demonstrate that the prion protein causes TSE.

BACTERIA ARE PROKARYOTES

Thousands of species of bacteria have been described. In contrast to viruses, which consist only of nucleic acid and protein, **bacteria** are cellular organisms. Bacteria are prokaryotes, and their cell structure is fundamentally different from the cells of other living organisms. For this reason some biologists have assigned them to their own kingdoms: Archaebacteria and Eubacteria. Other biologists, including most microbiologists, prefer to assign the bacteria to two domains: Archaea and Eubacteria.

Most bacterial cells are very small. Typically, their diameter ranges from 0.5 to 1.0 micrometer. Their cell volume is only about one-thousandth that of small eukaryotic cells, and their length is only about one-tenth. Most prokaryotes are unicellular, but some form colonies or filaments containing specialized cells.

Epulopiscium fishelsoni is an exception to the generalization that all bacteria are microscopic. This bacterium lives as a symbiont in the intestine of surgeonfish (Fig. 23–7). About 600 micrometers long and 80 micrometers wide, its volume is about a million times larger than that of a typical bacterium.

Bacteria have several common shapes

Although many species have irregular shapes, three common shapes are spherical, rod-shaped, and spiral (Fig. 23-8). Spherical bacteria, known as **cocci** (sing., *coccus*), occur singly in some species and in groups of independent cells in others. Cells may be grouped in twos (diplococci), in long chains (streptococci), or in irregular clumps that look like bunches of grapes (staphylococci). Rod-shaped bacteria, called **bacilli** (sing., *bacillus*), may occur as single rods or as long chains of rods. Some bacteria are helical. A bacterium shaped like a very short helix is called a **vibrio**. One that is a longer helix is known as a **spirillum** (plur., *spirilla*) if rigid, and as a **spirochete** if flexible. Shape is an important criterion in identifying bacterial species.

Prokaryotic cells lack membrane-bounded organelles

Recall that prokaryotic cells do not have membrane-bounded organelles typical of eukaryotic cells. Thus, bacterial cells do not have nuclei, mitochondria, chloroplasts, endoplasmic retic-

Figure 23–7 **The giant bacterium (*Epulopiscium*).** This bacterium, about a million times larger than a typical bacterium, is shown here with four paramecia (which are eukaryotes). *(Esther R. Angert and Norman R. Pace, Indiana University)*

1 μm

(a)   1.0 μm   (b)   3.0 μm   (c)   2.0 μm

Figure 23–8 **Common shapes of bacteria.** *(a)* SEM of *Micrococcus,* cocci bacteria. *(b)* SEM of *Salmonella,* bacilli bacteria. *(c)* SEM of *Spiroplasma,* spirilla bacteria. *(Visuals Unlimited/David M. Phillips)*

ula, Golgi complexes, or lysosomes (Fig. 23-9). The dense cytoplasm of the bacterial cell contains ribosomes and storage granules that hold glycogen, lipid, or phosphate compounds. Enzymes needed for metabolic activities may be located in the cytoplasm. Although the membranous organelles of eukaryotic cells are absent, in some bacterial cells the plasma membrane is extensively folded inward. Enzymes needed for cellular respiration and photosynthesis may be associated with the plasma membrane or its folds.

A cell wall typically covers the cell surface

Most prokaryotic cells have a cell wall surrounding the plasma membrane. The cell wall provides a rigid framework that supports the cell, maintains its shape, and keeps it from bursting under hypotonic conditions (see Chapter 5). Most bacteria seem to be adapted to hypotonic surroundings. When wall-less forms of bacteria are produced experimentally, they must be maintained in isotonic solutions to keep them from bursting. However, cell walls are of little help when a bacterium is in a hypertonic environment, as found in food preserved by means of a high sugar or salt content. That is why most bacteria grow poorly in jellies, jams, salted fish, and other foods preserved in these ways.

Bacteria were once classified with plants because they have cell walls. However, eukaryotic cell walls are homoplastic, not homologous, to bacterial cell walls; their structure and composition are different. Unlike plant cell walls, the bacterial cell wall is not made of cellulose. The eubacterial cell wall is composed of **peptidoglycan,** a complex polymer that consists of two unusual types of sugars (amino sugars) linked with short polypeptides. The sugars and peptides are linked to form a single macromolecule that surrounds the entire plasma membrane. The cell wall structure varies among species.

Differences in bacterial cell wall composition are of great interest to microbiologists and are important clinically. In 1888 the Danish physician Christian Gram developed the Gram staining procedure. Bacteria that absorb and retain crystal violet stain in the laboratory are referred to as **gram-positive,**

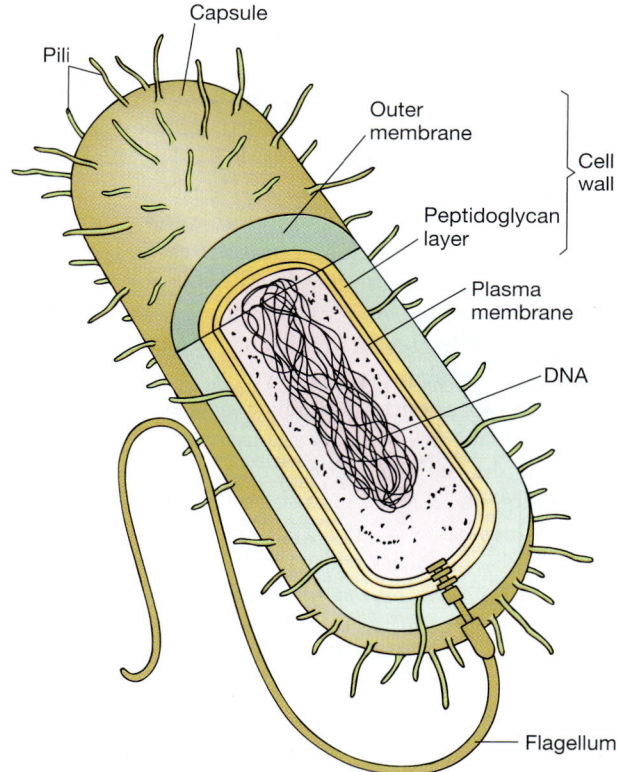

Figure 23–9 **Bacteria are prokaryotic cells.** This bacillus is a gram-negative bacterium (discussed in text). Note the absence of a nuclear envelope surrounding the bacterial DNA.

Figure 23-10 **Bacterial cell walls.** *(a)* In the gram-positive cell wall, many layers of peptidoglycan are held together by amino acids. *(b)* In the gram-negative cell wall, a thin peptidoglycan layer is covered by a thick outer membrane.

(a) Gram-positive cell wall

Thick peptidoglycan layer

Plasma membrane (inner membrane)

Transport protein

Polysaccharides

Lipoprotein

Outer membrane

Thin peptidoglycan layer

Plasma membrane

Transport protein

(b) Gram-negative cell wall

whereas those that do not retain the stain when rinsed with alcohol are **gram-negative.** The cell walls of gram-positive bacteria are very thick and consist primarily of peptidoglycan. The cell walls of gram-negative bacteria consist of two layers, a thin peptidoglycan wall and a thick outer membrane. The outer membrane resembles the plasma membrane but contains polysaccharides bonded to lipids (Fig. 23-10).

Distinguishing between gram-positive and gram-negative bacteria is important in treating certain diseases. For example, the antibiotic penicillin interferes with peptidoglycan synthesis, ultimately resulting in a fragile cell wall that cannot protect the cell (see Chapter 6). Predictably, penicillin works most effectively against gram-positive bacteria.

Some species of bacteria produce a **capsule** or slime layer that surrounds the cell wall. In free-living species, the capsule may provide the cell with added protection against phagocytosis (engulfment; see Chapter 5) by other microorganisms. In disease-causing bacteria, the capsule may protect against phagocytosis by the host's white blood cells. For example, the ability of *Streptococcus pneumoniae* to cause bacterial pneumonia depends on its capsule. A strain of *S. pneumoniae* that lacks a capsule does not cause the disease. Bacteria also use their capsules to attach to surfaces such as rocks, plant roots, or human teeth (where they cause dental plaque).

Some bacteria have hundreds of hairlike appendages known as **pili** (sing., *pilus*). These protein structures are organelles of attachment that help bacteria adhere to one another or to certain surfaces, such as the cells they infect. Some pili are involved in the transmission of DNA between bacteria.

Many types of bacteria are motile

Can you imagine trying to swim through molasses? Water has the same relative viscosity to bacteria that molasses has to humans. Most motile bacteria move by means of rotating **flagella.** The number and location of flagella are important in classification of some bacterial species.

Prokaryotic flagella are quite different than eukaryotic flagella (see Chapter 4). The bacterial flagellum is not composed of microtubules. It consists of three parts: a basal body, a hook, and a single filament (Fig. 23–11). The basal body is a complex structure that anchors the flagellum into the cell wall by disc-shaped plates. The curved hook connects the basal body to the long, hollow filament that extends into the outside environment. The basal body is a motor. The bacterium uses energy from ATP to pump protons out of the cell. Diffusion of these protons back into the cell powers the motor that spins the flagellum like a propeller. Thus, the flagellum produces a rotary motion that pushes the cell much as a propeller pushes a ship through the water.

Bacteria have a single DNA molecule

The genetic material of a bacterium lies in the cytoplasm and is not surrounded by a nuclear envelope. It is contained in a single, circular DNA molecule. If stretched out to its full length, this molecule would be about 1000 times longer than the cell itself. Unlike eukaryotic chromosomes, the bacterial DNA has little protein associated with it.

(a)

5 μm

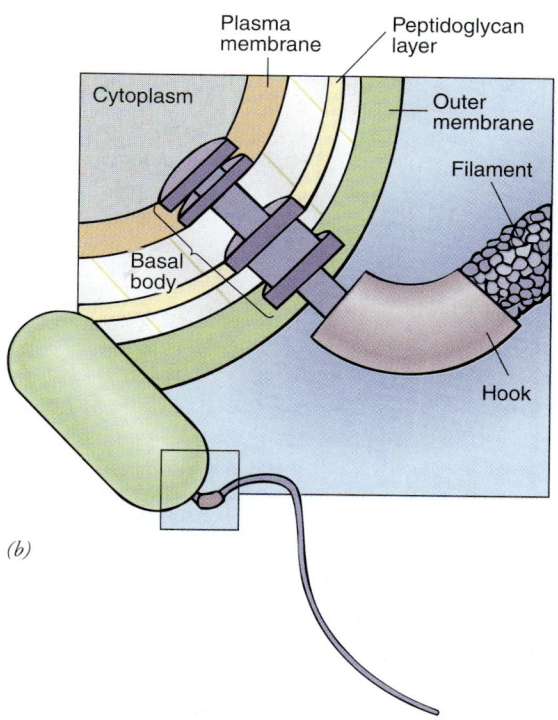

(b)

Figure 23–11 **Bacterial flagella.** *(a)* Phase contrast LM of *Lophotrichous spirillum,* a bacterium with a tuft of flagella at one end of the cell. *(b)* Structure of a bacterial flagellum. The bacterium uses energy from ATP to pump protons out of the cell. Diffusion of these protons into the cell powers the protein motor that spins the flagellum like a propeller. The motor consists of a series of protein rings that (1) anchor the flagellum to the cell wall and plasma membrane and (2) spin the hook and filament of the flagellum. *(a, E.C.S. Chan/ Visuals Unlimited)*

In addition to the genomic DNA, a small amount of genetic information may be present as **plasmids,** smaller circular fragments of DNA. Plasmids can replicate independently of the genomic DNA (see Chapter 14). Bacterial plasmids often have genes that code for catabolic enzymes, for genetic exchange, or for resistance to antibiotics.

Bacteria reproduce by binary fission

Bacteria reproduce asexually, generally by **binary fission,** a process in which one cell divides into two similar cells (see chapter opener). First the circular bacterial DNA replicates, and then a transverse wall is formed by an ingrowth of both the plasma membrane and the cell wall.

Binary fission occurs with remarkable speed; under ideal conditions some bacterial species divide in less than 20 minutes. At this rate, if nothing interfered, one bacterium would give rise to more than 1 billion bacteria within 10 hours! However, bacteria cannot reproduce at this rate for very long because their growth is eventually affected by lack of food or by the accumulation of waste products.

Bacteria can also reproduce asexually by budding or fragmentation. In **budding** a cell develops a bulge, or bud, that enlarges, matures, and eventually separates from the mother cell. In **fragmentation** walls develop within the cell, which then separates into several new cells.

Although sexual reproduction involving the fusion of gametes does not occur in bacteria, genetic material is sometimes exchanged between individuals. This exchange takes place by three different mechanisms: transformation, transduction, and conjugation.

1. In **transformation,** fragments of DNA released by a cell are taken in by another bacterial cell. Recall that this mechanism was used experimentally to demonstrate that genes can be transferred from one bacterium to another and that DNA is the chemical basis of heredity (see Chapter 11).
2. In the second process of gene transfer, **transduction,** a phage carries bacterial genes from one bacterial cell into another (Fig. 23–12).
3. In **conjugation,** two cells of different mating types come together, and genetic material is transferred from one to the other (Fig. 23–13). In contrast to transformation and transduction, conjugation involves contact between two cells.

Conjugation has been most extensively studied in the bacterium *Escherichia coli.* In the *E. coli* population there are donor cells (sometimes referred to as male cells) that have plasmids that can be transmitted to recipient (female) cells. A pilus on the donor cell recognizes the recipient cell and makes the first contact. A cytoplasmic bridge forms between the two cells, and DNA is transferred from donor to recipient cell.

Some bacteria form endospores

When the environment becomes unfavorable, such as when it gets very dry, the cells of some bacteria become dormant. Cells lose water, shrink slightly, and remain inactive until water is again available. Some other bacteria form dormant, extremely durable cells called **endospores.** After the endospore forms, the cell wall of the original cell lyses, releasing the endospore. Endospores can survive in very dry, hot, or frozen environments, or when food is scarce (Fig. 23–14). Some endospores

Step 1.
The DNA of a temperate phage penetrates the bacterial cell.

Step 2.
The phage DNA may become integrated with host cell DNA as a prophage.

Phage DNA with bacterial genes

Fragmented bacterial DNA

Step 3.
When the prophage becomes lytic, bacterial DNA is degraded and new phages are produced. The new phages may contain some bacterial DNA.

Step 4.
The bacterial cell lyses and releases many phages, which can then infect other cells.

Step 5.
A phage infects a new host cell.

Step 6.
Bacterial genes introduced into the new host cell are integrated into the host's DNA. They become a part of the bacterial DNA and are replicated along with it.

Figure 23–12 **Transduction.** In this process, a phage transfers bacterial DNA from one bacterium to another.

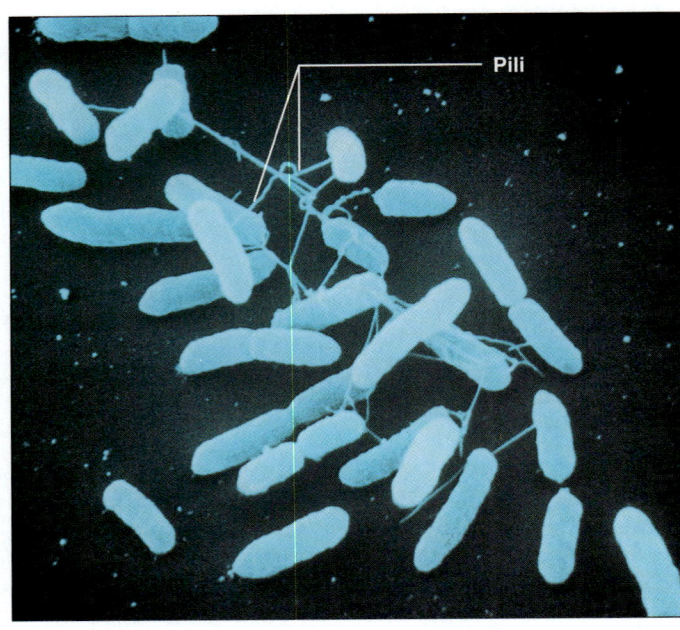

Pili

5 μm

Figure 23–13 **Conjugation.** SEM of *Serratia marcescens* bacteria, which are connected by pili. Plasmid DNA is transferred through cytoplasmic bridges (not shown). *(Manfred Kage/Peter Arnold, Inc.)*

2 μm

Figure 23–14 **Endospores.** Colorized TEM of *Clostridium tetani*, the bacterium that causes tetanus. Each bacterial cell contains one endospore, a resistant, dehydrated cell that develops within the original cell. *(Alfred Pasieka/Peter Arnold, Inc.)*

are so resistant that they can survive an hour or more of boiling, or centuries of freezing. When environmental conditions are again suitable for growth, the endospore germinates, forming an active, growing bacterial cell.

Endospores are not comparable to the reproductive spores of fungi and plants, and endospore formation is not really a kind of reproduction in bacteria. Only one endospore is formed per original cell, so the total number of individuals does not increase.

Metabolic diversity is evident among bacteria

A bacterial cell contains about 5000 different kinds of molecules. The functions of these molecules, how they interact, and how the bacterium synthesizes them from the nutrients it takes in are complex biochemical problems that have interested researchers for years. Much of the knowledge gained from studying these mechanisms in bacterial cells has been successfully applied to cells of humans and other organisms. The basic biochemical processes of all organisms are surprisingly similar.

Bacteria are either heterotrophs or autotrophs. Most bacteria are **heterotrophs** that must obtain organic compounds from other organisms. The majority of these heterotrophic bacteria are free-living **saprobes** (also called **saprotrophs**), organisms that get their nourishment from dead organic matter. Other heterotrophic bacteria obtain their nourishment from living organisms, in some cases harming them by causing diseases, and in other cases actually providing a beneficial service for their host. In fact, survival of most animals, including humans, depends on the microorganisms that inhabit them.

Autotrophic bacteria are either photosynthetic or chemosynthetic and are able to manufacture their own organic molecules from simple raw materials. **Photosynthetic autotrophs,** or simply **photoautotrophs,** obtain their energy from light. Chemosynthetic autotrophs, or **chemoautotrophs,** obtain energy by oxidizing inorganic chemicals.

Whether they are heterotrophs or autotrophs, most bacterial cells are **aerobic** (like animal and plant cells), requiring atmospheric oxygen for cellular respiration. Some bacteria are **facultative anaerobes,** meaning that they can use oxygen for cellular respiration if it is available but can also carry on metabolism anaerobically when necessary. Other bacteria are **obligate anaerobes** that can carry on energy-yielding metabolism only anaerobically. Certain obligate anaerobes are actually killed by even low concentrations of oxygen. Some bacteria respire with terminal electron acceptors other than oxygen, for example, sulfate (SO_4^{2-}) nitrate (NO_3^-), or iron (Fe^{2+})

ARCHAEA AND EUBACTERIA ARE FUNDAMENTALLY DISTINCT

Under a microscope, most bacteria appear similar in size and form. However, evidence from molecular biology, particularly differences in ribosomal RNA, has led microbiologists to conclude that ancient prokaryotes split into two lineages early in the history of life. Microbiologists classify the modern descendants of these two ancient lines in two domains: the **Archaea,** which include a group of prokaryotes that produce methane gas from simple carbon sources and two groups that live in extreme environments, and the **Eubacteria,** also called **Bacteria,** which comprise all other prokaryotes (Fig. 23–15). The Eubacteria include the bacteria most familiar to biologists, whereas the Archaea are less familiar. In this system, eukaryotes are classified in domain Eukarya.

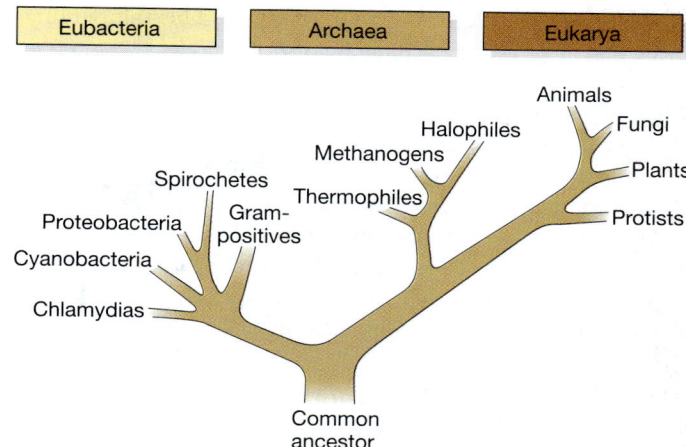

Figure 23–15 **Three domains.** Representative groups of each domain are shown. Gram-positive bacteria include actinomycetes, lactic acid bacteria, mycobacteria, streptococci, staphylococci, and clostridia. Spirochetes, proteobacteria, and chlamydias are gram-negative bacteria.

One of the most distinguishing features of the Archaea is the absence of peptidoglycan in their cell walls (Table 23–2). The lipids in their plasma membranes differ somewhat in structure (ether-linked rather than ester-linked) from those of Eubacteria and Eukarya. In some ways, the Archaea are more like the eukaryotes than the Eubacteria. Archaea do not have the simple RNA polymerase found in Eubacteria. Like eukaryotes, their transcription process is not sensitive to the antibiotic rifamycin. This antibiotic specifically inhibits the type

TABLE 23-2	Characteristics of the Three Domains		
Characteristic	Eubacteria	Archaea	Eukarya
Peptidoglycan in cell wall	Present	Absent	Absent
Nuclear envelope	Absent	Absent	Present
Membrane-bounded organelles	Absent	Absent	Present
Simple RNA polymerase	Present	Absent	Absent
70S ribosomes*	Present	Present	Absent
Flagellum (if present) has single filament	Yes	Yes	No
Transcription is sensitive to the antibiotic rifamycin	Yes	No	No

*The number 70S refers to the sedimentation coefficient (a measure of relative size) when centrifuged.

of RNA polymerase found in Eubacteria. The translational mechanisms of archaebacteria are also more like those of eukaryotes.

More than 1220 genera and 4000 species of prokaryotes have been classified. The editors of *Bergey's Manual of Determinative Bacteriology*[1], considered the definitive reference text by microbiologists, have divided prokaryotes into several groups based on their ribosomal RNA. Each group has distinctive nucleotide sequences in their ribosomal RNA. These sequences are known as **signature sequences.** Several representative groups of prokaryotes are described in Table 23–3.

The Archaea include methanogens, extreme halophiles, and extreme thermophiles

Many extant members of the Archaea survive in environments that are similar to conditions on early Earth. The **methanogens** are a large, diverse group that inhabit oxygen-free environments in sewage and swamps and are common in the digestive tracts of humans and other animals. They are strict anaerobes that produce methane gas from simple carbon compounds. The methanogens are important in recycling components of organic products of organisms that inhabit swamps.

Extreme halophiles are heterotrophs that live in saturated brine solutions such as salt ponds (Fig. 23–16). Some extreme halophiles capture the energy of sunlight with a purple pigment (bacteriorhodopsin) that is very similar to the pigment rhodopsin involved in animal vision. The process is different from photosynthesis carried out by plants or cyanobacteria (see Table 23–3).

Extreme thermophiles normally grow in hot (45°–110°C), sometimes acidic, environments. One species is found in the hot sulfur springs of Yellowstone Park at temperatures near 60°C and pH values of 1 to 2 (the pH of concentrated sulfuric acid). Others inhabit volcanic areas under the sea. One species, found in hot deep-sea vents on the sea floor, lives at temperatures from 80° to 110°C.

Eubacteria are the most familiar prokaryotes

The Eubacteria are widely distributed in the environment and are better known to microbiologists than the Archaea. Several groups of eubacteria are described in Table 23–3.

Bacteria are of great ecological importance

Bacteria are vital members of the biosphere. These microorganisms, especially actinomycetes and the myxobacteria, are the most numerous inhabitants of soil. As described in the

---

[1] *Bergey's Manual of Determinative Bacteriology, 9th ed.,* Williams & Wilkins, Baltimore, Philadelphia, 1994.

Figure 23–16 **Extreme halophiles.** These members of the Archaea thrive in salty environments. Seawater evaporating ponds near San Francisco Bay are colored pink, orange, and yellow from the large number of extreme halophiles that inhabit them. The salt that remains after the water has evaporated has commercial value. *(Helen E. Carr, Biological Photo Service)*

chapter introduction, bacteria are essential as decomposers and in recycling nutrients, including nitrogen, oxygen, carbon, phosphorus, sulfur, and many trace elements. (A discussion of the roles of bacteria in biogeochemical cycles is found in Chapter 53.)

Because nitrogen is constantly removed from the soil by plants and other natural processes, as well as by human activity, it must be continually added to the soil. Plant growth depends on the availability of usable nitrogen. Several types of bacteria transform atmospheric nitrogen to forms that can be used by plants (Fig. 23–17; also see *Making the Connection: Bacteria, Nitrogen Fixation, and Agriculture on page 500*).

Bacterial photosynthesis is ecologically very important, accounting for up to half of primary production in aquatic habitats. Like algae and plant cells, cyanobacteria carry out photosynthesis, using water as the electron source and generating oxygen. Other photosynthetic bacteria use compounds other than water, for example, hydrogen sulfide, as the electron source and do not produce oxygen.

Heterocysts

50 μm

Figure 23–17 **LM of *Anabaena*, a cyanobacterium that fixes nitrogen.** Nitrogen fixation takes place in the rounded cells, called heterocysts. *(Dennis Drenner)*

Some bacteria cause disease

A small percentage of bacterial species are important pathogens of plants and animals (Fig. 23–18). Other bacterial species have evolved along with eukaryotes and are interdependent with them. All plants and animals harbor a population of microorganisms that are considered normal microbiota—harmless symbiotic bacteria. In fact, the number of bacteria that normally inhabit the human body (approximately 700 trillion cells) exceeds the number of its own human cells (about 70 trillion cells). The presence of some of these bacterial populations prevents harmful microorganisms (including other bacteria) from flourishing. Some of the normal bacterial (and viral) inhabitants are opportunistic pathogens that can cause disease only under certain conditions. For example, if one's immune system is compromised, opportunistic bacteria may increase in number and cause disease.

Pathogenic microorganisms can enter the body in food, dust, droplets, or through wounds. Many diseases are transmitted by insect or animal bites. In order to cause disease, a pathogen must adhere to a specific cell type, multiply, and produce toxic substances. Adherence and multiplication can occur only when the pathogen competes successfully with the normal microbiota and counteracts the host's defenses against invasion.

Pathogens produce a variety of substances that increase their success. Some bacteria produce **exotoxins,** strong poisons that either are secreted from the cell or leak out when the bacterial cell is destroyed. The toxin, not the presence of the bacteria themselves, is responsible for the disease. As mentioned previously, diphtheria is caused by a gram-positive bacillus that is lysogenized by a phage. The diphtheria toxin kills cells and causes inflammation.

White blood cells

Bacteria

10 μm

Figure 23–18 **Pathogenic bacteria.** SEM of pus containing *Neisseria gonorrhea* (small pink spheres), the bacteria that cause gonorrhea, a common sexually transmitted disease. The larger cells are white blood cells. *(M.I. Walker/Science Source/Photo Researchers, Inc.)*

TABLE 23-3   Some Major Groups of Bacteria

Eubacteria with Gram-Negative Cell Walls

Proteobacteria   A large, very diverse group.

10.0 μm

SEM of *Escherichia coli* colony.   (David Scharf/Peter Arnold, Inc.)

Enterobacteria include decomposers that live on decaying plant matter, pathogens, and a variety of bacteria that inhabit humans. *Escherichia coli* inhabits the intestinal tracts of humans and other animals as part of the normal microbial population; however, certain strains can cause moderate to severe diarrhea. For example, in 1993 almost 500 people developed bloody diarrhea and three people died in the Pacific Northwest after ingesting hamburger meat contaminated with a new strain of *E. coli*. One species of *Salmonella* infects food and produces a toxin that causes a form of food poisoning; another species causes typhoid fever.

Vibrios are mainly marine; some are bioluminescent. One species is the causative agent of cholera.

Rhizobium species live symbiotically in root nodules of leguminous plants and convert atmospheric nitrogen to a form usable by plants (nitrogen fixation).

Pseudomonads are heterotrophs that produce nonphotosynthetic pigments. They cause disease in plants and animals, including humans.

Azotobacteria inhabit the soil and fix nitrogen under aerobic conditions. They form a resting cell termed a cyst that is resistant to drying.

Rickettsias are very small, rod-shaped bacteria. A few species are pathogenic to humans and other animals and are transmitted by arthropods through bites or contact with their excretions. Among these are typhus (transmitted by fleas and lice) and Rocky Mountain spotted fever (transmitted by ticks).

Myxobacteria (slime bacteria) secrete slime and glide or creep along. When nutrients are exhausted, these bacteria form masses that develop into stalked, multicellular reproductive structures called fruiting bodies. During this process, bacterial cells within the fruiting body enter a resting stage equivalent to spores. When conditions are favorable, the spores break open and the resting cells become active.

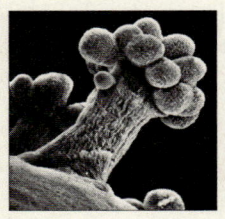

10 μm

Other medically important Proteobacteria include *Neisseria gonorrhoeae,* which causes gonorrhea, and *Legionella pneumophila,* which causes Legionnaires' disease.

SEM of fruiting body of the myxobacterium *Stigmatella aurantiaca*. Protective resting cells form within the fruiting bodies that are very resistant to heat and drying.   (From Grilicone, P. L., and Pangborn, J., Journal of Bacteriology 124:1558, 1975)

Other Gram-Negative Eubacterial Groups

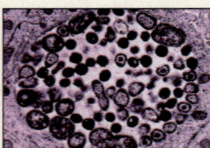

5 μm

TEM of *Chlamydia trachomatis* in oviduct cell.   (David M. Phillips/Visuals Unlimited)

Chlamydias lack peptidoglycan in their cell walls and do not depend on arthropod vectors for transmission. Although they do contain many enzymes and can carry on some metabolic processes, chlamydias are energy parasites, that is, completely dependent on their host for ATP. Chlamydias infect almost every species of bird and mammal. Perhaps 10% to 20% of the human population of the world is infected. Trachoma, the leading cause of blindness in the world, is caused by a strain of *Chlamydia;* sexually transmitted chlamydias are a major cause of pelvic inflammatory disease (PID) in women.

Spirochetes are spiral-shaped bacteria with flexible cell walls. They move by means of unique internal flagella called axial filaments. Some species are free-living and inhabit freshwater and marine habitats, whereas other species form symbiotic associations; a few are parasitic. The spirochete of greatest medical importance is *Treponema pallidum,* which causes syphilis. Lyme disease, a tick-borne disease of humans and some other animals, is caused by a spirochete belonging to the genus *Borrelia*.

Cyanobacteria, formerly known as blue-green algae, are gram-negative, photosynthetic bacteria that inhabit ponds, lakes, swimming pools, moist soil, dead logs, and the bark of trees. They contain chlorophyll *a* and use a photosynthetic process similar to that of plants and algae. The chloroplasts in algae and plants probably evolved from that of ancient cyanobacteria. Many species also fix nitrogen.

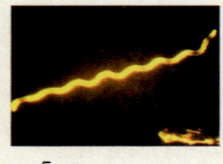

5 μm

LM of *Treponema pallidum,* the spirochete that causes syphilis.   (Charles W. Stratton/Visuals Unlimited/Science VU)

TABLE 23–3   Continued

Eubacteria with Gram-Positive Cell Walls

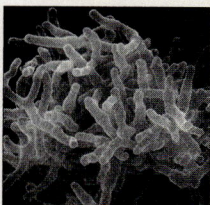

5 μm

SEM of *Actinomycetes naes-lundi,* a soil-dwelling bacterium, forms filamentous colonies. *(David M. Phillips/Visuals Unlimited)*

Actinomycetes superficially resemble fungi in that their cells remain together, forming branching filaments, and many produce moldlike spores. However, they have peptidoglycan in their cell walls, lack nuclear envelopes, and have other prokaryotic characteristics. Actinomycetes decompose organic materials in soil. Most members of this group are saprobes, and some are anaerobic. Several species of the genus *Streptomyces* produce antibiotics such as streptomycin, erythromycin, chloramphenicol, and the tetracyclines. In fact, most known antibiotics are derived from actinomycetes. Some actinomycetes cause serious lung disease or generalized infections in humans and other animals.

Lactic acid bacteria ferment sugar, producing lactic acid as the main end-product. They inhabit decomposing plant material, milk, and other dairy products. The characteristic taste of yogurt, acidophilus milk, pickles, sauerkraut, and green olives is due to the action of lactic acid bacteria. They are also among the normal inhabitants of the human mouth and vagina.

Mycobacteria are slender, irregular rods. They contain a waxy substance in their cell walls. One species causes tuberculosis; another causes leprosy.

Streptococci inhabit the mouth and digestive tract of humans and some other animals. Among the harmful species are those that cause "strep throat," dental caries, a form of pneumonia, scarlet fever, and rheumatic fever.

Staphylococci normally live in the nose and on the skin. They are opportunistic pathogens that cause disease when the immunity of the host is lowered. *Staphylococcus aureus* causes boils and skin infections (some extremely serious) and may infect wounds. Certain strains of *S. aureus* cause a form of food poisoning, and other strains cause toxic shock syndrome.

Clostridia are anaerobic. One species causes tetanus, another causes gas gangrene. *Clostridium botulinum* can cause botulism, an often fatal type of food poisoning. Botulism results from consuming foods, such as canned vegetables and smoked meats and fish, that have been inadequately sterilized; its endospores, which are resistant to heat, grow and produce the most potent toxin known. Approximately one microgram of toxin is enough to kill a human.

Eubacteria That Lack Cell Walls

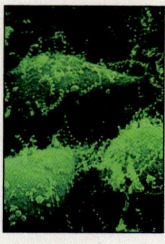

5 μm

Mycoplasmas lack cell walls. These bacteria inhabit soil and sewage; some are parasitic on plants or animals. Some inhabit human mucous membranes but do not generally cause disease; one species causes a mild type of bacterial pneumonia in humans.

SEM of *Mycoplasma* sp. on fibroblast cell. *(David M. Phillips/Visuals Unlimited)*

The Archaea: Bacteria with Cell Walls That Have No Peptidoglycan

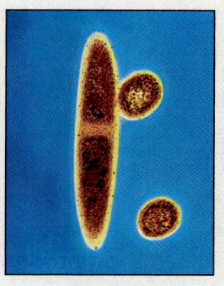

1 μm

Methanogens are anaerobes that produce methane gas from simple carbon compounds.

Extreme halophiles inhabit saturated salt solutions.

Extreme thermophiles grow at 70°C or higher; some thrive above 100°C.

TEM of a methanogen *(Methanospirillum hungatii)* undergoing cell division. Two other bacteria are visible in cross-section. Normally these bacteria are spiral-shaped. *(Dr. Kari Lounatmaa/Science Photo Library/Photo Researchers, Inc.)*

BACTERIA, NITROGEN FIXATION, AND AGRICULTURE

How does the mutualistic relationship between certain bacteria and plants benefit agriculture? Rhizobial bacteria (bacteria in the genus *Rhizobium*) form symbiotic associations with the roots of legumes, a large family of herbs, shrubs, and trees. Legumes include such important crops as peas, beans, lentils, soybeans, and peanuts, as well as clover and alfalfa, which are grown for livestock feed and to fertilize the soil.

Rhizobial bacteria are motile bacilli that inhabit the soil. After they infect the roots of a legume, the roots develop nodules (see Fig. 53–11*a*). The nodules consist of plant tissue in which the bacteria reside and fix nitrogen. Rhizobial bacteria and the roots of legumes form a mutualistic type of symbiotic relationship. In a mutualistic relationship both partners benefit (see Chapter 52). The

bacteria living in nodules supply the plant with all the nitrogen it requires, and the plant provides the bacteria with organic compounds, including sugar needed for cellular respiration.

Because legumes, like other plants, produce sugar by photosynthesis, a correlation exists between photosynthesis and nitrogen fixation. When a legume is photosynthesizing at a higher rate, it provides more sugar for its bacterial partners. The bacteria are then able to fix larger amounts of nitrogen. Plants without nodules must obtain nitrogen from the soil (see Chapter 34). Because many soils are deficient in nitrogen, legumes that have formed mutualistic associations with rhizobial bacteria have a decided advantage over other plants.

Botulism, a type of food poisoning that can lead to paralysis and sometimes death, results from ingestion of improperly canned food. Botulism is caused by an exotoxin released by the gram-positive bacterium, *Clostridium botulinum*. This exotoxin is so powerful that one gram could kill a million humans! Like many exotoxins, the one that causes botulism can be inactivated by heating. (Food must be heated to 80°C for 10 minutes, or boiled for 3 to 4 minutes.)

Endotoxins are not secreted by pathogens but are components of the cell walls of most gram-negative bacteria. These compounds affect the host only when the bacteria die and release them. Endotoxins bind to macrophages and stimulate them to release substances that cause fever and other symp-

toms of infection. Unlike exotoxins, which cause specific symptoms, all endotoxins appear to cause systemic symptoms such as fever. Endotoxins are not destroyed by heating.

Bacteria are used in many commercial processes

Many foods and beverages are produced with the help of microbial fermentation. Lactic acid bacteria are used in the production of acidophilus milk, yogurt, pickles, olives, and sauerkraut, and several types of bacteria are used to produce cheese (Fig. 23–19). Bacteria are involved in making fermented meats like salami and in the production of vinegar, soy sauce, chocolate, and certain B vitamins (B_{12} and riboflavin). Bacteria are also used in the production of citric acid, a compound added to candy and to most soft drinks. Some bacteria produce biocides that are used commercially to destroy insect pests.

Some types of microorganisms produce antibiotics, compounds that limit competition for nutrients by inhibiting or destroying other microorganisms. By the 1950s antibiotics had become important clinical tools that transformed the treatment of infectious disease. Today about 100 clinically useful antibiotics are available, and literally tons of antibiotics are produced annually. Pharmaceutical companies obtain most antibiotics from three groups of microorganisms: a large group of soil bacteria known as actinomycetes, gram-positive bacteria of the genus *Bacillus,* and molds (eukaryotes belonging to kingdom Fungi).

Because of their prolific reproduction rates, bacteria are ideal "factories" for the production of biomolecules. They have been genetically engineered (Chapter 14) to produce certain vaccines, human growth hormone, insulin, and many other clinically important compounds. About 50% of the insulin used to treat diabetics is derived from transgenic bacteria. Researchers are in the process of developing genetically engineered

5 µm

Figure 23–19 **LM of lactic acid bacteria (*Lactobacillus acidophilus*).** *(George J. Wilder/Visuals Unlimited)*

bacteria for production of many other medically useful products.

Bacteria are used in sewage treatment and to break down solid wastes in landfills. They are also being used in bioremediation, a process in which a contaminated site is exposed to microorganisms that break down the toxins, leaving behind harmless metabolic byproducts such as carbon dioxide and chlorides. To date, more than 1000 different species of bacteria and fungi have been used to clean up various forms of pollution.

SUMMARY WITH KEY TERMS

I. A **virus,** or **virion,** is a tiny particle consisting of a DNA or RNA genome surrounded by a **capsid** (protein coat).
  A. Viruses are subcellular particles that cannot metabolize on their own and are not considered to be truly living organisms.
  B. Viruses may be bits of nucleic acid that originally "escaped" from other organisms. They infect all types of organisms.
    1. **Phages (bacteriophages)** are viruses that infect bacteria.
    2. Many different viruses infect humans and other animals. Examples of human viral diseases include chickenpox, herpes simplex, infectious mononucleosis, mumps, warts, influenza, hepatitis, AIDS, as well as certain types of cancer. **Retroviruses,** such as HIV, use reverse transcriptase to transcribe their RNA genome into a DNA intermediate.
    3. Plant viruses cause serious agricultural losses. Viral diseases can be spread among plants by insect vectors.
  C. A viral reproductive cycle can be lytic or temperate.
    1. In a **lytic** cycle, the virus destroys the host cell. The five steps in a lytic cycle include: **attachment** to the host cell; **penetration** of viral nucleic acid into the host cell; **replication** of the viral nucleic acid; **assembly** of newly synthesized components into new viruses; and **release** from the host cell.
    2. **Temperate** viruses do not always destroy their hosts. In a **lysogenic** cycle the viral genome is replicated along with the host DNA.
      a. In some phages, the phage nucleic acid becomes integrated into the bacterial DNA; it is then called a **prophage.** Bacterial cells that carry prophages are **lysogenic cells.**
      b. In **lysogenic conversion,** bacterial cells containing certain temperate viruses exhibit new properties.
II. **Viroids** and **prions** are smaller than viruses.
  A. A viroid consists of a short strand of RNA with no protein coat.
  B. The prion appears to consist only of protein. Prions cause transmissible spongiform encephalopathies (TSEs).
III. Prokaryotes are divided into two groups: kingdom **Archaebacteria** and kingdom **Eubacteria.** Many biologists prefer classifying the bacteria into two domains: **Archaea** and **Eubacteria** (or simply, **Bacteria**).
  A. Prokaryotic cells do not have membrane-bounded organelles such as nuclei and mitochondria. Common shapes of bacterial cells include **coccus,** or spherical; **bacillus,** or rod-shaped; and spiral. Spiral bacteria include the **vibrio,** which is a short helix; **spirillum,** a longer, rigid helix; and **spirochete,** a longer, flexible helix.
  B. Most eubacteria have cell walls composed of **peptidoglycan.** The walls of **gram-positive** bacteria are very thick and consist mainly of peptidoglycan. The cell walls of **gram-negative** bacteria consist of a thin peptidoglycan layer and a thick outer membrane resembling the plasma membrane. Some species produce a **capsule** that surrounds the cell wall.
  C. Some bacteria have **pili,** protein organelles that extend out from the cell and help bacteria adhere to one another or to certain other surfaces.
  D. Bacterial **flagella** are structurally different from eukaryotic flagella; each flagellum consists of a basal body, hook, and filament. Unlike eukaryotic flagella, they produce a rotary motion.
  E. The genetic material of a prokaryote is a single circular DNA molecule. **Plasmids** may also be present.
  F. Bacteria reproduce asexually by **binary fission, budding,** or **fragmentation.**
  G. Genetic material may be exchanged by **transformation, transduction,** or **conjugation.**
  H. Some bacteria form dormant, durable cells called **endospores.**
  I. Bacteria are metabolically diverse; some are **heterotrophs,** whereas others are **autotrophs.**
    1. The majority of heterotrophic bacteria are free-living **saprobes,** organisms that obtain nourishment from dead organic matter.
    2. Autotrophs may be **photoautotrophs** or **chemoautotrophs.**
    3. Most bacteria are aerobic, but some are **facultative anaerobes;** others are **obligate anaerobes.**
IV. The cell walls of Archaea do not have peptidoglycan, and their translational mechanisms more closely resemble eukaryotic than other prokaryotic mechanisms. The three main groups of archaebacteria are the **methanogens,** the **extreme halophiles,** and the **extreme thermophiles.**
  A. The methanogens produce methane gas from simple carbon compounds. They inhabit anaerobic environments such as marshes, marine sediments, and the digestive tracts of animals.
  B. The extreme halophiles inhabit saturated salt solutions.
  C. The extreme thermophiles can inhabit environments at 70°C or higher. Some can even thrive at temperatures above 100°C.
V. See Table 23–3 to review some of the major groups of Eubacteria.
VI. Bacteria play essential ecological roles as decomposers and are important in recycling nutrients.
VII. Many bacteria are symbiotic with other organisms; some are important **pathogens** of plants and animals. Some pathogenic bacteria produce **exotoxins;** others produce **endotoxins.**

POST-TEST

1. Pathogens (a) are agents that cause disease (b) include most bacteria (c) include most cyanobacteria (d) include many autotrophs (e) three of the preceding answers are correct
2. The genome of a virus consists of (a) DNA (b) RNA (c) prions (d) DNA and RNA (e) DNA or RNA.
3. The capsid of a virus (a) consists of protein (b) consists of nucleic acid (c) is almost always helical (d) is a simple carbohydrate (e) answers b and c only are correct
4. Viruses that kill the host cell are (a) lysogenic (b) lytic (c) viroids (d) prophages (e) temperate.
5. The correct sequence in viral reproduction is (a) attachment, penetration, assembly, replication, release (b) penetration, absorption, assembly, replication, release (c) attachment, penetration, replication, assembly, release (d) release, assembly, penetration, replication, attachment (e) attachment, penetration, replication, release, assembly.
6. In lysogenic conversion (a) bacterial cells may exhibit new properties

(b) the host cell dies (c) prions sometimes convert to viroids (d) reverse transcriptase transcribes DNA into RNA (e) prion proteins convert to infective agents

7. Peptidoglycan is a chemical compound found in the cell walls of (a) most Eubacteria (b) most Archaea (c) prions (d) most viroids (e) Eukarya

8. Retroviruses (a) are DNA viruses (b) are RNA viruses (c) use reverse transcriptase to transcribe DNA into RNA (d) use prion protein to transcribe RNA (e) answers b and c are correct

9. Bacterial flagella (a) are homologous with eukaryotic flagella (b) are used in classifying some bacterial species (c) consist of a basal body and nine pairs of microtubules (d) are important in transduction (e) are gram-positive

10. In conjugation (a) two bacterial cells of different mating types come together, and genetic material is transferred from one to another (b) a bacterial cell develops a bulge that enlarges and eventually separates from the mother cell (c) fragments of DNA released by a broken cell are taken in by another bacterial cell (d) a phage carries bacterial genes from one bacterial cell into another (e) walls develop in the cell, which then divides into six new cells

11. Endospores (a) are formed by some viruses (b) are extremely durable cells (c) are comparable to the reproductive spores of fungi and plants (d) cause fever and other symptoms in the host (e) answers b and c are correct

12. The majority of heterotrophic bacteria are (a) free-living saprobes (b) photoautotrophs (c) chemoautotrophs (d) facultative anaerobes (e) obligate anaerobes

13. The Archaea include (a) bacteria that produce methane from carbon dioxide and hydrogen (b) thermophiles (c) halophiles (d) all of the gram-positive bacteria (e) answers a, b, and c are correct

REVIEW QUESTIONS

1. What characteristics does a virus share with a living cell? What characteristics of life are absent in a virus?
2. List the steps in a lytic cycle. Draw diagrams to illustrate your answer.
3. Contrast the sequence of events in a lysogenic cycle with that of a lytic cycle.
4. What are the distinguishing characteristics of viroids? Of prions?
5. What are the differences between the Archaea and the Eubacteria?
6. Contrast the cell wall of a gram-positive bacterium with that of a gram-negative bacterium.
7. Using Table 23–3 as a guide, give the distinguishing characteristics of each of the groups of bacteria described, and give examples of each group.
8. Label the diagram. Use Figure 23–9 to check your answers.

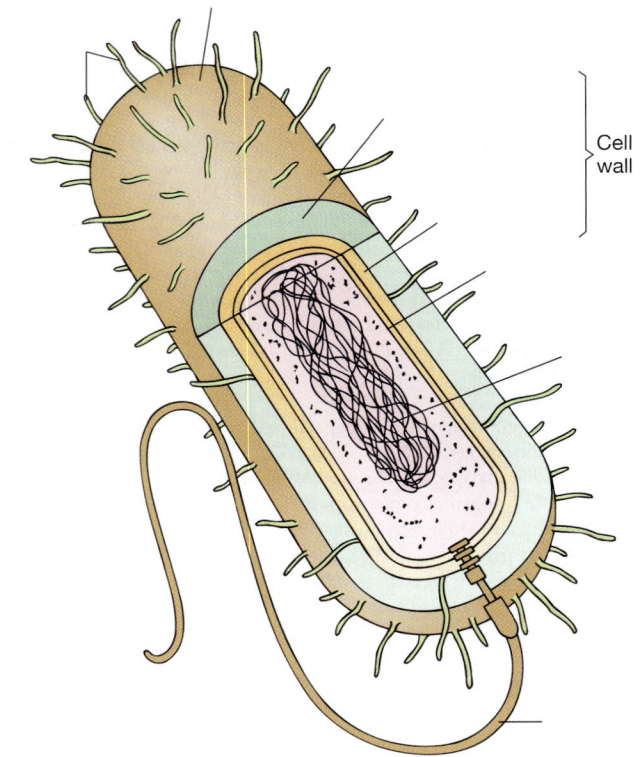

Cell wall

YOU MAKE THE CONNECTION

1. Historically, biologists thought that viruses, because of their simple structure, evolved before cellular organisms. Based on what you have learned about viruses, present an argument against this hypothesis.
2. Imagine that you discover a new microorganism. After careful study you determine that it should be classified in the domain Archaea. What characteristics might lead you to this classification?
3. How might life on planet Earth be different if bacteria had never evolved?

RECOMMENDED READINGS

Angert, E.R., K.D. Clements, and N.R. Pace. "The Largest Bacterium." *Nature,* 18 Mar. 1993. Discovery of a novel bacterium so large that it can be seen *without* a microscope.

Bessen, R.A. "Neurodegenerative Prion Diseases." *Science & Medicine,* Vol. 3, No. 5, Sept./Oct. 1996. A discussion of prions and their interspecies transmission.

Caldwell, M. "Prokaryotes at the Gate." *Discover,* Aug. 1994. An interesting account of bacterial resistance to antibiotics.

Canby, T.Y. "Bacteria: Teaching Old Bugs New Tricks." *National Geographic,* Aug. 1993. Considers the industrial potential of bacteria, from cleaning up toxic waste to degrading oil spills.

Carmichael, W.W. "The Toxins of Cyanobacteria." *Scientific American,* Vol. 270, No. 1, Jan. 1994. Many cyanobacteria produce unusual toxins that are deadly but potentially valuable.

Finlay, B.B. "Bacterial Virulence Factors." *Scientific American,* Vol. 272, No. 5, May/Jun. 1995. A fascinating discussion of the mechanisms that make pathogenic bacteria successful.

Hively, W. "Life Beyond Boiling." *Discover,* May 1993. Research discoveries about bacteria that thrive at temperatures well above 212°F (100°C). Most of these bacteria live in hot springs at the bottom of the ocean.

Krieg, A.M. "Human Endogenous Retroviruses." *Science & Medicine,* Vol. 4, No. 2, Mar./Apr. 1997. The genomes of humans and other vertebrates include retroviral sequences, some of which can express viral proteins.

Losick, R. and D. Kaiser. "Why and How Bacteria Communicate." *Scientific American,* Vol. 276, No. 2, Feb. 1997. An overview of the chemical signals bacteria send to one another and to their hosts.

Perry, J.J., and J.T. Staley. *Microbiology: Dynamics and Diversity.* Saunders College Publishing, Philadelphia, 1997. An excellent introduction to microbiology.

Prusiner, S.B. "The Prion Diseases." *Scientific American,* Vol. 272, No. 1, Jan. 1995. Infectious agents, smaller and simpler than viruses, may be responsible for a number of disorders.

Prusiner, S.B. "Prion Diseases and the BSE Crisis." *Science,* Vol. 278, 10 Oct. 1997. A discussion of the mechanisms by which prion proteins may become infective agents; includes a discussion of prevention.

Woese, C.R., O. Kandler, and M.L. Wheelis. "Towards a Natural System of Organisms: Proposal for the Domains Archaea, Bacteria, and Eucarya." *Proceedings of the National Academy of Science,* Vol. 87, Jun. 1990. Proposal of a classification system consisting of three domains.

Zimmer, C. "Triumph of the Archaea." *Discover,* Feb. 1995. Discussion of a molecular approach to the study of archaebacteria.

● Visit our website at **http://www.saunderscollege.com/lifesci/titles.html** and click on Solomon/Berg/Martin Biology for links to chapter-related resources on the World Wide Web.

CHAPTER 24

The Protist Kingdom

In Chapter 22 you learned that the kingdom Protista was first proposed in 1866 to include bacteria and other microorganisms that differ from plants and animals. When Robert Whittaker recommended the five-kingdom system of classification in 1969, only unicellular eukaryotic organisms were placed in kingdom Protista. The boundaries of this kingdom have expanded somewhat since then, although there is still no universal agreement among biologists about what constitutes a protist.

Kingdom Protista currently consists of a vast assortment of primarily aquatic eukaryotic organisms whose diverse body forms, types of reproduction, modes of nutrition, and lifestyles make them difficult to characterize. **Protists** are unicellular, colonial, or simple multicellular organisms that possess a eukaryotic cell organization. The word *protist,* from Greek, meaning "the very first," reflects the idea that protists were the first eukaryotes to evolve (discussed later in the chapter).

Eukaryotic cells, the unifying feature of protists, are common to complex multicellular organisms belonging to the three other eukaryotic kingdoms (Fungi, Animalia, and Plantae) but clearly differentiate protists from members of the prokaryotic kingdoms (Eubacteria and Archaebacteria). Recall from Chapter 4 that, unlike prokaryotic cells, eukaryotic cells possess nuclei and other membrane-bounded organelles such as mitochondria and plastids, 9 + 2 flagella, and multiple chromosomes in which DNA is complexed with proteins. Sexual reproduction, meiosis, and mitosis are also characteristic of eukaryotes.

Size varies considerably within the protist kingdom, from microscopic protozoa to giant kelps, which are brown algae that can reach 75 meters (250 feet) in length. Although most protists, like the dinoflagellates (*Ceratium* sp.) shown in the LM, are unicellular, some form **colonies** (loosely connected groups of cells), some are **coenocytic** (consisting of a multinucleate mass of cytoplasm), and some are multicellular. Unlike animals, fungi, and plants, most multicellular protists have relatively simple body forms without specialized tissues.

(from Freshwater Algae: Their Microscopic World Explored. Biopress, 1995. Photo by Hilda Canter-Lund)

Because of their huge numbers, members of kingdom Protista are important to the natural balance of the living world. Protists are an important source of food for other organisms, and photosynthetic protists also supply oxygen to aquatic and terrestrial ecosystems. Certain protists are economically important, while others cause diseases. Consideration of all protist phyla is beyond the scope of this text, but we will discuss sixteen representative phyla that include animal-like heterotrophs (protozoa), plantlike autotrophs (algae, including seaweeds), and fungus-like heterotrophs (slime molds and water molds).

1. Characterize the features common to the members of kingdom Protista.
2. Discuss in general terms the diversity inherent in the protist kingdom, including modes of nutrition, body forms, motility, and reproduction.
3. Briefly describe and compare these representative protozoa: amoebas, foraminiferans, actinopods, zooflagellates, ciliates, and apicomplexans.
4. Briefly describe and compare these representative algae: euglenoids, dinoflagellates, diatoms, golden algae, brown algae, green algae, and red algae.

5. Briefly describe and compare these representative fungus-like protists: plasmodial slime molds, cellular slime molds, and water molds.
6. Explain how multicellularity may have arisen within the protists, using the volvocine line as an example.
7. Discuss the controversies surrounding the evolutionary relationships both among the protists and between the protists and other eukaryotic kingdoms.

PROTISTS HAVE EVOLVED A VARIETY OF CELL STRUCTURES, ECOLOGICAL ROLES, AND LIFE HISTORIES

Size and structural complexity are not the only variable features of protists. During the course of evolutionary history, organisms in the kingdom Protista have evolved diversity in their (a) means of locomotion, (b) ways of obtaining nutrients, (c) interactions with other organisms, (d) habitat, and (e) modes of reproduction.

Protists, most of which are motile at some point in their life cycle, have various means of locomotion. They may move by pushing out cytoplasmic extensions (**pseudopodia**) as an amoeba does, by flexing individual cells, by gliding over surfaces, by waving **cilia** (short hairlike structures), or by lashing **flagella** (long, whiplike structures). Some protists use a combination of two or more means of locomotion, for example, both flagella and amoeboid movement.

Methods of obtaining nutrients differ widely in kingdom Protista. Most of the algae are autotrophic and photosynthesize as plants do. Some heterotrophic protists obtain their nutrients by absorption, as fungi do, whereas others resemble animals in that they ingest food. Some protists switch their modes of nutrition and are autotrophic at certain times and heterotrophic at others.

Although many protists are free-living, others form symbiotic associations with various organisms. These intimate associations range from *mutualism,* a more or less equal partnership where both partners benefit, to *parasitism,* where one partner (the parasite) lives on or in another (the host) and is metabolically dependent on it (see Chapter 52). Some parasitic protists are important pathogens (disease-causing agents) of plants or animals. Specific examples of symbiotic associations involving protists are described throughout this chapter.

Most protists are aquatic and live in the ocean or in freshwater ponds, lakes, and streams. They make up most of the **plankton,** the floating, often microscopic organisms that inhabit surface waters and are the base of the food web in aquatic

ecosystems. Other aquatic protists attach to rocks or other surfaces in the water. Even parasitic protists are aquatic because they live in the watery environments of other organisms' body fluids. Terrestrial protists are restricted to damp places such as soil, cracks in bark, and leaf litter.

Reproduction is varied in the kingdom Protista. Almost all protists reproduce asexually, and many also reproduce sexually, often by **syngamy,** the union of gametes. However, most protists do not develop multicellular reproductive organs, nor do they form embryos the way more complex organisms do.

Based on the diversity in their structures, ecological lifestyles, and reproduction, most biologists regard the protist kingdom as a polyphyletic group of organisms; that is, protists probably do not share a single common ancestor. Some biologists think as many as 50 phyla would be needed to reflect natural relationships within the protists! If a cladist were classifying these organisms based on evidence for a shared evolutionary history (see Chapter 22), many protist phyla would be converted to kingdom status, and there would be many more than six kingdoms. Fortunately, recent molecular analyses, such as comparison of ribosomal RNA sequences, are starting to clarify relationships among the various protist phyla and among protists and the other kingdoms (Fig. 24–1).

PROTOZOA ARE ANIMAL-LIKE PROTISTS

The name **protozoa** (from the Latin, meaning "first animals"; sing., *protozoon*) was originally given to animal-like organisms that are not multicellular. The term *protozoa* is used today as an informal designation for those protists that ingest food (as animals do). Protozoa are a polyphyletic group, and their evolutionary relationships are continually being reevaluated as additional evidence becomes available. In this chapter we consider six groups of protozoa: amoebas, foraminiferans, actinopods, zooflagellates, ciliates, and apicomplexans (see Table 24–1, p. 508).

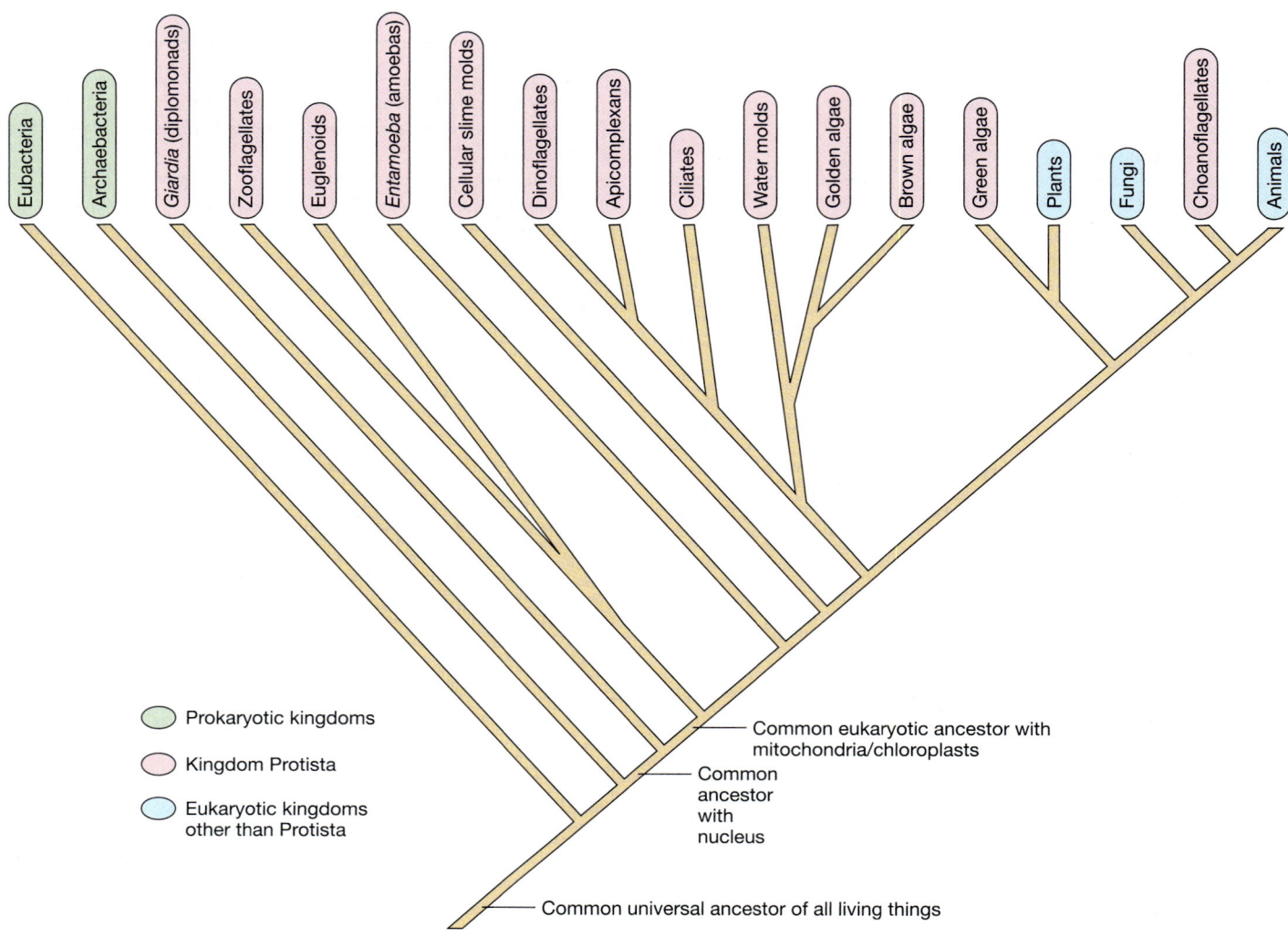

Prokaryotic kingdoms

Kingdom Protista

Eukaryotic kingdoms other than Protista

Common eukaryotic ancestor with mitochondria/chloroplasts

Common ancestor with nucleus

Common universal ancestor of all living things

Figure 24–1 **One interpretation of evolutionary relationships among certain protists.** This evolutionary tree is based on a comparison of ribosomal RNA data for several protist phyla as well as the other five kingdoms. Foraminiferans, actinopods, diatoms, red algae, and plasmodial slime molds are not shown. Biologists are not in complete agreement about the details of the protist family tree, and there are other interpretations of the protist lineage than the one depicted here.

Amoebas move by forming pseudopodia

Amoebas (phylum Rhizopoda) are unicellular organisms found in soil, fresh water, and the ocean. Because of the extreme flexibility of their outer plasma membrane, many members of this group have an indefinite body structure and continually change shape as they move. (The word *amoeba* is derived from a Greek word meaning "change.") An amoeba moves by pushing out temporary cytoplasmic projections called pseudopodia (sing., *pseudopodium,* meaning "false foot") from the surface of the cell. More cytoplasm flows into the pseudopodia, enlarging them until all the cytoplasm has entered and the organism as a whole has moved. Pseudopodia are also used to engulf and capture food by surrounding and forming a vacuole around it (Fig. 24–2). Food particles are di-

gested when the food vacuole fuses with a lysosome containing digestive enzymes. Digested materials are absorbed from the food vacuole, which gradually shrinks as it empties. Amoebas reproduce asexually by cell division after mitotic division of the nucleus; sexual reproduction has not been observed.

Parasitic amoebas include *Entamoeba histolytica,* which causes amoebic dysentery, a serious human intestinal disease characterized by severe diarrhea and the formation of large ulcers in the intestinal wall. In especially severe cases, the organism spreads from the large intestine and causes abscesses in the liver, lungs, or brain. According to the World Health Organization, about 48 million people were newly infected with this amoeba in 1996, and 70,000 people died. *Entamoeba histolytica* is transmitted as **cysts** in contaminated drinking water. (A cyst is a thick-walled, resistant, resting stage in the life

Figure 24-2 **Amoebas.** LM of a giant amoeba (*Chaos carolinense*), a single-celled protist that moves and feeds by means of pseudopodia, which are shown surrounding and ingesting a colonial green alga. (*Michael Abbey/Photo Researchers, Inc.*)

history of some protists.) Other amoebas, like *Acanthamoeba*, are usually free-living but can produce opportunistic infections such as eye infections in contact lens users.

Foraminiferans extend cytoplasmic projections through tests

Almost all **foraminiferans** (phylum Foraminifera) are marine organisms that produce shells, or **tests** (Fig. 24–3*a*). The ocean contains enormous numbers of foraminiferans, which secrete

chalky, many-chambered tests with pores through which cytoplasmic projections can be extended. The group gets its phylum name from this characteristic, as *foraminifera* is derived from Latin words that mean "bearing openings." The cytoplasmic projections form a sticky, interconnected net that entangles prey.

Dead foraminiferans sink to the bottom of the ocean, where their tests form a gray mud that is gradually transformed into chalk. With geological uplifting, these chalk formations can become part of the land, like the white cliffs of Dover in England (Fig. 24–3*b*). (The white cliffs of Dover are composed of the remains of a variety of carbonate organisms, not only foraminiferans.) Because foraminiferan tests are often found in rock layers covering oil deposits, geologists involved in oil exploration look for foraminiferan tests in rock strata.

Actinopods project slender axopods

Actinopods (phylum Actinopoda) have long, filamentous cytoplasmic projections called **axopods** that protrude through pores in their shells (Fig. 24–4). Each axopod is strengthened by a cluster of microtubules. Unicellular algae and other prey become entangled in these axopods and are engulfed outside the main body of the actinopod; cytoplasmic streaming carries the prey back within the shell. Many actinopods contain algal **endosymbionts** that provide them with the products of photosynthesis. (An endosymbiont lives inside the body of the organism with which it has formed a close relationship.)

Some actinopods secrete elaborate and beautiful shells made of silica. When actinopods die, their skeletons settle and become mud on the ocean floor; eventually they are compressed into sedimentary rock.

(a)

500 µm

(b)

Figure 24-3 **Foraminiferans.** These protists with shells, or tests, live primarily in marine environments. *(a)* SEM of a foraminiferan test. Note the pores through which cytoplasm can extrude. *(b)* The white cliffs of Dover, England, consist largely of the tests of foraminiferans. (*a, Biophoto Associates; b, Lynn McLaren/Photo Researchers, Inc.*)

TABLE 24-1   Representative Phyla in the Protist Kingdom

Common Name	Phylum	Morphology	Locomotion	Photosynthetic Pigments	Energy Reserves	Special features
Protozoa						
Amoebas	Rhizopoda	Unicellular, no definite shape	Pseudopodia	—	—	Some have shells (tests)
Foraminiferans	Foraminifera	Unicellular	Cytoplasmic projections	—	—	Pore-studded shells (tests)
Actinopods	Actinopoda	Unicellular	Some produce flagellated reproductive cells	—	—	Axopods protrude through pores in shell
Zooflagellates	Zoomastigina	Unicellular; some colonial	One to many flagella; some amoeboid	—	—	Symbiotic forms often highly specialized
Ciliates	Ciliophora	Unicellular	Cilia	—	—	Macronuclei and micronuclei
Apicomplexans	Apicomplexa	Unicellular	Move by flexing	—	—	Parasitic; develop resistant spores
Algae						
Euglenoids	Euglenophyta	Unicellular	Two flagella (one very short)	Chlorophylls *a* and *b;* carotenoids	Paramylon	Flexible outer covering (pellicle)
Dinoflagellates	Dinoflagellata	Unicellular; some colonial	Two flagella	Chlorophylls *a* and *c;* carotenoids, including fucoxanthin	Oils or polysaccharides	Many covered with cellulose plates
Diatoms	Bacillariophyta	Unicellular; some colonial	Most nonmotile; some glide over secreted slime	Chlorophylls *a* and *c;* carotenoids, including fucoxanthin	Oils or carbohydrates	Silica in shell

100 μm

Zooflagellates move by means of flagella

Zooflagellates (phylum Zoomastigina) are mostly unicellular (a few are colonial) organisms with spherical or elongated bodies, a single central nucleus, and from one to many long, whiplike flagella that enable them to move. Zooflagellates move rapidly, pulling themselves forward by lashing flexible flagella that are usually located at the anterior (front) end. Some zooflagellates are also amoeboid and engulf food by forming pseudopodia. Other zooflagellates ingest food by means of a definite "mouth," or *oral groove,* and "throat," or *cytopharynx.*

Zooflagellates are heterotrophic and obtain their food either by ingesting living or dead organisms or by absorbing nutrients from dead or decomposing organic matter. They may

◀ **Figure 24–4** **Actinopods.** LM of a living actinopod from the Red Sea. Note the many slender axopods that project from the cell. *(Robert Brons/Biological Photo Service)*

TABLE 24-1   Continued

Common Name	Phylum	Morphology	Locomotion	Photosynthetic Pigments	Energy Reserves	Special features
Golden algae	Chrysophyta	Unicellular; some colonial	Two flagella	Chlorophylls *a* and *c;* carotenoids, including fucoxanthin	Oils or carbohydrates	Covered with tiny scales of silica or calcium carbonate
Brown algae	Phaeophyta	Multicellular	Usually two flagella on reproductive cells	Chlorophylls *a* and *c;* carotenoids, including fucoxanthin	Laminarin	Kelps have bodies with blades, stipes, and holdfasts
Green algae	Chlorophyta	Unicellular; coenocytic; colonial; multicellular	Most flagellated at some stage in life; some entirely nonmotile	Chlorophylls *a* and *b;* carotenoids	Starch	Reproduction highly variable
Red algae	Rhodophyta	Most multicellular; some unicellular	No flagellated stages	Chlorophyll *a;* carotenoids; phycocyanin; phycoerythrin	Floridean starch	Some reef builders
Fungus-like protists						
Plasmodial slime molds	Myxomycota	Multinucleate plasmodium	Cytoplasmic streaming; flagellated or amoeboid reproductive cells	—	—	Reproduce by spores formed in sporangia
Cellular slime molds	Acrasiomycota	Unicellular feeding stage; multicellular (slug and fruiting body)	Pseudopodia (for single cells); cytoplasmic streaming (for multicellular forms)	—	—	Aggregation of cells signaled by cyclic AMP
Water molds	Oomycota	Coenocytic mycelium	Biflagellate zoospores	—	—	Cellulose and/or chitin in cell walls

be free-living or symbionts. For example, trichonymphs, complex, specialized zooflagellates with many flagella, live in the guts of termites (Fig. 24–5*a*). These zooflagellates possess the enzymes to digest cellulose in the wood that termites eat, and both the termite and the zooflagellates obtain their nutrients from this source. Termites would starve to death without their flagellated endosymbionts, as would the trichonymphs, which have the ability to digest but not to ingest wood.[1]

Some parasitic zooflagellates cause disease. For example, the zooflagellate *Trypanosoma* is a human parasite that causes African sleeping sickness (Fig. 24–5*b*). It is transmitted by the bite of infected tsetse flies. Early symptoms include recurring attacks of fever. Later, when the trypanosomes have invaded the central nervous system, infected individuals have difficulty speaking or walking; if untreated, African sleeping sickness can cause death.

Giardia intestinalis is a parasitic zooflagellate that causes backpackers' diarrhea, a common infection among campers and hikers, particularly in the mountains of the western United States. *Giardia* is eliminated as a resistant cyst in the feces of many vertebrate animals. These cysts are a common contaminant in mountain streams. Campers and hikers become infected when they drink or rinse dishes in the "clean" mountain water. In a heavy infection, much of the wall of the small intestine is coated with these zooflagellates, which interfere with the absorption of digested nutrients and cause weight loss, abdominal pain, and diarrhea.

Choanoflagellates (collared zooflagellates) are one of the classes of zooflagellates in the phylum Zoomastigina. These sessile marine and freshwater zooflagellates are attached by a stalk, and their single flagellum is surrounded by a delicate col-

---

[1]The zooflagellate endosymbionts of termites in turn possess endosymbionts, bacteria that reside within the zooflagellates. These bacteria, rather than the zooflagellates, may produce the enzymes that digest cellulose.

(a)

|————| 50 μm

Red blood cells

Trypanosome

Flagellum

(b)

|————| 25 μm

Flagellum

Collar of microvilli

Cell

Stalk

(c)

Figure 24–5 Zooflagellates. *(a)* LM of a complex zooflagellate (*Trichonympha* sp.) that lives in the gut of wood-eating termites and digests the cellulose of the wood particles eaten by its host, from which it obtains sugar for itself and its host. *Trichonympha* has hundreds of flagella. *(b)* LM of the zooflagellate (*Trypanosoma gambiense*) that causes sleeping sickness, among red blood cells in a human blood smear. *(c)* Choanoflagellates obtain food by waving their flagella, causing water currents to carry small particles of food into the collar of microvilli. Shown is a colonial form.   *(a, M.A. Abbey/Visuals Unlimited; b, Biophoto Associates/Photo Researchers, Inc.)*

lar of microvilli (Fig. 24–5c). They are of special interest because of their striking resemblance to collar cells in sponges (see Fig. 28–6). Some biologists think choanoflagellates are related to the ancestor of animals.

Ciliates use cilia for locomotion

Ciliates (phylum Ciliophora) are unicellular organisms with a flexible outer covering called a **pellicle** that gives them a definite but changeable shape. In *Paramecium,* the surface of the cell is covered with several thousand fine, short, hairlike organelles, called cilia, that extend through pores in the outer covering and permit movement (Fig. 24–6a,b). The cilia beat in such a precisely coordinated fashion that the organism can not only go forward but can also back up and turn around. Just under the pellicle, many ciliates possess numerous small **trichocysts,** organelles that discharge filaments thought to aid in trapping and holding prey.

Not all ciliates are motile. Some sessile forms are stalked, and others, although capable of some swimming, are more likely to remain attached to a rock or other surface at one spot (Fig. 24–6c). Their cilia set up water currents that draw food toward them.

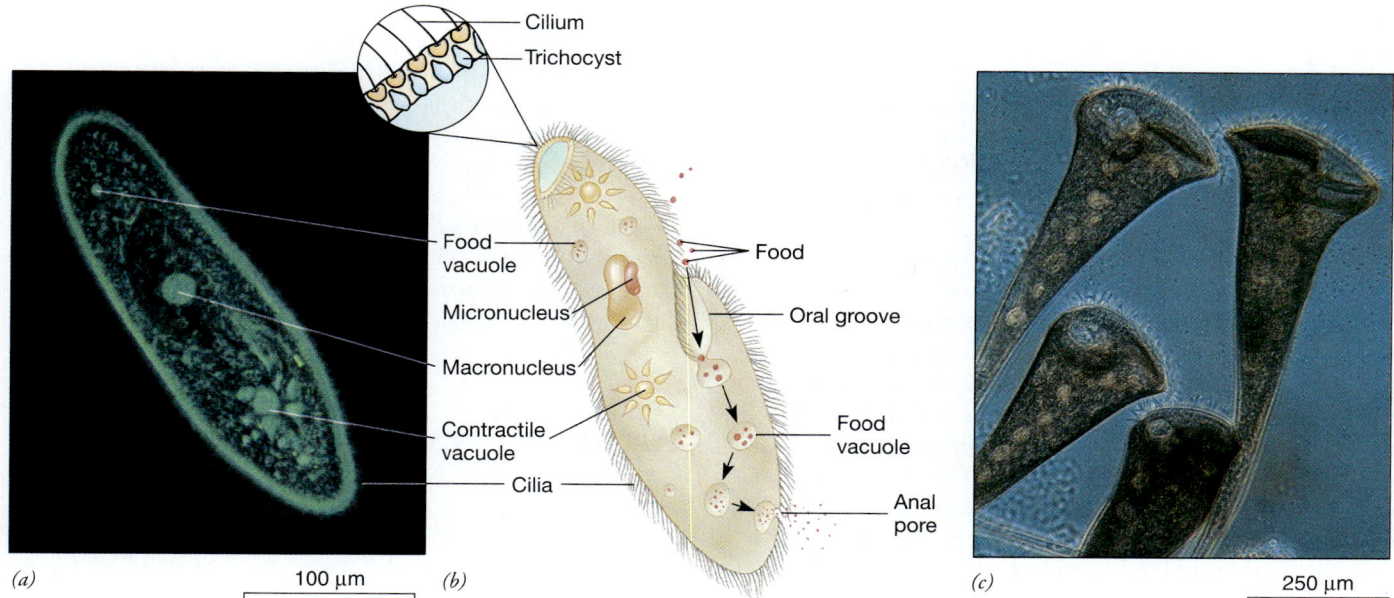

Cilium

Trichocyst

Food vacuole

Food

Micronucleus

Oral groove

Macronucleus

Contractile vacuole

Food vacuole

Cilia

Anal pore

(a)

100 µm

(b)

(c)

250 µm

Figure 24–6 **Ciliates.** *(a)* Note the complex cell structure seen in this LM of *Paramecium* sp., a freshwater ciliate. Like many ciliates, *Paramecium* has multiple nuclei—a macronucleus and one or more smaller micronuclei. *(b)* Food particles are swept into *Paramecium*'s ciliated oral groove and incorporated into food vacuoles. Lysosomes fuse with the food vacuoles, and the food is digested and absorbed; undigested wastes are eliminated through the anal pore. *Paramecium* absorbs water by osmosis from its freshwater surroundings, but it does not swell because contractile vacuoles fill with excess water and then contract to discharge it into the environment. *(c)* LM of *Stentor*, a sessile ciliate. Note the numerous cilia that direct food particles into its funnel-like "mouth." *(a, M. Abbey/Photo Researchers, Inc.; c, Eric Gravé/Photo Researchers, Inc.)*

Most ciliates ingest bacteria or other tiny protists; their cilia draw the food into a simple opening in some species and into a funnel-like oral groove in others. A vacuole forms around the food at the end of the opening, and the food is digested. Water regulation in freshwater ciliates is controlled by special organelles called **contractile vacuoles.** Being hypertonic to their environment, freshwater ciliates continually take in water by osmosis; the contractile vacuole continually expels excess water from the ciliate back to the environment.

Ciliates differ from other protozoa in having two kinds of nuclei: one or more small, diploid **micronuclei** that function in the sexual process, and a larger, polyploid **macronucleus** that controls cell metabolism and growth. Most ciliates are capable of a sexual process called **conjugation** where two individuals come together and exchange genetic material. Each ciliate species is divided into several different mating types. Individuals with different mating types are identical in appearance but genetically different in terms of sexual compatibility. Because there are no physical differences between the mating types, it is not appropriate to refer to them as "male" and "female."

During conjugation in *Paramecium,* two individuals of different mating types press their oral surfaces together. Within each individual the macronucleus disintegrates and the micronucleus undergoes meiosis, forming four haploid nuclei.

Three of these degenerate, leaving one. This nucleus then divides mitotically to form two identical haploid nuclei. One of these remains within the cell and the other nucleus crosses through the oral region into the other individual, where it fuses with the haploid nucleus in that cell. Thus, a single act of conjugation yields two cross fertilizations as each cell fertilizes the other. This results in two "new" cells that are genetically identical to each other but different from what they were before conjugation. Actual cell division need not follow immediately after conjugation. Cell division is a complex process involving more than simply splitting in half, because complex organelles must be duplicated. In addition, after the new micronucleus divides, a new macronucleus develops from one of the micronuclei.

Apicomplexans are spore-forming parasites of animals

Apicomplexans (phylum Apicomplexa) are a large group of parasitic protozoa, some of which cause serious diseases such as malaria in humans. They are thought to have evolved from parasitic dinoflagellates living in the intestines of marine invertebrates (dinoflagellates are discussed later in this chapter). Apicomplexans lack specific structures for locomotion but move by flexing. At some stage in their lives, they develop a

spore, a small infective agent transmitted to the next host. Many apicomplexans spend part of their life cycle in one host species and part in a different host species.

Malaria, which is caused by an apicomplexan, is the world's most serious infectious disease. According to the World Health Organization, approximately 300 to 500 million people currently have malaria, and 1.5 to 2.7 million people, many of them children, die each year from it. Although the disease was described for centuries in Chinese, Greek, Arabic, and Roman writings, its causal agent and mode of transmission were not identified until the end of the 19th century. *Plasmodium,* the apicomplexan that causes malaria, enters human blood through the bite of an infected female *Anopheles* mosquito (Fig. 24–7). *Plasmodium* first enters liver cells and then red blood cells, where it multiplies. When each infected red blood cell bursts, many new parasites are released. The released parasites infect new red blood cells, and the process is repeated. The simultaneous bursting of millions of red cells causes the symptoms of malaria: a chill, followed by high fever caused by toxic substances that are released and affect other organs of the body.

Malaria is currently recurring in many countries where it had been under control for decades. Formerly effective control methods—antimalarial drugs and pesticides—have lost much of their effectiveness. Chloroquine and several other antimalarial drugs are taken prophylactically to prevent malaria, but *Plasmodium* has evolved resistance to several of these in many areas, necessitating the use of a combination of drugs. Pesticides are used to control *Plasmodium's* vector, the mosquito, but mosquitoes have evolved resistance to many pesticides. New antimalarial drugs and several possible vaccines against malaria are currently being tested.

ALGAE ARE PLANTLIKE PROTISTS

Algae (sing., *alga*) are an informal group of mostly photosynthetic protists that range in size from unicellular, microscopic forms to large, multicellular seaweeds. (The word *alga* is derived from a Latin word for "seaweed.")

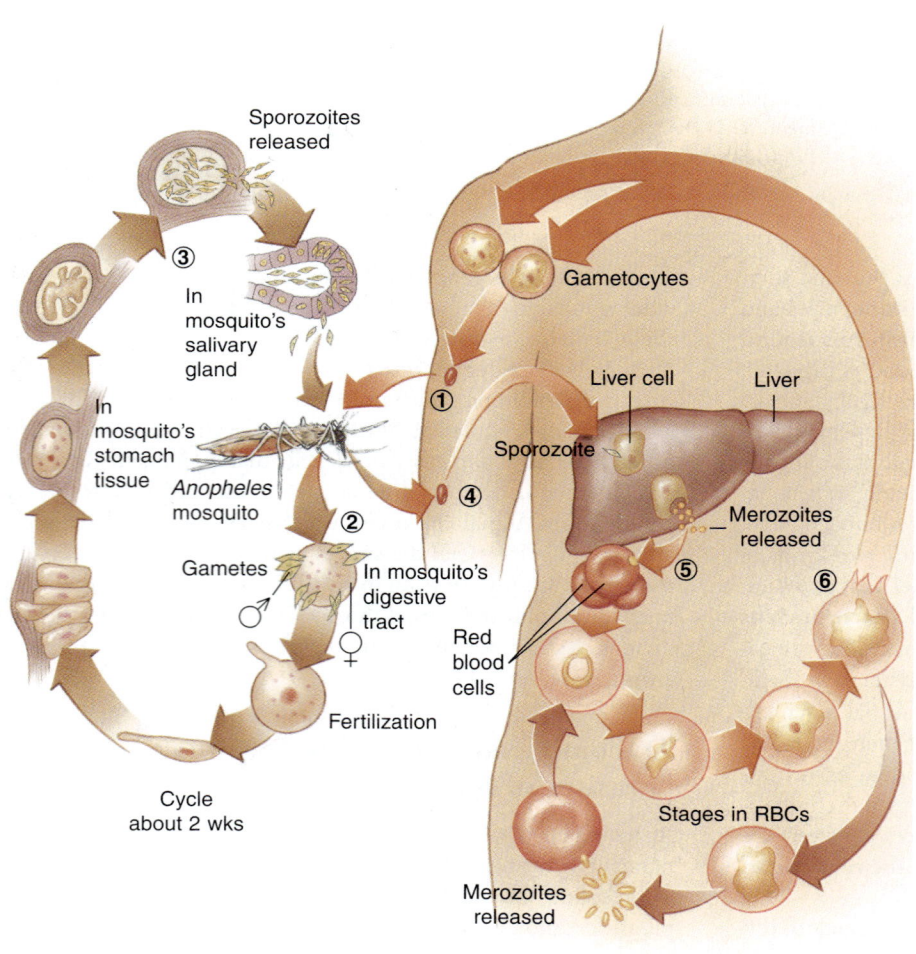

Sporozoites released

③ In mosquito's salivary gland

In mosquito's stomach tissue

Anopheles mosquito

② In mosquito's digestive tract

Gametes ♂ ♀

Fertilization

Cycle about 2 wks

① Sporozoite — Liver cell — Liver

④ Merozoites released

Gametocytes

⑤ Red blood cells

⑥ Stages in RBCs

Merozoites released

Cycle about 2 days

Figure 24–7 The life cycle of *Plasmodium* sp., the apicomplexan that causes malaria. (1) A female *Anopheles* mosquito bites an infected person and obtains gametocytes along with human blood. (2) In the mosquito's digestive tract, the gametocytes develop into gametes and fertilization occurs. (3) The zygote embeds in the mosquito's stomach lining and produces sporozoites (spores), which are released and migrate to the salivary gland. (4) The mosquito bites an uninfected human and transmits sporozoites to the human's blood. (5) The sporozoites enter liver cells and divide to produce merozoites that infect red blood cells. (6) In the blood cells, merozoites divide to form more merozoites, which infect more red blood cells. Some merozoites form gametocytes, which can be transmitted to the next mosquito that bites that human, and the process is repeated.

Flagellum
Eye spot
Contractile vacuole
Chloroplast
Nucleus
Paramylon body (stored food)
Pellicle

Mitochondria (indistinguishable in photo)
Nucleolus
Chromatin

(a)   25 μm   *(b)*

Figure 24-8 Euglenoids.
(a) Note the complex cell structure shown in this LM of a unicellular, flagellated euglenoid (*Euglena* sp.). *(b)* The eyespot is thought to shield a light detector at the base of the long flagellum, thereby helping *Euglena* move to light of an appropriate intensity. *Euglena*'s pellicle is flexible and enables it to change shape easily. *(Biophoto Associates)*

Although most algae are photosynthetic like plants, they are not considered plants because they lack many plant structures, such as roots, stems, and leaves. Algae lack a cuticle, which is a waxy covering over the aerial parts of plants that reduces water loss. When actively growing, algae are restricted to damp or wet environments, such as the ocean; freshwater ponds, lakes, and streams; hot springs; polar ice; alpine snow; moist soil, trees, and rocks; and the bodies of certain animals including sloths, sea anemones, corals, and worms. Also, most algae do not have multicellular **gametangia** (sing. *gametangium;* reproductive structures in which gametes are produced); algal gametangia generally are formed from single cells.

In addition to green chlorophyll *a* and yellow and orange **carotenoids,** which are photosynthetic pigments found in all algae (and plants), different algal phyla possess a variety of other pigments that are also important in photosynthesis. Classification into phyla is largely based on their pigment composition and what kinds of materials they produce to store energy reserves (fuel molecules). Other characteristics used to classify algae include their cell wall composition, the number and placement of flagella, and chloroplast structure. We will consider seven groups of algae: euglenoids, dinoflagellates, diatoms, golden algae, brown algae, green algae, and red algae (see Table 24–1).

Euglenoids are unique freshwater unicellular flagellates

Most **euglenoids** (phylum Euglenophyta) are unicellular flagellates, and about one-third of them are photosynthetic (Fig. 24–8). They generally possess two flagella, one long and whip-like and one so short that it does not extend outside the cell. Some euglenoids, such as *Euglena* sp., change shape continually as they move through the water because their pellicle (outer covering of protein) is flexible rather than rigid. Euglenoids reproduce asexually by longitudinal cell division; none has ever been observed to reproduce sexually.

Euglenoids have at various times been classified in the plant kingdom (with the algae) and in the animal kingdom (when protozoa were considered animals). Based on molecular data, euglenoids are thought to be closely related to zooflagellates. We include euglenoids in our discussion of algal protists rather than the discussion of zooflagellates because many species contain chloroplasts and photosynthesize. They have chlorophyll *a*, chlorophyll *b*, and carotenoids, which are the same pigments found in green algae and plants.[2] Their energy reserves are stored as *paramylon,* a polysaccharide. Some photosynthetic euglenoids lose their chlorophyll when grown in the dark and obtain their nutrients heterotrophically by ingesting organic matter. Other species of euglenoids are always colorless and heterotrophic.

Euglenoids inhabit freshwater ponds and puddles, particularly those with large concentrations of organic material. For that reason they are used as indicator species of organic pollution. Some euglenoids are also found in marine waters and mud flats.

[2]Although the euglenoids have the same pigments as the green algae and plants, they are not thought to be closely related to either group.

(a) 25 μm *(b)*

(c)

Figure 24–9 **Dinoflagellates.** *(a)* SEM of *Protoperidinium* sp., a unicellular, photosynthetic, bi-flagellate dinoflagellate. Note the cellulose plates that encase the unicellular body. The two flagella (not visible) are located in grooves. *(b)* A species of *Gonyaulax*, showing the two flagella located in grooves. *(c)* Aerial shot of a red tide around Matushima Island, Japan. The water is colored by the presence of billions of dinoflagellates. *(a, Courtesy of T.K. Maugel, University of Maryland; c, Suisan Aviastion Company)*

Most dinoflagellates are a part of marine plankton

One of the most unusual protist phyla is that of the **dinoflagellates** (phylum Dinoflagellata). Most dinoflagellates are unicellular, although a few are colonial. Their cells are often covered with shells of interlocking cellulose plates impregnated with silicates (Fig. 24–9*a,b*). The typical dinoflagellate has two flagella: one flagellum is wrapped around a transverse groove in the center of the cell like a belt, and the other is located in a longitudinal groove (perpendicular to the transverse groove) and projects behind the cell. The undulation of these flagella propels the dinoflagellate through the water like a spinning top. Indeed, the dinoflagellates' name is derived from the Greek *dinos,* meaning "whirling." Many marine dinoflagellates are bioluminescent.

Most dinoflagellates are photosynthetic and possess the photosynthetic pigments chlorophyll *a*, chlorophyll *c*, and carotenoids, including **fucoxanthin,** a special yellow-brown carotenoid. However, a number are colorless; some of these ingest other microorganisms for food. Dinoflagellates usually store energy reserves as oils or polysaccharides.

Many dinoflagellates are endosymbionts that reside in the bodies of marine invertebrates such as jellyfish, corals, and mollusks. These symbiotic dinoflagellates lack cellulose plates and flagella and are called **zooxanthellae.** Zooxanthellae photosynthesize and provide carbohydrates for their invertebrate partners. The contribution of zooxanthellae to the productivity of coral reefs is substantial. Other dinoflagellates that are endosymbionts lack pigments and do not photosynthesize; these heterotrophs are parasites that live off their hosts.

Reproduction in the dinoflagellates is primarily asexual, by longitudinal cell division, although a few species have been reported to reproduce sexually. The dinoflagellate nucleus is distinctive because the chromosomes are permanently condensed and always evident. Meiosis and mitosis are unusual because the nuclear envelope remains intact throughout cell division and the spindle is located *outside* the nucleus. (The chromosomes do not make direct contact with the spindle microtubules; instead, the chromosomes appear to be attached to the nuclear envelope, and the spindle separates the new nuclei from each other.)

Ecologically, dinoflagellates are one of the most important groups of producers in marine ecosystems. A few dinoflagellates are known to have occasional population explosions, or blooms. These blooms frequently color the water orange, red, or brown and are known as **red tides** (Fig. 24–9*c*). It is not known what environmental conditions initiate dinoflagellate blooms, but they are more common in the warm waters of late summer. Many experts think that human-produced coastal pollution triggers red tides, presumably by providing nutrients to the dinoflagellates. Some of the dinoflagellate species that form red tides produce a toxin that attacks the nervous systems of fishes, leading to massive fish kills. Birds such as cormorants suffer and sometimes die when exposed to the toxin from eating the dead fish. Scientists also determined that a red tide of the dinoflagellate *Gymnodinium breve* was responsible for the deaths of 158 manatees on Florida's gulf coast in 1996.

In August 1997 more than 30,000 fish with open, bleeding sores died in several tributaries of the Chesapeake Bay. Water skiers, fishermen, and scientists that came into contact with the contaminated water developed breathing difficulty, skin rashes, and temporary memory loss. This event, which received national attention, was caused by a population explosion of toxic dinoflagellates that are relatively new to science. *Pfiesteria piscicida* was the first dinoflagellate of this type to be scientifically named and identified, in 1991, by Joann Burkholder at North Carolina State University. *Pfiesteria* and similar dinoflagellates have complex life cycles with about 24 stages, including dormant cysts, nontoxic amoeboid forms, and both nontoxic and toxic flagellated forms. The algal blooms have been circumstantially linked to nitrogen and phosphorus pollution in shallow estuaries near hog and chicken operations and municipal sewage treatment plants. When *Pfiesteria* and similar dinoflagellates change to toxic forms, they release a poison that eats into the fish's skin, causing open, bleeding sores. The dinoflagellates feed on the fish tissues and blood and then quickly convert to nontoxic forms.

Humans sometimes get paralytic shellfish poisoning by eating filter-feeding mollusks, such as oysters, mussels, or clams, that have fed on certain dinoflagellates. Paralytic shellfish poisoning causes respiratory problems in humans who eat the contaminated shellfish; death from respiratory failure occasionally occurs. The dinoflagellates do not appear to harm the shellfish.

Diatoms have shells composed of two parts

Most **diatoms** (phylum Bacillariophyta) are unicellular, although a few exist as colonies. The cell wall of each diatom consists of two shells that overlap where they fit together, much like a petri dish. Silica is deposited in the shell, and this glasslike material is laid down in intricate patterns (Fig. 24–10*a*). There are two basic groups of diatoms, those with radial symmetry (wheel-shaped) and those with bilateral symmetry (boat-shaped or needle-shaped). Although some diatoms are part of the floating plankton, others live on rocks and sediments, where they move by gliding. This gliding movement is facilitated by the secretion of a slimy material from a small groove along the shell.

(a) 100 μm

(b)

Figure 24–10 **Diatoms.** *(a)* LM of diatoms, unicellular algae with shells that contain silica. These algae have strikingly beautiful patterns on their symmetrical shells. *(b)* Asexual reproduction in diatoms. As cell division occurs, each new cell retains half of the original shell. The other half of the shell is synthesized, always to fit *inside* the original half. As a result, one of the new cells is smaller than the other. *(a, The Stock Market/Phillip Harrington)*

(a)

25 μm

(b)

1 μm

Figure 24–11 **Golden algae.** *(a)* LM of a colonial, freshwater golden alga (*Dinobryon* sp.). As flagella beat, water currents are produced that carry bacteria into the transparent casing surrounding each cell of the colony. These bacteria are ingested and supplement the food that *Dinobryon* produces by photosynthesis. *(b)* SEM of a coccolithophorid (*Emiliania huxleyi*), a golden alga that is an important component of nanoplankton. Coccolithophorids are covered by tiny, overlapping scales of calcium carbonate. *(a, J.R. Waaland/Biological Photo Service; b, Dr. Elizabeth Venrick/Scripps Institution of Oceanography)*

Diatoms contain the photosynthetic pigments chlorophyll *a*, chlorophyll *c*, and carotenoids, including fucoxanthin; their pigment composition gives them a yellow or brown color. Energy reserves are stored as oils or carbohydrates.

Diatoms most often reproduce asexually by cell division. When a diatom divides, the two halves of its shell separate, and each becomes the larger half of a new diatom shell (Fig. 24–10*b*). Because the glass shell cannot grow, some diatom cells get progressively smaller with each succeeding generation. When a diatom is a small fraction of its original size, sexual reproduction is triggered, with the production of shell-less gametes. Sexual reproduction restores the diatom to its original size because the resulting zygote grows substantially before producing a new shell.

Diatoms are common in fresh water, but they are especially abundant in relatively cool ocean water. They are major producers in aquatic ecosystems because of their extremely large numbers. When diatoms die, their shells trickle down to the ocean floor and accumulate in layers that eventually become sedimentary rock. After millions of years, some of these deposits have been exposed on land by geological upheaval. Called diatomaceous earth, these deposits are mined and used as filtering, insulating, and soundproofing material. As a filtering agent, diatomaceous earth is used to refine raw sugar and to process vegetable oils. Because of its abrasive properties, diatomaceous earth is a common ingredient in scouring powders and metal polishes; it is no longer added to most toothpastes because it is too abrasive for tooth enamel. The intricately detailed diatom shells are often used to test microscope resolution down to 1 μm.

Most golden algae are flagellated, unicellular organisms

Golden algae (phylum Chrysophyta) are a complex group found in both freshwater and marine environments. Most species are biflagellated, unicellular organisms, although some are colonial (Fig. 24–11*a*). A few lack flagella and are similar to amoebas in appearance except that they contain chloroplasts. Cells may be covered by tiny scales of either silica or calcium carbonate. Reproduction in golden algae is primarily asexual and involves the production of flagellated, motile spores called **zoospores.**

Most golden algae are photosynthetic and produce the same pigments as diatoms: chlorophyll *a*, chlorophyll *c*, and carotenoids, including fucoxanthin. The pigment composition of golden algae gives them a golden or golden brown color. As in diatoms, energy reserves are stored as oils or carbohydrates. A few species ingest bacteria and other particles of food. Ecologically, golden algae are an important group of producers in marine environments. They comprise a significant portion of the ocean's **nanoplankton,** extremely minute algae that are major producers because of their great abundance.

Classification of golden algae is controversial. Some biologists lump diatoms and golden algae in a single phylum, whereas others think they should be classified as brown algae (discussed shortly). At the other extreme, some biologists divide the golden algae into two phyla by placing many of the marine species, such as **coccolithophorids** (Fig. 24–11*b*), in a separate phylum.

Blade

Stipe

Holdfast

(a)

(b)

Figure 24–12 **Brown algae.** *(a) Laminaria* sp. is a representative brown alga. Note its blade, stipe, and holdfast, all of which are visible in this photograph because this individual was removed from the water. Species of *Laminaria* are widely distributed on rocky coastlines of temperate and polar seas. *(b)* A kelp bed off the coast of California. These underwater forests are ecologically important, supporting large numbers of aquatic organisms. *(a, J.R. Waaland/Biological Photo Service; b, Richard Herrmann)*

Brown algae are multicellular seaweeds

Brown algae (phylum Phaeophyta) include the giants of the protist kingdom. All brown algae are multicellular and range in size from a few centimeters (an inch or so) to approximately 75 m (250 ft) in length. Their body forms may be branched filaments, tufts, "fleshy ropes," or thick, flattened branches. The largest brown algae, called kelps, are tough and leathery in appearance; many kelps possess leaflike **blades,** stemlike **stipes,** and rootlike anchoring **holdfasts** (Fig. 24–12*a*). They often have gas-filled floats *(bladders)* that provide buoyancy. (The blades, stipes, and holdfasts of brown algae are not homologous to the leaves, stems, and roots of plants. Brown algae and plants arose from different unicellular ancestors.)

Brown algae are photosynthetic and possess chlorophyll *a*, chlorophyll *c*, and carotenoids, including fucoxanthin, in their chloroplasts. The main energy storage reserve in brown algae is a carbohydrate called *laminarin.*

Reproduction is varied and complex in the brown algae. Their reproductive cells, both asexual zoospores and sexual gametes, are usually flagellated. Most have life cycles with an **al-**

ternation of generations, in which they spend a portion of their life cycles as haploid organisms and a portion as diploid organisms (see Fig. 9–12*c*).

Brown algae are commercially important for several reasons. Their cell walls contain a polysaccharide called algin that possibly helps cement the cell walls of adjacent cells together. Algin is used as a thickening and stabilizing agent in ice creams, marshmallows, toothpastes, shaving creams, hair sprays, and hand lotions. Brown algae are an important human food, particularly in East Asian countries, and they are rich sources of certain vitamins and of minerals such as iodine. Brown algae are one source of the antiseptic tincture of iodine.

Brown algae are common in cooler marine waters, especially along rocky coastlines, where they can be found mainly in the intertidal zone or relatively shallow offshore waters. Kelps form extensive underwater "forests" called kelp beds (Fig. 24–12*b*). They are essential in that ecosystem for two reasons: they are the primary food producers and they provide habitats for many marine invertebrates, fish, and mammals. The diversity of life supported by kelp beds rivals that found in coral reefs. There is also an extensive population of floating brown

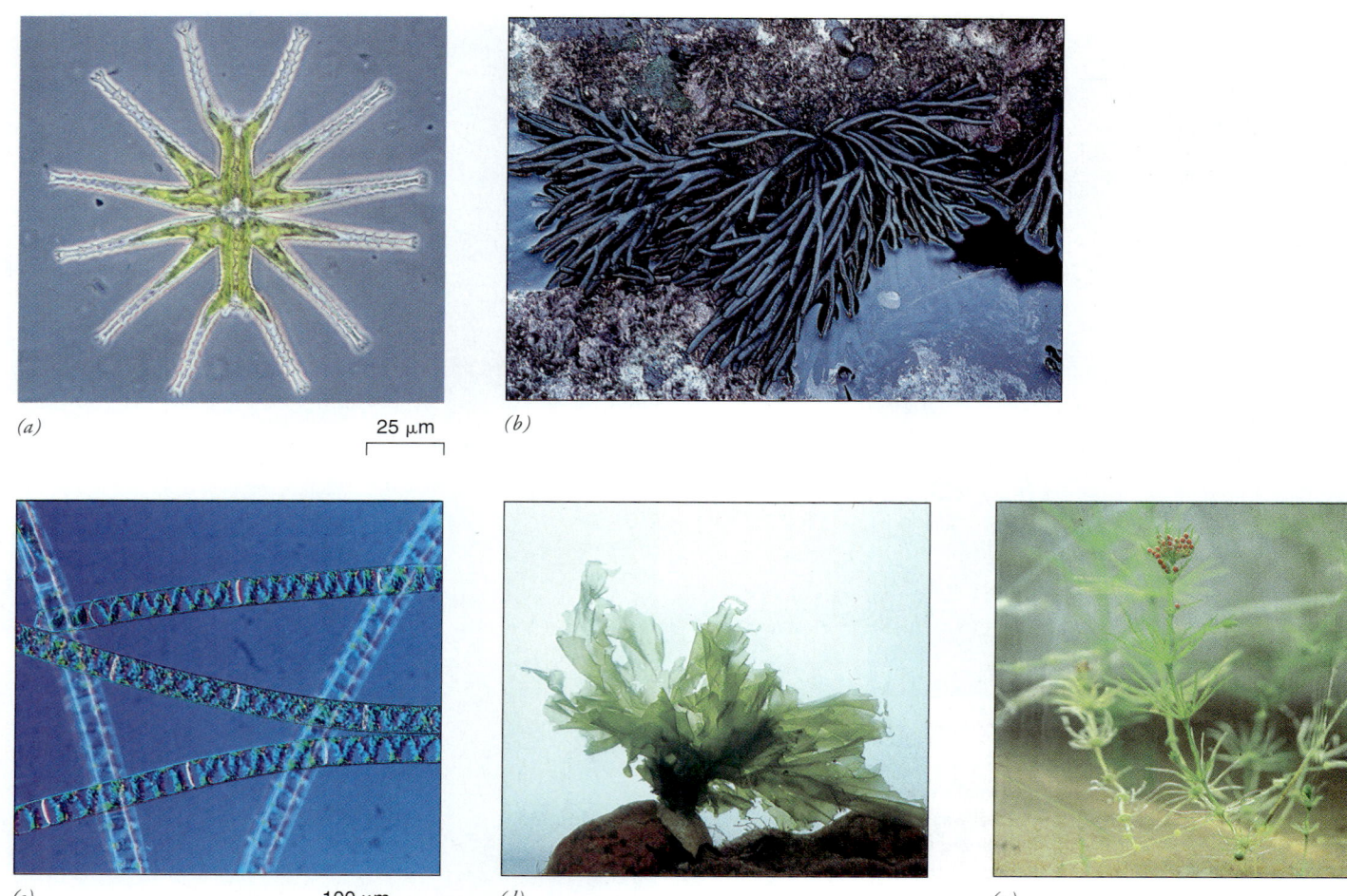

(a)   25 μm

(c)   100 μm   (d)   (e)

Figure 24 – 13   Green algae. *(a)* LM of a widely distributed desmid (*Micrasterias* sp.), a unicellular green alga with mirror-image halves. *(b)* Dead man's fingers (*Codium* sp.) exposed during low tide off Big Sur, California. A green alga like *Codium* is coenocytic. *(c)* Spiral-shaped chloroplasts visible in this LM are characteristic of *Spirogyra* sp., a widely distributed, multicellular green alga with a filamentous body form. *(d)* Some multicellular green algae are sheetlike. The thin, leaflike form has given *Ulva* sp. its common name, "sea lettuce." (e) *Chara* sp, a green alga commonly called a stonewort, is closely related to plants. *Chara* is widely distributed in fresh water, where it grows to 30 cm (1 ft) or more.   *(a, Biophoto Associates; b, David J. Wrobel/Biological Photo Service; c, Brian Parker/Tom Stack & Associates; d, J.M. Kingsbury; e, James W. Perry)*

algae in a central area of the North Atlantic Ocean called the Sargasso Sea, named for the brown alga *Sargassum*. (The Sargasso Sea is not greatly affected by the surface ocean currents rotating around the margins of the North Atlantic, and so the floating *Sargassum* remains there.)

Green algae share many similarities with plants

Green algae (phylum Chlorophyta) have pigments, energy reserve products, and cell walls that are chemically identical to those of plants. Green algae are photosynthetic, with chlorophyll *a*, chlorophyll *b*, and carotenoids present in chloroplasts

of a wide variety of shapes. Their main energy reserves are stored as starch. Most green algae possess cell walls with cellulose, although some lack walls. Because of these and other similarities, it is generally accepted that plants arose from ancestral green algae. Taxonomy of the green algae is currently under study, and research in ultrastructure (cell structure studied with the aid of electron microscopy), molecular biology, and biochemistry is providing insights into this extremely diverse group.

Green algae exhibit a variety of body types, from single cells to colonial forms, to coenocytic algae, to multicellular filaments and sheets (Fig. 24–13). The multicellular forms do not have cells differentiated into tissues, a characteristic that

ASEXUAL REPRODUCTION (by mitosis)

Zoospores

(n)

Gamete production (by mitosis)

SEXUAL REPRODUCTION

Meiosis

Haploid

Diploid

Fertilization (syngamy)

Isogamy

from a different strain

2n

Zygote

Figure 24–14 **Life cycle of *Chlamydomonas* sp.** *Chlamydomonas,* a haploid green alga with two mating types, (+) and (−), that are visually indistinguishable, is an example of isogamous sexual reproduction. Both mating types reproduce asexually by mitosis; only the (−) strain is shown. During sexual reproduction, a (+) gamete fuses with a (−) gamete, forming a diploid zygote. Meiosis occurs, and four haploid cells emerge, two (+) and two (−).

separates them from plants. Most green algae are flagellated or produce flagellated cells during their life history, although a few are totally nonmotile.

Reproduction in the green algae is as varied as their body forms. Both sexual and asexual reproduction occur, and many green algae have life cycles with an alternation of generations. Asexual reproduction may be by cell division in single cells or by fragmentation in multicellular forms. Many green algae produce spores asexually by mitosis; if these spores are flagellated and motile, they are called zoospores (Fig. 24–14). Sexual reproduction in the green algae involves the formation of gametes in unicellular gametangia. Three types of sexual reproduction—isogamous, anisogamous, and oogamous—are recognized in green algae. If the two flagellated gametes that fuse are identical in size and appearance, sexual reproduction is said to be **isogamous** (Fig. 24–14). **Anisogamous** sexual reproduction involves the fusion of two flagellated gametes of different sizes (Fig. 24–15). Some green algae are **oogamous** and produce a nonmotile egg and a flagellated male gamete. In addition to sexual reproduction by the fusion of gametes, some green algae exchange genetic information by a form of conjugation, in which the genetic material of one cell passes into a recipient cell (Fig. 24–16).

Green algae can be found in both aquatic and terrestrial environments. Aquatic green algae primarily inhabit fresh water, although there are a number of marine species. Terrestrial green algae are restricted to damp soil, cracks in tree bark, and other moist places. Many of the green algae are symbionts with other organisms; some live as endosymbionts in body cells of invertebrates, and a few grow together with fungi as "dual organisms" called lichens (see Chapter 25). Regardless of where they live, green algae are ecologically important as the base of the food web.

Red algae do not produce motile cells

The vast majority of **red algae** (phylum Rhodophyta) are multicellular organisms, although there are a few unicellular species. The multicellular body form of red algae is commonly composed of complex, interwoven filaments that are delicate and feathery (Fig. 24–17*a*); a few red algae are flattened sheets of cells (Fig. 24–17*b*). Most multicellular red algae attach to rocks or other substrates by a basal holdfast. Reproduction in the red algae is remarkably complex, with an alternation of sexual and asexual stages. Although sexual reproduction is com-

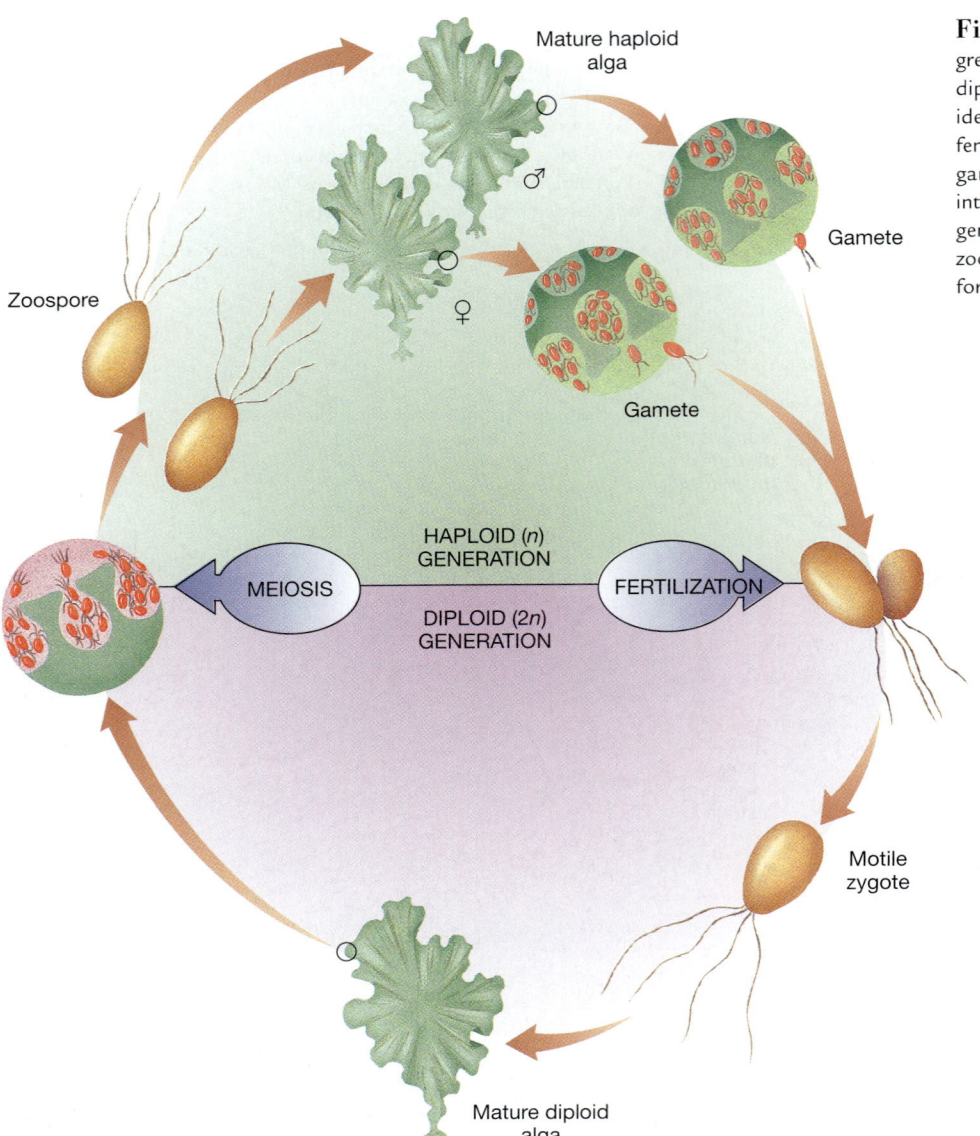

Figure 24–15 **Life cycle of *Ulva* sp.** The green alga *Ulva* alternates between haploid and diploid multicellular generations, which are identical in overall appearance. The male and female haploid algae produce anisogamous gametes that fuse and subsequently develop into the diploid alga. Special cells in the diploid generation undergo meiosis to form haploid zoospores that develop directly into haploid forms, and the cycle continues.

Labels on figure:
Mature haploid alga
Zoospore
Gamete
HAPLOID (*n*) GENERATION
MEIOSIS
DIPLOID (*2n*) GENERATION
FERTILIZATION
Gamete
Motile zygote
Mature diploid alga

mon, at no stage in the life history of red algae do any flagellated cells develop.

The chloroplasts of red algae contain **phycoerythrin,** a red pigment, and **phycocyanin,** a blue pigment, in addition to chlorophyll *a* and carotenoids. Their energy reserves are stored as *floridean starch,* a polysaccharide similar to glycogen. The red algae have the same pigment composition as the cyanobacteria (a group of photosynthetic eubacteria; see Chapter 23).

The cell walls of red algae often contain thick, sticky polysaccharides that are of commercial value. For example, agar is a polysaccharide extracted from certain red algae and used as a food thickener and culture medium (a substance on which to grow microorganisms and propagate some plants, such as orchids). Another polysaccharide extracted from red algae, car-

rageenan, is a food additive used to stabilize chocolate milk and to provide a thick, creamy texture to ice creams and other soft, processed foods. (A stabilizer keeps the texture uniform by preventing different ingredients from settling out.) Carrageenan is also used to stabilize paints and cosmetics. Red algae are a source of vitamins (particularly A and C) and minerals for humans, particularly in Japan and other East Asian countries where they are eaten fresh, dried, or toasted in such traditional foods as sushi and nori.

The red algae are primarily found in warm tropical ocean waters, although a few species occur in fresh water and in soil. Some red algae incorporate calcium carbonate in their cell walls from the ocean waters (Fig. 24–17c). These coralline red algae are extremely important in building "coral" reefs and are possibly more important than coral animals in this process.

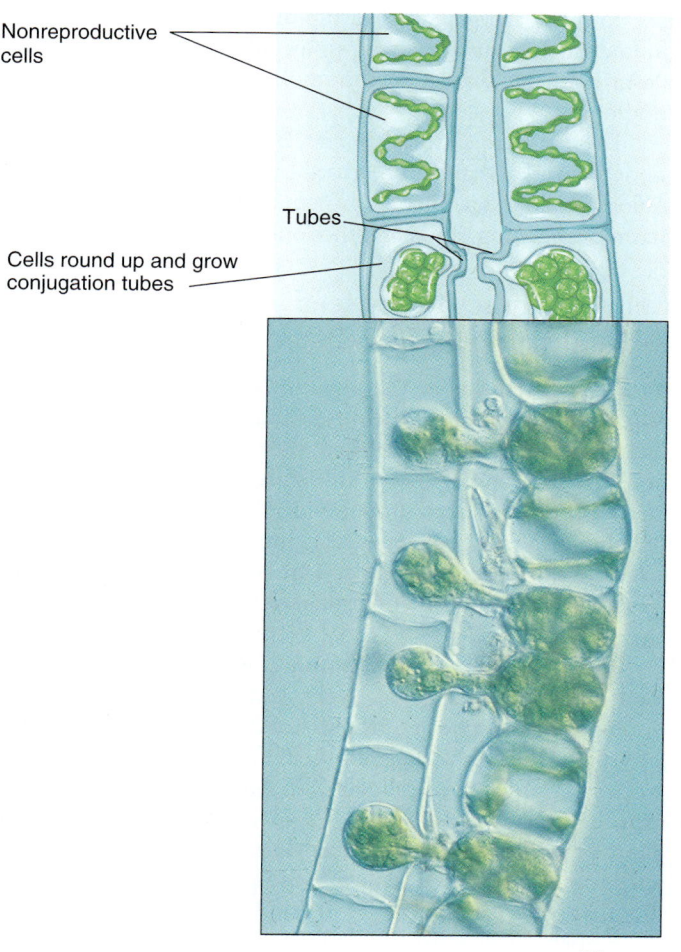

Nonreproductive cells

Tubes

Cells round up and grow conjugation tubes

25 μm

◀ **Figure 24–16** **Conjugation in *Spirogyra* sp.** *(Top)* Filaments of two different mating types of *Spirogyra*, a haploid green alga, line up, and conjugation tubes grow between cells of the two filaments. *(LM on bottom)* The contents of one cell passes into the other through the conjugation tube. The two protoplasts (cells without their walls) then fuse and form a diploid zygote. Following a period of dormancy, the zygote undergoes meiosis, restoring the haploid condition. *(Biophoto Associates)*

SLIME MOLDS AND WATER MOLDS ARE FUNGUS-LIKE PROTISTS

Some protists superficially resemble fungi in that they are not photosynthetic, and some have bodies formed of threadlike structures called *hyphae*. However, fungus-like protists are not fungi for several reasons. Many of these protists have centrioles and produce cellulose as a major component of their cell walls, whereas fungi lack centrioles and have cell walls of chitin. We consider three groups of fungus-like protists: plasmodial slime molds, cellular slime molds, and water molds (see Table 24–1).

Plasmodial slime molds feed as multinucleate plasmodia

The feeding stage of a **plasmodial slime mold** (phylum Myxomycota) is a **plasmodium,** a multinucleate mass of cytoplasm that can grow to 30 cm (1 ft) in diameter (Fig. 24–18*a*). The

(a)

(b)

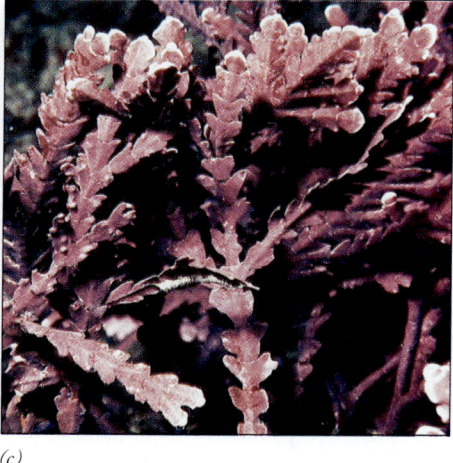
(c)

Figure 24–17 **Red algae.** *(a) Polysiphonia* sp., which is widely distributed throughout the world, has a highly branched body of interwoven filaments. *(b) Rhodymenia* sp., which is common in the Northern Hemisphere, has a flattened, branching body. *(c) Bossiella* sp., which is found in the Pacific Ocean, is a coralline red alga with a hard, brittle body encrusted with calcium carbonate. *(a, Philip Sze/Visuals Unlimited; b, Richard H. Gross; c, D. Gotshall/Visuals Unlimited)*

(a)

(b)

250 μm

Figure 24–18 **The plasmodial slime mold** *Physarum poly-cephalum.* *(a)* The plasmodium of *P. polycephalum* is brightly pigmented. This naked mass of cytoplasm is multinucleate and feeds on bacteria and other microorganisms. *(b)* The reproductive structures of *P. polycephalum* are stalked sporangia. *(a, b, R. Calentine/CBR Images)*

plasmodium, which is slimy in appearance, streams over damp, decaying logs and leaf litter, often forming a network of channels that covers a larger surface area. As it creeps along, it ingests bacteria, yeasts, spores, and decaying organic matter.

When the food supply dwindles or there is insufficient moisture, the plasmodium crawls to an exposed surface and initiates reproduction. Stalked structures of intricate complexity and beauty usually form from the drying plasmodium (Fig. 24–18*b*). Within these structures, called **sporangia,** meiosis produces haploid **spores** that are extremely resistant to adverse environmental conditions.

When conditions become favorable again, the spores germinate, and a haploid reproductive cell emerges from each. This haploid cell may be biflagellate (called a *swarm cell*) or amoeboid (called a *myxamoeba*), depending on how wet the environment is; flagellated cells form in wet conditions. Swarm cells and myxamoebas act as gametes. Eventually two gametes

fuse to form a zygote with a diploid nucleus. The resultant diploid nucleus divides many times by mitosis, but the cytoplasm does not divide, so the result is a multinucleate plasmodium.

The plasmodial slime mold *Physarum polycephalum* is a model organism that has been used to study many fundamental biological processes, such as growth and differentiation, cytoplasmic streaming, and the function of the cytoskeleton.

Cellular slime molds feed as single amoeboid cells

Although organisms in the phylum Acrasiomycota are called **cellular slime molds,** their resemblance to the plasmodial slime molds is superficial. Indeed, they have much closer affinities with amoebas, as shown in Figure 24–1. During its feeding stage, each cellular slime mold is an individual amoeboid cell that behaves as a separate, solitary organism. Each cell creeps over rotting logs and soil or swims in fresh water, ingesting bacteria and other particles of food as it goes. Each cell has a haploid nucleus.

When moisture or food becomes inadequate, certain cells send out a chemical signal, cyclic AMP (see Fig. 3–25), that causes them to aggregate by the hundreds or thousands (Fig. 24–19*a*). During this stage the cells creep about for short distances as a single multicellular unit, called a **pseudoplasmodium,** or "slug" (Fig. 24–19*b*). Each cell of the slug retains its plasma membrane and individual identity. Eventually the slug settles and constructs a stalked fruiting body, which is a structure that bears spores (Fig. 24–19*c*). After being released, each spore opens, a single haploid amoeboid cell emerges, and the cycle repeats. This reproductive cycle is asexual, although sexual reproduction has been observed occasionally. There are no flagellated stages for most cellular slime molds.

The cellular slime mold *Dictyostelium discoideum* is a model organism for the study of cell differentiation. Its biology has been studied intensively, particularly as it relates to signal molecules, such as cyclic AMP, which are found in many organisms in addition to the cellular slime molds.

Water molds produce flagellated reproductive cells

Water molds (phylum Oomycota) were once classified as fungi because of their superficial resemblance. Both water molds and fungi have a body, termed a **mycelium,** that grows over organic material, digesting it and then absorbing the predigested nutrients. The threadlike **hyphae** that make up the mycelium in water molds are coenocytic; there are no cross walls, and the body is like one giant multinucleate cell. The cell walls of water molds may be composed of cellulose (as in plants), chitin (as in fungi), or both. Most biologists classify the water molds as protists rather than as fungi.

When food is plentiful and environmental conditions are favorable, water molds reproduce asexually (Fig. 24–20). A hyphal tip swells and a cross wall is formed, separating the hyphal tip from the rest of the mycelium. Within this structure,

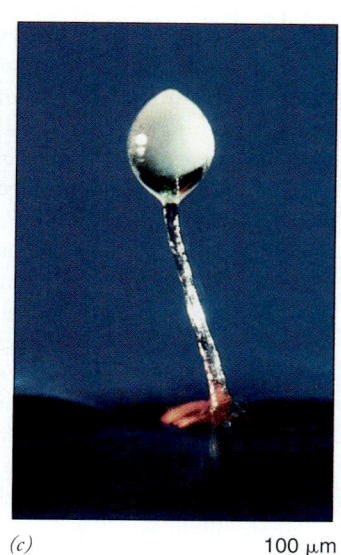

(a)   50 μm   (b)   50 μm

Figure 24–19   **The cellular slime mold *Dictyostelium discoideum*. (a)** Aggregation of amoeboid cells. **(b)** The aggregation organizes into a migrating pseudoplasmodium, called a slug, that eventually forms a stalked fruiting body. **(c)** The mature fruiting body releases spores, each of which opens to liberate an amoeboid cell.   *(a, b, c, Cabisco/Visuals Unlimited)*

(c)   100 μm

Zoosporangium

Zoospores

Primary zoospore (pear-shaped)

Encysted primary zoospore

ASEXUAL REPRODUCTION Haploid

Secondary zoospore (bean-shaped)

Encysted secondary zoospore

Germination of the zoospore

Mycelium

Antheridium (♂)

Sperm nucleus

SEXUAL REPRODUCTION

Enlarged hyphae

Oogonium (♀)

Egg nucleus

Germination of the oospore

Haploid

Diploid

MEIOSIS

FERTILIZATION

Oospores (develop from zygotes)

Figure 24–20   **Life cycle of *Saprolegnia* sp.** *Saprolegnia*, a water mold, reproduces both sexually by antheridia and oogonia, and asexually by zoosporangia.

tiny biflagellate zoospores form, each of which can develop into a new mycelium. When environmental conditions worsen, water molds initiate sexual reproduction. After fusion of male and female nuclei, a thick-walled **oospore** develops from the zygote. Water molds often overwinter as oospores.

Some water molds have played infamous roles in human history. For example, the Irish potato famine of the 19th century was precipitated by the water mold *Phytophthora infestans,* which causes late blight of potatoes. Because of several rainy, cool summers in Ireland in the 1840s, the water mold multiplied unchecked, causing potato tubers to rot in the fields. Since potatoes were the staple of the Irish peasants' diet, many people starved to death. Estimates of the number of deaths that resulted from the outbreak of this plant disease range from 250,000 to more than 1 million. The famine prompted a mass migration out of Ireland to such countries as the United States.

Late blight is still a problem today. Beginning in the late 1980s and 1990s, new strains of the water mold that causes it were reported in many states in the United States. The new strains have also appeared in parts of Canada, northern Europe, South America, Japan, South Korea, the Middle East, and Africa. These water mold strains are resistant to metalaxyl, the fungicide usually used to control the disease. Because today most people eat a varied diet, late blight is not expected to cause a famine, but it costs potato growers millions of dollars in lost crops.

THE EARLIEST EUKARYOTES WERE PROTISTS

Protists are thought to have been the first eukaryotic cells to evolve from ancestral prokaryotes. They appeared in the fossil record 1.9 to 2.1 billion years ago. A few protists with hardened shells, for example, diatoms and foraminiferans, produced abundant fossils. However, most ancient protists did not leave extensive fossil records because their bodies were too soft to leave permanent traces. Therefore, evolutionary studies of protists are based primarily on comparisons of present-day organisms, which contain many clues about their evolutionary history. Some of the most useful data for evolutionary interpretations come from electron microscopy, biochemistry, and molecular biology (see section on molecular comparisons among organisms in Chapter 17).

From an evolutionary perspective, one of the most interesting groups of modern-day protists are the **diplomonads,** which are zooflagellates that may have one of the more ancient lineages of extant (living) protists. Comparative ribosomal RNA sequencing has revealed that diplomonads such as *Giardia* (discussed earlier in the chapter) are more closely related to prokaryotes than are any other protists (see Fig. 24–1). Interestingly, *Giardia*'s cell structure—it possesses two haploid nuclei—suggests a partial explanation of how diploid eukaryotes may have arisen from haploid prokaryotes (Fig. 24–21).

2 μm

Figure 24–21 *Giardia intestinalis.* Colorized TEM of *Giardia* reveals two nuclei; this suggests that *Giardia*'s ancestors may have been an intermediate stage in the evolution of diploidy in eukaryotic life cycles. *Giardia intestinalis* is a parasite that can live in the small intestines of many vertebrates, including humans. *(E. White/Visuals Unlimited)*

Some biologists hypothesize that the first eukaryotes possessed a single haploid nucleus. Most eukaryotes today possess a single diploid nucleus that is formed at some stage in the life cycle when two haploid nuclei fuse. Perhaps the ancestors of *Giardia* represent an intermediate stage in eukaryotic evolution when cells each possessed two haploid nuclei but fusion had not yet occurred:

Haploid prokaryote → eukaryote with single haploid nucleus → eukaryote with two haploid nuclei → eukaryote with single diploid nucleus

Giardia also lacks mitochondria, which suggests that the nucleus may have evolved earlier than other membrane-bounded organelles such as mitochondria. Some molecular data dispute this idea, however. *Giardia* sp. contains a gene that codes for a protein thought to be associated with mitochondria in other organisms. This information suggests that an ancestor of *Giardia* may have possessed mitochondria, which were somehow lost at a later time during its evolutionary history.

How did mitochondria and other membrane-bounded organelles arise? The **endosymbiont theory,** popularized since the 1960s by Lynn Margulis, suggests that certain eukaryotic organelles (mitochondria, chloroplasts, and perhaps even cilia and flagella) arose from symbiotic relationships between larger prokaryotes and smaller prokaryotes that lived within them (see Fig. 20–6). Chloroplasts are thought to have originated from photosynthetic prokaryotes, and mitochondria from aerobic bacteria.

(a)   2.5 μm   *(b)*   50 μm

Figure 24–22   **Evolution of multicellularity.** Multicellularity may have evolved in this sequence: unicellular → colonial → multicellular organisms. *(a)* SEM of *Chlamydomonas,* a biflagellated unicellular green alga. *(b)* LM of *Volvox,* a colonial green alga composed of up to 50,000 *Chlamydomonas*-like cells. New colonies can be observed inside the parental colony, which eventually breaks apart to release them.   (a, *Courtesy of T.K. Maugel/University of Maryland; b, James W. Richardson/CBR Images*)

Multicellularity arose several times in the protist kingdom

Until recently, the timing of the origin of multicellularity during the Precambrian era has been uncertain, largely because of the scarcity of fossils. In the mid-1990s, however, hundreds of fossils of multicellular organisms were discovered in the Tuan-shanzi Formation in north China, which has been dated at 1.7 billion years old. The fossils, which range up to about 1 cm in length and resemble multicellular algae, with leaflike blades, stipes, and holdfasts, suggest that multicellularity arose at least 1.7 billion years ago.

Green algae, red algae, and brown algae are extant protist phyla with multicellular species. However, multicellular green algae have more in common with unicellular green algae than with other multicellular protists. Similarly, multicellular forms of both red and brown algae have little in common with each other or with the multicellular green algae. Because these groups are so different, it is likely that multicellularity arose several times within the kingdom Protista. That is, the multicellular green algae, red algae, and brown algae probably had different unicellular ancestors.

Studying the protists living today provides clues about how multicellularity may have arisen. Colonies, for example, may have been transitional stages between unicellular organisms and multicellular organisms. *Chlamydomonas* is a unicellular green alga that uses two flagella for motility (Fig. 24–22*a*). The green algae also include a number of colonial species that consist of attached *Chlamydomonas*-like cells. For example, *Gonium* is a colonial green alga that consists of four *Chlamydomonas*-like cells, and *Pandorina* is a colonial organism composed of 16 to 32 *Chlamydomonas*-like cells. The largest colonies of *Chlamydomonas*-like cells—from 1000 to 50,000 cells—are in the genus *Volvox* (Fig. 24–22*b*). Specialization in cell structure and function occurs in the larger colonies. *Volvox* sp., which has some division of labor among the cells (for example, only some cells are capable of carrying out reproduction), approaches multicellularity in complexity.

The **volvocine line** (the series of *Chlamydomonas*-like organisms just discussed) is an evolutionary dead end; that is, it did not give rise to multicellular green algae. However, the trend in increasing colony size and cell differentiation within the volvocine line indicates one possible way that multicellularity may have originated:

Single cells → colonies → multicellular organisms

I. Kingdom Protista is composed of "simple" eukaryotic organisms, most of which are unicellular and live in aquatic environments.
   A. **Protists** range in size from microscopic single cells to multicellular organisms 75m long.
   B. Protists obtain their nutrients autotrophically or heterotrophically.
   C. Protists may be free-living or symbiotic, with symbiotic relationships ranging from mutualism to parasitism.
   D. Many protists reproduce both sexually and asexually; others reproduce only asexually.
   E. Protists have various means of locomotion, including pseudopodia, flagella, and cilia. A few are nonmotile.

II. **Protozoa** are heterotrophic, animal-like protists.
   A. **Amoebas** move and obtain food using cytoplasmic extensions called **pseudopodia.**
   B. **Foraminiferans** secrete many-chambered **tests** (shells) with pores through which cytoplasmic projections extend to move and obtain food.
   C. **Actinopods** obtain food by means of **axopods,** slender cytoplasmic projections that extend through pores in their skeletons.
   D. **Zooflagellates** are heterotrophic protozoa that move by means of whiplike **flagella.**
   E. **Ciliates** move by hairlike **cilia,** have **micronuclei** (for sexual reproduction) and **macronuclei** (for controlling cell metabolism and growth), and undergo complex reproduction.
   F. **Apicomplexans** are parasites that produce **spores** and are nonmotile. An apicomplexan causes malaria.

III. **Algae** are autotrophic, plantlike protists.
   A. **Euglenoids** are unicellular, flagellated algae. Many are not photosynthetic.
   B. **Dinoflagellates** are mostly unicellular, biflagellated, photosynthetic organisms of great ecological importance as major producers in marine ecosystems. Some dinoflagellates produce toxic blooms known as red tides.
   C. **Diatoms,** which are major producers in aquatic ecosystems, are mostly unicellular, with shells containing silica.

   D. **Golden algae** are mostly unicellular, biflagellated freshwater and marine algae that are of ecological importance as a major component of the ocean's nanoplankton.
   E. **Brown algae** are multicellular seaweeds that are ecologically important in cooler ocean waters. They produce flagellated cells during their complex reproductive cycles.
   F. **Green algae** exhibit a wide diversity in size, structural complexity, and reproduction. They share many similarities with plants, and it is thought that ancestral green algae gave rise to plants.
   G. **Red algae,** which are mostly multicellular seaweeds, are ecologically important in warm tropical ocean waters. They lack motile cells and have complex reproductive cycles.

IV. Fungus-like protists were originally classified with the fungi but have several protist features.
   A. The feeding stage of **plasmodial slime molds** is a multinucleate **plasmodium.** Reproduction is by spores.
   B. **Cellular slime molds** feed as individual amoeboid cells. They reproduce by aggregating into a **pseudoplasmodium** (slug), then forming asexual spores.
   C. **Water molds** have a **coenocytic mycelium.** They reproduce asexually by forming biflagellate **zoospores** and sexually by forming **oospores.**

V. Protists originated about 2.1 billion years ago and were the first eukaryotes.
   A. Based on RNA sequencing data and cell ultrastructure studies, it is thought that **diplomonads,** which are a group of zooflagellates, may have one of the more ancient lineages of extant protists.
   B. Some of the earliest fossils of multicellular organisms, dated at 1.7 billion years, were discovered in the Tuanshanzi Formation in north China. The fossils resemble multicellular algae.
   C. Multicellularity probably evolved several times within the kingdom Protista, possibly by following a trend in greater complexity from single cells to colonies to multicellular organisms, as represented by the **volvocine line.**

1. Amoebas move and obtain food by means of (a) pseudopodia (b) flagella (c) cilia (d) gametangia (e) trichocysts
2. Foraminiferans (a) are endosymbionts in many marine invertebrates (b) were responsible for the Irish potato famine in the 19th century (c) secrete many-chambered tests with pores through which cytoplasmic extensions project (d) possess numerous trichocysts that may aid in trapping and holding prey (e) contain phycocyanin and phycoerythrin, pigments found in no other protist group
3. Unicellular protozoa that are free-living or parasitic, move by means of flagella, and do not photosynthesize are called (a) euglenoids (b) dinoflagellates (c) myxamoebas (d) zooflagellates (e) apicomplexans
4. *Paramecium* and other ciliates often display a sexual phenomenon called (a) oogamy (b) conjugation (c) anisogamy (d) red tide (e) macronucleus
5. Parasitic protozoa that form spores at some stage in their life belong to which group? (a) actinopods (b) ciliates (c) coccolithophorids (d) apicomplexans (e) dinoflagellates
6. Algae characterized by two flagella, one wrapped around the center of the cell like a belt and the other projecting behind the cell, are known as (a) actinopods (b) ciliates (c) coccolithophorids (d) apicomplexans (e) dinoflagellates

7. Photosynthetic protists with shells composed of two halves that fit together like a petri dish are known as (a) golden algae (b) diatoms (c) euglenoids (d) brown algae (e) foraminiferans
8. Chlorophyll *a*, chlorophyll *b*, and carotenoids are found in (a) green algae, red algae, and plants (b) green algae, euglenoids, and plants (c) brown algae, green algae, and golden algae (d) brown algae, diatoms, and golden algae (e) green algae, euglenoids, and diatoms
9. Which protists have photosynthetic pigments similar to those of the cyanobacteria? (a) golden algae (b) diatoms (c) euglenoids (d) brown algae (e) red algae
10. The multicellular bodies of _____ are differentiated into blades, stipes, holdfasts, and gas-filled floats. (a) golden algae (b) diatoms (c) euglenoids (d) kelps (e) green algae
11. The feeding stage of plasmodial slime molds is a multinucleate (a) plasmodium (b) pseudoplasmodium (c) pseudopodium (d) gametangium (e) mycelium
12. Water molds reproduce asexually by forming biflagellate _____, and sexually by forming _____. (a) oospores; holdfasts (b) zoospores; zooxanthellae (c) zoospores; oospores (d) holdfasts; isogametes (e) oospores; isogametes

REVIEW QUESTIONS

1. What are the features of a typical protist? Why are protists so difficult to classify?
2. Why are protists important to humans? How are they important ecologically?
3. What are the three main informal groups of protists? Describe and give at least three examples from each group.
4. Distinguish among amoebas, foraminiferans, and actinopods, and among zooflagellates, ciliates, and apicomplexans.
5. Distinguish among dinoflagellates, diatoms, golden algae, and euglenoids, and among green algae, red algae, and brown algae.
6. Distinguish among plasmodial slime molds, cellular slime molds, and water molds. What features do they share?
7. How is the structure of each of the following types of protists related to its way of life? (a) zooflagellates (b) green algae (c) water molds
8. Use the volvocine line to explain how multicellularity may have arisen within the protists.
9. Discuss the uncertainties surrounding the evolutionary relationships between the protists and the other eukaryotic kingdoms.
10. Label the diagram. Use Fig. 24–6*b* to check your answers.

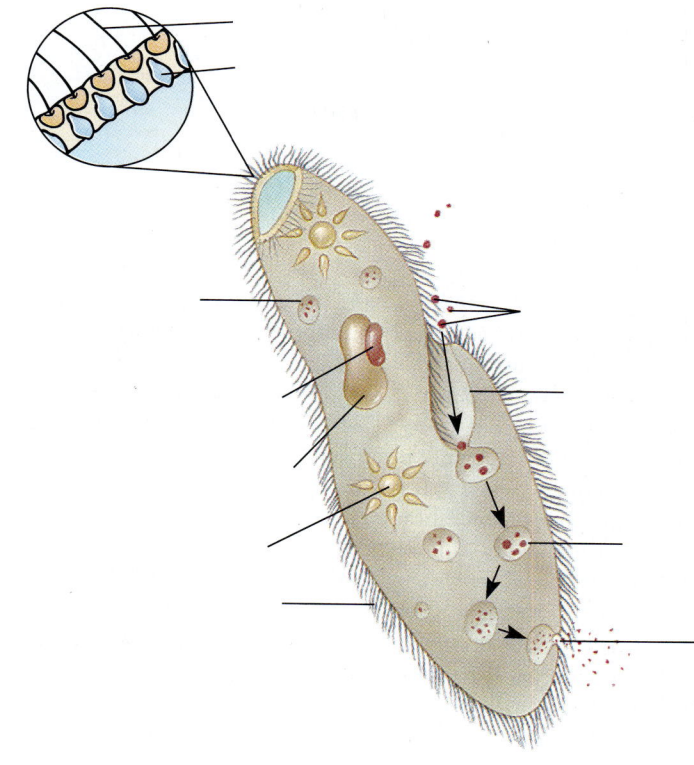

YOU MAKE THE CONNECTION

1. Some biologists still classify the protozoa as animals, the algae as plants, and the water molds as fungi. Explain their rationale. Why do most biologists classify these organisms as protists?

2. Molecular analyses of DNA sequences in plant chloroplasts and in free-living cyanobacteria are compellingly similar. How do these data support the endosymbiont theory?

RECOMMENDED READINGS

Broad, W.J. "A Spate of Red Tides Menaces Coastal Seas." *New York Times*, 27 Aug., 1996. Red tides have proliferated in recent years, possibly the result of increased pollution.

Fry, W.E. and S.B. Goodwin. "Resurgence of the Irish Potato Famine Fungus." *BioScience*, Vol. 47, No. 6, Jun. 1997. Fungicide-resistant strains of the water mold that causes late blight are causing problems worldwide.

Kunzig, R. "Invisible Garden." *Discover*, April 1990. Although they are invisible, microscopic algae are ecologically important to the biosphere.

Margulis, L., and R. Gurrero. "Kingdoms in Turmoil." *New Scientist*, 23 Mar., 1991. Highlights the widely differing interpretations of protist taxonomy.

Mlot, C. "The Rise in Toxic Tides." *Science News*, Vol. 152, 27 Sept., 1997. Discusses the problems stemming from an increase in harmful algal blooms over the past two decades.

Pelley, J. "What Is Causing Toxic Algal Blooms?" *Environmental Science and Technology*, 1 Jan., 1998. Examines possible links between pollution and blooms of *Pfiesteria*-like organisms.

Richardson, L.L. "Remote Sensing of Algal Bloom Dynamics." *BioScience*, Vol. 46, No. 7, Jul./Aug. 1996. Satellite sensors are increasingly being used to measure algal pigments as a way to study regional aquatic ecosystems.

Sharnoff, S.D. "Beauties from a Beast: Woodland Jekyll and Hydes." *Smithsonian*, July 1991. Beautiful photographs of the sporangia of slime molds.

● Visit our website at **http://www.saunderscollege.com/lifesci/titles.html** and click on Solomon/Berg/Martin Biology for links to chapter-related resources on the World Wide Web.

CHAPTER 25

Kingdom Fungi

Mushrooms, morels, and truffles, delights of the gourmet, have much in common with the black mold that forms on stale bread and the mildew that collects on damp shower curtains. These life forms belong to the kingdom Fungi, a diverse group of more than 81,000 known species, most of which are terrestrial. Although they vary strikingly in size and shape, all **fungi** (sing. *fungus*) are eukaryotes; their cells contain membrane-bounded nuclei, mitochondria, and other organelles. Like the cells of bacteria, certain protists, and plants, fungal cells are enclosed by cell walls during at least some stage in their life cycle. Fungal cell walls, however, have a different chemical composition than cell walls of other organisms. In most fungi, the cell wall is composed of complex carbohydrates, including **chitin,** a polymer that consists of subunits of a nitrogen-containing sugar (see Fig. 3–10). Chitin, which is coincidentally a component of the external skeletons of insects and other arthropods, is far more resistant to breakdown by microorganisms than are cellulose and lignin, which make up plant cell walls.

Fungi lack chlorophyll and are nonphotosynthetic. As heterotrophs, they cannot produce their own organic materials from a simple carbon source (carbon dioxide) but instead obtain preformed carbon molecules produced by other organisms. Fungi, however, do not ingest food as animals do, but secrete digestive enzymes and then absorb the predigested food (as small organic molecules) through their cell walls and plasma membranes. They obtain their nutrients from dead organic matter (as decomposers) or from other living organisms (as parasites). As decomposers, fungi, along with the bacteria (see Chapter 23), play an important ecological role.

Fungi grow best in moist habitats, but they are found universally wherever organic material is available. They require moisture to grow, and they can obtain water from a humid atmosphere as well as from the medium on which they live. When the environment becomes extremely dry, fungi survive by going into a resting stage or by producing spores that are resistant to desiccation (drying out). The rounded earthstar *(Geastrum saccatum)* shown in the photograph, for example, releases a puff of microscopic spores after the sac wall is hit by a raindrop. This fungus is common in leaf litter under trees.

Although the optimum pH for most species is about 5.6, different fungi can tolerate and grow in environments where the pH ranges from 2 to 9. Many fungi are less sensitive to high osmotic pressures than are bacteria. As a result, they can grow in concentrated salt solutions, or in sugar solutions such

(Jeff Lepore/Photo Researchers, Inc.)

as jelly, that discourage or prevent bacterial growth. Fungi may also thrive over a wide temperature range. Even refrigerated food is not immune to fungal invasion.

Fungi were classified originally in the plant kingdom, but biologists today recognize that they are not plants, and recent studies suggest that fungi are related more closely to animals than to plants. Because fungi are distinct from plants, animals, and protists in many ways, they are assigned to a separate kingdom.

AFTER YOU HAVE STUDIED THIS CHAPTER YOU SHOULD BE ABLE TO

1. Describe the distinguishing characteristics of the kingdom Fungi.
2. Describe the body plan of most fungi.
3. Trace the fate of a fungal spore that lands in an appropriate location, such as an overripe peach, and describe conditions that permit fungal growth.
4. Explain why biologists think chytridiomycetes may have been the earliest fungal group to have evolved from the flagellated protist hypothesized as the common ancestor of fungi.
5. List distinguishing characteristics, describe a typical life cycle,

and give examples of each of the following fungal groups: chytridiomycetes, zygomycetes, ascomycetes, basidiomycetes, and imperfect fungi.
6. Characterize the unique nature of a lichen.
7. Explain the ecological significance of fungi as decomposers.
8. Describe the special ecological role of mycorrhizae.
9. Summarize the various uses of fungi.
10. Identify at least three fungal diseases of plants and three of animals.

MOST FUNGI HAVE A FILAMENTOUS BODY PLAN

The vegetative (nonreproductive) body plan of most fungi consists of long, branched, threadlike filaments called **hyphae** (sing., *hypha*) (Fig. 25–1*a,b*). Hyphae form a tangled mass or tissue-like aggregation known as a **mycelium** (pl., *mycelia*). The cobweb-like mold sometimes seen on bread is the mycelium of a fungus. What is not seen is the extensive mycelium that grows down into the bread.

Some hyphae are **coenocytic;** that is, they are not divided into individual cells and are instead an elongated, multinucleated giant cell (Fig. 25–1*c*). Other hyphae are divided by cross walls called **septa** (sing., *septum*) into individual cells containing one or more nuclei (Fig. 25–1*d,e*). The septa of many sep-

tate fungi are perforated by a pore that may be large enough to permit organelles to flow from cell to cell. Cytoplasm flows within the hypha, providing a system of internal transport.

MOST FUNGI REPRODUCE BY SPORES

Most fungi reproduce by means of microscopic **spores,** which are nonmotile reproductive cells dispersed by wind, water, or animals. Spores are usually produced on specialized aerial hyphae or in fruiting structures. This arrangement permits the spores to be more easily dispersed. The aerial hyphae of some fungi form large, complex reproductive structures, called **fruiting bodies,** in which spores are produced. The familiar part of a mushroom or toadstool is a large fruiting body. We do

Hyphae

(a)

(b)

25 µm

(c)

(d)

(e)

Figure 25–1 **The fungus body plan.** *(a)* A fungal mycelium growing on agar in a culture dish. In nature, fungal mycelia are rarely so symmetrical. *(b)* SEM of a mycelium, which consists of a mass of threadlike hyphae. *(c)* A coenocytic hypha. *(d)* A hypha divided into cells by septa; each cell is monokaryotic (has one nucleus). In some taxa the septa are perforated (as shown), permitting cytoplasm to stream from one cell to another. *(e)* A septate hypha in which each cell is dikaryotic (has two genetically distinct nuclei). *(a, Dennis Drenner; b, G.T. Cole, University of Texas/BPS)*

Spore

Hypha

Mycelium

Figure 25–2 Germination of a spore to form a mycelium.

When a fungal spore comes into contact with an appropriate food source, perhaps an overripe peach that has fallen to the ground, it germinates and begins to grow (Fig. 25–2). A threadlike hypha emerges from the tiny spore and grows, branching frequently. Soon a mycelium infiltrates the peach, degrading its organic compounds to small molecules that the fungus can absorb. Fungi are extremely efficient at converting nutrients into new cell material. If excessive amounts of nutrients are available, fungi are able to store them, usually as lipid droplets or glycogen, in the mycelium.

FUNGI HAVE EVOLVED A VARIETY OF SPORES AND FRUITING BODIES

Classification of fungi traditionally was based mainly on the characteristics of their sexual spores and fruiting bodies, but molecular data, such as comparative DNA and RNA sequences, are playing an increasingly important role in determining relationships among fungi. Most biologists assign these diverse organisms to five phyla: Chytridiomycota, Zygomycota, Ascomycota, Basidiomycota, and Deuteromycota. Table 25–1 summarizes fungal classification. Slime molds and water molds were originally classified as fungi, but are now generally considered protists (see Chapter 24).

Chytridiomycetes are the most primitive fungi

Until recently, **chytridiomycetes** (phylum Chytridiomycota), also called *chytrids,* were considered by most biologists to be fungus-like protists similar in many respects to the water molds (see Chapter 24). However, molecular comparisons, particularly of DNA and RNA sequences, have provided compelling evidence that the approximately 750 species of chytridiomycetes are closely allied to fungi and not closely related to

not normally see the bulk of the fungus, a nearly invisible mycelium buried out of sight in the rotting material or soil on which it grows.

Fungi may produce spores either sexually or asexually. Unlike most animal and plant cells, fungal cells usually contain haploid nuclei. In sexual reproduction, the hyphae of two genetically compatible mating types come together, and their cytoplasm and nuclei fuse, forming a diploid cell known as a **zygote.**[1] In two fungal groups, the ascomycetes and basidiomycetes, the hyphae fuse, but the two different nuclei do not fuse immediately; rather, they remain separate within the fungal cytoplasm. Hyphae that contain two genetically distinct, sexually compatible nuclei within each cell are **dikaryotic,** a condition that is described as $n + n$ rather than $2n$ because there are two separate haploid nuclei. Hyphae that contain only one nucleus per cell are said to be **monokaryotic.**

[1] In certain fungi (*Allomyces,* for example), zygotes are typically formed by the union of two haploid gametes rather than two haploid nuclei in compatible hyphae.

TABLE 25–1	Phyla of Kingdom Fungi		
Phylum	Common Types	Axexual Reproduction	Sexual Reproduction
Chytridiomycota (chytridiomycetes)	*Allomyces*	Zoospores	Flagellated gametes in some chytrids
Zygomycota (zygomycetes)	Black bread mold	Nonmotile spores form in a sporangium	Zygospores
Ascomycota (ascomycetes or sac fungi)	Yeasts, powdery mildews, molds, morels, truffles	Conidia pinch off from conidiophores	Ascospores
Basidiomycota (basidiomycetes or club fungi)	Mushrooms, bracket fungi, puffballs, rusts, smuts	Uncommon	Basidiospores
Deuteromycota (deuteromycetes or imperfect fungi)	Molds	Conidia	Sexual stage not observed

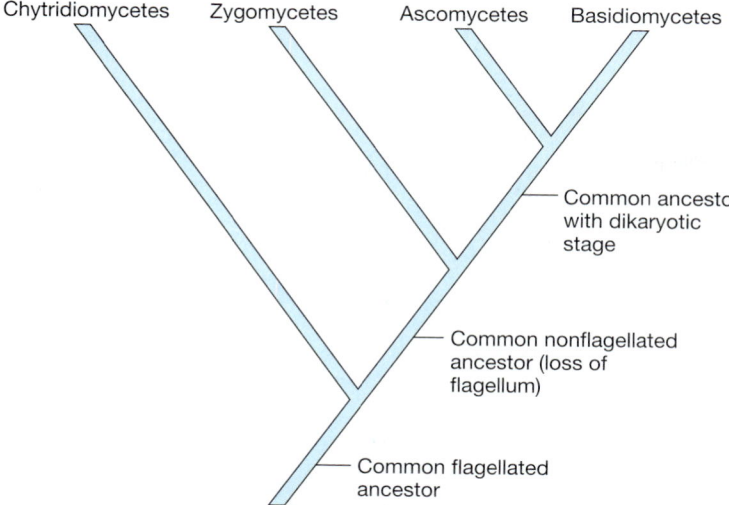

Chytridiomycetes   Zygomycetes   Ascomycetes   Basidiomycetes

Common ancestor with dikaryotic stage

Common nonflagellated ancestor (loss of flagellum)

Common flagellated ancestor

Figure 25–3   **Fungal evolution.** This diagram shows phylogenetic relationships among living fungi, based on current scientific evidence.

water molds or other fungus-like protists. The molecular evidence also indicates that chytridiomycetes, which are small, relatively simple aquatic fungi, are the most primitive group of fungi living today; *primitive* in this context implies that chytridiomycetes were probably the earliest fungal group to evolve from the ancient flagellated protist that has been hypothesized as the common ancestor of all fungi. Chytridiomycetes, which produce flagellated cells at some stage in their life history, have presumably retained this feature of their protist ancestor. All other fungal phyla are nonflagellated and apparently lost the ability to produce motile cells at some point in their evolutionary past, perhaps during the transition from aquatic to terrestrial habitats (Fig. 25–3).

Chytridiomycetes are parasites or decomposers that are found principally in fresh water, although a few species occur on land (in moist environments) and in marine water. Like other fungi, chytridiomycetes have cell walls containing chitin and store their food reserves as glycogen. Their motile cells possess a single, posterior flagellum. They reproduce both sexually and asexually.

One common chytridiomycete is *Allomyces,* which has an unusual life cycle compared to most fungi because it has an **alternation of generations,** spending part of its life as a haploid (*n*) **thallus** and part as a diploid (*2n*) thallus (Fig. 25–4). (The term *thallus* is used to describe the simple body plan of certain algae, fungi, or plants.) The haploid and diploid thalli are similar in appearance and consist of a stout trunklike part with slender branches. The haploid thallus bears male and female **gametangia,** which are structures in which gametes are formed, at the tips of the branches. When a flagellated male gamete fuses with a slightly larger, flagellated female gamete, the resulting zygote develops into a diploid thallus. The diploid thallus bears two kinds of spore cases, zoosporangia and resting sporangia. Zoosporangia produce flagellated diploid **zoospores** that may settle down and develop into new diploid thalli. Meiosis occurs within resting sporangia to form haploid zoospores, each of which has the potential to settle down and develop into a haploid thallus.

Zygomycetes reproduce sexually by forming zygospores

About 800 species of **zygomycetes** (phylum Zygomycota) produce sexual spores, called **zygospores.** Their hyphae are coenocytic; that is, they lack regularly spaced septa. Septa do form, however, to separate the hyphae from reproductive structures. Most zygomycetes are decomposers that live in the soil on decaying plant or animal matter, although some are parasites of plants and animals and others are mycorrhizal fungi (discussed shortly).

One common zygomycete is the black bread mold, *Rhizopus stolonifer,* a decomposer that breaks down bread and other foods (Fig. 25–5). If preservatives are not added, bread left at room temperature often becomes covered with a black, fuzzy growth in a few days. Bread becomes moldy when a spore falls on it and then germinates and grows into a mycelium. Hyphae penetrate the bread and absorb nutrients. Eventually, certain hyphae grow upward and develop spore sacs called **sporangia** (sing., *sporangium*) at their tips. Clusters of black asexual spores develop within each sporangium and are released when the delicate sporangium ruptures. The spores give the black bread mold its characteristic color.

Sexual reproduction in the black bread mold occurs when the hyphae of two different mating types, designated as plus (+) and minus (−), grow into contact with one another. The bread mold is **heterothallic,** meaning that an individual fungal hypha is self-sterile and mates only with a hypha of a different mating type. That is, sexual reproduction occurs only between a member of a (+) strain and one of a (−) strain, not between members of two (+) strains or members of two (−)

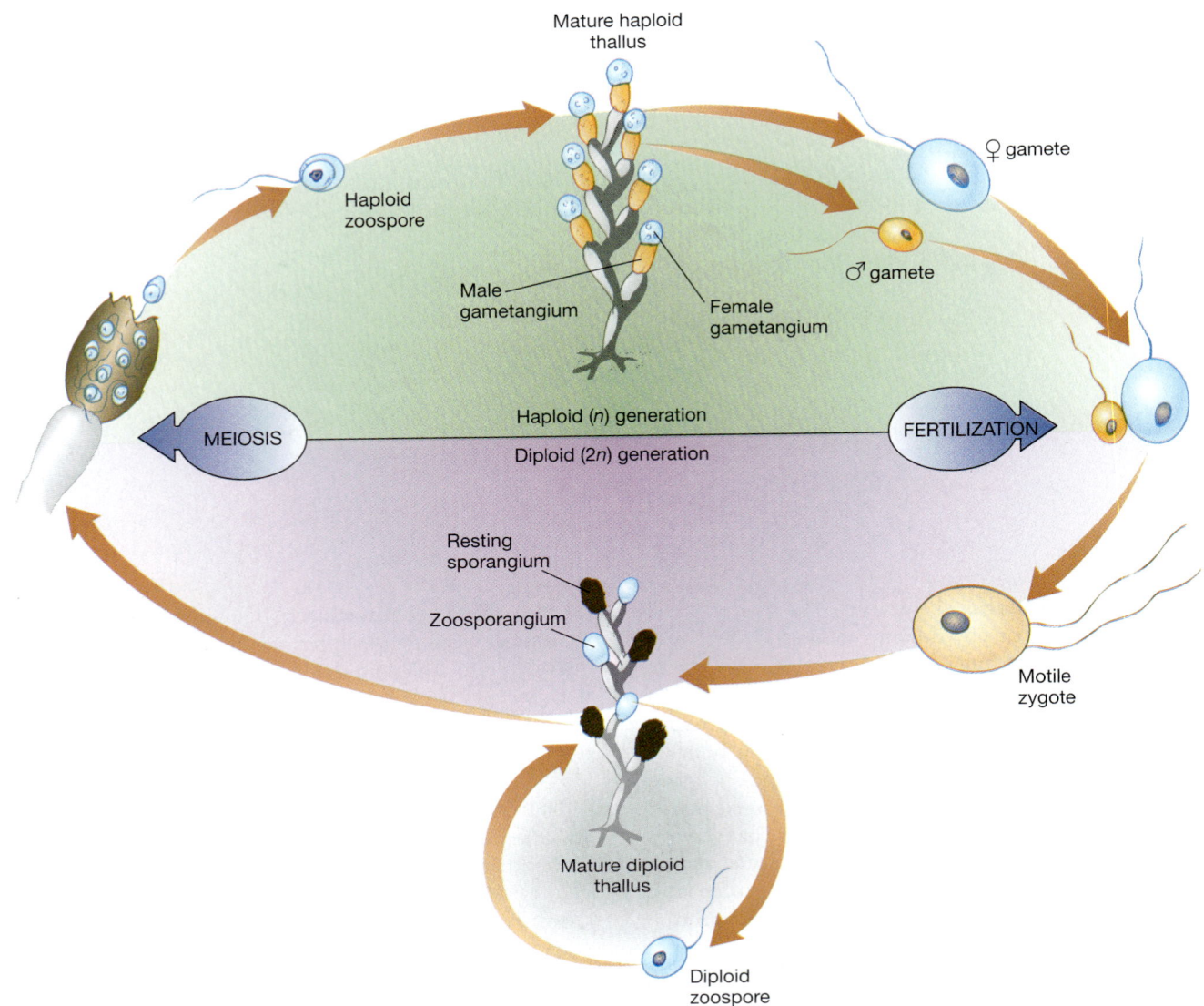

Mature haploid
thallus

♀ gamete

♂ gamete

Haploid
zoospore

Male
gametangium

Female
gametangium

MEIOSIS

Haploid (n) generation

Diploid (2n) generation

FERTILIZATION

Resting
sporangium

Zoosporangium

Motile
zygote

Mature diploid
thallus

Diploid
zoospore

Figure 25–4 **The life cycle of *Allomyces arbuscula*.** *(Top)* *Allomyces* alternates between haploid and diploid stages, which are similar in appearance. The haploid thallus gives rise to male and female gametes that fuse and subsequently develop into the diploid thallus. Meiosis occurs in the diploid resting sporangia to form haploid zoospores that develop into haploid thalli, and the cycle continues. *(Bottom)* *Allomyces* also reproduces asexually by forming diploid zoospores.

strains. Because there are no physical differences between the two mating types, it is not appropriate to refer to them as "male" and "female."

When hyphae of opposite mating types grow in close proximity, hormones are produced that cause the tips of the hyphae to come together and form specialized sexual structures known as gametangia. These structures then unite, and the (+) and (−) nuclei fuse to form a diploid nucleus, the zygote. A zygospore develops, consisting of a thick protective covering around the zygote. The zygospore may lie dormant for several months and can survive desiccation and extreme temperatures. Meiosis probably occurs at or just before germination of the zygospore. When the zygospore germinates, an aerial

hypha develops with a haploid sporangium at the tip. Mitosis within the sporangium produces haploid spores that are released and that may germinate to form new hyphae. Only the zygote of a black bread mold is diploid; all of the hyphae and the asexual spores are haploid.

Ascomycetes reproduce sexually by forming ascospores

Ascomycetes (phylum Ascomycota) comprise a large group of fungi consisting of about 30,000 described species. Ascomycetes are sometimes referred to as *sac fungi* because their sexual spores are produced in microscopic sacs called **asci**

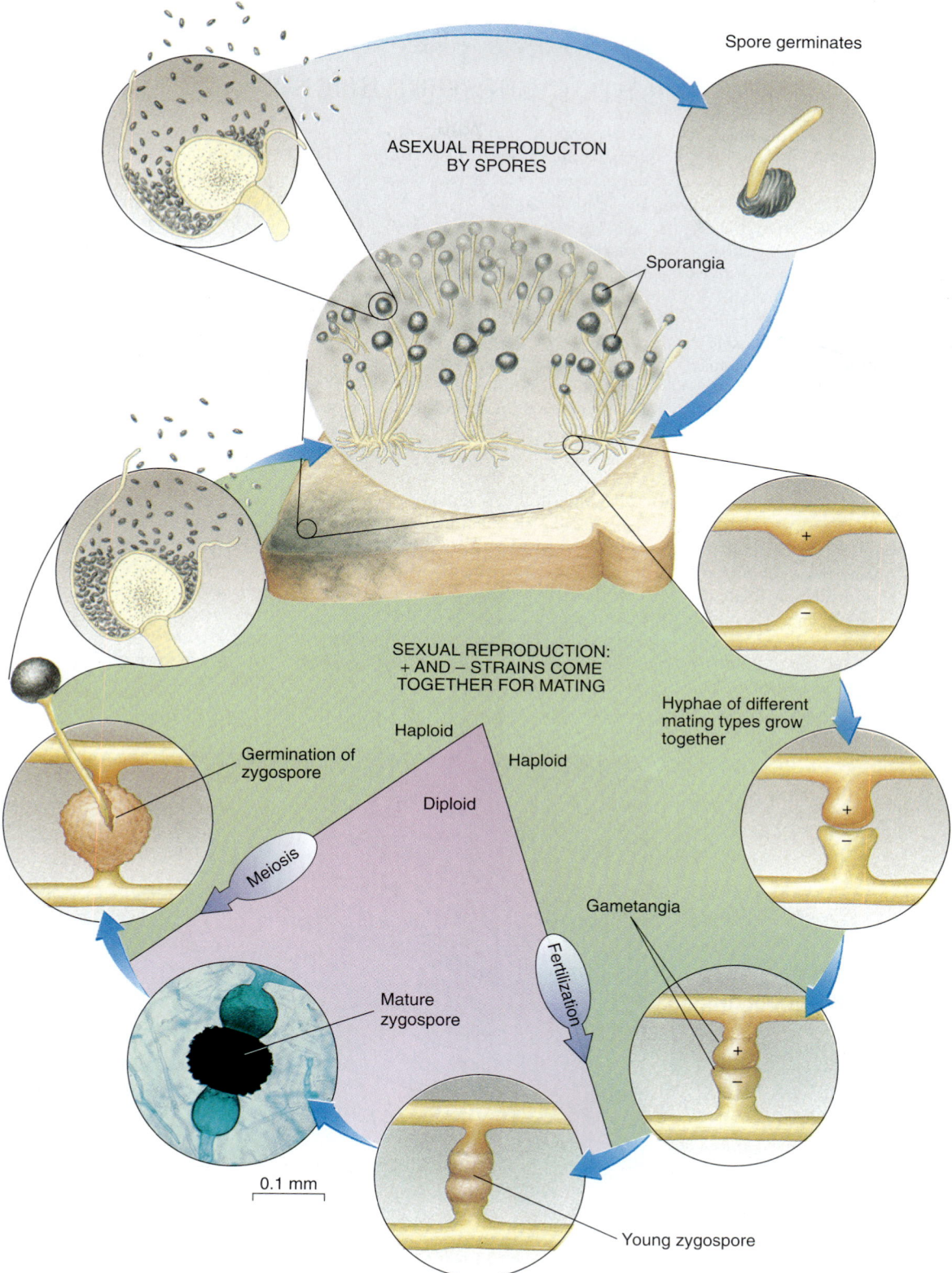

Spore germinates

ASEXUAL REPRODUCTON
BY SPORES

Sporangia

SEXUAL REPRODUCTION:
+ AND − STRAINS COME
TOGETHER FOR MATING

Hyphae of different
mating types grow
together

Germination of
zygospore

Haploid

Haploid

Diploid

Meiosis

Fertilization

Gametangia

Mature
zygospore

0.1 mm

Young zygospore

Figure 25–5 **The life cycle of the black bread mold** *(Rhizopus stolonifer)*. *(Top)* Asexual reproduction involves the formation of haploid spores. *(Bottom)* Sexual reproduction takes place only between different mating types, designated (+) and (−). After their hyphae meet and the nuclei fuse,
a zygospore develops *(inset, stained LM)*. *(Inset, Ed Reschke)*

FIGHTING CHESTNUT BLIGHT

Prior to the twentieth century, the American chestnut *(Castanea dentata)* was a very important tree, both ecologically and economically, in eastern forests. Throughout its natural range, which extended from Maine to Georgia and west to Illinois, the chestnut tree provided food and shelter for forest animals, including squirrels, birds, and insects. The tree grew rapidly, forming a straight, tall trunk that provided valuable, decay-resistant wood for house foundations and interior trim, furniture, railroad ties, and fence posts. The wood was also an important source of tannin, which is a chemical used to convert animal hides into leather. The edible nuts of the American chestnut were harvested, roasted, and eaten directly or ground into flour.

Early in this century the chestnut blight fungus *(Cryphonectria parasitica),* an ascomycete, was accidentally imported on diseased oriental chestnuts and quickly attacked native chestnuts, which had no resistance to the fungus. First identified in New York in 1904, the blight had killed or damaged several billion mature trees—almost every North American chestnut tree throughout its entire natural range—by the late 1940s (see figure).

A spore of the fungus enters a chestnut tree through a wound or crack in the bark, and the hyphae grow rapidly, attacking the inner bark. When the trunk is girdled (completely encircled) by the fungus, the portion of the plant above the injury dies. The fungus does not kill the roots, however, and they periodically produce new shoots, only to be killed back by the blight once again. The spores are carried by wind, water, and animals from tree to tree.

Although the American chestnut is on the brink of extinction, there are hopes for its eventual recovery. Some biologists have bred the American chestnut with the Chinese chestnut, a related species that is resistant to chestnut blight. With careful selective breeding over many years, it may be possible to develop a disease-resistant variety of American chestnut.

Some biologists are approaching the problem in a different way, by trying to isolate strains of the chestnut blight fungus that are less virulent (less deadly). A less virulent strain of the chestnut blight fungus was first identified in Europe in the 1950s; some of the trees there have been able to recover from the disease. More recently, a less virulent strain has appeared in the United States, and, even more encouraging, it appears that the less virulent strain is slowly spreading, replacing the more deadly strain. It is not known *why* the less virulent strain is replacing the more virulent one.

Biologists determined that the reduced virulence of the chestnut blight fungus is caused by the presence of a virus. They genetically engineered a synthetic form of the virus and introduced it into the virulent strain of the chestnut blight fungus, which subsequently lost its virulence. They plan to treat infected chestnut trees with the benign fungal strain, which they hope will spread rapidly and eventually replace the virulent strain.

American chestnuts. These massive trees, which were photographed in the late 1800s in the Great Smoky Mountains, and almost all other American chestnuts were killed by the chestnut blight fungus during the first part of the 20th century. *(Inset)* Chestnuts that sprout from the roots do not grow very large before they are killed back by the chestnut blight. *(Photo courtesy of the American Chestnut Foundation; inset, Marion Lobstein)*

(sing., *ascus*). Their hyphae usually have septa, but these cross walls have pores so that cytoplasm is continuous from one compartment to another.

The diverse ascomycetes include most yeasts; the powdery mildews; most of the blue-green, pink, and brown molds that cause food to spoil; decomposer cup fungi; and the edible morels and truffles. Some ascomycetes cause serious plant diseases such as Dutch elm disease, ergot disease on rye, powdery

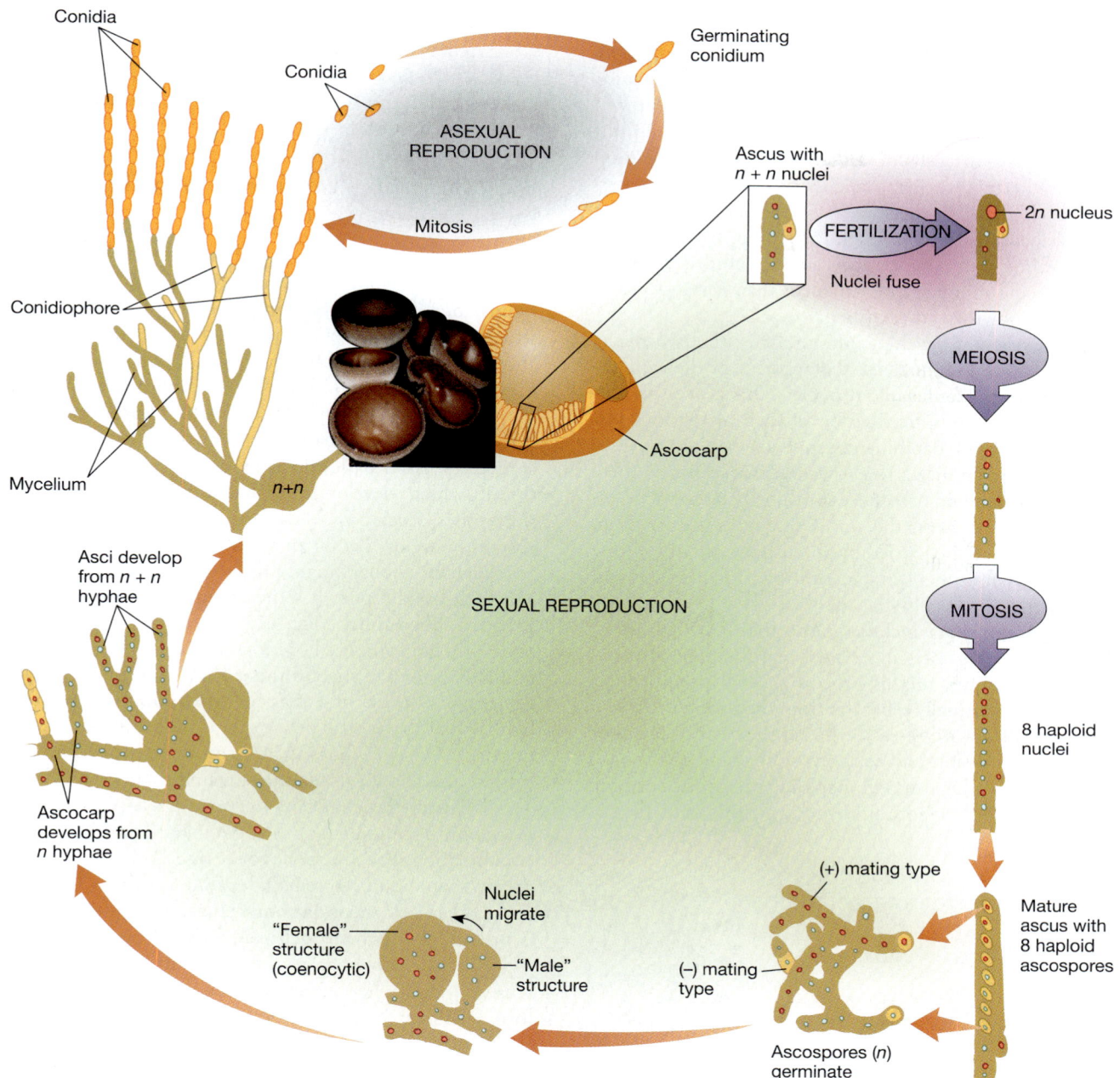

Figure 25–6 The life cycle of ascomycetes. *(Top)* Asexual reproduction involves the formation of haploid conidia. *(Bottom)* Sexual reproduction involves the fusion of two haploid hyphae of different mating types to form dikaryotic (*n* + *n*) hyphae from which asci develop. In each ascus the two nuclei fuse, followed by meiosis and mitosis to produce eight ascospores. The asci form the inner layer of a fruiting body called the ascocarp. Shown are ascocarps of the rainforest cup fungus (*Cookeina sulcipes*). *(James L. Castner)*

mildew on fruits and ornamental plants, and chestnut blight (see *Focus On: Fighting Chestnut Blight*).

In most ascomycetes, asexual reproduction involves production of spores called **conidia** (sing., *conidium*), which are formed at the tips of certain specialized hyphae known as **conidiophores** (from the Greek, meaning "dust-bearers") (Fig. 25–6, top). Conidia are a means of rapidly propagating new

mycelia when environmental conditions are favorable. Conidia occur in various shapes, sizes, and colors in different species. The color of the conidia is responsible for the characteristic brown, blue-green, pink, or other tints of many of these molds.

Some species of ascomycetes are heterothallic and have different mating strains; others are **homothallic,** which means that they are self-fertile and have the ability to mate with themselves. In both heterothallic and homothallic ascomycetes, sexual reproduction takes place after two hyphae (or, in some species, gametangia) grow together and their cytoplasm mingles (Fig. 25–6, bottom). Within this fused structure, the two haploid nuclei, one from each parent hypha, pair but do not fuse. New hyphae, the cells of which are dikaryotic, develop from the fused structure and branch repeatedly until the hyphal tips reach the site where asci will be produced. As the many sac-shaped asci develop, each containing two dissimilar nuclei (one from each parent), they are surrounded by intertwining haploid (monokaryotic) hyphae that develop into a fruiting body known as an **ascocarp.**

Within a cell that develops into an ascus, the two nuclei fuse and form a diploid nucleus, the zygote that then undergoes meiosis to form four haploid nuclei. This process is usually followed by one mitotic division of each of the four nuclei, resulting in eight haploid nuclei. Each haploid nucleus becomes incorporated into a heavy-walled **ascospore,** so there are usually eight haploid ascospores within the ascus. The ascospores are released when the tip of the ascus breaks open. Individual ascospores are carried by air currents, often for long distances. If one lands in a suitable location, it germinates and forms a new mycelium.

Phylum Ascomycota includes more than 300 species of unicellular **yeasts** (Fig. 25–7). Asexual reproduction of yeasts is mainly by **budding;** in this process a small protuberance (bud) grows and eventually separates from the parent cell. Each bud can grow into a new yeast cell. Some yeasts also reproduce asexually by fission and sexually by forming ascospores. During sexual reproduction, two haploid yeasts fuse, forming a diploid zygote. The zygote undergoes meiosis, and the re-

10 μm

Figure 25–7 **LM of baker's yeast (*Saccharomyces* sp.).** These unicellular ascomycetes possess mitochondria and carry out aerobic respiration when oxygen is present. In the absence of oxygen, yeasts carry out alcohol fermentation, which is used to produce alcohol and make bread rise. *(Dwight R. Kuhn)*

sulting haploid nuclei are incorporated into ascospores. These spores remain enclosed for a time within the original diploid cell wall, which corresponds to an ascus. Yeasts are essential in making bread and fermenting alcoholic beverages (discussed later in this chapter).

Basidiomycetes reproduce sexually by forming basidiospores

The 25,000 or more species of **basidiomycetes** (phylum Basidiomycota) include the most familiar of the fungi: the mushrooms, bracket fungi, and puffballs (Fig. 25–8). Some destructive plant parasites of important crops, such as wheat rust and corn smut, are also basidiomycetes. Basidiomycetes, sometimes called *club fungi,* derive their name from the fact that they develop microscopic club-shaped **basidia** (sing., *basidium*), structures comparable in function to the asci of ascomycetes. Each basidium is an enlarged hyphal cell, on the tip of which develop four **basidiospores.** Note that basidiospores develop on the *outside* of a basidium, whereas ascospores develop *within* an ascus.

Each individual fungus produces millions of basidiospores, and each basidiospore has the potential to give rise to a new **primary mycelium** (Fig. 25–9). Hyphae of a primary mycelium are composed of monokaryotic cells. The mycelium of a basidiomycete such as the cultivated mushroom, *Agaricus campestris,* consists of a mass of white, branching, threadlike hyphae that live mostly below ground. The hyphae are divided into cells by septa, but, as in ascomycetes, the septa are perforated and allow cytoplasmic streaming between cells.

When in the course of its growth a hypha of a primary mycelium encounters another monokaryotic hypha of a different mating type, the two hyphae fuse. As in the ascomycetes, the two haploid nuclei remain separate within each cell. In this way a dikaryotic **secondary mycelium** with dikaryotic hyphae is produced, in which each cell contains two haploid nuclei.

The *n + n* hyphae of the secondary mycelium grow extensively and, when environmental conditions are favorable, form compact masses, called buttons, along the mycelium. Each button grows into a fruiting body that we call a mushroom. A mushroom, which consists of a stalk and a cap, is more formally referred to as a **basidiocarp.** Each basidiocarp consists of intertwined hyphae that are matted together. The lower surface of the cap usually consists of many thin perpendicular plates called **gills** that radiate from the stalk to the edge of the cap (Fig. 25–10*a*). On the gills of the mushroom, haploid nuclei of the dikaryotic cells within the young basidia fuse to form diploid zygotes. These are the only diploid cells that form during a basidiomycete's life history. Meiosis then takes place, forming four haploid nuclei that move to the outer edge of the basidium. Finger-like extensions of the basidium develop, into which the nuclei and some cytoplasm move; each of these extensions becomes a basidiospore. A septum forms that separates the basidiospore from the rest of the basidium by a delicate stalk that breaks when the basidiospore is forcibly discharged (Fig. 25–10*b*).

(a)

(b)

(c)

(d)

Figure 25–8 **Basidiomycete fruiting bodies.** *(a)* Basidia line the gills of the jack-o'-lantern mushroom *(Omphalotus olearius)*, a poisonous species whose gills glow in the dark. *(b)* A giant puffball *(Calvatia gigantea)* in White Plains, New York. At maturity, a dried-out puffball often has a pore through which the basidiospores are discharged as a puff of dust. *(c)* The stinkhorn *(Phallus ravenelii)* has a foul smell that attracts flies, which help disperse the slimy mass of basidiospores. *(d)* Bracket fungi *(Fomes* sp.*)* grow on both dead and living trees, producing shelflike fruiting bodies. Basidiospores are produced in pores located underneath each shelf. *(a, Dennis Drenner; b, Ed Kanze/Dembinsky Photo Associates; c, Richard D. Poe/Visuals Unlimited; d, Richard H. Gross)*

Imperfect fungi have no known sexual stage

About 25,000 species of fungi have been assigned to a *form phylum* referred to as the **deuteromycetes** (phylum Deuteromycota). Members of a form phylum are similar to one another in certain respects but probably do not share a common ancestry; that is, the group is polyphyletic (see Chapter 22). Deuteromycetes, also known as **imperfect fungi** because in many of them no sexual stage has been observed at any point during their life cycle, are classified together simply as a matter of convenience. Should further study reveal a sexual stage, these species will be reassigned to a different phylum.

Increasingly, comparing DNA and/or RNA sequences among various species helps biologists determine relationships between imperfect fungi and their sexually reproducing relatives. Most imperfect fungi reproduce only by means of conidia and are closely related to the ascomycetes. A few appear to be more closely related to the basidiomycetes.

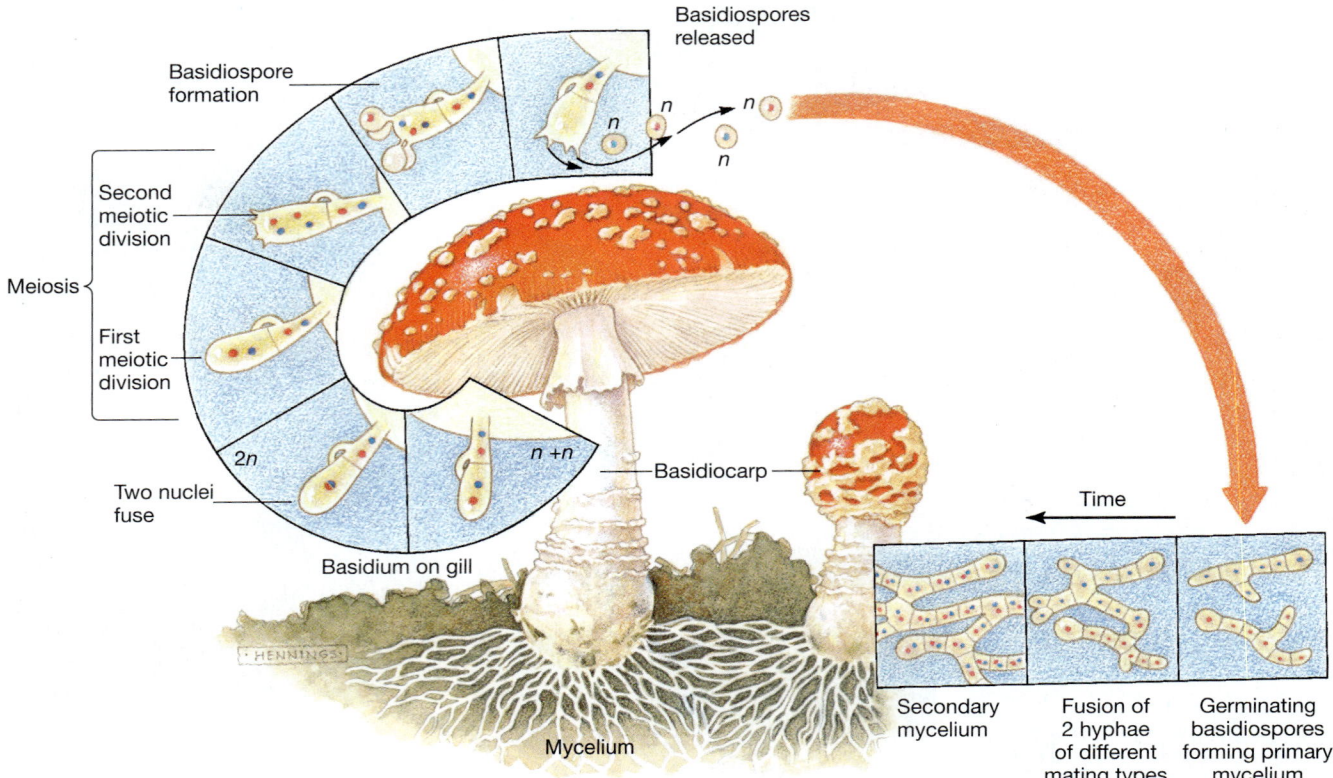

Figure 25-9 **The life cycle of a typical basidiomycete.** Sexual reproduction involves the fusion of two haploid hyphae of different mating types to form dikaryotic ($n + n$) hyphae from which basidia develop. In each basidium the two nuclei fuse, followed by meiosis to produce four basidiospores. In basidiomycetes in which the fruiting body is a mushroom, the basidia line the outer surfaces of the gills.

LICHENS ARE DUAL "ORGANISMS"

Although a **lichen** looks like a single organism, it is actually a symbiotic association between two organisms: a *phototroph* (a photosynthetic organism) and a fungus (Fig. 25–11*a*). (A symbiotic association is an intimate relationship between organisms of different species.) About 13,500 kinds of lichens have been described.

The phototrophic component of a lichen is either a green alga or a cyanobacterium. The fungus is most often an ascomycete, although in some tropical lichens the fungal partner is a basidiomycete. Most of the phototrophic organisms found in lichens also occur as free-living species in nature, but the fungal components are generally found only as a part of the lichen.

In the laboratory the fungal and phototrophic components can be isolated and grown separately in appropriate culture media. The phototroph grows more rapidly when separated, whereas the fungus grows more slowly and requires many complex carbohydrates. Neither organism resembles a lichen in appearance when grown separately. The phototroph and fungus can be reassembled as a lichen thallus, but only if they are placed in a culture medium under conditions that cannot support either of them independently.

What is the nature of this partnership? The lichen was originally considered a definitive example of mutualism, a symbiotic relationship that is beneficial to both species. The phototroph carries on photosynthesis, producing food for both members of the lichen, but it is unclear how the phototroph benefits from the relationship. It has been suggested that the phototroph obtains water and minerals from the fungus as well as protection against desiccation. More recently some biologists have suggested that the lichen partnership is not really a case of mutualism but one of controlled parasitism of the phototroph by the fungus.

Lichens typically possess one of three different growth forms (Fig. 25–11*b*). *Crustose lichens* are flat and grow tightly against their substrate (the surface they are growing on); *foliose lichens* are also flat, but they have leaflike lobes and are not so tightly pressed to the substrate; *fruticose lichens* grow erect and have many branches.

Able to tolerate extremes of temperature and moisture, lichens grow in almost all terrestrial environments except polluted cities. They exist farther north than any plants of the arc-

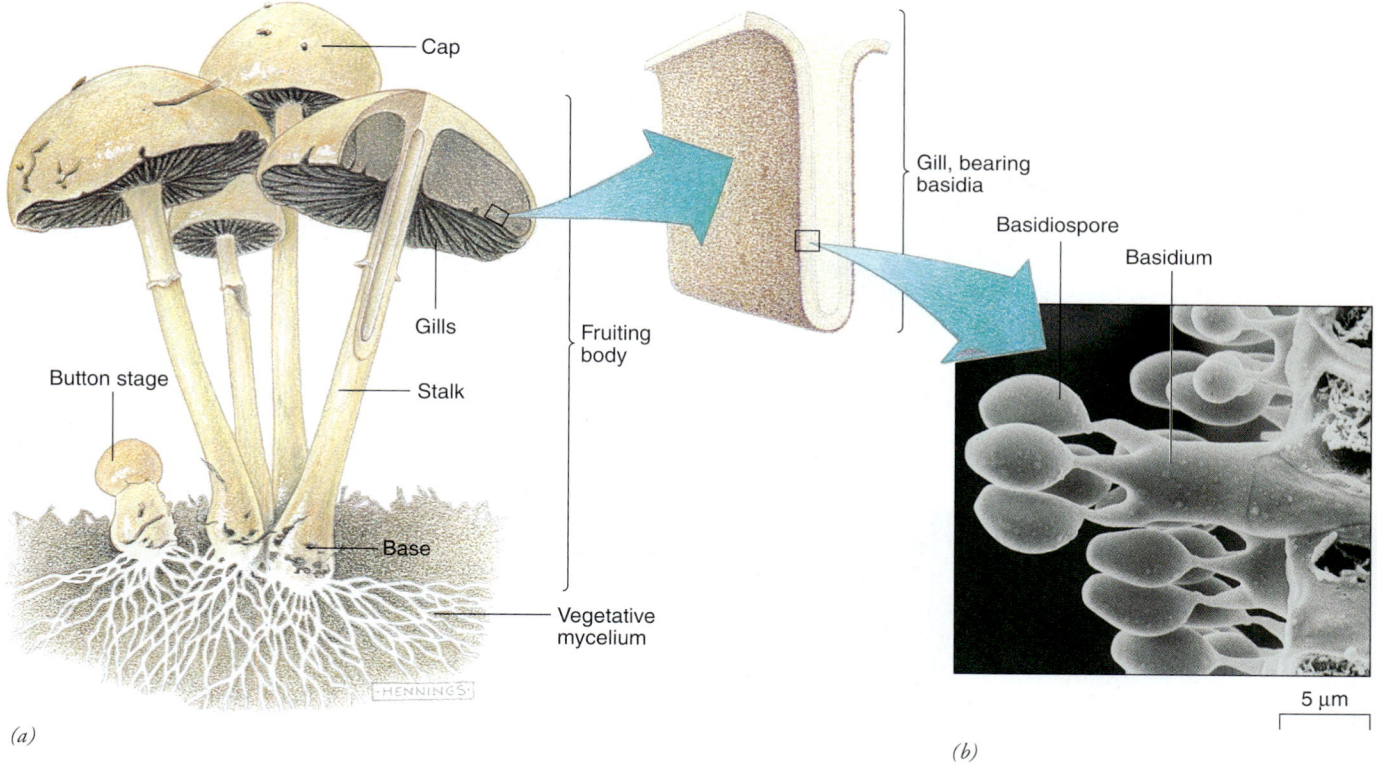

(a)

(b)

5 μm

Figure 25–10 **Details of mushroom basidiocarps.** *(a)* Interwoven hyphae from the vegetative mycelium form the basidiocarp commonly called a mushroom. Numerous basidia are borne along the gills. *(b)* SEM of a basidium. Each basidium produces four basidiospores. *(Biophoto Associates)*

Labels in figure: Cap, Gills, Stalk, Button stage, Base, Fruiting body, Vegetative mycelium, Gill, bearing basidia, Basidiospore, Basidium, HENNINGS

tic region and are equally at home in the steaming equatorial rain forest. They grow on tree bark, leaves, and exposed rock surfaces, from solidified lava to tombstones. In fact, lichens are often the first organisms to inhabit rocky areas. Lichen growth in these areas plays an important role in the formation of soil from rock because they gradually etch tiny cracks in rock (see Chapter 52). This process sets the stage for further disintegration of the rock by wind and rain.

Reindeer mosses of the arctic region, which serve as the main source of food for migrating herds of caribou, are not mosses but lichens. Some lichens produce colored pigments. One of them, orchil, is used to dye woolens, and another, litmus, is widely used in chemistry laboratories as an acid-base (pH) indicator.

Lichens vary greatly in size. Some are almost invisible, whereas others, like the reindeer mosses, may cover kilometers of land with an ankle-deep growth. Growth proceeds slowly; the radius of a lichen may increase by less than one millimeter each year. Some mature lichens are thought to be thousands of years old.

Lichens absorb minerals mainly from the air, rainwater, and the surface on which they grow. They cannot excrete the elements they absorb, and perhaps for this reason they are extremely sensitive to toxic compounds. A reduction in lichen growth has been used as a sensitive indicator of air pollution,

particularly sulfur dioxide. In 1997, for example, Italian researchers were able to establish a relationship between lung cancer and air pollution by comparing the locations of low lichen biodiversity (and therefore of air pollution) with the locations of lung cancer deaths in young male residents. The return of lichens to an area indicates an improvement in air quality. (Recall from Chapter 1 that air pollution killed lichens on tree trunks in England, thereby affecting the evolution of peppered moths.)

Lichens reproduce mainly by asexual means, usually by fragmentation, a process in which special dispersal units of the lichen, called **soredia,** break off and, if they land on a suitable surface, establish themselves as new lichens. Soredia contain cells of both partners. In some lichens, the fungus produces ascospores, which may be dispersed by wind and find an appropriate algal partner only by chance.

FUNGI ARE ECOLOGICALLY IMPORTANT

Fungi make important contributions to the ecological balance of our world. Like bacteria, most fungi are **saprotrophs** (also called *saprobes*), decomposers that absorb nutrients from organic wastes and dead organisms. Unlike bacteria, many fun-

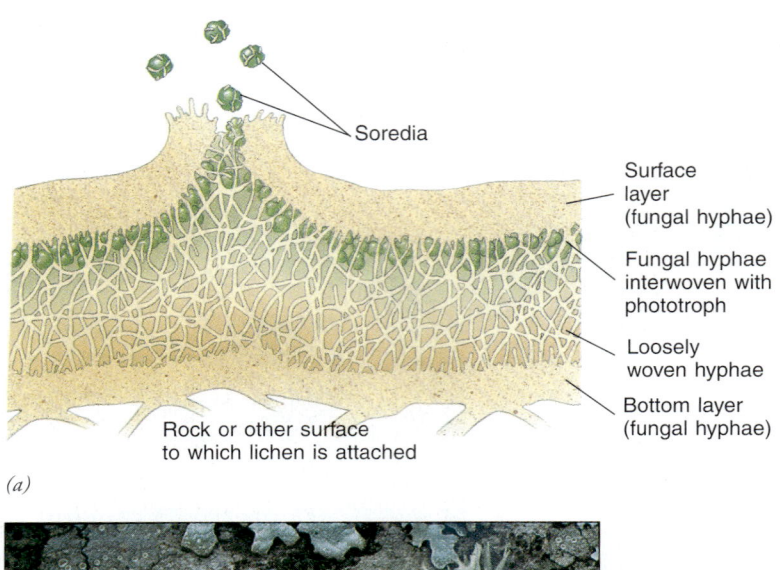

Soredia

Surface
layer
(fungal hyphae)

Fungal hyphae
interwoven with
phototroph

Loosely
woven hyphae

Bottom layer
(fungal hyphae)

Rock or other surface
to which lichen is attached

(a)

Figure 25-11 **Lichens.** *(a)* This cross section of a typical lichen shows distinct layers. The soredium (pl., *soredia*), an asexual reproductive structure, is composed of clusters of algal or cyanobacterial cells enclosed by fungal hyphae. *(b)* Lichens vary in color, shape, and overall appearance. Three growth forms—crustose, foliose, and fruticose—are shown. Crustose lichens grow tightly attached to tree trunks, rocks, and some other surfaces. Foliose lichens are leaflike in appearance, whereas fruticose lichens are branching and shrublike. *(Fred M. Rhoades)*

Fruticose lichen (*Ramalina*)

Crustose lichens (*Bacidia, Lecanora*)

Foliose lichen (*Parmelia*)

(b)

(a)   *(b)*

Figure 25-12 **Western red cedar *(Thuja plicata)* seedlings respond to mycorrhizae.** *(a)* Control plants grown in low phosphorus in the absence of the fungus. *(b)* These seedlings were grown under conditions identical to the control, except that their roots have formed mycorrhizal associations. *(a, b, Courtesy of Randy Molina, U.S. Forest Service)*

FUNGI, FLOWERING PLANTS, AND MIMICRY

Do fungi ever resemble flowers? One of the most fascinating phenomena relating to living organisms is mimicry, in which, during the course of evolution, one organism comes to resemble another organism or an inanimate object (see Chapter 52). In 1993 the journal *Nature* reported an unusual example of mimicry that actually requires *two* different organisms, a fungus and a plant, to produce the deception.* The fungus *(Puccinia monoica)* is a plant parasite that causes a rust disease in rock cress *(Arabis holboellii)*, a plant widely distributed across the northern part of North America.

When *Puccinia* infects rock cress, it doesn't kill the plant immediately, but it does change the plant's growth pattern. Infected plants grow much taller than normal and produce a cluster of leaves at the top of the plant. The fungal mycelium, which is bright yellow, grows over and covers these leaves, giving them the appearance of buttercup flowers (see figure). Even botany students have been fooled by the remarkable resemblance. The fungal mimicry is not only visual. The fungus also secretes a sugary solution and produces a strong scent, imitating the nectar and aroma of flowers (see Chapter 35).

What is the advantage of this elaborate mimicry? As you might guess, the answer involves the fungus' reproductive cycle. In the life cycles of flowers and *Puccinia monoica*, insects play the key role of increasing the chances of successful reproduction. The same characteristics that attract insects to real flowers also attract them to the fungal imitation.

In the case of flowers, the insects transfer pollen grains to the female part of another flower. In the case of *Puccinia*, a basidiomycete, sexual reproduction requires the union of nuclei from two different mating types. Insects, particularly flies, are attracted to the fungal "flowers," where they eat the sugary syrup. As the insects feed, pieces of the fungus cling to their bodies. Flying from one *Puccinia* "flower" to another, the insects distribute complementary mating types from fungus to fungus over a broad range. Thus, for the fungus, flower mimicry leads to better chances for its successful reproduction.

*Roy, B.A. "Floral Mimicry Induced by a Plant Pathogen." *Nature,* Vol. 362, 4 Mar. 1993.

A fungus *(Puccinia monoica)* that parasitizes rock cress *(Arabis holboellii)*. As a result of this parasitic interaction, rock cress changes its growth habit and produces, with the fungus' help, fake flowers. The "flowers" are so realistic that they attract insect pollinators, which help the fungus (instead of the plant) reproduce. *(Photograph by B.A. Roy, reprinted by permission from Nature, Vol. 362, cover and page 57)*

---

gal saprotrophs are able to degrade cellulose and lignin, the main components of plant cell walls. When fungi degrade wastes and dead organisms, water, carbon (as CO_2), and mineral components of organic compounds are released, and these elements are recycled (see Chapter 53). Without this continuous decomposition, essential nutrients would soon become locked up in huge mounds of dead animals, feces, branches, logs, and leaves. The nutrients would be unavailable for use by new generations of organisms, and life would cease.

Although most fungi are saprotrophs, others form symbiotic relationships of various kinds. Some fungi are **parasites,** organisms that live in or on other organisms and are harmful to their hosts. Parasitic fungi absorb food from the living bodies of their hosts. (See *Making the Connection: Fungi, Flowering Plants, and Mimicry* for an unusual example of a fungal parasite of a plant.)

Some types of fungi form mutualistic relationships with other organisms. **Mycorrhizae** (from Greek words meaning "fungus-roots") are mutualistic relationships between fungi and the roots of plants (Fig. 25–12; also see Fig. 34–11). Such relationships occur in more than 90% of all plant families. The mycorrhizal fungus benefits the plant by decomposing organic material in the soil and providing water and minerals such as phosphorus to the plant. At the same time, the roots supply sugars, amino acids, and other organic substances to the fungus.

TABLE 25-2 Total Plant Dry Mass of Two Grass Species Grown in Prairie Soil with and without Mycorrhizal Fungi*

	Andropogon gerardii		Sorghastrum nutans	
	With mycorrhizae	With mycorrhizae suppressed**	With mycorrhizae	With mycorrhizae suppressed**
Total dry mass (g) per plastic container (40 × 52 × 32 cm)	88.84	65.60	120.88	61.61

*Source: Wilson, G.W.T., and D.C. Hartnett. "Effects of Mycorrhizae on Plant Growth and Dynamics in Experimental Tallgrass Prairie Microcosms." *American Journal of Botany*, Vol. 84, No. 4, 1997.

**Soil was treated with a fungicide to suppress mycorrhizal fungi.

The importance of mycorrhizae first became evident when horticulturalists observed that orchids do not grow unless an appropriate fungus lives with them. Similarly, it has been shown that many forest trees such as pines decline and eventually die from mineral deficiencies when transplanted to mineral-rich grassland soils that lack the appropriate mycorrhizal fungi. When forest soil containing the appropriate fungi or their spores is added to the soil around these trees, they quickly resume normal growth. Similar results were reported in 1997 for tallgrass prairie plant species with and without mycorrhizal fungi (Table 25–2).

FUNGI ARE ECONOMICALLY IMPORTANT

The same powerful digestive enzymes that enable fungi to decompose wastes and dead organisms also permit them to reduce wood, fiber, and food to their basic components with great efficiency. Thus, various fungi cause incalculable damage to stored goods and building materials each year. Bracket fungi, for example, cause enormous losses by decaying wood, both in living trees and in stored lumber.

Fungi cause economic gains as well as losses. People eat them and grow them to make various chemicals. At the other extreme, some fungi are harmful from a human perspective because they cause diseases in humans and other animals and are the most destructive disease-causing organisms of plants. Their activities cost billions of dollars in agricultural damage yearly.

Fungi provide beverages and food

The capability of yeasts to produce ethyl alcohol and carbon dioxide from sugars such as glucose by fermentation is exploited to make wine, beer, and other fermented beverages, and also to make bread. Wine is produced when yeasts ferment fruit sugars, and beer results when yeasts ferment sugars derived from starch in grains (usually barley). During the process of making bread, carbon dioxide produced by yeast becomes trapped in dough as bubbles, causing the dough to rise;

this is what gives leavened bread its light texture. Both the carbon dioxide and the alcohol produced by the yeast escape during baking (Fig. 25–13).

The unique flavor of cheeses such as Roquefort, Brie, Gorgonzola, and Camembert is produced by the action of species

Figure 25–13 Foods and beverages produced in part by fungi. Yeasts ferment sugars from fruits to make wine or from grains to make beer, producing ethyl alcohol. That same process produces the carbon dioxide bubbles responsible for making bread rise. The bluish areas in the cheese are patches of mycelium. *(Raymond Tschoepe)*

(a)

(b)

Figure 25–14 **Edible ascomycetes.** *(a)* The common morel *(Morchella esculenta)* and *(b)* the truffle *(Tuber* sp.) are expensive gourmet treats. Both are ascocarps that produce ascospores. Truffles are subterranean ascocarps that people find with the help of trained dogs or pigs. Here truffles are shown whole and sectioned. *(a, Richard Shiell/Dembinsky Photo Associates; b, John D. Cunningham/Visuals Unlimited)*

of *Penicillium* (Fig. 25–13). *Penicillium roquefortii,* for example, is found in caves near the French village of Roquefort; only cheeses produced in this area can be called Roquefort cheese. In Roquefort and certain other cheeses, the blue mycelium of the fungus is visible in the cheese.

Aspergillus tamarii and other imperfect fungi are used in the Orient to produce soy sauce by fermenting soybeans with the fungi for three or more months.[2] Soy sauce enriches other foods with more than just its special flavor. It also adds vital amino acids from both the soybeans and the fungi themselves to supplement the low-protein rice diet.

Among the basidiomycetes, there are some 200 kinds of edible mushrooms and about 70 species of poisonous ones, sometimes called toadstools. Some edible mushrooms are cultivated commercially. In fact, more than 350,000 metric tons (about 780 million pounds) are produced each year in the United States alone. The common mushroom *(Agaricus bisporus)* is the only fungal species grown extensively for food, although several exotic mushrooms, such as oyster, shiitake, portobello, and straw mushrooms, are becoming increasingly popular. Morels, which superficially resemble mushrooms, and truffles, which produce underground fruiting bodies, are ascomycetes (Fig. 25–14). These gourmet delights are now being cultivated as mycorrhizal fungi on the roots of tree seedlings.

Edible and poisonous mushrooms can look very much alike and may even belong to the same genus. There is no simple way to tell them apart; they must be identified by an expert. Some of the most poisonous mushrooms belong to the genus *Amanita* (Fig. 25–15). Toxic species of this genus have been appropriately called such names as "destroying angel"

(Amanita virosa) and "death cap" *(Amanita phalloides).* Eating a single mushroom of either species can be fatal.

Ingestion of certain species of mushrooms causes intoxi-

Figure 25–15 **The destroying angel *(Amanita virosa).*** This extremely poisonous mushroom is distinguished, as are other amanitas, by the ring of tissue around its stalk and the underground cup from which the stalk protrudes. About 50 grams (2 ounces) of this mushroom could kill an adult man. Initial symptoms include vomiting, diarrhea, and cramps. If the poisoning goes untreated, liver and kidney failure occur, resulting in death in five to ten days. *(James W. Richardson/CBR Images)*

[2]In the United States soy sauce is often made by adding flavoring to salt water rather than soaking fermented soybeans.

cation and hallucinations. The sacred mushrooms of the Aztecs, *Conocybe* and *Psilocybe,* are still used in religious ceremonies by Native Americans of Central America for their hallucinogenic properties. The chemical ingredient psilocybin, related to lysergic acid diethylamide (LSD), is responsible for the trances and visions experienced by those who eat these mushrooms. Ingestion of psychoactive mushrooms is not recommended, because negative reactions vary considerably, from mild indigestion, sweating, and heart palpitations, to death. In addition, the possession and use of such mushrooms are illegal in the United States.

Fungi produce useful drugs and chemicals

In 1928 Alexander Fleming noticed that one of his petri dishes containing a bacterial culture was contaminated by mold. The bacteria were not growing in the vicinity of the mold, leading Fleming to the conclusion that the mold was releasing some substance harmful to them. Within a decade or so of Fleming's discovery, penicillin produced by *Penicillium notatum* was purified and used in treating bacterial infections. Penicillin is still among the most widely used and effective antibiotics. Other drugs derived from fungi include the antibiotic griseofulvin (used clinically to inhibit the growth of fungi), lovastatin (used to lower blood cholesterol levels), and cyclosporine (used to suppress immune responses in patients receiving organ transplants).

The ascomycete *Claviceps purpurea* infects the flowers of rye plants and other cereals. It produces a structure called an *ergot* where a seed would normally form in the grain head. When livestock eat this grain or when humans eat bread made from ergot-contaminated rye flour, they may be poisoned by the extremely toxic substances in the ergot. These substances may cause nervous spasms, convulsions, psychotic delusions, and even gangrene. This condition, called ergotism, was known as St. Anthony's fire during the Middle Ages, when it occurred often. In the year 994, for example, an epidemic of St. Anthony's fire caused more than 40,000 deaths. In 1722, the cavalry of Czar Peter the Great was felled by ergotism on the eve of the battle for the conquest of Turkey. This was one of several recorded times that a fungus changed the course of human history. Lysergic acid, one of the constituents of ergot, is an intermediate in the synthesis of LSD. Some of the compounds produced by ergot are now used clinically in small quantities as drugs to induce labor, to stop uterine bleeding, to treat high blood pressure, and to relieve one type of migraine headache.

Some fungi are grown commercially to produce citric acid and other industrial chemicals. Also, biologists are using recombinant DNA techniques to manipulate yeasts and certain filamentous fungi in order to produce important biological molecules such as hormones.

Fungi cause many important plant diseases

Fungi are responsible for many serious plant diseases, including epidemic diseases that spread rapidly and often result in

Figure 25–16   How a fungus parasitizes a plant. In this example, the hypha enters the leaf through a stoma and then grows, branching extensively through the internal air spaces and penetrating plant cells with specialized hyphal extensions called haustoria.

complete crop failure. All plants are apparently susceptible to some fungal infection. Damage may be localized in certain tissues or structures of the plant, or the disease may be systemic and spread throughout the entire plant. Fungal infections may cause stunting of plant parts or of the entire plant, they may cause growths like warts, or they may kill the plant.

A plant often becomes infected after hyphae enter through stomata (pores) in the leaf (Fig. 25–16) or stem or through wounds in the plant body. Alternatively, the fungus may produce an enzyme called cutinase that dissolves the waxy cuticle that covers the surface of leaves and stems. After dissolving the cuticle, the fungus easily invades the plant tissues. As the mycelium grows, it may remain mainly between the plant cells or it may penetrate the cells. Parasitic fungi often produce special hyphal branches called **haustoria** (sing., *haustorium*) that penetrate the host cells and obtain nourishment from the cytoplasm.

Some important plant diseases caused by ascomycetes are powdery mildew; chestnut blight; Dutch elm disease; apple scab; and brown rot, which attacks cherries, peaches, plums, and apricots (Fig. 25–17*a*). Diseases caused by basidiomycetes include smuts and rusts that attack various plants, for example, corn, wheat, oats, and other grains (Fig. 25–17*b*). Some of these parasites, such as the stem rust of wheat, have complex life cycles that involve two or more different host plants and the production of several kinds of spores. For example, wheat rust must infect a barberry plant at one stage in its life cycle. Since this fact was discovered, the eradication of barberry plants in wheat-growing regions has reduced infection with wheat rust. Wheat rust has not been eliminated by eradication of the barberry, however, because the fungus overwinters on wheat at the southern end of the Grain Belt and forms asexual spores. During the spring, wind blows these spores for hundreds of miles, reinfecting northern areas of the United States and Canada.

(a)

(b)

Figure 25–17 **Fungi that cause plant diseases.** *(a)* Brown rot of peaches is caused by *Monilinia fruticola*, an ascomycete. *(b)* Corn smut on an ear of sweet corn is caused by *Ustilago maydis*, a basidiomycete. *(a, Kathy Merrifield/ Photo Researchers, Inc.; b, Runk/Schoenberger, from Grant Heilman)*

Certain imperfect fungi also cause plant diseases. Examples include verticillium wilt on potatoes, which is caused by the deuteromycete *Verticillium* sp., and bean anthracnose, which is caused by the deuteromycete *Colletotrichum lindemuthianum.*

Fungi cause certain animal diseases

Some fungi cause superficial infections in which only the skin, hair, or nails are infected. Ringworm and athlete's foot are examples of superficial fungal infections; both are caused by imperfect fungi. Candidiasis, a yeast infection of mucous membranes of the mouth, throat, or vagina, is among the most common fungal infections; it is also caused by an imperfect fungus.

Other fungi cause systemic infections that infect internal tissues and organs and may spread through many regions of the body. Histoplasmosis, for example, is a serious infection of the lungs caused by inhaling spores of a fungus common in soil contaminated with bird feces. Most people in the eastern and midwestern parts of the United States have been exposed to this fungus at some time, and an estimated 40 million American have had mild infections by this pathogen (disease-causing organism). Fortunately, the infection usually stays in the lungs and is of short duration, but if the infection spreads into the bloodstream and from there into the heart, brain, eyes, kidneys, or other parts of the body, it can be serious and sometimes fatal.

Most pathogenic fungi are opportunists that cause infections only when the body's immunity is lowered. For example, the deuteromycete *Aspergillus fumigatus* does not usually cause disease in humans but does cause aspergillosis in people with defective immune systems, such as AIDS patients. During the course of this disease, the fungus can invade the lungs, heart, brain, kidneys, and other vital organs and can cause death. Other patients at high risk of acquiring life-threatening fungal infections include those with cancer and those with organ transplants.

Some fungi produce poisonous compounds collectively called **mycotoxins.** A few species of *Aspergillus,* for example, produce potent mycotoxins called *aflatoxins* that harm the liver and are known carcinogens. Foods on which aflatoxin-producing fungi commonly grow include peanuts, pecans, corn, and other grains. Other foods that may contain traces of aflatoxins include animal products such as milk, eggs, and meat (from animals that consumed feed contaminated by aflatoxin). It is impossible to avoid aflatoxin in the diet, but exposure should be minimized as much as possible. Any human food or animal forage product that has become moldy should be suspected of aflatoxin contamination and discarded.

I. **Fungi** are eukaryotes with cell walls composed of **chitin.**
   A. Fungi lack chlorophyll and are heterotrophic; they absorb predigested food molecules.
   B. A fungus may be unicellular (**yeast**), coenocytic (**mold**), or multicellular (**mold**).
      1. The structure of a multicellular fungus consists of long, branched **hyphae** that form a **mycelium.**
      2. In the **zygomycetes,** the hyphae are **coenocytic** (undivided by septa).
      3. In other fungi, perforated **septa** are present that divide the hyphae into individual cells.
   C. Most fungi reproduce both sexually and asexually by means of **spores.** When a fungal spore lands in a suitable spot, it germinates and begins to grow.
      1. Some hyphae infiltrate the substrate and digest its organic compounds.
      2. Spores are produced on aerial hyphae.
II. Fungi are classified into phyla based largely on their modes of sexual reproduction.
   A. **Chytridiomycetes** are simple aquatic fungi that produce motile cells with single, posterior flagella for both sexual (gametes) and asexual reproduction (**zoospores**).
      1. *Allomyces* is a representative of this group.
      2. Chytridiomycetes may have been the earliest fungal group to evolve from the ancient flagellated protist hypothesized as the common ancestor of all fungi.
   B. **Zygomycetes** produce both asexual spores and sexual spores (**zygospores**). The black bread mold is a representative of this group.
   C. **Ascomycetes** produce asexual spores called **conidia;** sexual spores called **ascospores** are produced in **asci.** Ascomycetes include yeasts, cup fungi, morels, truffles, and pink, brown, and blue-green molds.
   D. **Basidiomycetes** produce sexual spores called **basidiospores** on the outside of a **basidium;** basidia develop on the surface of **gills** in mushrooms, a type of **basidiocarp.** Basidiomycetes include mushrooms, puffballs, bracket fungi, rusts, and smuts.
   E. A sexual stage has not been observed in **imperfect fungi** (**deuteromycetes**). Most reproduce asexually by forming conidia. Members of this group include *Aspergillus tamarii* (used to produce soy sauce), species of *Penicillium,* and fungi that cause certain human fungal infections.
   F. A **lichen** is a symbiotic combination of a fungus and a phototroph (an alga or cyanobacterium); the association is one of controlled parasitism. Lichens have three main growth forms: crustose, foliose, and fruticose.
III. Fungi are ecologically significant.
   A. Many fungi are **saprotrophs** that break down organic compounds.
   B. **Mycorrhizae** are mutualistic relationships between fungi and the roots of plants. The fungus supplies minerals to the plant, and the plant secretes organic compounds needed by the fungus.
   C. Lichens play an important role in soil formation.
IV. Fungi have both positive and negative economic importance.
   A. Mushrooms, morels, and truffles are used as food; yeasts are vital in the production of beer, wine, and bread; certain fungi are used to produce cheeses and soy sauce.
   B. Fungi are used to make penicillin and other antibiotics; ergot is used to produce certain drugs; other fungi make citric acid and many other industrial chemicals.
   C. Fungi cause many plant diseases, including wheat rust, Dutch elm disease, and chestnut blight; they cause human diseases such as ringworm, athlete's foot, candidiasis, and histoplasmosis.

POST-TEST

1. Which of the following fungi does NOT have a mycelium? (a) black bread mold (b) yeast (c) rainforest cup fungus (d) cultivated mushroom (e) *Penicillium*
2. With the exception of chytridiomycetes, fungi are generally disseminated by (a) water currents (b) fragmentation of hyphae (c) soredia (d) airborne spores (e) flagellated zoospores
3. Which statement is NOT true of the chytridiomycetes? (a) they are simple aquatic fungi (b) they produce motile cells with single, posterior flagella (c) they have both sexual and asexual reproduction (d) the black bread mold is a representative of this group (e) they are the most primitive group of fungi
4. Which statement is NOT true of the zygomycetes? (a) they are simple, aquatic fungi (b) their sexual spores are called zygospores (c) they have both sexual and asexual reproduction (d) the black bread mold is a representative of this group (e) they have coenocytic hyphae
5. Which statement is NOT true of the ascomycetes? (a) their sexual spores are produced in asci (b) they produce motile cells with single, posterior flagella (c) they have both sexual and asexual reproduction (d) their asexual spores are called conidia (e) some species cause serious plant diseases
6. Which statement is NOT true of the basidiomycetes? (a) they have a diploid thallus that produces zoospores (b) their sexual spores are called basidiospores (c) they produce a secondary mycelium with $n + n$ hyphae (d) mushrooms, bracket fungi, and puffballs are examples of this group (e) basidiomycetes include both edible and poisonous species
7. Which statement is NOT true of deuteromycetes? (a) they have a diploid thallus that produces zoospores (b) they are also known as imperfect fungi (c) they have both sexual and asexual reproduction (d) their asexual spores are called conidia (e) both a and c are not true
8. The familiar portion of a mushroom is actually a large fruiting body called as a(an) _____. (a) ascocarp (b) basidium (c) basidiocarp (d) gametangium (e) ascus
9. A(an) _____ is a symbiotic association between a phototroph and a fungus. (a) soredium (b) basidiomycete (c) lichen (d) haustorium (e) saprotroph
10. Which characteristic is true of *all* fungi? (a) saprotrophic (b) parasitic (c) nonflagellated (d) pathogenic (e) heterotrophic
11. Mutualistic relationships between fungi and the roots of plants are called (a) lichens (b) mycorrhizae (c) deuteromycetes (d) haustoria (e) conidiophores

REVIEW QUESTIONS

1. What characteristics distinguish fungi from other organisms?
2. How does the body of a typical yeast differ from that of a mold?
3. What is the ecological importance of saprotrophic fungi? Of lichens? Of mycorrhizae?
4. Which fungal phylum is thought to be the most primitive? What evidence supports your answer?
5. Diagram the life cycle of the black bread mold.
6. Distinguish among each of the following: (a) ascocarp, ascus, and as-cospore (b) basidiocarp, basidium, and basidiospore (c) conidium and as-cospore (d) ascus and basidium (e) sporangium and conidium.
7. Some dictionaries erroneously define a morel as a type of mushroom. Why isn't a morel a mushroom?
8. Briefly describe three important fungal diseases of plants and three fungal diseases of humans.
9. Label the following diagram. Use Figure 25–10 to check your answers.

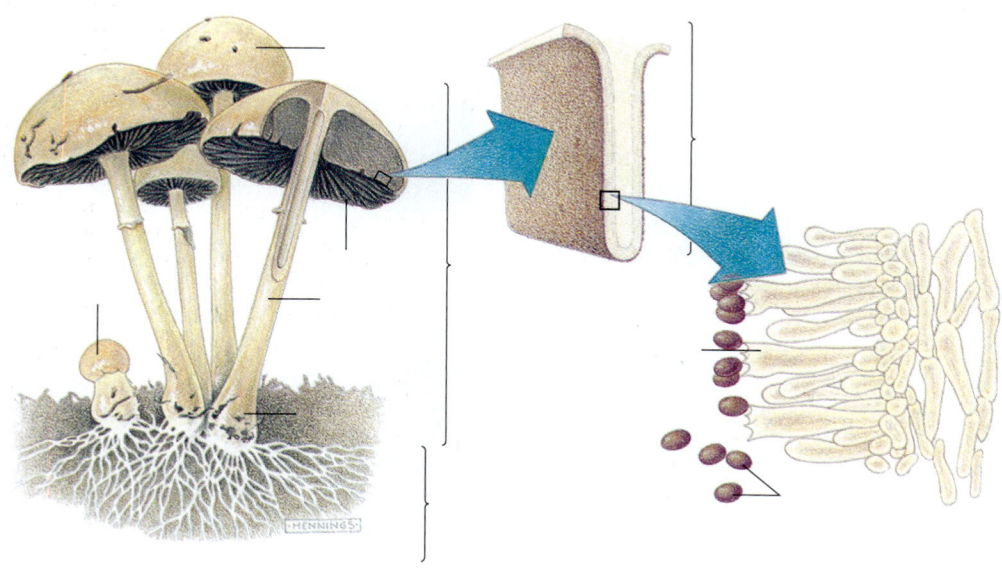

YOU MAKE THE CONNECTION

1. How are the life cycles of the marine alga *Ulva* (see Chapter 24) and *Allomyces* similar?
2. What measures can you suggest to prevent bread from becoming moldy?
3. If you do not see any mushrooms in your lawn, can you conclude that no fungi live there? Why or why not?
4. Biologists have discovered that many mycorrhizal fungi are sensitive to a low pH. What human-caused environmental problem might prove catastrophic for these fungi? How might this problem affect their plant partners?

RECOMMENDED READINGS

Alexopoulos, C.J., C.W. Mims, and M. Blackwell. *Introductory Mycology,* 4th ed. John Wiley and Sons, Inc., New York, 1996. A modern mycology textbook that provides comprehensive information on the biology of the fungi.

The Audubon Society Field Guide to North American Mushrooms, Alfred A. Knopf, New York, 8th Printing, 1992. A guide to common fungi with color photographs of each species discussed.

Cislaghi, C., and P.L. Nimis, "Lichens, Air Pollution, and Lung Cancer." *Nature,* Vol. 387, 29 May 1997. The authors demonstrated a relationship between lung cancer and air pollution by comparing biodiversity maps of pollution-sensitive lichens with mortality maps in northeastern Italy.

Gould, S.J. "Fungal Forgery." *Natural History,* Sept. 1993. A fascinating account of a fungus that mimics flowers.

Lewis, R. "A New Place for Fungi?" *BioScience* Vol. 44, No. 6, Jun. 1994. Molecular evidence indicates that fungi are more closely related to animals than to plants.

Lipske, M. "A New Gold Rush Packs the Woods in Central Oregon." *Smithsonian,* Jan. 1994. Mushroom gatherers are flocking to forests in the Northwest.

Radetsky, P. "The Yeast Within." *Discover,* Mar. 1994. The discovery that baker's yeast can be induced to grow like a filamentous mold has far-reaching medical implications.

Sharnoff, S.D. "Lichens: More Than Meets the Eye." *National Geographic,* Vol. 191, No. 2, Feb. 1997. Beautiful photographs accompany this essay on the ecological and economic value of lichens.

Walsh, R. "Seeking the Truffle." *Natural History,* Jan. 1996. A fascinating account of the history of truffles in France.

● Visit our website at **http://www.saunderscollege.com/lifesci/titles.html** and click on Solomon/Berg/Martin Biology for links to chapter-related resources on the World Wide Web.

CHAPTER 26

The Plant Kingdom: Seedless Plants

About 440 million years ago planet Earth would have seemed a most inhospitable place because, although life abounded in the oceans, it did not yet exist in abundance on land. The oceans were filled with vast numbers of fish, mollusks, and crustaceans as well as countless microscopic algae, and the water along rocky coastlines was home to large seaweeds. Occasionally, perhaps, an animal would crawl out of the water onto land, but it never stayed there permanently because there was little to eat on land, not a single blade of grass, no fruit, and no seeds.

During the next 30 million years, a time corresponding roughly to the Silurian period of the Paleozoic era in geological time, plants appeared in abundance and colonized the land. Where did they come from? Although plants living today exhibit great diversity in size, form, and habitat, they are all thought to have evolved from a common ancestor, an ancient green alga. Biologists infer this relationship because modern green algae share a number of biochemical and metabolic traits with modern plants. Both green algae and plants contain the same photosynthetic pigments: chlorophylls *a* and *b* and accessory pigments, the yellow and orange carotenoids, including xanthophylls (yellow pigments) and carotenes (orange pigments). Also, both store their excess carbohydrates as starch and possess cellulose as a major component of their cell walls. In addition, certain details of cell division, including the formation of a cell plate during cytokinesis (see Chapter 9), are shared by plants and some green algae.

Recent ultrastructural and molecular data indicate that plants probably descended from a group of green algae called **charophytes** (see Fig. 24–13*e*). Molecular comparisons, particularly of DNA and RNA sequences, have provided compelling evidence that charophytes are closely allied to plants. These data include comparisons among plants and various green algae, including charophytes, of chloroplast DNA sequences, of certain nuclear DNA sequences, and of ribosomal RNA sequences. In each case, the closest match occurs between charophytes and plants, indicating that modern charophytes and plants probably shared a common charophyte ancestor.

Today the plant kingdom comprises hundreds of thousands of species that live in varied habitats, from frozen Arctic tundra to lush tropical rain forests to harsh deserts to moist stream banks, as shown by the mosses on rocks in Marron Creek, which is found in Snowmass Wilderness, Colorado. Plants are complex multicellular organisms that range in size from minute, almost microscopic duckweeds to massive giant sequoias, some of the largest organisms that have ever lived.

(Sydney Karp/Photo/Nats, Inc.)

AFTER YOU HAVE STUDIED THIS CHAPTER YOU SHOULD BE ABLE TO

1. Discuss some of the environmental challenges of living on land and describe several adaptations that plants possess to meet these challenges.
2. Name the protist group from which plants are thought to have descended and describe supporting evidence.
3. Summarize the features that distinguish bryophytes from green algae and from other plants.
4. Name and briefly describe the three phyla of bryophytes.
5. Diagram the life cycle of mosses and compare their gameto-phyte and sporophyte generations.
6. Discuss the features that distinguish ferns and fern allies from algae and from other plants.
7. Name and briefly describe the four phyla of seedless vascular plants.
8. Diagram the life cycle of ferns and compare their sporophyte and gametophyte generations.
9. Compare the generalized life cycles of homosporous and heterosporous plants.

PLANTS HAVE ADAPTED TO LIFE ON LAND

What are some of the features of plants that have permitted them to colonize so many different environments? One important difference between plants and algae is that the aerial portion of a plant is covered by a waxy **cuticle.** A cuticle is essential for existence on land because it helps prevent the desiccation, or drying out, of plant tissues by evaporation. Plants are rooted in the ground and, unlike animals, cannot move to wetter areas during dry spells; therefore, a cuticle is critical to a land plant's survival.

Plants obtain the carbon they need for photosynthesis from the atmosphere as carbon dioxide (CO_2). In order for CO_2 to be fixed into organic molecules such as sugar, it must first diffuse into the chloroplasts that are inside green plant cells. Because the external surfaces of leaves and stems are covered by a waxy cuticle, however, gas exchange through the cuticle between the atmosphere and the insides of cells is negligible. To facilitate gas exchange, tiny pores called **stomata** (sing., *stoma*) dot the surfaces of leaves and stems; algae lack stomata.

Most plants possess multicellular sex organs called **gametangia** (sing., *gametangium*), while the gametangia of algae are single-celled (Fig. 26–1). Each plant gametangium has a layer of sterile (nonreproductive) cells that surrounds and protects the delicate gametes (eggs and sperm cells) it produces.

In plants, after fertilization occurs, the fertilized egg develops into a multicellular **embryo** within the female gametangium. Thus, the embryo is protected during its development. In algae, the fertilized egg develops away from its gametangium; in some algae, the gametes are released before fertilization, whereas in others the fertilized egg is released.

THERE ARE FOUR MAJOR GROUPS OF PLANTS

The plant kingdom consists of four major groups of plants: bryophytes, seedless vascular plants, gymnosperms, and flowering plants (Table 26–1 and Fig. 26–2). The mosses and other bryophytes are small nonvascular plants that lack a specialized vascular, or conducting, system to transport dissolved nutrients, water, and essential minerals throughout the plant body. In the absence of such a system, bryophytes rely on diffusion and osmosis to obtain needed materials. This reliance means that bryophytes are restricted in size; if they were much larger, some of their cells could not obtain the necessary materials in sufficient quantities. Bryophytes are seedless plants that reproduce and disperse primarily via haploid spores. Recent fossil evidence indicates that bryophytes may have been some of the earliest plants to colonize land (discussed later in the chapter).

(a)

(b)

(c)

(d)

Figure 26–1 **Generalized reproductive structures of algae and plants.** *(a,b)* In algae, gametangia are generally single-celled. When the gametes are released, the only thing remaining is the wall of the original cell. *(c,d)* In plants, the gametangia are multicellular, but only the inner cells become gametes. The gametangium is surrounded by a protective layer of sterile cells.

The other three groups of plants—ferns, gymnosperms, and flowering plants—possess vascular tissues: **xylem** for water and mineral conduction, and **phloem** for conduction of dissolved organic molecules such as sugar. A key step in the evolution of vascular plants was the ability to produce **lignin,** a strengthening polymer found in the walls of cells that function for support and conduction. The stiffening property of lignin enabled plants to grow tall and dominate the landscape. The successful occupation of the land by plants in turn made the evolution of terrestrial animals possible by providing them with both habitat and food (see *Making the Connection: Meeting the Environmental Challenges of Living on Land*).

Ferns and their allies (whisk ferns, horsetails, and club mosses) are seedless vascular plants that, like the bryophytes, reproduce and disperse primarily via spores. Seedless vascular plants arose and diversified during the Silurian and Devonian periods of the Paleozoic era, between 420 and 360 million years ago (discussed later in the chapter). Ferns and fern allies extend back more than 420 million years and were of consid-

TABLE 26-1	The Plant Kingdom

Nonvascular plants with a dominant gametophyte generation (bryophytes)
    Phylum Bryophyta (mosses)
    Phylum Hepaticophyta (liverworts)
    Phylum Anthocerotophyta (hornworts)

Vascular plants with a dominant sporophyte generation
  Seedless plants
    Phylum Pterophyta (ferns)
    Phylum Psilotophyta (whisk ferns)
    Phylum Sphenophyta (horsetails)
    Phylum Lycophyta (club mosses)
  Seed plants
    Plants with naked seeds (gymnosperms)
      Phylum Coniferophyta (conifers)
      Phylum Cycadophyta (cycads)
      Phylum Ginkgophyta (ginkgoes)
      Phylum Gnetophyta (gnetophytes)
    Seeds enclosed within a fruit
      Phylum Anthophyta (angiosperms or flowering plants)
        Class Dicotyledones (dicots)
        Class Monocotyledones (monocots)

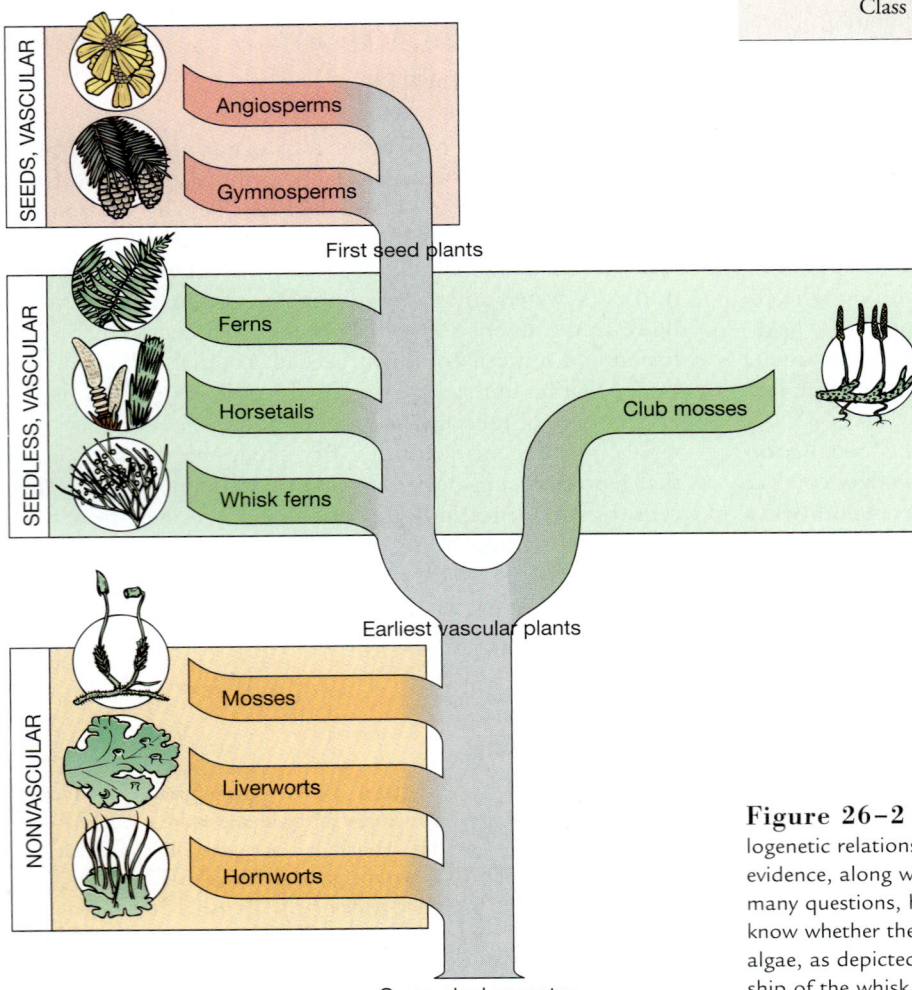

Figure 26-2 **Plant evolution.** This diagram shows possible phylogenetic relationships among living plants, based on current scientific evidence, along with several highlights in plant evolution. There are many questions, however, that remain to be resolved. We do not know whether the nonvascular plants descended directly from green algae, as depicted, or from early vascular plants. The exact relationship of the whisk ferns to other seedless vascular plants is also not clear.

MEETING THE ENVIRONMENTAL CHALLENGES OF LIVING ON LAND

How have terrestrial organisms met the environmental challenges of living on land? Life began in the oceans, but many life forms have since adapted to terrestrial life in a sea of air. Every single organism living on land has to meet the same environmental challenges: obtaining enough water; preventing excessive water loss; obtaining sufficient energy; and, in temperate and polar regions, tolerating widely varying temperature extremes. How those challenges are met varies from one organism to another and in large part explains the diversity of life encountered on land today. Let us compare how vertebrates (animals with backbones) and plants meet several terrestrial challenges.

1. **Obtaining enough water.** Animals are motile; they walk, slither, fly, run, or crawl to water sources. This requires not only the ability to move (skeletal and muscular systems) but also the ability to sense the presence of water (a nervous system). Plants have adapted in a much different way to this challenge. They have roots that not only anchor the plant in the soil but also absorb water and essential dissolved minerals.

2. **Preventing excessive water loss.** The outer layers of terrestrial vertebrates and plants protect the moist inner tissues from drying out. Vertebrates that are adapted to living on land have an accumulation of a water-insoluble protein called keratin in their epithelial cells. Keratin is particularly thick in reptiles, where it forms scales that greatly reduce water loss by evaporation. Plants possess a water-insoluble, waxy coating called a cuticle over their epidermal cells. Most plants that are adapted to moister habitats may have a very thin layer of wax, whereas those adapted to drier environments often possess a thick, crusty cuticle. Many desert plants also possess a reduced surface area, particularly of leaves, that minimizes water loss.

3. **Obtaining sufficient energy.** Animals are heterotrophs that eat plants or other animals that eat plants. Almost all plants are autotrophs and must absorb enough sunlight for effective photosynthesis. Some plants obtain adequate sunlight by growing tall (to maximize light interception); this adaptive approach required the evolution of strong supporting cells such as fibers, which contain the strengthening polymer lignin, because plants lack skeletal systems for support. Other plants have adapted to lower light intensities and so are able to grow in the shade of larger plants, albeit more slowly.

4. **Tolerating widely varying temperature extremes.** Air temperature varies greatly, particularly in temperate and polar regions. Many animals avoid hot temperatures by resting in the shade or by burrowing in the ground during the day when the temperatures are high; these animals become active at night when it is cooler. Sweat glands in mammalian skin produce sweat that cools the body by evaporation. Plants also rely on evaporative cooling; although they don't produce sweat, plants lose large quantities of water through tiny surface pores called stomata. As the water evaporates, enough heat leaves the plants to make them as much as 10°C cooler. To prevent overheating, desert plants reflect light from silvery colored leaves or stems, or shade themselves with pleated stems.

    Vertebrates deal with the cold temperatures of winter in several ways. Mammalian hair and bird feathers trap air next to the skin's surface, thereby providing insulation and allowing the body to conserve heat. Some animals avoid colder temperatures by migrating to warmer climates for the winter; whereas others avoid the cold by passing the winter in a dormant state called hibernation. Many plants also spend winter in a dormant state. The aerial parts of some plants die during the winter, but the underground parts remain alive; the following spring, they resume metabolic activity and develop new aerial shoots. Many trees are deciduous, that is, they shed their leaves for the duration of their dormancy. Shedding leaves is actually an adaptation to the "dryness" of winter. Roots cannot absorb water from ground that is cold or frozen. By shedding its leaves, the plant reduces water loss during the cold winter months when obtaining water from the soil is impossible.

erable importance as Earth's dominant plants in past ages. Fossil evidence indicates that many species of these plants were the size of immense trees. Many ferns and most fern allies are extinct today; a few small representatives of the ancient groups survive.

The gymnosperms are vascular plants that reproduce by forming seeds (see Chapter 27). Gymnosperms produce seeds borne exposed (unprotected) on a stem or in a cone. Plants with seeds as their primary means of reproduction and dis-

persal first appeared about 360 million years ago, at the end of the Devonian period. These early seed plants diversified into many varied species of gymnosperms.

The most recent plant group to appear is the flowering plants, or angiosperms, which arose during the early Cretaceous period of the Mesozoic era about 130 million years ago. Like gymnosperms, flowering plants reproduce by forming seeds. Flowering plants, however, produce seeds enclosed within a fruit.

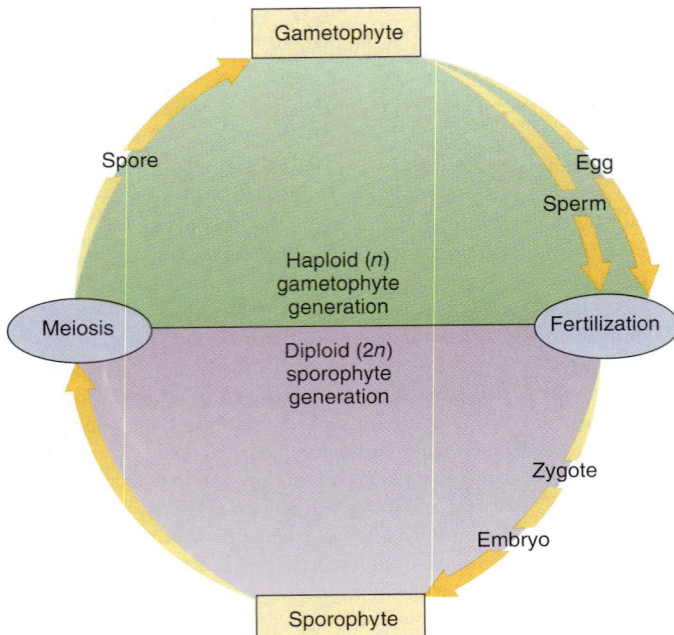

Figure 26-3 **The basic plant life cycle.** Plants alternate generations, spending part of the cycle in a haploid gametophyte stage and part in a diploid sporophyte stage. All plants have modifications of this cycle.

THE PLANT LIFE CYCLE ALTERNATES HAPLOID AND DIPLOID GENERATIONS

Plants have a clearly defined **alternation of generations** in which they spend part of their lives in a multicellular haploid stage and part in a multicellular diploid stage[1] (Fig. 26-3). The haploid portion of the life cycle is called the **gametophyte generation** because it gives rise to haploid gametes by mitosis. When two gametes fuse, the diploid portion of the life cycle, called the **sporophyte generation,** begins. The sporophyte generation produces haploid spores by the process of meiosis; these spores represent the first stage in the gametophyte generation.

Let us examine alternation of generations more closely. The haploid gametophytes produce antheridia (male gametangia), in which sperm cells form, and/or archegonia (female gametangia), each bearing a single egg (Fig. 26-4). Sperm cells reach the female gametangium in a variety of ways, and one sperm cell fertilizes the egg to form a zygote, or fertilized egg.

The diploid zygote is the first stage in the sporophyte generation. The zygote divides by mitosis and develops into a multicellular embryo. Embryo development takes place within the

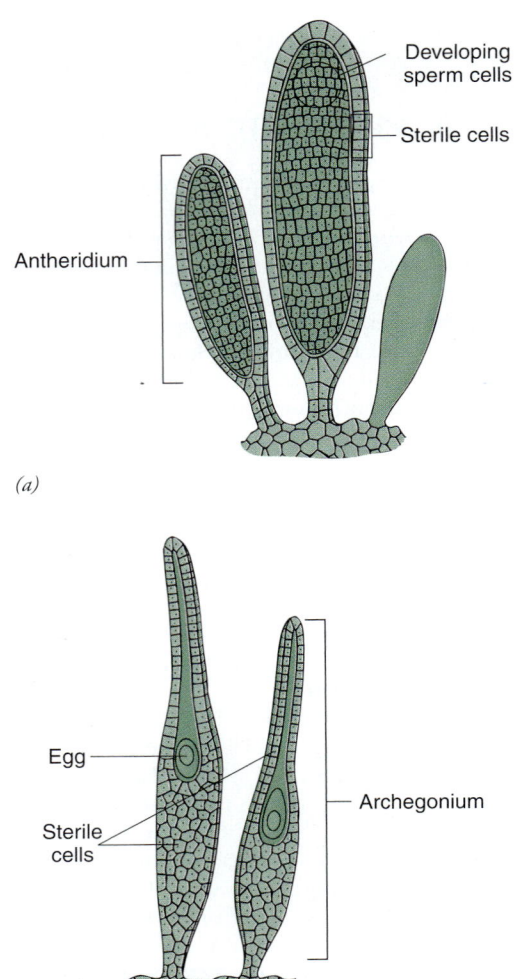

Figure 26-4 **Plant gametangia.** *(a)* Each antheridium, the male gametangium, produces numerous sperm cells. *(b)* Each archegonium, the female gametangium, produces a single egg. Shown are generalized moss gametangia.

archegonium; thus, during its development, the embryo is protected. Eventually the embryo matures into a sporophyte. The sporophyte has special cells called *sporogenous cells* (spore-producing cells, also called *spore mother cells*) that divide by meiosis to form haploid spores. All plant spores are produced by meiosis, in contrast to algae and fungi, which may produce spores by meiosis or mitosis.

The spores represent the first stage in the gametophyte generation. Each spore divides by mitosis to produce a multicellular gametophyte, and the cycle continues. Plants therefore have an alternation of generations, alternating between a haploid gametophyte generation and a diploid sporophyte generation.

[1]For convenience we limit our discussion to plants that are not polyploid, although polyploidy is very common in the plant kingdom. We therefore use the terms diploid and $2n$ (and haploid and n) interchangeably, although these terms are not actually synonymous.

TABLE 26-2 A Comparison of Major Groups of Seedless Plants

Plant Group	Dominant Stage of Life Cycle	Representative Genera
Nonvascular; reproduce by spores		
Mosses	Gametophyte: leafy plant	*Polytrichum, Sphagnum*
Liverworts	Gametophyte: thalloid or leafy plant	*Marchantia, Porella*
Hornworts	Gametophyte: thalloid plant	*Anthoceros*
Vascular; reproduce by spores		
Ferns	Sporophyte: roots, rhizomes, and leaves (megaphylls)	*Pteridium, Polypodium*
Whisk ferns	Sporophyte: rhizomes and erect stems; no true roots or leaves	*Psilotum, Tmesipteris*
Horsetails	Sporophyte: roots, rhizomes, erect stems, and leaves (reduced megaphylls)	*Equisetum*
Club mosses	Sporophyte: roots, rhizomes, erect stems, and leaves (microphylls)	*Lycopodium, Selaginella*

MOSSES AND OTHER BRYOPHYTES ARE NONVASCULAR PLANTS

The **bryophytes** (from the Greek words meaning "moss plant") comprise over 15,000 species of mosses, liverworts, and hornworts; bryophytes are the only nonvascular plants (Table 26–2). Because they are nonvascular and have no means for extensive internal transport of water, dissolved sugar, and essential minerals, bryophytes are typically quite small. They generally require a moist environment for active growth and reproduction, but some bryophytes tolerate dry areas. Although the three groups of bryophytes differ in many ways and may or may not be closely related, their life cycles are similar.

The bryophytes are divided into three distinct phyla: mosses (phylum Bryophyta), liverworts (phylum Hepaticophyta), and hornworts (phylum Anthocerotophyta).[2] These three groups of plants differ in many ways and may or may not be closely related. They are usually studied together, however, because they lack vascular tissues and have similar life cycles.

Moss gametophytes are small leafy shoots

Mosses (phylum Bryophyta), with about 9000 species, usually live in dense colonies or beds (Fig. 26–5a). Each individual plant has tiny hairlike absorptive structures, called *rhizoids,* and

[2]Until 1993 plants were classified into divisions rather than phyla. The International Botanical Congress, however, has approved the use of either the phylum or division designation for plants.

an upright, stemlike structure that bears leaflike blades. Because mosses lack vascular tissues, they do not possess true roots, stems, or leaves; that is, the moss structures are not homologous to roots, stems, or leaves in vascular plants. Some moss species possess water-conducting cells and sugar-conducting cells, although these cells are not as specialized or as effective as the conducting cells of vascular plants.

An alternation of generations is clear in the life cycle of mosses (Fig. 26–6). The leafy green moss gametophyte bears its gametangia at the top of the plant. Many moss species have separate sexes: male plants that bear antheridia and female plants that bear archegonia. Other moss species produce antheridia and archegonia on the same plant.

For fertilization to occur, one of the sperm cells must fertilize the egg within the archegonium. Sperm cells, which are flagellated, are transported from antheridium to archegonium by flowing water, such as splashing rain droplets. A raindrop lands on the top of a male gametophyte, and sperm cells are released into it from the antheridia. When another raindrop lands on the male plant, it may splash the sperm-laden droplet into the air and onto the top of a nearby female plant. Or insects may touch the sperm-laden fluid and inadvertently carry it for considerable distances. Once in a film of water on the female moss, a sperm cell swims into the archegonium, which secretes sucrose to attract and guide the sperm cells, and fuses with the egg.

The diploid zygote, formed as a result of fertilization, grows by mitosis into a multicellular embryo that develops into a mature moss sporophyte. This sporophyte grows out of the top of the female gametophyte, remaining attached and nutritionally dependent on the gametophyte throughout its existence (Fig. 26–7). The sporophyte is initially green and pho-

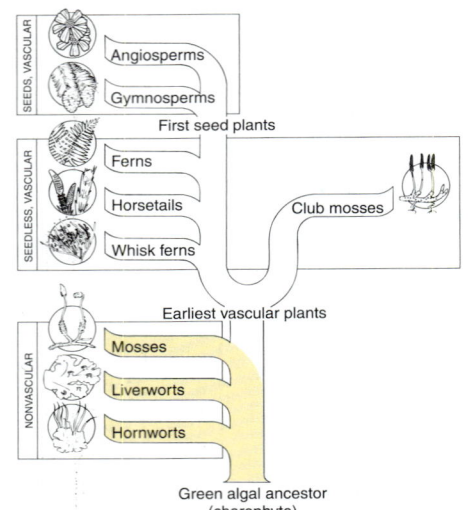

Figure 26–5 **Mosses, liverworts, and hornworts.** *(a)* A closeup of moss gametophytes, which grow in dense clusters. *(b)* The gametophyte of many liverworts is characterized by flattened, ribbon-like lobes. The common liverwort (*Marchantia polymorpha*) thallus produces gemmae cups. Each gemmae cup contains tiny reproductive bodies, called gemmae, that are dispersed by splashing raindrops and can then grow into new liverwort thalli. *(c)* The gametophyte with mature sporophytes (the "horns" projecting out of the gametophyte) of the common hornwort (*Anthoceros natans*). *(a, Rod Planck/Dembinsky Photo Associates; b, Carlyn Iverson; c, Robert A. Ross)*

(a)

(b)

(c)

tosynthetic, but becomes a golden brown at maturity. It is composed of three main parts: a *foot,* which anchors the sporophyte to the gametophyte and absorbs minerals and nutrients from it; a *seta,* or stalk; and a *capsule,* which contains sporogenous cells (spore mother cells).

The sporogenous cells undergo meiosis to form haploid spores. When the spores are mature, the capsule opens to release the spores. These microscopic cells are transported by wind or rain. If a moss spore lands in a suitable spot, it germinates and grows into a filament of cells called a **protonema.** The protonema, which superficially resembles a filamentous green alga, forms buds, each of which grows into a leafy green gametophyte, and the life cycle continues.

The haploid gametophyte generation is considered the dominant generation in mosses because it is capable of living independently of the diploid sporophyte. In contrast, the moss sporophyte is attached to and nutritionally dependent on the gametophyte.

Mosses make up an inconspicuous but significant part of

their environment. They play an important role in forming soil (see Chapter 52). Because they grow tightly packed together in dense colonies, mosses hold the soil in place and help prevent erosion.

Commercially, the most important mosses are the peat mosses in the genus *Sphagnum.* One of the distinctive features of *Sphagnum* "leaves" is the presence of many large dead cells that are able to absorb and hold water. This feature makes peat moss particularly beneficial as a soil conditioner. When added to sandy soils, for example, peat moss helps to absorb and retain moisture. In some countries such as Ireland and Scotland, layers of dead peat moss that have accumulated for hundreds of years are extracted from peat bogs, dried, and burned for fuel.

The name "moss" is often commonly misused to refer to plants that are not truly mosses. For example, reindeer moss is a lichen that is a dominant form of vegetation in the Arctic tundra, Spanish moss is a flowering plant, and club moss (discussed later in this chapter) is a relative of ferns.

Leafy gametophytes

Development of the gametophyte

Antheridia at the tip of the gametophyte shoot

Spore germinates

Protonema

Archegonia at the tip of the gametophyte shoot

Sperm released

Antheridium with sperm cells

Spores released

Archegonium with egg

Sperm cell

HAPLOID GAMETOPHYTE GENERATION

MEIOSIS

FERTILIZATION

DIPLOID SPOROPHYTE GENERATION

Zygote

Sporogenous cells that undergo meiosis

Calyptra

Capsule

Embryo

Sporophyte

Development of the sporophyte

Figure 26–6 The life cycle of mosses. The gametophyte generation, represented by leafy green plants, is dominant in the moss life cycle. After sexual reproduction takes place in mosses, the sporophyte grows out of the gametophyte. Mosses require water as a transport medium for sperm cells during fertilization. *(Carlyn Iverson)*

Liverwort gametophytes are either thalloid or leafy

Liverworts (phylum Hepaticophyta) comprise about 6000 species of nonvascular plants with a dominant gametophyte generation, but the gametophytes of some liverworts are quite different from those of mosses. Their body form is often a flattened, lobed, leaflike **thallus,** as in the common liverwort *(Marchantia polymorpha)* (Fig. 26–5b). Liverworts are so named because the lobes of their thalli superficially resemble the lobes of the human liver; *wort* is derived from the Old English word *wyrt,* meaning "plant." On the underside of the

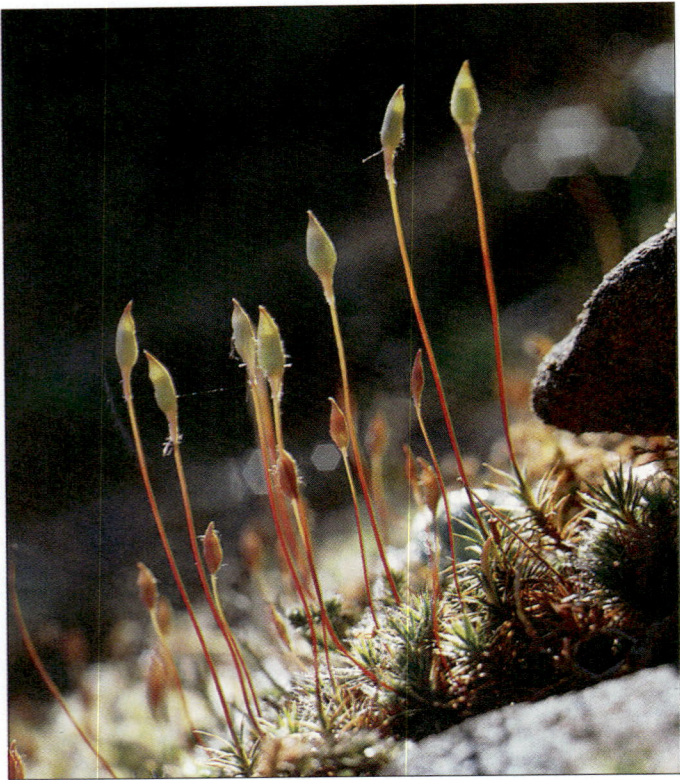

Figure 26-7 **Moss sporophytes.** The moss sporophytes grow out of the top of the gametophytes. Spores are produced within the capsule at the tip of each sporophyte. Shown is the haircap moss *(Polytrichum commune),* which is drought resistant and grows well in direct sunlight. *(David Cavagnaro)*

nected the individual lobes dies. Both of these mechanisms of reproduction, gemmae and thallus branching, are asexual because they do not involve fusion of gametes.

Hornwort gametophytes are inconspicuous thalloid plants

Hornworts (phylum Anthocerotophyta) are a small group of about 100 species of bryophytes whose gametophytes superficially resemble those of the thalloid liverworts. Hornworts are found in disturbed habitats such as fallow fields and roadsides.

Hornworts may or may not be closely related to other bryophytes. For example, their cell structure, particularly the presence of a single large chloroplast in each cell, is more like certain algal cells than it is like plant cells. In contrast, mosses, liverworts, and other plants have many disk-shaped chloroplasts per cell.

In hornworts, as for example, the common hornwort *(Anthoceros natans),* archegonia and antheridia are embedded in the gametophyte thallus so that they are not visible as in liverworts such as *Marchantia*. After fertilization and development, the needle-like sporophyte projects out of the gametophyte thallus, forming a spike or "horn"—hence the name *hornwort* (Fig. 26–5c). A single gametophyte often produces multiple sporophytes. Meiosis occurs within each sporangium (pl., *sporangia*), or spore case, and spores are formed. The sporangium splits open from the top to release the spores; each spore has the potential to give rise to a new gametophyte thallus. One unique feature of hornworts is that the sporophyte, unlike those of mosses and liverworts, continues to grow from its base for the remainder of the gametophyte's life.

Bryophytes are used for experimental studies

Scientists use certain bryophytes as experimental models to study many fundamental aspects of plant biology, including genetics, growth and development, plant ecology, plant hormones, and photoperiodism (plant responses to varying periods of night and day length; see Chapter 36). As experimental organisms, bryophytes are easy to grow on artificial media and do not require much space because they are so small.

THE EVOLUTION OF BRYOPHYTES IS OBSCURE

Although all plants probably descended from ancestral green algae, mosses and other bryophytes are probably not in a direct evolutionary path to the vascular plants. That is, the general consensus is that vascular plants did not have bryophyte ancestors. Some fossil evidence indicates that mosses and other bryophytes are ancient plants, perhaps representing an evolutionary sideline that arose from ancestral green algae. Alternatively, bryophytes may have evolved from early vascular plants by becoming simpler and losing their vascular tissues. The fos-

liverwort thallus are hairlike rhizoids that anchor the plant to the soil. Other liverworts have a leafy appearance rather than a lobed thallus and superficially resemble mosses, with "leaves," "stems," and rhizoids. As with other bryophytes, liverworts are small, generally inconspicuous plants that are largely restricted to damp environments.

Liverworts reproduce both sexually and asexually (Fig. 26–8). Their sexual reproduction involves the production of archegonia and antheridia on the haploid gametophyte. Their life cycle is basically the same as that of mosses, although some of the structures look quite different. The liverwort sporophyte, which is usually somewhat spherical, is attached to the gametophyte, as in mosses.

Some liverworts reproduce asexually by forming tiny balls of tissue called *gemmae* (sing., *gemma*), which are borne in a saucer-shaped structure, the gemmae cup, directly on the liverwort thallus. Splashing raindrops and small animals aid in the dispersal of gemmae. When a gemma lands in a suitable place, it grows into a new liverwort thallus. Liverworts may also reproduce asexually by thallus branching and growth. The individual thallus lobes elongate, and each becomes a separate plant when the older part of the thallus that originally con-

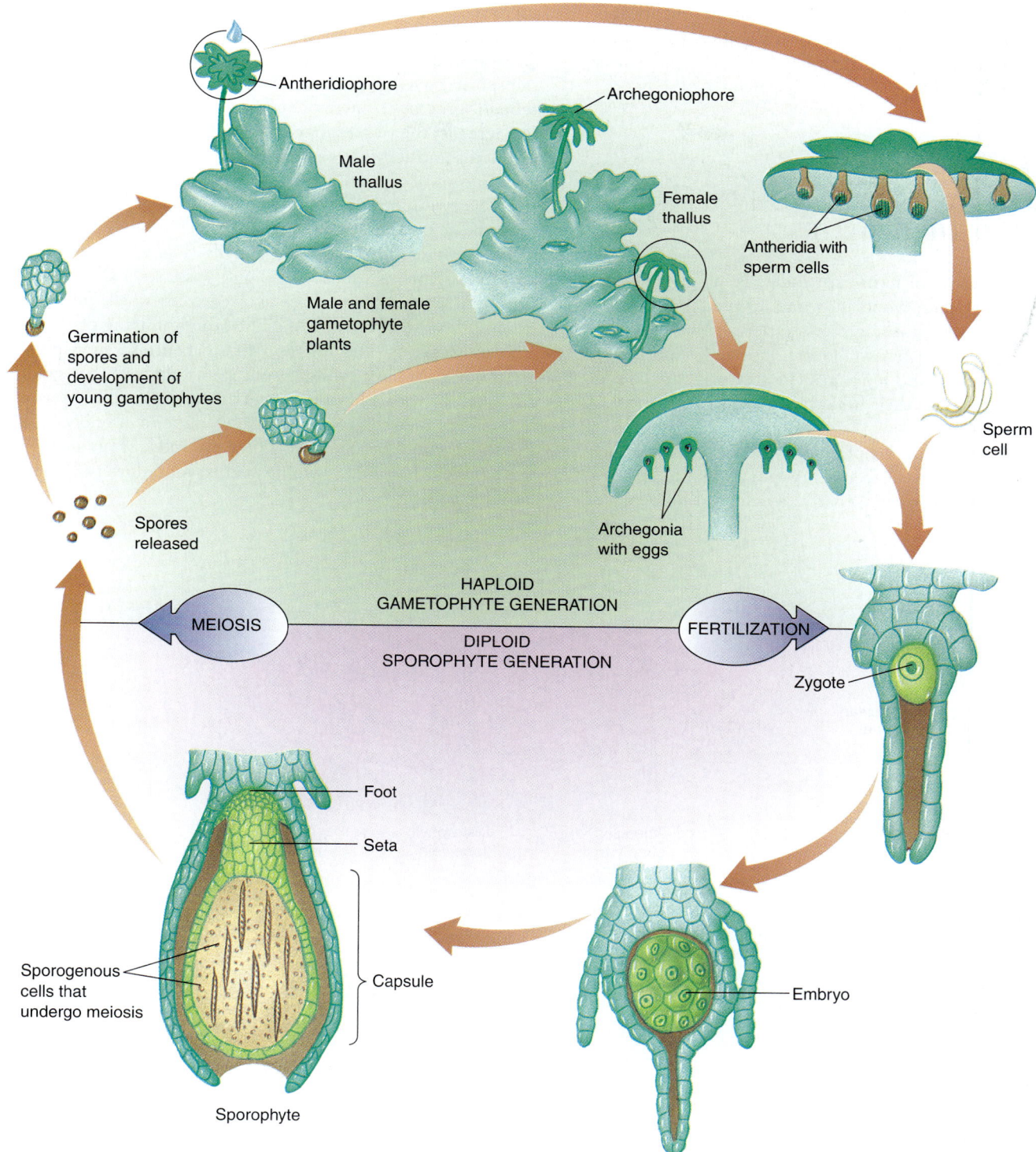

Figure 26–8 **The life cycle of the common liverwort (Marchantia polymorpha).** The dominant generation is the gametophyte, represented by ribbon-like male and female thalli.

Labels in figure:
Antheridiophore
Archegoniophore
Male thallus
Female thallus
Antheridia with sperm cells
Germination of spores and development of young gametophytes
Male and female gametophyte plants
Sperm cell
Spores released
Archegonia with eggs
HAPLOID GAMETOPHYTE GENERATION
MEIOSIS
DIPLOID SPOROPHYTE GENERATION
FERTILIZATION
Zygote
Foot
Seta
Capsule
Sporogenous cells that undergo meiosis
Embryo
Sporophyte

sil record of ancient bryophytes is incomplete, consisting mostly of spores and small tissue fragments, and can be interpreted in different ways. As a result, it does not provide a definite answer on bryophyte evolution.

In 1995 an exciting fossil discovery was reported in the journal *Nature:* a complete fossil plant about 400 million years old. This fossil, obtained from a coal deposit in the United Kingdom, resembles modern-day liverworts in many respects, but its spores are virtually identical to those found in 460-million-year-old rocks. This discovery is significant because it indicates that liverwort-like plants may have been the earliest plants to colonize land.

SEEDLESS VASCULAR PLANTS INCLUDE FERNS AND THEIR ALLIES

About 11,000 species of ferns exist today. Ferns are especially common in temperate woodlands and tropical rain forests, where they are found in the greatest variety. Three groups of vascular plants—whisk ferns (about 12 species), club mosses (about 1000 species), and horsetails (15 species)—are considered fern allies because their life cycles are similar to those of ferns (Table 26–2).

The most important adaptation found in ferns and their allies, though absent in algae and bryophytes, is the presence of specialized vascular tissues, that is, xylem and phloem, for support and conduction. This system of conduction enables vascular plants to achieve larger sizes than the bryophytes do because water, dissolved minerals, and dissolved sugar can be transported over great distances to all parts of the plant. Although ferns in temperate environments are relatively small, tree ferns in the tropics may grow to heights of 18 m (60 ft). The ferns and fern allies all have true stems with vascular tissues, and most also have true roots and leaves.

The evolution of the leaf as the main organ of photosynthesis has been studied extensively. There are two basic types of leaves: microphylls and megaphylls (Fig. 26–9). The **microphyll**, which is usually small and possesses a single vascular strand, is thought to have evolved from small, projecting extensions of stem tissue. Only one group of living plants, the club mosses, possesses microphylls. In contrast, **megaphylls** are thought to have evolved from stem branches that gradually filled in with additional tissue to form most leaves as we know them today. Megaphylls possess more than one vascular strand, as would be expected if they evolved from branch systems. Ferns, horsetails, gymnosperms, and flowering plants have megaphylls.

(a) MICROPHYLL EVOLUTION

(b) MEGAPHYLL EVOLUTION

Figure 26–9 Microphylls and megaphylls. *(a)* Microphylls probably originated as outgrowths (enations) of stem tissue that later developed a single, vascular strand. *(b)* Megaphylls, which are more complex and have multiple veins, probably evolved from the evolutionary modification of side branches. Webbing is the evolutionary process in which the spaces between close branches become filled with chlorophyll-containing cells.

(a)   (b)   (c)

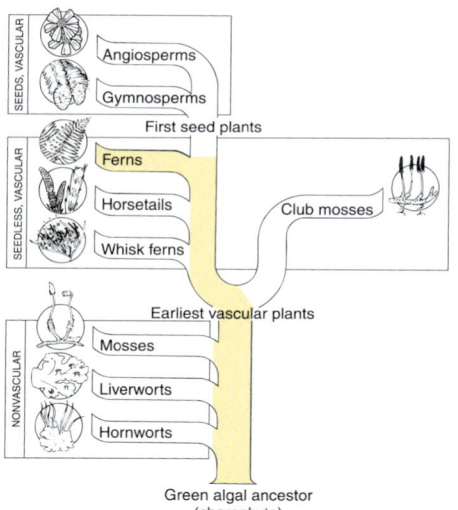

Figure 26–10 **Ferns.** *(a)* The Christmas fern *(Polystichum acrostichoides)* is green at Christmas, making it a popular holiday decoration. This fern was photographed in the Great Smoky Mountains. *(b)* The mosquito fern *(Azolla caroliniana)* is a free-floating aquatic fern that does not resemble "typical" ferns. It sometimes grows so densely across ponds that it reportedly smothers mosquito larvae. *(c)* The staghorn fern *(Platycerium bifurcatum)* is native to Australian rain forests and is widely cultivated elsewhere. In nature the staghorn fern is an epiphyte and grows attached to tree trunks but derives no nourishment from them. *(a, Ed Reschke; b, W. Ormerod/Visuals Unlimited; c, Carlyn Iverson)*

Ferns have a dominant sporophyte generation

Most of the 11,000 species of **ferns** (phylum Pterophyta) are terrestrial, although a few have adapted to aquatic habitats (Fig. 26–10). Ferns range from the tropics to the Arctic Circle, with most species living in tropical rain forests where they perch high in the branches of trees. In temperate regions, ferns commonly inhabit swamps, marshes, moist woodlands, and stream banks, although some species can be found in fields, rocky crevices on cliffs or mountains, or even deserts. The most common fern species throughout the world is bracken *(Pteridium aquilinum),* a rugged, coarse, weedy plant that grows well on poor soil and is uncommon in the moist habitats favored by other ferns.

The life cycle of ferns involves a clearly defined alternation of generations (Fig. 26–11). The ferns grown as houseplants (such as Boston fern, maidenhair fern, and staghorn fern) represent the larger, more conspicuous sporophyte generation. The fern sporophyte is composed of a horizontal underground stem, or *rhizome,* that bears leaves, called *fronds,* and true roots. As each young frond first emerges from the

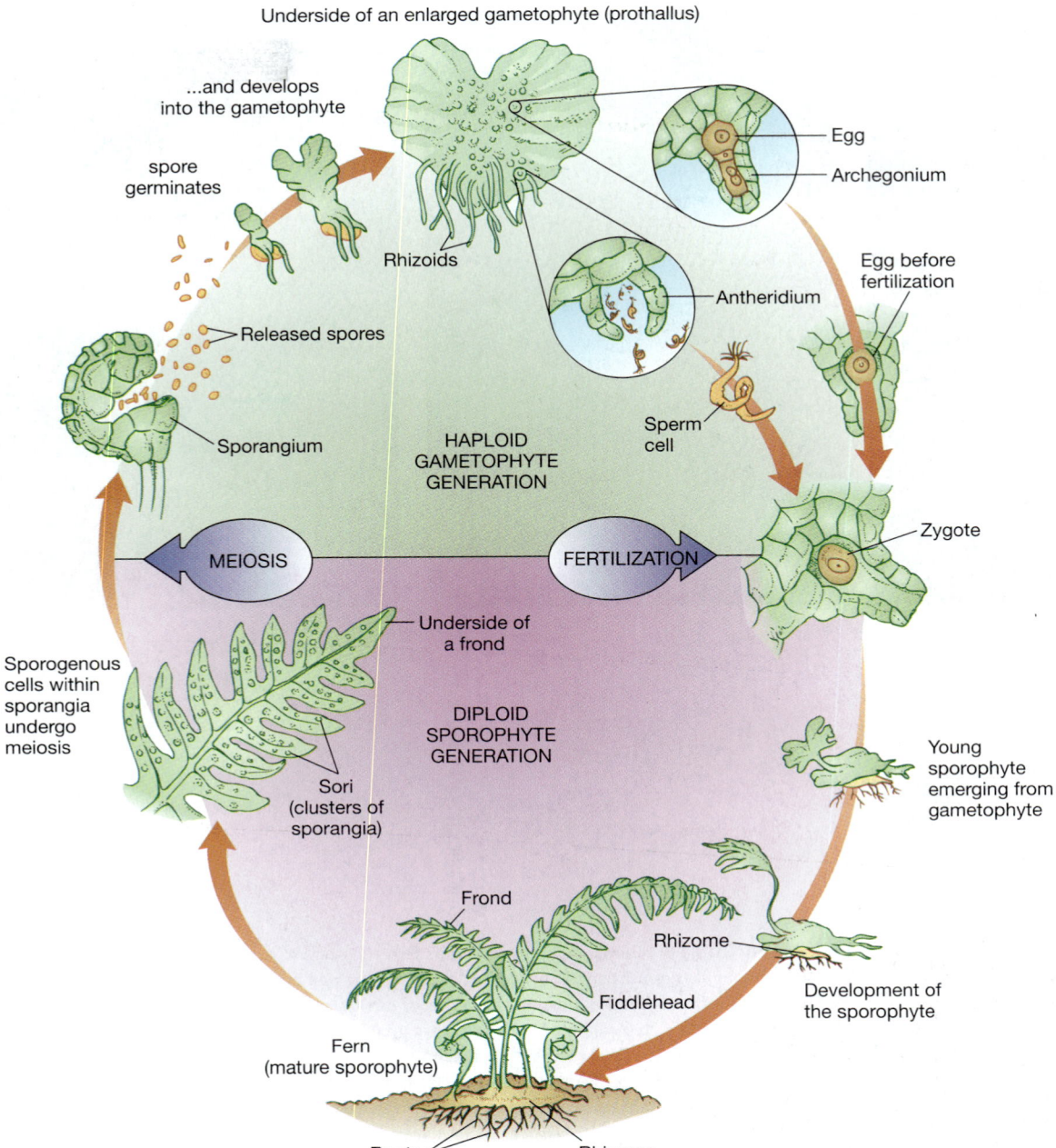

Underside of an enlarged gametophyte (prothallus)

...and develops into the gametophyte

spore germinates

Egg
Archegonium

Rhizoids

Antheridium

Egg before fertilization

Released spores

Sperm cell

HAPLOID GAMETOPHYTE GENERATION

Sporangium

Zygote

MEIOSIS

FERTILIZATION

Underside of a frond

Sporogenous cells within sporangia undergo meiosis

DIPLOID SPOROPHYTE GENERATION

Young sporophyte emerging from gametophyte

Sori (clusters of sporangia)

Frond

Rhizome

Fiddlehead

Development of the sporophyte

Fern (mature sporophyte)

Roots

Rhizome

Figure 26-11 **The life cycle of ferns.** Note the clearly defined alternation of generations between the gametophyte (prothallus) and sporophyte (leafy plant) generations. Fertilization in ferns requires water as a transport medium for sperm cells.

ground, it is tightly coiled and resembles the top of a violin, resulting in the name *fiddlehead* (Fig. 26–12). As fiddleheads grow, they unroll and expand to form fronds. Fern fronds are usually compound (that is, the blade is divided into several leaflets), with the leaflets forming beautifully complex leaves. Fronds, roots, and rhizomes all contain vascular tissues.

Spore production usually occurs in certain areas on the

fronds, which develop sporangia in which sporogenous cells (spore mother cells) undergo meiosis to form haploid spores. The sporangia are frequently borne in clusters, called **sori** (sing., *sorus*), on the fronds. When the sporangia burst open and the spores are discharged, they may germinate and grow by mitosis into gametophytes.

The mature fern gametophyte, which bears no resem-

Figure 26–12 **Fiddleheads.** The tightly coiled young fronds, or fiddleheads, are characeristic of ferns. *(Marion Lobstein)*

Archegonium

Rhizoids

500 μm

Figure 26–13 **Fern *(Woodwardia virginica)* prothallus.** The dark spots near the notch of the "heart" are archegonia; no antheridia are visible. *(Carolina Biological Supply Company/Phototake NYC)*

blance to the sporophyte, is a tiny (about the size of half of one of your fingernails), green, often heart-shaped structure that grows flat against the ground. Called a **prothallus** (pl., *prothalli*), the fern gametophyte lacks vascular tissues and has tiny hairlike absorptive rhizoids to anchor it (Fig. 26–13). The prothallus usually produces both archegonia and antheridia on its underside. Each archegonium contains a single egg, whereas numerous sperm cells are produced in each antheridium.

Although ferns are considered more advanced than mosses because they possess vascular tissues, they have retained a primitive fertilization technique: the use of water as a transport medium. The flagellated sperm cells swim to the neck of an archegonium through a thin film of water on the ground underneath the prothallus. After one of the sperm cells fertilizes the egg, a diploid zygote grows by mitosis into a multicellular embryo. At this stage in its life, the sporophyte embryo is attached to and dependent on the gametophyte, but as the embryo matures, the prothallus withers and dies, and the sporophyte becomes free-living.

The fern life cycle alternates between the dominant, diploid sporophyte with its rhizome, roots, and fronds, and the haploid gametophyte (prothallus). The sporophyte generation is dominant not only because it is larger than the gametophyte but also because it persists for an extended period of time (most fern sporophytes are perennials), whereas the gametophyte dies soon after reproducing.

Whisk ferns are the simplest vascular plants

Only about 12 species of **whisk ferns** (phylum Psilotophyta) exist today, and the fossil record contains several extinct species. All are relatively simple in structure and lack true roots and leaves but possess vascularized stems. *Psilotum nudum,* a representative whisk fern, has both a horizontal underground rhizome and vertical aerial stems (Fig. 26–14*a*). Whenever the stem forks or branches, it always divides into two equal halves. Botanists consider this **dichotomous** branching a primitive characteristic. In contrast, when most plant stems branch, one stem is more vigorous and becomes the main trunk.

The upright stems of *Psilotum* are green and are the main organs of photosynthesis. Tiny round sporangia, borne directly on the erect, aerial stems, contain sporogenous cells that undergo meiosis to form haploid spores. After being dispersed, the spores germinate to form haploid prothalli. The prothalli of whisk ferns are difficult to study because they grow underground. They are nonphotosynthetic, owing to their subterranean location, and they apparently have a symbiotic relationship with mycorrhizal fungi (see Chapter 25) that provide them with sugar and minerals.

Most species of whisk ferns are extinct, and the few surviving species are found mainly in the tropics and subtropics. Although whisk ferns do not closely resemble ferns in appearance, they are considered fern allies because of similarities in their life cycles.

Whisk ferns have been carefully studied in recent years, but botanists disagree about how to interpret their structures. Many botanists consider whisk ferns to be surviving representatives of very primitive vascular plants (see discussion of rhyniophytes later in the chapter). Other botanists think whisk

Sporangia

(a)

Strobilus

(b)

Strobilus

(c)

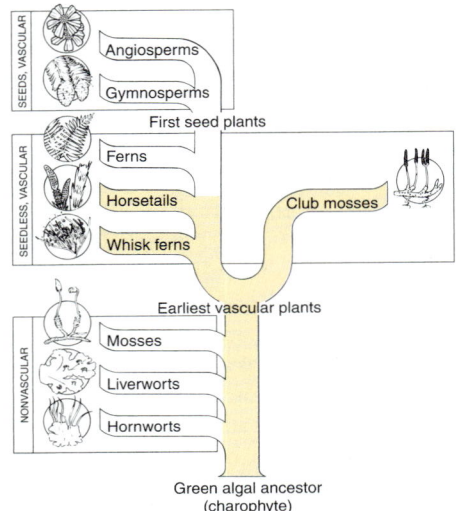

Figure 26–14 Fern allies. *(a)* The sporophyte of *Psilotum nudum,* a whisk fern. The stem is the main organ of photosynthesis in this rootless, leafless, vascular plant. Sporangia, which are initially green but turn yellow as they mature, are borne on short lateral branches directly on the stems. *(b) Equisetum telematia,* a horsetail with a wide distribution in Eurasia, Africa, and North America, has unbranched fertile shoots bearing conelike strobili and separate, highly branched vegetative (nonreproductive) shoots. In some horsetail species, both fertile and vegetative shoots are unbranched. *(c)* The sporophyte of *Lycopodium* sp., a club moss, has small, scalelike leaves (microphylls) that are evergreen. Spores are produced in sporangia on fertile leaves clustered in a conelike strobilus (as shown) or, in other species, scattered along the stem. *(a, John Arnaldi; b, J. Robert Waaland/Biological Photo Service; c, Ed Reschke)*

ANCIENT PLANTS AND COAL FORMATION

Our industrial society depends on energy from fossil fuels that formed from the remains of ancient organisms. One of our most important fossil fuels is coal, which is burned to produce electricity and to manufacture items made of steel and iron. Although coal is mined as a mineral, it is not an inorganic mineral like gold or aluminum, but an organic material formed from the remains of ancient vascular plants, particularly those of the Carboniferous period (approximately 300 million years ago). Five main groups of plants contributed to coal formation. Three were seedless vascular plants: the club mosses, horsetails, and ferns. The other two important groups were seed plants: seed ferns (now extinct) and primitive gymnosperms.

It is hard to imagine that relatives of the small, relatively inconspicuous club mosses, ferns, and horsetails of today could have been so significant in forming vast beds of coal. However, many of the members of these groups that existed during the Carboniferous period were giants compared with their modern counterparts and formed immense forests (see Fig. 20–10).

The climate during the Carboniferous period was warm, moist, and mild. Plants in most locations could grow year-round because of the favorable conditions. Forests of these plants often occurred in low-lying, swampy areas that were periodically flooded when the sea level rose. When the sea level receded, these plants would become reestablished.

When these large plants died or were blown over during storms, they decomposed incompletely because they were covered by swamp water. (The anaerobic con-

ditions of the water prevented wood-rotting fungi from decomposing the plants, and anaerobic bacteria do not decompose wood rapidly.) Thus, over time the partially decomposed plant material accumulated and consolidated.

Layers of sediment formed over the plant material each time the water level rose and flooded the low-lying swamps. With time, heat and pressure built up in these accumulated layers and converted the plant material to coal and the sediment layers to sedimentary rock. Much later, geological upheavals raised the layers of coal and sedimentary rock. For example, coal is found in seams (layers) in the Appalachian Mountains. The various grades of coal (lignite, bituminous, and anthracite) were formed as a result of the different temperatures and pressures to which the layers were exposed.

---

ferns are highly modified relatives of ferns. Recent molecular data, including comparisons of nucleotide sequences of ribosomal RNA, chloroplast DNA, and mitochondrial DNA in living species, support the hypothesis that the whisk ferns are more closely related to ferns than to other seedless vascular plants.

Horsetails have hollow, jointed stems

About 300 million years ago the **horsetails** (phylum Sphenophyta) were among the dominant plants and grew as large as modern trees (Fig 26–15). These ancient horsetails are still significant today because they contributed to Earth's vast coal deposits (see *Focus On: Ancient Plants and Coal Formation*). The few surviving horsetails, about 15 species in the genus *Equisetum,* grow mostly in wet, marshy habitats and are small (less than 1.3 m, or 4 ft, tall) but extremely distinctive (Fig. 26–14*b*). They are widely distributed on every continent except Australia.

Horsetails have true roots, stems (both rhizomes and erect aerial stems), and small leaves. The hollow, jointed stems are impregnated with silica, which gives them a gritty texture. Small leaves, interpreted as reduced megaphylls, are fused in

Figure 26–15 **Reconstruction of *Calamites* sp.** This ancient horsetail was the size of a small tree (to 20 m, or about 65 ft). *Calamites* had an underground rhizome where roots and aerial shoots originated.

Figure 26–16 Reconstruction of *Lepidodendron* sp. This ancient club moss was the size of a large tree (to 40 m, or about 130 ft). Numerous fossils of *Lepidodendron* were preserved in coal deposits, particularly in Great Britain and the central United States.

whorls at each node (the area on the stem where leaves attach). The green stem is the main organ of photosynthesis. Horsetails are so-named because certain vegetative (nonreproductive) stems have whorls of branches that give the appearance of a bushy horse's tail. In pioneer days horsetails were called "scouring rushes" and were used to scrub out pots and pans along stream banks.

Each reproductive branch of a horsetail bears a terminal conelike **strobilus** (pl., *strobili*). The strobilus is composed of several stalked, umbrella-like structures, each of which bears five to ten sporangia in a circle around a common axis.

The horsetail life cycle is similar in many respects to the fern life cycle. In horsetails, as in ferns, the sporophyte is the conspicuous plant, whereas the gametophyte is a minute, lobed thallus ranging in width from the size of a pinhead to about 1 cm (less than 0.5 in) across. The sporophyte and gametophytes are both photosynthetic and nutritionally independent at maturity. Like ferns, horsetails require water as a medium for flagellated sperm cells to swim to the egg.

Club mosses are small plants with rhizomes and short, erect branches

Like horsetails, **club mosses** (phylum Lycophyta) were important plants millions of years ago, when species that are now extinct often attained great size (Fig. 26–16). These large, treelike plants, like the ancient horsetails, were major contributors

to our present-day coal deposits. The 1000 or so species of club mosses living today, such as *Lycopodium* sp. (Fig. 26–14*c*), are small (less than 25 cm, or 10 in, tall), attractive plants commonly found in temperate woodlands. They possess true roots; both rhizomes and erect aerial stems; and small, scale-like leaves (microphylls). Sporangia are borne on fertile leaves that are either clustered in conelike strobili at the tips of stems or scattered in fertile areas along the stem axes. Club mosses are evergreen and often fashioned into Christmas wreaths and other decorations. In some areas they are endangered from overharvesting.

The fact that common names can be misleading in biology is vividly evident in this group of plants. The most common names for the phylum Lycophyta are "club mosses" and "ground pines," yet these plants are neither mosses nor pines and are most closely allied to the ferns.

More advanced plants are less dependent on water as a transport medium for reproductive cells

Many algae produce flagellated reproductive cells, both spores and sperm cells, that can swim through the water. Although reproduction by flagellated spores and sperm cells is an advantage in aquatic environments, it may be detrimental on land, particularly in locations where extended dry periods occur. In such terrestrial sites, the production of nonmotile, airborne spores and sperm cells may be more advantageous. Thus,

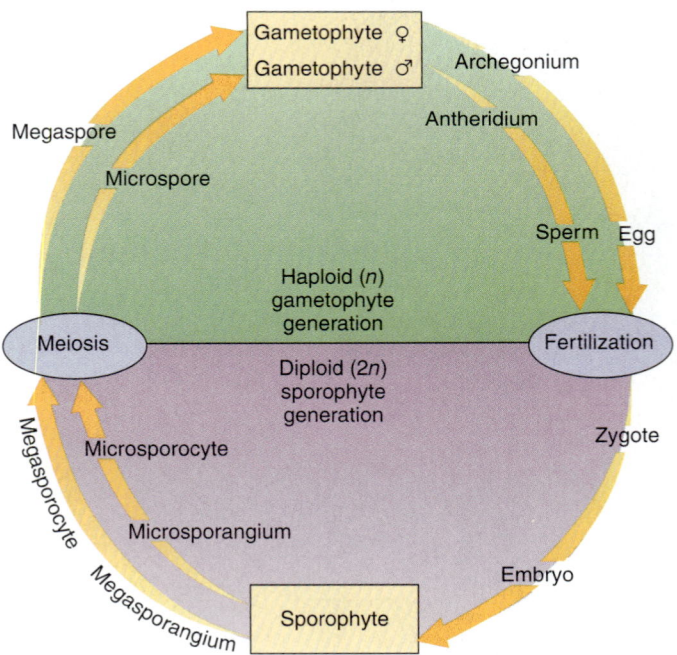

Figure 26–17 **The basic life cycle of heterosporous plants.** Two types of spores, microspores and megaspores, are produced during the life cycle of heterosporous plants.

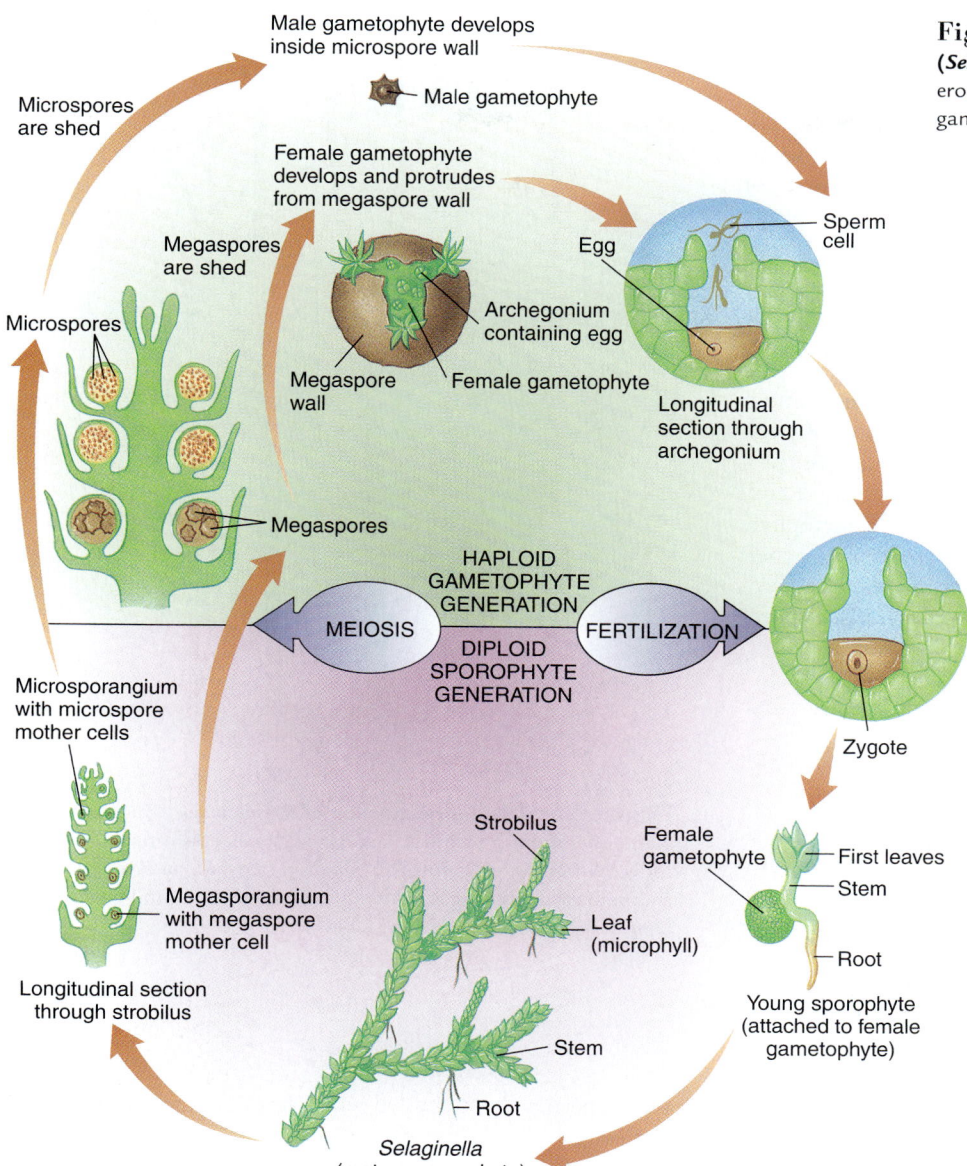

Figure 26–18 The life cycle of spike moss (*Selaginella* sp.). Because spike moss is heterosporous, it produces separate male and female gametophytes.

**Figure labels:**

Male gametophyte develops inside microspore wall

Male gametophyte

Microspores are shed

Microspores

Female gametophyte develops and protrudes from megaspore wall

Megaspores are shed

Egg

Sperm cell

Archegonium containing egg

Megaspore wall

Female gametophyte

Longitudinal section through archegonium

Megaspores

HAPLOID GAMETOPHYTE GENERATION

MEIOSIS

FERTILIZATION

DIPLOID SPOROPHYTE GENERATION

Microsporangium with microspore mother cells

Zygote

Strobilus

Female gametophyte

First leaves

Stem

Root

Leaf (microphyll)

Megasporangium with megaspore mother cell

Young sporophyte (attached to female gametophyte)

Longitudinal section through strobilus

Stem

Root

Selaginella (mature sporophyte)

---

a general survey of algae and plants shows that algae have motile spores and sperm cells; the relatively primitive mosses and ferns have nonmotile spores but have retained motile sperm cells; and the more advanced gymnosperms and flowering plants (see Chapter 27) have nonmotile spores and sperm cells.

Some ferns and club mosses are heterosporous

In the life cycles examined thus far, plants produce only one type of spore as a result of meiosis. This condition, known as **homospory,** is characteristic of bryophytes, horsetails, whisk ferns, and most ferns and club mosses. However, certain ferns and club mosses exhibit **heterospory,** in which they produce two different types of spores: microspores and megaspores. Fig-

ure 26–17 illustrates the generalized life cycle of a heterosporous plant.

Spike moss (*Selaginella* sp.), a small, delicate club moss, is an example of a heterosporous plant (Fig. 26–18). Each strobilus usually bears two kinds of sporangia: microsporangia and megasporangia. *Microsporangia* are sporangia that produce *microsporocytes* (also called *microspore mother cells*), which undergo meiosis to form microscopic, haploid **microspores.** Each microspore can develop into a male gametophyte that produces sperm cells within antheridia. *Megasporangia* in the *Selaginella* strobilus produce *megasporocytes* (also called *megaspore mother cells*). When megasporocytes undergo meiosis, they form haploid **megaspores,** each of which can develop into a female gametophyte that produces eggs in archegonia. In *Selaginella,* the development of male gametophytes from microspores and of female gametophytes from megaspores occurs within their respective spores. As a result, the male and female gametophytes are not truly free-living, unlike the gametophytes of other seedless vascular plants.

Heterospory was a significant development in plant evolution because it was the forerunner of the evolution of seeds. Heterospory is found in the two most successful groups of plants existing today, the gymnosperms and the flowering plants, both of which produce seeds (see Chapter 27).

Seedless vascular plants are used for experimental studies

Many seedless vascular plants are used as experimental models to study certain aspects of plant biology, such as physiology, growth, and development. Some ferns, for example, produce haploid sporophytes and diploid gametophytes. (Recall that the sporophyte is normally diploid, and the gametophyte is normally haploid.) Research involving alternation of generations implicates environmental conditions, rather than chromosome number, in the expression of the characteristic sporophyte and gametophyte morphology.

SEEDLESS VASCULAR PLANTS AROSE MORE THAN 420 MILLION YEARS AGO

Currently, the oldest known megafossils of early vascular plants are from mid-Silurian (420 million years ago) deposits in Europe. (*Megafossils* are fossilized roots, stems, leaves, and reproductive structures.) Megafossils of several kinds of small, seedless vascular plants have also been discovered from Silurian deposits in Bolivia, Australia, and northwestern China. Microscopic spores of early vascular plants appear in the fossil record earlier than megafossils, suggesting that even older megafossils of simple vascular plants remain to be discovered.

The oldest known vascular plants are assigned to phylum Rhyniophyta which, according to the fossil record, arose some 420 million years ago and became extinct about 380 million years ago. *Rhynia* sp. is an example of an early vascular plant

Figure 26–19   **Reconstruction of *Rhynia major*.** This leafless plant, one of Earth's earliest vascular plants, is now extinct. Fossils show that *Rhynia,* which probably lived in marshes, had rhizoids, dichotomously branching rhizomes and upright stems, and sporangia. *Rhynia major* was about 50 cm (20 in) tall.

that superficially resembled whisk ferns in that it consisted of leafless upright stems that branched dichotomously from an underground rhizome (Fig. 26–19). *Rhynia* lacked roots, although it possessed absorptive rhizoids. Sporangia formed at the ends of some of its branches.

SUMMARY WITH KEY TERMS

I. Plants are complex multicellular organisms that obtain energy by photosynthesis. Recent ultrastructural and molecular data indicate that plants probably arose from a group of green algae called charophytes.
   A. Plants and green algae have similar biochemical characteristics: the same photosynthetic pigments, cell wall components, and carbohydrate storage material.
   B. Plants and green algae share similarities in certain fundamental processes like cell division.
II. The colonization of land by plants required the evolution of a number of anatomical, physiological, and reproductive adaptations.
   A. Plants possess a waxy **cuticle** to protect against water loss and **stomata** for gas exchange needed for photosynthesis.
   B. An evolutionary trend in plants has been toward a larger, more dominant **sporophyte generation** and a smaller, less dominant **gametophyte generation**.
   C. Most plants produce multicellular **gametangia** with a protective jacket of sterile cells surrounding the gametes. **Antheridia** produce sperm cells, and **archegonia** produce eggs.
   D. Mosses and ferns, although adapted to life on land, have motile sperm cells and require water as a transport medium for fertilization.
   E. Vascular plants possess **xylem** to conduct water and dissolved minerals and **phloem** to conduct dissolved sugar.
III. Plant life cycles have an **alternation of generations** in which they spend part of their life cycle as a multicellular haploid gametophyte and part as a multicellular diploid sporophyte.
   A. The gametophyte produces haploid gametes by mitosis.

B. These gametes fuse to form a diploid zygote during fertilization.

C. The first stage in the sporophyte generation is the zygote, which develops into a multicellular **embryo** that is protected and nourished by the gametophyte.

D. The mature sporophyte produces sporogenous cells (spore mother cells) that undergo meiosis to form haploid spores, which are the first stage in the gametophyte generation.

IV. **Mosses** and other **bryophytes** have several adaptations that green algae lack, including the possession of a cuticle, stomata, and multicellular gametangia. They are nonvascular (lacking xylem and phloem).

A. Bryophytes are the only plants with a dominant gametophyte generation. Their sporophytes remain permanently attached and nutritionally dependent on the gametophyte.

B. Moss gametophytes are leafy plants that grow from a filamentous **protonema.**

C. **Liverwort** gametophytes are either leafy or **thalloid.**

D. **Hornworts** have thalloid gametophytes.

V. Two types of leaves evolved in the plant kingdom.

A. **Microphylls,** small leaves thought to have evolved from lateral projections of stem tissue, are characteristic of club mosses.

B. **Megaphylls,** leaves that probably evolved from branch systems, are characteristic of all vascular plants other than whisk ferns, which lack leaves, and club mosses.

VI. **Ferns** and fern allies have several adaptations that algae and bryophytes lack, including vascular tissues and a dominant sporophyte generation.

A. Ferns are the largest and most diverse group of seedless vascular plants.

1. Fern sporophytes have roots, rhizomes, and leaves that are megaphylls. Their leaves, called fronds, bear sporangia that produce haploid spores.

2. The fern gametophyte, called a **prothallus,** develops from a haploid spore and bears both archegonia and antheridia.

3. Reproduction in ferns depends on water as a transport medium for their motile sperm cells.

B. Sporophytes of **whisk ferns** consist of dichotomously branching rhizomes and erect stems; they lack true roots and leaves.

C. **Horsetail** sporophytes have roots, rhizomes, aerial stems that are hollow and jointed, and leaves that are reduced megaphylls.

D. Sporophytes of **club mosses** consist of roots, rhizomes, erect branches, and leaves that are microphylls.

VII. Vascular plants are either homosporous or heterosporous.

A. **Homospory,** the production of one kind of spore, is characteristic of bryophytes, whisk ferns, horsetails, most club mosses, and most ferns.

B. **Heterospory,** the production of two kinds of spores (microspores and megaspores), occurs in certain club mosses and ferns and in all seed plants. The evolution of heterospory was an essential step in the evolution of seeds.

1. **Microspores** give rise to male gametophytes that produce sperm cells.

2. **Megaspores** give rise to female gametophytes that produce eggs.

VIII. The fossil record indicates that several lines of vascular plants had evolved by the mid-Silurian period about 420 million years ago.

A. *Rhynia* sp. was one of the earliest vascular plants.

B. The ferns and fern allies were Earth's dominant plants in past ages.

C. Coal formed largely from the prehistoric remains of ancient ferns, club mosses, and horsetails. These plants existed during the Carboniferous period, approximately 300 million years ago.

P O S T - T E S T

1. The bryophytes (a) include mosses, liverworts, and hornworts (b) include whisk ferns, horsetails, and club mosses (c) are small plants that lack a vascular system (d) both a and c (e) both b and c

2. The waxy layer that covers aerial parts of plants is the (a) cuticle (b) archegonium (c) protonema (d) stoma (e) thallus

3. A strengthening compound found in cell walls of vascular plants is the (a) xanthophyll (b) lignin (c) cutin (d) cellulose (e) carotenoid

4. Stomata (a) help prevent desiccation of plant tissues (b) transport water and minerals through plant tissues (c) allow gas exchange for photosynthesis (d) strengthen cell walls (e) produce male gametes

5. The female gametangium, or _____, produces an egg; the male gametangium, or _____, produces sperm cells. (a) antheridium; archegonium (b) archegonium; megaphyll (c) megasporangium; antheridium (d) archegonium; antheridium (e) megasporangium; megaphyll

6. Liverworts and hornworts share life cycle similarities with (a) ferns (b) mosses (c) horsetails (d) club mosses (e) both a and c

7. The leafy green moss plant (a) is the gametophyte generation (b) is the sporophyte generation (c) is called a protonema (d) contains cells with single large chloroplasts (e) both b and c

8. Seedless vascular plants possess _____ to conduct water and dissolved minerals and _____ to conduct dissolved sugar. (a) cuticle; xylem (b) phloem; stoma (c) phloem; xylem (d) stoma; cuticle (e) xylem; phloem

9. Whisk ferns, horsetails, and club mosses share life cycle similarities with (a) ferns (b) mosses (c) hornworts (d) liverworts (e) b, c, and d

10. A(an) _____ is a leaf that arose from a branch system. (a) antheridium (b) microphyll (c) megaphyll (d) sorus (e) microspore

11. These plants have vascularized stems, but lack true roots and leaves (a) mosses (b) club mosses (c) horsetails (d) whisk ferns (e) hornworts

12. These plants have hollow, jointed stems that are impregnated with silica (a) mosses (b) club mosses (c) horsetails (d) whisk ferns (e) hornworts

R E V I E W   Q U E S T I O N S

1. What are the most important environmental challenges that plants face living on land, and what adaptations do they possess to meet these challenges?

2. Plants are thought to have descended from which group of protists? What kinds of evidence support this idea?

3. Define alternation of generations, and distinguish between gametophyte and sporophyte generations.

4. Sketch the generalized life cycle of a moss and label: gametophyte, antheridia, archegonia, sperm cells, egg cell, fertilization, zygote, embryo, sporophyte, capsule, spore mother cells, meiosis, spores, and protonema.

5. How are mosses, liverworts, and hornworts similar? How is each group distinctive?

6. State the adaptations of bryophytes that algae lack. What adaptations do ferns have that both algae and bryophytes lack?

7. Distinguish between megaphylls and microphylls.

8. Sketch the generalized life cycle of a fern and label: sporophyte, frond, rhizome, roots, sorus, sporangium, meiosis, spores, gametophyte (prothallus), antheridium, archegonium, sperm cells, egg cell, fertilization, and zygote.

9. Name the three groups of plants known as fern allies. What features do these plants share with ferns? How is each group distinctive?

10. How does heterospory modify the life cycle?

11. Label the diagram on p. 568. Use Fig. 26–3 to check your answers.

Haploid (*n*)
gametophyte
generation

Diploid (2*n*)
sporophyte
generation

YOU MAKE THE CONNECTION

1. Which group probably colonized the land first, plants or animals? Explain.
2. How might the following trends in plant evolution be adaptive to living on land?
   a. Dependence on water for fertilization → no need for water as a transport medium
   b. Dominant gametophyte generation → dominant sporophyte generation
   c. Homospory → heterospory

RECOMMENDED READINGS

Berg, L.R. *Introductory Botany: Plants, People, and the Environment.* Saunders College Publishing, Philadelphia, 1997. An introduction to general botany with an environmental emphasis.

Camus, J.M., A.C. Jermy, and B.A. Thomas. *A World of Ferns.* Natural History Museum Publications, London, 1991. An outstanding book on ferns.

Graham, L.E. *Origin of Land Plants.* Wiley, New York, 1993. The algal relatives of plants provide insights into plant evolution in a terrestrial environment.

Kenrick, P. and P.R. Crane. "The Origin and Early Evolution of Plants on Land." *Nature,* Vol. 389, 4 Sept. 1997. This review article discusses how recent fossil discoveries and advances in plant systematics have added to our understanding of plant evolution.

Levanthes, L.E. "Mysteries of the Bog." *National Geographic,* Vol. 171, No. 3, Mar. 1987. Discusses peat bogs and the mosses that form them.

Niklas, K.J. "One Giant Step for Life." *Natural History,* Jun. 1994. An overview of the transition from water to land during the evolution of plants.

Pearce, F. "Peat Bogs Hold Bulk of Britain's Carbon." *New Scientist,* Vol. 144, 19 Nov. 1994. A strong case for the preservation of Scottish peat bogs, which contain far more carbon than would be stored in equivalent areas of forest. Because stored carbon is not in the atmosphere as CO_2, it helps prevent global warming.

● Visit our website at **http://www.saunderscollege.com/lifesci/titles.html** and click on Solomon/Berg/Martin Biology for links to chapter-related resources on the World Wide Web.

CHAPTER 27

The Plant Kingdom: Seed Plants

Our discussion of plants in Chapter 26 focused on bryophytes and on ferns and their allies, all of which reproduce by means of *spores,* which are haploid reproductive units that give rise to gametophytes. Although the most successful and widespread plants also produce spores, their primary means of reproduction and dispersal is by **seeds,** each of which consists of an embryonic sporophyte and nutritive tissue surrounded by a protective coat. Seeds develop from the female gametophyte and its associated tissues. The two groups of the seed plants, gymnosperms and angiosperms (flowering plants), exhibit the greatest evolutionary complexity in the plant kingdom and are the dominant plants in most terrestrial environments.

Seeds, such as the lupine (*Lupinus* sp.) seeds in the photograph, are reproductively superior to spores for three main reasons. First, a seed contains a multicellular, well developed young plant with embryonic root, stem, and one or more leaves already formed, whereas a spore is a single cell. Second, a seed contains an abundant food supply. After germination, the plant embryo is nourished by food stored in the seed until it becomes self-sufficient. Because a spore is a single cell, few food reserves exist for the plant that develops from a spore. Third, a seed is protected by a resistant seed coat. Like spores, seeds can live for extended periods of time at reduced rates of metabolism, germinating when conditions become favorable.

Seeds and seed plants have been intimately connected with the development of human civilization. From prehistoric times, early humans collected and used seeds for food. The food stored in seeds is a concentrated source of proteins, oils, carbohydrates, and vitamins, which are nourishing for humans as well as for germinating plants. Also, seeds are easy to store (pro-

(David Cavagnaro)

vided they are kept dry); this has allowed humans to collect them during times of plenty and save them for times of need. Few other foods can be stored as conveniently or for as long. Although most seeds that humans consume are produced by flowering plants, seeds of certain gymnosperms, the piñon pine *(Pinus edulis),* for example, are edible. They are usually sold as "pine nuts."

AFTER YOU HAVE STUDIED THIS CHAPTER YOU SHOULD BE ABLE TO

1. Compare the features of seeds with those of spores and discuss the advantages of plants that reproduce primarily by seeds rather than by spores.
2. Trace the steps in the life cycle of a pine and compare its sporophyte and gametophyte generations.
3. Summarize the features that distinguish gymnosperms from other plants.
4. Name and briefly describe the four phyla of gymnosperms.
5. Summarize the features that distinguish flowering plants from other plants.

6. Diagram a generalized life cycle of a flowering plant and describe double fertilization.
7. Contrast dicots and monocots, the two classes of flowering plants.
8. Discuss the evolutionary adaptations of flowering plants.
9. Trace the evolution of gymnosperms from seedless vascular plants and that of flowering plants from gymnosperms.

GYMNOSPERMS AND FLOWERING PLANTS BEAR SEEDS

In seed plants, each seed develops from an **ovule,** which is a megasporangium and its enclosed structures, following fertilization of an egg cell within. Seed plants are divided into two groups based on whether or not their ovules are protected. The two groups of seed plants are the **gymnosperms** and the **angiosperms** (see Table 27–1). The word *gymnosperm* is adapted from the Greek for "naked seed." Gymnosperms produce seeds that are totally exposed or borne on the scales of cones. In other words, the ovules of gymnosperms are unprotected. Pine, spruce, fir, hemlock, and *Ginkgo* are examples of gymnosperms.

The Greek expression from which the term *angiosperm* is derived translates as "seed enclosed in a vessel or case." Angiosperms are flowering plants that produce their seeds within a fruit. Thus the ovules of angiosperms are protected. Flowering plants are extremely diverse and include corn, oaks, water lilies, cacti, apples, grasses, palms, and buttercups.

Both gymnosperms and flowering plants possess vascular tissues: xylem for the conduction of water and dissolved minerals, and phloem for the conduction of dissolved sugar. Both

have life cycles with an alternation of generations, that is, they spend a portion of their lives in the diploid sporophyte stage and a portion in the haploid gametophyte stage. The sporophyte generation is the dominant stage in each group, and the gametophyte generation is significantly reduced in size and entirely dependent on the sporophyte generation. Unlike the plants we have considered so far (bryophytes and ferns and most of their allies), the gymnosperms and flowering plants do not have free-living gametophytes; instead the gametophyte is attached to and nutritionally dependent on the sporophyte generation (see Chapter 26). All gymnosperms and flowering plants are heterosporous and produce two types of spores, microspores and megaspores.

GYMNOSPERMS ARE THE "NAKED SEED" PLANTS

The gymnosperms include some of the most interesting members of the plant kingdom, including a number of record holders. For example, a giant sequoia *(Sequoiadendron giganteum)*

TABLE 27-1 A Comparison of Gymnosperms and Flowering Plants

Characteristic	Gymnosperms	Angiosperms
Growth habit	Woody trees and shrubs	Woody or herbaceous
Conducting cells in xylem	Tracheids*	Vessel elements and tracheids*
Reproductive structures	Cones (usually)	Flowers
Pollen grain transfer	Wind	Animals or wind
Fertilization	Egg and sperm → zygote; double fertilization in gnetophytes	Double fertilization: egg and sperm → zygote; 2 polar nuclei and sperm → endosperm
Seeds	Exposed or borne on scales of cones	Enclosed within fruit

*See Chapter 31.

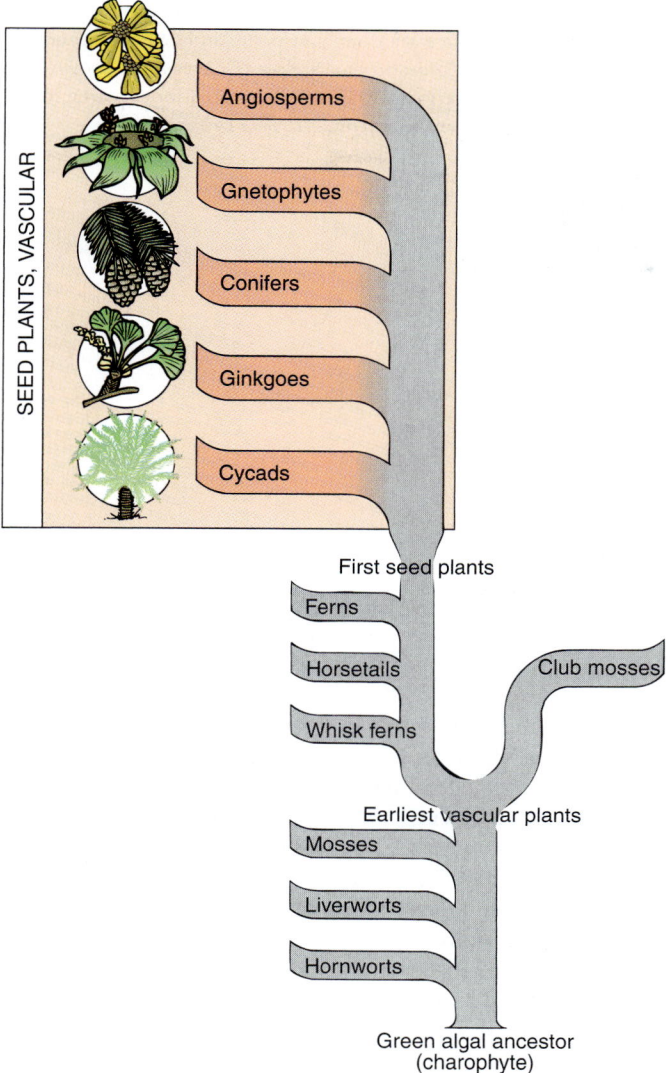

Figure 27–1 **Possible evolutionary relationships among plants.** The seed plants, which are the focus of this chapter, are the five phyla at the top of the tree. Four of these phyla (cycads, ginkgoes, conifers, and gnetophytes) are gymnosperms. The angiosperms, or flowering plants, have more living species than all other plant phyla combined.

cycads (phylum Cycadophyta), represent evolutionary remnants of groups that were more significant in the past. The fourth phylum of gymnosperms, the gnetophytes (phylum Gnetophyta), is a collection of some unusual plants that share certain traits not found in other gymnosperms.

Conifers are woody plants that produce seeds in cones

The **conifers** (phylum Coniferophyta), which include pines, spruces, hemlocks, and firs, are the most familiar group of gymnosperms (Fig. 27–2). They are woody trees or shrubs that produce annual additions of secondary tissues (wood and bark; see Chapter 33); there are no herbaceous (nonwoody) conifers. The wood (secondary xylem) is composed of tracheids, which are long, tapering cells with pits through which water and dissolved minerals move from one cell to another.

Many conifers produce **resin,** a viscous, clear or translucent substance consisting of several organic compounds that may protect the plant from attack by fungi or insects. The resin collects in resin ducts, tubelike cavities that extend throughout the roots, stems, and leaves. Resin is produced and secreted by cells lining the resin ducts.

Most conifers have leaves (megaphylls; see Chapter 26) called **needles** that are commonly long and narrow, tough, and leathery (Fig. 27–3). Pines bear clusters of two to five needles, depending on the species. In a few conifers such as American arborvitae *(Thuja occidentalis),* however, the leaves are scalelike and cover the stem. Most conifers are evergreen and bear their leaves throughout the year. Only a few, such as the dawn redwood, larch, and bald cypress, are deciduous and shed their needles at the end of the growing season.

Most conifers are **monoecious,** which means that they have separate male and female reproductive parts in different locations on the same plant. These reproductive parts are generally borne in strobili that are commonly called cones, hence the name *conifer,* which means "bears cones."

The approximately 550 species of conifers occupy extensive areas, ranging from the Arctic to the tropics, and are the dominant vegetation in the forested regions of Alaska, Canada, northern Europe, and Siberia. In addition, they are important in the Southern Hemisphere, particularly in wet, mountainous areas of temperate and tropical regions in South America, Australia, New Zealand, and Malaysia. Southwestern China, with more than 60 species of conifers, has the greatest regional diversity of conifer species on Earth. California, New Caledonia (an island west of Australia), southeastern China, and Japan also have considerable diversity of conifer species.

Ecologically, conifers contribute food and shelter to animals and other organisms, and their roots hold the soil in place and help prevent soil erosion. Humans use conifers for their wood (for building materials as well as paper products), medicine (for example, the anticancer drug taxol from the Pacific yew), turpentine, and resins. Because of their attractive appearance, conifers are grown for landscape design and Christmas trees.

known as the General Sherman Tree, located in Sequoia National Park in California, is one of the world's most massive organisms. It is 82 m (272 ft) tall and has a girth of 23.7 m (79 ft) measured 1.5 m (5 ft) above ground level. Another gymnosperm, a coast redwood *(Sequoia sempervirens),* is probably the world's tallest tree, measuring almost 117 m (385 ft) in height. One of the oldest living trees, a bristlecone pine *(Pinus aristata)* in the White Mountains of California, has been determined by tree ring analysis to be 4900 years old!

Gymnosperms are usually classified into four phyla, which represent four different evolutionary lines (Fig. 27–1). Numbering 550 species, the largest phylum of gymnosperms is the phylum Coniferophyta, commonly called conifers. Two phyla of gymnosperms, the ginkgoes (phylum Ginkgophyta) and the

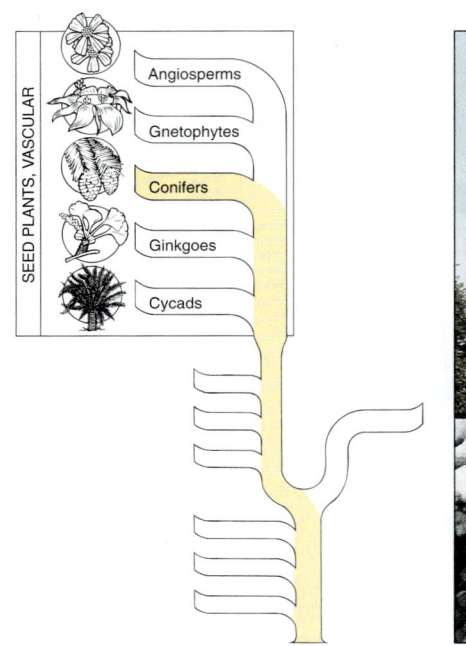

SEED PLANTS, VASCULAR

Angiosperms

Gnetophytes

Conifers

Ginkgoes

Cycads

Figure 27-2 **A representative conifer.**
Conifers are dominant plants in northern latitudes.
The Colorado blue spruce *(Picea pungens)* shown is
an evergreen that is well adapted for surviving in
cold environments. *(Dennis Drenner)*

Pines represent a typical conifer life cycle

The genus *Pinus,* by far the largest genus in the conifers, consists of about 100 species. A pine tree is a mature sporophyte. Pine is heterosporous and therefore produces microspores and megaspores, in separate cones (Fig. 27–4).[1] Male cones, usually 1 cm or less in length, are smaller than female cones and are generally produced on the lower branches each spring (Fig. 27–5). The more familiar, woody female cones, which are on the tree year-round, are usually found on the upper branches

[1]It might be helpful to review alternation of generations in Chapter 26 before studying the pine life cycle.

of the tree and bear seeds after reproduction. Female cones vary considerably in size. The sugar pine *(Pinus lambertiana)* of California produces the world's longest female cones, which reach lengths of 60 cm (2 ft).

Each male cone (also called a pollen cone) is composed of **sporophylls,** leaflike structures that bear sporangia at their bases. At the base of each sporophyll are two microsporangia, which contain numerous microsporocytes (also called microspore mother cells). Each microsporocyte undergoes meiosis to form four haploid microspores. Microspores then develop into extremely reduced male gametophytes. Each immature male gametophyte, also called a **pollen grain,** consists of four cells, two of which—a generative cell and a tube cell—are involved in reproduction. The other two cells soon degenerate. Two large air sacs on each pollen grain provide

(a) White pine
(Pinus strobus)

(b) American arborvitae
(Thuja occidentalis)

Figure 27-3 **Leaf variation in conifers.** *(a)* In white pine *(Pinus strobus),* leaves are long, slender needles that occur in clusters of five. *(b)* In American arborvitae *(Thuja occidentalis),* leaves are small and scalelike *(see inset).*

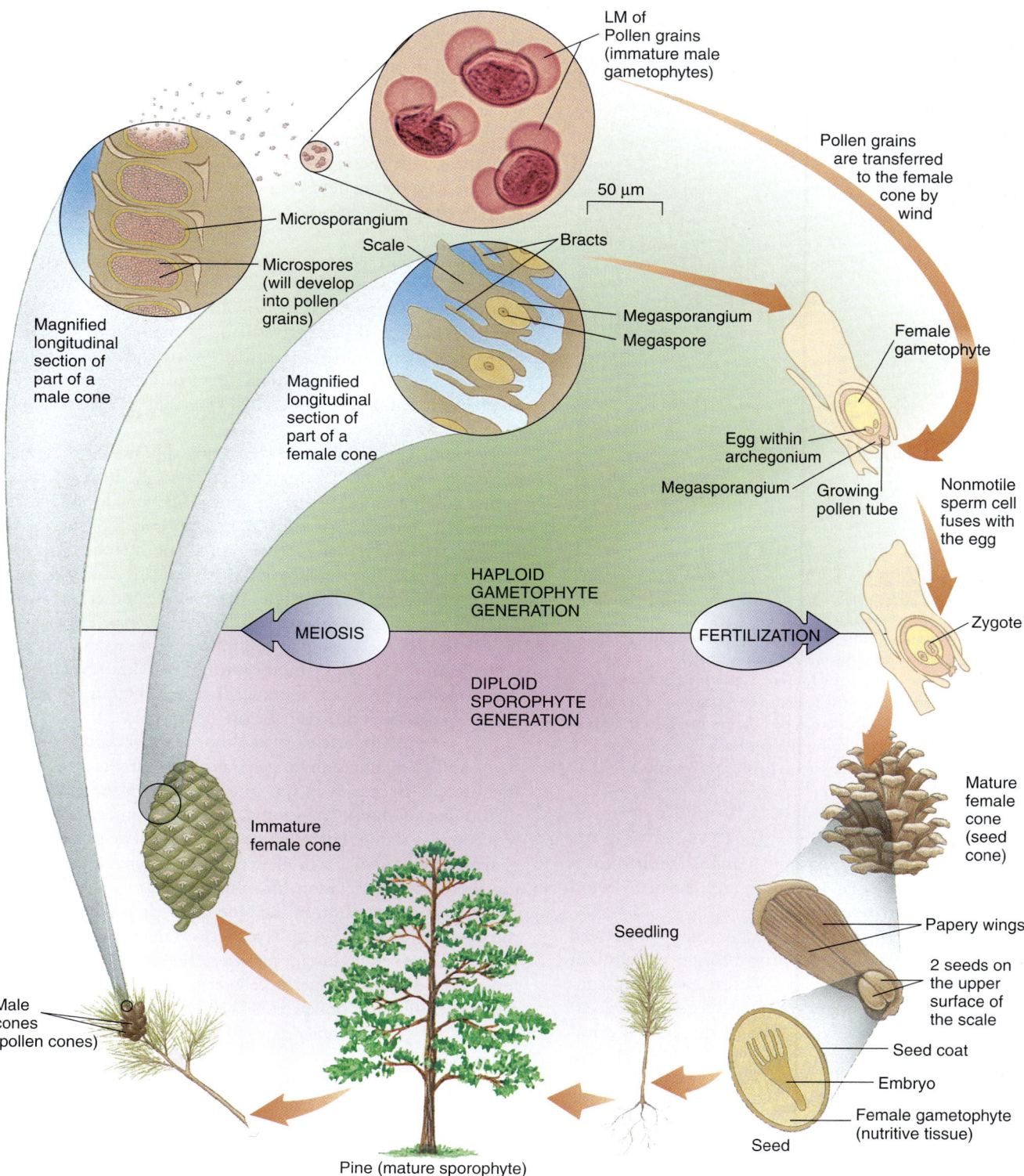

LM of Pollen grains (immature male gametophytes)

Pollen grains are transferred to the female cone by wind

Microsporangium

Scale

Bracts

50 μm

Microspores (will develop into pollen grains)

Magnified longitudinal section of part of a male cone

Megasporangium

Megaspore

Female gametophyte

Magnified longitudinal section of part of a female cone

Egg within archegonium

Megasporangium

Growing pollen tube

Nonmotile sperm cell fuses with the egg

HAPLOID GAMETOPHYTE GENERATION

MEIOSIS

FERTILIZATION

Zygote

DIPLOID SPOROPHYTE GENERATION

Immature female cone

Mature female cone (seed cone)

Papery wings

2 seeds on the upper surface of the scale

Seedling

Seed coat

Embryo

Female gametophyte (nutritive tissue)

Male cones (pollen cones)

Seed

Pine (mature sporophyte)

Figure 27–4 **The life cycle of pine.** One major advantage of gymnosperms over the seedless vascular plants is the production of wind-borne pollen grains *(top)*. Pines and other gymnosperms do not depend on water as a transport medium for sperm cells. *(Manfred Kage/Peter Arnold, Inc.)*

Figure 27–5 Male and female cones in lodgepole pine **(Pinus contorta).** *(Bottom)* Clusters of golden male cones produce copious amounts of pollen grains in the spring. *(Top)* Mature, dark brown female cones have opened to shed their seeds. The location of female cones above male cones facilitates cross pollination between different trees, because it is unlikely that wind will blow pollen grains directly upward to female cones on the same tree. *(Walt Anderson/Visuals Unlimited)*

buoyancy for wind dissemination. Pollen grains are shed from male cones in great numbers, and some are carried by wind currents to the immature female cones.

The woody scales of the female cones (also called seed cones) are considered by many botanists to be sporophylls that bear megasporangia at their bases. Within each megasporangium, meiosis of a megasporocyte (megaspore mother cell) produces four haploid megaspores. One of these divides mitotically, developing into the female gametophyte, which produces an egg within each of several archegonia. The other three megaspores are nonfunctional and soon degenerate. When the ovule is ready to receive pollen, it produces a sticky droplet at the opening where the pollen grains land. **Pollination,** the transfer of pollen to the female cones, occurs in the spring during a period of a week or ten days, after which the pollen cones wither and drop off the tree.

One of the many pollen grains that adhere to the sticky female cone grows a **pollen tube,** an outgrowth that digests its way through the megasporangium to the egg within the archegonium. The tube cell, which is involved in pollen tube growth, and the generative cell enter the pollen tube. The generative cell divides and forms a stalk cell and a body cell; the body cell later divides and forms two nonmotile sperm cells. When the pollen tube reaches the female gametophyte, it discharges the two sperm cells near the egg. One of these sperm cells fuses with the egg, in a process known as **fertilization,** to form a zygote, or fertilized egg, which subsequently grows into a young pine embryo in the seed. The other sperm cell degenerates.

The developing embryo, which consists of an embryonic root and an embryonic shoot with several cotyledons (embryonic leaves), is embedded in haploid female gametophyte tissue that becomes the nutritive tissue in the mature pine seed. The embryo and nutritive tissue are surrounded by a tough, protective seed coat that extends out to form a thin papery wing at one end of the seed. This wing enables the seed to be dispersed by air currents.

A long time elapses between the appearance of female cones on a tree and the maturation of seeds in those cones. When pollination occurs in the spring, the female cone is still immature, and meiosis of the megasporocytes (megaspore mother cells) has not yet occurred. During the following year, the female reproductive tissue gradually matures, and eggs form within archegonia. Meanwhile, the pollen tube grows slowly through the megasporangium to the archegonia. Fertilization occurs during the spring of the following year, and the embryo begins to develop. Seed maturation takes several additional months, although some seeds remain within the female cones for a number of years before being shed.

In the pine life cycle, the sporophyte generation is dominant, and the gametophyte generation is restricted in size to microscopic structures in the cones. Although the female gametophyte produces archegonia, the male gametophyte is so reduced that it does not produce antheridia. The gametophyte generation in pines, as in all seed plants, depends totally on the sporophyte generation for nourishment.

A major adaptation in the pine life cycle is elimination of the need for external water as a sperm transport medium. Instead, pine pollen grains are carried to female cones by air currents, and nonmotile sperm cells accomplish fertilization by moving through a pollen tube to the egg. Pine and other conifers are plants whose reproduction is totally adapted for life on land.

Cycads have seed cones and compound leaves

Cycads (phylum Cycadophyta) were very important during the Triassic period, which began approximately 248 million years ago and is sometimes referred to as the "Age of Cycads." Most species are now extinct, and the few surviving cycads, about 140 species, are tropical and subtropical plants with stout, trunklike stems and compound leaves that resemble

Figure 27-6 **Cycads.** Cycads are tropical gymnosperms with a palmlike appearance. Note the immense seed cones on this South African cycad. *(Walter H. Hodge/Peter Arnold, Inc.)*

those of palms or tree ferns (Fig. 27–6). Many cycads are endangered, primarily because they are popular as ornamentals and are gathered from the wild and sold to collectors.

Cycad reproduction is similar to that in pines except that cycads are **dioecious** and therefore have seed cones on female plants and pollen cones on male plants. Their seed structure is most like that of the earliest seeds found in the fossil record. Cycads have also retained the primitive feature of motile sperm cells, each of which possesses many hairlike flagella. These motile sperm cells are a vestige retained from the ancestors of cycads, in which sperm cells swam from antheridia to archegonia. Cycad pollen grains, however, are carried by air or possibly insects to the female plants; there the pollen grain germinates and grows a pollen tube down which the sperm cells swim to get to the egg. In other words, despite having motile sperm cells, cycads do not need water as a transport medium for fertilization.

Ginkgo biloba is the only living species in its phylum

Ginkgoes (phylum Ginkgophyta) are represented by a single living species, the maidenhair tree *(Ginkgo biloba)* (Fig. 27–7). A native of eastern China, it has been found growing in the wild in only two locations, and it is likely that it would have become extinct had it not been cultivated for centuries in Chinese monasteries. *Ginkgo* is the oldest genus (and species) of living trees. Fossil ginkgoes 200 million years old have been discovered that are nearly identical to the modern-day ginkgo.

Ginkgo is commonly planted in North America and Europe today, particularly in parks and along city streets, because it is hardy and somewhat resistant to air pollution. Its leaves are deciduous and turn a beautiful yellow before being shed in the fall.

Like cycads, ginkgoes are dioecious, with separate male and female trees. They have flagellated sperm cells, an evolutionary vestige that is not required because ginkgoes produce airborne pollen grains. Ginkgo seeds are completely exposed rather than occurring within cones. Male trees are typically planted because the female trees bear seeds whose fleshy outer coverings give off a foul odor that smells like rancid butter. In China and Japan, where people eat the seeds, the female trees are more common.

Ginkgo has been an important medicinal plant for centuries. Extracts from the leaves may enhance neurological functioning by increasing blood flow to the brain. Several studies suggest that ginkgo may improve memory in elderly people suffering from Alzheimer's disease; no data suggest that it improves memory in young people, however.

Gnetophytes include three unusual genera

The **gnetophytes** (phylum Gnetophyta) consist of about 70 species in three diverse genera *(Gnetum, Ephedra, Welwitschia)*. Gnetophytes share certain features that make them clearly more advanced than the rest of the gymnosperms. For example, gnetophytes have more efficient water-conducting cells, called vessel elements, in their xylem (see Chapter 31). Flow-

(a)

(b)

Figure 27–7 **The ginkgo, or maidenhair tree (Gingko biloba).** *(a)* A young male ginkgo tree. *(b)* Close-up of a branch from a female ginkgo tree, showing the exposed seeds and the distinctive, fan-shaped leaves. *(a, Carlyn Iverson; b, Marion Lobstein)*

ering plants also have vessel elements in their xylem, but no gymnosperms except the gnetophytes do. Also, the cone clusters produced by some of the gnetophytes resemble flower clusters, and certain details in their life cycles resemble those of flowering plants.

The genus *Gnetum* contains tropical vines, shrubs, and trees with broad leaves that resemble those of flowering plants (Fig. 27–8*a*). Species in the genus *Ephedra* include many shrubs and vines found in deserts and other dry temperate and tropical regions. Some *Ephedra* species resemble horsetails in that they possess jointed green stems with tiny leaves (Fig. 27–8*b*). Commonly called joint fir, *Ephedra* has been used medicinally for centuries. An Asiatic *Ephedra* is the source of ephedrine, which stimulates the heart and raises blood pressure. Ephedrine is sold over the counter in weight-control medications and herbal energy-boosters; several deaths have been reported from chronic use or overdose of products containing ephedrine.

The third gnetophyte genus, *Welwitschia,* contains a single species found in deserts of southwestern Africa (Fig. 27–8*c*). Most of *Welwitschia's* body—a long taproot—grows underground. Its short, wide stem forms a shallow disk, up to 0.9 m (3 ft) in diameter, from which two ribbon-like leaves extend. These two leaves continue to grow from the stem throughout the plant's life, but their ends are usually broken and torn by the wind, giving the appearance of numerous

leaves. Each leaf grows to about 2 m (6.6 ft) in length. When *Welwitschia* reproduces, it forms cones around the edge of its disklike stem.

FLOWERING PLANTS PRODUCE FLOWERS, FRUITS, AND SEEDS

Flowering plants or **angiosperms** (phylum Anthophyta)[2] are the most successful plants today, surpassing even the gymnosperms in importance. They have adapted to almost every habitat and, with at least 235,000 species, are Earth's dominant plants. Flowering plants come in a wide variety of sizes and forms, from herbaceous violets to massive eucalyptus trees. Some flowering plants—tulips and roses, for example—have large, conspicuous flowers, whereas others, such as grasses and oaks, produce flowers that are small and inconspicuous.

Flowering plants are vascular plants that reproduce sexually by forming flowers, and, after a unique double fertilization process that we will discuss shortly, seeds within fruits.

[2]Some botanists prefer to classify flowering plants in phylum Magnoliophyta instead of phylum Anthophyta. The International Botanical Congress has not yet standardized phylum names, although the matter is currently under study.

(a)

(b)

(c)

Figure 27–8 **Gnetophytes.** *(a)* The leaves of *Gnetum gnemon* resemble those of flowering plants. The exposed seeds are red to yellow when ripe. *(b)* A male joint fir (*Ephedra* sp.) has pollen cones clustered at the nodes. Most species of *Ephedra* are dioecious and have separate male and female plants. Nineteenth century pioneers used species native to desert areas in the American Southwest to make a beverage called Mormon tea. *(c)* *Welwitschia* is a gnetophyte that is native to deserts in southwestern Africa. The two leaves of this extremely unusual plant grow from the edges of a short, wide stem and become tattered and torn over time. *(a, John D. Cunningham/Visuals Unlimited; b, David Cavagnaro; c, Patti Murray)*

The fruit protects the developing seeds and often aids in their dispersal (see Chapter 35). Flowering plants possess efficient water-conducting cells called vessel elements in their xylem, and efficient sugar-conducting cells called sieve tube members in their phloem (see Chapter 31).

Flowering plants are extremely important to humans because our survival as a species literally depends on them. All of our major food crops are flowering plants, including cereal crops such as rice, wheat, corn, and barley. Woody flowering plants such as oak, cherry, and walnut provide valuable lumber. Flowering plants give us fibers like cotton and linen and medicines like digitalis and codeine. Products as diverse as rubber, tobacco, coffee, chocolate, and aromatic oils for perfumes come from flowering plants. Economic botany is the subdiscipline of botany that deals with plants of economic importance, most of which are flowering plants.

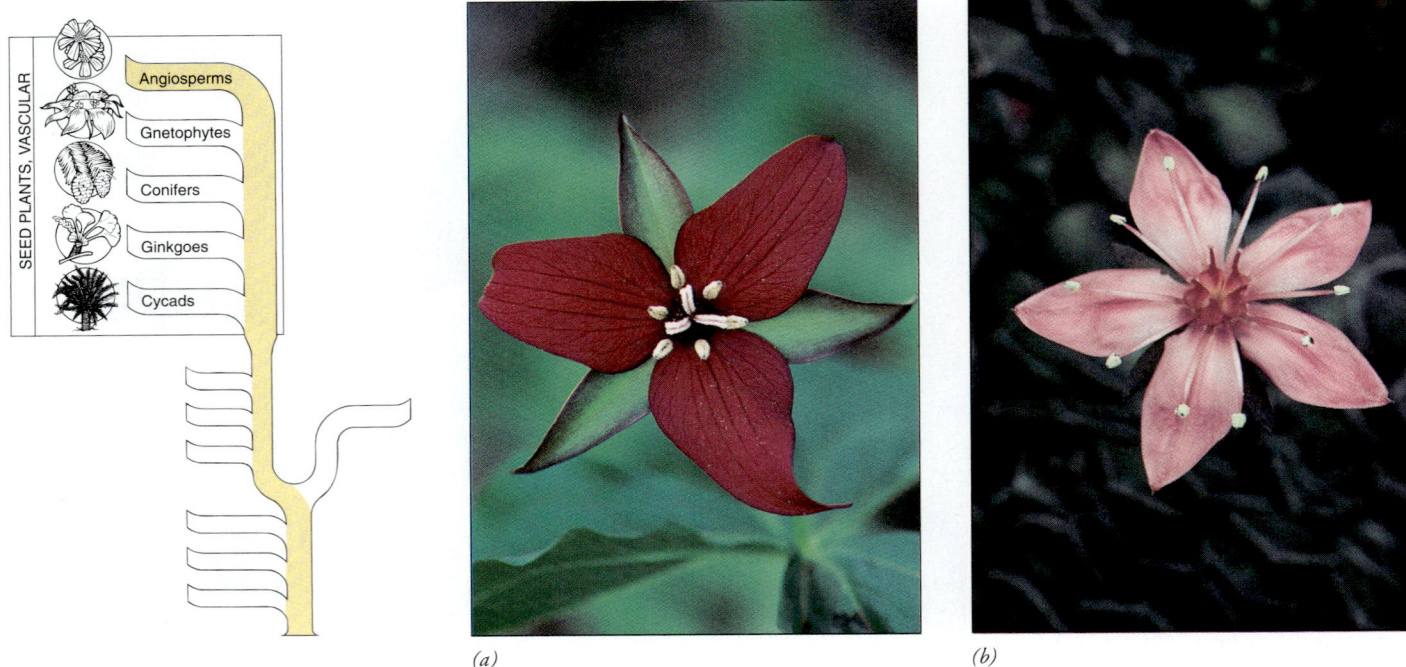

Figure 27–9 **Monocots and dicots.** *(a)* Most monocots such as this *Trillium erectum* have their floral parts in threes. Note the three green sepals, three red petals, six stamens, and three stigmas (the compound pistil is composed of three fused carpels). *(b)* Most dicots such as this *Tacitus* sp. have their floral parts in fours or fives. Note the five petals, ten stamens, and five separate pistils. *(a, John Gerlach/Tom Stack & Associates; b, Richard H. Gross)*

Monocots and dicots are the two classes of flowering plants

Phylum Anthophyta is divided into two classes, the monocots (class Monocotyledones) and dicots (class Dicotyledones)[3] (Fig. 27–9). Monocots include palms, grasses, orchids, irises, onions, and lilies. The dicot class includes oaks, roses, mustards, cacti, blueberries, and sunflowers. Dicots are more diverse and include many more species (at least 170,000) than the monocots (at least 65,000).

The cladistic analysis of recent molecular evidence suggests that the dicots are not a monophyletic group but instead are paraphyletic. Recall from Chapter 22 that a paraphyletic group contains a common ancestor and some, but not all, of its descendants. According to molecular evidence such as DNA sequence comparisons, some dicots—the magnolias and laurels—are more closely related to the monocots than to other dicots. Despite these data, botanists currently keep the classes Monocotyledones and Dicotyledones for convenience. Table 27–2 provides a comparison of some of the features of the two classes.

Monocots are mostly herbaceous plants with long, narrow leaves that have parallel veins (the main leaf veins run parallel to one another). The flower parts of monocot flowers usually occur in threes or multiples of three. For example, a flower might have three sepals, three petals, six stamens, and a compound pistil consisting of three fused carpels (these flower parts are discussed shortly). Monocot seeds have a single **cotyledon,** or embryonic seed leaf, and **endosperm,** nutritive tissue, is usually present in the mature seed.

Dicots may be herbaceous (for example, a tomato plant) or woody (for example, a hickory tree). Their leaves vary in shape but usually are broader than monocot leaves, with netted veins (branched veins resembling a net). Flower parts usually occur in fours or fives or multiples thereof. Two cotyledons are present in dicot seeds, and endosperm is usually absent in the mature seed, having been absorbed by the two cotyledons.

Flowers are involved in sexual reproduction

Flowers are reproductive shoots usually composed of four parts —sepals, petals, stamens, and carpels—arranged in whorls (circles) on the end of a flower stalk, or **peduncle** (Fig. 27–10). The peduncle may terminate in a single flower or a cluster of flowers known as an **inflorescence.** The tip of the peduncle enlarges to form a **receptacle** that bears some or all of the flower parts.

Botanists hypothesize that the flower is a modified branch in which the leaves evolved into sepals, petals, stamens, and

[3]Some botanists prefer to classify monocots in class Liliopsida and dicots in class Magnoliopsida. The International Botanical Congress has not yet standardized class names, although the matter is currently under study.

TABLE 27-2   Distinguishing Features of Dicots and Monocots

Feature	Dicots	Monocots
Flower parts	Usually in fours or fives	Usually in threes
Leaf venation	Usually netted	Usually parallel
Vascular bundles in stem cross section	Arranged in a circle (ring)	Usually scattered or more complex arrangement
Roots	Taproot system	Fibrous root system
Seeds	Two cotyledons	One cotyledon

carpels. That is, sepals and petals are thought by many botanists to be sterile modified leaves, while stamens and carpels are thought to be fertile modified leaves. (Recall from Chapter 19 that evolutionary novelties, such as the flower, originate as modifications of preexisting structures.)

All four floral parts are important in the reproductive process, but only the stamens (the "male" organs) and carpels (the "female" organs) produce gametes. A flower that has all four parts is said to be **complete,** whereas an **incomplete** flower lacks one or more of these four parts. A flower with both stamens and carpels is said to be **perfect,** whereas an **imperfect** flower has stamens or carpels, but not both.

Sepals, which make up the lowermost and outermost whorl on a floral shoot, are leaflike in appearance and often green (Fig. 27–11a). Sepals cover and protect the other flower parts when the flower is a bud. As the blossom opens, the sepals fold back to reveal the more conspicuous petals. The collective term for all the sepals of a flower is the **calyx.**

The whorl just above the sepals consists of **petals,** which are broad, flat, and thin (like sepals and leaves) but vary in shape and are frequently brightly colored. Petals play an important role in attracting animal pollinators to the flower (see Chapter 35). Sometimes petals are fused to form a tube (for example, trumpet honeysuckle flowers, Fig. 27–11b) or other floral shape (for example, snapdragons). The petals of a flower are referred to collectively as the **corolla.**

Just inside the petals is a whorl of **stamens** (Fig. 27–11c). Each stamen is composed of a thin stalk, called a **filament,** and a saclike **anther,** where meiosis occurs to form microspores that develop into pollen grains. Each pollen grain produces two cells surrounded by a tough outer wall. One cell divides to form two male gametes, or sperm cells, and the other pro-

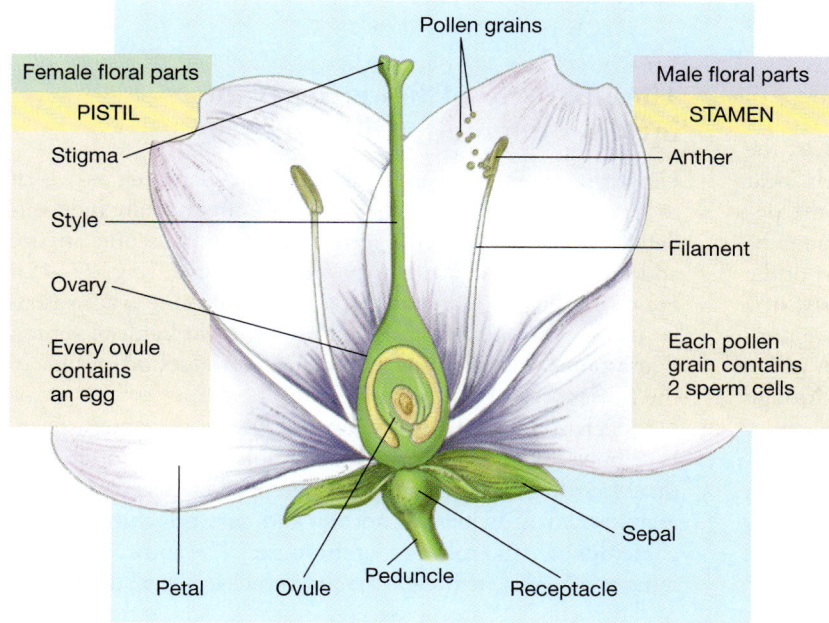

Figure 27–10   Floral structure. This cutaway view of a "typical" flower shows the details of basic floral structure. This flower is both a complete and a perfect flower. Not all flowers possess all these structures.

Petals Sepals

(a)

(c)

(b)

Figure 27–11 **The four parts of a flower.** *(a)* The leaflike sepals of a rose *(Rosa* sp.) bud cover and protect the inner flower parts. *(b)* The five petals of each trumpet honeysuckle *(Lonicera semper-virens)* flower are fused to form a tubular corolla. *(c)* A twinleaf *(Jeffersonia diphylla)* flower has eight stamens. Note the compound pistil with its rounded green ovary in the center of the flower. *(a, James Mauseth, University of Texas; b, James L. Castner; c, Marion Lobstein)*

duces a pollen tube through which the sperm cells travel to reach the ovule.

In the center of most flowers is one or more **carpels,** the "female" reproductive organs (Fig. 27–11*c*). Carpels bear ovules, which, as you may recall, are structures with the potential to develop into seeds. The carpels of a flower may be separate or fused into a single structure. The female part of the flower is also referred to as a **pistil.** A pistil may consist of a single carpel (a *simple* pistil) or a group of fused carpels (a *compound* pistil). Each pistil has three sections: a **stigma,** on which the pollen grain lands; a **style,** a necklike structure through which the pollen tube grows; and an **ovary,** an enlarged structure that contains one or more ovules. Each young ovule contains a female gametophyte that forms one female gamete (an egg) and two polar nuclei. The egg and polar nuclei participate directly in fertilization, and, following fertilization, the ovule develops into a seed and the ovary into a fruit.

The life cycle of flowering plants includes double fertilization

Flowering plants have an alternation of generations in which the sporophyte generation is larger and nutritionally independent and the gametophyte generation is microscopic in size and nutritionally dependent on the sporophyte (Fig. 27–12). Flowering plants, like gymnosperms and certain other vascular plants, are heterosporous and produce two kinds of spores: microspores and megaspores. Sexual reproduction occurs in the flower.

Each young ovule within an ovary contains a megasporo-cyte (megaspore mother cell) that undergoes meiosis to produce four haploid megaspores. Three of these usually disintegrate, and one divides mitotically and develops into a female gametophyte, also called an **embryo sac.** The embryo sac contains seven cells with eight haploid nuclei. Six of these cells,

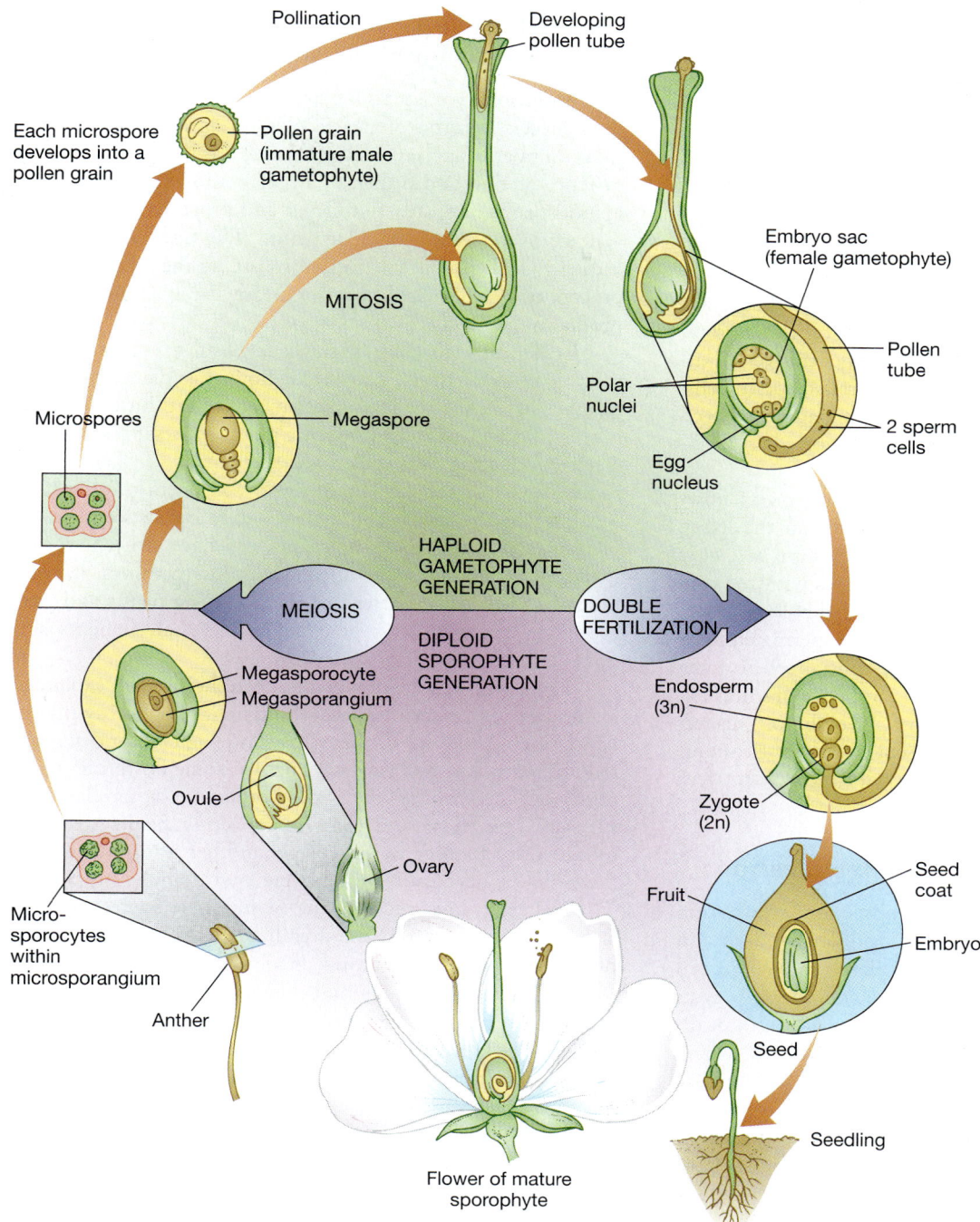

Figure 27-12 **The life cycle of flowering plants.** A significant feature of the flowering plant life cycle is double fertilization, in which one sperm cell unites with the egg, forming a zygote, and the other sperm cell unites with the two polar nuclei, forming a triploid cell that gives rise to endosperm.

including the egg cell, contain a single nucleus each, and a central cell has two nuclei, called **polar nuclei.** The egg and the central cell with two polar nuclei are directly involved in fertilization; the other five cells in the embryo sac apparently have no direct role in the fertilization process and disintegrate. It has been hypothesized, however, that as the synergids (the two cells closely associated with the egg) disintegrate, they release chemicals that may affect the direction of pollen tube growth.

Each pollen sac, or microsporangium, of the anther contains numerous microsporocytes (microspore mother cells), each of which undergoes meiosis to form four haploid microspores. Every microspore develops into an immature male gametophyte, also called a pollen grain. Pollen grains are extremely small; each consists of two cells, the tube cell and the generative cell.

The anther sacs split open and begin to shed pollen. Pollen grains are transferred to the stigma by a variety of agents, including wind, water, insects, and other animal pollinators (see Chapter 35). If compatible with the stigma, the pollen grain germinates; that is, the tube cell forms a pollen tube that grows down the style and into the ovary. Next, the generative cell divides to form two nonmotile male gametes called sperm cells. The sperm cells move down the pollen tube and are discharged into the embryo sac. Both sperm cells are involved in fertilization.

Something happens during sexual reproduction in flowering plants that does not occur anywhere else in the living world. When the two sperm cells enter the embryo sac, *both* participate in fertilization. One sperm cell fuses with the egg, forming a zygote that grows by mitosis and develops into a multicellular embryo in the seed. The second sperm cell fuses with the two haploid polar nuclei of the central cell to form a triploid ($3n$) cell that grows by mitosis and develops into endosperm, a nutrient tissue that nourishes the embryo. This fertilization process, which involves two separate nuclear fusions, is called **double fertilization** and is, with two exceptions, unique to flowering plants.[4]

Seeds and fruits develop after fertilization

As a result of double fertilization and subsequent growth and development, each seed contains a young plant embryo and nutritive tissue (the endosperm), both of which are surrounded by a protective seed coat. In monocots the endosperm persists and is the main source of food in the mature seed. In most dicots the endosperm nourishes the developing embryo, which subsequently stores food in its cotyledons.

As a seed develops from an ovule following fertilization, the ovary wall surrounding it enlarges dramatically and develops into a **fruit.** In some instances, other tissues associated with the ovary also enlarge to form the fruit. Fruits serve two purposes: to protect the developing seeds from desiccation as they grow and mature and to aid in the dispersal of seeds (see Chapter 35). For example, dandelion fruits have feathery plumes that enable the entire fruit to be carried by air currents. Once a seed lands in a suitable place, it may germinate and develop into a mature sporophyte that produces flowers, and the life cycle continues as described.

Flowering plants have many adaptations that account for their success

The evolutionary adaptations of flowering plants account for their success in terms of their ecological dominance and their great number of species. Seed production as the primary means of reproduction and dispersal, an adaptation shared with the gymnosperms, is clearly significant and provides a definite advantage over seedless vascular plants. The presence of closed carpels, which results in fruits surrounding the seeds, and the process of double fertilization increase the likelihood of reproductive success.

In addition to their highly successful reproduction involving flowers, fruits, and seeds, flowering plants possess a number of distinctive features that have contributed to their success. Recall that most flowering plants possess very efficient water-conducting cells, called vessel elements, in their xylem, in addition to tracheids. In contrast, the xylem of almost all seedless vascular plants and gymnosperms consists exclusively of tracheids. Most flowering plants also have efficient carbohydrate-conducting cells, called sieve tube members, in their phloem. With the exception of the flowering plants and gnetophytes, vascular plants generally lack vessel elements and sieve tube members.

The leaves of flowering plants, with their broad, expanded blades, are very efficient at absorbing light for photosynthesis. Abscission (shedding) of these leaves during cold or dry spells reduces water loss and thus has enabled some flowering plants to expand into habitats that would otherwise be too harsh for survival. The stems and roots of flowering plants are often modified for food or water storage, another feature that helps flowering plants to survive in severe environments.

Probably most crucial to the evolutionary success of flowering plants, however, is the overall adaptability of the sporophyte generation. As a group, flowering plants readily adapt to new habitats and changing environments. This adaptability is evident in the great diversity exhibited by the species of flowering plants. For example, the cactus is remarkably well adapted to desert environments. Its stem stores water; its leaves (spines) have a reduced surface area available for transpiration and may also protect against thirsty herbivorous animals; and its thick, waxy cuticle reduces water loss. On the other hand, the water lily is well adapted for wet environments, in part because it has air channels that provide adequate oxygen to stems and roots living in oxygen-deficient water and mud.

THE FOSSIL RECORD PROVIDES CLUES ABOUT THE EVOLUTION OF SEED PLANTS

One of the groups that descended from ancestral seedless vascular plants was the **progymnosperms,** all of which are now

---

[4]Double fertilization was reported in the gymnosperms *Ephedra nevadensis* (in 1990) and *Gnetum gnemon* (in 1996). This process differs from double fertilization in flowering plants in that an additional zygote, rather than endosperm, is produced. The second zygote later disintegrates.

extinct (Fig. 27–13a). Progymnosperms had two derived features that their immediate ancestors lacked: leaves that were megaphylls and woody tissue, that is, secondary xylem, similar to that of modern gymnosperms. Progymnosperms retained a primitive feature of their ancestors, however: they reproduced by spores, not seeds.

Fossils of several progymnosperms with reproductive structures intermediate between those of spore plants and seed plants have been discovered. For example, the evolution of microspores into pollen grains and of megasporangia into ovules (seed-producing structures) can be traced in fossil progymnosperms. Plants producing seeds appeared during the late Devonian period, more than 360 million years ago. The fossil record indicates that different groups of seed plants apparently arose independently several times.

As mentioned previously, fossilized remains of ginkgoes are found in 200-million-year-old rocks, and other groups of gymnosperms were well established by 160 to 120 million years ago. Although the gymnosperms are an ancient group, some questions persist about the exact pathways of gymnosperm evolution. The fossil record indicates that progymnosperms probably gave rise to conifers and to another group of extinct plants called **seed ferns,** which were seed-bearing woody plants with fernlike leaves (Fig. 27–13b). The seed ferns in turn probably gave rise to cycads and possibly ginkgoes, as well as to several gymnosperm groups that are now extinct. The origin of gnetophytes remains unclear.

Flowering plants are the most recent group of plants to evolve. The fossil record, although incomplete, indicates that flowering plants probably descended from gymnosperms. By the middle of the Jurassic period, approximately 180 million years ago, several gymnosperm lines existed with features reminiscent of flowering plants. Among other traits, these derived gymnosperms possessed leaves with broad, expanded blades and the first closed carpels (which enclose the ovules). It is also evident that beetles were visiting these plants, and biologists have suggested that perhaps this relationship was the beginning of coevolution (mutual adaptation) between plants and their animal pollinators.

One important task facing paleobotanists (biologists who study fossil plants) is determining which of the ancient gymnosperms are in the direct line of evolution leading to the flowering plants. Based on structural data, most botanists think that flowering plants arose only once, that is, that there is only one line of evolution from the gymnosperms to the flowering plants. The gnetophytes are the gymnosperm group that many botanists consider the closest living relatives of flowering plants; this conclusion is supported by both structural simi-

(a)

(b)

Figure 27–13 **Evolution of seed plants.** *(a) Archaeopteris,* a progymnosperm that existed about 360 million years ago, had a number of structural features in common with modern seed plants but did not produce seeds. *(b) Emplectopteris,* a seed fern that had seeds attached to leaves that were similar to those of modern-day ferns.

larities and molecular data. Gnetophytes are similar to angiosperms in that they possess vessels, lack archegonia, possess flower-like compound strobili, and undergo double fertilization.

The oldest definitive trace of flowering plants to appear in the fossil record is of pollen grains in Cretaceous rocks some 130 million years old, whereas the oldest fossilized flowers are about 120 million years old. (However, a fossil flower was tentatively identified in 130-million-year-old clay rocks in southern England in 1996.)

Based on the fossil record as well as both structural and molecular data of living angiosperms, the first flowering plants are thought to have been small, weedy shrubs or herbaceous plants that were adapted to the cool, dry climates of upland habitats. If this is the case, it explains why there are few fossils of early angiosperms: the environment in which they evolved was not favorable for their preservation as fossils.

Botanists hypothesize that the rapid diversification of angiosperms did not occur until early flowering plants had invaded lowland regions. By 90 million years ago, during the late Cretaceous period, flowering plants had diversified and had begun to replace gymnosperms as Earth's dominant plants. Fossils of flowering plant leaves, stems, flowers, fruits, and seeds are numerous and diverse. They outnumber fossils of gymnosperms and ferns in late Cretaceous deposits, indicating the rapid success of flowering plants once they appeared (Fig. 27–14). Many angiosperm species apparently arose as a result of changes in chromosome number.

Figure 27–14 **A fossil flower.** The fossilized flower of *Archaeanthus linnenbergeri,* an extinct plant that lived during the Cretaceous period, about 95 to 98 million years ago. The scars on the reproductive axis (receptacle) may have been where stamens, petals, and sepals were originally attached but abscised (fell off). Many spirally arranged pistils were still attached at the time this flower was fossilized. *(Courtesy of D. Dilcher)*

SUMMARY WITH KEY TERMS

I. **Seeds** are the primary means of reproduction and dispersal of gymnosperms and angiosperms (flowering plants). Each seed contains a well developed embryonic sporophyte and a food supply, surrounded by a protective seed coat.

II. **Gymnosperms** are vascular plants with seeds that are totally exposed or borne on the scales of cones.
   A. Gymnosperms produce wind-borne **pollen grains,** a feature that seedless vascular plants lack.
   B. There are four phyla of gymnosperms.
      1. **Conifers,** which are the largest phylum of gymnosperms, are woody plants that bear **needles** (leaves that are usually evergreen) and produce seeds in cones. Most conifers are **monoecious.**
         a. Male cones produce pollen grains (immature male gametophytes) that are carried by air currents to female cones.
         b. Female cones produce female gametophytes within **ovules** (megasporangia).
         c. Following fertilization, the zygote develops into an embryo that is encased inside a seed that is adapted for wind dispersal.
      2. **Cycads** are palmlike or fernlike in appearance. They are **dioecious** but reproduce in a manner similar to pines. There are relatively few living members of this once-large phylum.
      3. *Ginkgo biloba,* the only surviving species in its phylum, is a deciduous, dioecious tree. Female **ginkgoes** produce fleshy seeds directly on branches.
      4. **Gnetophytes** share a number of traits with angiosperms, such as efficient water-conducting cells in their xylem.

III. **Flowering plants (angiosperms)** constitute the phylum of vascular plants that produce seeds enclosed within a **fruit.** They are the most diverse and most successful group of plants.
   A. Flowering plants have several specialized features that help account for their success.
      1. The flower, which may contain **sepals, petals, stamens,** and **carpels,** functions in sexual reproduction.
      2. **Double fertilization,** which results in the formation of a diploid zygote and triploid **endosperm,** is characteristic of flowering plants.
      3. Unlike those of gymnosperms, the ovules of flowering plants are enclosed within an **ovary.** After fertilization, the ovules become seeds, and the ovary develops into a fruit.
      4. Flowering plants possess efficient water-conducting cells, called vessel elements, in their xylem and efficient carbohydrate-conducting cells, called sieve tube members, in their phloem.
      5. Various flowering plants use wind, water, insects, or other animals to transfer pollen grains.
   B. There are two classes of flowering plants.
      1. Most **monocots** have floral parts in multiples of three, and their seeds each contain one **cotyledon.** The nutritive tissue in their mature seeds is endosperm.
      2. **Dicots** usually have floral parts in multiples of four or five, and their seeds each contain two cotyledons. The nutritive organs in their mature seeds are usually the cotyledons, which have absorbed the nutrients in the endosperm.

IV. Seed plants arose from seedless vascular plants.
   A. **Progymnosperms** were seedless vascular plants that had megaphylls and "modern" woody tissue.
      1. Progymnosperms probably gave rise to conifers.

2. Progymnosperms probably gave rise to **seed ferns** as well, which in turn probably gave rise to cycads and possibly ginkgoes.
B. The evolution of gnetophytes is unclear, but they are thought to be the closest living relatives of flowering plants.
C. Flowering plants probably descended from ancient gymnosperms that had specialized features, such as leaves with broad, expanded blades and closed carpels.
1. Flowering plants probably arose only once.
2. The first flowering plants were probably dicots that were weedy shrubs or small herbaceous plants.

POST-TEST

1. Conifers, cycads, ginkgoes, and gnetophytes are collectively called (a) fern allies (b) gymnosperms (c) angiosperms (d) dicots (e) seedless vascular plants
2. Most conifers are _____ having male and female reproductive parts at different locations on the same plant. (a) incomplete (b) imperfect (c) monoecious (d) dioecious (e) perfect
3. The immature male gametophytes of pine are called (a) ovules (b) stamens (c) seed cones (d) pollen grains (e) polar nuclei
4. The transfer of pollen grains from the male to the female reproductive structure is known as (a) pollination (b) fertilization (c) embryo sac development (d) seed development (e) fruit development
5. Motile sperm cells are found as vestiges in these two gymnosperm groups: (a) monocots, dicots (b) gnetophytes, conifers (c) gnetophytes, flowering plants (d) cycads, conifers (e) cycads, ginkgoes
6. More than _____ species of flowering plants have been identified. (a) 235 (b) 2350 (c) 23,500 (d) 235,000 (e) 2,350,000
7. This class of flowering plants includes the palms, grasses, and orchids. (a) dicots (b) gnetophytes (c) cycads (d) monocots (e) conifers
8. The pistil has three sections: (a) stigma, style, and anther (b) anther, filament, and ovule (c) stigma, style, and ovary (d) ovary, ovule, and sepal (e) corolla, stamen, and sepal
9. A simple pistil consists of a single: (a) calyx (b) carpel (c) ovule (d) filament (e) petal
10. A flower that lacks stamens is said to be both _____ and _____. (a) complete; imperfect (b) incomplete; perfect (c) complete; perfect (d) incomplete; imperfect
11. After fertilization, the _____ develops into a fruit and the _____ develops into a seed. (a) ovary; ovule (b) polar nuclei; ovule (c) ovary; endosperm (d) ovule; ovary (e) ovule; polar nuclei
12. The female gametophyte in flowering plants is also called the (a) polar nuclei (b) anther (c) embryo sac (d) endosperm (e) sporophyll

REVIEW QUESTIONS

1. Why are seeds such a significant evolutionary innovation?
2. What distinguishing features do cycads, ginkgoes, and gnetophytes share with conifers?
3. How are the vegetative adaptations of flowering plants different from those of gymnosperms?
4. How does the gymnosperm life cycle differ from that of flowering plants?
5. What are the two classes of flowering plants, and how can one distinguish between them?
6. How does pollination occur in the gymnosperms? In flowering plants?
7. How does fertilization differ in gymnosperms and flowering plants?

YOU MAKE THE CONNECTION

1. How are cones and flowers alike? How are they different? (*Hint:* Your answer should consider microspores/megaspores and seeds.)
2. How do the life cycles of seedless plants (see Chapter 26) and seed plants differ? In what fundamental way are they alike?
3. Describe the evolutionary changes that occurred as ancient seedless vascular plants evolved into gymnosperms, and as ancient gymnosperms evolved into flowering plants.
4. In contrast to the cones of gymnosperms, which are either male or female, most flowers contain both male and female reproductive structures. Explain how bisexual flowers might be advantageous to flowering plants.

RECOMMENDED READINGS

Berg, L.R. *Introductory Botany: Plants, People, and the Environment.* Saunders College Publishing, Philadelphia, 1997. An introduction to general botany that stresses environmental links between plants and humans.

Heywood, V.H. *Flowering Plants of the World.* Oxford University Press, New York, 1993. This beautifully illustrated guide describes more than 300 families of angiosperms, including their economic uses.

Holden, C. "Ancient Trees Down Under." *Science,* 20 Jan. 1995. A conifer species thought to be extinct for the past 50 million years was recently discovered in Australia.

Mlot, C. "Early Flowering Tree Rediscovered." *Science News,* Vol. 152, 2 Aug. 1997. A rare flowering tree with many primitive features, such as lack of vessels in the xylem, has been found in a rain forest in Madagascar.

Palmer, D. "First Flowers Emerge from Triassic Mud." *New Scientist,* 29 Jan. 1994. 200-million-year-old rocks may contain the oldest traces of flowering plant pollen.

Poinar, G.O. "Still Life in Amber." *The Sciences,* Vol. 34, No. 2, 1993. The fossilized remains of many insects and other small animals have been preserved in amber, which is the fossilized resin of prehistoric gymnosperms.

Wolf, T.H. "The Object at Hand." *Smithsonian,* Sept. 1990. In this issue this regular column highlights the dawn redwood, a gymnosperm that was thought to have been extinct for millions of years but was found living in China.

● Visit our website at **http://www.saunderscollege.com/lifesci/titles.html** and click on Solomon/Berg/Martin Biology for links to chapter-related resources on the World Wide Web.

CHAPTER 28

The Animal Kingdom: Animals Without a Coelom

Biologists have described more than a million species of animals, and several million more probably remain to be discovered and classified. Although most animal species are readily recognizable as animals, the identity of some others is less obvious. Early naturalists thought that sponges were plants because they did not move from place to place. Some people still mistake certain marine animals, such as the purple tube sponge *(Callyspongia vaginalis)* and corals shown in the photograph, for plants. Moving from place to place is not a distinguishing qualification for being classified as an animal.

Animals are so diverse that for almost any definition, we can find exceptions. Still, the following characteristics describe most animals:

1. Animals are multicellular eukaryotes.
2. Cells that make up the animal body are specialized to perform specific functions. In all but the simplest animals, cells are organized to form tissues, and tissues are organized to form organs. In simple, small animals, life processes such as gas exchange, circulation of materials, and waste disposal take place by diffusion. In large, complex animals, specialized organ systems and mechanisms have evolved that carry on these life processes.
3. Animals are **heterotrophs.** They ingest their food first and then digest it inside the body, usually within a digestive system.
4. Most animals are capable of locomotion at some time during their life cycle. Some animals (the sponges, for example) move about as larvae (immature forms) but are **sessile** (firmly attached to the ground or some other surface) as adults.
5. Most animals have nervous systems and muscle systems that enable them to respond rapidly to stimuli in their environment.
6. Most animals are diploid organisms that reproduce sexually, with large, nonmotile eggs and small, flagellated sperm. A haploid sperm unites with a haploid egg, forming a zygote (fertilized egg) that undergoes **cleavage,** a series of mitotic cell divisions. During cleavage the zygote is converted to a hollow ball of cells called a **blastula.** Cells of the blastula undergo **gastrulation**, a process of forming and segregating specific layers of tissue, called germ layers. Some animals develop directly into adults. Others develop into a **larva,** a sexually immature form that may look very differ-

(Charles V. Angelo/Photo Researchers Inc.)

ent from the adult. Larvae then undergo **metamorphosis,** a developmental process that converts the immature animal into a juvenile that can then grow into an adult.

Most biologists divide **kingdom Animalia** into about 35 phyla. The animals most familiar to us, dogs, birds, fishes, frogs, snakes, are **vertebrates** (a subphylum of phylum Chordata). A vertebrate is an animal with a backbone (vertebral column). You may be surprised to learn that vertebrates account for fewer than 5% of the species in the animal kingdom. The majority of animals are the less familiar **invertebrates,** animals without backbones. The invertebrates include such diverse animals as sponges, jellyfish, worms, mollusks, insects, crustaceans, and sea stars.

1. List several characteristics common to most animals.
2. Describe the ecological roles and distribution of animals and compare the advantages and disadvantages of life in the sea, in fresh water, and on land.
3. Justify the classification and relationships of animal phyla on the basis of symmetry, type of body cavity, and pattern of development (that is, protostomes and deuterostomes).
4. Identify distinguishing characteristics of phyla Porifera, Cnidaria, Ctenophora, Platyhelminthes, Nemertea, Nematoda, and Rotifera and classify a given animal into the appropriate phylum.

5. Summarize the life cycles of the following parasites: *Ascaris,* tapeworm, hookworm, and trichina worm. Identify several adaptations that these animals possess for their parasitic lifestyles.
6. Describe the adaptive advantages of the following characteristics: bilateral symmetry, multiple tissue layers, cephalization, a motile life stage, a digestive cavity with two openings, and hermaphroditism.

ANIMALS INHABIT MOST ENVIRONMENTS

As consumers, all animals depend on producers for their raw materials, energy, and oxygen. They also depend on decomposers for recycling nutrients.

Animals are distributed in virtually every environment of planet Earth. They probably evolved during the Precambrian in shallow, marine environments (see Chapter 20). Members of most animal phyla still inhabit such environments. Of the three major environments, salt water, fresh water, and land, the sea is the most hospitable. Most salt water is isotonic to the tissue fluids of most animals, so fluid and salt balance can be more easily maintained. The buoyancy of sea water helps support its inhabitants, and the temperature is relatively stable due to the large volume of water. **Plankton,** the animals and protists that are suspended in water and float with its movement, provides a ready source of food.

Life in the sea has certain disadvantages. Although the continuous motion of water brings nutrients to animals and washes their wastes away, they must be able to cope with the water's movements and the currents that could sweep them away. Squids, fishes, and marine mammals are strong swimmers and can usually direct their movements and maintain their location. However, most invertebrates and young vertebrates are unable to swim strongly, and they have adapted in many different ways to the tides and currents. Some sessile animals attach permanently to a stable structure such as a rock. Others burrow in the sand and silt that cover the sea bottom. Many invertebrates have adapted by maintaining a small body size and becoming part of the plankton. They survive because while they are tossed about, their food supply continues to surround them.

Far fewer kinds of animals make their homes in fresh water than in the sea. Fresh water is hypotonic to the tissue fluids of animals, so water tends to move into the animal by osmosis. Freshwater species must have mechanisms for removing excess water while retaining salts. This osmoregulation requires an expenditure of energy. Fresh water offers a much less constant environment than sea water and generally contains less food. Oxygen content and temperature vary, and turbidity (due

to sediment suspended in the water) and even water volumes fluctuate.

Living on land is even more difficult than living in fresh water. Desiccation is a serious threat because water is constantly lost by evaporation and is often difficult to replace. A variety of adaptations found in terrestrial species reduce water loss. The temperature extremes of terrestrial habitats also present challenges. Only a few animal groups, most notably representatives of the arthropods (insects, spiders) and some vertebrate groups, have successfully made their homes on land.

BODY STRUCTURE AND PATTERN OF DEVELOPMENT ARE USEFUL IN RECONSTRUCTING PHYLOGENY

Biologists have evidence to support the hypothesis that animals evolved from protists, probably from colonial flagellates. Although the relationships among the various animal phyla remain a matter of debate, a few of the more widely held hypotheses are presented in this section.

Biologists divide animals into two groups: **parazoa** (*para,* "alongside" and *zoa,* "animals") and **eumetazoa** (*eu,* "true"; *meta,* "later"; *zoa,* "animals"). The sponges (phylum Porifera) are classified as parazoa. Although sponges are multicellular animals and can be large, they function much like colonial, unicellular protozoa. Most cells of sponges are extremely versatile and can change form and function; they are not organized into tissues. Sponges are so different from other animals that biologists think that they are not directly ancestral to any other animal group. Other animals have true tissues and are classified as eumetazoa.

Animals can be classified according to body symmetry

Symmetry refers to the arrangement of body structures in relation to some axis of the body. Sponges have an interesting variety of shapes. Most are not symmetrical, so that when cut

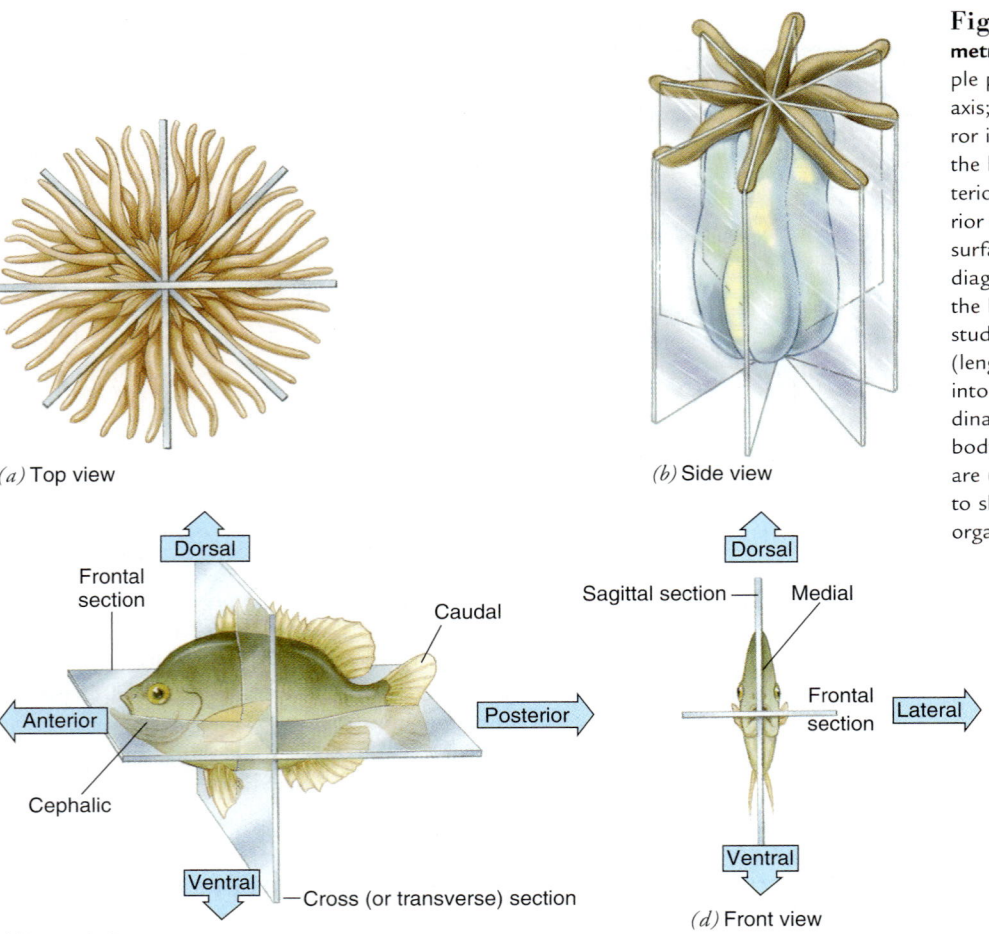

(a) Top view

(b) Side view

(c) Lateral view

Dorsal
Frontal section
Caudal
Anterior
Posterior
Cephalic
Ventral
Cross (or transverse) section

(d) Front view

Dorsal
Sagittal section
Medial
Frontal section
Lateral
Ventral

Figure 28-1 Radial and bilateral symmetry. *(a)* and *(b)* In radial symmetry, multiple planes can be drawn through the central axis; each divides the organism into two mirror images. *(c)* and *(d)* In bilateral symmetry, the head end of the animal is typically its anterior end, and the opposite end is its posterior end. The back of the animal is its dorsal surface, and the belly is its ventral surface. The diagrams also illustrate various ways in which the body can be sectioned (cut) in order to study its internal structure. A sagittal cut (lengthwise vertical cut) divides the animal into right and left parts. A frontal, or longitudinal, cut (lengthwise horizontal) divides the body into dorsal and ventral parts. Sections are used in illustrations throughout this book to show relationships among tissues and organs.

in half the two halves are not similar to one another. Most other animals exhibit either radial or bilateral body symmetry. Members of phylum Cnidaria (jellyfish, sea anemones, and their relatives) and adult echinoderms (sea stars and their relatives) have **radial symmetry.** In radial symmetry, the body has the general form of a wheel or cylinder, and similar structures are regularly arranged as spokes from a central axis (Fig. 28–1*a*). Multiple planes can be drawn through the central axis, each dividing the organism into two mirror images. An animal with radial symmetry receives stimuli equally from all directions in the environment. Many radially symmetrical animals actually have modified radial symmetry. For example, sea anemones and ctenophores (comb jellies) have **biradial symmetry** in which parts of the body have become specialized so that only two planes can divide the body into similar halves.

Most animals are **bilaterally symmetrical,** at least in their larval stages. A bilaterally symmetrical animal can be divided through only one plane (which passes through the midline of the body) to produce roughly equivalent right and left halves that are mirror images (Fig. 28–1*d*). Bilateral symmetry is considered an adaptation to locomotion. The front end of the animal typically has a head, where nervous system and sense organs are concentrated. This end receives most environmental

stimuli and generally moves into the environment first. The rear end of the animal may be equipped with a tail for swimming, or it may just be carried along.

To locate body structures in bilaterally symmetrical animals, it is helpful to define some basic terms and directions. The back surface of an animal is its **dorsal** surface; the underside (belly) is its **ventral** surface. **Anterior** means toward the front (head end) of the animal; **posterior,** or **caudal,** means toward the back (tail end). A structure is said to be **medial** if it is located toward the midline of the body and **lateral** if it is toward one side of the body; for example, the human ear is lateral to the nose. The term **cephalic** (or **superior**) refers to the head end of the body. (The term **inferior** is used in human anatomy to mean located below some point of reference, or toward the feet.)

A bilaterally symmetrical animal has three axes, each at right angles to the other two: an anterior-posterior axis extending from head to tail; a dorsoventral axis extending from back to belly; and a left-right axis extending from side to side. We can distinguish three planes or sections that divide the body into specific parts. A **sagittal plane** divides the body into right and left parts. A sagittal plane passes from anterior to posterior and from dorsal to ventral. A **frontal** plane divides a bilateral body into dorsal and ventral parts. A **transverse section,** or **cross section,** cuts at right angles to the body axis and separates anterior and posterior parts.

Animals can be grouped according to type of body cavity

The structures of most animals develop from three embryonic tissue layers, called **germ layers.** The outer germ layer, called the **ectoderm,** gives rise to the outer covering of the body and to the nervous system (if the animal has one). The inner layer, or **endoderm,** forms the lining of the digestive tube and other digestive organs. Cnidarians and ctenophores only develop these two germ layers and are referred to as **diploblastic.** All of the other eumetazoa are **triploblastic.** They develop a middle layer, called **mesoderm,** that gives rise to most other body structures, including muscles, bones, and circulatory system (when they are present).

A widely held system for grouping triploblastic animals is based on the presence and type of **body cavity,** a fluid-filled space between the body wall and the digestive tube. The flatworms are triploblastic but have a solid body, that is, they have no body cavity. They are referred to as **acoelomates** (*a*, "without"; *coelom,* "cavity") (Fig. 28–2).

Generally animals with a body cavity have a **tube-within-a-tube body plan.** The body wall, which forms the outer tube, is covered with tissue that develops from ectoderm. Tissue derived from endoderm lines the inner tube—the digestive tube, or gut—which has an opening at each end: the mouth and the anus. Beneath the ectoderm, the outer tube may be lined with tissue derived from mesoderm. The space that exists between the two tubes is the body cavity. If the body cavity is not completely lined with mesoderm it is called a **pseudocoelom** ("false coelom"). Animals with a pseudocoelom, like roundworms and rotifers, are referred to as **pseudocoelomates.**

In still more complex animals, the body cavity is completely lined with mesoderm. Such a body cavity is a true **coelom.** Only animals with true coeloms are referred to as **coelomates.** The tree shown in Figure 28–3 indicates one view of the relationships among the major phyla of animals based on their body cavity types.

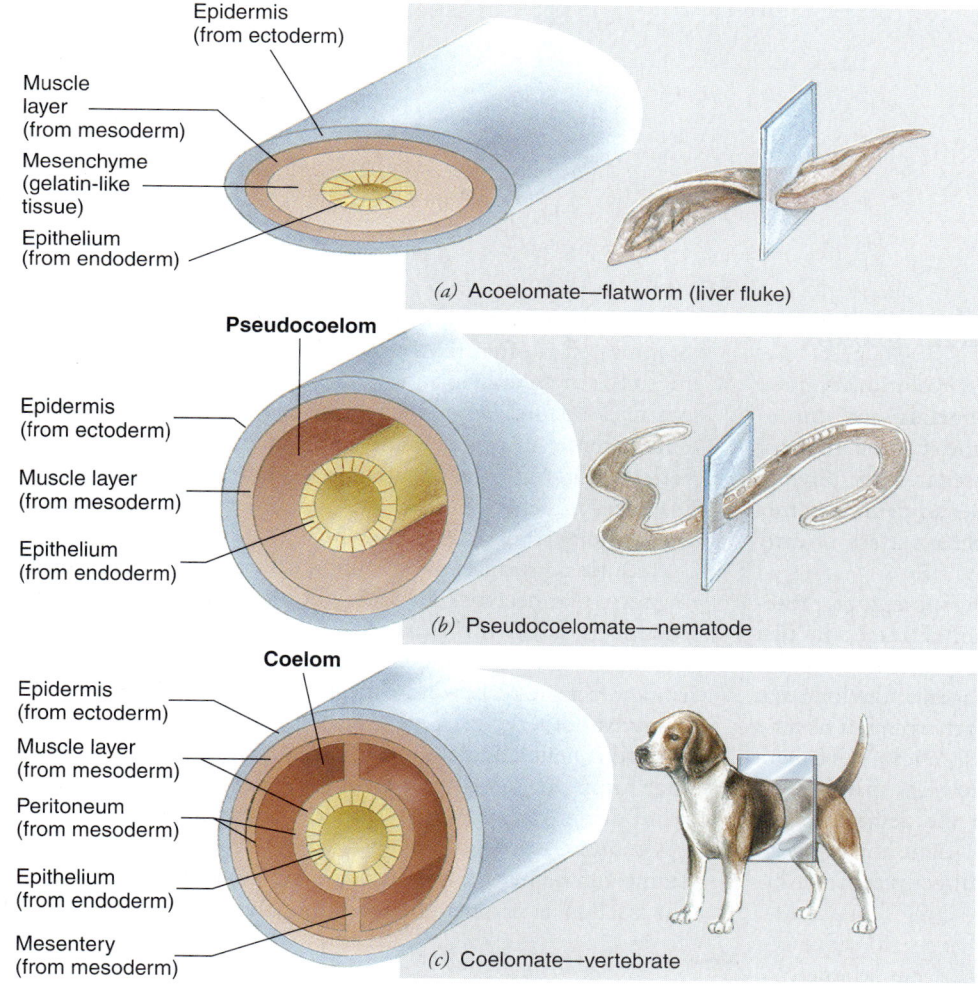

Figure 28–2 Three basic body plans in triploblastic animals. The germ layer from which each tissue was derived is indicated in parentheses. *(a)* An acoelomate animal has no body cavity. *(b)* A pseudocoelomate animal has a body cavity that is not completely lined with mesoderm. *(c)* In a coelomate animal the body cavity, or coelom, is completely lined with tissue that develops from mesoderm.

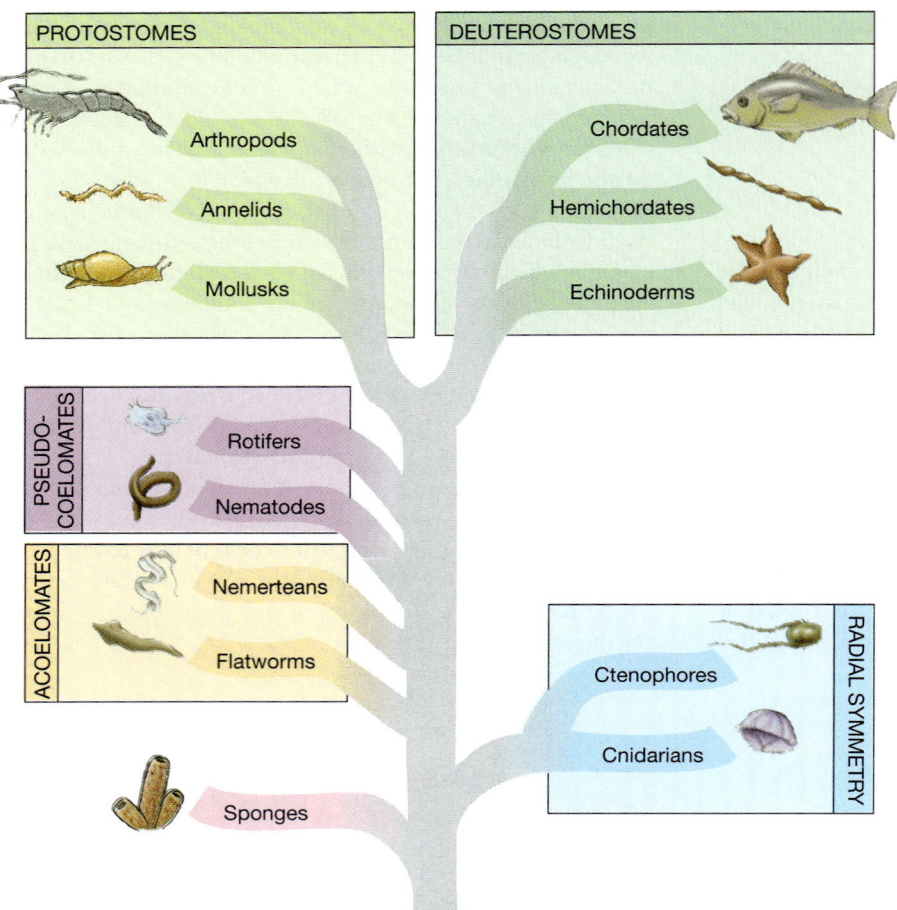

PROTOSTOMES

- Arthropods
- Annelids
- Mollusks

DEUTEROSTOMES

- Chordates
- Hemichordates
- Echinoderms

PSEUDO-COELOMATES

- Rotifers
- Nematodes

ACOELOMATES

- Nemerteans
- Flatworms

RADIAL SYMMETRY

- Ctenophores
- Cnidarians

Sponges

Figure 28-3 This phylogenetic tree illustrates some hypothetical evolutionary relationships. Flatworms and nemerteans have a solid body and so are referred to as acoelomate. Nematodes and rotifers are pseudocoelomates. Most other bilateral animals and the echinoderms have a true coelom. Two main evolutionary branches of coelomates are protostomes (mollusks, annelids, and arthropods) and deuterostomes (echinoderms, hemichordates, and chordates).

Coelomate animals form two main groups

Animals with a true coelom form two main evolutionary lines: **protostomes,** which include mollusks, annelids, and arthropods, and **deuterostomes,** which include the echinoderms (for example, sea stars and sea urchins) and chordates (the phylum that includes the vertebrates). Protostomes and deuterostomes are distinguished by basic differences in their pattern of early development.

An important difference in the development of protostomes and deuterostomes is the pattern of cleavage, the first several cell divisions of the embryo. In many protostomes, the early cell divisions are diagonal to the polar axis (the long axis of the egg), resulting in a somewhat spiral arrangement of cells; any one cell is located between the two cells above or below it (Fig. 28–4). This pattern of division is known as **spiral cleavage.** In **radial cleavage,** characteristic of the deuterostomes, the early divisions are either parallel or at right angles to the polar axis. The resulting cells are located directly above or below one another.

In the protostomes, the developmental fate of each embryonic cell is often fixed very early. For example, if the first four cells of an annelid embryo are separated, each cell develops into only a fixed quarter of the larva; this is referred to as **determinate cleavage.** In deuterostomes, cleavage is usually **indeterminate.** If the first four cells of a sea star embryo, for instance, are separated, each cell is capable of forming a complete, though small, larva.

During gastrulation a group of cells moves inward, forming a sac that becomes the embryonic gut. The opening to the outside is called the **blastopore.** In most protostomes, the blastopore develops into the mouth. The word protostome comes from Greek words meaning "first, the mouth." In deuterostomes the blastopore does not give rise to the mouth. Instead it generally develops into the anus. A second opening that forms later in development gives rise to the mouth. The word deuterostome means "second, the mouth."

Another, though less reliable, difference between protostome and deuterostome development is the manner in which the coelom is formed. In most protostomes, the mesoderm

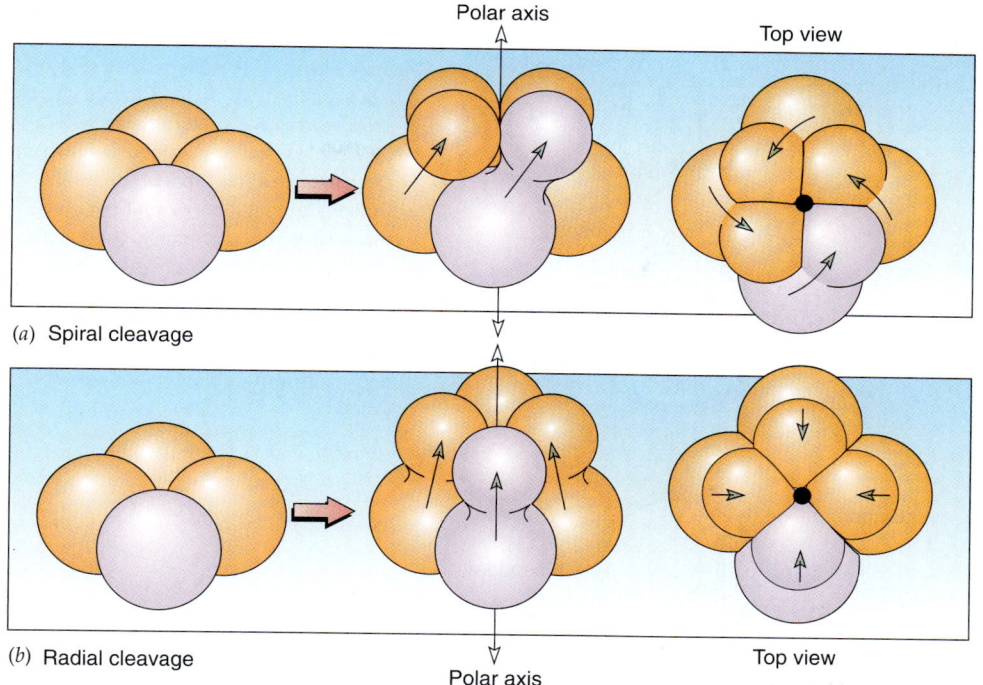

Figure 28-4 **Spiral and radial cleavage.** *(a)* Spiral cleavage is characteristic of protostomes. Note the spiral arrangement of the cells. *(b)* In radial cleavage, characteristic of deuterostome embryos, the early divisions are either parallel to the polar axis or at right angles to it. The cells are stacked in layers. The pattern of cleavage can be appreciated by comparing the position of the purple cells in *(a)* and *(b)*.

Polar axis

Top view

(a) Spiral cleavage

(b) Radial cleavage

Polar axis

Top view

splits, and the split widens into a cavity that becomes the coelom (Fig. 28–5). This method of coelom formation is known as **schizocoely,** and, for this reason, these protostomes are sometimes called **schizocoelomates.** In many deuterostomes, the mesoderm forms as "outpocketings" of the developing gut. These outpocketings eventually pinch off and form pouches; the cavity within the pouches becomes the coelom. This type of coelom formation is called **enterocoely,** and these animals are sometimes referred to as **enterocoelomates**.

Now that we have briefly discussed body structure and patterns of development in animals, we survey representative animal phyla. The remainder of this chapter discusses animals without a coelom: sponges, cnidarians, comb jellies, flatworms, ribbon worms, roundworms, and rotifers.

SPONGES HAVE UNIQUE FLAGELLATED CELLS

About 9000 species of sponges have been identified and assigned to phylum **Porifera.** The name *Porifera,* meaning "to have pores," aptly describes the sponges, whose bodies are perforated by tiny holes. Sponges are aquatic, mainly marine, animals that range in size from 1 to 200 cm (0.4–79 in) in height. Many are asymmetrical, but they vary in shape from flat, encrusting growths to balls, cups, fans, or vases. Living sponges

may be brightly colored—green, orange, red, yellow, blue, or purple—or they may be white or drab (Fig. 28–6). Some species are inhabited by symbiotic bacteria or algae that give them color.

Sponges are thought to have evolved from choanoflagellates. Recall from Chapter 24 that these protozoa have a single flagellum surrounded by a collar of microvilli. Sponges are the only animals with **choanocytes,** or **collar cells,** flagellated cells that are strikingly similar to the choanoflagellates. In the sense that they apparently did not give rise to any other animal group, sponges seem to represent a dead end in evolution. Of course, sponges themselves continue to evolve as they are subjected to continual selective pressure from their environment.

Biologists divide sponges into three main classes on the basis of the type of skeleton they secrete. Members of class **Calcarea** secrete a chalky skeleton composed of small calcium carbonate spikes, or **spicules.** Members of class **Hexactinellida,** the glass sponges, have a skeleton made of six-rayed spicules containing silicon. Most sponges belong to class **Demospongiae,** characterized by variable skeletons: some are made of a fibrous protein material known as *spongin,* others contain spicules of silicon, and most have a combination of both.

In a simple sponge, water enters through hundreds of tiny pores *(ostia),* passes into the central cavity, or **spongocoel** (not a digestive cavity), and flows out through the sponge's open end, the **osculum.** In most types of sponges, the body wall is

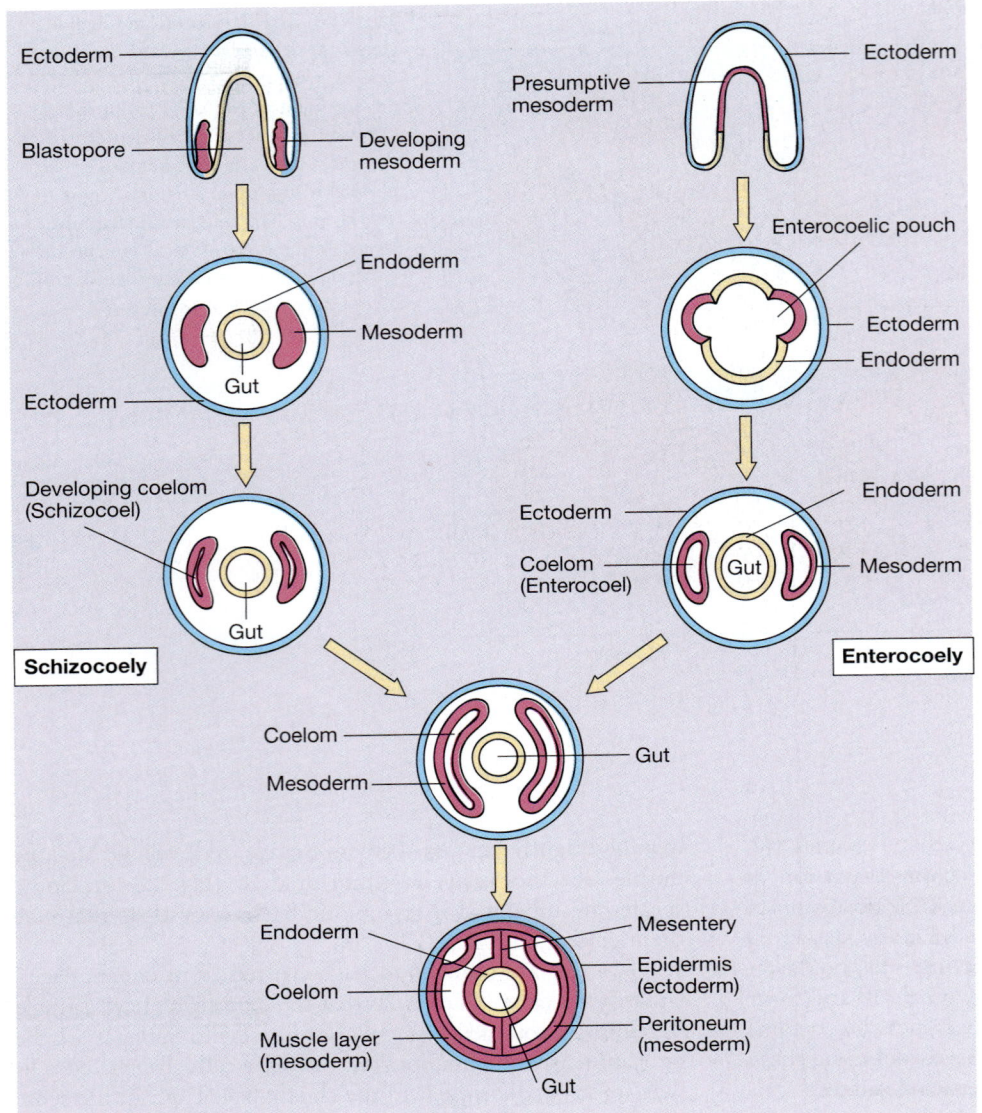

Figure 28-5 Coelom formation.
The coelom originates in the embryo from blocks of mesoderm that split off from each side of the embryonic gut. In protostomes the coelom typically forms by a process called schizocoely in which the mesoderm *(red)* splits. The split widens forming a cavity that becomes the coelom. In enterocoely, characteristic of deuterostomes, the mesoderm outpockets from the gut, forming pouches. The cavity within these pouches becomes the coelom. Ectoderm is shown in blue, endoderm in yellow.

extensively folded, and there are complicated systems of canals that provide increased surface area for food capture.

Although sponges are multicellular, their cells are loosely associated and do not form definite tissues. However, a division of labor exists among the several types of cells that make up the sponge, with certain cells specializing in nutrition, support, contraction, or reproduction. Epidermal cells form the outer layer of the sponge and line the canals. Specialized tube-like cells, called *porocytes,* form the pores of a simple sponge. These cells regulate the diameter of the pores by contracting.

The collar cells, which make up the inner layer of certain sponges, create the water current that brings food and oxygen to the cells and carries away carbon dioxide and other wastes. These cells also trap and phagocytize food particles. Each of these cells is equipped with a tiny collar that surrounds the base of the flagellum. The collar is an extension of the plasma membrane and consists of microvilli. Together, the collar cells of some complex sponges can pump a volume of water equal to the volume of the sponge each minute!

Between the outer and inner cell layers of the sponge body is a gelatin-like layer, the *mesohyl,* supported by skeletal spicules. Amoeboid cells, which wander about in this layer, are important in digestion and food transport. Other amoeboid cells in the mesohyl secrete the spicules.

Sponge larvae are flagellated and able to swim about. However, the adult sponge remains attached to some solid object and is incapable of locomotion. Sponges are *suspension feeders,* adapted for trapping and eating whatever food the water brings to them. As water circulates through the body, food is trapped along the sticky collars of the choanocytes. Food particles are either digested within the collar cell or transferred to an amoeboid cell for digestion and transport of nutrients to epidermal

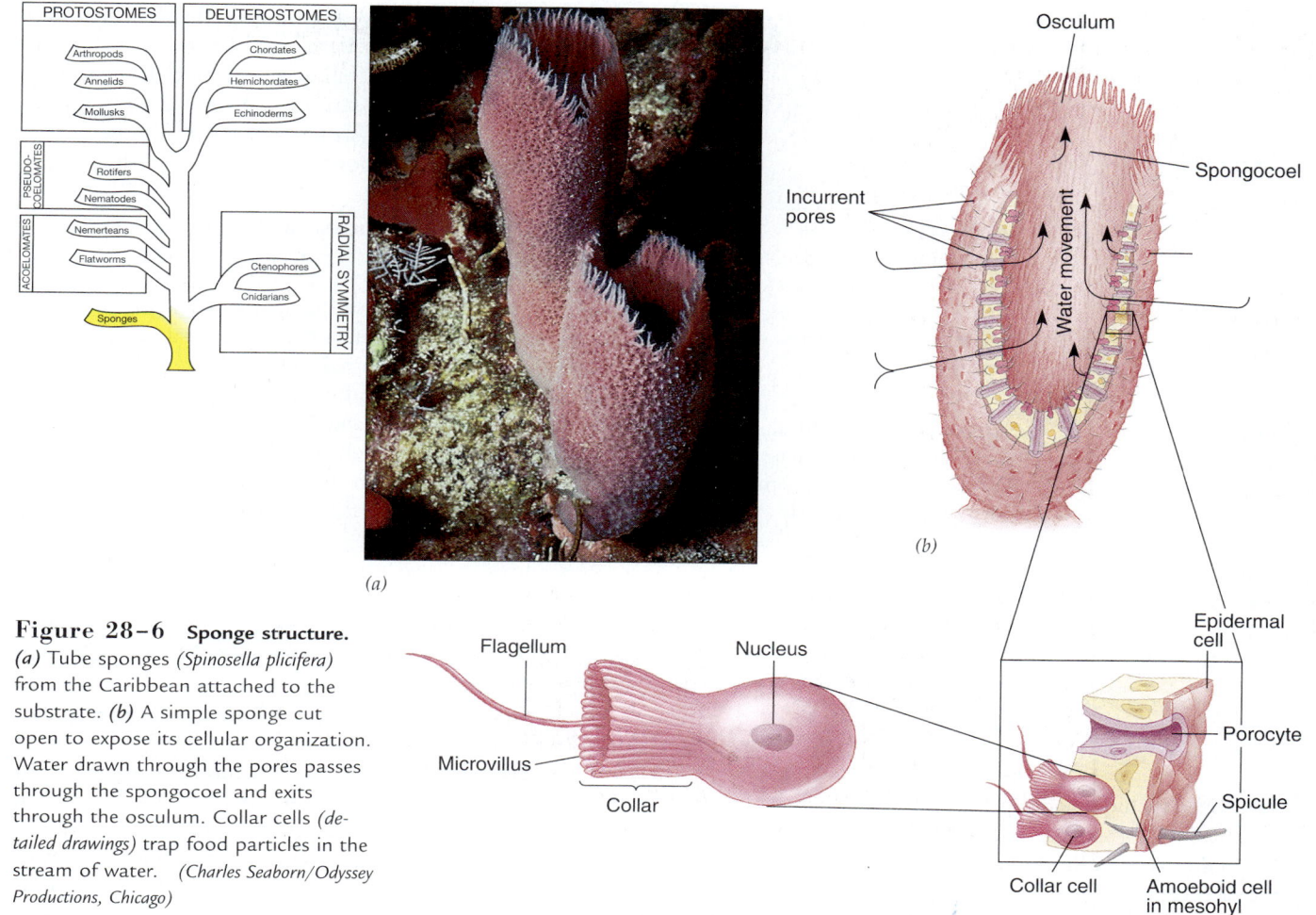

Figure 28–6 **Sponge structure.**
(a) Tube sponges (*Spinosella plicifera*) from the Caribbean attached to the substrate. **(b)** A simple sponge cut open to expose its cellular organization. Water drawn through the pores passes through the spongocoel and exits through the osculum. Collar cells (*detailed drawings*) trap food particles in the stream of water. (*Charles Seaborn/Odyssey Productions, Chicago*)

cells. Undigested food passes out through the osculum and is simply eliminated into the water.

Gas exchange and excretion of wastes depend on diffusion into and out of individual cells. Although cells of the sponge are irritable and can react to stimuli, sponges do not have specialized nerve cells and so cannot react as a whole. Behavior appears limited to basic metabolic necessities such as capturing food and regulating the flow of water through the body.

Sponges can reproduce asexually. A small fragment or bud may break free from the parent sponge and give rise to a new sponge. Such fragments may attach to the parent sponge, forming a colony. Sponges also reproduce sexually. Most sponges are **hermaphroditic,** meaning that the same individual can produce both eggs and sperm. Some of the amoeboid cells develop into sperm cells, others into egg cells. However, hermaphroditic sponges usually produce eggs and sperm at different times, and they cross-fertilize with other sponges. Mature sperm are released into the water and are taken in by other sponges. Fertilization and early development take place within the jelly-like mesohyl. Embryos eventually move into

the spongocoel and leave the parent along with the stream of outflowing water. After swimming about for a while, a larva finds a solid object, attaches to it, and settles down to a sessile life.

Sponges possess a remarkable ability to repair themselves when injured and to regenerate lost parts. If the cells of a sponge are separated from one another in the laboratory, they recognize one another and their place in the whole and reaggregate, forming a complete sponge again.

CNIDARIANS POSSESS RADIAL SYMMETRY AND UNIQUE STINGING CELLS

Most of the 10,000 or so species of phylum **Cnidaria** (pronounced "ni-dah′-ree-ah") are marine. These animals get their name from specialized cells, called **cnidocytes** (from a Greek word meaning "sea nettles"), that contain stinging organelles. Cnidarians are divided into three main classes (Table 28–1).

TABLE 28-1	Major Classes of Phylum Cnidaria
Class and Representative Animals	**Characteristics**
Hydrozoa *Hydra* *Obelia* Portuguese man-of-war	Mainly marine, but some freshwater species; alternation of polyp and medusa stages in most species (polyp form only in *Hydra*); some form colonies.
Scyphozoa Jellyfish	Mainly marine; typically inhabit coastal water, free-swimming medusa most prominent form; polyp stage often reduced.
Anthozoa Sea anemones Corals Sea fans	Marine; solitary or colonial polyps; in most no medusa stage; gastrovascular cavity divided by partitions into chambers, increasing area for digestion; sessile.

Class **Hydrozoa** includes hydras, hydroids, such as *Obelia,* and the Portuguese man-of-war; class **Scyphozoa** includes jellyfish; and class **Anthozoa** includes sea anemones and corals (Fig. 28–7). Evidence suggests that the hydrozoans are ancestral to the other two groups. Some cnidarians live a solitary existence, whereas many others, such as corals, form colonies. Some colonies, for example, the Portuguese man-of-war, consist of many individuals.

Basically, the cnidarian body is radially symmetrical and is organized as a hollow sac with the mouth and surrounding tentacles located at one end. The mouth leads into the digestive cavity, called the **gastrovascular cavity.** The mouth is the only opening into the gastrovascular cavity and so must serve for both ingestion of food and expulsion of wastes.

Much more highly organized than the sponge, the cnidarian is diploblastic; that is, it has two definite tissue layers. The ectoderm gives rise to the outer **epidermis,** a protective layer. The endoderm gives rise to the inner **gastrodermis** which functions in digestion. These layers are separated by a gelatin-like **mesoglea.**

Cnidarians have two body shapes, the polyp and the medusa. The **polyp** form, represented by *Hydra,* resembles an upside-down, slightly elongated jellyfish. Some cnidarians have the polyp shape during one stage of their life cycle and the **medusa** (pl., *medusae*), or jellyfish, form during another stage. In the medusa, the mouth is located in the lower concave, or *oral,* surface; the convex upper surface is the *aboral* surface. The Portuguese man-of-war and some other cnidarians consist of many individuals, some of which are polyps and others medusae.

The first true nerve cells in the animal kingdom are found in the cnidarians, but these nerve cells are not organized to form a brain or nerve cord. Rather, the nerve cells form irregular **nerve nets** connecting sensory cells in the body wall to contractile and gland cells. An impulse set up in one part of the body passes in all directions more or less equally.

Class Hydrozoa includes solitary and colonial forms

Although not really typical, the cnidarian most often studied by beginning biology students is the tiny, solitary *Hydra* found in freshwater ponds. To the naked eye *Hydra* looks like a bit of frayed string (Fig. 28–8). Because it has a remarkable ability to regenerate, *Hydra* is named after the multiheaded monster of Greek mythology which was able to grow two new heads for each head cut off. When *Hydra* is cut into several pieces, each piece may regrow all the missing parts and become a whole animal.

Hydra lives in fresh water and is typically attached to a rock, aquatic plant, or detritus by a disk of cells at its base. At the other end, the mouth connects the gastrovascular cavity with the outside. The mouth is surrounded by tentacles that are used for feeding.

The hydra's body consists of an outer, protective epidermis and an inner gastrodermis that functions in digestion. Both layers have cells specialized to contract (however, they are not muscle cells). Contractile fibers in the epidermal cells run lengthwise, and those in the gastrodermis run circularly. These two sets of contractile cells, act on the water-filled gastrovascular cavity, which forms a **hydrostatic skeleton** that supports the body and allows movement. By the contraction of one set of contractile cells or the other, the hydra can shorten, lengthen, or bend its body.

Cnidocytes are located mainly in the epidermis, especially on the tentacles. The cnidocytes contain stinging "thread capsules," or **nematocysts** (Fig. 28–9). When stimulated, the nematocysts release a coiled, hollow thread. Some types of nematocyst threads are sticky. Others are long and coil around prey. A third type bears barbs or spines and can inject a protein toxin that paralyzes prey animals such as small crustaceans. Each cnidocyte has a small, projecting trigger (cnidocil) on its outer surface. Stimuli such as touch or chemicals dissolved in the water cause the nematocyst to fire its thread.

Captured prey are pushed into the mouth by the tentacles. Digestion begins in the gastrovascular cavity. Partially digested fragments are taken up by the gastrodermal cells, and digestion is completed within food vacuoles in these cells.

Gas exchange and excretion occur by diffusion. The body wall of a hydra is thin enough so that no cell is far from the surface. The motion of the body as it stretches and shortens helps circulate the contents of the gastrovascular cavity.

Hydras reproduce asexually by budding during periods when environmental conditions are optimal. However, they differentiate as males and females in the fall or when pond water becomes stagnant. Females develop an ovary that produces

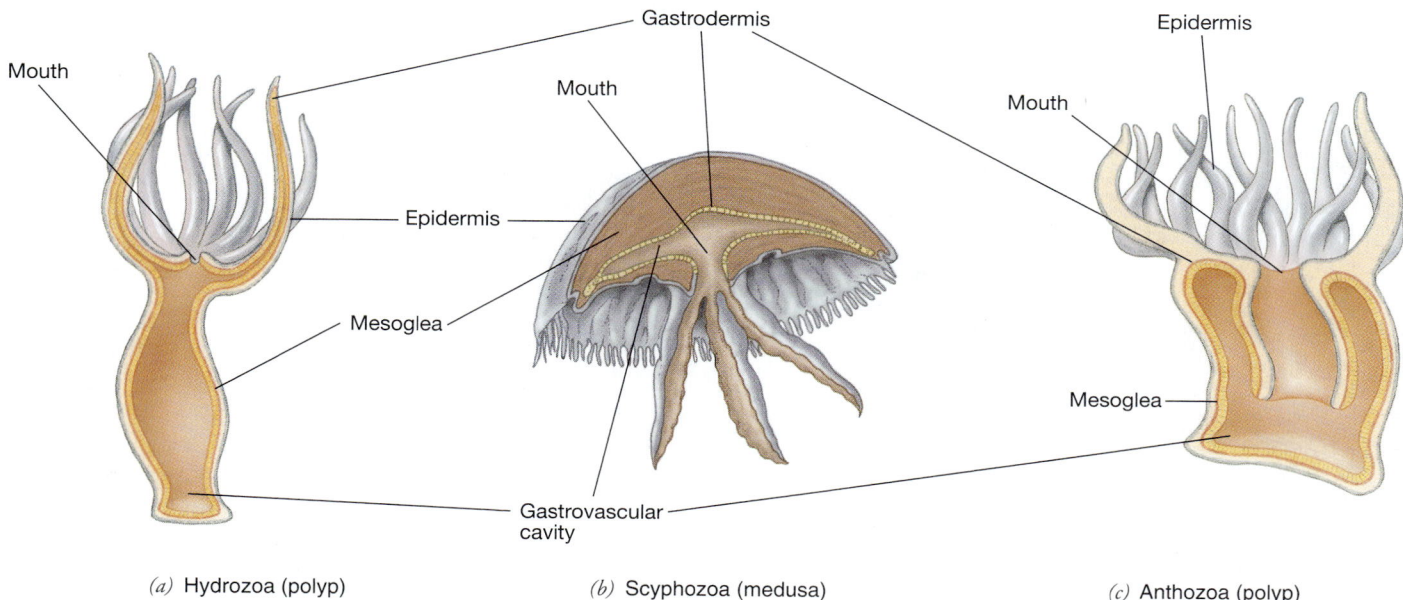

Mouth

Gastrodermis

Mouth

Epidermis

Mouth

Epidermis

Mesoglea

Mesoglea

Gastrovascular cavity

(a) Hydrozoa (polyp)

(b) Scyphozoa (medusa)

(c) Anthozoa (polyp)

Figure 28–7 **Polyp and medusa body forms.** *(a)* This hydrozoan, *(Gonothyraea loveni)* forms a colony of polyps. The drawing illustrates a single polyp. *(b)* The sea nettle *(Chrysaora fuscescens)*, like other jellyfish, uses its tentacles equipped with cnidocytes to capture small animals (zooplankton) suspended in the water, and to carry this food to the mouth. *(c)* Coral polyps *(Montastrea cavernosa)* extended for feeding. *(a, Robert Brons/Biological Photo Service; b, Brian Parker/Tom Stack & Associates; c, Mike Bacon/Tom Stack & Associates)*

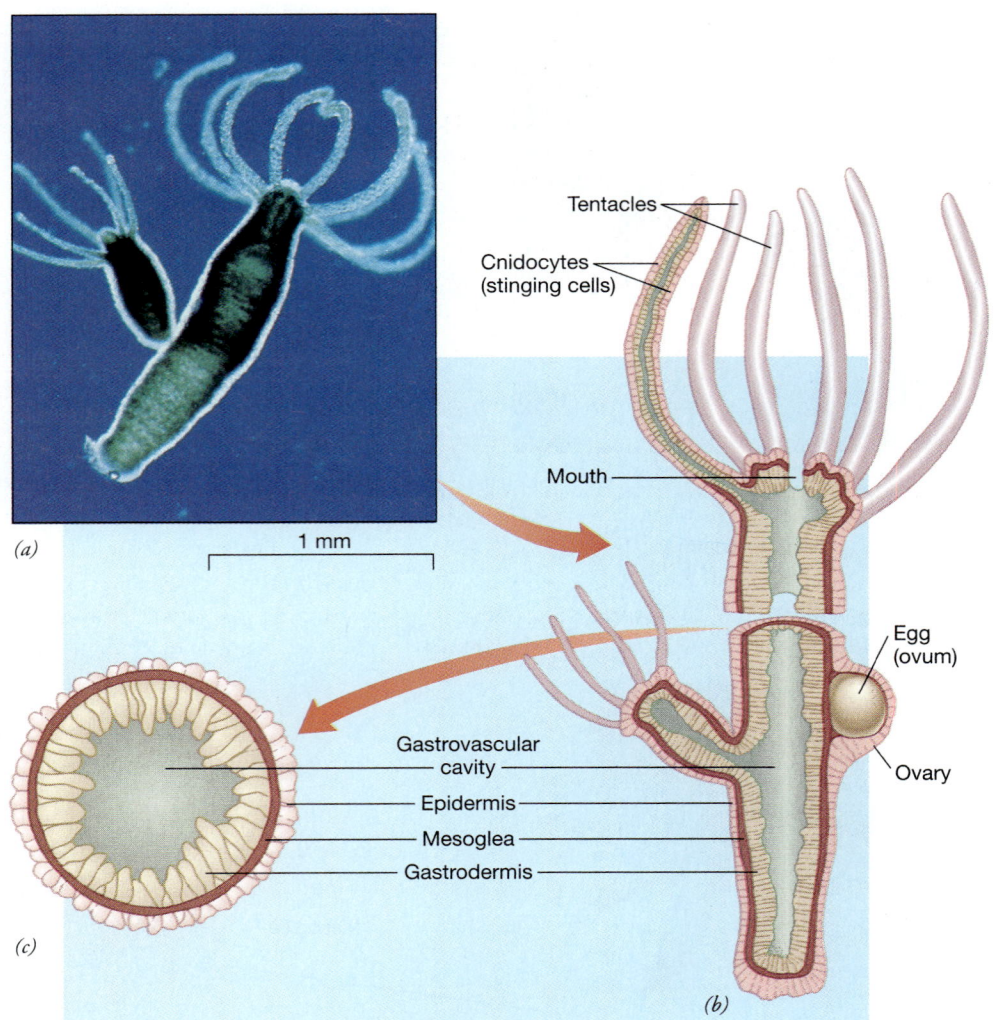

(a)

1 mm

Tentacles

Cnidocytes
(stinging cells)

Mouth

Egg
(ovum)

Ovary

Gastrovascular
cavity

Epidermis

Mesoglea

Gastrodermis

(c)

(b)

Figure 28-8 The basic cnidarian body plan. *(a)* *Hydra viridis* with a large bud. When the bud separates from its parent, it becomes an independent individual. *(b)* Hydra cut longitudinally to show its internal structure. Asexual reproduction by budding is represented on the left; sexual reproduction is represented by the ovary on the right. Male hydras develop testes that produce sperm. *(c)* Cross section through the body of a hydra. *(Biophoto Associates/Photo Researchers, Inc.)*

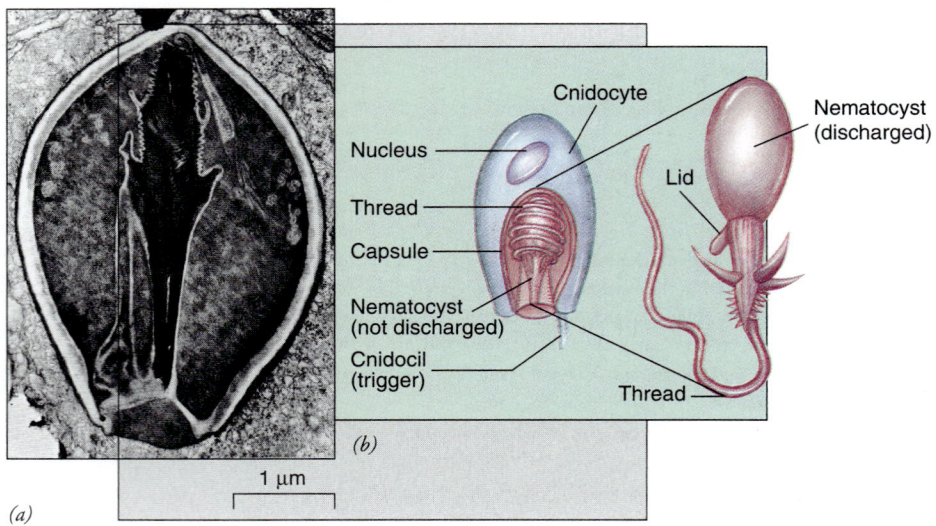

Cnidocyte

Nucleus

Thread

Capsule

Nematocyst
(not discharged)

Cnidocil
(trigger)

Nematocyst
(discharged)

Lid

Thread

(b)

1 μm

(a)

Figure 28-9 Nematocysts. These thread capsules are found within cnidocytes, cnidarian stinging cells. *(a)* TEM of an undischarged nematocyst of *Hydra* *(sagittal section)*. *(b)* When the cnidocil senses contact with an object, the nematocyst discharges, ejecting a thread that may entangle or penetrate the prey. Some nematocysts secrete a toxic substance that immobilizes the prey. *(a, G.B. Chapman, Cornell University Medical College)* a single egg, and males form a testis that produces sperm. After fertilization, the zygote may become covered with a shell. It leaves the parent and remains within the protective shell throughout the winter.

Many hydrozoans form colonies consisting of hundreds or thousands of individuals. A colony begins with a single polyp that reproduces asexually by budding. However, instead of separating from the parent, the bud remains attached and continues to form additional buds. Several types of individuals may arise in the same colony, some specialized for feeding, some for reproduction, and others for defense.

Some marine cnidarians are remarkable for an alternation of sexual and asexual stages. This alternation of stages differs from the alternation of generations in plants in that both sexual and asexual forms are diploid; only sperm and eggs are haploid. The life cycle of the colonial marine hydrozoan *Obelia* illustrates alternation of sexual and asexual stages (Fig. 28–10). In this polyp colony, the asexual stage consists of two types of polyps: those specialized for feeding and those for reproduction. Free-swimming male and female medusae bud off from the reproductive polyps. These medusae eventually produce

sperm and eggs, and fertilization takes place. The zygote develops into a ciliated swimming larva called a **planula.** The larva attaches to some solid object and begins to form a new generation of polyps by asexual reproduction.

The Portuguese man-of-war, *Physalia,* superficially resembles a jellyfish but is actually a hydrozoan colony of polyps and medusae. An iridescent purple, gas-filled sac helps maintain the animal's position in the water. *Physalia's* long tentacles may hang down for several meters below the float, Its cnidocytes are capable of paralyzing a large fish and can severely wound a human swimmer.

The medusa stage is dominant among the jellyfish

Among the jellyfish, members of class Scyphozoa, the medusa is the predominant body form. Scyphozoan medusae are generally larger than hydrozoan medusae, and they possess a thick, viscous mesoglea that gives firmness to the body. In scyphozoans, the polyp stage is small and inconspicuous or may even be absent. The largest jellyfish, *Cyanea,* may be more than 2 m

(a)

250 µm

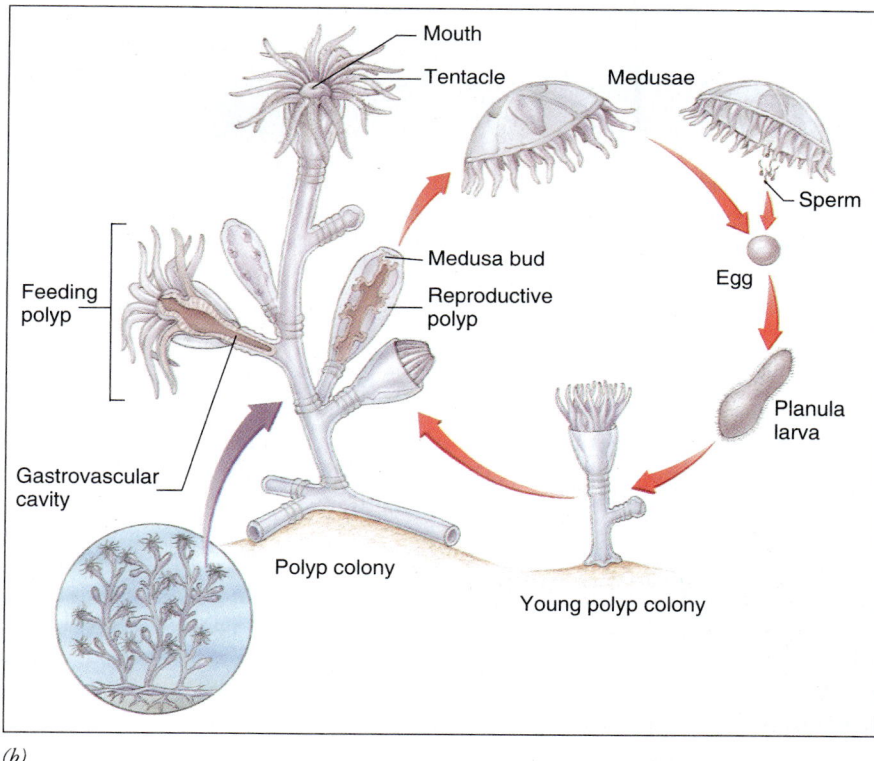

(b)

Figure 28–10 *Obelia,* **a marine colonial hydrozoan.** *(a)* LM of *Obelia.* Some polyps have tentacles and are specialized for feeding whereas others are specialized for reproduction. *(b)* Life cycle of *Obelia.* Reproductive polyps give rise asexually to medusae. The free-swimming medusae reproduce sexually, and the zygote develops into a planula larva. The larva develops into a polyp which forms a new colony. The colony grows as new polyps bud and remain attached. *(David M. Phillips/Visuals Unlimited)*

CORAL REEFS AND ENVIRONMENTAL ISSUES

Does environmental stress cause coral bleaching? Although biologists have known about coral bleaching for more than 75 years, bleaching has recently become widespread. The phenomenon of coral bleaching, the stress-induced loss of the colorful symbiotic algae that inhabit coral cells, has been the focus of recent research. Without their algae, coral become malnourished and die.

The causes of coral bleaching are not well understood. Suspected environmental factors include pollution, changes in salinity, disease, increased ultraviolet radiation (associated with the destruc-

Bleached coral *(Oculina patagonica)* from the Mediterranean coast of Israel. *(Courtesy of A. Kushmaro, Y. Loya, M. Fine, and E. Rosenberg. Bacterial infection and coral bleaching. Nature: 380:396, Apr. 1996. Photo by A. Shoob)*

tion of the ozone layer), and unusually high or low temperatures. Some scientists think that one of the most important factors in recent years has been abnormally high water temperature, possibly caused by global warming (see Chapter 55). Healthy coral thrives in a narrow temperature range. An increase of only one or two degrees above the normal summer maximum temperature can cause widespread coral mortality.

In 1996 a team of researchers at Tel Aviv University reported that a bacterial infection caused bleaching of a coral species *(Oculina patagonica;* shown in the photograph) that inhabits the Mediterranean coast of Israel. These investigators isolated and cultured the bacteria which they identified as belonging to the genus *Vibrio* (Chapter 23). The researchers demonstrated that coral experimentally infected with the bacteria became bleached. Based on their findings, the researchers suggested that an increase in seawater temperature lowers the resistance of the coral to infection. Some marine biologists think that the bacteria are opportunistic and infect the coral only when it is stressed by environmental conditions.

Coral bleaching appears to be a response to environmental stress, but some biologists suggest that bleaching may also be a mechanism that allows a different algal partner to establish residence within the coral. The new partnership may be more resistant to environmental stress.

Several international monitoring projects are gathering data that are needed to help us understand coral destruction and take action to protect these ecologically important and beautiful ecosystems. Once destroyed, it is difficult to reclaim them because of the amount of time needed to form new reefs. For example, the major reef builder in Florida and Caribbean waters, a species known as star coral *(Montastrea annularis),* requires about 100 years to form a reef just one meter high.

(6.5 ft) in diameter and have tentacles 30 m (98 ft) long. These orange and blue "monsters," among the largest invertebrates, are dangerous to swimmers in the North Atlantic Ocean.

Anthozoans have no medusa stage

The anthozoans—sea anemones and corals—have either individual or colonial polyps, but no free-swimming medusa stage. The polyp produces eggs and sperm, and the fertilized egg develops into a small ciliated planula. This larval form may swim to a new location before attaching to develop into a polyp.

Anthozoans differ from hydrozoans in that the gastrovascular cavity is partially divided into a number of connected chambers by a series of vertical partitions. The partitions increase the surface area for digestion, enabling an anemone to digest an animal as large as a crab or bass. Although corals can capture prey, many tropical species depend for nutrition on

photosynthetic algae (zooxanthellae) that live within cells lining the coral's digestive cavity (see Chapter 24). The relationship between coral and zooxanthellae is symbiotic and mutually beneficial. The algae provide the coral with oxygen and with carbon and nitrogen compounds. In exchange, the coral supplies the algae with waste products such as ammonia, from which the algae make nitrogenous compounds for both partners (see Chapter 52).

In warm, shallow seas, much of the bottom is covered with coral or anemones, most of them brightly colored. Coral communities (reefs) are among the most productive of all ecosystems, rivaling the tropical rain forests in species diversity. A single reef can serve as home for more than 3000 species of fishes and other marine organisms, and an estimated one-fourth of all marine species depend on them. Reef organisms form complex food webs. Some organisms attach to the reef, while others find shelter within its crevices. Many terrestrial organisms also benefit from coral reefs, which form and maintain the foundation of thousands of islands. By providing a barrier against waves, reefs also protect shorelines against storms and erosion (see Chapter 54).

The reefs and atolls of the South Pacific are the remains of billions of microscopic, cup-shaped calcareous skeletons, secreted during past ages by coral colonies and by coralline algae. Living colonies occur only in the uppermost regions of such reefs, adding their own skeletons to the forming rock. Living coral reefs are made up of colonies of millions of corals and by certain algae (mainly coralline red algae). These algae and the zooxanthellae that live within the coral cells contribute to the reef's brilliant colors.

Human activity, including overfishing, mining reefs for building materials, polluting coastal waters with industrial chemicals, and smothering coral with the silt that washes downstream from clearcut forests, threatens these important ecosystems. Studies indicate that many coral reefs, especially those in coastal waters, have suffered serious damage during the past several years (see *Making the Connection: Coral Reefs and Environmental Issues*).

COMB JELLIES MOVE BY CILIA

The 100 or so species of comb jellies, members of phylum **Ctenophora,** are all marine. They are fragile, luminescent animals that may be as small as a pea or larger than a tomato. Ctenophores are biradially symmetrical, meaning that you could obtain equal halves by cutting through the body (oral-aboral) axis in two different ways. Their body plan is somewhat similar to that of a medusa. They consist of two cell layers separated by a thick jelly-like mesoglea.

The outer surface of a ctenophore bears eight rows of cilia, resembling combs (Fig. 28–11). The coordinated beating of the cilia in these combs moves the animal through the water. A sense organ functions in balance and helps the animal orient itself. Some ctenophores have two tentacles. They do not have the stinging nematocysts characteristic of the cnidarians. However, their tentacles are equipped with adhesive glue cells that trap prey.

FLATWORMS ARE BILATERAL ACOELOMATES

Members of phylum **Platyhelminthes,** the **flatworms,** are flat, elongated, acoelomate animals that exhibit bilateral symmetry. At present the 20,000 species are divided into four classes. Class **Turbellaria** comprises the free-living flatworms, including planarians and their relatives. Classes **Trematoda** and **Monogenea** include the flukes, which are either internal or external parasites. Class **Cestoda** includes the tapeworms, which as adults are intestinal parasites of vertebrates (Table 28–2).

Some important characteristics of this phylum follow:

1. **Bilateral symmetry** and **cephalization.** Along with their symmetry, flatworms have definite anterior and posterior ends. An animal with a front end ("head") generally moves forward. With a concentration of sense organs in the part of the body that first meets its environment, the animal can find food or detect an enemy quickly. A rudimentary head, the beginnings of **cephalization,** is evident in flatworms.
2. **Three definite tissue layers** (triploblastic). In addition to an outer epidermis, derived from ectoderm, and an inner endodermis (epithelium), derived from endoderm, the flatworm has a middle tissue layer that develops from mesoderm.
3. **Well developed organs.** The flatworms are the simplest animals that have well developed organs, functional structures made of two or more kinds of tissue. Among their organs are a muscular pharynx for taking in food, a simple brain, eyespots and other sensory organs in the head, a sys-

Figure 28–11 Ctenophore (comb jelly). Ctenophores are free-swimming, bioluminescent hermaphrodites capable of self-fertilization. The sea gooseberry (*Pleurobrachia* sp.) has long aboral tentacles with adhesive cells used to capture prey. *(David Wrobel/Biological Photo Service)*

TABLE 28-2	Classes of Phylum Platyhelminthes
Class and Representative Animals	**Characteristics**
Turbellaria Planarians	Free-living flatworms; mainly marine; body covered by ciliated epidermis; typically carnivorous; prey on tiny invertebrates.
Trematoda and Monogenea Flukes	All parasites with a wide range of vertebrate and invertebrate hosts; may require intermediate hosts; adults have suckers for attachment to host.
Cestoda Tapeworms	Parasites of vertebrates; complex life cycle usually with one or two intermediate hosts; larval host may be invertebrate; tapeworms typically have suckers and sometimes hooks for attachment to host; eggs produced within proglottids, which are shed; no digestive or nervous systems.

tem for excreting wastes and maintaining fluid balance, and complex reproductive organs.

4. **A simple nervous system.** The simple brain typically consists of two masses of nervous tissue, called **ganglia,** in the head region. In many species, the ganglia are connected to two nerve cords that extend the length of the body. A series of nerves connects the cords like the rungs of a ladder.

5. **Protonephridia,** structures that function in osmoregulation and metabolic waste disposal (excretion).

6. **A gastrovascular cavity** in most species. As in the cnidarians, the digestive system has only one opening, the mouth, but the digestive system is often extensively branched.

Flatworms have no organs for circulation or gas exchange. These functions depend largely on diffusion.

Parasitic flatworms—flukes and tapeworms—are highly adapted to and modified for their parasitic life style. They have suckers or hooks for holding onto their hosts. The bodies of those that live in digestive tracts are resistant to the digestive enzymes secreted by their hosts. Many have complicated life cycles and produce large numbers of eggs. Other adaptations include the loss of structures such as sense organs and the digestive system.

Class Turbellaria includes planarians

Members of class Turbellaria are free-living flatworms. Most are marine, but many inhabit freshwater habitats, and a few tropical forms are terrestrial. **Planarians** are turbellarian flat-

worms found in ponds and quiet streams throughout the world. The common American planarian *Dugesia* is about 15 mm (0.6 in) long, with what appear to be crossed eyes and flapping "ears" called **auricles** (Fig. 28–12). The auricles actually serve as organs of chemoreception, important in locating food.

Planarians are carnivorous, trapping small animals in a mucous secretion. The digestive system consists of a single opening (the mouth), a tubelike **pharynx** (the first portion of the digestive tube), and a branched gastrovascular cavity. A planarian can project its pharynx outward through its mouth, using it to ingest its prey. Extracellular digestion takes place in the gastrovascular cavity by enzymes secreted by gland cells. Digestion is completed after the nutrients have been absorbed into individual cells. Undigested food is eliminated through the mouth. The long, highly branched gastrovascular cavity helps to distribute food to all parts of the body, so that each cell can receive nutrients by diffusion.

A planarian's flattened body ensures that gases can reach all of its cells by diffusion. Excretion also takes place mainly by diffusion, but some metabolic wastes are excreted by the protonephridia. These structures function primarily in fluid balance, or osmoregulation. Protonephridia are blind tubules that end in **flame bulbs,** collecting cells equipped with cilia. The beating of the cilia channels waste into the system of tubules and eventually out of the body through excretory pores.

Planarians are capable of learning. Memory is not localized within the ganglia but appears to be retained throughout the nervous system. Planarians can reproduce either asexually or sexually. In asexual reproduction, an individual constricts in the middle and divides into two planarians. Each regenerates its missing parts. Sexually, these animals are hermaphroditic. During the warm months of the year, each is equipped with a complete set of male and female organs. Two planarians come together in copulation and exchange sperm cells so that their eggs are cross-fertilized.

Flukes parasitize other animals

Although their body plan resembles that of the free-living flatworms, specialized adaptations make the **flukes,** members of classes Trematoda and Monogenea, successful parasites. For example, most adult flukes have structures, such as hooks and suckers, for attachment to the host. Flukes also have extremely complex and prolific reproductive organs (Fig. 28–13).

Flukes that are parasitic in humans include blood flukes, widespread in tropical areas of the world, and liver flukes, common in Asia, particularly in areas where human feces are used for fertilizing crops. Blood flukes of the genus *Schistosoma* infect about 200 million people who live in tropical areas. Both blood flukes and liver flukes go through complicated life cycles involving a number of different forms, alternation of sexual and asexual stages, and parasitism on one or more intermediate hosts (snails and fishes for example) (Fig. 28–14). The aquatic snails that serve as intermediate hosts thrive in ponds, rice paddies, and marshy areas that form when dams are built.

Figure 28–12 **The common planarian, *Dugesia*.** *(a)* Internal structure. *(b)* LM of a living planarian *(Dugesia dorotocephala)*. Note the prominent auricles, used to locate food. *(c)* Cross section through a planarian. *(b, T.E. Adams/Peter Arnold, Inc.)*

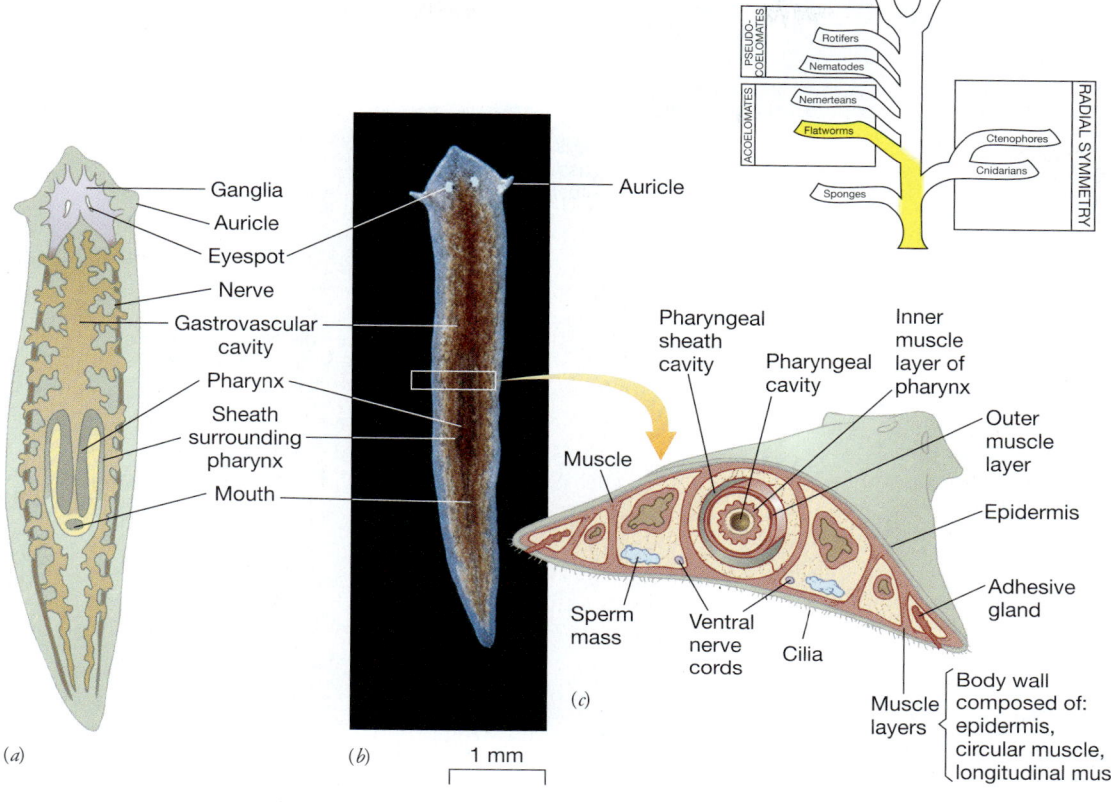

Tapeworms inhabit the intestines of vertebrates

Adult members of the more than 5000 different species of class Cestoda live as parasites in the intestines of probably every kind of vertebrate, including humans. Tapeworms are long, flat, ribbon-like animals strikingly specialized for their parasitic mode of life. Among their many adaptations are suckers and sometimes hooks on the "head," or **scolex,** that enable the parasite to attach to the host's intestine (Fig. 28–15).

The reproductive adaptations and abilities of tapeworms are extraordinary. The body of the tapeworm consists of a long chain of segments called **proglottids.** Each proglottid is an entire reproductive machine equipped with both male and female reproductive organs and containing up to 100,000 eggs. Because an adult tapeworm may have as many as 2000 seg-

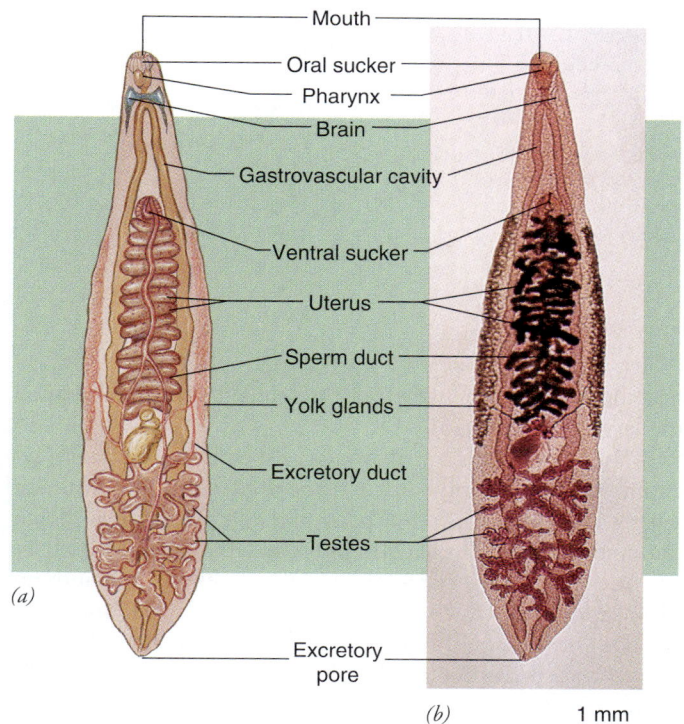

Figure 28–13 **Structure of a fluke.** *(a)* The liver fluke's oral sucker and well developed reproductive system are adaptations to its parasitic life style. *(b)* LM of human liver fluke *(Clonorchis sinensis)*. *(b, Carolina Biological Supply Company/Phototake)*

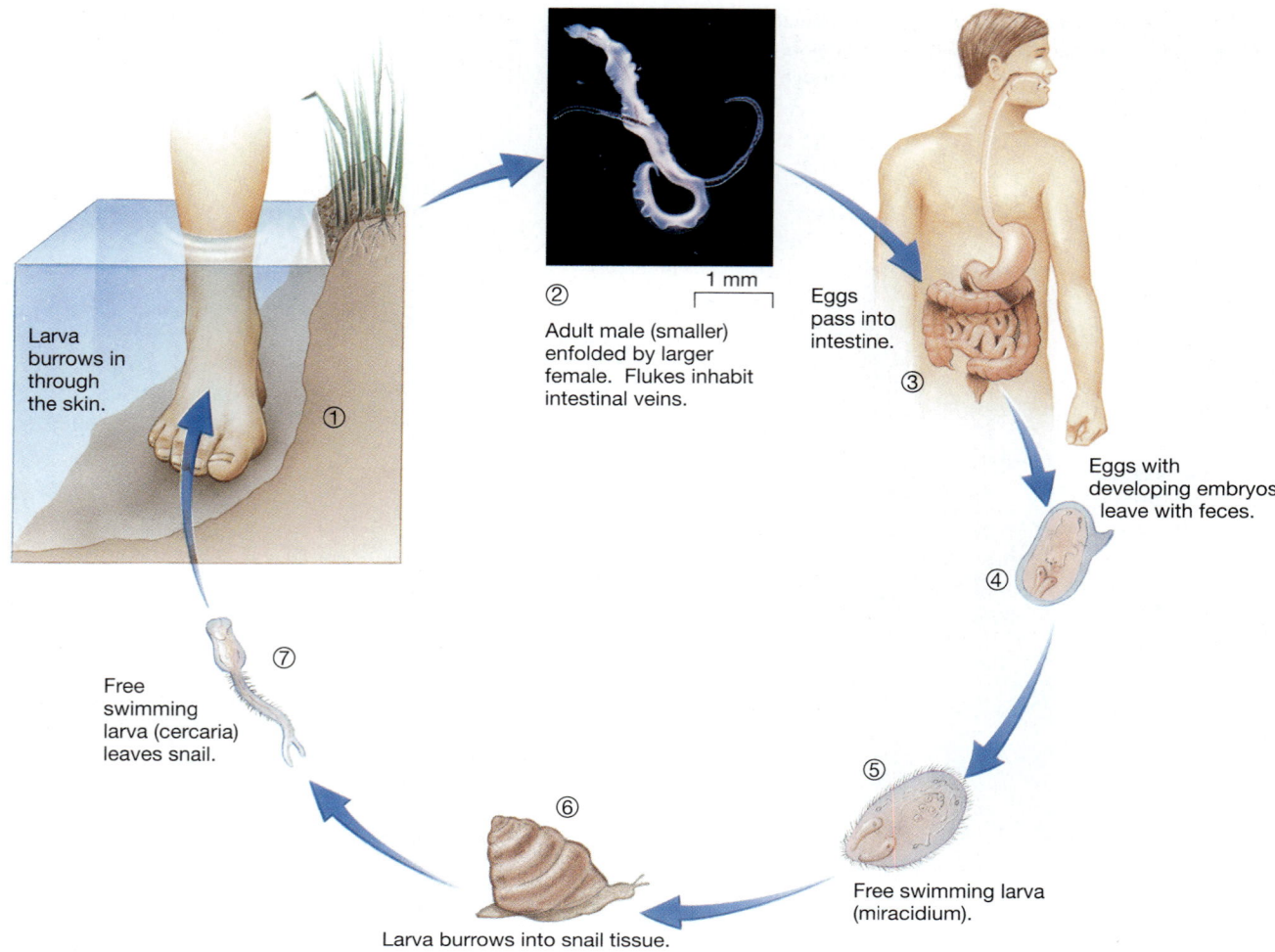

① Larva burrows in through the skin.

② 1 mm

Adult male (smaller) enfolded by larger female. Flukes inhabit intestinal veins.

Eggs pass into intestine. ③

④ Eggs with developing embryos leave with feces.

⑤ Free swimming larva (miracidium).

⑥ Larva burrows into snail tissue.

⑦ Free swimming larva (cercaria) leaves snail.

Figure 28–14 **Life cycle of the blood fluke (*Schistosoma* sp.).** (1) Humans are one of the two hosts. After burrowing through the skin, larvae make their way to the circulatory system. (2) The LM shows an adult male and female. They remain permanently paired for life. During reproduction, which takes place in intestinal veins, the female holds the male in a long groove. (3) Eggs pass into the intestine. (4) Eggs containing developing embryos leave the body with the feces. (5) If they find their way to fresh water, the eggs hatch releasing free-swimming larvae (miracidia). (6) To survive, the larvae must enter a second host, an aquatic snail. After burrowing into the tissues of the snail, larvae develop into a form that reproduces asexually. This process greatly increases the number of larvae. (7) Finally, fork-tailed larvae (cercariae) develop and leave the snail. (*Centers for Disease Control and Prevention, Atlanta, Georgia/Biological Photo Service*)

ments, its reproductive potential is staggering. A single tapeworm can produce 600 million eggs in a year. Proglottids farthest from the tapeworm's head contain the ripest eggs; these segments are shed from the host's body along with the feces.

The tapeworm has no mouth or digestive system. Digested food from the host is absorbed across the worm's body wall. The tapeworm also lacks well developed sense organs. Some tapeworms have complex life cycles, spending their larval stage within the body of an intermediate host and their adult life within the body of a different, final host. Let us consider the life cycle of the beef tapeworm, so named because humans can

become infected when they eat undercooked beef containing the larvae (Fig. 28–16).

The microscopic tapeworm larva spends part of its life cycle encysted within the muscle tissue of cattle. When a human ingests raw or rare infected beef, digestive juices break down the cyst, releasing the larva. Soon the larva attaches itself to the intestinal lining and within a few weeks matures into an adult tapeworm, which may grow to a length of about 15 m (50 ft). The parasite reproduces sexually within the human intestine and sheds proglottids filled with zygotes. Once established within a human host, the tapeworm makes itself very

Figure 28–15 **Scolex of the small tapeworm (*Acanthrocirrus retrisrostris*).** This tapeworm reaches maturity in the intestines of wading birds that eat barnacles. This false-color SEM shows the piston-like rostellum, which can be withdrawn into the head or thrust out and buried in the host's tissue. Beneath the rostellum, two of the four powerful suckers are visible. *(Cath Ellis/Science Photo Library/Photo Researchers, Inc.)*

250 µm

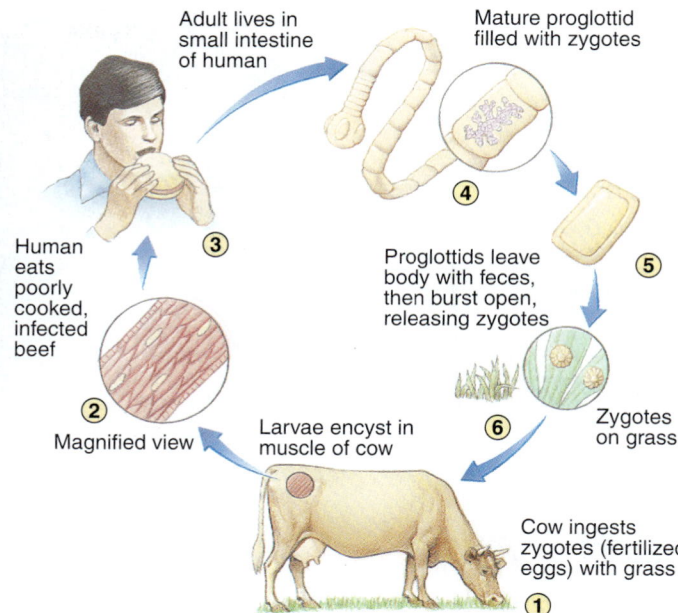

Figure 28–16 **Life cycle of the beef tapeworm.** (1) Cattle ingest zygotes with grass. Larvae hatch and encyst in the muscle of a cow or steer. (2) Magnified view of muscle containing encysted larvae. (3) When a human eats poorly cooked, infected beef, digestive juices dissolve the cyst, freeing the larva. (4) The adult tapeworm lives in the human's small intestine. Zygotes are produced within proglottids. (5) Mature proglottids leave the body with the feces and burst open, releasing zygotes. (6) Magnified view of zygotes on grass.

much at home and may remain for the rest of its life, as long as ten years. A person infected with a tapeworm may suffer pain or discomfort, increased appetite, weight loss, and other symptoms, or may be totally unaware of its presence.

In order for the life cycle of the tapeworm to continue, its eggs must be ingested by an intermediate host, in this case, a cow or steer. This requirement explains why we are not completely overrun by tapeworms and why a tapeworm must produce millions of eggs to ensure that at least a few survive. When cattle eat grass or other foods contaminated with human feces, eggs may be ingested. The eggs hatch in the cattle's intestines, and the larvae make their way into muscle. There they encyst and remain until released by a final host, perhaps a human eating very rare steak.

Two other tapeworms that infect humans are the pork tapeworm, found in undercooked, infected pork, and the fish tapeworm, found in raw or undercooked, infected fish. Like most parasites tapeworms tend to be species-specific; that is each can infect only certain specific species. For example, the beef tapeworm can only spend its adult life in a human host.

PHYLUM NEMERTEA IS CONSIDERED AN EVOLUTIONARY LANDMARK

Phylum **Nemertea,** the **ribbon worms,** is a relatively small group (about 900 species) of free-living animals (Fig. 28–17). Almost all are marine, although a few inhabit fresh water or damp soil. Nemerteans have long narrow bodies, either cylindrical or flattened, generally ranging in length from 5 cm (2 in) to about 2 m (6.5 ft), although some are much longer. Some are a vivid orange, red, or green, with black or colored stripes.

Their most remarkable organ, the **proboscis,** from which they get one of their common names *proboscis worms,* is a long, hollow, muscular tube that can be everted from the anterior end of the body and used in seizing food or in defense. The proboscis secretes mucus, which is helpful in catching and trapping prey. Although the nemerteans are functionally acoelomate, the chamber surrounding the proboscis is a true coelomic space derived like the coelom in coelomate protostomes.

Phylum Nemertea is of considerable evolutionary interest

(a)

Figure 28–17 **Nemerteans (proboscis worms).** *(a)* Proboscis worm *(Lineus)* from Panama, Pacific. *(b)* Lateral view of a typical nemertean. Note the complete digestive tract that extends from mouth to anus, giving this animal a tube-within-a-tube body plan. *(Kjell B. Sandved)*

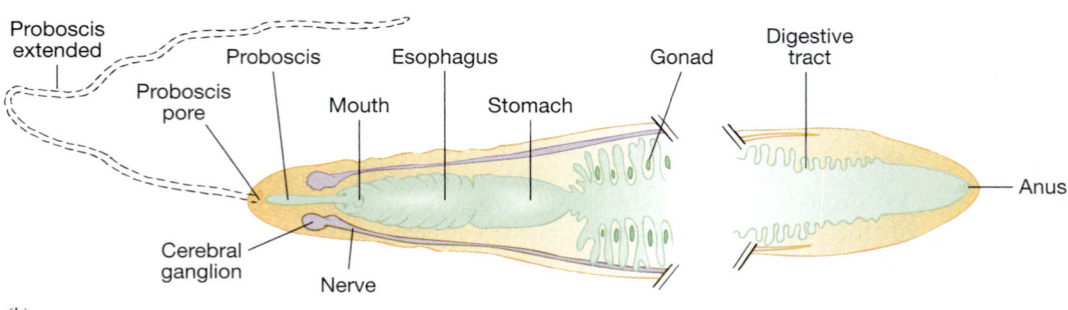

(b)

to biologists because its members have a circulatory system and a tube-within-a-tube body plan. The digestive tract is a complete tube, with a mouth at one end for taking in food and an anus at the other for eliminating undigested food. Recall that in most cnidarians and flatworms, food enters and wastes leave through the same opening.

In nemerteans the digestive and circulatory functions are separated. The primitive circulatory system consists simply of blood vessels, muscular tubes extending the length of the body and connected by transverse vessels. Nemerteans have no heart to pump the blood. The blood is circulated through the vessels by movements of the body and contractions of the muscular blood vessels. The development of a circulatory system for internal transport of materials provides an alternative to diffusion. The circulatory system and other advances that have evolved in the nemerteans permit these animals to be larger and more active than the flatworms.

ROUNDWORMS ARE OF GREAT ECOLOGICAL IMPORTANCE

Members of phylum **Nematoda,** the **roundworms,** play key roles in decomposition and nutrient recycling. Nematodes are numerous and widely distributed in soil and in marine and freshwater sediments. More than 12,000 species have been identified. A spadeful of soil may contain more than a million of these mainly microscopic worms, which thrash around coiling and uncoiling. Contraction of longitudinal muscles in the body wall produces these movements.

Although most nematodes are free living (Fig. 28–18), others are important parasites in plants and animals. More than 30 roundworms, including hookworms, pinworms, trichina worms, filarial worms, and the intestinal roundworm *Ascaris,* are human parasites. *Caenorhabditis elegans,* a free-living nematode, is an important research organism for biologists studying the genetic control of development (see Chapter 16).

The elongated, cylindrical, threadlike nematode body is pointed at both ends and covered with a tough **cuticle** (Fig. 28–19). Secreted by the underlying epidermis, the cuticle enables nematodes to resist desiccation, permitting them to inhabit dry soils and even deserts. Beneath the epidermis is a layer of longitudinal muscles. No circular muscles are present in the body wall.

Nematodes have a fluid-filled pseudocoelom that serves as a hydrostatic skeleton that transmits the force of muscle contraction to the enclosed fluid. Movement of fluid in the pseudocoelom is also important in nutrient transport and distribution. Like the ribbon worms, the nematodes exhibit bilateral symmetry, a complete digestive tract, three definite tis-

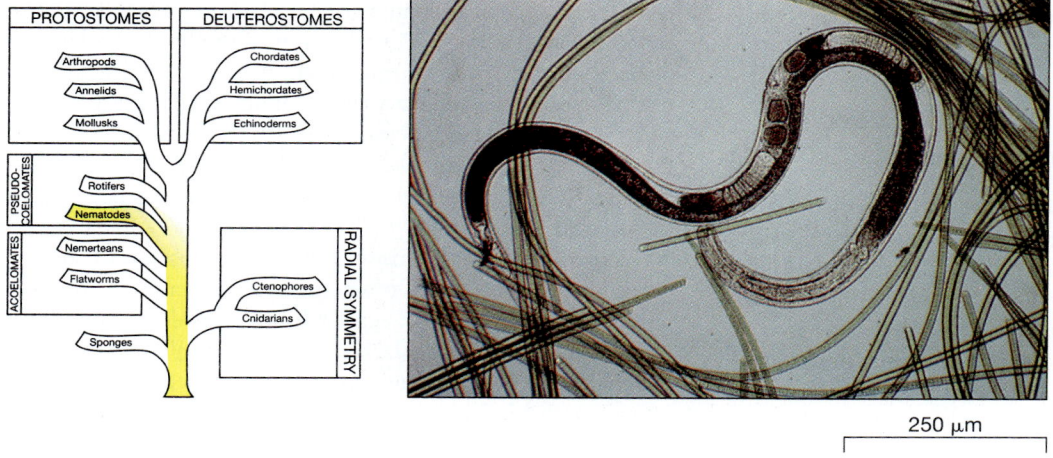

Figure 28–18 **LM of a free-living nematode.** This aquatic nematode is shown among the cyanobacteria *Oscillatoria*, which it eats. *(T.E. Adams/Visuals Unlimited)*

Mouth

Pharynx

Dorsal nerve

Muscle of pharynx wall

Excretory canal

Excretory gland

Ovary

Pseudocoelom

Excretory canal

Uterus

Intestine

Epidermis

Cuticle

Muscle layer

Vulva

(a)

(b)

Ventral nerve

Anus

Figure 28–19 **The roundworm *Ascaris*.** *(a)* Longitudinal section. Note the complete digestive tract that extends from mouth to anus. *(b)* Cross section through *Ascaris* shows the tube-within-a-tube body plan. The protective cuticle that covers the body helps the animal resist drying.

sue layers, and definite organ systems; however, they lack specific circulatory structures. The sexes are usually separate, and the male is generally smaller than the female.

Ascaris is a parasitic roundworm

A common intestinal parasite of humans, *Ascaris* is a white worm about 25 cm (10 in) long. *Ascaris* spends its adult life in the human intestine, where it ingests partly digested food. Like the tapeworm, it must devote a great deal of effort to reproduction in order to ensure survival of its species. The sexes are separate, and copulation takes place within the host. A mature female may produce as many as 200,000 eggs a day.

Ascaris eggs leave the human body with the feces. Where sanitation is poor, they find their way onto the soil. In many parts of the world, human wastes are used as fertilizer, a practice that encourages the survival of *Ascaris* and many other parasites. People become infected when they ingest *Ascaris* eggs on unwashed vegetables or fruit, or from hands dirty with contaminated soil.

The eggs hatch in the intestine, and the larvae then take a remarkable journey through the body before settling in the small intestine. To begin the journey, larvae burrow through the intestinal wall into blood vessels or lymph vessels. Then they are carried in blood vessels through the heart to the lungs. There, they break through into the air sacs and move up the air passageways to the throat, where they are swallowed. Finally, the larvae pass through the stomach and into the intestine, where they settle, feed, and mature. During their migration, the larvae damage the lungs and other tissues. When worms settle in the small intestine, they may cause abdominal pain, allergic reactions, or malnutrition. Sometimes a tangled mass of these worms blocks the intestine.

Several other parasitic roundworms infect humans

The life cycle of a human **hookworm** resembles that of *Ascaris*. Adult worms, which are less than 1.5 m (0.6 in) long, live in the human intestine. They lay eggs, which pass out of the body with the feces. Larvae hatch and feed on bacteria in the soil. After a period of maturation, they become infective. When a potential host walks barefoot, the microscopic larvae bore through the skin and enter the blood. They migrate through the body and find their way to the intestine, where they mature. Hookworms can cause serious tissue damage and loss of blood.

The **trichina** worm can live inside a variety of animals including pigs, rats, and bears. Humans typically become infected by eating undercooked, infected meat. Adult trichina worms live in the small intestine of the host. The females pro-duce larvae, which migrate through the body to skeletal muscle, where they encyst. Continuation of the life cycle depends on ingestion by another mammal. Because humans are not normally eaten, trichina larvae are not liberated from their cysts and eventually die. The cysts become calcified and remain in the muscles, causing stiffness and discomfort. Migrating larvae and adults cause other symptoms. No cure has been found for trichina infection.

Pinworms are the most common worms found in children. The tiny eggs are often ingested by eating with dirty hands contaminated with eggs. Adult worms, less than 1.3 cm (0.5 in) long, live in the large intestine. Female pinworms often migrate to the anal region at night to deposit their eggs. Irritation and itching caused by this practice induce scratching. Eggs may be distributed in the air and are in this way scattered throughout the house. Mild infestations may go unnoticed, but those more serious may result in discomfort, irritation, and injury to the intestinal wall.

ROTIFERS HAVE A CROWN OF CILIA

Among the more obscure invertebrates are the "wheel animals" of phylum **Rotifera**. Although no larger than many protozoa, these aquatic, microscopic animals are multicellular. More than 1800 species have been described. Most inhabit fresh water, but some live in marine environments or damp soil. Rotifers have a characteristic crown of cilia on their anterior end. The cilia beat rapidly during swimming and feeding, giving the appearance of a spinning wheel (Fig. 28–20 on p. 610).

Rotifers are pseudocoelomates with a complete digestive tract. They feed on tiny organisms suspended in the stream of water drawn into the mouth by the action of the cilia. A muscular organ posterior to the mouth grinds the food. Rotifers have a nervous system with a "brain" and sense organs, including eyespots. Protonephridia with flame cells remove excess water from the body and may also excrete metabolic wastes.

Like some other pseudocoelomate animals, rotifers are "cell constant": each member of a given species is composed of exactly the same number of cells. Indeed, each part of the body is made of a precisely fixed number of cells arranged in a characteristic pattern. Cell division does not take place after embryonic development, and mitosis cannot be induced; growth and repair are not possible. One of the challenging problems of biological research is discovering the difference between such nondividing cells and the dividing cells of other animals. Do you think rotifers could develop cancer?

Several small pseudocoelomate phyla are not discussed in this chapter. The main phyla that are discussed here are compared in Table 28–3.

TABLE 28-3	Comparison of Some Acoelomate Invertebrate Phyla				
	Porifera (pore bearers)	Cnidaria	Platyhelminthes (flatworms)	Nemertea	Nematoda (roundworms)
Representative Animals	Sponges	Hydras Jellyfish Corals	Planarians Flukes Tapeworms	Proboscis worms	Ascarids Hookworms Nematodes
Level of Organization and Body Plan	Multicellular; cells loosely arranged	Tissues	Triploblastic; organs; acoelomate	Triploblastic; organ systems; acoelomate	Triploblastic; organ systems; pseudocoelomate
Symmetry	None or radial	Radial	Bilateral; rudimentary head	Bilateral	Bilateral
Digestion	Intracellular	Gastrovascular cavity with only one opening; extra- and intracellular digestion	Gastrovascular cavity with only one opening; extra- and intracellular digestion	Complete digestive tract with mouth and anus; extra- and intracellular digestion	Complete digestive tract with mouth and anus; extra- and intracellular digestion
Circulation	Diffusion	Diffusion	Diffusion	At least two pulsating longitudinal blood vessels; no heart; blood cells with hemoglobin	Diffusion
Gas Exchange	Diffusion	Diffusion	Diffusion	Diffusion	Diffusion
Waste Disposal	Diffusion	Diffusion	Protonephridia; flame cells and ducts	Lateral excretory canals with flame cells	Excretory canals; most with unique excretory cells
Nervous System	Irritability of cytoplasm	Nerve net; no centralization of nerve tissue	Simple brain; two nerve cords; ladder type system; simple sense organs	Simple brain; nerve cords; cross nerves; simple sense organs	Simple brain; dorsal and ventral nerve cords; simple sense organs
Reproduction	Asexual, by budding; sexual, most are hermaphroditic	Asexual, by budding; sexual, sexes usually separate	Asexual, by fission; sexual, hermaphroditic, but cross-fertilization in most species	Asexual, by fragmentation; sexual, sexes separate	Sexual, sexes separate
Support and Movement	Support by spicules of calcium carbonate, silica, or spongin; contractile cells around pores	Support by mesoglea, calcareous skeletons (coral), or by fluid in gastrovascular cavity that serves as hydrostatic skeleton; contractile cells	Support by its tissues; well developed muscle tissue	Support by its tissues; locomotion by muscles or cilia	Support by tough cuticle; fluid in pseudocoelom serves as hydrostatic skeleton; longitudinal muscles in body wall
Environment and Lifestyle	Aquatic, mainly marine; ciliated swimming larvae; adults attach; suspension feeders	Aquatic, mainly marine; some float or swim, others sessile; polyp and medusa forms; some form colonies; capture food with cnidocytes, tentacles	Aquatic, some terrestrial in damp areas; many are carnivores; flukes and tapeworms are all parasites	Mainly marine; mainly carnivores; use proboscis for capturing food and defense	Widely distributed in the soil, saltwater, and fresh water; carnivores, scavengers, parasites

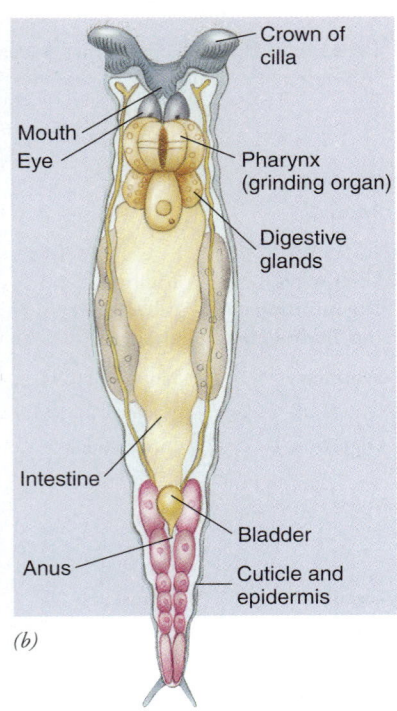

Figure 28–20 **Rotifers (wheel animals).** (*a*) LM of an Antarctic rotifer (*Philodina gregaria*) that survives the winter by forming a cyst. It reproduces in great numbers, sometimes coloring lake water red. (*b*) Longitudinal section showing rotifer anatomy. The motion of its cilia draws particles of food such as algae into the mouth. (*a, John Walsh/Science Photo Library/Photo Researchers, Inc.*)

(*a*)

100 mm

(*b*)

SUMMARY WITH KEY TERMS

I. Members of **kingdom Animalia** are eukaryotic, multicellular, heterotrophic organisms with cells specialized to perform specific functions. They generally are capable of locomotion at some time during their life cycle, can reproduce sexually, and can respond adaptively to external stimuli.

  A. In sexual reproduction, sperm and egg unite to form a zygote that undergoes **cleavage.** Multiple cell divisions result in the development of a hollow ball of cells, a **blastula.** The blastula undergoes **gastrulation,** forming embryonic tissues.

  B. Most animals are **invertebrates,** animals without backbones. **Vertebrates** are a subphylum of phylum Chordata.

II. Animals are consumers that inhabit marine environments, fresh water, and land. However, only a few groups have successfully adapted to terrestrial environments.

III. Body structure and developmental pattern provide clues to animal phylogeny.

  A. Cnidarians and adult echinoderms have **radial symmetry;** most other animals are **bilaterally symmetrical,** at least in their larval stages. In bilaterally symmetrical animals, the location of body structures can be defined by such relative directional terms as: **dorsal** or **ventral; anterior** or **posterior** (or **caudal**); **medial** or **lateral;** and **cephalic.**

  B. **Triploblastic** animals can be classified as **acoelomate** (no body cavity), **pseudocoelomate** (body cavity not completely lined with mesoderm), or **coelomate** (body cavity completely lined with mesoderm).

    1. Most structures develop from embryonic tissues called **germ layers.** The germ layers include the outer **ectoderm** that gives rise to the body covering and the nervous system; the inner **endoderm** that lines the gut and other digestive organs; and a middle **meso-**

derm that gives rise to most other body structures.

    2. A true **coelom** is a body cavity that is completely lined with mesoderm.

    3. Two evolutionary branches of coelomates are **protostomes** (mollusks, annelids, arthropods) and **deuterostomes** (echinoderms, hemichordates, and chordates). In protostomes the blastopore develops into the mouth; in deuterostomes it does not; instead it typically becomes the anus.

IV. Phylum **Porifera** consists of the sponges, animals characterized by flagellated **collar cells (choanocytes).**

  A. Sponges are divided into three classes on the basis of the type of skeleton they secrete.

  B. The sponge body is a sac with tiny openings through which water enters; a central cavity (**spongocoel**); and an open end, or **osculum,** through which water exits.

V. Phylum **Cnidaria,** which includes the hydras, jellyfish, and corals, is characterized by radial symmetry, two tissue layers, and **cnidocytes,** cells that contain stinging organelles (**nematocysts**).

  A. Phylum Cnidaria includes three main classes. Class **Hydrozoa** includes hydras, hydroids, and the Portuguese man-of-war; class **Scyphozoa** comprises the jellyfish; and class **Anthozoa** includes sea anemones and corals.

  B. In many types of cnidarians, the life cycle includes a sessile **polyp** stage and a free-swimming **medusa** stage.

  C. The **gastrovascular cavity** has a single opening that serves as both mouth and anus.

  D. Nerve cells form irregular, nondirectional **nerve nets** that connect sensory cells with contractile and gland cells.

VI. Phylum **Ctenophora** consists of the comb jellies: fragile, luminescent, biradially symmetrical marine predators.

VII. Phylum **Platyhelminthes,** the flatworms, are acoelomate animals characterized by bilateral symmetry, **cephalization,** three definite tissue layers, and well developed organs. Flatworms are **hermaphroditic;** a single animal produces both sperm and eggs. Phylum Platyhelminthes includes four classes.

   A. Class **Turbellaria** is comprised of free-living flatworms, including **planarians.** Classes **Trematoda** and **Monogenea** include the parasitic **flukes,** and class **Cestoda** comprises the parasitic tapeworms.

   B. Flatworms have a ladder-type nervous system, typically consisting of sense organs and a simple brain consisting of two **ganglia** connected to two nerve cords that extend the length of the body.

   C. Flatworms have **protonephridia,** organs that function in osmoregulation and disposal of metabolic wastes.

   D. The parasitic flukes and tapeworms typically have suckers or hooks for holding on to their hosts; they have complicated reproductive systems and life cycles.

VIII. Members of phylum **Nemertea** (ribbon worms) have a tube-within-a-tube body plan, a complete digestive tract with mouth and anus, and a separate circulatory system. The **proboscis** is a muscular tube used in capturing food and in defense.

IX. Phylum **Nematoda,** the **roundworms,** includes species of great ecological importance. Some species are parasitic in plants and animals.

   A. Nematodes are pseudocoelomates with bilateral symmetry, three tissue layers, and a complete digestive tract.

   B. The nematode body is covered by a tough **cuticle** that helps prevent dessication.

   C. Parasitic nematodes in humans include *Ascaris,* **hookworms, trichina worms,** and **pinworms.**

X. Members of phylum **Rotifera** (wheel animals) are characterized by a crown of cilia. Rotifers are aquatic, pseudocoelomate, microscopic animals that exhibit cell constancy (each individual of a species has the same number of cells).

POST-TEST

1. All invertebrates are (a) animals with vertebral columns (b) acoelomates (c) deuterostomes (d) animals without backbones (e) two of the preceding answers are correct

2. Radial symmetry is characteristic of (a) protostomes (b) acoelomates (c) deuterostomes (d) cnidarians (e) nematodes

3. The auricles of a planarian are _____ to its cerebral ganglia (a) medial (b) cephalic (c) caudal (d) anterior (e) lateral

4. The germ layer that gives rise to the outer covering of the body and the nervous system is the (a) cuticle (b) ectoderm (c) contractile layer (d) endoderm (e) mesoderm

5. A true coelom is completely lined with (a) cuticle (b) ectoderm (c) contractile layer (d) endoderm (e) mesoderm

6. Collar cells are most characteristic of phylum (a) Porifera (b) Cnidaria (c) Platyhelminthes (d) Nematoda (e) Rotifera

7. Rudimentary cephalization, three tissue layers, and protonephridia characterize phylum (a) Porifera (b) Cnidaria (c) Platyhelminthes (d) Nematoda (e) Rotifera

8. Tapeworms are classified in phylum (a) Porifera (b) Cnidaria (c) Platyhelminthes (d) Nematoda (e) Rotifera

9. Which of the following is NOT characteristic of proboscis worms? (a) pseudocoelomate (b) tube-within-a-tube body plan (c) complete digestive tube (d) muscular tube for capturing food (e) circulatory system

10. Which of the following is NOT a nematode? (a) hookworm (b) fluke (c) *Ascaris* (d) trichina worm (e) pinworm

11. Corals (a) are coelomates (b) have cnidocytes (c) lack a polyp stage (d) have a tough cuticle (e) have protonephridia

12. Which of the following is NOT an adaptation of parasitic life? (a) production of a few well protected eggs (b) hooks (c) suckers (d) reduced digestive system (e) intermediate host

REVIEW QUESTIONS

1. For centuries sponges were classified as plants. Justify their current classification as animals.

2. Why is the sea a more hospitable environment for many animals than the land or fresh water?

3. As compared to phylum Cnidaria, what advances do members of phylum Platyhelminthes exhibit? In what ways are these animals alike?

4. What are the advantages of bilateral symmetry and cephalization?

5. Identify the phyla (from among those studied in this chapter) that have the following characteristics: (a) radial symmetry (b) protonephridia (c) acoelomate (d) pseudocoelomate (e) alternation of sexual and asexual stages (f) cnidocytes (g) nerve net (h) digestive tract with an opening at each end (i) circulatory system

6. Describe the alternation of stages exhibited by *Obelia.*

7. How do flatworms survive without specialized structures for gas exchange and internal transport of materials?

8. What special adaptations do tapeworms have for their parasitic mode of life?

9. Describe the life cycles of the following animals: (a) beef tapeworm (b) *Ascaris;* (c) hookworm

10. Label the diagram on p. 610. Use Fig. 28–19 to check your answers.

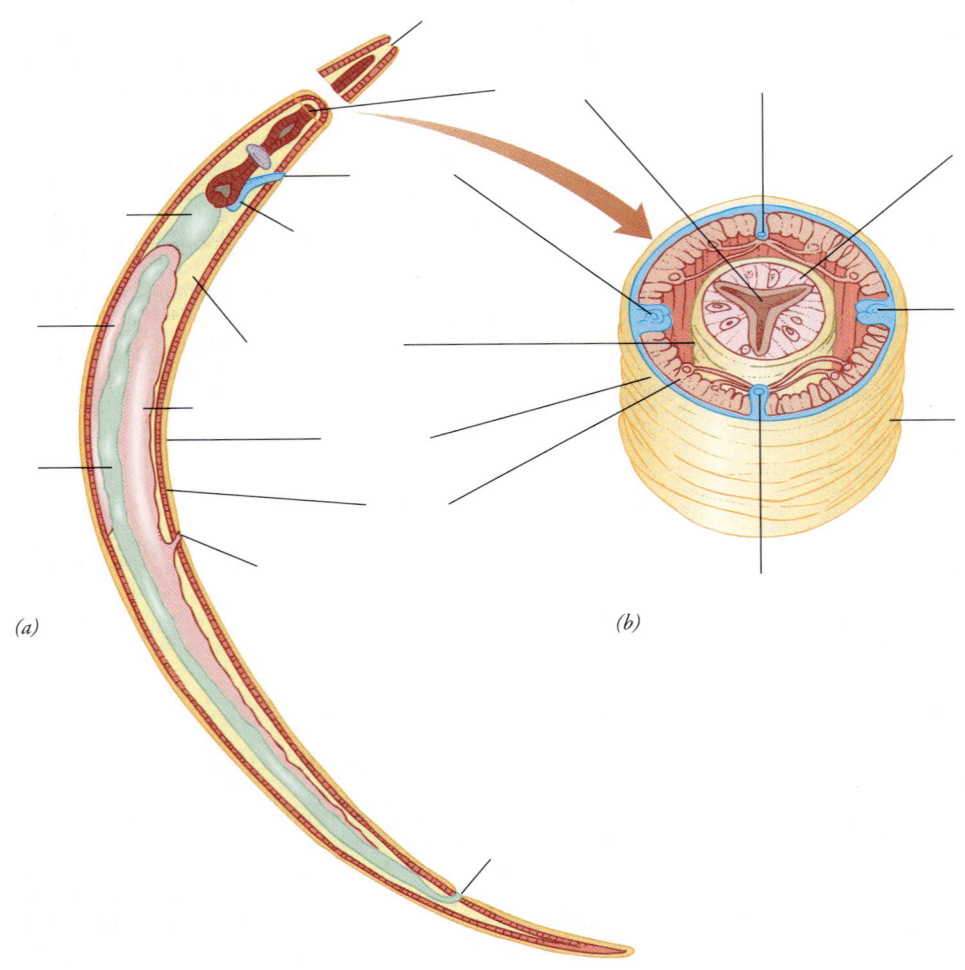

(a)

(b)

YOU MAKE THE CONNECTION

1. Every evolutionary adaptation has both benefits and costs. Explain this concept in terms of each of the following: (a) cephalization (b) circulatory system (c) hermaphroditism (d) pseudocoelom
2. Several international monitoring projects are gathering data needed to help us understand coral reef destruction. Why is it important to take action to protect coral reefs?
3. Imagine that you discover a new animal in a rain forest. How would you decide to which phylum to assign it? What are some characteristics that might contribute to your decision?

RECOMMENDED READINGS

Brown, B.E., and J.C. Ogden. "Coral Bleaching." *Scientific American,* Jan. 1993. Environmental stressors can cause permanent damage to coral reefs.

Brusca, R.C., and G.J. Brusca. *Invertebrates.* Sinauer Associates, Inc., Sunderland, Massachusetts, 1990. The authors have combined taxonomic and systems approaches in this comprehensive survey of the invertebrate phyla.

Fabricius, K.E., Y. Benayahu, and A. Genin. "Herbivory in Asymbiotic Soft Corals." *Science,* Vol. 268, 7 Apr. 1995. Unlike most corals, a group of soft corals feed on phytoplankton.

Knoll, A.H. "End of the Proterozoic Eon." *Scientific American,* Vol. 265, No. 4, Oct. 1991. A discussion of environmental change that may have led to the evolution of large animals.

Morse, A.N.C., and D.E. Morse. "Flypapers for Coral and Other Planktonic Larvae." *BioScience,* Vol. 46, No. 4, Apr. 1996. Coral and other larvae recognize specific chemical signals that induce attachment and metamorphosis. Some of the chemical compounds involved may prove important to commercial aquaculture or medicine.

Ruppert, E.E., and R.D. Barnes. *Invertebrate Zoology,* 6th ed. Saunders College Publishing, Philadelphia, 1994. This comprehensive textbook discusses the life processes of each invertebrate phylum.

Travis, J. "Bleaching Power: Marine Bacteria Rout Coral's Colorful Algae." *Science News,* Vol. 149, 15 Jun. 1996. An interesting account of a research project suggesting that bacteria cause coral bleaching.

• Visit our website at **http://www.saunderscollege.com/lifesci/titles.html** and click on Solomon/Berg/Martin Biology for links to chapter-related resources on the World Wide Web.

CHAPTER 29

The Animal Kingdom: The Coelomate Protostomes

What does an earthworm have in common with a clam, a spider, and a butterfly? All of these animals have a coelom, and all are protostomes. The coelomate protostomes include the annelids, mollusks, and arthropods, as well as several smaller, related phyla. As explained in Chapter 28, the **coelom** is a space completely lined by mesoderm that lies between the digestive tube and the outer body wall. Coelomate protostomes have a complete digestive tract with separate mouth and anus, and most have well developed circulatory, excretory, and nervous systems.

Animals with a coelom, and to a lesser extent those with a pseudocoelom, have certain advantages over those lacking a body cavity. The coelom serves as a fluid-filled space that protects internal organs by cushioning them. The coelom permits a separation between the muscles of the body wall and those in the wall of the digestive tract. It permits organs to develop and move independently of the outer wall of the body. For example, the digestive tube can move food along independently of body movements. And the pumping action of the heart would not be possible without the surrounding space provided by the coelom. The coelom also provides space for the gonads to develop. During the breeding season of many animals such as birds, the gonads enlarge as they fill with ripe gametes.

As an enclosed compartment (or series of compartments) of fluid under pressure, the coelom can serve as a **hydrostatic skeleton,** which provides shape to the body of soft animals. In many coelomates, like the bristleworm *(Hermodice)* shown here, a hydrostatic skeleton provides a firm structure against which surrounding muscles can contract. In some animals,

(Marty Snyderman/Visuals Unlimited)

fluid within the coelom helps transport materials such as food, oxygen, and wastes. Cells bathed by the coelomic fluid can exchange materials with it. The cells receive nutrients and oxygen from the coelomic fluid and excrete wastes into it. Some coelomates have excretory structures that remove wastes directly from the coelomic fluid.

AFTER YOU HAVE STUDIED THIS CHAPTER YOU SHOULD BE ABLE TO

1. Identify several advantages of having a coelom.
2. Identify challenges associated with terrestrial living and describe adaptations that enable terrestrial animals to meet those challenges.
3. Describe the distinguishing characteristics of mollusks, annelids, and arthropods and properly classify an animal that belongs to any of these phyla.
4. Describe the classes of mollusks discussed and give examples of animals that belong to each.

5. Describe and give examples of each of the three classes of annelids discussed.
6. Distinguish among the subphyla and classes of arthropods and give examples of animals that belong to each group.
7. Discuss factors that have contributed to the great biological success of insects.

LIFE ON LAND REQUIRES MANY ADAPTATIONS

Based on the fossil record, many biologists hypothesize that the first air-breathing land animals were millipede-like arthropods that came ashore in the Silurian period about 450 million years ago. The first land vertebrates, the amphibians, did not appear until the latter part of the Devonian period, about 30 million years later.

Many modern invertebrate coelomates still inhabit the sea. The earthworm is a terrestrial animal, but most annelids are marine. A few snails inhabit the land, but most mollusks also live in the sea. Among the arthropods, most crustaceans (crabs, lobsters, and their relatives) and the merostomes (horseshoe crabs) are also marine forms. However, certain modern arthropods, including the insects and spiders, are very successful terrestrial animals.

The chief problem facing all terrestrial organisms is that of drying out in the absence of a surrounding watery medium. A body covering adapted to minimize fluid loss helps solve this problem in many land animals. Location of the respiratory surface deep within the animal also helps prevent fluid loss. Thus, while gills are typically located externally, lungs and tracheal tubes (found in insects) are internal.

Another problem associated with life on land is supporting the body against the pull of gravity in the absence of the buoyant effect of water. Some animals, such as earthworms, do not face this challenge because they have small bodies. Larger animals generally need some sort of supporting skeleton. Arthropods and most mollusks have a tough **exoskeleton,** a supporting armor that covers the body. Vertebrates have an **endoskeleton,** a supporting framework within the body.

Reproduction on land poses still another challenge. Many aquatic forms shed their gametes in the water, where fertilization occurs. The surrounding water serves as an effective shock absorber, protecting the delicate embryos as they develop. Some land animals, including most amphibians, return to the water for reproduction; their larval forms develop in the water. Earthworms, land snails, insects, reptiles, birds, and mammals engage in internal fertilization. They transfer sperm from the body of the male directly into the body of the female by

copulation. The sperm are surrounded by a watery medium. Another important adaptation to reproduction on land is the tough, protective shell that surrounds the eggs of many species. Secreted by the females, this shell protects the developing embryo from drying out. An alternative adaptation for terrestrial reproduction is the development of the embryo within the moist body of the mother.

Now that we have discussed adaptations of terrestrial animals, we survey those animals that have a protostome pattern of early development: mollusks, annelids, and arthropods.

MOLLUSKS HAVE A FOOT, VISCERAL MASS, AND MANTLE

Mollusks are among the best known of the invertebrates. Most of us have walked along the seashore collecting their shells. Phylum **Mollusca** includes clams, oysters, octopods, snails, slugs, and the largest of all the invertebrates, the giant squid, which averages 9 to 16 m (about 30 to 53 ft) in length, including its tentacles. More than 50,000 living species and 35,000 fossil species (second only to the arthropods in number) have been described. Representative mollusks are illustrated in Figure 29–1. Four of the eight recognized classes are discussed in this chapter and are listed in Table 29–1.

▶ **Figure 29–1 The molluskan body plan.** *(a)* Chitons are sluggish marine animals with shells composed of eight overlapping plates. This sea cradle chiton *(Tonicella lineata)* inhabits coastal waters off the Pacific Northwest. *(b)* The broad, flat foot of the gastropod is an adaptation to its mobile lifestyle. The terrestrial banded garden snail *(Cepaea nemoralis)* is widely distributed. *(c)* The compressed body of the horseneck clam *(Tresus capax)* is adapted for burrowing in the mud. *(d)* The squid body is streamlined for swimming. To avoid being seen by potential predators, the squid can change color to blend with its background. This squid *(Sepioteuthis lessoniana)* is native to Hawaii. *(a, Kjell B. Sandved/Visuals Unlimited; b, William E. Ferguson; c, Tom McHugh/ Photo Researchers, Inc.; d, Mike Severns/Tom Stack & Associates)*

(a) Polyplacophora

Digestive tract

Shell

Foot

(b) Gastropoda

(c) Bivalvia

Foot

Digestive tract

Shell

(d) Cephalopoda

TABLE 29-1	Major Classes of Phylum Mollusca
Class and Representative Animals	**Characteristics**
Polyplacophora Chitons	Primitive marine animals with shell consisting of eight separate transverse plates; head reduced; broad foot used for locomotion.
Gastropoda Snails Slugs Nudibranchs	Marine, freshwater, or terrestrial; coiled shell in some species; torsion of visceral mass; well developed head with tentacles and eyes.
Bivalvia Clams Oysters Mussels	Marine and freshwater; body laterally compressed; two-part shell hinged dorsally; hatchet-shaped foot; suspension feeders.
Cephalopoda Squids Octopods	Marine; predatory; foot modified into tentacles, usually bearing suckers; well developed eyes; closed circulatory system.

Although most mollusks are marine, some snails and clams live in fresh water, and many species of snails and slugs inhabit the land. Mollusks probably evolved early in the protostome clade soon after the evolution of the coelom but before the origin of the segmented body that is characteristic of annelids and arthropods. Although mollusks vary widely in outward appearance, most share certain basic characteristics:

1. A soft body, usually covered by a dorsal shell composed mainly of calcium carbonate.
2. A broad, flat, muscular **foot,** located ventrally, which is used for locomotion.
3. The body organs (viscera) concentrated as a **visceral mass** located above the foot.
4. A **mantle,** a thin sheet of tissue that covers the visceral mass and usually contains glands that secrete a shell. The mantle generally overhangs the visceral mass, forming a mantle cavity that contains gills and other structures.
5. A rasplike structure called the **radula,** which is a belt of teeth in the mouth region. (The radula is not present in clams, their relatives, or in other suspension feeders.)
6. The coelom is generally reduced to small compartments around certain organs, including the heart and excretory organs (metanephridia). The main body cavity is typically a **hemocoel,** a space containing blood (see discussion of open circulatory system that follows).

All of the organ systems typical of complex animals are present in mollusks. The digestive system is a tube, often coiled, consisting of a mouth, buccal cavity (mouth cavity), esophagus, stomach, intestine, and anus. The radula, located within the buccal cavity, can be projected out of the mouth and used to scrape particles of food from the surface of rocks or the ocean floor. Sometimes the radula is used to drill a hole in the shell of a prey animal or to tear off pieces of a plant.

Most mollusks have an **open circulatory system** in which the blood, called **hemolymph,** bathes the tissues directly. In mollusks, the heart pumps blood into a single blood vessel, the aorta, which may branch into other vessels. Eventually blood flows into a network of large spaces called **sinuses,** where the tissues are bathed directly; this network makes up the hemocoel, or blood cavity. From the sinuses, blood drains into vessels that conduct it to the gills, where it is recharged with oxygen. From the gills, the blood returns to the heart. Thus, blood flow in a mollusk follows the pattern:

heart $\longrightarrow$ aorta $\longrightarrow$ smaller blood vessels $\longrightarrow$
blood sinuses $\longrightarrow$ blood vessels to gills $\longrightarrow$ heart

In open circulatory systems, blood pressure tends to be low, and tissues are not very efficiently oxygenated. However, most mollusks are slow-moving animals with low metabolic rates, and this type of circulatory system is adequate. In the active cephalopods (the class that includes the squids and octopods), the circulatory system is closed. In a **closed circulatory system,** blood flows through a complete circuit of blood vessels.

Large, paired **metanephridia** are excretory structures characteristic of mollusks. Each metanephridium is a ciliated tubule with a funnel at one end that removes wastes from the fluid in the coelom. Wastes are discharged through the other end of the tubule, which leads to an excretory pore. Open at both ends, metanephridia are quite different from the protonephridia of flatworms, which are blind tubules that open only to the exterior.

Most marine mollusks pass through one or more larval stages. The first larval stage is typically a **trochophore larva,** a free-swimming, ciliated, top-shaped larva characteristic of mollusks and annelids (Fig. 29–2). In many mollusks (gastropods and bivalves), the trochophore larva develops into a **veliger larva,** which has a shell, foot, and mantle. The veliger larva is unique to the mollusks.

Chitons may be similar to ancestral mollusks

Class **Polyplacophora** (meaning "many plates") comprises the **chitons,** sluggish marine animals with flattened bodies (Fig. 29–1a). Their most distinctive feature is a shell composed of eight separate but overlapping dorsal plates. The head is reduced and there are no eyes or tentacles.

The chiton inhabits rocky intertidal zones, using its broad, flat foot to move and to hold firmly onto rocks. By pressing

(a)

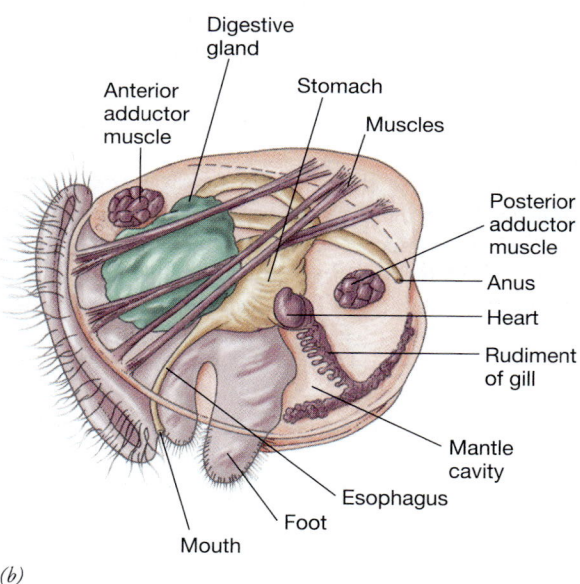

(b)

Figure 29–2 Larval stages of marine mollusks. (a) A trochophore larva, the first larval stage of a marine mollusk, is also characteristic of annelids. Just above the mouth, there is a characteristic band of ciliated cells that function as a swimming organ and, in some species, collects suspended food particles. (b) Typically, the trochophore larva develops into a bilateral veliger larva with foot, shell, and mantle.

its mantle against the substratum and lifting the inner edge of the mantle, the chiton can produce a partial vacuum. The resulting suction enables the animal to adhere powerfully to its perch. Using its radula for grazing, the chiton scrapes algae and other small organisms off rocks and shells.

(a)

(b)

Figure 29–3 Gastropods. (a) A rough keyhole limpet (*Diodora aspera*) escapes the arms of a seastar by exuding slippery mucus from its mantle. (b) Porter's nudibranch (*Chromodoris porterae*) feeding on a lightbulb tunicate (*Clavelina huntsmani*) (a, Norbert Wu/Peter Arnold, Inc.; b, Bruce Watkins/Animals Animals)

Gastropods are the largest group of mollusks

Class **Gastropoda,** which includes the snails and slugs and their relatives, is the largest and most diverse group of mollusks (Fig. 29–3). In fact, gastropods comprise the second largest class in the animal kingdom, second only to insects. Most gastropods inhabit marine waters, but others make their homes in brackish water, fresh water, or on land. We think of snails as having a single, spirally coiled shell into which they can withdraw the body, and many do. However, other gastropods, such as limpets, have shells like flattened dunce caps. Still others, such as garden slugs and the beautiful marine snails known as **nudibranchs,** have no shell at all.

Many gastropods have a well developed head with tentacles. Two simple eyes may be located on stalks that extend from

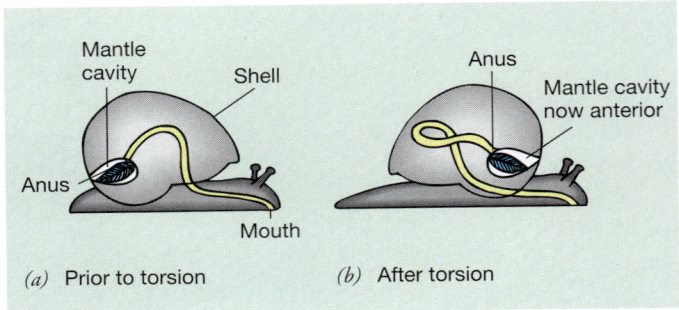

Mantle
cavity
Shell
Anus
Mantle cavity
now anterior
Anus
Mouth
(a) Prior to torsion
(b) After torsion

Figure 29–4 **Embryonic torsion in a gastropod.** As the bilateral larva develops, the visceral mass twists 180 degrees relative to the head.

Bivalves typically burrow in the mud

Class **Bivalvia** includes the clams, oysters, mussels, scallops and their relatives. The soft body of members of class Bivalvia is laterally compressed and completely enclosed by a two-part shell that hinges dorsally and opens ventrally (Fig. 29–5). This arrangement allows the hatchet-shaped foot to protrude ventrally for locomotion and for burrowing in the mud. The two parts, or *valves,* of the shell are connected by an elastic ligament. Stretching of the ligament opens the shell. Large, strong adductor muscles attached to the shell permit the animal to close its shell.

The inner pearly layer of the bivalve shell is made of calcium carbonate secreted in thin sheets by the epithelial cells of the mantle. Known as *mother-of-pearl,* this material is valued for making jewelry and buttons. Should a bit of foreign matter lodge between the shell and the mantle, the epithelial cells covering the mantle may be stimulated to secrete concentric layers of calcium carbonate around the intruding particle. Many species of oysters and clams form pearls in this way.

The bivalve nervous system has three pairs of ganglia, two pairs of nerve cords, and a variety of sense organs. Pigmented eyespots *(ocelli)* may be present along the edge of the mantle, enabling the animal to detect changes in light intensity that could be caused by the shadow of a predator. In scallops and in some oysters and clams, the ocelli are well developed, permitting formation of images.

Some bivalves, such as oysters, attach permanently to the substratum. Others, like clams, burrow slowly through rock or wood, seeking protected dwellings. The shipworm, *Teredo,* which damages dock pilings and other marine installations, is just looking for a home. A few bivalves, such as scallops, swim rapidly by clapping their two shells together with the contraction of a large adductor muscle (the part of the scallop that is eaten by humans).

the head. The broad, flat foot is used for creeping. Most land snails do not have gills. Instead, the mantle is highly vascularized and functions as a lung. These garden snails and slugs are described as **pulmonate** (meaning "having a lung").

A unique feature of gastropods is **torsion,** a twisting of the visceral mass. (This twisting is unrelated to the coiling of the shell.) As the bilateral larva develops, one side of the visceral mass grows more rapidly than the other side. This uneven growth results in rotation of the visceral mass. The visceral mass and mantle twist permanently up to 180 degrees, relative to the head. As a result, the digestive tract becomes somewhat U-shaped, and the anus comes to lie above the head and gill (Fig. 29–4). Subsequent growth is dorsal and usually in a spiral coil. Torsion limits space in the body, and typically the gill, metanephridium (kidney), and gonad are absent on one side. Some biologists hypothesize that torsion is an adaptation that protects the head by allowing it to enter the shell first during withdrawal. Without torsion, the foot would be withdrawn first.

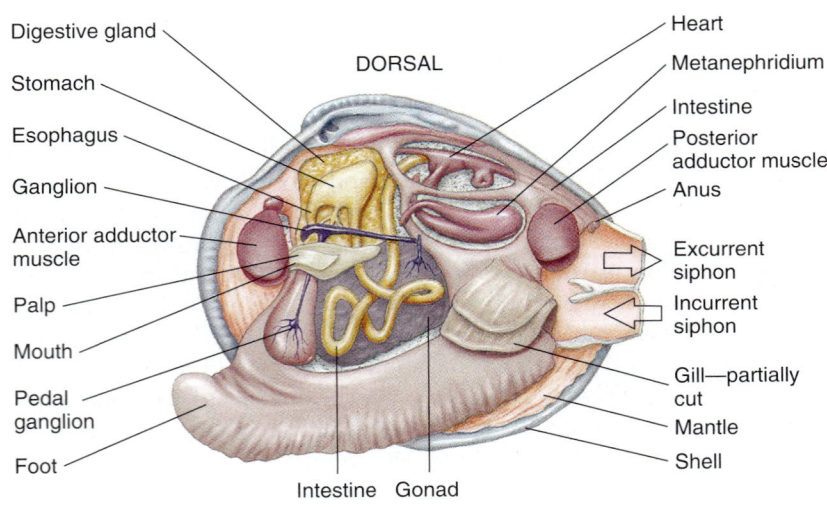

Digestive gland
Stomach
Esophagus
Ganglion
Anterior adductor muscle
Palp
Mouth
Pedal ganglion
Foot

DORSAL

Heart
Metanephridium
Intestine
Posterior adductor muscle
Anus
Excurrent siphon
Incurrent siphon
Gill—partially cut
Mantle
Shell

Intestine   Gonad
VENTRAL

Figure 29–5 **Internal anatomy of a clam.** The two shells of a bivalve hinge dorsally and open ventrally.

Clams and oysters are suspension feeders that trap food particles in sea water. They take sea water in through an extension of the mantle called the *incurrent siphon*. Water leaves by way of an excurrent siphon. As the water passes over the gills, food particles in the water are trapped in mucus secreted by the gills. Cilia move the food to the mouth. An oyster can filter more than 30 L (about 32 quarts) of sea water per hour. As suspension feeders, bivalves have no radula, and indeed they are the only group of mollusks that lack this structure.

Most bivalves have two distinct sexes. In some marine and nearly all freshwater bivalves, sperm are shed into the water and fertilize the eggs within the mantle cavity of the female. In these species, the female also carries her young within the mantle cavity, and development of the young takes place among the female's gill filaments. A marine bivalve typically develops as a trochophore larva, which later becomes a veliger larva. Some freshwater species have a modified larval stage, called a *glochidium*. In some species, glochidia spend several weeks as parasites on the gills or fins of fishes.

Cephalopods are active, predatory animals

In contrast to most other mollusks, members of the class **Cephalopoda** (meaning "head-feet") are fast-swimming predatory animals. The cephalopod mouth is surrounded by tentacles, or arms: 10 in squids, 8 in octopods, and as many as 90 in the nautilus. The large head has well developed eyes that form images. Although they develop differently, the eyes are structurally similar to vertebrate eyes and function in much the same way. The octopus has no shell, and the shell of the squid is greatly reduced and is located inside the body.

Nautilus has a coiled shell consisting of many chambers built up over time. Each year the animal lives in the newest and largest chamber of the series. *Nautilus* secretes a mixture of gases similar to air into the other chambers. By regulating the amount of gas in the chambers, the animal can control its depth in the water.

The tentacles of squids, octopods, and cuttlefish are covered with suckers for seizing and holding prey. In addition to a radula, the mouth is equipped with two strong, horny beaks used to kill prey and tear it to bits. The mantle is thick and muscular and fitted with a funnel-like **siphon.** By filling the cavity with water and ejecting it through the siphon, the cephalopod achieves forceful jet propulsion. Squids and cuttlefish have streamlined bodies adapted for efficient swimming.

Besides its speed, the cephalopod has two other important adaptations that enable it to escape from its predators, which include certain whales and moray eels. One is its ability to confuse the enemy by rapidly changing colors. By expanding and contracting *chromatophores,* cells in the skin that contain pigment granules, the cephalopod can display an impressive variety of mottled colors. Another defense mechanism is its **ink sac,** which produces a thick, black liquid that is released in a dark cloud when the animal is alarmed. While its enemy pauses, temporarily blinded and confused, the cephalopod easily escapes. The ink has been shown to inactivate the chemi-cal receptors of some predators, rendering them incapable of detecting their prey.

The octopus feeds on crabs and other arthropods, catching and killing them with a poisonous secretion of its salivary glands. During the day, the octopus usually hides among the rocks; in the evening, it emerges to hunt for food. Its motion is incredibly fluid, giving little hint of the considerable strength in its eight arms.

Small octopods survive well in aquaria and have been studied extensively. Because they have a relatively high degree of intelligence and can make associations among stimuli, they have been used as models for studying learning and memory. Their highly adaptable behavior more closely resembles that of the vertebrates than the more stereotypic patterns of behavior seen in other invertebrates (see Chapter 50).

ANNELIDS ARE SEGMENTED WORMS

Members of phylum **Annelida,** the segmented worms, have bilateral symmetry and a tubular body that may be partitioned into more than 100 ringlike segments. This phylum, composed of about 15,000 species, includes three main classes: the polychaetes, a group of marine and freshwater worms; the earthworms; and the leeches (Table 29–2 and Fig. 29–6).

The term *Annelida* (meaning "ringed") refers to the series of rings, or segments, that make up the annelid body. Both the body wall and many of the internal organs are segmented. Some structures, such as the digestive tract and certain nerves, extend the length of the body, passing through successive segments. Other structures, such as excretory organs, are repeated

TABLE 29-2	Main Classes of Phylum Annelida
Class and Representative Animals	**Characteristics**
Polychaeta Sandworms Tubeworms	Mainly marine; each segment bears a pair of parapodia with many setae; well developed head; separate sexes; trochophore larva.
Oligochaeta Earthworms	Terrestrial and freshwater worms; few setae per segment; lack well developed head; hermaphroditic.
Hirudinea Leeches	Most are blood-sucking parasites that inhabit fresh water; appendages and setae absent; prominent muscular suckers.

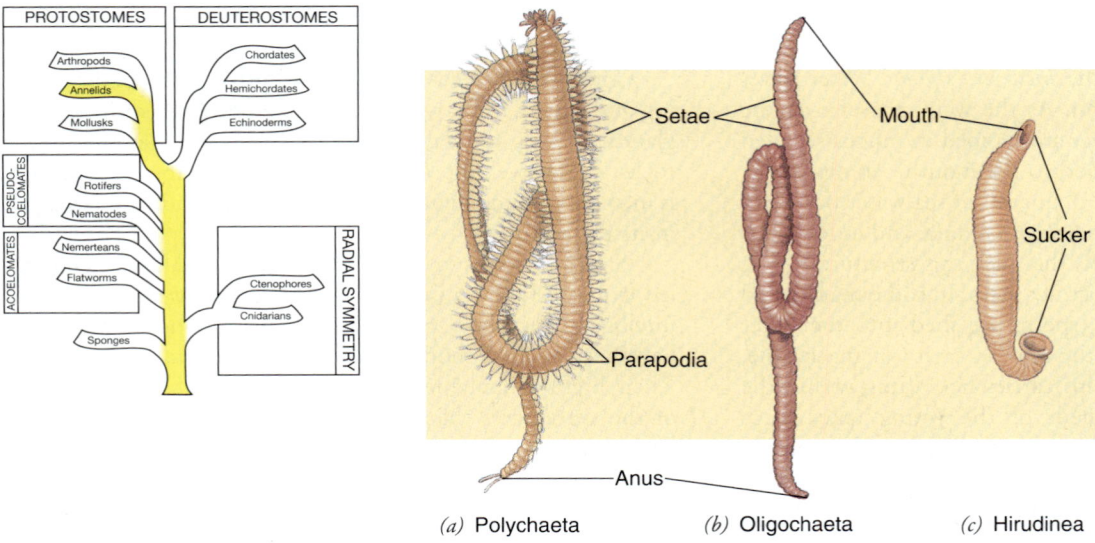

Figure 29–6 **Three major classes of annelids.** *(a)* Polychaetes are marine worms with paddle-shaped parapodia. *(b)* Oligochaetes, which include the earthworms, inhabit fresh water and moist terrestrial areas. *(c)* Many leeches, members of class Hirudinea, are parasites that feed on blood.

in each segment. In polychaetes and earthworms, segments are separated from one another internally by transverse partitions called **septa.**

One advantage of segmentation is that it facilitates locomotion. The coelom is divided into segments, and each segment has its own muscles. This arrangement allows the animal to elongate one part of its body while shortening another part. The annelid's hydrostatic skeleton is important in movement. In polychaetes and earthworms, bristle-like structures called **setae** (sing., *seta*), located on each segment, anchor the worm to the ground. To move along, the worm contracts its muscles (alternating contraction of longitudinal and circular muscles).

The annelid nervous system generally consists of a simple brain composed of a pair of ganglia, and ventral nerve cord. A pair of ganglia and lateral nerves are repeated in each segment. Annelids have a large, well developed coelom, a closed circulatory system, and a complete digestive tract extending from mouth to anus. Respiration is *cutaneous,* that is, through the skin, or by gills. Typically, a pair of metanephridia (excretory tubules) is found in each segment (described in Chapter 46).

Most polychaetes are marine worms with parapodia

Class **Polychaeta** includes marine worms that swim freely in the sea, burrow in the mud near the shore, or live in tubes secreted by the animal or made by cementing bits of shell and sand together with mucus (Fig. 29–7). Each body segment typically bears a pair of paddle-shaped appendages called **parapodia** (sing., *parapodium*) that function in locomotion and in gas exchange. These fleshy structures bear many stiff setae (the name *Polychaeta* means "many bristles"). Most polychaetes have

a well developed head bearing eyes and antennae. The head may also be equipped with tentacles and palps (feelers). Polychaetes develop from free-swimming trochophore larvae similar to those of mollusks.

Many polychaete species have evolved behavioral patterns that ensure fertilization. By responding to certain rhythmic variations, or cycles, in the environment, nearly all of the females and males of a given species release their gametes into the water at the same time. For example, more than 90% of reef-dwelling *Palolo* worms of the South Pacific shed their eggs and sperm within a 2-hour period on one night of the year. In this animal the seasonal rhythm limits the reproductive period, which is related to the lunar cycle and tides, to October or November. The posterior portion of the *Palolo* worm, loaded with eggs or sperm, breaks off from the rest of the body. It comes to the surface and bursts, releasing its gametes. Local islanders eagerly await this annual event when they can gather up great numbers of the swarming polychaetes, which they bake or eat raw.

Earthworms help maintain fertile soil

The 3000 or so species of the class **Oligochaeta** are found almost exclusively in fresh water and in moist terrestrial habitats. These worms lack parapodia, have few bristles per segment (the name *Oligochaeta* means "few bristles"), and lack a well developed head. All oligochaetes are hermaphroditic.

Lumbricus terrestris, the common earthworm, is about 20 cm (8 in) long. Its body is divided into more than 100 segments separated externally by grooves that indicate the internal position of the septa (Fig. 29–8). The earthworm's body is somewhat protected from drying by a thin, transparent cuticle, secreted by the cells of the epidermis. Mucus secreted by

(a)

(b)

Figure 29–7 **Polychaete annelids.** *(a)* The West Indies fireworm *(Hermodice carunculata)* is a crawling polychaete with poisonous setae. *(b)* The Christmas tree worm *(Spirobranchus giganteus)* photographed in a Florida coral reef, has a crown of sensory structures. *(a, Charles Seaborn/Odyssey Productions, Chicago; b, James H. Carmichael/Coastal Creations)*

Clitellum

Intestine Gizzard Crop Esophagus Pharynx

Dorsal vessel
Ventral vessel

Muscles — Longitudinal
Circular

Septa

Nerve cord Cerebral ganglia Mouth

(a)

Intestine Dorsal blood vessel

Metanephridia

Ventral blood vessel

Coelom

Nerve cord

(b)

Figure 29–8 **Earthworm structure.** *(a)* The internal structure has been exposed at the anterior end of an earthworm. *(b)* Cross section of an earthworm.

gland cells of the epidermis forms an additional protective layer over the body surface. The body wall has an outer layer of circular muscles and an inner layer of longitudinal muscles.

An earthworm literally eats its way through the soil, ingesting its own weight in soil and decaying vegetation every 24 hours. As the earthworm moves about, the soil is turned, aerated, and enriched by its nitrogenous wastes. This is how earthworms enhance the formation and maintenance of fertile soil. The earthworm's meal of soil, containing nutritious decaying vegetation, is processed in its complex digestive system.

Food is swallowed through the muscular **pharynx** and passes through the **esophagus,** which may be modified to form a thin-walled **crop** where food is stored and a thick-walled, muscular **gizzard.** In the gizzard food is ground to bits as it is moved against sand grains consumed along with the meal. The rest of the digestive system is a long, straight **intestine** where food is chemically digested and absorbed. The surface area of the intestine is increased by a dorsal, longitudinal fold, the *typhlosole.* Wastes pass out of the intestine to the exterior through the **anus.**

The efficient, closed circulatory system consists of two main blood vessels that extend longitudinally. The dorsal blood vessel, just above the digestive tract, collects blood from vessels in the segments, then contracts to pump the blood anteriorly. In the region of the esophagus, five pairs of blood vessels propel blood from the dorsal to the ventral blood vessel. Located just below the digestive tract, the ventral blood vessel carries blood posteriorly. Small vessels branch from it and deliver blood to the various structures in each segment, as well as to the body wall. Within these structures blood flows through tiny capillaries before returning to the dorsal blood vessel.

Gas exchange takes place through the moist skin. Oxygen is transported by the respiratory pigment hemoglobin in the blood plasma (the liquid part of the blood). The excretory system consists of paired metanephridia, repeated in almost every segment of the body. The ciliated funnel of each metanephridium opens into the next anterior coelomic cavity.

The nervous system consists of a pair of **cerebral ganglia** that serve as a brain, just above the pharynx, and a subpharyngeal ganglion, just below the pharynx. A ring of nerve fibers connects these ganglia. From the lower ganglion a ventral nerve cord extends beneath the digestive tract to the posterior end of the body. In each segment along the nerve cord, there is a pair of fused segmental ganglia. Nerves extend laterally from the segmental ganglia to the muscles and other structures of that segment. The segmental ganglia coordinate the contraction of the muscles of the body wall, allowing the worm to creep along.

Like other oligochaetes, earthworms are hermaphroditic. During copulation, the worms exchange sperm. Two worms, headed in opposite directions, press their ventral surfaces together. These surfaces become glued together by the thick mucous secretions of each worm's *clitellum,* a thickened ring of epidermis. Sperm are then transferred and stored in the *seminal receptacles* (small sacs) of the other worm. The worms then separate. A few days later each clitellum secretes a membranous cocoon containing a sticky fluid. As the cocoon is slipped forward, eggs are laid into it. Sperm are added as the cocoon passes over the seminal receptacles. As the cocoon slips free over the worm's head, its openings constrict so that a spindle-shaped capsule is formed. The fertilized eggs develop into tiny worms within this capsule. This complex reproductive pattern is an adaptation to terrestrial life; the cocoon protects the delicate gametes and young worms from drying out.

Many leeches are blood-sucking parasites

Most leeches, members of class **Hirudinea,** inhabit fresh water, but some live in the sea or in moist areas on land. About 75% of the known species of leeches are blood-sucking parasites. Some leeches are nonparasitic predators that capture small invertebrates such as earthworms or snails. Leeches differ from other annelids in having neither setae nor parapodia.

Leeches have muscular suckers at both anterior and posterior ends. Most parasitic leeches attach themselves to a vertebrate host, bite through the skin, and suck out a quantity of blood, which is stored in pouches in the digestive tract. *Hirudin,* an anticoagulant secreted by glands in the crop, ensures leeches a full meal of blood. Some leeches need feed only about twice each year because they digest their food slowly over several months.

Leeches have been used since ancient times for drawing blood from areas swollen by poisonous stings and bites. During the 19th century they were widely used to remove "bad blood," thought to be the cause of many diseases. Leeches have been used in modern medicine to remove excess fluid and blood that accumulates within body tissues as a result of injury, disease, or surgery (Fig. 29–9). The leech attaches its sucker near the site of injury, makes an incision, and deposits hirudin. The hirudin prevents the blood from clotting and dissolves already existing clots. In 30 minutes, a leech can suck out as much as ten times its own weight in blood.

ARTHROPODS ARE CHARACTERIZED BY JOINTED APPENDAGES AND AN EXOSKELETON OF CHITIN

Biologists consider **arthropods** the most diverse and biologically successful group of animals (Fig. 29–10). Phylum **Arthropoda** includes about one million species, and arthropods live in a greater range of habitats than the members of any other phylum. The following adaptations are major contributions to their success:

1. The arthropod body, like that of the annelid, is **segmented.** Groups of segments may be specialized (by virtue of their shape, muscles, or the appendages they bear) to perform particular functions.

(a)   500 μm   (b)

Figure 29-9 **Leeches.** *(a)* LM of a blood-sucking leech *(Helobdella stagnalis)* that feeds on mammals. The digestive tract *(dark area)* of its swollen body is filled with ingested blood. *(b)* The medicinal leech, *Hirudo medicinalis,* is used to treat hematoma, an accumulation of blood within body tissues that results from injury or disease. The leech releases hirudin, an anticoagulant that prevents the blood from clotting, and dissolves preexisting clots. *(a, Visuals Unlimited/T.E. Adams; b, St. Bartholomew's Hospital/Science Photo Library/Photo Researchers, Inc.)*

2. A hard **exoskeleton,** composed of chitin and protein, covers the entire body and appendages. The exoskeleton serves as a coat of armor protecting against predators and against excessive loss of moisture. It also gives support to the underlying soft tissues. Distinct muscle bands attach to the inner surface of the exoskeleton and operate the joints of the body and appendages. The exoskeleton has an important disadvantage, however. As the arthropod grows, it periodically outgrows this nonliving shell. The process of shedding an old exoskeleton and growing a larger one is known as **molting.** The shed exoskeleton represents a net metabolic loss, and molting also leaves the arthropod temporarily vulnerable to predators.

3. **Paired, jointed appendages,** from which this phylum gets its name (*arthropod* means "jointed foot") are modified for many different functions. They serve as swimming paddles, walking legs, mouth parts, sensory structures, or accessory reproductive organs for transferring sperm.

4. The nervous system, which resembles that of the annelids, consists of a "brain" (cerebral ganglia) and ventral nerve cord with ganglia. In some arthropods, successive ganglia may fuse together. Arthropods have a variety of very effective sense organs. Many have organs of hearing, and antennae sensitive to touch and chemicals. Most insects and many crustaceans have **compound eyes** composed of many light-sensitive units called **ommatidia** (see Chapter 41). The compound eye can form an image and is especially adapted for detecting movement.

5. Arthropods have specialized respiratory systems. Most aquatic arthropods have gills that function in gas exchange. In contrast, many terrestrial forms have a system of internal branching air tubes called **tracheae,** or **tracheal tubes.** Other terrestrial arthropods have plate-like **book lungs.**

Arthropods have an open circulatory system. A dorsal, tubular heart pumps hemolymph into a dorsal artery, which may branch into smaller arteries. From the arteries, hemolymph flows into large spaces that collectively make up the hemocoel. Hemolymph in the hemocoel bathes the tissues directly. Eventually hemolymph reenters the heart through openings, called **ostia,** in its walls. The coelom is small and filled chiefly by the organs of the reproductive system. The arthropod digestive system is a tube similar to that of the earthworm. Excretory structures vary somewhat from class to class.

Arthropod evolution and classification are controversial

Arthropods represent the pinnacle of evolutionary development among the protostomes. Many biologists think that arthropods and annelids shared a common ancestor. The relationship between arthropods and annelids is evident in their basic body plans. Like annelids, arthropods are segmented, at least as embryos. The segments develop in the same way in both phyla. However, in contrast to most annelids, each arthropod has a fixed number of segments that remains the same throughout life.

(a)

(b)

(c)

Figure 29–10 **Arthropods.** *(a)* The emperor scorpion *(Pandinis imperator).* Probably among the first terrestrial arthropods, scorpions have an elongated, segmented abdomen ending with a poisonous stinger. *(b)* This sponge crab *(Criptodromia octodenta),* photographed in Australia, wears a living sponge on its back as camouflage. *(c)* Green lacewings *(Chrysopoda* sp.) are seldom attacked by predators because they have an unpleasant odor. *(a, Carlyn Iverson; b, Fred Baverdam/Peter Arnold, Inc.; c, Stephen Dalton/Photo Researchers, Inc.)*

Segmentation is important from an evolutionary perspective because it provides the opportunity for specialization of body regions. In many annelids the individual segments are almost all alike, but in many segmented animals (arthropods and chordates), different segments and groups of segments are specialized to perform different functions. In some groups, the specialization is so pronounced that the basic segmentation of the body plan may not be apparent. (For example, in humans, although segmentation of the body is not obvious, muscles and nerves are segmentally organized.)

The basic plan of the nervous system is similar in both annelids and arthropods. A ventral nerve cord extends from a dorsal, anterior brain, and ganglia are present in each segment.

Some biologists hotly debate the majority view that the arthropods are a monophyletic group, asserting that they are polyphyletic. Some propose that the chelicerates (horseshoe crabs, spiders, ticks) and crustaceans (lobsters, crabs, and their relatives) evolved from a marine ancestor and that the uniramians (insects) evolved from a terrestrial ancestor. According to this polyphyletic view, basic arthropod features such as jointed appendages and an exoskeleton of chitin evolved independently at least twice. Some biologists who hold this view suggest that uniramians evolved from members of phylum **Onychophora,** wormlike animals that inhabit humid tropical

ANIMAL EVOLUTION AND MOLECULAR BIOLOGY

Taxonomists are using molecular methods to search for clues about arthropod evolution. Some biologists have viewed the onychophorans, known as velvet worms, as the "missing link" between the annelids and some groups of arthropods. As described in this chapter, onychophorans have some annelid and some arthropod characteristics. They are internally segmented like annelids but with an open circulatory system and tracheal tubes like arthropods. Also like arthropods, they have antennae and grow by molting.

Taxonomists have used molecular methods to study the evolutionary position of the onychophorans. Investigators compared sequences from 12S ribosomal RNA for onychophorans, annelids, mollusks, arthropods, echinoderms, and chordates. Their results suggest that onychophorans are not a link between the annelids and arthropods. Rather, onychophorans may have evolved from the arthropods and be part of the arthropod clade. The data suggest that they may be related to the chelicerates and crustaceans. Are similarities at the molecular level better indicators of evolutionary relationships than anatomical characters? A synthesis of structural and molecular data may suggest different possibilities. A decision on the evolutionary place of the onychophorans must await additional studies.

rain forests. Onychophorans, also known as velvet worms, have some annelid and some arthropod features (Fig. 29–11; also see *Making the Connection: Animal Evolution and Molecular Biology*). Like annelids, they are internally segmented, and many organs are duplicated serially. However, the jaws are derived from appendages, as in arthropods. Also in common with arthropods, onychophorans have an open circulatory system and a respiratory system consisting of tracheal tubes.

In this book we present arthropod classification based on the monophyletic view and accordingly assign the living arthropods to three subphyla: Chelicerata, Crustacea, and Uniramia (Tables 29–3 and 29–4). Subphylum **Chelicerata** includes the horseshoe crabs and arachnids (spiders, scorpions, ticks, mites). Chelicerates, the only arthropods without antennae, have fanglike feeding appendages called **chelicerae** (singular, *chelicera*).

Members of subphylum **Crustacea** (lobsters, crabs, shrimp, and barnacles) and members of subphylum **Uniramia** (insects, centipedes, and millipedes) have jawlike **mandibles** instead of chelicerae. Other than the extinct trilobites (discussed next), the crustaceans are the only group to have **biramous appendages,** that is, appendages with two jointed branches arising from their base. They are also the only group to have two pairs of antennae. Members of subphylum **Uniramia** have **uniramous** (unbranched) **appendages** and a single pair of antennae.

Trilobites were early arthropods

Among the earliest arthropods to evolve, **trilobites** inhabited shallow Paleozoic seas more than 600 million years ago. These arthropods, which have been extinct for 250 million years, lived on the sea bottom and dug into the mud. Most ranged from 3 to 10 cm, but a few reached almost a meter in length.

Covered by a hard, segmented exoskeleton, the trilobite body was a flattened oval divided into three parts: an anterior head bearing a pair of antennae and a pair of compound eyes; a thorax; and a posterior abdomen (Fig. 29–12). At right angles to these divisions, two dorsal grooves extended the length of the animal, dividing the body into a median lobe and two lateral lobes. (The name *trilobite* derives from this division of the body into three longitudinal parts.) Each segment had a pair of segmented biramous appendages, consisting of an inner walking leg and an outer branch with gills.

The trilobites may have given rise to the chelicerates, which also lived in Paleozoic waters. Only a few species of marine chelicerates, most notably the horseshoe crabs, exist today. Fossil evidence suggests that some extinct marine che-

Figure 29–11 **Onychophorans (velvet worms).** *Peripatus* sp. with young. Note the soft, sluglike body and the series of nonjointed legs. Onychophorans have features of both arthropods and annelids. *(Thomas C. Boydean/Visuals Unlimited)*

TABLE 29-3 Arthropod Subphyla

Subphylum and Selected Classes	Subphylum Characteristics
Subphylum Chelicerata Class Merostomata (horseshoe crabs) Class Arachnida (spiders, scorpions, ticks, mites)	First pair of appendages are the chelicerae used to manipulate food; body consists of cephalothorax and abdomen; no mandibles; no antennae.
Subphylum Crustacea Class Malacostraca (lobsters, crabs, shrimp)	Biramous appendages; mandibles; two pairs of antennae.
Subphylum Uniramia Class Insecta (grasshoppers, roaches) Class Chilopoda (centipedes) Class Diplopoda (millipedes)	Uniramous appendages; mandibles; single pair of antennae.

licerates invaded fresh water and may have given rise to the arachnids. The early arachnids were aquatic. The first terrestrial arachnids appeared more than 360 million years ago (in the Devonian period) and most living arachnids are terrestrial.

Chelicerates have no antennae

Subphylum Chelicerata includes the merostomes (horseshoe crabs) and the arachnids. The chelicerate body consists of a cephalothorax (fused head and thorax) and an abdomen. These animals have no antennae and no chewing mandibles (Table 29–3). Instead, the first pair of appendages, located immediately anterior to the mouth, are the chelicerae. The second pair of appendages are the **pedipalps.** The chelicerae and pedipalps are modified to perform different functions in various groups, including manipulation of food, locomotion, defense, and copulation. Posterior to the pedipalps are four pairs of legs specialized for walking.

TABLE 29-4 General Characteristics of the Principal Arthropod Groups

	Arachnida (About 60,000 species)	Crustacea (About 32,000 species)	Insecta (About 800,000 species)	Chilopoda (About 3000 species)	Diplopoda (About 7500 species)
Main Habitat	Mainly terrestrial	Marine or fresh water; a few are terrestrial	Mainly terrestrial	Terrestrial	Terrestrial
Body Divisions	Cephalothorax and abdomen	Head, thorax, and abdomen	Head, thorax, and abdomen	Head with segmented body	Head with segmented body
Gas Exchange	Book lungs or tracheae	Gills	Tracheae	Tracheae	Tracheae
Appendages	Uniramous	Biramous	Uniramous	Uniramous	Uniramous
Antennae	None	2 pairs	1 pair	1 pair	1 pair
Mouth parts	Chelicerae, pedipalps	Mandibles, 2 pairs of maxillae (for food handling)	Mandibles, maxillae	Mandibles, maxillae	Mandibles, maxillae
Legs	4 pairs on cephalothorax	1 pair per segment or less	3 pairs on thorax	1 pair per segment	Usually 2 pairs per segment
Development	Direct, except mites and ticks	Usually larval stages (nauplius larva)	Usually larval stages; most with complete metamorphosis	Direct	Direct

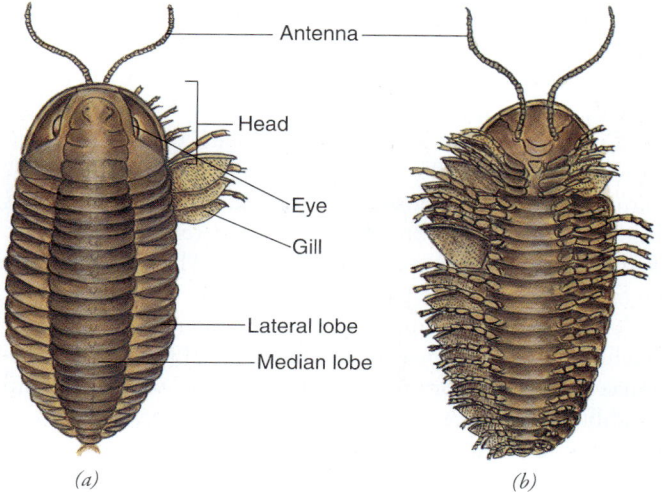

Figure 29–12 **Trilobites.** These extinct marine arthropods are considered the most primitive members of the phylum. *(a)* Dorsal view of a trilobite. *(b)* Ventral view.

Horseshoe crabs are the only living merostomes

Almost all of the merostomes are extinct. Only the horseshoe crabs have survived, essentially unchanged for 350 million years or more. *Limulus polyphemus,* the species common along the shore of North America, is horseshoe-shaped (Fig. 29–13*a*). Its long, spikelike tail is used in locomotion, not for defense or offense. Horseshoe crabs feed on mollusks, worms, and other invertebrates that they find on the sandy ocean floor.

Most arachnids are carnivores

Arachnids include spiders, scorpions, ticks, harvestmen (daddy long-legs), and mites (Fig. 29–13*b,c*). Most of the 60,000 or so species are carnivorous and prey on insects and other small arthropods. The arachnid body consists of a cephalothorax and abdomen, and there are six pairs of jointed appendages. In spiders, the first pair, the chelicerae, are fanglike structures used to penetrate prey. In some spider species, the chelicerae are used to inject poison into the prey. The second pair of ap-

(a)

(b)

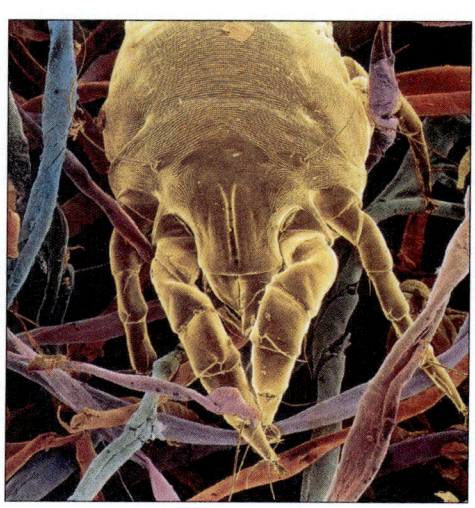

(c)

Figure 29–13 **Subphylum Chelicerata.** *(a)* The only living merostomes are a few closely related species of horseshoe crabs. Seasonally, horseshoe crabs *(Limulus polyphemus)* return to beaches for mating. In this photograph, males are competing for a female. *(b)* The venom of the black widow spider is a neurotoxin. A female black widow *(Latrodectus mactans)* is shown here on her web. *(c)* SEM of house-dust mite *(Dermatophagoides* sp.*)*, a common inhabitant of homes. This mite has been implicated in house-dust allergies. *(a, Milton H. Tierney, Jr./Visuals Unlimited; b, Steve Maslowski/Visuals Unlimited; c, Photo Researchers, Inc.)*

pendages, the pedipalps, are used by spiders to hold and chew food. In some species the pedipalps are modified as sense organs for tasting the food or for reproduction (for sperm transfer and courtship displays). In scorpions, the very large pedipalps are used as pincers for capturing prey. The other four pairs of appendages are used for walking.

Typically, spiders have eight eyes arranged in two rows of four in each row along the anterior dorsal edge of the exoskeleton. The eyes detect movement and locate objects. Jumping spiders are capable of forming a relatively sharp image.

Gas exchange in arachnids takes place either by tracheal tubes, book lungs, or both. A book lung consists of 15 to 20 plates, like pages of a book, that contain tiny blood vessels. Air enters the body through abdominal slits and circulates between the plates. As air passes over the blood vessels in the plates, oxygen diffuses into the blood and carbon dioxide diffuses out of the blood into the air. As many as four pairs of book lungs may be present, providing an extensive surface area for gas exchange.

The spider has glands in its abdomen that secrete silk, an elastic protein that is spun into fibers by organs called spinnerets. The silk is liquid as it emerges from the spinnerets but hardens after it leaves the body. Spiders use silk to build nests, to encase their eggs in a cocoon, and, in some species, to trap prey in a web. Many spiders lay down a silken dragline as they venture forth. The dragline serves as a safety line and also as a means of communication between members of a species. From a dragline another spider can determine the sex and maturity level of the spinner.

Although spiders have poison glands useful in capturing prey, only a few produce poison that is toxic to humans. The most widely distributed poisonous spider in the United States is the black widow (*Latrodectus mactans;* Fig. 29–13*b*). Its poison is a neurotoxin that interferes with transmission of messages from nerves to muscles. The brown recluse *(Loxosceles reclusa)* is smaller than the black widow and has a violin-shaped dorsal stripe on its back. Its venom destroys the tissues surrounding the bite and, like the venom of the black widow, can occasionally be fatal. Although painful, spider bites cause fewer than five fatalities per year in the United States.

Mites and ticks are among the most serious arthropod nuisances. They eat crops, infest livestock and pets, and inhabit our own bodies. Many live unnoticed, due to their small size, but others cause disease. Certain mites cause mange in dogs and other domestic animals. Chiggers (red bugs), the larval form of red mites, attach themselves to the skin and secrete an irritating digestive fluid that may cause itchy red welts. Larger than mites, ticks are parasites on dogs and many other animals. They can transmit diseases such as Rocky Mountain spotted fever, Texas cattle fever, relapsing fever, and Lyme disease.

Crustaceans are vital members of marine food webs

Subphylum Crustacea includes the lobsters, crabs, shrimp, and their relatives. Many crustaceans are primary consumers of algae and detritus. Countless billions of microscopic crustaceans contribute to marine **plankton,** the free-floating, mainly microscopic organisms found in the upper layers of the ocean. These crustaceans are food for many fishes and other marine animals, such as certain baleen whales.

Crustacean fossils have been dated back to the Cambrian period (more than 505 million years ago), but neither their origin nor their relationship to other arthropod subphyla is clear. A distinctive feature of crustaceans is the *nauplius larva,* which is often the first stage after hatching. This larva has only the most anterior three pairs of appendages.

Crustaceans are also characterized by mandibles, biramous appendages, and two pairs of antennae (Fig. 29–14). Their antennae serve as sensory organs for touch and taste. The hard mandibles, which are the third pair of appendages, are located on each side of the ventral mouth and are used for biting and grinding food. Posterior to the mandibles are two pairs of appendages, the first and second **maxillae,** used for manipulating and holding food. Several other pairs of appendages are present. Usually five pairs are modified for walking. Others may be specialized for swimming, sperm transmission, carrying eggs and young, or sensation.

As the only group of arthropods that are primarily aquatic, crustaceans usually have gills for gas exchange. Two large **antennal glands** located in the head remove metabolic wastes from the blood and body fluids. Wastes are excreted through ducts opening at the base of each antenna. Most adult crustaceans have compound eyes, and many crustaceans have **statocysts,** sense organs that detect the pull of gravity.

Crustaceans characteristically have separate sexes. During copulation, the male uses specialized appendages to transfer sperm into the female. The fertilized eggs are usually carried on some part of the body. The newly hatched animals may resemble the adults, or they may pass by successive molts through a series of larval stages before they develop the adult body. The lobster, for example, molts seven times during its first summer; after each molt it gets larger and more closely resembles the adult. After it becomes a small adult, additional molts allow for growth.

Barnacles, the only sessile crustaceans, differ markedly in their external anatomy from other members of the subphylum. They are marine suspension feeders that secrete complex limestone cups within which they live. The larvae of barnacles are free-swimming forms that go through several molts. They eventually become sessile and develop into the adult form. The barnacle was described by the 19th century naturalist Louis Agassiz as "nothing more than a little shrimplike animal standing on its head in a limestone house and kicking food into its mouth." The bane of marine boaters, barnacles can proliferate on ship bottoms in great numbers. Their presence may reduce the speed of a ship by more than 30%.

Isopods are mainly tiny (5 to 15 mm in length) marine crustaceans that inhabit the ocean floor. However, this group includes some familiar terrestrial animals: pillbugs and sowbugs. The mainly microscopic *copepods* are the largest group of crustaceans. Marine copepods are the most numerous component of plankton.

The largest order of crustaceans, *Decapoda,* contains more

(a)

(b)

(c)

Figure 29–14 **Subphylum Crustacea.**
(a) Goose barnacles *(Pollicipes polymerus)* are stalked barnacles that occur in large numbers on intertidal rocks along the West Coast of the United States. *(b)* Antarctic krill *(Euphausia superba)* are shrimplike crustaceans that are an important component of marine plankton. *(c)* Broken-back shrimp from Monterey Bay. *(a, Fred Bavendam/Peter Arnold, Inc.; b, Flip Nicklin/Minden Pictures; c, Frans Lanting/Minden Pictures)*

than 10,000 species of lobsters, crayfish, crabs, and shrimp. Most decapods are marine, but a few, such as the crayfish, certain shrimp, and a few crabs, live in fresh water. The crustaceans in general and the decapods in particular show striking specialization and differentiation of parts in the various regions of the animal. In the lobster, the appendages in the different parts of the body differ markedly in form and function (Fig. 29–15).

The five segments of the lobster's head and the eight segments of its thorax are fused into a cephalothorax, which is covered on the top and sides by a shield, the **carapace,** composed of chitin impregnated with calcium salts. The two pairs of antennae serve as chemoreceptors and tactile sense organs. The mandibles are short and heavy, with opposing surfaces used in grinding and biting food. Behind the mandibles are two pairs of accessory feeding appendages, the first and sec-

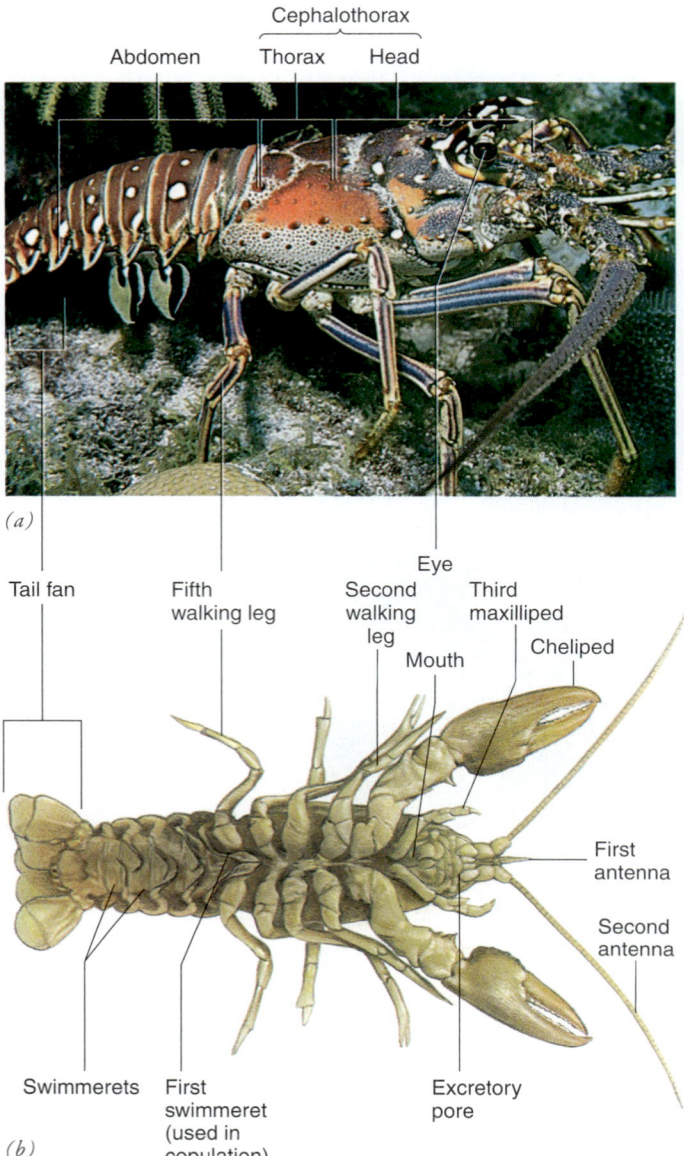

Cephalothorax

Abdomen | Thorax | Head

(a)

Tail fan | Fifth walking leg | Second walking leg | Third maxilliped | Eye | Mouth | Cheliped

First antenna

Second antenna

Swimmerets | First swimmeret (used in copulation) | Excretory pore

(b)

Figure 29–15 **Structure of the lobster.** *(a)* Like other decapods, the spiny lobster *(Panulirus argus)* has five pairs of walking legs. The first pair of walking legs is modified as chelipeds (large claws); *(b)* Ventral view of a lobster. Note the variety of specialized appendages. *(a, Larry Lipsky/Tom Stack & Associates)*

ond maxillae. The appendages of the first three segments of the thorax are the maxillipeds, which aid in chopping up food and passing it to the mouth. The fourth segment of the thorax has a pair of large **chelipeds,** or pinching claws. The last four thoracic segments bear **walking legs.**

The appendages of the first abdominal segment are part of the reproductive system and function in the male as sperm-transferring structures. The following four abdominal segments bear paired **swimmerets,** small paddle-like structures used by some decapods for swimming and by the females of all species for holding eggs. Each branch of the sixth abdominal ap-

pendages (called *uropods*) consists of a large flattened structure. Together with the flattened posterior end of the abdomen (the *telson*), they form a fan-shaped structure used for swimming backwards.

Subphylum Uniramia includes the insects, centipedes, and millipedes

The insects, centipedes, and millipedes are grouped together in subphylum Uniramia because they all have uniramous (unbranched) appendages, similar mouthparts, and a single pair of antennae (rather than two pairs as in crustaceans).

Class Insecta is the most successful group of animals

With more than 800,000 described species, the class **Insecta** is the most successful group of animals on our planet in terms of diversity, geographic distribution, number of species, and number of individuals[1] (Table 29–5). More species of insects have been identified than of all other classes of animals combined. What they lack in size, insects make up in sheer numbers. If we could weigh all the insects in the world, their weight would exceed that of all of the remaining terrestrial animals. Insects have an extraordinary ability to adapt to changes in the environment. Although primarily terrestrial animals, some species live in fresh water, a few are truly marine, and others inhabit the shore between the tides.

The earliest fossil insects—primitive, wingless species—date back to the Devonian period more than 360 million years ago. Insect fossils from the Carboniferous period more than 286 million years ago include both wingless and primitive winged species. Cockroaches, mayflies, and cicadas are among the insects that have survived relatively unchanged from the Carboniferous period to the present day.

We can describe an insect as an **articulated** (jointed), **tracheated** (having tracheal tubes for gas exchange), **hexapod** (having six feet). The insect body consists of three distinct parts: head, thorax, and abdomen (Fig. 29–16 on p. 631). Three pairs of legs emerge from the adult thorax, and insects in many orders have one or two pairs of wings. One pair of antennae protrudes from the head, and the sense organs include both simple and compound eyes. Their complex mouthparts may be adapted for piercing, chewing, sucking, or lapping.

The tracheal system of insects has made an important contribution to their diversity. Air enters the tracheal tubes (also called tracheae) through tiny openings, called **spiracles,** in the body wall. Oxygen passes directly to the internal organs. This effective oxygen delivery system permits the insects to have the high metabolic rate necessary for activities such as flight, even though they have an open circulatory system.

Excretion is accomplished by two or more **Malpighian tubules,** which receive metabolic wastes from the blood, concentrate the wastes, and discharge them into the intestine.

[1]The nematodes rival the insects in number of individuals.

(Text continues on page 631)

TABLE 29-5   Some Orders of Insects

Order, Number of Species, and Examples	Representative Member	Name	Some Characteristics
Insects with No Metamorphosis (Egg → immature form → adult)			
Thysanura (320) Silverfish Bristletails		Silverfish *Lepisma saccharina*	No wings; biting-chewing mouthparts; 2–3 "tails" extend from posterior tip of abdomen; inhabit dead leaves; eat starch in books
Insects with Incomplete Metamorphosis (Egg → nymph → adult)			
Odonata (5,000) Dragonflies Damselflies		Damselfly *Ischnura* sp.	Two pairs of long, membranous wings; chewing mouthparts; large, compound eyes; active predators
Orthoptera (20,000) Grasshoppers Crickets		Fork-tailed bush katydid *Scudderia furcata*	Forewings leathery, hindwings membranous; chewing mouthparts; most herbivorous, some cause crop damage; some predatory
Blattodea (3,700) Cockroaches		American cockroach *Periplaneta americana*	When wings present, forewings leathery, hindwings membranous; chewing mouthparts; legs adapted for running
Isoptera (2,000) Termites		Eastern subterranean termite *Reticulitermes flavipes*	2 pairs of wings, or none; wings shed by sexual forms after mating; chewing mouthparts; social insects, form large colonies; eat wood
Anoplura (500) Sucking lice		Human body louse *Pediculus humanus*	No wings; piercing-sucking mouthparts; ectoparasites of mammals; head louse and crab louse are human parasites; vectors of typhus fever

TABLE 29–5 Continued

Order, Number of Species, and Examples	Representative Member	Name	Some Characteristics
Insects with Incomplete Metamorphosis *(continued)* **Hemiptera** (35,000) (true bugs) Chinch bugs Bed bugs Water striders		Chinch bug *Blissus leucopterus*	Hindwings membranous; forewings smaller; piercing-sucking mouthparts form beak; most herbivorous; some parasitic
Homoptera (33,000) Aphids Leafhoppers Cicadas Scale insects	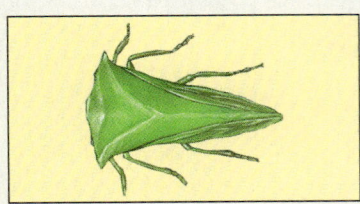	Buffalo treehopper *Stictocephala bubalus*	2 pairs membranous wings; piercing-sucking mouthparts form beak; some infect plants; others are vectors of plant diseases
Insects with Complete Metamorphosis (Egg → larva → pupa → adult) **Lepidoptera** (120,000) Moths Butterflies		Luna moth *Aetias luna*	Usually 2 pairs of membranous, colorful, scaled wings; sucking mouthparts; larvae are wormlike caterpillars that eat plants; adults suck flower nectar; important pollinators
Diptera (150,000) Houseflies Mosquitos		Deerfly *Chrysops vittatus*	Only forewings functional in flying; hindwings small, knoblike halteres; mouthparts usually adapted for sucking (and piercing in some); larvae are maggots or wrigglers and may damage domestic animals or food; adults may transmit disease such as sleeping sickness or yellow fever
Siphonaptera (1,750) Fleas		Dog flea *Ctenocephalides canis*	No wings; piercing-sucking mouthparts; legs adapted for clinging and jumping; parasites on birds and mammals; vectors of bubonic plague and typhus

Order, Number of Species, and Examples	Representative Member	Name	Some Characteristics
Insects with Complete Metamorphosis (continued)			
Coleoptera (300,000) Beetles Weevils		Colorado potato beetle *Leptinotarsa decemlineata*	Forewings modified as protective coverings for membranous hindwings (which are sometimes absent); chewing mouthparts; largest order of insects; most herbivorous; some aquatic
Hymenoptera (130,000) Ants Bees Wasps		Bald-faced hornet *Vespula maculata*	Usually 2 pairs of membranous wings; mouthparts may be modified for sucking or lapping nectar; many are social insects; some sting

Unique to terrestrial arthropods, Malpighian tubules perform the extremely important function of conserving water (discussed further in Chapter 46).

The sexes are separate, and fertilization takes place internally. Several molts occur during development. Primitive insects with no wings, such as silverfish, have direct, or simple, development: the young hatch as juveniles that resemble the adult form. Other insects, such as grasshoppers and cockroaches, undergo **incomplete metamorphosis** in which the egg gives rise to a nymph that resembles the adult in many ways but lacks functional wings and reproductive structures.

Most insects, including bees, butterflies, and fleas, undergo **complete metamorphosis** with four distinct stages in the life cycle: **egg, larva, pupa,** and **adult.** The wormlike larva does not look at all like the adult. For example, caterpillars have an entirely different body form from butterflies. Typically, an insect spends most of its life as a larva. Eventually the larva stops feeding, molts, and enters a pupal stage, usually within a protective cocoon or underground burrow. The pupa does not feed and cannot defend itself. Energy reserves stored during the larval stage are spent remodeling its body. When it emerges, it is equipped with functional wings and reproductive organs.

Certain species of bees, ants, and termites exist as colonies or societies made up of several different types of individuals, each adapted for some particular function (Fig. 29–17). The members of some insect societies communicate with each other by "dances" and by means of **pheromones,** substances secreted to the external environment. Social insects and their communication are discussed in Chapter 50.

Many adaptations contribute to the biological success of the insects

What are the secrets of insect success? One important feature is the tough exoskeleton, which protects insects from predators and helps to prevent water loss by evaporation. At the

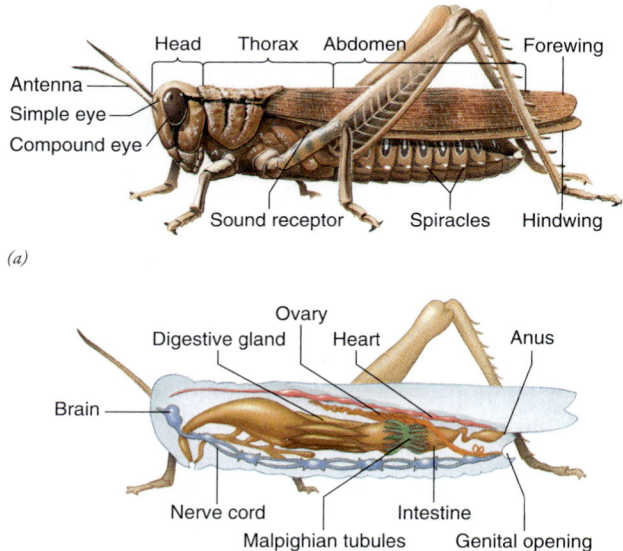

(a)

(b)

Figure 29–16   Structure of the grasshopper. *(a)* External anatomy. Note the three pairs of segmented legs. *(b)* Internal anatomy.

Figure 29–17 Queen and worker termites (*Nasutitermes* sp.). The queen, with an enlarged abdomen, occupies the center of the royal chamber. Most of the individuals are workers. A few soldiers with red heads and reduced mandibles can be seen. *(James L. Castner)*

same time, the segmented exoskeleton permits mobility and flexibility. The insect body plan has been modified and specialized in so many ways that insects are adapted to a remarkable number of lifestyles. Segmentation permits regional specialization of the body. For example, cephalization and highly developed sense organs are important in both offense and defense. The jointed appendages are specialized for various types of feeding, sensory functions, and different types of locomotion (walking, jumping, swimming). For example, grasshopper mouthparts are adapted for biting and chewing leaves. Moth and butterfly mouthparts are adapted for sucking nectar from flowers, and mosquito mouthparts are adapted for sucking blood. Aphids and leafhoppers have mouthparts specialized to pierce plants and feed on plant juices.

Another reason for insect success is the ability to fly. Unlike other invertebrates, which creep slowly along on (or under) the ground, many insects fly rapidly through the air. Their wings and small size facilitate their wide distribution and, in many instances, their immediate survival. For example, when a pond dries up, adult aquatic insects can fly to another habitat.

The reproductive capacity of insects is amazing. Under ideal conditions, the fruit fly *Drosophila* can produce 25 generations in one year! Insect eggs are protected by a thick membrane, and several eggs may, in addition, be enclosed in a protective egg case. Typically, 50 to 100 eggs are laid at a time. Metamorphosis divides the insect life cycle into different stages. This strategy reduces intraspecific competition by placing larval forms in different lifestyles so that they do not compete with adults for food or habitats.

Insects have many interesting adaptations for offense and defense. We are all familiar with the stingers of bees and wasps, which are specialized egg-laying structures (ovipositors). Some insects "play dead," by remaining motionless. Many have cryptic coloration, blending beautifully with the background of their habitat; some look like dead twigs or leaves (see Fig. 52–4*a*). Some mimic poisonous insects, deriving protection from this resemblance (see Fig. 52–5).

Insects communicate by tactile, auditory, visual, or chemical signals. For example, certain ant species use pheromones to mark trails and to warn of danger. Other insects use pheromones to attract a mate.

Insects have a great impact on humans

Not all insects compete with us for food or cause us to scratch, swell up, or recoil from their presence. Bees, wasps, beetles, and many other insects pollinate flowers of crops and fruit trees. Some insects destroy other insects that are harmful. For example, dragonflies eat mosquitos. And some organic farmers and home gardeners purchase ladybird beetles, adept at ridding plants of aphids and other insect pests. Insects are important members of many food webs. Many birds, mammals, amphibians, reptiles, and some fishes depend on insects for food. Many beetles and the larvae (maggots) of flies are detritus feeders; they break down dead plants and animals and their wastes, permitting nutrients to be recycled.

Many insect products are useful to us. Bees produce honey as well as beeswax, which is used in making candles, lubricants, and other products. Shellac is made from lac, a substance given off by certain scale insects that feed on the sap of trees. And the labor of silkworms provides us with beautiful fabric.

On the negative side, billions of dollars worth of crops are destroyed each year by insect pests. Whole buildings may be destroyed by termites, and clothing can be damaged by moths. Fire ants not only inflict painful stings but cause farmers serious economic loss because of their large mounds, which damage mowers and other farm equipment. Mounds also reduce grazing land because livestock quickly learn to avoid them.

Blood-sucking flies, screw worms, lice, fleas, and other insects annoy and transmit disease in both humans and domestic animals. Mosquitos are vectors of malaria and yellow fever. Body lice may carry the typhus rickettsia, and houseflies sometimes transmit typhoid fever and dysentery. Tsetse flies transmit African sleeping sickness, and fleas may be vectors of bubonic plague.

Classes Chilopoda and Diplopoda include centipedes and millipedes

Members of class **Chilopoda** are called *centipedes* ("hundred-legged"), and members of class **Diplopoda** are known as *millipedes* ("thousand-legged"). These animals are all terrestrial and are typically found beneath stones or wood in the soil in both temperate and tropical regions.

Centipedes and millipedes are similar in having a head and an elongated trunk with many segments, each bearing uniramous legs (Fig. 29–18). The centipedes have one pair of legs on each segment behind the head. Most centipedes do not have enough legs to merit their name, the most common number is 30 or so, although a few species have 100 or more. The legs of centipedes are long, enabling them to run rapidly. Centipedes are carnivorous and feed on other animals, mostly insects. Larger centipedes will eat snakes, mice, and frogs. The

(a)

(b)

Figure 29–18 **Chilopods and diplopods.** *(a)* Centipede (*Lithobius* sp.), a member of class Chilopoda. Centipedes have one pair of uniramous appendages per segment. *(b)* Millipede *(Diplopoda pachydesmus),* a member of the class Diplopoda. Millipedes have two pairs of uniramous appendages per segment. *(a, Dwight Kuhn; b, John R. MacGregor/Peter Arnold, Inc.)*

TABLE 29–6   Comparison of Characteristics of Some Coelomate Protostome Phyla*

	Mollusca (50,000 species)	Annelida (segmented worms) (15,000 species)	Arthropoda (joint-footed animals) (over 1 million species)
Representative Animals	Clams Snails Squids	Earthworms Leeches Marine worms	Crustaceans Insects Spiders
Circulation	Open system (closed in cephalopods)	Closed system	Open system
Gas Exchange	Gills and mantle	Diffusion through moist skin; oxygen circulated by blood	Tracheae in insects; gills in crustaceans; book lungs or tracheae in spider group
Waste Disposal	Metanephridia	Pair of metanephridia in each segment	Malphighian tubules in insects; antennal (green) glands in crustaceans
Nervous System	Simple brain (cerebral ganglia); simple sense organs	Simple brain; ventral nerve cord; simple sense organs	Brain; ventral nerve cord; well developed sense organs
Reproduction	Sexual	Sexual	Sexual
Support and Movement	Most have hydrostatic skelton; most have shell and ventral foot for locomotion	Fluid-filled coelom serves as hydrostatic skeleton; well developed muscles in body wall	Tough exoskeleton; jointed appendages (some have wings); well developed muscles
Environment and Lifestyle	Mainly marine, some inhabit fresh water or are terrestrial; herbivores, carnivores, scavengers, or suspension feeders	Marine, freshwater, terrestrial; herbivores, carnivores, scavengers, suspension feeders	Most diverse group in habitat and lifestyle; marine, freshwater, and terrestrial; herbivores, carnivores, scavengers

*Members of these phyla are at the organ system level of organization, have bilateral symmetry, and have a complete digestive tract.

prey is captured and killed with poison claws located just behind the head on the first trunk segment.

Millipedes have two pairs of legs on most body segments. Diplopods are not as agile as chilopods, and most species can crawl only slowly over the ground, although they can powerfully force their way through earth and rotting wood. Millipedes are generally herbivorous and feed on both living and dead vegetation.

The three main phyla discussed in this chapter are compared in Table 29–6.

SUMMARY WITH KEY TERMS

I. The coelomate protostomes include the mollusks, annelids, and arthropods, as well as several minor phyla.

II. The **coelom** protects internal organs and permits a separation between body wall and digestive tract, allowing the digestive tract to move food along independently of body movements. It also provides space for the gonads to develop.

III. Terrestrial animals require a body covering that prevents fluid loss; some sort of skeleton that withstands the pull of gravity; and reproductive adaptations, such as internal fertilization, and either development within the mother's body or shells to prevent the developing embryo from drying out.

IV. Members of phylum **Mollusca** are soft-bodied animals usually covered by a shell.
   A. Mollusks have a ventral **foot** for locomotion and a **mantle** that covers the **visceral mass**, a concentration of body organs.
   B. Mollusks have an **open circulatory system** with the exception of cephalopods, which have a **closed circulatory system.** Mollusks have excretory tubules called **metanephridia.** A rasplike **radula** functions as a scraper in feeding in all groups except the bivalves, which are suspension feeders.
   C. Typically, marine mollusks have a **trochophore larva** stage; in many marine gastropods and bivalves, the trochophore larva develops into a **veliger larva.**
   D. Class **Polyplacophora** includes the sluggish marine chitons, which have shells that consist of eight overlapping plates.
   E. Class **Gastropoda,** the largest and most successful group of mollusks, includes the snails, slugs, limpets, and their relatives. In gastropods, the body undergoes **torsion,** a twisting of the visceral mass. The shell (when present) is coiled.
   F. Class **Bivalvia** includes the aquatic clams, scallops, and oysters, suspension feeders enclosed by a two-part shell that is hinged dorsally.
   G. Class **Cephalopoda** includes the squids, octopods, and nautilus. Cephalopods are active, predatory swimmers. Tentacles surround the mouth, located in the large head.

V. Phylum **Annelida,** the segmented worms, includes many aquatic worms, earthworms, and leeches.
   A. Annelids have conspicuously long bodies with **segmentation** both internally and externally; their large compartmentalized coelom serves as a **hydrostatic skeleton.**
   B. Class **Polychaeta** consists of marine worms characterized by bristled **parapodia,** used for locomotion. The parapodia bear many **setae.**
   C. Class **Oligochaeta,** the earthworms, is characterized by a few short setae per segment. The body is divided into more than 100 segments separated internally by **septa.**
   D. Class **Hirudinea,** the leeches, is characterized by the absence of setae and appendages. Parasitic leeches are equipped with suckers for holding on to their host.

VI. Phylum **Arthropoda** is composed of segmented animals with paired, **jointed appendages** and an armor-like **exoskeleton** of chitin. **Molting** is necessary for the arthropod to grow.
   A. Arthropods have an open circulatory system with a dorsal heart that pumps **hemolymph.**
   B. Aquatic forms have gills; terrestrial forms have either **tracheae** or **book lungs.**
   C. Many biologists consider the arthropods a monophyletic group, but some argue that they are polyphyletic. Some biologists have suggested that phylum **Onychophora** may be a link between annelids and arthropods, but RNA sequencing data suggest that the onychophorans evolved from the arthropods.
   D. The **trilobites** are extinct marine arthropods covered by a hard, segmented shell.
   E. Subphylum **Chelicerata** includes the **merostomes** (horseshoe crabs) and the **arachnids** (spiders, mites, and their relatives).
      1. The chelicerate body consists of a cephalothorax and abdomen; there are six pairs of jointed appendages, of which four pairs serve as legs.
      2. The first pair of appendages are **chelicerae,** the second pair are **pedipalps.** These appendages may be adapted for manipulation of food, locomotion, defense, or copulation. Chelicerates have no antennae and no mandibles.
   F. Subphylum **Crustacea** includes lobsters, crabs, pillbugs, and barnacles.
      1. The crustacean body consists of a cephalothorax and abdomen; typically, five pairs of walking legs are present.
      2. Crustaceans have two pairs of **antennae** that sense taste and touch. The third appendages are **mandibles** used for chewing. Two pairs of **maxillae,** posterior to the mandibles, are used for manipulating and holding food.
   G. Subphylum **Uniramia** includes class Insecta, class Chilopoda, and class Diplopoda; members of this subphylum have unbranched appendages and a single pair of antennae.
      1. An insect is an **articulated, tracheated hexapod;** its body consists of head, thorax, and abdomen.
      2. Insects have a system of tracheae for gas exchange and **Malpighian tubules** for excretion.
      3. The biological success of the insects results from many adaptations, including the versatile exoskeleton, segmentation, specialized jointed appendages, ability to fly, highly developed sense organs, **metamorphosis** (which reduces intraspecific competition), effective reproductive strategies, effective mechanisms for defense and offense, and the ability to communicate.
      4. The centipedes have one pair of legs per body segment, whereas the millipedes have two pairs of legs per body segment. Centipedes are carnivorous, whereas millipedes are typically herbivorous.

POST-TEST

1. Which of the following belong to phylum Annelida? (a) gastropods (b) polychaetes (c) uniramians (d) nudibranchs (e) cephalopods
2. Which of the following belong to phylum Mollusca? (a) gastropods

(b) polychaetes (c) chelicerates (d) crustaceans (e) gastropods and crustaceans
3. Which of the following belong to subphylum Uniramia? (a) insects

(b) polychaetes (c) crustaceans (d) nudibranchs (e) chelicerates

4. In mollusks the first larval stage is typically a (a) pupa (b) veliger larva (c) trochophore larva (d) chiton (e) nymph

5. The correct sequence in insect complete metamorphosis is (a) egg → immature form → adult (b) egg → trochophore larva → veliger larva → adult (c) egg → pupa → larva → adult (d) egg → larva → pupa → adult (e) adult → larva → egg → pupa

6. Which of the following is characteristic of members of phylum Annelida? (a) open circulatory system (b) hemocoel (c) segmentation (d) mantle (e) mandibles

7. Which of the following characteristics are associated with phylum Mollusca? (a) mandibles (b) visceral mass (c) pedipalps (d) chelipeds (e) setae

8. Which of the following is characteristic of insects? (a) gills (b) six legs (c) metanephridia (d) eight legs (e) two pairs of antennae

9. Which of the following is NOT characteristic of arthropods? (a) ex-

oskeleton (b) veliger larva (c) paired, jointed appendages (d) chitin (e) segmentation

10. Biramous appendages are characteristic of (a) crustaceans (b) insects (c) centipedes (d) millipedes (e) spiders

11. Which of the following is NOT associated with insect success? (a) radula (b) metamorphosis (c) ability to fly (d) small size (e) cryptic coloration

12. Torsion in mollusks (a) is characteristic of bivalves (b) is a twisting of the visceral mass (c) involves coiling of the molluscan shell (d) begins in the adult stage (e) depends on action of parapodia

13. Spiders are likely to have (a) mandibles and maxillae (b) four pairs of legs on the abdomen (c) one pair of antennae (d) biramous appendages (e) chelicerae and pedipalps

14. Trilobites (a) were early mollusks (b) are members of phylum Onychophora (c) are characterized by parapodia and setae (d) were early arthropods (e) are an evolutionary link between annelids and arthropods

REVIEW QUESTIONS

1. Identify characteristics that mollusks and annelids share. In what ways are these animals different? What do their characteristics suggest about their evolutionary relationship?

2. Give two distinguishing characteristics for each of the following: (a) mollusks; (b) annelids; (c) arthropods

3. What are the advantages of each of the following? (a) presence of a coelom (b) the arthropod exoskeleton (c) segmentation with specialization

4. Contrast the lifestyles of a gastropod and a cephalopod. Identify adaptations that have evolved in each for its particular lifestyle.

5. Describe (a) a trochophore larva (b) a veliger larva (c) an insect nymph

6. Describe five adaptations that have contributed to insect success.

7. Distinguish between insects and spiders.

8. What are the distinguishing features of each of the arthropod subphyla? Identify animals that belong to each group.

9. Contrast the four molluscan classes discussed.

10. Compare complete and incomplete metamorphosis in insects.

11. Label the diagram. Use Figure 29–16b as a reference.

YOU MAKE THE CONNECTION

1. Discuss the idea that every evolutionary adaptation has both advantages and disadvantages, using each of the following characteristics as an example. (a) presence of a coelom (b) the arthropod exoskeleton (c) segmentation (d) complete metamorphosis (e) a complete digestive tract

2. Hypothesize a benefit that might explain why oysters secrete calcium carbonate layers around foreign particles.

3. Insects that undergo complete metamorphosis outnumber those that do not by more than ten to one. Hypothesize an explanation.

RECOMMENDED READINGS

Berenbaum, M.R. *Bugs in the System: Insects and Their Impact on Human Affairs.* Addison-Wesley, Reading, MA, 1995. A readable book on insects that describes their life histories, physiology, behavior, and relationships with humans.

Chadwick, D.H. "Planet of the Beetles." *National Geographic,* Vol. 193, No. 3, Mar. 1998. This article highlights the remarkable diversity of beetles, which make up about 25% of all known animal species.

Holldobler, B., and Wilson, E.O. *Journey to the Ants: A Story of Scientific Exploration.* The Belknap Press of Harvard University Press, Cambridge, 1994. Everything you ever wanted to know about ants, with emphasis on their biological success.

Knowlten, N. "A Tale of Two Seas." *Natural History,* Jun. 1994. A discussion of reproductive isolation and shrimp diversity.

Morell, V. "Life on a Grain of Sand." *Discover,* Apr. 1995. An interesting look at some unusual marine animals.

Natural History, Vol. 104, No. 3, Mar. 1995. This issue features an interesting collection of articles devoted to spiders.

Zimmer, C. "See How They Run." *Discover,* Vol. 15, No. 9, Sept. 1994. A look at how roaches, ants, centipedes, and other arthropods move.

● Visit our website at **http://www.saunderscollege.com/lifesci/titles.html** and click on Solomon/Berg/Martin Biology for links to chapter-related resources on the World Wide Web.

CHAPTER 30

The Animal Kingdom: The Deuterostomes

What does a sea star have in common with a fish, frog, hawk, or human? It may seem strange to group the **echinoderms**—the sea stars, sea urchins, and sand dollars—with the **chordates,** the phylum to which the vertebrates (animals with a backbone) belong. Although these animals look and behave very differently, evidence suggests that echinoderms and chordates arose on the same evolutionary line. They are both **deuterostomes,** the second main branch of the animal kingdom.

Recall from Chapter 28 that deuterostomes share similarities in their patterns of development. They are characterized by radial, rather than spiral, cleavage. Their cleavage is generally indeterminate, which means that the fate of their cells is fixed later in development than is the case in protostomes. In deuterostomes, the mesoderm typically develops from paired pouches that pocket out from the primitive gut. The coelom generally forms from cavities within the mesodermal out-pocketings.

Biologists include two less familiar phyla as deuterostomes: **hemichordates,** a small group of wormlike marine animals, and **chaetognaths,** or arrowworms, a group of about 50 species found in marine plankton. The larvae of echinoderms and hemichordates are very similar. They are bilateral and have a characteristic ring of cilia surrounding the mouth. This similarity suggests that these phyla evolved from a common ancestor. The hemichordates, like chordates, have pharyngeal gill slits and a dorsal nerve cord.

Although their evolutionary place is controversial, some biologists also include the **lophophorate phyla**—Phoronida, Ectoprocta, and Brachiopoda—as deuterostomes. These relatively obscure phyla are described in Table 30–1. With a few exceptions, lophophorates are marine animals that are adapted for life on the ocean bottom. The **lophophore,** a ciliated ring of tentacles that surrounds the mouth, is specialized for capturing suspended particles in the water.

In this chapter we focus on echinoderms, such as the pencil urchin (*Heterocentrus mammillatus*) photographed in Hawaii, and chordates, represented in the photograph by the fishes, the clown wrasse (*Coris gailmard*) and Moorish idol (*Zanclus cornutus*). The largest chordate subphylum is Verte-

(Ed Robinson/Tom Stack & Associates)

brata, animals with backbones (vertebral columns). The **vertebrates** are the animals with which we are most familiar—fishes, amphibians, reptiles, birds, and mammals. Fossil evidence of the first chordates, and perhaps also the first vertebrates, has been found in rocks from the Cambrian period more than 505 million years old.

Recent evidence suggests that all the invertebrates have a comparable number of genes and that all of the vertebrates have roughly double that number. Peter W. H. Holland, an English molecular zoologist at the University of Reading, has suggested that more than 500 million years ago a random mutation in an ancestral chordate resulted in a doubling of the chromosomes. Natural selection acting on mutations in the duplicated genes, some of which play critical roles in embryonic development, provided the information needed to develop a new body design, including a vertebrate-type head.

1. Discuss the relationship of the echinoderms and chordates, giving specific reasons for grouping them together as deuterostomes.
2. Describe the minor phyla listed in Table 30–1: lophophorate phyla, chaeognaths, and hemichordates.
3. Describe the distinguishing characteristics of the echinoderms.
4. Describe and give examples of each of the five main classes of echinoderms.
5. Distinguish among the subphyla of phylum Chordata and describe the characteristics that they have in common; describe the characteristics of subphylum Vertebrata.
6. Distinguish among the classes of vertebrates and assign a given vertebrate to the correct class.
7. Trace the evolution of vertebrates according to current theory.
8. Identify adaptations that reptiles and other terrestrial vertebrates have made to life on land.
9. Contrast monotremes, marsupials, and placental mammals and give examples of members that belong to each group.
10. Identify the major orders of placental mammals and give examples of animals that belong to each order.

ECHINODERMS ARE CHARACTERIZED BY THEIR WATER VASCULAR SYSTEM

All of the members of phylum **Echinodermata** inhabit marine environments. They are found in the oceans at all depths. About 7000 living and more than 13,000 extinct species have been identified. The echinoderms probably evolved during the early Cambrian period and achieved maximum diversity by the middle of the Paleozoic era, about 400 million years ago. By the beginning of the Mesozoic era 248 million years ago, they had declined, leaving six principal groups that have survived to the present day (Fig. 30–1): class Crinoidea, sea lilies and feather stars; class Asteroidea, sea stars; class Ophiuroidea, basket stars and brittle stars; class Echinoidea, sea urchins and sand dollars; class Holothuroidea, sea cucumbers; class Concentricycloidea, sea daisies.

The echinoderms are in many ways unique in the animal kingdom. Echinoderm larvae are bilaterally symmetrical, ciliated, and free-swimming. During development the body form reorganizes, and the adult exhibits **pentaradial symmetry,** in which the body is arranged in five parts around a central axis. Echinoderms have an **endoskeleton** (internal skeleton) consisting of $CaCO_3$ plates. The endoskeleton often bears spines that project outward; the name *Echinodermata,* meaning "spiny-skinned," reflects this trait. The endoskeleton is covered by a thin, ciliated epidermis.

Echinoderms are characterized by a unique, hydraulic **water vascular system,** a network of fluid-filled canals that functions in locomotion, feeding, and gas exchange. Branches of the water vascular system lead to numerous tiny **tube feet** that extend when filled with fluid. Each tube foot receives fluid from the main system of canals. A rounded muscular sac, or **ampulla,** at the base of the foot, stores fluid and uses it to operate the tube foot. The tube foot is separated from other parts of the system by a valve. When the valve shuts, the ampulla contracts, forcing fluid into the tube foot so that it extends. At the bottom of the foot is a suction-type structure that presses against and adheres to whatever surface the tube foot is on.

Echinoderms have a well developed coelom, and the coelomic fluid transports materials. Although its structure varies in different groups, the complete digestive system is the most prominent body system. A variety of respiratory structures are found in the various classes. No excretory organs are present. The nervous system is simple, generally consisting of nerve rings with radiating nerves about the mouth. Echinoderms have no brain. The sexes are usually separate, and eggs and sperm are generally released into the water, where fertilization takes place.

Members of class Crinoidea are suspension feeders

Class **Crinoidea,** the oldest class of living echinoderms, includes the feather stars and the sea lilies (Fig. 30–1*a*). Although a great many extinct crinoids are known, there are relatively few living species. The feather stars are motile crinoids, although they often remain in the same location for long periods of time. Sea lilies are sessile and remain attached to the ocean floor by a stalk.

Crinoids remove suspended food from the water. In all other echinoderms, the mouth is located on the underside of the disk toward the substratum, but in crinoids the **oral** (mouth) **surface** is on the upper side of the disk. A number of branched, feathery arms also extend upward. Along the feathery arms are numerous tube feet that are shaped like small tentacles and coated with mucus that traps microscopic organisms.

Many members of class Asteroidea capture prey

Sea stars, or starfish, are members of class **Asteroidea** (Fig. 30–1*b*). Their bodies consist of a central disk from which radiate from 5 to more than 20 arms, or rays (Fig. 30–2). The undersurface of each arm is equipped with hundreds of pairs

TABLE 30-1 Some Less Common Deuterostome Phyla

Phylum		Characteristics
Lophophorate phyla		A lophophore, a ciliated ring of tentacles that surrounds the mouth. Bottom dwellers. Suspension feeders. U-shaped gut.
Phoronids	**Lophophores project from the tubes of these phoronids (Phoronis vancouverensis).** These animals inhabit the Pacific Coast of the United States. *(Gary R. Robinson/Visuals Unlimited)*	Tube-dwelling marine worms.
Ectoprocts (Bryozoans)	**A colony of spiral, tufted bryozoans (Bugula turrita) with lophophores extended for feeding.** *(H. W. Pratt/Biological Photo Service)*	Sessile colonies produced by asexual budding. Moss-like appearance. Mainly marine. Lophophore can be retracted.
Brachiopods	**Northern lamp shells (Tetratulina septentrionalis).** *(Fred Bavendam/ Peter Arnold, Inc.)*	Known as lamp shells. Solitary, marine. Body enclosed between two shells (superficially resemble bivalve mollusks like clams).
Chaetognatha (Arrow worms)	**Arrow worm (Sagitta macrocephela).** *(Peter Parks/Oxford Scientific Films/Animals Animals)*	Marine. Make up part of plankton. Predatory carnivores.
Hemichordata (Acorn worms)	**Acorn worm (Saccoglossus kowalewski).** *(C. R. Wyttenbach, University of Kansas/Biological Photo Service)*	Sedentary, wormlike, marine bottom dwellers. Some species construct mucus-lined burrows in the mud. Have pharyngeal slits and dorsal nerve cord.

of tube feet. The mouth lies in the center of the underside of the disk. The endoskeleton consists of a series of calcareous plates that permit some movement in the arms. Delicate dermal gills, small extensions of the body wall, carry on gas exchange. Around the base of the skin gills, tiny pincer-like structures *(pedicellariae)* keep the surface of the animal free of debris.

Most sea stars are carnivorous predators and scavengers that feed on crustaceans, mollusks, annelids, and even other echinoderms. Occasionally they catch small fish. The sea star's water vascular system does not permit rapid movement, so its prey are usually slow-moving or stationary animals such as clams.

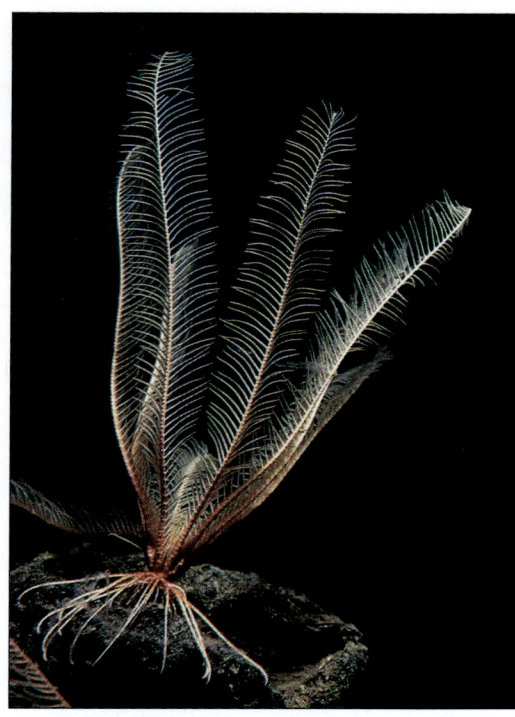

Figure 30-1 **Echinoderms.** *(a)* Feather stars use their slender, jointed appendages to cling to the surface of a rock or coral reef. They are able to creep and often swim away to escape predators. *(b)* Orange and red sea star *(Fromia monilis)* on bubble coral. *(c)* Daisy brittle star *(Ophiopholis aculeata)* photographed in Muscongus Bay, Maine. *(d)* With their flattened, circular bodies, sand dollars (such as *Derdraser excentricus* shown here) are adapted for burrowing in the ocean floor. *(e)* A sea cucumber *(Thelonota* sp.), raises its body to spawn. *(a,d, D. J. Wrobel, Monterey Bay Aquarium/Biological Photo Service; b, Marc Chamberlain/ Tony Stone Images; c, Robert Dunne/Photo Researchers, Inc.; e, Peter Scoones/Seaphot, Ltd.)*

(a) Crinoidea

(b) Asteroidea

(c) Ophiuroidea

(d) Echinoidea

(e) Holothuroidea

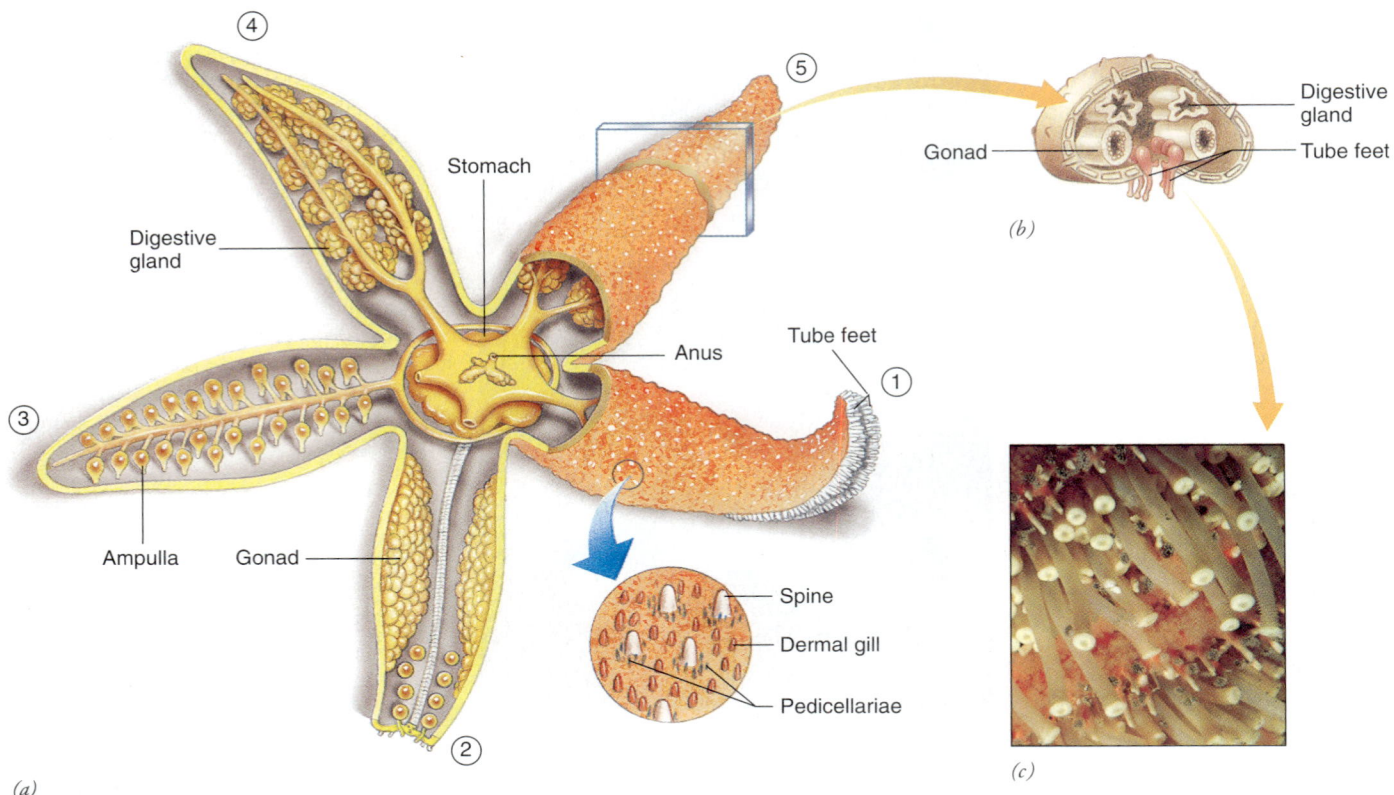

(a)

Figure 30–2 **Body plan of a sea star.** *(a)* A sea star viewed from above, with its arms in various stages of dissection. Similar structures are present in each arm. The two-part stomach is in the central disc with the anus on the aboral (upper) surface and the mouth beneath on the oral surface. (1) Upper surface with magnified detail. The end is turned up to show the tube feet on the lower surface. (2) The arm is dissected to show well-developed gonads. (3) Upper body and digestive glands have been removed, exposing the ampullae of some of the hundreds of tube feet *(magnified view)*. (4) Other organs have been removed to show the digestive glands. (5) Upper surface. *(b)* Cross section through arm and tube feet. *(c)* LM of tube feet of a sea star. *(c, Charles Seaborn/ Odyssey Productions, Chicago)*

To attack a clam or other bivalve mollusk, the sea star mounts it and assumes a humped position as it straddles the edge opposite the hinge (Fig. 30–3). Then, holding itself in position with its tube feet, the sea star slides its thin, flexible stomach out through its mouth and between the closed, or slightly gaping, valves (shell parts) of the clam. The sea star secretes enzymes that digest the soft parts of the clam to the consistency of a thick soup while the clam is still in its own shell. The partly digested meal passes into the sea star body, where it is further digested by enzymes secreted from glands located in each arm.

The circulatory system in sea stars is poorly developed and probably of little help in circulating materials. Instead, this function is assumed by the coelomic fluid, which fills the large coelom and bathes the internal tissues. Metabolic wastes pass to the outside by diffusion across the tube feet and dermal gills. The nervous system consists of a ring of nervous tissue encircling the mouth and a nerve cord extending from this ring into each arm.

Figure 30–3 **Painted sea star *(Orthiasterias koehleri)* attacking a clam.** The sea star inserts its thin, everted stomach between the valves of the clam and begins to digest the clam while it is still in its shell. *(Richard Chesher/Seaphot, Ltd.)*

Class Ophiuroidea is the largest class of echinoderms

Basket stars and brittle stars (serpent stars), members of class **Ophiuroidea,** are the largest group of echinoderms, both in number of species and in number of individuals (Fig. 30–1*c*). These animals resemble sea stars in that their bodies also consist of a central disk with arms, but the arms are long and slender and more sharply set off from the central disk. Ophiuroids can move more rapidly than sea stars, using their arms to perform rowing or even swimming movements. Their tube feet lack suckers and are not used in locomotion. They are used to collect and handle food and may also serve a sensory function, perhaps that of smell or taste.

Members of class Echinoidea have moveable spines

Sea urchins and sand dollars, the animals that comprise class **Echinoidea,** have no arms (Fig. 30–1*d*). Their skeletal plates are flattened and fused to form a solid shell called a test. The flattened body of the sand dollar is adapted for burrowing in the sand, where it feeds on tiny organic particles. Sand dollars have smaller spines than do sea urchins.

The sea urchin body is covered with spines that in some species can penetrate flesh and are difficult to remove. So threatening are these spines that swimmers on tropical beaches are often cautioned to wear shoes when venturing offshore, where these living pincushions may be found in abundance. Sea urchins use their tube feet for moving and their moveable spines for pushing themselves along. Many sea urchins graze on algae, scraping the sea floor with their calcareous teeth.

Members of class Holothuroidea are elongated, sluggish animals

Sea cucumbers, members of class **Holothuroidea,** are appropriately named, for some species are about the size and shape of a cucumber. The elongated sea cucumber body is a flexible, muscular sac (Fig. 30–1*e*). The mouth is usually surrounded by a circle of tentacles that are modified tube feet. Another characteristic of sea cucumbers is the reduction of the endoskeleton to microscopic plates. The circulatory system is more highly developed than that of other echinoderms and functions to transport oxygen and perhaps nutrients as well.

Sea cucumbers are sluggish animals that usually live on the bottom of the sea, sometimes burrowing in the mud. Some graze with their tentacles, while others stretch their branched tentacles out in the water and wait for dinner to float by. Algae and other morsels are trapped in mucus along the tentacles.

An interesting habit of some sea cucumbers is evisceration, in which the digestive tract, respiratory structures, and gonads are ejected from the body, usually when environmental conditions are unfavorable. When conditions improve, the lost parts are regenerated. Even more curious is the fact that when certain sea cucumbers are irritated or attacked, they direct their rear end toward the enemy and shoot red tubules out of their anus! These unusual weapons are sticky, and the attacking animal may become hopelessly entangled. Some of these tubules release a toxic substance.

Members of class Concentricycloidea have a unique water vascular system

First discovered in 1983, sea daisies inhabit bacteria-rich wood sunk in deep water. These small (less than 1 cm in diameter), disk-shaped echinoderms appear to have two ring canals with the tube feet projecting from the outer one. Class Concentricycloidea was established to accomodate these interesting echinoderms.

CHORDATE CHARACTERS INCLUDE A NOTOCHORD, DORSAL TUBULAR NERVE CORD, AND PHARYNGEAL GILL SLITS DURING SOME TIME IN THE LIFE CYCLE

The phylum of animals to which humans belong, Chordata, is divided into three subphyla: subphylum **Urochordata,** which consists of marine animals called tunicates; subphylum **Cephalochordata,** composed of marine animals called lancelets; and subphylum **Vertebrata,** animals with backbones (Fig. 30–4).

Chordates are coelomate animals with bilateral symmetry, a tube-within-a-tube body plan, and three well developed germ layers (Fig. 30–5). Typically, they have an endoskeleton and a closed circulatory system with a ventral heart. Four characteristics distinguish the chordates:

1. All chordates have a **notochord** during some time in their life cycle. The notochord is a dorsal longitudinal rod that is firm, but flexible, and supports the body.
2. At some time in their life cycle, chordates have a **dorsal tubular nerve cord.** The chordate nerve cord is different from the nerve cord of most other animals in that it is located dorsally rather than ventrally, is hollow rather than solid, and is single rather than double.
3. Chordates have **pharyngeal gill slits** (also called *pharyngeal slits*) during some time in their life cycle. In the embryo, a series of alternating branchial (gill) arches and grooves develop in the body wall in the pharyngeal (throat) region. Pharyngeal pouches extend laterally from the anterior portion of the digestive tract toward the grooves (described in more detail in Chapter 49).

   Early chordates, like many living chordates, may have been suspension feeders. The arrangement of pharyngeal pouches and slits permitted them to take water in through the mouth, concentrate small particles of food in the gut, and let the water escape from the body through the slits. The pharyngeal slits evolved into gill slits that function in

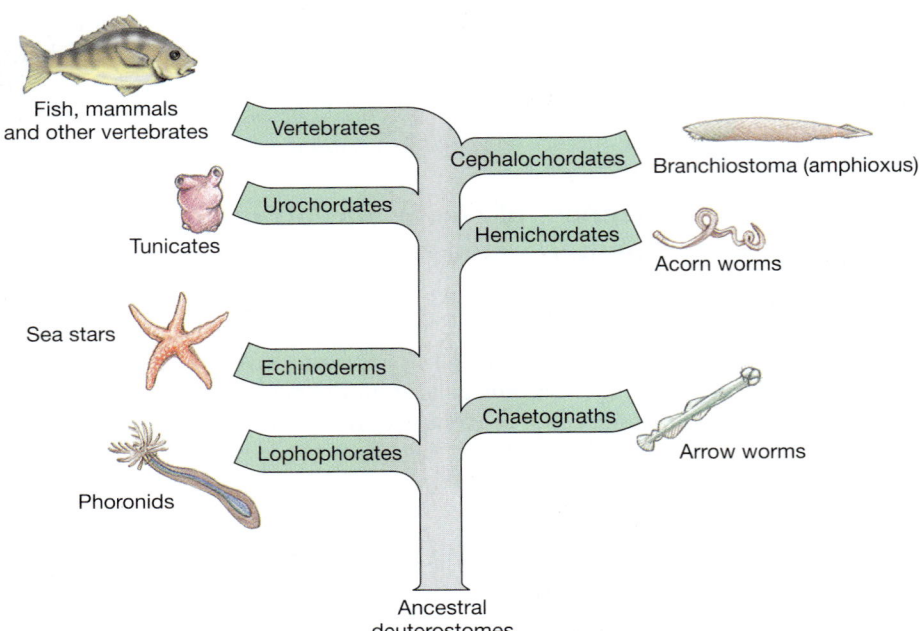

Figure 30–4 **Chordate evolution.** This diagram shows possible phylogenetic relationships among deuterostomes, including chordates, based on current scientific evidence.

Fish, mammals and other vertebrates

Vertebrates

Cephalochordates

Branchiostoma (amphioxus)

Urochordates

Tunicates

Hemichordates

Acorn worms

Sea stars

Echinoderms

Chaetognaths

Lophophorates

Arrow worms

Phoronids

Ancestral deuterostomes

respiration in fishes. In the embryos of terrestrial animals they develop into structures more suitable for life on land.

4. Most chordates have a **muscular postanal tail,** an appendage that extends posterior to the anus.

No clear fossil record of the chordate ancestors exists, but they were probably small, soft-bodied animals. A lancelet-like animal, *Pikaia,* has been found in the Burgess Shale of British Columbia, Canada. These rocks, which date back to the Cambrian period, have been a rich source of fossils. In 1995 scientists reported finding an earlier chordate (which they named *Yunnanozoon*) in an early Cambrian fossil site in China. Whether this fossil is truly a chordate or whether it is a hemichordate has been a matter of debate.

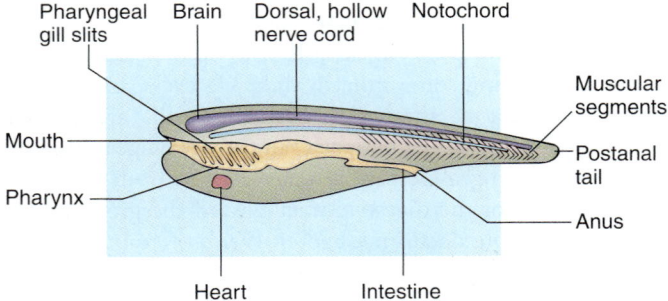

Pharyngeal gill slits   Brain   Dorsal, hollow nerve cord   Notochord

Muscular segments

Mouth

Postanal tail

Pharynx

Anus

Heart   Intestine

Figure 30–5   **Generalized chordate body plan.** Note the notochord; dorsal, hollow nerve cord; pharyngeal gill slits; and postanal tail.

Tunicates are sessile, marine animals

The **tunicates,** which comprise subphylum Urochordata, include the sea squirts, or *ascidians,* and their relatives. Larval tunicates have typical chordate characteristics and superficially resemble tadpoles. The expanded body has a pharynx with slits, and the long muscular tail contains a notochord and a dorsal, hollow nerve cord. Some tunicates *(appendicularians)* are common members of the zooplankton that retain their chordate features and ability to swim about.

Most tunicates are sea squirts (class **Ascidiacea**). In sea squirts, the larva swims for a time, then attaches itself to the sea bottom and loses its tail, notochord, and much of its nervous system. Adult sea squirts are barrel-shaped, sessile marine animals unlike other chordates. Indeed, they are often mistaken for sponges or cnidarians (Fig. 30–6). Only the pharyngeal slits (and the structure of its larva) suggest that the sea squirt is a chordate.

Adult tunicates develop a protective covering, or tunic, that may be soft and transparent or quite leathery. Curiously, the tunic is composed of a carbohydrate much like cellulose. The tunic has two openings: the incurrent siphon, through which water and food enter, and the excurrent siphon, through which water, waste products, and gametes pass to the outside. Sea squirts get their name from their practice of forcefully expelling a stream of water from the excurrent siphon when irritated.

Tunicates are suspension feeders that remove plankton from the stream of water passing through the pharynx. Food particles are trapped in mucus secreted by cells of the *endostyle,* a groove that extends the length of the pharynx. Ciliated cells

(c)

(a)

Incurrent siphon

Ganglion

Oral tentacles

Excurrent siphon

Pharynx with gill slits

Endostyle

Tunic

Intestine

Atrium

Ovary

Esophagus

Testis

Digestive gland

Heart

Stomach

(b)

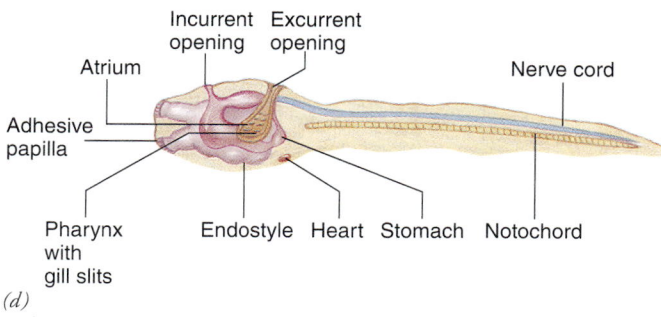

Incurrent opening   Excurrent opening

Atrium

Nerve cord

Adhesive papilla

Pharynx with gill slits

Endostyle   Heart   Stomach   Notochord

(d)

Figure 30–6 **Tunicate body plan.** *(a)* Incurrent *(top)* and excurrent *(side)* siphons of a sea peach tunicate *(Halocynthia auratium).* *(b)* Lateral view of an adult tunicate. The red arrows represent the flow of water, and the blue arrows represent the path of food. The stomach, intestine, and other visceral organs are embedded in the mantle. *(c)* Swimming larval stage of a tunicate, *Distaplia occidentalis.* *(d)* Internal structure of a larval tunicate *(lateral view).* *(a, Robert Shupak; c, Richard A. Cloney/University of Washington)*

within the endostyle move the stream of food-laden mucus into the esophagus. Much of the water entering the pharynx passes out through the pharyngeal slits into an **atrium** (chamber) and is discharged through the excurrent siphon.

Some species of tunicates form large colonies in which members may share a common tunic and excurrent siphon. Colonial forms often reproduce asexually by budding. Sexual forms are usually hermaphroditic.

Lancelets exhibit basic chordate characters

Subphylum Cephalochordata consists of the **lancelets,** translucent, fish-shaped animals, 5 to 10 cm (approximately 2 to 4 in) long and pointed at both ends. Lancelets are widely distributed in shallow seas, either swimming freely or burrowing in the sand near the low-tide line. In some parts of the world, lancelets are an important source of food. One Chinese fishery reports an annual catch of 35 tons (about 1 billion lancelets).

Chordate characteristics are highly developed in lancelets. The notochord extends from the tip of the head (hence the

name *Cephalochordata*) to the tip of the tail. Many pairs of slits are evident in the large pharyngeal region, and a hollow, dorsal nerve cord extends the entire length of the animal (Fig. 30–7). The most common genus, *Branchiostoma* (commonly known as amphioxus), exhibits the basic chordate characteristics so well that it is usually the first animal studied in comparative chordate anatomy courses. Although superficially similar to fishes, lancelets have a far simpler body plan. They do not have paired fins, jaws, sense organs, a heart, or a well defined brain.

Like the tunicates, lancelets use their cilia to draw a current of water into the mouth and then strain out microscopic organisms. Food particles are trapped in mucus in the pharynx and are then carried back to the intestine.

Water passes through the pharyngeal slits into the atrium, a chamber with a ventral opening (the *atriopore*) just anterior to the anus. Metabolic wastes are excreted by segmentally arranged, ciliated protonephridia that open into the atrium. In contrast to other invertebrates, the blood flows anteriorly in the ventral vessel and posteriorly in the dorsal vessel. This circulatory pattern is similar to that of fishes.

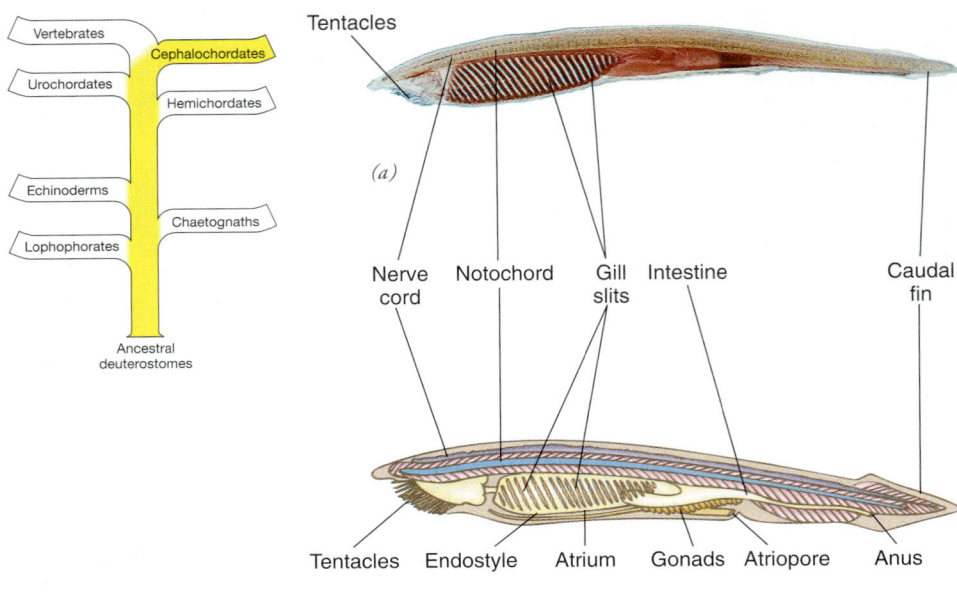

Tentacles

(a)

Nerve cord   Notochord   Gill slits   Intestine   Caudal fin

Tentacles   Endostyle   Atrium   Gonads   Atriopore   Anus

(b)

Figure 30–7   Cephalochordate body structure. *(a)* Photograph of a lancelet *Branchiostoma* (amphioxus). Note the prominent pharyngeal gill slits. *(b)* Longitudinal section showing internal structure.   *(John D. Cunningham/Visuals Unlimited)*

The success of the vertebrates is linked to the evolution of key adaptations

The vertebrates, members of subphylum Vertebrata, are distinguished from other chordates in having a backbone, or **vertebral column,** that forms the skeletal axis of the body. This flexible support develops around the notochord, and in most species it largely replaces the notochord during embryonic development. The vertebral column consists of cartilaginous or bony segments called **vertebrae.** Dorsal projections of the vertebrae enclose the nerve cord along its length. Anterior to the vertebral column, a cartilaginous or bony **cranium,** or braincase, encloses and protects the brain, the enlarged anterior end of the nerve cord.

The cranium and vertebral column are part of the **endoskeleton.** In contrast to the nonliving exoskeleton of many invertebrates, the vertebrate endoskeleton is a living tissue that grows with the animal. Two pairs of appendages are present in most vertebrates. The fins of fishes are appendages that stabilize them in the water. Paired pectoral and pelvic fins are also used in steering. As vertebrates moved onto the land, jointed appendages that facilitated locomotion evolved from lobed fins.

Recall that in invertebrates there is an evolutionary trend toward **cephalization,** the concentration of nerve cells and sense organs in a definite head. Vertebrate evolution is characterized by *pronounced* cephalization. The brain became larger and more elaborate, and its various regions became specialized to perform different functions. Ten or 12 pairs of cranial nerves emerge from the brain and extend to various organs of the body. Vertebrates have well developed sense organs concentrated in the head: eyes; ears that serve as organs of balance and, in some vertebrates, for hearing as well; and organs of smell and taste.

Vertebrates have a closed circulatory system with a ventral heart, paired kidneys that regulate fluid balance, and a complete digestive tract with large digestive glands (the liver and pancreas). The sexes are typically separate.

With about 48,000 species, the vertebrates are less diverse and much less numerous than the insects, but rival them in their adaptation to an enormous variety of lifestyles. Traditionally, extant vertebrates have been divided into three classes of fishes, referred to as superclass **Pisces,** and four classes of land vertebrates, superclass **Tetrapoda** (Fig. 30–8 and Table 30–2). In this classification scheme, the classes of fishes include class **Agnatha,** the jawless fishes such as lampreys; class **Chondrichthyes,** the sharks and rays with cartilaginous skeletons; and class **Osteichthyes,** the bony fishes. The four-legged land vertebrates, or tetrapods, are grouped in class **Amphibia,** the frogs, toads, and salamanders; class **Reptilia,** the lizards, snakes, turtles, and alligators; class **Aves,** the birds; and class **Mammalia,** the mammals. Not all the tetrapods have four legs (e.g., the snakes), but all evolved from four-legged ancestors. Nor do all tetrapods now live entirely on land (e.g., sea turtles, penguins, whales, seals), but all of these aquatic forms evolved from terrestrial ancestors.

Recently, taxonomists have divided fishes into six classes, resulting in ten classes of vertebrates. Although, as with most taxonomic schemes, there is lack of agreement among biologists, the newer classification scheme is presented along with the more traditional approach.

The jawless fishes represent the earliest vertebrates

Toothlike fossils, known as *conodonts,* discovered in Cambrian rocks, are thought to represent the earliest vertebrate, an eel-shaped animal. More is known about the **ostracoderms,** small,

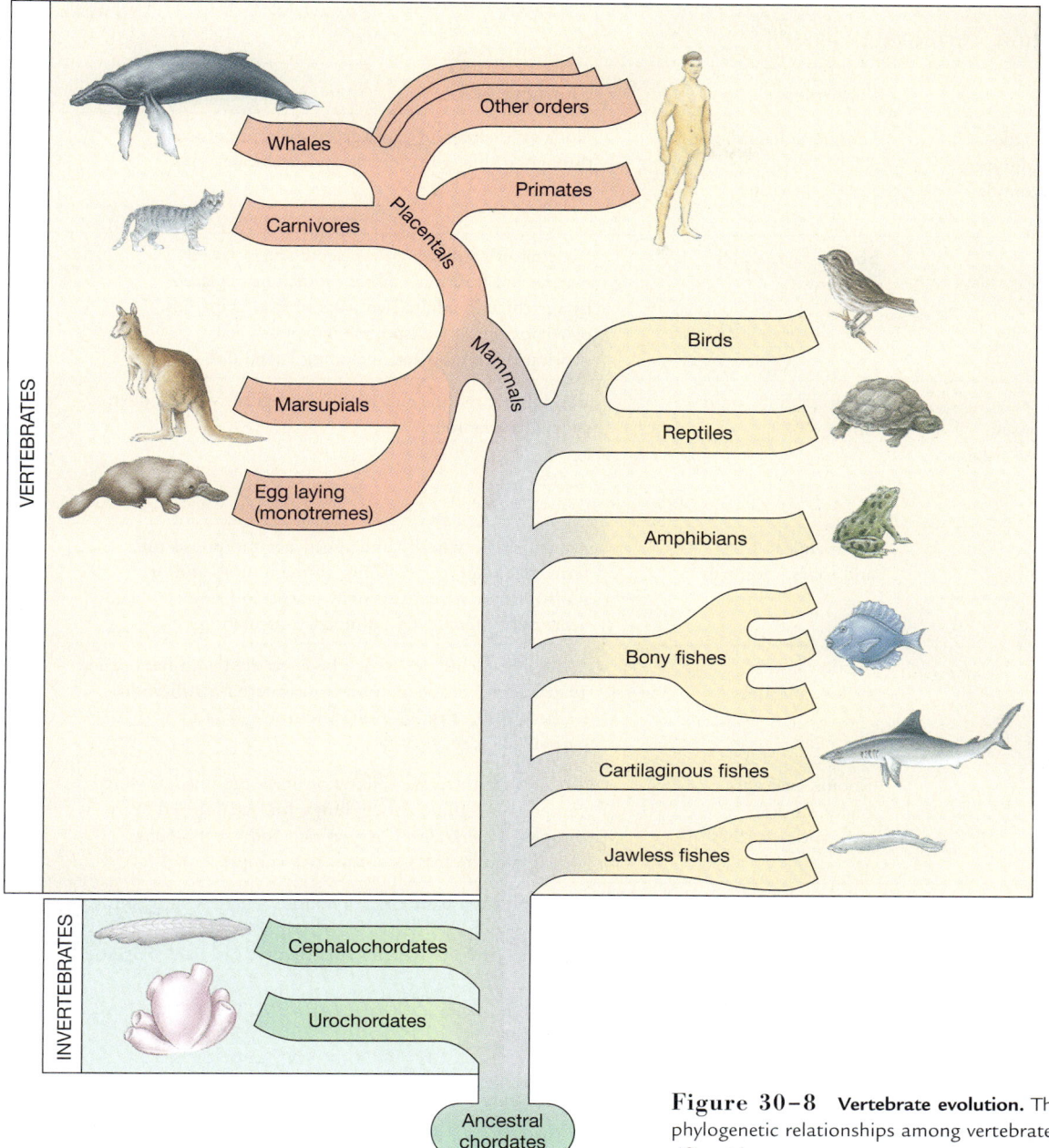

Figure 30–8 Vertebrate evolution. This diagram shows possible phylogenetic relationships among vertebrates, based on current scientific evidence.

armored, jawless fishes that lived on the bottom and strained their food from the water (see Fig. 20–9). Their heads were protected by thick bony plates, and their trunks and tails were covered with thick scales. Most ostracoderms had poorly developed fins. Fragments of ostracoderm scales have been found in rocks from the Cambrian period, but most ostracoderm fossils date back to the Ordovician and Silurian periods. By the Silurian and Devonian periods, ostracoderms had radiated extensively and were mainly freshwater fishes.

Although the ostracoderms became extinct by the end of the Devonian period, they are survived by their descendants, the lampreys and hagfishes historically classified in class **Ag-**

natha (*a,* "without"; *gnathos,* "jaw"). These eel-shaped animals, up to 1 m long, are supported by a cartilaginous skeleton. Their smooth skin lacks scales, and they have neither jaws nor paired fins. Some systematists have made Agnatha a superclass and have divided agnathans into two extant classes: class *Myxinoidea,* the hagfishes, and class *Petromyzontida,* the lampreys.

Hagfishes are marine scavengers that burrow for worms and other invertebrates or prey on dead and disabled fishes. Hagfishes differ from lampreys in not having a larval stage (at least none has been discovered). These fishes are of interest to medical researchers because, in addition to a heart, they have several other contractile structures that help circulate blood.

TABLE 30-2 Extant Vertebrate Classes

Class	Examples	Characteristics
Agnatha (Some biologists divide this class into 2 classes: Myxinoidea, the hagfishes, and Petromyzontida, the lampreys.)	Lampreys, hagfishes	No jaws; skeleton of cartilage; notochord persists throughout life; gills; marine and freshwater.
Chondrichthyes	Sharks, rays, skates, chimeras	Skeleton of cartilage; gills; placoid scales; jawed marine and freshwater fishes; notochord replaced by vertebrae in adult. Two pairs of fins; oviparous, ovoviviparous, a few species viviparous; well developed sense organs, including lateral line system.
Osteichthyes (Some biologists divide bony fishes into 3 classes: Actinopterygii, the ray-finned fishes; Actinistia, lobe-finned fishes; and Dipnoi, lungfishes)	Perch, salmon, trout, tuna	Bony fishes; marine and fresh water; gills; swim bladder; generally oviparous.
Amphibia	Salamanders, frogs and toads, caecilians	Aquatic larva typically undergoes metamorphosis into terrestrial adult; gas exchange through lungs and/or moist skin; heart consists of two atria and single ventricle; systemic and pulmonary circulation.
Reptilia	Turtles, lizards, snakes, alligators	Tetrapods; mainly terrestrial; body covered with hard scales; adapted for reproduction on land (internal fertilization, leathery shell, amnion); lungs; ventricle of heart partially divided.
Aves	Robins, pelicans, eagles, ducks, penguins, ostriches	Tetrapods with feathers; anterior limbs modified as wings; compact streamlined body; lungs; four-chambered heart; complete separation of oxygen-rich and oxygen-poor blood; endotherms; vocal calls and complex songs.
Mammalia	Monotremes (platypus); marsupials (kangaroos); placentals (whales, humans)	Tetrapods with hair; females nourish young with mammary glands; diaphragm moves air in and out of lungs; highly developed nervous system; endotherms; most are viviparous.

These additional "hearts," which help maintain blood pressure, are regulated by a compound called *eptatretin.* Researchers are studying eptatretin in the hope that it could be used to treat certain types of abnormal heart rhythms in humans.

Most lampreys live in fresh water. Some spend their adult lives in the ocean and return to fresh water to reproduce. Many species of adult lampreys are parasites on other fishes (Fig. 30–9). Adult parasitic lampreys have a circular sucking disk around the mouth, which is located on the ventral side of the anterior end of the body. Using this disk to attach to a fish, the lamprey bores through the skin of its host with horny teeth on the disk and tongue. Then the lamprey injects an anticoagulant into its host and sucks blood and soft tissues.

Adult lampreys leave the ocean or lake and swim upstream to spawn. They build a nest, a shallow depression in the gravelly stream bed, into which they shed eggs and sperm. After spawning they die. The fertilized eggs develop into larvae, which drift downstream to a pool and live as suspension feeders in burrows in the muddy bottom for three to seven years. They then undergo a metamorphosis, become adult lampreys, and migrate back to the ocean or lake. In lampreys the notochord persists throughout life and is not replaced by vertebrae.

The earliest jawed fishes are now extinct

Fossil evidence suggests that, during the late Silurian and Devonian periods, fishes evolved with jaws and finlike paired appendages. The first two classes of jawed fishes were the now extinct **acanthodians,** armored fishes with paired spines and pectoral and pelvic fins, and **placoderms,** armored fishes with paired fins (Fig. 30–10).

(a)

(b)

(a) Climatius

(b) Dinichthys

Figure 30–10 Early jawed fishes. Acanthodians and placoderms flourished in the Devonian period. *(a)* Climatius, a spiny-skinned acanthodian with large fin spines and five pairs of accessory fins between the pectoral and pelvic pairs. *(b)* Dinichthys, a giant placoderm that grew to a length of 9 m (30 ft). Its head and thorax were covered by bony armor, but the rest of the body and tail were naked.

The evolution of jaws from a portion of the gill arch skeleton and the development of fins enabled fishes to change from suspension-feeding bottom dwellers to active predators. This change afforded many new opportunities for finding food. The success of the jawed vertebrates may have contributed to the extinction of the ostracoderms.

Members of class Chondrichthyes are cartilaginous fishes

The ostracoderms and early jawed fishes inhabited mainly fresh water. Only a few ventured into the ocean. Members of class Chondrichthyes, the cartilaginous fishes, evolved as successful marine forms in the Devonian period. This class includes the sharks, rays, and skates (Fig. 30–11). Most species have remained as ocean dwellers, but a few have secondarily returned to a freshwater habitat. Except for the whales, the sharks are the largest living vertebrates. The whale shark *(Rhincodon)* may exceed 15 m (49 ft) in length, making it the largest fish.

Figure 30–11 Cartilaginous fishes (class Chondrichthyes). *(a)* Blue-spotted sting ray *(Taeniura lymma)*. Sting rays typically feed on shellfish and bottom-dwelling fishes. *(b)* The great white shark *(Carcharodon carcharias)*, photographed in Australia, is considered the most dangerous shark to humans. This shark is actually white only on its ventral aspect; the rest of the body is brownish-gray or bluish-gray. *(a, Jeffrey L. Rotman/Peter Arnold, Inc.; b, Kelvin Aitken/Peter Arnold, Inc.)*

(a)

(b)

Most rays and skates are sluggish, flattened creatures that live partly buried in the sand. Their enormous pectoral fins propel them along the bottom, where they feed on mussels and clams. The sting ray has a whiplike tail with a barbed spine at its base that can inflict a painful wound. The electric ray has electric organs on either side of the head. These modified muscles can discharge enough electric current (up to 2500 watts) to stun fairly large fishes, as well as human swimmers.

The chondrichthyes retain their cartilaginous embryonic skeleton. Although this skeleton is not replaced by bone, it may be strengthened by a deposit of calcium salts. All chondrichthyes have paired jaws and two pairs of fins. The skin contains **placoid scales.** Each scale is a toothlike structure composed of an outer layer of enamel and an inner layer of dentine (Fig. 30–12). The lining of the mouth contains larger, but essentially similar, scales that serve as teeth. The teeth of other vertebrates are homologous with these scales. Shark teeth are embedded in the flesh and not attached to the jawbones; new teeth develop continuously in rows behind the functional teeth and migrate forward to replace any that are lost.

The shark body is adapted for swimming. Lift is provided by body shape, fins, and by swimming swiftly. The shark stores a great deal of oil in the large liver. In some sharks the liver accounts for up to 30% of the body weight, and most of that weight is due to stored oil. Fats and oils decrease the overall density of fishes and contribute to buoyancy. Even so, the shark body is denser than water, so sharks tend to sink unless they are actively swimming.

Most sharks are streamlined predators that swim actively and catch and eat other fishes as well as crustaceans and mollusks. The largest sharks and rays, like the largest whales, are suspension feeders that strain plankton from the water. They gulp water through the mouth. Then, as the water passes through the pharynx and out the gill slits, food particles are trapped in a sieve-like structure.

Predatory sharks are attracted to blood, so a wounded swimmer or a skin diver towing speared fish is a target. However, although books and films portray sharks as monstrous enemies, most do not go out of their way to attack humans. In fact, of the approximately 350 known shark species, fewer than 30 have been known to attack humans.

The shark has a complex brain, and a spinal cord protected by vertebrae. Their well developed sense organs very effectively locate prey in the water. Sharks may detect other

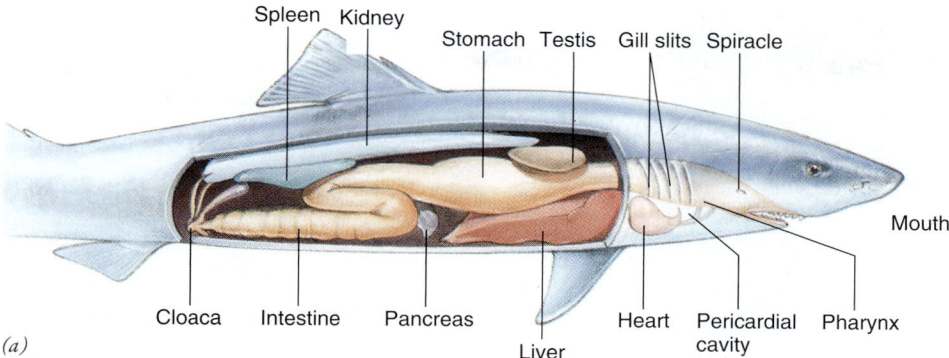

Spleen Kidney Stomach Testis Gill slits Spiracle

Mouth

Cloaca Intestine Pancreas Liver Heart Pericardial cavity Pharynx

(a)

Figure 30–12 **Sharks.** *(a)* Internal structure of a shark. *(b)* Structure of a placoid scale.

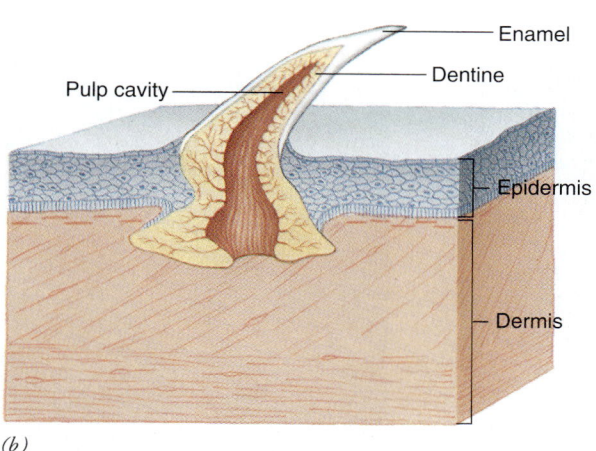

Enamel

Dentine

Pulp cavity

Epidermis

Dermis

(b)

animals electrically before sensing them by sight or smell. **Electroreceptors** on the shark's head can sense weak electric currents generated by the muscular activity of animals. The **lateral line organ,** found in all fishes, is a groove along each side of the body with many tiny openings to the outside. Sensory cells in the lateral line organ are sensitive to waves and other motion in the water, alerting the shark to the presence of predator or prey (see Fig. 41–3).

Cartilaginous fishes have five to seven pairs of gills. A current of water enters the mouth and passes over the gills and out the gill slits, constantly providing the fish with a fresh supply of dissolved oxygen. Sharks that actively swim depend on their motion to enhance gas exchange. Sharks that spend time on the ocean floor, and rays and skates, use muscles of the jaw and pharynx to pump water over their gills.

The digestive tract of sharks consists of the mouth cavity; a long pharynx leading to the stomach; a short, straight intestine; and a **cloaca,** which opens on the underside of the body and is characteristic of many vertebrates. The liver and pancreas discharge digestive juices into the intestine. The cloaca receives digestive wastes, as well as metabolic wastes from the urinary system. In females, the cloaca also serves as a reproductive organ.

The sexes are separate, and fertilization is internal. In the mature male, each pelvic fin has a slender, grooved section, known as a **clasper,** used to transfer sperm into the female's cloaca. The eggs are fertilized in the upper part of the female's oviducts. Part of the oviduct is modified as a shell gland, which secretes a protective coat around the egg.

Skates and some species of sharks are **oviparous;** that is, they lay eggs. Many species of sharks, however, are **ovoviviparous,** meaning that their young are enclosed in eggs that are incubated within the mother's body. During development, the young depend on stored yolk for their nourishment, rather than on transfer of materials from the mother. The young are born after hatching from the eggs.

A few species of sharks are **viviparous.** Not only do the embryos develop within the uterus, but much of their nourishment is delivered to them by the mother's blood. This is accomplished by transfer of nutrients between the blood vessels in the lining of the uterus and the yolk sac surrounding each embryo.

Members of class Osteichthyes are bony fishes

Class Osteichthyes includes more than 24,000 living species of freshwater and saltwater bony fishes, of many shapes and colors (Fig. 30–13). Bony fishes range in size from the Philippine goby, which is only about 10 mm (0.4 in) long, to the ocean sunfish (or *Mola;* see Fig. 19–12c), which may reach 4 m and weigh about 1500 kg (about 3300 lb). Although bony fishes appear earlier in the fossil record than cartilaginous fishes, both groups may have evolved about the same time, during the Devonian period. The two groups share many characteristics (such as continuous tooth replacement), but they also differ in important ways.

Most osteichthyes are characterized by a bony skeleton with many vertebrae (Fig. 30–14). Bone has advantages over cartilage: it provides excellent support and serves as a very effective storage site for calcium. The osteichthyes body is covered with overlapping, bony dermal scales. Most species have flexible median and paired fins, supported by long rays made of cartilage or bone. A lateral bony flap, the **operculum,** extends posteriorly from the head and protects the gills.

Unlike most sharks, bony fishes are generally oviparous.

Figure 30–13 Bony fishes (class Osteichthyes). (a) The porcupinefish *(Diodon hystrix)* can swallow air or water and inflate its body, a strategy to discourage potential predators (photographed in the Virgin Islands). **(b)** Schooling goggle-eyes *(Priacanthus hamrur)*, off Lizard Island, Great Barrier Reef, Australia. **(c)** The parrotfish *(Scarus gibbus)* feeds on coral, grinds it in its digestive tract, and extracts the corraline algae. The fish eliminates a fine white sand. These fishes contribute to white sand beaches in many parts of the world. Parrotfish begin life as females and later become males. **(d)** The leaf-like extensions of the body wall of this Australian leafy sea dragon *(Phyllapteryx taeniolatus)* help camouflage it in surrounding kelp. *(a, David Hall/Photo Researchers, Inc.; b, Fred Bavendam/Peter Arnold, Inc.; c, Jeffrey L. Rotman/Peter Arnold, Inc.; d, Norbert Wu/Peter Arnold, Inc.)*

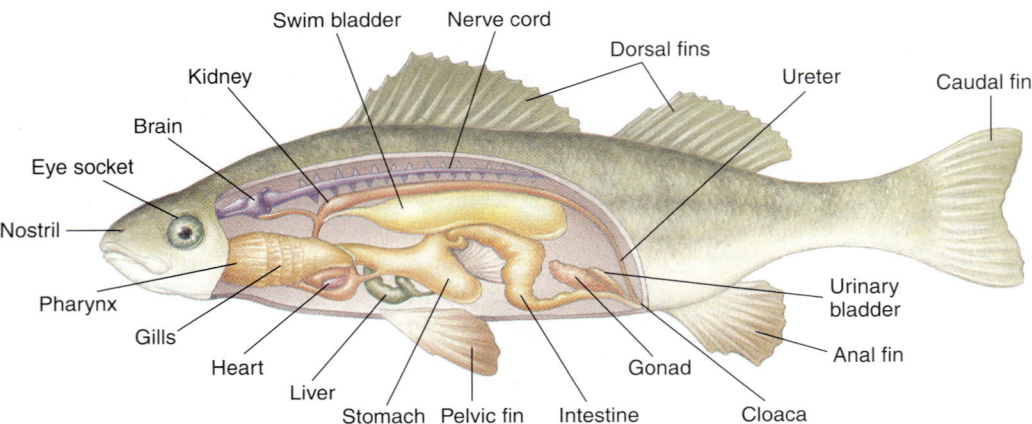

Figure 30–14 Perch, a representative bony fish. The swim bladder is a hydrostatic organ that enables the fish to change the density of its body, and remain stationary at a given depth. Pectoral *(not shown)* and pelvic fins are paired.

They lay an impressive number of eggs that they fertilize externally. The ocean sunfish, for example, lays over 300 million eggs! Of course, most of the eggs and young become food for other animals. The probability of survival is increased by certain behavioral adaptations. For example, many species of fishes build nests for their eggs and even watch over and protect them.

The ray-finned fishes gave rise to modern bony fishes

During the Devonian period, the bony fishes diverged into two major groups: the **ray-finned fishes** *(actinopterygians)* and the **lobe-finned fishes** *(sarcopterygians).* The ancestors of the ray-finned fishes are thought to have had lungs. The ray-finned fishes gave rise to most modern bony fishes. Some systematists have replaced class Osteichthyes with three classes: class *Actinopterygii,* which includes the ray-finned fishes; class *Actinistia,* the lobe-finned fishes; and class *Dipnoi,* the lungfishes (Fig. 30–15).

The ray-finned fishes underwent two important adaptive radiations (see Chapter 20). The first gave rise during the late Paleozoic era to a group of fishes that are now mostly extinct. The second radiation began during the early Mesozoic era and gave rise to the modern bony fishes. The lungs became modified as a **swim bladder,** an air sac that helps regulate buoyancy. By regulating gas exchange between the blood and the swim bladder, a fish can control the amount of gas in the swim bladder, changing the overall density of its body. This ability allows a bony fish, in contrast to a shark, to hover at a given depth of water without much muscular effort.

Descendants of the lobe-finned fishes moved onto the land

The lobe-finned fishes are characterized by fleshy, lobed fins and lungs. Early lobe-finned fishes (sarcopterygians) evolved along two separate lines: lungfishes (class Dipnoi) and lobe-finned fishes (class Actinistia). Three genera of lungfishes have survived to the present day in the rivers of tropical Africa, Australia, and South America. The lobe-finned fishes, considered the ancestors of the land vertebrates, were almost extinct by the end of the Paleozoic era. Only one species *(Latimeria chaluminae),* classified as a **coelacanth,** survives in the deep waters off the east coast of Africa near the Comoro Islands (Fig. 30–16). Nearly 2 m (about 6 ft) long, these giant "living fossils" are nocturnal predators on other fishes. Based on similarity of their hemoglobin, *Latimeria* appears to be more closely related to frogs than are other fishes.

During the Devonian period, frequent seasonal droughts caused swamps to become stagnant or even to dry up completely. Devonian lobe-finned fishes had a tremendous advantage for survival under those conditions: they were strikingly preadapted for survival on land. They had lungs for breathing air and their sturdy, fleshy fins may have allowed them to "walk" along in shallow water. These fins could support the fish's weight, enabling it to emerge onto dry land and make its way to another pond or stream.

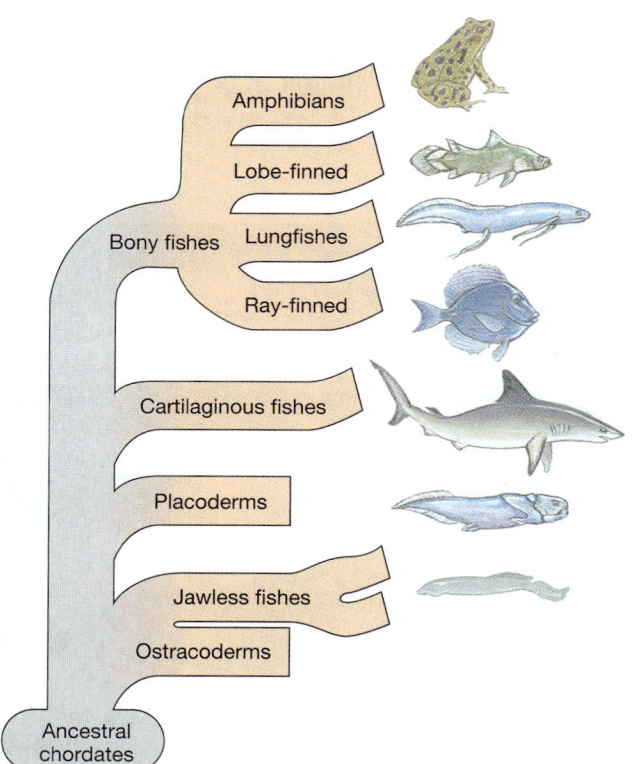

Figure 30–15 **Hypothetical evolutionary relationships of fishes and amphibians.** During the Devonian period, bony fishes diverged into the ray-finned fishes, lungfishes, and lobe-finned fishes. The lobe-finned fishes are thought to have given rise to the amphibians.

Figure 30–16 **Coelacanth.** Ancestors of this lobe-finned fish probably gave rise to the amphibians. The paired fins indicate the basic plan of a jointed series of bones that could evolve into the limbs of a terrestrial vertebrate. Living coelacanths *(Latimeria)* are difficult to observe because they inhabit deep ocean waters, and when brought to the surface, they do not survive the change in pressure. *(Estate of Dr. J. Metzner/Peter Arnold, Inc.)*

Figure 30–17 *An artist's conception of labyrinthodonts.* These early amphibians existed about 150 million years ago, from the late Devonian to the early Jurassic. *(Photograph by Logan, courtesy of Department of Library Services, American Museum of Natural History)*

The ability to move about, however awkwardly, on dry land gave these animals access to new food sources. Terrestrial plants were already established, and terrestrial insects and arachnids were rapidly evolving. A vertebrate that could survive on land had less competition for food. Laying eggs on land, away from the many ocean predators, also increased their chances for successful reproduction. Natural selection favored those individuals best adapted for making their way on land, leading to the evolution of the amphibians and, later, the reptiles.

Amphibians were the first successful land vertebrates

The first successful **tetrapods,** or land vertebrates, were the **labyrinthodonts** (Fig. 30–17), clumsy, salamander-like animals with short necks and heavy, muscular tails. These ancient members of class Amphibia somewhat resembled their ancestors, the lobe-finned fishes. However, they had limbs strong enough to support the weight of their bodies on land. The largest labyrinthodonts were the size of crocodiles. They flourished during the late Paleozoic and early Mesozoic eras, then became extinct. It is probable that the labyrinthodonts gave rise to other primitive amphibians, to frogs and salamanders, and to the earliest reptiles, the **cotylosaurs.**

Modern amphibians are classified in three orders: order **Urodela** ("visible tail") includes the salamanders, mudpuppies, and newts, all animals with long tails; order **Anura** ("no tail") is made up of the frogs and toads, with legs adapted for hopping; and order **Apoda** ("no feet") contains the wormlike caecilians (Fig. 30–18). Although some adult amphibians are quite successful as land animals and can live in dry environments, most return to the water to reproduce. Eggs and sperm are generally released in the water.

The embryos of frogs and toads develop into larvae called **tadpoles.** These larvae have tails and gills, and most feed on aquatic plants. After a time, the tadpole undergoes metamorphosis. The gills and gill slits disappear, the tail is resorbed, and limbs emerge. The digestive tract shortens, and food preference shifts from plant material to a carnivorous diet; the mouth widens; a tongue develops; the tympanic membrane

(a)

(b)

Figure 30–18 *Modern amphibians (class Amphibia). (a)* This red dart frog *(Dendrobates pumioio)* is a poison arrow frog. *(b)* The red eft newt *(Notophthalmus viridescens)* is mainly aquatic, but spends one to three years as a pre-adult in a moist terrestrial environment. *(a, Gerald and Buff Corsi/Tom Stack; b, The Stock Market/Roy Morsch)* (ear drum) and eyelids appear; and the eye lens changes shape. Many biochemical changes also accompany the transformation from a completely aquatic life to an amphibious one.

Amphibian metamorphosis is under hormonal control and is regulated by hormones secreted by the thyroid gland. Amphibians undergo a single, although complex, change from larva to adult. Several salamanders, such as the mudpuppy *Necturus*, do not undergo complete metamorphosis; they retain many larval characteristics even when sexually mature adults. Recall from Chapter 19 that this is an example of paedomorphosis (see Fig. 19–13). This type of development permits these salamanders to remain aquatic rather than having to compete on land.

The coloration of amphibians may conceal them in their habitat or may be very bright and striking. Many of the brightly colored species are poisonous (see Fig. 30—18*a*). Their distinctive colors warn predators that they are not encountering an ordinary amphibian. Some frogs can camouflage themselves by changing color.

Adult amphibians do not depend solely on their primitive lungs for the exchange of respiratory gases. Their moist, glandular skin, which lacks scales and is plentifully supplied with blood vessels, also serves as a respiratory surface. The numerous mucous glands within the skin help to keep the body surface moist, which is important in gas exchange. The mucus also makes the animal slippery, facilitating its escape from predators. Some amphibians have glands in their skin that secrete poisonous substances harmful to predators.

The amphibian heart is divided into three chambers: two **atria** receive blood, and a single **ventricle** pumps it into the arteries. A double circuit of blood vessels keeps oxygen-rich and oxygen-poor blood partially separate. Blood passes through the **systemic circulation** to the various tissues and organs of the body. Then, after returning to the heart, it is directed through the **pulmonary circulation** to the lungs and skin, where it is recharged with oxygen. The oxygen-rich blood returns to the heart to be pumped out into the systemic circulation again. The comparative anatomy of the heart and circulation of various vertebrate classes is discussed in Chapter 42.

Reptiles were once the dominant land animals

Reptiles, which descended from ancestral amphibians, have adaptations that make them completely terrestrial. Evolution of the **amnion,** a membrane that forms a fluid-filled sac around the embryo, was an important step in adapting to life on land. The amnion keeps the embryo moist, permitting independence from a watery external environment. Terrestrial vertebrates—reptiles, birds, and mammals—are referred to as **amniotes** because their embryos are enclosed by an amnion. Amniotes are thought to be a monophyletic group, that is, they have a common ancestor that was itself an amniote (Fig. 30–19).

In addition to the amnion, amniotes have three other extraembryonic (not part of the developing body itself) membranes that protect them, store nutrients *(yolk sac),* carry on gas exchange *(chorion and allantois),* and store wastes *(allantois).* Development of the extraembryonic membranes is discussed in Chapter 49.

The earliest reptiles are thought to have somewhat resembled lizards. By the late Carboniferous period, about 290 million years ago, amniotes had diverged into two groups. One of these groups evolved into the mammals; a second gave rise to all of the other reptiles and to the birds.

Many more extinct reptiles than living species are known. The Mesozoic era, which ended about 65 million years ago, is known as the Age of Reptiles. During that time reptiles were the dominant terrestrial animals (see Chapter 20). They had radiated into an impressive variety of ecological lifestyles (Fig. 20–11). Some were able to fly, others became marine, and many filled terrestrial habitats. Some of the dinosaurs were among the largest land animals to have ever lived. Some dinosaurs apparently traveled in social groups and took care of their young.

The reptiles were the dominant land animals for almost 200 million years. Then, toward the end of the Mesozoic era, many, including all the dinosaurs, disappeared from the fossil record. In fact, more than half of all animal species became extinct at that time (see Chapter 20).

Many reptilian characters are adaptations to terrestrial life

Most members of class Reptilia, which includes turtles, lizards, snakes, and alligators, inhabit the land and do not need to return to water to reproduce. A variety of adaptations make the reptilian lifestyle possible. The female secretes a protective leathery shell around the egg, which helps prevent the developing embryo from drying out. Because sperm cannot penetrate this shell, fertilization occurs within the body of the female before the shell is added. In this process of internal fertilization, the male uses a copulatory organ to transfer sperm into the female reproductive tract.

As the embryo develops within the protective shell, an amnion forms and surrounds the embryo. The amnion secretes fluid, providing the embryo with its own private "pond." In addition to keeping the embryo moist, the amniotic fluid serves as a shock absorber, cushioning the egg.

The hard, dry, horny scales that protect the reptilian body from drying are another adaptation to life on land. This scaly protective armor, which also helps protect the reptile from predators, is shed periodically.

The dry reptilian skin cannot serve as an organ for gas exchange. Reptilian lungs are better developed than the saclike lungs of amphibians. Divided into many chambers, the reptilian lung provides a greatly increased surface area for gas exchange. Most reptiles have a three-chambered heart that is more efficient than the amphibian heart. The ventricle has a partition, though incomplete, that separates oxygen-rich and

Present

CENOZOIC	Tertiary	Mammals / Turtles / Snakes and Lizards / Crocodiles/Alligators / Birds
MESOZOIC	Cretaceous	THERAPSIDS (Mammal-like reptiles) / Plesiosaurs (swimming reptiles) / Ichthyosaurs (swimming reptiles) / Dinosaurs (Ornithischia) / Dinosaurs (Saurischia) / Pterosaurs (flying reptiles)
	Jurassic	
	Triassic	
PALEOZOIC	Permian	DIAPSIDS
	Carboniferous	ANAPSIDS / SYNAPSIDS

Stem
Reptiles

Figure 30–19 **Phylogeny of the amniotes.** *Some proposed evolutionary relationships among extinct and present-day reptiles, birds, and mammals.*

oxygen-poor blood, facilitating oxygenation of body tissues. In crocodiles the partition is complete, and the heart has four chambers.

Another adaptation to life on land is a method of metabolic waste disposal that conserves water. Much of the fluid filtered from the blood by the kidneys is reabsorbed in the kidney tubules and urinary bladder. In aquatic animals, nitrogenous wastes from protein and nucleic acid metabolism are excreted as ammonia. Because it is quite toxic, large amounts of water are needed to dilute it. Reptiles (and birds, as well as most terrestrial arthropods) convert ammonia to uric acid, which is much less toxic and can be excreted as relatively insoluble crystals. These solid crystals do not require much water for their excretion, and so water is conserved.

Like fishes and amphibians, reptiles generally lack metabolic mechanisms for regulating body temperature. They are **ectothermic,** meaning that their body temperature fluctuates with the temperature of the surrounding environment. Some reptiles have behavioral adaptations that enable them to maintain a body temperature higher than that of their environment. You may have observed a lizard basking in the sun, which raises its body temperature and so increases its metabolic rate. This permits the lizard to hunt actively for food. When the body of a reptile is cold, the metabolic rate is low and the animal tends to be sluggish. Ectothermia may explain why reptiles are more successful in warm than in cold climates.

Although a few species of turtles, tortoises, and lizards are herbivores (animals that eat vegetation), most reptiles are carnivores (animals that eat meat). Their paired limbs, usually with five toes, are well adapted for running and climbing, and their well developed sense organs enable them to locate prey.

Modern reptiles are assigned to three main orders

The extant reptiles are assigned to three main orders: order **Chelonia** includes turtles, terrapins, and tortoises; order **Squamata** is composed of lizards, snakes, iguanas, and geckos; and order **Crocodilia** contains crocodiles, alligators, caimans, and gavials (Fig. 30–20).

Members of order Chelonia are enclosed in a protective shell made up of bony plates overlaid by horny scales. Some terrestrial species can withdraw their heads and legs completely into their shells. The size of adult turtles ranges from about 8 cm (3 in) long to that of the great marine species, which measure more than 2 m (6.5 ft) in length and may weigh 450 kg (almost 1000 lb). The land species are usually referred to as tortoises, whereas the aquatic forms are called turtles; freshwater types are sometimes called terrapins.

Lizards and snakes are the most common of the modern reptiles. These animals have rows of scales that overlap like shingles on a roof, forming a continuous flexible armor that may be shed periodically. Lizards range in size from certain geckos, which may weigh as little as 1 g (0.035 oz), to the Ko-

(a) Order Chelonia

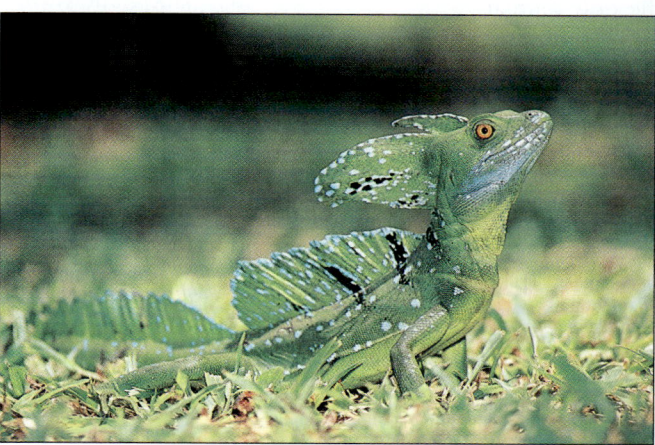

(b) Order Squamata

Figure 30–20 **Modern reptiles (class Reptilia).** *(a)* The paddlelike appendages of this green turtle *(Chelonia mydas)*, are adapted for swimming. *(b)* This basilisk lizard *(Basiliscus plumifrons)*, photographed in Costa Rica, appears to be thermoregulating. *(c)* This Nile crocodile *(Crocodilus niloticus)* is in the process of emerging from its leathery egg. *(a, Carlyn Iverson; b, Y. Lefevre/Peter Arnold, Inc.; c, Frans Lanting/Minden Pictures)*

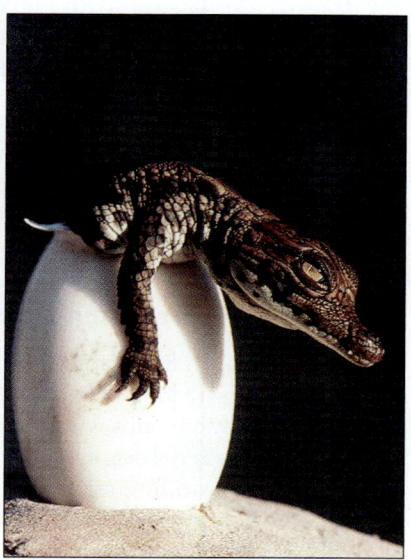

(c) Order Crocodilia

modo dragon of Indonesia, which may weigh 100 kg (220 lbs). Their body sizes and shapes vary greatly. Some, like the glass snake, which is really a lizard, are legless.

Snakes are characterized by a flexible, loosely jointed jaw structure that permits them to swallow animals larger than the diameter of their own jaws. Snakes lack legs, and their bodies are elongated. Their eyes, covered by a transparent cuticle, do not have moveable eyelids. Also absent are an external ear opening, a tympanic membrane (ear drum), and a middle ear cavity.

The forked tongue of a snake, which often darts quickly from its mouth, is used as an accessory sensory organ for touch and smell. Chemicals from the ground or air adhere to the tongue. The tip is then projected into a sense organ located in the roof of the mouth that detects odors. Pit vipers and some boas also have a prominent **sensory pit** on each side of the head that enables them to detect heat. These sense organs permit them to locate and capture small nocturnal mammals.

Some snakes, for example, king snakes, pythons, and boa constrictors, capture their prey by rapidly wrapping themselves around the animal and squeezing it so that it cannot breathe. Others have fangs, which are hollow teeth connected to venom glands in the mouth. When the snake bites, the venom is pumped through the fangs into the prey's body. Some snake venoms cause the breakdown of red blood cells; others, such as that of the coral snake, are neurotoxins that interfere with nerve function. Venomous snakes of the United States include rattlesnakes, copperheads, cottonmouths, and coral snakes. All except the coral snakes are pit vipers.

Three groups of crocodilians are: (1) the crocodiles of Africa, Asia, and America; (2) the alligators of the southern United States and China, plus the caimans of Central America; and (3) the gavials of Southeast Asia. Most species live in swamps, in rivers, or along sea coasts, feeding on various kinds of animals. Crocodiles are the largest living reptiles; some exceed 7 m (23 ft) in length. The crocodile can be distinguished from the alligator or caiman by its long, slender snout and by the large fourth tooth on the bottom jaw that is visible when the mouth is closed.

Birds are adapted for flight

Birds, which comprise class Aves, are the only animals with feathers (Fig. 30–21). Thought to have evolved from reptilian scales, feathers are very light, yet flexible and strong. They protect the body, decrease water and heat loss, and aid in flight by presenting a flat surface to the air.

Early birds had reptilian characteristics

Birds are thought to have evolved from saurischian dinosaurs, many of which were long-tailed animals that moved about on two feet and had forelimbs with three clawed fingers (see Fig. 20–12). Feathers may have first evolved as an adaptation that conserved body heat. Bird ancestors became endothermic and more active. According to one hypothesis, these animals ran along the ground, using their feathered forelimbs to swat insects, which they then ate. Another hypothesis holds that bird ancestors climbed trees and used their feathered forelimbs for gliding.

Although the bones of birds are fragile and disintegrate quickly, a few fossils of early birds have been found. The first birds looked very much like reptiles. They had teeth (which modern birds lack), a long tail, and bones with thick walls. Unlike reptiles, their jaws were elongated into beaks, and they had feathers and wings.

One of the earliest known birds, **Archaeopteryx** (meaning "ancient wing"), was about the size of a pigeon and had rather feeble wings. Its jawbones were armed with reptilian-type teeth, and it had a long reptilian tail covered with feathers. Each of its wings was equipped with three claw-bearing digits (Fig. 30–22). Several specimens of this genus have been found in the Jurassic limestone of Bavaria, which was laid down about 150 million years ago.

Cretaceous rocks have yielded fossils of other early birds. *Hesperornis*, which lived in North America, was a toothed, aquatic diving bird with powerful hind legs and vestigial wings. *Ichthyornis* was a toothed, flying bird about the size of a sea gull. In 1990 a fossil found in the rocky remnants of an ancient lake in China was described as the earliest known example of a bird with modern flying ability. This bird apparently had the adaptations necessary for living in trees. From the Tertiary period onward, the fossil record of birds shows an absence of teeth and progressive changes leading to the modern birds.

Modern birds are a very successful group

Even modern birds possess some characteristics in common with the reptiles. For example, they lay eggs and have reptilian-type scales on their legs. About 9000 species of birds have been described, and they have been classified into 30 orders. Birds inhabit a wide variety of habitats and can be found on all of the continents, most islands, and even the open sea. The largest living birds are the ostriches of Africa, which may be up to 2 m (6.5 ft) tall and weigh 136 kg (300 lb), and the great condors of the Americas, with wingspans of up to 3 m (10 ft). The smallest known bird is Helena's hummingbird of Cuba, with a length of less than 6 cm (2 in) and a weight of less than 4 g (0.14 oz).

The anterior limbs of birds are wings, usually modified for flight. The posterior limbs are modified for walking, swimming, or perching. Not all birds fly. Some, such as penguins, have small, flipper-like wings used in swimming. Others, such as the ostrich and cassowary, have only vestigial wings, but well developed legs.

In addition to feathers and wings, birds have many other adaptations for flight. Their bodies are compact and streamlined, and the fusion of many bones gives them the rigidity needed for flying. Their bones are strong but very light; many are hollow, containing large air spaces. The avian jaw is light,

(a)   *(b)*

Figure 30–21 **Modern birds (class Aves).** Although they are diverse, birds are highly adapted for flight and are structurally similar. *(a)* A male peacock *(Pavo cristatus)* impressively displaying his feathers. *(b)* This Atlantic puffin *(Fratercula arctica)* has caught several fish. *(c)* Cattle egrets *(Bubulcus ibis)* photographed in Kenya. *(a, Frans Lanting/ Minden Pictures; b, Gary Meszaros/Dembinsky Photo Associates; c, D. W. Fawcett)*

(c)

and, instead of teeth, there is a light, horny beak. The breastbone is broad and flat for the attachment of the large flight muscles.

Birds have efficient lungs with thin-walled extensions, called air sacs, that occupy spaces between the internal organs and within certain bones. Like mammals, birds have a four-chambered heart and a double circuit of blood flow. Blood delivers oxygen to the tissues and then is recharged with oxygen in the lungs before being pumped out into the systemic circulation again. The very effective respiratory and circulatory systems provide the cells with enough oxygen to permit a high metabolic rate, which is necessary for the tremendous muscular activity that flying requires. Some of the heat generated by metabolic activities is used to maintain a constant body temperature. Birds and mammals are **endotherms.** Their ability to maintain a constant body temperature permits birds to remain active in cold climates.

Birds excrete nitrogenous wastes mainly as semisolid uric acid. Because they lack a urinary bladder, these solid wastes are delivered into the cloaca. They leave the body with the feces, which are dropped frequently. This adaptive mechanism helps to maintain a light body weight.

Birds have become adapted to a variety of environments, and various species have very different types of beaks, feet, wings, tails, and behavioral patterns. Bills are specifically adapted for the type of food the bird eats. Although all birds must eat frequently (because they have a high metabolic rate and do not store much fat), the choice of food varies widely

Figure 30–22 *Archaeopteryx,* **a very early bird.** This reconstruction represents the hypothesis that *Archaeopteryx* was a climbing animal that had at least some ability to use its wings and feathers for gliding. Other hypotheses suggest that *Archaeopteryx* remained mainly on the ground, using its wings to trap small insects and its feathers for insulation. *(From a painting by Rudolph Freund, courtesy of Carnegie Museum of Natural History)*

heavily on vision, and their eyes are relatively larger than those of other vertebrates. Hearing is also well developed.

In striking contrast to the relatively silent reptiles, birds are very vocal. Most have short, simple calls that signal danger or influence feeding, flocking, or interaction between parent and young. Songs are usually more complex than calls and are performed mainly by males. Songs are related to reproduction, attracting and keeping a mate, and claiming and defending territory.

One of the most fascinating aspects of bird behavior is the annual migration that many species make. Some birds, such as the golden plover and Arctic tern, fly from Alaska to Patagonia, South America, and back each year, covering perhaps 40,250 km (25,000 mi) en route. Migration and navigation are discussed in Chapter 50.

Beautiful and striking colors are found among birds. Color is due partly to pigments deposited during the development of the feathers and partly to reflection and refraction of light of certain wavelengths. Many birds, especially females, are protectively colored by their plumage (see Chapter 52). Brighter colors are often assumed by the male during the breeding season to help in attracting a mate.

Mammals have hair and mammary glands

Distinguishing features of mammals (class Mammalia) include hair, which insulates and protects the body; **mammary glands,** which produce milk for the young; and the differentiation of teeth into incisors, canines, premolars, and molars. A muscular **diaphragm** helps to move air into and out of the lungs. Like birds, mammals are endotherms. The process of maintaining a constant body temperature is enhanced by the covering of insulating hair, by the four-chambered heart, and by separate pulmonary and systemic circulations. Red blood cells without nuclei serve as excellent oxygen transporters.

Contributing significantly to the success of the mammals, the complex nervous system is more highly developed than in any other group of animals. The cerebrum is especially large and complex, with an outer gray region called the cerebral cortex.

Fertilization is always internal, and, except for the primitive monotremes that lay eggs, mammals are viviparous. Most mammals develop a **placenta,** an organ of exchange between developing embryo and mother, through which the embryo receives its nourishment and oxygen and rids its blood of wastes. By carrying their developing young internally, mammals avoid the hazards of having their eggs consumed by predators. By nourishing the young and caring for them, the parents offer both protection and an "education" on how to obtain food and avoid being eaten.

The limbs of mammals are variously adapted for walking, running, climbing, swimming, burrowing, or flying. In four-legged mammals, the limbs are more directly under the body than they are in extant reptiles, which contributes to speed and agility. Life processes of mammals are discussed in detail in Part 7.

among species. Most birds eat energy-rich foods such as seeds, fruits, worms, mollusks, or arthropods. Warblers and some other species eat mainly insects. Owls and hawks eat rodents, rabbits, and other small mammals. Vultures feed on dead animals. Pelicans, gulls, terns, and kingfishers catch fishes. Some hawks catch snakes and lizards.

An interesting feature of the bird digestive system is the **crop,** an expanded, saclike portion of the digestive tract below the esophagus, in which food is temporarily stored. The stomach is divided into a **proventriculus,** which secretes gastric juices, and a thick, muscular **gizzard,** which grinds food. The bird swallows small bits of gravel that act as "teeth" in the gizzard, mechanically breaking down food.

Birds have a well developed nervous system with a brain that is proportionately larger than that of reptiles. Birds rely

Figure 30–23 Therapsid *(Lycaenops).* The therapsids were mammal-like reptiles. *Lycaenops* lived in the late Permian period in South Africa. *(Painting by John C. Germann, Department of Library Services, American Museum of Natural History)*

Many adaptations permitted the early mammals to compete for a place on our planet. Perhaps one of the most important was their skill at being inconspicuous. They were **arboreal** (tree-dwelling) and **nocturnal** (active at night), searching for food (mainly insects and plant material, and perhaps reptile eggs) at night while the reptiles were inactive. This lifestyle is suggested by the large eye sockets seen in fossil species, indicating that they possessed the large eyes characteristic of present-day nocturnal mammals.

As many reptiles died out, the mammals adapted to their abandoned niches (lifestyles). During this time, the flowering plants, including many trees, underwent adaptive radiation, providing new habitats, sources of food, and protection from predators. Larger forms and numerous varieties of mammals evolved. During the early Cenozoic era (more than 55 million years ago), the mammals underwent adaptive radiation, becoming widely distributed and adapted to an impressive variety of ecological lifestyles.

Early mammals were small, endothermic animals

Mammals are thought to have evolved from a group of reptiles called **therapsids** (Fig. 30–23) during the Triassic period some 200 million years ago. The therapsids were doglike carnivores with differentiated teeth (a mammalian trait) and legs adapted for running. The fossil record indicates that the early mammals were small, about the size of a mouse or shrew. Some may have been endothermic, and some may have had fur.

How did the mammals manage to coexist with the reptiles during the 160 million years or so that reptiles ruled Earth?

Modern mammals are assigned to three subclasses

By the end of the Cretaceous period, three main groups of mammals had evolved. Today mammals inhabit virtually every corner of Earth; they are found on land, in fresh and salt water, and in the air. They range in size from the tiny pigmy shrew, weighing about 2.5 g (less than 0.1 oz), to the blue whale, which may weigh more than 90,000 kg (88 tons) and is thought to be one of the largest animals that has ever lived.

Modern mammals are classified in three subclasses: **Prototheria** includes the egg-laying mammals, also called monotremes; **Metatheria** includes the marsupials, or pouched mammals; and **Eutheria** includes the placental mammals.

The **monotremes** are the only living order of subclass Prototheria. One genus includes the duck-billed platypus *(Ornithorhynchus)* and a second, the spiny anteater or echidna *(Tachyglossus)* (Fig. 30–24; see also Figs. 1–17 and 22–5).

Figure 30–24 Monotremes. The spiny anteater *(Tachyglossus aculeatus)* is an egg-laying mammal. *(Tom McHugh/Photo Researchers, Inc.)*

(a)

(b)

Figure 30–25 **Marsupials.** *(a)* Eastern gray kangaroo (*Macropus giganteus*) with joey. The kangaroo is native to Australia. *(b)* A kangaroo soon after birth. Marsupials are born still in an embryonic state and continue to develop in the safety of the marsupium (pouch). *(a, John Cancalosi/Peter Arnold, Inc.; b, Robert Anderson, reprinted with permission of Hubbard Scientific Company)*

These animals are found in Australia and Tasmania; the spiny anteater is also found in New Guinea. The females lay eggs that may be carried in a pouch on the abdomen or kept warm in a nest. When the young hatch, they are nourished with milk from the mammary glands. As their name suggests, spiny anteaters feed on ants, which they catch with their long, sticky tongues.

The duck-billed platypus lives in burrows along river banks. It has webbed feet and a flat, beaver-type tail, which aids in swimming. The duck-billed platypus preys on freshwater invertebrates.

Marsupials include pouched mammals such as kangaroos and opossums. Embryos begin their development in the mother's uterus, where they are nourished by fluid and yolk. After a few weeks, still in a very undeveloped stage, the young are born. In many species, the young crawl to the **marsupium** (pouch), where they complete their development. The young marsupial attaches its mouth to a mammary gland nipple and is nourished by its mother's milk (Fig. 30–25).

At one time, marsupials probably inhabited much of the world, but they were largely replaced by placental mammals. Now marsupials are found mainly in Australia and in Central and South America. The only marsupial species that ranges into the United States is the house-cat sized Virginia opossum (*Didelphis virginiana*). Australia became geographically isolated from the rest of the world before placental mammals reached it, and there the marsupials remained the dominant mammals (see Chapter 52). They underwent adaptive radiation, paralleling the evolution of placental mammals elsewhere. Thus, in Australia and adjacent islands we find marsupials that correspond to North American placental wolves, bears, rats, moles, flying squirrels, and even cats (Fig. 30–26).

Most familiar to us are **placental mammals,** characterized by development of a placenta, an organ of exchange between the developing embryo and its mother (Fig. 30–27). The placenta forms from both embryonic membranes and the maternal uterine wall. In it the blood vessels of the embryo come very close to the blood vessels of the mother, so materials can be exchanged by diffusion. (The two circulations do not normally mix.) The placenta enables the young to remain within the mother's body until embryonic development is complete.

Placental mammals are born at a more mature stage than are marsupials. Indeed, among some species the young can walk around and begin to interact with other members of the group within a few minutes after birth. Extant placental mammals are classified into about 16 orders. A brief summary of some of these orders is given in Table 30–3. The echinoderm and chordate phyla are compared in Table 30–4.

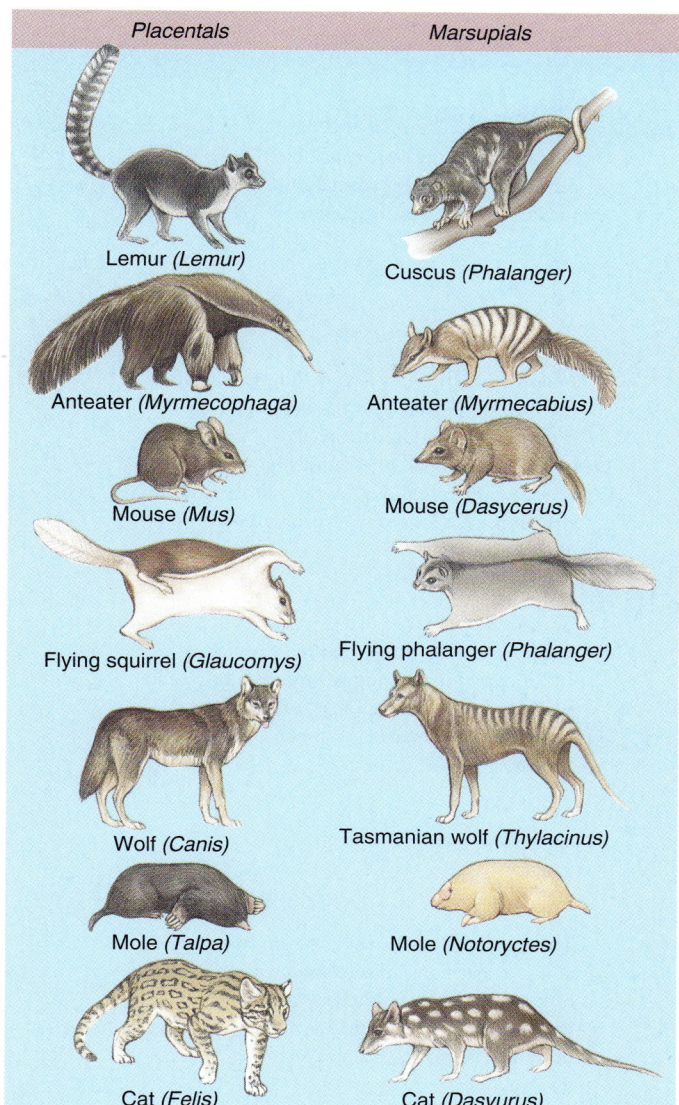

Placentals | Marsupials

Lemur (Lemur) — Cuscus (Phalanger)

Anteater (Myrmecophaga) — Anteater (Myrmecabius)

Mouse (Mus) — Mouse (Dasycerus)

Flying squirrel (Glaucomys) — Flying phalanger (Phalanger)

Wolf (Canis) — Tasmanian wolf (Thylacinus)

Mole (Talpa) — Mole (Notoryctes)

Cat (Felis) — Cat (Dasyurus)

▲ **Figure 30–26 Convergent evolution in placental and marsupial mammals.** For each mammal with a given niche in one group, there is a counterpart in the other group. Their similarities include both lifestyles and structural features.

(a)

(b)

▶ **Figure 30–27 Placental mammals.** *(a)* Wildebeest *(Connochaetes taurinus)* photographed in Tanzania. The wildebeest, the dominant plains antelope in many areas of eastern and southern Africa, migrates long distances during the dry season in search of food and water. *(b)* Humpback whales *(Megaptera novaeangliae)* exhibiting a rare double breach. *(c)* Polar bears *(Ursus maritimus)*, the largest living land carnivores, feed mainly on seals.   *(a, McMurray Photography; b, James D. Watt/Animals Animals; c, Michio Hoshino/Minden Pictures)*

(c)

TABLE 30-3 Some Orders of Living Placental Mammals

Order and Examples			Some Characteristics
Insectivora Moles, hedgehogs, and shrews		African hedgehog	Nocturnal; eat insects; considered most primitive placental mammals. Shrew is the smallest living mammal; some weigh less than 5 grams.
Chiroptera Bats		Bat	Adapted for flying; a fold of skin extends from the elongated fingers to the body and legs, forming a wing. Guided in flight by a type of biological sonar: they emit high-frequency squeaks and are guided by the echoes from obstructions. Eat insects and fruit, or suck blood of other animals.
Carnivora Cats, dogs, wolves, foxes, bears, otters, mink, weasels, skunks	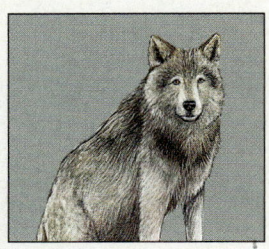	Wolf	Carnivores with sharp, pointed canine teeth and molars for shearing. Keen sense of smell; complex social interactions. Among fastest, strongest, and smartest animals.
Edentata Sloths, anteaters, armadillos		Nine-banded armadillo	Teeth reduced or no teeth. Sloths are sluggish animals that hang upside down from branches; often protectively colored by green algae that grow on their hair. Armadillos are protected by bony plates; eat insects and small invertebrates.
Rodentia Squirrels, beavers, rats, mice, hamsters, porcupines, guinea pigs		Flying squirrel	Gnawing animals with chisel-like incisors. As they gnaw, teeth are worn down, and so must grow continually.
Lagomorpha Rabbits, hares, pikas		Pika	Like rodents, have chisel-like incisors. Typically have long hind legs adapted for jumping. Many have long ears.

TABLE 30-3 (Continued)

Order and Examples			Some Characteristics
Primates Lemurs, monkeys, apes, humans		Ring-tailed lemur	Highly developed brains and eyes. Nails instead of claws. Opposable thumb. Eyes directed forward. Omnivores. Most species arboreal. (Primate evolution is discussed in Chapter 21.)
Perissodactyla Horses, zebras, tapirs, rhinoceroses		Tapir	Herbivores. Hoofed with an odd number of digits per foot; one or three toes. Teeth adapted for chewing. Usually large animals with long legs. (Hoofed mammals are referred to as ungulates.)
Artiodactyla Cattle, sheep, pigs, deer, giraffes		American elk	Hoofed with even number of digits per foot; most have two toes, some have four. Most have antlers or horns. Herbivores; most are ruminants that chew a cud and have a series of stomachs in which bacteria that digest cellulose are incubated; this contributes to their success as herbivores.
Proboscidea Elephants		African elephant	Largest land animals; weigh up to 7 tons; large head; broad ears; long, muscular trunk (proboscis) that is very flexible. Thick, loose skin characteristic. The two upper incisors are elongated as tusks. This order includes the extinct mastodons and wooly mammoths.
Sirenia Sea cows, manatees		Manatee	Herbivorous, aquatic mammals with finlike forelimbs and no hind limbs. They are probably the basis for most tales about mermaids.

TABLE 30-3   (Continued)

Order and Examples			Some Characteristics
Cetacea Whales, dolphins, porpoises		Humpback whale	Adapted for aquatic life with fish-shaped body and broad, paddle-like forelimbs (flippers). Posterior limbs absent. Many have thick layer of fat (blubber) under the skin. Some are suspension feeders. Mate and bear their young in the water; suckle young. Very intelligent.
Pinnipedia Seals, sea lions, walruses		Sea lions	Marine; limbs adapted as flippers for swimming. Carnivores; eat fish.

TABLE 30-4   Comparison of Echinoderms and Chordates*

	Echinodermata (Spiny-skinned animals) (6000 species)	Chordata (47,000 species)
Representative animals	Sea stars, sea urchins, sand dollars	Tunicates, lancelets, vertebrates
Body symmetry	Embryo: bilateral; adult: pentaradial	Bilateral
Gas exchange	Skin; gills	Gills or lungs; skin in some
Circulation	Open system; reduced	Closed system; ventral heart
Fluid balance/Waste disposal	Diffusion	Kidneys; also lungs, skin, gills
Nervous system	Nerve rings; no brain	Dorsal nerve cord with brain at anterior end
Reproduction	Sexual; sexes almost always separate	Sexual; sexes separate
Support and movement	Endoskeleton bearing spines; muscles; tube feet	Notochord; endoskeleton of cartilage and/or bone; well developed muscles
Environment and lifestyle	Marine; mainly carnivores	Diverse habitats and lifestyles; herbivores, carnivores, omnivores, scavengers, suspension feeders
Other characteristics	Water vascular system	Notochord; pharyngeal gill slits; postanal tail

*Members of these phyla are at the organ system level of organization and have a complete digestive tract.

I. **Deuterostomes** include **echinoderms, chordates, hemichordates,** and **chaetognaths.** Some biologists also classify the **lophophorate** phyla as deuterostomes.

II. Phylum **Echinodermata** includes marine animals with a spiny "skin," **water vascular system, tube feet,** and **endoskeleton.** The larvae have bilateral symmetry; most of the adults exhibit **pentaradial symmetry.**

   A. Class **Crinoidea** includes the sea lilies and feather stars. In these echinoderms, the **oral surface** is turned upward; some are sessile.

   B. Class **Asteroidea** consists of the sea stars, echinoderms with a central disk from which radiate five or more arms.

   C. Class **Ophiuroidea** includes the brittle stars, which resemble asteroids but have longer, more slender arms that are set off more distinctly from the central disk.

   D. Class **Echinoidea** includes the sea urchins and sand dollars, animals that lack arms; they have a solid shell and are covered with spines.

   E. Class **Holothuroidea** consists of sea cucumbers, animals with elongated flexible bodies; the mouth is surrounded by a circle of modified tube feet that serve as tentacles.

III. Phylum Chordata consists of three subphyla: **Urochordata, Cephalochordata,** and **Vertebrata.** At some time in its life cycle, a chordate has a **notochord,** a **dorsal tubular nerve cord,** and **pharyngeal gill slits.** Chordates are also characterized by a **postanal tail.**

   A. The **tunicates,** which belong to subphylum Urochordata, are sessile, suspension-feeding marine animals with tunics. Larvae have typical chordate characteristics and are free-swimming.

   B. Subphylum Cephalochordata consists of the **lancelets,** small, segmented, fishlike animals that exhibit chordate characteristics.

   C. Subphylum Vertebrata includes animals with a **vertebral column** forming the chief skeletal axis of the body. Vertebrates also have a **cranium,** pronounced cephalization, a complex brain, muscles attached to the endoskeleton for movement, and two pairs of appendages.

      1. Class **Agnatha,** the jawless fishes, includes the lampreys and hagfishes. Some systematists now divide the jawless fishes into two classes: Myxinoidea (hagfishes) and Petromyzontida (lampreys).

      2. Descendants of the **ostracoderms,** early jawless fishes that are now extinct, are thought to have given rise to the modern jawed fishes. Early jawed fishes included the **acanthodians** and **placoderms.**

      3. Class **Chondrichthyes,** the cartilaginous fishes, includes the sharks, rays, and skates. These fishes have jaws, two pairs of fins, and **placoid scales.** Skates and some species of sharks are **oviparous,** meaning that they lay eggs. Many species of sharks are **ovoviviparous;** their young are enclosed by eggs that are incubated in the mother's body. A few shark species are **viviparous;** the young develop in the mother's uterus and are nourished by transfer of nutrients from the mother's blood.

      4. Class **Osteichthyes,** the bony fishes, includes freshwater and saltwater fishes. The bony fishes and cartilaginous fishes are thought to have evolved at about the same time. Some biologists now divide Osteichthyes into three classes: Actinopterygii, ray-finned fishes; Actinistia, lobe-finned fishes; and Dipnoi, lungfishes.

         a. During the Devonian, bony fishes gave rise to two evolutionary lines: the **ray-finned fishes** and the **lobe-finned fishes.**

         b. The ray-finned fishes gave rise to the modern fishes. In these fishes, the lungs have been modified as a **swim bladder,** an air sac for regulating buoyancy.

         c. The lobe-finned fishes are thought to have given rise to the lungfishes and the group of lobe-finned fishes represented today by the **coelacanth.** Lobe-finned fishes were apparently preadapted for life on land, and they gave rise to the **tetrapods,** or land vertebrates.

      5. The first successful tetrapods were the now-extinct **labyrinthodonts.** These animals are thought to have been ancestors of the amphibians.

      6. Modern **amphibians** include salamanders, frogs and toads, and wormlike caecilians.

         a. Most amphibians return to the water to reproduce. Frog embryos develop into **tadpoles,** which undergo metamorphosis to become adults.

         b. Amphibians use their moist skin as well as lungs for gas exchange. They have a three-chambered heart and **systemic** and **pulmonary circulations.**

      7. Class **Reptilia** includes turtles, lizards, snakes, and alligators. The earliest reptiles were the **cotylosaurs.**

         a. Reptiles are **amniotes;** most are terrestrial.

         b. Fertilization is internal; most reptiles secrete a leathery protective shell around the egg; the embryo develops protective membranes, including an **amnion** that keeps it moist.

         c. A reptile has a dry skin with horny scales, lungs with many chambers, and a three-chambered heart with some separation of oxygen-rich and oxygen-poor blood.

         d. Reptiles dominated Earth during the Mesozoic era; then, toward the end of the Cretaceous period, many reptiles, including all of the dinosaurs, became extinct.

      8. Birds (class **Aves**) have many adaptations for flight, including feathers, wings, and light, hollow bones containing air spaces.

         a. Birds have a four-chambered heart, very efficient lungs, a high metabolic rate, and a constant body temperature; they excrete solid metabolic wastes (uric acid).

         b. Birds have a well developed nervous system and excellent vision and hearing.

         c. Birds communicate with simple calls and complex songs, as well as with color and behavior.

      9. **Mammals** have hair, **mammary glands,** and differentiated teeth. They maintain a constant body temperature and have a highly developed nervous system and a muscular **diaphragm.**

         a. **Monotremes,** mammals that lay eggs, include the duck-billed platypus and spiny anteaters.

         b. **Marsupials** include pouched mammals, e.g., kangaroos and opossums. The young are born in an embryonic stage and complete their development in the **marsupium,** where they are nourished with milk from the mammary glands.

         c. **Placental mammals** are characterized by an organ of exchange, the **placenta,** that develops between the embryo and the mother. Both oxygen and nutrients diffuse across the placenta from mother to embryo, permitting development to take place within the uterus. Living placental mammals are classified into about 16 orders.

1. Which of the following is NOT a deuterostome? (a) echinoderm (b) chordate (c) hemichordate (d) arthropod (e) lophophorate
2. Which of the following belongs to subphylum Vertebrata? (a) lophophorate (b) lamprey (c) lancelet (d) tunicate (e) echinoid
3. Which of the following are found in tunicates? (a) notochord (b) tube feet (c) anal gill slits (d) two pairs of appendages (e) vertebral column
4. Which of the following was an early jawed fish? (a) placoderm (b) crinoid (c) cotylosaur (d) labyrinthodont (e) ostracoderm.
5. An evolutionary sequence subscribed to by many biologists is (a) lancelet → ancestral chordate → tunicate (b) lobe-finned fish → labyrinthodont → modern amphibian (c) ancestral chordate → turtle→ cotylosaur→ amniote (d) therapsid → placental mammal → monotreme (e) dinosaur → *Archaeopteryx* → therapsid → marsupial.
6. Which of the following characteristics is NOT associated with sea stars? (a) tube feet (b) water vascular system (c) central disk with five or more arms (d) spiny skeleton (e) notochord
7. Which of the following characteristics is associated with amphibians? (a) mantle (b) placoid scales (c) three-chambered heart (d) swim bladder (e) amnion
8. Which of the following is NOT characteristic of class Aves? (a) feathers (b) ectothermic (c) amnion (d) high metabolic rate (e) reptilian–like scales on legs
9. Which of the following is NOT true of the duck-billed platypus? (a) classified as a marsupial (b) lays eggs and carries them in a pouch (c) notochord (d) pharyngeal gill slits (e) predator of freshwater invertebrates
10. Which of the following is NOT a placental mammal? (a) shrew (b) human (c) dolphin (d) opossum (e) bat
11. Which of the following is true of mammals? (a) all have placentas (b) evolved from saurischian dinosaurs (c) muscular diaphragm (d) most are ovoviviparous (e) evolved during the Devonian period
12. The vertebral column (a) is characteristic of all chordates (b) forms the skeletal axis of the chordate body (c) may consist of cartilaginous or bony segments (d) is well developed in lancelets (e) three of the preceding answers are correct
13. A shark is characterized by (a) placoid scales (b) bony skeleton (c) water vascular system (d) amnion (e) muscular diaphragm
14. Reptiles (a) have dry, scaly skin (b) are amniotes (c) are thought to have given rise to the birds and mammals (d) answers a, b, and c are correct (e) answers a and c only are correct

REVIEW QUESTIONS

1. Echinoderms and chordates are both deuterostomes. What are some characteristics that they have in common? What other phyla are classified as deuterostomes?
2. What are four principal distinguishing characteristics of a chordate? How are these evident in a tunicate larva? In an adult tunicate? In a lancelet? In a human?
3. What characteristics distinguish the vertebrates from the rest of the chordates?
4. How do lampreys and hagfishes differ from other fishes? Of what economic importance are agnathans?
5. Compare the skins of sharks, frogs, snakes, and mammals.
6. Identify organisms that possess each of the following, and give the location and function of each structure: (a) swim bladder (b) placenta (c) operculum (d) amnion (e) marsupium
7. Classify each of the following animals, giving the phylum, subphylum, class (and order if you can): (a) human (b) turtle (c) lamprey (d) *Branchiostoma* (amphioxus) (e) shark (f) whale (g) frog (h) pelican (i) bat
8. Which vertebrate groups maintain a constant body temperature? How do they accomplish this? Why is it advantageous?
9. From an evolutionary perspective, what is the significance of each of the following: (a) coelacanths; (b) placoderms; (c) labyrinthodonts (d) therapsids (e) *Archaeopteryx*
10. Label the diagram. Refer to Figure 30–5 to check your answers.

YOU MAKE THE CONNECTION

1. What is the function of gills? In general terms, how do they work? Why do you suppose aquatic mammals do not have them?
2. Some paleontologists consider monotremes to be therapsid reptiles rather than mammals. Give arguments for and against this position.
3. Which are more specialized, birds or mammals? Explain your answer.
4. Imagine that you discover an interesting new animal. You find that it has a dorsal tubular nerve cord, a cranium, moist skin, and a heart with two atria and a ventricle. How would you classify the animal? Explain each step in your decision.

RECOMMENDED READINGS

Blaustein, A.R. and David B. Wake. "The Puzzle of Declining Amphibian Populations," *Scientific American,* Vol. 272, No. 4, Apr. 1995. A look at contributing causes, such as habitat destruction, to decrease in amphibian populations.

Diamond, J. "Eat Dirt," *Discover,* Vol. 19, No. 2, Feb. 1998. A discussion of adaptations of parrots, one of the most successful groups of birds.

McClanahan, L.L., Ruibal, R. and V.H. Shoemaker. "Frogs and Toads in the Desert," *Scientific American,* Vol. 270, No. 3, Mar. 1994. A discussion of the adaptations that permit certain amphibians to inhabit desert areas.

Monastersky, R. "Jump-Start for the Vertebrates." *Science News,* Vol. 149, Feb. 3, 1996. Interesting discussion of the role of chromosome doubling in the evolution of the vertebrates.

Rismiller, P.D., and R. S. Seymour. "The Echidna." *Scientific American,* Vol. 264, No. 2, Feb. 1991. A discussion of the natural history and reproductive behavior of the spiny anteater, a mammal that lays eggs.

Zimmer, C. "Coming onto the Land." *Discover,* Vol. 16, No. 6, Jun. 1995. An interesting discussion of the evolution of terrestrial vertebrates.

● Visit our website at **http://www.saunderscollege.com/lifesci/titles.html** and click on Solomon/Berg/Martin Biology for links to chapter-related resources on the World Wide Web.

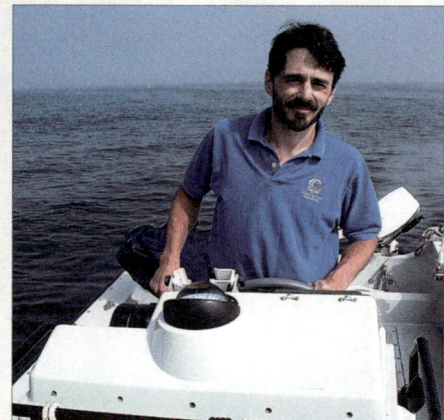

Aquarist/Nature Photographer

DAVID WROBEL

Dave Wrobel majored in biology and chemistry at State University of New York at Albany in upstate New York. He received his undergraduate degree in 1978. He entered the graduate program in Zoology and Ecology at Duke University and obtained his Master's Degree in 1981. His thesis involved the unusual life cycle of water fleas, important organisms in the food web of freshwater systems. Always interested in photography, he then worked as an electron microscopist where he learned technical aspects of photography. After moving to California, he obtained his scuba certification and took courses in underwater photography. Today, Dave's position as an aquarist of the jellyfish displays at the Monterey Bay Aquarium combines his knowledge of biological diversity and physiological systems with his love of nature and underwater photography. His photos have appeared in many magazines, as well as in zoology and biology textbooks. He is also co-author of the field guide, *Pacific Coast Pelagic Invertebrates: A Guide to the Common Gelatinous Animals.*

As an undergraduate, did you know what career you wanted to pursue?

I ended up at the Monterey Bay Aquarium through a very roundabout route. When I got out of college, I never could have guessed that I'd end up here.

As a college junior, I did a research project at Albany State on the behavioral differences between two species of salamanders. I've always had a love of animals, particularly the slimy kind. It was a small part of a broader research project run by

Professor Bob Jaeger. Working with him was an important part of my last couple of years as an undergraduate. I decided, based on that experience, that I wanted to go to graduate school to study zoology.

What brought you to California?

I took a job as an electron microscopist at the Duke Medical Center because I was interested in photography. Eventually, I got tired of being a technician, so I moved to California with the idea of going back to graduate school.

Then, I heard about the Brooks Institute in Santa Barbara, a well-known photography school. I thought I would try photography as a full time career. I took their underwater photography program in the summer of 1983. Just prior to that, I had received my scuba certification. I did a lot of diving that summer with the class.

How did you come to the Aquarium in Monterey?

In the Fall of 1983, the Monterey Bay Aquarium was new. I heard about it at the Brooks Institute. After doing a little research, I decided to move to Monterey. I did not really have any aquarium experience so at first I volunteered as a diver at the Aquarium, collecting, cleaning, and maintaining the larger exhibits. I also began working in the husbandry department where I currently work.

What are your current responsibilities as an aquarist?

I am now in charge of maintaining the jellyfish exhibits for the Aquarium. That involves acquiring the animals either by having them shipped or by collecting them, and then keeping them healthy in their displays. The Monterey Bay Aquarium is known for the jellyfish and their special tank, with its special water circulation system. The 2000 gallon jellyfish tank is the biggest jellyfish tank in the world. It's a lot of routine maintenance, particularly in the first few hours of any day, cleaning off the algae that could be growing on the surfaces and anything else that might be obstructing

the view. We prefer that the adult jellies do not produce their polyps in the display tank because it causes maintenance nightmares. Therefore, the reproductive process has to be closely watched. I also have to make sure they are getting a proper diet; therefore feeding often takes up a fair chunk of time.

Have you ever had disease problems with the jellies?

The time to be very vigilant is when we collect wild jellies. They may be carrying commensal organisms, creatures that are hitching rides, like small crabs or amphipods. We have to pick those off. They may not necessarily be parasitic, but they can be harmful to the jelly when they start gnawing on its delicate tissue.

What fascinates you about jellies?

I have come to appreciate their beauty and their mystery—the fact that they are here one minute and gone the next. For reasons that are not clearly understood, jellyfish can appear suddenly in huge swarms, and then just as suddenly seem to disappear. They just may be farther off shore or in deeper water. It's their drifting lifestyle that interests me.

What is the diversity of the jellyfish at the Aquarium?

We have 10 to 12 species on display, depending on their availability. We have sea nettles, which are big cnidarian jellies that have bells that are a foot and a half in diameter, and oral arms and tentacles that often trail ten or more feet. They are highly conspicuous, easily recognized as jellyfish. Most jellies, however, are fairly small and inconspicuous. We display some of those, certain hydromedusae, which are small, transparent, and camouflaged in their natural environment by their transparency.

How does your biology major help you as an aquarist and nature photographer?

A biology degree is a prerequisite for the aquarium job. It helps me understand the differences between jellies and all other animals that we have here. I have a deeper understanding of feeding biology and the reproductive biology that helps me maintain the exhibits. Any nature photographer who understands the relationships among organisms will take better pictures. Knowing some aspect of an animal's feeding behavior, camouflage strategy or defensive posture will make them more accessible photographic subjects. The photos will also be more eye-catching.

Structure and Life Processes in Plants

Crimson clover and lupine add color to a field in Del Norte Redwoods National Park, California. (Ron Thomas/FPG International)

CHAPTER 31

Plant Structure, Growth, and Differentiation

About 90 percent of the approximately 262,000 species of plants are flowering plants, which are vascular plants characterized by flowers, double fertilization, endosperm, and seeds enclosed within fruits (see Chapter 27). Because flowering plants comprise the largest, most successful group of plants, they are the focus of much of this chapter and of Chapters 32 to 36. Let us begin our study of the structure and functions of plants by examining a familiar species, the oak, and its fruit, the acorn.

Squirrels often temporarily store acorns in holes in the ground to provide winter food. Many of these are never retrieved, and so a new oak often begins its life after a squirrel has planted an acorn. The seed within the acorn first absorbs water from the surrounding soil. Then germination occurs as a root emerges and works its way down into the soil, absorbing additional water and anchoring the young plant. A miniature shoot begins to grow upward and breaks through the soil. At this point the young basket oak *(Quercus prinus)* shown in the photograph is called a seedling because it has just emerged from the seed and because it is still dependent on the food supply stored within the seed.

The stem continues to elongate, and small leaves develop, expand, and begin to photosynthesize. The young plant is now independently established: it is anchored in the ground, it absorbs water and dissolved nutrient minerals from the soil, and it uses energy from organic molecules, such as glucose, that it has produced by photosynthesis.

Smaller roots branch off the original root, and the tips of its shoot produce clusters of buds, dormant structures that develop into branches the following spring. The young tree grows taller, always by growth at the tips of its branches; the roots likewise elongate at their tips. As it ages, the tree also increases in girth (thickness) by forming additional wood and bark. This growth occurs along the sides of the more mature stems and roots. Thus, the oak progressively accumulates more wood and bark, more stem and root tissues, and more leaves.

Growth and expansion of both the root and shoot systems continue throughout the life of the oak. As in all plants, growth is flexible and dynamic, enabling the oak to respond to its environment. For example, the young oak may grow very slowly for several years, particularly if it is shaded by mature trees surrounding it. When conditions change, perhaps when an older tree nearby dies and falls to the ground, thereby permitting

(Runk/Schoenberger from Grant Heilman)

direct sunlight to reach the oak, the tiny tree is poised for rapid growth.

Oaks reproduce by forming separate male and female flowers in spring. The minute male flowers are more conspicuous than the female flowers because they are found together on slender, drooping, caterpillar-shaped catkins (clusters); in contrast, the tiny female flowers are often solitary. After pollen from the male flower is blown by the wind to the female flower, a sperm cell fertilizes the egg, and an acorn develops, partially enclosed by a scaly cup. First green in color, the acorn fruit turns a deep brown as it matures. Within the acorn is a seed containing an embryo of a miniature oak along with a supply of stored food. Thus the cycle of life repeats itself as squirrels busily collect these acorns and cache them for later retrieval.

In this chapter we examine the external structure of the flowering plant body; the organization of its cells, tissues, and tissue systems; and how it grows.

1. Discuss structural and functional aspects of the vascular plant body, including the nutrient and water-absorbing root system and the photosynthesizing shoot system.
2. Describe the ground tissue system (parenchyma tissue, collenchyma tissue, and sclerenchyma tissue).
3. Describe the structure and function of the vascular tissue system (xylem and phloem).
4. Describe the dermal tissue system (epidermis and periderm).
5. Discuss what is meant by growth in plants, and relate how it differs from growth in animals.
6. Distinguish between primary and secondary growth.

PLANTS EXHIBIT SIMILARITY AND DIVERSITY IN STRUCTURE AND LIFE SPAN

Remarkable variety is represented in the 235,000 or so species of flowering plants that live in and are adapted to the many environments offered by our planet. Yet all of these—from desert cacti with enormously swollen stems, to cattails partly submerged in marshes, to orchids growing in the uppermost branches of lush tropical rainforest trees—are recognizable as plants. From the tiny floating water-meal *(Wolffia microscopica),* the smallest flowering plant known, to Australian gum trees *(Eucalyptus* spp.), some of Earth's tallest trees, almost all plants have the same basic body plan, which consists of roots, stems, and leaves (Fig. 31–1).

Most plants are either herbaceous or woody. *Herbaceous* plants are nonwoody plants whose aerial parts (stems and leaves) die back to the ground at the end of the growing season. In contrast, the aerial parts of *woody* plants (trees and shrubs) persist. Botanically speaking, woody plants produce hard, lignified secondary tissues (i.e., cell walls contain lignin), and herbaceous plants do not. (The production of secondary tissues is discussed later in the chapter.)

Annuals are herbaceous plants (such as corn, geranium, and marigold) that grow, reproduce, and die in one year or less. Other herbaceous plants (such as carrot, Queen Anne's lace, cabbage, and foxglove) are **biennials** and take two years

(a)

(b)

Figure 31–1 **Size variation in flowering plants.** *(a)* Duckweeds (Spirodela sp.) are tiny floating aquatic plants about 1 cm (3/8 in) long. If you look closely at the frog's body, you'll see minute green dots. Each is a water-meal (*Wolffia* sp.) plant. These tiny floating herbs, about 1.5 mm (1/16 in) long, are the smallest known flowering plants. *(b)* This mountain ash (*Eucalyptus regnans*) was photographed in Bushy Park, Australia. Although most mountain ashes grow to 75 m (246 ft), some specimens have been measured at heights of 100 m (328 ft) and are the world's tallest flowering plants.   *(a, Carlyn Iverson; b, Joyce Photographers/Photo Researchers, Inc.)*

to complete their life cycles before dying. During their first season biennials produce extra carbohydrates, which they store and use during their second year when they typically form flowers and reproduce. **Perennials** are herbaceous and woody plants that have the potential to live for more than two years. In temperate climates, the aerial (above-ground) stems of herbaceous perennials such as iris, rhubarb, onion, and asparagus die back each winter. Their underground parts (roots and underground stems) become dormant during the winter and send out new growth each spring. (In dormancy, an organism reduces its metabolic state to a minimum level to survive unfavorable conditions.) Likewise, in certain tropical climates with pronounced wet and dry seasons, the aerial parts of herbaceous perennials die back and the underground parts become dormant during the unfavorable dry season. Other tropical plants, such as orchids, are herbaceous perennials that grow year-round.

All woody plants are perennials, and some of them live for hundreds or even thousands of years. In temperate climates, the above-ground stems of woody plants become dormant during the winter. Most temperate woody perennials are **deciduous,** that is, they shed their leaves before winter and produce new stems with new leaves the following spring. Other woody perennials are **evergreen** and shed their leaves over a long time period, so that some leaves are always present. Because they have permanent woody stems that are the starting points for new growth the following season, many trees attain massive sizes.

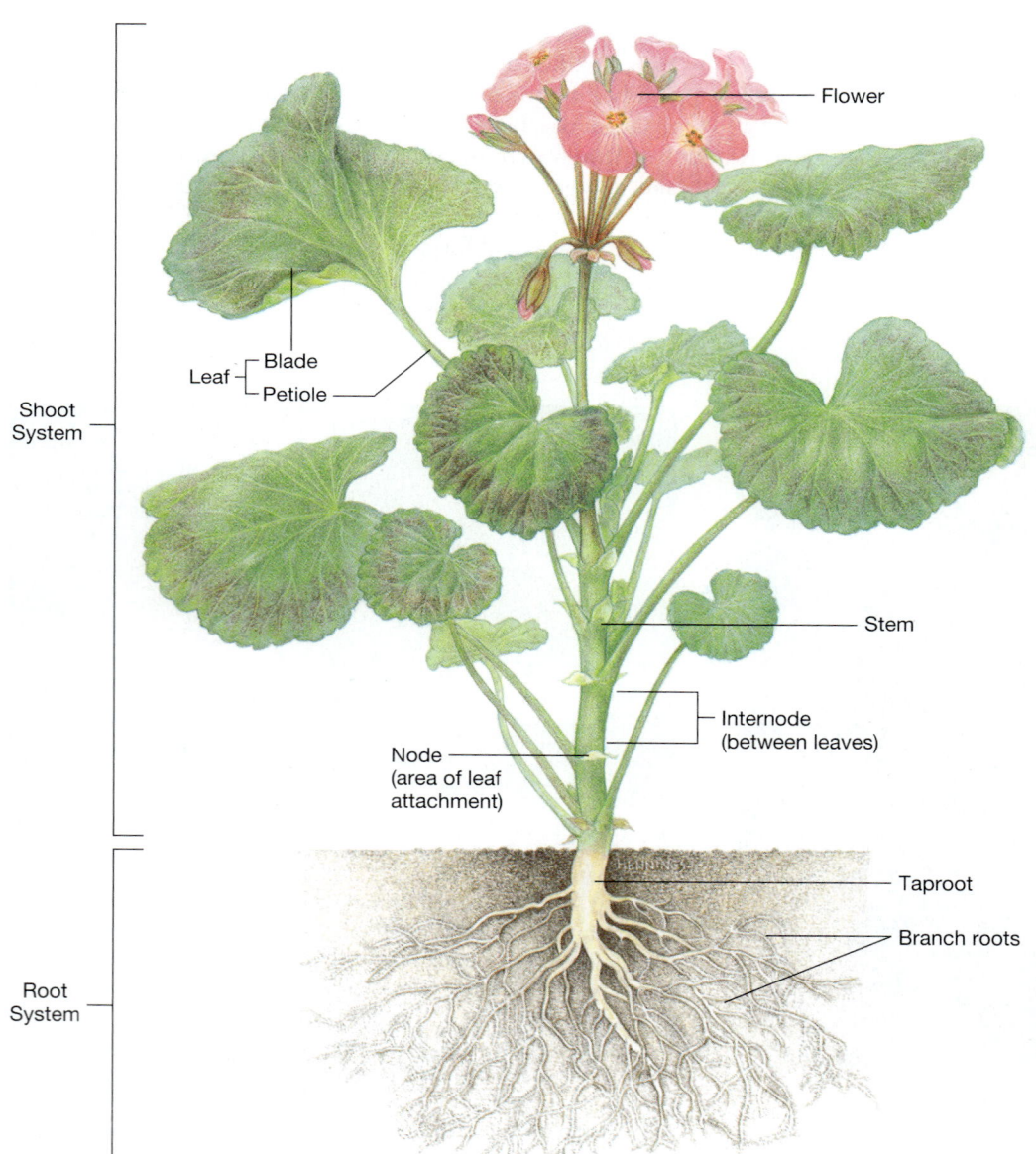

Figure 31–2 The geranium plant body. The plant body, which consists of a root system and a shoot system, possesses all of the necessary parts to survive and reproduce.

ROOTS, STEMS, LEAVES, FLOWERS, AND FRUITS MAKE UP THE PLANT BODY

The plant body of flowering plants (and other vascular plants) is usually organized into a root system and a shoot system (Fig. 31–2). The **root system** is generally the below-ground portion. The above-ground portion, the **shoot system,** generally consists of a vertical stem that bears leaves, and, in flowering plants, flowers and fruits that contain seeds.

Each plant typically grows in two different environments: the dark, moist soil and the illuminated, relatively dry air. Plants usually have both roots and shoots because they require resources from both environments. Thus, roots branch extensively through the soil, forming a network that anchors the plant firmly in place and absorbs water and dissolved nutrient minerals from the soil. Leaves, the flattened organs of photosynthesis, are attached more or less regularly on the stem, where they absorb the sun's light and atmospheric CO_2 used in photosynthesis.

THE PLANT BODY IS COMPOSED OF CELLS AND TISSUES

As in other organisms, the basic structural and functional unit of plants is the cell. During the course of evolution, plants have developed a diversity of cell types, each specialized for particular functions.

Like animal cells, plant cells are organized into tissues. A **tissue** is a group of cells that form a structural and functional unit. Some plant tissues, known as *simple tissues,* are composed of only one kind of cell, whereas other plant tissues, *complex tissues,* have two or more kinds of cells.

In vascular plants, tissues are organized into three tissue systems, each of which extends throughout the plant body. Each tissue system contains two or more kinds of tissues (Table 31–1). Most of the plant body is composed of the **ground tissue system,** which consists of three tissues that exhibit a variety of functions, including photosynthesis, storage, and support. The **vascular tissue system,** an intricate conducting system that extends throughout the plant body, is responsible for conduction of various substances, including water, dissolved nutrient minerals, and food (dissolved sugar). It also functions in strengthening and supporting the plant. The **dermal tissue system** provides a covering for the plant body.

Roots, stems, leaves, flower parts, and fruits are referred to as **organs** because each is composed of several different tissues. The tissue systems of different plant organs form an interconnected network throughout the plant. For example, the vascular tissue of a leaf is continuous with the vascular tissue of the stem to which it is attached, and the vascular tissue of the stem is continuous with the vascular tissue of the root.

The ground tissue system is composed of three simple tissues

The bulk of an herbaceous plant is its ground tissue system, which is composed of three tissues: parenchyma, collenchyma, and sclerenchyma (Table 31–2). These tissues can be distin-

---

TABLE 31–1   Tissue Systems, Tissues, and Cell Types of Flowering Plants

Tissue System	Tissue	Cell Types	Cell Geometry of Tissue System
Ground tissue system	Parenchyma tissue Collenchyma tissue Sclerenchyma tissue	Parenchyma cells Collenchyma cells Sclerenchyma cells (sclereids or fibers)	10–100 μm, relatively constant diameter
Vascular tissue system	Xylem	Tracheids Vessel elements Parenchyma cells Fibers	Very elongated and highly differentiated
	Phloem	Sieve tube members Companion cells Parenchyma cells Fibers	
Dermal tissue system	Epidermis	Parenchyma cells Guard cells Trichomes	Unusual asymmetrical shapes, flattened
	Periderm	Cork cells Cork cambium cells Cork parenchyma	

TABLE 31-2   Cell Types in the Ground Tissue System

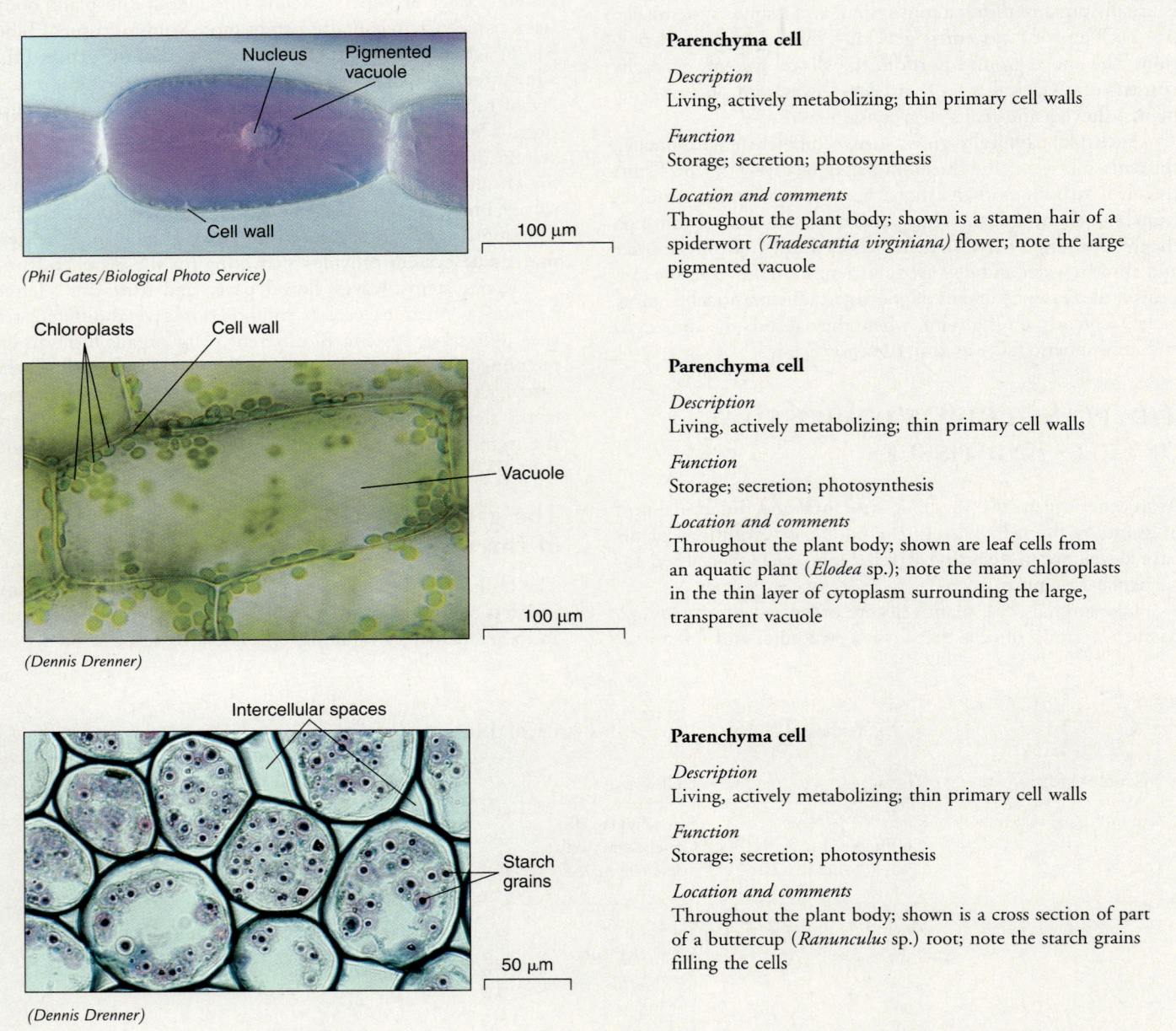

Parenchyma cell

Description
Living, actively metabolizing; thin primary cell walls

Function
Storage; secretion; photosynthesis

Location and comments
Throughout the plant body; shown is a stamen hair of a spiderwort *(Tradescantia virginiana)* flower; note the large pigmented vacuole

Labels on first image: Nucleus, Pigmented vacuole, Cell wall, 100 μm
(Phil Gates/Biological Photo Service)

Parenchyma cell

Description
Living, actively metabolizing; thin primary cell walls

Function
Storage; secretion; photosynthesis

Location and comments
Throughout the plant body; shown are leaf cells from an aquatic plant *(Elodea* sp.); note the many chloroplasts in the thin layer of cytoplasm surrounding the large, transparent vacuole

Labels on second image: Chloroplasts, Cell wall, Vacuole, 100 μm
(Dennis Drenner)

Parenchyma cell

Description
Living, actively metabolizing; thin primary cell walls

Function
Storage; secretion; photosynthesis

Location and comments
Throughout the plant body; shown is a cross section of part of a buttercup *(Ranunculus* sp.) root; note the starch grains filling the cells

Labels on third image: Intercellular spaces, Starch grains, 50 μm
(Dennis Drenner)

guished by their cell wall structures. Recall that plant cells are surrounded by a cell wall that provides structural support (see Chapter 4). A growing plant cell secretes a thin *primary cell wall*, which stretches and expands as the cell increases in size. After the cell stops growing, it sometimes secretes a thick, strong *secondary cell wall*, which is deposited *inside* the primary cell wall—that is, between the primary cell wall and the plasma membrane. (See *Making the Connection: Relating Cell Wall*

Chemistry to Different Cell Types for a discussion of how parenchyma, collenchyma, and sclerenchyma can be distinguished by their cell wall chemistry.)

Parenchyma cells have thin primary walls

Parenchyma tissue, a simple tissue composed of parenchyma cells, is found throughout the plant body and is the most

TABLE 31-2 (Continued)

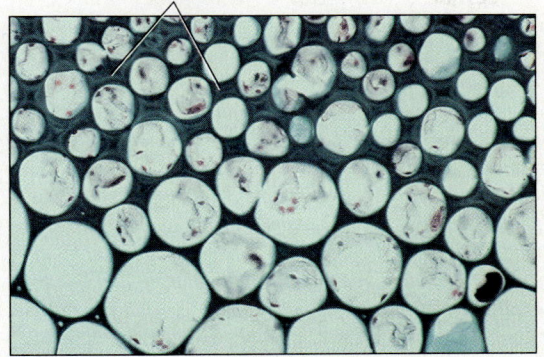

Primary cell walls are thickened in corners

50 μm

(James Mauseth, University of Texas)

Collenchyma cell

Description
Living; unevenly thickened primary cell walls

Function
Support

Location and comments
Just under stem epidermis; shown is a cross section of a water lily (*Nymphaea* sp.) petiole (leaf stalk); note the unevenly thickened cell walls that are especially thick in the corners, making the cell contents assume a spherical shape

Nucleus   Secondary cell walls   Cytoplasm   Pits

50 μm

(James Mauseth, University of Texas)

Sclereid

Description
May be living or dead at maturity; thick secondary cell walls; lacks secondary wall at pits

Function
Strength; sclereid-rich tissue is hard and inflexible

Location and comments
Shells of walnuts and coconuts; pits of cherries and peaches; shown are living sclereids in a section through a cherry (*Prunus avium*) pit

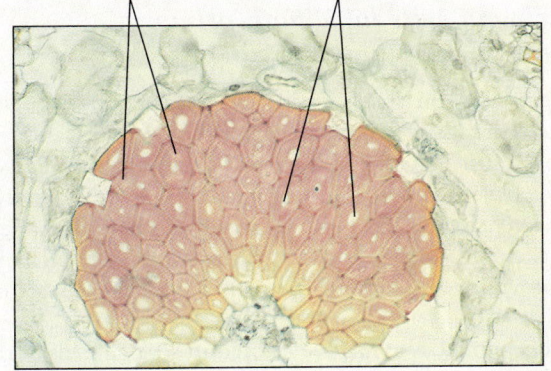

Thick secondary cell walls   Cavity where protoplast was when these cells were alive

50 μm

(James Mauseth, University of Texas)

Fiber

Description
Often dead at maturity; thick secondary cell walls; fewer pits than in sclereids

Function
Support; provides strength and elasticity

Location and comments
Throughout the plant body; common in stems and certain leaves; shown is a cross section through a clump of fibers from an *Agave* sp. leaf

RELATING CELL WALL CHEMISTRY TO DIFFERENT CELL TYPES

Can parenchyma, collenchyma, and sclerenchyma cells be distinguished by the chemistry of their cell walls? To answer this question, we must first examine the main chemical components of plant cell walls.

Cell walls may contain cellulose, hemicellulose, pectin, and lignin. **Cellulose,** the most abundant polymer in the world, accounts for about 40% to 60% of the dry weight of plant cell walls. As discussed in Chapter 3, cellulose is a polymer of glucose units joined by β-1,4 bonds. Each cellulose molecule consists of thousands of glucose subunits joined to form a flat, ribbon-like chain. From 40 to 70 of these chains lie parallel to one another and connect by hydrogen bonding to form a **cellulose microfibril,** a strong, tiny strand visible under the electron microscope (see Fig. 3–9*a*).

Cellulose microfibrils are cemented together by hemicelluloses and pectins. **Hemicelluloses** are a group of polysaccharides that vary in their chemical composition from one species to another. Despite their name, the chemical structure of the hemicelluloses is significantly different from that of cellulose. **Pectin,** another ce-

menting polysaccharide, is less variable in its monomer composition than are the hemicelluloses.

An important component of secondary plant cell walls, particularly those of wood, is **lignin** (see Chapter 26). Comprising as much as 35% of the dry weight of the secondary cell wall, lignin is a strengthening polymer composed of complex monomers derived from certain amino acids. The chemical structure of lignin has not been completely determined at this time.

Having examined the main chemicals in plant cell walls, we can now generalize about the cell chemistry of parenchyma, collenchyma, and sclerenchyma cells. The thin primary cell walls of parenchyma cells contain predominantly cellulose, although they also contain hemicelluloses and pectin. Both parenchyma and collenchyma cells have primary cell walls, but their walls are chemically distinct because the thickened areas of collenchyma walls contain large quantities of pectin, in addition to cellulose and hemicelluloses. The thick secondary walls of sclerenchyma cells are chemically different because they are rich in lignin, in addition to cellulose, hemicelluloses, and pectin.

common type of cell and tissue. The soft parts of a plant, such as the edible part of an apple or a potato, consist largely of parenchyma.

Parenchyma cells perform a number of important functions for plants, such as photosynthesis, storage, and secretion. Parenchyma cells that function in photosynthesis contain green chloroplasts, whereas nonphotosynthetic parenchyma cells lack chloroplasts and are often colorless. Materials stored in parenchyma cells include starch grains, oil droplets, water, and salts (sometimes visible as crystals). The various functions of parenchyma require that they be living, metabolizing cells.

Parenchyma cells have the ability to differentiate into other kinds of cells, particularly when a plant has been injured. If xylem (water-conducting) cells are severed, for example, adjacent parenchyma cells may divide and differentiate into new xylem cells within a few days. (Recall from Chapter 16 that it is possible to induce certain plant cells to become the equivalent of embryonic cells that can then differentiate into specialized cells.)

Collenchyma cells have unevenly thickened primary walls

Collenchyma tissue, a simple plant tissue composed of collenchyma cells, is an extremely flexible structural tissue that provides much of the support in soft, nonwoody plant organs. Support is a crucial function in plants, in part because it allows them to grow upward, thus enabling them to compete with other plants for available sunlight in a plant-crowded area.

Plants lack the bony skeletal system that is typical of many animals; instead, the plant body is supported by individual cells, including collenchyma cells.

Collenchyma cells, which are usually elongated, are alive at maturity. Their primary cell walls are unevenly thickened and are especially thick in the corners. Collenchyma is not found uniformly throughout the plant and often occurs as long strands near stem surfaces and along leaf veins. The "strings" in a celery stalk, for example, consist of collenchyma.

Sclerenchyma cells have both primary walls and thick secondary walls

Another simple plant tissue specialized for structural support is **sclerenchyma** tissue, whose cells have both primary and secondary cell walls. The root of the word *sclerenchyma* is derived from a Greek word *(sclero)* meaning "hard." The secondary cell walls of sclerenchyma cells become strong and hard due to extreme thickening. At functional maturity, when sclerenchyma tissue is providing support for the plant body, its cells are often dead.

Sclerenchyma tissue may be located in several areas of the plant body. Two types of sclerenchyma cells are sclereids and fibers. **Sclereids** are short, cubical cells common in the shells of nuts and in the pits of stone fruits such as cherries and peaches. Pears owe their slightly gritty texture to the presence of sclereids. **Fibers,** which are long, tapered cells that often occur in patches or clumps, are particularly abundant in the wood and inner bark of flowering plants.

The vascular tissue system consists of two complex tissues

The vascular tissue system, which is embedded in the ground tissue, transports needed materials throughout the plant via two complex tissues: xylem (pronounced zye'-lem) and phloem (pronounced flo'-em) (Table 31–3). Both xylem and phloem are continuous throughout the plant body. (Chapter 33 discusses the mechanisms of transport in xylem and phloem.)

The conducting cells in xylem are tracheids and vessel elements

Xylem conducts water and dissolved nutrient minerals from the roots to the stems and leaves and provides structural support. In flowering plants, xylem is a complex tissue composed of four different cell types: tracheids, vessel elements, parenchyma cells, and fibers. Two of the four cell types found in xylem, the **tracheids** and **vessel elements,** actually conduct water and dissolved nutrient minerals. In addition to these cells, xylem also contains parenchyma cells, known as *xylem parenchyma,* that perform storage functions, and fibers that provide support.

Tracheids and vessel elements are highly specialized for conduction. When mature, both cell types are dead and therefore hollow; only their cell walls remain. Tracheids, the chief water-conducting cells in gymnosperms (such as pine) and seedless vascular plants (such as ferns), are long, tapering cells located in patches or clumps. Water is conducted upward, from roots to shoots, passing from one tracheid into another through *pits,* thin areas in the tracheids' cell walls where a secondary wall did not form.

In addition to a relatively few tracheids, flowering plants possess extremely efficient water-conducting cells called vessel elements. The cell diameters of vessel elements are usually greater than those of tracheids. Vessel elements are hollow, but unlike tracheids, the end walls either have holes, known as *perforations,* or the end walls may be entirely dissolved away. Vessel elements are stacked one on top of the other, and water is conducted readily from one vessel element into the next. A stack of vessel elements, called a *vessel,* resembles a miniature water pipe. Vessel elements also have pits that permit the lateral transport (sideways movement) of water from one vessel to another.

Sieve tube members are the conducting cells of phloem

Phloem conducts food materials, that is, carbohydrates formed in photosynthesis, throughout the plant and provides structural support. In flowering plants, phloem is a complex tissue composed of four different cell types: sieve tube members, companion cells, fibers, and phloem parenchyma. Fibers are frequently extensive in the phloem of herbaceous plants, providing additional structural support for the plant body.

Food materials are conducted *in solution,* that is, dissolved in water, through the **sieve tube members,** which are among the most specialized cells in nature. Sieve tube members are joined end-on-end to form long sieve tubes. The cells' end walls, called *sieve plates,* have a series of holes through which cytoplasm extends from one sieve tube member into the next. Sieve tube members are living at maturity, but many of their organelles, including the nucleus, vacuole, mitochondria, and ribosomes, disintegrate or shrink as they mature.

Sieve tube members are among the few eukaryotic cells that can function without nuclei. These cells typically live for less than a year. There are, however, notable exceptions: certain palms have sieve tube members that have remained alive approximately 100 years!

Adjacent to each sieve tube member is a **companion cell** that assists in the functioning of the sieve tube member. The companion cell is a living cell, complete with a nucleus. This nucleus is thought to direct the activities of both the companion cell and sieve tube member. Numerous **plasmodesmata**—cytoplasmic connections through which cytoplasm extends from one cell to another (see Chapter 5)—occur between a companion cell and its sieve tube member. The companion cell plays an essential role in moving sugar into the sieve tube members for transport to other parts of the plant.

The dermal tissue system consists of two complex tissues

The dermal tissue system, the epidermis and periderm, provides a protective covering over plant parts (Table 31–4). In herbaceous plants, the dermal tissue system is a single layer of cells called the epidermis. Woody plants initially produce an epidermis, but it splits apart as the plant increases in girth as it produces additional woody tissues. Periderm, a tissue several to many cell layers thick, forms under the epidermis to provide a new protective tissue. Periderm, which replaces the epidermis in the stems and roots of older woody plants, composes the outer bark.

Epidermis is the outermost layer of cells of a herbaceous plant

The **epidermis** is a complex tissue composed mostly of parenchyma cells with scattered guard cells and outgrowths called trichomes (discussed shortly). In most plants, the epidermis consists of a single layer of flattened cells. Epidermal parenchyma cells generally contain no chloroplasts and are therefore transparent, allowing light to penetrate into the interior tissues of stems and leaves. In both stems and leaves, photosynthetic tissues lie *beneath* the epidermis.

An important requirement of the aerial parts (stems and leaves) of a plant is the ability to control water loss. Epidermal cells of aerial parts secrete a waxy layer called a **cuticle** over the surface of their exterior walls; this wax layer greatly restricts the loss of water from plant surfaces.

Although the cuticle is extremely efficient at preventing most water loss through epidermal cells, it also prevents the carbon dioxide required for photosynthesis from diffusing

TABLE 31-3 Selected Cell Types in the Vascular Tissue System

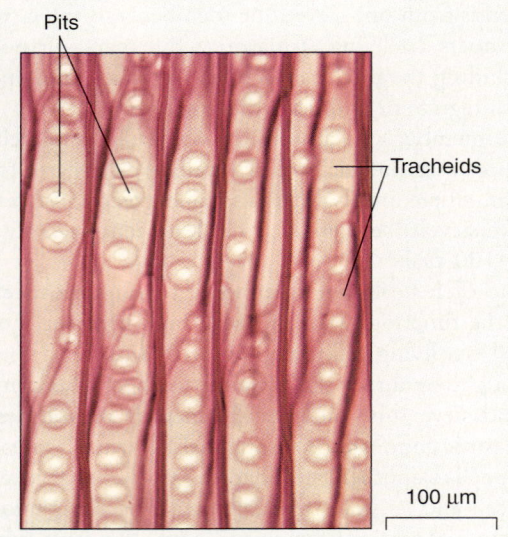

Pits

Tracheids

100 μm

(John D. Cunningham/Visuals Unlimited)

Tracheid

Description
Dead at maturity; lacks secondary wall at pits

Function
Conduction of water and minerals

Location and comments
Occurs in clumps in xylem throughout plant body; shown is a longitudinal section of tracheids from white pine *(Pinus strobus)* wood

Fibers

Vessel element dotted with numerous pits

Perforation

Vessel element dotted with numerous pits

100 μm

(James Mauseth, University of Texas)

Vessel element

Description
Dead at maturity; end walls have perforations; lacks secondary wall at pits

Function
Conduction of water and minerals

Location and comments
Xylem throughout plant body; vessel elements are more efficient than tracheids in conduction; shown is a longitudinal section of two vessel elements from an unknown woody dicot

from the atmosphere into the leaf or stem. The diffusion of carbon dioxide is facilitated by **stomata** (singular, *stoma*). Stomata are tiny pores in the epidermis surrounded by two cells called **guard cells.** Many gases, including carbon dioxide, oxygen, and water vapor, pass through the stomata by diffusion. Stomata are generally open during the day when photosynthesis is occurring, and the loss of water that also takes place when stomata are open provides some evaporative cooling. During the night, stomata usually close. During drought conditions, the need to conserve water overrides the need to cool the leaves and exchange gases. Thus, during a drought, the stomata close in the daytime. Stomata are discussed in greater detail in Chapter 32.

The epidermis may also contain special outgrowths, or hairs, called **trichomes,** which occur in many sizes and shapes and have a variety of functions. Root hairs are simple, unbranched trichomes that increase the surface area of the root epidermis (which comes into contact with the soil) for more effective water and mineral absorption. Plants that can tolerate salty environments such as the seashore often have specialized trichomes on their leaves to remove excess salt that accumulates in the plant. The presence of trichomes on the aerial

TABLE 31-3 (Continued)

Sieve tube member

Sieve plate

Sieve tube member

50 μm

(James Mauseth, University of Texas)

Sieve tube member

Description
Living but lacks nucleus and other organelles at maturity; end walls are sieve plates

Function
Conduction of sugar in solution

Location and comments
Phloem throughout plant body; shown is a longitudinal section through a clump of sieve tube members in a squash (*Cucurbita* sp.) petiole (leaf stalk)

Sieve tube members

Companion cells

Sieve plate

50 μm

(J. Robert Waaland/Biological Photo Service)

Companion cell

Description
Living; has cytoplasmic connections with sieve tube member

Function
Assists in moving sugars into and out of sieve tube member

Location and comments
Phloem throughout plant body; shown is phloem from a squash (*Cucurbita* sp.) petiole (leaf stalk) in cross section

parts of desert plants may increase the reflection of light off the plants, thereby keeping the internal tissues cooler and decreasing water loss. Other trichomes have a protective function. For example, the trichomes on stinging nettle leaves and stems contain irritating substances that may discourage herbivorous animals from eating the plant.

Epidermis is replaced by periderm in woody plants

As a woody plant begins to increase in girth, its epidermis is sloughed off and replaced by **periderm.** Periderm forms the outer bark of older stems and roots. It is a complex tissue composed mainly of cork cells and cork parenchyma cells. **Cork cells** are dead at maturity, and their walls are heavily coated with a waterproof substance called *suberin,* which helps to reduce water loss. **Cork parenchyma** cells function primarily in storage.

PLANTS EXHIBIT LOCALIZED GROWTH AT MERISTEMS

Growth is a complex phenomenon involving three different processes: cell division, cell elongation (the lengthening of a cell), and cell differentiation. Cell division is an essential part of growth that results in an increase in the number of cells. These new cells elongate as the cytoplasm grows and the vacuole fills with water, which exerts pressure on the cell wall and causes it to expand. In an onion root cell, the vacuole increases in size by some 30 to 150 times during elongation. Plant cells also **differentiate,** or specialize, into the various cell types just discussed. These cell types constitute the mature plant body and perform the various functions required in a multicellular organism. Although differentiation does not contribute to an increase in size, it is considered an important aspect of growth because it is essential for tissue formation.

TABLE 31–4 Selected Cell Types in the Dermal Tissue System

Epidermis   Cuticle

Collenchyma

100 μm

(James Mauseth, University of Texas)

Epidermal cell (ground parenchyma)

Description
Living parenchyma cell with thin primary wall; outer wall often thicker and covered by a noncellular waxy layer (cuticle)

Function
Protective covering over surface of plant body; helps reduce water loss

Location and comments
Epidermis is usually one cell thick; shown is a cross section through epidermis of ivy *(Hedera helix)* stem

Guard cell
Stoma
Epidermal cell
Guard cell

25 μm

(Dwight R. Kuhn)

Guard cell

Description
Chloroplast-containing cell that occurs in pairs; pair changes shape to open and close stomatal pore

Function
Opens and closes stomatal pore

Location and comments
Epidermis of stems and leaves; shown is epidermis of spiderwort *(Tradescantia virginiana)* leaf

One difference between plants and animals is the *location* of growth. When a young animal is growing, all parts of its body grow, although not necessarily at the same rate. However, when plants grow, their cells divide only in specific areas, called **meristems,** which are composed of cells that do not differentiate. Meristematic cells retain the ability to divide by mitosis, a trait that many differentiated cells lose. The persistence of mitotically active meristems means that plants, unlike most animals, retain the capability for growth throughout their entire life spans.

Two kinds of meristematic growth may occur in plants. **Primary growth** refers to an increase in stem and root length. All plants exhibit primary growth, which produces the entire plant body in herbaceous plants and the young, soft shoots and roots in woody trees and shrubs. **Secondary growth** refers to an increase in the girth of a plant. For the most part, only gymnosperms and woody dicots have secondary growth.[1] Tis-

sues produced by secondary growth comprise the wood and bark, which make up most of the bulk of trees and shrubs. A few annuals, geranium and sunflower, for example, have limited secondary growth despite the fact that they lack obvious wood and bark tissues. (See *Making the Connection: Plant Life History Strategies* for a brief discussion of the relative ecological advantages of being either a woody perennial or an herbaceous annual.)

Primary growth takes place at apical meristems

Primary growth occurs as a result of the activity of **apical meristems,** areas located at the tips of roots and within the buds of stems. Such growth is evident when a root tip (Fig. 31–3) is examined. The root tip is covered by a protective layer of cells called a root cap. Directly behind the root cap is the root apical meristem, which consists of meristematic cells. Meristematic cells, which are quite small and cubical in shape, remain small because they are continually dividing.

Further from the root tip, just behind the area of cell division, is an area of cell elongation where the cells are no longer

[1]Recall from Chapter 27 that, on the basis of structural features, flowering plants are divided into two groups, informally called dicots and monocots. Oak, sycamore, ash, cherry, apple, and maple are examples of woody dicots, whereas bean, daisy, and snapdragon are examples of herbaceous dicots. Palm, corn, bluegrass, lily, and tulip are examples of monocots.

TABLE 31–4 (Continued)

Trichome

Description
Hair or other epidermal outgrowth; may be single-celled or multicellular; occurs in variety of sizes and shapes

Function
Varied: absorption; secretion; excretion; protection; reduces water loss

Location and comments
Epidermis; shown is an SEM of a nettle *(Solanum carolinense)* leaf, which has trichomes that break off inside the skin of animals and inject irritating substances that cause stinging sensation

Trichomes

Epidermis

500 μm

(Biophoto Associates)

Cork cell

Description
Dead at maturity; cell walls are impregnated with waterproof materials

Function
Reduces water loss and prevents disease-causing organisms from penetrating

Location and comments
Produced in large numbers; cork often forms just under the epidermis; replaces epidermis in older stems and roots; shown is a cross section through periderm of a geranium *(Pelargonium sp.)* stem

Remnants of epidermis

Cork cells

Cork cambium

Cork parenchyma

Periderm

50 μm

(Dennis Drenner)

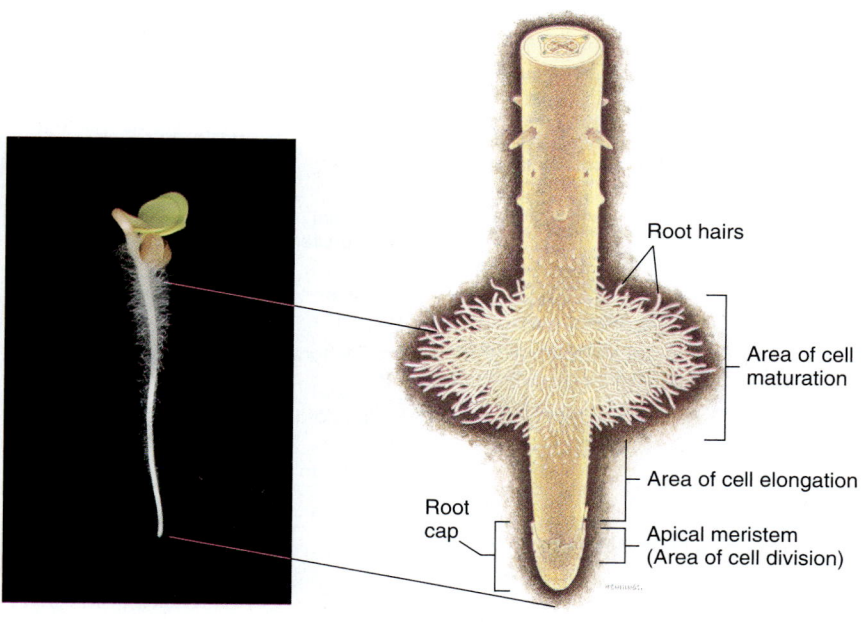

Root hairs

Area of cell maturation

Area of cell elongation

Root cap

Apical meristem (Area of cell division)

Figure 31–3 A root tip. The root apical meristem (where cells divide and thus increase in number) is protected by a root cap. Farther from the tip is an area of cell elongation, where cells enlarge and begin to differentiate. The area of cell maturation has fully mature, differentiated cells. Note the young root hairs in this area and on the root of a young radish *(Raphanus sativus)* seedling *(left).   (Dennis Drenner)*

PLANT LIFE HISTORY STRATEGIES

Under what conditions is it more favorable for individuals of a species to be long-lived or short-lived? Woody perennials often live for hundreds of years, whereas some herbaceous annuals may live for only a few weeks or months. Such characteristic features of an organism's life cycle, particularly as they relate to its survival and successful reproduction, are known as **life history strategies.** Biologists try to understand the relative advantages of each life history strategy. It appears that in some environments a longer life span is advantageous, whereas in others a shorter life span actually increases a species' chances for reproductive success.

When an environment is relatively favorable, it is filled with plants competing for available space. Because such an environment is so crowded, it has few open spots in which new plants can become established. When a plant dies, the empty area is quickly filled by another plant, but not necessarily by the same species as before. Thus, an adult perennial survives well in such a habitat, but young plants, whether perennials or annuals, do not. A plant with a long life span thrives in this type of environment because it can occupy a piece of soil and continue to produce seeds for many

years. In a tropical rain forest, for example, competition prevents most young plants from becoming established; hence woody perennials predominate.

In a relatively unfavorable environment, many possible sites are usually available. This type of environment is not crowded, and young plants usually don't have to compete against large, fully established plants. Here, smaller, short-lived plants have the reproductive advantage. These plants are opportunists: they grow and mature quickly during the brief periods when environmental conditions are most favorable. As a result, all of their resources are directed into producing as many seeds as possible before dying. In deserts following a rainy spell, for example, annuals are more prevalent than woody perennials.

Thus, each species has its own characteristic life history strategy, with some plants adapted to variable environments and others adapted to stable environments. The longer life span characteristic of woody perennials is just one of several successful life history plans. We return to life history strategies in our discussion of population ecology (see Chapter 51).

---

dividing but instead are growing longer, pushing the root tip deeper into the soil. Some differentiation also occurs in the area of cell elongation, and immature tissues, such as differentiating xylem and phloem, become evident. The tissue systems continue to develop and differentiate into primary tissues (epidermis, xylem, phloem, and ground tissues) of the adult plant. Further from the tip, following the area of cell elongation, the cells have completely differentiated and are fully mature. Root hairs are evident in this area.

A stem bud, the terminal bud, for example, is quite different in appearance from a root tip (Fig. 31–4). A dome of tiny, regularly arranged meristematic cells—the stem apical meristem—is located within every bud. *Leaf primordia* (developing leaves) and *bud primordia* (developing buds) arise from the stem apical meristem. The tiny leaf primordia tend to cover and protect the stem apical meristem. As the cells formed by the stem apical meristem elongate, the stem apical meristem is pushed upward. Subsequent cell divisions produce

Coleus

Older leaf

Leaf primordia

Apical meristem

Older leaf

Trichome

Bud primordium

100 μm

Figure 31–4 **A terminal bud.** LM of a longitudinal section through a terminal bud of coleus (*Coleus* sp.) showing the stem apical meristem, leaf primordia, and bud primordia. *(James Mauseth, University of Texas)*

additional stem tissue and cause new leaf and bud primordia to appear. Further from the stem tip, the immature cells differentiate into the three tissue systems of the mature plant body.

Secondary growth takes place at lateral meristems

In addition to primary growth, trees and shrubs have secondary growth. These plants increase in length by primary growth and increase in girth by secondary growth. The increase in girth, which occurs in areas that are no longer elongating, is due to cell divisions that take place in **lateral meristems,** areas extending the entire length of the stems and roots, except at the tips. Two lateral meristems, the vascular cambium and the cork cambium, are responsible for secondary growth, which is the formation of secondary tissues: secondary xylem, secondary phloem, and periderm (Fig. 31–5).

The **vascular cambium** is a layer of meristematic cells that forms a long, thin cylinder within the stem and root trunk. It is located between the wood and bark of a woody plant. Cells of the vascular cambium divide, adding more cells to the wood (secondary xylem) and inner bark (secondary phloem).

The **cork cambium** is composed of a thin cylinder or irregular arrangement of meristematic cells and is located in the outer bark. Cells of the cork cambium divide and form the cork cells and cork parenchyma cells that make up the periderm (discussed previously).

We are now ready to give a more precise definition of bark. **Bark,** the outermost covering over woody stems and roots,

Outer bark (periderm)
Inner bark (secondary phloem)
Bark
Wood (secondary xylem)
Vascular cambium

Figure 31–5 **Secondary growth.** The vascular cambium, a thin layer of cells sandwiched between the wood and bark, produces the secondary vascular tissues: the wood, which is secondary xylem, and the inner bark, which is secondary phloem. The cork cambium produces the periderm, the outer bark tissue that replaces the epidermis in the secondary plant body.

consists of all plant tissues located outside the vascular cambium. Bark has two regions, a living inner bark composed of secondary phloem and a mostly dead outer bark composed of periderm. A more comprehensive discussion of secondary growth is given in Chapters 33 and 34.

SUMMARY WITH KEY TERMS

I. The vascular plant body typically consists of a root system and a shoot system.
  A. The **root system** is generally underground and obtains water and dissolved nutrient minerals for the plant. Roots also anchor the plant firmly in place.
  B. The **shoot system** is generally aerial and obtains sunlight and exchanges gases such as carbon dioxide and oxygen.
    1. The shoot system consists of a vertical stem that bears leaves (the main organs of photosynthesis) and reproductive structures (in flowering plants, flowers and fruits).
    2. Buds (undeveloped embryonic shoots) develop on stems.
II. The plant body is composed of three tissue systems: ground, vascular, and dermal.
  A. The **ground tissue system** consists of three tissues with a variety of functions.
    1. **Parenchyma** tissue is composed of living parenchyma cells that possess thin primary cell walls. Functions of parenchyma tissue include photosynthesis, storage, and secretion.
    2. **Collenchyma** tissue is composed of collenchyma cells with unevenly thickened primary cell walls. This tissue provides flexible structural support.
    3. **Sclerenchyma** tissue is composed of sclerenchyma cells (**sclereids** or **fibers**) that have both primary and secondary cell walls. Scle-

renchyma cells are often dead at maturity, but they provide structural support.
  B. The **vascular tissue system** conducts materials throughout the plant body and provides strength and support.
    1. **Xylem** is a complex tissue that conducts water and dissolved nutrient minerals. The actual conducting cells of xylem are **tracheids** and **vessel elements.**
    2. **Phloem** is a complex tissue that conducts sugar in solution. **Sieve tube members** are the conducting cells of phloem; they are assisted by **companion cells.**
  C. The **dermal tissue system** is the outer protective covering of the plant body.
    1. The **epidermis** is a complex tissue that covers the herbaceous plant body.
      a. The epidermis that covers aerial parts secretes a layer of wax, called the **cuticle,** that reduces water loss.
      b. Gas exchange between the interior of the shoot system and the surrounding atmosphere occurs through **stomata.**
    2. The **periderm** is a complex tissue that covers the woody parts of the plant body in woody plants.
  D. Although separate organs (roots, stems, leaves, flower parts, and fruits) exist in the plant, many tissues are integrated throughout the plant body, providing continuity from organ to organ.

III. Growth in plants is localized in specific regions, called **meristems,** and involves cell division, cell elongation, and cell differentiation.
   A. **Primary growth** is an increase in stem and root length.
      1. Primary growth occurs in all plants.
      2. Primary growth results from the activity of **apical meristems** that are localized at the tips of roots and the buds of stems.
   B. **Secondary growth** is an increase in stem and root girth (thickness).

   1. In addition to primary growth, woody plants also have secondary growth.
   2. Secondary growth is localized, typically occurring in long cylinders of active growth throughout the length of older stems and roots.
   3. The two **lateral meristems** responsible for secondary growth are the **vascular cambium** and the **cork cambium.**

POST-TEST

1. Plants that complete their life cycles in one year are called _____; those that complete them in two years are _____; and those that live year after year are _____. (a) annuals; perennials; biennials (b) biennials; annuals; perennials (c) annuals; biennials; perennials (d) perennials; annuals; biennials (e) perennials; biennials; annuals

2. Most of the plant body consists of the _____ tissue system. (a) ground (b) vascular (c) periderm (d) dermal (e) cortex

3. Storage, secretion, and photosynthesis are the functions of (a) collenchyma (b) vessel elements (c) lateral meristems (d) sclerenchyma (e) parenchyma

4. The two simple tissues that are specialized for support are (a) parenchyma and collenchyma (b) collenchyma and sclerenchyma (c) sclerenchyma and parenchyma (d) parenchyma and xylem (e) xylem and phloem

5. Conduction of water and minerals in the xylem occurs in vessel elements and (a) sieve tube members (b) tracheids (c) collenchyma (d) cork cells (e) phloem

6. Conduction of sugar in solution in the sieve tube members is aided by (a) cork cells (b) sclerenchyma (c) parenchyma (d) guard cells (e) companion cells

7. The outer covering of plants with primary growth is _____, whereas plants with secondary growth are covered by _____. (a) cuticle; cork parenchyma (b) periderm; phloem (c) epidermis; periderm (d) epidermis; collenchyma (e) cellulose; lignin

8. The noncellular layer of wax secreted by the epidermis over its surface is called (a) lignin (b) cuticle (c) periderm (d) cellulose (e) trichome

9. Tiny pores known as _____ dot the surface of the epidermis of leaves and stems; each pore is bordered by two _____. (a) stomata; guard cells (b) stomata; fibers (c) sieve tube members; companion cells (d) sclereids; guard cells (e) cuticle; guard cells

10. Localized areas within the plant body where cell divisions occur are known as (a) organs (b) fibers (c) meristems (d) cork parenchyma (e) stomata

11. Primary growth, an increase in the length of a plant, occurs at a(an) (a) cork cambium (b) apical meristem (c) vascular cambium (d) both a and c (e) a, b, and c are correct

12. The two lateral meristems responsible for secondary growth are the (a) cork cambium and apical meristem (b) apical meristem and cork parenchyma (c) vascular cambium and apical meristem (d) vascular cambium and cork cambium (e) cork cambium and cork parenchyma

REVIEW QUESTIONS

1. What are some of the functions of roots? Of shoots?
2. What are the three tissue systems in plants? Describe the functions of each.
3. Compare the cellular structures and functions of parenchyma, collenchyma, and sclerenchyma tissues.
4. What are the functions of xylem and phloem?
5. Compare and contrast epidermis and periderm.
6. What is the role of plant meristems?
7. Distinguish between primary and secondary growth. Between apical and lateral meristems.
8. Identify each type of tissue in the figure. Use Tables 31–2, 31–3, and 31–4 to check your answers.

(John D. Cunningham/Visuals Unlimited)

(Dennis Drenner)

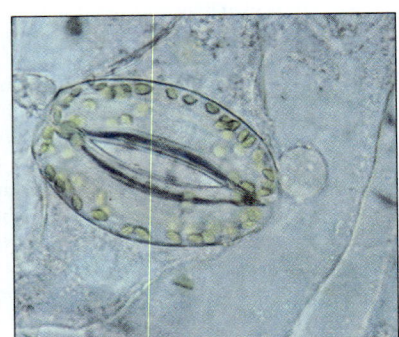

(Dwight R. Kuhn)

1. Grasses have a special meristem situated at the base of the leaves. Relate this information to what you know about mowing the lawn.
2. A couple carved a heart with their initials into a tree trunk four feet above ground level; the tree was 25 feet tall at the time. Twenty years later the tree was 50 feet tall. How far above the ground were the initials? Explain your answer.
3. Sclerenchyma in plants is the functional equivalent of bone in humans (i.e., both sclerenchyma and bone provide support). However, sclerenchyma is dead, while bone is living tissue. What are some of the advantages of a plant having dead support cells? Can you think of any disadvantage?

RECOMMENDED READINGS

Bell, A.D. *Plant Form: An Illustrated Guide to Flowering Plant Morphology.* Oxford University Press, Oxford, 1991. The external features of plants are examined in this beautifully illustrated book.

Berg, L.R. *Introductory Botany: Plants, People, and the Environment.* Saunders College Publishing, Philadelphia, 1997. A general botany text with an environmental emphasis.

Pimm, S.L. "In Search of Perennial Solutions." *Nature,* Vol. 389, 11 Sep. 1997. This short article highlights the environmental problems caused by traditional agriculture, which makes use of annual crops. The possibility of developing and farming perennial crops is explored.

Rensberger, B. "Getting to the Root of Plant Growth." *Washington Post,* Science Section, 13 Jul. 1992. Why roots generally grow downward and shoots upward.

● Visit our website at **http://www.saunderscollege.com/lifesci/titles.html** and click on Solomon/Berg/Martin Biology for links to chapter-related resources on the World Wide Web.

CHAPTER 32

Leaf Structure and Function

Suppose for a moment that you are taking a course in engineering and have been asked to design an efficient solar collector capable of converting the radiant energy it collects into chemical energy of organic molecules. Where would you start? It might be helpful to check library books and periodicals to see how solar collectors/energy converters have been designed in the past. In this instance, it would also be wise to ask a biologist, or even a biology student like yourself, whether anything comparable exists in nature. The answer, of course, is yes: plants have organs that are effective solar collectors and energy converters. They are called *leaves.*

Plants allocate many resources into the production of leaves. According to John E. Dale, a University of Edinburgh botanist who specializes in leaf development, a large maple tree annually produces 46.5 square meters (500 square feet) of leaves, which weigh more than 113 kilograms (250 pounds). The metabolic cost of producing so many leaves is high, but leaves are essential to the survival of the tree.

Leaves gather the sunlight necessary for **photosynthesis,** the biological process that converts radiant energy into the chemical energy of sugars. Plants then use these molecules as starting materials to synthesize all other organic compounds and as fuel to provide energy for their metabolic processes. During a single summer, the leaves of the red maple *(Acer rubrum)* shown here will "fix" about 454 kilograms (1000 pounds) of carbon dioxide into sugars and other organic compounds. Leaves are the main photosynthetic organs of most plants, and the structure of a leaf is superbly adapted for its primary function of photosynthesis. Most leaves are thin and flat, a shape that allows maximum absorption of light energy and the efficient internal diffusion of gases such as CO_2 and O_2. As a result of their ordered arrangement on the stem, leaves efficiently catch the sun's rays. The leaves of plants form an intricate green mosaic, bathed in sunlight and atmospheric gases.

In addition to meeting the requirements of photosynthesis, the structure of foliage leaves is also adapted to prevent excessive water loss. Leaves possess several features that help reduce or control water loss, the most important of which is a thin, transparent layer of wax that covers the leaf surface. However, structural adaptations are compromises between competing needs. Thus, some features that optimize photosynthesis actually *promote* water loss. For example, plants have minute

(David Sieren/Visuals Unlimited)

pores that allow an exchange of gases for photosynthesis, but these tiny openings also allow water to escape into the atmosphere as water vapor. Thus leaf structure represents a trade-off between photosynthesis and water conservation.

LEARNING OBJECTIVES

AFTER YOU HAVE STUDIED THIS CHAPTER YOU SHOULD BE ABLE TO

1. Describe the major tissues of the leaf: epidermis, mesophyll, xylem, and phloem.
2. Compare leaf anatomy in dicots and monocots.
3. Relate leaf structure to its function of photosynthesis.
4. Outline the physiological changes that accompany stomatal opening and closing.

5. Discuss transpiration and its effects on plants.
6. Define leaf abscission, explain why it occurs, and describe the physiological and anatomical changes that precede it.
7. List five examples of modified leaves and give the function of each.

FOLIAGE LEAVES VARY GREATLY IN EXTERNAL FORM

Leaves are the most variable plant organ, so much so that plant biologists had to develop specific terminology to describe their shapes, margins (edges), vein patterns, and the way they attach to stems. Because each leaf is characteristic of the plant species on which it grows, many plants can be identified by their leaves alone. Leaves may be round, needle-like, scalelike, cylindrical, heart-shaped, fan-shaped, or thin and narrow. They vary in size from those of the raffia palm *(Raphia ruffia)*, whose leaves often grow to more than 20 m (65 ft) long, to those of

water-meal (*Wolffia* sp.), whose leaves are so small that 16 of them laid end-to-end equal 2.5 cm (1 in) (see Fig. 31–1*a*).

The typical leaf is composed of two parts, a blade and a petiole. The broad, flat portion of a leaf is the **blade;** the stalk that attaches the blade to the stem is the **petiole.** Some leaves also have **stipules,** which are leaflike outgrowths usually present in pairs at the base of the petiole (Fig. 32–1). Some leaves do not have petioles or stipules.

Leaves may be *simple* (having a single blade) or *compound* (having a blade divided into two or more leaflets) (Fig. 32–2*a*). Sometimes it is difficult to tell whether a plant has formed one compound leaf or a small stem bearing several simple leaves. One easy way to determine if a plant has simple or compound leaves is to look for axillary buds, so-called because each develops in a leaf *axil* (the angle between the stem and petiole). Axillary buds form at the base of a leaf, whether it is simple or compound. However, axillary buds never develop at the base of leaflets. Also, the leaflets of a compound leaf lie in a single plane (you can lay a compound leaf flat on a table), whereas simple leaves usually are not arranged in one plane on a stem.

Leaves are arranged on a stem in one of three possible ways (Fig. 32–2*b*). Plants such as beeches and walnuts have an *alternate leaf arrangement,* with one leaf at each node. (A *node* is the area of the stem where one or more leaves are attached.) In an *opposite leaf arrangement,* as occurs in maples and ashes, two leaves grow at each node. In a *whorled leaf arrangement,* as occurs in catalpa, three or more leaves grow at each node.

Leaf blades may possess *parallel venation,* in which the primary veins run approximately parallel to one another (generally characteristic of monocots[1]), or *netted venation,* in which veins are branched in such a way that they resemble a net (generally characteristic of dicots; Fig. 32–2*c*). Netted veins can be *palmately netted,* in which several major veins radiate out from one point (see Fig. 32–1), or *pinnately netted,* in which major veins branch off along the entire length of one main vein.

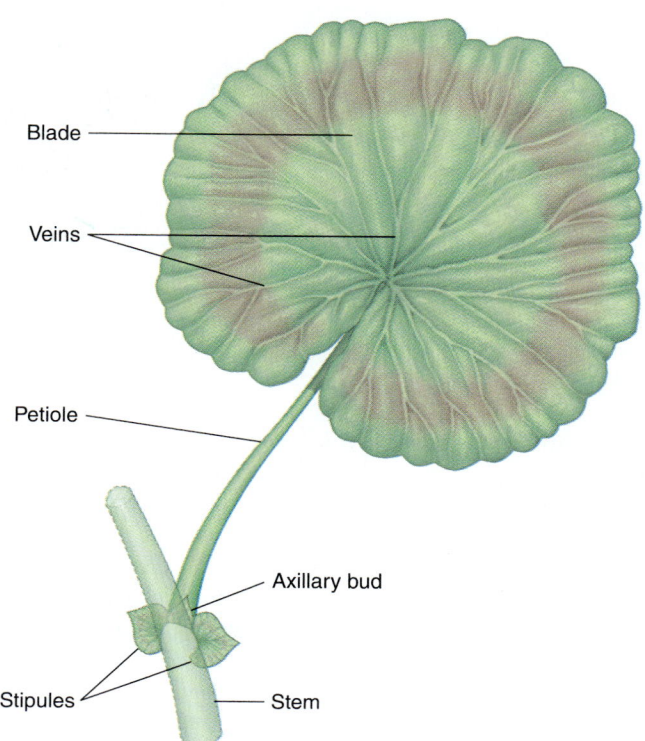

Figure 32–1 **A typical leaf.** A geranium leaf consists of a blade, a petiole, and two stipules at the base of the leaf. Note the axillary bud in the leaf axil.

Labels: Blade, Veins, Petiole, Axillary bud, Stipules, Stem

[1]Recall that flowering plants, the focus of this chapter, are divided into two groups, informally called dicots and monocots, based on a number of structural features (see Chapter 27). Dicots include such well known plants as beans, petunias, oaks, cherry trees, roses, and snapdragons; examples of monocots include corn, lilies, grasses, palms, tulips, orchids, and bananas.

(a) Simple/compound leaves

Simple

California white oak
(Quercus lobata)

Pinnately compound

White ash
(Fraxinus americana)

Palmately compound

Ohio buckeye
(Aesculus glabı

(b) Leaf arrangements on stems

Alternate

American beech
(Fagus grandifolia)

Opposite

Sugar maple
(Acer saccharum)

Whorled

Southern catalpa
(Catalpa bignonioides)

(c) Venation patterns

Parallel

Bermuda grass
(Cynodon dactylon)

Pinnately
netted

Black willow
(Salix nigra)

Palmately
netted

Sweetgum
*(Liquidambar
styraciflua)*

Figure 32–2 **Leaf morphology.** *(a)* Simple, pinnately compound, and palmately compound leaves. *(b)* Leaf arrangement may be alternate, opposite, or whorled, depending on the number of leaves at each node. *(c)* Venation patterns include parallel, which is characteristic of monocot leaves, pinnately netted, and palmately netted. All leaves shown are woody dicot trees from North America, except bermuda grass, which is a herbaceous monocot native to Europe and Asia.

EPIDERMIS, MESOPHYLL, XYLEM, AND PHLOEM ARE THE MAJOR LEAF TISSUES

The leaf is a complex organ composed of several tissues, all of which are organized to optimize photosynthesis (Fig. 32–3). Because the leaf blade has upper and lower surfaces, it is covered by two epidermal layers, the **upper epidermis** and the **lower epidermis.** Most cells in these layers are living parenchyma cells (see Chapter 31) that lack chloroplasts and are relatively transparent. One interesting feature of leaf epidermal cells is that the cell wall facing toward the outside of the leaf is somewhat thicker than the cell wall facing inward. This extra thickness may provide the plant with additional protection against injury or water loss.

Because leaves have such a large surface area exposed to the atmosphere, water loss by evaporation from the leaf's surface is unavoidable. However, epidermal cells secrete a waxy layer, the **cuticle,** that reduces water loss from their exterior walls (see Table 31–4). The cuticle, which consists primarily of a waxy substance called **cutin,** varies in thickness in different plants, in part due to environmental conditions. As one might expect, the leaves of plants adapted to hot, dry climates have extremely thick cuticles. Furthermore, a leaf's exposed (and warmer) upper epidermis generally has a thicker cuticle than its shaded (and cooler) lower epidermis.

The epidermis of many leaves is covered with various hair-like structures called **trichomes** (see Table 31–4). Some leaves, such as those of the popular cultivated plant called lamb's ear

Figure 32–3 Tissues in a typical leaf blade. The blade is covered by an upper and lower epidermis. The photosynthetic parenchyma tissue, called mesophyll, is often arranged into palisade and spongy layers. Veins branch throughout the mesophyll.

Palisade mesophyll

Vein (vascular bundle)

Spongy mesophyll

Stoma

Upper epidermis

Cuticle

Bundle sheath

Xylem

Phloem

Lower epidermis

Stoma

Guard cells

(*Stachys byzantina*), have so many trichomes that they actually feel fuzzy (Fig. 32–4). Trichomes may have several functions. Trichomes of some plants help reduce water loss from the leaf surface by retaining a layer of moist air next to the leaf and by reflecting sunlight, thereby protecting the plant from overheating. Some trichomes secrete stinging irritants for protection from herbivores, and a leaf covered with trichomes is difficult for an insect to walk over or eat. Other trichomes excrete excess salts absorbed from a salty soil.

The leaf epidermis is typically covered with minute openings, or **stomata** (sing., *stoma*), for gas exchange. Each stoma is flanked by two specialized epidermal cells called **guard cells** (Fig. 32–5). The guard cells are responsible for opening and closing the stoma. Guard cells are usually the only epidermal cells with chloroplasts, but the functional significance of this is not clear. Stomata are especially numerous on the lower epidermis of horizontally oriented leaves (an average of about 100 stomata per square millimeter), and in many species are lo-

Figure 32–4 Trichomes on leaves of lamb's ear (Stachys byzantina). This popular garden plant has so many trichomes that the leaves look and feel like soft felt. *(Alan L. Detrick/Photo Researchers, Inc.)*

(a)

25 μm

(b)

25 μm

Figure 32–5 **Stomata.** LM of *(a)* an open stoma and *(b)* a closed stoma from the leaf epidermis of wandering jew (*Zebrina pendula*). Also note the chloroplasts present in the guard cells. *(a, b, Dwight R. Kuhn)*

cated *only* on the lower surface. The lower epidermis of apple (*Malus sylvestris*) leaves, for example, has an average of 387 stomata per square millimeter, whereas the upper epidermis has none. This adaptation reduces water loss because stomata on the lower epidermis are shielded from direct sunlight and are therefore cooler than those on the upper epidermis. On the other hand, leaves of aquatic plants such as water lilies have stomata only on the upper epidermis.

Guard cells may be associated with special epidermal cells called **subsidiary cells** that are structurally different from all other epidermal cells. Subsidiary cells provide a reservoir of water and ions that move into and out of the guard cells as they change shape during stomatal opening and closing (discussed later in this chapter).

The photosynthetic parenchyma tissue of the leaf, called the **mesophyll** (from the Greek *meso,* "the middle of," and *phyll,* "leaf"), is sandwiched between the upper epidermis and the lower epidermis. Mesophyll cells, which are parenchyma cells packed with chloroplasts, are loosely arranged with many air spaces between them that facilitate gas exchange. These intercellular air spaces account for as much as 70% of the leaf's volume. In many plants, the mesophyll is divided into two sublayers. Toward the upper epidermis, the columnar cells are stacked closely together in a **palisade** layer. In the lower portion, they are more loosely and irregularly arranged in a **spongy** layer. The two layers have different functions. Palisade mesophyll is the main site of photosynthesis in the leaf. Although photosynthesis also occurs in the spongy mesophyll, its primary function is to allow for diffusion of gases (particularly CO_2) within the leaf.

Palisade mesophyll may be further organized into one, two, three, or even more rows of cells. The presence of additional layers of palisade mesophyll is at least partly an adaptation to environmental conditions. Leaves exposed to direct sunlight are thicker because they contain more layers of palisade mesophyll than do shaded leaves on the same plant. In direct sunlight, the light is strong enough to effectively penetrate multiple layers of palisade mesophyll, allowing all layers to photosynthesize efficiently.

The **veins,** or **vascular bundles,** of a leaf extend through the mesophyll. Branching is extensive, and no mesophyll cell is more than two or three cells away from a vein. Therefore, movement of needed resources between mesophyll cells and veins is not limited by the slow process of diffusion. Each vein contains two types of vascular tissue: xylem and phloem (see Chapter 31). **Xylem,** which conducts water and dissolved nutrient minerals, is usually located in the upper part of a vein, toward the upper epidermis, whereas **phloem,** which conducts dissolved sugars, is usually confined to the lower part of a vein.

Veins are usually surrounded by one or more layers of nonvascular cells that make up the **bundle sheath.** Bundle sheaths are composed of parenchyma or sclerenchyma cells (see Chapter 31). Frequently, the bundle sheath has support columns, called **bundle sheath extensions,** that extend through the mesophyll from the upper epidermis to the lower epidermis (Fig. 32–6). Bundle sheath extensions may be composed of parenchyma, collenchyma, or sclerenchyma cells.

Leaf structure differs in dicots and monocots

A dicot leaf is usually composed of a broad, flattened blade and a petiole. As mentioned previously, dicot leaves typically have netted venation. In contrast, monocot leaves often lack a petiole; they are narrow and often wrap around the stem, forming a sheath. Parallel venation is characteristic of monocot leaves.

Dicots and certain monocots also differ in internal leaf anatomy (Fig. 32–7). While most dicots and monocots have both palisade and spongy layers, some monocots (corn and other grasses) do not have mesophyll differentiated into dis-

Upper epidermis
Bundle sheath extension
Midvein
Bundle sheath
Bundle sheath extension
Lower epidermis

250 μm

Figure 32–6   A bundle sheath extension. LM of a wheat *(Triticum aestivum)* midvein cross section, showing bundle sheath extensions to both the upper epidermis and lower epidermis.   *(Phil Gates/ Biological Photo Service)*

▶ **Figure 32–7   Dicot and monocot leaf cross sections.** *(a)* LM of leaf of privet *(Ligustrum vulgare)*, a dicot, reveals a mesophyll with distinct palisade and spongy sections. *(b)* LM of leaf of corn *(Zea mays)*, a monocot, reveals that it lacks distinct regions of palisade and spongy mesophyll.   *(a, Ed Reschke; b, Dwight R. Kuhn)*

Midvein
Upper epidermis
Palisade mesophyll
Lengthwise view of vein
Spongy mesophyll
Lower epidermis
Xylem
Phloem
(a)
250 μm

Midvein
Upper epidermis
Mesophyll
Lower epidermis
Xylem
Phloem
(b)
25 μm

Dicots and some monocots

Open

Closed

Certain monocots (grasses)

Open

Subsidiary cells

Guard cells

Closed

Figure 32–8 **Variation in guard cells.** Guard cells of dicots and many monocots are bean-shaped. Other monocot guard cells (those of grasses, reeds, and sedges) are narrow in the center and thicker at each end. Guard cells are often associated with special epidermal cells called subsidiary cells.

tinct palisade and spongy layers. Because dicots have netted veins, a cross section of a dicot blade often shows veins in both cross-sectional and lengthwise views. In a cross section of a monocot leaf, on the other hand, the parallel venation pattern produces evenly spaced veins, all of which are in cross section.

Differences between the guard cells in dicot and certain monocot leaves also occur (Fig. 32–8). The guard cells of dicots and many monocots are shaped like kidney beans. Other monocot leaves (those of grasses, reeds, and sedges) have guard cells shaped like dumbbells. These structural differences affect how the cells swell or shrink to open or close the stoma.

LEAF STRUCTURE IS RELATED TO FUNCTION

The primary function of leaves is to collect radiant energy and convert it to the chemical energy stored in the bonds of organic molecules such as glucose. This process, called photosynthesis, was examined in detail in Chapter 8. During photosynthesis, plants take relatively simple molecules (carbon dioxide and water) and convert them into sugars. Oxygen is given off as a byproduct. The sugar formed during photosynthesis is used by the plant in two ways. First, it is broken down by aerobic respiration to release the chemical energy stored in

its bonds for other cellular activities (see Chapter 7). Second, sugar molecules provide the cell with basic building materials. The cell modifies sugars, converting them into all the organic compounds used by the plant (starch and cellulose, amino acids, lipids, nucleic acids, etc.).

How is leaf structure related to its primary function of photosynthesis? The epidermis of a leaf is relatively transparent and allows light to penetrate to the interior of the leaf where the photosynthetic tissue, the mesophyll, is located. Stomata, which dot the leaf surfaces, permit the exchange of gases between the atmosphere and the leaf's internal tissues. Carbon dioxide, a raw material of photosynthesis, diffuses into the leaf through stomata, and the oxygen produced during photosynthesis diffuses rapidly out of the leaf through stomata. Stomata also permit other gases, including air pollutants, to enter the leaf (see *Focus On: The Effects of Air Pollution on Leaves*).

Water required for photosynthesis is obtained from the soil and transported in the xylem to the leaf, where it moistens the surfaces of mesophyll cells. The loose arrangement of the mesophyll tissue, with air spaces between cells, allows for rapid diffusion of carbon dioxide to the mesophyll cell surfaces; there it dissolves in a film of water before diffusing into the cells.

The veins not only supply the photosynthetic tissue with water and minerals (from the roots, by way of the xylem), but

THE EFFECTS OF AIR POLLUTION ON LEAVES

The air we breathe is often dirty and contaminated with many pollutants, particularly in urban areas. Air pollution consists of gases, liquids, or solids present in the atmosphere in levels high enough to harm humans and other organisms, as well as nonliving materials. Although air pollutants can come from natural sources, as, for example, a lightning-caused fire or a volcanic eruption, human activities make a major contribution to global air pollution. Motor vehicles and industry are the two main human sources of air pollution.

All parts of a plant can be damaged by air pollution, but leaves are particularly susceptible because of their structure and function. The thin blade provides a large surface area that comes into contact with the surrounding air. The thousands of tiny stomatal pores that dot the epidermis and allow gas exchange with the atmosphere also permit pollutants to diffuse into the leaf. Just as lungs, the organs of gas exchange in humans, often are affected by air pollution, so too the leaves of a plant are most affected.

Trees provide a dramatic demonstration of the effect of air pollution on biological longevity. According to American Forests (formerly the American Forestry Association), the average life span for a tree living in a rural setting, where air pollution is generally low, is 150 years. Even in the most ideal urban setting, a tree typically lives only 60 years, and in a normal city

setting the average life span is 32 years. Trees living in a downtown setting, where air quality is lowest, live only 7 years on average. It should be noted, however, that air pollution is not the only cause of short life span in urban trees. Other factors include inadequate room for root growth, lack of water, and polluted runoff from streets and sidewalks.

Many studies have shown that the overall productivity of crop plants is also reduced by high levels of most forms of air pollution. The worst pollutant in terms of yield loss is ozone, a toxic gas produced when sunlight catalyzes a reaction between pollutants emitted by motor vehicles and industries. In plants, ozone inhibits photosynthesis because it damages the mesophyll cells, probably by altering the permeability of their cell membranes. Exposure to low levels of air pollution often causes a decline in photosynthesis without any other symptoms of injury. Lesions on leaves and other obvious symptoms appear at much higher levels of air pollution.

When air pollution is combined with other environmental stresses (such as low winter temperatures, prolonged droughts, insects, and bacterial, fungal, and viral diseases), it can cause plants to decline and die. More than half of the red spruce trees in the mountains of the northeastern United States have died since the mid-1970s, and sugar maples in eastern Canada

and the United States are also dying. Many still-living trees are exhibiting symptoms of **forest decline,** characterized by gradual deterioration and often death of trees. The general symptoms of forest decline are reduced vigor and growth, but some plants exhibit specific symptoms, such as yellowing of needles in conifers. Forest decline is more pronounced at higher elevations, possibly because most trees growing at high elevations are at the limit of their normal range and are therefore more susceptible to wind and low temperatures.

Many factors can interact to decrease the health of trees, and no single factor accounts for the recent instances of forest decline. Several human-induced air pollutants have been implicated, including acid rain (see Fig. 2–18), ozone, and toxic heavy metals such as lead, cadmium, and copper. These pollutants are produced by power plants, ore smelters, refineries, and motor vehicles. Insects and weather factors such as drought and severe winters may also be important. To complicate matters further, the actual causes of forest decline may vary from one tree species to another and from one location to another. Thus, forest decline appears to result from the combination of multiple stresses. When one or more stresses weaken a tree, then an additional stress may be enough to cause its death.

---

also carry (in the phloem) dissolved sugar to all parts of the plant. Bundle sheaths and bundle sheath extensions associated with the veins provide additional support to prevent the leaf, which is structurally weak because of the large amount of air space in the mesophyll, from collapsing under its own weight.

The leaves of each type of plant help it survive in the environment to which it is adapted

The environment to which a particular plant is adapted is reflected in its leaf structure. Although both aquatic plants and those adapted to dry conditions perform photosynthesis and have the same basic leaf anatomy, their leaves are modified to enable them to survive different environmental conditions. The leaves of water lilies, for example, have petioles long

enough to allow the blade to float on the water's surface (Fig. 32–9). These petioles and other submerged parts have an internal system of air ducts; oxygen moves through these ducts from the floating leaves to the underwater roots and stems, which live in a poorly aerated environment.

The leaves of conifers, an important group of woody trees and shrubs that includes pine, spruce, fir, redwood, and cedar, are waxy needles.[2] Most conifers are evergreen, which means they produce and lose leaves throughout the year rather than during certain seasons. Conifers dominate a large portion of

---

[2]As discussed in Chapter 27, conifers are gymnosperms, one of the two groups of seed plants (the other group is the flowering plants, or angiosperms). Unlike flowering plants, whose seeds are enclosed in fruits, conifers bear "naked" seeds on the scales of female cones.

Figure 32–9 **Water lily (*Nymphaea* sp.) leaves.** This unusual view of water lily leaves shows their petioles as well as their blades, which float on the water's surface. One of water lily's adaptations for an aquatic lifestyle is that the length of its petioles depends on the depth of the water. *(Frans Lanting/Minden Pictures)*

Blade   Petiole

Earth's land area, particularly in northern forests and mountains. Their needles have structural adaptations that help them survive winter, the driest part of the year. (Winter is arid even in areas of heavy snows because roots cannot absorb water from soil when the soil temperature is low.) Indeed, many of the structural features of needles are also found in many desert plants.

Figure 32–10 shows a cross section of a pine needle. Note that the needle is somewhat thickened rather than thin and bladelike. The needle's relative thickness, which results in less surface area exposed to the air, reduces water loss. Other features that help conserve water include the thick, waxy cuticle and sunken stomata; these permit gas exchange while minimizing water loss. Thus, needles help conifers tolerate the dry

(low relative humidity) winds that occur during winter. With the warming of spring, soil water again becomes available, and the needles quickly resume photosynthesis.

STOMATAL OPENING AND CLOSING ARE DUE TO CHANGES IN GUARD CELL TURGIDITY

Stomata are adjustable pores that are usually open during the day when carbon dioxide is required for photosynthesis and closed at night when photosynthesis is shut down (see the section on CAM photosynthesis in Chapter 8 for an interesting exception). The opening and closing of stomata are controlled by changes in the shape of the two guard cells that surround each pore. When water moves into guard cells from surrounding cells, they become turgid (swollen), and the inner cell walls bend outward at the center, producing a pore. When water leaves the guard cells, they become flaccid (limp) and collapse against one another, closing the pore.

Although the opening and closing of stomata are triggered by light or darkness, other environmental factors are also involved, including carbon dioxide concentration. A low concentration of CO_2 in the leaf induces stomata to open even in the dark. The effects of light and CO_2 concentration on stomatal opening are interrelated. Photosynthesis, which occurs in the presence of light, reduces the internal concentration of CO_2 in the leaf, triggering stomatal opening. Another environmental factor that affects stomatal opening and closing is dehydration (water stress). During a prolonged drought, stomata will remain closed even during the day. Stomatal opening and closing are also under hormonal control (see Chapter 36).

The opening and closing of stomata also appear to be under the control of an internal biological clock that in some way measures time. For example, plants placed in continual dark-

Epidermis and cuticle

Resin duct

Endodermis

Xylem ⎤
        ⎬ Vascular bundle
Phloem ⎦

Mesophyll cell (photosynthetic parenchyma cell)

Sunken stoma

250 µm

Figure 32–10 **LM of a pine (*Pinus* sp.) needle in cross section.** The thick, waxy cuticle and sunken stomata are two structural adaptations that enable the pine tree to retain its needles throughout the winter. *(John D. Cunningham/Visuals Unlimited)* ness persist in opening and closing their stomata at more or less the same time each day. Such biological rhythms that follow an approximate 24-hour cycle are known as **circadian rhythms.** Other examples of circadian rhythms are provided in Chapter 36.

Stomata open and close in response to the movement of H⁺ and K⁺ across the guard cells' plasma membranes

Data from numerous experiments and observations suggest that the movement of H^+ and K^+ across the plasma membranes of guard cells explains the opening and closing of stomata (Fig. 32–11). Light, particularly blue light, triggers the movement of potassium ions (K^+) into the guard cells from subsidiary cells or ordinary epidermal cells. This movement of potassium ions, which has been experimentally measured by patch clamp techniques (see Chapter 5), occurs by active transport through specific ion channels in the guard cells' plasma membranes and requires ATP. The ATP supplies energy to pump protons (H^+) out of the guard cells; the H^+ is formed when malic acid produced in the chloroplasts of the guard cells ionizes. The removal of H^+ from the guard cells produces an electrochemical gradient that can drive the uptake of potassium ions through specific K^+ channels. This ion movement is an example of a linked cotransport system (see Chapter 5). The K^+ accumulates in the vacuoles of the guard cells.

The active uptake of K^+ in the guard cells increases the solute concentration in the vacuoles. As a result, water enters the guard cells from surrounding epidermal cells by osmosis. (Recall from the discussion of osmosis in Chapter 5 that when a cell has a solute concentration greater than that of surrounding cells, water flows into the cell.) The increased turgidity of the guard cells changes their shape, and the pore opens:

Light → proton pump moves H^+ out of guard cells → K^+ actively transported into guard cells → water diffuses into guard cells → guard cells change shape and pore appears

In the late afternoon or early evening, stomata close, possibly by a reversal of the opening process. The potassium ions are pumped out of the guard cells into the surrounding epidermal cells. Water leaves the guard cells by osmosis, the cells lose their turgidity, and the pore closes as they collapse. Stomatal closure, however, may not be an *exact* reversal of stomatal opening. There is some evidence that Ca^{2+} triggers stomatal closure but inhibits stomatal opening. The actual mechanism whereby Ca^{2+} exerts this effect is under investigation.

As in other plant responses to light, stomata vary in their sensitivity to different colors, that is, different wavelengths, of light. Stomatal opening is most pronounced in blue and, to a lesser extent, in red light. Also, dim blue light induces stomatal opening, whereas dim red light does not. Any plant re-

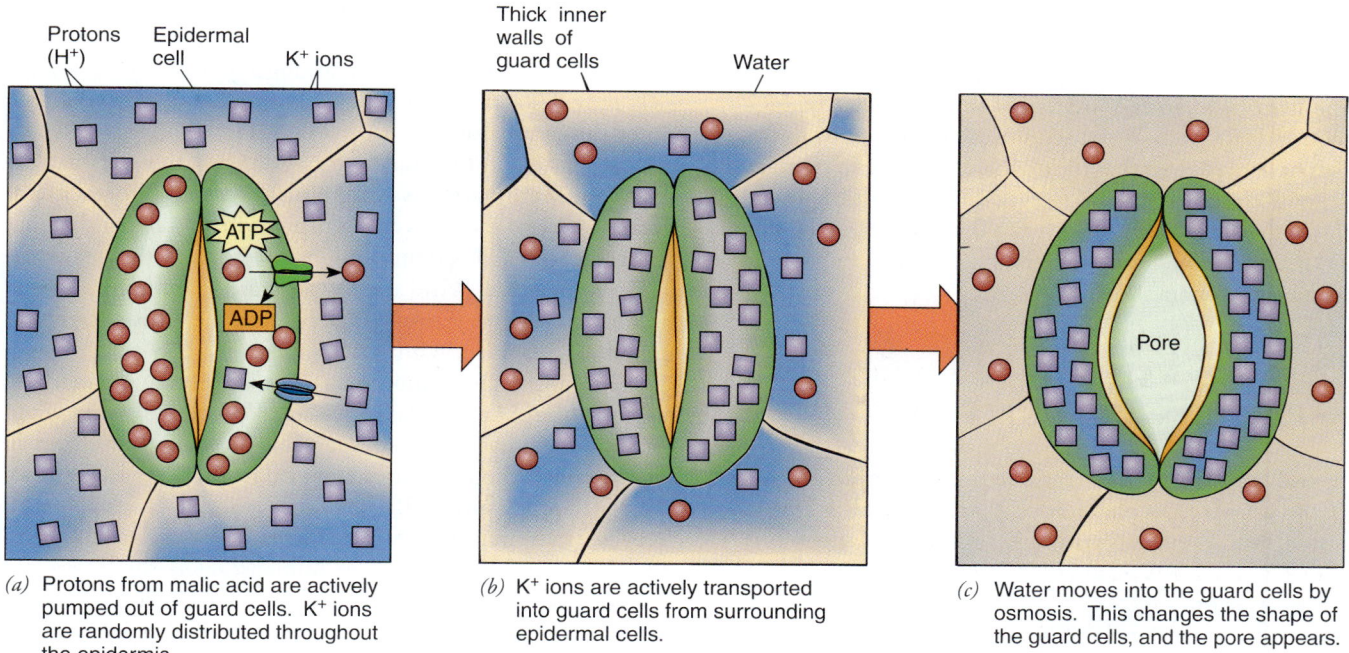

(a) Protons from malic acid are actively pumped out of guard cells. K^+ ions are randomly distributed throughout the epidermis.

(b) K^+ ions are actively transported into guard cells from surrounding epidermal cells.

(c) Water moves into the guard cells by osmosis. This changes the shape of the guard cells, and the pore appears.

Figure 32–11 **The mechanism of stomatal opening and closing.** Evidence suggests that stomatal opening is caused by the movement of *(a)* H^+ out of the guard cells, followed by the movement of *(b)* K^+ and then *(c)* water into the guard cells. For simplicity, the H^+ and K^+ linked cotransport system is shown in the right guard cell of *(a)* only.

TREES, TRANSPIRATION, AND CLIMATE

Do trees influence an area's climate, that is, the average weather conditions that occur in that area over a period of years? Factors that determine climate include temperature and precipitation, both of which are influenced by transpiration. Transpiration by trees, for example, influences the local temperature of forests. If you have ever walked into a forest on a hot summer day, you have probably noticed that the air is cooler and moister there than outside the forest. Transpired water has a cooling effect, not only for the plant but also for the local area.

Transpiration is an important part of the **hydrologic cycle** (see Chapter 53), in which water cycles from the ocean and land to the atmosphere, and then back to the ocean and land. As a result of transpiration, water evaporates from leaves and stems to form clouds in the atmosphere. Thus, transpiration eventually results in precipitation. As you might expect, forest trees release substantial amounts of moisture into the air by transpiration. Researchers have determined that at least half the rain that falls in the Amazon rain-forest basin is recycled again and again by transpiration and precipitation.

Climate modelers who simulate deforestation in the Amazon rain forest also have determined that less moisture is carried into the deforested basin by trade winds; they therefore predicted that rainfall would decline and droughts become more common in that region. This prediction has been observed and verified where large

tracts of Amazon rain forest have been burned to grow crops; often people have observed droughts in areas that formerly had rainfall almost every day of the year. The explanation for less rainfall in a deforested region is that the heat carried into the atmosphere by transpiring trees accelerates the moist air movement upward, producing a low pressure area below it that pulls air in from over the Atlantic Ocean. (Air is not pulled in from over the Pacific Ocean because the Andes Mountains act as a barrier.) Similar declines in precipitation have also been observed in arid regions such as Australia, where extensive clearing of eucalyptus woodlands between 1950 and 1980 was followed by a 20% decline in winter rainfall.

Although the trees/rain connection holds in many areas, the relationship between trees and climate is not as simple as this brief discussion indicates. For example, climate modelers have determined that deforestation in India could *increase* precipitation there. The reason for the difference between the Amazon basin and India is related to geography. As you know, India is surrounded on three sides by ocean. During the day, the land heats up more than the surrounding ocean water. The heated land surface causes warm air masses to rise over the land, which pulls in moist air masses from the ocean. It is thought that the deforested land will heat up more, thereby magnifying this effect. Thus, complex interactions occur among ocean, atmosphere, land, and plants to determine regional climate.

sponse to light must involve a pigment, a molecule that absorbs the light prior to the induction of a particular biological response. Data such as the responses of stomata to different colors of light suggest that the pigment involved in stomatal opening and closing is yellow (yellow pigments strongly absorb blue light and weakly absorb red light). The yellow pigment is thought to be located in the guard cells, probably in their plasma membranes.

LEAVES LOSE WATER BY TRANSPIRATION AND GUTTATION

Despite leaf adaptations such as the cuticle, approximately 99% of the water that a plant absorbs from the soil is lost by evaporation from the leaves and, to a lesser extent, the stems. Loss of water vapor by evaporation from aerial plant parts is called **transpiration.**

The cuticle is extremely effective in reducing water loss from transpiration. It is estimated that only 1% to 3% of the water lost from a plant transpires directly through the cuticle. Most transpiration occurs through open stomata. The numerous stomatal pores that are so effective in gas exchange for photosynthesis also provide openings through which water vapor

escapes. In addition, the loose arrangement of the mesophyll cells provides a large surface area within the leaf from which water can evaporate.

Several environmental factors influence the rate of transpiration. More water is lost from plant surfaces at higher temperatures. Light increases the transpiration rate, in part because it triggers stomatal opening and in part because it increases the leaf's temperature. Wind and dry air increase transpiration, but humid air *decreases* transpiration because the air is already saturated, or nearly so, with water vapor.

Although transpiration may seem wasteful, particularly to farmers in arid lands, it is an essential process that has adaptive value. Transpiration is responsible for water movement in plants, and without it water would not reach the leaves from the soil (see Chapter 33). The large amount of water plants lose by transpiration may provide some additional benefits. Transpiration, like sweating in humans, cools the leaves and stems. When water passes from a liquid state to a vapor, it absorbs a great deal of heat. When the water molecules leave the plant as water vapor, they carry this heat with them. Thus, the cooling effect of transpiration may prevent the plant from overheating, particularly in direct sunlight. On a hot summer day, for example, the internal temperature of leaves is measurably lower than that of the surrounding air.

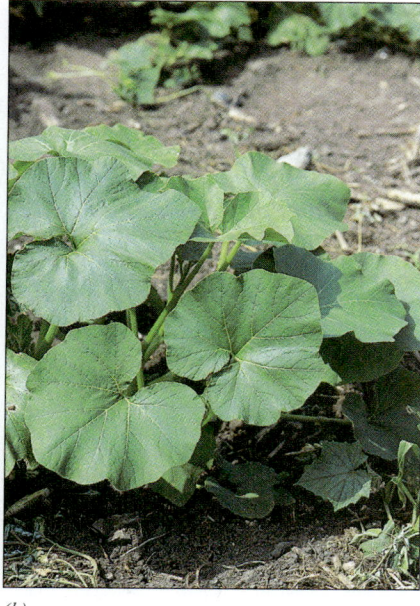

(a)   (b)

Figure 32–12 **Temporary wilting in squash (*Cucurbita pepo*) leaves.** *(a)* In the late afternoon of a hot day and *(b)* recovery the following morning. Note that wilting helps reduce the surface area from which transpiration occurs. During the night while transpiration is negligible, the plants recover by absorbing water from the soil. *(Carlyn Iverson)*

A second benefit of transpiration is that it provides the plant with sufficient essential minerals. The water a plant transpires is initially absorbed from the soil, where it is not present as pure water but rather as a dilute solution of dissolved mineral salts. The water and dissolved nutrient minerals are then transported in the xylem throughout the plant, including its leaves. Water moves from the plant to the atmosphere during transpiration, but minerals do not. Many of these minerals are required for the plant's growth. It has been suggested that transpiration enables a plant to take in sufficient water to provide enough essential minerals and that plants cannot satisfy their mineral requirements if the transpiration rate is not high enough.

There is no doubt, however, that under certain circumstances excessive transpiration can be harmful to a plant. On hot summer days, plants frequently lose more water by transpiration than they can take in from the soil. Their cells experience a loss of turgor, and the plant wilts (Fig. 32–12). If a plant is able to recover overnight, because of the combination of negligible transpiration (recall that stomata are closed) and absorption of water from the soil, the plant is said to have experienced *temporary wilting*. Most plants recover from temporary wilting with no ill effects. In cases of prolonged drought, however, the soil may not contain sufficient moisture to permit recovery from wilting. A plant that cannot recover is said to be *permanently wilted* and will die.

Some plants exude liquid water

Many leaves have special structures through which liquid water is literally forced out. This loss of liquid water, known as **guttation,** occurs when transpiration is negligible and available soil moisture is high. Guttation typically occurs at night because the stomata are closed, but water continues to move into the roots by osmosis. People sometimes think erroneously that the early morning water droplets on leaf margins are dew (water condensation from the air) rather than guttation (Fig. 32–13). (The mechanism for guttation is discussed in Chapter 33.)

Figure 32–13 **Guttation in strawberry (*Fragaria* sp.) leaves.** Many people mistake guttation for early morning dew. *(Ed Reschke/Peter Arnold, Inc.)*

LEAF ABSCISSION ALLOWS PLANTS TO SURVIVE AN UNFAVORABLE SEASON

All trees shed leaves; many conifers, for example, shed their needles year-round. The leaves of deciduous plants, however, turn color and **abscise,** or fall off, once a year—as winter approaches in temperate climates or before the dry period in tropical climates with pronounced wet and dry seasons. In temperate forests, for example, most woody plants with broad leaves shed their leaves in order to survive the low temperatures of winter. During winter, the plant's metabolism, including its photosynthetic machinery, slows down or halts temporarily.

Another reason for abscission is related to a plant's water requirements, which become critical during the physiological drought of winter. As mentioned previously, as the ground chills, absorption of water by the roots is inhibited. When the ground freezes, *no* absorption occurs. If the broad leaves were to stay on the plant during the winter, the plant would continue to lose water by transpiration but would be unable to replace it with water absorbed from the soil.

Leaf abscission is a complex process that involves many physiological changes, all of which are initiated and orchestrated by changing levels of plant hormones (see Chapter 36). Briefly, the process is this: As autumn approaches, sugars, amino acids, and many essential minerals (such as nitrogen, phosphorus, and possibly potassium) are mobilized and transported from the leaves into the woody tissues. Chlorophyll breaks down, allowing the orange carotenes and yellow xanthophylls, some of the accessory pigments in the chloroplasts of leaf cells, to become evident. Recall from Chapter 8 that accessory pigments are always present in the leaf but are masked by the green of the chlorophyll. In addition, red water-soluble pigments called anthocyanins may be synthesized and stored in the vacuoles of leaf cells in some species; their function is unknown. The various combinations of these pigments are responsible for the brilliant colors found in autumn landscapes in temperate climates.

In many leaves, abscission occurs at an abscission zone near the base of the petiole

The area where a petiole detaches from the stem is structurally different from surrounding tissues. This area, called the **abscission zone,** is composed primarily of thin-walled parenchyma cells and, because it contains few fibers, is anatomically weak (Fig. 32–14). A protective layer of cork cells develops on the stem side of the abscission zone. These cells have a waxy, waterproof material impregnated in their walls. Enzymes then dissolve the *middle lamella* (the "cement" that holds the primary cell walls of adjacent cells together) in the abscission zone. Once this process is completed, nothing holds the leaf to the stem but a few xylem cells. A sudden breeze is enough to make the final break, and the leaf detaches. The protective layer of cork remains, sealing off the area and forming a leaf scar.

Figure 32–14 **Abscission zone.** This LM of a longitudinal section through a silver maple (*Acer saccharinum*) branch shows the abscission zone at the base of a petiole. The abscission zone is an area of thin-walled cells at the petiole's base where the leaf will separate from the stem. An axillary bud with its protective bud scales is also evident above the petiole. *(James Mauseth, University of Texas)*

MODIFIED LEAVES MAY HAVE VARIED FUNCTIONS

Although photosynthesis is the main function of leaves, certain leaves possess special modifications for other functions. Some plants have leaves specialized for protection. **Spines,** modified leaves that are hard and pointed, may be found on many desert plants like cacti (Fig. 32–15*a*). In the cactus, the main organ of photosynthesis is the stem rather than the leaf. Spines discourage animals from eating the succulent stem tissue.

Vines are climbing plants whose stems cannot support their own weight, and so they often possess **tendrils** that help keep the vine attached to the structure on which it is growing. The tendrils of many vines, such as peas, cucumbers, and squash, are specialized leaves (Fig. 32–15*b*). However, some tendrils, such as those of ivy, Virginia creeper, and grape, are specialized stems.

(a) *(b)* *(c)*

(d) *(e)*

Figure 32–15 **Leaf modifications.** *(a)* The leaves of cacti such as this barrel cactus *(Ferocactus sp.)* are modified as spines for protection. *(b)* Tendrils, which wind around objects and aid vines in climbing, may be modified leaves or stems. Shown are tendrils of bur cucumber *(Echinocystis lobata)*, which are modified leaves. *(c)* Overlapping bud scales are leaves modified to protect buds. Shown here are a terminal bud and two axillary buds of a maple *(Acer sp.)* twig. *(d)* The leaves of bulbs such as the onion *(Allium cepa)* are fleshy for storage of food materials and water. *(e)* Some plants have thick, succulent leaves modified for water storage as well as photosynthesis. The succulent leaves of the string-of-beads plant *(Senecio rowleyanus)* are spherical to minimize surface area, thereby conserving water. *(a, e, James Mauseth, University of Texas; b, Stephen P. Parker/Photo Researchers, Inc.; c, d, Dennis Drenner)*

The winter buds of a dormant woody plant are covered by **bud scales,** modified leaves that protect the delicate meristematic tissue of the bud from being injured and drying out (Fig. 32–15c).

Leaves may also be modified for storage of water or food. For example, a **bulb** is a short underground stem to which large, fleshy leaves are attached (Fig. 32–15d). Onions and tulips form bulbs. Many plants adapted to arid conditions, for example, jade plant *(Crassula arborescens)*, medicinal aloe *(Aloe barbadensis)*, and string-of-beads *(Senecio rowleyanus)*, have succulent leaves for water storage (Fig. 32–15e). These leaves are usually green and function in photosynthesis as well.

Modified leaves of insectivorous plants capture insects

Insectivorous plants are plants that capture insects. Most insectivorous plants grow in poor soil that is deficient in certain essential minerals, particularly nitrogen. These plants meet some of their mineral requirements by digesting insects and other small animals. The leaves of insectivorous plants are adapted to attract, capture, and digest their animal prey.

Some insectivorous plants have passive traps. The leaves of a pitcher plant, for example, are shaped so that rainwater collects and forms a reservoir that also contains acid secreted by the plant (Fig. 32–16). Some pitchers are quite large; in the tropics, for example, pitcher plants may be large enough to hold one liter (approximately one quart) or more of liquid. An insect attracted by the odor or nectar of the pitcher may lean over the edge and fall in. Although it may make repeated attempts to escape, the insect is prevented from crawling out by the slippery sides and the rows of stiff hairs that point downward around the lip of the pitcher. The insect eventually drowns, and part of its body eventually disintegrates and is absorbed. Interestingly, although most insects are killed in pitcher plants, the larvae of several insects (flies, midges, and mosquitoes), as well as a large community of microorganisms, actually live inside the pitchers. All three insect species obtain their food from the insect carcasses, and the pitcher plant digests what remains. It is not known how these insects survive the acidic environment inside the pitcher.

The Venus flytrap is an insectivorous plant with active traps. Its leaves resemble tiny bear traps (see Fig. 1–4). Each side of the leaf contains three small hairs. If an insect alights and brushes against two of the hairs, or against the same hair twice in quick succession, the trap springs shut with amazing rapidity. (The mechanism whereby the leaves shut is discussed further in Chapter 36.) After the insect has died and been digested, the trap reopens and the indigestible remains fall out.

Figure 32–16   **A common pitcher plant (Sarracenia purpurea).** This species is widely distributed in acidic bogs and marshes in eastern North America. Young pitchers are green but turn red as they age. The insectivorous pitcher plant has leaves modified to form a water-collecting pitcher that drowns its prey. Note the dead beetle in the "pitcher." *(Bill Lea/Dembinsky Photo Associates)*

SUMMARY WITH KEY TERMS

I. Leaves exhibit variation in shape and form.
  A. Leaves typically consist of a broad, flat **blade** and a stalklike **petiole.** Some leaves also possess small, leaflike outgrowths from the base called **stipules.**
  B. Leaves may be simple or compound.
  C. Leaf arrangement on a stem may be alternate, opposite, or whorled.
  D. Leaves may have parallel or netted (either pinnately netted or palmately netted) venation.
II. Leaf structure is adapted for its primary function of **photosynthesis.**
  A. Most leaves have a broad, flattened blade that is quite efficient in collecting the sun's radiant energy.
  B. The transparent **epidermis** allows light to penetrate into the mesophyll, where photosynthesis occurs.
    1. **Stomata** are small pores in the epidermis that permit gas exchange needed for photosynthesis and **transpiration** needed to obtain water and minerals from the soil. Each pore is surrounded by two **guard cells** that are often associated with special epidermal cells called **subsidiary cells.** Subsidiary cells provide a reservoir of water and ions that move into and out of the guard cells as they change shape during stomatal opening and closing.
    2. The waxy **cuticle** coats the epidermis, enabling the plant to survive the dry conditions of a terrestrial existence.
  C. The **mesophyll** tissue contains air spaces that permit rapid diffusion of carbon dioxide and water into, and oxygen out of, mesophyll cells.
  D. Leaf **veins** have **xylem** to conduct water and essential minerals to the leaf, and **phloem** to conduct sugar produced by photosynthesis to the rest of the plant.
III. Monocot and dicot leaves can be distinguished based on their external and internal structures.
  A. Monocot leaves have parallel venation, whereas dicot leaves have netted venation.
  B. Some monocots (corn and other grasses) do not have mesophyll differentiated into distinct palisade and spongy layers. Some monocots (grasses, reeds, and sedges) have guard cells shaped like dumbbells.
IV. Stomata generally open during the day and close at night.
  A. The movement of H^+ and K^+ across the plasma membranes of

guard cells explains the opening and closing process.

1. Light, particularly blue light, initiates a proton pump in the guard cells' plasma membranes that pumps H^+ out of the guard cells. The H^+ is produced by the ionization of malic acid, which is produced in the chloroplasts of guard cells.
2. The electrochemical gradient established by the proton pump drives the active transport of potassium ions into the guard cells.
3. The resulting osmotic movement of water into the guard cells causes them to become turgid, forming a pore.
4. Stomata close when potassium ions leave the guard cells, causing water to flow out by osmosis. As the guard cells become less turgid and collapse, the pore closes.

B. A number of factors affect stomatal opening and closing, including light or darkness, CO_2 concentration, water stress, and the plant's circadian rhythm.

V. **Transpiration** is the loss of water vapor from aerial parts of plants.
A. It occurs primarily through the stomata.
B. The rate of transpiration is affected by environmental factors like temperature, wind, and relative humidity.
C. Transpiration appears to be both beneficial and harmful to the plant (the CO_2 requirement for photosynthesis/water conservation trade-off).

VI. **Guttation** is the release of liquid water from leaves of some plants that occurs through special structures when transpiration is negligible and available soil moisture is high.

VII. Leaf **abscission** is a complex process involving physiological and anatomical changes prior to leaf fall.

VIII. Leaves may be modified for functions other than photosynthesis.
A. **Spines** are leaves adapted to provide protection.
B. Some **tendrils** are leaves modified for grasping and holding onto other structures (to support weak stems).
C. **Bud scales** are leaves modified to protect delicate meristematic tissue or dormant buds.
D. **Bulbs** are short underground stems with fleshy leaves specialized for storage.
E. Insectivorous plants have leaves modified to trap insects.

POST-TEST

1. The photosynthetic tissue in the middle of the leaf is called (a) cutin (b) mesophyll (c) the abscission zone (d) subsidiary mesophyll (e) palisade and spongy stomata
2. Gas exchange occurs through microscopic pores formed by two (a) subsidiary cells (b) abscission cells (c) mesophyll cells (d) guard cells (e) stipules
3. Most stomata are usually located in the _____ of the leaf. (a) upper epidermis (b) lower epidermis (c) cuticle (d) spongy mesophyll (e) palisade mesophyll
4. The thin, noncellular layer of wax secreted by the epidermis of leaves is the (a) stoma (b) subsidiary cell (c) trichome (d) bundle sheath (e) cuticle
5. The _____ encircles a vein. (a) palisade mesophyll (b) guard cell (c) bundle sheath (d) blade (e) cuticle
6. The _____ of a leaf vein transports water and dissolved nutrient minerals, whereas the _____ transports sugars produced by the leaf during photosynthesis. (a) xylem; phloem (b) xylem; bundle sheath (c) phloem; xylem (d) phloem; trichome (e) cuticle; bundle sheath
7. Most of the water that a plant absorbs from the soil is lost by the process of (a) guttation (b) circadian rhythm (c) abscission (d) transpiration (e) photosynthesis
8. When transpiration is negligible, plants such as grasses exude excess water by (a) guttation (b) circadian rhythm (c) abscission (d) pumping H^+ out of and K^+ into guard cells (e) photosynthesis
9. Place the following events of stomatal opening in correct order.
   (1) proton pump moves H^+ out of guard cells
   (2) guard cells change shape and pore appears
   (3) leaf is exposed to light
   (4) water diffuses into guard cells
   (5) K^+ actively transported into guard cells
   (a) 1-5-4-2-3 (b) 5-3-4-2-1 (c) 1-3-4-2-5 (d) 3-5-1-2-4 (e) 3-1-5-4-2
10. The seasonal detachment of leaves is known as (a) forest decline (b) transpiration (c) abscission (d) guttation (e) dormancy
11. Modified leaves that enable a stem to climb are called _____, whereas modified leaves that cover the winter buds of a dormant woody plant are called _____. (a) spines; bud scales (b) bud scales; tendrils (c) tendrils; bud scales (d) tendrils; spines (e) insectivorous; spines
12. Leaves have a trade-off, or compromise, between photosynthesis and transpiration, which results from (a) the numerous stomatal pores that provide gas exchange for photosynthesis and openings through which water vapor escapes (b) the secretion of a waxy layer, the cuticle, that reduces water loss (c) blue light triggering an influx of potassium ions (K^+) into the guard cells (d) the abscission of leaves of deciduous plants as winter approaches in temperate climates (e) the stomata being closed at night, although water continues to move into the roots by osmosis

REVIEW QUESTIONS

1. Give the general equation for photosynthesis (see Chapter 8), and discuss how the leaf is organized to deliver the raw materials and remove the products of photosynthesis.
2. How is leaf structure related to both photosynthesis and transpiration?
3. Relate the series of physiological changes that occur in guard cells during stomatal opening and closing.
4. Discuss at least two ways that, during the course of evolution, certain leaves became adapted to conserve water.
5. How do environmental factors (sunlight, temperature, humidity, wind) influence the rate of transpiration? How does the environment influence stomatal opening and closing?
6. Why do many temperate woody plants lose their leaves in autumn?
7. Discuss the specialized features of the leaves of insectivorous plants.
8. Label the diagram. Use Fig. 32–3 to check your answers.

YOU MAKE THE CONNECTION

1. Suppose you are asked to observe a micrograph of a leaf cross section and distinguish between the upper and lower epidermis. How would you make this decision?
2. Given that (1) xylem is located toward the upper epidermis in leaf veins while phloem is toward the lower epidermis, and (2) the vascular tissue of a leaf is continuous with that of the stem, suggest one possible arrangement of vascular tissues in the stem that might account for the arrangement of vascular tissue in the leaf.
3. What might be some of the advantages of a plant having a few very large leaves? Disadvantages? What might be some advantages of having many very small leaves? Disadvantages? How would your answer differ along a moisture gradient, from a humid environment to a desert?

RECOMMENDED READINGS

Baskin, Y. *The Work of Nature: How the Diversity of Life Sustains Us.* Island Press, Washington, D.C., 1997. Chapter 8, "Climate and Atmosphere," examines some of the connections between regional climate and trees.

Beerling, D.J. and C.K. Kelly. "Stomatal Density Responses of Temperate Woodland Plants over the Past Seven Decades of CO_2 Increase: A Comparison of Salisbury (1927) with Contemporary Data." *American Journal of Botany,* Vol. 84, No. 11, 1997. Some plants have responded to changes in the concentration of atmospheric CO_2 by adjusting their stomatal density.

Berg, L.R. *Introductory Botany: Plants, People, and the Environment.* Saunders College Publishing, Philadelphia, 1997. A general botany text with an environmental emphasis.

Dale, J.E. "How Do Leaves Grow?" *BioScience,* Vol. 42, No. 6, Jun. 1992. Leaf research is "branching out" in new directions, away from describing leaf variations and toward understanding changes that occur during leaf development.

Dale, J.E. "Power Plants." *The Sciences,* Sep./Oct. 1994. What is currently known about how a leaf primordium, which starts out as a mound of cells on the stem apical meristem, develops into a fully expanded leaf.

Raloff, J. "When Nitrate Reigns," *Science News,* Vol. 147, No. 6, 11 Feb. 1995. This article dispels the myth that nitrates and other forms of nitrogenous air pollution are good for trees because they provide fertilizer. In actuality, an excess of such nitrogen compounds promotes forest decline.

Smith, W.K., T.C. Vogelmann, E.H. DeLucia, D.T. Bell, and K.A. Shepherd. "Leaf Form and Photosynthesis." *BioScience,* Vol. 47, No. 11, Dec. 1997. Leaf structure is strongly linked to internal light and CO_2 levels.

Vogel, S. "When Leaves Save the Tree." *Natural History,* Sep. 1993. In high winds, leaves roll or clump together to minimize drag.

Zimmer, C. "The Processing Plant." *Discover,* Sep. 1995. Discusses the unique ecosystem and food chain inside a pitcher plant, which involves three insect species as well as the plant.

● Visit our website at **http://www.saunderscollege.com/lifesci/titles.html** and click on Solomon/Berg/Martin Biology for links to chapter-related resources on the World Wide Web.

CHAPTER 33

Stems and Plant Transport

A vegetative (not sexually reproductive) vascular plant has three parts: roots, leaves, and stems. As discussed in Chapter 31, roots serve to anchor the plant and absorb materials from the soil, whereas leaves are primarily for photosynthesis, converting radiant energy into the chemical energy of carbohydrate molecules. Stems, the focus of this chapter, link a plant's roots to its leaves and are usually located above ground, although many plants have underground stems. Stems exhibit varied forms, ranging from ropelike vines to massive tree trunks. They can be either herbaceous (consisting of soft, nonwoody tissues) or woody (with extensive hard tissues of wood and bark).

Stems perform three main functions in plants. First, stems of most species support leaves and reproductive structures. The upright position of most stems and the arrangement of the leaves on them allow each leaf to absorb light for use in photosynthesis. Reproductive structures (flowers and fruits) are located on stems in areas accessible to insects, birds, and air currents, which transfer pollen from flower to flower and help disperse seeds and fruits.

Second, stems provide internal transport. They conduct water and dissolved nutrient minerals from the roots, where these materials are absorbed from the soil, to leaves and other plant parts. Stems also conduct the sugar produced in leaves by photosynthesis to roots and other parts of the plant. Remember, however, that stems are not the only plant organ that conducts materials. The vascular system is continuous throughout all parts of a plant, and conduction occurs in roots, stems, leaves, and reproductive structures.

Third, stems produce new living tissue. They continue to grow throughout a plant's life, producing buds that develop into stems with new leaves and/or reproductive structures. In

(Frans Lanting/Minden Pictures)

addition to the main functions of support, conduction, and production of new stem tissues, stems of some species are modified for asexual reproduction (see Chapter 35) or, if green, to manufacture sugar by photosynthesis. Also, some stems are specialized to store starch, as in these baobab (*Adansonia digitata*) trees, which are native to Africa, Madagascar, and Australia. Baobab trees store large volumes of water and starch in their massive trunks.

AFTER YOU HAVE STUDIED THIS CHAPTER YOU SHOULD BE ABLE TO

1. Describe three functions of stems.
2. Label cross sections of herbaceous dicot and monocot stems and describe the functions of each tissue.
3. Outline the transition from primary growth to secondary growth in a woody stem, name the two lateral meristems, and describe the tissues that arise from each.
4. Describe the pathway of water movement in plants.

5. Discuss tension-cohesion and root pressure as mechanisms to explain the rise of water and dissolved nutrient minerals in xylem.
6. Describe the pathway of sugar translocation in plants.
7. Discuss the pressure-flow hypothesis of sugar translocation in phloem.

A WOODY TWIG DEMONSTRATES EXTERNAL STEM STRUCTURE

Although stems exhibit great variation in structure and growth, they all have **buds,** which are undeveloped embryonic shoots (see Fig. 31–4). A **terminal bud** is the embryonic shoot located at the tip of a stem. When a terminal bud is dormant (that is, not actively growing), its apical meristem is covered and protected by an outer protective layer of **bud scales,** which are modified leaves (see Fig. 32–15c). **Axillary buds,** also called **lateral buds,** are located in the *axils* of a plant's leaves (see Fig. 32–1). An axil is the upper angle between a leaf and the stem to which it is attached. When terminal and axillary buds grow, they form stems that bear leaves and/or flowers. The area on a stem where each leaf is attached is called a **node,** and the region between two successive nodes is an **internode.**

A woody twig of a deciduous tree that has shed its leaves can be used to demonstrate certain stem structures (Fig. 33–1). The terminal bud is covered by bud scales that protect its delicate apical meristem during dormancy. When the bud resumes growth, the bud scales covering the terminal bud fall off, leaving **bud scale scars** on the stem where the bud scales were attached. Because temperate-zone plants form terminal buds at the end of each year's growing season, the number of sets of bud scale scars on a twig indicates its age. A **leaf scar** shows where each leaf was attached on the stem; the vascular (conducting) tissue that extends from the stem out into the leaf forms **bundle scars** within a leaf scar. Axillary buds may be found above the leaf scars. Also, the bark of a woody twig has **lenticels,** sites of loosely arranged cells that allow oxygen to diffuse into the interior of the woody stem. Lenticels look like tiny specks on the bark of a twig.

STEMS ORIGINATE AND DEVELOP AT MERISTEMS

Recall from Chapter 31 that plants have two different types of growth. Primary growth is an increase in the length of a plant and occurs at **apical meristems** located at the tips of stems and roots. Secondary growth is an increase in the girth (thick-

ness) of a plant due to the activity of **lateral meristems** located within stems and roots. The new tissues formed by the lateral meristems are called *secondary tissues* to distinguish them from *primary tissues* produced by apical meristems.

All plants have primary growth; some plants have both primary and secondary growth. Recall that stems with only primary growth are herbaceous, while those with both primary

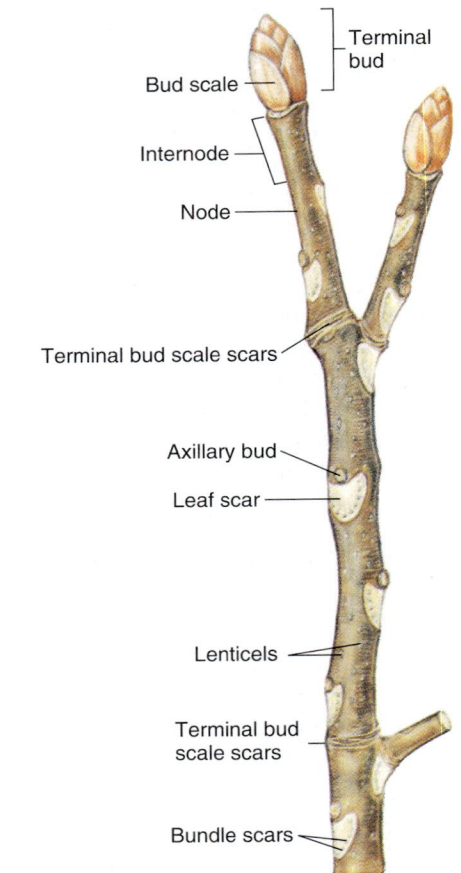

Figure 33–1 **External structure of a woody twig in its winter condition.** The age of a woody twig can be determined by the number of sets of bud scale scars (don't count side branches). How old is this twig?

and secondary growth are woody.[1] A woody plant increases in length by primary growth at the tips of its stems and roots, while its older stems and roots further back from the tips increase in girth by secondary growth. In other words, at the same time that secondary growth is adding wood and bark (thereby causing the stem to thicken), primary growth (which increases the length of the stem) continues.

HERBACEOUS DICOT AND MONOCOT STEMS DIFFER IN INTERNAL STRUCTURE

Although considerable structural variation exists in stems, they all possess an outer protective covering (epidermis or periderm), one or more types of ground tissue, and vascular tissues (xylem and phloem). Let us first consider the structure of herbaceous dicot stems and then of monocot stems.

Vascular bundles of herbaceous dicot stems are arranged in a circle in cross section

A young sunflower stem is a representative herbaceous dicot stem that exhibits primary growth (Fig. 33–2). Its outer covering, the **epidermis,** provides protection in herbaceous stems, as it does in leaves and herbaceous roots. The epidermis is covered by a cuticle, a waxy layer of *cutin* that also covers the leaf

[1]Many herbaceous stems (such as those of geranium and sunflower) also have a limited amount of secondary growth.

epidermis (see Chapter 32) and reduces water loss from the stem surface.

Inside the epidermis is the **cortex,** a cylinder of cells that may contain parenchyma, collenchyma, and sclerenchyma cells (see Chapter 31). As might be expected from the various types of cells that it contains, the cortex in herbaceous dicot stems can have several functions, such as photosynthesis, storage, and support. If a stem is green, photosynthesis occurs in chloroplasts of cortical parenchyma cells. Parenchyma in the cortex also stores starch (in amyloplasts) and crystals (in vacuoles). Collenchyma and sclerenchyma in the cortex provide strength and structural support for the stem.

The vascular tissues provide conduction and support. In herbaceous dicot stems, the vascular tissues are located in bundles that, when viewed in cross section, are arranged in a circle. However, viewed lengthwise, these bundles extend as long strands throughout the length of a stem and are continuous with vascular tissues of both roots and leaves.

Each vascular bundle contains both **xylem,** which transports water and dissolved nutrient minerals from roots to leaves, and **phloem,** which transports dissolved sugar. Xylem is located on the inner side of the vascular bundle, and phloem is found toward the outside. Sandwiched between xylem and phloem is a single layer of cells called the **vascular cambium,** a lateral meristem responsible for secondary growth (discussed shortly).

Because most stems grow through the air and support the plant body, they are much stronger than roots. The thick walls of tracheids and vessel elements in xylem help support the plant. Fibers also occur in both xylem and phloem, although

(a) 500 µm

(b) 250 µm

Epidermis
Cortex
Phloem fiber cap
Phloem
Vascular cambium
Xylem
Vascular bundle

Vascular bundles
Pith
Cortex
Epidermis

Figure 33–2 **LMs of a herbaceous dicot stem.** *(a)* Cross section of a sunflower *(Helianthus annuus)* stem. Note the vascular bundles arranged in a circle around a central core of pith. *(b)* Close-up of two vascular bundles. Xylem is located toward a stem's interior, and phloem is toward the exterior. Each vascular bundle is "capped" by a batch of fibers for additional support. *(a and b, Ed Reschke)* they are usually more extensive in phloem. These fibers add considerable strength to the herbaceous stem. In sunflowers and certain other herbaceous dicot stems, phloem contains a cluster of fibers toward the outside of the vascular bundle, called a **phloem fiber cap,** that helps strengthen the stem. The phloem fiber cap is not present in all herbaceous dicot stems.

The **pith** at the center of the herbaceous dicot stem is composed of large, thin-walled parenchyma cells that function primarily in storage. Due to the arrangement of the vascular tissues in bundles, there is no distinct separation of cortex and pith between the vascular bundles. The areas of parenchyma between the vascular bundles are often referred to as **pith rays.**

Vascular bundles are scattered throughout monocot stems

Monocot stems, as exemplified by the herbaceous stem of corn, are covered by an epidermis with its waxy cuticle. As in herbaceous dicot stems, the vascular tissues run in strands throughout the length of a stem. In cross section the vascular bundles contain xylem toward the inside and phloem toward the outside. In contrast to herbaceous dicots, however, vascular bundles of monocots are not arranged in a circle but instead are scattered throughout the stem (Fig. 33–3). Each vascular bundle is enclosed in a bundle sheath of supporting sclerenchyma cells. The monocot stem does not have distinct areas of cortex and pith. The **ground tissue** in which the vascular tissues are embedded performs the same functions as cortex and pith in herbaceous dicot stems.

The lateral meristems, vascular cambium and cork cambium, that give rise to secondary growth do not occur in monocot stems. Monocots have primary growth only and do not produce wood and bark. Although some treelike monocots such as palms attain considerable size, they do so by a modified form of primary growth in which parenchyma cells divide and enlarge. Stems of some monocots like bamboo and palm contain a great deal of sclerenchyma tissue, which makes them extremely hard.

WOODY PLANTS HAVE STEMS WITH SECONDARY GROWTH

Woody plants undergo secondary growth, an increase in the girth of stems and roots. Secondary growth occurs as a result of the activity of two lateral meristems: vascular cambium and cork cambium. Among flowering plants, only woody dicots (such as apple, hickory, and maple) have secondary growth. Cone-bearing gymnosperms (such as pine, juniper, and spruce) also have secondary growth. Table 33–1 summarizes the relationships of meristems and tissues in a woody dicot stem.

Cells in **vascular cambium** divide and produce two conducting and supporting tissues: secondary xylem (wood) to replace primary xylem, and secondary phloem (inner bark) to replace primary phloem. Primary xylem and primary phloem are not able to transport materials indefinitely and so are replaced if the plant is to survive long-term. Cells of the second

(a)   500 μm

(b)   100 μm

Figure 33–3 **LMs of a monocot stem.** *(a)* Cross section of a corn *(Zea mays)* stem shows vascular bundles scattered throughout ground tissue. *(b)* Close-up of a vascular bundle. The air space is the site where the first xylem elements were formed and later disintegrated. The entire bundle is enclosed in a bundle sheath of sclerenchyma for additional support. *(a and b, Ed Reschke)*

VINES, EVOLUTIONARY ADAPTATIONS, AND ECOLOGY

How are vines adapted to their environment? Vines are weak-stemmed plants that depend on other plants for support. Because vines do not expend many resources to produce structurally strong stems, they are able to grow extremely rapidly, often ascending through the shaded understory to the sunlit canopy of forest trees where they produce luxuriant foliage.

Vines have a variety of adaptations that fit their climbing lifestyle. As newly germinated plants that are rooted in the soil, many vine seedlings grow *away* from sunlight rather than toward it. In growing toward the darkest part of its environment, a vine seedling usually encounters a large tree that it then ascends. Along their stems, woody vines (known as *lianas*) often produce special roots with adhesive pads that stick to the bark of the host tree. Herbaceous vines frequently have tendrils, modified leaves or stems that wrap around supports (see Chapter 32), while other vines are *twiners*, with stems that spiral around their host as they ascend it.

Vines are most numerous and diverse in tropical forests, particularly tropical rain forests (see Chapter 54), where both herbaceous vines and lianas abound. Lianas that have grown into the upper canopy often grow from the branches of one forest tree to another, connecting the tops of the trees and providing a walkway for many of the canopy's animal residents. These and other vines provide nectar and fruit for many tree-dwelling animals.

Temperate forests, boreal (far northern) forests, and island forests have far fewer vines than do tropical rain forests. It is thought that vines are less common in temperate and boreal forests because they are less resistant to droughts and fires than are the trees that grow there. Vines in tropical rain forests generally need not adapt to long droughts or forest fires. Some botanists have suggested that vines are relatively uncommon on islands because their seeds, which are dispersed by wind, cannot reach islands to colonize them.

---

lateral meristem, **cork cambium,** divide and produce cork cells and cork parenchyma. Cork cambium and the tissues it produces are collectively referred to as **periderm** (outer bark), which functions as a replacement for the epidermis (see the LM in Table 31–4).

Vascular cambium gives rise to secondary xylem and secondary phloem

Primary tissues in woody dicot stems are organized similarly to those in herbaceous dicot stems, with the vascular cambium a thin layer of cells sandwiched between xylem and phloem in the vascular bundles. Once secondary growth begins, however, the internal structure of a stem changes considerably (Fig. 33–4). Although vascular cambium is not initially a continuous cylinder of cells (because the vascular bundles are separated by pith rays), it becomes continuous when production of secondary tissues begins. This continuity develops because certain parenchyma cells in each pith ray retain the ability to divide. These cells connect to vascular cambium cells in each vascular bundle, thus forming a complete ring of vascular cambium.

Cells in the vascular cambium divide and produce cells in two directions. The cells formed from the dividing vascular cambium are located either *inside* the ring of vascular cam-

---

TABLE 33–1   Development in a Woody Dicot Stem

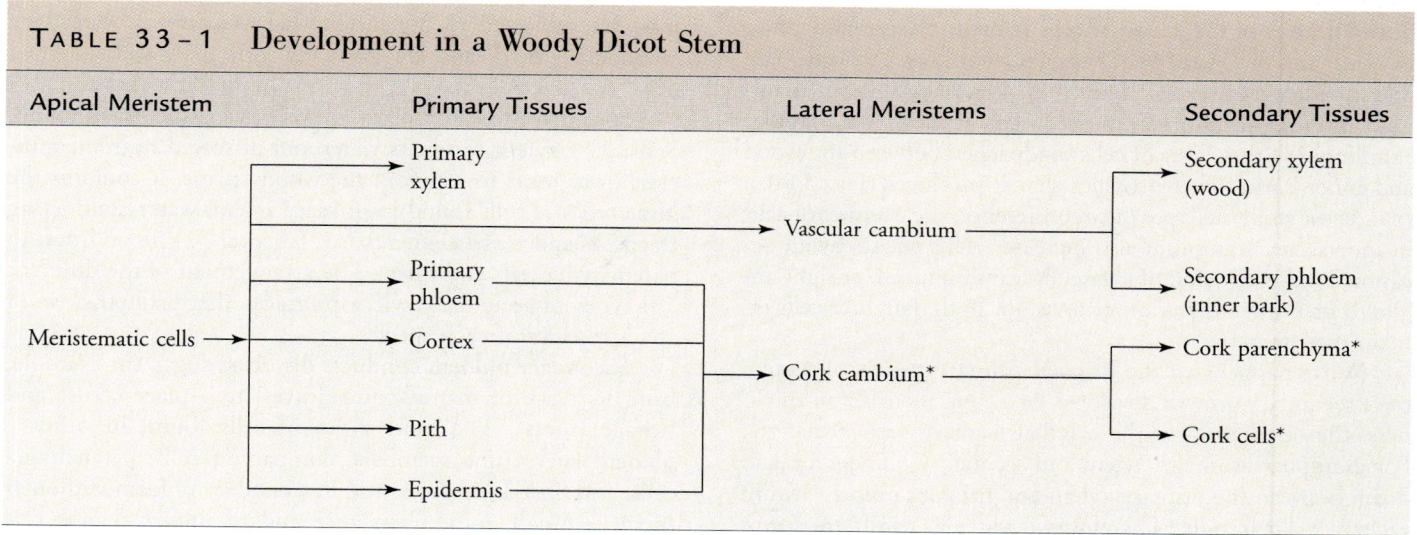

Apical Meristem	Primary Tissues	Lateral Meristems	Secondary Tissues
Meristematic cells	Primary xylem	Vascular cambium	Secondary xylem (wood)
	Primary phloem		Secondary phloem (inner bark)
	Cortex	Cork cambium*	Cork parenchyma*
	Pith		Cork cells*
	Epidermis		

*Collectively known as periderm (outer bark).

(a)

Figure 33–4 Development of secondary growth in a dicot stem. Vascular cambium and the tissues it produces are shown in cross section; cork cambium is not depicted. *(a)* A herbaceous dicot stem. Vascular cambium is sandwiched between primary xylem and primary phloem in each vascular bundle. At the onset of secondary growth, vascular cambium arises in the parenchyma between the vascular bundles (that is, in the pith rays), forming a cylinder of meristematic tissue that appears as a gray circle in cross section. *(b)* Vascular cambium begins to divide, forming secondary xylem on the inside and secondary phloem on the outside. Note that the primary xylem and primary phloem in the original vascular bundles are separated during secondary growth. *(c)* A young woody stem. Vascular cambium produces significantly more secondary xylem than secondary phloem.

(b)

(c)

bium (to become secondary xylem, or wood), or *outside* it (to become secondary phloem, or inner bark) (Fig. 33–5). When a cell in the vascular cambium divides tangentially (inward or outward), one of the daughter cells remains meristematic; that is, it remains as a part of the vascular cambium. The other cell may divide again several times, but eventually it stops dividing and develops into mature secondary tissue. Thus, vascular cambium is a thin layer of cells sandwiched between the wood and inner bark, the two tissues that it produces (Fig. 33–6).

As the stem increases in circumference, the number of cells in the vascular cambium also increases. This occurs by an occasional radial division of a vascular cambium cell, at right angles to its normal direction of division. Both daughter cells remain meristematic.

What happens to the original primary tissues of a stem once secondary growth develops? As a stem increases in thickness, the orientation of the original primary tissues changes. For example, secondary xylem and secondary phloem are laid down between the primary xylem and primary phloem within each vascular bundle. Therefore, as vascular cambium forms secondary tissues, the primary xylem and primary phloem in each vascular bundle become separated from one another (Fig. 33–7). The primary tissues located outside the cylinder of secondary growth (that is, primary phloem, cortex, and epidermis) are subjected to the mechanical pressures produced by secondary growth and are gradually torn apart and sloughed off.

Secondary tissues replace the primary tissues in function. Secondary xylem conducts water and dissolved nutrient minerals from roots to leaves in the woody plant. It contains the same types of cells found in primary xylem: water-conducting tracheids and vessel elements (see Chapter 31), in addition to parenchyma cells and fibers. The arrangement of the different cell types in secondary xylem produces the distinctive wood characteristics of each species.

Secondary phloem conducts dissolved sugar, for example, from its place of manufacture (leaves) to a place of use and storage (roots). The same types of cells found in primary phloem (sieve tube members, companion cells, parenchyma cells, and fibers) are also found in secondary phloem, although there are usually more fibers in secondary phloem than in primary phloem.

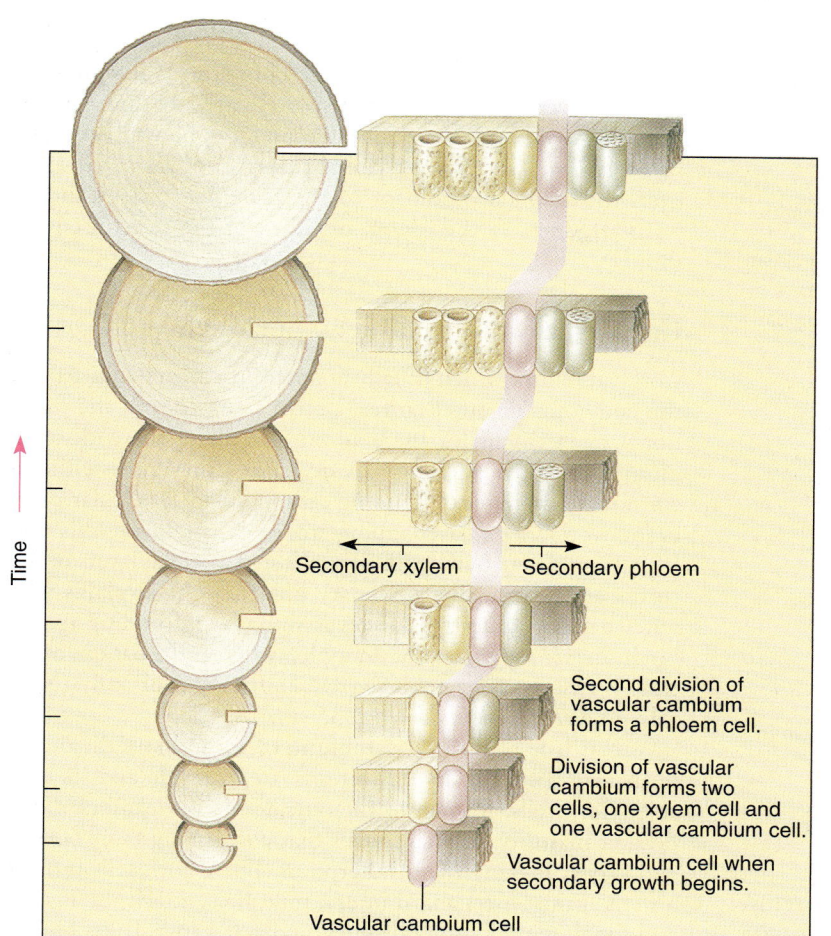

Figure 33-5 **Development of secondary xylem and secondary phloem.** To study the figure, which shows a radial view of a dividing vascular cambium cell, start at the bottom and move up. Note that vascular cambium divides in two directions, forming secondary xylem to the inside and secondary phloem to the outside. These cells differentiate to form the mature cell types associated with xylem and phloem. As secondary xylem accumulates, vascular cambium "moves" outward, and the woody stem increases in diameter.

Time

Secondary xylem ← → Secondary phloem

Second division of vascular cambium forms a phloem cell.

Division of vascular cambium forms two cells, one xylem cell and one vascular cambium cell.

Vascular cambium cell when secondary growth begins.

Vascular cambium cell

Pith

Primary xylem

Annual ring of second xylem

Secondary xylem (wood)

Vascular cambium

Secondary phloem

Primary phloem and cortex

Cork parenchyma

Cork cambium

Ray

Cork cells sloughing off

500 μm

Figure 33-6 **LM of a three-year-old basswood *(Tilia americana)* stem in cross section.** Note the location of the vascular cambium between the secondary xylem (wood) and secondary phloem (inner bark). *(Carolina Biological Supply Company/Phototake-NYC)*

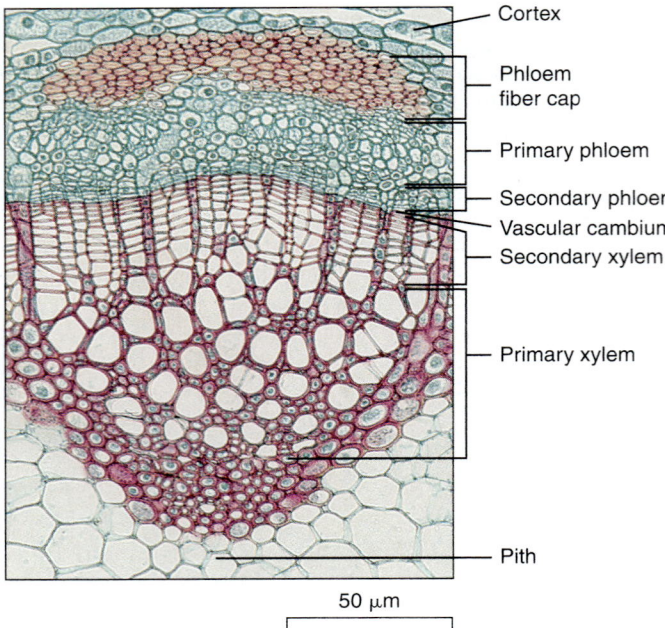

Cortex
Phloem fiber cap
Primary phloem
Secondary phloem
Vascular cambium
Secondary xylem
Primary xylem
Pith

50 μm

Figure 33–7 **LM of part of a magnolia (*Magnolia* sp.) stem in cross section.** Note that the vascular bundle has been split apart by secondary growth. Compare this vascular bundle with the one in Fig. 33-2*b*, which has primary growth only. *(Dennis Drenner)*

Whereas secondary xylem and secondary phloem transport water, minerals, and sugar vertically throughout the woody plant body, materials must also move horizontally, that is, laterally. Lateral movement occurs through **rays,** which are chains of parenchyma cells that radiate out from the center of the woody stem or root (see Fig. 33–6). Rays, which are often continuous from the secondary xylem to the secondary

phloem, are formed by the vascular cambium. Water and dissolved nutrient minerals are transported laterally through rays from the secondary xylem to the secondary phloem. Likewise, rays form pathways for the lateral transport of dissolved sugar from the secondary phloem to the secondary xylem and of waste products to the center, or heart, of the tree (discussed shortly).

Cork cambium produces periderm

Cork cambium, which usually arises from parenchyma cells in the epidermis or outer cortex, produces **periderm,** the functional replacement for the epidermis. Cells of the cork cambium retain their ability to divide. Cork cambium is either a continuous cylinder of dividing cells (similar to vascular cambium) or a series of overlapping arcs of meristematic cells that form from parenchyma cells in successively deeper layers of the cortex and, eventually, secondary phloem. This explains why outer bark of some tree species is smooth and peeling (paper birch, *Betula papyrifera,* for example), whereas the bark in other trees is fissured (bur oak, *Quercus macrocarpa*) or scaly (Norway pine, *Pinus resinosa*) (Fig. 33–8).

As is true of vascular cambium, cork cambium divides to form new tissues in two directions: to its inside and its outside. Cork cells, formed to the outside of cork cambium, are dead at maturity and have walls that contain layers of *suberin* and waxes, making them waterproof. These cork cells protect a stem against mechanical injury, mild fires, attacks by insects and fungi, temperature extremes, and water loss. To its inside, cork cambium sometimes forms cork parenchyma cells that store water and starch granules. Cork parenchyma is only one to several cells thick, much thinner than the cork cell layer.

Cork cells are impermeable to water and gases, yet the living internal cells of the woody stem require oxygen and must

(a)   *(b)*   *(c)*

Figure 33–8 **Variation in bark.** *(a)* Paper birch (*Betula papyrifera*) has a smooth, peeling bark. *(b)* Bur oak (*Quercus macrocarpa*) bark is deeply fissured. *(c)* Bark from Norway pine (*Pinus resinosa*) is scaly. *(a,b,c, Carlyn Iverson)*

Lenticel

Cork cells

Cork cambium and cork parenchyma

200 μm

Figure 33–9 **LM of stem periderm, showing a lenticel.** The epidermis has ruptured due to the proliferation of loosely arranged cork cells in the lenticel. Lenticels permit gas exchange through the periderm. *(James Mauseth, University of Texas)*

be able to exchange gases with the surrounding atmosphere. As a stem thickens from secondary growth, the epidermis, including stomata that allowed gas exchange for the herbaceous stem, dies. Stomata are replaced by lenticels, which permit gas exchange (Fig. 33–9).

Common terms associated with wood are based on plant structure

If you've ever examined different types of lumber, you may have noticed that some trees have wood with two different colors (Fig. 33–10). The functional secondary xylem, that is, the part that conducts water and dissolved nutrient minerals, is the *sapwood,* a thin layer of younger, lighter colored wood that is closest to the bark. *Heartwood,* the older wood in the center of the tree, is typically a brownish red color. A microscopic examination of heartwood reveals that its vessels and tracheids are plugged up with pigments, tannins, gums, resins, and other materials. Therefore, heartwood no longer functions in conduction but instead functions as a storage site for waste products. Heartwood is denser than sapwood and therefore provides structural support for trees. Some evidence suggests that heartwood is also more resistant to decay.

Almost everyone has heard of hardwood and softwood. Botanically speaking, *hardwood* is the wood of flowering plants and *softwood* is the wood of conifers (cone-bearing gymnosperms). The wood of pine and other conifers typically lacks fibers (with their thick secondary cell walls) and vessel elements; the conducting cells in gymnosperms are tracheids. These cell differences generally make conifer wood softer than the wood of flowering plants, although there is a substantial variation from one species to another. The balsa tree, for example, is a "hardwood" whose extremely light, soft wood is used to fashion airplane models.

Woody plants that grow in temperate climates where there is a growing period (during spring and summer) and a dormant period (during winter) exhibit *annual rings,* concentric circles found in cross sections of wood (Fig. 33–10). To determine the age of a woody stem in the temperate zone, simply count the annual rings. In the tropics, environmental conditions, particularly seasonal or year-round precipitation patterns, determine the presence or absence of rings, and rings are not a reliable method of determining the ages of most tropical trees.

Examination of annual rings with a magnifying lens reveals no actual "ring," or line, separating one year's growth from the next. The appearance of a ring in cross section is due

Heartwood

Sapwood

Figure 33–10 **Heartwood and sapwood.** The wood of older trees consists of a dense, central heartwood and an outer layer of sapwood. The sapwood is the functioning xylem that conducts water and dissolved nutrient minerals. *(Carlyn Iverson)*

to differences in cell size and cell wall thickness between secondary xylem formed at the end of the preceding year's growth and that formed at the beginning of the following year's growth. In the spring, when water is plentiful, wood formed by vascular cambium has large-diameter conducting cells (tracheids and vessel elements) and few fibers and is appropriately called *springwood* or *early wood*. As summer progresses and water becomes less plentiful, the wood formed, known as *summerwood* or *late wood*, has narrower conducting cells and many fibers. It is this difference in cell size between the summerwood of one year and the springwood of the following year that gives the appearance of rings (Fig. 33–11). A great deal of information about climate in past times can be learned from the study of annual rings of ancient trees (see *Making the Connection: Tree Ring Analysis and Global Climate*).

As a woody stem increases in girth over the years, the branches that it bears grow along with it as long as they are alive. If a branch dies, it no longer continues to grow with the stem. In time, as the stem increases in girth, it surrounds the base of the dead branch. The basal portion of an embedded dead branch is called a *knot*. It is possible for a knot to contain bark as well as wood. The presence of knots in wood reduces its commercial value, except for ornamental purposes. Some plants, such as knotty pine, are valued for their high production of knots.

TRANSPORT IN PLANTS OCCURS IN XYLEM AND PHLOEM

Roots obtain water and dissolved nutrient minerals from the soil. Once inside roots, these materials are transported upward to stems, leaves, flowers, fruits, and seeds. Furthermore, sugar molecules manufactured in leaves by photosynthesis are transported in solution (that is, dissolved in water) throughout the plant, including into the subterranean roots. Water and dissolved nutrient minerals are transported from roots to other parts of the plant in xylem, whereas dissolved sugar is **translocated** in phloem.

Xylem transport and phloem translocation do not resemble the movement of materials in animals, because in plants nothing *circulates* in a system of vessels. Water and minerals, transported in xylem, travel in one direction only (upward), whereas translocation of dissolved sugar may occur upward or downward in separate phloem cells. Also, xylem transport and phloem translocation differ from internal circulation in animals because movement in both xylem and phloem is driven largely by natural physical processes rather than by a pumping organ, or heart.

How, exactly, do materials travel in the continuous system of the plant's vascular tissues? We first examine water and its movement through the plant, and later we discuss the translocation of dissolved sugar.

WATER AND MINERALS ARE TRANSPORTED IN XYLEM

Water initially moves horizontally into roots from the soil, passing through several tissues until it reaches xylem. Once the water moves into the tracheids and vessel elements of root xylem, it travels upward through a continuous network of these hollow, dead cells from root to stem to leaf. Dissolved nutrient

Cross section of 3-year-old *Tilia* stem

Secondary phloem

Vascular cambium

Summerwood

Annual ring of xylem

Springwood

Summerwood of preceding year

100 μm

Figure 33–11 LM of a portion of a basswood *(Tilia americana)* stem cross section. One annual ring, or growth increment, is shown. Note the differences in cell size between the vessel elements of springwood and summerwood. *(Dennis Drenner)*

TREE RING ANALYSIS AND GLOBAL CLIMATE

In temperate climates, the age of a tree can be determined by counting the number of annual rings. Other useful information can be determined by analyzing tree rings as well. For example, the size of each ring varies depending on local weather conditions, including precipitation and temperature. Sometimes the variation in tree rings can be attributed to a single environmental factor, and similar patterns appear in the rings of different tree species over a large geographical area. For example, trees in the southwestern United States have similar ring patterns due to variations in the amount of annual precipitation. Years with adequate precipitation produce wider rings of growth, whereas years of drought produce much narrower rings.

It is possible to study ring sequences going back several thousand years. First, a *master chronology,* a complete sample of rings dating back as far as possible, is developed *(see figure).* A small core of wood is bored out of the trunk of an old living tree to obtain a sample of rings. The oldest rings (those toward the center of the tree) are matched with the youngest rings (those toward the outside) of an older tree or even an old piece of wood from a house. A master chronology of the area is obtained by using successively older and older sections of wood, even those found in prehistoric dwellings, and overlapping their matching ring sequences. The longest master chronology is of bristlecone pines in the western United States; it goes back almost 9000 years.

Dendrochronology, the study of both visible and microscopic details of tree rings, has been used extensively in several fields. Tree ring analysis has been extremely useful in dating prehistoric sites of Native Americans in the Southwest. For example, the Cliff Palace in the Mesa Verde National Park dates back to the year 1073 A.D. Tree ring analysis indicates that an extended drought forced the original inhabitants to abandon their homes. Tree ring analysis is also useful in other disciplines, including ecology (to study changes in a forest community over time), environmental science (to study the effects of air pollution on tree growth), and geology (to date earthquakes and volcanic eruptions).

Climatologists are increasingly using tree ring data to study past climatic patterns. Annual ring widths of certain tree species that grow at high elevations are sensitive to yearly temperature variations; the rings of these trees are wider in warm years and narrower in cool years. By studying tree rings across long time sequences, the natural pattern of global temperature fluctuations can be determined. This information is particularly important because of concerns about human influence on global climate. Scientists generally agree that the Earth has warmed in recent decades, and there is little doubt that human production of "greenhouse gases" such as carbon dioxide has contributed to this warming (see Chapter 55). Researchers, however, are not sure how much of the recent warming is due to human influence, as opposed to natural climate variability. It is thought that tree ring analysis will help answer this vital question. Scientists in nine European countries, for example, are cooperating in a massive tree ring analysis to construct an annual history of temperatures across northern Europe and Asia since the end of the last Ice Age, about 10,000 years ago. Local diving clubs are obtaining samples of well preserved logs that are thousands of years old from river and lake sediments. Scientists will use these and other samples, including living trees, to piece together a 10,000-year master chronology. When that is accomplished (sometime before the year 2005), it should be possible to determine conclusively if today's climate is distinctly different from the natural climate patterns of the past.

Outer bark

Secondary phloem

Vascular cambium

Annual ring

Summerwood

Springwood

Resin duct

Pith

1950

Outermost ring is year when tree was cut.

1939

1932

Other dates obtained by counting rings inward from bark.

1926

1920

Matching and overlapping older and older wood sections extends dates back in time.

Tree ring dating. A master chronology is developed using progressively older pieces of wood from the same geographical area. By matching the rings of a wood sample of unknown age to the master chronology, the age of the sample can be accurately determined. Ring matching, once a painstakingly tedious job, is usually done by computer today.

TABLE 33-2	Xylem and Phloem Transport Rates in Selected Plants*	
Plant	Maximum Rate in Xylem (cm/hr)	Maximum Rate in Phloem (cm/hr)
Conifer	120	48
Woody dicot	4,400	120
Herbaceous dicot/monocot	6,000	168–660
Herbaceous vine	15,000	72

*Xylem and phloem rates are from different plants within each general group and should be used for comparative purposes only. Adapted from Mauseth, J.D. *Botany: An Introduction to Plant Biology*, 2nd ed. Saunders College Publishing, Philadelphia, 1995.

minerals are carried along passively in the water. The plant does not expend any energy of its own to transport water, which moves as a result of natural physical processes. The transport of xylem sap is the most rapid of any movement of materials in plants (Table 33–2).

How does water move to the tops of plants? It is either pushed up from the bottom of the plant or pulled up to the top of the plant. Although both mechanisms exist, current evidence indicates that most water is transported through xylem by being *pulled* to the top of the plant.

Water movement can be explained by a difference in water potential

In order to understand how water moves, it is helpful to introduce **water potential,** which is defined as the free energy of water. Water potential is important in plant physiology because it is a measure of a cell's ability to absorb water by osmosis (see Chapter 5). Water potential also provides a measure of water's tendency to evaporate from cells.

The water potential of pure water is conventionally set at 0 megapascals[2] because it cannot be measured directly. It is possible to measure differences in the free energy of water molecules in different situations, however. When solutes are dissolved in water, the free energy of water decreases.[3] This means that dissolved solutes lower the water potential to a negative number. *Water moves from a region of higher (less negative) water potential to a region of lower (more negative) water potential.*

The water potential of the soil varies, depending on how much water it contains. When a soil is extremely dry, its wa-

[2]A megapascal (MPa) is a unit of pressure equal to about 10 atmospheres, or 145.1 pounds per square inch.

[3]Solutes induce the formation of hydration layers, in which water molecules surround polar molecules and ions, keeping them in solution by preventing them from coming together. The association of water molecules in hydration layers reduces their motion, thereby decreasing their free energy.

ter potential is very low (very negative). When a soil is moister, its water potential is higher, although it is still a negative number because dissolved nutrient minerals are present in dilute concentrations.

The water potential in root cells is also negative owing to the presence of dissolved solutes. Roots contain more dissolved materials than does soil water, however, unless the soil is extremely dry. This means that under normal conditions the water potential of the root is more negative than the water potential of the soil. Thus, water moves by osmosis from the soil into the root.

Tension-cohesion pulls water up a stem

According to the **tension-cohesion model,** also known as the **transpiration-cohesion model,** water is pulled up the plant as a result of a *tension* produced at the top of the plant (Fig. 33–12). This tension, which resembles that produced when drinking a liquid through a straw, is caused by the evaporative pull of transpiration. Recall from Chapter 32 that **transpiration** is the evaporation of water vapor from plants. Most water loss from transpiration takes place through the numerous microscopic pores (stomata) present on leaf and stem surfaces. The tension extends from leaves, where most transpiration occurs, down the stems and into the roots. It draws water up stem xylem to leaf cells that have lost water as a result of transpiration, and pulls water from root xylem into stem xylem. As water is pulled upward, additional water from the soil is drawn into the roots. Thus, the pathway of water movement is as follows:

Soil → root tissues (epidermis, cortex, etc.) → root xylem → stem xylem → leaf xylem → leaf mesophyll → atmosphere

This upward pulling of water is only possible as long as there is an unbroken column of water in xylem throughout the plant. Water forms an unbroken column in xylem because of the *cohesiveness* of water molecules. Recall from Chapter 2 that water molecules are cohesive, that is, strongly attracted to one another, because of *hydrogen bonding*. In addition, the *adhesion* of water to the walls of xylem cells (also the result of hydrogen bonding) is an important factor in maintaining an unbroken column of water. Thus, the cohesive and adhesive properties of water enable it to form an unbroken column that can be pulled up through the xylem.

The movement of water in xylem due to the tension-cohesion mechanism can be explained in terms of water potential. The atmosphere has an extremely negative water potential. For example, air with a relative humidity of 50% has a water potential of −100 MPa; even moist air at a relative humidity of 90% has a negative water potential of −13 MPa. Thus, there is a water potential gradient from the least negative (the soil) up through the plant to the most negative (the atmosphere). This gradient literally pulls the water from the soil up through the plant.

Coleus

Figure 33–12 **The tension-cohesion model.** This model hypothesizes that water vapor diffuses from the surfaces of leaf mesophyll cells to the drier atmosphere through stomata *(top)*. This produces a tension that pulls water out of leaf xylem toward the mesophyll cells. *(middle)* The cohesion of the water molecules, caused by hydrogen bonding, allows unbroken columns of water to be pulled up the narrow vessels and tracheids of stem xylem. *(bottom)* This in turn pulls water up root xylem, forming a continuous column of water from root xylem to stem xylem to leaf xylem. As water moves upward in the root, it produces a pull that causes soil water to diffuse into the root.

Is tension-cohesion powerful enough to explain the rise of water in the tallest plants? Plant biologists have calculated that the tension produced by transpiration is strong enough to pull water upward 150 meters (500 feet) in tubes the diameter of xylem vessels. Because the tallest trees are no more than about 117 meters (375 feet) high, the tension-cohesion model easily accounts for the transport of water. Currently, most botanists consider the tension-cohesion model to be the dominant mechanism of xylem transport in most plants.

Although the tension-cohesion model was proposed more than 100 years ago to explain water transport in xylem, conclusive experimental evidence to support this mechanism was first obtained in 1995 by two research groups working independently. Both groups demonstrated that large negative pressures exist in xylem and that the water potential gradients in root, stem, and leaf xylem are adequate to explain the observed movement of water.

Root pressure pushes water from the root up a stem

In the less important mechanism for water transport, known as **root pressure**, water that moves into a plant's roots from the soil is *pushed* up through xylem toward the top of the plant. Root pressure occurs because nutrient mineral ions that are actively absorbed from the soil at night are pumped into the xylem, decreasing its water potential. Water then moves into xylem cells from surrounding root cells. In turn, water moves into roots by osmosis because of the difference in water po-

tential between the soil and root cells. The accumulation of water in root tissues produces a positive pressure (as high as +2MPa) that forces the water up through the xylem.

Guttation, a phenomenon in which liquid water is forced out through special openings in the leaves (see Chapter 32), is a manifestation of root pressure. However, root pressure is not strong enough to explain the rise of water to the tops of coastal redwoods and other tall trees. Root pressure exerts an influence in smaller plants, particularly in the spring when the soil is quite wet, but it clearly does not cause water to rise 100 meters or more in the tallest plants. Furthermore, root pressure does not occur to any appreciable extent in summer (when water is often not plentiful in soil), yet the movement of water is greatest during hot summer days.

SUGAR IN SOLUTION IS TRANSLOCATED IN PHLOEM

The sugar produced during photosynthesis is converted into sucrose (common table sugar), a disaccharide composed of one molecule of glucose and one of fructose (see Fig. 3–7*b*), before being loaded into phloem and translocated to the rest of the plant. Sucrose is the predominant photosynthetic product carried in phloem. Translocation in phloem sap is not as swift-moving as xylem transport (Table 33–2).

Fluid within phloem tissue moves both upward and downward. Sucrose is translocated in individual sieve tubes from a *source,* an area of excess sugar supply (usually a leaf), to a *sink,* an area of storage (as insoluble starch) or of sugar use such as roots, apical meristems, fruits, or seeds.

The pressure-flow hypothesis explains translocation in phloem

Translocation of dissolved sugar in phloem is explained by the **pressure-flow hypothesis,** which postulates that dissolved sugar moves in phloem by means of a pressure gradient (that is, a difference in pressure). The pressure gradient exists between the source, where the sugar is loaded into phloem, and the sink, where the sugar is removed from phloem.

At the source, the dissolved sucrose is moved from a leaf's mesophyll cells, where it was manufactured, into the sieve tube members of phloem. This movement occurs by active transport, a process that requires ATP (Fig. 33–13). The ATP supplies energy to pump protons out of the sieve tube members, producing a proton gradient that drives the uptake of sugar through specific channels by the cotransport of protons back into the sieve tube members (an example of a linked cotransport system, discussed in Chapter 5). The sugar therefore accumulates in the sieve tube member. The increase in dissolved sugars in the sieve tube member at the source—a concentration that is two to three times greater than in surrounding cells—decreases (makes more negative) the water potential of that cell. As a result, water moves by osmosis into the sieve tubes, increasing the hydrostatic pressure inside them. Thus, phloem loading at the source is as follows:

Proton pump moves H^+ out of sieve tube member → sugar is actively transported into sieve tube member → water diffuses into sieve tube member → hydrostatic pressure increases

At its destination (the sink), sugar is unloaded by various mechanisms, both active and passive, from the sieve tube members. With the loss of sugar, the water potential in the sieve tube members at the sink increases (becomes less negative). Therefore, water moves out of the sieve tubes by osmosis and into surrounding cells. This decreases the hydrostatic pressure

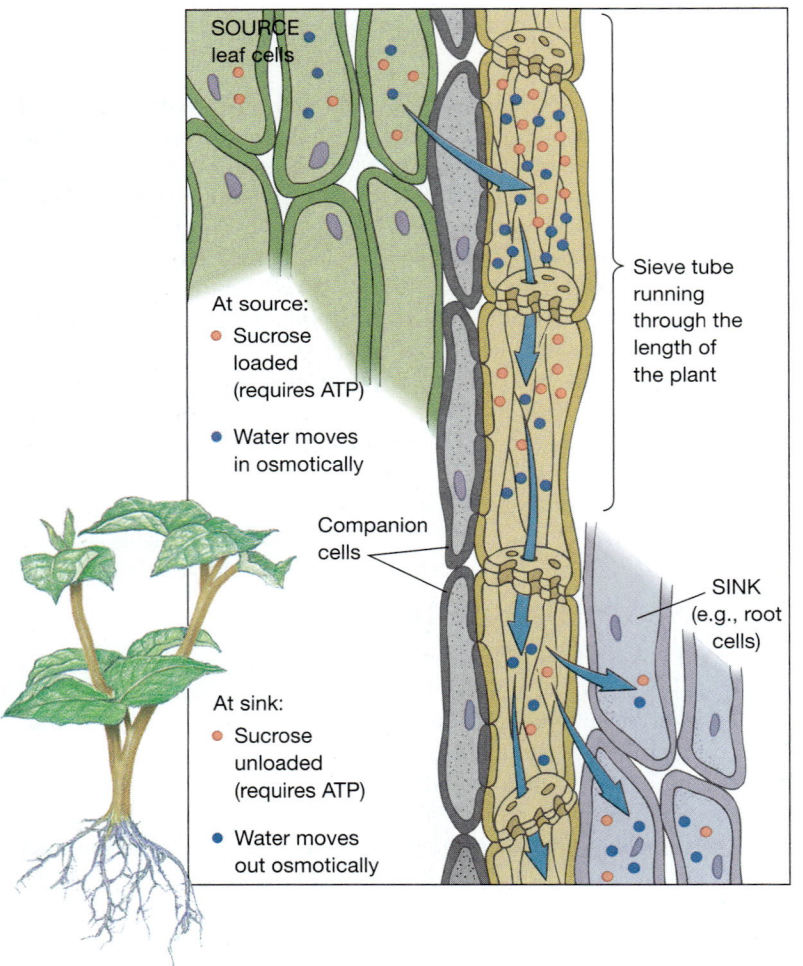

SOURCE
leaf cells

At source:
- Sucrose loaded (requires ATP)
- Water moves in osmotically

Sieve tube running through the length of the plant

Companion cells

SINK (e.g., root cells)

At sink:
- Sucrose unloaded (requires ATP)
- Water moves out osmotically

Figure 33–13 **The pressure-flow hypothesis.** Sugar is actively loaded into the sieve tube member at the source. As a result, water moves into the sieve tube member by osmosis. At the sink, the sugar is actively or passively unloaded, and water leaves the sieve tube member, again by osmosis. The pressure gradient from source to sink causes translocation from the area of higher hydrostatic pressure (the source) to the area of lower hydrostatic pressure (the sink).

(a)

(b)

25 µm

Figure 33–14 **Aphids used to study translocation in phloem.** *(a)* Mature aphid, a tiny insect about 3 to 6 mm in length, feeding on a stem. *(b)* LM of phloem cells, showing a sieve tube member that has been penetrated by the aphid mouthpart. *(a, Dwight Kuhn; b, M.H. Zimmerman,* Science, *Vol. 133, pp. 73–79 [Fig. 4], 13 Jan. 1961. Copyright 1999 by the American Association for the Advancement of Science)*

inside the sieve tubes at the sink. Thus, phloem unloading at the sink is as follows:

> Sugar is transported out of sieve tube member → water diffuses out of sieve tube member → hydrostatic pressure decreases

Thus, the pressure-flow hypothesis explains the movement of dissolved sugar in phloem by means of a pressure gradient. It is the difference in sugar concentrations between the source and the sink that causes translocation in phloem as water and dissolved sugar flow along the pressure gradient. This pressure gradient pushes the sugar solution through phloem much as water is forced through a hose.

The actual translocation of dissolved sugar in phloem does not require metabolic energy. However, the loading of sugar at the source and the active unloading of sugar at the sink require energy derived from ATP to move the sugar across cell membranes by active transport.

Although the pressure-flow hypothesis adequately explains current data on phloem translocation, much remains to be learned about this complex process. Phloem translocation is difficult to study in plants. Because phloem cells are under pressure, cutting into phloem to observe it releases the pres-

sure and causes the contents of the sieve tube members (the phloem sap) to exude and mix with the contents of other severed cells that are also unavoidably cut. In the 1950s botanists developed a unique research tool to avoid contaminating the phloem sap: aphids, which are small insects that insert their mouthparts into phloem sieve tubes for feeding (Fig. 33–14). The pressure in the punctured phloem drives the sugar solution through the aphid's mouthpart and into its digestive system. When the aphid's mouthpart is severed from its body by a laser beam, the sugar solution continues to flow through the mouthpart at a rate proportional to the pressure in phloem. This rate can be measured, and the effects on phloem transport of different environmental conditions—varying light intensities, darkness, or mineral deficiencies, for example—can be ascertained.

The identity and proportions of translocated substances can also be determined using severed aphid mouthparts. This technique has verified that in most plant species the sugar sucrose is the main or only carbohydrate transported in phloem; however, some species transport other sugars, such as raffinose, or sugar alcohols, such as sorbitol. Additional substances transported in phloem include certain plant hormones, ATP, amino acids, K^+ and other inorganic ions, and disease-causing plant viruses.

SUMMARY WITH KEY TERMS

I. The main functions of stems are support, conduction, and production of new living tissues.

II. Woody twigs demonstrate the external structure of stems.

A. **Buds** are undeveloped embryonic shoots. A **terminal bud** is located at the tip of a stem, whereas **axillary buds (lateral buds)** are located in leaf axils.

B. A dormant bud is covered and protected by **bud scales.** When the bud resumes growth, bud scales covering the bud fall off, leaving **bud scale scars.**

C. The area on a stem where each leaf is attached is called a **node,** and the region of a stem between two successive nodes is an **internode.**

D. A **leaf scar** shows where each leaf was attached to the stem. The areas within a leaf scar where the vascular tissue extended from the stem to the leaf are called **bundle scars.**

E. **Lenticels** are sites of loosely arranged cells that allow oxygen to diffuse into the interior of a woody stem.

III. Herbaceous stems possess an epidermis, vascular tissue, and either cortex and pith or ground tissue.

A. The **epidermis** is a protective layer covered by a water-conserving cuticle. Stomata permit gas exchange.

B. **Xylem** conducts water and dissolved nutrient minerals, and **phloem** conducts dissolved sugar and small quantities of other substances, such as hormones, ATP, amino acids, inorganic ions such as K^+, and viruses.

C. The **cortex, pith,** and **ground tissue** function primarily for storage.

IV. Although herbaceous stems all have the same basic tissues, their arrangement varies considerably.

A. Herbaceous dicot stems have the vascular bundles arranged in a circle (in cross section) and have a distinct cortex and pith.

B. Monocot stems have vascular bundles scattered in ground tissue.

V. Secondary growth occurs in some flowering plants (woody dicots) and in all cone-bearing gymnosperms.

A. **Vascular cambium** develops between the primary xylem and the primary phloem. Vascular cambium produces secondary xylem (wood) to the inside and secondary phloem (inner bark) to the outside.

B. **Cork cambium** arises near a stem's surface.

1. Cork cambium produces **periderm,** which consists of cork parenchyma to the inside and cork cells to the outside.

2. Cork cells are the functional replacement for epidermis in the woody stem. Cork parenchyma functions primarily for storage in the woody stem.

C. **Rays** are chains of parenchyma cells that radiate out from the center of a stem and form pathways for the lateral movement of materials between the secondary xylem and the secondary phloem.

VI. Water and dissolved nutrient minerals move upward in xylem from roots to stems and leaves.

A. **Water potential** is a measure of the free energy of water.

1. Pure water has a water potential of 0 megapascals, whereas water with dissolved solutes has a negative water potential.

2. Water moves from an area of higher (less negative) water potential to an area of lower (more negative) water potential.

B. The **tension-cohesion model** explains the rise of water in even the largest plants.

1. The evaporative pull of **transpiration** causes tension at the top of the plant. This tension is the result of a water potential gradient that ranges from the slightly negative water potentials in the soil and roots to the very negative water potentials in the atmosphere.

2. As a result of the gradient in water potentials, water moves from the soil to root xylem to stem xylem to leaf xylem to mesophyll cells to the atmosphere.

3. As a consequence of the cohesive and adhesive properties of water, the column of water pulled up through the plant remains unbroken.

C. **Root pressure,** caused by the movement of water into roots from the soil as a result of the active absorption of nutrient mineral ions from the soil, helps explain the rise of water in small plants, particularly when the soil is wet. Root pressure pushes water up through xylem.

VII. Dissolved sugar is **translocated** upward or downward in phloem.

A. Sucrose is the predominant sugar translocated in phloem.

B. The movement of materials in phloem is explained by the **pressure-flow hypothesis.**

1. Sugar is actively loaded (requiring ATP) into the sieve tubes at the source.

   a. The ATP supplies energy to pump protons out of the sieve tube members.

   b. The proton gradient drives the uptake of sugar by the cotransport of protons back into the sieve tube members.

   c. The sugar therefore accumulates in the sieve tube member, causing the movement of water into the sieve tubes by osmosis.

2. Sugar is actively (requiring ATP) and passively (not requiring ATP) unloaded from the sieve tubes at the sink. As a result, water leaves the sieve tubes by osmosis, decreasing the hydrostatic pressure inside the sieve tubes.

3. The flow of materials between source and sink is driven by the hydrostatic pressure gradient produced by water entering phloem at the source and water leaving phloem at the sink.

POST-TEST

1. All stems have undeveloped embryonic shoots called (a) lenticels (b) buds (c) lianas (d) phloem fiber caps (e) periderm

2. The tissue in monocot stems in which the vascular tissues are embedded is (a) cork cambium (b) cortex (c) ground tissue (d) pith (e) phloem

3. The protective outer layer of cells covering herbaceous stems is the (a) periderm (b) cork cambium (c) lateral meristem (d) epidermis (e) bud scale

4. Ground tissue in monocot stems performs the same functions as _____ and _____ in herbaceous dicot stems (a) phloem; xylem (b) cork cambium; vascular cambium (c) epidermis; periderm (d) primary xylem; secondary xylem (e) cortex; pith

5. The two lateral meristems responsible for secondary growth are (a) phloem and xylem (b) cork cambium and vascular cambium (c) epidermis and periderm (d) primary xylem and secondary xylem (e) cortex and pith

6. Cork cambium and the tissues it produces are collectively called (a) periderm (b) lenticels (c) cortex (d) epidermis (e) wood

7. Horizontal movement of materials in woody plants occurs in (a) bud scales (b) cortex (c) rays (d) lenticels (e) pith rays

8. Water potential is (a) the formation of a proton gradient across a cell membrane (b) the transport of a watery solution of sugar in phloem (c) the transport of water in both xylem and phloem (d) the removal of sucrose at the sink, causing water to move out of the sieve tubes (e) the free energy of water in a particular situation

9. Which of the following mechanisms of water movement in xylem does NOT generate sufficient force to explain the rise of water to the tops of the tallest trees? (a) pressure-flow hypothesis (b) tension-cohesion (c) root pressure (d) active transport of potassium into guard cells (e) transpiration

10. According to which of the following mechanisms of water movement in xylem is a tension produced at the top of the plant by the evaporative pull of transpiration and the cohesive and adhesive properties of water? (a) pressure-flow (b) tension-cohesion (c) root pressure (d) active transport of potassium into guard cells (e) guttation-transpiration

11. According to which of the following mechanisms of phloem transport is dissolved sugar moved in phloem by means of a pressure gradient that exists between the source and the sink? (a) pressure-flow (b) tension-cohesion (c) root pressure (d) active transport of potassium into guard cells (e) guttation-transpiration hypothesis

12. How does increasing solute concentration affect water potential? (a) water potential becomes more positive (b) water potential becomes more negative (c) water potential becomes more positive under certain conditions and more negative under other conditions (d) water potential is not affected by solute concentration

REVIEW QUESTIONS

1. List several functions of stems and describe the tissue(s) responsible for each.
2. If you are examining a cross section of a herbaceous stem of a flowering plant, how would you determine whether it is a dicot or a monocot?
3. What happens to the primary tissues of a stem when secondary growth occurs?
4. When a strip of bark is peeled off a tree branch, what tissues are usually removed?
5. Distinguish between vascular cambium and cork cambium and describe the tissues that arise from each.
6. Name and briefly describe the model used to explain the rise of water in the tallest trees.
7. Describe root pressure. What are its limitations?
8. Describe the pressure-flow hypothesis of sugar movement in phloem, including the activities at source and sink.
9. Label the various tissues of this herbaceous dicot stem. Give at least one function for each of the tissues. Use Fig. 33–2a to check your answers.

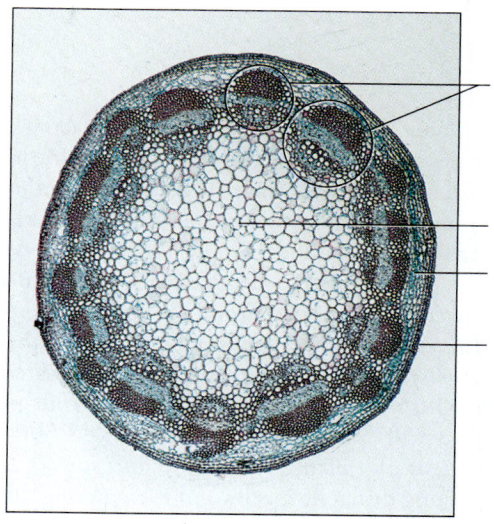

YOU MAKE THE CONNECTION

1. When secondary growth is initiated, certain cells become meristematic and begin to divide. Could a mature tracheid ever do this? A sieve tube member? Why or why not?
2. Why does the wood of many tropical trees lack annual rings?

3. Why is hardwood more desirable than softwood for making furniture? Explain your answer based on the structural differences between hardwood and softwood.

RECOMMENDED READINGS

Berg, L.R. *Introductory Botany: Plants, People, and the Environment.* Saunders College Publishing, Philadelphia, 1997. A general botany text with an environmental emphasis.

Moffett, M. "These Plants Claw and Strangle Their Way to the Top." *Smithsonian,* Sep. 1993. Spectacular photographs accompany this article on vines that grow in tropical rain forests.

Niklas, K.J. "How to Build a Tree." *Natural History,* Feb. 1996. The ability to make woody tissues has evolved independently many times during the history of plants.

Pearce, F. "Lure of the Rings." *New Scientist,* Vol. 152, 14 Dec. 1996. The analysis of tree rings produced over thousands of years may determine the extent of human influence on global climate.

Renfrew, C. "Kings, Tree Rings, and the Old World." *Nature,* Vol. 381, 27 Jun. 1996. This perspective of a research paper in the same issue explains how tree rings have been used to date archaeological sites in the eastern Mediterranean region, from 2220 to 718 B.C.

Scuderi, L.A. "A 2000-Year Tree Ring Record of Annual Temperatures in the Sierra Nevada Mountains." *Science,* Vol. 259, 5 Mar. 1993. How tree rings were used to reconstruct the mean temperatures in the Sierra Nevada of California.

Thybony, S., "Dead Trees Tell Tales." *National Wildlife,* Vol. 25, No. 2, Aug./Sep. 1987. An enjoyable account of the potential uses of tree ring analysis.

● Visit our website at **http://www.saunderscollege.com/lifesci/titles.html** and click on Solomon/Berg/Martin Biology for links to chapter-related resources on the World Wide Web.

CHAPTER 34

Roots and Mineral Nutrition

I n Chapters 32 and 33 we discussed the aerial vegetative structures of a plant: the leaves and stems. In this chapter we turn to the third major vegetative organ: the roots. Branching underground root systems are often more extensive than a plant's above-ground parts. The roots of a corn plant, for example, may grow to a depth of 2.5 m (about 8 ft) and spread outward 1.2 m (4 ft) from the stem. Desert-dwelling tamarisk (*Tamarix* sp.) trees reportedly have roots that grow to a depth of 50 m (163 ft) to tap underground water. The total root length, not counting root hairs, of a four-month-old rye *(Secale cereale)* plant was found to exceed 500 km (310 mi)! The extent of a plant's root depth and spread varies considerably among different species and even among different individuals in the same species. Soil conditions, discussed in this chapter, greatly affect the extent of root growth.

Because roots are usually underground and out of sight, people do not always appreciate the important functions that they perform. First, as anyone who has ever pulled weeds can attest, roots anchor a plant securely in the soil. A plant needs a solid foundation from which to grow. Firm anchorage is also essential to a plant's survival so that the stem remains upright, enabling leaves to absorb sunlight effectively.

Second, roots absorb water and dissolved nutrient mineral salts such as nitrates, phosphates, and sulfates, which are necessary for the synthesis of important organic molecules. These dissolved nutrient minerals are then transported throughout the plant in the xylem.

Storage is the third main function performed by many roots, and the carrots *(Daucus carota)* shown in the photograph, along with other root crops, are important sources of human food. Surplus sugars produced in the leaves by photosynthesis are transported in the phloem to the roots for storage (usually as starch or sucrose) until needed. Carrot roots have an extensive phloem for this purpose. Although roots use some photosynthetic products for their own respiratory needs, most are stored and later transported out of the roots for use by the plant. Both *taproots* (carrots, beets, radishes, and turnips) and *fibrous roots* (sweet potatoes and yams) may be modified for storage. Plants with storage taproots are usually biennials (see Chapter 31) that, as part of the strategy to survive winter, store their food reserves in the root during the first year's growth and use these reserves to reproduce during the second year's growth. Other plants, particularly those living in arid regions, possess storage roots adapted to store water.

(R. Calentine/Visuals Unlimited)

In certain species, roots are modified for functions other than anchorage, absorption, conduction, and storage. Roots specialized to perform unusual functions are discussed later in this chapter.

1. Describe the functions of roots and how their structure correlates with these functions.
2. Distinguish between the structures of roots and stems.
3. Label cross sections of a primary dicot root and a monocot root and describe the functions of each tissue.
4. Trace the pathway of water from the soil through the various root tissues.
5. Discuss the structure of roots with secondary growth and describe how secondary tissues form.
6. Describe several roots that are modified to perform unusual functions.
7. List the five components of soil and give the ecological significance of each.
8. Describe how roots absorb positively charged nutrient mineral ions by the process of cation exchange.
9. Describe the factors involved in soil formation.
10. Distinguish between plant macronutrients and micronutrients and give a physiological role for each essential element.

THERE ARE TWO BASIC TYPES OF ROOT SYSTEMS

Two types of root systems, a taproot system and a fibrous root system, may develop (Fig. 34–1). A **taproot** system consists of one main root (formed from the seedling's enlarging *radicle,* or embryonic root) with many smaller lateral roots coming out of it. Lateral roots often initially occur in regular rows along the length of the main root. Taproots are characteristic of many dicots and gymnosperms. A dandelion is a good example of a common herbaceous plant with a taproot system. A few trees, hickory, for example, retain their taproots, which become quite massive as the plants age. Most mature trees, however, do not retain their taproots and have root systems that consist of large, shallow lateral roots from which other roots branch off and grow downward.

A **fibrous root** system has several to many roots of the same size developing from the end of the stem, with smaller lateral roots branching off these roots. Fibrous root systems form in plants that have a short-lived embryonic root. The roots first originate from the base of the embryonic root and later from stem tissue. Because the main roots of a fibrous root system do not arise from preexisting roots, but rather from the stem, they are said to be **adventitious.** Adventitious organs occur in an unusual location, such as roots that develop on a stem or buds that develop on roots. Onions, crabgrass, and other monocots have fibrous root systems.

Taproot and fibrous root systems are adapted to obtain water in different sections of the soil. Taproot systems often extend down into the soil to obtain water located deep underground, whereas fibrous root systems, which are located relatively close to the surface of the soil, are adapted to obtain rainwater from a larger area as it drains into the soil.

(a)   *(b)*

Figure 34–1 Root systems.
(a) The roots of a fibrous root system are adventitious and develop from stem tissue. *(b)* A taproot system develops from the embryonic root in the seed.

Figure 34–2 **Structures unique to roots.** *(a)* LM of an oak (*Quercus.* sp.) root tip showing its root cap. The root apical meristem is protected by the root cap. *(b)* Root hairs on a radish seedling. Each delicate hair is an extension of a single cell of the root epidermis. Root hairs increase the surface area of the root in contact with the soil.   *(a, Runk/Rannels/Grant Heilman Photography; b, Dennis Drenner)*

ROOTS POSSESS ROOT CAPS AND ROOT HAIRS

Because of the need to adapt to the soil environment instead of the atmospheric environment, roots have several structures, such as root caps and root hairs, that stems lack. Although stems and leaves have various types of hairs, they are distinct from root hairs in structure and function.

Each root tip is covered by a **root cap,** a protective thimble-like layer many cells thick that covers the delicate root apical meristem (Fig. 34–2*a*; also see Fig. 31–3). As the root grows, pushing its way through the soil, parenchyma cells of the root cap are sloughed off by the frictional resistance of the soil particles and replaced by new cells formed by the root apical meristem. The root cap also appears to be involved in orienting the root so that it grows downward (see discussion of gravitropism in Chapter 36). When a root cap is removed, the root apical meristem grows a new cap. However, until the root cap has regenerated, the root grows randomly rather than in the direction of gravity.

Root hairs are short-lived extensions of single epidermal cells located just behind the growing root tip. Root hairs continually form in the area of cell maturation closest to the root tip to replace those that are dying off at the more mature end of the root hair zone (Fig. 34–2*b*; also see Fig. 31–3). Each root hair is short (typically less than 1 cm, or 0.4 in, in length), but they are quite numerous. Root hairs greatly increase the absorptive capacity of roots by increasing their surface area in contact with moist soil. Soil particles are coated with a microscopically thin layer of water in which minerals are dissolved. The root hairs establish an intimate contact with soil particles, which allows absorption of much of the water and minerals.

Unlike stems, roots lack nodes and internodes and do not usually produce leaves or buds. While herbaceous roots have certain primary tissues (such as epidermis, xylem, phloem, and parenchyma of the cortex and pith) found in herbaceous stems, these tissues are arranged quite differently. Table 34–1 summarizes the major differences between roots and stems in herbaceous dicots.

TABLE 34–1   General Differences Between Herbaceous Dicot Roots and Stems*

Roots	Stems
No nodes and internodes	Nodes and internodes
No leaves or buds	Leaves and buds
Nonphotosynthetic	Photosynthetic
No pith	Pith
No cuticle	Cuticle
Root cap	No cap
Root hairs	Trichomes
Pericycle	No pericycle
Endodermis	No endodermis
Lateral roots form internally from the pericycle	Branches form externally from lateral buds

*Some exceptions to these general differences exist.

THE ARRANGEMENT OF VASCULAR TISSUES DISTINGUISHES HERBACEOUS DICOT ROOTS AND MONOCOT ROOTS

Although considerable variation exists in herbaceous dicot and monocot roots, they all possess an outer protective covering (epidermis or periderm), a cortex for storage of starch and other organic molecules, and vascular tissues for conduction. Let us first consider the structure of herbaceous dicot roots.

In most herbaceous dicot roots, the central core of vascular tissue lacks pith

The buttercup root is a representative dicot root with primary growth (Fig. 34–3). Like other parts of this herbaceous dicot, its roots are covered by a single layer of protective tissue, the **epidermis.** The root hairs are a modification of the root epidermis that enables it to absorb more water from the soil. The root epidermis does not secrete a thick, waxy cuticle in the region of root hairs because this layer would impede the absorption of water from the soil. Both the lack of a cuticle and the presence of root hairs increase absorption. (*Why* water moves from the soil into the root was explained in Chapter 33.)

Most of the water that enters the root moves along the cell walls rather than entering the cells. One of the major components of cell walls is cellulose, which absorbs water like a sponge. An example of the absorptive properties of cellulose is found in cotton balls, which are almost pure cellulose.

The **cortex,** which is primarily composed of loosely packed parenchyma cells, comprises the bulk of a herbaceous dicot root. Roots usually lack supporting collenchyma cells, probably because the soil supports the root, although they may develop some sclerenchyma (another supporting tissue; see Chapter 31) as they age. The primary function of the root cortex is storage. A microscopic examination of the parenchyma cells that form the cortex often reveals numerous amyloplasts, which store starch. Starch, an insoluble carbohydrate composed of glucose subunits, is the most common form of stored energy in plants. When used at a later time, these reserves provide energy for such activities as growth following winter or cell replacement following an injury.

The large intercellular (between cells) spaces, a common feature of the root cortex, provide a pathway for water uptake and allow for aeration of the root. The oxygen that root cells need for aerobic respiration (see Chapter 7) diffuses from air spaces in the soil into the intercellular spaces of the cortex and from there into the cells of the root.

The inner layer of the cortex, the **endodermis,** regulates the movement of nutrient minerals that enter the xylem in the root's interior. Structurally, the endodermis differs from the rest of the cortex. Endodermal cells fit snugly against each other, and each has a special bandlike region, called a **Casparian strip** (Fig. 34–4), on its radial (side) and transverse (upper and lower) walls. If you compare the endodermis to a cylinder constructed of bricks, endodermal cells correspond to the bricks, and the Casparian strips correspond to the mortar between them. Casparian strips contain *suberin,* a fatty material that is waterproof. (Recall from Chapter 33 that suberin is also the waterproof material in cork cell walls.)

The water and dissolved nutrient minerals that enter the root cortex from the epidermis move in solution along two pathways, the apoplast and symplast. The **apoplast** follows along the interconnected porous cell walls, whereas the **symplast** is the route from one cell's cytoplasm to the next through plasmodesmata (Fig. 34–5). Until the endodermis is reached, most of the water and dissolved nutrient minerals have not

(a)

250 μm

Cortex cells filled with amyloplasts

Endodermis cell

Pericycle cell

Phloem cell

Xylem vessel elements

(b)

25 μm

Figure 34–3 **LMs of cross sections of a buttercup (*Ranunculus* sp.) root.** *(a)* Cortex comprises the bulk of the root. *(b)* Close-up of the root's stele. Note the solid core of vascular tissues. *(a, b, Ed Reschke)*

Figure 34–4 The endodermis controls the root's mineral uptake. Note the Casparian strip around the radial and transverse walls.

passed through a plasma membrane or entered the cytoplasm of a root cell (i.e., most have traveled along the apoplast). However, the waterproof Casparian strip on the radial and transverse walls of the endodermal cells prevents water and minerals from continuing along the cell walls. Water enters by osmosis, whereas minerals enter the endodermal cells by passing through carrier proteins in their plasma membranes. Thus, the endodermis is responsible for controlling the movement of

minerals into the root, even though it is an internal cell layer, and minerals must pass through the epidermis and cortex to reach it.

Dissolved nutrient mineral ions are actively transported through carrier proteins in the plasma membranes of endodermal cells (see Chapter 5). In active transport, the mineral ions move *against* the concentration gradient, that is, from an area of *low* concentration in the soil solution to an area of *high*

Apoplast: interconnected cell wall spaces

Water and dissolved nutrient minerals

Symplast: interconnected cytoplasm of living cells

Figure 34–5 Pathways of water and dissolved nutrient minerals that enter the root. Water and dissolved nutrient minerals travel from cell to cell along the interconnected porous cell walls (the apoplast) or from one cell's cytoplasm to another through plasmodesmata (the symplast). On reaching the endodermis, water and minerals can only continue to move into the root's center if they pass through a plasma membrane and enter an endodermal cell. The Casparian strip blocks the passage of water and minerals between the endodermal cells.

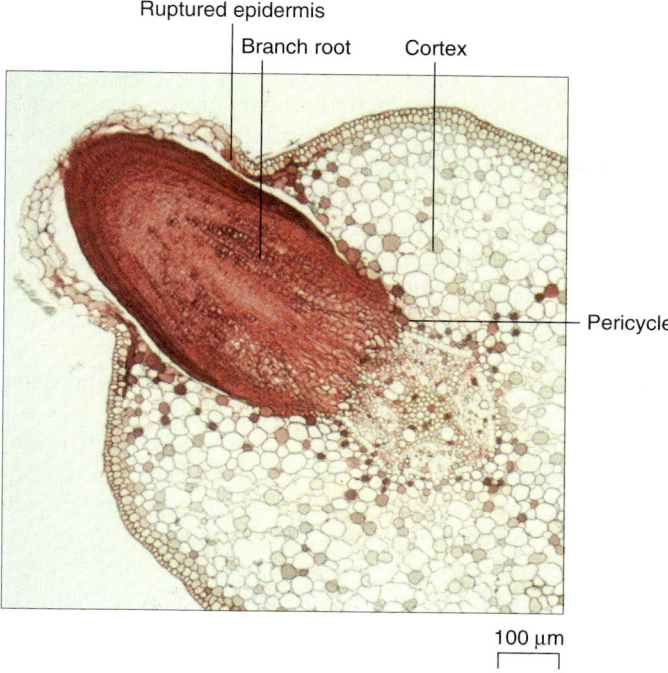

Figure 34–6 **LM of a multicellular lateral root.** Lateral roots originate at the pericycle. *(James Mauseth, University of Texas)*

100 μm

concentration of that mineral in the plant's cells. One of many reasons that root cells require sugar and oxygen for aerobic respiration is that this active transport requires the expenditure of energy, usually in the form of ATP. From the endodermis, water and nutrient mineral ions enter the root xylem (precisely how this is done is not known) and are conducted to the rest of the plant.

Just inside the endodermis is a single layer of cells called the **pericycle,** which gives rise to multicellular lateral roots, also called branch roots (Fig. 34–6). The pericycle is composed of parenchyma cells that remain meristematic. Lateral roots originate when cells in a portion of the pericycle start dividing. As it grows, the lateral root pushes through several layers of root tissue (endodermis, cortex, and epidermis) before entering the soil. Each lateral root has all the structures and features—root cap, root hairs, epidermis, cortex, endodermis, pericycle, xylem, and phloem—of the larger root from which it emerges. In addition to producing lateral roots, the pericycle is involved in forming the lateral meristems that produce secondary growth in woody roots (discussed in the next section).

At the center of a dicot primary root is the **stele,** a central cylinder of vascular tissues (see Fig. 34–3*b*). **Xylem,** the center-most tissue, often has two, three, four, or more extensions, or "xylem arms." **Phloem** is located in patches between the xylem arms. The xylem and phloem of the root have the same functions as in the rest of the plant: water and dissolved nutrient minerals are conducted in xylem, and dissolved sugar (sucrose) is conducted in phloem.

After passing through the endodermal cells, water enters the root xylem, often at one of the xylem arms. Up to this point the pathway of water has been horizontal from the soil into the center of the root:

Root hair → epidermis → cortex → endodermis → pericycle → root xylem

Once water enters the xylem, it is transported upward through root xylem into stem xylem and throughout the rest of the plant (see Chapter 33).

One pathway of phloem conduction is from the leaves, where sugar is made by photosynthesis, to the root, where sugar is used for the growth and maintenance of root tissues or stored, usually as starch. Another pathway of phloem conduction is from the root, where sugar is stored as starch, to other parts of the plant, where sugar is used for growth and maintenance of tissues. The same kinds of cells found in stem xylem and phloem are found in root xylem and phloem, except that roots typically have fewer fibers for support. The **vascular cambium,** which gives rise to secondary tissues, is sandwiched between the xylem and phloem. Because it possesses an inner core of vascular tissue, the primary dicot root lacks **pith,** a ground tissue found in the centers of many stems and roots.

Xylem does not form the central tissue in some monocot roots

Although monocot roots exhibit considerable variation in structure when compared to dicot roots, they possess the same basic structure. Starting at the outside of a typical monocot root, there is epidermis, then cortex, endodermis, and pericycle (Fig. 34–7). Unlike herbaceous dicot roots, the xylem in a

Figure 34–7 **LM of a cross section of a greenbriar (*Smilax*) root.** As in herbaceous dicot roots, the cortex is extensive. *(Dennis Drenner)*

250 μm

monocot root does not form a solid cylinder in the center. Instead, the phloem and xylem are located in separate alternating bundles arranged around the central pith, which is composed of parenchyma cells.

Because the vast majority of monocots do not have secondary growth, no vascular cambium exists in monocot roots. Despite their lack of secondary growth, long-lived monocots, such as palms, often have thickened roots produced by a modified form of primary growth in which parenchyma cells in the cortex divide and enlarge.

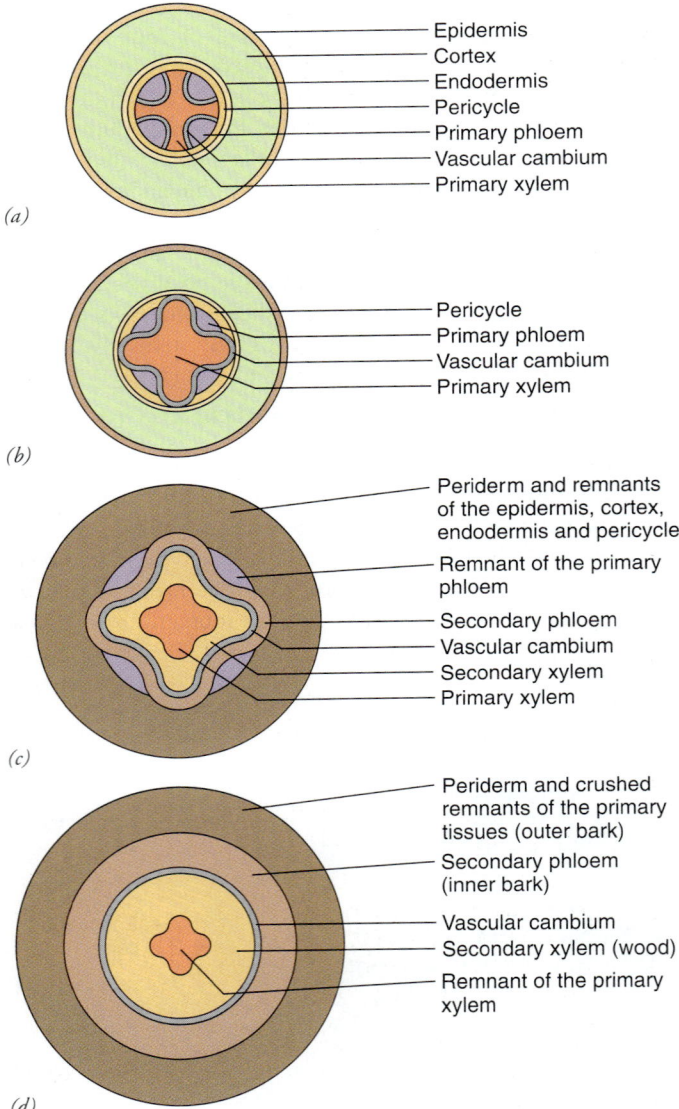

(a)
- Epidermis
- Cortex
- Endodermis
- Pericycle
- Primary phloem
- Vascular cambium
- Primary xylem

(b)
- Pericycle
- Primary phloem
- Vascular cambium
- Primary xylem

(c)
- Periderm and remnants of the epidermis, cortex, endodermis and pericycle
- Remnant of the primary phloem
- Secondary phloem
- Vascular cambium
- Secondary xylem
- Primary xylem

(d)
- Periderm and crushed remnants of the primary tissues (outer bark)
- Secondary phloem (inner bark)
- Vascular cambium
- Secondary xylem (wood)
- Remnant of the primary xylem

Figure 34–8 Development of secondary vascular tissues in a primary root. *(a)* The tissues in a primary root. *(b)* At the onset of secondary growth, the vascular cambium extends out to the pericycle, forming a continuous, noncircular loop. *(c)* The vascular cambium produces secondary xylem to its inside and secondary phloem to its outside. Note that the primary phloem is pushed outward. *(d)* Over time, the ring of vascular cambium gradually becomes circular. As it continues to divide, the epidermis, cortex, and primary phloem found in the outer bark are torn apart.

WOODY PLANTS HAVE ROOTS WITH SECONDARY GROWTH

Plants that produce stems with secondary growth also produce roots with secondary growth. Recall from Chapter 33 that these plants, gymnosperms and woody dicots, have primary growth at apical meristems, while secondary growth occurs at lateral meristems. The production of secondary tissues occurs some distance back from the root tips and is the result of the activity of the same two lateral meristems found in woody stems: the vascular cambium and the cork cambium. Major roots of trees are often massive and possess both wood and bark. In temperate climates, the wood of both roots and stems exhibits annual rings in cross section.

Before secondary growth starts in a root, the vascular cambium is sandwiched between the primary xylem and the primary phloem (Fig. 34–8). At the onset of secondary growth, the vascular cambium extends out to the pericycle. As the vascular cambium divides to produce secondary tissues, it eventually forms a cylinder of vascular cambium that continues to divide, producing secondary xylem (wood) to the inside and secondary phloem (inner bark) to the outside. The root increases in girth (thickness), and the vascular cambium continues to move outward.

The epidermis, cortex, endodermis, and primary phloem are gradually torn apart as the root increases in girth. The root epidermis is replaced by periderm, composed of cork cells and cork parenchyma, both produced by the cork cambium (Table 31–4 shows a LM of periderm). The cork cambium in the root initially arises from regions in the pericycle.

SOME ROOTS ARE SPECIALIZED FOR UNUSUAL FUNCTIONS

Adventitious roots often arise from the nodes of stems. Many aerial adventitious roots are adapted for functions other than anchorage, absorption, conduction, or storage. **Prop roots** are adventitious roots that develop from branches or a vertical stem and that grow downward into the soil to help support the plant in an upright position (Fig. 34–9*a,b*). Prop roots are more

▶ **Figure 34–9 Specialized roots.** *(a,b)* Prop roots are adventitious roots that provide additional support. *(a)* Prop roots in corn *(Zea mays)* arise near the base of the stem. *(b)* The banyan tree *(Ficus benghalensis)* has prop roots that develop from branches. *(c)* The moth orchid *(Phalaenopsis* hybrid) has photosynthetic aerial roots. *(d)* LM of a parasitized juniper *(Juniperus* sp.) branch showing a parasitic mistletoe haustorium that obtains water from the host on which it lives. *(e)* Black mangrove *(Avicennia germinans)* produces pneumatophores that possibly provide oxygen for roots buried in anaerobic (oxygen-deficient) mud. *(f)* Plants that produce corms or bulbs often have contractile roots. During successive seasons, contractile roots pull the corm or bulb deeper into the soil.   *(a, Dennis Drenner; b, James Mauseth, University of Texas; c, John Arnaldi ; d, Courtesy of C. Calvin, Portland State University; e, Nada Pecnik/Visuals Unlimited; f, Courtesy of Judith Jernstedt, University of California, Davis)*

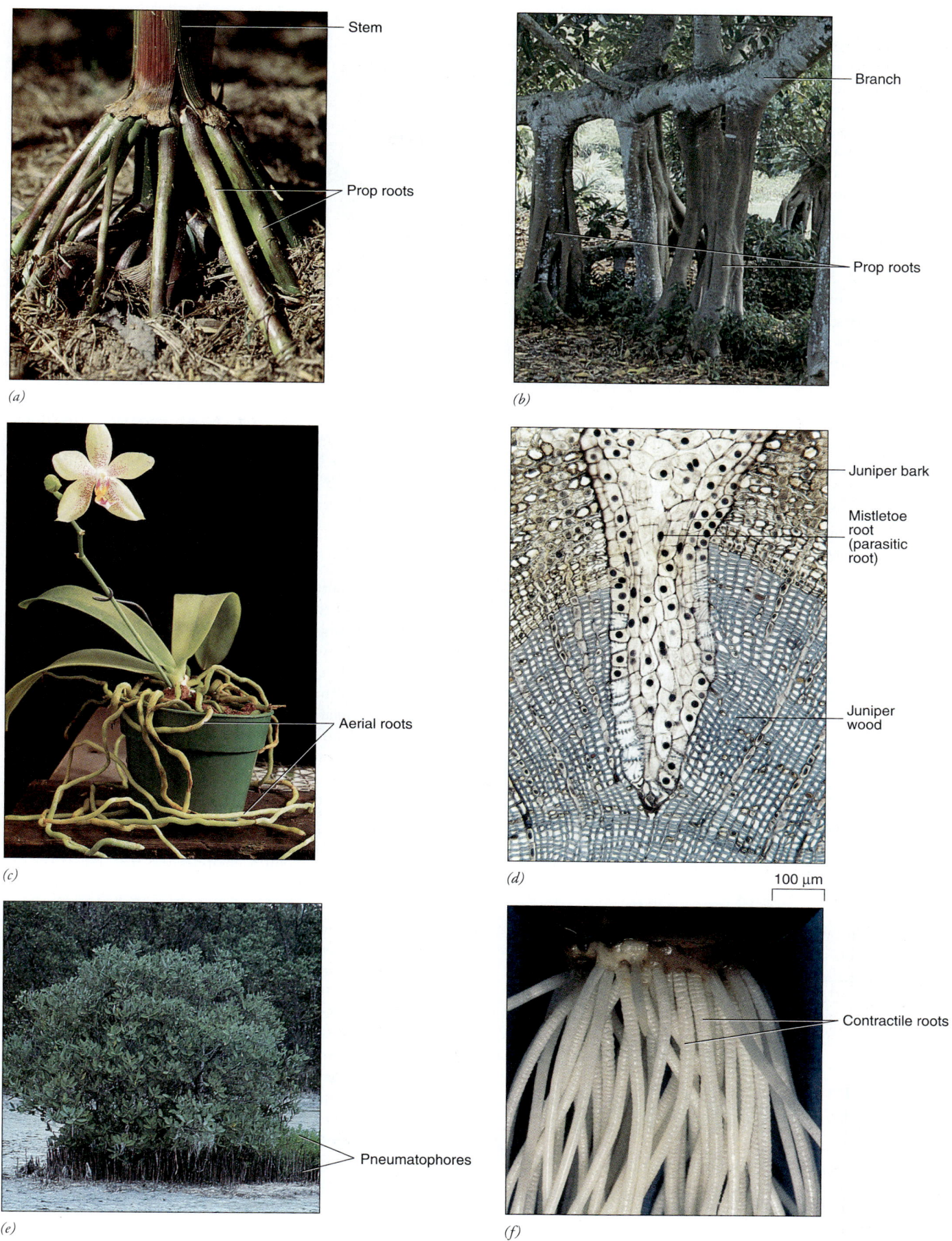

(a)

Stem

Prop roots

(b)

Branch

Prop roots

(c)

Aerial roots

(d)

Juniper bark

Mistletoe root (parasitic root)

Juniper wood

100 μm

(e)

Pneumatophores

(f)

Contractile roots

common in monocots than in dicots. Corn and sorghum, both monocots, are herbaceous plants that produce prop roots. Many tropical and subtropical dicot trees, such as red mangrove *(Rhizophora mangle)* and banyan *(Ficus benghalensis)*, also produce prop roots.

Epiphytes (plants that grow attached to other plants) and climbing plants have aerial roots that anchor the plant to the bark, branch, or other surface on which it grows. Some epiphytes have aerial roots specialized for functions other than anchorage. Certain epiphytic orchids, for example, have photosynthetic roots (Fig. 34–9*c*). Epiphytic roots may absorb moisture as well. Some parasitic epiphytes such as mistletoe have modified roots that penetrate the host plant tissues and absorb water (Fig. 34–9*d*). These roots, called **haustoria** (sing., *haustorium*), consist of parenchyma cells and vessel elements; a haustorium lacks many typical root structures such as a root cap, root apical meristem, and cortex. Another plant that starts its life as an epiphyte is the strangler fig, which produces long roots that eventually reach the ground and anchor the plant (now a tree rather than an epiphyte) in the soil. The tree that the strangler fig originally grew on is often killed as the strangler fig grows around it, competing with it for light and other resources and crushing its secondary phloem.

Other aerial roots have additional functions. In swampy or tidal environments where the soil is flooded or waterlogged, roots often grow upward until they are above the high-tide level. Even though roots live in the soil, they still require oxygen for aerobic respiration. Flooded soils are depleted of oxygen, so these aerial "breathing" roots, known as **pneumatophores,** may assist in getting oxygen to the submerged roots. Pneumatophores have a well developed system of internal air spaces that is continuous with the submerged parts of the root, presumably allowing gas exchange. Black mangrove, white mangrove, and bald cypress are examples of plants with pneumatophores (Fig. 34–9*e*).

Plants that produce corms or bulbs (underground stems or buds specialized for asexual reproduction; see Chapter 35) often have wiry **contractile roots** in addition to their "normal" roots. The contractile roots grow into the soil and then contract (the cortical cells shorten or totally collapse), thus pulling the corm or bulb deeper into the soil (Fig. 34–9*f*). Contractile roots are necessary for corms because each succeeding year's growth is *on top of* the preceding year's growth. As a result, corms tend to move upward in the soil over time. Without contractile roots they would eventually be exposed at the soil's surface. Contractile roots are more common in monocots, but certain dicots and ferns also possess them.

ROOTS FORM RELATIONSHIPS WITH OTHER SPECIES

As roots of certain trees grow through the soil, they sometimes encounter roots of other trees of the same or different species. When this occurs, they may grow together to form a natural

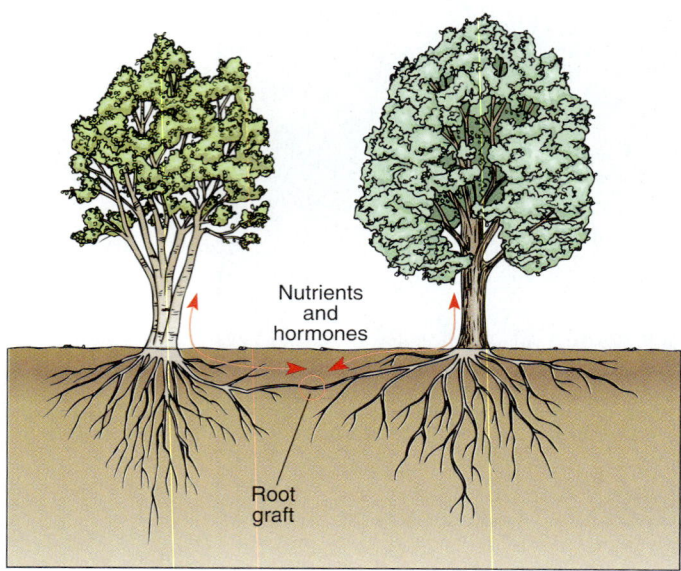

Figure 34–10 **Natural root grafts.** Root grafts often occur between two different trees, such as a birch *(left)* and a maple *(right)*, allowing for the exchange of nutrients and hormones.

graft, as, for example, between a birch and a maple (Fig. 34–10). Because their vascular tissues are connected in the graft, dissolved sugars and other materials such as hormones pass between the two trees. Root grafts have been observed in more than 160 tree species.

The roots of most plant species form *mutualistic* (mutually beneficial) relationships with certain soil fungi (see Chapters 25 and 52). These subterranean associations, known as **mycorrhizae,** permit the transfer of materials (such as sugars) from roots to fungus. At the same time, essential minerals such as phosphorus move from fungus to roots. The threadlike body of the fungal partner extends into the soil, extracting minerals well beyond the reach of the plant's roots. In some mycorrhizae, the fungal mycelium encircles the root like a sheath, whereas in others, the fungus penetrates root cells (Fig. 34–11). The relationship is considered mutually beneficial because when mycorrhizae are not present, neither the fungus nor the plant grows as well (see Fig. 25–12). Recent evidence indicates that mycorrhizae also promote species diversity in forest communities (see *On the Cutting Edge: Mycorrhizae and Carbon Exchange between Plant Species in a Forest Ecosystem*).

The roots of some plants—clover, peas, and soybeans, for example—form an association with certain nitrogen-fixing bacteria. **Nodules** (swellings) that house millions of the bacteria develop on the roots (see Fig. 53–11*a*). Like mycorrhizae, the association between nitrogen-fixing bacteria and the roots of plants is mutually beneficial. Bacteria receive the photosynthetic products from plants while helping them to meet their nitrogen requirements by producing excess ammonia (NH_3).

Figure 34–11 **Mycorrhizae.** (*a*) LM of ectomycorrhizae, fungal associations that form a sheath around the root. The fungal hyphae penetrate the root between cortical cells but do not enter the cells. (*b*) LM of endomycorrhizae, fungal associations in which the fungal hyphae penetrate root cortical cells and form branched haustoria within the cells to aid in delivering and receiving nutrients. Roots of the majority of vascular plant species are colonized by endomycorrhizae. *(a, Robert Knauft/Biology Media/Photo Researchers, Inc.; b, Cabisco/Visuals Unlimited)*

SOIL ANCHORS PLANTS AND PROVIDES WATER AND MINERALS

We now examine the soil environment in which most roots live. Soil is a relatively thin layer of Earth's crust that has been modified by the natural actions of weather, wind, water, and organisms. It is easy to take soil for granted. We walk on and over it throughout our lives, but rarely stop to think about how important it is to our survival. Vast numbers and kinds of organisms colonize soil and depend on it for shelter and food. Most plants anchor themselves in soil, and from it they receive water and essential minerals. Most of the 16 different elements essential for plant growth are obtained directly from the soil (these elements are discussed later in this chapter). Most plants cannot survive on their own without soil, and because we depend on plants for our food, humans could not exist without soil, either.

Most soils are formed from rock (called parent material) that is gradually broken down into smaller and smaller particles by biological, chemical, and physical **weathering.** Two important factors that work together in the weathering of rock are climate and organisms. When plant roots and other organisms living in the soil respire, they produce carbon dioxide, CO_2, which diffuses into the soil and reacts with soil wa-

ter to form carbonic acid, H_2CO_3. Soil organisms such as lichens[1] also produce other kinds of acids. These acids etch tiny cracks, or fissures, in the rock surface; water then seeps into these cracks. If the parent material is located in a temperate climate, the alternate freezing and thawing of the water during winter causes the cracks to enlarge, breaking off small pieces of rock. Small plants can then become established and send their roots into the larger cracks, fracturing the rock further.

Topography, a region's surface features—including the presence or absence of mountains and valleys—is also involved in soil formation. Steep slopes often have little or no soil on them because the soil and rock are continually transported down the slopes by gravity; runoff from precipitation tends to amplify erosion on steep slopes. Moderate slopes, on the other hand, may encourage the formation of deep soils.

The disintegration of solid rock into finer and finer mineral particles and the accumulation of organic material (discussed shortly) in the soil take an extremely long time, usually thousands of years. Soil forms continually as the weathering

---

[1]A lichen is a dual organism composed of a fungus and a phototroph (photosynthetic organism). See Chapter 25 for further discussion of lichens.

Mycorrhizae and Carbon Exchange Between Plant Species in a Forest Ecosystem

Hypothesis: The exchange of carbon compounds between tree species connected by mycorrhizal fungi can be measured and is influenced by field conditions such as relative light intensity.

Method: Use isotope tracers to measure the transfer of photosynthetically produced carbon compounds between same-age trees of different species (Douglas fir and paper birch) connected by mycorrhizal fungi. Compare the net transfer between healthy trees exposed to full sunlight and less healthy trees growing in deep shade.

Results: Carbon transfer between Douglas fir and paper birch is bidirectional, but net transfer is always from paper birch to Douglas fir. Transfer of carbon compounds is greatest when paper birches are grown in full sun and Douglas firs are shaded.

Conclusion: Douglas firs receive a net gain of carbon compounds from paper birches by way of mycorrhizal fungi connections between the two trees. The magnitude of carbon transfer is affected by relative light intensity.

Biologists have studied mycorrhizae for many years, but most have used a reductionist approach (see Chapter 1) in which the fungus and plant are isolated from the natural environment and grown together in a pot. Many important aspects of the fungus-plant relationship have been determined using this approach, but other questions about the ecological significance of mycorrhizae in natural communities remained unanswered.

In 1997 Suzanne Simard of the British Columbia Ministry of Forests and her colleagues from Oregon State University, Okanagan University College in British Columbia, and the U. S. Department of Agriculture reported one of the first field studies involving mycorrhizae in a forest community.* This experiment demonstrated that in natural ecosystems the subterranean relationship between plants and fungi is more complex than originally supposed. Simard et al. studied ectomycorrhizae, whose fungal hyphae ensheathe the root and then extend into the soil. These fungi are not host-specific; many ectomycorrhizal fungus species can colonize a single plant species, and each fungus species can form mycorrhizal relationships with many plant species. For example, as many as ten fungal species form mycorrhizal relationships with both paper birch (*Betula papyrifera*) and Douglas fir (*Pseudotsuga menziesii*). The mycelia of these fungi often connect the different plant species, and materials such as phosphorus, nitrate, and carbon compounds pass from one plant to another by way of the mycorrhizal fungus.

The work of Simard et al. is unique because they were able to quantify the amount of exchange between two plant species. They exposed young (two-year-old seedlings in one experiment and three-year-old seedlings in another) birch and Douglas fir trees growing close together in the forest to *reciprocal isotope labeling*. One plant received ^{14}C-labeled CO_2 and the other received ^{13}C-labeled CO_2. (The appropriate isotope was injected into a sealed bag that enclosed each plant.) After a short time to allow for the accumulation of carbon compounds produced by photosynthesis, the plants were harvested. It was hypothesized that if carbon compounds were not being transferred from one plant to the other by their fungal connection,

then each plant would contain carbon compounds labeled with either ^{14}C or ^{13}C, but not both. Each plant was evaluated for the presence of ^{14}C- and ^{13}C-labeled carbon-containing molecules, and each contained both isotopes. These data indicate that the fungal connections linking the two plant species had transferred materials between them in both directions.

Although the movement of carbon compounds was bidirectional, Douglas fir received more carbon compounds from paper birch than paper birch received from Douglas fir. Perhaps even more interesting, the *net gain* of carbon compounds from paper birch to Douglas fir was influenced by relative light intensity. When Douglas firs were grown in deep shade by covering them with tents four weeks before the labeling experiment, the amount of carbon transferred from the paper birch to the shaded Douglas fir increased dramatically. On average, the paper birches provided 6% of their total photosynthetic production of carbon compounds to Douglas firs growing in deep shade.

Why does a shaded Douglas fir receive more carbon compounds from paper birch than does a Douglas fir grown in full sunlight? Simard suggests that having mycorrhizal relationships with two strong, vigorously growing trees benefits the fungus more than such connections with a strong tree and a weaker one. (The weaker tree is the one growing in shade because it cannot produce enough energy-storing, carbon compounds to grow vigorously.) Thus, the fungus may control the net transfer of carbon compounds between plants to optimize the overall health of their mutualistic partners. An alternative explanation is that there may be a pressure-flow (see Chapter 33) type of transfer from source (paper birch) to sink (Douglas fir). This transfer would be caused by a physical mechanism and would not be regulated by the mycorrhizal fungus. Additional research will have to be done to determine the cause of the differential transfer of carbon compounds.

Simard's results have also raised ecological questions. To what extent do mycorrhizal relationships promote forest diversity? Do mycorrhizae influence the diversity of plant communities other than forests? How do Simard's results affect our understanding of the importance of competition among plant species? (Chapter 52 examines the role of competition, including whether competition is the main ecological determinant of community structure.)

*Simard, S.W., D.A. Perry, M.D. Jones, D.D. Myrold, D.M. Durall, and R. Molina. "Net Transfer of Carbon Between Ectomycorrhizal Tree Species in the Field." *Nature,* Vol. 388, 7 Aug. 1997.

of parent material beneath already-formed soil continues to add new soil. Soil thickness varies from a thin film to more than 3 m (10 ft) deep.

Soil is composed of inorganic minerals, organic matter, organisms, air, and water

The five distinct components of soil—inorganic mineral particles, organic matter, organisms, water, and air—all interact with one another. Some minerals, for example, continually cycle from the soil to organisms, which use them in their biological processes. When the organisms die, they are decomposed by bacteria and other soil organisms, returning the minerals to the soil.

The inorganic mineral particles, which come from weathered rock, constitute most of what we call soil. Because different rocks are composed of different minerals, soils vary in chemical composition. Also, soils formed from the same kind of parent material may not develop in the same way because other factors, such as weather, topography, and organisms, differ.

The texture, or structural characteristic, of a soil is determined by the percentages (by weight) of the different-sized inorganic mineral particles—sand, silt, and clay—that it contains. The size assignments for sand, silt, and clay give soil scientists a way to classify soil texture. Particles larger than 2 mm in diameter, called gravel or stones, are not considered soil particles because they do not have any direct value to plants. The largest soil particles are called *sand* (0.02 to 2 mm in diameter), the medium-sized particles are called *silt* (0.002 to 0.02 mm in diameter), and the smallest particles are called *clay* (less than 0.002 mm in diameter). Sand particles are large enough to be seen easily with the eye; silt particles (about the size of flour particles) are barely visible with the eye; and most individual clay particles are too small to be seen with an ordinary light microscope; they can only be seen under an electron microscope.

The clay component of a soil is particularly important in determining many of its characteristics, in part because clay particles have the greatest surface area of all soil particles. If the surface areas of about 450 g (1 lb) of clay particles were laid out side by side, they would occupy 1 hectare (2.5 acres). Each clay particle has negative electrical charges on its outer surface that attract and reversibly bind cations (positively charged mineral ions). Many cations, such as potassium (K^+) and magnesium (Mg^{2+}), are essential for plant growth and are retained in the soil by clay particles. Roots secrete protons (H^+), which are exchanged for other positively charged mineral ions in a process known as **cation exchange.** These "freed" ions and the water that forms a film around the soil particles are absorbed by the plant's roots (Fig. 34–12). In contrast, anions (negatively charged mineral ions) are usually not held as tightly in the soil and are often washed out of the root zone.

Soil always contains a mixture of different-sized particles, but the proportions vary from one soil to another. A *loam*,

Figure 34–12 **Cation exchange.** Negatively charged clay particles bind to positively charged nutrient mineral cations, holding them in the soil. Roots secrete protons (H^+), which are exchanged for the cations, facilitating their absorption.

which is an ideal agricultural soil, has an optimum combination of different soil particle sizes: it contains approximately 40% each of sand and silt, and about 20% of clay. Generally speaking, the larger particles provide aeration while the smaller ones bind together into clumps and hold nutrient minerals and water. Soils with larger proportions of sand are not as desirable for most plants because they do not hold water and mineral ions well; plants grown in such soils are more susceptible to drought and mineral deficiencies. Soils with larger proportions of clay are also not desirable for most plants because they provide poor drainage and often do not provide enough oxygen. Clay soils used in agriculture tend to get compacted, which reduces the number of soil spaces that can be filled by water and air.

A soil's organic matter consists of the wastes and remains of soil organisms

The organic matter in soil is composed of litter (dead leaves and branches on the soil's surface), droppings (animal dung), and the dead remains of plants, animals, and microorganisms in various stages of decomposition. Organic matter is decomposed by microorganisms, particularly bacteria and fungi, that inhabit the soil. During decomposition, essential nutrient mineral ions are released into the soil, where they may be bound by soil particles or absorbed by plants. Organic matter increases the soil's water-holding capacity by acting much like a sponge. For this reason gardeners often add organic matter to soils, especially sandy soils, which are naturally low in organic matter.

Figure 34–13 Humus. Humus is partially decomposed organic material that comes primarily from plant and animal remains. Soil rich in humus has a loose, somewhat spongy structure with several properties, such as increased water-holding capacity, that are beneficial for plants and other organisms living in it. *(USDA/Natural Resources Conservation Service)*

The partly decayed organic portion of the soil is referred to as **humus** (Fig. 34–13). Humus serves much the same functions as clay in holding cations as well as water.

The organisms living in the soil form a complex ecosystem

A single teaspoon of fertile agricultural soil may contain millions of living microorganisms such as bacteria, fungi, algae, protozoa, and microscopic worms. Many other organisms also colonize soil, including earthworms, insects, plant roots, and animals such as moles, snakes, and groundhogs (Fig. 34–14). Most numerous in soil are bacteria, which number in the hundreds of millions per gram of soil.

Worms are some of the most important organisms living in soil. Earthworms, probably one of the most familiar soil inhabitants, ingest soil and obtain energy and raw materials by digesting humus. *Castings,* bits of soil that have passed through the gut of an earthworm, are deposited on the soil surface. In this way, nutrient minerals from deeper layers are brought to upper layers. Earthworm tunnels serve to aerate the soil, and the worms' waste products and corpses add organic material to the soil.

Ants live in the soil in enormous numbers, constructing tunnels and chambers that help to aerate it. Members of soil-dwelling ant colonies forage on the surface for bits of food, which they carry back to their nests. Not all of this food is eaten, however, and its eventual decomposition helps increase the organic matter in the soil.

About 30% to 60% of soil volume is composed of pore spaces

Soil has numerous pore spaces of different sizes around and among the soil particles. Pore spaces are filled with varying proportions of air and water (Fig. 34–15), both of which are necessary to produce a moist but aerated soil that sustains plants and other soil-dwelling organisms. Generally speaking, water is held in the smaller pores, while air is found in the larger pores. After a prolonged rain, almost all of the pore spaces may be filled with water, but water drains rapidly from the larger pore spaces, drawing air from the atmosphere into those spaces.

Soil air contains the same gases as atmospheric air, although they are usually present in different proportions. As a result of aerobic respiration by soil organisms, there is usually less oxygen and more carbon dioxide in soil air than in atmospheric air. (Recall from Chapter 7 that aerobic respiration uses oxygen and produces carbon dioxide.) Among the important gases in soil are oxygen (O_2), required by soil organisms for aerobic respiration; nitrogen (N_2), used by nitrogen-fixing bacteria; and carbon dioxide (CO_2). As mentioned earlier in the chapter, carbon dioxide, a product of aerobic respiration, dissolves in water to form carbonic acid, a weak acid that accelerates the weathering process during soil formation.

Soil water originates as precipitation, which drains downward, or as groundwater (water stored in porous underground rock), which rises upward from the water table (the uppermost level of groundwater). Soil water contains low concentrations of dissolved nutrient minerals that enter the roots of plants when they absorb water. Water not bound to soil particles or absorbed by roots percolates (moves down) through soil, carrying dissolved nutrient minerals with it. The removal of dissolved materials from soil by percolating water is called **leaching**. Some anions completely leach out of the soil because they are so soluble that they migrate all the way down to the groundwater. (Recall that not all soluble materials leach out of the soil because soil particles, particularly clay particles, bind them.) It is also possible for water to move *upward* through the soil and carry dissolved materials with it.

Soil pH affects soil characteristics and plant growth

As discussed in Chapter 2, soil acidity is measured using the pH scale, which runs from 0 (extremely acidic) through 7 (neutral) to 14 (extremely alkaline). The pH of most soils ranges from 4 to 8, but some soils are outside this range. The soil of the Pygmy Forest in Mendocino County, California, is extremely acidic (pH 2.8–3.9). At the other extreme, certain soils in Death Valley, California, have a pH of 8.5.

Plants are affected by soil pH partly because the solubility of certain minerals varies with differences in pH. Soluble minerals can be absorbed by the plant, whereas insoluble forms cannot. At a low pH, for example, the aluminum and manganese in soil water are more soluble and are sometimes ab-

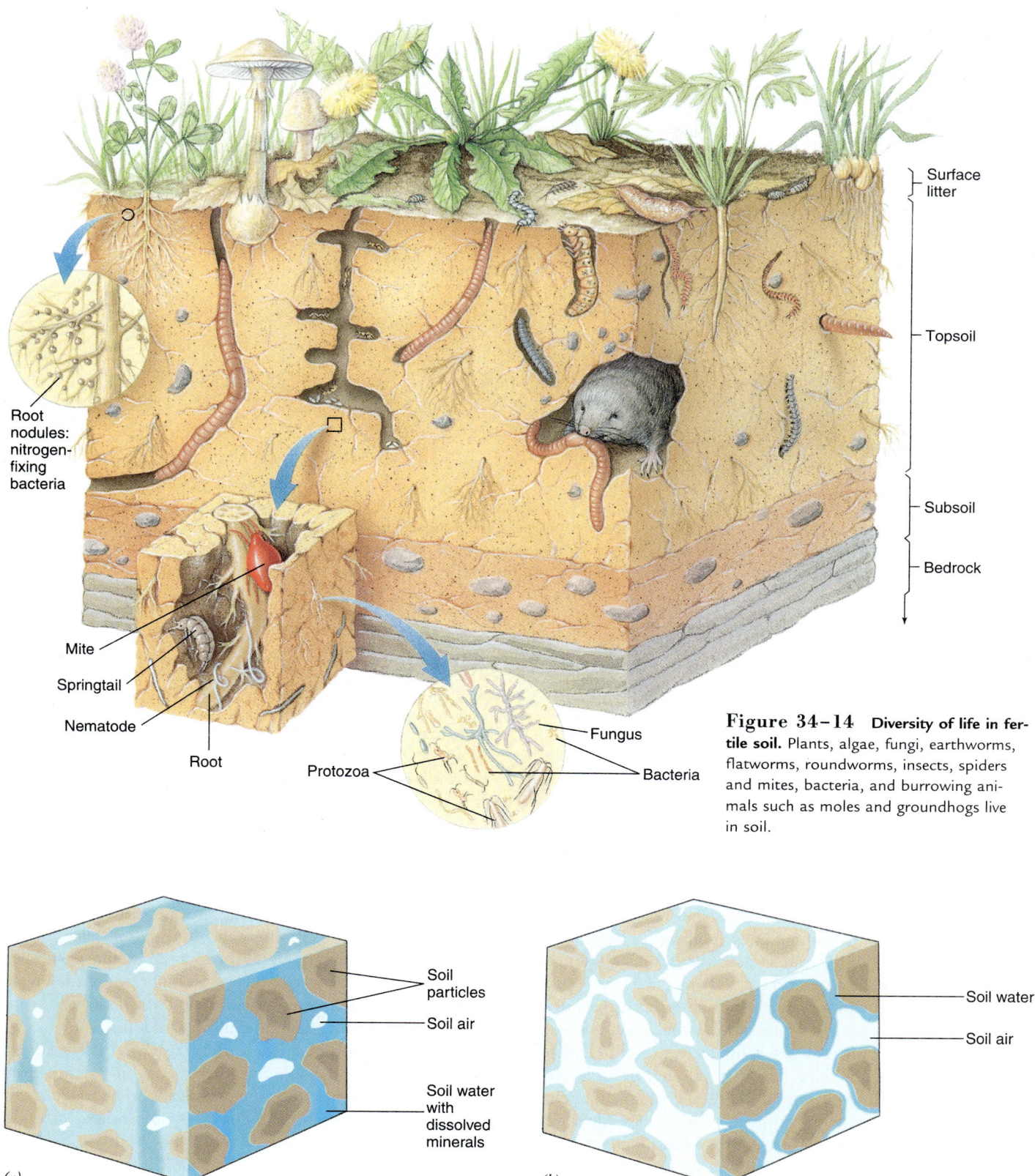

Root
nodules:
nitrogen-
fixing
bacteria

Mite

Springtail

Nematode

Root

Protozoa

Fungus

Bacteria

Surface
litter

Topsoil

Subsoil

Bedrock

Figure 34–14 Diversity of life in fertile soil. Plants, algae, fungi, earthworms, flatworms, roundworms, insects, spiders and mites, bacteria, and burrowing animals such as moles and groundhogs live in soil.

Soil
particles

Soil air

Soil water
with
dissolved
minerals

Soil water

Soil air

(a)

(b)

Figure 34–15 Pore space, soil air, and water. *(a)* In a wet soil, most of the pore space is filled with water. *(b)* In a dry soil, a thin film of water is tightly bound to soil particles, and most of the pore space is occupied by air.

Figure 34–16 **How acid alters soil chemistry.** *(a)* In normal soil, positively charged nutrient mineral ions are attracted to the negatively charged soil particles. *(b)* In acidified soil, hydrogen ions displace the cations. Aluminum ions released when the soil becomes acidified also adhere to soil particles.

sorbed by the roots in toxic concentrations. Certain mineral salts essential for plant growth, such as calcium phosphate, become less soluble and thus less available to plants at a higher pH.

Soil pH also affects the leaching of nutrients. An acidic soil has less ability to bind positively charged ions to it (Fig. 34–16). As a result, certain minerals essential for plant growth, such as potassium (K^+), are leached more readily from acidic soil. The optimum soil pH for most plant growth is 6.0 to 7.0, because most essential nutrient minerals are available to plants in that pH range.

Soil pH affects plants, but it is also, in turn, influenced by plants and other organisms. Cation exchange performed by roots decreases the pH of soil, as does the decomposition of humus and the cellular respiration of soil organisms. **Acid precipitation,** a type of air pollution in which sulfuric and nitric acids produced by human activities fall to the ground as acid rain, sleet, snow, or fog, can seriously decrease soil pH. Acid precipitation is one of several factors implicated in **forest decline,** the gradual deterioration and, often, death of trees that has been observed in many European and North American forests. Forest decline may be partly the result of soil changes caused by acid precipitation. A 1989 study of Central European forests that have experienced forest decline, for example, found a strong correlation between forest damage and soil chemistry altered by acid precipitation.

SOIL PROVIDES MOST OF THE MINERALS FOUND IN PLANTS

More than 90 naturally occurring elements exist on Earth, and more than 60 of these, including elements as common as carbon and as rare as gold, have been found in plant tissues. Not all of these elements are considered essential for plant growth, however.

Sixteen elements have been found essential for plant growth (Table 34–2). Nine of these are required in fairly large quantities (greater than 0.05% dry weight) and are therefore known as **macronutrients.** These include carbon, hydrogen, oxygen, nitrogen, phosphorus, potassium, sulfur, calcium, and magnesium. The remaining seven **micronutrients** are needed in trace amounts (less than 0.05% dry weight) for normal plant growth and development. These include iron, boron, manganese, copper, molybdenum, chlorine, and zinc.

Four of the 16 elements—carbon, oxygen, hydrogen, and nitrogen—come directly or indirectly from soil water or from gases in the atmosphere. Carbon is obtained from carbon dioxide (CO_2) in the atmosphere during photosynthesis. Oxygen is obtained from atmospheric oxygen (O_2) and water (H_2O). Water also supplies hydrogen to the plant. Plants absorb their nitrogen from the soil as ions of nitrogen salts, but the nitrogen in nitrogen salts ultimately comes from atmospheric nitrogen (see *Making the Connection: Soil Nitrogen and the*

SOIL NITROGEN AND THE INTERDEPENDENCE OF ORGANISMS

How does the requirement of nitrogen by organisms show the interdependence of life? Because nitrogen is an essential part of biologically important molecules such as proteins and nucleic acids, all organisms must have nitrogen in order to survive. Animals, including humans, cannot use nitrogen as atmospheric N_2 or as inorganic ions (nitrate, NO_3^-, and ammonium, NH_4^+), and so they get their nitrogen from proteins and other nitrogen-containing compounds in the foods they eat. Ultimately, this nitrogen is traced back to plant sources.

Plants obtain their nitrogen as nitrate or ammonium ions from the soil. But where do the NO_3^- and NH_4^+ come from?

Nitrogen is converted into those forms from atmospheric N_2 by billions upon billions of individual bacteria of just a few species (see Chapter 23). Thus, the nitrogen supply of the entire biosphere, including humans, comes from a few, at first glance insignificant, species. Without these bacteria, the supply of nitrogen in a form suitable for plants would be rapidly depleted, and plants would die. With the demise of plants, animals, including humans, would be unable to survive. We say more about the crucial role of microorganisms in the nitrogen cycle in Chapter 53.

Interdependence of Organisms). The remaining 12 essential elements are obtained from the soil as dissolved nutrient mineral ions. Their ultimate source is the parent material from which the soil was formed.

Carbon, hydrogen, and oxygen are found as part of the structure of all biologically important molecules, including lipids, carbohydrates, nucleic acids, and proteins. Nitrogen is part of proteins, nucleic acids, and chlorophyll. Phosphorus is critical for plants because it is found in nucleic acids, phospholipids (an essential part of cell membranes), and energy

transfer molecules such as ATP. Calcium plays a key structural role as a component of the middle lamella (the cementing layer between cell walls of adjacent plant cells). Calcium has also been implicated in a number of physiological roles in plants, such as altering membrane permeability. Magnesium is part of the chlorophyll molecule. Sulfur is essential because it is found in certain amino acids and vitamins.

Potassium, which plants use in fairly substantial amounts, is not found in a specific compound or group of compounds in plant cells. Instead, it remains as free K^+ and plays a key

---

TABLE 34-2 Functions of Essential Elements

Element	Taken Up As	Major Functions
Carbon	CO_2	Component of carbohydrate, lipid, protein, and nucleic acid molecules
Hydrogen	H_2O	Component of carbohydrate, lipid, protein, and nucleic acid molecules
Oxygen	CO_2, H_2O	Component of carbohydrate, lipid, protein, and nucleic acid molecules
Nitrogen	NO_3^-, NH_4^+	Component of proteins, nucleic acids, chlorophyll, certain coenzymes
Phosphorus	HPO_4^-, $H_2PO_4^{2-}$	In nucleic acids, phospholipids, ATP (energy transfer compound)
Calcium	Ca^{2+}	In cell walls; involved in membrane permeability; enzyme activator
Magnesium	Mg^{2+}	In chlorophyll; enzyme activator in carbohydrate metabolism
Sulfur	SO_4^{2-}	In certain amino acids and vitamins
Potassium	K^+	Osmosis and ionic balance; opening and closing of stomata; enzyme activator (for 40+ enzymes)
Chlorine	Cl^-	Ionic balance; involved in photosynthesis
Iron	Fe^{2+}, Fe^{3+}	Part of enzymes and electron transport molecules involved in photosynthesis, respiration, and nitrogen fixation
Manganese	Mn^{2+}	Part of enzymes involved in respiration and nitrogen metabolism; required for photosynthesis
Copper	Cu^+, Cu^{2+}	Part of enzymes involved in photosynthesis
Zinc	Zn^{2+}	Part of enzymes involved in respiration and nitrogen metabolism
Molybdenum	MoO_4^{2-}	Part of enzymes involved in nitrogen metabolism
Boron	$H_2BO_3^-$	Exact role unclear; involved in membrane transport and calcium utilization

physiological role in maintaining the turgidity of cells because it is osmotically active. The presence of K^+ in cytoplasm causes the cell to have a greater solute concentration than surrounding cells. As a result, water passes through the plasma membrane into the cell by osmosis. Potassium is also involved in the opening and closing of stomata (see Chapter 32).

Another element that has a role in maintaining turgidity of cells is chlorine. In addition to its osmotic role, the chloride (Cl^-) ion, which is present in trace amounts in plants, is essential for photosynthesis.

Five of the micronutrients (iron, manganese, copper, zinc, and molybdenum) are associated with various plant enzymes, often as enzyme activators, and are involved in certain enzymatic reactions. While the role of boron in plants is unclear, some data suggest that it is involved in the transport of carbohydrates across cell membranes and affects calcium utilization.

In addition to the sixteen essential elements, several other elements have been demonstrated to be essential for specific plants. Nickel is involved in enzymatic reactions in nitrogen-fixing legumes such as peas and beans; sodium is probably essential for some plants adapted to a salty soil, some desert plants, sugar beets, bluegrass, and plants with C_4 photosynthesis; silicon enhances the growth of various grasses. After further evaluation, one or more of these elements may be added to the list of essential elements.

How do biologists determine whether an element is essential?

It is impossible to conduct mineral nutrition experiments by growing plants in soil because soil is too complex and contains too many elements. Thus, one of the most useful methods to test whether or not an element is essential is **hydroponics,** which is the growing of plants in aerated water to which nutrient mineral salts have been added. Hydroponics also has commercial applications in addition to its scientific use (Fig. 34–17).

If biologists suspect that a particular element is essential for plant growth, they grow plants in a nutrient solution that contains all known essential elements *except* the one in question. If plants grown in the absence of that element are unable to develop normally or to complete their life cycle, the element may be essential. Additional criteria are used to confirm whether an element is essential. For example, it must be demonstrated that the element has a direct effect on the plant's metabolism, and that the element is essential for a wide variety of plants.

SOIL CAN BE DAMAGED BY HUMAN MISMANAGEMENT

Soil is a valuable natural resource on which humans depend for food. Many human activities generate or aggravate soil problems, including mineral depletion, soil erosion, and accumulation of salt.

Mineral depletion occurs in soils that are farmed

In a natural ecosystem, the essential minerals removed from the soil by plants are returned when the plants or the animals that eat them die and decompose. An agricultural system disrupts this pattern. Crops, which contain minerals, are removed from the cycle when they are harvested. As a result, they cannot decay and release their nutrients back into the soil. Thus, over time, soil that is farmed eventually loses its fertility, that is, its ability to produce abundant crops. In similar manner, homeowners often mow their lawns and remove the clippings, preventing decomposition and cycling of minerals that were in the grass blades.

Plant growth is usually limited by the essential material (water, sunlight, or some essential element) in shortest supply. This phenomenon is sometimes called the concept of **limiting factors.** The three elements that are most often limiting factors for plants are nitrogen, phosphorus, and potassium. In

Figure 34–17 *Hydroponically grown lettuce.* Lettuce is one of several hydroponic crops now grown commercially. Because they are not anchored in soil, the plants are supported by the white boards. A chemically defined liquid solution of nutrient minerals trickles over their roots, which also have access to atmospheric oxygen. *(Hank Morgan/Photo Researchers, Inc.)*

order to sustain the productivity of agricultural soils, fertilizers that contain nutrients are periodically added to depleted soils to replace the minerals that limit plant growth.

The two main types of fertilizers are organic and inorganic. *Organic fertilizers* include such natural materials as cow manure, crop residues, bone meal, blood, and compost. Green manure, a special type of organic fertilizer, is actually a crop that is planted and deliberately plowed into the soil to decompose instead of being harvested. Frequently the plant grown as green manure has nitrogen-fixing bacteria living in root nodules, thereby increasing the amount of nitrogen in the soil. Organic fertilizers are complex, and their exact compositions vary. The mineral nutrients in them become available to plants only as the organic material decomposes. For that reason, organic fertilizers are slow-acting and long-lasting.

Inorganic fertilizers are manufactured from chemical compounds, and their exact compositions are known. Because they are soluble, they are immediately available to plants. Inorganic fertilizers are available in the soil for only a short period of time (relative to organic fertilizers), because they quickly leach away. Most inorganic fertilizers contain the three elements (nitrogen, phosphorus, and potassium) that are usually the limiting factors in plant growth. The numbers on fertilizer bags (for example, 10–20–20) tell the relative percentage concentrations of each of these three elements (N, P, K).

Soil erosion is the loss of soil from the land

Water, wind, ice, and other agents cause **soil erosion,** the wearing away or removal of soil from the land. Water and wind are particularly effective in removing soil: rainfall loosens soil particles that can then be transported away by moving water (Fig. 34–18), while wind loosens soil and blows it away, particularly if the soil is exposed and dry.

Soil erosion is a national and international problem that does not make the headlines very often. To get a feeling for how serious the problem is, consider that approximately 4.0 billion metric tons (4.4 billion tons) of topsoil are lost from U. S. croplands and pasturelands as a result of soil erosion *each year.* The U. S. Department of Agriculture estimates that approximately one-fifth of U. S. cropland is vulnerable to soil erosion damage. Erosion causes a loss in soil fertility because essential minerals and organic matter are also removed. As a result of these losses, the productivity of eroded agricultural soils declines, and more fertilizer must be used to replace the nutrients lost to erosion.

Humans often accelerate soil erosion through poor soil management practices. Agriculture is not the only culprit, since the removal of natural plant communities during surface mining and during the construction of roads and buildings also accelerates erosion. In addition, some logging practices, such as clear-cutting large forested areas for lumber and pulpwood, cause severe erosion.

Soil erosion has an impact on other natural resources. Sediment that enters streams, rivers, and lakes degrades water quality and fish habitats. Pesticides and fertilizer residues may be present in sediment, adding additional pollutants to the wa-

Figure 34–18 **Soil erosion in an open field.** Precipitation can cause gullies to enlarge rapidly because they provide channels for the runoff of water. *(USDA/Natural Resources Conservation Service)*

ter. Also, when forests are removed within the watershed of a hydroelectric power facility, accelerated soil erosion can cause the reservoir behind the dam to fill in with sediment much faster than usual. This process results in a loss of electricity productivity at that facility.

Sufficient plant cover reduces the amount of soil erosion: leaves and stems cushion the impact of rainfall, and roots help to hold the soil in place. Although soil erosion is a natural process, abundant plant cover makes it negligible in many natural ecosystems.

Salt accumulates in soil that is improperly irrigated

Although irrigation improves the agricultural productivity of arid and semiarid lands, it can also cause salt to accumulate in the soil, a process called **salinization.** In a natural scenario, as a result of precipitation runoff, rivers carry dissolved salts away. Irrigation water, however, normally soaks into the soil and does not run off the land into rivers, so when it evaporates, the salt remains behind and accumulates in the soil. Salty soil results in a decline in productivity and in extreme cases renders the soil completely unfit for crop production.

Most plants cannot obtain all the water they need from salty soil because a water balance problem exists: water moves by osmosis *out* of plant roots and into the salty soil. Obviously, most plants cannot survive under these conditions (see Fig. 5–13). Plant species that thrive in saline soils have special adaptations that enable them to tolerate the high amount of salt. Black mangroves, for example, excrete excess salt through their leaves. Most crops, unless they have been genetically selected to tolerate high salt, are not productive in saline soil.

I. Anchorage, absorption, conduction, and storage are the main functions of roots.

II. Two types of root systems exist in plants.
  A. A **taproot** system has one main root from which many smaller lateral roots extend.
  B. A **fibrous root** system has several to many **adventitious roots** of the same size developing from the end of the stem. Smaller lateral roots extend from these adventitious roots.

III. Roots differ structurally from stems in several ways.
  A. Each root tip is covered by a **root cap,** a protective layer that covers the delicate root apical meristem and may orient the root so that it grows downward.
  B. **Root hairs,** short-lived extensions of epidermal cells, increase the surface area of the root in contact with the moist soil.
  C. Roots lack nodes and internodes and do not usually produce leaves or buds.

IV. Primary roots possess an epidermis, ground tissues (cortex and in certain roots, pith), and vascular tissues (xylem and phloem).
  A. **Epidermis** protects the root, and its root hairs aid in absorption of water and dissolved nutrient minerals.
  B. **Cortex** consists of parenchyma cells that often store starch. The water and dissolved nutrient minerals move through the cortex along one of two pathways, the **apoplast** (along the interconnected porous cell walls) or the **symplast** (from one cell's cytoplasm to the next through plasmodesmata).
  C. **Endodermis,** the innermost layer of the cortex, regulates the movement of nutrient minerals into the root xylem.
    1. Cells of the endodermis possess a **Casparian strip** around their radial and transverse walls that is impermeable to water and dissolved nutrient minerals.
    2. Nutrient minerals are actively transported through membrane carrier proteins into the cytoplasm of the endodermal cells.
  D. **Pericycle** gives rise to lateral roots and lateral meristems.
  E. **Xylem** conducts water and dissolved nutrient minerals; **phloem** conducts dissolved sugar.

V. Roots vary in internal structure.
  A. Xylem of herbaceous dicot roots forms a solid core in the center of the root. In contrast, the center of monocot roots often consists of **pith** surrounded by a ring of alternating bundles of xylem and phloem.

  B. Monocot roots lack a **vascular cambium** and therefore do not have secondary growth.
  C. Roots of gymnosperms and woody dicots develop secondary tissues (wood and bark).

VI. Some roots are modified for specialized functions.
  A. **Prop roots** develop from branches or from a vertical stem and grow downward into the soil to help support the plant in an upright position.
  B. **Pneumatophores** are aerial "breathing" roots that may assist in getting oxygen to the submerged roots.
  C. Certain epiphytes have roots modified to photosynthesize, absorb moisture, or, if parasitic, penetrate host tissues and absorb nutrients.
  D. Corms and bulbs often have **contractile roots** that grow into the soil and then contract, thereby pulling the corm or bulb deeper into the soil.

VII. Roots often form associations with other species.
  A. A root **graft** is a natural union between the roots of trees belonging to the same or different species.
  B. **Mycorrhizae** are mutually beneficial associations between roots and soil fungi.
  C. Root **nodules** are swellings that develop on certain roots and house millions of nitrogen-fixing bacteria.

VIII. Soil is the complex material in which roots grow.
  A. Factors influencing soil formation include parent material, climate, organisms, the passage of time, and topography.
  B. Soil is composed of inorganic minerals, organic matter, organisms, air, and water.

IX. Plants require 16 essential elements for normal growth.
  A. Nine elements are **macronutrients**: carbon, oxygen, hydrogen, nitrogen, potassium, phosphorus, sulfur, magnesium, and calcium.
  B. Seven elements are **micronutrients**: iron, boron, manganese, copper, zinc, molybdenum, and chlorine.
  C. Plants obtain positively charged nutrient mineral ions from clay particles in the soil by **cation exchange,** in which roots secrete protons (H^+), which are exchanged for other positively charged mineral ions, freeing them to be absorbed by roots.

P O S T - T E S T

1. One main root, formed from the enlarging embryonic root, with many smaller lateral roots coming out of it is a (an) (a) fibrous root system (b) adventitious root system (c) taproot system (d) contractile root system (e) prop root system

2. Roots produced at unusual places on the plant are (a) fibrous (b) adventitious (c) taproots (d) contractile (e) mycorrhizae

3. Plants with corms often have _____ roots that pull the bulb deeper into the ground. (a) fibrous (b) adventitious (c) tap (d) contractile (e) prop

4. Certain plants adapted to flooded soil produce aerial "breathing" roots known as (a) fibrous roots (b) pneumatophores (c) mycorrhizae (d) contractile roots (e) prop roots

5. Unlike stems, roots produce (a) nodes and internodes (b) root caps and internodes (c) axillary buds and root hairs (d) terminal buds and axillary buds (e) root caps and root hairs

6. The waterproof region around the radial and transverse walls of endodermal cells is the: (a) Casparian strip (b) pericycle (c) apoplast (d) symplast (e) pneumatophore

7. Plants obtain positively charged nutrient mineral ions from clay particles in the soil by cation exchange, in which: (a) roots passively absorb the positively charged mineral ions that they require (b) mineral ions flow freely along porous cell walls (c) roots secrete protons (H^+), which free other positively charged mineral ions to be absorbed by roots (d) the Casparian strip effectively blocks the passage of water and nutrient mineral ions along the endodermal cell wall (e) a well developed system of internal air spaces in the root allows both gas exchange and cation exchange

8. The cell layer from which lateral roots originate is (a) epidermis (b) cortex (c) endodermis (d) pericycle (e) vascular cambium

9. The center of a herbaceous dicot root is composed of _____ ,

whereas the center of a monocot root is composed of _____ . (a) pith; cortex (b) xylem; phloem (c) phloem; xylem (d) xylem; pith (e) pith; xylem

10. Mutually beneficial associations between certain soil fungi and the roots of most plant species are called (a) mycorrhizae (b) pneumatophores (c) nodules (d) steles (e) humus

11. The technique of growing plants in aerated water with dissolved nutrient mineral salts is known as: (a) hydration (b) hydroponics (c) hydrophilic (d) hydrostatic (e) hydrolysis

12. Carbon, hydrogen, oxygen, nitrogen, phosphorus, sulfur, magnesium, calcium, and potassium are collectively known as (a) micronutrients (b) microvilli (c) micronuclei (d) macronuclei (e) macronutrients

REVIEW QUESTIONS

1. List several functions of roots and describe the tissue(s) or cell layer(s) responsible for each function.
2. How would you distinguish between a root hair and a small lateral root?
3. If you were examining a cross section of a primary root of a flowering plant, how would you determine whether it was a dicot or a monocot?
4. Trace the pathway of water from the soil through the various tissues in a herbaceous dicot root.
5. How does a herbaceous dicot root develop secondary tissues?
6. Distinguish among the following specialized roots: (a) storage root (b) prop root (c) aerial "breathing" root (d) photosynthetic root (e) contractile root.
7. List the five components of soil and describe how each is important to plants.
8. Explain how weathering processes convert rock into soil.
9. Distinguish between macronutrients and micronutrients.
10. Label the various tissues of this herbaceous dicot root. Use Fig. 34–3 to check your answers.

YOU MAKE THE CONNECTION

1. A mesquite root is found penetrating a mine shaft about 46 meters (150 feet) below the surface of the soil. How might you determine when the root first grew into the shaft? (*Hint:* mesquite is a woody plant.)
2. A barrel cactus that is 2 ft tall and 1 ft in diameter has roots over 10 ft long. However, all of the plant's roots are found in the soil at a depth of 2 to 6 in. What possible adaptive value does such a shallow root system confer on a desert plant?
3. You are given a plant structure that was found growing in the soil and are asked to determine whether it is a root or an underground stem. How would you identify the plant part without a microscope? With a microscope?

4. How would you design an experiment to determine whether gold is essential for plant growth? What would you use for an experimental control?
5. Why does overwatering a plant often kill it?
6. Explain why, once secondary growth has occurred, that portion of the root is no longer involved in absorption. Where does absorption of water and dissolved nutrient minerals occur in plants that have roots with secondary growth?

RECOMMENDED READINGS

Baskin, Y. *The Work of Nature: How the Diversity of Life Sustains Us.* Island Press, Washington, D.C., 1997. This excellent book, written by a distinguished science writer, examines threats posed by the loss of biodiversity, including how changes in soil-dwelling species have affected soil fertility.

Berg, L.R. *Introductory Botany: Plants, People, and the Environment.* Saunders College Publishing, Philadelphia, 1997. A general botany text with an environmental emphasis.

Brown, J.C., and V.D. Jolley. "Plant Metabolic Responses to Iron-Deficiency Stress." *BioScience,* Vol. 39, No. 8, Sep. 1989. An in-depth discussion of the factors that make iron in the soil available to roots.

Feldman, L.J. "The Habits of Roots." *BioScience,* Vol. 38, No. 9, Oct. 1988.

How roots interact with their soil environment.

Moore, P. "Upwardly Mobile Roots." *Nature,* Vol. 341, 21 Sep. 1989. An essay on unusual root behaviors, including roots that grow out of the soil and up the trunk of a tree to obtain nutrients leaching from the forest canopy.

Wolkomir, R. "Unearthing Secrets Locked Deep Inside Each Fistful of Soil." *Smithsonian,* Mar. 1997. Highlights the research of scientists at the National Soil Tilth Lab in Ames, Iowa.

Zimmer, C. "The Web Below." *Discover,* Nov. 1997. This article highlights the research of Simard et al., which was featured in this chapter's *On the Cutting Edge* box.

● Visit our website at **http://www.saunderscollege.com/lifesci/titles.html** and click on Solomon/Berg/Martin Biology for links to chapter-related resources on the World Wide Web.

CHAPTER 35

Reproduction in Flowering Plants

In Chapter 31 you learned that flowering plants, which include at least 235,000 species, are the largest, most successful group of plants. One reason for the success of flowering plants, or angiosperms, is their ability to reproduce both sexually and asexually. You may have admired flowers, such as the painted trillium *(Trillium undulatum)* shown in the photograph, for their various colors and shapes as well as their often pleasing fragrances. The biological function of flowers, however, is sexual reproduction. Their colors, shapes, and fragrances are adaptations that increase the likelihood that pollen grains, which produce sperm cells, are carried from plant to plant.

As in most organisms, sexual reproduction in plants includes meiosis and the fusion of reproductive cells—egg and sperm cells, collectively called gametes. The union of gametes, which is called *fertilization,* occurs within the flower's ovary. After fertilization, flowering plants produce seeds inside fruits.

The offspring of parents that reproduce sexually exhibit considerable genetic variation. They may resemble one of the parent plants, both of the parents, or neither of the parents. Sexual reproduction offers the advantage of new gene combinations, not found in either parent, that might make an individual plant better suited to its environment. These new gene combinations largely result from the independent assortment of chromosomes that occurs during meiosis before the production of both egg and sperm cells. (How this variation occurs, that is, the details of independent assortment and genetic recombination, is discussed in Chapters 9 and 10.)

Most flowering plants reproduce asexually to some extent. Asexual reproduction does not usually involve the formation of flowers, seeds, and fruits. Instead, offspring generally form when a vegetative part of an existing plant expands, grows, and then becomes separated from the rest of the plant, often by the death of tissues. This part subsequently grows to form a complete, independent plant.

Flowering plants have many methods of asexual reproduction, most of which involve modified vegetative organs such as stems, roots, and leaves. In particular, many modified stems, such as rhizomes, tubers, corms, and stolons, reproduce asexually. Roots that form suckers are an important means of asexual reproduction in other plants. Because asexual reproduction requires only one parent, and no meiosis or fusion of

(Skip Moody/Dembinsky Photo Associates)

gametes occurs, the offspring of asexual reproduction are virtually genetically identical to each other and to the parent plant from which they came.[1]

In this chapter we examine various aspects of both sexual and asexual reproduction in flowering plants, including floral adaptations that are important in pollination, seed and fruit structure and dispersal, and several kinds of asexual reproduction. We conclude with a discussion of the evolutionary advantages and disadvantages of sexual and asexual reproduction.

---

[1]Although offspring of asexual reproduction are generally considered to be genetically uniform, somatic mutations can result in some variability among asexually derived offspring.

AFTER YOU HAVE STUDIED THIS CHAPTER YOU SHOULD BE ABLE TO

1. State the differences between sexual and asexual reproduction and discuss the evolutionary advantages and disadvantages of each.
2. Label the parts of a flower on a diagram, and describe the functions of each part.
3. Relate where eggs and pollen grains are formed within the flower and distinguish between pollination and fertilization.
4. Compare the evolutionary adaptations that characterize flowers pollinated in different ways (by insects, birds, bats, and wind).
5. Define coevolution and give examples of ways that plants and

their animal pollinators have affected one another's evolution.
6. Trace the stages of embryo development in flowering plants and list and define the main parts of seeds.
7. Distinguish among simple, aggregate, multiple, and accessory fruits; give examples of each type; and cite several different methods of seed and fruit dispersal.
8. Explain how the following structures may be used to propagate plants asexually: rhizomes, tubers, stolons, corms, bulbs, plantlets, suckers. Explain how apomixis can occur.

THE FLOWERING PLANT LIFE CYCLE ALTERNATES BETWEEN A CONSPICUOUS SPOROPHYTE AND A REDUCED GAMETOPHYTE

In Chapters 26 and 27 you learned that angiosperms and other plants have a cyclic **alternation of generations** in which they spend a portion of their lives in a multicellular haploid stage and a portion in a multicellular diploid stage. The haploid portion, called the **gametophyte generation,** gives rise to gametes by mitosis. When two gametes fuse, the diploid portion of the life cycle, called the **sporophyte generation,** begins. The sporophyte generation produces haploid spores by meiosis. Each spore has the potential to give rise to a gametophyte plant, and the cycle continues.

In flowering plants the diploid sporophyte generation is larger and nutritionally independent, and the haploid gametophyte generation, which is located in the flower, is microscopic in size and nutritionally dependent on the sporophyte. We revisit alternation of generations in flowering plants after our discussion of the specialized reproductive structures called flowers. It may be helpful for you to review Fig. 27–12, which shows the main stages in the flowering plant life cycle.

FLOWERS ARE INVOLVED IN SEXUAL REPRODUCTION

Flowers are reproductive shoots usually composed of four kinds of organs—sepals, petals, stamens, and carpels—arranged in whorls (circles) on the end of a flower stalk (Fig. 35–1a). In flowers with all four organs, the normal order of whorls from the flower's periphery to the center (or from the flower's base upward) is:

Sepals → petals → stamens → carpels

The tip of the stalk enlarges to form a **receptacle** on which some or all of the flower parts are borne. All four floral parts are important in the reproductive process, but only the stamens (the "male" organs) and carpels (the "female" organs) participate directly in sexual reproduction; sepals and petals are sterile.

Sepals, which constitute the lowest and outermost whorl on a floral shoot, cover and protect the flower parts when the flower is a bud. Sepals are leaflike in shape and form, and are often green. Some sepals, such as those in lily flowers, resemble petals. As the blossom opens from a bud, the sepals fold back to reveal the more conspicuous petals. The collective term for all the sepals of a flower is **calyx.**

The whorl just above the sepals consists of **petals,** which are broad, flat, and thin (like sepals and leaves) but tremendously varied in shape and frequently brightly colored, attracting pollinators. As we will see shortly, petals play an important role in ensuring that sexual reproduction will occur. Sometimes petals are fused to form a tube or other floral shape. The collective term for all the petals of a flower is **corolla.**

Just inside the petals are the **stamens,** the male reproductive organs. Each stamen is composed of a thin stalk, called a **filament,** at the top of which is an **anther,** a saclike structure in which **pollen grains** form (Fig. 35–1b). For sexual reproduction to occur, pollen grains must be transferred from the anther to the female reproductive structure (the carpel), usually of another flower of the same species. Each pollen grain produces two cells surrounded by a tough outer wall. One cell divides mitotically to form two nonflagellated male gametes, known as sperm cells, and the other produces a **pollen tube** through which the sperm cells travel to reach the ovule.

In the center or top of most flowers are one or more **carpels,** the female reproductive organs. Carpels bear **ovules,** which are structures with the potential to develop into seeds. The carpels of a flower may be separate or fused together into a single structure. The female part of the flower, often referred to as a **pistil,** may consist of a single carpel (a *simple pistil*) or a group of fused carpels (a *compound pistil*). Each pistil has three sections: a **stigma,** on which the pollen grains land; a **style,** a necklike structure through which the pollen tube

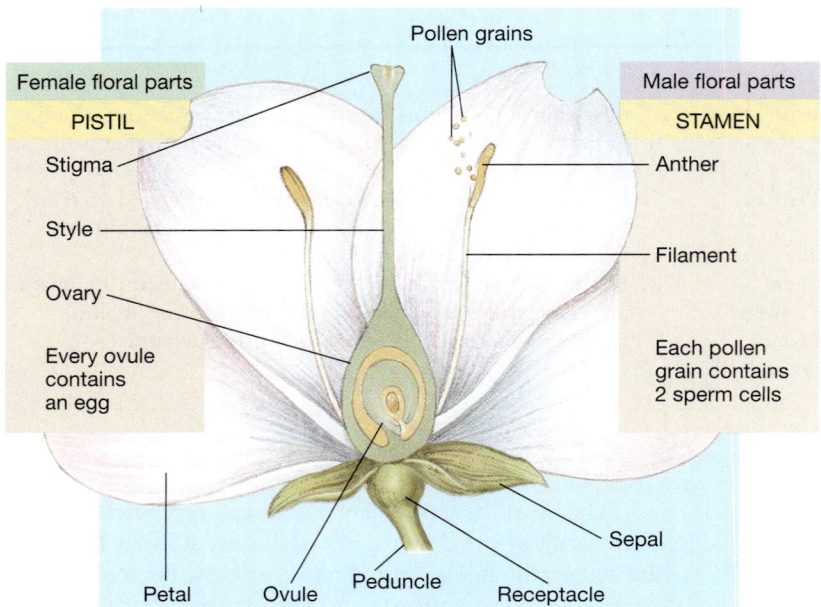

Figure 35–1 **Structure of a flower.** *(a)* Cutaway view of a "typical" flower. The pistil in this flower consists of a single carpel; other pistils are composed of several to many fused carpels. *(b)* Pollen grains develop within sacs in the anther. *(c)* An embryo sac, with its egg and polar nuclei, forms within the ovule. The ovary depicted here contains a single ovule.

(a)

(b)

(c)

grows; and an **ovary,** a juglike structure that contains one or more ovules (Fig. 35–1*c*). Each ovule contains an **embryo sac** that forms one female gamete (an egg) and two **polar nuclei.** The egg and both polar nuclei participate directly in fertilization. (See *Making the Connection: Flowers and Evolution* for a brief discussion of the evolution of sepals, petals, stamens, and carpels.)

Female gametophytes occur in the ovary, male gametophytes in the anther

Before we proceed further, it might be helpful to relate the stages in alternation of generations to floral structure. As discussed in Chapters 26 and 27, angiosperms and certain other seed plants are heterosporous and produce two kinds of spores, megaspores and microspores (Fig. 35–2).

Each young ovule within an ovary contains a diploid cell, the megasporocyte, which undergoes meiosis to produce four haploid cells. Three of these usually disintegrate, and the fourth, the megaspore, divides mitotically to produce a multicellular female gametophyte, also called an embryo sac. The embryo sac, which is embedded in the ovule, contains seven cells with eight haploid nuclei. Six of these cells, including the egg cell, contain a single nucleus each; a large central cell has two nuclei, called polar nuclei.

Pollen sacs within the anther contain numerous diploid cells called microsporocytes, each of which undergoes meiosis to produce four haploid cells called microspores. Each microspore divides mitotically to produce an immature male gametophyte, also called a pollen grain, that consists of two cells, the tube cell and the generative cell. The pollen grain becomes mature when its generative cell divides to form two nonmotile sperm cells.

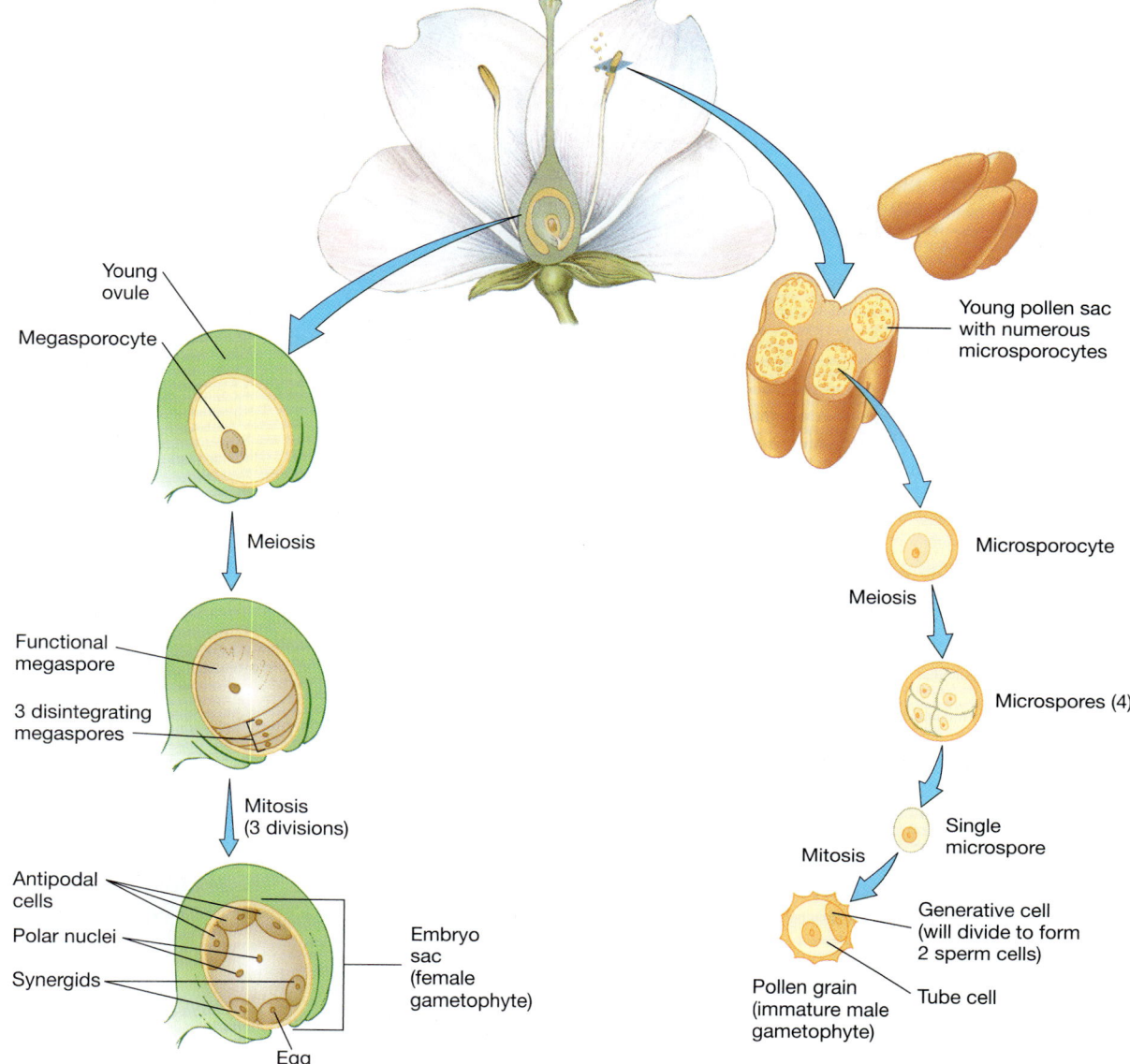

Figure 35–2 **Development of female and male gametophytes.** *(Left side)* The female gametophyte, or embryo sac, develops within the ovule. *(Right side)* The immature male gametophytes, or pollen grains, develop within pollen sacs. Each male gametophyte becomes mature when its generative cell divides mitotically to produce two sperm cells.

POLLINATION IS THE FIRST STEP TOWARD FERTILIZATION

Before fertilization can occur, pollen grains must travel from the anther (where they form) to the stigma. The transfer of pollen grains from anther to stigma is known as **pollination.** Plants are *self-pollinated* if pollination occurs within the same flower or a different flower on the same plant. When pollen grains are transferred to a flower on another plant, we say the plant is *cross-pollinated*. Flowering plants accomplish pollina-

tion in a variety of ways. Some flowers are pollinated by beetles, bees, flies, butterflies, moths, wasps, or other insects. Other animals such as birds, bats, snails, and small rodents also pollinate plants. Wind is the agent of pollination for certain flowers, while a few aquatic flowers are pollinated by water.

Flowering plants and their animal pollinators have coevolved

Flowers pollinated by animals have various features to attract the pollinators, including showy petals (a visual attractant) and

FLOWERS AND EVOLUTION

How did flowers evolve? Recall from Chapter 19 that in evolution new structures or organs often originate by modification of previously existing structures or organs. Much scientific evidence supports the classical interpretation that the four organs of a flower arose from highly modified leaves. This evidence includes comparisons of the arrangement of vascular tissues in both flowers and leafy stems and of the development of floral parts and leaves.

Sepals are the most leaflike of the four floral organs, and botanists generally agree that sepals are specialized leaves. Although petals of many flowering plant species are leaflike in appearance, botanists generally view petals as modified stamens that became sterile and leaflike. Cultivated roses and camellias provide evidence of this hypothesis; in some varieties the stamens have been transformed into petals, forming showy flowers with large numbers of petals.

The origin of stamens and carpels from leaves is supported by the remarkably leaflike stamens and carpels of certain tropical trees and other species. Consider, for example, the carpel of *Drimys,* a genus of evergreen trees and shrubs native to Southeast Asia, Australia, and South America. This carpel resembles a leaf folded inward along the midrib, thereby enclosing the ovules, and joined along the entire length of the leaf's margin *(see figure).* The fundamental question is whether these leaflike stamens and carpels are primitive organs that were conserved (retained) during the course of evolution or are highly specialized organs that do not resemble early stamens and carpels. Many botanists who have studied this question have concluded that stamens and carpels are probably derived from leaves. The origin of stamens and carpels from highly modified leaves is not accepted by all botanists, however.

During the course of more than 120 million years of angio-

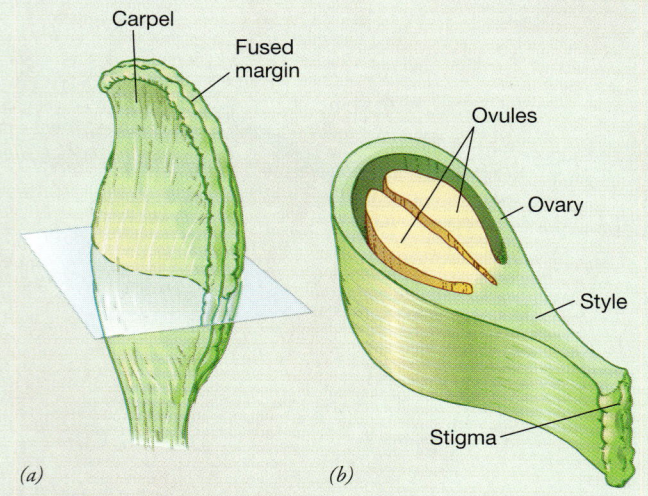

The carpel of *Drimys piperita*. *(a)* The carpel resembles a folded leaf in which the ovules borne on its upper surface are enclosed. *(b)* A cross section of the carpel, cut along the dashed line in part *(a)*.

sperm evolution, flower structure diversified as floral organs fused together or became reduced in size or number. These changes led to greater complexity in floral structure in some species and to greater simplicity in other species. Interpreting the floral structures of so many different angiosperm species is sometimes difficult, but it is important because correct interpretations are essential to devising a classification scheme that is phylogenetic (see Chapter 22).

---

scent (an olfactory attractant) (Fig. 35–3). One of the rewards for the animal pollinator is food. Nectar is a sugary solution produced by some flowers in special floral glands called nectaries and used as an energy-rich food by pollinators. Pollen grains are also a protein-rich food for many animals. As they move from flower to flower searching for food, pollinators inadvertently carry pollen grains on their body parts, thus facilitating sexual reproduction in plants.

Biologists estimate that about 70% of all flowering plant species are pollinated by insects. Bees are particularly important as pollinators of crop plants. About 30% of human food comes from crops pollinated by bees. Plants pollinated by insects often have blue or yellow petals. The insect eye does not perceive color in the same way that the human eye does. Most insects see well in the violet, blue, and yellow range of visible light but do not perceive red as a distinct color. Consequently, flowers pollinated by insects are not usually red. Insects can also see in the ultraviolet range, wavelengths that are invisible to the human eye. Insects see ultraviolet radiation as a color

called *bee's purple.* Many flowers have dramatic markings that may or may not be visible to humans but that direct insects to the center of the flower where the pollen grains and nectar are located (Fig. 35–4).

Insects have a well developed sense of smell, and many insect-pollinated flowers have a strong scent that may be pleasant or foul to humans. The carrion plant, for example, is pollinated by flies and smells like the rotting flesh in which flies lay their eggs. As flies move from one reeking flower to another looking for a place to lay their eggs, they transfer pollen grains.

Birds such as hummingbirds are important pollinators (Fig. 35–5a). Flowers pollinated by birds are usually red, orange, or yellow because birds see well in this range of visible light. Because birds do not have a strong sense of smell, bird-pollinated flowers usually lack a scent.

Bats, which feed at night and do not see well, are important pollinators, particularly in the tropics where they are most abundant (Fig. 35–5b). Bat-pollinated flowers are night-

Figure 35–3 **Visualizing floral scent.** Devil's tongue produces a floral odor that is disagreeable to humans. The odor-causing chemicals react with HCl released from the filter paper, producing a visible precipitate. Devil's tongue is pollinated by beetles and flies. *(Courtesy of J.M. Patt and B.J.D. Meeuse)*

(a)

(b)

Figure 35–4 **Ultraviolet markings on insect-pollinated flowers.** *(a)* A flower as seen by the human eye is solid yellow. *(b)* The same flower viewed under ultraviolet radiation provides clues about how the insect eye perceives it. The light blue portions of the petals appear purple to a bee, whereas the dark blue inner parts appear yellow. These differences in coloration draw attention to the center of the flower, where the pollen grains and nectar are located. *(a, b, Thomas Eisner)*

blooming and have dull white petals and a strong scent, usually of fermented fruit. Nectar-feeding bats are attracted to the flowers by the scent; they lap up the nectar with their long, extendible tongues. As they move from flower to flower, they transfer pollen grains.

Animal pollinators and the plants they pollinate have had such close, interdependent relationships over time that they have affected the evolution of certain physical and behavioral features in each other. The term coevolution describes such mutual adaptation, in which two different species interact so closely that they become increasingly adapted to one another.

During the time that plants were coevolving specialized features such as petals, scent, and nectar to attract pollinators, animal pollinators coevolved specialized body parts and behaviors that adapted them to aid pollination as they obtain nectar and pollen grains as a reward. For example, coevolution has selected for the bumblebees' hairy bodies, which catch and hold the sticky pollen grains for transport from one flower to another. Coevolution may also have led to the long, curved beaks of certain honeycreepers, Hawaiian birds that insert their beaks into tubular flowers to obtain nectar (Fig. 35–6). The long, tubular corolla of the flowers that honeycreepers visit also came about through coevolution.

Animal behavior has also coevolved, sometimes in bizarre ways. The flowers of certain orchids (*Ophrys* sp.), for example, resemble female wasps in coloring and shape. These flowers also secrete a scent similar to that produced by female wasps, and the males are irresistibly attracted to it. The resemblance between *Ophrys* flowers and female wasps is so strong that male wasps mount the flowers and attempt to copulate with them. During this misdirected activity, known as *pseudocopulation,* a pollen sac usually attaches to the back of the wasp. When the frustrated wasp departs and attempts to copulate with another orchid flower, pollen grains are transferred to that flower.

(a)

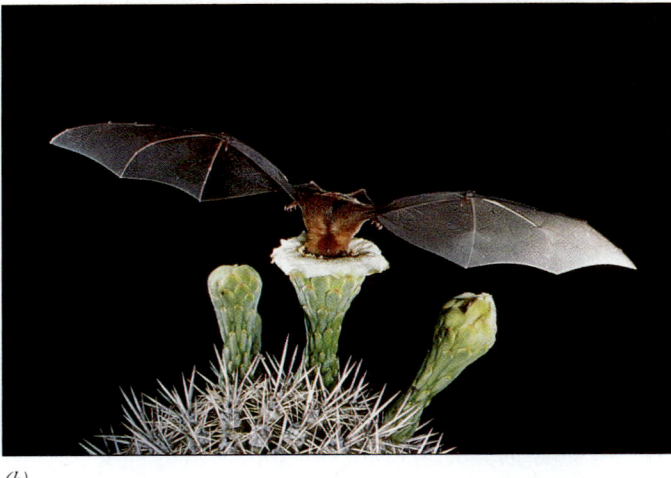

(b)

Figure 35–5 **Animal pollinators.** *(a)* A ruby-throated hummingbird obtains nectar from a trumpet vine flower. The pollen grains on the bird's feathers will be carried to the next plant. *(b)* A lesser long-nosed bat obtains nectar from a cardon cactus flower. *(a, Dan Dempster/Dembinsky Photo Associates; b, Merlin Tuttle/Bat Conservation International/Photo Researchers, Inc.)*

Figure 35–6 **Coevolution between flowering plants and animal pollinators.** The 'i'iwi, one of the Hawaiian honeycreepers, is thought to possess its gracefully curved bill in order to sip nectar from flowers of the lobelia. The 'i'iwi bill fits perfectly into the long, tubular lobelia flowers. Both bird and flower coevolved during the thousands of years of this relationship. Interestingly, lobelias have become rare during the 20th century, largely as a result of grazing by cows and feral goats. About 25% of lobelioid species have become extinct. The 'i'iwi now feeds largely on the petal-less flowers of the ohia tree, and the 'i'iwi bill appears to be slowly adapting to this change in feeding preference. A comparison of the bills of 'i'iwi museum specimens collected in 1902 with the bills of live birds captured in the 1990s shows that 'i'iwi bills are now about 3% shorter than they were in 1902.

Some flowering plants depend on wind to disperse pollen

Some flowering plants, such as grasses, ragweed, maples, and oaks, are pollinated by wind and produce many small, inconspicuous flowers (Fig. 35–7). Wind-pollinated plants do not produce large, colorful petals, scent, or nectar. Some have large, feathery stigmas, presumably to trap wind-borne pollen grains. Because wind pollination is a hit-or-miss affair, the likelihood of a particular pollen grain landing on a stigma of the same species of flower is slim. Wind-pollinated plants produce large quantities of pollen grains, which increases the likelihood that some grains will land on the appropriate stigma.

FERTILIZATION IS FOLLOWED BY SEED AND FRUIT DEVELOPMENT

Once pollen grains have been transferred from anther to stigma, the tube cell, one of the two cells in the pollen grain, grows a thin pollen tube down through the style and into an ovule in the ovary. The second cell within the pollen grain divides to form two male gametes (the sperm cells), which move down the pollen tube and enter the ovule (Fig. 35–8).

A unique double fertilization process occurs in flowering plants

The egg within the ovule unites with one of the sperm cells, forming a zygote (fertilized egg) that will develop into an embryonic plant contained in a seed. The two polar nuclei in the central cell of the ovule fuse with the second sperm cell to form the first cell of the triploid ($3n$) **endosperm,** the tissue with

Figure 35-7 Wind pollination. Each cluster of male oak flowers dangles from a tree branch and sheds a shower of pollen when the wind blows. These flowers lack petals. *(Dr. Jeremy Burgess/Science Photo Library/Photo Researchers, Inc.)*

Anther

Pollen grains

Generative cell

Tube cell

Flower

Style

Pollen grain

Ovary

Ovule (containing embryo sac)

Figure 35-8 Pollination and pollen tube growth. After pollination, a pollen tube grows through the style to the ovule in the ovary. Two sperm cells move down the pollen tube and enter the ovule, where they participate in double fertilization.

Stigma

Pollen tube

Pollen tube grows and enters ovule

2 polar nuclei

Embryo sac

Egg

Sperm cells

Endosperm

Zygote

Double fertilization occurs

SEED BANKS

In order to preserve older, more diverse varieties of plants, many countries are collecting plant **germplasm,** which is any plant material that may be used in breeding. Germplasm includes seeds, plants, and plant tissues of traditional crop varieties. The International Plant Genetics Resource Institute in Rome, Italy, is the scientific organization that oversees plant germplasm collections worldwide. More than 700 seed collections, called *seed banks,* exist around the world and collectively hold about 2.5 million samples of thousands of different kinds of plants *(see figure).* The U. S. National Plant Germplasm System in Fort Collins, Colorado, stores seeds of about 250,000 different species and varieties.

Most seed banks help to preserve the genetic variation within different varieties of crops. Farmers typically discontinue planting local varieties when newer, improved varieties become available. The newer varieties have desirable genetic characteristics, such as a greater yield, for example, but the local, discarded varieties also contain valuable genes. Each local variety's characteristic combination of genes gives it distinctive nutritional value, size, color, flavor, resistance to disease, and adaptability to different climates and soil types. Maintaining the genetic diversity present in local crop varieties will help preserve genes that we may need in the future. The gene combinations of local varieties are potentially valuable to agricultural breeders because they can be transferred to other varieties, either by traditional breeding methods or by genetic engineering.

Seed banks offer the advantage of storing a large amount of live plant genetic material in a very small space. Seeds stored in seed banks are safe from habitat destruction, climate changes, and general neglect. There have even been some instances of using seeds from seed banks to reintroduce a plant species that has become extinct in the wild.

There are some disadvantages to seed banks. Seeds do not remain alive indefinitely and must be germinated periodically so that new ones can be collected. In addition, accidents such as fires or power failures can result in the permanent loss of the genetic diversity represented in a seed bank. Also, many types of plants, such as avocados and coconuts, for example, cannot be stored as seeds. The seeds of these plants do not tolerate being dried out, which is a necessary step before they can be frozen. (If 20% or more moisture remains in a frozen seed, it will probably die.) Some seeds cannot be stored successfully because they only remain viable for a short time—a few months or even just a few days.

Perhaps the most important disadvantage of seed banks is that plants stored in this manner remain stagnant in an evolutionary sense. Removed from their natural habitats, they are no longer subject to the forces of natural selection. As a result, they may be less fit for survival when they are reintroduced into the wild.

Despite their shortcomings, seed banks are increasingly viewed as an important way to safeguard seed for future generations. Other international efforts to preserve plant

Seed storage. These small vials and packets of seeds are stored in the seed bank in Svalbard, Norway. *(Courtesy of Nordiska Genbanken, Alnarp, Svenge)*

genetic diversity are also being planned and implemented. For example, some farmers may soon be paid to set aside some of their land for cultivating local varieties of crops, thereby preserving the genetic diversity in agriculturally important plants.

---

nutritive and hormonal functions that surrounds the developing embryonic plant in a seed. This process, in which two separate cell fusions occur, is called **double fertilization.** It is, with two exceptions, unique to flowering plants. (As mentioned in Chapter 27, a type of double fertilization has been reported in two gymnosperm species, *Ephedra nevadensis* and *Gnetum gnemon.*)

After double fertilization has occurred, the ovule develops into a seed, and the ovary surrounding it develops into a fruit. (Double fertilization and the flowering plant life cycle are discussed in greater detail in Chapter 27.)

Embryonic development in seeds is orderly and predictable

Flowering plants produce a young plant embryo complete with stored nutrients in a compact package, the **seed,** that develops from the ovule after fertilization. The nutrients in seeds are not only used by germinating plant embryos but also consumed by animals, including humans (see *Focus On: Seed Banks*). Development of the embryo and endosperm following fertilization is possible because of the constant flow of nutrients into the developing seed from the parent plant.

Cell divisions of the fertilized egg to form a multicellular embryo proceed in a variety of ways in flowering plants. The following description is of dicot embryonic development; monocot embryonic development is identical in the early stages.

The two cells formed as a result of the first division of the fertilized egg establish polarity, or direction, in the embryo. The bottom cell (toward the outside of the ovule) typically develops into a **suspensor,** which is a multicellular structure that anchors the embryo and aids in nutrient uptake from the endosperm. The top cell (toward the inside of the ovule) develops into the actual embryo.

Initially, the top cell divides to form a short chain of cells, called a *proembryo* (Fig. 35–9). As cell division continues, a small ball of cells, often called a *globular embryo,* develops. Cells begin to develop into specialized tissues during this stage. When the dicot embryo starts to develop its two cotyledons

(seed leaves), it has two lobes and resembles a heart; this is often called the *heart stage.* During the *torpedo stage,* the embryo continues to grow as the cotyledons elongate. As the embryo enlarges, it often curves back on itself and crushes the suspensor beyond recognition.

The mature seed contains an embryonic plant and storage materials

A mature seed contains an embryonic plant and food (stored in either the cotyledons or endosperm), surrounded by a tough, protective **seed coat.** The seeds in turn are enclosed within a fruit.

The mature embryo within the seed consists of a short embryonic root, or **radicle;** an embryonic shoot; and one or two seed leaves, or **cotyledons.** Monocots have a single cotyle-

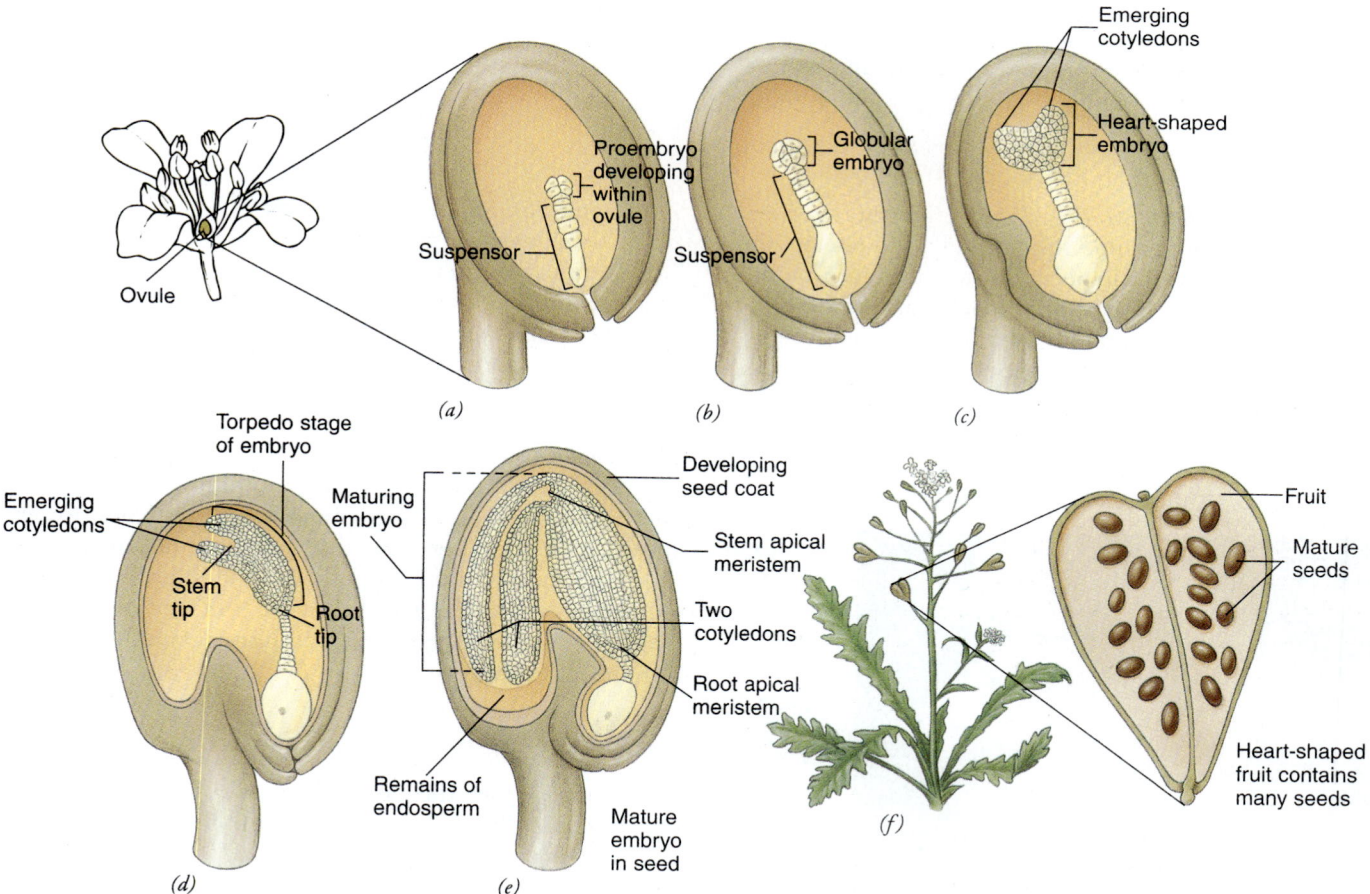

Figure 35–9 **Embryonic development in shepherd's purse (Capsella bursa-pastoris). (a)** The proembryo is the earliest multicellular stage of the embryo (shown outside the ovary). **(b)** As cell division continues, the embryo becomes a ball of cells, called the globular stage. **(c)** As the two cotyledons begin to emerge, the embryo is shaped like a heart. **(d)** The cotyledons continue to elongate, forming the torpedo stage. **(e)** A mature embryo within the seed. The food originally stored in the endosperm has been almost completely depleted during embryonic growth and development. Most of the food for the embryonic plant is in its cotyledons. **(f)** A longitudinal section through a heart-shaped fruit of shepherd's purse reveals numerous tiny seeds, each containing a mature embryo.

SEED SIZE AND SURVIVAL STRATEGIES

How large should seeds be? In flowering plants seed size varies considerably, from the microscopic, dustlike seeds of orchids to the giant seeds of the double coconut *(Lodoicea seychellarum),* which weigh as much as 27 kg (almost 60 lb). Despite this variation among different species, seed size is a remarkably constant trait within a species.

Assuming that a given plant species invests a fixed amount of its energy in reproduction, is it more advantageous to produce a large number of small seeds, or a few big ones? By observing the seed sizes that predominate in different environments, biologists have concluded that in some environments, a smaller seed appears to be advantageous, whereas in others, a larger seed may be better.

For example, plants that grow in widely scattered open sites (such as old fields) usually produce smaller seeds, perhaps because they can be more easily dispersed over large areas than can larger seeds. On the other hand, wide dispersal is probably less important for plants adapted to densely vegetated areas such as forests. These plants generally produce bigger seeds with an ample food reserve that may confer a greater likelihood of becoming successfully estab-

lished in a shaded environment. The stored energy may allow the young seedling to grow tall enough to reach adequate sunlight for photosynthesis.

Other ecological factors are associated with seed size. Larger seeds are typical of many plants that live in arid habitats, possibly because the food stored in a large seed allows a young seedling to establish an extensive root system quickly, thereby enabling it to survive the dry climate. Island plants also produce larger seeds than similar species on the nearby mainland. In this case, biologists hypothesize that large seeds are less likely to be widely dispersed and therefore less likely to fall into the ocean, where chances of survival are slim. On the other hand, seeds that remain dormant in the soil for a long period of time tend to be smaller; the exact reason for this is not known.

Despite many observations that associate seed size with specific environments, the general principles concerning the adaptive advantages of large seeds versus small seeds remain to be determined.

---

don, whereas dicots have two (Fig. 35–10). The short portion of the embryonic shoot connecting the radicle to one or two cotyledons is known as the **hypocotyl.** The shoot apex, or terminal bud, located above the point of attachment of the cotyledon(s), is the **plumule.** After the radicle, hypocotyl, cotyledon(s), and plumule have formed, the young plant's development is arrested (by dessication or dormancy, discussed

in Chapter 36). When conditions are right for continuation of the developmental program, the seed germinates.

Because the embryonic plant is nonphotosynthetic, it must have nutrients during germination until it becomes photosynthetic and, therefore, self-sufficient. The cotyledons of many plants function as storage organs and become large, thick, and fleshy as they absorb the food reserves (starches, oils, and proteins) initially produced as endosperm. Seeds that store nutrients in cotyledons have little or no endosperm at maturity. Examples of such seeds are peas, beans, squashes, sunflowers, and peanuts. Other plants, wheat and corn, for example, have thin cotyledons that function primarily to help the young plant digest and absorb food stored in the endosperm. (See *Making the Connection: Seed Size and Survival Strategies.* Also see Chapter 36 for a discussion of seed germination and early growth.)

Fruits are mature, ripened ovaries

After double fertilization takes place within the ovule, the ovule develops into a seed (just described), and the ovary surrounding it develops into a **fruit.** For example, a pea pod is a fruit, and the peas within it are seeds. A fruit may contain one or more seeds; some orchid fruits contain several thousand to a few million seeds! Fruits provide protection for the enclosed seeds and sometimes aid in their dispersal.

There are several types of fruits; their differences result from variations in the structure or arrangement of the flowers from which they were formed. The four basic types of fruits are simple fruits, aggregate fruits, multiple fruits, and accessory fruits (Fig. 35–11).

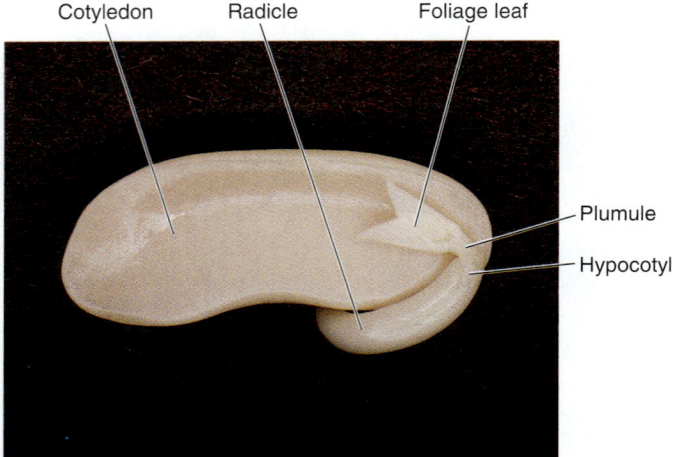

Cotyledon   Radicle   Foliage leaf

Plumule

Hypocotyl

Figure 35–10 **Seed structure.** A bean seed has been dissected—its seed coat and one of its two cotyledons were removed—to show the radicle, hypocotyl, plumule, cotyledon, and foliage leaf produced at the shoot apex. This particular seed had begun germinating, so its radicle is larger than in an ungerminated seed. *(James Mauseth, University of Texas)*

Berry-
A simple, fleshy fruit in which the fruit wall is soft throughout.

Seed

Tomato (*Lycopersicon lycopersicum*)

Drupe-
A simple, fleshy fruit in which the inner wall of the fruit is hard and stony (the pit).

Single seed inside pit

Peach (*Prunus persica*)

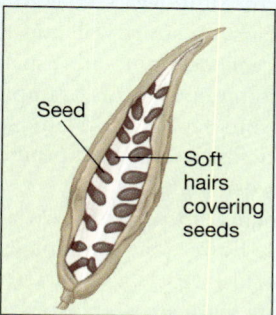

Follicle-
A simple, dry fruit that splits open along one suture to release its seeds.

Seed

Soft hairs covering seeds

Milkweed (*Asclepias syriaca*)

Legume-
A simple, dry fruit that splits open along two sutures to release its seeds.

Seed

Seed

Green bean (*Phaseolus vulgaris*)

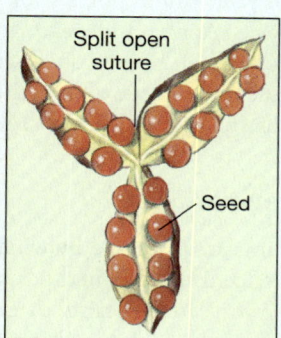

Capsule-
A simple, dry fruit that splits open along three or more sutures or pores to release its seeds.

Split open suture

Seed

Iris (*Iris* sp.)

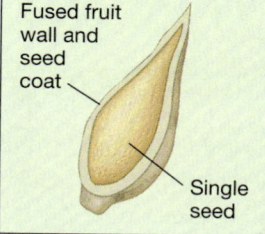

Grain-
A simple, dry fruit in which the fruit wall is fused to the seed coat.

Fused fruit wall and seed coat

Single seed

Wheat (*Triticum* sp.)

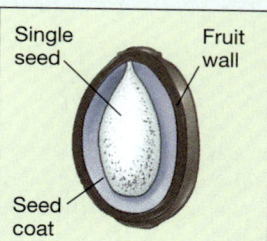

Achene-
A simple, dry fruit in which the fruit wall is separate from the seed coat.

Single seed

Fruit wall

Seed coat

Sunflower (*Helianthus annuus*)

Nut-
A simple, dry fruit that has a stony wall, is usually large, and does not split open at maturity.

Woody fruit wall

Cup of fused bracts

Single seed

Oak (*Quercus* sp.)

Aggregate fruit-
A fruit that develops from a single flower with several to many pistils (i.e., carpels are not fused into a single pistil).

Seed

Blackberry (*Rubens* sp.)

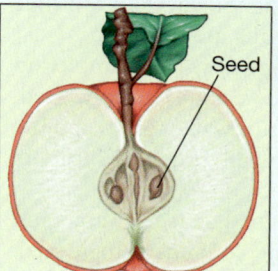

Multiple fruit-
A fruit that develops from the ovaries of a group of flowers.

Seed

Mulberry (*Morus* sp.)

Accessory fruit-
A fruit composed primarily of tissue (such as the receptacle) other than ovary tissue.

Seed

Apple (*Malus sylvestris*)

Figure 35–11 **Fruit types.** Fruits are botanically classified into four groups—simple, aggregate, multiple, and accessory fruits—based on structure and mechanism of seed dispersal.

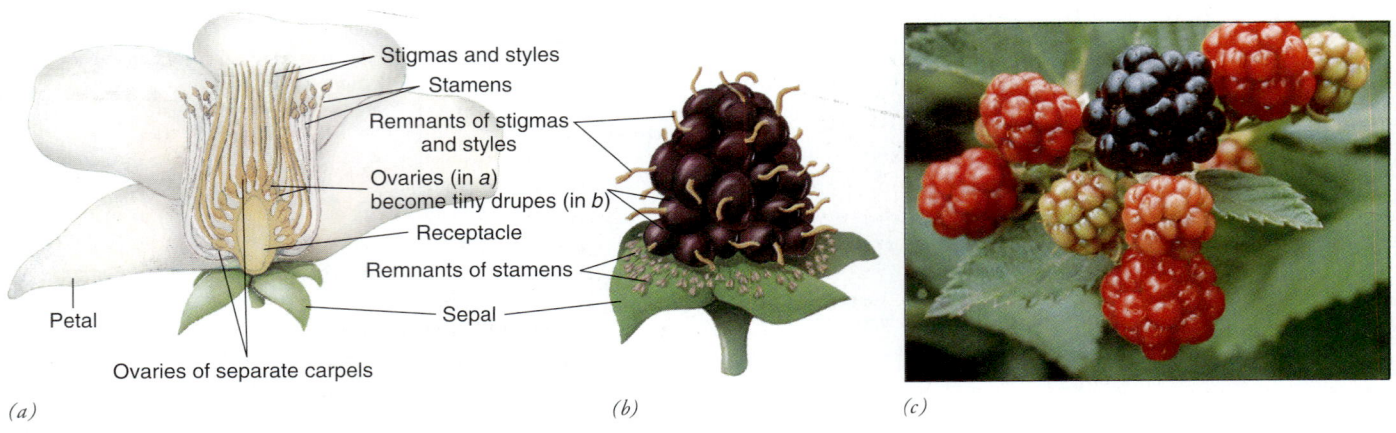

Figure 35–12 **Aggregate fruit.** *(a)* Cutaway view of a blackberry flower, showing the many separate carpels in the center of the flower. *(b)* A developing blackberry fruit is an aggregate of tiny drupes. The little "hairs" on the blackberry are remnants of stigmas and styles. *(c)* Developing fruits at various stages of maturity. *(c, Dennis Drenner)*

Labels in figure (a):
- Stigmas and styles
- Stamens
- Petal
- Ovaries of separate carpels

Labels in figure (b):
- Remnants of stigmas and styles
- Ovaries (in *a*) become tiny drupes (in *b*)
- Receptacle
- Remnants of stamens
- Sepal

(a)　　　　　*(b)*　　　　　*(c)*

Most fruits are simple fruits. A **simple fruit** develops from a single pistil (which may consist of a single carpel or several fused carpels). At maturity, simple fruits may be fleshy or dry. Two examples of simple, fleshy fruits are berries and drupes. A **berry** is a fleshy fruit that has soft tissues throughout and contains few to many seeds; a blueberry is a berry, as are grapes, cranberries, bananas, and tomatoes. Many so-called berries do not fit the botanical definition. Strawberries, raspberries, and mulberries, for example, are not berries; these three non-berries will be discussed shortly.

A **drupe** is a simple, fleshy (or fibrous) fruit that contains a hard, stony pit surrounding a single seed. Examples of drupes include peaches, plums, olives, avocados, and almonds. The almond shell is actually the stony pit, which remains after the rest of the fruit has been removed.

Many simple fruits are dry at maturity; some of these split open, usually along seams, called *sutures,* to release their seeds. A milkweed pod is an example of a **follicle,** a simple, dry fruit that splits open along one suture to release its seeds. A **legume** is a simple, dry fruit that splits open along two sutures. Pea pods are legumes, as are green beans, although both are generally harvested before the fruit has dried out and split open. Pea seeds are usually removed from the fruit and consumed, whereas in green beans the entire fruit and seeds are eaten. A **capsule** is a simple, dry fruit that splits open along multiple sutures or pores. Poppy and cotton fruits are capsules.

Other simple, dry fruits, such as **grains,** for example, do not split open at maturity. Each grain contains a single seed. Because the seed coat is fused to the fruit wall, a grain appears to be a seed rather than a fruit. Kernels of corn and wheat are fruits of this type.

An **achene** is similar to a grain in that it is simple and dry, does not split open at maturity, and contains a single seed. However, the seed coat of an achene is not fused to the fruit wall. Instead, the single seed is attached to the fruit wall at one point only, permitting an achene to be separated from its seed.

The sunflower fruit is an example of an achene. One can peel off the fruit wall (the shell) to reveal the sunflower seed within.

Nuts are simple, dry fruits that have a stony wall and do not split open at maturity. Unlike achenes, nuts are usually large and are often derived from a compound pistil. Examples of nuts include chestnuts, acorns, and hazelnuts. Many so-called nuts do not fit the botanical definition. Shelled peanuts and Brazil nuts, for example, are seeds, not nuts.

Aggregate fruits are a second main type of fruit. An **aggregate fruit** is formed from a single flower that contains several to many separate (free) carpels (Fig. 35–12). After fertilization, each ovary from each individual carpel enlarges. As they enlarge, the ovaries may fuse together to form a single fruit. Raspberries, blackberries, and magnolia fruits are examples of aggregate fruits.

A third type is the **multiple fruit,** formed from the ovaries of many flowers that grow in close proximity on a common floral stalk. The ovary from each flower fuses with nearby ovaries as it develops and enlarges after fertilization. Pineapples, figs, and mulberries are multiple fruits (Fig. 35–13).

Accessory fruits are the fourth type. They differ from other fruits in that other plant tissues in addition to ovary tissue make up the fruit. For example, the edible portion of a strawberry is the red, fleshy receptacle. Apples and pears are also accessory fruits; the outer part of each of these fruits is an enlarged *floral tube,* consisting of receptacle tissue along with portions of the calyx, that surrounds the ovary (Fig. 35–14).

Seed dispersal is highly varied

The seeds and fruits of flowering plants are dispersed by wind, animals, water, and explosive dehiscence. Effective methods of seed dispersal have made it possible for certain plants to expand their geographical range. In some cases, the seed is the actual agent of dispersal, whereas in others, the fruit performs this role. In tumbleweeds, such as Russian thistle, the entire

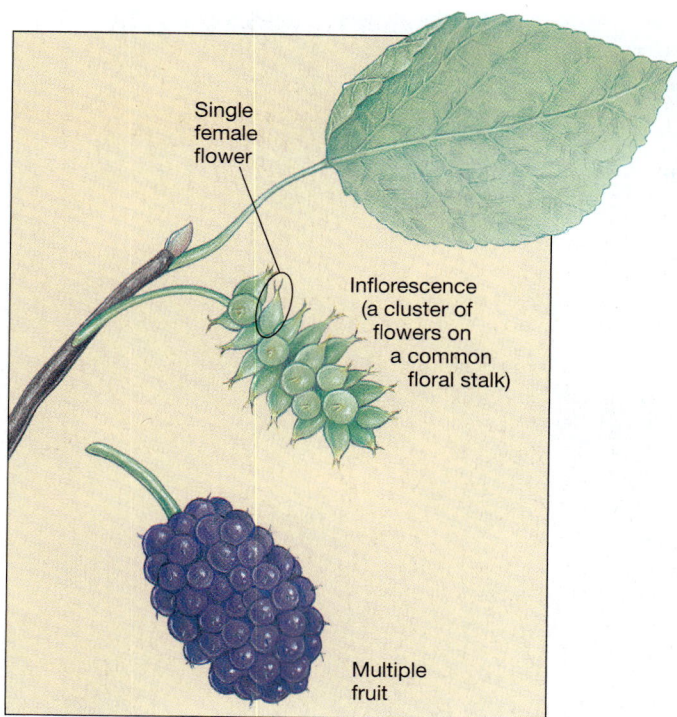

Figure 35–13 Multiple fruit. Mulberry is formed from the ovaries of many flowers that fused to become a multiple fruit. Mulberry flowers are imperfect and contain either stamens or pistils. The inflorescence of female flowers from which the mulberry fruit develops is also shown.

Labels in figure: Single female flower; Inflorescence (a cluster of flowers on a common floral stalk); Multiple fruit

plant is the agent of dispersal because it detaches and blows across the ground, scattering seeds as it bumps along. Tumbleweeds are lightweight but have a great deal of wind resistance and are sometimes blown many kilometers by the wind.

Wind is responsible for seed dispersal in many plants. Plants such as maple have winged fruits adapted for wind dispersal. Light, feathery plumes are other structures that allow seeds or fruits to be transported by wind, often for considerable distances. Both dandelion fruits and milkweed seeds (Fig. 35–15a) have this type of adaptation.

Some plants have special structures that aid in dispersal of their seeds and fruits by animals. The spines and barbs of burdock burs and similar fruits catch in animal fur and are dispersed as the animal moves about (Fig. 35–15b). Fleshy, edible fruits are also adapted for animal dispersal. As these fruits are eaten, the seeds are either discarded or swallowed. Many seeds that are swallowed have thick seed coats and are not digested, but instead are passed through the digestive tract and deposited in the animal's feces some distance from the parent plant. In fact, some seeds will not germinate unless they have passed through an animal's digestive tract; it is likely that the animal's digestive juices facilitate germination by helping to break down the seed coat. (Interestingly, a 1994 paper published in the journal *Ecology* reported that some edible fruits contain chemicals that function as laxatives to *speed* the passage of seeds through the animal's digestive tract. The longer these seeds spend in the digestive tract, the less likely they are to germinate.)

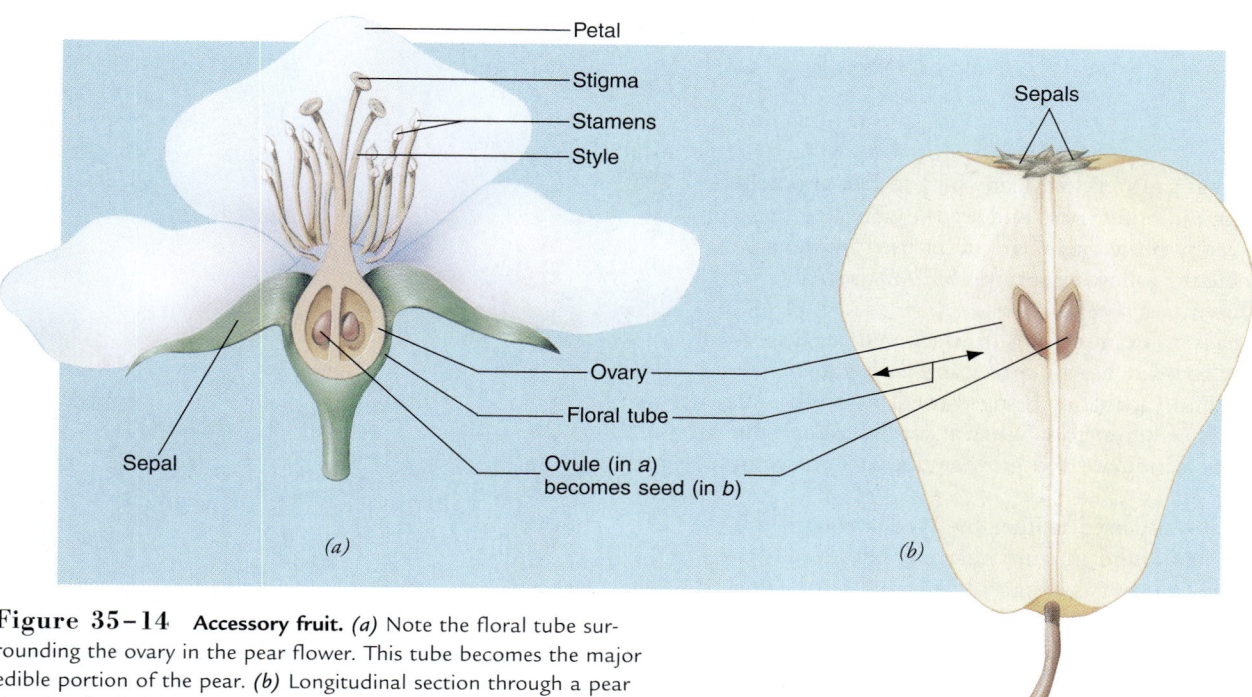

(a)   (b)

Labels in figure: Petal; Stigma; Stamens; Style; Sepals; Sepal; Ovary; Floral tube; Ovule (in a) becomes seed (in b)

Figure 35–14 Accessory fruit. *(a)* Note the floral tube surrounding the ovary in the pear flower. This tube becomes the major edible portion of the pear. *(b)* Longitudinal section through a pear showing the fruit tissue, which is derived from both the floral tube and the ovary.

(a)

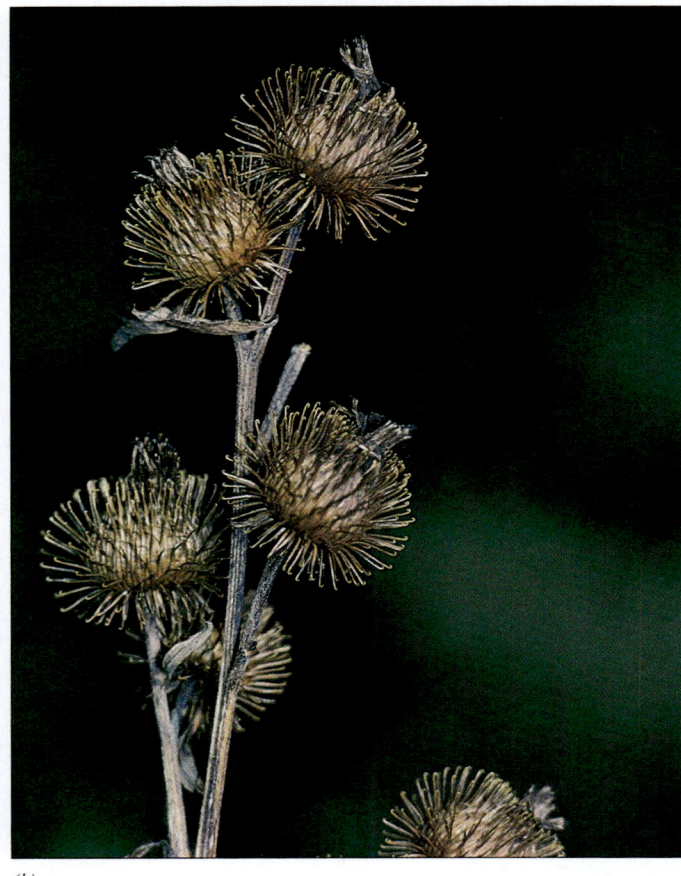

(b)

Figure 35-15 **Seed (and fruit) dispersal.** *(a)* The feathery plumes of milkweed seeds make them buoyant for dispersal by wind. *(b)* Burdock burs (the hooked fruits) are carried from the parent plant after they become matted in bird feathers, mammal fur, or human clothing. *(a, John Serrao/Visuals Unlimited; b, DPA/Dembinsky Photo Associates)*

Animals such as squirrels and many bird species also help to disperse acorns and other fruits and seeds by burying them for winter use. Many buried seeds are never used by the animal and germinate the following spring. (See *Making the Connection: Seed Dispersal and Ants.*)

The coconut is an example of a fruit adapted for dispersal by water. The coconut has air spaces and corky floats that make it buoyant and capable of being carried by ocean currents for thousands of kilometers. When it washes ashore, the seed within it may germinate and grow into a coconut palm tree.

Some seeds are dispersed neither by wind, animals, nor water. Such seeds are found in fruits that use *explosive dehiscence,* in which the fruit bursts open suddenly and quite often violently, forcibly discharging its seeds (Fig. 35–16). Pressures due to differences in turgor or to drying out cause these fruits to burst open. The fruits of plants such as touch-me-not and bitter cress split open so explosively that seeds are scattered a meter or more.

(a) (b)

Figure 35-16 **Explosive dehiscence in bitter cress (*Cardamine pratensis*).** *(a)* An intact fruit before it has opened. *(b)* The fruit splits open with explosive force, propelling the seeds some distance from the plant.

MAKING THE CONNECTION

SEED DISPERSAL AND ANTS

How does a plant species disperse its seeds to places where they can successfully germinate and grow? As discussed in this chapter, plants possess a variety of dispersal methods that increase the chances of seeds landing in suitable locations. Regardless of how they are dispersed, however, most seeds land in places that are unsuitable for growth, or are eaten and destroyed by animals such as mice and squirrels shortly after being dispersed.

Some plants have an increased seed survival because of a dispersal method that insures that seeds are buried underground. Such seeds are less likely to be eaten by animals. The role of burying seeds is performed for many plants by ants, which collect the seeds and take them underground to their nests. Ants disperse and bury seeds for hundreds of different plant species in almost every terrestrial environment, from northern coniferous forests to tropical rain forests to deserts.

Both ants and flowering plants benefit from their association. The ants ensure the reproductive success of the plants whose seeds they bury, and the plants supply food to the ants. A seed that is collected and taken underground by ants often contains a special structure called an *elaiosome*, or *oil body*, that protrudes from the seed *(see figure)*. Elaiosomes are a nutritious food for ants, which carry seeds underground before removing the elaiosome. Once an elaiosome is removed from a seed, the ants discard the undamaged seed in an underground refuse pile, which happens to be rich in organic material (such as ant droppings and dead ants) and contains the minerals required by young seedlings. Thus, ants not only bury the seeds away from animals that might eat them, but also place the seeds in rich soil that is ideal for germination and seedling growth.

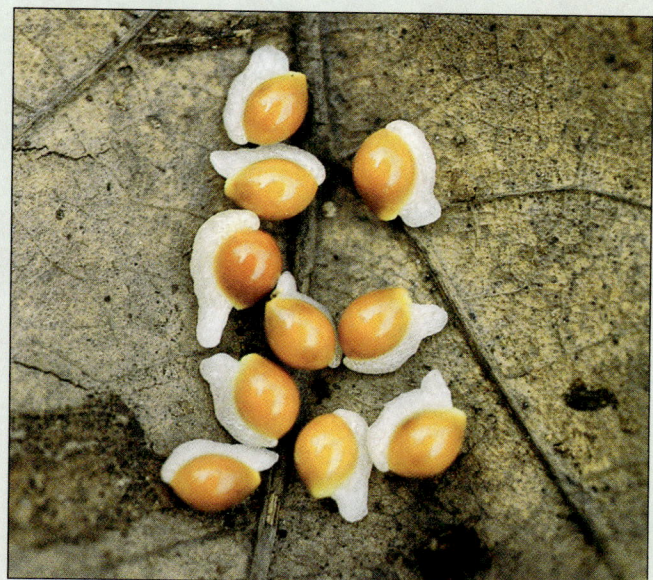

Bloodroot (*Sanguinaria canadensis*) seeds. The brown part is the seed proper, and the white part is the elaiosome, or oil body. The seeds have been placed on an oak leaf to indicate scale. *(Marion Lobstein)*

ASEXUAL REPRODUCTION IN FLOWERING PLANTS MAY INVOLVE MODIFIED STEMS, LEAVES, OR ROOTS

Flowering plants have many kinds of asexual reproduction, a number of which involve modified stems (rhizomes, tubers, bulbs, corms, and stolons). A **rhizome** is a horizontal underground stem that may or may not be fleshy. Fleshiness indicates that the rhizome is used for storing food materials such as starch (Fig. 35–17*a*). Although rhizomes resemble roots, they are really stems, as indicated by the presence of scalelike leaves, buds, nodes, and internodes. Rhizomes frequently branch in different directions. Over time, the old portion of the rhizome dies, and two branches eventually separate to become two distinct plants. Irises, bamboos, ginger, and many grasses are examples of plants that reproduce asexually by forming rhizomes.

Some rhizomes produce greatly thickened ends called **tubers,** which are fleshy underground stems enlarged for food storage. When the attachment between a tuber and its parent plant breaks, often as a result of the death of the parent plant, the tuber grows into a separate plant. White potatoes and elephant's ear (*Caladium* sp.) are examples of plants that produce tubers (Fig. 35–17*b*). The "eyes" of a potato are actually axillary buds, evidence that the tuber is an underground stem rather than a storage root like sweet potatoes or carrots.

A **bulb** is a modified underground bud in which fleshy storage leaves are attached to a short stem (Fig. 35–18*a*). A bulb is globose (round) and covered by paper-like bulb scales, which are modified leaves. It frequently forms axillary buds that develop into small daughter bulbs (bulblets). These new bulbs are initially attached to the parent bulb, but when the parent bulb dies and rots away, each daughter bulb can become established as a separate plant. Lilies, tulips, onions, and daffodils are some of the plants that form bulbs.

A **corm** is a very short, erect underground stem that superficially resembles a bulb (Fig. 35–18*b*). Unlike the bulb, whose food is stored in underground leaves, the corm's storage organ is a thickened underground stem covered by papery

(a)

(b)

Rhizome
Tuber
Roots

Rhizome
Root

Figure 35–17 **Rhizomes and tubers.** *(a)* Irises have horizontal underground stems called rhizomes. New aerial shoots arise from buds that develop on the rhizome. *(b)* Potato plants form rhizomes, which enlarge into tubers (the potatoes) at the ends.   *(Carlyn Iverson)*

scales (modified leaves). Axillary buds that give rise to new corms frequently arise; the death of the parent corm separates these daughter corms, which then become established as separate plants. Familiar garden plants that produce corms include crocus, gladiolus, and cyclamen.

Stolons, or **runners,** are horizontal above-ground stems that grow along the surface and are characterized by long internodes (Fig. 35–19). Buds develop along the stolon, and each bud gives rise to a new shoot that roots in the ground. When the stolon dies, the daughter plants live separately. The strawberry plant produces stolons.

Some plants are capable of forming **plantlets** (small plants) along their leaf margins. *Kalanchoe* sp., commonly called "mother of thousands," has meristematic tissue that gives rise to an individual plantlet at each notch in the leaf (Fig. 35–20). When these plantlets attain a certain size, they drop to the ground, root, and grow.

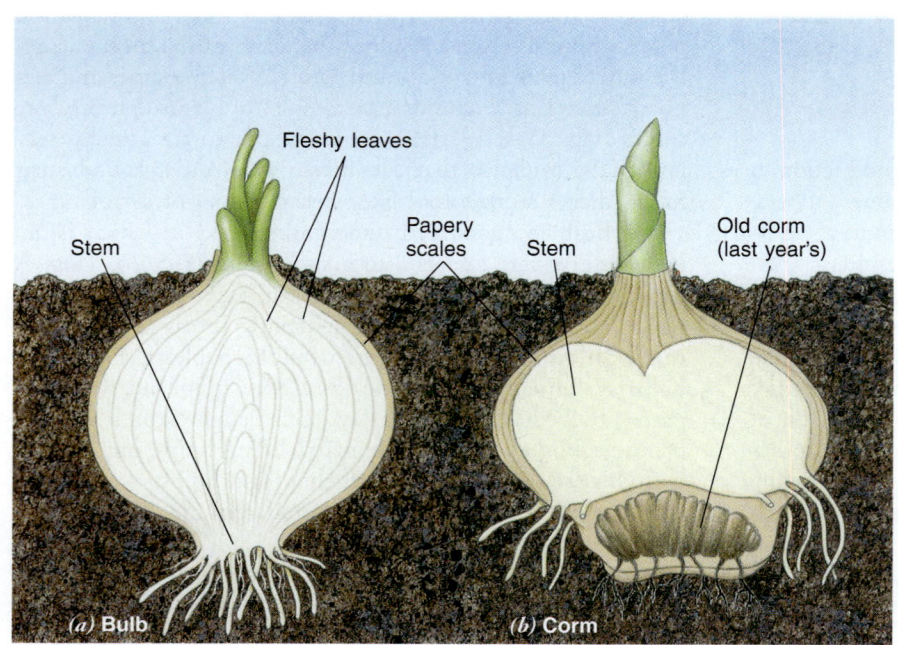

Fleshy leaves

Stem

Papery scales

Stem

Old corm (last year's)

(a) Bulb

(b) Corm

Figure 35–18 **Bulbs and corms.** *(a)* A bulb is a short underground stem to which overlapping, fleshy leaves are attached; most of the bulb consists of leaves. *(b)* A corm is an underground stem that is almost entirely stem tissue surrounded by a few papery scales.

Figure 35–19 **Stolons.** Strawberries reproduce asexually by forming stolons, or runners.

Figure 35–21 **Suckers.** A grove of aspen trees is often descended from a single tree that reproduced asexually by forming suckers. Because all the trees in the grove are genetically identical, their responses to the environment are uniform. In spring they break dormancy simultaneously, and in the fall their leaves turn color at the same time. *(Sharon Cummings/Dembinsky Photo Associates)*

Some plants reproduce asexually by producing **suckers,** above-ground stems that develop from adventitious buds on the roots. Each sucker grows additional roots and becomes an independent plant when the parent plant dies. Examples of plants that form suckers include black locust, pear, apple, cherry, blackberry, and aspen (Fig. 35–21). A quaking aspen *(Populus tremuloides)* colony in the Wasatch Mountains of Utah contains at least 47,000 tree trunks formed from suckers that can be traced back to a single individual; this massive "organism" occupies almost 43 hectares (106 acres). Some weeds, such as field bindweed, for example, can produce many suckers. These plants are difficult to control, because pulling the plant

out of the soil seldom removes all of the roots, which can grow as deep as 3 m (10 ft). In fact, in response to wounding, the roots produce additional suckers, which can be a considerable nuisance.

Apomixis is the production of seeds without the sexual process

Sometimes flowering plants produce embryos in seeds without meiosis and the fusion of gametes. This asexual process is known as **apomixis.** For example, an embryo may develop from a diploid cell in the ovule rather than from a diploid zygote that forms from the union of two haploid gametes. Seed production by apomixis is a form of asexual reproduction; because there is no fusion of gametes, the embryo is virtually genetically identical to the maternal genotype. However, the advantage of apomixis over other methods of asexual reproduction is that the seeds and fruits produced by apomixis can be dispersed by methods associated with sexual reproduction. Apomixis occurs in various species of more than 40 angiosperm families. Examples of plants that reproduce by apomixis include dandelions, citrus trees, blackberries, garlic, and certain grasses.

SEXUAL AND ASEXUAL REPRODUCTION HAVE DIFFERENT FUNCTIONS

Sexual and asexual reproduction are suited for different environmental circumstances. As you know, sexual reproduction results in offspring that are genetically different from the

Figure 35–20 **Plantlets.** The "mother of thousands" (*Kalanchoe* sp.) produces plantlets along the margins of its leaves. The young plantlets will drop off and root in the ground. *(Dennis Drenner)*

parents, that is, the parental genotypes are not preserved in the offspring. This genetic diversity of offspring may be selectively advantageous, particularly in an unstable, or changing, environment. If a plant species that reproduces sexually (and is therefore genetically diverse) is exposed to increasing annual temperatures as a result of global warming, for example, some of the individuals may be more fit than either the parents or other offspring. If, on the other hand, all members of a genetically identical species are sensitive to increasing temperatures, they might all die, and the species would become extinct. The genetic diversity that results from sexual reproduction may also permit the individuals of a species to exploit new environments, thereby expanding their range.

You have seen that asexual reproduction results in offspring that are virtually genetically identical to the parent, that is, the parental genotype is preserved. Assuming that the parent is well adapted to its environment (that is, has a favorable combination of alleles), this genetic similarity may be selectively advantageous if the environment remains stable (unchanging) for several generations. None of the offspring of asexual reproduction are more fit than the parent, but neither are any of them less fit.

Despite the apparent advantages of asexual reproduction, most plant species whose reproduction is primarily asexual occasionally reproduce sexually. Even in a stable environment, plants are exposed to changing selective pressures, such as changes in the number and kinds of predators and parasites, the availability of food, competition from other species, and climate. Sexual reproduction permits species whose reproduction is primarily asexual to increase their genetic variability so at least some individuals are adapted to the changing selective pressures in a stable environment.

Sexual reproduction has some disadvantages

Although the genetic diversity produced by sexual reproduction is advantageous to a species' survival, sexual reproduction is a "costly" form of reproduction. In sexual reproduction, both males and females are required, and their gametes have to meet in order for reproduction to occur. The many adaptations of flowers for different modes of pollination represent one of the costs of sexual reproduction.

Sexual reproduction produces some individuals with genotypes that are well adapted to the environment, but it also produces some individuals that are less well adapted to the environment. Therefore, sexual reproduction is usually accompanied by high death rates among offspring, particularly when selective pressures are strong. As discussed in Chapter 17, however, this aspect of sexual reproduction is an important part of evolution by natural selection.

Every biological process involves trade-offs, and sexual reproduction is no exception. Although sexual reproduction has its costs, the adaptive advantages of sexual reproduction clearly outweigh any disadvantages.

SUMMARY WITH KEY TERMS

I. The offspring produced by sexual reproduction are genetically variable.
  A. Sexual reproduction occurs in the flower. A flower may contain sepals, petals, stamens, and carpels (pistils).
    1. **Sepals** cover and protect the flower parts when the flower is a bud.
    2. **Petals** play an important role in attracting animal pollinators to the flower.
    3. **Stamens** produce pollen grains.
    4. The **carpel** is the female reproductive unit. A **pistil** may consist of a single carpel or a group of fused carpels. Each pistil has three sections: a **stigma,** on which the pollen grains land; a **style,** through which the pollen tube grows; and an **ovary** that contains one or more **ovules.**
  B. Each pollen grain contains two cells. One generates two sperm cells, and the other produces a **pollen tube** through which the sperm cells will reach the ovule.
  C. An egg and two **polar nuclei** are formed in the ovule. Both egg and polar nuclei participate directly in fertilization.
II. **Pollination** is the transfer of pollen grains from **anther** to stigma.
  A. Some plants rely on animals to transfer pollen grains.
    1. **Coevolution** occurs when two different organisms (such as flowering plants and their animal pollinators) form an interdependent relationship and affect the course of one another's evolution.
    2. Flowers pollinated by insects are often yellow or blue and possess a scent.

    3. Bird-pollinated flowers are often yellow, orange, or red and do not have a strong scent.
    4. Bat-pollinated flowers often have dusky white petals and possess a scent.
  B. Plants pollinated by wind often have smaller petals or lack petals altogether and do not produce a scent or nectar; wind-pollinated flowers make copious amounts of pollen.
III. After pollination, fertilization (fusion of gametes) occurs.
  A. A pollen tube grows down the style into the ovary, and the two sperm cells travel down the tube into the ovule.
  B. Flowering plants have **double fertilization.**
    1. In the ovule, the egg fuses with one sperm cell, forming a zygote (fertilized egg) that eventually develops into a multicellular embryo in the seed.
    2. The two polar nuclei fuse with the second sperm cell, forming a nutritive tissue called **endosperm.**
IV. The seed and fruit develop as a result of successful fertilization.
  A. A dicot embryo develops in the seed in an orderly fashion, from proembryo to globular embryo to the heart stage to the torpedo stage. The mature flowering plant embryo consists of a **radicle,** a **hypocotyl, cotyledons** (one in monocots or two in dicots), and a **plumule.**
  B. A mature **seed** contains both a young embryo and nutritive tissue (stored in the endosperm or cotyledons) for use during germination.
  C. Seeds are enclosed within **fruits,** which are mature, ripened ovaries.

1. **Simple fruits** develop from a single pistil that consists of one carpel or several fused carpels. Some simple fruits **(berries, drupes)** are fleshy at maturity, whereas others **(follicles, legumes, capsules, grains, achenes, nuts)** are dry.
2. **Aggregate fruits** develop from a single flower with many separate ovaries.
3. **Multiple fruits** develop from the ovaries of many flowers growing in close proximity on a common axis.
4. In **accessory fruits,** the major part of the fruit consists of tissue other than ovary tissue.

D. Seeds and fruits are adapted for various means of dispersal, including animals, wind, water, and explosive dehiscence.

V. Asexual reproduction involves the formation of offspring without the fusion of gametes. The offspring are virtually genetically identical to the single parent plant.
   A. **Rhizomes, tubers, bulbs, corms,** and **stolons** are stems specialized for asexual reproduction.
   B. Some leaves have meristematic tissue along their margins and give rise to **plantlets.**
   C. Roots may develop adventitious buds that develop into **suckers.** Suckers produce additional roots and may give rise to new plants.

D. **Apomixis** is the production of seeds and fruits without sexual reproduction.
VI. Sexual and asexual reproduction have different functions.
   A. The parental genotypes are not preserved in the offspring of sexual reproduction.
      1. Genetic diversity among offspring produced by sexual reproduction may be selectively advantageous, particularly in an unstable, or changing, environment.
      2. Genetic diversity may also permit individuals to exploit new environments.
      3. However, sexual reproduction is costly because both male and female gametes have to meet.
   B. The parental genotype is preserved in asexual reproduction.
      1. Genetic similarity may be selectively advantageous if the environment remains stable (unchanging) for several generations.
      2. In asexual reproduction all individuals have the potential to produce offspring.
      3. Despite the apparent advantages of asexual reproduction, most plant species whose reproduction is primarily asexual occasionally reproduce sexually, thereby increasing their genetic variability.

POST-TEST

1. The pistil is composed of (a) stigma, style, and stamen (b) anther and filament (c) sepal and petal (d) stigma, style, and ovary (e) radicle, hypocotyl, and plumule
2. The petals of a flower are collectively called a(an) (a) calyx (b) capsule (c) carpel (d) cotyledon (e) corolla
3. The transfer of pollen grains from anther to stigma is known as (a) fertilization (b) double fertilization (c) pollination (d) germination (e) apomixis
4. The observation that long tubular flowers are pollinated by insects with long mouthparts, whereas short flowers are pollinated by insects with short mouthparts, is explained by (a) coevolution (b) germination (c) double fertilization (d) apomixis (e) explosive dehiscence
5. The process of _____ in flowering plants involves one sperm cell fusing with an egg cell and one sperm cell fusing with two polar nuclei. (a) coevolution (b) germination (c) double fertilization (d) apomixis (e) pollination
6. The nutritive tissue in the seeds of flowering plants that is formed from the union of a sperm cell and two polar nuclei is called the (a) plumule (b) endosperm (c) cotyledon (d) hypocotyl (e) radicle
7. The _____ is a multicellular structure that anchors the embryo

and aids in nutrient uptake from the endosperm (a) proembryo (b) heart stage (c) suspensor (d) cotyledon (e) torpedo stage
8. After fertilization the ovule develops into a(an) _____, and the ovary into a(an) _____ (a) fruit; seed (b) seed; fruit (c) calyx; corolla (d) corolla; calyx (e) follicle; legume
9. _____ fruits form from many ovaries of a single flower, whereas _____ fruits develop from the ovaries of many separate flowers (a) multiple; accessory (b) simple; accessory (c) aggregate; multiple (d) accessory; aggregate (e) simple; multiple
10. Apples, strawberries, and pears are examples of what kind of fruit? (a) accessory (b) simple (c) multiple (d) aggregate (e) legume
11. A horizontal, underground stem that bears leaves and buds and is often specialized for asexual reproduction is called a (a) stolon (b) bulb (c) corm (d) rhizome (e) tuber
12. Place the following events in the correct order: (1) pollen tube grows into ovule (2) insect lands on flower to drink nectar (3) embryo develops within the seed (4) fertilization occurs (5) pollen carried by insect drops onto stigma (a) 2–5–1–4–3 (b) 1–4–2–5–3 (c) 3–2–5–1–4 (d) 5–1–3–4–2 (e) 2–5–4–3–1

REVIEW QUESTIONS

1. Distinguish between sexual and asexual reproduction, including the advantages and disadvantages of each.
2. What is the difference between pollination and fertilization? Which process occurs first?
3. Describe a "typical" insect-pollinated flower.
4. What is coevolution?
5. Put the following stages of embryonic development in order and briefly describe each: torpedo stage, globular stage, proembryo, and heart stage.

6. Distinguish among simple, aggregate, multiple, and accessory fruits, and give examples of each.
7. Explain some of the features possessed by seeds and fruits that are dispersed by animals.
8. List and describe four modified stems that are involved in asexual reproduction.
9. Identify the fruit types on the next page. Use Figure 35–11 to check your answers.

(a) (b) (c) (d) (e) (f)

(g) (h) (i) (j) (k)

YOU MAKE THE CONNECTION

1. How might the presence of sori on the fronds of many fern species be used as indirect evidence for the evolution of stamens and carpels from leaves?
2. Draw pictures to show the kinds of flowers that might form simple, aggregate, multiple, and accessory fruits.
3. Could seed dispersal by ants be considered an example of coevolution? Why or why not?
4. Based on what you have learned in this chapter, speculate whether it is more likely that offspring of asexual reproduction develop in close proximity to or widely dispersed from the parent plant. Explain your reasoning. How could you design an experiment to test your hypothesis?
5. Which type of reproduction, sexual or asexual, might be more beneficial in the following circumstances, and why? (a) a perennial (plant that lives more than two years) in a stable environment; (b) an annual (plant that lives one year) in a rapidly changing environment; (c) a plant adapted to an extremely narrow climate range.

RECOMMENDED READINGS

Berg, L.R. *Introductory Botany: Plants, People, and the Environment.* Saunders College Publishing, Philadelphia, 1997. A general botany text with an environmental emphasis.

Cox, P.A. "Water-pollinated Plants." *Scientific American,* Vol. 269, No. 4, Oct. 1993. Some aquatic plants rely on water currents to transfer pollen to female flowers.

Fleming, T.H. "Cardon and the Night Visitors." *Natural History,* Oct. 1994. Many cacti are pollinated by bats.

Handel, S.N., and A.J. Beattie. "Seed Dispersal by Ants." *Scientific American,* Vol. 263, No. 2, Aug. 1990. How plants induce ants to disperse seeds without harming them.

Heinrich, B. "Of Bedouins, Beetles, and Blooms." *Natural History,* May 1994. In the Judean Desert, competition among various flowers for animal pollinators has resulted in convergent evolution. Many flower species have red petals (even buttercups, which are yellow elsewhere) and are pollinated by unusual beetles that can see red.

Kaerns, C.A., and D.W. Inouye. "Pollinators, Flowering Plants, and Conservation Biology." *BioScience,* Vol. 47, No. 5, May 1997. Threats to animal pollinators, such as habitat alteration and pesticides, also affect the well-being of the plants that require pollinators.

Meyerowitz, E.M. "The Genetics of Flower Development." *Scientific American,* Vol. 271, No. 5, Nov. 1994. Studies of normal and mutant *Arabidopsis* flowers have shown it possible to predict how mutations will alter the structure of flowers.

Moore, P.D. "Not Only Gone with the Wind." *Nature,* Vol. 35, No. 14, May 1992. Some plant species have two methods of seed dispersal instead of one. Violets, for example, have explosive dehiscence as their primary means of dispersal and ants for secondary dispersal after the seeds have been scattered.

Tanksley, S.D., and S.R. McCouch. "Seed Banks and Molecular Maps: Unlocking Genetic Potential from the Wild." *Nature,* Vol. 277, 22 Aug. 1997. This article discusses how we may use the genetic potential of seed banks to benefit society.

● Visit our website at **http://www.saunderscollege.com/lifesci/titles.html** and click on Solomon/Berg/Martin Biology for links to chapter-related resources on the World Wide Web.

CHAPTER 36

Growth Responses and Regulation of Growth

As you learned in Chapter 16, the ultimate control of plant growth and development is genetic. If the genes required for development of a particular trait, such as the shape of the leaf, the color of the flower, or the type of root system, are not present, that characteristic does not develop. When a particular gene is present, its expression, that is, how it exhibits itself as an observable feature of an organism, is determined by a variety of factors, including signals from other genes and from the environment. The location of a cell in the young plant body, for example, has a profound effect on gene expression during development. Experiments suggest that chemical signals from adjacent cells may help the cell "perceive" its location within the plant body. Each cell's spatial environment, therefore, helps to determine what that cell ultimately becomes.

Environmental cues, such as changing day length and variations in precipitation and temperature, also exert an important influence on gene expression, as they do on all aspects of plant growth and development. The initiation of sexual reproduction is often under environmental control, particularly in temperate latitudes, and plants switch from vegetative to reproductive growth upon receiving the appropriate signals from the environment. These black-eyed Susans *(Rudbeckia hirta),* for example, produce flowers in response to the shortening nights of spring and early summer; black-eyed Susans typically flower between June and October in North America. Such control is important for the plant's survival because the timing of sexual reproduction is critical for reproductive success: all flowering plants in temperate climates must flower and form seeds before the onset of winter induces **dormancy**, which is a temporary state of reduced physiological activity. Many detect changes in the relative amounts of daylight and darkness that accompany the changing seasons and flower in response to those changes. Other plants have temperature requirements that induce sexual reproduction.

In this chapter we consider the role of plant **hormones**, organic chemical compounds produced in one part of a plant and transported to another part, where they elicit some kind of physiological response. The production of hormones, which is under both genetic and environmental control, regulates many aspects of plant growth and development. Growth and development include **germination** (when a seed sprouts) and the growth of seedlings (young plants that develop from

(Dwight Kuhn)

germinating seeds) into mature plants. A plant's responses to changes in various aspects of its environment, including temperature, light, gravity, and touch, are also a part of growth and development.

AFTER YOU HAVE STUDIED THIS CHAPTER YOU SHOULD BE ABLE TO

1. Summarize the influence of environmental factors on the germination of seeds.
2. Discuss genetic and environmental factors that affect plant growth and development.
3. Explain how flowering is induced by varying amounts of light and darkness and describe the role of phytochrome in flowering.
4. Distinguish between a nastic movement and a tropism and describe phototropism, gravitropism, thigmotropism, and heliotropism.

5. Define circadian rhythm and give an example.
6. List several different ways each of the following hormones affects plant growth and development: auxins, gibberellins, cytokinins, ethylene, and abscisic acid.
7. Describe the hormone regulation of each of the following: seed dormancy, germination, apical dominance, stem elongation, fruit ripening, and leaf abscission.

EXTERNAL AND INTERNAL FACTORS AFFECT GERMINATION AND EARLY GROWTH

In Chapter 35 you learned how pollination and fertilization are followed by seed and fruit development in flowering plants. Each seed develops from an ovule and contains an embryonic plant and food to provide nourishment for the embryo during germination. A mature seed, that is, a seed in which the embryo is fully developed, is often dormant and may not germinate immediately even if growing conditions are ideal.

Numerous factors influence whether or not a seed germinates. Many of these are environmental cues, including the presence of water and oxygen, proper temperature, and sometimes the presence of light penetrating the soil surface. No seed germinates, for example, unless it has absorbed water. The embryo in a mature seed is dehydrated, and a watery environment is necessary for active metabolism. When a seed germinates, its metabolic machinery is turned on, and numerous materials are synthesized and others degraded. Therefore, water is an absolute requirement for germination.

The absorption of water by a dry seed is known as **imbibition**. As a seed imbibes water, it often swells to several times its original, dry size (Fig. 36–1). Cells imbibe water by adsorption of water onto and into colloidal materials such as cellulose, pectin, and starches within the seed. Water is attracted to and bound to these materials by adhesion, the attraction between unlike materials (see Chapter 2).

Germination and subsequent growth also require a great deal of energy. Because young plants obtain this energy by converting the energy of food molecules stored in the endosperm or cotyledons to ATP by aerobic respiration, much oxygen is usually needed during germination. Some plants, such as rice, can respire anaerobically during the early stages of germination and seedling growth. This enables rice plants to grow and become established in flooded soil, an environment that would suffocate most young plants.

Temperature is another environmental factor that affects germination. Each species has an optimal, or ideal, temperature at which the germination percentage is highest. For most

plants, the optimal germination temperature falls between 25° and 30° C (77° and 86° F). Some seeds, such as those of apples, require prolonged exposure to low temperatures before their seeds are able to germinate at any temperature. Some of the environmental factors that are needed for germination help ensure the survival of the young plant. The requirement of a prolonged low temperature period ensures that seeds adapted to temperate climates germinate in the spring rather than in the winter. Some plants, especially those with tiny seeds, such as lettuce, require light for germination. A light requirement ensures that a tiny seed germinates only if it is close to the surface of the soil. If such a seed were to germinate several inches below the soil surface, it might not have enough food reserves to grow to the surface. On the other hand, if this light-dependent seed remains dormant until the soil is disturbed and it is brought to the surface, it has a much greater likelihood of survival.

In certain seeds, internal factors, which are under genetic control, prevent germination even when all external conditions are favorable. Many seeds are dormant either because the embryo is immature and must develop further or because certain chemicals are present. The presence of such chemical inhibitors

Figure 36–1   Imbibition.   Dry seeds must imbibe water before they germinate. Pinto bean seeds before imbibition *(left)* and after *(right)*.   *(Marion Lobstein)*

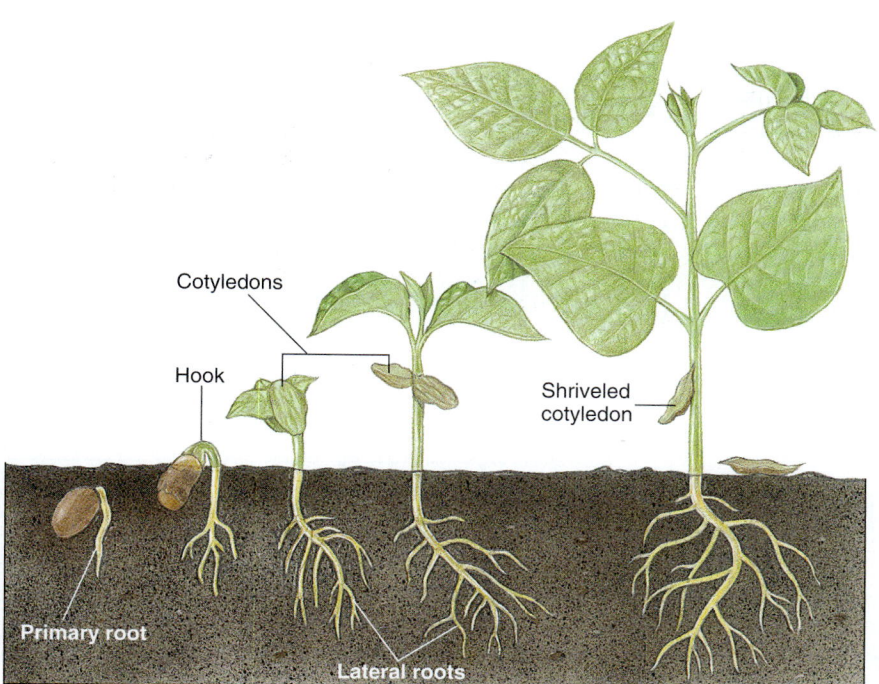

Figure 36–2 Germination and growth of a dicot seedling. Shown is soybean, a dicot. Note the hook in the stem, which protects the delicate stem tip as it moves up through the soil. Once the shoot has emerged from the soil, the hook straightens.

Cotyledons

Hook

Shriveled cotyledon

Primary root

Lateral roots

helps ensure the survival of the plant. The seeds of many desert plants, for example, often contain high levels of abscisic acid, which inhibits germination (discussed later in this chapter). Abscisic acid is washed out only when rainfall is sufficient to support the plant's growth after the seed germinates. Some seeds, such as legumes, have extremely hard, thick seed coats that prevent water and oxygen from entering, thereby inducing dormancy. *Scarification*, the process of scratching or nicking the seed coat (physically with a knife or chemically with an acid) before sowing it, induces germination in these plants. In nature scarification occurs when these seeds pass through the digestive tracts of animals or when the seed coats are partially digested by bacteria.

Dicots and monocots exhibit characteristic patterns of early growth

Once conditions are right for germination, the first part of the plant to emerge from the seed is the radicle, or embryonic root. As the root grows and forces its way through the soil, it encounters considerable friction from soil particles. The delicate apical meristem of the root tip is protected by a root cap (see Chapter 34).

The shoot is next to emerge from the seed. Stem tips are not protected by a structure comparable to a root cap, but plants have ways to protect the delicate stem tip as it grows through the soil to the surface. The stem of a bean seedling (a dicot), for instance, curves over to form a hook so that the stem tip and cotyledons are actually *pulled up* through the soil (Fig. 36–2). Corn and other grasses (monocots) have a special sheath of cells called a **coleoptile** that surrounds and protects the young shoot (Fig. 36–3). First, the coleoptile pushes up

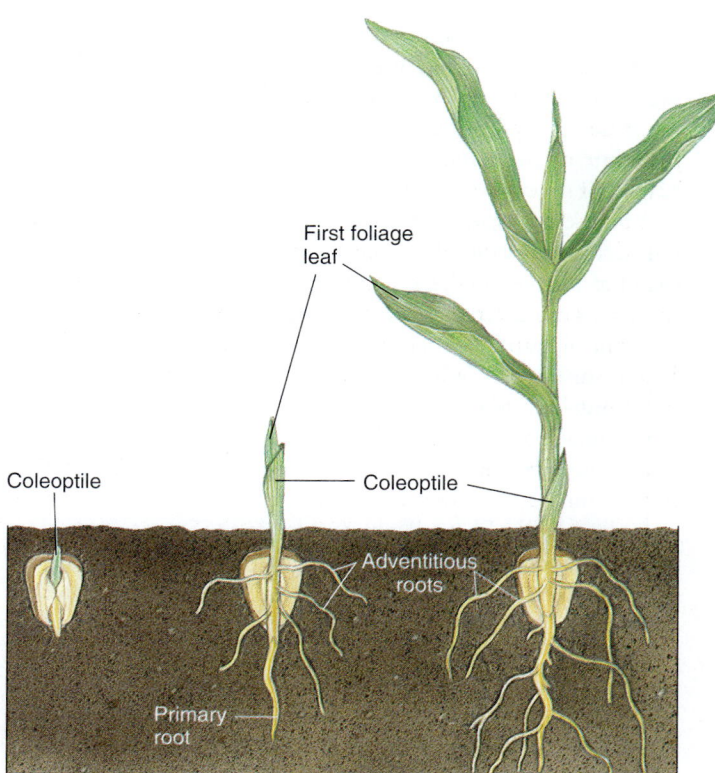

First foliage leaf

Coleoptile

Coleoptile

Adventitious roots

Primary root

Figure 36–3 Germination and growth of a monocot seedling. Shown is corn, a monocot. Note the coleoptile, a sheath of cells that emerges first from the soil. The shoot and leaves grow up through the middle of the coleoptile.

through the soil, and then the leaves and stem grow through the tip of the coleoptile.

Certain parts of a plant grow throughout its life. This **indeterminate growth**, the ability to grow indefinitely, is characteristic of stems and roots, both of which arise from apical meristems. Hypothetically, stems and roots could continue to grow forever. Other parts of a plant, such as leaves and flowers, have **determinate growth**; that is, they stop growing after reaching a certain size. The size of each of these structures varies from species to species and from individual to individual, depending on the plant's genetic programming and on environmental conditions, such as availability of sunlight, water, and essential minerals.

ENVIRONMENTAL CUES AFFECT FLOWERING AND OTHER PLANT RESPONSES

Photoperiodism is any response of a plant to the relative lengths of daylight and darkness. Initiation of flowering is one of several physiological activities that are photoperiodic in many plants. Plants are classified into four main groups—short-day, long-day, intermediate-day, and day-neutral—on the basis of how photoperiodism affects their flowering.

Short-day plants (also called **long-night plants**) flower when the night length is equal to or greater than some critical period (Fig. 36–4a). The minimum critical night length varies considerably from one plant species to another, but falls between 12 and 14 hours for many. The initiation of flowering in short-day plants is not due to the shorter period of daylight but to the long, uninterrupted period of darkness. Examples of short-day plants are chrysanthemum, aster, ragweed, and poinsettia, which typically flower in later summer or fall. Poinsettias, for example, typically initiate flower buds in early October in the Northern Hemisphere and flower about eight to ten weeks later; hence, their traditional association with Christmas. Short-day plants can detect the lengthening nights of late summer or fall, and they flower at that time.

Long-day plants (also called **short-night plants**) flower when the night length is equal to or less than some critical period (Fig. 36–4b). Plants such as clover, spinach, black-eyed Susan, and lettuce flower in late spring or summer and are long-day plants. These plants can detect the shortening nights of spring and early summer, and they flower at that time.

Intermediate-day plants flower when they are exposed to days and nights of intermediate length. Sugarcane and several other grasses are intermediate-day plants. These plants do not flower when day length is either too long or too short.

Some plants, called **day-neutral plants**, do not initiate flowering in response to seasonal changes in the period of daylight and darkness but instead respond to some other type of stimulus, external or internal. Tomato, dandelion, string bean, grape, cucumber, corn, and pansy are examples of day-neutral plants. Many of these plants originated in the tropics where

(a) Short-day plants

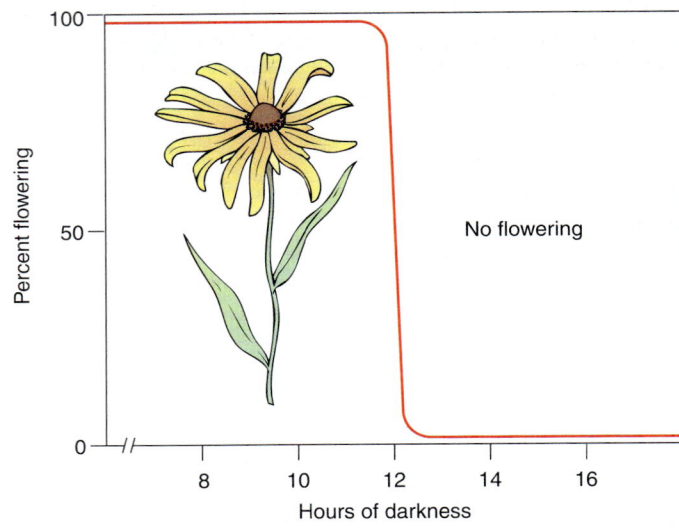

(b) Long-day plants

Figure 36–4 Photoperiodic responses in short-day and long-day plants. *(a)* Short-day plants, such as chrysanthemum, flower when the night length is equal to or exceeds a certain critical length. *(b)* Long-day plants, such as black-eyed Susan, flower when the night length is equal to or less than a certain critical length. The critical length, arbitrarily chosen as 12 hours in both *(a)* and *(b)*, varies among different species.

day length does not vary appreciably during the year. (In contrast, short-day, long-day, and intermediate-day plants are temperate, or mid-latitude, species.)

Phytochrome detects day length

For a plant or any organism to have a biological response to light, it must contain a light-sensitive substance, called a *photoreceptor*, to absorb the light. The photoreceptor for photoperiodism and several other light-initiated plant responses is a family of about five, blue-green pigments, each coded by a different gene, and collectively called **phytochrome**. Phytochrome is present in cells of all vascular plants examined so far. Each phytochrome molecule exists in two forms and readily converts from one form to the other after absorption of light of specific wavelengths. One form, designated P_r (for *r*ed-absorbing *p*hytochrome), strongly absorbs red light with a relatively short wavelength (660 nm). In the process, the shape of the molecule changes to the second form of phytochrome, P_{fr}, so designated because it absorbs *far-r*ed light, which is red light with a relatively long wavelength (730 nm) (Fig. 36–5). When P_{fr} absorbs far-red light, it reverts back to the original form, P_r. The P_{fr} form of phytochrome is less stable than the P_r form, so P_{fr} also reverts slowly to P_r in the dark. P_{fr} is the form of phytochrome that triggers or inhibits physiological responses such as flowering.

What does a pigment that absorbs red light and far-red light have to do with daylight and darkness? Sunlight is composed of the entire spectrum of visible light, in addition to ultraviolet and infrared radiation. Because sunlight contains more red than far-red light, however, when a plant is exposed to sunlight, its level of P_{fr} increases. During the night, its level of P_{fr} slowly decreases as P_{fr} reverts to P_r.

In short-day plants, the P_{fr} form of phytochrome *inhibits* flowering, so these plants need long nights in order to flower. The long period of darkness allows the P_{fr} to revert back to P_r, so the plant has some minimum time during the 24-hour period with no P_{fr} present. This absence of P_{fr} initiates flowering in short-day plants. Biologists have experimented with short-day plants by growing them under a short-day/long-night regimen and interrupting the night with a short burst of red light (Fig. 36–6). Exposure to red light for a few minutes in the middle of the night prevents flowering in short-day plants. This effect occurs because the brief exposure to red light converts some of the phytochrome from the P_r form to the P_{fr} form. Therefore, the plant does not have a sufficient period of nighttime without any P_{fr}.

In long-day plants, the active form of phytochrome, P_{fr}, *induces* flowering. The long days cause these plants to produce predominantly P_{fr}. During the short nights, some P_{fr} changes to P_r, but because the night is short, the plant still has a sufficient level of P_{fr} present to flower.

Plant biologists are puzzled by the observation that P_{fr} inhibits flowering in short-day plants but induces flowering in long-day plants. Why different plants vary in their responses to P_{fr} is unknown at this time, although some plant biologists hypothesize that the different members of the phytochrome family (recall that there are five different phytochrome molecules) are involved in these varying responses. Nevertheless, the importance of phytochrome to plants cannot be overemphasized. Timing of day length and darkness is the most reliable way for plants to measure the change from one season to the next. This measurement is crucial for their survival, particularly in environments where the climate goes through a regular, annual pattern of favorable and unfavorable seasons.

Phytochrome is involved in many other responses to light

Phytochrome is involved in the light requirement that some seeds have for germination. Seeds with a light requirement must be exposed to red light. Exposure to red light converts P_r to P_{fr}, and germination occurs. Other physiological functions under the influence of phytochrome include sleep movements in leaves (discussed shortly); shoot dormancy; leaf abscission (see Chapter 32); pigment formation in flowers, fruits, and leaves; and shade avoidance (see *Making the Connection: Phytochrome and Competition Among Shade-Avoiding Plants for Sunlight*).

The molecular mechanism by which phytochrome changes gene expression is unknown

Biologists know that phytochrome monitors light exposure and causes changes in gene expression that enable a plant to optimize its growth and development in response to changing environmental conditions. Each phytochrome molecule consists of a protein attached to a light-absorbing photoreceptor. It is thought that the absorption of light by the photoreceptor portion of phytochrome elicits a change in the shape of the larger protein component. Scientists do not know how phytochrome relays its sensory information to the genes that it regulates, however. At least one **signal transduction** pathway is involved, and several have been hypothesized, but definitive evidence

Red light (660 nm)

Inactive form

P_r

Active form

P_{fr}

Far-red light (730 nm)

Physiological response (such as flowering)

or dark reversion

Figure 36–5 Phytochrome. This pigment occurs in two forms, designated P_r and P_{fr}, and readily converts from one form to the other. Red light (660 nm) converts P_r to P_{fr}, and far-red light (730 nm) converts P_{fr} to P_r.

Figure 36–6 **Experimental evidence of photoperiodic responses of short-day and long-day plants.** Note that the short-day plant does not flower when exposed to a long night (16 hours of darkness) interrupted with a brief flash of light. This same treatment induces the long-day plant to flower.

Day lengths

| Short nights 24 hrs | Long nights | Long nights with interruption |

Short-day plant: Vegetative growth only | Flowers | Vegetative growth only

Long-day plant: Flowers | Vegetative growth only | Flowers

is currently lacking. (See *Making the Connection: Information Transfer Across the Plasma Membrane* in Chapter 5 for a discussion of signal transduction.)

Temperature may affect reproduction

Certain plants have a temperature requirement that must be met if they are to flower. The promotion of flowering by exposure to low temperature for a period of time is known as **vernalization**. Although the exact temperature and amount of time required varies among species, vernalization typically occurs between 0° and 10° C (32° to 50° F). The part of the plant that must be exposed to low temperature varies. Seeds of some plants must be moistened and exposed to low temperature for a period of several weeks in order for flowering to occur after the seeds germinate and grow. For other plants, the stem apical meristem has a low-temperature requirement. In some plants the requirement of a low-temperature period is absolute, meaning that they will not flower unless they have been vernalized. Other plants flower sooner if exposed to low temperatures, but will still flower at a later date if they are not exposed to low temperatures.

Examples of plants with a low-temperature requirement include annuals such as winter wheat, which grow, reproduce, and die in one year, and biennials such as carrots, which take two years to complete their life cycles. Winter wheat is planted in the fall and germinates at that time. The stem apical meristems of the young seedlings are exposed to low temperatures during the winter and subsequently flower after resuming growth the following spring. Winter wheat is a particularly suitable crop in the Midwest because it can be harvested early, before the summer drought.

Carrots and other biennials grow vegetatively (that is, their leaves, stems, and roots grow) the first year and store surplus food in their roots. If the roots are not harvested, the plants flower and reproduce sexually during the second year, after their apical meristems are exposed to the low temperatures of winter. Carrots growing in a warm environment and not exposed to low temperatures continue vegetative growth and do not initiate sexual reproduction.

An external stimulus to which a plant responds, such as low temperature, may be moderated and influenced by internal conditions, such as hormone levels in the plant. It is possible, for example, to eliminate the low-temperature require-

PHYTOCHROME AND COMPETITION AMONG SHADE-AVOIDING PLANTS FOR SUNLIGHT

Can plants sense the presence of nearby plants? The answer is yes: they not only detect the presence of their neighbors, but they also react to nearby plants by changing the way they grow and develop. Most plants compete for light, a response known as **shade avoidance**, in which plants tend to grow taller when closely surrounded by other plants.

The environmental factor that triggers shade avoidance has been recognized by botanists since the 1970s. Although plants do not possess real eyes, they are able to use their phytochrome "eye" to perceive changes in the ratio of red to far-red light that result from the presence of nearby plants. The leaves of neighboring plants absorb much more red light than far-red light. (Recall from Chapter 8 that the green pigment chlorophyll used in photosynthe-

sis strongly absorbs red light.) In a densely plant-populated area, the ratio of red light to far-red light (r/fr) decreases because more far-red light is reflected from leaves of nearby plants. The phytochrome "eye," sensing more far-red light, triggers a series of responses that cause the shade-avoiding plant, which is adapted to high-light environments, to grow taller.

When a plant is using many of its resources for stem elongation, it has fewer resources to allocate for new leaves and branches, storage tissues, or reproductive tissues. However, for a shade-avoiding plant that is shaded by its neighbors, a rapid increase in stem length is advantageous because once this plant is taller than its neighbors, it obtains a larger share of unfiltered sunlight. We discuss other aspects of competition in Chapters 51 and 52.

---

ment for flowering in biennials by treating the plants with gibberellin, a plant hormone that is discussed later in this chapter.

A BIOLOGICAL CLOCK INFLUENCES MANY PLANT RESPONSES

Most organisms, including plants, animals, fungi, eukaryotic microorganisms, and many prokaryotes, appear to have an internal timer, or biological clock, that approximates a 24-hour cycle. These internal cycles, known as **circadian rhythms** (from the Latin *circum*, "around," and *diurn*, "daily"), help the organism respond to the time of day. In contrast, photoperiodism enables a plant to detect time of year.

In the absence of external cues, circadian rhythms repeat every 20 to 30 hours. In nature, the rising and setting of the sun reset the biological clock so that the cycle repeats every 24 hours. For many plants, the photoreceptor portion of phytochrome has been implicated as the light-absorbing group involved in resetting the biological clock. Additionally, certain amino acid sequences of the protein portion of phytochrome are homologous with amino acid sequences of clock proteins in fruit flies, fungi, mammals, and bacteria; this molecular evidence strongly supports the circadian clock role of phytochrome.

One example of a circadian rhythm in plants is the opening and closing of stomata, independent of light and darkness. Plants placed in continual darkness for extended periods continue to open and close their stomata on an approximate 24-hour cycle. **Sleep movements** observed in the common bean and other plants are another example of a circadian rhythm in plants (Fig. 36–7). During the day, bean leaves are horizon-

tal, possibly for optimal light absorption, but at night the leaves fold down or up, a movement that orients them perpendicular to their daytime position. The biological significance of sleep movements is unknown at this time. These movements persist in the absence of cues; if bean plants are placed in continual darkness or continual light, sleep movements continue on an approximate 24-hour cycle. The first plant with a mutant biological clock gene was reported in 1995 (see *On the Cutting Edge: Circadian Clock Mutants in* Arabidopsis).

Why do plants and other organisms exhibit circadian rhythms? Predictable environmental changes, such as sunrise and sunset, occur during the course of each 24-hour period. These predictable changes may be important to an individual organism, causing it to change its behavior (in the case of animals) or its physiological activities. It is thought that circadian rhythms help an organism to synchronize repeated daily activities so that they occur at the appropriate time each day. If, for example, a firefly flashes its light on and off at the wrong time of day, it will not find a mate. Likewise, if an insect-pollinated flower does not open at the time of day that pollinating insects are foraging for food, reproduction will be unsuccessful.

CHANGES IN TURGOR CAN INDUCE NASTIC MOVEMENTS

The sensitive plant (*Mimosa pudica*) dramatically folds its leaves and droops in response to touch (or to an electrical, chemical, or thermal stimulus) (Fig. 36–8). The response, which typically occurs in a few seconds, spreads throughout the plant even if only one leaflet is initially stimulated. When a sensitive plant is touched, an electrical impulse moves down

(Text continued on page 769.)

(a)

(b)

Figure 36–7 **Sleep movements in a bean (Phaseolus vulgaris) seedling.** *(a)* Leaf position at noon. *(b)* Leaf position at midnight. It is not known why some plants exhibit sleep movements, but they occur on an approximate 24-hour cycle, regardless of the amount of light or darkness to which they are exposed. *(a, b, Dennis Drenner)*

(a)

(b)

(c)

Figure 36–8 **Nastic movements in the sensitive plant (Mimosa pudica).** The plant *(a)* before and *(b)* several seconds after being touched. Note how the compound leaves have folded and drooped. *(c)* How the drooping occurs. Pulvini are located in three areas—at the base of each leaflet, at the base of each cluster of leaflets, and at the base of each leaf. Only changes in the pulvini at the base of leaflets is shown. *(Top right)* Section through two leaflets, showing their pulvini when the leaf is undisturbed. *(Bottom right)* Section through the two leaflets, showing how a loss of turgor produces the folding of the leaves. *(a, b, Dennis Drenner)*

Circadian Clock Mutants in *Arabidopsis*

HYPOTHESIS: It is possible to identify and study plants with mutant circadian clock genes.

METHOD: Develop a line of transgenic *Arabidopsis* plants in which the promoter region of an *Arabidopsis* gene known to be influenced by a circadian clock is spliced to the luciferase gene from fireflies.

RESULTS: When exposed to a chemical called luciferin, wild-type (normal) transgenic plants that contain the luciferase gene glow during the day and do not glow at night. Transgenic plants that are circadian clock mutants glow at unpredictable times.

CONCLUSION: Plants with mutant circadian clocks can be identified and studied.

Circadian clock genes have been identified and cloned in organisms as diverse as cyanobacteria, fungi, fruit flies, and mice, but until the mid-1990s such genes in plants eluded biologists. To identify circadian clock genes, researchers typically study mutant individuals with abnormal internal rhythms, such as, a fungus that develops reproductive structures at the "wrong" time of day. However, plant rhythms are generally less obvious than those in other organisms, and mutants have therefore been extremely difficult to identify. (Imagine how daunting it would be, for example, to monitor stomata from plants to determine if any are opening or closing at unpredictable times.)

In 1995 a research team led by Steve Kay of the National Science Foundation Center for Biological Timing at the University of Virginia reported that they had identified circadian clock mutants in seedlings of *Arabidopsis*, a diminutive plant from the mustard family (see Chapter 16).* One of the *Arabidopsis* genes known to exhibit a circadian rhythm is *cab*, the gene that codes for chlorophyll-*a/b* binding protein. The *cab* gene is actively transcribed during the day but switched off at night. As do other genes that follow a circadian rhythm, the *cab* gene continues this rhythm on an approximate 24-hour cycle when *Arabidopsis* plants are placed in continual light.

A transgenic line of *Arabidopsis* plants was developed in which the *cab* promoter (a regulatory sequence; see Chapter 13) was spliced to the luciferase gene from fireflies. The luciferase gene codes for an enzyme that catalyzes a reaction causing the chemical luciferin to glow. It was hoped that linking the luciferase gene to the promoter of *cab*, which is influenced by as-yet-unidentified circadian clock genes, would cause the luciferase gene to be expressed during the day but not at night.

The results were highly informative. When *Arabidopsis* seedlings containing the recombinant DNA were sprayed with luciferin, they glowed only during the day *(see figure)*. All that remained was to screen the seedlings for individuals that glowed either earlier or later than normal. Altogether, 26 mutant seedlings,

*Millar, A.J., I.A. Carre, and S.A. Kay. "Circadian Clock Mutants in *Arabidopsis* Identified by Luciferase Imaging." *Science*, Vol. 267, 24 Feb. 1995.

representing at least 21 separate mutations, were identified. When various circadian rhythms were examined in these mutants, it was found that their leaf movements were timed differently than those of wild-type plants. Other circadian rhythms were normal, however, which may indicate that *Arabidopsis* contains more than one circadian clock gene or that there is a "central" clock with separate links to different physiological processes.

Biologists are currently making a concerted effort to identify and clone the circadian clock gene in *Arabidopsis*. After that goal is reached, the gene can be compared with known circadian clock genes in other organisms to determine how similar they are. Biologists also want to learn how the wild-type circadian clock gene works.

Transgenic *Arabidopsis* seedlings glow after being sprayed with luciferin. The timing of their bioluminescence helped researchers identify circadian clock mutants. (The intensity of light produced by the glowing seedlings was digitally transformed into color. White, pink, and red indicate the strongest intensities.) *(Courtesy of Steve A. Kay)*

the leaf to special cells housed in an organ at the base of each leaflet, each cluster of leaflets, and each petiole, called the **pulvinus** (pl., *pulvini*). The pulvinus is a somewhat swollen joint that acts as a hinge. When the electrical signal reaches cells in

the pulvinus, it induces a chemical signal that increases membrane permeability to certain ions. A loss of *turgor* occurs in certain pulvinus cells as potassium ions exit through the now-permeable plasma membrane, causing water to leave the cells by osmosis (see Chapter 5 for a discussion of turgor pressure). The sudden change in turgor causes the leaf movement. Such **nastic movements** occur in response to external stimuli, but the direction of movement is predetermined and is independent of the direction of the stimulus. Nastic movements are temporary and reversible. The movement of potassium ions and water back into the pulvinus cells causes the plant part to return to its original position, although recovery takes several to many minutes longer than the original movement.

The Venus flytrap *(Dionaea muscipula),* which snaps shut after an insect lands and brushes against two of its three trigger hairs, provides an impressive example of nastic movements. Several competing hypotheses seek to explain the mechanism by which the Venus flytrap leaf closes. One hypothesis is that leaf closure in the Venus flytrap is similar to that of the sensitive plant. According to this hypothesis, an electrical signal, which moves much more rapidly than in the sensitive plant, induces a chemical signal that causes a movement of potassium ions out of certain cells, followed by the exit of water; the loss of turgor causes the leaf to snap shut.

A TROPISM IS DIRECTIONAL GROWTH IN RESPONSE TO AN EXTERNAL STIMULUS

A plant may respond to an external stimulus, such as light, gravity, or touch, by directional growth. Such a directional growth response, called a **tropism**, results in a change in the position of a plant part. Tropisms are irreversible and may be positive or negative, depending on whether the plant grows toward the stimulus (a positive tropism) or away from it (a neg-

ative tropism). Tropisms are under hormonal control, which is discussed later in this chapter.

Phototropism is the directional growth of a plant caused by light (Fig. 36–9). Most growing shoot tips exhibit positive phototropism by bending (growing) toward light, something you may have observed if you place houseplants near a sunny window. This growth response increases the likelihood that stems and leaves will receive adequate light for photosynthesis. The bending response of phototropism is triggered by blue light with wavelengths less than 500 nanometers. The photoreceptor that absorbs blue light and triggers the phototropic response is thought to be a yellow pigment. However, it has never been isolated and biochemically identified because it is present in such low concentrations in the cell. Ongoing research may determine the signal-transduction pathway by which blue light triggers phototropism.

Growth in response to the direction of gravity is called **gravitropism**. Most stem tips exhibit negative gravitropism by growing away from the center of Earth, whereas most root tips exhibit positive gravitropism (Fig. 36–10). The root cap is the site of gravity perception in roots. Special cells in the root cap possess starch-containing **amyloplasts** that collect toward the bottoms of the cells in response to gravity, and for many years it was thought that these amyloplasts initiated the gravitropic response (Fig. 36–11). If the root is moved, as when a potted plant is laid on its side, the amyloplasts tumble to a new position, always settling in the direction of gravity. The gravitropic response occurs shortly thereafter and involves the hormone auxin (discussed later in the chapter).

Despite the movement of amyloplasts in response to gravity, their role in gravitropism has been questioned. Botanists at Michigan State University identified a mutant *Arabidopsis* plant that lacks amyloplasts in its root cap cells. When placed on their sides, roots of these mutant plants still responded gravitropically, indicating that roots do not necessarily need amyloplasts to respond to gravity. Ongoing research may help clarify how roots perceive gravity.

Thigmotropism is growth in response to a mechanical stimulus, such as contact with a solid object. The twining or curling growth of tendrils or stems, which helps attach a climbing plant such as a vine to some type of support, is an example of thigmotropism (see Fig. 32–15b).

Heliotropism, also called **solar tracking**, is the ability of leaves or flowers of certain plants, such as sunflower, soybean, and cotton, to follow the sun's movement across the sky (Fig. 36–12). Frequently, the leaves of such plants arrange themselves so that they are perpendicular to the sun's rays, regardless of the time of day or the sun's position in the sky. This positioning allows for maximal light absorption. Many solar trackers have pulvini at the bases of their petioles. Changes in turgor in the cells of the pulvinus help position the leaf in its proper orientation relative to the sun. Like phototropism, heliotropism is triggered by blue light.

Tropisms in plants may also be caused by other stimuli in the environment such as water, temperature, chemicals, and oxygen. Because tropisms are growth responses, they cause permanent changes in the positions of plant parts.

Figure 36–9 **Phototropism in corn *(Zea mays)* seedlings.** Stems grow in the direction of light and therefore exhibit positive phototropism. The bending toward light is caused by greater elongation on the shaded side of each stem than on the lighted side. *(Runk/Schoenberger/Grant Heilman)*

(a)

(b)

(c)

(d)

Figure 36–10 **Gravitropism in corn (Zea mays) seedlings.** *(a)* Two corn seeds were germinated at the same time. The plant on the left served as a control. *(b)* The plant on the right was turned on its side on day 3. *(c)* In one hour, the root and shoot tips had curved. *(d)* In 24 hours, the new root growth was downward (positive gravitropism), and the new shoot growth was upward (negative gravitropism). Bending of both root and stem was caused by greater elongation on one side of the stem or root than on the other. *(Dennis Drenner)*

Root apical meristem

Root cap

(a)   100 μm

Amyloplasts

(b)   25 μm

Figure 36–11 **Amyloplasts in root cap cells.** *(a)* LM of a root cap of a root that is growing downward, in the direction of gravity. *(b)* LM at a higher magnification shows amyloplasts, which have collected on the bottoms of the cells. *(a, b, Courtesy of R. Moore and E. McClellan, Baylor University)*

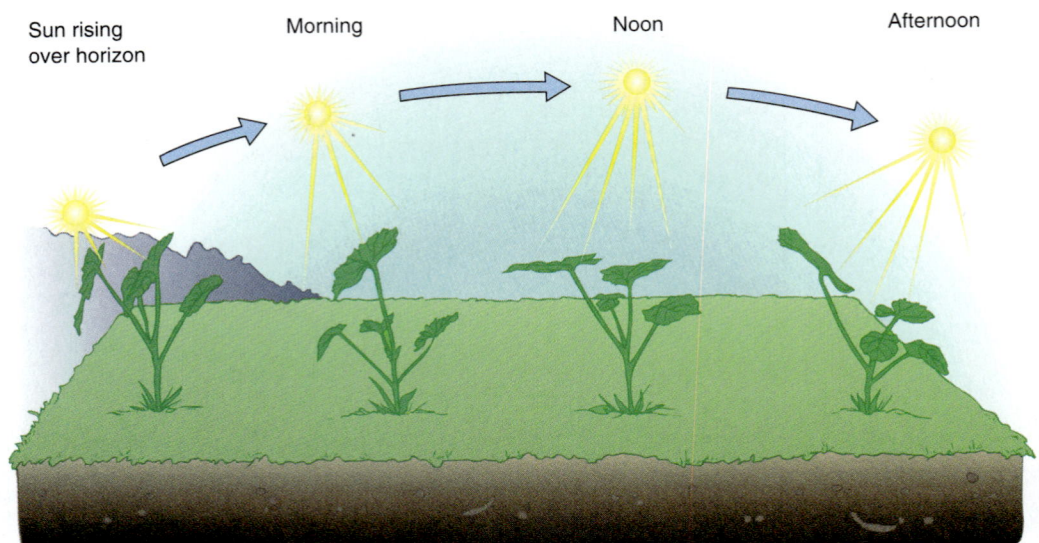

Figure 36–12 **Heliotropism.** Note that the leaves are oriented so that they are perpendicular to the sun's rays throughout the day, thereby maximizing the amount of light absorbed.

Sun rising over horizon   Morning   Noon   Afternoon

HORMONES ARE CHEMICAL MESSENGERS THAT REGULATE GROWTH AND DEVELOPMENT

As mentioned earlier in the chapter, a *hormone* is an organic chemical produced in one part of a plant and transported to another part, where it elicits some kind of physiological response. The study of plant hormones and their effects is very challenging and difficult because hormones are effective in extremely small amounts, and each hormone elicits many different responses. The latter feature contrasts with the actions of animal hormones, which are usually very specific. In addition, the effects of different plant hormones overlap, so it is difficult to determine which hormone, if any, is the primary cause of a particular response. Moreover, plant hormones may stimulate a certain response at one concentration and inhibit that same response at a different concentration.

Plant and animal hormones are similar in their basic mechanism of action. Like animal hormones, plant hormones bind to specific receptor proteins in or on the cells of the target tissue. Each receptor has a three-dimensional shape that only binds with one kind of hormone molecule. Sometimes this binding triggers the production of a **second messenger**, an in-

(a)

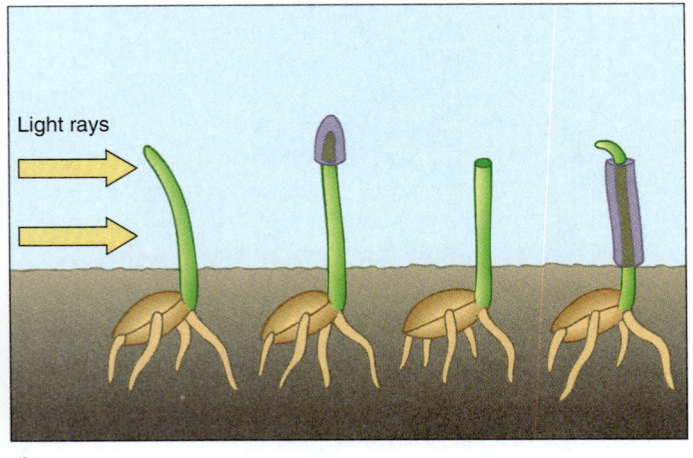

Light rays

(b)

Figure 36–13 **The Darwins' phototropism experiments with coleoptiles of canary grass seedlings.** *(a)* Some plants were uncovered, some were covered only at the tip, some had the tip removed, and some were covered everywhere but at the tip. *(b)* After exposure to light coming from one direction, the uncovered plants and the plants with uncovered tips grew toward the light. The plants with tips removed or covered did not bend toward light.

TABLE 36-1   The Five Classes of Plant Hormones

Hormone	Chemical Structure	Site of Production	Primary Method of Translocation
Auxin (IAA)		Shoot apical meristem, young leaves, seeds	Polar transport (tip toward base) in parenchyma cells
Gibberellin (GA₃)		Young leaves, root and shoot apical meristems, embryo in seed	Xylem and phloem
Cytokinin (Zeatin)		Roots	Xylem
Ethylene		Stem nodes, ripening fruit, damaged or senescing tissue	Thought to be diffusion
Abscisic acid		Older leaves, root cap, stem	Vascular tissue

tracellular signal that affects the function of the cell. Ions such as Ca^{2+} may serve as second messengers in many plant cells. Once the Ca^{2+} concentration in the cell increases, Ca^{2+} may bind to and activate certain enzymes. Other plant hormones may function by influencing gene expression, that is, transcription and/or translation. (Chapter 47 discusses second messengers as they relate to animal hormones.)

Biologists have identified five major classes of plant hormones: auxins, gibberellins, cytokinins, ethylene, and abscisic acid (Table 36–1). These hormones and others regulate the growth and development of a plant.

Auxin promotes cell elongation

Charles Darwin, best known for developing the theory of natural selection to explain evolution, also provided the first evidence for the existence of auxin. The experiments that Dar-

win and his son Francis performed in the 1870s involved positive phototropism, the directional growth of plants toward light. The plants they used were newly germinated canary grass seedlings (Fig. 36–13). As in all grasses, the first part of a canary grass seedling to emerge from the soil is the coleoptile, a protective sheath that encircles the stem. When coleoptiles are exposed to light from only one direction, they bend toward the light. The bending occurs close to the tip of the coleoptile.

The Darwins tried to influence this bending in several ways. For example, they covered the tip of the coleoptile as soon as it emerged from the soil. Even though they covered that part of the coleoptile above where the bend would be expected to occur, the plants treated in this manner did not bend. On other plants, they removed the coleoptile tip and found that bending did not occur. When the bottom of the coleoptile was shielded from the light, however, the coleoptile did

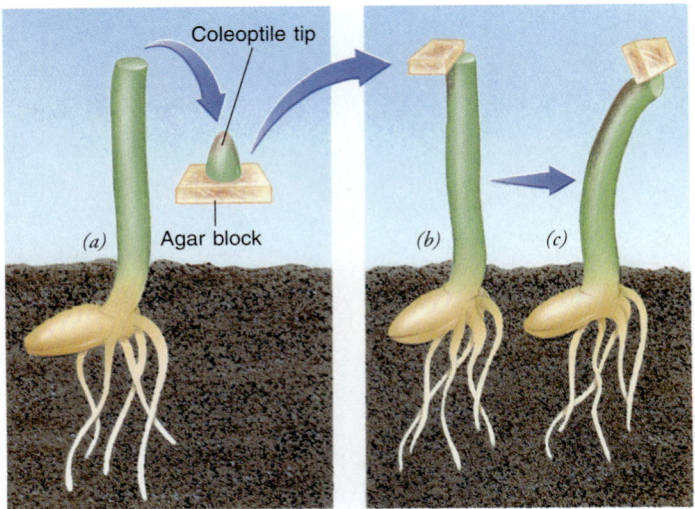

Figure 36–14 **Isolating auxin from coleoptiles.** *(a)* Coleoptile tips were placed on agar blocks for a period of time. *(b)* The agar block was transferred to a decapitated coleoptile. It was placed off-center, and the coleoptile was left in continual darkness. *(c)* The coleoptile bent. This indicated that a chemical had been transferred from the original coleoptile tip to the agar block and from there to one side of the decapitated coleoptile, causing that side to elongate.

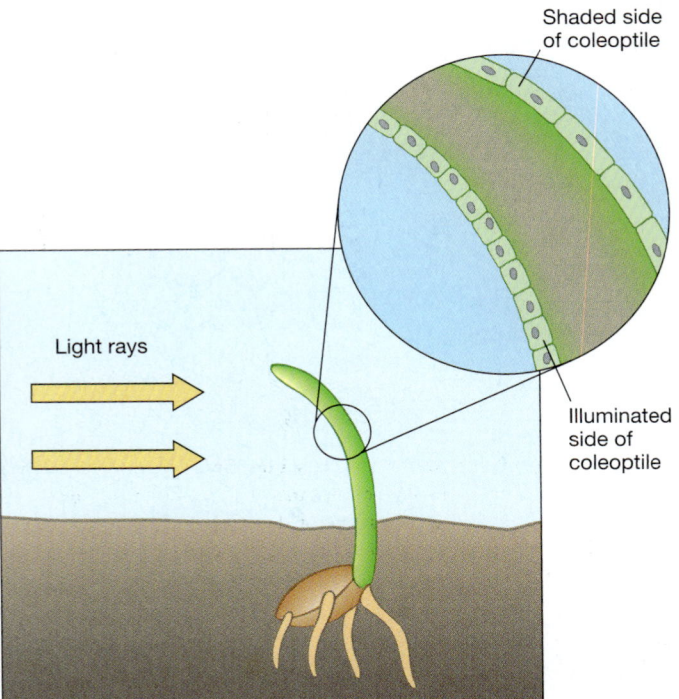

Figure 36–15 **Phototropism and the unequal distribution of auxin.** Auxin travels down the side of the stem or coleoptile *away* from the light, causing cells on the shaded side to elongate. Therefore, the stem or coleoptile bends toward light.

bend toward light. From these experiments, the Darwins concluded that "some influence is transmitted from the upper to the lower part, causing it to bend."

In the 1920s Frits Went, a young Dutch scientist, isolated the phototropic hormone from oat coleoptiles. He removed the coleoptile tips and placed them on tiny blocks of agar for a period of time. When he put one of these agar blocks squarely on a decapitated coleoptile, normal growth resumed. When he placed one of these agar blocks to one side of the tip of a decapitated coleoptile in the dark, bending occurred (Fig. 36–14). This indicated that the substance had diffused from the coleoptile tip into the agar, and later from the agar into the decapitated coleoptile. Went named this substance **auxin** (from the Greek *aux,* "enlarge" or "increase"). The purification and elucidation of auxin's chemical structure were accomplished in the early 1940s by a research team led by Kenneth Thimann at the California Institute of Technology.

Auxin is a group of natural and artificial plant hormones, the most important of which is **indoleacetic acid (IAA).** The movement of auxin in the plant is said to be *polar,* that is, it moves downward from its site of production, usually the shoot apical meristem. Young leaves and seeds are also sites of auxin production.

Auxin's most characteristic action is promotion of cell elongation, an effect it apparently exerts by acidification of cell walls, which increases the plasticity of the cell walls so they expand under the force of the cell's internal turgor pressure. Auxin's effect on cell elongation also provides an explanation for phototropism. When a plant is exposed to a light from only one direction, the auxin migrates laterally to the shaded side of the stem before moving down the stem by polar transport.

When auxin concentration reaches a certain level on the shaded side of the stem, the cells there elongate more than the cells on the light side, and the stem bends toward the light (Fig. 36–15). Auxin is involved in gravitropism, thigmotropism, and possibly heliotropism as well.

Auxin exerts other effects on plants. For example, some plants tend to branch out very little when they grow. Growth in these plants occurs almost exclusively from the apical meristem rather than from axillary buds, which do not develop as long as the terminal bud is present. Such plants are said to exhibit **apical dominance,** the inhibition of axillary buds by the apical meristem. In plants with strong apical dominance, it appears that auxin produced in the apical meristem inhibits axillary buds near the apical meristem from developing into actively growing shoots. When the apical meristem is pinched off, the auxin source is removed, and axillary buds grow to form branches (Fig. 36–16). Apical dominance is often quickly reestablished, however, as one branch begins to inhibit the growth of others. Recent evidence indicates that other hormones (ethylene and cytokinin, both discussed shortly) are also involved in apical dominance. As with other physiological activities, the changing ratios of these hormones may be the factor responsible for apical dominance.

Auxin produced by developing seeds stimulates the development of the fruit. When auxin is applied to certain flowers in which fertilization has not occurred (and, therefore, in which seeds are not developing), the ovary enlarges and de-

(a)   (b)

Figure 36–16 Auxin and axillary bud development. *(a)* When the tip (stem apical meristem) of a plant with strong apical dominance is intact, its axillary buds do not develop. *(b)* The stem tip has been removed. Because no auxin is moving down from the tip, axillary buds develop into branches.

Figure 36–17 Auxin and root development on honeysuckle (*Lonicera fragrantissima*) cuttings. *(Left)* Many adventitious roots developed on a honeysuckle cutting placed in a solution with a high concentration of synthetic auxin. *(Middle)* Fewer roots developed in a lower auxin concentration. *(Right)* The cutting placed in water (no auxin) served as a control and did not form roots in the same time period. *(Joe Eakes, Color Advantage/Visuals Unlimited)*

velops into a seedless fruit. Seedless tomatoes have been produced in this manner.[1] Auxin is not the only hormone involved in fruit development, however.

Some manufactured, or synthetic, auxins have been made that have structures similar to indoleacetic acid. The synthetic auxin naphthalene acetic acid is used to stimulate root development on stem cuttings for asexual propagation, particularly of woody plants with horticultural importance (Fig. 36–17). The synthetic auxins 2,4-D and 2,4,5-T are used as selective herbicides (weed killers). The compounds 2,4-D and 2,4,5-T kill plants with broad leaves but, for reasons not completely understood at this time, do not kill grasses. Both herbicides are similar in structure to IAA and disrupt the plants' normal growth processes. Both cause exaggerated growth in some plant parts and growth inhibition in others. Because many of the world's most important crops are grasses (for example, wheat, corn, and rice), both 2,4-D and 2,4,5-T can be used to kill broadleaf weeds that compete with these crops. The use of 2,4,5-T is no longer allowed in the United States, however, because of its association with dioxins, a group of extremely toxic compounds formed as by-products during the manufacture of 2,4,5-T.

A gene that codes for an enzyme that regulates auxin has been isolated

In 1994 biologists at Michigan State University isolated the first gene known to be involved in regulating auxin. Called the *iaglu* gene, it codes for an enzyme that changes auxin from its active to its inactive form. The *iaglu* gene has been found in many different plants, and biologists suspect it may be present in all plants.

[1]Not all seedless fruits are produced by treatment with auxin. In Thompson seedless grapes, for example, fertilization occurs but the embryos abort, and therefore the seeds fail to develop.

The *iaglu* gene has received a great deal of attention because it may be possible to use it to modify plant growth by genetic engineering. For example, by inserting extra copies of the *iaglu* gene into a plant, researchers hypothesize that smaller amounts of active auxin would be available and, as a result, the plant would be shorter and bushier. By inserting *reversed* copies of the *iaglu* gene (copies that code for mRNA that is complementary to normal mRNA), researchers think that larger amounts of active auxin would be available and the plant would be taller:

Reversed copy of *iaglu* gene is inserted → plant makes reversed copy of RNA by transcription → reversed RNA binds by complementary base-pairing to normal mRNA produced by transcription of normal *iaglu* gene → translation is blocked and *iaglu* enzyme is not produced → more active auxin is present in plant → plant grows taller

Ultimately, biologists may be able to alter growth by manipulating hormones in specific parts of the plant in order to make larger fruits or larger clusters of flowers, for instance. How easily plant growth can be manipulated using the *iaglu* gene remains to be seen. Other as-yet-unknown genes are probably involved as well in controlling auxin production and regulation, in which case manipulating the *iaglu* gene by itself may not be particularly effective.

Gibberellins promote stem elongation

In the 1920s, a Japanese biologist was studying a disease of rice in which the young rice seedlings grew extremely tall and spindly, fell over, and died. The cause of the disease, dubbed the "foolish seedling" disease, was a fungus (*Gibberella fujikuroi*) that produces a chemical substance named **gibberellin**. Not until after World War II did scientists in Europe and North

(a)

(b)

Figure 36–18 Gibberellin and stem elongation. *(a)* An experiment testing the effects of gibberellin on normal and dwarf corn *(Zea mays)* plants shows that dwarf plants respond to gibberellin much more dramatically than normal plants. (This dwarf variety is a mutant with a single recessive gene that impairs gibberellin metabolism.) From left to right: dwarf, untreated; dwarf, treated with gibberellin; normal, treated with gibberellin; normal, untreated. *(b)* Like many biennials, Indian blanket *(Gaillardia pulchella)* grows as a rosette, which is a circular cluster of leaves close to the ground, during its first year *(left)* and then bolts when it initiates flowering in the second year *(right)*. *(a, Courtesy of B.O. Phinney, University of California, Los Angeles; b, Robert E. Lyons)*

America learn of the exciting work done by the Japanese. During the 1960s studies in the United States showed that gibberellins are produced by healthy plants as well as by the fungus that causes the "foolish seedling" disease. Gibberellins were found to be hormones involved in many normal plant functions. The symptoms of the "foolish seedling" disease were caused by an abnormally high gibberellin concentration in the plant tissue (because both plant and fungus were producing gibberellin). Currently, more than 80 naturally occurring gibberellins are known, although it is thought that many of these are inactive precursors; there are no synthetic gibberellins.

As in the "foolish seedling" disease, gibberellins promote stem elongation in many plants. When gibberellin is applied to a plant, particularly to certain dwarf varieties, this elongation may be spectacular. Some corn and pea plants that are dwarf as a result of one or more mutations grow to a normal height when treated with gibberellin (Fig. 36–18*a*). Gibberellins are also involved in **bolting**, the rapid elongation of a floral stalk that occurs naturally in many plants when they

initiate flowering (Fig. 36–18*b*). In all these cases, gibberellins cause stem elongation by stimulating cells to divide as well as elongate. The actual mechanism of cell elongation appears to be different from that caused by auxin, however. Recall that IAA-induced cell elongation involves the acidification of the cell wall; in gibberellin-induced cell elongation, cell-wall acidification does not occur.

Gibberellins are involved in several reproductive processes in plants. They stimulate flowering, particularly in long-day plants. In addition, they can substitute for the low temperature that biennials require before the initiation of flowering. If gibberellins are applied to biennials during their first year of growth, flowering occurs without the period of low temperature. Gibberellins, like auxin, affect the development of fruits. Gibberellins are applied to several varieties of grapes to produce larger fruits.

Gibberellins are also involved in the germination of seeds in many plants. In a classic experiment involving the germination of barley seeds, the release of gibberellin from the

CELL AND TISSUE CULTURE

Cells can be isolated from certain plants and grown in a chemically defined, sterile nutrient medium. In initial experiments with such cultures, plant cells could be kept alive, but they did not divide. It was later discovered that the addition of certain natural materials such as the liquid endosperm of coconut, also known as coconut milk, induced cells to divide in culture. By the late 1950s, plant cells from a variety of sources could be cultured successfully, dividing to produce a mass of undifferentiated cells, or **callus**.

In 1958, F. C. Steward, a plant physiologist at Cornell University, succeeded in generating an entire carrot plant from a single callus cell derived from a carrot root (see Fig. 16–3). This demonstrated conclusively that each plant cell contains a genetic blueprint for all features of an entire organism. His work also showed that an entire plant can be grown from a single cell, provided the proper genes are expressed at the appropriate times.

Since Steward's pioneering work, many plants have been successfully cultured using a variety of cell sources. Plants have been regenerated from different tissues, organ explants (excised organs or parts such as root apical meristems or young embryos), and single cells. Under certain conditions, the genes that control embryonic development are expressed in plant tissue culture, and all of the plant's embryonic stages in the seed can be observed in their normal progression *(see figure)*.

Cell and tissue culture techniques are used to help answer many fundamental questions involving growth and development in plants (see Chapter 16). These techniques also have great practical potential. Using tissue culture, it is possible to regenerate large numbers of genetically identical plants from cells of a single, genetically superior plant; this has been done for many different kinds of plants, from orchids and African violets to coastal redwoods. It is also possible to alter the genetic composition of a cell while it is in culture and then have these changes expressed in the whole plant during regeneration. This provides a valuable tool to genetic engineers who wish to introduce desirable new traits into crop species (see Chapter 14). Research involving plant cell and tissue culture is one of the exciting and fruitful areas of biological research today.

5 mm

Callus   Embryos   Cotyledons

Tiny embryoids bud from callus. Compare these embryoids with the one depicted in Fig. 35–9d. *(Courtesy of Dennis Gray, University of Florida)*

embryo was shown to trigger the synthesis of the enzyme *α*-amylase, which digests starch in the endosperm. As a result, glucose is available for absorption by the embryo. Although the formation of enzymes to mobilize starch reserves occurs in many types of seeds, control of enzyme formation by gibberellin appears to be restricted to seeds of cereals and other grasses. In addition to mobilizing food reserves in newly germinated grass seeds, the application of gibberellins can substitute for low-temperature or light requirements for germination in plants such as lettuce, oats, and tobacco.

Cytokinins promote cell division

During the 1940s and 1950s, researchers were trying to find substances that might induce plant cells to divide in **tissue culture**, a technique in which cells are isolated from plants and grown in a nutrient medium (see *Focus On: Cell and Tissue Culture*). It was discovered that cells would not divide without a substance found in coconut milk. Because coconut milk has a complex chemical composition, the division-inducing substance was not chemically identified for some time. Finally, an

Figure 36–19 Auxin-cytokinin interactions in tobacco tissue culture. Varying amounts of auxin and cytokinin in the culture media produce different growth responses. *(a)* The initial explant is a small piece of sterile tissue from the pith of a tobacco stem, which is placed on a nutrient agar medium. *(b)* Nutrient agar containing 2.0 mg of auxin per liter and 0.2 mg of cytokinin per liter caused cells to divide and form a clump of undifferentiated tissue called a callus. *(c)* Root growth is stimulated by a medium with a high ratio of auxin to cytokinin. *(d)* Shoot growth is stimulated by a medium containing a low ratio of auxin to cytokinin.

active substance was isolated from aged DNA from herring sperm. It was called **cytokinin** because it induces cell division, or cytokinesis. In 1963 the first naturally occurring cytokinin, zeatin, was identified from corn, and since that time similar molecules have been identified from other plants; several synthetic cytokinins have also been synthesized. Little is known, however, about the natural biosynthesis of cytokinin, which is structurally similar to adenine (a part of DNA and RNA molecules), or about its mechanism of action.

Cytokinins promote cell division and differentiation of young, relatively unspecialized cells into mature, more specialized cells in intact plants. They are also a required ingredient in any plant tissue culture medium and must be present in order for cells to divide. In tissue culture, cytokinins interact with auxin during the formation of plant organs such as roots and stems (Fig. 36–19). For example, in tobacco tissue culture, a low ratio of auxin to cytokinin induces shoots to form, whereas a high ratio induces roots.

Cytokinins and auxin also interact in the control of apical dominance. Here their relationship is antagonistic: auxin inhibits the growth of axillary buds, and cytokinin promotes their growth.

One very interesting effect of cytokinins on plant cells is to delay the aging process, known as **senescence**. Plant cells, like all living cells, go through a natural aging process. This process is accelerated in cells of plant parts that are cut, such as flower stems. It is thought that plants must have a continual supply of cytokinins from the roots. Cut stems, of course, lose their source of cytokinins and therefore age and die rapidly.

Figure 36–20 Cytokinin synthesis and delay of senescence. The tobacco *(Nicotiana tabacum)* plant on the left was genetically engineered to produce additional cytokinin as it aged, whereas the tobacco plant of the same age on the right was not genetically engineered and served as a control. Note the extensive senescence and death of older leaves on the control plant. *(Courtesy of Dr. Richard M. Amasino, University of Wisconsin)*

When cytokinins are sprayed on leaves of a cut stem, they remain green, while any leaves that are not sprayed turn yellow and die.

Despite their involvement in many aspects of plant growth and development, cytokinins currently have few commercial applications other than plant tissue culture. However, in 1995, molecular biologists at the University of Wisconsin combined a promoter from a gene that is activated during normal senescence with a gene that encodes an enzyme involved in cytokinin synthesis. The leaves of transgenic tobacco plants that contained this recombinant DNA produced more cytokinin and therefore lived longer and continued to photosynthesize (Fig. 36–20). Some biologists think this system has the potential to increase the longevity and productivity of certain crops into which it is engineered.

Ethylene stimulates abscission and fruit ripening

During the early 20th century, scientists had observed the gas **ethylene** had several effects on plant growth, but it wasn't until 1934 that plants were demonstrated to produce ethylene. Many diverse plant processes are influenced by this natural plant hormone. Ethylene inhibits cell elongation, promotes the germination of seeds, promotes apical dominance, and is involved in plant responses to wounding or invasion by disease-causing microorganisms.

Ethylene also has a major role in many aspects of senescence, including fruit ripening. As a fruit ripens, it produces ethylene, which triggers an acceleration of the ripening process. This induces the fruit to produce more ethylene, which further accelerates ripening. The expression "One rotten apple spoils the barrel" is true. A rotten apple is one that is overripe and produces large amounts of ethylene, which diffuses and triggers the ripening process in nearby apples. Ethylene is used commercially to promote the uniform ripening of bananas. Bananas are picked while green and shipped to their destination, where they are exposed to ethylene before they are delivered to grocery stores.

Plants growing in a natural environment are subjected to rain, hail, wind, and contact with passing animals. All of these forms of mechanical stress alter the growth and development of plants, making them shorter and stockier than plants grown in a greenhouse. Such developmental responses to mechanical stimuli, known as **thigmomorphogenesis**, are regulated by ethylene. Plants that are mechanically disturbed produce additional ethylene, which in turn inhibits stem elongation and enhances cell wall thickening in supporting tissues such as collenchyma and sclerenchyma. These changes are adaptive because shorter, thicker stems are less likely to be damaged by mechanical stresses.

Ethylene has been implicated as the hormone that induces leaf abscission. However, abscission is actually influenced by two antagonistic plant hormones, ethylene and auxin. As a leaf ages (when autumn approaches, for deciduous trees in temperate climates), the level of auxin in the leaf decreases. Concurrently, cells in the abscission layer at the base of the petiole (where the leaf will break away from the stem) begin producing ethylene. To further complicate the process, cytokinins are possibly involved in abscission. Cytokinins, like auxin, decrease in concentration as leaf tissue ages.

Abscisic acid promotes bud and seed dormancy

Abscisic acid was discovered in 1963 by two different research teams. Its name is an unfortunate choice because abscisic acid is primarily involved in dormancy and does not induce abscission in most plants. Abscisic acid is an environmental stress hormone that promotes changes in plant tissues that are stressed, or exposed to unfavorable conditions such as freezing, high salt levels, and droughts. (Recall that ethylene also

affects plant responses to certain stresses, such as mechanical stress and wounding.)

The effect of abscisic acid on plants suffering from water stress is understood best. The level of abscisic acid increases dramatically in the leaves of plants that are exposed to severe drought conditions. The high level of abscisic acid in the leaves triggers the closing of stomata, which saves the water that would normally be lost by transpiration, thereby increasing the plant's likelihood of survival.

The onset of winter is also a type of stress on plants. A winter adaptation that involves abscisic acid is dormancy in seeds. Many seeds have high levels of abscisic acid in their tissues and are therefore unable to germinate until the abscisic acid washes out. In a corn mutant unable to synthesize abscisic acid, the seeds are not dormant. Instead, they germinate as soon as the embryos are mature, even while attached to the ear (Fig. 36–21).

Abscisic acid is not the only hormone involved in seed dormancy. For example, addition of gibberellin reverses the effects of dormancy. In seeds, the level of abscisic acid decreases during the winter, and the level of gibberellin increases. Cytokinins have also been implicated in breaking dormancy. Once again we see that a single physiological activity such as

Figure 36–21 Abscisic acid and seed dormancy. Inability to produce abscisic acid can prevent seed dormancy in corn *(Zea mays).* Some of the kernels have germinated while still on the ear, producing roots *(arrows). (Courtesy of M.G. Neuffer)*

TABLE 36-2   Some of the Interactions Among Plant Hormones

Physiological Activity	Auxin	Gibberellin	Cytokinin	Ethylene	Abscisic Acid	Other Factors for Some Plants
Seed germination		Promotes germination	?	Promotes germination	Inhibits germination	Cold requirement, light requirement, scarification
Growth of seedling into mature plant	Cell elongation, organogenesis,* promotes xylem and phloem differentiation	Cell division and elongation, promotes transition of embryonic root to adult	Cell division and differentiation, promotes chloroplast development, organogenesis*	Inhibits root and stem elongation		Light requirement
Apical dominance	Inhibits axillary bud development		Promotes axillary bud development	?		
Initiation of reproduction (flowering)		Stimulates flowering in some plants, can stimulate differentiation of male structures	?	Stimulates flowering in some plants, can stimulate differentiation of female structures		Cold requirement, photoperiod requirement
Fruit development and ripening	Fruit development, delays ripening	Fruit development		Promotes ripening		Light requirement (for pigment formation)
Leaf senescence and abscission	Inhibits abscission at moderate concentrations		Inhibits senescence and abscission	Promotes senescence and abscission		Light requirement
Winter dormancy of plant		Breaks dormancy	?	Promotes dormancy	?	Light requirement
Seed dormancy		Breaks dormancy	Breaks dormancy		Promotes dormancy	

*In plant tissue culture.

seed dormancy may be controlled in plants by the interaction of several hormones. The plant's actual response may result from changing ratios of hormones rather than the effect of each individual hormone. Table 36–2 summarizes some of the interactions among the five main groups of plant hormones during growth and development.

OTHER CHEMICALS HAVE BEEN IMPLICATED IN PLANT GROWTH AND DEVELOPMENT

In addition to the five major classes of plant hormones, several chemicals are implicated in certain specific aspects of plant growth and development. Four such chemical regulators are polyamines, polypeptides, oligosaccharins, and salicylic acid. Not much is currently known about how these chemicals function at the molecular level, but each is the focus of ongoing research.

Polyamines, organic molecules with two or more amino ($-NH_2$) groups, affect a variety of physiological activities such as cell division, fruit development, and senescence. Polyamines are not considered plant hormones because (1) they are present in high concentrations (about 1000 times greater than hormones such as auxin), and (2) they are not transported extensively through the plant. Polyamines may function by influencing gene expression; they appear to increase both transcription of DNA and translation of mRNA.

Although many animal hormones are known to be polypeptides, the first *plant polypeptide* with hormonal properties was not isolated until 1991. Known as *systemin*, this polypeptide is transported systemically throughout the plant in response to wounding by insects. It appears to stimulate a natural defense mechanism at extremely low concentrations, as low as one part per trillion. Systemin may trigger the plant to produce protease inhibitors, molecules that disrupt insect digestion, thereby curbing leaf damage done by caterpillars and other herbivorous insects. The discovery of systemin caused a flurry of research in search of additional polypeptide regulators in plants, and several additional plant polypeptides were discovered in the mid-1990s.

Oligosaccharins are cell-wall carbohydrate fragments consisting of short, branched chains of sugar residues. They are present in extremely small quantities in cells and active at much lower concentrations (100 to 1000 times lower) than hormones such as auxin. Different oligosaccharins appear to have distinct functions. Some trigger the production of *phytoalexins* (from the Greek *phyto*, "plant", and *alexi*, "to ward off"), antibiotics that kill plant pathogens, particularly fungi. Other oligosaccharins inhibit flowering and induce vegetative growth. One oligosaccharin has been implicated in a negative feedback loop that regulates the effects of auxin; the presence of auxin apparently stimulates synthesis of this oligosaccharin, which in turn lessens the effect of auxin on cell elongation.

Salicylic acid was first extracted from willow (*Salix* sp.) bark and is the active ingredient in aspirin (acetylsalicylic acid). More recently, salicylic acid has been shown to help defend plants against insect pests and pathogens such as viruses. When a plant is under attack, the concentration of salicylic acid increases, and it spreads systemically throughout the plant. It is thought that salicylic acid binds to a cell receptor, thereby switching on genes that code for proteins that fight infection and promote wound healing. In 1993 the receptor to which salicylic acid binds was identified as catalase, an enzyme that catalyzes the degradation of hydrogen peroxide in eukaryotic cells. The binding of salicylic acid to catalase apparently starts a chain of events that, while still hypothetical, is supported by current data:

Salicylic acid binds to catalase → catalase is inhibited → hydrogen peroxide level increases in the cell → specific proteins (transcription factors?) become activated → specific genes are turned on and expressed → natural defense response occurs.

An additional role of salicylic acid was reported in 1997 by Vladimir Shulaev and colleagues at Rutgers University. They found that tobacco plants infected with tobacco mosaic virus released into the air a volatile form of salicylic acid, known as methyl salicylate or oil of wintergreen. When nearby healthy plants received the airborne chemical signal, they began synthesizing antiviral proteins that enhanced their resistance to the virus.

Although steroid hormones are known to have crucial roles in animals, their roles in plants have been unclear. A team of biologists at the Salk Institute reported in 1996 that they identified *Arabidopsis* plants with genetic mutations in the gene that codes for an enzyme involved in steroid synthesis. Interestingly, this enzyme is similar in sequence to one involved in the synthesis of the male hormone testosterone in mammals. The mutant plants had growth defects in certain light-regulated developmental steps, but these defects were reversed by the application of the plant steroid *brassinolide*. It therefore appears that brassinolide, long suspected of being a plant hormone, may be involved in the signal transduction pathway of light-mediated development.

Florigen is a hypothetical hormone

Experiments in which different tobacco species are grafted together indicate that both flower-promoting and flower-inhibiting substances may exist. *Nicotiana silvestris* is a long-day tobacco plant, and a variety of *N. tabacum* is a day-neutral tobacco plant. When a long-day tobacco is grafted to a day-neutral tobacco and exposed to short nights, both plants flower (Fig. 36–22). The day-neutral tobacco plant flowers

Graft

Long-day induction

Day-neutral plant grafted to long-day plant

Both plants flower

Figure 36–22 **Evidence for the existence of a flower-promoting substance.** When a long-day tobacco (*Nicotiana silvestris*) is grafted to a day-neutral tobacco (*N. tabacum*) and both plants are exposed to a long-day/short-night regimen, they both flower. The day-neutral plant flowers sooner than it normally would, presumably because a flower-promoting substance passes from the long-day plant to the day-neutral one through the graft.

sooner than it normally would. It has been suggested that a flower-promoting substance may be induced in the long-day tobacco and transported to the day-neutral tobacco through the graft union, causing the day-neutral tobacco to flower sooner than expected. The hypothetical flower-promoting substance is known as **florigen**.

When a long-day tobacco is grafted to a day-neutral tobacco and exposed to long nights, neither plant flowers. As long as these conditions continue, the day-neutral plants do not flower even when they would normally do so. In this case, it has been suggested that the long-day tobacco may produce a flower-inhibiting substance that is transported to the day-neutral tobacco through the graft union, preventing the day-neutral tobacco from flowering. The many attempts that have been made to isolate and chemically characterize substances that are clearly identifiable as flower promoters (florigen) or flower inhibitors have been unsuccessful.

SUMMARY WITH KEY TERMS

I. The location of a cell in the young plant body, that is, each cell's spatial environment, affects gene expression during development by causing some genes in that cell to be turned off and others to be turned on.

II. **Germination** is affected by both internal and external factors.
   A. External environmental factors that may affect germination include requirements for oxygen, water, temperature, and light.
   B. Internal factors affecting whether or not a seed germinates include the maturity of the embryo, the presence or absence of chemical inhibitors, and the presence or absence of hard, thick seed coats.

III. Plant growth and development is controlled not only by internal genetic factors but also by factors, such as changing day length, in the physical environment.

IV. Many plants flower in response to specific cues from the environment.
   A. **Photoperiodism** is the response of plants to the duration and timing of light and dark.
      1. In many plants, flowering is a photoperiodic response; some are **short-day plants**, some are **long-day plants**, and others are **intermediate-day plants**. In **day-neutral plants**, flowering is not affected by photoperiod.
      2. The photoreceptor in photoperiodism is **phytochrome**, a blue-green pigment with two forms, P_r and P_{fr}, named by the wavelength of light they absorb.
   B. **Vernalization** is the promotion of flowering by exposure of seeds or stem apical meristems to low temperatures.

V. **Circadian rhythms** are regular periodicities in growth or activities of a plant or other organism that approximate the 24-hour day and are reset by the rising and setting of the sun.

VI. Nastic movements and tropisms are the two kinds of plant movements that occur in response to external stimuli.
   A. **Nastic movements**, which are temporary and reversible, occur in response to external stimuli, but the direction of movement is independent of the direction of the stimulus.
   B. **Tropisms** are directional growth responses.
      1. **Phototropism** is plant growth in response to the direction of light.
      2. **Gravitropism** is plant growth in response to the influence of gravity.
      3. **Thigmotropism** is plant growth in response to contact with a solid object.
      4. **Heliotropism** is the ability of leaves or flowers to track the sun across the sky.

VII. Plants respond to **hormones**, chemical messengers that regulate plant growth and development.
   A. Hormones are effective in extremely small concentrations.
   B. The functions of some plant hormones overlap.
   C. Many physiological activities of plants may be due to the interactions of several hormones rather than to the effect of a single hormone.
   D. There are five major classes of plant hormones.
      1. **Auxin** is involved in cell elongation, tropisms, **apical dominance**, and fruit development.
      2. **Gibberellins** are involved in stem elongation, flowering, and germination.
      3. **Cytokinins** promote cell division and differentiation, delay **senescence**, and interact with auxin and ethylene in apical dominance.
      4. **Ethylene** has a role in the ripening of fruits, apical dominance, leaf abscission, wound response, **thigmomorphogenesis**, and senescence.
      5. **Abscisic acid** is an environmental stress hormone involved in seed dormancy and in stomatal closure due to water stress.
   E. Several other chemicals are implicated in specific aspects of plant growth and development.
      1. Polyamines are organic molecules with two or more amino ($-NH_2$) groups that affect a variety of plant processes, possibly by influencing gene expression.
      2. Systemin, a plant polypeptide with hormonal properties, stimulates a natural defense mechanism in which the plant produces molecules that disrupt insect digestion.
      3. Oligosaccharins, sugar residues that are fragments of cell wall carbohydrates, have a variety of functions and exert their effects by binding to membrane receptors and altering gene expression.
      4. Salicylic acid helps defend plants against pathogens and insect pests. It may bind to a cell receptor, thereby switching on genes that fight infection and promote wound healing. A volatile form of salicylic acid, known as methyl salicylate, may serve as an airborne chemical signal from virus-infected plants to healthy ones.
      5. The plant steroid brassinolide may be involved in the signal transduction pathway of light-mediated development.
      6. Grafting experiments indicate the possible existence of both flower-promoting (**florigen**) and flower-inhibiting substances, neither of which have been successfully isolated.

POST-TEST

1. A plant's response to the relative amounts of daylight and darkness is known as (a) apical dominance (b) bolting (c) gravitropism (d) photoperiodism (e) phototropism

2. P_{fr}, the active form of _____, is formed when red light is absorbed. (a) gibberellin (b) phytochrome (c) far-red light (d) phototropism (e) cytokinin

3. The promotion of flowering by the exposure of seeds or stem apical meristems to a low-temperature treatment is known as (a) thigomo-

morphogenesis (b) bolting (c) gravitropism (d) photoperiodism (e) vernalization

4. In _____, leaves or other plant organs track the sun's movement across the sky. (a) heliotropism (b) thigmotropism (c) phototropism (d) bolting (e) photoperiodism

5. The growth of a plant due to the direction of light is called _____ , whereas the twining of tendrils is an example of _____ (a) heliotropism; gravitropism (b) photoperiodism; nastic movements (c) phototropism; gravitropism (d) phototropism; thigmotropism (e) photoperiodism; thigmotropism

6. A synthetic _____ known as 2,4-D is used as a selective herbicide. (a) auxin (b) gibberellin (c) cytokinin (d) ethylene (e) abscisic acid

7. Research on a fungal disease of rice provided the first clues about this plant hormone. (a) auxin (b) gibberellin (c) cytokinin (d) ethylene (e) abscisic acid

8. This plant hormone interacts with auxin during the formation of plant organs in tissue culture. (a) florigen (b) gibberellin (c) cytokinin (d) ethylene (e) abscisic acid

9. The plant hormone _____ delays senescence, whereas the plant hormone _____ promotes it. (a) cytokinin; auxin (b) auxin; cytokinin (c) cytokinin; ethylene (d) abscisic acid; ethylene (e) gibberellin; auxin

10. The stress hormone that helps plants respond to drought is (a) auxin (b) gibberellin (c) cytokinin (d) ethylene (e) abscisic acid

11. This hormone promotes seed dormancy. (a) auxin (b) gibberellin (c) cytokinin (d) ethylene (e) abscisic acid

REVIEW QUESTIONS

1. What factors influence the germination of seeds? Explain the role that each of these factors plays in germination and discuss why the germinating seed responds the way it does.

2. Why are plant growth and development so sensitive to environmental cues?

3. What is phytochrome? Describe its role in flowering. List several other roles of phytochrome.

4. Distinguish between photoperiodism and circadian rhythms.

5. Distinguish between nastic movements and tropisms.

6. How is auxin involved in phototropism?

7. Discuss the interactions of various hormones involved in each of the following physiological processes: (a) seed germination, (b) stem elongation, (c) fruit ripening, (d) leaf abscission, (e) seed dormancy.

YOU MAKE THE CONNECTION

1. Predict whether flowering would be expected to occur in the following situations. Explain each answer. (a) A short-day plant is exposed to 15 hours of daylight and 9 hours of darkness. (b) A short-day plant is exposed to 9 hours of daylight and 15 hours of darkness. (c) A short-day plant is exposed to 9 hours of daylight and 15 hours of darkness, with a 10-minute exposure to red light in the middle of the night.

2. If you transplanted a short-day plant to the tropics, would it flower? Explain your answer.

3. Are sleep movements an example of a nastic movement or a tropism? Explain your answer.

4. What biological advantages are conferred on a plant whose stems are positively phototropic and whose roots are positively gravitropic?

RECOMMENDED READINGS

Berg, L.R. *Introductory Botany: Plants, People, and the Environment.* Saunders College Publishing, Philadelphia, 1997. A general botany text with an environmental emphasis.

Chapin, F.S., III. "Integrated Responses of Plants to Stress." *BioScience*, Vol. 41, Jan. 1991. Plants respond to many environmental stresses by altered growth.

Evans, M.L., R. Moore, and K.H. Hasenstein. "How Roots Respond to Gravity." *Scientific American*, Vol. 255, No. 6, Dec. 1986. A closer look at gravitropism.

Jones, M. "Keeping in Time." *New Scientist*, Vol. 149, 16 Mar. 1996. An excellent overview of circadian rhythms and internal biological clocks in diverse organisms.

"Leaves with Clocks," in "Breakthroughs" section of *Discover*, Sep. 1993. Sleep movements are another example of circadian rhythms in plants.

Mores, P.B., and N.H. Chua, "Light Switches and Plant Genes." *Scientific American*, Vol. 258, No. 4, Apr. 1988. How an environmental cue is related to gene expression.

Raloff, J. "When Tomatoes See Red." *Science News*, Vol. 152, 13 Dec. 1997. Scientists have discovered that using different colors of plastic mulches to cover the soil results in different plant growth responses.

- Visit our website at **http://saunderscollege.com/lifesci/titles.html** and click on Solomon/Berg/Martin Biology for links to chapter-related resources on the World Wide Web.

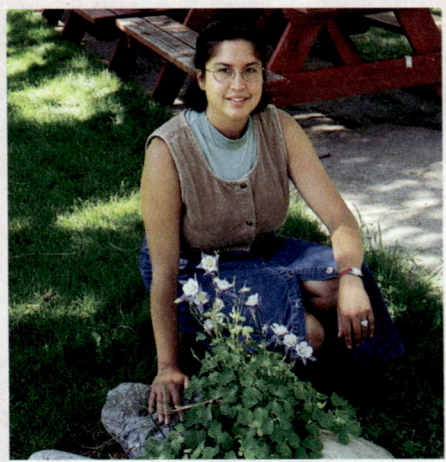

La Titia Taylor is the administrative head of the Agricultural Division in the Natural Resources Department for the Southern Ute Indian Tribe. She grew up on the Southwest Colorado Ute reservation and is one of 1300 tribal members. La Titia received her B.S. in Biology from the University of Central Oklahoma in 1989, and her Master's Degree in Botany in 1992 from Southwest Texas State University. Her intention was to become a research botanist. However, economic circumstances drew her back to her family in the Southern Ute Indian Tribe. There she became involved with an agricultural extension program that educates tribal members in agriculture and horticulture. She was appointed division head two years ago. La Titia oversees the soil conservation program, the noxious weed program, a drought relief program, and a custom farm program. She also runs the Youth Garden, a 4-H program for children offered in the summertime.

What do you do in your position?

My current position as Agriculture Division Head of the Natural Resources Department is an administrative one. I have 15 employees in the winter but 22 in the summer, due to the summer months' custom farm program. The custom farm program provides a service to the tribal families at a reduced cost for agriculture and related practices such as plowing fields, baling hay, cutting irrigation ditches for those members who can't afford the farming equipment. I oversee at least three additional programs. I also run the 4-H youth garden.

Agricultural Director

LA TITIA TAYLOR

Tell me about your undergraduate education.

I was a troubled youth, into drugs and alcohol. When I graduated from high school I was reading at the fifth grade level. In college I turned my life around. My education was paid for by the tribe. When I was a kid, I loved looking at plants so I decided to study biology, more specifically, botany. After two years at Christian College (now called Oklahoma Christian University), I transferred to the University of Central Oklahoma. I studied under Dr. Darrin Keck at Christian College and Professor Joe Vaughn at the University of Central Oklahoma.

Did you go right away to graduate school at Southwest Texas State?

Yes I did, on a recommendation from Prof. Vaughn. He introduced me to Dr. Edward Schneider, who was the Dean of Science at Southwest Texas State University. For my assistantship as a Master's candidate, I helped Dr. Schneider with his water lily collection. During my studies for my Master's thesis on the four species of the *Barclaya*, a water lily, I was also mentored by Dr. Paula Williamson.

What kinds of crops in particular do the Southern Utes grow?

We emphasize crops for cattle and horses, like alfalfa, hay, and grass hay. I have been telling our people for five years that specialty crops are important. We need to start looking into vegetables, crops for human consumption. We live in Durango, where there are many wealthy people who are interested in herbs and crops grown by Native Americans. Expensive plants like asparagus grow well here. It could be quite a successful tribal enterprise. I think some of the Tribal Council members are starting to listen to me.

Is drought a concern?

We had a problem with drought a few years ago. It killed many grass pastures. It was too dry in the winter. In the winter months, if there is enough moisture in the soil, the water actually keeps the soil warmer. However, if there is just dry soil, then the roots will freeze and die. This is what happened in our pastures. So, we implemented the drought relief program to help the tribal members. We have about 60 tribal families that are interested in agriculture because they have the farms and ranches. During the drought, about half of them accessed a $100,000 cost-share program. We cost-shared on feed and vitamins.

What other kinds of programs have you implemented?

I now manage the soil conservation program. We also have a noxious weed program. Our custom farm program operation is under the Agriculture Division, too. We are trying to make custom farming more of an enterprise, where it's going to make money. In the noxious weed program, we are trying to control most weeds by providing tribal members with herbicides, spraying equipment, and information on their proper use. We would like to "eradicate" the weeds, but the best you can do is control them. There are about three species of knap weed. There are also toxic weeds, such as water hemlock, in the waterways. They are often ingested by the livestock. The noxious weed program is very successful.

Do you interact with federal government departments?

We have an EPA (Environmental Protection Agency) office, which the tribe funds along with EPA grants. Primarily they do air and water quality control through air, water, and soil sampling. We also have a USDA program, the Natural Resources Conservation Service. We pushed for this representation and got it. We work closely with the conservationist in that office.

What do you hope your work will give to the tribe?

I hope the children of our tribe will become interested in their own education, especially in agriculture, biology, and other natural resource sciences. The tribal members need to become more aware of their natural resources and to preserve the land for its beauty and for agricultural practices, so that it's not taken over by weeds, or turned into oil and gas sites.

Structure and Life Processes in Animals

The leopard *(Pantera pardus)*, a large, spotted cat with a heavy build and short powerful limbs, is a formidable predator that can leap out of a tree onto its prey. Animals are extremely diverse with a wide variety of size, form, and function. Each animal is adapted to its habitat and to the role it plays in the biosphere. *(Gerald Lacz/Peter Arnold, Inc.)*

CHAPTER 37

The Animal Body: Introduction to Structure and Function

Animals can grow to large sizes because they are composed of many cells. The size of a single cell is limited by the ratio of its surface area (plasma membrane) to its volume (Chapter 4). The plasma membrane needs to be large enough relative to the cell's volume to permit passage of materials into and out of the cell, thereby maintaining conditions necessary for life. In a multicellular animal, each cell has a large enough surface to volume ratio to effectively regulate its internal environment. Individual cells can live and die and be replaced while the organism continues to maintain itself and thrive.

The number of cells, not their individual sizes, is responsible for the size of an animal. The cells of the Yacare caiman *(Caiman yacare)* in the photograph and the flambeau butterfly *(Dryas julia)* on its head are all about the same size. The caiman is larger because its genes code for a larger number of cells.

Consider how different the caiman and the butterfly are, not only in size, but also in body form and lifestyle. Multicellularity permits the specialization of cells. In a unicellular organism, such as a bacterium or a protist, the single cell must carry on all the activities necessary for life. In a multicellular organism, such as the caiman, cells specialize to perform specific tasks. A **tissue** consists of a group of closely associated, similar cells that carry out specific functions. Animal tissues are classified as epithelial, connective, muscle, or nervous tissue. Classification of tissues is based on their origin as well as on their structure. Each kind of tissue (and subtissue) is composed of cells with characteristic sizes, shapes, and arrangements. Some tissues are specialized to transport materials, whereas others contract, enabling the organism to move. Still others secrete hormones that regulate metabolic processes.

Tissues associate to form **organs** such as the heart or stomach. Groups of tissues and organs form the **organ systems** of the body. In this unit we discuss how cells associate with one another and how tissues, organs, and organ systems perform specialized functions such as circulating blood, hearing, or digestion.

Cells, tissues, organs, and organ systems work together to maintain appropriate conditions in the body. The tendency to

(Fritz Polking/Dembinsky Photo Associates)

maintain a relatively constant internal environment is called **homeostasis,** and the processes that accomplish the task are **homeostatic mechanisms.** For example, in mammals nervous, endocrine, and circulatory systems work together to regulate body temperature. In the next several chapters we will discuss the organ systems of a complex animal and focus on how they work together to maintain homeostasis of the organism as a whole.

AFTER YOU HAVE STUDIED THIS CHAPTER YOU SHOULD BE ABLE TO

1. Discuss the advantages of multicellularity.
2. Define tissue, organ, and organ system.
3. Compare the general structure and function of the four principal kinds of animal tissues: epithelial, connective, muscle, and nervous tissues.
4. Describe the main types of epithelial tissue and give their functions.
5. Compare the main types of connective tissue and summarize their functions.
6. Contrast the three types of muscle tissue and their functions.
7. Identify the cells that make up nervous tissue and give their functions.
8. List and briefly describe the organ systems of a complex animal.
9. Define homeostasis and discuss how each organ system helps maintain homeostasis.
10. Describe strategies used by ectotherms and endotherms to maintain body temperature.
11. Contrast the cells of a malignant neoplasm with those of normal tissue and discuss current thinking about the mechanisms, causes, and treatment of cancer.

EPITHELIAL TISSUES COVER THE BODY AND LINE ITS CAVITIES

Epithelial tissue (also called **epithelium**) covers body surfaces and lines cavities. It forms the outer layer of the skin and the linings of the digestive, respiratory, excretory, and reproductive tracts. Epithelial tissue consists of cells fitted tightly together to form a continuous layer, or sheet, of cells. One surface of the sheet is typically exposed because it lines a cavity, such as the lumen of the intestine, or covers the body (outer layer of the skin). The other surface of an epithelial layer is attached to the underlying tissue by a noncellular **basement membrane** composed of tiny fibers and nonliving polysaccharide material produced by the epithelial cells. Table 37–1 illustrates the main types of epithelial tissue, indicates their locations in the body, and describes their functions.

Epithelial tissues perform a wide variety of functions, including protection, absorption, secretion, and sensation. The epithelial layer of the skin, the **epidermis,** covers the entire body and protects it from mechanical injury, chemicals, bacteria, and fluid loss. The epithelial tissue lining the digestive tract absorbs nutrients and water into the body. Some epithelial cells are organized into **glands** that secrete cell products like hormones, enzymes, or sweat. Other epithelial cells are specialized as sensory receptors that receive information from the environment. For example, epithelial cells in taste buds and in the nose are specialized as chemical receptors.

Everything that enters or leaves the body must cross at least one layer of epithelium. Food taken into the mouth and swallowed is not really "inside" the body until it is absorbed through the epithelium of the gut and enters the blood. To a large extent, the permeabilities of the various epithelia regulate the exchange of substances between the different parts of the body, as well as between the organism and the external environment.

Many epithelial membranes, such as the epidermis of the skin, are subjected to continuous wear and tear. As outer cells are sloughed off, they must be replaced by new ones from below. Such epithelial tissues generally have a rapid rate of cell division that continuously produces new cells to replace those lost.

Three types of epithelial cells can be distinguished on the basis of shape (Table 37–1). **Squamous** epithelial cells are thin, flattened cells shaped like flagstones. **Cuboidal** epithelial cells are short cylinders that from the side appear cube-shaped, resembling dice. Actually, each cell is typically hexagonal in cross-section, making it an eight-sided polyhedron. **Columnar** epithelial cells look like columns or cylinders when viewed from the side. The nucleus is usually located near the base of the cell. Viewed from above or in cross section, these cells often appear hexagonal. On its free surface, a columnar epithelial cell may have cilia that beat in a coordinated way, moving materials over the tissue surface. Most of the upper respiratory tract is lined with ciliated columnar epithelium that moves particles of dust and other foreign material away from the lungs.

Epithelial tissue may be simple, stratified, or pseudostratified. **Simple epithelium** is composed of one layer of cells. It is usually located where substances are secreted, excreted, or absorbed or where materials diffuse between compartments, for example, lining the kidney tubules. **Stratified** epithelium, composed of two or more layers, is found where protection is required. For example, it makes up the outer layer of the skin and lines the mouth of humans and other vertebrates. The cells of **pseudostratified epithelium** falsely appear to be layered. Although all of its cells rest on a basement membrane, not every cell extends to the exposed surface of the tissue. This arrangement gives the impression of two or more cell layers. Some of the respiratory passageways are lined with pseudostratified epithelium equipped with cilia.

The linings of blood and lymph vessels have a different embryonic origin than "true" epithelium, and they are referred to as **endothelium.** Structurally though, the cells of these linings look like typical epithelial cells.

A gland consists of one or more epithelial cells specialized to produce and secrete a product such as sweat, milk, mucus, wax, saliva, hormones, or enzymes (Fig. 37–1). Epithelial tis-

TABLE 37-1   Epithelial Tissues

Nuclei

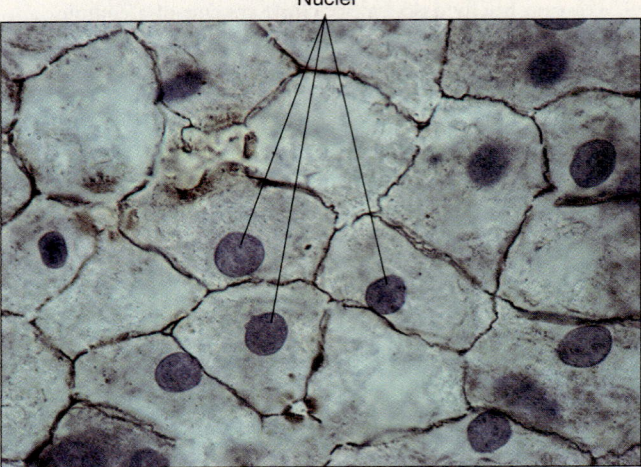

LM of simple squamous epithelium.   *(Ed Reschke)*

Simple squamous epithelium

Main Locations
Air sacs of lungs; lining of blood vessels

Functions
Passage of materials where little or no protection is needed and where diffusion is major form of transport

Description and Comments
Cells are flat and arranged as single layer

25 µm

Nuclei of cuboidal epithelial cells   Lumen of tubule

LM of simple cuboidal epithelium.   *(Ed Reschke)*

Simple cuboidal epithelium

Main Locations
Linings of kidney tubules; gland ducts

Functions
Secretion and absorption

Description and Comments
Single layer of cells; photograph shows cross section through tubules; from the side each cell looks like short cylinder; some have microvilli for absorption

25 µm

Goblet cell   Nuclei of columnar cells

LM of simple columnar epithelium.   *(Ed Reschke)*

Simple columnar epithelium

Main Locations
Linings of much of digestive tract and upper part of respiratory tract

Functions
Secretion, especially of mucus; absorption; protection; movement of mucous layer

Description and Comments
Single layer of columnar cells; sometimes with enclosed secretory vesicles (in goblet cells); highly developed Golgi complex; often ciliated

25 µm

TABLE 37-1 *continued*

LM of stratified squamous epithelium. *(Ed Reschke)*

Stratified squamous epithelium

Main Locations

Skin; mouth lining; vaginal lining

Functions

Protection only; little or no absorption or transit of materials; outer layer continuously sloughed off and replaced from below

Description and Comments

Several layers of cells, with only the lower ones columnar and metabolically active; division of lower cells causes older ones to be pushed upward toward surface, becoming flatter as they move

50 μm

Basement membrane   Cilia

LM of pseudostratified columnar epithelium, ciliated. *(Ed Reschke)*

Pseudostratified epithelium

Main Locations

Some respiratory passages; ducts of many glands

Functions

Secretion; protection; movement of mucus

Description and Comments

Ciliated, mucus-secreting, or with microvilli; comparable in many ways to columnar epithelium except that not all cells are the same height. Thus, though all cells contact the same basement membrane, the tissue appears stratified

25 μm

sue lining the cavities and passageways of the body typically contains some specialized mucus-secreting cells called **goblet cells.** The mucus lubricates these surfaces and facilitates the movement of materials.

Glands can be classified as exocrine or endocrine. **Exocrine glands,** like goblet cells and sweat glands, secrete their products onto a free epithelial surface, typically through a duct (tube). All **endocrine glands** lack ducts. Endocrine glands release their products, called **hormones,** into the **interstitial fluid** (tissue fluid) or blood; hormones may be transported to other parts of the body by the circulatory system. (Endocrine glands are discussed in Chapter 47.)

An **epithelial membrane** consists of a sheet of epithelial tissue and a layer of underlying connective tissue. Types of ep-

ithelial membranes include mucous membranes and serous membranes. A **mucous membrane,** or mucosa, lines a body cavity that opens to the outside of the body, such as the digestive or respiratory tract. The epithelial layer secretes mucus that lubricates the tissue and protects it from drying.

A **serous membrane** lines a body cavity that does not open to the outside of the body. It consists of simple squamous epithelium over a thin layer of loose connective tissue. This type of membrane secretes fluid into the cavity it lines. Some serous membranes with which you may be familiar are the pleural membranes lining the pleural cavities around the lungs, the pericardial membranes lining the pericardial cavity around the heart, and the peritoneum lining the abdominal cavity.

Cilia

Unicellular glands (goblet cells)

Basement membrane

(a) Goblet cells

Skin

(b) Sweat gland

(c) Compound gland

Figure 37–1 **Glands.** A gland consists of one or more epithelial cells. *(a)* Goblet cells are unicellular glands that secrete mucus. *(b)* Sweat glands are simple tubular glands with coiled tubes. Their walls are constructed of simple cuboidal epithelium. *(c)* Compound glands, such as the parotid salivary glands, have branched ducts.

CONNECTIVE TISSUES JOIN AND SUPPORT OTHER BODY STRUCTURES

Almost every organ in the body has a framework of **connective tissue** that supports and cushions it. Cartilage and bone are examples of connective tissues that support the vertebrate body and protect organs such as the heart and lungs. Blood is a connective tissue that transports materials, thereby connecting distant parts of the body. Adipose tissue provides a protective cushion and stores fat.

There are many kinds of connective tissues and many systems for classifying them. Some of the main types of connective tissue are: (1) loose and dense connective tissues; (2) elastic connective tissue; (3) reticular connective tissue; (4) adipose tissue; (5) cartilage; (6) bone; and (7) blood, lymph, and tissues that produce blood cells. The tissues vary widely in their structural details and in the functions they perform (Table 37–2 on pp. 792–793).

Typically, connective tissues contain relatively few cells; these are embedded in an extensive **intercellular substance** consisting of threadlike, microscopic **fibers** scattered throughout a **matrix,** a thin gel composed of polysaccharides secreted by the cells. The cells of different kinds of connective tissues differ in their shapes and structures and in the kinds of fibers and matrices they secrete. The nature and function of each kind of connective tissue are determined in part by the structure and properties of the intercellular substance.

Connective tissue contains collagen, elastic, and reticular fibers

Connective tissue typically contains three types of fibers: collagen, elastic, and reticular. **Collagen fibers,** the most numerous type, are composed of collagens, the most abundant proteins in the body. Collagen is a very tough material. (Meat is tough because of its collagen content.) The tensile strength (ability to be stretched without tearing) of collagen fibers is comparable to that of steel. Collagen fibers are wavy and flexible, allowing them to remain intact when tissue is stretched. When treated with hot water, collagen is converted into gelatin, a soluble protein.

Elastic fibers branch and fuse to form networks. They can be stretched by a force and then (like a stretched rubber band) return to their original size and shape when the force is removed. Elastic fibers, composed of the protein elastin, are an important component of structures that must stretch.

Reticular fibers are very small branched fibers that form delicate networks not visible in ordinary stained slides; however, they become apparent when tissue is stained with silver. Reticular fibers are composed of collagen and some glycoprotein. The framework of many organs such as the liver and lymph nodes consists of reticular fibers.

Connective tissue contains specialized cells

Fibroblasts are connective tissue cells that produce the fibers, as well as the protein and carbohydrate complexes, of the matrix. Fibroblasts release protein components that become arranged to form the characteristic fibers. These cells are especially active in developing tissue and are important in healing wounds. As tissues mature, the number of fibroblasts decreases and they become less active.

Macrophages, the scavenger cells of the body, commonly wander through connective tissues, cleaning up cellular debris and phagocytizing foreign matter, including bacteria. Among the other types of cells seen in connective tissues are *adipose* (fat) *cells; mast cells,* which release histamine during allergic reactions; and *plasma cells,* which secrete antibodies.

Loose connective tissue is widely distributed

Loose connective tissue (also called *areolar tissue*) is the most widely distributed connective tissue in the body. Found as a thin filling between body parts, it serves as a reservoir for fluid and salts. Nerves, blood vessels, and muscles are wrapped in this tissue. Together with adipose tissue, loose connective tissue forms the subcutaneous (below the skin) layer that attaches skin to the muscles and other structures beneath. Loose connective tissue consists of fibers running in all directions though a semifluid matrix. Its flexibility permits the parts it connects to move.

Dense connective tissue consists mainly of fibers

Dense connective tissue, found in the lower layer (dermis) of the skin, is very strong, though somewhat less flexible than loose connective tissue. Collagen fibers predominate. In **irregular dense connective tissue,** the collagen fibers are arranged in bundles distributed in all directions through the tissue.

In **regular dense connective tissue,** collagen bundles are arranged in a definite pattern, making the tissue greatly resistant to stress. Tendons, the cords that connect muscles to bones, and ligaments, the cables that connect bones to one another, consist of regular dense connective tissue.

Elastic tissue is found in structures that must expand

Elastic connective tissue consists mainly of bundles of parallel elastic fibers. This tissue type is found in structures that must expand and then return to their original size, such as lung tissue and the walls of large arteries.

Reticular connective tissue provides support

Reticular connective tissue is composed mainly of interlacing reticular fibers. It forms a supporting framework in many organs, including the liver, spleen, and lymph nodes.

Adipose tissue stores energy

Adipose tissue is rich with fat cells that store fat and release it when fuel is needed for cellular respiration. Adipose tissue is found in the subcutaneous layer and in tissue that cushions internal organs. An immature fat cell is somewhat star-shaped. As fat droplets accumulate within the cytoplasm, the cell assumes a more rounded appearance (Fig. 37–2). Fat droplets merge with one another, eventually forming a single large drop of fat that occupies most of the volume of the mature fat-storing cell. The cytoplasm and its organelles are pushed to the cell edges, where a bulge is typically formed by the nucleus. In cross section a fat cell looks like a ring (the cytoplasm) with a single gemstone (the nucleus).

When you study a section of adipose tissue through a microscope, it may remind you of chicken wire. The rings of cytoplasm are the "wire," and the large spaces indicate where fat drops existed before they were dissolved by chemicals used to prepare the tissue. These spaces may cause the cells to collapse, giving the tissue a wrinkled appearance.

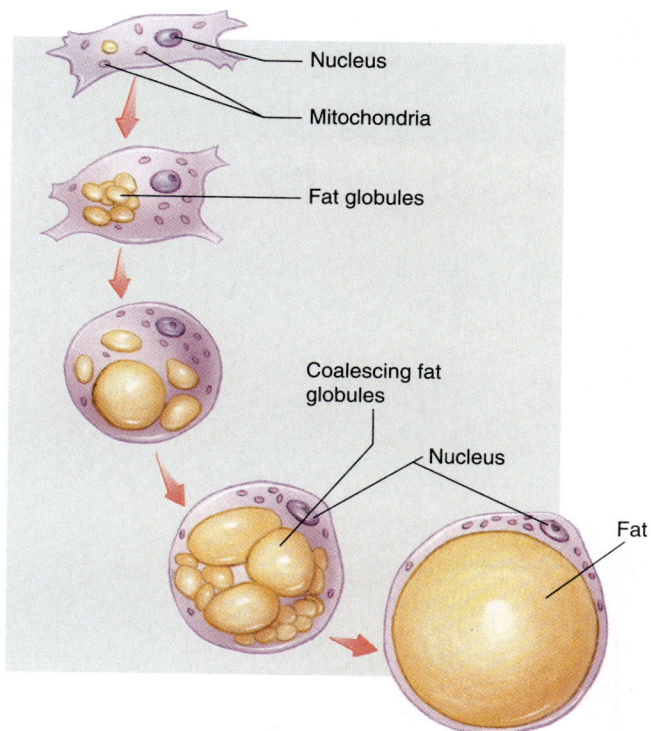

Figure 37–2 **Development of a fat cell.** As fat droplets accumulate in the cytoplasm, they coalesce to form a very large globule of fat. Such a fat globule may occupy most of the cell, pushing the cytoplasm and the organelles to the periphery. Fat cells make up adipose tissue. See Table 37–2 for a LM of fat cells filled with fat.

Cartilage and bone provide support

The supporting skeleton of a vertebrate is composed of cartilage, or of both cartilage and bone. Recall that **cartilage** is the supporting skeleton in the embryonic stages of all vertebrates, but is largely replaced in the adult by bone in all but class Chondrichthyes (sharks and rays). In humans, cartilage forms the supporting structure of the external ear, the supporting rings in the walls of the respiratory passageways, the tip of the nose, the ends of some bones, and the discs that serve as cushions between our vertebrae.

Cartilage is firm yet elastic. Its cells, called **chondrocytes,** secrete a hard, rubbery matrix around themselves. Chondrocytes also secrete collagen fibers, which become embedded in the matrix and strengthen it. Chondrocytes eventually come to lie, singly or in groups of two or four, in small cavities in the matrix called **lacunae** (Table 37–2). Chondrocytes remain alive and are nourished by nutrients and oxygen that diffuse through the matrix. Cartilage tissue lacks nerves, lymph vessels, and blood vessels.

Bone, the main vertebrate skeletal tissue, is similar to cartilage in that it consists mostly of matrix material containing lacunae. The bone cells, called **osteocytes,** secrete and main-

(Text continues on page 795.)

TABLE 37–2   Connective Tissues

Collagen fibers   Nuclei of fibroblasts

50 μm

LM of loose connective tissue.   *(Ed Reschke)*

Loose (areolar) connective tissue
Main Locations
Everywhere that support must be combined with elasticity, such as subcutaneous layer

Functions
Support; reservoir for fluid and salts

Description and Comments
Fibers produced by fibroblast cells embedded in semifluid matrix and mixed with miscellaneous other cells

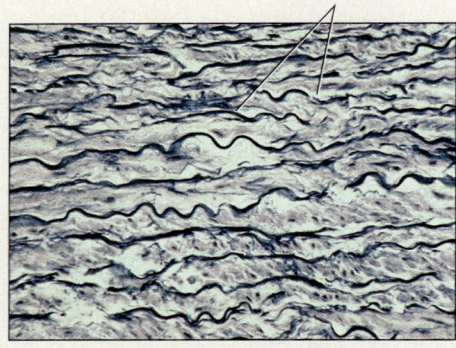

25 μm

LM of dense connective tissue.   *(Dennis Drenner)*

Dense connective tissue
Main Locations
Tendons; many ligaments; dermis of skin

Functions
Support; transmission of mechanical forces

Description and Comments
Collagen fibers may be regularly or irregularly arranged

Elastic fibers

50 μm

LM of elastic connective tissue.   *(Ed Reschke)*

Elastic connective tissue
Main Locations
Structures that must both expand and return to their original size, such as lung tissue and large arteries

Functions
Confers elasticity

Description and Comments
Branching elastic fibers interspersed with fibroblasts

Reticular fibers

50 μm

LM of reticular connective tissue.   *(Ed Reschke)*

Reticular connective tissue
Main Locations
Framework of liver; lymph nodes; spleen

Functions
Support

Description and Comments
Consists of interlacing reticular fibers

TABLE 37-2 *continued*

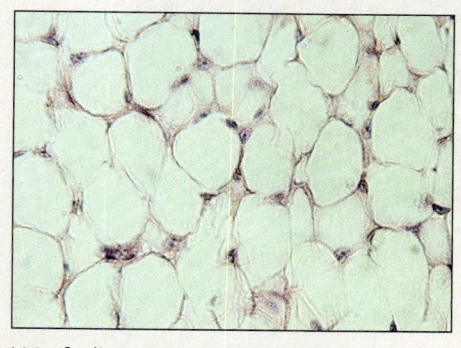

LM of adipose tissue. *(Dennis Drenner)*

Adipose tissue

Main Locations

Subcutaneous layer; pads around certain internal organs

Functions

Food storage; insulation; support of such organs as mammary glands, kidneys

Description and Comments

Fat cells are star-shaped at first; fat droplets accumulate until typical ring-shaped cells are produced

Chondrocytes   Lacuna   Intercellular substance

LM of cartilage. *(Ed Reschke)*

Cartilage

Main Locations

Supporting skeletons in sharks and rays; ends of bones in mammals, and some other vertebrates; supporting rings in walls of some respiratory tubes; tip of nose; external ear

Functions

Flexible support and reduction of friction in bearing surfaces

Description and Comments

Cells (chondrocytes) separated from one another by intercellular substance; cells occupy lacunae

Lacunae   Haversian canal   Matrix

LM of bone. *(Dennis Drenner)*

Bone

Main Locations

Forms skeletal structure in most vertebrates

Functions

Support and protection of internal organs; calcium reservoir; skeletal muscles attach to bones

Description and Comments

Osteocytes in lacunae; in compact bone, lacunae arranged in concentric circles surrounding Haversian canals

Red blood cells   White blood cells

LM of blood. *(Ed Reschke)*

Blood

Main Locations

Within heart and blood vessels of circulatory system

Functions

Transports oxygen, nutrients, wastes, and other materials

Description and Comments

Consists of cells dispersed in fluid intercellular substance

Lacunae

Haversian
canal

Matrix

Haversian
canal

(a)

(b)

Blood vessel

Cytoplasmic
extensions

Osteon

(d)

Spongy
bone

Compact
bone

Blood
vessel

(c)

Lacuna

Osteocyte

Cytoplasmic
extensions

Matrix

(e)

Figure 37–3 **Bone.** *(a)* The human skeleton consists mainly
of bone. *(b)* A bone is cut open, exposing its internal structure.
(c) Blood vessels and nerves run through the Haversian canal within
each osteon of compact bone. LM of osteon is shown above. *(d, e)*
The bone matrix is rigid and hard. Osteocytes become trapped within
lacunae but communicate with one another by way of cytoplasmic extensions that extend through tiny canals. *(Dennis Drenner)*

tain the matrix (Fig. 37–3). Unlike cartilage, however, bone is a highly vascular tissue with a substantial blood supply. Osteocytes communicate with one another and with capillaries by tiny channels (canaliculi) that contain long cytoplasmic extensions of the osteocytes.

Diffusion alone would not provide sufficient nourishment for the osteocytes. This is because bone matrix consists not only of collagen, mucopolysaccharides, and other organic materials, but also of hydroxyapatite crystals, composed mainly of calcium phosphate. Diffusion through this substance is very slow.

A typical bone has an outer layer of *compact bone* surrounding a filling of *spongy bone*. Compact bone consists of spindle-shaped units called **osteons.** Within each osteon, osteocytes are arranged in concentric layers called *lamellae,* which are formed by the matrix. In turn, the lamellae surround central microscopic channels known as **Haversian canals,** through which capillaries and nerves pass (Fig. 37–3*d*). Thus, each osteon consists of a central blood vessel, surrounded by lamellae, and osteocytes.

Bones are amazingly light and strong. Calcium salts of bone render the matrix very hard, and collagen prevents the bony matrix from being overly brittle. Most bones have a large central **marrow cavity** that may contain a spongy tissue called marrow. Yellow marrow consists mainly of fat. Red marrow is the connective tissue in which blood cells are produced.

Bone is a dynamic, living tissue that gradually changes its shape and internal architecture in response to normal growth processes and physical stress. In growth and remodeling, large multinucleated cells, called **osteoclasts,** as well as the osteocytes themselves, help sculpt bone by dissolving and removing parts of the bony substance.

Blood and lymph are circulating tissues

Blood and **lymph** are circulating tissues that help other parts of the body communicate and interact. Like other connective tissues, they consist of specialized cells dispersed in an intercellular substance. Blood and lymph are discussed in Chapter 42.

In mammals, blood is composed of red blood cells, white blood cells, and platelets suspended within **plasma,** the liquid, noncellular part of the blood. Plasma consists of water, proteins, salts, and a variety of substances such as hormones that it transports from one part of the body to another.

The **red blood cells** *(erythrocytes)* of humans and other vertebrates contain the red respiratory pigment hemoglobin, which transports oxygen. The red blood cells of most mammals are flattened, biconcave discs that lack nuclei (Table 37–2). Those of other vertebrates are oval and have nuclei. Human blood contains five main types of **white blood cells** *(leukocytes),* each with distinct size, shape, structure, and functions. The white blood cells are an important line of defense against disease-causing microorganisms. **Platelets** are small fragments broken off from large cells in the bone marrow. In complex vertebrates they play a key role in blood clotting.

MUSCLE TISSUE IS SPECIALIZED TO CONTRACT

The movements of most animals result from the contraction of the elongated, cylindrical, or spindle-shaped cells of **muscle tissue.** Each muscle cell is referred to as a **muscle fiber** because of its length. A muscle fiber contains many thin, longitudinal, parallel contractile units called **myofibrils.** Two proteins, **myosin** and **actin,** are the chief components of myofibrils, and play a key role in contraction of muscle fibers.

Vertebrates have three types of muscle tissue: smooth, skeletal, and cardiac (Table 37–3). **Smooth muscle** occurs in the walls of the digestive tract, uterus, blood vessels, and certain other internal organs. Each spindle-shaped fiber contains a single nucleus.

Skeletal muscle makes up the large muscle masses attached to the bones of the body. Skeletal muscle fibers are very long—up to 2 or 3 cm (about 1 in). Each skeletal muscle fiber has many nuclei, a consequence of its formation from the fusion of several embryonic cells. The nuclei of skeletal muscle fibers are also unusual in their position. They lie just under the plasma membrane, which frees the entire central part of the skeletal muscle fiber for the **myofibrils.** This arrangement appears to be an adaptation that increases the efficiency of contraction. (Muscle contraction will be discussed in Chapter 38.) Whereas skeletal muscle fibers are generally under voluntary control, cardiac and smooth muscle fibers are normally not regulated at will.

Light microscopy reveals that both skeletal and cardiac fibers have alternating light and dark transverse stripes, or **striations,** that change their relative sizes during contraction. Striated muscle fibers can contract rapidly but cannot remain contracted for a long period of time. They must relax and rest momentarily before contracting again.

Cardiac muscle is the main tissue of the heart. The fibers of cardiac muscle are joined end-to-end, and they branch and rejoin, forming complex networks. One or two nuclei are found within each fiber. A characteristic feature of cardiac muscle tissue is the presence of *intercalated discs,* specialized junctions where the fibers join.

NERVOUS TISSUE CONTROLS MUSCLES AND GLANDS

Nervous tissue is composed of **neurons,** cells specialized for transmitting electrochemical nerve impulses, and **glial cells,** cells that support and nourish the neurons (Fig. 37–4). Certain neurons receive signals from the external or internal environment and transmit them to the spinal cord and brain. Other neurons relay, process, or store information. Still others transmit signals from the brain and spinal cord to the muscles, glands, and other organs of the body. Neurons communicate at junctions called **synapses. A nerve** consists of a great many neurons bound together by connective tissue.

TABLE 37-3   Muscle Tissues

	Skeletal	Smooth	Cardiac
Location	Attached to skeleton	Walls of stomach, intestines, etc.	Walls of heart
Type of Control	Voluntary	Involuntary	Involuntary
Shape of Fibers	Elongated, cylindrical, blunt ends	Elongated, spindle-shaped, pointed ends	Elongated, cylindrical fibers that branch and fuse
Striations	Present	Absent	Present
Number of Nuclei per Fiber	Many	One	One or two
Position of Nuclei	Peripheral	Central	Central
Speed of Contraction	Most rapid	Slowest	Intermediate; varies
Ability to Remain Contracted	Least	Greatest	Intermediate

(a) Skeletal muscle fibers   (b) Smooth muscle fibers   (c) Cardiac muscle fibers

A typical neuron has an enlarged **cell body** containing the nucleus, and two types of cytoplasmic extensions (see Chapter 39). **Dendrites** are cytoplasmic extensions specialized for receiving impulses and transmitting them to the cell body. The single **axon** transmits impulses away from the cell body. Axons are usually long and smooth but may give off an occasional branch. They typically end in a group of fine branches.

Axons range in length from a millimeter or two to over a meter. Those extending from the spinal cord down the arm or leg in a human, for example, may be a meter or more in length.

In this chapter, we have focused on normal tissues. For a discussion of some abnormal tissues, see *Focus On: Unwelcome Tissues: Cancers.*

COMPLEX ANIMALS HAVE ORGANS AND ORGAN SYSTEMS

Although an animal organ may be composed predominantly of one type of tissue, other types are needed to provide support, protection, and a blood supply, and to allow transmission of nerve impulses. For example, the heart consists mainly of cardiac muscle tissue, but it is lined and covered by endothelium, is regulated by nervous tissue, and contains blood vessels composed of smooth muscle and connective tissue.

Several tissues and organs may work together, performing a specialized set of functions. Such an organized group of structures makes up an organ system. In complex animals we can identify ten major organ systems that work together to make up the organism (Fig. 37–5). The main organ systems of complex animals include the **integumentary, skeletal, muscle, nervous, circulatory, digestive, respiratory, urinary, endocrine,** and **reproductive systems.** Table 37–4 summarizes their principal organs and functions.

100 µm

Figure 37–4   LM of nervous tissue. Neurons transmit impulses. Glial cells support and nourish neurons. *(Ed Reschke)*

(Text continues on p. 800.)

UNWELCOME TISSUES: CANCERS

In this chapter we have focused on normal tissues. A **neoplasm** ("new growth"), or **tumor,** is an abnormal mass of cells (also see Chapter 16). A neoplasm may be benign or malignant (cancerous). A **benign** ("kind") tumor tends to grow slowly, and its cells stay together. Because benign tumors form masses with distinct borders, they can usually be removed surgically.

A **malignant** ("wicked") **neoplasm,** or **cancer,** usually grows much more rapidly and invasively than a benign tumor. In fact, two basic defects in behavior that characterize most cancer cells are rapid multiplication and abnormal relations with neighboring cells. Unlike normal cells, which respect one another's boundaries and form tissues in an orderly, organized manner, cancer cells grow helter-skelter upon one another and infiltrate normal tissues. They are apparently no longer able to receive or respond appropriately to signals from surrounding cells; communication is lacking.

When a cell that has transformed into a cancer cell multiplies, all the cells derived from it are also abnormal. Unlike the cells of benign tumors, cancer cells do not retain normal structural features. Cancers that develop from connective tissues or muscle are referred to as **sarcomas.** Those that originate in epithelial tissue are called **carcinomas.** Most human cancers are epithelial.

Death from cancer almost always results from **metastasis,** a migration of cancer cells through blood or lymph channels to distant parts of the body. Once there, they multiply, forming new malignant neoplasms that can interfere with the normal functions of the tissues being invaded. Cancer often spreads so rapidly and extensively that surgeons are unable to locate or remove all the malignant masses.

Studies suggest that many neoplasms grow to several millimeters in diameter and then enter a dormant stage, which may last for months or even years. At some point, cells of the neoplasm release a chemical substance that stimulates nearby blood vessels to develop new capillaries that grow into the abnormal mass of cells. Nourished by its new blood supply, the neoplasm may begin to grow rapidly. Newly formed blood vessels have leaky walls and so are an important route for metastasis. Malignant cells enter the blood through these walls and are transported to new sites.

Cancers appear to arise when a cell accumulates multiple mutations that act together to prevent normal control of cell division. **Proto-oncogenes** are normal genes

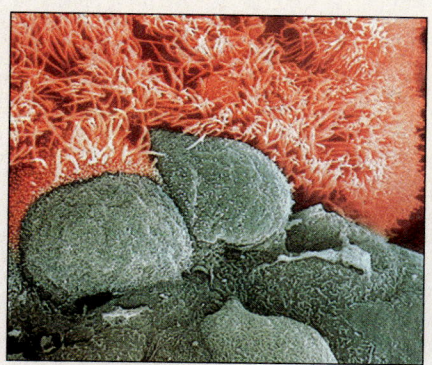

10 μm

SEM of lung cancer. Cancer cells multiply rapidly and invade normal tissues, interfering with normal function. The bronchial passageway is lined with ciliated cells. The cilia *(orange)* sweep dust and other foreign particles away from the lungs. Cancer cells *(green)* invade the bronchial wall and crowd out normal cells lining the bronchial passageway. *(Boehringer Ingelheim International GmbH, Photo by Lennart Nilsson,* The Incredible Machine, *National Geographic Society.)*

found in all cells, that are important in controlling growth and development (Chapter 16). Mutations in proto-oncogenes can convert them to **oncogenes,** which can transform normal cells into cancer cells. More than 60 oncogenes have been identified.

Mutations can also inactivate the **tumor-suppressor genes** (anti-oncogenes) that normally inhibit proliferation of transformed cells. A tumor-suppressor gene that has been extensively studied is known as **P53.** Since its discovery in 1979, more than 5000 studies have been published about this gene, which appears to prevent the cell from copying damaged DNA. Sometimes it causes mutated cells to self-destruct. Mutations in the P53 gene have been found in at least 50% of all human cancer cells. If the P53 gene is damaged or absent, the cell may become malignant. Cancers associated with P53 malfunction appear to be very aggressive, and investigators have suggested that the mutated gene may stimulate cell division instead of inhibiting it. In some cases, the P53 gene is not mutated but is somehow prevented from doing its job. In about 30% of sarcomas, a protein repressor called Mdm-2 binds with P53, rendering it inactive.

Alleles of some genes appear to affect an individual's level of tolerance to carcinogens (cancer-producing agents). More than 80% of cancer cases are thought to be trig-

gered by carcinogens in the environment.

Cancer is the second greatest cause of death in the United States. One in three persons in the United States gets cancer at some time in his or her life, and two out of three cancer patients die within five years of diagnosis. Currently, the key to survival is early diagnosis and treatment with some combination of surgery, hormonal treatment, radiation therapy, and drugs that suppress mitosis, such as chemotherapy. In patients with a functional P53 gene, radiation treatment and some forms of chemotherapy (e.g., Tamoxifen) appear to work by activating the gene. Investigators are developing several approaches to P53-based cancer treatment. For example, viruses can be used to deliver functional P53 genes into malignant neoplasms.

Because cancer is actually an entire family of closely related diseases (there are more than 100 distinct varieties), it is probable that no single cure exists.

Cancer researchers are developing a variety of new strategies for preventing metastasis and more effective treatments for cancer patients whose tumors have already metastasized. For example, a group of anti-angiogenesis compounds (such as angiostatin and endostatin) have been developed that block the development of new blood vessels that nourish neoplasms. First identified by Judah Folkman at Harvard University's Children's Hospital in Boston, these compounds have been shown to shrink cancers in mice. They are currently being clinically tested on human cancers. Another approach that appears promising involves the use of immune therapy in which customized cancer vaccines destroy neoplasms.

Risk of developing cancer can be decreased by following these recommendations:

1. Do not smoke or use tobacco. Smoking is responsible for more than 80% of lung cancer cases.
2. Avoid prolonged exposure to the sun. When in the sun, use sunscreen or sun block. Exposure to the sun is responsible for almost all of the 400,000 cases of skin cancer reported each year.
3. Increase the fiber content of your diet and avoid high-fat, smoked, salt-cured, and nitrite-cured foods.
4. Avoid unnecessary exposure to x rays.
5. Women should examine their breasts each month, have regular mammograms, and obtain annual Pap tests. Men should regularly examine their testes and should have prostate examinations yearly after age 50.

Hair

Skin

Fingernails

(1) THE INTEGUMEN-TARY SYSTEM consists of the skin and the structures such as nails and hair that are derived from it. This system protects the body, helps to regulate body temperature, and receives stimuli such as pressure, pain, and temperature.

Toenails

(2) THE SKELETAL SYSTEM consists of bones and cartilage. This system helps to support and protect the body.

(3) THE MUSCULAR SYSTEM consists of the large skeletal muscles that enable us to move, as well as the cardiac muscle of the heart and the smooth muscle of the internal organs.

Brain

Nerves

Spinal cord

(4) THE NER-VOUS SYSTEM consists of the brain, spinal cord, sense organs, and nerves. This is the principal regulatory system.

Pineal
Thyroid
Parathyroids
Thymus

Hypothalamus
Pituitary

Adrenals

Pancreas (islets)

Ovaries

Testes

(5) THE ENDO-CRINE SYSTEM consists of the ductless glands that release hormones. It works with the nervous system in regulating metabolic activities.

Arteries

Heart

Veins

(6a) THE CIRCU-LATORY SYSTEM includes the heart and blood vessels. Transports materials; defends body against disease organisms.

Figure 37-5 The ten principal organ systems of the human body.

Thymus
Thoracic duct
Lymph node
Spleen
Lymph vessels

(6b) THE LYMPHATIC SYSTEM is a subsystem of the circulatory system; it returns excess tissue fluid to the blood and defends the body against disease.

Nasal cavity
Pharynx (throat)
Lungs
Oral cavity (mouth)
Larynx (voice box)
Trachea (windpipe)
Bronchus
Diaphragm

(7) THE RESPIRATORY SYSTEM consists of the lungs and air passageways. This system supplies oxygen to the blood and excretes carbon dioxide.

Pharynx
Oral cavity
Salivary glands
Esophagus
Liver
Stomach
Gallbladder
Pancreas
Small intestine
Large intestine
Rectum
Anus

(8) THE DIGESTIVE SYSTEM consists of the digestive tract and glands that secrete digestive juices into the digestive tract. This system mechanically and enzymatically breaks down food; functions in nutrient absorption; eliminates wastes.

Kidney
Ureter
Urinary bladder
Urethra

(9) THE URINARY SYSTEM is the main excretory system of the body and helps to regulate blood chemistry. The kidneys remove wastes and excess materials from the blood and produce urine.

Uterine tube
Ovary
Uterus
Vagina

Prostate gland
Vas deferens
Penis
Testis

(10) MALE AND FEMALE REPRODUCTIVE SYSTEMS. Each reproductive system consists of gonads and associated structures. The reproductive system maintains the sexual characteristics and passes on genes to the next generation.

TABLE 37-4 The Mammalian Organ Systems and Their Functions

System	Components	Functions	Homeostatic Ability
Integumentary	Skin, hair, nails, sweat glands	Covers and protects body	Sweat glands help control body temperature; as a barrier, the skin helps maintain a steady state
Skeletal	Bones, cartilage, ligaments	Supports and protects body; provides for movement and locomotion; stores calcium	Helps maintain constant calcium level in blood
Muscular	Skeletal muscle; cardiac muscle; smooth muscle	Moves parts of skeleton; provides locomotion; moves internal materials	Ensures vital functions requiring movement, e.g., cardiac muscle circulates the blood
Digestive	Mouth, esophagus, stomach, intestines, liver, pancreas, salivary glands	Ingests and digests foods; absorbs nutrients into blood	Maintains adequate supplies of fuel molecules and building materials
Circulatory	Heart, blood vessels, blood; lymph and lymph structures (lymphatic system is a subsystem of the circulatory system)	Transports materials from one part of body to another; defends body against disease organisms	Transports oxygen, nutrients, hormones, wastes; maintains water and ionic balance of tissues
Respiratory	Lungs, trachea, and other air passageways	Exchanges gases between blood and external environment	Maintains adequate blood oxygen content and helps regulate blood pH; eliminates carbon dioxide
Urinary	Kidney, bladder, and associated ducts	Excretes metabolic wastes; removes excessive substances from blood	Helps regulate volume and composition of blood and body fluids
Nervous	Nerves and sense organs; brain and spinal cord	Receives stimuli from external and internal environment; conducts impulses; integrates activities of other systems	Principal regulatory system
Endocrine	Ductless glands (e.g., pituitary, adrenal, thyroid) and tissues that secrete hormones	Regulates blood chemistry and many body functions	In conjunction with nervous system, regulates metabolic activities and blood levels of various substances
Reproductive	Testes, ovaries, and associated structures	Sexual reproduction	Maintains sexual characteristics

Consider the digestive system as an example of an organ system. Its organs include the mouth, esophagus, stomach, small and large intestines, liver, pancreas, and salivary glands. The digestive system processes food, reducing it to small molecular components. The products of digestion are absorbed into the blood, which transports them to all the cells in the body.

ORGAN SYSTEMS WORK TOGETHER TO MAINTAIN HOMEOSTASIS

In a complex animal, billions of cells are organized to form tissues, organs, and organ systems. The organism functions effectively, in large part because very precise control mechanisms

STRESS AND HOMEOSTASIS

How does stress affect the body? Stressors are changes in the internal or external environment that disturb homeostasis, causing stress. Examples of external stressors are heat, cold, noise, abnormal atmospheric pressure, and lack of oxygen. Internal stressors include changes in blood pressure, pH, and salt concentration, as well as high and low blood-sugar levels. Many stressors occur routinely. The body responds automatically by activating homeostatic mechanisms that expertly manage the stress. These mechanisms often involve the nervous and endocrine systems. Other stressors are more severe and may cause serious disruption of homeostasis.

Under conditions of stress, cells produce **stress proteins.** These proteins belong to a group of proteins known as **chaperones** that bind to polypeptide segments and stabilize their correct conformation. **Heat shock proteins** are stress proteins that are produced when cells are stressed by elevated temperature (42°C or higher). Recall that heat denatures proteins. Heat shock proteins have been conserved in evolution. In one group of these proteins, 50% of the amino acid sequence is the same in both *E. coli* and humans.

When homeostatic mechanisms are unable to restore the steady (normal) state, stress may cause malfunction that leads to disease or even death. In fact, we can view death as the failure of some homeostatic mechanism. In Chapter 47 we will discuss the role of the adrenal glands in regulating homeostatic adjustments that help the body effectively cope with stress.

---

maintain homeostasis. If the organism is to survive and function, the composition of the fluids that bathe its cells must be carefully regulated. An appropriate concentration of nutrients, oxygen and other gases, ions, and compounds needed for metabolism must be available at all times. In addition, internal temperature and pressure must be maintained within relatively narrow limits.

Homeostasis is a basic concept in physiology. First coined by the physiologist Walter Cannon, the word *homeostasis* is derived from the Greek *homoios,* meaning "same," and *stasis,* "standing." Actually, the internal environment never really stays the same. Homeostasis is always being challenged by **stressors,** changes in the internal or external environment that affect normal conditions within the body, causing **stress.** Homeostatic mechanisms interact continuously to manage stress, maintaining the internal environment within the physiological limits that support life (see *Making the Connection: Stress and Homeostasis*). All the organ systems participate in these regulatory mechanisms, but most of them are controlled by the nervous and endocrine systems. During our study of organ systems, we will discuss numerous ways in which organ systems interact to maintain the steady state of the organism.

How do homeostatic mechanisms work? Many are **feedback systems,** sometimes called biofeedback systems. Such a system consists of a cycle of events in which information about a change (e.g., a change in temperature) is fed back into the system so that the regulator (the temperature-regulating center in the brain) can control the process (temperature regulation). The desired condition (normal body temperature) is referred to as the **set point.** When body temperature becomes too high or too low, the change serves as input, triggering the regulator to counteract the change. The regulator activates mechanisms that bring the system back to the set point. The

return to normal temperature signals the temperature-regulating center to "shut off" the homeostatic mechanisms.

In this type of feedback system, the response counteracts the inappropriate change, thus restoring the steady state. This is a **negative feedback system,** because the response of the regulator is opposite (negative) to the output (Fig. 37–6*a*). Most homeostatic mechanisms in the body are negative feedback systems. When some condition varies too far from the steady state (either too high or too low), a control system using negative feedback brings the condition back to the steady state.

There are a few **positive feedback systems** in the body. In these systems a deviation from the steady state sets off a series of changes that intensify (rather than reverse) the changes. Therefore, although many positive feedback systems are beneficial, they do not maintain homeostasis. A positive feedback cycle operates during the birth of a baby. As the baby's head pushes against the opening of the uterus (cervix), a reflex action causes the uterus to contract. The contraction forces the head against the cervix again, resulting in another contraction, and the positive feedback cycle is repeated again and again until the baby is born. Some positive feedback sequences, such as those that deepen circulatory shock following severe hemorrhage, can lead to disruption of steady states and even to death.

Regulation of blood-sugar level provides a good example of homeostatic mechanisms at work (Fig. 37–6*b*). When you wake up in the morning your blood-sugar (glucose) level is about 90 mg of glucose per 100 mL of blood. Perhaps you eat a big breakfast that includes pastry or a doughnut. Many of the starches and sugars in your breakfast are digested to glucose. The glucose is then absorbed into the circulatory system, causing the blood-sugar level to rise.

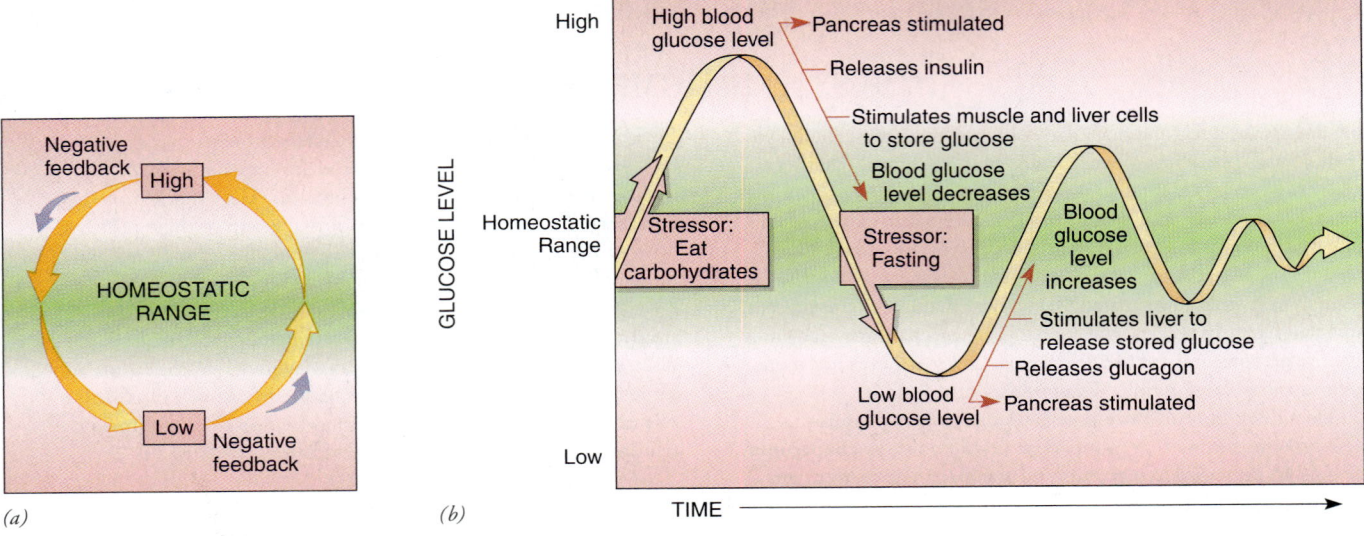

Figure 37–6 **Negative feedback mechanisms maintain homeostasis.** *(a)* In negative feedback the response of the regulator is opposite to the output. *(b)* Regulation of blood-sugar concentration by negative feedback.

An increase in blood-sugar level stimulates the pancreas to release the hormone insulin. This hormone causes the body cells to take up glucose from the blood, and stimulates the liver and muscle cells to store glucose (as glycogen). As a result, the glucose level in the blood decreases and returns to the normal fasting level of 90 mg/100 mL.

After several hours, when the glucose level of the blood begins to fall below the normal level, the pancreas releases the hormone glucagon. This hormone raises the blood-sugar level by stimulating the liver cells to slowly convert glycogen to glucose and thereby release their stored glucose. In this way, insulin and glucagon act in seesaw fashion to maintain a steady state of glucose in the blood.

MANY ANIMALS THERMOREGULATE

Most animals cannot regulate their body temperature and passively conform to the changing temperature of their environment. Other animals can thermoregulate. **Thermoregulation** is the ability to maintain body temperature within certain limits, even when the temperature of the environment is very different. Some animals, for example, snowshoe hares, snowy owls, and weasels, are able to survive in cold arctic regions. Others, such as the Cape ground squirrel, are adapted to hot tropical climates. Although some animals can survive at temperature extremes, most can survive only within moderate temperature ranges. In fact, each animal species has an optimal temperature range.

Animals have a number of structural, behavioral, and physiological strategies for thermoregulation. They produce heat as a byproduct of metabolic activities. Insulating feathers or hair are structures that decrease heat loss to the environment. Heat can be gained from the environment by behavioral strategies such as basking in the sun. Heat can also be transferred to the environment, for example, by **evaporation,** the conversion of a liquid, such as sweat, to water vapor. When sweat evaporates from the surface of the body or when a dog pants, the vaporizing water molecules transfer heat from the body to the surroundings.

Ectotherms absorb heat from their surroundings

Most animals are **ectotherms,** which means that their body temperature depends to a large extent on heat from the surrounding environment. The metabolic rate of an ectotherm tends to change with the weather. Many ectotherms use structural and behavioral strategies to adjust body temperature. For example, lizards bask in the sun, orienting their bodies to expose the maximum surface to the sun's rays (Fig. 37–7). Other behavioral strategies for regulating temperature include hibernation and migration.

An advantage of ectothermy is the absence of direct metabolic cost. Most of the heat for thermoregulation comes from the sun. As a result, ectotherms have a much lower daily energy expenditure than do endotherms, and more of the energy in their food can be converted to growth and reproduction. One of the disadvantages of ectothermy is that activity may be limited by daily and seasonal temperature conditions.

(a)

Figure 37–7 **Behavioral adaptations for thermoregulation.**
(a) This marine iguana *(Amblyrhynochos cristatus)* is an ectotherm that increases its body temperature by sunning itself. *(b)* The body temperature of this baby sooty tern of Hawaii, an endotherm, is reduced as heat leaves the body through its open mouth. *(a, Breck P. Kent/ Animals Animals; b, Frans Lanting/Minden Pictures)*

(b)

Endotherms derive heat from metabolic processes

Birds and mammals, as well as some species of fish (e.g., tuna) and insects, are **endotherms,** which means that they have homeostatic mechanisms that maintain body temperature despite changes in the external temperature. Endotherms have structural, behavioral, and physiological mechanisms for maintaining body temperature. For example, the insulating feathers of birds, hair of mammals, and chitin hairs or fur of some insects are structures that facilitate thermoregulation. The Cape ground squirrel uses its tail to shade its body from the direct rays of the sun. Endotherms have homeostatic mechanisms for regulating heat production and regulating heat exchange with the environment. Most of their body heat comes from their own metabolic processes (see *Making the Connection: Electron Transport and Heat,* Chapter 7). Receptors located mainly in the hypothalamus and spinal cord regulate temperature. In cold weather they can increase heat production or decrease heat loss. Heat production can be increased by contracting muscles. Heat from metabolic activities can be increased either directly, or indirectly by the action of hormones (e.g., thyroid hormones) that increase metabolic rate.

Some insects use a combination of behavior and metabolic heat production to regulate body temperature. The "furry"

body of the moth helps conserve body heat. When a moth prepares for flight, it contracts its flight muscles with little movement of its wings. The heat generated enables the moth to sustain the intense metabolic activity needed for flight.

In humans, receptors in the skin, the hypothalamus of the brain, and certain other areas are sensitive to body temperature changes (Fig. 37–8). Information about body temperature is sent to the temperature-regulating center in the hypothalamus. Nerves signal muscles to shiver, or allow us to move muscles voluntarily to increase body temperature. When body temperature increases, nerves increase the activity of sweat glands.

The nervous system also helps regulate body temperature by dilating blood vessels in the skin when we are hot. Increased blood flow to the skin brings body heat to the surface. The skin acts as a heat radiator, allowing heat to leave the body. When we are cold, blood vessels in the skin constrict, reducing heat loss.

Important advantages of endothermy include increased rate of enzyme activity and constant body temperature. Endotherms can carry out their activities even in low winter temperatures. However, endotherms are disadvantaged by the high energy cost of thermoregulation during times when they are inactive. We must maintain our body temperature even when we are asleep.

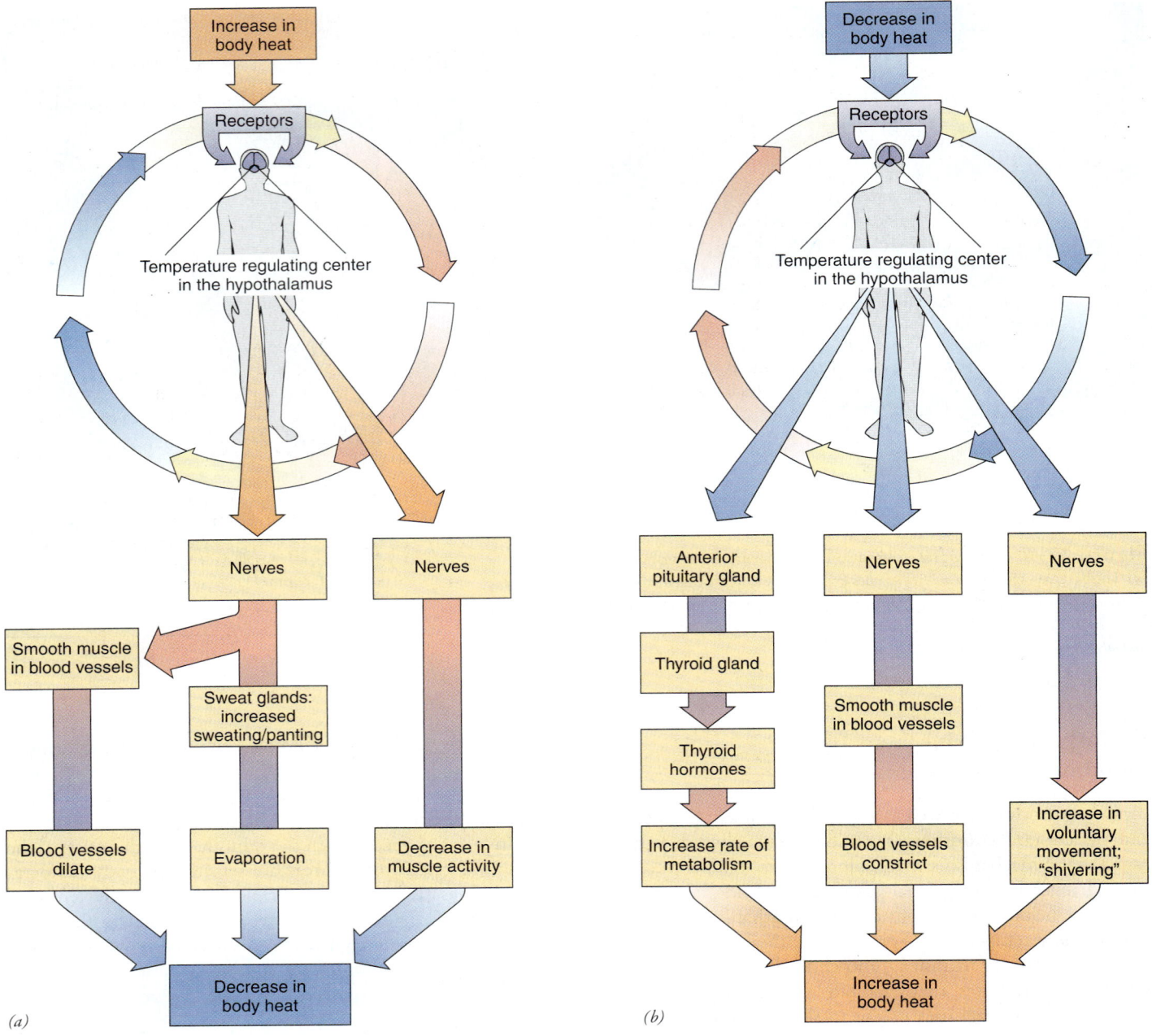

Figure 37–8 Regulation of temperature in the human body. *(a)* Mechanisms that restore homeostasis when body temperature increases. *(b)* Mechanisms that restore homeostasis when body temperature decreases.

SUMMARY WITH KEY TERMS

I. Multicellular organisms can be much larger and more diverse than unicellular ones, and their cells can specialize, performing specific functions.

II. A **tissue** consists of a group of similarly specialized cells that associate to perform one or more functions. Animal tissues are classified as epithelial, connective, muscular, or nervous.

    A. **Epithelial tissue (epithelium)** may form a continuous layer, or sheet, of cells covering a body surface or lining a body cavity. One surface of an epithelial layer is attached to the underlying tissue by a **base-**

ment membrane, consisting of fibers and polysaccharides made by the epithelial cells.

    1. Epithelial tissue functions in protection, absorption, secretion, or sensation.

    2. Epithelial cells may be **squamous, cuboidal,** or **columnar** in shape.

    3. Epithelial tissue may be **simple, stratified,** or **pseudostratified** (summarized in Table 37–1).

4. Some epithelial tissue is specialized to form **glands. Goblet cells** are unicellular glands that secrete mucus. Goblet cells are **exocrine glands** that secrete their product through a duct onto an exposed epithelial surface. In contrast, **endocrine glands** release hormones into the **interstitial fluid** or blood.
5. An **epithelial membrane** consists of a sheet of epithelial tissue and a layer of underlying connective tissue. A **mucous membrane** lines a cavity that opens to the outside of the body. A **serous membrane** lines a body cavity that does not open to the outside.

B. **Connective tissue** joins other tissues of the body, supports the body and its organs, and protects underlying organs. Connective tissue consists of relatively few cells separated by **intercellular substance,** which is composed of **fibers** scattered through a **matrix.**
1. The intercellular substance contains **collagen fibers, elastic fibers,** and **reticular fibers.** Connective tissue contains specialized cells such as **fibroblasts** and **macrophages.**
2. Some main types of connective tissue are **loose connective tissue, dense connective tissue, elastic connective tissue, reticular connective tissue, adipose tissue, cartilage, bone,** and **blood** (summarized in Table 37–2).
3. Cartilage cells, called **chondrocytes,** are located in **lacunae,** small cavities in the cartilage matrix.
4. **Osteocytes** secrete and maintain the matrix of bone. As bone is remodeled, osteocytes and **osteoclasts** dissolve and remove parts of the bony substance. Compact bone consists of **osteons,** which are spindle-shaped units that consist of a central blood vessel that runs through a **Haversian canal,** surrounded by lamellae, concentric layers containing osteocytes.

C. **Muscle tissue** is composed of cells specialized to contract. Each cell is an elongated muscle fiber containing many contractile units called **myofibrils.**
1. **Skeletal muscle** is striated and under voluntary control.
2. **Cardiac muscle** is striated; its contraction is involuntary.
3. **Smooth muscle** contracts involuntarily. It is responsible for movement of food through the digestive tract and for movement of other body organs.

D. **Nervous tissue** is composed of **neurons,** cells specialized for conducting impulses, and **glial cells,** which are supporting cells.

III. Tissues and **organs** work together, forming **organ systems.** In complex animals, ten principal organ systems work together, making up the living organism. Among these are the **integumentary system, endocrine system, nervous system,** and **skeletal system** (summarized in Table 37–4).

IV. **Homeostasis** is the body's automatic tendency to maintain a constant internal environment, or steady state. It is maintained by **negative feedback** systems in which the response of the regulator counteracts the change.

V. Many animals are capable of **thermoregulation,** the ability to maintain body temperature within certain limits even when the temperature of the environment changes.
A. In **ectotherms,** body temperature depends to a large extent on the temperature of the environment. However, many ectotherms have structural and behavioral strategies for regulating body temperature.
B. In addition to structural and behavioral strategies for maintaining body temperature, **endotherms** have homeostatic mechanisms for regulating heat generated by metabolic activities and for regulating heat exchange with the environment.

P O S T - T E S T

1. A group of closely associated cells that carry out specific functions is a (an) (a) colony (b) tissue (c) organ system (d) organelle (e) homeostatic mechanism
2. Epithelial cells that produce and secrete a product into a duct form a (an) (a) endocrine gland (b) exocrine gland (c) pseudostratified membrane (d) answers a, b, and c are correct (e) answers a and c only are correct
3. A serous membrane (a) lines a body cavity that does not open to the outside of the body (b) consists of loose connective tissue (c) has a framework of reticular connective tissue (d) answers a, b, and c are correct (e) answers a and c only are correct
4. The most numerous type of fibers in connective tissue is (a) reticular (b) myofibrils (c) collagen (d) elastic (e) glial
5. Dense connective tissue is likely to be found in (a) the outer layer of skin (b) tendons (c) bone (d) lungs (e) the framework of the lymph nodes
6. Chondrocytes are most likely to be found in (a) the outer layer of skin (b) bone (c) cartilage (d) lungs (e) blood
7. Tissue that contains fibroblasts and a great deal of intercellular substance is (a) connective tissue (b) muscle tissue (c) nervous tissue (d) pseudostratified epithelium (e) two of the preceding answers are correct
8. Muscle that is striated and involuntary is (a) connective (b) smooth (c) skeletal (d) pseudostratified (e) cardiac
9. Glial cells are most characteristic of (a) muscle that is striated and voluntary (b) cardiac muscle tissue (c) nervous tissue (d) stratified epithelium (e) loose connective tissue
10. Tissue that has a hard, rubbery matrix and lacks blood vessels is (a) integumentary tissue (b) bone (c) nervous tissue (d) stratified epithelium (e) cartilage
11. The contractile elements in muscle tissue are (a) myofibrils (b) elastic (c) collagen (d) lacunae (e) reticular
12. Which system consists of glands that secrete hormones? (a) integumentary (b) skeletal (c) nervous (d) endocrine (e) exocrine
13. Which system has the homeostatic function of helping to maintain constant calcium level in the blood? (a) integumentary (b) skeletal (c) nervous (d) reproductive (e) exocrine
14. Which system has the homeostatic function of helping to regulate volume and composition of blood and body fluids? (a) integumentary (b) muscular (c) reproductive (d) urinary (e) exocrine
15. Many homeostatic functions are maintained by (a) negative feedback systems (b) positive feedback systems (c) set points (d) stressors (e) exocrine glands
16. An ectotherm (a) may use behavioral strategies to help adjust body temperature (b) has a variety of homeostatic mechanisms to regulate body temperature (c) depends on sensors in the hypothalamus to regulate temperature (d) has a higher rate of enzyme activity than a typical endotherm (e) must expend more energy on thermoregulation than an endotherm

R E V I E W   Q U E S T I O N S

1. What advantages do multicellular organisms have over unicellular organisms? Can you think of any disadvantages?
2. What are the functions of epithelial tissues? How are these tissues adapted to carry out these functions?
3. What is the structure of bone? Of adipose tissue? Of loose connective tissue? How is each adapted to carry out its special functions?
4. Compare the properties of the three types of muscle.
5. How is the neuron uniquely adapted for its function?

(a) *(b)* *(c)*

(d) *(e)*

(Ed Reschke)

6. What kinds of tissues would you expect to find in the following organs: (a) lung (b) heart c) intestine (d) salivary glands?

7. List the principal organ systems found in a complex animal and give the functions of each.

8. How do the cells of a malignant neoplasm differ from those of a normal tissue? What is metastasis?

9. Identify each type of tissue in the figures above. Refer to Tables 37–1 and 37–2 to check your answers.

YOU MAKE THE CONNECTION

1. Imagine that all of the epithelium in a complex animal, such as a human, suddenly disappeared. What effects might this have on the body and its ability to function?

2. What would connective tissue be like if it had no intercellular substance?

What effect would the absence of intercellular substance have on the body?

3. A high concentration of carbon dioxide in the blood and interstitial fluid results in more rapid breathing. Explain this observation in terms of homeostasis.

RECOMMENDED READINGS

Engel, J. "Biochemistry: Versatile Collagens in Invertebrates," *Science,* Vol. 277, 19 Sep. 1997. An interesting discussion of the structure and versatility of collagen.

Fackelman, K. "Immune Attack on Cancer," *Science News,* Vol. 153, 13 Jun. 1998. An overview of the development of immune therapy in treating cancer.

Marshall, E. "Cancer: The Roadblocks to Angiogenesis Blockers," *Science,* Vol. 280, 15 May 1998. A brief summary of the status of a group of an-

titumor agents that block the development of new blood vessels.

Rhoades, R., and R. Pflanzer. *Human Physiology,* 3rd ed. Saunders College Publishing, Philadelphia, 1996. A readable presentation of how the human body works.

Ruben, J.A., et al. "The Metabolic Status of Some Late Cretaceous Dinosaurs," *Science,* Vol. 273, 30 Aug. 1996. The cross-sectional area of nasal passages was used to assess the metabolic rates of three theropod dinosaurs.

● Visit our website at **http://www.saunderscollege.com/lifesci/titles.html** and click on Solomon/Berg/Martin Biology for links to chapter-related resources on the World Wide Web.

CHAPTER 38

Protection, Support, and Movement: Skin, Skeleton, and Muscle

In our survey of the animal kingdom (Chapters 28 to 30), we made many references to protective coverings, movement, and skeletal systems in terms of adaptation of various animal groups. In Chapter 37 we described animal tissues, organs, and systems, laying the foundation for the more detailed discussions in this and later chapters. Here we focus on epithelial coverings, skeleton, and muscle—systems that are closely interrelated in function and significance. We compare these systems in several animal groups and then focus on their structures and functions in the human body.

In addition to protecting underlying tissues, the **epithelial coverings** of animals may be specialized to exchange gases, excrete wastes, regulate temperature, or secrete substances such as mucus or poison. For example, the striped land snail *(Helicella candicans)* in the photograph glides through a slime track consisting of mucus secreted by the epithelial covering of its foot.

In many animals a **skeletal system** also functions to protect the body. Whether it is a fluid-filled compartment, a shell, or a system of bones, the skeletal system provides support for the body and typically protects internal organs. The **muscular system** and skeletal system work together. Muscles responsible for locomotion are anchored to the skeleton, which gives them something firm on which to act. In humans and most other vertebrates, bones serve as levers that transmit the force necessary to move various parts of the body.

Some animals run; some jump; some fly. Others remain rooted to one spot, sweeping their surroundings with tentacles. Many contain internal circulating fluids, pumped by hearts and contained by hollow vessels that maintain their pressure with gentle squeezing. Most have digestive systems that push food along with peristaltic contractions. In all of these cases, each action is powered by muscle, a tissue specialized to contract.

(Y.Momatiuk/Photo Researchers, Inc.)

1. Compare the structure and function of the external epithelium of invertebrates and vertebrates and identify the principal derivatives of vertebrate skin.
2. Compare the advantages and disadvantages of different types of skeletal systems, including the hydrostatic skeleton, exoskeleton, and endoskeleton.
3. Identify the main divisions of the vertebrate skeleton and the bones that make up each division.
4. Describe the structure of a typical long bone and differentiate between endochondral and intramembranous bone development.

5. Describe the macroscopic and microscopic structures of skeletal muscle.
6. List, in sequence, the events that take place during muscle contraction.
7. Compare the roles of glycogen, creatine phosphate, and ATP in providing energy for muscle contraction.
8. Describe the antagonistic action of muscles.
9. Summarize the functional relationship between skeletal and muscle tissues.

EPITHELIAL COVERINGS PROTECT THE BODY

Epithelial tissue covers all external and internal surfaces of the animal body. The epithelial covering forms a protective shield around the body.

Invertebrate epithelium may function in secretion or gas exchange

In invertebrates, the external epithelium protects the body and may also be specialized for secretion, absorption, or gas exchange. Epithelial cells may be modified as sensory cells that are selectively sensitive to light, chemical stimuli, or mechanical stimuli such as contact or pressure.

In many species, the epithelium contains secretory cells that produce a protective cuticle or secrete lubricants or adhesives. In some species, secretory cells release odorous secretions that are used for communication among members of the species or for marking trails. In other species, the cells produce poisonous secretions that are used for offense or defense. In earthworms, a lubricating mucous secretion serves as a moist slime that promotes efficient diffusion of gases across the body wall and also reduces friction during movement through the soil.

An epithelial secretion may be limited to a particular region of the body surface. The foot of the gastropod mollusk, for example, releases a mucous secretion, producing a slime track along which the snail glides. Certain insects such as weaver ants release epithelial secretions as fine, very strong threads that are used to construct nests. The spinning glands of spiders develop from epithelial cells, and lepidopteran insects such as butterflies and moths synthesize silk from amino acids in silk-forming glands.

The vertebrate skin functions in protection and temperature regulation

The **integumentary system** of vertebrates includes the skin and structures that develop from it. In many fishes, in the African ant-eating pangolin (a mammal), and in some reptiles, skin has developed into a set of scales formidable enough to be considered armor. Even human skin has considerable strength. Structures derived from skin include fingernails and toenails, hair, sweat glands, oil (sebaceous) glands, and several types of sensory receptors that give us the ability to feel pressure, temperature, and pain. The skin of mammals contains mammary glands, specialized in females for secretion of milk.

Oil glands in human skin empty via short ducts into hair follicles. A **hair follicle** is the part of a hair below the skin surface, together with its epithelial and connective tissue coverings. Oil glands secrete a substance called **sebum,** a complex mixture of fats and waxes. In humans these glands are especially numerous on the face and scalp. The oil secreted keeps the hair moist and pliable and prevents the skin from drying and cracking. Sebum also contains substances that inhibit the growth of harmful bacteria. (At puberty, excessive sebum, produced in response to increased levels of sex hormones, may fill the glands and follicles, producing acne.)

In humans (and some other mammals) the skin helps maintain body temperature (see Chapter 37). About 2.5 million sweat glands secrete sweat, and its evaporation from the surface of the skin lowers the body temperature. Constriction and dilation of capillaries in the skin are also part of homeostatic mechanisms for regulating body temperature.

Derivatives of skin differ considerably among vertebrates. Fish have bony or toothlike scales. Amphibians have naked skin covered with mucus and some species are equipped with poison glands. Reptiles have epidermal scales, mammals have hair, and birds have feathers that provide even more effective insulation than fur. Skin and its derivatives are often brilliantly colored in connection with courtship rituals, territorial displays, and other kinds of communication. The human blush pales alongside the spectacular displays of animals such as peacocks.

The epidermis is a waterproof protective barrier

The outer layer of skin, the **epidermis,** is the interface between the delicate tissues within the vertebrate body and the hostile outside environment. The epidermis consists of several strata,

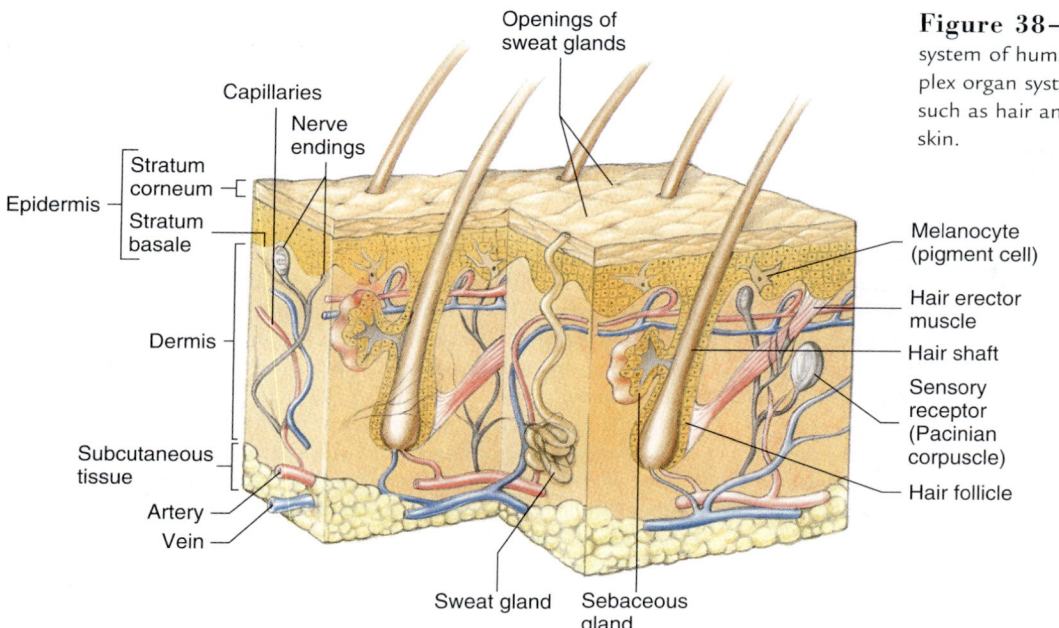

Figure 38–1   **Human skin.** The integumentary system of humans and other mammals is a complex organ system consisting of skin and structures such as hair and nails that are derived from the skin.

or sublayers. The deepest is **stratum basale,** and the most superficial is **stratum corneum** (Fig. 38–1). Pigment cells in stratum basale and in the dermis produce **melanin,** a pigment that contributes to the color of the skin (see *Making the Connection: The Sun, Ultraviolet Radiation, and Skin*). In stratum basale, cells divide and are pushed outward as other cells are produced below them. The epidermal cells mature as they move toward the skin surface. Because most vertebrates have no capillaries in the epidermis, the maturing cells receive less and less nourishment, which diminishes their metabolic activity.

As they move toward the body surface, epidermal cells manufacture **keratin,** an elaborately coiled protein that gives the skin considerable mechanical strength and flexibility. Keratin is quite insoluble and serves as a diffusion barrier for the body surface. As epidermal cells move through stratum corneum, they die. When they reach the outer surface of the skin, they wear off and must be continuously replaced.

The dermis contains blood vessels and other structures

Beneath the epidermis lies the **dermis** (Fig. 38–1), composed of a dense, fibrous connective tissue made up mainly of collagen fibers. Collagen imparts strength and flexibility to the skin. Sweat glands and hair follicles are embedded in the dermis (Fig. 38–2). The dermis also contains blood vessels that nourish the skin, and sensory receptors for touch, pain, and temperature. Mammalian skin rests on a layer of **subcutaneous tissue** composed mainly of adipose tissue that insulates the body from outside temperature extremes.

250 μm

Figure 38–2   **SEM of human skin showing a hair follicle.** Part of the hair follicle was torn during preparation of the tissue. *(Courtesy of Dr. Karen A. Holbrook)*

THE SUN, ULTRAVIOLET RADIATION, AND SKIN

How does ultraviolet (UV) radiation, the short, invisible rays from the sun, affect the skin? Exposure to UV is necessary for the body to make vitamin D. However, too much can cause sunburn and skin damage and can eventually result in wrinkling of the skin, cataracts, and skin cancer. Three bands of UV radiation have been identified: UVA, UVB, and UVC. UVA contributes to wrinkling, sunburn, and skin cancer. UVB is stronger than UVA and is the most common cause of sunburn; it contributes to cataracts and skin cancer and damages the immune system. UVC is the strongest and most dangerous type of UV radiation.

The amount of UV radiation that reaches the Earth's surface depends on the time of year, altitude, distance from the equator, and cloud cover. UV radiation is most intense near the equator, at high altitudes, and around noon on a clear summer day. The amount of UV radiation that reaches us is also affected by the ozone layer, a layer of gas in the stratosphere that absorbs incoming UVB and UVC radiation. Scientists warn that the ozone layer is being destroyed by pollutants such as chlorofluorocarbons, called CFCs. These compounds are used as propellants for aerosol cans, as coolants in air conditioners and refrigerants (e.g., Freon), and as foam for insulation and packaging (e.g., Styrofoam). In 1985 scientists discovered a large hole in the ozone layer over Antarctica. A smaller hole has been found over the Arctic, allowing more UV radiation to reach the surface of Earth. Stratospheric ozone levels have been decreasing worldwide for several decades. Despite international agreements to reduce CFC production, depletion of the ozone layer continues. CFCs are stable compounds that will continue to damage the ozone layer for many years after they are no longer used. In addition, some damaging compounds are not covered by the agreements. (Stratospheric ozone depletion is discussed in greater detail in Chapter 55.)

Exposure to ultraviolet rays causes the epidermis to thicken and stimulates pigment cells in the skin to produce melanin at an increased rate. An increase in melanin causes the skin to become darker. Melanin is an important protective screen against the sun

because it absorbs some of the harmful ultraviolet rays. The suntan so prized by sun worshipers is actually a sign that the skin has been exposed to too much ultraviolet radiation. When the melanin is not able to absorb all the ultraviolet rays, the skin becomes inflamed, or sunburned. Because dark-skinned people have more melanin, they suffer less sunburn, wrinkling, and skin cancer. Ultraviolet radiation damages DNA, causing mutations that lead to malignant transformation (see Chapter 16, *Focus On: Oncogenes and Cancer*).

Skin cancer is on the rise. The incidence of malignant melanoma, a type of skin cancer, is increasing faster than any other type of cancer. Some forms of malignant melanoma spread rapidly through the body and may cause death within a few months after diagnosis. Most cases of skin cancer are caused by excessive, chronic exposure to UV radiation.

Most skin cancer can be prevented by protecting the body from sunlight and other forms of ultraviolet radiation. Tanning machines expose the body to ultraviolet rays and so also pose risks. Those who choose to expose themselves to sunlight can protect their skin by limiting the time they spend in the sun, and by applying effective sun screens. Dermatologists suggest that everyone wear sun screen on exposed skin every day. Consumer organizations have persuaded the National Weather Service to include a UV Index as part of the daily weather forecast. The index indicates the time it would take for a fair-skinned North American to burn when exposed to the sun at noon.

New Zealand, southern Australia, parts of the United States, and certain other geographical areas are reporting significant increases in harmful radiation. Evidence suggests that continuing depletion of the ozone shield around our planet will cause 1 million extra cases of skin cancer each year and that about 30,000 of these victims will die. The highest incidences of skin cancer have been reported in Australia, New Zealand, Hawaii, and certain areas in the United States. About 75% of Australians of European ancestry will have at least one skin cancer removed in their lifetime.

SKELETONS FUNCTION IN LOCOMOTION, PROTECTION, AND SUPPORT

Some animals do not have hard skeletons. In them muscle may act on a thickened epidermis or on a fluid-filled body cavity. In more complex animals, muscles act on hard structures such as chitin or bone that transmit and transform muscle contraction into the variety of motions animals use. In addition to its function in locomotion, the skeleton supports the body and protects the internal organs.

In echinoderms and chordates the skeleton is internal, in the form of plates or shafts of calcium-impregnated tissue (such as cartilage or bone). But in most animals, the skeleton is not a living tissue but a lifeless deposit atop the epidermis, a shell or **exoskeleton.**

In hydrostatic skeletons, body fluids are used to transmit force

Imagine an elongated balloon full of water. If you were to pull on it, it would lengthen and become thinner. It would do the same if you squeezed it. Conversely, if you pushed the ends toward the center, it would shorten and thicken. Many animals, including cnidarians and annelids, have a **hydrostatic skeleton** that works something like a balloon filled with water. Fluid in a closed compartment of the body is held under pressure. Contracting muscles push against the tube of fluid. Because fluids cannot be compressed, the force is transmitted through the fluid, causing change in shape and movement of the body.

In *Hydra* and other cnidarians, cells of the two body lay-

Longitudinal
contractile fibers
of epidermal layer

Circular
contractile fibers
of gastrodermis

(a)　　　　　　　　　*(b)*

(c)

Figure 38–3　**Hydrostatic skeleton of *Hydra*.** The longitudinally arranged contractile cells are antagonistic to the cells arranged in circles around the body axis. *(a)* Contraction of the circular contractile fibers elongates the body. *(b)* Contraction of the longitudinal fibers shortens the body. *(c) Hydra* in motion. The body of this *Hydra* sp. appears to be shortening as it moves in on its prey, a water flea (*Daphnia* sp.). *(c, Dwight R. Kuhn/DRK Photo)*

ers are capable of contraction. The contractile cells in the outer epidermal layer are arranged longitudinally, whereas the contractile cells of the inner layer (the gastrodermis) are arranged circularly around the central body axis (Fig. 38–3). The two groups of cells work in **antagonistic** fashion: what one can do, the other can undo. When the epidermal (longitudinal) layer contracts, the hydra shortens. Because of the fluid in the gastrovascular cavity, force is transmitted so that the hydra thickens as well. On the other hand, when the inner (circular) layer contracts, the hydra thins, and its fluid contents force it to lengthen.

Mechanically, the hydra is a bag of fluid. The fluid acts as a hydrostatic skeleton because it transmits force when the contractile cells contract against it. (Although technically not a closed compartment, the gastrovascular cavity can function as a hydrostatic skeleton because its opening is small.) Hydrostatic skeletons permit only crude mass movements of the body or its appendages. Delicate movements are difficult because force tends to be transmitted equally in all directions throughout the entire fluid-filled body of the animal. For example, it is not easy for the hydra to thicken one part of its body while thinning another.

The more sophisticated hydrostatic skeleton of the annelid worm enables it to be more versatile in movement. An earthworm's body consists of a series of segments divided by transverse partitions, or septa (Chapter 29). The septa isolate portions of the body cavity and its contained fluid, permitting the hydrostatic skeletons of each segment to be largely independent of one another. Thus, contraction of the circular muscle in the elongating anterior end need not interfere with the ac-

tion of the longitudinal muscle in the segments of the posterior end.

Some examples of hydrostatic skeletons occur even in complex invertebrates equipped with shells or endoskeletons, and in vertebrates with endoskeletons of cartilage or bone. Among mollusks, for example, the clam extends and anchors its foot by a hydrostatic blood pressure mechanism similar to that used by earthworms. Sea stars and sea urchins move their tube feet by an ingenious version of the hydrostatic skeleton (Chapter 30). And even the human penis becomes erect and stiff because of the turgidity of pressurized blood in its cavernous spaces.

Mollusks and arthropods have nonliving exoskeletons

In both mollusks and arthropods, the exoskeleton is a nonliving product of the epidermal cells. In mollusks, the exoskeleton provides protection, a retreat used in emergencies, with the bulk of the naked, tasty body exposed at other times.

Exoskeletons of arthropods serve not only to protect but also to transmit forces. In this respect they are comparable to the skeletons of vertebrates. Although the arthropod exoskeleton is a continuous, one-piece sheath covering the entire body, it varies greatly in thickness and flexibility. Large, thick, inflexible plates are separated from one another by thin, flexible joints arranged segmentally. Enough joints are provided to make the arthropod's body as flexible as those of many vertebrates. The exoskeleton is also extensively modified to form

Figure 38–4 **Molting.** A greengrocer cicada requires 13 years underground to mature. It then emerges from the soil, climbs a tree, and molts prior to reproducing. *(Judy Davidson/Science Photo Library/Photo Researchers, Inc.)*

Figure 38–5 **The echinoderm endoskeleton.** The endoskeleton provides support and protection. As in other echinoderms, the sea urchin endoskeleton is composed of spicules and plates of nonliving calcium salts embedded in tissues of the body wall. *(Brian Parker/ Tom Stack & Associates)*

specialized tools or weapons or is otherwise adapted to a vast variety of lifestyles.

A disadvantage of the rigid arthropod exoskeleton is that it interferes with growth. To accommodate growth, an arthropod must **molt,** that is, shed its exoskeleton and replace it with a new, larger one (Fig. 38–4). During molting the animal is weak and vulnerable to predators.

Internal skeletons are capable of growth

Endoskeletons, or internal skeletons, are well developed in echinoderms and chordates. Composed of living tissue, the endoskeleton grows along with the animal as a whole. The echinoderm endoskeleton consists of spicules and plates of calcium salts embedded in the tissue of the body walls, that is, beneath an epidermis that covers the body. This endoskeleton forms what amounts to an internal shell that provides support and protection (Fig. 38–5). Many echinoderm endoskeletons bear spines that project to the outer surface.

The internal skeletons of vertebrates provide support and protection, and transmit forces. Members of class Chondrichthyes (sharks and rays) have skeletons composed of cartilage, but in most vertebrates the skeleton consists mainly of bone. Although the skeletons of adult vertebrates vary, their bones have been shown to be homologous. For example, the

bones of the human middle ear originate in the embryo in the same way that the gill arches of fishes do.

The vertebrate skeleton has two main divisions: the axial and appendicular skeletons. The **axial skeleton,** located along the central axis of the body, consists of the skull, vertebral column, ribs, and sternum (breastbone). The **appendicular skeleton** consists of the bones of the limbs (arms and legs) plus the bones making up the shoulder (pectoral) girdle and most of the hip (pelvic) girdle; these girdles connect the limbs to the axial skeleton (Fig. 38–6).

The **skull,** the bony framework of the head, consists of the cranial and facial bones. In the human, eight cranial bones enclose the brain, and 14 bones make up the facial portion of the skull. Several cranial bones that are single in the adult human result from the fusion of two or more bones that are separate in the embryo or even in the newborn.

The vertebrate spine, or **vertebral column,** supports the body and bears its weight. In humans it consists of 24 **vertebrae** and two fused bones, the **sacrum** and **coccyx.** The vertebral column consists of the **cervical** (neck) region, with 7 vertebrae; the **thoracic** (chest) region, with 12 vertebrae; the **lumbar** (back) region, with 5 vertebrae; the **sacral** (pelvic) region, with 5 fused vertebrae; and the **coccygeal** region, also composed of fused vertebrae.

Although vertebrae differ in size and shape in various regions of the vertebral column, a typical vertebra consists of a bony **centrum** (central portion), which bears most of the body weight, and a dorsal ring of bone called the **neural arch,** which

Figure 38-6 **The human skeletal system.** *(a)* Bones of the axial skeleton, anterior view. *(b)* Bones of the appendicular skeleton, anterior view.

surrounds and protects the delicate spinal cord. Vertebrae may also have projections for the attachment of ribs and muscles and for articulating (joining) with neighboring vertebrae. The first cervical vertebra, the **atlas** (named for the mythical Greek who held the world on his shoulders), has rounded depressions on its upper surface into which fit two projections from the base of the skull, allowing the head to nod up and down. The second cervical vertebra, the **axis,** serves as a pivot for rotation of the atlas and skull, permitting the head to move from side to side.

The **rib cage** is a bony "basket" formed by the sternum (breastbone), thoracic vertebrae, and, in mammals, 12 pairs of ribs. (Males and females have the same number of ribs.) The rib cage protects the internal organs of the chest, including the heart and lungs. It also supports the chest wall, preventing it from collapsing as the diaphragm contracts with each breath. Each pair of ribs is attached dorsally to a separate vertebra. Of the 12 pairs of ribs in the human, the first seven are attached ventrally to the sternum; the next three are attached indirectly by cartilages; and the last two, the "floating ribs," have no attachments to the sternum.

The **pectoral girdle** consists of two collarbones, or **clavicles,** and two shoulder blades, or **scapulas.** The **pelvic girdle** consists of a pair of large bones, each composed of three fused hipbones. Whereas the pelvic girdle is securely fused to the vertebral column, the pectoral girdle is loosely and flexibly attached to it by muscles.

Each human limb consists of 30 bones and terminates in five **digits,** the fingers and toes. The more specialized appendages of other tetrapods may be characterized by four digits (as in the pig), three (as in the rhinoceros), two (as in the camel), or one (as in the horse).

Great apes and humans have a highly specialized feature: the opposable thumb. (Great apes also have an opposable big toe. Although the human big toe is not opposable, it is similar enough in structure to the thumb that it can be surgically substituted.) The opposable thumb can be readily wrapped around objects, such as a tree limb, in climbing, and it is especially useful in grasping and manipulating objects.

In humans, the upper limbs are not generally used for locomotion as they are in other mammals, including the great apes. The combination of opposable thumbs and upright posture enables us to use our hands to write, shape, build, and use weapons. These abilities have permitted the human species to change its environment to a greater extent than any other organism in the history of the Earth.

A typical long bone consists of compact and spongy bone

The radius, one of the two bones of the forearm, is a typical long bone (Fig. 38–7). Its numerous muscle attachments are arranged in such a way that the bone rotates about its long axis and also operates as a lever amplifying the motion generated by the muscles. By themselves, muscles cannot shorten enough to produce large movements of the body parts to which they are attached.

Like other bones, the radius is covered by a connective tissue membrane, the **periosteum,** to which muscle tendons and ligaments attach. The periosteum is capable of producing fresh layers of bone and thus increasing the bone's diameter. The main shaft of a long bone is its **diaphysis;** each expanded end is called an **epiphysis.** In children, a disk of cartilage, the **metaphysis,** is found between the epiphyses and the diaphysis. Metaphyses are growth centers that disappear at maturity, be-

coming vague **epiphyseal lines.** Long bones contain a central cavity that contains **bone marrow.** Yellow marrow consists mainly of a fatty connective tissue; the red marrow found in certain bones produces blood cells.

The radius has a thin outer shell of **compact bone,** which is very dense and hard. Compact bone is found primarily near the surfaces of a bone, where it provides great strength. Compact bone consists of interlocking spindle-shaped units called osteons, or Haversian systems (see Fig. 37–3). Within an osteon, **osteocytes** (bone cells) are found in small cavities called **lacunae** (sing., *lacuna*). The lacunae are arranged in concentric circles around central **Haversian canals.** Blood vessels that nourish the bone tissue pass through the Haversian canals. Osteocytes are connected by threadlike extensions of their cytoplasm that extend through narrow channels called canaliculi.

Interior to the thin shell of compact bone is a filling of **spongy bone** (also called *cancellous bone*). Spongy bone, which consists of a network of thin strands of bone, provides most of the bone's mechanical strength. The spaces within spongy bone are filled with bone marrow.

Bones are remodeled throughout life

During fetal development, bones form in two ways. Long bones, such as the radius, develop from cartilage templates in a process called **endochondral** bone development. A bone begins to ossify in its diaphysis, and secondary sites of bone production develop in the epiphyses. The part of the bone between the ossified regions can grow. Eventually the ossified regions fuse. In contrast, other bones, including the flat outer bones of the skull, develop from a noncartilage, connective tissue scaffold. This process is known as **intramembranous bone development.**

Osteoblasts are bone-building cells. They secrete the protein collagen, which forms the strong fibers of bone. The compound hydroxyapatite, composed mainly of calcium phosphate, is present in the interstitial fluid. It automatically crystallizes around the collagen fibers, forming the hard matrix of bone. As the matrix forms around the osteoblasts, they become isolated within the lacunae. The trapped osteoblasts are then referred to as osteocytes.

Little was known about the biochemistry of osteoblasts until 1994 when developmental geneticist Gerard Karsentry, at the M. D. Anderson Cancer Center of the University of Texas, Houston, identified genes in the mouse that code for osteocalcin, a protein that is part of the bone matrix. Another developmental geneticist, Patricia Ducy, working with Karsentry, then found the gene that switches the osteocalcin gene on in osteoblasts. In 1997 Ducy reported that when she spliced this regulatory gene, called CBFA-1, into mouse skin cells, the cells developed into osteoblasts and produced osteocalcin. Based on these and other studies, some researchers suggest that the CBFA-1 gene may be a key regulator that stimulates preosteoblasts to differentiate into osteoblasts. Investigators are hopeful that understanding the mechanisms of bone development could lead to treatment of bone diseases. For example,

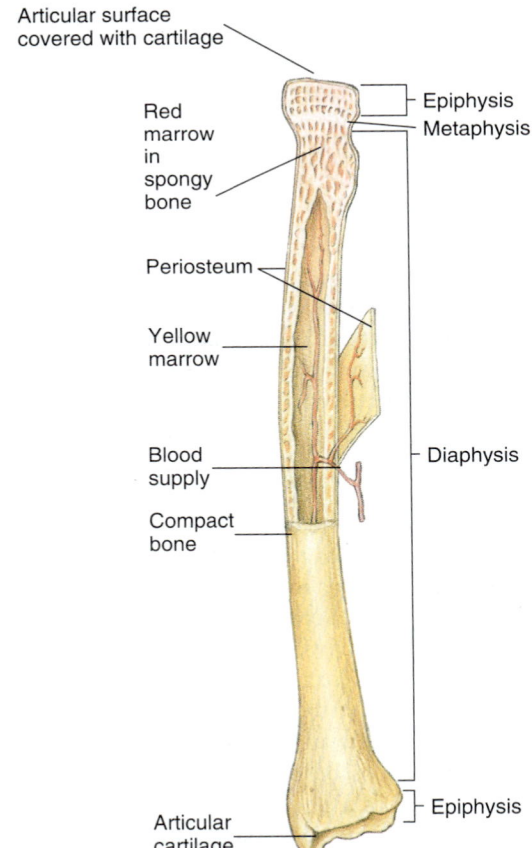

Articular surface
covered with cartilage

Red
marrow
in
spongy
bone

Periosteum

Yellow
marrow

Blood
supply

Compact
bone

Articular
cartilage

Epiphysis
Metaphysis

Diaphysis

Epiphysis

Figure 38–7   A typical long bone. A long bone has a thin shell of compact bone and a filling of spongy bone containing marrow.

osteoporosis, a progressive degenerative bone disease, may be caused by the body's decreasing ability to produce new osteoblasts as aging occurs. The CBFA-1 gene might someday be used to stimulate an increase in bone formation, thus slowing the disease process.

Bones are modeled during growth and remodeled continuously throughout life in response to physical stresses and other changing demands. As muscles develop in response to physical activity, the bones to which they are attached thicken and become stronger. As a bone grows, tissue is removed from its interior, especially from the walls of the marrow cavity. **Osteoclasts** are very large cells that break down bone in a process referred to as bone resorption. The osteoclasts move about, secreting enzymes that digest bone. Osteoclasts and osteoblasts are synergistic; together they shape bones. In humans, bone is replaced as many as ten times during the course of an average lifetime.

Joints are junctions between bones

Joints, or articulations, are the sites of junction of two or more bones of the skeleton. Joints facilitate flexibility and movement. At the joint, the outer surface of each bone consists of articular cartilages. One way to classify joints is according to the degree of movement they allow. The *sutures* found between bones of the skull are **immovable joints.** In a suture, bones are held together by a thin layer of dense fibrous connective tissue, which may be replaced by bone in the adult. **Slightly movable joints,** found between vertebrae, are made of cartilage and help absorb shock.

Most joints are **freely movable joints.** Each is enclosed by a joint capsule composed of connective tissue and lined with a membrane that secretes a lubricant called **synovial fluid.** This thick fluid reduces friction during movement and also absorbs shock. The joint capsule is typically reinforced by **ligaments,** bands of fibrous connective tissue that connect bones and limit movement at the joint.

Joints wear down with time and use. In osteoarthritis, a common joint disorder, cartilage repair does not keep up with degeneration, and the articular cartilage wears out. In rheumatoid arthritis, the synovial membrane thickens and becomes inflamed. Synovial fluid accumulates, causing pressure, pain, stiffness, and progressive deformity and loss of function.

MUSCLE TISSUE GENERATES MOVEMENT IN MOST ANIMALS

All eukaryotic cells contain the contractile protein **actin,** the major component of microfilaments. Actin is important in many cell processes, including amoeboid movement and attachment of cells to surfaces. In most cells, actin is functionally associated with the contractile protein **myosin.** Actin and myosin are most highly organized in **muscle** cells.

We have discussed the contractile cells of the hydrostatic skeleton of *Hydra* and other cnidarians. In flatworms and most other animal groups, muscle occurs as a specialized tissue organized into definite layers, or straplike bands. Skeletal and smooth muscle are found among some invertebrate phyla. The three types of vertebrate muscle—skeletal, smooth, and cardiac—are compared in Table 37–3.

Bivalve mollusks, such as scallops, have two sets of muscles for opening and closing the shell. Smooth muscle, which is capable of slow, sustained contraction, can be used to keep the two shells tightly closed for long periods, even days or weeks at a time. Striated muscle, which contracts rapidly, is used for swimming and to shut the shell quickly when the mollusk is threatened.

Arthropod muscles are striated, even in the walls of the digestive tract. Insect flight muscles contract more rapidly than any other known muscle, up to 1000 contractions per second (see *Making the Connection: Muscles and Insect Flight*). Throughout the animal kingdom, muscle tissue generates the mechanical forces and motion necessary for locomotion, manipulation of objects, circulation of blood, movement of food through the digestive tract, and many other processes.

A vertebrate muscle may consist of thousands of muscle fibers

In vertebrates, each skeletal muscle may be considered an organ. Its elongated cells, referred to as **muscle fibers,** are organized in bundles, called **fascicles,** that are wrapped by connective tissue. The biceps in your arm, for example, consists of thousands of individual muscle fibers and their connective tissue coverings.

Each striated muscle fiber is a long, cylindrical cell with many nuclei (Fig. 38–8). The plasma membrane (known as the *sarcolemma* in a muscle fiber) has multiple inward extensions that form a set of **T tubules** (transverse tubules). The cytoplasm of a muscle fiber is referred to as **sarcoplasm,** and the endoplasmic reticulum as **sarcoplasmic reticulum.**

Threadlike structures called **myofibrils** run lengthwise through the muscle fiber. They are composed of even smaller structures, the **myofilaments** or simply **filaments.** There are two types of myofilaments: myosin and actin filaments. **Myosin filaments** are thick myofilaments consisting mainly of the protein myosin. The thin **actin filaments** consist mostly of the protein actin; they also contain the proteins **tropomyosin** and the **troponin complex** that regulate the actin filament's interaction with myosin filaments.

Myosin and actin filaments are organized into repeating units called **sarcomeres,** the basic units of muscle contraction. Each sarcomere consists of overlapping myosin and actin filaments. The filaments overlap lengthwise in the muscle fibers, producing the pattern of transverse bands, or striations, characteristic of striated muscle (Figs. 38–8 and 38–9). The bands are designated by the letters A, H, and I. Sarcomeres are joined at their ends by an interweaving of filaments called the Z line. Hundreds of sarcomeres connected end-to-end make up a myofibril.

MUSCLES AND INSECT FLIGHT

How do the structure and function of insect flight muscles permit insects to fly? Flight is an adaptation that has contributed to the great biological success of insects (Chapter 29). In many flying insects, the striated flight muscles are attached not directly to the wings but to the flexible portions of the exoskeleton that articulate with the wings. Each contraction of the muscles produces a dimpling of the exoskeleton in association with a downstroke and sometimes, depending on the exact arrangement of the muscles, on the upstroke as well. When the dimple springs back into its resting position, the muscles attached to it are stretched. The stretching immediately initiates another contraction, and the cycle is repeated.

The deformation of the cuticle is transmitted as a force to the wings which beat—so fast that we may perceive the sound as a musical tone. In the common blowfly, for instance, the wings may beat at a frequency of 120 cycles *per second*. Yet, in the same blowfly, the neurons that innervate those furiously contracting flight muscles are delivering impulses to them at the astonishingly low frequency of three per second. The mechanical properties of the musculoskeletal arrangement provide the stimuli for contraction, by stretching the muscle fibers at a high frequency. But the nerve impulses are needed to maintain it.

Flight muscles must be kept at appropriate operating temperature if they are to function. You have probably noticed the constant twitching of the wings of such insects as wasps even when they are crawling instead of flying. Probably this behavior is necessary to keep the temperature of the flight muscles high for instant readiness. You may also have noticed that the bodies of many moths are quite furry. Moth fur serves the same function as fur in a mammal, to conserve body heat. When the moth awakens and prepares for flight, it shivers its flight muscles at a low frequency to warm them up, constricting its abdominal blood vessel to keep the heat in its thorax. Gradually the frequency of the shivering increases until, at a critical moment, the moth spreads its wings and hums off into the darkness.

Insect flight muscles in action have a very high metabolic rate, perhaps the highest of any tissue anywhere. Accordingly, these muscles contain more mitochondria than any known variety of muscle, and they are elaborately infiltrated with tiny air-filled tracheae that carry oxygen directly to each fiber.

Many insects have special adaptations to rid the body of the excess heat produced by the flight muscles. The rapidly flying sphinx moth, for example, has in its abdomen what amounts to a radiator—a great blood vessel that carries heat from the thorax, where it is generated, and emits it into the cool of the night.

Just how insects fly has been an aerodynamic mystery. Insect flight involves much more than just flapping the wings up and down. The flapping motion of the insect wing changes direction and speed, and upstrokes alternate with downstrokes at very high rates. At each shift of stroke, the wing rotates about its long axis, and tilts to just the right angle for the new direction of motion.

Somehow insects are able to create lift that is 20 or more times their body weight. In 1996 Charles P. Ellington and his colleagues at the University of Cambridge in England reported that as insects flap their wings downward, some of the air flowing over the wings rolls up along the front edge forming a vortex (like a whirlpool) that becomes larger as it moves along toward the tip of the wing. The vortex generates a low pressure region above the wing that sucks the wing upward. This generates an extra lifting force.

Elephant hawk moth *(Deilephila elpenor)* in flight. *(Stephen Dalton/ Photo Researchers, Inc.)*

Contraction occurs when actin and myosin filaments slide past each other

Muscle contraction occurs when the sarcomeres, and thus the muscle fibers, shorten. This shortening takes place when the actin and myosin filaments slide past each other, increasing their overlap. This theory of muscle contraction, known as the **sliding filament model,** was developed in the 1950s by two British biologists, Hugh Huxley and Andrew Huxley. It is im-portant to understand that neither the actin nor myosin filaments change in length. The length of the muscle shortens as the filaments increase their overlap. You might think of an extension ladder. The overall ladder length changes as the ends get closer or farther apart, but the length of each ladder section stays the same.

In a **motor unit,** a motor neuron (a neuron that stimulates muscle) is functionally connected with an average of about

(Text continued on page 819.)

Figure 38–8 **Muscle structure.** *(a)* A muscle such as the biceps in the arm consists of many fascicles (bundles) of muscle fibers. *(b)* A fascicle wrapped in a connective tissue covering. *(c)* Part of a muscle fiber showing the structure of myofibrils. The Z lines mark the ends of the sarcomeres. *(d)* TEM of striated muscle. *(e)* LM showing striations. *(d, D.W. Fawcett; e, Ed Reschke)*

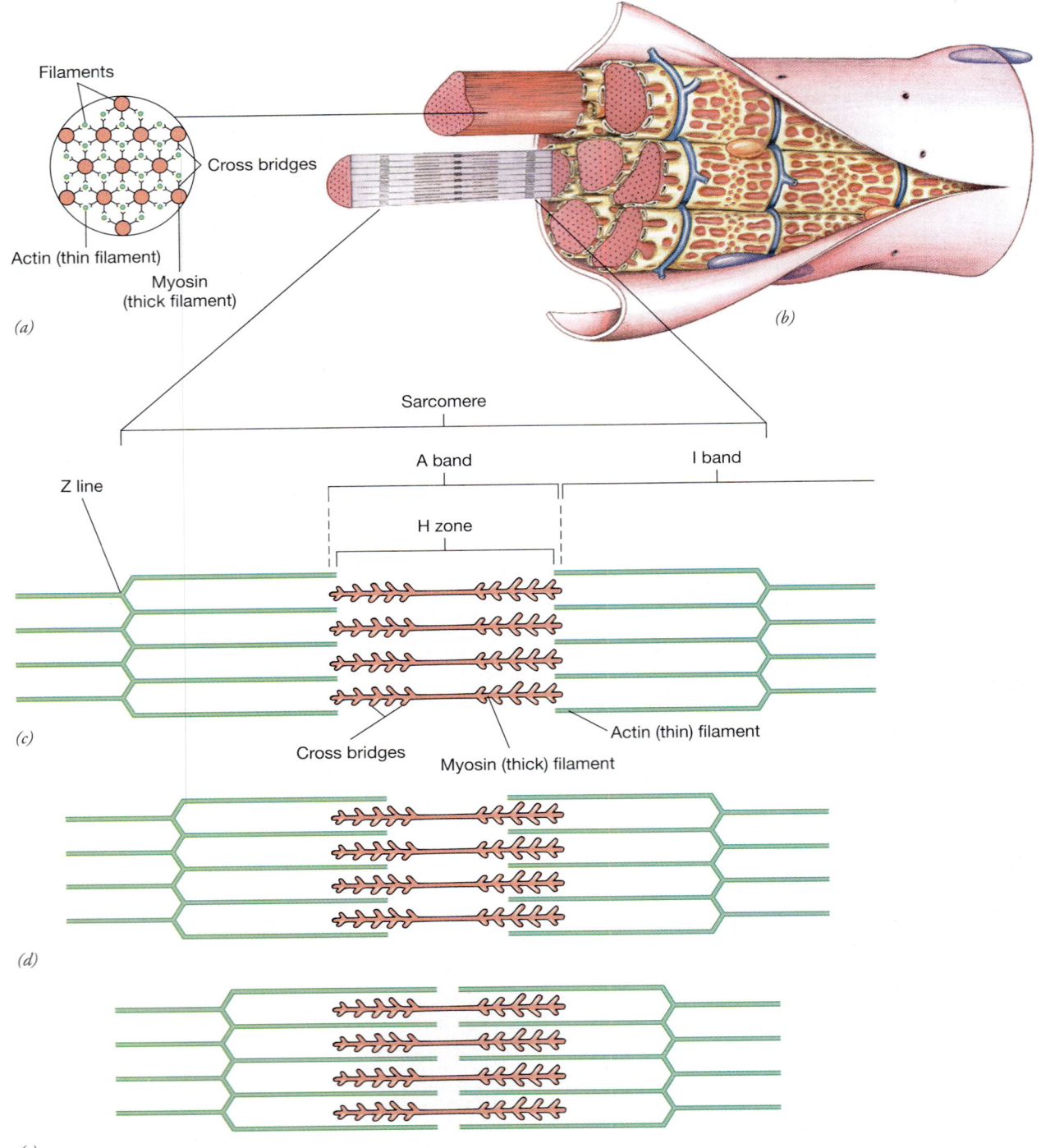

(a) Filaments

Cross bridges

Actin (thin filament)

Myosin
(thick filament)

(b)

Sarcomere

A band

I band

Z line

H zone

Cross bridges

Myosin (thick) filament

Actin (thin) filament

(c)

(d)

(e)

Figure 38−9   Muscle contraction. Myofibrils are threadlike structures that contain actin and myosin filaments. *(a)* Cross section of a myofibril shows the arrangement of actin and myosin filaments. *(b)* Part of a muscle fiber showing the location of the filaments. *(c)* The regular pattern of overlapping filaments gives skeletal and cardiac muscle their striated appearance. *(d)* During contraction actin filaments slide toward each other, increasing the amount of overlap between actin and myosin filaments. *(e)* As the sarcomeres become shorter, the muscle fiber becomes shorter.

Figure 38-10 **A motor unit.** A motor unit consists of a motor nerve fiber and the muscle fibers that it innervates. A motor unit typically includes about 150 muscle fibers, but some units have less than a dozen fibers, while others have several hundred. *(a)* The motor unit illustrated here shows only a single motor nerve fiber. *(b)* SEM of some of the fibers in a motor unit. Note how the neuron branches to innervate all of the cells in the motor unit. *(b, Don Fawcett/ Science Source/Photo Researchers, Inc.)*

150 muscle fibers (Fig. 38–10). When a motor neuron transmits a message, it releases the neurotransmitter **acetylcholine** into the synaptic cleft (the gap) between the motor neuron and each muscle fiber. Acetylcholine binds with receptors on each muscle fiber, causing an electrical change, called **depolarization,** along its sarcolemma. Depolarization may cause an electrical impulse, or **action potential,** to be generated.

In a muscle fiber, an action potential is a wave of depolarization that travels along the sarcolemma and into the system of T tubule membranes. The T tubules release a compound called inositol triphosphate (IP_3) that diffuses to the sarcoplasmic reticulum. In response, the sarcoplasmic reticulum releases stored calcium into the myofibrils.

Calcium ions bind to troponin molecules in the actin filaments, which causes the troponin to change shape. This change results in the troponin pushing the tropomyosin away from the active sites on the actin (Fig. 38–11). These sites are now exposed.

One end of each myosin molecule is folded into two globular structures called heads. The rounded heads of the myosin molecules extend away from the body of the myosin filament. Each myosin molecule also has a long tail that joins other myosin tails to form the thick filament. ATP is bound to the myosin when the muscle fiber is at rest (not contracting). Myosin is an ATPase, an enzyme that splits ATP. The ADP and phosphate initially remain attached to the myosin head.

1. Acetylcholine released by motor neuron combines with receptors on muscle fiber, causing depolarization and action potential.

2. Impulse spreads through T tubules, stimulating Ca^{2+} release from sarcoplasmic reticulum.

Figure 38–11 **Model of muscle contraction.** Cross bridges move the thin and thick filaments past each other. Parts of this model are still hypothetical.

ATP binding site

Myosin filament

Actin filament

Tropomyosin

Troponin

Active site

3. Ca^{2+} bind to troponin causing change in shape. Troponin pushes tropomyosin away, exposing active sites on actin filaments.

4. ATP is split. The myosin head, now cocked, binds to exposed active site, forming a cross bridge.

7. Actin-myosin complex binds ATP and myosin detaches from actin.

5. P_i and ADP released.

6. Cross bridge flexes and actin filament is pulled toward center of sarcomere.

Myosin heads (bound to ADP and P_i) bind to exposed active sites on the actin filament, forming **cross bridges** that link the myosin and actin filaments.

According to the currently accepted model of muscle contraction, the ADP and phosphate are released, causing the myosin to undergo a conformational change. The myosin bends about 45 degrees in a flexing motion that is the *power stroke* that pulls the actin filament along toward the center of the sarcomere. The actin-myosin complex then binds with a new ATP and myosin detaches from the actin. ATP must bind to the myosin head before the cross bridge can detach itself from the actin and begin a new cycle.

Energized once again, the myosin heads contact a second set of active sites, those closer to the end of the sarcomere. The process is repeated with a third set, and so on (Fig. 38–11). This series of stepping motions pulls the actin filaments toward the center of the sarcomere. Thus, when the cross bridges attach, move 45 degrees, detach, and then reattach farther along the actin filament, the muscle shortens. One way to visualize this process is to imagine the myosin heads engaging "hand-over-hand" in a kind of tug-of-war on the actin filaments. When many sarcomeres contract simultaneously, they produce the contraction of the muscle as a whole. The sequence of events in muscle contraction can be summarized as follows:

Motor neuron releases acetylcholine → combines with receptors on muscle fiber → action potential → impulse spreads through T tubules → Ca^{2+} released from sarcoplasmic reticulum → Ca^{2+} binds to troponin, causing shape change → troponin pushes tropomyosin away, exposing active sites on actin filaments → ATP (attached to myosin) is split → myosin head binds to exposed active site on actin filament, forming cross bridge → P_i and ADP released → cross bridge flexes and actin filament pulled toward center of sarcomere → actin-myosin complex binds ATP and myosin detaches from actin → if sufficient Ca^{2+}, sequence repeats with activation of myosin and splitting of ATP

After contracting, muscle fibers return to their resting state. Acetylcholine in the synaptic cleft is inactivated by the enzyme acetylcholinesterase. Calcium ions are pumped back into the sarcoplasmic reticulum by active transport, a process that requires ATP. Without calcium ions, the active sites on the actin filaments are again covered by the tropomyosin-troponin complex. The actin filaments slide back to their original position, and the muscle relaxes. This entire series of events happens in milliseconds.

Even when we are not moving, our muscles are in a state of partial contraction known as **muscle tone.** At any given moment some muscle fibers are contracted, stimulated by messages from nerve cells. Muscle tone is an unconscious process that helps keep muscles prepared for action. When the motor nerve to a muscle is cut, the muscle becomes limp (completely relaxed), or flaccid.

ATP powers muscle contraction

Muscle cells are often called on to perform strenuously, and so they must be provided with large amounts of energy. The immediate source of energy necessary for muscle contraction is ATP, needed not only for the pull exerted by the cross bridges but also for their release from each active site as they engage in their tug-of-war on the actin filaments. Rigor mortis, the temporary but very marked muscular rigidity that appears after death, results from ATP depletion following the cessation of cellular respiration after death.[1] Without ATP, the cross bridges that have been formed are not released.

Sufficient energy can be held in ATP molecules for only a few seconds of strenuous activity. Fortunately, muscle cells have a backup energy storage compound called **creatine phos-**phate that can be stockpiled. The energy stored in creatine phosphate is transferred to ATP as needed. But during vigorous exercise the supply of creatine phosphate is also short-lived. As ATP and creatine phosphate stores are depleted, muscle cells must replenish their supplies of these energy-rich compounds.

Fuel is stored in muscle fibers in the form of **glycogen,** a large polysaccharide formed from hundreds of glucose units. Stored glycogen is degraded, yielding glucose, which is then broken down in cellular respiration. When sufficient oxygen is available, enough energy is captured from the glucose to produce needed quantities of ATP and creatine phosphate.

During a burst of strenuous exercise, the circulatory system cannot deliver sufficient oxygen to keep up with the demand of the rapidly metabolizing muscle cells. This results in an **oxygen debt.** Under these conditions, muscle cells can break down fuel molecules anaerobically (without oxygen) for short periods. Lactic acid fermentation is a method of generating ATP anaerobically, but not in great quantity (Chapter 7). ATP depletion results in weaker contractions and muscle fatigue. Accumulation of the waste product lactic acid also contributes to muscle fatigue. Well conditioned athletes learn to tolerate the high levels of lactic acid generated during high performance activity. The period of rapid breathing that generally follows strenuous exercise pays back the oxygen debt.

The energy conversion of muscle contraction is relatively efficient compared to human-made machines. Recall from Chapter 7 that a steam power plant has an efficiency of about 35%. About 40% of the chemical energy of the glucose fuel is actually converted to mechanical work. The remaining energy is accounted for as heat, produced mainly by frictional forces within the muscle cell. That is why we get hot during hard physical labor, and shiver when we are cold. The muscle contractions involved in shivering are one way that animals produce heat to warm the body.

Skeletal muscle action depends on muscle pairs that work antagonistically

Skeletal muscles produce movements by pulling on **tendons,** tough cords of connective tissue that anchor muscles to bone. Tendons then pull on bones. Skeletal muscles, or their tendons, pass across a joint and are attached to the bones on each side of the joint. When the muscle contracts, it draws one bone toward or away from the bone with which it articulates.

Muscles can only pull; they cannot push. Muscles act **antagonistically** to one another, which means that the move-

---

[1]Rigor mortis does not persist indefinitely, however, for the entire contractile apparatus of the muscles eventually decomposes, restoring pliability. The phenomenon is temperature-dependent, so, given the prevailing temperature, a medical examiner can estimate the time of death of a cadaver from its degree of rigor mortis. Perhaps it should be said that rigor mortis by itself is not muscular contraction; it only tends to freeze the corpse in its position at the time of death. Thus, tales of corpses sitting up and pointing to their murderers, and otherwise carrying on posthumously may be entertaining but have no factual basis.

Biceps relaxes

Triceps contracts

Triceps relaxes

Biceps contracts

Extension

Flexion

(a)

(b)

Figure 38–12 **Muscle action.** The biceps and triceps muscles function antagonistically.

ment produced by one can be reversed by another. The biceps muscle, for example, flexes (bends) your arm, whereas contraction of the triceps muscle extends it once again (Fig. 38–12). Thus, the biceps and triceps work antagonistically.

The muscle that contracts to produce a particular action is known as the **agonist.** The muscle that produces the opposite movement is the **antagonist.** When the agonist is contracting, the antagonist is relaxed. Generally, movements are accomplished by groups of muscles working together, so there may be several agonists and several antagonists in any action. Note that muscles that are agonists in one movement may be antagonists in another. The superficial skeletal muscles of the human body are shown in Figures 38–13 and 38–14.

Smooth, cardiac, and skeletal muscle respond in specific ways

Each type of muscle is specialized for particular types of responses. **Smooth muscle** is not attached to bones, but forms tubes that squeeze like the muscle tissue in the body wall of the earthworm. Smooth muscle often contracts in response to simple stretching, and its contraction tends to be sustained. It is well adapted to performing such tasks as the regulation of blood pressure by sustained contraction of the walls of the ar-

terioles. Although smooth muscle contracts slowly, it shortens much more than striated muscle does; it squeezes superlatively.

Smooth muscle is specialized for its type of response. The fibers of smooth muscle tissue can function as a unit because they are connected by gap junctions (Chapter 5). These junctions permit electrical signals to pass rapidly from fiber to fiber. Although smooth muscle contraction is basically similar to contraction of skeletal muscle, the cross bridges in smooth muscle remain in the attached state longer. This translates into less ATP being required to maintain a high level of force.

Cardiac muscle contracts and relaxes in alternating rhythm, propelling blood with each contraction. Sustained contraction of cardiac muscle would be disastrous! Like smooth muscle fibers, cardiac muscle fibers are electrically coupled by gap junctions. Cardiac muscle (also like smooth muscle) can produce its own signals for contraction.

Skeletal muscle, when stimulated by a single brief electrical stimulus, contracts with a single quick contraction called a **simple twitch.** Simple twitches ordinarily do not occur except in laboratory experiments. In the normal animal, skeletal muscle receives a series of separate stimuli timed very close together. These do not produce a series of simple twitches, however, but a single, smooth, sustained contraction called **tetanus.** Depending on the identity and number of our muscle cells tetanically contracting, we might thread a needle, rock a baby, or run a mile.

Not all muscular activities are the same. Dancing and, even more so, typing require quick response rather than the long, sustained effort that might be appropriate in hauling a rope or maintaining posture. Two main types of muscle fibers are **white fibers** specialized for rapid response and **red (dark) fibers** specialized for slower, endurance activities. In many animals, entire muscles are specialized for quick or slow responses. In chickens, for instance, the white breast muscles are efficient for quick responses, because flight is an escape mechanism for chickens. On the other hand, they walk about on the ground all day, so the dark (red) meat of the leg and thigh is composed of muscle specialized for more sustained activity. Birds that fly have red breast muscles specialized to support sustained activity.

Like other vertebrates, humans possess white and red muscle fibers. White fibers obtain most of their energy from glycolysis; they have few mitochondria. White fibers can generate a great deal of power, but can sustain that activity for only a brief period. When their glycogen supply is depleted, they fatigue rapidly.

Red fibers are rich in **myoglobin,** a red pigment similar to hemoglobin. Myoglobin, like the hemoglobin in red blood cells, stores oxygen. Myoglobin enhances the rapid diffusion of oxygen during strenuous muscle exertion. Red fibers, which require a steady supply of oxygen, have many mitochondria and a rich supply of capillaries.

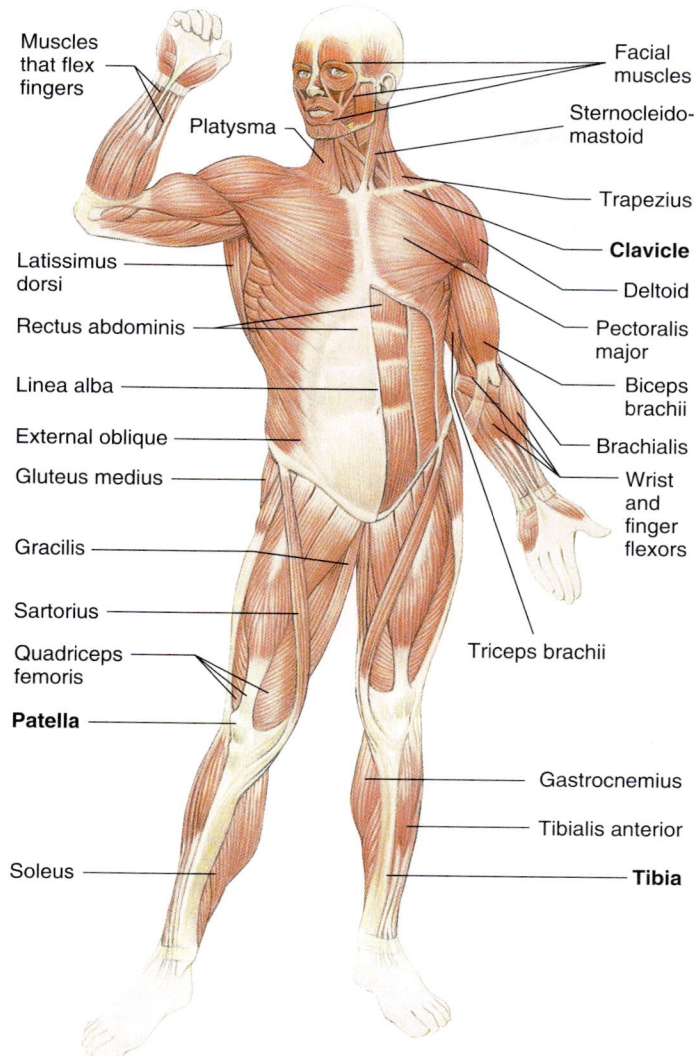

Figure 38–13 **Some superficial muscles of the human body.** Anterior view.

Labels (clockwise):
Muscles that flex fingers · Platysma · Facial muscles · Sternocleidomastoid · Trapezius · **Clavicle** · Deltoid · Pectoralis major · Biceps brachii · Brachialis · Wrist and finger flexors · Triceps brachii · Gastrocnemius · Tibialis anterior · **Tibia** · Soleus · **Patella** · Quadriceps femoris · Sartorius · Gracilis · Gluteus medius · External oblique · Linea alba · Rectus abdominis · Latissimus dorsi

Figure 38–14 **Some superficial muscles of the human body.** Posterior view.

Labels: Sternocleidomastoid · Trapezius · Deltoid · Triceps brachii · Brachioradialis · Muscles that flex fingers · Biceps brachii · Brachialis · Latissimus dorsi · External oblique · Gluteus maximus · Hamstring muscles · Gracilis · Semitendinosus · Biceps femoris · Semimembranosus · Gastrocnemius · Soleus · Achilles tendon · **Calcaneus**

There are two types of red fibers: **fast-twitch fibers** specialized for moderate endurance activities and **slow-twitch fibers** specialized for long, slow response. Fast-twitch fibers have a higher ATPase activity; their myosin can break down ATP more rapidly than slow-twitch fibers. As a result they contract and relax more rapidly. Slow twitch fibers function more slowly and are very resistant to fatigue.

The proportions of slow-twitch and fast-twitch fibers vary from person to person and from muscle to muscle in the same person. The relative proportions of the two appear to be genetically determined, and they influence the kind of athletic activity at which each of us has the greatest potential proficiency. Someone whose leg and thigh muscles contain a high proportion of fast-twitch fibers could, with proper training, become a good sprinter. An athlete with a greater proportion of slow-twitch fibers may be better suited to marathon activities. The proportions of the two kinds of fibers can be changed by appropriate training.

I. **Epithelial coverings** protect underlying tissues. They may be specialized for sensory or respiratory functions, to produce a protective cuticle, to secrete lubricants or adhesives, to produce odorous or poisonous secretions, or to produce threads for nests or webs.

II. The **integumentary system** of vertebrates includes the skin and structures that develop from it. In humans this system includes the skin, nails, hair, sweat glands, oil glands, and sensory receptors. The vertebrate skin protects, may help prevent dehydration, may be specialized for secretion or reception of stimuli, and may help regulate body temperature.

A. In human skin, cells in **stratum basale** of the **epidermis** divide and are pushed upward toward the skin surface. These cells mature, flatten, produce **keratin**, and eventually die. **Stratum corneum**, the most superficial layer of the epidermis, consists of dead cells filled with keratin.

B. The **dermis**, which consists of dense, fibrous connective tissue, rests on a layer of **subcutaneous tissue** composed largely of fat.

III. The **skeletal system** transmits mechanical forces generated by muscle and also supports and protects the body.

A. Many invertebrates have a **hydrostatic skeleton** in which fluid in a closed body compartment is used to transmit forces generated by contractile cells or muscle.

B. **Exoskeletons** are characteristic of mollusks and arthropods. The arthropod skeleton, composed mainly of chitin, is jointed for flexibility. This nonliving skeleton does not grow, making it necessary for arthropods to **molt** periodically.

IV. The **endoskeletons** of echinoderms and chordates are composed of living tissue and therefore are capable of growth.

A. The vertebrate skeleton consists of an **axial skeleton** and an **appendicular skeleton.**

1. The axial skeleton consists of **skull, vertebral column,** ribs, and sternum.

2. The appendicular skeleton consists of bones of the limbs, pectoral girdle, and pelvic girdle.

B. A typical long bone consists of a thin outer shell of **compact bone** surrounding the inner **spongy bone** and a central marrow cavity.

C. Long bones develop from cartilage templates during **endochondral bone formation.** Other bones, such as the flat bones of the skull, develop from a noncartilage connective tissue model by **intramembranous bone development.**

D. **Osteoblasts,** cells that produce bone, and **osteoclasts,** cells that break down bone, work together to shape bone.

E. **Joints** are junctions of two or more bones. The sutures of the skull are **immovable joints;** joints between vertebrae are **slightly movable joints.** A **freely movable joint** is enclosed by a joint capsule lined with a membrane that secretes **synovial fluid. Ligaments** are connective tissue bands that connect bones and limit movement at the joint.

V. All animals have the ability to move. A **muscular system** is found in most invertebrate phyla and in vertebrates. As muscle tissue contracts (shortens), it moves body parts by pulling on them. Three types of muscle are **skeletal, smooth,** and **cardiac muscle.**

A. A muscle such as the biceps is an organ made up of hundreds of muscle fibers. Each fiber is made up of threadlike **myofibrils** that are composed of smaller **myofilaments,** or simply **filaments.**

B. The striations of skeletal muscle fibers reflect the overlapping of their **actin filaments** and **myosin filaments.** A **sarcomere** is a contractile unit of actin (thin) and myosin (thick) filaments.

C. During muscle contraction, the thin filaments are pulled by the myosin filaments toward the center of the myofibril.

1. **Acetylcholine** released by a motor neuron combines with receptors on the surface of a muscle fiber. This may cause **depolarization** of the sarcolemma and transmission of an **action potential.**

2. The impulse spreads through the **T tubules,** which release inositol triphosphate (IP_3). This compound stimulates calcium ion release from the sarcoplasmic reticulum.

3. Calcium ions bind to **troponin** in the actin filaments, causing the troponin to change shape. Troponin pushes **tropomyosin** away from the active sites on the actin filaments.

4. ATP is bound to myosin when the muscle is not contracting. ATP is split, which puts the myosin head in a high energy state (it is "cocked"). Energized myosin heads bind to the exposed active sites on the actin filaments, forming **cross bridges** that link the myosin and actin filaments.

5. After myosin attaches to the actin filament, phosphate and ADP are released.

6. It is thought that this release causes the cross bridge to flex, producing the power stroke that pulls the actin filament toward the center of the sarcomere.

7. The actin-myosin complex binds ATP, and the myosin detaches from the actin. As long as calcium concentration remains elevated, the ATP is split (step 4), and the sequence of steps repeats. The myosin reattaches to new active sites so that the filaments are pulled past one another and the muscle continues to shorten.

D. ATP is the immediate source of energy for muscle contraction. The energy from ATP hydrolysis provides the energy to "cock" the myosin, and it also detaches the actin and myosin after movement has taken place. Muscle tissue has an intermediate energy storage compound, **creatine phosphate. Glycogen** is the fuel stored in muscle fibers.

E. **Muscle tone** is the state of partial contraction characteristic of muscles. Muscles act by pulling on **tendons,** connective tissue cords that attach muscles to bones. Muscles act **antagonistically** to one another. The muscle that produces a particular action is the **agonist.** The muscle that produces the opposite movement is the **antagonist.**

F. When activated by a brief electrical stimulus, skeletal muscle responds with a **simple twitch.** Typically skeletal muscle is stimulated with a series of separate stimuli timed close together and responds with a smooth, sustained contraction called **tetanus.**

G. **White fibers** are specialized for rapid response. **Red fibers,** which are rich in **myoglobin,** are specialized for slower, longer responses. There are two types of red fibers: **fast-twitch fibers** specialized for moderate endurance responses, and **slow-twitch fibers,** specialized for long, slow responses.

POST-TEST

1. The vertebrate skin consists of (a) outer epidermis, inner hypodermis (b) outer epidermis, inner endoskeleton (c) outer endodermis, inner epidermis (d) outer epidermis, inner dermis (e) outer subcutaneous layer, inner dermis

2. Cells actively divide in (a) stratum basale (b) stratum corneum (c) stratum dermis (d) layer with cells that contain keratin (e) two of the preceding answers are correct

3. An endoskeleton (a) is typically composed of dead tissue (b) is characterized by fluid in a closed compartment (c) is typical of echinoderms (d) is typical of arthropods (e) requires the animal to molt

4. Which of the following is NOT part of the axial skeleton? (a) skull (b) vertebral column (c) pelvic girdle (d) atlas (e) sacrum

5. The thin outer shell of a long bone is composed of (a) compact bone (b) spongy bone (c) epiphyses (d) cancellous bone (e) mainly chondrocytes

6. Which of the following connects bones? (a) tendons (b) ligaments (c) osteoclasts (d) synovial membranes (e) smooth fibers

7. In endochondral bone formation (a) osteoclasts produce bone (b) joints connect fibers (c) the skeleton consists of cartilage (d) bones develop from cartilage templates (e) bones form in noncartilage connective tissue

8. All animals have (a) muscles (b) actin (c) bones (d) endoskeleton or exoskeleton (e) dermis

9. An energy storage compound that can be stockpiled in muscle cells is (a) creatine phosphate (b) ATP (c) troponin (d) myosin (e) two of the preceding answers are correct

10. Which sequence most accurately describes events of muscle contraction? (a) acetylcholine released → action potential → calcium ions stimulate process that leads to exposure of active sites → myosin binds to actin, forming cross bridges → ATP is split/cross bridges flex → myosin filament shortens (b) acetylcholine released → action potential → calcium ions stimulate process that leads to exposure of active sites → ATP is split/cross bridges flex → myosin binds to actin → actin filament shortens (c) acetylcholine released → action potential → calcium ions stimulate process that leads to exposure of active sites → myosin binds to actin, forming cross bridges → ATP is split/cross bridges flex → filaments slide past each other/muscle fiber shortens (d) troponin released → action potential → calcium ions stimulate process that leads to exposure of active sites → myosin binds to actin, forming cross bridges → ATP is split/cross bridges flex → muscle fiber shortens (e) acetylcholine released → action potential → calcium ions form cross bridges → acetylcholine binds to actin → ATP is formed/cross bridges flex → actin filament shortens

11. When skeletal muscle is stimulated with a series of separate stimuli timed close together (a) it responds with a smooth, sustained contraction called tetanus (b) a simple twitch occurs (c) white fibers respond (d) red fibers respond (e) muscle tone occurs.

12. Glycogen (a) is produced by actin (b) is depleted within one second of strenuous activity (c) is a form of long term energy storage (d) is synthesized when cross bridges form (e) when depleted is responsible for oxygen debt

REVIEW QUESTIONS

1. Compare the external epithelium of invertebrates and vertebrates.
2. What properties does keratin give human skin?
3. Describe a hydrostatic skeleton. What functions does it perform? How do the septa in the annelid worm contribute to the versatility of its hydrostatic skeleton?
4. What are some disadvantages of an exoskeleton? Some advantages?
5. Describe the divisions of the human skeleton.
6. Draw a typical long bone, such as the radius, and label its parts.
7. What are the functions of osteoblasts and osteoclasts? Why is it important that bones be continuously remodeled?
8. Describe a skeletal muscle fiber and compare its two types of filaments.
9. Outline the sequence of events thought to occur when a muscle fiber contracts, beginning with the release of acetylcholine and including cross bridge action.
10. What is the role of ATP in muscle contraction? What are the functions of creatine phosphate and glycogen?
11. Label the diagram. Use Fig. 38–8 to check your answers.

YOU MAKE THE CONNECTION

1. Compare the arthropod exoskeleton with the vertebrate endoskeleton. What are some benefits and some disadvantages of each type of skeleton?
2. What are some examples of hydrostatic support in plants?
3. Why is it important that a muscle be able to shift functions, sometimes acting as an agonist and at other times acting as an antagonist?

RECOMMENDED READINGS

Dickman, S. "No Bones About a Genetic Switch for Bone Growth." *Science,* Vol. 276, 6 Jun. 1997. A brief account of recent research on the genetics of bone development.

Fackelman, K. "Melanoma Madness: The Scientific Flap over Sunscreens and Skin Cancer." *Science News,* Vol. 153 6 Jun. 1998. A reasoned discussion of the costs and benefits of using sunscreen.

Hadley, N.F. "The Arthropod Cuticle." *Scientific American,* Vol. 255, No. 1, Jul. 1986. The arthropod exoskeleton is largely responsible for the adaptive success of the phylum. The author discusses the properties that enable the exoskeleton to provide protection and support.

Mayor, M.B., and J. Collier. "The Technology of Hip Replacement." *Scientific American Science and Medicine,* May/Jun. 1994. An interesting account of technology for replacing damaged joints.

Zimmer, C. "See How They Run." *Discover,* Sep. 1994. A discussion of animal locomotion.

● Visit our website at **http://www.saunderscollege.com/lifesci/titles.html** and click on Solomon/Berg/Martin Biology for links to chapter-related resources on the World Wide Web.

CHAPTER 39

Neural Control: Neurons

An organism's ability to survive and to maintain homeostasis depends largely on how effectively it responds to internal signals like hunger or lowered blood pressure and to external signals like changes in temperature or the presence of predators. Changes within the body or in the outside world that can be detected by an organism are called **stimuli.** In all animals except the sponges, responses to stimuli depend on the activities of networks of nerve cells, or **neurons.** These cells are specialized for transmitting electrical and chemical signals. In most animals, neurons and supporting cells are organized as a **nervous system** that functions like a computer, taking information in, integrating it, and responding. Just how animals respond to stimuli depends on how their neurons are organized and connected to one another. A single neuron in the vertebrate brain may be functionally connected to thousands of other neurons.

Neurobiology is one of the most exciting areas of biological research. Recently, much effort has been focused on **neurotransmitters,** the chemical messengers used by neurons to signal other neurons, and on the receptors that bind with the neurotransmitters. Mood states are influenced by the concentration of certain neurotransmitters, and disorders such as schizophrenia and depression are related to abnormal amounts. Many antidepressant medications act by selectively inhibiting the reuptake of the neurotransmitter serotonin by neurons in the brain. This action results in a greater concentration of serotonin, leading to an elevation in mood.

Another active area of research is the role of **glial cells** in the nervous system. These cells support and protect the neurons and have many regulatory functions. The immunofluorescent LM shows a tangle of neurons (*green*) and a large microglial cell (*red*) within brain tissue. This type of glial cell is

(Nancy Kedersha/UCLA/Science Photo Library/Photo Researchers, Inc.)

a phagocytic (scavenger) cell that ingests bacteria and other foreign matter.

In most animals, the endocrine system works with the nervous system to regulate many behaviors and physiological processes. The endocrine system generally provides relatively slow and long-lasting regulation, whereas the nervous system typically permits more rapid, but brief, responses.

AFTER YOU HAVE STUDIED THIS CHAPTER YOU SHOULD BE ABLE TO

1. Trace the flow of information through the nervous system. Describe the four processes involved in responding to a stimulus: reception, transmission, integration, and response.
2. Draw a typical neuron, label its parts (including myelin and cellular sheaths) and give the function of each.
3. Summarize the process by which an impulse is transmitted along a neuron.
4. Explain the ionic basis of the resting membrane potential and relate the conduction of an action potential to changes in ion distribution.

5. Compare continuous conduction with saltatory conduction.
6. Trace the events that take place in synaptic transmission and draw diagrams to support your description.
7. Identify the neurotransmitters described in the chapter and describe mechanisms for inactivating them.
8. Identify factors that affect speed of neural transmission.
9. Describe how a postsynaptic neuron integrates incoming stimuli and "decides" whether to fire.
10. Draw diagrams of neural circuits showing convergence, divergence, and reverberation and explain why each is important.

THE NERVOUS SYSTEM IS FUNCTIONALLY ORGANIZED

Thousands of stimuli constantly bombard an animal, and its survival depends on identifying and either ignoring or responding appropriately to these stimuli. In most animals, appropriate response to a stimulus involves four processes: reception, transmission, integration, and action by muscles or glands. **Reception,** the process of detecting a stimulus, is the job of the neurons and of specialized sense organs such as eyes and ears. **Transmission** is the process of sending messages along a neuron, from one neuron to another or from a neuron to a muscle or gland. In vertebrates, a neural message is transmitted from a receptor to the **central nervous system (CNS),** which consists of the brain and spinal cord. Neurons that transmit information to the CNS are called **sensory neurons,** also known as **afferent neurons.**

Afferent neurons generally transmit information to **interneurons** (sometimes referred to as **association neurons**) that *integrate* input and output. **Integration** involves sorting and interpreting incoming sensory information and determining the appropriate response. Neural messages are transmitted from the CNS by **efferent neurons,** or **motor neurons,** to **effectors** (muscles and glands). The **action by effectors** is the actual response to the stimulus (Fig. 39–1). Sensory receptors and afferent and efferent neurons are part of the **peripheral nervous system (PNS).**

NEURONS AND GLIAL CELLS ARE THE CELLS OF THE NERVOUS SYSTEM

The functional unit of the nervous system is the neuron, a cell specialized to receive and send information. The neuron works by producing and transmitting electrical signals called **nerve impulses,** or **action potentials,** and by synthesizing and re-

leasing neurotransmitters. A second cell type unique to the nervous system is the glial cell, which provides support and protection for the neurons.

Glial cells provide metabolic and structural support

Sometimes glial cells are referred to collectively as the **neuroglia,** which literally means "nerve glue." Vertebrates have three main types of glial cells in the CNS: microglia, astrocytes, and oligodendrocytes. **Microglia** are found near blood vessels in the nervous system. When an injury or infection occurs, microglial cells migrate into the CNS, perhaps from the bone marrow. These phagocytic cells remove cellular debris. Recent studies suggest that they may play a role in the development of prion diseases (see Chapter 23).

Astrocytes are star-shaped glial cells that have a variety of functions. Some are phagocytic and remove invading microorganisms and debris from the nervous tissue. Others help regulate the concentration of potassium ions in the extracellular fluid of nervous tissue. Still other astrocytes regulate the concentration of neurotransmitters.

Oligodendrocytes are glial cells that envelop neurons in the CNS, forming insulating sheaths around them. This covering consists of myelin, a white, fatty substance found in the plasma membrane of the glial cell. Because it is an excellent electrical insulator, its presence speeds the transmission of neural impulses. In **multiple sclerosis,** a neurological disease that affects about 300,000 people in the United States, patches of myelin deteriorate at irregular intervals along axons and are replaced by scar tissue. This damage interferes with conduction of neural impulses, and the victim suffers loss of coordination, tremor, and partial or complete paralysis of parts of the body. The cause of multiple sclerosis has been a mystery, but some evidence suggests that it is an autoimmune disease, in which the body attacks its own tissue, in this case neurons and glial cells (autoimmune diseases are discussed in Chapter 43).

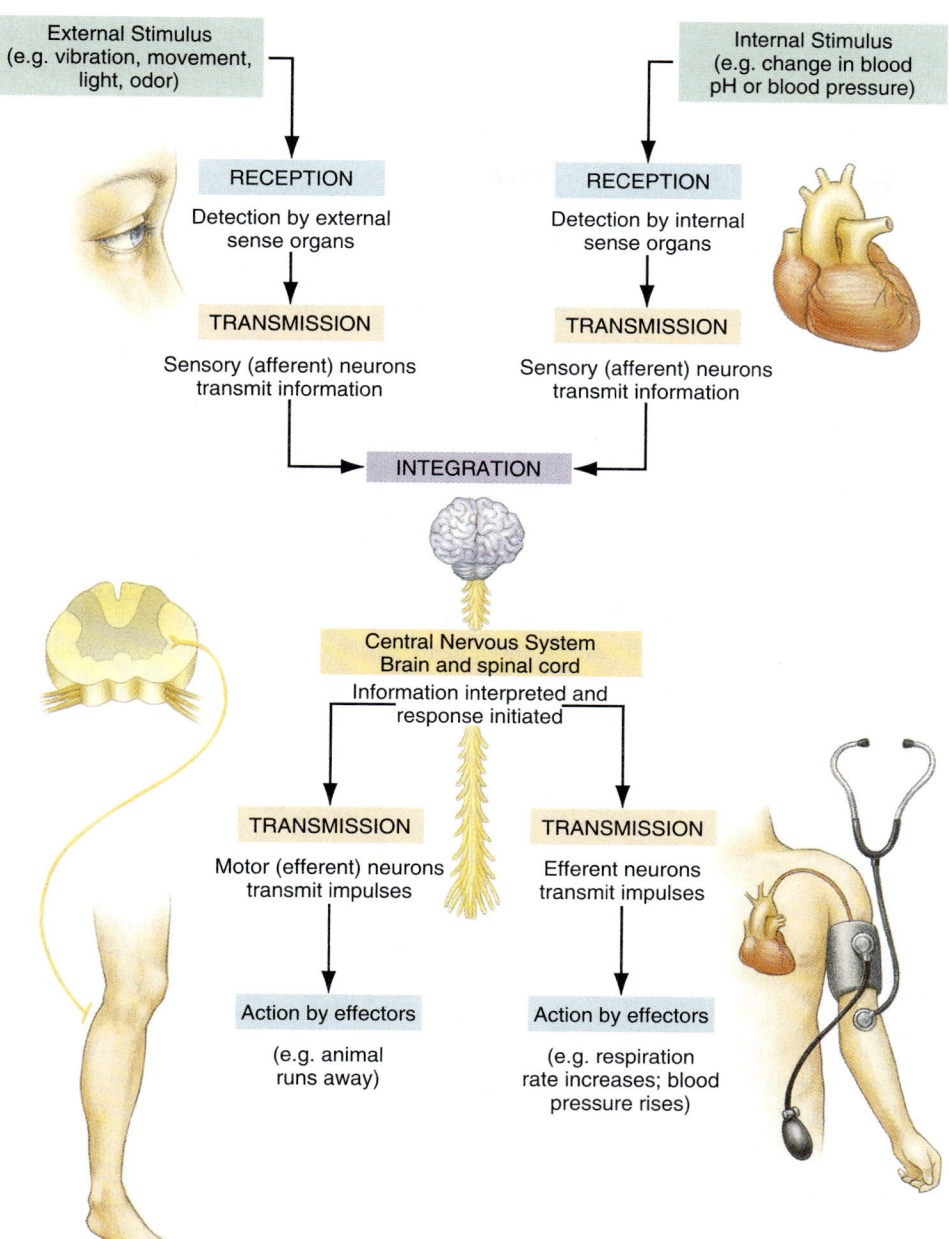

Figure 39–1 **Response to a stimulus.**
Whether a stimulus originates in the outside world or inside the body, information must be received, transmitted to the central nervous system, integrated, then transmitted to the muscles or glands that carry out some action.

External Stimulus (e.g. vibration, movement, light, odor)

Internal Stimulus (e.g. change in blood pH or blood pressure)

RECEPTION
Detection by external sense organs

RECEPTION
Detection by internal sense organs

TRANSMISSION
Sensory (afferent) neurons transmit information

TRANSMISSION
Sensory (afferent) neurons transmit information

INTEGRATION

Central Nervous System
Brain and spinal cord
Information interpreted and response initiated

TRANSMISSION
Motor (efferent) neurons transmit impulses

TRANSMISSION
Efferent neurons transmit impulses

Action by effectors
(e.g. animal runs away)

Action by effectors
(e.g. respiration rate increases; blood pressure rises)

In vertebrates, **Schwann cells,** another type of glial cell, are located outside the central nervous system. Schwann cells form sheaths around some neurons.

A typical neuron consists of a cell body, dendrites, and an axon

Highly specialized to receive stimuli and transmit messages in the form of electrical impulses, the neuron is distinguished from all other cells by its long cytoplasmic extensions, or processes. Examine the structure of a common type of neuron, the multipolar neuron in Figure 39–2.

The largest portion of the neuron, the **cell body,** contains the nucleus, the bulk of the cytoplasm, and most of the organelles. Two types of cytoplasmic extensions project from the cell body of a multipolar neuron: Numerous dendrites extend from one end, and a long, single axon projects from the opposite end.

Dendrites are typically short, highly branched processes specialized to receive stimuli and send signals to the cell body. The cell body integrates incoming signals.

Although microscopic in diameter, an **axon** may be 1m (over 1 yd) or more in length and may divide, forming branches called **axon collaterals.** The axon conducts nerve impulses

Figure 39–2 **Structure of a multipolar neuron.** The cell body contains most of the organelles. Many dendrites and a single axon extend from the cell body. Schwann cells form a myelin sheath around the axon.

Labels (top to bottom, left and right): Dendrites covered with dendritic spines · Cell body · Nucleus · Nodes of Ranvier · Collateral branch · Synaptic terminals · Axon · Schwann cell · Myelin sheath · Terminal branches

away from the cell body to another neuron, or to a muscle or gland. At its end the axon divides forming many **terminal branches** that end in **synaptic terminals.** The synaptic terminals release **neurotransmitters,** chemicals that transmit signals from one neuron to another, or from neuron to effector. The junction between a synaptic terminal and another neuron (or effector) is called a **synapse.** Typically, a small space exists between the membranes of these two cells.

In vertebrates, the axons of many neurons outside the CNS are surrounded by a series of Schwann cells that form an insulating covering, the **myelin sheath.** Gaps in the myelin sheath, called **nodes of Ranvier,** occur between successive Schwann cells. At these points the axon is not insulated with myelin. Axons more than 2 μm in diameter have myelin sheaths and are described as *myelinated.* Those of smaller diameter are generally unmyelinated.

A **nerve** consists of hundreds or even thousands of axons wrapped together in connective tissue (Fig. 39–3). We can compare a nerve to a telephone cable. The individual axons correspond to the wires that run through the cable, and the sheaths and connective tissue coverings correspond to the insulation. Within the CNS, bundles of axons are referred to as **tracts** or **pathways** rather than nerves. Outside the CNS, the cell bodies of neurons are usually grouped together in masses

called **ganglia** (sing., *ganglion*). Inside the CNS, collections of cell bodies are generally referred to as *nuclei* rather than ganglia.

NEURONS USE ELECTRICAL SIGNALS TO TRANSMIT INFORMATION

Most animal cells have a difference in electrical charge across the plasma membrane—a more negative electrical charge inside the cell compared with the electrical charge of the interstitial fluid outside. The plasma membrane is said to be electrically **polarized,** meaning that one side, or pole, has a different charge from the other side. When electrical charges are separated in this way, a potential energy difference exists across the membrane. This difference in electrical charge across the plasma membrane is referred to as the **membrane potential** or **resting potential.** Should the charges be permitted to come together, they have the *potential* to do work. Thus, the cell can be thought of as a biological battery. The ability of certain excitable cells, such as neurons and muscle cells, to send signals depends on the resting potential.

The neuron membrane has a sizable resting potential

The resting potential may be expressed in units called *millivolts* (mV). (A millivolt equals one-thousandth of a volt.) Voltage is the force that causes charged particles to flow between two points. Like other cells that can produce action potentials, the neuron has a large resting potential, about 70 mV. By convention this is expressed as -70 mV because the inner surface of the plasma membrane is negatively charged relative to the interstitial fluid (Fig. 39–4).

We can measure the potential across the membrane by placing one electrode, insulated except at the tip, inside the cell, and a second electrode outside the cell. The two electrodes are connected through a voltmeter or oscilloscope that measures the charge separation across the membrane. If we place both electrodes on the outside surface of the neuron, no potential difference between them is registered. All points on the same side of the membrane are at the same potential.

How does the membrane potential develop? In a resting neuron—one not transmitting an impulse—there is a slight excess of positive ions outside the plasma membrane and a slight excess of negative ions inside the membrane. The distribution of ions inside neurons and in the interstitial fluid surrounding them is similar to that of most other cells in the body. The K^+ concentration is about ten times greater inside than outside the cell. In contrast, the Na^+ concentration is about ten times greater outside than inside.

This distribution of ions across the plasma membrane is brought about by the action of ion pumps, selective ion channels, and gates. In vertebrate neurons (and skeletal muscle fibers), the resting membrane potential depends mainly on the

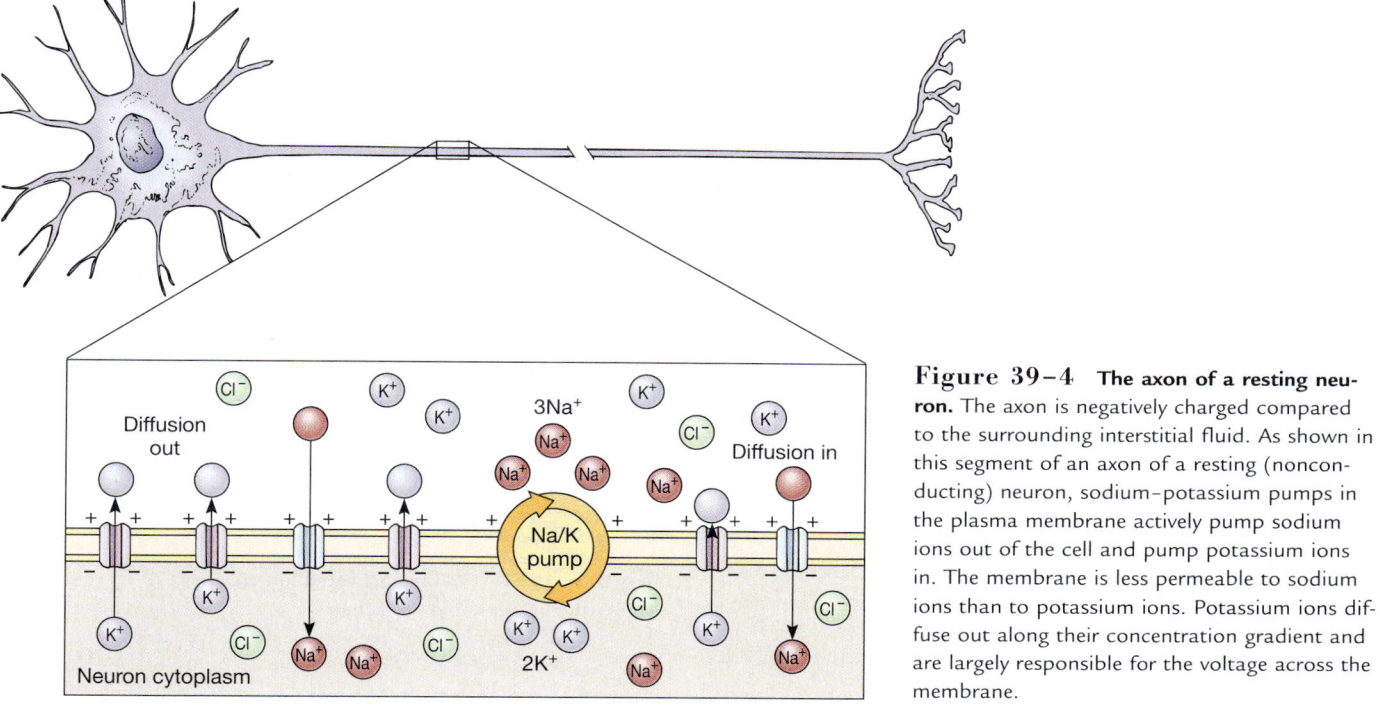

Figure 39–3 **Structure of a nerve.** *(a)* A nerve consists of bundles of axons held together by connective tissue. The cell bodies belonging to the axons of a nerve are grouped together in a ganglion. *(b)* SEM showing a cross section through a myelinated afferent nerve of a bullfrog. *(b, E.R. Lewis/Biological Photo Service)*

Figure 39–4 **The axon of a resting neuron.** The axon is negatively charged compared to the surrounding interstitial fluid. As shown in this segment of an axon of a resting (nonconducting) neuron, sodium–potassium pumps in the plasma membrane actively pump sodium ions out of the cell and pump potassium ions in. The membrane is less permeable to sodium ions than to potassium ions. Potassium ions diffuse out along their concentration gradient and are largely responsible for the voltage across the membrane.

diffusion of ions down their electrochemical gradients. These gradients are maintained by ion pumping. The neuron plasma membrane has very efficient **sodium-potassium pumps** that actively transport sodium ions out of the cell and potassium ions into the cell (see Fig. 5–15). Both sodium and potassium ions are pumped against a concentration gradient, and these pumps require ATP. For every three sodium ions pumped out of the cell, two potassium ions are pumped in. Thus, more positive ions are pumped out than in.

Ions also cross the membrane by diffusion through membrane proteins that form **ion channels.** Net movement of ions occurs from an area of higher concentration to one of lower concentration. Typically, these channels allow only specific types of ions to pass. The protein that makes up the ion channel may have charged regions that act as gates. When these gates are open the channel is activated; however when the gates are closed, ions cannot pass through the membrane.

Neurons have three types of ion channels: passive ion channels, voltage-activated channels, and chemically activated ion channels. **Passive ion channels** permit the passage of specific ions such as Na^+, K^+, Cl^-, and Ca^{2+}. These channels are generally open (Fig. 39–5). **Voltage-activated ion channels** (also called **voltage-gated channels**) are kept closed by gates that open in response to changes in voltage (Fig. 39–6). **Chemically activated ion channels** are found mainly on the dendrites and cell body. (These channels will be discussed in a later section.)

Potassium channels are the most common type of passive ion channel in the plasma membrane, and neurons are more permeable to potassium than to other types of ions. In fact, in the resting neuron the membrane is up to 100 times more permeable to K^+ ions than to Na^+ ions. Sodium ions pumped out of the neuron cannot easily pass back into the cell, but potassium ions pumped into the neuron can diffuse out.

The sodium-potassium pumps maintain a higher concentration of K^+ inside the cell than outside. Potassium ions leak out through passive ion channels following their concentration gradient. As these positively charged ions diffuse out of the neuron, the cell becomes more negatively charged. A steady state, known as the *equilibrium potential,* is reached when the K^+ outflow equals the inward flow of K^+ ions. At this point a potential of about -70 mV has developed across the membrane; this is the resting potential. Although the resting potential is primarily established by the K^+ ion gradient, the influx of chloride ions contributes slightly. The plasma membrane is permeable to negatively charged chloride ions. These ions accumulate along the inside of the plasma membrane, contributing to the negative charge.

The nerve impulse is an action potential

Neurons are highly excitable cells. They have the ability to respond to stimuli and to convert stimuli into nerve impulses. An electrical, chemical, or mechanical stimulus may alter the resting potential by increasing the membrane's permeability to sodium. If the neuron membrane is only slightly stimulated,

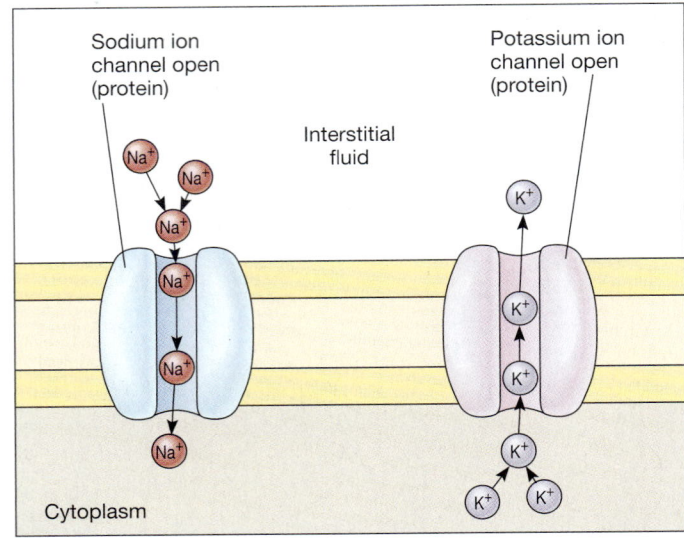

Figure 39–5 **Passive ion channels.** Proteins in the plasma membrane form specific passive ion channels. Ions flow through these channels down their concentration gradient.

only a local disturbance may occur in the membrane. If the stimulus is strong enough, however, the neuron fires a nerve impulse, or **action potential,** an electrical excitation that travels rapidly down the axon into the synaptic terminals.

Specific voltage-activated ion channels in the plasma membrane of the axon and cell body open when they detect a change in the membrane potential. These channels have been studied

Figure 39–6 **Voltage-activated ion channels.** *(a)* In the resting state, voltage–activated Na^+ channels are closed. *(b)* When the voltage reaches threshold level, the voltage-activated gates open, allowing Na^+ to pass into the cell. After a certain amount of time elapses, gates close, inactivating the channels.

using the *patch clamp technique,* also known as single-channel recording (see *Making the Connection: The Patch Clamp Technique in the Study of Ion Channels* in Chapter 5). Using this technique, biologists can measure currents across a very small segment of the plasma membrane rather than across the entire membrane. The patch of membrane measured is so small that it may contain a single channel. Investigators have established that when the voltage reaches a certain critical point, the **threshold level,** the protein making up the channel changes shape so that the gates to the channel open. When the gates open, specific ions can pass through the channel.

The membrane of the neuron can depolarize by about 15 mV, that is, to a resting potential of about −55 mV, without actually initiating an impulse. However, when the extent of depolarization is greater than −55 mV, the threshold level has been exceeded. At that point activating gates on voltage-activated sodium ion channels open, and Na$^+$ ions diffuse into the cell, moving from an area of higher concentration to an area of lower concentration. After a certain period of time, inactivating gates at the cytoplasmic end of the channel close the channels. Their closing depends on time rather than on voltage. Voltage-gated potassium ion channels also open when the threshold level is reached. They open more slowly and remain open until the resting potential has been restored. Although not all voltage-activated channels in a neuron have the precisely same threshold, when some channels open, depolarization increases, leading to the opening of more channels by a positive feedback mechanism.

As the action potential is produced, the neuron membrane quickly reaches zero potential and even overshoots to +35 mV or more, as a momentary reversal in polarity takes place. The sharp rise and fall of the action potential is referred to as a spike. Figure 39–7 illustrates an action potential that has been recorded by placing one electrode inside an axon and one just outside.

Once initiated, the action potential is self-propagating. The action potential is an electrical current of sufficient strength to induce collapse of the resting potential in the adjacent area of the membrane. The depolarization in one area causes adjacent voltage-activated ion channels to open, allowing sodium ions to enter (Fig. 39–8). This action causes the next adjacent voltage-activated ion channels to open, permitting sodium ions to enter in that area, and the process is repeated like a chain reaction until the end of the axon is reached. Thus, a **wave of depolarization** travels down the length of the axon at a velocity and amplitude that are constant for each type of neuron (Fig. 39–9). Conduction of a neural impulse is sometimes compared to burning a trail of gunpowder. Once the gunpowder is ignited at one end of the trail, the flame moves steadily to the other end by igniting the powder particles ahead of it.

By the time the action potential moves a few millimeters down the axon, the membrane over which it has just passed begins to **repolarize.** After a certain time, the sodium gates close, and the membrane again becomes impermeable to sodium. Gates in voltage-activated potassium ion channels

Figure 39–7 **Action potential.** An action potential can be recorded by placing one electrode inside the cell and one outside the plasma membrane. When the axon depolarizes to about −55 mV, an action potential is generated. (The numerical values vary for different nerve cells.)

open at this time, allowing potassium to leak out of the neuron. This decrease in K$^+$ returns the interior of the membrane to its relatively negative state, repolarizing the membrane. This entire mechanism—depolarization and then repolarization—can take place in less than 1 millisecond.

Even though repolarization takes place very quickly, the redistribution of sodium and potassium to normal resting conditions requires more time. Resting conditions are reestablished when the sodium-potassium pump actively transports excess sodium out of the cell. It should be clear that as the wave of depolarization moves down the membrane of the neuron, the normal polarized state is quickly reestablished behind it. We might imagine the action potential as a ring-shaped zone of negative charge traveling just beneath the plasma membrane from one end of the axon to the other.

During the millisecond or so in which it is depolarized, the axon membrane is in an **absolute refractory period:** it cannot transmit another action potential no matter how great a stimulus is applied. This is because the voltage-activated Na$^+$ channels are inactivated. Until their gates are reset, they cannot be reopened. When enough Na$^+$ channel gates have

Figure 39–8 Action of voltage-activated ion channels. *(a)* In the resting state, voltage-activated ion channels are closed. *(b)* At threshold, gates open and Na⁺ ions enter the neuron, causing further depolarization. *(c)* After a certain time, the Na⁺ channels close. Repolarization occurs as voltage-activated K⁺ channels open, permitting K⁺ to leave the cell. *(d)* Resting conditions are restored.

(a) Resting state. Na⁺ channels closed.

(b) Depolarization. Na⁺ channels open. Na⁺ ions enter cell; inside of cell becomes positive.

(c) Repolarization. Na⁺ channels close, K⁺ channels open; K⁺ moves out of cell, restoring negative charge to inside of cell.

(d) Return to resting potential, Na⁺ and K⁺ channels close.

been reset, the neuron enters a **relative refractory period** that lasts for a few additional milliseconds. During this period, the axon can transmit impulses, but the threshold is higher. Even with the limits imposed by their refractory periods, most neurons can transmit several hundred impulses per second.

Neuron diameter affects speed of conduction

Compared with the speed of an electrical current or the speed of light, a nerve impulse travels rather slowly. Most axons transmit impulses at about 1 to 10 meters per second. The smooth, progressive impulse transmission just described, called **continuous conduction**, occurs in unmyelinated neurons. In unmyelinated axons, the speed of transmission is proportional to the diameter of the axon. Axons with larger diameters transmit more rapidly than those with smaller diameters. This is because an axon with a large diameter presents less resistance to the ions flowing along its length.

Squids and certain other invertebrates have giant axons, up to 1 mm in diameter, that permit them to respond rapidly when escaping from predators. Neurobiologists have taken advantage of these large neurons in their studies of neural transmission. For example, in a series of experiments in the 1940s, pioneering researchers Alan Hodgkin and Andrew Huxley inserted electrodes into large, squid axons. By measuring voltage changes as they varied ion concentrations, they demonstrated that passage of Na⁺ ions into the neuron and K⁺ ions out of the neuron resulted in an action potential.

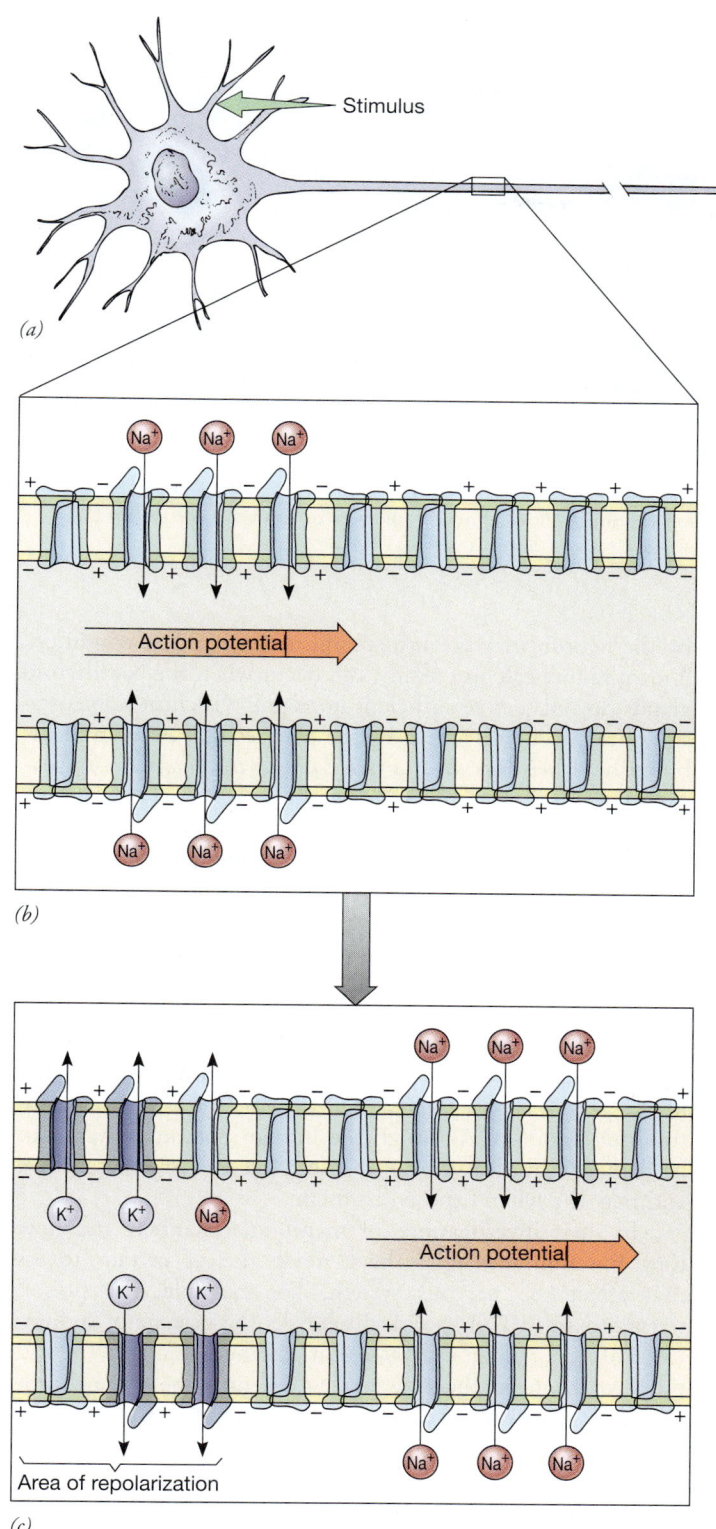

(a)

(b)

Action potential

(c)

Area of repolarization

Figure 39–9 **Transmission of an action potential along the axon.** *(a)* The dendrites (or cell body) of a neuron are stimulated sufficiently to depolarize the membrane to its firing level. *(b)* and *(c)* An action potential is transmitted as a wave of depolarization that travels down the axon. At the region of depolarization, sodium ions diffuse into the cell. As the action potential progresses along the axon, repolarization occurs quickly behind it.

makes direct contact with the surrounding interstitial fluid only at the nodes, and voltage-activated sodium and potassium ion channels are concentrated there. Ion movement across the membrane occurs only at the nodes. Because the ion activity at the active node depolarizes the next node along the axon, the action potential jumps along the axon from one node of Ranvier to the next (Fig. 39–10). This type of impulse transmission is known as **saltatory conduction** (from the Latin word *saltus,* which means "to leap").

In myelinated neurons, the distance between successive nodes of Ranvier affects speed of transmission: the farther apart the nodes, the faster the axon conducts. Using saltatory conduction, a myelinated axon can conduct an impulse up to 50 times faster than the fastest unmyelinated axon. Saltatory conduction has another advantage over continuous conduction: it requires less energy. Current flows only at the nodes, so fewer sodium and potassium ions are displaced. As a result, the sodium-potassium pumps do not have to expend as much ATP to reestablish resting conditions each time an impulse is conducted.

The neuron obeys an all-or-none law

Any stimulus too weak to depolarize the neuron to threshold level cannot fire the neuron. It merely sets up a local response that fades and dies within a few millimeters from the point of stimulus. Only a stimulus strong enough to depolarize the neuron to its critical threshold level results in transmission of an impulse along the axon. The neuron is said to obey an **all-or-none law** because it either transmits an action potential or it does not. No variation exists in the strength of a single impulse.

Saltatory conduction is rapid

In vertebrates, another strategy has evolved that speeds transmission—myelinated neurons. Myelin acts as an effective electrical insulator around the axon except at the nodes of Ranvier, which are not myelinated. The axon plasma membrane

Figure 39–11 **Japanese pufferfish *(Fugu rubripes).*** A toxin, tetrodotoxin, isolated from this fish binds to voltage-activated sodium ion channels, allowing neurobiologists to study them. *(Jeffrey L. Rotman/Peter Arnold, Inc.)*

Figure 39–10 **Saltatory conduction.** The impulse leaps along from one node of Ranvier to the next.

If the all-or-none law is valid, how can we explain differences in the levels of intensity of sensations? After all, we have no difficulty distinguishing between the pain of a severe toothache and that of a minor cut on the arm. This apparent inconsistency is explained by the fact that intensity of sensation depends on the *number of neurons stimulated and on their frequency of discharge.* For example, suppose you burn your hand. The larger the area burned, the more pain receptors are stimulated and the more neurons are depolarized. Also, the stronger the stimulus, the greater the number of action potentials each neuron transmits per unit of time.

The threshold level for different types of stimuli varies with the type of neuron. For example, neurons in the retina of the eye have lower thresholds to light stimuli than do other types of neurons. Pain receptors have low thresholds to strong mechanical stimuli, such as pricking the skin with a needle.

Certain substances affect excitability

Any substance that increases the permeability of the membrane to sodium ions causes the neuron to become more excitable than normal. Substances that decrease the permeability of the membrane to sodium ions make the neuron less excitable.

Calcium balance is essential to normal neural function. Calcium ions are thought to bind to the proteins that make up the sodium channels. The positively charged calcium ions affect the channel proteins, increasing the voltage necessary to open the gates. When calcium ions are too numerous, neurons are less excitable and more difficult to fire.

In contrast, when insufficient numbers of calcium ions are present, the sodium gates apparently fail to close completely between action potentials. As a result, sodium ions leak into the cell. This makes the potential less negative, bringing the neuron closer to firing. The neuron fires more easily and sometimes even spontaneously. As a result, the muscle innervated

by the neuron may go into spasm, or **tetany.** The condition, known as low-calcium tetany, can occur when the parathyroid glands do not secrete sufficient hormone. Calcium homeostasis appears to be disrupted by the plaques that develop in the brain in Alzheimer's disease (see *Making the Connection: Neurons, Genes, and Alzheimer's Disease*).

Many narcotics and anesthetics block conduction of nerve impulses. Local anesthetics such as cocaine, procaine (Novocain), and lidocaine (Xylocaine) affect the voltage-activated sodium ion channels, decreasing the permeability of the neuron to sodium. Excitability may be so reduced that the neuron cannot transmit an impulse through the anesthetized region.

DDT and other chlorinated hydrocarbon pesticides interfere with the action of the sodium-potassium pumps. When nerves are poisoned by such substances, they are unable to transmit impulses. Although the human nervous system can be damaged by these poisons, insects are even more sensitive and may die when exposed to them.

In their investigations of membrane channels and how they function, biologists have made use of certain toxins that affect the nervous system. For example, the poison tetrodotoxin (TTX) specifically blocks the passage of sodium ions through voltage-activated ion channels. Using TTX and other toxins, researchers were able to identify the channel protein. TTX was first isolated from the Japanese pufferfish *(Fugu rubripes),* which is eaten as a delicacy (Fig. 39–11). In small amounts TTX tingles the taste buds, but in a larger dose it can prevent breathing. (The Japanese government operates a certification program for pufferfish chefs!)

NEURONS SIGNAL OTHER CELLS ACROSS SYNAPSES

A synapse is a junction between two neurons or between a neuron and an effector. The synapse between a neuron and a muscle cell is referred to as a **neuromuscular junction** or **mo-**

NEURONS, GENES, AND ALZHEIMER'S DISEASE

Researchers are integrating their knowledge of gene function, proteins, and neurons to solve the mystery of *Alzheimer's disease (AD).* This progressive, degenerative brain disorder afflicts more than 4.5 million people in the United States, making it the fourth leading cause of death. Although AD strikes some individuals in midlife, more than 90% of Alzheimer's patients develop the disease after age 65. In fact, it is the leading cause of senile dementia, which is the loss of memory, judgment, and the ability to reason that we often associate with aging.

In Alzheimer's disease, cells are lost in certain parts of the brain, including the cerebral cortex and hippocampus, areas that are important in thinking and remembering. Neurons that secrete the neurotransmitter acetylcholine are especially affected.

Clues to untangling the mysteries of Alzheimer's disease are leading researchers in many directions. Some are working to develop a simple diagnostic test. To date, the only way to definitively diagnose this disease is by post-mortem examination of brain tissue. Other neurobiologists are studying prevention, or looking for causes and factors that influence the progression of the disease. Still others are developing drugs and other treatments that may slow or stop the disease process.

By studying the brains of individuals who have died, investigators have demonstrated that two of the abnormalities that develop in brain tissue as we age, senile plaques and neurofibrillary tangles, are especially characteristic of AD. These abnormal developments damage brain cells, leading to deterioration in general function. Researchers are investigating the biochemistry and genetic basis of both plaques and tangles for clues to the causes and cures of Alzheimer's disease.

Senile plaques are clusters of abnormal neurons and glial cells. Investigators have demonstrated that these plaques have a central core consisting of a peptide called **beta amyloid.** The precursor (beta–APP) of this protein fragment is a large transmembrane protein coded by a gene located on chromosome 21. Normal brain cells make a soluble form of beta amyloid peptide. Alzheimer's disease may develop when an imbalance in brain metabolism results in an insoluble peptide form, which leads to plaque formation. The plaques are thought to disrupt calcium homeostasis, which in turn leads to neural malfunction and brain cell death.

Neurofibrillary tangles consist of abnormal accumulations of certain cytoskeletal proteins in the neuron cytoplasm. One of the proteins involved, called **tau,** normally stimulates the protein tubulin to form microtubules. When too many phosphates attach to tau, it can no longer adhere to microtubules. Instead, tau molecules join with one another to form fibrous deposits that make up the neurofibrillary tangles. Neurons that have such tangles form fewer synapses with other neurons.

Other abnormal deposits, AMY plaques, were identified in 1997. These are extracellular plaques so-named because they somewhat resemble the amyloid plaques. Researchers debate whether the plaques and tangles *cause* death of brain cells or whether they are an *effect,* developing during the disease process as neurons are damaged and destroyed.

In 1993 neurologist Allen Roses at Duke University Medical Center discovered a genetic clue to the Alzheimer's disease mystery.

We all have a gene that codes for a protein called apolipoprotein E, or Apo-E, which helps transport cholesterol in the blood. Three different forms of this protein are known. The common form, Apo-E3, binds to tau and may inhibit phosphate-bonding. Apo-E3 may also be important in transporting beta amyloid to cells for processing. The Apo-E4 form of the protein (which differs in only one amino acid) does not inhibit phosphate-bonding.

Roses reported that individuals who are homozygous for the Apo-E4 allele (that is, have two copies of the gene that codes for the Apo-E4 form of the protein) are eight times more likely to develop AD than are individuals with the more common form of the gene, Apo-E3. However, Apo-E4 gene has been found to be a risk factor only for patients who develop the disease prior to age 70, only a small percent of the cases.

In 1997 W. Davis Parker, Jr. of the University of Virginia Medical Center reported a genetic link to the more common form of AD. His team identified mutations in two mitochondrial genes that code for cyctochrome *c* oxidase, an enzyme important in the mitochondrial electron transport chain. Again, it is not yet clear whether the finding is a cause or a consequence of the disease.

Another debate surrounds the disease process. Some investigators have proposed that AD may be a lifelong process. One startling study demonstrated that a person's writing early in life can predict the disease. Young adults whose writings had a lower density of ideas were more likely to develop the disease. Another approach to detecting development of AD is the use of PET (positron emission tomography) scans to study glucose uptake in the brain. Uptake of less glucose in certain areas of the brains of AD patients indicates that neurons in these areas may be damaged. Would you want to know if you had risk factors for AD? When reliable predictive tests are developed, will they be required for certain jobs?

More women than men suffer from AD. However, in two long-term studies, older women who took estrogen after menopause were about 60% less likely to develop AD. More long-term studies are needed, and researchers are having difficulty recruiting subjects. Many women are already convinced that estrogen can protect them from AD and are unwilling to participate in a study in which they may be given a placebo for several years.

The damaged neurons characteristic of AD have been found to have receptors for nerve growth factor (NGF), a protein known to promote development and survival of neurons. Researchers hypothesized that they could stop the disease process if they could provide NGF to affected cells. To test this hypothesis, they needed an experimental animal with AD. Transgenic mice have now been produced that model AD, and investigators are studying the effects of introducing NGF produced from recombinant DNA into areas of the brain with degenerating neurons. However, NGF is only one of a group of nerve growth factors (neurotrophins), and investigators have much to learn about their biochemical pathways, interactions, and effects before they can develop effective treatment strategies.

More than a dozen drugs designed to slow the progress of Alzheimer's disease are in clinical trials. One of the challenges of developing a drug has been getting the medication through the blood-brain barrier (see Chapter 40). Current research on the causes and cures of Alzheimer's disease is providing new insights into the metabolism of neurons and the function of the nervous system.

TABLE 39-1   Some Neurotransmitters

Substance	Where Secreted	Comments
Acetylcholine	Neuromuscular junctions; autonomic system; CNS*	Inactivated by cholinesterase; excitatory at skeletal muscle receptors; may be excitatory or inhibitory at other synapses
Biogenic amines		
Norepinephrine	Autonomic system; CNS	Inactivated slowly by monoamine oxidase (MAO); mainly inactivated by reuptake; norepinephrine level in brain affects mood
Dopamine	CNS	Thought to affect motor function; may be involved in schizophrenia; amount reduced in Parkinson's disease.
Serotonin (5-hydroxytryptamine, 5-HT)	CNS	May play role in sleep; LSD antagonizes serotonin; generally inhibitory
Amino acids		
GABA (gamma-aminobutyric acid)	CNS	Acts as inhibitor; may play role in pain perception
Neuropeptides		
Endorphins	CNS and pituitary gland	Morphine-like properties; suppress pain; may help regulate cell growth; linked to learning and memory
Enkephalins	CNS and digestive tract	Inhibit pain impulses by inhibiting release of substance P; bind to same receptors in brain as morphine
Substance P	CNS; sensory nerves; intestine	Transmits pain impulses from pain receptors into CNS

*These and other structures listed in this table are discussed in Chapter 40.

tor end plate. A neuron that *terminates* at a specific synapse is referred to as a **presynaptic neuron,** while a neuron that *begins* at a synapse is known as a **postsynaptic neuron.** Note that these terms are relative to a specific synapse. A neuron that is postsynaptic with respect to one synapse may be presynaptic to the next synapse in the sequence.

Signals across synapses can be electrical or chemical

Based on how pre- and postsynaptic neurons communicate, two types of synapses have been identified: **electrical synapses** and **chemical synapses.** In electrical synapses, the presynaptic and postsynaptic neurons occur very close together (within 2 nm of one another) and form gap junctions (see Chapter 5). The two cells are connected by a protein channel. Electrical synapses allow the passage of ions from one cell to another, permitting an impulse to be directly and rapidly transmitted from presynaptic to postsynaptic neuron. The escape responses of many invertebrates and vertebrates involve electrical synapses. For example, the "tail-flick" escape response of the crayfish involves giant axons in the nerve cord that form electrical synapses with large motor axons that then signal muscles in the abdomen.

Most synapses are chemical synapses. Pre- and postsynaptic cells are separated by a space, the **synaptic cleft,** more than 20 nm wide (less than one-millionth of an inch). Because depolarization is a property of the plasma membrane, when an action potential reaches the end of the axon it is unable to jump the gap. The electrical signal must be converted into a chemical one. Neurotransmitters are the chemical messengers that conduct the neural signal across the synapse and bind to chemically activated ion channels in the membrane of the postsynaptic neuron. When a postsynaptic neuron reaches threshold level, it transmits an action potential.

When considering speed of conduction through a sequence of neurons, the number of chemical synapses must be taken into account, because each time an impulse is conducted from one neuron to another there is a slight synaptic delay

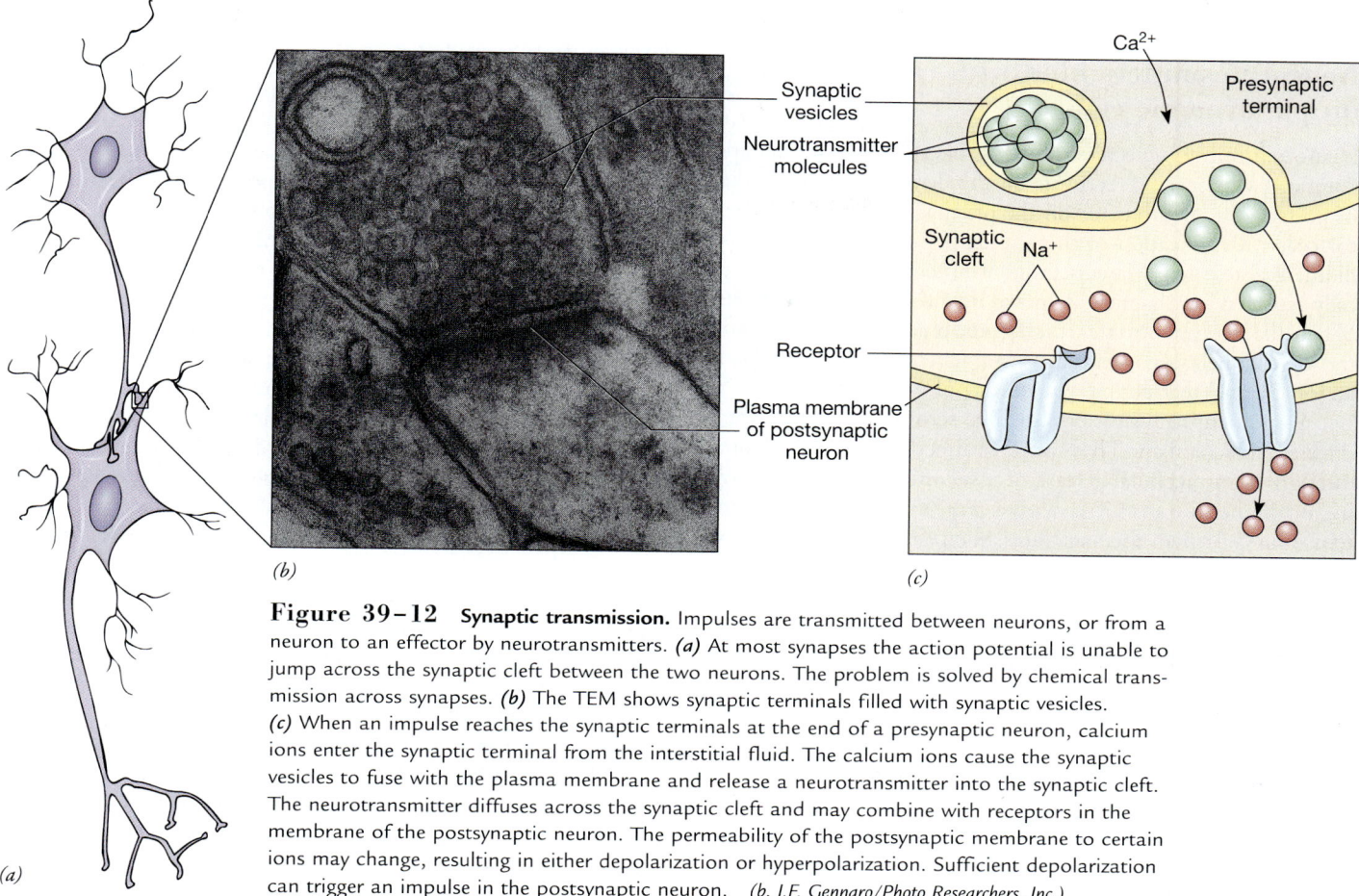

(b)

(a)

(c)

Figure 39–12 **Synaptic transmission.** Impulses are transmitted between neurons, or from a neuron to an effector by neurotransmitters. *(a)* At most synapses the action potential is unable to jump across the synaptic cleft between the two neurons. The problem is solved by chemical transmission across synapses. *(b)* The TEM shows synaptic terminals filled with synaptic vesicles. *(c)* When an impulse reaches the synaptic terminals at the end of a presynaptic neuron, calcium ions enter the synaptic terminal from the interstitial fluid. The calcium ions cause the synaptic vesicles to fuse with the plasma membrane and release a neurotransmitter into the synaptic cleft. The neurotransmitter diffuses across the synaptic cleft and may combine with receptors in the membrane of the postsynaptic neuron. The permeability of the postsynaptic membrane to certain ions may change, resulting in either depolarization or hyperpolarization. Sufficient depolarization can trigger an impulse in the postsynaptic neuron. *(b, J.F. Gennaro/Photo Researchers, Inc.)*

(about 0.5 millisecond). This delay is the time required for the release of neurotransmitter, its diffusion, and its binding to postsynaptic membrane receptors.

Neurons signal other cells with neurotransmitters

More than 40 different chemical compounds are now known or suspected to function as neurotransmitters (Table 39–1). In Chapter 38 we discussed how **acetylcholine,** released from motor neurons, triggers muscle contraction. Acetylcholine is also released by some neurons in the brain and in the autonomic nervous system (see Chapter 40). Cells that release this neurotransmitter are referred to as **cholinergic neurons.**

Neurons that release **norepinephrine** are called **adrenergic neurons.** Norepinephrine and the neurotransmitters **serotonin** and **dopamine** belong to a class of compounds called **biogenic amines,** or **catecholamines.** Biogenic amines affect mood, and their imbalance has been linked to several mental disorders including major depression, attention deficit disorder (ADD), and schizophrenia. Antidepressants and many other mood-affecting drugs work by altering the levels of biogenic amines in the brain.

Gamma-aminobutyric acid (GABA), an amino acid, inhibits neurons in the brain and spinal cord. The actions of GABA are enhanced by drugs, such as the benzodiazepines and

barbiturates, used to treat anxiety. The opiates **enkephalin** and **beta-endorphin** are neuropeptides that bind to opioid receptors and block pain signals. Opiates modulate the effect of other neurotransmitters. (Pain perception is discussed in Chapter 41.) **Nitric oxide (NO),** a signaling molecule that has been the focus of much recent research, may transmit information in the direction opposite that of neurotransmitters.

Many neurotransmitters, including acetylcholine, the biogenic amines, and GABA, are small molecules that act rapidly. These low molecular weight neurotransmitters are synthesized within synaptic terminals. Mitochondria provide the ATP required for this synthesis. Enzymes needed for neurotransmitter synthesis are produced in the cell body and transported down the axon to the synaptic terminals. Higher molecular weight neuropeptide neurotransmitters, such as the opiates, are synthesized in the cell body and transported down the axon to the synaptic terminals.

Neurotransmitters are stored in the synaptic terminals within small membrane-bounded sacs called **synaptic vesicles.** Each time an action potential reaches a synaptic terminal, the resulting change in membrane potential activates voltage-sensitive calcium channels. Calcium ions from the surrounding interstitial fluid then pass into the synaptic terminal. The Ca^{2+} ions cause synaptic vesicles to fuse with the presynaptic membrane and release neurotransmitter molecules into the synaptic cleft by exocytosis (Fig. 39–12).

Neurotransmitters bind with receptors on postsynaptic cells

Neurotransmitter molecules diffuse across the synaptic cleft and combine with specific receptors on the dendrites or cell bodies of postsynaptic neurons (or on the plasma membranes of effector cells). Identification of these receptors is an area of intense biomedical research. Many neurotransmitter receptors are chemically activated ion channels as known **ligand-gated ion channels.** When the neurotransmitter, the ligand, binds with the receptor, the ion channel opens. The acetylcholine receptor, for example, is an ion channel that permits the passage of Na^+ and K^+.

Some neurotransmitters, such as serotonin, operate via a different mechanism. They work indirectly through the production or intracellular release of a **second messenger.** Binding of the neurotransmitter with a receptor activates a G protein. The G protein then activates an enzyme, such as adenylyl cyclase, in the postsynaptic membrane. Adenylyl cyclase converts ATP to **cyclic AMP (cAMP),** which acts as a second messenger (see *Making the Connection: Information Transfer across the Plasma Membrane* in Chapter 5). Cyclic AMP activates a protein kinase that phosphorylates a protein that closes K^+ channels.

If repolarization of a postsynaptic cell is to occur quickly, any excess neurotransmitter in the synaptic cleft must be removed. Some neurotransmitters are inactivated by enzymes. For example, excess acetylcholine is degraded into its chemical components, choline and acetate, by the enzyme acetylcholinesterase. Other neurotransmitters, for example, the biogenic amines, are actively transported back into the synaptic terminals, a process known as **reuptake.** These neurotransmitters are repackaged in vesicles and recycled. Many drugs inhibit the reuptake of neurotransmitters. For example, many antidepressants work by inhibiting the reuptake of serotonin, thus increasing its concentration in the synaptic cleft. Cocaine inhibits the reuptake of dopamine.

Neurotransmitter receptors can send excitatory or inhibitory signals

Depending on the type of postsynaptic receptor with which it combines, the same neurotransmitter can have different effects. For example, acetylcholine has an excitatory effect on skeletal muscle. It opens sodium ion channels, which increases the permeability of the muscle fiber membrane to Na^+. The influx of Na^+ depolarizes the membrane, leading to contraction. In contrast, acetylcholine has an inhibitory effect on cardiac muscle, resulting in a decreased heart rate. A postsynaptic neuron may have receptors for more than one type of neurotransmitter. Indeed, some of its receptors may be excitatory and some may be inhibitory.

When neurotransmitter molecules bind to receptors, they directly or indirectly affect ion channels. The resulting redistribution of ions changes the electrical potential of the membrane. The membrane may become depolarized, or it may become **hyperpolarized,** in which case the membrane potential becomes more negative. If sufficiently intense, a local depolarization can set off an action potential.

When neurotransmitter molecules combine with a receptor that opens sodium channels, the resulting influx of sodium ions partially depolarizes the membrane. A change in membrane potential that brings the neuron closer to firing is called an **excitatory postsynaptic potential (EPSP)** (Fig. 39–13a). Let us say that sufficient sodium ions enter to change the membrane potential, from −70 mV to −60 mV. The membrane would be only −5 mV away from threshold. Under such conditions, a relatively weak stimulus can cause the neuron to fire.

Recall that some neurotransmitters act indirectly to close K^+ channels. When these channels are closed, K^+ cannot diffuse out of the cell. As they accumulate, the inner surface of the plasma membrane becomes more positive, leading to depolarization.

Some neurotransmitter-receptor combinations *hyperpolarize* the postsynaptic membrane. Because such an action takes the neuron farther away from the firing level, a potential change in this direction is called an **inhibitory postsynaptic potential (IPSP)** (Fig. 39–13b). For example, if the membrane potential changes from −70 mV to −80 mV, the membrane is farther away from threshold and a stronger stimulus will be required to fire the neuron.

Like an EPSP, an IPSP can be produced in several ways. Binding of GABA to a $GABA_B$ receptor opens K^+ channels. As K^+ moves out, the neuron becomes more negative, hyperpolarizing the membrane. Activated $GABA_A$ receptors produce IPSPs by opening Cl^- channels. In this case, the influx of negative ions hyperpolarizes the membrane.

When certain types of receptors on the postsynaptic neuron are activated, the reactivity of the neuron can be affected for long periods of time, even years. Neurotransmitters that cause such long-term effects are sometimes referred to as modulators. They appear to work by affecting genes or activating enzymes that increase or decrease the number of receptors. These mechanisms will be discussed further in Chapter 40 in connection with the process of learning.

Graded potentials vary in magnitude

Each EPSP or IPSP is a local response in the neuron membrane. Such responses are referred to as **graded potentials** because they vary in magnitude depending on the strength of the stimulus applied. A local change in potential causes a flow of ions. The greater the change in potential, the greater the flow of ions. Such a local response can function as a signal only over a very short distance, because it fades out within a few millimeters of its point of origin. However, graded potentials can be added together, resulting in action potentials.

One EPSP is usually too weak to trigger an action potential by itself. Its effect is *subliminal,* that is, below threshold level. Even though subthreshold EPSPs do not produce an action potential, they do affect the membrane potential. EPSPs may be added together in a process known as **summation. Temporal summation** occurs when repeated stimuli

Figure 39–13 Comparison of *(a)* an excitatory postsynaptic potential (EPSP) and *(b)* an inhibitory postsynaptic potential (IPSP).

cause new EPSPs to develop before previous EPSPs have decayed. By summation of several EPSPs, the neuron may be brought to the critical firing level. When several synaptic terminals release neurotransmitter simultaneously, the postsynaptic neuron is stimulated at several places at once. This effect, called **spatial summation,** can also bring the postsynaptic neuron to the threshold level.

NEURAL IMPULSES MUST BE INTEGRATED

Neural integration is the process of summing, or integrating, incoming signals. Each neuron may synapse with hundreds of other neurons. Indeed, as much as 40% of a postsynaptic neuron's dendritic surface and cell body may be covered by thousands of synaptic terminals of presynaptic neurons. It is the

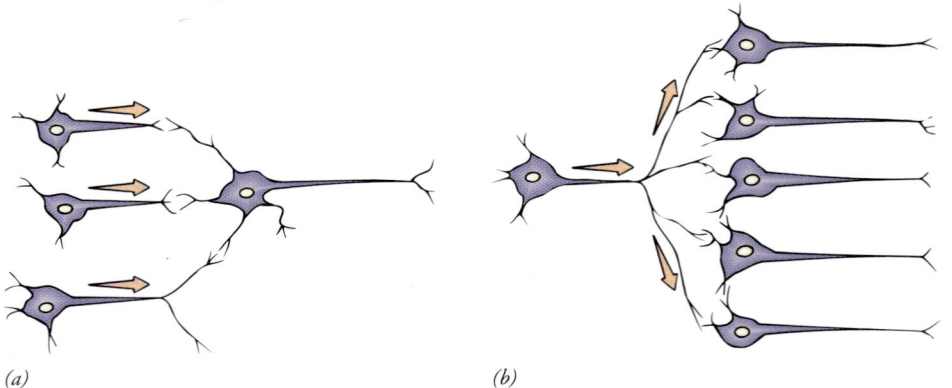

Figure 39–14 **Neural circuits.** *(a)* Convergence of neural input. Several presynaptic neurons synapse with one postsynaptic neuron. This organization permits one neuron to receive signals from many sources. *(b)* Divergence of neural output. A single presynaptic neuron synapses with several postsynaptic neurons. This organization allows one neuron to communicate with many others.

(a)

(b)

job of the dendrites and the cell body of every neuron to integrate the hundreds of messages that continually bombard them.

EPSPs and IPSPs are produced continually in postsynaptic neurons, and IPSPs cancel the effects of some of the EPSPs. It is important to remember that each EPSP and IPSP is not an all-or-none response. Rather, each is a local response (it does not travel like an action potential) that may be added to or subtracted from other EPSPs and IPSPs. As the neuron membrane continuously updates its molecular tabulations, the neuron may be inhibited or brought to threshold level. This mechanism provides for integration of hundreds of signals (EPSPs and IPSPs) before an all-or-none action potential is actually transmitted along the axon of a postsynaptic neuron. Local responses permit the neuron and the entire nervous system a far greater range of response than would be the case if every EPSP generated an action potential.

Where does neural integration take place? Every neuron sorts through (on a molecular level) the hundreds and thousands of bits of information continually bombarding it. In vertebrates more than 90% of the neurons in the body are located in the CNS, so most neural integration takes place there, within the brain and spinal cord. These neurons are responsible for making most of the "decisions." In the next chapter the brain and spinal cord will be examined in some detail.

NEURONS ARE ORGANIZED INTO CIRCUITS

The CNS contains millions of neurons, but it is not just a tangled mass of nerve cells. Its neurons are organized into separate **neural networks,** and within each network the neurons are arranged in specific pathways, or **neural circuits.** Although each network has some special characteristics, the neural circuits in all of the networks share many organizational features. For example, convergence and divergence are probably characteristic of all of them.

In **convergence,** a single neuron is controlled by converging signals from two or more presynaptic neurons (Fig.

39–14*a*). An interneuron in the spinal cord, for instance, may receive converging information from sensory neurons entering the cord, from neurons bringing information from various parts of the brain, and from neurons coming from different levels of the spinal cord. Information from all these sources has to be integrated before an action potential can be sent and an appropriate motor neuron stimulated. Convergence is an important mechanism by which the CNS can integrate the information that impinges on it from various sources.

In **divergence,** a single presynaptic neuron stimulates many postsynaptic neurons (Fig. 39–14*b*). Each presynaptic neuron may branch and synapse with as many as 25,000 or more different postsynaptic neurons. For example, a single neuron transmitting an impulse from the motor area of the brain may synapse with hundreds of interneurons in the spinal cord,

(a)

(b)

Figure 39–15 **A reverberating circuit.** A neuron may fire multiple times. *(a)* A simple reverberating circuit in which an axon collateral of the second neuron turns back on its own dendrites, so the neuron continues to stimulate itself. *(b)* In this neural circuit an axon collateral of the second neuron synapses with an interneuron. The interneuron synapses with the first neuron in the sequence. New impulses are triggered again and again in the first neuron, causing reverberation.

and each of these may diverge in turn, so that hundreds of muscle fibers may be stimulated.

Another important type of neural circuit is the **reverberating circuit,** a neural pathway arranged so that an axon collateral synapses with an interneuron (Fig. 39–15). The interneuron synapses with a neuron in a sequence that can send new impulses through the circuit. This is an example of positive feedback. New impulses can be generated again and again until the synapses fatigue (from depletion of neurotransmitter) or until stopped by some sort of inhibition. Reverberating circuits are thought to be important in rhythmic breathing, mental alertness, and perhaps short-term memory.

SUMMARY WITH KEY TERMS

I. Information flow through the nervous system typically follows this sequence: **reception** of information; **transmission** to interneurons in the CNS via **afferent (sensory) neurons; integration** by interneurons in the CNS; transmission from the CNS via **efferent neurons;** and **action by effectors,** the muscles and glands

II. **Glial cells** and **neurons** are the two types of cells characteristic of neural tissue.
  A. Glial cells are supporting cells.
  B. Neurons are specialized to receive stimuli and transmit electrical and chemical signals.
    1. A typical neuron consists of a **cell body** with many branched **dendrites** and a single long **axon.** Branches at the end of the axon give rise to **synaptic terminals.**
    2. Many axons are surrounded by a **myelin sheath.** Outside the central nervous system, the myelin sheath is produced by **Schwann cells.**
    3. A **nerve** consists of several hundred axons wrapped in connective tissue; a **ganglion** is a mass of neuron cell bodies.

III. Neurons transmit electrical impulses.
  A. A neuron that is not transmitting an impulse has a **membrane potential,** or **resting potential.** This results from differences in concentrations of specific ions across the membrane.
    1. **Sodium-potassium pumps** continuously transport sodium ions out of the neuron and potassium ions in.
    2. Ions pass through specific **passive ion channels** by facilitated diffusion. K^+ leaks out more readily than Na^+ can leak in. Cl^- ions accumulate along the inner surface of the plasma membrane.
    3. In the resting neuron, the inner surface of the plasma membrane is negatively charged compared to the outside; the membrane is polarized. A potential difference of about -70 mV exists across the membrane.
  B. Excitatory stimuli open **voltage-activated ion channels** in the plasma membrane. Na^+ enters and K^+ leaves the neuron through specific channels, altering the membrane potential toward **depolarization.** Inhibitory stimuli **hyperpolarize** the membrane.
  C. When the membrane potential exceeds **threshold level,** an **action potential** is generated.
    1. The action potential is a wave of depolarization that moves down the axon.

2. The action potential obeys an **all-or-none law** that states that no variation exists in the strength of a single impulse; the membrane potential either exceeds threshold level and transmits an action potential or it does not.
    3. As the action potential moves down the axon, **repolarization** occurs behind it.
      a. The axon enters an **absolute refractory period** during depolarization.
      b. When enough gates controlling Na^+ channels have been reset, the neuron enters a **relative refractory period.**
  D. **Saltatory conduction** takes place in myelinated neurons. In this type of transmission, depolarization skips along the axon from one **node of Ranvier** to the next—sites where the axon is not covered and Na^+ channels are concentrated.
  E. Excitability of a neuron can be affected by calcium concentration and by certain substances such as local anesthetics and pesticides.

IV. The junction between two neurons or between a neuron and effector is a **synapse.** Synaptic transmission is generally chemical and depends on release of a neurotransmitter from vesicles in the synaptic terminals of the presynaptic axon.
  A. Some of the more studied neurotransmitters include **acetylcholine;** the **biogenic amines,** including **norepinephrine, serotonin,** and **dopamine; GABA,** a widespread inhibitory neurotransmitter; and the neuropeptides, which include **beta-endorphin.**
  B. Neurotransmitter diffuses across the **synaptic cleft** and combines with specific receptors on the postsynaptic neuron.
  C. Many neurotransmitter receptors are proteins that form **ligand-gated ion channels.** Others work through a **second messenger** such as **cyclic AMP.**
  D. Receptor activation can cause either an **excitatory postsynaptic potential (EPSP)** or an **inhibitory postsynaptic potential (IPSP),** depending on the type of receptor.

V. Each EPSP or IPSP is a **graded potential** that varies in magnitude depending on the strength of the stimulus applied. The mechanism of neural integration is **summation:** the process of adding and subtracting incoming signals.

VI. Complex **neural circuits** are possible because of **reverberating circuits** and neural associations such as **convergence** and **divergence.**

POST-TEST

1. Summing incoming neural signals is part of (a) reception (b) transmission (c) integration (d) action by effectors (e) afferent neuron transmission
2. Which of the following are phagocytic and remove debris from nervous tissue? (a) neurons (b) neurotransmitters (c) certain glial cells (d) effectors (e) afferent neurons
3. Most of the neuron cytoplasm is located in the (a) cell body (b) axon (c) dendrites (d) synaptic terminals (e) nodes
4. Nerve impulses are transmitted away from the cell body by the (a) ganglia (b) axon (c) dendrites (d) synaptic terminals (e) nodes

5. The correct sequence in response to a stimulus is (a) reception→efferent neuron→CNS→afferent neuron (b) reception→transmission→integration→transmission→action by an effector (c) afferent neuron→reception in CNS→efferent neuron→integration (d) reception→integration→effector→transmission (e) afferent neuron→efferent neuron→CNS→effector action
6. The myelin sheath is produced by the (a) axon (b) neuron cell body (c) dendrites (d) Schwann cells (e) synaptic terminals
7. Neurotransmitters are released by the (a) axon (b) neuron cell body (c) dendrites (d) Schwann cells (e) synaptic terminals

8. Which of the following contribute to the resting potential of a neuron? (a) sodium-potassium pumps (b) ion channels (c) differences in concentration of ions across the membrane (d) answers a, b, and c are correct (e) answers b and c only are correct

9. Which of the following occur first when voltage reaches the threshold level? (a) gates of certain voltage-activated ion channels open (b) K^+ channels close (c) the membrane hyperpolarizes (d) neurotransmitter is released (e) neurotransmitter reuptake takes place at the synapse

10. Saltatory conduction (a) requires more energy than continuous conduction (b) occurs in unmyelinated neurons (c) occurs when the action potential jumps from one node of Ranvier to the next (d) slows transmission of an impulse (e) depends on the presence of GABA

11. Acetylcholine (a) is a biogenic amine (b) is recycled (c) is released by motor neurons and by some neurons in the brain (d) works through G proteins (e) binds to opioid receptors

12. Some neurotransmitter receptors (a) are voltage-activated ion channels (b) permit influx of chloride ions, leading to depolarization of the membrane (c) work through a second messenger (d) inhibit reuptake of the neurotransmitter (e) three of the preceding answers are correct

13. IPSPs (a) excite presynaptic neurons (b) excite postsynaptic neurons (c) cancel the effects of some EPSPs (d) release large amounts of neurotransmitters (e) are released by postsynaptic cell bodies

14. A presynaptic neuron in the cerebrum synapses with hundreds of other neurons. This is an example of (a) convergence (b) divergence (c) summation (d) a reverberating circuit (e) graded potential

REVIEW QUESTIONS

1. Distinguish between a neuron and a nerve.
2. Imagine that you are swimming and you suddenly spot a shark fin moving in your direction. What processes must take place within your nervous system before you can make your escape?
3. Describe the sequence of events that occurs when a neuron exceeds threshold level.
4. Describe the functions of the following: (a) myelin (b) voltage-activated ion channels (c) glial cells (d) dendrites (e) synaptic terminals
5. What is meant by the resting potential of a neuron? How do sodium-potassium pumps and ion channels contribute to the resting potential?
6. What is an action potential? How does it differ from a graded potential?
7. Contrast saltatory conduction with conduction in an unmyelinated neuron.
8. How is neural function affected by the presence of too much calcium? Too little calcium?
9. Describe the functions and mechanisms of action of the following substances: (a) acetylcholine (b) acetylcholinesterase; (c) norepinephrine (d) GABA
10. Contrast convergence and divergence.
11. What is summation?
12. Label the diagram. Use Fig. 39–2 to check your answers.

YOU MAKE THE CONNECTION

1. Discuss several specific ways that the nervous system helps maintain homeostasis.
2. Stimulant drugs such as amphetamines increase the activity of the nervous system. Propose two mechanisms involving synaptic transmission that could explain the action of such drugs.
3. Develop a hypothesis to explain the fact that acetylcholine has an excitatory effect on skeletal muscle but an inhibitory effect on cardiac muscle.

4. Investigators have genetically engineered cells to produce acetylcholine, dopamine, GABA, and other neurotransmitters. In what ways might these cells be useful in neurobiological research? How might they be useful in clinical medicine?
5. Identify some of the ethical challenges posed by neurobiological research. (*Hint:* Read *Making the Connection: Neurons, Genes, and Alzheimer's Disease*)

RECOMMENDED READINGS

Fackelman, K. "Forecasting Alzheimer's Disease." *Science News,* Vol. 149, 18 May, 1996. Brain scans and samples of writing may predict Alzheimer's disease.

Fuller, R.W. "Neural Functions of Serotonin." *Scientific American Science & Medicine,* Vol. 2, No. 4, Jul./Aug. 1995. The neurotransmitter serotonin is released by neurons in many parts of the brain, and many drugs, including Prozac, work by modifying serotonin levels.

Kandel, E.R., J.H. Schwartz, and T.M. Jessell. *Essentials of Neural Science and Behavior.* Appleton & Lange, Norwalk, Connecticut, 1995. An excellent introduction to neurobiology.

Matthews, G.G. *Neurobiology: Molecules, Cells, and Systems.* Blackwell Science, Malden, MA, 1998. An excellent and very readable discussion of selected topics in neurobiology.

Neher, E., and B. Sakmann. "The Patch Clamp Technique." *Scientific American,* Vol. 266, No. 3, Mar. 1992. The 1991 Nobel Prize–winning authors discuss their technique for isolating ion channels in cell membranes.

Roses, D. "Apolipoprotein E and Alzheimer Disease." *Scientific American Science & Medicine,* Vol. 2, No. 5, Sep./Oct. 1995. A discussion of genes linked to Alzheimer Disease.

Seeman, P. "Dopamine Receptors and Psychosis." *Scientific American Science & Medicine,* Vol. 2, No. 5, Sep./Oct. 1995. Actions of dopamine receptors affect mental health, and these receptors offer a target for pharmacological intervention.

Tobin, A.J., and R.E. Morel. *Asking About Cells.* Saunders College Publishing, Philadelphia, 1997. Chapter 23 provides an excellent introduction to neurons.

● Visit our website at **http://www.saunderscollege.com/lifesci/titles.html** and click on Solomon/Berg/Martin Biology for links to chapter-related resources on the World Wide Web.

CHAPTER 40

Neural Regulation: Nervous Systems

A frog ejects its tongue with lightning speed to capture a fly, a rabbit escapes from a predator, and a human learns biology. In a complex nervous system millions of neurons work together to produce effective responses to stimuli in the external environment. The nervous system also regulates heart rate, breathing, and hundreds of other internal activities. Just how neurons interact has been the focus of extensive research, and we are only beginning to understand the mechanisms that permit the complex functions of the nervous system.

Animals can change their behavior based on experience. Even animals with very simple nervous systems can learn to repeat behaviors associated with reward and avoid behaviors that cause pain. Such changes in behavior reflect **neural plasticity,** the ability of the nervous system to modify its own structure and function. Many areas of the brain that were once thought to be "hard-wired" are now known to be flexible and capable of change. Familiar examples of neural plasticity include learning to walk, skate, ride a bicycle, or hit a baseball. At first you were probably clumsy, but with practice, your performance became smoother and more precise. The ability to learn information and motor skills and to remember them depends on neural plasticity. But what are the mechanisms by which neurons signal one another to carry out such diverse activities as movement and learning?

Because interneuron signaling takes place at synapses, these junctions between neurons have been the focus of intense research. Neural plasticity may, in fact, be **synaptic plasticity,** as evidenced by the observation that changes that take place during learning and remembering appear to take place at synapses. Working on a molecular level, neurobiologists are studying neurotransmitter action on membrane receptors. They are asking questions more quickly than they can answer them. How is synaptic activity depressed, and how is it enhanced? Do neurons change the contacts they make with one another? Do they make connections with new neurons? What mechanisms are activated by neurotransmitter-receptor binding? These mechanisms are known to include complex intracellular signaling systems that involve second messengers, gene activation, and protein synthesis. Just how they work is still a mystery.

Neurobiologists use a variety of methods to study the mechanisms of neural function. For example, improved imag-

(Volker Steger/Peter Arnold, Inc.)

ing methods have revolutionized the study of the brain. Functional magnetic resonance imaging (fMRI) has provided investigators with a window through which to observe brain function. fMRI allows neurophysiologists to study the responses of neural networks in the brain while an individual is performing a task, such as speaking. During performance of the task, an area of the brain becomes active, and flow of oxygenated blood to that area increases. In the image shown here, red indicates areas of greatest activation during speech; yellow indicates medium activation. fMRI can detect changes that take place in response to a very brief stimulus. For example, a visual stimulus lasting only 30 milliseconds stimulates brain activation that can be detected by fMRI.

In this chapter we compare various animal nervous systems. We then examine the structure and function of the vertebrate nervous system, with emphasis on the function of the human brain. We explore some of the frontiers of neurophysiology such as the cellular mechanisms of memory and learning, focusing on synaptic events and intracellular signals.

AFTER YOU HAVE STUDIED THIS CHAPTER YOU SHOULD BE ABLE TO

1. Contrast nerve nets and radial nervous systems with bilateral nervous systems.
2. Compare the vertebrate nervous system with a bilateral invertebrate nervous system.
3. Trace the development of the principal vertebrate brain regions from the forebrain, midbrain, and hindbrain, and compare the brains of fish, amphibians, reptiles, birds, and mammals.
4. Describe the functions and structure of the vertebrate spinal cord.
5. Locate the following parts of the human brain and give the functions of each: medulla, pons, midbrain, thalamus, hypothalamus, cerebellum, and cerebrum. Include a description of the functional areas of the cerebrum.
6. Describe the sleep-wake cycle and contrast REM and non-REM sleep.
7. Describe the actions of the limbic system.

8. Summarize neurophysiological changes that take place during various types of learning. Cite experimental evidence to support your descriptions.
9. Cite experimental evidence linking environmental stimuli with changes in the brain and with learning and motor abilities.
10. Describe the organization of the vertebrate nervous system, comparing the central nervous system with the peripheral nervous system and comparing the somatic system with the autonomic system.
11. Contrast the sympathetic and parasympathetic divisions of the autonomic system, giving examples of the effects of these systems on specific organs such as the heart and intestine.
12. Discuss the biological actions and effects on mood of the following types of drugs: alcohol, antidepressants, barbiturates, antianxiety drugs, antipsychotic drugs, opiates, stimulants, hallucinogens, and marijuana.

MANY INVERTEBRATES HAVE COMPLEX NERVOUS SYSTEMS

An animal's lifestyle is closely related to the organization and complexity of its nervous system. *Hydra* and other cnidarians have a **nerve net** consisting of neurons scattered throughout the body with no central control organ. Electrical signals are sent from neuron to neuron in more than one direction (Fig. 40–1). Responses may involve large parts of the body. An advantage of the nerve net is that the cnidarian can respond effectively to predator or prey approaching from any direction. *Hydra* can respond by discharging nematocysts (stinging struc-

tures) and coordinating the movements of its tentacles to capture food. Some cnidarians have two or more nerve nets. In some jellyfish, a nerve net that transmits slowly coordinates movement of the tentacles, while another, which is faster, coordinates swimming.

The **radial nervous system** of the sea star and other echinoderms (Chapter 30) is a modified nerve net. This system exhibits some degree of selective organization of neurons into more than a diffuse network. It consists of a nerve ring around the mouth, from which a large radial nerve extends into each arm. Branches of these nerves, which form a network somewhat similar to the nerve net of *Hydra*, coordinate movement of the animal.

Bilaterally symmetrical animals generally have a more complex nervous system than those with radial symmetry. A bilaterally symmetrical animal typically moves forward, head first. With sense organs concentrated at the front of the body, the animal can detect an enemy quickly or sense food in time to capture it. We can identify the following trends in the evolution of nervous systems:

1. **Increased number of nerve cells.**
2. **Concentration of nerve cells,** forming masses of tissue that become **ganglia** and **brain,** and thick cords of tissue that become **nerve cords** and **nerves.**
3. **Specialization of function.** For example, transmission of nerve impulses in one direction requires both **afferent nerves,** which conduct impulses toward a central nervous system (CNS), and **efferent nerves,** which transmit impulses away from the CNS and to the **effectors** (muscles and glands). Certain parts of the central nervous system are typically specialized to perform specific functions, and distinct structural and functional regions can be identified.
4. An **increased number of association neurons and more complex synaptic contacts** permit greater integration of

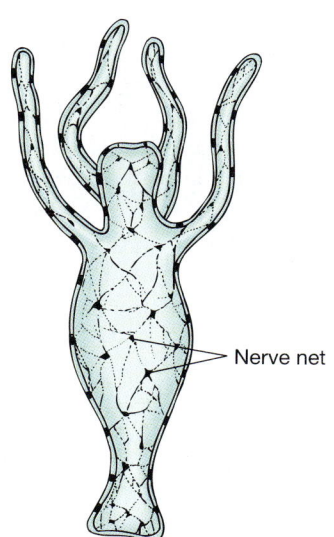

Figure 40–1 *Hydra's* **nerve net.** *Hydra* and other cnidarians have a network of neurons with no central control organ.

Nerve net

incoming messages, provide a greater range of responses, and allow more precision in responses.

5. **Cephalization, or formation of a head.** A bilaterally symmetrical animal generally moves in a forward direction. Concentration of sense organs at the front end of the body allows the animal to detect an enemy quickly enough to escape or to see or smell food in time to capture it. Response can be rapid if sense organs are linked by short pathways to decision-making nerve cells nearby. Therefore, nerve cells are typically concentrated in the head region and organized to form ganglia or a brain.

In planarian flatworms, the head region contains concentrations of nerve cells referred to as **cerebral ganglia** (Fig. 40–2). These serve as a primitive brain and exert some measure of control over the rest of the nervous system. Typically, two solid ventral longitudinal nerve cords extend from the ganglia to the posterior end of the body. Transverse nerves connect the two nerve cords and connect the brain with the eyespots. This arrangement is referred to as a ladder-type nervous system.

Annelids and arthropods typically have a ventral nerve cord (Fig. 40–3). The cell bodies of many of the neurons are massed into ganglia. Afferent and efferent neurons are located in lateral nerves that link the ganglia with muscles and other body structures. If an earthworm's brain is removed, the animal can move almost as well as before. However, when it bumps into an obstacle, it persists in a futile effort to move forward rather than turning aside. Its brain is necessary for the earthworm to respond adaptively to environmental change.

The cerebral ganglia of some arthropods are different than those of annelids in that they have specific functional regions. These areas are specialized for integrating information transmitted to the ganglia from sense organs.

Mollusks typically have at least three pairs of ganglia: (1) *cerebral ganglia,* which serve as a coordinating center for complex reflexes and have a motor function, are found dorsal to the esophagus; (2) *visceral ganglia,* which control shell opening and closing, are distributed among the organs; and (3) *pedal ganglia,* which control the movement of the foot, are found in the foot. The visceral and pedal ganglia are connected to the cerebral ganglia by nerve cords.

In cephalopod mollusks such as the octopus, neurons are concentrated in a central region (Fig. 40–4). Ganglia are massed in a ring surrounding the esophagus, making up a brain

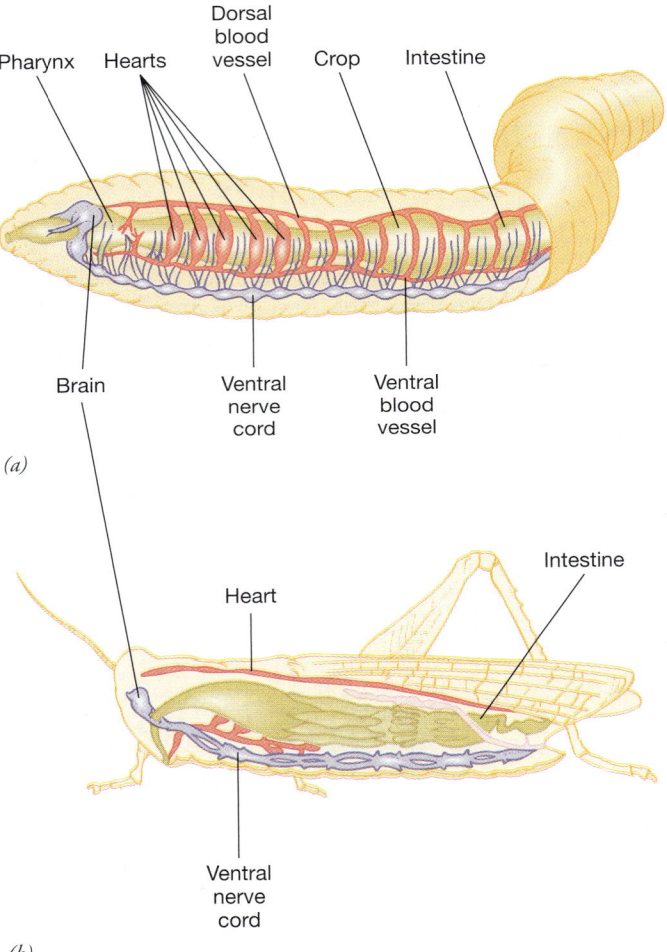

(a)

(b)

Figure 40–3 The annelid and arthropod nervous systems.
(a) Like other annelids, the earthworm nervous system includes a dorsal anterior brain and one or more ventral nerve cords. Cell bodies of the neurons are located in ganglia that are connected by the ventral nerve cord. *(b)* In the insect nervous system, the brain is connected to a ventral nerve cord. The brain is more specialized, and there is less segmentation than in annelids.

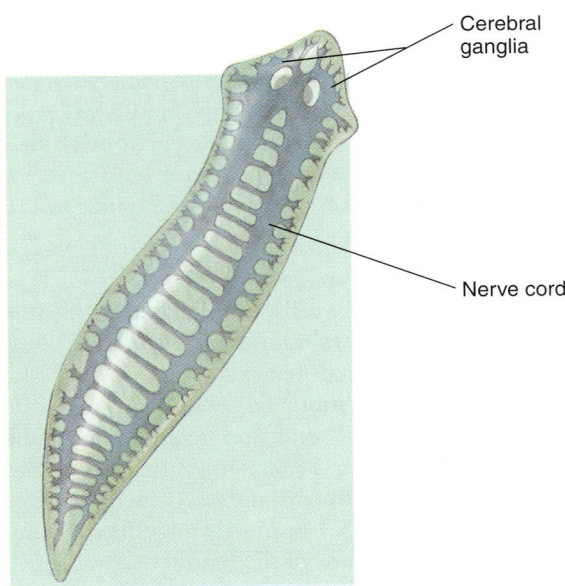

Figure 40–2 Ladder-type nervous system of flatworms. Cerebral ganglia in the head region serve as a simple brain. The longitudinal nerve cords are connected by transverse nerves.

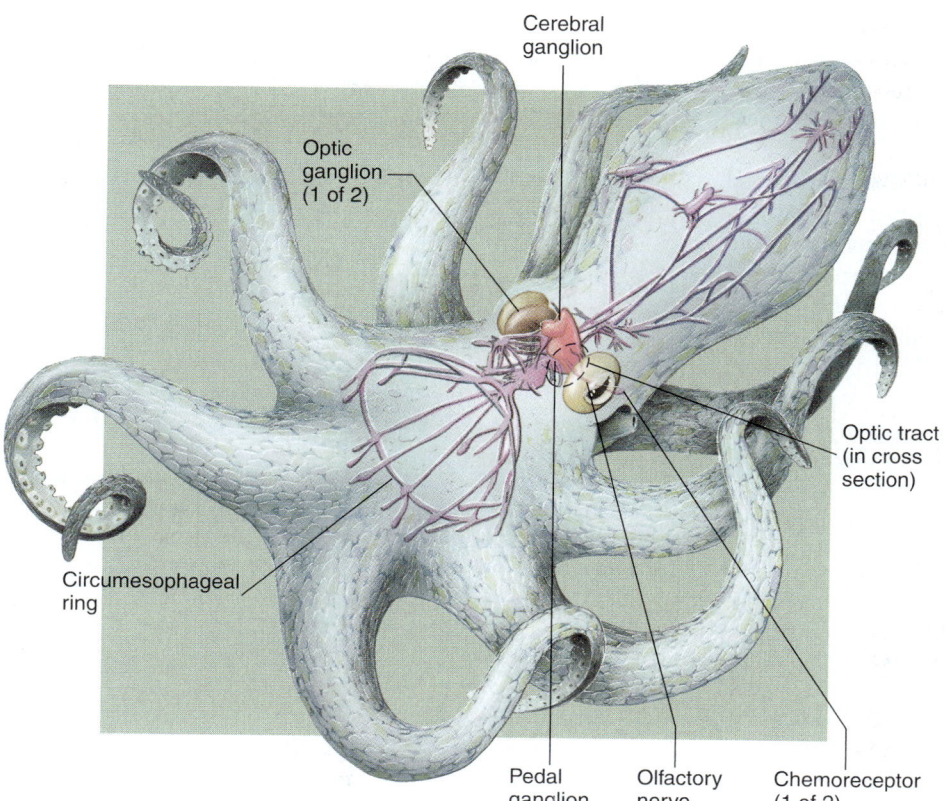

Cerebral
ganglion

Optic
ganglion
(1 of 2)

Optic tract
(in cross
section)

Circumesophageal
ring

Pedal
ganglion

Olfactory
nerve

Chemoreceptor
(1 of 2)

Figure 40–4 **The cephalopod nervous system.** Millions of neurons are concentrated in ganglia.

that contains about 168 million nerve cells. This complex nervous system, which includes well developed sense organs, is correlated with the active, predatory lifestyle of these animals. With its highly developed nervous system, the octopus is capable of considerable learning and can be taught quite complex tasks. In fact, the octopus is considered to be among the most intelligent invertebrates.

THE VERTEBRATE NERVOUS SYSTEM HAS TWO MAIN DIVISIONS: CNS AND PNS

An animal's range of possible responses depends in large part on the number of its neurons and how they are organized in the nervous system. As animal groups evolved, nervous systems became increasingly complex. The vertebrate nervous system has two main divisions: the **central nervous system (CNS)** and the **peripheral nervous system (PNS)** (Table 40–1). The CNS consists of a complex brain that is continuous with the dorsal tubular spinal cord. Serving as central control, these organs integrate incoming information and determine appropriate responses.

The PNS is made up of the sensory receptors (e.g., tactile, auditory, and visual receptors) and the nerves, which are the communication lines. Various parts of the body are linked to the brain by cranial nerves and to the spinal cord by spinal

---

TABLE 40–1   Divisions of the Vertebrate Nervous System

Central nervous system (CNS)
   Brain
   Spinal cord

Peripheral nervous system (PNS)
   Somatic portion
   1. Receptors
   2. Afferent (sensory) nerves—transmit information from receptors to CNS
   3. Efferent nerves—transmit information from CNS to skeletal muscles
   Autonomic portion
   1. Receptors
   2. Afferent (sensory) nerves—transmit information from receptors in internal organs to CNS
   3. Efferent nerves—transmit information from CNS to glands and involuntary muscle in organs
      a. Sympathetic nerves—generally stimulate activity that results in mobilization of energy (e.g., speeds heartbeat)
      b. Parasympathetic nerves—action results in energy conservation or restoration (e.g., slows heart but speeds digestion)

nerves. Afferent neurons in these nerves continuously inform the CNS of changing conditions. Then efferent neurons transmit the "decisions" of the CNS to appropriate muscles and glands, which make the adjustments needed to maintain homeostasis.

For convenience the PNS can be subdivided into **somatic** and **autonomic** portions. Most of the receptors and nerves concerned with changes in the external environment are somatic. Those that regulate the internal environment are autonomic. Both systems have afferent nerves, which transmit messages from receptors to the CNS, and efferent nerves, which transmit information back from the CNS to the structures that must respond. In the autonomic system there are two kinds of efferent pathways: **sympathetic** and **parasympathetic** nerves. Sympathetic nerves generally stimulate activities that mobilize energy, for example, they stimulate the heart to contract more rapidly. Parasympathetic nerves stimulate activities that conserve or restore energy, for example, they cause the heart to decrease its rate of contraction.

Figure 40-5 **Early development of the vertebrate nervous system.** Early in the development of the vertebrate embryo, the anterior end of the neural tube differentiates into the forebrain, midbrain, and hindbrain. These primary divisions subdivide and eventually give rise to specific structures of the adult brain (see Table 40-2).

THE EVOLUTION OF THE VERTEBRATE BRAIN IS MARKED BY INCREASING COMPLEXITY

All vertebrates, from fishes to mammals, have the same basic brain structure. Different parts of the brain are specialized in the various vertebrate classes, and the evolutionary trend is toward increasing complexity, especially of the cerebrum and cerebellum.

In the early vertebrate embryo, the brain and spinal cord differentiate from a single tube of tissue, the **neural tube.** Anteriorly, the tube expands and develops into the brain. Posteriorly, the tube becomes the spinal cord. Brain and spinal cord remain continuous, and their cavities communicate. As the brain begins to differentiate, three bulges become visible: the hindbrain, midbrain, and forebrain (Fig. 40-5).

The hindbrain develops into the medulla, pons, and cerebellum

The **hindbrain** subdivides to form the metencephalon, which gives rise to the **cerebellum** and **pons,** and the myelencephalon, which gives rise to the **medulla.** The medulla, pons, and midbrain make up the **brain stem,** the elongated portion of the brain that looks like a stalk holding up the cerebrum.

The medulla, the most posterior part of the brain, is continuous with the spinal cord. Its cavity, the **fourth ventricle,** is continuous with the central canal of the spinal cord and with a channel that runs through the midbrain. The walls of the medulla are thick and made up largely of nerve tracts (bundles of axons) that connect the spinal cord with various parts of the brain. In complex vertebrates, the medulla contains discrete nuclei (masses of nerve cell bodies) that serve as vital cen-

ters, regulating respiration, heartbeat, and blood pressure. Other reflex centers in the medulla regulate such activities as swallowing, coughing, and vomiting.

The size and shape of the cerebellum vary among the vertebrate classes (Fig. 40-6). Development of the cerebellum in different animals is roughly correlated with the extent and complexity of muscular activity. In some fishes, birds, and mammals, the cerebellum is highly developed, whereas it tends to be small in agnathans, amphibians, and reptiles. The cerebellum coordinates muscle activity and is responsible for muscle tone, posture, and equilibrium.

Injury or removal of the cerebellum results in impaired muscle coordination. A bird without a cerebellum cannot fly, and its wings thrash about jerkily. When the human cerebellum is injured by a blow or by disease, muscular movements are uncoordinated. Any activity requiring delicate coordination, such as threading a needle, is very difficult, if not impossible.

In mammals, a large mass of fibers known as the pons forms a bulge on the anterior surface of the brain stem. The pons is a bridge connecting the spinal cord and medulla with upper parts of the brain. It contains centers that help regulate respiration and nuclei that relay impulses from the cerebrum to the cerebellum.

The midbrain is prominent in fishes and amphibians

In fishes and amphibians, the **midbrain** is the most prominent part of the brain, serving as the main association area. It receives incoming sensory information, integrates it, and sends decisions to appropriate motor nerves. The dorsal portion of

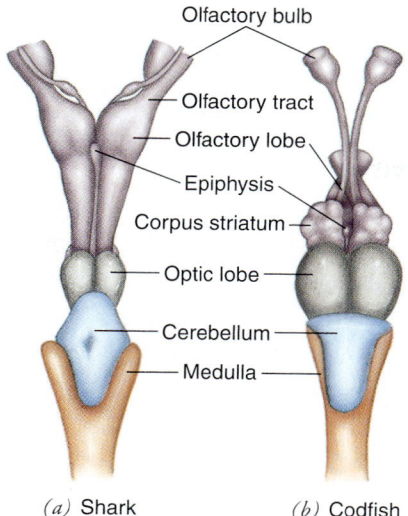

(a) Shark   (b) Codfish

Labels for (a)/(b):
- Olfactory bulb
- Olfactory tract
- Olfactory lobe
- Epiphysis
- Corpus striatum
- Optic lobe
- Cerebellum
- Medulla

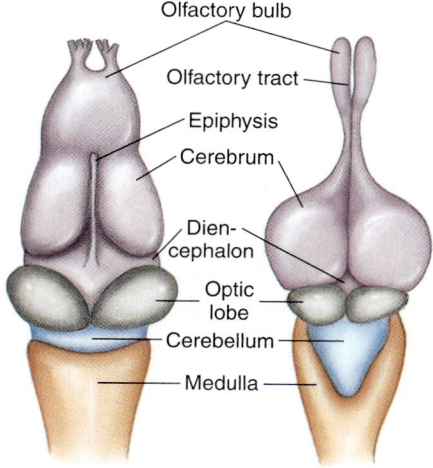

(c) Amphibian (frog)   (d) Reptile (alligator)

Labels for (c)/(d):
- Olfactory bulb
- Olfactory tract
- Epiphysis
- Cerebrum
- Dien-cephalon
- Optic lobe
- Cerebellum
- Medulla

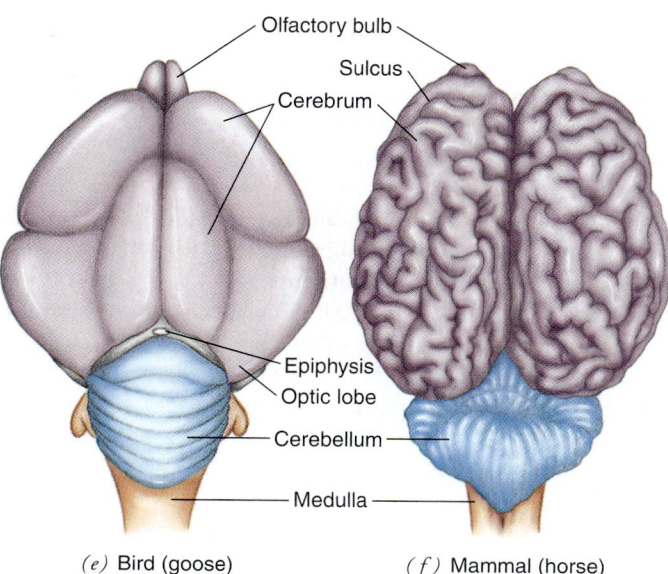

(e) Bird (goose)   (f) Mammal (horse)

Labels for (e)/(f):
- Olfactory bulb
- Sulcus
- Cerebrum
- Epiphysis
- Optic lobe
- Cerebellum
- Medulla

Figure 40–6 **Evolution of the vertebrate brain.** Comparison of the brains of members of six vertebrate classes reveals basic similarities and evolutionary trends. Note that different parts of the brain are specialized in the various groups. For example, the large olfactory lobes in the shark brain *(a)* are essential to this predator's highly developed sense of smell. *(b* through *f)* During the course of evolution, the cerebrum and cerebellum have become larger and more complex. In mammals *(f)*, the cerebrum is the most prominent part of the brain; the cerebral cortex, the thin outer layer of the cerebrum, is highly convoluted (folded), which greatly increases its surface area.

the midbrain is differentiated to some extent. For example, the optic lobes are specialized for visual interpretations.

In reptiles, birds, and mammals, many of the functions of the optic lobes are assumed by the cerebrum, which develops from the forebrain. In mammals, the midbrain consists of the **superior colliculi,** centers for visual reflexes such as pupil constriction, and the **inferior colliculi,** centers for certain auditory reflexes. The mammalian midbrain also contains a center (the red nucleus) that helps maintain muscle tone and posture.

The forebrain gives rise to the thalamus, hypothalamus, and cerebrum

As indicated in Table 40–2, the **forebrain** subdivides to form the **telencephalon** and **diencephalon.** The telencephalon gives rise to the **cerebrum,** and the diencephalon to the **thalamus** and **hypothalamus.** The lateral ventricles (also called the first and second ventricles) are located within the cerebrum. Each lateral ventricle is connected with the third ventricle (within the diencephalon) by way of a channel.

In all vertebrate classes, the thalamus is a relay center for motor and sensory messages. In mammals, all sensory messages (except those from the olfactory receptors) are delivered to the thalamus before they are relayed to the sensory areas of the cerebrum.

The hypothalamus, which lies below the thalamus, forms the floor of the third ventricle. The hypothalamus contains olfactory centers and is the principal integration center for the regulation of the viscera (internal organs). It provides input to centers in the medulla and spinal cord that regulate activities such as heart rate, respiration, and digestive system function. In reptiles, birds, and mammals, the hypothalamus controls body temperature. It also regulates appetite and water balance and is involved in emotional and sexual responses. As will be discussed in Chapter 47, the hypothalamus links the nervous and endocrine systems and produces certain hormones.

The telencephalon gives rise to the cerebrum and, in most vertebrate groups, the olfactory bulbs. These structures are important in the chemical sense of smell—the dominant sense in most aquatic and terrestrial vertebrates. In fact, much of brain development in vertebrates appears to be focused on the integration of olfactory information. In fish and amphibians, the cerebrum is almost entirely devoted to this function.

TABLE 40-2 Differentiation of CNS Structures

Early Embryonic Divisions	Subdivisions	Derivatives in Adult	Cavity
Brain			
Forebrain	Telencephalon	Cerebrum	Lateral ventricles (first and second ventricles)
	Diencephalon	Thalamus, hypothalamus, epiphysis (pineal body)	Third ventricle
Midbrain	Midbrain (Mesencephalon)	Optic lobes in fish and amphibians; superior and inferior colliculi	Cerebral aqueduct
Hindbrain	Metencephalon	Cerebellum, pons	
	Myelencephalon	Medulla	Fourth ventricle
Spinal cord		Spinal cord	Central canal

Birds are an exception among the vertebrates in that their sense of smell is generally poorly developed. A part of their cerebrum, the *corpus striatum,* however, is highly developed. This structure is thought to control the innate, stereotypical, yet complex action patterns characteristic of birds. Just above the corpus striatum is a region thought to govern learning.

In most vertebrates, the cerebrum is divided into right and left **hemispheres.** Most of the cerebrum is made of **white matter,** which consists mainly of myelinated axons that connect various parts of the brain. In mammals and most reptiles, a layer of **gray matter,** the **cerebral cortex,** makes up the outer portion of the cerebrum. Gray matter contains cell bodies and dendrites.

Certain reptiles and all mammals have a type of cerebral cortex, the **neopallium,** not found in less complex vertebrates. It serves as an association area—a region that links sensory and motor functions and is responsible for higher functions such as learning. The neopallium is very extensive in mammals, making up the bulk of the cerebrum. In humans, about 90% of the cerebral cortex is neopallium, consisting of six distinct cell layers.

In mammals, the cerebrum is the most prominent part of the brain. During embryonic development, it expands and grows backwards, covering many other brain structures. The cerebrum is responsible for many of the functions performed by other parts of the brain in other vertebrates. In addition, it has many complex association functions that are lacking in reptiles, amphibians, and fish.

In small or simple mammals, the cerebral cortex may be smooth. However, in large complex mammals, the surface area is greatly expanded by numerous folds called **convolutions.** The furrows between them are called **sulci** (sing., *sulcus*) if shallow and **fissures** if deep. The number of folds (not the size of the brain) has been associated with complexity of brain function.

THE HUMAN CENTRAL NERVOUS SYSTEM IS REMARKABLY COMPLEX

The soft, fragile human brain and spinal cord are well protected. Encased within bone, they are covered by three layers of connective tissue, the **meninges.** The three meningeal layers are the tough, outer **dura mater,** the middle **arachnoid,** and the thin, vascular **pia mater,** which adheres closely to the tissue of the brain and spinal cord (Fig. 40–7). Meningitis is a disease in which these coverings become infected and inflamed.

Between the arachnoid and the pia mater is the subarachnoid space, which contains **cerebrospinal fluid (CSF).** The fluid is produced by special networks of capillaries, collectively called the **choroid plexus,** that *project* (extend) from the pia mater into the ventricles. The choroid plexus extracts nutrients from the blood and adds them to the CSF. Together the choroid plexus and the arachnoid serve as a barrier between blood and CSF. They prevent harmful substances in the blood from entering the brain.

CSF is a shock-absorbing fluid that cushions the brain and spinal cord against mechanical injury. This fluid also serves as a medium for exchange of nutrients and waste products between the blood and brain. CSF circulates down through the ventricles and passes into the subarachnoid space surrounding the brain and spinal cord. It is then reabsorbed into large blood sinuses within the dura mater.

The spinal cord transmits impulses to and from the brain

The tubular **spinal cord** extends from the base of the brain to the level of the second lumbar vertebra. A cross section through the spinal cord reveals a small **central canal** surrounded by an

Dural sinus (superior sagittal sinus)

Skin of scalp
Skull
Dura mater
Arachnoid
Subarachnoid space
Pia mater
Cortex of brain
Brain (cerebrum)

(a)

Skin
Skull
Blood sinus
Pia mater
Brain
Lateral ventricle
Interventricular foramen
Third ventricle

Dura mater
Choroid plexus
Cerebral aqueduct
Arachnoid
Subarachnoid space
Fourth ventricle
Choroid plexus

Spinal cord

(b)

Figure 40–7 Protection of the brain and spinal cord. The CNS is well protected by the skull and meninges, and by cerebrospinal fluid. *(a)* Frontal section through the superior part of the brain. Note the large blood sinus shown between two layers of the dura mater. Blood leaving the brain flows into such sinuses and then circulates to the large jugular veins in the neck. *(b)* The cerebrospinal fluid cushions the brain and spinal cord. Produced by the choroid plexi in the walls of the ventricles, this fluid circulates through the ventricles and subarachnoid space. It is continuously produced and continuously reabsorbed into the blood of the dural sinuses.

area of gray matter shaped somewhat like the letter **H** (Fig. 40 –8). The gray matter is composed of large masses of cell bodies, dendrites, unmyelinated axons, glial cells, and blood vessels and is subdivided into sections called **horns.** The white matter, found outside the gray matter, consists of myelinated axons arranged in bundles called **tracts** or **pathways.** Long ascending tracts conduct impulses up the cord to the brain. For example, the spinothalamic tracts in the anterior and lateral columns of the white matter conduct pain and temperature information from sensory neurons in the skin. The pyramidal tracts are descending tracts that convey impulses from the cerebrum to motor nerves at various levels in the cord. The spinal nerves are described in a later section of this chapter.

In addition to transmitting impulses to and from the brain, the spinal cord controls many reflex activities. A **reflex action** is a relatively fixed response pattern to a simple stimulus. The response is predictable and automatic, not requiring conscious thought. Many of the activities of the body, such as breathing, are regulated by reflex actions.

Although most reflex actions are much more complex, let us consider a **withdrawal reflex,** in which a neural circuit consisting of only three types of neurons is needed to carry out a response to a stimulus (Fig. 40–9). Suppose you touch a hot stove. Almost instantly, and even before you are consciously aware of what has happened, you jerk your hand away. In this brief instant a message has been carried by a sensory neuron from pain receptors in the skin to the spinal cord. In the tissue of the spinal cord, the message is transmitted from the sensory neuron to an association neuron. Finally, the message is transmitted to an appropriate motor neuron, which conducts it to groups of muscles that respond by contracting and moving the limb toward the midline of the body, moving the hand away from the stove. Actually, many neurons in sensory, association, and motor nerves participate in such a reaction, and complicated switching is involved. Generally, we are not even consciously aware that all these responding muscles exist.

Quite probably, at the same time that the reflex pathway is activated, a message is also sent up the spinal cord to the conscious areas of the brain. As you withdraw your hand from the hot stove, you become aware of what has happened and feel the pain. This awareness, however, is not part of the reflex response.

Recent research suggests that contrary to long accepted dogma, the spinal cord has plasticity and is capable of training. For example, sensory feedback from walking exercise may increase the strength of neural connections in the spinal cord.

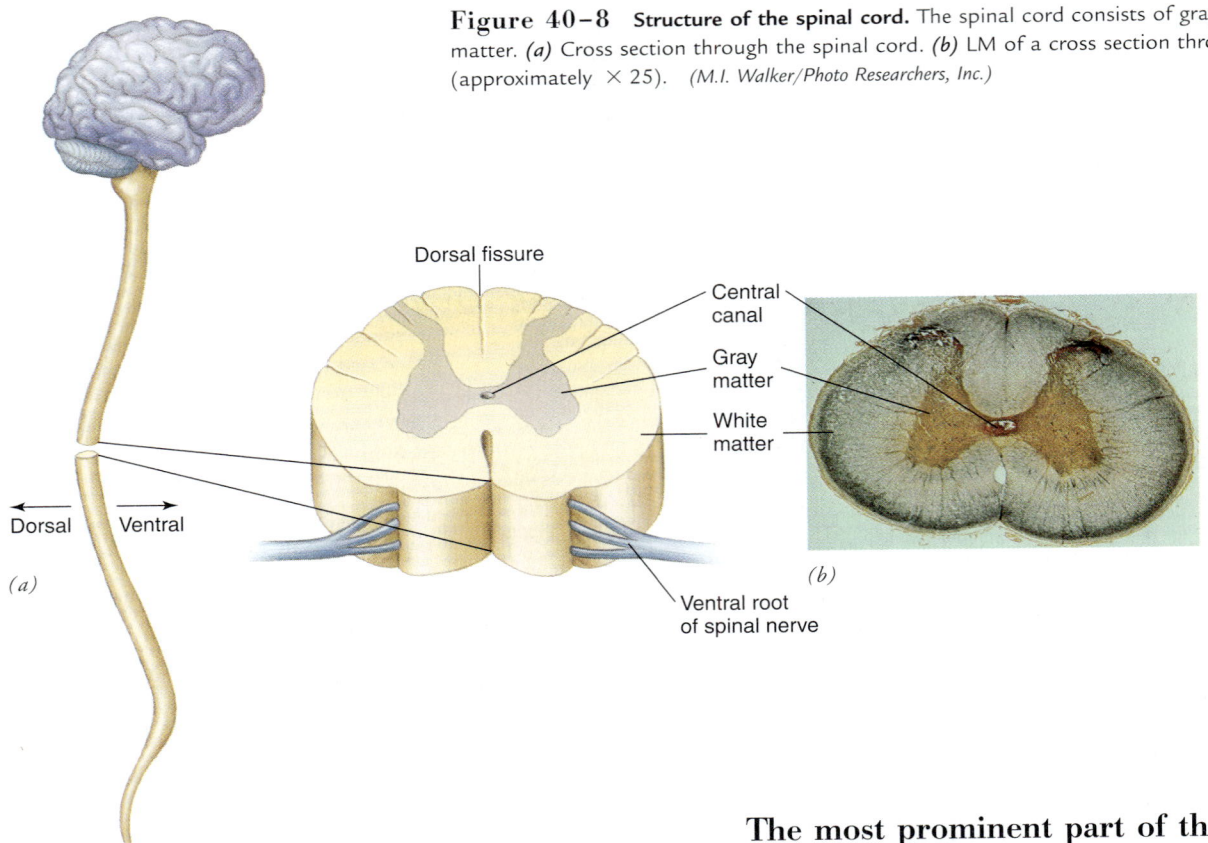

Dorsal fissure

Central canal

Gray matter

White matter

Dorsal Ventral

(a)

(b)

Ventral root of spinal nerve

Such findings raise hope that some victims of spinal cord injury (some 200,000 patients in the United States alone) could potentially be helped to develop at least some limited ability to walk again.

The most prominent part of the human brain is the cerebrum

The structure and functions of the main parts of the human brain are summarized in Table 40–3. The human brain is illustrated in Figures 40–10 and 40–11. As in other mammals, the human cerebral cortex is functionally divided into three areas: (1) **sensory areas** that receive incoming signals from the

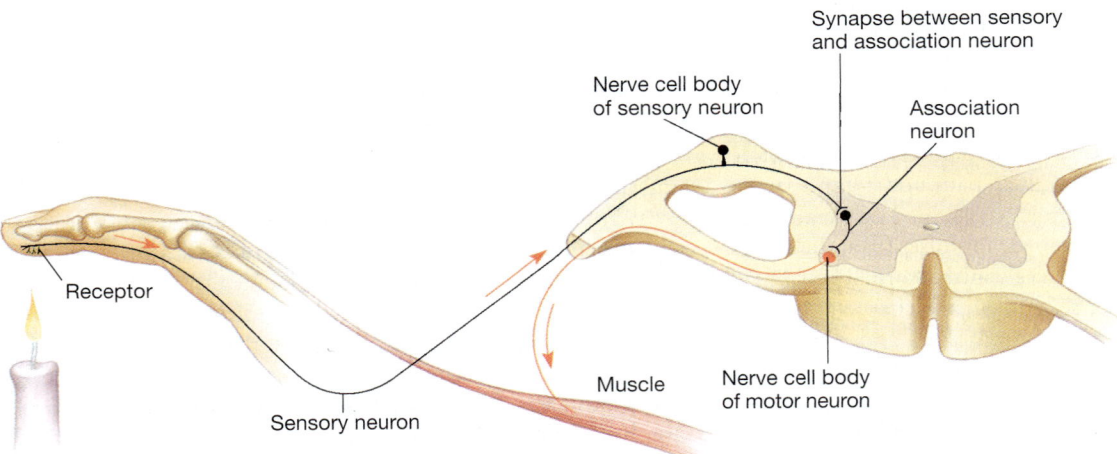

Synapse between sensory and association neuron

Nerve cell body of sensory neuron

Association neuron

Receptor

Muscle

Nerve cell body of motor neuron

Sensory neuron

Figure 40–9 **Withdrawal reflex.** A sensory neuron transmits impulses from the receptor to the central nervous system, where it synapses with an association neuron. Then an appropriate motor neuron *(shown in red)* transmits impulses to the muscles that move the hand away from the flame (the actual response).

TABLE 40-3 The Brain

Structure	Description	Function
Brain stem		
Medulla	Continuous with spinal cord; primarily made up of nerves passing from spinal cord to rest of brain	Contains vital centers (clusters of neuron cell bodies) that control heartbeat, respiration, and blood pressure; contains centers that control swallowing, coughing, vomiting
Pons	Forms bulge on anterior surface of brain stem	Connects various parts of brain with one another; contains respiratory center
Midbrain	Just above pons; largest part of brain in fishes and amphibians; in humans most of its functions are assumed by cerebrum	Center for visual and auditory reflexes (e.g., pupil reflex, blinking, adjusting ear to volume of sound)
Thalamus	At top of brain stem	Main sensory relay center for conducting information between spinal cord and cerebrum. Neurons in thalamus sort and interpret all incoming sensory information (except olfaction) before relaying messages to appropriate neurons in cerebrum
Hypothalamus	Just below thalamus; pituitary gland is connected to hypothalamus by stalk of neural tissue	Contains centers for control of body temperature, appetite, fat metabolism, and certain emotions; regulates pituitary gland
Cerebellum	Second largest division of brain	Reflex center for muscular coordination and refinement of movements; when it is injured, performance of voluntary movements is uncoordinated and clumsy
Cerebrum	Largest, most prominent part of human brain; longitudinal fissure divides cerebrum into right and left hemispheres, each divided into lobes: frontal, parietal, temporal, and occipital lobes	Center of intellect, memory, consciousness, and language; also controls sensation and motor functions
Cerebral cortex (outer gray matter)	Arranged into convolutions (folds) that increase surface area; functionally, cerebral cortex is divided into:	
	1. Motor cortex	Controls movement of voluntary muscles
	2. Sensory cortex	Receives incoming information from eyes, ears, pressure and touch receptors, etc.
	3. Association cortex	Site of intellect, memory, language, and emotion; interprets incoming sensory information
White matter	Consists of myelinated axons of neurons that connect various regions of brain; these axons are arranged into bundles (tracts)	Connects: 1. Neurons within same hemisphere 2. Right and left hemispheres 3. Cerebrum with other parts of brain and spinal cord

sense organs; (2) **motor areas** that control voluntary movement; and (3) **association areas** that link the sensory and motor areas and are responsible for thought, learning, language, memory, judgment, and personality.

Investigators have been able to map the cerebral cortex, locating the areas responsible for different functions. The **occipital lobes** contain the visual centers. Stimulation of these areas, even by a blow on the back of the head, causes the sensation of light; their removal causes blindness. The centers for hearing are located in the **temporal lobes** of the brain above the ear; stimulation by a blow causes a sensation of noise. Removal of both auditory areas causes deafness. Removal of one does not cause deafness in one ear but rather produces a decrease in the auditory acuity of both ears.

Cerebellum   Parietal   Brainstem
             lobe
                         Central
                         sulcus

                         Frontal
                         lobe

Occipital
lobe

                         Prefrontal
                         area

Temporal
lobe

Medulla

(b)

Figure 40–10   The human brain. *(a)* Photograph of human brain, lateral view. The diencephalon and part of the brain stem are covered by the cerebrum. *(b)* Lateral view of the human brain showing the lobes of the cerebrum.   *(Fred Hossler/Visuals Unlimited)*

A groove called the **central sulcus** crosses the top of each hemisphere from medial to lateral edge. This groove partially separates the **frontal lobes** from the **parietal lobes.** The **primary motor areas** in the frontal lobes control the skeletal muscles. The **primary sensory areas** in the parietal lobes receive information regarding heat, cold, touch, and pressure from sense organs in the skin.

The size of the motor area in the brain for any given part of the body is proportional not to the amount of muscle but to the complexity of movement involved. Predictably, areas that control the hands and face are relatively large (Fig. 40–12). A similar relationship exists between the sensory area and the sensitivity of the region of the skin from which it receives impulses. In connections between the body and the

brain, not only do the fibers cross so that one side of the brain controls the opposite side of the body, but another "reversal" makes the uppermost part of the cortex control the lower limbs of the body.

When all the areas of known function are plotted, they cover almost all of the rat's cortex, a large part of the dog's, a moderate amount of the monkey's, and only a small part of the total surface of the human cortex. The remaining cortical areas are the association areas. Somehow the association regions integrate all the diverse impulses reaching the brain into a meaningful unit so that an appropriate response is made. When disease or accident destroys the functioning of one or more association areas, the ability to recognize certain kinds of symbols may be lost. For example, the names of objects may be forgotten, although their functions are remembered and understood.

The white matter of the cerebrum lies beneath the cerebral cortex. Nerve fibers of the white matter connect the cortical areas with one another and with other parts of the nervous system. A large band of white matter, the **corpus callosum,** connects the right and left hemispheres (Figure 40–11).

Deep within the white matter of the cerebrum lie the **basal ganglia,** paired groups of nuclei (gray matter). These nuclei play an important role in coordination of movement. The basal ganglia send signals to the **substantia nigra,** which is located in the midbrain, and they receive inputs back from the substantia nigra. The neurons that project to the basal ganglia produce dopamine (see *Making the Connection: Dopamine and Motor Function*). Other neurons from the substantia nigra signal nuclei in the thalamus that in turn relay information to the motor cortex. The substantia nigra neurons that signal the thalamus release the neurotransmitter GABA. Just how these areas work together to coordinate motor function is not yet fully understood.

The brain integrates information

Most of the integration of the nervous system takes place within the cerebral cortex. The cerebral cortex integrates information about such diverse activities as arousal and sleep, emotion, and information processing.

Brain activity cycles in a sleep-wake pattern

Brain activity can be studied by measuring and recording the electrical potentials, or "brain waves," generated by thousands of active neurons in various parts of the brain. This electrical activity can be recorded by a device known as an electroencephalograph. To obtain a recording of this electrical activity, called an **electroencephalogram (EEG),** electrodes are taped to different parts of the scalp, and the activity of the cerebral cortex is measured. The EEG shows that the brain is continuously active. The most regular indication of activity, **alpha waves,** comes mainly from the visual areas in the occipital lobes

Figure 40-11 **Midsagittal section through the human brain.** Half of the brain has been cut away, exposing structures normally covered by the cerebrum. Compare the diagram **(a)** with the photograph of the human brain **(b)**. *(b, Science Pictures Limited/Science Photo Library/Photo Researchers, Inc.)*

when the person being tested is resting quietly with eyes closed. Alpha waves occur rhythmically at the rate of about ten per second (Fig. 40–13).

When the eyes are opened, alpha waves disappear and are replaced by more rapid, irregular waves. As you are reading this biology text, your brain is emitting **beta waves.** These have a fast-frequency rhythm characteristic of heightened mental activity such as information processing. During sleep, the brain emits waves with a lower frequency and a higher amplitude. The slow, large waves associated with certain stages of sleep are called **delta waves.** During dreaming flurries of irregular waves occur.

Certain brain diseases change the pattern of brain waves. Individuals with epilepsy, for example, exhibit a distinctive, recognizable, abnormal wave pattern. The location of a brain tumor or the site of brain damage caused by a blow to the head can sometimes be determined by noting the part of the brain that shows abnormal waves.

The **reticular activating system (RAS)** is a complex neural pathway within the brain stem and thalamus. Sometimes referred to as an arousal system, the RAS receives messages from neurons in the spinal cord and from many other parts of the nervous system and communicates with the cerebral cortex by complex neural circuits. The RAS maintains **consciousness**—a cognitive system that produces awareness of sensations and memories. The extent of RAS activity helps determine the state of alertness. When the RAS bombards the cerebral cortex with stimuli, you feel alert and are able to focus your attention on specific thoughts. If the RAS is severely damaged, the victim may pass into a deep, permanent coma.

Sleep is an alteration of consciousness during which there is decreased electrical activity of the cerebral cortex and from which a person can be aroused. Two main stages of sleep are recognized: **non-REM** and **REM.** *REM* is an acronym for *rapid eye movements.* During non-REM sleep, sometimes called normal sleep, metabolic rate decreases, breathing slows, and

Figure 40–12 Functional areas of the brain. *(a)* Several sensory, motor, and association areas are shown. *(b)* A cross section through the primary motor area showing which area of cerebral cortex controls each body part. The figure shown here, known as a motor homunculus ("little person"), is proportioned to reflect the amount of cerebral cortex that controls each body part. Note that more cortical tissue is devoted to controlling those body structures capable of skilled, complex movement.

Labels (a):
Central sulcus
General sensory area
Visual association area
Auditory area
Auditory association area
Primary visual area
Broca's speech area
Primary motor area
Premotor area
Frontal lobe
(a)

Labels (b):
Knee
Hip
Trunk
Shoulder
Elbow
Wrist
Hand
Little finger
Ring finger
Middle finger
Index finger
Thumb
Neck
Brow
Eyelid and eyeball
Ankle
Toes
Face
Lips
Jaw
Tongue
Swallowing
Vocalization
Salivation
Chewing
(b)

blood pressure decreases. Delta waves, thought to be generated spontaneously by the cerebral cortex when it is not driven by impulses from other parts of the brain, are characteristic of non-REM sleep.

Every 90 minutes or so, a sleeping person enters the REM stage for a time. During this stage, which accounts for about one-fourth of total sleep time, the eyes move about rapidly beneath the closed but fluttering lids. Brain waves change to a desynchronized pattern of beta waves. Sleep researchers claim that everyone dreams, especially during REM sleep. In 1998, Allen Braun of the National Institutes of Health and his colleagues reported in *Science* that they found differences in blood flow through the brain during different stages of sleep. Comparing positive emission tomography (PET) scans of sleeping subjects, they found that during REM sleep, blood flow in the frontal lobes was reduced, whereas blood flow increased in areas that produce visual scenes and emotion. Other investigators have suggested that during dreams, norepinephrine release

within the RAS increases, which generates stimulating impulses that are fed into the cerebral cortex.

The hypothalamus and brain stem regulate the sleep-wake cycle. Using rats, investigators have demonstrated that stimulation of the *preoptic area* of the hypothalamus induces non-REM sleep. The preoptic nucleus is located near the *suprachiasmatic nucleus,* which is thought to be one of the most important of the body's biological clocks. The suprachiasmatic nucleus receives information about the duration of light and dark from the retina of the eye and apparently transmits this information to the preoptic nucleus. The fatigue and decreased physical and mental performance referred to as jet lag occurs when we fly to a different time zone where the body is no longer synchronized with the light-dark cycle.

The raphe nuclei in the brain stem (lower pons and medulla) also appear to be important in producing REM sleep. Many of the neurons that project from the raphe nuclei release serotonin, a neurotransmitter involved in sleep. During

DOPAMINE AND MOTOR FUNCTION

How does the neurotransmitter dopamine affect motor function? The function of dopamine was discovered through an interesting series of somewhat unrelated events. During the mid-1950s, the drug reserpine became popular as a major tranquilizer for mental patients. Then, in 1959, investigators noticed that some patients taking reserpine experienced distressing side effects, such as muscle rigidity and persistent tremors (shaking). These symptoms were very similar to those seen in patients with Parkinson's disease, a disorder in which movement is slow, shaky, and difficult. Victims of Parkinson's disease have a shuffling gait and suffer from tremors even when they are not attempting to move. This observation led to studies showing that reserpine greatly reduces the amount of dopamine in two of the basal ganglia within the white matter of the cerebrum. Investigators then discovered that patients with Parkinson's disease have only about 50% of the normal amount of dopamine in their basal ganglia.

Attempts to administer dopamine to these patients were not successful because dopamine cannot penetrate the blood-brain barrier. However, L-dopa, a substance from which dopamine is synthesized in the body, does penetrate the blood-brain barrier. Its use has dramatically relieved the symptoms of Parkinson's disease in many patients. Dopamine is thought to restore balance between inhibition and excitation of neurons involved in motor function.

Dopamine depletion may occur with aging. As a result, even healthy persons experience changes in motor abilities as they age. Body movements and even reflexes slow, and movement becomes more difficult. Studies suggest that treatment with L-dopa may be helpful. Investigators have had some success using animal models of Parkinson's disease to study the effects of transplanting dopamine-producing neurons directly into the brain.

REM sleep, neurons in the brain stem release norepinephrine, which may stimulate heightened activity in certain regions of the brain. The hormone *melatonin* may also be involved in inducing sleep.

The sleep-wake cycle may be affected by fatigue of the RAS after many hours of activity. Sleep centers are then activated, and their neurons release serotonin. After sufficient rest, the inhibitory neurons of the sleep centers become less excitable, while the excitatory neurons of the RAS become more excitable.

Although birds, mammals, and many other animals have a sleep-wake cycle, neurophysiologists do not know why sleep is necessary. When a person stays awake for unusually long periods, fatigue and irritability result, and even routine tasks cannot be performed well.

Not only is non-REM sleep required, but REM sleep is apparently also necessary. In sleep deprivation experiments performed with human volunteers, lack of REM sleep makes subjects anxious and irritable. When the subjects are permitted to sleep normally again, they spend more time than usual in the REM stage for a period. Many types of drugs, for example amphetamines, alter sleep patterns and affect the amount of REM sleep. Certain drugs that induce sleep may increase total sleeping time but decrease REM time. When a person stops taking such a drug, several weeks may be required before normal sleep patterns are reestablished.

The limbic system affects emotional aspects of behavior

The **limbic system,** an action system of the brain, consists of certain structures of the cerebrum and hypothalamus, as well as several nuclei in the midbrain. The limbic system affects the emotional aspects of behavior, evaluates rewards, and is important in motivation. This system also plays a role in sexual behavior, biological rhythms, and autonomic responses. Stimulation of certain areas of the limbic system in an experimental animal results in increased general activity and may cause fighting behavior or extreme rage.

Figure 40–13 **Electroencephalograms showing electrical activity in the brain.** The regular waves characteristic of the relaxed state are called alpha waves. During mental activity, rapid, low-amplitude beta waves are dominant. Recordings made in various stages of sleep show theta and delta waves, which have a lower frequency and greater amplitude.

Investigators first discovered that the limbic system is an important motivational system when they implanted electrodes in certain areas of the brains of laboratory animals. They found that a rat may press a lever that stimulates the so-called reward area as many as 15,000 times per hour. The rat will forego food and water in favor of self-stimulation until it drops from exhaustion. One of the areas that makes up this motivational system is the substantia nigra in the midbrain. When something that feels good happens, neurons in this area release dopamine. Drugs such as cocaine and amphetamines block the reuptake of dopamine, leading to an increased rate of self-stimulation.

Until 1990 neurophysiologists thought that dopamine functioned directly to produce feelings of pleasure and euphoria. Since that time investigators have reported conflicting data about dopamine's role, and its function is a matter of controversy. Some neurophysiologists now think that dopamine acts to call our attention to surprising events or to those that predict rewards. Dopamine may motivate us to act when something important is happening. This view may help us understand the mechanisms of such disorders as attention deficit disorder (ADD) and schizophrenia, which involve abnormal amounts of dopamine in the brain. Such individuals have difficulty filtering out sensory stimuli. The presence of excess dopamine might drive them to divert their attention to so many sensory stimuli that they have difficulty focusing on work.

Using PET scans, Dr. Mark George and his team at the National Institute of Mental Health in Bethesda, Maryland, showed that areas outside the limbic system are also involved in emotion. In 1995 George reported that when his subjects were happy, activity decreased in the temporal-parietal region of the cortex, an area involved in complex planning. During sadness, the left prefrontal area of the cortex, and the amygdala, a pair of structures within the limbic system, were activated. In subjects with clinical depression, the left prefrontal area appeared to be shut down and subjects felt emotionally numb.

Learning involves the storage of information and its retrieval

Learning is a relatively long-lasting change in behavior resulting from experience. Laboratory experiments have shown that members of every animal phylum can learn. In order for learning to occur, we must be able to remember what we experience. **Memory** is the storage of information and the ability to recall it.

Information processing At least two stages of memory are recognized: short-term, and long-term. At any moment you are bombarded with thousands of bits of sensory information. At this very moment, your eyes are receiving information about the words on this page, the objects around you, and the intensity of the light in the room. At the same time you may be hearing a variety of sounds—music, your friends talking in the next room, the hum of an air conditioner. Your olfactory epithelium may sense cologne or the smell of coffee. Sensory receptors in your hands may be receiving information regarding the weight and position of your book. Most sensory stimuli are not important to remember and so are filtered out.

When we focus our attention on certain bits of sensory information or other information of which we become aware, that information may be registered as **short-term memory.** In order for sensory stimuli to be recognized, we must relate them to past experience or knowledge, and this requires further processing. We recognize patterns and begin to encode information. Typically, we can hold only about seven chunks of information (a chunk is some unit such as a word, syllable, or number) at a time in short-term memory.

Short-term memory allows us to recall information for a few minutes. Usually when we look up a phone number, for example, we remember it only long enough to dial. Should we need the same number an hour later, most of us would have to look it up again. One hypothesis of short-term memory suggests that it is based on reverberating circuits (see Fig. 39–15). A memory circuit may continue to reverberate for several minutes until it fatigues or until new signals are received that interfere with the old.

When information is selected for long-term storage, the brain rehearses the material and then stores it in **long-term memory** in association with similar memories. Several minutes are required for a memory to become consolidated within long-term memory. Should a person suffer a brain concussion or undergo electroshock therapy, for example, memory of what happened immediately prior to the incident may be completely lost. This is known as *retrograde amnesia.* When the hippocampus is damaged, short-term memory is unimpaired and the patient can still recall information stored in the past. However, new short-term memories can no longer be converted to long-term memories.

Retrieval of information stored in long-term memory is of considerable interest, especially to students. Some researchers think that once information is consolidated in long-term memory, it remains within the brain permanently. The challenge is to find it when you need it. When you seem to forget something, the problem may be that you have not effectively searched for the memory. Information retrieval can be improved by careful storage. One method is to form strong associations between items when they are being stored.

Researchers have worked with the brains of experimental animals for years without finding specific regions where information is stored. Some forms of learning take place in association areas within lower brain regions, the thalamus, for example. Even very simple animals like flatworms that completely lack a cerebral cortex are capable of some types of learning.

When large areas of the mammalian cerebral cortex are destroyed, information is lost somewhat in proportion to the extent of lost tissue. However, no specific area can be labeled the "memory bank." Rather, memories appear to be stored within the many sensory areas of the brain. For example, visual memories may be stored in the visual centers of the occipital lobes, and auditory memories may be stored in the temporal lobes. Memories are thought to be integrated in many areas of the brain, including association areas of the cerebral cortex, parts of the limbic system (the amygdala and the hippocampus), and the thalamus and hypothalamus.

Wernicke's area in the temporal lobe has been identified as a very important association area for complex thought processes. This area is an important center for language function involved in the recognition and interpretation of words. Neurons within the association areas form highly complex pathways that permit complicated reverberation.

Neurophysiological changes during learning Short-term memory involves brief changes in neurotransmitter receptors linked by second messengers to ion channels in postsynaptic neurons. The second messenger cyclic AMP is thought to facilitate short-term memory by activating a specific protein kinase. By phosphorylating certain molecules, this protein kinase affects specific ion channels.

Long-term memory involves slower, but longer lasting, changes in postsynaptic neurons. Activated receptors are linked to G proteins, and cyclic AMP acts as a second messenger. A high level of cyclic AMP activates a protein kinase, which enters the nucleus, leading to gene activation and protein synthesis. In this process, protein kinase phosphorylates a regulatory protein (a transcription factor) known as CREB (for cyclic AMP response element binding protein). CREB then turns on the transcription process of certain genes. CREB has been shown to be a central memory molecule in many animals including fruit flies and mice.

In some types of learning, changes take place in presynaptic terminals or postsynaptic neurons that permanently enhance or inhibit the transmission of impulses. In some cases, specific neurons may become more sensitive to neurotransmitter.

Unraveling the mechanisms that change the strength of synaptic connections appears to be an important key to understanding the process of learning. Neurons typically transmit action potentials in bursts, and the amount of neurotransmitter released by each action potential may increase or decrease. An increase, referred to as **synaptic enhancement,** is thought to occur as a result of calcium ion accumulation inside the presynaptic terminal. One type of synaptic enhancement, **synaptic facilitation,** can occur after a single action potential. However, synaptic facilitation lasts for only milliseconds. **Potentiation** (which means to strengthen or make more potent) is a longer form of synaptic enhancement that can last for several minutes. It occurs when a presynaptic neuron continues to transmit action potentials at a high rate for a minute or longer.

For more than 30 years, neurophysiologists have studied learning in *Aplysia,* a marine snail (a nudibranch) (Fig. 40–14). Because its nervous system is relatively simple and because the cellular mechanisms of neural functioning have been conserved during evolution, *Aplysia* has proved to be a good

(a)

(b) Dorsal view

(c)

Figure 40–14 **A model for studying learning.** *(a)* Neurobiologists have taken advantage of the relatively simple nervous system of the marine snail *Aplysia* sp. to study learning. *(b)* Structure of *Aplysia.* Seawater flows in through the slightly open mantle, bathes the gills, then exits through the siphon. *(c)* The neural circuit for the withdrawal reflex in *Aplysia.* Sensory neurons that innervate the siphon are sensitive to touch. These neurons project directly to the motor neurons of the gill muscles. They also project indirectly to the gill muscles by way of interneurons. When the siphon is touched, the siphon and gill are withdrawn into the protective mantle. *(Randy Morse/Tom Stack & Associates)* model of memory and learning in other animals, including mammals.

An important focus of study has been the gill-withdrawal reflex, a relatively simple neural circuit that helps protect the animal from predators. *Aplysia's* gill is located in its mantle (a fold of tissue), which is normally slightly open allowing sea water to bathe the gill. The water exits through its siphon. When *Aplysia's* exposed siphon is touched, a reflex action is initiated resulting in withdrawal of the siphon and gill into the mantle. This behavior is a reflex action that can be modified by experience.

The neural pathway includes sensory neurons that synapse with motor neurons. When a sensory neuron is stimulated, a message is transmitted to a motor neuron that stimulates the gill to withdraw. Interneurons synapse with sensory and motor neurons; they can be activated by the sensory neurons and then signal the motor neurons. Using this model, neurophysiologists have been able to study several neural processes, including habituation, sensitization, and classical conditioning.

Habituation is the decrease in response that occurs after repeated exposure to a harmless stimulus. This process permits animals to disregard constant stimuli. When *Aplysia's* siphon is repeatedly stimulated, the gill-withdrawal reflex habituates and the gill is no longer withdrawn when the siphon is touched. By studying the mechanism of habituation in this marine snail, investigators learned how habituation comes about in many other animals. They found that, with repeated stimulation, sensory neurons release less neurotransmitter, resulting in a decreased number of action potentials in the motor neurons. The decrease in neurotransmitter release was traced to inactivation of calcium channels in the presynaptic terminals. This leads to a reduction in the number of calcium ions entering the presynaptic terminals in response to repeated action potentials. (Recall from Chapter 39 that the influx of calcium ions stimulates neurotransmitter release.)

Sensitization results in an increased response after experience of an unpleasant stimulus. Stimulating *Aplysia's* tail with a series of electric shocks sensitizes the gill-withdrawal reflex.

Figure 40–15 Sensitization of the gill-withdrawal reflex in the marine snail *Aplysia*. *(a)* Unpleasant stimuli applied to the tail activate facilitatory interneurons. These interneurons signal the presynaptic terminals of siphon sensory neurons. After sensitization, the duration of the presynaptic action potential increases, and larger excitatory postsynaptic potentials are produced. *(b)* A model for the events associated with sensitization of the gill withdrawal reflex. The axon terminals of the facilitatory interneuron release the neurotransmitter serotonin which binds with receptors in the membrane of the presynaptic axon terminals of the sensory neuron. The activated receptor activates two G proteins. One activates adenylyl cyclase which increases cyclic AMP. The cAMP stimulates protein kinase A which closes K^+ channels. This action promotes passage of Ca^{2+} into the neuron. The second G protein activates protein kinase C that may potentiate neurotransmitter release by mobilizing reserve neurotransmitter vesicles.

This increased behavioral response was studied by Eric Kandel and his team at Columbia University. Sensory neurons in the tail synapse on facilitatory interneurons that end on the siphon sensory neuron and on interneurons (Fig. 40–15*a*). Note that these facilitatory interneurons synapse on the synaptic terminals of siphon sensory neurons. Sensitization depends on facilitation (enhancement) of neurotransmitter release. Kandel reported that when these interneurons are stimulated, they release serotonin (Fig. 40–15*b*). The serotonin initiates a complex series of reactions involving G-proteins and cyclic AMP. Cyclic AMP activates protein kinase A, which inactivates potassium channels. These channels are important in repolarization of the sensory neuron. Their inactivation provides a longer time for the axon terminal to remain depolarized. As a result more calcium enters the axon terminal and more neurotransmitter is released. This stimulates a stronger response in the postsynaptic motor neuron. Thus, sensitization depends on an increase in neurotransmitter released by the sensory neuron.

Classical conditioning is a form of learning in which an association is formed between some normal response to a normal stimulus, called the unconditioned stimulus (US) and a new stimulus, the conditioned stimulus (CS). The animal learns the association, and the conditioned stimulus then elicits the same response, now called the conditioned response (CR). (Classical conditioning and other types of learning are described in Chapter 50.) *Aplysia's* gill-withdrawal response can be classically conditioned (Fig. 40–16). A strong electric shock applied to the tail causes a strong unconditioned withdrawal response. In contrast, when the siphon is lightly touched, only a mild response occurs. When the light touch is paired with the electric shock several times, the snail *learns* to respond to a light touch (CS) to the siphon with a strong withdrawal response (CR).

Classical conditioning is similar to sensitization in that interneurons release serotonin. However, sensitization normally lasts only a few minutes, whereas conditioning can last for weeks, months, or years. One difference may be that in conditioning the sensory neuron is already active when it is signaled by the facilitatory interneuron. If a sensory neuron is already active when serotonin binds with it, it is strongly affected and becomes conditioned. When it fires again, it releases an increased amount of neurotransmitter, resulting in a strong behavioral response. The cellular mechanism is thought to involve a G protein, cyclic AMP, and a protein kinase. The protein kinase may activate genes and protein synthesis.

In 1973 T.V.P. Bliss and T.J. Lomo reported that repeated electrical stimulation of neurons in the rabbit brain caused a functional change at synapses. This resulted in a persistent, increased response of postsynaptic neurons to an action potential, a response known as **long-term potentiation,** or **LTP.** LTP appears to be an important mechanism of learning. Is classical conditioning an example of LTP? In 1997 Geoffrey Murphy and David Glanzman reported that classical conditioning of *Aplysia's* siphon-withdrawal reflex is mediated by LTP (see *On the Cutting Edge: The Neurophysiology of Learning*).

Figure 40–16 **Classical conditioning of a withdrawal response in the marine snail *Aplysia*.** An electric shock to the tail is the unconditioned stimulus (US). A sensory neuron from the tail is stimulated and signals a facilitatory interneuron. The interneuron releases neurotransmitter, which signals the motor neurons to withdraw the gill (the unconditioned response or UR). A light touch to the siphon is the conditioned stimulus (CS). During conditioning these stimuli (the electric shock and the light touch) are paired. Neurotransmitter from the interneuron then stimulates the already activated sensory neuron from the siphon. The sensory neuron is strongly affected, and the next time it fires it releases a large amount of neurotransmitter. This causes a response as strong as that elicited by the original shock stimulus to the tail. Thus, the light touch (the CS) now elicits an intense response that withdraws the gill (the conditioned response).

The Neurophysiology of Learning

HYPOTHESIS: Classical conditioning of the siphon-withdrawal reflex in *Aplysia* depends on NMDA (N-methyl-D-aspartate) receptors that are involved in LTP (long-term potentiation).

METHOD: Cultured *Aplysia* neurons were used as an in vitro model system to study a simple type of learning: the classical conditioning of the siphon-withdrawal reflex. The response of neurons exposed to an antagonist of the NMDA receptor was compared to the response of control neurons.

RESULTS: The NMDA receptor antagonist blocked conditioning in the model system.

CONCLUSION: Because conditioning of the siphon-withdrawal response requires normally functioning NMDA receptors, this learning process appears to involve LTP.

Memory and learning depend on synaptic plasticity, which involves complex biochemical processes, including actions of neurotransmitters and postsynaptic receptors. Recent studies suggest that long-term potentiation, or LTP, is an important mechanism in learning.

Investigators have determined that strong stimulation of a presynaptic neuron strengthens the synapse. The same strong stimulation also strengthens weakly stimulated synapses that are activated at the same time. The strong stimulation leads to strong depolarization of the postsynaptic neuron. In turn, this depolarization triggers LTP. Just what is the relationship of LTP to classical conditioning? And what are the cellular mechanisms involved?

In 1997 Geoffrey Murphy and David Glanzman reported the results of their studies of classical conditioning of *Aplysia* in the journal *Science*.[*] They tested the hypothesis that classical conditioning of the siphon-withdrawal reflex (Fig. 40–16) depends on NMDA (N-methyl-D-aspartate) receptors, which appear to be important in LTP. NMDA receptors belong to a class of receptors known as ionotropic glutamate receptors. These receptors respond to the neurotransmitter glutamate by opening ion channels that allow Na^+ and K^+ to diffuse through the plasma membrane. Several types of ionotropic glutamate receptors are known. An important difference among them is that some types actively bind NMDA, an artificial ligand. (Recall that a ligand is any molecule that binds to a specific receptor.)

NMDA receptors appear to require both glutamate and glycine and, under some conditions, these receptors allow the passage of Ca^{2+} as well as Na^+ and K^+. When the postsynaptic neuron is at its resting potential, the ion channel is blocked by Mg^{2+}

*Murphy, G.G., and D.L. Glanzman. "Mediation of Classical Conditioning in *Aplysia californica* by Long-Term Potentiation of Sensorimotor Synapses." *Science*, Vol. 278, 17 Oct. 1997.

ions. The channel can open only after the neuron is depolarized and both glutamate and glycine are bound to the receptor. Because the NMDA receptor requires both ligand binding and a particular voltage, it is a *doubly gated receptor*.

Murphy and Glanzman conditioned cultured neurons in the presence of APV (DL-2-amino-5-phosphonovalerate), an antagonist of the NMDA receptor. Their results indicated that the antagonist blocks conditioning. The antagonist did not affect facilitation of the neural pathway in control groups that were not being conditioned. Murphy and Glanzman's study provides evidence that LTP can be produced by mechanisms that involve NMDA receptors and that such mechanisms are involved in classical conditioning in *Aplysia*.

Other studies have demonstrated that NMDA receptors are important in LTP in mammals; when these receptors are blocked, LTP does not occur. S. Tonegawa at the Massachusetts Institute of Technology and Eric Kandel at Columbia University have reported that when they eliminated NMDA receptors and a protein kinase known to be important for LTP from the hippocampus, LTP was disrupted and mice were not able to learn a water maze.

One model for the neurophysiological basis of LTP is illustrated in the figure. A presynaptic neuron releases glutamate, which combines with non-NMDA receptors. The postsynaptic neuron is depolarized and Mg^{2+} moves from the NMDA receptors, unblocking them. When the NMDA receptor channel opens, Ca^{2+} moves into the cell. The Ca^{2+} appears to be an important trigger for LTP.

Calcium ions act as a second messenger initiating long-term changes. One effect of Ca^{2+} may be to stimulate release of a **retrograde signal** (one that moves backward) from the *post*synaptic neuron that would signal the *pre*synaptic neuron. This signal would enhance neurotransmitter release by the presynaptic neuron, thus strengthening the relationship between the two neurons. Some in-

---

Effects of experience Many studies have demonstrated neural plasticity, the ability to change in response to environmental stimuli, in rats and other laboratory animals exposed to enriched environments. In contrast to rats housed in standard cages and provided with the basic necessities, those exposed to enriched environments are given toys and other stimulating objects, as well as the opportunity to socially interact with other rats. Animals reared in a complex environment ex-

hibit increased synaptic contacts and may be able to process and remember information more quickly than animals not given such advantages. In 1997, Gerd Kempermann and his colleagues reported in the journal *Nature* that mice reared in an enriched environment developed significantly greater numbers of neurons in the hippocampus. These investigators also reported that the experimental mice learned a maze significantly faster than controls.

A proposed model for the mechanism of LTP. *(a)* A single stimulus results in glutamate release from a presynaptic neuron. The glutamate acts mainly on non-NMDA receptors. NMDA receptors are blocked by Mg^{2+}. *(b)* A series of action potentials activates NMDA receptors. Calcium ions pass through the ion channel and enter the cell and then activate a second messenger system. A retrograde signal is produced that signals the presynaptic neuron to release more glutamate.

vestigators think that the retrograde signal is a soluble gas, perhaps nitric oxide. Calcium ions act on the enzyme NO synthase, which produces nitric oxide. This gas passes easily through cell membranes, and, because it is very reactive, it only lasts a few seconds. Nitric oxide may also signal nearby neurons, thus recruiting more neurons in the learning process. Nitric oxide has been identified as a signal in several types of tissues and can activate other signaling molecules on the cell surface.

As this brief glimpse of the cellular processes involved in synaptic plasticity indicates, the neurophysiology of learning and memory are extremely complex, and it is likely that many different mechanisms for LTP will be discovered.

Early environmental stimulation can also enhance the development of motor areas in the brain. For example, the brains of rats encouraged to exercise become slightly heavier than those of control animals. Characteristic changes occur within the cerebellum, including the development of larger dendrites.

Apparently, during early life, certain critical or sensitive periods of nervous system development occur that are influenced by environmental stimuli. For example, when the eyes of young mice first open, neurons in the visual cortex develop large numbers of dendritic spines (structures on which synaptic contact takes place). If the animals are kept in the dark and deprived of visual stimuli, fewer dendritic spines form. If the mice are exposed to light later in life, some new dendritic spines form, but never the number that develop in a mouse reared in a normal environment.

Studies linking the development of the brain with environmental experience indicate that early stimulation is important for the sensory, motor, and intellectual development of children. Such studies have led to the rapidly expanding educational toy market and to widespread acceptance of early education programs. Researchers have also suggested that continuing environmental stimulation is needed to maintain the status of the cerebral cortex in later life.

THE PERIPHERAL NERVOUS SYSTEM INCLUDES SOMATIC AND AUTONOMIC SUBDIVISIONS

The **peripheral nervous system (PNS)** consists of the sensory receptors, the nerves that link these receptors with the CNS, and the nerves that link the CNS with the effectors (muscles and glands). The portion of the PNS that helps the body respond to changes in the external environment is the somatic system. The nerves and receptors that maintain homeostasis despite internal changes make up the autonomic nervous system.

The somatic system helps the body adjust to the external environment

The **somatic nervous system** includes the receptors that react to changes in the external environment, the sensory neurons that keep the CNS informed of those changes, and the motor neurons that adjust the positions of the skeletal muscles, maintaining the body's well-being. In mammals, 12 pairs of **cranial nerves** emerge from the brain. They transmit information regarding the senses of smell, sight, hearing, and taste from the special sensory receptors, and information from the general

	TABLE 40-4	The Cranial Nerves of Mammals	
Number	Name	Origin of Sensory Fibers	Effector Innervated by Motor Fibers
I	Olfactory	Olfactory epithelium of nose (smell)	None
II	Optic	Retina of eye (vision)	None
III	Oculomotor	Proprioceptors* of eyeball muscles (muscle sense)	Muscles that move eyeball (with IV and VI); muscles that change shape of lens; muscles that constrict pupil
IV	Trochlear	Proprioceptors of eyeball muscles	Other muscles that move eyeball
V	Trigeminal	Teeth and skin of face	Some muscles used in chewing
VI	Abducens	Proprioceptors of eyeball muscles	Other muscles that move eyeball
VII	Facial	Taste buds of anterior part of tongue	Muscles used for facial expression; submaxillary and sublingual salivary glands
VIII	Auditory (vestibulocochlear)	Cochlea (hearing) and semicircular canals (senses of movement, balance, and rotation)	None
IX	Glossopharyngeal	Taste buds of posterior third of tongue and lining of pharynx	Parotid salivary gland; muscles of pharynx used in swallowing
X	Vagus	Nerve endings in many of the internal organs; lungs, stomach, aorta, larynx	Parasympathetic fibers to heart, stomach, small intestine, larynx, esophagus, and other organs
XI	Spinal accessory	Muscles of shoulder	Muscles of neck and shoulder
XII	Hypoglossal	Muscles of tongue	Muscles of tongue

*Proprioceptors are receptors located in muscles, tendons, or joints that provide information about body position and movement (see Chapter 41).

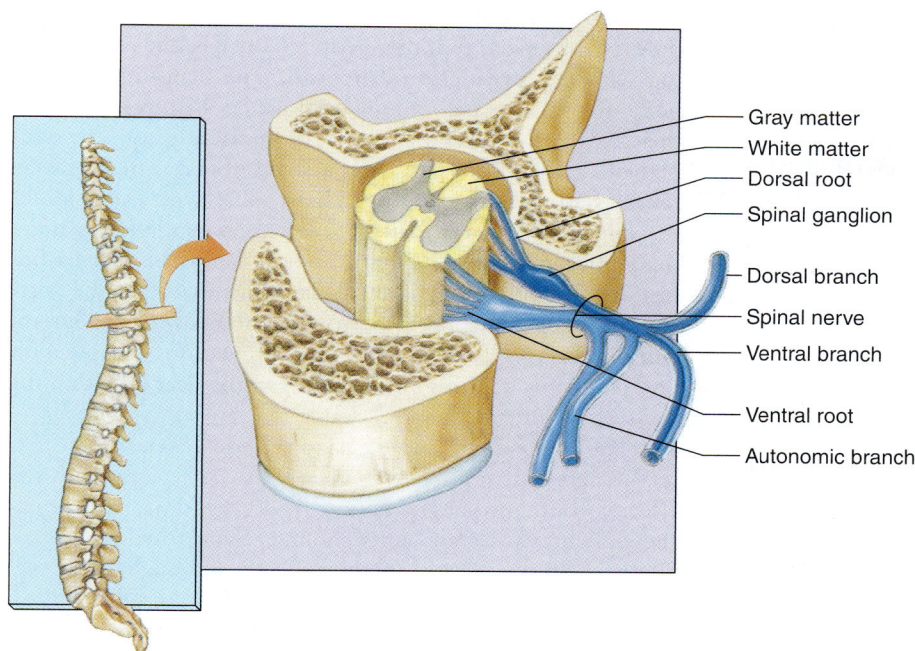

Figure 40–17 **Spinal nerve.** Dorsal and ventral roots emerge from the spinal cord and join to form a spinal nerve. Each spinal nerve divides into several branches.

Gray matter
White matter
Dorsal root
Spinal ganglion

Dorsal branch
Spinal nerve
Ventral branch

Ventral root
Autonomic branch

sensory receptors, especially in the head region. For example, cranial nerve II, the optic nerve, transmits signals from the retina of the eye to the brain. The cranial nerves also bring orders from the CNS to the voluntary muscles that control movements of the eyes, face, mouth, tongue, pharynx, and larynx. Cranial nerve VII, the facial nerve, for example, transmits signals to the muscles used in facial expression and to the salivary glands. The names of the cranial nerves and the structures they innervate are given in Table 40–4.

In humans, 31 pairs of spinal nerves emerge from the spinal cord. Named for the general region of the vertebral column from which they originate, they comprise eight pairs of cervical spinal nerves, twelve pairs of thoracic, five pairs of lumbar, five pairs of sacral, and one pair of coccygeal spinal nerves.

Each spinal nerve has a dorsal root and a ventral root (Fig. 40–17). The dorsal root consists of afferent fibers that transmit information from the sensory receptors to the spinal cord. Just before the dorsal root joins with the cord, it is marked by a swelling, the spinal ganglion, which consists of the cell bodies of the sensory neurons. The ventral root is a grouping of the efferent fibers leaving the cord en route to the muscles and glands. Cell bodies of the motor neurons occur within the gray matter of the cord.

Peripheral to the junction of the dorsal and ventral roots, each spinal nerve divides into branches. The dorsal branch serves the skin and muscles of the back. The ventral branch serves the skin and muscles of the sides and ventral part of the body. The autonomic branch innervates the viscera. The ventral branches of several spinal nerves form tangled networks called plexi (sing., *plexus*). Within a plexus, the fibers of a spinal nerve may separate and then regroup with fibers that origi-

nated in other spinal nerves. Thus, nerves emerging from a plexus consist of neurons that originated in several different spinal nerves.

The autonomic system regulates the internal environment

The autonomic system helps to maintain homeostasis in the internal environment. For instance, it regulates the heart rate and helps maintain a constant body temperature. The autonomic system works automatically and without voluntary input. Its effectors are smooth muscle, cardiac muscle, and glands. Like the somatic system, it is functionally organized into reflex pathways. Receptors within the viscera relay information via afferent nerves to the CNS. The information is integrated at various levels. Then the decision is transmitted along efferent nerves to the appropriate muscles or glands.

Afferent and efferent neurons of the autonomic system are found within cranial or spinal nerves. For example, afferent fibers of cranial nerve X, the vagus nerve, transmit signals from many organs of the chest and upper abdomen to the CNS. Its efferent fibers transmit signals from the brain to the heart, stomach, small intestine, and several other organs.

The efferent portion of the autonomic system is subdivided into **sympathetic** and **parasympathetic systems.** In general, the sympathetic nerves operate to stimulate organs and to mobilize energy, especially in response to stress, whereas the parasympathetic nerves influence organs to conserve and restore energy, particularly during quiet, calm activities. Many organs are innervated by both types of nerves, which act on

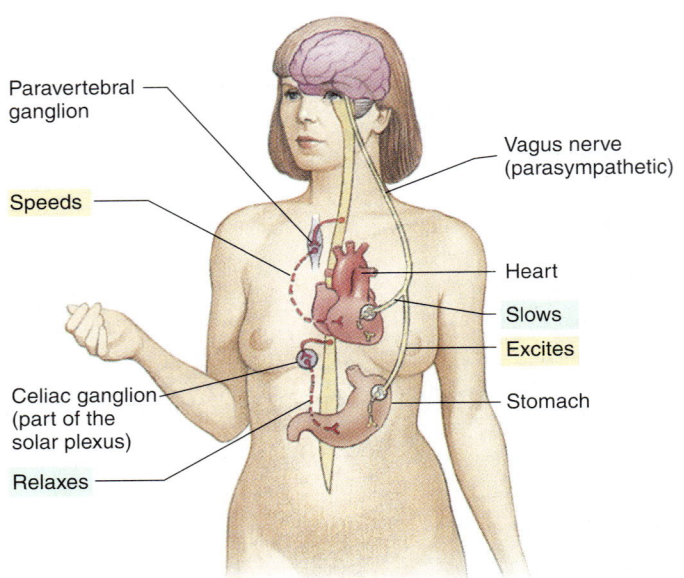

Paravertebral ganglion

Speeds

Celiac ganglion (part of the solar plexus)

Relaxes

Vagus nerve (parasympathetic)

Heart

Slows

Excites

Stomach

Figure 40–18 **Dual innervation of the heart and stomach.**
Many organs are innervated by both sympathetic and parasympathetic nerves. A sympathetic nerve increases the heart rate whereas the vagus nerve, a parasympathetic nerve, slows the heartbeat. In contrast, sympathetic nerves slow contractions of the stomach and intestine while the vagus nerve stimulates contractions of the digestive organs. Sympathetic nerves are shown in red; postganglionic fibers are shown as dotted lines.

the organ in a complementary way (Figs. 40–18 and 40–19; Table 40–5). For example, the heart rate is slowed by impulses from its parasympathetic nerve fibers and speeded up by messages from its sympathetic nerve supply.

Instead of using a single efferent neuron, as in the somatic system, the autonomic system uses a relay of two neurons between the CNS and the effector. The first neuron, called the **preganglionic neuron,** has a cell body and dendrites within the CNS. Its axon, part of a peripheral nerve, ends by synapsing with a **postganglionic neuron.** The dendrites and cell body of the postganglionic neuron are in a ganglion outside the CNS. Its axon terminates near or on the effector. The sympathetic ganglia are paired, and a chain of them, the **paravertebral sympathetic ganglion chain,** occurs on each side of the spinal cord from the neck to the abdomen. Some sympathetic preganglionic neurons do not end in these ganglia but instead pass on to ganglia in the abdomen, close to the aorta and its major branches. These ganglia are known as **collateral ganglia.** Parasympathetic preganglionic neurons synapse with postganglionic neurons in **terminal ganglia** near or within the walls of the organs they innervate.

The sympathetic and parasympathetic systems also differ in the neurotransmitters they release at the synapse with the effector. Both preganglionic and postganglionic parasympathetic neurons secrete acetylcholine. Sympathetic postganglionic neurons release norepinephrine (although their preganglionic neurons secrete acetylcholine). Table 40–6 compares the actions of the sympathetic and parasympathetic systems on selected effectors.

The autonomic system got its name from the original belief that it was independent of the CNS, that is, autonomous. Physiologists have shown that this is not so and that the hypothalamus and many other parts of the CNS help to regulate the autonomic system. Although the autonomic system usually functions automatically, its activities can be consciously influenced. Biofeedback provides a person with visual or auditory evidence concerning the status of an autonomic body function. For example, a tone may be sounded when blood pressure reaches a desirable level. Using such techniques, subjects have learned to control autonomic activities such as blood pressure, brain wave pattern, heart rate, and blood-sugar level. Even certain abnormal heart rhythms can be consciously modified.

TABLE 40–5	Comparison of Sympathetic and Parasympathetic Systems	
Characteristic	Sympathetic System	Parasympathetic System
General effect	Prepares body to cope with stressful situations	Restores body to resting state after stressful situation; actively maintains normal body functions
Extent of effect	Widespread throughout body	Localized
Neurotransmitter released at synapse with effector	Norepinephrine (usually)	Acetylcholine
Duration of effect	Lasting	Brief
Outflow from CNS	Thoracic and lumbar nerves from spinal cord	Cranial nerves and sacral nerves from spinal cord
Location of ganglia	Chain and collateral ganglia	Terminal ganglia
Number of postganglionic fibers with which each preganglionic fiber synapses	Many	Few

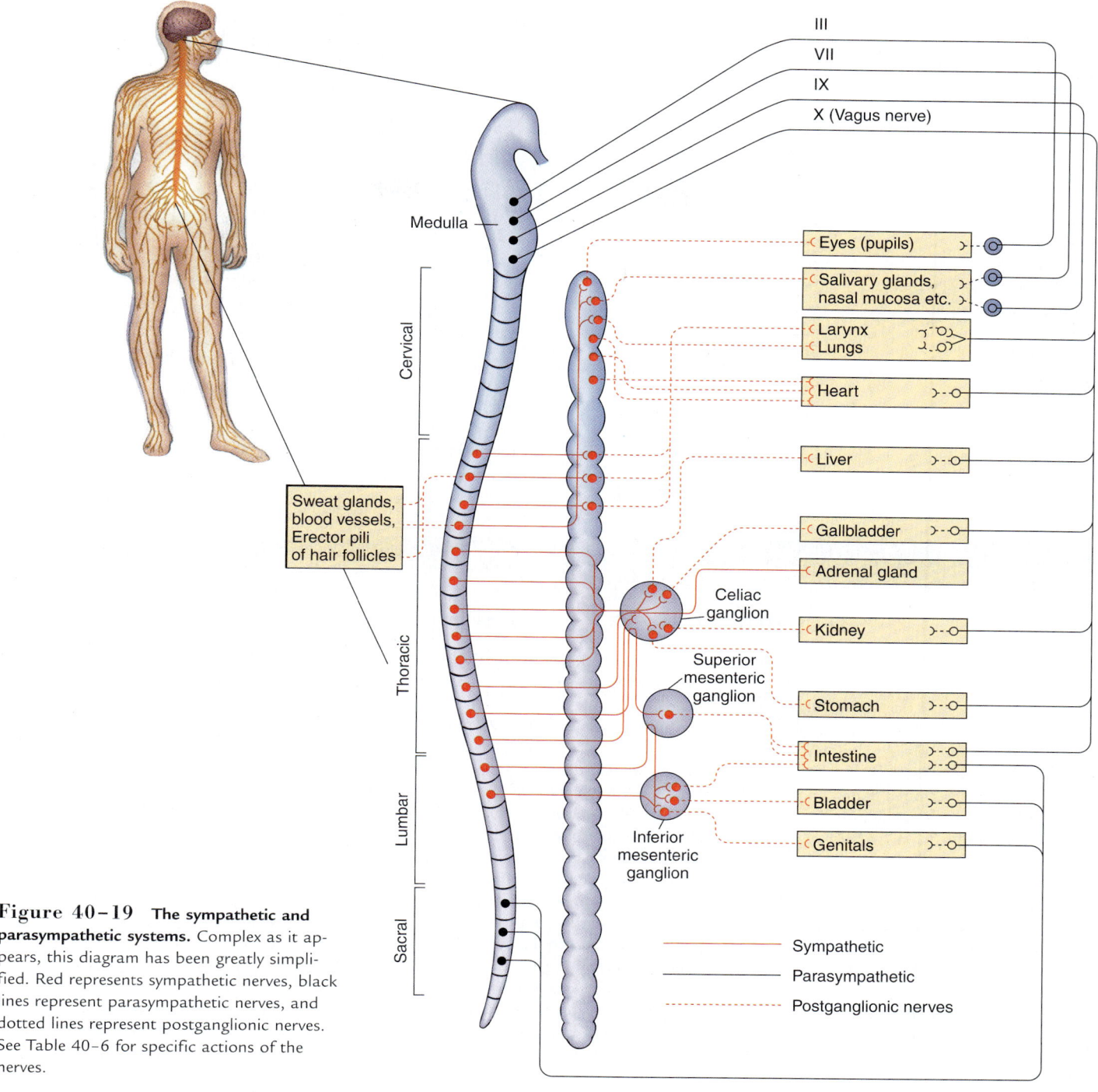

Figure 40-19 The sympathetic and parasympathetic systems. Complex as it appears, this diagram has been greatly simplified. Red represents sympathetic nerves, black lines represent parasympathetic nerves, and dotted lines represent postganglionic nerves. See Table 40-6 for specific actions of the nerves.

Labels within figure:
III
VII
IX
X (Vagus nerve)
Medulla
Cervical
Thoracic
Lumbar
Sacral
Sweat glands, blood vessels, Erector pili of hair follicles
Eyes (pupils)
Salivary glands, nasal mucosa etc.
Larynx
Lungs
Heart
Liver
Gallbladder
Adrenal gland
Kidney
Stomach
Intestine
Bladder
Genitals
Celiac ganglion
Superior mesenteric ganglion
Inferior mesenteric ganglion

— Sympathetic
— Parasympathetic
- - - Postganglionic nerves

MANY DRUGS AFFECT THE NERVOUS SYSTEM

About 25% of all prescribed drugs are taken to alter psychological conditions, and almost all the commonly abused drugs affect mood. Many of them act by changing the levels of neurotransmitters within the brain. In particular, levels of norepinephrine, serotonin, and dopamine are thought to influence mood. For example, when excessive amounts of norepineph-

rine are released in the reticular activating system, we feel stimulated and energetic, whereas low concentrations of this neurotransmitter reduce anxiety. Table 40-7 lists several commonly used and abused drugs and gives their effects. Also see *Focus On: Alcohol: The Most Abused of All Drugs* and *Focus On: Crack Cocaine.*

Habitual use of almost any mood-altering drug can result in **psychological dependence,** in which the user becomes emotionally dependent on the drug. When deprived of it, the

TABLE 40-6 Comparison of Sympathetic and Parasympathetic Actions on Selected Effectors*

Effector	Sympathetic Action	Parasympathetic Action
Heart	Increases rate and strength of contraction	Decreases rate; no direct effect on strength of contraction
Bronchial tubes	Dilates	Constricts
Iris of eye	Dilates pupil	Constricts pupil
Sex organs (male)	Constricts blood vessels; ejaculation	Dilates blood vessels; erection
Blood vessels	Generally constricts	No innervation for many
Sweat glands	Stimulates	No innervation
Intestine	Inhibits motility	Stimulates motility and secretion
Liver	Stimulates glycogenolysis (conversion of glycogen to glucose)	No effect
Adipose tissue	Stimulates free fatty acid release from fat cells	No effect
Adrenal medulla	Stimulates secretion of epinephrine and norepinephrine	No effect
Salivary glands	Stimulates thick, viscous secretion	Stimulates a profuse, watery secretion

*Refer to Figure 40–19 as you study this table. Notice that many other examples could be added to this list.

ALCOHOL: THE MOST ABUSED OF ALL DRUGS

After tobacco, alcohol is the leading cause of premature death in the United States. It is linked to more than 100,000 deaths every year and costs our society more than $100 billion annually. According to the pollster Louis Harris, there are 28 million alcoholics in the United States, and about one in three homes includes someone with a serious drinking problem. Alcohol abuse is not limited to adults. About 4.6 million adolescents, or nearly one of every three high school students, experience negative consequences from alcohol use, including difficulty with parents, poor performance at school, and trouble with the law.

Alcohol abuse results in physiological, psychological, and social impairment for the abuser and has serious negative consequences for family, friends, and society. Alcohol abuse has been linked to the following:

- More than 50% of all traffic fatalities.
- More than 50% of violent crimes, and more than 60% of child abuse and spouse abuse cases.
- More than 50% of suicides.
- More than 15,000 babies born each year with serious birth defects because their mothers drank alcohol during pregnancy. Fetal alcohol syndrome is the leading cause of preventable mental retardation in the United States.
- Recent studies suggest that as few as three

drinks per week increase breast cancer risk by 50%.

A single drink, that is, 12 ounces of beer, 5 ounces of wine, or 1.5 ounces of 80-proof liquor, results in a blood alcohol concentration of approximately 20 mg/dL (milligrams per deciliter). This represents about 0.5 ounce of pure alcohol in the blood. Alcohol accumulates in the blood because absorption occurs more rapidly than do oxidation and excretion. Every cell in the body can take in alcohol from the blood. At first the drinker may feel stimulated. But alcohol actually causes depression of the central nervous system.

As blood alcohol concentration rises, information processing, judgment, memory, sensory perception, and motor coordination all become progressively impaired. Depression and drowsiness generally occur. Contrary to popular belief, alcohol decreases sexual performance in males. Some individuals become loud, angry, or violent. In most states, a blood alcohol concentration of 100 mg/dL (or 0.10) legally defines driving while intoxicated (DWI). A 150-pound man typically reaches this level by drinking two or three beers in an hour.

Alcohol metabolism occurs at a fixed rate in the liver so only time, not coffee, will decrease its effects. Because alcohol inhibits water reabsorption in the kidneys, more fluid is excreted as urine than is consumed.

This results in dehydration that, together with low blood sugar level, may cause the stupor produced by excessive drinking.

In chronic drinkers, cells of the central nervous system adapt to the presence of the drug. This causes tolerance (more and more alcohol is needed to experience the same effect) and physical dependence. Abrupt withdrawal can result in sleep disturbances, severe anxiety, tremors, seizures, hallucinations, and psychoses.

The *A1* allele of the D2 dopamine receptor gene increases the risk for alcoholism. Individuals with this allele have fewer D2 dopamine receptors in their brain. Alcohol, as well as cocaine, heroin, amphetamines, and nicotine, enhances dopamine activity in the mesolimbic circuits of the brain. Individuals with the *A1* allele may compensate for their low dopamine levels by using alcohol and other drugs. However, the mechanisms underlying alcohol addiction are far more complex. Evidence suggests that many neurotransmitter systems are involved, including glutamate, serotonin, GABA, and opioid peptide systems.

Treatment for alcohol problems includes various forms of psychotherapy, including relapse prevention therapy in which individuals are encouraged not to consider lapses from abstinence as failure. The group support offered by Alcoholics Anonymous (AA) has proved effective for many struggling with alcohol abuse.

CRACK COCAINE

With the exception of alcoholics, the majority of persons seeking treatment for drug abuse are now crack cocaine addicts. Cocaine use by teenagers alone has increased about 400% during the past ten years, involving an estimated 2 million. Crack is a very concentrated and extremely powerful form of cocaine—five to ten times more addictive than other forms of cocaine. This drug is produced in illegal, makeshift labs by converting powdered cocaine into small "rocks" that are up to 80% pure cocaine. Crack is smoked in pipes or added to tobacco or marijuana cigarettes.

Use of crack results in an intense, brief high beginning in 4 to 6 seconds and lasting for 5 to 7 minutes. Physiologically, crack stimulates a massive release of norepinephrine and dopamine in the brain and blocks their reuptake by postsynaptic neurons. Excitation of the sympathetic nervous system occurs, and users report experiencing feelings of self-confidence, power, and euphoria. As the neurotransmitters are depleted, the high is followed by a "crash," a

period of deep depression. The abuser experiences an intense craving for another crack "hit" in order to get more stimulation.

Some abusers spend days smoking crack without stopping to eat or sleep. Although a vial of rocks can be obtained for about $20, many abusers develop habits that cost hundreds of dollars a week. Supporting an expensive drug habit leads many abusers to prostitution, drug dealing, and other forms of crime.

Cocaine addicts report problems with memory loss, fatigue, depression, insomnia, paranoia, loss of sexual drive, violent behavior, and attempts at suicide. Crack can cause respiratory problems, brain seizures, cardiac arrest, and elevation of blood pressure that leads to stroke. Many users have suffered fatal reactions to impurities in the drug or have died as a result of accidents related to its use.

The effects of cocaine on the developing brain has been the focus of extensive investigation. Pediatrician Ira Chasnoff of the University of Illinois College of Medi-

cine in Chicago reported that children whose mothers used cocaine during pregnancy had significantly more behavioral problems. They were more anxious, aggressive, impulsive, and had difficulty paying attention and staying focused. They were also depressed. Linda Mayes of Yale University reported that by three to six months of age infants exposed to cocaine prenatally were more irritable than control infants. By the time they were about a year old, these children had difficulty focusing. Animal studies suggest that when the brain is exposed to high levels of dopamine and serotonin, the brain attempts to compensate and permanent changes occur that may affect behavior.

No pharmaceutical agent has been developed to treat cocaine overdose or addiction. Addiction is a chronic, relapsing disorder with genetic, physiological, behavioral, and societal components. Effective treatments will need to address all of these aspects of the disorder.

Cocaine use decreases brain activity as shown in these PET scans. Brain activity was measured by how fast glucose was used by brain cells. (A radioactive analogue of glucose was injected.) Brain activity is indicated by color; the warmest colors (red and orange) indicate the highest activity (see color bar). When a cocaine abuser recieved cocaine, he experienced euphoria accompanied by a dramatic decrease in activity throughout the brain (right). Compare this image to the placebo conditions (left) in which the same individual was injected with saline. The decrease in brain activity is seen as a shift from red to yellow. Opiates produce a similar effect. *(Steven Grant and Edythe London, Brain Imaging Center, National Institute on Drug Abuse)*

user craves the feeling of euphoria (well-being) that the drug induces. Some drugs induce **tolerance** after several days or weeks. This means that response to the drug decreases, and greater amounts are required to obtain the desired effect. Tolerance often occurs because the liver cells are stimulated to produce more of the enzymes that metabolize and inactivate the drug. It can also occur due to a decrease in the number of

postsynaptic receptors that bind with the drug (a mechanism known as downregulation).

Drug addiction is a serious societal problem that involves compulsive use of a drug despite negative health consequences and negative effects on the ability to function socially and occupationally. Use of some drugs, such as heroin, tobacco, alcohol, and barbiturates, may also result in physical depen-

(*Text continues on p. 874*)

TABLE 40-7   Effects of Some Commonly Used Drugs

Name of Drug	Effect on Mood	Actions on Body	Side Effects/Dangers Associated with Abuse
Antidepressants			
Tricyclic antidepressants (e.g., Elavil, Anafranil)	Elevate mood; relieve depression; also used to treat obsessive-compulsive behavior	Most block reuptake of biogenic amines (especially norepinephrine) increasing their concentration in the synapses	Can cause sedation, weight gain, sexual dysfunction
Selective serotonin reuptake inhibitors (SSRIs) (Prozac, Zoloft, Effexor)	Relieve depression; used to treat obsessive-compulsive behavior	Block serotonin reuptake	Nausea, headache, insomnia, anxiety
MAO inhibitors	Relieve depression	Block enzymatic breakdown of biogenic amines, increasing their concentration in the synapses	Liver toxicity, excessive CNS stimulation; overdose may affect blood pressure, cause hallucinations
Anti-anxiety drugs			
Benzodiazepines (e.g., Xanax, Valium, Librium)	Sedation; induce sleep	Bind to GABA receptors on post-synaptic neuron; increase effectiveness of GABA in opening chloride channels, causing hyperpolarization; cause relaxation of skeletal muscles	Drowsiness, confusion; psychological dependence, addiction; effects additive with alcohol
Barbiturates "Downers" (e.g., Phenobarbital)	Sedation; induce sleep	Bind to receptor adjacent to GABA receptor on chloride channels; chloride channels open so more chloride ions move in, causing hyperpolarization	Drowsiness, confusion; psychological dependence, tolerance, addiction; overdose can cause severe CNS depression, resulting in coma and death (especially lethal in combination with alcohol)
Antipsychotic medications			
Phenothiazines (e.g., Thorazine, Mellaril, Stelazine)	Relieve some symptoms of schizophrenia; reduce impulsive and aggressive behavior	Block dopamine receptors	Muscle spasms, shuffling gait, and other symptoms of Parkinson's disease
Newer antipsychotic medications (e.g., Clozaril, Risperdal)	Similar to phenothiazines, but also appear to increase motivation	Block dopamine receptors and several other neurotransmitter receptors	Fewer side effects than with the phenothiazines
Narcotic analgesics			
Opiates (e.g., morphine, codeine, Demerol, heroin)	Euphoria; sedation; relieve pain	Mimic actions of endorphins; bind to opiate receptors	Depress respiration; constrict pupils; impair coordination; tolerance, psychological dependence, addiction; convulsions, death from overdose

TABLE 40-7 continued

Name of Drug	Effect on Mood	Actions on Body	Side Effects/Dangers Associated with Abuse
Cocaine	Euphoria; excitation followed by depression	CNS stimulation followed by depression; autonomic stimulation; dilation of pupils; local anesthesia; stimulates release and inhibits reuptake of norepinephrine and dopamine	Mental impairment, convulsions, hallucinations, unconsciousness; death from overdose
Marijuana	Euphoria	Impairs coordination; impairs depth perception and alters sense of timing; impairs short-term memory (probably by decreasing acetylcholine levels in the hippocampus); inflames eyes; causes peripheral vasodilation	In large doses, sensory distortions, hallucinations; evidence of lowered sperm counts and testosterone (male hormone) levels
Amphetamines "Uppers," "pep pills," "crystal meth" (e.g., Dexedrine)	Euphoria, stimulation, hyperactivity	Stimulate release of dopamine and norepinephrine; enhance flow of impulses in RAS; increase heart rate; raise blood pressure; dilate pupils	Tolerance, possible physical dependence, hallucinations; increased blood pressure; psychotic episodes; death from overdose
LSD (lysergic acid diethylamide)	Overexcitation; sensory distortions, hallucinations	Alters levels of transmitters in brain; potent CNS stimulator; dilates pupils, sometimes unequally; increases heart rate; raises blood pressure	Irrational behavior
Methaqualone (e.g., Quaalude, Sopor)	Hypnotic	Depresses CNS; depresses certain spinal reflexes	Tolerance, physical dependence; convulsions, death
Caffeine	Increases mental alertness; decreases fatigue and drowsiness	Acts on cerebral cortex; relaxes smooth muscle; stimulates cardiac and skeletal muscle; increases urine volume	Very large doses stimulate centers in the medulla (may slow the heart); toxic doses may cause convulsions
Nicotine	Lessens psychological tension	Stimulates sympathetic nervous system; stimulates synthesis of lipid in arterial wall	Tolerance, physical dependence; stimulates development of atherosclerosis
Alcohol	Euphoria, relaxation, release of inhibitions	Depresses CNS; impairs vision, coordination, judgment; lengthens reaction time	Physical dependence, damage to pancreas, liver cirrhosis, brain damage

dence, in which physiological changes occur that make the user dependent on the drug. For example, when heroin or alcohol are withheld, the addict suffers characteristic withdrawal symptoms. Physical addiction can occur when a drug, for example morphine, has components similar to substances that body cells normally manufacture on their own. Some highly addictive drugs such as crack cocaine and methamphetamine do not cause serious withdrawal symptoms.

The neurophysiological mechanisms for drug addiction involve a network of neurons that release dopamine. These neurons are located in the midbrain and give rise to a dopamine system that extends into behavioral control centers in the limbic system. Facilitation of neurons in this pathway has been demonstrated in nicotine, heroin, amphetamine, and cocaine addiction.

SUMMARY WITH KEY TERMS

I. Nerve nets and radial nervous systems are typical of radially symmetrical invertebrates. Bilateral nervous systems are characteristic of bilaterally symmetrical animals.
   A. Cnidarians have a **nerve net** consisting of nerve cells scattered throughout the body; no central nervous system is present.
   B. Echinoderms typically have a **radial nervous system** consisting of a nerve ring and nerves that extend to various parts of the body.
   C. In a **bilateral nervous system,** nerve cells concentrate to form nerves, nerve cords, **ganglia,** and **brain;** typically, sense organs are concentrated in the head region. An increase in number of neurons, especially association neurons, permits a wide range of responses.
      1. In planarian flatworms the nervous system includes **cerebral ganglia** and, typically, two solid ventral nerve cords connected by transverse nerves.
      2. Annelids and arthropods typically have a ventral nerve cord and numerous ganglia. The cerebral ganglia of arthropods have specialized regions.
      3. Mollusks typically have at least three pairs of ganglia. Octopods and other cephalopod mollusks have the most complex invertebrate nervous systems.
II. The vertebrate nervous system consists of the **central nervous system (CNS)** and **peripheral nervous system (PNS).**
   A. The CNS consists of the brain and dorsal, tubular **spinal cord.**
   B. The PNS consists of sensory receptors and nerves. The PNS consists of the **somatic system** and the **autonomic system.** Two types of efferent pathways in the autonomic system are the **sympathetic system** and the **parasympathetic system.**
III. In the vertebrate embryo, the brain and spinal cord arise from the **neural tube.** The anterior end of the tube differentiates into **forebrain, midbrain,** and **hindbrain.**
   A. The hindbrain subdivides into the metencephalon and myelencephalon.
      1. The myelencephalon develops into the **medulla,** which contains vital centers and other reflex centers. The **fourth ventricle,** the cavity of the medulla, communicates with the **central canal** of the spinal cord.
      2. The metencephalon gives rise to the **cerebellum,** which is responsible for muscle tone, posture, and equilibrium, and to the **pons,** which connects various parts of the brain.
   B. The midbrain is the largest part of the brain in fishes and amphibians. It is their main association area, linking sensory input and motor output. In reptiles, birds, and mammals, the midbrain serves as a center for visual and auditory reflexes.
   C. The medulla, pons, and midbrain make up the **brain stem.**
   D. The forebrain differentiates to form the **diencephalon** and telencephalon.
      1. The diencephalon develops into the **thalamus** and **hypothalamus.**
         a. The thalamus is a relay center for motor and sensory information.
         b. The hypothalamus controls autonomic functions; links nervous and endocrine systems; controls temperature, appetite,

and fluid balance; and is involved in some emotional and sexual responses.
      2. The telencephalon develops into the **cerebrum** and **olfactory bulbs.**
         a. In most vertebrates the cerebrum is divided into right and left **hemispheres.**
         b. In fishes and amphibians, the cerebrum mainly integrates incoming sensory information.
         c. In birds, the corpus striatum controls stereotypical but complex behavior patterns, and another part of the cerebrum governs learning.
         d. In mammals, the **neopallium** accounts for a large part of the **cerebral cortex,** and the cerebrum has complex association functions.
IV. The human brain and spinal cord are protected by bone and three **meninges**—the **dura mater, arachnoid,** and **pia mater;** brain and spinal cord are cushioned by **cerebrospinal fluid.**
   A. The spinal cord transmits impulses to and from the brain and controls many **reflex actions.**
      1. The spinal cord consists of ascending **tracts,** which transmit information to the brain, and descending tracts, which transmit information from the brain. Its gray matter contains nuclei that serve as reflex centers.
      2. A **withdrawal reflex** involves receptors; sensory, association, and motor neurons; and effectors.
   B. The human cerebral cortex consists of **gray matter,** which form folds or **convolutions.** Deep furrows between the folds are called **fissures.**
      1. The cerebrum is functionally divided into motor **areas** that control voluntary movement; **sensory areas** that receive incoming sensory information; and **association areas** that link sensory and motor areas and are responsible for learning, language, thought, and judgment.
      2. The cerebrum consists of lobes including the **frontal lobes, parietal lobes, temporal lobes,** and **occipital lobes.**
      3. The **white matter** of the cerebrum lies beneath the cerebral cortex. The **corpus callosum,** a large band of white matter, connects right and left hemispheres. The **basal ganglia,** a cluster of nuclei within the white matter, are important centers for motor function.
   C. Brain activity cycles in a sleep-wake pattern that is regulated by the hypothalamus and brain stem.
      1. **Alpha wave** patterns are characteristic of relaxed states, **beta wave** patterns of heightened mental activity, and **theta** and **delta waves** of non-REM sleep.
      2. The **reticular activating system (RAS)** is an arousal system; its neurons filter sensory input, selecting which information is transmitted to the cerebrum.
      3. Electrical activity of the cerebral cortex and metabolic rate slows during **non-REM sleep. REM sleep** is characterized by dreaming.
      4. The body's main biological clock (suprachiasmatic nucleus) receives information about light and dark and apparently transmits

it to other nuclei that regulate sleep. When sleep centers are activated, they are thought to release serotonin.

D. The **limbic system** affects the emotional aspects of behavior, motivation, sexual behavior, autonomic responses, and biological rhythms.

E. **Learning** is a long-lasting change in behavior that results from experience. **Memory** is the storage of knowledge and the ability to retrieve it.

1. **Short-term memory** allows us to recall information, like a telephone number, for a few minutes. Information can be transferred from short-term memory to **long-term memory.**

2. Memories appear to be stored throughout the cerebrum; they are integrated in many areas of the brain, including the association areas of the cerebral cortex and parts of the limbic system.

3. Understanding synaptic plasticity may be the key to understanding learning. An increase in neurotransmitter released by a presynaptic neuron as a result of an action potential is **synaptic facilitation.** This type of synaptic enhancement lasts less than a second. Long-term memory storage may involve gene activation and long-term functional changes at synapses. Known as **long term potentiation (LTP),** such changes involve increased sensitivity to an action potential by a postsynaptic neuron.

4. Investigators have used the marine snail *Aplysia* as a model of memory and learning in animals.

5. **Habituation** is the decrease in response that occurs after repeated exposure to a harmless stimulus. Habituation results from inactivation of calcium channels in presynaptic terminals in response to repeated action potentials. This results in a decrease in neurotransmitter released by the presynaptic neurons.

6. **Sensitization** results in an increased response after experience of an unpleasant stimulus. It depends on an increase in neurotransmitter released by sensory (presynaptic) neurons.

7. **Classical conditioning** is a form of learning in which an association forms between an unconditioned stimulus (US) and a conditioned stimulus (CS). The CS then elicits a conditioned response (CR). The cellular mechanism for conditioning may involve a second messenger, protein kinase, gene activation, and protein synthesis.

8. The physical structure and chemistry of the brain can be altered by environmental experience.

V. The peripheral nervous system consists of sensory receptors and nerves, including the **cranial nerves** and **spinal nerves** and their branches.

VI. The autonomic system regulates the internal activities of the body.

A. The sympathetic system permits the body to respond to stressful situations.

B. The parasympathetic system influences organs to conserve and restore energy.

C. Many organs are innervated by sympathetic and parasympathetic nerves, which function in a complementary way.

VII. Some of the types of drugs that affect the nervous system are amphetamines, antianxiety drugs, barbiturates, antipsychotic drugs, antidepressants, narcotic analgesics, and alcohol.

A. Many drugs alter mood by increasing or decreasing the concentrations of specific neurotransmitters within the brain.

B. Habitual use of mood-altering drugs can result in **psychological dependence** or **addiction.** Many drugs induce **tolerance,** in which the body's response to the drug decreases so that greater amounts are needed to obtain the desired effect.

POST-TEST

1. A radially symmetrical animal is likely to have (a) a forebrain (b) a nerve net (c) a cerebral ganglia (d) a ventral nerve cord (e) answers a and b only are correct

2. In vertebrate embryos the brain arises from the (a) spinal cord (b) sympathetic nervous system (c) parasympathetic nervous system (d) neural tube (e) forebrain

3. Which part of the brain maintains posture, muscle tone, and equilibrium? (a) cerebrum (b) medulla (c) cerebellum (d) neopallium (e) thalamus

4. Which part of the brain controls autonomic functions and regulates body temperature? (a) cerebrum (b) hypothalamus (c) cerebellum (d) pons (e) thalamus

5. The correct sequence for neural transmission in a withdrawal reflex is (a) reception → motor neuron → association neuron in CNS → afferent neuron (b) reception → sensory neuron → motor neuron → association neuron in CNS (c) afferent neuron → reception → efferent neuron → motor neuron (d) reception → sensory neuron → association neuron in CNS → motor neuron (e) sensory neuron → CNS → afferent neuron

6. Association areas in the human brain are concentrated in the (a) cerebral cortex (b) medulla (c) ventricle (d) corpus callosum (e) meninges

7. The human brain is protected by (a) meninges (b) cerebrospinal fluid (c) skull bones (d) answers a, b, and c are correct (e) answers a and c only are correct

8. Which of the following is NOT a function of the spinal cord? (a) controls many reflex actions (b) transmits information to the brain (c) transmits information from the brain (d) functions as a biological clock (e) controls the withdrawal reflex

9. The most prominent part of the amphibian brain is the (a) midbrain (b) medulla (c) cerebellum (d) neopallium (e) cerebrum

10. The visual centers are located in the (a) parietal lobes (b) thalamus (c) occipital lobes (d) limbic lobes (e) frontal lobes

11. As you are answering these questions, what is the predominant type of wave that your brain is emitting? (a) REM (b) alpha (c) theta (d) delta (e) beta

12. Long term potentiation (LTP) (a) is mainly the responsibility of the tenth cranial nerve (b) is a functional change at the synapse (c) is associated with short term memory (d) depends on sympathetic nerves (e) answers a and c are correct

13. The heart rate is slowed by (a) sympathetic nerves (b) parasympathetic nerves (c) corpus callosum (d) answers a, b, and c are correct (e) answers b and c only are correct

14. After taking a mood-altering drug for several weeks, a patient notices that it no longer works as effectively. This is an example of (a) psychological dependence (b) withdrawal (c) addiction (d) tolerance (e) neurotransmitter increases

REVIEW QUESTIONS

1. Contrast the nervous system of a planarian flatworm with that of *Hydra*.
2. What are some characteristics of the nervous system in a bilaterally symmetrical animal? What are some advantages of this type of system?
3. Compare the flatworm nervous system with that of a vertebrate. In each

case, what is the relationship between the type of nervous system and the animal's lifestyle?
4. Identify the parts of the adult vertebrate brain derived from the embryonic forebrain, midbrain, and hindbrain.

5. Give the function of each of the following structures in the human brain: (a) medulla (b) midbrain (c) cerebellum (d) thalamus (e) hypothalamus (f) cerebrum.
6. Compare the midbrain and cerebrum of the fish with those of the mammal.
7. Describe the protective coverings of the human CNS and give the function of the cerebrospinal fluid.
8. How is information processed? How might short-term memory be adaptive? Where are memories stored?
9. What mechanisms are involved in habituation? Sensitization? Classical conditioning? What is long term potentiation (LTP)?
10. Contrast the functions of the CNS with those of the PNS.
11. Compare the functions of the somatic and autonomic systems.
12. Contrast the structure and function of the sympathetic system with those of the parasympathetic system.
13. Describe how each of the following drugs affects the CNS: (a) alcohol (b) antipsychotic drugs (c) antidepressants (d) amphetamines.
14. Label the diagram. Use Figure 40–11a to check your answer.

YOU MAKE THE CONNECTION

1. What general trends can you identify in the evolution of the vertebrate brain?
2. In what way does electrical activity of the brain reflect your state of consciousness? Of what use might it be to learn to control your brain wave patterns?
3. Imagine that you have just become a parent. What kind of things could you do to enhance the development of your child's intellectual abilities?
4. Hypothesize a possible relationship between inheritance and intelligence based on gene activation during information processing.

RECOMMENDED READINGS

Gazzaniga, M.S. "The Split Brain Revisited." *Scientific American,* Vol. 279, No. 1, Jul. 1998. Studies on brain organization provide insight into the workings of the mind.

Ho, D.Y., and R.M. Sapolsky. "Gene Therapy for the Nervous System." *Scientific American,* Vol. 276, No. 6, Jun. 1997. Damage from degenerative neurological disease may some day be treated by inserting genes into brain cells.

Karni, A. "Adult Cortical Plasticity and Reorganization." *Science and Medicine,* Vol. 4, No. 1, Jan./Feb. 1997. Studies indicate that the adult brain retains an important degree of plasticity.

Landry, D.W. "Immunotherapy for Cocaine Addiction." *Scientific American,* Vol. 276, No. 2, Feb. 1997. A discussion of cocaine addiction and a possible future treatment.

Maudgil, D.D., and S.D. Shorvon. "Locating the Epileptogenic Focus by MRI." *Science and Medicine,* Vol. 4, No. 5, Sep./Oct. 1997. New imaging techniques provide valuable clinical information regarding lesions in the brain related to epilepsy.

Noble, E.P. "The Gene That Rewards Alcoholism." *Science and Medicine,* Vol. 3, No. 2, Mar./Apr. 1996. A gene that codes for a dopamine receptor has been linked to alcoholism.

Science, Vol. 278 3 Oct. 1997. A series of articles on "Frontiers in Neuroscience: The Science of Substance Abuse." Discussion of the current view of drug addiction and what we know about neurophysiological mechanisms underlying addiction.

Youdim, M.B.H., and P. Riederer. "Understanding Parkinson's Disease." *Scientific American,* Vol. 276, No. 1, Jan. 1997. A discussion of the role of free radicals in the etiology of Parkinson's disease.

● Visit our website at **http://www.saunderscollege.com/lifesci/titles.html** and click on Solomon/Berg/Martin Biology for links to chapter-related resources on the World Wide Web.

CHAPTER 41

Sensory Reception

Bats, dolphins, and a few other vertebrates can detect distant objects by **echolocation,** sometimes called biosonar. Echolocation works something like radar but uses sound rather than radio waves. For example, a bat emits high-pitched sounds that bounce off of objects in its path and echo back. The bat responds instantly to the echo, deftly avoiding obstacles or capturing prey. The California leaf-nosed bat *(Macrotus californicus)* in the photo has located an insect. The ability to detect sounds emitted by some foraging bats has evolved in certain moths and other insect prey. Some species of bats have adapted, in turn, by emitting very high frequency sounds that their prey cannot detect.

You may have heard dolphins emit clicking sounds. Humans can hear only a limited number of the wide range of sound frequencies that dolphins emit. Using echolocation, dolphins can determine the size, shape, texture, and density of an object, as well as its location.

Sensory receptors are structures that detect information about changes in the internal or external environment. By informing an animal about its environment, sensory receptors are the link between the animal and the outside world. The kinds of sense organs an animal possesses determine just how it perceives its surroundings. We humans live in a world of rich colors, numerous shapes, and varied sounds. But we cannot hear the high-pitched whistles that are audible to dogs and cats, or the ultrasonic echoes by which bats navigate. Neither do we ordinarily recognize our friends by their distinctive odors. And although vision is our dominant and most refined sense, we are blind to the ultraviolet hues that light up the world for insects.

Sensory receptors consist of neuron endings or specialized cells in close contact with neurons. These receptors **transduce** (convert) the energy of the stimulus to electrical signals, the information currency of the nervous system. The transduction mechanisms that couple the stimulus with the opening or closing of ion channels in the plasma membrane of sensory receptors is the focus of a great deal of research. When a sensory receptor is depolarized, an action potential may be initiated in a sensory neuron. As we discussed in Chapter 40, sensory neurons transmit information from receptors to the central nervous system (CNS).

(Merlin D. Tuttle/Bat Conservation International/Photo Researchers, Inc.)

Sensory receptors, along with other types of cells, make up the familiar complex **sense organs:** eyes, ears, nose, and taste buds. A human taste bud, for example, consists of modified epithelial cells that detect chemicals dissolved in saliva. In addition to the five senses of sight, hearing, smell, taste, and touch, neurobiologists recognize balance as a sense. They view touch as a compound sense that allows us to detect pressure, pain, and temperature. In this chapter, we also consider receptors that enable us to sense muscle tension and joint position.

AFTER YOU HAVE STUDIED THIS CHAPTER YOU SHOULD BE ABLE TO

1. Compare exteroceptors, proprioceptors, and interoceptors, and explain the importance of each group.
2. Distinguish among the five kinds of sensory receptors that are classified according to the types of energy to which they respond and give examples of each type of sense organ.
3. Describe how a sensory receptor functions: define energy transduction, receptor potential, and adaptation as part of your answer.
4. Describe the following mechanoreceptors: tactile receptors, statocysts, lateral line organs, and proprioceptors.
5. Compare the function of the saccule and utricle with that of the semicircular canals in maintaining equilibrium.
6. Trace the path taken by sound waves through the structures of the ear and explain how the organ of Corti functions as an auditory receptor.

7. Describe the structure and function of the receptors of taste and smell.
8. Relate the presence of thermoreceptors to the lifestyles of animals that have them.
9. Contrast simple eyes, compound eyes, and vertebrate eyes.
10. Label the structures of the vertebrate eye on a diagram and give the functions of each of its accessory structures.
11. Describe the events that take place in human vision from the time light passes through the cornea to the integration of visual information in the brain. (Compare the two types of photoreceptors and describe the signal transduction pathway as part of your explanation.)

SENSORY RECEPTORS CAN BE CLASSIFIED ACCORDING TO SOURCE OR TYPE OF STIMULUS

Exteroceptors receive stimuli from the outside environment, enabling an animal to know and explore the world, search for food, find and attract a mate, recognize friends, and detect enemies. **Proprioceptors** are sensory receptors within muscles, tendons, and joints that enable the animal to perceive the positions of its arms, legs, head, and other body parts, along with the orientation of its body as a whole. With the help of proprioceptors, humans can get dressed or eat even in the dark.

Interoceptors are sensory receptors within body organs that detect changes in pH, osmotic pressure, body temperature, and the chemical composition of the blood. We are not usually conscious of messages sent to the CNS by these receptors as they work continuously to maintain homeostasis. We become aware of their activity when they enable us to perceive such diverse internal conditions as thirst, hunger, nausea, pain, and orgasm.

Sensory receptors can also be classified according to the *types* of stimuli to which they respond (Table 41–1). **Mechanoreceptors** respond to mechanical energy—touch, pressure, gravity, stretching, and movement. **Chemoreceptors** respond to certain chemical compounds, and **photoreceptors** detect light energy. **Thermoreceptors** respond to heat and cold. Some fish have well developed **electroreceptors,** which detect electrical energy.

RECEPTOR CELLS PRODUCE RECEPTOR POTENTIALS

Sensory receptors absorb energy, transduce (convert) that energy into electrical energy, and produce a **receptor potential,** a depolarization or hyperpolarization of the membrane. In its

capacity as a detector, or sensor, a sensory receptor absorbs a small amount of energy from the environment. Each kind of sensory receptor is especially sensitive to one particular form of energy. For example, photoreceptor cells (rods and cones) in the eye absorb light energy, while temperature receptors respond to infrared thermal energy (heat). Receptor cells are remarkably sensitive to appropriate stimuli. The photoreceptors of the eye are stimulated by an extremely faint beam of light, the taste receptors by a minute amount of acetic acid in vinegar, and the olfactory receptors by a few molecules of vanilla.

When unstimulated, a sensory receptor maintains a resting potential, that is, a difference in charge between the inside and the outside of the cell. When the receptor cell is stimulated and its membrane potential changes, the permeability of its plasma membrane is altered. Specific types of ion channels open or close. If the difference in charge increases, the receptor becomes hyperpolarized. If the potential decreases, the receptor becomes depolarized. The change in membrane potential is the receptor potential.

The receptor potential, the state of depolarization caused by a stimulus, is a graded response that spreads relatively slowly down the dendrite, fading as it goes. If the sensory neuron of which the receptor is a part is sufficiently depolarized to reach its threshold level, an action potential is generated (see Fig. 39–7). The action potential travels along the axon to the CNS. In summary:

Stimulus (e.g., light energy) → transduction into electrical energy → receptor potential → action potential

The sensory receptor performs three important functions: It (1) detects a stimulus in the environment by absorbing energy; (2) converts the energy of the stimulus into electrical energy, the process known as transduction; and (3) produces a receptor potential that may result in the transmission of in-

TABLE 41-1   Classification of Receptors by Stimuli to Which They Respond

Type of Receptor	Examples	Effective Stimuli

Mechanoreceptors	Tactile receptors
	    Pacinian corpuscles
	    Meissner's corpuscles
		Touch, pressure

	Proprioceptors
	    Muscle spindles
	    Golgi tendon organs
	    Joint receptors
		Movement, body position
		Muscle contraction
		Stretch of a tendon
		Movement in ligaments

	Lateral line organs in fish	Waves, currents in water
	Statocysts in invertebrates	Gravity

	Labyrinth of vertebrate ear
	    Saccule and utricle
	    Semicircular canals
	    Hair cells in the organ of Corti in the cochlea
		Gravity; linear acceleration
		Angular acceleration
		Pressure waves (sound)

Chemoreceptors	Taste buds; olfactory epithelium	Specific chemical compounds
Thermoreceptors	Temperature receptors in blood-sucking insects and ticks; pit organs in pit vipers; nerve endings and receptors in skins and tongues of many animals	Heat
Electroreceptors	Organs in skin of some fishes	Electrical currents in water
Photoreceptors	Eyespots; ommatidia of arthropods; rods and cones in retinas of	Light energy

formation to the CNS by an action potential. With minor variations, this is how all receptors operate.

The intensity of a stimulus is coded by the frequency of action potentials fired by sensory neurons during the stimulus. A strong stimulus causes a greater depolarization of the receptor membrane and causes the sensory neuron to fire action potentials with greater frequency than would a weak stimulus. This occurs because, like an EPSP (see Chapter 39), the receptor potential is a *graded response*, that is, it can vary in magnitude. In contrast, according to the *all-or-none law*, the amplitude (size) of each action potential does not vary, and therefore has no relation to the magnitude of the stimulus. (Some receptors work by causing a decrease in the frequency of action potentials.)

Sensation depends on transmission of a coded message

How do you know whether you are seeing a blue sky, tasting a doughnut, or hearing a note played by a piano? All action potentials are qualitatively the same. Light of the wavelength 450 nanometers (blue), sugar molecules (sweet), and sound waves of 440 hertz (A above middle C) all cause transmission of similar action potentials. Our ability to differentiate stimuli depends on both the sensory receptor itself and on the brain. We can distinguish the color of a blue sky from the scent of cologne; a sweet taste from a light breeze; or the sound of a piano from the heat of the sun because cells of each sensory receptor are connected to specific neurons in particular parts of the brain. Because a receptor normally responds to only one type of stimulus, for example, light or sound, the brain interprets a message arriving from a particular receptor as meaning that a certain type of stimulus occurred (e.g., a flash of color).

Sensation takes place in the brain. Interpretation of the message and the type of sensation depends on which association neurons receive the message. The rods and cones of the eye do not see. When stimulated, they send a message to the brain that interprets the signals and translates them into a rainbow, an elephant, or a child. Artificial (e.g., electrical) stimulation of brain centers can also result in sensation. On the other hand, many sensory messages never give rise to sensations at all. For example, certain chemoreceptors sense internal changes in the body but never stir our consciousness.

When stimulated, a sensory receptor initiates what might be considered a coded message, composed of action potentials transmitted by nerve fibers. This coded message is later decoded in the brain. Impulses from the sensory receptor may differ in (1) the total number of fibers transmitting, (2) the specific fibers carrying action potentials, (3) the total number of action potentials passing over a given fiber, and (4) the frequency of the action potentials passing over a given fiber. For example, the difference in sound intensity between the gentle rustling of leaves and a clap of thunder depends on the number of neurons transmitting action potentials, as well as the frequency of the action potentials transmitted by each neuron.

Just how the sensory receptor initiates different codes and how the brain analyzes and interprets them to produce various sensations are not completely understood.

Sensory receptors adapt to stimuli

Many sensory receptors do not continue to respond at the initial rate, even if the stimulus continues at the same intensity. With time, the frequency of action potentials in the sensory neuron decreases. This diminishing response to a continued, constant stimulus is called **sensory adaptation.** Sensory adaptation occurs for two reasons. First, during a sustained stimulus, the receptor produces a smaller receptor potential resulting in a lower frequency of action potentials in the sensory neurons. Second, changes take place at synapses in the neural pathway activated by the receptor. For example, the release of neurotransmitter from a presynaptic terminal may decrease in response to a series of action potentials.

Some receptors, such as those for pain or cold, adapt so slowly that they continue to trigger action potentials as long as the stimulus persists. Receptors that adapt slowly, or not at all, are called **tonic receptors.** Other receptors, called **phasic receptors,** adapt rapidly, permitting an animal to ignore persistent unpleasant or unimportant stimuli. For example, when you first pull on a pair of tight jeans, your pressure receptors let you know that you are being squished, and you may feel uncomfortable. Soon, though, these receptors adapt, and you hardly notice the sensation of the tight fit. In the same way, we quickly adapt to odors that at first seem to assault our senses. Sensory adaptation enables an animal to discriminate between unimportant background stimuli that can be ignored and new or important stimuli that require attention.

MECHANORECEPTORS RESPOND TO TOUCH, PRESSURE, GRAVITY, STRETCH, OR MOVEMENT

Mechanoreceptors are activated when they change shape as a result of being mechanically pushed or pulled. Some mechanoreceptors enable an organism to maintain its body position with respect to gravity (for us, head up and feet down; for a dog, dorsal side up and ventral side down; for a tree sloth, ventral side up and dorsal side down). Other mechanoreceptors are concerned with maintaining postural relations—the position of one part of the body with respect to another. This information is essential for all forms of locomotion and for all coordinated and skilled movements, from spinning a cocoon to completing a reverse one-and-a-half dive with twist.

Mechanoreceptors provide information about the shape, texture, weight, and topographical relations of objects in the external environment. Certain mechanoreceptors affect the operation of internal organs. For example, they supply information about the presence of food in the stomach, feces in the rectum, urine in the bladder, and a fetus in the uterus.

Touch receptors are located in the skin

The simplest mechanoreceptors are free nerve endings in the skin. These nerve endings detect touch, pressure, and pain when stimulated by objects that contact the body surface. Thousands of more specialized tactile (touch) receptors are also located in the skin (Fig. 41–1). Merkel discs sense touch and pressure. They adapt slowly, permitting us to know that an object continues to touch the skin. Three types of mechanoreceptors in the skin have endings that are encapsulated: Meissner corpuscles, Ruffini corpuscles, and Pacinian corpuscles. Meissner corpuscles are sensitive to light touch and pressure and adapt quickly to a sustained stimulus. Ruffini corpuscles, which adapt very slowly, inform us of heavy and continuous touch and pressure.

The **Pacinian corpuscle** is sensitive to deep pressure that causes rapid movement of the tissues. It is especially sensitive to stimuli that vibrate. The Pacinian corpuscle consists of a nerve ending surrounded by concentric connective tissue layers interspersed with fluid. Compression causes displacement of the layers, which stimulates the axon. Even though the displacement is maintained under steady compression, the receptor potential rapidly falls to zero, and action potentials cease—an excellent example of sensory adaptation. The Pacinian corpuscle is a phasic receptor that is stimulated only when there is movement of the tissue. The transduction mechanisms that couple touch or pressure on the skin with the opening or closing of ion channels in the plasma membrane of the receptor cell are not yet known.

In many invertebrates as well as vertebrates, tactile receptors lie at the base of a hair or bristle. Tactile hairs may sense the body's orientation in space with respect to gravity. They may also detect air and water vibrations and contact with other objects. The tactile receptor is stimulated indirectly when the hair is bent or displaced. A receptor potential develops, and a few action potentials may be generated. This type of receptor is a phasic receptor that responds only when the hair is moving. Even though the hair may be maintained in a displaced position, the receptor is not stimulated unless there is motion.

Many invertebrates have gravity receptors called statocysts

Every organism is oriented in a characteristic way with respect to gravity. When displaced from this normal position, it quickly adjusts its body to reassume normal orientation. In complex animals, receptors must continually send information to the CNS regarding the position and movements of the body.

Many invertebrates have gravity receptors called **statocysts.** A statocyst is basically an infolding of the epidermis

(a)

(b)   500 μm

Figure 41–1   **Sensory reception in human skin.** *(a)* This diagrammatic section through the human skin illustrates several mechanoreceptors. Merkel disks, Meissner corpuscles, and Pacinian corpuscles respond to touch and pressure. Ruffini corpuscles are sensitive to pressure. Hair follicle receptors respond to displacement of the hair. Also shown are free nerve endings that respond to pain; *(b)* Pacinian corpuscle, a deep pressure receptor. *(c)* The transduction mechanism that couples touch or pressure on the skin to the opening of ion channels in the receptor membrane is not yet understood. It is thought that sodium ion channels open, leading to depolarization. *(Ed Reschke)*

(c) Membrane of the Pacinian corpuscle

lined with receptor cells, called sensory hair cells, equipped with sensory hairs (not true hairs) (Fig. 41–2). The cavity contains one or more **statoliths,** tiny granules of loose sand grains or calcium carbonate, held together by an adhesive material secreted by cells of the statocyst. Normally the particles are pulled downward by gravity and stimulate the hair cells. When the animal changes position, the position of the statolith also changes, stimulating different hair cells. Each sensory cell responds maximally when the animal is at a particular position with respect to gravity. The mechanical displacement results in receptor potentials and action potentials that inform the CNS of the change in position. By "knowing" which hair cells are firing, the animal knows where "down" is and so can correct any abnormal orientation.

In a classic experiment, the function of the statocyst was demonstrated by substituting iron filings for sand grains in the statocysts of crayfish. The force of gravity was overcome by holding magnets above the animals: the iron filings were attracted upward toward the magnets, and the crayfish began to swim upside down in response to the new information provided by their gravity receptors.

Lateral line organs supplement vision in fish

Lateral line organs of fish and aquatic amphibians detect vibrations in the water. They are thought to supplement vision by informing the animal of obstacles in its way and of moving objects such as prey, enemies, and schooling mates. Researchers have suggested that sharks occasionally attack human swimmers because humans may produce vibrations that are similar to those produced by an injured fish.

Typically the lateral line organ consists of a long canal running the length of the body and continuing into the head (Fig.

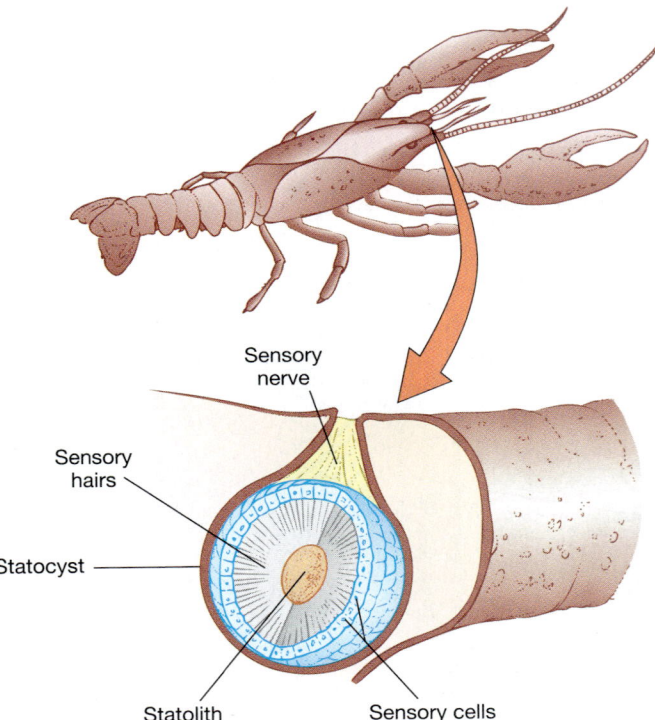

Figure 41–2 *A statocyst.* Many invertebrates have statocysts, receptors that sense gravitational force and provide information about orientation of the body with respect to gravity.

Sensory nerve

Sensory hairs

Statocyst

Statolith

Sensory cells

Proprioceptors help coordinate muscle movement

Proprioceptors are mechanoreceptors that continually respond to tension and movement in muscles and joints. Vertebrates have three main types of proprioceptors: **muscle spindles,** which detect muscle movement (Fig. 41–4); **Golgi tendon organs,** which determine stretch in the tendons that attach muscle to bone; and **joint receptors,** which detect movement in ligaments. These are tonic sensory receptors rather than phasic receptors. The receptor potential is maintained (though not at constant magnitude) as long as the stimulus is present, and action potentials continue to be generated. Thus, information about position is continuously supplied.

By means of these sensory receptors, we can carry out activities such as dressing or playing the piano even with our eyes closed. Impulses from the proprioceptors are important in ensuring the harmonious contractions of the distinct muscles involved in a single movement; without such receptors, complicated, skillful acts would be impossible. These organs are also important in maintaining balance. Proprioceptors are probably more numerous and more continually active than any of the other sensory receptors, although we are less aware of them than of most of the others.

The mammalian muscle spindle, one of the more versatile stretch receptors, helps maintain muscle tone. It consists of a bundle of specialized muscle fibers, in the center of which is a region encircled by sensory nerve endings. These neurons continue to transmit signals for a prolonged period of time, in proportion to the degree of stretch. Some of the nerve endings also exhibit a strong dynamic response to an increase in length, but only while the length is actually increasing.

The vestibular apparatus maintains equilibrium

When we think of the ear, we think of hearing. However, in vertebrates its main function is to help maintain equilibrium.

41–3). The canal is lined with sensory hair cells that have hair-like cilia. The tips of the hairs are enclosed by a **cupula,** a mass of gelatinous material secreted by the hair cells. Waves, currents, and even slight movement in the water cause vibrations in the lateral line organ. The water moves the cupula, causing the cilia to bend. This results in changes in the membrane potential of the hair cell, and a message may be dispatched to the CNS.

(a)   *(b)*   *(c)*

Figure 41–3 **Lateral line organ.** *(a)* The lateral line organ is a canal that extends the length of the body. *(b)* The canal has numerous openings to the outside environment. *(c)* The receptor cells have sensory hairs (cilia and stereocilia) that are enclosed by a gelatinous cupula. The receptor cells respond to waves, currents, or other disturbances in the water, informing the fish of obstacles or moving objects.

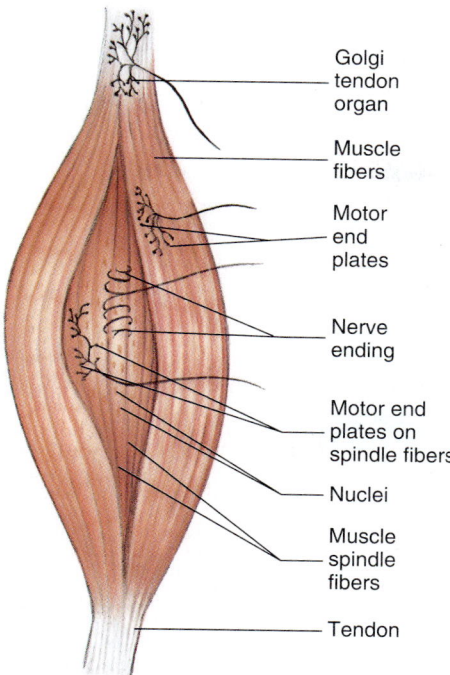

Figure 41–4 **Proprioceptors.** These sensory receptors respond to changes in movement, tension, and position in muscles and joints. Muscle spindles detect muscle movement. A muscle spindle consists of an elongated bundle of specialized muscle fibers. Golgi tendon organs signal force in muscles and their associated tendons.

Although many vertebrates do not have outer or middle ears, all have inner ears.

The inner ear consists of a complicated group of interconnected canals and sacs, referred to as the **labyrinth,** which includes a membranous labyrinth that fits inside a bony labyrinth. In mammals the membranous labyrinth consists of two saclike chambers, the **saccule** and the **utricle,** and three **semicircular canals.** Collectively, the saccule, utricle, and semicircular canals are referred to as the **vestibular apparatus** (Fig. 41–5). Destruction of these organs leads to a considerable loss of the sense of equilibrium. A pigeon whose vestibular apparatus has been destroyed cannot fly but in time can relearn how to maintain equilibrium using visual stimuli. Proprioceptors, cells sensitive to pressure in the soles of the feet, and vision all contribute to equilibrium in the human.

The saccule and utricle house gravity detectors in the form of small calcium carbonate ear stones called **otoliths** (Fig. 41–6). The sensory cells of these structures are similar to those of the lateral line organ. Each consists of a group of hair cells surrounded at their tips by a gelatinous cupula. The hair cells in the saccule and utricle lie in different planes. The surface of a vertebrate hair cell typically has a single long cilium and a number of shorter **stereocilia,** which are microvilli that contain actin filaments. The stereocilia increase in length from one side of the hair cell to the other.

Normally, the pull of gravity causes the otoliths to press against the stereocilia, stimulating them to initiate impulses that are sent to the brain by way of sensory nerve fibers at their bases. When the head is tilted or in linear acceleration (an increase in speed when the body is moving in a straight line), otoliths press on the stereocilia of different cells, deflecting them. Deflection of the cilia toward the longest cilium depolarizes the hair cell. Deflection in the opposite direction hyperpolarizes the hair cell. Thus, depending on the direction of movement, the hair cells release more or less neurotransmitter. The brain interprets the neural messages so that the animal is aware of its position relative to the ground regardless of the position of its head.

Information about turning movements, referred to as angular acceleration, is provided by the three semicircular canals.

Figure 41–5 **The human inner ear.** The saccule, utricle, and semicircular canals make up the vestibular apparatus. The membranous labyrinth is exposed. Because this is a posterior view, the utricle and saccule can be seen.

Figure 41-6 Function of the saccule and utricle in maintaining posture. The saccule and utricle sense linear acceleration, allowing us to know our position relative to the ground. Compare the positions of the otoliths and hairs in *(a)* with those in *(b)*. Changes in head position cause the force of gravity to distort the cupula, which in turn distorts the cilia of the hair cells. The hair cells respond by sending impulses along the vestibular nerve (part of the auditory nerve) to the brain.

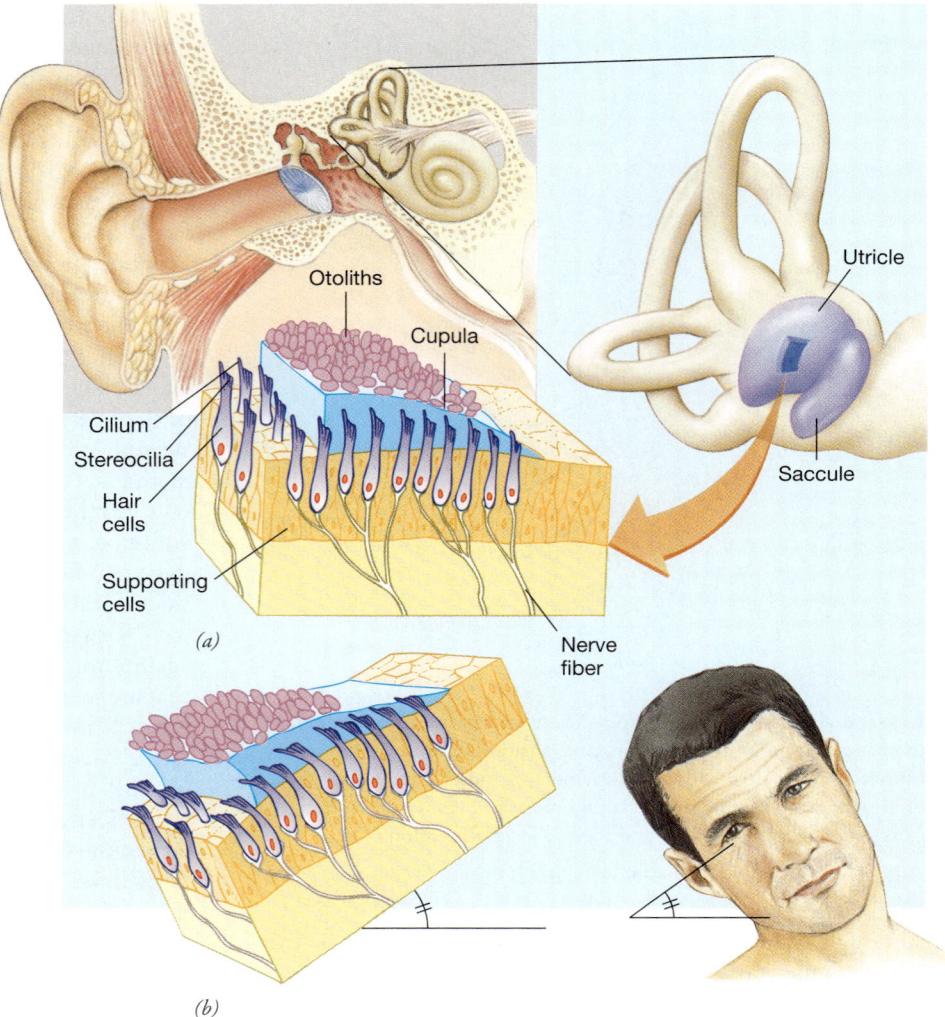

Each canal, a hollow ring connected with the utricle, lies at right angles to the other two and is filled with fluid called **endolymph.** At one of the openings of each canal into the utricle is a small, bulblike enlargement, the **ampulla.** Within each ampulla lies a clump of hair cells called a **crista** (pl., *cristae*), similar to the groups of hair cells in the utricle and saccule but lacking otoliths. The stereocilia of the hair cells of the cristae are stimulated by movements of the endolymph in the canals (Fig. 41–7).

When the head is turned, there is a lag in the movement of the fluid within the canals. The cilia move in relation to the fluid and are stimulated by its flow. This stimulation produces not only the consciousness of rotation but also certain reflex movements in response to it. These reflexes cause the eyes and head to move in a direction opposite that of the original rotation. Because the three canals are in three different planes, head movement in any direction stimulates fluid movement in at least one of the canals.

We humans are used to movements in the horizontal plane, but not to vertical movements (parallel to the long axis of the upright body). The motion of an elevator or of a ship pitching in a rough sea stimulates the semicircular canals in an unusual way and may cause seasickness or motion sickness, with resultant nausea or vomiting. When a seasick person lies down, the movement stimulates the semicircular canals in a more familiar way, and nausea is less likely to occur.

Auditory receptors are located in the cochlea

Many arthropods and most vertebrates have sound receptors, but for many of them hearing does not seem to be a sensory priority. It is important in tetrapods, however, and both birds and mammals have a highly developed sense of hearing. Their auditory receptors, located in the **cochlea** of the inner ear, contain mechanoreceptor hair cells that detect pressure waves.

The cochlea is a spiral tube that resembles a snail's shell (Figs. 41–8 and 41–9). If the cochlea were uncoiled, it would be seen to consist of three canals separated from each other by thin membranes and coming almost to a point at the apex. Two of these canals, or ducts, the **vestibular canal** (also known as the scala vestibuli) and the **tympanic canal** (scala tympani),

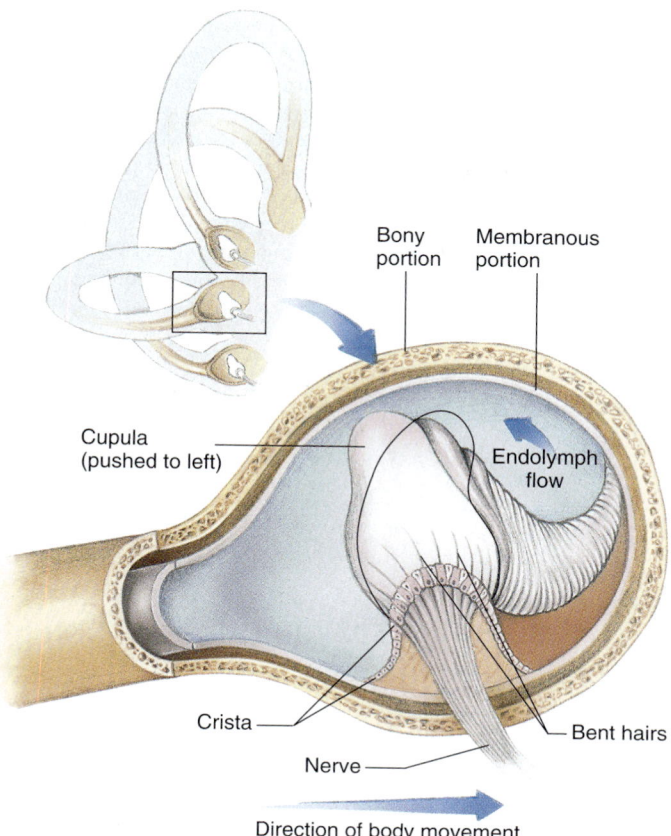

Bony portion

Membranous portion

Cupula (pushed to left)

Endolymph flow

Crista

Bent hairs

Nerve

Direction of body movement

Figure 41–7 **Semicircular canals and equilibrium.** When the head changes its rate of rotation, endolymph within the ampulla of the semicircular canal distorts the cupula. The hair cells of the cupula are bent, increasing the frequency of action potentials in sensory neurons. Information is transmitted to the brain via the vestibular nerve.

are connected at the apex of the cochlea and are filled with a fluid known as **perilymph.** The middle canal, the **cochlear duct** (scala media), is filled with endolymph and contains the auditory organ, the **organ of Corti.**

Each organ of Corti contains about 18,000 hair cells arranged in rows that extend the length of the coiled cochlea. Each cell is equipped with stereocilia that extend into the cochlear duct. The hair cells rest on the **basilar membrane,** which separates the cochlear duct from the tympanic canal. Overhanging and in contact with the hair cells of the organ of Corti is another membrane, the **tectorial membrane.** Movement of the stereocilia of the hair cells initiates impulses in the fibers of the cochlear nerve.

In terrestrial vertebrates sound waves in the air are transformed into pressure waves in the cochlear fluid. In the human ear, for example, sound waves pass through the **external auditory canal** and cause the **tympanic membrane,** or eardrum (the membrane separating the outer ear and the middle ear), to vibrate. The vibrations are transmitted across the middle ear by three tiny bones, the **malleus, incus,** and **stapes** (or hammer, anvil, and stirrup, so called because of their shapes). The malleus is in contact with the eardrum, and the stapes is in contact with the membrane covering the opening of the inner ear, called the **oval window.** These bones act as three interconnected levers that help amplify the vibrations. A very small movement in the malleus causes a larger movement in the incus and a very large movement in the stapes. The vibrations pass through the oval window to the fluid in the vestibular canal. If sound waves were conducted directly from air to the oval window, much energy would be lost. The middle ear functions to couple sound waves in the air with the pressure waves conducted through the fluid in the cochlea.

Auditory bones

Malleus   Incus   Stapes

Semicircular canals

Oval window

Vestibular nerve

Cochlear nerve

Cochlea

Vestibule

Round window

Tympanic membrane

Eustachian tube

External ear canal

Figure 41–8 **Structure of human ear.** The ear is adapted to direct sound waves from outside the body to receptors in the inner ear.

Figure 41–9 **The cochlea, organ of hearing.** (a) Cross section through the cochlea showing the cochlear nerve. (b) Enlarged view of the organ of Corti resting on the basilar membrane and covered by the tectorial membrane. (c) How the organ of Corti works. Vibrations transmitted by the malleus, incus, and stapes set the fluid in the tympanic canal in motion. These vibrations are transmitted to the basilar membrane, and, as the membrane vibrates, hair cells of the organ of Corti rub against the overlying tectorial membrane. This stimulation causes the hair cells to depolarize, resulting in action potentials in the sensory neurons of the cochlear nerve, a branch of the auditory nerve.

Cochlear nerve division of the vestibulocochlear (VIII) nerve

Organ of Corti

(a)

Oval window
Cochlear duct
Vestibular membrane

Walls of membranous labyrinth

Vestibular canal (contains perilymph)

Cochlear duct (contains endolymph)
Tectorial membrane
Organ of Corti
Basilar membrane

(b)

Tympanic canal (contains perilymph)

Hair cell
Cochlear nerve
Basilar membrane

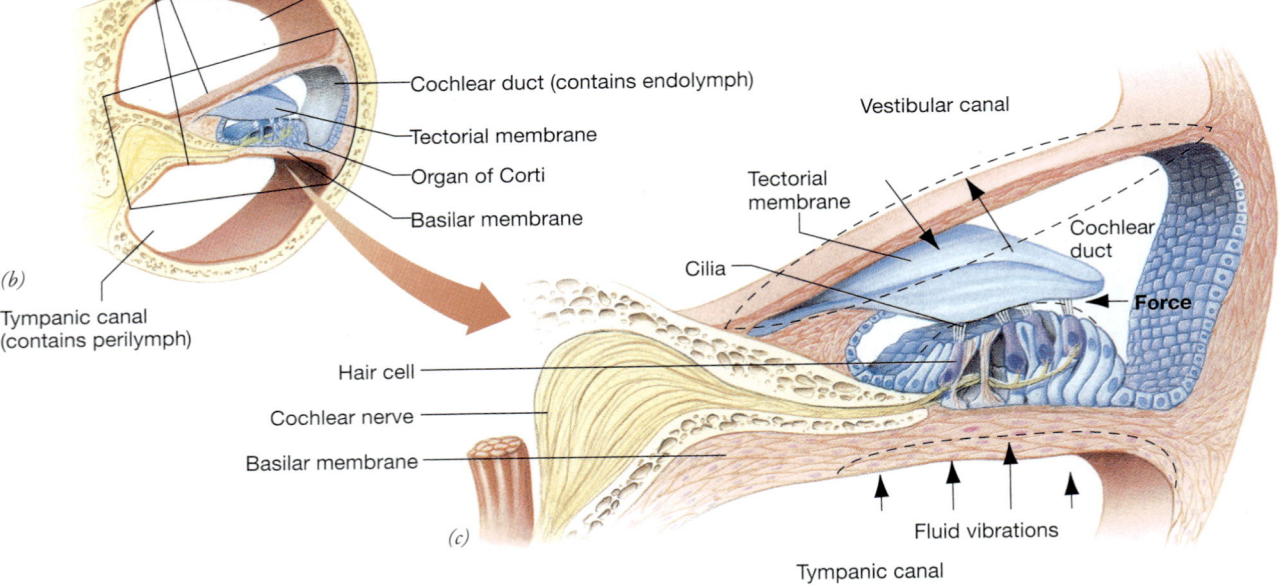

Vestibular canal
Tectorial membrane
Cilia
Cochlear duct
Force
Fluid vibrations
Tympanic canal
(c)

Because liquids cannot be compressed, the oval window could not cause movement of the fluid in the vestibular duct if there were not an escape valve for the pressure. This is provided by the **round window** at the end of the tympanic canal. The pressure wave presses on the membranes separating the three ducts, is transmitted to the tympanic canal, and causes a bulging of the round window. The movements of the basilar membrane produced by these pulsations cause the stereocilia of the organ of Corti to rub against the overlying tecto-

rial membrane. This stimulation initiates nerve impulses in the cochlear nerve. We can summarize the sequence of events involved in hearing as follows:

Sound waves enter external auditory canal→tympanic membrane vibrates→malleus, incus, and stapes amplify vibrations→oval window vibrates→vibrations are conducted through fluid→basilar membrane vibrates→hair cells in the organ of Corti are stimulated→cochlear nerve transmits impulses to brain

Sounds differ in pitch, loudness, and tone quality. **Pitch** depends on frequency of sound waves, or number of vibrations per second, and is expressed as hertz (Hz). Low-frequency vibrations result in the sensation of low pitch, whereas high-frequency vibrations result in the sensation of high pitch. Frequencies greater than 60 Hz result in unequal vibration along the length of the basilar membrane. Sounds of a given frequency set up resonance waves in the cochlear fluid that cause a particular section of the basilar membrane to vibrate. High frequencies are detected by hair cells located near the base of the basilar membrane, whereas low frequencies are sensed by hair cells near the apex of the basilar membrane. The brain infers the pitch of a sound from the particular hair cells that are stimulated.

Loud sounds cause resonance waves of greater amplitude (height). The hair cells are more intensely stimulated, and the cochlear nerve then transmits a greater number of impulses per second. Variations in the quality of sound, such as those evident when an oboe, a cornet, and a violin play the same note, depend on the number and kinds of overtones, or harmonics, produced. These provide stimulation to different hair cells in addition to the main stimulation common to all three instruments. Thus, differences in tone quality are recognized in the *pattern* of the hair cells stimulated.

The human ear is equipped to register sound frequencies between about 20 and 20,000 Hz although individuals vary greatly. Dogs and some other animals can hear sounds of much higher frequencies. The human ear is more sensitive to sounds between 1000 and 4000 Hz than to higher or lower ones. Within this intermediate range, the ear is extremely sensitive. In fact, when the energy of audible sound waves is compared with the energy of visible light waves, the ear is ten times more sensitive than the eye.

Deafness may be caused by injury to or malformation of either the sound-transmitting mechanism of the outer, middle, or inner ear, or the sound-perceiving mechanism of the inner ear. Intense sound can injure the organ of Corti. People continually exposed to heavily amplified music or workers subjected to loud, high-pitched noises over a period of years often become deaf to high tones because the cells near the base of the organ of Corti become damaged.

CHEMORECEPTORS ARE ASSOCIATED WITH THE SENSES OF TASTE AND SMELL

Two highly sensitive chemoreceptive systems are the senses of **taste** (gustation) and **smell** (olfaction). These senses allow us to detect chemical substances in food, water, and in the air. Unlike most neurons, both taste and smell receptors are continuously regenerated. Chemoreception is also an important method of intraspecific communication as discussed in *Making the Connection: Chemoreception and Animal Behavior.*

Taste buds detect dissolved food molecules

The organs of taste in mammals are **taste buds,** located in the mouth. In humans, they are found mainly in tiny elevations, or papillae, on the tongue. Each of the thousands of taste buds is an oval epithelial capsule containing about 100 taste receptor cells interspersed with supporting cells (Fig. 41–10). The plasma membrane at the tip of each taste receptor cell has microvilli that extend into a taste pore on the surface of the tongue, where they are bathed in saliva. The taste receptors detect chemical substances dissolved in saliva. Certain molecules, for example, those perceived as sweet, activate a signal transduction process involving a G protein. Adenylyl cyclase activity increases, elevating cAMP levels. A protein kinase is activated that phosphorylates and closes K^+ channels. This decrease in K^+ permeability sets up a depolarizing receptor potential. Action potentials are then generated in sensory neurons that synapse with the taste receptor cell. One sensory neuron can innervate several taste buds.

Traditionally, four basic tastes have been recognized: sweet, sour, salty, and bitter. Although the greatest sensitivity to each of these tastes occurs in a given area of the human tongue (Fig. 41–10), not all papillae are restricted to a single category of taste. Flavor depends on the four basic tastes in combination with smell, texture, and temperature. Smell affects flavor because odors pass from the mouth to the nasal chamber via the internal nares. No doubt you have observed that when your nose is congested, food seems to have little "taste." The taste buds are not affected, but the blockage of nasal passages severely reduces the participation of olfactory reception in the composite sensation of flavor.

An awareness of the genetic component of taste can be traced to 1931 when Arthur L. Fox, a chemist at the du Pont Company synthesized a compound called phenylthiocarbamide (PTC). Some PTC blew into the air and was inhaled by a colleague who experienced it as very bitter. Because Fox himself could not taste it, he became intrigued and asked other people to taste it. Fox found that about 25% of people were nontasters. Everyone else experienced PTC as bitter. Further research showed that the ability to taste PTC is inherited as a dominant trait.

In the 1970s Linda Bartoshuk of Yale University continued taste research using a chemical compound called PROP (6-*n*-propylthiouracil). She found that some tasters were even more sensitive to the bitter taste of PROP. Subsequent studies have suggested that about 25% of the U. S. population are supertasters, 50% are regular tasters, and 25% are nontasters. In 1997, Adam Drewnowski at the University of Michigan and his team reported that supertasters avoid broccoli, Brussels sprouts, cabbage, and many other vegetables and fruits that have a bitter taste. These foods contain flavonoids and other compounds that are thought to protect against cancer. Continuing investigation of taste preferences and their nutritional consequences may lead to more effective approaches to improving nutrition and to a better understanding of the relative importance of genetics and learning in food selection.

(a)

(b)

50 µm

(c)

Figure 41–10 **Taste buds.** Taste receptors are located mainly on the surface of the tongue. *(a)* The surface of the tongue, showing distribution of taste buds that are sensitive to sweet, bitter, sour, and salt. A single taste receptor may respond to more than one category of taste. *(b)* LM of a taste bud, which consists of an epithelial capsule containing several taste receptors. *(c)* A sugar molecule activates a signal transduction process. (1) The sugar binds with a receptor in the plasma membrane of a taste receptor. (2) A G protein is activated. (3) Adenylyl cyclase is activated. (4) ATP is converted to cyclic AMP. (5) Cyclic AMP activates a protein kinase. (6) Action of the protein kinase closes K^+ channels. *(b, Ed Reschke)*

The olfactory epithelium is responsible for the sense of smell

Many animals rely on the sense of smell to find food, identify predators, and find mates. In terrestrial vertebrates, **olfaction,** the detection of odors, occurs in the nasal epithelium. In humans, the **olfactory epithelium** is found in the roof of the nasal cavity (Fig. 41–11). It contains about 100 million olfactory receptor cells with ciliated tips. The cilia extend into a layer of mucus on the epithelial surface of the nasal passageway. Receptor molecules on the cilia bind with compounds dissolved in the mucus. The other end of each olfactory receptor cell is an axon that projects directly to the brain. These axons make up the olfactory nerve (the first cranial nerve) which extends to the olfactory bulb in the brain. From there information is transmitted to the olfactory cortex, then to the limbic system and finally to other areas of the cortex by way of the thalamus.

When a molecule binds with a receptor on the cilia of an olfactory receptor cell, a signal transduction process is initiated. A G protein is activated, leading to the synthesis of cyclic AMP, which opens gated channels in the plasma membrane. These channels permit Na^+ and other cations to enter the cell, causing a depolarization, which is the receptor potential. The number of odorous molecules determines the intensity, which in turn determines the magnitude of the receptor potential.

Humans can detect at least seven main groups of odors: camphor, musk, floral, peppermint, ethereal, pungent, and putrid. About 1000 genes code for 1000 types of olfactory receptors. Each odor consists of several component chemical

(a)

(b)

(c)

Figure 41–11 **Olfactory epithelium.** *(a)* Smell depends on thousands of chemoreceptor cells in the roof of the nasal cavity. *(b)* The receptor cells are neurons located in the olfactory epithelium. *(c)* An odor molecule binds to a receptor in the plasma membrane of an olfactory cell. This activates a G protein and adenylyl cyclase. ATP is converted to cyclic AMP, which opens gated channels. Sodium ions enter the cell, leading to depolarization.

MAKING THE CONNECTION

CHEMORECEPTION AND ANIMAL BEHAVIOR

How does detection of chemical compounds affect the interactions of animals? Many species communicate with chemical messengers called **pheromones,** small volatile molecules that are secreted into the environment. Animals use pheromones to communicate danger, ownership of territory, and availability for mating. Insects use specific chemical compounds to communicate, defend themselves against predators, and recognize specific foods. Ants mark their trails with pheromones. Female moths release pheromones that attract males.

Many vertebrates, for example dogs and wolves, use chemical secretions to mark territory. Among some mammals, an ovulating female (one physiologically ready to mate) releases pheromones as part of her vaginal secretion. When these chemical odors are detected by males, they increase sexual interest. A recently impregnated female mouse aborts if she detects a particular pheromone in the urine of a strange adult male (one other than the male with

which she mated). How might this be adaptive?

Studies of rodents have shown that pheromones are detected, not by the olfactory epithelium, but by the **vomeronasal organ,** which is also located in the wall of the nasal cavity. The pheromone binds to a neuron receptor and triggers an action potential. The neural message is not transmitted through olfactory pathways. Instead of being transmitted to higher cognitive centers in the brain, the message is conveyed to the amygdala and hypothalamus, structures that regulate emotional responses and certain endocrine processes. About 100 genes that are thought to code for pheromone receptors have been identified in the mouse and rat. These receptors initiate signal transduction processes that involve G proteins. When neurons of the vomeronasal system are damaged in virgin mice, they do not mate.

The extent to which humans are affected by pheromones is currently under study.

groups, and each type of receptor may bind with a particular component. The combination of receptors activated determines what odor we perceive. We are capable of perceiving about 10,000 scents.

The olfactory receptors respond to remarkably small amounts of a substance. For example, ionone, the synthetic substitute for the odor of violets, can be detected by most people when it is present in a concentration of only one part to more than 30 billion parts of air. Despite its sensitivity, smell is perhaps the sense that adapts most quickly. The olfactory receptors adapt about 50% in the first second or so after stimulation, so even offensively odorous air may seem odorless after only a few minutes.

THERMORECEPTORS ARE SENSITIVE TO HEAT

Heat is another form of radiant energy to which organisms respond. Although not much is known about their specific thermoreceptors, many invertebrates are sensitive to changes in temperature. Mosquitoes, ticks, and other blood-sucking arthropods use thermoreception in their search for an endothermic host. Some have temperature receptors on their antennae that are sensitive to changes of less than 0.5°C. At least two types of snakes, pit vipers and boas, use thermoreceptors to locate their prey (Fig. 41–12).

In mammals, which are endothermic, free nerve endings and specialized receptors in the skin and tongue detect temperature changes in the outside environment. In humans, changes in temperature are detected by at least three types of receptors: cold receptors, warmth receptors, and pain receptors (for temperature extremes). (For a discussion of pain, see *Focus On: Pain Perception.*) Thermoreceptors in the hypothalamus detect internal changes in temperature and receive and integrate information from thermoreceptors on the body surface. The hypothalamus then initiates homeostatic mechanisms that ensure a constant body temperature.

ELECTRORECEPTORS DETECT ELECTRICAL CURRENTS IN WATER

Some predatory species of sharks, rays, and bony fishes can detect the electrical fields generated in the water by the muscle activity of their prey. Their electroreceptors are modified hair cells that may be associated with the lateral line organ. In addition to their lateral line system, sharks have electroreceptors, called *ampullae of Lorenzini,* on their heads that may also be used to locate prey. Some electroreceptors are sensitive enough to detect the Earth's magnetic field.

Several groups of fishes have electric organs, specialized muscle or nerve cells that produce external electrical fields. In

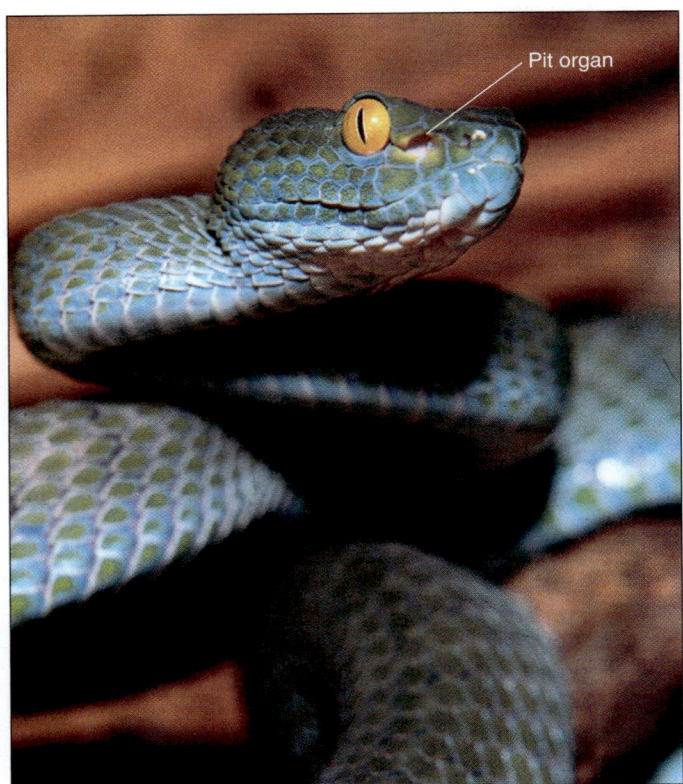

Figure 41–12 **Thermoreception.** The pit organ of this bamboo viper *(Trimersurus stejnegeri)* is a sense organ located between each eye and nostril. The pit organ of many snakes can detect the heat from an endothermic animal up to a distance of 1 to 2 m. *(Zig Leszczynski/ Animals Animals)*

species that produce a weak current, electric organs may help in orientation. This is particularly useful in murky water, where visibility and olfaction are poor. Electroreception also appears to be important in communication, for example, in recognition of a potential mate. Males have a different frequency of discharge than females. A few fishes, such as electric eels or electric rays, have electric organs in their heads capable of delivering powerful shocks that stun prey or predators.

PHOTORECEPTORS USE PIGMENTS TO ABSORB LIGHT

Most animals have photoreceptors that use pigments to absorb light energy. **Rhodopsins** are the photopigments found in the eyes of cephalopod mollusks, arthropods, and vertebrates. Light energy striking a light-sensitive receptor cell containing these pigments triggers chemical changes in the pigment molecules. As a result of this transduction, the receptor cell may transmit a nerve impulse.

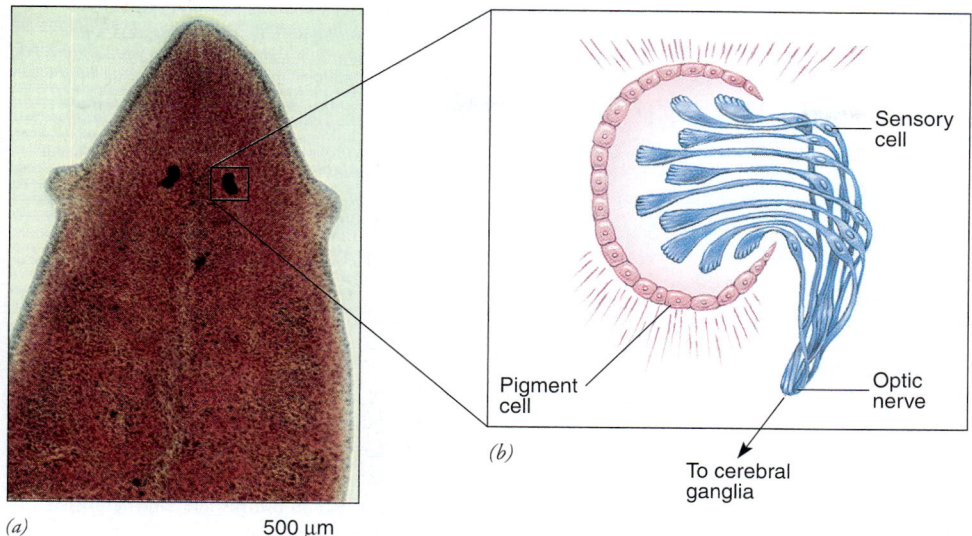

Figure 41–13 **Eyespots.** The simplest light-sensitive structures are the ocelli, or eyespots, found in certain invertebrates. *(a)* Planarian worm *(Planaria agilis)* showing eyespots. *(b)* Structure of the eyespot of a planarian worm. *(Terry Ashley/Tom Stack & Associates)*

Sensory cell

Pigment cell

Optic nerve

(b)

To cerebral ganglia

(a)

500 μm

Eyespots, simple eyes, and compound eyes are found among invertebrates

The simplest light-sensitive structures in animals are found in certain cnidarians and flatworms (Fig. 41–13). They are **eyespots,** called **ocelli,** that detect light but do not see objects. Eyespots are often bowl-shaped clusters of light-sensitive cells within the epidermis. They may detect the direction of the source of light and distinguish light intensity.

Effective image formation requires a more complex **eye,** usually with a lens. A lens is a structure that concentrates light on a group of photoreceptors. Vision also requires a brain that can interpret the action potentials generated by the photoreceptors. The brain must integrate information about movement, brightness, location, position, and shape of the visual stimulus.

Two fundamentally different types of eyes evolved: the camera eye of vertebrates and some mollusks (squids and octopods), and the compound eye of arthropods.

Compound eyes found in crustaceans and insects are structurally and functionally different from vertebrate eyes (Fig. 41–14 on page 894). The surface of a compound eye appears faceted, which means having many faces, like a diamond. Each **facet** is the convex cornea of one of the eye's visual units, called **ommatidia** (sing., *ommatidium*). The number of ommatidia varies with the species. For example, each eye of certain crustaceans has only 20 ommatidia, whereas the eye of a dragonfly has as many as 28,000.

The optical part of each ommatidium includes a biconvex **lens** and a **crystalline cone.** These structures focus light onto photoreceptor cells, called **retinular cells.** These cells have a light-sensitive membrane made up of microvilli containing rhodopsin. The membranes of adjacent retinular cells may fuse, forming a rod-shaped **rhabdome** which is sensitive to light.

Compound eyes do not perceive form well. Although the lens system of each ommatidium is adequate to focus a small inverted image, there is little evidence that they are actually perceived as images by the organism. However, all the ommatidia together do produce a composite image, or **mosaic** picture. Each ommatidium, in gathering a point of light from a narrow sector of the visual field, is in fact sampling a mean intensity from that sector. All of these points of light taken together form a mosaic picture.

To appreciate the nature of this mosaic picture, we need only look at a newspaper photograph through a magnifying glass; it is a mosaic of many dots of different intensities. The clarity and definition of the picture depend on how many dots there are per unit area—the more dots, the better the picture. So it is with the compound eye. The image perceived by the animal is probably much better in quality than might be suspected from the structure of the compound eye. The nervous system of an insect is apparently capable of image processing similar to that employed to improve the quality of photographs sent to Earth by robot spacecraft.

Arthropod eyes usually adapt to different intensities of light. A sheath of pigmented cells envelops each ommatidium, and screening pigments are present in cells called iris cells and in retinular cells. In nocturnal and crepuscular (active at dusk) insects and many crustaceans, pigment is capable of migrating proximally and distally. When the pigment is in the proximal position, each ommatidium is shielded from its neighbor, and only light entering directly along its axis can stimulate the receptors. When the pigment is in the distal position, light striking at any angle can pass through several ommatidia and stimulate many retinal units. As a result, sensitivity is increased in

PAIN PERCEPTION

Pain is a signal that protects us from harmful stimuli. In the human body, pain receptors are dendrites of certain sensory neurons found in almost every tissue. One type of pain receptor responds to strong mechanical stimulation. Another responds to temperatures above 45°C.

When stimulated, pain receptors send a message through sensory neurons to the spinal cord. The message is transmitted to the opposite side of the spinal cord and then sent upward to the thalamus, where pain perception begins *(figure)*. From the thalamus, impulses are sent into the parietal lobes of the cerebrum. At that time the individual becomes fully aware of the pain and can evaluate the situation. How threatening is the stimulus? How intense is the pain? What is the most adaptive response? From the thalamus, messages are also sent to a region in the limbic system, which is the brain's emotional center.

Pain perception is colored by emotional experience. Pain can be inhibited or facilitated at many levels in the nervous system. The intensity of one's pain perception depends on the particular situation and on how one has learned to deal with pain. A child with a bruised knee may emotionally heighten the feeling of pain, whereas a professional fighter may virtually ignore a long series of well delivered blows.

The brain locates pain on the basis of past experience. Generally, pain at the body surface is accurately projected back to the injured area. For example, when you step on a nail, your brain perceives the pain and then projects it back to the injured foot, so that you feel pain at the site of puncture. Artificial stimulation of the leg nerves may produce a sensation of pain in the foot even though the foot is untouched. In fact, for years after an amputation, a patient may feel **phantom pain** in the missing limb. This occurs because when the severed nerve is stimulated and sends a message to the brain, the brain "remembers" the nerve as it originally was—connected to the missing limb.

Most internal organs are poorly supplied with pain receptors. Pain is often not projected back to the organ that is stimulated. Instead it is referred to an area just under the skin that may be some distance from the organ involved. The area to which the pain is referred generally is connected to nerve fibers from the same level of the spinal cord as the organ involved.

A person with angina who feels heart pain in the left arm is experiencing **referred pain.** Neurons from both the heart and the arm converge on the same neurons in the central nervous system. The brain interprets the incoming message as coming from the body surface because that is the more common origin. When pain is felt both at the site of the distress and as referred pain, it may seem to spread, or radiate, from the organ to the superficial area.

The physiology of pain is an important area of research. Sensory neurons that transmit pain impulses to the spinal cord release the peptide neurotransmitter **substance P.** Pain can be perceived by the brain only in the presence of substance P. The body has a variety of mechanisms for analgesia (pain control). Neurophysiologists have long known that opiates, such as morphine, relieve pain. (Morphine and morphine-derivatives are widely used clinically to control pain. For example, patient-controlled morphine pumps are routinely used in many hospitals.) Opiates work by blocking the release of substance P. More than ten opiates have now been discovered in the brain, spinal cord, and pituitary gland. Among the more important ones are **beta-endorphin** (for "endogenous morphine-like"), **enkephalins,** and **dynorphin.** These endogenous opiates work by inhibiting the release of substance P. These compounds are more effective than morphine, and several are currently being investigated as potential analgesic (pain-killing) drugs.

In the spinal cord, certain neurons that release serotonin act on interneurons that synapse with pain neurons. These interneurons release enkephalin, which is thought to cause presynaptic inhibition of incoming pain-transmitting neurons. Enkephalin may block calcium channels, preventing release of substance P. In this way the body can block pain signals when

---

dim light, and the eye is protected from excessive stimulation in bright light. Pigment migration is under neural control in insects and under hormonal control in crustaceans. In some species it follows a daily rhythm.

Although the compound eye can form only coarse images, it compensates by being able to follow flickers to higher frequencies. Flies are able to detect up to about 265 flickers per second. In contrast, the human eye can detect only 45 to 53 flickers per second; for us, flickering lights fuse above these values, so we see light provided by an ordinary bulb as steady and the movement in motion pictures as smooth. To an insect, both room lighting and motion pictures must flicker horribly. The advantage of the insect's high critical flicker fusion threshold is that it permits immediate detection of even slight movement by prey or enemy. The compound eye is an important adaptation to the arthropod's way of life.

Compound eyes differ from our eyes in another respect. They are sensitive to wavelengths of light in the range from red to ultraviolet (UV). Accordingly, an insect can see UV light well, and its world of color is very different from ours. Because flowers reflect UV light to various degrees, flowers that appear identically colored to us may appear strikingly dissimilar to insects (see Fig. 41–14d; see also Fig. 35–4).

Vertebrate eyes form sharp images

The position of the eyes in the front of the head of humans and certain other vertebrates permits both eyes to be focused they first enter the spinal cord.

Many areas of the brain have opiate receptors, so pain signals that reach the brain can be blocked by beta-endorphin, enkephalins, and dynorphins. The neurotransmitter GABA (gamma aminobutyric acid) may inhibit release of substance P in some areas of the brain.

Some neurobiologists think that endorphins may explain the mechanism of action of acupuncture. There is some evidence that acupuncture needles stimulate nerves deep within the muscles, which in turn stimulate the pituitary gland and parts of the brain to release endorphins.

You may have observed that rubbing the skin around an injury can alleviate pain. Activating certain large sensory neurons that deliver messages about touch or pressure can stimulate the interneurons in the spinal cord that release enkephalin. Acupuncture may also stimulate these large sensory neurons.

Clinical methods have been developed for relieving pain by electrical stimulation of large sensory nerve fibers. Stimulation of the skin over a painful area with electrodes has successfully relieved pain in some patients. This procedure is called transcutaneous electrical nerve stimulation (TENS). Electrodes can be implanted in the brain, allowing the patient to control chronic pain by stimulating the release of endorphins.

on the same object (Fig. 41–15). The overlap in information they receive results in the same visual information striking the two retinas (light-sensitive areas) at the same time. This **binocular vision** is an important factor in judging distance and depth. Variations of eye position in other vertebrates offer different advantages. For example, the eyes of grazing animals are positioned laterally, enabling them to spot a predator approaching from behind.

The vertebrate eye can be compared to a camera. An adjustable lens can be focused for different distances, and a diaphragm, called the **iris,** regulates the size of the light opening, called the *pupil* (Fig. 41–16). The **retina** corresponds to the light-sensitive film used in a camera. Outside the retina is the **choroid layer,** a sheet of cells filled with black pigment

that absorbs extra light and prevents internally reflected light from blurring the image. (Cameras are also black on the inside.) The choroid is rich in blood vessels that supply the retina.

The outer coat of the eyeball, called the **sclera,** is a tough, opaque, curved sheet of connective tissue that protects the inner structures and helps to maintain the rigidity of the eyeball. On the front surface of the eye, this sheet becomes the thinner, transparent **cornea,** through which light enters. The cornea serves as a fixed lens.

The lens of the eye is a transparent, elastic ball just behind the iris. It bends the light rays coming in and brings them to a focus on the retina. The lens is aided by the curved surface of the cornea and by the refractive properties (ability to bend light rays) of the liquids inside the eyeball. The **an-**

Facets
Ommatidia
Optic ganglion
Optic nerve

(b)

Lens
Crystalline cone
Crystalline cone stalk
Iris cell with screening pigment
Rhabdome
Retinular cell

(a)

1 mm

(d)

(c)

Figure 41–14 **Compound eyes.** *(a)* The Mediterranean fruit fly *(Ceratitis capitata)* has prominent compound eyes. *(b)* Structure of the compound eye showing several ommatidia. The eye registers changes in light and shade, permitting the animal to detect movement. *(c)* Structure of an ommatidium. The rhabdome is the light-sensitive core of the ommatidium. *(d)* A bee eye's view of a flower. This photo was taken using an optical device to achieve an approximate simulation of how the bee might see this flower. However, the bee would see UV light rather than red, and circles would be vertically elongated elipses. *(a, David Scharf/Peter Arnold, Inc.; d, Courtesy of J. Gould, Princeton)*

terior cavity between the cornea and the lens is filled with a watery substance, the **aqueous fluid.** The larger **posterior cavity** between the lens and the retina is filled with a more viscous fluid, the **vitreous body.** Both fluids are important in maintaining the shape of the eyeball by providing an internal fluid pressure.

At its anterior margin, the choroid is thick and projects medially into the eyeball to form the **ciliary body,** which consists of ciliary processes and the ciliary muscle. The ciliary processes are glandlike folds that project toward the lens and secrete the aqueous fluid.

We focus a camera by changing the distance between the lens and the film. The eye has the power of **accommodation,** the ability to change focus for near or far vision by changing the shape of the lens (Fig. 41–17). This is accomplished by the **ciliary muscle,** a part of the ciliary body. To focus on objects that are near, the ciliary muscle contracts, causing the elastic lens to assume a rounder shape. To focus on more distant objects, the ciliary muscle relaxes and the lens assumes a flattened (ovoid) shape.

The most common disorders of vision are nearsightedness, farsightedness, and astigmatism (Fig. 41–18). In *nearsighted-*

(a)

(b)

Figure 41–15 **Position of the eyes in various vertebrates.** *(a)* The eyes of the zebra are positioned laterally, enabling the animal to see on both sides. Even while grazing, it can spot a predator approaching from behind. *(b)* Like many other nocturnal animals, the owl monkey *(Aotus evingatus)* has large eyes. Its eyes are positioned at the front of the head, and it has binocular vision, permitting it to judge distances. *(c)* The orbits (bony cavities that contain the eyeballs) of the hippopotamus are elevated, enabling the animal to see even when most of its head is underwater. *(a, Diane Blell/Peter Arnold, Inc.; b, Stephen Dalton/Animals Animals; c, Frans Lanting/Minden Pictures)*

(c)

ness *(myopia)*, the eyeball is elongated. The light rays converge at a point in front of the retina and are diverging again when they reach it. This results in a blurred image. Concave lenses correct for the nearsighted condition by bringing the light rays to a focus at a point farther back. In a *farsighted* eye, the eyeball is too short and the retina too close to the lens. Light rays strike the retina before they have converged, again resulting in a blurred image. Convex lenses correct for the farsighted condition by causing the light rays to converge farther forward.

In *astigmatism* the cornea is curved unequally in different planes, so that light rays in one plane are focused at a different point from those in another plane. To correct for astigmatism, lenses are ground unequally to compensate for the unequal curvature of the cornea.

The amount of light entering the eye is regulated by the iris, a ring of smooth muscle that appears as blue, green, gray, or brown depending on the amount and nature of pigment present. The iris is composed of two mutually antagonistic sets

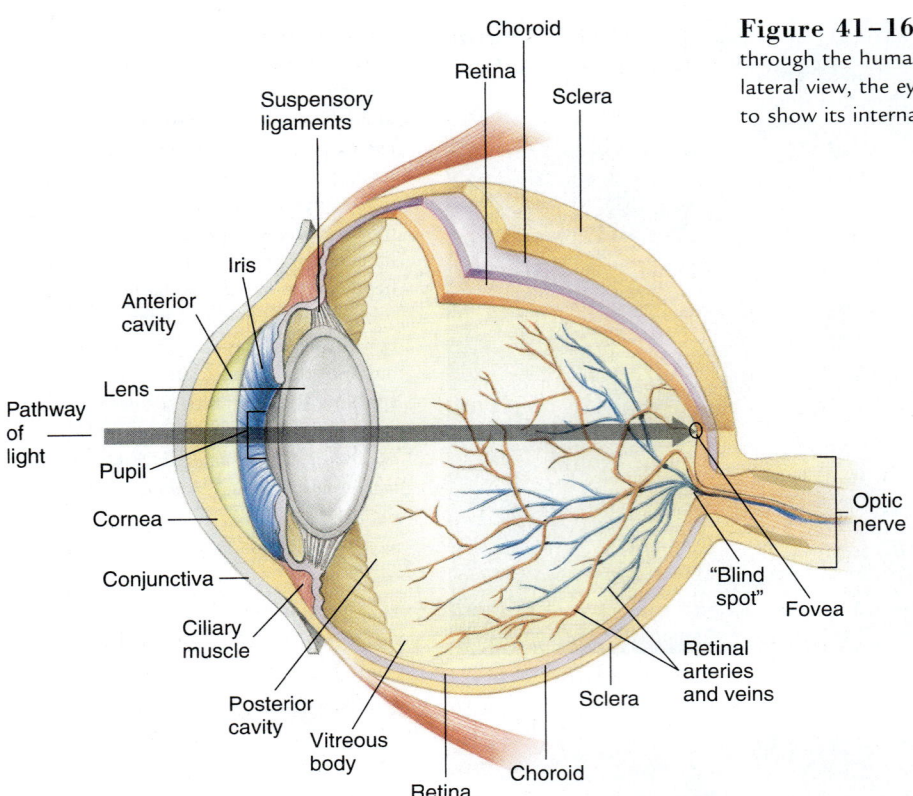

Choroid
Retina
Sclera
Suspensory
ligaments

Figure 41−16 **Structure of the human eye.** Light passes through the human eye to photoreceptor cells in the retina. In this lateral view, the eye is shown partly sectioned along the sagittal plane to show its internal structures.

Iris
Anterior
cavity
Lens
Pathway
of
light
Pupil
Cornea
Conjunctiva
Ciliary
muscle
Posterior
cavity
Vitreous
body
Retina

Optic
nerve
"Blind
spot"
Fovea
Retinal
arteries
and veins
Sclera
Choroid

Figure 41−17 **Focusing on near and far objects.**

Ciliary muscles contract,
lens "rounds up"
Retina
Lens
Fovea
Cornea
Optic
nerve
Suspensory ligaments

(a) Near vision (accommodation)

Ciliary muscles
relax and lens is
pulled to a flatter shape

(b) Distant vision

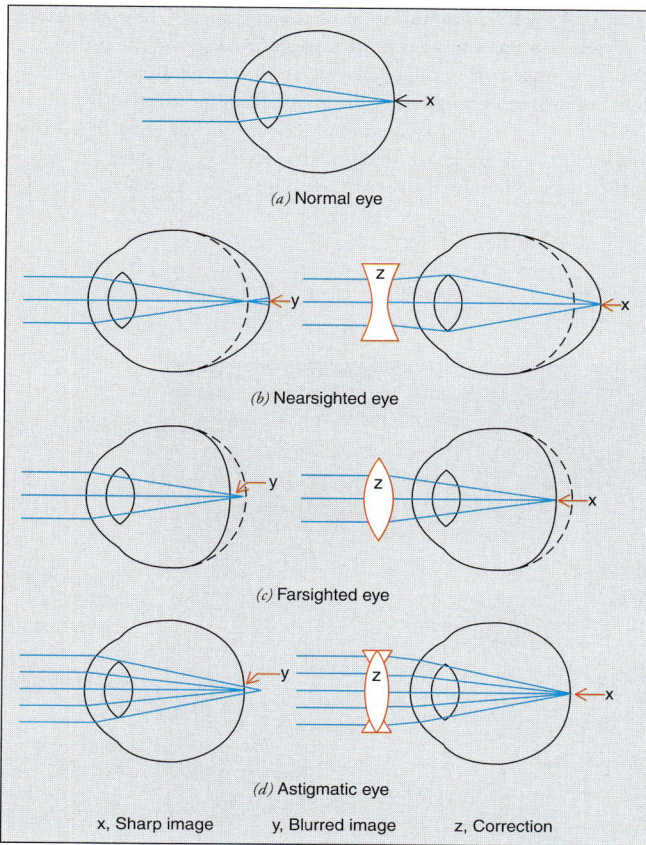

Figure 41–18 **Vision disorders.** Common abnormalities of the eye result in defects in vision. *(a)* In the normal eye, parallel light rays coming from a point in space are focused as a point on the retina. *(b)* In the nearsighted eye, the eyeball is elongated. Parallel light rays are brought to a focus in front of the retina (on dotted line, which represents the position of the retina in a normal eye). A blurred image is formed on the retina. Nearsightedness is corrected by placing a concave lens in front of the eye, which diverges the light rays so that they are focused on the retina. *(c)* In the farsighted eye, the eyeball is shortened and light rays are focused behind the retina. A convex lens converges the light rays. *(d)* In astigmatism, light rays passing through one part of the lens are focused on the retina, while light rays passing through another area of the lens are not focused on the retina. A cylindrical lens corrects this by bending light rays passing through only certain parts of the cornea.

of muscle fibers. One set is arranged circularly and contracts to *decrease* the size of the pupil. The other is arranged radially and contracts to *increase* the size of the pupil.

Each eye has six muscles that extend from the surface of the eyeball to various points in the bony socket. These muscles enable the eye as a whole to move and be oriented in a given direction. Cranial nerves innervate the muscles in such a way that the eyes normally move together and focus on the same area.

The retina contains light-sensitive rods and cones

The light-sensitive structure in the vertebrate eye is the retina, which lines the posterior two-thirds of the eyeball, covering the choroid. The retina, which is composed of ten layers, contains the photoreceptor cells called, according to their shapes, **rods** and **cones.** The human eye has about 125 million rods and 6.5 million cones. Rods function in dim light, allowing us to detect shape and movement. They are not sensitive to colors. Because the rods are more numerous in the periphery of the retina, you can see an object better in dim light if you look slightly to one side of it (allowing the image to fall on the rods).

Cones respond to light at higher levels of intensity, for example daylight, and they allow us to perceive fine detail. Cones are responsible for color vision; they are differentially sensitive to different frequencies (colors) of light. The cones are most concentrated in the **fovea,** a small depressed area in the center of the retina. The fovea is the region of sharpest vision because it has the greatest density of receptor cells and because the retina is thinner in that area.

Light must pass through several layers of connecting neurons in the retina to reach the rods and cones (Fig. 41–19). The retina consists of five main types of neurons. (1) Photoreceptors (rods and cones) synapse on (2) **bipolar cells** which make synaptic contact with (3) **ganglion cells.** Two types of lateral interneurons are the (4) **horizontal cells** that receive information from the photoreceptor cells and send it to bipolar cells and (5) **amacrine cells** that receive messages from the bipolar cells and send signals to ganglion cells or back to the bipolar cells (Fig. 41–20).

The axons of the ganglion cells extend across the surface of the retina and unite to form the **optic nerve.** The area where the optic nerve passes out of the eyeball, the optic disk, is known as the "blind spot"; because it lacks rods and cones, images falling on it cannot be perceived. A simplified summary of the visual pathway follows:

Light passes through cornea→through aqueous fluid→through lens→through vitreous body→image forms on retina (photoreceptor cells)→impulses in bipolar cells→impulses in ganglion cells→optic nerve transmits nerve impulses to thalamus→integration by visual areas of the cerebral cortex

A chemical change in rhodopsin leads to the response of a rod to light

How is light converted into the neural signals that transmit information about environmental stimuli into pictures in the brain? Rod cells are so sensitive to light that they can respond to a single photon. Rhodopsin in the rod cells and some very closely related photopigments in the cone cells are responsible for the ability to see. Rhodopsin consists of opsin, a large protein that is chemically joined with **retinal,** an aldehyde of vi-

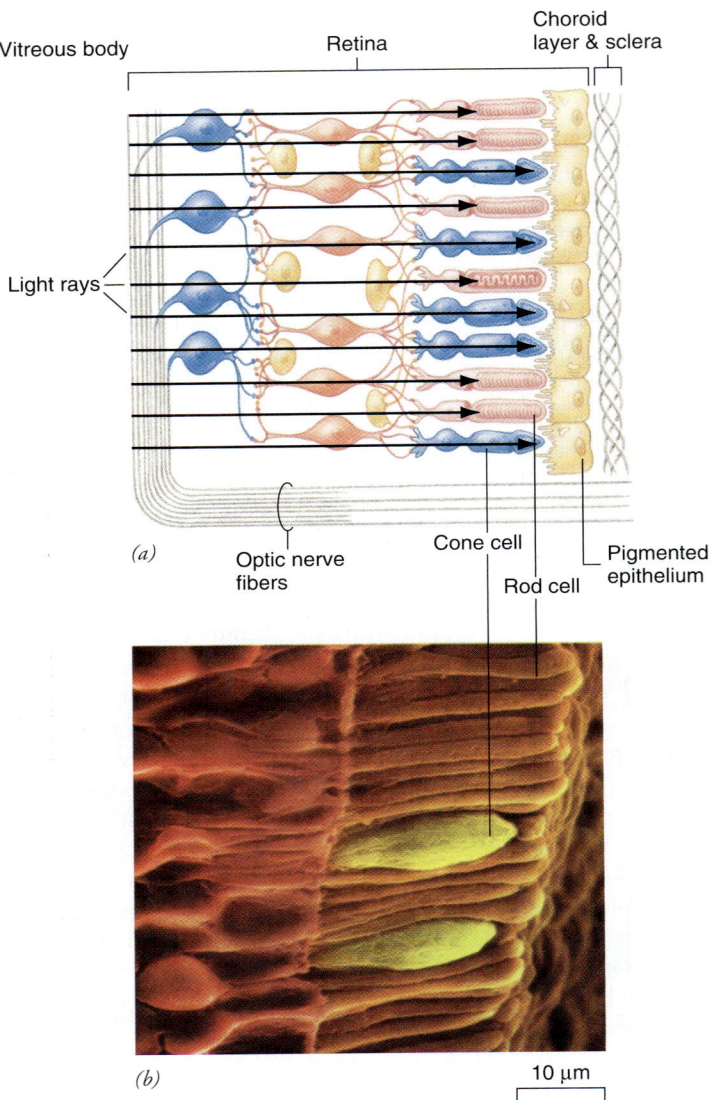

Vitreous body | Retina | Choroid layer & sclera

Light rays

Optic nerve fibers

Cone cell

Rod cell

Pigmented epithelium

(a)

(b)   10 μm

Figure 41–19 **Organization of the retina.** *(a)* The elaborate interconnections among the various layers of neurons in the retina allow them to interact and to influence one another. The rods and cones are the photoreceptor cells. *(b)* Rods *(red elongated structures)* and two cones *(shorter, thicker, yellow structures)* are seen in this SEM. The elongated rods permit us to see shape and movement, whereas the shorter cones allow us to view our world in color. *(c)* LM of the retina. *(b, Lennart Nilsson, from* The Incredible Machine, *p. 279; c, Manfred Kage/Peter Arnold, Inc.)*

Sclera

Choroid

Retinal pigment epithelium

Photoreceptor layer

Outer nuclear area

Inner nuclear layer

Ganglion cell layer

Retinal nerve fiber layer

Internal limiting membrane

(c)   50 μm

tamin A (see *Making the Connection: Molecules that Absorb Light* in Chapter 3). Two isomers of retinal exist: the *cis* form, which is folded and the *trans* form, which is straight.

When it is dark, retinal binds to opsin in the *cis* form. Cyclic GMP (cyclic guanosine monophosphate, a molecule similar to cyclic AMP) opens nonspecific channels that permit passage of Na⁺ and other cations into the rod cell (Fig. 41–21). This depolarizes the rod cell, and it releases glutamate, an inhibitory neurotransmitter. The glutamate hyperpolarizes the membrane of the bipolar cell so that it does not transmit messages. Note that the photoreceptor is different from other neurons in that the ion channels in its membrane are normally open; it is depolarized and continually releases inhibitory neurotransmitter. Another unusual characteristic of photoreceptors, and also bipolar cells, is that they do not produce action potentials. Their release of neurotransmitter is graded, regulated by the extent of depolarization.

When light strikes rhodopsin, it transforms *cis*-retinal to *trans*-retinal. This change in shape causes rhodopsin to change shape and to break down into its components, opsin and retinal. This is the light-dependent process in vision. Rhodopsin is part of a signal transduction pathway. When it changes shape it binds with a G protein called **transducin.** Transducin activates an enzyme that hydrolyzes cyclic GMP to GMP, thereby reducing the concentration of cyclic GMP. When the amount of cyclic GMP decreases, the Na⁺ channels begin to close. Fewer cations pass into the rod cell, and it becomes more negative, or hyperpolarized. The rod cell then releases less glutamate. Thus, light causes a decrease in neural signals from the rod cells.

The release of glutamate normally hyperpolarizes the membrane of the bipolar cell, so a decrease in glutamate release results in its depolarization. The depolarized bipolar cell increases its release of excitatory neurotransmitter, which typ-

Figure 41–20   **Neural pathway in the retina.** The photoreceptor cells are located in the back of the retina. They synapse with bipolar and horizontal cells. The bipolar cells synapse with ganglion cells and amacrine cells. Axons of ganglion cells make up the optic nerve.

(a) In dark, the rod cell is depolarized

(b) In light, the rod cell becomes hyperpolarized

Figure 41–21   **The biochemistry of vision.** (a) In the dark, sodium channels are open and the rod cell is depolarized. It releases the inhibitory neurotransmitter glutamate. (b) Light causes rhodopsin to change shape, activating a signal transduction pathway in which activation of a G protein leads to hyperpolarization. See text for further explanation.

ically stimulates the ganglion cell. (Some ganglion cells decrease their firing rate in response to stimulation.)

We have seen that photoreceptor cells synapse on bipolar cells, which synapse on ganglion cells. Ganglion cells can be inhibited by horizontal cells. Ganglion cells are activated or inhibited depending on which photoreceptor cells have been stimulated by light. Each ganglion cell has a receptive field, the area on which light must strike for the neuron to be stimulated.

Color vision depends on three different types of cones

Three different types of cones, commonly referred to as blue, green, and red cones, function in color vision. Each contains a slightly different photopigment. Although the retinal portion of the pigment molecule is the same as in rhodopsin, the opsin protein differs slightly in each type of photoreceptor. Each type of cone can respond to light within a considerable range of wavelengths but is named for the wavelength its pigment responds to most strongly. For example, red light can be absorbed by all three types of cones, but those cones most sensitive to red act as red receptors. By comparing the relative responses of the three types of cones, the brain can detect colors of intermediate wavelengths. Color blindness occurs when one or more of the three types of cones is absent. This is usually an inherited X-linked condition (see Chapter 10, Fig. 10–15).

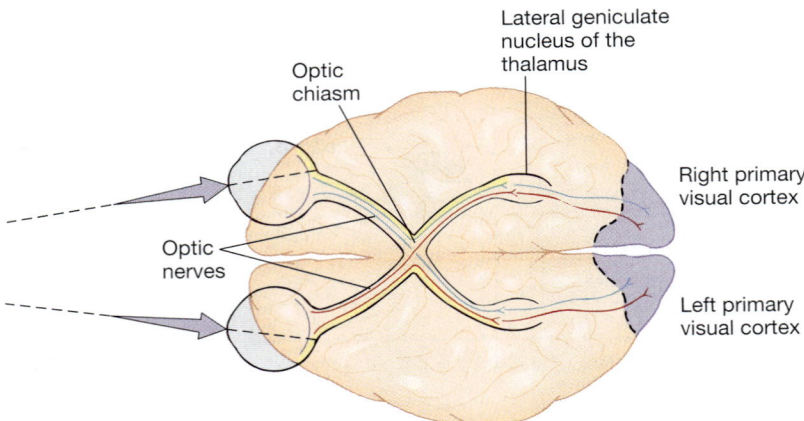

Figure 41–22   Neural pathway for transmission of visual information.

Optic chiasm

Lateral geniculate nucleus of the thalamus

Right primary visual cortex

Optic nerves

Left primary visual cortex

Integration of visual information begins in the retina

The size, intensity, and location of light stimuli determine initial processing in the retina. The pattern of neuron firing in the retina appears to be very important. Ganglion cells detect the characteristics of a visual image. Each ganglion cell transports information about specific types of visual stimuli such as color, brightness, or motion.

Axons of ganglion cells form the optic nerves that transmit information to the brain by way of complex, encoded signals. The optic nerves cross in the floor of the hypothalamus, forming an X-shaped structure, the **optic chiasm** (Fig. 41–22). Some of the axons of the optic nerves cross over and then extend to the opposite side of the brain. Axons of the optic nerves end in the **lateral geniculate nuclei** of the thalamus. From there, neurons convey information to the **primary visual cortex** in the occipital lobe of the cerebrum. The lateral geniculate nucleus controls which information is sent to the visual cortex. Neurons in the reticular activating system are

involved in this integration. Information can be transmitted from the primary visual cortex to other cortical areas for further processing.

Different types of ganglion cells receive information about different aspects of a visual stimulus. As a result, information from each point in the retina is transmitted in parallel by several different types of ganglion cells. These cells project to different kinds of neurons in the lateral geniculate nucleus and primary visual cortex.

Neurobiologists have not yet discovered all of the mechanisms by which the brain makes sense out of the visual information it receives. We do know that a large part of the association areas of the cerebrum are involved in integrating visual input. The neurons of the visual cortex are organized as a map of the external visual field. They are also organized into columns. Each column of neurons responds to light entering either the right or left eye. Columns are also organized by orientation of light. Neurons in a column are most stimulated by light with the same orientation but different locations in the retina.

SUMMARY WITH KEY TERMS

I. **Sensory receptors** are specialized to respond to specific energy stimuli in the environment.
   A. Sensory receptors may be neuron endings or specialized **receptor cells** in close contact with neurons.
   B. **Sense organs** are composed of sensory receptors and accessory cells.

II. **Exteroceptors** receive information from the outside world. **Proprioceptors,** sensory receptors within muscles, tendons, and joints, enable the animal to perceive orientation of the body and the positions of its parts. **Interoceptors,** sensory receptors within body organs, help maintain homeostasis.

III. Based on the types of stimuli to which they respond, sensory receptors also may be classified as **mechanoreceptors, chemoreceptors, photoreceptors, thermoreceptors,** or **electroreceptors.**

IV. Receptor cells absorb energy, transduce that energy into electrical energy, and produce **receptor potentials,** which are depolarizations, or hyperpolarizations of the membrane.

V. **Sensory adaptation** to a continuous stimulus results in decreased response to that stimulus.

A. Receptors that adapt slowly or not at all are **tonic receptors.**
B. Receptors that adapt rapidly are **phasic receptors.**

VI. Mechanoreceptors respond to touch, pressure, gravity, stretch, or movement.
   A. The **tactile receptors** in the skin respond to mechanical displacement of hairs or to displacement of the receptor cells themselves. For example, the **Pacinian corpuscle** responds to touch and pressure.
   B. **Statocysts** are gravity receptors found in many invertebrates.
   C. **Lateral line organs** supplement vision in fish and some amphibians by informing the animal of moving objects or objects in its path.
   D. **Muscle spindles, Golgi tendon organs,** and **joint receptors** continually respond to tension and movement in the muscles and joints.
   E. The vertebrate **inner ear** consists of a **labyrinth** of fluid-filled chambers and canals. The **vestibular apparatus** in the upper part of the labyrinth consists of the **saccule, utricle,** and **semicircular canals.**
      1. The saccule and utricle contain **otoliths** that change position when the head is tilted or when the body is moving in a straight

line. The hair cells stimulated by otoliths send impulses to the brain, enabling the animal to perceive the direction of gravity.

    2. The semicircular canals inform the brain about turning movements. The **cristae** within each **ampulla** are stimulated by movements of the **endolymph,** a fluid that fills each canal.

  F. In birds and mammals, the **organ of Corti** within the **cochlea** contains auditory receptors.

    1. Sound waves pass through the external auditory canal, cause the eardrum to vibrate, and are amplified and transmitted through the middle ear by the **malleus, incus,** and **stapes.**

    2. Vibrations pass through the **oval window** to fluid within the vestibular duct. Pressure waves press on the membranes that separate the three ducts of the cochlea.

    3. The bulging of the **round window** serves as an escape valve for the pressure. The pressure waves cause movements of the **basilar membrane.** These movements stimulate the hair cells of the organ of Corti by rubbing them against the overlying **tectorial membrane.**

    4. Nerve impulses are initiated in the dendrites of neurons that lie at the base of each hair cell, and neural impulses are transmitted by the cochlear nerve to the brain.

VII. Chemoreceptors include receptors for taste and smell.

  A. Taste receptors are specialized epithelial cells in **taste buds.**

  B. The **olfactory epithelium** contains specialized olfactory cells with axons that extend to the brain as fibers of the olfactory nerves. When a molecule binds with a receptor on an olfactory receptor cell, a signal transduction process involving a G protein is initiated, leading to depolarization.

VIII. Thermoreceptors are important in endothermic animals because they provide cues about body temperature. In some invertebrates they are used to locate an endothermic host.

  IX. Electroreceptors detect electrical currents in water and are used by predatory fishes to detect prey.

  X. Photoreceptors in **eyespots,** or **ocelli,** detect light, but do not form images effectively. Effective image formation and interpretation are called vision.

  A. The **compound eye** of insects and crustaceans consists of visual units called **ommatidia,** which collectively produce a mosaic image. Each ommatidium has a transparent **lens** and a crystalline cone that focus light onto receptor cells, called retinular cells**.**

  B. In the human eye, light enters through the cornea, is focused by the lens, and produces an image on the **retina.** The **iris** regulates the amount of light that can enter.

  C. The retina contains the photoreceptor cells: **rods** that function in dim light and form images in black and white; and **cones** that function in bright light and permit color vision.

  D. In addition to the rods and cones, the retina contains **bipolar cells, ganglion cells, horizontal cells,** and **amacrine cells.** Axons of the ganglion cells make up the **optic nerve.** Horizontal and amacrine cells are interneurons that integrate information in the retina.

  E. When light strikes **rhodopsin** in the rod cells, the retinal portion of the rhodopsin molecule changes shape and initiates a signal transduction process that involves a G protein. The membrane becomes hyperpolarized and less inhibitory neurotransmitter is released. Bipolar and ganglion cells are affected.

  F. The optic nerves transmit information to the **lateral geniculate nuclei** in the thalamus. From there neurons project to the **primary visual cortex** and then to integration centers in the cerebral cortex.

POST-TEST

1. A sensory receptor (a) detects a stimulus in the environment (b) converts energy of the stimulus into chemical energy (c) hyperpolarizes in response to neurotransmitter release (d) answers a, b, and c are correct (e) answers a and c only

2. Which sequence most accurately describes how a sensory receptor works? (a) stimulus→transduction into chemical energy→release of neurotransmitter→action potential (b) stimulus→transduction into electrical energy→receptor potential→action potential (c) conversion of a stimulus into a receptor potential→sensory adaptation→action potential (d) receptor potential→neurotransmitter release→action potential (e) stimulus→transduction into electrical energy→action potential→sensory adaptation

3. Interoceptors (a) help maintain homeostasis (b) are found in muscles and joints (c) include photoreceptors (d) answers a, b, and c are correct (e) answers b and c only

4. A sensory receptor that rapidly decreases its response to a stimulus is an example of a (a) sensory adaptation (b) receptor potential (c) tonic receptor (d) answers a, b, and c are correct (e) answers a and c only

5. A Pacinian corpuscle (a) is a mechanoreceptor (b) is a tactile receptor (c) responds to pressure (d) answers a, b, and c are correct (e) answers a and c only

6. Which of the following are located in the inner ear? (a) vestibular apparatus (b) cochlea (c) malleus, incus, and stapes (d) answers a, b, and c are correct (e) answers a and b only

7. The auditory receptors are (a) cochlea (b) organ of Corti (c) rods and cones (d) answers a, b, and c are correct (e) answers a and b only

8. Which of the following are associated with informing the brain about turning movements? (a) ampulla (b) saccule (c) endolymph (d) answers a, b, and c are correct (e) answers a and c only

9. Which sequence most accurately describes events involved in hearing? (a) tympanic membrane vibrates→bones in middle ear amplify and transmit vibrations→vibrations conducted to ampullae→cochlear nerve transmits impulses to organ of Corti (b) external auditory canal vibrates→bones in middle ear amplify and conduct vibrations→vibrations conducted to basilar membrane→organ of Corti vibrates (c) tympanic membrane vibrates→bones in middle ear amplify and conduct vibrations→vibrations conducted to basilar membrane→cochlear nerve transmits impulses to organ of Corti (d) tympanic membrane vibrates→bones in middle ear amplify and transmit vibrations→vibrations conducted to basilar membrane→hair cells in organ of Corti stimulated→cochlear nerve transmits impulses to brain (e) tympanic membrane vibrates→vibrations conducted to tectorial membrane→cochlear nerve transmits impulses to organ of Corti

10. Which sequence most accurately describes the visual pathway? (a) cornea→rod→ganglion cell→bipolar cell→optic nerve (b) cornea→lens→ganglion cell→rod→optic nerve (c) cornea→rod→bipolar cell→ganglion cell→optic nerve (d) cornea→lens→rod→horizontal cell→amacrine cell→optic nerve (e) cornea→rod→lens→bipolar cell→optic nerve

11. Cones (a) are most concentrated in the ganglion area (b) are more numerous than the rods (c) are responsible for bright-light vision (d) answers a, b, and c are correct (e) answers b and c only

12. Rhodopsin (a) is concentrated in the bipolar cells (b) binds with a G protein (c) changes shape when the membrane of the cone is depolarized (d) answers a, b, and c are correct (e) answers b and c only

13. The lateral geniculate nuclei are most closely associated with (a) taste (b) olfaction (c) hearing (d) vision (e) pain

REVIEW QUESTIONS

1. Contrast mechanoreceptors with chemoreceptors.
2. Which sensory receptors permit us to perform actions such as getting dressed or finding our way into bed in the dark? Explain how they work.
3. What is the physiological basis of seasickness?
4. Draw a sequence diagram summarizing the process of hearing.
5. Discuss the mechanism by which the sensory cells of the ear are stimulated by sound waves.
6. What are otoliths, and what is their role in maintaining equilibrium?
7. Draw a diagram of the human retina, labeling all parts. Diagram the neural pathway in the retina.
8. What is the function of rhodopsin? Describe the signal transduction pathway triggered by rhodopsin.
9. How does the human eye adjust to near and far vision?
10. Contrast the function of the insect's compound eye with that of the vertebrate eye.
11. Explain the statement "Vision happens mainly in the brain."
12. Label the diagrams. Use Figures 41–8 and 41–16 to check your answers.

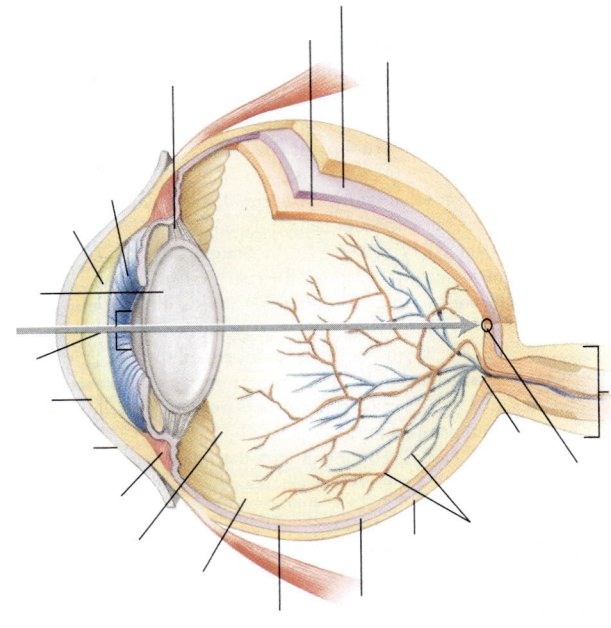

YOU MAKE THE CONNECTION

1. If all neurons transmit the same type of message, how do we know the difference between sound and light? How are we able to distinguish between an intense pain and a mild one? How are these discriminations adaptive?

2. Connoisseurs can recognize many varieties of cheese or wine by "tasting." How can this be, when there are only four types of taste receptors? How might it be adaptive for organisms to have only four or five basic tastes?

RECOMMENDED READINGS

Axel, R. "The Molecular Logic of Smell." *Scientific American,* Vol. 173, No. 4, Oct. 1995. An overview of the reception and integration of olfactory information.

Berne, R.M., and Levy, M.N. *Principles of Physiology,* 2nd ed.. Mosby, St. Louis, 1996. Chapters 7 and 8 are well written introductions to the general sensory system and special senses.

DISCOVER, Vol. 14, No. 6, Jun. 1993. Special Issue: "The Mystery of Sense." Several articles discussing vision, hearing, touch, smell, and taste.

Fackelmann, K. "The Bitter Truth: Do Some People Inherit a Distaste for Broccoli?" *Science News,* Vol. 152, Jul. 1997. Twenty-five percent of the U.S. population are supertasters who are especially sensitive to the bitter taste of certain vegetables.

Kirchner, W.H., and W.F. Towne. "The Sensory Basis of the Honeybee's Dance Language." *Scientific American,* Vol. 270, No. 6, Jun. 1994. Bees can hear the sounds produced by the dances they use to communicate.

Long, M.E. "The Sense of Sight." *National Geographic,* Vol. 182, No. 5, Nov. 1992. A beautifully illustrated discussion of vision and new technology for restoring sight.

Matthews, G.G. *Neurobiology: Molecules, Cells, and Systems.* Blackwell Scientific, Malden, MA, 1998. In Part IV the author very clearly discusses the sensory systems in some detail.

Rhoades, R., and Pflanzer, R. *Human Physiology,* 3rd ed.. Saunders College Publishing, Philadelphia, 1996. Chapter 8 is an excellent overview of sensory systems.

Travis, J.T. "Eye-Opening Gene: How Many Times Did Eyes Arise?" *Science News,* Vol. 151, May 1997. An interesting discussion of the controversy surrounding the evolution of eyes.

● Visit our website at **http://www.saunderscollege.com/lifesci/titles.html** and click on Solomon/Berg/Martin Biology for links to chapter-related resources on the World Wide Web.

CHAPTER 42

Internal Transport

Most cells require a continuous supply of nutrients and oxygen and removal of waste products. In very small animals, these metabolic needs can be met by simple **diffusion,** the net movement of particles from a region of higher concentration to a region of lower concentration, resulting from random motion. A molecule can diffuse 1 micrometer (μm) in less than 1 millisecond (msec), so over microscopic distances, diffusion is adequate. In invertebrates that are only a few cells thick, diffusion is an effective mechanism for distributing materials to and from their cells. No specialized circulatory structures are present in sponges, cnidarians, ctenophores, flatworms, or nematodes. As in all animals, the fluid between the cells, called **interstitial fluid,** or tissue fluid, bathes the cells and provides a medium for diffusion of oxygen and nutrients.

The time that diffusion requires increases with the square of the distance over which diffusion occurs. A cell that is 10 μm away from its oxygen (or nutrient) supply can receive oxygen by diffusion in about 50 msec, but a cell that is one mm away from its oxygen supply would have to wait several minutes and could not survive if it had to depend on diffusion alone. In animals that are many cells thick, specialized **circulatory systems** transport oxygen, nutrients, hormones, and other materials to the interstitial fluid surrounding all of the cells and remove metabolic wastes. A circulatory system reduces the diffusion distance that needed materials must travel. Typically, a circulatory system interacts with every other organ system in the body.

A circulatory system typically consists of the following:

1. Blood, a connective tissue consisting of cells and cell fragments dispersed in fluid known as plasma
2. A pumping organ, generally a heart
3. A system of blood vessels or spaces through which the blood circulates

The human circulatory system, also known as the **cardiovascular system,** is the focus of extensive research because cardiovascular disease is the number one cause of death in the United States and most other industrial societies. The PET scan shown here visualizes the left ventricle of the human heart. The scan measures metabolic activity based on glucose levels in the cardiac muscle tissue, providing information about the condition of the heart. (The color progression from least to most active is blue, green, yellow, red.)

(Ted Horowitz/The Stock Market)

A major risk factor for cardiovascular disease is elevated levels of cholesterol and low density lipoprotein (LDL) in the blood. In contrast, high density lipoprotein (HDL) appears to play a protective role, by removing excess cholesterol from the blood and tissues. Cells in the liver and certain other organs somehow bind with the HDL, remove the cholesterol, and use it in synthesizing needed compounds.

In 1997 molecular biologist Monty Krieger of the Massachusetts Institute of Technology and his research team reported that they had identified a gene in mice that codes for an HDL receptor. When these investigators knocked out the mouse gene, cholesterol levels more than doubled. Discovering receptors and other mechanisms involved in lipid transport and metabolism may provide targets for drugs and other treatments for cardiovascular disease.

AFTER YOU HAVE STUDIED THIS CHAPTER YOU SHOULD BE ABLE TO

1. Compare and contrast internal transport in animals with no circulatory system, those with an open circulatory system, and those with a closed circulatory system.
2. Relate structural adaptations of the vertebrate circulatory system to each function it performs.
3. Compare the structure and functions of red blood cells, white blood cells, and platelets.
4. Summarize in proper sequence the events involved in blood clotting.
5. Compare the structure and function of different types of blood vessels, including arteries, arterioles, capillaries, and veins.
6. Trace the evolution of the vertebrate heart from fish to mammal.
7. Describe the structure and function of the human heart and

label a diagram of the heart. Include a description of cardiac muscle and of the heart's conduction system.
8. Trace the events of the cardiac cycle and relate normal heart sounds to the events of this cycle.
9. Define cardiac output, describe how it is regulated, and identify factors that affect it.
10. Identify factors that determine and regulate blood pressure and compare blood pressure in different types of blood vessels.
11. Trace a drop of blood through the pulmonary and systemic circulations, naming in sequence each structure through which it passes.
12. Identify the risk factors for cardiovascular disease, trace the progress of atherosclerosis, and summarize its possible complications, including angina pectoris and myocardial infarction.
13. Describe the structure and functions of the lymphatic system.

SOME INVERTEBRATES HAVE NO CIRCULATORY SYSTEM

As indicated in the chapter introduction, many small, aquatic invertebrates have no circulatory system. In cnidarians, the central gastrovascular cavity (as its name implies) serves as a circulatory organ as well as a digestive organ (Fig. 42–1). The animal's tentacles capture prey and deliver it through the mouth into the cavity, where digestion occurs. The digested nutrients then pass into the cells lining the cavity and through them to cells of the outer layer. As the animal stretches and contracts, movements of the body stir up the contents of the gastrovascular cavity and help distribute nutrients to all parts of the body.

The flattened body of the flatworm permits effective gas exchange by diffusion. Its branched intestine brings nutrients close to all of the cells. As in cnidarians, circulation is aided by contractions of the muscles of the body wall, which agitate the intestinal fluid and the tissue fluid. The branching excretory system of planarians provides for internal transport of metabolic wastes that are then expelled from the body.

Fluid in the body cavity of nematodes and other pseudocoelomate animals helps to circulate materials. Nutrients, oxygen, and wastes dissolve in this fluid and diffuse through it to and from the individual cells of the body. Body movements of the animal result in movement of the fluid, facilitating distribution of these materials to all parts of the body.

MANY INVERTEBRATES HAVE AN OPEN CIRCULATORY SYSTEM

Arthropods and most mollusks have an **open circulatory system,** in which the heart pumps blood into vessels that have open ends. Blood and interstitial fluid are not distinguishable;

they are collectively referred to as **hemolymph.** Hemolymph spills out of the open ends of the blood vessels, filling large spaces, called sinuses, that make up the **hemocoel** (blood cavity) which is not part of the coelom. (In arthropods and mollusks the coelom is reduced.) The hemolymph bathes the cells of the body directly. Blood reenters the circulatory system through openings in the heart (in arthropods) or through open-ended vessels that lead to the gills (in mollusks).

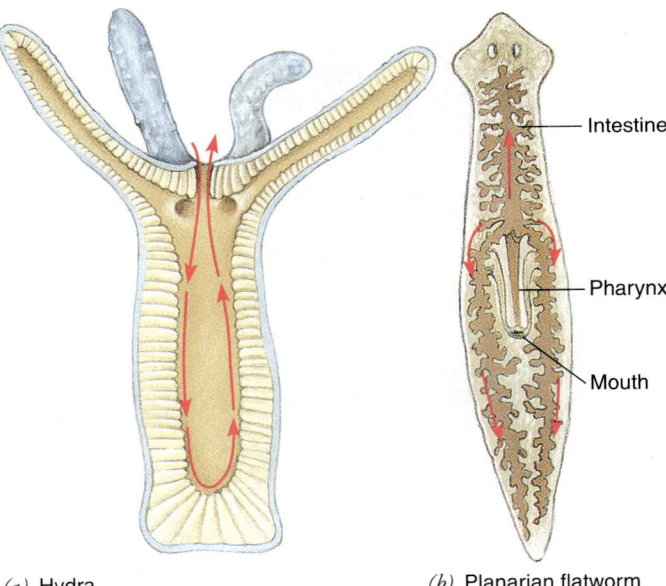

(a) Hydra   *(b)* Planarian flatworm

Figure 42–1 **Invertebrates with no circulatory system.** *(a)* In *Hydra* and other cnidarians, oxygen and nutrients circulate through the gastrovascular cavity and come in contact with the inner layer of body cells. These materials can diffuse the short distance to the outer layer of cells. *(b)* In planarian flatworms, the branched intestine circulates nutrients and oxygen within close proximity to the body cells.

In the open circulatory system of most mollusks, the heart typically consists of three chambers: two atria and a ventricle. The atria receive hemolymph from the gills. The ventricle pumps oxygen-rich hemolymph to the tissues. Blood vessels conducting hemolymph from the heart open into large sinuses, enabling the hemolymph to bathe the body cells. From the hemocoel, hemolymph passes into vessels that lead to the gills. There it is recharged with oxygen and passes into blood vessels that return it to the heart.

Some mollusks, as well as arthropods, have a hemolymph pigment, **hemocyanin,** that contains copper. The copper ions bind with oxygen. When oxygenated, hemocyanin is blue and imparts a bluish color to the hemolymph of these animals (the original blue bloods!).

In arthropods, a tubular heart pumps hemolymph into blood vessels (arteries) that deliver it to the sinuses of the hemocoel (Fig. 42–2). Hemolymph then circulates through the hemocoel, eventually returning to the pericardial cavity surrounding the heart. Hemolymph enters the heart through tiny openings that are equipped with valves to prevent backflow. Some insects have accessory "hearts," modified blood vessels that help pump hemolymph through the extremities, particularly the wings. The rate of hemolymph circulation increases when the insect moves. Thus, when an animal is active and most in need of nutrients for fuel, its own movement ensures effective circulation. An open circulatory system cannot provide enough oxygen to maintain the active lifestyle of insects. Indeed, insect hemolymph mainly distributes nutrients and hormones. Oxygen diffuses directly to the cells through a system of air tubes (tracheae) that make up the respiratory system.

SOME INVERTEBRATES HAVE A CLOSED CIRCULATORY SYSTEM

Annelids, some mollusks (cephalopods), echinoderms, and vertebrates have a **closed circulatory system.** In them, blood flows through a continuous circuit of blood vessels. The walls of the smallest blood vessels, the capillaries, are thin enough to permit diffusion of gases, nutrients, and wastes between blood in the vessels and the interstitial fluid that bathes the cells.

A rudimentary closed circulatory system is found in the proboscis worms (phylum Nemertea). This system consists of a complete network of blood vessels but no heart. Blood flow depends on movements of the animal and on contractions in the walls of the large blood vessels.

Earthworms and other annelids have a complex, closed circulatory system (Fig. 42–3). Two main blood vessels extend lengthwise in the body. The ventral vessel conducts blood posteriorly, and the dorsal vessel conducts blood anteriorly. Dorsal and ventral vessels are connected by lateral vessels in every segment. Branches of the lateral vessels deliver blood to the surface, where it is oxygenated. In the anterior part of the worm, five pairs of contractile blood vessels (sometimes referred to as "hearts") connect dorsal and ventral vessels. Contractions of these paired vessels and of the dorsal vessel, as well

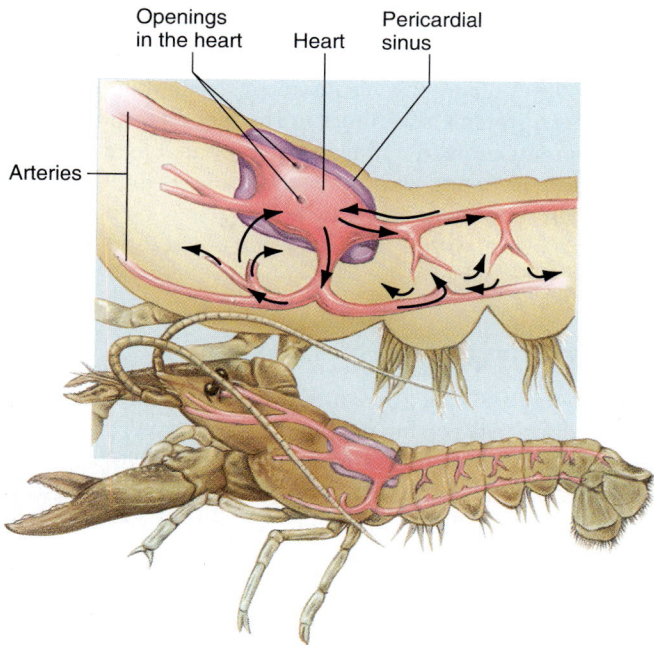

Figure 42–2 **Open circulatory system of the crayfish.** The heart pumps the blood into arteries, which end in sinuses of the hemocoel. Hemolymph circulates through the hemocoel and returns to the heart through openings in the heart wall.

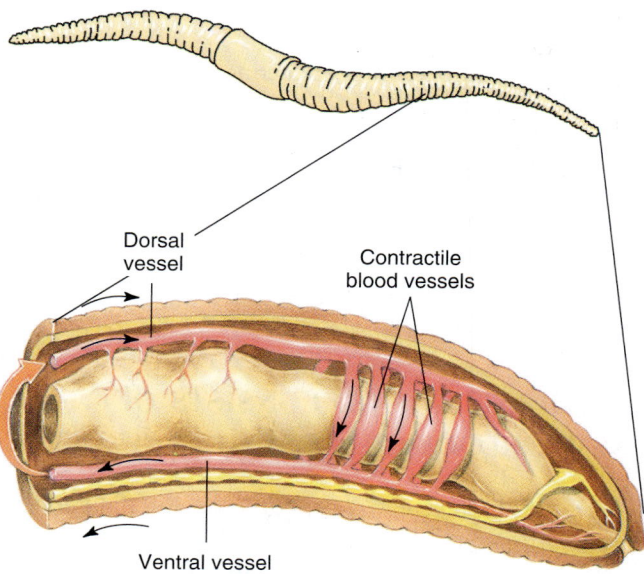

Figure 42–3 **Closed circulatory system of the earthworm.** Blood circulates through a continuous system of blood vessels. Five pairs of contractile blood vessels deliver blood from the dorsal vessel to the ventral vessel.

as contraction of the muscles of the body wall, circulate the blood. Earthworms have **hemoglobin,** the same red pigment that transports oxygen in vertebrate blood. However, their hemoglobin is not contained within red blood cells but is dissolved in the blood plasma.

Although other mollusks have an open circulatory system, the fast-moving cephalopods, such as the squid and octopus, require a more efficient means of internal transport. They have a closed system made even more effective by accessory "hearts" at the base of the gills, which speed the passage of blood through the gills.

THE CLOSED CIRCULATORY SYSTEM OF VERTEBRATES IS ADAPTED FOR SEVERAL FUNCTIONS

The circulatory system is basically similar in all vertebrates, from fishes, frogs, and reptiles to birds and mammals. All have a ventral, muscular heart that pumps blood into a closed system of blood vessels. The vertebrate circulatory system consists of heart, blood vessels, blood, lymph, lymph vessels, and associated organs such as the thymus, spleen, and liver. The tiniest blood vessels, **capillaries,** have very thin walls that permit exchange of materials between blood and interstitial fluid.

The vertebrate circulatory system performs several functions:

1. Transports nutrients from the digestive system and from storage depots to each cell of the body
2. Transports oxygen from respiratory structures (gills or lungs) to the cells of the body
3. Transports metabolic wastes from each cell to organs that excrete them
4. Transports hormones from endocrine glands to target tissues
5. Helps maintain fluid balance
6. Defends the body against invading microorganisms
7. Helps distribute metabolic heat within the body, which helps maintain a constant body temperature in endothermic animals
8. Helps maintain appropriate pH

VERTEBRATE BLOOD CONSISTS OF PLASMA, BLOOD CELLS, AND PLATELETS

In vertebrates, **blood** consists of a pale yellowish fluid, known as **plasma,** in which red blood cells, white blood cells, and platelets are suspended (Fig. 42–4; Table 42–1). In humans the total circulating blood volume is about 8% of the body weight—5.6 L (6 qt) in a 70-kg (154-lb) person. About 55% of the blood volume is plasma. The remaining 45% is made up of blood cells and platelets. Because cells and platelets are heavier than plasma, they can be separated from it by cen-

trifugation. Plasma does not separate from blood cells in the body because the blood is constantly mixed as it circulates in the blood vessels.

Plasma is the fluid component of blood

Plasma is composed of water (about 92%), proteins (about 7%), salts, and a variety of materials being transported, such as dissolved gases, nutrients, wastes, and hormones. Plasma is in dynamic equilibrium with the interstitial fluid bathing the cells and with the intracellular fluid. As blood passes through the capillaries, substances constantly move into and out of the plasma. Changes in its composition initiate responses on the part of one or more organs of the body to restore homeostasis.

Plasma contains several kinds of **plasma proteins,** each with specific properties and functions: fibrinogen; alpha, beta, and gamma globulins; and albumin. Fibrinogen is one of the proteins involved in the clotting process. When the proteins involved in blood clotting have been removed from the plasma, the remaining liquid is called **serum.** Alpha globulins include certain hormones and proteins that transport hormones; prothrombin, a protein involved in blood clotting; and high density lipoproteins (HDL) which transport fats and cholesterol (see Chapter 45). Beta globulins include other lipoproteins that transport fats and cholesterol, as well as proteins that transport certain vitamins and minerals. The gamma globulin fraction contains many types of antibodies that provide immunity to diseases such as measles and infectious hepatitis. Purified human gamma globulin is sometimes used to treat certain diseases or to reduce the possibility of contracting a disease.

Plasma proteins, especially albumins and globulins, help regulate the distribution of fluid between plasma and interstitial fluid. Too large to pass readily through the walls of blood vessels, these proteins contribute to the blood's osmotic pressure, important in maintaining an appropriate blood volume. Plasma proteins (along with the hemoglobin in the red blood cells) are also important acid-base buffers. They help keep the pH of the blood within a narrow range—near its normal, slightly alkaline pH of 7.4.

Red blood cells transport oxygen

Erythrocytes, also called **red blood cells (RBCs),** are highly specialized for transporting oxygen. In most vertebrates except mammals, circulating RBCs have nuclei. For example, birds have large, oval, nucleated RBCs. In mammals, the nucleus is ejected from the RBC as the cell develops. Each mammalian RBC is a flexible, biconcave disc, 7 to 8 μm in diameter and 1 to 2 μm thick. An internal elastic framework maintains the disc shape and permits the cell to bend and twist as it passes through blood vessels even smaller than its own diameter. Its biconcave shape provides a high ratio of surface area to volume, allowing efficient diffusion of oxygen and carbon dioxide into and out of the cell. In an adult human, about 30 trillion RBCs circulate in the blood, approximately 5 million per μL.

Figure 42–4 **Composition of vertebrate blood.** Blood consists of a fluid plasma in which white blood cells, red blood cells, and platelets are suspended.

10 to 14 µm
Neutrophil

15 to 20 µm
Monocyte

10 to 14 µm
Eosinophil

8 to 10 µm
Lymphocyte

10 to 14 µm
Basophil

Erythrocytes are produced within the red bone marrow of certain bones: vertebrae, ribs, breastbone, skull bones, and long bones. As an RBC develops, it produces great quantities of **hemoglobin,** the oxygen-transporting pigment that gives vertebrate blood its red color. (Oxygen transport is discussed in Chapter 44.) The life span of a human RBC is about 120 days. As blood circulates through the liver and spleen, phagocytic cells remove worn-out RBCs from the circulation. These RBCs are then disassembled, and some of their components are recycled. In the human body, more than 2.4 million RBCs are destroyed every second, so an equal number must be produced in the bone marrow to replace them. Red blood cell production is regulated by the hormone **erythropoietin,** which is released by the kidneys in response to a decrease in oxygen.

Anemia is a deficiency in hemoglobin (often accompanied by a decrease in the number of RBCs). When hemoglobin is insufficient, the amount of oxygen transported is inadequate to supply the body's needs. An anemic person may complain of feeling weak and may become easily fatigued. Three general causes of anemia are (1) loss of blood due to hemorrhage or internal bleeding; (2) decreased production of hemoglobin or red blood cells as in iron-deficiency anemia; and (3) increased rate of RBC destruction—the **hemolytic anemias,** like sickle cell anemia (see Chapter 15).

White blood cells defend the body against disease organisms

The **leukocytes,** or **white blood cells (WBCs),** are specialized to defend the body against harmful bacteria and other microorganisms. Leukocytes are amoeba-like cells capable of independent movement. Some types routinely slip through the walls of blood vessels and enter the tissues. Human blood contains five kinds of leukocytes, which may be classified as either granular or agranular (Fig. 42–4).

The **granular leukocytes** are manufactured in the red bone marrow. These cells are characterized by large lobed nuclei and distinctive granules in their cytoplasm. The three varieties of granular leukocytes are the neutrophils, eosinophils, and basophils. **Neutrophils,** the principal phagocytic cells in the blood, are especially adept at seeking out and ingesting bacteria. They also phagocytize dead cells, a clean-up task especially demanding after injury or infection. Most of the granules in neutrophils contain enzymes that digest ingested material.

Eosinophils have large granules that stain bright red with eosin, an acidic dye. The lysosomes of these WBCs contain enzymes such as oxidases and peroxidases, suggesting that they

TABLE 42-1 Cellular Components of Blood

	Normal Range	Function	Pathology
Red blood cells	Male: 4.2–5.4 million/μL Female: 3.6–5.0 million/μL	Oxygen transport; carbon dioxide transport	Too few: anemia Too many: polycythemia
Platelets	150,000–400,000/μL	Essential for clotting	Clotting malfunctions; bleeding; easy bruising
White blood cells (total)	5000–10,000/μL		
Neutrophils	(about 60% of WBCs)	Phagocytosis	Too many: may be due to bacterial infection, inflammation, leukemia (myelogenous)
Eosinophils	1–3% of WBCs	Play some role in allergic response	Too many: may result from allergic reaction, parasitic infestation
Basophils	1% of WBCs	May play role in prevention of clotting in body	
Lymphocytes	25–35% of WBCs	Produce antibodies; destroy foreign cells	Atypical lymphocytes present in infectious mononucleosis; too many may be due to leukemia (lymphocytic), certain viral infections
Monocytes	6% of WBCs	Differentiate to form macrophages	May increase in monocytic leukemia and fungal infections

function in detoxifying foreign proteins and other substances. Eosinophils increase in number during allergic reactions and during parasitic (e.g., tapeworm) infestations. **Basophils** exhibit deep blue granules when stained with basic dyes. Like eosinophils, these cells are thought to play a role in allergic reactions. Basophils do not contain lysosomes. Granules in their cytoplasm contain **histamine,** a substance that dilates blood vessels and makes capillaries more permeable. Basophils release histamine in injured tissues and in allergic responses. Other basophil granules contain the anticlotting chemical **heparin,** an anticoagulant that helps prevent blood from clotting inappropriately within the blood vessels.

Agranular leukocytes are manufactured in the red bone marrow. These cells lack large distinctive granules, and their nuclei are rounded or kidney-shaped. Two types of agranular leukocytes are lymphocytes and monocytes. Some **lymphocytes** are specialized to produce antibodies, whereas others attack foreign invaders such as bacteria or viruses directly. Just how they manage these feats is discussed in the next chapter.

Monocytes are the largest WBCs, reaching 20 μm in diameter. After circulating in the blood for about 24 hours, a monocyte leaves the circulation and completes its development in the tissues. The monocyte greatly enlarges and becomes a **macrophage,** a giant scavenger cell. All of the macrophages found in the tissues develop in this way. Macrophages voraciously engulf bacteria, dead cells, and debris.

In human blood there are normally about 7000 WBCs per μL of blood (only one for every 700 RBCs). During bac-

terial infections the number may rise sharply, so that a WBC count is a useful diagnostic tool. The proportion of each kind of WBC is determined by a differential WBC count. The normal distribution of leukocytes is indicated in Table 42–1.

Leukemia is a form of cancer in which any one of the various kinds of white cells multiplies rapidly within the bone marrow. Many of these cells do not mature, and their large numbers crowd out developing RBCs and platelets, leading to anemia and impaired clotting. A common cause of death from leukemia is internal hemorrhaging, especially in the brain. Another frequent cause of death is infection; although the white cell count may rise dramatically, the cells are immature and abnormal and cannot defend the body against disease organisms. *Acute leukemias* have a rapid onset and are characterized by large numbers of immature WBCs, whereas *chronic leukemias* have a more gradual onset and are marked by large numbers of mature WBCs. No cure for leukemia has been discovered, but radiation treatment and therapy with antimitotic drugs (chemotherapy) can induce partial or complete remissions lasting 15 years or longer in some patients.

Platelets function in blood clotting

In most vertebrates other than mammals, the blood contains small, oval cells called **thrombocytes,** which have nuclei. In mammals, thrombocytes are tiny spherical or disc-shaped bits of cytoplasm that lack nuclei. They are usually referred to as **platelets.** About 300,000 platelets per μL are present in hu-

① Injury to blood vessel

Blood flow

② Wall of vessel contracts

Blood flow decreases

③ Platelets adhere to collagen fibers of damaged vessel wall.

Platelet plug

Blood flow decreases

Figure 42–5 **Blood clotting.** Platelets and a variety of clotting factors are important in blood clotting. The SEM *(false-color)* of part of a blood clot shows red blood cells enmeshed in a network of fibrin. *(Lennart Nilsson, Boehringer Ingelheim International, GmbH)*

④ More permanent clot forms

Flow ceases

Damaged cells and platelets release substances that activate clotting factors.

Series of reactions involving clotting factors in the blood. → Prothrombin activator

Prothrombin → (Ca²⁺) → Thrombin

Fibrinogen → (Ca²⁺) → Fibrin threads (clot) (Red blood cells trapped among fibrin threads strengthen clot)

5 μm

man blood. Platelets are pinched off from very large cells (megakaryocytes) in the bone marrow. Thus, a platelet is not a whole cell but a fragment of cytoplasm enclosed by a membrane.

Platelets play an important role in blood clotting. When a blood vessel is cut, it constricts, reducing loss of blood. Platelets stick to the rough, cut edges of the vessel, physically patching the break in the wall. As platelets begin to gather, they release substances that attract other platelets. The platelets become sticky and adhere to collagen fibers in the blood vessel wall. Within about 5 minutes after injury, they form a platelet plug, or temporary clot.

At the same time that the temporary clot forms, a stronger, more permanent clot begins to develop. More than 30 different chemical substances interact in this very complex process. The series of reactions that leads to clotting is triggered when one of the clotting factors in the blood is activated by contact with the injured tissue. In **hemophilia,** one of the clotting factors is absent as a result of an inherited genetic mutation (see Chapter 15). The clotting process is summarized in Figure 42–5.

Prothrombin, a plasma protein manufactured in the liver, requires vitamin K for its production. In the presence of clotting factors, calcium ions, and compounds released from platelets, prothrombin is converted to **thrombin.** Then thrombin catalyzes the conversion of the soluble plasma protein **fibrinogen** to an insoluble protein, **fibrin.** Once formed, fibrin polymerizes, producing long threads that stick to the damaged surface of the blood vessel and form the webbing of the clot. These threads trap blood cells and platelets, which help strengthen the clot.

VERTEBRATES HAVE THREE MAIN TYPES OF BLOOD VESSELS

The vertebrate circulatory system includes three main types of blood vessels: arteries, capillaries, and veins (Fig. 42–6). An **artery** carries blood away from a heart chamber, toward other tissues. When an artery enters an organ, it divides into many smaller branches called **arterioles.** The arterioles deliver blood into the microscopic **capillaries.** After blood courses through an organ, capillaries eventually merge to form **veins** that transport it back toward the heart. Because this basic plan is similar in all vertebrates, much can be learned about the human circulatory system by dissecting an animal such as a shark or a frog. A blood vessel wall has three layers (Fig. 42–6*b*). The

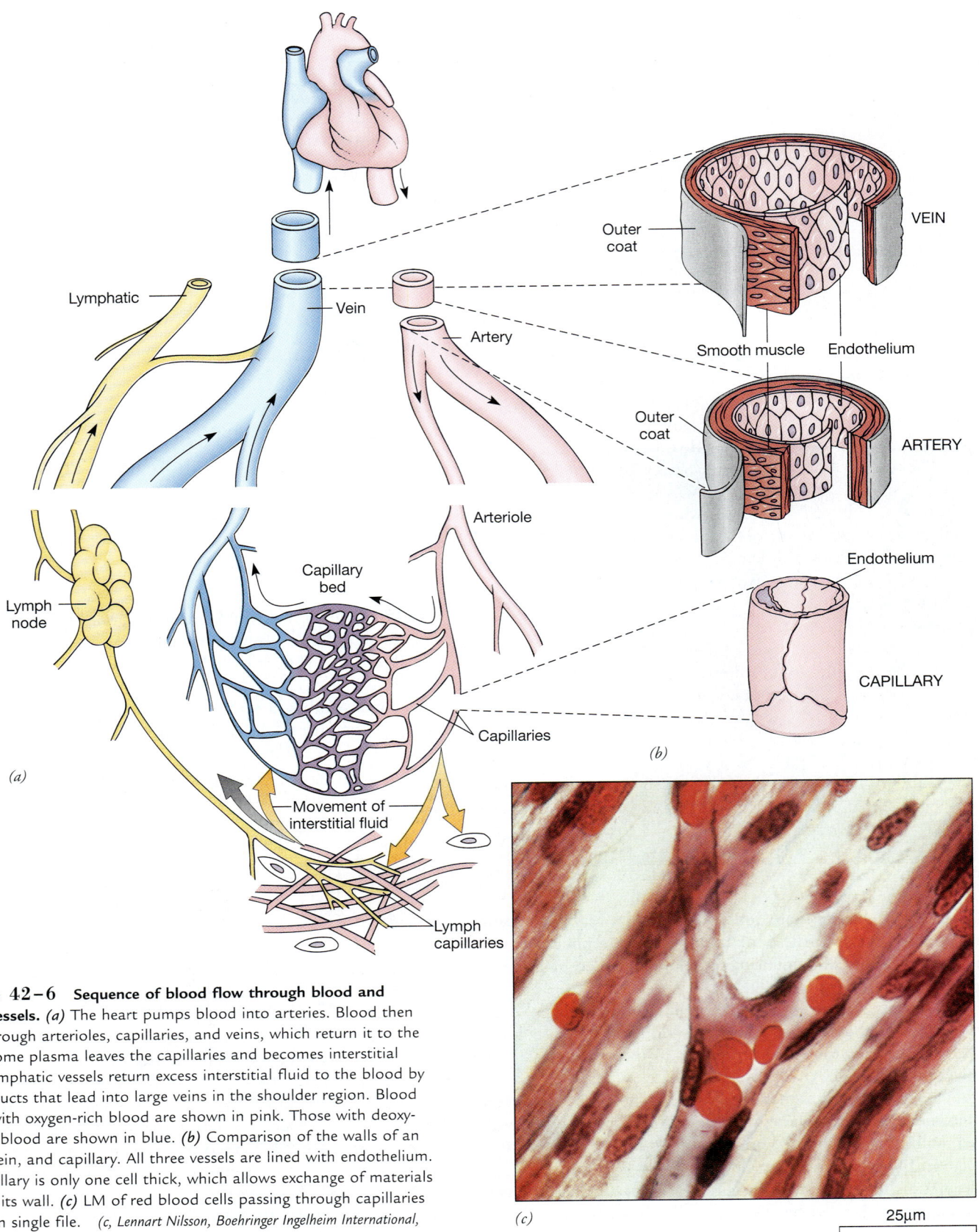

Figure 42–6 Sequence of blood flow through blood and lymph vessels. *(a)* The heart pumps blood into arteries. Blood then flows through arterioles, capillaries, and veins, which return it to the heart. Some plasma leaves the capillaries and becomes interstitial fluid. Lymphatic vessels return excess interstitial fluid to the blood by way of ducts that lead into large veins in the shoulder region. Blood vessels with oxygen-rich blood are shown in pink. Those with deoxygenated blood are shown in blue. *(b)* Comparison of the walls of an artery, vein, and capillary. All three vessels are lined with endothelium. The capillary is only one cell thick, which allows exchange of materials through its wall. *(c)* LM of red blood cells passing through capillaries almost in single file.   *(c, Lennart Nilsson, Boehringer Ingelheim International, GmbH)*

Labels in figure: Lymphatic · Vein · Artery · Lymph node · Arteriole · Capillary bed · Capillaries · Movement of interstitial fluid · Lymph capillaries · Outer coat · Smooth muscle · Endothelium · VEIN · ARTERY · Endothelium · CAPILLARY · 25μm

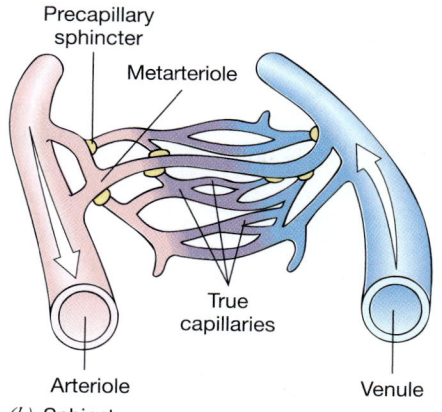

Precapillary sphincter
Metarteriole
True capillaries
Arteriole
Venule

(a) Sphincters closed

Precapillary sphincter
Metarteriole
True capillaries
Arteriole
Venule

(b) Sphincters open

Figure 42–7 **Blood flow through a capillary network.** As a tissue becomes active, the pattern of blood flow through its capillary networks changes. *(a)* When a tissue is inactive, only its metarterioles are open. *(b)* When the tissue becomes active, decreased oxygen tension in the tissue brings about relaxation of the precapillary sphincters, and the capillaries open. This increases the blood supply and the delivery of oxygen to the active tissue.

innermost layer, which lines the blood vessel, consists mainly of **endothelium,** a tissue that resembles squamous epithelium (Chapter 37). The middle layer consists of connective tissue and smooth muscle cells, and the outer coat is composed of connective tissue rich in elastic and collagen fibers.

The thick walls of arteries and veins prevent gases and nutrients from passing through. Materials are exchanged between the blood and interstitial fluid bathing the cells through the capillary walls, which are only one cell thick. Capillary networks in the body are so extensive that at least one of these tiny vessels is located close to every cell in the body. The total length of all capillaries in the body has been estimated to be more than 60,000 miles!

Smooth muscle in the arteriole wall can constrict (**vasoconstriction**) or relax (**vasodilation**), changing the radius of the arteriole. Such changes help maintain appropriate blood pressure and can help control the volume of blood passing to a particular tissue. Changes in blood flow are regulated by the nervous system in response to the metabolic needs of the tissue, as well as by the demands of the body as a whole. For example, when a tissue is metabolizing rapidly, it needs a greater supply of nutrients and oxygen. During exercise, arterioles within the muscles dilate, increasing by more than tenfold the amount of blood flowing to the muscle cells.

If all the blood vessels were dilated at the same time, there would not be sufficient blood to fill them completely. Normally the liver, kidneys, and brain receive the lion's share of blood. However, if an emergency suddenly occurred requiring rapid action, the blood would be rerouted quickly in favor of the heart and muscles. This would enable rapid, effective action. At such a time the digestive system and kidneys can do with less blood, for they are not critical in responding to the crisis.

The small vessels that directly link arterioles with venules (small veins) are **metarterioles.** The so-called true capillaries branch off from the metarterioles and then rejoin them (Fig. 42–7). True capillaries also interconnect with one another. Wherever a capillary branches from a metarteriole, a smooth muscle cell called a precapillary sphincter is present. Precapillary sphincters open and close continuously, directing blood first to one and then to another section of tissue. These sphincters also (along with the smooth muscle in the walls of arter-

ies and arterioles) regulate the blood supply to each organ and its subdivisions.

EACH VERTEBRATE CLASS EXHIBITS DIFFERENT ADAPTATIONS OF THE HEART AND CIRCULATION

The vertebrate circulatory system became modified in the course of evolution as the site of gas exchange shifted from gills to lungs and as vertebrates became active, endothermic animals with higher metabolic rates. The vertebrate heart has one or two **atria** (sing., *atrium*), chambers that receive blood returning from the tissues, and one or two **ventricles** that pump blood into the arteries (Fig. 42–8). Additional chambers are present in some classes.

Although it contains only one atrium and one ventricle, the fish heart technically has four chambers. A thin-walled *sinus venosus* receives blood returning from the tissues and pumps it into the atrium. The atrium then pumps blood into the ventricle. Next, the ventricle pumps blood into an elastic *conus arteriosus*, which does not contract.

In fishes, blood flows through a single circuit. Blood returning to the heart is poor in oxygen. It is oxygenated as it passes through capillaries in the gills. From the gills, blood circulates to other organs of the body. After blood circulates through capillaries in the gills its pressure is low, so blood leaving the gills passes very slowly to the other organs. Circulation is helped by the movements of the fish while it is swimming. This single-circuit, low-pressure circulatory system permits only a low metabolic rate in fishes.

In amphibians, blood flows through a double circuit: the **pulmonary circulation** and the **systemic circulation.** Oxygen-rich and oxygen-poor blood are kept somewhat separate. The amphibian heart has two atria and one ventricle. A sinus venosus collects oxygen-poor blood returning from the veins and pumps it into the right atrium. Blood returning from the lungs passes directly into the left atrium. Both atria pump into the single ventricle, but oxygen-poor blood is pumped out of the ventricle before oxygen-rich blood enters it. Blood passes into an artery (the conus arteriosus) equipped with a fold that helps keep the blood separate. Much of the oxygen-poor blood

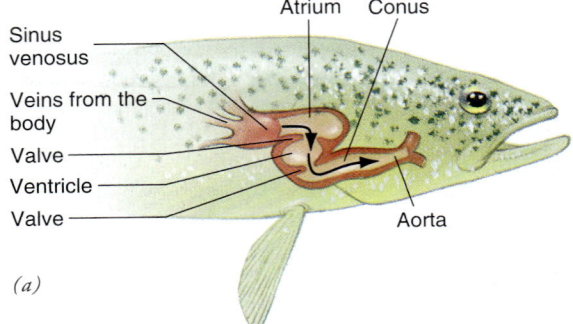

Atrium Conus

Sinus venosus

Veins from the body

Valve

Ventricle

Valve

Aorta

(a)

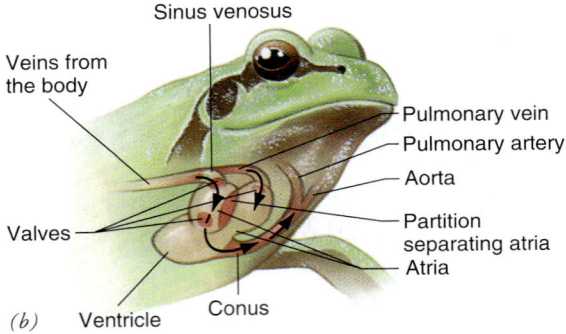

Sinus venosus

Veins from the body

Pulmonary vein

Pulmonary artery

Aorta

Partition separating atria

Atria

Valves

Conus

Ventricle

(b)

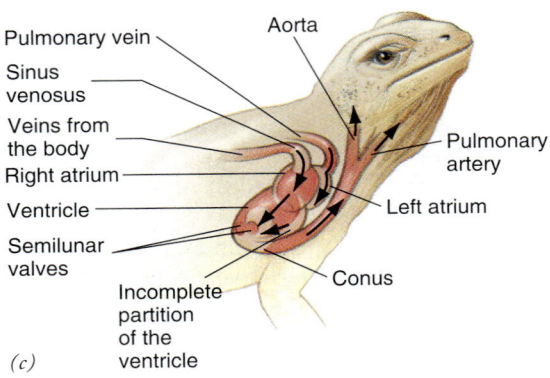

Pulmonary vein

Aorta

Sinus venosus

Veins from the body

Right atrium

Ventricle

Semilunar valves

Incomplete partition of the ventricle

Pulmonary artery

Left atrium

Conus

(c)

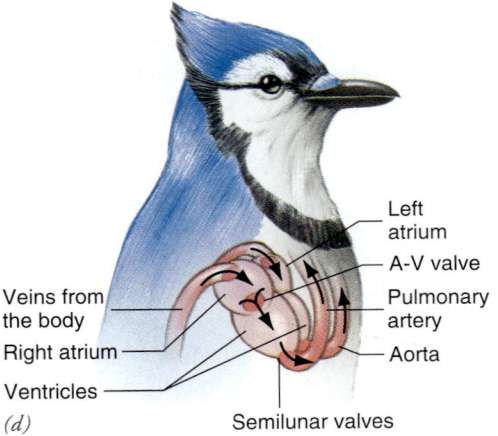

Left atrium

A-V valve

Veins from the body

Pulmonary artery

Right atrium

Aorta

Ventricles

Semilunar valves

(d)

Figure 42–8 Evolution of the vertebrate heart. Through evolution, the heart has become well adapted for the way of life of each class of vertebrates. *(a)* The single atrium and ventricle of the fish heart is part of a single circuit of blood flow. *(b)* In amphibians blood flows through a double circuit, and the heart consists of two atria and one ventricle. *(c)* The reptilian heart has two atria and two ventricles, but in all but the crocodiles and alligators, the wall separating the ventricles is incomplete so that blood from the right and left chambers mixes to some extent. *(d)* Birds and mammals have two atria and two ventricles, and blood rich in oxygen is kept completely separate from oxygen-poor blood.

is directed into the pulmonary circulation, which delivers it to the lungs and skin where it is recharged with oxygen. Oxygen-rich blood is delivered by the systemic circulation into arteries that conduct it to the various tissues of the body.

Most reptiles also have a double circuit of blood flow, made more efficient by a wall that partly divides the ventricle. Although some mixing of oxygen-rich and oxygen-poor blood occurs, it is minimized by the timing of contractions of the left and right sides of the heart and by pressure differences. In crocodiles, the wall between the ventricles is complete, so the heart consists of two atria and two ventricles. Thus, a four-chambered heart is first seen among the reptiles.

Unlike birds and mammals, amphibians and reptiles do not ventilate their lungs continuously. Therefore, it would be inefficient to pump blood through the pulmonary circulation continuously. The shunts between the two sides of the heart allow the blood to be distributed to the lungs as needed.

In all birds and mammals, the wall between the ventricles is complete, preventing the mixing of oxygen-rich blood in the left chamber with oxygen-poor blood in the right chamber. The conus arteriosus has split and become the base of the **aorta** (the largest artery) and pulmonary artery. No sinus venosus is present as a separate chamber, but a vestige remains as the sinoatrial node (the pacemaker).

Complete separation of the right and left sides of the heart makes it necessary for blood to pass through the heart twice each time it tours the body. As a result, it is possible to maintain higher blood pressures, and materials are delivered to the tissues rapidly and efficiently. Because the blood of birds and mammals contains more oxygen per unit volume and circulates more rapidly than in other vertebrates, the tissues receive more oxygen. As a result, birds and mammals can maintain a higher metabolic rate and a constant, high body temperature even in cold surroundings.

The pattern of blood circulation in birds and mammals may be summarized as follows:

Veins (conduct blood from organs) → right atrium → right ventricle → pulmonary arteries → capillaries in the lungs → pulmonary veins → left atrium → left ventricle → aorta → arteries (conduct blood to organs) → arterioles → capillaries

THE HUMAN HEART IS WELL ADAPTED FOR PUMPING BLOOD

Not much bigger than a fist and weighing less than a pound, the human **heart** is a remarkable organ that beats about 2.5 billion times in an average lifetime, pumping about 300 million L (80 million gal) of blood (Fig. 42–9). To meet the body's

Left subclavian artery

Brachiocephalic veins

Left common carotid artery

Brachiocephalic artery

ARCH OF AORTA

Superior vena cava

Right pulmonary artery

Left pulmonary artery

ASCENDING AORTA

Right pulmonary veins

Left pulmonary veins

Right coronary artery

Pulmonary artery

Auricle of right atrium

Auricle of left atrium

RIGHT ATRIUM

Descending branch of left
coronary artery

RIGHT VENTRICLE

LEFT VENTRICLE

Inferior vena cava

DESCENDING AORTA

(a)

Aorta

Pulmonary arteries

Superior vena cava

Left pulmonary veins

Right pulmonary veins

Circumflex artery

RIGHT ATRIUM

Great cardiac vein

Inferior vena cava

LEFT ATRIUM

Coronary sinus

Right coronary artery

LEFT VENTRICLE

RIGHT VENTRICLE

(b)

Figure 42-9 **External view of the human heart.** (a) Anterior view. Note the coronary blood vessels that bring blood to and from the heart muscle itself. (b) Posterior view.

Figure 42–10 Section through the human heart. Identify the right and left atria which receive blood, and the right and left ventricles which pump blood into the arteries. Arrows indicate the pattern of blood flow.

Superior vena cava

Right pulmonary arteries

Pulmonary valve

Right atrium

Pulmonary veins

Tricuspid valve

Right ventricle

Inferior vena cava

Aorta

Left pulmonary arteries

Pulmonary artery

Pulmonary veins

Left atrium

Mitral valve

Aortic semilunar valve

Chordae tendineae ("heart strings")

Papillary muscles

Left ventricle

Aorta

changing needs, the heart can vary its output from 5 to more than 20 L of blood per minute.

The heart is a hollow, muscular organ located in the chest cavity directly under the breastbone. Enclosing it is a tough connective tissue sac, the **pericardium.** The inner surface of the pericardium and the outer surface of the heart are covered by a smooth layer of endothelium. Between these two surfaces is a small **pericardial cavity** filled with fluid, which reduces friction to a minimum as the heart beats.

The wall of the heart is composed mainly of cardiac muscle attached to a framework of collagen fibers. The right atrium and ventricle are separated from the left atrium and ventricle by a wall, or **septum.** Between the atria the wall is known as the **interatrial septum;** between the ventricles it is the **interventricular septum.** A shallow depression, the **fossa ovalis,** on the interatrial septum marks the place where an opening, the **foramen ovale,** was located in the fetal heart. In the fetus, the foramen ovale permits the blood to flow directly from right to left atrium so that very little passes to the nonfunctional lungs. At the upper surface of each atrium lies a small muscular pouch called an **auricle.**

To prevent blood from flowing backward, the heart is equipped with valves that close automatically (Fig. 42–10). The valve between the right atrium and right ventricle is called the **right atrioventricular (AV) valve,** or **tricuspid valve.** The **left AV valve** (between the left atrium and left ventricle) is referred to as the **mitral valve,** or **bicuspid valve.** The AV valves are held in place by stout cords, or "heart-strings," the **chordae tendineae.** These cords attach the valves to the papillary muscles that project from the walls of the ventricles.

When blood returning from the tissues fills the atria, blood pressure on the AV valves forces them to open into the ventricles, which then fill with blood. As the ventricles contract,

blood is forced back against the AV valves, pushing them closed. Contraction of the papillary muscles and tensing of the chordae tendineae prevent the valves from opening backward into the atria. These valves are like swinging doors that can open in only one direction.

Semilunar valves (named for their flaps, which are shaped like half-moons) guard the exits from the heart. The semilunar valve between the left ventricle and the aorta is the **aortic valve,** and the one between the right ventricle and the pulmonary artery is the **pulmonary valve.** When blood passes out of the ventricles, the flaps of the semilunar valves are pushed aside and offer no resistance to blood flow. But when the ventricles are relaxing and filling with blood from the atria, the blood pressure in the arteries is higher than that in the ventricles. Blood then fills the pouches of the valves, stretching them across the artery so that blood cannot flow back into the ventricle.

Valve deformities are sometimes present at birth, or they may result from certain diseases such as rheumatic fever or syphilis. Inflammation and scarring may narrow the blood passageway by thickening the valves. Sometimes erosion of the valve tissues prevents the flaps from closing tightly, causing blood to leak backward and reducing the efficiency of the heartbeat. Diseased or deformed valves can be surgically repaired or replaced with artificial valves.

Each heartbeat is initiated by a pacemaker

Horror films frequently feature a scene in which a heart cut out of a human body continues to beat. Scriptwriters of these tales actually have some factual basis for their gruesome fantasies, for when carefully removed from the body, the heart does continue to beat for many hours if kept in a nutritive, oxygenated fluid. This is possible because the contractions of cardiac muscle begin within the muscle itself and can occur independently of any nerve supply.

At their ends cardiac muscle cells are joined by dense bands called **intercalated discs** (Fig. 42–11 *b* and *c*). Each disc is a type of gap junction (see Chapter 5) in which two cells are connected through pores. This type of junction is of great physiological importance because it offers very little resistance to the passage of an action potential. Ions move easily through the gap junctions, allowing the entire atrial (or ventricular) muscle mass to contract as one giant cell.

Compared to skeletal muscle with action potentials that typically last 1 to 2 msec, the action potentials of cardiac muscle are much longer, several hundred milliseconds. Voltage-dependent calcium ion channels open during depolarization of cardiac muscle fibers. Entrance of the Ca^{2+} contributes to the longer depolarization time. Another contributing factor is the presence of a kind of potassium channel that stays open when the cell is at its resting potential but closes during depolarization. This decreases permeability of the membrane to K^+ ions, which also lengthens depolarization. From patch-clamp experiments on isolated cardiac muscle fibers, investigators have found that spontaneous contraction occurs as a result of the combination of a slow decrease in potassium permeability and a slow increase in sodium and calcium permeability.

A specialized conduction system ensures that the heart beats in a regular and effective rhythm. Each beat is initiated by the **pacemaker**, called the **sinoatrial (SA) node** (Fig. 42–11*a*). The SA node is a small mass of specialized cardiac muscle in the posterior wall of the right atrium near the opening of a large vein, the superior vena cava. The action potential in the SA node is triggered mainly by the opening of Ca^{2+} channels. Ends of the SA node fibers fuse with surrounding ordinary atrial muscle fibers so that each action potential spreads through both atria, producing atrial contraction.

One group of atrial muscle fibers conducts the action potential directly to the **atrioventricular (AV) node,** located in the right atrium along the lower part of the septum. Here transmission is delayed briefly, permitting the atria to complete their contraction before the ventricles begin to contract. From the AV node the action potential spreads into specialized muscle fibers called **Purkinje fibers.** These large fibers make up the **atrioventricular (AV) bundle.** The AV bundle divides, sending branches into each ventricle. When an impulse reaches the ends of the Purkinje fibers, it spreads through the ordinary cardiac muscle fibers of the ventricles.

Each minute the heart beats about 70 times. One complete heartbeat takes about 0.8 second and is referred to as a **cardiac cycle.** That portion of the cycle in which contraction occurs is known as **systole;** the period of relaxation is **diastole.** Figure 42–12 shows the sequence of events that occur during one cardiac cycle.

You can measure your heart rate by placing a finger over the radial artery in your wrist or the carotid artery in your neck

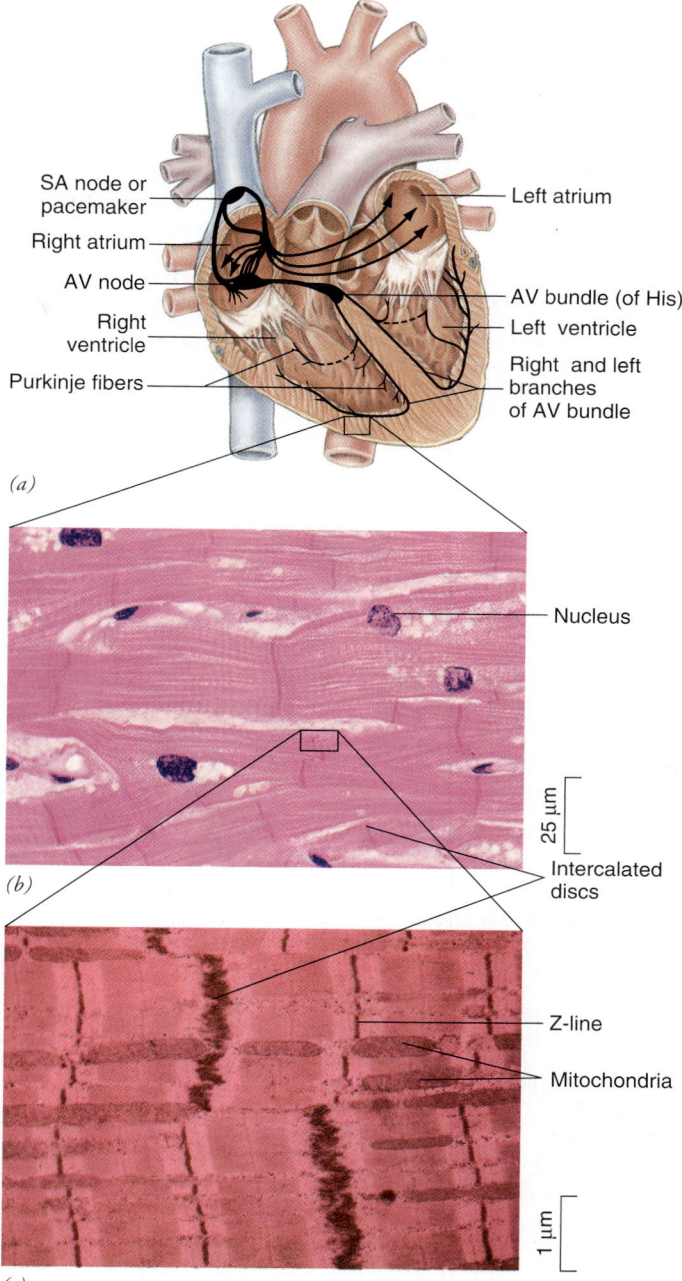

(a)

(b)

(c)

Figure 42–11 **Conduction system of the heart.** (*a*) The SA node initiates each heartbeat. The action potential spreads through the muscle fibers of the atria, producing atrial contraction. Transmission is briefly delayed at the AV node before the action potential spreads through specialized muscle fibers into the ventricles. (*b*) LM of cardiac muscle. (*c*) TEM of cardiac muscle. (*b, Ed Reschke; c, Don Fawcett/Visuals Unlimited*)

and counting the pulsations for one minute. Arterial pulse is the alternate expansion and recoil of an artery. Each time the left ventricle pumps blood into the aorta, the elastic wall of the aorta expands to accommodate the blood. This expansion moves in a wave down the aorta and the arteries that branch from it. When this pressure wave passes, the elastic arterial wall snaps back to its normal size.

Superior vena cava — Aorta — Pulmonary artery

Right atrium —
Tricuspid valve —
Inferior vena cava —

— Semilunar valves
— Pulmonary vein
— Left atrium
— Mitral valve

(a) Atrial systole: Atria contract, pushing blood through the open tricuspid and mitral valves into the ventricles. Semilunar valves are closed.

Right ventricle — Left ventricle

(e) Period of falling pressure: Blood flows from veins into the relaxed atria. Tricuspid and mitral valves open when pressure in the ventricles falls below that in the atria; blood then flows into the ventricles.

Heart sounds

(b) Beginning of ventricular systole: Ventricles contract, pressure within the ventricles increases and closes tricuspid and mitral valves, causing the first heart sound.

(d) Beginning of ventricular diastole: pressure in the relaxing ventricles drops below that in the arteries. Semilunar valves snap shut, causing the second heart sound.

(c) Period of rising pressure: Semilunar valves open when pressure in the ventricle exceeds that in the arteries. Blood spurts into the aorta and pulmonary artery.

Figure 42–12 The cardiac cycle. The cycle comprises contraction of both atria followed by both ventricles. Arrows indicate the direction of blood flow; dotted lines indicate the change in size as contraction occurs.

Two main heart sounds can be distinguished

When you listen to the heartbeat with a stethoscope you can hear two main heart sounds, "lub-dup," which repeat rhythmically. These sounds result from the closure of the heart valves. When the valves close they causes turbulence in blood flow that sets up vibrations in the walls of the heart chambers. The first heart sound, lub, is low-pitched, not very loud, and fairly long-lasting. It is caused mainly by the closing of the AV (mitral and tricuspid) valves and marks the beginning of ventricular systole. The lub sound is quickly followed by the higher-pitched, louder, sharper, and shorter dup sound. Heard almost as a quick snap, the dup marks the closing of the semilunar valves and the beginning of ventricular diastole.

The quality of these sounds tells a discerning physician much about the state of the valves. When the semilunar valves are injured, a soft hissing noise ("lub-shhh") is heard in place of the normal sound. This is known as a **heart murmur** and

may be caused by any condition that prevents the valves from closing tightly, permitting blood to flow backward into the ventricles during diastole.

The electrical activity of the heart can be recorded

As each wave of contraction spreads through the heart, electrical currents flow into the tissues surrounding the heart and onto the body surface. By placing electrodes on the body surface on opposite sides of the heart, the electrical activity can be amplified and recorded either by an oscilloscope or an electrocardiograph. The written record produced is called an **electrocardiogram** (**ECG** or **EKG**).

An ECG begins with a **P wave,** which is caused by the firing of the SA node and depolarization of the atrial muscle (Fig. 42–13). Then a **QRS complex** appears, reflecting the firing of the AV node and depolarization of the ventricles. The

T wave results from repolarization of the ventricles. The heart then repeats its pattern of electrical impulses, generating a new P wave, QRS complex, and T wave.

Abnormalities in the ECG indicate disorders in the heart or its rhythm. One class of disorders that can be diagnosed with the help of the ECG is **heart block.** In this condition, transmission of an impulse is delayed or blocked at some point in the conduction system. **Artificial pacemakers** can be implanted in patients with severe heart block. A pacemaker is implanted beneath the skin, and its electrodes are connected to the heart. This device provides continuous rhythmic impulses that avoid the block and drive the heartbeat.

Cardiac output varies with the body's need

The **cardiac output** (CO) is the volume of blood pumped by the left ventricle into the aorta in one minute. The volume of blood pumped by one ventricle during one beat is called the **stroke volume.** By multiplying the stroke volume by the number of times the left ventricle beats per minute, the cardiac output can be computed. For example, in a resting adult the heart may beat about 72 times per minute and pump about 70 mL of blood with each contraction.

$$CO = \text{stroke volume} \times \text{heart rate (number of ventricular contractions per min)}$$
$$= 70 \text{ mL/stroke} \times 72 \text{ strokes/min}$$
$$= 5040 \text{ mL/min (about 5 L/min)}$$

The cardiac output varies dramatically with the changing needs of the body. During stress or heavy exercise, the normal heart can increase its cardiac output fourfold to fivefold, so that 20 to 30 liters of blood can be pumped per minute. Cardiac output varies with changes in either stroke volume or heart rate.

Stroke volume depends on venous return

Stroke volume depends mainly on venous return, the amount of blood delivered to the heart by the veins. According to **Starling's law of the heart,** the greater the amount of blood delivered to the heart by the veins, the more blood the heart pumps (within physiological limits). When extra amounts of blood fill the heart chambers, the cardiac muscle fibers are stretched to a greater extent and contract with greater force, pumping a larger volume of blood into the arteries. This increase in stroke volume increases the cardiac output (Fig. 42–14).

The release of norepinephrine by sympathetic nerves also increases the force of contraction of the cardiac muscle fibers. Epinephrine released into the blood by the adrenal glands during stress has a similar effect on the heart muscle. When the force of contraction increases, the stroke volume increases, and this in turn increases cardiac output.

Heart rate is regulated by the nervous system

Although the heart is capable of beating independently, its rate is, in fact, carefully regulated by the nervous system and endocrine system (Fig. 42–14). Sensory receptors in the walls of certain blood vessels and heart chambers are sensitive to changes in blood pressure. When stimulated, they send messages to **cardiac centers** in the medulla of the brain. These cardiac centers maintain control over two sets of autonomic nerves that pass to the SA node. Parasympathetic and sympathetic nerves have opposite effects on heart rate (Chapter 40). Parasympathetic nerves release acetylcholine, which slows the heart. Acetylcholine slows the rate of depolarization by in-

(a)

(b)

Figure 42–13 Electrocardiograms. *(a)* Tracing from a normal heart. The P wave corresponds to the contraction of the atria, the QRS complex to the contraction of the ventricles, and the T wave to the relaxation of the ventricles. *(b)* Tracing from a patient with atrial fibrillation. The individual muscle fibers of the atria twitch rapidly and independently. There is no regular atrial contraction and no P wave. The ventricles beat independently and irregularly, causing the QRS wave to appear at irregular intervals. *(Courtesy of Dr. Lewis Dexter, Peter Bent Brigham Hospital, Boston, Mass.)*

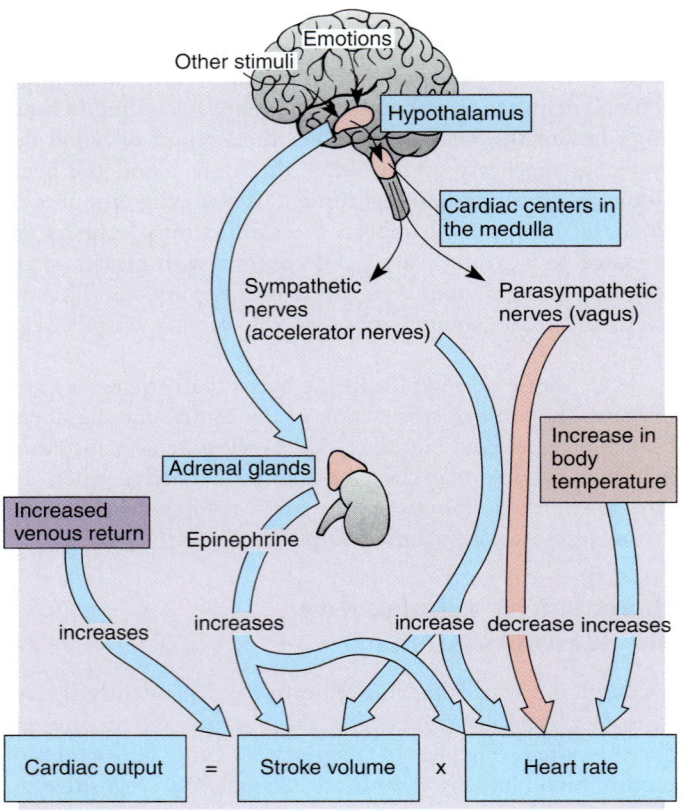

Figure 42-14 Some factors that influence cardiac output.

creasing the membrane's permeability to potassium (Fig. 42-15a). Sympathetic nerves release norepinephrine, which speeds the heart rate and increases the strength of contraction. Norepinephrine stimulates calcium ion channel opening during depolarization (Fig. 42-15b). Both norepinephrine and acetylcholine act indirectly on ion channels. They activate a signal transduction process involving a G protein. Norepinephrine binds to beta-adrenergic receptors. These receptors are targeted by *beta blockers,* drugs used clinically in the treatment of hypertension and other types of heart disease.

In response to physical and emotional stressors, the adrenal glands release epinephrine and norepinephrine, which speed the heart. An elevated body temperature also increases heart

▼ **Figure 42-15** **Actions of sympathetic and parasympathetic neurons on cardiac muscle cells.** Neurotransmitters released by sympathetic and parasympathetic neurons initiate a signal transduction process. **(a)** Parasympathetic neurons release acetylcholine (1), which binds with receptors on the plasma membrane of cardiac muscle (2). The receptor then activates a G protein (3), which can bind with a K^+ channel, causing the channel to open (4). K^+ leaves the cell (5). The membrane is hyperpolarized, and the rate of action potentials slows. **(b)** Sympathetic neurons release norepinephrine (1), which binds with receptors on the plasma membrane of cardiac muscle (2). The receptor then activates a G protein (3), which activates adenylyl cyclase (4). This enzyme converts ATP to cyclic AMP, which then activates a protein kinase (5). The protein kinase phosphorylates calcium ion channels so that they open more easily when the neuron is depolarized (6). Action potentials occur more rapidly.

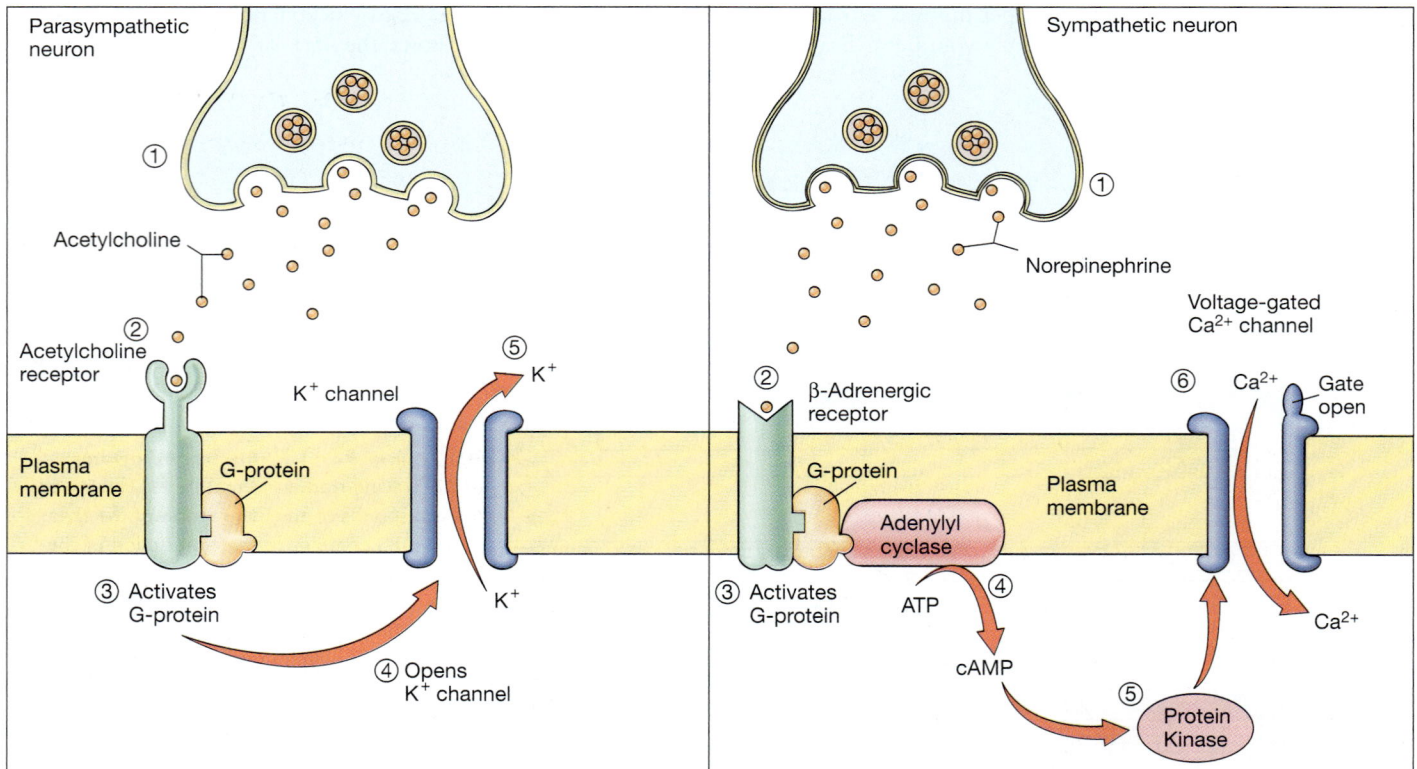

(a) Parasympathetic action on cardiac muscle

(b) Sympathetic action on cardiac muscle

rate. During fever, the heart may beat more than 100 times per minute. As you might expect, heart rate decreases when body temperature is lowered. This is why a patient's temperature may be deliberately lowered during heart surgery.

BLOOD PRESSURE DEPENDS ON BLOOD FLOW AND RESISTANCE TO BLOOD FLOW

Blood pressure is the force exerted by the blood against the inner walls of the blood vessels. It is determined by cardiac output, blood volume, and by the resistance to blood flow (Fig. 42–16a). When cardiac output increases, blood flow increases, causing a rise in blood pressure. When cardiac output decreases, blood flow decreases, causing a fall in blood pressure. The volume of blood flowing through the system also affects blood pressure. If blood volume is reduced by hemorrhage or by chronic bleeding, the blood pressure drops. On the other

hand, an increase in blood volume results in an increase in blood pressure. For example, a high dietary intake of salt causes water retention. This results in an increase of blood volume and leads to higher blood pressure.

Blood flow is impeded by resistance; when the resistance to flow increases, blood pressure rises. **Peripheral resistance** is the resistance to blood flow caused by viscosity of the blood and by friction between the blood and the wall of the blood vessel. In the blood of a healthy person, viscosity remains fairly constant and is only a minor factor influencing changes in blood pressure. More important is the friction between the blood and the wall of the blood vessel. The length and diameter of a blood vessel determine the amount of surface area in contact with the blood. The length of a blood vessel does not change, but the diameter, especially of an arteriole, does. A small change in the diameter of a blood vessel causes a big change in blood pressure.

Blood pressure in arteries rises during systole and falls during diastole. Normal blood pressure (measured in the upper arm) for a young male adult is about 120/80 mm of mercury, abbreviated mm Hg, as measured by the sphygmomanometer. Systolic pressure is indicated by the numerator, diastolic by the denominator.

When the diastolic pressure consistently measures more than 95 mm Hg, the patient may be suffering from high blood pressure, or **hypertension.** In hypertension, there is usually increased vascular resistance, especially in the arterioles and small arteries. The heart's workload increases because it must pump against this greater resistance. If this condition persists, the left ventricle enlarges and may begin to deteriorate in function. Heredity, aging, and obesity appear to be important in the development of hypertension.

Blood pressure is highest in arteries

As you might guess, blood pressure is greatest in the large arteries, decreasing as blood flows away from the heart and through the smaller arteries and capillaries (Fig. 42–16b). By the time blood enters the veins, its pressure is very low, even approaching zero. Flow rate can be maintained in veins at low pressure because they are low resistance vessels. Their diameter is larger than that of corresponding arteries, and there is little smooth muscle in their walls. Flow of blood through veins depends on several factors, including muscular movement, which compresses veins. Most veins larger than 2 mm (0.08 in) in diameter that conduct blood against the force of gravity are equipped with valves to prevent backflow (Fig. 42–17). Such valves usually consist of two cusps formed by inward extensions of the vein wall.

When a person stands perfectly still for a long time, as when a soldier stands at attention or a store clerk stands at a cash register, blood tends to pool in the veins, which when fully distended, can accept no more blood from the capillaries. Pressure in the capillaries increases, and large amounts of plasma are forced out of the circulation through the thin capillary walls. Within just a few minutes, as much as 20% of the blood volume can be lost from the circulation—with drastic

Figure 42–16 **Blood pressure.** *(a)* Blood pressure depends on blood flow and resistance to that flow. A variety of factors can affect blood flow and resistance. *(b)* Blood pressure in different types of blood vessels. Systolic and diastolic variations in arterial blood pressures are shown. Note that the venous pressure drops below zero (below atmospheric pressure) near the heart.

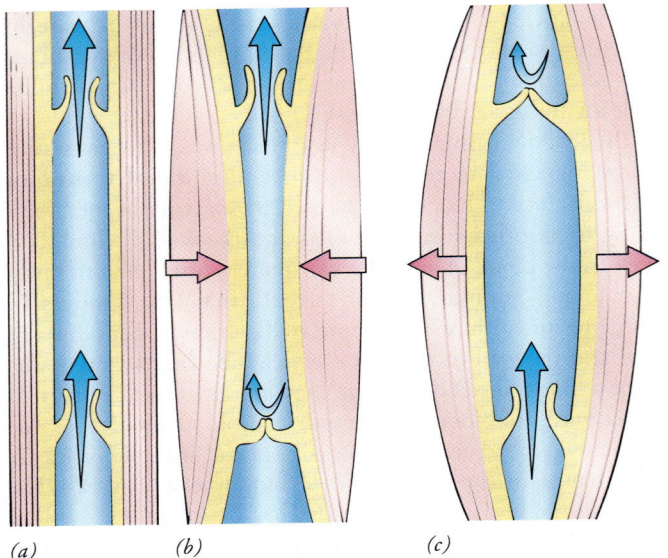

(a)   *(b)*   *(c)*

Figure 42–17 **Venous blood flow.** Contraction of skeletal muscles helps move blood through the veins. *(a)* Resting condition. *(b)* Muscles contract and bulge, compressing veins and forcing blood toward the heart. The lower valve prevents backflow. *(c)* Muscles relax, and the vein expands and fills with blood from below. The upper valve prevents backflow.

effect. Arterial blood pressure falls dramatically, reducing blood flow to the brain. Sometimes the resulting lack of oxygen in the brain causes fainting, a protective response. Lying in a prone position increases blood supply to the brain. In fact, lifting a person who has fainted to an upright position can result in circulatory shock and even death.

Blood pressure is carefully regulated

Each time you get up from a horizontal position, changes occur in your blood pressure. Several complex mechanisms interact to maintain normal blood pressure so that you do not faint when you get out of bed each morning or change position during the day. When blood pressure falls, sympathetic nerves to the blood vessels stimulate vasoconstriction so that pressure rises again.

The **baroreceptors** present in the walls of certain arteries and in the heart wall are sensitive to changes in blood pressure. When an increase in blood pressure stretches the baroreceptors, messages are sent to the cardiac and vasomotor centers in the medulla of the brain. The cardiac center stimulates parasympathetic nerves that slow the heart, lowering blood pressure. The vasomotor center inhibits sympathetic nerves that constrict arterioles; this action causes vasodilation, which lowers blood pressure. These neural reflexes continuously work in this complementary way to maintain blood pressure within normal limits.

Hormones are also involved in regulating blood pressure. In response to low blood pressure, the kidneys release **renin.** This enzyme stimulates formation of **angiotensins** from a

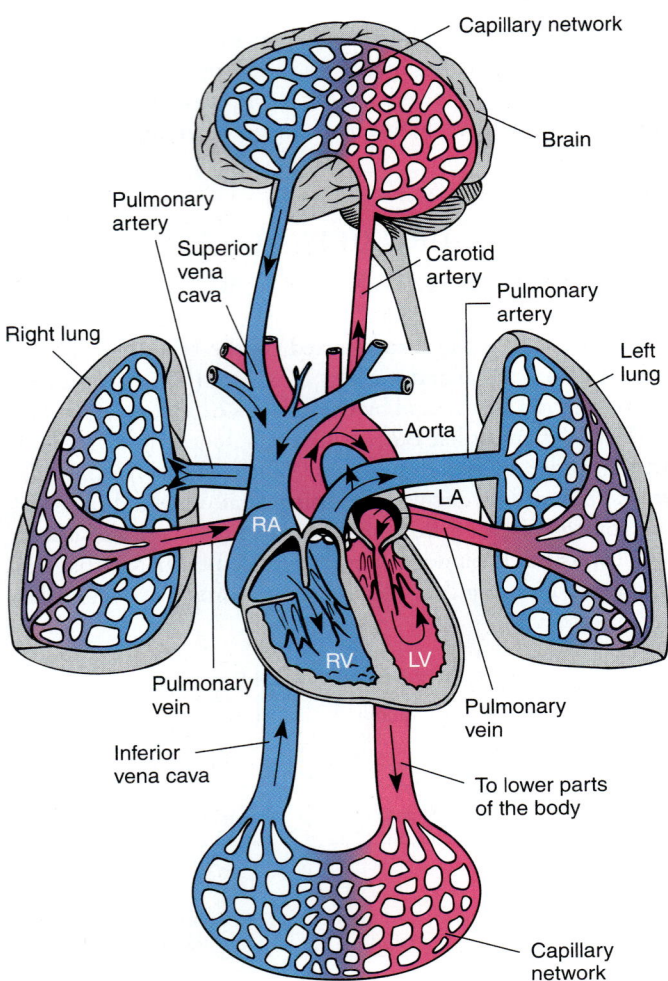

Figure 42–18 **Systemic and pulmonary circulation.** In this highly simplified diagram, red represents oxygen-rich blood; blue represents oxygen-poor blood.

plasma protein. The angiotensins are a group of hormones that act as powerful vasoconstrictors. Angiotensins also act indirectly to maintain blood pressure by increasing the synthesis and release of the hormone **aldosterone.** Aldosterone increases the retention of sodium ions by the kidneys, resulting in greater fluid retention and increased blood volume (Chapter 46).

BLOOD IS PUMPED THROUGH PULMONARY AND SYSTEMIC CIRCUITS

One of the main jobs of the circulation is to bring oxygen to all the cells of the body. In mammals and birds, blood is charged with oxygen in the lungs. As in amphibians and reptiles, birds and mammals have a double circuit of blood vessels: (1) the pulmonary circulation, which connects the heart and lungs; and (2) the systemic circulation, which connects the heart with all of the tissues of the body. This general pattern of circulation may be traced in Figure 42–18.

The pulmonary circulation oxygenates the blood

Blood from the tissues returns to the right atrium of the heart, partly depleted of its oxygen supply. This oxygen-poor blood, loaded with carbon dioxide, is pumped by the right ventricle into the pulmonary circulation. As it emerges from the heart, the large pulmonary trunk branches to form the two pulmonary arteries, one going to each lung. These are the only arteries in the body that carry oxygen-poor blood.

In the lungs the pulmonary arteries branch into smaller and smaller vessels, which finally give rise to extensive networks of pulmonary capillaries that surround the air sacs of the lungs. As blood circulates through the pulmonary capillaries, carbon dioxide diffuses out of the blood and into the air sacs. Oxygen from the air sacs diffuses into the blood so that, by the time blood enters the pulmonary veins leading back to the left atrium of the heart, it is charged with oxygen. Pulmonary veins are the only veins in the body that carry blood rich in oxygen.

In summary, blood flows through the pulmonary circulation in the following sequence:

Right atrium → right ventricle → pulmonary arteries → pulmonary capillaries (in lungs) → pulmonary veins → left atrium

The systemic circulation delivers blood to the tissues

Blood entering the systemic circulation is pumped by the left ventricle into the **aorta,** the largest artery of the body. Arteries that branch off from the aorta conduct blood to all regions of the body. Some of the principal branches include the **coronary arteries** to the heart wall itself, the **carotid arteries** to the brain, the **subclavian arteries** to the shoulder region, the **mesenteric artery** to the intestine, the **renal arteries** to the kidneys, and the **iliac arteries** to the legs (Fig. 42–19). Each of these arteries gives rise to smaller and smaller vessels, somewhat like branches of a tree that divide until they form tiny twigs. Eventually blood flows into the capillary network within each tissue or organ.

Blood returning from the capillary networks within the brain passes through the **jugular veins.** Blood from the shoulders and arms drains into the **subclavian veins.** These veins and others returning blood from the upper portion of the body merge to form a very large vein that empties blood into the right atrium. In humans this vein is called the **superior vena cava. Renal veins** from the kidneys, **iliac veins** from the lower limbs, **hepatic veins** from the liver, and other veins from the lower portion of the body return blood to the **inferior vena cava,** which delivers blood to the right atrium.

Figure 42–19 **Blood circulation through some of the principal arteries and veins of the human body.** Blood vessels carrying oxygen-rich blood are red; those carrying oxygen-poor blood are blue.

Carotid arteries
Jugular veins
Right subclavian artery
Superior vena cava
Axillary artery
Right lung
Liver
Renal vein
Inferior vena cava
Common iliac vein
Femoral vein

Left subclavian vein
Aortic arch
Left pulmonary artery
Left pulmonary vein
Left ventricle
Right ventricle
Renal artery
Kidney
Inferior mesenteric artery
Abdominal aorta
Common iliac artery
External iliac artery
Femoral artery

CARDIOVASCULAR DISEASE

Cardiovascular disease is the number one cause of death in the United States and in most other industrial societies. Most often death results from some complication of **atherosclerosis,** a disease in which the arteries narrow as a result of lipid deposits in their walls. Although it can affect almost any artery, the disease most often develops in the aorta and in the coronary and cerebral arteries. When it occurs in the cerebral arteries, it can cause vascular deficits and lead to a **cerebrovascular accident (CVA),** commonly referred to as a stroke.

Several major modifiable risk factors for cardiovascular disease have been identified:

1. **Elevated cholesterol levels** in the blood, often associated with diets rich in total calories, total fats, saturated fats, and cholesterol.
2. **Hypertension.** The higher the blood pressure, the greater the risk.
3. **Cigarette smoking.** The risk of developing atherosclerosis is two to six times greater in smokers than in nonsmokers and is directly proportional to the number of cigarettes smoked daily. Components of cigarette smoke damage the endothelial lining of blood vessels, leading to atherosclerosis.
4. **Diabetes mellitus,** an endocrine disorder in which glucose is not metabolized nor-

mally. The body shifts to fat metabolism and there is a marked increase of circulating lipids, leading to atherosclerosis.
5. **Physical inactivity.** One in four adults in the United States has a sedentary lifestyle and does not engage in regular exercise.
6. **Obesity.** Can affect cholesterol levels and increase the risk of hypertension and diabetes.

The risk of developing cardiovascular disease also increases with age. Estrogen hormones are thought to offer some protection in women until after menopause, when the concentration of these hormones decreases. Other probable risk factors currently being studied are hereditary predisposition, stress and behavior patterns, and dietary factors.

In atherosclerosis, the endothelial lining of the arterial wall is damaged. The underlying smooth muscle then proliferates, thickening the arterial wall. Lipids in the blood are deposited in the smooth muscle cells of the arterial wall. These cells proliferate, and the inner lining thickens. More lipid, especially cholesterol from low-density lipoproteins, accumulates in the wall, and fibrous tissue forms around the deposits. Eventually calcium is deposited there, contributing to the slow formation of hard plaque that further narrows the diameter of the artery. As the plaque develops, arteries lose their ability to stretch when filled

with blood, and they become progressively occluded (blocked), as shown in the figure. Less blood can be delivered to the tissues, which then may become **ischemic,** meaning that they are lacking in blood. Under these conditions the tissue is deprived of an adequate oxygen and nutrient supply.

When a coronary artery becomes narrowed, **ischemic heart disease** can develop. Sufficient oxygen may reach the heart tissue during normal activity, but the increased need for oxygen during exercise or emotional stress results in the pain known as **angina pectoris.** Persons with this condition may carry nitroglycerin pills for use during an attack. This drug dilates veins, reducing venous return. Cardiac output is lowered so that the heart is not working so hard and requires less oxygen. Nitroglycerin also dilates the coronary arteries slightly, allowing more blood to reach the heart muscle. Calcium ion channel blockers slow the heart by inhibiting the passage of calcium into cardiac muscle fibers. They also dilate the coronary arteries.

Myocardial infarction (MI), popularly referred to as heart attack, is a very serious, often fatal, consequence of ischemic heart disease. MI often results from a sudden decrease in coronary blood supply. The portion of cardiac muscle deprived of oxygen dies within a few minutes and is then referred to as an **infarct.** MI is the leading

---

As an example of blood circulation through the systemic circuit, let us trace a drop of blood from the heart to the right leg and back to the heart:

Left atrium → left ventricle → aorta → right common iliac artery → smaller arteries in leg → capillaries in leg → small veins in leg → common iliac vein → inferior vena cava → right atrium

The coronary circulation delivers blood to the heart

The heart muscle is not nourished by the blood within its chambers because its walls are too thick for nutrients and oxygen to diffuse through them to reach all of the muscle fibers.

Instead, the cardiac muscle is supplied by **coronary arteries** branching from the aorta at the point where that vessel leaves the heart. These arteries branch, giving rise to a network of blood vessels within the wall of the heart. Nutrients and gases are exchanged through coronary capillaries. Blood from these capillaries flows into **coronary veins,** which join to form a large vein, the **coronary sinus.** The coronary sinus empties directly into the right atrium; it does not join either of the venae cavae.

When one of the coronary arteries is blocked, the cells in the area of the heart muscle that is served by that artery die as a result of deprivation of oxygen and nutrients, and the affected muscle stops contracting. If a sufficient amount of cardiac muscle is affected, the heart may stop beating entirely. This is a common cause of "heart attack" and is often the result of atherosclerosis (see *Focus On: Cardiovascular Disease*).

(a)

500 µm

(b)

500 µm

cause of death and disability in the United States. Just what triggers the sudden decrease in blood supply is a matter of some debate.

Platelets may adhere to the roughened wall of an artery and initiate clotting. This can produce a **thrombus,** a clot that forms within a blood vessel or within the heart. If a thrombus blocks a sizable branch of a coronary artery, blood flow to a portion of heart muscle is impeded or completely halted. When such blockage, referred to as a coronary occlusion, prevents blood flow to a large region of cardiac muscle, the heart may stop beating, that is, **cardiac arrest** may occur, and death can follow within moments. If only a small region of the heart is affected, however, the heart may continue to function. Cells in the re-

gion deprived of oxygen die and are replaced by scar tissue.

An episode of ischemia may trigger a fatal arrhythmia such as **ventricular fibrillation,** a condition in which the ventricles contract very rapidly without actually pumping blood. The pulse may stop, and blood pressure may fall precipitously. Ventricular fibrillation has been linked with about 65 percent of cardiac arrests. The only effective treatment for fibrillation is **defibrillation** with electric shock. The shock appears to depolarize every muscle fiber in the heart so that its timing mechanism can reset.

Patients with progressive cardiovascular disease can be treated with coronary bypass surgery in which veins from another location in the patient's body are grafted

around occluded coronary arteries. The newly positioned blood vessels restore adequate blood flow to the affected area. Another procedure, coronary angioplasty, involves inserting a small balloon into an occluded coronary artery. When the balloon is inflated, the plaque in the arterial wall is broken up, increasing the diameter of the vessel. Other new approaches to treating cardiovascular disease include gene therapy aimed at production of a protein that would dissolve thrombi, as well as metabolic interventions that would shift fatty acid metabolism to glucose metabolism.

Four arteries deliver blood to the brain

Four arteries, two internal carotid arteries and two vertebral arteries (branches of the subclavian arteries), deliver blood to the brain. At the base of the brain, branches of these arteries form an arterial circuit called the **circle of Willis.** In the event that one of the arteries serving the brain becomes blocked or injured in some way, this arterial circuit helps ensure blood delivery to the brain cells via other vessels. Blood from the brain returns to the superior vena cava by way of the internal jugular veins at either side of the neck.

The hepatic portal system delivers nutrients to the liver

Blood almost always travels from artery to capillary to vein to the heart. An exception to this sequence occurs in the **hepatic**

portal system, which delivers blood rich in nutrients to the liver. Blood is conducted to the small intestine by the superior mesenteric artery. Then, as it flows through capillaries within the wall of the intestine, blood picks up glucose, amino acids, and other nutrients. This blood passes into the mesenteric vein and then into the **hepatic portal vein.** Instead of going directly back to the heart (as most veins do), the hepatic portal vein delivers nutrients to the liver.

Within the liver, the hepatic portal vein gives rise to an extensive network of tiny blood sinuses. As blood courses through the hepatic sinuses, liver cells remove nutrients and store them. Eventually liver sinuses merge to form hepatic veins, which deliver blood to the inferior vena cava. The hepatic portal vein contains blood that, although laden with food materials, has already given up some of its oxygen to the cells of the intestinal wall. Oxygen-rich blood is supplied to the liver by the hepatic artery.

THE LYMPHATIC SYSTEM IS AN ACCESSORY CIRCULATORY SYSTEM

In addition to the blood circulatory system, vertebrates have an accessory circulatory system, the **lymphatic system** (Fig. 42–20), which has three important functions: (1) to collect and return interstitial fluid to the blood; (2) to defend the body against disease organisms by way of immune mechanisms; and (3) to absorb lipids from the digestive tract. In this section we focus on the first function. Immunity is discussed in Chapter 43, and lipid absorption is discussed in Chapter 45.

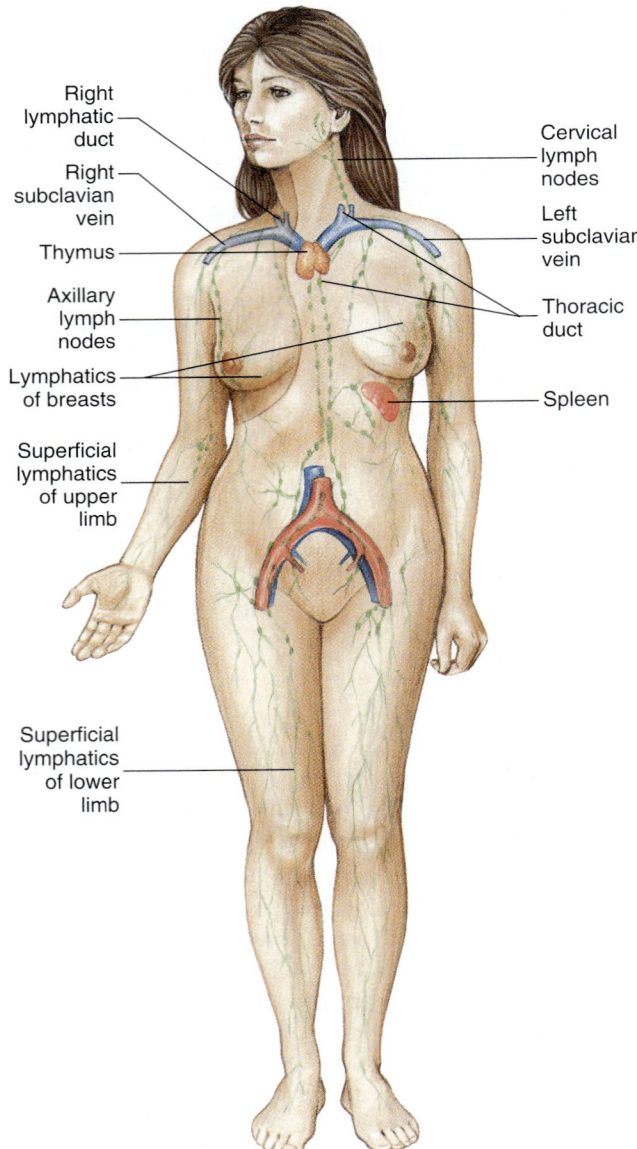

Figure 42–20 **Human lymphatic system.** Note that while the lymphatic vessels extend into most tissues of the body, the lymph nodes are clustered in certain regions. The right lymphatic duct drains lymph from the upper right quadrant of the body. The thoracic duct drains lymph from other regions of the body.

The lymphatic system consists of lymphatic vessels and lymph tissue

The lymphatic system consists of (1) an extensive network of **lymphatic vessels** that conduct **lymph,** the clear, watery fluid formed from interstitial fluid, and (2) **lymph tissue,** a type of connective tissue with large numbers of lymphocytes. Lymph tissue is organized into small masses of tissue called **lymph nodes** and **lymph nodules.** The tonsils, thymus gland, and spleen, which consist mainly of lymph tissue, are also part of the lymphatic system.

Tiny "dead-end" capillaries of the lymphatic system extend into almost all of the tissues of the body (Fig. 42–21). Lymph capillaries join to form larger lymphatics (lymph veins). There are no lymph arteries.

Interstitial fluid enters lymph capillaries and then is referred to as lymph. The lymph is conveyed into lymphatics. At certain locations the lymphatics empty into lymph nodes, where phagocytes filter out bacteria and other harmful materials from the lymph. The lymph then flows into lymphatics that conduct it away from the lymph node. Lymphatics from all over the body conduct lymph toward the shoulder region. These vessels join the circulatory system at the base of the subclavian veins by way of ducts: the **thoracic duct** on the left side and the **right lymphatic duct** on the right.

Tonsils are masses of lymph tissue under the lining of the oral cavity and throat. (When enlarged, the pharyngeal tonsils in back of the nose are called **adenoids.**) Tonsils help protect the respiratory system from infection by destroying bacteria and other foreign matter that enter the body through the mouth or nose. Unfortunately, tonsils are sometimes overcome by invading bacteria, and become the site of frequent infection themselves.

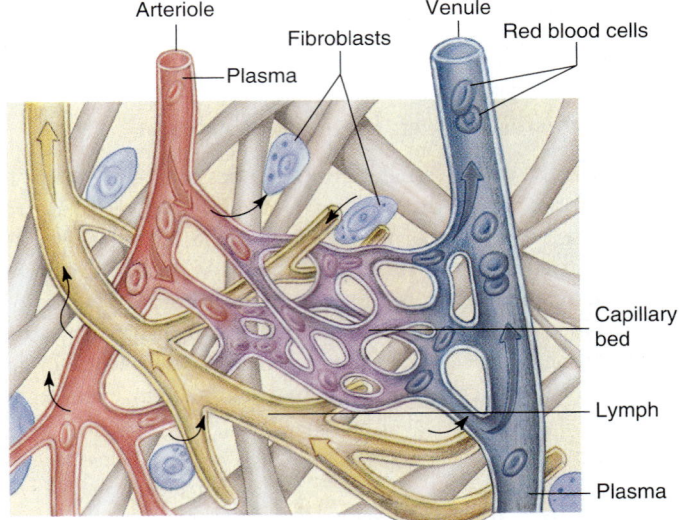

Figure 42–21 **Lymph capillaries.** Lymph capillaries drain excess interstitial fluid from the tissues. Note that blood capillaries are connected to vessels at both ends, whereas lymph capillaries, shown in yellow, are "dead-end streets." The arrows indicate direction of flow.

Some nonmammalian vertebrates such as the frog have lymph "hearts," which pulsate and squeeze lymph along. However, in mammals the walls of the lymph vessels themselves pulsate. Valves within the lymph vessels prevent the lymph from flowing backward. When muscles contract or when arteries pulsate, pressure on the lymph vessels increases lymph flow. The rate at which lymph flows is slow and variable, and the total lymph flow is about 100 mL per hour—very much slower than the 5 L per min of blood flowing in the vascular system.

The lymphatic system plays an important role in fluid homeostasis

When blood enters a capillary network it is under rather high pressure, so some plasma is forced out of the capillaries and into the tissues. Once it leaves the blood vessels, this fluid is called interstitial fluid, or tissue fluid. It contains no red blood cells or platelets and only a few white blood cells. Its protein content is about one fourth of that found in plasma because proteins are too large to pass easily through capillary walls. Smaller molecules dissolved in the plasma do pass out with the fluid leaving the blood vessels. Thus, interstitial fluid contains glucose, amino acids, other nutrients, and oxygen, as well as a variety of salts. This nourishing fluid bathes all the cells of the body.

The main force pushing plasma out of the blood is hydrostatic pressure, that is, the blood pressure against the capillary wall, caused by the beating of the heart (Fig. 42–22). The osmotic pressure of the interstitial fluid adds to the filtration pressure. The principal opposing force is the osmotic

pressure of the blood (also called colloid osmotic pressure, or oncotic pressure), which restrains fluid loss from the capillary.

At the venous ends of the capillaries the blood pressure is much lower, and the osmotic pressure of the blood draws fluid back into the capillary. However, not as much fluid is absorbed back into the circulation as is filtered out. Furthermore, protein does not return effectively into the venous capillaries and instead tends to accumulate in the interstitial fluid. These potential problems are so serious that without the lymphatic system, fluid balance in the body would be significantly disturbed within a few hours, and death would occur within about 24 hours. The lymphatic system preserves fluid balance by collecting about 10% of the interstitial fluid and the protein that accumulates in it.

The walls of the lymph capillaries are composed of endothelial cells that overlap slightly. When interstitial fluid accumulates, it presses against these cells, pushing them inward like tiny swinging doors that can swing in only one direction. As fluid accumulates within the lymph capillary, these cell doors are pushed closed.

Obstruction of the lymphatic vessels causes **edema,** swelling that results from excessive accumulation of interstitial fluid. Lymphatic vessels can be blocked as a result of injury, inflammation, surgery, or parasitic infection. For example, when a breast is removed (mastectomy) because of cancer, lymph nodes in the underarm region may also be removed in an effort to prevent the spread of cancer cells. The disrupted lymph circulation causes the patient's arm to swell. New lymph vessels develop within a few months, and the swelling slowly subsides.

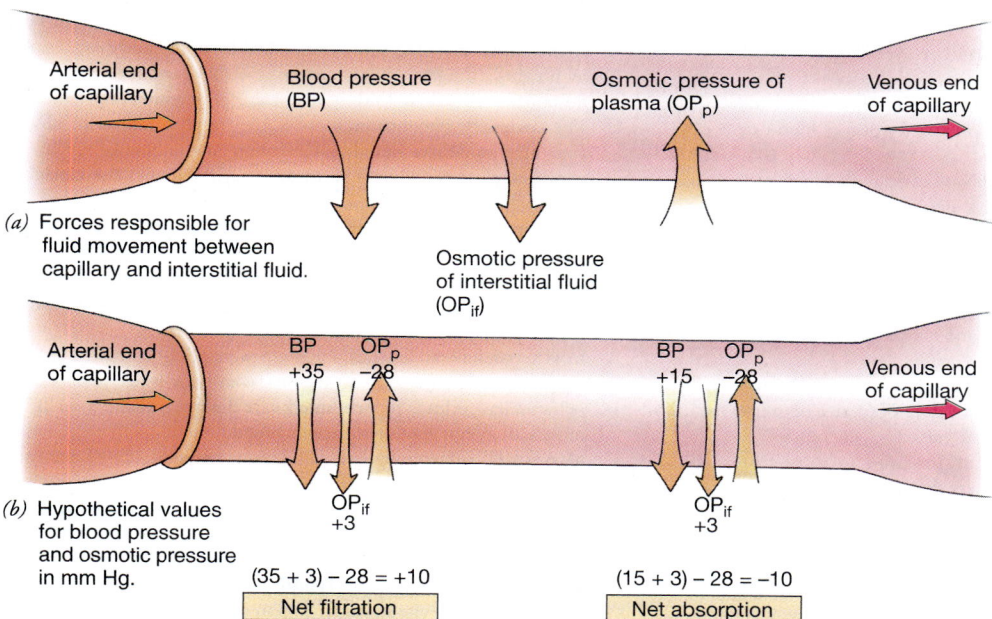

(a) Forces responsible for fluid movement between capillary and interstitial fluid.

(b) Hypothetical values for blood pressure and osmotic pressure in mm Hg.

$(35 + 3) - 28 = +10$ Net filtration

$(15 + 3) - 28 = -10$ Net absorption

Figure 42–22 **Fluid movement between blood and interstitial fluid.** *(a)* Blood pressure (hydrostatic pressure) and osmotic pressures are responsible for fluid movement and, thus, exchange of dissolved materials between blood and interstitial fluid. At the arterial end of a capillary, blood pressure forces plasma out of the capillary. The osmotic pressure of the blood is an opposing force acting to draw fluid into the blood. Osmotic pressure of the interstitial fluid contributes to the net filtration pressure but does not change much between arterial and venous ends of the capillary. At the venous end of the capillary, fluid enters the blood because blood pressure is much lower. *(b)* The numbers given are hypothetical and represent mm Hg. Net filtration is the total pressure moving fluid out of the capillary. Net absorption refers to the total pressure drawing fluid into the capillary.

I. Small, simple invertebrates, such as sponges, cnidarians, flatworms, and nematodes depend on diffusion for internal transport. Larger animals require a specialized **circulatory system,** which typically consists of **blood,** a **heart,** and a system of blood vessels or spaces through which blood circulates. In all animals **interstitial fluid,** the fluid between cells, is important in bringing oxygen and nutrients into contact with cells.

II. Arthropods and most mollusks have an **open circulatory system** in which blood flows into a **hemocoel,** bathing the tissues directly.

III. Some invertebrates and all vertebrates have a **closed circulatory system** in which blood flows through a continuous circuit of blood vessels.

IV. The vertebrate circulatory system consists of a muscular heart that pumps blood into a system of **arteries, capillaries,** and **veins.** This system transports nutrients, oxygen, wastes, and hormones; helps maintain fluid balance, appropriate pH, and body temperature; and defends the body against disease. The human circulatory system is often referred to as the **cardiovascular system.**

V. Vertebrate blood consists of liquid **plasma** in which **red blood cells (erythrocytes), white blood cells (leukocytes),** and **platelets** are suspended.
   A. Plasma consists of water, salts, substances in transport, and **plasma proteins,** including **albumins, globulins,** and **fibrinogen.**
   B. Red blood cells transport oxygen and carbon dioxide.
      1. In vertebrates, red blood cells produce large quantities of **hemoglobin,** a red pigment that binds with oxygen.
      2. A deficiency in hemoglobin is known as **anemia.**
   C. White blood cells defend the body against disease organisms. **Lymphocytes** and **monocytes** are agranular white blood cells; **neutrophils, eosinophils,** and **basophils** are granular white blood cells.
   D. Platelets patch damaged blood vessels and release substances essential for blood clotting. During the clotting process thrombin catalyzes the conversion of fibrinogen to an insoluble protein, **fibrin.**

VI. Arteries carry blood away from the heart chambers; veins return blood to the heart chambers.
   A. **Arterioles** constrict, called **vasoconstriction,** and dilate, called **vasodilation,** to regulate blood pressure and distribution of blood to the tissues.
   B. Capillaries are the thin-walled exchange vessels through which materials are transferred between the blood and tissues.

VII. The vertebrate heart consists of one or two **atria,** which receive blood, and one or two **ventricles,** which pump blood into the arteries.
   A. The four-chambered hearts of birds and mammals separate oxygen-rich blood from oxygen-poor blood.
   B. The heart is enclosed by a **pericardium** and is equipped with valves that prevent backflow of blood.
      1. The valve between right atrium and ventricle is the **right atrioventricular (AV) valve,** or **tricuspid valve.** The valve between left atrium and ventricle is the **mitral valve.**
      2. **Semilunar valves** guard the exits from the heart.
   C. Cardiac muscle fibers are joined by **intercalated discs.**
   D. The **sinoatrial (SA) node,** or pacemaker, initiates each heartbeat. A specialized electrical conduction system ensures that the heart beats in a coordinated manner.
   E. One complete heartbeat makes up a **cardiac cycle.** Contraction occurs during **systole.** The period of relaxation is **diastole.**
   F. An **electrocardiogram (ECG or EKG)** begins with a **P wave** caused by depolarization of the SA node and atrial muscle. Then, a **QRS complex** reflects the action potential spreading through the ventricles. The **T wave** occurs during repolarization of the ventricles.
   G. **Cardiac output** equals **stroke volume** times heart rate.
      1. Stroke volume depends on **venous return** and on neural messages and hormones, especially epinephrine and norepinephrine.
      2. According to **Starling's law of the heart,** the more blood delivered to the heart by the veins, the more blood the heart pumps.
      3. Heart rate is regulated mainly by the nervous system and is influenced by hormones and body temperature.
      4. The heart rate can be measured by counting the pulse. **Arterial pulse** is the alternate expansion and recoil of an artery.

VIII. **Blood pressure** is the force exerted by the blood against the inner walls of the blood vessel.
   A. Blood pressure is determined by cardiac output, blood volume, and resistance to blood flow. **Peripheral resistance** is the resistance to blood flow caused by the viscosity of the blood and by friction between the blood and the wall of the blood vessel.
   B. Blood pressure is greatest in the arteries and decreases as blood flows through the capillaries.
   C. **Baroreceptors** sensitive to changes in blood pressure send messages to the cardiac and vasomotor centers in the medulla of the brain. When informed of an increase in blood pressure, the cardiac center stimulates parasympathetic nerves that slow the heart rate, and the vasomotor center inhibits sympathetic nerves that constrict blood vessels. These actions reduce blood pressure.
   D. **Angiotensins** are hormones that raise blood pressure. **Aldosterone** helps regulate salt excretion, which affects blood volume and blood pressure.

IX. The **pulmonary circulation** connects heart and lungs; the **systemic circulation** connects the heart and the tissues.
   A. In the pulmonary circulation, the right ventricle pumps blood into the **pulmonary arteries,** one going to each lung. Blood circulates through pulmonary capillaries in the lung and then is conducted to the left atrium by a **pulmonary vein.**
   B. In the systemic circulation, the left ventricle pumps blood into the **aorta,** which branches into arteries leading to the body organs. After flowing through capillary networks within various organs, blood flows into veins that conduct it to the right atrium.
   C. The **coronary circulation** supplies the heart muscle with blood.
   D. Four arteries deliver blood to the brain. At the base of the brain, branches of these arteries form an arterial circuit, the **circle of Willis.**
   E. The **hepatic portal system** circulates nutrient-rich blood through the liver.

X. Modifiable risk factors for **cardiovascular disease** include elevated cholesterol levels, hypertension, cigarette smoking, diabetes mellitus, physical inactivity, and obesity.
   A. In **atherosclerosis,** arteries become progressively narrowed as they thicken in response to lipid deposition in their walls.
   B. Atherosclerosis leads to **ischemic heart disease,** in which the heart muscle does not receive sufficient blood.
   C. **Myocardial infarction** (MI) is a serious consequence of ischemic heart disease.

XI. The **lymphatic system** collects interstitial fluid, also called **tissue fluid,** and returns it to the blood. It plays an important role in homeostasis of fluids.
   A. **Lymphatic vessels** conduct **lymph,** a clear fluid formed from interstitial fluid, to the shoulder region and return it to the circulatory system.
   B. **Lymph nodes** are small masses of lymph tissue that filter the lymph, removing bacteria and harmful materials.

1. An open circulatory system (a) is found in flatworms (b) typically includes a hemocoel (c) has a continuous circuit of vessels with openings in the capillaries (d) is characteristic of vertebrates (e) is typically found in animals with a two-chambered heart

2. Which of the following is NOT a function of the vertebrate circulatory system? (a) helps maintain appropriate pH (b) transports nutrients, oxygen, and metabolic wastes (c) helps maintain fluid balance (d) produces hemocyanin (e) provides internal defense

3. Lipoproteins (a) are mainly transported in granular leukocytes (b) transport cholesterol (c) have been linked to clotting disorders (d) are associated with platelets (e) are stored in red blood cells

4. Which of the following are most closely associated with oxygen transport? (a) red blood cells (b) platelets (c) neutrophils (d) basophils (e) lymphocytes

5. Which of the following are most closely associated with blood clotting? (a) red blood cells (b) platelets (c) neutrophils (d) basophils (e) lymphocytes.

6. In blood clotting (a) thrombin → prothrombin; fibrinogen → fibrin (b) prothrombin → thrombin; fibrin → fibrinogen (c) prothrombin → thrombin; fibrinogen → fibrin (d) clotting factors → platelets; thrombin → fibrinogen (e) prothrombin → thrombin; fibrinogen → platelets

7. Blood vessels that carry blood away from the heart are (a) arteries (b) arterioles (c) veins (d) capillaries (e) arterioles and arteries

8. Arterioles (a) help regulate blood pressure (b) help regulate distribution of blood to the tissues (c) deliver blood to arteries (d) answers a, b, and c are correct (e) answers a and b only

9. Which sequence most accurately describes a sequence of blood flow? (a) right atrium → right ventricle → pulmonary artery (b) right atrium → left atrium → left ventricle → aorta (c) left atrium → left ventricle → pulmonary artery (d) left ventricle → left atrium → aorta (e) right atrium → right ventricle → aorta

10. Which sequence most accurately describes a sequence of blood flow? (a) pulmonary vein → pulmonary artery → right atrium (b) pulmonary artery → left atrium → left ventricle (c) pulmonary artery → pulmonary capillaries → pulmonary vein → left atrium (d) left ventricle → aorta → pulmonary artery (e) pulmonary artery → pulmonary capillaries → pulmonary vein → right atrium

11. A cardiac cycle (a) consists of one ventricular heart beat (b) includes a systole (c) equals stroke volume times heart rate (d) includes a diastole (e) includes a systole and a diastole

12. Blood pressure is determined by (a) cardiac output (b) peripheral resistance (c) blood volume (d) answers a, b, and c are correct (e) answers b and c only

13. Lymph forms from (a) interstitial fluid (b) blood serum (c) plasma combined with protein (d) fluid released by lymph nodes (e) angiotensins

14. The valve between the right atrium and right ventricle is the (a) mitral valve (b) semilunar valve (c) tricuspid valve (d) pulmonary valve (e) aortic valve

15. Atherosclerosis (a) is associated with thickening of arteries and veins (b) is thought to be caused by high concentrations of LDL (c) can lead to ischemic heart disease (d) answers a, b, and c are correct (e) answers b and c only

R E V I E W   Q U E S T I O N S

1. Compare how nutrients and oxygen are transported to the body cells in a hydra, planarian, earthworm, insect, and frog.
2. Contrast an open with a closed circulatory system.
3. Describe five functions of the vertebrate circulatory system.
4. List the functions of the main groups of plasma proteins.
5. Contrast the structure and functions of red and white blood cells.
6. Summarize the process by which blood clots.
7. Compare the functions of arteries, capillaries, and veins. Why are arterioles important in maintaining homeostasis?
8. Compare and contrast the hearts of a fish, amphibian, reptile, and bird.
9. Define cardiac output and describe factors that influence it.
10. Describe the conduction system of the heart and how the heart is regulated. Include a description of the actions of acetylcholine and norepinephrine.
11. What is the relationship between blood pressure and peripheral resistance? What mechanisms regulate blood pressure?
12. Trace the path of a red blood cell: (a) from the inferior vena cava to the aorta; and (b) from the renal vein to the renal artery.
13. What is the function of the hepatic portal system? How does its sequence of blood vessels differ from that in most other circulatory routes?
14. List five modifiable risk factors associated with the development of cardiovascular disease. Describe the disease process in atherosclerosis, and explain the association between atherosclerosis and ischemic heart disease? Describe myocardial infarction.
15. What is the role of lipoproteins in the development of cardiovascular disease? How does exercise affect lipoprotein concentration?
16. What is the relationship between plasma, interstitial fluid, and lymph? How does the lymph system help maintain fluid balance?
17. Blood pressure is low in capillaries. Explain how this helps to retain fluid in the circulation.
18. Label the diagram. Use Fig. 42–9 to check your answers.

YOU MAKE THE CONNECTION

1. How do the changes that take place during exercise affect the cardiovascular system? Compare the heart function of an adult who does not exercise to that of an athlete.
2. Cartilage lacks blood and lymph vessels. How do you imagine its cells are nourished? What effect, if any, would a lack of blood vessels have on cartilage healing after an injury?

3. How is the heart of the fish specifically adapted to its lifestyle?
4. When the nerves to the heart are cut, the heart rate increases to about 100 contractions per minute. What does this indicate about the regulation of the heart rate?

RECOMMENDED READINGS

Eisenberg, M.S. "Defibrillation: The Spark of Life." *Scientific American,* Vol. 278, No. 6, Jun. 1998. The author discusses the history of using electricity to restart the heart and compares defibrillation to rebooting a computer that has suddenly frozen.

Harken, A.H. "Surgical Treatment of Cardiac Arrhythmias." *Scientific American,* Vol. 269, No. 1, Jul. 1993. A surgical procedure is used to correct dangerously rapid heartbeats.

Lopaschuk, G.D., and Stanley, W.C. "Manipulation of Energy Metabolism in the Heart." *Science & Medicine,* Vol. 4, No. 6, Nov./Dec. 1997. Shifting energy substrate preference away from fatty acid metabolism and toward glucose use appears to be an effective new approach to treating cardiovascular disease.

Nabel, E.G., "Cardiovascular Gene Therapy." *Science & Medicine,* Vol. 4, No. 4, Jul./Aug. 1997. Clinical trials of gene therapy to treat cardiovascular disease are in progress.

NIH Consensus Development Panel on Physical Activity and Cardiovascular Health. "Physical Activity and Cardiovascular Health." *Journal of the American Medical Association,* Vol. 276, No. 3, 17 Jul. 1996. An assessment of the relationship between physical activity and cardiac health.

Stamler, J., and Neaton, J.D. "Benefits of Lower Cholesterol." *Scientific American: Science & Medicine,* Vol. 1, No. 2, May/Jun. 1994. Recent declines in death rate from heart disease appear to be related to lifestyle changes.

● Visit our website at **http://www.saunderscollege.com/lifesci/titles.html** and click on Solomon/Berg/Martin Biology for links to chapter-related resources on the World Wide Web.

CHAPTER 43

Internal Defense

Animals have internal defense mechanisms to protect them against disease-causing organisms that enter the body with air, food, and water, and through wounds in the skin. Disease-causing microorganisms, including certain viruses, bacteria, fungi, and protozoa, are referred to as **pathogens.** Internal defense depends on the animal's ability to distinguish between *self* and *nonself.* Such recognition is possible because organisms are biochemically unique. Cells have surface proteins different from those on the cells of other species or even other members of the same species. An animal recognizes its own cells and can identify those of other animals as foreign.

Pathogens manufacture macromolecules that the body usually identifies as foreign. A single bacterium may have from 10 to more than 1000 distinct macromolecules on its surface. Pathogens may also secrete macromolecules, some of which are toxic to most organisms. When a pathogen invades an animal, its distinctive macromolecules stimulate the animal's defense mechanisms.

The term *immune* is derived from a Latin word meaning "safe." **Immunology,** the study of internal defense mechanisms, is one of the most rapidly changing and exciting fields of biomedical research today. An **immune response** involves recognition of foreign macromolecules and a response aimed at eliminating them. Immune responses depend on communication among cells, or **cell signaling.** As discussed in previous chapters, an important aspect of cell signaling is signal transduction, the conversion of an extracellular signal into a series of intracellular events.

Two main types of immune responses are nonspecific and specific immune responses. **Nonspecific defense mechanisms,** also called **innate immune responses,** provide general protection against pathogens. These mechanisms prevent most pathogens from entering the body and rapidly destroy those pathogens that do penetrate the outer defenses. For example, the cuticle or skin provides a barrier to pathogens that come in contact with an animal's body. Phagocytosis of invading bacteria is another example of a nonspecific defense mechanism. Innate immune responses can be activated by the chemical properties of the foreign agent.

Specific defense mechanisms are tailor-made to combat specific macromolecules associated with each pathogen. Specific immune responses are also referred to as **acquired** or **adaptive immune responses.** Specific immune responses are directed to the particular type of foreign substance or pathogen

(NIBSC/Science Photo Library/Photo Researchers, Inc.)

that has gained entrance to an animal's body. Any molecule that can be specifically recognized as foreign by cells of the immune system is called an **antigen.** Many macromolecules, including proteins, RNA, DNA, and some carbohydrates, are antigens. An important specific defense mechanism is the production of **antibodies,** highly specific proteins that recognize and bind to specific antigens. In complex animals, specific defense mechanisms include immunological memory, the capacity to respond more effectively the second time foreign molecules invade the body.

Sometimes the immune system malfunctions and attacks the tissues of the body as though they were pathogens. At other times, the body may be overcome by pathogens. HIV (human immunodeficiency virus), the pathogen that causes AIDS, infects T cells (also called T lymphocytes), an important component of the immune system. The T cell *(green)* in the color-enhanced SEM shown here is infected with HIV *(red).*

AFTER YOU HAVE STUDIED THIS CHAPTER YOU SHOULD BE ABLE TO

1. Compare in general terms internal defense mechanisms of invertebrates and vertebrates.
2. Distinguish between specific and nonspecific defense mechanisms and describe nonspecific mechanisms including physical barriers, cytokines, complement, inflammation, and phagocytosis.
3. Draw the basic structure of an antibody and describe how antibodies recognize antigens.
4. Describe the principal cells of the immune system and contrast T and B cells with respect to their development and function.
5. Summarize the mechanisms of antibody-mediated immunity, including the effects of antigen-antibody complexes on pathogens; include a discussion of the complement system.
6. Describe the mechanisms of cell-mediated immunity, including the development of memory cells.

7. Contrast a secondary with a primary immune response.
8. Compare active and passive immunity, giving examples of each.
9. Summarize the immunological basis of graft rejection and explain how the effects of graft rejection can be minimized.
10. Describe how the body defends itself against cancer cells.
11. Explain the immunological basis of allergy and briefly describe the events that occur during (1) a hayfever response, and (2) systemic anaphylaxis.
12. Describe the immunological basis of autoimmune diseases and list possible causes.
13. Describe the cause of AIDS, risk factors and progress of the disease, and the difficulties encountered in developing a vaccine.

INVERTEBRATES HAVE MAINLY NONSPECIFIC INTERNAL DEFENSE MECHANISMS

All invertebrate species that have been studied demonstrate the ability to distinguish between self and nonself. Invertebrates make nonspecific defense responses such as phagocytosis and the inflammatory response. Most invertebrates can also demonstrate some specificity in their immune responses.

Sponge cells have specific glycoproteins on their surfaces that enable them to distinguish between self and nonself. When cells of two different species are mixed together, they reassort according to species. When two different species of sponges are forced to grow in contact with each other, tissue is destroyed along the region of contact. Cnidarians also reject grafted tissue and destroy foreign tissue.

Invertebrates with a coelom have amoeba-like phagocytes that engulf and destroy bacteria and other foreign matter. Many coelomate invertebrates also have substances in the hemolymph that kill bacteria, inactivate cells of some pathogens, and cause some foreign cells to clump. In mollusks, these hemolymph substances enhance phagocytosis by the phagocytes.

Certain cnidarians (e.g., corals) and arthropods (e.g., insects) have some specific immune mechanisms and immunological memory. In them, and in some echinoderms and simple chordates, the body appears to remember antigens for a short period of time; thus the body can respond more effectively when the same pathogens are encountered again. Echinoderms and tunicates are the simplest animals known to have differentiated white blood cells that perform limited immune functions.

VERTEBRATES LAUNCH NONSPECIFIC AND SPECIFIC IMMUNE RESPONSES

Like invertebrates, vertebrates protect themselves against pathogens with both nonspecific and specific defense mechanisms (Fig. 43–1). More sophisticated specific defense mechanisms are possible because vertebrates have a specialized lymphatic system (Chapter 42). Only vertebrates have **lymphocytes,** white blood cells specialized to carry out immune responses. In the discussion that follows, we will focus on the human immune system, with references to those of other vertebrates.

VERTEBRATE NONSPECIFIC DEFENSE MECHANISMS INCLUDE MECHANICAL AND CHEMICAL BARRIERS

An animal's first line of defense against pathogens is its outer covering. For example, the intact human skin presents both a mechanical and a chemical barrier to microorganisms. Sweat and sebum contain chemicals that destroy certain types of bacteria. **Lysozyme,** an enzyme found in many tissues, and in tears and other body fluids, attacks the cell walls of many gram-positive bacteria.

Microorganisms that enter with food are usually destroyed by the acid secretions and enzymes of the stomach. Pathogens that enter the body with inhaled air may be filtered out by hairs in the nose or trapped in the sticky mucous lining of the respiratory passageways. There, they may be destroyed by phagocytes.

Figure 43–1 **Nonspecific and specific defense mechanisms.** Nonspecific (innate) defense mechanisms include physical barriers that prevent entrance of pathogens, physiological mechanisms that destroy pathogens that manage to cross the barriers, phagocytosis, and inflammation. Cytokines released by cells during inflammation signal specific (adaptive) defense mechanisms. Specific defense mechanisms include cell-mediated immunity and antibody-mediated immunity.

Cytokines are important in nonspecific and specific immune responses

When pathogens breach the body's external barriers, other defense responses are activated. Cells of the immune system secrete a remarkable number of regulatory proteins known as **cytokines.** Cytokines are important in signaling cells during immune responses. (They are also important in regulating other important biological processes such as cell growth, repair, and cell activation.) Like hormones, cytokines can act on the cells that produce them, can regulate the activity of nearby cells, and in some cases can modify their functions. Three groups of cytokines that will be introduced here are interferons, interleukins, and tumor necrosis factors.

When infected by viruses or other intracellular parasites (some types of bacteria, fungi, and protozoa), certain cells respond by secreting cytokines called **interferons.** Type I interferons are produced either by **macrophages** (interferon alpha) or by the fibroblasts of connective tissues (interferon beta). (Recall from Chapter 42 that macrophages are large phagocytic cells.) Type I interferons inhibit viral replication and also activate cells called natural killer cells that have antiviral actions. Viruses produced in cells exposed to Type I interferons are not very effective at infecting other cells. Another group, Type II interferons (interferon gamma), produced by part of the specific immune system, enhance the activities of other immune cells. This group of interferons can stimulate macrophages to destroy tumor cells and host cells that have been infected by viruses.

Since their discovery in 1957, interferons have been the focus of much research. Recombinant DNA techniques are now used to produce large quantities of some interferons. The U.S. Food and Drug Administration (FDA) has approved in-terferons for treating several diseases, including hepatitis B and hepatitis C, genital warts, a type of leukemia, a type of multiple sclerosis, and AIDS-related Kaposi's sarcoma. Interferons are being tested in clinical trials for treatment of HIV infection and several types of cancer.

Other types of cytokines involved in nonspecific defense mechanisms include some of the interleukins. **Interleukins** are cytokines secreted mainly by macrophages and lymphocytes. They are numbered according to their order of discovery. This diverse group of proteins regulates interactions between lymphocytes and other cells of the body. Therefore, some interleukins have widespread effects. For example, during infection, **interleukin-1 (IL-1)** can reset the body's thermostat in the hypothalamus, resulting in fever and its symptoms. Just as there are overlaps between the functions of nonspecific and specific defense mechanisms, the cytokines of these subsystems also overlap. For example, cytokines produced by nonspecific cells such as macrophages can activate specific lymphocytes.

Tumor necrosis factors (TNF) are cytokines that are secreted by macrophages (TNF alpha) and by lymphocytes called T cells (TNF beta). TNF is important in mediating inflammation. TNF also kills tumor cells, offering promise in terms of immunotherapy for cancer patients. In addition, TNF can stimulate immune cells to initiate an inflammatory response. Sometimes infection by gram-negative bacteria, such as *Salmonella typhi,* results in the release of large amounts of TNF and other cytokines. This can lead to septic shock, a potentially lethal condition that may involve high fever and malfunction of the circulatory system. Thus, cytokines can sometimes have harmful effects. Cytokines that are more closely associated with specific immune responses will be discussed later in the chapter.

Complement leads to destruction of pathogens

Complement, so-named because it *complements* the action of other defense mechanisms, consists of more than 20 proteins present in plasma and other body fluids. Normally, complement proteins are inactive until the body is exposed to an antigen. Certain pathogens activate the complement system directly. In other cases, the binding of an antigen and antibody stimulate activation. Complement activation involves a cascade of reactions with each component acting on the next in the series. Proteins of the complement system then work to destroy pathogens.

Complement proteins can be activated against any antigen, and their action is nonspecific. Activated complement proteins have four main actions: (1) certain complement proteins lyse the pathogen cell wall; (2) others coat pathogens, making them less "slippery" so that phagocytes (macrophages and neutrophils) can phagocytize them more easily, a process known as opsonization; (3) some complement proteins attract white blood cells to the site of infection; (4) some complement proteins increase inflammation by stimulating release of **histamine** and other compounds that dilate blood vessels and increase capillary permeability.

Inflammation is a protective mechanism

The **inflammatory response (inflammation)** is the body's reaction to pathogen invasion or physical injury (Fig. 43–2). Inflammation is regulated by proteins in the plasma, by cytokines, by substances released by platelets, by certain white blood cells (basophils), and by **mast cells,** large connective tissue cells filled with distinctive granules. *Bradykinin,* a peptide in the plasma, dilates blood vessels and increases capillary permeability. Platelets, basophils, and mast cells release histamine and serotonin, compounds that dilate blood vessels in the affected area and increase capillary permeability. Blood flow increases to the infected region, bringing great numbers of phagocytic cells such as neutrophils (a type of white blood cell; see Chapter 42). The increased blood flow makes the skin feel warm, and makes skin that contains little pigment appear red. Phagocytes migrate out of the capillaries and into the infected tissues.

Increased capillary permeability allows fluid and antibodies to leave the circulation and enter the tissues. As the volume of interstitial fluid increases, **edema** (swelling) occurs. The edema (and also certain enzymes in the plasma) causes the pain characteristic of inflammation. Thus, the clinical characteristics of inflammation are *heat, edema, pain,* and possible *redness.*

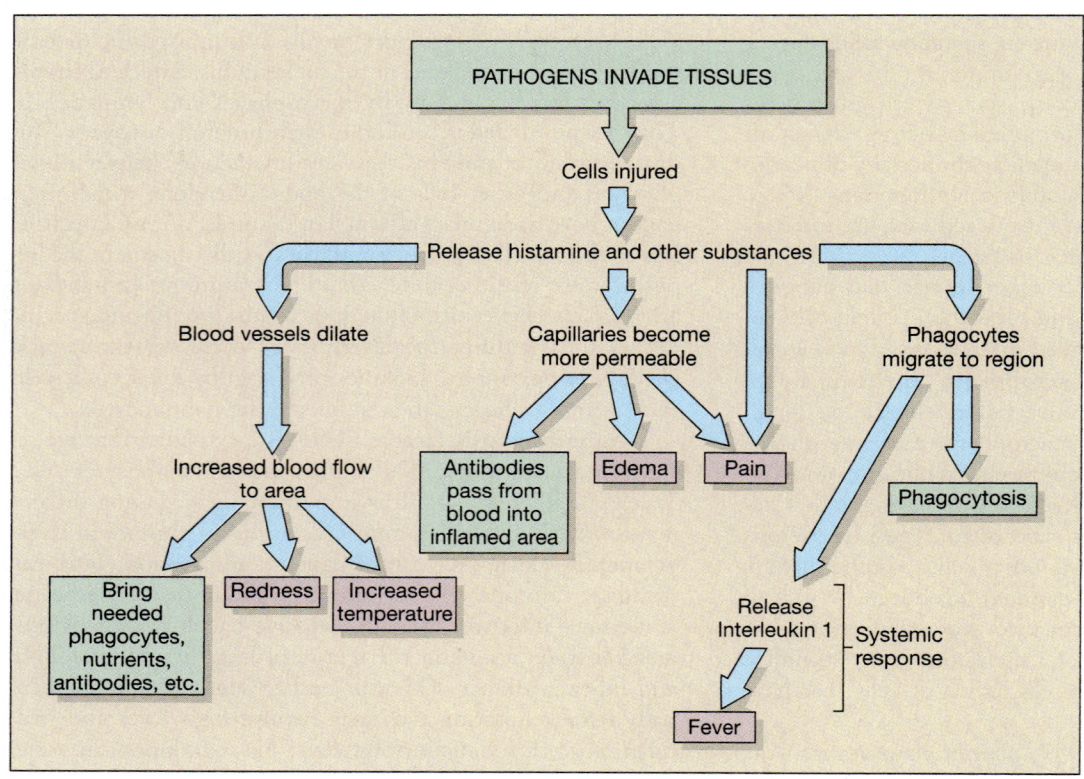

Figure 43–2 **Inflammation.** During inflammation, protective immune mechanisms are localized at the area of infection. Inflammation provides the conditions for phagocytic cells, antibodies, and other needed compounds to enter the tissue where injury or pathogen invasion is taking place.

(a)

5 μm

(b)

1 μm

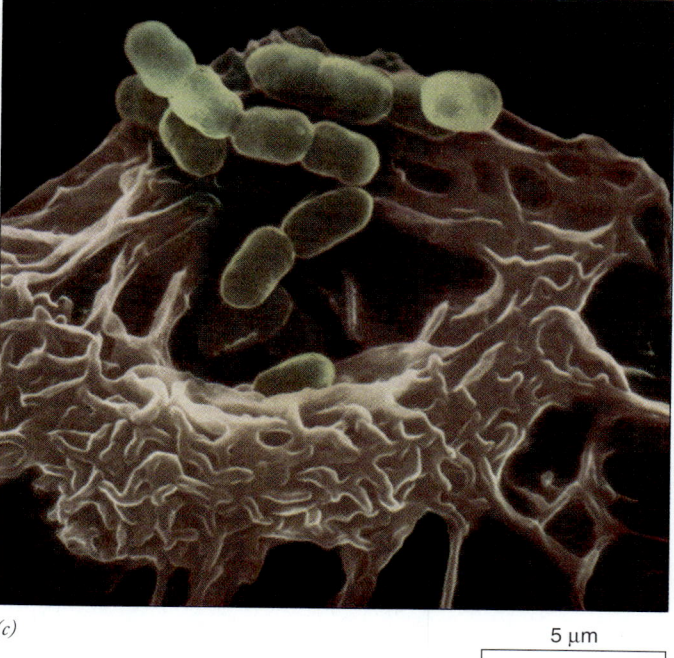

(c)

5 μm

Figure 43–3 **False-color SEMs of macrophages, remarkably efficient warriors.** *(a)* A macrophage extends a pseudopod toward an invading *E. coli* bacterium that is already multiplying. *(b)* The bacterium is trapped within the engulfing pseudopod. *(c)* The macrophage takes in the trapped bacteria along with its own plasma membrane. The macrophage plasma membrane will seal over the bacteria, and powerful lysosomal enzymes will destroy them. *(Lennart Nilsson, Boehringer Ingelheim International GmbH)*

Although inflammation is often a local response, sometimes the entire body is involved. **Fever** is a common clinical symptom of widespread inflammatory response. Macrophages and certain other cells release compounds, such as the cytokine interleukin-1 (IL-1), that reset the body's thermostat in the hypothalamus, resulting in fever. Prostaglandins (an important group of compounds derived from fatty acids) are also involved in this resetting process.

Fever helps the body fight infection. The increased body temperature interferes with the growth and replication of some microorganisms, and may kill some pathogens. Fever also causes lysosomes to break down, destroying cells infected by viruses. In addition, increased temperature promotes the activity of certain lymphocytes (T cells) and the production of antibodies, and increases phagocytosis. A short-term, low fever helps speed recovery.

Phagocytes destroy pathogens

One of the main functions of inflammation appears to be increased phagocytosis (see Chapter 5). A neutrophil can phagocytize 20 or so bacteria before it becomes inactivated (perhaps by leaking lysosomal enzymes) and dies (see Fig. 5–17b). A macrophage can phagocytize about 100 bacteria during its lifespan.

Can bacteria counteract the phagocyte's attack? Certain bacteria release enzymes that destroy the membranes of the phagocyte's lysosomes. The powerful lysosomal enzymes then spill out into the cytoplasm and may destroy the phagocyte. Some bacteria have cell walls or capsules that resist the action of lysosomal enzymes.

Some macrophages wander through the body's tissues, phagocytizing foreign matter (including bacteria) and, when appropriate, releasing antiviral agents (Fig. 43–3). Others stay in one place and destroy bacteria that pass by. For example, air sacs in the lungs contain large numbers of macrophages that destroy foreign matter entering with inhaled air.

SPECIFIC DEFENSE MECHANISMS INCLUDE ANTIBODY-MEDIATED AND CELL-MEDIATED IMMUNITY

While nonspecific defense mechanisms are destroying pathogens and preventing the spread of infection, the body is also mobilizing its specific defense mechanisms. Several days are required to activate specific immune responses, but, once in gear, these mechanisms are extremely effective. Two main types of specific immunity are antibody-mediated immunity and cell-mediated immunity.

Cells of the immune system include phagocytes and lymphocytes

Specific defense mechanisms depend on two main groups of white blood cells: phagocytes and lymphocytes. Phagocytes include neutrophils and macrophages. The main warriors in specific defense mechanisms are the lymphocytes. Lymphocytes develop and mature in primary lymph organs, including the bone marrow and the thymus gland. Secondary lymph organs, where large numbers of lymphocytes reside, include the spleen, lymph nodes, and other lymphatic tissues strategically positioned throughout the body.

Three main types of lymphocytes are **T lymphocytes,** or **T cells; B lymphocytes,** or **B cells;** and **natural killer (NK) cells.** NK cells kill virally infected cells and tumor cells. In **antibody-mediated immunity,** B cells mature into **plasma cells** that produce specific antibodies. Activation of B cells is a complex process that requires participation of a type of T cells, called **helper T cells.** In **cell-mediated immunity,** T cells kill body cells infected by invading pathogens and also attack cells altered by mutation (cancer cells).

Although T and B cells have different functions and life histories, they are similar in appearance when viewed with a light microscope. Sophisticated techniques such as fluorescence microscopy, however, demonstrate that the B and T cells can be differentiated by their unique cell surface macromolecules. They also tend to locate in (or "home" to) separate regions of the spleen, lymph nodes, and other lymph tissues.

Macrophages are important in nonspecific and specific defense mechanisms

When a macrophage ingests a bacterium, most, but not all, of the bacterial antigens are degraded by lysososomal enzymes. Fragments of the foreign antigens associate with a certain type of self-molecule (MHC, class II) and are then displayed on the surface of the macrophage. Thus, the macrophage is an **antigen-presenting cell (APC)** that displays foreign antigens as well as its own surface proteins. The foreign antigen-self-molecule combination activates certain T cells.

Macrophages secrete about 100 different compounds, including interferons and enzymes that destroy bacteria. When macrophages are stimulated by bacteria, they secrete interleukins, which activate B cells and certain T cells. Interleukins also promote a general response to injury, causing fever and activating other mechanisms that defend the body against invasion.

Natural killer cells attack cancer cells

Natural killer cells are large, granular lymphocytes that originate in the bone marrow. At first, immunologists thought that NK cells functioned only against tumor cells. However, studies have shown that these cells are active against a wide variety of targets, including cells infected with viruses, some bacteria, and some fungi.

NK cells destroy target cells by both nonspecific and specific (antibody requiring) processes. NK cells release cytokines (including interferon gamma, IL-2, IL-12, and TNF) as well as enzymes known as perforins and granzymes that destroy target cells.

NK activity is stimulated by several cytokines, including interferon gamma. When NK levels are high, resistance to certain cancers may be increased. Psychological stressors are thought to decrease NK cell activity and thus enhance tumor growth.

B cells are responsible for antibody-mediated immunity

Millions of B cells are produced in the bone marrow daily. B cells mature in the fetal liver and in the adult bone marrow (from which their name is derived). Each B cell is genetically programmed to encode a glycoprotein receptor that binds with a specific type of antigen. When a B cell comes into contact with the antigen that binds to its receptors, it divides rapidly, forming a clone of identical cells. These B cells differentiate into **plasma cells,** which produce antibody, a soluble form of the receptor molecule that can be secreted. A plasma cell can produce more than 10 million molecules of antibody per hour! Antibody binds to the antigen that originally activated the B cells.

Some activated B cells do not become plasma cells, but instead become **memory B cells.** Memory cells continue to produce small amounts of antibody after an infection has been overcome.

T cells are responsible for cellular immunity

T cells originate from stem cells in the bone marrow (Fig. 43–4). On their way to the lymph tissues, the future T cells stop off in the **thymus gland** for processing. (The "T" in T cells stands for *thymus-derived.*) The thymus makes T cells immunocompetent, that is, capable of immunological response (discussed in a later section). As T cells move through the thy-

Figure 43–4 **Cells of the immune system.** These cells have different developmental histories and different functions. They interact by way of complex signaling.

mus, they divide many times and develop specific surface proteins with distinctive receptor sites. Only the T cells possessing specific receptors are selected to divide. This is a form of **positive selection.**

T cells that react to self-antigens undergo **apoptosis,** or programmed cell death. This is a form of **negative selection.** Immunologists estimate that more than 90% of developing T cells are negatively selected. The remaining T cells differentiate and leave the thymus to take up residence in other lymph tissues or to launch immune responses in infected tissues. By selecting only appropriate T cells, the thymus gland ensures that T cells can distinguish between the body's own antigens and foreign antigens.

The differentiation of the majority of T cells within the thymus is thought to take place just before birth and during the first few months of postnatal life. If the thymus is removed before this processing takes place, an animal is not able to develop cellular immunity. If the thymus is removed after that time, cellular immunity does not seem to be as seriously impaired.

T cells are distinguished by the **T cell antigen receptor (TCR).** Several types and subtypes of T cells have been identified. One group, known as **CD8 T cells** because they have a surface marker designated CD8, includes **cytotoxic T cells (T_c),** also known as *killer T cells.* Cytotoxic T cells recognize and destroy cells with foreign antigens on their surfaces. Among their target cells are virus-infected cells, cancer cells, and foreign tissue grafts. T_c cells kill their target cells by releasing a variety of cytokines and enzymes that lyse cells.

Another category of T cells are **helper T cells (T_h),** also known as **CD4 T cells** because they have a surface marker designated CD4. Helper T cells secrete substances that activate or enhance immune responses. Two subsets of helper T cells have been identified: T helper 1 (Th1) cells and T helper 2 (Th2) cells. They secrete different types of cytokines and have different functions. T helper 1 cells mainly promote cell-mediated immune responses. T helper 2 cells stimulate B cells to divide and produce antibodies, and thus function mainly in antibody-mediated immunity. The balance between T helper 1 and T helper 2 cells appears to be important in effective response to infection. If the relative proportion of Th1 and Th2 cells could be shifted in diseases such as AIDS, a more positive outcome might occur.

Both helper T cells and cytotoxic T cells can suppress immune responses (discussed later in this chapter). Both types of T cells also include memory T cells.

T cells produce many cytokines important in immune responses. Some of the cytokines affect T cell development, others B cell development, and still others influence the action of macrophages. T helper 1 cells produce interleukin-2, or IL-2, which stimulates development of NK cells, T cells, and B cells. Efforts have been made to treat people with depressed immune systems with IL-2 in order to increase their responses to cancer. Unfortunately, this type of therapy tends to be fairly toxic to the patient. T helper 2 cells, B cells, and macrophages produce interleukin-4, or IL-4. Whereas IL-2 enhances cell-mediated immunity, IL-4 tends to enhance antibody-mediated immunity. Another cytokine, IL-12, is thought to enhance Th-1 action and therefore IL-2, as well as the action of cytotoxic T cells. Such interactions of cytokine activity provide mechanisms for checks and balances in this very complex system. Investigators continue to identify cytokines produced by T cells and other cells of the immune system.

The major histocompatibility complex permits recognition of self

The ability of the vertebrate immune system to distinguish self from nonself depends largely on a group of cell surface proteins, known as **MHC antigens.** These antigens are coded for by a set of closely linked genes known as the **major histocompatibility complex (MHC).** In humans, the MHC is called the **HLA** (human leukocyte antigen) group. These genes are polymorphic (variable). Within the population there are multiple alleles for each locus (sometimes more than 40 alleles for a given gene). As a result, the cell surface proteins for which they code are generally different in each individual. With so many possible combinations, no two people, except identical twins, are likely to have all of the same MHC proteins on their cells. The more closely related two individuals are, the more MHC genes they have in common. Thus, the MHC proteins of an individual are a biochemical "fingerprint."

The MHC is divided into three groups of genes that code for distinct sets of proteins. These proteins differ in terms of tissue distribution and chemical structure. MHC class I antigens are found on most nucleated cells and are important in distinguishing between self and nonself. They bind foreign antigens produced within cells (for example, by viruses or by foreign tissue grafts), forming molecular complexes that are displayed on the cell surface. These foreign antigen-MHC complexes are recognized by cytotoxic T cells.

MHC class II antigens are found primarily on cells of the immune system, particularly B cells, macrophages, some T cells, and **dendritic cells,** specialized cells located throughout the body, especially in the spleen and lymph nodes. MHC class II antigens regulate the interactions among T cells, B cells, and antigen-presenting cells. MHC class II antigens bind peptide fragments of proteins that have entered the cell via foreign sources such as bacteria and have been degraded. The foreign antigen-MHC complex is displayed on the cell surface and stimulates helper T cells. MHC class III proteins include components of the complement system.

ANTIBODY-MEDIATED IMMUNITY IS A CHEMICAL WARFARE MECHANISM

B cells are responsible for antibody-mediated immunity (also called humoral immunity). A given B cell can produce many copies of one specific antibody. Recall that antibody molecules serve as cell surface receptors that combine with antigens. Only a B cell displaying a matching receptor on its surface can bind a particular antigen. This binding activates the B cell.

In most cases, activation of B cells is a complex process that involves dendritic cells or macrophages, and helper T cells (Fig. 43–5). Dendritic cells capture antigens in the tissues and migrate to the lymph nodes or spleen where they present antigen fragments to helper T cells. Macrophages display fragments of antigens from pathogens they have engulfed. The foreign antigen forms a complex with MHC class II molecules of the macrophage. This foreign antigen-MHC complex is then displayed on the cell surface.

When a macrophage (or other APC) displaying a foreign antigen-MHC complex contacts a helper T cell with complementary T cell receptors, a complicated interaction occurs. Multiple chemical signals are sent back and forth between cells. For example, the macrophage secretes interleukins, such as IL-1, that activate helper T cells. T cells do not recognize an antigen that is presented alone. Helper T cells require that the antigen be presented as part of a foreign antigen-MHC class II complex on the surface of an APC.

The antibody receptor of a B cell binds with complementary antigen. Inside the B cell, the antigen is degraded into peptide fragments. The B cell then displays the peptide fragments together with MHC protein class II on its surface. An activated helper T cells binds with the foreign antigen-MHC complex on the B cell. Thus, the B cell can serve as an APC to T cells. An activated helper T cell then releases interleukins, which, together with antigen, activate the B cell.

Bacterium
Antigen
Lysosome
Antigen-MHC complex
APC

(a) An APC degrades antigen and displays it in combination with MHC class II.

Bacterium
Antigen
B cell

A B cell independently interacts with foreign antigen on the surface of a bacterium.

Cytokines
APC
Helper T cell
Receptor

(b) Helper T cell is activated (when its receptors combine with foreign antigen-MHC complex)

Foreign antigen
Helper T cell
B cell
Cytokines
Cell division

(c) B cell is activated by a helper T cell

Figure 43–5 **B cell activation.** Antigen presentation and helper T cells are required. *(a)* An antigen presenting cell (APC), such as a macrophage, takes in a bacterium or foreign antigen, breaks it down, and presents foreign antigens on its surface in combination with MHC antigens. B cells can combine with specific antigens. However, B cells are typically not active until stimulated by helper T cells. *(b)* Helper T cells are activated when their receptors combine with foreign antigen-MHC complex on an APC. Complex signaling takes place via cytokines secreted by both cells. *(c)* The activated helper T cell secretes cytokines that can activate a B cell. The B cell then divides, forming a clone of cells.

Once activated, a B cell increases in size. It then divides by mitosis, giving rise to a clone of identical cells (Fig. 43–6). This cell division in response to a specific antigen is known as **clonal selection.** This concept is based on the work of Frank Macfarlane Burnet, David Talmadge, and Niels Jerne which in the late 1950s led to the **clonal selection theory.** According to this theory, each lymphocyte and its clone make antibodies specific to one specific antigen. The specificity of the clone is determined before the B cell encounters the antigen.

Some cells of the B cell clone mature into plasma cells that secrete the type of antibody specific to the antigen. Unlike T cells, most plasma cells do not leave the lymph nodes. Only the antibodies they secrete pass out of the lymph tissues and make their way via the lymph and blood to the infected area. This sequence is summarized as follows:

Pathogen (bearing foreign antigens) invades body → APC phagocytizes pathogen → foreign antigen-MHC complex displayed on APC surface → helper T cell binds with foreign antigen-MHC complex → activated helper T cell interacts with a B cell that displays the same complex → B cell activated → clone of B cells produced → B cells differentiate → plasma cells secrete antibody → antibodies form complexes with pathogen → antibody complexes trigger processes leading to pathogen destruction

Some activated B cells do not differentiate into plasma cells, but instead become **memory B cells.** In these cells, a "survival gene" is activated that permits them to prevent the programmed death (apoptosis) that is the eventual fate of plasma cells. Memory cells continue to live and produce small

APC

Antigen

An APC presents foreign antigen-MHC complex on surface

Helper T cell

A helper T cell binds with complex and secretes cytokines

B cell

B cell

B cell

Combination of cytokines and antigens activates B cells

Bacterium

Activated B cell

Activated B cell increases in size and divides by mitosis

Cytokines

B cells

Clone of competent B cells is produced

B cells differentiate into plasma cells and memory B cells

Plasma cells secrete antibodies

Memory B cells

Plasma cells secrete specific antibodies

Antibodies transported via lymph and blood to infected region

Bacteria with antigen

Site of infection

Antigen-antibody complex

Antibodies combine with antigens on the surface of the pathogen to form antigen-antibody complexes.

Figure 43–6 Antibody-mediated immunity. A B cell becomes activated when it binds with a specific antigen and an activated helper T cell releases cytokines. Once activated, the B cell divides, producing a clone. Many of these cells differentiate and become plasma cells which secrete antibodies. The plasma cells remain in the lymph tissues, but the antibodies are transported to the site of infection by the blood or lymph. The antigen-antibody complexes that form destroy pathogens. Some of the B cells become memory cells that continue to secrete small amounts of antibody for years after the infection is over.

amounts of antibody long after an infection has been overcome. This antibody, part of the gamma globulin fraction of the plasma, becomes part of the body's arsenal of chemical weapons. Should the same pathogen enter the body again, this circulating antibody immediately targets it for destruction. At the same time, specific memory cells are stimulated to divide, producing new clones of the appropriate plasma cells.

A typical antibody consists of four polypeptide chains

An antibody molecule has two main functions: It combines with antigen and it activates processes that destroy the antigen that binds to it. For example, the antibody may stimulate phagocytosis. Note that the antibody does not destroy the antigen directly. Rather, it *labels* the antigen for destruction. The part of an antibody that binds an antigen is the **Fab** portion. The part that interacts with cells of the immune system is its **Fc** portion. Many cells of the immune system have Fc receptors.

How do antibodies, also called **immunoglobulins** or **Ig**, "recognize" antigens? In an antigen that is a protein, specific sequences of amino acids make up an **antigenic determinant,** or **epitope** (Fig. 43–7). These amino acids give part of the antigen molecule a specific shape that can be recognized by an antibody or T-cell receptor. Usually, an antigen has many different antigenic determinants on its surface. Some have hundreds. These antigenic determinants may differ from one another, so several different kinds of antibodies can combine with a single antigen.

A typical antibody is a Y-shaped molecule in which the two arms of the Y (the Fab portion) bind with antigen (Fig. 43–7). This shape permits the antibody to combine with two antigen molecules and allows formation of **antigen-antibody complexes.** While the arms of the Y bind to antigen, the tail of the Y, the Fc portion, interacts with cells of the immune system, such as phagocytes, or binds with molecules of the complement system.

The antibody molecule consists of four polypeptide chains: two identical long chains called heavy chains, and two identical short chains called light chains (Fig. 43–7). Each chain has a constant segment, a junctional segment, and a variable segment. In the **constant segment,** or **C region,** of the heavy chains, the amino acid sequence is constant within a particular immunoglobulin class. The C region may be thought of as the handle portion of a door key. Like the elongated part of a key that slides into a lock, the amino acid sequence of the **junctional segment,** or **J region,** is somewhat variable. Finally, like the pattern of bumps and notches at the end of a key, the **variable segment,** or **V region,** has a unique amino acid sequence. In B-cell receptors, the variable region of the immunoglobulin protrudes from the B cell, whereas the constant region anchors the molecule to the cell. (See *Making the Connection: Immunity and Genetics* for a discussion of antibody diversity.)

(a)

(b)

Figure 43–7   Structure and functions of antibodies. Antibodies combine with antigens, forming antigen-antibody complexes. *(a)* The antibody molecule is composed of two light chains and two heavy chains, joined together by disulfide bonds. The constant and variable regions of the chains are indicated. The Fab part of the antibody binds to antigen. The Fc part of the molecule binds with cells of the immune system. *(b)* Antigen-antibody complexes directly inactivate pathogens and increase phagocytosis. They also activate the complement system.

The V region is the part of the key that is unique for a specific antigen (the lock). At its variable regions, the antibody folds three-dimensionally, assuming a shape that enables it to combine with a specific antigen. When they meet, antigen and

(Text continues on p. 942)

IMMUNITY AND GENETICS

How can the immune system recognize every possible antigen, even those produced by newly mutated viruses never before encountered during the evolution of our species? Do our cells contain millions of separate antibody genes, each coding for an antibody with a different specificity? Although each human cell has a large amount of DNA, it is not enough to provide a different gene to code for each specific antibody molecule.

Recombinant DNA technology (see Chapter 14) has allowed researchers to make direct comparisons between the DNA of the B cells that produce antibodies and that of other cells of the body. They have found that separate DNA segments code for different regions of the heavy and light chains of an antibody (see Fig. 43–7). During the development of a B cell, the DNA segments are shuffled and then joined, making one combined gene. By shuffling the gene segments in this way, the number of potential combinations is enormous! Millions of different types of B (and T) cells are produced. By chance, one of

those cells may produce just the right antibody to destroy a virus that invades the body. You create diverse combinations of things in your everyday life. A familiar example is when you make your own ice cream sundae. Imagine the varied combinations that are possible using ten types of ice cream, six types of sauce, and 15 kinds of toppings.

Additional sources of antibody diversity are known. For example, the DNA of the mature B lymphocytes that codes for the variable regions of immunoglobulins mutates very readily. These mutations produce genes that code for slightly different antibodies.

We have used antibody diversity here as an example, but similar genetic mechanisms account for the diversity of T cell receptors. Although we may actually use only a relatively few types of antibodies or T cells in a lifetime, the remarkable diversity of the immune system prepares it to attack most potentially harmful antigens that may invade the body.

Antibody diversity. DNA rearrangement occurs during production of an antibody. In undifferentiated cells, gene segments are present for a number of different variable (V) regions, for one or more junction (J) regions, and for one or more different constant (C) regions. During differentiation, the segments are rearranged. A gene segment extending from the end of one of the V segments to the beginning of one of the J segments may be deleted. This produces a gene that can be transcribed. The RNA transcript is processed to remove introns, and the mRNA produced is translated. Each of the polypeptide chains of an antibody molecule is produced in this way. This diagram is greatly simplified.

(a)

(b)

Figure 43–8 **Antigen-antibody complex.** The components of an antigen-antibody complex fit together as shown in this computer-simulation of their molecular structure. The antigen lysozyme is shown in green, the heavy chain of the antibody is shown in blue, and the light chain in yellow. *(a)* A portion of the antigenic determinant, shown in red, fits into a groove in the antibody molecule. *(b)* The antigen-antibody complex has been pulled apart to show its structure.

antibody fit together somewhat like a lock and key. They must fit in just the right way, though not perfectly, for the antibody to be effective (Fig. 43–8). A given antibody can bind with different strengths, or **affinities,** to different antigens. In the course of an immune response, stronger (higher affinity) antibodies are generated.

Antibodies are grouped in five classes

Antibodies are grouped in five classes defined by antigenic differences in the constant region of the heavy chains. Using the abbreviation Ig for immunoglobulin, the classes are designated IgG, IgM, IgA, IgD, and IgE. In humans, about 75% of the antibodies belong to the **IgG** class; these are part of the gamma globulin fraction of the plasma. IgG and **IgM** defend against many pathogens carried in the blood, including bacteria, viruses, and some fungi. IgG and IgM antibodies interact with macrophages and activate the complement system.

IgA, present in mucus, tears, saliva, and milk, prevents viruses and bacteria from attaching to epithelial surfaces. This immunoglobulin defends against inhaled or ingested pathogens. IgA is strategically secreted into the respiratory passageways, digestive tract, urinary tract, and reproductive tract.

IgD has a low concentration in the plasma. Along with IgM, it is an important immunoglobulin on the B cell surface. IgD helps activate B cells following antigen binding. Although **IgE** has a low plasma concentration, it is important because it can bind to mast cells, connective tissue cells that contain potent mediators, such as histamine. When an antigen binds to IgE on a mast cell, these mediators are released. Histamine triggers many allergy symptoms. IgE is also responsible for immunity to invading parasitic worms.

The binding of antibody to antigen activates other defense mechanisms

Antibodies mark a pathogen as foreign by combining with an antigen on its surface. Generally, several antibodies bind with several antigens, creating a mass of clumped antigen-antibody complexes. The combination of antigen and antibody activates several defense mechanisms:

1. The antigen-antibody complex may inactivate the pathogen or its toxin. For example, when an antibody attaches to the surface of a virus, the virus may lose its ability to attach to a host cell.

2. The antigen-antibody complex stimulates phagocytic cells to ingest the pathogen.
3. Antibodies of the IgG and IgM groups work mainly through the complement system (discussed earlier in this chapter). When antibodies combine with a specific antigen on a pathogen, complement proteins destroy the pathogens. IgG antibodies have an Fc part of the molecule that binds Fc receptors that are expressed on most immune cells. When the Fc part of an antibody molecule that has bound to a pathogen binds with Fc receptors on a phagocyte, the pathogen is more easily destroyed.

CELL-MEDIATED IMMUNITY PROVIDES CELLULAR WARRIORS

The T cells and macrophages are responsible for cell-mediated immunity (Fig. 43–9). They destroy cells infected with viruses and cells that have been altered in some way, such as cancer cells. They also destroy foreign grafts such as transplanted kidneys. There are thousands of different populations of T cells. Each T cell has more than 50,000 identical receptors (TCRs) that bind to one specific type of antigen. Like B cells, T cells are clonal.

How do T cells know which cells to attack? T cells do not recognize antigens unless they are presented properly. For example, when a virus infects a cell, some of the viral protein is broken down to peptides and displayed with MHC class I molecules on the cell surface. T cell receptors can react with such foreign antigen-MHC complexes. Only T cells with receptors that bind to the specific antigen presented together with the MHC receptor become activated. Generally, fewer than 1 in 10,000 T cells can respond. Helper T cell activation is thought to be necessary for activating many aspects of the immune response.

Once activated, a T cell increases in size and gives rise to a clone of helper T cells, cytotoxic T cells, and memory cells (Fig. 43–9). Cytotoxic T cells make up the cellular infantry. They leave the lymph nodes and make their way to the infected area. These killer cells can destroy a target cell within seconds after contact.

After a cytotoxic T cell combines with antigen on the surface of the target cell, it secretes granules. These granules contain proteins that destroy the target cell. After releasing cytotoxic substances, the T cell disengages itself from its victim cell and seeks out a new target. This sequence is summarized as follows:

Virus invades body cell → foreign antigen-MHC class I complex displayed on cell surface → specific T cell activated by this complex → clone of T cells produced; some become cytotoxic T cells → cytotoxic T cells migrate to area of infection → cytotoxic T cells release proteins that destroy target cells

Helper T cells and macrophages at the site of infection secrete interleukins that help regulate immune function. For example, some interleukins stimulate activated T cells to divide. Others enhance inflammation, attracting great numbers of macrophages to the site of infection. Experimental evidence suggests that both helper and cytotoxic T cells can, under some conditions, suppress immune responses. They may act by destroying antigen-presenting cells.

LONG-TERM IMMUNITY DEPENDS ON IMMUNOLOGICAL MEMORY

Memory B and memory T cells are responsible for long-term immunity. They enable the body to launch more effective immune responses and they also permit immunization against disease.

A secondary immune response is more effective than a primary response

The first exposure to an antigen stimulates a **primary response.** Injection of an antigen into an animal causes specific antibodies to appear in the blood plasma in 3 to 14 days. After injection of the antigen, there is a brief latent period, during which the antigen is recognized and appropriate lymphocytes begin to form clones. Then there is a logarithmic phase, during which the antibody concentration rises rapidly for several days until it reaches a peak (Fig. 43–10). IgM is the principal antibody synthesized during the primary response. Finally, there is a decline phase, during which the antibody concentration decreases to a very low level.

A second injection of the same antigen, even years later, results in a **secondary response** (Fig. 43–10). Because memory cells bearing antibodies to that antigen may persist for many years, the secondary response is generally much more rapid than the primary response, with a shorter latent period. Much less antigen is necessary to stimulate a secondary response than a primary response, and more antibodies are produced. The affinity of antibodies is generally much higher. The predominant antibody in a secondary response is IgG.

The body's ability to launch a rapid, effective response during a second encounter with an antigen explains why we do not usually suffer from the same infectious disease several times. Most persons contract measles or chicken pox, for example, only once. When exposed a second time, the immune system responds quickly, destroying the pathogens before they have time to multiply and cause symptoms of the disease. Booster shots of vaccine are given in order to elicit a secondary response, thus reinforcing immunological memory.

You may wonder, then, how a person can get influenza (the flu) or a cold more than once. Unfortunately, there are many varieties of these diseases, each caused by a virus with slightly different antigens. For example, more than 100 different viruses cause the common cold, and new varieties of

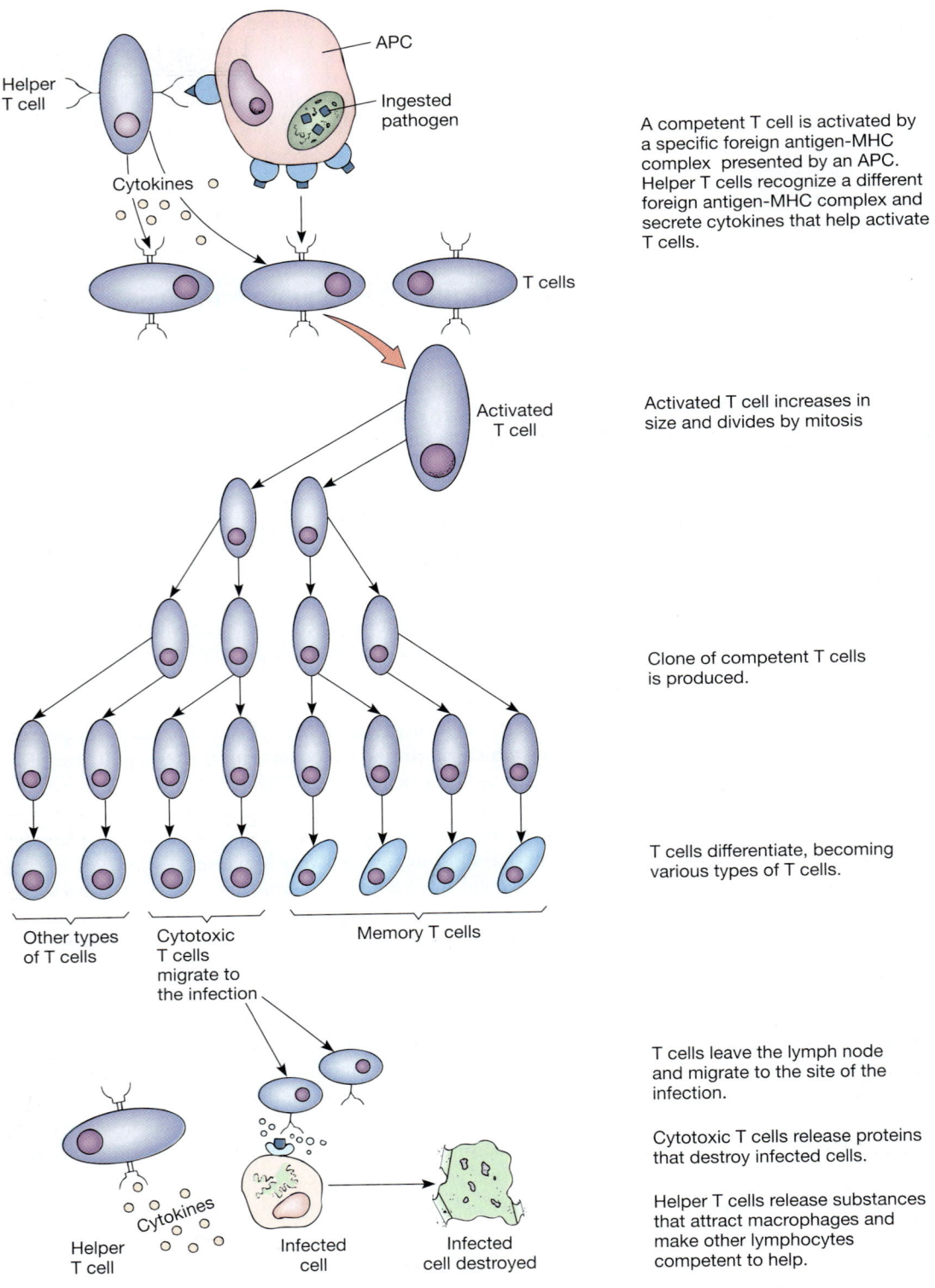

APC

Helper T cell

Ingested pathogen

Cytokines

T cells

A competent T cell is activated by a specific foreign antigen-MHC complex presented by an APC. Helper T cells recognize a different foreign antigen-MHC complex and secrete cytokines that help activate T cells.

Activated T cell

Activated T cell increases in size and divides by mitosis

Clone of competent T cells is produced.

T cells differentiate, becoming various types of T cells.

Other types of T cells

Cytotoxic T cells migrate to the infection

Memory T cells

T cells leave the lymph node and migrate to the site of the infection.

Cytotoxic T cells release proteins that destroy infected cells.

Helper T cells release substances that attract macrophages and make other lymphocytes competent to help.

Helper T cell

Cytokines

Infected cell

Infected cell destroyed

Figure 43–9   Cell-mediated immunity. When activated by a foreign antigen-MHC complex presented by an APC and by cytokines, a T cell divides, giving rise to a clone of cells. Some of these cells become cytotoxic T cells, which migrate to the site of infection and release proteins that destroy invading pathogens. Other cells become helper T cells, suppressor cells, or memory T cells.

cold and flu virus evolve continuously by mutation (a survival mechanism for them), which may result in changes in their surface antigens. Even a slight change may prevent recognition by memory cells. Because the immune system is so specific, each different antigen is treated by the body as a new immunological challenge.

Active immunity can be induced with immunization

We have been considering **active immunity,** immunity developed following exposure to antigens. When you have chicken pox as a young child, for example, you develop immunity that protects you from contracting it again. Active immunity can be *naturally* or *artificially* induced (Table 43–1). If someone with chicken pox sneezes near you and you contract the disease, you develop active immunity naturally. Active immunity can also be artificially induced by **immunization,** that is, by exposure to a **vaccine.**

The term vaccination is derived from the first vaccine which was prepared in 1796 by Edward Jenner against vaccinia, the cowpox virus. This vaccine provided humans immunity to smallpox, a deadly disease. Jenner had no knowledge of microorganisms or of immunology, and it remained for Louis Pasteur 100 years later to begin to develop scientific methods for preparing vaccines. Pasteur showed that inoculations with preparations of attenuated (weakened) pathogens could be used to develop immunity against the virulent form of the pathogen. However, it was not until 20th century advances in immunology, for example, Burnet's clonal selection theory in 1957 and the discovery of T and B cells in 1965, that a modern understanding of vaccines was developed. Effective vaccination depends on stimulating the body to launch an immune response against the antigens contained in the vac-

Figure 43–10 **Immunological memory.** Antigen 1 was injected at day 0, and the immune response was assessed by measuring antibody levels to the antigen. At week 4, the primary response had subsided. Antigen 1 was injected again, together with a new protein, antigen 2. The secondary response was greater and more rapid than the primary response. It was also specific to antigen 1. A primary response was made to the newly encountered antigen 2.

cine. Memory cells develop, and future encounters with the same pathogen are dealt with swiftly.

Effective vaccines are prepared in a number of ways. A virus may be attenuated by successive passage through cells of nonhuman hosts. In the process, mutations adapt the pathogen to the nonhuman host so that it can no longer cause disease in humans. This is how Sabin polio vaccine and measles vac-

TABLE 43–1   Active and Passive Immunity

Type of Immunity	When Developed	Development of Memory Cells	Duration of Immunity
Active			
Naturally induced	After pathogens enter the body through natural encounters (e.g., person with measles sneezes on you)	Yes	Many years
Artificially induced	After immunization with a vaccine	Yes	Many years
Passive			
Naturally induced	After transfer of antibodies from mother to developing baby	No	Few months
Artificially induced	After injection with gamma globulin	No	Few months

cine are produced. Whooping cough and typhoid fever vaccines are made from killed pathogens that still have the necessary antigens to stimulate an immune response. Tetanus and botulism vaccines are made from toxins secreted by the respective pathogens. The toxin is altered so that it can no longer destroy tissues, but its antigenic determinants are still intact.

Most vaccines consist of the entire pathogen, attenuated or killed, or of a protein from the pathogen. Researchers are investigating a number of approaches that would reduce potential side effects. For example, they are designing vaccines consisting of synthetic peptides that are only a small part of the antigen. Another new approach is development of DNA vaccines, made from a part of the pathogen's genetic material that codes for antigens. When any vaccine is introduced into the body, the immune system actively develops clones of cells, produces antibodies, and develops memory cells.

In **passive immunity,** an individual is given antibodies actively produced by another organism. The serum or gamma globulin that contains these antibodies can be obtained from humans or animals. Animal sera are less desirable because their nonhuman proteins can themselves act as antigens, stimulating an immune response that may result in an illness known as serum sickness.

Passive immunity is borrowed immunity, and its effects are not lasting. It is used to boost the body's defense temporarily against a particular disease. For example, when someone is diagnosed with hepatitis A, a form of viral hepatitis that can be spread through contaminated food or water, persons at risk of infection can be injected with gamma globulin containing antibodies to the hepatitis pathogen. However, such gamma globulin injections offer protection for only a few weeks. Because the body has not actively launched an immune response, it has no memory cells and cannot produce antibodies to the pathogen. Once the injected antibodies are broken down, the immunity disappears.

Pregnant women confer natural passive immunity on their developing babies by manufacturing antibodies for them. These maternal antibodies, of the IgG class, pass through the placenta (the organ of exchange between mother and developing child) and provide the fetus and newborn infant with a defense system until its own immune system matures. Babies who are breastfed continue to receive immunoglobulins, particularly IgA, in their milk.

MONOCLONAL ANTIBODIES ARE HIGHLY SPECIFIC

In the 1970s Cesar Milstein and Georges Kohler at the Laboratory of Molecular Biology in Cambridge, England, developed **monoclonal antibodies,** identical antibodies produced by cells cloned from a single cell. To produce monoclonal antibodies in the laboratory, mice are injected with the antigen of interest, for example, antigens from human cancer cells. Af-ter the mice have produced antibodies to the antigen, their B cells are collected. These cells survive in culture for only a few generations. In contrast, cancer cells can live and divide in tissue culture indefinitely. The B cells can be suspended in a culture medium together with myeloma (a type of cancer) cells from other mice, and the cells can be induced to fuse. The hybrid cells, called **hybridomas,** have properties of the two "parent" cells. They can be cultured indefinitely (a cancer cell property) and continue to secrete antibodies (a B cell property).

Researchers select hybrid cells that are manufacturing the specific antibody needed and then clone them in a separate cell culture. Cells of this clone secrete large amounts of the specific antibody, thus, the name monoclonal antibodies. Each type of monoclonal antibody is specific for a single antigenic determinant.

Because of their purity and specificity, monoclonal antibodies have proven to be invaluable tools in modern biology. For example, a researcher may want to be able to detect a particular molecule even when it is present in very small amounts in a mixture. The reaction of that molecule (the antigen) with a specific monoclonal antibody makes its presence known. Monoclonal antibodies are used in similar ways in various diagnostic tests. For example, the commonly used home pregnancy tests make use of a monoclonal antibody that is specific for hCG, a hormone produced by a developing human embryo (see Chapters 48 and 49). Therapeutic uses for monoclonal antibodies, especially for cancer, will be discussed shortly.

GRAFT REJECTION IS AN IMMUNE RESPONSE AGAINST TRANSPLANTED TISSUE

Skin can be successfully transplanted from one part of the same body to another or from one identical twin to another. However, when skin is taken from one person and grafted onto the body of a nontwin, it is rejected and sloughs off. Why? Recall that tissues from the same individual or from identical twins have identical MHC alleles and thus the same MHC antigens. Such tissues are compatible.

Because there are many alleles for each of the MHC genes, it is difficult to find identical matches. When a tissue or organ is taken from a donor and transplanted to the body of a nontwin host, several of the MHC antigens are likely to be different. The host's immune system regards the graft as foreign and launches an immune response called **graft rejection.** T cells attack the transplanted tissue and can destroy it within a week (Fig. 43–11).

Before transplants are performed, tissues from the patient and from potential donors must be typed and matched as closely as possible. Cell typing is somewhat similar to blood typing but is more complex. If all of the MHC antigens are

Figure 43–11 Graft rejection.

matched, the graft has about a 95% chance of surviving the first year. Unfortunately, not many persons are lucky enough to have an identical twin to supply spare parts, so perfect matches are difficult to find. Furthermore, some parts such as the heart cannot be spared. Most organs to be transplanted, therefore, are removed from unrelated donors, often from patients who have just died.

To prevent graft rejection in less compatible matches, physicians use drugs that suppress the immune system. Unfortunately, while these drugs reduce graft rejection, they also make the transplant patient more vulnerable to pneumonia or other infections and increase the risk of certain types of tumor growths. If the patient can survive the first few months, dosages of these drugs can be reduced. Researchers are developing specific immunosuppression techniques that will act only on the specific lymphocyte clones that cause graft rejection.

A few immunologically privileged locations exist in which foreign tissue is accepted by a host. For example, corneal transplants are highly successful because the cornea has almost no blood or lymphatic vessels associated with it and so is out of reach of most lymphocytes. Furthermore, antigens in the corneal graft probably would not find their way into the circulatory system and so would not stimulate an immune response.

Within the United States alone, thousands of patients are in need of organ transplants. Because the number of human donors does not meet this need, investigators are developing techniques for transplanting animal tissues and organs to humans, a procedure known as *xenotransplantation.* Challenges include differences in organ size between the animal donor and the human recipient, risks of transmitting diseases, and graft rejection. Researchers are developing methods for genetically engineering pigs and other animals so that they do not produce antigens that stimulate immune responses in human recipients. These animals could then be cloned and used as donors of hearts, kidneys, and other organs.

THE BODY DEFENDS ITSELF AGAINST CANCER

A few normal cells may be transformed into precancer cells daily in each of us in response to radiation, certain viruses, or chemical carcinogens in the environment. Because they are abnormal cells, some of their surface proteins may be different from those of normal body cells. Such proteins may act as antigens, stimulating an immune response that typically destroys these abnormal cells. Until recently, immunologists thought that, in this way, the immune system provided surveillance against malignancy. Investigators now think that the mechanisms involved are more complex. For example, evidence suggests that immune responses may be more effective against cancers caused by oncogenic viruses than against other types of cancer. Research findings also suggest that rather than having unique antigens, cancer cells may produce much larger quantities of certain antigens than are produced by other cells.

Although many components of the immune system help defend against cancer cells, NK cells and cytotoxic T cells appear to be most important (Fig. 43–12). T cells produce interleukins, which attract macrophages and NK cells and activate them. The T cells also produce interferons, which have an antitumor effect. The macrophages themselves produce factors, including TNF (tumor necrosis factor), that inhibit tumor growth.

Immunologists have discovered that cancer cells can also destroy lymphocytes. Some T cells have a surface molecule referred to as Fas ligand (FasL). When FasL binds with FAS, a receptor on many cell surfaces, the cell self-destructs. In 1996 Peter Galle, of the University of Heidelberg in Germany, and his colleagues reported that human liver cancer cells can produce FasL and use it to kill T cells.

When monoclonal antibodies were first discovered, they were hailed as a promising new treatment for cancer. When injected into the same patient whose cancer cells were used to stimulate their production, monoclonal antibodies may destroy the cancer cells. Monoclonal antibodies can be tagged with toxic drugs that are then delivered specifically to the cancer cells. Such targeted drug therapy would be expected to cause fewer side effects than current types of chemotherapy, which damage normal cells as well as cancer cells. Although such techniques have been very successful in animal studies, investigators have encountered serious problems in applying these techniques to humans. For example, the human immune system responds strongly to the injected mouse antibodies. Despite setbacks spanning more than 20 years, monoclonal antibodies are emerging as effective therapies for a few cancers. Immunologists are developing new approaches that will make the injected antibodies more like human antibodies, so that they are less likely to elicit an immune response. Another strategy is to target antigens that are growth factors needed by the cancer cells. Other investigators are targeting signals that induce blood vessel formation (angiogenesis) in tumors.

(a)   10 μm

(b)   10 μm

(c)   10 μm

Figure 43–12   **SEMs showing cancer cell destruction.**
(a) An army of cytotoxic T cells surrounds a large cancer cell. The
T cells recognize the cancer cell as nonself because it displays altered
or unique antigens on its surface. *(b)* Some of the cytotoxic T cells
elongate as they chemically attack the cancer cell, breaking down its
plasma membrane. *(c)* The cancer cell has been destroyed. Only a
collapsed fibrous cytoskeleton remains.   *(Lennart Nilsson, Boehringer
Ingelheim International GmbH)*

MALFUNCTION OF THE IMMUNE SYSTEM CAN CAUSE ALLERGY AND DISEASE

The immune system is extremely complex and so presents
many opportunities for mechanisms to malfunction. **Hyper-sensitivity** refers to an exaggerated, damaging immune response to an antigen that is normally harmless. Inappropri-
ately directed or ineffective immune responses can also result
in disease states. The absence or failure of normal function of
the immune system can result in **immunodeficiency disease,**
a condition that causes increased susceptibility to infection.
Worldwide, the leading cause of immunodeficiency in children
is protein malnutrition. Lack of protein causes a decrease in T
cell numbers, resulting in increased risk for opportunistic infections. AIDS and other infections can also cause immunodeficiency.

Allergic reactions are directed against ordinary environmental antigens

About 20% of the population of the United States is plagued by an allergic disorder such as allergic asthma or hayfever. A predisposition toward these disorders appears to be inherited. In **allergic reactions,** hypersensitivity results in the manufacture of antibodies against mild antigens, called **allergens,** that normally do not stimulate an immune response. Common environmental agents such as house-dust mites or pollen can trigger allergic reactions in some individuals. Allergic reactions are referred to as Type I hypersensitivity. In many kinds of allergic reactions, distinctive IgE immunoglobulins are produced.

Let us examine a common allergic reaction, a hayfever response to ragweed pollen (Fig. 43–13). The first step is *sensitization.* Macrophages degrade the allergen and present fragments of it to T cells. The activated T cells then stimulate B cells to become plasma cells and produce IgE. These antibodies attach to receptors on mast cells. Each IgE molecule attaches to a receptor by its C region end, leaving the V region end free to combine with the ragweed pollen allergen.

The second step is *activation of mast cells.* When a sensitized, allergic person inhales the microscopic pollen, allergen molecules rapidly attach to the IgE on mast cells. This binding of allergen with IgE antibody stimulates the mast cell to release granules filled with chemicals like histamine and serotonin that cause inflammation (Fig. 43–14). These substances cause blood vessels to dilate and capillaries to become more permeable, leading to edema and redness. Such responses cause the victims' nasal passages to become swollen and irritated. Their noses run, they sneeze, their eyes water, and they feel generally uncomfortable.

A third step may occur in which *the allergic response is prolonged.* Chemical compounds released by the mast cells lure certain white blood cells to leave the circulation and migrate to the inflamed area. These cells then release compounds that damage tissue and prolong the allergic reaction.

Allergens on pollen → plasma cells sensitized → allergen-specific IgE released → IgE combines with mast cell receptors → bound IgE combines with allergen → mast cells release granules containing histamine and other chemicals → allergic symptoms

In **allergic asthma,** an allergen-IgE response occurs in the bronchioles of the lungs. Mast cells release substances that cause smooth muscle to contract. The airways in the lungs

1. Previous exposure to pollen results in plasma cells making pollen-specific IgE.

2. IgE combines with mast cell receptors in the lining of the nasal passages.

3. Pollen is inhaled.

4. Allergen combines with variable region of the IgE on surface of sensitized mast cell.

5. Mast cell releases histamine and other chemicals.

6. This release causes
   -Increased vasodilation
   -Increased capillary permeability resulting in...

   ...edema, redness, constriction of the respiratory passageways.

Pollen grains

Nasal mucosa

Plasma cell

IgE

Soluble antigens

Mast cell

Histamine and other compounds

Hay fever symptoms

Figure 43–13 **Allergic response.** Pollen causes a common type of allergic response in many people.

Figure 43–14 **SEM showing release of granules by mast cells.** When an allergen combines with IgE bound to a mast cell receptor, the mast cell explosively releases granules filled with histamine and other compounds that cause the symptoms of an allergic response. *(Lennart Nilsson, Boehringer Ingelheim International GmbH)*

sometimes constrict for several hours, making breathing difficult.

Certain foods or drugs act as allergens in some persons, causing a reaction in the walls of the digestive tract that leads to discomfort and diarrhea. The allergen may be absorbed and cause mast cells to release granules elsewhere in the body. When the allergen-IgE reaction takes place in the skin, the histamine released by mast cells causes swollen red welts known as **hives.**

Systemic anaphylaxis is a dangerous allergic reaction that can occur when a person develops an allergy to a specific drug such as penicillin, to compounds in the venom injected by a stinging insect, or even to certain foods. Within minutes after the substance enters the body, a widespread allergic reaction takes place. Mast cells release large amounts of histamine and other compounds into the circulation. These compounds cause extreme vasodilation and permeability. So much plasma may be lost from the blood that circulatory shock and death can occur within a few minutes.

The symptoms of allergic reactions are often treated with **antihistamines,** drugs that block the effects of histamines. These drugs compete for the same receptor sites on cells targeted by histamine. When the antihistamine combines with the receptor, it prevents the histamine from binding and thus prevents its harmful effects. Antihistamines are useful clinically

in relieving the symptoms of hives and hayfever. They are not completely effective because mast cells release substances other than histamines that also cause allergic symptoms.

In serious allergic disorders, patients are sometimes given a form of immunotherapy known as desensitization. Small amounts of the allergen are injected weekly over a period of months or years. Just how this treatment works is not completely understood, but production of IgG antibody to the allergen or inducing T cell tolerance may be involved.

In an autoimmune disease the body attacks its own tissues

During lymphocyte development, complex mechanisms establish self-tolerance (self recognition) so that these cells do not attack tissues of their own body. However, some lymphocytes remain that have the potential to be **autoreactive,** that is to launch an immune response against self-issues. Such autoreactivity can lead to a form of hypersensitivity known as **autoimmune disease** in which T cells react immunologically against self. Some of the diseases that result from such failure in self-tolerance are rheumatoid arthritis, multiple sclerosis, systemic lupus erythematosus (SLE), insulin-dependent diabetes, psoriasis, and scleroderma.

In autoimmune diseases, antibodies and T cells attack the body's own tissues (see *On the Cutting Edge: Using Gene Therapy to Treat Autoimmune Disease*). In rheumatoid arthritis, T cells in the area of inflammation produce IL-15, an interleukin that promotes inflammation. In multiple sclerosis, antibodies attack the glial cells that produce the myelin sheath surrounding neurons in the brain and spinal cord. Magnetic resonance images (MRIs) show the absence of myelin sheaths around axons that are normally myelinated. Patients generally suffer weakness and visual problems, and become progressively more disabled. Genetic risk factors play a role in multiple sclerosis. For example, when one identical twin has multiple sclerosis, the other twin has about a 30% chance of contracting the disease also.

Studies indicate that viral or bacterial infection often precedes the onset of an autoimmune disease. Some pathogens have evolved a tactic known as molecular mimicry. They trick the body by producing molecules that look like self-molecules. For example, an adenovirus that causes respiratory and intestinal illness produces a peptide that mimics myelin protein. When the body launches responses to the adenovirus peptide, it may also begin to attack the similar self-molecule, myelin.

AIDS is an immune disease caused by a retrovirus

Human immunodeficiency virus (HIV) was first isolated in 1983 and was shown to be the cause of **acquired immune deficiency syndrome (AIDS)** in 1984. HIV, a retrovirus, has probably been studied more than any other virus. (Recall that a retrovirus is an RNA virus that uses its RNA as a template to make DNA with the help of reverse transcriptase.) Several

Using Gene Therapy to Treat Autoimmune Disease

HYPOTHESIS: Autoreactive T cells can be genetically modified to deliver therapeutic agents to a region of inflammation caused by an autoimmune response.

METHOD: Researchers used autoreactive T cells to treat mice that can be used as a model for multiple sclerosis (MS), an autoimmune disease. These T cells had been genetically altered to express an anti-inflammatory cytokine in the area of inflammation.

RESULTS: The genetically modified T cells migrated to the area of inflammation caused by the autoimmune response and inhibited the progression of the disease.

CONCLUSION: Genetically modified T cells can be used as vectors to deliver anti-inflammatory agents to treat an autoimmune disease.

Immunologists have recognized that the ability of T cells to migrate to areas of inflammation and to target specific antigens makes them attractive candidates for delivering therapeutic agents to treat autoimmune disease. One subset of helper T cells (Th1) promotes inflammation by expressing inflammatory cytokines such as interleukin 2 (IL-2) and tumor necrosis factor alpha. Another helper T cell subset (Th2) expresses anti-inflammatory cytokines IL-4 and IL-10. Could autoreactive T cells (those that act against an individual's own tissues) be genetically engineered to deliver anti-inflammatory cytokines to areas of the body inflamed by autoimmune disease?

Immunologists Peter Mathisen, Vincent Tuohy, and their colleagues at the Lerner Research Institute at The Cleveland Clinic Foundation, genetically modified autoreactive T cells in an effort to develop a treatment for experimental allergic encephalomyelitis (EAE) in mice, a model for the human autoimmune disease multiple sclerosis (MS).* The inflammation characteristic of EAE ap-

pears to be mediated, at least in part, by Th1 cells that react to myelin proteins in the central nervous system (CNS). These cells produce cytokines that cause inflammation and trigger events that result in deterioration of myelin (demyelination) of neurons in the CNS. Mathisen and Tuohy used autoreactive memory T cells as vectors for delivering anti-inflammatory cytokines to areas of inflammation associated with EAE.

Autoreactive memory T cells were genetically modified to express an anti-inflammatory cytokine, IL-10, only when activated by the presence of a particular myelin protein (the autoantigen). This was accomplished by placing the IL-10 cDNA under the control of an IL-2 promoter region that could be induced only by myelin. In this way production of IL-10 could be limited to the part of the mouse body affected by the autoimmune disease. These researchers demonstrated that the modified T cells could both inhibit the onset of the disease and slow its course after the inflammatory process and neurological symptoms had developed. A significant correlation was found between treatment with the genetically engineered T cells and decreased symptoms (i.e., decreased demyelination).

Using genetically modified T cells as vectors in treating autoimmune disease appears to be safer than using viral vectors because immunologists have found that viral vectors can themselves trigger inflammation. Gene therapy is also being intensively investigated as a treatment for immune deficiency diseases (gene therapy for SCID is discussed in Chapter 15).

*Mathisen, P.M., and V.K. Tuohy. "Gene Therapy in the Treatment of Autoimmune Disease." *Immunology Today,* Vol. 19, No. 3, Mar. 1998. Mathisen, P.M., Yu, M., Johnson, J.M., Drazba, J.A., and V.K. Tuohy. "Prevention and Treatment of Experimental Autoimmune Encephalomyelitis (EAE) with Genetically Modified Autoreactive T-Cells." *Journal of Allergy and Clinical Immunology,* Vol. 99, No. 1, Part 2, supplement, Jan. 1997, p. 1975.

---

different strains of the virus are known, and some may be more virulent than others. HIV damages the immune system by destroying helper T cells (Fig. 43–15). As a consequence of this immunosuppression, AIDS patients usually die from rare forms of cancer, pneumonia, and other opportunistic infections.

AIDS is a deadly disease currently spreading through the world's human population at an alarming rate. Epidemiologists estimate that more than 30 million people worldwide are now infected with the AIDS virus. These numbers may reflect only a small percentage of the actual number of individuals infected; an estimated 6 million new infections occur each year. More than 12 million people have died from AIDS since the beginning of the epidemic.

HIV is transmitted mainly during sexual intercourse with an infected person or by direct exposure to infected blood or blood products. This virus is not spread by casual contact. People do not contract the virus by hugging, casual kissing, or using the same bathroom facilities. Friends and family members who live with AIDS patients are not more likely to get the virus.

Most persons currently infected in the United States are males who engage in homosexual and bisexual behavior and individuals who use intravenous drugs. New infections are increasing most rapidly among women and teenagers who contract the virus through heterosexual contact. Heterosexual contact with infected individuals accounts for more than one-third of HIV infections in women and an increasing number of cases

(a)

1 μm

(b)

0.25 μm

(c)

0.25 μm

(d)

0.1 μm

Figure 43–15 **SEMs of HIV infection of a helper T cell.**
(a) HIV virus particles *(blue)*, which cause AIDS, attack a helper T cell *(light green)*. *(b)* HIV-1 virus particles budding from the ends of the branched microvilli of a T cell. *(c)* An even higher magnification of virus particles budding from a "bleb" (a cytoplasmic extension broader than a microvillus). Note the pentagonal symmetry often evident in biological branching and flowering structures. *(d)* HIV-1 virus particles at extremely high magnification. The surfaces are grainy and the outlines slightly blurred because the preparation was coated with coarse-grained heavy metal (palladium) salt.   *(Lennart Nilsson, Boehringer Ingelheim International GmbH)*

in both men and women. Use of a latex condom during sexual intercourse provides some protection against the virus. Use of a spermicide containing nonoxynol-9 is thought to provide additional protection.

An estimated 10% of AIDS patients are children born to infected mothers, but recent drug treatment protocols have reduced the rate of HIV transmission between mother and fetus. Effective blood-screening procedures have been developed to safeguard blood bank supplies, markedly reducing the risk of infection from blood transfusion in developed countries.

Response to exposure with HIV and progression of AIDS may depend on a combination of genetic and environmental factors. Some evidence suggests that susceptibility is affected by psychosocial factors, including personality variables and coping styles.

When HIV infects the body, an immune response may be launched. About 15% of infected individuals experience mild flulike symptoms (fever and aching muscles) for a week or so. Most others have no symptoms. However, some cells infected by HIV are not destroyed, and the virus may continue to replicate slowly for many years. After a time, a progression of symptoms occurs, including swollen lymph glands, night sweats,

fever, and weight loss (Fig. 43–16). In about one-third of AIDS patients, the virus infects the nervous system, causing AIDS dementia complex. These patients exhibit progressive cognitive, motor, and behavioral dysfunction that typically ends in coma and death. The patient may develop serious opportunistic infections and malignancies may develop. Kaposi's sarcoma is an endothelial cell tumor that causes purplish spots on the skin.

HIV enters a host cell by attaching a protein on its outer envelope to CD4, a protein present on the surface of several types of immune system cells. Helper T cells (CD4 T cells) appear to be the main target of the virus (chapter opening figure and Fig. 43–15). HIV destroys helper T cells and over time causes a dramatic decrease in their numbers. When the helper T cell population is depressed, the ability to resist infection is severely impaired, making the AIDS patient more susceptible to cancer and other opportunistic infections.

Researchers throughout the world are searching for drugs that will successfully combat the AIDS virus. Because HIV often infects the central nervous system, an effective drug must cross the blood-brain barrier. **AZT** (azidothymidine, a deoxyribonucleotide analog), the first drug developed to treat

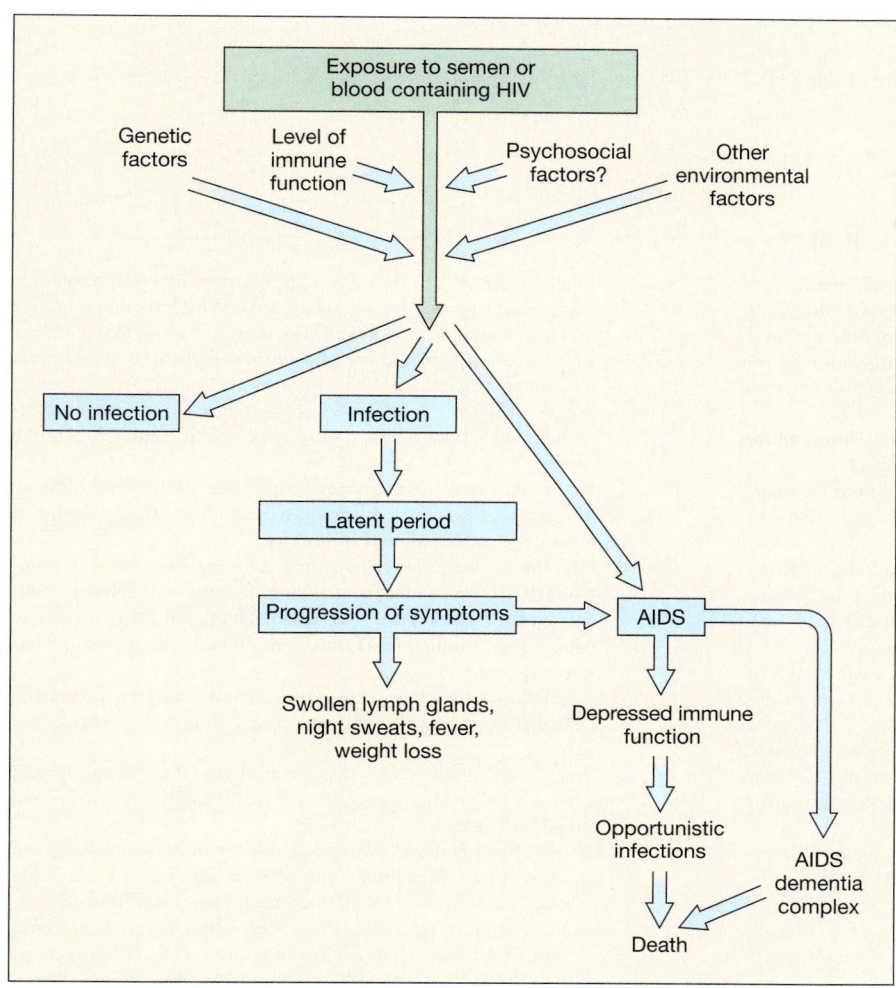

Figure 43–16 **HIV infection.** Exposure to semen or blood that contains HIV can lead to AIDS. Although some exposed individuals apparently do not become infected, the risk increases with multiple exposures. Many factors apparently determine whether a person exposed to the AIDS virus contracts the disease.

HIV infection, can prolong the period prior to the onset of AIDS symptoms. AZT blocks HIV replication by inhibiting the action of reverse transcriptase, the enzyme used by the retrovirus to synthesize DNA. Without producing DNA, the virus cannot incorporate itself into the host cell's DNA. Unfortunately, the reverse transcriptase used by the virus to synthesize DNA makes many mistakes: about one in every 2000 nucleotides it incorporates is incorrect. By mutating in this way, the virus has developed strains that are resistant to AZT.

Protease inhibitors are a more recently developed group of anti-HIV drugs. These drugs block the viral enzyme protease resulting in viral copies that cannot infect new cells. Protease inhibitors can be used in combination with AZT and other anti-HIV drugs. Triple combination treatment has been very effective for many AIDS patients, preventing opportunistic infections and prolonging life. In fact, combination treatment has led to a recent decline in AIDS incidence and mortality in the United States. Unfortunately, people in many parts of the world cannot afford these expensive drugs, and in those poor areas the AIDS epidemic continues to grow.

In January 1998 Ralph Nachman and his colleagues at New York Hospital–Cornell Medical Center reported that a glycoprotein (thrombospondin) found in saliva can block the growth of laboratory strains of HIV. This finding explains why only small amounts of HIV are found in the saliva of an HIV positive person. Nachman suggested that this glycoprotein could be used in condoms and other products to block HIV transmission.

Worldwide containment of the AIDS epidemic will require an effective vaccine that prevents the spread of HIV. Developing such a vaccine has been a most daunting challenge for immunologists. Because of its high rate of mutations, new viral strains with new antigens evolve quickly. A vaccine would not be effective against new antigens and so would quickly become obsolete. An effective vaccine would also have to overcome the problem created by viral destruction of key cells needed to mount an immune response. Other barriers to the development of a vaccine include the absence of an effective animal model for AIDS and the ethical and practical difficulties associated with using human volunteers to test the vaccine.

While immunologists work to develop a successful vaccine and more effective drugs to treat infected patients, massive educational programs are being developed to slow the spread of HIV. Educating the public that having multiple sexual partners increases the risk of AIDS and teaching sexually active individuals the importance of "safe" sex may help to slow the epidemic. Some have suggested that public health facilities offer free condoms to those who are sexually active and free sterile hypodermic needles to those addicted to drugs. The cost of these measures would be far less than the cost of medical care for increasing numbers of AIDS patients and the toll in human suffering.

SUMMARY WITH KEY TERMS

I. The body defends itself against **pathogens** and other foreign agents. The study of internal defense mechanisms is called **immunology.** An **immune response** involves recognizing foreign macromolecules and mounting a response aimed at eliminating them. Thus, immune responses depend on the ability of an organism to distinguish between self and nonself.
   A. **Nonspecific defense mechanisms,** also called **innate immune responses,** provide general protection against pathogens.
   B. **Specific defense mechanisms,** also known as **acquired** or **adaptive immune responses,** target specific macromolecules associated with a pathogen.
   C. An **antigen** is a substance specifically recognized as foreign by cells of the immune system. **Antibodies** are highly specific proteins produced by animals in response to antigens. Antibodies bind with antigens.

II. Many invertebrates depend mainly on nonspecific defense mechanisms such as physical barriers (cuticle, skin, mucous membranes) and phagocytosis.

III. Vertebrate nonspecific defense mechanisms include physical barriers, such as the skin and mucous lining of the respiratory and digestive tracts. Should pathogens break through these first line defenses, other nonspecific defense mechanisms are activated.
   A. **Cytokines** are regulatory proteins that mediate interactions between cells.
      1. **Interferons** inhibit viral replication and enhance activities of immune cells.
      2. **Interleukins** help regulate various cells of the immune system and also affect a variety of other cells in the body.
      3. **Tumor necrosis factors (TNF)** can kill tumor cells and can stim-

ulate immune cells to initiate an **inflammatory response.**
   B. Some **complement** proteins lyse the cell wall of the pathogen. Some coat the pathogen, enhancing phagocytosis, while others attract white blood cells to the site of infection. Still others increase the inflammatory response.
   C. When pathogens invade tissues, they trigger an inflammatory response, which brings needed phagocytic cells and antibodies to the infected area.
   D. Neutrophils and **macrophages** phagocytize and destroy bacteria.

IV. Vertebrate specific defense mechanisms include **antibody-mediated immunity** and **cell-mediated immunity.**
   A. Cells of the immune system include: **phagocytes** such as **neutrophils** and macrophages; and **lymphocytes—T cells, B cells,** and **natural killer cells (NK cells). Dendritic cells,** present in many tissues throughout the body, also function as part of the immune system.
   B. Macrophages, B cells, and dendritic cells are **antigen-presenting cells** that display foreign antigens as well as their own surface proteins.
   C. Natural killer cells (NK cells) use both specific and nonspecific methods to destroy tumor cells and cells infected with viruses and other pathogens.
   D. B cells (B lymphocytes) are responsible for antibody-mediated immunity. B cells differentiate into **plasma cells** which produce antibodies. Some activated B cells become **memory B cells** which continue to produce antibodies after the infection has been overcome.
   E. T cells (T lymphocytes), distinguished by the **T cell antigen receptor (TCR),** are responsible for cellular immunity. The **thymus gland** confers immunological competence on T cells by making

them capable of distinguishing between self and nonself. Among the several types of T cells are the **cytotoxic T cells (CD8 T cells)**, **helper T cells (CD4 T cells)**, and **memory T cells.**

   F. Immune responses depend on protein antigens coded for by a group of genes known as the **major histocompatibility complex (MHC).**

V. In antibody-mediated immunity, B cells are activated when they combine with antigen. Activation requires an antigen-presenting cell (such as a macrophage) that has a foreign antigen-MHC complex displayed on its surface. A helper T cell that secretes interleukins is also needed.

   A. Activated B cells multiply, giving rise to clones of cells.

   B. Some B cells differentiate to become plasma cells, which secrete specific antibodies. Others become memory B cells.

   C. Antibodies are highly specific proteins, also called **immunoglobulins (Ig),** that are produced in response to specific antigens.

      1. A typical antibody is Y-shaped; the two arms of the Y combine with antigen. An antibody molecule consists of four polypeptide chains: two identical heavy chains and two shorter chains, called light chains. Each chain has a C region (constant segment), J region (junctional segment), and V region (variable segment).

      2. Antibodies are grouped in five classes determined by differences in the constant regions of their heavy polypeptide chains. **IgG** and **IgM** defend against pathogens in the blood. **IgA** prevents pathogens from attaching to epithelial surfaces. **IgD** helps activate B cells after they bind with antigens. **IgE** is important in allergic reactions.

      3. A given antibody can bind with different strengths, or **affinities,** to different antigens. Antibodies with higher affinity are produced during an immune response.

   D. An antibody combines with a specific antigen to form an **antigen-antibody complex,** which may inactivate the pathogen, stimulate phagocytosis, or activate the complement system.

VI. In cell-mediated immunity, specific T cells are activated by helper T cells and by a foreign antigen-MHC complex on a cell surface.

   A. Activated T cells multiply, giving rise to a clone.

   B. Some T cells differentiate to become cytotoxic T cells, which migrate to the site of infection and chemically destroy cells infected with viruses.

   C. Some activated T cells remain in the lymph nodes as memory T cells; others become helper T cells.

VII. Memory B and memory T cells are responsible for long-term immunity.

   A. The first exposure to an antigen stimulates a **primary response.** Second exposure to the same antigen evokes a **secondary immune response,** which is more rapid and more intense than the primary response.

   B. **Active immunity** develops as a result of exposure to antigens; it may occur naturally after recovery from a disease or be artificially induced by immunization with a **vaccine. Passive immunity** is a temporary condition that develops when an individual receives antibodies produced by another person or animal.

   C. **Monoclonal antibodies** are identical antibodies produced by cells cloned from a single cell. They are important tools in biological research.

   D. Transplanted tissues have MHC antigens that stimulate **graft rejection,** an immune response in which T cells destroy the transplant.

   E. Normally, the immune system destroys precancer cells when they arise; diseases such as cancer develop when this immune mechanism fails to operate effectively. **Monoclonal antibodies** can be produced that are specific for a particular cancer antigen; these antibodies may prove effective therapies for certain types of cancer.

VIII. Malfunction of the immune system can lead to allergy, autoimmune disease, or immunodeficiency.

   A. In an **allergic reaction,** an **allergen** can stimulate production of IgE, which combines with the receptors on **mast cells.** When the allergen combines with the IgE, the mast cells release **histamine** and other substances that cause symptoms of allergy such as inflammation. **Systemic anaphylaxis** is a rapid, widespread allergic reaction that can lead to death.

   B. In **autoimmune diseases,** the body reacts immunologically against its own tissues.

   C. **Acquired immune deficiency syndrome (AIDS)** is caused by the retrovirus known as **human immunodeficiency virus (HIV).** HIV damages the immune system by destroying helper T cells. The ability to resist infection is severely impaired, putting the patient at risk for opportunistic infections.

POST-TEST

1. A substance that is recognized as foreign by cells of the immune system is a(an) (a) antibody (b) antigen (c) immunoglobulin (d) interferon (e) cytokine

2. Nonspecific (innate) defense mechanisms include (a) physical barriers like the skin (b) antigen-antibody complexes (c) immunoglobulin action (d) complement and memory T cells (e) interferon and memory B cells

3. Invertebrate defense mechanisms include (a) phagocytosis (b) nonspecific defense mechanisms (c) ability to distinguish between self and nonself (d) answers a, b, and c are correct (e) answers b and c only

4. Cytokines (a) are regulatory nucleic acids (b) prevent the inflammatory response (c) include interferons and interleukins (d) are immunoglobulins (e) include complement proteins

5. Which of the following is NOT an action of complement? (a) enhance phagocytosis (b) enhance inflammatory response (c) coat pathogens (d) stimulate histamine release (e) stimulate allergen release

6. Which of the following cells are antigen-presenting cells? (a) NK cells (b) plasma cells (c) macrophages (d) memory B cells (e) memory T cells

7. Which of the following cells are especially adept at destroying tumor cells? (a) NK cells (b) plasma cells (c) neutrophils (d) cytotoxic B cells (e) mast cells

8. Which of the following cells become immunologically competent after processing in the thymus gland? (a) NK cells (b) T cells (c) macrophages (d) B cells (e) plasma cells

9. Cells that have a surface marker called CD4 are (a) NK cells (b) cytotoxic T cells (c) helper T cells (d) B cells (e) suppressor cells

10. The major histocompatibility complex (MHC) (a) codes for a group of cell surface proteins (b) codes for proteins that form complexes with certain cytokines (c) is important mainly in allergic reactions (d) suppresses complement release from macrophages (e) is Y-shaped

11. Which sequence most accurately describes antibody-mediated immunity?
    1. B cell divides and gives rise to clone
    2. antibodies produced
    3. cells differentiate, forming plasma cells
    4. activated helper T cell interacts with B cell displaying same antigen complex
    5. B cell activated
    (a) 1→2→3→4→5 (b) 3→2→1→4→5 (c) 4→5→3→4→5 (d) 4→5→1→3→2 (e) 4→3→1→2→5

12. A typical antibody (a) has a Y shape (b) has four identical heavy chains and four identical light chains (c) has IgG and IgD components (d) suppresses allergic reactions (e) attacks cancer cells

13. Immunoglobulin A (a) is important in allergic reactions (b) combines with mast cells (c) prevents pathogens from attaching to epithelial surfaces (d) is found mainly on B cell surfaces (e) is found mainly on T cells

14. When a person is exposed to the same antigen a second time, the response (a) is called a secondary immune response (b) is more rapid (c) is less intense (d) answers a, b, and c are correct (e) answers a and b only

15. Graft rejection (a) is an example of passive immunity (b) occurs in mild form after immunization (c) generates monoclonal antibodies (d) does not occur when tissue is transplanted from one identical twin to the other (e) is initiated by the thymus gland

16. In an allergic reaction (a) the body is immunodeficient (b) allergen binds with IgE (c) helper T cells release histamine (d) allergen stimulates graft rejection (e) mast cells are deactivated

17. HIV (a) is a retrovirus (b) destroys helper T cells (c) causes acquired immune deficiency syndrome (d) answers a, b, and c are correct (e) none of the preceding answers is correct

REVIEW QUESTIONS

1. How does the body distinguish between self and nonself? What evidence is there that invertebrates are capable of making this distinction?

2. Contrast specific and nonspecific defense mechanisms. Which type confronts invading pathogens immediately? In what ways do the two types of processes interact?

3. How does inflammation help to restore homeostasis?

4. Give two specific ways in which cell-mediated and antibody-mediated immune responses are similar and two ways in which they are different.

5. Describe how antibodies work against pathogens.

6. John is immunized against measles. Jack contracts measles from a playmate in nursery school because his parents failed to have him immunized. Compare the immune responses of the two children. Five years later, John and Jack are playing together when Judy, who is coming down with measles, sneezes on both of them. Compare their immune responses.

7. Why is passive immunity temporary?

8. What is graft rejection? What is the immunological basis for it?

9. List the immunological events that take place in a common type of allergic reaction such as hayfever.

10. What is an autoimmune disease? Give two examples.

11. Specificity, diversity, and memory are key features of the immune system. Giving specific examples, explain how each of these features is important.

YOU MAKE THE CONNECTION

1. Imagine that you are a researcher developing new HIV treatments. What approaches might you take? What public policy decisions would you recommend that might help slow the spread of AIDS while new treatments or vaccines are being developed?

2. Macrophages can be selectively destroyed in the body by the administration of a certain chemical. What would be the effects of such a loss of macrophages? Which do you think would have a greater effect on the immune system, loss of macrophages or loss of B cells?

3. What are the advantages of having MHC antigens? Disadvantages? What do you think would be the consequences of not having them?

RECOMMENDED READINGS

Abraham, S.N. "Discovering the Benign Traits of the Mast Cell," *Science & Medicine,* Vol. 4, No. 5, Sep./Oct. 1997. A discussion of the role of mast cells in human responses.

Behar, S.M., and S.A. Porcelli. "The Immunology of Inflammatory Arthritis." *Science & Medicine,* Vol. 4, No. 1, Jan./Feb. 1997. A discussion of an abnormal immune response.

Black, P.H. "Psychoneuroimmunology: Brain and Immunity." *Scientific American: Science & Medicine,* Vol. 2, No. 6, Nov./Dec. 1995. A discussion of what is known about the relationship between stress and immunosuppression.

Collins, T. "Adhesion Molecules in Leukocyte Emigration." *Scientific American: Science & Medicine,* Vol. 2, No. 6, Nov./Dec. 1995. How white blood cells leave the circulation and enter the tissues in search of foreign molecules.

Condra, J.H. and E.A. Emini. "Preventing HIV-1 Drug Resistance." *Science & Medicine,* Vol. 4, No. 1, Jan./Feb. 1995. A discussion of the mechanisms that confer drug resistance on HIV.

Eisenbarth, G.S., and D. Bellgrau. "Autoimmunity." *Scientific American: Science & Medicine,* Vol. 1, No. 2, May/Jun. 1994. A discussion of autoimmune disease.

Langer, R., and J.P. Vacanti. "Artificial Organs." *Scientific American,* Vol. 273, No. 3, Sep. 1995. A preview of possible 21st century technologies for fabricating artificial organs.

Lanz, R.P., Cooper, D.K.C. and W.L. Chick. "Xenotransplantation." *Scientific American,* Vol. 277, No. 1, Jul. 1997. An overview of current research on organ transplantation from pig and other animal donors to human recipients.

Levine, A.J. "Tumor Suppressor Genes." *Scientific American: Science & Medicine,* Jan./Feb. 1995. An article that makes important connections between genetics, immunology, virology, molecular biology, and epidemiology.

Nowak, M.A., and A.J. McMichael. "How HIV Defeats the Immune System." *Scientific American,* Vol. 273, No. 2, Aug. 1995. A discussion of the hypothesis that continued mutation of the HIV virus allows it to eventually overcome the body's defenses.

"The Pathogenesis of AIDS: A Special Report." *Science and Medicine,* Vol. 4, No. 2, Mar./Apr. 1997. Through the process of testing new drugs, investigators are learning about the role of selection pressure on emergence of viral mutations. Researchers are also learning more about the role of cytokines in inducing or suppressing HIV.

Roitt, I., J. Brostoff, and D. Male. *Immunology,* 5th ed. Mosby, London, 1998. An excellent immunology textbook.

Science, Vol. 280, No. 5371, 19 Jun. 1998. A special issue on AIDS research.

Scientific American, Vol. 275, No. 3, Sep. 1996. A single-topic issue on causes, immunology, mechanisms, prevention, and therapy of cancer.

Streit, W.J., and C.A. Kincaid-Colton, "The Brain's Immune System." *Scientific American,* Vol. 273, No. 5, Nov. 1995. Microglia, a type of glial cells, have immunological properties; however, sometimes they damage neurons.

● Visit our website at **http://www.saunderscollege.com/lifesci/titles.html** and click on Solomon/Berg/Martin Biology for links to chapter-related resources on the World Wide Web.

CHAPTER 44

Gas Exchange

Most animal cells require a continuous supply of oxygen for cellular respiration. Some cells, such as mammalian brain cells, may be damaged beyond repair if their oxygen supply is cut off for only a few minutes. Animal cells must also rid themselves of carbon dioxide. The exchange of gases between an organism and its environment is known as **respiration.** Two phases of respiration are organismic and cellular respiration. During **organismic respiration,** oxygen from the environment is taken up by the animal and delivered to its individual cells. At the same time, carbon dioxide generated during cellular respiration is excreted into the environment. In **aerobic cellular respiration,** which takes place in mitochondria, oxygen is necessary for the citric acid cycle to proceed. Oxygen serves as the final electron acceptor in the mitochondrial electron transport chain (see Chapter 7). Carbon dioxide is produced as a metabolic waste product of cellular respiration.

In small, aquatic organisms such as sponges, hydras, and flatworms, gas exchange occurs entirely by simple **diffusion,** the passive movement of particles (atoms, ions, or molecules) from a region of higher concentration to a region of lower concentration, that is, down a concentration gradient. Most cells are in direct contact with the environment. Dissolved oxygen from the surrounding water diffuses into the cells, while carbon dioxide diffuses out of the cells and into the water. No specialized respiratory structures are needed.

Oxygen diffuses through tissues slowly, however. In an organism more than about 1 mm thick, oxygen cannot diffuse quickly enough through layers of cells to support life. Specialized respiratory structures such as gills or lungs are required to deliver oxygen to the cells or to a transport system and to facilitate the excretion of carbon dioxide. Such respiratory systems, as well as circulatory systems in many animals, provide the efficient intake and transport of oxygen necessary to support high metabolic rates.

If the air or water supplying oxygen to the cells can be continuously renewed, more oxygen will be available. For this reason animals carry on **ventilation;** that is, they actively move air or water over their respiratory surfaces. Sponges do this by setting up a current of water through the channels of their bodies by means of flagella. Most fishes gulp water, which then passes over their gills. Terrestrial vertebrates, such as the white-tailed deer *(Odocoileus virginianus)* shown here, breathe air; the diaphragm and other muscles move air in and out of the lungs.

(Skip Moody/Dembinsky Photo Associates)

1. Compare the advantages and disadvantages of gas exchange in air with those in water.
2. Describe the following adaptations for gas exchange: the body surface, tracheal tubes, gills, and lungs.
3. Identify two respiratory pigments and describe their function.
4. Trace the passage of oxygen through the human respiratory system from nostrils to alveoli.
5. Summarize the mechanics and the regulation of breathing in humans. (Include the role of chemoreceptors.)
6. Describe how oxygen and carbon dioxide are exchanged in the lungs and in the tissues.

7. Explain the role of hemoglobin in oxygen transport and identify factors that determine and influence the oxygen-hemoglobin dissociation curve.
8. Summarize the mechanisms by which carbon dioxide is transported in the blood.
9. Describe the physiological effects of hyperventilation and of sudden decompression when surfacing too quickly from a deep-sea dive.
10. Describe the defense mechanisms that protect the lungs and the effects on the respiratory system of breathing polluted air.

RESPIRATORY STRUCTURES ARE ADAPTED FOR GAS EXCHANGE IN AIR OR WATER

Gills are adapted for gas exchange in water, while the tracheal tubes of insects and lungs of vertebrates are respiratory structures adapted for gas exchange in air. Respiratory surfaces must be kept moist to prevent drying out, and oxygen and carbon dioxide are dissolved in the fluid that bathes the cells of these surfaces. Whether an animal makes its home on land or water, gas exchange takes place across a moist surface.

Animals that carry out gas exchange in water require no special mechanisms for maintaining moisture. In contrast, animals that respire in air struggle continuously with water loss. Adaptations have evolved that keep respiratory surfaces moist and minimize dessication. For example, the lungs of air-breathing vertebrates are located deep within the body, not exposed like gills. Air is humidified and brought to body temperature as it passes through the upper respiratory passageways, and expired air must again pass through these airways (thus allowing for retention of water) before leaving the body. These adaptations help protect the lungs from the drying and cooling effects of air.

Gas exchange in air has certain advantages over gas exchange in water. Compared to water, air contains a much higher concentration of molecular oxygen. Oxygen also diffuses much faster through air than through water. Another advantage is that air is not salty like seawater, so air-breathers do not have to cope with diffusion of ions into their body fluids along with oxygen. As a result, they have an easier time maintaining appropriate internal ion concentrations.

FOUR TYPES OF SURFACES HAVE EVOLVED FOR GAS EXCHANGE

Not only must respiratory structures be moist, they must also have thin walls through which diffusion can easily occur. Respiratory structures are generally richly supplied with blood ves-

sels to facilitate transport and exchange of respiratory gases. Four main types of respiratory surfaces used by animals are: the animal's own body surface, tracheal tubes, gills, and lungs (Fig. 44–1).

The body surface may be adapted for gas exchange

Gas exchange occurs through the entire body surface in many animals, including nudibranch mollusks, most annelids, and a few vertebrates. All of these animals are small, with a high surface-to-volume ratio. They also have a low metabolic rate that requires smaller quantities of oxygen per cell. In aquatic animals, the body surface is kept moist by the surrounding water. In terrestrial animals, the body secretes fluids that keep its surface moist. Many animals that exchange gases across the body surface also have gills or lungs.

How does an animal such as an earthworm exchange gases across its body surface? Gland cells in the epidermis secrete mucus, which keeps the body surface moist while also offering protection. Oxygen, present in air pockets in the loose soil that the earthworm inhabits, dissolves in the mucus. Then, the oxygen diffuses through the body wall and into blood circulating in a network of capillaries just beneath the outer cell layer. Carbon dioxide is transported by the blood to the body surface; from there it diffuses out into the environment.

Tracheal tube systems of arthropods deliver air directly to the cells

In insects and some other arthropods (e.g., chilopods, diplopods, some mites, and some spiders), the respiratory system consists of a network of **tracheal tubes** (Fig. 44–2). Air enters the tracheal tubes through a series of up to 20 tiny openings called **spiracles** along the body surface. In some insects, especially large, active insects, muscles help ventilate the tracheae by pumping air in and out of the spiracles. For example, the grasshopper draws air in through the first four pairs of spiracles when the abdomen expands. Then the abdomen contracts, forcing air out through the last six pairs of spiracles.

Figure 44–1 **Adaptations for gas exchange.** *(a)* Some animals exchange gases through the body surface. *(b)* Insects and some other arthropods exchange gases through a system of tracheal tubes, or tracheae. *(c)* Many aquatic animals have gills for gas exchange. *(d)* Lungs are adaptations for terrestrial gas exchange.

Once inside the body, the air passes through the branching tracheal tubes, which extend to all parts of the animal. The tracheal tubes terminate in microscopic, fluid-filled tracheoles. Gases are exchanged between this fluid and the body cells. The tracheal system supplies oxygen needed to support the high metabolic rates characteristic of many insects. The relatively inefficient open circulatory system of arthropods cannot supply these active animals with sufficient oxygen.

Gills of aquatic animals are respiratory surfaces

Found mainly in aquatic animals, **gills** are moist, thin structures that extend from the body surface. They are supported by the buoyancy of water but tend to collapse in air. In many animals, the outer surface of the gills is exposed to water, whereas the inner side is in close contact with networks of blood vessels.

Sea stars and sea urchins have **dermal gills** that project from the body wall. Their ciliated epidermal cells ventilate the gills by beating a stream of water over them. Gases are exchanged between the water and the coelomic fluid inside the body by diffusion through the gills.

Various types of gills are found in some annelids, aquatic mollusks, crustaceans, fishes, and amphibians. Molluskan gills

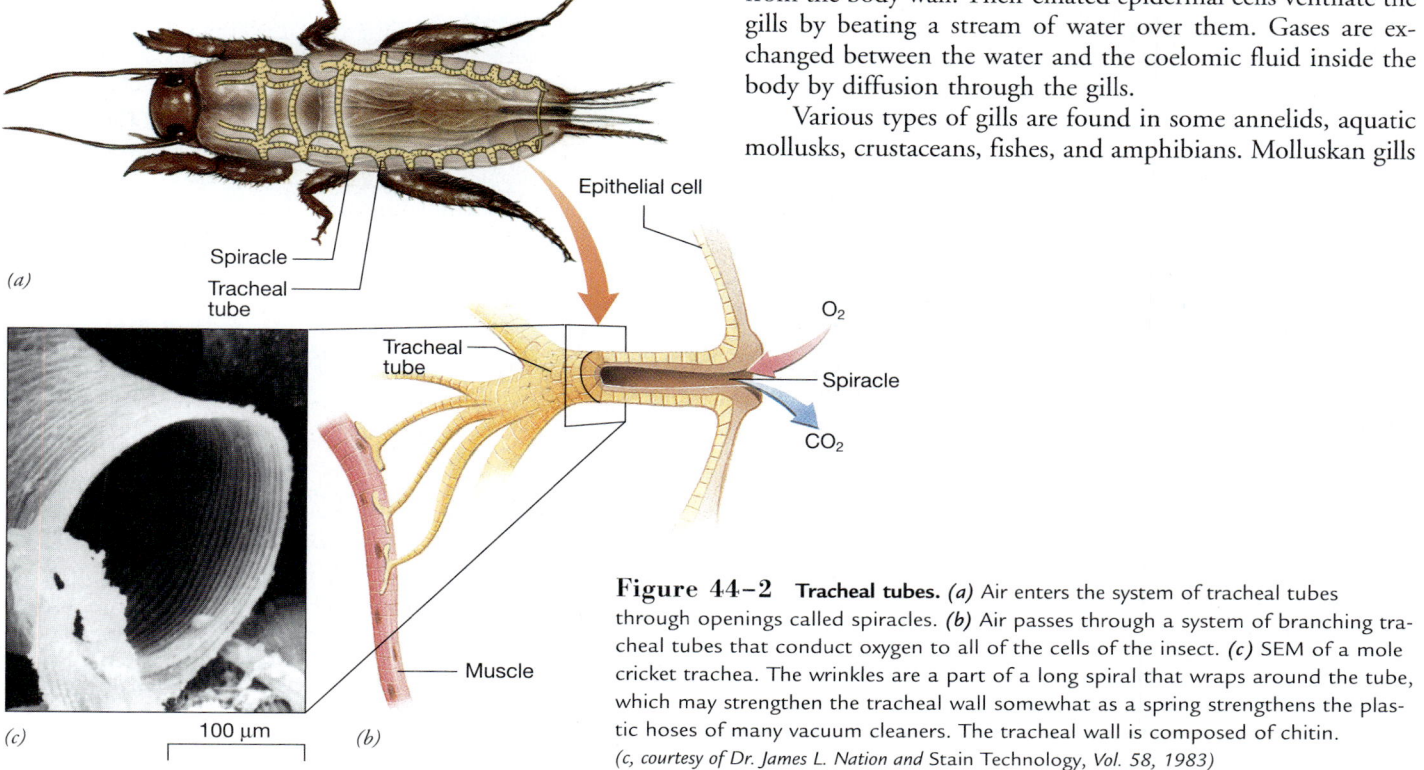

Figure 44–2 **Tracheal tubes.** *(a)* Air enters the system of tracheal tubes through openings called spiracles. *(b)* Air passes through a system of branching tracheal tubes that conduct oxygen to all of the cells of the insect. *(c)* SEM of a mole cricket trachea. The wrinkles are a part of a long spiral that wraps around the tube, which may strengthen the tracheal wall somewhat as a spring strengthens the plastic hoses of many vacuum cleaners. The tracheal wall is composed of chitin. *(c, courtesy of Dr. James L. Nation and* Stain Technology, *Vol. 58, 1983)*

are folded, providing a large surface for respiration. In clams and other bivalve mollusks and in simple chordates, gills may also be adapted for trapping and sorting food. Rhythmic beating of cilia draws water over the gill area, and food is filtered out of the water as gases are exchanged. In mollusks, gas exchange also takes place through the mantle.

In chordates, gills are usually internal. A series of slits perforates the pharynx, and the gills are located along the edges of these gill slits. In bony fishes, the fragile gills are protected by an external bony plate, the **operculum.** In some fish, movements of the jaw and operculum help pump water rich in oxygen through the mouth and across the gills. The water exits through the gill slits.

Each gill in the bony fish consists of many **filaments,** which provide an extensive surface for gas exchange (Fig. 44–3).

The filaments extend out into the water, which continuously flows over them. A capillary network delivers blood to the gill filaments, facilitating diffusion of oxygen and carbon dioxide between blood and water. The very impressive efficiency of this system is possible because blood flows in a direction opposite to the movement of the water. This arrangement, referred to as a **countercurrent exchange system,** maximizes the difference in oxygen concentration between blood and water throughout the area where the two remain in contact.

If blood and water flowed in the *same* direction, that is, *concurrent exchange,* the difference between the oxygen concentrations in blood (low) and water (high) would be very large initially and very small at the end. The oxygen concentration in the water would decrease as that in the blood increased. When the concentrations in the two fluids became equal, an

Figure 44–3 **Gills in bony fish.** *(a)* The gills are located under a bony plate, the operculum, which has been removed in this side view. The gills form the lateral wall of the pharyngeal cavity. *(b)* Gills of the salmon. *(c)* Each gill consists of a cartilaginous gill arch to which two rows of leaflike gill filaments are attached. As water flows past the gill filaments, blood circulates within them. *(d)* Each gill filament has many smaller extensions rich in capillaries. Blood entering the capillaries is deficient in oxygen. The blood flows through the capillaries in a direction opposite to that taken by the water. This countercurrent exchange system efficiently charges the blood with oxygen. *(e)* If the system were concurrent, that is if blood flowed through the capillaries in the same direction as the flow of the water, much less of the oxygen dissolved in the water could diffuse into the blood. *(b, G.I. Bernard/Animals Animals)*

equilibrium would be reached and the net diffusion of oxygen would stop. Only about 50% of the oxygen dissolved in the water could diffuse into the blood.

In the countercurrent exchange system, however, blood low in oxygen comes in contact with water that is partly oxygen-depleted. Then, as the blood flowing through the capillaries becomes more and more oxygen-rich, it comes in contact with water with a progressively higher concentration of oxygen. Thus, all along the capillaries, the diffusion gradient favors passage of oxygen from the water into the gill. A high rate of diffusion is maintained, ensuring that a very high percentage (more than 80%) of the available oxygen in the water diffuses into the blood.

Oxygen and carbon dioxide do not interfere with one another's diffusion, and they simultaneously diffuse in opposite directions. This is because oxygen is more concentrated outside the gills than within, but carbon dioxide is more concentrated inside the gills than outside. Thus, the same countercurrent exchange mechanism that ensures efficient inflow of oxygen also results in equally efficient outflow of carbon dioxide.

Terrestrial vertebrates exchange gases through lungs

Lungs are respiratory structures that develop as ingrowths of the body surface or from the wall of a body cavity such as the pharynx. For example, the **book lungs** of spiders are enclosed in an inpocketing of the abdominal wall. These lungs consist of a series of thin parallel plates of tissue (like the pages of a book) filled with hemolymph (see Fig. 44–1). The plates of tissue are separated by air spaces that receive oxygen from the outside environment through a spiracle. A different type of lung evolved in land snails and slugs. These terrestrial mollusks lack gills. Gas exchange takes place through a lung, which is a vascularized region of the mantle.

Fossil evidence suggests that early lobe-finned fishes had lungs somewhat similar to those of modern lungfish. The three extant genera of lungfish live in the headwaters of the Nile, in the Amazon, and in certain Australian rivers. Streams inhabited by these fishes dry up during seasonal droughts. During these dry periods, lungfish remain in the mud of the stream, exchanging gases by means of their lungs. These fishes are also equipped with gills, which they use when swimming. The African lungfish uses both its gills and its lungs throughout the year.

Remains of early lobe-finned fishes occur extensively in the fossil record. Those found in Devonian strata are thought to be similar to the ancestors of amphibians, and numerous amphibian fossils occur in adjacent ancient strata. Some paleontologists hypothesize that the evolution of lungs occured as an adaptation to the periodic droughts that occurred in Devonian times, and that lungs or lunglike structures may have been present in all early bony fishes. Most modern bony fishes have no lungs, but nearly all of them do possess homologous **swim bladders** (see Chapter 30). By adjusting the amount of gas in its swim bladder, the fish can control its buoyancy.

Some amphibians do not have lungs. Among plethodontid (lungless) salamanders, for example, all gas exchange takes place in the pharynx, or across the thin, wet skin. However, even though they depend mainly on their body surface for gas exchange, most amphibians have lungs (Fig. 44–4). The lungs of mud puppies are two long, simple sacs richly supplied by capillaries. Frogs and toads have ridges containing connective tissue on the inside of the lungs, thereby increasing the respiratory surface somewhat.

The lungs of most reptiles are rather simple sacs, with only some folding of the wall to increase the surface for gas exchange. Gas exchange is not very effective and does not supply sufficient oxygen to sustain long periods of activity. In some lizards, turtles, and crocodiles, the lungs are somewhat more complex with subdivisions that give them a spongy texture. In 1997 physiologist John A. Ruben examined fossilized lung remains from a dinosaur (a therapod) and reported in *Science* that these lungs were similar to those of crocodiles. Arguing against the prevailing view, Ruben held that that dinosaurs could not have been endothermic, because their respiratory systems were too inefficient. He also challenged the widely accepted hypothesis that dinosaurs gave rise to birds (see Chapter 20), arguing that dinosaur lungs could not have evolved into the highly efficient lungs of modern birds.

Whatever their common ancestor, birds do have the most efficient respiratory system of any living vertebrate. Very active,

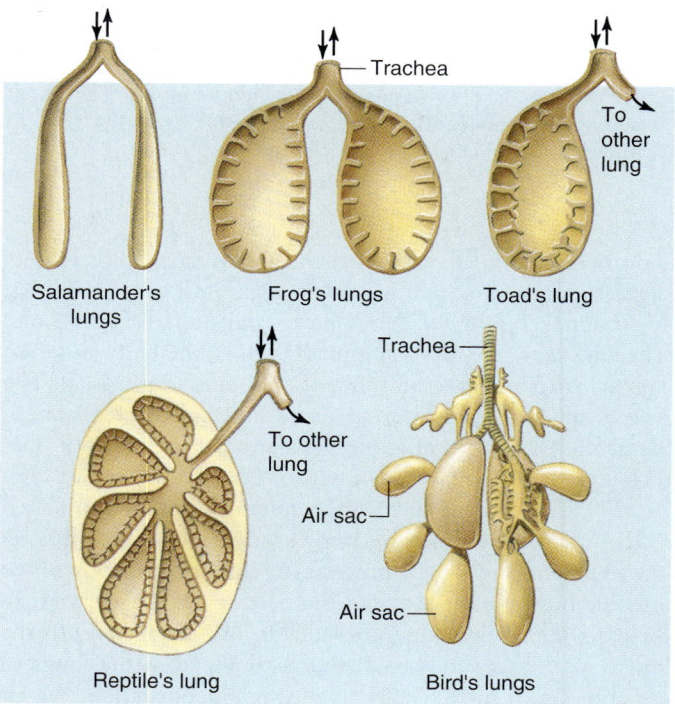

Figure 44–4 **Comparison of vertebrate lungs.** The surface area of the lung has increased during vertebrate evolution. Salamander lungs are simple sacs. Other amphibians and reptiles have lungs with small ridges or folds that help increase surface area. Birds have an elaborate system of lungs and air sacs. Mammalian lungs have millions of alveoli that increase the surface available for gas exchange.

(a)

Air

(b) Inhale   (c) Exhale   (d) Inhale   (e) Exhale

Figure 44–5 **Function of bird lungs.** The bird has a two-cycle breathing process that supports a one-way flow of air. *(a)* The bird respiratory system includes lungs and extensions called air sacs. *(b)* As the bird inhales, air flows into the posterior air sacs. *(c)* As it exhales, air is forced into the expanding lungs. *(d)* At the second inhalation, air from the first breath moves into the anterior air sacs. *(e)* Finally, at the second exhalation, air from the first inhalation leaves the body. Air from the second inhalation is not shown.

endothermic animals with high metabolic rates, birds require large amounts of oxygen to sustain flight and other activities. Their small, bright red lungs have extensions (usually nine) called **air sacs,** which reach into all parts of the body and even connect with air spaces in some of the bones (Fig. 44–5). The air sacs act as bellows drawing air into the system. Collapse of the air sacs during expiration (exhalation) forces air out. Gas exchange does not take place across the walls of the air sacs.

The high-performance bird respiratory system is arranged so that air flows in one direction through the lungs and is renewed during a two-cycle process. Air entering the body passes into the posterior air sacs and the part of the lungs closest to these air sacs. When the bird exhales, that air flows into the lungs. At the second breath, the air flows from the lungs to the anterior air sacs. Finally, at the second exhalation, the air leaves the body. Thus, a bird gets fresh air across its lungs through both inhalation and exhalation.

Bird lungs have tiny, thin-walled tubes, the **parabronchi,** which are open at both ends. Gas exchange takes place across the walls of these tubes. Chickens and other weak-flying birds have about 400 parabronchi per lung compared to pigeons and

other strong-flying birds which have about 1800 parabronchi per lung. The direction of blood flow in the lungs is opposite that of air flow through the parabronchi. This arrangement, somewhat similar to the countercurrent exchange in the gills of fishes, increases the amount of oxygen that enters the blood. However, in birds there is a **crosscurrent,** rather than countercurrent, arrangement; the capillaries are oriented at right angles to the parabronchi, rather than along their length.

The lungs of mammals are very complex and have an enormous surface area. Using the human respiratory system as an example, we examine gas exchange in mammals in later sections of this chapter.

RESPIRATORY PIGMENTS INCREASE CAPACITY FOR OXYGEN TRANSPORT

Complex animals have **respiratory pigments** that combine reversibly with oxygen and greatly increase the capacity of blood to transport it. For example, the hemoglobin in human blood increases its capacity to transport oxygen by about 75 times. Oxygen enters the pulmonary capillaries and combines with hemoglobin in the red blood cells. Then, as blood circulates through tissues where the oxygen concentration is low, hemoglobin releases oxygen, which diffuses out of the blood and into the tissue cells.

Hemoglobin is the respiratory pigment characteristic of vertebrates. It is also present in many invertebrate species including annelids, nematodes, mollusks, and arthropods. In some of these animals, the hemoglobin is dispersed in the plasma rather than confined to blood cells.

Hemoglobin is actually a general name for a group of related compounds, all of which consist of an iron-porphyrin, or heme, group bound to a protein known as a globin. The protein portion of the molecule varies in size, amino acid composition, and physical properties among various species. When combined with oxygen, hemoglobin is bright red; without oxygen, it appears dark red, imparting a purplish color to venous blood.

Hemocyanins, another type of respiratory pigment, are copper-containing proteins found in many species of mollusks and arthropods. These do not have a heme (porphyrin) group. When oxygen is combined with the copper, the compound appears blue. Without oxygen, it is colorless. Hemocyanins are dispersed in the blood rather than confined within cells.

THE HUMAN RESPIRATORY SYSTEM IS TYPICAL OF AIR-BREATHING VERTEBRATES

The respiratory system in humans and other air-breathing vertebrates consists of a series of tubes through which air passes on its journey from the nostrils to the lungs and back (Fig. 44–6).

Figure 44–6 **The human respiratory system.** The muscular diaphragm forms the floor of the thoracic cavity. An internal view of one lung illustrates a portion of its extensive system of air passageways. The respiratory centers in the brain regulate the rate of respiration.

Respiratory centers

Pharynx

Esophagus

Space occupied by heart

Right lung

Diaphragm

Sinuses

Nasal cavity

Tongue

Epiglottis

Larynx

Trachea

Bronchioles

Bronchus

Left lung

The airway conducts air into the lungs

A breath of air enters the body through the **nostrils** and flows through the **nasal cavities.** As air passes through the nose, it is filtered, moistened, and brought to body temperature. The nasal cavities are lined with a moist, ciliated epithelium that is rich in blood vessels. Inhaled dirt, bacteria, and other foreign particles are trapped in the stream of mucus that is produced by cells within the epithelium and pushed along toward the throat by the cilia. In this way, foreign particles are delivered to the digestive system, which can more effectively dispose of such materials than can the delicate lungs. A person normally swallows more than a pint of nasal mucus each day, more during an infection or allergic reaction.

The back of the nasal cavities is continuous with the throat region, or **pharynx.** Air finds its way into the pharynx whether one breathes through the nose or mouth. An opening in the floor of the pharynx leads into the **larynx,** sometimes called the "Adam's apple." Because the larynx contains the vocal cords, it is also referred to as the voice box. Cartilage embedded in its wall prevents the larynx from collapsing and makes it hard to the touch when felt through the neck.

During swallowing, a flap of tissue called the **epiglottis** automatically closes off the larynx so that food and liquid enter the esophagus rather than the lower airway. Should this defense mechanism fail and foreign matter enter the sensitive larynx, a **cough reflex** is initiated, expelling the material. Despite these mechanisms, choking sometimes occurs.

From the larynx, air passes into the **trachea,** or windpipe, which is kept from collapsing by rings of cartilage in its wall. The trachea divides into two branches, the **bronchi** (sing., *bronchus*), each of which connects to a lung.

Both trachea and bronchi are lined by a mucous membrane containing ciliated cells. Many medium-sized particles that have escaped the cleansing mechanisms of nose and larynx are trapped here. Mucus that contains these particles is constantly beaten upward by the cilia to the pharynx, where it is periodically swallowed. This mechanism, functioning as a cilia-propelled mucous elevator, helps keep foreign material out of the lungs.

Gas exchange occurs in the alveoli of the lungs

The lungs are large, paired, spongy organs occupying the thoracic (chest) cavity. The right lung is divided into three lobes, the left lung into two lobes. Each lung is covered with a **pleural membrane,** which forms a continuous sac that encloses the lung and becomes the lining of the thoracic cavity. The pleural cavity is the space between the pleural membranes covering the lung and the pleural membrane lining the thoracic cavity. A film of fluid in the pleural cavity provides lubrication between the lungs and the chest wall.

Because the lung consists largely of air tubes and elastic tissue, it is a spongy, elastic organ with a very large internal

Alveolus

Bronchiole

Capillary

Macrophage

Red blood cells

Capillaries

Epithelial cell of the adjacent alveolus

Epithelial cell of the wall of the alveolus

Alveolus

Alveolus

(a)

Alveolus

(b)   50 μm

Wall of alveolus   Wall of capillary   Red blood cell

(c)   1 μm

Figure 44–7 **Structure of alveoli.** *(a)* Gas exchange takes place across the thin wall of the alveolus. Note that the alveolar wall consists of extremely thin squamous epithelium. Extensive capillary networks lie between the walls of the alveoli. *(b)* SEM showing the capillary network surrounding a portion of several alveoli. *(c)* LM of a portion of a capillary and the wall of an alveolus. The dark structure extending through the capillary is a portion of a red blood cell. The wall of the alveolus is visible just above the wall of the capillary. Notice the very short distance oxygen must diffuse to get from the air within the alveolus to the red blood cells that will transport it to the body tissues.   *(b, Kessel, R.G. and R.H. Kardon, Tissues and Organs: A Text-Atlas of Scanning Electron Microscopy. San Fransisco, W.H. Freeman Co., 1979; c, Courtesy of Drs. Peter Gehr, Marianne Bachofen, and Ewald R. Wiebel)*

surface area for gas exchange. Inside the lungs the bronchi branch, becoming smaller and more numerous. These branches give rise to more than one million tiny **bronchioles** in each lung. Each bronchiole ends in a cluster of tiny air sacs, the **alveoli** (sing., *alveolus*) (Fig. 44–7). The human lung contains more than 300 million alveoli with an internal surface area the approximate size of a tennis court. Each alveolus is lined by an extremely thin, single layer of epithelial cells. Gases diffuse freely through the wall of the alveolus and into the capillaries that surround it. Only two thin cell layers, the epithelia of the alveolar wall and the capillary wall, separate the air in the alveolus from the blood.

In summary, the sequence of structures through which air passes after it enters the body is:

Nostrils ⟶ nasal cavities ⟶ pharynx ⟶ larynx ⟶ trachea ⟶ bronchi ⟶ bronchioles ⟶ alveoli

Ventilation is accomplished by breathing

Breathing is the mechanical process of moving air from the environment into the lungs and of expelling air from the lungs. Inhaling air is referred to as **inspiration;** exhaling air is **expiration.** A resting adult breathes about 14 times each minute.

The thoracic cavity is closed so that no air can enter except through the trachea. (When the chest wall is punctured, for example, by a gunshot wound, air enters the pleural space and the lung collapses.)

During inspiration, the volume of the thoracic cavity is increased by the contraction of the **diaphragm,** the dome-shaped muscle that forms the floor of the thoracic cavity. When the diaphragm contracts, it moves downward, increasing the volume of the thoracic cavity (Fig. 44–8). During forced inspiration, when a large volume of air is inspired, the *external intercostal muscles* contract as well. This action moves the ribs upward, thereby increasing the volume of the thoracic cavity. Because the lungs adhere to the walls of the thoracic cavity, when the volume of the thoracic cavity increases, the space within each lung also increases. The air in the lungs now has more space in which to move about, and the pressure of the air in the lungs falls by 2 or 3 mm Hg below the pressure of the air outside the body. As a result of this pressure difference, air from the outside rushes in through the respiratory passageways and fills the lungs until the two pressures are equal once again.

Expiration occurs when the diaphragm relaxes. The volume of the chest cavity decreases, increasing the pressure in the lungs to 2–3 mm Hg above atmospheric pressure. The millions of distended air sacs deflate, expelling the air that was inhaled. The pressure returns to normal, and the lung is ready for another inspiration. Thus, in inspiration, the millions of alveoli fill with air like so many tiny balloons. Then during expiration, the air rushes out of the alveoli, partially deflating them. During exercise or forced expiration, muscles of the abdominal wall and the *internal intercostal muscles* contract, pushing the diaphragm up and the ribs down.

The quantity of respired air can be measured

The amount of air moved into and out of the lungs with each normal resting breath is called the **tidal volume.** The normal tidal volume is about 500 mL. The **vital capacity** is the maximum amount of air a person can exhale after filling the lungs to the maximum extent.

Vital capacity is greater than tidal volume because the lungs are not completely emptied of stale air and filled with fresh air with each breath. The volume of air that remains in the lungs at the end of a normal expiration is the **residual capacity.** Table 44–1 shows the percentages of oxygen and carbon dioxide present in exhaled (alveolar) air compared with inhaled air. Because carbon dioxide is produced during cellular respiration, more of this gas—100 times more, in fact—enters the alveoli from the blood than is present in air inhaled from the environment.

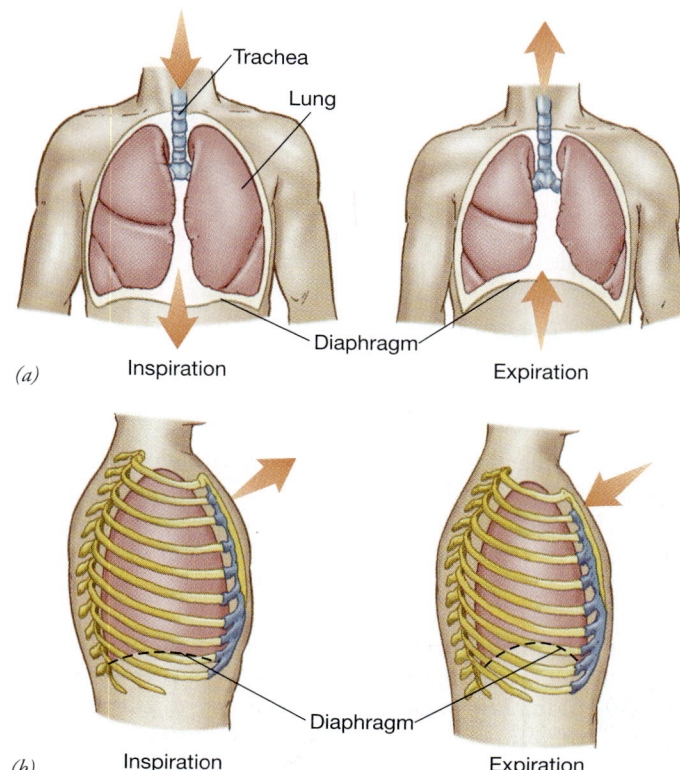

Figure 44–8 **Mechanics of breathing.** *(a)* Changes in the position of the diaphragm in expiration and inspiration result in changes in the volume of the thoracic cavity. During inspiration, the diaphragm contracts, increasing the volume of the thoracic cavity. When the volume of the thoracic cavity increases, air moves into the lungs. During expiration, the diaphragm relaxes, decreasing the volume of the thoracic cavity. *(b)* During forced inspiration, the elevation of the front ends of the ribs by the external intercostal muscles causes an increase in the front-to-back dimension of the chest and a corresponding increase in the volume of the thoracic cavity. During expiration, the rib cage drops, decreasing the volume of the thoracic cavity.

Gas exchange takes place in the alveoli

The respiratory system delivers oxygen to the alveoli, but if oxygen were to remain in the lungs, all the other body cells would soon die. The vital link between alveolus and body cell is the circulatory system. Each alveolus serves as a tiny depot from which oxygen is loaded into blood brought close to the alveolar air by capillaries (Fig. 44–9).

Oxygen molecules efficiently pass by simple diffusion from the alveoli, where they are more concentrated, into the blood in the pulmonary capillaries, where they are less concentrated. At the same time, carbon dioxide moves from the blood, where it is more concentrated, to the alveoli, where it is less concentrated. Each gas diffuses through the single layer of cells lining the alveoli and the single layer of cells lining the capillaries.

Cellular respiration results in the continuous production of carbon dioxide and utilization of oxygen. Consequently, the concentration of oxygen in the cells is lower than in the cap-

TABLE 44-1 Composition of Inhaled Air Compared with that of Exhaled Air

	% Oxygen (O_2)	% Carbon Dioxide (CO_2)	% Nitrogen (N_2)
Inhaled air* (atmospheric air)	20.9	0.04	79
Exhaled air (alveolar air)	14.0	5.60	79

*As indicated, the body uses up about one-third of the inhaled oxygen. The percentage of CO_2 increases more than 100-fold because it is produced during cellular respiration.

illaries entering the tissues, and the concentration of carbon dioxide is higher in the cells than in the capillaries. Thus, as blood circulates through capillaries of a tissue such as brain or muscle, oxygen moves by simple diffusion from the blood to the cells, while carbon dioxide moves from the cells into the blood.

The factor that determines the direction and rate of diffusion is the pressure or tension of the particular gas. According to **Dalton's law of partial pressures,** in a mixture of gases the total pressure of the mixture is the sum of the pressures of the individual gases. Each gas exerts, independently of the others, a **partial pressure**—the same pressure it would exert if it were present alone. At sea level, the barometric pressure (the pressure of Earth's atmosphere) is able to support a column of mercury (Hg) 760 mm high. Because oxygen makes up about 21% of the atmosphere, oxygen's share of that pressure is 0.21 × 760 = 160 mm Hg. Thus, 160 mm Hg is the partial pressure of atmospheric O_2, abbreviated P_{O_2}. In contrast the partial pressure of atmospheric CO_2 is 0.3 mm Hg, abbreviated P_{CO_2}.

Fick's law of diffusion explains that the amount of oxygen or carbon dioxide that diffuses across the membrane of an alveolus depends on the differences in partial pressure on the two sides of the membrane and also on the surface area of the membrane. The greater the difference in pressure and the larger the surface area, the faster the gas will diffuse.

Gas exchange takes place in the tissues

The partial pressure of oxygen in arterial blood is about 100 mm Hg. The P_{O_2} in the tissues is still lower, ranging from 0 to 40 mm Hg. Consequently, oxygen diffuses out of the capillaries and into the tissues. Not all of the oxygen leaves the blood, however. The blood passes through the tissue capillaries too rapidly for equilibrium to be reached. As a result, the partial pressure of oxygen in venous blood returning to the lungs is about 40 mm Hg. Thus, expired air has had only about one third of its oxygen removed and can be breathed over again —a good thing for those in need of mouth-to-mouth resuscitation!

Oxygen is transported in combination with hemoglobin

At rest, the cells of the human body use about 250 mL of oxygen per minute, or about 300 L every 24 hours. With exercise or work this rate may increase as much as 10- or 15-fold. If oxygen were simply dissolved in plasma, blood would have to circulate through the body at a rate of 180 L per minute to supply enough oxygen to the cells at rest. This is because, as we have seen, oxygen is not very soluble in blood plasma. Actually, the blood of a human at rest circulates at a rate of about 5 L per minute and supplies all of the oxygen the cells need. Why are only 5 L per minute rather than 180 L per minute required?

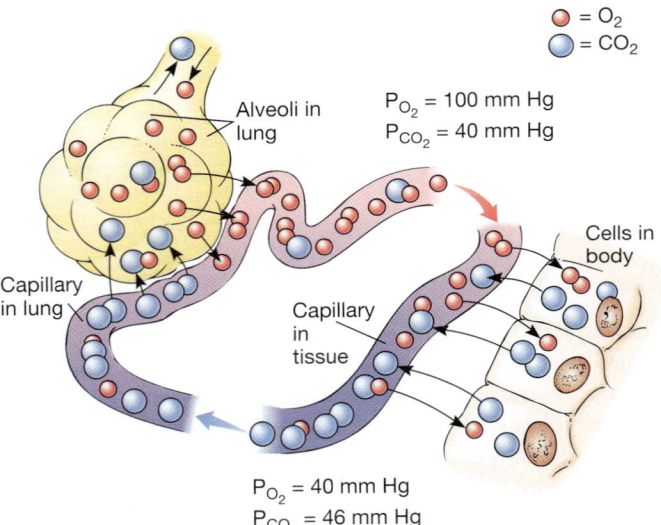

Figure 44-9 **Gas exchange in the lungs and tissues.** The concentration of oxygen is greater in the alveoli than in the pulmonary capillaries, so oxygen diffuses from the alveoli into the blood. Carbon dioxide is more concentrated in the blood than in the alveoli, so it diffuses out of the capillaries and into the alveoli. In the tissues, oxygen is more concentrated in the blood than in the body cells; it diffuses out of the capillary into the cells. Carbon dioxide is more concentrated in the cells, so it diffuses out of the cells and moves into the blood. Note the differences in partial pressures of oxygen and carbon dioxide before and after gases are exchanged in the tissues.

The answer is hemoglobin, the respiratory pigment in red blood cells. Hemoglobin transports about 97% of the oxygen. Only about 3% is dissolved in the plasma. Plasma in equilibrium with alveolar air can take up only 0.25 mL of oxygen per 100 mL, but the properties of hemoglobin permit whole blood to carry some 20 mL of oxygen per 100 mL. The protein portion of hemoglobin is composed of four peptide chains, typically two α and two β chains, each attached to a heme (porphyrin) ring (see Fig. 3–21). An iron atom is bound in the center of each heme ring.

Hemoglobin has the remarkable property of forming a weak chemical bond with oxygen. An oxygen molecule can attach to the iron atom in each heme. In the lung (or gill), oxygen diffuses into the red blood cells and combines with hemoglobin (Hb) to form **oxyhemoglobin (HbO$_2$).** Because the chemical bond formed between the oxygen and the hemoglobin is weak, the reaction is readily reversible. In the body tissues, the reaction proceeds to the left, releasing oxygen.

$$Hb + O_2 \rightleftharpoons HbO_2$$

The maximum amount of oxygen that can be transported by hemoglobin is called the **oxygen carrying capacity.** The actual amount of oxygen bound to hemoglobin is the **oxygen content.** The ratio of O$_2$ content to O$_2$ capacity is the **percent O$_2$ saturation** of the hemoglobin. The percent saturation is highest in the pulmonary capillaries, where the concentration of oxygen is greatest. In the capillaries of the tissues, where there is less oxygen, the oxyhemoglobin dissociates, releasing oxygen. There, the percent saturation of hemoglobin is correspondingly lower. The **oxygen-hemoglobin dissociation curve** shown in Figure 44–10a illustrates this relationship. As oxygen concentration increases, there is a progressive increase in the percentage of hemoglobin that is combined with oxygen. The ability of oxygen to combine with hemoglobin and be released from oxyhemoglobin is influenced by several factors in addition to percent O$_2$ saturation; these include pH, carbon dioxide concentration, and temperature.

Carbon dioxide formed in respiring tissue reacts with water in the plasma to form carbonic acid, H$_2$CO$_3$. In this way an increase in the carbon dioxide concentration increases the acidity (lowers the pH) of the blood. Oxyhemoglobin dissociates its oxygen more readily in an acidic environment than in an environment with normal pH. Lactic acid released from active muscles also lowers the pH of the blood and has a similar effect on the oxygen-hemoglobin dissociation curve. Displacement of the oxygen-hemoglobin dissociation curve by a change in pH is known as the **Bohr effect** (Fig. 44–10b).

Some carbon dioxide is transported by the hemoglobin molecule. Although it attaches to the hemoglobin molecule in a different way and at a different site than oxygen, the attachment of a carbon dioxide molecule causes the release of an oxygen molecule from the hemoglobin. The effect of carbon dioxide concentration on the oxygen-hemoglobin dissociation curve is important. In the capillaries of the lungs (or gills in fishes), carbon dioxide concentration is relatively low and oxygen concentration is high, so oxygen combines with

(a) Normal oxygen-hemoglobin dissociation curve

(b) Effect of pH

Figure 44–10 **Oxygen-hemoglobin dissociation curves.**
(a) Normal curve showing the relationship between the partial pressure of oxygen (*horizontal axis*) and the percent saturation of hemoglobin (*vertical axis*). Oxygen binds to hemoglobin in the lungs and dissociates from hemoglobin in the tissues. **(b)** Effect of pH on the oxygen-hemoglobin curve (Bohr effect). The normal pH of human blood is 7.4. Find the location on the horizontal axis where the partial pressure of oxygen is 40 and follow the line up through the curves. Notice how the saturation of hemoglobin with oxygen differs among the three curves, even though the partial pressure of oxygen is the same. A shift to the right indicates decreased ability of the hemoglobin to bind oxygen, so oxygen is unloaded. A shift to the left represents an increased ability to bind oxygen. For example, during muscular activity the pH decreases, and more oxygen is unloaded and available for the muscle cells.

a very high percentage of hemoglobin. In the capillaries of the tissues, carbon dioxide concentration is high and oxygen concentration is low, so oxygen is readily released from the hemoglobin.

Carbon dioxide is transported mainly as bicarbonate ions

Carbon dioxide is transported in the blood in three ways. About 7 to 10% of carbon dioxide is dissolved in the plasma. Another 20% enters the red blood cells and combines with hemoglobin, forming carbaminohemoglobin. Because the bond between the hemoglobin and carbon dioxide is very weak, the reaction is readily reversible. Most of the carbon dioxide (about 70%) is transported in the plasma as **bicarbonate ions (HCO_3^-).**

Carbon dioxide combines with water to form carbonic acid. This reaction is catalyzed in red blood cells by the enzyme **carbonic anhydrase.** The carbonic acid dissociates, forming hydrogen ions and bicarbonate ions.

$$CO_2 + H_2O \xrightarrow{\text{Carbonic anhydrase}} H_2CO_3 \longrightarrow H^+ + HCO_3^-$$

Carbon dioxide   Water   Carbonic acid   Hydrogen ion   Bicarbonate ion

Most of the hydrogen ions released from carbonic acid combine with hemoglobin, which is a very effective buffer. Many of the bicarbonate ions diffuse into the plasma. Chloride ions diffuse into the red blood cells to replace the negative charges of bicarbonate ions, a process known as the **chloride shift.** As CO_2 diffuses out of the alveolar capillaries, the resulting lower CO_2 concentration causes the reversal of the reaction sequence.

Any condition (such as pneumonia) that interferes with the removal of carbon dioxide by the lungs can lead to respiratory acidosis. In this condition there is an increased concentration of carbonic acid and bicarbonate ions in the blood. Although the pH of the blood is not actually acidic, it is lower than normal.

Breathing is regulated by respiratory centers in the brain

Breathing is a rhythmic, involuntary process regulated by **respiratory centers** in the brain stem (see Fig. 44–6). Neurons in the dorsal region of the medulla regulate the basic rhythm of breathing. These neurons send a burst of impulses to the diaphragm and external intercostal muscles, causing them to contract. After several seconds, these neurons become inactive, the muscles relax, and expiration occurs. Respiratory centers in the pons help control the transition from inspiration to expiration. These centers can stimulate or inhibit the medullary respiratory centers.

The cycle of activity and inactivity repeats itself so that at rest we breathe about 14 times per minute. A group of neurons in the ventral region of the medulla becomes active only when we need to breath forcefully. Overdose of certain med-

ications such as barbiturates depresses the respiratory centers and may lead to respiratory failure.

The basic rhythm of respiration can be altered in response to changing needs of the body. When you are engaged in a strenuous game of tennis, you require more oxygen than when studying biology. During exercise the rate of aerobic cellular respiration increases, producing more carbon dioxide. The body must dispose of this carbon dioxide through increased ventilation. Carbon dioxide concentration is the most important chemical stimulus for regulating the rate of respiration. Specialized **chemoreceptors** in the medulla and in the walls of the aorta and carotid arteries are sensitive to changes in arterial carbon dioxide concentration. When stimulated they send impulses to the respiratory centers, which leads to an increase in breathing rate.

The chemoreceptors in the walls of the aorta, called *aortic bodies,* and those in the walls of the carotid arteries, the *carotid bodies,* are sensitive to changes in hydrogen ion concentration and oxygen concentration, as well as to carbon dioxide levels. Recall that an increase in carbon dioxide concentration results in an increase in hydrogen ions from carbonic acid, lowering the pH of the blood. Even a slight decrease in pH stimulates these chemoreceptors, leading to a faster breathing rate. As carbon dioxide is removed by the lungs, the hydrogen ion concentration in the blood and other body fluids decreases, and homeostasis is maintained.

Interestingly, oxygen concentration generally does not play an important role in regulating respiration. Only if the partial pressure of oxygen falls markedly do the chemoreceptors in the aorta and carotid arteries become stimulated to send messages to the respiratory centers.

Although breathing is an involuntary process, the action of the respiratory centers can be consciously influenced for a short time by stimulating or inhibiting them. For example, you can inhibit respiration by holding your breath. You cannot hold your breath indefinitely, however, because eventually you feel a strong urge to breathe. Even if you were able to ignore this, you would eventually pass out, and breathing would resume.

Individuals who have stopped breathing because of drowning, smoke inhalation, electric shock, or cardiac arrest can sometimes be sustained by mouth-to-mouth resuscitation until their own breathing reflexes are initiated again. **Cardiopulmonary resuscitation (CPR)** is a method for aiding victims who have suffered respiratory and cardiac arrest. CPR must be started immediately, because irreversible brain damage may occur within about 4 minutes of respiratory arrest. A number of organizations offer training in CPR. Its ABCs are: clear Airway, restore Breathing, and restore Circulation by using external cardiac compression.

Hyperventilation reduces carbon dioxide concentration

Underwater swimmers and some Asian pearl divers voluntarily **hyperventilate** before going under water. By taking a series of deep inhalations and exhalations, they "blow off" CO_2, markedly reducing the carbon dioxide content of the alveolar air and of the blood. As a result, it takes longer before the urge to breathe becomes irresistible.

When hyperventilation is continued for a long period, dizziness and sometimes unconsciousness may occur. This is because a certain concentration of carbon dioxide is needed in the blood to maintain normal blood pressure. (This mechanism operates by way of the vasoconstrictor center in the brain, which maintains the muscle tone of blood vessel walls.) Furthermore, if divers hold their breath too long, the low concentration of oxygen may result in unconsciousness and drowning.

High flying or deep diving can disrupt homeostasis

The barometric pressure decreases at progressively higher altitudes. Because the concentration of oxygen in the air remains at 21%, the partial pressure of oxygen decreases along with the barometric pressure. At an altitude of 6000 m (19,500 ft), the barometric pressure is about 350 mm Hg, the partial pressure of oxygen is about 75 mm Hg, and the hemoglobin in arterial blood is about 70% saturated with oxygen. At 10,000 m (33,000 ft) the barometric pressure is about 225 mm Hg, the partial pressure of oxygen is 50 mm Hg, and arterial oxygen saturation is only 20%. Thus, getting sufficient oxygen from the air becomes an ever-increasing problem at higher altitudes.

When a person moves to a high altitude, the body adjusts over a period of time by producing a greater number of red blood cells. In a person breathing pure oxygen at 10,000 m, the oxygen would have a partial pressure of 225 mm Hg, and the hemoglobin would be almost fully saturated with oxygen. Above 13,000 m, however, barometric pressure is so low that even breathing pure oxygen does not permit complete oxygen saturation of arterial hemoglobin.

A person becomes unconscious when the arterial oxygen saturation falls to between 40% and 50%. This level is reached at about 7000 m (23,000 ft) when the person is breathing air, or 14,500 m (47,100 ft) when pure oxygen is used. All high-flying jets have airtight cabins pressurized to the equivalent of the barometric pressure at an altitude of about 2000 m.

Hypoxia, a deficiency of oxygen, results in drowsiness, mental fatigue, headache, and sometimes euphoria. The ability to think and make judgments is impaired, as is the ability to perform tasks requiring coordination. If a jet flying at 11,700 m (over 38,000 ft) underwent sudden decompression, the pilot would lose consciousness in about 30 sec and become comatose in about 1 min.

In addition to the problems of hypoxia, a rapid decrease in barometric pressure can cause **decompression sickness** (commonly known as the "bends" because those suffering from it bend over in pain). Whenever the barometric pressure drops below the total pressure of all gases dissolved in the blood and other body fluids, the dissolved gases tend to come out of solution and form gas bubbles. A familiar example of such bub-

bling occurs each time you uncap a bottle of soda, thus reducing the pressure in the bottle. The carbon dioxide is released from solution and bubbles out into the air. In the body, nitrogen has a low solubility in blood and tissues. When it comes out of solution, the bubbles formed may damage tissues and block capillaries, interfering with blood flow. The clinical effects of decompression sickness are pain, dizziness, paralysis, unconsciousness, and even death.

Decompression sickness is even more common in scuba diving than in high-altitude flying. As a diver descends, the surrounding pressure increases tremendously—1 atmosphere (the atmospheric pressure at sea level which equals 760 mm Hg) for each 10 m. To prevent the collapse of the lungs, a diver must be supplied with air under pressure, thereby exposing the lungs to very high alveolar gas pressures.

At sea level an adult human has about 1 L of nitrogen dissolved in the body, with about half in the fat and half in the body fluids. After a diver's body has been saturated with nitrogen at a depth of 100 m (325 ft) the body fluids contain about 10 L of nitrogen. To prevent this nitrogen from rapidly bubbling out of solution and causing decompression sickness, the diver must be brought to the surface gradually, with stops at certain levels on the way up. This allows the nitrogen to be expelled slowly through the lungs.

Some mammals can spend rather long periods of time in the ocean depths without coming up for air (see *Focus On: Adaptations of Diving Mammals*).

BREATHING POLLUTED AIR DAMAGES THE RESPIRATORY SYSTEM

Several defense mechanisms protect the delicate lungs from the harmful substances we breathe (Fig. 44–11). The hair around the nostrils, the ciliated mucous lining in the nose and phar-

Figure 44–11 **Urban air pollution.** Industry spews tons of pollutants into the atmosphere. Air pollution in urban areas such as Copsa Mica (built by the Romanian dictator Nicolae Ceausescu as a "model" industrial city) contributes to respiratory disorders. *(Florent Flipper/Unicorn Stock Photos)*

ADAPTATIONS OF DIVING MAMMALS

Dolphins, whales, seals, beavers, and several other air-breathing mammals have structural and physiological adaptations that permit them to dive for food or to elude their enemies. With their streamlined bodies and forelimbs modified as fins or flippers, diving mammals perform impressive aquatic feats. The Weddell seal *(Leptonychotoes weddelli)* can swim under the ice at a depth of 596 m (1968 ft) for more than an hour without coming up for air. The enormous northern elephant seal *(Mirounga angustirostris)* which measures about 5 m (16 to 18 ft) in length and weighs 2 to 4 tons can plunge even deeper. A female elephant seal carried a depth recorder to a depth of more than 1500 m (about 4000 ft) and stayed beneath the surface 2 hours.

Turtles and birds that dive depend on oxygen stored in their lungs. Diving mammals do not take in extra air before a dive. In fact, seals exhale before they dive. With less air in their lungs they are less buoyant. Their lungs collapse at about 50 to 70 m into their dive and then reinflate as they ascend. This means that their lungs do not function for most of the dive. These adaptations are thought to reduce the chance of decompression sickness, because with less air in the lungs there is less nitrogen in the blood to dissolve during the dive.

Physiological adaptations, including

Northern elephant seal *(Mirounga angustirostris)* with depth recorder. *(Frank Balthis Photography)*

ways to distribute and store oxygen, permit some mammals to dive deeply and remain under water for long periods. Seals have about twice the volume of blood, relative to their body weight, as nondiving mammals. Diving mammals also have high concentrations of **myoglobin,** an oxygen-binding pigment similar to hemoglobin, which stores oxygen in muscles. These animals have up to ten times more myoglobin than terrestrial mammals. The very large spleen typical of many diving mammals is thought to store oxygen-rich red blood cells. Under

pressure, the spleen is squeezed and releases these red blood cells into the circulation.

When a mammal dives to its limit, a group of physiological mechanisms known collectively as the **diving reflex** are activated. Metabolic rate decreases by about 20%, which conserves oxygen. Breathing stops and bradycardia (slowing of the heart rate) occurs. The heart rate may decrease to one-tenth of the normal rate, reducing the body's consumption of oxygen and energy. Blood is redistributed; skin, muscles, digestive organs, and other internal organs can survive with less oxygen and receive less blood while an animal is submerged.

The diving reflex is present to some extent in humans, where it may act as a protective mechanism during birth, when an infant may be deprived of oxygen for several minutes. Many cases of near-drownings, especially of young children, have been documented in which the victim had been submerged for as long as 45 minutes in very cold water before being rescued and resuscitated. In many of these survivors there was no apparent brain damage. The shock of the icy water slows the heart rate, increases blood pressure, and shunts the blood to the internal organs of the body that most need oxygen (blood flow in the arms and legs decreases). Metabolic rate decreases so that less oxygen is required.

---

ynx, and the cilia-mucus elevator of the trachea and bronchi serve to trap foreign particles in inspired air. One of the body's most rapid defense responses to breathing dirty air is **bronchial constriction.** In this process, the bronchial tubes narrow, increasing the chance that inhaled particles will land on the sticky mucous lining. Unfortunately, bronchial constriction increases airway constriction so that less air can pass through to the lungs, thus decreasing the amount of oxygen available to body cells. Fifteen puffs on a cigarette during a five-minute period increases airway resistance as much as threefold, and this added resistance to breathing lasts more than 30 minutes. Chain smokers and those who breathe heavily polluted air may remain in a state of chronic bronchial constriction.

Neither the smallest bronchioles nor the alveoli are equipped with mucus or ciliated cells. Foreign particles that get through other respiratory defenses and find their way into the alveoli may be engulfed by macrophages. The macrophages may then accumulate in the lymph tissue of the lungs. Lung

tissues of chronic smokers and those who work in dirty fossil-fuel-burning industries contain large blackened areas where carbon particles have been deposited (Fig. 44–12).

Continued insult to the respiratory system results in disease. Chronic bronchitis and emphysema are **chronic obstructive pulmonary diseases** that have been linked to smoking and breathing polluted air. More than 75% of patients with chronic bronchitis have a history of heavy cigarette smoking (see *Focus On: The Effects of Smoking*). In **chronic bronchitis,** irritation from inhaled pollutants causes the bronchial tubes to secrete too much mucus. Ciliated cells, damaged by the pollutants, cannot effectively clear the mucus and trapped particles from the airways. The body resorts to coughing in an attempt to clear the airways. The bronchioles become constricted and inflamed, and the patient is short of breath.

Victims of chronic bronchitis often develop **pulmonary emphysema,** a disease most common in cigarette smokers. In this disorder, alveoli lose their elasticity, and walls between

THE EFFECTS OF SMOKING

Smoking is the single most preventable cause of death in our society. Tobacco smoke is an important risk factor for cardiovascular disease and by far the most important risk factor for lung cancer, the most common lethal cancer in the United States and in the world. Cigarette smoke is a "portable" air pollutant. Smokers move about, exhaling tobacco smoke into the air we all must breathe. This environmental tobacco smoke has been linked to death from lung cancer of about 3000 nonsmokers each year in the United States.

Nicotine has been shown to be highly addictive. Mechanisms leading to nicotine addiction are the focus of ongoing research. For example, Gaetano Di Chiara and his colleagues at the University of Cagliari in Italy injected nicotine into the veins of rats and then studied the effects on the brain. In 1996 these researchers reported in *Nature* that nicotine, like morphine, amphetamines, and cocaine, increases dopamine concentration and activates cells in the base of the forebrain, in an area called the nucleus accumbens. This area helps integrate emotion, and Di Chiara suggested that dopamine may facilitate learning an association between the pleasurable effects of the drug and other stimuli such as the smell of smoke.

Cigarette smoke contains more than 4700 chemical compounds, including compounds that damage the inner lining of blood vessels, leading to development of atherosclerosis, and more than 40 known cancer-causing agents. In 1996, biologist Mikhail Denissenko and his colleagues reported in *Science* that they had demonstrated a direct link between a carcinogen in cigarette smoke and human cancer. The mechanism involves benzo α pyrene, one of the most potent mutagens and carcinogens known. Denissenko's group used polymerase chain reaction techniques to show that this carcinogen causes mutations in the P53 tumor suppressor gene. This gene is mutated in about 60% of lung cancers and also in many other types of cancer. (See *Focus on Unwelcome Tissues: Cancers* in Chapter 37.)

Some facts about smoking:

- The life of a 30-year-old who smokes 15 cigarettes a day is shortened by an average of more than five years.

- If you smoke more than one pack per day, you are about 20 times more likely to develop lung cancer than is a nonsmoker. According to the American Cancer Society, cigarette smoking causes more than 75% of all lung cancer deaths.
- If you smoke, you double your chances of dying from cardiovascular disease.
- If you smoke, you are 20 times more likely to develop chronic bronchitis and emphysema than is a nonsmoker.
- If you smoke, you have about 5% less oxygen circulating in your blood (because carbon monoxide binds to hemoglobin) than does a nonsmoker.
- If you smoke when you are pregnant, your baby will weigh about 6 ounces less at birth, and there is double the risk of miscarriage, stillbirth, and infant death.
- Infants whose parents smoke have double the risk of contracting pneumonia or bronchitis in their first year of life.
- Workers who smoke one or more packs of cigarettes per day are absent from their jobs because of illness 33% more often than are nonsmokers.
- When smokers quit smoking, their risk of dying from chronic pulmonary disease, cardiovascular disease, or cancer decreases. (Precise changes in risk depend on the number of years the person smoked, the number of cigarettes smoked per day, the age of starting to smoke, and the number of years since quitting.)
- Nicotine replacement with gum, patches, or nasal spray has been shown to be effective as an aid to smoking cessation with many individuals, especially when used in conjunction with behavioral therapy. More recently certain antidepressants have been used to reduce the craving for nicotine.

Almost every American who takes up smoking is a teenager — 3000 every day, more than 1 million every year. The prevalence of smoking among adolescents increased significantly during the 1990s. Ten percent of children who begin smoking start by the fourth grade, and nearly two-thirds start by the tenth grade. In 1996 the U. S. Food and Drug Administration (FDA) took jurisdiction over cigarettes and smokeless tobacco. This action was taken, in part, because the scientific community had accepted the evidence that nicotine is addictive and that consumers use it for pharmacological purposes. The FDA then developed the Children's Tobacco Rule, an effort to reduce tobacco use by children by regulating advertisements that appeal to children and restricting access to tobacco products. Just how effective the campaign to reduce child nicotine addiction will be will depend on how serious our nation is about solving this public health problem.

(a)

(b)

Effects of cigarette smoking. (a) Lungs and major bronchi of a nonsmoker. (b) Lungs and heart of a cigarette smoker. The dark spots in the lung tissue are particles of carbon, tar, and other substances that passed through the respiratory defenses and lodged in the lungs. (a, b, Martin Rotker/Taurus Photos)

Alveoli

(a)   100 μm

Alveoli have been
destroyed in this area

(b)   100 μm

Figure 44–12 Comparison of normal and diseased lung tissue. *(a)* LM of normal lung tissue. *(b)* LM of lung tissue with accumulated carbon particles. Despite the body's defenses, when we inhale smoky, polluted air, especially over a long period of time, dirt particles do enter the lung tissue and some remain lodged there. *(a, b, Alfred Paseika/Taurus Photos)*

adjacent alveoli are destroyed. The surface area of the lung is so reduced that gas exchange is seriously impaired. Air is not expelled effectively, and stale air accumulates in the lungs. The emphysema victim struggles for every breath, and still the body does not get enough oxygen. To compensate, the right ventricle of the heart pumps harder and becomes enlarged. Emphysema patients frequently die of heart failure.

Cigarette smoking is also the main cause of lung cancer. More than 40 of the compounds in the tar of tobacco smoke have been shown to cause cancer. These carcinogenic substances irritate the cells lining the respiratory passages and alter their metabolic balance. Normal cells are transformed into cancer cells, which may multiply rapidly and invade surrounding tissues.

SUMMARY WITH KEY TERMS

I. **Respiration** is the process of gas exchange.
   A. During **organismic respiration** oxygen from the environment is taken up by the animal and delivered to its cells, while carbon dioxide is excreted into the environment.
   B. **Aerobic cellular respiration** is the process by which cells generate ATP through a series of redox reactions in which oxygen serves as the final electron acceptor. Carbon dioxide is produced as a waste product. Aerobic cellular respiration takes place in mitochondria.

II. Gas exchange in air has advantages and disadvantages compared with gas exchange in water.
   A. Air contains a higher concentration of molecular oxygen than water.
   B. Oxygen diffuses more rapidly through air than through water.
   C. Dessication is a problem for terrestrial animals; these animals have adaptations that protect their respiratory surfaces from drying.

III. In small aquatic animals gas exchange takes place by diffusion, with no specialized respiratory structures required. Large animals have specialized respiratory structures such as gills, tracheal tubes, or lungs. Animals carry on **ventilation,** that is, they actively move air or water over their respiratory surfaces.
   A. Many invertebrates, including nudibranch mollusks, most annelids and small arthropods, and some vertebrates, exchange gases across the body surface.
   B. In insects and some other arthropods, air enters a network of **tracheal tubes,** or **tracheae,** through openings, called **spiracles,** along the body surface. Tracheal tubes branch and extend to all regions of the body.
   C. **Gills** are moist, thin projections of the body surface found mainly in aquatic animals.

   1. In chordates, gills are usually internal, located along the edges of the gill slits. In bony fishes the gills are protected by an **operculum.**
   2. In bony fishes, a **countercurrent exchange system** maximizes diffusion of oxygen into the blood and diffusion of carbon dioxide out of the blood.
   D. Terrestrial vertebrates have **lungs** with some means of ventilating them.
   1. Most fishes do not have lungs, but have a homologous **swim bladder** that permits the fish to control its buoyancy.
   2. Amphibians and reptiles have lungs with only some ridges or folds that increase surface area.
   3. In birds, the lungs have extensions called **air sacs** that act as bellows, drawing air into the system. Air flows in one direction: From the outside into the posterior air sacs, to the lung, through the anterior air sacs and then out of the body. Gas exchange takes place through the walls of the **parabronchi** in the lungs. A **crosscurrent** arrangement, in which blood flow is at right angles to the parabronchi, increases the amount of oxygen that enters the blood.

IV. Respiratory pigments, such as **hemoglobin** and **hemocyanins,** increase the capacity of blood to transport oxygen.

V. The mammalian respiratory system includes the lungs and a system of airways. In humans and other mammals a breath of air passes in sequence through the **nostrils, nasal cavities, pharynx, larynx, trachea, bronchi, bronchioles,** and **alveoli.** Each lung occupies a pleural cavity and is covered with a **pleural membrane.**
   A. During breathing, the **diaphragm** and intercostal muscles contract, expanding the chest cavity. The membranous walls of the lungs move

outward along with the chest walls, decreasing the pressure within the lungs. Air from outside the body rushes in through the air passageways and fills the lungs until the pressure equals atmospheric pressure.

B. **Tidal volume** is the amount of air moved into and out of the lungs with each normal breath. **Vital capacity** is the maximum volume that can be exhaled after the lungs are filled to the maximum extent. The volume of air that remains in the lungs at the end of a normal expiration is the **residual capacity.**

C. Oxygen and carbon dioxide are exchanged between alveoli and blood by diffusion.
1. The pressure of a particular gas determines its direction and rate of diffusion.
2. **Dalton's law of partial pressures** explains that in a mixture of gases, the total pressure of the mixture is the sum of the pressures of the individual gases. Thus, each gas in a mixture of gases exerts a **partial pressure,** the same pressure it would exert if it were present alone. The partial pressure of atmospheric oxygen, P_{O_2}, is 160 mm Hg at sea level.
3. According to **Fick's law of diffusion,** the greater the difference in pressure on the two sides of a membrane and the larger the surface area, the faster the gas will diffuse across the membrane.

D. About 97% of the oxygen in the blood is transported as **oxyhemoglobin (HbO$_2$).**
1. The maximum amount of oxygen that can be transported by hemoglobin is the **oxygen carrying capacity.**
2. The actual amount of oxygen bound to hemoglobin is the **oxygen content.**
3. The **percent O$_2$ saturation,** the ratio of O$_2$ content to O$_2$ carrying capacity, is highest in pulmonary capillaries, where oxygen concentration is greatest.
4. The **oxygen-hemoglobin dissociation curve** shows that as oxygen concentration increases, there is a progressive increase in the amount of hemoglobin that combines with oxygen. The curve is affected by pH, temperature, and CO$_2$ concentration.
5. Due to lowered pH caused by carbonic acid, oxyhemoglobin dissociates more readily as carbon dioxide concentration increases. This is the **Bohr effect.**

E. About 70% of the carbon dioxide in the blood is transported as bicarbonate ions. About 20% combines with hemoglobin, and another 10% is dissolved in plasma.
1. Carbon dioxide combines with water to form carbonic acid; the reaction is catalyzed by **carbonic anhydrase.** Carbonic acid dissociates, forming bicarbonate ions (HCO$_3^-$) and hydrogen ions (H$^+$).
2. Hemoglobin combines with H$^+$, buffering the blood. Many bicarbonate ions diffuse into the plasma and are replaced by Cl$^-$ ions; this exchange is known as the **chloride shift.**

F. **Respiratory centers** in the medulla and pons regulate respiration. These centers are stimulated by **chemoreceptors** sensitive to an increase in carbon dioxide concentration, an increase in hydrogen ions, and to very low oxygen concentration.

G. **Hyperventilation** reduces the concentration of carbon dioxide in the alveolar air and in the blood.

H. As altitude increases, barometric pressure (the pressure of Earth's atmosphere) decreases and less oxygen enters the blood. This situation can lead to **hypoxia,** a deficiency of oxygen, which can lead to loss of consciousness and death. In addition to hypoxia, rapid decrease in barometric pressure can cause **decompression sickness,** a condition especially common among divers who ascend too rapidly.

VI. Inhaling polluted air results in **bronchial constriction,** increased mucous secretion, damage to ciliated cells, and coughing. Breathing polluted air or inhaling cigarette smoke can lead to **chronic obstructive pulmonary diseases** such as **chronic bronchitis** and **pulmonary emphysema,** or to lung cancer.

VII. Physiological adaptations, including ways to distribute and store oxygen, permit some mammals to dive deeply and remain under water for long periods. Diving mammals have high concentrations of **myoglobin,** a pigment that binds oxygen, in their muscles. The **diving reflex** is a group of physiological mechanisms, such as decrease in metabolic rate, that are activated when a mammal dives to its limit.

POST-TEST

1. Breathing is an example of (a) countercurrent exchange (b) cellular respiration (c) ventilation (d) answers a, b, and c are correct (e) answers a and c only

2. Which of the following is a benefit of gas exchange in air compared with water? (a) higher concentration of molecular oxygen (b) oxygen diffuses more slowly in air (c) no energy required for ventilation (d) moist respiratory surface not needed (e) better physiological mechanisms to cope with ion diffusion

3. Which of the following adaptations for gas exchange are most typically found in insects? (a) lungs (b) tracheal tubes (tracheae) (c) parabronchi (d) air sacs (e) gills

4. Which of the following are accurately matched? (a) bony fish/operculum (b) insect/alveoli (c) bird/spiracles (d) mammal/filaments (e) shark/dermal gills

5. Tracheal tubes (tracheae) (a) are typically found in mollusks (b) are highly vascular (c) branch and extend to all the cells (d) are characteristic of many mammals (e) end in book lungs

6. The most efficient vertebrate respiratory system is found in (a) amphibians (b) birds (c) reptiles (d) mammals (e) humans

7. In a bird the correct sequence for a breath of air would be (a) anterior air sacs → posterior air sacs → lung (b) posterior air sacs → lung → anterior air sacs (c) parabronchi → posterior air sacs→anterior air sacs (d) posterior air sacs → alveoli → anterior air sacs (e) posterior air sacs → capillaries → cells.

8. Respiratory pigments (a) combine reversibly with oxygen (b) are found only in vertebrates (c) all have a heme (porphyrin) group that combines with oxygen (d) diffuse into the air sacs (e) attach to the alveolar wall

9. Which sequence most accurately describes air flow in the human respiratory system? (a) pharynx → bronchus → trachea → alveolus (b) pharynx → parabronchi → alveoli → bronchioles (c) bronchus → trachea → larynx → lung (d) larynx → trachea → bronchus → bronchiole (e) trachea → larynx → bronchus → alveolus

10. The amount of air moved in and out of the lungs with each normal resting breath is the (a) vital capacity (b) residual capacity (c) vital volume (d) partial pressure (e) tidal volume

11. The greater the difference in pressure and the larger the surface area, the faster a gas will diffuse. This is explained by (a) Dalton's law (b) Fick's law (c) the percent saturation (d) the Bohr effect (e) the oxygen-hemoglobin dissociation curve

12. Oxygen in the blood is transported mainly (a) in combination with hemoglobin (b) as bicarbonate ions (c) as carbonic acid (d) dissolved in plasma (e) combined with carbon dioxide

13. The concentration of which of the following substances is most important in regulating the rate of respiration? (a) chloride ions (b) oxygen (c) bicarbonate ions (d) nitrogen (e) carbon dioxide

14. When a diver ascends too rapidly (a) bronchial constriction occurs (b) a diving reflex is activated (c) nitrogen rapidly bubbles out of solution in the body fluids (d) nitrogen hypoxia occurs (e) carbon dioxide bubbles damage the alveoli

15. Pulmonary emphysema is (a) a chronic obstructive pulmonary disease (b) common in cigarette smokers (c) characterized by loss of elasticity of the alveolar walls (d) a, b, and c are correct (e) a and b only

REVIEW QUESTIONS

1. Why are specialized respiratory structures necessary in a tadpole but not in a flatworm?
2. Compare ventilation in a sponge with ventilation in a human.
3. Compare gas exchange in the following animals: (a) earthworm (b) grasshopper (c) fish (d) frog (e) bird
4. What is the function of respiratory pigments? How do they work?
5. Why does alveolar air differ in composition from atmospheric air? Explain.
6. What physiological mechanisms bring about an increase in rate and depth of breathing during exercise? Why is such an increase necessary?
7. Under what conditions might it be advantageous for a fish to have lungs as well as gills? What function do the "lungs" of modern fishes serve?
8. How does the countercurrent exchange system increase the efficiency of gas exchange between a fish's gills and blood?
9. What mechanisms does the human respiratory system have for getting rid of inhaled dirt? What happens when so much dirty air is inhaled that these mechanisms cannot function effectively?
10. What is meant by the percent O_2 saturation of hemoglobin? What factors affect the dissociation of oxyhemoglobin?
11. Describe the physiological effects of each of the following: hyperventilation, sudden decompression at 12,000 m altitude, and surfacing too quickly from a scuba dive.
12. Label the diagram. Use Figure 44–6 to check your answers.

YOU MAKE THE CONNECTION

1. What problems would be faced by a terrestrial animal possessing gills instead of lungs?
2. Aquatic mammals such as whales and dolphins use lungs rather than gills for gas exchange. Propose a hypothesis to explain this.
3. What are the advantages of having millions of alveoli rather than a pair of simple, balloon-like lungs?

RECOMMENDED READINGS

Bartecchi, C.E., T.D. MacKenzie, and R.W. Schrier. "The Global Tobacco Epidemic." *Scientific American*, Vol. 272, No. 5, May 1995. An overview of the worldwide increase in cigarette smoking and of the health effects of tobacco.

Gostin, L.O., P.S. Arno, and A.M. Brandt. "FDA Regulation of Tobacco Advertising and Youth Smoking." *Journal of the American Medical Association*, Vol. 277, No. 5, 5 Feb. 1997. An overview of U.S. policy regarding regulation of tobacco and smoking by minors.

Kessler, D.A., P.S. Barnett, A. Witt, M.R. Zeller, J.R. Mande, and W.B. Schultz. "The Legal and Scientific Basis for FDA's Assertion of Jurisdiction Over Cigarettes and Smokeless Tobacco." *Journal of the American Medical Association*, Vol. 277, No. 5, 5 Feb. 1997. An overview of an important shift in regulation of tobacco based on recognition of the addictive qualities of nicotine and on the dangers of secondary smoke.

Kooyman, G.L., and P.J. Ponganis. "The Challenges of Diving to Depth." *American Scientist*, Vol. 85, No. 6, Nov./Dec. 1997. Diving birds and mammals have adaptations for conserving oxygen and responding to pressure.

McAuliffe, K. "Elephant Seals, The Champion Divers of the Deep." *Smithsonian*, Vol. 26, No. 6, Sept. 1995. An overview of research findings about diving mammals.

Moon, R.E., R.D. Vann, and P.B. Bennett. "The Physiology of Decompression Illness." *Scientific American*, Vol. 272, No. 2, Aug. 1995. A discussion of some aspects of decompression sickness.

Ruben, J.A., T.D. Jones, N.R. Geist, and W.J. Hillenius. "Lung Structure and Ventilation in Theropod Dinosaurs and Early Birds." *Science*, Vol. 278, 14 Nov. 1997. A bold challenge to the prevailing hypotheses that dinosaurs were endothermic and that they gave rise to birds.

● Visit our website at **http://www.saunderscollege.com/lifesci/titles.html** and click on Solomon/Berg/Martin Biology for links to chapter-related resources on the World Wide Web.

CHAPTER 45

Processing Food and Nutrition

Giraffes eat leaves, antelope graze, frogs and lions capture other animals. Sponges and baleen whales feed on food particles suspended in the water. Although their choices of food and feeding mechanisms are diverse, all animals are **heterotrophs,** organisms that must obtain their energy and nourishment from the organic molecules manufactured by other organisms. What they eat, how they obtain food, and how they process and use food are the focus of this chapter.

Nutrients are substances in food that are used as energy sources to run the systems of the body, as ingredients to make compounds for metabolic processes, and as building blocks in the growth and repair of tissues. Obtaining nutrients is of such vital importance that both individual organisms and ecosystems are structured around the central theme of **nutrition,** the process of taking in and using food. An organism's body plan and its lifestyle are adapted to its particular mode of obtaining food, as illustrated by the variety of predators and scavengers shown gathered around a dead elephant (photographed in East Africa). For example, spotted hyenas *(Crocuta crocuta),* which are formidable predators as well as scavengers, have massive jaws that permit them to consume an entire elephant carcass including bones and hide. Few nutrients are wasted.

With only slight variations, all animals require the same basic nutrients: minerals, vitamins, carbohydrates, lipids, and proteins. Carbohydrates, lipids, and proteins can all be used as energy sources. Eating too much of any of these nutrients can result in weight gain, whereas eating too few nutrients or an unbalanced diet can result in malnutrition and death. Malnutrition in humans, particularly protein deficiency, is a serious health problem in many parts of the world.

Most animals have a specialized digestive system that processes the food they eat. After foods are selected and obtained, they are taken into the digestive cavity. **Ingestion** generally includes taking the food into the mouth and swallowing it. The process of breaking down food is called **digestion.** Because animals eat the macromolecules tailor-made by and for other organisms, they must break down these molecules and refashion them for their own needs. For example, the hyena cannot incorporate the proteins from the elephant carcass directly into its own cells. It must *mechanically digest* its food, and then *chemically digest* it by enzymatic hydrolysis (degrading it with the addition of water; see Chapter 3). In this way proteins are broken down into their component amino acids.

(McMurray Photography)

Amino acids pass through the lining of the digestive tract and into the blood by **absorption.** Like other nutrients, amino acids are transported to the body cells by the circulatory system. In the cells they can be used to synthesize hyena proteins. Food that is not digested and absorbed is discharged from the body. This process is called **egestion** in simple animals and **elimination** in more complex animals.

Food processing and nutrition are active areas of research. For example, researchers investigating the regulation of digestion are discovering a number of peptide messengers in the digestive tract. Nutritionists are studying a vast variety of plant compounds that appear to be important in maintaining health. And the food industry continues to search for new fat and sugar substitutes.

AFTER YOU HAVE STUDIED THIS CHAPTER YOU SHOULD BE ABLE TO

1. Compare food processing, including ingestion, digestion, absorption, and egestion or elimination, of an animal (such as *Hydra*) that has a single-opening digestive system and an animal (such as a vertebrate) that has a digestive system with two openings.
2. Identify on a diagram or model each of the structures of the human digestive system described in this chapter and give the function of each.
3. Trace the pathway traveled by an ingested meal in the human digestive system and describe the step-by-step digestion of carbohydrate, protein, and lipid; summarize the functions of the accessory digestive glands of humans and other terrestrial vertebrates.
4. Describe the structural adaptations that increase the surface area of the digestive tract.
5. Compare lipid absorption with absorption of other nutrients.
6. Trace the fate of glucose, lipids, and amino acids after their absorption, and discuss their roles in the body.
7. Discuss the roles of vitamins and minerals, and distinguish between water-soluble and fat-soluble vitamins.
8. Contrast basal metabolic rate with total metabolic rate; write the basic energy equation for maintaining body weight, and describe the consequences of altering it in either direction.
9. In general terms, describe the problem of world food supply relative to world population, and describe the effects of malnutrition.
10. Summarize the challenges encountered in obtaining adequate amounts of amino acids in a vegetarian diet and describe how a nutritionally balanced vegetarian diet could be planned.

ANIMALS ARE ADAPTED TO THEIR MODE OF NUTRITION

Animals can be classified as herbivores, carnivores, or omnivores on the basis of the type of food they typically eat (Fig. 45–1). Animals that feed directly on producers are **herbivores,** or primary consumers. Animals cannot digest cellulose of plant cell walls, and many adaptations have evolved for extracting nutrients from the plant material they eat. For example, many herbivores have a symbiotic relationship with bacteria that inhabit their digestive tracts. In exchange for food and shelter, the bacteria break down cellulose cell walls, allowing the host's digestive enzymes access to the nutrients within the plant cells. Termites, cows, and horses are among the herbivores that enjoy such symbiotic relationships.

In the cud-chewing ruminants (cattle, sheep, deer, giraffes), the stomach is divided into four chambers. Bacteria inhabiting the first two chambers digest cellulose, splitting some of it into sugars, which are then used by the host and the bacteria themselves. The bacteria produce fatty acids during their metabolism, some of which are absorbed by the animal and serve as an important energy source. Food that is not sufficiently chewed, called cud, is regurgitated into the animal's mouth and chewed again.

Many herbivores eat great quantities of food. Grasshoppers, locusts, elephants, and cattle, for example, all spend a major part of their lives eating. Most of what they eat is not efficiently digested and is eliminated from the body almost unchanged as waste. However, they eat large enough quantities to provide the nourishment they need.

Herbivores are sometimes eaten by flesh-eating **carnivores,** which may also eat one another. Carnivores (secondary and higher level consumers in ecosystems) are adapted for capturing and killing prey. Some carnivores seize their victims and swallow them alive and whole (Fig. 45–1*d*). Others paralyze, crush, or shred their prey before ingesting it. Carnivorous mammals have well developed canine teeth for stabbing their prey during combat. The digestive juice of the stomach breaks down proteins, and because meat is more easily digested than plant food, their digestive tracts are shorter than those of herbivores.

Omnivores, such as bears and humans, consume both plants and animals. Earthworms ingest large amounts of soil containing both animal and plant material. The blue whale, the largest animal, is a filter feeder that strains out tiny algae and animals as it swims. Omnivores often possess adaptations that permit them to distinguish among a wide range of smells and tastes and thereby select a variety of foods.

Animals can also be classified according to the mechanisms they use to feed. Many animals, including carnivores, ingest large pieces of food. Adaptations for this type of feeding include claws, fangs, poison glands, tentacles, and teeth. The beaks of birds and the teeth of many vertebrates are specialized for cutting, tearing, or chewing food.

Many omnivores are **suspension feeders** that remove suspended food particles from the water. Some, like bivalve mollusks, filter water. Many expose a sticky, mucus-coated surface to flowing water; suspended particles adhere to the surface. For example, some echinoderms have tentacles coated with mucus. Baleen whales use rows of hard plates (baleen) suspended from the roof of the mouth to filter small crustaceans.

Some animals feed on fluids by piercing and sucking. Mosquitos have highly adapted structures for piercing skin and sucking blood. Birds that feed on pollen and nectar have long bills and tongues. The shape, size, and curve of the beak may be specialized for feeding on a particular type of flower. Bats that feed on nectar have a long tongue and reduced dentition.

(a)   *(b)*

(c)   *(d)*

Figure 45–1   **Adaptations for obtaining and processing food.** *(a)* The impressively long "snout" of the herbivorous acorn weevil (*Circulio* sp.) is adapted both for feeding and for making a hole in the acorn through which it deposits an egg. When it has hatched, the larva feeds on the contents of the acorn seed. *(b)* The giant panda's (*Ailuropoda melanoleuca*) large, flat teeth and well developed jaws and jaw muscles are adaptations for grinding high-fiber plant food. *(c)* The mouth of the carnivorous long-nose butterfly fish (*Forcipiger longirostris*) is adapted for extracting small worms and crustaceans from tight spots in coral reefs. *(d)* This carnivorous snake (*Dromicus* sp.) is strangling a lava lizard (*Tropiduris* sp.).   *(a, Darwin Dale/Photo Researchers, Inc.; b, Tom McHugh/Photo Researchers, Inc.; c, Zig Leszczynski/ Animals Animals; d, Frans Lanting/Minden Pictures)*

SOME INVERTEBRATES HAVE A DIGESTIVE CAVITY WITH A SINGLE OPENING

The simplest invertebrates, sponges, have no digestive system at all. Sponges obtain food by filtering microscopic organisms from the surrounding water. Individual cells phagocytize the food particles, and digestion is *intracellular* within food vacuoles. Wastes are egested into the water that continuously circulates through the sponge body.

Cnidarians (such as hydras and jellyfish) and flatworms have a **gastrovascular cavity,** a central digestive cavity with only a single opening. Cnidarians capture small aquatic animals with the help of their stinging cells and tentacles (Fig. 45–2a). The mouth opens into the gastrovascular cavity. Cells lining this digestive cavity secrete enzymes that break down proteins. Digestion continues *intracellularly* within food vacuoles, and digested nutrients diffuse into other cells. Undigested food particles are egested through the mouth by contraction of the body.

Free-living flatworms begin to digest their prey even before ingesting it. They extend the pharynx out through their mouth and secrete digestive enzymes onto the prey (Fig. 45–2b). When ingested, the food enters the branched gastrovascular cavity where enzymes continue to digest it. Partly digested food fragments are then phagocytized by cells lining the gastrovascular cavity, and digestion is completed within food vacuoles. As in cnidarians, the flatworm digestive cavity has only one opening, so undigested wastes are egested through the mouth.

MOST ANIMAL DIGESTIVE SYSTEMS HAVE TWO OPENINGS

Most invertebrates, and all vertebrates, have a tube-within-a-tube body plan. The body wall forms the outer tube. The inner tube is a digestive tract with two openings, referred to as a complete digestive system (Fig. 45–3). Food enters through the mouth and undigested food is eliminated through the anus. **Motility** refers to the mixing and propulsive movements of the digestive tract. The propulsive activity characteristic of most regions of the digestive tract is **peristalsis,** waves of muscular contraction that push the food in one direction. More food can be taken in while previously eaten food is being digested and absorbed farther down the digestive tract.

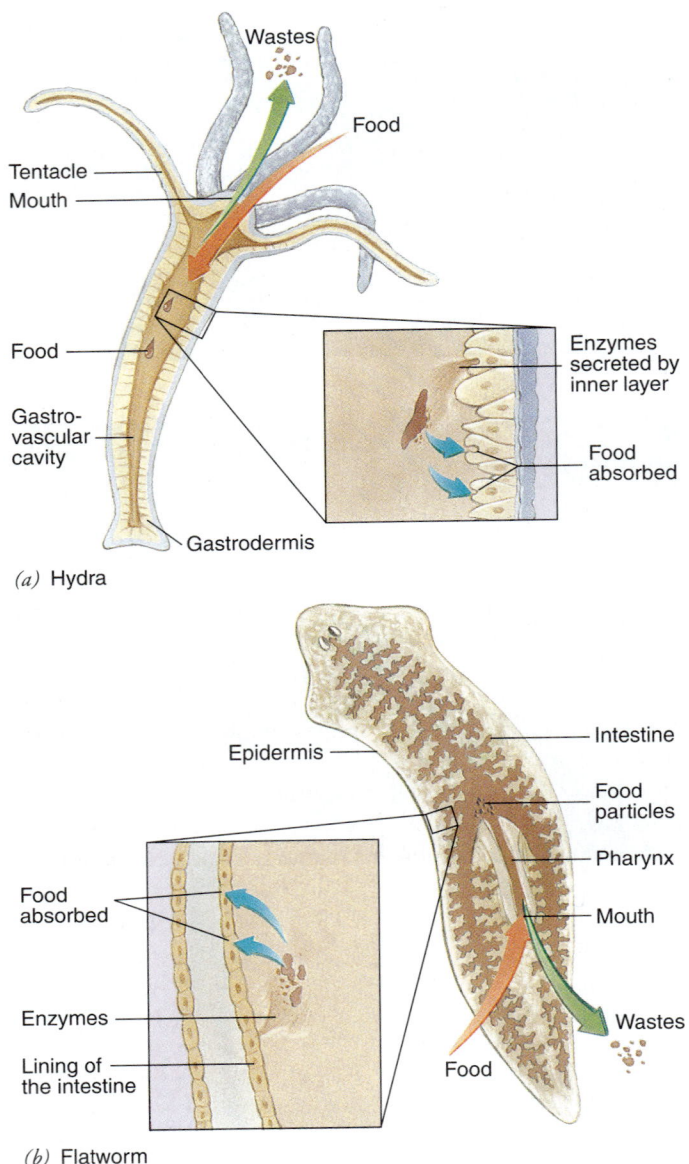

Figure 45–2 **Simple invertebrate digestive systems.** (a) Hydras and (b) planarians have digestive tracts with a single opening that serves as both mouth and anus.

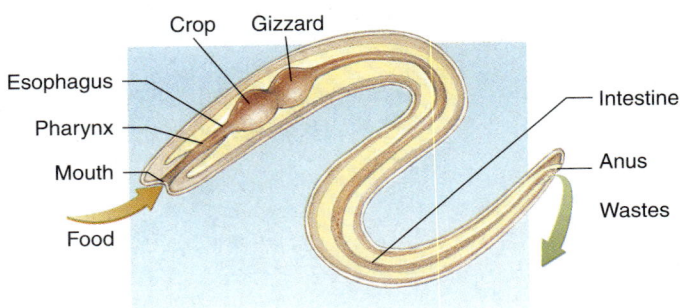

Figure 45–3 **Digestive tract with two openings.** The earthworm, like most complex animals, has a complete digestive tract extending from mouth to anus. Various regions of the digestive tract are specialized to perform different food processing functions.

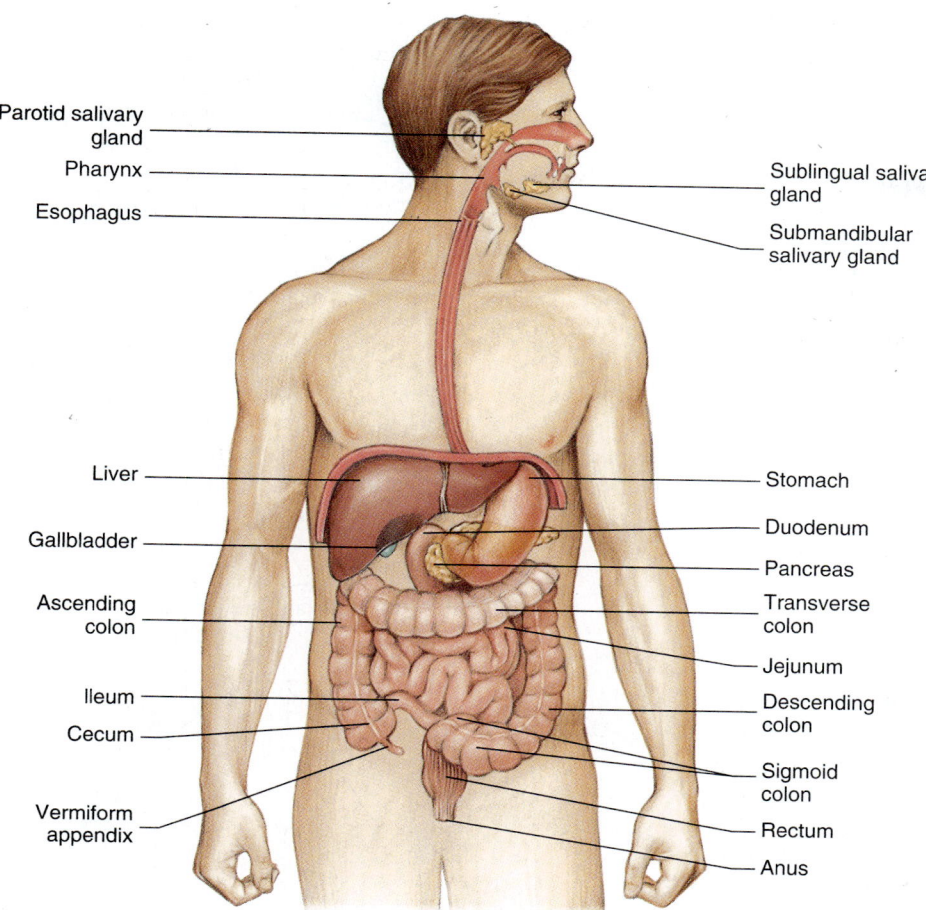

Parotid salivary gland
Pharynx
Esophagus
Sublingual salivary gland
Submandibular salivary gland
Liver
Stomach
Gallbladder
Duodenum
Pancreas
Ascending colon
Transverse colon
Jejunum
Ileum
Descending colon
Cecum
Sigmoid colon
Vermiform appendix
Rectum
Anus

Figure 45-4 **Human digestive tract.** The human digestive tract is a long, coiled tube extending from mouth to anus. The small intestine consists of the duodenum, jejunum, and ileum. The large intestine includes the cecum, colon, rectum, and anus. Locate the three types of accessory glands.

In a digestive tract with two openings, various regions of the tube are specialized to perform specific functions. For example, in the vertebrate digestive tract, food passes in sequence through the following specialized regions:

Mouth ⟶ pharynx (throat) ⟶ esophagus ⟶ stomach ⟶ small intestine ⟶ large intestine ⟶ anus

All vertebrates have accessory glands that secrete digestive juices into the digestive tract. These include the liver, the pancreas, and, in terrestrial vertebrates, the salivary glands.

THE VERTEBRATE DIGESTIVE SYSTEM IS HIGHLY SPECIALIZED FOR PROCESSING FOOD

Various regions of the vertebrate digestive tract are specialized to perform specific functions (Fig. 45–4). The wall of the digestive tract is composed of four layers. Although they vary somewhat in structure in various regions, the layers are basically similar throughout the digestive tract (Fig. 45–5).

The **mucosa,** a layer of epithelial tissue and underlying connective tissue, lines the *lumen* (inner space) of the digestive tract. In the stomach and intestine, the mucosa is greatly folded to increase the secreting and absorbing surface. Surrounding the mucosa is the **submucosa,** a connective tissue layer rich in blood vessels, lymphatic vessels, and nerves.

Surrounding the submucosa is a **muscle layer** consisting of two sublayers of smooth muscle. In the inner sublayer the muscle fibers are arranged circularly around the digestive tube. In the outer sublayer the muscle fibers are arranged longitudinally. The outer connective tissue coat of the digestive tract is the **adventitia.** Below the level of the diaphragm, the adventitia becomes the **visceral peritoneum.** By various folds it is connected to the **parietal peritoneum,** a sheet of connective tissue that lines the walls of the abdominal and pelvic cavities. The visceral and parietal peritonea enclose part of the coelom called the **peritoneal cavity.** Inflammation of the peritoneum, called **peritonitis,** can be very serious because infection can spread along the peritoneum to most of the abdominal organs.

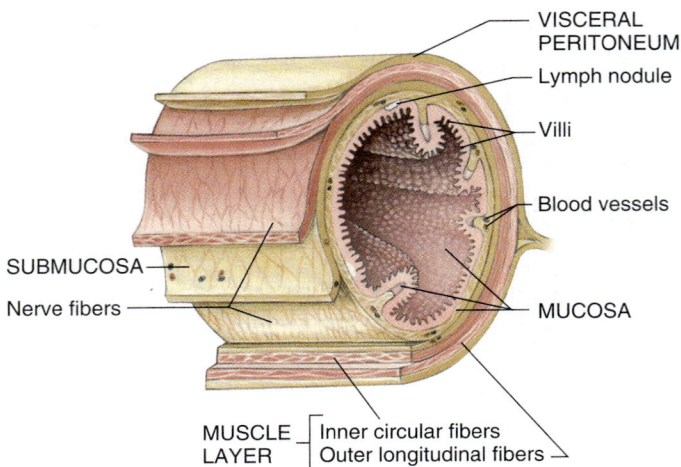

Figure 45-5 **Wall of the digestive tract.** From inside out, the layers of the wall are the mucosa, submucosa, muscle layer, and visceral peritoneum.

Food processing begins in the mouth

Imagine that you have just taken a big bite of a hamburger. The mouth is specialized for ingestion and for beginning the digestive process. Mechanical digestion begins as you bite, grind, and chew the meat and bun with your teeth. Unlike the simple, pointed teeth of fish, amphibians, and reptiles, the teeth of mammals vary in size and shape and are specialized to perform specific functions. The chisel-shaped **incisors** are used for biting, while the long, pointed **canines** are adapted for tearing food (Fig. 45-6). The flattened surfaces of the **premolars** and **molars** are specialized for crushing and grinding.

Each tooth is covered by **enamel,** the hardest substance in the body (Fig. 45-7). Most of the tooth consists of **dentin,** which resembles bone in composition and hardness. Beneath the dentin is the **pulp cavity,** a soft connective tissue containing blood and lymph vessels, and nerves.

While food is being mechanically disassembled by the teeth, it is also moistened by saliva. Some of its molecules dissolve, enabling you to taste the food. Recall from Chapter 41 that taste buds are located on the tongue and other surfaces of the mouth. Three pairs of **salivary glands** secrete about a liter of saliva into the mouth cavity each day. Saliva contains an enzyme, **salivary amylase,** which initiates the chemical digestion of starch into sugar.

The pharynx and esophagus conduct food to the stomach

After the bite of food has been chewed and fashioned into a lump called a **bolus,** it is swallowed, that is, moved through the **pharynx** and into the **esophagus.** The pharynx, or throat, is a muscular tube that serves as the hallway of the respiratory system as well as the digestive system. During swallowing, the

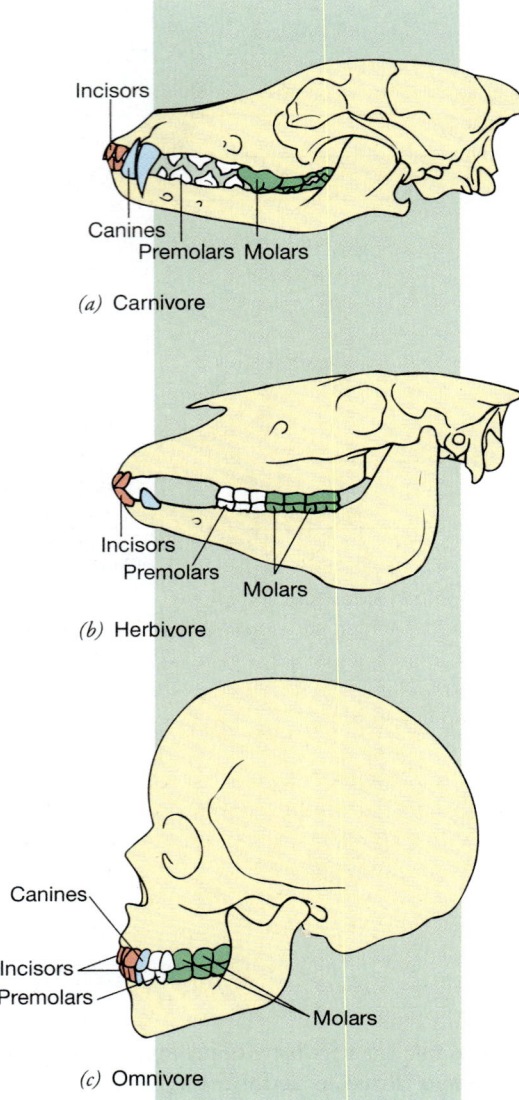

(a) Carnivore

(b) Herbivore

(c) Omnivore

Figure 45-6 **Teeth and diet.** *(a)* Skull of a coyote showing the pointed incisors and canines, adaptations for ripping flesh. *(b)* In contrast, herbivores, such as horses, have incisors (and sometimes canines) adapted for cutting off bits of vegetation. Canines are absent in some herbivores. The broad, ridged surfaces of the molars are adapted for grinding plant material. *(c)* The teeth of omnivores, such as humans, are adapted for chewing a variety of foods.

opening to the airway is closed by a small flap of tissue, the **epiglottis.**

Waves of muscular contraction, called peristalsis, sweep the bolus through the pharynx and esophagus toward the stomach (Fig. 45-8). Circular muscle fibers in the wall of the esophagus contract around the top of the bolus, pushing it downward. Almost at the same time, longitudinal muscles around the bottom of the bolus and below it contract, shortening the tube.

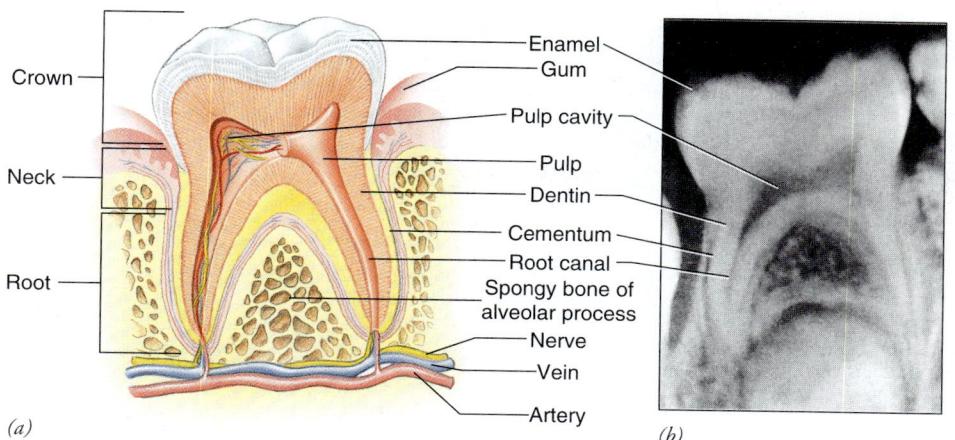

Figure 45–7 **Tooth structure.**
(a) Sagittal section through a human lower molar showing the crown, neck, and root. *(b)* X ray of a healthy tooth.

When the body is in an upright position, gravity helps to move the food through the esophagus, but gravity is not essential. Astronauts are able to eat in its absence, and even if you are standing on your head, food will reach your stomach.

Food is mechanically and enzymatically digested in the stomach

The entrance to the large, muscular **stomach** is normally closed by a ring of muscle at the lower end of the esophagus. When a peristaltic wave passes down the esophagus, the muscle relaxes, permitting the bolus to enter the stomach (Fig. 45–9). When empty, the stomach is collapsed and shaped almost like a hotdog. Folds of the stomach wall, called **rugae,** give the inner lining a wrinkled appearance. As food enters, the rugae gradually smooth out, expanding the capacity of the stomach to more than a liter.

The stomach is lined with a simple, columnar epithelium that secretes large amounts of mucus. Tiny pits mark the entrances to the millions of **gastric glands,** which extend deep into the stomach wall. **Parietal cells** in the gastric glands secrete hydrochloric acid and a substance known as intrinsic factor, needed for adequate absorption of vitamin B_{12}. **Chief cells** in the gastric glands secrete **pepsinogen,** an inactive enzyme precursor. When pepsinogen comes in contact with the acidic gastric juice in the stomach, it is converted to **pepsin,** the main digestive enzyme of the stomach. Pepsin hydrolyzes proteins, converting them to short polypeptides.

Several protective mechanisms prevent the gastric juice from digesting the wall of the stomach. Cells of the gastric mucosa secrete an alkaline mucus that coats the stomach wall and neutralizes the acidity of the gastric juice along the lining. In addition, the epithelial cells of the lining fit tightly together, preventing gastric juice from leaking between them and into the tissue beneath. If some of the epithelial cells become damaged, they are quickly replaced. In fact, about a half million of these cells are shed and replaced every minute!

These mechanisms sometimes malfunction and part of the stomach lining is digested, leaving an open sore or **peptic** **ulcer.** Such ulcers often occur in the duodenum and sometimes in the lower part of the esophagus. The bacterium *Helicobacter pylori* has been implicated as a causative factor in ulcers. *H. pylori* infects the mucus-secreting cells of the stomach lining, resulting in a decrease in protective mucus that may lead to peptic ulcers or cancer. *H. pylori* infection responds to antibiotic therapy.

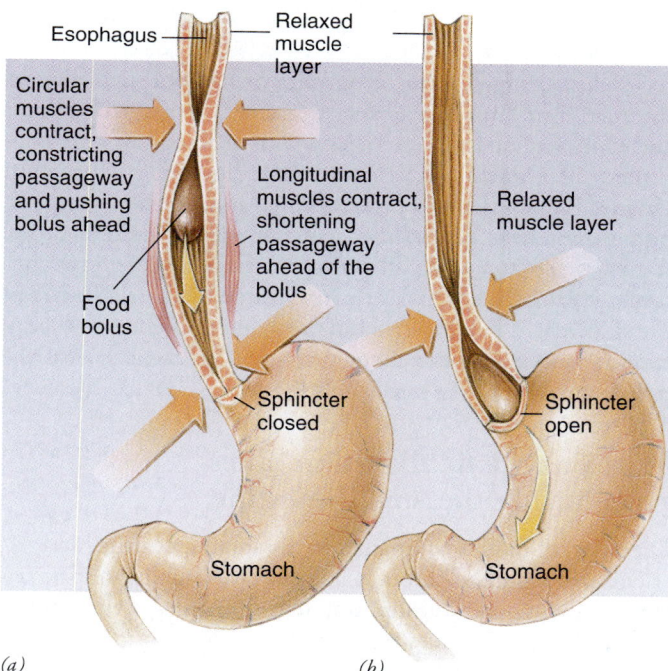

Figure 45–8 **Peristalsis.** Food is moved through the digestive tract by waves of muscular contractions known as peristalsis. *(a)* A bolus is moved down through the esophagus by peristaltic contractions. *(b)* When the sphincter (ring of muscle) at the entrance of the stomach opens, food enters the stomach.

Visceral peritoneum
Esophagus
Sphincter
Circular muscle layer
Longitudinal muscle layer
Oblique muscle layer
Pyloric sphincter

Nucleus
Chief cell
Parietal cell

Duodenum
Rugae

(a)

Openings into gastric glands
Epithelium

Gastric glands

Surface epithelium

Parietal cells

Gastric glands

Lymph nodule

Chief cells

(b)
Gastric mucosa

Gastric glands

Figure 45–9 **Structure of the stomach.** From the esophagus, food enters the stomach, where it is mechanically and enzymatically digested. *(a)* The wall of the stomach has been progressively removed to show muscle layers and rugae. *(b)* Stomach lining and gastric glands.

What changes occur in our bite of hamburger during its three- to four-hour stay in the stomach? The stomach churns and chemically degrades the food so that it assumes the consistency of a thick soup; this partially digested food is called **chyme.** Protein digestion then begins, and much of the hamburger protein is degraded to polypeptides. Digestion of the starch in the bun to small polysaccharides and maltose continues until salivary amylase is inactivated by the acidic pH of the stomach. When digestion in the stomach is complete, peristaltic waves propel the chyme through the stomach exit, the **pylorus,** and into the small intestine.

Most enzymatic digestion takes place inside the small intestine

Digestion of food is completed in the **small intestine,** and nutrients are absorbed through its wall. The small intestine has three regions: the **duodenum, jejunum,** and **ileum.** Most chemical digestion takes place in the duodenum, the first portion of the small intestine, not in the stomach as is commonly believed. Bile from the liver and enzymes from the pancreas are released into the duodenum and act on the chyme. Then enzymes produced by the epithelial cells lining the duodenum catalyze the final steps in the digestion of the major types of nutrients.

The lining of the small intestine appears velvety because of its millions of tiny finger-like projections, the intestinal **villi** (sing., *villus*) (Fig. 45–10). The villi increase the surface area of the small intestine for digestion and absorption of nutrients. The intestinal surface is further expanded by thousands of microvilli, folds of cytoplasm on the exposed surface of the simple columnar epithelial cells of the villi. About 600 microvilli protrude from the surface of each cell, giving the epithelial lining a fuzzy appearance (described as a brush border) when viewed with the electron microscope.

If the intestinal lining were smooth like the inside of a water pipe, food would zip right through the intestine, and many valuable nutrients would not be absorbed. Folds in the wall of the intestine, the villi, and microvilli together increase the surface area of the small intestine by about 600 times. If we could unfold and spread out the lining of the small intestine of an adult human, its surface would approximate the size of a tennis court.

The liver secretes bile, which mechanically digests fats

Just under the diaphragm lies the **liver,** the largest and also one of the most complex organs in the body (Fig. 45–11). A single liver cell can carry on more than 500 separate, special-

Figure 45–10 **Villi and microvilli.** The inner wall of the small intestine is studded with villi and tiny openings that lead into the intestinal glands. *(a)* SEM of a cross section of the small intestine. *(b)* Microscopic view of a small portion of the intestinal wall. Some of the villi have been opened to show the blood and lymph vessels within. *(c)* SEM of the surface of an epithelial cell from the lining of the small intestine, showing microvilli. The epithelium has been cut vertically, allowing the microvilli to be viewed from the side as well as from above. *(a, G. Shih-R. Kessel/Visuals Unlimited; c, Courtesy of J.D. Hoskings, W.G. Henk, and Y.Z. Abdelbaki, from the* American Journal of Veterinary Research, *Vol. 43, No. 10)*

ized metabolic activities. The liver's food processing functions include the following:

1. Secretes **bile,** which is important in the mechanical digestion of fats.
2. Helps maintain homeostasis by removing or adding nutrients to the blood.
3. Converts excess glucose to glycogen, and stores it.
4. Converts excess amino acids to fatty acids and urea.
5. Stores iron and certain vitamins.
6. Detoxifies alcohol and other drugs and poisons that enter the body.

Bile consists of water, bile salts, bile pigments, cholesterol, salts, and lecithin (a phospholipid). Bile produced in the liver is stored in the pear-shaped **gallbladder.** The gallbladder con-

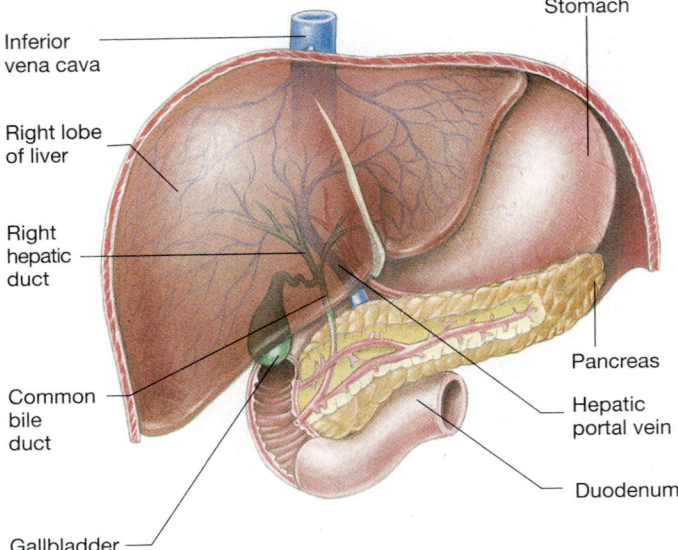

Inferior vena cava

Right lobe of liver

Right hepatic duct

Common bile duct

Gallbladder

Stomach

Pancreas

Hepatic portal vein

Duodenum

Figure 45–11   The liver and pancreas. The gallbladder stores bile from the liver. Note the ducts that conduct bile to the gallbladder and the duodenum.

centrates the bile and releases it into the duodenum as needed. Bile mechanically digests fats by a detergent-like action that decreases the surface tension of fat particles. This action permits the fat molecules to disperse so they can be attacked by lipase (fat-digesting enzymes). The dispersion of fat globules by bile is called **emulsification.** Bile contains no digestive enzymes and so does not enzymatically digest food.

The pancreas secretes digestive enzymes

The **pancreas** is an elongated gland that secretes both digestive enzymes and hormones that help regulate the level of glucose in the blood. Among its enzymes are **trypsin** and **chymotrypsin,** which digest polypeptides to dipeptides; **pancreatic lipase,** which degrades neutral fats; **pancreatic amylase,** which breaks down almost all types of carbohydrates, except cellulose, to disaccharides; and **ribonuclease** and **deoxyribonuclease,** which split the nucleic acids ribonucleic acid (RNA) and deoxyribonucleic acid (DNA) to free nucleotides.

Enzymatic digestion occurs as food moves through the digestive tract

Chyme moves through the digestive tract by peristalsis, mixing contractions, and motions of the villi. As chyme passes through the digestive tract, nutrients in the chyme come into contact with enzymes that digest them.

Carbohydrates are digested to monosaccharides

Polysaccharides, such as starch and glycogen, are important components of the food ingested by humans and most other animals. The glucose units of these large molecules are connected by glycosidic bonds linking carbon 4 (or 6) of one glucose molecule with carbon 1 of the adjacent glucose molecule. These bonds are hydrolyzed by **amylases,** enzymes that digest polysaccharides to the disaccharide maltose (Table 45–1). Although the amylase of the digestive tract can split the α-glycosidic linkages present in starch and glycogen, they cannot split the β-glycosidic linkages present in cellulose (see Figures 3–8 and 3–9 in Chapter 3).

Amylase cannot split the bond between the two glucose units of maltose. Enzymes produced by the cells lining the small intestine break down disaccharides such as maltose to monosaccharides. **Maltase,** for example, splits maltose into two glucose molecules. Hydrolysis occurs while the disaccharides are being absorbed through the epithelium.

Proteins are digested to amino acids

Several kinds of proteolytic enzymes are secreted into the digestive tract (Table 45–1). Each breaks peptide bonds at one or more specific locations in a polypeptide chain. Trypsin, secreted in an inactive form by the pancreas, is activated by an enzyme called enterokinase. The trypsin then activates chymotrypsin and carboxypeptidase, as well as additional trypsin. Pepsin, trypsin, and chymotrypsin break certain internal peptide bonds of proteins and polypeptides. Carboxypeptidase removes amino acids with free terminal carboxyl groups from the end of polypeptide chains. **Dipeptidases** released by the duodenum then split the remaining small peptides to amino acids.

Fats are digested to fatty acids and monoacylglycerols

Lipids are usually ingested as large masses of triacylglycerols (also called triglycerides). They are digested mainly within the duodenum by pancreatic lipase (Table 45–1). Like many other proteins, lipase is water-soluble, but its substrates are not. Thus, the enzyme can attack only the fat molecules at the surface of a mass of fat. The bile salts act like detergents to reduce the surface tension of fats. Their action breaks large masses of fat into smaller droplets. This greatly increases the surface area of fat exposed to the action of pancreatic lipase and so increases the rate of lipid digestion.

Conditions in the intestine are usually not optimal for the complete hydrolysis of lipids to glycerol and fatty acids. The products of lipid digestion therefore include monoacylglycerols (monoglycerides) and diacylglycerols (diglycerides) as well as glycerol and fatty acids. Undigested triacylglycerols remain as well, and some of these are absorbed without digestion.

Nerves and hormones regulate digestion

Most digestive enzymes are produced only when food is present in the digestive tract. Salivary gland secretion is controlled entirely by the nervous system, but secretion of other digestive juices is regulated by both nerves and hormones. The wall

TABLE 45-1   Summary of Digestion

Location	Source of Enzyme	Digestive Process*
Carbohydrate digestion		
Mouth	Salivary glands	Polysaccharides (e.g., starch) $\xrightarrow{\text{Salivary amylase}}$ Maltose + Small polysaccharides
Stomach		Action continues until salivary amylase is inactivated by acidic pH
Small intestine	Pancreas	Undigested polysaccharides $\xrightarrow{\text{Pancreatic amylase}}$ Maltose and small polysaccharides
	Intestine	Disaccharides hydrolyzed to monosaccharides as follows:
		Maltose (malt sugar) $\xrightarrow{\text{Maltase}}$ Glucose + Glucose
		Sucrose (table sugar) $\xrightarrow{\text{Sucrase}}$ Glucose + Fructose
		Lactose (milk sugar) $\xrightarrow{\text{Lactase}}$ Glucose + Galactose
Protein digestion		
Stomach	Stomach (gastric glands)	Protein $\xrightarrow{\text{Pepsin}}$ Short polypeptides
Small intestine	Pancreas	Polypeptides A—A—A—A—A | A—A—A—A—A $\xrightarrow{\text{Trypsin, Chymotrypsin}}$ Polypeptides + Dipeptides A—A—A A—A
		Polypeptides A—A—A—A $\xrightarrow{\text{Carboxypeptidase}}$ Peptides and free amino acids A—A—A A A
	Small intestine	Peptides A—A—A + Dipeptides A—A $\xrightarrow{\text{Peptidases, Dipeptidases}}$ Free amino acids A A A
Lipid digestion		
Small intestine	Liver	Glob of fat $\xrightarrow{\text{Bile salts}}$ Emulsified fat (individual triacylglycerols)
	Pancreas	Triacylglycerol $\xrightarrow{\text{Pancreatic lipase}}$ Fatty acids + Glycerol

* ○ = monosaccharide; ⊑ = triacylglycerol; ⌒ = fatty acid; A = amino acid

TABLE 45-2   Hormonal Control of Digestion

Hormone	Source	Target Tissue	Actions	Factors that Stimulate Release
Gastrin	Stomach (mucosa)	Stomach (gastric glands)	Stimulates gastric glands to secrete pepsinogen	Distention of the stomach by food; certain substances such as partially digested proteins and caffeine
Secretin	Duodenum (mucosa)	Pancreas Liver	Signals secretion of sodium bicarbonate Stimulates bile secretion	Acidic chyme acting on mucosa of duodenum
Cholecystokinin (CCK)	Duodenum (mucosa)	Pancreas Gallbladder	Stimulates release of digestive enzymes Stimulates emptying of bile	Presence of fatty acids and partially digested proteins in duodenum
Gastric inhibitory peptide (GIP)	Duodenum (mucosa)	Stomach	Decreases stomach churning, thus slowing emptying	Presence of fatty acids or glucose in duodenum

of the digestive tract contains dense networks of neurons. This so-called enteric nervous system continues to regulate many motor and secretory activities of the digestive system even if sympathetic and parasympathetic nerves to these organs are cut. Many neuropeptides present in the brain are also released by neurons in the digestive tract and help regulate digestion. For example, substance P stimulates smooth muscle contraction of the digestive tract, while enkephalin inhibits motor activity.

At least four hormones, **gastrin, secretin, cholecystokinin (CCK),** and **gastric inhibitory peptide (GIP),** help regulate the digestive system (Table 45–2). All of these hormones are polypeptides secreted by endocrine cells in the mucosa of certain regions of the digestive tract. Investigators are studying several other messenger peptides that appear to be important in regulating digestive activity.

As an example of the regulation of the digestive system, consider the secretion of gastric juice. Seeing, smelling, tasting, or even thinking about food causes the brain to send neural messages to the gastric glands in the stomach, stimulating them to secrete. In addition, when food distends the stomach, stretch receptors send neural messages to the medulla. The medulla then sends messages to endocrine cells in the stomach wall that secrete the hormone gastrin. Although gastrin is absorbed into the blood, it returns to the stomach, where it stimulates gastric juice release, gastric emptying, and intestinal motility.

Absorption takes place mainly through the villi of the small intestine

Only a few substances—water, simple sugars, salts, alcohol, and certain drugs—are small enough to be absorbed through the stomach wall. Absorption of nutrients is primarily the job of the intestinal villi. As illustrated in Figure 45–10, the wall of a villus consists of a single layer of epithelial cells. Inside each villus is a network of capillaries and a central lymph vessel, called a lacteal.

To reach the blood (or lymph), a nutrient molecule must pass through an epithelial cell of the intestinal lining and through a cell of the blood or lymph vessel lining. Absorption occurs by a combination of simple diffusion, facilitated diffusion, and active transport. Because glucose and amino acids cannot diffuse through the intestinal lining, they must be absorbed by active transport. Absorption of these nutrients is coupled with the active transport of sodium (see Chapter 5). Fructose is absorbed by facilitated diffusion.

Amino acids and glucose are transported directly to the liver by the **hepatic portal vein.** In the liver this vein divides into a vast network of tiny blood vessels similar to capillaries. As the nutrient-rich blood moves slowly through the liver, nutrients and certain toxic substances are removed from the circulation.

The products of lipid digestion are absorbed by a different process and different route (Fig. 45–12). Fatty acids and monoacylglycerols combine with bile salts to form soluble complexes called **micelles.** The micelles serve as a reservoir for fatty acids and monoacylglycerols. Only small amounts of free fatty acids and monoacylglycerols are present in the intestine. As these molecules are absorbed by diffusion (they are soluble in the lipid of the plasma membrane), they are immediately replaced by molecules from the micelles. This process greatly facilitates absorption of fats. As monoacylglycerols and fatty acids leave a micelle, new fatty acids and monoacylglycerols combine with it. Micelles also dissolve cholesterol and fat-soluble vitamins, facilitating their absorption.

1 Glycerol, fatty acids, and monoacylglycerols in micelles

Monoacylglycerol

Fatty acid

Glycerol

Micelles free to pick-up more fatty acids

Epithelial cell of the intestinal mucosa

2 Glycerol, fatty acids, and monoacylglycerols absorbed by diffusion

Diffusion

3 Within epithelial cell fatty acids and glycerol resynthesized into triacylglycerols

4 Triacylglycerols packaged with cholesterol and phospholipids as chylomicrons

Chylomicron

P — Phospholipids

C — Cholesterol

Triacylglycerol

5 Chylomicrons pass out of cell into lacteal by diffusion

Diffusion

6 Chylomicrons transported via lymph system to blood; triacylglycerols enter fat cells for storage

Through lymphatic system to blood

Figure 45–12 **Lipid absorption.** Fatty acids and monoacylglycerols combine with bile salts in the intestine to form micelles, which transport the fatty substances to the epithelial cells lining the intestine. In the epithelial cells, fatty acids and glycerol combine to form triacylglycerols. These molecules, along with phospholipids and cholesterol, form protein-covered globules called chylomicrons, which are transported to the blood by the lymphatic system.

Once free fatty acids and monoacylglycerols enter an epithelial cell in the intestinal lining, they are reassembled as triacylglycerols in the smooth endoplasmic reticulum. The triacylglycerols are packaged into droplets, which become larger as absorbed cholesterol and phospholipids are added. The droplets are then covered with a thin coat of protein. These protein-covered fat droplets, called **chylomicrons,** diffuse out of the epithelial cell and into the lacteal (lymph vessel) of the villus.

Chylomicrons are transported by the lymph to the subclavian veins where the lymph and its contents enter the blood. About 90% of absorbed fat enters the blood circulation in this indirect way. The rest, mainly short-chain fatty acids such as those in butter, are absorbed directly into the blood. After a meal rich in fats, the great number of chylomicrons in the blood may give the plasma a turbid, milky appearance for a few hours.

Most of the nutrients in the chyme are absorbed by the time it reaches the end of the small intestine. What is left (mainly waste) passes through a sphincter, the **ileocecal valve,** into the large intestine.

The large intestine eliminates waste

Undigested material, such as the cellulose of plant foods, along with unabsorbed chyme, passes into the **large intestine** (Fig. 45–4). Although only about 1.3 m (about 4 ft) long, this part of the digestive tract is referred to as "large" because its diameter is greater than that of the small intestine. The small intestine joins the large intestine about 7 cm (2.8 in) from the end of the large intestine, forming a blind pouch, the **cecum.** The **vermiform appendix** projects from the end of the cecum. (Appendicitis is an inflammation of the appendix.) The functions of the cecum and appendix in humans are not known. They are generally considered vestigial organs. Herbivores such as rabbits have a large, functional cecum that holds food while bacteria digest its cellulose.

From the cecum to the **rectum** (the last portion of the large intestine), the large intestine is known as the **colon.** The regions of the large intestine are the cecum; ascending colon; transverse colon; descending colon; sigmoid colon; rectum; and anus, the opening for the elimination of wastes.

As the chyme passes slowly through the large intestine, water and sodium are absorbed from it, and it gradually assumes the consistency of normal feces. Bacteria inhabiting the large intestine are nourished by the last remnants of the meal and benefit their host by producing vitamin K and certain B vitamins that can be absorbed and used.

A distinction should be made between elimination and excretion. *Elimination* is the process of getting rid of digestive wastes—materials that have not been absorbed from the digestive tract and did not participate in metabolic activities. *Excretion* refers to the process of getting rid of *metabolic wastes* and in mammals is mainly the function of the kidneys and lungs. The large intestine, however, does excrete bile pigments.

When chyme passes through the intestine too rapidly, **defecation** (expulsion of feces) becomes more frequent, and the feces are watery. This condition, called diarrhea, may be caused by anxiety, certain foods, or by disease organisms that irritate the intestinal lining. Prolonged diarrhea results in loss of water and salts and leads to dehydration—a serious condition, especially in infants. Constipation results when chyme passes through the intestine too slowly. Because more water than usual is removed from the chyme, the feces may be hard and dry. Constipation is often caused by a diet deficient in fiber.

In Western countries, cancer of the colon and rectum accounts for more new cases each year than do any other cancers except lung cancer (Fig. 45–13). Studies indicate that a high incidence of this disease is found in populations whose diets are very low in fiber and high in animal protein, fat, and refined carbohydrate. It has been suggested that diets low in fiber result in less frequent defecation, allowing prolonged contact between the mucous membrane of the colon and such carcinogens as nitrites (used as preservatives) in foods.

ADEQUATE AMOUNTS OF NUTRIENTS ARE NECESSARY TO SUPPORT METABOLIC PROCESSES

All animals require carbohydrates, lipids, proteins, vitamins, and minerals. Although not considered a nutrient in a strict sense, water is a necessary dietary component. Sufficient fluid must be ingested to replace that lost in urine, sweat, feces, and breath. In addition, nutritionists are investigating a host of plant compounds, called phytochemicals, that appear to be important in nutrition.

Adequate amounts of essential nutrients are necessary for metabolic processes. Recall from Chapter 1 that metabolism refers to all of the chemical processes that take place in the body. Metabolic processes include anabolism and catabolism. Anabolism refers to synthetic processes such as the production of proteins. Catabolism includes breakdown processes such as hydrolysis. Nutritionists measure the energy value of food in kilocalories, or simply Calories. A Calorie, spelled with a capital C, is a **kilocalorie (kcal),** defined as the amount of heat required to raise the temperature of a kilogram of water 1 degree Celsius (see *Making the Connection: Energy, Work, and Heat* in Chapter 6).

Carbohydrates are a major energy source in the omnivore diet

Sugars and starches are the principal sources of energy in the ordinary human diet. However, they are not considered essential nutrients, because the body can obtain sufficient energy from a mixture of proteins and fats. In the average American diet, carbohydrates provide about 48% of the kcal ingested daily.

Most carbohydrates are ingested in the form of starch and cellulose, both polysaccharides. (You may want to review the discussion of carbohydrates in Chapter 3.) Nutritionists refer to polysaccharides as **complex carbohydrates.** Foods rich in complex carbohydrates include rice, potatoes, corn, and other cereal grains. These are the least expensive foods, and for this

Figure 45–13 Colon cancer. Cancer is evident as a mass that projects into the lumen of the colon in this radiographic view of the large intestine. The large intestine has been filled with a suspension of barium sulfate, which makes irregularities in the wall visible.

reason the proportion of carbohydrate in a family's diet often reflects economic status. Very poor people subsist on diets that are almost exclusively carbohydrate, while the more affluent enjoy the more costly protein-rich foods such as meat and dairy products.

Nutritionists suggest that we increase our consumption of complex carbohydrates and fiber by eating more fruits, vegetables, and whole grains. **Fiber** is mainly a complex mixture of cellulose and other indigestible carbohydrates. The U.S. diet is low in fiber due to low intake of fruits and vegetables and use of refined flour. Increasing fiber in the diet may decrease the risk of colon cancer. Fiber may also stimulate the feeling of being satisfied (satiety) with the amount of food intake and thus could be useful in treating obesity.

When an excess of carbohydrate-rich food is eaten, the liver cells may become fully packed with glycogen and still have more glucose entering them. Liver cells then convert excess glucose to fatty acids and glycerol. These compounds are converted to triacylglycerols and sent to the fat depots of the body for storage.

Lipids are used as an energy source and to make biological molecules

Cells use ingested lipids as fuel, as components of cell membranes, and to make lipid compounds such as steroid hormones

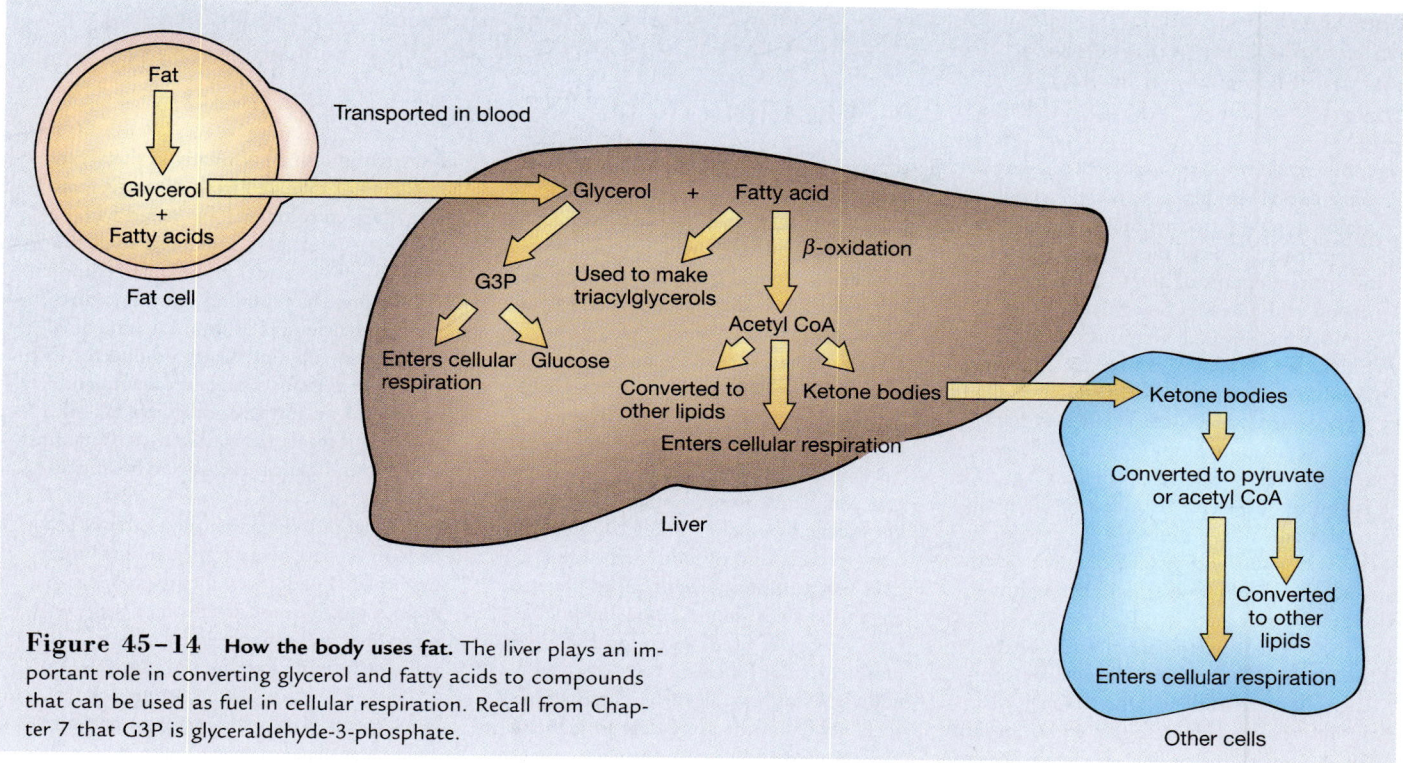

Figure 45–14 **How the body uses fat.** The liver plays an important role in converting glycerol and fatty acids to compounds that can be used as fuel in cellular respiration. Recall from Chapter 7 that G3P is glyceraldehyde-3-phosphate.

and bile salts. Lipids account for about 34% of the kcal in the average U.S. diet. Three polyunsaturated fatty acids (linoleic, linolenic, and arachidonic acids) are essential fatty acids that must be obtained in the human diet. Given these and sufficient nonlipid nutrients, the body can make all of the lipid compounds (including fats, cholesterol, phospholipids, and prostaglandins) that it needs.

About 98% of the lipids in the diet are ingested in the form of triacylglycerols (triglycerides). (Recall from Chapter 3 that a triacylglycerol is a glycerol molecule chemically combined with three fatty acids; see Fig. 3–11*b*.) Triacylglycerols may be saturated, that is, fully loaded with hydrogen atoms, or their fatty acids may be monounsaturated (containing one double bond) or polyunsaturated (containing two or more double bonds).

Generally, animal foods are rich in both saturated fats and cholesterol, while plant foods contain unsaturated fats and no cholesterol. Commonly used polyunsaturated vegetable oils are corn, soy, cottonseed, and safflower oils. Olive, canola, and peanut oils contain large amounts of monounsaturated fats. Butter contains mainly saturated fats.

The average U.S. diet provides about 700 mg of cholesterol each day, whereas about 300 mg is the recommended maximum. High cholesterol sources include egg yolks, butter, and meat. The body is not dependent on dietary sources for cholesterol because it is able to synthesize cholesterol from other nutrients. In fact, dietary intake of saturated fats can increase cholesterol level markedly.

When needed, stored fats are hydrolyzed to fatty acids and released into the blood. Before these fatty acids can be used by cells as fuel, they must be broken down into smaller compounds and combined with coenzyme A to form molecules of acetyl coenzyme A (Fig. 45–14). Acetyl coenzyme A enters the citric acid cycle (see Chapter 7). This transformation is accomplished in the liver by a process known as β-oxidation.

For transport to the cells, acetyl coenzyme A is converted into one of three types of *ketone bodies* (four-carbon ketones). Normally, the level of ketone bodies in the blood is low, but in certain abnormal conditions, such as starvation and diabetes mellitus, fat metabolism is tremendously increased. Ketone bodies are then produced so rapidly that their level in the blood becomes excessive, which may cause the blood to become too acidic. Such disruption of normal pH balance can lead to death. (How the Inuit, who live on extremely high-fat diets, manage to maintain acid-base homeostasis is something of a mystery.)

Proteins serve as enzymes and as structural components of cells

Proteins are essential building blocks of cells, serve as enzymes, and are also used to make needed substances such as hemoglobin and myosin. Protein consumption is an index of a country's (or an individual's) economic status, because high-quality protein tends to be the most expensive and least available of the nutrients.

VEGETARIAN DIETS

Most of the world's population depends almost entirely on plants, especially cereal grains (typically rice, wheat, or corn) as the staple food. None of these foods contains adequate amounts of all of the essential amino acids. Besides being deficient in some of the essential amino acids, plant foods contain a lower percentage of protein than do animal foods. Meat contains about 25% protein, whereas even the new high-yield grains contain only 5% to 13% protein. What protein is available in plant food is also less digestible than that found in animal foods. Because most of the protein is encased within indigestible cellulose cell walls, much of it passes right through the digestive tract.

Despite these potential nutritional problems, more and more people are turning to vegetarian diets. One reason is that in meat-based diets we ingest a great deal of fat along with the protein, leading to health problems. Another reason is that meats are becoming increasingly expensive because they are ecologically expensive to produce. About 21 kg of protein in grain, for example, is required to produce just 1 kg of beef

protein. If the human population of our planet continues to expand at a much greater rate than does our food production, more grain will be diverted for human food and less for animal feed. The price of meat will continue to soar and may become unaffordable for many of us.

A vegetarian diet can be nutritionally balanced provided one takes into account the nutritional risks (especially in growing children) associated with non-meat diets. An important guideline is to select proteins that complement one another. This requires knowledge of which amino acids are deficient in each kind of food. Since the body does not store amino acids, all of the essential amino acids should be ingested daily. For example, if rice is eaten for dinner, and black beans for lunch the next day, the body will not have all the essential amino acids needed at the same time to manufacture proteins. If beans and rice are eaten together, however, all of the needed amino acids are provided, because one food contributes what the other lacks. (Rice is limited in the amino acid lysine but is high in methionine; beans are low in methionine,

but high in lysine.) Similarly, if dairy products are not excluded from the vegetarian diet, then macaroni can be paired with cheese, or cereal with milk, and all of the essential amino acids can be obtained. Other concerns regarding vegetarian diets include deficiencies of iron, calcium, zinc, and certain vitamins. Dairy products, which are a major source of calcium and vitamin D (in fortified milk), are excluded in certain (vegan) vegetarian diets. Vitamin B_{12} is found almost exclusively in animal products.

Balanced vegetarian diets offer health benefits in part because they are low in saturated fat. In addition, increases in legume, grain, vegetable, and fruit intake provide more fiber and phytochemicals, as well as more of certain vitamins and minerals. Studies of Seventh Day Adventists, a religious group that excludes animal products from their diets, compared with non–Seventh Day Adventists living in the same area indicate that the incidence of cardiovascular disease is reduced by about 50%. Other studies have shown that vegetarians have a lower risk for certain types of cancer.

The recommended daily adult intake of protein is about 56 g—only about an eighth of a pound (dry weight). In the United States and other developed countries, most people eat about twice as much protein as they require. In other parts of the world, protein poverty is one of the most pressing health problems. Millions of humans suffer from poor health, disease, and even death as a consequence of protein malnutrition.

Ingested proteins are degraded in the digestive tract to amino acids. Of the 20 or so amino acids important in nutrition, approximately eight (nine in children) cannot be synthesized by humans at all, or at least not in sufficient quantities to meet the body's needs. These, which must be provided in the diet, are referred to as **essential amino acids.**

Complete proteins, those that contain the most appropriate distribution of amino acids for human nutrition, are found in eggs, milk, meat, and fish. Some foods, such as gelatin or soybeans, contain a high proportion of protein. However, they either do not contain all of the essential amino acids, or they do not contain them in proper nutritional proportions. Most plant proteins are deficient in one or more essential amino acids (See *Focus On: Vegetarian Diets*).

Amino acids circulating in the blood can be taken up by cells and used for the synthesis of proteins. Excess amino acids are removed from the circulation by the liver. In the liver cells, these are deaminated; that is, the amino group is removed (Fig. 45–15). During deamination, ammonia forms from the amino group. Ammonia, which is toxic at high concentrations, is converted to urea and excreted from the body.

The remaining carbon chain of the amino acid (called a keto acid) may be converted into carbohydrate or lipid and used as fuel or stored. Thus, even people who eat high-protein diets can gain weight if they eat too much.

Vitamins are organic compounds essential for normal metabolism

Vitamins are organic compounds required in the diet in relatively small amounts for normal biochemical functioning. Many are components of coenzymes (see Chapter 6). Vitamins may be divided into two main groups. **Fat-soluble vitamins,** those that can be dissolved in fat, include vitamins A, D, E, and K. **Water-soluble vitamins** are the B and C vitamins.

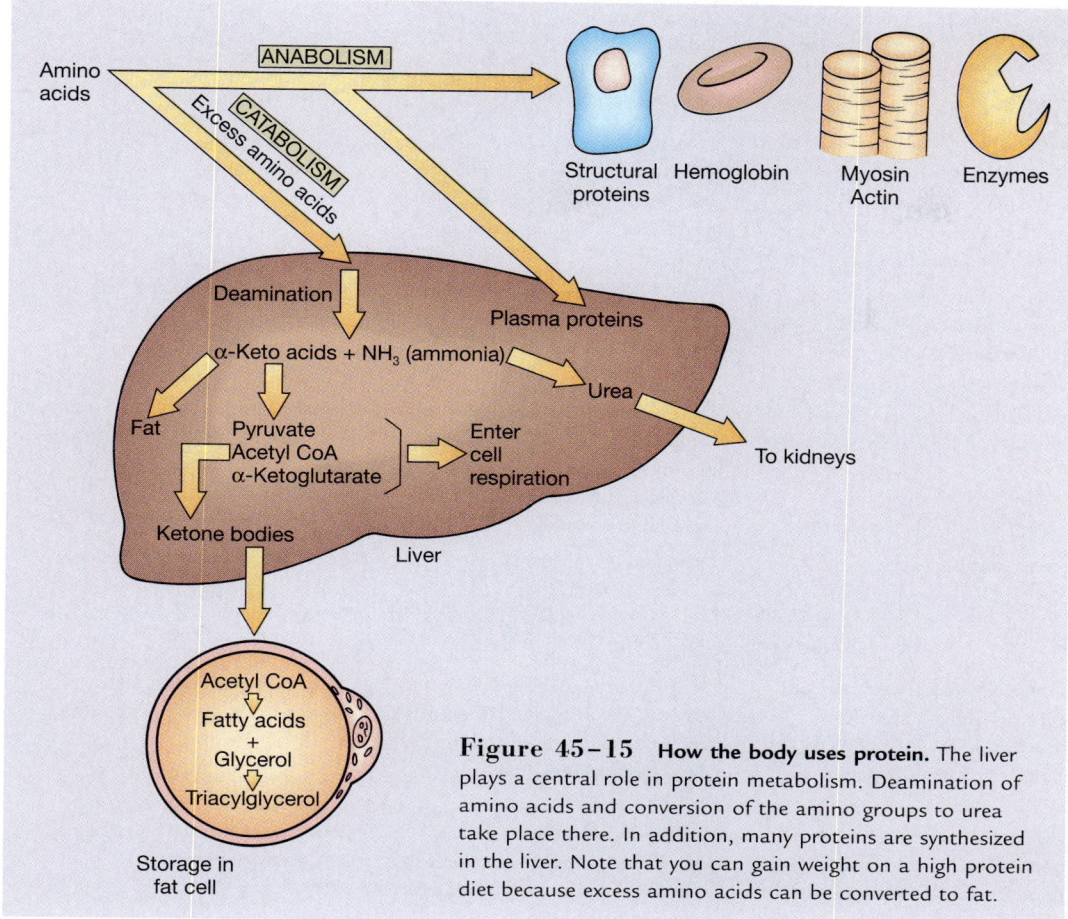

Figure 45–15   How the body uses protein. The liver plays a central role in protein metabolism. Deamination of amino acids and conversion of the amino groups to urea take place there. In addition, many proteins are synthesized in the liver. Note that you can gain weight on a high protein diet because excess amino acids can be converted to fat.

Table 45–3 provides the sources, functions, and consequences of deficiency for most of the vitamins.

Health professionals debate the advisability of taking large amounts of certain specific vitamins, such as vitamin C to prevent colds, or vitamin E to protect against vascular disease. Some studies suggest that vitamin A (found in yellow and green vegetables) and vitamin C (found in citrus fruit and tomatoes) may help protect against certain forms of cancer. We do not yet understand all of the biochemical roles played by vitamins, or the interactions among various vitamins and other nutrients.

We do know that large overdoses of vitamins, like vitamin deficiency, can be harmful. Moderate overdoses of the B and C vitamins are excreted in the urine, but surpluses of the fat-soluble vitamins are not easily excreted and can accumulate to harmful levels.

Minerals are inorganic nutrients

Minerals are inorganic nutrients ingested in the form of salts dissolved in food and water (Table 45–4). Essential minerals required in amounts of 100 mg or more daily include sodium, chloride, potassium, calcium, phosphorus, magnesium, and sulfur. Several others, such as iron, copper, iodide, fluoride, and selenium, are **trace elements,** minerals that are required in amounts of less than 100 mg per day.

Minerals are needed as components of body tissues and fluids. Salt content (about 0.9% in plasma) is vital in maintaining the fluid balance of the body, and since salts are lost from the body daily in sweat, urine, and feces, they must be replaced by dietary intake. Sodium chloride (common table salt) is the salt needed in largest quantity in blood and other body fluids. A deficiency results in dehydration.

Iron is the mineral most likely to be deficient in the diet. In fact, iron deficiency is one of the most widespread nutritional problems in the world. In most developing countries, an estimated two-thirds of children and women of childbearing age suffer from iron deficiency. In the United States, Europe, and Japan, 10% to 20% of women of childbearing age have this deficiency.

Phytochemicals play important roles in maintaining health

Results of hundreds of studies indicate that diets rich in fruit and vegetables lower the incidence of cancer and heart disease. Yet, nutritionists estimate that in the United States fewer than 10% of people include adequate amounts of fruit and vege-

TABLE 45-3 The Vitamins

Vitamins and U.S. RDA*	Actions	Effect of Deficiency	Sources
Fat-soluble			
Vitamin A, retinol 5000 IU†	Converted to retinal; essential for normal vision; essential for normal growth and differentiation of cells; reproduction; immunity	Growth retardation; night blindness; worldwide 500,000 preschool children are blinded by vitamin A deficiency each year.	Liver, fortified milk, yellow and green vegetables such as carrots and broccoli
Vitamin D, calciferol 400 IU	Promotes calcium and phosphorus absorption from digestive tract; essential to normal growth and maintenance of bone	Bone deformities: rickets in children; osteomalacia in adults	Fish oils, egg yolk, fortified milk, butter, margarine
Vitamin E, tocopherols 30 IU	Antioxidant; protects unsaturated fatty acids and cell membranes	Increased catabolism of unsaturated fatty acids, so that not enough are available for maintenance of cell membranes; prevention of normal growth; nerve damage	Vegetable oils, nuts; leafy greens
Vitamin K, probably about 1 mg	Essential for blood clotting	Prolonged blood clotting time	Normally supplied by intestinal bacteria; leafy greens, legumes
Water-soluble			
Vitamin C, ascorbic acid 60 mg	Collagen synthesis; antioxidant; needed for synthesis of some hormones and neurotransmitters; important in immune function	Scurvy (wounds heal very slowly and scars become weak and split open; capillaries become fragile; bone does not grow or heal properly)	Citrus fruit, strawberries, tomatoes, leafy vegetables, cabbage
B-complex vitamins			
Vitamin B$_1$, thiamine 1.5 mg	Active form is a coenzyme in many enzyme systems; important in carbohydrate and amino acid metabolism	Beriberi (weakened heart muscle, enlarged right side of heart, nervous system and digestive tract disorders); common in alcoholics	Liver, yeast, whole and enriched grains, meat, green leafy vegetables
Vitamin B$_2$, riboflavin 1.7 mg	Used to make coenzymes (e.g., FAD) essential in cellular respiration	Dermatitis, inflammation and cracking at corners of mouth; confusion	Liver, milk, eggs, green leafy vegetables, enriched grains
Niacin 20 mg	Component of important coenzymes (NAD$^+$ and NADP$^+$); essential to cellular respiration	Pellagra (dermatitis, diarrhea, mental symptoms, muscular weakness, fatigue)	Liver, chicken, tuna, milk, green leafy vegetables, enriched grains
Vitamin B$_6$, pyridoxine 2 mg	Derivative is coenzyme in many reactions in amino acid metabolism	Dermatitis; digestive tract disturbances; convulsions	Meat, whole grains, legumes, green leafy vegetables
Pantothenic acid 10 mg	Constituent of coenzyme A (important in cellular metabolism)	Deficiency extremely rare	Meat, whole grains, legumes
Folic acid 0.4 mg	Coenzyme needed for nucleic acid synthesis and for maturation of red blood cells	A type of anemia; certain birth defects; increased risk of cardiovascular disease; deficiency in alcoholics, smokers, and pregnant women	Liver, legumes, dark green leafy vegetables, orange juice
Biotin 0.3 mg	Coenzyme important in metabolism	—	Produced by intestinal bacteria; liver, chocolate, egg yolk
Vitamin B$_{12}$ 6 micrograms (0.006 mg)	Coenzyme important in metabolism; contains cobalt	A type of anemia	Liver, meat, fish

*RDA is the recommended dietary allowance, established by the Food and Nutrition Board of the National Research Council, to maintain good nutrition for healthy persons.

†International Unit: the amount that produces a specific biological effect and is internationally accepted as a measure of the activity of the substance.

TABLE 45-4   Some Important Minerals and Their Functions

Mineral	Functions	Sources; Comments
Calcium	Component of bones and teeth; essential for normal blood clotting; needed for normal muscle and nerve function	Milk and other dairy products, fish, green leafy vegetables; bones serve as calcium reservoir
Phosphorus	Performs more functions than any other mineral. Structural component of bone; component of ATP, DNA, RNA, and phospholipids	Meat, dairy products, cereals
Sulfur	Component of many proteins and vitamins	High-protein foods such as meat, fish, legumes, nuts
Potassium	Principal positive ion within cells; influences muscle contraction and nerve excitability	Fruit, vegetables, grains
Sodium	Principal positive ion in interstitial fluid; important in fluid balance; neural transmission	Many foods, table salt; too much ingested in average American diet; excessive amounts may contribute to high blood pressure
Chloride	Principal negative ion of interstitial fluid; important in fluid balance and in acid-base balance	Many foods; table salt
Magnesium	Needed for normal muscle and nerve function	Nuts; whole grains; green, leafy vegetables
Copper	Component of enzyme needed for melanin synthesis; component of many other enzymes; essential for hemoglobin synthesis	Liver, eggs, fish, whole wheat flour, beans
Iodide	Component of thyroid hormones (hormones that increase metabolic rate). Deficiency results in goiter (abnormal enlargement of thyroid gland)	Seafood, iodized salt, vegetables grown in iodine-rich soils
Manganese	Necessary to activate arginase, an enzyme essential for urea formation; activates many other enzymes	Whole-grain cereals, egg yolks, green vegetables; poorly absorbed from intestine
Iron	Component of hemoglobin, myoglobin, important respiratory enzymes (cytochromes), and other enzymes essential to oxygen transport and cellular respiration. Deficiency results in anemia and may impair cognitive function	Mineral most likely to be deficient in diet. Good sources: meat (especially liver), nuts, egg yolk, legumes, dried fruit
Fluoride	Component of bones and teeth; makes teeth resistant to decay; excess causes tooth mottling	Fish; in areas where it does not occur naturally, fluoride may be added to municipal water supplies (fluoridation)
Zinc	Cofactor for at least 70 enzymes; helps regulate synthesis of certain proteins; needed for growth and repair of tissues; deficiency may impair cognitive function	Meat, milk, yogurt, some seafood
Selenium	Antioxidant (part of a peroxidase that breaks down peroxides)	Seafood, eggs, liver, garlic, mushrooms

tables in their diet. A diet that includes all of the essential nutrients does not provide the same health benefits as one rich in fruit and vegetables. The missing ingredients appear to be **phytochemicals,** compounds found in plants. Nutritionists are just beginning to intensively investigate these chemicals. Except for carotenoids, yellow-orange pigments that the body can convert into vitamin A, phytochemicals have not been established as essential nutrients.

Among the important phytochemicals are the *flavonoids,* water-soluble pigments in fruits, vegetables, grains, flowers, and seeds, that function as *antioxidants* (see *Making the Connection: Antioxidants and Cell Function*). More than 4000

ANTIOXIDANTS AND CELL FUNCTION

What are antioxidants, and why are they an important part of a healthy diet? Normal cell processes that require oxygen produce highly reactive molecules (oxidants) such as free radicals, peroxides, and superoxides. Free radicals are also generated by ionizing radiation, tobacco smoke, and other forms of air pollution.

Oxidants can damage DNA, proteins, and unsaturated fatty acids by snatching electrons. Damage to DNA can cause mutations that lead to cancer. Injury to unsaturated fatty acids can result in damage to cell membranes. Free radicals are thought to contribute to atherosclerosis by causing oxidation of LDL cholesterol. Oxidative damage to the body over the years contributes to the aging process.

Cells have **antioxidants** that destroy free radicals and other reactive molecules. Antioxidants in tissues include enzymes such as superoxide dismutase, catalase, peroxidase, and glutathione. Their action requires minerals such as selenium, zinc, manganese, and copper. Certain vitamins—vitamin C, vitamin E, and beta-carotene—have strong antioxidant activity. Vitamins A and E protect cell membranes from free radicals. In addition, a variety of phytochemicals, compounds found in plants, are potent antioxidants *(see text)*.

Chocolate contains phytochemicals, known as flavonoid phenols, that have potent antioxidant activity. Tea contains flavonoids of the catechin group that act synergistically with vitamins E and C. In mice exposed to nitrosamines (cancer-causing compounds formed from certain food preservatives), green tea reduced lung cancer by 45%. In a study of more than 35,000 women who drank more than two cups of black tea per day, risk for urinary tract cancer was reduced by 60% and cancers of the gastrointestinal tract by 32%.

Many antioxidants act synergistically. For example, vitamin E helps prevent selenium deficiency. Investigators have not yet identified or studied many phytochemicals, and we do not yet know how much we need of which ones. We do not know whether antioxidant supplements are useful. Meanwhile nutritionists recommend that we increase the antioxidants in our diet by eating fruit, vegetables, and other foods high in antioxidants.

---

flavonoids have been identified. Indoles and isocyanates are phytochemicals that enhance the synthesis of enzymes that detoxify cancer-causing agents. Cruciferous vegetables—cabbage, brussel sprouts, cauliflower, and their relatives—are rich in flavonoids, indoles, and isocyanates. Studies indicate that diets rich in these vegetables reduce the risk of several types of cancer, including colon cancer.

For hundreds of years garlic has been used medicinally. Nutritionists are just beginning to understand how its phytochemicals may protect against heart disease and cancer. Onions, green tea, and soy products also contain phytochemicals that apparently contribute to health. In Asian countries where diets are low in fat and high in soy and green tea, the incidence of breast, prostate, and colorectal cancer is low.

ENERGY METABOLISM IS BALANCED WHEN ENERGY INPUT EQUALS ENERGY OUTPUT

The amount of energy liberated by the body per unit time is a measure of the **metabolic rate.** Much of the energy expended by the body is ultimately converted to heat. Metabolic rate may be expressed either in kcal of heat energy expended per day or as a percentage above or below a standard normal level.

The **basal metabolic rate** (**BMR**) is the rate at which the body releases heat as a result of breaking down fuel molecules. BMR is the body's basic cost of living, that is, the rate of energy used during resting conditions. This energy is required to maintain body functions such as heart contraction, breathing, and kidney function. An individual's **total metabolic rate** is the sum of his or her BMR and the energy used to carry on all daily activities. For example, a laborer has a greater total metabolic rate than does an executive whose job requirements do not include a substantial amount of movement and who does not exercise regularly.

An average-sized man who does not engage in any exercise program and who sits at a desk all day expends about 2000 kcal daily. If the food he eats each day also contains about 2000 kcal, he will be in a state of energy balance; that is, his energy input will equal his energy output. This is an extremely important concept, because body weight remains constant when:

$$\text{Energy input} = \text{Energy output}$$

When energy output is greater than energy input, stored fat is burned, and body weight decreases. On the other hand, people gain weight when they take in more energy in food than they expend in daily activity, in other words, when:

$$\text{Energy input} > \text{Energy output}$$

Obesity is a serious nutritional problem

Obesity, the excess accumulation of body fat, is a serious form of malnutrition that has become a problem of epidemic proportions in affluent societies. According to a 1998 report of the World Health Organization, more than 50% of U.S. adults are overweight (33%) or clinically obese (25%). An estimated 20% of U.S. children and adolescents are obese. Several large studies (including the well known Framingham study) suggest that an overweight person is at greater risk for heart disease, diabetes, and other ailments. In the Framingham study of more than 2000 men, those who were 20% overweight had a significantly greater mortality rate from all causes. According to many researchers, obesity contributes to about 300,000 deaths annually in the United States, and is the second leading preventable cause of death (second only to smoking). Recently, some researchers have suggested that lack of physical activity may be a confounding factor in many studies, and being physically unfit may be the most important factor in higher mortality.

Body mass index (BMI) is now used worldwide as a measure of body size. BMI is an index of weight in relation to height. It is calculated by dividing the square of the weight (kg^2) by height (m). The English equivalent is 4.89 times the weight (in lb) divided by the square of the height (in ft). A person is considered obese if the BMI is 30 or more. Each of us appears to have a **set point,** or steady state, around which body weight is regulated. When BMI decreases below an individual's set point, energy conserving mechanisms are activated and energy expenditure decreases.

Researchers have identified human gene mutations that affect body weight, suggesting that we inherit the risk for obesity (see *On the Cutting Edge: In Search of a Cure for Obesity*). In fact, some investigators estimate that 40 to 70 percent of the factors involved in obesity are inherited. Obesity can result from an increase in the size of fat cells or from an increase in the number of fat cells, or both. The number of fat cells in the adult is apparently determined mainly by the amount of fat stored during infancy and childhood. When we are overfed early in life, abnormally large numbers of fat cells are formed. Later in life, these fat cells may be fully stocked with excess lipids or may shrink in size, but they are always there. People with such increased numbers of fat cells are thought to be at greater risk for obesity than are those with normal numbers.

Whatever the underlying causes, overeating and/or underexercising appear to be the routes to obesity. For every 9.3 kcal of excess food taken into the body, about 1 g of fat is stored. (An excess of about 140 kcal, less than a typical candy bar, per day for a month results in a 1-lb weight gain.)

Malnutrition can cause serious health problems

While millions of people eat too much, about 840 million humans do not have enough to eat, or do not eat a balanced diet. Individuals suffering from malnutrition are weak, easily fatigued, and highly susceptible to infection. Essential amino acids, iron, calcium, and vitamin A are commonly deficient nutrients.

Of all the required nutrients, essential amino acids are the ones most often deficient in the diet. Millions of people suffer from poor health and a lowered resistance to disease because of protein deficiency. Children's physical and mental development are retarded when the essential building blocks of cells are not provided in the diet. Because their bodies cannot manufacture antibodies (which are proteins) and cells needed to fight infection, common childhood diseases, such as measles, whooping cough, and chicken pox, are often fatal in children suffering from protein malnutrition.

In young children, severe protein malnutrition results in the condition known as **kwashiorkor.** This term is an African word that means "first-second." It refers to the situation in which a first child is displaced from its mother's breast when a younger sibling is born. The older child is placed on a diet of starchy cereal or cassava that is deficient in protein. Growth becomes stunted, muscles are wasted, edema develops (as displayed by a swollen belly), the child becomes apathetic and anemic, and metabolism is impaired (Fig. 45–16). Without essential amino acids, digestive enzymes cannot be manufactured, so what little protein is ingested cannot be digested.

Figure 45–16 **Protein deficiency.** Millions of children suffer from kwashiorkor, a disease caused by severe protein deficiency. Note the characteristic swollen belly, which results from fluid imbalance. *(P. Pittet/United Nations Food and Agricultural Organization)*

In Search of a Cure for Obesity

HYPOTHESIS: Genetic variations contribute to human obesity; a gene product, leptin, is important in regulating body weight and energy balance in humans.

METHOD: Leptin genomic DNA from two severely obese children was cloned using PCR techniques and sequenced.

RESULTS: A homozygous frameshift mutation involving the deletion of a single guanine nucleotide in the gene for leptin was identified.

CONCLUSION: A mutation in the gene coding for leptin leads to obesity in humans.

In the 1950s a spontaneous recessive mutation occurred in a strain of laboratory mice that caused the mutant mice to become grossly obese. The increase in adipose tissue in these mice is part of a syndrome that parallels morbid obesity in humans, a condition in which an individual's body weight is 100 lb or more (45.5 kg) above normal. In 1994, Jeffrey M. Friedman's research team at Rockefeller University isolated the gene (*ob* gene) that, when mutated, was resposible for the obese phenotype. Mice with the mutated gene apparently lack some weight-regulating substance.

In 1995, three different research teams reported in *Science* that injections of the *ob* gene product, later named **leptin**, resulted in weight loss by obese and non-obese mice. All three groups transferred the *ob* gene into bacteria, which then made large quantities of the protein. When leptin was injected into grossly obese mice, their appetites decreased and their energy use increased, resulting in weight loss, mainly from loss of body fat. Friedman's team fed the same diet to obese mice who were injected with leptin and to a control group of obese mice who did not receive leptin. The mice in the treated group lost 50% more weight than the untreated animals (see figure). Frank Collins' research group at Amgen showed

that treated mutant mice became more active and that their metabolic rate increased. Arthur Campfield's team at Hoffman-La Roche Inc. reported that leptin acts directly on neural networks in the brain. They suggested that leptin is a hormone produced in adipose tissue; it acts on centers in the brain responsible for regulating feeding behavior and energy metabolism.

Several additional genes and a variety of compounds have also been identified that affect body weight in mice. The *diabetes (db)* gene encodes the leptin receptor. Mice with a *db* mutation produce leptin but cannot respond to it. Leptin appears to be produced by fat cells in proportion to body fat mass. A decrease in body fat results in a decrease in leptin, which leads to increased food intake. In contrast, an increase in body fat results in increased secretion of leptin, leading to a decrease in food intake. Receptors for leptin occur in the hypothalamus. **Neuropeptide Y (NPY),** a neurotransmitter produced in the hypothalamus, increases appetite and slows metabolism when leptin levels and food intake are low as during dieting or starvation. A group of peptides called **melanocortins** are also involved in regulation of body weight. Their receptors appear to decrease appetite in response to increased fat stores.

SUMMARY WITH KEY TERMS

I. **Nutrition** is the process of taking in and using food. Animals are **heterotrophs**; they must obtain their **nutrients** from the organic molecules manufactured by other organisms.

II. Processing food includes **ingestion, digestion, absorption** of nutrients, and **egestion** or **elimination** of wastes.

III. An organism's body plan and lifestyle are adapted to its mode of nutrition. Animals can be classified according to the type of food they typically eat and according to their mechanisms for obtaining food.
   A. **Herbivores** feed directly on producers.
   B. **Carnivores** are adapted for capturing and killing prey.
   C. **Omnivores** eat both plants and animals.
   D. Some animals eat large pieces of food. Others are **suspension feeders** that trap or filter food from water or air surrounding them. Still others suck fluids from other organisms.

IV. The simplest invertebrates, the sponges, have no digestive system; they digest food intracellularly. Most other animals have organs or systems specialized for processing food.

   A. In cnidarians and flatworms, food is digested in the **gastrovascular cavity.** This cavity has only one opening that serves as both mouth and anus. Cnidarians have intracellular and extracellular digestion.
   B. In more complex invertebrates and in all vertebrates, the digestive tract is a complete tube with an opening at each end. Digestion is mainly extracellular.

V. In vertebrates, various parts of the digestive tract are specialized to perform specific functions. Food passes in sequence through the mouth, pharynx, esophagus, stomach, small intestine, large intestine, and anus.
   A. **Motility** refers to the mixing and propulsive movements of the digestive tract. The propulsive activity characteristic of most regions of the digestive tract is **peristalsis,** waves of muscular contraction that push the food along.
   B. The wall of the digestive tract consists of four layers. The **mucosa,** a layer of epithelial and connective tissue, lines the lumen. The **submucosa** consists of connective tissue rich in blood and lymph

Mutant mice before and after treatment with leptin. *(Courtesy of John Sholtis/The Rockefeller University, New York, NY)*

Human obesity appears to be very complex. Research findings reported in the journal *Nature* in 1997 by Carl T. Montague of the University of Cambridge and his colleagues provide evidence that genetic variations contribute to human obesity and that leptin is important in regulating body weight and energy balance in humans.* They cloned the gene encoding leptin from two severely obese children and from normal controls. The researchers identified a frameshift mutation involving the deletion of a single guanine nucleotide in the leptin gene. The Cambridge team has also identified a mutation in a second gene involved in human weight control. Just how leptin affects human body weight is not known, but congenital deficiency of leptin causes similar effects to those seen in leptin-deficient mice. However, overweight people typically have high leptin levels in their blood, and the problem does not appear to be in the leptin receptors. Some researchers think that NPY or some other compound that interacts with leptin may be a key.

The mechanisms of energy balance and body weight in humans involve several hormones and other regulatory molecules. Recent studies indicate that the hormones leptin and insulin are both secreted in response to the amount of adipose tissue. During negative energy balance, the body's adipose tissue decreases, resulting in a decrease in leptin and insulin secretion. In response, Neuropeptide Y (and probably other yet unknown factors) are released and act to restore energy homeostasis. In contrast when energy balance is positive, the amount of adipose tissue increases, and insulin and leptin secretion increase. Catabolic activities increase, favoring a return to energy homeostasis. Corticotropin-releasing hormone from the hypothalamus also appears to be involved.

Unraveling the genetic and biochemical mechanisms underlying obesity provides clues for the development of pharmacological treatments for obesity. Pharmaceutical companies are investigating a variety of targets for weight-loss drugs. For example, melanocortin receptors are G protein-coupled receptors that are easily blocked. A drug that targeted these receptors might decrease appetite. Other new targets are *uncoupling proteins* that may uncouple metabolism and energy production (see *Making the Connection: Electron Transport and Heat* in Chapter 7). As investigators continue to uncover clues to the puzzle of human obesity, effective treatments will likely be developed.

*Montague, C.T., et al. "Congenital Leptin Deficiency Is Associated with Severe Early-Onset Obesity in Humans." *Nature*, Vol. 387, 26 Jun. 1997. *Science*, Vol. 290, 29 May 1998 includes several articles on weight regulation.

---

vessels and nerves. The **muscle layer** consists of an inner sublayer of circularly arranged muscle fibers and an outer longitudinal sublayer. The outer connective tissue coat is the **adventitia.** Below the level of the diaphragm, the adventitia is called the **visceral peritoneum.** The visceral peritoneum is connected by folds to the **parietal peritoneum,** a sheet of connective tissue that lines the walls of abdominal and pelvic cavities. The visceral and parietal peritonea enclose part of the coelom, the **peritoneal cavity**.

C. Mechanical digestion and enzymatic digestion of carbohydrates begin in the mouth.
1. Mammalian teeth include **incisors** for biting, **canines** for tearing food, **premolars** and **molars** for crushing and grinding. Each tooth is covered by a hard **enamel.** Most of a tooth consists of **dentin** which surrounds a **pulp cavity** containing nerves and blood and lymph vessels.
2. Three pairs of **salivary glands** secrete saliva which contains **salivary amylase,** an enzyme that digests starch.

D. As food is swallowed, it is propelled through the **pharynx** and **esophagus.** A **bolus** of food is moved along through the digestive tract by peristalsis.

E. In the **stomach,** food is mechanically digested by vigorous churning, and proteins are enzymatically digested by the action of **pepsin** in the gastric juice. **Rugae** are folds in the stomach wall that expand as the stomach fills with food. **Gastric glands** secrete hydrochloric acid and pepsinogen, the precursor of pepsin.

F. After several hours, a soup of partly digested food, called **chyme,** leaves the stomach through the **pylorus** and enters the small intestine. The surface area of the small intestine is greatly expanded by folds in its wall, by the intestinal **villi,** and by microvilli. Most enzymatic digestion takes place in the **duodenum,** which receives secretions from the liver and pancreas and produces several digestive enzymes of its own.

G. The **liver** produces **bile,** which emulsifies fats.

H. The **pancreas** releases enzymes that digest protein, lipid, and carbohydrate, as well as RNA and DNA. **Trypsin** and **chymotrypsin** digest polypeptides to dipeptides. **Pancreatic lipase** degrades neutral fats. **Pancreatic amylase** digests complex carbohydrates.

I. Activities of the digestive system are regulated by both nerves and hormones. In addition to sympathetic and parasympathetic nerves, digestive activity is regulated by networks of neurons within the intestinal wall. Among the hormones secreted by the digestive tract are **gastrin, secretin, cholecystokinin (CCK),** and **gastric inhibitory peptide (GIP).**

J. Nutrients in chyme are enzymatically digested as they move through the digestive tract.
  1. Polysaccharides are digested to maltose by salivary and pancreatic amylases. Maltase in the small intestine splits maltose into glucose, the main product of carbohydrate digestion.
  2. Proteins are split by pepsin in the stomach and by proteolytic enzymes in the pancreatic juice. The peptides and dipeptides produced are then split by dipeptidases. The end products of protein digestion are amino acids.
  3. Lipids are emulsified by bile salts and then hydrolyzed by pancreatic lipase.

K. Nutrients are absorbed through the thin walls of the intestinal villi. Amino acids and glucose are transported to the liver by the **hepatic portal vein.** Fatty acids and monoacylglycerols combine with bile salts to form soluble complexes called **micelles.** Fatty acids and monacylglycerols leave a micelle and enter an epithelial cell in the intestinal lining. There they are reassembled into triacylglycerols and are packaged into droplets. The droplets include cholesterol and phospholipids and are covered by a thin coat of protein. These **chylomicrons** are transported by the lymphatic system to the blood circulation.

L. The **large intestine,** which consists of the **cecum, colon, rectum,** and **anus,** is responsible for the elimination of undigested wastes. It also incubates bacteria that produce vitamin K and certain B vitamins.

VI. For a balanced diet, humans and other animals require carbohydrates, lipids, proteins, vitamins, and minerals.
  A. Most carbohydrates are ingested in the form of polysaccharides—starch and cellulose. Polysaccharides are referred to as **complex carbohydrates. Fiber** is mainly a mixture of cellulose and other indigestible carbohydrates.
    1. Carbohydrates are used mainly as fuel.
    2. Glucose concentration in the blood is carefully regulated. Excess glucose is stored as glycogen and can also be converted to fat.
  B. Lipids are used as fuel, as components of cell membranes, and to synthesize steroid hormones and other lipid substances.
    1. Most lipids are ingested in the form of triacyglycerols.
    2. Fatty acids are converted to molecules of acetyl coenzyme A and used as fuel. Excess fatty acids are converted to triacylglycerol and stored as fat.
  C. Proteins serve as enzymes and are essential structural components of cells.
    1. The best distribution of **essential amino acids** is found in the complete proteins of animal foods.
    2. Excess amino acids are deaminated by liver cells. Amino groups are converted to urea and excreted in urine; the remaining keto acids are converted to carbohydrate and used as fuel or converted to lipid and stored in fat cells.
  D. **Vitamins** are organic compounds required in small amounts for many biochemical processes. Many serve as components of coenzymes. **Fat-soluble vitamins** include vitamins A, D, E, and K. **Water-soluble vitamins** are the B and C vitamins.
  E. **Minerals** are inorganic nutrients ingested as salts dissolved in food and water. **Trace elements** are minerals required in amounts less than 100 mg per day.

VII. **Basal metabolic rate (BMR)** is the body's cost of metabolic living.
  A. **Total metabolic rate** is the BMR plus the energy used to carry on daily activities.
  B. When energy (kcal) input equals energy output, body weight remains constant.

VIII. **Obesity** is a serious nutritional problem in which an excess amount of fat accumulates in adipose tissues. A person gains weight by taking in more energy, in the form of kilocalories, than is expended in activity.

IX. Millions of people suffer from malnutrition. Essential amino acids are the nutrients most often deficient in the diet.

POST-TEST

1. The process of taking in and using food is (a) nutrition (b) chemical digestion (c) egestion (d) ingestion (e) absorption
2. Animals that typically feed on producers are (a) herbivores (b) secondary consumers (c) animals with gastrovascular cavities (d) carnivores (e) secondary consumers and carnivores
3. Teeth adapted for crushing and grinding are (a) incisors (b) canines (c) molars (d) incisors and premolars (e) canines and molars
4. The layer of tissue that lines the lumen of the digestive tract is the (a) adventitia (b) visceral peritoneum (c) parietal peritoneum (d) mucosa (e) submucosa
5. Which of the following are accessory digestive glands? (a) salivary glands (b) pancreas (c) liver (d) answers a, b, and c are correct (e) answers a and b only
6. Which of the following is the correct sequence? (a) pharynx → stomach → esophagus (b) stomach → large intestine → small intestine (c) esophagus → pharynx → stomach (d) cecum → colon → duodenum (e) pharynx → esophagus → stomach
7. Amylase is produced by the (a) liver (b) pancreas (c) salivary glands (d) liver and pancreas (e) pancreas and salivary glands
8. Pepsin is produced by the (a) liver (b) stomach (c) pancreas (d) duodenum (e) salivary glands
9. Which sequence most accurately describes the digestion of protein? (a) polypeptide → monoacylglycerol → amino acids (b) polypeptide → dipeptides → amino acids (c) protein → amylose → amino acid (d) protein → emulsified peptide → fatty acids and glycerol (e) protein → triacylglycerol → fatty acids and glycerol
10. The surface area of the small intestine is increased by (a) folds in its wall (b) villi (c) microvilli (d) a, b, and c are correct (e) b and c only
11. Lipids are transported from the intestine to the cells of the body by (a) chylomicrons (b) micelles (c) rugae (d) villi (e) leptin
12. Most vitamins are (a) inorganic compounds (b) components of coenzymes (c) used as fuel (d) electrolytes (e) required by herbivores and carnivores but not omnivores
13. When energy input is greater than energy output (a) weight loss occurs (b) weight remains stable (c) weight gain occurs (d) leptin prevents weight gain (e) the *ob* gene is activated
14. A hormone that stimulates gastric glands to secrete pepsinogen is (a) secretin (b) gastric inhibitory peptide (c) gastrin (d) cholecystokinin (CCK) (e) substance P
15. Neuropeptide Y (a) increases appetite (b) increases metabolism (c) is encoded by the mutant *ob* gene (d) is a hormone produced by the pancreas (e) decreases bile production

REVIEW QUESTIONS

1. If you were presented with an unfamiliar animal and asked to determine its nutritional lifestyle, how would you go about this task?
2. What are the advantages of a two-opening digestive tract compared to one with a single-opening?
3. How are digestive structures and methods of processing food in sponges, hydras, and flatworms adapted to each group's lifestyle? Give specific examples.
4. Trace a bite of food through the human digestive tract, listing each structure through which it passes.
5. Give the functions of the three types of accessory glands that secrete digestive juices in vertebrates. Identify their secretions.
6. The inner lining of the digestive tract is not smooth like the inside of a water pipe. Why is this advantageous? What structures increase its surface area?
7. Summarize the step-by-step digestion of (a) carbohydrates (b) lipids (c) proteins
8. What happens to ingested cellulose in humans? Why?
9. Draw and label an intestinal villus and explain how its structure is adapted to its function.
10. How does the absorption of fat differ from the absorption of glucose?
11. List the nutrients that must be included in a balanced diet. What are some of the difficulties in planning a nutritionally balanced vegetarian diet?
12. Why, specifically, is each of the following essential? (a) iron (b) calcium (c) iodine (d) vitamin A (e) vitamin K (f) essential amino acids
13. Draw a diagram to illustrate the fate of each of the following in the body: (a) glucose (b) absorbed amino acids (c) absorbed fat
14. Write an equation to describe energy balance and explain what happens when the equation is altered in either direction.
15. Label the diagram. Use Figure 45–4 to check your answers.

YOU MAKE THE CONNECTION

1. A high percentage of adults suffer from gastrointestinal discomfort, including cramps and diarrhea, when they drink milk. What do you think could cause this condition, known as lactose intolerance?
2. Design an experiment to demonstrate the nutritional need for the B vitamin pyridoxine.
3. Why are proteolytic enzymes produced in an inactive form?
4. Investigators are unraveling the biochemistry of obesity in mice. How might their work help us understand human obesity? Give reasons to support your response.

RECOMMENDED READINGS

Borek, C. "Antioxidants and Cancer." *Science and Medicine,* Vol. 4, No. 6. Nov./Dec. 1997. Discussion of the importance of supplemental antioxidant vitamins and minerals.

Brown, L.R. et al. "Tracing the Prospect of Food Scarcity." *State of the World 1997.* Norton, New York, 1997. A Worldwatch Institute Report on environmental problems and progress. Includes an analysis of our ability to provide adequate food for our increasing population.

"Genetics of Colon Cancer: A Special Report." *Science and Medicine,* Vol. 4, No. 4, Jul./Aug. 1997. Cumulative events, including mutation of oncogenes and inactivation of tumor suppressor genes, appear to cause the progression of colon cancer.

Science, Vol. 280, No. 5368, 29 May 1998. Several articles on "Regulation of Body Weight." These articles discuss current research on obesity and eating disorders.

Smolin, L.A., and M.B. Grosvenor. *Nutrition: Science and Applications,* 2nd ed. Saunders College Publishing, Philadelphia, 1997. A readable introduction to nutritional science.

Weigle, D.S. and J.L. Kuijper. "Mouse Models of Human Obesity." *Science and Medicine,* Vol. 4, No. 3. May/Jun. 1997. Genetic variations in signaling molecules involved in regulating body composition account for the inherited portion of body fat.

● Visit our website at **http://www.saunderscollege.com/lifesci/titles.html** and click on Solomon/Berg/Martin Biology for links to chapter-related resources on the World Wide Web.

CHAPTER 46

Osmoregulation and Disposal of Metabolic Wastes

(McMurray Photography)

Water, the most abundant molecule in the cell and on Earth, shapes life and the distribution of organisms on our planet. Water is the medium in which most metabolic reactions take place, and the osmotic and ionic composition of an animal's intracellular and **extracellular fluids** (the interstitial fluid, blood, and other fluid outside of the cells) must be maintained within homeostatic limits. The body fluids of many marine invertebrates and many vertebrates have the same osmotic concentration as seawater. The *ionic composition* of the extracellular fluid is also similar to sea water in many invertebrates, but not in most vertebrates. Natural selection has resulted in the evolution of a variety of homeostatic mechanisms that regulate the volume and composition of fluids in the internal environment. In this chapter we will discuss some of these mechanisms.

Many small animals live in the ocean and obtain their food and oxygen directly from the sea water that surrounds them; they release waste products into the surrounding sea water. In larger, more complex animals and most terrestrial animals the extracellular fluid serves as an internal sea. Blood plasma, composed mainly of water, transports nutrients, gases, waste products, and other materials throughout the animal body. Interstitial fluid forms from the blood plasma and bathes all of the cells of the body. Excess water evaporates from the body surface or is excreted by specialized structures.

Terrestrial animals have a continuous need to conserve water. Its loss from the body must be carefully regulated, and water that is lost must be replaced. Water is ingested with food and drink and is also produced in some metabolic reactions. Most animals require a dependable source of water with which to replenish the body fluids. The zebra, wildebeest, and variety of birds, including flamingos, shown here have gathered to drink at a lake in Ngorongoro Crater in Tanzania. During annual East African dry seasons, hundreds of thousands of animals migrate long distances in search of water.

Two processes that maintain homeostasis of fluids in humans and other animals are osmoregulation and excretion of metabolic wastes. **Osmoregulation** is the active regulation of osmotic pressure of body fluids to keep them from becoming too dilute or too concentrated. **Excretion** is the process of ridding the body of metabolic wastes, including excess water. Efficient **excretory systems** have evolved that function in osmoregulation and in disposal of metabolic wastes, excess water and ions, and harmful substances. As we will discuss, hormones are important signaling molecules in these regulatory processes.

AFTER YOU HAVE STUDIED THIS CHAPTER YOU SHOULD BE ABLE TO

1. Relate the principal functions of excretory systems to specific osmoregulatory challenges posed by various environments.
2. Contrast the advantages and disadvantages of excreting ammonia, uric acid, or urea.
3. Compare nephridial organs and Malpighian tubules as osmoregulatory organs.
4. Relate the function of the vertebrate kidney to the success of vertebrates in a wide variety of habitats.
5. Compare the adaptations that freshwater fishes have evolved to solve their challenges of osmoregulation with those of marine bony fishes, sharks, and marine mammals.

6. Label on a diagram the organs of the mammalian urinary system and give the functions of each.
7. Identify on a diagram the principal parts of a nephron (including associated blood vessels) and give the functions of each structure.
8. Trace a drop of filtrate from Bowman's capsule to its release from the body as urine.
9. Describe the hormonal regulation of fluid balance by antidiuretic hormone (ADH), aldosterone, and atrial natriuretic peptide (ANP).

EXCRETORY SYSTEMS HELP MAINTAIN HOMEOSTASIS

Excretory systems maintain homeostasis by selectively adjusting the concentrations of salts and other substances in blood and other body fluids. Typically, an excretory system collects fluid, generally from the blood or interstitial fluid. It then adjusts the composition of this fluid by selectively returning needed substances to the body fluid. Finally, the adjusted excretory product containing excess or potentially toxic substances is released from the body. In many animals, a major excretory product is **urine,** a watery solution of metabolic wastes and other organic and inorganic substances.

The terms *excretion* and *elimination* are sometimes confused (Fig. 46–1). Food materials that have not been absorbed are eliminated from the body in the feces. Such substances never participated in the organism's metabolism or entered body cells, but merely passed through the digestive system.

METABOLIC WASTE PRODUCTS INCLUDE WATER, CARBON DIOXIDE, AND NITROGENOUS WASTES

Metabolic wastes must be excreted so that they do not accumulate and reach concentrations that would disrupt homeostasis. The principal metabolic waste products in most animals are water, carbon dioxide, and nitrogenous (nitrogen-containing) wastes. Carbon dioxide is excreted mainly by respiratory structures (see Chapter 44). Excess water is also lost from respiratory surfaces in terrestrial animals. Excretory organs, such as kidneys, remove and excrete most of the water and nitrogenous wastes.

Nitrogenous wastes include ammonia, uric acid, and urea. Recall that amino acids and nucleic acids contain nitrogen. During the breakdown of amino acids, the nitrogen-containing amino group is removed (in a process known as deamina-

tion) and converted to ammonia (Fig. 46–2). However, ammonia is highly toxic. Some aquatic animals excrete it into the surrounding water before it can build up to toxic concentrations in their tissues. A few terrestrial animals, including some terrestrial snails and wood lice, vent it directly into the air. But in many organisms, humans included, ammonia is converted to some less toxic nitrogenous waste such as uric acid or urea.

Uric acid is produced both from ammonia and by the breakdown of nucleotides from nucleic acids. Uric acid is insoluble in water and forms crystals that can be excreted as a crystalline paste with little fluid loss. This is an important water-conserving adaptation in many terrestrial animals, including insects, certain reptiles, and birds. In birds, the absence of a urinary bladder and the frequent excretion of uric acid as part of the feces contribute to the light body weight

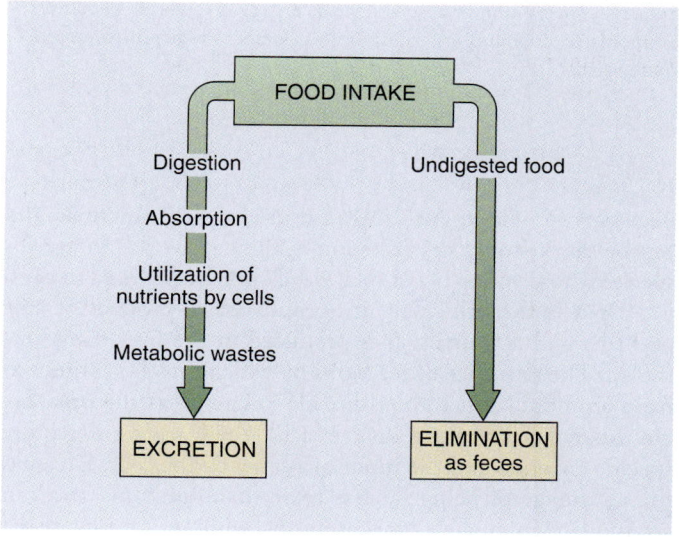

Figure 46–1   Excretion versus elimination. Excretion is the disposal of metabolic wastes. Elimination is the ejection of undigested and unabsorbed food from the body.

Figure 46–2 **Formation of nitrogenous wastes.** Deamination of amino acids and metabolism of nucleic acids produce nitrogenous wastes. Ammonia is the first metabolic product of deamination. In many animals, ammonia is converted to urea via the urea cycle. Other animals convert ammonia to uric acid. Energy is required to convert ammonia to urea and uric acid, but less water is required to excrete these wastes.

necessary for flight. Also, because uric acid is not toxic and can be safely stored, its excretion is an adaptive advantage for species whose young begin their development enclosed in eggs.

Urea is the principal nitrogenous waste product of amphibians and mammals. It is produced in the liver from ammonia. The sequence of reactions by which urea is synthesized from ammonia and carbon dioxide is known as the **urea cycle.** Like the formation of uric acid, these reactions require specific enzymes and the input of energy by the cells. Urea has the advantage of being far less toxic than ammonia and can accumulate in higher concentrations without causing tissue damage; thus, it can be excreted in more concentrated form. Because urea is highly soluble, it is dissolved in water. More water is needed to excrete urea than to excrete uric acid.

INVERTEBRATES HAVE ADAPTATIONS FOR OSMOREGULATION AND METABOLIC WASTE DISPOSAL

The sea is a stable environment and its salt concentration does not vary much. The body fluids of most marine invertebrates are in osmotic equilibrium with the surrounding sea water. These animals are known as **osmoconformers** because the concentration of their body fluids varies along with changes in the sea water. Even so, many osmoconforming marine animals do regulate some ions in their body fluids.

Coastal habitats, such as estuaries that contain brackish water, are much less stable environments than is the open sea. Salt concentrations change frequently with shifting tides. Many invertebrates (as well as vertebrates) that inhabit these environments are **osmoregulators** that maintain an optimal salt concentration in their tissues regardless of changes in the salt concentration of their surroundings.

In a coastal environment where fresh water enters the sea, the water may have a lower salt concentration than the body fluids of the animal. Water osmotically moves into the body, and salt diffuses out. An animal adapted to this environment has excretory structures that actively remove the excess water. Crabs and some other animals have cells in their gills that remove salts from the surrounding water and transport them into the body fluids. Certain polychaete worms and the blue crab are among the animals that can be osmoconformers or osmoregulators depending on environmental conditions.

Marine sponges and cnidarians need no specialized excretory structures. Their wastes pass by diffusion from their cells to the external environment and are washed away by water currents. They expend little or no energy to excrete wastes. When water becomes stagnant and currents do not wash away wastes, aquatic environments, such as coral reefs, are damaged by their accumulation.

The fluid concentration of terrestrial animals is higher than that of the air around them. They tend to lose water by evaporation from the body surface and from respiratory surfaces and may also lose water as wastes are excreted. As animals moved onto the land, natural selection favored the evolution of structures and processes that conserve water.

Nephridial organs are specialized for osmoregulation and/or excretion

A **nephridial organ,** or nephridium, is a specialized excretory structure that has evolved in many invertebrates. It consists of simple or branching tubes that usually open to the outside of the body through excretory pores, called nephridiopores.

In flatworms and nemerteans, metabolic wastes pass through the body surface by diffusion. However, these animals have specialized osmoregulatory nephridial organs, called **protonephridia,** composed of tubules with no internal openings. Their enlarged blind ends consist of **flame cells** with brushes of cilia, so named because their constant motion reminded

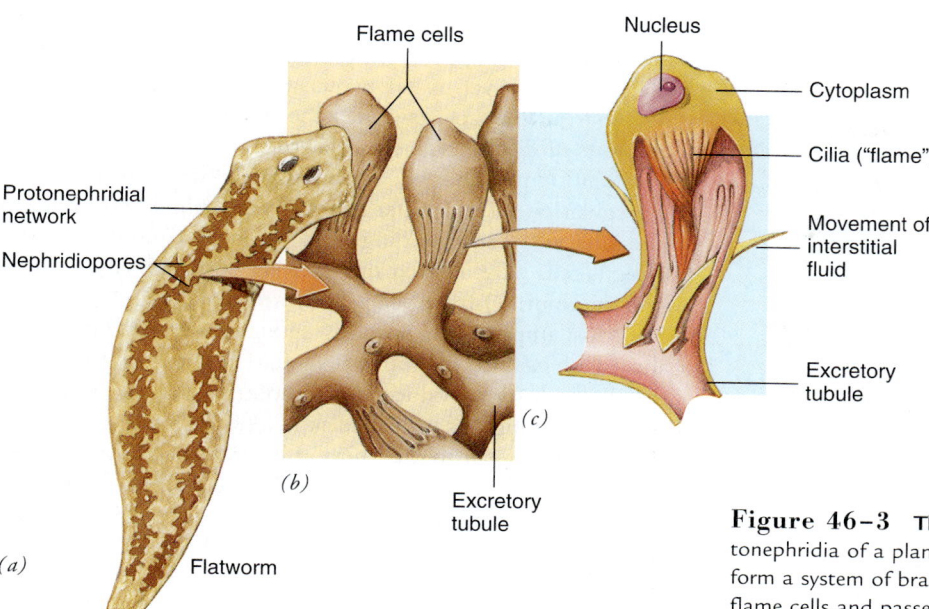

Figure 46–3 **The protonephridia of a flatworm.** *(a)* The protonephridia of a planarian, which function mainly in osmoregulation, form a system of branching tubules. *(b)* Interstitial fluid enters the flame cells and passes through a series of tubules. Excess fluid leaves the body through nephridiopores. *(c)* A single flame cell.

early biologists of flickering flames (Fig. 46–3). The flame cells lie in the interstitial fluid that bathes the body cells. Fluid enters the flame cells, and the beating of the cilia propels the fluid into the tubules. Excess fluid leaves the body through nephridiopores. A variety of other invertebrates, including rotifers, some annelids, and lancelets, also have protonephridia.

Most annelids, as well as mollusks, have more complex nephridial organs called **metanephridia.** Each segment of an earthworm has a pair of metanephridia. A metanephridium is a tubule open at both ends. The inner end opens into the coelom as a ciliated funnel (Fig. 46–4), and the outer end opens to the outside through a nephridiopore. Around each tubule is a network of capillaries. Fluid from the coelom passes into the tubule, bringing with it whatever it contains—glucose, salts, or wastes. As fluid moves through the tubule, needed materials (such as water and glucose) are removed from the fluid by the tubule and are reabsorbed by the capillaries, leaving the wastes behind. In this way, urine is produced that contains concentrated wastes.

Malpighian tubules are an important adaptation for conserving water in insects

The excretory system of insects and spiders consists of **Malpighian tubules** (Fig. 46–5). Two hundred to several hundred tubules may be present, depending on the species. Malpighian tubules are slender extensions of the gut wall. Their blind ends lie in the hemocoel (blood cavity) and are bathed in hemolymph. Cells of the tubule wall actively transport uric acid, potassium ions, and some other substances from the hemolymph into the tubule lumen. Other solutes and water follow by diffusion. The Malpighian tubules empty into the gut.

Water, some salts, and other solutes are reabsorbed into the hemolymph by a specialized epithelium in the rectum.

Uric acid, the major waste product, is excreted as a semidry paste with a minimum of water. Because the insect excretory system conserves body fluids, it has contributed significantly to the success of insects in terrestrial environments.

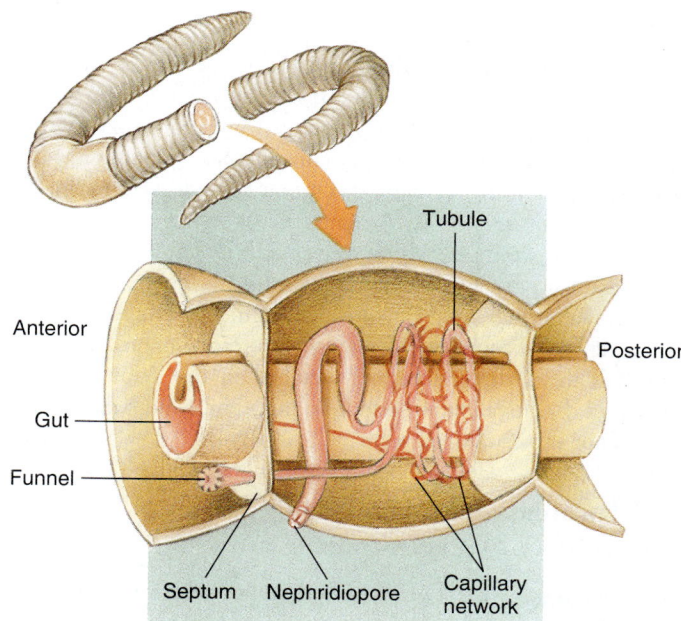

Figure 46–4 **A metanephridium of an earthworm.** Each metanephridium consists of a ciliated funnel opening into the coelom, a coiled tubule, and a nephridiopore opening to the outside. This three-dimensional internal view shows parts of three segments.

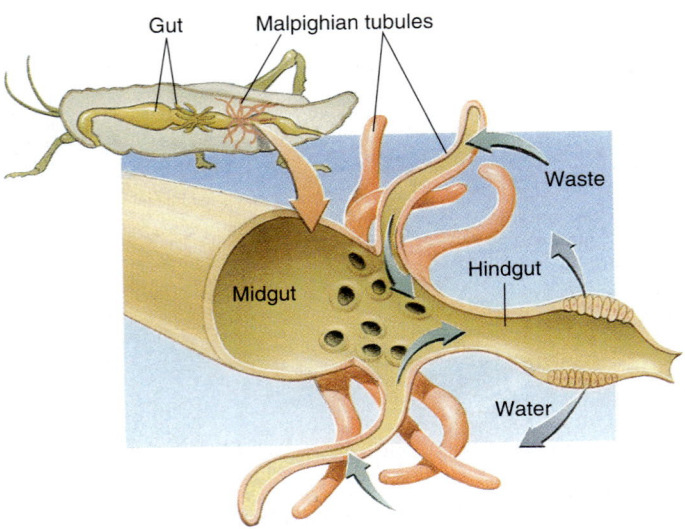

Figure 46–5 **The Malpighian tubules of an insect.** The slender Malpighian tubules have blind ends that extend into the hemocoel. Their cells transfer uric acid and some ions from the hemolymph to the cavity of the tubule. Water follows by diffusion. The wastes are discharged into the gut. The epithelium in the rectum actively reabsorbs most of the water and needed ions.

THE KIDNEY IS THE KEY VERTEBRATE ORGAN OF OSMOREGULATION AND EXCRETION

Vertebrates live successfully in a wide range of habitats—in fresh water, the ocean, tidal regions, and on land, even in extreme environments such as deserts. In response to the requirements of these diverse environments, vertebrates have evolved adaptations for regulating their salt and water content and for excreting wastes. An extreme example is the desert-dwelling kangaroo rat, which must carefully conserve water. It obtains most of its water from its own metabolism, and its kidneys are so efficient that it loses little fluid as urine.

The main osmoregulatory and excretory organ in vertebrates is the **kidney.** In most vertebrates, the skin, lungs or gills, and digestive system also help maintain fluid balance and dispose of metabolic wastes.

Freshwater vertebrates must rid themselves of excess water

As fishes began to move into freshwater habitats about 460 million years ago, there was strong selection for the evolution of adaptations for effective osmoregulation. Because the body fluids of freshwater animals have higher salt concentrations and are thus hypertonic to their surroundings, water passes into them osmotically. As a result, they are in constant danger of becoming waterlogged. Freshwater fishes are covered by scales and a mucous secretion that retards the passage of water into the body. However, water enters through the gills. The kidneys of these fishes have become adapted to excrete large amounts of dilute urine (Fig. 46–6a).

Water entry, though, is only part of the challenge of osmoregulation in freshwater fishes. These animals also tend to lose salts to the surrounding fresh water. To compensate, special gill cells have evolved that actively transport salts (mainly sodium chloride) from the water into the body.

Most amphibians are at least semiaquatic, and their mechanisms of osmoregulation are similar to those of freshwater fishes. They, too, produce large amounts of dilute urine. For example, through its urine and skin, a frog can lose an amount of water equivalent to one-third of its body weight in one day. Active transport of salt inward by special cells in the skin compensates for loss of salt through skin and urine.

Marine vertebrates must replace lost fluid

Freshwater fishes have adapted very successfully to their aquatic habitats. Evolution of body fluids more dilute than seawater is one of their chief adaptations. Thus, when some freshwater fishes returned to the sea about 200 million years ago, their blood and body fluids were less salty than (hypotonic to) their surroundings. They tended to lose water osmotically and to take in salt. To compensate for fluid loss, many marine bony fishes drink sea water (Fig. 46–6b). They retain the water and excrete salt by the action of specialized cells in their gills. Very little urine is excreted by the kidneys; the nephrons (microscopic units of the kidney that produce urine) have only small or no capillary clusters (glomeruli) that filter the blood and produce urine in other vertebrates.

Marine cartilaginous fishes (sharks and rays) have different osmoregulatory adaptations that allow them to tolerate the salt concentrations of their environment. These animals accumulate and tolerate urea (Fig. 46–6c). Their tissues are adapted to function at concentrations of urea that would be toxic to most other animals. The high urea concentration makes the body fluids slightly hypertonic to sea water, resulting in a net inflow of water. Their well developed kidneys excrete a large volume of urine. Excess salt is excreted by the kidneys and, in many species, by a rectal gland.

The heads of certain reptiles and marine birds have salt glands that excrete salt that has entered the body with ingested seawater. Salt glands are usually inactive; they function only in response to osmotic stress. Thus, only when seawater or salty food is ingested do the salt glands excrete a fluid laden with sodium and chloride.

Whales, dolphins, and other marine mammals ingest seawater along with their food. Their kidneys produce a very concentrated urine, much saltier than seawater. This is an important physiological adaptation, especially for marine carnivores. The high-protein diet of these animals results in production of large amounts of urea, which must be excreted in the urine or, in some cases, by special accessory salt glands.

(a) Freshwater fish

(b) Marine bony fish

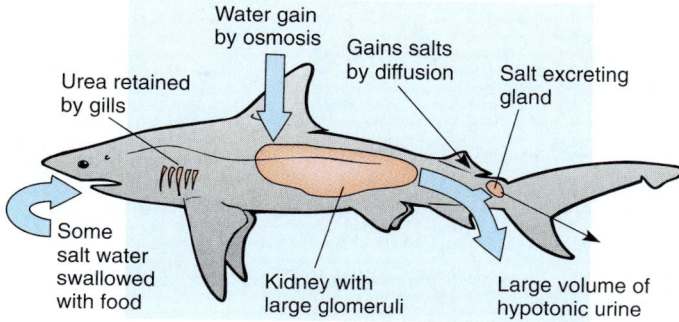

(c) Cartilaginous fish (shark)

Figure 46–6 Osmoregulation in fishes. *(a)* Freshwater fishes live in a hypotonic medium, so water continuously enters the body by osmosis, and salts diffuse out. These fishes excrete large quantities of dilute urine and actively absorb salts through the gills. *(b)* Marine fishes live in a hypertonic medium and therefore lose water by osmosis. They gain salts from the water they drink and by diffusion. To compensate, the fish drinks water, excretes the salt, and produces a small volume of urine. *(c)* In contrast, the shark accumulates urea in high enough concentration that its tissues become hypertonic to the surrounding medium. As a result, water enters its body by osmosis. A large quantity of dilute urine is excreted.

The mammalian kidney helps maintain homeostasis

In mammals, the kidneys are the principal excretory organs. They are responsible for the excretion of most nitrogenous wastes and for helping to maintain fluid balance by adjusting the salt and water content of the urine. As in other terrestrial vertebrates, the lungs, skin, and digestive system are also important in mammalian osmoregulation and waste disposal (Fig. 46–7). Most carbon dioxide and a great deal of water are excreted by the lungs. Although primarily concerned with the regulation of body temperature, the sweat glands of humans and some other mammals excrete 5% to 10% of all metabolic wastes.

Most of the bile pigments produced by the breakdown of red blood cells are normally excreted by the liver into the intestine. From the intestine they then pass out of the body with the feces. The liver also produces both urea and uric acid, which are transported by the blood to the kidneys.

THE KIDNEYS, URINARY BLADDER, AND THEIR DUCTS MAKE UP THE URINARY SYSTEM

The mammalian **urinary system** consists of the kidneys, the urinary bladder, and associated ducts. The overall structure of the human urinary system is shown in Figure 46–8. Located just below the diaphragm in the "small of the back," the kidneys look like a pair of giant, dark-red lima beans, each about the size of a fist. Each kidney is covered by a connective tissue capsule. The outer portion of the kidney is called the **renal cortex;** the inner portion is the **renal medulla** (Fig. 46–9). The renal medulla contains a number (usually eight to ten) of cone-shaped structures called **renal pyramids.** The tip of each pyramid is a **renal papilla.** Each papilla has several pores, the openings of **collecting ducts.**

As urine is produced, it flows from collecting ducts through a renal papilla and into the **renal pelvis,** a funnel-shaped chamber. Urine then flows into one of the paired **ureters,** ducts that connect the kidney with the **urinary bladder.** The urinary bladder is a remarkable organ capable of holding (with practice) up to 800 mL (about a pint and a half) of urine. Emptying the bladder changes it from the size of a small melon to that of a pecan. This feat is made possible by the smooth muscle and special *transitional epithelium* of the bladder wall, which is capable of great shrinkage and stretching.

During **urination,** urine is released from the bladder and flows through the **urethra,** a duct leading to the outside of the body. In the male, the urethra is lengthy and passes through the penis. Semen, as well as urine, is transported through the male urethra. In the female, the urethra is short and transports only urine. Its opening to the outside is just above the opening of the vagina. The length of the male urethra discourages

(Text continues on page 1008.)

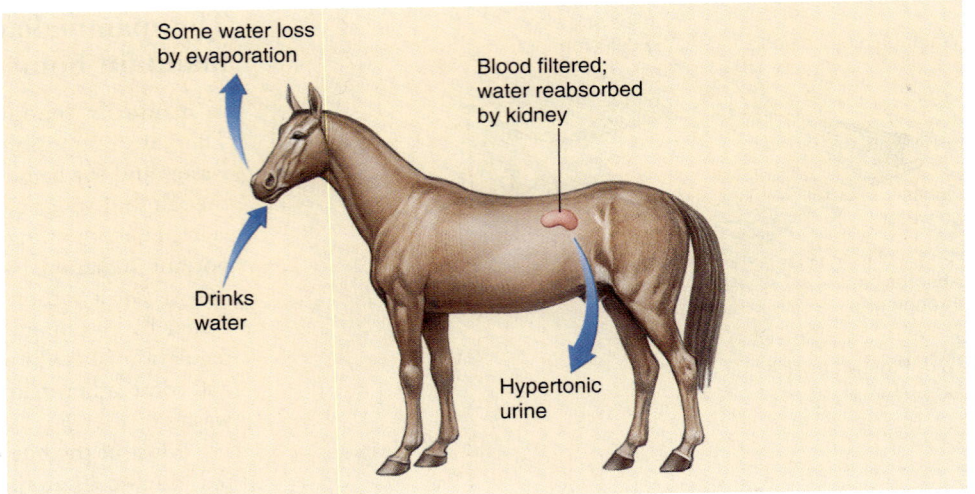

Some water loss by evaporation

Blood filtered; water reabsorbed by kidney

Drinks water

Hypertonic urine

(a)

LIVER

ALL CELLS

Wastes produced

Hb (hemoglobin) breakdown

Cellular respiration

Breakdown of nucleic acids

Deamination of amino acids

Wastes

Uric acid

Bile pigments

Water

Carbon dioxide

Urea

Organs of excretion

KIDNEY

DIGESTIVE SYSTEM

SKIN

LUNGS

Excretion

Urine

Feces

Sweat

Exhaled air containing
water vapor   carbon dioxide

(b)

Figure 46–7 **Excretory organs in terrestrial vertebrates.** *(a)* The vertebrate kidney conserves water by reabsorbing it. *(b)* Disposal of metabolic wastes in humans and other terrestrial mammals. To conserve water, a small amount of hypertonic urine is produced. Nitrogenous wastes are produced by the liver and transported to the kidneys. All cells produce carbon dioxide and some water during cellular respiration.

Adrenal gland

Left renal artery

Hilus

Left kidney

Abdominal aorta

Right and left ureters

Urinary bladder

External urethral orifice

Right kidney

Right renal vein

Renal pelvis

Inferior vena cava

Ureteral orifices

Urethra

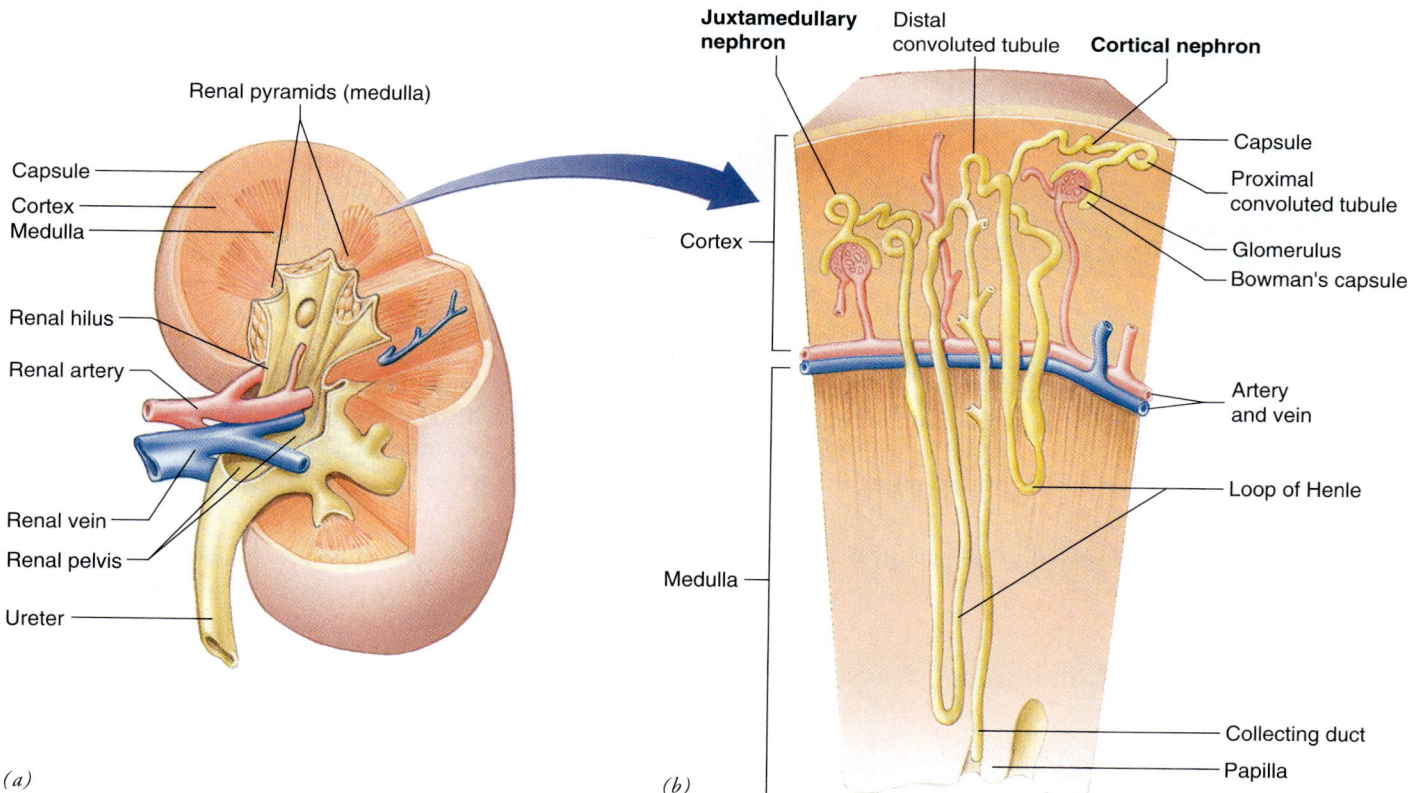

Renal pyramids (medulla)

Capsule

Cortex

Medulla

Renal hilus

Renal artery

Renal vein

Renal pelvis

Ureter

Juxtamedullary nephron

Distal convoluted tubule

Cortical nephron

Capsule

Proximal convoluted tubule

Glomerulus

Bowman's capsule

Artery and vein

Loop of Henle

Collecting duct

Papilla

Cortex

Medulla

(a)

(b)

Figure 46-9 **Structure of the kidney.** *(a)* The kidney is covered by a fibrous capsule. The outer region of the kidney is the cortex; the inner region is the medulla. When urine is produced, it flows into the renal pelvis and leaves the kidney through the ureter. The renal artery delivers blood to the kidney; the renal vein drains blood away from the kidney. *(b)* Longitudinal section showing the location of the two main types of nephrons, the juxtamedullary nephron and the cortical nephron.

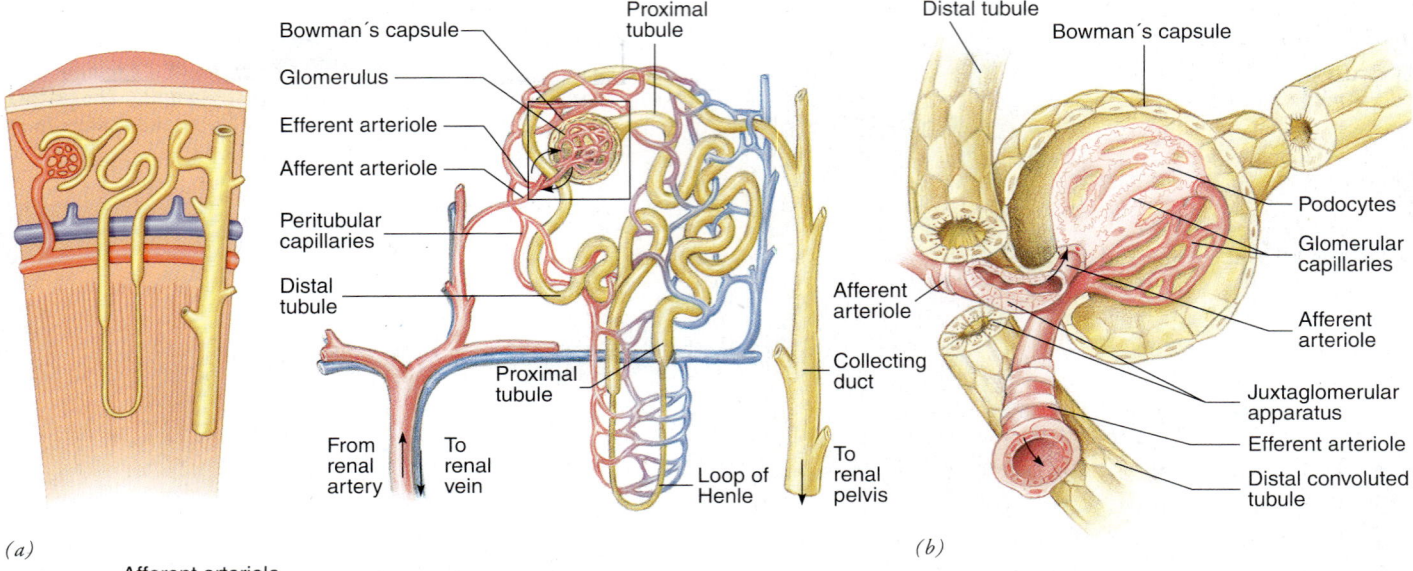

Bowman's capsule
Proximal tubule
Glomerulus
Efferent arteriole
Afferent arteriole
Peritubular capillaries
Distal tubule
Proximal tubule
From renal artery   To renal vein
Loop of Henle
Collecting duct
To renal pelvis

(a)

Distal tubule
Bowman's capsule
Podocytes
Glomerular capillaries
Afferent arteriole
Afferent arteriole
Juxtaglomerular apparatus
Efferent arteriole
Distal convoluted tubule

(b)

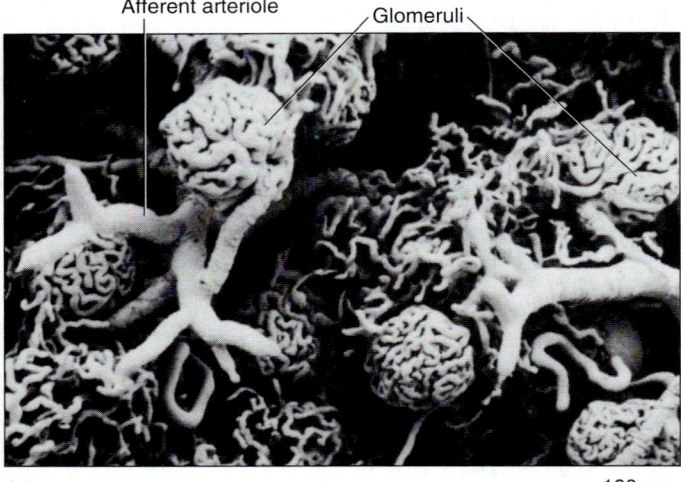

Afferent arteriole   Glomeruli

(c)   100 µm

Figure 46–10 **Nephron structure.** *(a)* The location and the basic structure of a nephron. Urine forms by filtration of the blood in the glomeruli and by adjustment of the filtrate as it passes through the tubules that drain the glomeruli. *(b)* Detailed view of Bowman's capsule. Note that the distal convoluted tubule is actually adjacent to the afferent arteriole. The juxtaglomerular apparatus (discussed later in the chapter) is a small group of cells located in the walls of the tubule and arterioles. *(c)* Low-power SEM of a portion of the kidney cortex, showing glomeruli and associated blood vessels. *(c, CNRI/ Science Photo Library/Photo Researchers, Inc.)*

bacterial invasions of the bladder. This length difference helps explain why bladder infections are more common in females than in males.

In summary, urine flows through the following structures:

Kidney (through renal pelvis) ⟶ ureter ⟶ urinary bladder ⟶ urethra

The nephron is the functional unit of the kidney

Each kidney consists of more than one million functional units called **nephrons.** A nephron consists of a cuplike **Bowman's capsule** connected to a long, partially coiled **renal tubule** (Figs. 46–9 and 46–10). Positioned within Bowman's capsule is a cluster of capillaries known as a **glomerulus.**

Three main regions of the renal tubule are the **proximal convoluted tubule,** which conducts the filtrate from Bow-

man's capsule; the **loop of Henle,** an elongated, hairpin-shaped portion; and the **distal convoluted tubule,** which conducts the filtrate to a collecting duct. Thus, filtrate passes through the following structures:

Bowman's capsule ⟶ proximal convoluted tubule ⟶ loop of Henle ⟶ distal convoluted tubule ⟶ collecting duct

The human kidney has two types of nephrons: the more numerous (85%) cortical nephrons and the more internal juxtamedullary nephrons (see Fig. 46–9*b*). **Cortical nephrons** have relatively small glomeruli and are located almost entirely within the cortex or outer medulla. **Juxtamedullary nephrons** have large glomeruli, and their very long loops of Henle extend deep into the medulla. The loop of Henle consists of a *descending loop* that receives filtrate from the proximal convoluted tubule and an *ascending loop,* through which the filtrate passes on its way to the distal convoluted tubule. The juxtamedullary nephrons contribute to the ability of the mammalian kidney to concentrate urine. Excretion of urine that is hypertonic to body fluids is an important mechanism for conserving water.

Blood is delivered to the kidney by the renal artery. Small branches of the renal artery give rise to **afferent arterioles.** (Afferent means "to carry toward.") An afferent arteriole conducts blood into the capillaries that make up each glomerulus. As blood flows through the glomerulus, some of the plasma is forced into Bowman's capsule.

You may recall that in a usual circulatory route, capillaries deliver blood into veins. Circulation in the kidneys is an exception because blood flowing from the glomerular capillaries next passes into an **efferent arteriole.** Each efferent arteriole conducts blood *away* from a glomerulus. The efferent arteriole delivers blood to a second capillary network, the **peritubular capillaries** surrounding the renal tubule.

As blood flows through the first set of capillaries, those of the glomerulus, it is filtered. The peritubular capillaries receive materials returned to the blood by the renal tubule. Blood from the peritubular capillaries enters small veins that eventually lead to the renal vein. In summary, blood circulates through the kidney in the following sequence:

Renal artery ⟶ afferent arteriole ⟶ capillaries of glomerulus ⟶ efferent arteriole ⟶ peritubular capillaries ⟶ small veins ⟶ renal vein

Urine is produced by filtration, reabsorption, and secretion

Urine is produced by a combination of three processes: filtration, reabsorption, and tubular secretion (Fig. 46–11).

Filtration is not selective with regard to ions and small molecules

Blood flows through the glomerular capillaries under high pressure, forcing more than 10% of the plasma out of the capillaries and into Bowman's capsule. **Filtration** is somewhat similar to the mechanism whereby interstitial fluid is formed as blood flows through other capillary networks in the body. However, blood flow through glomerular capillaries is at much higher pressure so more plasma is filtered in the kidney.

Several factors contribute to filtration. First, the hydrostatic pressure in the glomerular capillaries is higher than in other capillaries. This high pressure is mainly due to the high resistance to outflow presented by the efferent arteriole, which is smaller in diameter than the afferent arteriole. A second factor that contributes to the large amount of **glomerular filtrate** is the large surface area for filtration provided by the highly coiled glomerular capillaries. A third factor is the great permeability of the glomerular capillaries. Numerous small pores (fenestrations) are present between the endothelial cells that form their walls, making the glomerular capillaries more porous than typical capillaries.

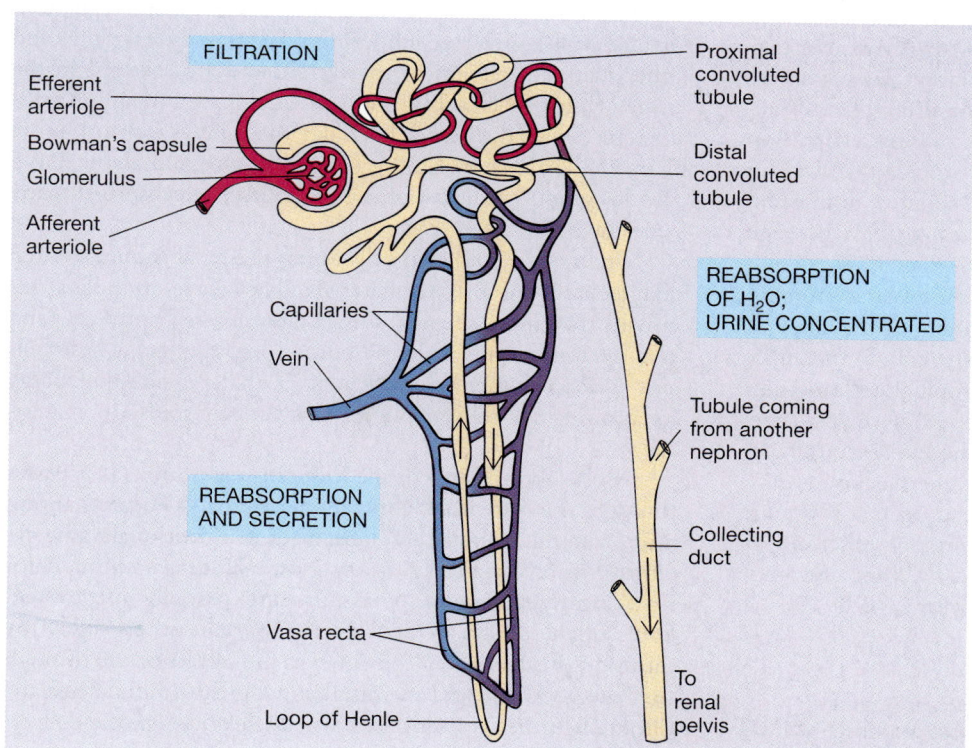

Figure 46–11 Sites of filtration, reabsorption, and secretion.

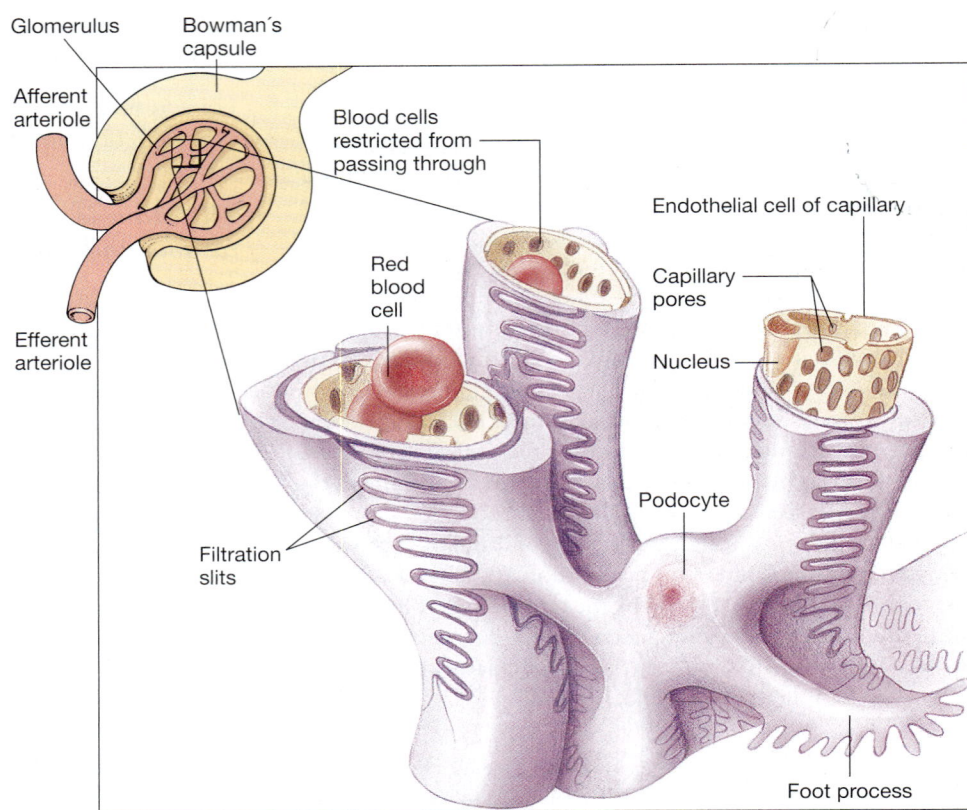

Figure 46–12 Filtration membrane of the kidney. Consisting of podocytes and glomerular capillary walls, this filtration barrier is highly permeable to water and small molecules but restricts the passage of blood cells and large molecules.

Glomerulus
Bowman's capsule
Afferent arteriole
Efferent arteriole
Blood cells restricted from passing through
Red blood cell
Filtration slits
Endothelial cell of capillary
Capillary pores
Nucleus
Podocyte
Foot process

The wall of Bowman's capsule in contact with the capillaries consists of specialized epithelial cells called **podocytes.** These cells have numerous cytoplasmic extensions called foot processes that cover most of the surfaces of the glomerular capillaries (Fig. 46–12). Foot processes of adjacent podocytes are separated by narrow gaps called **filtration slits.** The permeable glomerular capillary walls and podocytes form a **filtration membrane** that permits fluid and small solutes dissolved in the plasma, such as glucose, amino acids, sodium, potassium, chloride, bicarbonate, other salts, and urea, to pass through and become part of the filtrate. This filtration membrane holds back blood cells, platelets, and most of the plasma proteins.

The total volume of blood passing through the kidneys is about 1200 mL per minute, or about one fourth of the entire cardiac output. The plasma passing through the glomerulus loses more than 10% of its volume to the glomerular filtrate; the rest leaves the glomerulus through the efferent arteriole. The normal glomerular filtration rate amounts to about 180 L (about 45 gal) each 24 hours. This is four and a half times the amount of fluid in the entire body! Common sense tells us that urine could not be excreted at that rate. Within a few moments, dehydration would become life-threatening.

Reabsorption is highly selective

The threat to homeostasis posed by the vast amounts of fluid filtered by the kidneys is avoided by **reabsorption.** About 99%

of the filtrate is reabsorbed into the blood through the renal tubules, leaving only about 1.5 L to be excreted as urine during a 24 hour period. Reabsorption permits precise regulation of blood chemistry by the kidneys. Wastes, excess salts, and other materials remain in the filtrate and are excreted in the urine, while needed substances such as glucose and amino acids are returned to the blood. Each day the tubules reabsorb more than 178 L of water, 1200 g (2.6 lb) of salt, and about 250 g (0.5 lb) of glucose. Most of this, of course, is reabsorbed many times over.

The simple epithelial cells lining the renal tubule are well adapted for reabsorbing materials. They have abundant microvilli that increase their surface area for reabsorption (and give the inner lining a "brush border" appearance). These cells also contain numerous mitochondria that provide the energy for running the cellular pumps that actively transport materials.

Most (about 65%) of the filtrate is reabsorbed as it passes through the proximal convoluted tubule. Glucose, amino acids, vitamins, and other substances of nutritional value are entirely reabsorbed there. Many ions, including sodium, chloride, bicarbonate, and potassium, are partially reabsorbed. Some of these ions are actively transported; others follow by diffusion. Reabsorption continues as the filtrate passes through the loop of Henle and the distal convoluted tubule. Then the filtrate is further concentrated as it passes through the collecting duct that leads to the renal pelvis.

Normally, substances that are useful to the body, such as glucose or amino acids, are reabsorbed from the renal tubules. If the concentration of a particular substance in the blood is high, however, the tubules may not be able to reabsorb all of it. The maximum rate at which a substance can be reabsorbed is called its **tubular transport maximum (Tm).** For example, the Tm for glucose averages 320 mg per minute for an adult human. Normally, the tubular load of glucose is only about 125 mg per minute, so almost all of it is reabsorbed. However, if glucose is filtered in excess of the Tm, that excess will not be reabsorbed but will instead pass into the urine.

Each substance that has a Tm also has a **renal threshold** concentration in the plasma. When a substance exceeds its renal threshold, the portion not reabsorbed is excreted in the urine. In a person with uncontrolled diabetes mellitus, the concentration of glucose in the blood exceeds its renal threshold (about 150 mg of glucose per 100 mL of blood), so glucose is excreted in the urine. Its presence there is evidence of this disorder (Chapter 47).

Some substances are actively secreted from the blood into the filtrate

Tubular **secretion** is the passage of substances across the tubule epithelium in a direction opposite to that of reabsorption. Secretion occurs mainly in the region of the distal convoluted tubule. Potassium, hydrogen, and ammonium ions are secreted into the filtrate. Certain drugs, such as penicillin, are also removed from the blood by secretion.

Secretion of hydrogen ions, an important homeostatic mechanism for regulating the pH of the blood, takes place through the formation of carbonic acid. Carbon dioxide, which diffuses from the blood into the cells of the distal tubules and collecting ducts, combines with water to form carbonic acid. This acid then dissociates, forming hydrogen ions and bicarbonate ions. When the blood becomes too acidic, more hydrogen ions are secreted into the urine.

$$CO_2 + H_2O \rightleftharpoons H_2CO_3 \rightleftharpoons H^+ + HCO_3^-$$

Potassium secretion is also an important homeostatic mechanism. When potassium concentration is too high, nerve impulses are not effectively transmitted and the strength of muscle contraction decreases. The heart can be weakened and even fail. When potassium ions are too highly concentrated, they are secreted from the blood into the renal tubules and then are excreted in the urine. Secretion results partly from a direct effect of the potassium ions on the tubules. In addition, a high potassium ion concentration in the blood stimulates the adrenal cortex to increase its output of the hormone **aldosterone,** which further stimulates secretion of potassium.

Urine becomes concentrated as it passes through the renal tubule

A salt concentration gradient is established, in part, by reabsorption of salt from various regions of the renal tubule.

Sodium ions are actively transported out of the proximal tubule, and water follows osmotically. The loop of Henle is specialized to produce a high concentration of sodium chloride in the medulla. This is important in maintaining a highly hypertonic interstitial fluid in the medulla near the bottom of the loop, which in turn permits the kidneys to produce a concentrated urine. The walls of the descending loop of Henle are relatively permeable to water but relatively impermeable to sodium and urea. There is a high concentration of sodium in the interstitial fluid, so as the filtrate passes down the loop of Henle, water moves out by osmosis. This concentrates the filtrate inside the loop of Henle (Fig. 46–13).

At the turn of the loop of Henle, the walls become more permeable to salt and less permeable to water. As the concentrated filtrate moves up the ascending portion of the loop of Henle, salt diffuses out into the interstitial fluid. This contributes to the high salt concentration in the interstitial fluid in the medulla of the kidney surrounding the loop of Henle. Further along the ascending part of the loop of Henle, sodium is actively transported out of the tubule.

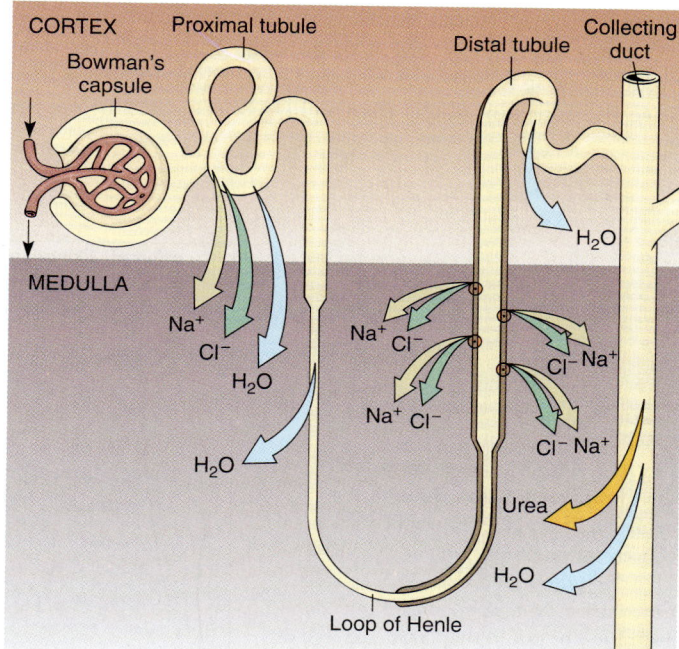

Figure 46–13 **Movement of water, ions, and urea through the renal tubule and collecting duct.** Water passes out of the descending loop of Henle, leaving a more concentrated filtrate inside. Salt moves out of the ascending loop. Chloride pumps transport chloride out into the interstitial fluid, and sodium follows. The saltier the interstitial fluid becomes, the more water moves out of the descending loop. This leaves a concentrated filtrate inside, so more salt passes out. Note that this is a positive feedback system. Urea also moves out into the interstitial fluid through the collecting ducts. Water from the collecting ducts moves out osmotically into this hypertonic interstitial fluid. The heavy outline along the ascending loop indicates that this region is relatively impermeable to water.

Because water passes out of the descending portion of the loop of Henle, the filtrate at the bottom of the loop has a high salt concentration. However, because salt (but not water) is removed in the ascending portion, by the time the filtrate moves through the distal tubule, it is isotonic (or even hypotonic) to blood. The filtrate passes from the renal tubule into a larger collecting duct that eventually empties into the renal pelvis.

Note that there is a *counterflow* of fluid through the two limbs of the loop of Henle. Filtrate passing down through the descending portion of the loop is flowing in a direction opposite the filtrate moving upward through the ascending loop. The filtrate is concentrated as it moves down the descending portion of the loop and diluted as it moves up the ascending part of the loop. This countercurrent mechanism helps maintain a high salt concentration in the interstitial fluid of the medulla. The hypertonic interstitial fluid draws water osmotically from the filtrate in the collecting ducts.

The inner medullary collecting ducts are permeable to urea, allowing the concentrated urea in the filtrate to diffuse out into the interstitial fluid. This urea contributes to the high solute concentration of the inner medulla and so helps in the process of concentrating urine (Fig. 46–14).

The collecting ducts are routed to pass through the zone of very salty interstitial fluid. As the filtrate moves down the collecting duct, water passes osmotically into the interstitial fluid, where it is collected by capillaries. So much water may leave the collecting ducts that highly concentrated urine can be produced. Because it has a low concentration of water, a hypertonic urine conserves water.

Some of the water that diffuses from the filtrate into the interstitial fluid is removed by capillaries known as the **vasa recta** and carried off in the venous drainage of the kidney. The vasa recta are long, looped extensions of the efferent arterioles of the juxtamedullary nephrons. They extend deep into the medulla, only to negotiate a hairpin curve and return to the cortical venous drainage of the kidney.

Blood flows in opposite directions in the ascending and descending regions of the vasa recta, just as filtrate flows in opposite directions in the ascending and descending portions of the loop of Henle. As a consequence of this countercurrent flow, much of the salt and urea that enter the blood leave again; the solute concentration of the blood leaving the vasa recta is only slightly higher than that of the blood entering. This mechanism helps maintain the high solute concentration of the interstitial fluid in the renal medulla.

Urine is composed of water, nitrogenous wastes, and salts

By the time the filtrate reaches the renal pelvis, its composition has been precisely adjusted. Useful materials have been returned to the blood by reabsorption. Wastes and excess materials that entered by filtration or secretion have been retained

Figure 46–14 **Concentration of the filtrate as it moves through the nephron.** This figure shows the relative concentration of ions, mainly Na$^+$ and Cl$^-$, during formation of a very concentrated urine. Numbers indicate the concentration of salt in the filtrate expressed in milliosmols per liter. (A milliosmol is a unit of osmotic pressure equal to one-thousandth gram molecular weight of a substance divided by the number of ions into which a substance dissociates in one liter of solution.) The more hypertonic a solution is, the higher its osmotic pressure. The very hypertonic interstitial fluid near the renal pelvis draws water osmotically from the filtrate in the collecting ducts. The heavy outline along the ascending loop indicates that this region is relatively impermeable to water.

KIDNEY DISEASE

Our kidneys excrete metabolic wastes and help regulate the volume and composition of body fluids. Their vital function is compromised in more than 13 million people in the United States who suffer from kidney disease. In fact, kidney disease ranks fourth in prevalence among major human diseases in the United States. Kidney function can be impaired by infections, tumors, kidney stones, shock, circulatory disease, or exposure to substances such as mercury, lead, or carbon tetrachloride. One of the most common kidney diseases, glomerulonephritis, is actually a large number of related chronic diseases in which the filtering units of the kidney (the glomeruli) are damaged. This damage is thought to result from an autoimmune response.

In chronic kidney disease, a progressive loss of function may eventually reach the stage of kidney failure. A decrease in the rate of filtration occurs, causing loss of homeostatic water and salt balance and an inability to effectively excrete urea and other nitrogenous wastes. Retention of water causes edema (swelling). As the concentration of hydrogen ions increases, body fluids become too acidic (acidosis). Nitrogenous wastes accumulate in the blood and tissues, causing a condition called uremia. If untreated, acidosis and uremia can cause coma and, eventually, death.

Chronic kidney failure is treated by kidney dialysis or by kidney transplant.

More than 200,000 individuals in the United States alone undergo kidney dialysis. In dialysis, solutes are exchanged by diffusion across a selectively permeable membrane between solutions of different compositions. A kidney dialysis machine can be used to temporarily restore appropriate blood solute balance to a patient whose kidneys are not functioning.

In one type of dialysis (hemodialysis), blood circulates from one of the patient's arteries through the dialysis machine and is then returned through a vein. A plastic tube is surgically inserted into both an artery and a vein in the patient's arm or leg. These tubes are then connected to a circuit of plastic tubing from a dialysis machine. The patient's blood flows through the tubing, which is permeable to certain substances such as nitrogenous wastes and ions, but not permeable to proteins and blood cells. The tubing is immersed in a solution containing most of the normal blood plasma components in their normal proportions. Nitrogenous wastes diffuse down their concentration gradient from the patient's blood through minute pores in the tubing and into the surrounding solution in the dialysis machine. As the blood circulates repeatedly through the tubing in the machine, dialysis continues, eventually adjusting most of the values of the patient's blood chemistry to normal ranges.

Although much improved by recent engineering advances, mechanical dialysis is expensive, clumsy, and inconvenient. Patients typically undergo dialysis three times a week, and the procedure takes 4 to 6 hours. This type of dialysis can also produce serious side effects such as osteoporosis ("brittle bone syndrome," caused by bone calcium loss).

Another dialysis technique, continuous ambulatory peritoneal dialysis (CAPD), uses the patient's own peritoneum (the lining of the abdominal cavity), which is a selectively permeable multicellular membrane. A plastic bag containing dialysis fluid is attached to the patient's abdominal cavity, and the fluid is allowed to run into the abdominal cavity. After about 30 minutes, the fluid is withdrawn into the bag and discarded. This process is repeated about three times each day. This type of dialysis is much more convenient but poses the threat of peritonitis if bacteria enter the body cavity with the dialysis fluid.

Long-term dialysis is not as desirable for the patient as a functioning kidney. With a successful kidney transplant, a patient can live a more normal life with far less long-term expense. At present, more than two-thirds of kidney transplants are successful for several years, although physicians must routinely treat the problems of graft rejection (see Chapter 43). Many recipients of kidney transplants have survived for more than 20 years.

by the tubules. The adjusted filtrate, called **urine,** is composed of approximately 96% water, 2.5% nitrogenous wastes (mainly urea), 1.5% salts, and traces of other substances, such as bile pigments, that may contribute to the characteristic color and odor.

Healthy urine is sterile and has been used to wash battlefield wounds when clean water was not available. However, urine swiftly decomposes when exposed to bacterial action, forming ammonia and other products. It is the ammonia that produces the diaper rash of infants.

The composition of urine yields many clues to body function and malfunction. **Urinalysis,** the physical, chemical, and microscopic examination of urine, is a very important diagnostic tool that has been used to monitor diabetes mellitus and many other disorders. Urinalysis is also extensively used in drug testing because breakdown products of some drugs can be identified in the urine for several weeks. (See *Focus On: Treating Kidney Disease.*)

Kidney function is regulated by hormones

Several hormones interact to regulate urine volume and concentration, and investigators think that additional hormones that help maintain fluid homeostasis remain to be identified

TABLE 46–1 Hormonal Control of Kidney Function

Hormone	Source	Target Tissue	Actions	Factors That Stimulate Release
Antidiuretic hormone (ADH)	Produced in hypothalamus; released by posterior pituitary gland	Collecting ducts	Increases permeability of the collecting ducts to water, increasing reabsorption and decreasing water excretion	Low fluid intake decreases blood volume and increases osmotic pressure of blood; receptors in hypothalamus stimulate posterior pituitary
Aldosterone	Adrenal glands (cortex)	Distal tubules and collecting ducts	Increases sodium reabsorption by distal tubules and collecting ducts	Angiotensin II
Angiotensin II	Produced from blood protein; renin necessary	Blood vessels and adrenal glands	Constricts blood vessels which raises blood pressure; stimulates aldosterone secretion	Decrease in blood pressure causes renin secretion
Atrial natriuretic peptide (ANP)	Atrium of heart	Afferent arterioles; collecting ducts	Dilates afferent arterioles; inhibits sodium reabsorption by collecting ducts; inhibits aldosterone secretion	Stretching of atria due to increased blood volume

(Table 46–1). The amount of urine produced depends on the body's need to retain or rid itself of water. We have seen that salt reabsorption in the loops of Henle establishes a very salty interstitial fluid that draws water osmotically from the collecting ducts. Permeability of the collecting ducts to water is regulated by **antidiuretic hormone (ADH).** When the body needs to conserve water, the posterior pituitary gland increases its release of ADH (Fig. 46–15). This hormone makes the collecting ducts more permeable to water so that more water is reabsorbed and a small volume of concentrated urine is produced.

Secretion of ADH is stimulated by special receptors in the hypothalamus. When fluid intake is low, the body begins to dehydrate, causing the blood volume to decrease. As blood volume decreases, the concentration of salts dissolved in the blood becomes greater, causing an increase in osmotic pressure. Receptors in the hypothalamus are sensitive to this osmotic change and stimulate the posterior lobe of the pituitary to release more ADH. A thirst center in the hypothalamus also responds to dehydration, stimulating an increase in fluid intake.

When one drinks a great deal of water, the blood becomes diluted and its osmotic pressure falls. Release of ADH by the pituitary gland decreases, lessening the amount of water reabsorbed from the collecting ducts. A large volume of dilute urine is produced.

Occasionally, the pituitary gland malfunctions and does not produce sufficient ADH. The resulting condition is called *diabetes insipidus* (not to be confused with the more common

disorder, diabetes mellitus). This condition can also result from an acquired insensitivity of the kidney to ADH. In diabetes insipidus, water is not efficiently reabsorbed from the ducts, so a large volume of urine is produced. A person with severe diabetes insipidus may excrete up to 25 quarts of urine each day, a serious loss of water to the body. The affected individual becomes dehydrated and must drink almost continually to offset fluid loss. Diabetes insipidus can often be controlled by injections of ADH or by use of an ADH nasal spray.

Sodium is the most abundant extracellular ion, accounting for about 90% of all positive ions outside cells in the extracellular fluid. Sodium concentration is precisely regulated by the actions of several hormones. **Aldosterone,** which is secreted by the cortex of the adrenal glands, stimulates the distal tubules and collecting ducts to increase sodium reabsorption. When the adrenal glands of experimental animals are removed, too much sodium is excreted, leading to serious depletion of the extracellular fluid.

Aldosterone secretion can be stimulated by a decrease in blood pressure (which is caused by a decrease in volume of blood and interstitial fluid). When blood pressure falls, cells of the **juxtaglomerular apparatus** secrete the enzyme **renin** and activate the **renin-angiotensin-aldosterone pathway.** The juxtaglomerular apparatus is a small group of cells located in the region where the renal tubule contacts the afferent and efferent arterioles (see Fig. 46–10b). Renin acts on a plasma protein (angiotensinogen), converting it to angiotensin. An enzyme from the lungs converts angiotensin into its active form,

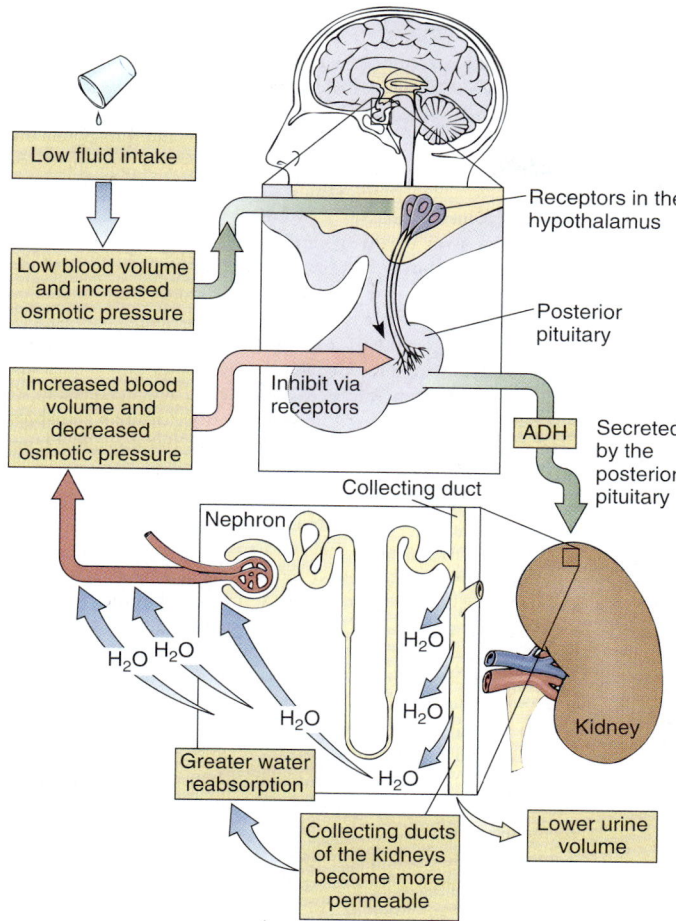

Figure 46–15 **Regulation of urine volume by antidiuretic hormone (ADH).** When the body is dehydrated, the hormone ADH increases the permeability of the collecting ducts to water. More water is reabsorbed, and only a small volume of concentrated urine is produced.

angiotensin II, a peptide hormone that stimulates aldosterone secretion.

Angiotensin II not only increases the synthesis and release of aldosterone, but also raises blood pressure directly by constricting blood vessels. Angiotensin II also stimulates sodium reabsorption by the proximal convoluted tubules and may stimulate the posterior pituitary to release ADH. All of these actions help restore extracellular fluid volume and normal blood pressure.

> Decrease in blood volume ⟶ decrease in blood pressure ⟶ cells of juxtaglomerular apparatus secrete renin ⟶ angiotensinogen ⟶ angiotensin ⟶ angiotensin II ⟶ constricts blood vessels and stimulates aldosterone secretion ⟶ aldosterone increases sodium reabsorption ⟶ blood pressure increases

Atrial natriuretic peptide (ANP), a hormone produced by the heart, increases sodium excretion and decreases blood pressure. In 1981 Adolpho de Bold and his research team at Queen's University in Ontario reported that when they injected homogenized rat atria into rats, they observed a dramatic increase in excretion of sodium and water by the kidney. During the next few years, the amino acid sequence of ANP was determined, and it was chemically synthesized. Rat, mouse, and human genes for ANP were then cloned, and their nucleotide sequences were determined. ANP is stored in granules in atrial muscle cells and is released into the circulation in response to stretching of the atria by increased blood volume.

ANP dilates afferent arterioles, thereby increasing glomerular filtration rate. It inhibits sodium reabsorption by the collecting ducts directly and also indirectly by inhibiting secretion of aldosterone. ANP acts on the adrenal cortex to inhibit aldosterone secretion and also reduces plasma aldosterone concentration by inhibiting release of renin. These actions of ANP lower blood volume and blood pressure. The renin-angiotensin system and ANP work antagonistically in regulating fluid balance, salt (electrolyte) balance, and blood pressure.

> Increase in blood volume ⟶ increase in blood pressure ⟶ atria of heart stretched ⟶ atria release ANP ⟶ directly inhibits sodium reabsorption and inhibits aldosterone secretion, which also inhibits sodium reabsorption ⟶ decrease in blood volume ⟶ blood pressure decreases

SUMMARY WITH KEY TERMS

I. **Osmoregulation** is the active regulation of osmotic pressure of body fluids so that homeostasis is maintained. **Excretion** is the process of ridding the body of metabolic wastes, including excess water. **Excretory systems** help maintain homeostasis by regulating the concentration of body fluids through osmoregulation and excretion of metabolic wastes.

II. The principal waste products of animal metabolism are water; carbon dioxide; and **nitrogenous wastes,** including **ammonia, urea,** and **uric acid. Urine** is a watery solution of metabolic wastes and other organic and inorganic substances.

III. Most marine invertebrates are **osmoconformers;** the salt concentration of their body fluids varies along with changes in the sea water. Some marine invertebrates, especially those inhabiting coastal habitats, are **osmoregulators;** they maintain an optimal salt concentration despite changes in salinity of their surroundings.

A. Invertebrate mechanisms of osmoregulation and waste disposal include the **protonephridia** of flatworms, **metanephridia** of annelids, and **Malpighian tubules** of insects.

B. The enlarged blind end of a protonephridium is a **flame cell** with brushes of cilia that propel fluid into a system of tubules. Fluid leaves the body through nephridiopores.

   C. A metanephridium is a tubule open at both ends. Fluid from the coelom enters the tubule through a ciliated funnel. As fluid moves through the tubule, needed materials are reabsorbed by capillaries that surround the tubule.

   D. Malpighian tubules are extensions of the insect gut wall. Their blind ends lie in the hemocoel. Cells of the tubule actively transport uric acid and some other substances from the hemolymph into the tubule, and water follows by diffusion. The contents of the tubule pass into the gut, and water and some solutes are reabsorbed by a specialized epithelium in the rectum.

IV. The vertebrate **kidney,** which functions in excretion and osmoregulation, is vital in maintaining homeostasis.

   A. Aquatic vertebrates have the continuous challenge of osmoregulation. Freshwater fishes take in water osmotically; they excrete a large volume of dilute urine.

   B. Marine bony fishes lose water osmotically. They compensate by drinking sea water and excreting salt through their gills; only a small volume of urine is produced.

   C. Marine cartilaginous fishes retain large amounts of urea, which enables them to take in water osmotically through the gills. This water can be used to excrete a hypotonic urine.

   D. Other aquatic vertebrates have specific adaptations for dealing with osmoregulation. Marine birds and reptiles, for example, have salt glands that excrete excess salt.

V. The **urinary system** is the principal excretory system in humans and other vertebrates. The kidney is the key organ of the urinary system.

   A. In mammals, the kidneys produce urine, which passes through the **ureters** to the **urinary bladder** for storage. During urination, the urine is released from the body through the **urethra.**

   B. The outer portion of each kidney is the **renal cortex;** the inner portion is the **renal medulla.** The renal medulla contains 8 to 10 **renal pyramids.** The tip of each pyramid is a **renal papilla.** As urine is produced it flows into **collecting ducts,** which empty through a renal papilla into a funnel-shaped chamber, the **renal pelvis.**

   C. Each **nephron** consists of a cluster of capillaries, called a **glomerulus,** surrounded by a **Bowman's capsule** that opens into a long, coiled renal tubule. The **renal tubule** consists of a **proximal convoluted tubule, loop of Henle,** and **distal convoluted tubule.**

   D. **Cortical nephrons** have small glomeruli and are located almost entirely within the cortex or outer medulla. **Juxtamedullary nephrons** have large glomeruli and long loops of Henle that extend deep into the medulla. These nephrons are important in concentrating urine.

   E. Small branches of the **renal artery** deliver blood to **afferent arterioles** that conduct blood to the glomerular capillaries. From the glomerular capillaries, blood flows into an **efferent arteriole** that delivers blood into a second set of capillaries, the **peritubular capil-**

laries that surround the renal tubule. Blood leaves the kidney through the **renal vein.**

   F. Urine formation is accomplished by the **filtration** of plasma, **reabsorption** of needed materials, and **secretion** of a few substances such as potassium and hydrogen ions into the renal tubule.

   1. Plasma filters out of the glomerular capillaries and into Bowman's capsule. The permeable walls of the capillaries and specialized epithelial cells, called **podocytes,** that make up the inner wall of Bowman's capsule, serve as a **filtration membrane.** Filtration is nonselective with regard to small molecules; glucose and other needed materials, as well as metabolic wastes, become part of the filtrate.

   2. About 99% of the filtrate is reabsorbed from the renal tubules into the blood. Reabsorption is a highly selective process that returns usable materials to the blood but leaves wastes and excesses of other substances to be excreted in the urine. The maximum rate at which a substance can be reabsorbed is its **tubular transport maximum (Tm).** Each substance that has a Tm also has a **renal threshold** concentration in the plasma.

   3. In secretion, certain substances and drugs are actively transported into the renal tubule to become part of the urine.

   4. The ability to produce a concentrated urine depends on a high salt and urea concentration in the interstitial fluid of the kidney medulla. The interstitial fluid in the medulla has a salt concentration gradient in which the salt is most concentrated around the bottom of the loop of Henle. This gradient is maintained, in part, by salt reabsorption from various parts of the renal tubule. There is a counterflow of fluid through the two limbs of the loop of Henle; filtrate becomes concentrated as it moves down the descending loop and diluted as it moves up the ascending loop.

   5. Water is drawn by osmosis from the filtrate as it passes through the collecting ducts. This permits the concentration of urine in the collecting ducts.

   6. Some of the water that diffuses from the filtrate into the interstitial fluid is removed by a system of capillaries known as the **vasa recta.**

   7. Urine consists of excess water, nitrogenous wastes, excess salts, and other substances not needed by the body.

   G. Urine volume is regulated by **antidiuretic hormone (ADH)** which is released by the posterior lobe of the pituitary gland in response to an increase in osmotic concentration of the blood (caused by dehydration). ADH increases the permeability of the collecting ducts. As a result, more water is reabsorbed and only a small volume of urine is produced.

   H. **Aldosterone** and **atrial natriuretic peptide (ANP)** regulate sodium reabsorption. When blood pressure decreases, cells of the **juxtaglomerular apparatus** secrete the enzyme **renin,** which activates a pathway leading to production of **angiotensin II,** a hormone that increases release of aldosterone.

POST-TEST

1. The process that maintains homeostasis of body fluids, keeping them from becoming too dilute or too concentrated is called (a) excretion (b) elimination (c) osmoregulation (d) glomerular filtration (e) tubular secretion

2. The main nitrogenous waste product of insects and birds is (a) urea (b) uric acid (c) ammonia (d) carbon dioxide (e) nitrate

3. The main nitrogenous waste product of amphibians and mammals is (a) urea (b) uric acid (c) ammonia (d) carbon dioxide (e) nitrate

4. Osmoconformers (a) maintain an optimal salt concentration despite fluctuations in salt concentration of their surroundings (b) include many animals that inhabit coastal habitats (c) experience variation in concentration of their body fluids along with changes in salinity of the seawater

(d) typically have cells in their gills that remove salts from the surrounding water (e) two of the preceding answers are correct

5. Which of the following is NOT a correct pair (a) protonephridia/flatworm (b) metanephridia/annelid (c) flame cell/flatworm (d) Malpighian tubule/mollusk (e) kidney/vertebrate

6. To compensate for fluid loss, many marine bony fishes (a) accumulate urea (b) have glands that excrete glucose (c) eat a low-protein diet (d) excrete a large volume of hypertonic urine (e) drink sea water

7. Which sequence is accurate? (a) kidney ⟶ ureter ⟶ urinary bladder ⟶ urethra (b) kidney ⟶ urethra ⟶ urinary bladder ⟶ ureter (c) ureter ⟶ kidney ⟶ urinary bladder ⟶ urethra (d) urethra ⟶ kidney ⟶ urinary bladder ⟶ ureter (e) kidney ⟶ collecting duct ⟶ urinary bladder ⟶ urethra

8. Which sequence is accurate? (a) proximal convoluted tubule ⟶ Bowman's capsule ⟶ loop of Henle ⟶ distal convoluted tubule ⟶ collecting duct (b) Bowman's capsule ⟶ proximal convoluted tubule ⟶ loop of Henle ⟶ distal convoluted tubule ⟶ collecting duct (c) Bowman's capsule ⟶ renal papilla ⟶ distal convoluted tubule ⟶ collecting duct (d) distal convoluted tubule ⟶ Bowman's capsule ⟶ loop of Henle ⟶ proximal convoluted tubule ⟶ collecting duct (e) Bowman's capsule ⟶ distal convoluted tubule ⟶ loop of Henle ⟶ proximal convoluted tubule ⟶ collecting duct

9. Which sequence is accurate? (a) renal artery ⟶ efferent arteriole ⟶ capillaries of glomerulus ⟶ afferent arteriole (b) efferent arteriole ⟶ renal artery ⟶ capillaries of glomerulus ⟶ afferent arteriole (c) renal vein ⟶ efferent arteriole ⟶ capillaries of glomerulus ⟶ afferent arteriole (d) renal artery ⟶ efferent arteriole ⟶ peritubular capillaries ⟶ renal vein (e) renal artery ⟶ afferent arteriole ⟶ capillaries of glomerulus ⟶ efferent arteriole

10. Which of the following do NOT contribute to the process of filtration? (a) high hydrostatic pressure in glomerular capillaries (b) large surface area for filtration (c) permeability of glomerular capillaries (d) active transport by epithelial cells lining renal tubule (e) podocytes

11. Tubular transport maximum refers to (a) the maximum concentration of a substance in the plasma that can be reabsorbed by the kidney (b) the most rapid rate at which urine can be transported through the ureter (c) the maximum rate at which a substance can be reabsorbed from the filtrate in the renal tubules (d) the maximum rate at which a substance can pass through the loop of Henle (e) the maximum rate at which a substance can be secreted into the filtrate

12. Which of the following does NOT contribute to the high salt concentration in the interstitial fluid in the medulla of the kidney? (a) reabsorption of salt from various regions of Bowman's capsule (b) diffusion of salt from the ascending portion of the loop of Henle (c) active transport of sodium from the upper part of the ascending loop of Henle (d) counterflow of fluid through the two limbs of the loop of Henle (e) diffusion of urea out of the collecting duct

13. Which of the following is NOT normally present in urine? (a) urea (b) glucose (c) salts (d) water (e) traces of bile pigments

14. Which is NOT true of ADH? (a) released by posterior pituitary (b) increases water reabsorption (c) secretion increases when osmotic pressure in body increases (d) increases urine volume (e) secretion decreases when you drink a lot of water

15. Aldosterone (a) is released by posterior pituitary (b) decreases sodium reabsorption (c) secretion is stimulated by an increase in blood pressure (d) is an enzyme that converts angiotensin into angiotensin II (e) secretion increases in response to angiotensin II

REVIEW QUESTIONS

1. Compare osmoregulation in flatworms with that in insects.
2. What type of osmoregulatory challenge is faced by marine fishes? By freshwater fishes? What adaptations have evolved that meet these challenges?
3. What are the principal types of nitrogenous wastes?
4. Name the structure(s) in the mammalian body associated with each of the following: (a) urea formation (b) urine formation (c) temporary storage of urine (d) conduction of urine out of the body
5. Which part of the nephron is associated with the following? (a) filtration (b) reabsorption (c) secretion
6. Contrast filtration, reabsorption, and secretion.
7. List the sequence of blood vessels through which a drop of blood passes as it is conducted to and from a nephron.
8. How is urine volume regulated? Explain. Why must victims of untreated diabetes insipidus drink great quantities of water?
9. What are the actions of the renin-angiotensin-aldosterone pathway?
10. Describe two treatment strategies for kidney failure. What are the advantages of each?

YOU MAKE THE CONNECTION

1. The number of protonephridia in a planarian is related to the salinity of its environment. Planaria inhabiting slightly salty water develop fewer protonephridia, but the number quickly increases when the concentration of salt in the environment is lowered. Explain why.
2. What types of osmoregulatory challenges do humans experience? Explain. What mechanisms do we have to solve these challenges?
3. Why is glucose normally not present in urine? Why is it present in diabetes mellitus? Why do you suppose diabetics experience an increased output of urine?
4. The kangaroo rat's diet consists of dry seeds, and it drinks no water. Speculate about the adaptations this animal needs in order to survive.

RECOMMENDED READINGS

McClanahan, L.L., R. Ruibal, and V.H. Shoemaker. "Frogs and Toads in Deserts." *Scientific American,* Vol. 270, No. 3, Mar. 1994. A discussion of some unusual adaptations for life in the desert.

Rhoades, R. and R. Pflanzer. *Human Physiology,* 3/ed. Saunders College Publishing, Philadelphia, 1996. Chapters 23 to 25 focus on the kidney and its homeostatic functions.

● Visit our website at **http://www.saunderscollege.com/lifesci/titles.html** and click on Solomon/Berg/Martin Biology for links to chapter-related resources on the World Wide Web.

CHAPTER 47

Endocrine Regulation

A caterpillar becomes a butterfly. A crustacean such as the juvenile Puget Sound king crab *(Lopholithodes mandtii)* in the photograph, changes color to blend with its background. A young girl develops into a woman. An adult copes with chronic stress. These physiological processes and many other adjustments of metabolism, fluid balance, growth, and reproduction are regulated by the **endocrine system.** This system works closely with the nervous system to maintain the steady state of the body. **Endocrinology,** the study of endocrine activity, is a very active and exciting field of biomedical research.

The endocrine system is a diverse collection of cells, tissues, and organs, including specialized **endocrine glands** that produce and secrete **hormones,** chemical messengers responsible for the regulation of many body processes. The term *hormone* is derived from a Greek word meaning "to excite." Hormones excite, or stimulate, changes in specific tissues.

Endocrine cells secrete hormones into the interstitial fluid or blood. Hormones are typically transported by the blood, and produce a characteristic response only after they reach a **target tissue** and bind with specific receptors. Target tissues may be another endocrine gland or an entirely different type of organ, such as a bone or the kidney. Often the target tissue is located far from the endocrine gland. For example, the vertebrate thyroid gland secretes hormones that stimulate metabolism in tissues throughout the body.

Modern endocrinology has its roots in experiments performed by German physiologist A.A. Berthold in the 1840s. Berthold removed the testes from young roosters and observed that their combs (a male secondary sex characteristic) did not grow as large as those in normal roosters. He then transplanted testes into some of the birds and observed that the combs grew back to normal size.

Berthold's methods became a model for subsequent studies in endocrinology and are still used by researchers today. To test the hypothesis that a particular tissue is endocrine, an investigator first removes the tissue. A first question is, does removal of the tissue produce deficiency symptoms in the experimental animal? Investigators then replace the suspected endocrine tissue by transplanting similar tissue from another animal. They often transplant the new tissue to a different location in the body in order to determine whether the effects depend on a signal that moves throughout the body in the blood. As with Berthold's roosters, the changes induced by removing the tissue should be reversed by replacing it.

Investigators extract the suspected compound from the endocrine tissue of one animal and inject it into an experimen-

(Shark Song/M. Kazmers/Dembinsky Photo Associates)

tal animal from which the tissue producing the compound has been removed. Deficiency symptoms should be relieved by replacing the suspected hormone. Researchers then isolate the active compound and determine its chemical structure. Finally, the compound is synthesized in the laboratory and injected into experimental animals. If its effects are those predicted, the researchers have data to support their hypothesis.

Using such procedures, endocrinologists have identified about ten discrete endocrine glands. More recently investigators have identified specialized cells in the digestive tract, heart, kidneys, and many other parts of the body that also release hormones. In addition, some neurons release hormones. As a result of these discoveries, the scope of endocrinology has been broadened to include the production and actions of chemical messengers produced by a wide variety of organs, tissues, and cells. One of the "hottest" current areas of research in endocrinology is the study of the mechanisms of hormone action, which includes characterizing receptors and identifying the molecules involved in signal transduction. In this chapter we will discuss the actions of a variety of hormones and will examine how overproduction or deficiency of various hormones interferes with normal functioning.

AFTER YOU HAVE STUDIED THIS CHAPTER YOU SHOULD BE ABLE TO

1. Describe the sources, general chemical structure, transport, and actions of classical hormones and of related chemical signals.
2. Compare the mechanisms of action of steroid and protein-type hormones; include the role of second messengers, such as cyclic AMP.
3. Summarize the roles of hormones in invertebrates and describe the hormonal regulation of insect development.
4. Identify the principal vertebrate endocrine glands, locate them in the body, and list the hormones secreted by each.
5. Summarize the regulation of endocrine glands by negative feedback mechanisms, giving specific examples.
6. Describe the mechanisms by which the hypothalamus serves as

an important link between nervous and endocrine systems.
7. Compare the functions of the posterior and anterior lobes of the pituitary; describe the actions of their hormones.
8. Describe the actions of growth hormone on growth and metabolism and contrast the consequences of hyposecretion and hypersecretion.
9. Define the actions of the thyroid hormones, their regulation, and the effects of thyroid malfunction.
10. Contrast the actions of insulin and glucagon and describe the disorders associated with the malfunction of the islet cells of the pancreas.
11. Describe the actions of the adrenal glands, including their role in helping the body cope with stress.

CELLS COMMUNICATE BY CHEMICAL SIGNALS

Endocrine glands lack ducts; they differ from **exocrine glands** (such as sweat glands and gastric glands) that release their secretions into ducts. Endocrine glands secrete their hormones into the surrounding interstitial fluid or into the blood. Typically, hormones diffuse into capillaries and are transported by the blood to target tissues. In addition to the discrete classical endocrine glands, specialized cells in many tissues and organs of the body also release hormones or hormone-like substances. Certain neurons, known as **neuroendocrine cells,** are an important link between the nervous and endocrine systems. Neuroendocrine cells produce **neurohormones** that are transported down axons and released into the interstitial fluid. They typically diffuse into capillaries and are transported by the blood (Fig. 47–1). Invertebrate endocrine systems are largely neuroendocrine. In vertebrates, the hypothalamus produces several neurohormones that link the nervous system with the pituitary gland, a very important endocrine gland.

The complexity of animal physiology challenges simplistic definitions. As new chemical signals and their modes of action have been discovered, the old definition of a hormone as a substance secreted by an endocrine gland and transported by the blood has become inadequate. In **autocrine regulation,** a hormone, or other regulator, acts on the very cells that produce it. The female hormone estrogen, which functions as a classical hormone, may also exert an autocrine effect that stimulates additional estrogen secretion. Moreover, certain reproductive hormones (androgens) that are indisputably hormones are typically transported by the blood but, under some conditions, also diffuse through the interstitial fluid and act on nearby target cells. This type of local regulation is called **paracrine regulation.** Other paracrine regulators include local chemical mediators such as histamine, growth factors, and

prostaglandins. Recall from Chapter 43 that **histamine** is stored in mast cells and is released in response to allergic reactions, injury, or infection. Histamine causes blood vessels to dilate and capillaries to become more permeable. **Growth factors** are typically peptides that stimulate cell division and normal development. Growth factors have autocrine or paracrine effects, and their mechanisms of action are similar to those of peptide hormones.

Prostaglandins are modified fatty acids released by many different organs, including the lungs, liver, digestive tract, and certain reproductive organs. Although present in very small quantities, prostaglandins affect a wide range of body processes. They are sometimes referred to as **local hormones** because they are paracrine regulators that act on cells in their immediate vicinity. Prostaglandins modify cyclic AMP levels, and they interact with other hormones to regulate various metabolic activities.

About 16 prostaglandins are known, and they have different actions on different tissues. Some reduce blood pressure, whereas others raise it. Various prostaglandins dilate the bronchial passageways, inhibit gastric secretion, stimulate contraction of the uterus, affect nerve function, cause inflammation, and affect blood clotting. Those synthesized in the temperature-regulating center of the hypothalamus cause fever. In fact, nonsteroidal anti-inflammatory drugs such as aspirin and ibuprofen reduce fever and decrease inflammation by inhibiting prostaglandin synthesis.

Because prostaglandins are involved in the regulation of so many metabolic processes, they have great potential for a variety of clinical uses. At present, prostaglandins are used to induce labor in pregnant women, to induce abortion, and to promote the healing of ulcers in the stomach and duodenum. Their use as a birth control drug is being investigated. Prostaglandins may someday be used to treat a wide variety of illnesses, including asthma, arthritis, kidney disease, certain cardiovascular disorders, and some forms of cancer.

Figure 47–1 **Some types of endocrine signaling.** *(a)* In classical endocrine signaling, endocrine cells release hormones that are transported to target cells by the blood. *(b)* In neuroendocrine signaling, neurons release neurohormones, which may be transported by blood or may diffuse through interstitial fluid. *(c)* In paracrine regulation hormones diffuse through the interstitial fluid and act on nearby target cells. *(d)* In autocrine regulation a hormone acts on the very cells that produce it.

Pheromones, another type of chemical messenger, are produced by animals for communication with other animals of the same species. Because pheromones are generally produced by exocrine glands and do not regulate metabolic activities within the animal that produces them, most biologists do not classify them as hormones. Their role in regulating behavior is discussed in Chapter 50.

HORMONES CAN BE ASSIGNED TO FOUR CHEMICAL GROUPS

Although hormones are chemically diverse, they generally belong to one of four different chemical groups: (1) fatty acid derivatives, (2) steroids, (3) amino acid derivatives, or (4) peptides or proteins (Figs. 47–2).

The juvenile hormones of insects and prostaglandins are **fatty acid derivatives** (Fig. 47–2*a*). Prostaglandins are synthesized from arachidonic acid, a 20-carbon fatty acid. A prostaglandin has a five-carbon ring in its structure.

The molting hormone (also called ecdysone) of insects is a **steroid hormone** (Fig. 47–2*b*). In vertebrates, the adrenal cortex, testis, ovary, and placenta secrete steroid hormones synthesized from cholesterol. An example is cortisol, a hormone that stimulates liver cells to produce glucose from other nutrients. Progesterone, one of the female sex hormones, is the precursor of the steroid hormones produced by the adrenal cortex, as well as of the male hormone testosterone and the female hormone estradiol. Synthetic hormones known as anabolic steroids are sometimes abused by athletes (see *Making the Connection: Anabolic Steroids, Physical Endurance, and Drug Abuse*).

Chemically, the simplest hormones are **amino acid derivatives** (Fig. 47–2*c*). The thyroid hormones (T_3 and T_4) are synthesized from the amino acid tyrosine and iodide. Epinephrine and norepinephrine, hormones produced by the medulla of the adrenal gland, are also derived from tyrosine. Melatonin is synthesized from the amino acid tryptophan.

The water-soluble **peptide hormones** are the largest hormone group. Oxytocin and antidiuretic hormone (ADH), produced by neuroendocrine cells in the hypothalamus, are short peptides composed of nine amino acids (Fig. 47–2*d*). Seven of the amino acids are identical in the two hormones, but the actions of these hormones are quite different. The hormones glucagon, secretin, adrenocorticotropic hormone (ACTH),

Figure 47-2 Major chemical groups of hormones. *(a)* The juvenile hormone of insects and prostaglandins are derived from fatty acids. *(b)* The molting hormone of insects, hormones produced by the mammalian adrenal cortex, such as cortisol, and reproductive hormones, such as estradiol, are steroid hormones. *(c)* Norepinephrine and epinephrine produced by the adrenal medulla, and the thyroid hormones are derived from amino acids. Note the presence of iodine in the thyroid hormones. *(d)* Two peptide hormones are produced in the hypothalamus and secreted by the posterior lobe of the pituitary gland. Oxytocin and antidiuretic hormone are both small peptides containing nine amino acids. Note that the structures of these hormones differ by only two amino acids.

and calcitonin are somewhat longer peptides consisting of about 30 amino acids. Insulin, secreted by the islets of Langerhans in the pancreas, is a small protein consisting of two peptide chains joined by disulfide bonds (see Fig. 3–18). Growth hormone, thyroid-stimulating hormone, and the gonadotropic hormones, all secreted by the anterior lobe of the pituitary gland, are large proteins with molecular weights of 25,000 or more.

ANABOLIC STEROIDS, PHYSICAL ENDURANCE, AND DRUG ABUSE

Why do an estimated one million individuals in the United States, one-half of them adolescents, abuse a group of synthetic hormones known as anabolic steroids? These synthetic androgens (male reproductive hormones) were developed in the 1930s to prevent muscle atrophy in patients with diseases that prevented them from moving about. In the 1950s, anabolic steroids became popular with professional athletes, who used them to increase muscle mass, physical strength, endurance, and aggressiveness. In truth, their athletic performance was probably enhanced, at least in part, by drug-induced euphoria and increased enthusiasm for training.

As with other hormones, the concentration of steroids circulating in the body is precisely regulated. Use of anabolic steroids interferes with normal physiological processes. Even during short-term use and at relatively low doses, anabolic steroids have a significant effect on mood and behavior. At higher doses, users experience disturbed thought processes, forgetfulness, confusion, and often find themselves easily distracted. "Steroid rage" refers to the mood swings, unpredictable rage, increased aggressiveness, and irrational behavior exhibited by many users. Anabolic steroids damage the liver and increase LDL concentration, raising the risk of cardiovascular disease. In adolescents, anabolic hormones stunt growth by prematurely closing the growth plates in bones. Abuse of these hormones can shrink the testes, leading to sterility. These drugs remain in the body for a long time. Their metabolites (breakdown products) can be detected in the urine for up to six months.

When their serious side effects became known in the 1960s, anabolic steroid use became controversial, and in 1973 the Olympic Committee banned their use. They are now prohibited worldwide by amateur and professional sports organizations. However, according to the U.S. Drug Enforcement Administration, a multimillion-dollar black market exists for these synthetic hormones. They are both injected and taken in pill form. The typical anabolic steroid user is a male (95%) athlete (65%), most often a football player, weight lifter, or wrestler. Surprisingly, though, about 10% of male high school students have used anabolic steroids and about one-third of these students are not even on a high school team. They use the hormone only to change their physical appearance—to pump up their muscles ("bulk up")—and increase endurance. Many anabolic steroid users have difficulty realistically perceiving their body images. They remain unhappy even after dramatic increases in muscle mass.

HORMONE SECRETION IS REGULATED BY NEGATIVE FEEDBACK MECHANISMS

Although present in minute amounts, more than 50 different hormones may be circulating in the blood of vertebrates at any time. Steroid hormones and thyroid hormones are transported bound to plasma proteins or to specific carrier proteins. Most peptide hormones circulate without transport proteins. Hormone molecules are continuously removed from the circulation by target tissues. They are also removed by the liver, which inactivates some hormones, and by the kidneys, which excrete them.

Hormone secretion is typically regulated by **negative feedback mechanisms.** Information about the amount of hormone or of some other substance in the blood or interstitial fluid is fed back to the gland, which then responds to restore homeostasis. The parathyroid glands, located in the neck of tetrapod vertebrates, provide a good example of how negative feedback works.

The parathyroid glands regulate the calcium concentration of the blood. When the calcium concentration is not within homeostatic limits, nerves and muscles cannot function properly. For example, when insufficient amounts of calcium ions are present, neurons can fire spontaneously, causing muscle spasms. When calcium concentration varies too far from the steady state (either too high or too low), negative feedback mechanisms bring the condition back to the steady state. A decrease in the calcium concentration in the plasma signals the parathyroid glands to release more parathyroid hormone (Fig. 47–3). This hormone increases the concentration of calcium in the blood. (The details of parathyroid action are discussed later in this chapter.) When the calcium concentration rises above normal limits, the parathyroid glands slow their output of hormone. Both responses are negative feedback mechanisms. An increase in calcium concentration results in decreased release of parathyroid hormone, whereas a decrease in calcium leads to increased hormone secretion. In each case, the response counteracts the inappropriate change, thus restoring the steady state.

HORMONES COMBINE WITH SPECIFIC RECEPTORS IN TARGET CELLS

A hormone may pass through many tissues seemingly "unnoticed" until it reaches its target tissue. How does the target tissue recognize a hormone? Specific receptors in or on the cells of the target tissues recognize the hormone and bind with it. Researchers have determined that hormone receptors are large proteins or glycoproteins. The receptor site is similar to a lock, and the hormones are similar to different keys. Only the hormone that fits the lock, the specific receptor, can influence the metabolic machinery of the cell. Receptors are responsible for the specificity of the endocrine system.

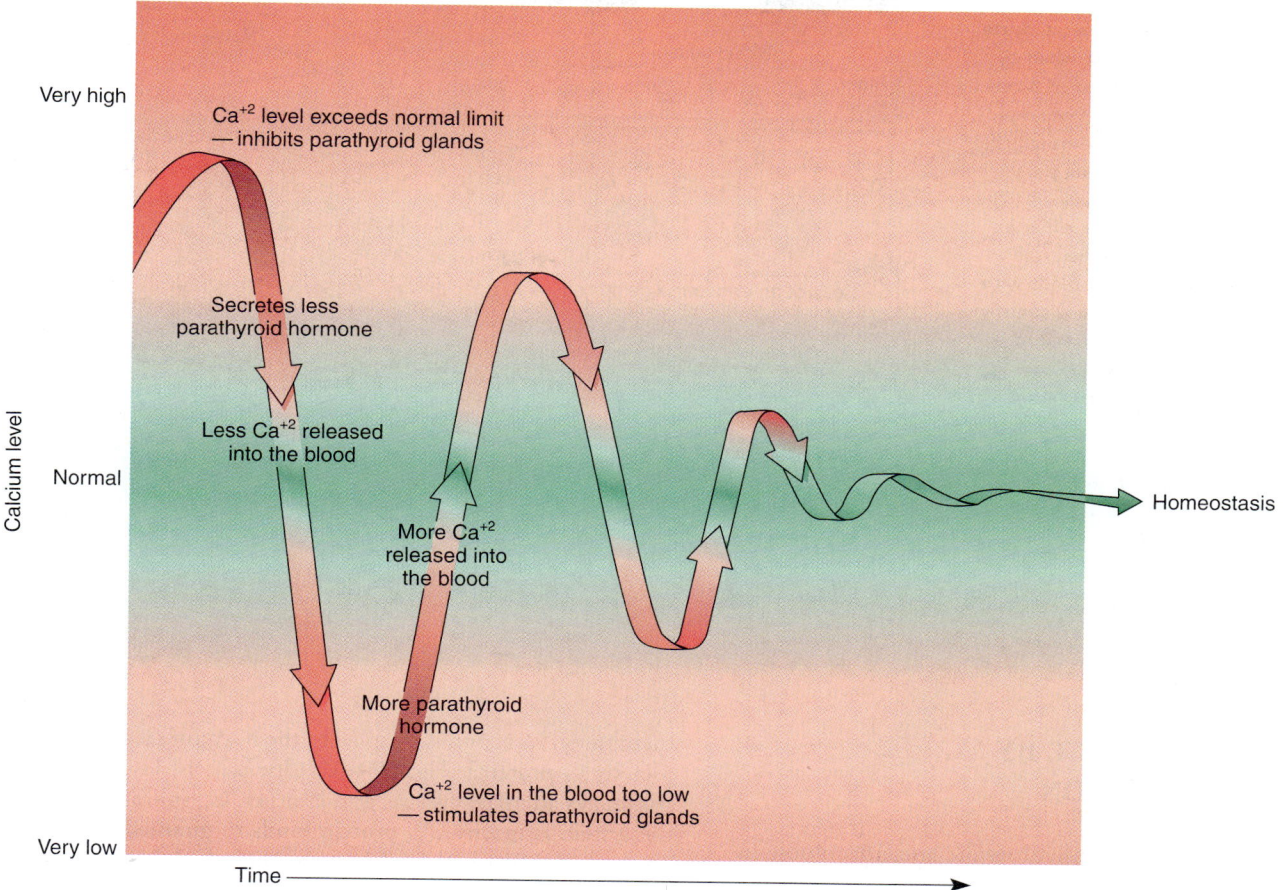

Figure 47–3 **Regulation by negative feedback.** Should calcium exceed its normal concentration in the blood, the parathyroid glands are inhibited and slow their release of parathyroid hormone. When the calcium concentration in the blood falls below normal levels, the parathyroid glands are stimulated to release more parathyroid hormone. This hormone acts on target tissues that increase the calcium level in the blood, thus restoring homeostasis.

Receptors are continuously synthesized and degraded. They can increase or decrease their numbers by **receptor up-regulation** and **receptor down-regulation.** Some hormones *up-regulate,* or stimulate synthesis of their own receptors, thus amplifying their effect on the cell. Other hormones *down-regulate,* or decrease the number of their own receptors. This mechanism suppresses the sensitivity of target cells to the hormone.

Many hormones act by **signal transduction** (Chapter 5). The transmembrane receptors act as signal transducers that convert an extracellular hormone signal into an intracellular signal that affects some intracellular process. The hormone, which serves as the first messenger, may relay information to a **second messenger,** or intracellular signal. The second messenger then relays information to effector molecules that carry out the action. Some hormones activate a series of molecular events that comprise a *signaling cascade.*

Several types of hormones may be involved in regulating the metabolic activities of a particular type of cell. In fact, many hormones produce a synergistic effect in which the presence of one hormone enhances the effects of another.

SOME HORMONES ENTER THE CELL AND ACTIVATE GENES

Steroid hormones and thyroid hormones are relatively small, lipid-soluble (hydrophobic) molecules that easily pass through the plasma membrane of the target cell. Specific protein receptors in the cytoplasm or in the nucleus bind with the hormone to form a hormone-receptor complex (Fig. 47–4). This complex combines with specific acceptor sites on the nuclear DNA. Their interaction turns specific genes on or off. The hormone-receptor complex either activates or represses transcription of messenger RNA molecules coding for specific proteins. These proteins produce the changes in structure or metabolic activity responsible for the effect of the hormone.

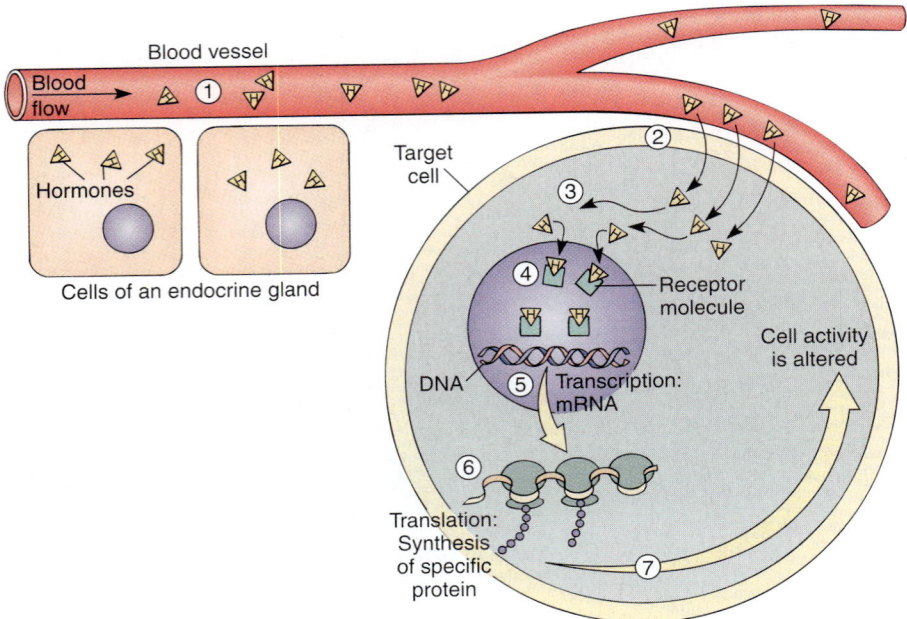

Figure 47–4 Steroid hormones activate genes. (1) Steroid hormones are secreted by an endocrine gland and transported to a target cell. (2) Steroid hormones are small, lipid-soluble molecules that pass freely through the plasma membrane. (3) The hormone passes through the cytoplasm to the nucleus. (4) The hormone combines with a receptor either in the cytoplasm or in the nucleus. Then, the steroid hormone-receptor complex binds with DNA. (5) This activates (or represses) specific genes, leading to mRNA transcription and (6) synthesis of specific proteins. The proteins cause the response recognized as the hormone's action (7).

Labels in figure: Blood vessel; Blood flow; ① Hormones; Cells of an endocrine gland; Target cell; ②; ③; ④ Receptor molecule; DNA; ⑤ Transcription: mRNA; Cell activity is altered; ⑥ Translation: Synthesis of specific protein; ⑦

MANY HORMONES ACT BY SIGNAL TRANSDUCTION

Because peptide hormones are hydrophilic and not soluble in the lipid layer of the plasma membrane, they do not enter the target cell. Instead, they initiate signal transduction. They bind with receptors on the plasma membrane, and the extracellular signal is converted into intracellular signals that affect the target cell (Fig. 47–5).

Two main types of cell surface hormone receptors are known: **enzyme-linked receptors** and **G-protein-linked receptors.** Most enzyme-linked receptors are transmembrane proteins with a hormone-binding site outside the cell and an enzyme site (typically a protein kinase) inside the cell. G-protein-linked receptors are transmembrane proteins that loop back and forth through the plasma membrane seven times. These receptors activate **G proteins,** a group of integral regulatory proteins. The G indicates that they bind **guanosine triphosphate (GTP),** which, like ATP, is an important molecule in energy reactions. G proteins are shuttles that transmit a signal between the receptor and a second messenger.

Cyclic AMP is the most common second messenger in animal cells

In the 1960s, Earl Sutherland identified **cyclic AMP (cAMP)** as a second messenger, and for his pioneering work he received a Nobel prize in 1971. When certain hormones bind with receptors on the extracellular surface of the plasma membrane, a membrane-bound enzyme, **adenylyl cyclase,** is activated. Adenylyl cyclase catalyzes the conversion of ATP to cAMP. This enzyme is located on the cytoplasmic side of the plasma

membrane. Its action is coupled to the hormone-receptor complex by G proteins (Fig. 47–6). One type of G protein, G_s, stimulates adenylyl cyclase, and another, G_i, inhibits it. When the system is inactive, G protein binds to **guanosine diphosphate,** or GDP, which is similar to ADP, the hydrolyzed form of ATP.

Most hormones bind to a stimulatory receptor, resulting in activation of adenylyl cyclase. After the receptor-hormone complex is formed, the G protein releases GDP and then binds to GTP. This binding produces a conformational change in the G protein that enables it to bind with and activate adenylyl cyclase. By coupling the hormone-receptor complex to

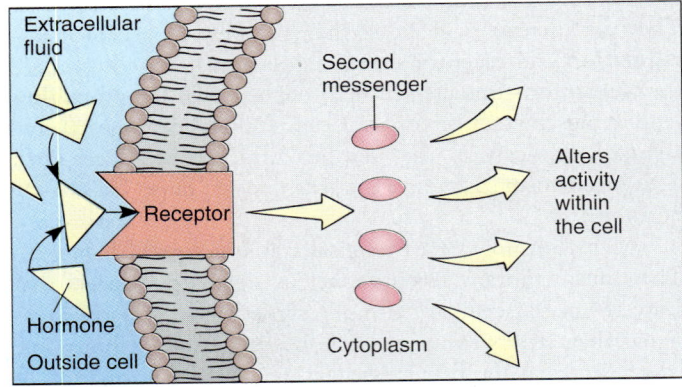

Labels in figure: Extracellular fluid; Second messenger; Receptor; Alters activity within the cell; Hormone; Outside cell; Cytoplasm

Figure 47–5 Action of hormones that have membrane receptors. Peptide hormones bind with receptors on the plasma membrane of a target cell. The receptor is a signal transducer that converts the hormone signal to an intracellular signal. The signal may be relayed by a second messenger. Typically, a sequence of several signaling molecules relays the message.

(a)

(b)

Figure 47-6 **Role of G proteins.** G proteins couple adenylyl cyclase activity with membrane receptors. *(a)* When a hormone binds to a receptor on the plasma membrane, the hormone-receptor complex binds to a G protein. *(b)* GDP on the G protein is replaced by GTP. The G protein undergoes a conformational change (change in shape), allowing it to bind with adenylyl cyclase. G_s activates adenylyl cyclase, which catalyzes the conversion of ATP to cAMP. G_i *(not shown)* inhibits adenylyl cyclase.

an enzyme that generates a signal, many second messenger molecules can be rapidly produced and the actions of the hormone are amplified.

Once activated, adenylyl cyclase catalyzes the conversion of ATP to cyclic AMP (Fig. 47–7). Cyclic AMP then activates one or more enzymes known as **protein kinases.** Each type of protein kinase phosphorylates (adds a phosphate group to) a specific protein. When the protein is phosphorylated, its function is altered, and it triggers a chain of reactions leading to some metabolic effect. Any increase in cAMP is temporary. Cyclic AMP is rapidly inactivated by enzymes known as **phosphodiesterases,** which convert it to AMP. Thus, the concentration of cAMP depends on the activity of both adenylyl cyclase that produces it and phosphodiesterase that breaks it down.

Protein kinases can inhibit or activate certain cellular proteins. Each protein kinase acts on a different type of protein. Because different types of protein kinases exist in each kind of target cell (and even within different organelles of the same target cell), protein kinases are able to produce a wide variety of responses. For example, activation of one type of protein kinase may have a metabolic effect on the cell, a second protein kinase may affect membrane permeability, and a third may activate genes (Fig. 47–8).

Calcium ions can act as second messengers

Certain receptors are linked to gated ion channels, for example calcium channels. When a hormone binds to the receptor, the calcium channel opens, allowing calcium ions to move into the cell. A G protein may be involved in coupling the hormone-receptor complex to the opening of the calcium channel (Fig. 47–9). The hormone-receptor complex can also in-

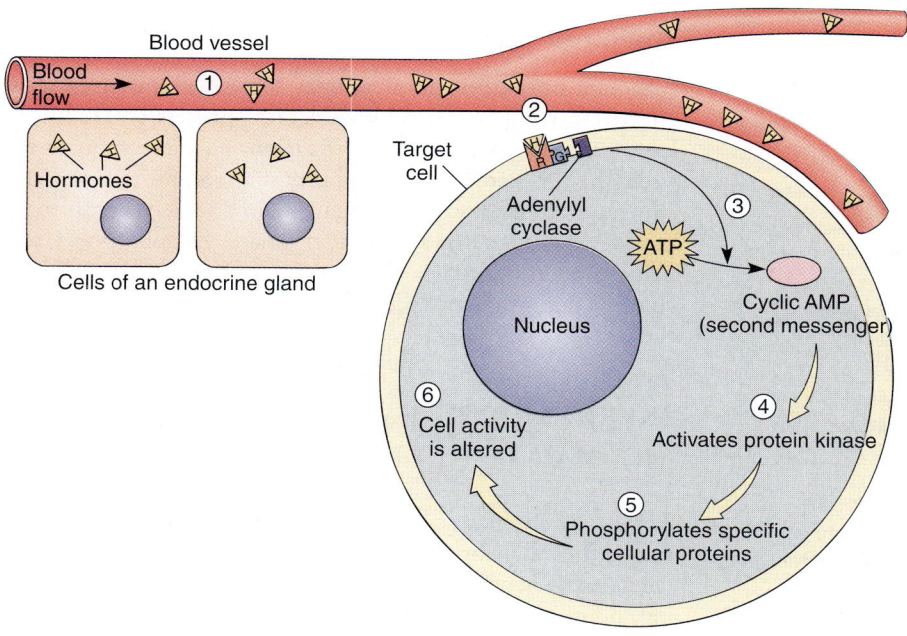

Figure 47-7 **Mechanism of action of hormones that work through second messengers.** (1) Peptide hormones are transported in the blood. (2) They bind with receptors in the plasma membrane of a target cell. (3) The hormone-receptor combination is coupled by a G protein to adenylyl cyclase. The adenylyl cyclase catalyzes the conversion of ATP to cyclic AMP, a second messenger. (4) Cyclic AMP then activates one or more enzymes that (5) change proteins. (6) These proteins alter the activity of the cell in some way.

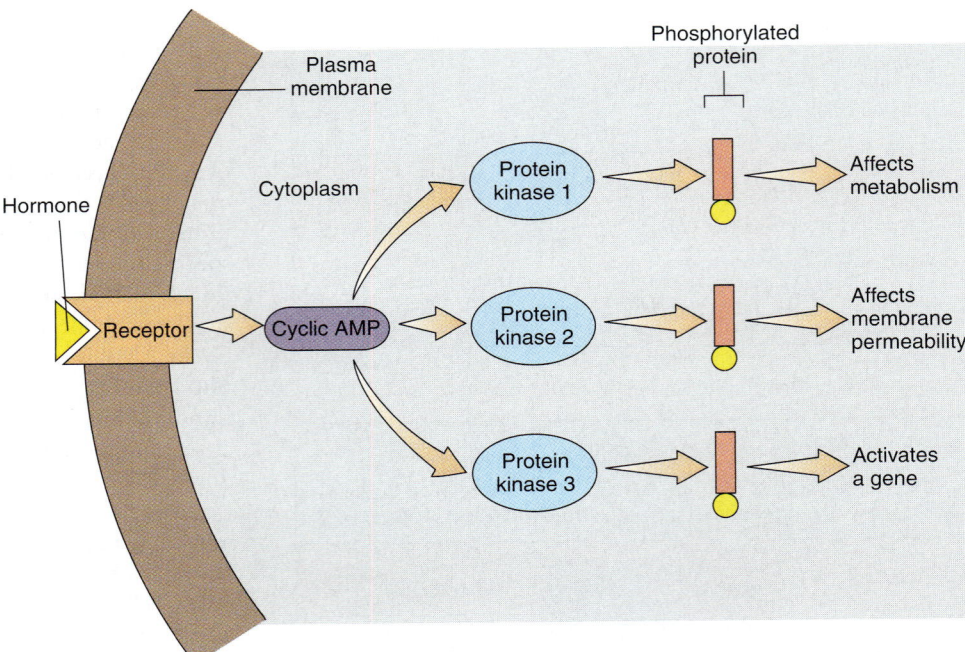

Figure 47–8 Effects of protein kinases. When hormone action increases the amount of cyclic AMP in the cell, protein kinases are activated. Protein kinases phosphorylate proteins, leading to a variety of effects.

crease the cellular concentration of calcium by releasing calcium ions stored in the endoplasmic reticulum. Once the calcium concentration in the cell is increased, calcium ions bind to **calmodulin,** a protein found in all eukaryotic cells that have been examined to date. The calmodulin changes shape and can then activate certain enzymes, including kinases. Cellular processes regulated by calmodulin include membrane phosphorylation, neurotransmitter release, and microtubule disassembly.

Figure 47–9 Calcium ions as second messengers. Certain receptors are linked by a G protein to gated calcium ion channels. When the gate opens, Ca^{2+} enters the cell and binds to calmodulin. The activated calmodulin activates kinases, enzymes that produce changes in the cell.

Phospholipid products can act as second messengers

Certain hormone-receptor complexes (e.g., antidiuretic hormone) activate a G protein that then activates the membrane-bound enzyme phospholipase C (Fig. 47–10). This enzyme splits a membrane protein (phosphotidylinositol 4,5-bisphosphate, or PIP_2) into two products, **inositol triphosphate (IP_3)** and **diacylglycerol (DAG),** that act as second messengers. IP_3 stimulates the endoplasmic reticulum to release calcium ions that then act as a third messenger. The calcium ions bind to calmodulin, and together with DAG they activate the enzyme protein kinase C, which phosphorylates specific proteins.

Hormone signals are amplified

Although hormones are present in very small amounts, they effectively regulate many physiological processes. This is in large part due to **signal amplification,** an increase in signal strength. For example, a single hormone-receptor complex can stimulate the production of many cAMP molecules. In turn, each cAMP can activate a protein kinase that phosphorylates many protein molecules. In this way, a single hormone molecule can activate many proteins.

INVERTEBRATE ENDOCRINE SYSTEMS ARE MAINLY NEUROENDOCRINE

Among invertebrates, hormones are secreted mainly by neurons rather than by endocrine glands. These neurohormones regulate regeneration in hydras, flatworms, and annelids; molting and metamorphosis in insects; color changes in crustaceans;

Figure 47–10 **Phospholipid products as second messengers.** Certain hormone-receptor complexes activate G proteins that activate phospholipase C. This enzyme splits a membrane lipid (PIP_2) forming two second messengers, inositol triphosphate (IP_3) and diacylglycerol (DAG). IP_3 stimulates the release of Ca^{2+} from the endoplasmic reticulum (ER). Ca^{2+} acts as a third messenger. Ions bind with calmodulin, and together with DAG they activate protein kinase C. The protein kinase phosphorylates certain proteins, leading to various changes in the cell.

and growth, gamete production, reproductive behavior, and metabolic rate in other groups. Trends in the evolution of invertebrate endocrine systems include a larger number of both neurohormones and hormones secreted by endocrine glands, and a greater role of hormones in physiological processes.

Color change in crustaceans is regulated by hormones

Crustaceans possess true (nonneural) endocrine glands, as well as masses of neuroendocrine cells. Hormonal regulation in crustaceans is complex and affects many activities, including molting, reproduction, heart rate, and metabolism. One of the most interesting and novel activities regulated by hormones is color change (see chapter opening photograph).

Pigment cells of crustaceans are located in the integument beneath the exoskeleton. Pigment may be black, yellow, red, white, or even blue. Color changes are produced by changes in the distribution of pigment granules within the cells. Distribution of pigment is regulated by neurohormones produced by neuroendocrine cells. When the pigment is concentrated near the center of a cell, its color is only minimally visible. However, when it is dispersed throughout the cell, the color shows to advantage. These pigments provide protective coloration. By appropriate concentration or dispersal of specific types of pigments, a crustacean can approximate the color of its background.

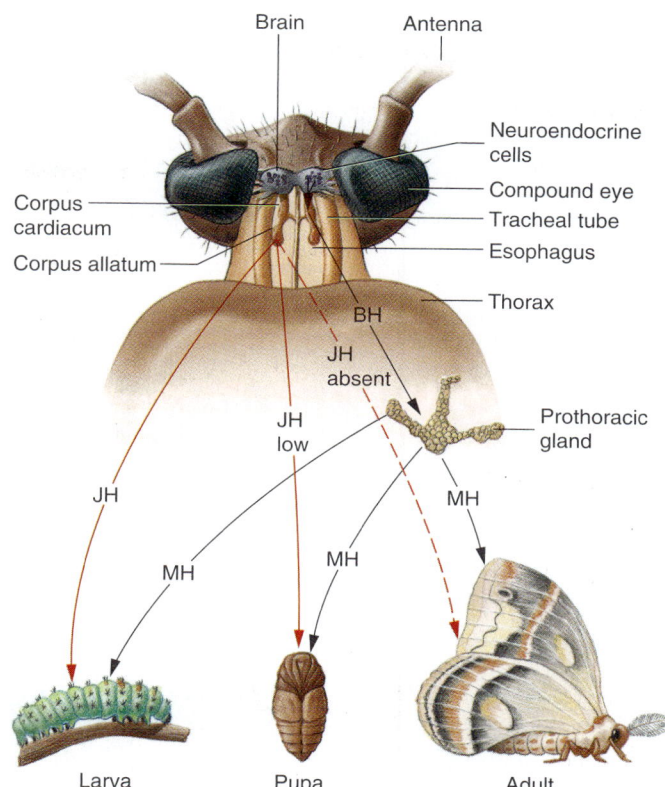

Figure 47–11 **Regulation of growth and molting in insects.** The dissection of an insect head shows the paired corpora allata and corpora cardiaca, as well as the brain. Neuroendocrine cells in the brain secrete the neurohormone BH, which is stored in the corpora cardiaca. When released, BH stimulates the prothoracic glands to secrete molting hormone (MH), which stimulates growth and molting. In the immature insect, the corpora allata secrete juvenile hormone (JH), which suppresses metamorphosis at each larval molt. Metamorphosis to the adult form occurs when molting hormone acts in the absence of juvenile hormone.

Insect development is regulated by hormones

Like crustaceans, insects have endocrine glands as well as neuroendocrine cells. The various hormones interact with one another to regulate reproduction, metabolism, growth, and development, including molting and morphogenesis.

Hormonal control of development in insects is complex and varies among the many species. Generally, some environmental factor (e.g., temperature change) affects neuroendocrine cells in the brain. When activated, these cells produce a neurohormone referred to as **brain hormone** (**BH,** or ecdysiotropin), which is transported down axons and stored in the paired **corpora cardiaca** (sing, *corpus cardiacum;* Fig. 47–11). When released from the corpora cardiaca, BH stimulates the **prothoracic glands,** endocrine glands in the prothorax, to produce **molting hormone (MH),** also called **ecdysone.** Molting hormone stimulates growth and molting.

TABLE 47–1   Some Endocrine Glands and Their Hormones*

Endocrine Gland and Hormone	Target Tissue	Principal Actions
Hypothalamus		
Releasing and inhibiting hormones	Anterior lobe of pituitary	Regulate secretion of hormones by the anterior pituitary
Hypothalamus (production) **Posterior lobe of pituitary** (storage and release)		
Oxytocin	Uterus	Stimulates contraction
	Mammary glands	Stimulates ejection of milk into ducts
Antidiuretic hormone (ADH)	Kidneys (collecting ducts)	Stimulates reabsorption of water; conserves water
Anterior lobe of pituitary		
Growth hormone (GH)	General	Stimulates production of insulin-like growth factors (IGFs); stimulates growth by promoting protein synthesis
Prolactin	Mammary glands	Stimulates milk production
Thyroid-stimulating hormone (TSH)	Thyroid gland	Stimulates secretion of thyroid hormones; stimulates increase in size of thyroid gland
Adrenocorticotropic hormone (ACTH)	Adrenal cortex	Stimulates secretion of adrenal cortical hormones
Gonadotropic hormones* (follicle-stimulating hormone, FSH; luteinizing hormone, LH)	Gonads	Stimulate gonad function and growth
Thyroid gland		
Thyroxine (T_4) and triiodothyronine (T_3)	General	Stimulate metabolic rate; essential to normal growth and development
Calcitonin	Bone	Lowers blood-calcium level by inhibiting Ca^{2+} release from bone
Parathyroid glands		
Parathyroid hormone	Bone, kidneys, digestive tract	Increases blood-calcium level by stimulating Ca^{2+} release from bone; stimulates calcium reabsorption by kidneys; activates vitamin D, which increases intestinal absorption of calcium

*The gonadotropic hormones (FSH and LH) and the ovaries and testes and their hormones are discussed in Chapter 48.

In the immature insect, paired endocrine glands called **corpora allata** (sing, *corpus allatum*) secrete **juvenile hormone (JH).** This hormone suppresses metamorphosis at each larval molt so that the insect increases in size while remaining in its immature state; after the molt, the insect is still in a larval stage. When the concentration of juvenile hormone decreases, metamorphosis occurs, and the insect is transformed into a pupa (see Chapter 29). In the absence of juvenile hormone, the pupa molts and becomes an adult. The secretory activity of the corpora allata is regulated by the nervous system, and the amount of juvenile hormone decreases with successive molts.

THE HYPOTHALAMUS LINKS THE VERTEBRATE NERVOUS AND ENDOCRINE SYSTEMS

Vertebrate hormones regulate such diverse activities as growth, development, reproduction, metabolic rate, fluid balance, blood homeostasis, and coping with stress. Most vertebrates have similar endocrine glands, though the actions of some hormones may be different in various groups. Table 47–1 gives the sources, target tissues, and physiological actions of some of the major vertebrate hormones. Many of these hormones

TABLE 47-1 continued

Endocrine Gland and Hormone	Target Tissue	Principal Actions
Islets of Langerhans of pancreas		
Insulin	General	Lowers glucose concentration in the blood by facilitating glucose uptake and utilization by cells; stimulates glycogen production; stimulates fat storage and protein synthesis
Glucagon	Liver, adipose tissue	Raises glucose concentration in the blood; stimulates glycogen breakdown; mobilizes fat
Adrenal medulla		
Epinephrine and norepinephrine	Muscle; cardiac muscle; blood vessels; liver; adipose tissue	Help body cope with stress; increase heart rate, blood pressure, metabolic rate; reroute blood; mobilize fat; raise blood-sugar level
Adrenal cortex		
Mineralocorticoids (aldosterone)	Kidney tubules	Maintain sodium and potassium balance; increase sodium reabsorption; increase potassium excretion
Glucocorticoids (cortisol)	General	Help body cope with long-term stress; raise blood-glucose level; mobilize fat
Pineal gland		
Melatonin	Hypothalamus	Important in biological rhythms; influences reproductive processes in hamsters and other animals; pigmentation in some vertebrates; may help control onset of puberty in humans
Ovary[†]		
Estrogens (estradiol)	General; uterus	Develop and maintain sex characteristics in female; stimulate growth of uterine lining
Progesterone	Uterus; breast	Stimulates development of uterine lining
Testis[†]		
Testosterone	General; reproductive structures	Develops and maintains sex characteristics of males; promotes spermatogenesis; responsible for adolescent growth spurt
Inhibin	Anterior lobe of pituitary	Inhibits FSH release in male

[†]For more detailed description see Table 48–1 and Table 48–2.

are regulated by the hypothalamus and pituitary gland. The principal human endocrine glands are illustrated in Figure 47–12.

Homeostasis depends on normal concentrations of hormones

When a disorder or disease process affects an endocrine gland, the rate of secretion may become abnormal. If **hyposecretion** (abnormally reduced output) occurs, target cells are deprived of needed stimulation. If **hypersecretion** (abnormally increased output) occurs, the target cells may be overstimulated.

In some endocrine disorders, an appropriate amount of hormone is secreted, but the target cell receptors do not function properly. As a result, the target cells may not be able to respond to the hormone. Any of these abnormalities leads to loss of homeostasis, resulting in predictable metabolic malfunctions and clinical symptoms (Table 47–2).

The hypothalamus regulates the pituitary gland

Most endocrine activity is controlled directly or indirectly by the hypothalamus, which links the nervous and endocrine

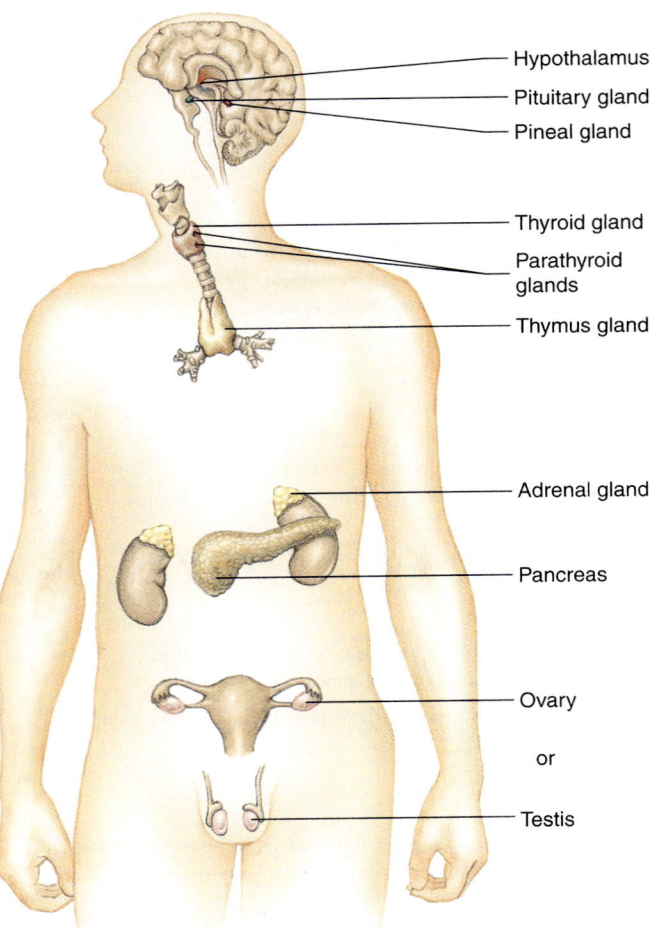

Hypothalamus
Pituitary gland
Pineal gland

Thyroid gland
Parathyroid glands
Thymus gland

Adrenal gland

Pancreas

Ovary

or

Testis

Figure 47–12 **The classical human endocrine glands.** The hormones produced by these glands and their actions are discussed in this chapter.

systems both anatomically and physiologically. The pituitary gland is connected to the hypothalamus by the pituitary stalk. In response to input from other areas of the brain and from hormones in the blood, neurons of the hypothalamus secrete neurohormones that regulate specific physiological processes. The hypothalamus also produces releasing hormones and inhibiting hormones that regulate the release of hormones by the pituitary gland.

Because its secretions control the activities of several other endocrine glands, the **pituitary gland** is known as the master gland of the body. Although the pituitary gland is only the size of a large pea and weighs only about 0.5 g (0.02 oz), it produces and secretes at least seven peptide hormones that exert far-reaching influence over body activities. The pituitary gland consists of two main lobes, the **anterior lobe** and the **posterior lobe.** In some vertebrates, an intermediate lobe secretes hormones that regulate skin color. The role, if any, of the intermediate lobe in humans is not known.

The posterior lobe of the pituitary gland releases hormones produced by the hypothalamus

The posterior lobe of the pituitary gland, or neurohypophysis, develops from brain tissue. This neuroendocrine organ *secretes* two peptide hormones, **oxytocin** and **antidiuretic hormone** (**ADH;** see Chapter 46). These hormones are actually *produced* by neuroendocrine cells in two distinct areas of the hypothalamus. Enclosed within vesicles, they are transported down axons that extend into the posterior lobe (Fig. 47–13). The vesicles are stored in these axon terminals until the neuron is stimulated. Then the hormones are released and diffuse into surrounding capillaries.

Oxytocin levels rise toward the end of pregnancy, stimulating the strong contractions of the uterus needed to expel a baby. Oxytocin is sometimes administered clinically (under the trade name Pitocin) to initiate or speed labor. After birth, when an infant sucks at its mother's breast, sensory neurons signal the pituitary to release oxytocin. The hormone stimulates contraction of cells surrounding the milk glands so that milk is let down into the ducts, from which it can be sucked by the infant. Because oxytocin also stimulates the uterus to contract, breast feeding promotes rapid recovery of the uterus to nonpregnant size.

The anterior lobe of the pituitary gland regulates growth and other endocrine glands

The anterior lobe of the pituitary develops from epithelial, rather than neural, cells. The anterior lobe functions like a classical endocrine gland in that it receives signals by way of the blood and releases its hormones into the blood. It secretes prolactin, growth hormone, and several **tropic hormones,** which stimulate other endocrine glands (Fig. 47–14). **Prolactin** stimulates the cells of the mammary glands to produce milk in a nursing mother.

The hypothalamus secretes several **releasing hormones** and **inhibiting hormones** that regulate the anterior lobe of the pituitary gland. These neurohormones enter capillaries and pass through special portal veins that connect the hypothalamus with the anterior lobe of the pituitary (Fig. 47–14). (These portal veins, like the hepatic portal vein, do not deliver blood to a larger vein directly but connect two sets of capillaries.) Within the anterior lobe of the pituitary, the portal veins divide into a second set of capillaries. The releasing and inhibiting hormones pass through the walls of these capillaries into the tissue of the anterior lobe, where each regulates the production and secretion of specific pituitary hormones.

Growth hormone stimulates protein synthesis

Small children measure themselves periodically against their parents, eagerly awaiting that time when they, too, will be "big." Whether one will be tall or short depends on many factors, including genes, diet, and hormonal balance.

TABLE 47-2 Consequences of Endocrine Malfunction

Hormone	Hyposecretion	Hypersecretion
Growth hormone	Pituitary dwarfism	Gigantism if malfunction occurs in childhood; acromegaly in adult
Thyroid hormones	Cretinism (in children); myxedema, a condition of pronounced adult hypothyroidism (metabolic rate is reduced by about 40%; patient feels tired and may be mentally slow); dietary iodine deficiency can lead to hyposecretion and goiter (abnormal enlargement of the thyroid gland; see figure)	Hyperthyroidism; increased metabolic rate, nervousness, irritability; goiter; can be caused by Grave's disease.
Parathyroid hormone	Spontaneous discharge of nerves; spasms; tetany; death	Weak, brittle bones; kidney stones
Insulin	Diabetes mellitus	Hypoglycemia
Hormones of adrenal cortex	Addison's disease (inability to cope with stress; sodium loss in urine may lead to shock)	Cushing's disease (edema gives face a full-moon appearance; fat is deposited about trunk; blood-glucose level rises; immune responses are depressed)

Goiter commonly results from iodine deficiency. (John Paul Kay/Peter Arnold, Inc.)

Growth hormone (**GH;** also called somatotropin) is referred to as an **anabolic hormone** because it promotes tissue growth. Many of the effects of GH on skeletal growth are indirect. GH stimulates the liver to produce peptides called **somatomedins,** including **insulin-like growth factors (IGFs).** These growth factors (1) promote the linear growth of the skeleton by stimulating growth of cartilage in growth areas of the bones, and (2) stimulate general tissue growth and increase in size of organs by promoting protein synthesis and other anabolic processes.

Growth hormone stimulates uptake of amino acids by the cells and promotes protein synthesis. GH also affects lipid and carbohydrate metabolism. In adipose tissue, GH decreases glucose uptake and promotes mobilization of fat, raising the level of free fatty acids in the blood. In this protein-sparing process, fatty acids become available for cells to use as fuel. How does this help to promote growth? Fat mobilization by GH is also important during fasting or periods of prolonged stress, when the blood-sugar level is low. Can you explain why?

Growth is affected by many factors

In adults as well as in growing children, GH is secreted in pulses throughout the day. Secretion of GH is regulated by both a **growth hormone–releasing hormone (GHRH)** and a **growth hormone–inhibiting hormone (GHIH,** also called

somatostatin) released by the hypothalamus. A high level of GH in the blood signals the hypothalamus to secrete the inhibiting hormone, and the pituitary release of GH slows. A low level of GH in the blood stimulates the hypothalamus to secrete the releasing hormone, which in turn stimulates the pituitary gland to release more GH. Many other factors, including blood-sugar level, decrease in amino acid concentration, and stress, influence GH secretion.

The age-old notions that children need plenty of sleep, a proper diet, and regular exercise in order to grow are supported by research. Secretion of GH increases during exercise, probably because rapid metabolism by muscle cells lowers the blood-sugar level. GH is secreted about 1 hour after the onset of deep sleep, and in a series of pulses 2 to 4 hours after a meal.

Emotional support is also necessary for proper growth. Growth may be retarded in children who are deprived of cuddling, playing, and other forms of nurture, even when their needs for food and shelter are met. In extreme cases, childhood stress can produce a form of retarded development known as psychosocial dwarfism. Some emotionally deprived children exhibit abnormal sleep patterns, which may be the basis for decreased secretion of GH.

Other hormones also influence growth. Thyroid hormones appear to be necessary for normal GH secretion and function and for normal tissue response to IGFs. Sex hormones

Figure 47–13 The relationship between the hypothalamus and posterior lobe of the pituitary gland. The hormones secreted by the posterior lobe of the pituitary are manufactured in neuroendocrine cells of the hypothalamus. The axons of these neurons extend down into the posterior lobe of the pituitary. Their hormones are packaged in vesicles that are transported through these axons and stored in their ends. When needed, the hormones are secreted into the blood and transported by the circulatory system.

Neuroendocrine cells

Brain

Skull

Hypothalamus

Anterior lobe of pituitary gland

Posterior lobe of pituitary gland

Pituitary stalk

Axons

Posterior lobe

Capillary

Vesicles containing hormone

Anterior lobe

Hormones

Antidiuretic hormone (ADH)

Oxytocin

Kidney tubules

Increases permeability

Increased water reabsorption

Uterus

Stimulates contraction

Mammary glands

Stimulates milk release

must be present for the adolescent growth spurt to occur. However, the presence of sex hormones eventually causes the growth centers within the long bones to ossify, so that further increase in height is impossible even when GH is present.

Inappropriate amounts of growth hormone secretion result in abnormal growth

Have you ever wondered why some people fail to grow normally? Some may be **pituitary dwarfs**—individuals whose pituitary glands do not produce sufficient growth hormone during childhood. Although miniature, a pituitary dwarf has normal intelligence and is usually well proportioned. If the

growth centers in the long bones are still active when this condition is diagnosed, it can be treated clinically by injection with human growth hormone, which can be synthesized by recombinant DNA technology. Growth problems may also result from the malfunction of other mechanisms, such as regulating hormones from the hypothalamus.

An individual may become abnormally tall when the anterior pituitary secretes excessive amounts of growth hormone during childhood. This condition is referred to as **gigantism.** If pituitary malfunction leads to hypersecretion of growth hormone during adulthood, the individual cannot grow taller. Instead, connective tissue thickens, and bones in the hands,

Figure 47–14 **The hypothalamus regulates the anterior lobe of the pituitary gland.** The hypothalamus secretes several specific releasing and inhibiting hormones that are transported to the anterior lobe of the pituitary gland by way of portal veins. These regulatory hormones stimulate or inhibit the release of specific hormones by cells of the anterior lobe. The anterior lobe secretes hormones that act on a wide variety of target tissues.

feet, and face may increase in diameter. This condition is known as **acromegaly,** which means "large extremities."

Thyroid hormones increase metabolic rate

The **thyroid gland** is located in the neck region, in front of the trachea and below the larynx (Fig. 47–12). Two of the **thyroid hormones, thyroxine,** also known as T_4 and **triiodothyronine,** or T_3 are synthesized from the amino acid tyrosine and from iodine. Thyroxine has four iodine atoms attached to each molecule; T_3 has three. In most target tissues, T_4 is converted to T_3, the more active form. We will describe the actions of calcitonin, another hormone secreted by the thyroid gland, when we discuss the parathyroid glands.

In vertebrates, thyroid hormones are essential for normal growth and development, and they increase the rate of metabolism in most body tissues. T_3 binds with its receptor in the nucleus of a target cell. The T_3-receptor complex induces or suppresses synthesis of specific enzymes and other proteins. Thyroid hormones help regulate the synthesis of proteins necessary for cellular differentiation. For example, tadpoles cannot develop into adult frogs without thyroxine.

Thyroid secretion is regulated by negative feedback mechanisms

The regulation of thyroid hormone secretion depends mainly on a negative feedback loop between the anterior pituitary and the thyroid gland (Fig. 47–15). When the concentration of thyroid hormones in the blood rises above normal, the anterior pituitary secretes less thyroid-stimulating hormone (TSH):

High concentration of thyroid hormones ⟶ anterior pituitary secretes less TSH ⟶ thyroid gland secretes less hormone ⟶ homeostasis

Too much thyroid hormone in the blood also affects the hypothalamus, inhibiting secretion of a TSH-releasing hormone, TRH. However, the hypothalamus is thought to exert its regulatory effects mainly in certain stressful situations, such as extreme weather change. Exposure to very cold weather may stimulate the hypothalamus to increase secretion of TSH-releasing hormone. This action raises body temperature through increased metabolic heat production. Increased body temperature has a negative feedback effect, limiting further secretion of thyroid hormones.

When the concentration of thyroid hormones decreases, the pituitary secretes more TSH. TSH acts by way of cyclic AMP to promote synthesis and secretion of thyroid hormones and also to promote increased size of the thyroid gland itself. The effect of TSH can be summarized as follows:

Low concentration of thyroid hormones ⟶ anterior pituitary secretes more TSH ⟶ thyroid gland secretes more hormone ⟶ homeostasis

Malfunction of the thyroid gland leads to specific disorders

Extreme hypothyroidism during infancy and childhood results in low metabolic rate and can lead to **cretinism,** a condition of retarded mental and physical development. When diagnosed early enough and treated with thyroid hormones, the effects of cretinism can be prevented.

An adult who feels like sleeping all of the time, has little energy, and is mentally slow or confused may be suffering from hypothyroidism. When there is almost no thyroid function, the basal metabolic rate is reduced by about 40% and the patient develops **myxedema,** characterized by a slowing down of

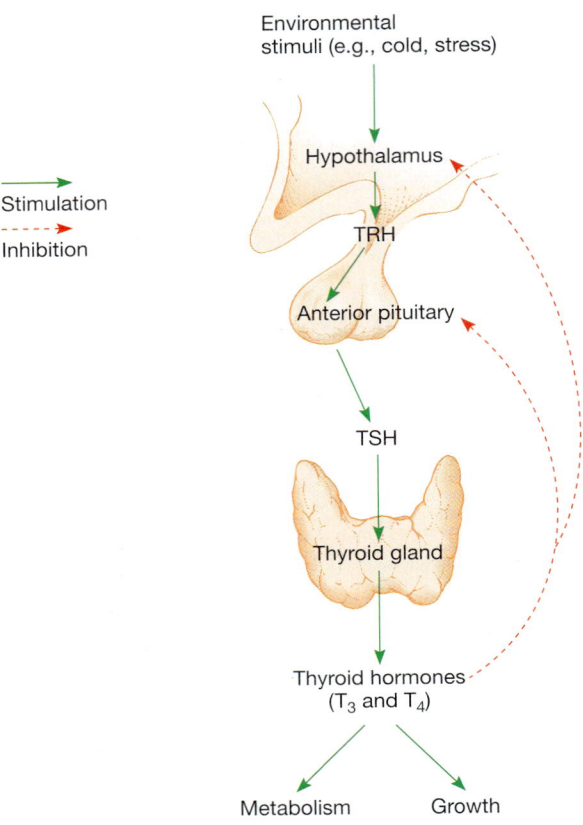

Figure 47–15 **Regulation of thyroid hormone secretion by negative feedback mechanisms.** When the concentration of thyroid hormones increases above normal levels, they signal the anterior pituitary and hypothalamus to limit TSH (thyroid stimulating hormone) production. Thus, thyroid hormones limit their own production by negative feedback. Green arrows indicate stimulation; red arrows indicate inhibition.

physical and mental activity. Hypothyroidism can be treated readily by oral administration of the missing hormone.

Hyperthyroidism does not cause abnormal growth but does increase metabolic rate by 60% or even more. This increase in metabolism results in the rapid use of nutrients, causing the individual to be hungry and to eat more. But this is not sufficient to meet the demands of the rapidly metabolizing cells, so individuals with this condition often lose weight. They also tend to be nervous, irritable, and emotionally unstable. The most common cause of hyperthyroidism is **Grave's disease,** an autoimmune disease. Abnormal antibodies bind to TSH receptors, activating them. This leads to increased production of thyroid hormones.

An abnormal enlargement of the thyroid gland, or **goiter,** may be associated with either hypersecretion or hyposecretion (see figure in Table 47–2). In Grave's disease, the abnormal antibodies that activate TSH receptors can cause the development of a goiter. One cause of hyposecretion is dietary iodine deficiency. Without iodine the gland cannot make thyroid hormones, so their concentration in the blood decreases. In compensation, the anterior pituitary secretes large amounts of

Figure 47–16 **Hormonal regulation of calcium homeostasis.** Parathyroid hormone (PTH) and calcitonin regulate calcium concentration in the blood and interstitial fluid. Green arrows indicate stimulation; red arrows indicate inhibition. *(a)* When calcium concentration decreases below normal limits, the parathyroid glands secrete PTH, which stimulates homeostatic mechanisms that restore appropriate calcium concentration. *(b)* When calcium concentration increases above normal limits, the parathyroid glands are inhibited. The thyroid gland is stimulated to secrete calcitonin, which inhibits Ca^{2+} release from bone and decreases Ca^{2+} reabsorption in the kidneys.

TSH, which stimulates growth of the thyroid gland, sometimes to gigantic proportions. However, enlargement of the gland cannot increase production of the hormones, because the needed ingredient is still missing. Seafood is a rich source of iodine, and iodine is also added to table salt as a nutritional supplement. In fact, thanks to iodized salt, goiter is no longer common in the United States and other developed countries. In other parts of the world, however, hundreds of thousands still suffer from this easily preventable disorder.

The parathyroid glands regulate calcium concentration

The **parathyroid glands** are typically embedded in the connective tissue surrounding the thyroid gland. The parathyroid glands secrete **parathyroid hormone (PTH),** a polypeptide that helps regulate the calcium level of the blood and interstitial fluid (Fig. 47–16). Parathyroid hormone acts by way of a membrane receptor and cAMP to stimulate calcium release from bones and to increase calcium reabsorption from the kidney tubules. It also activates vitamin D, which then increases the amount of calcium absorbed from the intestine.

Calcitonin, a peptide hormone secreted by the thyroid gland, works antagonistically to parathyroid hormone. When the concentration of calcium rises above homeostatic levels, calcitonin is released and rapidly inhibits removal of calcium from bone.

The islets of the pancreas regulate glucose concentration

In addition to secreting digestive enzymes (see Chapter 45), the pancreas is an important endocrine gland. Its hormones, **insulin** and **glucagon,** are secreted by cells that occur in little clusters, the **islets of Langerhans,** throughout the pancreas (Fig. 47–17). About one million islets are present in the human pancreas. They are composed mainly of **beta cells** which secrete insulin, and **alpha cells** which secrete glucagon.

Insulin lowers the concentration of glucose in the blood

When insulin binds to a receptor on a target cell plasma membrane, it activates the receptor, which is itself a tyrosine kinase. This enzyme can phosphorylate other insulin receptors, in-

Common bile duct

Pancreatic duct

Islet of Langerhans

100 µm

Figure 47–17 **LM of an islet of Langerhans.** Scattered throughout the pancreas, these clusters of endocrine cells secrete insulin and glucagon. *(Ed Reschke)*

creasing their activity. Other proteins, including a large protein called insulin receptor substrate (IRS), are also phosphorylated by the insulin receptor. Activation of IRS initiates a series of phosphorylations that result in activation of transport mechanisms in the plasma membrane and also affect DNA transcription.

Insulin is an anabolic hormone that promotes storage of fuel molecules. It stimulates cells of many tissues, including liver, muscle, and fat cells, to take up glucose from the blood by facilitated diffusion (see Chapter 5). Once glucose enters muscle cells, it is either used immediately as fuel or stored as glycogen. Insulin also inhibits liver cells from releasing glucose. Thus, insulin activity results in *lowering* the glucose level in the blood.

Insulin helps regulate fat and protein metabolism. This hormone reduces the use of fatty acids as fuel and instead stimulates their storage in adipose tissue. Insulin has an anabolic effect on protein metabolism, resulting in a net increase in protein. It promotes protein synthesis by stimulating the transport of certain amino acids into cells and by promoting transcription and translation.

Glucagon raises the concentration of glucose in the blood

The effects of **glucagon** are opposite to those of insulin. The main action of glucagon is to raise blood-glucose level. It does this by stimulating liver cells to convert glycogen to glucose, a process known as *glycogenolysis.* Glucagon also stimulates *gluconeogenesis,* the production of glucose from other metabolites. Glucagon mobilizes fatty acids and amino acids as well as glucose.

Insulin and glucagon secretion are regulated by glucose concentration

Insulin and glucagon secretion are directly controlled by the concentration of glucose in the blood (Fig. 47–18). After a meal, when the blood-glucose level rises as a result of intestinal absorption, beta cells are stimulated to increase insulin secretion. Then, as cells remove glucose from the blood, decreasing its concentration, insulin secretion decreases accordingly, as follows:

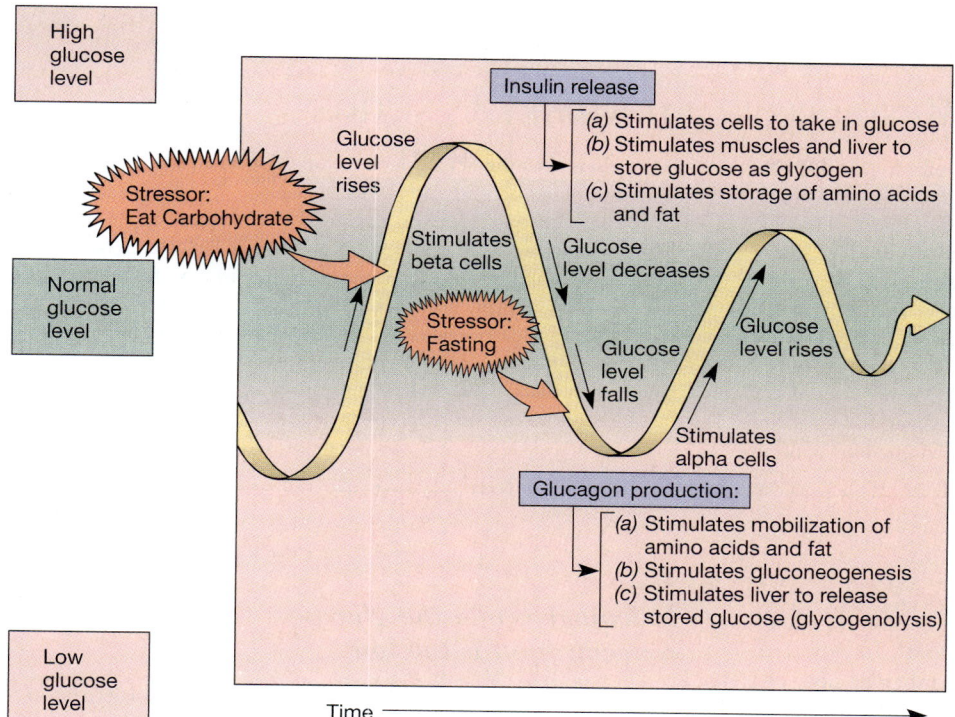

High glucose level

Normal glucose level

Low glucose level

Stressor: Eat Carbohydrate

Glucose level rises

Stimulates beta cells

Insulin release
(a) Stimulates cells to take in glucose
(b) Stimulates muscles and liver to store glucose as glycogen
(c) Stimulates storage of amino acids and fat

Glucose level decreases

Stressor: Fasting

Glucose level falls

Glucose level rises

Stimulates alpha cells

Glucagon production:
(a) Stimulates mobilization of amino acids and fat
(b) Stimulates gluconeogenesis
(c) Stimulates liver to release stored glucose (glycogenolysis)

Time

Figure 47–18 **Regulation of glucose concentration.** Insulin and glucagon work antagonistically to regulate blood-glucose levels.

Blood-glucose concentration too high ⟶ stimulates beta cells ⟶ insulin secretion increases ⟶ blood-glucose concentration decreases ⟶ homeostasis

When one has not eaten for several hours, the concentration of glucose in the blood begins to fall. When it falls from its normal fasting level of about 90 mg of glucose per 100 mL of blood to about 70 mg of glucose, the alpha cells of the islets increase their secretion of glucagon. Glucose is mobilized from storage in the liver cells, and blood-glucose concentration returns to normal:

Blood-glucose concentration too low ⟶ stimulates alpha cells ⟶ glucagon secretion increases ⟶ blood-glucose concentration increases ⟶ homeostasis

The alpha cells respond to the glucose concentration within their own cytoplasm, which reflects the blood-glucose level. When blood-glucose level is high, there is generally a high level of glucose within the alpha cells, and glucagon secretion is inhibited.

Note that these are negative feedback mechanisms and that insulin and glucagon work antagonistically to keep blood-glucose concentration within normal limits. When the glucose level rises, insulin release brings it back to normal; when it falls, glucagon acts to raise it again. The insulin-glucagon system is a powerful, fast-acting mechanism for keeping blood-glucose level within normal limits. Why do you think it is important to maintain a constant blood-glucose level? Recall that

brain cells depend on a continuous supply of glucose. They ordinarily are unable to use other nutrients as fuel. As we will discuss, several other hormones also affect blood-glucose concentration.

Diabetes mellitus is a serious disorder of carbohydrate metabolism

Diabetes mellitus, the most common endocrine disorder, is a worldwide health problem. Of the estimated 16 million diabetics in the United States, more than 40,000 die each year as a result of this disorder, making it one of the leading causes of death. Diabetes is a leading cause of blindness, kidney disorders, disease of small blood vessels, gangrene of the limbs, and various other malfunctions.

About 90% of diabetics have type 2 diabetes. This disorder develops gradually, usually in overweight persons over the age of 40. In type 2 diabetes, normal (or greater than normal) concentrations of insulin are present in the blood, but insulin receptors on target cells do not bind it, a condition known as **insulin resistance.** The causes of insulin resistance are not yet known, but there is a strong genetic component. Even if they are raised separately in very different environments, when one identical twin develops insulin resistance, the other one almost always does also. Many type 2 diabetics can keep their blood-glucose levels within normal range by managing their diets, weight loss, and regular exercise. When this treatment approach is not effective, type 2 diabetics are treated with oral hypoglycemic drugs that stimulate insulin secretion and promote the actions of insulin.

INSULIN, METABOLISM, AND KIDNEY FUNCTION

How does kidney function provide clues to insulin function? Insulin deficiency leads to disruption of carbohydrate, fat, and protein metabolism and to electrolyte imbalance. The concentration of glucose is so high that it exceeds the renal threshold: the tubules in the kidneys are unable to return all the glucose in the filtrate to the blood. As a result, glucose is excreted in the urine. The presence of glucose in the urine is a simple screening test for diabetes.

Insulin-dependent cells in a diabetic can take in only about 25% of the glucose they require for fuel. The body turns to fat and protein for energy. Increased fat metabolism increases the formation of compounds known as ketone bodies. These compounds build up in the blood, causing ketoacidosis, a condition in which the body fluids and blood become too acidic.

When the ketone level in the blood rises, ketones appear in the urine, another clinical indication of diabetes mellitus. When ketone bodies and glucose are excreted in the urine, water follows by osmosis; as a result, urine volume increases. The resulting dehydration causes the diabetic to feel continually thirsty. When ketones are excreted, they also take sodium, potassium, and some other cations with them. Loss of these ions in the urine causes electrolyte imbalance. If severe, ketoacidosis can lead to coma and death.

Insulin-dependent diabetes, referred to as type 1, usually develops before age 30, often during childhood. In type 1 diabetes there is a marked decrease in the number of beta cells in the pancreas, resulting in insulin deficiency. Daily insulin injections are needed to correct the carbohydrate imbalance that results. Studies suggest that type 1 diabetes is an autoimmune disease in which antibodies mark the beta cells for destruction. This disorder may be caused by a combination of genetic predisposition and infection by a virus. Patients with diabetes have a shortened life expectancy, in part, because impaired lipid metabolism can cause atherosclerotic disease (see Chapter 42).

Similar metabolic disturbances occur in both types of diabetes mellitus:

1. **Decreased use of glucose.** Because the cells of diabetics cannot take up glucose from the blood, it accumulates there, causing hyperglycemia (an abnormally high concentration of glucose in the blood). Instead of the normal fasting concentration of about 90 mg per 100 mL, the level typically exceeds 200 mg per 100 mL and may reach concentrations of more than 1000 mg per mL. When glucose concentration exceeds its renal threshold (about 150 mg of glucose per 100 mL of blood), glucose is excreted in the urine. The glucose exerts an osmotic effect that results in increased urine volume (see *Making the Connection: Insulin, Metabolism, and Kidney Function*).

2. **Increased fat mobilization.** Despite the large quantities of glucose in the blood, most cells cannot use it and must turn to other fuel sources. The absence of insulin promotes the mobilization of fat stores, providing nutrients for cellular respiration. But unfortunately, the blood lipid level may reach five times the normal level, leading to the development of atherosclerosis.

3. **Increased protein use.** Lack of insulin also results in increased protein breakdown relative to protein synthesis, so the untreated diabetic becomes thin and emaciated.

In hypoglycemia the glucose concentration is too low

Hypoglycemia, low blood-glucose concentration, sometimes occurs in people who later develop diabetes. It may be an overreaction by the islets to glucose challenge. Too much insulin is secreted in response to carbohydrate ingestion. About 3 hours after a meal, the blood-sugar concentration falls below normal, making the individual feel very drowsy. If this reaction is severe enough, the patient may become uncoordinated or even unconscious.

Serious hypoglycemia can develop if diabetics receive injections of too much insulin or if the islets, because of a tumor, secrete too much insulin. The blood-glucose concentration may then fall drastically, depriving brain cells of their needed fuel supply. These events can lead to **insulin shock,** a condition in which the patient may appear drunk, or may become unconscious, suffer convulsions, and even die.

The adrenal glands help the body cope with stress

The paired **adrenal glands** are small, yellow masses of tissue that lie in contact with the upper ends of the kidneys (Fig. 47–19). Each gland consists of a central portion, the **adrenal medulla,** and a larger outer section, the **adrenal cortex.** Although joined anatomically, the adrenal medulla and cortex develop from different types of tissue in the embryo and function as distinctly different glands. Both secrete hormones that help to regulate metabolism, and both help the body respond to stress.

The adrenal medulla initiates an alarm reaction

The adrenal medulla is coupled to the sympathetic nervous system. It develops from neural tissue, and its secretion is controlled by sympathetic nerves. This neuroendocrine gland se-

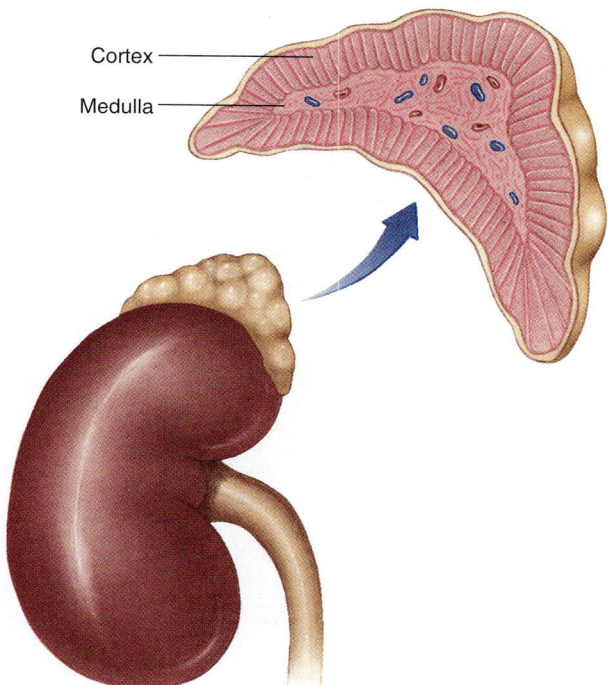

Cortex

Medulla

Figure 47–19 **Adrenal gland.** The adrenal glands are located above the kidneys. Each gland consists of a central adrenal medulla surrounded by an adrenal cortex.

cretes **epinephrine** (sometimes called adrenaline) and **norepinephrine** (noradrenaline). Chemically, these hormones are very similar; they belong to the chemical group known as **catecholamines.** Recall that norepinephrine also serves as a neurotransmitter released by sympathetic neurons and by some neurons in the central nervous system. Most of the hormone output of the adrenal medulla is epinephrine.

Under normal conditions, both epinephrine and norepinephrine are secreted continuously in small amounts. Their secretion is under neural control. When anxiety is aroused, messages are sent from the brain through sympathetic nerves to the adrenal medulla. Acetylcholine released by these neurons triggers the release of epinephrine and norepinephrine.

During a stressful situation, adrenal medullary hormones initiate an alarm reaction enabling you to think more quickly, fight harder, or run faster than usual. Metabolic rate increases by as much as 100%. Blood is rerouted to those organs essential for emergency action (Fig. 47–20). Blood vessels going to the brain, muscles, and heart are dilated, while those to the skin and kidneys are constricted. Constriction of blood vessels serving the skin has the added advantage of decreasing blood loss in case of hemorrhage (and explains the sudden paling that comes with fear or rage). At the same time, the heart beats faster. Thresholds in the reticular activating system of the brain are lowered, so you become more alert. Strength of muscle contraction increases. The adrenal medullary hormones also raise fatty acid and glucose levels in the blood, ensuring needed fuel for extra energy.

The adrenal cortex helps the body deal with chronic stress

All the hormones of the adrenal cortex are steroids synthesized from cholesterol, which in turn is made from acetyl coenzyme A. Although more than 30 types of steroids have been isolated from the adrenal cortex, this gland produces only three types of hormones in significant amounts: sex hormone precursors, mineralocorticoids, and glucocorticoids.

Sex hormone precursors, such as the androgen (masculinizing hormone) known as DHEA (dehydroepiandrosterone), are secreted by the adrenal cortex in both sexes. In the tissues they are converted to **testosterone,** the principal male sex hormone, and **estradiol,** the principal female sex hormone. In males, androgen production by the adrenal cortex is not significant because the testes produce testosterone. However, in females, the androgen produced by the adrenal cortex accounts for most of that circulating in the blood.

The principal **mineralocorticoid** is **aldosterone.** Recall from Chapter 46 that this hormone helps regulate fluid balance by regulating salt balance. In response to aldosterone, the kidneys reabsorb more sodium and excrete more potassium. As a result of increased sodium, extracellular fluid volume increases, which results in greater blood volume and elevated blood pressure.

When the adrenal glands do not produce enough aldosterone, large amounts of sodium are excreted in the urine. Water leaves the body with the sodium (because of osmotic pressure), and the blood volume may be so markedly reduced that the patient dies of low blood pressure.

Cortisol, also called hydrocortisone, accounts for about 95% of the **glucocorticoid** activity of the adrenal cortex. Cortisol helps ensure adequate fuel supplies for the cells when the body is under stress. Its principal action is to stimulate gluconeogenesis in liver cells. Cortisol helps provide nutrients for glucose production by stimulating transport of amino acids into liver cells (Fig. 47–20). It also promotes mobilization of fats so that fatty acids are available for conversion to glucose. These actions ensure that glucose and glycogen are produced in the liver, and the concentration of glucose in the blood rises. Thus, the adrenal cortex provides an important backup system for the adrenal medulla, ensuring glucose supplies when the body is under stress and in need of extra energy (see *Focus On: Coping with Stress*).

Almost any type of stress stimulates the hypothalamus to secrete **corticotropin-releasing factor, CRF.** This hormone stimulates the anterior pituitary to secrete **adrenocorticotropic hormone,** or **ACTH.** ACTH regulates both glucocorticoid and aldosterone secretion. ACTH is so potent that it can result in up to a 20-fold increase in cortisol secretion within minutes. When the body is not under stress, high levels of cortisol in the blood inhibit both CRF secretion by the hypothalamus and ACTH secretion by the pituitary.

Destruction of the adrenal cortex and the resulting decrease in aldosterone and cortisol secretion cause **Addison's disease.** Reduction in cortisol prevents the body from regu-

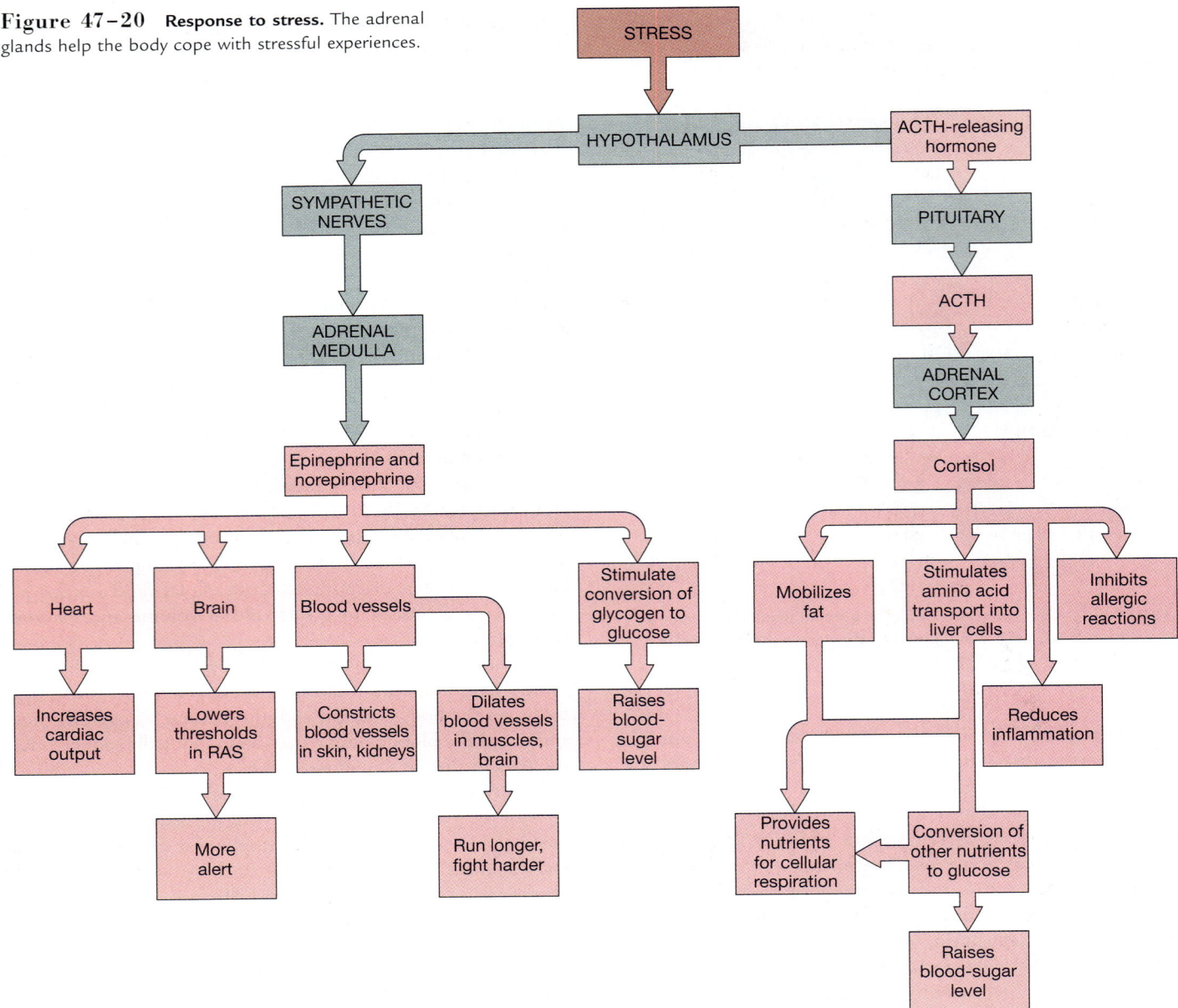

Figure 47–20 **Response to stress.** The adrenal glands help the body cope with stressful experiences.

lating the concentration of glucose in the blood because it cannot synthesize enough glucose. The cortisol-deficient patient also loses the ability to cope with stress. If cortisol levels are significantly depressed, even the stress of mild infections can cause death.

Glucocorticoids are used clinically to reduce inflammation in allergic reactions, infections, arthritis, and certain types of cancer. These hormones inhibit prostaglandin (which are mediators of inflammation) production by inducing the expression of an inhibitor of the enzyme phospholipase A. Glucocorticoids also reduce inflammation by decreasing the permeability of capillary membranes, thereby reducing swelling. In addition, they reduce the effects of histamine and so are used to treat allergic symptoms.

When used in large amounts over long periods of time, glucocorticoids can cause serious side effects. Although they help stabilize lysosome membranes so that they do not destroy tissues with their potent enzymes, the ability of lysosomes to degrade foreign molecules is also decreased. Glucocorticoids decrease interleukin-1 production, thereby blocking cell-mediated immunity and reducing the patient's ability to fight infections. Other side effects include ulcers, hypertension, diabetes mellitus, and atherosclerosis.

Abnormally large amounts of glucocorticoids, whether due to disease or drugs, can result in **Cushing's disease.** In this condition, fat is mobilized from the lower part of the body and deposited about the trunk. Edema gives the patient's face a full-moon appearance. Blood-glucose level rises to as much

COPING WITH STRESS

The breakup of a relationship, an infection, or the anxiety of taking a test when you are not prepared are all stressors that arouse the body to action. The brain and the adrenal glands work together to help the body cope effectively. Information is transferred by nerves and hormones to many tissues and organs. Neural messages from the brain stimulate the adrenal medulla to release epinephrine and norepinephrine that prepare the body for fight or flight.

During stress, the hypothalamus secretes corticotropin-releasing factor, which signals the anterior pituitary to secrete ACTH. The release of ACTH increases the secretion of glucocorticoids such as cortisol. These hormones adjust metabolism to meet the increased demands of the stressful situation (see Fig. 47–20).

Some forms of stress are short-lived. We react, quickly resolving the situation. Chronic stress, such as that from an unhappy marriage or psychologically toxic work situation, is harmful because of the side effects of long-term, elevated levels of glucocorticoids.

Chronic stress has also been shown to damage the brain. Studies indicate that when rodents and monkeys are subjected to prolonged stress, elevation of glucocorticoids leads to degeneration of neurons, especially in the hippocampus (a part of the brain involved in learning and remembering). Elevated concentration of glucocorticoids also impairs the capacity of neurons in the hippocampus to withstand physiological insult such as reduced blood flow or oxygen to the brain.

Individuals approach stressful situations in their lives differently. A stressor that may result in chronic, damaging levels of glucocorticoids in one person may be viewed as a challenge by another. One source of such individual differences may be variations in maternal care. Dong Liu and his research team studied the effects of variations in maternal care in rats during the first 10 days of life. In 1997 these investigators reported in *Science* that rats that were licked and groomed more by their mothers showed reduced ACTH and adrenocorticosteroid secretion in response to acute stress when they were older. Does early experience also determine how well humans cope with stress later in life?

We have no control over the maternal care we receive early in life, but we can learn to reduce our psychological and physiological responses to stressors by learning stress management techniques, such as meditation, visual imagery, progressive muscle relaxation, or self-hypnosis. Practicing a stress management technique can result in decreased activity of the sympathetic nervous system and reduced response to norepinephrine. For example, relaxation training has been shown to lower blood pressure in hypertensive patients, decrease the frequency of migraine headaches, and reduce chronic pain.

---

as 50% above normal, causing adrenal diabetes. If this condition persists for several months, the beta cells in the pancreas may "burn out" from secreting excessive amounts of insulin, leading to permanent diabetes mellitus. Reduction in protein synthesis causes weakness and decreases immune responses, so the patient often dies of infection.

Many other hormones are known

Many other tissues of the body secrete hormones. The **pineal gland,** located in the brain, produces a hormone called **melatonin,** which influences biological rhythms and the onset of sexual maturity. In humans, melatonin facilitates the onset of sleep. Exposure to light suppresses secretion of melatonin. An interesting study reported in 1998 in *Science* by Scott Campbell and Patricia Murphy provides evidence that human melatonin levels are affected by exposure to light at other parts of the body than the eyes. (These researchers applied light to various vascular parts of the body, including the back of the knees.)

Several hormones secreted by the digestive tract regulate digestive processes. The **thymus gland** produces **thymosin,** a hormone that plays a role in immune responses. **Atrial natriuretic factor (ANF),** secreted by the atrium of the heart, promotes sodium excretion and lowers blood pressure (Chapter 46). In Chapter 48 we discuss the principal reproductive hormones.

SUMMARY WITH KEY TERMS

I. The **endocrine system** consists of **endocrine glands,** cells, and tissues that secrete **hormones.** This system helps regulate many aspects of metabolism, fluid balance, growth, and reproduction. **Endocrinology** is the study of endocrine activity.

II. Hormones are an important type of chemical signal by which cells communicate with one another.

   A. Classically, hormones were defined as chemical messengers secreted by discrete endocrine glands, glands that lack ducts. Hormones are secreted into the interstitial fluid or blood and typically are transported by the blood. They bind with receptors on or in specific **target tissues.**

   B. In addition to the classical endocrine glands, many other cells, tissues, and organs secrete hormones and hormone-like substances. For example, **growth factors** are peptides that stimulate cell division and development. Hormones and other signaling molecules can be transported in axons or in interstitial fluid.

   C. **Neuroendocrine cells** secrete **neurohormones,** which are typically transported down axons and then secreted and transported by the blood.

   D. In **paracrine regulation,** a hormone (or other signal molecule) diffuses through interstitial fluid and acts on nearby target cells. **Prostaglandins** are a group of local hormones that help regulate

E. In **autocrine regulation**, a hormone (or other signal molecule) is secreted into the interstitial fluid and then acts back on the very cell that produced it.

III. Hormones can be assigned to four chemical groups: fatty acid derivatives, steroids, amino acid derivatives, or peptides and proteins.

A. Prostaglandins and the juvenile hormone of insects are derived from fatty acids.

B. Hormones secreted by the adrenal cortex, ovary, and testis, as well as the molting hormone of insects are **steroid hormones.**

C. Thyroid hormones and epinephrine are derived from amino acids.

D. Antidiuretic hormone (ADH) and glucagon are examples of peptide hormones. Insulin is a small protein.

IV. Hormone secretion is typically regulated by negative feedback control mechanisms.

V. Receptors on or in target cells determine the specificity of hormones. Hormones can stimulate synthesis of their own receptors, thus amplifying their effect on the cell. This mechanism is known as **receptor up-regulation.** In **receptor down-regulation,** hormones decrease the number of their own receptors, thereby suppressing the sensitivity of target cells to the hormone.

VI. Steroid hormones and thyroid hormones are hydrophobic molecules that pass through the plasma membrane and combine with receptors within the target cell; the hormone-receptor complex may activate or repress transcription of messenger RNA coding for specific proteins.

VII. Most hormones are hydrophilic and do not enter target cells. They combine with receptors on the plasma membrane of their target cells. Many hormones act via **signal transduction.** An extracellular hormone signal is transduced by the receptor into an intracellular signal.

A. Most peptide hormones are first messengers that carry out their actions by way of **second messengers,** such as **cyclic AMP (cAMP)** or calcium ions. The hormone-receptor complex activates a coupling molecule, typically a **G protein.** The G protein either stimulates or inhibits an enzyme that affects the second messenger. For example G proteins stimulate or inhibit **adenylyl cyclase,** the enzyme that catalyzes the conversion of ATP to cAMP. Second messengers typically act by stimulating the activity of protein kinases.

B. Protein kinases phosphorylate specific proteins that affect the activity of the cell.

C. Certain hormone-receptor complexes increase the concentration of calcium ions in the cell. Calcium ions bind with **calmodulin,** which activates certain enzymes.

D. **Inositol triphosphate (IP$_3$)** and **diacylglycerol (DAG)** are second messengers that increase calcium concentration and activate enzymes.

E. **Signal amplification** occurs as each hormone-receptor complex stimulates the production of many second messenger molecules. Second messengers, in turn, activate protein kinase molecules that can activate many protein molecules.

VIII. Many invertebrate hormones are secreted by neurons rather than by endocrine glands. These neurohormones help regulate regeneration, molting, metamorphosis, reproduction, and metabolism.

A. Pigment distribution in crustaceans is regulated by neurohormones.

B. Hormones control development in insects.

1. When stimulated by some environmental factor, neuroendocrine cells in the insect brain secrete **brain hormone (BH).**

2. BH stimulates the prothoracic glands to produce **molting hormone** (ecdysone), which stimulates growth and molting.

3. In the immature insect, the corpora allata secrete **juvenile hormone,** which suppresses metamorphosis at each larval molt. The amount of juvenile hormone decreases with successive molts.

IX. In vertebrates, hormones help regulate growth, reproduction, salt and fluid balance, and many aspects of metabolism.

X. Nervous and endocrine system regulation are integrated in the **hypothalamus,** which regulates the activity of the **pituitary gland.**

A. Endocrine disorders can result from **hyposecretion** (abnormally reduced output) or **hypersecretion** (abnormally increased output) of hormones.

B. The neurohormones **oxytocin** and **antidiuretic hormone (ADH)** are produced by the hypothalamus and released by the **posterior lobe** of the pituitary.

1. Oxytocin stimulates contraction of the uterus and stimulates ejection of milk by the mammary glands.

2. ADH stimulates reabsorption of water by the kidney tubules.

C. Secretion of hormones by the **anterior lobe** of the pituitary gland is regulated by **releasing hormones** and **inhibiting hormones** secreted by the hypothalamus.

D. The anterior lobe of the pituitary gland secretes **growth hormone, prolactin,** and several **tropic hormones** that stimulate other endocrine glands.

1. Growth hormone (GH) is an **anabolic hormone** that stimulates body growth by promoting protein synthesis. GH stimulates the liver to produce **somatomedins,** also called **insulin-like growth factors,** that promote skeletal growth and general tissue growth. Malfunctions in GH secretion can lead to **pituitary dwarfism, gigantism,** and **acromegaly.**

2. Prolactin stimulates the mammary glands to produce milk.

E. The **thyroid gland** secretes **thyroid hormones (thyroxine,** or **T$_4$,** and **triiodothyronine,** or **T$_3$)** that stimulate the rate of metabolism.

1. Regulation of thyroid secretion depends mainly on a feedback system between the anterior pituitary and the thyroid gland.

2. Hyposecretion of thyroxine during childhood may lead to **cretinism;** during adulthood it may result in **myxedema. Goiter,** an abnormal enlargement of the thyroid gland, is associated with both hyposecretion and hypersecretion. The most common cause of hyperthyroidism is **Grave's disease,** an autoimmune disease.

F. The **parathyroid glands** secrete **parathyroid hormone,** which regulates the calcium level in the blood.

1. Parathyroid hormone increases calcium concentration by stimulating calcium release from bones, increasing calcium reabsorption by kidney tubules, and increasing calcium reabsorption from the intestine.

2. **Calcitonin,** secreted by the thyroid gland, acts antagonistically to parathyroid hormone.

G. The **islets of Langerhans** in the pancreas secrete **insulin** and **glucagon.**

1. Insulin stimulates cells to take up glucose from the blood and so lowers blood-glucose concentration.

2. Glucagon raises blood-glucose concentration by stimulating conversion of glycogen to glucose (glycogenolysis) and by stimulating production of glucose from other nutrients (gluconeogenesis).

3. Insulin and glucagon secretion are regulated directly by blood-glucose levels.

4. In **diabetes mellitus,** either insulin deficiency or **insulin resistance** results in decreased utilization of glucose, increased fat mobilization, and increased protein utilization.

H. The **adrenal glands** secrete hormones that help the body cope with stress.

1. The **adrenal medulla** secretes **epinephrine** and **norepinephrine.**

2. The **adrenal cortex** secretes sex hormones; **mineralocorticoids,** such as **aldosterone;** and **glucocorticoids,** such as **cortisol.** Aldosterone increases the rate of sodium reabsorption and potassium excretion by the kidneys. Cortisol promotes gluconeogenesis.

3. The hormones of the adrenal medulla help the body respond to stress by increasing the heart rate and metabolic rate and the strength of muscle contraction and also by rerouting blood to those organs needed for fight or flight. The adrenal cortex ensures adequate fuel supplies for the rapidly metabolizing cells.

1. Which of the following is NOT true of endocrine glands? (a) they secrete hormones (b) they have ducts (c) their product is typically transported by the blood (d) they are typically regulated by negative feedback (e) when removed from an experimental animal, the animal exhibits symptoms of deficiency

2. A cell secretes a product that diffuses through the interstitial fluid and acts on nearby cells. This is an example of (a) neuroendocrine secretion (b) autocrine regulation (c) paracrine regulation (d) classical endocrine control (e) peptide hormone function

3. Paracrine regulators that are derived from fatty acids and are found in many different organs are (a) prostaglandins (b) histamine (c) growth factors (d) anabolic steroids (e) G proteins

4. Which of the following is (are) true of steroid hormones? (a) hydrophilic (b) secreted by the posterior pituitary (c) typically work through G protein and cyclic AMP (d) typically bind with receptor in nucleus and affect transcription (e) two of the preceding answers are correct

5. Which of the following is NOT a correct pair? (a) neurohormone; brain hormone (b) calcium; calmodulin (c) posterior lobe of pituitary; releasing hormone (d) anterior lobe of pituitary; growth hormone (e) hyposecretion; cretinism

6. Which of the following are second messengers? (a) hormone-receptor complex (b) calcium ions (c) inositol triphosphate (IP_3) (d) answers a, b, and c are correct (e) only answers b and c are correct

7. Growth hormone (a) is regulated mainly by calcium level (b) stimulates the liver to produce insulin-like growth factors (c) is a catabolic hormone (d) stimulates metabolic rate (e) signals the hypothalamus to produce a releasing hormone

8. Arrange the following events into an appropriate sequence. (1) high thyroid hormone concentration (2) anterior pituitary inhibited (3) homeostasis (4) lower level of TSH (5) thyroid gland secretes less thyroid hormone (a) 1, 2, 4, 5, 3 (b) 5, 4, 3, 2, 1 (c) 1, 2, 5, 4, 3 (d) 4, 5, 2, 3, 1 (e) 1, 4, 2, 5, 3

9. Arrange the following events into an appropriate sequence. (1) blood-glucose concentration increases (2) alpha cells in islets stimulated (3) homeostasis (4) low blood-glucose concentration (5) glucagon secretion increases (a) 1, 2, 3, 5, 4 (b) 5, 4, 2, 1, 3 (c) 1, 2, 5, 4, 3 (d) 4, 2, 5, 1, 3 (e) 4, 5, 1, 2, 3

10. Parathyroid hormone (a) increases glucose level in blood (b) helps body cope with stress (c) increases permeability of kidney tubules to water (d) promotes uptake of amino acids (e) increases calcium concentration in blood

11. An action of cortisol is (a) decreases glucose level in blood (b) helps body cope with stress (c) increases permeability of kidney tubules to water (d) promotes uptake of amino acids (e) increases calcium concentration in blood

12. Which of the following is NOT a correct pair? (a) thyroid gland; calcitonin (b) islets of Langerhans; glucagon (c) posterior lobe of pituitary; oxytocin (d) anterior lobe of pituitary; cortisol (e) adrenal medulla; epinephrine

13. Which of the following occur in diabetes mellitus? (a) decreased use of glucose (b) decreased fat metabolism (c) decreased protein use (d) increased concentration of TRH (e) two of the preceding answers

14. Aldosterone (a) is released by posterior pituitary (b) is an androgen (c) secretion is stimulated by an increase in TSH (d) is an enzyme that converts epinephrine to norepinephrine (e) increases sodium reabsorption

R E V I E W   Q U E S T I O N S

1. Define the term *hormone*. What are some of the important actions of hormones in invertebrates? In vertebrates?

2. How do hormones reach their target tissues? How do they "recognize" their target tissues?

3. What are the major chemical groups of hormones?

4. What is the role of receptors and second messengers in hormone action? Describe the mechanism of action of a hormone that uses (a) cyclic AMP as a second messenger (b) calcium ions as second messengers.

5. Why is the hypothalamus considered the link between the nervous and endocrine systems? What is the role of the anterior lobe of the pituitary? The posterior lobe?

6. Describe the actions of (a) prolactin (b) oxytocin (c) thyroid-stimulating hormone.

7. Draw a diagram to illustrate the regulation of (a) thyroid hormone secretion and (b) parathyroid hormone secretion.

8. Explain the hormonal basis for (a) acromegaly (b) pituitary dwarfism (c) cretinism (d) hypoglycemia (e) Cushing's disease.

9. Explain the antagonistic actions of insulin and glucagon in regulating blood-glucose level.

10. Describe several physiological disturbances caused by diabetes mellitus.

11. What are the actions of epinephrine and norepinephrine? How is the adrenal medulla regulated?

12. What types of hormones are released by the adrenal cortex, and what are the actions of each type?

13. Explain how the adrenal glands help the body respond to stress.

14. An injection of too much insulin may cause a diabetic to go into insulin shock, in which the patient may appear drunk or may become unconscious, suffer convulsions, and even die. From what you know about the actions of insulin, explain the physiological causes of insulin shock.

15. Label the diagram. Use Figure 47–12 to check your answer.

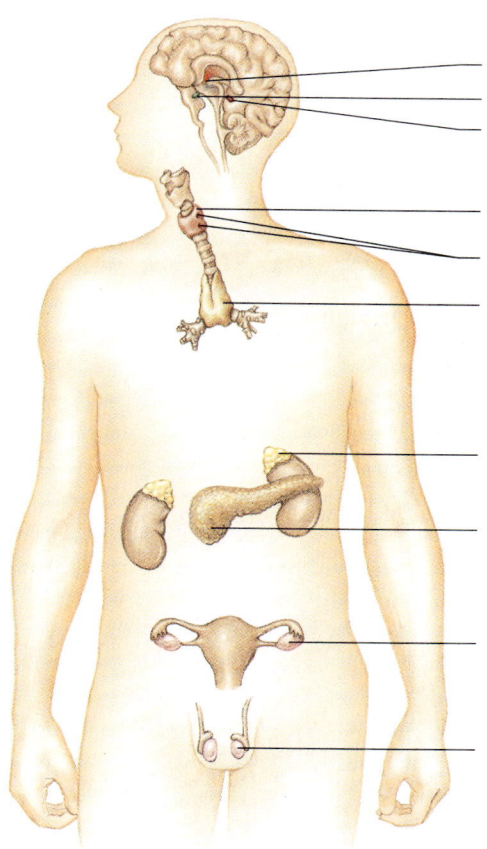

YOU MAKE THE CONNECTION

1. Human males have about the same amount of oxytocin circulating in their blood as do nonpregnant females who are not lactating, but its function in males is unknown. Hypothesize its function and design an ethical experiment to test your hypothesis. (*Hint:* Based on its effects on animal behavior, oxytocin has been referred to as the "nurturing hormone.")
2. How do receptors impart specificity within the endocrine system? What might be some advantages of having complex mechanisms for hormone action (e.g., second messengers)?
3. Why do you think it is important to maintain a constant blood-glucose level? Several hormones discussed in this chapter affect carbohydrate metabolism. Why is it important to have more than one? How do they interact?

RECOMMENDED READINGS

Lienhard, G.E., J. Slot, D.E. James, and M.M. Mueckler. "How Cells Absorb Glucose." *Scientific American,* Vol. 266, No. 1, Jan. 1992. A discussion of how insulin helps cells transport glucose.

Linder, M.E., and A.G. Gilman. "G Proteins." *Scientific American,* Vol. 267, No. 1, Jul. 1992. G proteins play an important role in the mechanism of hormone action.

Lowe, W.L., Jr. "Insulin-like Growth Factors." *Scientific American Science & Medicine,* Vol. 3, No. 2, Mar./Apr. 1996. A discussion of the process of establishing the role of IGFs and a description of their actions.

Youngren, J.F., and I.D. Goldfine. "The Molecular Basis of Insulin Resistance." *Science & Medicine,* Vol. 4, No. 3, May/Jun. 1997. Insulin resistance is a complex phenotype that may result from defects in insulin receptor signaling.

● Visit our website at **http://www.saunderscollege.com/lifesci/titles.html** and click on Solomon/Berg/Martin Biology for links to chapter-related resources on the World Wide Web.

CHAPTER 48

Reproduction

If there is one feature of a living organism that is unique, it is its ability to reproduce and perpetuate the species. The survival of each species requires that its members produce new individuals to replace those that die. At the molecular level, reproduction is a function of the capacity of nucleic acids to replicate themselves.

Many invertebrates can reproduce asexually. In **asexual reproduction** a single parent gives rise to two or more offspring that are genetically identical (except for mutation) to the parent. Asexual reproduction is an adaptation of sessile animals that cannot move about to search for mates. For animals that do move about, this method of reproduction is advantageous when the population density is low and mates are not readily available.

Sexual reproduction in animals involves the production and fusion of two types of **gametes** — sperm and eggs. Typically, as exemplified by the mating nudibranchs (*Chromodoris* sp.) photographed in Indonesia, a male parent contributes sperm and a female parent contributes an egg, or **ovum** (pl., *ova*). The sperm provides genes coding for some of the male parent's traits, and the egg contributes genes coding for some of the female parent's traits. When sperm and egg unite, a fertilized egg, or **zygote,** forms. The zygote develops into a new organism, similar to both parents but not identical to either.

Sexual reproduction has the biological advantage of promoting genetic variety among the members of a species. Each offspring is the product of a particular combination of genes contributed by both parents, rather than a genetic copy of a single individual. By making possible the recombination of the inherited traits of two parents, sexual reproduction gives rise to offspring that may be better able to survive than either parent. Sexual reproduction is particularly adaptive in an unstable, changing environment.

In many animals, reproduction involves remarkably complex structural, functional, and behavioral processes. In vertebrates, these processes are regulated by hormones secreted by the hypothalamus, pituitary gland, and gonads. This chapter summarizes some major features of animal reproductive processes and then focuses on human reproduction.

(1996 Bruce Watkins/ Animals Animals)

1. Compare asexual and sexual reproduction and compare external and internal fertilization.
2. Label the structures of the human male reproductive system on a diagram and describe the functions of each.
3. Trace the passage of sperm cells through the human male reproductive system from their origin in the seminiferous tubules to their expulsion from the body in the semen.
4. Describe the endocrine regulation of reproduction in the human male.
5. Label the structures of the human female reproductive system on a diagram and describe the functions of each.

6. Trace the development of a human ovum (egg) and its passage through the female reproductive system until it is fertilized.
7. Describe the endocrine regulation of reproduction in the human female and identify the important events of the menstrual cycle, such as ovulation and menstruation.
8. Summarize the process of human fertilization.
9. Compare the modes of action, effectiveness, advantages, and disadvantages of the methods of birth control in Table 48–3 and describe emergency contraception.
10. Identify common sexually transmitted diseases and describe their symptoms, effects, and treatments.

ASEXUAL REPRODUCTION IS COMMON AMONG SOME ANIMAL GROUPS

In asexual reproduction, a single parent may split, bud, or fragment to give rise to two or more offspring. Except for mutations, the offspring have hereditary traits identical to those of the parent. Sponges and cnidarians are among the animals that can reproduce by **budding.** A small part of the parent's body separates from the rest and develops into a new individual (Fig. 48–1). Sometimes the buds remain attached and become more-or-less independent members of a colony.

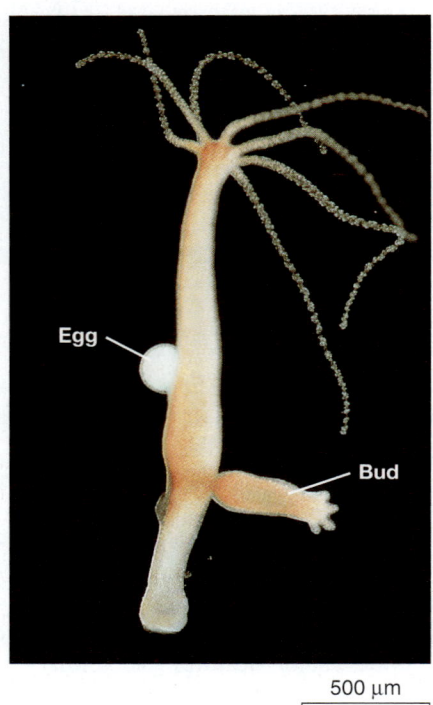

500 µm

Figure 48–1 **Asexual reproduction by budding.** A part of *Hydra's* body grows outward and then separates and develops into a new individual. The portion of the parent body that buds is not specialized exclusively for reproduction. The hydra shown here is also reproducing sexually as evidenced by the egg *(left)*. *(Richard Campbell/Biological Photo Service)*

Oyster farmers learned long ago that when they tried to kill sea stars by chopping them in half and throwing the pieces back into the sea, the number of sea stars preying on the oyster bed doubled! In flatworms, this ability to regenerate is part of a method of reproduction known as **fragmentation.** The body of the parent may break into several pieces. Each piece then regenerates the missing parts and develops into a whole animal.

Parthenogenesis ("virgin development") is a form of asexual reproduction in which an unfertilized egg develops into an adult animal. Parthenogenesis is known among many invertebrate and vertebrate groups, including some species of nematodes, gastropods, crustaceans, insects (especially honeybees and wasps), fishes, amphibians, and reptiles. Although a few species appear to reproduce solely by parthenogenesis, in most species episodes of parthenogenesis alternate with periods of sexual reproduction. Parthenogenesis may occur for several generations, followed at some point by sexual reproduction in which males develop, produce sperm, and mate with the females to fertilize their eggs. In some species, parthenogenesis appears to be an adaptation for survival in times of stress or serious population decline.

SEXUAL REPRODUCTION IS THE MOST COMMON TYPE OF ANIMAL REPRODUCTION

Most animals reproduce sexually by fusion of sperm and egg. The egg is typically large and nonmotile, with a store of nutrients that supports the development of the embryo. The sperm is usually small and motile, adapted to propel itself by beating its long, whiplike flagellum.

Many aquatic animals practice **external fertilization** in which the gametes meet outside the body (Fig. 48–2a). Mating partners usually release eggs and sperm into the water simultaneously. Gametes only live for a short time, and many are lost in the water; some are eaten by predators. However, so many gametes are released that sufficient numbers of sperm and egg cells do meet to perpetuate the species.

(a)

(b)

Figure 48–2 External and internal fertilization. *(a)* Like many aquatic animals, these spawning frogs *(Rana temporaria),* practice external fertilization. The female lays a mass of eggs, while the male mounts her and simultaneously deposits his sperm in the water. *(b)* Internal fertilization is practiced by aquatic reptiles, birds, and mammals and by most terrestrial animals, such as these lions *(Panthera leo).* *(a, Zig Leszczynski/Animals Animals; b, Fritz Polking/Dembinsky Photo Associates)*

In **internal fertilization,** matters are left less to chance. The male generally delivers sperm cells directly into the body of the female. Her moist tissues provide the watery medium required for the movement of sperm, and the gametes fuse inside the body. Most terrestrial animals, sharks, and aquatic reptiles, birds, and mammals practice internal fertilization (Fig. 48–2*b*).

Hermaphroditism is a form of sexual reproduction in which a single individual produces both eggs and sperm. A few hermaphrodites, such as the tapeworm, are capable of self-fertilization. Earthworms are more typical hermaphrodites. Two animals copulate, and mutual cross-fertilization occurs with each inseminating the other. In some hermaphroditic species, self-fertilization is prevented by the development of testes and ovaries at different times.

HUMAN REPRODUCTION: THE MALE PROVIDES SPERM

The human male, like other male mammals, has the reproductive role of producing sperm cells and delivering them into the female reproductive tract. The sperm that combines with an egg contributes its genes and determines the sex of the offspring. The male reproductive system is illustrated in Figure 48–3.

The testes produce sperm

In humans and other vertebrates, **spermatogenesis,** the process of sperm cell production, occurs in the paired male gonads, or **testes** (sing., *testis*). Spermatogenesis takes place within a vast tangle of hollow tubules, the **seminiferous tubules,** within each testis (Fig. 48–4). Spermatogenesis begins with undifferentiated cells, the **spermatogonia** in the walls of the tubules (Fig. 48–5).

The spermatogonia, which are diploid cells, divide by mitosis, producing more spermatogonia. Some enlarge and become **primary spermatocytes,** which undergo **meiosis,** producing haploid gametes. (You may want to review the discussion of meiosis in Chapter 9.) In many animals, gamete production occurs only in the spring or fall, but humans have no special breeding season. In the human adult male, spermatogenesis proceeds continuously, and millions of sperm are produced each day.

Each primary spermatocyte undergoes a first meiotic division, producing **secondary spermatocytes** (Fig. 48–6). In the second meiotic division, each secondary spermatocyte gives rise to two **spermatids.** Four spermatids are produced from the original primary spermatocyte. Each haploid spermatid differentiates into a mature sperm. The sequence is as follows:

Spermatogonium (diploid) ⟶ primary spermatocyte (diploid) ⟶ two secondary spermatocytes (haploid) ⟶ four spermatids (haploid) ⟶ four mature sperm (haploid)

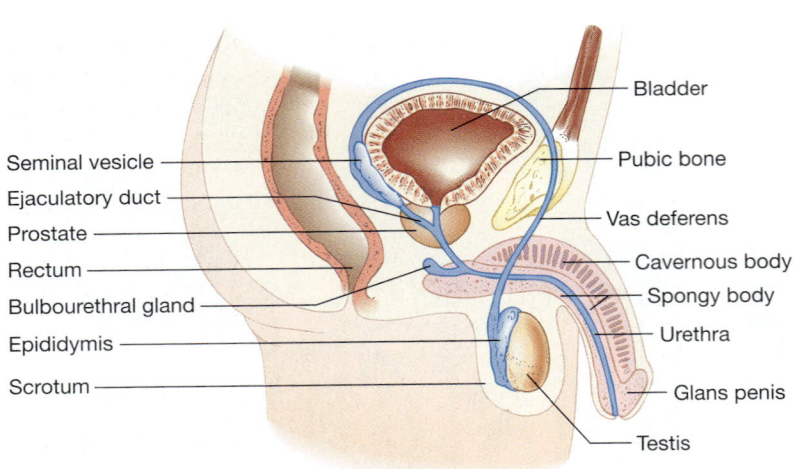

Figure 48-3 **Male reproductive system.** The scrotum, penis, and pelvic region of the human male are shown in sagittal section to illustrate their internal structure.

Labels for Figure 48-3:
Bladder
Seminal vesicle
Ejaculatory duct
Prostate
Rectum
Bulbourethral gland
Epididymis
Scrotum
Pubic bone
Vas deferens
Cavernous body
Spongy body
Urethra
Glans penis
Testis

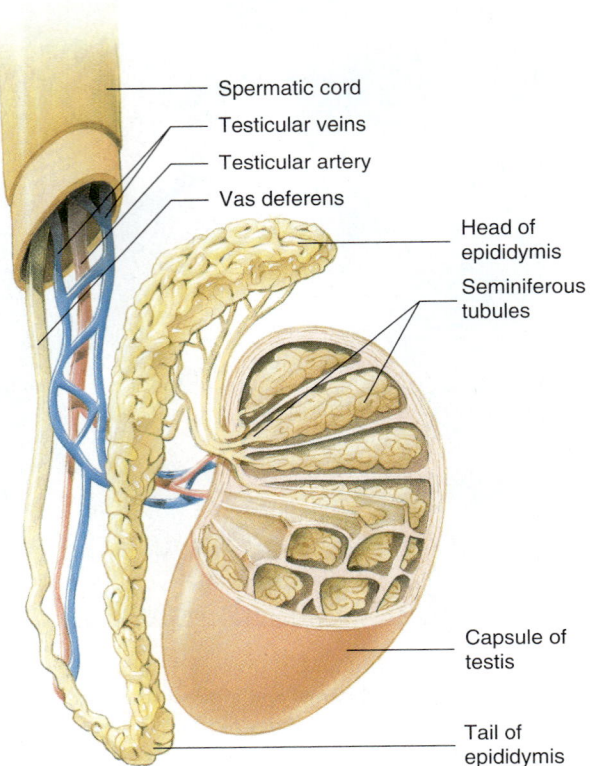

Figure 48-4 **Structure of testis, epididymis, and spermatic cord.** The organs are shown partly dissected and exposed. The testis is shown in sagittal section to illustrate the arrangement of the seminiferous tubules.

Labels for Figure 48-4:
Spermatic cord
Testicular veins
Testicular artery
Vas deferens
Head of epididymis
Seminiferous tubules
Capsule of testis
Tail of epididymis

(a)

100 μm

(b)

Figure 48-5 **Spermatogenesis in the seminiferous tubules.** *(a)* An SEM of a transverse section through a seminiferous tubule. Se, Sertoli cell; Sc, primary spermatocyte; Sg, spermatogonium. *(b)* Sperm cells can be seen in various stages of development. Can you identify the sequence of sperm cell differentiation? Note the large nutritive Sertoli cells. *(a, Custom Medical Stock Photo)*

Labels for Figure 48-5:
Mature sperm cells
Spermatids
Primary spermatocytes
Sertoli cell
Nucleus of a Sertoli cell
Spermatogonium
Wall of the seminiferous tubule

Figure 48–6   Spermatogenesis. A primary spermatocyte undergoes meiosis, giving rise to four spermatids. The spermatids differentiate, becoming mature sperm cells.

Each mature sperm consists of a head, a midpiece, and a flagellum (Fig. 48–7). The head consists of the nucleus and a cap, called an **acrosome,** that differentiates from the Golgi complex. The acrosome helps the sperm penetrate the egg. Mitochondria, located in the midpiece of the sperm, provide the energy for movement of the flagellum. The sperm flagellum has the typical eukaryotic 9 + 2 arrangement of microtubules. During its development, most of the sperm's cytoplasm is discarded and is phagocytized by the large nutritive **Sertoli cells** present within the seminiferous tubules (see *Making the Connection: Reproduction and Tight Junctions*).

Human sperm cells cannot develop at body temperature. Although the testes develop within the abdominal cavity of the male embryo, about two months before birth they descend into the **scrotum,** a skin-covered sac suspended from the groin. The scrotum serves as a cooling unit, maintaining sperm below body temperature. In rare cases, the testes do not descend. If this condition is not corrected (surgically or with hormone treatment), the seminiferous tubules eventually degenerate and the male becomes **sterile,** unable to produce offspring.

The scrotum is an outpocketing of the pelvic cavity and is connected to it by the **inguinal canals.** As they descend, the testes pull their blood vessels, nerves, and conducting tubes after them. The inguinal region is a weak place in the abdominal wall. Straining the abdominal muscles by lifting heavy ob-

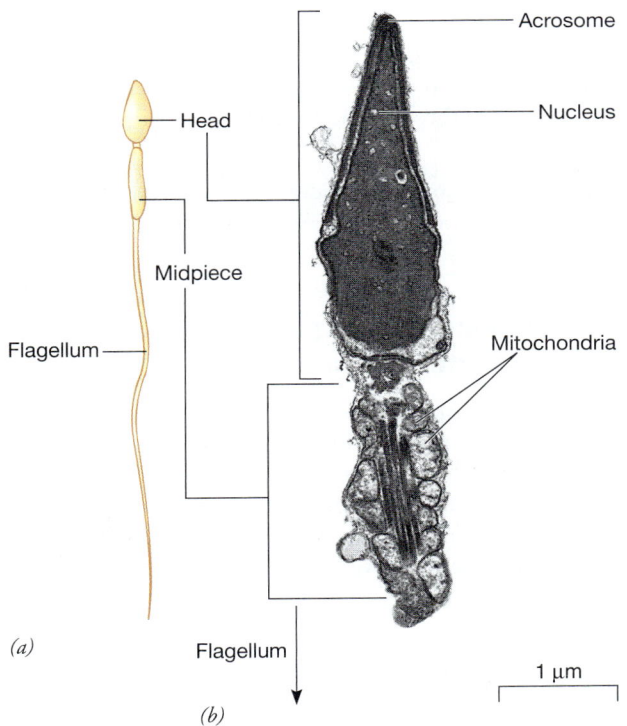

Figure 48–7   Structure of a mature sperm. *(a)* A mature sperm has a head, midpiece, and flagellum. *(b)* TEM of a human sperm cell. Mitochondria are visible in the midpiece.   *(b, Dr. Tony Brain/Science Photo Library/Photo Researchers, Inc.)*

REPRODUCTION AND TIGHT JUNCTIONS

Of what importance are tight junctions in the cells of the reproductive system? The Sertoli cells in the seminiferous tubules form a ring around the fluid-filled lumen of the tubule. They are joined to one another by tight junctions (see Chapter 5) and together form a blood-testis barrier. This barrier prevents the entrance into the tubule of harmful substances that could interfere with spermatogenesis. It also stops sperm from passing out of the tubule and into the blood, where they could stimulate an immune response.

In the female, the cells (known as granulosa cells) surrounding the developing follicle perform a corresponding function. They are connected by tight junctions that form a protective barrier around the developing ovum.

---

jects sometimes results in tearing the inguinal tissue. A loop of intestine can then bulge into the scrotum through the tear, a condition known as an inguinal hernia.

A series of ducts store and transport sperm

Sperm cells leave the seminiferous tubules of each testis through small tubules that empty into a larger coiled tube, the **epididymis.** There sperm complete their maturation and are stored.

During ejaculation, sperm pass from each epididymis into a sperm duct, the **vas deferens** (pl., *vasa deferentia*). The vas deferens extends from the scrotum through the inguinal canal and into the pelvic cavity. Each vas deferens empties into a short **ejaculatory duct,** which passes through the prostate gland and then opens into the urethra. The single urethra, which at different times conducts urine and semen, passes through the penis to the outside of the body. Thus, the sperm pass in sequence through the following structures:

Seminiferous tubules ⟶ epididymis ⟶ vas deferens ⟶ ejaculatory duct ⟶ urethra ⟶ released from body

The accessory glands produce the fluid portion of semen

As sperm are transported through the conducting tubes, they are mixed with secretions from three types of accessory glands. Approximately 3.5 mL of **semen** are ejaculated during sexual climax. Semen consists of about 200 million sperm cells suspended in the secretions of these glands.

The paired **seminal vesicles** secrete a nutritive fluid rich in fructose and prostaglandins into the vasa deferentia (Fig. 48–3). Nutrients in this secretion provide energy for the sperm after they are ejaculated. The single **prostate gland** secretes an alkaline fluid that may be important in neutralizing the acidic environment of the vagina and in increasing sperm cell motility. Prostaglandins in the semen stimulate contractions of the female uterus, which help move sperm up the female reproductive tract. The prostate gland is a common site of cancer in men over age 50. Its cause is not known, but prostate cancer is thought to be hormone-related.

During sexual arousal, the paired **bulbourethral glands,** located on each side of the urethra, release a mucous secretion. This fluid lubricates the penis, facilitating its penetration into the vagina.

A major cause of male infertility is insufficient sperm production. When sperm counts drop below 35 million per mL, fertility is impaired, and males with fewer than 20 million sperm per mL of semen are usually considered sterile. When a couple's attempts to produce a child are unsuccessful, a sperm count and analysis may be performed in a clinical laboratory. Sometimes semen is found to contain large numbers of abnormal sperm or, occasionally, no sperm at all.

In the United States in the 1970s an average healthy young man produced about 100 million sperm per mL of semen. Today that average has dropped to about 60 million. Although the cause of this decrease is not known, low sperm counts have been linked to a variety of environmental factors, including chronic marijuana use, alcohol abuse, and cigarette smoking. Studies show that men who smoke tobacco are more likely than nonsmokers to produce abnormal sperm. Exposure to industrial and environmental toxins such as DDT and PCBs (polychlorinated biphenyls) may also result in low sperm count and sterility. The use of anabolic steroids by athletes to accelerate muscle development can cause sterility in both males and females (See *Making the Connection: Anabolic Steroids, Physical Endurance, and Drug Abuse* in Chapter 47.)

The penis transfers sperm to the female

The **penis** is an erectile copulatory organ that delivers sperm into the female reproductive tract. It consists of a long shaft that enlarges to form an expanded tip, the **glans.** Part of the loose-fitting skin of the penis folds down and covers the proximal portion of the glans, forming a cuff called the **prepuce,** or foreskin. In the operation termed circumcision (commonly performed on male babies either for hygienic or religious reasons), the foreskin is removed.

Under the skin, the penis consists of three parallel columns of **erectile tissue,** two called the **cavernous bodies** and one

Urinary
bladder

Prostate gland

Orifices of
ejaculatory
ducts

Bulbourethral
gland

Openings of
bulbourethral
glands

Cavernous
bodies

Spongy body

Urethra

Glans penis

Prepuce

Urethral orifice

(a)

Dorsal veins
(dilated)

Dorsal artery
(constricted)

Cavernous
bodies

Spongy body

Urethra

Section of flaccid penis

Dorsal veins
(constricted)

Artery
(dilated)

Connective
tissue

Cavernous
bodies

Spongy body

Section of erect penis

(b)

Figure 48-8 **Internal structure of the penis.** *(a)* Longitudinal section through the prostate gland and penis. Note the three parallel columns of erectile tissue in the penis. *(b)* Cross section through a flaccid and an erect penis. Note that the erectile tissues of the cavernous and spongy bodies are engorged with blood in the erect penis.

known as the **spongy body** (Fig. 48–8). The spongy body surrounds the portion of the urethra that passes through the penis. When the male is sexually stimulated, nerve impulses cause the arteries of the penis to dilate. Blood rushes into the numerous blood vessels of the erectile tissue, causing the tissue to swell. This compresses veins that conduct blood away from the penis, slowing the outflow of blood. Thus, more blood enters the penis than can leave, causing the erectile tissue to become further engorged with blood. The penis becomes erect, that is, longer, larger in circumference, and firm. Although the human penis contains no bone, penis bones do occur in some other mammals, such as bats, rodents, and some primates.

The hypothalamus, pituitary, and testes interact to regulate male reproduction

When a boy is about ten years old the hypothalamus begins to secrete **gonadotropin-releasing hormone** (**GnRH;** Table 48–1). This hormone stimulates the anterior pituitary to secrete the gonadotropic hormones **follicle-stimulating hormone (FSH)** and **luteinizing hormone (LH),** also called **interstitial cell stimulating hormone (ICSH).** (GnRH has a greater effect on LH than on FSH, and investigators suspect that a separate FSH-releasing hormone may eventually be identified.) Both FSH and LH are glycoproteins that use cAMP as

a second messenger. LH stimulates the **interstitial cells,** lying between the seminiferous tubules in the testes, to secrete the steroid hormone **testosterone,** which is the principal **androgen,** or male sex hormone.

FSH, LH, and testosterone all appear to stimulate testosterone secretion as well as spermatogenesis. However, some studies suggest that, after puberty, LH and testosterone can maintain adequate levels of spermatogenesis. FSH is thought to have an indirect effect on spermatogenesis. A high concentration of testosterone in the testes is required for spermatogenesis. FSH stimulates Sertoli cells to produce **androgen-binding protein (ABP).** This regulatory protein binds to testosterone and concentrates it in the testes. Interestingly, developing sperm cells appear to lack receptors for sex steroids, so just how testosterone acts on them is not known.

Reproductive hormone concentrations are regulated by negative feedback mechanisms. Testosterone acts on both the hypothalamus and the pituitary gland to inhibit secretion of FSH and LH. Testosterone acts on the hypothalamus, decreasing its secretion of GnRH. It inhibits the anterior lobe of the pituitary by blocking the normal actions of GnRH on FSH and LH synthesis and release. Another negative feedback loop involves the peptide hormone called **inhibin,** which is secreted by Sertoli cells. (FSH stimulates inhibin secretion.) Inhibin is transported by the blood to the pituitary gland, where it in-

TABLE 48-1 Principal Male Reproductive Hormones

Endocrine Gland and Hormones	Principal Target Tissue	Principal Actions
Hypothalamus Gonadotropin-releasing hormone (GnRH)	Anterior pituitary	Stimulates release of FSH and LH
Anterior pituitary Follicle-stimulating hormone (FSH)	Testes	Stimulates development of seminiferous tubules; stimulates spermatogenesis
Luteinizing hormone (LH); also called interstitial cell-stimulating hormone (ICSH)	Testes	Stimulates interstitial cells to secrete testosterone and stimulates spermatogenesis
Testes Testosterone	General	*Before birth:* stimulates development of primary sex organs and descent of testes into scrotum *At puberty:* responsible for growth spurt; stimulates development of reproductive structures and secondary sex characteristics (male body build, growth of beard, deep voice, etc.) *In adult:* responsible for maintaining secondary sex characteristics; stimulates spermatogenesis
Inhibin	Anterior pituitary	Inhibits secretion of FSH

hibits FSH secretion (Fig. 48–9). **Activin,** still another hormone secreted by Sertoli cells, stimulates FSH secretion. From this brief overview, it should be evident that the endocrine regulation of reproductive function is extremely complex and is not yet completely understood.

In addition to stimulating spermatogenesis, testosterone affects almost every tissue in the body. It directly affects muscle and bone. Testosterone is responsible for the adolescent growth spurt in males at about age 13. This hormone stimulates growth of the reproductive organs and so is responsible for the male's **primary sex characteristics.** Testosterone is also responsible for the **secondary sex characteristics** that develop at puberty. These include growth of facial and body hair, muscle development, and the increase in vocal cord length and thickness that causes the voice to deepen.

What happens when testosterone is insufficient or absent? Insufficient testosterone results in sterility. If a male is **castrated,** that is, the testes are removed before puberty, he is deprived of testosterone and becomes a eunuch. He retains childlike sex organs and does not develop secondary sexual characteristics. If castration occurs after puberty, increased secretion of male hormones by the adrenal glands helps to maintain masculinity.

HUMAN REPRODUCTION: THE FEMALE PRODUCES GAMETES AND INCUBATES THE EMBRYO

The female reproductive system produces oocytes (immature gametes), receives the penis and sperm released from it during sexual intercourse, houses and nourishes the embryo during prenatal development, gives birth, and produces milk for the young (lactation). These processes are regulated and coordinated by the interaction of hormones secreted by the hypothalamus, pituitary gland, and ovaries. The principal organs of the female reproductive system are illustrated in Figures 48–10 and 48–11.

The ovaries produce gametes and sex hormones

Like the male gonads, the female gonads, or **ovaries,** produce both gametes and sex hormones. About the size and shape of large almonds, the ovaries are located close to the lateral walls of the pelvic cavity, and are held in position by several connective tissue ligaments. Internally, the ovary consists mainly

(a)

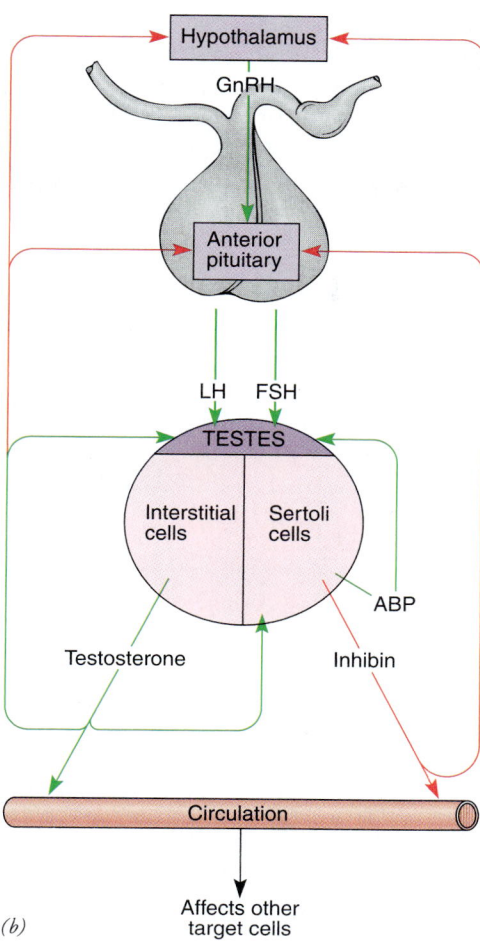

(b)

Figure 48–9 **Regulation of reproduction in the male.** The hypothalamus, anterior pituitary, and testes interact to regulate male reproductive function. *(a)* Overview of hormonal regulation. The hypothalamus secretes gonadotropin-releasing hormone (GnRH), which stimulates the anterior pituitary to secrete FSH and LH. These hormones stimulate the testes to produce testosterone and stimulate spermatogenesis. *(b)* Several negative feedback systems operate to regulate hormone level. Testosterone inhibits the production of GnRH by the hypothalamus and inhibits its action on the pituitary. Inhibin also inhibits FSH and LH secretion.

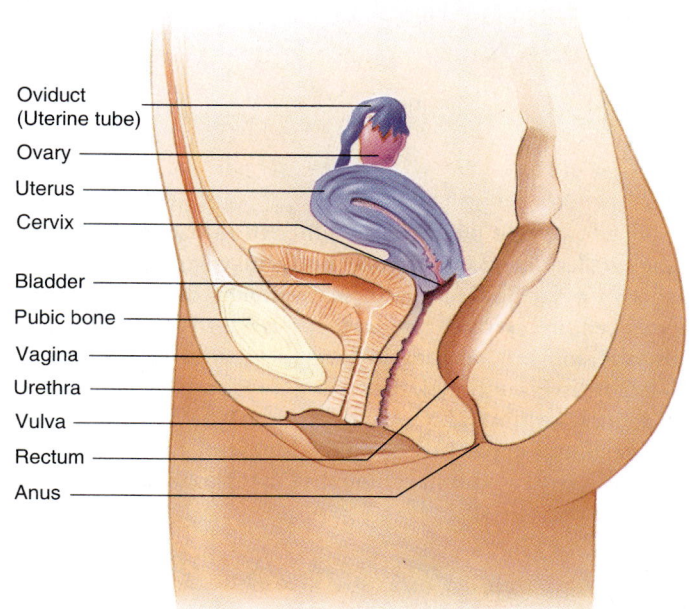

Figure 48–10 **Female reproductive system.** Midsagittal section through the female pelvis. Note the position of the uterus relative to the vagina.

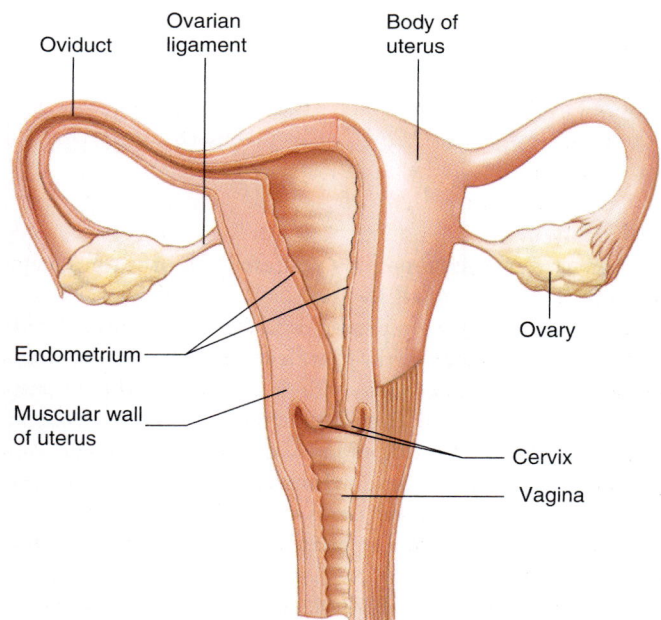

Figure 48–11 **Anterior view of the female reproductive system.** Some organs have been cut open to expose their internal structure. Connective tissue ligaments help hold the reproductive organs in place.

Figure 48-12 **Development of follicles in the ovary.**
(a) LM of a developing follicle. The secondary oocyte is surrounded by a layer of follicle cells that will be released along with it. The layer of follicle cells (called the corona radiata) acts as a barrier to sperm cells. *(b)* Follicles in various stages of development are scattered throughout the ovary. *(a, Biophoto Associates)*

Secondary oocyte

(a)

Antrum

500 μm

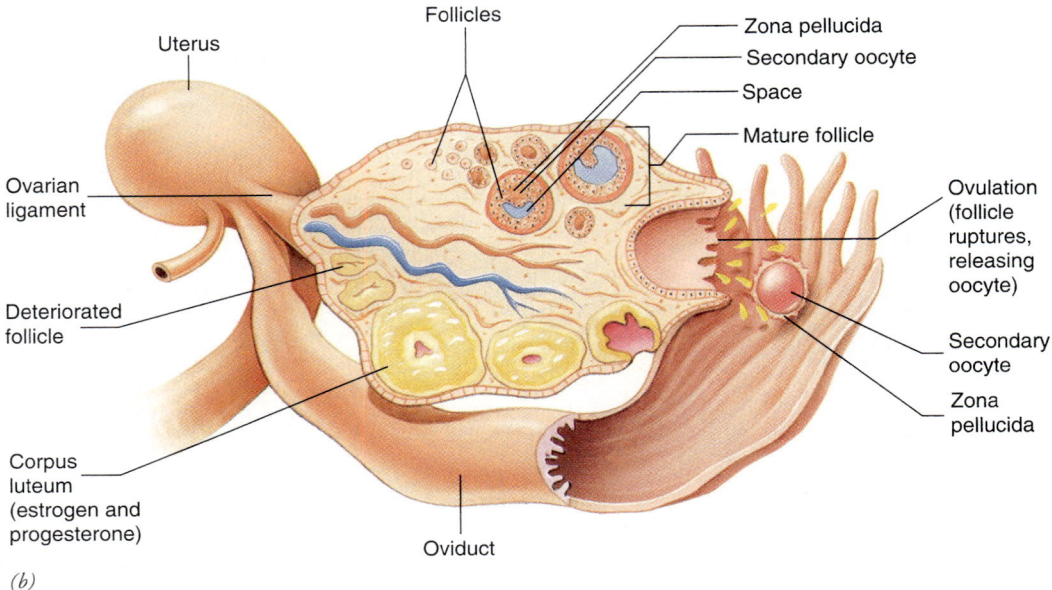

Uterus

Follicles

Zona pellucida

Secondary oocyte

Space

Mature follicle

Ovarian ligament

Ovulation (follicle ruptures, releasing oocyte)

Deteriorated follicle

Secondary oocyte

Zona pellucida

Corpus luteum (estrogen and progesterone)

Oviduct

(b)

of connective tissue containing scattered ova in various stages of maturation (Fig. 48–12).

The process of ovum production, called **oogenesis,** begins in the ovaries. Before birth, hundreds of thousands of **oogonia** are present in the ovaries. All of a female's gametes originate during embryonic development. No new oogonia are formed after birth. During prenatal development, the oogonia increase in size and become **primary oocytes.** By the time of birth, they are in the prophase of the first meiotic division. At this stage, they enter a resting phase that lasts throughout childhood and into adult life.

A primary oocyte and the cluster of cells surrounding it together make up a **follicle.** With the onset of puberty, a few follicles begin to mature each month in response to FSH secreted by the anterior pituitary gland. As the follicle grows, the primary oocyte completes its first meiotic division.

The two cells produced are different in size (Fig. 48–13). The smaller one, the first **polar body,** may later divide, forming two polar bodies, but these eventually disintegrate. The larger cell, the **secondary oocyte,** proceeds to the second meiotic division but remains in metaphase until it is fertilized. When meiosis continues, the second meiotic division gives rise to a single ovum and a second polar body. The polar bodies are small and apparently serve to dispose of unneeded chromosomes with a minimal amount of cytoplasm. The sequence is as follows:

Oogonium (diploid) ⟶ primary oocyte (diploid) ⟶
secondary oocyte + first polar body (both haploid) ⟶
ovum + second polar body (both haploid)

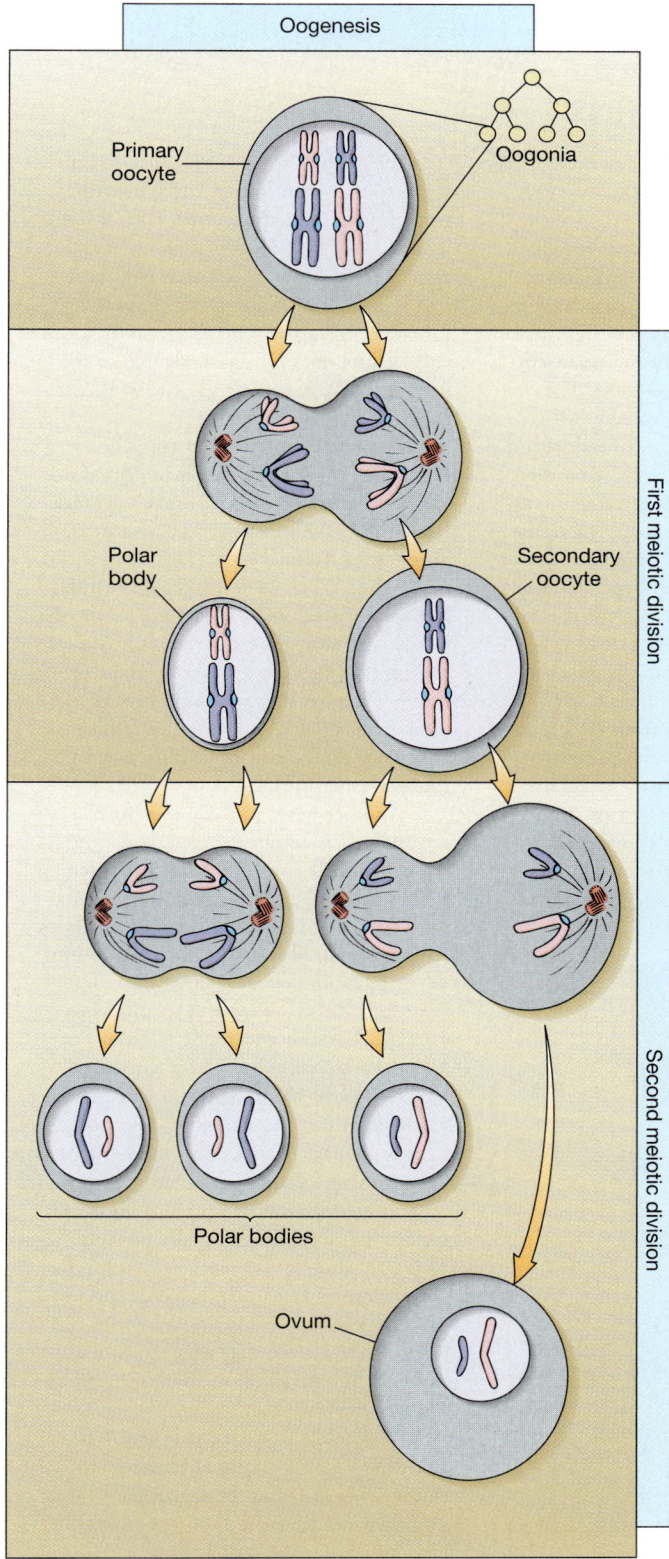

Figure 48–13 **Oogenesis.** Before birth oogonia divide many times by mitosis, then become primary oocytes that undergo meiosis. Only one functional ovum is produced from each primary oocyte. The other cells produced are polar bodies that degenerate. The second meiotic division is completed after fertilization.

Labels in figure:
- Oogenesis
- Oogonia
- Primary oocyte
- First meiotic division
- Polar body
- Secondary oocyte
- Second meiotic division
- Polar bodies
- Ovum

In the male, each primary spermatocyte gives rise to four sperm. In contrast, each primary oocyte generates only one ovum.

As an oocyte develops, it becomes separated from its surrounding follicle cells by a layer of glycoproteins called the **zona pellucida.** As the follicle develops, follicle cells secrete fluid, which collects in the space, called the **antrum,** between them (Fig. 48–12). The follicle cells also secrete **estrogens,** female sex hormones. Typically, only one follicle fully matures each month. Several others may develop for awhile and then deteriorate.

As a follicle matures, it moves closer to the surface of the ovary, eventually resembling a fluid-filled bulge on the ovarian surface. Follicle cells secrete proteolytic enzymes that break down a small area of the ovary wall. During **ovulation,** the secondary oocyte is ejected through the ovary wall and into the pelvic cavity. The portion of the follicle that remains in the ovary develops into the **corpus luteum,** a temporary endocrine gland that secretes estrogens and **progesterone.**

The oviducts transport the secondary oocyte

Almost immediately after ovulation, the secondary oocyte passes into the funnel-shaped opening of the **oviduct,** or **uterine tube** (also called the fallopian tube). Peristaltic contractions of the muscular wall of the oviduct and beating of the cilia in its lining help to move the secondary oocyte along toward the uterus. Fertilization takes place within the oviduct. If fertilization does not occur, the secondary oocyte degenerates there.

Scarring of the oviducts (for example, by sexually transmitted disease) can block the tubes so that the fertilized ovum cannot pass to the uterus. Sometimes partial constriction of the oviduct results in tubal pregnancy, in which the embryo begins to develop in the wall of the oviduct because it cannot progress to the uterus. Oviducts are not adapted to bear the burden of a developing embryo; thus, the oviduct and the embryo it contains must be surgically removed before it ruptures and endangers the life of the mother. Women with blocked oviducts may be infertile. They usually can produce ova and can incubate an embryo. However, they need clinical assistance in getting the ovum from the ovary to the uterus (see *Focus On: Novel Origins*).

The uterus incubates the embryo

The oviducts open into the upper corners of the pear-shaped **uterus** (Fig. 48–11). About the size of a fist, the uterus (or womb) occupies a central position in the pelvic cavity. It has thick walls of smooth muscle and an epithelial lining, the **endometrium,** that thickens each month in preparation for possible pregnancy.

If a secondary oocyte is fertilized, the tiny embryo finds its way into the uterus and implants in the endometrium. There it grows and develops, sustained by nutrients and oxygen delivered by surrounding maternal blood vessels. If fertil-

NOVEL ORIGINS

About 15% of married couples in the United States are affected by infertility, the inability of a couple to achieve conception after using no contraception for at least one year. About 30% of cases involve both male and female factors. Male inferility is often attributed to low sperm count. Among the common causes of female infertility are failure to ovulate, production of infertile eggs (common in older women), and oviduct scarring (often caused by pelvic inflammatory disease) that blocks the passage of the secondary oocyte to the uterus. Women with blocked oviducts can usually produce ova and incubate an embryo normally but require clinical assistance in getting the ovum from the ovary to the uterus.

In the United States, more than 3 million infertile couples consult health care professionals each year. Some are helped with conventional treatment, for example, with hormone therapy that regulates ovulation or with fertility drugs. But more than 40,000 couples need more sophisticated clinical help and turn to high-tech assisted reproductive techniques that have been developed through reproduction and human embryo research. At present these techniques are expensive and the success rate is low, less than one-third.

The most common assisted reproductive procedure is **artificial insemination,** in which a catheter is used to inject sperm directly into the cervix or uterus. Transfer into the uterus is called **intrauterine insemination (IUI).** More than 600,000 IUI procedures are performed annually with a success rate of about 10%. Artificial insemination is indicated when the male partner of a couple desiring a child is sterile or carries a genetic defect. If the male is infertile due to a low sperm count, his sperm can be concentrated, or alternatively, sperm from a donor can be used. Although the sperm donor usually remains anonymous to the couple, his genetic qualifications are screened by physicians.

With **in vitro fertilization (IVF),** an ovum is fertilized with sperm in the laboratory and then implanted in the woman's uterus. This procedure was first used in England in 1978 to help a couple who had tried unsuccessfuly for several years to have a child. Since that time, thousands of "test-tube babies" have been conceived in this way and born to previously infertile women. Typically, the woman takes a fertility drug that induces ovulation of several eggs. The eggs can be removed and fertilized with sperm. Embryos can be screened for chromosome or gene abnormalities, and healthy ones can be implanted in the uterus. In 1995, the success rate of in vitro fertilization was about 20%.

Gamete intrafallopian transfer (GIFT) is a technique in which eggs and sperm are inserted into a woman's oviduct. More than 4000 of these expensive procedures are performed each year with a success rate of about 28%. In **zygote intrafallopian transfer (ZIFT),** eggs are fertilized in the laboratory, and the resulting zygotes are inserted into the oviduct. In these procedures, the patient's own eggs may be used, or if she does not produce fertile eggs, they can be contributed by a donor **(oocyte donation).**

Another novel procedure is **host mothering.** An embryo is removed from its natural mother and implanted into a female substitute. The foster mother can support the developing embryo either until birth, or temporarily until it is implanted again into the original mother or another host. This technique has proved useful to animal breeders. For example, embryos from prize sheep can be temporarily implanted into rabbits for easy shipping by air and then implanted into a host mother sheep, perhaps of inferior quality. Host mothering has the advantage of allowing an animal with superior genetic traits to produce more offspring than would be naturally possible. This procedure is also used to increase the populations of certain endangered species *(see Figure).* Host mothering may someday be popular with women who can produce embryos but are either unable or unwilling to carry them to term.

Technology is available to freeze the gametes or embryos of many species, including humans, and then transplant them into their donors or into host mothers. Freezing eggs may become popular with young women not yet ready to become parents, but who want to preserve young eggs with lower risk for chromosome abnormality, and reimplant them at a later time.

Newborn bongo with its surrogate mother, an eland. As a young embryo, the bongo was transplanted into the eland's uterus, where it implanted and developed. Bongos are a rare and elusive species inhabiting dense forests in Africa. The larger and more common elands, members of the same genus, inhabit open areas.

ization does not occur during the monthly cycle, the endometrium sloughs off and is discharged in the process known as **menstruation.**

More than five million women in the United States are affected by **endometriosis,** a painful disorder in which fragments of the tissue lining the uterus migrate to other areas such as the oviducts or ovaries. Endometriosis causes scarring that can lead to infertility.

The lower portion of the uterus, called the **cervix,** extends slightly into the vagina. The cervix is a common site of cancer in women. Detection is usually possible by the routine Papanicolaou test (Pap smear) in which a few cells are scraped from the cervix during a regular gynecological examination and studied microscopically. When cervical cancer is detected at very early stages of malignancy, the chances that the patient can be cured are good.

The vagina receives sperm

The **vagina** is an elastic, muscular tube that extends from the uterus to the exterior of the body. The vagina serves as a receptacle for sperm during sexual intercourse and as part of the birth canal (Fig. 48–11).

The vulva are external genital structures

The external female sex organs, collectively known as the **vulva,** include liplike folds, the **labia minora,** that surround the vaginal and urethral openings (Fig. 48–14). External to the delicate labia minora are the thicker, hair-covered **labia majora.** Anteriorly, the labia minora merge to form the pre-

puce of the **clitoris,** a small erectile structure comparable to the male glans penis. Like the penis, the clitoris contains erectile tissue that becomes engorged with blood during sexual excitement. Rich in nerve endings, the clitoris is highly sensitive to touch, pressure, and temperature, and serves as a center of sexual sensation in the female.

The **mons pubis** is the mound of fatty tissue just above the clitoris at the junction of the thighs and torso. At puberty it becomes covered by coarse pubic hair. The **hymen** is a thin ring of tissue that forms a border around the entrance to the vagina.

The breasts function in lactation

Each breast is composed of 15 to 20 lobes of glandular tissue. The amount of adipose tissue around these lobes determines the size of the breasts and accounts for their softness. Gland cells are arranged in grapelike clusters called **alveoli** (Fig. 48–15). Ducts from each cluster join to form a single duct from each lobe, producing 15 to 20 tiny openings on the surface of each nipple. The breasts are the most common site of potentially deadly cancer in women (see *Focus On: Breast Cancer*).

Lactation is the production of milk for the nourishment of the young. During pregnancy, high concentrations of the female reproductive hormones, the estrogens and progesterone, stimulate the breasts to increase in size. For the first couple of days after childbirth, the mammary glands produce a fluid called **colostrum,** which contains protein and lactose but lit-

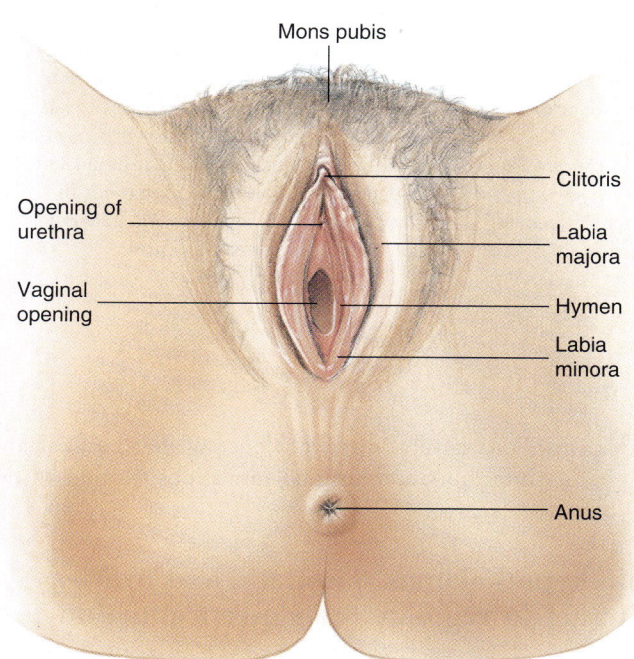

Figure 48–14 **External female genital structures.** Collectively these structures are referred to as the vulva.

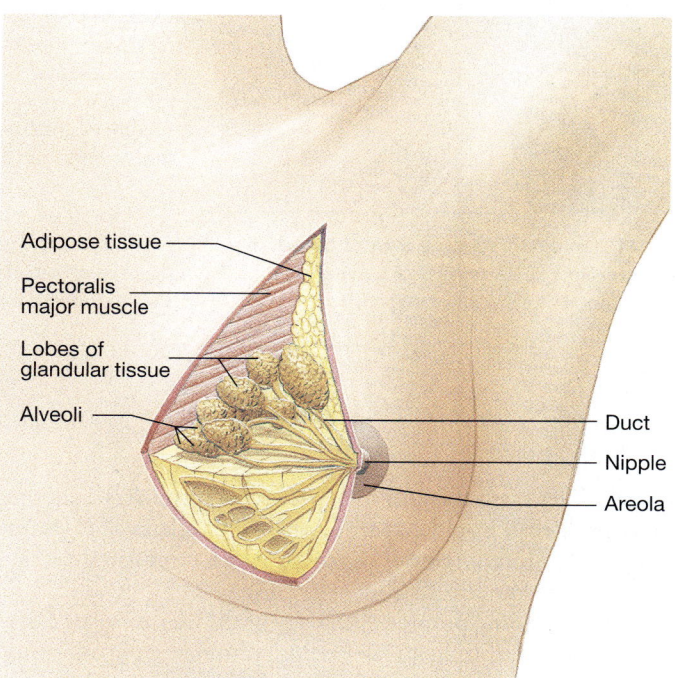

Figure 48–15 **Structure of the mature female breast.** The breast contains lobes of glandular tissue. The lobes consist of alveoli, clusters of gland cells.

BREAST CANCER

Breast cancer is the most common type of cancer among women other than skin cancer. It is a leading cause of cancer deaths in women, secondary only to lung cancer. The causes of breast cancer are not known, but there appears to be a higher risk in women with a family history of the disease. An estimated 5% of breast cancers are familial. Mutations of two tumor suppressor genes (BRCA1 and BRCA2) may predispose a woman to develop breast cancer.

Although no conclusive evidence yet exists, some investigators think that other risk factors include a high-fat diet, obesity, exposure to radiation, and exposure to certain chemicals. Smoking cigarettes increases a woman's risk of dying from breast cancer

Mammogram showing area of breast cancer. Note the extensive vascularization. *(Visuals Unlimited/SIU)*

by at least 25%. Women who smoke two packs or more of cigarettes a day have a 75% greater risk. Taking estrogen for five or more years as hormone replacement therapy after menopause appears to increase the risk of breast cancer.

About 50% of breast cancers begin in the upper, outer quadrant of the breast *(see photograph)*. As a malignant tumor grows, it may adhere to the deep tissue of the chest wall. Sometimes it extends to the skin, causing dimpling. Eventually the cancer spreads to the lymphatic system. About two-thirds of breast cancers have metastasized (spread) to the lymph nodes by the time they are first diagnosed. When diagnosis and treatment begin early, 80% of patients survive for five years, and 62% survive for ten years or longer. Untreated patients have a five-year survival rate of only 20%.

Mastectomy (surgical removal of the breast) and radiation treatment are common methods of treating breast cancer. Lumpectomy (surgical removal of only the affected portion of the breast) in conjunction with radiation treatment is thought to be as effective as mastectomy in some cases. Chemotherapy is useful in preventing metastasis, especially in premenopausal patients. A recent development in cancer treatment is the use of biological response modifiers, which include substances such as interferons, interleukins, and monoclonal antibodies (see Chapter 43).

About one-third of breast cancers are estrogen-dependent, that is, their growth depends on circulating estrogens. Removing the ovaries in patients with these tumors re-

lieves the symptoms and may cause remission of the disease for months or even years. Developing pharmacological agents that antagonize the action of estrogen receptors has been an important approach in the treatment of breast cancer. A challenge has been to inhibit estrogen in the breast while at the same time retaining its beneficial effects on bone, blood vessels, and the brain.

The synthetic drug tamoxifen blocks estrogen receptors, and is a potent chemotherapeutic agent for treating breast cancer patients. This drug appears to be effective in preventing cancer development, especially in patients who have had tumors removed surgically and are at high risk for recurrence of breast cancer. However, studies suggest that tamoxifen may increase risk of uterine cancer. The drug raloxifene is one member of a more recent generation of selective estrogen receptor modulators. Raloxifene has been used clinically to prevent osteoporosis, particularly in patients unable to take estrogen for hormone replacement therapy. Clinical studies suggest that raloxifene may prevent breast cancer without increasing the risk of uterine cancer.

Because early detection of breast cancer greatly increases the chances of cure and survival, campaigns have been launched to educate women on the importance of self-examination. Mammography, a soft-tissue radiological study of the breast, is helpful in detecting very small lesions that might not be identified by routine examination. In mammography, lesions show on an x-ray plate as areas of increased density.

---

tle fat. After birth the hormone **prolactin** stimulates milk production. Recall from Chapter 47 that when a baby suckles, the posterior pituitary releases **oxytocin** which stimulates ejection of milk into the breasts.

Breastfeeding promotes recovery of the uterus because oxytocin released during breastfeeding stimulates the uterus to contract to nonpregnant size. Breastfeeding offers advantages to the baby as well. It promotes a close bond between mother and child and provides milk tailored to the nutritional needs of the human infant. Breast milk contains antibodies, and

breastfed infants have a lower incidence of diarrhea, ear and respiratory infections, and hospital admissions than do bottle-fed babies.

The hypothalamus, pituitary, and ovaries interact to regulate female reproduction

As in the male, regulation of female reproduction is extremely complex, involving many hormones and other signal molecules. Here we provide a simplified overview of the hormonal

TABLE 48-2　Principal Female Reproductive Hormones

Endocrine Gland and Hormones	Principal Target Tissue	Principal Actions
Hypothalamus		
Gonadotropin-releasing hormone (GnRH)	Anterior pituitary	Stimulates release of FSH and LH
Anterior pituitary		
Follicle-stimulating hormone (FSH)	Ovary	Stimulates development of follicles and secretion of estrogen
Luteinizing hormone (LH)	Ovary	Stimulates ovulation and development of corpus luteum
Prolactin	Breast	Stimulates milk production (after breast has been prepared by estrogen and progesterone)
Posterior pituitary		
Oxytocin	Uterus	Stimulates contraction and stimulates prostaglandin release
	Mammary glands	Stimulates ejection of milk into ducts
Ovary		
Estrogens (estradiol)	General	Stimulate growth of sex organs at puberty and development of secondary sex characteristics (breast development, broadening of pelvis, distribution of fat and muscle)
	Reproductive structures	Induce maturation; stimulate monthly preparation of the endometrium for pregnancy; make cervical mucus thinner and more alkaline
Progesterone (secreted mainly by corpus luteum)	Uterus	Completes preparation of endometrium for pregnancy

interactions that have been worked out. Table 48–2 lists the actions of some of the principal female reproductive hormones. Like testosterone in the male, estrogens are responsible for growth of the sex organs at puberty, for body growth, and for the development of secondary sexual characteristics. In the female, these include the development of the breasts, the broadening of the pelvis, and the characteristic development and distribution of muscle and fat responsible for the shape of the female body.

Hormones of the hypothalamus, anterior pituitary, and ovaries regulate the **menstrual cycle,** the monthly sequence of events that prepares the body for possible pregnancy. The menstrual cycle runs its course every month from puberty until menopause occurs at about age 50. Although wide variations exist, a typical menstrual cycle is 28 days long (Fig. 48–16). The first day of the cycle is marked by the onset of menstruation, the monthly discharge through the vagina of blood and tissue from the endometrium. Ovulation occurs on about the 14th day of the cycle.

During the menstrual phase of the cycle, which lasts about five days, gonadotropin-releasing hormone (GnRH) is released from the hypothalamus. GnRH stimulates the anterior pituitary to release follicle-stimulating hormone (FSH) and luteinizing hormone (LH) (Fig. 48–17). FSH stimulates the initial development of a few follicles in the ovary. After a few days, only one follicle continues to develop. FSH also stimulates transcription of the gene that codes for an enzyme, *aromatase,* that is important in estrogen synthesis.

During the **preovulatory phase** (also called the follicular phase) of the menstrual cycle, the developing follicles secrete estrogens. Estrogens have an autocrine action (see Chapter 47) on the follicle cells that produce them. They increase the number of receptors for both estrogens and FSH on the follicle cells. They also promote further secretion of estrogens. Estrogens stimulate growth of the endometrium, which thickens and develops new blood vessels and glands. The rise in the concentration of estrogens in the blood signals the anterior pituitary to secrete LH. Increased concentration of LH in the blood

Figure 48-16 Menstrual cycle.
When fertilization does not occur, the menstrual cycle repeats itself about every 28 days. The events that take place within the ovary and uterus are precisely coordinated by hormones. Note that estrogen concentration is highest during the preovulatory phase, whereas progesterone concentration is highest during the postovulatory phase. Compare this illustration with Figure 48-17.

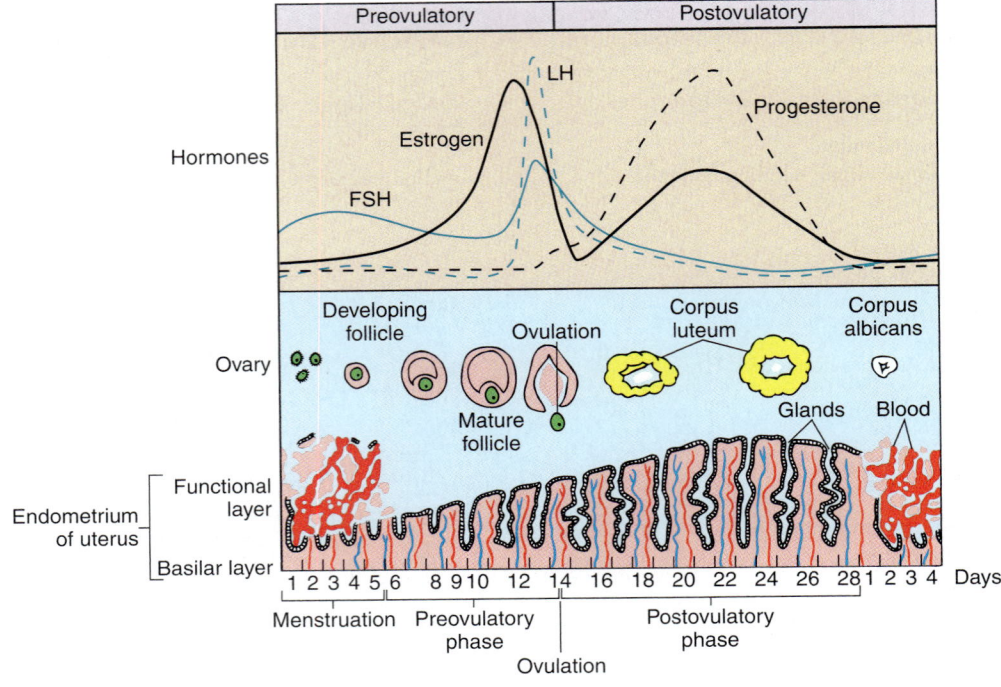

Figure 48-17 Regulation of reproduction in the female. Hormones from the hypothalamus, anterior pituitary, and ovary interact to regulate the menstrual cycle. (a) Hormonal interactions during the preovulatory phase. (b) Hormonal interactions during the postovulatory phase. Note that estrogens have a positive feedback effect on the hypothalamus and pituitary during the preovulatory phase but a negative feedback effect during the postovulatory phase. Red arrows indicate inhibition.

(a) Preovulatory phase

(b) Postovulatory phase

stimulates follicle cells to secrete small amounts of **progesterone.**

FSH and LH together stimulate the final maturation of the follicle. The rise in estrogens secreted by the developing follicle has a positive feedback effect, stimulating the anterior pituitary to secrete a surge of LH. This positive feedback takes place at both the hypothalamus and the pituitary. The surge of LH is necessary for the final maturation of the follicle and stimulates ovulation to occur at about day 14 of the cycle.

After the secondary oocyte has been ejected from the ovary, the **postovulatory phase** (also called the luteal phase) begins. LH stimulates development of the corpus luteum, which secretes a large amount of progesterone and a small amount of estrogens (Fig. 48–17b). These hormones stimulate the uterus to continue its preparation for pregnancy. Progesterone stimulates tiny glands in the endometrium to secrete a fluid rich in nutrients.

Although estrogens stimulate secretion of GnRH, FSH, and LH during the preovulatory phase, they inhibit secretion of these hormones during the postovulatory phase. The different effect is attributed to a change in the sensitivity of the hypothalamus to these hormones.

If the secondary oocyte is not fertilized, the corpus luteum begins to degenerate after about eight days. Although the mechanism responsible for corpus luteum degeneration is not completely understood, a decrease in LH may be a factor. The corpus luteum secretes prostaglandins, which are thought to contribute to its own demise. When the corpus luteum stops secreting progesterone and estrogens, the concentrations of these hormones in the blood fall markedly. As a result, small arteries in the endometrium constrict, reducing the oxygen supply. Menstruation begins as cells die and damaged arteries rupture and bleed. The low levels of estrogens and progesterone are insufficient to inhibit the anterior pituitary, and secretion of FSH and LH increases once again.

Premenstrual syndrome (PMS) is a condition experienced by some women starting several hours to ten days before menstruation and ending a few hours after onset of menstruation. Symptoms include fatigue, anxiety, depression, irritability, headache, edema, and skin eruptions. Its cause is not known.

If the secondary oocyte is fertilized, development begins as it makes its way to the uterus. The embryo begins to implant in the thick endometrium on about the seventh day after fertilization (Fig. 48–18). Membranes that develop around the embryo secrete **human chorionic gonadotropin (hCG),** a hormone that signals the mother's corpus luteum to continue to function.

SEXUAL RESPONSE INVOLVES PHYSIOLOGICAL CHANGES

During copulation, also called **coitus** or sexual intercourse in humans, the male deposits semen into the upper end of the vagina. The complex structures of the male and female reproductive systems, and the physiological, endocrine, and psychological processes associated with sexual activity, are adap-

Fertilization

Implantation

Changes in endometrium

Week 1 — Menstruation

Week 2

Ovulation

Week 3 — Embryo implants in uterus

Week 4 — No menstruation

Figure 48–18 Events following fertilization. The menstrual cycle is interrupted when pregnancy occurs. The corpus luteum does not degenerate, and menstruation does not take place. Instead, the wall of the uterus thickens even more, permitting the embryo to develop within it.

tations that promote the successful union of sperm with secondary oocyte, and the development of the resulting embryo.

Sexual stimulation results in two basic physiological responses: increased blood flow (vasocongestion) to reproductive structures and other tissues such as the skin, and increased muscle tension. During vasocongestion, erectile tissues within the penis and clitoris, as well as in other areas of the body, become engorged with blood.

Sexual response as described by Masters and Johnson includes four phases: **sexual excitement, plateau, orgasm,** and **resolution.** The *desire* to have sexual activity may be motivated by fantasies or thoughts about sex. This anticipation can lead to (physical) sexual excitement and a sense of sexual pleasure. Physiologically, the excitement phase involves vasocongestion and increased muscle tension. Before the penis can enter the vagina and function in coitus, it must be erect. Penile erection is the first male response to sexual excitement. In the female, vaginal lubrication is the first response to effective sexual stimulation. During the excitement phase, the vagina lengthens and expands in preparation for receiving the penis, the clitoris and breasts become vasocongested, and the nipples become erect.

If erotic stimulation continues, sexual excitement heightens to the plateau phase. Vasocongestion and muscle tension increase markedly. In both sexes, blood pressure increases and heart rate and breathing accelerate.

Coitus is usually initiated during the plateau phase. During coitus the penis creates friction as it is moved inward and outward in the vagina in actions referred to as pelvic thrusts. Physical and psychological sensations resulting from this friction (and from the entire intimate experience between the partners) may lead to orgasm, the climax of sexual excitement. In the female, stimulation of the clitoris is important in heightening the sexual excitement that leads to orgasm.

Although it lasts only a few seconds, orgasm is the phase of maximum sexual tension and its release. In both sexes, orgasm is marked by rhythmic contractions of the muscles of the pelvic floor and reproductive structures. These muscular contractions continue at about 0.8-second intervals for several seconds. After the first few contractions, their intensity decreases, and they become less regular and less frequent. Heart rate and respiration more than double, and blood pressure rises markedly, just before and during orgasm. In the male, orgasm is marked by the ejaculation of semen from the penis. No fluid ejaculation accompanies orgasm in the female. Orgasm is followed by the resolution phase, a state of well-being during which the body is restored to its unstimulated state.

Sexual dysfunction may be caused by psychological or biological factors. For example, chronic inability to sustain an erection, termed *erectile dysfunction* (formerly called impotence), is associated with a variety of physical causes and psychological issues. This disorder prevents effective coitus. *Vaginismus* is a condition in which, during sexual intercourse, a woman experiences painful involuntary spasms of the outer third of the vaginal muscles. Vaginismus is often associated with a history of sexual abuse.

FERTILIZATION IS THE FUSION OF SPERM AND EGG TO PRODUCE A ZYGOTE

In the process of **fertilization,** sperm and ovum fuse to produce a zygote. Fertilization and the subsequent establishment of pregnancy together are referred to as **conception.** When conditions in the vagina and cervix are favorable, sperm begin to arrive at the site of fertilization in the upper oviduct within 5 minutes after ejaculation. Contractions of the uterus and oviduct help transport the sperm. The sperms' own motility is probably most important in approaching and fertilizing the ovum. Sperm cannot fertilize a secondary oocyte until they have been in contact with the female reproductive system for several hours, a process known as *capacitation.* When a capacitated sperm encounters an egg, openings develop in the sperm acrosome, allowing enzymes to exit.

If only one sperm is needed to fertilize a secondary oocyte, why are millions involved in each act of coitus? For one thing, sperm movement is undirected, so that many lose their way. Others die as a result of unfavorable pH or phagocytosis by leukocytes and macrophages in the female tract. Only a few thousand succeed in traversing the correct oviduct and reaching the vicinity of the ovum. Additionally, large numbers of sperm may be necessary to penetrate the covering of follicle cells (the corona radiata) that surrounds the ovum. Each sperm releases small amounts of enzymes from its acrosome that help break down the cement-like substance holding the follicle cells together.

As soon as one sperm enters the ovum, a rapid electrical change occurs, followed by a slower chemical change in the plasma membrane of the ovum. These changes prevent the entrance of other sperm. As the fertilizing sperm enters the ovum, it usually loses its flagellum (Fig. 48–19). Sperm entry stimulates the ovum to complete its second meiotic division. The head of the haploid sperm then swells to form the male pronucleus and fuses with the female pronucleus, forming the diploid nucleus of the zygote. The process of fertilization is described in more detail in Chapter 49. (Also see *Focus on Novel Origins.*)

After ejaculation into the female reproductive tract, sperm remain alive and retain their ability to fertilize an ovum for only a few days. The ovum itself remains fertile for only about 24 hours after ovulation. A government study reported in 1995 in the *New England Journal of Medicine* provided evidence that conception is most probable when intercourse takes place on the day of ovulation or the five days preceding ovulation. Thus, in a very regular 28-day menstrual cycle, sexual intercourse on days 9 to 15 is most likely to result in fertilization. However, many women do not have regular menstrual cycles, and many factors can cause irregular cycles even in women who are generally regular.

In view of the many factors working against fertilization, it may seem remarkable that it ever occurs! Yet the frequency of coitus and the large number of sperm deposited at each ejaculation enable the human species not only to maintain itself, but to increase its numbers at an alarming rate.

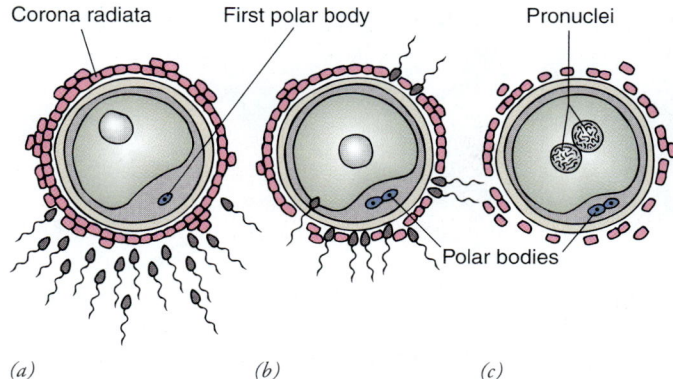

Corona radiata   First polar body   Pronuclei

(a)   *(b)*   *(c)*

Figure 48–19   Fertilization. *(a)* Each sperm is thought to release a small amount of an enzyme that helps to disperse the layer of
follicle cells (corona radiata) surrounding the ovum. *(b)* After a sperm
cell enters, the secondary oocyte completes its second meiotic division, producing an ovum and a polar body. *(c)* Pronuclei of sperm
and ovum combine, producing a zygote with the diploid number of
chromosomes. *(d)* SEM of human sperm cells surrounding a test
ovum. Sperm are being tested for viability.   *(d, David Scharf/Peter Arnold,
Inc.)*

HORMONES REGULATE THE BIRTH PROCESS

How does the body know when to eject the developing baby
from the mother's uterus? The mechanisms for terminating
pregnancy and bringing about the birth process, called **parturition,** are not fully understood. Some signal from the fetus
may be involved. Based on animal studies, some investigators
have proposed that the signal is cortisol because levels of this
hormone rise in the fetus before birth. Cortisol is thought to
help prepare the infant for independent life. For example,
it may stimulate maturation of the lungs and promote some
necessary circulatory changes. However, in humans, prostaglandins appear to be most important in initiating parturition.

Progesterone inhibits contractions of the uterus. The concentration of progesterone in the mother's blood decreases toward the end of pregnancy, removing this inhibition and setting the stage for the birth process. Toward the end of
pregnancy the muscle fibers of the uterine wall develop additional receptors for oxytocin, which plays an important role
during the birth process. As a result the uterus becomes more
sensitive to oxytocin, and uterine contractions become
stronger. In response to oxytocin, uterine muscle fibers secrete
prostaglandins, which through paracrine action also stimulate
uterine contraction.

A long series of involuntary contractions of the uterus are
experienced as **labor.** Labor, which may be divided into three
stages, begins when uterine contractions occur every 10 to 15
minutes. During the first stage, which typically lasts about 12

(d)   100 μm

hours, the contractions of the uterus move the fetus toward
the cervix, causing the cervix to *dilate* (open) to a maximum
diameter of about 10 cm. The cervix also becomes *effaced;* that
is, it flattens so that the fetal head can pass through. During
the first stage of labor the amnion usually ruptures, releasing
about a liter of amniotic fluid, which flows out through the
vagina.

During the second stage, which normally lasts between 20
minutes and an hour, the fetus passes through the cervix and
the vagina and is born, or "delivered" (Fig. 48–20). With each
uterine contraction the woman holds her breath and bears
down so that the fetus is expelled by the combined forces of
uterine contractions and contractions of abdominal wall muscles.

At birth, the baby is still connected to the placenta by the
umbilical cord. Contractions of the uterus squeeze much of
the fetal blood from the placenta into the infant. The cord is
tied and cut, separating the child from the mother. (The stump
of the **cord** gradually shrivels until nothing remains but the
scar, the **navel.**)

During the third stage of labor, which lasts 10 or 15 minutes after the birth, the placenta and the fetal membranes are

Figure 48–20 **Parturition.**
In about 95% of all human births the baby descends through the cervix and vagina in the head-down position. *(a)* The mother bears down hard with her abdominal muscles, helping to push the baby out. When the head fully appears, the physician or midwife can gently grasp it and guide the baby's entrance into the outside world. *(b)* Once the head has emerged, the rest of the body usually follows readily. The physician gently aspirates the mouth and pharynx to clear the upper airway of amniotic fluid, mucus, or blood. At this time the newborn usually takes its first breath. *(c)* The baby, still attached to the placenta by its umbilical cord, is presented to its mother. *(d)* During the third stage of labor, the placenta is delivered. *(Courtesy of Dan Atchinson)*

loosened from the lining of the uterus by another series of contractions and expelled. At this stage they are collectively called the **afterbirth** because these tissues are expelled from the vagina after the baby has been delivered.

During labor an obstetrician may administer oxytocin or prostaglandins to increase the contractions of the uterus, or may assist with special forceps or other techniques. In some women, the opening between the pelvic bones through which the vagina passes is too small to permit the passage of the baby. The baby must then be delivered by **Cesarean section,** an operation in which an incision is made in the abdominal wall and uterus.

BIRTH CONTROL METHODS ALLOW INDIVIDUALS TO CHOOSE

When a fertile, heterosexually active woman uses no form of birth control, her chances of becoming pregnant during the course of a year are about 90%. Any method for deliberately separating sexual intercourse from reproduction is considered **contraception** (literally, "against conception"). Since ancient times, humans have searched for effective contraceptive methods.

Although many couples worldwide agree that it is best to have babies by choice rather than by chance, the majority either are not educated about contraceptive methods or do not have contraceptives available to them. In underdeveloped countries, an estimated 54% of married women lack the means to limit family size. Studies indicate that many of these women would use modern birth control methods if they were available and affordable and if someone showed them how to use them.

More than half of the 7 million women who become pregnant each year in the United States did not intend to do so. Young people particularly lack the knowledge and means of protecting themselves from unwanted pregnancy. Only about one-third of sexually active teenagers consistently use birth control. Every year more than 1 million teenagers in the United States become pregnant, and thousands of girls aged 14 or younger have babies.

Modern science has developed a variety of contraceptives with a high percentage of reliability, but the ideal contraceptive has not yet been developed. Women in the United States currently have four highly effective, reversible methods of contraception: oral contraceptives, copper IUDs, progestin (a synthetic form of progesterone) implants (Norplant), and injectable progestin (DMPA or Depo-Provera). Norplant and Depo-Provera became available in the U.S. in the early 1990s. The only male method of contraception currently available is the condom.

Among the issues that must be considered in developing contraceptives are: safety, effectiveness, cost, convenience, and ease of use. Risks of cancer, birth defects in case the method fails and the woman becomes pregnant, permanent (or long-term) infertility, and side effects such as menstrual abnormalities must also be minimized.

Some researchers predict that the contraceptive methods of the 21st century will control regulatory peptides and the genes that code for them. For example, the genes that code for the pituitary hormones that stimulate the ovaries to release estrogens could be turned off selectively, or other hormone signals could be interrupted. Other approaches being studied are molecular interruption of fertilization, and contraceptive vaccines. Hormone therapy that will interfere with sperm production will require an estimated 15 to 20 years to develop. Within 25 years, current methods of contraception, abortion, and unwanted pregnancy may be replaced by more sophisticated molecular methods.

Some of the more common methods of birth control are described in the following paragraphs and in Table 48–3 (see also Fig. 48–21). Note that intrauterine devices (IUDs) as well as some types of oral contraceptives may not actually prevent fertilization; they probably destroy the embryo or prevent its implantation in the wall of the uterus.

Most hormone contraceptives prevent ovulation

Oral contraceptives, progestin implants (Norplant), and injectable progestin (DMPA, or Depo-Provera) are all hormone contraceptives. More than 80 million women worldwide (more than 8 million in the United States alone) use **oral contraceptives.** The most common preparations are combinations of progestin and synthetic estrogen. (Natural hormones are destroyed by the liver almost immediately, but synthetic ones are chemically modified so that they can be absorbed effectively and metabolized slowly.) In a typical regimen, a woman takes one pill each day for about three weeks. Then for one week she takes a sugar pill that allows menstruation to occur because of the withdrawal of the hormones. When taken correctly, these pills are about 99.7% effective in preventing pregnancy.

Most oral contraceptives prevent pregnancy by preventing ovulation. When postovulatory levels of ovarian hormones are maintained in the blood, the body is tricked into responding as though conception has occurred. The pituitary gland is inhibited and does not produce the surge of LH that stimulates ovulation. The chief advantage of oral contraceptives is their high rate of effectiveness.

Studies suggest that women over the age of 35 who smoke or have other risk factors, such as untreated hypertension, should not take oral contraceptives. Women in this category who take oral contraceptives have an increased risk of death from cardiovascular diseases such as stroke and myocardial infarction. The new low-dose oral contraceptive pills appear to be safe for nonsmokers up to the time of menopause. Oral contraceptives are linked to deaths in about 3 per 100,000 users. This compares favorably with the death rate of about 9 per 100,000 pregnancies (Table 48–4).

Depo-Provera (injectable progestin) prevents ovulation by suppressing anterior pituitary function. It is generally injected intramuscularly every three months. **Implantation of progestin** (Norplant) in the woman's body involves inserting several soft, flexible capsules under the skin of the upper arm. There the synthetic hormone is continuously released and is transported by the blood. The hormone inhibits ovulation and stimulates thickening of the cervical mucus (which makes it more difficult for the sperm to reach the egg). The capsules are effective for about five years, after which they must be replaced. Progestin implants are one of the most effective, reversible contraceptive procedures available. The failure rate is about 1 in 100. The most common side effect is irregular menstrual bleeding, which may last up to one year.

Use of the intrauterine device (IUD) has declined

The **intrauterine device (IUD)** is a small plastic loop or coil that must be inserted into the uterus by a medical professional. Once in place, it can be left in the uterus for up to ten years or until the woman wishes to conceive. IUDs are about 99% effective.

Figure 48–21 Some commonly used contraceptives. The only common male method of contraception is the condom. *(McMurray Photography)*

TABLE 48-3   Birth Control Methods

Method	Failure Rate*	Mode of Action	Advantages	Disadvantages
Oral contraceptives	0.3; 5	Inhibit ovulation; may also affect endometrium and cervical mucus and prevent implantation	Highly effective; regulate menstrual cycle	Minor discomfort in some women; possible thromboembolism, hypertension, heart disease in some users; possible increased risk of infertility; should not be used by women who smoke
Depo-Provera (DMPA)	About 1	Inhibits ovulation	Effective; long-lasting	Fertility may not return for 6–12 months after discontinued
Progesterone implantation (Norplant)	About 1	Inhibits ovulation	Effective; long-lasting	Irregular menstrual bleeding in some women
Intrauterine device (IUD)	1; 5	Not known; probably stimulates inflammatory response and prevents implantation	Provides continuous protection; highly effective	Cramps; increased menstrual flow; spontaneous expulsion; increased risk of pelvic inflammatory disease and infertility; not recommended for women who have not completed childbearing
Spermicides (foams, jellies, creams)	3; 20	Chemically kill sperm	No side effects (?); some kill HIV and other pathogens that cause STD	Some evidence linking spermicides to birth defects

*The lower figure is the failure rate of the method; the higher figure is the rate of method failure plus failure of the user to utilize the method correctly. Based on number of failures per 100 women who use the method per year in the United States.
†The failure rate is lower when the diaphragm is used together with spermicides.

The IUD's mode of action is not well understood. White blood cells, mobilized in response to the foreign body in the uterus, produce substances toxic to the fertilized ovum. There are two types of IUDs presently being used in the United States. One type contains progestin; the other (Copper-T) contains copper, which dissolves slowly in the uterine secretions and apparently interferes with embryo implantation. Disadvantages of the IUD include uterine cramping and bleeding, and increased risk of pelvic inflammatory disease (PID), described in Table 48–5.

Other common contraceptive methods include the diaphragm and condom

The **contraceptive diaphragm** mechanically blocks the passage of sperm from the vagina into the cervix. It is covered with spermicidal jelly or cream and inserted just prior to sexual intercourse.

The **condom** is also a mechanical method of birth control. The only contraceptive device currently sold for men, the condom provides a barrier that contains the semen so that sperm cannot enter the female tract. The condom is one of the only contraceptives that provides some protection against AIDS and other sexually transmitted diseases. Thinner, stronger condoms made of nonlatex material are being developed. A condom for women has been developed but is not widely used.

Emergency contraception is available

Oral contraceptives can be used as **morning-after** (post-coital) **pills,** that is, as emergency, after-the-fact contraception for rape victims and others who have had unprotected intercourse. These contraceptives, which are most commonly high doses of oral hormone contraceptives, change the endometrium so that the embryo cannot implant in the uterine wall. Taken up to

TABLE 48-3 continued

Method	Failure Rate*	Mode of action	Advantages	Disadvantages
Contraceptive diaphragm (with jelly)†	3; 14	Diaphragm mechanically blocks entrance to cervix; jelly is spermicidal	No side effects	Must be prescribed (and fitted) by physician; must be inserted prior to coitus and left in place for several hours after intercourse
Condom	2.6; 10	Mechanically prevents sperm from entering vagina	No side effects; some protection against STD, including AIDS	Interruption of foreplay to put it on; slightly decreased sensation for male; could break
Rhythm‡	13; 21	Abstinence during fertile period	No side effects (?)	Not very reliable
Douche	40	Flush semen from vagina	No side effects	Not reliable, sperm are beyond reach of douche in seconds
Withdrawal (coitus interruptus)	9; 22	Male withdraws penis from vagina prior to ejaculation	No side effects	Not reliable; sperm in the fluid secreted before ejaculation may be sufficient for conception
Sterilization				
Tubal ligation	0.04	Prevents ovum from leaving uterine tube	Most reliable method	Often not reversible
Vasectomy	0.15	Prevents sperm from leaving vas deferens	Most reliable method	Often not reversible
Chance (no contraception)	About 90			

‡There are several variations of the rhythm method. For those who use the calendar method alone, the failure rate is about 35. However, if the body temperature is taken daily and careful records are kept (temperature rises after ovulation), the failure rate can be reduced. When women use some method to help determine the time of ovulation and have intercourse *only* more than 48 hours *after* ovulation, the failure rate can be reduced to about 7.

72 hours after unprotected intercourse, they are about 75% effective in preventing pregnancy. Insertion of a Copper-T IUD within a week of unprotected intercourse is more than 99% effective in preventing pregnancy.

Emergency contraception is not an abortion. It does not appear to harm an embryo if it is mistakenly used in an already established pregnancy. After-the-fact contraception has not become popular in the United States because most women are unaware that anything can be done to prevent pregnancy after sex. Those who are aware may have difficulty obtaining a prescription within 72 hours. In 1997 David A. Kessler, Commissioner of the Food and Drug Administration, asserted that if physicians advocated and women used emergency contraception, an estimated 2 million unwanted pregnancies and thousands of abortions could be prevented.

TABLE 48-4 Deaths in the United States from Pregnancy and Childbirth and from Some Birth Control Methods

	Death Rate per 100,000
Pregnancy and childbirth	9
Oral contraception	3
IUD	0.5
Legal abortions—first trimester	1.9
Legal abortions—after first trimester (mainly therapeutic abortions)	12.5
Illegal abortions performed by medically untrained individuals	About 100

TABLE 48-5   Some Common Sexually Transmitted Diseases*

Disease and Causative Organism	Course of Disease	Treatment
Chlamydia (*Chlamydia trachomatis*, a bacterium)	Discharge and burning with urination, or may be asymptomatic; most common cause of nongonococcal urethritis in males; about 10% of male college students in the United States are infected	Antibiotics: Doxycyclin and azithromycin (Zithromax)
Gonorrhea (*Neisseria gonorrhoeae*, a gonococcus bacterium)	Bacterial toxin may produce redness and swelling at infection site; symptoms in males are painful urination and discharge of pus from penis; in about 60% of infected women no symptoms occur in initial states; can spread to epididymis (in males) or uterine tubes and ovaries (in females), causing sterility; can cause widespread pelvic or other infection, plus damage to heart valves, meninges (outer coverings of brain and spinal cord), and joints	Cephalosporin
Syphilis (*Treponema pallidum*, a spirochete bacterium)	Bacteria enter body through defect in skin near site of infection and spread throughout the body by lymphatic and circulatory routes; primary chancre (a small, painless ulcer) forms at site of initial infection and heals in about a month: highly infectious at this stage; secondary stage follows, in which a widespread rash and influenza-like symptoms may occur; scaly lesions may occur that teem with bacteria and are highly infectious; latent stage that follows can last 20 years; eventually, lesions called gummae may form, consuming parts of the body surface or damaging liver, bone, or spleen; serious brain damage may occur; death results in 5–10% of cases	Penicillin

Primary syphilitic chancre. This ulcer is typically the first symptom of syphilis. (*Custom Medical Stock Photo*)

Sterilization renders an individual incapable of producing offspring

Aside from total abstinence, **sterilization** is the only foolproof method of contraception. Sterilization is currently the most popular contraceptive method for couples in which the woman is over the age of 30.

Male sterilization is performed by vasectomy

An estimated 1 million **vasectomies** are performed each year in the United States. Using a local anesthetic, a small incision is made on each side of the scrotum. Then, each vas deferens is cut and its ends are tied or clipped so that they cannot grow back together (Fig. 48–22*a*). Because testosterone secretion and transport are not affected, vasectomy does not affect mas-

culinity. Sperm continue to be produced, though at a much slower rate, and are destroyed by macrophages in the testes. No change in the amount of semen ejaculated is noticed because sperm account for very little of the semen volume.

By surgically reuniting the ends of the vasa deferentia, surgeons can successfully reverse male sterilization in about 70% of attempts. Apparently, some sterilized men eventually develop antibodies against their own sperm and remain sterile even after their vasectomies have been surgically reversed. For this reason fertility rates are low (about 30%) for men who undergo vasectomy reversal after ten or more years.

An alternative to vasectomy reversal is the storage of frozen sperm in sperm banks. If the male should decide to father another child after he has been sterilized, he simply "withdraws" his sperm in order to artificially inseminate his mate. Sperm banks have been established throughout the United States. Not

TABLE 48–5   continued

Disease and Causative Organism	Course of Disease	Treatment
Genital herpes (herpes simplex type 2 virus)	Tiny, painful blisters appear on genitals; may develop into ulcers; influenza-like symptoms may occur; recurs periodically; threat to fetus or newborn infant; may predispose to cervical cancer in females	No effective cure; some drugs may shorten outbreaks or reduce severity of symptoms; new medication available that can be taken to prevent outbreak
Trichomoniasis (a protozoon)	Symptoms include itching, discharge, soreness; can be contracted from dirty toilet seats and towels; may be asymptomatic in males	Flagyl (an antibiotic)
"Yeast" infections (genital candidiasis; *Candida albicans*)	Irritation, soreness, discharge; especially common in females; may be asymptomatic in males	Antifungal drugs
Pelvic inflammatory disease (PID; primarily caused by gonorrhea or chlamydia)	Generalized infection of reproductive organs and pelvic cavity; usually chronic and difficult to treat; may lead to sterility (more than 15% of cases)	Antibiotics, surgical removal of affected organs
Genital warts (certain strains of human papilloma virus [HPV])	Warty growths may be present on the internal or external genitalia; associated with cervical cancer	Difficult to treat; antiviral drugs or injection with agents that destroy infected cells
Acquired immune deficiency syndrome (AIDS)* (caused by a retrovirus known as human immunodeficiency virus, HIV)	Influenza-like symptoms: swollen lymph glands, fever, night sweats, weight loss; decreased immunity, leading to pneumonia, rare forms of cancer	AZT, protease inhibitors, and a variety of experimental drugs

*AIDS was discussed in Chapter 43.

much is known yet about the effects of long-term sperm storage, and there may be an increased risk of genetic defects.

Female sterilization is by tubal ligation

An estimated one in four married women between the ages of 15 and 44 have chosen sterilization as a means of birth control. **Tubal ligation** is generally performed under general anesthesia by making an abdominal incision and using an instrument called a laparoscope to locate and cut the oviducts (Fig. 48–22b). The ends of the tubes are then tied or cauterized. As in the male, hormone balance and sexual performance are not affected by sterilization.

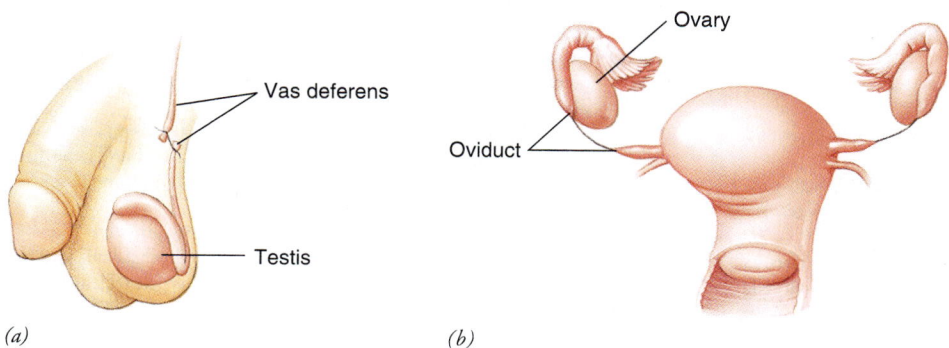

(a)   *(b)*

Figure 48–22 **Sterilization.** *(a)* In vasectomy, the vas deferens (sperm duct) on each side is cut and tied. *(b)* In tubal ligation, each oviduct is cut and tied so that ovum and sperm can no longer meet.

There are three types of abortion

Abortion is termination of pregnancy that results in the death of the embryo or fetus. Worldwide, an estimated 40 million deliberate abortions are performed each year (more than a million in the United States). Three types of abortions may be distinguished: spontaneous abortions, therapeutic abortions, and those performed as a means of birth control.

Spontaneous abortions (popularly called miscarriages) occur without intervention. Spontaneously aborted embryos are frequently abnormal. **Therapeutic abortions** are performed in order to maintain the health of the mother or when there is reason to suspect that the embryo is grossly abnormal. The third type of abortion, performed as a means of birth control, is the most controversial.

Most first-trimester abortions (those performed during the first three months of pregnancy) are performed by a suction method or by administering drugs that interrupt the pregnancy and induce labor. In the suction method of abortion, the cervix is dilated, and a suction aspirator is inserted into the uterus. The embryo and other products of conception are evacuated. The action of methotrexate, a drug used to treat cancer, is similar to that of the controversial French drug RU-486, known as mifepristone in the United States. (At the time of this writing mifepristone has been conditionally approved by the FDA but is still largely unavailable in the United States.) Both of these drugs interrupt pregnancy. Prostaglandins are then used to induce the uterine contractions that expel the embryo.

In pregnancies of more than 12 weeks, the method most commonly used is dilation and evacuation (D & E). The cervix is dilated, forceps are used to remove the fetus, and suction is used to aspirate the contents of the uterus. Drugs such as prostaglandins are being increasingly used to induce labor in second trimester abortions. After the first trimester, abortions are performed mainly because of fetal defects or maternal illness.

When abortions are performed during the first trimester by skilled medical personnel, the mortality rate in the United States is about 1.9 per 100,000. After the first trimester, the death rate rises to 12.5 per 100,000 (Table 48–4). The U.S. death rate from illegal abortions performed by medically untrained individuals is about 100 per 100,000. These statistics can be contrasted with the death rate from pregnancy and childbirth of about 9 per 100,000.

SEXUALLY TRANSMITTED DISEASES ARE SPREAD BY SEXUAL CONTACT

Sexually transmitted diseases (STD), also called venereal diseases (VD), are, next to the common cold, the most prevalent communicable diseases in the world. The World Health Organization has estimated that more than 250 million people are infected each year with **gonorrhea,** and more than 50 million are infected with **syphilis.** Currently, the most common STD in the United States is **chlamydia,** which is the most common cause of **pelvic inflammatory disease (PID).** According to Planned Parenthood, one out of four U.S. teenagers becomes infected with an STD by age 21. Some common sexually transmitted diseases are described in Table 48–5. (AIDS was discussed in Chapter 43.)

SUMMARY WITH KEY TERMS

I. In **asexual reproduction,** a single parent endows its offspring with a set of genes identical to its own (except for mutations). Asexual reproduction can take place by budding, fragmentation, or parthenogenesis.
  A. In **budding,** a part of the parent's body grows and separates from the rest of the body.
  B. In **fragmentation,** the parent's body may break into several pieces; each piece can develop into a new animal.
  C. In **parthenogenesis,** an unfertilized egg develops into an adult.
II. In **sexual reproduction,** offspring are produced by the fusion of two types of **gametes** (egg and sperm). When sperm and egg fuse, a fertilized egg, or **zygote,** forms.
  A. In **external fertilization,** mating partners typically release eggs and sperm into the water simultaneously.
  B. In **internal fertilization,** the male delivers sperm into the female's body.
  C. In **hermaphroditism,** a single individual produces both eggs and sperm.
III. The human male reproductive system includes the **testes,** which produce sperm and **testosterone;** a series of conducting ducts; accessory glands; and the **penis.**
  A. The testes, housed in the **scrotum,** contain the **seminiferous tubules,** where **spermatogenesis** takes place. The **interstitial cells** in the testes secrete testosterone. **Sertoli cells** are large cells that produce a fluid that nourishes sperm cells.

1. **Spermatogonia** divide by mitosis; some differentiate and become **primary spermatocytes** which undergo **meiosis.**
2. The first meiotic division produces two **secondary spermatocytes.** In the second meiotic division, each secondary spermatocyte yields two **spermatids.**
3. Each spermatid differentiates to form a mature sperm. The head of a sperm consists of the nucleus and a cap, or **acrosome,** containing enzymes.
  B. Sperm complete their maturation and are stored in the **epididymis** and **vas deferens.**
  C. During ejaculation, sperm pass from the vas deferens to the **ejaculatory duct** and then into the **urethra,** which passes through the penis.
  D. Each ejaculate of semen contains about 400 million sperm suspended in the secretions of the **seminal vesicles** and **prostate gland.** The **bulbourethral glands** release a mucous secretion.
  E. The penis consists of three columns of **erectile tissue,** two **cavernous bodies** and one **spongy body** which surrounds the urethra. When this tissue becomes engorged with blood, the penis becomes erect.
  F. The endocrine regulation of male reproduction involves the hypothalamus, pituitary gland, and testes. The hypothalamus secretes **gonadotropin-releasing hormone (GnRH)** which stimulates the anterior pituitary gland to secrete the gonadotropic hormones: **follicle-stimulating hormone (FSH)** and **luteinizing hormone**

(LH), also called **interstitial cell stimulating hormone (ICSH).**

1. FSH, LH, and testosterone directly or indirectly stimulate sperm production. LH stimulates the interstitial cells of the testes to produce testosterone.
2. FSH stimulates the Sertoli cells to produce (1) **androgen-binding protein (ABP),** which binds to testosterone and concentrates it; and (2) **inhibin,** a hormone that inhibits secretion of FSH.
3. Testosterone is responsible for establishing and maintaining male **primary sex characteristics** and **secondary sex characteristics.**

IV. The female reproductive system includes the **ovaries,** which produce oocytes and the steroid hormones **estrogens** and **progesterone;** the **oviducts (uterine tubes);** the **uterus** that incubates the developing embryo; the **vagina; vulva,** or external genital structures; and breasts.
   A. **Oogenesis** takes place in the ovaries. **Oogonia** differentiate into **primary oocytes.** A primary oocyte and the cluster of cells surrounding it make up a **follicle.**
      1. As the follicle grows, the primary oocyte undergoes the first meiotic division, giving rise to a **secondary oocyte** and a **polar body.** After **ovulation,** the **secondary oocyte** enters the oviduct, where it may be fertilized.
      2. The part of the follicle remaining in the ovary develops into a **corpus luteum.**
   B. The uterus serves as an incubator for the developing embryo. The epithelial lining of the uterus is the **endometrium.** The lower part of the uterus, the **cervix,** extends into the vagina.
   C. The vagina is the lower part of the birth canal and also receives the penis during sexual intercourse.
   D. The vulva include the labia majora, labia minora, clitoris, and mons pubis.
   E. The breasts function in **lactation.** Each breast consists of 15 to 20 lobes of glandular tissue. Gland cells are arranged in **alveoli.**
   F. The endocrine regulation of female reproduction involves the hypothalamus, pituitary gland, and ovaries. The first day of the **menstrual cycle** is marked by the beginning of menstrual bleeding. **Ovulation** occurs at about day 14 in a typical 28-day menstrual cycle. Events of the menstrual cycle are coordinated by gonadotropic and ovarian hormones.
      1. During the **preovulatory phase,** the GnRH from the hypothalamus stimulates the anterior lobe of the pituitary to secrete

FSH, which stimulates follicle development. The developing follicles release estrogens that stimulate development of the endometrium and signal the anterior pituitary to secrete LH. LH stimulates ovulation.
      2. During the **postovulatory phase,** LH promotes development of the corpus luteum. The corpus luteum secretes progesterone and estrogens that stimulate final preparation of the uterus for possible pregnancy. During the postovulatory phase, estrogens inhibit secretion of GnRH, FSH, and LH.
      3. If the secondary oocyte is fertilized, development begins and the tiny embryo implants in the uterus; membranes that develop around the embryo secrete **human chorionic gonadotropin (hCG),** a hormone that maintains the corpus luteum. If fertilization does not occur, the corpus luteum degenerates, concentrations of estrogens and progesterone in the blood fall, and menstruation occurs.
      4. Estrogens are responsible for the secondary female sex characteristics.

V. Vasocongestion and increased muscle tension are physiological responses to sexual stimulation. The phases of sexual response include **sexual excitement, plateau, orgasm,** and **resolution.**

VI. Human **fertilization** is the fusion of secondary oocyte and sperm to form a zygote. Fertilization and establishment of pregnancy together are called **conception.**

VII. Several hormones, including oxytocin and prostaglandins, regulate **parturition,** the birth process. Labor can be divided into three stages; the baby is delivered during the second stage.

VIII. Effective methods of **contraception** include hormonal methods (**oral contraceptives, Depo-Provera, implantation of progestin,** and **morning-after pills**); intrauterine devices; condoms; contraceptive **diaphragms;** and **sterilization** (**vasectomy** or **tubal ligation**). Emergency contraception can be used to prevent unwanted pregnancy after rape or unprotected intercourse.

IX. **Spontaneous abortions** (miscarriages) occur without intervention. **Therapeutic abortions** are performed to maintain the mother's health or when the embryo is thought to be grossly abnormal. Abortions are also performed as a means of birth control.

X. Among the common types of **sexually transmitted diseases (STD)** are **chlamydia, gonorrhea, syphilis, genital herpes, pelvic inflammatory disease (PID),** and **AIDS.**

POST-TEST

1. Which of the following is NOT an example of asexual reproduction? (a) budding (b) external fertilization (c) fragmentation (d) parthenogenesis (e) regeneration to form new individual from a fragment
2. Hermaphroditism (a) is a form of asexual reproduction (b) occurs when an unfertilized egg develops into an adult animal (c) can involve cross-fertilization between two animals (d) typically involves self-fertilization (e) typically involves only one animal
3. The seminiferous tubules (a) are the site of spermatogenesis (b) produce most of the seminal fluid (c) empty directly into the vas deferens (d) are located within the cavernous body (e) receive fluid from the bulbourethral glands
4. Which sequence is correct? (a) spermatogonium ⟶ spermatid ⟶ primary spermatocyte ⟶ secondary spermatocyte ⟶ sperm (b) spermatid ⟶ spermatogonium ⟶ primary spermatocyte ⟶ secondary spermatocyte⟶sperm (c) spermatid ⟶ primary spermatocyte ⟶ secondary spermatocyte ⟶ spermatogonium ⟶ sperm (d) spermatogonium ⟶ secondary spermatocyte ⟶ primary spermatocyte ⟶ spermatid ⟶ sperm (e) spermatogonium ⟶ primary spermatocyte ⟶ secondary spermatocyte ⟶ spermatid ⟶ sperm
5. Which sequence best describes passage of sperm? (a) seminiferous tubules ⟶ vas deferens ⟶ epididymis ⟶ ejaculatory duct ⟶ urethra

(b) seminiferous tubules ⟶ epididymis ⟶ vas deferens ⟶ ejaculatory duct ⟶ urethra (c) seminiferous tubules ⟶ prostate ⟶ epididymis ⟶ ejaculatory duct ⟶ urethra (d) seminiferous tubules ⟶ vas deferens ⟶ epididymis ⟶ seminal vesicles ⟶ urethra (e) seminiferous tubules ⟶ epididymis ⟶ vas deferens ⟶ prostate ⟶ seminal vesicle ⟶ urethra

6. Which of the following characteristics is NOT associated with testosterone? (a) maintains secondary sex characteristics (b) responsible for primary sex characteristics (c) principal androgen (d) protein hormone (e) necessary for spermatogenesis
7. Androgen-binding protein (ABP) (a) is secreted by Sertoli cells (b) is transported in the blood (c) inhibits secretion of FSH (d) inhibits spermatogenesis (e) two of the preceding answers are correct
8. Which of the following cells is haploid? (a) primary oocyte (b) oogonium (c) secondary oocyte (d) corpus luteum (e) follicle cell
9. The corpus luteum (a) is surrounded by ova (b) degenerates if fertilization occurs (c) develops in the preovulatory phase (d) is maintained by prostaglandins (e) serves as a temporary endocrine gland
10. After ovulation the secondary oocyte enters the (a) ovary (b) corpus luteum (c) cervix (d) oviduct (e) vagina
11. The endometrium (a) is the muscle layer of the uterus (b) is thickest during the preovulatory phase (c) is the site of embryo implantation

(d) lines the vagina (e) is directly affected by FSH

12. The hormone that reaches its highest level during the postovulatory phase is (a) progesterone (b) estrogens (c) FSH (d) LH (e) testosterone

13. Uterine contraction is strongly stimulated by (a) progesterone (b) FSH (c) LH (d) estrogens (e) oxytocin

14. Which of the following is NOT a hormonal type of contraception? (a) Depo-Provera (b) contraceptive diaphragm (c) injectable progestin (d) implantation of progestin (e) oral contraceptives

15. A primary chancre is a sign of (a) syphilis (b) gonorrhea (c) genital herpes (d) pelvic inflammatory disease (e) chlamydia

REVIEW QUESTIONS

1. Compare asexual with sexual reproduction, and give two specific examples of asexual reproduction.

2. Explain the physiological basis of erection of the penis.

3. Compare the functions of ovaries and testes.

4. Trace the passage of sperm from a seminiferous tubule through the male reproductive system until it leaves the male body during ejaculation. Assuming that ejaculation takes place within the vagina, trace the journey of the sperm until it meets the ovum.

5. What are the actions of testosterone? Give an overview of the endocrine regulation of male reproduction.

6. What is the function of the corpus luteum? Which hormone stimulates its development?

7. Give an overview of the endocrine regulation of female reproduction. What are specific actions of FSH and LH in the female? Of estrogens? Of progesterone?

8. Why are so many sperm produced in the male and so few ova produced in the female?

9. Which methods of birth control are most effective? Least effective? Describe emergency contraception.

10. Draw a diagram of the principal events of the menstrual cycle, including ovulation and menstruation. Indicate on which days of the cycle sexual intercourse would most likely result in pregnancy.

YOU MAKE THE CONNECTION

1. What are the biological advantages of hermaphroditism with self-fertilization? When cross-fertilization is necessary?

2. What would happen if ovulation occurred but no corpus luteum developed?

3. Why do you think a variety of effective methods of male contraception have not yet been developed? If you were a researcher, what are some approaches you might take to male contraception?

RECOMMENDED READINGS

Alexander, N.J. "Future Contraceptives." *Scientific American*, Vol. 273, No. 3, Sept. 1995. An interesting discussion of future contraceptive options, including some for males.

Davis, D.L., Axelrod, D., Osborne, M.P., and N.T. Telang. "Environmental Influences on Breast Cancer Risk." *Science & Medicine*, Vol. 4, No. 3, May/Jun. 1997. Exposure to estrogen is a major risk factor for breast cancer.

Duellman, W.E. "Reproductive Strategies of Frogs." *Scientific American*, Vol. 267, No. 1, Jul. 1992. In addition to the egg-to-tadpole progression, frogs have other reproductive strategies such as egg to froglet, egg brooding, and tadpoles in the mother's stomach.

Hoberman, J.M., and Yesalis, C.C. "The History of Synthetic Testosterone." *Scientific American*, Vol. 272, No. 2, Feb. 1995. The authors suggest that testosterone therapy could become common during the next few years.

Kuiper, G.G.J.M., Carlquist, M., and J. Gustafsson. "Estrogen Is a Male and Female Hormone." *Science & Medicine*, Vol. 5, No. 4, Jul./Aug. 1998. Discovery of a second type of estrogen receptor appears to be related to a new understanding of the actions of estrogens in males.

Osborne, C.K., Elledge, R.M., and S.A.W. Fuqua. "Estrogen Receptors in Breast Cancer Therapy." *Science & Medicine*, Vol. 3, No. 1, Jan./Feb. 1996.

Wright, K. "Human in the Age of Mechanical Reproduction." *Discover*, May 1998. An interesting discussion of non-coital conception.

● Visit our website at **http://www.saunderscollege.com/lifesci/titles.html** and click on Solomon/Berg/Martin Biology for links to chapter-related resources on the World Wide Web.

CHAPTER 49

Animal Development

Development includes all the changes that take place during the entire life of an individual. In this chapter, however, we focus mainly on the fertilization of the egg to form a zygote, and the subsequent development of the young animal before birth or hatching. Just how does a microscopic, single-celled zygote give rise to the bones, muscles, brain, and other structures of a complex animal? These are derived from a balanced combination of several fundamental processes: cell division and growth, determination and cell differentiation, and pattern formation and morphogenesis.

The zygote divides by mitosis, forming an **embryo** that subsequently undergoes an orderly sequence of cell divisions. Although the very early embryo usually does not grow, later cell divisions typically do contribute to growth. (Growth in animals occurs mainly by an increase in the number of cells, although in some cases growth may be due to an increase in cell size). However, by itself, cell division, which is under the control of a genetic program that interacts with environmental signals (see Chapter 9), would produce only a formless heap of similar cells.

As embryonic development proceeds, certain cells become biochemically and structurally specialized to carry out specific functions, through a process known as cell **differentiation**. In our discussion of the genetic control of development in Chapter 16, we examined how cell differentiation occurs through cell **determination,** a series of events in which cells become progressively committed to follow a particular differentiation pathway. We also discussed evidence supporting the principle of **nuclear equivalence,** which states that in most cases neither determination nor cell differentiation is accompanied by a loss of genetic information from the cell nucleus. That is, the nuclei of virtually all differentiated cells of an animal contain the same genetic information that was present in the zygote, but each cell type expresses a different subset of that information. Cell differentiation is therefore an expression of changes in the activity of specific genes, and this genetic activity in turn is influenced by a variety of factors inside and outside the cell. It is this **differential gene expression** that is responsible for variations in chemistry, behavior, and structure among cells. Through this process, an embryo develops into an organism composed of more than 200 types of cells, each specialized to perform specific functions.

Cell differentiation by itself does not explain development. The differentiated cells must become progressively organized, shaping the intricate pattern of tissues and organs that characterizes a multicellular animal. This development of form,

(Courtesy of Bonnie Ullmann, from Kimmel, *et al.* "Stages of the Embryonic Development of the Zebrafish." *Developmental Dynamics* 203: 253–310, Wiley-Liss, 1995.)

known as **morphogenesis,** proceeds through the process of **pattern formation,** a series of steps requiring signaling between cells, changes in the shapes of certain cells, precise cell migrations, interactions with the extracellular matrix, and even **apoptosis** (controlled death) of some cells.

In this chapter, we compare and contrast the sequences of developmental events in several different animals. These model organisms have been chosen by researchers because they have certain desirable characteristics. Some embryos, like the zebrafish *(Danio rerio)* in the photograph, are particularly easy to observe because they are transparent. It is also advantageous if investigators can study very large numbers of embryos, all developing synchronously. As you read, note the intimate interrelationships among the basic developmental processes, as well as the fundamental similarities among the early developmental events in the organisms featured.

1. Summarize the roles of the following in the development of an animal: cell growth and division; cell determination and cell differentiation; pattern formation and morphogenesis.
2. Summarize the genetic and physiological functions of fertilization and describe the four processes involved.
3. Trace the generalized pattern of early development of the embryo from zygote through early cleavage; formation of the morula, blastula, and gastrula; and early organ development.
4. Contrast early development, including cleavage and gastrulation, in the echinoderms (or in amphioxus), the amphibian, and the bird, paying particular attention to the importance of the amount and distribution of the yolk.
5. Summarize the fate of each of the germ layers.
6. Define organogenesis and trace the early development of the nervous system.
7. Trace the development of the extraembryonic membranes and placenta, giving the functions of each membrane.
8. Describe the general course of early human development, including fertilization, the fates of the trophoblast and inner cell mass, implantation, and the role of the placenta.
9. Contrast postnatal with prenatal life, describing several changes that occur at or shortly after birth that allow the neonate to live independently.
10. List specific steps that a pregnant woman can take to promote the well-being of her developing child; describe how the embryo can be affected by nutrients, drugs, cigarette smoking, pathogens, and ionizing radiation.
11. Describe some anatomical and physiological changes that occur with aging and discuss current hypotheses of aging.

FERTILIZATION HAS GENETIC AND PHYSIOLOGICAL CONSEQUENCES

In Chapter 48 we discussed **spermatogenesis** and **oogenesis,** the processes by which meiosis leads to the formation of haploid cells, which differentiate as sperm and eggs respectively. In **fertilization** a flagellated, motile sperm fuses with a much larger, immotile ovum to produce a **zygote,** or fertilized egg. Fertilization has two important genetic consequences: restoration of the diploid chromosome number, and in mammals and many other animals, determination of the sex of the offspring (see Chapter 10). Fertilization also has profound physiological effects, for it activates the egg by stimulating reactions that permit development to take place.

Fertilization involves four steps: First, the sperm contacts the egg and recognition occurs. Second, the sperm or sperm nucleus enters the egg. Third, the egg becomes activated, and certain developmental changes begin. Finally, relatively late in the activation program, the sperm and egg nuclei fuse. Unless otherwise stated, our discussion applies to sea urchins and other echinoderms such as sea stars, which have been studied intensively because they produce very large numbers of gametes, and because fertilization is external.

The first step in fertilization involves contact and recognition

Although eggs are immotile, they are active participants in the fertilization process. An egg is surrounded not only by a plasma membrane, but also by one or more external coverings that are important in fertilization. For example, as discussed in Chapter 48, a mammalian egg is enclosed by a thick acellular **zona pellucida,** which is surrounded by a layer of small cells derived from the follicle in which the egg developed.

The egg coverings not only facilitate fertilization by sperm of the same species but also serve as barriers to interspecific fertilization, a function that is particularly important in species that practice external fertilization. External to the plasma membrane of a sea urchin egg are two acellular layers that interact with sperm: a very thin **vitelline envelope** and, outside of this, a thick glycoprotein layer called the **jelly coat.** Sea urchin sperm become motile when they are released into the sea water, and their motility is increased when they reach the vicinity of a sea urchin egg. However, motility alone is not enough to ensure that actual contact with the egg will occur. In sea urchins, as well as certain fish and some cnidarians, the egg or one of its coverings secretes a chemotactic substance that attracts sperm of the same species. However, for a great many species it has not been possible to demonstrate the existence of a specific chemical that attracts the sperm to the egg.

A sea urchin sperm becomes activated when it contacts the jelly coat. It undergoes an **acrosome reaction,** in which the membranes surrounding the acrosome (the cap at the head of the sperm, see Fig. 48–7) fuse and pores in the membrane enlarge. Calcium ions from the sea water move into the acrosome, which then swells and begins to disorganize. The acrosome then releases proteolytic enzymes that digest a path through the jelly coat to the vitelline envelope of the egg. If the sperm and egg are of the same species, a species-specific protein known as **bindin,** located on the acrosome, adheres to species-specific bindin receptors located on the vitelline envelope.

Before a mammalian sperm can participate in fertilization, it must first undergo **capacitation,** a maturation process that occurs in the female reproductive tract. During capacitation, which in humans can take several hours, sperm become increasingly motile and capable of undergoing an acrosome reaction when they encounter an egg. There is evidence that fertilization requires the interaction between sperm and species-specific glycoproteins in the zona pellucida.

Sperm entry is regulated

Once acrosome-vitelline envelope contact occurs, enzymes dissolve a bit of the vitelline envelope in the area of the sperm head. The plasma membrane of the egg is covered with microvilli, several of which elongate to surround the head of the sperm. As this occurs, the plasma membranes of sperm and egg fuse, and a **fertilization cone** is formed which contracts to draw the sperm into the egg (Fig. 49–1).

As soon as one sperm enters the egg, two reactions occur that prevent the entry of additional sperm; these are known as the fast and slow blocks to polyspermy. In the fast block to polyspermy the egg plasma membrane becomes depolarized, preventing its fusion with additional sperm. An unfertilized egg is polarized, i.e., the cytoplasm is negatively charged relative to the outside, but within a few seconds after sperm fusion, ion channels in the plasma membrane open, permitting positively charged calcium ions to diffuse across the membrane and depolarize the egg.

The slow block to polyspermy, the **cortical reaction,** requires about one to several minutes to complete, but it is a complete block. As a result of the depolarization of the egg plasma membrane, calcium ions stored in the endoplasmic reticulum are released into the cytosol. This causes thousands of cortical granules, membrane-bounded vesicles in the egg cortex (region close to the plasma membrane), to release enzymes by exocytosis into the space between the plasma membrane and the vitelline envelope. Some of these enzymes dissolve the protein linking the two membranes, allowing the space to enlarge as additional substances released by the cortical granules raise the osmotic pressure, causing an influx of water from the surroundings. Thus the vitelline envelope becomes elevated away from the plasma membrane and forms the fertilization envelope, a hardened covering that prevents entry of additional sperm.

In mammals, a fertilization envelope does not form, but the enzymes released alter the sperm receptors on the zona pellucida so that no additional sperm can bind to them. In some species more than one sperm enters the egg, but only the first sperm fertilizes it.

Fertilization activates the egg

Release of calcium ions into the egg cytoplasm is necessary for the cortical reaction, and it also triggers the activation program, a series of metabolic changes within the egg. Aerobic respiration increases, certain maternal enzymes and other proteins become active, and within a few minutes after sperm entry, a burst of protein synthesis occurs.

In some species, an egg can be artificially activated without sperm penetration by swabbing it with blood and pricking the plasma membrane with a needle, by calcium injection, or by certain other treatments. These haploid eggs may go through some developmental stages parthenogenetically, i.e., without fertilization, but they are unable to complete development. Special mechanisms of egg activation have evolved in species that are naturally parthenogenetic (see Chapter 48).

Figure 49–1 **Fertilization.** A fertilization cone (FC) forms as a sperm enters a sea urchin egg. The TEM shows various sperm organelles in the process of being engulfed by the egg cytoplasm (SN, sperm nucleus; SM, sperm mitochondrion; SF, sperm flagellum). *(Frank J. Longo)*

Sperm and egg pronuclei fuse

After the sperm nucleus enters the egg through the fertilization cone, it is thought to be guided toward the egg nucleus by a system of microtubules that forms within the egg. The sperm nucleus swells, forming the male pronucleus, and the nucleus of the ovum becomes the female pronucleus. Then the haploid pronuclei fuse to form the diploid nucleus of the zygote, and DNA synthesis occurs in preparation for the first cell division. (Recall from Chapter 48 that at the time of fertilization a mammalian egg is actually a secondary oocyte, which must complete meiosis before an egg pronucleus can form.)

DURING CLEAVAGE THE ZYGOTE DIVIDES, GIVING RISE TO MANY CELLS

Despite its simple appearance, the zygote is **totipotent,** that is, it has the potential to give rise to all the cell types of the new individual. Because the ovum is very large compared with the sperm, the bulk of the zygote cytoplasm and organelles come from the ovum. However, the sperm and ovum usually contribute equal numbers of chromosomes.

(a) Unfertilized egg   100 µm

Nucleus

(b) Two-cell stage   50 µm

(c) Four-cell stage   50 µm

(d) 16-cell stage   50 µm

(e) Blastula   50 µm

Blastocoel

Archenteron   Blastopore

(f) Early gastrula   50 µm

Archenteron

Blastopore

(g) Middle gastrula   50 µm

Mouth

Anus

Stomach

(h) Sea star larva   50 µm

(i) Young sea star   1 mm

Figure 49–2   **LMs showing sea star development** . *(a)* The isolecithal egg has a small amount of uniformly distributed yolk. *(b-e)* The cleavage pattern is radial and holoblastic (the entire egg becomes partitioned into cells). *(f, g)* The three germ layers form during gastrulation. The blastopore is the opening into the developing gut cavity, the archenteron. The rudiments of organs are evident in the sea star larva *(h)* and the young sea star *(i)*. All views are side views with the animal pole at the top, except *(c)* and *(i)*, which are top views. *(Carolina Biological Supply Company/Phototake—NYC)*

Shortly after fertilization, the zygote undergoes **cleavage,** a series of rapid mitotic divisions with no period of growth during each cell cycle. For this reason, although the cell number increases, the embryo does not increase in size. The zygote initially divides to form a two-celled embryo. Then, each of these cells undergoes mitosis and divides, bringing the number of cells to four. Repeated divisions increase the number of cells, called **blastomeres,** that make up the embryo. At about the 32-cell stage the embryo is a solid ball of blastomeres called a **morula.** Eventually, anywhere from 64 to several hundred blastomeres form the **blastula,** which is usually a hollow ball with a fluid-filled cavity, the **blastocoel.**

The pattern of cleavage is affected by yolk

Many animal eggs contain **yolk,** a mixture of proteins, phospholipids, and fats that serves as food for the developing embryo. The amount and distribution of yolk vary among different animal groups, depending on the needs of the embryo. Mammalian eggs have very little yolk because the embryo obtains maternal nutritional support throughout most of its development, whereas the eggs of birds and reptiles must contain sufficient yolk to sustain the embryo until hatching. Echinoderm eggs typically need only enough yolk to nourish the embryo until it becomes a tiny larva capable of obtaining its own food.

Most invertebrates and simple chordates have **isolecithal** eggs with relatively small amounts of yolk uniformly distributed through the cytoplasm. Isolecithal eggs divide completely **(holoblastic cleavage).** Cleavage of these eggs can be radial or spiral. Radial cleavage is characteristic of deuterostomes, whereas spiral cleavage is characteristic of protostomes (see Chapter 28).

In **radial cleavage,** the first division is vertical and splits the egg into two equal cells. The second cleavage division, also vertical, is at right angles to the first division and separates the two cells into four equal cells. The third division is horizontal, at right angles to the other two, and separates the four cells into eight cells: four above and four below the third line of cleav-

Figure 49–3 **Cleavage in amphioxus.** As in the sea star, cleavage is holoblastic and radial. The embryos shown are viewed from the side. *(a)* Mature egg with polar body. *(b–e)* Two-, four-, eight-, and 16-cell stages. *(f)* Embryo cut open to show the blastocoel. *(g)* Blastula. *(h)* Blastula cut open. *(i)* Early gastrula showing beginning of invagination at vegetal pole. *(j)* Late gastrula. Invagination is completed and the blastopore has formed.

age. This pattern of radial cleavage occurs in echinoderms and in the simple chordate amphioxus (Figs. 49–2, *a–e* and 49–3).

In **spiral cleavage,** after the first two divisions, the plane of cytokinesis is diagonal to the polar axis (Fig. 49–4). This results in a spiral arrangement of cells, with each cell located above and between the two underlying cells. This pattern is typical of annelids and mollusks.

. Many vertebrate eggs are **telolecithal,** meaning that they have large amounts of yolk concentrated at one end of the cell, known as the **vegetal pole.** The opposite, more metabolically active, pole is the **animal pole.** The eggs of amphibians are moderately telolecithal (mesolecithal). Although cleavage is

Figure 49–4 **Spiral cleavage in an annelid embryo.** *a–f* are top views of the animal pole. The successive cleavage divisions occur in a spiral pattern as illustrated. *(g)* A typical trochophore larva. The upper half of the trochophore develops into the extreme anterior end of the adult worm; all the rest of the adult body develops from the lower half.

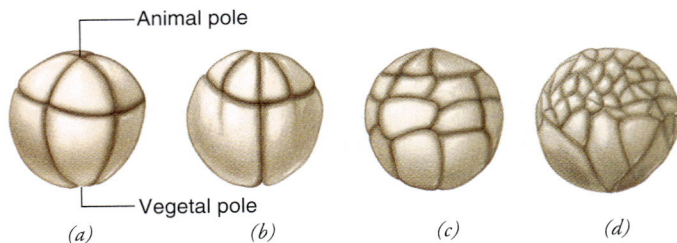

Figure 49–5 Cleavage pattern in a frog egg. *(a–d)* Although cleavage is holoblastic, the large amount of yolk concentrated in the vegetal hemisphere slows cleavage. As a result, fewer cells develop at the vegetal hemisphere than at the animal hemisphere. These embryos are viewed from the side.

radial and holoblastic, the divisions in the vegetal hemisphere are slowed by the presence of the inert yolk. As a result, the blastula consists of many small cells in the animal hemisphere, and fewer but larger cells in the vegetal hemisphere (Fig. 49–5). The blastocoel is displaced toward the animal pole.

The telolecithal eggs of reptiles and birds have very large amounts of yolk at the vegetal pole and only a small amount of cytoplasm concentrated at the animal pole. The yolk of such eggs never cleaves. Cell division is restricted to the **blastodisc,** the small disc of cytoplasm at the animal pole (Fig. 49–6); this type of cleavage is termed **meroblastic.** In birds and some reptiles, the blastomeres form two layers, separated by the blastocoel cavity: an upper epiblast and below that a thin layer of flat cells, the hypoblast.

Cleavage may distribute developmental determinants

The pattern of cleavage in a particular species depends not only on yolk, but on other factors as well. Recall from Chapter 16 that some organisms such as the roundworm *Caenorhabditis* have relatively rigid developmental patterns, known as **mosaic development.** This is largely a consequence of the unequal distribution of important materials in the cytoplasm of the zygote. Because the zygote cytoplasm is not homogeneous, cytoplasmic developmental determinants portioned out to each new cell during cleavage may be different. Such differences help determine the course of development. At the other extreme are mammals, which have zygotes with very homogenous cytoplasm. They exhibit highly **regulative development,** in which individual cells produced by the cleavage divisions are equivalent, allowing the embryo to develop as a self-regulating whole. Developmental patterns of most animals fall somewhere on a continuum between these two extremes.

In some species the distribution of developmental determinants in the unfertilized egg is maintained in the zygote. In others sperm penetration initiates rearrangement of the cytoplasm. For example, fertilization in the amphibian egg causes a shift of some of the cortical cytoplasm. These cytoplasmic movements can be easily followed because the egg cortex con-

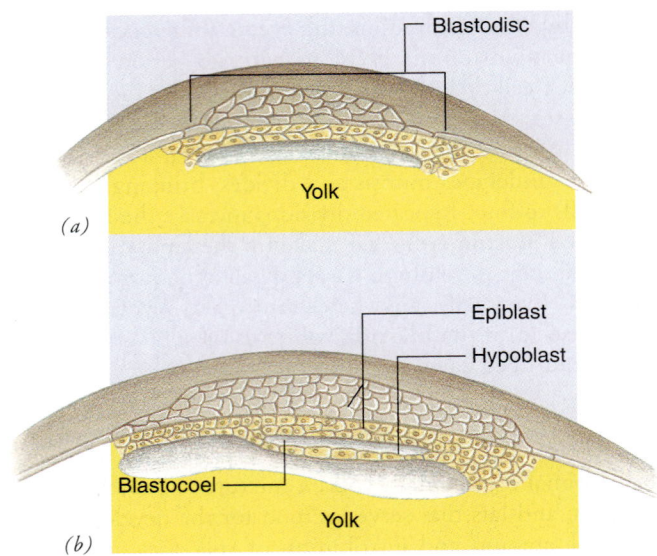

Figure 49–6 Cleavage in a bird embryo. Cleavage is meroblastic, i.e., it is restricted to the blastodisc, a small disk of cytoplasm on the upper surface of the egg yolk. *(a)* Early blastodisc formation. This cutaway view shows cells on the blastodisc surface, as well as in the interior. *(b)* The blastodisc splits into two tissue layers, an upper epiblast and a lower hypoblast, separated by the blastocoel.

tains dark pigment granules. As a result of this rearrangement, a crescent-shaped region of underlying lighter-colored (gray) cytoplasm becomes evident directly opposite the point on the cell where the sperm penetrated the egg. This **gray crescent** region is thought to contain growth factors and other developmental determinants. The first cleavage bisects the gray crescent, distributing half to each of the first two blastomeres; in this way the position of the gray crescent establishes the future right and left halves of the embryo. As cleavage continues, the gray crescent material becomes partitioned into certain blastomeres. Those that contain parts of the gray crescent eventually develop into the dorsal region of the embryo.

Experiments have confirmed the importance of determinants in the gray crescent to development. If the first two frog blastomeres are separated experimentally, each develops into a complete tadpole (Fig. 49–7). When the plane of the first division is altered so that the gray crescent is completely absent from one of the cells, that cell does not develop normally.

Cleavage provides building blocks for development

Cleavage partitions the zygote into many small cells that serve as basic building blocks. Their small size allows the cells to move about with relative ease, arranging themselves into the patterns necessary for further development. Each cell moves by ameboid motion. Surface proteins are important in helping cells to "recognize" one another and therefore in determining which ones adhere to form tissues.

Gray crescent

(a)

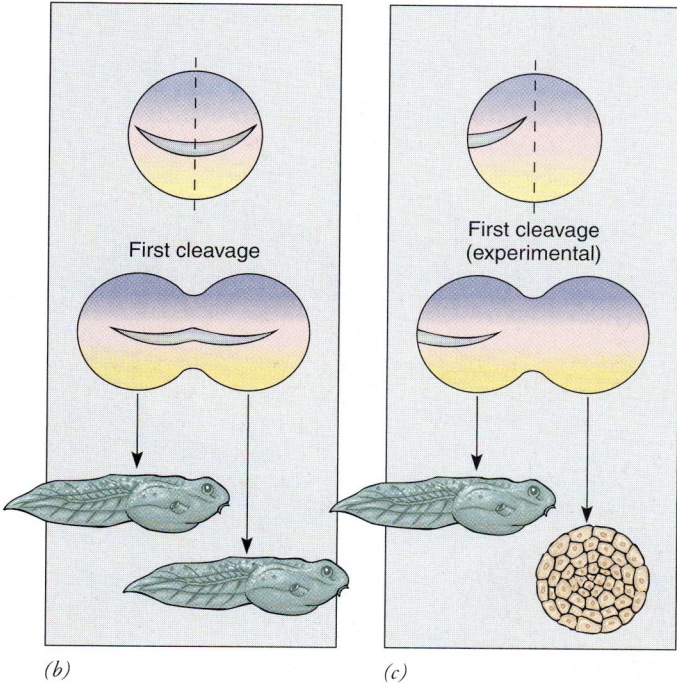

First cleavage

First cleavage (experimental)

(b)   *(c)*

Figure 49–7 Cytoplasmic determinants in frog development. *(a)* The position of the gray crescent in the frog zygote determines the main axes of the body. *(b)* The first division of the zygote partitions the gray crescent into the two daughter cells. If these cells are separated, each can develop into a tadpole. *(c)* If the plane of cleavage is changed experimentally so that only one cell receives the gray crescent, only that cell can develop into a tadpole.

THE GERM LAYERS FORM DURING GASTRULATION

The process by which the blastula becomes a three-layered embryo, or **gastrula,** is called **gastrulation.** Thus, early development proceeds through the following stages:

Zygote ⟶ early cleavage stages ⟶ morula ⟶ blastula ⟶ gastrula

During gastrulation, the embryo begins to take on a first approximation of its adult form as cells arrange themselves into three distinct **germ layers,** or embryonic tissue layers: the outermost layer, the ectoderm; the innermost, the endoderm; and

the mesoderm, which develops between them. Each of these layers will develop into specific parts of the embryo. As you study the illustrations, keep in mind that a conventional color-code accepted by developmental biologists is used to depict the germ layers: endoderm, yellow; mesoderm, red or pink; ectoderm, blue.

The pattern of gastrulation is affected by the amount of yolk

The simple type of gastrulation that occurs in echinoderms and in amphioxus is illustrated in Figures 49–2 *f–g* and 49–3 *i–j,* respectively. Gastrulation begins when a group of cells at the vegetal pole undergoes a series of changes in shape, causing that part of the blastula wall to first flatten and then bend inward (invaginate). The invaginated wall eventually meets the opposite wall, obliterating the blastocoel.

We can roughly demonstrate this type of gastrulation by pushing inward on the wall of a partly deflated rubber ball until it rests against the opposite wall. In a similar way the embryo is converted into a double-walled, cup-shaped structure. The new internal wall lines the **archenteron,** the newly formed cavity of the developing gut. The opening of the archenteron to the exterior, the **blastopore,** is the site of the future anus.

This simple type of gastrulation cannot occur in the amphibian embryo because the large yolk-laden cells in the vegetal half of the blastula would obstruct any inward movement at the vegetal pole. Instead, cells from the animal pole move down the surface of the embryo and when they reach the region derived from the gray crescent they begin to move into the interior. This inward movement is accomplished as the cells change shape, first becoming flask or bottle shaped (so that most of their mass is actually below the surface) and then sinking into the interior as they lose their remaining connections with other cells on the surface. This spot on the embryo's surface, referred to as the **dorsal lip of the blastopore,** is marked by a dimple, shaped like a **C** lying on its side (Fig. 49–8). As the process continues, the blastopore becomes ring-shaped as cells lateral, and then ventral, to the dorsal lip of the blastopore become involved in similar movements. The yolk-filled cells fill the space enclosed by the lips of the blastopore, forming the yolk plug.

The archenteron forms and is lined on all sides by cells that have moved in from the surface. At first, the archenteron is a narrow slit, but it gradually expands at the anterior end of the embryo, causing the blastocoel to become progressively smaller and to be eventually eliminated.

Although the details differ somewhat, gastrulation in the bird is basically similar to amphibian gastrulation. Epiblast cells migrate toward the midline to form a thickened cellular region, known as the **primitive streak,** that elongates and narrows as it develops. At its center is a narrow furrow, the **primitive groove.** The streak is a dynamic structure. The cells composing it constantly change as they migrate in from the epiblast, sink inward at the primitive groove, then move out

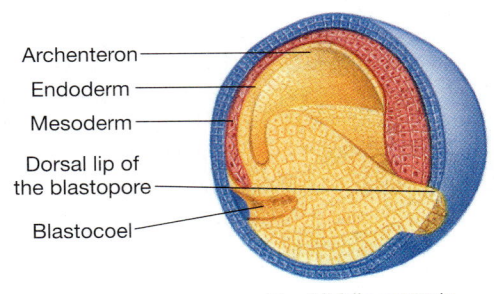

Figure 49–8 Gastrulation in a frog embryo. The embryos in the diagrams are cut in half longitudinally to show the formation of the germ layers. *(a)* Late blastula. *(b)* Early gastrula. *(c)* Middle gastrula. *(d)* Late gastrula. *(d)* and *(e)* Nervous system development begins with the formation of the neural plate.

laterally and anteriorly in the interior (Fig. 49–9). The primitive groove is the functional equivalent of the blastopore of the echinoderm, amphioxus, and amphibian embryos. However, a bird embryo contains no cavity that is homologous to the archenteron.

At the anterior end of the primitive streak, cells destined to form the notochord (supporting rod of mesodermal, cartilage-like cells) become concentrated in a thickened knot known as **Hensen's node.** These cells sink into the interior and then grow anteriorly just beneath the epiblast, forming a narrow extension from the node. Cells that will form other types of mesoderm move laterally and anteriorly from the primitive streak between the epiblast (which becomes ectoderm) and the hypoblast. Other cells form the endoderm, displacing the hypoblast cells and causing them to move laterally. These displaced cells will form part of the extraembryonic membranes, to be discussed in a later section.

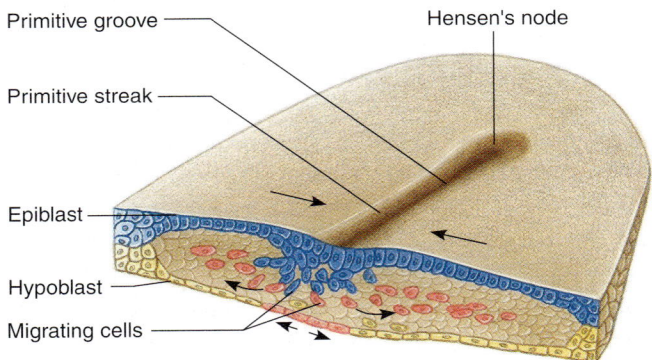

Figure 49–9 Gastrulation in birds. A three-layered embryo is formed as cells move toward the primitive streak, move inward at the primitive groove, and migrate laterally and forward in the interior.

ORGANOGENESIS BEGINS WITH THE DEVELOPMENT OF THE NERVOUS SYSTEM

In **organogenesis,** or organ formation, the cells of the three germ layers continue the processes of pattern formation that will lead to the formation of specific structures. The ectoderm eventually forms the outer layer of the skin and gives rise to the nervous system and sense organs. Tissues that eventually line the digestive tract, and organs that develop as outgrowths of the digestive tract (including the liver, pancreas, and lungs) are all of endodermal origin. Skeletal tissue, muscle, and the circulatory, excretory, and reproductive systems are all derived from mesoderm (Table 49–1; see Fig. 16–1).

The notochord, brain, and spinal cord are among the first organs to develop in the early vertebrate embryo (Figs. 49–8 *d* and *e,* and 49–10). First the notochord, which is mesodermal tissue, grows forward along the length of the embryo as a cylindrical rod of cells. The notochord, which serves as the flexible skeletal axis in all chordate embryos, eventually will be replaced by the vertebral column, although remnants will remain in the discs of cartilage between the vertebrae.

The developing notochord has yet another crucial function. Numerous experiments in which researchers have transplanted portions of the notochord mesoderm to other locations in the embryo have demonstrated that it is responsible for causing the overlying ectoderm to thicken and differentiate to form the precursor of the central nervous system, the **neural plate.** Such phenomena, in which certain cells stimulate or otherwise influence the differentiation of their neighbors, are examples of **induction** (see Chapter 16). Because it has been shown repeatedly that induction of the neural plate cells does not require direct cell-to-cell contact with the developing notochord cells, it is thought that this induction involves the diffusion of some type of signal molecule, although the signal itself has never been identified. The induction of the neural plate by the notochord is a good example of the importance of the position of a cell in relation to its cellular neighbors. That position is often critical in determining the fate of a cell because it determines its exposure to substances released from other cells.

The cells of the neural plate undergo changes in shape that cause the central cells of the neural plate to move downward and form a depression called the **neural groove;** the cells flanking the groove on each side form **neural folds.** Continued changes in cell shape bring the folds closer together until they meet and fuse, forming the **neural tube.** In this process, the hollow neural tube comes to lie beneath the surface. The ectoderm overlying it will form the outer layer of skin. The anterior portion of the neural tube grows and differentiates into the brain; the remainder of the tube develops into the spinal cord. The brain and the spinal cord remain hollow even in the adult; the ventricles of the brain and central canal of the spinal cord are persistent reminders of their embryonic origins.

Various motor nerves grow out of the developing brain and spinal cord, but sensory nerves have a separate origin.

TABLE 49–1	Fate of the Germ Layers Formed at Gastrulation

Ectoderm
  Nervous system and sense organs
  Outer layer of skin (epidermis) and its
    associated structures (nails, hair, etc.)
  Pituitary gland

Mesoderm
  Notochord
  Skeleton (bone and cartilage)
  Muscles
  Circulatory system
  Excretory system
  Reproductive system
  Inner layer of skin (dermis)
  Outer layers of digestive tube and of
    structures that develop from it, such as
    part of respiratory system

Endoderm
  Lining of digestive tube and of structures
    that develop from it, such as lining of
    the respiratory system

When the neural folds fuse to form the neural tube, bits of nervous tissue known as the **neural crest** arise from ectoderm in the approximate region of fusion on each side of the tube. Neural crest cells migrate downward from their original position and form the dorsal root ganglia of the spinal nerves and the postganglionic sympathetic neurons. From sensory cells in the dorsal root ganglia, dendrites grow out to the sense organs, and axons grow into the spinal cord. Neural crest cells migrate to various locations in the embryo. They give rise to parts of certain sense organs and nearly all pigment-forming cells in the body.

As the nervous system develops, other organs also begin to take shape. Blocks of mesoderm known as **somites** form on either side of the neural tube. These will give rise to the vertebrae, muscles and other components of the body axis. Mesodermal structures in addition to the skeleton and muscles include the kidneys and reproductive structures, as well as the circulatory organs. In fact the heart and blood vessels are among the first structures to take form, and they must function while still developing.

The development of the four-chambered heart of birds and mammals is a good example of the origin of a complex organ (Fig. 49–11; see Chapter 42). The heart originates through the fusion of paired blood vessels. Venous blood enters the single atrium, passes into the single ventricle, which is located above the atrium, and is then pumped into the embryo. During subsequent development, the atrium undergoes

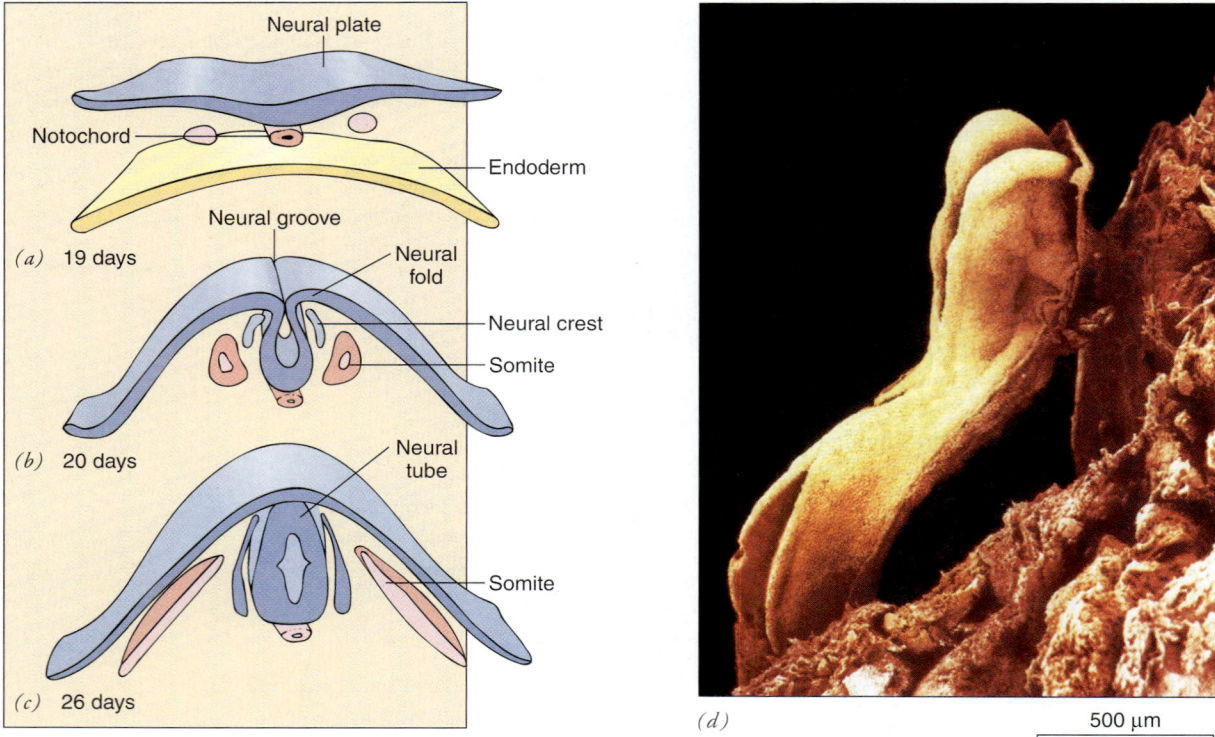

(a) 19 days

(b) 20 days

(c) 26 days

(d)

500 µm

Figure 49-10   **Development of the nervous system.** *(a–c)* Cross sections of human embryos. *(a)* Approximately 19 days. The neural plate has indented to form a shallow groove flanked by neural folds. *(b)* Approximately 20 days. The neural folds approach one another. The neural crest cells, derived from ectoderm, will migrate in the embryo and give rise to sensory neurons; the somites are blocks of mesoderm that will become the vertebrae and other segmented body parts. The endoderm is not shown. *(c)* Approximately 26 days. The neural folds have joined to form the hollow neural tube, which will give rise to the brain at the anterior end of the embryo and the spinal cord posteriorly. *(d)* The neural folds in the 20-day-old human embryo in this photograph have begun to fuse in the middle, but not in the region of the developing brain *(top)* nor in the posterior portion of the spinal cord.   *(d, Lennart Nilsson, Boehringer Ingelheim International GmbH, from* A Child Is Born, *Dell Publishing 1989 p. 76.)*

a process of torsion that brings it to a position above the ventricle, and partitions form that divide the atrium and the ventricle into right and left chambers.

The digestive tract is first formed as a separate foregut and hindgut as the body wall grows and folds, cutting them off

from the yolk sac as two simple tubes. (Fig. 49–12*a*). As the embryo grows, these tubes, which are lined with endoderm, grow and become greatly elongated. The liver, pancreas, and trachea originate as hollow, tubular outgrowths from the gut. As the trachea grows downward, it gives rise to the paired lung

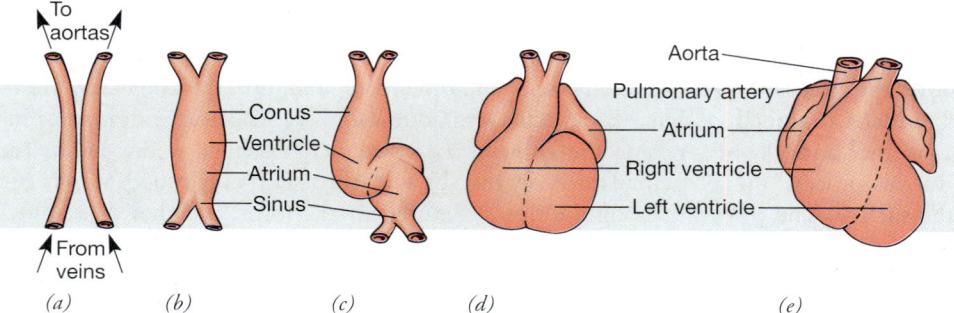

(a)   (b)   (c)   (d)   (e)

Figure 49-11   **Formation of the human heart.** The heart forms from the fusion of two blood vessels. These ventral views of successive stages in the development of the heart show that at first the heart is upside down; the end that receives blood from the veins is at the bottom. As it develops, the heart twists and turns to carry the atrium to a position above the ventricle, and the conus and sinus disappear. The chambers divide to form the four-chambered heart.

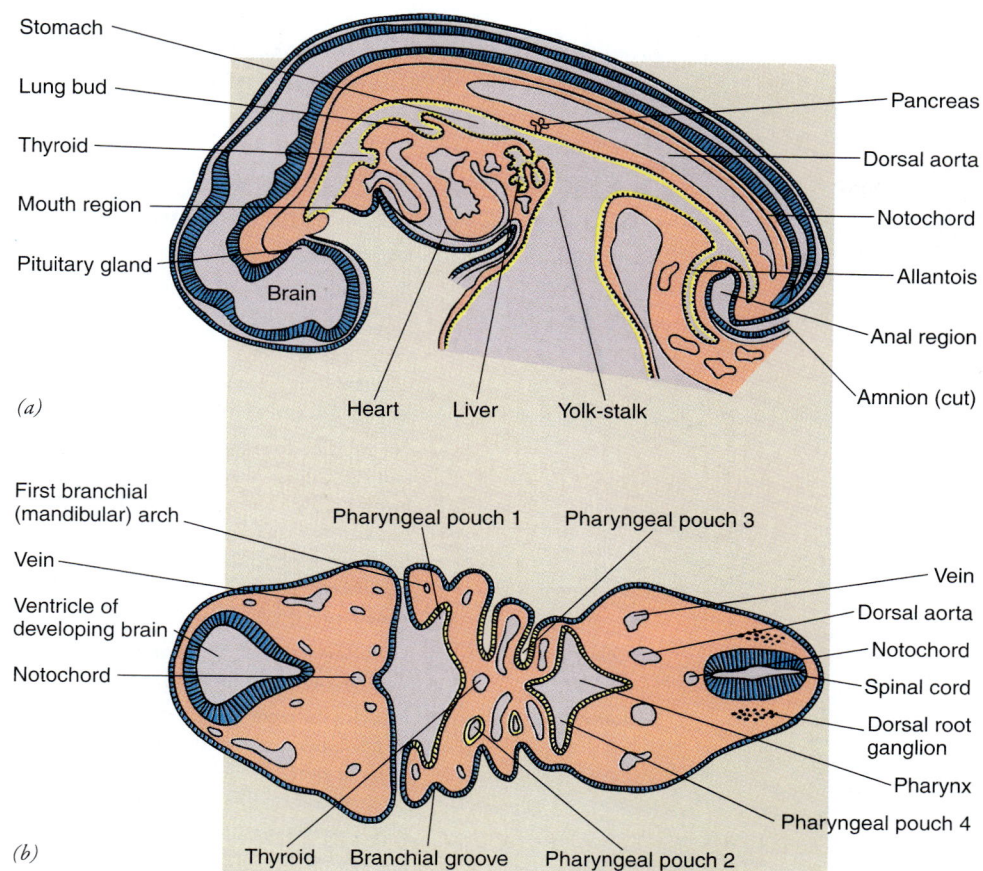

Stomach
Lung bud
Thyroid
Mouth region
Pituitary gland
Brain
Pancreas
Dorsal aorta
Notochord
Allantois
Anal region
Heart   Liver   Yolk-stalk
Amnion (cut)
(a)

First branchial (mandibular) arch
Vein
Ventricle of developing brain
Notochord
Pharyngeal pouch 1   Pharyngeal pouch 3
Vein
Dorsal aorta
Notochord
Spinal cord
Dorsal root ganglion
Pharynx
Pharyngeal pouch 4
Thyroid   Branchial groove   Pharyngeal pouch 2
(b)

Figure 49–12 **Early organ formation.** The diagrams illustrate some of the organs that form in a human embryo during the fifth week. *(a)* Sagittal section. Note that the liver, pancreas, and lungs develop as outpocketings from the digestive tract. *(b)* Cross section through the head and neck region. Note the branchial grooves, pharyngeal pouches, and branchial arches. The embryo is flexed so both the brain and spinal cord are present in the cross section.

buds, which develop into lungs. The most anterior part of the foregut becomes the pharynx. A series of small outpocketings of the pharynx, the **pharyngeal pouches,** bud out laterally (Fig. 49–12b). These pouches meet a corresponding set of inpocketings from the overlying ectoderm, the **branchial grooves.** The arches of tissue formed between the grooves are called **branchial arches.** They contain the rudimentary skeletal, neural, and vascular elements of the face, jaws, and neck.

In fishes and some amphibians, the pharyngeal grooves and branchial grooves meet and form a continuous passage from the pharynx to the outside; these are the gills slits, which function as respiratory organs. In terrestrial vertebrates, each branchial groove remains separated from the corresponding pharyngeal pouch by a thin membrane of tissue, and these structures give rise to organs more appropriate for life on land.

EXTRAEMBRYONIC MEMBRANES PROTECT AND NOURISH THE EMBRYO

Terrestrial vertebrates have four **extraembryonic membranes:** the chorion, amnion, allantois, and yolk sac (Fig. 49–13). Although they develop from the germ layers, these membranes are not part of the embryo itself and are discarded at hatching or birth. The extraembryonic membranes are adaptations

to the challenges of embryonic development on land. They protect the embryo, prevent it from drying out, and help in obtaining food and oxygen and eliminating wastes.

The outermost membrane, the **chorion,** encloses the entire embryo. Lying underneath the egg shell in birds and reptiles, it functions as the major organ of gas exchange. As we will see in the next section, its functions have been extended in humans and most other mammals.

Like the chorion, the **amnion** also encloses the entire embryo. The amniotic cavity, the space between the embryo and the amnion, becomes filled with amniotic fluid secreted by the membrane. Recall from Chapter 30 that terrestrial vertebrates are known as *amniotes* because their embryos develop within this pool of fluid. The amniotic fluid prevents the embryo from drying out and permits it a certain freedom of motion. The fluid also serves as a protective cushion that absorbs shocks and prevents the amniotic membrane from sticking to the embryo.

The **allantois** is an outgrowth of the developing digestive tract. In reptiles and birds, it stores nitrogenous wastes. In humans, the allantois is small and nonfunctional, except that its blood vessels contribute to the formation of umbilical vessels joining the embryo to the placenta.

In vertebrates with yolk-rich eggs, the **yolk sac** encloses the yolk, slowly digests it, and makes it available to the embryo. A yolk sac, connected to the embryo by a yolk stalk (see

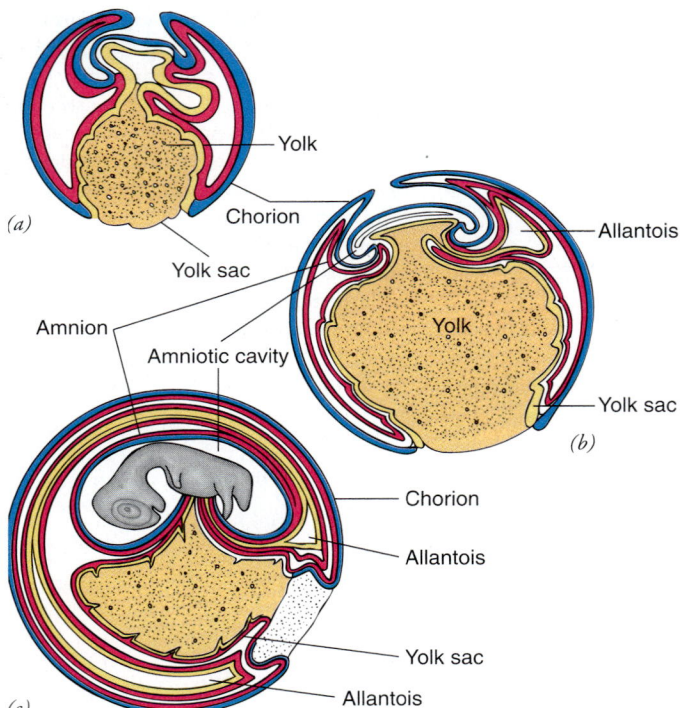

Figure 49–13 **Extraembryonic membranes.** The formation of the extraembryonic membranes of the chick is illustrated at *(a)* 4 days, *(b)* 5 days, and *(c)* 9 days of development. Each of the membranes develops from a combination of two germ layers. The chorion and amnion form from lateral folds of the ectoderm and mesoderm that extend over the embryo and fuse. The allantois develops from an outpocketing of the gut. The allantois, an elongated sac, and the yolk sac develop from endoderm and mesoderm.

Fig. 49–12*a*), develops even in mammalian embryos that have little or no yolk. Its walls serve as temporary centers for the formation of blood cells.

HUMAN PRENATAL DEVELOPMENT REQUIRES ABOUT 266 DAYS

The human **gestation period,** the duration of pregnancy, averages 266 days (38 weeks, or about 9 months) from the time of fertilization to the birth of the baby (Table 49–2). We will use this system of timing when discussing the events of early development. Because the time of fertilization is not easily marked, obstetricians usually time the pregnancy by beginning counting from the date of onset of the mother's last menstrual period; by this calculation an average pregnancy is 280 days (40 weeks).

Development begins in the oviduct

Fertilization occurs in the oviduct, and within 24 hours the human zygote has divided to become a two-celled embryo (Fig.

TABLE 49–2   **Some Important Developmental Events in the Human Embryo**

Time from Fertilization	Event
24 hours	Embryo reaches two-cell stage
3 days	Morula reaches uterus
7 days	Blastocyst begins to implant
2.5 weeks	Notochord and neural plate are formed; tissue that will give rise to heart is differentiating; blood cells are forming in yolk sac and chorion
3.5 weeks	Neural tube forming; primordial eye and ear visible; pharyngeal pouches forming; liver bud differentiating; respiratory system and thyroid gland just beginning to develop; heart tubes fuse, bend, and begin to beat; blood vessels are laid down
4 weeks	Limb buds appear; three primary divisions of brain forming
2 months	Muscles differentiating; embryo capable of movement; gonad distinguishable as testis or ovary; bones begin to ossify; cerebral cortex differentiating; principal blood vessels assume final positions
3 months	Sex can be determined by external inspection; notochord degenerates; lymph glands develop
4 months	Face begins to look human; lobes of cerebrum differentiate; eyes, ears, and nose look more "normal"
Third trimester	A covering of downy hair covers the fetus, then later is shed; neuron myelination begins; tremendous growth of body
266 days (from conception}	Birth

49–14). Cleavage continues as the embryo is propelled along the oviduct by ciliary action and muscular contraction.

When the embryo enters the uterus on about the fifth day of development, the zona pellucida (its surrounding coat) is dissolved. During the next few days, the embryo floats free in the uterine cavity, nourished by a nutritive fluid secreted by the glands of the uterus. Its cells arrange themselves, forming a blastula, which in mammals is called a **blastocyst** (Fig. 49–15). The outer layer of cells, the **trophoblast,** eventually forms the chorion and amnion that surround the embryo. A

(a)   50 µm

(b)   50 µm

(c)   50 µm

(d)   50 µm

Figure 49–14   **Cleavage in a human embryo.** *(a)* The human zygote contains the genetic instructions for producing a complete human. *(b)* Two-cell stage. *(c)* Eight-cell stage. *(d)* Cleavage continues, giving rise to a cluster of cells called the morula. *(Lennart Nilsson, from Being Born, 1992, pp. 14, 15, 17)*

little cluster of cells, the **inner cell mass,** projects into the cavity of the blastocyst. The inner cell mass gives rise to the embryo proper.

The embryo implants in the wall of the uterus

On about the seventh day of development the embryo begins the process of **implantation,** in which it becomes embedded in the endometrium of the uterus (see Fig. 49–15*a*). The trophoblast cells in contact with the uterine lining secrete enzymes that erode an area just large enough to accommodate the tiny embryo. Slowly the embryo works its way down into the underlying connective and vascular tissues, and the opening in the lining of the uterus repairs itself. All further development of the embryo takes place *within* the endometrium of the uterus.

The placenta is an organ of exchange

In placental mammals, the **placenta** is the organ of exchange between mother and embryo (Fig. 49–15*e*). The placenta provides nutrients and oxygen for the fetus and removes wastes, which the mother then excretes. In addition, the placenta is an endocrine organ that secretes hormones to maintain preg-

nancy. The placenta develops from both the chorion of the embryo and the uterine tissue of the mother. In early development, the chorion grows rapidly, invading the endometrium and forming finger-like projections known as chorionic villi. These structures are used in the chorionic villus sampling technique, which allows the prenatal detection of certain genetic disorders (see Chapter 15). The villi become vascularized (infiltrated with blood vessels) as the embryonic circulation develops.

As the human embryo grows, the **umbilical cord** develops and connects the embryo to the placenta (Fig. 49–15*e*). The umbilical cord contains the two umbilical arteries and the umbilical vein. The umbilical arteries connect the embryo to a vast network of capillaries developing within the villi. Blood from the villi returns to the embryo through the umbilical vein.

The placenta consists of the portion of the chorion that develops villi, together with the uterine tissue underlying the villi that contains maternal capillaries and small pools of maternal blood. The fetal blood in the capillaries of the chorionic villi comes in close contact with the mother's blood in the tissues between the villi. However, they are always separated by a membrane through which substances may diffuse or be actively transported. In addition, certain pathogens have evolved mechanisms that permit them to cross the placenta. However, *maternal and fetal blood do not normally mix in the placenta or any other place.*

Several hormones are produced by the placenta. From the time the embryo first begins to implant itself, its trophoblastic cells release **human chorionic gonadotropin (hCG),** which signals the corpus luteum that pregnancy has begun. In response, the corpus luteum increases in size and releases large amounts of progesterone and estrogens. These hormones stimulate continued development of the endometrium and placenta.

Figure 49-15 diagram panels:

(a) 7 days — labels: Trophoblast, Inner cell mass, Uterine epithelium, Inner cell mass, Trophoblast, Uterine blood vessel, Uterine gland

(b) 10 days — labels: Healing site of implantation, Yolk sac, Uterine epithelium, Chorion, Embryonic disc, Amniotic cavity, Maternal vessel

(c) 12 days — labels: Site of implantation, Epithelium of uterus, Embryonic disc

(d) 25 days — labels: Amniotic cavity, Amnion, Embryo, Yolk sac, Chorionic cavity, Chorionic villi (area of future placenta), Maternal blood

(e) 45 days — labels: Chorion, Amnion, Umbilical arteries and vein, Umbilical cord, Placenta, Amniotic cavity

Figure 49–15 **Implantation and early development in the uterus.** *(a)* About seven days after fertilization, the blastocyst drifts to an appropriate site along the uterine wall and begins to implant. The cells of the trophoblast proliferate and invade the endometrium. *(b)* About ten days after fertilization, the chorion has formed from the trophoblast. *(c)* This LM shows an implanted blastocyst at about 12 days after fertilization. *(d)* After 25 days, intimate relationships have been established between the embryo and the maternal blood vessels. Oxygen and nutrients from the maternal blood now satisfy the embryo's needs. Note the specialized region of the chorion that will soon become the placenta. *(e)* At about 45 days the embryo and its membranes together are about the size of a ping-pong ball, and the mother still may be unaware of her pregnancy. The amnion filled with amniotic fluid surrounds and cushions the embryo. The yolk sac has been incorporated into the umbilical cord. Blood circulation has been established through the umbilical cord to the placenta. *(c, Courtesy of Carnegie Institute of Washington)*

Without hCG, the corpus luteum would degenerate and the embryo would be aborted and flushed out with the menstrual flow. The woman would probably not even know that she had been pregnant. When the corpus luteum is removed before about the 11th week of pregnancy, the embryo is spontaneously aborted. After that time, the placenta itself produces enough progesterone and estrogens to maintain pregnancy.

Organ development begins during the first trimester

Gastrulation occurs during the second and third weeks of development. Then the notochord begins to form and induces

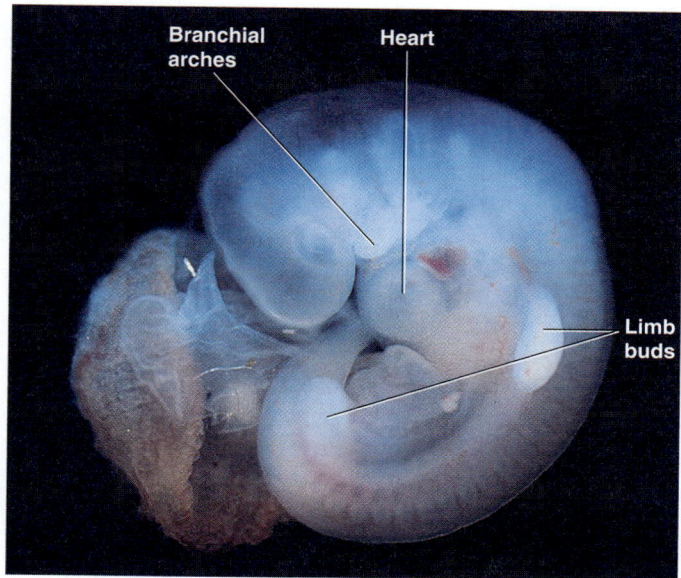

Figure 49–16 **A human embryo at 29 days.** Developing limb buds are evident and there is a distinct tail, which will regress during later stages. The heart can be seen below the head near the mouth of the embryo. Branchial arches appear as "double chins." At this stage the embryo is about 7mm (0.3 in) long. *(Lennart Nilsson, from* A Child Is Born, *Dell Publishing 1989)*

formation of the neural plate. The neural tube develops, and the forebrain, midbrain, and hindbrain are evident by the fifth week of development. A week or so later the forebrain begins to grow outward, forming the rudiments of the cerebral hemispheres.

The heart begins to develop, and after 3.5 weeks begins to beat spontaneously (see Fig. 49–11). Pharyngeal pouches, branchial grooves, and branchial arches form in the region of the developing pharynx. In the floor of the pharynx, a tube of cells grows downward to form the primordial trachea, which gives rise to the lung buds. The digestive system also gives rise to outgrowths that will develop into the liver, gallbladder, and pancreas. A thin tail becomes evident, but does not grow as rapidly as the rest of the body and so becomes inconspicuous by the end of the second month. Near the end of the fourth week the limb buds begin to differentiate; these eventually give rise to arms and legs (Fig. 49–16).

All the organs continue to develop during the second month (Fig. 49–17). Muscles develop, and the embryo becomes capable of movement. The brain begins to send impulses that regulate the functions of some organs, and a few simple reflexes are evident. After the first two months of development, the embryo is referred to as a **fetus** (Fig. 49–18).

By the end of the **first trimester** (the first three months of development) the fetus is about 56 mm (2.2 in) long and

(a)

(b)

Figure 49–17 **The second month of development.** The amnion is prominent as a transparent fluid-filled sac surrounding these embryos. *(a)* Human embryo at five and a half weeks, 1 cm (0.4 inch) long. Limb buds have lengthened, and the eyes have become more evident. The balloon-like object at the left, connected by a stalk, is the yolk sac. *(b)* In its seventh week of development, the embryo is 2 cm (0.8 in) long. The dark red object inside the embryo is the liver. The large fluffy mass in the lower right-hand corner of the photograph is part of the placenta. *(a, Petit Format/ Nestle/Photo Researchers, Inc.; b, Lennart Nilsson, from* A Child Is Born, *Dell Publishing 1989)*

Yolk sac

Uterine cavity

Amniotic cavity

Placenta

Cervix

Figure 49–18   **Human fetus at ten weeks.** Note the position of the fetus within the uterine wall.

weighs about 14 g (0.5 oz). Although small, it has a recognizably human appearance. The external genital structures have differentiated, indicating the sex of the fetus. Ears and eyes approach their final positions. Some of the skeleton becomes distinct, and the notochord has been replaced by the developing vertebral column. The fetus performs breathing movements, pumping amniotic fluid into and out of its lungs, and even makes sucking motions.

Development continues during the second and third trimesters

During the second trimester (months four through six), the fetal heart can be heard with a stethoscope. The fetus moves freely within the amniotic cavity, and during the fifth month the mother usually becomes aware of relatively weak fetal movements ("quickening").

The fetus grows rapidly during the final trimester (months seven through nine), and final differentiation of tissues and organs occurs. If born at 24 weeks (out of 40 weeks), the fetus has only about a 50% chance of surviving, even with the best of medical care, because its brain is not sufficiently developed to sustain vital functions such as rhythmic breathing, and because the kidneys and lungs are immature.

During the seventh month the cerebrum grows rapidly and develops convolutions. The grasping and sucking reflexes are evident, and the fetus may suck its thumb. Any infant born before 37 weeks of gestation is considered premature. However, if born after 30 weeks, the baby has a good chance of surviving. At birth the average full-term baby weighs about 3000 g (6.6 lb) and measures about 52 cm (20 in) in total length.

The neonate must adapt to its new environment

Important changes take place after birth. During prenatal (before birth) life, the fetus received both food and oxygen from the mother through the placenta. Now the newborn's own digestive and respiratory systems must function. Correlated with these changes are several major changes in the circulatory system.

Normally, the **neonate** (newborn infant) begins to breathe within a few seconds of birth and cries within half a minute. If anesthetics have been given to the mother, however, the fetus may also have been anesthetized, and breathing and other activities may be depressed. Some infants may not begin breathing until several minutes have passed. This is one of the reasons that many women request childbirth methods that minimize the use of medication.

The neonate's first breath is thought to be initiated by the accumulation of carbon dioxide in the blood after the umbilical cord is cut. The carbon dioxide stimulates the respiratory centers in the medulla. The resulting expansion of the lungs enlarges its blood vessels (which previously were partially collapsed). Blood from the right ventricle flows in increasing amounts through these larger pulmonary vessels. (During fetal life, blood bypasses the lungs in two ways: by flowing through an opening, the foramen ovale, which shunts blood from the right atrium to the left atrium, and by flowing through an arterial duct connecting the pulmonary artery and aorta. Both of these routes become closed off after birth.)

Environmental factors affect the embryo

We all know that the growth and development of babies are influenced by the food they eat, the air they breathe, the disease organisms that infect them, and the chemicals or drugs to which they are exposed. **Prenatal** development is also affected by these environmental influences. Life before birth is even more sensitive to environmental changes than it is in the fully formed baby. Although there is no direct mixing of maternal and fetal blood, diffusion and various other mechanisms allow many substances—nutrients, drugs, pathogens, and gases—to be transported across the placenta.

Table 49–3 describes some of the environmental influences on development. Some of these are **teratogens,** drugs or other substances that interfere with morphogenesis, causing malformations (Fig. 49–19). Many factors, such as smoking, alcohol use, and poor nutrition contribute to low birth weight, a condition responsible for a great number of infant deaths.

About 5% of newborns (more than 150,000 babies per year) in the United States have a defect of clinical significance. Such birth defects account for about 15% of deaths among newborns. Birth defects may be caused by genetic or environmental factors, or a combination of the two. Genetic factors were discussed in Chapter 15. In this section, we examine some environmental conditions that affect the well-being of the embryo.

TABLE 49–3 Environmental Influences on the Embryo

Factor	Example and Effect	Comment
Nutrition	Severe protein malnutrition doubles number of defects; fewer brain cells are produced, and learning ability may be permanently affected; deficiency of folic acid (a vitamin) linked to CNS defects such as spina bifida (open spine); low birth weight	Growth rate mainly determined by rate of net protein synthesis by embryo's cells
Medications	Many medications, even aspirin, affect development of the fetus	
Excessive vitamins	Vitamin D essential, but excessive amounts may result in form of mental retardation; an excess of vitamins A and K may also be harmful	Vitamin supplements are normally prescribed for pregnant women, but only the recommended dosage should be taken
Thalidomide	Thalidomide, marketed in Europe as a mild sedative, was responsible for serious malformations in more than 7000 babies born in the late 1950s in 20 countries; principal defect was phocomelia, a condition in which babies are born with extremely short limbs, often with no fingers or toes	This teratogenic drug interferes with the development of blood vessels and nerves; most hazardous when taken during fourth to fifth weeks, when limbs are developing; thalidomide has been approved for the treatment of leprosy, and clinical trials are being conducted to test its usefulness in the treatment of some of the complications of AIDS and cancer
Accutane (isotretinoin)	Accutane, a synthetic derivitive of vitamin A, is used for the treatment of a severe form of acne known as cystic acne. A woman who takes Accutane during pregnancy has a 1 in 5 chance of having a child with serious malformations of the brain (causing mental retardation), head and face, thymus gland (interfering with immune function), or heart	Although Accutane is marketed with severe restrictions to prevent pregnant women from taking this teratogenic drug, several children are born each year with birth defects attributable to Accutane exposure
Pathogens		
Rubella	Rubella (German measles) virus crosses placenta and infects embryo; interferes with normal metabolism and cell movements; causes syndrome that involves blinding cataracts, deafness, heart malformations, and mental retardation; risk is greatest (about 50%) when rubella is contracted during the first month of pregnancy; risk declines with each succeeding month	Rubella epidemic in the United States in 1963–1965 resulted in about 20,000 fetal deaths and 30,000 infants born with serious defects; immunization is available, but must be administered several months before conception
HIV	HIV can be transmitted from mother to baby before birth, during birth, or by breastfeeding	See discussion of AIDS in Chapter 43
Syphilis	Syphilis is transmitted to fetus in about 40% of infected women; fetus may die or be born with defects and congenital syphilis	Pregnant women are routinely tested for syphilis during prenatal examinations; most cases can be safely treated with antibiotics
Ionizing radiation	When mother is subjected to x rays or other forms of radiation during pregnancy, infant has a higher risk of birth defects and leukemia	Radiation was one of the earliest causes of birth defects to be recognized

(Table continues on page 1090)

TABLE 49-3   Continued

Factor	Example and Effect	Comment
Recreational and/or abused substances		
Alcohol	When a woman drinks heavily during pregnancy, the baby may be born with fetal alcohol syndrome (FAS), which includes certain physical deformities, and mental and physical retardation; low birth weight and structural abnormalities have been associated with as little as two drinks a day; some cases of hyperactivity and learning disabilities have occurred	Fetal alcohol syndrome is the leading cause of preventable mental retardation in the United States
Cigarette smoking	Cigarette smoking reduces the amount of oxygen available to the fetus because some of the maternal hemoglobin is combined with carbon monoxide; may slow growth and can cause subtle forms of damage	Mothers who smoke deliver babies with lower-than-average birth weights and have a higher incidence of spontaneous abortions, stillbirths, and neonatal deaths; studies also indicate a possible relationship between maternal smoking and slower intellectual development in offspring
Cocaine	Causes constriction of fetal arteries, resulting in retarded development and low birth weight; severe cases may be mentally retarded, have heart defects and other medical problems	Thousands of cocaine-addicted babies are born to mothers who use cocaine during pregnancy; cocaine users frequently abuse alcohol as well, so it is difficult to separate the effects
Heroin	High rates of mortality and prematurity; low birth weight	Infants that survive are born addicted and must be treated for weeks or months

Timing is important. Each developing structure has a critical period during which it is most susceptible to unfavorable conditions. Generally, this critical period occurs early in the development of the structure, when interference with cell movements or divisions may prevent formation of normal shape or size, resulting in permanent malformation. Because most structures form during the first three months of embryonic life, the embryo is most susceptible to environmental factors during this early period. During a substantial portion of this time, the mother may not even realize that she is pregnant and so may not take special precautions to minimize potentially dangerous influences.

Physicians are now able to diagnose some defects while the embryo is in the uterus. In some cases, treatment is possible before birth. Amniocentesis and chorionic villus sampling, discussed in Chapter 15, are techniques used to detect certain defects. Ultrasound imaging techniques are used to produce a type of image known as a **sonogram** (see Fig. 15–10). Such previews are helpful in diagnosing defects and also in determining the position of the fetus and whether a multiple birth is pending. Unlike imaging techniques that use radiation, the available evidence is that ultrasound is harmless to the fetus. New methods currently under development include special types of MRI (magnetic resonance imaging) and high resolution 3-D ultrasound imaging (Fig. 49–20).

More than one mechanism can lead to a multiple birth

Occasionally, the cells of the two-celled embryo separate and each cell develops into a complete organism. Or sometimes the inner cell mass subdivides, forming two groups of cells, each of which develops independently. Because these cells have identical sets of genes, the individuals formed are exactly alike —**monozygotic, or identical, twins.** Very rarely, the two inner cell masses do not completely separate and so give rise to **conjoined twins** (formerly known as Siamese twins.)

Dizygotic twins, also called **fraternal twins,** develop when two eggs are ovulated and each is fertilized by a different sperm. Each zygote has its own distinctive genetic endowment, so the individuals produced are not identical. They may not even be of the same sex. Similarly, triplets (and other multiple births) may be either identical or fraternal.

In the United States, twins are born once in about 80 births (about 30% of twins are monozygotic), triplets once in 80^2 (or 1 in 6400), and quadruplets once in 80^3 (or 1 in

(a)   *(b)*

Figure 49–19   **Effects of thalidomide.** Thalidomide is a potent teratogen in humans and other primates. When administered to the marmoset *(Callithrix jacchus)*, it produces a pattern of developmental defects similar to those found in humans. *(a)* Control marmoset fetus obtained from an untreated mother on day 125 of gestation. *(b)* Fetus of marmoset (same age as control) treated with 25mg/kg thalidomide from days 38 to 52 of gestation. The drug suppresses limb formation, perhaps by interfering with the formation of blood vessels or the function of certain nerves. *(Courtesy of Dr. W.G. McBride and P.H. Vardy, Foundation 41; from* Development, Growth and Differentiation, *Vol. 25, No. 4, 1983, pp. 361–373.)*

512,000). These statistics appear to be changing. Multiple births have been increasing with increased use of fertility-inducing agents (Chapter 48).

A family history of twinning increases the probability of having dizygotic twins. However, giving birth to monozygotic twins does not appear to be influenced by heredity, age of the mother, or other known factors. An estimated two-thirds of multiple pregnancies end in the birth of a single baby; the other embryo(s) may be absorbed within the first 10 weeks of pregnancy, or spontaneously aborted. As mentioned previously, ultrasonic imaging techniques used early in pregnancy can give valuable information regarding the presence of multiple embryos. This is medically important because multiple births are associated with higher mortality, which is largely a consequence of an increased risk of prematurity and low birth weight.

THE HUMAN LIFE CYCLE EXTENDS FROM FERTILIZATION TO DEATH

We have examined briefly the development of the embryo and fetus, the birth process, and the adjustments required of the neonate. The human life cycle then proceeds through the stages of infant, child, adolescent, young adult, middle-aged adult, and elderly adult.

Figure 49–20   **Three-dimensional ultrasound image of a human fetus.** Note the enhanced soft tissue detail. *(Courtesy of Advanced Technology Laboratories)*

Aging is not a uniform process

Development encompasses any biological change that takes place within an organism over time, including the changes commonly called **aging.** Changes during the aging process result in decreased function in the older organism. The declining capacities of the various systems in the human body, although most apparent in the elderly, may begin much earlier in life.

The aging process is far from uniform among different individuals or in various parts of the body. The systems of the body generally decline at different times and rates. On average, between the ages of 30 and 75 a man loses 64% of his taste buds, 44% of the glomeruli in his kidneys, and 37% of the axons in his spinal nerves. His nerve impulses are propagated at a 10% slower rate, the blood supply to his brain is 20% less, his glomerular filtration rate has decreased 31%, and the vital capacity of his lungs has declined 44%.

While marked improvements in medicine and public health have led to survival to an advanced age of a larger fraction of the total human population, there has been no corresponding increase in the maximum life expectancy. Although relatively little is known about the aging process itself, this is now an active field of scientific investigation.

Homeostatic response to stress decreases during aging

Recent research findings support the idea that most aging occurs because a combination of inheritance and environmental stress makes the individual less able to respond to additional stressors. One major question is whether there exists a genetic program that has evolved to cause aging to occur, or if genetic involvement in the process is more circumstantial. Most of the available evidence favors the latter view.

Certainly some genetically programmed developmental events do seem to be related to the aging process. Cells that normally stop dividing when they differentiate appear to be more subject to the changes of aging than are those that continue to divide throughout life. Furthermore, it has been hypothesized that certain parts of the body begin to fail to function normally because a genetic program causes key cells to stop dividing and replenishing themselves. In the model system known as **cellular aging,** normal human cells eventually lose their ability to divide when grown in culture. Furthermore, cells taken from an older person can divide fewer times than those from a younger person. Cellular aging appears to be related to the fact that most normal somatic cells of humans are genetically programmed to lose the ability to produce active telomerase, an enzyme that replicates the DNA of the end caps (telomeres) of the chromosomes (see Chapter 11, *On the Cutting Edge: Telomeres, Cellular Aging, and Cancer*). This loss may help protect the individual against the growth of tumors.

Genetically programmed cell death, apoptosis, is an essential developmental mechanism that may play a role in aging (Chapters 4 and 16). However, it is not thought that most aging is a natural consequence of the genetic program leading to apoptosis; instead, mistakes in the control of apoptosis may be involved in certain degenerative conditions, such as Alzheimer's disease or in some of the cell deaths that occur following a heart attack or stroke.

In general, researchers are investigating genes that affect how the body is maintained and repaired in the face of various stressors. Like other life processes, aging may be accelerated by certain environmental influences and may occur at different rates in different individuals because of inherited differences. Experimental evidence suggests that aging, at least in rats, can be delayed by severe caloric restriction. There is also evidence that premature aging can be precipitated by hormonal changes; by various malfunctions of the immune system, including autoimmune responses; by the accumulation of specific waste products within the cells; by changes in the molecular structure of macromolecules such as collagen; and by damage to DNA by continued exposure to cosmic radiation and x rays.

SUMMARY WITH KEY TERMS

I. Development proceeds as a balanced combination of growth, cell **determination** leading to cell **differentiation,** and **pattern formation** leading to **morphogenesis,** the development of form.

II. Cell differentiation is a consequence of **differential gene expression.**

III. Fertilization involves four processes:
   A. Contact and recognition occur between acellular egg coverings and sperm.
      1. The coverings of echinoderm eggs are the **vitelline envelope** and the **jelly coat;** a **zona pellucida** encloses the mammalian egg.
      2. Upon contact a sperm undergoes an **acrosome reaction,** which facilitates penetration of the coverings.
   B. Sperm entry is regulated to prevent interspecific fertilization and polyspermy. Sea urchin fertilization is followed by a fast block to polyspermy (depolarization of the plasma membrane) and a slow block to polyspermy (the **cortical reaction**).
   C. Fertilization activates the egg, triggering the events of early development.
   D. Fusion of sperm and egg nuclei restores the diploid condition.

IV. Development is regulated by the interaction of genes with cytoplasmic factors and environmental factors. **Mosaic development** depends heavily on the distribution of cytoplasmic determinants, whereas the embryo acts as a self-regulating whole in **regulative development.**

V. The zygote undergoes **cleavage,** a series of rapid cell divisions without a growth phase.
   A. Cleavage leads to the formation of a solid ball of cells (the **morula**) and then usually a hollow ball of cells (the **blastula**).
   B. The main effect of cleavage is to partition the zygote into many small cells. As cells divide, the distribution of materials in the cytoplasm influences development.
   C. The **isolecithal** eggs of most invertebrates and simple chordates have evenly distributed **yolk.** They undergo **holoblastic cleavage,** which involves division of the entire egg.
   D. In the moderately telolecithal eggs of amphibians, a concentration of yolk at the **vegetal pole** slows cleavage so that only a few large cells form there, compared to a large number of smaller cells at the **animal pole.**

E. The highly telolecithal eggs of reptiles and birds, with a large concentration of yolk at one end, undergo **meroblastic cleavage,** which is restricted to the **blastodisc.**

VI. In **gastrulation,** three **germ layers**—**ectoderm, mesoderm,** and **endoderm**—form; each gives rise to specific structures.
   A. In the sea star and in amphioxus, cells from the blastula wall invaginate and eventually meet the opposite wall; the new cavity formed, the forerunner of the digestive tube, is the **archenteron,** which has an opening to the exterior, the **blastopore.**
   B. In the amphibian, invagination at the vegetal pole is obstructed by large, yolk-laden cells; instead, cells from the animal pole move down over the yolk-rich cells and invaginate, forming the dorsal lip of the blastopore.
   C. In the bird, invagination occurs at the **primitive streak,** and no archenteron forms.

VII. **Organogenesis** is the process of organ development. One of the earliest events of organogenesis is the **induction** of nervous system development by the developing notochord. The brain and spinal cord develop from the **neural tube.**

VIII. In terrestrial vertebrates, four **extraembryonic membranes**—**chorion, amnion, allantois,** and **yolk sac**—protect the embryo and help it obtain food and oxygen and eliminate wastes. The amnion is a fluid-filled sac that surrounds the embryo and keeps it moist; it also acts as a shock absorber.

IX. Early human development follows a fairly typical vertebrate pattern.
   A. Cleavage takes place as the embryo is moved down the oviduct toward the uterus.
   B. In the uterus, the embryo develops into a **blastocyst** consisting of an outer **trophoblast,** which will give rise to the chorion and amnion, and an **inner cell mass,** which will become the embryo proper. The blastocyst undergoes **implantation** in the endometrium.
   C. After the first two months of development, the embryo is referred to as a **fetus.**
   D. In placental mammals, the **umbilical cord** connects the embryo to the **placenta,** the organ of exchange between the maternal and fetal circulation. The placenta is derived from the embryonic chorion and maternal tissue.

E. **Prenatal** development requires 266 days from time of fertilization. The **neonate** (newborn) must undergo rapid adaptations to independent life.

X. By controlling environmental factors such as exposure to **teratogens,** diet, smoking, and the intake of alcohol and drugs, a pregnant woman can help ensure the well-being of her unborn child. A **sonogram** (ultrasound image) can diagnose certain defects and also multiple pregnancy.

XI. **Monozygotic (identical) twins** arise from a single fertilized egg; **dizygotic (fraternal) twins** arise from fertilization of two different eggs.

XII. The **aging** process is marked by a decrease in homeostatic response to stress.

POST-TEST

1. The main function of the acrosome reaction is to (a) activate the egg (b) improve sperm motility (c) prevent interspecific fertilization (d) facilitate penetration of the egg coverings by the sperm (e) cause fusion of the sperm and egg pronuclei
2. The fast block to polyspermy in sea urchins (a) is a depolarization of the egg plasma membrane (b) requires exocytosis of the cortical granules (c) includes the elevation of the fertilization envelope (d) involves the hardening of the jelly coat (e) is a complete block
3. Place the following events of sea urchin fertilization in the proper sequence. (1) fusion of egg and sperm pronuclei (2) DNA synthesis (3) increased protein synthesis (4) release of calcium ions into the egg cytoplasm (a) 4-3-1-2 (b) 3-2-4-1 (c) 2-3-1-4 (d) 1-2-3-4 (e) 4-1-2-3
4. The cleavage divisions of a sea urchin embryo (a) occur in a spiral pattern (b) do not include DNA synthesis (c) do not include cytokinesis (d) are holoblastic (e) do not occur at the vegetal pole
5. Meroblastic cleavage is typical of embryos formed from _____ eggs. (a) moderately telolecithal (b) highly telolecithal (c) isolecithal (d) b and c (e) a, b, and c

6. The primitive groove of the bird embryo is the functional equivalent of the _____ in the amphibian embryo. (a) yolk plug (b) archenteron (c) blastocoel (d) gray crescent (e) blastopore
7. Which of the following are mismatched? (a) endoderm; lining of the digestive tube (b) ectoderm; circulatory system (c) mesoderm; notochord (d) mesoderm; reproductive system (e) ectoderm; sense organs
8. Which of the following has three germ layers? (a) morula (b) gastrula (c) blastula (d) blastocyst (e) trophoblast
9. An unidentified substance or substances released from the developing notochord cause the overlying ectoderm to form the neural plate. This phenomenon is known as (a) activation (b) determination (c) induction (d) implantation (e) mosaic development
10. Which of the following is composed of both fetal and maternal tissues? (a) umbilical cord (b) placenta (c) amnion (d) allantois (e) yolk sac
11. The embryo proper of a mammal develops from the (a) trophoblast (b) umbilical cord (c) inner cell mass (d) entire blastocyst (e) yolk sac

REVIEW QUESTIONS

1. What mechanisms ensure fertilization by only one sperm of the same species?
2. How do the mechanisms of fertilization ensure control of both quality (fertilization by a sperm of the same species) and quantity (fertilization by only one sperm)?
3. Contrast cleavage and gastrulation in sea stars, amphibians, and birds.
4. Trace the formation of the neural tube and explain how cell division, growth, cell differentiation, and morphogenesis interact in the process.

5. Give examples of adult structures that develop from each germ layer.
6. Why do terrestrial vertebrate embryos develop an amnion?
7. Describe human blastocyst formation and implantation.
8. What adaptations must the neonate make immediately after birth?
9. What are some of the nongenetic factors that influence development?
10. What steps can the pregnant woman take to help ensure the safety and well-being of her developing child?
11. Describe some of the changes that take place during the aging process.

YOU MAKE THE CONNECTION

1. What is the adaptive value of developing a placenta?
2. For almost 200 years, scientists debated whether an egg or sperm cell contains a completely formed, miniature human (preformation), or if structures develop gradually from a formless zygote (epigenesis). Relate these views to current concepts of development. Can you find any truth in the preformationist view?
3. Why are not all teratogenic medications banned?

RECOMMENDED READINGS

Browder, L.W., C.A. Erickson, and W.R. Jeffery. *Developmental Biology,* 3rd ed. Saunders College Publishing, Philadelphia, 1991. A excellent introduction to animal development.

Lithgow, G.J., and T.B. Kirkwood. "Mechanisms and Evolution of Aging." *Science,* Vol. 273, 5 Jul. 1996. This review, which provides a perspective on aging research, is one of a series of articles in the same issue devoted to "Patterns of Aging."

Roush, W. "A Womb with a View." *Science,* Vol. 278, 21 Nov. 1997. A discussion of new technological applications that are yielding improved images of developing embryos.

● Visit our website at **http://www.saunderscollege.com/lifesci/titles.html** and click on Solomon/Berg/Martin Biology for links to chapter-related resources on the World Wide Web.

CHAPTER 50

Animal Behavior

Suppose your professor provided you with a hypodermic syringe full of poison and instructed you to find a particular type of insect, one that you had never seen before and that was armed with active defenses. You then had to inject the ganglia of your victim's nervous system (about which you had been taught nothing) with just enough poison to paralyze but not kill it. You would have difficulty accomplishing these tasks, but a solitary wasp no larger than the first joint of your thumb does it all with elegance and surgical precision, without instruction.

The sand wasp *Philanthus* sp. captures a bee, stings it, and places the paralyzed insect in a burrow excavated in the sand (see photograph). She then lays an egg on her victim, which is devoured alive by the larva that hatches from that egg. From time to time, *Philanthus* returns to her hidden nest to reprovision it until the larva becomes a hibernating pupa in the fall. Her offspring will repeat this behavior, executing each step to perfection without ever having seen it done.

An animal's **behavior** is its response to stimuli in its environment, or more simply, what it does. A dog may wag its tail, a bird may sing, a butterfly may release a volatile sex attractant. Behavior is just as diverse as biological structure and just as characteristic of a given species as its anatomy or physiology. Structure, physiology, and behavior are adaptations that help define an organism and equip it for survival.

The capacity for behavior is inherited, but much inherited behavior can be modified by experience. Learning involves persistent changes in behavior that result from experience. In considering complex behaviors such as the reproductive behavior of *Philanthus,* we might wonder *why* she behaves as she does, and we might also be interested in *how* she accomplishes her task. Early investigators of animal behavior focused on *how* questions. These questions address **proximate causes,** immediate causes like the genetic, developmental, and physiological processes that permit the animal to carry out the particular behavior.

More recently, biologists have added the *why* perspective, asking questions that address the **ultimate causes.** These ques-

(Darryl T. Gwynne)

tions, which have evolutionary explanations, ask *why* the animal has evolved its proximate processes. Ultimate considerations address costs and benefits of behavior patterns. When studying ultimate causes we may ask what the adaptive value of a particular behavior might be. An understanding of behavior requires consideration of both proximate and ultimate causes.

AFTER YOU HAVE STUDIED THIS CHAPTER YOU SHOULD BE ABLE TO

1. Define behavior and explain how it is (a) adaptive, (b) homeostatic, and (c) flexible.
2. Describe the interactions of heredity, environment, and maturation in animal behavior.
3. Classify a learned behavior as an example of (a) classical conditioning, (b) operant conditioning, (c) habituation, or (d) insight learning.
4. Discuss the adaptive significance of imprinting.
5. Cite examples of biological rhythms, and suggest some of the mechanisms responsible for them.
6. Distinguish between directional orientation and navigation and discuss some biological advantages and disadvantages of migration.

7. Discuss the hypothesis that optimal foraging behavior is adaptive.
8. Describe an animal society and identify the potential advantages of cooperative behavior.
9. Summarize the modes of communication that animals employ.
10. Describe the concept of a dominance hierarchy, giving at least one example, and discuss its adaptive significance and social function.
11. Distinguish between home range and territory and describe costs and benefits of territoriality.
12. Describe different types of mating systems and the adaptive value of courtship behavior.
13. Define inclusive fitness and relate this concept to helping behavior.

MOST BEHAVIOR IS ADAPTIVE

Whether biologists study behavior in an animal's natural environment or in the laboratory, they must consider that what an animal does cannot be isolated from the way in which it lives. **Behavioral ecology** is the study of behavior in natural environments from the evolutionary perspective. For two decades, behavioral ecology has been the main approach of biologists that study animal behavior. Prior to the emergence of this approach the study of animal behavior was referred to as *ethology,* and this term is sometimes still used to refer to the overall study of animal behavior.

Behavioral ecologists address both the benefits and costs of specific behaviors. A behavior may help an organism obtain food or water, acquire and maintain territory in which to live, protect itself, or reproduce. The benefits typically contribute to **direct fitness,** which is measured by reproductive success. At the same time, behaviors involve costs. For example, while a parent hunts for food for its offspring, the young may be killed by predators. If the benefits are greater than the costs, the behavior is adaptive.

Behavior tends to be homeostatic. Certain responses may lead to the death of the individual while increasing the chance that copies of its genes will survive through the enhanced production or survival of its offspring or other relative. The ultimate function of a behavior is therefore to increase the probability that the genes of the *individual* animal will be passed to future generations. In this chapter, we consider how an animal's behavior contributes to its survival, allowing it to pass its genes on to its offspring.

BEHAVIOR DEPENDS ON THE INTERACTION OF GENES AND ENVIRONMENTAL FACTORS

Early investigators in the field of animal behavior debated about nature versus nurture, that is, the relative importance of genes compared to environmental experience. They defined **innate** (inborn) **behavior** (popularly referred to as instinct) as genetically programmed and **learned behavior** as behavior that has been modified in response to environmental experience. More recently, behavioral ecologists have recognized that no true dichotomy exists. All behavior has a genetic basis. Even the *capacity* for learned behavior is inherited. However, behavior is modified by the environment in which an animal lives; it is a product of the interaction between genetic capacity and environmental influences. Thus, behavior begins with an inherited framework that experience can modify.

We can think of a range of behavior from the more rigidly genetically programmed types, through those that, although they have a genetic component, are extensively developed through experience. The wasp *Philanthus,* discussed in the chapter introduction, efficiently carries out a complex, largely genetically programmed, sequence of behaviors. How to dig the burrow, how to cover it, how to kill the bees—these behaviors appear to be genetically determined. Yet some of her behavior is learned. There is no way that her ability to locate the burrow could be genetically programmed. Because a burrow can be dug only in a suitable spot, its location must be learned *after* it is dug. When *Philanthus* covers a nest with sand, she takes precise bearings on the location of the burrow by circling the area a few times before flying off again to hunt.

This behavior of *Philanthus* was studied by the Dutch investigator Niko Tinbergen. Tinbergen surrounded the wasp's burrow with a circle of pine cones, on which the wasp took her bearings (Fig. 50–1). Before she returned with another bee, Tinbergen moved or removed the pine cones. The wasp could not find her burrow without them. When Tinbergen moved the pine cones to an area where there was no burrow, the female wasp responded as though the burrow were there. When the investigator completely removed them, the female appeared to be very confused. Only when the experimenter restored the cones to their original location could the wasp find her burrow.

Studies of fruit fly courtship and mating have provided interesting examples of interaction between genes and behavior. A ritual consisting of a complex sequence of steps, almost like a dance, must occur before mating takes place. This courtship ritual involves an exchange of visual, auditory, tactile, and chemical signals between the male and female. J.B. Hall at Brandeis University and his colleagues have identified more than a dozen genes controlling these actions, suggesting that courtship behavior is inherited and preprogrammed. Despite this, the fruit fly has the capacity to learn from experience, a capacity which, of course, is also inherited. Male flies quickly lose interest in females that have already mated. After about 30 minutes, the inhibitory pheromones released by the mated female inhibit further courtship behavior. For several hours after this experience, the male refrains from courtship activity with any female.

The interaction between genes and environment has been studied in many vertebrates. Several species of the lovebird *(Agapornis)* differ not only in appearance but in behavior. One species uses its bill to transport small pieces of bark for building a nest. Another species tucks nest-building materials under its rump feathers. In 1962 William Dilger tested the hypothesis that these behaviors are genetically determined, by producing hybrids. These birds appeared confused. They attempted to tuck the material under their feathers, then tried to carry it in their beaks, repeating the pattern several times. Eventually, most of the birds carried the material in their bills, but it took them up to three years to perfect this behavior, and most of them continued to make futile attempts to tuck material into their feathers. These studies suggest that the method of transporting materials is inherited but somewhat flexible.

Figure 50–1 Niko Tinbergen's sand wasp experiment. Although the ability of *Philanthus* to learn is quite limited, it is adequate for most natural situations. When the ring of pine cones is moved from position *(a)* to position *(b)*, the *Philanthus* wasp behaves as if her nest were still located at the center because she learned its position in relation to the cones. The wasp responds to the arrangement of the cones, rather than the cones themselves, as shown by the substitution of a ring of stones for cones *(c)*. *(After Tinbergen)*

Behavior develops

Behavior involves all body systems, but it is influenced mainly by the nervous and endocrine systems. The capacity for behavior therefore depends on the genetic characteristics that govern the development and functions of these systems. Before an animal can exhibit any pattern of behavior, it must be physiologically ready to produce the behavior. For example, breeding behavior does not ordinarily occur among birds or most mammals unless certain concentrations of steroid sex hormones are present in their blood. A human baby cannot walk until its muscles and neurons are sufficiently developed. These states of physiological readiness are themselves produced by a continuous interaction with the environment. The level of sex hormones in a bird's blood may be determined by seasonal variations in day length. The baby's muscles develop with age, as well as experience.

Several factors influence the development of song in male white-crowned sparrows (generally only male songbirds sing a complex song). These birds exhibit considerable regional variation in their song. During early development, young sparrows normally hear adult males sing the distinctive song of their population. Days 10 to 50 are a critical period for learning the song. When he is several months old, a young male sparrow "practices" the song over several weeks until he eventually sings in the local "dialect."

In laboratory experiments, birds kept in isolation and deprived of the acoustic experience of hearing the song of mature males eventually sing a very poorly developed but recognizable white-crowned sparrow song. When a young white-crowned sparrow is permitted to interact socially with a strawberry finch (which belongs to a different genus of birds), it learns the song of the finch, rather than its own species-specific song. This occurs even if the white-crowned sparrow can hear the song of other sparrows, but does not interact socially. From these experiments, investigators have concluded that while the white-crowned sparrow is hatched equipped with a rough genetic pattern of its song, social and acoustical stimuli are both important in developing its ability to sing its specific song.

Many behavior patterns depend on motor programs

Many behaviors that we think of as automatic depend on coordinated sequences of muscle actions now referred to as **motor programs.** Some motor programs, for example walking in newborn gazelles, appear to be mainly innate. Others, such as walking in human infants, have a greater learned component.

A classic example of a motor program in vertebrates is egg-rolling in the European graylag goose. When an egg is removed from the nest of a goose and placed a few centimeters in front of her, she reaches out with her neck and pulls the egg back into the nest (Fig. 50–2). If, while the goose is rolling the egg back toward the nest, the egg veers off to the side, the goose steers it back on course. If the egg is quickly removed, the goose continues her head and neck movements as she persists in moving the now nonexistent egg back toward the nest. Once activated by a simple sensory stimulus, this behavior continues to completion regardless of sensory feedback. There is little flexibility. Ethologists called this behavior a **fixed action pattern (FAP).**

An FAP can be elicited by a **sign stimulus** or **releaser,** a simple signal that triggers a specific behavioral response. A wooden egg is a sign stimulus that elicits egg-rolling behavior in the graylag goose. Another classical example of a sign stimulus is the red stripe on the ventral surface of a male stickleback fish. The red stripe triggers aggressive behavior by a male whose territory is being invaded. Tinbergen found that crude models painted with a red belly were more likely to be attacked than more realistic models lacking the red belly (Fig. 50–3).

ANIMALS LEARN FROM EXPERIENCE

Learning is a change in behavior due to experience. The capacity to learn appropriate responses to new situations is adaptive, because learned behavior can be shaped to meet the needs of the changing environment that most animals experience.

Figure 50–2 **Egg-rolling behavior in the European graylag goose.** This behavior is a fixed action pattern (FAP). The goose reaches out by extending her neck and uses her bill to pull the egg back into the nest. If the investigator quickly removes the egg while the goose is in the process of reaching for it or pulling it back, she continues the FAP to completion, as though pulling the non-egg back to the nest.

Figure 50–3 **A sign stimulus triggers a fixed action pattern.**
A male stickleback fish *(a)* will not attack a realistic model of another male stickleback if it lacks a red belly *(b)*, but it will attack another model, however unrealistic, that has a red "belly" *(c)*. The aggressive behavior is triggered by the red sign stimulus, rather than by recognition based on a combination of features.

An animal habituates to irrelevant stimuli

Habituation is a type of learning in which an animal learns to ignore a repeated, irrelevant stimulus. Pigeons gathered in a city park learn by repeated harmless encounters that humans are no more dangerous to them than are cows to crows, and behave accordingly. This is to their advantage. A pigeon intolerant of people might waste energy by flying away each time a human approached and might not get enough to eat. Many African animals habituate to humans on photo safari and to the vans that transport them (Fig. 50–4). Urban humans become habituated to the noise of traffic. In fact, many urban dwellers report that they do not sleep well when they visit a quiet rural area.

Imprinting occurs during an early critical period

Anyone who has watched a mother duck with her ducklings must have wondered how she can keep track of such a horde of almost identical little creatures tumbling about in the grass, let alone distinguish them from those belonging to another duck (Fig. 50–5).Although she is capable of recognizing her offspring to an extent, basically they have the responsibility of keeping track of her. The survival of a duckling requires that it quickly establish a behavioral bond with its mother. This bond, which forms during a **critical period,** usually within a

Figure 50–4 **Habituation.** After repeated safe encounters with vans transporting humans on photo safari, many animals, including giraffes, zebras, lions, and elephants, in the Serengeti learn to ignore them. Elephants typically ignore the vans unless the driver provokes them by moving too close. In that event an elephant may challenge and even charge the van. *(McMurray Photography)*

few hours after birth (or hatching), develops through **imprinting.** Konrad Lorenz, an early investigator of this type of learning, discovered that a newly hatched bird imprints on the first moving object it sees—even a human or an inanimate object such as a colored sphere or light. Although the process of imprinting is genetically determined, the bird *learns* to respond to a particular animal or object.

Among many types of birds, especially ducks and geese, the older embryos are able to exchange calls with their nest

Figure 50–5 **Imprinting.** Parent-offspring bonds generally form very early. Some young animals follow the first moving object they encounter. Usually, the object is their mother, but under experimental conditions, young animals have imprinted on humans or other unnatural objects. *(J.H. Dick/VIREO)*

mates and parents through the porous eggshell. When they hatch, at least one parent is normally on hand, emitting the characteristic sounds with which the hatchlings are already familiar. If the parent moves, the chicks follow. This movement plus the sounds produce imprinting. During a brief critical period after hatching, the chicks learn the appearance of the parent.

Imprinting establishes the bond between parent and offspring among many mammals, as well as among birds. In many species, the mother also establishes a bond with her offspring during a critical period. The mother in some species of hoofed mammals, such as sheep, will accept her offspring for only a few hours after its birth. If they are kept apart past that time, the young are rejected. Normally, this behavior enables the mother to distinguish her own offspring from those of others, evidently by olfactory cues.

In classical conditioning, a reflex becomes associated with a new stimulus

In a type of learning called **classical conditioning,** an association is formed between some normal body function and a new stimulus. If you have observed dog or cat behavior, you know that the sound of a can opener at dinner time captivates a pet's attention. Ivan Pavlov, a Russian physiologist, discovered early in this century that if he rang a bell just before he fed a dog, the dog formed an association between the sound of the bell and the food. Eventually (Fig. 50–6), even when the bell was rung in the absence of food, the dog salivated. Pavlov called the physiologically meaningful stimulus (food, in this case) the *unconditioned stimulus.* The normally irrelevant stimulus (the bell) that became a substitute for it was the *conditioned stimulus.* Because a dog does not normally salivate at the sound of a bell, the association was clearly a learned one. It could also be forgotten. If the bell no longer signaled food, the dog eventually stopped responding to it. Pavlov called this process extinction. Many predators become conditioned to the scent or sound of potential prey.

In operant conditioning spontaneous behavior is reinforced

In **operant conditioning** (also called instrumental conditioning), the animal must do something in order to gain a reward **(positive reinforcement)** or avoid punishment. In a typical experiment, a rat is placed in a cage containing a movable bar. When random actions of the rat result in pressing the bar, a pellet of food rolls down a chute and is delivered to the rat. Thus, the rat is positively reinforced for pressing the bar. Eventually, the rat learns the association and presses the bar to obtain food.

In **negative reinforcement,** removal of a stimulus increases the probability that a behavior will occur. For example, a rat may be subjected to an unpleasant stimulus, such as a low-level electric shock. When the animal presses a bar, this negative reinforcer is removed.

When presented with food (the unconditioned stimulus), the dog begins to salivate.

Salivation

A bell (the conditioned stimulus) is rung whenever food is given to the dog. This is repeated a number of times so that an association between the food and the bell is formed.

Bell

Eventually the dog salivates at the sound of the bell alone.

Figure 50–6 **Classical conditioning.** Pavlov's experiment demonstrated that, through classical conditioning, dogs can learn to substitute a new stimulus (the conditioned stimulus) for one that was physiologically meaningful (the unconditioned stimulus).

Many variations of these techniques have been developed. A pigeon might be trained to peck at a lighted circle to obtain food, a chimpanzee might learn to perform some task in order to get tokens that can be exchanged for food, or children might learn to stay quietly in their seats at school to obtain the teacher's praise. Operant conditioning is probably the way that animals learn to perform complex tasks like perfecting feeding skills.

Operant conditioning plays a role in the development of some behaviors that appear to be genetically programmed. An example is the feeding behavior of gull chicks. Herring gull chicks peck the beaks of the parents, which regurgitate partially digested food for them. The chicks are attracted by two stimuli: the general appearance of the parent's beak with its elongated shape and distinctive red spot, and its downward movement as the parent lowers its head. Like the rat's chance pressing of the bar, this behavior is sufficiently functional to get the chicks their first meal, but they waste a lot of energy in pecking. Some pecks are off target and are therefore not rewarded. However, this pecking behavior becomes more efficient over time. Thus, a behavior that might appear to be entirely genetic may be improved by learning (Fig. 50–7).

Insight learning uses recalled events to solve new problems

Perhaps the most complex learning is **insight learning,** which is the ability to adapt past experiences that may involve different stimuli to solve a new problem. A dog can be placed in

Figure 50–7 **Operant conditioning.** With the experience of being positively reinforced for success, the pelican chick learns to be more accurate in begging for food from a parent. *(H. Cruickshank/VIREO)*

Figure 50–8 **Insight learning.** Confronted with the problem of reaching food hanging from the ceiling, the chimpanzee stacks boxes until it can climb and reach the food. What former experience might the chimp be applying to this new situation?

a blind alley that it must circumvent in order to reach a reward. The difficulty lies in the fact that the animal must move *away* from the reward in order to get to it. Typically, the dog flings itself at the barrier nearest the food. Eventually, by trial-and-error, the frustrated dog may find its way around the barrier and reach the reward.

In contrast to the dog, a chimpanzee placed in a similar situation is likely to make new associations between tasks it has learned before in order to solve the problem (Fig. 50–8). Primates are especially skilled at insight learning, but some other mammals and a few birds also seem to possess this ability to some degree.

Learning abilities are biased

Animals learn some things more easily than others. For instance, learning language comes very naturally to young humans. A child will learn the speech of her caretakers, even if no one deliberately instructs her. In general, learning biases reflect an animal's specialized mode of life.

The information most important to survival appears to be most easily learned. The same rat that may have taken a dozen trials to learn the artificial task of pushing a lever to get a reward learns in a single trial to avoid a food that has made it ill. Those who poison rats to get rid of them can readily appreciate the adaptive value of this learning ability. Such quick learning in response to an unpleasant experience forms the basis of warning coloration, which is found in many poisonous insects and brilliantly colored, but distasteful, bird eggs. Once made ill by such a meal, predators quickly learn to avoid them.

Play may be practice behavior

Perhaps you have watched a kitten pouncing on a dead leaf or practicing a carnivore neck bite or a hind-claw disemboweling stroke on a littermate without causing injury. Many animals, especially young birds and mammals, appear to be practicing adult patterns of behavior while they play (Fig. 50–9). They may improve their ability to escape, kill prey, or perform sexual behavior. Other hypotheses for the ultimate causes of play behavior include exercise, learning to coordinate movements, and learning social skills.

Some investigators have suggested that young animals play just to have "fun." Dolphins appear to play just for pleasure. One play activity that has been studied in bottlenose dolphins is swirling water with their fins, then blowing bubbles to produce rings and helices of air. Of course, there could be some other explanation, yet unknown to investigators.

Figure 50–9 **Cheetah cubs playing.** Play frequently appears to serve as a means of practicing behavior that will be used in earnest later in life, possibly in hunting, fighting for territory, or competing for mates. Play may be an example of operant conditioning in action. Photographed in Kenya. *(Peter Arnold, Inc./BIOS, M&C Denis-Huot)*

BIOLOGICAL RHYTHMS AFFECT BEHAVIOR

Many types of biological rhythms are known including daily, monthly, and annual rhythms. Among many animals, including humans, periods of activity and sleep, feeding and drinking, body temperature fluctuations, and secretion of some hormones have cycles that appear to follow an internal rhythm. Human body temperature, for example, follows a typical daily curve. These daily activities are referred to as **circadian** (meaning "approximately one day") **rhythms.** Such biological rhythms suggest that animals have **biological clocks** that are precisely adjusted or reset by environmental cues.

The behavior of many animals, like the activities of many plants (see Chapter 36), appears to be organized around circadian rhythms. **Diurnal** animals, such as honeybees and pigeons, are most active during the day. Most bats and moths are **nocturnal** animals, most active during the hours of darkness. **Crepuscular** animals, like many mosquitoes and fiddler crabs, are busiest at dawn or dusk, or both. Generally, there are ecological reasons for these patterns. If an animal's food is most plentiful in the early morning, for example, its cycle of activity must be regulated so that it becomes active shortly before dawn.

Some biological rhythms of animals reflect the **lunar** (moon) **cycle.** The most striking rhythms are those in marine organisms that are tuned to changes in tides and phases of the moon. For instance, a combination of tidal, lunar, and annual rhythms governs the reproductive behavior of the grunion, a small fish that lives off the Pacific coast of North America.

The grunion swarms from April through June on those three or four nights when the highest tides of the year occur.

At precisely the high point of the tide, the fishes squirm onto the beach and deposit eggs and sperm in the sand. They return to the sea in the next wave. By the time the next tide reaches that portion of the beach 15 days later, the young fishes have hatched in the damp sand and are ready to enter the sea. This synchronization may help protect fish eggs from aquatic predators.

Normally, an organism's metabolic processes and behavior are synchronized with the cyclic changes in its external environment. Its behavior anticipates these regular changes. The little fiddler crabs of marine beaches often emerge from their burrows at low tide to engage in social activities such as territorial disputes. They must return to their burrows before the tide returns to avoid being washed away. How do the crabs "know" that high tide is about to occur?

One might guess that the crabs recognize clues present in the seashore. However, when the crabs are isolated in the laboratory away from any known stimulus that could relate to time and tide, their characteristic behavioral rhythms persist.

Many biological rhythms appear to be regulated by *internal* timing mechanisms that serve as biological clocks. Current evidence suggests that most animals have no single biological clock. Instead, the interaction of a number of biochemical processes (possibly involving cell membranes) might be responsible for governing physiological and behavioral rhythms. The pineal gland is thought to play a role in the timing systems of birds, rats, humans, and some other vertebrates. Regions of the hypothalamus are also part of the biological clock in mammals. Although some parts of an organism may coordinate or dominate the function of the biological clock, it is likely that every cell has some type of timing mechanism.

MIGRATION INVOLVES INTERACTIONS AMONG BIOLOGICAL RHYTHMS, PHYSIOLOGY, AND ENVIRONMENT

Ruby-throated hummingbirds cross the vast reaches of the Gulf of Mexico twice each year, and the sooty tern travels across the entire South Atlantic from Africa to reach its tiny island breeding grounds south of Florida. Birds, butterflies, fishes, sea turtles, wildebeest, zebras, and whales are among the many animals that travel long distances. **Migration** is a periodic long-distance movement from one location to another. Many migrations involve astonishing feats of endurance and navigation.

Migration is an adaptation to environmental change

Why do animals migrate? Ultimate causes of migration apparently involve the advantages of moving from an area that seasonally becomes less hospitable to a region more likely to

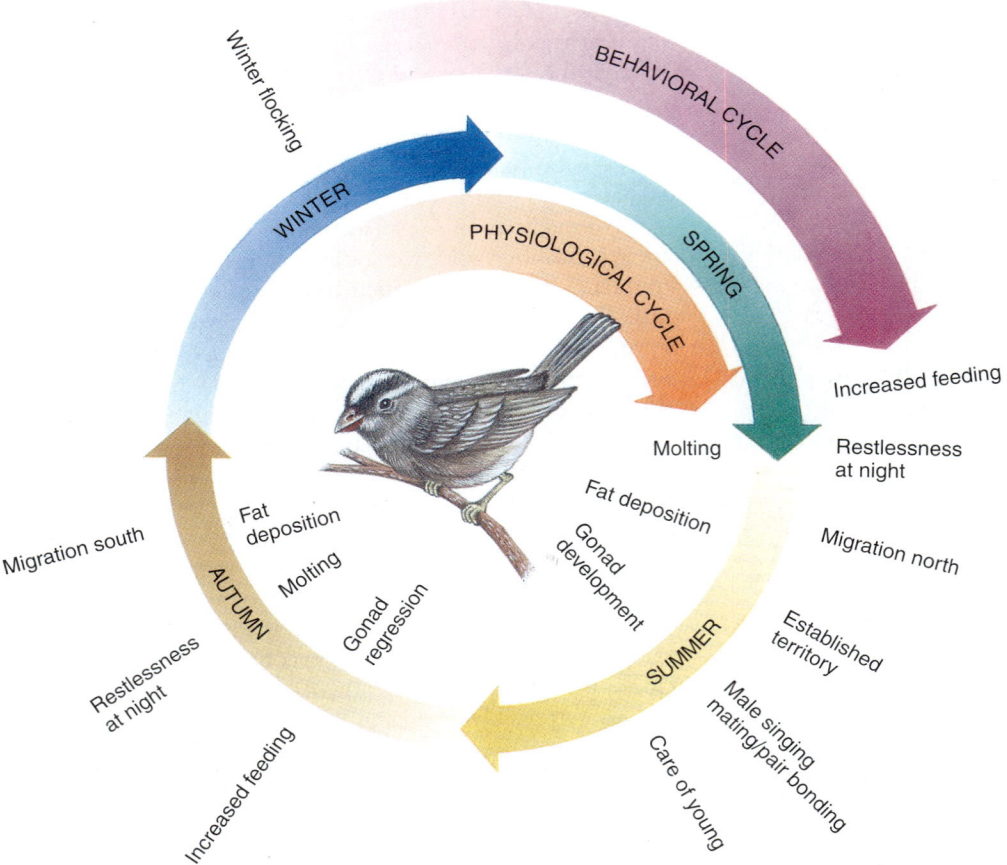

Figure 50–10 Seasonal changes in the physiology and behavior of the white-crowned sparrow.
Note the increased rate of feeding and then restlessness that precedes each period of migration. *(From Alcock, J., Animal Behavior: An Evolutionary Approach, 2nd ed., Sunderland, Mass., Sinauer Associates, 1979)*

support reproduction or survival. Seasonal changes include shifts in climate, availability of food resources, and safe nesting sites. For example, as winter approaches, many birds migrate to a warmer region.

Each year millions of monarch butterflies from Canada and the continental United States migrate to Mexico—a 2500-km journey (about 1500 mi) for some. Their dramatic annual migration appears to be related to the availability of milkweed plants on which females lay their eggs. In cold regions these plants die in late autumn and grow again with the warmer weather of spring. The wintering destinations of monarchs also seem to be determined by temperature and humidity. The investment of migrating monarch butterflies in their long journey increases their probability of reproducing. Benefits may include the opportunity to winter in an area that offers an abundant food supply, nonfreezing temperatures, and moist air.

The benefits of migration are not without cost. Many weeks may be spent each year on energy-demanding journeys. Some animals may become lost or die along the way from fatigue or predation. When in unfamiliar areas, migrating individuals are often at greater risk from predators. In recent years, human activities have interfered with migrations of many kinds

of animals. For example, after millions of years of biological success, survival of sea turtles is threatened by fishing and shrimping nets in which they get entangled. As a result, thousands of turtles drown each year.

Proximate causes of migration include signals from the environment

Some animals migrate when they mature. However, in many animals, certain environmental cues trigger physiological responses that lead to migration. In migratory birds, for example, the pineal gland senses changes in day length and then releases hormones that cause restless behavior. The birds show an increased readiness to fly, and to fly for longer periods of time (Fig. 50–10).

How do migrating animals find their way? **Directional orientation** refers to travel in a specific direction. To travel in a straight line toward a destination requires a sense of direction, or **compass sense.** Many animals use the sun to orient themselves. Because the sun appears to move across the sky each day, an animal must have a sense of time. Indeed animals appear to have biological clocks that regulate circadian rhythms.

Navigation is more complex, requiring both compass sense and **map sense,** which is an "awareness" of location. Navigation involves the use of cues to change direction when necessary to reach a specific destination. When navigating, an animal must integrate information about distance and time, as well as direction.

DNA tests confirm that loggerhead sea turtles that hatch on Florida beaches along the Atlantic swim hundreds of miles across the ocean to the Mediterranean Sea, an area rich in food. Several years later, those that survive to become adults mate, and the females navigate back, often to the same beach, to lay their eggs. Their journey requires both compass and map sense. Marine biologist Kenneth Lohmann fitted hatchling turtles with harnesses connected to a swivel arm in the center of a large tank (Fig. 50–11). A computer connected to the swivel arm recorded a record of their swimming movements. Manipulating light and magnetic fields, Lohmann demonstrated that both environmental cues are important in turtle migration. The turtles swam toward the east until Lohmann reversed the magnetic field. The turtles reversed their direction to swim toward the new "magnetic east," which was now actually west. Recent work suggests that young turtles use wave direction to help set magnetic direction preference. Some biologists suspect that turtles also use their sense of smell to guide them, particularly to guide the females to the very same beach to lay their eggs. Similarly, adult salmon use the unique odors of different streams to help find their way back to the same stream from which they hatched.

Birds and some other animals that navigate by day rely on the position of the sun; those that travel at night use the stars to guide them. Working in the 1950s, Franz and Eleonore Sauer hand-reared a number of whitethroats, a species of small European warbler. This ruled out the possibility that the parents had transmitted any information to their offspring. When (and only when) the birds could see the star patterns of the night sky, they attempted to fly in the appropriate migration direction for their species, a direction they had had no opportunity to learn.

Studies by Stephen Emlen in 1975 showed that young indigo buntings *(Passerina cyanea)* learn the constellations, using the position of the North Star as their reference point. (Other stars in the Northern Hemisphere appear to rotate about the North Star.) When Emlen rearranged the constellations in a planetarium sky, birds learned the altered patterns of stars and later attempted to fly in a direction consistent with the artificial pattern. These and other studies suggested that birds have a genetic ability to learn constellation patterns and use them to orient themselves during migration.

Investigators have long observed that some species of birds can navigate even when the sky is overcast and they cannot see the stars. They hypothesized that these birds use Earth's magnetic field to navigate. Studies of garden warblers *(Sylvia borin)* by P. Weindler and his colleagues in 1996 indicate that, as these birds make their way from central Europe to Africa each winter, they navigate both by the stars and by the Earth's magnetic field. These investigators found that warblers used the stars to determine the general direction of travel, then used magnetic information to refine and correct their course. Honeybees, some fishes, amphibians, sea turtles, and some other animals also appear to be sensitive to Earth's magnetic field, and use it as a guide.

EFFICIENT FORAGING BEHAVIOR CONTRIBUTES TO SURVIVAL

Feeding behavior, or **foraging,** involves locating and selecting food, as well as food gathering and food capture. Some behavioral ecologists study the costs and benefits of searching for and selecting certain types of food, as well as the mechanisms used to locate prey. For example, many camouflage strategies have evolved that make prey difficult to detect. As predators experience repeated success in locating a particular prey, they are thought to develop a *search image,* a constellation of cues that help them identify hidden prey.

Why do grizzly bears spend hours digging Arctic ground squirrels out of burrows, while ignoring larger prey such as caribou? It is energetically more efficient to dig for squirrels because the bears' efforts most probably will be rewarded, whereas caribou are more likely to escape, leaving the bears hungry. This is an example of **optimal foraging,** the most efficient way for an animal to obtain food. When animals maximize energy obtained per unit of foraging time, they may maximize reproductive success. Many factors, such as avoiding predators while foraging, must be considered in determining efficient or optimal strategies.

In habitats where the most prized food items are abundant and an animal does not have to travel far to obtain them, animals can afford to be very selective. In contrast, the optimal strategy in poor habitats, where it takes longer to find the best food items, is to select a greater variety of items. Animals

Figure 50–11 **Navigation by light and magnetic field.** To study turtle navigation, researcher Kenneth Lohmann harnessed leatherback turtles *(Dermochelys coriacea)* like this hatchling and wired them to a computer that recorded their swimming direction. *(Kenneth Lohmann)*

may learn to forage efficiently through operant conditioning, that is, by randomly trying various strategies and selecting the one associated with the most rewards (and the fewest hunger pangs at the end of the day).

Foraging is also affected by risk of predators. Guy Cowlishaw studied foraging behavior in a population of baboons *(Papio cynocephalus ursinus)*. He found that the baboons spent more time foraging in a habitat in which food was relatively scarce than in an area with more abundant food that had a high risk of predation by lions and leopards.

Lions live, and often forage, in social units called *prides*. A pride typically consists of a group of related adult females, their cubs, and unrelated males. Lion behavior has been extensively studied in Serengeti National Park, Tanzania, by Craig Packer and his research team. Packer has radio-collared at least one female in each of 21 prides and has tracked them for several years. During the season when prey is abundant, lions hunt wildebeest, gazelle, and zebra that have migrated into the area. When these herds migrate out of the area during the dry season, prey becomes scarce and lions feed mainly on warthog and Cape buffalo. Packer has reported that when prey is abundant, the size of the foraging group has little effect on daily food intake. Whether a lion hunts individually or in small (two to four females) or large (five to seven) groups, food is plentiful and is captured by individuals or groups of any size. However, during the season when prey is scarce, lions are more successful if they hunt alone or in large groups (Fig. 50–12). Despite this, Packer's data from radio-collared lions indicated that females in small prides typically forage in as large a group as they can, even though this approach decreases foraging efficiency. Thus, selection pressure appears to be stronger to protect cubs and defend territory against larger prides than to forage optimally (alone).

SOCIAL BEHAVIOR HAS BOTH BENEFITS AND COSTS

The mere presence of more than one individual does not mean that a behavior is social. Many factors of the physical environment bring animals together in **aggregations,** but whatever interaction they experience may be circumstantial. A light shining in the dark is a stimulus that draws large numbers of moths. The high humidity under a log may attract wood lice. Although these aggregations may have adaptive value, they are not truly social, because the organisms are not responding to one another.

We can define **social behavior** as the interaction of two or more animals, usually of the same species. Many animals benefit from living in groups. By cooperation and division of labor, some insects construct elaborate nests and raise young by mass-production methods. Schools of fishes tend to confuse predators, and so individuals within the school are less vulnerable to predators than a solitary fish would be. Zebras can more effectively protect themselves when they are in groups (Fig. 50–13). Zebra stripes appear to be a visual antipredator

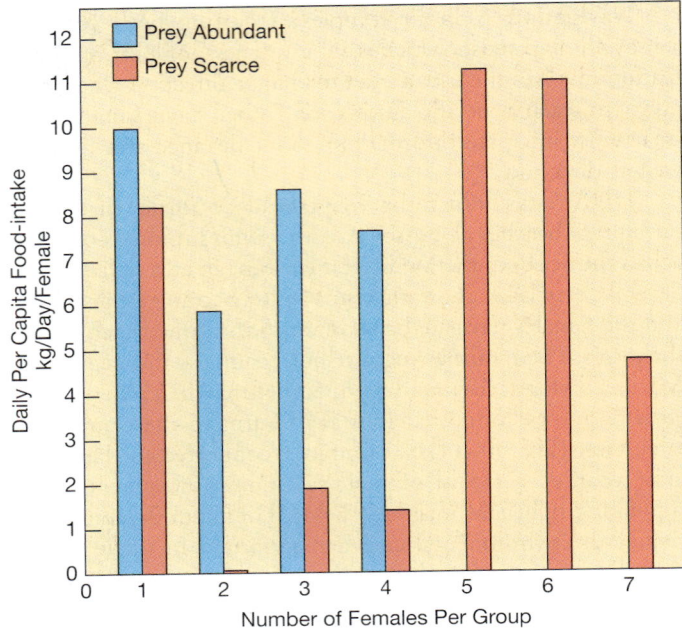

Figure 50–12 Optimal foraging and group size in lions. Researcher Craig Packer and his colleagues observed that during times of prey scarcity, optimal foraging was related to group size: Lions hunted most successfully alone or in large groups of five to seven. Observations of group size suggested that the benefits of optimal defense often outweighed the benefits of optimal foraging. *(Based on data of Packer et al)*

adaptation. When viewed from a distance, the stripes tend to visually break up the form of the animal so that individuals cannot be distinguished. A herd of zebras tends to confuse predators, whereas a solitary animal is easy prey to lions, cheetahs, or spotted hyenas.

Figure 50–13 Social behavior in zebras. The social unit consists of a fairly stable group of females with young and a dominant stallion. Being in a group of many striped zebras may be advantageous to individuals attempting to avoid predators (confusion effect). Photographed in East Africa. *(McMurray Photography)*

Social foraging is an adaptive, efficient strategy that is used routinely by many animal species. A pack of wolves has greater success in hunting than the individual wolves would have if hunting alone. Among some birds of prey, such as ospreys, group hunting results in locating prey more quickly.

Some species that engage in social behavior form societies. A **society** is an actively cooperating group of individuals belonging to the same species and often closely related. A hive of bees, a flock of birds, a pack of wolves, and a school of fish are examples of societies. Some societies are loosely organized, whereas others have complex structures. Characteristics of a highly organized society include cooperation and division of labor among animals of different sexes, age groups, or castes. A complex system of communication reinforces the organization of the society. The members tend to remain together and to resist attempts by outsiders to enter the group.

Social behavior offers benefits that increase the chances of perpetuating the genes that produce such behavior. However, social behavior also has certain costs. Living together means increased competition for food and habitats. Social interaction, as well as behaviors such as nesting and roosting, also increase the risk of attracting predators and of transmitting disease. Typically, for animals that exhibit social groupings, the benefits of the society outweigh the costs. For the many animal species that do not form social groups, the costs of forming a society outweigh the benefits.

COMMUNICATION IS NECESSARY FOR SOCIAL BEHAVIOR

One animal can influence the behavior of another only if the two of them can exchange mutually recognizable signals (Fig. 50–14). **Communication** is most evident when an animal performs an act that changes the behavior of another organism. Communication may be important in finding food, as in the elaborate dances of the honeybees. Animals may communicate to hold a group together, warn of danger, indicate social status, indicate willingness to accept or provide care, identify members of the same species, or indicate sexual maturity or readiness.

Animals communicate in a wide variety of ways

In animal communication, one animal signals auditory, visual, tactile, chemical, or electrical information to another. Orcas (*Orcinus orca;* formerly known as killer whales) communicate by sounds and songs. Members of a given pod (social group) have an average of 12 different calls, which vary in pitch and duration and reflect their "emotional" state. In many bird species, territorial males announce their presence and willingness to interact socially by singing. Along with their songs, many birds present visual displays.

Some animals communicate by scent. Antelopes and deer rub facial gland secretions on conspicuous objects in their

(a)

(b)

Figure 50–14 **Animal communication.** *(a)* A male spring peeper frog *(Hyla crucifer)* calling to locate a mate. *(b)* Communicating with language. Chimpanzee Tatu *(top)* is signing "food" to Washoe *(below)*. Washoe was the first chimp to learn American Sign Language from a human. She then taught other chimps how to sign. *(a, R. Lindholm/ Visuals Unlimited; b, April Ottey, Chimpanzee and Human Communication Institute, Central Washington University)*

vicinity and urinate on the ground. Dogs mark territory by frequent urination. Certain fishes (gymnotids) use electric pulses for navigation and communication, including territorial threat, in a fashion similar to bird vocalization. As sociobiologist Edward O. Wilson has said, "The fish, in effect, sing electrical songs."

Pheromones are chemical signals used in communication

Pheromones are chemical signals secreted by animals that convey information between members of a species. They are a simple, widespread means of communication. Most pheromones elicit a very specific, immediate, but transitory type of behavior. Others trigger hormonal activities that result in slow but long-lasting responses (Chapter 47). Some pheromones may act in both ways.

An advantage to pheromone communication is that relatively little energy is needed to synthesize the simple, but distinctive, organic compounds involved. Members of the same species have receptors that fit the molecular configuration of the pheromone; other species usually ignore it or do not detect it at all. Pheromones are effective in the dark, they can pass around obstacles, and they last for several hours or longer. Major disadvantages of pheromone communication are slow transmission and limited information content. Some animals compensate for the latter disadvantage by secreting different pheromones with different meanings.

Pheromones are important in attracting the opposite sex and in sex recognition in many species. Many female insects produce pheromones that attract males. We have taken advantage of some sex-attractant pheromones to help control such pests as gypsy moths by luring the males to traps baited with synthetic versions of female pheromone.

Some aspects of the vertebrate sexual cycle are affected by pheromones. When the odor of a male mouse is introduced among a group of females, the reproductive cycles of the female mice become synchronized. In some species of mice, the odor of a strange male, a sign of high population density, causes a newly impregnated female to abort. Among humans, it appears that some unconsciously perceived body odor is capable of synchronizing the menstrual cycles of women who associate closely (for instance, college roommates or cellmates in prison). Pheromones more strictly govern the reproduction of many social insects.

DOMINANCE HIERARCHIES ARE SOCIAL RANKINGS

In the spring, female paper-wasps awaken from hibernation and begin to build a nest together. During the early course of construction, a series of squabbles among the females takes place in which the combatants bite one another's bodies or legs. Finally, one of the wasps emerges as dominant. After that, she is rarely challenged. This queen wasp spends more and more time tending the nest and less and less time out foraging for herself. She takes the food she needs from the others as they return.

The queen then begins to take an interest in raising a family—her family. Because she is almost always in the nest, she is able to prevent other wasps from laying eggs in the brood cells by rushing at them, jaws agape. Because of her supreme dominance, the queen can bite any other wasp without serious risk of retaliation.

The other wasps of the nest are further organized into a **dominance hierarchy,** a ranking of status in which each wasp has more status than the wasps that are lower in rank. Wasps lower in the hierarchy are subordinate to those above them.

Queen $\longrightarrow$ Wasp A $\longrightarrow$ Wasp B $\longrightarrow$ ... Wasp J $\longrightarrow$ Wasp K

Dominance hierarchies suppress aggression

Once a dominance hierarchy is established, little or no time is wasted in fighting. When challenged, subordinate wasps exhibit submissive poses that, in turn, inhibit the queen's aggressive behavior. Consequently, few or no colony members are lost through wounds sustained in fighting one another. This ensures greater reproductive success for the colony.

Many factors affect dominance

In some animals, dominance is a function of aggressiveness, which is itself often influenced directly by sex hormones (Fig. 50–15). Among chickens, the rooster is the most dominant; as with most vertebrates, the hormone testosterone increases aggressiveness. If a hen receives testosterone injections, her place in the dominance hierarchy shifts upward. When male rhesus monkeys are dominant, their testosterone levels are much higher than when they have been defeated. Not only can estrogen sometimes reduce dominance and testosterone increase dominance, but dominance may even increase testosterone. It is not always easy to determine cause and effect. (Studies suggest that the effects of testosterone on human mood and behavior may be different.)

Establishing dominance is often strenuous and dangerous. Tremendous energy is often needed for the fights engaged in

Figure 50–15 Communicating dominance. Social animals use many signals to convey messages relating to social dominance. This baboon bares his teeth and screams in an unmistakable show of aggression, a signal that allows him to establish and maintain dominance. *(Gerald Lacz/Peter Arnold, Inc.)*

Figure 50–16 Home range. The massive Cape buffalo *(Syncerus caffer)* lives in herds of several hundred animals. Each herd, like this one photographed in the Serengeti, has a fairly constant home range. Buffalo protect herd members, especially calves. *(McMurray Photography)*

by some birds and many mammals, including rams and bull seals. Males may expend energy in wallowing, roaring, and leaping about. These behaviors appear to be a test of male quality. The male with the greatest endurance will likely gain dominance and will have the greatest opportunity to mate and to perpetuate his genes.

In many species, males and females have separate dominance systems. However, in many monogamous animals, especially birds, the female acquires the dominance status of her mate by virtue of their relationship and the male's willingness to defend his mate.

Like many fishes and some invertebrates, certain coral reef fishes (labrids) are capable of sex reversal. The largest, most dominant individual is always male, and the remaining fishes within his territory are all female. If the male dies or is removed, the most dominant female will become the new male. Should any harm come to him, the next-ranking female will undergo sex reversal, take charge, and protect the group territory. Other types of fishes exhibit the reverse behavior, in which the most dominant fish is always female. In these species, size is less important for aggressive defense than for maximal egg production.

MANY ANIMALS DEFEND A TERRITORY

Most animals have a **home range,** a geographical area that they seldom leave (Fig. 50–16). Because the animal has the opportunity to become familiar with everything in that range, it has an advantage over its competitors, predators, and prey in negotiating cover and finding food. Some, but not all, animals exhibit **territoriality.** They defend a **territory,** a portion of the home range, often against other individuals of their species and sometimes against individuals of other species (Fig. 50–17). Many species are territorial for only part of the year, often during the breeding season, but others are territorial throughout the year. Territoriality has been positively correlated with the availability of needed resources that occur in small areas that can be defended.

Figure 50–17 Territoriality in lions. Lions *(Panthera leo)* form territorial groups typically consisting of several females and their offspring and accompanied by a coalition of males, generally brothers who are unrelated to the females. Successful breeding appears to depend on having a territory. *(Steve McCutcheon/Visuals Unlimited)*

Territoriality is easily studied in birds. Typically, the male chooses a territory at the beginning of the breeding season. This behavior results from high sex hormone concentrations in his blood. The males of adjacent territories fight until territorial boundaries become established. Generally, the dominance of a male is directly associated with his nearness to the center of his territory. Thus, close to home he is like a lion, but when invading some other bird's territory, he may behave more like a lamb. The interplay of dominance values among territorial males may produce a neutral line at which neither is dominant.

Bird songs announce the existence of a territory and often serve as a substitute for violence. Furthermore, they announce to eligible females that a propertied male resides in the territory. Typically, male birds take up a conspicuous station, sing, and sometimes display striking patterns of coloration or aerial acrobatics to their neighbors, their rivals, and sometimes their mates.

Studies of lions in the Serengeti show that these animals engage in group territoriality. As discussed previously, a pride consists of several adult females, their dependent young, and a group of immigrant males. The females protect their young and defend the territory against invasion by foreign females. The males defend the territory against strange males. At night males patrol their turf, roaring fiercely to announce their presence.

The costs of territoriality include the time and energy expended in staking out and defending a territory and the risks involved in fighting for it. Benefits often include exclusive rights to food within the territory and greater reproductive success. The males of a lion pride father all the cubs born within that pride. Their ability to defend their territory from invasion by strange males ensures their reproductive success. Among many species, animals that fail to establish territories fail to reproduce. Territoriality also tends to reduce conflict among members of the same species and ensures efficient use of environmental resources by encouraging individuals to spread throughout a habitat.

Usually, territorial behavior is related to the specific lifestyle of the animal and to whatever aspect of its ecology is most critical to its reproductive success. For instance, sea birds may range over hundreds of square kilometers of open water but exhibit territorial behavior only at crowded nesting sites on an island. The nesting sites are their scarcest resource and the one for which competition is keenest.

SEXUAL SELECTION RESULTS IN REPRODUCTIVE ADVANTAGE

In many species individuals actively compete for mates. Typically in a population, there is an abundance of males competing for a limited number of receptive females. For a male, reproductive success depends on how many females he can impregnate. Females may have the opportunity to select a sexual partner from among several males. For a female, reproductive success depends on how many eggs she can produce during her reproductive lifetime, the quality of the sperm that fertilizes them, and on the survival of her offspring to reproductive age. An animal's reproductive success is a measure of its *direct fitness*. **Sexual selection,** a type of natural selection, occurs when individuals vary in their ability to compete for mates. This leads to the reproductive advantage that some individuals have over others of the same sex and species.

Dominance and ornamental traits influence mate choice

In many species, success of a male in dominance encounters with other males indicates his quality to the female, and she allows the victorious male to court her. Many studies confirm that males that rank higher in a dominance hierarchy mate more frequently than males that rank lower.

Many exceptions demonstrate the complexity of reproductive behavior among some species. For example, among baboons with a definite dominance hierarchy, males lower on the hierarchy copulated with females as frequently as males of higher status. Investigators observed, though, that dominant males copulated more frequently with females who were in estrus, their fertile period. In addition, some lower ranking males develop alternative strategies for attracting females. For instance, a male might win a female's interest by protecting her baby (even though it is not his own).

Among crickets and many other insect species, a courting male captures prey and offers this gift of food to a prospective mate. Studies show that the larger or higher quality the food offering, the better the chances the male will be accepted.

Females may select their mates based on ornamental displays. Male fishes are often brightly colored. Among many bird species, males exhibit bright colors and dramatic plumage, and male deer display elaborate antlers. Investigators suggest that the expression of ornamental traits may give the female important information about the male's physical condition. The size of the antlers of red deer (*Cervus elaphus*), for example, reflects proper nutrition and good health.

Among some species of insects, birds, and bats, males gather in a small display area called a **lek** where they compete for females. When a receptive female appears, males may excitedly display themselves and compete for her attention. Among some species, the female selects a male based on his location in the lek, rather than on his appearance. The dominant male may occupy a central position and be chosen by most of the females. The female selects a male, mates with him, and then leaves the lek. The male remains to woo other females.

Courtship rituals ensure that the male is indeed a male and is a member of the same species, and also provide the female further opportunity to evaluate him. In some species, courtship may also be necessary as a signal to trigger nest

(a)

(b)

Figure 50–18 **Courtship displays.** *(a)* The male great frigate-bird *(Frigata minor)* inflates his red throat sac in display as part of a courtship ritual. Photographed on Christmas Island, in the Pacific. *(b)* Egrets *(Egretta rufescens)* performing a mating dance. *(a, Sid Bart/Photo Researchers, Inc.; b, Arthur Morris/Visuals Unlimited)*

building or ovulation. Courtship rituals can last seconds or hours and often involve a series of fixed action patterns (Fig. 50–18). The first display by the male releases a counter behavior by the female. This, in turn, releases additional male behavior, and so on until the pair is physiologically ready for copulation. Specific cues enable courtship rituals to function as reproductive isolating mechanisms among species (see Chapter 19).

An extreme courtship ritual has been described for red-back spiders *(Latrodectus hasselti),* which are closely related to the black widow spider. During copulation, the small male spider positions himself above his larger mate's jaws. During 65% of matings, the finale of mating is that the female eats her suitor. The apparent explanation for this behavior is that the suicidal male is able to copulate for a longer period and thus fertilize more eggs than noncannibalized males and that the female is more likely to reject additional mates.

Sexual selection favors polygynous mating systems

In most species, males make little parental investment in their offspring, apart from providing sperm. Males ensure reproductive success by impregnating many females, increasing the probability that their genes will be propagated in multiple offspring. Thus, sexual selection favors male **polygyny,** mating with many females.

In the mating system known as **polyandry,** a female mates with several males. Benefits may include receiving gifts from several males or enlisting several males to help care for the young. Data collected at Jane Goodall's research center at Gombe National Park in Tanzania indicate that female chim-

panzees mate several hundred times. In addition to mating with males of their own group, many females slip off to neighboring communities when they are most fertile and copulate with less familiar males. It is possible that these sexual rendezvous protect against inbreeding. Investigators also suggest that mating with many males provides insurance against infanticide. Males do not aggress against infants of mothers with whom they have mated. Males in polyandrous species may accept their status because guarding their females may be costly, ineffective, or both. Females may be able to range over a wide area, easily escaping guarding behavior of males. (See *Focus On: Dominance Effects on Reproductive Success in Female Chimpanzees.*)

Polyandry and polygyny sometimes occur in the same species. After mating, a female giant water bug attaches a clutch of eggs to the back of her mate. He takes care of the eggs until they hatch. She then may mate with a different male and glue the new clutch of eggs to his back. However, if the male has more space for eggs, he may mate with another female.

Among some species, a male guards his partner after copulation to ensure that she will not copulate with another male. Mate guarding behavior is likely to occur when the female is receptive and has eggs that might be fertilized by another male. For example, dominant male African elephants guard a female only during the phase of estrus when she is most likely to have a fertile egg. Before that time, or later in estrus after mating has already occurred, younger male elephants of lesser social status can copulate with the female. One cost of mate guarding is the loss of opportunity for a dominant male to mate with other females.

Among some species males are **monogamous,** that is, they mate with a single partner during a breeding season.

DOMINANCE EFFECTS ON REPRODUCTIVE SUCCESS IN FEMALE CHIMPANZEES

Until recently dominance hierarchies were thought to be a "male thing" among chimpanzees. Now, data from a 35-year long field study of chimpanzees in Gombe National Park in Tanzania suggest otherwise. Anne Pusey, Jennifer Williams, and Jane Goodall reported in *Science* (August 1997) that dominant female chimpanzees are more successful reproductively than those with lower social status.

Adult male chimpanzees travel together over the home range of the social community, defending its borders. When challenged by a dominant male, a lower ranking male or female communicates recognition of its lower status by making pant-grunts. These hnn-hnn-hnn noises signal submission. An adult female chimpanzee spends about 65% of her time alone with her dependent offspring. Much of her time is spent foraging for food in a core area that may overlap with the core area of another adult. When she meets another female, the two adults often ignore each other. However, sometimes one female pant-grunts to the other.

The research team mapped female dominance hierarchies by recording the direction of pant-grunts between females. They found that the direction was reliably consistent. If female X pant-grunted to female A at any encounter, female X would again pant-grunt to female A the next month or the next year.

Chimpanzee (Pan triglodytes) with young. Photographed at Gombe Stream, Tanzania. *(Kennan Ward/DRK Photo)*

The team looked at the reproductive histories of the females and found that rank had a significant effect on reproductive success. Offspring of high-ranking mothers were much more likely to survive to age seven years than offspring of low-ranking mothers. In addition, daughters of high-ranking mothers reached sexual maturity as much as four years earlier. The investigators also found that high-ranking females live longer than low-ranking females.

How does dominance rank affect reproductive success? The research team observed that high-ranking females often grabbed and killed the infants of low-ranking mothers. Infanticide appeared to be a significant threat. The investigators found that female offspring of dominant mothers weighed more than those of low-ranking mothers, and higher weight correlated with reaching sexual maturity at an earlier age. They suggested that better nutrition might account for higher survival rates of high-ranking females and their young. Better nutrition may be a consequence of claiming and maintaining a core area with abundant, more nutritious fruit.

How do female chimpanzees achieve social status? Some females apparently become dominant through aggressive behavior. Others achieve high rank by virtue of their mother's status. Does social rank determine success, or does success determine social rank? Or is there another, yet unknown, factor responsible for both? Research continues at Gombe in the pursuit of understanding more about social relationships among our closest relatives.

Monogamy is a rare mating system among mammals, but common in birds. Male monogamy occurs when males need to guard their mates so that they do not mate with other males. This mating system also occurs when males are needed to protect and feed the young. For example, among wolves and many other carnivores, males defend the young and the food supply.

A **pair bond** is a stable relationship between two animals of the opposite sex that may ensure cooperative behavior in mating and the rearing of the young (Fig. 50–19). In an estimated 90% of bird species, a male pair-bonds with a female during the breeding season. The mechanisms involved in establishing and maintaining the pair bond are often remarkably detailed.

Many organisms care for their young

Care of the young is an important part of successful reproduction in many species (Fig. 50–20). The benefit of investment in parental care is the increased probability that each individual offspring will survive. The costs include a reduction in the number of offspring that can be produced and the risks taken in protecting the young from predators. Natural selection has favored parental care in species in which the benefits to offspring survival outweigh the costs of decreased opportunities to produce additional offspring.

Females of many vertebrate animals have more invested in gamete formation than males; they produce relatively few,

Figure 50–19 **Pair bond in birds.** Black-browed albatrosses *(Diomedea melanophris)* form durable mating pairs and may stay together for life. *(Johnny Johnson/DRK Photo)*

large eggs. Because of the time and energy spent producing eggs and carrying the developing embryo, the female has more to lose than the male if the young do not develop. Thus, females are more likely than males to brood eggs and young, and usually females invest more in parental care. Parental care is especially skewed toward the female in mammals because female mammals, not males, provide milk to nourish their young.

Investing time and effort in care of the young is usually less advantageous to a male (assuming that the female can handle the job by herself), for time spent in parenting is time lost from inseminating other females. In some species, it may not even be certain who fathered the offspring. Raising some other male's offspring reduces a male's ability to produce and care for his own and so is a genetic disadvantage. Thus, males may compromise mate chasing in favor of mate guarding.

In some situations, however, a male can benefit by helping to rear his own young or even those of a genetic relative. Receptive females may be scarce, breeding territories may be

(a)

(b)

(c)

Figure 50–20 **Caring for the young.** Parental investment in offspring increases the probability that the young will survive. *(a)* Adult robins invest energy in feeding their young. *(b)* Female crocodile transporting hatchlings in her mouth from their nest site to Lake St. Lucia. *(c)* Female cheetah *(Acinonyx jubatus)* stands watch as her cubs eat the Thompson's gazelle she has hunted and killed for them. Photographed in Masi Mara, East Africa. *(a, Dominique Braud/Dembinsky Photo Associates; b, M. Reardon/Photo Researchers, Inc.; c, McMurray Photography)* difficult to establish or guard, and gathering sufficient food may require more effort than one parent can provide. In some habitats, the young may need protection against predators or cannibalistic males of the same species.

Among many species of fishes, the male cares for the young. The ultimate cause for this behavior appears to be that the parenting costs to the males are less than they would be for the females. A male fish must have a territory in order to attract a female. While guarding his territory, the male guards the fertilized eggs as well. Having eggs also appears to make the male more attractive to potential mates. In contrast, when a female is caring for her eggs or young, she is not able to feed optimally. As a result her body size is smaller and her fertility is reduced.

ELABORATE SOCIETIES ARE FOUND AMONG THE SOCIAL INSECTS

Although many insects, such as tent caterpillars, which spin a communal nest, cooperate socially, the most elaborate insect societies are found among the bees, ants, wasps, and termites. (The first three of these all belong, not coincidentally, to the same order, Hymenoptera.) Insect societies are held together by a complex system of sign stimuli that are keyed to social interaction. As a result, they tend to be quite rigid.

The social organization of honeybees has been studied more extensively than that of any other social insect. Instructions for the honeybee society are inherited and preprogrammed, and the size and structure of the bee's nervous system permit only a limited range of variation of behavior. Yet these insects are not automatons. Within those limits, the complex bee society can respond flexibly to food and other stimuli in the environment.

A honeybee society generally consists of a single adult queen, up to 80,000 worker bees (all female), and, at certain times, a few males called drones that fertilize newly developed queens. The queen, the only female in the hive capable of reproduction, deposits about 1000 fertilized eggs per day in the wax cells of a comb.

Division of labor among worker bees is mostly determined by age. The youngest worker bees serve as nurses that nourish larval bees. After about a week as nurse bees, workers begin to produce wax and build and maintain the wax cells. Older workers are foragers, bringing home nectar and pollen. Most worker bees die at the ripe old age of 42 days.

The composition of a bee society is generally controlled by an antiqueen pheromone secreted by the queen. It inhibits the workers from raising a new queen and prevents development of ovaries in the workers (Fig. 50–21). If the queen dies, or if the colony becomes so large that the inhibiting effect of the pheromone is dissipated, the workers begin to feed some larvae special food that promotes their development into new queens.

Figure 50–21 **Maintaining a complex social structure.** A queen honeybee *(center)* surrounded by numerous workers. They constantly lick secretions from the queen bee. These secretions, transmitted throughout the hive, suppress the activity of the workers' ovaries. *(Treat Davidson/Photo Researchers, Inc.)*

The most sophisticated known mode of communication among bees is a stereotyped series of body movements referred to as a dance. When a honeybee scout locates a rich source of nectar, it apparently observes the angle between the food, hive, and the sun. The bee then communicates the direction and distance of the food source relative to the hive (Fig. 50–22). If the food supply is within about 50 m, the scout performs a **round dance,** which generally excites the other bees and causes them to fly short distances in all directions from the hive until they find the nectar. If the food is distant, however, the scout performs a **waggle dance,** which follows a figure-eight pattern.

In the 1940s Karl von Frisch pioneered studies in bee communication. He found that the waggle dance conveys information about both distance and direction. The bee typically performs the dance in the hive on the vertical surface of the comb. During the straight run part of the figure eight, the number and frequency of the waggles indicate the distance. The orientation of the movements indicates the direction of the food source in reference to the position of the sun (Fig. 50–22). For example, if the bee dances straight up, the nectar is located directly toward the sun. If the bee dances 40 degrees to the left of the vertical surface of the comb, the nectar is located 40 degrees to the left of a line between the hive and the sun.

Von Frisch thought that through their dances the bee scouts symbolically communicated information about food sources to the worker bees. Indeed, human observers can find food sources based on information conveyed by the dances. However, some investigators have been skeptical that bees could have such a complex system of communication. Perhaps

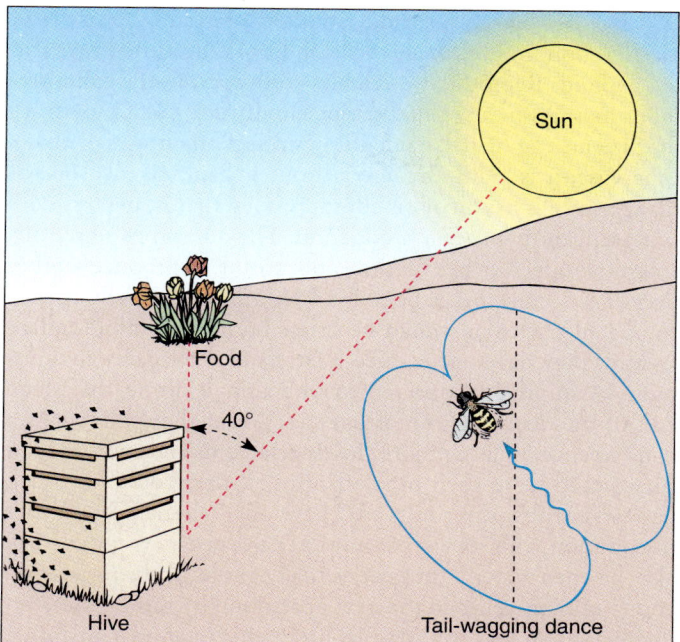

Figure 50-22 **The honeybee waggle dance.** The scout is waggling upward, indicating that the food is toward the sun, and inclined 40 degrees to the left, which reveals the angle of the food source relative to the sun.

the flower scent on the dancing bee might be the true clue to the source of the food, and the dance itself does not really communicate anything to the bees. Studies using mechanical bees (with no scent) have supported Von Frisch's conclusions. Current evidence suggests that bees can communicate in both ways—through their dances and also through scent.

VERTEBRATE SOCIETIES TEND TO BE RELATIVELY FLEXIBLE

Vertebrate societies usually contain nothing comparable to the physically and behaviorally specialized castes of honeybees or ants. An exception is the naked mole rat, a rodent that has a social structure closely resembling that of the social insects.

Although in some ways most vertebrate societies seem simpler than insect societies, they are also more flexible. Vertebrate societies share a great range and plasticity of potential behaviors and can effectively modify behavior to meet environmental challenges. The behavioral plasticity of vertebrates makes possible the transmission of culture in some bird and primate groups. Human society is based to a great extent on the symbolic transmission of culture.

ALTRUISTIC BEHAVIOR CAN BE EXPLAINED BY INCLUSIVE FITNESS

Sociobiology focuses on the evolution of social behavior through natural selection. In his landmark book, *Sociobiology: The New Synthesis* published in 1975, Edward O. Wilson combined principles of population genetics, evolution, and animal behavior to present a comprehensive view of the evolution of social behavior. Wilson's work influenced the development of behavioral ecology, and many of the concepts discussed in this chapter, such as paternal investment in care of the young and helping behavior are based on contributions made by sociobiologists.

From the sociobiological perspective, an organism and its adaptations, including its behavior, ensure that genes make more copies of themselves. The cells and tissues of the body support the functions of the reproductive system, and the reproductive system transmits genetic information to succeeding generations.

If an animal's primary mission is to perpetuate its genes, we must wonder why some animals appear to spend time and energy in helping others. We have discussed examples of cooperative behavior such as group hunting in which each animal in the group benefits. This type of helpful behavior is referred to as **mutualism.** In a type of mutualism known as **reciprocal altruism,** one animal helps another with no immediate benefit. However, at some later time the animal that was helped repays the debt.

In **altruistic behavior** there appears to be no payoff: one individual behaves in a way that seems to benefit others rather than itself, with no potential payoff. However, true altruistic behavior would not be adaptive because the animal's response decreases its reproductive success relative to individuals that do not exhibit the altruistic behavior. In 1964 William D. Hamilton offered a plausible explanation. He suggested that evolution does not distinguish between genes transmitted directly from parent to offspring and those transmitted indirectly through close relatives. According to Hamilton, natural selection favors animals that help a relative because the relative's offspring carries some of the helper's genes. This concept is known as **inclusive fitness** because it includes the genes an animal perpetuates in its kin as well as the genes it perpetuates in its own offspring.

Inclusive fitness can be measured by the **coefficient of relatedness,** the probability that two individuals inherit the same gene from a common ancestor. The coefficient of relatedness for two animals that are not related is 0, whereas between a parent and its offspring this coefficient would be 0.5, indicating that half of their genes are the same. The coefficient of relatedness for siblings is also 0.5; for first cousins it is 0.125. The higher the coefficient of relatedness, the more likely an animal will help a relative. DNA fingerprinting of lions in the Serengeti National Park and in the Ngorongoro Crater in Tanzania has confirmed that lion brothers that cooperate in prides

Figure 50–23 **Kin selection in prairie dogs.** Low-ranking members of this social rodent group act as sentries, risking their own lives by exposing themselves outside their burrows. However, in this way they protect their siblings and by so doing ensure that the genes they share in common will be perpetuated in the population. *(Wu Wal Ping/Bruce Coleman, Inc.)*

have a greater probability of perpetuating their genes than those that go off on their own. This is true even if the lion perpetuates his genes only by proxy through his brother. **Kin selection** is a form of natural selection that increases inclusive fitness through the breeding success of close relatives (Fig. 50–23).

Among some birds (e.g., Florida jays), nonreproducing individuals aid in the rearing of the young. Nests tended by these additional helpers as well as parents produce more young than do nests with the same number of eggs overseen only by parents. The nonreproducing helpers are close relatives who increase their own biological success by ensuring the successful, though indirect, perpetuation of their genes. Helpers may be prevented from producing their own offspring by such limiting factors as a shortage of mates or territories.

The bee society provides another interesting example of kin selection. In bees, males develop from unfertilized eggs and are haploid. Reproductive females store sperm cells from previous matings in a seminal receptacle. If they release sperm to fertilize an egg as it is laid, the resulting offspring is female; otherwise, it is male. Because a drone is haploid, each one of his sperm cells has *all* of his chromosomes; that is, meiosis does not occur during sperm production. The queen bee stores this sperm throughout her lifetime and uses it to produce worker bees. Thus, the worker bees of a hive are more closely related to one another than would be sisters born of a diploid father. Indeed, they have up to three-quarters of their genes in common. (Assuming the queen bred with a single drone, they share half of the queen's chromosomes and all of the drone's.) As a consequence, they are more closely related to one another than they would be to their own offspring, if they could have any. (A worker bee's offspring would have only one-half of its genes in common with its worker mother.) Because new queens will also be their sisters, they are actually more likely to pass on copies of their genes to the next generation by raising these individuals than they would be by producing their own offspring.

SOCIOBIOLOGY EXPLAINS HUMAN SOCIAL BEHAVIOR IN TERMS OF ADAPTATION

Like Darwin and many other biologists of the past, E. O. Wilson and other sociobiologists suggest that human behavior can be studied in evolutionary terms. Sociobiology is controversial, at least in part because of its possible ethical implications. This approach is sometimes viewed as denying that human behavior is flexible enough to permit substantial improvements in the quality of our social lives. Yet sociobiologists do not disagree with their critics that human behavior is flexible. At least some of the debate therefore seems to rest on the degree to which human behavior is genetically determined and the extent to which it can be modified.

As sociobiologists acknowledge, people can, through culture, change their way of life far more profoundly in a few years than could a hive of bees or a troop of baboons in hundreds of generations of genetic evolution. This capacity to make changes is indeed genetically determined, and that is a great gift. How we use it and what we accomplish with it is not a gift but a responsibility upon which our own well-being and the well-being of other species depend.

S U M M A R Y   W I T H   K E Y   T E R M S

I. **Behavior** is what an animal does; it consists of the responses of an organism to stimuli in its environment.
   A. **Proximate causes** of behavior are immediate causes such as genetic, developmental, and physiological processes that permit the animal to carry out a specific behavior. Proximate causes answer *how* questions.

   B. **Ultimate causes** are the evolutionary explanations for *why* a certain behavior occurs.

II. **Behavioral ecology** is the scientific study of behavior in natural environments from the evolutionary perspective. The overall study of animal behavior is referred to as ethology.
   A. Behavior has costs and benefits. Benefits contribute to **direct fit-**

ness, the reproductive success of the animal. When benefits outweigh costs, the behavior is adaptive.

B. Behavior tends to be adaptive and homeostatic.

III. Behavior results from the interaction of genes (**innate behavior**) and environmental factors. The capacity for behavior is inherited, but behavior is modified by environmental stimuli. **Learned behavior** is behavior that has been modified in response to environmental experience.

A. Before an organism can show any pattern of behavior, it must be physiologically ready to produce the behavior.

B. Walking and many other behaviors that we view as automatic are **motor programs,** coordinated sequences of muscle actions.

C. A **fixed action pattern (FAP)** is an automatic behavior that, once activated by a simple sensory stimulus, continues to completion regardless of sensory feedback. An FAP can be triggered by a specific unlearned **sign stimulus,** or **releaser.**

D. **Habituation** is a type of learning in which an animal learns to ignore a repeated, irrelevant stimulus.

E. **Imprinting** establishes a parent-offspring bond during a critical period early in development.

F. In **classical conditioning** an association is formed between some normal body function and a new stimulus. A predator can become classically conditioned to the scent of its prey.

G. In **operant conditioning** an animal learns a behavior in order to receive positive reinforcement or to avoid punishment.

H. **Insight learning** is the ability to adapt past experiences that may involve different stimuli to solve a new problem.

I. Information important to survival appears to be most easily learned.

J. Play may give a young animal a chance to acquire adult patterns of behavior.

IV. It is adaptive for an organism's metabolic processes and behavior to be synchronized with cyclical changes in the environment. **Diurnal** animals are most active during the day, **nocturnal** animals at night, and **crepuscular** animals at dawn and/or dusk.

A. In many species, physiological processes and activity follow **circadian rhythms.** Some biological rhythms reflect the **lunar cycle** or the changes in tides due to phases of the moon.

B. No single **biological clock** has been found. Biological rhythms are thought to be regulated by both internal timing mechanisms as well as external factors.

V. **Migration** is periodic long-distance travel from one location to another.

A. Ultimate causes of migration include the advantage of moving away from an area that seasonally becomes too cold, too dry, or depleted of food to a more hospitable area.

B. Proximate causes of migration include physiological responses that are triggered by environmental changes.

C. **Directional orientation** refers to travel in a specific direction and requires **compass sense,** a sense of direction. Many animals use the sun to orient themselves.

D. **Navigation** requires both compass sense and **map sense,** an awareness of location. Birds that navigate at night use the stars to guide them. Birds and many other animals also use Earth's magnetic field to navigate.

VI. **Optimal foraging,** the most efficient strategy for an animal to obtain food, enhances reproductive success.

VII. **Social behavior** is adaptive interaction, usually among members of the same species.

A. A **society** is a group of individuals of the same species that may work together in an adaptive manner.

B. Many animal societies are characterized by a means of communication, cooperation, division of labor, and a tendency to stay together.

VIII. Animal **communication** involves the exchange of mutually recognizable signals.

A. Communication signals can be auditory, visual, electrical, or chemical.

B. **Pheromones** are chemical signals that convey information between members of a species.

IX. A **dominance hierarchy** is a ranking of status within a group in which more dominant members are accorded benefits (such as food or mates) by subordinates, usually without overt aggressive behavior.

X. Organisms often inhabit a **home range,** a geographical area that they seldom leave, but do not necessarily defend.

A. A defended area within a home range is called a **territory,** and the defensive behavior is **territoriality.**

B. Some animals, like lions, engage in group territoriality.

XI. **Sexual selection,** a type of natural selection, occurs when individuals vary in their ability to compete for mates. Individuals with reproductive advantages are selected over others of the same sex and species.

A. Mate choice may be influenced by dominance, gifts, ornaments, and courtship displays. Males of some species gather in a **lek,** a small display area where they compete for females.

B. Courtship behavior ensures that the male is a member of the same species, and it permits the female to assess the quality of the male.

C. Sexual selection often favors male **polygyny,** a mating system in which a male mates with many females. In **polyandry,** a female mates with several males. **Monogamy,** mating with a single partner during a breeding season, is common among birds.

D. Among many species, males guard their mates, especially when the female is most fertile, to prevent other males from fertilizing her eggs.

E. A **pair bond** is a stable relationship between a male and a female that may involve cooperative behavior in mating and in rearing the young.

F. Parental care increases the probability that the offspring will survive. A high investment in parenting is typically less advantageous to the male than to the female.

XII. Insect societies tend to be rigid, with the role of the individual narrowly defined. Division of labor is mainly determined by age.

A. Bees communicate the direction and distance of a food source with a series of body movements referred to as a dance. They probably also receive information through scent.

B. A round dance indicates that the food source is within 50 m; a waggle dance communicates information about both distance and direction when the food is farther away.

XIII. Vertebrate societies are more flexible than insect societies. Symbolic transmission of culture is important in human societies.

XIV. Some animals appear to spend time and energy helping others.

A. In mutualism, animals engage in cooperative behavior (e.g., hunting) that benefits all. In a type of mutualism known as **reciprocal altruism,** the helper does not immediately benefit but is helped at a later time by the animal it helped.

B. In **altruistic behavior,** one individual appears to behave in a way that benefits others rather than itself.

C. The concept of **inclusive fitness** suggests that natural selection favors animals that help a relative because this is an indirect way to perpetuate some of the helper's own genes. Inclusive fitness can be measured by the **coefficient of relatedness,** the probability that two individuals inherit the same gene from a common ancestor.

D. **Kin selection** is a type of natural selection that increases inclusive fitness through successful reproduction of close relatives.

E. **Sociobiology** focuses on the evolution of social behavior through natural selection. Sociobiology offers an adaptation explanation for human social behavior.

1. The contemporary study of behavior in natural environments from the point of view of adaptation is (a) ecology (b) ethology (c) behavioral ecology (d) answers a, b, and c are correct (e) answers a and c only are correct

2. If you were interested in exploring why the graylag goose "rolls" a nonexistent egg toward her nest, you might explore (a) proximate causes (b) ultimate causes (c) pheromones (d) answers a, b, and c are correct (e) answers a and b only are correct.

3. The responses of an organism to signals from its environment are its (a) behavior (b) learned behavior (c) ultimate behavior (d) releasers (e) motor programs

4. A fixed action pattern (FAP) is elicited by a(an) (a) motor program (b) sign stimulus (c) releaser (d) answers a, b, and c are correct (e) answers b and c only are correct.

5. In operant conditioning, a typical sequence might be (a) animal performs a behavior → animal receives positive reinforcement → animal repeats behavior (b) wolf hears rustling in bushes → wolf pounces on rabbit → wolf associates rustling with food (c) sounds of traffic keep person awake → sounds persist/nothing harmful happens → person adjusts and sleeps despite traffic sounds (d) stimulus presented during critical period → animal continues to respond to this stimulus (e) child is given candy and then asked to clean his room → child continues to clean his room each time he is asked

6. A form of learning in which a young animal forms a strong attachment to a moving object (usually its parent) within a few hours of birth is (a) classical conditioning (b) operant conditioning (c) imprinting (d) insight learning (e) parental investment

7. An organism learns to ignore a repeated, irrelevant stimulus. This is (a) classical conditioning (b) operant conditioning (c) imprinting (d) insight learning (e) habituation

8. Salivation by a student when the noon bell rings is an example of (a) classical conditioning (b) operant conditioning (c) imprinting (d) insight learning (e) habituation

9. Animals that are most active at dawn or twilight are described as (a) circadian (b) lunar (c) crepuscular (d) nocturnal (e) diurnal.

10. Both compass sense and map sense are necessary for (a) navigation (b) migration (c) directional orientation (d) answers a, b, and c are correct (e) answers a and b only are correct

11. In optimal foraging (a) animals always hunt in social groups (b) an animal obtains food in the most efficient way (c) animals can be selective in areas where food is plentiful (d) answers a, b, and c are correct (e) answers b and c only are correct

12. Chemical signals that convey information between members of a species are (a) pheromones (b) hormones (c) neurotransmitters (d) leks (e) three of the preceding answers are correct

13. The benefits of territoriality include (a) rights to food within the territory (b) increased reproductive success (c) release of excess energy in staking out and defending the area (d) answers a, b, and c are correct (e) answers a and b only are correct

14. Sexual selection (a) is a form of natural selection (b) occurs when animals are very similar in their ability to compete for mates (c) results in animals that have lower direct fitness (d) occurs mainly among animals that practice polygyny (e) occurs mainly among animals that practice polyandry

15. Mate choice may be influenced by (a) ornamental displays such as antlers (b) dominance (c) competition in a lek (d) answers a, b, and c are correct (e) answers b and c only are correct

16. A stable relationship between two animals of the opposite sex that may involve cooperative behavior in mating and rearing the young is (a) typical in polyandry (b) a pair bond (c) characteristic of polygamous relationships (d) found only among mammals (e) extremely rare among non-humans

17. The round dance of the bee (a) indicates that food is close to the hive (b) communicates direction (c) results in bees flying long distances in all directions (d) indicates that food is distant from the hive (e) indicates both direction and height

18. Behavior that appears to have no payoff, that is, an individual appears to act to benefit others rather than itself is known as (a) mutualism (b) altruistic behavior (c) reciprocal altruism (d) inclusive fitness (e) helping behavior

19. Inclusive fitness (a) considers only the genes an animal transmits to its own offspring (b) can be measured by the coefficient of relatedness (c) explains territoriality (d) is a form of social behavior (e) is a form of reciprocal altruism

20. Kin selection (a) increases inclusive fitness (b) is a way of perpetuating genes of nonrelatives (c) accounts for some forms of migration (d) typically involves mate guarding (e) typically involves ornamental displays and use of a lek

R E V I E W Q U E S T I O N S

1. In what ways are the behaviors of *Philanthus,* the sand wasp, adaptive?

2. Why is it adaptive for some species to be diurnal but for others to be nocturnal or crepuscular? What behavioral (and other) traits would you expect to observe in species with diurnal, nocturnal, or crepuscular lifestyles?

3. Behavior capacity is inherited and is modified by learning. Give an example.

4. When Konrad Lorenz kept a greylag goose isolated from other geese for the first week of its life, the goose persisted in following him about in preference to other geese. How can this behavior be explained?

5. How does physiological readiness affect innate behavior? How does it affect learned behavior?

6. What distinguishes an organized society from a mere aggregation of organisms? Cite an example of an organized society and describe characteristics that qualify the society as organized.

7. How does an organism learn its place in a dominance hierarchy? What determines this place? What costs and benefits accrue to dominant individuals in a dominance hierarchy? To subordinate individuals?

8. What is territoriality? What functions does it seem to serve?

9. What is sexual selection? What are some factors that influence mate choice?

10. Under what conditions does sexual selection favor polygynous (or polyandrous) mating systems?

11. What are some advantages of courtship rituals?

12. Why do more females than males invest in caring for their offspring? Explain in terms of costs, benefits, and fitness.

13. Compare and contrast the dance of bees with human language. Compare the society of a social insect with human society.

14. Define inclusive fitness. How is kin selection used by sociobiologists to explain the evolution of altruistic behavior? What is the possible role of kin selection in the maintenance of insect and other animal societies?

YOU MAKE THE CONNECTION

1. What might be the adaptive value of sea turtle migration? Consider how this adaptive value has changed if, as a result of human activities, migration now puts sea turtles at greater risk than if they restricted their habitat to a single location. Discuss the possible evolutionary mechanisms by which the behavior of these species may (or may not) adapt to these environmental pressures. What conservation efforts should we take to ensure continued migration?

2. Using what you know about animal learning, propose experiments to test whether it might be possible to teach sea turtles "safer" reproductive behavior. As you do, discuss the following questions: What might be the broader, long-term ecological consequences of your proposals? Is it advis-able for humans to attempt to alter the behavior of another species? Is this an ethical way of preserving a species that has become endangered as a result of human activities?

3. Behavioral ecologists have demonstrated that young tiger salamanders can become cannibals and devour other salamanders. Investigators observed that the cannibal salamanders preferred eating unrelated salamanders over their own cousins and preferred eating cousins over their siblings. Explain this behavior based on what you have learned in this chapter.

4. What similarities between the transmission of information by symbolic language and by heredity can you think of? What are some differences?

RECOMMENDED READINGS

Alcock, J. *Animal Behavior,* 6th ed. Sinauer Associates, Sunderland, MA, 1997. An excellent, basic evolutionary approach to animal behavior.

Dugatkin, L.A. "The Evolution of Cooperation." *Bioscience,* Vol. 47, No. 6, Jun. 1997. The author discusses four evolutionary paths to cooperative behavior.

Greenspan, R.J. "Understanding the Genetic Construction of Behavior." *Scientific American,* Vol. 272, No. 4, Apr. 1995. Studies of courtship and mating in the fruit fly suggest how genes might affect complex behavior.

Hill, G.E. "Ornamental Traits as Indicators of Environmental Health." *BioScience,* Vol. 45, No. 1, Jan. 1995. A discussion of ornamental factors such as bright colors and dramatic plumages in mate selection.

Kirchner, W.H., and W.F. Towne. "The Sensory Basis of the Honeybee's Dance Language." *Scientific American,* Vol. 270, No. 6, Jun. 1994. A discussion of honeybee communication.

Lohmann, K.J. "How Sea Turtles Navigate." *Scientific American,* Vol. 266, No. 1, Jan. 1992. Biological compasses and maps guide sea turtles as they migrate across hundreds of miles of ocean.

Narins, P.M. "Frog Communication." *Scientific American,* Vol. 273, No. 2, Aug. 1995. Some frogs croak loudly to ensure that they are heard; others transmit vibrations by tapping on the ground or on plants.

Packer, C., and A.E. Pusey. "Divided We Fall: Cooperation Among Lions." *Scientific American,* Vol. 276, No. 5, May 1997. A discussion of the costs and benefits of cooperative behavior among lions.

Pfennig, D.W., and P.W. Sherman. "Kin Recognition." *Scientific American,* Vol. 272, No. 6, Jun. 1995. The ability of many animals to recognize their relatives may have a variety of survival advantages.

Preston-Mafham, K., and R. Preston-Mafham. "Mating Strategies of Spiders." *Scientific American,* Vol. 279, No. 5, Nov. 1998, Stunning photographs illustrate a brief discussion of mating rituals of some spiders.

Pool, R. "Putting Game Theory to the Test." *Science,* 17 Mar. 1995. A discussion of how game theory can be applied to animal behavior.

Pusey, A., J. Williams, and J. Goodall. "The Influence of Dominance Rank on the Reproductive Success of Female Chimpanzees." *Science,* Vol. 277, 8 Aug. 1997. Data from a 35-year field study of chimpanzee behavior.

Sherman, P.W. and J. Alcock (eds.) *Exploring Animal Behavior: Readings from American Scientist,* 2nd ed. Sinauer Assoc. Sunderland, MA, 1998. Compilation of three articles on animal behavior originally published in *American Scientist.*

● Visit our website at **http://www.saunderscollege.com/lifesci/titles.html** and click on Solomon/Berg/Martin Biology for links to chapter-related resources on the World Wide Web.

General Surgeon

HARRY P. MARSHALL, JR., M.D.

Harry Marshall is a general surgeon in the Department of Surgery at Georgetown University Medical Center in Washington, D.C. He received his B.S. in biology from Morehouse College in Atlanta, Georgia, in 1980 and his medical degree from Howard University in 1984. After an internship at Howard, he elected to repay his ROTC scholarship with four years in the Army. He entered as a general medical officer and, later, became a flight surgeon. In 1989, after 4 years of active duty in the Army, Harry began a five-year general surgery residency training program at Georgetown University Medical Center. He interrupted his residency to complete a one-year surgical critical care fellowship and two years of research at the University of Pittsburgh Medical Center (1992–1995). After completing his residency, he opted to remain at Georgetown, joining the Department of Surgery as an attending surgeon in 1997. His undergraduate biology background in anatomy and physiology was particularly helpful in enabling him to compete successfully for the residency appointment, and ultimately, in fulfilling his goal of becoming a general surgeon. Surgical training is long and sometimes grueling, but the rewards include significant personal satisfaction and the opportunity to help people overcome disease. Harry strongly believes that a doctor's responsibility extends beyond the patients, and includes the community as well. In addition to his surgical practice, Harry founded and runs an organization called M-PowerHouse, which seeks to prevent trauma and violence in the community by exposing inner city children to different career paths in an effort to set a more positive course for their future.

Why did you choose Morehouse College for your undergraduate work?

My high school counselor's son went to Morehouse. We were neighbors. The positive image he painted of the institution appealed to me. I was impressed by the concept of a large group of young, motivated, idealistic African-American men, with different interests but similar goals.

When you entered college, did you know what major interested you?

I wanted to be a surgeon early in life. When I was a senior in high school, we had to dissect the nervous system of a frog in an advanced biology class. As you can imagine, it's a small nervous system, but it was something I looked forward to every day. I dissected the nervous system completely, even out to the digits of the frog, with very basic laboratory instruments. It was a little more than the original assignment required. That's when I realized that using anatomy and physiology to help people was really what I wanted to do.

Were you a biology major on a premed track?

I was a biology major taking a premed curriculum. Unlike some large universities, Morehouse did not offer a premed degree.

You either majored in chemistry or biology or math. In addition to college semesters, I spent many summers doing some kind of medical school preparation. These Morehouse College summer programs reinforced the basic sciences—anatomy, physiology, and biochemistry, and also included some time management and reading courses. Morehouse is a small, all-male liberal arts college. When I was there, the total enrollment was only 1800. I graduated in 1980.

When you graduated, where did you apply to medical school?

I applied to a combination of large and small schools, and ended up going to Morehouse School of Medicine for the first two years. At that time, it was a brand new two-year program. After the two years of basic sciences at Morehouse, I got my medical degree from Howard University. I finished medical school in 1984, and then did a year of general surgery internship at Howard.

At that point, I opted to repay my ROTC scholarship by serving four years in the Army. I entered the Army as a general medical officer and later became a flight surgeon, which was very exciting. I was in the Army from 1985 until 1989.

Where were you stationed?

I was stationed in Ft. Stewart, Georgia; Camp Humphrey, Korea; and Ft. Rucker, Alabama. I moved around by design. I wanted to experience life. I wanted to travel in the Army, not stay in one place. When I was in the Army I was assigned to an infantry brigade, the 24th infantry.

Later, as a flight surgeon I was assigned to an army aviation post in the East, so I did not have hospital assignments per se.

What sorts of medical situations came your way as a flight surgeon?

The routine daily job was taking care of aviators. I was the primary care physician on call. The challenge came with trauma situations, involving unstable patients, who needed to be transported to a major medical facility. In such cases, the flight surgeon was the physician on whom they relied to stabilize and transport the patient.

Let's tally this up: four years of undergraduate college, four years of medical school, one year internship, four years of the Army, and that takes us to Georgetown. Is that right?

Yes, almost. I also did a one-year fellowship in critical care at the University of Pittsburgh and two years of research. This added another three years to my five-year general surgery residency. Surgery is one of the longest training programs; generally, it's a minimum of five years. By this time, I was getting older. I had to decide what kind of work I wanted to do for the rest of my life.

What is general surgery? Do you serve all the other departments?

General surgery is primarily surgery of the abdomen, and includes surgical oncology, trauma, critical care, breast, and endocrine surgery. There are many other "subspe-

cialty" areas in surgery, including neurosurgery, cardiothoracic surgery, plastic surgery, and orthopedics. Depending on what area of the body is involved, either the general surgeon or the subspecialty surgeon will operate. Occasionally, both the general surgeon and the subspecialty surgeon will operate together on the same patient. The general surgeons also take care of most of the post-operative patients.

I live in a low-income area by choice to provide an ongoing example: Here's how a doctor lives . . . it's attainable.

Did you have any idea what you were getting into when you made the decision to go into general surgery?

I knew that it was a long residency program, and the hours were extremely demanding. On any given day, I would arrive at the hospital at 6:00 AM and leave for home (on a good day) at 7:30 or 8:00 PM. I was on call about every third night,

and those nights were all spent in the hospital. A 12-hour day is considered a good day.

Now that you have completed your residency, how does your day look?

My hours are more flexible now but they are also more demanding. If my patient is not doing well, I cannot just go home at the end of the day. The responsibility is much more, but the time is more my own. The reward is in being able to see my patients do well.

I understand that being selected for a surgical residency is very competitive. How would you advise somebody to prepare? How did you do it?

Definitely, in terms of pursuing a medical career, a good foundation in the biomedical sciences is essential.

What undergraduate courses helped you the most?

Biochemistry, anatomy, and physiology were the most helpful.

How did you prepare for the MCAT exam?

Being a biology major really helps prepare for the MCATs. I also think anyone who is planning to take them should take a preparatory course. You do not need a science degree to get into medical school, but you need to take the core requirements, which are biology, organic chemistry, calculus, physics, and biochemistry.

Do most of your colleagues have a similar career history?

Minus the Army, most of my colleagues took a similar path. Generally, every university surgeon or physician will do some type of fellowship to bring some expertise to the table. I mentioned the critical care fellowship earlier, which I did at the University of Pittsburgh. It was an intensive care doctor fellowship, dealing with the care and management of critically ill patients.

Do you have an area of specialty?

Critical care and trauma are two of my areas of interest, but I consider myself primarily a general surgeon. My interest in trauma incorporates my commitment to providing health care to the disenfranchised. In addition, there is significant opportunity for preventive medicine in the form of workshops on violence prevention.

Your assistant told me that you run a clinic for the uninsured?

The hospital has a clinic every Wednesday for the uninsured. I inherited it as the most junior faculty member.

I also have a related injury prevention project, which is aimed at the prevention of youth violence. I started it in Pittsburgh in 1993, and it is based both in Pittsburgh and in Kenilworth Parkside here in Washington, D.C. It's called M-PowerHouse.

What do you do there?

We focus mainly on younger children from ages 8 to 15, and try to get them interested in various careers and activities outside their daily experience. These kids live in a low-income environment that is not particularly motivating. I would like to make the program larger, since right now it's still in its infancy. We're always looking for volunteers and support.

I live in Kenilworth Parkside here in D.C. In Pittsburgh, when I started this, I lived in Homewood, which is also a low-income area. What the project does for the community is to inject a sense of hope and possibility.

Did you come from a low-income neighborhood?

I grew up in a part of Northeast Washington, D.C., which was a middle- to low-income area in the early to mid-60s. We lived there until 1971. With the imple-mentation of the Equal Housing Opportunity Act, we were part of the great migration to the suburbs, and moved out to Rockville, Maryland. I think my empathy comes from having lived in both places. Back in the '60s, people of color did not move out of the city into certain neighborhoods, even if they had the means. At the same time, our inner city communities were stronger because we had people from all sorts of occupations living in one neighborhood. For example, my dad was a personnel officer at NIH (National Institutes of Health). When Equal Housing Opportunity happened, people who could move did move, and a chasm was created. I believe that those of us who moved out have to be the bridge builders.

I do not see how you would have much time to devote to M-PowerHouse?

I do not have that much time. I live in a low income area by choice because I believe the key is actually being there. The kids do not need you every minute of every day. But I provide them with an ongoing example: Here's how a doctor lives. I open up my life to them so it's not some unknown; it appears attainable.

Do you also teach?

We teach the medical residents every Tuesday and Friday mornings. The students seem smarter and smarter every year. It's amazing. You have to try to stay at least one step ahead of them. It makes you pay attention. It makes things interesting.

I am in the process of developing a surgical anatomy course for the residents, which is exciting. It's a course where we take the interns through some basic surgical procedures and basic surgical anatomy. Then, we have a cadaver course designed specifically for the senior residents.

What do you like most about being a surgeon?

I like being able literally to reach into someone, perform an operation that is curative or that relieves the ailment, close them up, and make them whole again.

Would you ever specialize down the road?

No, I have done all the specialization I am going to do. If I were younger, would I specialize? No. I like what I do. I like the variety. It suits me.

What do you see in a typical surgical day?

In a typical day, I do a laparoscopic cholecystecstomy, which is surgical removal of the gall bladder; an appendectomy for appendicitis; and a hernia repair.

How do you prepare yourself for these surgeries, the shifts from one area of the body to another?

My preparation is my years of training. In a good general surgery program, you will have done at least 900 to 1200 surgery cases over that time, covering all of these areas. Some of them will be subspecialty cases; the majority of them will be general surgery cases. And, of course, you review the cases. You meet the patients in the office, examine them, discuss their ailments, and their peri-operative courses. When they come in for surgery, you meet them in the waiting area, answer any questions, and assure them. After surgery, you speak with family members and with the patients. I generally follow patients through at least two post-operative visits. I see them 7 to 10 days after surgery. Then, depending on how they are doing, I will see them one or two more times. After that, I turn them back over to their primary care physician.

Where will you be in five years?

Overall, I would not change my general surgery practice. I would like M-PowerHouse to be larger in both Pittsburgh and Washington, D.C. I'd like to see it develop into an established, fully funded program.

Medicine to me is an extremely rewarding and noble profession. It is a privilege to have someone seek your expertise for their health problems. It's also a humbling experience because sometimes things go wrong. To anyone contemplating a surgical career, I think you will enjoy it more if it really wakes you up in the morning, and if you are driven to make people feel better. The rewards that make it worthwhile, like personal satisfaction, have to come from within you.

The Interactions of Life: Ecology

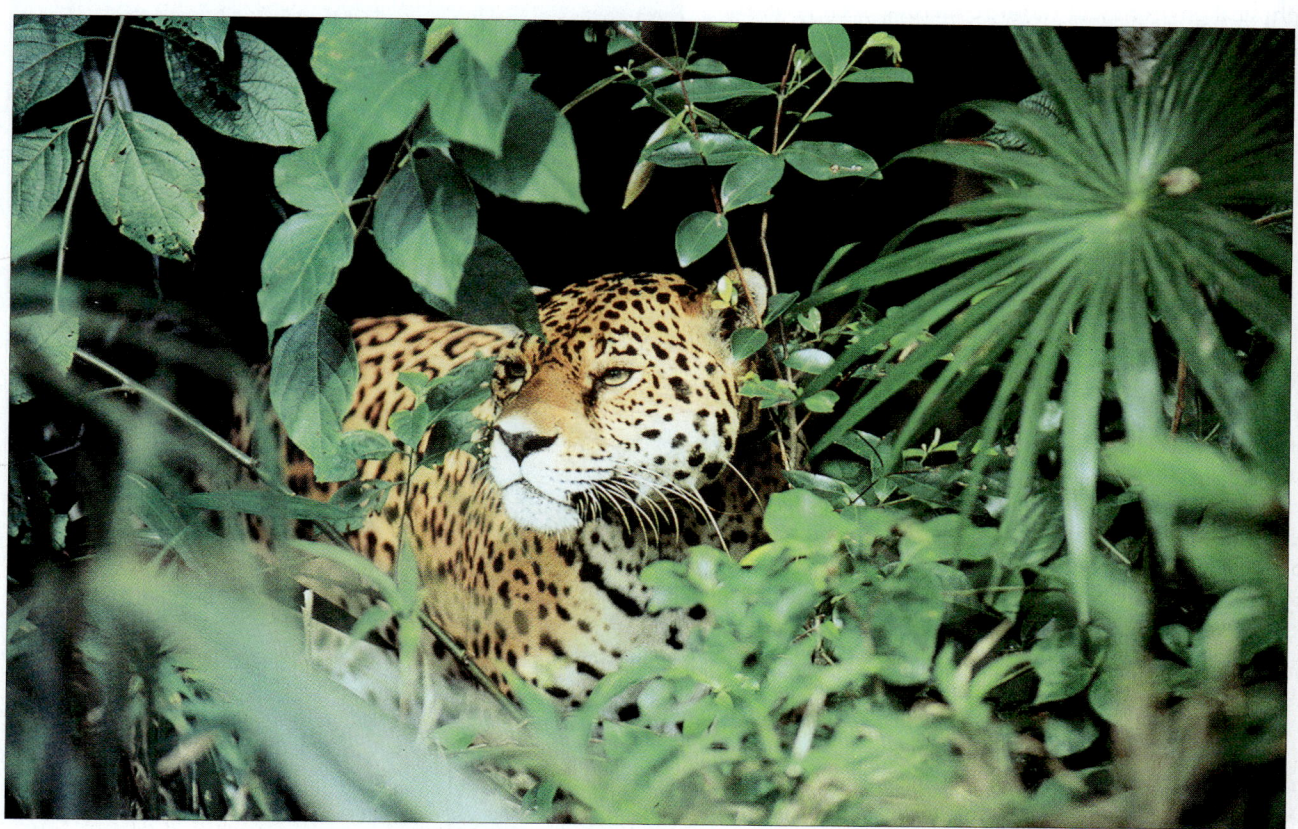

A jaguar rests in the vegetation of a Brazilian rain forest. The jaguar is inextricably connected to the plants surrounding it and to the physical environment in which it exists. (Jeanne Drake/Tony Stone Images)

CHAPTER 51

Introduction to Ecology: Population Ecology

Living organisms and the physical environment interact in an immense and complicated web of relationships. **Ecology** is the science that studies interactions among organisms (**biotic factors**), and between organisms and their nonliving, physical environment (**abiotic factors,** such as water, temperature, pH, wind, and chemical nutrients). Ecologists formulate hypotheses to explain such phenomena as the distribution and abundance of life on Earth, the ecological role of specific species, the interactions among species in communities, and the importance of ecosystems in maintaining the health of the biosphere. They then test these hypotheses empirically.

Ecology is the broadest field in biology, with explicit links to evolution and every other biological discipline. Its universality also encompasses subjects that are not traditionally part of biology. Earth science, geology, oceanography, climatology, and meteorology are extremely important to ecology, especially when ecologists examine the abiotic environment of planet Earth. Because humans are part of Earth's complex web of life, all of our activities have a bearing on other organisms. Even economics and politics have profound ecological implications.

The levels of biological organization that most interest ecologists are those including and above the level of the individual organism. An individual belongs to a **population,** a group consisting of members of the same species that live together in the same area at the same time. A population ecologist might study a population of polar bears or Mexican poppies (shown in the photograph) to see how individuals within it live and interact with each other, with other organisms in their community, and with their physical environment.

Populations are organized into communities. A **community** consists of all the populations of all of the different species that live and interact together within an area. A community ecologist might study how species interact with one another, including, for example, who eats whom, in a coral reef or in an alpine meadow community. An **ecosystem** encompasses a community in a specific area and links it to the abiotic environment. Thus, an ecosystem includes not only all the interactions among the living organisms of a community but also the interactions between the organisms and their abiotic environment. An ecosystem ecologist, for example, might examine how temperature, light, precipitation, and soil factors affect the organisms living in a desert or in a coastal bay. Ecosystem

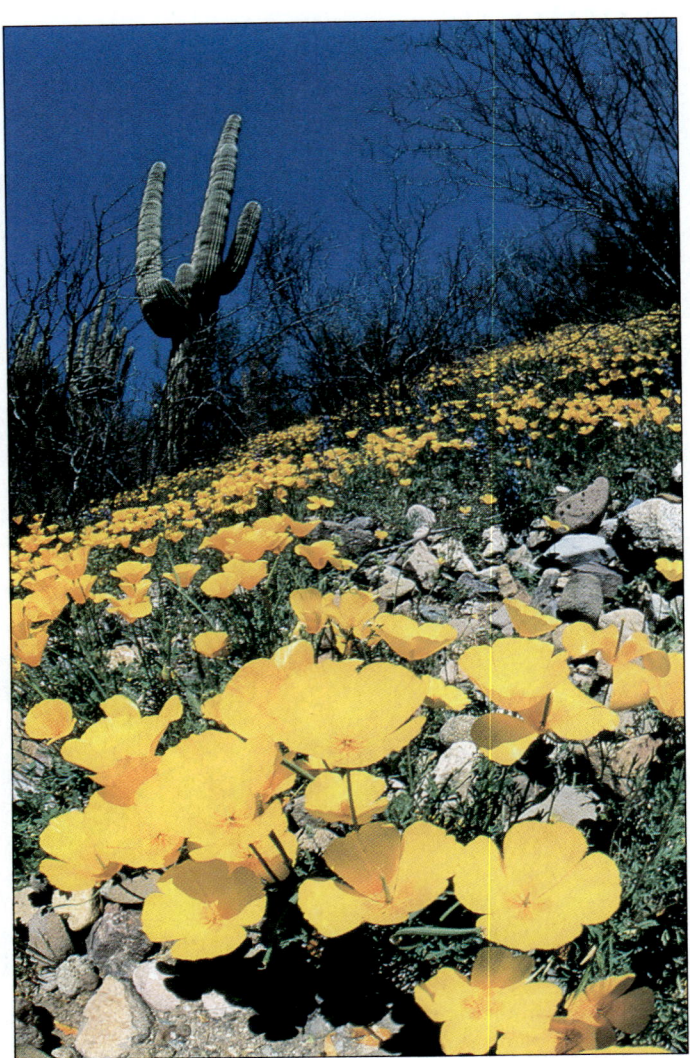

(Jim Brandenburg/Minden Pictures)

ecologists focus on such questions as how nutrients cycle through an ecosystem and how energy flows through food webs.

The **biosphere** is a global ecological system comprising all of Earth's communities. The organisms of the biosphere depend on one another and on Earth's physical environment—

its atmosphere, hydrosphere, and lithosphere. The *atmosphere* is the gaseous envelope surrounding Earth, the *hydrosphere* is Earth's supply of water (liquid and frozen, fresh and salty), and the *lithosphere* is the soil and rock of Earth's crust. Ecologists who study the biosphere examine the complex physical and chemical interrelationships among Earth's organisms, atmosphere, land, and water. The effect of deforestation (the clearance of forests for agriculture or other uses) on global carbon dioxide levels is an example of an ecological study at the biosphere level.

In this chapter we begin our study of ecological principles by focusing on the properties and dynamics of populations, followed by a discussion of the human population. Subsequent chapters examine the interactions among different populations within communities (Chapter 52), the dynamic exchanges between communities and their physical environments (Chapter 53), the characteristics of Earth's major biological ecosystems (Chapter 54), and some of the impacts of humans on the biosphere (Chapter 55).

LEARNING OBJECTIVES

AFTER YOU HAVE STUDIED THIS CHAPTER YOU SHOULD BE ABLE TO

1. Define ecology and distinguish among the following ecological levels: population, community, ecosystem, and biosphere.
2. Define population density and dispersion and describe the main types of population dispersion.
3. Explain the four factors that produce changes in population size and solve simple problems involving these changes.
4. Define biotic potential and carrying capacity and explain the differences between J-shaped and S-shaped growth curves.
5. Distinguish between density-dependent and density independent factors and give examples of each.
6. Distinguish between *K* strategies and *r* strategies and give an example of each.
7. Describe Type I, Type II, and Type III survivorship curves and explain which curve corresponds to *r* selection and which to *K* selection.
8. Summarize the history of human population growth.
9. Explain how developed and developing countries differ in population characteristics such as infant mortality rate, total fertility rate, replacement-level fertility, and age structure.
10. Distinguish between people overpopulation and consumption overpopulation.

THE POPULATION IS AN IMPORTANT UNIT OF ECOLOGICAL STUDY

Individuals of a given species are part of a larger organization known as a population. Populations exhibit characteristics distinctive from those of the individuals of which they are composed. Some of the features discussed in this chapter that are unique to populations are population density, population dispersion, birth and death rates, growth rates, and age structure.

Although communities are composed of all the populations of all the different species that live together within an area, populations have properties that communities lack. Populations, for example, share a common gene pool (see Chapter 18). As a result, allele frequency changes resulting from natural selection occur in populations. Natural selection, therefore, acts directly to produce adaptive changes in populations and only indirectly produces adaptive change at the community level.

Population ecology considers the number of individuals of a particular species that are found in an area, and how and why those numbers change (or remain fixed) over time. Population ecologists try to determine the processes common to all populations. They study how a population interacts with its environment, such as how individuals in a population compete for food or other resources, and how predation and other environmental pressures affect the population.

Additional aspects of populations that interest biologists are their reproductive success or failure (extinction), their evolution, their genetics, and how they affect the normal functioning of communities and ecosystems. Biologists in applied disciplines, such as forestry, agronomy, and wildlife management, must possess this understanding in order to manage populations of economic importance, such as forests, field crops, game animals, and fishes. An understanding of the population dynamics of endangered species plays a key role in efforts to prevent their slide to extinction.

DENSITY AND DISPERSION ARE IMPORTANT FEATURES OF POPULATIONS

Population size is only meaningful when the boundaries of that population are defined. Consider, for example, the difference between a thousand mice in 100 hectares (250 acres) and a thousand mice in one hectare (2.5 acres).

Often a population is too large to study in its entirety. Such a population is examined by sampling a part of it and then expressing the population in terms of density, such as, for example, the number of grass plants of a single species per square meter, the number of water fleas *(Daphnia)* per liter, or the number of cabbage aphids per square centimeter of cabbage leaf. **Population density** is the number of individuals of a species per unit of area or volume at a given time.

Different environments vary in the population density of any species of organism that they can support. This density may also vary in a single habitat (local environment) from season to season or year to year. For example, consider red grouse populations in the treeless moors of northwest Scotland at two locations only 2.5 km (1.5 mi) apart. At one location the population density remained stationary during a three-year period, but at the other site it almost doubled in the first two years and then declined to its initial density in the third year. The reason was likely a difference in habitat. The area where the population density increased initially and then decreased had been experimentally burned; young heather *(Calluna vulgaris)* shoots produced after the burn provided nutritious food for the red grouse. So population density may be determined in large part by external factors in the environment.

The individuals comprising the population often exhibit characteristic patterns of **dispersion,** or spacing relative to each other. Individuals may be spaced in a clumped, uniform, or random dispersion. Both clumped and uniform dispersion, being nonrandom, are sometimes referred to as *dispersion patterns.*

Perhaps the most common spacing is **clumped dispersion,** also called **aggregated distribution** and **patchiness,** which occurs when individuals are concentrated in specific parts of the habitat (Fig. 51–1*a*). Clumped dispersion often results from the patchy distribution of resources in the environment. It is also caused in animals by the presence of family groups and pairs, and in plants by limited seed dispersal or by asexual reproduction. An entire grove of aspen trees, for example, may originate asexually from a single plant (see Fig. 35–21). Clumped dispersion may sometimes be advantageous, as social animals derive many benefits from their association. Many fish species, for example, associate in dense schools for at least part of their life cycle, possibly because schooling may reduce the risk of predation for any particular individual. The many pairs of eyes of schooling fish tend to detect predators more effectively than a single pair of eyes of a single fish. When threatened, schooling fish clump together more closely, making it difficult for a predator to single out an individual.

Uniform dispersion occurs when individuals are more evenly spaced than would be expected from a random occupation of a given habitat (Fig. 51–1*b*). A nesting colony of seabirds, in which the birds are nesting in a relatively homogeneous environment and place their nests at a more-or-less equal distance from each other, is an example of uniform dispersion. In this case, uniform dispersion may occur as a result of nesting territoriality. Aggressive interactions among the nesting birds as they strike at one another from their nests cause each pair to place its nest just beyond the reach of nearby nesting birds. Uniform dispersion also occurs when competition

(a)

(b)

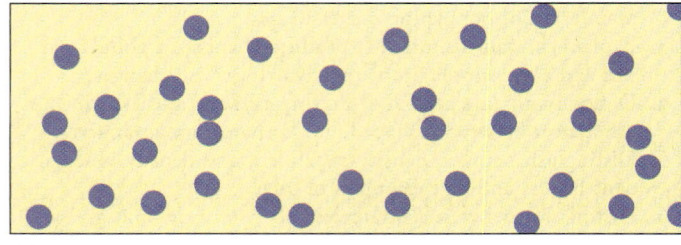

(c)

Figure 51–1 Dispersion of individuals within a population. *(a)* Clumped dispersion is often the result of the patchy distribution of resources in the environment. *(b)* Uniform dispersion is often caused by negative, or antagonistic, interactions among individuals of a population. *(c)* Random dispersion occurs when individuals distribute themselves with no regard for the positions of other individuals in the population.

among individuals is severe, when plants wage chemical warfare on one another by allelopathy (Fig. 51–2), or when animals establish feeding or mating territories.

Random dispersion occurs when individuals in a population are spaced in an unpredictable manner that is unrelated to the presence of others (Fig. 51–1*c*). Trees of the same species, for example, sometimes appear to be distributed randomly in a tropical rain forest. Of the three major types of dispersion, random dispersion is least common and hardest to observe in nature; some ecologists question the existence of random dispersion. Random dispersion may occur infrequently because important environmental factors affecting dispersion usually do not occur at random. Flour beetle larvae in a container of flour are randomly dispersed, for example, but their environment (flour) is unusually homogeneous.

Some populations have different spacing patterns at different ages. Competition for sunlight among same-aged sand pine in a Florida scrub community, for example, resulted in a change over time from either random or clumped dispersion

Figure 51–2 **Example of uniform dispersion.** The uniform dispersion of certain desert plants such as the creosote bush and the saltbush is probably due to allelopathy, the production of toxic substances that inhibit the growth of nearby plants. *(William E. Ferguson)*

when the plants were young to uniform dispersion when the plants were old. Sand pine is a fire-adapted plant with cones that do not release their seeds until they have been exposed to high temperatures (45° to 50°C or higher). As a result of seed dispersal and soil conditions following a fire, the seedlings grow back in dense stands that exhibit random or slightly clumped dispersion. Over time, however, many of the more crowded trees tend to die, resulting in uniform dispersion of the surviving trees (Table 51–1).

MATHEMATICAL MODELS DESCRIBE POPULATION GROWTH

Part of the scientific process is the discovery of common threads or patterns among separate observations. As mentioned previously, population ecologists wish to understand general processes that are shared by many different populations, so they develop mathematical models, equations that represent population dynamics. Population models are rarely perfect representations of a population, but models help to illuminate

complex processes. Moreover, mathematical modeling enhances the scientific process by providing a framework with which experimental population studies can be compared. As more knowledge accumulates from observations and experiments, the model is refined and made more precise.

Size of populations, whether they are sunflowers, elephants, or humans, changes over time. On a global scale, this change is ultimately due to two factors: **natality,** the rate at which organisms produce offspring (the birth rate), and **mortality,** the rate at which organisms die (the death rate). In humans the birth rate is usually expressed as the number of births per 1000 people per year and the death rate as the number of deaths per 1000 people per year.

To determine the rate of change in population size, we must also take into account the time interval involved, that is, the change in time. To express these changes in equations, we employ the Greek letter delta (Δ):

$$(1)\ \Delta N/\Delta t = b - d$$

where ΔN is the change in the number of individuals in the population, Δt the change in time, b the natality, and d the mortality. Thus, the rate of change, or **growth rate** (r) of a population is the birth rate minus the death rate ($r = b - d$).

A modification of this equation tells us the rate at which the population is growing at a particular instant in time (rather than the average growth rate during the entire period of study), that is, its instantaneous growth rate (dN/dt).[1] Using differential calculus, this growth rate can be expressed as:

$$(2)\ dN/dt = rN$$

where N is the number of individuals in the existing population, t the time, and r the growth rate.

Because $r = b - d$, if organisms in the population are born faster than they die, r is a positive value, and population size increases. If organisms in the population die faster than they are born, r is a negative value, and population size decreases. If r is equal to zero, births and deaths match, and population size is stationary despite continued reproduction.

[1]The symbol dN is the differential of N; it is not a product, nor should the d in dN be confused with the death rate, d. The same applies to dt.

TABLE 51–1	Dispersion in a Sand Pine Population in Florida*	
Tree Trunks Examined†	**Density (per square meter)**	**Dispersion**
All (alive and dead)	0.16	Random
Alive only	0.08	Uniform

*Adapted from Laessle, A.M. "Spacing and Competition in Natural Stands of Sand Pine."
Ecology, Vol. 46, 1965, pp. 65–72.
†Data were collected 51 years following a fire.

Migration affects the growth rate in some populations

In addition to the number of births and deaths, migration (movement from one region or country to another) must be considered when examining changes in populations on a *local* scale. There are two types of migration: immigration and emigration. **Immigration** occurs when individuals enter a population and thus increase its size. **Emigration** occurs when individuals leave a population and thus decrease its size. The growth rate of a local population must take into account birth rate (b), death rate (d), immigration (i), and emigration (e). The growth rate equals (the birth rate minus the death rate) plus (immigration minus emigration):

$$(3) \quad r = (b - d) + (i - e)$$

Each population has a characteristic biotic potential

The maximum rate at which an organism or population could increase under ideal conditions, when resources are abundant, is known as its **biotic potential** (r_{max}). Different species have different biotic potentials. A species' biotic potential is influenced by several factors, including the age at which reproduction begins, the fraction of the lifespan during which the organism is capable of reproducing, the number of reproductive periods per lifetime, and the number of offspring the organism is capable of producing during each period of reproduction. These factors, called life history characteristics, determine whether a particular species has a large or small biotic potential.

Generally, larger organisms such as blue whales and elephants have the smallest biotic potentials, whereas microorganisms have the greatest biotic potentials. Under ideal conditions (an environment with unlimited resources), certain bacteria can reproduce by binary fission every 20 minutes. At this rate of growth, a single bacterium would increase to a population of more than 1 billion in just 10 hours!

If we plot the population size versus time, the graph has a J shape that is characteristic of **exponential population growth,** the constant growth rate that occurs under optimal conditions (Fig. 51–3). When a population grows exponentially, the larger that population gets, the faster it grows.

Regardless of which organism we are considering, whenever a population is growing at its biotic potential, population size plotted versus time gives a curve of the same shape; the only variable is time. For example, it may take longer for an elephant population than for a bacterial population to reach a certain size, but both populations will always increase exponentially under "ideal conditions."

No population can increase exponentially indefinitely

Certain populations may exhibit exponential growth for short periods of time. Exponential growth has been experimentally demonstrated in bacterial and protist cultures and in certain

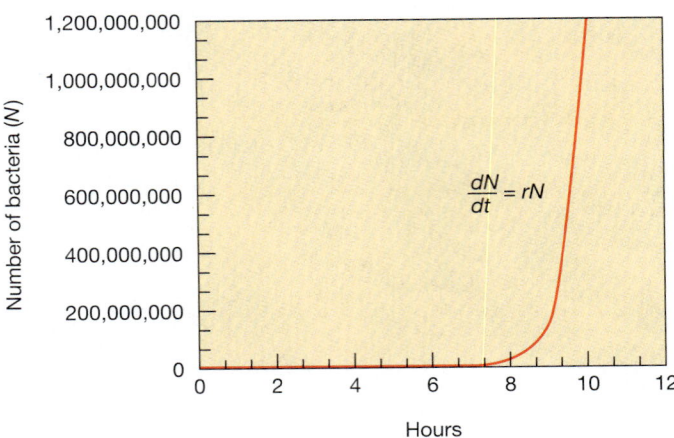

Figure 51–3 **Exponential population growth.** When bacteria divide every 20 minutes, their numbers increase exponentially. The curve of exponential population growth has a characteristic J shape. The ideal conditions under which bacteria or other organisms reproduce exponentially rarely occur in nature, and when these conditions do occur, they are of short duration.

insects. However, organisms cannot reproduce indefinitely at their biotic potential because the environment sets limits, which are collectively called **environmental resistance.** Using the preceding example, bacteria would never be able to reproduce unchecked for an indefinite period of time because they would eventually run out of food and living space, and poisonous wastes would accumulate in their vicinity. With crowding, they probably would also become more susceptible to parasites and predators. As their environment changed, their birth rate (b) would decline and their death rate (d) would increase due to such factors as shortages of food, increased predation, increased competition, and other environmental stresses. Conditions might worsen to a point where d would exceed b, and the population would decrease in size. The number of individuals in a population is controlled by the ability of the environment to support them; as the number of individuals in a population (N) increases, so does environmental resistance.

Over longer periods of time, the rate of population growth may decrease to nearly zero. This leveling out occurs at or near the limits of the environment to support the population. The **carrying capacity (K)** represents, in theory, the largest population that can be maintained for an indefinite period of time by a particular environment, assuming there are no changes in that environment. In nature, the carrying capacity is dynamic and changes in response to environmental changes.

When the size of a population regulated by environmental resistance is graphed over longer periods of time, the curve has a characteristic **S** shape that shows the population's initial exponential increase (note the curve's J shape at the start, when environmental resistance is low), followed by a leveling out as the carrying capacity of the environment is approached (Fig. 51–4). The **S**-shaped growth curve, also called **logistic population growth,** can be modeled by a modified growth equation called a logistic equation. It describes a population increasing from a small number of individuals to a larger number

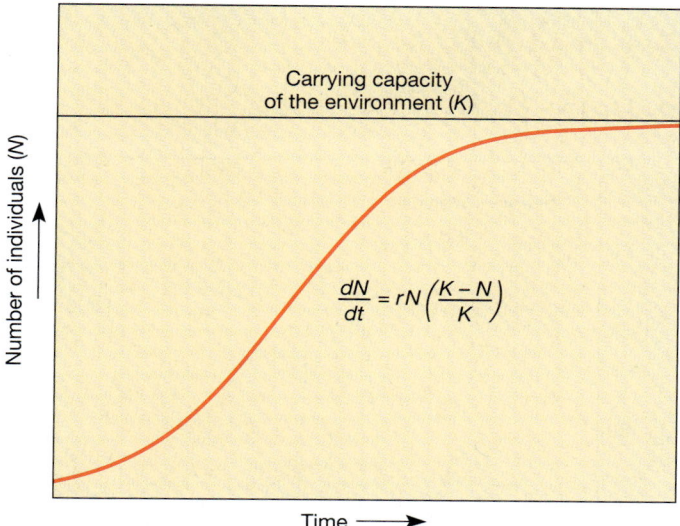

$$\frac{dN}{dt} = rN\left(\frac{K - N}{K}\right)$$

Figure 51–4 **Carrying capacity and logistic population growth.** In many laboratory studies, exponential population growth slows as the carrying capacity (K) of the environment is approached. The logistic model of population growth, when graphed, has a characteristic S-shaped curve.

The graph's vertical axis is labeled "Number of individuals (N)", the horizontal axis "Time", and a horizontal line at top is labeled "Carrying capacity of the environment (K)".

of individuals that are ultimately limited by a finite environment. The logistic equation takes the carrying capacity of the environment into account:[2]

$$(4)\quad dN/dt = rN[(K - N)/K]$$

Note that part of the equation is the same as equation (2). The added element $[(K - N)/K]$ reflects a decline in growth as a population size approaches its carrying capacity. When the number of organisms (N) is small, the rate of population growth is unchecked by the environment, because the expression $[(K - N)/K]$ has a value of almost 1. But as the population (N) begins to approach the carrying capacity (K), the growth rate declines because the value of $[(K - N)/K]$ approaches zero.

Although the S curve is a simplification of actual population changes over time, it appears to fit some populations that have been studied in the laboratory, as well as a few studied in nature. For example, G. F. Gause, a Russian ecologist who conducted experiments during the 1930s, grew a population of *Paramecium caudatum* in a test tube. He supplied a constant but limited amount of food daily and replenished the growth medium occasionally to eliminate the accumulation of metabolic wastes. Under these conditions, the population of *P. caudatum* increased exponentially at first. The paramecia became so numerous that the water was cloudy with them. But then their growth rate declined to zero, and the population size leveled off.

[2]The logistic model of population growth was developed to explain population growth in continually breeding populations. Similar models exist for populations that have specific breeding seasons.

A population rarely stabilizes at K (carrying capacity), but may temporarily rise higher than K. It will then experience a *population crash,* an abrupt decline from high to low population density. Such an abrupt change is commonly observed in bacterial cultures, zooplankton, and other populations whose resources are nonrenewable or inflexible and have been exhausted. The carrying capacity for reindeer, which live in cold northern habitats, is determined largely by the availability of winter forage. In 1910 a small herd of 26 reindeer was introduced on one of the Pribilof Islands of Alaska. The herd's population increased exponentially for about 25 years until there were approximately 2000 reindeer, many more than the island could support, particularly in winter. The reindeer overgrazed the vegetation until the plant life was almost wiped out. Then, in slightly over a decade, as reindeer died from starvation, the number of reindeer plunged to eight, one-third the size of the original introduced population. Recovery of subarctic and arctic vegetation after overgrazing by reindeer can take 15 to 20 years, during which time the carrying capacity for reindeer is greatly reduced.

POPULATION SIZE VARIES OVER TIME

The observation that small populations tend to grow and decline in size rapidly, whereas larger populations grow and decline in size less rapidly, suggests that certain mechanisms influence population size. Factors that affect population size fall into two categories: density-dependent factors (the "regulatory" mechanisms) and density-independent factors (the "nonregulatory" mechanisms). These two sets of factors vary in importance from one species to another, and in most cases probably interact simultaneously in complex ways to determine the size of a population.

Density-dependent factors regulate population size

If a change in population density alters how an environmental factor affects that population, then the environmental factor is said to be a **density-dependent factor.** As population density increases, density-dependent factors tend to slow population growth by causing an increase in death rate and/or a decrease in birth rate. The effect of these density-dependent factors on population growth becomes stronger as the population density increases; that is, density-dependent factors affect a larger *proportion* (not just a larger number) of the population as population size increases. Density-dependent factors can also exert an effect on population growth when population density declines; a decrease in population density results in enhancement of population growth by density-dependent factors. When population density is low, these factors either decrease the death rate and/or increase the birth rate. Density-dependent factors, then, tend to regulate a population at a relatively constant size that is near the carrying capacity of the

LONG-TERM PREDATOR-PREY DYNAMICS

Are there any long-term examples of predator-prey dynamics? Wildlife biologists have studied the populations of moose and wolves on Isle Royale since the 1950s. As discussed in this chapter, during the early 1900s a small herd of moose wandered across the ice of frozen Lake Superior to an island, Isle Royale. In the ensuing years, until 1949, moose became successfully established there, although the population sometimes experienced large oscillations. In 1949, a few Canadian wolves wandered across the frozen lake and discovered abundant moose prey on the island. The wolves also remained and became established.

Moose have been the wolves' primary winter prey on Isle Royale. Wolves hunt in packs by encircling a moose and trying to get it to run so they can attack it from behind. (The moose is the wolf's largest, most dangerous prey. A standing moose is more dangerous than one that is running because when standing, it can kick and slash its attackers with its hooves.)

Wildlife biologists, such as Rolf Peterson of Michigan Technological University, have studied the effects of both density-dependent and density-independent factors on the Isle Royale moose and wolf populations. They found that the moose and wolf populations fluctuated over the years *(see figure)*. Generally, as the population of wolves increased, the population of moose declined. Wolves tended to reduce the population of moose, but they did not eliminate them from the island. Studies indicate that wolves primarily feed on the very old and very young in the moose population. Healthy moose in their peak reproductive years are not eaten.

A new episode in the Isle Royale story began during the late 1980s and early 1990s. The wolf population plunged, from 50 animals in 1980 to a low of 12 animals in 1989, possibly the result of a deadly disease, canine parvovirus. (Analysis of their blood revealed the presence of antibodies to canine parvovirus, confirming that the wolves had been exposed to this disease.)

Fluctuations in the populations of wolves and moose on Isle Royale, 1959 to 1995. *(Rolf O. Peterson, Michigan Technological University)*

As expected, the moose population increased as the wolf population declined; in 1995 there were more than 2400 moose on Isle Royale, many more than the island's vegetation could support. The moose overgrazed the island, particularly mountain ash and aspen, their preferred food. Lack of food in combination with a particularly bad winter (1995 to 1996) caused hundreds of moose to die; only 500 moose were counted in the 1997 survey. A year later, in 1998, the wolf population plummeted to 14 individuals. Lack of vulnerable prey may have caused the wolf decline, but scientists must again check for another disease outbreak.

---

environment. (Keep in mind, however, that the carrying capacity of the environment frequently changes.)

Predation, disease, and competition are examples of density-dependent factors. As the density of a population increases, for example, it is logical that predators are more likely to find an individual of a given prey species. Also, when population density is high, it is reasonable that the members of a population encounter one another more frequently and that the chance of their transmitting infectious disease organisms increases.

Competition is an important density-dependent factor

As population density increases, so does competition for resources such as living space, food, cover, water, minerals, and sunlight. Eventually, it may reach the point where many members of a population fail to obtain the minimum amount of whatever resource is in shortest supply. This raises the death

rate and/or lowers the birth rate, inhibiting further population growth.

Competition occurs both within a given population (**intraspecific competition**) and among populations of different species (**interspecific competition**). We consider the effects of intraspecific competition here; interspecific competition is discussed in Chapter 52.

Individuals of the same species compete for a resource in limited supply by contest competition or by scramble competition. In **contest competition** certain dominant individuals obtain an adequate supply of the limited resource at the expense of other individuals in the population. In **scramble competition** all the individuals in a population "share" the limited resource equally so that at high population densities none of them obtains an adequate amount. The populations of species in which scramble competition operates often oscillate over time, and there is always a risk that the population size will drop to zero. In contrast, those species in which contest competition operates experience a relatively small drop in popula-

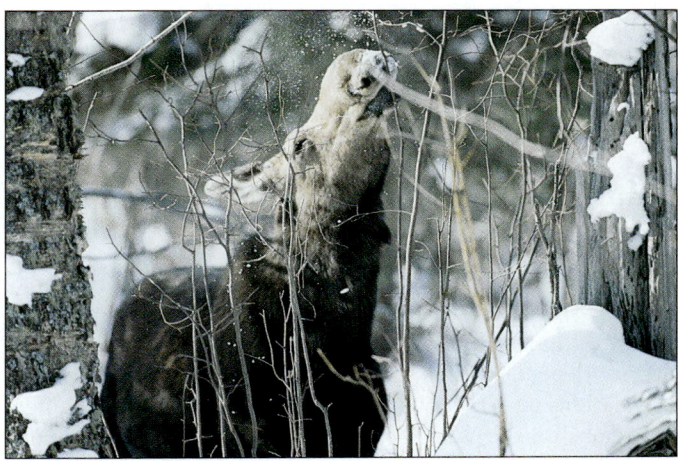

Figure 51–5 Scramble competition among moose on Isle Royale in Lake Superior. Individual moose compete with one another for food during the long, cold winter months. *(Rolf O. Peterson, Michigan Technological University)*

tion size, caused by the death of individuals that are unable to compete successfully.

Intraspecific competition among red grouse involves contest competition. When red grouse populations are low, the birds are less aggressive, and most young birds are able to establish a feeding territory (an area defended against other members of the same species). However, when the population is large, establishing a territory is very difficult because there are more birds than there are territories, and the birds are much more aggressive. Those birds without territories often die from predation or starvation. Thus, birds with territories use a larger share of the limited resource (the territory with its associated food and cover), whereas birds without territories cannot compete successfully.

The moose population on Isle Royale, the largest island in Lake Superior, provides a vivid example of scramble competition (Fig. 51–5). Isle Royale differs from most islands in that large mammals can walk to it when the lake freezes over in winter. The minimum distance to be walked is 24 kilometers (15 miles), however, so this movement has happened infrequently. Around 1900, some moose reached the island for the first time. By 1934 the moose population on the island had increased to about 3000 and had consumed almost all the edible vegetation. In the absence of this food resource, there was massive starvation in 1934. Over 60 years later, in 1996, a similar die-off claimed 80% of the moose after they had again increased to a high density. Thus, scramble competition for scarce resources can result in dramatic population oscillations. (See *Making the Connection: Long-term Predator-Prey Dynamics* for a further discussion of Isle Royale Moose.)

The effects of different density-dependent factors are difficult to assess in nature

Most studies of density dependence have been conducted in laboratory settings where all density-dependent factors except one are controlled experimentally. But populations in natural settings are exposed to a complex set of variables that continually change. As a result, in natural communities it can be difficult to evaluate the relative effects of different density-dependent factors. For example, few spiders occur on tropical islands inhabited by lizards, whereas larger numbers and more species of spiders are found on lizard-free islands.

Deciding to study these observations experimentally, Thomas Schoener and David Spiller selected 12 small Caribbean islands, 4 with lizards and 8 without; all contained web-building spiders. Schoener and Spiller introduced a small population of lizards onto four of the lizard-free islands (Fig. 51–6). At the end of a seven-year study period, spider population densities were higher in the lizard-free islands than in islands with lizards. Moreover, the islands without lizards had more species of spiders.

Figure 51–6 Interaction of density-dependent factors. D.A. Spiller and T.W. Schoener, two ecologists at the University of California at Davis, tested the effect of the presence of lizards on spider population density on Bahamian islands. Shown is one of the enclosures used for their study. *(Courtesy of Thomas W. Schoener)*

Even the results of this relatively simple experiment may be explained by a combination of two density-dependent factors: predation (lizards eat spiders) and interspecific competition (lizards compete with spiders for insect prey). In this case, the effects of the two factors in determining spider population size cannot be evaluated separately.

Density-independent factors limit population size

Any environmental factor that affects the size of a population but is not influenced by changes in population density is called a **density-independent factor.** Such factors are typically abiotic. For example, random weather events that reduce population size serve as density-independent factors. These often affect population density in unpredictable (random) ways. A severe blizzard, hurricane, or fire, for example, may cause extreme and irregular reductions in a vulnerable population and thus might be considered largely density-independent (Fig. 51–7).

Consider a density-independent factor that influences mosquito populations in arctic environments. These insects produce several generations per summer and achieve high population densities by the end of the season. A shortage of food does not seem to be a limiting factor for mosquitoes, nor is there any shortage of ponds in which to breed. What puts a stop to the skyrocketing mosquito population is winter. Not a single adult mosquito survives winter, and the entire population must grow afresh the next summer from the few eggs and hibernating larvae that survive. Thus, severe winter weather is a density-independent factor that affects arctic mosquito populations.

Density-independent and density-dependent factors are often interrelated. Social animals, for example, are often able to resist dangerous weather conditions by means of their collective behavior, as in the case of sheep huddling together in a snowstorm. In this case it appears that the greater the population density, the better their ability to resist the environmental pressure of a snowstorm.

Natural populations may fluctuate chaotically

An additional reason that populations fluctuate over time may be their inherent instability. Ecologists in recent years have tried to understand why, in the absence of any obvious density-dependent and density-independent factors, certain populations undergo wild and unpredictable changes. Dungeness crabs on the West Coast of the United States, for example, exhibit huge population swings.

Some ecologists have turned to the mathematical theory of chaos to help explain the state of flux displayed by some populations. **Chaos** is the tendency of a simple system to exhibit complex, erratic dynamics. Researchers from the University of California at Davis published the results of a computer model of Dungeness crab populations. The assumptions that they built into the model were simple: (1) Adults produce many larvae, most of which die from competition and predation. (2) After reproducing, the adults die. (3) Most of the surviving larvae do not migrate far away. The computer model predicted the crab population for the next 20,000 years. Despite the simple assumptions, the model showed occasional wild fluctuations that were independent of any environmental factors (Fig. 51–8). It therefore appears that a certain amount of chaos may be inherent in population dynamics. Therefore, not all populations that fluctuate in size do so because of explicit factors that can be studied experimentally.

SPECIES DIFFER IN REPRODUCTIVE STRATEGIES

Each organism has a lifestyle that is uniquely adapted to its reproductive pattern. Many years pass before a young magnolia tree flowers and produces seeds, whereas a poppy plant grows from seed, flowers, and dies in a single season. A mating pair of black-browed albatrosses produces a single chick every year, whereas a mating pair of gray-headed albatrosses produces a single chick biennially (every other year). Ecologists try to understand the adaptive consequences of these various life history strategies.

Imagine an organism possessing the "perfect" life history strategy that ensures reproduction at its biotic potential. In other words, this hypothetical organism produces the maximum number of offspring, and the majority of these offspring survive to reproductive maturity and then reproduce also. Such an organism would have to reach reproductive maturity soon after it was born. It would have to reproduce frequently throughout its long life and produce large numbers of offspring

Figure 51–7 Example of a density-independent factor. The effects of fire and other density-independent factors are not influenced by changes in population density. *(Doug Sokell/Tom Stack & Associates)*

Figure 51–8 Chaos in a computer model of population dynamics in Dungeness crabs. When the population was projected for 3000 years, it showed periods of stability as well as occasional wild fluctuations. For example, examine site #150 over time by visually scanning from the top to the bottom of the figure above the #150 site location. Two short vertical lines have been designated 1 and 2 on the figure. The area of the first short line shows one time period during the simulation when the population at site #150 did not fluctuate appreciably. The second short line shows the same location during a later time period when the population fluctuated wildly. *(Courtesy of Kevin Higgins and Alan Hastings)*

each time. Further, it would have to be able to care for all of its young in order to assure maximum survival.

In nature, such an organism does not exist because if an organism were to put all its energy into being a perfect "reproductive machine," it could not expend any energy toward ensuring its own survival. Animals must use energy to hunt for food, and plants need energy to grow taller than surrounding plants to obtain adequate sunlight. Natural selection, then, requires organisms to make trade-offs in the expenditure of energy. Organisms, if they are to be successful, must do what is required to survive as individuals and as populations (by reproducing). If they allocate all their energy for reproduction, none is available for the survival of the individual, and the individual dies. If they allocate all their energy for the individual, none is available for reproduction, and there are no future generations.

Each species has its own life history strategy—its own reproductive characteristics, body size, habitat requirements, and migration patterns—that is a series of trade-offs reflecting this "energetic compromise." Although a continuum of life histories exist, two main groups can be distinguished with respect to their reproductive strategies: *r*-selected species and *K*-selected species. Keep in mind as you read the following descriptions of *r* selection and *K* selection, however, that these concepts, while useful, probably oversimplify most life histories. Many species possess a combination of *r*-selected and *K*-selected traits, as well as traits that cannot be classified as either *r*-selected or *K*-selected.

Populations described by the concept of **r selection** have a strategy that has evolved through the natural selection of traits that lead to a high population growth rate. (Recall that *r* designates the growth rate. Because such organisms have a high *r*, they are known as **r strategists** or **r-selected species.**) Small body size, early maturity, short lifespan, large broods, and little or no parental care are typical of many *r* strategists, which are usually opportunists found in variable, temporary, or unpredictable environments where the probability of long-term survival is low.

Some of the best examples of *r* strategists are common weeds such as the dandelion (Fig. 51–9*a*). Dandelions are propagated by parachute-like fruits containing little seeds that are scattered widely by any strong breeze or wind. In temperate climates, any suitable disturbed habitat can be colonized by dandelions as soon as it becomes available. The seeds of some varieties of dandelions are produced by the asexual process of apomixis (see Chapter 35). After germinating, the young plant uses another form of asexual reproduction. Its long taproot splits longitudinally to form two roots, then four roots, and so on, so that natural clones of plants spread out from the original plant in genetically identical clumps.

Asexual reproduction has two key advantages for dandelions and some other *r* strategists. First, asexual reproduction in plants does not require insect pollinators or even other plants, both of which may be in short supply in unpredictable habitats. Second, as long as the particular kind of habitat for which an *r* strategist is adapted is common, asexual reproduction preserves a genotype adapted to that habitat.

In populations described by the concept of **K selection,** evolution has selected traits that maximize the chance of surviving in an environment where the number of individuals (*N*) is near *K*, the carrying capacity of the environment. Because populations of species with *K* selection are maintained at or near the carrying capacity, individuals have little need for a high reproductive rate. These organisms, called **K strategists** or **K-selected species,** do not produce large numbers of offspring. They characteristically have long lifespans with slow development, late reproduction, large body size, and a low reproductive rate. Animals that are *K* strategists typically invest in parental care of their young. *K* strategists are found in relatively constant or stable environments, where they have a high competitive ability.

Tawny owls (*Strix aluco*) are *K* strategists that pair-bond

(a)

(b)

Figure 51–9　**The reproductive strategies, *r* selection and *K* selection.** *(a)* Dandelions *(Taraxacum officinale)* are *r* strategists—annuals that mature early and produce many small seeds. A dandelion population fluctuates from year to year but rarely approaches the carrying capacity of its environment. *(b)* Tawny owls *(Strix aluco)* are *K* strategists and maintain a fairly constant population size at or near the carrying capacity. They mature slowly, delay reproduction, and have a relatively large body size.　*(a, Marion Lobstein; b, Stephan Dalton/Photo Researchers, Inc.)*

for life, with both members of a pair living and hunting in adjacent, well-defined territories (Fig. 51–9*b*). They regulate their reproduction in accordance with the resources, especially food, present in their territories. In an average year, 30% of the birds do not breed. If food supplies are more limited than initially indicated, many of those that do breed fail to incubate their eggs. Rarely do the owls lay the maximum number of eggs, and often they delay breeding until late in the season when the rodent populations on which they depend have become large. Thus, tawny owls behaviorally regulate their population size so that its number stays at or near the carrying capacity of the environment. Starvation, an indication that the tawny owl population has exceeded the carrying capacity, rarely occurs.

Survivorship is related to *r* and *K* selection

The various reproductive strategies that organisms possess, from *r* selection to *K* selection, are associated with different patterns of **survivorship**, the proportion of individuals in a population that survive to a particular age. Figure 51–10 is a graph of the three main survivorship curves that ecologists recognize as hypothetical constructs to explain survivorship over the lifespan. In Type I survivorship, as exemplified by humans, the young have a high probability of living, and the probability of survival decreases more rapidly with increasing age; that is, mortality is concentrated later in life. *K*-selected species are characterized by Type I survivorship. Humans and other *K*-selected species are long-lived organisms whose young have a high likelihood of survival.

In Type III survivorship, the probability of mortality is greatest early in life, and those individuals that survive early mortality have a high probability of survival; that is, the probability of survival increases with increasing age. Type III survivorship is characteristic of oysters and other *r*-selected species.

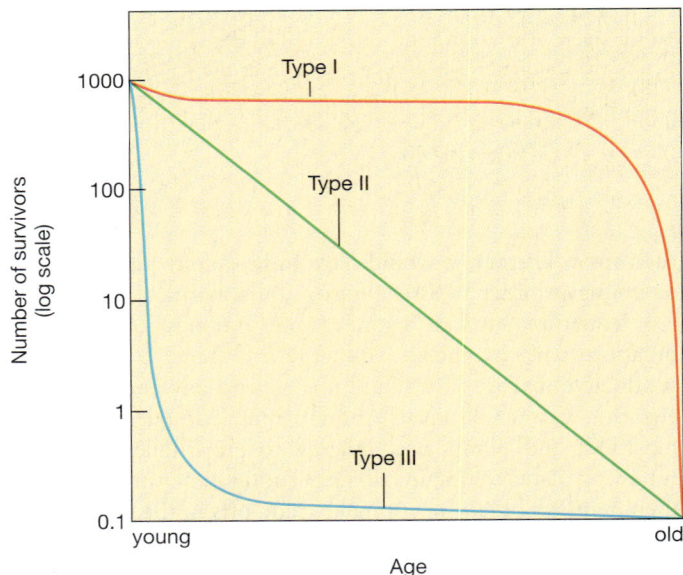

Figure 51–10　**Survivorship curves.** These curves represent the ideal survivorships of organisms in which mortality is greatest in old age (Type I), spread evenly across all age groups (Type II), and greatest among the young (Type III). The survivorship of most organisms can be compared to these curves.

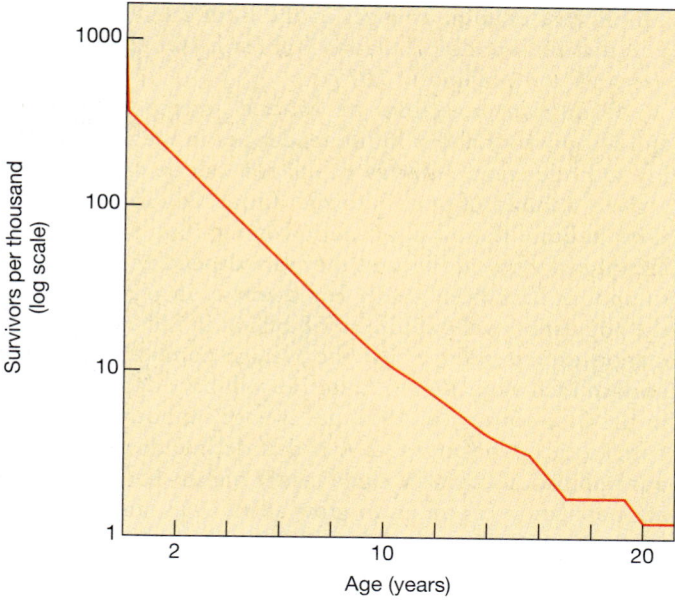

Figure 51–11 **Survivorship curve for a herring gull population.**
Herring gulls have Type III survivorship as chicks and Type II survivorship as adults. Data were collected from Kent Island, Maine, during a five-year period; baby gulls were banded to establish identity. *(After Paynter)*

Young oysters have three free-swimming larval stages before settling down and secreting a shell. These larvae are vulnerable to predation, and few survive to adulthood.

In Type II survivorship, intermediate between Types I and III, the probability of survival does not change with age. Probability of mortality is equally likely across all age groups, resulting in a linear decline in survivorship. This constancy probably results from essentially random events that cause death with little age bias. Certain annual plants and some lizards have Type II survivorship.

The three survivorship curves are generalizations, and few populations exactly fit one of the three. Some populations have one type of survivorship curve early in life and another type as adults. Populations of herring gulls, for example, start out with a Type III survivorship curve but develop a Type II curve as adults (Fig. 51–11). Note that most mortality occurs almost immediately after hatching, despite the protection and care given to the chicks by the parent bird. Herring gull chicks die from predation or attack by other herring gulls, inclement weather, infectious diseases, or starvation following death of the parent. Once the chicks become independent, their survivorship increases dramatically, and mortality occurs at about the same rate throughout their remaining lives. As a result, few or no herring gulls die from the degenerative diseases of "old age" that cause death in most humans.

THE PRINCIPLES OF POPULATION ECOLOGY APPLY TO HUMAN POPULATIONS

Now that we have examined some of the basic concepts of population ecology, we can apply those concepts to the human population. Examine Figure 51–12, which shows the world increase in the human population since the development of agriculture approximately 10,000 years ago. Now look back at Figure 51–3 and compare the two curves. The characteristic J curve of exponential population growth shown in Figure 51–12 reflects the decreasing amount of time it has taken to add each additional billion people to our numbers. It took thousands of years for the human population to reach 1 billion, a milestone that took place around 1800. It took 130 years to reach 2 billion (in 1930), 30 years to reach 3 billion (in 1960), 15 years to reach 4 billion (in 1975), and 12 years to reach 5 billion (in 1987). The population is projected to reach 6 billion in 1999.

Thomas Malthus (1766–1834), a British economist, was one of the first to recognize that the human population cannot continue to increase indefinitely (see Chapter 17). He pointed out that human population growth is not always desirable (a view contrary to the beliefs of his day and to those

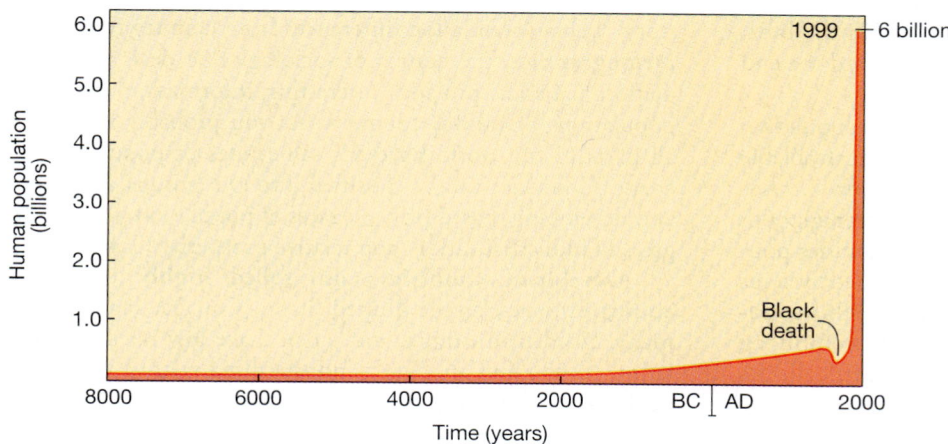

Figure 51–12 **Human population growth.** During the last 1000 years, the human population has been increasing nearly exponentially.

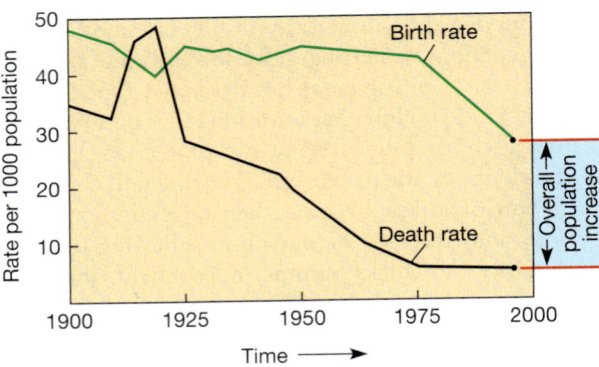

Figure 51–13 **Birth and death rates in Mexico, 1900 to 1995.** Both birth and death rates declined during the 20th century, but because the death rate declined much more than the birth rate, Mexico has experienced a high growth rate. (The high death rate prior to 1920 was caused by the Mexican Revolution.)

of many people even today) and that the human population is capable of increasing faster than the food supply. He maintained that the inevitable consequences of population growth are famine, disease, and war.

The world population was 5.93 billion in 1998 and increased by approximately 86 million from 1997 to 1998. This change was not caused by an increase in the birth rate *(b)*. In fact, the world birth rate has actually declined slightly during the past 200 years. The increase in growth rate is due instead to a large *decrease in the death rate (d),* which has occurred primarily because of greater food production, better medical care, and improved sanitation practices. For example, from 1920 to 1995, the death rate in Mexico fell from approximately 40 per 1000 individuals to 5 per 1000. During the same period the birth rate only dropped from approximately 40 per 1000 individuals to 27 per 1000 (Fig. 51–13).

The human population has reached a turning point. Although our numbers continue to increase, the growth rate *r* has declined over the past several years, from a peak in 1965 of about 2% per year to a 1998 growth rate of 1.4% per year. Population experts at the United Nations and the World Bank have projected that the growth rate will continue to slowly decrease until zero population growth is attained. It is projected that **zero population growth**—the point at which the birth rate equals the death rate (*r* = 0)— will occur toward the end of the 21st century.

The United Nations periodically publishes population projections for the 21st century. The latest U.N. data available forecast that the human population will be between 7.8 billion (their low projection) and 12.5 billion (their high projection) in the year 2050. Another group that examines population trends is the International Institute for Applied Systems Analysis (IIASA) in Vienna, Austria. The latest (1996) population projections of the IIASA were obtained by combining statistically the predictions of 12 population experts. IIASA also took into account future changes in the death rate due to such factors as starvation and disease. (Most population projections just examine changes in the birth rate.) IIASA projects that the world's population will reach 10 billion in 2050 and peak at 11 billion in 2075.

Population projections are "what if" exercises: given certain assumptions about future tendencies in natality, mortality, and migration, an area's population can be calculated for a given number of years into the future. Population projections indicate that changes are upcoming, but they must be interpreted with care because they vary depending on what assumptions have been made. For example, in projecting that the population will stabilize at 7.8 billion in 2050, U.N. population experts assume that the average number of children born to each woman in all countries will have declined to 1.7 in the 21st century. In 1998 the average number of children born to each woman was 2.9. If that decline does not occur, our population could be significantly higher. For example, if the population continues to grow at its 1998 rate, there will be almost 13 billion humans in 2050.

The main unknown factor in this population growth scenario is Earth's carrying capacity. No one knows how many humans can be supported, and projections and estimates vary widely depending on what assumptions are made about standard of living, resource consumption, and waste generation. If we want all people to have a high level of material well-being equivalent to the lifestyles common in developed countries, then Earth will clearly be able to support far fewer humans than if everyone lives just above the subsistence level.

It is also not clear what will happen to the human population if or when the carrying capacity is approached. Optimists suggest that the human population will stabilize because of a decrease in the birth rate. Some experts take a more pessimistic view and predict that the widespread degradation of our environment caused by our ever-expanding numbers will make Earth uninhabitable for humans and other species. These experts contend that a massive wave of human suffering and death will occur. Some experts think human population has already exceeded the carrying capacity of the environment.

Not all countries have the same growth rate

While world population size illustrates overall trends, they do not describe other important aspects of the human population story, such as population differences from country to country. **Demographics,** the branch of sociology that deals with population statistics, provides interesting information about the populations of various countries. As you probably know, not all parts of the world have the same rates of population increase. Countries can be classified into two groups, developed and developing, based on their rates of population growth, degrees of industrialization, and relative prosperity (Table 51–2).

Developed countries (also called **highly developed countries**), such as the United States, Canada, France, Germany, Sweden, Australia, and Japan, have low rates of population growth and are highly industrialized relative to the rest of the world. Developed countries have the lowest birth rates in the world. Indeed, some developed countries such as Ger-

TABLE 51-2 Comparison of 1998 Population Data in Developed and Developing Countries

	Developed	Developing	
	(Highly Developed) United States	(Moderately Developed) Brazil	(Less Developed) Ethiopia
Fertility rate	2.0	2.5	7.0
Doubling time at current rate	116 years	48 years	28 years
Infant mortality rate	7.0 per 1000	43 per 1000	128 per 1000
Life expectancy at birth	76 years	67 years	42 years
Per capita GNP (U.S. $; 1995)	$28,020	$4400	$100
Women using modern contraception	68%	70%	27%

many have birth rates just below that needed to sustain the population and are thus declining slightly in numbers. Highly developed countries also have low **infant mortality rates** (the number of infant deaths per 1000 live births). The infant mortality rate of the United States was 7.0 in 1998, for example, compared with a 1998 worldwide rate of 58. Highly developed countries also have longer life expectancies (76 years in the United States versus 66 years worldwide) and higher average per capita GNPs ($28,020 in the United States versus $5180 worldwide).[3]

Developing countries fall into two subcategories: moderately developed and less developed. Mexico, Turkey, Thailand, and most countries of South America are *moderately developed*. When compared with highly developed countries, the birth rates and infant mortality rates of moderately developed countries are higher. Moderately developed countries have a medium level of industrialization, and their average per capita GNPs are lower than those of highly developed countries. *Less developed* countries, which include Bangladesh, Niger, Ethiopia, Laos, and Cambodia, have the highest birth rates, the highest infant mortality rates, the lowest life expectancies, and the lowest average per capita GNPs in the world.

One way to represent the population growth of a country is to determine the **doubling time,** the amount of time it would take for its population to double in size, assuming that its current growth rate did not change. A look at a country's doubling time can identify it as a highly, moderately, or less developed country: the shorter the doubling time, the less developed the country. At 1998 growth rates, the doubling time is 19 years for Togo, 28 years for Ethiopia, 32 years for Mexico, 116 years for the United States, and 529 years for Belgium.

It is also instructive to examine **replacement-level fertility,** the number of children a couple must produce in order to "replace" themselves. Replacement-level fertility is usually given as 2.1 children in developed countries and 2.7 children in developing countries. The number is always greater than 2.0 because some children die before they reach reproductive age. Higher infant mortality rates are the main reason that replacement levels in developing countries are greater than in developed countries. The **total fertility rate,** the average number of children born to a woman during her lifetime, is 2.9 worldwide, which is well above replacement levels.

The population in many developing countries is beginning to approach stabilization. Fertility rates must decline in order for the population to stabilize (see Table 51–3 and note the general decline in total fertility rate from the 1960s to 1998 in selected developing countries). The total fertility rate in de-

TABLE 51-3 Fertility Changes in Selected Developing Countries

Country	Total Fertility Rate*	
	1960–65	1998
Afghanistan	7.0	6.1
Bangladesh	6.7	3.3
Brazil	6.2	2.5
China	5.9	1.8
Egypt	7.1	3.6
Guatemala	6.9	5.1
India	5.8	3.4
Kenya	8.1	4.5
Mexico	6.8	3.1
Nepal	5.9	4.6
Nigeria	6.9	6.5
Thailand	6.4	2.0

*Total fertility rate = average number of children born to each woman during her lifetime.

[3]GNP stands for gross national product, the total value of a nation's annual output in goods and services.

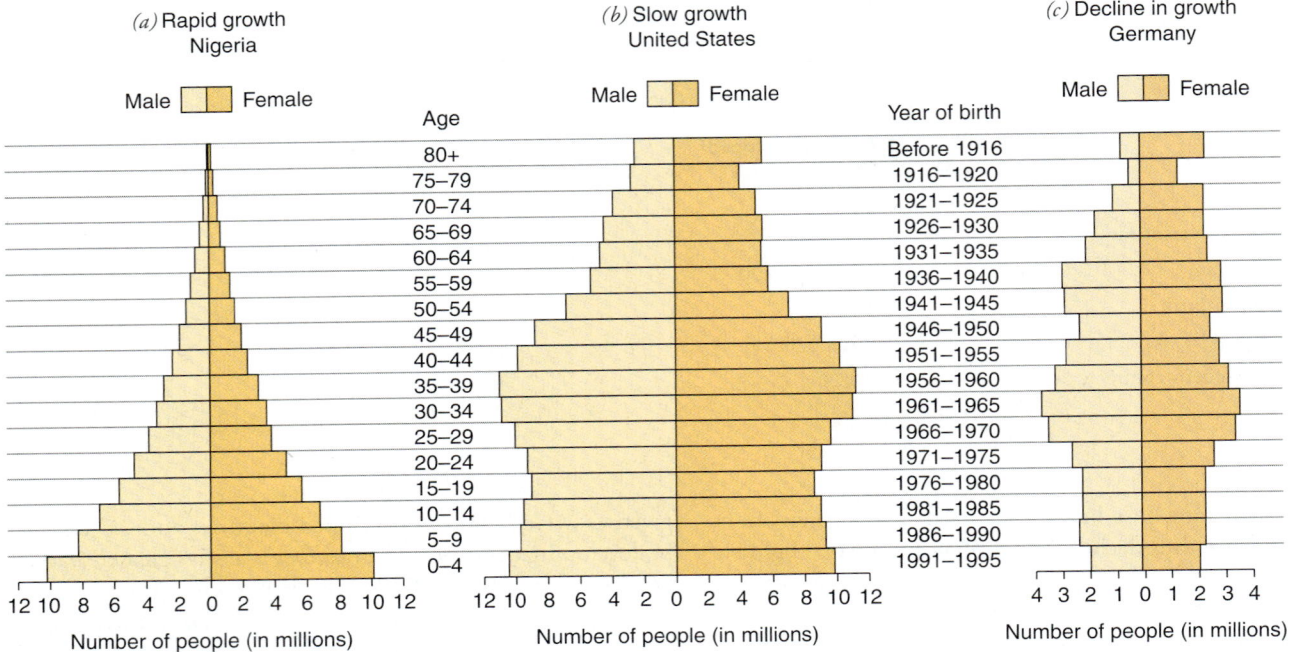

Figure 51-14 Age structure diagrams. These age structure diagrams for *(a)* Nigeria, *(b)* the United States, and *(c)* Germany indicate that less developed countries such as Nigeria have a greater percentage of young people than highly developed countries. As a result, less developed countries are projected to have greater population growth than highly developed countries.

veloping countries has decreased from an average of 6.1 children per woman in 1970 to 3.3 in 1998. In countries such as Brazil, Indonesia, and Mexico, fertility rates have declined by at least 25% in the past decade.[4]

The age structure of a country can be used to predict future population growth

In order to predict the future growth of a population, it is important to know its **age structure,** which is the percentages of the population at different ages. The number of males and females at each age, from birth to death, is represented in age structure diagrams.

The overall shape of an age structure diagram indicates whether the population is increasing, stationary, or shrinking. The age structure diagram of a country with a high growth rate (for example, Nigeria or Bolivia) is shaped like a pyramid (Fig. 51–14a). Because the largest percentage of the population is in the prereproductive age group (ages 0 to 14), the probability of future population growth is great. A strong *population growth momentum* exists because when all these children mature, they will become the parents of the next gener-

ation, and this group of parents will be larger than the previous group. Thus, even if the fertility rate of such a country declines to replacement level, the population will continue to grow.

In contrast, the more tapered bases of the age structure diagrams of countries with slowly growing, stationary, or declining populations indicate a smaller proportion of children, who will become the parents of the next generation (Fig. 51–14b and c). The age structure diagram of a stationary population, one that is neither growing nor shrinking, demonstrates that the number of people at prereproductive (0 to 14) and reproductive age (15 to 44) is approximately the same. Also, a larger percentage of the population is older, that is, postreproductive (45 and older), than in a rapidly increasing population. Many countries in Europe have stationary populations.

In a shrinking population, the prereproductive age group is *smaller* than either the reproductive or postreproductive groups. Germany, Russia, and Bulgaria are examples of countries with shrinking populations.

Worldwide, 32% of the human population is under age 15. When these people enter their reproductive years, they have the potential to cause a large increase in the growth rate. Even if the birth rate does not increase, the growth rate will increase simply because there are more people reproducing.

Most of the world population increase that has occurred since 1950 has taken place in developing countries as a result of the younger age structure and the higher-than-replacement-

[4]Although the fertility rates in these countries have declined, it should be remembered that they exceed replacement-level fertility. Consequently, the populations in these countries are still increasing.

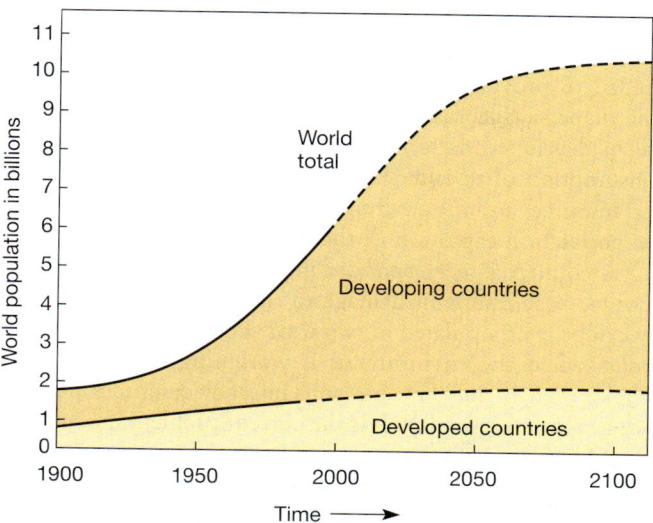

Figure 51–15 **Proportions of world population in developed and developing countries.** Much of the worldwide increase in population since 1950 has occurred in developing countries.

(a)

(b)

Figure 51–16 **People and natural resources.** *(a)* The rapidly increasing number of people in developing countries overwhelms their natural resources, even though individual resource requirements may be low. Shown is a typical Indian family, from Ahraura Village, India, with all their possessions. *(b)* People in developed countries consume a disproportionate share of natural resources. Shown is a typical American family from Pearland, Texas, with all their possessions. *(a,b, Peter Ginter/Material World)*

level fertility rates of their populations (Fig. 51–15). In 1950, 66.8% of the world's population was in the developing countries in Africa, Asia (minus Japan), and Latin America. Between 1950 and 1998, the world's population more than doubled, but most of that growth occurred in developing countries. As a reflection of this, in 1998 the people in developing countries had increased to 80.1% of the world's population. Because different age structures in populations imply different potential changes when we project to the future, most of the population increase that will occur during the next century will also take place in developing countries. As you know, these countries are least able to support such growth. By 2020 it is estimated that about 85% of the people in the world will live in developing countries.

Environmental degradation is related to population growth and resource consumption

The relationships among population growth, use of natural resources, and environmental degradation are complex, but we can make two useful generalizations. First, the resources that are essential to an individual's survival are relatively small, but a rapidly increasing number of people (as we see in developing countries) tends to overwhelm and deplete a country's soils, forests, and other natural resources (Fig. 51–16a). Second, in developed nations, individual resource demands are relatively large, far above requirements for survival. In order to satisfy their desires rather than their basic needs, people in more affluent nations exhaust resources and degrade the global environment through excessive consumption and "throwaway" lifestyles (Fig. 51–16b).

Rapid population growth can cause natural resources to be overexploited. For example, when fisheries are overharvested, there will be too few fish to serve as a food source. A similar problem arises when land that is inappropriate for farming, such as mountain slopes or some tropical rain forests, is used to grow crops. Although this practice may provide a short-term solution to the need for food, it does not work in the long run, because when these lands are cleared for farming, their agricultural productivity declines rapidly and severe environmental deterioration occurs.

The effects of population growth on natural resources are particularly critical in developing countries. The economic growth of developing countries is often tied to the exploitation of natural resources. These countries are faced with the difficult choice of exploiting natural resources to provide for their expanding populations in the short term or conserving those resources for future generations. (It is instructive to note that the economic growth and development of the United States and of other highly developed nations came about through the exploitation, and in some cases the destruction, of their natural resources. Moreover, highly developed countries have used and continue to use the resources of developing countries.)

Although resource issues are clearly related to population size (more people use more resources), an equally, if not more, important factor is a population's *resource consumption*. People in developed countries are conspicuous consumers; their use of resources is greatly out of proportion to their numbers. A single child born in a developed country such as the United States, for example, causes a greater impact on the environment and on resource depletion than do a dozen or more children born in a developing country. Many natural resources are needed to provide the automobiles, air conditioners, disposable diapers, videocassette recorders, and other "comforts" of life in developed nations. Thus, the disproportionately large consumption of resources by developed countries affects natural resources and the environment as much or more than does the population explosion in the developing world.

A country is *overpopulated* if the level of demand on its resource base results in damage to the environment. A country can be overpopulated in two ways. **People overpopulation** occurs when the environment is worsening from too many people, even if those people consume few resources per person. People overpopulation is the current problem in many developing nations. In contrast, **consumption overpopulation** occurs when each individual in a population consumes too large a share of resources, that is, more than is needed to survive. The effect of consumption overpopulation on the environment is the same as that of people overpopulation—pollution and degradation of the environment. Many affluent developed nations suffer from consumption overpopulation: developed nations represent only 20 percent of the world's population, yet they consume significantly more than half of its resources.

S U M M A R Y   W I T H   K E Y   T E R M S

I. **Ecology** is the study of all relationships among organisms and their abiotic environment.
   A. A **population** is all the members of a particular species that live together in the same area.
   B. A **community** is all the populations of different species living in the same area. An **ecosystem** is a community and its environment.
   C. The **biosphere** is the global ecological system that comprises all the communities on Earth. The biosphere concept includes interactions among all Earth's communities and the Earth's atmosphere, lithosphere, and hydrosphere.
II. **Population ecology** is the branch of biology that deals with the number of individuals of a particular species that are found in an area and how and why those numbers change over time. Populations have certain properties, such as birth rates and death rates, that individual organisms lack.
   A. **Population density** is the number of individuals of a species per unit of area or volume at a given time.
   B. Population **dispersion** (spacing) may be **random** (unpredictably spaced), **clumped** (clustered in specific parts of the habitat), or **uniform** (evenly spaced).
   C. Population size is affected by the number of births *(b)*, deaths *(d)*, immigrants *(i)*, and emigrants *(e)*.
      1. The **growth rate** *(r)* of a population is its rate of change in size.
      2. $r = b - d$ on a global scale (when migration is not a factor). Populations increase in size as long as the birth rate (**natality**) is greater than the death rate (**mortality**).
      3. $r = (b - d) + (i - e)$ for a local population (where migration is a factor).

III. **Biotic potential** is the maximum rate at which a species or population could increase in number under ideal conditions. Although certain populations exhibit **exponential population growth** for limited periods of time (the J-shaped curve), eventually the growth rate decreases to around zero or becomes negative.
   A. Population size is modified by limits set by the environment, which are collectively called **environmental resistance.**
   B. The **carrying capacity (*K*)** of the environment is the largest population that can be maintained for an indefinite time by a particular environment.
IV. **Logistic population growth,** when graphed, shows a characteristic S-shaped curve. It shows an initial lag phase (when the population is small), followed by an exponential phase, followed by a leveling phase as the carrying capacity of the environment is reached. Seldom do natural populations follow the logistic growth curve very closely.
V. Environmental factors limit population growth.
   A. **Density-dependent factors** regulate population growth by affecting a larger proportion of the population as population density rises. Predation, disease, and competition are examples.
   B. **Density-independent factors** limit population growth but are not influenced by changes in population density. Hurricanes and fires are examples.
VI. Each organism has its own life history strategy.
   A. An *r* **strategy** emphasizes a high growth rate. These organisms often have small body sizes, high reproductive rates, and short lifespans, and they typically inhabit variable environments.
   B. A *K* **strategy** results in maintenance of a population near the carrying capacity of the environment. These organisms often have large

body sizes, low reproductive rates, and long lifespans, and they typically inhabit stable environments.

C. There are three general types of **survivorship** curves. Type I survivorship, in which mortality is greatest in old age, is typical of *K*-selected species. Type III survivorship, in which mortality is greatest among the young, is typical of *r*-selected species. In Type II survivorship, mortality is spread evenly across all age groups.

VII. The principles of population ecology apply to humans.

A. The world population was 5.93 billion in 1998 and is increasing by more than 80 million a year.

1. Although our numbers continue to increase, the growth rate *(r)* has declined over the past several years, from a peak in 1965 of about 2% per year to a 1998 growth rate of 1.4% per year.

2. **Demographers,** scientists who study human population statistics, project that the world population will become stationary (*r* = 0, or **zero population growth**) by the end of the 21st century.

B. **Highly developed countries** have the lowest birth rates, lowest **infant mortality rates,** longest life expectancies, and highest per

capita GNPs. **Developing countries** have the highest birth rates, highest infant mortality rates, shortest life expectancies, and lowest per capita GNPs.

C. The **age structure** of a population greatly influences population dynamics. It is possible for a country to have **replacement-level fertility** and still experience population growth if the largest percentage of the population is in the prereproductive years.

VIII. There are two kinds of overpopulation, people overpopulation and consumption overpopulation.

A. **Developing countries** tend to have **people overpopulation,** in which population increase degrades the environment even though each individual uses few resources.

B. **Developed countries** have **consumption overpopulation,** in which each individual in a slow-growing or stationary population consumes a large share of resources, resulting in environmental degradation.

POST-TEST

1. Population _____ is the number of individuals of a species per unit of habitat area or volume at a given time. (a) dispersion (b) density (c) survivorship (d) age structure (e) demographics

2. Which of the following patterns of cars parked along a street is an example of uniform dispersion? (a) five cars parked next to one another in the middle, leaving two empty spaces at one end and three empty spaces at the other end (b) five cars parked in the pattern: car, empty space, car, empty space, etc. (c) five cars parked in no discernable pattern, sometimes having empty spaces on each side and sometimes parked next to another car.

3. This type of dispersion occurs when individuals are concentrated in certain parts of the habitat. (a) clumped (b) uniform (c) random (d) both a and b (e) both b and c

4. The birth rate, or _____, increases population size, whereas the death rate, or _____, decreases population size. (a) natality; demography (b) scramble competition; contest competition (c) mortality; natality (d) total fertility rate; mortality (e) natality; mortality

5. The maximum rate at which a population could increase under ideal conditions is known as its (a) total fertility rate (b) survivorship (c) biotic potential (d) doubling time (e) age structure

6. In a graph of population size versus time, a J-shaped curve is characteristic of _____ population growth. (a) exponential (b) logistic (c) zero (d) replacement-level (e) both a and b have J-shaped curves

7. Populations never remain at their biotic potential indefinitely because population size is modified by limits, collectively called (a) survivorship (b) contest competition (c) density-dependent factors (d) scramble competition (e) environmental resistance

8. The largest population that can be maintained by a particular environment for an indefinite period of time is known as the (a) biotic potential (b) exponential growth (c) growth rate (d) carrying capacity (e) density

9. Predation, disease, and competition are examples of _____ factors. (a) density-dependent (b) density-independent (c) survivorship (d) emigration (e) immigration

10. _____ competition occurs within a population and _____ competition occurs among populations of different species. (a) Interspecific; intraspecific (b) Intraspecific; interspecific (c) Type I survivorship; Type II survivorship (d) Contest; scramble (e) Scramble; contest

11. Organisms with _____ selection have high rates of natural increase, whereas organisms with _____ selection do not use environmental resources and energy to produce large numbers of offspring. (a) *K*; *r* (b) *r*; *K* (c) natality; mortality (d) interspecific; intraspecific (e) density-dependent; density-independent

12. A highly developed country has a (a) low doubling time (b) low infant mortality rate (c) high per capita GNP (d) both a and b are correct (e) a, b, and c are correct

REVIEW QUESTIONS

1. List and define the levels of biological organization that ecologists study.

2. Give several biological advantages for a clumped dispersion. What are the disadvantages?

3. Define each of the following and explain its effect on population size: (a) natality; (b) mortality; (c) immigration; (d) emigration

4. The 1998 population of the Netherlands was 15.7 million, and its land area is 990 square miles. The 1998 population of the United States was 270.2 million, and its land area is 3,615,200 square miles. Which country has the greater population density?

5. The population of India in 1998 was 988.7 million, and its growth rate was 1.9 percent per year. Calculate the 1999 population of India. Assuming that the 1998 growth rate remains constant, in what year will India's population be double that of its 1998 population?

6. The world population in 1998 was 5.93 billion, and its annual growth rate was 1.4 percent. If the 1998 birth rate was 23 per 1000 people, what was the 1998 death rate, expressed as number per 1000 people?

7. Explain the J-shaped and S-shaped population growth curves in terms of biotic potential and carrying capacity.

8. Give three examples each of density-dependent and density-independent factors that affect population growth.
9. Draw the three main types of survivorship curves and relate each to *r* selection and *K* selection.
10. What is replacement-level fertility? Why is it higher in developing countries than in developed countries?

11. Explain how a single child born in the United States can have a greater effect on the environment and natural resources than a dozen children born in Kenya.
12. Which of these age structure diagrams indicates a population that is growing? Explain your answer.

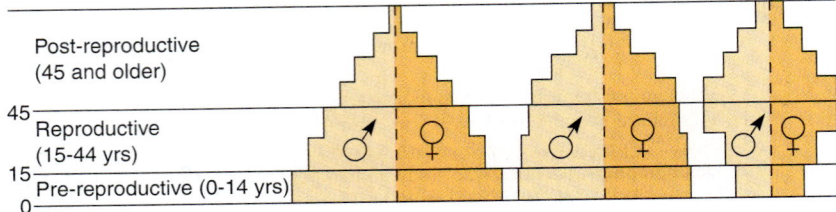

YOU MAKE THE CONNECTION

1. How might pigs at a trough be an example of scramble competition? Of contest competition?
2. Explain why the population size of a species that competes by contest competition is often near the carrying capacity, whereas the population size of a species that competes by scramble competition is often greater than or below the carrying capacity.
3. A female elephant bears a single offspring every two to four years. Based on this information, do you think elephants are *r*-selected or *K*-selected species? Which survivorship curve do you think is representative of elephants? Explain your answers.
4. In Bolivia, 42% of the population is younger than age 15, and 4% is older than 65. In Austria, 18% of the population is younger than 15, and 15% is older than 65. Which country will have the highest growth rate over the next two decades? Why?

RECOMMENDED READINGS

Cohen, J.E. "Population Growth and Earth's Human Carrying Capacity." *Science*, Vol. 269, 21 Jul. 1995. Explores the question of how many people Earth can support.

Line, L. "After Years of Gains, Wolf Population on Isle Royale Plummets." *The New York Times*, 21 Apr. 1998. Provides an update on the status of Isle Royale's moose and wolf populations.

Mitchell, J.D. "Before the Next Doubling." *World Watch*, Vol. 11, No. 1, Jan./Feb. 1998. The author argues that we must begin now to prevent the world population from doubling in the latter half of the 21st century.

Moffat, A.S. "Ecologists Look at the Big Picture." *Science*, Vol. 273, 13 Sep. 1996. This short article explains why experts have made such widely varying estimates of how many people the Earth can support.

Moore, P.D. "Feeding Patterns on Forest Floors." *Nature*, Vol. 390, 20 Nov. 1997. This article summarizes two reports from the *Journal of Ecology* on the subtle effects of two rainforest mammals (red howler monkeys and tapirs) on the dispersion patterns of certain rainforest plants.

Raloff, J. "The Human Numbers Crunch." *Science News*, Vol. 149, 22 Jun. 1996. An overview of the challenges that population experts think we will face during the next 50 years.

Raven, P.H., L.R. Berg, and G.B. Johnson. *Environment*, 2nd ed. Saunders College Publishing, Philadelphia, 1998. This environmental science text offers a detailed situation analysis of planet Earth, including how humans interact with and affect its physical and living systems.

1998 World Population Data Sheet, Population Reference Bureau, Washington, D.C. A chart that provides current population data for all countries. Includes birth rates, death rates, infant mortality rates, total fertility rates, and life expectancies, as well as other pertinent information.

● Visit our website at **http://www.saunderscollege.com/lifesci/titles.html** and click on Solomon/Berg/Martin Biology for links to chapter-related resources on the World Wide Web.

CHAPTER 52

Community Ecology

In Chapter 51 we examined the dynamics of single populations, including the ways individual populations change and what factors affect these changes. In the natural world, however, species do not exist as isolated populations. Rather, most populations are the interacting parts of a complex **community,** which consists of populations of different species that live and interact in a given environment.

The populations that compose a community interact with one another and are interdependent in a variety of ways. Species compete with one another for food, water, living space, and other resources. Organisms kill and eat other organisms. Species form symbiotic associations with one another. Each organism plays one of three main roles in community life: producer, consumer, or decomposer. In this photograph, for example, the trees are producers that manufacture complex organic molecules by photosynthesis, and the raccoon is a consumer that eats berries, nuts, bird eggs, insects, crayfish, rodents, and even carrion. Understanding the many interactions and interdependencies of organisms living together as a community is one of the goals of community ecologists.

Although each community forms a distinct living system, communities vary greatly in size, typically do not have precise boundaries, and are rarely completely isolated. Communities interact with and influence one another in countless ways, even when the interaction is not readily apparent. Furthermore, there are communities within communities. A forest is a community, but so is a rotting log in that forest. The log contains bacteria, fungi, slime molds, worms, insects, and perhaps even mice. The microorganisms living within the gut of a termite in the rotting log also form a community.

Organisms occupy an abiotic (nonliving) environment that is as essential to their lives as their interactions with other organisms. Minerals, air, water, and sunlight are just as much a part of a honeybee's environment, for example, as are the flowers that it pollinates and from which it takes nectar. A biological community and its abiotic environment together compose an **ecosystem.**

This chapter is concerned with understanding community structure and diversity by finding common patterns and processes in a wide variety of communities. Although the living community is emphasized in this chapter, communities and their abiotic environments are inseparably linked; the abiotic components of ecosystems are considered in Chapter 53.

(Gary Meszaros/Dembinsky Photo Associates)

AFTER YOU HAVE STUDIED THIS CHAPTER YOU SHOULD BE ABLE TO

1. Distinguish between a community and an ecosystem.
2. Define predation and describe the effects of natural selection on predator-prey relationships.
3. Explain symbiosis and distinguish among mutualism, commensalism, and parasitism.
4. Define competition and distinguish between interspecific and intraspecific competition.
5. Define ecological niche and distinguish between an organism's fundamental niche and its realized niche.
6. Summarize the concept of competitive exclusion.
7. Give several examples of limiting factors and discuss how they might affect an organism's ecological niche.

8. Summarize the main determinants of species diversity in a community and describe factors associated with high species diversity.
9. Define ecological succession and distinguish between primary and secondary succession.
10. Discuss the two views of the nature of communities: the organismic model and the individualistic model.
11. Define biogeography and explain the relationship between biogeography and community ecology.

COMMUNITIES CONTAIN AUTOTROPHS AND HETEROTROPHS

Sunlight is the source of energy that powers almost all life processes on Earth. **Primary producers,** also called *autotrophs* or, simply, *producers,* are organisms that make complex organic molecules from simple inorganic substances (generally carbon dioxide and water), usually using the energy of sunlight. In other words, most producers perform the process of photosynthesis. By incorporating the chemicals they manufacture into their own **biomass** (living material), producers become potential food resources for other organisms. While plants are the most significant producers on land, algae and cyanobacteria are important producers in aquatic environments. In rift-vent, hot spring communities deep in the ocean, nonphotosynthetic bacteria are the producers. These producers use chemical energy to power their metabolism (see *Focus On: Life without the Sun*).

All other organisms in a community are *heterotrophs* that extract energy from organic molecules produced by other organisms. **Consumers** are heterotrophs that obtain energy and body-building materials by feeding on organic molecules formed by other organisms. Consumers that eat only producers are called **primary consumers,** which usually means that they are exclusively **herbivores** (plant eaters). Deer and grasshoppers are examples of primary consumers. **Secondary consumers** eat primary consumers and include flesh-eating **carnivores,** which eat other animals. Lions and spiders are examples of carnivores. Other consumers, called **omnivores,** eat a variety of organisms, both plant and animal, and therefore function as both primary and secondary consumers. Bears, pigs, and humans are examples of omnivores.

Some consumers, called **detritus feeders,** or **detritivores,** consume **detritus,** which is dead organic matter that includes animal carcasses, leaf litter, and feces. Detritus feeders, such as snails, crabs, clams, and worms, are especially abundant in aquatic environments, where they burrow in the bottom muck and consume the organic matter that collects there. Earthworms are terrestrial detritus feeders, as are termites, beetles, snails, mites, and millipedes. Detritus feeders work together with microbial decomposers to destroy dead organisms and waste products. An earthworm, for example, actually eats its way through the soil, digesting much of the organic matter contained there.

Many consumers do not fit readily into a single category of herbivore, carnivore, omnivore, or detritivore. To some degree these organisms modify their food preferences as the need arises. Furthermore, some organisms change food preferences over their lifetime. Tadpoles, which eat algae and plant material, are primary consumers, but adult frogs, which eat earthworms, snails, insects, crayfish, fish, other frogs, and tadpoles, are secondary consumers.

Decomposers (also called **saprotrophs**) include microbial heterotrophs that break down organic material and use the decomposition products to supply themselves with energy. They typically release simple inorganic molecules, such as carbon dioxide and mineral salts, that can then be reused by producers. Most bacteria and fungi are important decomposers. Dead wood, for example, is invaded first by sugar-metabolizing fungi that consume the wood's simple carbohydrates, such as glucose and maltose. When these carbohydrates are exhausted, other fungi, often aided by termites (with symbiotic zooflagellates in their guts; see Chapter 24) complete the digestion of the wood by breaking down cellulose, a complex carbohydrate that is the main component of wood.

Producers and decomposers have indispensable roles in ecosystems. Producers provide both food and oxygen for the rest of the community. Decomposers are also necessary for the long-term survival of any community, because without them dead organisms and waste products would accumulate indefinitely. Without decomposers, essential elements such as potassium, nitrogen, and phosphorus would remain in dead organisms and hence be unavailable for use by new generations of organisms.

LIFE WITHOUT THE SUN

In 1977 an oceanographic expedition aboard the research submersible Alvin studied the Galapagos Rift, a deep cleft in the ocean floor off the coast of Ecuador. The expedition revealed, on the floor of the abyss, a series of hydrothermal vents where sea water apparently had penetrated and been heated by the hot rocks below. During its time within the Earth, the water had also been charged with mineral compounds, including hydrogen sulfide, H_2S, which is toxic to most species.

At the tremendous depths of greater than 2500 m (8200 ft) found in the Galapagos Rift, there is no light for photosynthesis. But hydrothermal vents support a rich and bizarre variety of life forms *(see figure)*, in contrast with the surrounding "desert" of the abyssal floor. Many of the species in these oases of life are not found in other habitats. For example, giant blood-red tube worms almost 3 m (10 ft) in length cluster in great numbers around the vents. Other animals around the hydrothermal vents include unique species of clams, crabs, barnacles, and mussels. Since 1977 such hydrothermal vent communities, some as large as football fields, have been identified at many other oceanic sites.

Scientists initially wondered what energy source sustains these organisms. Most deep-sea communities depend on the organic matter that drifts down from the surface waters; in other words, they rely on

energy ultimately derived from photosynthesis. Hydrothermal vent communities, however, are too densely clustered and too productive to be dependent on chance encounters with organic material from surface waters. Instead, the base of the food web in these aquatic oases is occupied by chemosynthetic autotrophic bacteria that possess enzymes that catalyze the oxidation of hydrogen sulfide to water and sulfur or sulfate. Such chemical reactions are exergonic (see Chapter 6) and provide the energy required to fix CO_2, which is dissolved in the water, into organic compounds. Many of the animals consume the bacteria directly by filter-feeding, but others, such as the giant tube worms, get their energy from bacteria that live in their tissues.

Scientists continue to generate questions about hydrothermal vent communities. How, for example, do the organisms find and colonize vents, which are widely scattered on the ocean floor? How have the inhabitants of these communities adapted to survive the harsh living conditions, including high pressure, high temperatures, and toxic chemicals? As vent community research continues, scientists hope to discover answers to these and other questions.

A hydrothermal vent community. Chemoautotrophic bacteria living in the tissues of these tube worms extract energy from hydrogen sulfide to manufacture organic compounds. Because these worms lack digestive systems, they depend on the organic compounds provided by the endosymbiotic bacteria, along with materials filtered from the surrounding water and digested extracellularly. Also visible in the photograph are some filter-feeding mollusks *(yellow)* and a crab *(white)*. *(D. Foster, Science VU-WHOI/Visuals Unlimited)*

ORGANISMS INTERACT IN A VARIETY OF WAYS

No species exists independently of other organisms. The primary producers, consumers, and decomposers of a community interact with one another in a variety of complex ways, and each forms associations with other organisms. Three main types of interactions occur among species in a community: predation, symbiosis, and competition. Table 52–1 summarizes these interactions as they relate to any two species.

Natural selection shapes both prey and predator

Predation is the consumption of one species, the *prey*, by another, the *predator*. It includes animals eating other animals,

as well as animals eating plants. Predation has resulted in an evolutionary "arms race," with both the evolution of predator strategies—more efficient ways to catch prey—and prey strategies—better ways to escape the predator. A predator that is more efficient at catching prey exerts a strong selective force on its prey, which over time may evolve some sort of countermeasure that reduces the probability of being captured. The countermeasure acquired by the prey in turn acts as a strong selective force on the predator. This type of interdependent evolution of two interacting species is known as **coevolution** (see Chapter 35).

Pursuit and ambush are two predator strategies

A brown pelican sights a fish while in flight. Less than two seconds after diving into the water, it has its catch (Fig.

Interaction	Effect on Species 1	Effect on Species 2
Predation of Species 2 by Species 1	Beneficial (+)	Harmful (−)
Symbiosis		
Mutualism of Species 1 and Species 2	Beneficial (+)	Beneficial (+)
Commensalism of Species 1 with Species 2	Beneficial (+)	No effect (0)
Parasitism By Species 1 on Species 2	Beneficial (+)	Harmful (−)
Competition between Species 1 and Species 2	Harmful (−)	Harmful (−)

TABLE 52-1 Ecological Interactions among Species

52–1). Orcas (formerly known as killer whales) hunt in packs and often herd salmon or tuna into a cove so that they are easier to catch. Any trait that increases hunting efficiency, such as the speed of a brown pelican or the cunning tactics of orcas, favors predators that pursue their prey. Because these carnivores must be able to process information quickly during the pursuit, their brains are generally larger relative to body size than those of the prey they pursue.

Ambush is another effective way to catch prey. Some small spiders are the same color as the flowers on which they hide. This camouflage keeps unwary insects that visit the flower for nectar from noticing the spider until it is too late. Predators that are able to attract prey are particularly effective at ambushing. For example, a diverse group of deep-sea fishes known as angler fish possess rodlike luminescent lures to attract food.

Figure 52–1 Pursuit of prey. A brown pelican alights in the water, having just caught a fish in its pouch. This effective predator strategy requires speed. *(SharkSong/M. Kazmers/Dembinsky Photo Associates)*

Chemical protection is an effective plant defense against herbivores

Plants cannot escape predators by fleeing, but they possess a number of adaptations that protect them from being eaten. The presence of spines, thorns, tough leathery leaves, or even thick wax on leaves discourages foraging herbivores from grazing. Other plants produce an array of protective chemicals that are unpalatable or even toxic to herbivores. Many of the active ingredients that discourage the foraging of herbivores affect hormone activity or nerve, muscle, liver, or kidney functions. Interestingly, many of the chemical defenses in plants are useful to humans. India's neem tree, for example, contains valuable chemicals that are effective against more than 100 species of herbivorous insects, mites, and nematodes. Nicotine from tobacco, pyrethrum from chrysanthemum, and rotenone from the derris plant are other examples of chemicals extracted and used as insecticides. Such plant-derived pesticides are called *botanicals*.

Milkweeds are an excellent example of the biochemical co-evolution between plants and herbivores (Fig. 52–2). Milkweeds produce alkaloids and cardiac glycosides, chemicals that are poisonous to all animals except for a small group of insects. During the course of evolution, these insects acquired the ability to either tolerate or metabolize the milkweed toxins. As a result, they can eat milkweeds without being poisoned. These insects avoid competition from other herbivorous insects because few others are able to tolerate milkweed toxins. Predators also learn to avoid these insects, which accumulate the toxins in their tissues and therefore become toxic themselves. An insect's bright **warning coloration** clearly announces its toxicity to predators that have learned to associate bright colors with illness. The black, white, and yellow-banded caterpillar of the monarch butterfly is an example of a milkweed feeder.

Animals possess a variety of defensive adaptations to avoid predators

Many animals, such as meadow voles and woodchucks, flee from predators by rapidly running to underground homes.

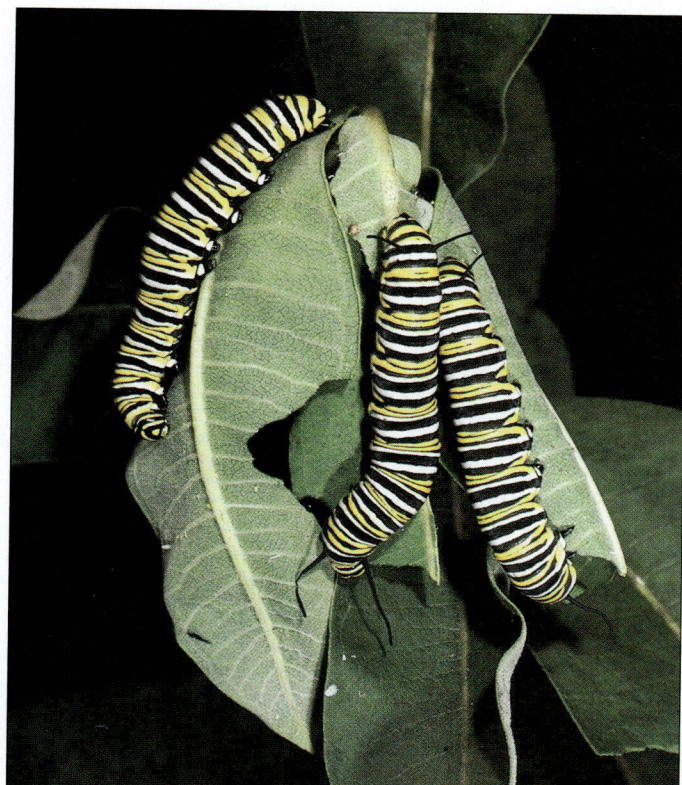

Figure 52-2 **Chemical plant defenses.** The common milkweed (*Asclepias syriaca*) is protected by its toxic chemicals. Its leaves are poisonous to most herbivores except monarch caterpillars *(shown)* and a few other insects. *(Patti Murray)*

Figure 52-3 **Chemical animal defenses.** The poison arrow frog (*Dendrobates tinctorius*) advertises its poisonous nature with its conspicuous coloring, warning away would-be predators. *(Michael Fogden/Animals Animals)*

Others have mechanical defenses, as, for example, the barbed quills of a porcupine and the shell of a pond turtle. Some animals live in groups, such as herds of antelope, colonies of honeybees, schools of anchovies, and flocks of pigeons. Because a group has so many eyes, ears, and noses watching, listening, and smelling for predators, this social behavior decreases the likelihood of a predator catching any one member unaware.

Chemical defenses are also common among animal prey. The South American poison arrow frog (*Dendrobates*) has poison glands in its skin. Its bright yellow and black warning coloration prompts avoidance by experienced predators (Fig. 52–3; also see Fig. 30–18*a*). Snakes and other animals that have tried once to eat a poisonous frog do not repeat their mistake! Other examples of warning coloration occur in the striped skunk, which sprays acrid chemicals from its anal glands, and the bombardier beetle, which spews harsh chemicals at potential predators (see Fig. 6–9).

Some animals have **cryptic coloration,** colors or markings that help them hide from predators by blending into their physical surroundings. Certain caterpillars resemble twigs so closely that you would never guess they were animals unless they moved (Fig. 52–4*a*). Pipefish are almost perfectly camouflaged in green eel grass (Fig. 52–4*b*). Such cryptic coloration has been preserved and accentuated by means of natural selection; these animals are less likely to be captured by predators and, therefore, more likely to live to maturity and produce offspring that also carry the genes for cryptic coloration.

Sometimes a defenseless species (a *mimic*) is protected from predation by its resemblance to a species that is dangerous in some way (a *model*). Such a strategy is known as **Batesian mimicry.** Many examples of this phenomenon exist. For example, a harmless moth may look so much like a bee or wasp that predators avoid it (Fig. 52–5). Even a biologist would hesitate to pick it up!

In **Müllerian mimicry,** different species, all of which are poisonous, harmful, or distasteful, resemble one another. Although their harmfulness protects them as individual species, their similarity in appearance works as an added advantage. Potential predators can more easily learn one common warning coloration. Viceroy and monarch butterflies are currently viewed as an example of Müllerian mimicry (see *Focus On: Batesian Butterflies Disproved*).

Symbiosis involves close association between species

Symbiosis is any intimate, long-term relationship or association between two or more species. The partners of a symbiotic relationship, called *symbionts,* may benefit from, be unaffected by, or be harmed by the relationship. The thousands,

(a)

(b)

Figure 52–4 **Cryptic coloration.** *(a)* Geometrid larvae are caterpillars that resemble twigs. Can you find the caterpillar? *(b)* The bay pipefish is thin, narrow, and green like the eel grass or algae in which it hides. Note its habit of holding its body in a position that resembles waving eel grass or algae. *(a, James L. Castner; b, Doug Wechsler)*

or perhaps even millions, of symbiotic associations are all products of coevolution. Symbiosis takes three forms: mutualism, commensalism, and parasitism.

In mutualism, benefits are shared

Mutualism is a symbiotic relationship in which both partners benefit. The association between nitrogen-fixing bacteria of the genus *Rhizobium* and legumes (plants such as peas, beans, and clover) is an example of mutualism. Nitrogen-fixing bacteria, which live inside nodules on the roots of legumes, supply the plants with all the nitrogen they require to manufacture such nitrogen-containing compounds as chlorophylls, proteins, and nucleic acids (see Fig. 53–3*a*). The legumes supply sugars and other energy-rich organic molecules to their bacterial symbionts.

Another example of mutualism is the association between reef-building animals and dinoflagellates called zooxanthellae (see Chapters 28 and 54). These symbiotic algae live inside cells of the coral, where they photosynthesize and provide the animal with carbon and nitrogen compounds as well as oxygen (Fig. 52–6). Zooxanthellae also stimulate corals, causing calcium carbonate skeletons to form around their bodies much faster when the algae are present. The coral, in turn, supplies its zooxanthellae with waste products such as ammonia, which the algae use to make nitrogen compounds for themselves and their partners.

Mycorrhizae are mutualistic associations between fungi and the roots of plants belonging to about 90% of all plant families. The fungus absorbs essential minerals, especially phosphorus, from the soil and provides them to the plant. In return, the plant provides the fungus with organic molecules produced by photosynthesis. Plants grow more vigorously in the presence of mycorrhizal fungi (see Figs. 25–12 and 34–11), and they are better able to tolerate environmental stresses such as drought and high soil temperatures.

A remarkable example of mutualism occurs in the rain forests of South and Central America between flowering vines and ants or wasps. Those vines that produce bright red flowers are usually pollinated by hummingbirds, which are rewarded with nectar from the blossoms. Some insects, however,

(a)

(b)

Figure 52–5 **Batesian mimicry.** In this example, a fly *(a)* is the mimic and a yellowjacket wasp *(b)* is the model. *(a, b, James L. Castner)*

BATESIAN BUTTERFLIES DISPROVED

The monarch butterfly (*Danaus plexippus*) is an attractive insect found throughout much of North America *(see figure, left side)*. As a caterpillar, it feeds exclusively on milkweed leaves. The milky white liquid produced by the milkweed plant contains poisons that apparently do not harm the insect, but remain in its tissues for life. When a young bird encounters and eats its first monarch butterfly, it sickens and vomits. Thereafter, the bird avoids eating the distinctively marked insect.

Many people confuse the viceroy butterfly (*Limenitis archippus*) *(see figure, right side)* with the monarch. The viceroy, which is found throughout most of North America, is approximately the same size, and the color and markings of its wings are almost identical to those of the monarch. As caterpillars, viceroys eat willow and poplar leaves, which apparently do not contain poisonous substances.

During the past century, it was thought that the viceroy butterfly was a tasty food for birds, but that its close resemblance to monarchs gave it some protection against being eaten. In other words, birds that had learned to associate the distinctive markings and coloration of the monarch butterfly with its bad taste tended to avoid viceroys because they were similarly marked. The viceroy butterfly was therefore considered a classic example of Batesian mimicry.

In the journal *Nature* in 1991, biologists David Ritland and Lincoln Brower of the University of Florida reported the results of an experiment that tested the long-held notion that birds like the taste of viceroys but avoid eating them because of their resemblance to monarchs. They removed the wings of different kinds of butterflies—monarchs, viceroys, and several tasty species—and fed the seemingly identical wingless bodies to red-winged blackbirds. The results were surprising: monarchs and viceroys were equally distasteful to the birds.

As a result of this work, biologists are reevaluating the evolutionary significance of different types of mimicry. Rather than being an example of Batesian mimicry, monarchs and viceroys now appear to be an example of Müllerian mimicry, in which two or more different species that are distasteful or poisonous have come to resemble one another during the course of evolution. This likeness provides an adaptive advantage because predators learn quickly to avoid all butterflies with the coloration and markings of monarchs and viceroys. As a result, fewer butterflies of either species die, and more individuals survive to reproduce.

The butterfly study provides us with a useful reminder about the process of science. Expansion of knowledge in science is an ongoing enterprise, and newly acquired evidence helps scientists to reevaluate current models or ideas. Thus, scientific knowledge and understanding is not static, but continually changing.

Müllerian mimicry. Monarch *(left)* and viceroy *(right)* butterflies may be an example of Müllerian mimicry, in which two or more poisonous, harmful, or distasteful organisms resemble each other. *(Thomas C. Emmel)*

Figure 52–6 Mutualism. The greenish-brown specks in these coral polyps are zooxanthellae, algae that live symbiotically within the coral's tissue and supply the coral with carbon and nitrogen compounds. In return, the coral provides its zooxanthellae with nitrogen in the form of ammonia. *(P. Parks-OSF/Animals Animals)*

Figure 52–7 **Commensalism.** Epiphytes, small plants that grow attached to the body of a tree, are an example of commensalism. *(Carlyn Iverson)*

Figure 52–8 **Parasitism.** Tiny tracheal mites live in the breathing tubes of honeybees and suck their blood, weakening and eventually killing them. *(USDA, Agricultural Research Service)*

try to rob the flower of its nectar without performing the important task of pollination. The ants or wasps patrol the plant and attack any would-be nectar robber; the insect guards are rewarded with nectar produced by the leaves.

Commensalism is taking without harming

Commensalism is a type of symbiosis in which one organism benefits and the other one is neither harmed nor helped. One example of commensalism is the relationship between two kinds of insects, silverfish and army ants. Certain types of silverfish move along in permanent association with the marching columns of army ants and share the food caught in their raids. The army ants derive no apparent benefit or harm from the silverfish.

Another example of commensalism is the relationship between a tropical tree and its epiphytes, which are smaller plants attached to the bark of its branches (Fig. 52–7). The epiphyte anchors itself to the tree but does not obtain nutrients or water directly from it. Living on the tree enables it to obtain adequate light, water (as rainfall dripping down the branches), and required minerals (washed out of the tree's leaves by rainfall). Thus, the epiphyte benefits from the association, whereas the tree is apparently unaffected.

Parasitism is taking at another's expense

Parasitism is a symbiotic relationship in which one member, the *parasite,* benefits, while the other, the *host,* is adversely affected. The parasite lives in or on its host, from which it obtains nourishment. A parasite rarely kills its host directly but may weaken it, rendering it more vulnerable to predators, competitors, or abiotic factors. Parasitism is a successful lifestyle; by one estimate, more than two-thirds of all species are parasites. More than 100 parasites, from ticks to tapeworms, live in or on the human species alone!

Since the 1980s, wild and domestic honeybees in the United States have been dying off. Although habitat loss and pesticide use have contributed to the problem, tracheal mites (Fig. 52–8) and larger varroa mites have been a major reason for the honeybee decline. The number of commercial colonies has fallen by about 50% during the past several decades. Honeybees pollinate up to $10 billion of apples, almonds, and other crops each year and produce about $250 million of honey, so their decline is a major threat to U.S. agriculture. Entomologists are searching for mite-resistant native pollinators to replace the honeybee.

When a parasite causes disease and sometimes the death of a host, it is called a **pathogen.** Crown gall disease, which is caused by a bacterial pathogen, occurs in many different kinds of plants, and results in millions of dollars of damage each year to ornamental and agricultural plants. Crown gall bacteria, which also live on organic debris in the soil, enter plants through small wounds such as those caused by insects. They cause galls (tumor-like growths), often at a plant's crown, that is, the area between the stem and roots at or near the surface of the ground. Although plants seldom die from crown gall disease, they weaken, grow more slowly, and often succumb to other pathogens.

Competition may be intraspecific or interspecific

Competition occurs when two or more individuals attempt to use an essential common resource such as food, water, shelter, living space, or sunlight. Because resources are often in limited supply in the environment, their use by one individual decreases the amount available to others. If a tree in a dense forest grows taller than surrounding trees, for example, it is able to absorb more of the incoming sunlight. Less sunlight is therefore available for nearby trees that are shaded by the taller tree.

Competition can occur among individuals within a population (**intraspecific competition**) or between different species (**interspecific competition**). Intraspecific competition was discussed in Chapter 51.

Ecologists traditionally assumed that competition is the most important determinant of both the number of species found in a community as well as the size of each population (recall the discussion of carrying capacity in Chapter 51). Today ecologists recognize that competition is only one of many interacting biotic and abiotic factors that affect community structure. Furthermore, competition is not always a straightforward, direct interaction. A variety of flowering plants, for example, live in a young pine forest and presumably compete with the conifers for such resources as soil moisture and soil minerals. Their relationship, however, is more complex than simple competition. The flowers produce nectar that is consumed by some insect species that also prey on needle-eating insects, thereby reducing the number of insects feeding on pines. It is therefore difficult to assess the overall effect of flowering plants on pines. If the flowering plants were removed from the community, would the pines grow faster because they were no longer competing for necessary resources? Or would pine growth be inhibited by the increased presence of needle-eating insects (caused by fewer omnivorous insects)?

Short-term experiments in which one competing plant species is removed from a forest community often have demonstrated an improved growth for the remaining species. However, very few studies have tested the long-term effects on forest species of the removal of one competing species. These long-term effects may be subtle, indirect, and difficult to ascertain; they may lower or negate the negative effects of competition for resources.

THE NICHE IS AN ORGANISM'S ROLE IN THE COMMUNITY

We have seen that a diverse assortment of organisms inhabit each community and that these organisms obtain nourishment in a variety of ways. We have also examined some of the ways that species interact to form interdependent relationships within the community. A biological description of an organism includes whether it is a producer, consumer, or decomposer; whether it is a predator and/or a prey for other predators; the kinds of symbiotic associations it forms; and the nature of its intraspecific and interspecific competitive interactions. Other details are needed, however, to provide a complete picture.

Every organism is thought to have its own ecological role within the structure and function of a community; we call this role its **ecological niche.** An ecological niche, which is difficult to define precisely, takes into account all biotic and abiotic aspects of the organism's existence: all of the physical, chemical, and biological factors that the organism needs to survive, remain healthy, and reproduce. Among other things, the niche includes the physical surroundings in which an organism lives (its **habitat**). An organism's niche also encompasses what it consumes, what consumes it, the organisms with which it competes, and how it is influenced by the abiotic components of its environment (e.g., light, temperature, and moisture). The niche, then, represents the totality of an organism's adaptations to its environment, its use of resources, and the lifestyle to which it is suited. Thus, a complete description of an organism's ecological niche involves numerous dimensions.

The ecological niche of an organism may be far broader in potentiality than in actuality. An organism is usually capable of using much more of its environment's resources or of living in a wider assortment of habitats than it actually does. The potential ecological niche of an organism is its **fundamental niche,** but various factors such as competition with other species may exclude it from part of this fundamental niche. Thus, the lifestyle that an organism actually pursues and the resources that it actually uses make up its **realized niche.**

An example may help to make this distinction clear. The green anole (*Anolis carolinesis*), a lizard native to Florida and other southeastern states, perches on trees, shrubs, walls, or fences during the day waiting for insect and spider prey (Fig. 52–9*a*). In the past these little lizards were widespread in Florida. Several years ago, however, a related species, the brown anole (*Anolis sagrei*), was introduced from Cuba into southern Florida and quickly became common (Fig. 52–9*b*). Suddenly the green anoles became rare, apparently driven out of their habitat by competition from the slightly larger brown anoles. Careful investigation disclosed, however, that green anoles were still around. They were now confined largely to the vegetation in wetlands and to the leafy crowns of trees, where they were less easily seen.

The habitat portion of the green anole's fundamental niche includes the trunks and crowns of trees, exterior house walls, and many other locations. Once the brown anoles became established in the green anole's habitat, the green anoles were driven from all but wetlands and tree crowns; as a result of interspecific competition, their realized niche became much smaller than it had been (Fig. 52–9*c, d*). Because communities consist of numerous species, many of which compete to some extent, the complex interactions among them produce each species' realized niche.

Competition between species with overlapping niches may lead to competitive exclusion

When two species are similar, as are the green and brown anoles, their fundamental niches may overlap. However, according to one school of thought, no two species can indefinitely occupy the same niche in the same community because competitive exclusion eventually occurs. In **competitive exclusion,** one species is excluded from a niche by another as a result of interspecific competition. Although it is possible for different species to compete for some necessary resource without being total competitors, two species with absolutely iden-

(a)

(b)

(c)

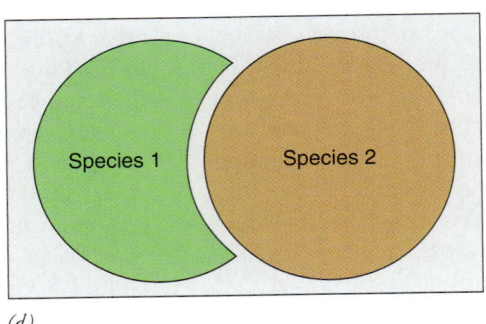
(d)

Figure 52–9 **Effect of competition on an organism's realized niche.** *(a)* The green anole is native to Florida. *(b)* The six-inch-long brown anole was introduced to Florida. *(c)* The fundamental niches of the two lizards overlap. Species 1 represents the green anole, and Species 2 represents the brown anole. *(d)* The brown anole out-competes the green anole in the area where their niches overlap, restricting the niche of the green anole. *(a, Ed Kanze/Dembinsky Photo Associates; b, Robert Clay/ Visuals Unlimited)*

tical ecological niches cannot coexist. Coexistence can occur, however, if the overlap between the two niches is reduced. In the lizard example, direct competition between the two species was reduced as the brown anole excluded the green anole from most of its former habitat.

The initial evidence that interspecific competition contributes to a species' realized niche came from a series of laboratory experiments by the Russian biologist G.F. Gause in 1934 (see descriptions of other experiments by Gause in Chapter 51). In one study Gause grew populations of two species of protozoa, *Paramecium aurelia* and the larger *P. caudatum* (Fig. 52–10). When grown in separate test tubes, the population of each species quickly increased to a high level and remained there for some time thereafter. When grown together, however, only *P. aurelia* thrived, while *P. caudatum* dwindled and eventually died out. Under different sets of culture conditions, *P. caudatum* prevailed over *P. aurelia*. Gause interpreted this to mean that although one set of conditions favored one species, a different set favored the other. Nonetheless, because both species were so similar, in time one or the other would eventually triumph at the other's demise.

Competition, then, has adverse effects on all species that use a limited resource, and may result in competitive exclusion of one or more species. It therefore follows that over time natural selection should favor individuals of each species that avoid or, at least, reduce competition by **resource partitioning,** differences in resource use among species. Evidence of re-source partitioning in animals is well documented and includes studies in tropical forests of Central and South America that demonstrate little overlap in the diets of fruit-eating birds, primates, and bats that coexist in the same habitat. Although fruits are the primary food for several hundred bird, primate, and bat species, the wide variety of fruits available have allowed fruit eaters to specialize, thereby reducing competition.

Resource partitioning also may include timing of feeding, location of feeding, nest sites, and other aspects of an organism's ecological niche. Robert MacArthur's study of five North American warbler species is a classic example of resource partitioning (Fig. 52–11). Although it initially appeared that their niches were nearly identical, MacArthur determined that individuals of each species spend most of their feeding time in different portions of the spruces and other conifer trees they frequent. They also move in different directions through the canopy, consume different combinations of insects, and nest at slightly different times.

Apparent contradictions to the competitive exclusion principle exist. In Florida, for example, native and introduced (non-native) fish seem to coexist in identical niches. Similarly, botanists have observed closely competitive plant species in the same location. Although such situations seem to contradict the concept of competitive exclusion, the realized niches of these organisms may differ in some way that biologists do not yet understand (as with the warblers before MacArthur studied them).

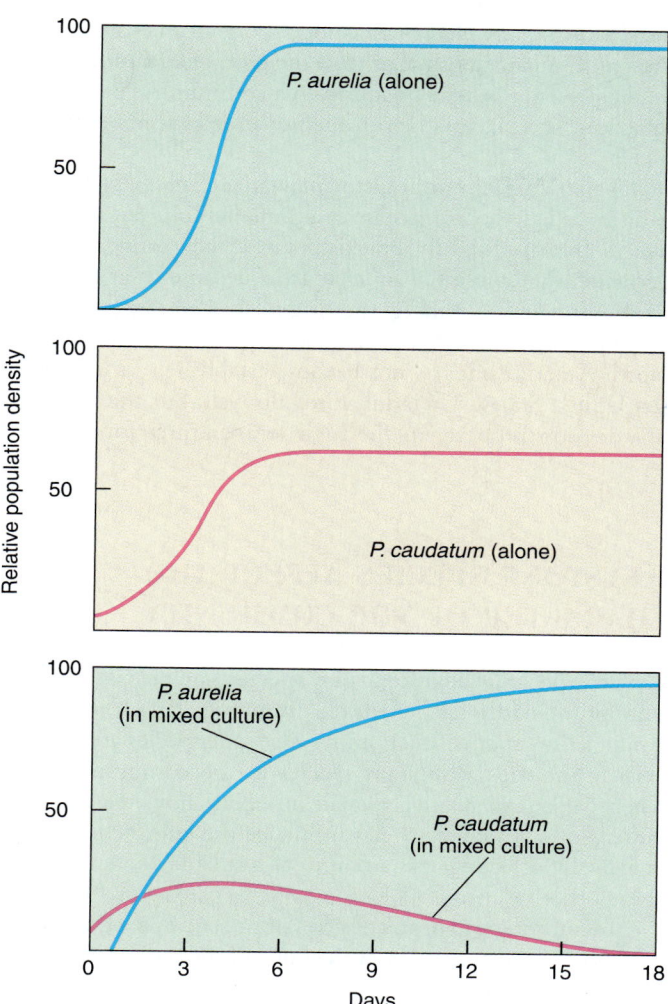

Figure 52–10 **Interspecific competition.** Competition between two species of *Paramecium* was studied by G.F. Gause. The top two graphs show how a population of each species of *Paramecium* flourishes when grown alone; the bottom graph shows how they grow together, in competition with each other. *(After Gause)*

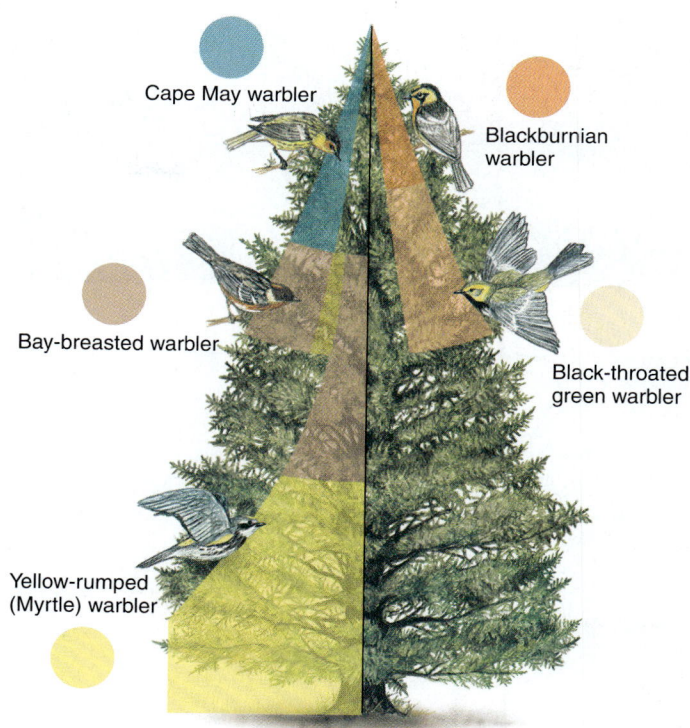

Figure 52–11 **Resource partitioning.** Each warbler species spends at least half its foraging time in the designated area of a spruce tree. *(After MacArthur)*

Character displacement reduces interspecific competition

Sometimes populations of two similar species occur both sympatrically and allopatrically (see Chapter 19). Where their geographical distributions overlap, the two species tend to differ more in their structural, ecological, and behavioral characteristics than they do where each occurs in separate geographical areas. Such divergence in traits in two species living in the same geographical area is known as **character displacement.** It is thought that character displacement reduces competition between two species, since their differences give them somewhat different ecological niches in the same environment.

There are several well documented examples of character displacement between two closely related species. For exam-

ple, the flowers of two species of *Solanum* in Mexico are quite similar in areas where either one or the other occurs, but in areas where their distributions overlap, the two species differ significantly in flower size and are pollinated by different kinds of bees. In other words, character displacement reduces interspecific competition, in this case for the same animal pollinator.

The bill sizes of Darwin's finches provide another example of character displacement (Fig. 52–12). On large islands in the Galapagos where the medium ground finch *(Geospiza fortis)* and the small ground finch *(G. fuliginosa)* occur sympatrically (together), their bill depths are distinctive: *G. fuliginosa* has a smaller bill depth that enables it to crack small seeds, whereas *G. fortis* has a larger bill depth that enables it to crack medium-sized seeds. However, *G. fortis* and *G. fuliginosa* are also found allopatrically (separately) on other islands. Where the two species live separately, bill depths tend to be the same intermediate size, perhaps because there is no competition from the other species.

Although these examples may be explained in terms of character displacement, it was only recently, in 1994, that an experiment was conducted to test the character displacement hypothesis (see *On the Cutting Edge: Experimental Evidence of Character Displacement*).

(a)

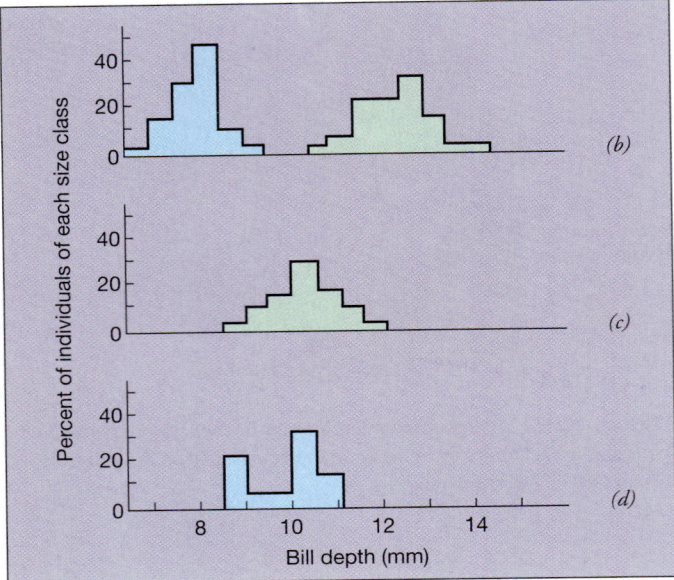

(b)

(c)

(d)

Figure 52–12 **Character displacement.** *(a)* Bill depth in two species of Galapagos Island finches, *Geospiza fuliginosa* and *G. fortis.* *(b)* Where the two species are found on the same island, *G. fuliginosa* *(blue)* has a smaller average bill depth than *G. fortis (green)*. Where they occur on separate islands *(c, d)*, the average bill depths of each are similar.

Limiting factors restrict an organism's ecological niche

The environmental factors that actually determine a species' ecological niche can be extremely difficult to identify. For this reason, the concept of ecological niche is largely abstract, although some of its dimensions can be experimentally determined. Any environmental variable that tends to restrict the ecological niche of an organism is called a **limiting factor.**

What factors actually determine an ecological niche? A niche is basically determined by the total of a species' structural, physiological, and behavioral adaptations. Such adaptations determine, for example, the tolerance of an organism for environmental extremes. If any feature of its environment lies outside the bounds of its tolerance, then the organism cannot live there. Just as you would not expect to find a cactus living in a pond, you would not expect water lilies in a desert.

Most of the limiting factors that have been studied are simple variables such as the soil's mineral content, temperature extremes, and precipitation amounts. Such investigations have

disclosed that any factor exceeding an organism's tolerance, or present in quantities smaller than the required minimum, limits the presence of that organism in a community. By their interaction, such factors help to define an organism's ecological niche.

A particular limiting factor may affect only part of an organism's life cycle. For instance, although adult blue crabs can live in fresh or slightly brackish water, they cannot become permanently established in such areas because their larvae require salt water. Similarly, the ring-necked pheasant, a popular game bird native to Eurasia, has been widely introduced in North America but has not become established in the southern United States. The adult birds do well, but the eggs cannot develop properly in the high temperatures found there.

KEYSTONE SPECIES AFFECT THE CHARACTER OF THE COMMUNITY

Certain species, called **keystone species,** are crucial in determining the nature of the entire community, that is, its species composition and its functioning. Keystone species are not necessarily the most abundant species in the community. Even when present in relatively small numbers, however, the individuals of a keystone species influence the entire community, usually by affecting the amount of available food, water, or some other resource.

Identifying and protecting keystone species is a crucial goal of conservation biology because if one of them disappears from a community, many other species in that community may also go extinct. Because fig trees produce a continuous crop of fruits, they appear to be keystone species in tropical rain forests of Central and South America. Monkeys, fruit-eating birds, bats, and other vertebrates of the forest do not normally consume large quantities of figs in their diets. During that time of the year when other fruits are less plentiful, however, fig trees become important in sustaining fruit-eating vertebrates. Should the fig trees disappear, most of the fruit-eating vertebrates would also be eliminated. In turn, should the fruit-eaters disappear, the spatial distribution of other fruit-bearing plants would become more limited because the fruit-eaters help disperse their seeds. Thus, protecting fig trees in such tropical rainforest ecosystems increases the likelihood that many other species will survive.

Another example of a keystone species is a top predator such as the gray wolf. Where wolves were hunted to extinction, for example, the populations of elk, deer, and other larger herbivores increased explosively. As these herbivores overgrazed the vegetation, many plant species that could not tolerate such grazing pressure disappeared. Smaller animals such as rodents, rabbits, and insects declined in number because the plants that they depended on for food became less abundant. The number of foxes, hawks, owls, and badgers that prey on these small animals decreased, as did the number of ravens, eagles, and other scavengers that eat wolf kill. Thus, the disappearance of

Experimental Evidence of Character Displacement

HYPOTHESIS: Interspecific competition in stickleback fish promotes character displacement.

METHOD: Place a species of stickleback fish that includes a range of phenotypes (the Cranby species) into two divided experimental ponds. Add individuals of a potential competitor species (the limnetic species) to one separated half of each pond.

RESULTS: In the presence of the limnetic species, the Cranby individuals that were least like the limnetic species were more likely to survive and grow.

CONCLUSION: Evidence was obtained of natural selection against phenotypes most similar to the competitor species. Such natural selection is consistent with the hypothesis of character displacement.

There are many examples in nature of similar species that are markedly different where they are found together, but very alike (yet distinguishable, at least by experts) where each is found alone. For example, two species of threespine stickleback fish, a limnetic species and a benthic species, occur in several lakes of coastal British Columbia. (*Limnetic* organisms live in the shallow, open water of a lake; *benthic* organisms live on the bottom.) Where they are found together, individuals belonging to the limnetic species are small, have narrow mouths, and eat plankton in the open water. Individuals belonging to the benthic species, on the other hand, are large, have wide mouths, and feed on invertebrates along the shoreline. In lakes where each species occurs alone, the fish are "generalists": they are intermediate in form and eat both plankton and invertebrates.

The hypothesis of character displacement, which has been advanced to explain these observations in threespine sticklebacks and many other organisms, had never been experimentally tested until recently. Dolph Schluter, a researcher at the University of British Columbia who had previously made extensive observations on natural threespine stickleback populations, designed just such an experiment.* He chose to study two species.† One was the limnetic species from a two-species lake. The other, referred to as the Cranby species, was from a nearby lake where it occurs alone. Prior to beginning his experiments, Schluter performed a series of genetic crosses in the laboratory to make the Cranby species, which exhibits the intermediate phenotype, more variable. After the crosses, the Cranby species contained a range of phenotypes, from benthic-like to intermediate to limnetic-like.

Schluter began his experiments by dividing two artificial ponds into separate halves. (Two ponds were used to provide a replication of the experiment.) He placed 1800 young Cranby individuals into each half-pond. To one of the halves of each pond he also added 1200 young limnetic individuals; these were the experimental halves. The half-ponds with only Cranby individuals were the control halves.

After 90 days, the fish were harvested and preserved. About 65 randomly chosen Cranby individuals from each group were sized (measured for length), and their phenotypes were noted. There were no significant differences in survival and growth among the three types of Cranby individuals when they were grown alone.

*Schluter, D. "Experimental Evidence That Competition Promotes Divergence in Adaptive Radiation." *Science,* Nov. 1994.
†Threespine stickleback fish were once thought to be a single species, *Gasterosteus aculeatus.* Although several species are now known to exist, all are currently lumped together in the *Gasterosteus aculeatus* complex because their taxonomic relationships have not been determined.

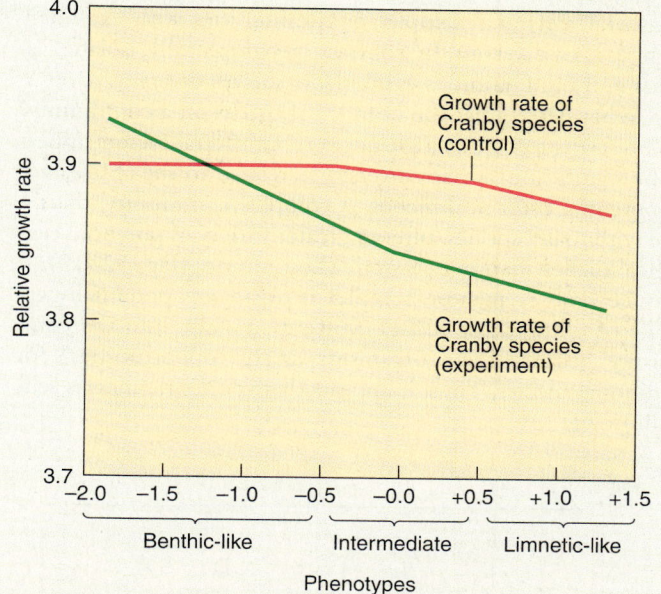

Experimental evidence of character displacement. A comparison of growth rates for the Cranby species alone *(control),* and in the presence of the limnetic species *(experiment).* The structural index reflects the range of Cranby phenotypes, from those most benthic-like (−2.0) to those most limnetic-like (+1.5). Values for relative growth rates were obtained statistically.   *(Courtesy of Dolph Schluter)*

However, in the presence of the limnetic species, Cranby individuals with a limnetic phenotype exhibited significantly lower rates of growth and survival than did the benthic and intermediate types *(see figure).*

Because only one generation was studied, this experiment investigated only natural selection, the first step of the process of character displacement. The next step, an evolutionary change in gene frequencies over successive generations, will require a long-term evaluation of several generations.

This experiment has been both hailed as a landmark and criticized for its design and analysis. For example, some have argued that the experimental half-ponds had much higher population densities (1800 plus 1200 fish) than the control half-ponds (only 1800 fish) and that this might have affected the results. Further refinements of the experimental design, including studies on multiple generations, are expected to resolve some of these concerns and to raise many interesting new questions.

the wolf resulted in ecosystems with considerably less biological diversity. The reintroduction of wolves to Yellowstone National Park in early 1995 has given biologists a unique opportunity to study the ecological impact of a keystone species.

COMMUNITIES VARY IN SPECIES RICHNESS

Species richness, the number of species within a community, varies greatly from one community to another and is influenced by many biotic and abiotic factors. Tropical rain forests and coral reefs are examples of communities with extremely high species diversity. In contrast, geographically isolated islands and mountaintops exhibit low species diversity.

What determines the number of species in a community? There seems to be no single answer, but several factors appear to be significant. Species diversity is related to the abundance of potential ecological niches. A complex community, that is, one with many species, offers a greater variety of potential ecological niches than does a simple community (Fig. 52–13). It may become even more complex if organisms potentially capable of filling those niches evolve or migrate into the community because these organisms create "opportunities" for additional species. Thus, it appears that species richness is self-reinforcing.

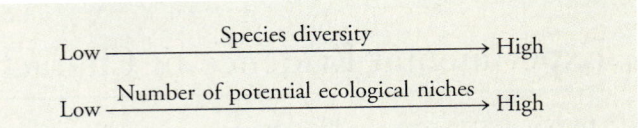

Species diversity is inversely related to the geographical isolation of a community. Isolated island communities tend to be much less diverse than are communities in similar environments on continents, in part because of the difficulty encountered by many species in reaching and successfully colonizing the island (the *distance effect;* Fig. 52–14). Also, sometimes species become locally extinct as a result of random events. In isolated habitats such as islands or mountaintops, locally extinct species are not readily replaced. Isolated areas are likely to be small and to possess fewer potential ecological niches.

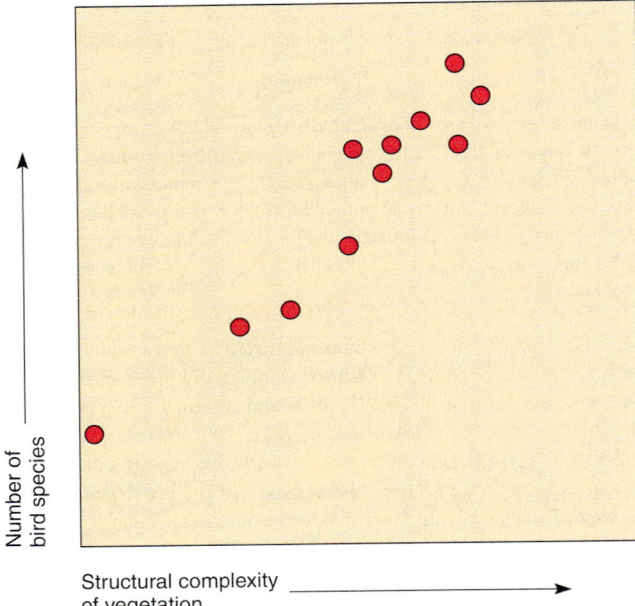

Figure 52–13 Effect of community complexity on species diversity. A community in which the vegetation is structurally complex—a forest, for example—provides birds with more kinds of food and hiding places than does a grassland. *(After MacArthur and MacArthur, 1961)*

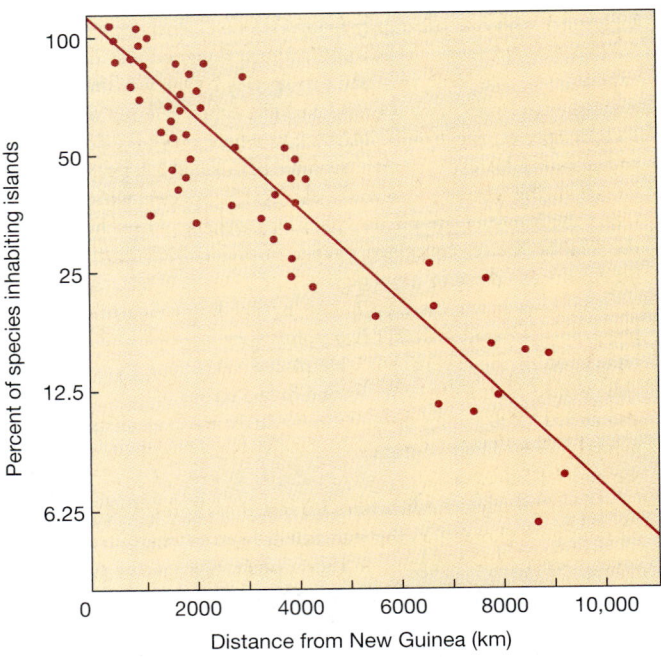

Figure 52–14 The distance effect. This graph shows that the percent of bird species on islands in the South Pacific is related to their distance from the large island of New Guinea, which is a source of colonizing species for these islands. Note that species diversity declines as the distance from New Guinea increases. *(After J.M. Diamond, Proceedings of the National Academy of Sciences, Vol. 69, 1972)*

Generally, species diversity is inversely related to the environmental stress of a habitat. Only those species capable of tolerating extreme conditions can live in an environmentally stressed community. Thus, the species diversity of a polluted stream is low compared to that of a nearby pristine stream. Similarly, the species diversity of high-latitude (further from the equator) communities with harsh climates is lower than that of lower-latitude (closer to the equator) communities with milder climates. Known as the *species richness–energy hypothesis,* it suggests that different latitudes affect species richness due to variations in solar energy. Greater energy, for example, may permit more species to coexist in a given region. Although the equatorial countries of Colombia, Ecuador, and Peru occupy only 2% of Earth's land, they contain an astonishing 45,000 native plant species. In contrast, the continental United States and Canada, with a significantly larger land area, possess a total of 19,000 native plant species. Ecuador alone contains more than 1300 native species of birds—twice as many as the United States and Canada combined.

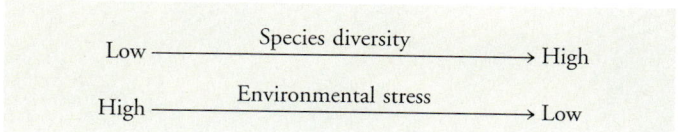

Species diversity is usually greater at the margins of distinct communities than it is in their centers. This is because **ecotones,** transitional zones where two or more communities meet, contain all or most of the ecological niches of the adjacent communities as well as some unique to the ecotone (see Chapter 54). This change in species composition produced at ecotones is known as the **edge effect.**

Species diversity is reduced when any one species enjoys a decided position of dominance within a community that enables it to appropriate a disproportionate share of available resources, thus crowding out, or outcompeting, other species. Ecologist James H. Brown of the University of New Mexico addressed species composition and diversity in experiments conducted from 1977 to 1994 in the Chihuahuan desert of southeastern Arizona. In one experiment, the removal of three dominant species, all kangaroo rats, from several plots resulted in an increased diversity of rodent species. This increase was ascribed both to lowered competition for food and also to an altered habitat, because the abundance of grasses increased dramatically after the removal of the kangaroo rats.

Species diversity is greatly affected by geological history. Tropical rain forests are thought to be old, stable communities that have undergone relatively few climatic changes through Earth's entire history. During this time, myriad species evolved in tropical rain forests, having experienced few or no abrupt climatic changes that might have led to their extinction. In contrast, glaciers have repeatedly altered temperate and arctic regions during Earth's history. An area recently vacated by glaciers has a low species diversity because few species have as yet had a chance to enter it and become established.

Species richness probably causes community stability

Ecologists have traditionally assumed that community stability, that is, the ability of a community to withstand disturbances, was a consequence of community complexity. That is, a community with considerable species diversity was thought to be more stable than a community with less species diversity. According to this view, the greater the species diversity, the less critically important any single species should be. With many possible interactions within the community, it appeared unlikely that any single disturbance could affect enough components of the system to make a significant difference in its functioning. Evidence for this hypothesis can be found in the fact that destructive outbreaks of pests are more common in cultivated fields, which are low-diversity communities, than in natural communities with greater species diversity. As another example, the almost complete loss of the American chestnut tree to the chestnut blight fungus had little ecological impact on the moderately diverse Appalachian woodlands of which it used to be a part.

Ongoing studies by David Tilman of the University of Minnesota and John Downing of the University of Iowa, however, have strengthened the link between species richness and community stability. In their initial study, reported in the journal *Nature* in 1994, they established and monitored 207 plots of Minnesota grasslands for seven years. During the study period, Minnesota's worst drought in 50 years occurred (1987–88). The biologists found that those plots with the greatest number of plant species lost less ground cover and recovered faster than species-poor plots. Additional work published in 1996 in the journal *Ecology* supported these conclusions and showed a similar effect of species richness on community stability during nondrought years.

Although species richness increases community stability, populations of individual species within a diverse community may vary significantly from year to year. It may seem paradoxical that instability within populations of individual species relates to the stability of the entire community. When one considers all the interactions among the organisms in a community, however, it is obvious that some species benefit at the expense of others; if one species declines in a given year, for example, other species that compete with it may flourish. Tilman's group is currently doing additional research on the relationship between species richness and other aspects of community stability, such as adaptability to predatory insects or diseases.

SUCCESSION IS COMMUNITY CHANGE OVER TIME

A mature community does not spring into existence full-blown but develops gradually, through a series of stages, each dominated by different organisms. The process of sequential community development over time, which involves species in one stage being replaced by different species, is called **succession.** An area is initially colonized by certain species that are replaced over time by others, which themselves may be replaced much later by still others.

Ecologists initially thought that succession inevitably led to a stable and persistent community, known as a *climax community*. But more recently, this traditional view has fallen out of favor. The apparent stability of a climax forest, for example, is probably the result of how long trees live relative to the human lifespan. Mature climax communities are not in a state of permanent equilibrium, but rather in a state of continual flux. A mature community changes in species composition and in the relative abundance of each species despite the fact that it retains a relatively uniform appearance to the untrained eye.

Succession is usually described in terms of the changes in the species composition of an area's vegetation, although each successional stage also has its own characteristic kinds of animals and other organisms. The time involved in ecological succession is on the order of tens, hundreds, or thousands of years, not the millions of years involved in the evolutionary time scale.

Sometimes a community develops in a "lifeless" environment

Primary succession is the change in species composition over time in a habitat that was not previously inhabited by organisms. No soil exists when primary succession begins. Bare rock surfaces, such as recently formed volcanic lava and rock scraped clean by glaciers, are examples of sites where primary succession might occur.

During primary succession on bare rock, the rock is eventually transformed into soil

Although the details vary from one site to another, in primary succession on bare rock one might first observe a community of lichens, which form through the symbiotic relationship between a fungus and an alga or cyanobacterium (Fig. 52–15*a*). Lichens are often the most important element in the **pioneer community,** which is the first community to appear in a succession. Lichens secrete acids that help to break the rock apart, which is how soil starts to form. Lichens also add valuable organic materials to the young soil. Over time, the lichen community may be replaced by mosses and drought-resistant ferns, followed in turn by tough grasses and herbs. Once sufficient soil has formed, grasses and herbs may be replaced by low shrubs (Fig. 52–15*b*), which in turn may be replaced by forest trees (Fig. 52–15*c*) in several distinct stages. Primary succession on bare rock, which may take hundreds or thousands of years, might proceed from a pioneer community to a forest community as follows:

Lichens ⟶ mosses ⟶ grasses ⟶ shrubs ⟶ trees

(a)

(b)

(c)

Figure 52–15 **Primary succession after the retreat of glaciers.** Although these photos were not taken in the same area, they show some of the stages of primary succession on glacial moraine (rocks, gravel, and sand deposited by a glacier). *(a)* The barren landscape exposed after the retreat of the glacier is initially colonized by lichens, then mosses. *(b)* At a later time, dwarf trees and shrubs (alder, cottonwood, and willow) colonize the area. *(c)* Still later, hemlocks and spruces dominate the community. *(a, c, Wolfgang Kaehler; b, Glenn N. Oliver/Visuals Unlimited)*

Figure 52–16 **Primary succession on sand dunes.** Empire Bluffs, Sleeping Bear Sand Dunes National Lakeshore, Michigan, shows several stages of primary succession on sand dunes. The first plants to colonize a sand dune are grasses *(foreground)*, followed by low-growing shrubs *(midground on left)*, and later by trees *(background)*. *(Joe Sroka/Dembinsky Photo Associates)*

Primary succession occurs on sand dunes

Some lake and ocean shores have extensive sand dunes that are deposited by wind and water. At first these dunes are blown about by the wind. The sand dune environment is severe, with high temperatures during the day and low temperatures at night. The sand may also be deficient in certain mineral nutrients needed by plants. As a result, few plants can tolerate the environmental conditions of a sand dune.

Grasses are common pioneer plants on sand dunes around the Great Lakes. As the grasses extend over the surface of a dune, their roots hold it in place, helping to stabilize it (Fig. 52–16). At this point, mat-forming shrubs may invade, further stabilizing the dune. Much later, shrubs may be replaced by pines, which years later are replaced by oaks. Because the soil fertility remains low, oaks are rarely replaced by other forest trees. A summary of how primary succession on sand dunes around the Great Lakes might proceed is:

Grasses ⟶ shrubs ⟶ pine trees ⟶ oak trees

Sometimes communities develop where a previous community existed

Secondary succession is the change in species composition over time in a habitat already substantially modified by a pre-existing community. Soil is already present at these sites. An open area caused by a forest fire or an abandoned agricultural field are common examples of sites where secondary succession occurs.

Secondary succession has been studied extensively in abandoned farmland, particularly in North Carolina. Although it takes more than 100 years for secondary succession to occur at a single site, it is possible for a single researcher to study old field succession in its entirety by observing different sites in the same area that have been undergoing succession for different amounts of time. By examining court records, the biologist can accurately determine when each field was abandoned.

Abandoned farmland in North Carolina is colonized by a predictable succession of communities. The first year after cultivation ceases, the field is dominated by crabgrass. During the second year, horseweed, a larger plant that outgrows crabgrass, is the dominant species. Horseweed does not dominate for more than one year, however, because decaying horseweed roots inhibit the growth of young horseweed seedlings. In addition, horseweed does not compete well with the other plants, such as broomsedge, ragweed, and aster, that appear and become established during the third year. Typically, broomsedge, which is drought-tolerant, outcompetes aster, which is not.

In 5 to 15 years, the dominant plants in an abandoned field are pines such as shortleaf pine and loblolly pine. Through the buildup of litter (pine needles and branches) on the soil, pines produce conditions that cause the earlier dominant plants to decline in importance. Over time, pines give up their dominance to hardwoods such as oaks. The replacement of pines by oaks depends primarily on the environmental alterations produced by the pines. The pine litter causes soil changes, such as an increase in water-holding capacity, that are necessary for young hardwood seedlings to become established. In addition, hardwood seedlings are more tolerant of shade than are pine seedlings. Secondary succession on abandoned farmland might proceed as follows:

Crabgrass ⟶ horseweed ⟶ perennial weeds ⟶ pine trees ⟶ hardwood trees

Animal life also changes during secondary succession

As secondary succession proceeds, a progression of animal life follows the changes in vegetation. Although a few animals, the short-tailed shrew, for example, are found in all stages of abandoned farmland succession in North Carolina, most animals appear with certain stages and disappear with later stages. During the crabgrass and weed stages of secondary succession, the habitat is characterized by open fields that support grasshoppers, meadow mice, cottontail rabbits, and birds such as grasshopper sparrows and meadowlarks. As young pine seedlings become established, animals of open fields give way to animals common in mixed herbaceous and shrubby habitats. Now white-tailed deer, white-footed mice, ruffed grouse, robins, and song sparrows are common, whereas grasshoppers, meadow mice, grasshopper sparrows, and meadow larks decline. As the pine seedlings grow into trees, animals of the forest replace those common in mixed herbaceous and shrubby habitats. Cottontail rabbits give way to red squirrels; and ruffed grouse, robins, and song sparrows are replaced by warblers and veeries. Thus, each stage of succession supports its own ecological niches that are occupied by characteristic animal life.

ECOLOGISTS CONTINUE TO STUDY COMMUNITY STRUCTURE

F. E. Clements, an ecologist who was professionally active during the first third of the 20th century, was struck by the worldwide uniformity of large tracts of vegetation, for example, tropical rain forests in South America, Africa, and Southeast Asia (see Chapter 54). He also noted that even though the species composition of a community in a particular habitat might be different from that of a community in a habitat with a similar climate elsewhere in the world, overall the components of the two communities were usually similar. He viewed communities as something like compound organisms, "superorganisms" whose member species cooperated with one another in a manner that resembled the cooperation of the parts of an individual organism's body. Clements' view was that a community went through certain stages of development, like those of an organism, and eventually reached an adult state; the developmental process was succession, and the adult state was the climax community. This cooperative view of the community, called the **organismic model,** stresses the interaction of the members, which tend to cluster in groups within discrete community boundaries (Fig. 52–17a).

Opponents of the organismic model, particularly H. A. Gleason, have held that biological interactions are less important in the production of communities than are environmental gradients (such as climate and soil) or even chance. Indeed, the concept of a community is questionable. It might be a classification category with no reality outside the minds of ecologists, reflecting little more than the tendency of organisms with similar environmental requirements to live in similar places. This school of thought, called the **individualistic model,** emphasizes species individuality, with each species having its own particular abiotic living requirements. It holds that communities are therefore not interdependent associations of organisms; rather, each species is independently distributed across a continuum of areas that meets its own individual requirements (Fig. 52–17b).

Studies testing the organismic and individualistic hypotheses of communities do not seem to support Clements' interactive concept. Instead, most studies favor the individualistic model. Tree species in Wisconsin forests, for example, are distributed in a continuum from wet to dry environments (Fig. 52–17c). Keep in mind, however, that support for the individualistic model does not mean that species do not interact or form important associations within a community.

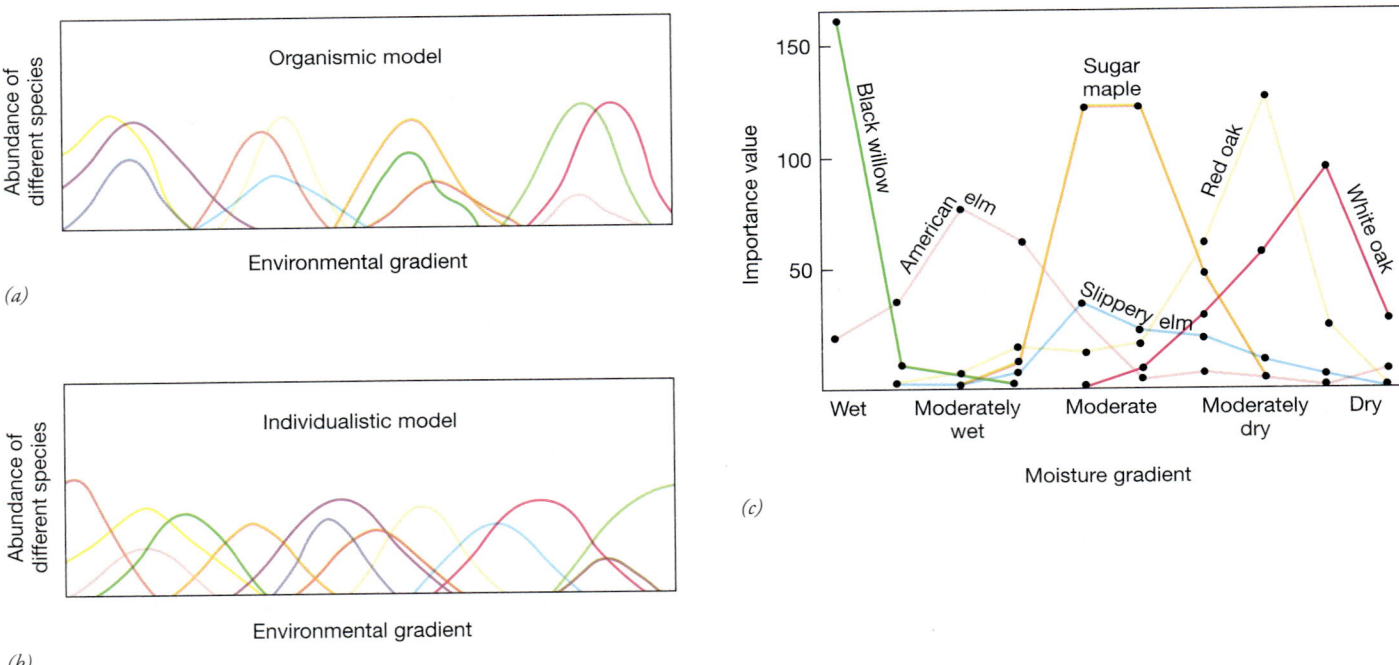

Figure 52–17 **Diagrammatic representations of species' distributions along an environmental gradient.** *(a)* The hypothesized organismic model. *(b)* The hypothesized individualistic model. *(c)* Tree species in Wisconsin forests are distributed along a moisture gradient. The "importance value" of each species in a given location combines three aspects (density, frequency, and size). *(Adapted from Curtis, 1959, p. 99, Copyright by the Regents of the University of Wisconsin)*

ANCIENT COMMUNITIES FORM A CONTINUUM WITH MODERN COMMUNITIES

You learned in Chapter 17 that the study of the geographical distribution of plants and animals is called **biogeography.** Biogeographers search for patterns in geographical distribution and try to explain how such patterns arose, including where populations originated, how they spread, and when. Biogeographers recognize that geological and climate changes such as mountain building, continental drift (see Chapter 19), and periods of extensive glaciation influence the distribution of species. Biogeography is linked to evolutionary history and provides insights into how organisms may have interacted in ancient communities. Studying biogeography helps us relate ancient communities to modern communities that currently exist in a given area.

One of the basic tenets of biogeography is that each species originated only once. The particular place where this occurred is known as the species' **center of origin.** The center of origin is not a single point but the distribution of the population when the new species originated. From its center of origin, each species spread until halted by a barrier of some kind, such as an ocean, desert, or mountain range; unfavorable climate; or the presence of organisms that compete with it for food or shelter.

Most plant and animal species have characteristic geographical distributions. The **range** of a particular species is that portion of the Earth in which it is found. The range of some species may be a relatively small area, as for example, wombats, which are found only in drier parts of Australia and nearby islands. Such localized, native species are said to be **endemic,** that is, they are not found anywhere else in the world. In contrast, some species have a nearly worldwide distribution and occur on more than one continent or throughout much of the ocean. Such species are said to be **cosmopolitan.**

One of the early observations of biogeographers is that the ranges of different species do not include everywhere that they could survive. Central Africa has elephants, gorillas, chimpanzees, lions, antelopes, umbrella trees, and guapiruvu trees, whereas areas in South America with a similar climate have none of these. These animals and plants, which originated in Africa after continental drift had already separated the supercontinent Pangaea into several land masses, could not expand their range into South America because the Atlantic Ocean was an impassible barrier. Likewise, the ocean was a barrier to South American monkeys, sloths, tapirs, balsa trees, snakewood trees, and many palms, none of which is found in Africa.

Land areas are divided into six biogeographical realms

As the various continents were explored and their organisms studied, biologists observed that the world could be divided into major blocks of vegetation, such as forests, grasslands, and deserts, and that these vegetation types corresponded to specific climates (see Chapter 54). The relationship between animal distribution, geography, and climate was not deduced until 1876 when Alfred Wallace, who also developed the same theory of evolution as Charles Darwin (see Chapter 17), divided Earth's land areas into six major biogeographical realms: the Palearctic, Nearctic, Neotropical, Ethiopian, Oriental, and Australian (Fig. 52–18). Each of the six biogeographical realms is separated from the others by a major barrier, such as a mountain range, desert, or ocean, that helps maintain each region's biological distinctiveness. Wallace's classification was quickly embraced by many biologists and is still considered valid, except that human activities, such as the intentional and unintentional introduction of foreign species, are contributing to a homogenization of the biogeographical realms (see Chapter 55).

Refer to Figure 52–18 as we briefly consider some of the characteristic animals in each realm. The *Nearctic* and *Palearctic realms* are more closely related than the other regions, especially in their northern parts where they share many animals such as wolves, hares, and caribou.

The *Neotropical realm* was almost completely isolated from the Nearctic realm and other land masses for almost all of the past 70 million years; during this time, many marsupial species evolved. The isthmus of Panama, which formed about 3 million years ago, linked North and South America and provided a route for animal migrations. Only three species, the opossum, armadillo, and porcupine, are descendants of animals that survived the northward migration from South America, but many species, such as the tapir and llama, are descendants of animals that survived the southward migration from North America. These species caused many of South America's marsupial species to go extinct.

The *Ethiopian realm,* which is separated from other land masses by the Sahara Desert, contains the most varied vertebrates of all six realms. Some overlap exists between the Ethiopian and *Oriental realms* because a land bridge with a moist climate linked Africa to Asia during the Miocene and Pliocene epochs (see Chapter 20). The Oriental realm has the fewest endemic species of all the tropical realms.

The *Australian realm* has not had a land connection with other regions for more than 85 million years. It has no placental mammals and instead is dominated by marsupials and monotremes, including the duck-billed platypus and the spiny anteater. Adaptive radiation of the marsupials during their long period of isolation led to species with ecological niches similar to those of placental mammals of other realms.

The six biogeographical realms are an example of classical biogeography. We now discuss an example of how contemporary research impacts both community ecology and biogeography.

Analysis of mitochondrial DNA is a useful biogeographical tool

Mitochondrial DNA data can provide insights about an area's biogeographical history. Consider Florida, a peninsula bor-

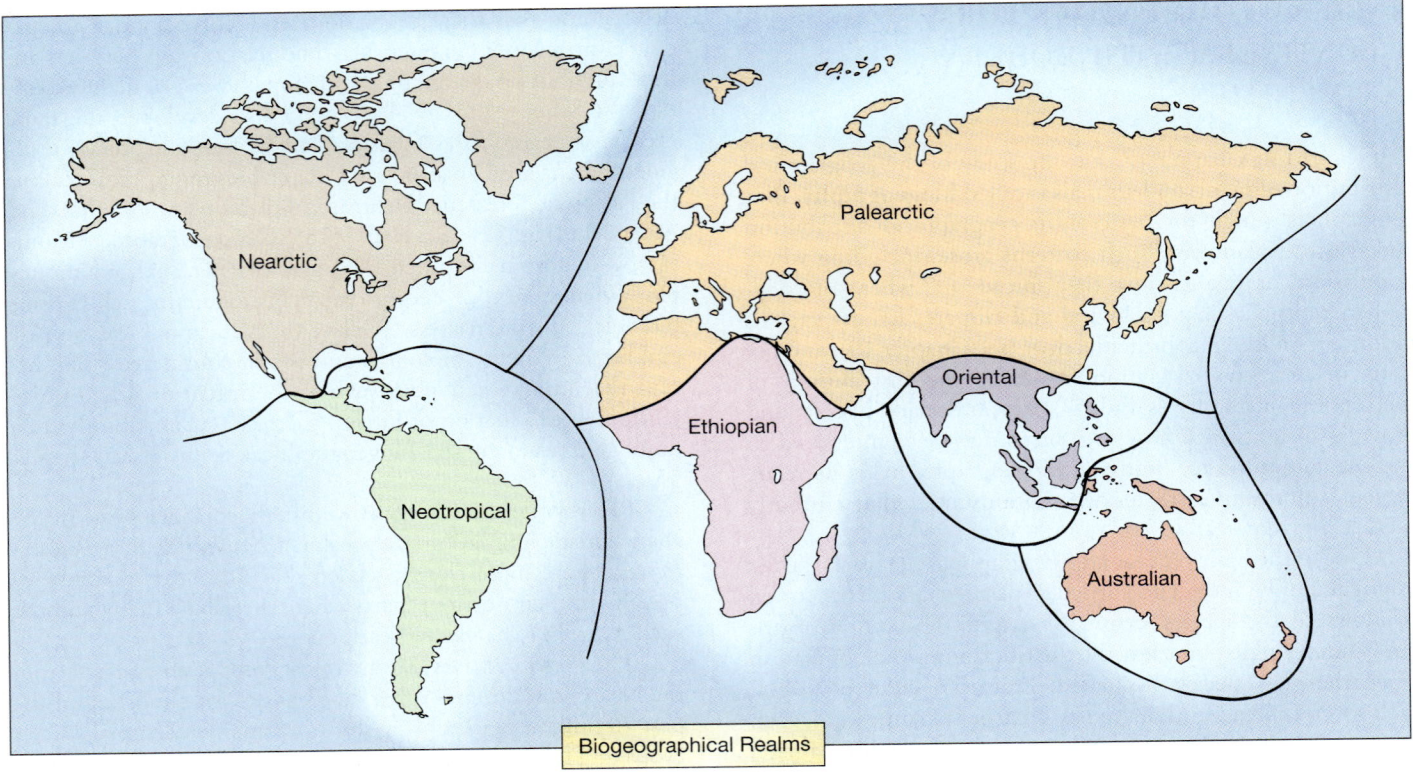

Figure 52–18 Wallace's biogeographical realms. These six biogeographical realms are characterized by the presence of certain unique species whose distributions are the direct outcome of their centers of origin, past migrations, and the barriers that they encountered.

dered by the Atlantic Ocean on the east and the Gulf of Mexico on the west. During the Pleistocene epoch, recurring periods of glaciation caused the sea level of the Atlantic Ocean and the Gulf of Mexico to lower as snow and ice accumulated on the land. Florida had a larger land area that experienced drier and cooler conditions during these periods of glaciation; when the glaciers receded, Florida's land area was smaller and its climate was moister and warmer. The ranges of Florida's plant and animal species presumably varied in response to these changes in climate, but exactly how their ranges were affected was not clear until mitochondrial DNA data helped reconstruct Florida's biogeographical history.

During the 1990s the mitochondrial DNA of many Florida fish, oyster, terrapin, crab, rodent, and bird species was analyzed and revealed that each species has distinctive eastern and western populations. These genetic patterns may be evidence that ancient climate changes in Florida produced some type of east–west barrier running the length of the state. The barrier effectively divided each species into smaller, geographically separate populations that diverged (evolved differences) from each other during their separation. (It should be noted that although mtDNA data indicate an east–west barrier, biogeographers have not yet deduced the nature of the barrier from Florida's climate and ecological history.)

SUMMARY WITH KEY TERMS

I. A biological **community** consists of a group of organisms of different species that interact and live together. A community and its abiotic environment compose an **ecosystem.**

II. The major roles of organisms in communities are those of autotrophs and heterotrophs.

   A. **Producers** are the photosynthetic autotrophs at the base of most food webs. They include plants and algae.

   B. **Consumers** are heterotrophs that feed on other organisms. **Primary consumers,** or **herbivores,** feed on plants; **secondary consumers,** or **carnivores,** feed on primary consumers. **Detritus feed-**

ers, or **detritivores,** feed on **detritus,** dead organic material. **Omnivores** eat a variety of foods.

   C. **Decomposers,** or **saprotrophs,** are microbial heterotrophs (bacteria and fungi) that recycle the components of dead organisms and organic wastes.

III. **Predation** is the consumption of one species (the prey) by another (the predator).

   A. During **coevolution** between predator and prey, the predator evolves more efficient ways to catch prey, and the prey evolves better ways to escape the predator.

B. Two effective predator strategies are pursuit and ambush.

C. Plants possess a number of adaptations that protect them from being eaten, including spines, thorns, tough leathery leaves, and protective chemicals that are unpalatable or toxic to herbivores.

D. Animals possess many strategies that help them avoid being killed and eaten.

    1. Many animals flee from predators, some have mechanical defenses, and some associate in groups.

    2. Some animals that possess chemical defenses also exhibit **warning coloration.**

    3. Some animals exhibit **cryptic coloration** that helps them hide from predators by blending into their surroundings.

    4. In **Batesian mimicry,** a harmless or edible species resembles another species that is dangerous in some way. Predators avoid the mimic as well as the model.

    5. In **Müllerian mimicry,** several different species, all of which are poisonous, harmful, or distasteful, resemble one another. Predators easily learn to avoid their common warning coloration.

IV. **Symbiosis** is any intimate or long-term association between two or more species.

A. In **mutualism,** both partners benefit.

B. In **commensalism,** one organism benefits and the other is unaffected.

C. In **parasitism,** one organism (the parasite) benefits while the other (the host) is harmed. Some parasites are **pathogens** that cause disease.

V. **Competition** occurs when two or more individuals attempt to use an essential common resource such as food, water, shelter, living space, or sunlight.

A. Competition can occur among individuals within a population (**intraspecific competition**) or between different species (**interspecific competition**).

B. Interactions among species competing for the same resources may be complex.

VI. The distinctive lifestyle and role of an organism in a community is its **ecological niche.** An organism's ecological niche takes into account all abiotic and biotic aspects of the organism's existence. An organism's **habitat** (where it lives) is one of the parameters used to describe the niche.

A. Organisms are potentially able to exploit more resources and play a broader role in the life of their community than they actually do.

    1. The potential ecological niche for an organism is its **fundamental niche,** whereas the niche it actually occupies is its **realized niche.**

    2. Interspecific competition is one of the chief biological determinants of a species' realized niche.

B. It is thought that **competitive exclusion** prevents two species from occupying the same niche in the same community for an indefinite period of time.

    1. In competitive exclusion, one species is excluded by another as a result of competition for a resource in limited supply (a **limiting factor**).

    2. An organism's limiting factors (such as the mineral content of soil, temperature extremes, and amount of precipitation) tend to restrict its realized niche.

    3. Some species reduce competition by **resource partitioning,** in which they evolve differences in resource use.

    4. Some species reduce competition by **character displacement,** in which differences in their structural, ecological, and behavioral characteristics evolve where their ranges overlap.

VII. Each community may have one or more **keystone species** that are crucial in determining the nature of the entire community. Identifying and protecting keystone species is a goal of conservation biology.

VIII. Community complexity is related to a variety of factors.

A. Community complexity is expressed in terms of **species richness,** the number of species within a community.

B. Species diversity is often great when there are many potential ecological niches, when a community is not isolated or severely stressed, when more energy is available (the species richness-energy hypothesis), in **ecotones,** and in communities with long histories without major disturbances.

IX. **Succession** is the orderly replacement of one community by another.

A. **Primary succession** occurs in an area that has not previously been inhabited (for example, bare rock or shifting sand dunes).

B. **Secondary succession** begins in an area where there was a preexisting community and well formed soil (for example, abandoned farmland).

X. There are two views of the nature of communities.

A. The **organismic model** views a community as a "superorganism" that goes through certain stages of development (succession) toward adulthood (climax). In this view, biological interactions are primarily responsible for species composition, and organisms are highly interdependent.

B. Most ecologists support the **individualistic model,** which challenges the concept of a highly interdependent community. According to this model, abiotic environmental factors are the primary determinants of species composition in a community, and organisms are largely independent of each other.

XI. **Biogeography** is the study of the geographical distribution of plants and animals, including where populations came from, how they got there, and when.

A. Each species originated only once, at its **center of origin.** From its center of origin, each species spread until halted by a physical, environmental, or biological barrier.

B. The **range** of a particular species is that portion of the Earth in which it is found.

C. Alfred Wallace divided Earth's land areas into six major biogeographical realms: the Palearctic, Nearctic, Neotropical, Ethiopian, Oriental, and Australian.

    1. Each realm has maintained its biological distinctiveness because it is separated from the others by a mountain range, desert, ocean, or other barrier.

    2. Today human activities are contributing to a homogenization of the biogeographical realms.

POST-TEST

1. An association of populations of different species living together in one area is a(an) (a) succession (b) ecological niche (c) ecotone (d) community (e) habitat

2. Ecologically speaking, grasses are classified as _____ and deer are classified as _____. (a) herbivores; carnivores (b) primary consumers; secondary consumers (c) detritivores; herbivores (d) primary producers; primary consumers (e) producers, secondary consumers

3. Monarch and viceroy butterflies are an example of (a) Batesian mimicry (b) character displacement (c) coevolution (d) Müllerian mimicry (e) cryptic coloration

4. A symbiotic association in which organisms are beneficial to one another is known as (a) predation (b) interspecific competition (c) intraspecific competition (d) commensalism (e) mutualism

5. An organism's _____ is the totality of its adaptations, its use of resources, and its lifestyle. (a) habitat (b) ecotone (c) ecological niche (d) competitive exclusion (e) coevolution

6. Competition from other species helps to determine an organism's (a) ecotone (b) fundamental niche (c) realized niche (d) limiting factor (e) ecosystem

7. "Complete competitors cannot coexist" is a statement of the principle

of (a) primary succession (b) limiting factors (c) Müllerian mimicry (d) competitive exclusion (e) character displacement

8. The _____ signifies that species diversity is greater where two communities meet than at the center of either community. (a) edge effect (b) fundamental niche (c) character displacement (d) realized niche (e) limiting factor

9. Primary succession occurs on (a) bare rock (b) newly cooled lava (c) abandoned farmland (d) both a and b are examples (e) a, b, and c are examples

10. The pioneer community of bare rock would be (a) ferns (b) grasses (c) lichens (d) shrubs (e) trees

11. The tendency for two similar species to differ from one another more markedly in areas where they occur together is known as (a) Müllerian

mimicry (b) Batesian mimicry (c) resource partitioning (d) competitive exclusion (e) character displacement

12. An unpalatable species demonstrates its threat to potential predators by displaying (a) character displacement (b) limiting factors (c) cryptic coloration (d) warning coloration (e) competitive exclusion

13. Support for the individualistic model of community structure includes (a) the decline of honeybees due to two species of parasitic mites (b) the identification of fig trees as a keystone species in tropical forests (c) the competitive exclusion of one *Paramecium* species by another (d) the distribution of trees along a moisture gradient in Wisconsin forests (e) the effects of the removal of a dominant rodent species from an Arizona desert

REVIEW QUESTIONS

1. Distinguish a community from an ecosystem.
2. Distinguish between a decomposer and a detritus feeder.
3. Describe how natural selection has affected predator-prey relationships.
4. List the three kinds of symbiosis, give an example of each, and state the effects of the symbiotic interaction on each of the two species involved.
5. Why is an organism's realized niche usually narrower, or more restricted, than its fundamental niche?

6. What is the principle of competitive exclusion?
7. Describe at least two factors that affect species diversity in a community.
8. Distinguish between primary and secondary succession and give an example of each.
9. Contrast the organismic and individualistic models of the nature of communities.
10. How is biogeography related to community ecology?

YOU MAKE THE CONNECTION

1. In what symbiotic relationships are humans involved?
2. Describe the ecological niche of humans. Do you think our realized niche has changed during the past 1000 years? Why or why not?
3. Which part of the stickleback experiment described in the Cutting Edge box exhibited directional selection? Explain your answer.

4. How is interspecific competition related to secondary succession on old fields?
5. Many plants that produce nodules for nitrogen-fixing bacteria are common on disturbed sites. Explain how these plants might simultaneously compete and cooperate with other plant species.

RECOMMENDED READINGS

Ackerman, J. "Parasites: Looking for a Free Lunch." *National Geographic,* Vol. 191, No. 10, Oct. 1997. This article includes some spectacular photographs of parasites, from protozoa to giant tapeworms.

Baskin, Y. "The Work of Nature." *Natural History,* Feb. 1997. The importance of species diversity (particularly keystone species) to community function.

MacDonald, I.R., and C. Fisher. "Life without Light." *National Geographic,* Vol. 190, No. 4, Oct. 1996. Examines biological diversity at cold seeps, natural leaks of oil and natural gas, in the depths of the Gulf of Mexico.

Mlot, C. "The Coyotes of Lamar Valley." *Science News,* Vol. 153, 31 Jan. 1998. The coyote is learning to adapt to the presence of wolves in Yellowstone National Park.

Ostfeld, R.S., C.G. Jones, and J.O. Wolff. "Of Mice and Mast: Ecological Connections in Eastern Deciduous Forests." *BioScience,* Vol. 46, No. 5, May 1996. Ecologists have unraveled a complex cycle of events that links Lyme disease epidemics and outbreaks of gypsy moths in deciduous forests.

Poulin, R., and A.S. Grutter. "Cleaning Symbioses: Proximate and Adaptive Explanations." *BioScience,* Vol. 46, No. 7, Jul./Aug. 1996. Biologists are rethinking the symbiotic relationships between larger fish and the smaller fish that clean them at cleaning stations in coral reefs.

Power, M.E., et al. "Challenges in the Quest for Keystones." *BioScience,* Vol. 46, No. 8, Sep. 1996. Examines the significance of keystone species and some of the difficulties associated with identifying them.

Raven, P.H., L.R. Berg, and G.B. Johnson. *Environment,* 2nd ed. Saunders College Publishing, Philadelphia, 1998. This environmental science text offers a detailed situation analysis of planet Earth, including how humans interact with and affect its physical and living systems.

Smith, H.R., and C.L. Remington. "Food Specificity in Interspecies Competition." *BioScience,* Vol. 46, No. 6, Jun. 1996. Compares the food preferences of various animals, including a consideration of resource partitioning.

Turner, M.G., V.H. Dale, and E.H. Everham III. "Fires, Hurricanes, and Volcanoes: Comparing Large Disturbances." *BioScience,* Vol. 47, No. 11, Dec. 1997. This article compares the recovery of natural communities affected by three large-scale disturbances: the 1980 eruption of Mount St. Helens, the 1988 Yellowstone fires, and Hurricane Hugo in 1989.

Wakelin, D. "Parasites and the Immune System: Conflict or Compromise?" *BioScience,* Vol. 47, No. 1, Jan. 1997. Parasites and their hosts must strike balances between the beneficial and harmful consequences of parasitism.

● Visit our website at **http://www.saunderscollege.com/lifesci/titles.html** and click on Solomon/Berg/Martin Biology for links to chapter-related resources on the World Wide Web.

CHAPTER 53

Ecosystems and the Biosphere

Almost completely isolated from everything in the Universe but sunlight, our planet Earth has often been compared to a vast spaceship whose life support system consists of the communities of organisms that inhabit it, plus energy from the sun. These organisms produce oxygen, transfer energy, and recycle nutrients with great efficiency. Yet none of these ecological processes would be possible without the abiotic (nonliving) environment of our spaceship Earth, here viewed from space. Much of the climate to which organisms have adapted is produced by the sun, which warms the planet, powers the hydrological cycle (causes precipitation), and drives ocean currents and atmospheric circulation patterns. The sun also supplies the energy that almost all organisms use to carry on life processes.

The science of ecology deals with the abiotic environment as well as with living organisms. Individual communities and their abiotic environments are **ecosystems.** An ecosystem encompasses all the interactions among organisms living together in a particular place, and among those organisms and their abiotic environment. Earth, which encompasses the **biosphere** (all of Earth's communities) and its interactions with Earth's water, soil, rock, and atmosphere, is the largest ecosystem.

This chapter will develop three key concepts about ecosystems. First, energy flow in ecosystems is linear. Energy moves through food webs of ecosystems in a one-way direction from the environment to organisms to the environment. Once energy has been used to do biological work for an organism, it is unavailable to other organisms because, as work is performed, the energy is changed to heat and given off into the cooler surroundings. Energy cannot be recycled and reused.

Second, matter, the material of which organisms are composed, moves in numerous cycles within an ecosystem and from one ecosystem to another, that is, from one organism to another, and from organisms to the abiotic environment and back again. All materials vital to life are continually recycled through ecosystems and so become available to new generations of organisms.

(NASA)

Third, the abiotic environment causes conditions that determine where and how successfully species live. We will briefly consider five aspects of the abiotic environment that have helped shape the biotic component of ecosystems: solar radiation, the atmosphere, the ocean, weather and climate, and fire.

1. Compare how matter and energy operate in ecosystems.
2. Summarize the concept of energy flow through a food web.
3. Draw and explain typical pyramids of numbers, biomass, and energy.
4. Distinguish between gross primary productivity and net primary productivity.
5. Diagram the carbon, nitrogen, phosphorus, and hydrological cycles.

6. Summarize the effects of solar energy on Earth's temperatures.
7. Discuss the roles of solar energy and the Coriolis effect in the production of global air and water flow patterns.
8. Define El Niño–Southern Oscillation (ENSO) and describe some of its effects on climate and on organisms.
9. Give three causes of regional precipitation differences.
10. Discuss the effects of fire on certain ecosystems.

THE FLOW OF ENERGY THROUGH ECOSYSTEMS IS LINEAR

The passage of energy in a one-way direction through an ecosystem is known as **energy flow.** Energy enters an ecosystem as radiant energy (sunlight), a tiny portion (far less than 1%) of which is trapped and used by producers during photosynthesis. The energy, now in chemical form, is stored in the bonds of organic (carbon-containing) molecules such as glucose. When these molecules are broken apart by cellular respiration, energy becomes available (in the form of ATP) to do work such as repairing tissues, producing body heat, or reproducing. As the work is accomplished, energy escapes the organisms and dissipates into the environment as heat. Ultimately, this heat energy radiates into space. Thus, once energy has been used by an organism, it is unavailable for reuse (Fig. 53–1; see also the discussion of the second law of thermodynamics in Chapter 6).

Energy flow describes who eats whom in ecosystems

In an ecosystem, energy flow occurs in **food chains,** in which energy from food passes from one organism to the next in a sequence. Producers form the beginning of the food chain by capturing the sun's energy through photosynthesis. Herbivores (and omnivores) eat plants, obtaining the chemical energy of the producers' molecules as well as building materials from which they construct their own tissues. Herbivores are in turn consumed by carnivores and omnivores, who reap the energy stored in the herbivores' molecules. Decomposers break down organic molecules in the remains (carcasses and body wastes) of all members of the food chain. (See *Making the Connection: Food Chains and Poisons in the Environment* for a discussion of how certain poisons pass through food chains.)

Simple food chains as just described rarely occur in nature, because few organisms eat just one kind, or are eaten by just one kind, of other organism. More typically, the flow of energy and materials through ecosystems takes place in accordance with a range of food choices for each organism involved. In an ecosystem of average complexity, hundreds of alternative pathways are possible. Thus, a **food web,** which is a complex of interconnected food chains in an ecosystem, is a more realistic model of the flow of energy and materials through ecosystems (Fig. 53–2).

Each level in a food web is called a **trophic level.** The first trophic level is formed by producers (organisms that photosynthesize), the second by primary consumers (herbivores), the third by secondary consumers (carnivores and omnivores), and so on (Fig. 53–3). (See *Making the Connection: Interactions Among Trophic Levels in a Food Web* for a consideration of both direct and indirect interactions among trophic levels.)

The most important thing to remember about energy flow in ecosystems is that it is linear, or one-way. That is, energy

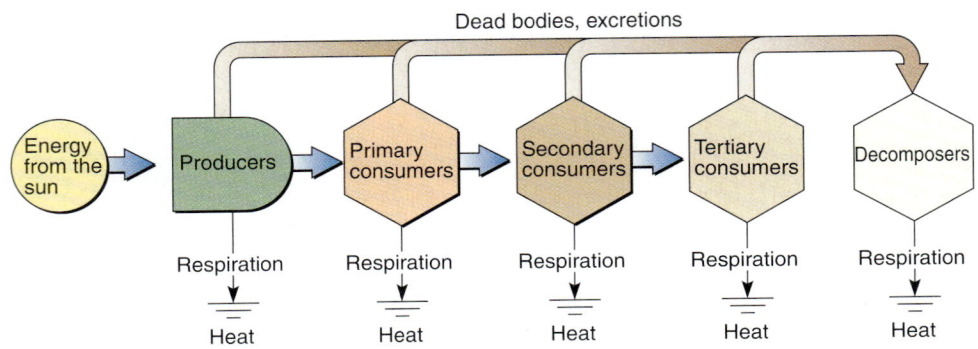

Figure 53–1 One-way energy flow through ecosystems. Energy enters ecosystems from an external source (the sun) and exits as heat loss. Much of the energy acquired by a given trophic level is used for metabolic purposes at that level and is therefore unavailable to the next trophic level.

FOOD CHAINS AND POISONS IN THE ENVIRONMENT

Certain toxic substances, including some pesticides, radioactive isotopes, heavy metals such as mercury, and industrial chemicals such as PCBs, can enter food chains and cause problems. The problem was first demonstrated by the effects of the pesticide DDT on some bird species. Falcons, pelicans, bald eagles, ospreys, and many other birds are very sensitive to traces of DDT in their tissues. One of the effects of DDT on these birds is that they lay eggs with extremely thin, fragile shells that usually break during incubation, causing the chicks' deaths. After 1972, the year DDT was banned in the United States, the reproductive success of many birds gradually improved.

The impact of DDT on birds is the result of three characteristics of DDT (and other toxins that cause problems in food webs): its persistence, bioaccumulation, and biological magnification. Some toxins are extremely stable and may take many years to be broken down into less toxic forms. The **persistence** of synthetic pesticides and industrial chemicals is a result of their novel chemical structures. Natural decomposers such as bacteria have not evolved ways to degrade these toxins, which therefore accumulate in the environment and increase in concentration as they pass from one trophic level to another.

When a persistent toxin is not metabolized (broken down) or excreted by an organism, it simply gets stored, often in fatty tissues. Over time, the organism may accumulate high concentrations of the toxin. The buildup of such a toxin in an organism's body is known as **bioaccumulation.**

Organisms at higher trophic levels tend to have greater concentrations of bioaccumulated toxins stored in their bodies than those at lower levels. The increase in concentration as the toxin passes through successive levels of the food web is known as **biological magnification.**

As an example of the concentrating characteristic of persistent toxins, consider a hypothetical food chain: algae → small fish → large fish → heron *(see figure)*. When a pesticide such as DDT is sprayed on plants, some of it gets into aquatic waterways; its concentration in water is extremely dilute, perhaps on the order of 0.00005 parts per million (ppm). The algae and aquatic plants, which take up and accumulate the toxin, contain a greater concentration of DDT, 0.04 ppm, than the waterway. Each small fish grazing on the algae and plants ingests and concentrates the pesticide in its tissues, up to 0.2 to 1.2 ppm. A large fish that eats many small fishes laced with pesticide ends up with a pesticide level of 1.0 to 2.0 ppm. The top carnivore in this example, a heron, may have a pesticide value ranging up to 76.0 ppm from eating contaminated fishes. Although this example involves a bird at the top of the food chain, it is important to recognize that *all* top carnivores, from fishes to humans, are at risk from biological magnification of persistent toxins.

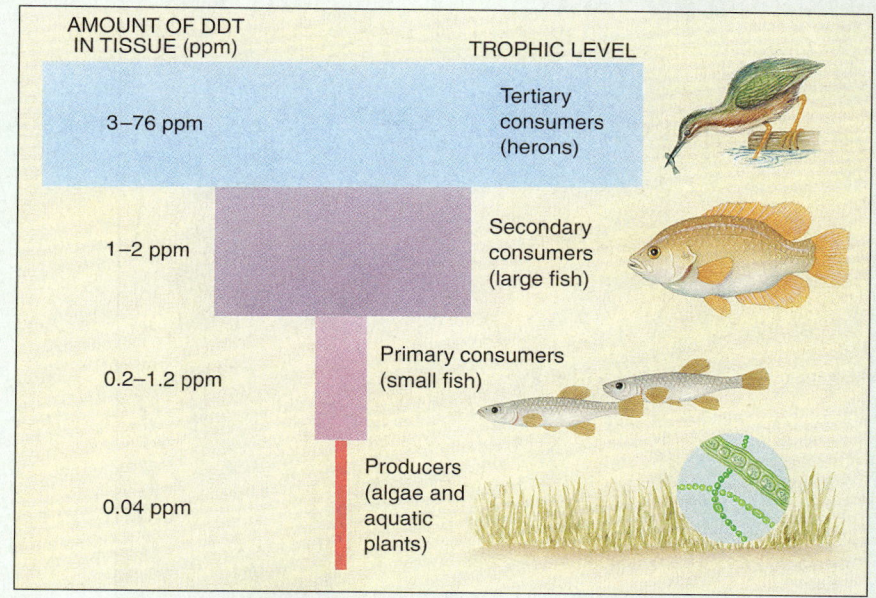

AMOUNT OF DDT IN TISSUE (ppm)	TROPHIC LEVEL
3–76 ppm	Tertiary consumers (herons)
1–2 ppm	Secondary consumers (large fish)
0.2–1.2 ppm	Primary consumers (small fish)
0.04 ppm	Producers (algae and aquatic plants)

Biological magnification of DDT in an aquatic ecosystem. Note how the level of DDT (expressed as parts per million) increases in the tissues of various organisms as DDT moves through the food chain from producers to consumers. The heron at the top of the food chain has approximately one million times more DDT in its tissues than the concentration of DDT in the water.

can move along a food web from one trophic level to the next trophic level as long as it is not used to do biological work. Once energy has been used by an organism, however, it is lost as heat and is unavailable to any other organism in the ecosystem.

Ecological pyramids illustrate how ecosystems work

Ecologists sometimes compare trophic levels by determining the number of organisms, the biomass, or the relative energy

INTERACTIONS AMONG TROPHIC LEVELS IN A FOOD WEB

Do food webs indicate relationships other than predator-prey interactions? It turns out that both negative and positive interactions occur in a food web. Because food webs are descriptions of "who eats whom," they indicate the negative effects that predators have on their prey. For example, consider a simple food chain: grass → field mouse → owl. The owl, which kills and eats mice, obviously exerts a negative effect on the mouse population; in like manner, field mice, which eat grass seeds, reduce the grass population.

One trophic level in a food web also influences other trophic levels to which it is not directly linked. Producers and top carnivores do not exert direct effects on one another, yet each is indirectly affected by the other. In our example, the owls help the producers by keeping the population of seed-eating mice under control. Likewise, the grasses benefit owls by supporting a population of mice on which the owl population feeds. These indirect interactions may be as important as direct, predator-prey interactions in food web dynamics.

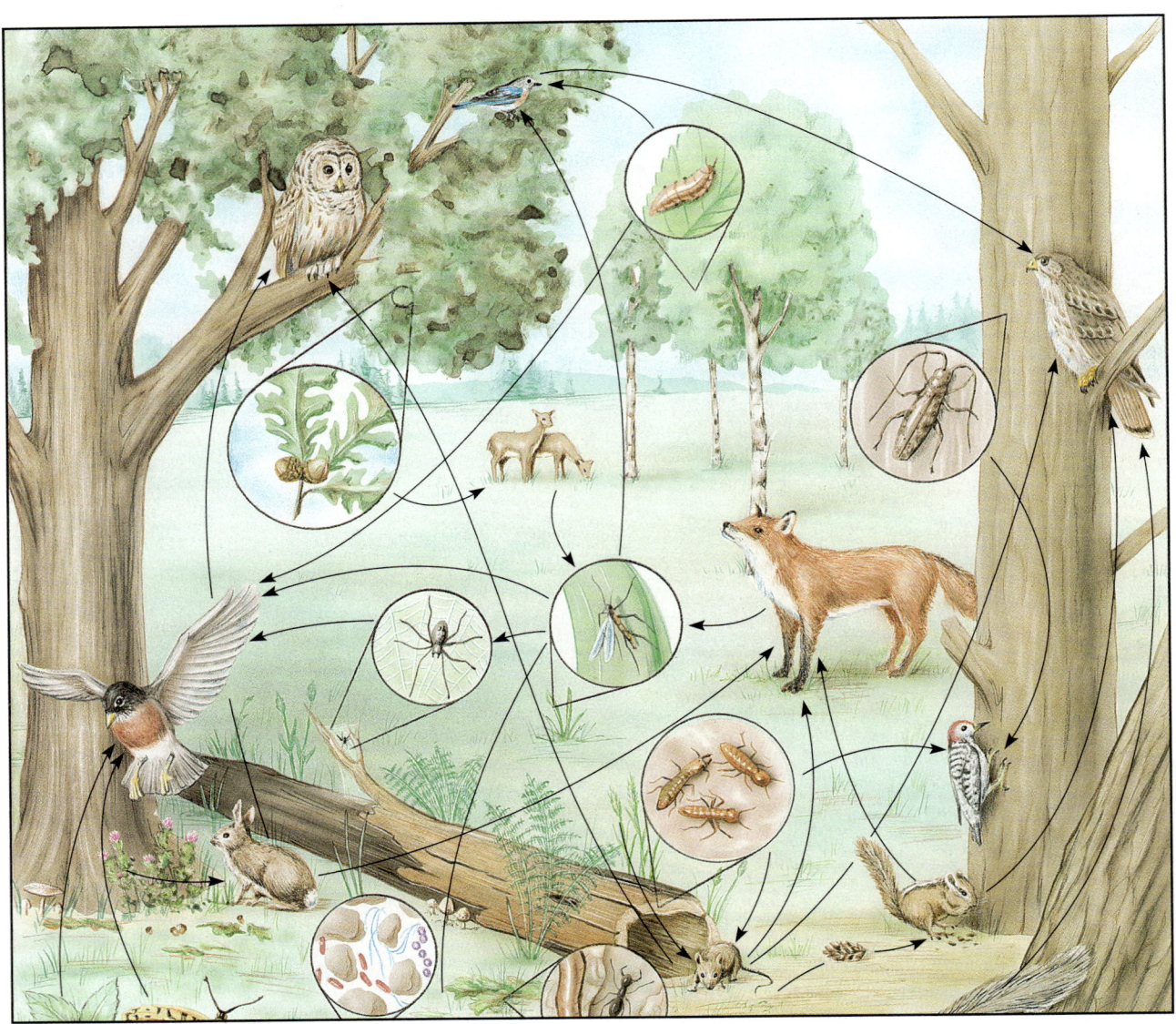

Figure 53–2   A deciduous forest food web. This diagram is greatly simplified compared to what actually happens in nature. Groups of species are lumped into single categories such as "spiders"; many species are not included; and numerous links in the web are not shown.

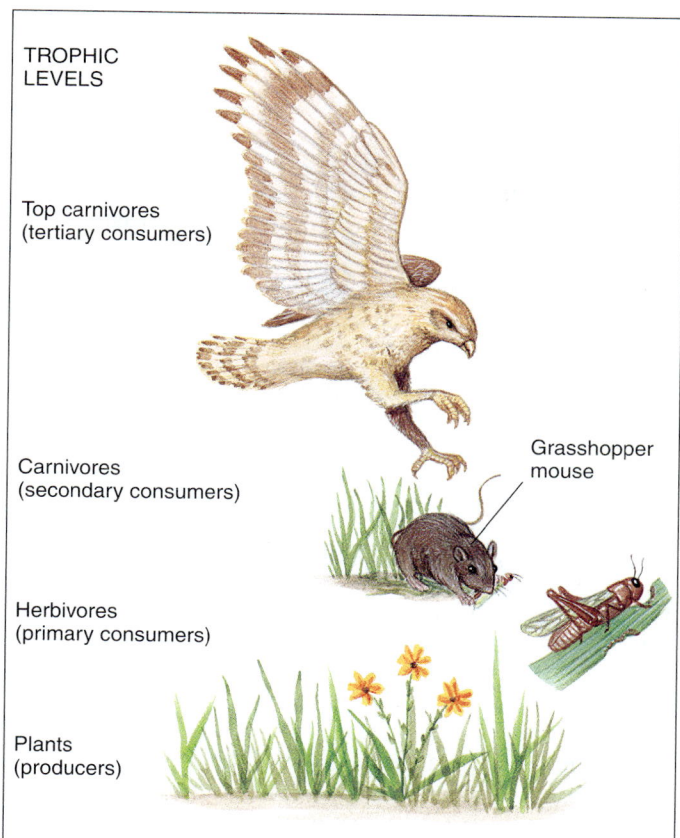

Figure 53–3 **Trophic levels.** Trophic levels, which show how chemical energy in food flows through ecosystems, help us simplify food webs by grouping organisms based on their position in the web. As stipulated by the second law of thermodynamics, most energy at each level is released into the environment as heat.

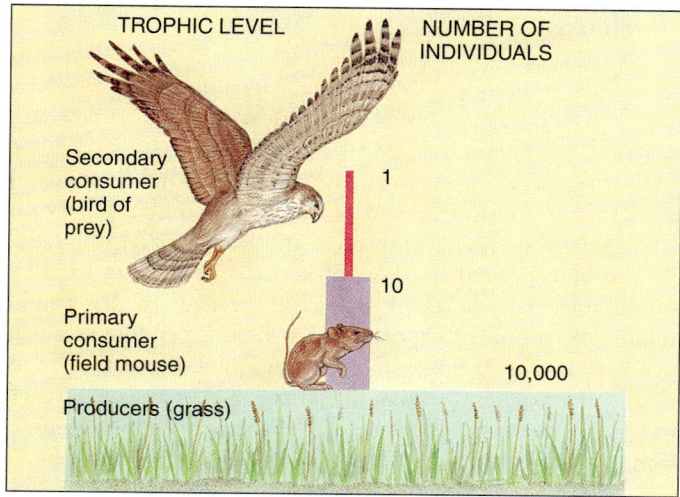

Figure 53–4 **A pyramid of numbers.** Typically, there are more producers than primary consumers, more primary consumers than secondary consumers, and so on. A pyramid of numbers is not as useful as other ecological pyramids because it provides no information about biomass or energy relationships between one trophic level and the next.

(a)

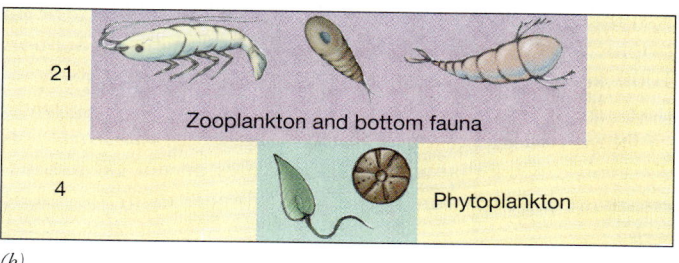

(b)

Figure 53–5 **Pyramids of biomass.** These pyramids are based on the biomass at each trophic level and generally have a pyramid shape with a large base and progressively smaller areas for each succeeding trophic level. *(a)* A pyramid of biomass for a hypothetical area of a temperate grassland. *(b)* An inverted biomass pyramid occurs when highly productive lower trophic levels experience high rates of turnover.

found at each level. This information is presented graphically as **ecological pyramids.**

A **pyramid of numbers** shows the number of organisms at each trophic level in a given ecosystem, with greater numbers illustrated by a wider pyramid (Fig. 53–4). In most pyramids of numbers, each successive trophic level is occupied by fewer organisms. Thus, the number of herbivores (such as zebras and wildebeests) is greater than the number of carnivores (such as lions). Inverted pyramids of numbers, in which higher trophic levels have more organisms than lower trophic levels, are often observed among decomposers, parasites, herbivorous insects, and similar organisms. One tree can provide food for thousands of leaf-eating insects, for example.

A **pyramid of biomass** illustrates the total biomass at each successive trophic level. **Biomass** is a quantitative estimate of the total mass, or amount, of living material; it indicates the amount of fixed energy at a particular time. Units of measure vary; biomass may be represented as total volume, dry weight, or live weight. Typically, these pyramids illustrate a progressive reduction of biomass in succeeding trophic levels (Fig. 53–5*a*).

Figure 53–6 A pyramid of energy. Energy flow, the functional basis of ecosystem structure, is represented by a hypothetical pyramid of energy. Note the substantial loss of usable energy from one trophic level to the next. (Energy in producers represents gross primary productivity.)

GROSS PRIMARY PRODUCTIVITY

Figure 53–7 Comparison of gross and net primary productivities. This pyramid of energy for a river ecosystem illustrates gross primary productivity and net primary productivity.

Assuming an average biomass reduction of about 90% for each trophic level,[1] 10,000 kg of grass should be able to support 1000 kg of grasshoppers, which in turn support 100 kg of frogs. By this logic, the biomass of frog-eaters (such as snakes) could only weigh, at most, about 10 kg. From this brief exercise, you can see that although carnivores do not eat producers, many producers are required to support them via a food web.

Occasionally we find an inverted pyramid of biomass in which the primary consumers outweigh the producers (Fig. 53–5b). In these instances the producers, which are usually unicellular algae that are short-lived and reproduce quickly, are consumed in large numbers by herbivores such as large zooplankton and fish. Thus, although at any point in time relatively few algae are present, the rate of biomass production of the primary consumers is much less than that of the producers.

A **pyramid of energy** indicates the energy content in the biomass of each trophic level. These pyramids show that less energy reaches each successive trophic level from the level beneath it because some of the energy at the lower level is used by those organisms to perform work, while some of it is lost (no biological process is 100% efficient) (Fig. 53–6). The second law of thermodynamics explains why there are few trophic levels. Food webs are short because of the dramatic reduction in energy content that occurs at each successive trophic level.

Ecosystems vary in productivity

The **gross primary productivity (GPP)** of an ecosystem is the rate at which energy is captured during photosynthesis.[2] Thus,

gross primary productivity is the total amount of photosynthetic energy captured in a given period of time. Of course, plants must respire to provide energy for their life processes, and cellular respiration acts as a drain on photosynthetic output. Energy that remains in plant tissues after cellular respiration has occurred is called **net primary productivity (NPP)**. That is, net primary productivity is the amount of biomass (the energy stored in plant tissues) found in excess of that broken down by a plant's cellular respiration. Net primary productivity represents the rate at which this organic matter is actually incorporated into plant tissues in order to produce growth (Fig. 53–7).

Net primary productivity = Gross primary productivity − Plant respiration
(Plant growth)   (Total photosynthesis)

Only the energy represented by net primary productivity is available for consumers, and of this energy only a portion is actually used by them. Both gross and net primary productivity can be expressed in terms of kilocalories (energy fixed by photosynthesis) per square meter per year, or of dry weight (grams of carbon incorporated into tissue) per square meter per year. In Figure 53–7, for example, the GPP of the plants in a river ecosystem is 20,810 kcal per m^2 per year. However, the plants use 11,977 kcal per m^2 per year in cellular respiration, leaving an NPP of 8833 kilocalories per m^2 per year.

Many factors may interact to determine productivity. Some plants (C$_4$ plants) are more efficient than others (C$_3$ plants) in fixing carbon (see Chapter 8). Environmental factors are also important. These include the availability of solar energy, mineral nutrients, and water; other climatic factors; the degree of maturity of the community; and the severity of hu-

---

[1] The 90% reduction in biomass is an approximation; actual biomass reduction varies widely in nature.

[2] Gross and net primary productivities are referred to as *primary* because plants occupy the first position in food webs.

man modification of the environment. Many of these factors are difficult to assess.

Ecologists use different methods to measure primary productivity, depending on whether gross or net primary productivity is being assessed. Methods also vary from one type of ecosystem to another. On land, for example, ecologists might cut, dry, and weigh plants at the end of a growing season to measure NPP. Dry weights are measured because the water content of different species varies considerably. This method would not be useful to measure NPP in the ocean, however, where the main producers are short-lived microscopic algae that are heavily grazed.

Ecosystems differ strikingly in their productivity (Fig. 53–8 and Table 53–1). On land, tropical rain forests have the highest productivity, probably due to the abundant rainfall, warm temperatures, and intense sunlight. As you might expect, tundra with its short, cool growing season and deserts with their lack of precipitation are the least productive terrestrial ecosystems. In ecosystems with comparable annual temperatures—temperate deciduous forest, temperate grassland, and temperate desert, for example—water availability (as annual precipitation) directly affects NPP. Availability of essential minerals such as nitrogen and phosphorus can also affect NPP.

TABLE 53–1	Net Primary Productivities (NPP) for Selected Ecosystems*
Ecosystem	Average NPP (grams dry matter/m²/year)
Swamp and marsh	3000
Tropical rain forest	2200
Temperate evergreen forest	1300
Temperate deciduous forest	1200
Savanna	900
Boreal (northern) forest	800
Woodland and shrubland	700
Agricultural land	650
Temperate grassland	600
Lake and stream	400
Arctic and alpine tundra	140
Desert and semidesert scrub	90
Extreme desert (rock, sand, ice)	3

*After Whittaker, R. H., and Likens, G. E. 1975.

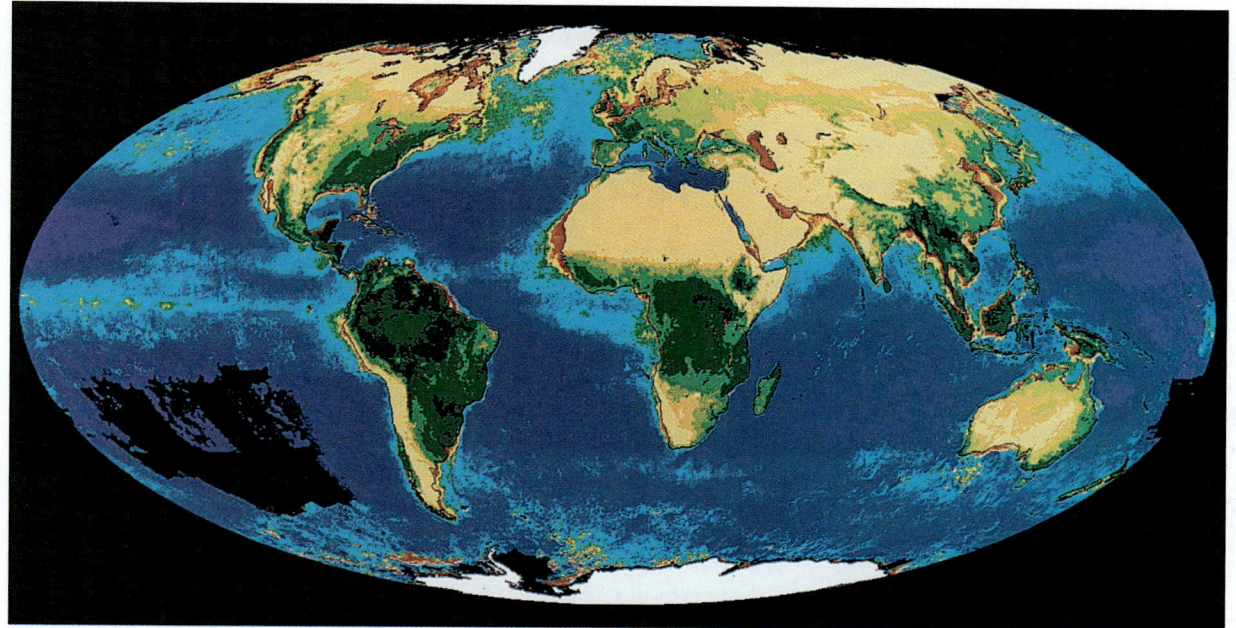

Figure 53–8 **Earth's primary productivity as measured by satellite.** On land, the most productive areas, such as tropical rain forests, are dark green, whereas the least productive ecosystems (deserts) are yellow. In the ocean and other aquatic ecosystems, the most productive regions are red, followed by orange, yellow, green, and blue (the least productive). Data are not available for the black areas. *(NASA/Goddard Space Flight Center)*

Wetlands (swamps and marshes), which connect terrestrial and aquatic environments, are extremely productive. The most productive aquatic ecosystems are algal beds, coral reefs, and estuaries. The lack of available mineral nutrients in the sunlit region of the open ocean makes this area extremely unproductive, equivalent to an aquatic desert. (Earth's major aquatic and terrestrial ecosystems are discussed in greater detail in Chapter 54.)

The relationship of productivity to biological diversity is complex

Conventional ecological theory once held that the more productive the ecosystem, the greater the biodiversity it supported. Now ecologists are seeing a recurring pattern worldwide: ecosystems are more diverse as productivity increases, but at some point diversity actually declines with increasing productivity. For example, the resource-poor depths of the Atlantic Ocean's abyssal plain have a higher species diversity than the productive shallow waters near the coasts, and intermediate depths exhibit the greatest diversity. Ecologists have little solid data to help explain the pattern, which holds true with rodents in Israel, birds in South America, and large mammals in Africa. Mathematical ecosystem models, however, suggest that a *patchy distribution of resources* reduces competition and allows the coexistence of a greater variety of organisms (see Chapter 51).

The bad news for global biodiversity is that humans are constantly enriching our environment, for example, with nitrogen and phosphorus inputs from fossil fuels, fertilizers, and livestock. This continual fertilization may make the Earth's ecosystems more and more productive, a shift that some ecologists think could cost the world a substantial loss of biodiversity. (Other factors that affect species diversity were discussed in Chapter 52.)

Humans consume an increasingly greater percentage of global primary productivity

It should come as no surprise that humans consume far more of the Earth's resources than any of the other millions of animal species. When both direct and indirect human impacts are accounted for, humans are estimated to use 25% of the global annual NPP and 40% of the annual NPP of terrestrial ecosystems. This disproportionate use of global productivity is crowding out other species and may contribute to the loss of many species, some potentially useful to humans, through extinction or genetic impoverishment. Clearly, at these levels of consumption and exploitation of the Earth's resources, explosive human population growth becomes a serious threat to the planet's ability to support its occupants.

MATTER CYCLES THROUGH ECOSYSTEMS

Matter cycles from the living world to the abiotic environment and back again. We call these cycles **biogeochemical cycles.** Earth is essentially a closed system (one into which matter does

not enter and from which matter cannot escape). The materials used by organisms cannot be "lost," although they can end up in locations outside the reach of organisms for a long period of time. Usually, however, materials are reused and often recycled both within and among ecosystems.

Four different biogeochemical cycles of matter—carbon, nitrogen, phosphorus, and water—represent all biogeochemical cycles. These four cycles are particularly important to organisms as they involve materials used to make the chemical components of cells. Carbon, nitrogen, and water have gaseous components and so cycle over large distances of the atmosphere with relative ease. Phosphorus, however, is an element that is never in a gaseous phase, and, as a result, only local cycling of phosphorus occurs easily.

CARBON DIOXIDE IS THE PIVOTAL MOLECULE OF THE CARBON CYCLE

Carbon must be available to organisms because proteins, nucleic acids, lipids, carbohydrates, and other molecules essential to life contain carbon. Carbon is present in the atmosphere as the gas carbon dioxide (CO_2), which makes up approximately 0.03% of the atmosphere. It is also present in the ocean and fresh water as dissolved carbon dioxide, as carbonate (CO_3^{2-}), and as bicarbonate (HCO_3^-); and in carbonate rocks such as limestone ($CaCO_3$). The global movement of carbon between the abiotic environment, including the atmosphere, and organisms is known as the **carbon cycle** (Fig. 53–9).

During photosynthesis, plants, algae, and cyanobacteria remove carbon dioxide from the air and *fix,* or incorporate, it into complex organic compounds such as glucose. Plants use much of the glucose to make cellulose, starch, amino acids, nucleic acids, and other compounds. Thus, photosynthesis incorporates carbon from the abiotic environment into the biological compounds of producers. Many of these compounds are used as fuel for cellular respiration by the producer that made them, by a consumer that eats the producer, or by a decomposer that breaks down the remains of the producer or consumer. Thus, the process of cellular respiration returns carbon dioxide to the atmosphere. A similar carbon cycle occurs in aquatic ecosystems between aquatic organisms and dissolved carbon dioxide in the water.

Sometimes the carbon in biological molecules is not recycled back to the abiotic environment for some time. For example, a large amount of carbon is stored in the wood of trees, where it may stay for several hundred years or even longer. In addition, millions of years ago vast coal beds formed from the bodies of ancient trees that were buried and subjected to anaerobic conditions before they had fully decayed. Similarly, the oils of unicellular marine organisms probably gave rise to the underground deposits of oil and natural gas that accumulated in the geological past. Coal, oil, and natural gas, called **fossil fuels** because they formed from the remains of ancient organisms, are vast depositories of carbon compounds, the end

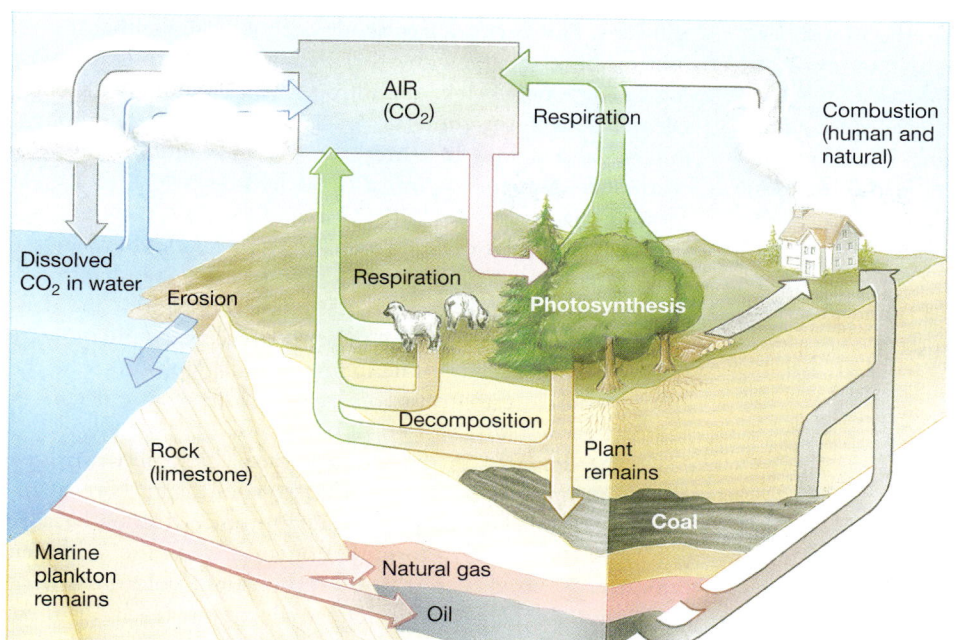

Figure 53–9 A simplified diagram of the carbon cycle. Carbon (as carbon dioxide) enters organisms when primary producers photosynthesize. Carbon returns to the environment by respiration, combustion, and erosion. The carbon in fossil fuels, limestone rock, and marine animal shells may take millions of years to cycle back to the biotic world.

products of photosynthesis that occurred millions of years ago. Fossil fuels are nonrenewable resources; that is, Earth has a finite, or limited, supply of them. Although fossil fuels are still being formed by natural processes today, they are forming too slowly to replace the fossil fuel reserve we are using.

The carbon in coal, oil, natural gas, and wood may be returned to the atmosphere by the process of burning, or combustion. In combustion, organic molecules are rapidly oxidized (combined with oxygen), converting them into carbon dioxide and water with an accompanying release of light and heat.

An even greater amount of carbon that is stored for millions of years is incorporated into the shells of marine organisms. When these organisms die, their shells sink to the ocean floor and are covered by sediments. The shells form seabed deposits thousands of meters thick that are eventually cemented together to form a sedimentary rock called limestone. Earth's crust is dynamically active, and over millions of years, sedimentary rock on the bottom of the sea floor may be lifted up to form land surfaces. (The summit of Mount Everest, for example, is composed of sedimentary rock.) When limestone is exposed by the process of geological uplift, it is slowly eroded by chemical and physical weathering processes. This returns carbon to the water and atmosphere, where it is available to participate in photosynthesis once again.

Thus, photosynthesis removes carbon from the abiotic environment and incorporates it into biological molecules, while cellular respiration, combustion, and erosion return carbon to the water and atmosphere of the abiotic environment. Table 53–2 provides some estimates of the global carbon budget, that is, the global quantities of carbon (expressed as billion metric tons) in organisms and in abiotic reservoirs such as the atmosphere and ocean.

Human activities have disturbed the global carbon budget

From the advent of the Industrial Revolution to the present, humans have burned increasing amounts of fossil fuels—coal, oil, and natural gas. This trend, along with a greater combustion of wood as a fuel and the burning of large sections of

TABLE 53–2	Some Estimates of the Global Carbon Budget
Total Carbon	**Amount of Carbon (in billion metric tons)**
Atmosphere (as CO_2)	755
Dissolved carbon in the ocean (90% as HCO_3^-)	38,000
Upper one meter of soil	1200
Terrestrial organisms (forests, etc.)	450 to 600
Marine organisms	3
Recoverable fossil fuels	10,000
Annual increase of atmospheric carbon (due primarily to burning of fossil fuels)	3

Adapted from Moore, B. "The Oceanic Sink for Excess Atmospheric Carbon Dioxide." In *Wastes in the Ocean*, Vol. 4. John Wiley, Chichester, England, 1985.

tropical forest, has released carbon dioxide into the atmosphere at a rate greater than producers and the ocean can absorb it from the atmosphere.

The slow and steady rise of CO_2 in the atmosphere may cause changes in climate, called global warming. Global warming could result in a rise in sea level, changes in precipitation patterns, death of forests, extinction of organisms, and problems for agriculture. It could force the displacement of millions of people, particularly from coastal areas. A more thorough discussion of increasing atmospheric CO_2 and potential impacts of global warming is found in Chapter 55.

BACTERIA ARE ESSENTIAL TO THE NITROGEN CYCLE

Nitrogen is crucial for all organisms because it is an essential part of proteins, nucleic acids, and chlorophyll. Because Earth's atmosphere is about 78% nitrogen gas (N_2), it would appear that there could be no possible shortage of nitrogen for or-

ganisms. However, molecular nitrogen is so stable that it does not readily combine with other elements. Therefore, N_2 must be broken apart before the nitrogen can combine with other elements to form proteins, nucleic acids, and chlorophyll. Chemical reactions that break up N_2 and combine nitrogen with such elements as oxygen and hydrogen require a great deal of energy.

The **nitrogen cycle,** in which nitrogen cycles between the abiotic environment and organisms, has five steps: nitrogen fixation, nitrification, assimilation, ammonification, and denitrification (Fig. 53–10). All these steps except assimilation are performed by bacteria. The following paragraphs describe the nitrogen cycle that occurs on land; a similar cycle occurs in aquatic ecosystems.

The first step in the nitrogen cycle, biological **nitrogen fixation,** involves conversion of gaseous nitrogen (N_2) to ammonia (NH_3). This process is called nitrogen fixation because nitrogen is *fixed* into a form that organisms can use. Nitrogen can also be fixed as nitrate by combustion, volcanic action, lightning discharges, and industrial means (each of these processes supplies enough energy to break up molecular ni-

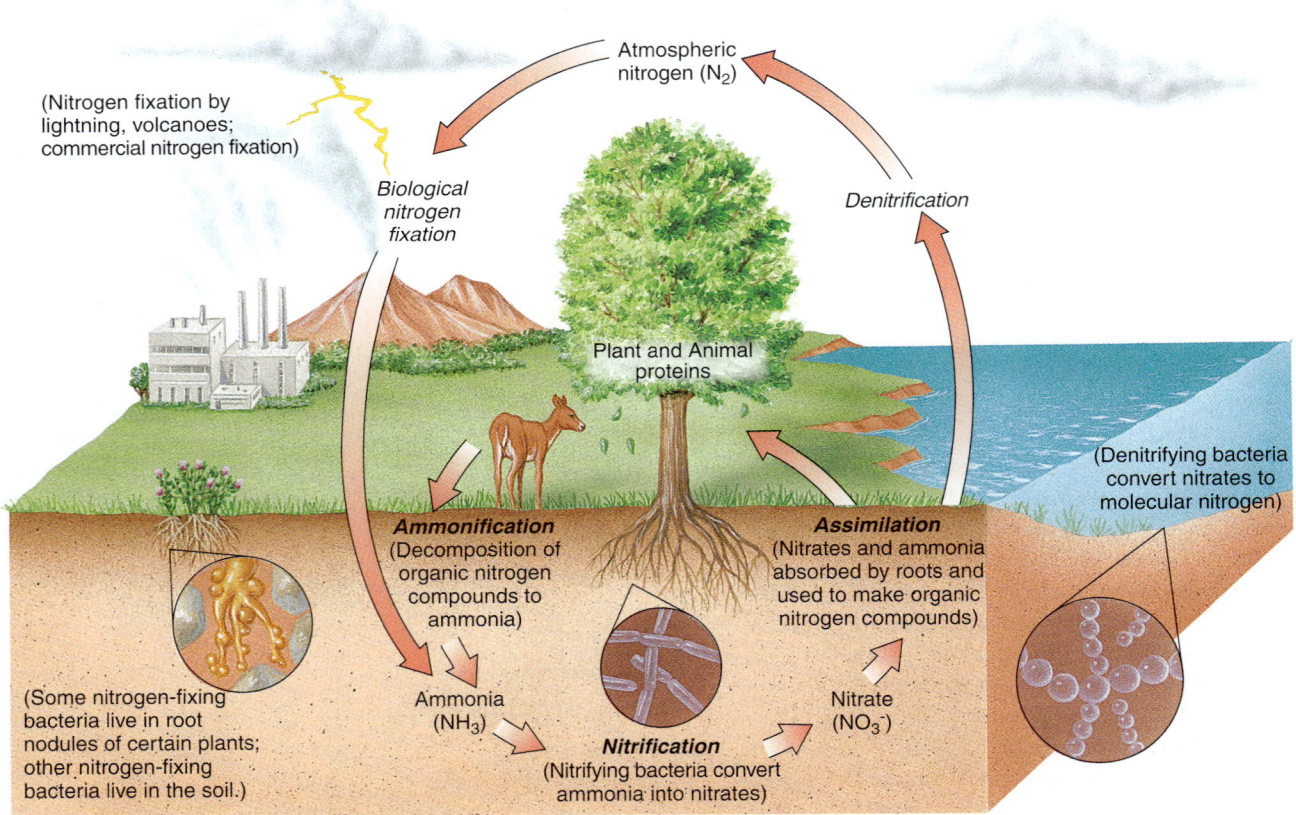

Figure 53–10 **A simplified diagram of the nitrogen cycle.** Nitrogen-fixing bacteria convert atmospheric nitrogen (N_2) into ammonia (NH_3). Ammonia is converted to nitrate (NO_3^-) by nitrifying bacteria in the soil. Plants assimilate nitrate and ammonia, producing proteins and nucleic acids in the process; then animals eat plant proteins and produce animal proteins. When organisms produce wastes or die, their nitrogen compounds are broken down by ammonifying bacteria, and ammonia is released. Nitrogen is returned to the atmosphere by denitrifying bacteria, which convert nitrate to molecular nitrogen.

(a)

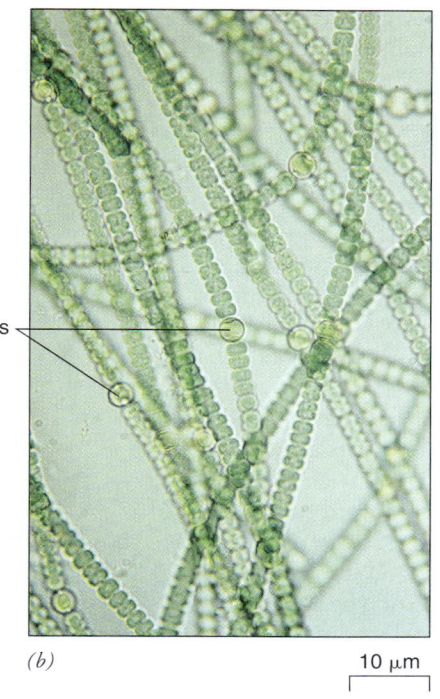

(b)

Heterocysts

10 μm

Figure 53–11 Nitrogen fixation. Nitrogenase, the enzyme involved in biological nitrogen fixation, only functions in the absence of oxygen. *(a)* Root nodules of a pea plant provide an oxygen-free environment for nitrogen-fixing *Rhizobium* bacteria that live in them. *(b)* LM of *Anabaena,* a cyanobacterium with distinctive oxygen-excluding heterocysts in which nitrogen fixation occurs. *(a, Hugh Spencer/Photo Researchers, Inc.; b, Dennis Drenner)*

trogen). Biological nitrogen fixation is carried out by nitrogen-fixing bacteria, including cyanobacteria, in soil and aquatic environments. Nitrogen-fixing bacteria employ an enzyme called **nitrogenase** to break up molecular nitrogen and combine it with hydrogen.

Because nitrogenase functions only in the absence of oxygen, the bacteria that fix nitrogen must insulate the enzyme from oxygen by some means. Some nitrogen-fixing bacteria live beneath layers of oxygen-excluding slime on the roots of a number of plants. Other important nitrogen-fixing bacteria, in the genus *Rhizobium,* live in special swellings, or **nodules** (Fig. 53–11*a*), on the roots of legumes such as beans and peas and some woody plants.

In aquatic environments, most nitrogen fixation is performed by cyanobacteria. Filamentous cyanobacteria have special oxygen-excluding cells called **heterocysts** that function to fix nitrogen (Fig. 53–11*b*). Some water ferns have cavities in which cyanobacteria live, in a manner comparable to the way *Rhizobium* lives in root nodules of legumes. Other cyanobacteria fix nitrogen in symbiotic association with cycads or some other terrestrial plants, or as the photosynthetic partner of certain lichens.

The reduction of nitrogen gas to ammonia by nitrogenase is a remarkable accomplishment by living organisms, achieved without the tremendous heat, pressure, and energy required to do the same thing during the manufacture of commercial fertilizers. Even so, nitrogen-fixing bacteria must consume the energy in 12 g of glucose (or the equivalent) to fix 1 g of nitrogen biologically.

The second step of the nitrogen cycle is **nitrification,** the conversion of ammonia (NH_3) to nitrate (NO_3^-). Nitrifica-

tion, a two-step process, is accomplished by soil bacteria. First, the soil bacteria *Nitrosomonas* and *Nitrococcus* convert ammonia to nitrite (NO_2^-). Then the soil bacterium *Nitrobacter* oxidizes nitrite to nitrate. The process of nitrification furnishes these bacteria, called nitrifying bacteria, with energy.

In the third step, called **assimilation,** roots absorb either ammonia (NH_3) or nitrate (NO_3^-) that was formed by nitrogen fixation and nitrification, and incorporate the nitrogen into proteins, nucleic acids, and chlorophyll. When animals consume plant tissues, they assimilate nitrogen by taking in plant nitrogen compounds and converting them to animal nitrogen compounds.

The fourth step is **ammonification,** which is the conversion of organic nitrogen compounds into ammonia. Ammonification begins when organisms produce nitrogen-containing wastes such as urea (in urine) and uric acid (in the wastes of birds). These substances, along with the nitrogen compounds in dead organisms, are decomposed, releasing the nitrogen into the abiotic environment as ammonia (NH_3). The bacteria that perform ammonification in both the soil and aquatic environments are called ammonifying bacteria. The ammonia produced by ammonification is available for the processes of nitrification and assimilation. As a matter of fact, most available nitrogen in the soil derives from the recycling of organic nitrogen by ammonification.

The fifth, and final, step of the nitrogen cycle is **denitrification,** which is the reduction of nitrate (NO_3^-) to gaseous nitrogen (N_2). Denitrifying bacteria reverse the action of nitrogen-fixing and nitrifying bacteria by returning nitrogen to the atmosphere as nitrogen gas. Denitrifying bacteria are anaerobic and therefore live and grow best where there is little or no free oxygen. For example, they are found deep in the soil near the water table, an environment that is nearly oxygen-free.

Human interaction with the nitrogen cycle can cause water pollution

Humans affect the nitrogen cycle by producing large quantities of nitrogen fertilizer, both ammonia and nitrate, from nitrogen gas. The increasing use of fertilizer has resulted in higher crop yields. In 1997, for example, 131 million tons of fertilizer were used worldwide; this amount is equal to 22.4 kg (49.4 lb) of fertilizer per person.

Although the production and use of fertilizer in itself is not harmful, the overuse of commercial fertilizers on the land can cause water quality problems. The amount of nitrate or ammonia in most aquatic ecosystems is in limited supply and therefore limits the growth of phytoplankton. Rain washes nitrate fertilizer into rivers and lakes, where it stimulates algal growth. As these algae die, their decomposition by aerobic bacteria robs the water of dissolved oxygen, which in turn causes other aquatic organisms, including many fishes, to die of suffocation.

Fertilizer on land → fertilizer in aquatic ecosystem → algal growth → decomposition of dead algae by aerobic decomposers → lower level of dissolved oxygen in water → death of aerobic aquatic organisms

Nitrates from fertilizers can also leach (dissolve and wash down) through the soil and contaminate groundwater. Many people who live in rural areas drink groundwater, and groundwater contaminated by nitrates is dangerous, particularly for infants and small children. The effects of nitrate contamination on the aquatic environment are discussed more thoroughly in Chapter 54.

THE PHOSPHORUS CYCLE LACKS A GASEOUS COMPONENT

In the **phosphorus cycle,** phosphorus, which does not exist in a gaseous state and therefore does not enter the atmosphere, cycles from the land to sediments in the ocean and back to the land (Fig. 53–12). As water runs over rocks containing phos-

Figure 53–12 **A simplified diagram of the phosphorus cycle.** Recycling of phosphorus (as phosphate) is slow because no form of phosphorus is gaseous. Phosphate that becomes part of marine sediments may take millions of years to solidify into rock, uplift as mountains, and erode again to become available to organisms.

phorus, it gradually erodes the surface and carries off inorganic phosphate (PO_4^{3-}) molecules.

The erosion of phosphorus rocks releases phosphate into the soil, where it is taken up by roots. Once in plant cells, phosphate is incorporated into a variety of biological molecules, including nucleic acids and the phospholipids that make up cell membranes. Animals obtain most of their required phosphorus from the food they eat, although in some places drinking water may contain a substantial amount of inorganic phosphate. Phosphate released by decomposers becomes part of the pool of inorganic phosphate in the soil that can be reused by plants.

Phosphorus cycles through aquatic ecosystems in much the same way that it does through terrestrial ecosystems. Dissolved phosphate enters aquatic ecosystems through absorption by algae and aquatic plants, which are consumed by plankton and larger organisms. These are in turn eaten by a variety of fin fish and shellfish. Ultimately, decomposers that break down wastes and dead organisms release inorganic phosphate into the water, making it available for use again by aquatic producers.

Some phosphate in the aquatic food web finds its way back to the land. A tiny fraction of fish and aquatic invertebrates are eaten by sea birds, which may defecate on the land where they roost. Guano, the manure of sea birds, contains large amounts of phosphate and nitrate. Once on land, these minerals may be absorbed by plant roots. The phosphate contained in guano may enter terrestrial food webs in this way, although the amounts involved are quite small.

Phosphate can be lost for varying time periods from biological cycles. Some phosphate is carried from the land by streams and rivers to the ocean, where it can be deposited on the sea floor and remain for millions of years. The geological process of uplift may someday expose these seafloor sediments as new land surfaces, from which phosphate will be once again eroded. Phosphate deposits are also mined for agricultural use in phosphate fertilizers.

Humans affect the natural cycling of phosphorus

You have seen that once phosphorus moves from terrestrial to aquatic ecosystems, its return to land is very slow. Certain human activities accelerate the long-term loss of phosphorus from terrestrial ecosystems. For example, corn grown in Iowa, which contains phosphate absorbed from the soil, may be used to fatten cattle in an Illinois feedlot. Part of the phosphate absorbed by the roots of the corn plants thus ends up in the feedlot wastes, much of which is used as fertilizer and may eventually wash into the Mississippi River. Beef from the Illinois cattle may be consumed by people living far away, in New York City, for instance, where more of the phosphate ends up in human wastes and is flushed down toilets into the New York City sewer system. Sewage treatment rarely removes phosphates, and so they cause water quality problems in rivers and lakes.

To compensate for the steady loss of phosphate from their land, farmers must add phosphate fertilizer to their fields. More than likely, that fertilizer is extracted from the large deposits of phosphate rock in Florida.

In natural terrestrial communities, very little phosphorus is lost from the cycle, but few communities today are in a natural state, that is, unaltered in some way by humans. Phosphorus loss from the soil is accelerated by land-denuding practices such as the clearcutting of timber and by erosion of agricultural and residential lands. For practical purposes, phosphorus that washes from the land into the ocean is permanently lost from the terrestrial phosphorus cycle, for most of it remains in the ocean for millions of years.

WATER MOVES AMONG THE OCEAN, LAND, AND ATMOSPHERE IN THE HYDROLOGICAL CYCLE

Life on planet Earth would be impossible without water, which makes up a substantial part of the mass of most organisms. All life forms, from bacteria to plants and animals, use water as a medium for chemical reactions as well as for the transport of materials within and among cells. (Recall from Chapter 2 that water has many unique properties that help shape the continents, moderate climate, and allow organisms to survive.)

Water continuously circulates from the ocean to the atmosphere to the land and back to the ocean. It provides a renewable supply of purified water for terrestrial organisms. This complex cycle, known as the **hydrological cycle,** results in a balance between water in the ocean, on the land, and in the atmosphere (Fig. 53–13). Water moves from the atmosphere to the land and ocean in the form of precipitation (rain, sleet, snow, or hail). When water evaporates from the ocean surface and from soil, streams, rivers, and lakes, it eventually condenses and forms clouds in the atmosphere. In addition, **transpiration,** the loss of water vapor from land plants, adds a considerable amount of water vapor to the atmosphere.

Water may evaporate from land and reenter the atmosphere directly. Alternatively, it may flow in rivers and streams to coastal estuaries, where fresh water meets the ocean. The movement of surface water from land to ocean is called *runoff,* and the area of land being drained by runoff is called a *watershed.* Water also percolates (seeps) downward in the soil to become *groundwater,* where it is trapped and held for a time. The underground caverns and porous layers of rock in which groundwater is stored are called *aquifers.* Groundwater may reside in the ground for hundreds to many thousands of years, but eventually it supplies water to the soil, to streams and rivers, to plants, and to the ocean. The removal by humans of more groundwater than can be recharged by precipitation or melting snow, called *aquifer depletion,* eliminates groundwater as a water resource.

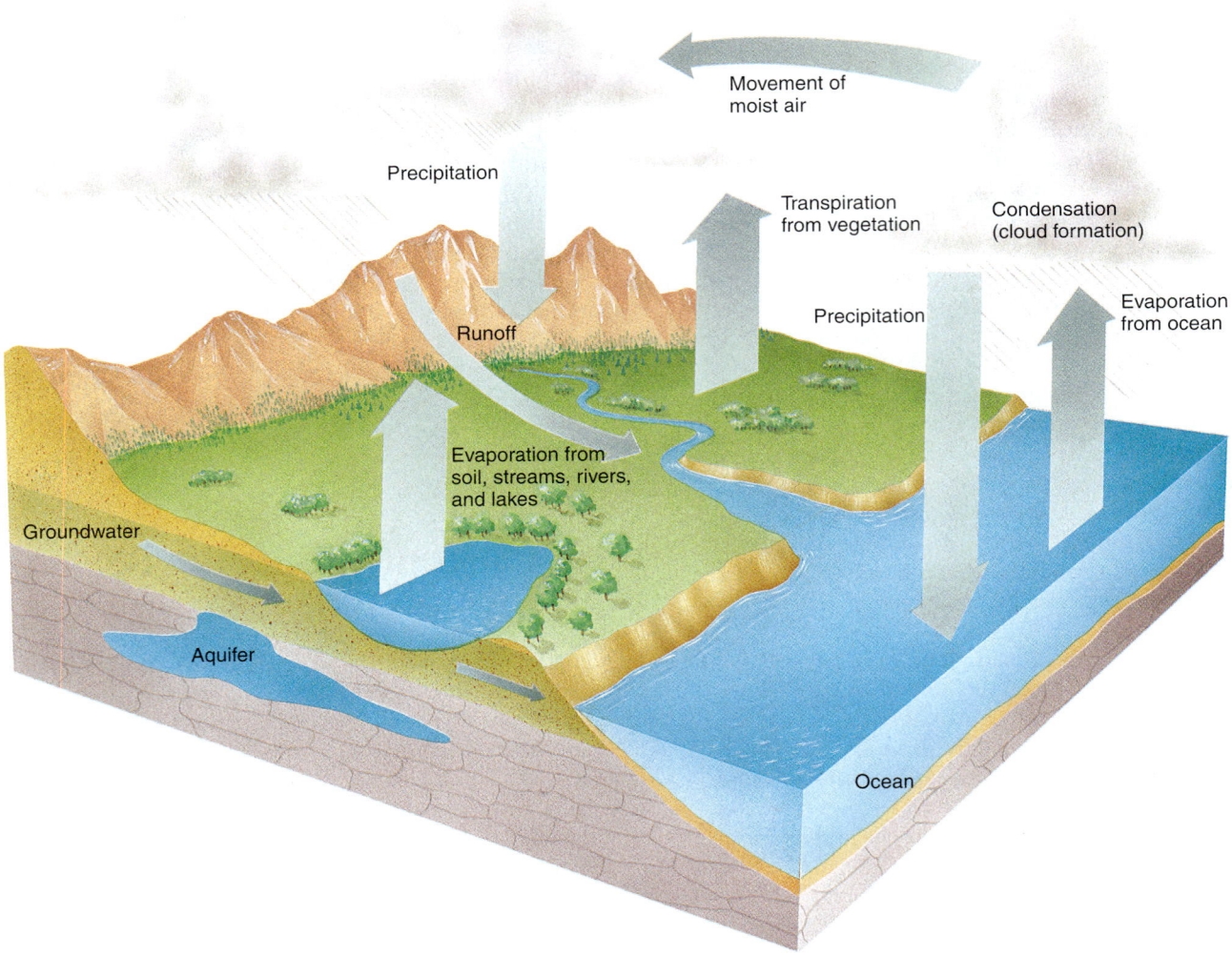

Figure 53–13 **A simplified diagram of the hydrological cycle.** Tremendous quantities of water are cycled annually between Earth and its atmosphere. An estimated 396,000 cubic km (95,000 cubic mi) enter the atmosphere each year. Approximately three-fourths of this water reenters the ocean directly as precipitation over water; the remainder falls on land. Although some water molecules are unavailable for thousands of years (locked up in polar ice or underground aquifers, for example), all water molecules eventually pass through the hydrological cycle.

Regardless of its physical form (solid, liquid, or vapor) or location, every molecule of water eventually moves through the hydrological cycle. As is true of the other cycles, water (in the form of glaciers and ice caps) can be lost from the cycle for thousands of years.

ABIOTIC FACTORS INFLUENCE WHERE AND HOW SUCCESSFULLY AN ORGANISM CAN SURVIVE

We have seen how organisms depend on the abiotic environment to supply energy and essential materials (in biogeochemical cycles). Abiotic factors such as solar radiation, the atmosphere, the ocean, weather and climate, and fire also affect

the distributions of organisms. For a given abiotic factor, each organism has an optimal range in which it can survive and reproduce. Water and temperature are probably the two abiotic factors that affect the distribution of most organisms. Temperature, for example, is the most important abiotic factor affecting the geographical range of certain aquatic insects. If the water temperature is too low, both metabolism and growth are limited, whereas if the water temperature is too high, metabolism increases but growth is suppressed.

THE SUN WARMS EARTH

The sun makes life on Earth possible. It warms the planet to habitable temperatures. Without the sun's energy, the temperature on planet Earth would approach absolute zero ($-273°$C)

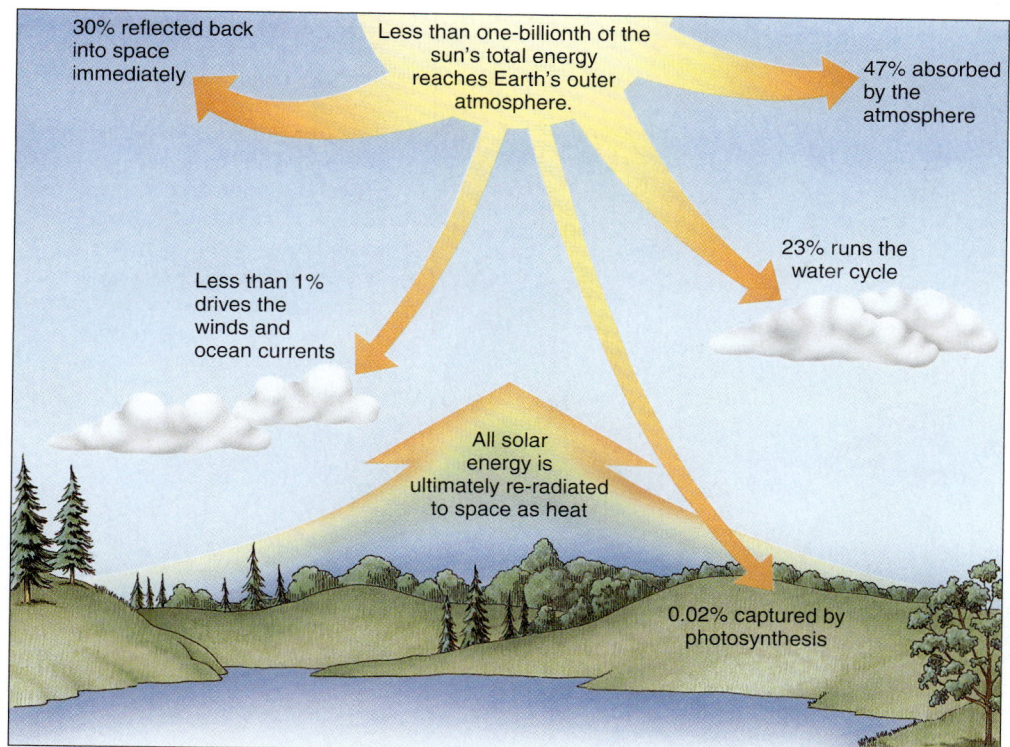

30% reflected back into space immediately

Less than one-billionth of the sun's total energy reaches Earth's outer atmosphere.

47% absorbed by the atmosphere

Less than 1% drives the winds and ocean currents

23% runs the water cycle

All solar energy is ultimately re-radiated to space as heat

0.02% captured by photosynthesis

Figure 53–14 **The fate of solar radiation that reaches Earth.** Most of the energy produced by the sun never reaches Earth. The solar energy that does reach Earth warms the planet's surface, drives the hydrological and other biogeochemical cycles, produces our climate, and powers almost all life through the process of photosynthesis.

and all water would be frozen, even in the ocean. The hydrological cycle, carbon cycle, and other biogeochemical cycles are powered by the sun, which also determines climate to a great extent. The sun's energy is captured by photosynthetic organisms and used to make organic compounds that are required by almost all forms of life. Most of our fuels, such as wood, oil, coal, and natural gas, represent solar energy captured by photosynthetic organisms. Without the sun, virtually all life on Earth would cease.

The sun's energy, which is the product of a massive nuclear fusion reaction, is emitted into space in the form of electromagnetic radiation, especially ultraviolet, visible, and infrared radiation. An infinitesimal portion of this energy, one-billionth of the sun's total production, strikes the atmosphere, and a minute part of this tiny trickle of energy operates the biosphere. On average, 30% of the solar radiation that falls on Earth is immediately reflected away by clouds and surfaces, especially snow, ice, and the ocean (Fig. 53–14). The remaining 70% is absorbed by Earth's surface and atmosphere and runs the water cycle, drives winds and ocean currents, powers photosynthesis, and warms the planet. Ultimately, however, all of this energy is lost by the continual radiation of longwave infrared (heat) energy into space. If heat gains were not exactly balanced by losses, the Earth would heat up or cool down.

Temperature changes with latitude

The most significant local variations in Earth's temperature are produced because the sun's energy does not uniformly reach all places. A combination of our planet's roughly spherical shape and the tilted angle of its axis produces a great deal of variation in the exposure of the surface to the energy delivered by sunlight. The sun's rays strike nearly vertically near the equator, concentrating the energy and producing warmer temperatures. Near the poles the sun's rays strike more obliquely and, as a result, are spread over a larger surface area. Also, rays of light entering the atmosphere obliquely near the poles must pass through a deeper envelope of air than those entering near the equator. This causes more of the sun's energy to be scattered and reflected back into space, which further lowers temperatures near the poles. Thus, the solar energy that reaches polar regions is less concentrated and produces lower temperatures.

Temperature changes with season

Seasons are determined by two main factors: Earth's inclination on its axis (the more important factor), and distance from the sun, which varies during the year. Since Earth's inclination on its axis is always the same (23.5 degrees), during half of the year (March 21 to September 22) the Northern Hemisphere tilts toward the sun, concentrating the sunlight and making the days longer (Fig. 53–15). During the other half of the year (September 22 to March 21) the Northern Hemisphere tilts away from the sun, giving it a lower concentration of sunlight and shorter days. The orientation of the Southern Hemisphere is just the opposite.

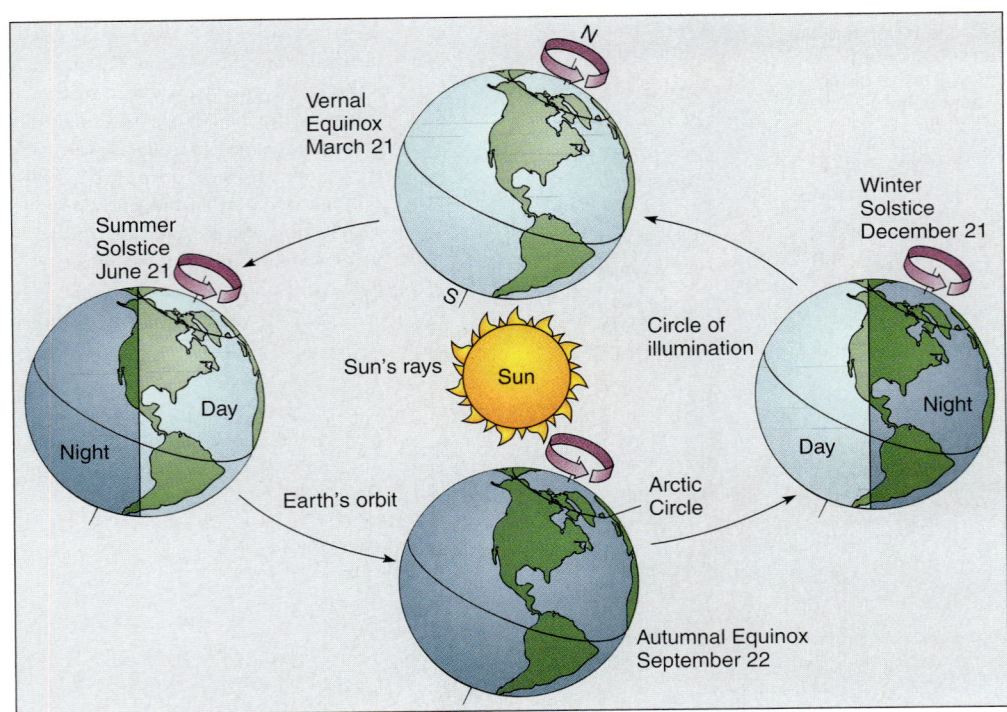

Figure 53–15 Seasonal changes in temperature. Earth's inclination on its axis remains the same as it travels around the sun. Thus, the sun's rays hit the Northern Hemisphere obliquely during winter months and more directly during summer months. In the Southern Hemisphere, the sun's rays are oblique during their winter, which corresponds to our summer. At the equator, the sun's rays are approximately vertical on March 21 and September 22.

THE ATMOSPHERE CONTAINS SEVERAL GASES ESSENTIAL TO ORGANISMS

The atmosphere is an invisible layer of gases that envelops the Earth. Oxygen (21%) and nitrogen (78%) are the predominant gases in the atmosphere, accounting for about 99% of dry air; other gases, including argon, carbon dioxide, neon, and helium, make up the remaining 1%. In addition, water vapor and particles such as dust, pollen, and microorganisms are present. Atmospheric oxygen is essential to plants, animals, and other organisms that respire aerobically, whereas carbon dioxide is required by plants and other photosynthetic organisms.

The atmosphere performs several ecologically important functions. It protects the surface from most of the sun's ultraviolet radiation and x rays and from lethal amounts of cosmic rays from space. Without this shielding by the atmosphere, life as we know it would cease. While the atmosphere protects Earth from high-energy radiation, it allows visible light and some infrared radiation to penetrate, and they warm the surface and the lower atmosphere. This interaction between the atmosphere and solar energy is responsible for weather and climate.

Organisms depend on the atmosphere, but they also help to maintain and, in certain instances, modify its composition. For example, atmospheric oxygen is thought to have increased to its present level as a result of billions of years of photosynthesis. The level is maintained, in part, by a balance between oxygen-producing photosynthesis and oxygen-using aerobic respiration.

Global atmospheric circulation is driven by the sun

In large measure, differences in temperature caused by variations in the amount of solar energy at different locations drive the circulation of the atmosphere. The warm surface near the equator heats the air with which it comes into contact, causing this air to expand and rise. As the warm air rises, it flows away from the equator, cools, and sinks again. Much of it recirculates back to the same areas it left, but the remainder flows toward the poles, where eventually it is chilled. Similar upward movements of warm air and its subsequent flow toward the poles occur at higher latitudes (further from the equator) as well (Fig. 53–16). As air cools, it sinks and flows toward the equator, generally beneath the sheets of warm air flowing toward the pole at the same time. The constant motion of air transfers heat from the equator toward the poles, and as the air returns, it cools the land over which it passes. This constant turnover does not equalize temperatures over Earth's surface, but it does moderate them.

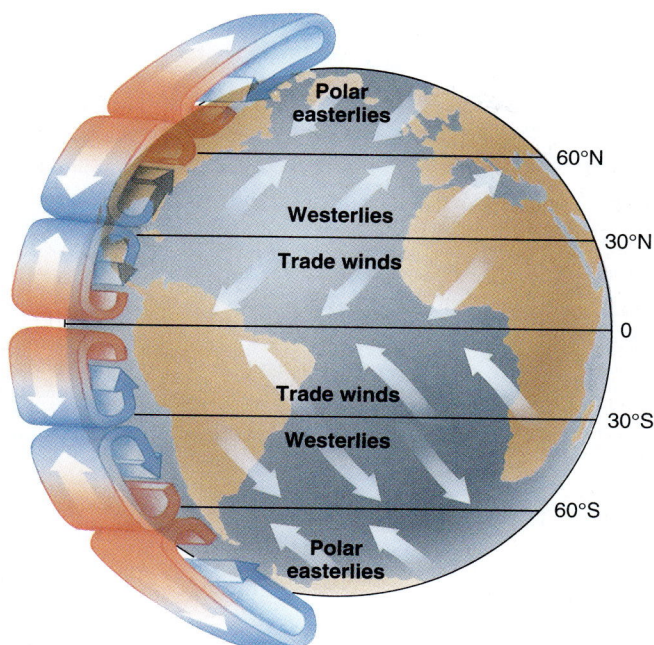

Figure 53–16 **Atmospheric circulation.** The greatest solar energy input occurs at the equator, heating air most strongly in that area. The air rises and travels poleward *(left)* but is cooled in the process so that much of it descends again around 30 degrees latitude in both hemispheres. At higher latitudes the patterns of air movement are more complex.

The atmosphere exhibits complex horizontal movements

In addition to global circulation patterns, the atmosphere exhibits complex horizontal movements that are commonly referred to as **winds.** The nature of wind, with its turbulent gusts, eddies, and lulls, is complex and difficult to understand or predict. It results in part from differences in atmospheric pressure and from the rotation of Earth. Air pressure is variable and changes with altitude, temperature, and humidity. Winds tend to blow from areas of high atmospheric pressure to areas of low pressure; the greater the difference between the high and low pressure areas, the stronger the wind.

Earth's rotation influences the direction that wind blows. Because Earth rotates from west to east, wind swerves to the right in the Northern Hemisphere and to the left in the Southern Hemisphere. This tendency of moving air to be deflected from its path by Earth's rotation is known as the **Coriolis effect.** The Coriolis effect can be visualized by imagining that you and a friend are sitting about 10 ft apart on a merry-go-round that is turning clockwise (Fig. 53–17). Suppose you throw a ball directly at your friend. By the time the ball arrives, your friend is no longer in that spot. The ball will have apparently swerved far to the left. This is how the Coriolis ef-

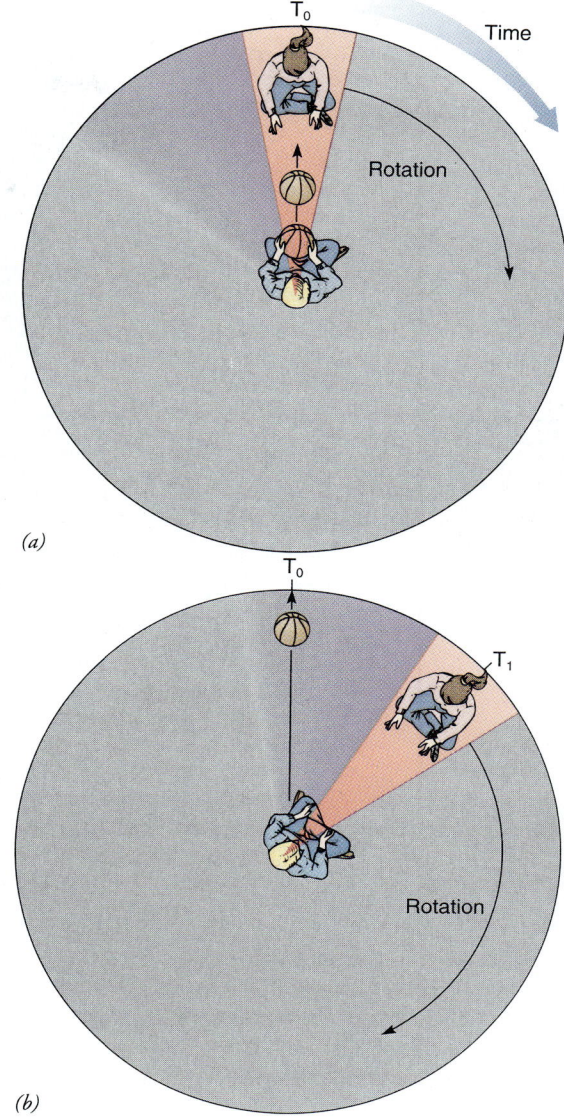

Figure 53–17 **The Coriolis effect as demonstrated by a merry-go-round.** The center of the merry-go-round *(shown from above)* is analogous to the South Pole, and the outer edge to the equator. *(a)* If you throw a ball to a friend at time zero (T_0), when the merry-go-round is rotating clockwise, the ball at time T_1 *(b)* appears to curve to the left instead of going straight. Because of Earth's rotation, winds appear to curve to the left in the Southern Hemisphere (and to the right in the Northern Hemisphere).

fect works in the Southern Hemisphere. To visualize how the Coriolis effect works in the Northern Hemisphere, imagine that you and your friend are sitting on the same merry-go-round, only this time it is moving counterclockwise. Now when you throw the ball, it will appear to swerve far to the right of your friend.

Figure 53–18 **Major surface ocean currents.** The basic pattern of ocean currents is caused largely by the action of winds. The clockwise flow in the Northern Hemisphere and counterclockwise flow in the Southern Hemisphere result partly from the Coriolis effect.

THE GLOBAL OCEAN COVERS MOST OF EARTH'S SURFACE

The global ocean is a huge body of salt water that surrounds the continents and covers almost three-fourths of Earth's surface. It is a single, continuous body of water, but geographers divide it into four sections separated by the continents: the Pacific, Atlantic, Indian, and Arctic Oceans. The Pacific Ocean is the largest by far: it covers one-third of Earth's surface and contains more than half of Earth's water.

Surface ocean currents are driven by winds

The persistent prevailing winds blowing over the ocean produce mass movements of surface ocean water known as **ocean currents** (Fig. 53–18); circular ocean currents generated by the prevailing winds are called *gyres*. The paths that surface ocean currents travel are partly caused by the Coriolis effect. Earth's rotation from west to east causes surface ocean currents to swerve to the right in the Northern Hemisphere, producing a clockwise gyre of water currents. In the Southern Hemisphere, ocean currents swerve to the left, producing a counterclockwise gyre.

The ocean interacts with the atmosphere

The ocean and the atmosphere are strongly linked, with wind from the atmosphere affecting the ocean currents, and heat from the ocean affecting atmospheric circulation. One of the best examples of the interaction between ocean and atmosphere is the **El Niño–Southern Oscillation (ENSO)** event. ENSO is a periodic warming of surface waters of the tropical East Pacific that alters both oceanic and atmospheric circulation patterns and results in unusual weather in areas far from the tropical Pacific. Normally, westward-blowing trade winds restrict the warmest waters to the western Pacific (near Australia). Every three to seven years, however, the trade winds weaken and the warm water mass expands eastward to South America, raising surface temperatures in the East Pacific to 3° to 5° C above average. Ocean currents, which normally flow westward in this area, slow down, stop altogether, or even reverse and go eastward. The phenomena is called *El Niño* (Spanish for "the child") because the warming usually reaches the fishing grounds off Peru just before Christmas.

An ENSO event changes biological productivity in parts of the ocean. The warmer sea surface temperatures and accompanying changes in ocean circulation patterns off the west coast of South America prevent colder, nutrient-laden deeper waters from **upwelling** (coming to the surface). This results in a severe drop in the populations of anchovies and many other marine fishes. For example, during the 1982 to 1983 ENSO, one of the worst ever recorded, the anchovy population decreased by 99%, aided, in part, by overfishing. Other species, such as shrimp and scallops, thrive during an ENSO event. Along the Pacific coast of North America, ENSO shifts the distribution of tropical fishes northward and even affects how the salmon run in Alaska.

TEMPERATURE AND PRECIPITATION ARE ASPECTS OF CLIMATE THAT PROFOUNDLY AFFECT ORGANISMS

Weather refers to the conditions in the atmosphere at a given place and time; it includes temperature, atmospheric pressure, precipitation, cloudiness, humidity, and wind. Weather changes from one hour to the next and from one day to the next.

Climate comprises the average weather conditions plus extremes (records) that occur in a given place over a period of years. The two most important factors that help determine an area's climate are temperature (both average temperature and temperature extremes) and precipitation (both average precipitation and seasonal distribution). Other climatic factors include wind, humidity, fog, cloud cover, and lightning-caused wildfires.

Day-to-day variations, day-to-night variations, and seasonal variations in climatic factors are also important dimensions of climate that affect organisms. Temperature, precipitation, and other aspects of climate are influenced by latitude, elevation, topography, vegetation, distance from the ocean or other large bodies of water, and location on a continent or other land mass. Unlike weather, which changes rapidly, climate generally changes slowly, over hundreds or thousands of years.

Earth has many different climates, and because they are relatively constant for many years, organisms have adapted to them. The wide variety of organisms on Earth evolved in part because of the large number of different climates, ranging from cold, snow-covered, polar climates to hot, tropical climates where it rains almost every day.

Air and water movements and surface features affect precipitation patterns

Precipitation refers to any form of water, such as rain, snow, sleet, and hail, that falls to the surface from the atmosphere. Precipitation varies from one location to another and has a profound effect on the distribution and kinds of organisms.

Differences in precipitation depend on several factors. The heavy-rainfall areas of the tropics result mainly from the uplifting of moisture-laden air. High surface-water temperatures (recall the enormous amount of solar energy striking the equator) cause the evaporation of vast quantities of water from tropical parts of the ocean. Prevailing winds blow the resulting moist warmer air over land masses. Heating of air by land surfaces warmed by the sun causes moist air to rise. As it rises, the air cools and its moisture content decreases (less moisture is present in cool air than warm air because there is less heat to keep water in a vapor state). When air reaches its saturation point (when it cannot hold any additional water vapor), clouds form and water may be released as precipitation. The air eventually returns to the surface on both sides of the equator between the Tropics of Cancer and Capricorn (latitudes 23.5 degrees north and south, respectively). By then, most of the moisture has precipitated so that dry air returns to the equator. This dry air makes little biological difference over the ocean, but its lack of moisture produces some of the great subtropical deserts, such as the Sahara.

Air may also be dried during long journeys over land masses. Near the windward (side from which the prevailing wind blows) coasts of continents, rainfall may be heavy. However, in the temperate zones—the areas between the tropical and polar zones—continental interiors are usually dry because they are far from the ocean that replenishes water vapor in the air passing over it.

Moisture is removed from humid air by mountains, which force air to rise. As it gains altitude, the air cools, clouds form, and precipitation typically occurs—primarily on the windward slopes of the mountains. As the air mass moves down on the other side of the mountain, it is warmed and clouds evaporate, thereby lessening the chance of precipitation of any remaining moisture. This situation exists on the west coast of North America, where precipitation falls on the western slopes of mountains that are close to the coast. The dry lands on the sides of the mountains away from the prevailing wind (in this case, east of the mountain range) are called **rain shadows** (Fig. 53–19).

Figure 53–19 **Rain shadow.** When wind blows moist air over a mountain range, precipitation occurs on the windward side of the mountain, causing a dry rain shadow on the leeward side. Such a rain shadow occurs in Washington state east of the Cascade Mountains. The west side of Olympic National Park receives more than 500 cm of precipitation annually, whereas the east side gets 40–50 cm.

Microclimates are local variations in climate

Differences in elevation, in the steepness and direction of slopes, and therefore in exposure to sunlight and prevailing winds may produce local variations in climate known as **microclimates,** which can be quite different from their overall surroundings. Patches of sun and shade on a forest floor, for example, produce a variety of microclimates for plants, animals, and microorganisms living there. The microclimate of an organism's habitat is of primary importance because that is the climate an organism actually experiences and must cope with. (Keep in mind, however, that microclimates are largely affected by the regional climate in which they are embedded.)

Sometimes organisms modify their own microclimate. For instance, trees modify the local climate within a forest so that in summer the temperature is usually lower, and the relative humidity greater, than outside the forest. The temperature and humidity beneath the litter of the forest floor differ still more; in the summer the microclimate of this area is considerably cooler and moister than the surrounding forest. As another example, many desert-dwelling animals burrow to avoid surface climatic conditions that would kill them in minutes. The cooler daytime microclimate in their burrows permits them to survive until night, when the surface cools off and they can come out to forage or hunt.

FIRES ARE A COMMON DISTURBANCE IN SOME AREAS

Wildfires are an important ecological force in many geographical areas. Those areas most prone to wildfires have wet seasons followed by dry seasons. Vegetation that grows and accumulates during the wet season dries out enough during the dry season to burn easily. When lightning hits vegetation or ground litter, it ignites the dry organic material, and a fire spreads through the area.

Fires have several effects on organisms. First, combustion frees the minerals that were locked in dry organic matter. The ashes remaining after a fire are rich in potassium, phosphorus, calcium, and other minerals essential for plant growth. Thus with the arrival of precipitation, vegetation generally flourishes following a fire. Second, fire removes plant cover and exposes the soil. This change stimulates the germination and establishment of seeds requiring bare soil, as well as encourages the growth of shade-intolerant plants. Sometimes, however, heavy precipitation and/or high winds may strike an area before the vegetative cover is reestablished. In these cases, accelerated erosion may remove valuable nutrients from the site.

Some plants are fire-adapted. Grasses and some shrubs, for example, have underground stems and buds that are typically unaffected by a fire sweeping over them. After the aerial parts have been killed by fire, the underground parts send up new sprouts. The availability of nutrients in the ash speed the regrowth of these plants. Fire-adapted trees such as bur oak and ponderosa pine have a thick bark that is resistant to fire. (In contrast, fire-sensitive trees, such as many hardwoods, have a thin bark.) Certain pines such as jack and lodgepole pine depend on fire for successful reproduction, because the heat of the fire opens the cones so that the seeds can be released.

Fires were a part of the natural environment long before humans appeared, and many terrestrial ecosystems have adapted to it. African savanna, California chaparral, North American grasslands, and ponderosa pine forests of the western United States are some fire-adapted ecosystems (see Chapter 54). For example, fire helps maintain grasses as the

Figure 53–20 **Fire as a tool of ecological management.** Here, a controlled burn helps maintain a ponderosa pine *(Pinus ponderosa)* stand in Oregon. *(Joan Landsberg, USDA Forest Service)*

dominant vegetation in grasslands by removing fire-sensitive hardwood trees.

The influence of fire on plants became even more pronounced once humans evolved. Because humans deliberately and accidentally set fires, fire became a more common occurrence. Humans set fires for many reasons: to provide the grasses and shrubs that many game animals require; to clear the land for agriculture and human development; and to reduce enemy cover in times of war.

Humans also try to prevent fires, and sometimes this effort can have disastrous consequences. When fire is excluded from a fire-adapted ecosystem, organic litter accumulates. As a result, when a fire does occur, it is much more destructive. The deadly fire in Colorado during the summer of 1994, which claimed the lives of 14 firefighters, was blamed in part on decades of suppressing fires in the region. Prevention of fire also converts grassland to woody vegetation and facilitates the invasion of fire-sensitive trees into fire-adapted forests.

Controlled burning is a tool of ecological management in which the plant litter is deliberately burned before it has accumulated to dangerous levels (Fig. 53–20). Controlled burns are also used to suppress fire-sensitive trees, thereby maintaining the natural fire-adapted ecosystem.

SUMMARY WITH KEY TERMS

I. **Energy flow** through an ecosystem is in a linear direction, from the sun to producer to consumer to decomposer. Much of this energy is converted to heat as it moves from one organism to another and is therefore unusable by organisms occupying the next higher trophic level.
   A. Trophic relationships may be expressed as **food chains** or, more realistically, as **food webs,** which show the many alternative pathways that energy may take among the producers, consumers, and decomposers of an ecosystem.
   B. **Ecological pyramids** typically express the progressive reduction in numbers of organisms, biomass, and energy found in successively higher trophic levels.
   C. **Gross primary productivity** of an ecosystem is the rate at which energy captured by photosynthesis accumulates as biomass. **Net primary productivity** is the energy that remains (as biomass) after plants carry out cellular respiration.
      1. The two most productive terrestrial ecosystems are wetlands (swamps and marshes) and tropical rain forests.
      2. The two least productive terrestrial ecosystems are tundra and desert.

II. **Biogeochemical cycles** are the cycling of matter from the abiotic environment to organisms and then back to the abiotic environment.
   A. Carbon dioxide is the important gas of the **carbon cycle.**
      1. Carbon enters plants, algae, and cyanobacteria as CO_2, which is incorporated into organic molecules by photosynthesis.
      2. Cellular respiration, combustion, and weathering return CO_2 to the atmosphere, making it available for producers again.
   B. The **nitrogen cycle** has five steps.
      1. Biological **nitrogen fixation** is the conversion of nitrogen gas to ammonia, an important form of nitrogen for certain plants.
      2. **Nitrification** is the biological conversion of ammonia to nitrate, an important form of nitrogen used by plants.
      3. **Assimilation** is the biological conversion of nitrates or ammonia to proteins, chlorophyll, and other nitrogen-containing compounds by plants. The conversion of plant proteins into animal proteins is also considered assimilation.
      4. **Ammonification** is the biological conversion of organic nitrogen to ammonia.
      5. **Denitrification** is the biological conversion of nitrate to nitrogen gas.
   C. The **phosphorus cycle** has no biologically important gaseous compounds.
      1. Phosphorus erodes from rock as inorganic phosphate and is absorbed from the soil by the roots of plants. Plants incorporate phosphorus into such compounds as nucleic acids and phospholipids.
      2. Animals obtain the phosphorus they need from their diets. Decomposers release inorganic phosphate into the environment.

      3. Phosphorus can be lost from biological cycles for millions of years when it washes into the ocean and is subsequently deposited in sea beds.
   D. The **hydrological cycle,** which continually renews the supply of water so essential to life, involves an exchange of water between the land, the atmosphere, and organisms.
      1. Water enters the atmosphere by evaporation and **transpiration** and leaves the atmosphere as precipitation.
      2. On land, water filters through the ground or runs off to lakes, rivers, and the ocean. The underground caverns and porous layers of rock in which groundwater is stored are called aquifers. The movement of surface water from land to ocean is called runoff.

III. The unique planetary environment of Earth makes life possible. Sunlight is the primary (almost sole) source of energy available to the biosphere.
   A. Of the solar energy that reaches Earth, 30% is immediately reflected away and the remaining 70% is absorbed by the atmosphere and surface.
   B. Ultimately, all absorbed solar energy is reradiated into space as infrared (heat) radiation.
   C. A combination of Earth's roughly spherical shape and the tilted angle of its axis concentrates solar energy at the equator and dilutes it at the poles.
      1. The tropics are therefore hotter and less variable in climate than are temperate and polar areas.
      2. Seasons are determined by two main factors: the inclination of Earth's axis (the more important factor) and the distance from the sun, which varies during the year.

IV. The atmosphere protects the surface from most of the sun's ultraviolet radiation and x rays, and from lethal amounts of cosmic rays from space. At the same time, visible light and some infrared radiation penetrate to warm the surface and the lower part of the atmosphere.
   A. Atmospheric heat transferred from the equator to the poles produces both a movement of warm air toward the poles and of cool air toward the equator, thus moderating the extremes of global climate.
   B. Winds result in part from differences in atmospheric pressure and from the **Coriolis effect.**

V. The global ocean is a single, continuous body of water that surrounds and covers almost three-fourths of Earth's surface.
   A. Surface **ocean currents** result in part from prevailing winds and the Coriolis effect.
   B. The **El Niño–Southern Oscillation (ENSO)** event alters both ocean and atmospheric currents, results in unusual weather patterns, and has significant ecological effects.

VI. An area's **climate** comprises the average weather conditions plus extremes that occur there over a period of years.
   A. The two most important factors that help determine an area's climate

are temperature (both average temperature and temperature extremes) and precipitation (both average precipitation and seasonal distribution).

B. Precipitation is greatest where warm air passes over the ocean, absorbing moisture, and then cools, such as when humid air is forced upward by mountains.

C. Deserts develop in the **rain shadows** of mountain ranges or in continental interiors.

D. **Microclimates,** which are local variations in climate, produce a variety of climatic conditions for organisms in a given habitat.

VII. Many ecosystems, such as savanna, chaparral, grasslands, and certain forests, contain fire-adapted organisms.

POST-TEST

1. A community and its abiotic environment best defines a(an) (a) biogeochemical cycle (b) biosphere (c) ecosystem (d) food web (e) trophic level

2. The movement of matter is _____ in ecosystems, and the movement of energy is _____. (a) linear; linear (b) linear; cyclic (c) cyclic; cyclic (d) cyclic; linear

3. A complex of interconnected food chains in an ecosystem is called a(an): (a) ecosystem (b) pyramid of numbers (c) pyramid of biomass (d) biosphere (e) food web

4. The quantitative estimate of the total amount of living material is called (a) biomass (b) energy flow (c) gross primary productivity (d) plant respiration

5. Which of the following equations shows the relationship between gross and net primary productivities? (a) GPP = NPP − photosynthesis (b) NPP = GPP − photosynthesis (c) GPP = NPP − plant respiration (d) NPP = GPP − plant respiration (e) NPP = GPP − animal respiration

6. Which of the following processes increase(s) the amount of atmospheric carbon in the carbon cycle? (a) photosynthesis (b) cellular respiration (c) combustion (d) both a and c (e) both b and c

7. In the nitrogen cycle, gaseous nitrogen is converted to ammonia by (a) nitrogen fixation (b) nitrification (c) assimilation (d) ammonification (e) denitrification

8. The conversion of ammonia to nitrate, known as _____, is a two-step process performed by soil bacteria. (a) nitrogen fixation (b) nitrification (c) assimilation (d) ammonification (e) denitrification

9. This biogeochemical cycle does not have a gaseous component but cycles from the land to sediments in the ocean and back to the land. (a) carbon cycle (b) nitrogen cycle (c) phosphorus cycle (d) hydrological cycle (e) neither a nor b have a gaseous component

10. The _____, which results from the rotation of Earth, displaces the paths of atmospheric and oceanic currents to the right (Northern Hemisphere) and the left (Southern Hemisphere). (a) upwelling (b) prevailing wind (c) ocean current (d) El Niño–Southern Oscillation (e) Coriolis effect

11. The periodic warming of surface waters of the tropical East Pacific that alters both oceanic and atmospheric circulation patterns is known as (a) upwelling (b) prevailing wind (c) ocean current (d) El Niño–Southern Oscillation (e) Coriolis effect

12. A mountain range may produce a downwind arid (a) upwelling (b) rain shadow (c) ocean current (d) microclimate (e) ecological pyramid

REVIEW QUESTIONS

1. Describe energy flow through a food web.
2. What are trophic levels and how are they related to ecological pyramids?
3. Define gross primary productivity (GPP) and net primary productivity (NPP).
4. Why is the cycling of matter essential to the long-term continuance of life?
5. Diagram the carbon cycle, including the following processes: photosynthesis, cellular respiration, combustion, and erosion.
6. List and describe the five steps in the nitrogen cycle.
7. What basic forces determine the circulation of the atmosphere? Describe the general directions of atmospheric circulation.
8. What basic forces produce the main ocean currents? Describe the general directions of these currents.
9. What are some of the factors that produce regional differences in precipitation?
10. Label the diagram showing energy flow through ecosystems. Use Figure 53–1 to check your answers.

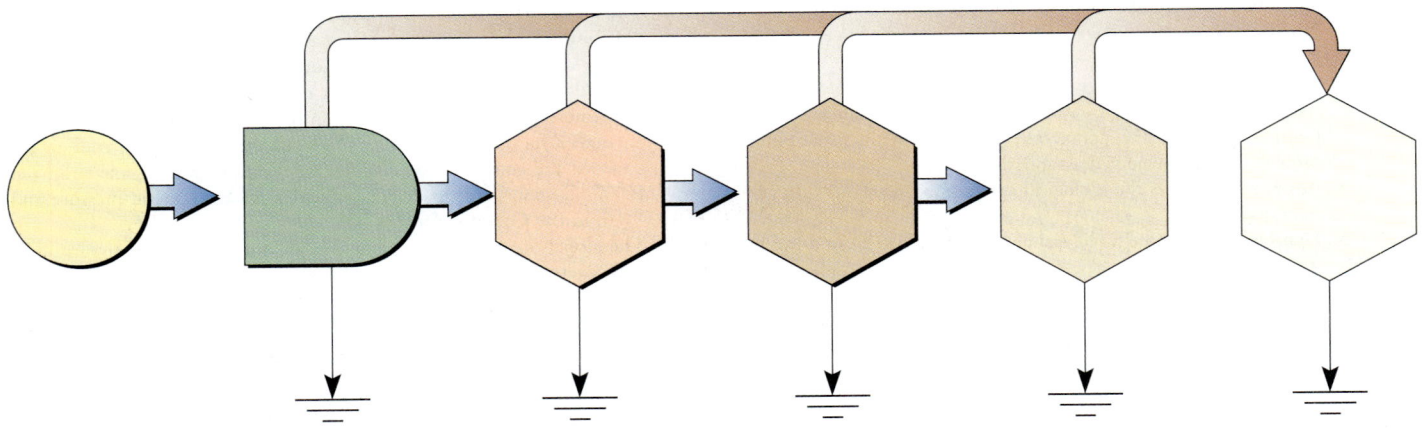

YOU MAKE THE CONNECTION

1. Describe the simplest stable ecosystem that you can imagine.

2. Would the microclimate of an ant be the same as that of an elephant living in the same area? Why or why not?

RECOMMENDED READINGS

Asner, G.P., T.R. Seastedt, and A.R. Townsend. "The Decoupling of Terrestrial Carbon and Nitrogen Cycles." *BioScience*, Vol. 47, No. 4, Apr. 1997. Human use of land, such as conversion of forest to agricultural land, and fertilizer production are changing natural links between the carbon and nitrogen cycles.

Brewer, R. *The Science of Ecology*, 2nd ed. Saunders College Publishing, Philadelphia, 1994. A readable general textbook on the principles of ecology, including ecosystem ecology.

Caraco, N.F. "Disturbance of the Phosphorus Cycle: A Case of Indirect Effects of Human Activity." *Tree*, Vol. 8, No. 2, Feb. 1993. Humans have affected the phosphorus cycle both directly and indirectly.

Chahine, M.T. "The Hydrologic Cycle and Its Influence on Climate." *Nature*, Vol. 359, 1 Oct. 1992. A review of our current understanding of how the hydrological cycle affects global climate.

Raven, P.H., L.R. Berg, and G.B. Johnson. *Environment*, 2nd ed. Saunders College Publishing, Philadelphia, 1998. This environmental science text offers a detailed situation analysis of planet Earth, including how humans interact with and affect its physical and living systems.

Schoen, D. "Primary Productivity: The Link to Global Health." *BioScience*, Vol. 47, No. 8, Sep. 1997. Examines an important question that scientists are currently addressing: how terrestrial primary productivity will respond to climate change.

Smil, V. "Global Population and the Nitrogen Cycle." *Scientific American*, Vol. 277, No. 1, Jul. 1997. Humans have had a global impact on the nitrogen cycle, largely due to our increasing production of nitrogen fertilizer to provide food for a growing human population.

Trefil, J. "Nitrogen." *Smithsonian*, Oct. 1997. An excellent overview of the nitrogen cycle and human-produced nitrogen pollution.

● Visit our website at **http://www.saunderscollege.com/lifesci/titles.html** and click on Solomon/Berg/Martin Biology for links to chapter-related resources on the World Wide Web.

CHAPTER 54

Ecology and the Geography of Life

In Chapter 53 you learned that climate, particularly temperature and precipitation, provides Earth with many different terrestrial environments. In each major kind of climate, a distinctive vegetation develops. Tropical rainforest plants, for example, are associated with warm, humid climates. The species within tropical rain forests may vary depending on their locations, but the overall appearance of each rain forest is much the same. You can easily recognize certain areas of South America, Africa, and Southeast Asia as tropical rain forests, although each is unique in its species composition. All tropical rain forests have enough characteristics in common to be collectively described as a single large terrestrial ecosystem, or *biome.*

In this chapter we examine some of the features of Earth's major biomes. These include the major biomes in North America: tundra, taiga, temperate rain forest, temperate deciduous forest, temperate grassland, chaparral, and desert. Tropical rain forest and savanna, which are important biomes of tropical regions, are also discussed.

We then turn our attention to aquatic ecosystems and examine the main features of both freshwater and saltwater habitats. As on land, the physical factors of aquatic ecosystems largely determine which organisms characterize each. However, the producers are often not as evident in aquatic ecosystems as they are in biomes. In the photograph, you cannot see the microscopic algae that form the base of the food web in the open ocean. The school of fast-swimming horse-eye jacks *(Caranx latus)* pictured here feed on a wide variety of invertebrates and fishes, which in turn consume smaller animals that graze on the algae.

Freshwater ecosystems include streams and rivers (flowing-water ecosystems), ponds and lakes (standing-water ecosystems), and marshes and swamps (freshwater wetlands). Estuaries are aquatic ecosystems that occur where fresh water and salt water meet and include salt marshes and mangrove forests. The ocean, which occupies about 70 percent of Earth's surface, consists of several major aquatic ecosystems that vary as much or more than the biomes on land. These include the challenging environment dominated by the tides (the intertidal zone), the ocean floor (the benthic environment), shallow water close to shore (the neritic province), and the deep ocean (the oceanic province).

(Larry Lipske/DRK Photo)

AFTER YOU HAVE STUDIED THIS CHAPTER YOU SHOULD BE ABLE TO

1. Define biome and briefly describe the nine major terrestrial biomes, giving attention to the climate, soil, and characteristic plants and animals of each.
2. Describe at least one human effect on each of the biomes described.
3. Explain the similarities in the changes in vegetation observed with increasing altitude and increasing latitude.
4. Explain the important environmental factors that affect aquatic ecosystems.

5. Distinguish among plankton, nekton, and benthos.
6. Briefly describe the various freshwater, estuarine, and marine ecosystems, giving attention to the environmental characteristics and representative organisms of each.
7. Describe at least one human effect on each of the aquatic ecosystems described.
8. Describe thermal stratification in a temperate lake.
9. Define and describe some of the features of an ecotone.

NATURAL SELECTION ADAPTS EACH ORGANISM TO ITS ENVIRONMENT

Natural selection determines an organism's ability to survive and reproduce in a given environment (see Chapter 17). In natural selection, both **abiotic** (nonliving) and **biotic** (living) factors act to eliminate the least fit individuals in a given population. Over time, each succeeding generation of organisms that live in each biome or major aquatic ecosystem becomes better adapted to local environmental conditions.

To demonstrate how each organism is well adapted to its environment, consider the black-tailed prairie dog *(Cynomys ludovicianus),* one of several prairie dog species. These small rodents live in the grasslands of western North America, from southern Canada to northern Mexico. Black-tailed prairie dogs are not found in temperate grasslands on other continents, although they could presumably survive there. A species occurs at a given location either because it evolved or migrated there. Black-tailed prairie dogs evolved in North America and were prevented from migrating to similar grasslands by biological and geographical barriers.

Black-tailed prairie dogs are superbly adapted to their environment. Their teeth and digestive tracts are modified to eat and easily digest the seeds and leaves of grasses that grow in great profusion on the Great Plains. They have an enlarged pair of incisors, for example, that are used to nibble grasses and gnaw through seeds. Prairie dogs also eat insects and underground roots and tubers.

Black-tailed prairie dogs have many behavioral adaptations that protect them from predators, such as coyotes, eagles, hawks, badgers, and ferrets. For example, prairie dogs are social animals and live in large colonies of about 500 individuals; the eyes of every individual in the colony watch for potential danger, and when they see it, prairie dogs call out to warn the rest of the colony. In fact, they are called "dogs" not because they are related to dogs (they aren't), but because their warning calls are similar to a dog's bark.

Black-tailed prairie dogs are burrowing animals (Fig. 54–1). When danger approaches—a coyote, perhaps—each prairie dog dives into its underground home. Their burrows have at least two openings and consist of an elaborate network of long tunnels with several chambers. One room may serve as a nursery for the young, one as a sleeping chamber, one as a toilet chamber, and one, which is close to an entrance, as a listening chamber. The burrow entrances are surrounded by high piles of excavated soil that help prevent flooding during rainstorms.

Black-tailed prairie dogs survive winter by hibernating in their burrows. Their metabolism slows, and they exist on the stored fat in their bodies. They do not go into deep hibernation, however, and may come out of their burrows to look for food when the weather warms.

Figure 54–1 **A black-tailed prairie dog.** Prairie dogs never wander far from their burrows, which they use to escape from predators. Photographed in North Dakota. *(Steve Kaufman/Peter Arnold, Inc.)*

Some ranchers view prairie dogs as a nuisance because they dig burrows in fields and pastures. Ranchers fear that their livestock will step into the burrow entrances and injure their legs. (Horses and cattle rarely step into the entrances, however.) Ranchers also think that prairie dogs compete with livestock for grasses. Despite extensive trapping and poisoning, however, these resourceful animals persist.

Why aren't prairie dogs found in other environments, such as deserts or deciduous forests? Natural selection has adapted prairie dogs to the environmental conditions found in temperate grasslands, and prairie dogs can usually tolerate most environmental extremes in that biome. In adapting a population to local conditions, however, natural selection may limit its successful dispersal to and establishment in other places with different environmental conditions.

Like prairie dogs, each species has evolved structural, behavioral, and physiological features that help adapt it to its own particular environmental and lifestyle. These features often limit its ability to survive and reproduce elsewhere where different conditions prevail, however. As we examine Earth's major biomes and aquatic ecosystems, including the species characteristic of each, think about the variety of adaptations that natural selection has produced in organisms in response to their particular environments.

BIOMES ARE LARGELY DETERMINED BY CLIMATE

A **biome** is a large, relatively distinct terrestrial region characterized by similar climate, soil, plants, and animals regardless of where it occurs. Because it is so large in area, a biome encompasses a number of interacting ecosystems. In terrestrial ecology, a biome is considered the next level of ecological organization above that of ecosystem.

A biome's boundaries are largely determined by invisible climate barriers, with temperature and precipitation being most important (Fig. 54–2; climate was discussed in Chapter

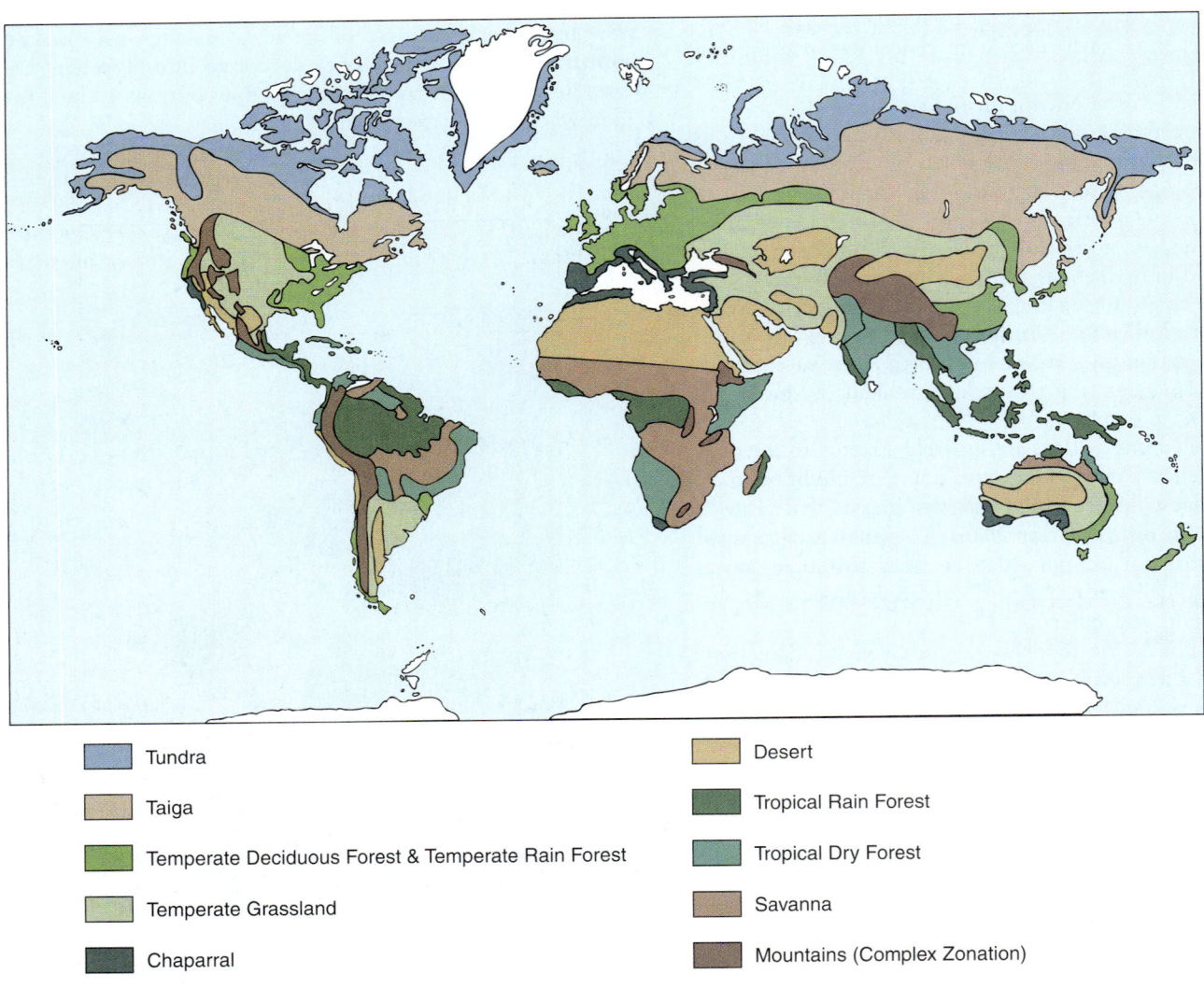

Tundra		Desert	
Taiga		Tropical Rain Forest	
Temperate Deciduous Forest & Temperate Rain Forest		Tropical Dry Forest	
Temperate Grassland		Savanna	
Chaparral		Mountains (Complex Zonation)	

Figure 54–2 The world's major biomes. This highly simplified diagram shows sharp boundaries between biomes. Biomes actually intergrade at their boundaries. *(Adapted from Odum, 1971)*

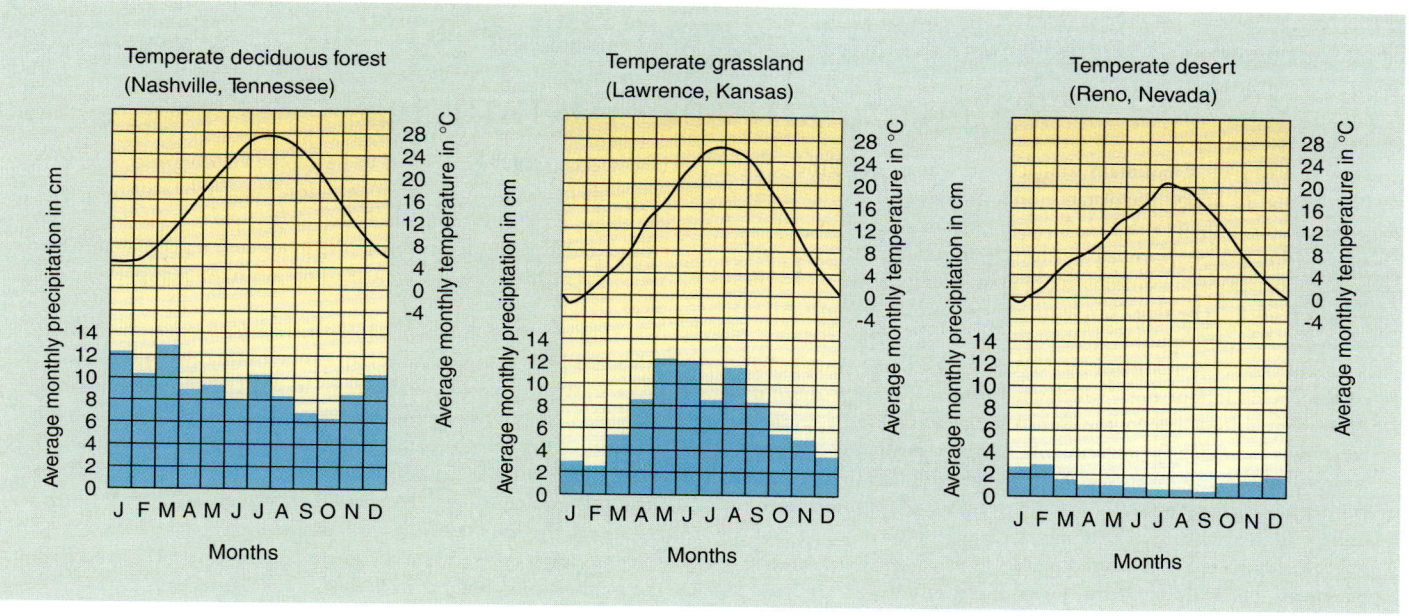

Figure 54-3 **Significance of precipitation in temperate biomes.** Average monthly temperature *(black)* and precipitation *(blue)* are given for three temperate biomes in North America. Although temperature is approximately the same in all three locations, precipitation varies a great deal, resulting in deciduous forest where precipitation is plentiful, grassland where it is less plentiful and more seasonal, and desert where it is quite low.

53). Near the poles, temperature is generally the overriding climate factor, whereas in temperate and tropical regions, precipitation becomes more significant than temperature (Fig. 54–3). Light is relatively plentiful in biomes, except in certain environments such as the rainforest floor. Other abiotic factors to which biomes are sensitive include temperature extremes as well as rapid temperature changes, fires, floods, droughts, and strong winds. Nine major biomes are discussed: tundra, taiga, temperate rain forest, temperate deciduous forest, temperate grassland, chaparral, desert, savanna, and tropical rain forest.

Tundra is the cold, boggy plains of the far north

Tundra (also called **arctic tundra**) occurs in extreme northern latitudes wherever snow melts seasonally (Fig. 54–4; also see Fig. 54–2). The Southern Hemisphere has no equivalent of the arctic tundra because it has no land in the proper latitudes. A similar ecosystem located in the higher elevations of mountains, above the tree line, is called **alpine tundra** to distinguish it from arctic tundra (see *Making the Connection: Comparing Altitudes and Latitudes*).

Arctic tundra has long, harsh winters and extremely short summers. Although the growing season, with its warmer temperatures, is as short as 50 days (depending on location), the days are long. Above the Arctic Circle the sun does not set at all for many days in midsummer, although the amount of light

at midnight is only one-tenth that at noon. There is little precipitation (10 to 25 cm, or 4 to 10 in, per year) over much of the tundra, with most of it falling during summer months.

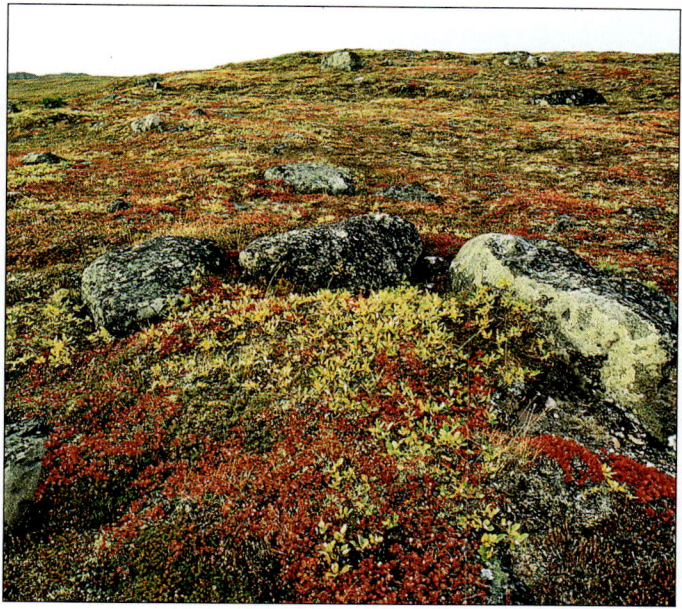

Figure 54-4 **Arctic tundra in Northwest Territories, Canada.** Because of its short growing season and permafrost, only small, hardy plants grow in the northernmost biome that encircles the Arctic Ocean. Photograph was taken in autumn. *(Eastcott/Momatiuk/ Earth Scenes)*

COMPARING ALTITUDES AND LATITUDES

How is hiking up a mountain similar to traveling toward the North Pole? Although the distance up a mountain is shorter than an excursion to the North Pole, the same major vegetation patterns are encountered *(see figure)*. This altitude-latitude similarity occurs because the temperature drops going up a mountain just as it does on a trip north. The types of plants growing on the mountain change as the temperatures change.

The base of a mountain in Colorado, for example, might be covered by deciduous trees, which shed their leaves every autumn. Above that altitude, where the climate is colder and more severe, a coniferous *subalpine forest* resembling northern taiga grows. Spruces and firs are the dominant trees here. Higher still, at the tree line, stunted pines cling to life. Above the tree line, where the climate is quite cold, a kind of tundra occurs, with vegetation composed of grasses, sedges, and small tufted plants, most of which are hardy perennials. Some alpine plants, for example, buttercups, are low-

land species that have adapted to the alpine environment, whereas other plants, for example, mountain douglasia, live exclusively in the mountains. This tundra is called *alpine tundra* to distinguish it from arctic tundra. At the top of the mountain, a permanent ice or snow cap might be found, similar to the nearly lifeless polar land areas.

Important environmental differences exist between high altitudes and high latitudes, however. Alpine tundra typically lacks permafrost and receives more precipitation than does arctic tundra. Also, high elevations of temperate mountains do not have the great extremes of day length that are associated with the changing seasons in high-latitude biomes. Furthermore, the intensity of solar radiation is greater at high elevations than at high latitudes. At high elevations, the sun's rays pass through less atmosphere, which results in greater exposure to ultraviolet radiation (less UV is filtered out by the atmosphere) than occurs at high latitudes.

ALTITUDE ZONES

High
Snow/ice

Alpine tundra

Subalpine coniferous forest

Low
Deciduous forest

LATITUDE ZONES

North Pole
Polar region

Arctic tundra

Northern taiga

Mid-latitudes
Temperate forest

A comparison of altitude and latitude zones. The cooler temperatures at higher elevations of a mountain produce a series of ecosystems similar to those encountered when going from the equator toward the North Pole.

Tundra soils tend to be geologically young, since most were formed only after the last Ice Age.[1] These soils are usually nutrient-poor and have little organic litter (dead leaves and stems, animal droppings, and remains of organisms) in the uppermost layer of soil. Although the soil surface melts during the summer, tundra has a layer of permanently frozen ground called **permafrost** that varies in depth and thickness. Because permafrost interferes with drainage, the thawed upper zone of soil is usually waterlogged during the summer. Permafrost limits the depth to which roots can penetrate, thereby preventing the establishment of most woody species. Limited precipitation, combined with low temperatures, flat topography (surface features), and permafrost, produces a landscape of broad shallow lakes, sluggish streams, and bogs.

Low species diversity and low primary productivity characterize tundra. Few plant species occur, but individual species often exist in great numbers. Tundra is dominated by mosses, lichens (such as reindeer moss), grasses, and grasslike sedges. No readily recognizable trees or shrubs grow except in sheltered locations, although dwarf willows, dwarf birches, and other dwarf trees are common. Tundra plants seldom grow taller than 30 cm (12 in) in open areas.

Year-round animal life of the tundra includes lemmings, voles, weasels, arctic foxes, gray wolves, snowshoe hares, ptarmigan, snowy owls, and musk-oxen. In the summer, caribou migrate north to the tundra to graze on sedges, grasses, and dwarf willow. Dozens of bird species also migrate north in summer to nest and feed on abundant insects. Mosquitoes, blackflies, and deerflies survive the winter as eggs or pupae and occur in great numbers during summer weeks. There are no reptiles or amphibians.

Tundra regenerates quite slowly after it has been disturbed. Even casual use by hikers can cause damage. Long-lasting injury, likely to persist for hundreds of years, has been done to large portions of the arctic tundra as a result of oil exploration and military use.

Taiga is an evergreen forest of the north

Just south of the tundra is the **taiga,** or **boreal forest,** which stretches across both North America and Eurasia and covers approximately 11% of Earth's land (Fig. 54–5; also see Fig. 54–2). A biome comparable to the taiga is not found in the Southern Hemisphere because it has no land at the corresponding latitudes. Winters are extremely cold and severe, although not as harsh as in the tundra. The growing season of the boreal forest is somewhat longer than that of the tundra. Taiga receives little precipitation, perhaps 50 cm (20 in) per year, and its soil is typically acidic, mineral-poor, and characterized by a deep layer of partly decomposed pine and spruce needles at the surface. Permafrost is patchy and, where found, is often deep underneath the soil. Taiga contains numerous

Figure 54–5   Taiga in Yukon, Canada. Taiga is coniferous forest that occurs in cold regions of the Northern Hemisphere adjacent to the tundra.   *(Beth Davidow/Visuals Unlimited)*

ponds and lakes in water-filled depressions that were dug by grinding ice sheets during the last Ice Age.

Black and white spruces, balsam fir, eastern larch, and other conifers (cone-bearing evergreens) clearly dominate the taiga, but **deciduous** trees such as aspen or birch, which shed their leaves in autumn, may form striking stands. Conifers have many drought-resistant adaptations, such as needle-like leaves with a minimal surface area to reduce water loss (see Chapter 32). Such an adaptation enables conifers to withstand the "drought" of the northern winter months, when roots cannot absorb water because the ground is frozen. Natural selection also favors conifers in the taiga because, being evergreen, they can resume photosynthesis as soon as warmer temperatures occur.

Animal life of the boreal forest includes some larger species such as caribou (which migrate from the tundra to the taiga for winter), wolves, bears, and moose. However, most animal life is medium-sized to small, and includes rodents, rabbits, and fur-bearing predators such as lynx, sable, and mink. Most species of birds are seasonally abundant but migrate to warmer climates for winter. Insects are numerous, but there are few amphibians and reptiles except in the southern taiga.

Most of the taiga is not suitable for agriculture because of its short growing season and mineral-poor soil. However, the boreal forest yields vast quantities of lumber and pulpwood (for making paper products), plus animal furs and other forest products. Currently, the boreal forests are the world's primary source of industrial wood and wood fiber. Beginning in the 1980s, extensive logging of certain boreal forests has

[1]Glacier ice, which occupied about 29% of Earth's land during the last Ice Age, began retreating about 17,000 years ago. Today, glacier ice occupies about 10% of the land surface.

occurred. The annual loss of boreal forests is estimated to encompass an area twice as large as the Amazonian rain forests of Brazil. About one million hectares (2.5 million acres) of Canadian forests are logged annually, and extensive tracts of Siberian forests in Russia are harvested, although exact estimates are unavailable.

Boreal forests are harvested primarily by clearcut logging, in which all trees are removed from an area. The land is then allowed to reseed and regenerate itself naturally or is planted with a few specific varieties of trees. Clearcutting is ecologically unsound because it destroys biological habitats and increases soil erosion, particularly on sloping land.

Temperate rain forest is characterized by cool weather, dense fog, and high precipitation

A coniferous **temperate rain forest** occurs on the northwest coast of North America. Similar vegetation exists in southeastern Australia and in southern South America (see Fig. 54–2). Annual precipitation in this biome is high, from 200 to 380 cm (80 to 152 in), and this is augmented by condensation of water from dense coastal fogs. The proximity of temperate rain forest to the coastline moderates the temperature so that seasonal fluctuation is narrow; winters are mild and summers are cool. Temperate rain forest has a relatively nutrient-poor soil, although its organic content may be high. (Cool temperatures slow the activity of bacteria and fungi. Thus, needles and large fallen branches and trunks accumulate on the ground as litter that takes many years to decay and release nutrients back to the soil.)

The dominant vegetation type in the North American temperate rain forest is large evergreen trees such as western hemlock, Douglas fir, Sitka spruce, and western red cedar. Temperate rain forest is rich in epiphytic vegetation, which are smaller plants that grow nonparasitically on the trunks and branches of large trees (Fig. 54–6). Epiphytes in this biome are mainly mosses, lichens, and ferns, all of which also carpet the ground. Deciduous shrubs such as vine maple grow wherever a break in the overlying canopy occurs. Squirrels, woodrats, mule deer, elk, numerous bird species, and several species of reptiles and amphibians are common temperate rain-forest animals.

Temperate rain forest is one of the richest wood producers in the world, supplying us with lumber and pulpwood. It is also one of the most complex ecosystems in terms of species diversity. Care must be taken to avoid overharvesting original old-growth forest, however, because such an ecosystem takes hundreds of years to develop. The logging industry typically harvests old-growth forest and replants the area with a monoculture (a single species) of trees that it can harvest in 40- to 100-year cycles. Thus, the old-growth forest ecosystem, once harvested, never has a chance to redevelop. A small fraction of the original, old-growth temperate rain forest in Washington, Oregon, and northern California remains untouched. At current logging rates, all unprotected old-growth forests in Washington and Oregon will be gone by 2023.

Figure 54–6   Temperate rain forest in Olympic National Park, Washington. Temperate rain forest is characterized by high amounts of precipitation. Note the epiphytes hanging from the branches of coniferous trees.   *(Terry Donnelly/Dembinsky Photo Associates)*

Temperate deciduous forest has a canopy of broad-leaf trees

Seasonality (hot summers and cold winters) is characteristic of **temperate deciduous forest,** which occurs in temperate areas where precipitation ranges from about 75 to 126 cm (30 to 50 in) annually (see Fig. 54–2). Typically, the soil of a temperate deciduous forest consists of a topsoil rich in organic material, and a deep, clay-rich lower layer. As organic materials decay, mineral ions are released. If they are not absorbed by the roots of living trees, these ions leach into the clay, where they may be retained.

Temperate deciduous forests of the northeastern and middle eastern United States are dominated by broad-leaf hardwood trees, such as oak, hickory, maple, and beech, that lose their foliage annually (Fig. 54–7). In southern reaches, the number of broad-leaf evergreen trees, such as magnolia, increases.

Temperate deciduous forest originally contained a variety of large mammals such as puma, wolves, bison, and other species now regionally extinct, plus deer, bears, many small mammals, and birds. Both reptiles and amphibians abounded, together with a denser and more varied insect life than exists today.

Figure 54–7 **Temperate deciduous forest in Pennsylvania.** Photograph was taken in autumn. The broad-leaf trees that dominate the temperate deciduous forest shed their leaves before winter. *(Barbara Miller/Biological Photo Service)*

As in Europe, much of the original temperate deciduous forest of North America was removed by logging and land clearing for farms, tree plantations, and cities. Where it has been allowed to regenerate, temperate deciduous forest is often in a seminatural state, that is, highly modified by humans for recreation, livestock foraging, timber harvest, and other uses. Although these returning forests do not have the biological diversity of virgin stands, many forest organisms have successfully become reestablished.

Worldwide, temperate deciduous forest was among the first biomes to be converted to agricultural use. In Europe and Asia, for example, many soils that originally supported temperate deciduous forest have been cultivated by traditional agricultural methods for thousands of years without substantial loss in fertility. During the 20th century, however, intensive agricultural practices have resulted in the widespread degradation of agricultural land. The first global assessment of soil conditions was released by the U.N. Environment Program in 1992. It reported that 1.96 billion hectares (4.84 billion acres) of soil—an area equal to 17% of the Earth's total vegetated surface area—have been degraded since the end of World War II. Soil degradation is primarily attributed to poor agricultural practices, overgrazing, and deforestation.

Temperate grasslands occur in areas of moderate precipitation

Summers are hot, winters are cold, and rainfall is often uncertain in **temperate grasslands** (see Fig. 54–2). Annual precipitation averages 25 to 75 cm (10 to 30 in). In grasslands with less precipitation, minerals tend to accumulate in a marked layer just below the topsoil. These minerals tend to leach out of the soil in areas with more precipitation. Grassland soil contains considerable organic material because aboveground portions of many grasses die off each winter and contribute to the organic content of the soil, while the roots and rhizomes (underground stems) survive underground. The roots that eventually die also contribute to the soil's organic material. Many grasses are sod formers: their roots and rhizomes form a thick, continuous underground mat.

Moist temperate grasslands, also known as *tallgrass prairies,* occur in the United States in Iowa, western Minnesota, eastern Nebraska, and parts of other Midwestern states. Although few trees grow except near rivers and streams, grasses grow in great profusion in the thick, rich soil (Fig. 54–8). Formerly, certain species of grass grew as tall as a person on horseback, and the land was covered with herds of grazing animals, particularly bison. The principal predators were wolves, although in sparser, drier areas their place was taken by coyotes. Smaller fauna included prairie dogs and their predators (foxes, black-footed ferrets, and various birds of prey), grouse, reptiles (such as snakes and lizards), and great numbers of insects.

Shortgrass prairies are temperate grasslands that receive less precipitation than the moister grasslands just described but more precipitation than deserts. In the United States, short-grass prairies occur in the eastern half of Montana, the western half of South Dakota, and parts of other Midwestern states. Shortgrass prairies are dominated by grasses that grow knee

Figure 54–8 **Bison grazing on temperate grassland in Oklahoma.** This tallgrass prairie is a preserve owned by the Nature Conservancy. Like other moist temperate grasslands, it is mostly treeless but contains a profusion of grasses and other herbaceous flowering plants. *(Harvey Payne)*

high or lower. The plants grow in less abundance than in the moister grasslands, and occasionally some bare soil is exposed. Native grasses of shortgrass prairies are drought-resistant.

The North American grassland, particularly the tallgrass prairie, was so well suited to agriculture that little of it remains. More than 90% has vanished under the plow, and the remainder is so fragmented that almost nowhere can we see even an approximation of what European settlers saw when they migrated into the Midwest. Today, the tallgrass prairie is considered North America's rarest biome. It is not surprising that the American Midwest, the Ukraine, and other moist temperate grasslands became the breadbaskets of the world, because they provide ideal growing conditions for crops such as corn and wheat, which are also grasses.

Chaparral is a thicket of evergreen shrubs and small trees

Some temperate environments have mild winters with abundant rainfall, combined with extremely dry summers. Such Mediterranean climates, as they are called, occur not only in the area around the Mediterranean Sea but also in California, western Australia, portions of Chile, and South Africa (see Fig. 54–2).[2] In southern California this environment is known as **chaparral.** Chaparral soil is thin and infertile. Frequent fires occur naturally in this environment, particularly in late summer and autumn (see section on fires in Chapter 53).

Chaparral vegetation looks strikingly similar in different areas of the world, even though the individual species are quite different. Chaparral is usually dominated by a dense growth of evergreen shrubs and may contain drought-resistant pine or scrub oak trees (Fig. 54–9). During the rainy winter season the landscape may be lush and green, but the plants lie dormant during the hot, dry summer. Trees and shrubs often have hard, small, leathery leaves that resist water loss. Many plants are also fire-adapted and grow best in the months following a fire. Such growth is possible because fire releases minerals that were tied up in above-ground parts of plants that burned. The underground parts of some plants and the seeds of many others are not killed by the fire, however, and with the new availability of essential minerals, plants sprout vigorously during winter rains. Mule deer, wood rats, chipmunks, lizards, and many species of birds are common animals of the chaparral.

Fires occur at irregular intervals in California chaparral vegetation. They are often quite costly because they consume expensive homes built on the hilly chaparral landscape. Unfortunately, efforts to control naturally occurring fires sometimes backfire. Denser, thicker vegetation tends to accumulate when periodic fires are prevented; then, when a fire does occur, it is much more severe. Removing the chaparral vegetation, whose roots hold the soil in place, can also cause problems: witness the mud flows that sometimes occur during winter rains in these areas.

[2]This vegetation type is known as maquis in the Mediterranean region, mallee scrub in Australia, matorral in Chile, and Cape scrub in Africa.

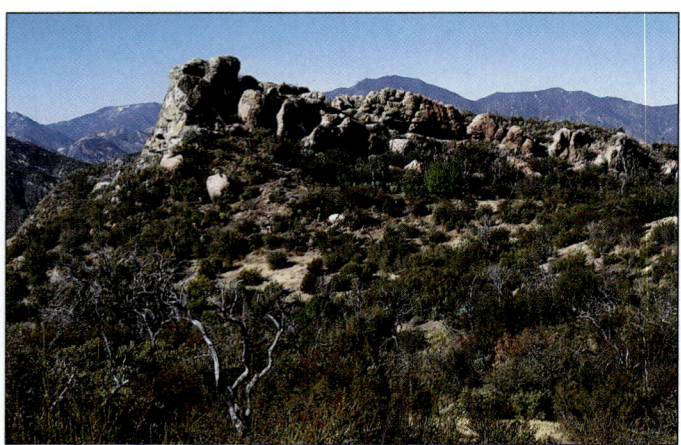

Figure 54–9 **Chaparral in the Santa Lucia Mountains, California.** Chaparral, which consists primarily of drought-resistant evergreen shrubs, develops where hot, dry summers alternate with mild, rainy winters. *(Edward Ely/Biological Photo Service)*

Deserts are arid ecosystems

Deserts are dry areas found in temperate *(cold deserts)* and subtropical or tropical regions *(warm deserts)* (see Fig. 54–2). The low water vapor content of the desert atmosphere leads to daily temperature extremes of heat and cold, so that a major change in temperature occurs in a single 24-hour period. Deserts vary greatly depending on the amount of precipitation they receive, which is generally less than 25 cm (10 in) per year. A few deserts are so dry that virtually no plant life occurs in them. As a result of sparse vegetation, desert soil is low in organic material but often high in mineral content, particularly the salts NaCl, $CaCO_3$, and $CaSO_4$. In some regions, such as areas of Utah and Nevada, the naturally occurring concentration of certain soil minerals reaches toxic levels for many plants.

Desert vegetation includes both perennials and, after a rain, flowering annuals. Plants in North American deserts include cacti, yuccas, Joshua trees, and sagebrushes (Fig. 54–10). Desert plants tend to have reduced leaves or no leaves, an adaptation that conserves water. In cacti such as the giant saguaro, for example, photosynthesis is carried out by the stem, which also expands accordion-style to store water; the leaves are modified into spines, which discourage herbivores. Other desert plants shed their leaves for most of the year, growing only during the brief moist season. Desert plants are noted for **allelopathy,** an adaptation in which toxic substances secreted by roots or shed leaves inhibit the establishment of competing plants nearby (see Figure 51–2). Many desert plants possess defensive spines, thorns, or toxins to resist the heavy grazing pressure often experienced in this food- and water-deficient environment.

Desert animals tend to be small. During the heat of the day they remain under cover or return to shelter periodically, whereas at night they come out to forage or hunt. In addition to desert-adapted insects, there are many specialized desert rep-

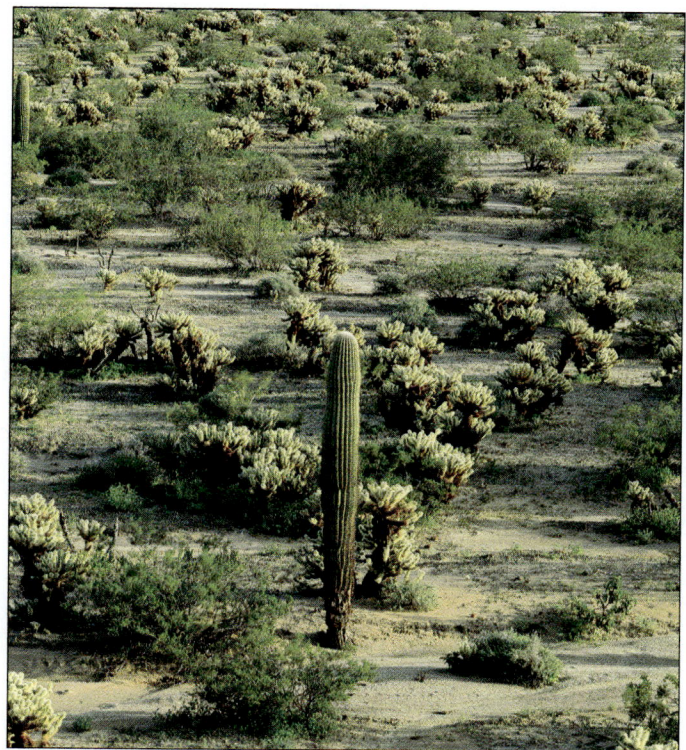

Figure 54–10 Desert in Arizona. Desert inhabitants are strikingly adapted to the demands of their environment. The warmer deserts of North America, such as the Sonoran Desert shown here, are characterized by summer rainfall. The Sonoran Desert, found in parts of Arizona, California, and Mexico, contains many species of cacti, including the large, treelike saguaro, which grows 15 to 18 m (50 to 60 ft). *(Willard Clay/Dembinsky Photo Associates)*

tiles (such as lizards, tortoises, and snakes) and a few desert-adapted amphibians (frogs and toads). Mammals include such rodents as the American kangaroo rat, which does not need to drink water but can subsist solely on the water content of its food (primarily seeds and insects). American deserts are also home to jack-rabbits, while Australian deserts have kangaroos. Carnivores like the African fennec fox and some birds of prey, especially owls, live on rodents and rabbits. During the driest months of the year, many desert insects, amphibians, reptiles, and mammals tunnel underground, where they remain inactive; this period of dormancy is known as *aestivation.*

American deserts have been altered by humans in several ways. Offroad vehicles damage desert vegetation, which sometimes takes years to recover, and certain cacti and desert tortoises are rare as a result of poaching. Also, houses, factories, golf courses, resorts, and farms built in desert areas require vast quantities of water, which typically must be imported from distant areas. Increased groundwater consumption by many desert communities has caused groundwater levels to drop. Aquifer depletion in U.S. deserts is particularly critical in southern Arizona and southwestern New Mexico.

Savanna is a tropical grassland with scattered trees

The **savanna** biome is a tropical grassland with widely scattered clumps of low trees (Fig. 54–11). Savanna is found in areas of relatively low or seasonal rainfall with prolonged dry periods (see Fig. 54–2). Temperatures in savannas vary little throughout the year, and seasons are regulated by precipitation, not by temperature as they are in temperate grasslands. Annual precipitation is 85 to 150 cm (34 to 60 in). Savanna

Figure 54–11 Savanna in Tanzania. Savanna is tropical grassland with widely scattered trees. African savanna formerly supported large herds of grazing animals and their predators; these are swiftly vanishing under pressure from pastoral and agricultural land use. *(Carlyn Iverson)*

soil is low in essential mineral nutrients, in part because it is strongly leached. Savanna soil is often rich in aluminum, which resists leaching, and in places the aluminum reaches levels that are toxic to many plants. Although the African savanna is best known, savanna also occurs in South America, western India, and northern Australia.

Savanna is characterized by wide expanses of grasses interrupted by occasional trees such as *Acacia,* which bristles with thorns that provide protection against herbivores. Both trees and grasses have fire-adapted features such as extensive underground root systems that survive the periodic fires that sweep through savanna.

The world's greatest assemblage of hoofed mammals occurs in the African savanna. Here live great herds of herbivores, including wildebeests, antelopes, giraffes, zebras, and elephants. Large predators, such as lions and hyenas, kill and scavenge the herds. In areas of seasonally varying rainfall, the herds and their predators may migrate annually.

Savanna is rapidly being converted to rangeland for cattle and other domesticated animals, which are replacing the big herds of wild animals. The problem is particularly acute in Africa, which has the most rapidly growing human population of any continent. In some places severe overgrazing by domestic animals has contributed to the conversion of marginal savanna into desert, a process known as **desertification.**

There are two types of forests in tropical areas

There are two types of tropical forests, tropical dry forests and tropical rain forests. **Tropical dry forests** occur in regions subjected to a wet season and a dry season (usually two to three months each year). Annual precipitation is 150 to 200 cm (60 to 80 in). During the dry season, many tropical trees shed their leaves and remain dormant, much as temperate trees do during the winter. India, Kenya, Zimbabwe, Egypt, and Brazil are some of the countries that have tropical dry forests (see Fig. 54–2).

Tropical rain forests occur where temperatures are warm throughout the year and precipitation occurs almost daily (see Fig. 54–2). The annual precipitation of tropical rain forests is 200 to 450 cm (80 to 180 in). Much of this precipitation comes from locally recycled water that enters the atmosphere by transpiration (see Chapter 32) by the forest's own trees.

Tropical rain forests are often located in areas with ancient, highly weathered, mineral-poor soil. Little organic matter accumulates in such soils. Because temperatures are high and soil moisture is abundant year round, decay organisms and detritus-feeding ants and termites decompose organic litter quite rapidly. Mineral nutrients from decomposing materials are quickly absorbed by vast networks of roots and mycorrhizae. Thus, minerals of tropical rain forests are tied up in the vegetation rather than in the soil.

Tropical rain forest is very productive despite the scarcity of mineral nutrients in the soil. Its plants capture considerable

energy by photosynthesis, stimulated by abundant solar energy and precipitation.

Of all the biomes, the tropical rain forest is unexcelled in species diversity. A person can often travel for 0.4 km (0.25 mi) without encountering two members of the same tree species. Local factors affecting rainforest species composition include varying soil fertility and topography (valleys have more plant diversity than hills, for example).

Most trees of tropical rain forests are evergreen flowering plants (Fig. 54–12*a*). Their roots are often shallow and concentrated near the surface in a mat only a few centimeters (an inch or so) thick. The root mat catches and absorbs almost all mineral nutrients released from leaves and litter by decay processes. Swollen bases or braces called *buttresses* hold the trees upright and aid in the extensive distribution of the shallow roots (Fig. 54–12*b*).

A fully developed rain forest has three or more distinct stories of vegetation. The topmost story consists of the crowns of occasional very tall trees, some 50 m (164 ft) or more in height, that are exposed to direct sunlight. The middle story, which reaches a height of 30 to 40 m (100 to 130 ft), forms a continuous canopy of leaves overhead that lets in little sunlight for the support of the sparse understory. Only 2 to 3% of the light bathing the forest canopy reaches the forest understory. The understory itself consists of both smaller plants specialized for life in shade, and seedlings of taller trees. Vegetation of tropical rain forests is not dense at ground level except near stream banks or where a fallen tree has opened the canopy.

Tropical rainforest trees support extensive epiphytic communities of smaller plants such as orchids and bromeliads. Although epiphytes grow in crotches of branches, on bark, or even on the leaves of their hosts, they only use their host trees for physical support, not for nourishment.

Because little light penetrates to the understory, many plants living there are adapted to climb already-established host trees rather than to invest their meager photosynthetic resources in the dead cellulose tissues of their own trunks. Lianas (woody tropical vines), some as thick as a human thigh, twist up through the branches of tropical rainforest trees (Fig. 54–12*c*; also see *Making the Connection: Vines, Evolutionary Adaptations, and Ecology* in Chapter 33).

Not counting bacteria and other soil-dwelling organisms, about 90% of tropical rainforest organisms live in the upper canopy (Fig. 54–13). Rainforest animals include the most abundant and varied insect, reptile, and amphibian fauna on Earth. Birds, too, are diverse, with some specialized to consume fruits (parrots, for example) and others to consume nectar (hummingbirds and sunbirds, for example). Most rainforest mammals, such as sloths and monkeys, live only in the trees and never climb down to the ground surface, although some large ground-dwelling mammals, including elephants, are also found in rain forests.

Unless strong conservation measures are initiated soon, human population growth and agricultural and industrial expansion in tropical countries will spell the end of tropical rain

(a)

(b)

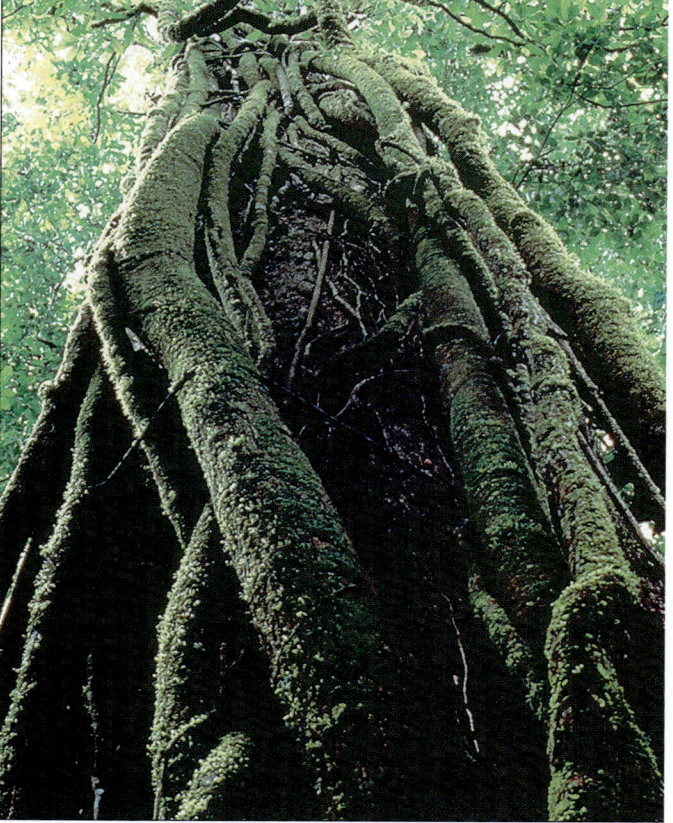

(c)

Figure 54–12 **Tropical rain forest.** *(a)* A broad view of tropical rainforest vegetation along a riverbank in Southeast Asia. Except at riverbanks, tropical rain forest has a closed canopy that admits little light to the forest floor. *(b)* Tropical rainforest trees typically possess buttress roots that support them in the shallow, often wet soil. Shown are buttress roots on Australian banyan *(Ficus macrophylla)*. *(c)* Thick, epiphyte-covered lianas grow into the canopy using a tree trunk for support. This photo was taken in Costa Rica. *(a, Frans Lanting/Minden Pictures; b, John Arnaldi; c, Mark Moffett/Minden Pictures)*

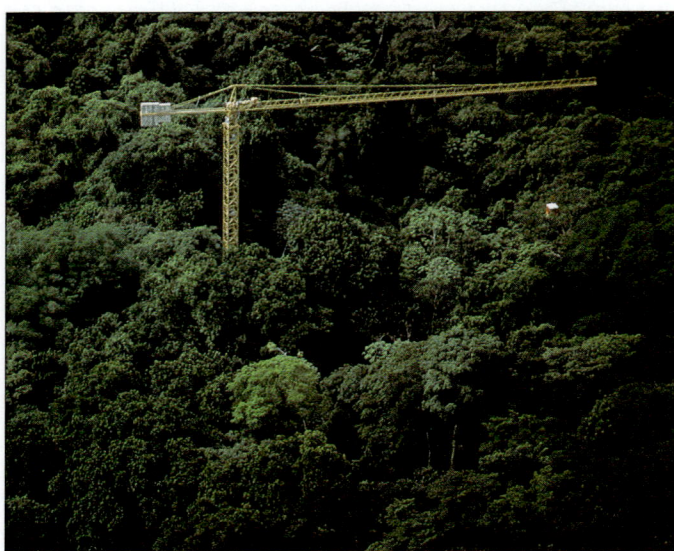

Figure 54–13 **Studying the rainforest canopy.** This construction crane was erected in a tropical rain forest in Panama to study rainforest organisms without harming them. Before cranes were used, biologists were unable to perform many experiments in the upper canopy. *(Photo by N. Guerre/STRI)*

forests by the middle of the next century. Biologists know that many rainforest species will become extinct before they have even been identified and scientifically described. Tropical rainforest destruction is discussed in detail in Chapter 55.

AQUATIC ECOSYSTEMS OCCUPY MOST OF EARTH'S SURFACE

Not surprisingly, the abiotic factors that help determine an aquatic life zone's boundaries differ significantly from those of terrestrial ecosystems. **Salinity** (the concentration of dissolved salts, such as sodium chloride, in a body of water) affects the kinds of organisms present in aquatic ecosystems, as does the amount of dissolved oxygen. Water greatly interferes with the penetration of light, so floating aquatic organisms that photosynthesize must remain near the water's surface, and vegetation attached to the bottom can grow only in shallow water. In addition, low levels of essential mineral nutrients often limit the number and distribution of organisms in certain aquatic environments. Other abiotic determinants of species composition in aquatic ecosystems include temperature, pH, and presence or absence of waves and currents.

Aquatic ecosystems contain three main ecological categories of organisms: free-floating plankton, strongly swimming nekton, and bottom-dwelling benthos. **Plankton** are usually small or microscopic organisms that are relatively feeble swimmers. For the most part, they are carried about at the mercy of currents and waves. They are unable to swim far horizontally, but some species are capable of large vertical migrations

and are found at different depths of water at different times of the day or at different seasons. Plankton are generally subdivided into two major categories: phytoplankton and zooplankton. **Phytoplankton** (photosynthetic cyanobacteria and free-floating algae) are producers that form the base of most aquatic food webs. **Zooplankton** are nonphotosynthetic organisms that include protozoa, tiny crustaceans, and the larval stages of many animals. **Nekton** are larger, more strongly swimming organisms such as fishes, turtles, and whales. **Benthos** are bottom-dwelling organisms that fix themselves to one spot (sponges, oysters, and barnacles), burrow into the sand (many worms and echinoderms), or simply walk or swim about on the bottom (crawfish, aquatic insect larvae, and brittle stars).

FRESHWATER ECOSYSTEMS ARE CLOSELY LINKED TO LAND AND MARINE ECOSYSTEMS

Freshwater ecosystems include streams and rivers (flowing-water ecosystems), ponds and lakes (standing-water ecosystems), and marshes and swamps (freshwater wetlands). Each type of freshwater ecosystem is distinguished by its own specific abiotic conditions and characteristic organisms.

Although freshwater ecosystems occupy a relatively small portion (about 2%) of Earth's surface, they have an important role in the hydrologic cycle: they assist in recycling precipitation that flows as surface runoff to the ocean (see Chapter 53). Large bodies of fresh water also help moderate daily and seasonal temperature fluctuations on nearby land. Freshwater habitats also provide homes for large numbers of species.

Streams and rivers are flowing-water ecosystems

Many different conditions exist along the length of a stream or river. The nature of a **flowing-water ecosystem** changes greatly from its source (where it begins) to its mouth (where it empties into another body of water). For example, certain parts of the stream may be shaded by surrounding forest, whereas other parts may be exposed to direct sunlight. Headwater streams (small streams that are the sources of a river) are usually shallow, clear, cold, swiftly flowing, and highly oxygenated. In contrast, rivers downstream from the headwaters are wider and deeper, cloudy (i.e., they contain suspended particulates), not as cold, slower flowing, and less oxygenated.

The kinds of organisms found in flowing-water ecosystems vary greatly from one stream to another, depending primarily on the strength of the current. In streams with fast currents, inhabitants may have adaptations such as suckers to attach themselves to rocks so they are not swept away. Alternatively, they may have flattened bodies that enable them to slip under or between rocks, or they may be streamlined and muscular enough to swim in the current (Fig. 54–14). Or-

TERRESTRIAL-AQUATIC LINKAGES—THE DETRITUS CONNECTION

Does leaf litter from a wooded streambank have any effect on a forest stream ecosystem? One of the long-held assumptions of stream ecology is that a forest provides a stream flowing through it with an important input of energy in the form of detritus (nonliving organic matter). Until recently, however, little direct evidence was available to support this idea. In 1997, J.B. Wallace and others from the University of Georgia and Virginia Polytechnic Institute and State University published results of a three-year study that link terrestrial litter to the organisms in a forest stream.*

Wallace and coworkers excluded detritus from 180 m of a stream for three years by erecting an overhead canopy and lateral fence. They used a nearby "unmanipulated" stream as a control in order to minimize the possibility that the changes they observed were due to natural environmental fluctuations instead of their experimental manipulations. They compared the changes in two different habitats in the stream: where the streambed was gravel or sand; and where it was moss-covered bedrock.

Compared to the unmanipulated stream, the abundance, biomass, and secondary production of aquatic invertebrates declined in those portions of the experimental stream where the streambed was gravel or sand (figure). The decline was observed throughout the food web, from detritivores (animals that eat detritus) to the predators of detritivores. Thus, Wallace's experiment demonstrates how a single change, reduction in the input of leaf litter, propagates through a food web.

Interestingly, excluding terrestrial inputs from the moss-covered portion of the experimental stream caused few changes in the invertebrate populations. Apparently, different food webs exist in different stream habitats, despite the fact that they are in close proximity to each another.

Wallace's work has important implications in the field of restoration ecology. Many natural and human-induced disturbances, such as logging, land-use change, and grazing, cause multiple changes in aquatic ecosystems, including a reduction in the supply of leaf litter to stream ecosystems. Wallace's experiment indicates that in order to restore the biological diversity in a stream ecosystem, inputs of terrestrial detritus must be established.

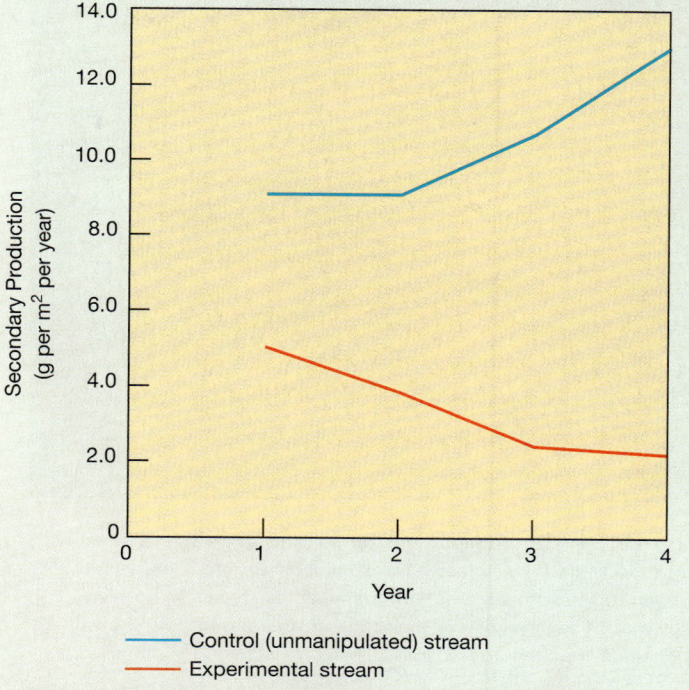

Annual secondary productivity of benthic organisms in gravel and sand portions of control (unmanipulated) and experimental streams. Secondary production is the assimilation of organic matter by the consumers (in this example, the detritivores and predators) in the stream ecosystem. It is measured in grams per square meter per year. Year 1 measurements were taken before litter was excluded from the experimental stream; litter exclusion occurred during years 2, 3, and 4. (Wallace, J.B., S.L. Eggert, J.L. Meyer, and J.R. Webster. "Multiple Trophic Levels in a Forest Stream Linked to Terrestrial Litter Inputs," Science, Vol. 277, 4 Jul. 1997)

*Wallace, J.B., S.L. Eggert, J.L. Meyer, and J.R. Webster. "Multiple Trophic Levels of a Forest Stream Linked to Terrestrial Litter Inputs." Science, Vol. 277, 4 Jul. 1997.

ganisms in large, slow-moving streams and rivers do not need such adaptations, although they are typically streamlined (as are most aquatic organisms) to lessen resistance during movement through water. Where current is slow, plants and animals of the headwaters are replaced by those characteristic of ponds and lakes.

Unlike other freshwater ecosystems, streams and rivers depend on land for much of their energy. In headwater streams, up to 99% of the energy input comes from detritus (dead organic material such as leaves) carried from the land into streams and rivers by wind or surface runoff (see *Making the Connection: Terrestrial-Aquatic Linkages—The Detritus Connection*). Downstream, rivers contain more producers, and therefore

have a slightly lower dependence on detritus as a source of energy than do the headwaters.

Human activities have several adverse impacts on rivers and streams, including water pollution and the effects of dams, which are built to contain the water of rivers or streams. Dams change the nature of flowing-water ecosystems, both upstream and downstream from the dam location. A dam causes water to back up, flooding large areas of land and forming a reservoir, which destroys terrestrial habitat. Below the dam, the once-powerful river is reduced to a relative trickle, altering ecosystems. For example, the Glen Canyon Dam, built in 1963, has profoundly affected the Colorado River in the Grand Canyon. Prior to the dam's construction, powerful spring

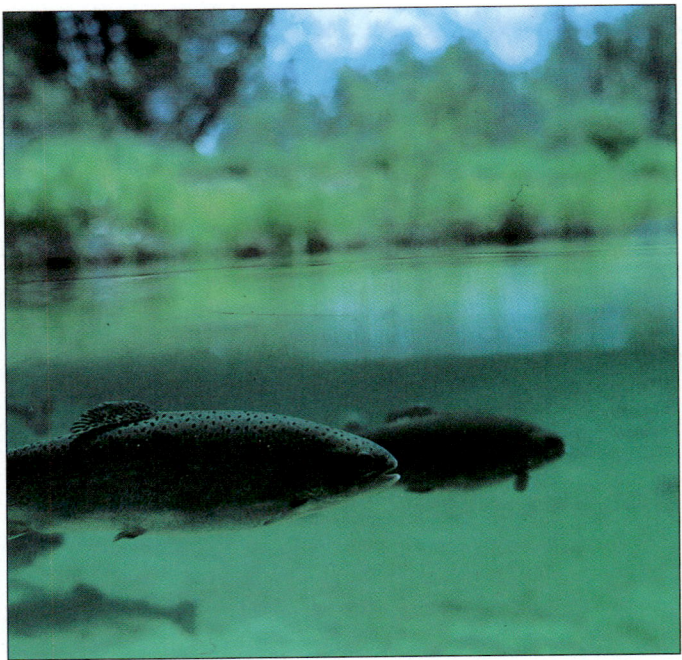

Figure 54–14 **Adaptations for streams and rivers.** The brown trout *(Salmo trutta)* is an active, streamlined fish that is muscular enough to reside in streams with many rapids. Typically, however, brown trout spend much of their time in stream pools, as shown. *(Jack Fields/Photo Researchers, Inc.)*

floods used to deposit beaches and sandbars that provided nesting sites for birds and shallow waters for breeding fishes. The regulated flow of water since the Glen Canyon Dam was built has changed the ecosystem to the detriment of some of the Grand Canyon's wildlife. The Bureau of Reclamation tried to rectify some of the changes that have occurred to the river by releasing an additional 117 billion gallons of water during a

one-week period in the spring of 1996. The experimental flood rebuilt some 50 beaches and sandbars that had disappeared since 1963 and enlarged most of the existing ones. It also partially restored fish-spawning habitats.

Pollution alters the physical environment and changes the biotic component downstream from the pollution source. For example, fertilizer runoff from Midwestern fields in such states as Iowa, Wisconsin, and Illinois eventually finds its way into the Mississippi River and, from there, into the Gulf of Mexico. These nutrients have created a huge "dead zone" in the Gulf of Mexico that is nearly as large as the state of New Jersey—some 17,500 square km (7000 square mi) in area. Only anaerobic bacteria exist in the dead zone because the water does not contain enough dissolved oxygen to support fishes, shrimp, or other aquatic organisms. This condition, known as **hypoxia,** occurs when algae grow rapidly due to the presence of nutrients in the water. When these algae die, they sink to the bottom and are decomposed by bacteria, whose metabolic activities deplete the water of dissolved oxygen, leaving too little for other aquatic life. Hypoxia occurs in many coastal areas where large rivers empty into the ocean.

Ponds and lakes are standing-water ecosystems

Standing-water ecosystems are characterized by zonation. A large lake has three basic layers or zones: the littoral, limnetic, and profundal zones (Fig. 54–15). Smaller lakes and ponds typically lack a profundal zone. The **littoral zone** is a shallow water area along the shore of a lake or pond. It includes rooted, emergent vegetation, such as cattails and burreeds, plus several deeper dwelling aquatic plants and algae. The littoral zone is the most highly productive zone of the lake, that is, photosynthesis is greatest here, in part because light is abundant and because the littoral zone receives nutrient inputs from surrounding land that stimulate the growth of plants and algae.

Figure 54–15 **Zonation in a large lake.** A lake is a standing-water ecosystem surrounded by land. Vegetation around the lake is not drawn to scale.

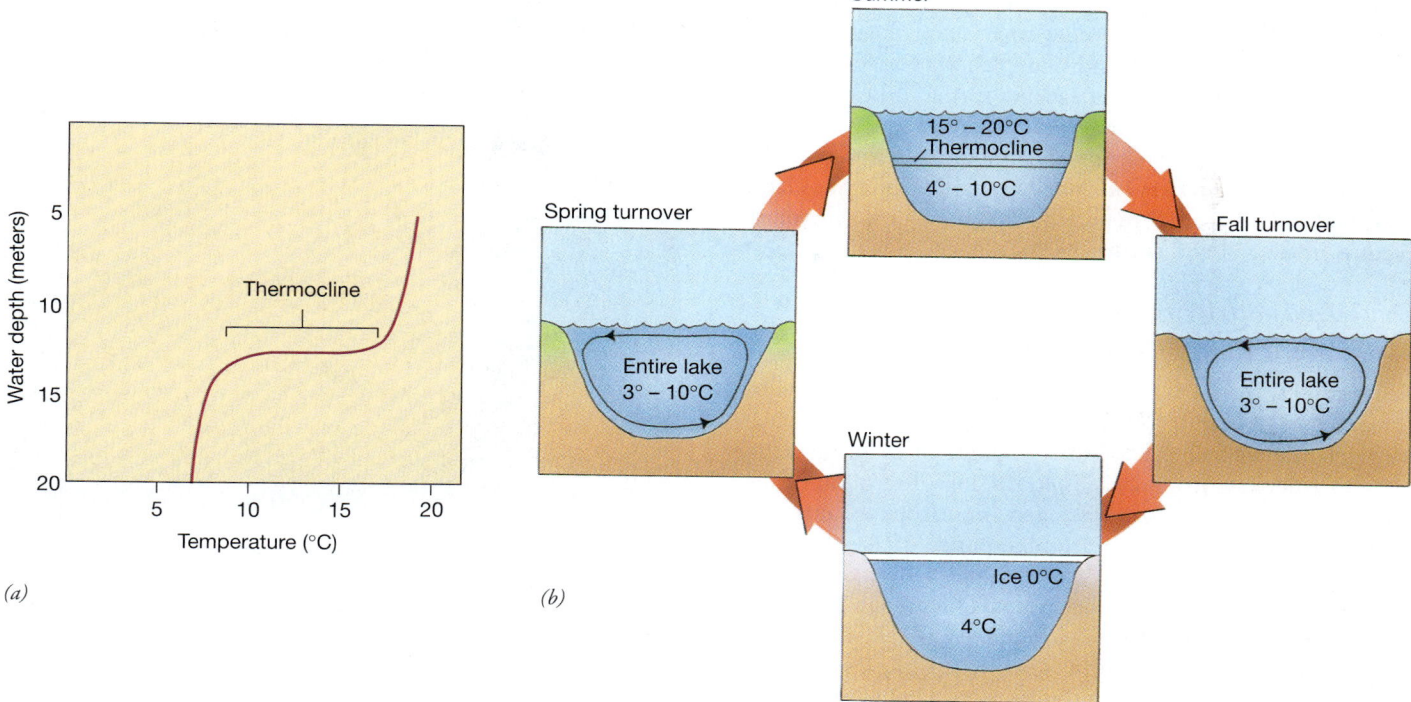

Figure 54–16 **Thermal stratification in a temperate lake.** *(a)* Temperature differences by depth during the summer. There is an abrupt temperature transition, called the thermocline. *(b)* During fall and spring turnovers, a mixing of upper and lower layers of water brings oxygen to the oxygen-depleted depths of the lake and minerals to the mineral-deficient surface waters.

Animals of the littoral zone include frogs and their tadpoles, turtles, worms, crayfish and other crustaceans, insect larvae, and many fishes such as perch, carp, and bass.

The **limnetic zone** is the open water away from the littoral zone; it extends down as far as sunlight penetrates. The main organisms of the limnetic zone are microscopic phytoplankton and zooplankton. Larger fishes also spend some of their time in the limnetic zone, although they may visit the littoral zone to feed and reproduce. Owing to its depth, less vegetation grows here than in the littoral zone.

Beneath the limnetic zone of a large lake is the **profundal zone.** Because light does not penetrate effectively to this depth, plants and algae do not live here. Food drifts into the profundal zone from the littoral and limnetic zones. Bacteria decompose dead plants and animals that reach the profundal zone, liberating minerals. These minerals are not effectively recycled because no photosynthetic organisms are present to absorb them and incorporate them into the food web. As a result, the profundal zone tends to be both mineral-rich and oxygen-deficient, with few forms of life occupying it other than anaerobic bacteria.

Thermal stratification occurs in temperate lakes

The marked layering of large temperate lakes caused by light penetration is accentuated by **thermal stratification,** in which the temperature changes sharply with depth (Fig. 54–16*a*). Thermal stratification occurs because the summer sunlight penetrates and warms surface waters, making them less dense.[3] In summer, cool (and therefore more dense) water remains at the lake bottom and is separated from warm (and therefore less dense) water above by an abrupt temperature transition called the **thermocline.**

In temperate lakes, falling temperatures in autumn cause a mixing of the lake waters called the **fall turnover** (Fig. 54–16*b*). (Since little seasonal temperature variation occurs in the tropics, such turnovers are uncommon there.) As surface water cools, its density increases, and eventually it sinks and displaces the less dense, warmer, mineral-rich water beneath. Warmer water then rises to the surface where it, in turn, cools and sinks. This process of cooling and sinking continues until the lake reaches a uniform temperature throughout.

When winter comes, surface water cools to below 4°C, its temperature of greatest density, and, if it is cold enough, ice forms. Ice, which forms at 0°C, is less dense than cold water; thus, ice forms on the surface, and the water on the lake bottom is warmer than the ice.

In the spring, a **spring turnover** occurs as ice melts and surface water reaches 4°C. Surface water again sinks to the bot-

[3]Recall that the density of water is greatest at 4°C; both above and below this temperature, water is less dense.

tom, and bottom water returns to the surface. As summer arrives, thermal stratification occurs once again.

The mixing of deeper, nutrient-rich water with surface, nutrient-poor water during the fall and spring turnovers brings essential minerals to the surface and oxygenated water to the bottom. The sudden presence of large amounts of essential minerals in surface waters encourages the development of large algal and cyanobacterial populations, which form temporary blooms in the fall and spring.

Increased nutrients, supplied by human activities, stimulate algal growth

Enrichment, the fertilization of a body of water, is caused by the presence of high levels of plant and algal nutrients such as nitrogen and phosphorus. Excess amounts of these nutrients get into waterways from sewage and from fertilizer runoff from lawns and fields. A pond or lake that is enriched is said to be **eutrophic.** The water in a eutrophic pond or lake is cloudy because of the presence of vast numbers of algae and cyanobacteria that are supported by the nutrients. Although eutrophic lakes contain large populations of aquatic animals, the species composition is different than in unenriched lakes. For example, an unenriched lake in the northeastern United States may contain pike, sturgeon, and whitefish in the deeper, colder part of the lake where there is a higher concentration of dissolved oxygen. In eutrophic lakes, on the other hand, the deeper, colder levels of water are depleted of dissolved oxygen because of the greater amount of decomposition on the lake floor. Therefore, fishes such as pike, sturgeon, and whitefish die out and are replaced by fishes, such as catfish and carp, that can tolerate lower concentrations of dissolved oxygen.

Eutrophication is reversible and has declined sharply in North America since the 1970s with the passage of legislation to limit the phosphate content of detergents and with the construction of better sewage treatment plants. According to the 1992 National Water Quality Inventory Report to Congress, the leading source of water quality impairment today is agriculture. Fertilizer runoff, as well as animal wastes and plant residues in waterways, still cause enrichment problems.

Freshwater wetlands are transitional between aquatic and terrestrial ecosystems

Freshwater wetlands, usually covered by shallow water for at least part of the year, have characteristic soils and water-tolerant vegetation. They include marshes, in which grasslike plants dominate, and swamps, in which woody trees or shrubs dominate (Fig. 54–17). Freshwater wetlands also include hardwood bottomland forests (lowlands along streams and rivers that are periodically flooded), prairie potholes (small, shallow ponds that formed when glacial ice melted at the end of the last Ice Age), and peat moss bogs (peat-accumulating wetlands where sphagnum moss dominates).

Wetland plants, which are highly productive, provide enough food to support a wide variety of organisms. Wetlands

Figure 54–17 A freshwater swamp in northeast Texas. Freshwater swamps are inland areas permanently saturated or covered by water and dominated by trees, such as bald cypress *(shown)*. The water surface is covered by a floating carpet of tiny aquatic plants. *(Gregory G. Dimijian/ Photo Researchers, Inc.)*

are valued as a wildlife habitat for migratory waterfowl and many other bird species, beaver, otter, muskrat, and game fishes. Wetlands help control flooding by acting as holding areas for excess water when rivers flood their banks. The floodwater stored in wetlands then drains slowly back into the rivers, providing a steady flow of water throughout the year. Wetlands also serve as groundwater recharging areas. One of their most important roles is to help cleanse and purify water by trapping and holding pollutants in the flooded soil.

At one time wetlands were thought of as wastelands, areas to be filled in or drained so that farms, housing developments, and industrial plants could be built on them. Wetlands are also breeding places for mosquitoes and therefore were viewed as a menace to public health. Although the crucial environmental services that wetlands provide are widely recognized today, wetlands are still threatened by agriculture, pollution, engineering (dams), and urban and suburban development. In the United States, wetlands have been steadily shrinking by an estimated 105,000 hectares (260,000 acres) per year since the mid-1980s. In the contiguous 48 states, only 38 million hectares (95 million acres) of wetlands remain of the more than 81 million hectares (200 million acres) that originally existed.

ESTUARIES OCCUR WHERE FRESH WATER AND SALT WATER MEET

Where the ocean meets the land there may be one of several kinds of ecosystems: a rocky shore, a sandy beach, an intertidal mud flat, or a tidal estuary. An **estuary** is a coastal body

Figure 54-18 A salt marsh at Assateague Island National Seashore, Maryland. Cordgrass *(Spartina)* is the dominant vegetation. *(Michael Gadomski/Earth Scenes)*

Figure 54-19 Red mangroves *(Rhizophora mangle)*. The stiltlike roots that support the tree grow into deeper water as well as into the mudflat exposed by low tide. Many animals live in the complex root system of mangrove forests. This photo was taken along the coast of Florida, near Miami. *(Patti Murray/Earth Scenes)*

of water, partly surrounded by land, with access to the open ocean and a large supply of fresh water from rivers. It usually contains **salt marshes,** shallow swampy areas dominated by grasses in which tides and precipitation cause salinity to fluctuate between that of sea water and that of fresh water (Fig. 54–18). Many estuaries undergo marked variations in temperature, salinity, and other physical properties in the course of a year. To survive, estuarine organisms must have a wide tolerance to such changes.

Estuaries are among the most fertile ecosystems in the world, often having much greater productivities than the adjacent ocean or freshwater river. This high productivity is brought about by (1) the action of tides, which promote a rapid circulation of nutrients and help remove waste products; (2) the transport of nutrients from land into streams and rivers that empty into the estuary; (3) the high level of light penetrating the shallow water; and (4) the presence of many plants, which provide an extensive photosynthetic carpet and whose roots and stems also mechanically trap detritus, forming the basis of detritus food webs. Most commercially important fin fishes and shellfish spend their larval stages in estuaries among the protective tangle of decaying stems.

Mangrove forests, the tropical equivalent of salt marshes, cover perhaps 70% of tropical coastlines. Like salt marshes, mangrove forests provide valuable environmental services. They are breeding grounds and nurseries for several commercially important deepwater fish species, and their roots stabilize the sediments, thereby preventing coastal erosion and providing a barrier against the ocean during storms (Fig. 54–19).

Estuaries and other coastal areas are threatened by population growth and economic development

Like freshwater wetlands, salt marshes have often appeared to be worthless, empty stretches of land to uninformed people. As a result, they have been used as dumps, becoming severely polluted, or have been filled with dredged bottom material to form land for residential, commercial, and industrial development. A large part of the estuarine environment has been lost in this way, along with many of its benefits: wildlife habitat, sediment and pollution trapping, and flood control.

Mangroves are under assault from coastal development and unsustainable logging. Some countries, such as the Philippines, Bangladesh, and Guinea-Bissau, have cut down 70% or more of their mangrove forests.

Some 3.8 billion people—about two-thirds of the world's population—live within 150 km (93 mi) of a coastline. Population experts project that three-fourths of all humans, perhaps as many as 6.4 billion, will live in that area by 2025. Many of the world's largest cities are situated in coastal areas, and these cities are currently growing more rapidly than noncoastal cities. If the world's natural coastal areas, including estuaries, are not to become urban sprawl or continuous strips of tourist resorts during the next century, coastal management strategies must be developed that take into account projections of human population growth and distribution.

MARINE ECOSYSTEMS DOMINATE EARTH'S SURFACE

Although oceans and lakes are comparable in many ways, they have many differences. Depths of even the deepest lakes do not approach those of the oceanic abysses, which are extremely deep areas that extend more than 6 km (3.6 mi) below the sunlit surface. Oceans are profoundly influenced by tides and currents. Gravitational pulls of both sun and moon produce two tides a day throughout the oceans, but the height of those tides varies with the phases of the moon (full moons and new moons cause the highest tides), the season, and the local topography.

The immense marine environment is subdivided into several zones: the intertidal zone, the benthic (ocean floor) environment, and the pelagic (ocean water) environment (Fig. 54-20). The pelagic environment is in turn divided into two provinces—the neritic province and the oceanic province.

The intertidal zone is transitional between land and ocean

The shoreline area between low and high tide is called the **intertidal zone.** Although high levels of light and nutrients, together with an abundance of oxygen, make the intertidal zone a biologically productive environment, it is also a stressful one. If an intertidal beach is sandy, inhabitants must contend with a constantly shifting environment that threatens to engulf them and gives them scant protection against wave action. Consequently, most sand-dwelling organisms, such as mole crabs, are continual and active burrowers. Because they are able to follow the tides up and down the beach, most do not have any notable adaptations to survive drying or exposure.

A rocky shore provides a fine anchorage for seaweeds and invertebrate animals. However, it is exposed to constant wave action when immersed during high tides and to drying and temperature changes when exposed to the air during low tides. A typical rocky-shore inhabitant has some way of sealing in moisture, perhaps by closing its shell, if it has one, plus a powerful means of anchoring itself to rocks. Mussels, for example, have horny, threadlike anchors, and barnacles have special cement glands. Rocky-shore intertidal algae (seaweeds) usually have thick, gummy polysaccharide coats, which dry out slowly when exposed, and flexible bodies not easily broken by wave action (Fig. 54–21). Some rocky-shore community inhabitants hide in burrows or crevices at low tide.

Coral reefs are an important part of the benthic environment

The **benthic environment** is the ocean floor. It consists of sediments (mostly sand and mud) in which many marine animals, such as worms and clams, burrow. Bacteria are common in marine sediments, but until the 1990s scientists assumed that bacteria did not extend very deep into the sediments. In the 1990s, however, living bacteria were found in ocean sediments more than 500 m (1625 ft) below the ocean floor at several different sites in the Pacific Ocean.

Particularly productive benthic communities in shallow ocean waters include kelp forests, seaweed beds, and coral reefs. We restrict our discussion here to coral reefs. (See Chapter 24 for a discussion of kelp forests.)

The living portions of coral reefs must grow in shallow waters where light penetrates. Many coral reefs are composed principally of red coralline algae that require light for photosynthesis. Coral animals also require light for the large number of symbiotic dinoflagellates, known as **zooxanthellae,** that live and photosynthesize in their tissues (see Fig. 52–6). Although species of coral without zooxanthellae exist, only those with zooxanthellae build reefs. In addition to obtaining food from the zooxanthellae living inside them, coral animals capture food at night by using their stinging tentacles to paralyze small animals that drift nearby.

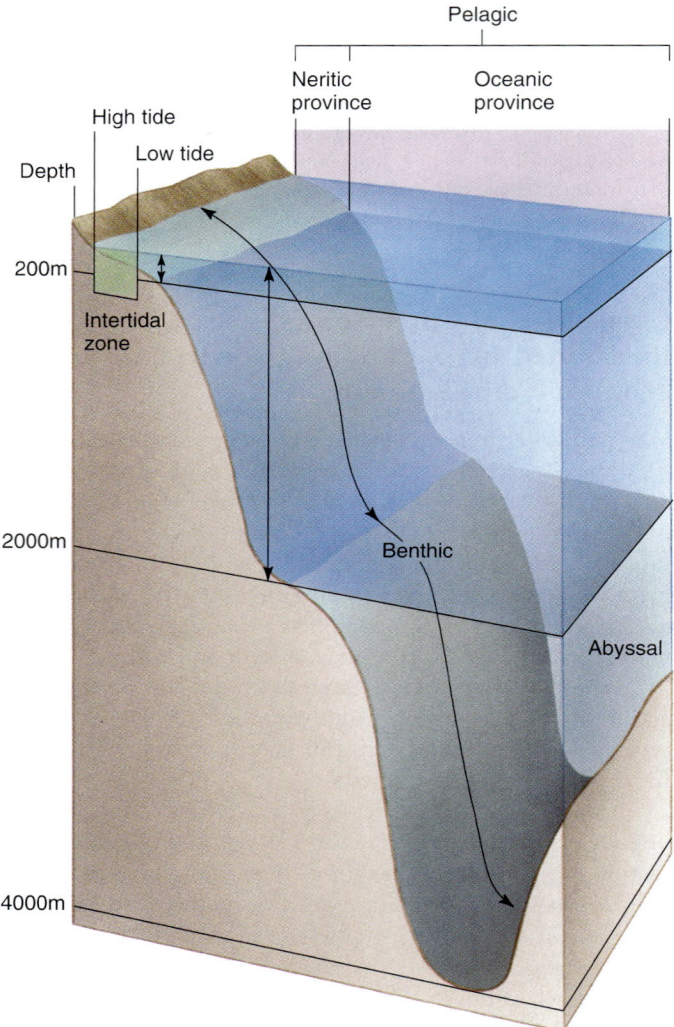

Figure 54–20 Zonation in the ocean. The ocean has three main life zones: the intertidal zone, the benthic environment, and the pelagic environment. The pelagic environment consists of the neritic and oceanic provinces.

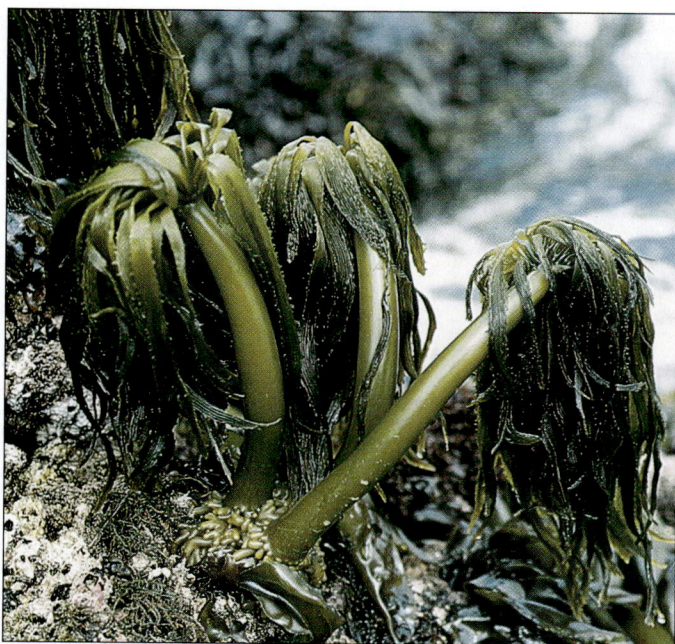

Figure 54–21 Sea palms *(Postelsia)* in a rocky intertidal zone. These sturdy seaweeds, which are common on the rocky Pacific coast from Vancouver Island to California, are exposed at low tide. Their bases are firmly attached to the rocky substrate, enabling them to withstand pounding wave action. *(William E. Ferguson)*

Coral reefs grow slowly in warm, shallow water, as coral organisms build on the calcareous remains of countless organisms before them. The waters in which coral reefs are found are often poor in nutrients. Other factors favor high productivity, however, including the presence of symbiotic zooxanthellae, favorable temperature, and plenty of sunlight.

Coral reef ecosystems are the most diverse of all marine environments and contain hundreds of species of fishes and invertebrates, such as giant clams, sea urchins, sea stars, sponges, brittle stars, sea fans, and shrimp (Fig. 54–22*a*). The Great Barrier Reef of Australia, for example, occupies only 0.1% of the ocean's surface, but 8% of the world's fish species live there. Many are brightly colored, which advertises the fact that they are poisonous. The complex multitude of relationships and interactions that occur at coral reefs is comparable only to the tropical rain forest among terrestrial ecosystems. As in the rain forest, competition is intense, particularly for light and space to grow.

Many unusual relationships occur at coral reefs. Certain tiny fishes, for example, swim over larger fishes and even inside their mouths to remove potentially harmful parasites (Fig. 54–22*b*). Fishes sometimes line up at these cleaning stations, waiting their turn to be serviced. As another example, *Podillopora* corals have a remarkable defense mechanism that protects them from coral browsers such as the crown-of-thorns sea star (Fig. 54–22*c*). Tiny crabs live among the branches of the coral. When a sea star crawls onto the coral, the crabs swarm over it, nipping off its tube feet and eventually killing it if it does not retreat! (Since the late 1960s, population explosions

of the crown-of-thorns sea stars have occurred on many coral reefs, causing extensive devastation of the coral. In most cases, the number of sea stars has eventually declined to normal levels. It is not known what causes these outbreaks, including the role of factors such as pollution.)

Coral reefs suffer from many human-related problems, including damage from divers, pollution, oil spills, boat groundings, fishing with dynamite or cyanide, and coral bleaching. Marine biologists are especially concerned about the more than one million scuba divers and snorkelers that visit coral reefs each year. Most divers and snorkelers are unaware of how vulnerable coral reefs and their associated organisms are to human interactions. Many reef dwellers are killed or injured simply by touching or squeezing them. When a diver or snorkeler accidently kicks or grabs the reef, pieces break off. Divers also stir up the bottom sediments, which suffocate the coral animals. Coral bleaching, another serious problem, is discussed in Chapter 28 in *Making the Connection: Coral Reefs and Environmental Issues.*

The neritic province consists of shallow waters close to shore

The **neritic province** is open ocean that overlies the continental shelves, that is, the ocean floor from the shoreline to a depth of 200 m (650 ft). Organisms that live in the neritic province are all floaters or swimmers. The upper reaches of the neritic province make up the **photic region,** which extends from the surface to a depth of approximately 100 m. Enough light penetrates the photic region to support photosynthesis.

Large numbers of phytoplankton, particularly diatoms in cooler waters and dinoflagellates in warmer waters, produce food by photosynthesis and are thus the base of food webs (Fig. 54–23). Zooplankton (including tiny crustaceans, jellyfish, comb jellies, protists such as foraminiferans, and larvae of barnacles, sea urchins, worms, and crabs) feed on phytoplankton. Zooplankton are consumed by plankton-eating nekton such as herring, sardines, squid, and baleen whales. These in turn become prey for carnivorous nekton such as sharks, tuna, dolphin, and toothed whales. Nekton are mostly confined to the shallower neritic waters (less than 60 m, or 195 ft, deep) because that is where their food is.

The oceanic province comprises most of the ocean

The average depth of the world's oceans is 4000 m (2.4 mi). The **oceanic province** is that part of the open ocean that covers the deep ocean basin, that is, the ocean floor at depths greater than 200 m. It is the largest marine environment and contains about 75% of the ocean's water. The oceanic province is characterized by cold temperatures, high hydrostatic pressure, and an absence of sunlight; these environmental conditions are uniform throughout the year.

Most organisms of the oceanic province depend on **marine snow,** organic debris that drifts down into the **aphotic** ("without light") **region** from the upper, lighted regions. Or-

(a)

(b)

(c)

ganisms of this little-known realm are filter-feeders, scavengers, or predators. Many are invertebrates, some of which attain great sizes. The giant squid, for example, measures up to 18 m (59 ft) in length, including its tentacles. Fishes of the oceanic province are strikingly adapted to darkness and food scarcity. For example, the gulper eel has huge jaws that enable it to swallow large prey (Fig. 54–24). (An organism that encounters food infrequently needs to eat as much as possible when it has the chance.) Many animals of the oceanic province have illuminated organs that enable them to see one another for mating or to capture food. Adapted to drifting or slow swimming, they are often characterized by reduced bone and muscle mass.

Life exists not only in the deepest ocean waters but also along volcanic ridges in the ocean floor. These ocean vents support a diverse marine community of clams, mussels, anemones, barnacles, crabs, worms, and fishes. Hydrothermal vent com-

munities exist in total darkness and are exposed to both temperature and pressure extremes. (See *Focus On: Life without the Sun* in Chapter 52 for additional discussion of hydrothermal vent communities.)

Overfishing is a threat to the neritic and oceanic provinces

In late 1994 the National Marine Fisheries Service of the U.S. Commerce Department closed Georges Bank, a 16,500 square km (6600 square mi) area off the coast of New England in the North Atlantic Ocean. The moratorium on fishing the Georges Bank, once one of the world's richest fishing grounds, was declared because catches of cod, haddock, and yellowtail flounder had been steadily declining there for the past ten years due to overfishing. *Overfishing* means that fish are harvested faster than they can replace themselves. Many economically impor-

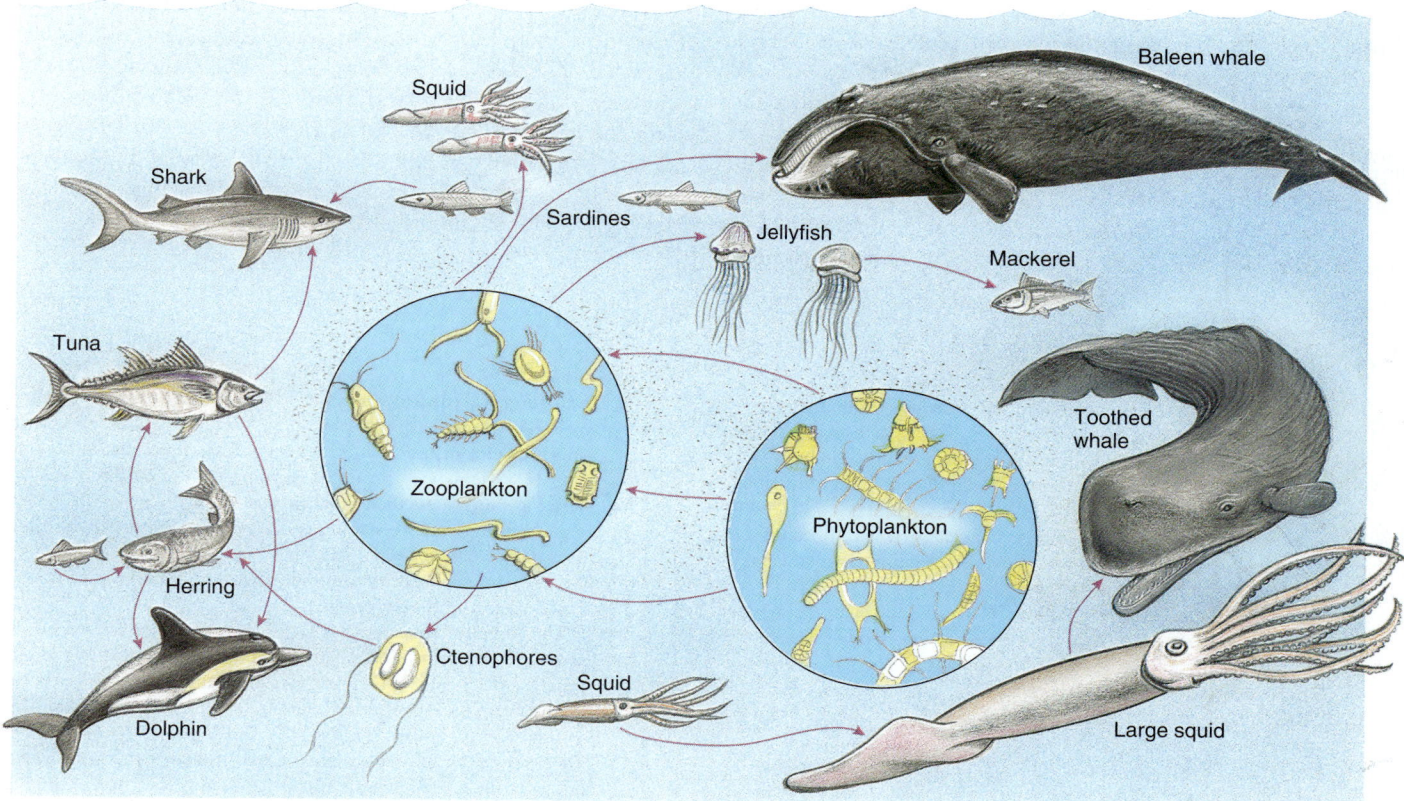

Figure 54–23 **Food web in the neritic province.** Phytoplankton are consumed by zooplankton, which in turn are consumed by many animals, including jellyfish, ctenophores, herring, and baleen whales.

tant fishes have reached **commercial extinction** in the Georges Bank because their numbers are so low that they are unprofitable to harvest. As a result, thousands of people lost their jobs. No one knows for sure how long it will take fish populations in the Georges Bank to recover, but estimates range to 20 years. As we go to press, there is no sign of a recovery.

The collapse of the Georges Bank Fishery is not unique. The Alaskan salmon fishery declined during the 1960s and 1970s, the Peruvian anchovy fishery collapsed in the 1970s, and the Newfoundland cod fishery along the Grand Banks closed in the 1990s. Fish shortages are occurring all over the globe as a result of overfishing. According to the National Marine Fisheries Service, 40% of commercially important species have been overfished in U.S. waters. The problem is particularly acute in the Pacific Northwest, near Nova Scotia, and the Gulf of Mexico.

ECOSYSTEMS INTERACT WITH ONE ANOTHER

We have discussed the various terrestrial biomes and aquatic ecosystems as if they were distinct and separate entities, but they intergrade with one another at their boundaries. You learned in Chapter 52 that the transition zone where two communities or biomes meet and intergrade is called an **ecotone.** Ecotones range in size from quite small, such as the area where

Figure 54–24 **A gulper eel from the oceanic province.** This deepwater fish uses its "trap-door" jaws to swallow prey as large as itself. The tail, of which only a portion is shown, makes up most of the length of a gulper eel's body. Gulper eels grow to 1.8 m (6 ft). *(Courtesy of Bruce H. Robison, Monterey Bay Aquarium)*

an agricultural field meets a woodland or where a stream flows through a forest, to continental in scope. For example, at the border between tundra and taiga, an extensive ecotone occurs that consists of tundra vegetation interspersed with small, scattered conifers. Such ecotones provide habitat diversity and are often populated by a greater variety and density of organisms than either adjacent ecosystem.

Ecologists who study ecotones look for adaptations that enable organisms to survive there. They also examine the re-lationship between biological diversity and ecotones, and how ecotones change over time. Long-term studies of ecotones have revealed that they are far from static. The ecotone boundary between desert and semiarid grassland in southern New Mexico, for example, has moved during the past 50 years as the desert ecosystem has expanded into the grassland. Many scientists think that ecotones will show the first measurable responses to global climate change.

SUMMARY WITH KEY TERMS

I. A **biome** is a large, relatively distinct terrestrial region with characteristic climate, soil, plants, and animals. Because it is so large in area, a biome encompasses a number of interacting ecosystems. Temperature and precipitation are important abiotic factors that influence biome distribution.
   A. **Tundra,** the northernmost biome, is characterized by a frozen layer of subsoil (**permafrost**), and low-growing vegetation that is adapted to extreme cold and a short growing season.
   B. The **taiga,** or **boreal forest,** lies south of the tundra and is dominated by coniferous trees that are adapted to the cold winters, short growing season, and acidic, mineral-poor soil.
   C. **Temperate rain forest,** such as occurs on the northwest coast of North America, receives high precipitation and is dominated by large conifers.
   D. **Temperate deciduous forest,** which occurs where precipitation is relatively high and soils are rich in organic matter, is dominated by broad-leaf trees that lose their leaves seasonally.
   E. **Temperate grassland** typically possesses a deep, mineral-rich soil and has moderate but uncertain precipitation.
   F. The **chaparral** is characterized by thickets of small-leaf shrubs and trees and a climate of wet, mild winters and dry summers.
   G. **Desert,** found in both temperate (cold deserts) and subtropical or tropical regions (warm deserts) with low levels of precipitation, possesses communities whose organisms have specialized water-conserving adaptations.
   H. Tropical grassland, called **savanna,** has widely scattered trees interspersed with grassy areas. It occurs in tropical areas with low or seasonal rainfall.
   I. **Tropical rain forest** is characterized by mineral-poor soil and high rainfall that is evenly distributed throughout the year. Tropical rain forest has a high species diversity and high productivity.
II. In aquatic ecosystems, important environmental factors include **salinity,** the amount of dissolved oxygen, and the availability of light for photosynthesis.
   A. Aquatic life is ecologically divided into **plankton** (free-floating), **nekton** (strongly swimming), and **benthos** (bottom-dwelling).
   B. **Phytoplankton** are photosynthetic algae and cyanobacteria that form the base of the food web in most aquatic communities. **Zooplankton** are nonphotosynthetic organisms that include protozoa, tiny crustaceans, and the larval stages of many animals.
III. Freshwater ecosystems include flowing-water ecosystems (streams and rivers), standing water ecosystems (ponds and lakes), and freshwater wetlands.

   A. In **flowing-water ecosystems** the water flows in a current. Flowing-water ecosystems have few phytoplankton and depend on detritus from the land for much of their energy. The organisms in flowing-water ecosystems vary greatly depending on the current, which is swifter in headwaters than downstream.
   B. Large **standing-water ecosystems** (freshwater lakes) are divided into zones on the basis of water depth.
      1. The marginal **littoral zone** contains both emergent vegetation and algae and is very productive.
      2. The **limnetic zone** is open water away from the shore that extends as far down as sunlight penetrates. Organisms in the limnetic zone include phytoplankton, zooplankton, and larger fishes.
      3. The deep, dark **profundal zone** holds little life other than bacterial decomposers.
   C. **Freshwater wetlands,** lands that are transitional between freshwater and terrestrial ecosystems, are usually covered at least part of the year by shallow water and have characteristic soils and vegetation. They are highly productive areas that perform many environmental services.
IV. An **estuary** is a coastal body of water, partly surrounded by land, with access to the ocean and a large supply of fresh water from rivers.
   A. Estuaries are very productive, in part because they receive a high input of nutrients from the adjacent land.
   B. One of the important roles of estuaries is to provide a nursery for the young of many aquatic organisms.
V. Four important marine environments are the intertidal zone, the benthic environment, the neritic province, and the oceanic province.
   A. The **intertidal zone** is the shoreline area between low and high tides. It is a very productive area. Organisms of the intertidal zone possess adaptations to resist wave action and the extremes of being covered by water (high tide) and exposed to air (low tide).
   B. The **benthic environment** is the ocean floor. Coral reefs are particularly important benthic communities in shallow ocean waters because they have high species diversity and are very productive.
   C. The **neritic province** is open ocean from the shoreline to a depth of 200 m. Organisms that live in the neritic province are all floaters or swimmers. Phytoplankton in the **photic region** are the base of the food web.
   D. The **oceanic province** is that part of the open ocean that is deeper than 200 m. The uniform environment is one of darkness, cold temperature, and high pressure. Animal inhabitants of the oceanic province are either predators or scavengers that subsist on **marine snow,** detritus that drifts down from other areas of the ocean.

POST-TEST

1. The northernmost biome, known as _____ , typically has little precipitation, a short growing season, and permafrost. (a) chaparral (b) taiga (c) tundra (d) northern deciduous forest (e) boreal forest
2. South of tundra is the _____ , which consists of coniferous forests

with many lakes. (a) chaparral (b) taiga (c) alpine tundra (d) northern deciduous forest (e) permafrost
3. These forests of the northeastern and middle eastern United States are dominated by broad-leaf hardwood trees that lose their foliage annually.

(a) temperate deciduous forest (b) tropical deciduous forest (c) northern coniferous forest (d) temperate rain forest (e) tropical rain forest

4. The thickest, richest soil in the world occurs in (a) temperate rain forest (b) tropical rain forest (c) savanna (d) temperate grassland (e) chaparral

5. This biome is characterized by a thicket of evergreen shrubs and small trees found in areas with Mediterranean climates. (a) temperate rain forest (b) tropical rain forest (c) savanna (d) temperate grassland (e) chaparral

6. This biome is a tropical grassland interspersed with widely spaced trees. (a) temperate rain forest (b) tropical rain forest (c) savanna (d) temperate grassland (e) chaparral

7. This biome has the greatest species diversity. (a) temperate rain forest (b) tropical rain forest (c) savanna (d) temperate grassland (e) chaparral

8. Organisms in aquatic environments fall into three categories: free-floating _____, strongly swimming _____, and bottom-dwelling _____. (a) nekton; benthos; plankton (b) nekton; plankton; benthos (c) plankton; benthos; nekton (d) plankton; nekton; benthos (e) benthos; nekton; plankton

9. Temperate-zone lakes are thermally stratified, with warm and cold layers separated by a transitional (a) aphotic region (b) thermocline (c) barrier reef (d) ecotone (e) littoral zone

10. Emergent vegetation grows in the _____ zone of freshwater lakes. (a) littoral (b) limnetic (c) profundal (d) neritic (e) intertidal

11. This coastal body of water has access to the open ocean and a large supply of fresh water from rivers. (a) intertidal zone (b) estuary (c) freshwater wetland (d) neritic province (e) standing-water ecosystem

12. The _____ is open ocean from the shoreline to a depth of 200 m. (a) benthic environment (b) intertidal zone (c) neritic province (d) oceanic province (e) aphotic region

13. The transition zone where two ecosystems or biomes meet and intergrade is called a(an): (a) biosphere (b) aphotic region (c) thermocline (d) ecosystem (e) ecotone

REVIEW QUESTIONS

1. What climate and soil factors produce the major biomes?
2. Describe representative organisms of these forest biomes: (a) taiga, (b) temperate deciduous forest, (c) temperate rain forest, (d) tropical rain forest.
3. In which biome do you live? If your biome does not match the description given in this text, explain the discrepancy.
4. Compare tundra with desert and temperate grassland with savanna.
5. Distinguish among plankton, nekton, and benthos, giving examples of each.
6. What environmental factors are most important in determining the adaptations of organisms that live in aquatic environments?
7. Distinguish between freshwater wetlands and estuaries, and between flowing-water and standing-water ecosystems.
8. List and briefly describe the four main marine environments.
9. Which aquatic ecosystem is often compared to tropical rain forests? Why?

YOU MAKE THE CONNECTION

1. When a black-tailed prairie dog or other small animal dies, other prairie dogs bury it. Develop a hypothesis to explain how this behavior may be adaptive. How would you test this hypothesis?
2. In which biomes would migration be most common? Hibernation? Aestivation? Explain your answers.
3. Develop a hypothesis to explain why there are no amphibians or reptiles in the tundra. How would you test your hypothesis?
4. Develop a hypothesis to explain why animals adapted to the desert are usually small. How would you test your hypothesis?
5. Why do most of the animals of the tropical rain forest live in trees?
6. What would happen to the organisms in a river with a fast current if a dam were built? Would there be any differences in habitat if the dam was upstream or downstream of the organisms in question? Explain your answers.

RECOMMENDED READINGS

Allen, W.H. "Travel Across the Treetops." *BioScience,* Vol. 46, No. 11, Dec. 1996. How construction cranes help ecologists study tropical forests.

Broad, W.J. "Fantastic World Found in Sea's Middle Depths." *New York Times,* 26 Sep. 1994. The Monterey Bay Aquarium Research Institute is exploring the little-known realm of life in the ocean's dark middle depths.

Chadwick, D.H. "Roots of the Sky." *National Geographic,* Vol. 184, No. 4, Oct. 1993. Wildlife flourishes in the patchy remnants of the North American prairie.

Johnson, B.L., W.B. Richardson, and T.J. Naimo. "Past, Present, and Future Concepts in Large River Ecology." *BioScience,* Vol. 45, No. 3, Mar. 1995. Provides an overview of what is known about large river ecology. This issue of *BioScience* contains a number of articles about flowing-water ecosystems.

Kappel-Smith, D. "Fickle Desert Blooms: Opulent One Year, No-shows the Next." *Smithsonian,* Mar. 1995. Vegetation varies widely in deserts from year to year, depending on local conditions.

"Life in the Deep." *Discover,* Nov. 1993. A photo essay of several unusual organisms adapted to various parts of the marine environment.

Madigan, M.T., and B.L. Marrs. "Extremophiles." *Scientific American,* Vol. 276, No. 4, Apr. 1997. Some environments thought to be too harsh for life (such as sea ice, deep sea vents, acid springs, salt lakes) contain thriving populations of microorganisms.

Pennisi, E. "Brighter Prospects for the World's Coral Reefs?" *Science,* Vol. 277, 25 Jul. 1997. Although the many threats to coral reefs are serious, a growing body of evidence suggests that when nations clean up pollution and restrict reef fishing, the coral reefs rebound.

Raven, P.H., L.R. Berg, and G.B. Johnson. *Environment,* 2nd ed. Saunders College Publishing, Philadelphia, 1998. This environmental science text offers a detailed situation analysis of planet Earth, including how humans interact with and affect its physical and living systems.

Rützler, K., and I.C. Feller. "Caribbean Mangrove Swamps." *Scientific American,* Vol. 247, Vol. 3, Mar. 1996. Discusses the diverse organisms that live around mangrove trees in Belize.

● Visit our website at **http://www.saunderscollege.com/lifesci/titles.html** and click on Solomon/Berg/Martin Biology for links to chapter-related resources on the World Wide Web.

CHAPTER 55

Humans in the Environment

The human species *(Homo sapiens)* has been present on Earth for about 800,000 years (see Chapter 21), which is a brief span of time compared with the age of our planet (some 4.6 billion years). Despite our relatively short tenure on Earth, our biological impact has been unparalleled. Our numbers have increased dramatically—the human population reached 6 billion in 1999—and we have expanded our biological range, moving into almost every habitat on Earth.

Wherever we have gone, we have altered the environment and shaped it to meet our needs, as demonstrated by the photograph of pedestrians and traffic on a street in Shanghai, China. In only a few generations we have transformed the face of Earth, placed a great strain on Earth's resources and resilience, and profoundly affected other organisms. Thus, the impact of humans on the environment merits special study in biology, not merely because we ourselves are humans but also because our impact on the rest of the biosphere has been so extensive.

Humans do not live alone on Earth, nor are we above the laws of nature; our actions have consequences. We are part of the same biosphere as all other organisms and are totally dependent on it and them for survival. Thus, an important goal is to avoid upsetting the biological systems that support us.

Many environmental concerns exist today, too many to be considered in a single chapter. The rapidly expanding human population underlies and exacerbates all environmental problems. The increasing population is placing a nonsustainable stress on the environment, as humans consume ever-increasing quantities of food and water, use more and more energy and raw materials, and produce enormous amounts of waste and pollution. Because the human population crisis was discussed in Chapter 51, we conclude this text by focusing our attention on four serious environmental issues that affect the biosphere: declining biological diversity, deforestation, global warming, and ozone depletion in the stratosphere.

Organisms are important natural resources that are not fully appreciated. Biological diversity contributes toward a sustainable environment, that is, one that provides a life support system that enables humans as well as other species to survive. The current reduction in biological diversity caused by the extinction of many species results in the increased instability of ecosystems and in lost opportunities and lost solutions to future problems.

(Jeff Greenberg/Photo Researchers, Inc.)

The world's forests provide many environmental services (habitat for organisms, watershed protection, and prevention of soil erosion, to name a few) as well as commercially important timber and numerous recreational opportunities. The greatest problem facing forests today is deforestation, the temporary or permanent removal of forest. Deforestation, in turn, contributes to a loss of biological diversity.

Production of atmospheric pollutants that trap solar heat in the atmosphere will probably affect Earth's climate during the 21st century. Global warming and the accompanying changes in precipitation patterns could alter food production, destroy forests, reduce biological diversity, and submerge coastal areas.

The chemical destruction of stratospheric ozone by gaseous air pollutants may permit large amounts of solar ultraviolet radiation to penetrate to Earth's surface. Living organisms, including humans, will be harmed by increased exposure to ultraviolet radiation.

AFTER YOU HAVE STUDIED THIS CHAPTER YOU SHOULD BE ABLE TO

1. Distinguish among threatened species, endangered species, and extinct species.
2. Discuss at least four causes of declining biological diversity and identify the most important cause.
3. Define conservation biology and compare in situ and ex situ conservation measures.
4. Describe the benefits and shortcomings of the following laws and agreements: the Endangered Species Act; the World Conservation Strategy; the Convention on International Trade in Endangered Species of Wild Flora and Fauna.
5. Discuss the ecological benefits of forests and describe the consequences of deforestation.
6. State at least three reasons that forests are disappearing today.
7. Name at least three greenhouse gases and explain how greenhouse gases contribute to global warming.
8. Describe how global warming may affect sea level, precipitation patterns, organisms (including humans), and food production.
9. Give examples of ways to prevent, mitigate, and adapt to global warming.
10. Distinguish between surface ozone and stratospheric ozone.
11. Cite the causes and potential effects of ozone destruction in the stratosphere.

SPECIES ARE DISAPPEARING AT AN ALARMING RATE

Extinction, the death of a species, occurs when the last individual member of a species dies (see Chapter 19). A natural biological process, extinction has been greatly accelerated by human activities. The burgeoning human population, which was considered in detail in Chapter 51, has forced us to spread into almost all areas of Earth. Whenever humans invade an area, the habitats of many plants and animals are disrupted or destroyed, which can contribute to their extinction. For example, the dusky seaside sparrow, a small bird that was found only in the marshes of the St. Johns River in Florida, became extinct in 1987, largely because of human destruction of its habitat (Fig. 55–1).

Biological diversity, also called **biodiversity,** is the variety of living organisms and of the ecosystems in which they live. Biological diversity includes the number of different species *(species diversity)*, the genetic variety within a species *(genetic diversity)*, and the variety of interactions within and among ecosystems *(ecosystem diversity)*.

Biological diversity is currently decreasing at an alarming rate. It is likely that thousands of species will be eliminated within the next few decades. As many as one-fourth of Earth's plant families[1] may be extinct by the end of the 21st century, and countless animal species that depend on those plants for food and habitat will probably also become extinct.

Some biologists fear that we are entering the greatest period of mass extinction in Earth's history, but the current situation differs from previous periods of mass extinction in several respects. First, its cause is directly attributable to human activities. Second, it is occurring in a tremendously compressed period of time (just a few decades as opposed to hundreds of thousands of years), much faster than rates of speciation (or replacement). Perhaps even more sobering, larger numbers of plant species are becoming extinct today than in previous mass extinctions. Because plants are the base of terrestrial food webs, extinction of animals that depend on plants cannot be far behind.

Figure 55–1 **Extinction.** The dusky seaside sparrow became extinct in 1987, largely owing to human destruction of its habitat in Florida. *(U. S. Fish and Wildlife Service)*

[1]Recall that a family consists of a number of related genera, each of which consists of a number of related species. When a family becomes extinct, all the species of all the genera comprising that family cease to exist.

FOREST FRAGMENTATION AND DECLINING BIRD POPULATIONS

How is habitat fragmentation related to declining bird populations? Evidence is mounting that the populations of many birds that were once abundant have been dropping steadily for more than a decade. Across the North American continent, population declines have been noted in every major group of birds. In the eastern United States, for example, 70% of neotropical bird species show declining numbers. *Neotropical birds* spend the winter in Central and South America and the Caribbean and then migrate north to breed in the United States and Canada during the summer. Cerulean warblers, olive-sided flycatchers, yellow-billed cuckoos, and rose-breasted grosbeaks are examples of declining neotropical birds.

Neotropical birds are faced with changing environments in both their winter and summer homes, and loss of habitat is the main reason for their decline. These migratory birds are under stress from the burning of tropical rain forests in Central and South America, as well as the fragmenting of forested areas in North America to accommodate suburban development, agriculture, and logging.

Several studies have shown that fragmentation of forests increases the likelihood of reproductive failure for neotropical birds, many of which are insect-eaters that nest only in forest interiors. As a result of forest fragmentation, their nests are more likely to be located near a *forest edge,* the boundary between the forest and surrounding farmlands or residential neighborhoods, rather than deep in the forest. Nests that are within 100 m (328 ft) of a forest edge are more vulnerable to predation by small mammalian predators such as raccoons and both domestic and feral cats (feral cats are domestic cats that have gone wild). Blue jays, American crows, and other egg-eating birds do most of their hunting along the forest edge.

Nest parasitism by brown-headed cowbirds is also more significant along the forest edge. Female cowbirds lay their eggs in the nests of other bird species. In a 1989 to 1993 study that monitored 5000 nests of neotropical birds, the majority of nests were parasitized in areas with less than 55% forest cover, and in many cases the nests contained more cowbird eggs than eggs that belonged there. Nest parasitism causes reproductive failure because the host parents provide food for the larger, more aggressive cowbird babies, while their own young starve.

Cowbirds do not nest or feed in intact forests. Before Europeans settled in North America, cowbirds followed the migratory herds of bison across the Great Plains, foraging on seeds and insects stirred up by the bison. With the settlement of the continent by Europeans, many forests were cut to provide agricultural lands, and the cowbird expanded its range into fields and cattle pastures. The cowbirds began parasitizing new bird species that had not yet evolved ways to resist them. (Some birds, such as robins, gray catbirds, and Baltimore orioles, reject cowbird eggs.)

According to the United States' Endangered Species Act, a species is designated as **endangered** when its numbers are so severely reduced that it is in imminent danger of extinction throughout all or a significant part of its range. Unless humans intervene, an endangered species will probably become extinct. When extinction is less imminent but the population of a particular species is quite low, the species is said to be **threatened.** A threatened species is likely to become endangered in the foreseeable future, throughout all or a significant part of its range. Designation of endangered or threatened status for a species already represents a decline in biological diversity, since it implies severely diminished genetic diversity. Endangered and threatened species are at greater risk of extinction than species with greater genetic variability, because long-term survival and evolution depend on genetic diversity (see *Making the Connection: Cheetahs, Genetics, and Ecology* in Chapter 18).

Human activities contribute to declining biological diversity

Species become endangered and extinct for a variety of reasons, including the destruction or modification of habitats and the production of pollution. Humans also upset the delicate balance of organisms in a given area by introducing new, exotic species or by controlling native pests or predators. Illegal hunting and uncontrolled commercial harvesting are also factors.

The majority of species facing extinction today are endangered because of destruction of natural habitats. Building roads, parking lots, and buildings; clearing forests to grow crops or graze domestic animals; and logging forests for timber all take their toll on natural habitats. Draining marshes converts aquatic habitats to terrestrial ones, while building dams floods terrestrial habitats. Habitat destruction threatens the survival of species because most organisms require a particular type of environment, and habitat destruction reduces their biological range and ability to survive.

Humans often leave small, isolated patches of natural landscape that are completely surrounded by roads, fields, and buildings. Like a land mass that is surrounded by water, an isolated habitat that is surrounded by an expanse of unsuitable territory is referred to as an *island* (recall the discussion of isolated island communities in Chapter 52). Species from the "developed" landscape may intrude into the island, and species that prefer the island habitat may occur in greatly reduced numbers or even disappear altogether. As a result, habitat fragments often support only a fraction of the species found in the original, unaltered environment. The effects of habitat fragmentation are supported by extensive scientific evidence (see *Making the Connection: Forest Fragmentation and Declining Bird Populations*).

THE GREAT LAKES

The Great Lakes of North America formed about 10,500 years ago when the melting waters of retreating glaciers drained into the lake basins that were carved from river valleys by the glaciers. The Great Lakes, which are connected to one another, collectively hold about one-fifth of the world's fresh water.

More than 33 million people live in the Great Lakes watershed, an area that is home to agriculture, trade, industry, and tourism. Important industrial cities, such as Duluth, Milwaukee, Chicago, Cleveland, Erie, Buffalo, and Toronto, are located along the lake shorelines. At least 38 million people obtain their drinking water from the Great Lakes.

During the 1960s, pollution in the Great Lakes became a highly visible problem, particularly in Lake Erie.* Thousands of toxic chemicals polluted the lakes. Eutrophication (enrichment; see Chapter 54) was pronounced, bacterial counts were high enough to be a health hazard, and fish kills were common. Birth defects, such as missing brains, internal organs located outside the body, and deformed feet and wings,

were observed in almost 50% of the animal species studied.

As of the late-1990s, the Great Lakes are in better shape than they have been in recent memory. The U. S. EPA and Environment Canada concluded in the *State of the Great Lakes 1995* that water quality and human health are slowly improving in the Great Lakes region. Levels of the pesticide DDT in women's breast milk have declined since 1967, for example, as have levels of industrial chemicals called PCBs in trout, Coho salmon, and herring gulls. Some animal populations, such as double-crested cormorants and bald eagles, have rebounded.

Many problems remain, however. Forty-three shoreline areas are so polluted that they are designated "areas of concern" by the U. S. EPA and Environment Canada. Zebra mussels, sea lampreys, and more than 130 other exotic species have proliferated and threaten native species. Zebra mussels clog water-intake pipes for utilities and industries. Sea lampreys prey on trout and other large fishes.† Shoreline development continues to encroach on natural areas.

Persistent toxic compounds remain in the lakes, much of them coming from air pollution or from contaminated lake sediments. Mercury contamination, for example, is on the increase, primarily from coal-fired power plants and hazardous waste incinerators. There is evidence that the presence of certain toxic chemicals causes hormonal changes that are linked to reproductive failures, abnormal development, and abnormal behaviors in some fishes, birds, and mammals. Because persistent toxic chemicals have accumulated in food webs (see Chapter 53), fish consumption advisories are issued warning people who eat Great Lakes' fish of potential health problems.

*Lake Erie is the most polluted of the Great Lakes because it has the largest human population on its shores, and it is the shallowest of the Great Lakes and therefore contains the smallest volume of water.

†The possibility of sea lamprey control is more promising since the recent development of a chemical that sterilizes male sea lampreys. Male sea lampreys are trapped, sterilized, and released, after which they mate with females, who subsequently do not bear offspring.

Even habitats left undisturbed and in their natural state are indirectly modified by human activities that produce acid precipitation and other forms of pollution. Acid precipitation has contributed to the decline of large stands of forest trees and to the biological death of many freshwater lakes, for example, in the Adirondack Mountains and in Nova Scotia. The production of other types of pollutants also adversely affects organisms (see *Focus On: The Great Lakes*). Such pollutants include industrial and agricultural chemicals, organic pollutants from sewage, acid wastes seeping from mines, and thermal pollution from the heated waste water of industrial plants.

Biotic pollution, the introduction of a foreign, or exotic, species into an area where it is not native, often upsets the balance among the organisms living in that area. The foreign species may compete with native species for food or habitat, or may prey on them. Generally, an introduced competitor or predator causes a greater negative effect on local organisms than do native competitors or predators. (Most introduced species lack natural agents, such as diseases, predators, and competitors, that would otherwise control them. Also without a shared evolutionary history, most native species typically are less equipped to cope with introduced species.) Although exotic species may be introduced into new areas by natural

means, humans are usually responsible for such introductions, either knowingly or accidentally.

One of North America's greatest biological threats is the zebra mussel, a native of the Caspian Sea that was probably introduced through ballast water flushed into the Great Lakes by a foreign ship in 1985 or 1986. Since then, the tiny freshwater mussel, which clusters in extraordinary densities, has massed on hulls of boats, piers, buoys, water intake systems and, most damaging of all, on native clam and mussel shells. The zebra mussel's strong appetite for algae, phytoplankton, and zooplankton is also cutting into the food supply of native fishes, mussels, and clams, threatening their survival. By the mid-1990s the zebra mussel had progressed from the Great Lakes into the Mississippi River. It is found as far south as New Orleans, as far north as Duluth, Minnesota, and as far east as the Hudson River in New York. Some scientists hypothesize that within the next decade it will have colonized all freshwater bodies in North America that meet its wide-ranging ecological requirements. Since the zebra mussel invaded North America, an estimated $5 billion dollars have been spent to repair damage such as clogged pipes (Fig. 55–2).

Islands are particularly susceptible to the introduction of exotic species. For example, Abingdon Island, one of the Gala-

Figure 55–2 Zebra mussels clog a pipe. The zebra mussel has caused billions of dollars in damage in addition to displacing native clams and mussels. *(Illinois Department of Conservation)*

Figure 55–3 Illegal commercial hunting. The Kenyan government destroys elephant tusks confiscated from poachers. Burning ensures that such illegally obtained wildlife trophies will never be sold on the black market. *(Steve Turner/Animals Animals)*

pagos Islands off the coast of South America, was home to an **endemic** (found nowhere else) giant tortoise. In 1957 several fishermen introduced goats to Abingdon Islands, and within five years the Abingdon tortoise was extinct. The goats, with no natural predators on the island, had greatly increased in number and had eaten the tortoises' food. In Hawaii, the introduction of mouplan sheep has imperiled both the mamane tree (because the sheep eat it) and a species of honeycreeper, an endemic bird that relies on the tree for food.

Sometimes species become endangered or extinct as a result of deliberate efforts to eradicate or control their numbers, often because they prey on game animals or livestock. Populations of large predators such as the wolf, mountain lion, and grizzly bear have been decimated by ranchers, hunters, and government agents. Predators of game animals and livestock are not the only animals vulnerable to human control efforts. Some animals are killed because their lifestyles cause problems for humans. The Carolina parakeet, a beautiful green, red, and yellow bird endemic to the southeastern United States, was extinct by 1920, exterminated by farmers because it ate fruit from their trees. Prairie dogs and pocket gophers were poisoned and trapped so extensively by ranchers and farmers that, between 1900 and 1960, they disappeared from about 98% of their original range. As a result of sharply decreased numbers of prairie dogs, the black-footed ferret, a natural predator of these animals, became endangered. By the winter of 1985–1986,

only ten ferrets were known to exist, four in Meeteetse, Wyoming, and six in captivity. A successful captive breeding program enabled biologists to release black-footed ferrets to the Wyoming prairie, beginning in 1991.

In addition to the control of predators and pests, humans hunt for three other reasons. *Commercial hunting* is the killing of animals for profit, for example by selling their fur; *sport hunting* is the killing of animals for recreation; and *subsistence hunting* is the killing of animals for food. Subsistence hunting has caused the extinction of certain species in the past but is not a major cause today, mainly because so few human groups still rely on subsistence hunting for their food supply. Sport hunting was also a factor in the extinction of certain animals in the past but is now strictly controlled in most countries. The passenger pigeon was one of the most common birds in North America in the early 1800s, but a century of overhunting (both commercial and sport) resulted in its extinction in the early 1900s.

Figure 55-4 Illegal commercial harvesting. These hyacinth macaws (*Anodorhynchus hyacinthus*) were seized in French Guiana in South America as part of the illegal animal trade there. *(Jany Sauvanet/NHPA)*

Illegal commercial hunting (poaching) continues to endanger a number of larger animals such as the tiger, cheetah, and snow leopard, whose beautiful furs are quite valuable. Rhinoceroses are slaughtered for their horns (used for dagger handles in the Middle East and as a medicine and an aphrodisiac in Asia) and bears for their gallbladders (used in Asian medicine to treat ailments ranging from indigestion to hemorrhoids). Although these animals are protected by law, demand for their products on the black market has promoted illegal hunting, particularly in impoverished countries where sale of contraband products can support a family for months. When illegally obtained wildlife trophies are confiscated by police or park rangers, they are destroyed (Fig. 55-3).

In contrast to commercial hunting, in which target organisms are killed, *commercial harvest* removes live organisms from the wild. Most organisms that are commercially harvested end up in pet stores. Several million birds are commercially harvested each year for the pet trade, but, unfortunately, many die in transit, and many more die from improper treatment in their owners' homes. At least 40 parrot species are now threatened or endangered, in part because of commercial harvest. Although it is illegal to capture endangered animals from the wild, a thriving black market exists, mainly because collectors in the United States, Europe, and Japan will pay extremely large sums for rare tropical birds (Figure 55-4). Imperial Amazon macaws, for example, fetch up to $20,000 each. Income from the sale of these organisms more than compensates for the risks of fines and prison if caught.

Animals are not the only organisms threatened by commercial harvest. A number of unique or rare plants have been collected from the wild so extensively that they are now classified as endangered. These include certain carnivorous plants, cacti, and orchids.

Where is the problem of declining biological diversity greatest?

Declining biological diversity is a concern throughout the United States but is most serious in the states of California, Hawaii, Texas, and Florida, where many ecosystems are deteriorating rapidly due to human population growth and land development. As serious as declining biological diversity is in the United States, it is even more serious abroad. Perhaps 40% of the world's species are concentrated in tropical rain forests, areas that are increasingly threatened by habitat destruction. This means that a relatively few countries, primarily developing nations, hold most of the biological diversity that is so ecologically and economically important to the entire world. The situation is complicated by the fact that these countries are least able to afford the protective measures needed to maintain biological diversity. International cooperation will clearly be needed to preserve our biological heritage.

Conservation biology addresses the issue of declining biological diversity

Conservation biology is the study and protection of biological diversity. It includes two types of efforts to save organisms from extinction: in situ and ex situ. **In situ conservation,** which includes the establishment of parks and reserves, concentrates on preserving biological diversity in the wild. A high priority of in situ conservation is the identification and protection of sites with a great deal of diversity. With increasing demands on land, however, in situ conservation cannot guarantee the preservation of all types of biological diversity. **Ex situ conservation** involves conserving biological diversity in human-controlled settings. Breeding captive species in zoos and seed storage of genetically diverse crops (see *Focus On: Seed Banks* in Chapter 35) are examples of ex situ conservation.

In situ conservation is the best way to preserve biological diversity

Many nations appreciate the need to protect their biological heritage and for this reason have set aside areas for wildlife habitats. Such natural ecosystems offer the best strategy for the long-term protection and preservation of biological diversity. There are currently more than 3000 national parks, marine sanctuaries, refuges, forests, and other protected areas throughout the world. Some of these areas have been chosen to protect specific endangered species. The first such refuge, established in 1903 at Pelican Island, Florida, was set aside to protect the brown pelican. Today the National Wildlife Refuge System of the United States includes more than 500 refuges; although the bulk of the protected land is in Alaska, refuges exist in all 50 states.

Many protected areas have multiple uses. National parks may serve recreational needs, for example, whereas national forests may be open for logging, grazing, and farming operations. The mineral rights to many refuges are privately owned, and some wildlife refuges have had oil, gas, and other mineral development. Hunting is allowed in more than half of the wildlife refuges in the United States, and military exercises are conducted in several. Such uses often conflict with the goal of preserving biodiversity.

Certain parts of the world are critically short of protected areas. Protected areas are urgently needed in tropical rain forests, the tropical grasslands and savannas of Brazil and Australia, and dry forests that are widely scattered around the world. Desert organisms are underprotected in northern Africa and Argentina, and the species of many islands and lakes also need protection.

Restoring damaged or destroyed habitats is the goal of restoration ecology

Disturbed lands can be reclaimed and converted into natural ecosystems with high levels of biological diversity. In certain cases, an area is restored to its original species composition, although in most cases, the area is rehabilitated by establishing at least some of the original species. Ecological restoration is a complicated process that requires advance planning and study of site conditions, including native and exotic species currently existing there. During the restoration process, native species typically are reintroduced, whereas exotic species are eliminated as much as possible.

One of the oldest and most famous examples of ecological restoration has been carried out since the 1930s by the University of Wisconsin–Madison Arboretum (Fig. 55–5). Several distinct natural communities have been carefully developed on damaged agricultural land. These communities include a tallgrass prairie, a xeric (dry) prairie, and several types of pine and maple forests native to Wisconsin.

Restoration ecology establishes wildlife habitats and also provides additional benefits, such as the regeneration of soil damaged by agriculture or mining. The disadvantages of restoration ecology include the time and expense required to restore an area. Nonetheless, restoration ecology is an impor-

(a)

(b)

Figure 55–5 **Restoring damaged lands.** The University of Wisconsin–Madison Arboretum has pioneered restoration ecology. *(a)* The restoration of the prairie was at an early stage in November, 1935. The men are digging holes to plant prairie grass sod. *(b)* The prairie as it looks today. This picture was taken at approximately the same location as the 1935 photograph. *(Courtesy of Virginia Kline, University of Wisconsin–Madison Arboretum)* tant aspect of conservation biology, as it is thought that restoration will deter many extinctions.

Ex situ conservation is used in an attempt to save species on the brink of extinction

Zoos, aquaria, and botanical gardens often attempt to save certain endangered species from extinction. Eggs may be collected from nature, for example, or the remaining few animals may be captured and bred in zoos and other research facilities. Special techniques such as artificial insemination and host mothering are used to increase the number of offspring. Sperm is collected from a suitable male of a rare species and is used to impregnate a female (perhaps located in another zoo in a different city or even in another country). This is an example of **artificial insemination.** In another approach, a female of a rare species is treated with fertility drugs, which cause her to produce multiple eggs. Some of these eggs are collected, fertilized with sperm, and surgically implanted into females of a related but less rare species, who later give birth to offspring of the rare species. This is an example of **host mothering,** also called **embryo transfer.** (Host mothering is also discussed in *Focus On: Novel Origins* in Chapter 48.)

A few spectacular successes have occurred in captive breeding programs, in which large enough numbers of a species have been produced to reestablish small populations in the wild. Conservation efforts, for example, have helped the bald eagle make a remarkable comeback. In the mid-1970s, the first eagles bred in captivity were released to the wild. In addition to captive breeding programs, biologists also remove eagle eggs from their nests in the wild, raise the baby eagles in wildlife refuges, and then return them to the wild. (Removal of eggs actually helps increase the number of eagles because nesting eagles often lay more eggs to replace those that were removed.) Other efforts also improved the bald eagle's chances, including protection of nesting sites and the banning of chlorinated pesticides (see *Making the Connection: Food Chains and Poisons in the Environment* in Chapter 53). As a result of continuing efforts on many fronts, the number of nesting pairs in the contiguous United States increased from 417 in 1963 to 4452 in 1994. In 1994 the bald eagle was removed from the endangered list and transferred to the less critical threatened list.

Attempting to save a species on the brink of extinction is usually extremely expensive; therefore, only a small proportion of endangered species can be saved. Moreover, zoos, aquaria, and botanical gardens do not have the space to try to save all endangered species. This means that conservation biologists must prioritize which species to attempt to save. Zoos have traditionally focused on large, charismatic animals because the public is more interested in them. Such conservation efforts ignore millions of less glamorous but ecologically important species. However, zoos, aquaria, and botanical gardens serve a useful purpose by educating the public about the value of conservation biology. Clearly, it is more cost effective to protect and maintain natural habitats so that species do not become endangered in the first place.

Figure 55-6 **Biological diversity and habitat size.** When a protected area is set aside, it is important to know what the minimum size of that area must be so that it will not be affected by encroaching species from surrounding areas. Shown are 1 hectare and 10 hectare plots that are part of the Minimum Critical Size of Ecosystems Project. *(R.O. Bierregaard)*

Conservation organizations are essential to conservation biology

Conservation organizations are an important part of the effort to maintain biological diversity. They help to educate policy makers and the public about the importance of biological diversity. In certain instances they serve as catalysts by galvanizing public support for important biodiversity preservation efforts. They also provide financial support for conservation projects, from basic research to the purchase of land that is a critical habitat for a particular organism or group of organisms.

The World Conservation Union (IUCN)[2] assists countries with hundreds of conservation biology projects. It and other conservation organizations are currently assessing how effective established wildlife refuges are in maintaining biological diversity. For example, the Minimum Critical Size of Ecosystems Project, a long-term study of the effects of fragmentation on Amazonian rain forest, is being conducted in Brazil by the World Wildlife Fund and Brazil's National Institute for Amazon Research (Figure 55-6). In addition, IUCN and the World Wildlife Fund[3] have identified major conservation priorities by determining which biomes and aquatic ecosystems are not represented by protected areas. IUCN

[2]Formerly called the International Union for Conservation of Nature and Natural Resources, the World Conservation Union still goes by the abbreviation IUCN.

[3]Outside of the United States, Canada, and Australia, the World Wildlife Fund is known as the World Wide Fund for Nature.

TABLE 55-1 Organisms Listed as Endangered or Threatened in the United States, 1996

Type of Organism	Number of Endangered Species	Number of Threatened Species
Mammals	55	9
Birds	74	16
Reptiles	14	19
Amphibians	7	5
Fishes	65	40
Snails	15	7
Clams	51	6
Crustaceans	14	3
Insects	20	9
Spiders	5	0
Flowering plants	405	90
Conifers	2	0
Ferns and other plants	26	2

maintains a data bank on the status of the world's species; its material about organisms and habitats is published in *The IUCN Red Data Books.*

The Endangered Species Act provides some legal protection for species and habitats

In 1973 the **Endangered Species Act** was passed in the United States, authorizing the U. S. Fish and Wildlife Service to protect endangered and threatened species in the United States and abroad. Many other countries now have similar legislation.

Since the passage of the Endangered Species Act, 955 species in the United States have been listed as endangered or threatened (Table 55–1). The act makes it illegal to sell or buy any product made from an endangered or threatened species. It further requires officials of the U. S. Fish and Wildlife Service to select critical habitats and design a recovery plan for each species listed. For example, the 1995 reintroduction of wolves to Yellowstone National Park is part of that species' recovery plan.

The Endangered Species Act, which was amended in 1982, 1985, and 1988, is scheduled to be reauthorized by Congress sometime in the late 1990s. It is considered one of the strongest pieces of environmental legislation in the United States, in part because species are designated as endangered or threatened entirely on biological grounds; currently, economic considerations cannot influence the designation of endangered or threatened species. The Endangered Species Act has also been one of the most controversial pieces of environmental legislation because it has interfered with some federally funded development projects.

Some critics view the Endangered Species Act as an impediment to economic progress. To protect the habitat of the northern spotted owl and 40 other endangered species, for example, the timber industry has been blocked from logging old-growth forests (2000 year old trees) in certain federal timberlands of the Pacific Northwest. Those who defend the Endangered Species Act point out that, of 34,000 past cases of endangered species versus "development," only 21 cases could not be resolved by some sort of compromise. When the black-footed ferret was reintroduced on the Wyoming prairie, for example, it was classified as an "experimental, nonessential species" so that its reintroduction would not block ranching and mining in the area. Thus, the ferret release program obtained the support of local landowners, support deemed crucial to the ferret's survival in nature.

Defenders of the Endangered Species Act agree that it is not perfect. Relatively few endangered species have recovered enough to be reclassified as threatened or delisted (removed from protection), although the U. S. Fish and Wildlife Service, in a report to Congress in 1994, said that 41% of listed species are stable or improving. The law is geared more to saving a few popular or unique endangered species rather than the much larger number of less glamorous species that perform valuable ecosystem services. Yet it is organisms such as fungi and insects that play a central role in ecosystems and contribute most to their functioning. Conservation biologists would like to see the Endangered Species Act strengthened to preserve whole ecosystems and maintain complete biological diversity rather than attempting to save individual endangered species. This approach offers protection to many declining species rather than to a single species.

International agreements help protect species and habitats

The exploitation of endangered species can be somewhat controlled through legislation. At the international level, 128 countries participate in the Convention on International Trade in Endangered Species of Wild Flora and Fauna (CITES), which bans hunting, capturing, and selling of endangered or threatened species. Unfortunately, enforcement of CITES varies from country to country, and even where enforcement exists, the penalties are not very severe. As a result, illegal trade in rare, commercially valuable species continues.

The **World Conservation Strategy,** a plan designed to conserve biological diversity worldwide, was formulated in 1980 by the IUCN, the World Wildlife Fund, and the U. N. Environment Programme. In addition to conserving biological diversity, the World Conservation Strategy seeks to preserve the vital ecosystem processes on which all life depends for survival and to develop sustainable uses of organisms and the ecosystems that they comprise. The most recent version of the World Conservation Strategy, published in 1991, also focuses on stabilizing the human population in *Caring for the Earth: A Strategy for Sustainable Living.*

The biological diversity treaty produced by the 1992 Earth Summit has been ratified by more than 150 nations and is now considered binding. Under the conditions of the treaty,

Figure 55–7 **Deforestation.** Aerial view of a clearcut forest in Washington. *(Gary Braasch/Tony Stone Images)*

each signatory nation must inventory its own biodiversity and develop a **national conservation strategy,** a detailed plan for managing and preserving the biological diversity of that specific country.

DEFORESTATION IS OCCURRING AT AN UNPRECEDENTED RATE

The most serious problem facing the world's forests is **deforestation,** which is defined as the temporary or permanent clearance of forests for agriculture or other uses (Fig. 55–7). When forests are destroyed, they no longer make valuable contributions to the environment or to the people who depend on them.

Deforestation results in increased soil erosion and thus decreased soil fertility. Increased sedimentation of waterways, caused by soil erosion, harms aquatic ecosystems by reducing light penetration, covering aquatic organisms, bringing insoluble toxic pollutants into the water, and filling in waterways. Uncontrolled soil erosion, particularly on steep deforested slopes, can cause mud flows that endanger human lives and property and can affect the production of hydroelectric power as silt builds up behind dams. In drier areas, deforestation can lead to the formation of deserts.

Deforestation also contributes to the loss of biological diversity. In particular, many tropical species have limited ranges within a forest, so they are especially vulnerable to habitat destruction or modification. Wildlife in temperate areas, particularly migratory birds and butterflies, also suffer because of tropical deforestation.

Forests, particularly on hillsides and mountains, provide nearby lowlands with some protection from floods by trapping and absorbing precipitation. When a forest is cut down, the watershed cannot hold water nearly as well, and the total amount of surface runoff flowing into rivers and streams increases. This not only causes soil erosion, but puts lowland areas at extreme risk of flooding and mudslides.

Regional and global climate changes are thought to be induced by deforestation. Transpiring trees release substantial amounts of moisture into the air. This moisture falls back to the surface in the hydrologic cycle (see Chapter 53). When a large forest is removed, rainfall may decline, droughts may become common in that region, and temperatures may rise slightly (because there is less evaporative cooling due to transpiration). Deforestation also may contribute to an increase in global temperature (discussed shortly) by causing a release of stored carbon into the atmosphere as carbon dioxide, which causes air to retain heat.

Where and why are forests disappearing?

During the past 1000 years, deciduous forests in temperate areas were largely cleared for housing and agriculture. Today, however, deforestation in the tropics is occurring much more rapidly and over a much larger area. Most of the remaining undisturbed tropical rain forests, which occur in the Amazon and Congo River basins of South America and Africa, are being cleared and burned at a rate unprecedented in human history. Tropical rain forests are also being destroyed at an extremely rapid rate in southern Asia, Indonesia, Central America, and the Philippines.

Exact figures on rates of tropical forest destruction are unavailable. However in the early 1990s, the Food and Agriculture Organization (FAO) of the United Nations released its second global assessment of tropical deforestation. A total of 87 tropical countries were evaluated in this study. The FAO estimated an average annual loss in forests of 0.9% per year from 1981 to 1990, and some areas, such as West Africa, experienced losses estimated at greater than 2% per year. If this average rate of deforestation—which represents a worldwide annual loss of 16.9 million hectares (41.8 million acres)—con-

Figure 55–8 **Clearing the forest.** Tropical rain forest in Brazil is burned to provide agricultural land. This type of cultivation is known as slash-and-burn agriculture. *(Dr. Nigel Smith/Earth Scenes)*

Figure 55–9 **Wood for fuel.** A woman in India lights a wood cooking fire. Half of the people in the world use wood fires to cook food. *(Inga Spence/Tom Stack & Associates)*

tinues, tropical forests will be all but gone by the second half of the 21st century.

Keeping in mind that tropical deforestation is a complex problem, three agents are thought to be the most immediate causes of deforestation in tropical rain forests: subsistence agriculture, commercial logging, and cattle ranching. Other reasons for the destruction of tropical forests include the development of hydroelectric power, which inundates large areas of forest; mining, particularly when ore smelters burn charcoal produced from rainforest trees; and plantation-style agriculture of crops such as citrus fruits and bananas. The main cause of deforestation in tropical dry forests is fuel wood consumption.

Subsistence agriculture, in which food is produced by a family to feed itself, accounts for perhaps 60% of tropical deforestation. In many developing countries where tropical rain forests are located, the majority of people do not own the land on which they live and work. For example, in Brazil only 5% of the farmers own 70% of the land. Most subsistence farmers have no place to go except into the forest, which they clear to grow food.

Subsistence farmers often follow loggers' access roads until they find a suitable spot. They first cut down the trees and allow them to dry, then they burn the area (Fig. 55–8) and plant crops immediately after burning. This is known as **slash-and-burn agriculture.** Yields from the first crop are often quite high because the nutrients that had been in the trees (see Chapter 54) are now available in the soil. However, soil productivity declines rapidly, and subsequent crops are poor. In a few years the farmer must move to a new part of the forest and repeat the process. Cattle ranchers often claim the abandoned

land for grazing, because land that is not rich enough to support crops can still support livestock.

Slash-and-burn agriculture done on a small scale with periods of 20 to 100 years between cycles is sustainable. The forest regrows rapidly after a few years of farming. But when millions of people try to obtain a living in this way, the land is not allowed to lie uncultivated long enough to recover.

About 20% of tropical deforestation is the result of commercial logging, and vast tracts of tropical rain forests, particularly in Southeast Asia, are being harvested for export abroad. Most tropical countries allow commercial logging to proceed at a much faster rate than is sustainable, because it supplies them with much-needed revenues. In the final analysis, tropical deforestation does not contribute to economic development; rather, it reduces or destroys the value of an important natural resource.

Approximately 12% of tropical rainforest destruction is carried out to provide open rangelands for cattle. Cattle ranching is particularly important in Central America. Much of the beef raised on these ranches, which are often owned by foreign companies, is exported to fast-food restaurant chains in North America and Europe. After the forests are cleared, cattle can graze on the land for up to perhaps 20 years, after which time the soil fertility is depleted. When this occurs, shrubby plants, known as scrub savanna, take over the range.

Perhaps half of the wood consumed worldwide is used in developing countries for heating and cooking fuel (Fig. 55–9). Fuel wood consumption is a greater concern in dry tropical forests—tropical areas subjected to a wet season followed by a prolonged dry season—than it is in humid forests. For example, about 340,000 hectares (840,000 acres) of India's dry tropical forests disappear each year, mostly for firewood consumption.

Boreal forests are also being destroyed

Although tropical forests have been depleted extensively during the second half of the 20th century, they are not the only forests being destroyed. Extensive deforestation of certain boreal (northern) forests began in the late 1980s and early 1990s. Boreal forests cover about 11% of the Earth's land area and occur in Alaska, Canada, Scandinavia, and northern Russia.

Boreal forests are currently the primary source of the world's industrial wood and wood fiber. Boreal forests are being harvested primarily by clearcutting, in which all trees are removed from an area. Timber companies prefer clearcutting, which is ecologically unsound, because it is the most cost-effective way to harvest trees. The annual loss of boreal forests has been estimated to encompass an area twice as large as the Amazonian rain forests of Brazil. About 1 million hectares (2.5 million acres) of Canadian forests are logged annually, and extensive tracts of Siberian forests in Russia are harvested, although exact estimates are unavailable. Alaska's boreal forests are also at risk because the U. S. government may increase logging on public lands in the future.

CERTAIN ATMOSPHERIC POLLUTANTS MAY AFFECT EARTH'S CLIMATE

Earth's average surface temperature in 1997, the latest data available as we go to press, edged to a record high and beat the previous records set in 1990 and 1995. The warmest half-decade on record was 1991 through 1995, and the second warmest was 1986 through 1990. The last two decades have been the warmest this century (Fig. 55–10). Reliable records of Earth's surface temperature have been kept only since the mid-nineteenth century. Other evidence that suggests an increase in global temperature includes a slight increase in sea level, the retreat of glaciers worldwide, and the increased incidence of extreme weather events in certain regions.

In December 1995 the U. N. Intergovernmental Panel on Climate Change (IPCC) reported that human-produced air pollutants have probably played a role in recent climate change. This report, compiled by a group of climate experts, is the most definitive scientific statement about global warming released to date. Based on studies by numerous scientists, the IPCC projected a 1° to 3.5° C increase in global temperature by the year 2100, although the warming will not be uniform from region to region. Thus, Earth may become warmer during the 21st century than it has been for hundreds of thousands of years.

Most climate experts agree with the IPCC's assessment that the warming trend has already begun. Although most sci-

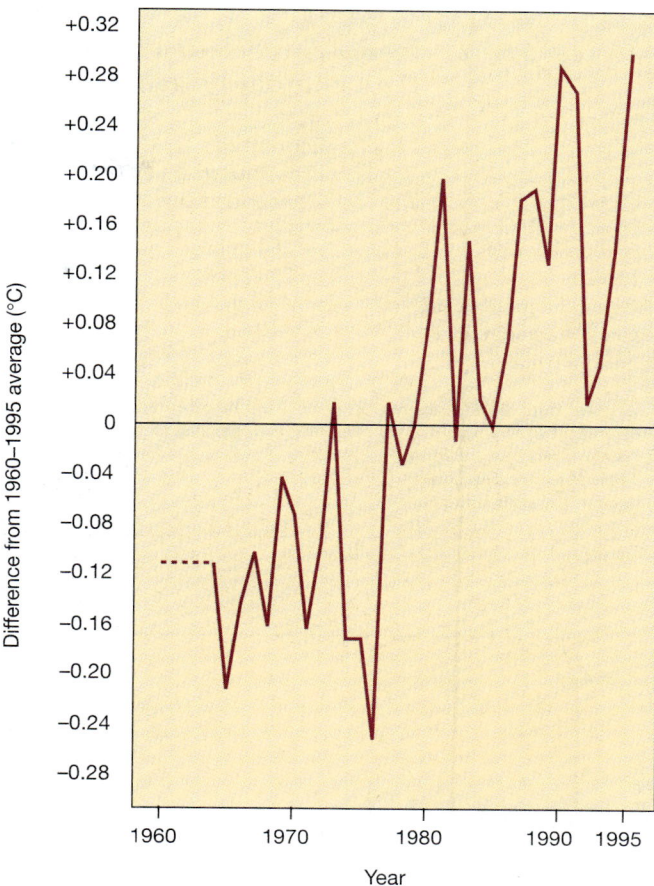

Figure 55–10 **The global warming trend.** Global temperature at Earth's surface, presented as the difference from the average 1960 to 1995 global temperature (°C). The horizontal line at 0 represents the 1960 to 1995 average, which was 15.09°C.

entists agree that the world will continue to warm, they disagree over how rapidly the warming will proceed, how severe it will be, and where it will be most pronounced. Recall that uncertainty and debate are part of the scientific process and that scientists can never claim to know a "final answer." As a result of these uncertainties about global warming, many people, including policy makers, are confused about what should be done and use this uncertainty as a justification to do nothing. Yet the stakes are quite high because human-induced global warming has the potential to disrupt Earth's climate and affect the economies of Earth's nations for a very long time.

Greenhouse gases probably cause global warming

Carbon dioxide (CO_2) and certain other trace gases, including methane (CH_4), surface ozone (O_3),[4] nitrous oxide (N_2O),

[4]Surface ozone (more precisely, ozone in the troposphere) is a greenhouse gas as well as a component of photochemical smog. Ozone in the upper atmosphere, the stratosphere, provides an important planetary service that is discussed later in this chapter.

TABLE 55-2	Increase in Selected Atmospheric Greenhouse Gases, Pre-industrial Revolution to 1994		
Gas	Estimated Pre-industrial Concentration	1994 Concentration	% Increase
Carbon Dioxide	280 ppm*	359 ppm	28.2
Methane	700 ppb†	1666 ppb	138.0
Chlorofluorocarbon-12	0 ppt‡	509 ppt	509.0
Chlorofluorocarbon-11	0 ppt	261 ppt	261.0
Nitrous oxide	285 ppb	309 ppb	8.4

Adapted from *World Resources 1996–1997* and based on multiple sources (World Resources Institute, United Nations Environment Programme, United Nations Development Programme, and World Bank).
*ppm = parts per million
†ppb = parts per billion
‡ppt = parts per trillion

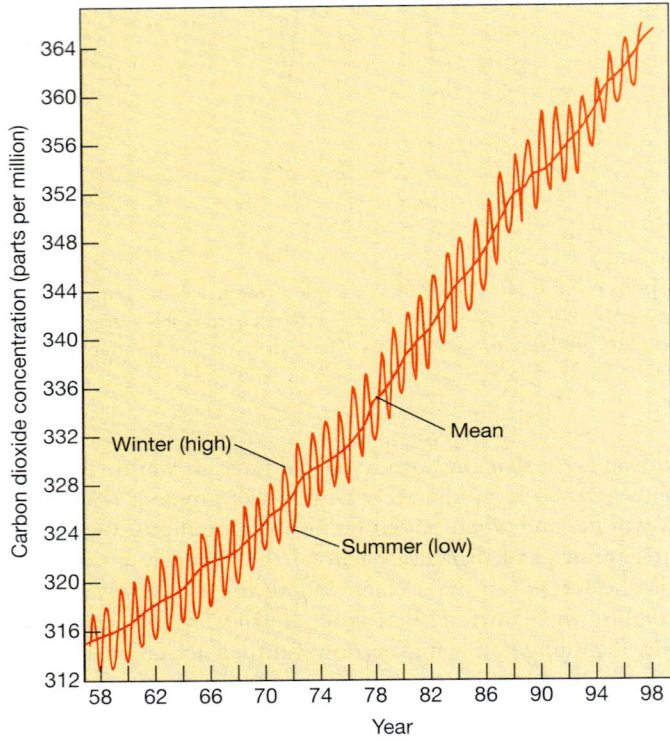

Figure 55–11 **Concentration of carbon dioxide in the atmosphere.** These measurements were taken at the Mauna Loa Observatory, far from urban areas where CO_2 is emitted by factories, power plants, and motor vehicles. The seasonal fluctuation each year corresponds to winter (a high level of CO_2) and summer (a lower level of CO_2) and is caused by greater photosynthesis in summer. *(Dave Keeling and Tom Whorf, Scripps Institution of Oceanography, LaJolla, California)*

and chlorofluorocarbons (CFCs), are accumulating in the atmosphere as a result of human activities (Table 55–2). The concentration of atmospheric CO_2 has increased from about 280 parts per million (ppm) approximately 200 years ago (before the Industrial Revolution began) to 361 ppm in 1995 (Fig. 55–11). Atmospheric CO_2 is still increasing, as are levels of the other trace gases associated with global warming. For example, every time you drive your car, the combustion of gasoline releases CO_2 and N_2O and triggers the production of surface O_3; a single gallon of gasoline burns to produce almost 5.5 pounds of CO_2. Every day additional CO_2 is released by the burning of tropical rain forests. One way that CFCs get into the atmosphere is from old, leaking refrigerators and air conditioners. Decomposition of organic material by anaerobic bacteria in moist places as varied as landfills and the intestinal tracts of cattle is a major source of CH_4.

Global warming occurs because these gases slow the loss of infrared radiation, that is, heat, from the atmosphere to space, thereby warming the atmosphere (Fig. 55–12). Some heat from the atmosphere is transferred to the ocean and raises its temperature as well. As the atmosphere and ocean warm, the overall global temperature rises. Because the atmospheric retention of infrared radiation by CO_2 and other gases is analogous to the way glass traps heat in a greenhouse, the global warming produced in this manner is known as the **greenhouse effect.** Gases that retain heat in the atmosphere are called **greenhouse gases.**

What are the potential effects of global warming?

We now consider some of the potential effects of global warming, including changes in sea level; changes in precipitation patterns; effects on organisms, including humans; and effects on agriculture.

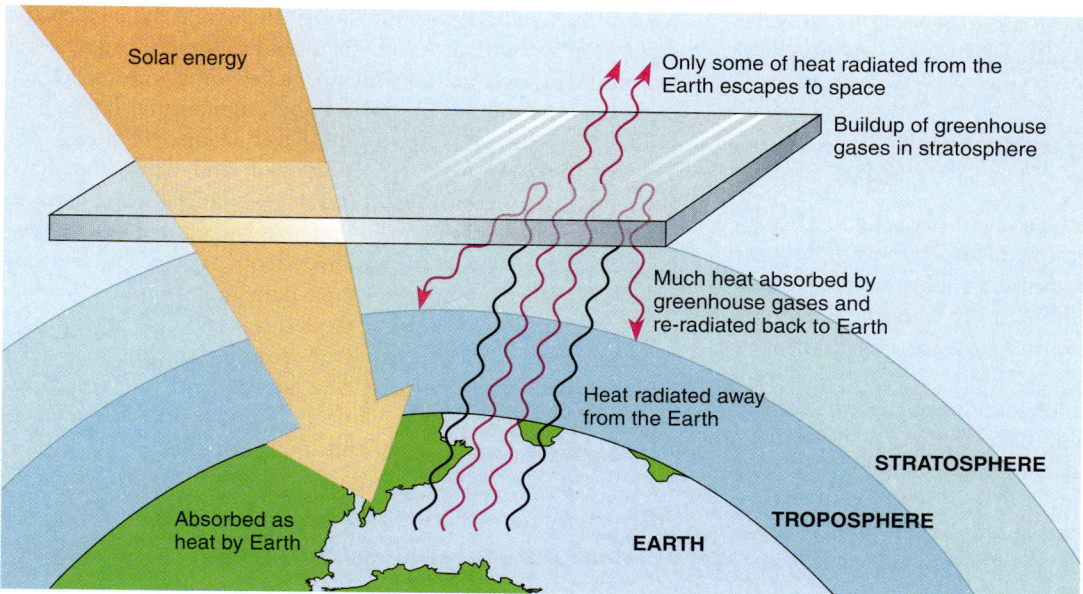

Figure 55–12 **Greenhouse effect.** The buildup of carbon dioxide and other greenhouse gases causes the greenhouse effect, which results in global warming. The increased level of greenhouse gases in the stratosphere is depicted in this diagram as a sheet of glass to emphasize that in this situation the greenhouse gases function much like the glass walls and roof of a greenhouse.

Labels in figure:
- Solar energy
- Only some of heat radiated from the Earth escapes to space
- Buildup of greenhouse gases in stratosphere
- Much heat absorbed by greenhouse gases and re-radiated back to Earth
- Heat radiated away from the Earth
- STRATOSPHERE
- TROPOSPHERE
- EARTH
- Absorbed as heat by Earth

With global warming, the sea level is expected to rise

If the overall temperature of Earth increases by just a few degrees, there could be a major thawing of glaciers and the polar ice caps. In January 1995, for example, a huge chunk of the Larsen Ice Shelf broke away from the Antarctic Peninsula. (Antarctic ice shelves are thick sheets of floating ice that are fed mainly by glaciers that flow off the land.) This breakup coincided with a trend of atmospheric warming in the Antarctic during the past 50 years. A 16-year study from 1978 to 1994 showed that the area of ice-covered ocean in the Arctic has also retreated. In addition to sea-level rise caused by the retreat of glaciers and thawing of polar ice, the sea level will probably rise because of thermal expansion of the ocean. (Water expands when it is warmer.)

During the past century, the sea level has risen by about 18 cm (7 in), most of it due to thermal expansion. The IPCC estimates that the sea level will rise by an additional 50 cm (20 in) by 2100. Such a rise in sea level would flood low-lying coastal areas. Because of climate changes, coastal areas that are not inundated will be more likely to suffer damage from coastal erosion and from weather events such as hurricanes, which cause storm surges. (Storm surges are temporary rises in the sea level caused by strong winds pushing ocean waters against the shore.) These likely effects are certainly a cause for concern, particularly since about two-thirds of the world's population lives within 150 km (93 mi) of a coastline.

With global warming, precipitation patterns may change

Precipitation patterns that occurred thousands of years ago when the Earth was warmer have been used to develop computer simulations of weather changes as global warming occurs. These simulations indicate that changes in precipitation patterns could affect the availability and quality of fresh water in many locations. It is projected that areas that are currently arid or semiarid will have the most troublesome water shortages as the climate changes. At the same time, heavier snow and rainstorms may cause flooding in other areas. The frequency and intensity of storms may also increase. These changes are expected because as the atmosphere warms, more water evaporates, which in turn releases more energy into the atmosphere (recall the discussion of water's heat of vaporization in Chapter 2). This additional energy generates more powerful storms.

In 1995 climatologists at the National Climatic Data Center in Asheville, North Carolina, analyzed temperature and precipitation records for the past 100 years. They determined that more frequent and more severe heat waves, droughts, and snow and rainstorms had occurred from 1980 to 1994 than during the previous 85 years. They then compared this trend to computer simulations of global warming and concluded that there is a 90 to 95% likelihood that recent precipitation extremes are the result of the increase in greenhouse gases in the atmosphere. The property insurance industry takes the possibility of global warming seriously because they will be directly affected by property damages caused by extreme weather events.

With global warming, the ranges of organisms will change

Biologists have started to examine some of the potential consequences of global warming on organisms. For example, one team of biologists placed infrared heaters above test plots in a Rocky Mountain meadow to mimic future warming. Changes in the kinds of plants growing in the plots were noted in just a few months.

Other researchers determined that populations of zooplankton in the California Current have declined by 80% since 1951, apparently because the current has warmed slightly. (The California Current flows from Oregon southward along the California coast.) The decline in zooplankton has affected the entire ecosystem's food web. Populations of seabirds and plankton-eating fishes, for example, have also declined.

The first scientific evidence that a species has shifted its geographical range in response to climate change was published in 1996. Camille Parmesan, a biologist at the University of California at Santa Barbara, compared the current geographical range of a western butterfly (the Edith's checkerspot butterfly) to previously recorded observations. She found that the butterfly has shifted its range northward by about 160 km (about 100 mi).

Each species reacts to changes in temperature differently. Some species will undoubtedly become extinct, particularly those with narrow temperature requirements, those confined to small reserves or parks, and those living in fragile ecosystems, whereas other species may survive in greatly reduced numbers and ranges. Ecosystems considered most vulnerable to species loss in the short term are polar seas; coral reefs; mountains, particularly alpine tundra; coastal wetlands; tundra; taiga; and temperate forests.

Some species may be able to migrate to new environments or adapt to the changing conditions in their present habitats. Also, some species may be unaffected by global warming, whereas others may emerge from it as winners, with greatly expanded numbers and ranges. Those considered most likely to prosper include weeds, pests, and disease-carrying organisms, all of which are generalists that are already common in many different environments.

In terms of human health, the increase in atmospheric CO_2 and the resulting more frequent and more severe heat waves during summer months will probably cause an increase in the number of heat-related illnesses and deaths. Human health will also be indirectly affected, as for example, if disease carriers such as mosquitoes expand their ranges into the newly warm areas and spread malaria, dengue, yellow fever, and viral encephalitis.

Global warming will probably affect agriculture

Global warming may increase problems for agriculture, which is already beset with the challenge of providing enough food for a hungry world without doing irreparable damage to the environment. Several studies document that the rising sea level will inundate river deltas, which have been important since ancient times as some of the world's best agricultural lands. The Nile River (Egypt), Mississippi River (United States), and Yangtze River (China) are examples of river deltas that have been studied. Certain agricultural pests and disease-causing organisms will probably proliferate. As mentioned earlier, global warming may also increase the frequency and duration of droughts and, in some areas, crop-damaging floods.

Current global warming models forecast that agricultural productivity will increase in some areas and decline in others. Some climate experts predict that tropical and subtropical regions, where many of the world's poorest people live, will be hardest hit by declining agricultural productivity.

Many actions have been suggested to deal with global warming

Even if we were to immediately stop polluting the atmosphere with greenhouse gases, Earth would still undergo climate change to some extent because of the greenhouse gases that have already accumulated during the past 100 years. The amount and severity of global warming depend on how much additional greenhouse gas emissions we add to the atmosphere.

There are three strategies to manage global warming: prevention, mitigation, and adaptation. *Prevention* of global warming involves developing ways to prevent the buildup of greenhouse gases in the atmosphere. It is the ultimate and best solution because it is permanent. *Mitigation,* which involves ways to moderate or postpone global warming, gives us time to pursue other, more permanent solutions. It also gives us time to understand more fully how global warming operates so we can avoid some of its worst consequences. *Adaptation* is responding to changes brought about by global warming.[5] Developing adaptive strategies to climate change implies an assumption that global warming is unavoidable.

We can prevent the buildup of greenhouse gases in the atmosphere

The international community recognizes that it must stabilize CO_2 emissions.[6] More than 165 nations, including the United States, have signed the climate change treaty developed by the 1992 U. N. Conference on Environment and Development. Its ultimate goal is to stabilize greenhouse gas concentrations in the atmosphere at levels low enough to prevent dangerous human influences on the climate. At the 1996 U. N. Climate Change Convention held in Geneva, Switzerland, developed countries agreed to establish legally binding timetables to cut

[5]*Adaptation* used in this context should not be confused with biological adaptation, which is evolutionary modification that improves the chances of survival and reproductive success.

[6]Other greenhouse gases will also have to be addressed, but carbon dioxide is the focus here because it is produced in the greatest quantities and has the largest total effect of all the greenhouse gases.

emissions of greenhouse gases, beginning in the year 2000. The timetables were decided at a meeting of representatives from 160 countries held in Kyoto, Japan, in December, 1997, but will not go into effect until at least 55% of the countries involved have ratified the Kyoto Protocol, as the agreement is called. Although each nation has its own individual goal, developed countries agreed collectively to reduce emissions of carbon dioxide, methane, and nitrous oxide by 5.2% below their 1990 levels by the year 2012.

Despite international resolve to reduce CO_2 emissions, which would mainly be accomplished by reducing our dependence on fossil fuels, there is much discussion about whether we should respond until we are more certain that global warming will occur and how fast we should respond. Should we postpone reducing CO_2 emissions until new technologies, including alternative sources of energy, have been fully developed? Or will the environmental costs of postponement be too high? The scientific uncertainties associated with global warming make it difficult to know how we should proceed, but many scientists do not think the potential seriousness of the problem can be ignored until more data are available.

We can mitigate global warming to some extent

One of the most effective ways to mitigate global warming involves forests. As you know, atmospheric CO_2 is removed from the air by actively growing forests, which incorporate carbon into leaves, stems, and roots by photosynthesis. On the other hand, deforestation releases CO_2 into the atmosphere as trees are burned or decomposed. We can mitigate global warming by planting and maintaining new forests and by protecting existing forests.

Increasing the energy efficiency of automobiles and appliances, which would reduce the output of CO_2, would also help mitigate global warming. This could be accomplished by establishing regulatory programs that increase minimum energy efficiency standards and by negotiating agreements with industry. Energy-pricing strategies, such as carbon taxes and the elimination of energy subsidies, are examples of other policies that could help mitigate global warming. Most experts think that by using existing technologies and developing such policies, greenhouse gas emissions can be significantly reduced with little cost to society.

We can adapt to the reality of global warming

Because the overwhelming majority of climate experts think that human-induced global warming is inevitable, government planners and social scientists from many countries are developing a number of strategies to help us adapt to global warming. One of the most pressing issues is rising sea level. People living in coastal areas could be moved inland, away from the dangers of storm surges. This solution would have high societal and economic costs, however. An alternative, also extremely expensive, is the construction of dikes and levees to protect coastal land. Rivers and canals that spill into the ocean would have to be channeled to protect fresh water and agricultural lands from salt water intrusion.

We also need to adapt to shifting agricultural zones. Many temperate countries are in the process of evaluating semitropical crops to determine the best ones to substitute for traditional crops. Drought-resistant strains of trees are being developed by large lumber companies now, because the trees planted today will be harvested in the middle of the 21st century, when global warming may already be well advanced.

STRATOSPHERIC OZONE CONTINUES TO DECLINE

Ozone (O_3) is a form of oxygen that is a human-made pollutant in the lower atmosphere but a naturally produced, essential part of the stratosphere (see Fig. 20–5). The **stratosphere,** which encircles our planet some 10 to 45 km (6 to 28 mi) above the surface, contains a layer of ozone that shields the surface from much of the ultraviolet radiation coming from the sun (Fig. 55–13). If ozone disappeared from the stratosphere, Earth would become unlivable for most forms of life. Ozone in the lower atmosphere is converted back to oxygen in a few days and so does not replenish the ozone that has been depleted in the stratosphere.

Ozone depletion, which is a lowered concentration of ozone in the stratosphere, was first observed and reported over Antarctica in 1985. This thinning of the ozone layer, which lasts for a couple of months, is popularly referred to as the ozone "hole." There, ozone levels decrease by as much as 67% during the Antarctic spring each year. The hole now encompasses about 23.3 million square km (9 million square mi), an

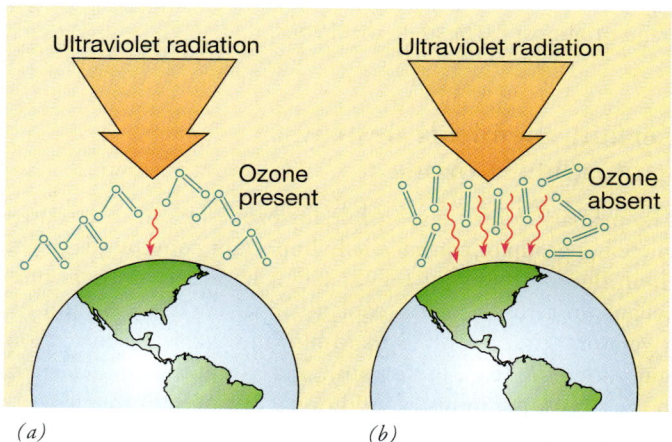

Figure 55–13 **Ultraviolet radiation and the ozone layer.** *(a)* Stratospheric ozone absorbs ultraviolet radiation, effectively shielding Earth's surface. *(b)* When stratospheric ozone is less abundant, more high-energy ultraviolet radiation penetrates the atmosphere to the surface, where it harms organisms. (For simplicity, part *b* shows an absence of ozone. In actuality, about 40% of the normal ozone level may be present in the ozone "hole.")

Figure 55–14 The ozone hole. A computer-generated image of part of the Southern Hemisphere on October 17, 1994, reveals the ozone "hole" (black and purple areas) over Antarctica and the tip of South America. Normal ozone levels are shown in yellow, orange, and red. The ozone hole is not stationary but moves as a result of air currents. *(Courtesy NASA)*

thousands of laboratory measurements, atmospheric observations, and calculations by computer models. In 1995 the Nobel Prize in chemistry was awarded to Sherwood Rowland, Mario Molina, and Paul Crutzen, the scientists who first explained the connection between the thinning ozone layer and chemicals such as CFCs. This Nobel Prize was the first ever given for work in environmental science.

CFCs and similar compounds slowly drift up to the stratosphere and become widely dispersed. Once there, ultraviolet radiation breaks them down, releasing chlorine. Similarly, bromine is released by the breakdown of halons and methyl bromide. Under certain conditions, chlorine or bromine is capable of attacking ozone, converting it into molecular oxygen. The chlorine or bromine is not altered by this process, and as a result, chlorine or bromine can break down many thousands of ozone molecules.

The hole in the ozone layer that was discovered over Antarctica occurs annually between September and November, when the **circumpolar vortex,** a mass of cold air, circulates around the southern polar region, in effect isolating it from the warmer air in the rest of the world. The cold causes polar stratospheric clouds to form; these clouds contain ice crystals to which chlorine, bromine, and other chemicals adhere, enabling them to attack ozone. When the circumpolar vortex breaks up, the ozone-depleted air spreads northward, diluting ozone levels in the stratosphere over South America, New Zealand, and Australia.

area about as large as North America (Fig. 55–14). A smaller hole has been detected in the stratospheric ozone layer over the Arctic. In addition, worldwide levels of stratospheric ozone have been decreasing for several decades. Ozone levels over Europe and North America, for example, have dropped by an average of almost 10% since the 1950s.

Certain chemicals destroy stratospheric ozone

The primary culprit responsible for ozone loss in the stratosphere is a group of commercially important compounds called chlorofluorocarbons, or CFCs. CFCs have been used as propellants in aerosol cans, as coolants in air conditioners and refrigerators (e.g., freon), as foam-blowing agents for insulation and packaging (e.g., Styrofoam), and as solvents and cleaners in the electronics industry. Additional compounds that also attack ozone include halons (found in many fire extinguishers); methyl bromide (a pesticide used as a soil and crop fumigant); methyl chloroform (an industrial solvent); and carbon tetrachloride (used in many industrial processes, including the manufacture of pesticides and dyes).

The scientific evidence linking CFCs and other human-made compounds to stratospheric ozone destruction includes

Ozone depletion harms organisms

With depletion of the ozone layer, more ultraviolet radiation reaches the surface. A study conducted in Toronto from 1989 to 1993, for example, showed that wintertime levels of UV-B increased by more than 5% each year as a result of lower ozone levels. (UV-B is one of three types of UV radiation; see *Making the Connection: The Sun, Ultraviolet Radiation, and Skin* in Chapter 38.) Excessive exposure to ultraviolet radiation is linked to a number of human health problems, including cataracts, skin cancer, and a weakened immune system.

Much scientific evidence also documents crop damage from exposure to high levels of ultraviolet radiation. However, these experiments indicate what could happen if the ozone layer is significantly destroyed and UV levels increase dramatically. No scientific evidence exists that plants are currently suffering from increased exposure to present levels of UV-B.

Biologists are concerned that the ozone hole over Antarctica could damage plankton that forms the base of the food web for the surrounding ocean. A 1992 study confirmed that increased ultraviolet radiation is penetrating surface waters around Antarctica and that productivity of Antarctic phytoplankton may have declined by at least 6% to 12% as a result. Also, a possible link exists between the widespread decline of amphibian populations and increased UV radiation (see *Making the Connection: Increased Ultraviolet Radiation and Declining Amphibian Populations*).

INCREASED ULTRAVIOLET RADIATION AND DECLINING AMPHIBIAN POPULATIONS

Is there a link between observed increases in worldwide ultraviolet radiation and declining amphibian populations? Amphibians such as frogs, toads, and salamanders have bare and extremely permeable skin and lay gelatinous and unprotected eggs. Since the 1970s, many of the world's frog populations have dwindled. For example, at least 14 species of Australian rainforest frogs have become endangered or extinct since 1980. Likewise, populations of all seven native species of frogs and toads in Yosemite National Park have declined.

Biologists are beginning to unravel some clues to these mysterious declines. One possible culprit appears to be increased UV radiation caused by ozone thinning. Dr. Andrew Blaustein and colleagues at Oregon State University reported in 1994 the results of their investigations on the effects of UV radiation on egg survival of some amphibian species. Amphibians possess photolyase, an enzyme that allows them to repair DNA damage caused by natural UV radiation. In the study, species suffering declines appear to be restricted in their ability to repair such cellular damage. The researchers exposed the eggs of three species to natural radiation. Egg survival was high for the Pacific tree frog, which had the greatest photolyase activity and is not in decline. Egg survival was much lower (only 45 to 65%) for the Western toad and Cascades frog but increased dramatically when eggs where shielded from UV radiation.

Scientists caution that all amphibian declines are not caused by the same threat and are likely the result of multiple effects such as habitat loss, diseases, invasive species, and air and water pollution. However, biologists increasingly perceive amphibians to be **bellwether species,** that is, organisms that provide an early warning of environmental damage.

International cooperation will prevent significant depletion of the ozone layer

In 1978 the United States, the world's largest user of CFCs, banned the use of CFC propellants in products such as antiperspirants and hair sprays. Although this ban was a step in the right direction, it did not solve the problem. Most nations did not follow suit, and besides, propellants represent only the tip of the iceberg in terms of CFC use.

In 1987, representatives from a number of countries met in Montreal to sign the Montreal Protocol, an agreement to reduce CFC production by 50% by 1998. Since 1987, more than 160 countries signed an agreement to phase out all use of CFCs by 2000. Despite these agreements, the news about the ozone layer has continually worsened since 1990. As a result, some nations, including the United States, took even stricter measures to limit CFC production.

Industrial companies that manufacture CFCs quickly developed substitutes, such as hydrofluorocarbons (HFCs) and hydrochlorofluorocarbons (HCFCs). HFCs do not attack ozone, although they are potent greenhouse gases. HCFCs attack ozone but are not as destructive as the chemicals they are replacing. Although production of HFCs and HCFCs is increasing rapidly, these chemicals are transitional substances that will only be used until industry develops nonfluorocarbon substitutes.

Production of CFCs, carbon tetrachloride, and methyl chloroform was completely phased out in the United States and other developed countries in 1996 except for a relatively small amount exported to developing countries. Existing stockpiles can be used after the deadline, however. Developing countries are on a different timetable and will phase out CFC use by the year 2006. Currently, methyl bromide use is scheduled to be phased out in 2005 in developed countries (the United States will ban production in 2001), and HCFCs in 2030.

Unfortunately, CFCs are extremely stable, and those being used today will continue to deplete stratospheric ozone for 50 to 150 years. Scientists expect the Antarctic ozone hole, for example, to reappear each year until around 2050. However, initial gains have been made. In 1996, measurements of ozone-destroying chemicals in the lower atmosphere showed they had peaked and were beginning to decline in concentration. It will take several years for this trend to become significant in the stratosphere. Ozone improvement directly attributable to the Montreal Protocol and its amendments is not expected to be measurable until 2010. (The increase in ozone must be several percent higher than natural fluctuations in global ozone levels before it can be attributed to the Montreal Protocol.)

ENVIRONMENTAL PROBLEMS ARE INTERRELATED

Several connections have been pointed out among the four environmental problems discussed in this chapter. For example, deforestation, global warming, and ozone depletion will likely cause future reductions in biological diversity. Similarly, any environmental problem you can think of, even if not discussed in this chapter, is related to other environmental concerns.

Most of all, however, environmental problems are connected to overpopulation. We have seen that the rate of human population growth is greatest in developing countries but that developed nations are overpopulated as well, in the sense that they have a high per capita consumption of resources (see Chapter 51). *Consumption overpopulation,* which is the disproportionately large consumption of resources by people in developed countries, affects the environment as much as the population explosion in the developing world does.

As living organisms, humans share much in common with the fate of other life forms on the planet. We are not immune to the environmental damage we have produced. We differ from other organisms, however, in our capacity to reflect on the consequences of our actions and to alter our behavior accordingly. Humans, both individually and collectively, are able to bring about change. The widespread substitution of constructive change for destructive change is the key to any hope of ensuring the biosphere's survival in its current form.

SUMMARY WITH KEY TERMS

I. **Biological diversity,** the number and variety of organisms, is decreasing worldwide.
   A. A species becomes **extinct** when its last individual member dies.
      1. A species whose severely reduced numbers put it in imminent danger of extinction is called an **endangered species.**
      2. When extinction is less imminent but the population is quite small, a species is called a **threatened species.**
   B. Human activities that contribute to a reduction in biological diversity include habitat loss and disturbance, pollution, introduction of foreign species (**biotic pollution**), pest and predator control, hunting, and commercial harvest. Of these, habitat loss is the most significant.
   C. **Conservation biology,** the study and protection of biological diversity, includes in situ and ex situ conservation.
      1. Efforts to preserve biological diversity in the wild are known as **in situ conservation.**
      2. **Ex situ conservation,** such as captive breeding, occurs in human-controlled settings.

II. Forests provide us with many ecological benefits, including watershed protection, soil erosion prevention, climate moderation, and wildlife habitat.
   A. The greatest problem facing forests today is **deforestation,** the temporary or permanent clearance of forests for agriculture or other uses.
   B. In the tropics, forests are destroyed to provide farmers with temporary agricultural land; produce timber, particularly for developed nations; provide open rangeland for cattle; and supply fuel wood.
   C. Extensive deforestation of certain boreal forests began in the late 1980s and 1990s to provide industrial wood and wood fiber.

III. Carbon dioxide and certain other gases cause the **greenhouse effect,** in which the atmosphere retains heat and warms the Earth's surface.
   A. Increased levels of CO_2 and other **greenhouse gases** in the atmosphere are causing concerns about a major global warming that may occur during the 21st century.
      1. The combustion of fossil fuels produces pollutants, especially CO_2, nitrous oxide, and surface ozone.
      2. Other greenhouse gases are methane and chlorofluorocarbons.
   B. Global warming may cause a rise in sea level, changes in precipitation patterns, the extinction of many species, and problems for agriculture. It could result in the displacement of millions of people.
   C. The challenges of global warming can be met by prevention (stop polluting the air with greenhouse gases), mitigation (slow down the rate of global warming), and/or adaptation (make adjustments to live with global warming).

IV. The **ozone (O_3)** layer in the **stratosphere** helps shield Earth's surface from damaging ultraviolet radiation.
   A. The total amount of ozone in the stratosphere is declining (i.e., **ozone depletion**), and large ozone holes develop over Antarctica and the Arctic each year.
   B. The attack on the ozone layer is caused by chlorofluorocarbons and similar chlorine- and bromine-containing compounds.
   C. CFCs peaked in the lower atmosphere in 1996 and will peak in the stratosphere a few years later. Ozone improvement due to the Montreal Protocol will probably not be measurable until 2010.

POST-TEST

1. Which of the following statements about extinction is NOT true? (a) extinction is the permanent loss of a species (b) extinction is a natural biological process (c) once a species is extinct, it can never reappear (d) human activities have little impact on extinctions (e) as many as one-fourth of plant families may be extinct by the end of the 21st century

2. An endangered species (a) is severely reduced in number (b) is in imminent danger of becoming extinct (c) usually does not have reduced genetic variability (d) is not in imminent danger of extinction (e) both a and b are true

3. The most important reason for declining biological diversity is (a) air pollution (b) introduction of foreign species (c) habitat destruction (d) commercial hunting (e) commercial harvesting

4. In situ conservation (a) includes breeding captive species in zoos (b) includes seed storage of genetically diverse crops (c) concentrates on preserving biological diversity in the wild (d) focuses exclusively on large, charismatic animals (e) both a and b are true

5. In _____ , sperm from a male is used to artificially impregnate a female. (a) host mothering (b) artificial insemination (c) in situ conservation (d) biotic pollution

6. When forests are destroyed (a) soil fertility increases (b) soil erosion decreases (c) some forest organisms decline in number (d) the danger of flooding in nearby lowlands declines (e) carbon dioxide is removed from the atmosphere and stored in forest plants

7. About 60% of tropical deforestation is the result of (a) commercial logging (b) cattle ranching (c) hydroelectric dams (d) mining (e) subsistence agriculture

8. Which of the following gases contributes to both global warming and thinning of the ozone layer? (a) CO_2 (b) CH_4 (c) surface O_3 (d) CFCs (e) N_2O

9. The first scientific evidence that a species had shifted its geographical range in response to climate change involved a(an) (a) flowering plant (b) butterfly (c) mosquito (d) coral reef (e) soil-dwelling fungus

10. What gas is a human-made pollutant in the lower (surface) atmosphere but a natural and beneficial gas in the stratosphere? (a) CO_2 (b) CH_4 (c) O_3 (d) CFCs (e) N_2O

11. Where is a stratospheric ozone depletion most pronounced? (a) over Antarctica (b) over the equator (c) over South America (d) over North America and Europe (e) over Alaska and Siberia

12. The Montreal Protocol is an agreement to (a) reduce greenhouse gas emissions (b) reduce CFC production (c) design a recovery plan for each endangered species (d) curtail deforestation (e) prevent the selling of products made from endangered or threatened species

REVIEW QUESTIONS

1. Which organism is more likely to become extinct, an endangered species or a threatened species? Explain your answer.
2. How does habitat destruction contribute to declining biological diversity?
3. Which type of conservation measure, in situ or ex situ, will help the greatest number of species? Why?
4. State at least three environmental benefits that forests provide.
5. Discuss two reasons for tropical deforestation.
6. Describe the greenhouse effect and list three or more greenhouse gases.
7. What are some of the significant problems that may be caused by global warming during the 21st century?
8. Discuss and give an example of each of the three approaches (prevention, mitigation, and adaptation) for dealing with global warming.
9. What environmental service does the stratospheric ozone layer provide?
10. Describe how ozone depletion occurs and discuss some of the consequences of the thinning ozone layer.

YOU MAKE THE CONNECTION

1. If most of the world's biological diversity were to disappear, how would it affect your life?
2. Because new species will eventually evolve to replace those that humans are driving to extinction, why is declining biological diversity such a threat to us?
3. Why does the most recent version of the World Conservation Strategy include stabilizing the human population as one of its goals?
4. Why might captive breeding programs that reintroduce species into natural environments be doomed to failure?
5. If you were given the task of developing a policy for the United States to deal with global climate change during the next 50 years, would you stress prevention, mitigation, or adaptation? Explain your answer.
6. It has been suggested that the wisest way to "use" fossil fuels would be to leave them in the ground. How would this affect global warming? Energy supplies?
7. This statement was overheard in an elevator: "CFCs cannot be the cause of ozone depletion over Antarctica because there are no refrigerators in Antarctica." Criticize the reasoning behind this statement.
8. It has been suggested that a more descriptive name for *Homo sapiens* would be "*Homo dangerous*." Explain this specific epithet, based on what you have learned in this chapter.

RECOMMENDED READINGS

Acharya, A. "Plundering the Boreal Forests." *World Watch,* May/Jun. 1995. Describes current massive deforestation in Canada and Siberia.

Easter-Pilcher, A. "Implementing the Endangered Species Act." *BioScience,* Vol. 46, No. 5, May 1996. An evaluation of how species are determined to be endangered or threatened.

Flanagan, R. "Engineering a Cooler Planet." *Earth,* Oct. 1996. Examines several possible ways to mitigate global warming.

Mahlman, J.D. "Uncertainties in Projections of Human-Caused Climate Warming." *Science,* Vol. 278, 21 Nov. 1997. The author, a respected climatologist, lists the known "facts," virtually certain projections, probable projections, and incorrect projections about human-induced climate warming.

Monastersky, R. "Health in the Hot Zone." *Science News,* Vol. 149, 6 Apr. 1996. Scientists are beginning to assess how global warming may affect human health.

Myers, N. "Environmental Services of Biodiversity." *Proceedings of the National Academy of Sciences,* Vol. 93, Apr. 1996. Examines the ecological and economic importance of biological diversity.

Parson, E.A., and O. Greene. "The Complex Chemistry of the International Ozone Agreements." *Environment,* Vol. 37, No. 2, Mar. 1995. Provides a chronology of ozone-protecting agreements and examines implementation by various countries.

Pitelka, L.F., and the Plant Migration Workshop Group. "Plant Migration and Climate Change." *American Scientist,* Vol. 85, Sep./Oct. 1997. Predicting plant migrations that might occur in response to climate change is difficult.

Raven, P.H., L.R. Berg, and G.B. Johnson. *Environment,* 2nd ed., Saunders College Publishing, Philadelphia, 1998. This environmental science text offers a detailed situation analysis of planet Earth, including how humans interact with and affect its physical and living systems.

Robinson, S.K. "Nest Gains, Nest Losses." *Natural History,* Jul. 1996. Includes a fascinating account of how humans enabled the cowbird to significantly expand its range at the expense of other bird species.

Serrill, M.S. "Ghosts of the Forests." *Time,* Nov. 1997. An overview of the many problems, from pollution to clearcutting, that face the world's forests.

Stein, B.A., and S.R. Flack. "Conservation Priorities: The State of U.S. Plants and Animals." *Environment,* Vol. 39, No. 4, May 1997. The Nature Conservancy surveys the condition of 20,439 species.

Vitousek, P.M., C.M. D'Antonio, L.L. Loope, and R. Westbrooks, "Biological Invasions as Global Environmental Change." *American Scientist,* Vol. 84, Sep./Oct. 1996. The introduction of exotic species harms ecosystems, threatens human health, and is economically disastrous.

Wheelwright, J. "The Berry and the Poison." *Smithsonian,* Dec. 1996. Examines the benefits and dangers of methyl bromide, a pesticide that attacks the ozone layer in the stratosphere.

White, R. "Climate Science and National Interests." *Issues in Science and Technology,* Fall 1996. This respected climate scientist examines the political complexities of responding to the threat of climate warming.

● Visit our website at **http://www.saunderscollege.com/lifesci/titles.html** and click on Solomon/Berg/Martin Biology for links to chapter-related resources on the World Wide Web.

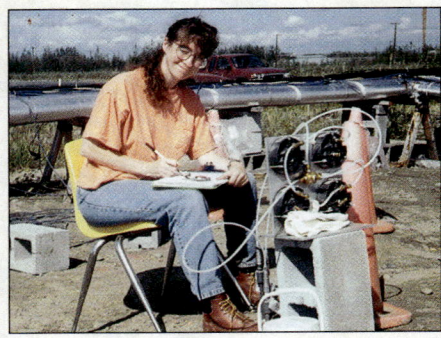

Bioremediation Specialist

ANDREA LEESON

Andrea Leeson received her B.S. in biology at Eastern Kentucky University with the long-term plan of doing environmental work as a career. She then obtained her Ph.D. in bioremediation from Johns Hopkins University and started working soon after as a bioremediation specialist at Battelle Memorial Institute in Columbus, Ohio. Battelle is a non-profit organization involved in industrial, medical, and environmental research and its practical applications to technology. As a bioremediation specialist, she applies her combined education in biology and engineering to developing innovative techniques for cleaning up hazardous waste sites for government agencies such as the Air Force and the Environmental Protection Agency. For the last two years, in addition to her other responsibilities, she has been managing editor on a new bioremediation journal and co-chair of a symposium in the field, both of which are very exciting to her.

What exactly is bioremediation? What does your job entail?

Essentially, bioremediation is the use of an organism (mainly microorganisms) to clean up hazardous waste sites. Common sites are created by degreasers, leaking of underground fuel tanks or pipes, and fire training pits. In our field research, we look at cleaning up sites through new ways of enhancing the natural microbial activity that already exists there. Primarily, I work on petroleum-contaminated sites but I work with chlorinated compounds as well.

How do you use bacteria and other microorganisms to clean up a site? How is your biology background used?

We use bacteria at a site by exploiting their natural ability to use contaminants as a source of carbon and/or energy. We supply them with whatever nutrients they need—usually oxygen to break down contaminants. My biology background is necessary to understand what is needed for the bacteria to grow, while the engineering background is necessary to understand how to supply the nutrients and design the systems. The major technology I use for supplying oxygen is called bioventing and air sparging, which is injecting air into the ground. Therefore, you have to be able to design an air injection system.

What undergraduate courses have helped you in your career?

The most important courses for this field were the microbiology courses. Those were essential. They gave me an advantage. In environmental engineering you have people coming in from a wide mix of fields, mostly engineering. One needs the engineering aspects to know how to do the technologies; however, coming in with a microbiology and biology background really helped me understand the basic biology that was taking place.

Did you ever have to make the decision between becoming an engineering major or a biology major?

I knew I wanted to be in the biological sciences when I started. I did not have the option of an engineering degree at Eastern Kentucky University but I probably would not have chosen engineering even if it had been available. My interest in engineering developed after I had been in college a few years. When I started college, I knew I wanted to do an environmental type of program. Biology gave me a more solid base than other environmentally oriented majors that were offered at the time.

Was there anything of a personal nature that attracted you to the environmental program?

I think it was always an interest of mine. My parents brought me up to appreciate the environment. We were outdoors-oriented. Also, when I was a kid in the early 70s, the big movement toward environmental awareness was just beginning. They had just formed the EPA. So, it hit me at an impressionable age.

Was there any experience in your undergraduate years that was pivotal to your later career interest and job search?

While getting my Bachelor's degree, I did a cooperative research program with Oak Ridge National Laboratory, Environmental Resources Division. This experience probably made the biggest impact on me: I was able to do a bit of environmental research while still an undergraduate.

What was the focus of your research?

I performed toxicity tests by exposing algae species to various contaminants. This work involved field studies and that's probably what influenced me the most. Later, when I had finished my Ph.D. and had been in the lab for years, I became eager to do field work again. So, I looked for companies that were doing field research in environmental engineering or environmental sciences.

Where did you do your Ph.D? How long is the program there?

I was at Johns Hopkins. It took me four and a half years, which is about average.

What was your area of research at Johns Hopkins?

It was bioremediation. I was specifically looking at aerobic biodegradation of chlorinated compounds.

Was that in the environmental engineering program?

Yes. Many environmental engineering programs are incorporated into civil engineering departments. Johns Hopkins was unusual at the time in that they had a department of environmental engineering —actually geography and environmental engineering. These departments are fairly common today. The department was, and still is, strong in areas like wastewater treatment, drinking water treatment, and hazardous waste treatment—hazardous waste being any substance from a man-made source that is harmful to human health or to the environment. Those are the three main areas for an environmental engineer.

What is the difference between environmental engineering and bioremediation?

Bioremediation falls under the broad category of environmental engineering. Environmental engineering started out as civil engineering or sanitary engineering, involving wastewater or drinking water treat-

ment. It started to spin off into its own specialty in the 1970s. Then, when there was heightened environmental awareness, more environmental emphasis was placed on treatment of hazardous waste. Until relatively recently, the only way to remove hazardous contaminants from soil was to excavate the soil from an entire field and incinerate it. As a result of research, other

---

Petroleum compounds are natural, so there's been time for the bacteria to evolve the ability to metabolize them.

---

techniques, including using microorganisms to degrade organic contaminants, are available to clean up contaminated soil.

Does bioremediation only deal with hazardous wastes?

Primarily, this is true. However, even at a wastewater treatment plant, where microorganisms are used to degrade sewage, the process is basically bioremediation.

How has this field changed in the last few years?

It used to be that bioremediation was not very well accepted as a treatment for contaminated sites, because it was new and was considered experimental. You had to use tried and true methods for remedia-

tion, such as incineration (already mentioned). In the last few years that has been changing. Bioremediation is much more commonly applied as a means of cleaning up a site. It's no longer just research in the field.

How long does it take to clean up a site?

That varies depending on how much contamination is present, what the soil type is like, and what kind of contaminant is involved. Generally, compounds with lighter molecular weights such as benzene (a component of some petroleum products) can be bioremediated within two years. If you have heavier, multi-carbon compounds, we often say anywhere from two to ten years.

What kinds of microbes other than bacteria are used in bioremediation?

It's mostly bacteria, but research is being done with certain fungi and algae. Those studies are still in the early stages. Fungi look especially promising.

What kinds of bacteria in particular are used?

What I prefer, and what is very common, is to work with whatever natural bacteria are already at the site and have already adapted to breaking down whatever contaminant is there. For example, petroleum compounds are natural, so there has been plenty of time for the bacteria to evolve the ability to metabolize them. In some instances, genetic engineering has been used to enhance the degrading abilities of a bacterial species. Those enhanced bacteria are then introduced into the environment. I personally am not involved with genetic engineering. When you introduce bacteria into the environment that do not naturally belong there, even if you have enhanced them to degrade your contaminants well, they often have difficulty surviving. I think genetic engineering offers promise for some of the more stubborn contaminants.

Are there types of contamination other than petroleum and chlorinated compounds that would call for your involvement?

There is some work being done with inorganics, using bacteria, even plant species, to bioaccumulate the inorganics. Bioaccumulation is used for remediating some of the heavy metals, such as mercury and lead. Bacteria or plants take the heavy metals up into their cellular structure. However, the living form cannot do anything with the heavy metals. We can then filter off the bacteria or harvest the plants, leaving the water or soil cleaner.

How involved is the government in what you do?

The government is our biggest client. We work with the Air Force and Navy a lot, as well as the EPA (Environmental Protection Agency). We assist them in researching the best ways to remediate the different sites they have. Petroleum compounds are the most common kind of contamination, followed by chlorinated contaminants. Many military installations use chlorinated solvents as degreasers. In the past, equipment may have been sprayed down with a solvent, generating runoff that contaminated the soil.

So, you are not typically called in by private organizations?

It does not happen often. That's largely because of the nature of Battelle, which is more of a research institute. Occasionally, we do work with industry, but a lot of companies have their own research departments.

Please explain the new bioremediation techniques that you are using now.

One of the biggest processes we are using is *bioventing*. From there we have moved into *air sparging*, which is very similar to bioventing, except we are injecting air directly into the groundwater rather than

into the soil to try to remediate the groundwater. The other technique is *bioslurping*, which is the removal of the portion of a fuel, like gasoline or jet fuel, which does not dissolve into water. Often there is spilled petroleum, which will work its way through the soil until it hits groundwater. It does not penetrate the groundwater because it's lighter than water. So, there are instances where you can use a technique such as bioslurping to remove the fuel, because much of the fuel is floating on top of the groundwater.

What kinds of equipment do you use for these two techniques?

Air sparging is quite simple. You just need some kind of air compressor to inject the air. The monitoring is also not very complicated. You put in small screens to collect water samples and air samples from the soil. We do have some field instrumentation that we use to measure oxygen in the field, but many samples are sent to an analytical laboratory. Bioslurping is a little more complicated. We have a liquid ring pump that can pull out the contaminated groundwater, which then has to go through an oil-water separator. We also send groundwater through a filter box to reduce emulsions. So, there is a little more equipment involved. Most of the equipment is nothing that we really built; it's all existing pieces that we put to a different use.

What do you like best about your job?

I like getting out to the field and doing the field work. I enjoy putting in wells and seeing how these things work once we get them up and going.

Do you do much traveling then?

Yes, I do a fair amount. I usually have one or two trips a month. I usually go for the first few days to set up the team and identify the equipment that needs to be used.

How do you keep up to date in your field?

I guess it's easier for me to keep current because I am editing *Bioremediation Journal* and running a symposium. With the journal, I get new bioremediation manuscripts in every week. So I know what people are doing right away. The bioremedia-

tion symposium, called the International Symposium on In Situ and On Site Bioremediation, is so well represented, we feel that the newest research is coming out of it. My co-chair and I read through every abstract. Last year there were over 800 abstracts. So, it's very easy to keep up that way.

Who are these symposium attendees? Do they belong to a particular professional association?

No, they represent all aspects of the bioremediation field. The symposium has received much interest, so you have people from academia, the government, and industry who attend. About two thousand people attend.

Do you see any pattern emerging in the material that's being submitted to you, anything new to the field?

We see a lot more research on chlorinated compounds like degreasers, and explosives, like TNT. I think this is because they are more difficult to remediate. Actually, in the journal we see more laboratory and small-scale studies than full research. This is probably due to funding issues. It's more expensive to go out into the field, so we do not see as many of those studies. I personally would like to see more.

What advice would you give someone considering your career?

Since my bachelor's was in biology, I always felt very strong in the biological aspects of my job. I probably would have taken more groundwater courses like hydrogeology. More math, such as differential equations and fluid mechanics, would have helped. Bioremediation is such a broad field, you really need a little bit of everything. Probably the hydrogeology is one of the best courses you could have for this field if you already have a biology major.

What is next for you?

I cannot really say. I am quite satisfied where I am. I really enjoy doing the journal and the symposium. These are the two big pluses for me, because I am in the middle of all the new research and because I like writing and editing. I suppose, if anything, I would like to teach at a university. However, that is in the future.

CHAPTER 1

1. a   2. a   3. e   4. c   5. a   6. b   7. a
8. c   9. c   10. b   11. c   12. a   13. b   14. b
15. d   16. b   17. b

CHAPTER 2

1. c   2. d   3. e   4. a   5. b   6. c   7. d
8. a   9. e   10. a   11. d

CHAPTER 3

1. a   2. d   3. b   4. e   5. e   6. d   7. c
8. a   9. d   10. a   11. e   12. c   13. b

CHAPTER 4

1. b   2. a   3. d   4. d   5. d   6. a   7. e
8. d   9. a   10. a   11. b   12. a   13. a

CHAPTER 5

1. d   2. c   3. e   4. c   5. a   6. e   7. a
8. b   9. b   10. b   11. e   12. c   13. e

CHAPTER 6

1. a   2. c   3. c   4. d   5. b   6. d   7. b
8. d   9. a   10. e   11. c   12. b   13. d   14. a

CHAPTER 7

1. c   2. a   3. d   4. e   5. a   6. a   7. b
8. c   9. a   10. a   11. e   12. b   13. c

CHAPTER 8

1. a   2. d   3. c   4. a   5. a   6. c   7. e
8. e   9. a   10. c   11. b   12. c   13. a   14. c

CHAPTER 9

1. d   2. a   3. d   4. e   5. c   6. a   7. e
8. d   9. b   10. d   11. c   12. b

CHAPTER 10

1. c   2. d   3. b   4. e   5. c   6. d   7. a
8. e   9. e   10. c   11. b   12. a

Genetics Problems

1. a. all yellow   b. $\frac{1}{2}$ yellow: $\frac{1}{2}$ green   c. all yellow
d. $\frac{3}{4}$ yellow: $\frac{1}{4}$ green

2. 150

3. The short-winged condition is recessive. Both parents are heterozygous.

4. The blue-eyed man is homozygous recessive; his brown-eyed wife is heterozygous.

5. Repeated matings of the roan bull and the white cow will yield an approximate 1:1 ratio of roan to white offspring. Repeated matings among roan offspring will yield red, roan, and white offspring in an approximate ratio of 1:2:1. The mating of two red individuals will yield only red offspring.

6. There are 36 possible outcomes when a pair of dice is rolled. There are six ways of obtaining a seven: 1,6; 6,1; 2,5; 5,2; 3,4; and 4,3. There are five ways of rolling a six: 1,5; 5,1; 2,4; 4,2; and 3,3. There are also five ways of rolling an eight: 2,6; 6,2; 3,5; 5,3; and 4,4.

7. The genotype of the brown, spotted rabbit is $bbSS$. The genotype of the black, solid rabbit is $BBss$. An F_1 rabbit would be black, spotted ($BbSs$). The F_2 is expected to be $\frac{9}{16}$ black, spotted ($B\_S\_$), $\frac{3}{16}$ black, solid (B_ss), $\frac{3}{16}$ brown, spotted ($bbS\_$), and $\frac{1}{16}$ brown, solid ($bbss$).

8. The F_1 offspring are expected to be black, short-haired. There is a $\frac{1}{16}$ chance of a brown and tan, long-haired offspring in the F_2 generation.

9. pleiotropic.

10. The rooster is *PpRr*; hen A is *PpRR*; hen B is *Pprr*; and hen C is *PPRR*.

11. a. Both parents are heterozygous (for a single locus). b. One parent is heterozygous and the other homozygous recessive (for a single locus). c. Both parents are heterozygous (for two loci.) d. One parent is heterozygous and the other homozygous (for two loci).

12. The expected types of F$_2$ plants are: $\frac{1}{16}$ of the plants produce 4-pound fruits; $\frac{1}{4}$ of the plants produce 3.5-pound fruits; $\frac{3}{8}$ of the plants produce 3-pound fruits; $\frac{1}{4}$ of the plants produce 2.5-pound fruits; and $\frac{1}{16}$ of the plants produce 2-pound fruits.

13. All males are barred, and all females are nonbarred, thus allowing the sex of the chicks to be determined from their phenotypes.

14. This is a two-point test cross involving linked loci. The parental classes of offspring are *Aabb* and *aaBb*; the recombinant classes are *AaBb* and *aabb*. There is 4.6% recombination, which corresponds to 4.6 map units between the loci.

15. If genes B and C are 10 map units apart, gene A is in the middle. If genes B and C are 2 map units apart, gene C is in the middle.

CHAPTER 11

1. d   2. a   3. e   4. a   5. d   6. c   7. e
8. b   9. a   10. c

CHAPTER 12

1. d   2. c   3. b   4. a   5. e   6. c   7. a
8. b   9. c   10. e

CHAPTER 13

1. e   2. c   3. a   4. c   5. a   6. b   7. c
8. a   9. b   10. c

CHAPTER 14

1. a   2. b   3. e   4. d   5. a   6. b   7. a
8. b   9. c   10. a   11. b

CHAPTER 15

1. a   2. b   3. c   4. c   5. a   6. e   7. d
8. a   9. e   10. d   11. c   12. e   13. e

CHAPTER 16

1. c   2. d   3. a   4. b   5. a   6. c   7. b
8. e   9. c   10. b

CHAPTER 17

1. d   2. a   3. c   4. c   5. d   6. b   7. a
8. e   9. b   10. b   11. b

CHAPTER 18

1. b   2. b   3. b   4. c   5. d   6. a   7. e
8. e   9. d   10. b   11. c   12. b

Review Questions

1. 0.6   2. 0.25; 0.5   3. 0.8   4. 0.4; 0.6   5. yes
6. no; if it were in genetic equilibrium, the genotype frequency of *Aa* would be $2pq = 0.48$ 7. frequency of $t = 0.55$; frequency of $T = 0.45$ 8. 0.3   9. $T = 0.6$; $t = 0.4$ 10. $T = 0.8$; $t = 0.2$ 11. 0.42

CHAPTER 19

1. e   2. a   3. a   4. c   5. d   6. b   7. c
8. b   9. e   10. c   11. d

CHAPTER 20

1. a   2. c   3. b   4. c   5. d   6. d   7. a
8. b   9. b   10. c

CHAPTER 21

1. a   2. e   3. d   4. b   5. b   6. e   7. c
8. a   9. b   10. b   11. d

CHAPTER 22

1. b   2. c   3. a   4. d   5. b   6. b   7. e
8. c   9. a   10. a   11. a   12. a   13. b

CHAPTER 23

1. a   2. e   3. a   4. b   5. c   6. a   7. a
8. b   9. b   10. a   11. b   12. a   13. e

CHAPTER 24

1. a   2. c   3. d   4. b   5. d   6. e   7. b
8. b   9. e   10. d   11. a   12. c

Chapter 25

1. b   2. d   3. d   4. a   5. b   6. a   7. e
8. c   9. c   10. e   11. b

Chapter 26

1. d   2. a   3. b   4. c   5. d   6. b   7. a
8. e   9. a   10. c   11. d   12. c

Chapter 27

1. b   2. c   3. d   4. a   5. e   6. d   7. d
8. c   9. b   10. d   11. a   12. c

Chapter 28

1. d   2. d   3. e   4. b   5. e   6. a   7. c
8. c   9. a   10. b   11. b   12. a

Chapter 29

1. b   2. a   3. a   4. c   5. d   6. c   7. b
8. b   9. b   10. a   11. a   12. b   13. e   14. d

Chapter 30

1. d   2. b   3. a   4. a   5. b   6. e   7. c
8. b   9. a   10. d   11. c   12. c   13. a   14. d

Chapter 31

1. c   2. a   3. e   4. b   5. b   6. e   7. c
8. b   9. a   10. c   11. b   12. d

Chapter 32

1. b   2. d   3. b   4. e   5. c   6. a   7. d
8. a   9. e   10. c   11. c   12. a

Chapter 33

1. b   2. c   3. d   4. e   5. b   6. a   7. c
8. e   9. c   10. b   11. a   12. b

Chapter 34

1. c   2. b   3. d   4. b   5. e   6. a   7. c
8. d   9. d   10. a   11. b   12. e

Chapter 35

1. d   2. e   3. c   4. a   5. c   6. b   7. c
8. b   9. c   10. a   11. d   12. a

Chapter 36

1. d   2. b   3. e   4. a   5. d   6. a   7. b
8. c   9. c   10. e   11. e

Chapter 37

1. b   2. b   3. a   4. c   5. b   6. c   7. a
8. e   9. c   10. e   11. a   12. d   13. b   14. d
15. a   16. a

Chapter 38

1. d   2. a   3. c   4. c   5. a   6. b   7. d
8. b   9. a   10. c   11. a   12. c

Chapter 39

1. c   2. c   3. a   4. b   5. b   6. d   7. e
8. d   9. a   10. c   11. c   12. c   13. c   14. b

Chapter 40

1. b   2. d   3. c   4. b   5. d   6. a   7. d
8. d   9. a   10. c   11. e   12. b   13. b   14. d

Chapter 41

1. a   2. b   3. a   4. a   5. d   6. e   7. e
8. e   9. d   10. c   11. c   12. b   13. d

Chapter 42

1. b   2. d   3. b   4. a   5. b   6. c   7. e
8. e   9. a   10. c   11. e   12. d   13. a   14. c
15. d

CHAPTER 43

1. b   2. a   3. d   4. c   5. e   6. c   7. a
8. b   9. c   10. a   11. d   12. a   13. c   14. e
15. d   16. b   17. d

CHAPTER 44

1. c   2. a   3. b   4. a   5. c   6. b   7. b
8. a   9. d   10. e   11. b   12. a   13. e   14. c
15. d

CHAPTER 45

1. a   2. a   3. c   4. d   5. d   6. e   7. e
8. b   9. b   10. d   11. a   12. b   13. c   14. c
15. a

CHAPTER 46

1. c   2. b   3. a   4. c   5. d   6. e   7. a
8. b   9. e   10. d   11. c   12. a   13. b   14. d
15. e

CHAPTER 47

1. b   2. c   3. a   4. d   5. c   6. e   7. b
8. a   9. d   10. e   11. b   12. d   13. a   14. e

CHAPTER 48

1. b   2. c   3. a   4. e   5. b   6. d   7. a
8. c   9. e   10. d   11. c   12. a   13. e   14. b
15. a

CHAPTER 49

1. d   2. a   3. a   4. d   5. b   6. e   7. b
8. b   9. c   10. b   11. c   12. d

CHAPTER 50

1. c   2. b   3. a   4. e   5. a   6. c   7. e
8. a   9. c   10. a   11. e   12. a   13. e   14. a
15. d   16. b   17. a   18. b   19. b   20. a

CHAPTER 51

1. b   2. b   3. a   4. e   5. c   6. a   7. e
8. d   9. a   10. b   11. b   12. e

CHAPTER 52

1. d   2. d   3. d   4. e   5. c   6. c   7. d
8. a   9. d   10. c   11. e   12. d   13. d

CHAPTER 53

1. c   2. d   3. e   4. a   5. d   6. e   7. a
8. b   9. c   10. e   11. d   12. b

CHAPTER 54

1. c   2. b   3. a   4. d   5. e   6. c   7. b
8. d   9. b   10. a   11. b   12. c   13. e

CHAPTER 55

1. d   2. e   3. c   4. c   5. b   6. c   7. e
8. d   9. b   10. c   11. a   12. b

Appendix C

The Classification of Organisms

The system of cataloging organisms used in this book is described in Chapter 1 and in Part 5. In this edition of *Biology*, we use a six-kingdom classification: Eubacteria, Archaebacteria, Protista, Fungi, Plantae, and Animalia. Prokaryotic organisms, distinguished from eukaryotes by their smaller ribosomes and absence of a nuclear envelope and other membranous organelles, are classified in kingdoms Eubacteria and Archaebacteria. Because many biologists prefer the three-domain approach, we also discuss the domains: **Archaea** (which corresponds to kingdom Archaebacteria), **Eubacteria** (or simply **Bacteria**), and **Eukarya** (eukaryotes). We have omitted many groups, especially extinct ones, in order to simplify the vast number of diverse categories of living organisms and their relationships to one another. Note that we have omitted the viruses from this survey, because they do not fit into any of the six kingdoms.

KINGDOM EUBACTERIA

Very large, diverse group of prokaryotic organisms (lack nuclear envelopes, mitochondria, and other membranous organelles; ribosomes smaller than in eukaryotes). Typically unicellular, but some form colonies or filaments. Mainly heterotrophic, but some groups are photosynthetic or chemosynthetic. Reproduction is primarily asexual by fission. Bacteria are nonmotile or move by beating flagella. When present, flagella are solid (rather than the 9 + 2 type typical of eukaryotes). More than 10,000 species.

Bacterial nomenclature and taxonomic practices are controversial and changing. In one system, eubacteria are classified in 11 phyla that fall into three main groups: (1) bacteria with a gram-negative type of cell wall, (2) bacteria with a gram-positive type of cell wall, and (3) bacteria with no cell walls (mycoplasmas). See Chapter 23, Table 23–3.

Gram-negative bacteria. Thin cell wall containing peptidoglycan. Include *proteobacteria, chlamydias, spirochetes, cyanobacteria.*

Gram-positive bacteria. Thick cell wall of peptidoglycan. All are nonphotosynthetic, and many produce spores. Include *actinomycetes, lactic acid bacteria, mycobacteria, streptococci, staphylococci, clostridia.*

Bacteria that lack a rigid cell wall. Extremely small bacteria bounded by a plasma membrane. *Mycoplasmas.*

KINGDOM ARCHAEBACTERIA

Prokaryotes with cell walls lacking peptidoglycan. Distinguished by their ribosomal RNA, lipid structure, and specific enzymes. Archaebacteria are found in extreme environments—hot springs, sea vents, dry and salty seashores, boiling mud, and near ash-ejecting volcanoes. Three main groups (considered phyla by some microbiologists) are *methanogens, extreme halophiles,* and *extreme thermophiles.*

Methanogens. Anaerobes that produce methane gas from simple carbon compounds.

Extreme Halophiles. Inhabit saturated salt solutions.

Extreme Thermophiles. Grow at 70°C or higher, some thrive above the boiling point.

KINGDOM PROTISTA

Primarily unicellular or simple multicellular eukaryotic organisms that do not form tissues and that exhibit relatively little division of labor. Most modes of nutrition occur in this kingdom. Life cycles may include both sexually and asexually reproducing phases and may be extremely complex, especially in parasitic forms. Locomotion is by cilia, flagella, amoeboid movement, or by other means. Flagella and cilia have 9 + 2 structure.

Animal-like protists

Phylum Zoomastigina. *Flagellates.* Single cells that move by means of flagella. Some free-living; many symbiotic; some pathogenic. Reproduction usually asexual by binary fission.

Phylum Rhizopoda. *Amoebas.* Shelled or naked unicellular protists whose movement is associated with pseudopods.

Phylum Foraminifera. *Foraminiferans.* Unicellular protists that produce calcareous tests (shells) with pores through which cytoplasmic projections extend, forming a sticky net to entangle prey.

Phylum Actinopoda. *Actinopods.* Unicellular protists that produce axopods (long, filamentous cytoplasmic projections) that protrude through pores in their siliceous shells.

Phylum Ciliophora. *Ciliates.* Unicellular protists that move by means of cilia. Reproduction is asexual by binary fission or sexual by conjugation. About 7200 species.

Phylum Apicomplexa. *Apicomplexans.* Parasitic unicellular protists that lack specific structures for locomotion. At some stage in life cycle, they develop spores (small infective agents). Some pathogenic. About 3900 species.

Plant-like protists

Phylum Dinoflagellata. *Dinoflagellates.* Unicellular (some colonial), photosynthetic, biflagellate. Cell walls, composed of overlapping cell plates, contain cellulose. Contain chlorophylls *a* and *c*, and carotenoids, including fucoxanthin. About 2100 species.

Phylum Bacillariophyta. *Diatoms.* Unicellular (some colonial), photosynthetic. Most nonmotile, but some move by gliding. Cell walls composed of silica rather than cellulose. Contain chlorophylls *a* and *c*, and carotenoids, including fucoxanthin. About 5600 species.

Phylum Chrysophyta. *Golden algae.* Unicellular (some colonial), photosynthetic, biflagellate (some lack flagella). Cells covered by tiny scales of either silica or calcium carbonate. Contain chlorophylls *a* and *c*, and carotenoids, including fucoxanthin. About 500 species.

Phylum Euglenophyta. *Euglenoids.* Unicellular, photosynthetic, two flagella (one of them very short). Flexible outer covering. Contain chlorophylls *a* and *b*, and carotenoids. About 1000 species.

Phylum Chlorophyta. *Green algae.* Unicellular, colonial, siphonous, and multicellular forms. Some motile and flagellated. Photosynthetic; contain chlorophylls *a* and *b*, and carotenoids. About 7000 species.

Phylum Rhodophyta. *Red algae.* Most multicellular (some unicellular), mainly marine. Some (coralline algae) have bodies impregnated with calcium carbonate. No motile cells. Photosynthetic; contain chlorophyll *a*, carotenoids, phycocyanin, and phycoerythrin. About 4000 species.

Phylum Phaeophyta. *Brown algae.* Multicellular, often quite large (kelps). Photosynthetic; contain chlorophylls *a* and *c*, and carotenoids, including fucoxanthin. Biflagellate reproductive cells. About 1500 species.

Fungus-like protists

Phylum Myxomycota. *Plasmodial slime molds.* Spend part of life cycle as a thin, streaming, multinucleate plasmodium that creeps along on decaying leaves or wood. Flagellated or amoeboid reproductive cells; form spores in sporangia. About 500 species.

Phylum Acrasiomycota. *Cellular slime molds.* Vegetative (nonreproductive) form unicellular; move by pseudopods. Amoeba-like cells aggregate to form a multicellular pseudoplasmodium that eventually develops into a fruiting body that bears spores. About 70 species.

Phylum Oomycota. *Water molds.* Consist of branched, coenocytic mycelia. Cellulose and/or chitin in cell walls. Produce biflagellate asexual spores. Sexual stage involves production of oospores. Some parasitic. About 580 species.

KINGDOM FUNGI

All eukaryotic, mainly multicellular organisms that are heterotrophic with saprotrophic or parasitic nutrition. Body form often a mycelium, and cell walls consist of chitin. No flagellated stages except in chytrids. Reproduce by means of spores, which may be produced sexually or asexually. Cells usually haploid or dikaryotic with brief diploid period following fertilization.

Phylum Chytridiomycota. *Chytrids.* Parasites and decomposers found principally in fresh water. Motile cells (gametes and zoospores) contain a single, posterior flagellum. Reproduce both sexually and asexually. About 1000 species.

Phylum Zygomycota. *Zygomycetes.* Produce sexual resting spores called zygospores, and asexual spores in a sporangium. Hyphae are coenocytic. Many are heterothallic (two mating types). About 800 species.

Phylum Ascomycota. *Sac fungi.* Sexual reproduction involves formation of ascospores in little sacs called asci. Asexual reproduction involves production of spores called conidia, which pinch off from conidiophores. Hyphae usually have perforated septa. About 30,000 species.

Phylum Basidiomycota. *Club fungi.* Sexual reproduction involves formation of basidiospores on a basidium. Asexual reproduction uncommon. Heterothallic. Hyphae usually have perforated septa. About 25,000 species.

Phylum Deuteromycota. *Imperfect fungi.* Sexual stage has not been observed. Most reproduce only by conidia. About 25,000 species.

KINGDOM PLANTAE

Multicellular eukaryotic organisms with differentiated tissues and organs. Cell walls contain cellulose. Cells frequently contain large vacuoles; photosynthetic pigments in plastids. Photosynthetic pigments are chlorophylls *a* and *b*, and carotenoids. Nonmotile. Reproduce both asexually and sexually, with alternation of gametophyte (*n*) and sporophyte (2*n*) generations.

Phylum Bryophyta. *Mosses.* Nonvascular plants that lack xylem and phloem. Marked alternation of generations with dominant gametophyte generation. Motile sperm. The gametophytes generally form a dense green mat consisting of individual plants. About 9000 species.

Phylum Hepatophyta. *Liverworts.* Nonvascular plants that lack xylem and phloem. Marked alternation of generations with dominant gametophyte generation. Motile sperm. The gametophytes of certain species have a flat, liver-like thallus; other species are more mosslike in appearance. About 6000 species.

Phylum Anthocerotophyta. *Hornworts.* Nonvascular plants that lack xylem and phloem. Marked alternation of generations with dominant gametophyte generation. Motile sperm. The gametophyte is a small, flat green thallus with scalloped edges. Spores are produced on an erect, hornlike stalk. About 100 species.

Phylum Pterophyta. *Ferns.* Vascular plants with a dominant sporophyte generation. Generally homosporous. Gametophyte is free-living and photosynthetic. Reproduce by spores. Motile sperm. About 11,000 species.

Phylum Psilotophyta. *Whisk ferns.* Vascular plants with a dominant sporophyte generation. Homosporous. Stem is distinctive because it branches dichotomously; plant lacks true roots and leaves. The gametophyte is subterranean and nonphotosynthetic and forms a mycorrhizal relationship with a fungus. Motile sperm. About 12 species.

Phylum Sphenophyta. *Horsetails.* Vascular plants with hollow, jointed stems and reduced, scalelike leaves. Although modern representatives are small, some extinct species were treelike. Homosporous. Gametophyte a tiny photosynthetic plant. Motile sperm. About 15 species.

Phylum Lycophyta. *Club mosses.* Sporophyte plants are vascular with branching rhizomes and upright stems that bear microphylls. Although modern representatives are small, some extinct species were treelike. Some homosporous, others heterosporous. Motile sperm. About 1000 species.

Phylum Coniferophyta. *Conifers.* Heterosporous vascular plants with woody tissues (trees and shrubs) and needle-shaped leaves. Most are evergreen. Seeds are usually borne naked on the surface of cone scales. Nutritive tissue in the seed is haploid female gametophyte tissue. Nonmotile sperm. About 550 species.

Phylum Cycadophyta. *Cycads.* Heterosporous, vascular, dioecious plants that are small and shrubby or larger and palmlike. Produce naked seeds in conspicuous cones. Flagellated sperm. About 140 species.

Phylum Ginkgophyta. *Ginkgo.* Broad-leaved deciduous trees that bear naked seeds directly on branches. Dioecious. Contain vascular tissues. Flagellated sperm. The ginkgo tree is the only living representative. One species.

Phylum Gnetophyta. *Gnetophytes.* Woody shrubs, vines, or small trees that bear naked seeds in cones. Contain vascular tissues. Possess many features similar to flowering plants. About 70 species.

Phylum Anthophyta. *Flowering plants or angiosperms.* Largest, most successful group of plants. Heterosporous; dominant sporophytes with extremely reduced gametophytes. Contain vascular tissues. Bear flowers, fruits, and seeds (enclosed in a fruit; seeds contain endosperm as nutritive tissue). Double fertilization. About 235,000 species.

KINGDOM ANIMALIA

Multicellular eukaryotic heterotrophs with differentiated cells. In most animals, cells are organized to form tissues, tissues are organized to form organs; and tissues and organs form specialized body systems that carry on specific functions. Most animals have a well developed nervous system and can respond rapidly to changes in their environment. Most are capable of locomotion during some time in their life cycle. Most animals reproduce sexually with large nonmotile eggs and flagellated sperm.

Phylum Porifera. *Sponges.* Mainly marine. Body bears many pores through which water circulates. Food is filtered from the water by collar cells (choanocytes). Solitary or colonial. Asexual reproduction by budding; external sexual reproduction in which sperm are released and swim to internal egg. Larva is motile. About 9000 species.

Phylum Cnidaria. *Hydras, jellyfish, sea anemones, corals.* Radial symmetry. Tentacles surrounding mouth. Stinging cells (cnidocytes) contain stinging structures called nematocysts. Polyp and medusa forms. Planula larva. Solitary or colonial. Marine, with a few freshwater forms. About 10,000 species.

Phylum Ctenophora. *Comb jellies.* Biradial symmetry. Free-swimming; marine. Two tentacles and eight longitudinal rows of cilia resembling combs; animal moves by means of these bands of cilia. About 100 species.

Acoelomates

No body cavity; region between body wall and internal organs filled with tissue.

Phylum Platyhelminthes. *Flatworms.* Planarians are free-living; flukes and tapeworms are parasitic. Body dorsoventrally flattened; cephalization; three tissue layers. Simple nervous system with ganglia in head region. Excretory organs are protonephridia with flame cells. About 20,000 species.

Phylum Nemertea. *Proboscis worms* (also called ribbon worms). Long, dorsoventrally flattened body with complex proboscis used for defense and for capturing prey. Simplest animal to have definite organ systems. Complete digestive tract. Circulatory system with blood. Functionally acoelomate, but do have a small true coelom in the proboscis. About 900 species.

Pseudocoelomates

Body cavity not completely lined with mesoderm. Complete digestive tract extending from mouth to anus.

Phylum Nematoda. *Roundworms. Ascaris,* hookworms, pinworms. Slender, elongated, cylindrical worms; covered with cuticle. Free-living and parasitic forms. About 12,000 species.

Phylum Rotifera. *Wheel animals.* Microscopic, wormlike animals. Anterior end has ciliated crown that looks like a wheel when the cilia beat. Posterior end tapers to a foot. Constant number of cells. About 1800 species.

Coelomates

These animals have a true coelom, a body cavity completely lined with mesoderm.

Protostomes

Coelomates with spiral, determinate cleavage; mouth typically develops from blastopore.

Phylum Mollusca. *Snails, clams, squids, octopods.* Unsegmented, soft-bodied animals usually covered by a dorsal shell. Have a ventral, muscular foot. Most organs located above foot in visceral mass. A shell-secreting mantle covers the visceral mass and forms a mantle cavity, which contains gills. Trochophore and/or veliger larva. About 50,000 species.

Phylum Annelida. *Segmented worms.* Polychaetes, earthworms, leeches. Both body wall and internal organs are segmented. Body segments separated by septa. Some have nonjointed appendages.

Setae used in locomotion. Closed circulatory system; metanephridia; specialized regions of digestive tract. Trochophore larva. About 15,000 species.

Phylum Arthropoda. *Arachnids (spiders, mites, ticks), crustaceans (lobsters, crabs, shrimp), insects, centipedes, millipedes.* Segmented animals with paired, jointed appendages and a hard exoskeleton made of chitin. Open circulatory system with dorsal heart. Hemocoel occupies most of body cavity, and coelom is reduced, More than 1 million species.

Deuterostomes

Coelomates with radial, indeterminate cleavage. Blastopore develops into anus, and mouth forms from a second opening.

Phylum Phoronida. One of the lophophorate phyla. Tube-dwelling marine worms. About 15 species.

Phylum Ectoprocta. *Bryozoans.* One of the lophophorate phyla. Mainly marine; sessile colonies produced by asexual budding. About 4500 species.

Phylum Brachiopoda. *Lamp shells.* One of the lophophorate phyla. Marine; body enclosed between two shells. About 300 species.

Phylum Echinodermata. *Sea stars, sea urchins, sand dollars, sea cucumbers.* Marine animals that have pentaradial symmetry as adults but bilateral symmetry as larvae. Endoskeleton of small, calcareous plates. Water vascular system; tube feet for locomotion. About 7000 species.

Phylum Chaetognatha. *Arrow worms.* Marine, make up part of plankton. Ciliated. Pharyngeal gill slits. About 100 species.

Phylum Hemichordata. *Acorn worms.* Marine animals with an anterior muscular proboscis, connected by a collar region to a long wormlike body. The larval form resembles an echinoderm larva. About 85 species.

Phylum Chordata. *Subphylum Urochordata (tunicates), subphylum Cephalochordata (lancelets), subphylum Vertebrata (fishes, amphibians, reptiles, birds, mammals).* Notochord; pharyngeal gill slits; dorsal, tubular nerve cord, and postanal tail present at some time in life cycle. About 50,000 species.

Appendix D
Careers in Biology

The following organizations, professional associations, and World Wide Web sites provide career information upon request.

GENERAL

American Association for the Advancement of Science

1200 New York Avenue, NW
Washington, DC 20005
202-326-6400
website: http://www.aaas.org
email: info@aaas.org

American Institute of Biological Sciences

730 11th Street, NW
Washington, DC 20001-4521
202-628-1500
website: http://www.aibs.org/careers.html
email: cgabriel@aibs.org

Association for Tropical Biology

Department of Botany MRC-166
National Museum of Natural History
Smithsonian Institute
Washington, DC 20560
202-357-3392
(fax) 202-786-2563
website: http://atb.botany.ufl.edu
email: kressj@nmnh.si.edu

Association for Women in Science (AWIS)

1200 New York Avenue, NW
Suite 650
Washington, DC 20005
202-326-8940
(fax) 202-326-8960
website: http://www.awis.org
email: awis@awis.org

Biology Careers Page

website: http://www.furman.edu/~snyder/careers/careers.html

Bio Online

website: http://www.bio.com
email: jobs@bio.com

Bioscience Links

website: http://golgi.harvard.edu/biopages/all.html

Federation of American Societies for Experimental Biology

9650 Rockville Pike
Bethesda, MD 20814-3998
301-530-7020
(fax) 301-571-0699
website: http://www.faseb.org
email: careers@faseb.org

JobHunt: Science, Engineering, and Medicine

website: http://www.job-hunt.org/science.html

National Academy of Science, National Academy of Engineering, Institute of Medicine, National Research Council: A Career Planning Center

website: http://www2.nas.edu/cpc/index.html

Pursuit

Pursuing careers in science, engineering, and mathematics for persons with disabilities
website: http://pursuit.rehab.uiuc.edu/pursuit/homepage.html

Women in Technology International (WITI)

4641 Burnet Avenue
Sherman Oaks, CA 91403
818-990-6705
(fax) 818-906-3299
website: http://www.witi.com

AGRICULTURE/AGRONOMY

Careers in Agriculture and Agri-Food

website: http://www.cfa-fca.ca/careers/index1.html

American Society of Agronomy

667 South Segoe Road
Madison, WI 53711
608-273-8080 (ext. 314)
website: http://www.agronomy.org
email: Lmalison@agronomy.org (Leann Malison)

ART AND COMMUNICATIONS

Association of Medical Illustrators

1819 Peachtree Street, N.E., Suite 620
Atlanta, GA 30309
404-350-7900
(fax) 404-351-3348
website: www.medical-illustrators.org
email: assnhq@mindspring.com

Biological Photographic Association

1819 Peachtree Street N.E., Suite 620
Atlanta, Georgia 30309
404-351-6300
(fax) 404-351-3348
website: http://www.thebpa.org
email: assnhq@atl.mindspring.com

Medical Library Association

Suite 300, Six North Michigan Avenue
Chicago, IL 60602-4805
312-419-9094
(fax) 312-419-8950
website: http://www.mlanet.org
email: mlapd1@mlahq.org

National Association of Science Writers

P.O. Box 294
Greenlawn, NY 11740
516-757-5664
(fax) 516-757-0069
website: http://www.nasw.org/
email: diane@nasw.org

BIOCHEMISTRY

American Society for Biochemistry and Molecular Biology (ASBMB)

9650 Rockville Pike
Bethesda, Maryland 20814-3996
website: http://www.biophysics.org/asbmb/asbmb.html
email: aps@scisoc.org

BIOMEDICAL ENGINEERING

Biomedical Engineering Society

P.O. Box 2399
Culver City, CA 90231
310-618-9322
(fax) 310-618-1333
website: http://mecca.mecca.org/BME/BMES/society/bmeshm.html
email: bmes@netcom.com

BIOMEDICAL SCIENCE

Career Resources and Employment Services for Biomedical Scientists

website: http://www.wam.umd.edu/~garlandc/
Jobs%20HTMLs/Resources.html

Employment Links for the Biomedical Scientist

website: http://www.his.com/~graeme/employ.html

BIOPHYSICS

The Biophysical Society

9650 Rockville Pike
Bethesda, MD 20814
301-530-7114
(fax) 301-530-7133
website: http://www.biophysics.org/biophys/society/biohome.htm
email: society@biophysics.faseb.org

BOTANY

American Phytopathological Society

3340 Pilot Knob Road
St. Paul, MN 55121-2097
612-454-7250
fax: (612) 454-0766
website: http://www.scisoc.org

American Society of Plant Physiologists

15501 Monona Drive
Rockville, MD 20855-2768
301-251-0560
(fax) 301-279-2996
website: http://www.aspp.org/
email: bhyps@access.digex.net

Botanical Society of America

1735 Neil Avenue
Columbus, OH 43210-1293
614-292-3519
website: http://www.botany.org

Mycological Society of America

P.O. Box 1897
Lawrence, Kansas 66044-8897
913-843-1235
(fax) 913-843-1274
website: http://www.erin.utoronto.ca/~w3msa/
email: allenpressam@delphi.com

Phycological Society of America

P.O. Box 1897
Lawrence, Kansas 66044-8897
913-843-1235
(fax) 913-843-1274
website: http://jupiter.phy.ohiou.edu/psa
email: allenpressam@delphi.com

CELL BIOLOGY

American Society for Cell Biology (ASCB)

9650 Rockville Pike
Bethesda, Maryland 20814-3992
301-530-7153
(fax) 301-530-7139
website: http://www.faseb.org/ascb
email: ascbinfo@ascb.org

Cell & Molecular Biology Online

website: http://www.cellbio.com

DENTISTRY

American Association of Dental Schools

202-667-9433
(fax) 202-667-0642
website: http://www.aads.jhu.edu
email: aads@aads.jhu.edu

American Dental Association

211 E. Chicago Avenue
Chicago, Illinois 60611
312-440-2500
(fax) 312-440-2800
website: http://www.ada.org
email: publicinfo@ada.org

EDUCATION

National Association of Biology Teachers (NABT)

11250 Roger Bacon Drive #19
Reston, VA 22090
703-471-1134
(fax) 703-435-5582
website: http://www.nabt.org
email: NABTer@aol.com

ENTOMOLOGY

Entomological Society of America

9301 Annapolis Road
Lanham, MD 20706-3115
301-731-4535
(fax) 301-731-4538
website: http://www.entsoc.org
email: esa@entsoc.org

ENVIRONMENT/ECOLOGY

Ecological Society of America

2010 Massachusetts Avenue, NW
Suite 400
Washington, DC 20036
202-833-8773
(fax) 202-833-8775
website: http://www.sdsc.edu/projects/ESA/esa.htm
email: esahq@esa.org

Environmental Careers Organization (ECO)

179 South Street
Boston, MA 02111
617-426-4375
(fax) 617-423-0998
website: http://www.eco.org

Midcontinent Ecological Science Center: Explore Science Careers

Midcontinent Ecological Science Center
4512 McMurry Avenue
Fort Collins, Colorado 80525-3400
970-226-9100
(fax) 970-226-9230
website: http://www.mesc.nbs.gov/science-careers.html

National Wildlife Federation

8925 Leesburg Pike
Vienna, VA 22184
703-790-4000
website: http://www.nwf.org/
email: jobopp@nwf.org

Nature Conservancy

International Headquarters
1815 North Lynn Street
Arlington, Virginia 22209
703-841-5300
website: http://www.tnc.org

ENVIRONMENTAL LAW

Environmental Defense Fund

257 Park Avenue South
New York, NY 10010
1-800-684-3322
website: http://www.edf.org

Environmental Law Institute

1616 P Street, NW, Suite 200
Washington, DC 20036
202-939-3800
(fax) 202-939-3868
website: http://www.eli.org/

FORENSICS

American Academy of Forensic Sciences

The Forensic Sciences Foundation, Inc.
P.O. Box 669
Colorado Springs, CO 80901-0669
719-636-1100
(fax) 719-636-1993
website: http://www.aafs.org/
email: Membship@aafs.org

GENETICS

Genetics Society of America

9650 Rockville Pike
Bethesda, MD 20814-3889
301-571-1825
website: http://www.faseb.org/genetics/gsa/gsamenu.htm
email: estrass@genetics.faseb.org

National Society of Genetic Counselors

NSGC Executive Office
233 Canterbury Drive
Wallingford, PA 19086-6617
610-872-7608
website: http://members.aol.com/nsgcweb/nsgchome.htm
email: nsgc@aol.com

Careers in Biotechnology

website: http://www.gene.com:80/ae/AB/CC/index.html

MARINE BIOLOGY

American Society of Limnology and Oceanography

website: http://aslo.org

Careers & Jobs in Marine Biology & Oceanography

website: http://www-marine.stanford.edu/HMS web/careers.html

Scripps Research Institute

10550 North Torrey Pines Road
La Jolla, California 92037
619-784-1000
website: http://www.scripps.edu

Virginia Institute of Marine Science

P.O. Box 1346
Gloucester Point, Virginia 23062-1346
804-684-7000
website: http://www.vims.edu/adv/ed/careers/

Woods Hole Oceanographic Institute

Woods Hole Oceanographic Institution
Information Office
Co-op Building, MS #16
Woods Hole, MA 02543
508-289-2252 or 508-289-2100
(fax) 508-457-2180
website: http://www.whoi.edu/
email: information@whoi.edu

MATHEMATICAL BIOLOGY (BIOINFORMATICS)

The Society for Mathematical Biology

http://www.smb.org/index.html

MEDICINE

American Academy of Family Physicians

8880 Ward Parkway
Kansas City, MO 64114
(816) 333-9700
website: http://www.aafp.org/
email: fp@aafp.org

American Academy of Pediatrics

141 Northwest Point Boulevard
Elk Grove Village, IL 60007-1098
847-228-5005
(fax) 847-228-5097
website: http://www.aap.org
email: pedscareer@aap.org

American Association of Immunologists

9650 Rockville Pike
Bethesda, Maryland 20814-3994
301-530-7178
(fax) 301-571-1816
website: http://www.scienceXchange.com/aai/
email: infoaai@aai.faseb.org

American Medical Association

515 North State Street
Chicago, IL 60610
312-464-5000
website: http://www.ama-assn.org

Clinical Immunology Society

611 E. Wells Street
Milwaukee, WI 53202
414-224-8095

Research! America (Medical Research)

908 King Street, Suite 400E
Alexandria, VA 22314
703-739-2577
website: http://www.researchamerica.org/
email: researcham@aol.com

MICROBIOLOGY

American Society for Microbiology

1325 Massachusetts Avenue, NW
Washington, DC 20005
202-942-9283
(fax) 202-942-9329
website: http://www.asmusa.org/
email: FellowshipsCareerInformation@asmusa.org

MICROSCOPY

Microscopy Society of America

435 North Michigan Avenue, Suite 1717
Chicago, IL 60611-4067
312-644-1527
(fax) 312-644-8557
website: http://www.msa.microscopy.com
email: BusinessOffice@MSA.Microscopy.Com

NURSING

American Nurses' Association

600 Maryland Avenue, SW
Suite 100W
Washington, DC 20024-2571
202-651-7000
(fax) 202-651-7001
website: www.nursingworld.org

National Association for Practical Nurse Education and Service

1400 Spring Street
Suite 300
Silver Spring, MD 20910
301-588-2491
website: napnes@bellatlantic.net

NUTRITION

American Dietetic Association

216 West Jackson Boulevard
Chicago, Illinois 60606-6995
312-899-0040
(fax) 312-899-1979
website: http://www.eatright.org/careers.html
email: network@eatright.org

Institute of Food Technologists

Professional Development Department
Institute of Food Technologists
221 North LaSalle Street, Suite 300
Chicago, IL 60601
312-782-8424
(fax) 312-782-0045

PHARMACEUTICALS

American Association of Colleges of Pharmacy

1426 Prince Street
Alexandria, VA 22314-2841
703-729-2330
(fax) 703-836-8982
website: http://www.aacp.org/
email: angieaacp@aol.com

American Pharmaceutical Association

2215 Constitution Avenue, NW
Washington, DC 20037
202-628-4410
(fax) 202-783-2351
website: http://www.aoa.dhhs.gov/aoa/dir/49.html

PHYSICAL THERAPY

American Physical Therapy Association

1111 North Fairfax Street
Alexandria, VA 22314
703-684-2782
(fax) 703-684-7343

PHYSIOLOGY

American Physiological Society

9650 Rockville Pike
Bethesda, MD 20814
301-530-7160
website: http://www.biophysics.org/aps/
email: mfrank@aps.faseb.org

PSYCHIATRY/PSYCHOLOGY

American Psychiatric Association

Division of Public Affairs
1400 K Street, NW
Washington, DC 20005
202-682-6000
website: http://www.psych.org
email: apa@psych.org

American Psychological Association

750 First Street, NE
Washington, DC 20002
202-336-5500
website: http://www.apa.org/
email: education@apa.org

PUBLIC HEALTH

American Public Health Association

1015 15th Street, NW
Washington, DC 20005-2605
202-789-5600
(fax) 202-789-5661
website: http://www.apha.org

TAXONOMY

American Society of Plant Taxonomists

website: http://www.csd1.tamu.edu/FLORA/aspt/aspthome.htm

Association of Systematic Collections

1725 K Street, NW, Suite 601
Washington, DC 20006-1401
(202) 835-7334
website: http://www.ascoll.org/
email: asc@ascoll.org

ZOOLOGY

American Association of Zoo Keepers

Topeka Zoological Park
635 S.W. Gage Boulevard
Topeka, KS 66606
913-272-5821 or 913-273-1980

American Society of Mammalogists

Monte L. Bean Life Science Museum
Brigham Young University
Provo, UT 84602-0200
website: http://wkuwebl.wku.edu/~asm/

American Zoo and Aquarium Association

Executive Office and Conservation Center
7090-D Old Georgetown Road
Bethesda, MD 20814-2493
website: http://www.aza.org

Society for Integrative and Comparative Biology (SICB) (formerly: American Society of Zoologists)

401 North Michigan Avenue
Chicago, IL 60611-4267
312-527-6697
(fax) 312-321-3700
website: http://www.sicb.org/
email: sicb@sba.com

Appendix E
Understanding Biological Terms

Your task of mastering new terms will be greatly simplified if you learn to dissect each new word. Many terms can be divided into a prefix, the part of the word that precedes the main root, the word root itself, and often a suffix, a word ending that may add to or modify the meaning of the root. As you progress in your study of biology, you will learn to recognize the more common prefixes, word roots, and suffixes. Such recognition will help you analyze new terms so that you can more readily determine their meaning and will also help you remember them.

Prefixes

a-, ab- from, away, apart (abduct, move away from the midline of the body)

a-, an-, un- less, lack, not (asymmetrical, not symmetrical)

ad- (also **af-, ag-, an-, ap-**) to, toward (adduct, move toward the midline of the body)

allo- different (allometric growth, different rates of growth for different parts of the body during development)

ambi- both sides (ambidextrous, able to use either hand)

andro- a man (androecium, the male portion of a flower)

anis- unequal (anisogamy, sexual reproduction in which the gametes are of unequal sizes)

ante- forward, before (anteflexion, bending forward)

anti- against (antibody, proteins that have the capacity to react against foreign substances in the body)

auto- self (autotroph, organism that manufactures its own food)

bi- two (biennial, a plant that takes two years to complete its life cycle)

bio- life (biology, the study of life)

circum-, circ- around (circumcision, a cutting around)

co-, con- with, together (congenital, existing with or before birth)

contra- against (contraception, against conception)

cyt- cell (cytology, the study of cells)

di- two (disaccharide, a compound made of two sugar molecules chemically combined)

dis- apart (dissect, cut apart)

ecto- outside (ectoplasm, outer layer of cytoplasm)

end-, endo- within, inner (endoplasmic reticulum, a network of membranes found within the cytoplasm)

epi- on, upon (epidermis, upon the dermis)

ex-, e-, ef- out from, out of (extension, a straightening out)

extra- outside, beyond (extraembryonic membrane, a membrane that encircles and protects the embryo)

gravi- heavy (gravitropism, growth of a plant in response to gravity)

hemi- half (cerebral hemisphere, lateral half of the cerebrum)

hetero- other, different (heterozygous, having unlike members of a gene pair)

homo-, hom- same (homologous, corresponding in structure; homozygous, having identical members of a gene pair)

hyper- excessive, above normal (hypersecretion, excessive secretion)

hypo- under, below, deficient (hypotonic, a solution whose osmotic pressure is less than that of a solution with which it is compared)

in-, im- not (incomplete flower, a flower that does not have one or more of the four main parts)

inter- between, among (interstitial, situated between parts)

intra- within (intracellular, within the cell)

iso- equal, like (isotonic, equal osmotic concentration)

macro- large (macronucleus, a large, polyploid nucleus found in ciliates)

mal- bad, abnormal (malnutrition, poor nutrition)

mega- large, great (megakaryocyte, giant cell of bone marrow)

meso- middle (mesoderm, middle tissue layer of the animal embryo)

meta- after, beyond (metaphase, the stage of mitosis after prophase)

micro- small (microscope, instrument for viewing small objects)

mono- one (monocot, a group of flowering plants with one cotyledon, or seed leaf, in the seed)

oligo- small, few, scant (oligotrophic lake, a lake deficient in nutrients and organisms)

oo- egg (oocyte, developing egg cell)

paedo- a child (paedomorphosis, the preservation of a juvenile characteristic in an adult)

para- near, beside, beyond (paracentral, near the center)

peri- around (pericardial membrane, membrane that surrounds the heart)

photo- light (phototropism, growth of a plant in response to the direction of light)

poly- many, much, multiple, complex (polysaccharide, a carbohydrate composed of many simple sugars)

post- after, behind (postnatal, after birth)

pre- before (prenatal, before birth)

pseudo- false (pseudopod, a temporary protrusion of a cell, i.e., "false foot")

retro- backward (retroperitoneal, located behind the peritoneum)

semi- half (semilunar, half-moon)

sub- under (subcutaneous tissue, tissue immediately under the skin)

super, supra- above (suprarenal, above the kidney)

sym- with, together (sympatric speciation, evolution of a new species within the same geographical region as the parent species)

syn- with, together (syndrome, a group of symptoms that occur together and characterize a disease)

trans- across, beyond (transport, carry across)

Suffixes

-able, -ible able (viable, able to live)

-ad used in anatomy to form adverbs of direction (cephalad, toward the head)

-asis, -asia, -esis condition or state of (euthanasia, state of "good death")

-cide kill, destroy (biocide, substance that kills living things)

-emia condition of blood (anemia, a blood condition in which there is a lack of red blood cells)

-gen something produced or generated or something that produces or generates (pathogen, an organism that produces disease)

-gram record, write (electrocardiogram, a record of the electrical activity of the heart)

-graph record, write (electrocardiograph, an instrument for recording the electrical activity of the heart)

-ic adjective-forming suffix that means *of* or *pertaining to* (ophthalmic, of or pertaining to the eye)

-itis inflammation of (appendicitis, inflammation of the appendix)

-logy study or science of (cytology, study of cells)

-oid like, in the form of (thyroid, in the form of a shield)

-oma tumor (carcinoma, a malignant tumor)

-osis indicates disease (psychosis, a mental disease)

-pathy disease (dermopathy, disease of the skin)

-phyll leaf (mesophyll, the middle tissue of the leaf)

-scope instrument for viewing or observing (microscope, instrument for viewing small objects)

Some Common Word Roots

abscis cut off (abscission, the falling off of leaves or other plant parts)

angi, angio vessel (angiosperm, plants that produce seeds enclosed within a fruit or "vessel")

apic tip, apex (apical meristem, area of cell division located at the tips of plant stems and roots)

arthr joint (arthropods, invertebrate animals with jointed legs and segmented bodies)

aux grow, enlarge (auxin, a plant hormone involved in growth and development)

bi, bio life (biology, study of life)

blast a formative cell, germ layer (osteoblast, cell that gives rise to bone cells)

brachi arm (brachial artery, blood vessel that supplies the arm)

bry grow, swell (embryo, an organism in the early stages of development)

cardi heart (cardiac, pertaining to the heart)

carot carrot (carotene, a yellow, orange, or red pigment in plants)

cephal head (cephalad, toward the head)

cerebr brain (cerebral, pertaining to the brain)

cervic, cervix neck (cervical, pertaining to the neck)

chlor green (chlorophyll, a green pigment found in plants)

chondr cartilage (chondrocyte, a cartilage cell)

chrom color (chromosome, deeply staining body in nucleus)

cili small hair (cilium, a short, fine cytoplasmic hair projecting from the surface of a cell)

coleo a sheath (coleoptile, a protective sheath that encircles the stem in grass seedlings)

conjug joined together (conjugation, a sexual phenomenon in certain protists)

cran skull (cranial, pertaining to the skull)

cyt cell (cytology, study of cells)

decid falling off (deciduous, a plant that sheds its leaves at the end of the growing season)

dehis split (dehiscent fruit, a fruit that splits open at maturity)

derm skin (dermatology, study of the skin)

ecol dwelling, house (ecology, the study of organisms in relation to their environment, i.e., "their house")

enter intestine (enterobacteria, a group of bacteria that include species that inhabit the intestines of humans and other animals)

evol to unroll (evolution, descent with modification, or gradual directional change)

fil a thread (filament, the thin stalk of the stamen in flowers)

gamet a wife or husband (gametangium, the part of a plant or protist that produces reproductive cells)

gastr stomach (gastrointestinal tract, the digestive tract)

glyc, glyco sweet, sugar (glycogen, storage form of glucose)

gon seed (gonad, an organ that produces gametes)

gutt a drop (guttation, loss of water as liquid "drops" from plants)

gymn naked (gymnosperm, a plant that produces seeds that are not enclosed with a fruit, i.e., "naked")

hem blood (hemoglobin, the pigment of red blood cells)

hepat liver (hepatic, of or pertaining to the liver)

hist tissue (histology, study of tissues)

hom, homeo same, unchanging, steady (homeostasis, reaching a steady state)

hydr water (hydrolysis, a breakdown reaction involving water)

leuk white (leukocyte, white blood cell)

menin membrane (meninges, the three membranes that envelop the brain and spinal cord)

morph form (morphogenesis, development of body form)

my, myo muscle (myocardium, muscle layer of the heart)

myc a fungus (mycelium, the vegetative body of a fungus)

nephr kidney (nephron, microscopic unit of the kidney)

neur, nerv nerve (neuromuscular, involving both the nerves and muscles)

occiput back part of the head (occipital, back region of the head)

ost bone (osteology, study of bones)

path disease (pathologist, one who studies disease processes)

ped, pod foot (bipedal, walking on two feet)

pell skin (pellicle, a flexible covering over the body of certain protists)

phag eat (phagocytosis, process by which certain cells ingest particles and foreign matter)

phil love (hydrophilic, a substance that attracts, i.e., "loves," water)

phloe bark of a tree (phloem, food-conducting tissue in plants that corresponds to bark in woody plants)

phyt plant (xerophyte, a plant adapted to xeric, or dry, conditions)

plankt wandering (plankton, microscopic aquatic protists that float or drift passively)

rhiz root (rhizome, a horizontal, underground stem that superficially resembles a root)

scler hard (sclerenchyma, cells that provide strength and support in the plant body)

sipho a tube (siphonous, a type of tubular body form found in certain algae)

som body (chromosome, deeply staining body in the nucleus)

sor heap (sorus, a cluster or "heap" of sporangia in a fern)

spor seed (spore, a reproductive cell that gives rise to individual offspring in plants and protists)

stom a mouth (stoma, a small pore, i.e., "mouth," in the epidermis of plants)

thigm a touch (thigmotropism, plant growth in response to touch)

thromb clot (thrombus, a clot within a blood vessel)

tropi turn (thigmotropism, growth of a plant in response to contact with a solid object, as when a tendril "turns" or wraps around a wire fence)

visc pertaining to an internal organ or body cavity (viscera, internal organs)

xanth yellow (xanthophyll, a yellowish pigment found in plants)

xyl wood (xylem, water-conducting tissue in plant, the "wood" of woody plants)

zoo an animal (zoology, the science of animals)

The biological sciences use a great many abbreviations and with good reason. Many technical terms in biology and biological chemistry are both long and difficult to pronounce. Yet it can be difficult for beginners, when confronted with something like NADPH or EPSP, to understand the reference. Here are some of the common abbreviations used in biology for your ready reference.

A Adenine
ABA Abscisic acid
ACTH Adrenocorticotropic hormone
ADA Adenosine deaminase
ADH Antidiuretic hormone
ADP Adenosine diphosphate
AIDS Acquired immune deficiency syndrome
AMP Adenosine monophosphate
amu Atomic mass unit (dalton)
APC Antigen presenting cell
ATP Adenosine triphosphate
AV node or valve Atrioventricular node or valve (of heart)
B lymphocyte or B cell Lymphocyte responsible for antibody-mediated immunity
BH Brain hormone (of insects)
BMR Basal metabolic rate
C Cytosine
C$_3$ Three-carbon pathway for carbon fixation (Calvin cycle)
C$_4$ Four-carbon pathway for carbon fixation (Hatch-Slack pathway)
CAM Crassulacean acid metabolism
cAMP Cyclic adenosine monophosphate
CAP Catabolite gene activator protein
CD4 T cells Helper T cells (T$_h$); have a surface marker designated CD4
CD8 T cells T cells with a surface marker designated CD8; include cytotoxic T cells
cDNA Complementary deoxyribonucleic acid
CFTR Cystic fibrosis transmembrane conductance regulator
CITES The Convention on International Trade in Endangered Species of Wild Flora and Fauna
CFCs Chlorofluorocarbons
CNS Central nervous system
CoA Coenzyme A
COPD Chronic obstructive pulmonary disease
CP Creatine phosphate
CR Conditioned response
CS Conditioned stimulus
CPR Cardiopulmonary resuscitation

CSF Cerebrospinal fluid
CVS Cardiovascular system
DAG Diacylglycerol
DNA Deoxyribonucleic acid
E$_A$ Activation energy (of an enzyme)
ECG Electrocardiogram
ECM Extracellular matrix
EEG Electroencephalogram
EKG Electrocardiogram
EM Electron microscope or micrograph
ENSO El Niño–Southern Oscillation
EPSP Excitatory postsynaptic potential (of a neuron)
ER Endoplasmic reticulum
F$_1$ First filial generation
F$_2$ Second filial generation
Fab portion The part of an antibody that binds to an antigen
Factor VIII Blood clotting factor (absent in hemophiliacs)
FAD/FADH$_2$ Flavin adenine dinucleotide (oxidized and reduced forms, respectively)
FAP Fixed action pattern
Fc portion The part of an antibody that interacts with cells of the immune system
FSH Follicle-stimulating hormone
G Guanine
G$_1$ phase First gap phase (of the cell cycle)
G$_2$ phase Second gap phase (of the cell cycle)
G3P Glyceraldehyde-3-phosphate
G protein Cell signaling molecule that requires GTP
GA$_3$ Gibberellin
GABA Gaba-aminobutyric acid
GH Growth hormone (somatotropin)
GnRH Gonadotropin-releasing hormone
GTP Guanosine triphosphate
HCFCs Hydrochlorofluorocarbons
hCG Human chorionic gonadotropin
HDL High density lipoprotein
HFCs Hydrofluorocarbons
hGH Human growth hormone
HIV Human immunodeficiency virus
HLA Human leukocyte antigen
IAA Indole acetic acid (natural auxin)
Ig Immunoglobulin, as in IgA, IgG, etc.
IGF Insulin-like growth factor
IP$_3$ Inositol triphosphate
IPCC United Nations Intergovernmental Panel on Climate Change
IPSP Inhibitory postsynaptic potential
IUCN World Conservation Union

IUD Intrauterine device
JH Juvenile hormone (of insects)
kb Kilobase
LDH Lactic dehydrogenase enzyme
LDL Low density lipoprotein
LH Luteinizing hormone
LM Light microscope or micrograph
LSD Lysergic acid diethylamide
LTP Long-term potentiation
MAO Monoamine oxidase
MAPs Microtubule-associated proteins
MHC Major histocompatibility complex
MI Myocardial infarction
MPF Mitosis-promoting factor
MRI Magnetic resonance imaging
mRNA Messenger RNA
mtDNA Mitochondrial DNA
MTOC Microtubule organizing center
9 + 2 structure Cilium or flagellum (of a eukaryote)
9 × 3 structure Centriole or basal body (of a eukaryote)
n, 2n The chromosome number of a gamete and of a zygote, respectively
NAD$^+$/NADH Nicotinamide adenine dinucleotide (oxidized and reduced forms, respectively)
NADP$^+$/NADPH Nicotinamide adenine dinucleotide phosphate (oxidized and reduced forms, respectively)
NK cell Natural killer cell
P generation Parental generation
P53 A tumor suppressor gene
P680 Reaction center of photosystem II
P700 Reaction center of photosystem I
PABA Para-aminobenzoic acid
PCR Polymerase chain reaction
PEP Phosphoenolpyruvate
P$_{fr}$ Phytochrome (form that absorbs far red light)
PGA Phosphoglycerate
PID Pelvic inflammatory disease

PKU Phenylketonuria
PNS Peripheral nervous system
pre-mRNA Precursor messenger RNA (in eukaryotes)
P$_r$ Phytochrome (form that absorbs red light)
PTH Parathyroid hormone
RAS Reticular activating system
RBC Red blood cell (erythrocyte)
REM sleep Rapid eye movement sleep
RFLP Restriction fragment length polymorphism
RNA Ribonucleic acid
rRNA Ribosomal RNA
Rubisco Ribulose bisphosphate carboxylase/oxygenase
RuBP Ribulose bisphosphate
S phase DNA synthetic phase (of the cell cycle)
SA node Sinoatrial node (of heart)
SCID Severe combined immune deficiency
SEM Scanning electron microscope or micrograph
snRNP Small nuclear ribonucleoprotein complex
STD Sexually transmitted disease
T Thymine
T lymphocyte or T cell Lymphocyte responsible for cell-mediated immunity
T$_c$ lymphocyte Cytotoxic T cell
T$_h$ lymphocyte Helper T cell
TATA box Base sequence in eukaryotic promoter
TCA cycle Tricarboxylic acid cycle (synonym for citric acid cycle)
TCR T cell antigen receptor
TEM Transmission electron microscope or micrograph
Tm Tubular transport maximum
TNF Tumor necrosis factor
tRNA Transfer RNA
U Uracil
UPE Upstream promoter element
UR Unconditioned response
US Unconditioned stimulus
UV light Ultraviolet light
WBC White blood cell (leukocyte)

Glossary

abiotic factors   Elements of the nonliving, physical environment that affect a particular organism. Compare with *biotic factors.*

abscisic acid (ab-sis´ik)   A plant hormone involved in dormancy and responses to stress.

abscission (ab-sizh´en)   The normal (usually seasonal) fall of leaves or other plant parts, such as fruits or flowers.

abscission layer   The area at the base of the petiole where the leaf will break away from the stem. Also known as abscission zone.

absorption (ab-sorp´shun)   The movement of nutrients and other substances through the wall of the digestive tract and into the blood or lymph.

absorption spectrum   A graph of the amount of light at specific wavelengths that has been absorbed as light passes through a substance. Each type of molecule has a characteristic absorption spectrum. Compare with *action spectrum.*

accessory fruit   A fruit composed primarily of tissue other than ovary tissue, e.g., apple, pear. Compare with *aggregate, simple,* and *multiple fruits.*

acetyl coenzyme A (acetyl CoA) (as´uh-teel)   A key intermediate compound in metabolism; consists of a two-carbon acetyl group covalently bonded to coenzyme A.

acetyl group   A two-carbon group derived from acetic acid (acetate).

acetylcholine (ah″see-til-koh´leen)   A common neurotransmitter released by cholinergic neurons, including motor neurons.

achene (a-keen´)   A simple, dry fruit with one seed in which the fruit wall is separate from the seed coat, e.g., sunflower fruit.

acid   A substance that is a hydrogen ion (proton) donor; acids unite with bases to form salts. Compare with *base.*

acidic solution   A solution in which the concentration of hydrogen ions (H^+) exceeds the concentration of hydroxide ions (OH^-). An acidic solution has a pH less than 7. Compare with *basic solution* and *neutral solution.*

acid precipitation   Precipitation that is acidic as a result of both sulfur and nitrogen oxides forming acids when they react with water in the atmosphere.

acoelomate (a-seel´oh-mate)   Animal lacking a body cavity (coelom). Compare with *coelomate* and *pseudocoelomate.*

acquired immune deficiency syndrome (AIDS)   A serious, potentially fatal disease caused by the human immunodeficiency virus (HIV).

acromegaly (ak″roh-meg´ah-lee)   A condition characterized by overgrowth of the extremities of the skeleton, fingers, toes, jaws, and nose. It may be produced by excessive secretion of growth hormone by the anterior pituitary gland.

acrosome reaction (ak´roh-sohm)   A series of events in which the acrosome, a caplike structure covering the head of a sperm cell, releases proteolytic (protein-digesting) enzymes and undergoes other changes that permit the sperm to penetrate the outer coverings of the egg.

actin (ak´tin)   The protein of which microfilaments are composed. Actin, together with the protein myosin, is responsible for muscle contraction.

actinopods (ak-tin´o-podz)   Protozoa characterized by axopods that protrude through pores in their shells.

action potential   The brief change in electrical activity developed across the plasma membrane of a muscle or nerve cell during activity; a neural impulse.

action spectrum   A graph of the effectiveness of light at specific wavelengths in promoting a light-requiring reaction. Compare with *absorption spectrum.*

activation energy (E_A)   The kinetic energy required to initiate a chemical reaction.

activator protein   A transcription factor that acts as a positive regulator, stimulating transcription when bound to DNA. Compare with *repressor protein.*

active site   Specific region of an enzyme (generally near the surface) that accepts one or more substrates and catalyzes a chemical reaction. Compare with *allosteric site.*

active transport   All forms of transport of a substance across a membrane that do not rely on the potential energy of a concentration gradient for the substance being transported, and therefore require an additional energy source (often ATP); includes carrier-mediated active transport, endocytosis, and exocytosis. Compare with *diffusion* and *facilitated diffusion.*

adaptation   (1) Evolutionary modification that improves an organism's chances of survival and reproductive success; (2) A decline in the response of a receptor subjected to repeated or prolonged stimulation.

adaptive radiation   The evolution of a large number of related species from an unspecialized ancestral organism.

adaptive zone   A new ecological opportunity that was not exploited by an ancestral organism; used by evolutionary biologists to explain the ecological paths along which different taxa evolve.

addiction   Physical dependence on a drug, generally based on physiological changes that take place in response to the drug; when the drug is withheld, the addict may suffer characteristic withdrawal symptoms.

adenine (ad´eh-neen)   A nitrogenous purine base that is a component of nucleic acids and ATP.

adenosine triphosphate (ATP) (a-den′oh-seen)   An organic compound containing adenine, ribose, and three phosphate groups; of prime importance for energy transfers in cells.

adipose tissue (ad′i-pohs)   Tissue in which fat is stored.

adrenal cortex (ah-dree′-nul kor′teks)   The outer region of each adrenal gland; secretes steroid hormones, including mineralocorticoids and glucocorticoids.

adrenal glands (ah-dree′nul)   Paired endocrine glands, one located just superior to each kidney; secrete hormones that help regulate metabolism and help the body cope with stress.

adrenal medulla (ah-dree′nul meh-dull′uh)   The inner region of each adrenal gland; secretes epinephrine and norepinephrine.

adrenergic neuron (ad-ren-er′jik)   A neuron that releases norepinephrine or epinephrine as a neurotransmitter. Compare with *cholinergic neuron.*

adventitious root (ad″ven-tish′us)   A root that arises in an unusual position on a plant, such as from a stem or leaf.

aerobe   Organism that grows or metabolizes only in the presence of molecular oxygen. Compare with *anaerobe.*

aerobic (air-oh′bik)   Growing or metabolizing only in the presence of molecular oxygen. Compare with *anaerobic.*

aerobic respiration   See *respiration.*

afferent (af′fer-ent)   Leading toward some point of reference. Compare with *efferent.*

age structure   The percentage of a population, including the number of individuals of each sex, at different ages. Age structure diagrams represent the number of males and females at each age in the population.

aggregate fruit   A fruit that develops from a single flower with many separate carpels, e.g., raspberry. Compare with *simple, accessory,* and *multiple fruits.*

aggregated distribution   See *clumped dispersion.*

agnathans (ag-na′thanz)   Jawless fishes; historical class of vertebrates, including lampreys, hagfishes, and many extinct forms.

albinism (al′bih-niz-em)   A hereditary inability to form melanin pigment, resulting in light coloration.

AIDS   See *acquired immune deficiency syndrome.*

albumin (al-bew′min)   A class of protein found in most animal tissues; a fraction of plasma proteins.

aldehyde   An organic molecule containing a carbonyl group bonded to at least one hydrogen atom. Compare with *ketone.*

aldosterone (al-dos′tur-ohn)   Steroid hormone produced by the vertebrate adrenal cortex; regulates excretion of sodium and other ions.

algae (al′gee) (sing. *alga*)   An informal group of unicellular, or simple multicellular, photosynthetic protists that are important producers in aquatic ecosystems; includes dinoflagellates, diatoms, euglenoids, golden algae, green algae, red algae, and brown algae.

allantois (a-lan′toe-iss)   One of the extraembryonic membranes of reptiles, birds, and mammals; most of the allantois is detached at hatching or birth.

allele frequency   The percentage of a specific allele out of the total of all alleles of a given locus in the population.

alleles (al-leels′)   Genes governing variation of the same character that occupy corresponding positions (loci) on homologous chromosomes; alternative forms of a gene.

allelopathy (uh-leel′uh-path″ee)   An adaptation in which toxic substances secreted by roots or shed leaves inhibit the establishment of competing plants nearby.

allergen   A substance that stimulates an allergic reaction.

allergy   A hypersensitivity to some substance in the environment, manifested as hay fever, skin rash, asthma, food allergies, etc.

all-or-none law   The principle that neurons transmit an impulse in a similar way no matter how weak or strong the stimulus is; the neuron either transmits an action potential (all) or does not (none).

allopatric speciation (al-oh-pa′trik)   Speciation that occurs when one population becomes geographically separated from the rest of the species and subsequently evolves. Compare with *sympatric speciation.*

allopolyploid (al″oh-pol′ee-ploid)   A polyploid whose chromosomes are derived from two different species.

allosteric regulators   Substances that affect protein function by binding to allosteric sites.

allosteric site (al-oh-steer′ik)   (1) A regulatory site located on a protein that is separate from the functional site. The binding of a specific regulator to the allosteric site alters the conformation and function of the protein; (2) A site on an enzyme other than the active site, to which a specific substance (other than the normal substrate) can bind, thereby changing the shape and activity of the enzyme.

alpha (α) helix   A type of secondary structure of a polypeptide chain consisting of a regular coiled structure, maintained by hydrogen bonds. Compare with *beta (β)-pleated sheet.*

alpine tundra   An ecosystem located in the higher elevations of mountains, above the tree line and below the snow line. Compare with *tundra.*

alternation of generations   A type of life cycle characteristic of plants and a few algae and fungi in which they spend part of their life in a multicellular *n* gametophyte stage and part in a multicellular 2*n* sporophyte stage. The gametophyte develops from a spore and produces gametes; the sporophyte develops from a zygote and produces spores.

altruistic behavior   Behavior in which one individual helps another, seemingly at its own risk or expense.

alveolus (al-vee′o-lus) (pl. *alveoli*)   (1) An air sac of the lung through which gas exchange with the blood takes place; (2) A saclike unit of some glands.

amino acid (uh-mee′no)   An organic compound containing an amino group ($-NH_2$) and a carboxyl group ($-COOH$); may be joined by peptide bonds to form the polypeptide chains of protein molecules.

amino group   A weakly basic functional group; abbreviated $-NH_2$.

aminoacyl-tRNA (uh-mee″no-as′seel)   Molecule consisting of an amino acid covalently linked to a transfer RNA.

aminoacyl-tRNA synthetase   One of a family of enzymes, each responsible for covalently linking an amino acid to its specific transfer RNA.

ammonification (uh-moe″nuh-fah-kay′shun)   The conversion of nitrogen-containing organic compounds to ammonia (NH_3) by certain soil bacteria (ammonifying bacteria); part of the nitrogen cycle.

amniocentesis (am″nee-oh-sen-tee′sis)   Sampling of the amniotic fluid surrounding a fetus in order to obtain information about its development and genetic makeup. Compare with *chorionic villus sampling.*

amnion (am′nee-on)   An extraembryonic membrane that forms a fluid-filled sac for the protection of the developing embryo.

amniotes Terrestrial vertebrates: reptiles, birds, and mammals; animals whose embryos are enclosed by an amnion.

amoeba (a-mee'ba) A unicellular protozoon that moves by means of pseudopodia.

amphibians Members of vertebrate class that includes salamanders, frogs, and caecilians.

amphipathic molecule (am"fih-pa'thik) Molecule containing both hydrophobic and hydrophilic regions.

ampulla Any small saclike extension, e.g., the expanded structure at the end of each semicircular canal of the ear.

amylase (am'-uh-laze) Starch-digesting enzyme, e.g., human salivary amylase or pancreatic amylase.

amyloplasts Colorless plastids (leukoplasts) specialized for the storage of large amounts of starch in cells of roots and tubers.

anabolic steroids Synthetic androgens that increase muscle mass, physical strength, endurance, and aggressiveness.

anabolism (an-ab'oh-lizm) The aspect of metabolism in which simpler substances are combined to form more complex substances, resulting in the storage of energy, the production of new cellular materials, and growth. Compare with *catabolism*.

anaerobe Organism that grows or metabolizes only in the absence of molecular oxygen. See *facultative anaerobe* and *obligate anaerobe*. Compare with *aerobe*.

anaerobic (an"air-oh'bik) Growing or metabolizing only in the absence of molecular oxygen. Compare with *aerobic*.

anaerobic respiration See *respiration*.

anaphase (an'uh-faze) The stage of mitosis, and of meiosis I and II, in which the chromosomes move to opposite poles of the cell; anaphase occurs after metaphase and before telophase.

anaphylaxis (an"uh-fih-lak'sis) An acute allergic reaction following sensitization to a foreign substance or other substance.

ancestral characters See *shared ancestral characters*.

androgen (an'dro-jen) Any substance that possesses masculinizing properties, such as a sex hormone. See *testosterone*.

anemia (uh-nee'mee-uh) A deficiency of hemoglobin or red blood cells.

aneuploidy (an'you-ploy-dee) Any chromosomal aberration in which there are either extra or missing copies of certain chromosomes.

angiosperms (an'jee-oh-spermz") The traditional name for flowering plants, a very large (about 235,000 species), diverse phylum of plants that form flowers for sexual reproduction and produce seeds enclosed in fruits; include monocots and dicots.

angiotensin (an-jee-o-ten'sin) A hormone formed by the action of renin and found in blood; stimulates aldosterone secretion by the renal cortex.

animal pole The non-yolky, metabolically active pole of a vertebrate or echinoderm egg. Compare with *vegetal pole*.

anion (an'eye-on) A particle with one or more units of negative charge, such as a chloride ion (Cl^-) or hydroxide ion (OH^-). Compare with *cation*.

anisogamy (an"eye-sog'uh-me) Sexual reproduction involving motile gametes of similar form but dissimilar size. Compare with *isogamy* and *oogamy*.

annelid (an'eh-lid) Member of phylum Annelida; segmented worm such as earthworm.

annual plant A plant that completes its entire life cycle in one year or less. Compare with *perennial* and *biennial*.

antenna complex The currently accepted arrangement of chlorophyll and accessory pigment molecules into units in the thylakoid membranes of photoautotrophic eukaryotes.

antennae (sing. *antenna*) Sensory structures characteristic of some arthropod groups.

anterior Toward the head end of a bilaterally symmetrical animal. Compare with *posterior*.

anther (an'thur) The part of the stamen in flowers that produces microspores and, ultimately, pollen grains.

antheridium (an"thur-id'ee-im) (pl. *antheridia*) In plants, the multicellular male gametangium (sex organ) that produces sperm cells. Compare with *archegonium*.

anthropoid (an'thra-poid) A member of a suborder of primates that includes monkeys, apes, and humans.

antibody (an-tih-bod'ee) A specific protein (immunoglobulin) that recognizes and binds to specific antigens; produced by plasma cells.

antibody-mediated immunity A type of specific defense mechanism in which B cells differentiate into plasma cells and produce antibodies that bind with foreign antigens, leading to the destruction of pathogens.

anticodon (an'ty-koh"don) A sequence of three nucleotides in transfer RNA that is complementary to, and combines with, the three nucleotide codon on messenger RNA, thus helping to specify the addition of a particular amino acid to the end of a growing polypeptide.

antidiuretic hormone (ADH) (an"ty-dy-uh-ret'ik) A hormone secreted by the posterior lobe of the pituitary that controls the rate of water reabsorption by the kidney.

antigen (an'tih-jen) Any molecule, usually a protein or large carbohydrate, that can be specifically recognized as foreign by cells of the immune system.

antigen-antibody complex The combination of antigen and antibody molecules.

anti-oncogene A gene (also known as a tumor suppressor gene) whose normal role is to block cell division in response to certain growth inhibiting factors; when mutated, may contribute to the formation of a cancer cell. Compare with *oncogene*.

anus (ay'nus) The distal end and outlet of the digestive tract.

aorta (ay-or'tah) The largest and main systemic artery of the vertebrate body; arises from the left ventricle and branches to distribute blood to all parts of the body except the lungs.

aphotic region (ay-fote'ik) The lower layer of the ocean (deeper than 100 meters or so) where light does not penetrate.

apical dominance (ape'ih-kl) The inhibition of lateral buds by a shoot tip.

apical meristem (mehr'ih-stem) An area of dividing tissue, located at the tip of a shoot or root, that gives rise to primary tissues; apical meristems cause an increase in the length of the plant body. Compare with *lateral meristem*.

apicomplexans A group of parasitic protozoa that lack structures for locomotion and that produce spores as infective agents; malaria is caused by an apicomplexan.

apoenzyme (ap"oh-en'zime) Protein portion of an enzyme; requires the presence of a specific coenzyme to become a complete functional enzyme.

apomixis (ap"uh-mix'us) A type of reproduction in which fruits and seeds are formed asexually.

apoplast A continuum consisting of the interconnected, porous plant cell walls. Compare with *symplast*.

apoptosis Programmed cell death; apoptosis is a normal part of development and maintenance.

arachnids (ah-rack'nids) Eight-legged arthropods such as spiders, scorpions, ticks, and mites.

archaebacteria (ar″kuh-bak-teer′ee-uh) Prokaryotic organisms with a number of features, such as the absence of peptidoglycan in their cell walls, that set them apart from the rest of the bacteria. Compare with *eubacteria*.

archaic *Homo sapiens* Regionally diverse descendants of *H. erectus* that lived in Africa, Asia, and Europe from about 800,000 to 100,000 years ago; considered by some paleoanthropologists to be a separate species, *H. heidelbergensis*.

archegonium (ar″ke-go′nee-um) (pl. *archegonia*) In plants, the multicellular female gametangium (sex organ) that contains an egg. Compare with *antheridium*.

archenteron (ar-ken′ter-on) The central cavity of the gastrula stage of embryonic development, which is lined with endoderm; primitive digestive system.

arctic tundra See *tundra*.

Ardipithecus ramidus The earliest known hominid; an australopithecine that lived about 4.4 million years ago. See *australopithecines*.

arterial pulse See *pulse, arterial*.

arteriole (ar-teer′ee-ole) A very small artery. Vasoconstriction and vasodilation of arterioles help regulate blood pressure.

artery A thick-walled blood vessel that carries blood away from a heart chamber and toward the body organs. Compare with *vein*.

arthropod (ar′throh-pod) Invertebrate that belongs to phylum Arthropoda; characterized by a hard exoskeleton, a segmented body, and paired, jointed appendages.

artificial insemination The impregnation of a female by artificially introducing sperm from a male.

artificial selection Selection by humans of traits that are desirable in plants or animals, and breeding only those individuals that possess the desired traits.

ascocarp (ass′koh-karp) The fruiting body of an ascomycete.

ascomycete (ass″koh-my′seat) Member of a phylum of fungi characterized by the production of nonmotile asexual conidia and sexual ascospores.

ascospore (ass′koh-spor) One of a set of sexual spores, usually eight, contained in a special spore case (an ascus) of an ascomycete.

ascus (ass′kus) A saclike spore case in ascomycetes that contains sexual spores called ascospores.

asexual reproduction Reproduction in which there is no fusion of gametes and in which the genetic makeup of parent and of offspring is usually identical. Compare with *sexual reproduction*.

assimilation (of nitrogen) The conversion of inorganic nitrogen (nitrate, NO_3^-, or ammonia, NH_3) to the organic molecules of living things; part of the nitrogen cycle.

association neuron See *interneuron*.

association areas Areas of the brain that link sensory and motor areas; responsible for thought, learning, memory, language abilities, judgment, and personality.

assortative mating Sexual reproduction in which individuals pair nonrandomly, i.e., select mates on the basis of phenotype.

asters Clusters of microtubules radiating out from the poles in dividing cells that have centrioles; present in most animal cells and in cells of certain plants, but not in the cells of flowering plants or of most gymnosperms.

atherosclerosis (ath″ur-oh-skle-row′sis) A progressive disease in which lipid deposits accumulate in the inner lining of arteries, leading eventually to impaired circulation and heart disease.

atom The smallest quantity of an element that can retain the chemical properties of that element.

atomic mass The total number of protons and neutrons in an atom; expressed in atomic mass units or daltons.

atomic mass unit (amu) The approximate mass of a proton or neutron; also called dalton.

atomic number The number of protons in the atomic nucleus of an atom, which uniquely identifies the element to which the atom corresponds.

ATP See *adenosine triphosphate*.

ATP synthase Large enzyme complex that catalyzes the formation of ATP from ADP and inorganic phosphate by chemiosmosis; contains a transmembrane channel through which protons diffuse down a concentration gradient; located in the inner mitochondrial membrane, the thylakoid membrane of chloroplasts, and the plasma membrane of bacteria.

atrial natriuretic peptide (ANP) A hormone released by the atrium of the heart; helps regulate sodium excretion and lowers blood pressure.

atrioventricular (AV) node (ay″tree-oh-ven-trik′you-lur) Mass of specialized cardiac tissue that receives an impulse from the sinoatrial node (pacemaker) and conducts it to the ventricles.

atrioventricular (AV) valve (of the heart) A valve between each atrium and its ventricle that prevents backflow of blood. The right AV valve is the tricuspid valve, the left AV valve is the mitral valve.

atrium (of the heart) (ay′tree-um) A heart chamber that receives blood from the veins.

australopithecines Early hominids that lived between about 4.4 and 1.25 million years ago, based on fossil evidence. Includes several species in two genera, *Ardipithecus* and *Australopithecus*.

Australopithecus afarensis Hominids that lived between about 3.6 and 3.0 million years ago, e.g., Lucy, discovered at Hadar, Ethiopia, in 1974. *A. afarensis* may have arisen from *A. anamensis*.

Australopithecus africanus Hominids that lived between about 3.0 and 2.5 million years ago. *A. africanus* may have arisen from *A. afarensis*.

Australopithecus anamensis Hominids that lived between about 3.9 and 4.2 million years ago. May have arisen from *Ardipithecus ramidus;* had an upright posture and was bipedal.

autocrine regulation A type of regulation in which a signal molecule (e.g., a hormone) is secreted into interstitial fluid and then acts on the cells that produce it. Compare with *paracrine regulation*.

autogenous model The idea that eukaryotes arose from prokaryotes by the proliferation of internal membranes, derived from the plasma membrane, to form cellular compartments. Compare with *endosymbiont theory*.

autoimmune disease (aw″toh-ih-mune′) A disease in which the body produces antibodies against its own cells or tissues.

autonomic nervous system (aw-tuh-nom′ik) The portion of the peripheral nervous system that controls the visceral functions of the body, e.g., regulates smooth muscle, cardiac muscle, and glands, thereby helping to maintain homeostasis. Its divisions are the sympathetic and parasympathetic nervous systems. Compare with *somatic nervous system*.

autoradiography Method for detection of radioactive decay; radiation causes the appearance of dark silver grains in special x-ray film.

autosome (aw′toh-sohm) A chromosome other than the sex (X and Y) chromosomes.

autotroph (aw'toh-trof)   Organism that synthesizes complex organic compounds from simple inorganic raw materials; also called producer or primary producer. Compare with *heterotroph*. See *chemoautotroph* and *photoautotroph*.

auxin (awk'sin)   A plant hormone involved in various aspects of growth and development, such as stem elongation, apical dominance, and root formation on cuttings, e.g., indole acetic acid (IAA).

avirulent   Unable to cause disease in a host. Compare with *virulent*.

Avogadro's number   The number of units (6.02×10^{23}) present in one mole of any substance.

axillary bud   A bud in the axil of a leaf. Compare with *terminal bud*.

axon (aks'on)   The long extension of the neuron that transmits nerve impulses away from the cell body. Compare with *dendrite*.

axopods (aks'o-podz)   Long, filamentous cytoplasmic projections characteristic of actinopods.

B cell (B lymphocyte)   A type of white blood cell responsible for antibody-mediated immunity. When stimulated, B cells differentiate to become plasma cells that produce antibodies. Compare with *T cell*.

bacillus (bah-sill'us) (pl. *bacilli*)   Rod-shaped bacterium. Compare with *coccus, spirillum, vibrio,* and *spirochete*.

background extinction   The continuous, low-level extinction of species that has occurred throughout much of the history of life. Compare with *mass extinction*.

bacteria (bak-teer'ee-uh)   General term for two groups of unicellular, prokaryotic microorganisms, the archaebacteria and eubacteria. Most bacteria are decomposers, but some are parasites and others are autotrophs.

bacteriophage (bak-teer'ee-oh-fayj)   Virus that can infect a bacterium (literally, "bacteria eater"). Also called phage.

balanced polymorphism (pol"ee-mor'fizm)   The presence in a population of two or more genetic variants that are maintained in a stable frequency over several generations.

bark   The outermost covering over woody stems and roots; consists of all plant tissues located outside the vascular cambium.

baroreceptors (bare"oh-ree-sep'torz)   Receptors within certain blood vessels that are stimulated by changes in blood pressure.

Barr body   A condensed and inactivated X-chromosome appearing as a distinctive dense spot in the nucleus of certain cells of female mammals.

basal body (bay'sl)   Structure involved in the organization and anchorage of a cilium or flagellum. Structurally similar to a centriole; each is in the form of a cylinder composed of nine triplets of microtubules (9×3 structure).

basal metabolic rate (BMR)   The amount of energy expended by the body at resting conditions, when no food is being digested and no voluntary muscular work is being performed.

base   A substance that is a hydrogen ion (proton) acceptor; bases unite with acids to form salts. Compare with *acid*.

base (in nucleotide)   See *purines* and *pyrimidines*.

basic solution   A solution in which the concentration of hydroxide ions (OH^-) exceeds the concentration of hydrogen ions (H^+). A basic solution has pH greater than 7. Compare with *acidic solution* and *neutral solution*.

base-substitution mutation   A change in one base pair in DNA. See *missense mutation* and *nonsense mutation*.

basidiocarp (ba-sid'e-o-karp)   The fruiting body of a basidiomycete, e.g., a mushroom.

basidiomycete (ba-sid"e-o-my'seat)   Member of a phylum of fungi characterized by the production of sexual basidiospores.

basidiospore (ba-sid'e-o-spor)   One of a set of sexual spores, usually four, borne on a basidium of a basidiomycete.

basidium (ba-sid'ee-um)   The clublike spore-producing organ of basidiomycetes that bears sexual spores called basidiospores.

basilar membrane   The multicellular tissue in the inner ear that separates the cochlear duct from the tympanic canal; the sensory cells of the organ of Corti rest on this membrane.

Batesian mimicry (bate'see-un mim'ih-kree)   The resemblance of a harmless or palatable species to one that is dangerous, unpalatable, or poisonous. Compare with *Müllerian mimicry*.

behavioral ecology   The scientific study of behavior in natural environments from the evolutionary perspective.

behavioral isolation   A reproductive isolating mechanism (prezygotic) in which reproduction between similar species is prevented because each group possesses its own characteristic courtship behavior; also called sexual isolation.

bellwether species   An organism that provides an early warning of environmental damage. Examples include lichens, which are very sensitive to air pollution, and amphibians, which are sensitive to a wide variety of environmental stressors.

benthos (ben'thos)   Bottom-dwelling sea organisms that fix themselves to one spot, burrow into the sediment, or simply walk about on the ocean floor.

berry   A simple, fleshy fruit in which the fruit wall is soft throughout, e.g., tomato, banana, grape.

beta (β) oxidation   Process by which fatty acids are converted to acetyl CoA before entry into the citric acid cycle.

beta (β)-pleated sheet   A regular, folded, sheetlike structure resulting from hydrogen bonding between two different polypeptide chains or two regions of the same polypeptide chain. Compare with *alpha (α) helix*.

biennial plant (by-en'ee-ul)   A plant that takes two years to complete its life cycle. Compare with *annual* and *perennial*.

bilateral symmetry   A body shape with right and left halves that are approximately mirror images of one another. Compare with *radial symmetry*.

bile   The fluid secreted by the liver; emulsifies fats.

binary fission (by'nare-ee fish'un)   Equal division of a cell or organism into two; a type of asexual reproduction.

binomial system of nomenclature (by-nome'ee-ul)   System of naming a species by the combination of the genus name and a specific epithet.

bioaccumulation   The buildup of a persistent toxic substance, such as certain pesticides, in an organism's body.

biodiversity   See *biological diversity*.

biogenic amines   A class of neurotransmitters that includes norepinephrine, serotonin, and dopamine.

biogeochemical cycle (bye"o-jee"o-kem'e-kl)   Process by which matter cycles from the living world to the nonliving, physical environment and back again, e.g., the carbon cycle, the nitrogen cycle, and the phosphorus cycle.

biogeography   The study of the past and present geographical distributions of organisms.

biological clocks   Mechanisms by which activities of organisms are adapted to regularly recurring changes in the environment.

biological diversity   The number and variety of living organisms; biological diversity is considered at all levels, including genetic diversity within a species, species diversity, and ecosystem diversity. Also called biodiversity.

biological magnification   The increased concentration of toxic chemicals, such as PCBs, heavy metals, and certain pesticides, in the tissues of organisms at higher trophic levels in food webs.

biological species concept   See *species*.

biomass (bye'o-mas)   A quantitative estimate of the total mass, or amount, of living material in a particular ecosystem.

biome (by'ohm)   A large, relatively distinct terrestrial region characterized by a similar climate, soil, plants, and animals, regardless of where it occurs on Earth.

biosphere   All of Earth's living organisms.

biotic factors   Elements of the living world that affect a particular organism, that is, its relationships with other organisms. Compare with *abiotic factors*.

biotic potential   The maximum rate at which an organism or population could increase when environmental conditions are optimal.

bipedal   Walking on two feet.

biramous appendages   Appendages with two jointed branches at their ends; characteristic of crustaceans.

bivalent (by-vale'ent or biv'ah-lent)   See *tetrad*.

blade   (1) The thin, expanded part of a leaf; (2) The flat, leaflike structure of certain multicellular algae.

blastocoel (blas'toh-seel)   The fluid-filled cavity of a blastula.

blastocyst   See *blastula*.

blastodisc   Small disc of cytoplasm at the animal pole of a reptile or bird egg; cleavage is restricted to the blastodisc (meroblastic cleavage).

blastopore (blas'toh-pore)   Primitive opening into the body cavity of an early embryo that may become the mouth (in protostomes) or anus (in deuterostomes) of the adult organism.

blastula (blas'tew-lah)   In animal development, a hollow ball of cells produced by cleavage of a fertilized ovum. Known as a blastocyst in mammalian development.

blood   A fluid, circulating connective tissue that transports nutrients and other materials through the body of many types of animals. In vertebrates, consists of red blood cells, white blood cells, and platelets suspended in a fluid component called plasma.

blood pressure   The force exerted by blood against the inner walls of the blood vessels.

Bohr effect   Increased oxyhemoglobin dissociation due to lowered pH; occurs as carbon dioxide concentration increases.

bolting   Production of a tall flower stalk by a plant that grows vegetatively as a rosette (growth habit with a short stem and a circular cluster of leaves).

bond energy   The energy required to break a particular chemical bond.

bone tissue   Principal vertebrate skeletal tissue; a type of connective tissue.

boreal forest (bor'ee-uhl)   See *taiga*.

bottleneck   A sudden decrease in a population size due to adverse environmental factors; may result in genetic drift; also called genetic bottleneck or population bottleneck.

Bowman's capsule   Double-walled sac of cells that surrounds the glomerulus of each nephron.

brachiopods (bray'kee-oh-pods)   Phylum of marine invertebrate deuterostomes possessing a pair of shells, and internally, a pair of coiled arms with ciliated tentacles; one of the lophophorate phyla.

brain   A concentration of nervous tissue that controls neural function; in vertebrates, the anterior, enlarged portion of the central nervous system.

brain stem   The part of the vertebrate brain that includes the medulla, pons, midbrain, thalamus, and hypothalamus.

branchial   Pertaining to the gills or gill region.

bronchiole (bronk'ee-ole)   Air duct in the lung that branches from a bronchus; divides to form air sacs (alveoli).

bronchus (bronk'us) (pl. *bronchi*)   One of the branches of the trachea and its immediate branches within the lung.

brown alga   One of a phylum of predominantly marine algae that are multicellular and contain the pigments chlorophyll *a* and *c,* and carotenoids, including fucoxanthin.

bryophytes (bry'oh-fites)   Nonvascular plants including mosses, liverworts, and hornworts.

bud   An undeveloped shoot that can develop into flowers, stems, or leaves. Buds are enclosed in bud scales.

bud scale   A modified leaf that covers and protects a dormant bud.

bud scale scar   Scar on a twig left when a bud scale abscises from the terminal bud.

budding   Asexual reproduction in which a small part of the parent's body separates from the rest and develops into a new individual. Characteristic of yeasts and certain other organisms.

buffer   A substance in a solution that tends to lessen the change in hydrogen ion concentration (pH) that otherwise would be produced by adding an acid or base.

bulb   A globose, fleshy, underground bud that consists of a short stem with fleshy leaves, e.g., onion.

bundle scar   Marks on a leaf scar left when vascular bundles of the petiole break during leaf abscission.

bundle sheath cells   Tightly packed cells that form a sheath around the veins of a leaf.

bundle sheath extension   Support cells that extend from the bundle sheath of a leaf vein toward the upper and/or lower epidermis.

C_3 plant   Plant that carries out carbon fixation solely by the Calvin cycle. Compare with *C_4 plant* and *CAM plant*.

C_4 plant   Plant that fixes carbon initially by the Hatch-Slack pathway, in which the reaction of CO_2 with phosphoenolpyruvate is catalyzed by PEP carboxylase in leaf mesophyll cells; the products are transferred to the bundle sheath cells, where the Calvin cycle takes place. Compare with *C_3 plant* and *CAM plant*.

calcitonin (kal-sih-toh'nin)   A hormone secreted by the thyroid gland that rapidly lowers the calcium content in the blood.

callus (kal'us)   Undifferentiated tissue formed on an explant (excised tissue or organ) in plant tissue culture.

calmodulin   A calcium-binding protein; when bound it alters the activity of certain enzymes or transport proteins.

calorie   The amount of heat energy required to raise the temperature of 1 g of water 1°C; equivalent to 4.184 joules. Compare with *kilocalorie*.

Calvin cycle   Cyclic series of reactions in the chloroplast stroma in photosynthesis; fixes carbon dioxide and produces carbohydrate. See *C_3 plant*.

calyx (kay'liks)   The collective term for the sepals of a flower.

cambium   See *lateral meristems*.

Cambrian explosion   The earliest geological period of the Paleozoic era, from about 570 to 505 million years ago, during which many new animal groups appeared in the fossil record.

CAM plant Plant that carries out crassulacean acid metabolism; carbon is initially fixed into organic acids at night in the reaction of CO_2 and phosphoenolpyruvate, catalyzed by PEP carboxylase; during the day the acids break down to yield CO_2, which enters the Calvin cycle. Compare with *C_3 plant* and *C_4 plant*.

cAMP See *cyclic AMP*.

cancer cells See *malignant*.

CAP See *catabolite gene activator protein*.

capillaries (kap′i-lare-eez) Microscopic blood vessels in the tissues that permit exchange of materials between cells and blood.

capillary action The ability of water to move in small diameter tubes as a consequence of its cohesive and adhesive properties.

capping See *mRNA cap*.

capsid Protein coat surrounding the nucleic acid of a virus.

capsule (1) The portion of the moss sporophyte that contains spores; (2) A simple, dry, dehiscent fruit that opens along many sutures or pores to release seeds; (3) A gelatinous coat that surrounds some bacteria.

carbohydrate Compound containing carbon, hydrogen, and oxygen, in the approximate ratio of $C:2H:O$, e.g., sugars, starch, and cellulose.

carbon cycle The worldwide circulation of carbon from the abiotic environment into living things and back into the abiotic environment.

carbon fixation reactions Reduction reactions of photosynthesis in which carbon from carbon dioxide becomes incorporated into organic molecules, leading to the production of carbohydrate; requires ATP and NADPH.

carbonyl group A polar functional group consisting of a carbon attached to an oxygen by a double bond; found in aldehydes and ketones.

carboxyl group A weakly acidic functional group; abbreviated —COOH.

carcinogen (kar-sin′oh-jen) An agent that causes cancer or accelerates its development.

cardiac cycle One complete heart beat.

cardiac muscle Involuntary, striated type of muscle found in the vertebrate heart. Compare with *smooth muscle* and *skeletal muscle*.

cardiac output The volume of blood pumped by the heart per unit of time.

cardiovascular disease Disease of the heart or blood vessels; the leading cause of death in most industrial societies.

carnivore (kar′ni-vor) An animal that feeds on other animals; flesh-eater; also called secondary consumer.

carotenoids (ka-rot′n-oidz) A group of yellow to orange plant pigments synthesized from isoprene subunits; include carotenes and xanthophylls.

carpel (kar′pul) The female reproductive unit of a flower; carpels bear ovules. Compare with *pistil*.

carrier-mediated active transport Transport across a membrane of a substance from a region of low concentration to a region of high concentration; requires both a transport protein with a binding site for the specific substance and an energy source (often ATP).

carrier-mediated transport Any form of transport across a membrane that uses a membrane-bound transport protein with a binding site for a specific substance; includes both facilitated diffusion and carrier-mediated active transport.

carrying capacity The largest population that a particular habitat can support and sustain for an indefinite period of time, assuming there are no changes in the environment.

cartilage Flexible skeletal tissue of vertebrates; a type of connective tissue.

Casparian strip (kas-pare′ee-un) A band of waterproof material around the radial and transverse walls of endodermal root cells.

catabolism The aspect of metabolism in which complex substances are broken down to form simpler substances; catabolic reactions are particularly important in releasing chemical energy stored by the cell. Compare with *anabolism*.

catabolite gene activator protein (CAP) A positively acting regulator that becomes active when bound to cAMP; active CAP stimulates transcription of the lactose operon and other operons that code for enzymes used in catabolic pathways. Also known as cyclic AMP receptor protein (CRP).

catalyst (kat′ah-list) A substance that increases the speed at which a chemical reaction occurs without being used up in the reaction. Enzymes are biological catalysts.

cation A particle with one or more units of positive charge, such as a hydrogen ion (H^+) or calcium ion (Ca^{2+}). Compare with *anion*.

cDNA library A collection of recombinant plasmids that contain complementary DNA (cDNA) copies of mRNA templates. The cDNA, which lacks introns, is synthesized by reverse transcriptase. Compare with *genomic DNA library*.

cell The basic structural and functional unit of life, which consists of living material bounded by a membrane.

cell cycle Cyclic series of events in the life of a dividing eukaryotic cell; consists of mitosis, cytokinesis, and the stages of interphase. The time required to complete one cell cycle is the generation time.

cell determination See *determination*.

cell differentiation See *differentiation*.

cell fractionation Technique used to separate the components of cells by subjecting them to centrifugal force. See *differential centrifugation* and *density gradient centrifugation*.

cell plate Structure that forms during cytokinesis in plants, separating the two daughter cells produced by mitosis.

cell signaling Mechanisms of communication between cells. Cells can signal one another with secreted signaling molecules, or a signaling molecule on one cell can combine with a receptor on another cell. Examples include the synaptic signaling of neurons and endocrine signaling. See *signal transduction*.

cell theory The theory that the cell is the basic unit of life, of which all living things are composed, and that all cells are derived from preexisting cells.

cell wall Structure outside the plasma membrane of certain cells; may contain cellulose (plant cells), chitin (most fungal cells), peptidoglycan and/or lipopolysaccharide (most bacterial cells), or other material.

cellular respiration See *respiration*.

cellular slime mold A phylum of fungus-like protists whose feeding stage consists of unicellular, amoeboid organisms that aggregate to form a pseudoplasmodium during reproduction.

cellulose (sel′yoo-lohs) A structural polysaccharide composed of beta glucose subunits; the main constituent of plant primary cell walls.

Cenozoic era A geological era that began 65 million years ago and extends to the present time.

center of origin   The geographical area where a given species originated.

central nervous system (CNS)   In vertebrates, the brain and spinal cord. Compare with *peripheral nervous system (PNS)*.

centrifuge   Device used to separate cells or their components by subjecting them to centrifugal force.

centriole (sen′tree-ohl)   One of a pair of small, cylindrical organelles lying at right angles to each other near the nucleus in the cytoplasm of animal cells and certain protist and plant cells; each centriole is in the form of a cylinder composed of nine triplets of microtubules (9×3 structure).

centromere (sen′tro-meer)   Specialized constricted region of a chromatid; contains the kinetochore. In cells at prophase and metaphase, sister chromatids are joined in the vicinity of their centromeres.

cephalization   The evolution of a head; the concentration of nervous tissue and sense organs at the front end of the animal.

cephalochordates   Members of the chordate subphylum that includes the lancelets.

cerebellum (ser-eh-bel′um)   A convoluted subdivision of the vertebrate brain concerned with the coordination of muscular movements, muscle tone, and balance.

cerebral cortex (ser-ee′brul kor′tex)   The outer layer of the cerebrum composed of gray matter and consisting mainly of nerve cell bodies.

cerebrospinal fluid (CSF)   The fluid that bathes the central nervous system of vertebrates.

cerebrum (ser-ee′brum)   Large, convoluted subdivision of the vertebrate brain; in humans, it functions as the center for learning, voluntary movement, and interpretation of sensation.

chaos   The tendency of a simple system to exhibit complex, erratic dynamics; used by some ecologists to explain the state of flux displayed by some populations.

chaparral (shap″uh-ral′)   A biome with a Mediterranean climate (mild, moist winters and hot, dry summers). Chaparral vegetation is characterized by drought-resistant, small-leaved evergreen shrubs and small trees.

chaperones   See *molecular chaperones*.

character displacement   The tendency for two similar species to diverge (become more different) in areas where their ranges overlap; reduces interspecific competition.

chelicerae (keh-lis′er-ee)   The first pair of appendages in certain arthropods; clawlike appendages located immediately anterior to the mouth and used to manipulate food into the mouth.

chemical bond   A force of attraction between atoms in a compound. See *covalent bond, hydrogen bond,* and *ionic bond*.

chemical compound   Two or more elements combined in a fixed ratio.

chemical equilibrium   See *dynamic equilibrium*.

chemical evolution   The origin of life from nonliving matter.

chemical formula   A representation of the composition of a compound; the elements are indicated by chemical symbols with subscripts to indicate their ratios. See *molecular formula* and *structural formula*.

chemical symbol   Abbreviation for an element; usually the first letter (or first and second letters) of the English or Latin name.

chemiosmosis   Process by which phosphorylation of ADP to form ATP is coupled to the transfer of electrons down an electron transport chain; the electron transport chain powers proton pumps that produce a proton gradient across the membrane; ATP is formed as protons diffuse through transmembrane channels in ATP synthase.

chemoautotroph (kee″moh-aw′toh-trof)   Organism that obtains energy from inorganic compounds and synthesizes organic compounds from inorganic raw materials; includes some bacteria. Compare with *photoautotroph*.

chemoheterotroph (kee″moh-het′ur-oh-trof)   Organism that uses organic compounds as a source of energy and carbon; includes animals, fungi, and many bacteria. Compare with *photoheterotroph*.

chemoreceptor (kee″moh-ree-sep′tor)   A sensory receptor that responds to chemical stimuli.

chiasma (ky-az′muh) (pl. *chiasmata*)   An X-shaped site in a tetrad (bivalent) usually marking the location where homologous (nonsister) chromatids previously underwent crossing-over.

chimera   Organism composed of two or more kinds of genetically dissimilar cells.

chitin (ky′tin)   A nitrogen-containing structural polysaccharide that forms the exoskeleton of insects and the cell walls of many fungi.

chlorophyll (klor′oh-fil)   A group of light-trapping green pigments found in most photosynthetic organisms.

chloroplasts (klor′oh-plastz)   Membranous organelles that are the sites of photosynthesis in eukaryotes; occur in some plant and algal cells.

cholinergic neuron (kohl″in-air′jik)   A nerve cell that secretes acetylcholine as a neurotransmitter. Compare with *adrenergic neuron*.

chondrichthyes   The class of cartilaginous fishes; includes the sharks, rays, and skates.

chondrocytes   Cartilage cells.

chordates (kor′dates)   Deuterostome animals that belong to phylum Chordata; at some time in their lives, possess a cartilaginous, dorsal skeletal structure called a notochord; a dorsal, tubular nerve cord; pharyngeal gill grooves; and a postanal tail.

chorion (kor′ee-on)   An extraembryonic membrane in reptiles, birds, and mammals that forms an outer cover around the embryo, and in mammals contributes to the formation of the placenta.

chorionic villus sampling (CVS) (kor″ee-on′ik)   Study of extraembryonic cells that are genetically identical to the cells of an embryo, making it possible to assess its genetic makeup. Compare with *amniocentesis*.

chromatid (kroh′mah-tid)   One of the two identical halves of a duplicated chromosome; the two chromatids that make up a chromosome are referred to as sister chromatids.

chromatin (kro′mah-tin)   The complex of DNA and protein that makes up eukaryotic chromosomes.

chromoplasts   Pigment-containing plastids; usually found in flowers and fruits.

chromosomes (kro′moh-soms)   Structures in the cell nucleus that are composed of chromatin and contain the genes. The chromosomes become visible under the microscope as distinct structures during cell division.

chylomicrons (kie-low-my′kronz)   Protein-covered fat droplets produced in the intestinal cells; they enter the lymphatic system and are transported to the blood.

chytrid   See *chytridiomycete*.

chytridiomycete (ki-trid″e-o-my′seat)   Member of a phylum of fungi characterized by the production of flagellated cells at some stage in their life history. Also called chytrid.

ciliate (sil′e-ate)   A unicellular protozoon covered by many short cilia.

cilium (sil'ee-um) (pl. *cilia*) One of many short, hairlike structures that project from the surface of some eukaryotic cells and are used for locomotion or movement of materials across the cell surface; structurally like eukaryotic flagella, including two central single microtubules surrounded by a cylinder of nine double microtubules (9 + 2 structure), covered by a plasma membrane.

circadian rhythm (sir-kay'dee-un) An internal rhythm that approximates the 24-hour day. Circadian rhythms are found in plants, animals, and eukaryotic microorganisms.

circulatory system The body system that functions in internal transport and protects the body from disease.

citrate (citric acid) A six-carbon organic acid.

citric acid cycle Series of chemical reactions in aerobic respiration in which acetyl coenzyme A is completely degraded to carbon dioxide and water with the release of metabolic energy that is used to produce ATP; also known as the Krebs cycle and the tricarboxylic acid (TCA) cycle.

clade A taxon containing a common ancestor and all the taxa descended from it; a monophyletic group.

cladistics An approach to classification based on recency of common ancestry rather than degree of structural similarity. Also called phylogenetic systematics. Compare with *phenetics* and *classical evolutionary taxonomy.*

cladogram A branching diagram that illustrates taxonomic relationships based on the principles of cladistics.

class A taxonomic category made up of related orders.

classical conditioning A type of learning in which an association is formed between some normal response to a stimulus and a new stimulus, after which the new stimulus elicits the response.

classical evolutionary taxonomy An approach to classification that considers both evolutionary relationships and the extent of divergence that has occurred since a group branched from an ancestral group. Compare with *cladistics* and *phenetics.*

cleavage Series of mitotic cell divisions, without growth, that converts the zygote to a multicellular blastula.

cline Gradual change in phenotype and genotype frequencies among contiguous populations that is the result of an environmental gradient.

clitoris (klit'o-ris) A small, erectile structure at the anterior part of the vulva in female mammals; homologous to the male penis.

cloaca (klow-a'ka) An exit chamber in some animals that receives digestive wastes and urine; may also serve as an exit for gametes.

clonal selection Lymphocyte activation in which a specific antigen causes activation, cell division, and differentiation only in cells that express receptors with which the antigen can bind.

clone (1) A population of cells descended by mitotic division from a single ancestral cell; (2) A population of genetically identical organisms asexually propagated from a single individual.

closed circulatory system A type of circulatory system in which the blood flows through a continuous circuit of blood vessels; characteristic of annelids, cephalopods, and vertebrates. Compare with *open circulatory system.*

closed system An entity that does not exchange energy or matter with its surroundings. Compare with *open system.*

club mosses A phylum of seedless vascular plants with a life cycle similar to ferns.

clumped dispersion The spatial distribution pattern of a population in which individuals are more concentrated in specific parts of the habitat. Also called aggregated distribution and patchiness. Compare with *random dispersion* and *uniform dispersion.*

cnidarians (ni-dah'ree-anz) Phylum of animals that have stinging cells called cnidocytes, two tissue layers, and radial symmetry; include hydras and jellyfish.

cnidocytes Stinging cells characteristic of cnidarians.

coated pit A depression in the plasma membrane, the cytosolic side of which is coated with the protein clathrin; important in receptor-mediated endocytosis.

cochlea (koke'lee-ah) The structure of the inner ear of mammals that contains the auditory receptors (organ of Corti).

coccus (kok'us) (pl. *cocci*) A bacterium with a spherical shape. Compare with *bacillus, spirillum, vibrio,* and *spirochete.*

codon (koh'don) A triplet of mRNA bases that specifies an amino acid, a start signal, or a signal to terminate the polypeptide.

codominance (koh"dom'in-ants) Condition in which two alleles of a locus are expressed in a heterozygote.

coelacanths A genus of lobe-finned fish that have survived to the present day.

coelom (see'lum) The main body cavity of most animals; a true coelom is lined with mesoderm. Compare with *pseudocoelom.*

coelomate (seel'oh-mate) Animal possessing a true coelom. Compare with *acoelomate* and *pseudocoelomate.*

coenocyte (see'no-site) An organism consisting of a multinucleate cell, i.e., the nuclei are not separated from one another by septa.

coenzyme (koh-en'zime) An organic cofactor for an enzyme; generally participates in the reaction by transferring some component, such as electrons or part of a substrate molecule.

coenzyme A (CoA) Organic cofactor responsible for transferring groups derived from organic acids.

coevolution The interdependent evolution of two or more species that occurs as a result of their interactions over a long period of time.

cofactor A nonprotein substance needed by an enzyme for normal activity; some cofactors are inorganic (usually metal ions); others are organic (coenzymes).

cohesive Having the property of sticking together.

colchicine A drug that blocks the division of eukaryotic cells by binding to tubulin subunits, which make up the microtubules that comprise the major component of the mitotic spindle.

coleoptile (kol-ee-op'tile) A protective sheath that encloses the young stem in certain monocots.

collagen (kol'ah-gen) Protein in connective tissues.

collecting duct A tube in the kidney that receives filtrate from several nephrons and conducts it to the renal pelvis.

collenchyma (kol-en'kih-mah) Living cells with moderately but unevenly thickened primary cell walls; collenchyma cells help support the herbaceous plant body.

colon (koe'lon) Portion of the large intestine between the cecum and rectum.

commensalism (kuh-men'sul-iz-m) A type of symbiosis in which one organism benefits and the other one is neither harmed nor helped. Compare with *mutualism* and *parasitism.*

community An association of different species living together in a defined habitat with some degree of interdependence. Compare with *ecosystem.*

compact bone Dense, hard bone tissue found mainly near the surfaces of a bone.

companion cell A cell in the phloem of flowering plants that is responsible for loading and unloading sugar into the sieve tube member for translocation.

competition The interaction among two or more individuals that attempt to use the same resource, such as food, water, sunlight, or living space. See *contest* and *scramble competition*. See *interspecific* and *intraspecific competition*.

competitive exclusion The concept that no two species with identical living requirements can occupy the same ecological niche indefinitely. Eventually, one species will be excluded by the other as a result of interspecific competition for a resource in limited supply.

competitive inhibitor A substance that binds to the active site of an enzyme, thus lowering the rate of the reaction catalyzed by the enzyme. Compare with *noncompetitive inhibitor*.

complement A group of proteins in blood and other body fluids that are activated by an antigen-antibody complex.

complementary DNA (cDNA) DNA synthesized by reverse transcriptase, using RNA as a template.

complete flower A flower that possesses all four parts: sepals, petals, stamens, and carpels. Compare with *incomplete flower*.

compound eye An eye, such as that of an insect, composed of many light-sensitive units called ommatidia.

concentration gradient A difference in the concentration of a substance from one point to another, as for example, across a cell membrane.

condensation synthesis Reaction in which two monomers are combined covalently through the removal of the equivalent of a water molecule. Compare with *hydrolysis*.

cone (1) In botany, a reproductive structure in many gymnosperms that produces either microspores or megaspores; (2) In zoology, one of the conical photoreceptive cells of the retina that is particularly sensitive to bright light and, by distinguishing light of various wavelengths, mediates color vision. Compare with *rod*.

conidiophore (kah-nid'e-o-for") A specialized hypha that bears conidia.

conidium (kah-nid'e-um) (pl. *conidia*) An asexual spore that is usually formed at the tip of a specialized hypha called a conidiophore.

conifer (kon'ih-fur) Any of a large phylum of gymnosperms that are woody trees and shrubs with needle-like, mostly evergreen, leaves and with seeds in cones.

conjugation (kon"jew-gay'shun) (1) A sexual phenomenon in certain protists that involves exchange or fusion of a cell with another cell; (2) A mechanism for DNA exchange in bacteria that involves cell-to-cell contact.

connective tissue Animal tissue consisting mostly of intercellular substance (fibers scattered through a matrix) in which the cells are embedded, e.g., bone.

consanguineous mating A mating between close relatives.

conservation biology The study and protection of biological diversity; includes in situ and ex situ conservation.

constitutive gene A gene that is constantly transcribed.

consumer See *heterotroph*.

consumption overpopulation A situation in which each individual in a human population consumes too large a share of resources; results in pollution, environmental degradation, and resource depletion. Compare with *people overpopulation*.

contest competition Intraspecific competition in which certain dominant individuals obtain an adequate supply of the limited resource at the expense of other individuals in the population. Compare with *scramble competition*.

continental drift The theory that continents were once joined together and later split and drifted apart.

contraception Any method used to intentionally prevent pregnancy.

contractile root (kun-trak'til) A specialized type of root that contracts and pulls a bulb or corm deeper into the soil.

contractile vacuole A membrane-bounded organelle that is found in certain freshwater protists, such as *Paramecium,* and that appears to have an osmoregulatory function; it periodically fills with water, then contracts to expel the contents into the surroundings.

control group In a scientific experiment, a group in which the experimental variable is kept constant. The control provides a standard of comparison used to verify the results of the experiment.

controlled mating A mating in which the genotypes of the parents are known.

convergent circuit (kun-vur'jent) A neural pathway in which a postsynaptic neuron is controlled by signals coming from two or more presynaptic neurons. Compare with *divergent circuit*.

convergent evolution (kun-vur'jent) The independent evolution of structural or functional similarity in two or more distantly related species, usually as a result of adaptations to similar environments.

corepressor Substance that binds to a repressor protein, converting it to its active form, which is capable of preventing transcription.

Coriolis effect (kor"e-o'lis) The tendency of moving air or water to be deflected from its path to the right in the Northern Hemisphere and to the left in the Southern Hemisphere. Caused by the direction of Earth's rotation.

cork cambium (kam'bee-um) A lateral meristem that produces cork cells and cork parenchyma; cork cambium and the tissues it produces make up the outer bark of a woody plant. Compare with *vascular cambium*.

cork cell A cell in the bark that is produced outwardly by the cork cambium; cork cells are dead at maturity and function for protection and reduction of water loss.

cork parenchyma (par-en'kih-mah) One or more layers of parenchyma cells produced inwardly by the cork cambium.

corm A short, thickened underground stem specialized for food storage and asexual reproduction, e.g., crocus, gladiolus.

cornea (kor'nee-ah) Transparent covering of an eye.

corolla (kor-ohl'ah) A collective term for the petals of a flower.

corpus callosum (kah-loh'sum) In mammals, a large bundle of nerve fibers interconnecting the two cerebral hemispheres.

corpus luteum (loo'tee"um) The temporary endocrine tissue in the ovary that develops from the ruptured follicle after ovulation; secretes progesterone and estrogen.

cortex (kor'tex) (1) The outer part of an organ, such as the cortex of the kidney (compare with *medulla*); (2) The tissue between the epidermis and vascular tissue in the stems and roots of many herbaceous plants.

cortical reaction Process occurring after fertilization that prevents additional sperm from entering the egg; also known as the slow block to polyspermy.

cosmopolitan species Species that have a nearly worldwide distribution and occur on more than one continent or throughout much of the ocean. Compare with *endemic species*.

cotransport The active transport of a substance from a region of low concentration to a region of high concentration by coupling its transport to the transport of a substance down its concentration gradient.

cotyledon (kot"uh-lee'dun) The seed leaf of a plant embryo, which may contain food stored for germination.

cotylosaurs The first reptiles; also known as stem reptiles.

countercurrent exchange system A biological mechanism that enables maximum exchange between two fluids. The two fluids must be flowing in opposite directions and have a concentration gradient between them.

coupled reactions Set of reactions in which an exergonic reaction provides the free energy required to drive an endergonic reaction; energy coupling generally occurs through a common intermediate.

covalent bond Chemical bond involving shared pairs of electrons; may be single, double, or triple (with one, two, or three shared pairs of electrons, respectively). Compare with *ionic bond* and *hydrogen bond*.

covalent compound A compound in which atoms are held together by covalent bonds; covalent compounds consist of molecules. Compare with *ionic compound*.

cranial nerves Ten to twelve pairs of nerves in vertebrates that emerge directly from the brain.

cranium The bony framework that protects the brain in vertebrates.

crassulacean acid metabolism See *CAM plant*.

creatine phosphate Energy-storing compound found in muscle cells.

cretinism (kree'tin-izm) A chronic condition due to lack of thyroid secretion during fetal development and early childhood; results in retarded physical and mental development if untreated.

cri-du-chat A human genetic disease caused by loss of part of the short arm of chromosome 5 and characterized by mental retardation, a cry that sounds like a kitten mewing, and death in infancy or childhood.

cristae (kris'tee) (sing. *crista*) Shelflike or finger-like inward projections of the inner membrane of a mitochondrion.

Cro-Magnons Prehistoric humans (*Homo sapiens*) with modern features (tall, erect, lacking a heavy brow) who lived in Europe some 30,000 years ago.

cross bridges The connections of myosin with actin that link thick and thin filaments in muscle fibers.

crossing-over The breaking and rejoining of homologous (nonsister) chromatids during early meiotic prophase I that results in an exchange of genetic material.

CRP See *catabolite gene activator protein*.

cryptic coloration Colors or markings that help some organisms hide from predators by blending into their physical surroundings. Compare with *warning coloration*.

ctenophores (ten'oh-forz) Phylum of marine animals (comb jellies) whose bodies consist of two layers of cells enclosing a gelatinous mass. The outer surface is covered with comblike rows of cilia, by which the animal moves.

cuticle (kew'tih-kl) (1) A noncellular, waxy covering over the epidermis of the aerial parts of plants that reduces water loss; (2) The outer covering of some animals, such as roundworms.

cyanobacteria (sy-an"oh-bak-teer'ee-uh) Prokaryotic photosynthetic microorganisms that possess chlorophyll and produce oxygen during photosynthesis. Formerly known as blue-green algae.

cycad (sih'kad) Any of a phylum of gymnosperms that live mainly in tropical and semitropical regions and have stout stems (to 20 m in height) and fernlike leaves.

cyclic AMP (cAMP) A form of adenosine monophosphate in which the phosphate is part of a ring-shaped structure; acts as a regulatory molecule and second messenger in organisms ranging from bacteria to humans.

cyclic AMP receptor protein See *catabolite gene activator protein*.

cyclic photophosphorylation In photosynthesis, a series of reactions involving Photosystem I; ATP is formed by chemiosmosis, but O_2 and NADPH are not produced. Compare with *noncyclic photophosphorylation*.

cyclins Regulatory proteins whose levels oscillate during the cell cycle; activate cyclin-dependent protein kinases.

cystic fibrosis A genetic disease with an autosomal recessive inheritance pattern; characterized by secretion of abnormally thick mucus, particularly in the respiratory and digestive systems.

cytochromes (sy'toh-kromz) Iron-containing heme proteins of an electron transport system.

cytokines Regulatory proteins released by cells of the immune system. See *interferons* and *interleukins*.

cytokinesis (sy"toh-kih-nee'sis) Stage of cell division in which the cytoplasm divides to form two daughter cells.

cytokinin (sy"toh-kih'nin) A plant hormone involved in various aspects of plant growth and development, such as cell division and delay of senescence.

cytoplasm The cell contents with the exception of the nucleus.

cytosine A nitrogenous pyrimidine base that is a component of nucleic acids.

cytoskeleton Dynamic internal network of protein fibers that includes microfilaments, intermediate filaments, and microtubules.

cytosol Fluid component of the cytoplasm in which the organelles are suspended.

cytotoxic T cell T lymphocyte that destroys cancer cells and other pathogenic cells on contact. Also known as killer T cell.

dalton See *atomic mass unit (amu)*.

day-neutral plant A plant whose flowering is not controlled by variations in day length that occur with changing seasons. Compare with *long-day, short-day,* and *intermediate-day plants*.

deamination (dee-am-ih-nay'shun) Removal of an amino group ($-NH_2$) from an amino acid or other organic compound.

decarboxylation Reaction in which a molecule of CO_2 is removed from a carboxyl group of an organic acid.

deciduous Term describing a plant that sheds leaves or other structures at regular intervals; e.g., during autumn. Compare with *evergreen*.

decomposers Microbial heterotrophs that break down dead organic material and use the decomposition products as a source of energy. Also called saprotrophs or saprobes.

deductive reasoning Reasoning that operates from generalities to specifics and can make relationships among data more apparent. Compare with *inductive reasoning*. See *hypothetico-deductive approach*.

deforestation The temporary or permanent removal of forest for agriculture or other uses.

dehydrogenation (dee-hy"dro-jen-ay'shun) A form of oxidation in which hydrogen atoms are removed from a molecule.

deletion (1) A chromosome abnormality in which part of a chromosome is missing, e.g., cri-du-chat; (2) The loss of one or more base pairs from DNA, which can result in a frameshift mutation.

demographics The branch of sociology that deals with human population statistics, such as density and distribution.

denature (dee-nay′ture) To alter the physical properties and three-dimensional structure of a protein, nucleic acid, or other macromolecule by treating it with excess heat, strong acids, or strong bases.

dendrite (den′drite) A branch of a neuron that receives and conducts nerve impulses toward the cell body. Compare with *axon*.

dendritic cells A set of immune cells present in many tissues that capture antigens and present them to T cells.

dendrochronology (den″dro-kruh-naal′uh-gee) A method of dating using the annual rings of trees.

denitrification (dee-nie″tra-fuh-kay′shun) The conversion of nitrate (NO_3^-) to nitrogen gas (N_2) by certain bacteria (denitrifying bacteria) in the soil; part of the nitrogen cycle.

dense connective tissue A type of tissue that may be irregular, as in the dermis of the skin, or regular, as in tendons.

density-dependent factor An environmental factor whose effects on a population change as population density changes; density-dependent factors tend to retard population growth as population density increases and enhance population growth as population density decreases. Compare with *density-independent factor*.

density gradient centrifugation A technique whereby organelles or large molecules can be separated on the basis of differences in density. See *differential centrifugation*.

density-independent factor An environmental factor that affects the size of a population but is not influenced by changes in population density. Compare with *density-dependent factor*.

deoxyribonucleic acid (DNA) Double-stranded nucleic acid; contains genetic information coded in specific sequences of its constituent nucleotides.

deoxyribose Pentose sugar lacking a hydroxyl (—OH) group on carbon-2′; a constituent of DNA.

depolarization (dee-pol″ar-ih-zay′shun) A decrease in the charge difference across a plasma membrane; may result in an action potential in a neuron or muscle cell.

derived characters See *shared derived characters*.

dermal tissue system The tissue that forms the outer covering over a plant; the epidermis or periderm.

dermis (dur′mis) The layer of dense connective tissue beneath the epidermis in the skin of vertebrates.

desert A temperate or tropical biome in which lack of precipitation limits plant growth.

desertification Degradation of once-fertile land into nonproductive desert; caused partly by soil erosion, deforestation, and overgrazing by domestic animals.

desmosomes (dez′moh-somz) Button-like plaques, present on two opposing cell surfaces, that hold the cells together by means of protein filaments that span the intercellular space.

determinate growth Growth of limited duration, as for example, in flowers and leaves. Compare with *indeterminate growth*.

determination The developmental process by which one or more cells become progressively committed to a particular fate. Also called cell determination. Determination usually leads to differentiation.

detritivore (duh-try′tuh-vore) An organism, such as an earthworm or crab, that consumes fragments of dead organisms; also called detritus feeder.

detritus (duh-try′tus) Organic debris from decomposing organisms.

detritus feeder See *detritivore*.

deuteromycetes (doot″er-o-my′seats) An artificial grouping of fungi characterized by the absence of sexual reproduction but usually having other traits similar to ascomycetes; also called imperfect fungi.

deuterostome (doo′ter-oh-stome) Major division of the animal kingdom in which the anus develops from the blastopore; includes the echinoderms and chordates. Compare with *protostome*.

diabetes mellitus (mel′i-tus) An endocrine disorder caused by insulin deficiency.

diacylglycerol (DAG) (di″as-il-glis′er-ol) A neutral fat consisting of a glycerol combined chemically with two fatty acids; also called diglyceride. Can act as a second messenger that increases calcium concentration and activates enzymes.

dialysis The diffusion of certain solutes across a selectively permeable membrane.

diaphragm In mammals, the muscular floor of the chest cavity; contracts during inhalation, expanding the chest cavity.

diastole (di-as′toh-lee) Phase of the cardiac cycle in which the heart is relaxed. Compare with *systole*.

diatom (die′eh-tom″) A usually unicellular alga that is covered by an ornate, siliceous shell consisting of two overlapping halves; an important component of plankton in both marine and fresh waters.

dichotomous (di-kaut′uh-mus) A type of branching in which one part always divides into two more or less equal parts.

dicot (dy′kot) One of the two classes of flowering plants; dicot seeds contain two cotyledons, or seed leaves. Compare with *monocot*.

differential centrifugation Separation of cellular particles according to their mass, size, or density. In differential centrifugation the supernatant is spun at successively higher revolutions per minute. See *density gradient centrifugation*.

differential gene expression The expression of different subsets of genes at different times and in different cells during development.

differentiation (dif″ah-ren-she-ay′shun) Development toward a more mature state; a process changing a young, relatively unspecialized cell to a more specialized cell. Also called cell differentiation.

diffusion Net movement of particles (atoms, molecules, or ions) from a region of higher concentration to a region of lower concentration (i.e., down a concentration gradient), resulting from random motion. Compare with *facilitated diffusion* and *active transport*.

digestion The breakdown of food to smaller molecules.

diglyceride See *diacylglycerol*.

dihybrid cross (dy-hy′brid) A genetic cross that takes into account the behavior of alleles of two loci. Compare with *monohybrid cross*.

dikaryotic (dy-kare-ee-ot′ik) Condition of having two nuclei per cell (i.e., $n + n$), characteristic of certain fungal hyphae. Compare with *monokaryotic*.

dimer An association of two monomers.

dinoflagellate (dy″noh-flaj′eh-late) A unicellular, biflagellate, typically marine alga that is an important component of plankton; usually photosynthetic.

dioecious (dy-ee′shus) Having male and female reproductive structures on separate plants; compare with *monoecious*.

dipeptide A compound consisting of two amino acids linked by a peptide bond.

diploid (dip'loyd)   The condition of having two sets of chromosomes per nucleus. Compare with *haploid* and *polyploid*.

directional selection   The gradual replacement of one phenotype with another due to environmental change that favors phenotypes at one of the extremes of the normal distribution. Compare with *stabilizing selection* and *disruptive selection*.

disaccharide (dy-sak'ah-ride)   A sugar produced by covalently linking two monosaccharides.

dispersion   The pattern of distribution in space of the individuals of a population relative to their neighbors; may be clumped, random, or uniform.

disruptive selection   A special type of directional selection in which changes in the environment favor two or more variant phenotypes at the expense of the mean. Compare with *stabilizing selection* and *directional selection*.

distal   Remote; farther from the point of reference. Compare with *proximal*.

distal convoluted tubule   The part of the renal tubule that extends from the loop of Henle to the collecting duct. Compare with *proximal convoluted tubule*.

divergent circuit   A neural pathway in which a presynaptic neuron stimulates many postsynaptic neurons. Compare with *convergent circuit*.

diving reflex   A group of physiological mechanisms, such as decrease in metabolic rate, that are activated when a mammal dives to its limit.

division   A taxonomic category below that of kingdom, comparable to a phylum; often used in classifying plants, fungi, and certain protists.

dizygotic twins   Twins that arise from the separate fertilization of two eggs; commonly known as fraternal twins. Compare with *monozygotic twins*.

DNA   See *deoxyribonucleic acid*.

DNA-dependent RNA polymerase   See *RNA polymerase*.

DNA ligase   Enzyme that catalyzes the joining of the 5′ and 3′ ends of two DNA fragments; essential in DNA replication and used in recombinant DNA technology.

DNA methylation   A process in which gene inactivation is perpetuated by enzymes that add methyl groups to DNA.

DNA polymerases   Family of enzymes that catalyze the synthesis of DNA from a DNA template, by adding nucleotides to a growing 3′ end.

DNA provirus   Double-stranded DNA molecule that is an intermediate in the life cycle of an RNA tumor virus (retrovirus).

DNA replication   The process by which DNA is duplicated; ordinarily a semiconservative process in which a double helix gives rise to two double helices, each with an "old" strand and a newly synthesized strand.

DNA sequencing   Procedure by which the sequence of nucleotides in DNA is determined.

domain   (1) A structural and functional region of a protein; (2) A taxonomic category that includes one or more kingdoms.

dominance hierarchy   A linear "pecking order" into which animals in a population may organize according to status; regulates aggressive behavior within the population.

dominant allele (al-leel')   An allele that is always expressed when it is present, regardless of whether it is homozygous or heterozygous. Compare with *recessive allele*.

dopamine   A neurotransmitter of the biogenic amine group.

dormancy   A temporary period of arrested growth in plants or plant parts such as spores, seeds, bulbs, and buds.

dorsal (dor'sl)   Toward the uppermost surface or back of an animal. Compare with *ventral*.

dosage compensation   Genetic mechanism by which the expression of X-linked genes is made equivalent in XX females and XY males.

double fertilization   A process in the flowering plant life cycle in which there are two fertilizations; one fertilization results in the formation of a zygote, while the second results in the formation of endosperm.

doubling time   The amount of time it takes for a population to double in size, assuming that its current rate of increase does not change.

Down syndrome   An inherited condition in which individuals have abnormalities of the face, eyelids, tongue, and other parts of the body, and are physically and mentally retarded; usually results from trisomy of chromosome 21.

drupe (droop)   A simple, fleshy fruit in which the inner wall of the fruit is hard and stony, e.g., peach, cherry.

duodenum (doo″o-dee′num)   The portion of the small intestine into which the contents of the stomach first enter.

duplication   An abnormality in which a set of chromosomes contains more than one copy of a particular chromosomal segment; the translocation form of Down syndrome is an example.

dynamic equilibrium   The condition of a chemical reaction when the rate of change in one direction is exactly the same as the rate of change in the opposite direction, i.e., the concentrations of the reactants and products are not changing, and the difference in free energy between reactants and products is zero; also called chemical equilibrium.

ecdysone (ek'dih-sone)   See *molting hormone*.

echinoderms (eh-kine'oh-derms)   Phylum of spiny-skinned marine deuterostome invertebrates characterized by a water vascular system and tube feet; include sea stars, sea urchins, and sea cucumbers.

ecological niche   See *niche*.

ecological pyramid   A graphical representation of the relative energy value at each trophic level. See *pyramid of biomass, pyramid of energy,* and *pyramid of numbers*.

ecological succession   See *succession*.

ecology (ee-kol'uh-jee)   A discipline of biology that studies the interrelations among living things and their environments.

ecosystem (ee'koh-sis-tem)   The interacting system that encompasses a community and its nonliving, physical environment. Compare with *community*.

ecotone   The transition zone where two communities or biomes meet and intergrade.

ectoderm (ek'toh-derm)   The outer germ layer of the early embryo; gives rise to the skin and nervous system. Compare with *mesoderm* and *endoderm*.

ectotherm   An animal whose temperature fluctuates with that of the environment; may use behavioral adaptations to regulate temperature; sometimes referred to as cold-blooded. Compare with *endotherm*.

edge effect   The ecological phenomenon in which ecotones between adjacent communities often contain a greater number of species or greater population densities of certain species than either adjacent community.

effector   A muscle or gland that contracts or secretes in direct response to nerve impulses.

efferent (ef'fur-ent)   Leading away from some point of reference. Compare with *afferent*.

ejaculation (ee-jak″yoo-lay′shun)   A sudden expulsion, as in the ejection of semen from the penis.

electrolyte   Substance that dissociates into ions when dissolved in water; the resulting solution can conduct an electrical current.

electron   A particle with one unit of negative charge and negligible mass, located outside the atomic nucleus. Compare with *neutron* and *proton*.

electron configuration   The arrangement of the electrons around the atom. In a Bohr model the electron configuration is depicted as a series of concentric circles.

electron microscope   Microscope capable of producing high resolution, highly magnified images through the use of an electron beam (rather than light). Transmission electron microscopes (TEM) produce images of thin sections; scanning electron microscopes (SEM) produce images of surfaces.

electron shell   Group of orbitals of electrons with similar energies.

electron transport system   A series of chemical reactions during which hydrogens or their electrons are passed along an electron transport chain from one acceptor molecule to another, with the release of energy.

electronegativity   A measure of an atom's attraction for electrons.

electroreceptor   A receptor that responds to electrical stimuli.

element   A substance that cannot be changed to a simpler substance by a normal chemical reaction.

elimination   Ejection of undigested food from the body. Compare with *excretion*.

El Niño–Southern Oscillation (ENSO) (el nee′nyo)   A recurring climatic phenomenon that involves a surge of warm water in the Pacific Ocean and unusual weather patterns elsewhere in the world.

elongation (in protein synthesis)   Cyclic process by which amino acids are added one by one to a growing polypeptide chain. See *initiation* and *termination*.

embryo (em′bree-oh)   (1) A young organism before it emerges from the egg, seed, or body of its mother; (2) Developing human until the end of the second month, after which it is referred to as a fetus; (3) In plants, the young sporophyte produced following fertilization and subsequent development of the zygote.

embryo sac   The female gametophyte generation in flowering plants.

embryo transfer   See *host mothering*.

emigration   A type of migration in which individuals leave a population and thus decrease its size. Compare with *immigration*.

enantiomers (en-an′tee-oh-merz)   Two isomeric chemical compounds that are mirror images.

endangered species   A species whose numbers are so severely reduced that it is in imminent danger of extinction throughout all or part of its range. Compare with *threatened species*.

Endangered Species Act   A U.S. law that authorizes the Fish and Wildlife Service to protect from extinction all endangered and threatened species in the United States and abroad.

endemic species   Localized, native species that are not found anywhere else in the world. Compare with *cosmopolitan species*.

endergonic reaction (end′er-gon″ik)   Nonspontaneous reaction; a reaction requiring a net input of free energy. Compare with *exergonic reaction*.

endocrine gland (en′doh-crin)   Gland that secretes hormones directly into the blood or tissue fluid instead of into ducts. Compare with *exocrine gland*.

endocrine system   Body system that helps regulate metabolic activities; consists of ductless glands and tissues that secrete hormones.

endocytosis (en″doh-sy-toh′sis)   The active transport of substances into the cell by the formation of invaginated regions of the plasma membrane that pinch off and become cytoplasmic vesicles. Compare with *exocytosis*.

endoderm (en′doh-derm)   The inner germ layer of the early embryo; becomes the lining of the digestive tract and the structures that develop from the digestive tract—liver, lungs, and pancreas. Compare with *ectoderm* and *mesoderm*.

endodermis (en″doh-der′mis)   The innermost layer of the plant root cortex. Endodermal cells have a waterproof Casparian strip around their radial and transverse walls that ensures that water and minerals can enter the root xylem only by passing through the endoderm cells.

endolymph (en′doh-limf)   The fluid of the membranous labyrinth and cochlear duct of the ear.

endomembrane system   See *internal membrane system*.

endometrium (en″doh-mee′tree-um)   Uterine lining.

endoplasmic reticulum (ER) (en′doh-plaz″mik reh-tik′yoo-lum)   Interconnected network of internal membranes in eukaryotic cells enclosing a compartment, the ER lumen. Rough ER has ribosomes attached to the cytosolic surface; smooth ER, a site of lipid biosynthesis, lacks ribosomes.

endorphins (en-dor′finz)   Polypeptides released by certain brain and visceral neurons; their action is mimicked by opiate alkaloids. Compare with *enkephalins*.

endoskeleton (en″doh-skel′eh-ton)   Bony and/or cartilaginous structures within the body that provide support. Compare with *exoskeleton*.

endosperm (en′doh-sperm)   The $3n$ nutritive tissue that is formed at some point in the development of all angiosperm seeds.

endospore   A resting cell formed by certain bacteria; highly resistant to heat, radiation, and disinfectants.

endosymbiont (en″doe-sim′bee-ont)   An organism that lives inside the body of another kind of organism. Endosymbionts may benefit their host (mutualism) or harm their host (parasitism).

endosymbiont theory   Theory that certain organelles such as mitochondria and chloroplasts originated as symbiotic prokaryotes that lived inside other, free-living, prokaryotic cells. Compare with *autogenous model*.

endothelium (en-doh-theel′ee-um)   The tissue that lines the cavities of the heart, blood vessels, and lymph vessels.

endotherm (en′doh-therm)   An animal that uses metabolic energy to maintain a constant body temperature despite variations in environmental temperature; e.g., birds and mammals. Compare with *ectotherm*.

endotoxin   A poisonous substance in the cell walls of gram-negative bacteria. Compare with *exotoxin*.

energy   The capacity to do work; can be expressed in kilojoules or kilocalories.

energy of activation   See *activation energy*.

enhancers   Regulatory elements that can be located long distances away from the actual coding regions of a gene.

enkephalins (en-kef′ah-linz)   Polypeptides released by certain brain neurons that seem to function in pain perception; their action is mimicked by opiate alkaloids. Compare with *endorphins*.

enterocoely (en′ter-oh-seely)   Process by which the coelom forms as a cavity within mesoderm produced by outpocketings of the primitive gut (archenteron); characteristic of many deuterostomes. Compare with *schizocoely*.

enthalpy   The total potential energy of a system; sometimes referred to as the heat content of the system.

entropy (en'trop-ee)   Disorderliness; a quantitative measure of the amount of the random, disordered energy that is unavailable to do work.

environmental resistance   Unfavorable environmental conditions, such as crowding, that prevent organisms from reproducing indefinitely at their biotic potential.

enzyme (en'zime)   An organic catalyst (usually a protein) that accelerates a specific chemical reaction by lowering the activation energy required for that reaction.

enzyme-substrate complex   The temporary association between enzyme and substrate that forms during the course of a catalyzed reaction; also called ES complex.

eosinophil (ee-oh-sin'oh-fil)   A type of white blood cell whose cytoplasmic granules absorb acidic stains.

epidermis (ep-ih-dur'mis)   (1) An outer layer of cells that covers the body of plants and animals and functions primarily for protection; (2) The outer layer of vertebrate skin.

epididymis (ep-ih-did'ih-mis) (pl. *epididymides*)   A coiled tube that receives sperm from the testis and conveys it to the vas deferens.

epiglottis   A thin, flexible structure that guards the entrance to the larynx, preventing food from entering the airway during swallowing.

epinephrine (ep-ih-nef'rin)   Hormone produced by the adrenal medulla; stimulates the sympathetic nervous system.

epistasis (ep'ih-sta-sis)   Condition in which certain alleles of one locus can alter the expression of alleles of a different locus.

epithelial tissue (ep-ih-theel'ee-al)   The type of animal tissue that covers body surfaces, lines body cavities, and forms glands; also called epithelium.

epoch   The smallest unit of geological time; a subdivision of a period.

equilibrium   See *dynamic equilibrium*.

era   One of the main divisions of geological time; eras are subdivided into periods.

erythroblastosis fetalis (i-rith-rho-blast-oh'sis fi-tal'is)   Serious condition in which Rh$^+$ red blood cells (which bear antigen D) of a fetus are destroyed by maternal anti-D antibodies.

erythrocyte (er-eeth'roh-site)   Vertebrate red blood cell.

ES complex   See *enzyme-substrate complex*.

essential nutrient   A nutrient that must be provided in the diet because the body cannot make it or cannot make it in sufficient quantities to meet nutritional needs, e.g., essential amino acids and essential fatty acids.

ester linkage   Covalent linkage formed by the reaction of a carboxyl group and a hydroxyl group, with the removal of the equivalent of a water molecule; the linkage includes an oxygen atom bonded to a carbonyl group.

estrogens (es'troh-jens)   Female sex hormones produced by the ovary; promote the development and maintenance of female reproductive structures and of secondary sexual characteristics.

estuary (es'choo-wear-ee)   A coastal body of water that connects to an ocean, in which fresh water from the land mixes with salt water.

ethology (ee-thol'oh-jee)   The study of animal behavior under natural conditions from the point of view of adaptation.

ethyl alcohol   A two-carbon alcohol.

ethylene (eth'ih-leen)   A gaseous plant hormone involved in various aspects of plant growth and development, such as leaf abscission and fruit ripening.

eubacteria (yoo"bak-teer'ee-ah)   Prokaryotes other than the archaebacteria.

euchromatin (yoo-croh'mah-tin)   Loosely coiled chromatin that is generally capable of transcription. Compare with *heterochromatin*.

euglenoids (yoo-glee'noids)   A group of mostly freshwater, flagellated, unicellular algae that move by means of an anterior flagellum and are usually photosynthetic.

eukaryote (yoo"kar'ee-ote)   Organism whose cells possess nuclei and other membrane-bounded organelles. Compare with *prokaryote*.

eustachian tube (yoo-stay'shee-un)   The auditory tube passing between the middle ear cavity and the pharynx in vertebrates; permits the equalization of pressure on the tympanic membrane.

eutrophic lake   A lake enriched with nutrients such as nitrate and phosphate and consequently overgrown with plants or algae.

evergreen   Shedding leaves over a long time period, so that some leaves are always present. Compare with *deciduous*.

evolution   Any cumulative genetic changes in a population from generation to generation. Evolution leads to differences in populations and explains the origin of all of the organisms that exist today or have ever existed.

excitatory postsynaptic potential (EPSP)   A change in membrane potential that brings a neuron closer to the firing level. Compare with *inhibitory postsynaptic potential (IPSP)*.

excretion (ek-skree'shun)   The discharge from the body of a waste product of metabolism (not to be confused with the elimination of undigested food materials). Compare with *elimination*.

excretory system   The body system in animals that functions to discharge metabolic wastes.

exergonic reaction (ex'er-gon"ik)   A reaction characterized by a release of free energy. Also called spontaneous reaction. Compare with *endergonic reaction*.

exocrine gland (ex'oh-crin)   Gland that excretes its products through a duct that opens onto a free surface such as the skin (e.g., sweat glands). Compare with *endocrine gland*.

exocytosis (ex"oh-sy-toh'sis)   The active transport of materials out of the cell by fusion of cytoplasmic vesicles with the plasma membrane. Compare with *endocytosis*.

exon   (1) A protein-coding region of a eukaryotic gene; (2) The RNA transcribed from such a region. Compare with *intron*.

exoskeleton (ex"oh-skel'eh-ton)   An external skeleton, such as the shell of mollusks or outer covering of arthropods; provides protection and sites of attachment for muscles. Compare with *endoskeleton*.

exotoxin   A poisonous substance released by certain bacteria. Compare with *endotoxin*.

exponential population growth   Population growth that occurs at a constant rate of increase over a period of time. When the increase in number versus time is plotted on a graph, exponential growth produces a characteristic J-shaped curve. Compare with *logistic population growth*.

ex situ conservation   Conservation efforts that involve conserving biological diversity in human-controlled settings, such as zoos. Compare with *in situ conservation*.

exteroceptor (ex'tur-oh-sep"tor)   One of the sense organs that receives sensory stimuli from the outside world, such as the eyes or touch receptors. Compare with *interoceptor*.

extinction The elimination of a species; occurs when the last individual member of a species dies.

extracellular matrix (ECM) A network of proteins and carbohydrates that surrounds many animal cells.

extraembryonic membranes Multicellular membranous structures that develop from the germ layers of a terrestrial vertebrate embryo but are not part of the embryo itself. See *chorion, amnion, allantois,* and *yolk sac.*

F_1 generation (first filial generation) The first generation of hybrid offspring resulting from a cross between parents from two different true-breeding lines.

F_2 generation (second filial generation) The offspring of the F_1 generation.

facilitated diffusion The passive transport of ions or molecules by a specific carrier protein in a membrane. As in simple diffusion, net transport is down a concentration gradient, and no additional energy has to be supplied. Compare with *diffusion* and *active transport.*

facilitation A process in which a neuron is brought close to the threshold level by stimulation from various presynaptic neurons.

facultative anaerobe Organism capable of carrying out aerobic respiration but able to switch to fermentation when oxygen is unavailable; e.g., yeast. Compare with *obligate anaerobe.*

FAD/FADH₂ Oxidized and reduced forms, respectively, of flavin adenine dinucleotide; coenzyme that transfers electrons (as hydrogen) in metabolism, including cellular respiration.

fallopian tube See *oviduct.*

family A taxonomic category made up of related genera.

fatty acid An organic acid containing a long hydrocarbon chain, with no double bonds *(saturated fatty acid),* one double bond *(monounsaturated fatty acid),* or two or more double bonds *(polyunsaturated fatty acid);* fatty acids are components of neutral fats and phospholipids.

feedback inhibition Type of enzyme regulation in which the accumulation of the product of a reaction inhibits an earlier reaction in the sequence.

fermentation Anaerobic process by which ATP is produced by a series of redox reactions in which organic compounds serve both as electron donors and terminal electron acceptors.

fern One of a phylum of seedless vascular plants that reproduce by spores produced in sporangia usually borne on leaves in sori; ferns have an alternation of generations between the dominant sporophyte (a fern plant) and the gametophyte (a prothallus).

fertilization Fusion of two *n* gametes; results in the formation of a *2n* zygote.

fetus The unborn human offspring from the third month of pregnancy to birth.

fiber (1) In plants, a type of sclerenchyma; fibers are long, tapered cells with thick walls. Compare with *sclereid;* (2) In animals, an elongated cell such as a muscle or nerve cell.

fibrin An insoluble protein formed from the plasma protein fibrinogen during blood clotting.

fibrous root system A root system in plants that has many roots that are similar in length and thickness. Compare with *taproot system.*

Fick's law of diffusion States that the rate of diffusion of a substance across a membrane is directly proportional to the surface area and to the difference in pressure between the two sides.

filament In flowering plants, the thin stalk of a stamen; the filament bears an anther at its tip.

first law of thermodynamics The law of conservation of energy, which states that the total energy of any closed system (any object plus its surroundings, i.e., the universe) remains constant. Compare with *second law of thermodynamics.*

fitness A measure of the ability of an organism, owing to its genotype, to compete successfully and make a genetic contribution to subsequent generations.

fixed action pattern (FAP) An innate behavior triggered by a sign stimulus.

flagellum (flah-jel'um) (pl. *flagella*) Long, whiplike, movable structure extending from the cell and used in locomotion. (1) Eukaryote flagella are composed of two central single microtubules surrounded by a cylinder of nine, double microtubules (9 + 2 structure), all covered by a plasma membrane. (2) Prokaryote flagella are filaments rotated by special structures located in the plasma membrane and cell wall.

flame cells Collecting cells that have cilia; part of the osmoregulatory system of flatworms.

flavin adenine dinucleotide See *FAD/FADH₂.*

flowering plants See *angiosperms.*

flowing-water ecosystem A river or stream ecosystem.

fluid-mosaic model The currently accepted model of the plasma membrane and other cell membranes, in which protein molecules float in a phospholipid bilayer.

fluorescence Emission of light of a longer wavelength (lower energy) than the light originally absorbed.

follicle (fol'i-kl) (1) A simple, dry, dehiscent fruit that splits open at maturity along one suture to liberate the seeds; (2) A small sac of cells in the mammalian ovary that contains a maturing egg; (3) The pocket in the skin from which a hair grows.

follicle-stimulating hormone (FSH) A gonadotropic hormone secreted by the anterior lobe of the pituitary gland; stimulates follicle development in the ovaries of females and sperm production in the testes of males.

food chain The series of organisms through which energy flows in an ecosystem. Each organism in the series eats or decomposes the preceding organism in the chain. See *food web.*

food web A complex interconnection of all of the food chains in an ecosystem.

foramen magnum The opening in the vertebrate skull through which the spinal cord passes.

foraminiferan (for"am-in-if'er-an) A marine protozoon that produces a shell, or test, that encloses an amoeboid body.

forest decline A gradual deterioration (and often death) of many trees in a forest; may be caused by a combination of factors, such as acid precipitation, toxic heavy metals, and surface-level ozone.

fossil Parts or traces of an ancient organism usually preserved in rock.

fossil fuel Combustible deposits in Earth's crust that are composed of the remnants of prehistoric organisms that existed millions of years ago, e.g., oil, natural gas, and coal.

founder cell A cell from which a particular cell lineage is derived.

founder effect Genetic drift that results from a small population colonizing a new area.

fovea (foe'vee-ah) The area of sharpest vision in the retina; cone cells are concentrated here.

fragile site A weak point at a specific location on a chromosome where part of a chromatid appears to be attached to the rest of the chromosome by a thin thread of DNA.

fragile X syndrome A human genetic disorder caused by a fragile site that occurs near the tip on the X chromosome; effects range from mild learning disabilities to severe mental retardation and hyperactivity.

frameshift mutation Mutation that results when one or two nucleotide pairs are inserted into or deleted from the DNA. The change causes the mRNA transcribed from the mutated DNA to have an altered reading frame such that all codons downstream from the mutation are changed.

free energy The maximum amount of energy available to do work under the conditions of a biochemical reaction.

frequency-dependent selection Selection in which the relative fitness of different genotypes is related to how frequently they occur in the population.

freshwater wetlands Land that is transitional between freshwater and terrestrial ecosystems and is covered with water for at least part of the year; e.g., marshes and swamps.

frontal lobes In mammals, the anterior part of the cerebrum.

fruit In flowering plants, a mature, ripened ovary. Fruits contain seeds and usually provide seed protection and dispersal.

fruiting body A multicellular structure that contains the sexual spores of certain fungi; refers to the ascocarp of an ascomycete and the basidiocarp of a basidiomycete.

fucoxanthin (few″koh-zan′thin) The brown carotenoid pigment found in brown algae, golden algae, diatoms, and dinoflagellates.

functional group A group of atoms that confers distinctive properties on an organic molecule (or region of a molecule) to which it is attached, e.g., hydroxyl, carbonyl, carboxyl, amino, phosphate, and sulfhydryl groups.

fundamental niche The potential ecological niche that an organism could occupy if there were no competition from other species. Compare with *realized niche.*

fungus (pl. *fungi*) A heterotrophic eukaryote with chitinous cell walls and a body usually in the form of a mycelium of branched, threadlike hyphae. Most fungi are decomposers; some are parasitic.

G protein One of a group of proteins that bind GTP and are involved in the transfer of signals across the plasma membrane.

G_1 phase The first gap phase within the interphase stage of the cell cycle; G_1 occurs before DNA synthesis (S phase) begins. Compare with *S* and *G_2 phases.*

G_2 phase Second gap phase within the interphase stage of the cell cycle; G_2 occurs after DNA synthesis (S phase) and before mitosis. Compare with *S* and *G_1 phases.*

gallbladder A small sac that stores bile.

gametangium (gam″uh-tan′gee-um) Special multicellular or unicellular structure of plants, protists, and fungi in which gametes are formed.

gamete (gam′eet) A sex cell; in plants and animals, an egg or sperm. In sexual reproduction, the union of gametes results in the formation of a zygote. The chromosome number of a gamete is designated *n*. Species that are not polyploid have haploid gametes and diploid zygotes.

gametic isolation (gam-ee′tik) A prezygotic reproductive isolating mechanism in which sexual reproduction between two closely related species cannot occur because of chemical differences in the gametes.

gametogenesis The process of gamete formation. See *spermatogenesis* and *oogenesis.*

gametophyte generation (gam-ee′toh-fite) The *n*, gamete-producing stage in the life cycle of a plant. Compare with *sporophyte generation.*

ganglion (gang′glee-on) (pl. *ganglia*) A mass of neuron cell bodies.

gap junction Structure consisting of specialized regions of the plasma membrane of two adjacent cells; contains numerous pores that allow the passage of certain small molecules and ions between them.

gastrin (gas′trin) A hormone released by the stomach mucosa; stimulates the gastric glands to secrete pepsinogen.

gastrovascular cavity A central digestive cavity with a single opening that functions as both mouth and anus; characteristic of cnidarians and flatworms.

gastrula (gas′troo-lah) A three-layered embryo formed by the process of gastrulation.

gastrulation (gas′troo-lay″shun) Process in embryonic development during which the three germ layers (ectoderm, mesoderm, and endoderm) form.

gel electrophoresis Procedure by which proteins or nucleic acids can be separated on the basis of size and charge as they migrate through a gel in an electrical field.

gene A segment of DNA that serves as a unit of hereditary information; includes a transcribable DNA sequence (plus associated sequences regulating its transcription) that yields a protein or RNA product with a specific function. Most eukaryotic genes are in chromosomes.

gene amplification Process by which multiple copies of a gene are produced by selective replication, thus allowing for increased synthesis of the gene product.

gene flow The movement of alleles between local populations due to the migration of individuals; can have significant evolutionary consequences.

gene locus See *locus.*

gene pool All the alleles of all the genes present in a freely interbreeding population.

generation time The time required for the completion of one cell cycle.

gene therapy Any of a variety of methods designed to correct a disease or alleviate its symptoms through the introduction of genes into the affected person's cells.

genetic bottleneck See *bottleneck.*

genetic drift A random change in allele frequency in a small breeding population.

genetic engineering Manipulation of genes, often through recombinant DNA technology.

genetic equilibrium The condition of a population that is not undergoing evolutionary change.

genetic polymorphism (pol″ee-mor′fizm) The presence in a population of two or more alleles for a given gene locus.

genetic probe A single-stranded nucleic acid (either DNA or RNA) that can be used to identify a complementary sequence by hydrogen-bonding to it.

genetic recombination See *recombination, genetic.*

genome (jee′nome) Originally, all the genetic material in a cell or organism. The term is used more than one way, depending on context: e.g., an organism's haploid genome is all the DNA contained in one haploid set of its chromosomes, and its mitochondrial genome is all the DNA in a mitochondrion.

genomic DNA library A collection of recombinant plasmids in which all the DNA in the genome is represented. Compare with *cDNA library.*

genotype (jeen′oh-type) The genetic makeup of an individual. Compare with *phenotype.*

genus (jee'nus)   A taxonomic category made up of related species.

germination   Resumption of growth of an embryo or spore; occurs when a seed or spore sprouts.

germ layers   In animals, three embryonic tissue layers: endoderm, mesoderm, and ectoderm.

germ line cell   In animals, a cell that is part of the line of cells that will ultimately undergo meiosis to form gametes. Compare with *somatic cell.*

germplasm   Any plant or animal material that may be used in breeding; includes seeds, plants, and plant tissues of traditional crop varieties and the sperm and eggs of traditional livestock breeds.

gibberellin (jib″ur-el′lin)   A plant hormone involved in many aspects of plant growth and development, such as stem elongation, flowering, and seed germination.

gills   (1) The respiratory organs characteristic of many aquatic animals, usually thin-walled projections from the body surface or from some part of the digestive tract; (2) The spore-bearing, platelike structures under the caps of mushrooms.

ginkgo (ging′ko)   Member of an ancient gymnosperm group that consists of a single living representative *(Ginkgo biloba),* a hardy, deciduous tree with broad, fan-shaped leaves and naked, fleshy seeds (on female trees).

gland   See *endocrine gland* and *exocrine gland.*

glial cells (glee′ul)   In nervous tissue, cells that support and nourish neurons.

globulin (glob′yoo-lin)   One of a class of proteins in blood plasma, some of which (gamma globulins) function as antibodies.

glomerulus (glom-air′yoo-lus)   The cluster of capillaries at the proximal end of a nephron; the glomerulus is surrounded by Bowman's capsule.

glucagon (gloo′kah-gahn)   A pancreatic hormone that stimulates glycogen breakdown, thereby increasing the concentration of glucose in the blood. Compare with *insulin.*

glucose   A hexose aldehyde sugar that is central to many metabolic processes.

glyceraldehyde-3-phosphate (G3P)   Phosphorylated 3-carbon compound that is an important intermediate in glycolysis and in the Calvin cycle.

glycerol   A three-carbon alcohol with a hydroxyl group on each carbon; a component of neutral fats and phospholipids.

glycocalyx (gly″koh-kay′lix)   A coating on the outside of an animal cell, formed by the polysaccharide portions of glycoproteins and glycolipids associated with the plasma membrane.

glycogen (gly′koh-jen)   The principal storage polysaccharide in animal cells; formed from glucose and stored primarily in the liver and, to a lesser extent, in muscle cells.

glycolipid   A lipid with covalently attached carbohydrates.

glycolysis (gly-kol′ih-sis)   The first stage of cellular respiration, literally the "splitting of sugar." The metabolic conversion of glucose into pyruvate, accompanied by the production of ATP.

glycoprotein (gly′koh-pro-teen)   A protein with covalently attached carbohydrates.

glycosidic linkage   Covalent linkage joining two sugars; includes an oxygen atom bonded to a carbon of each sugar.

glyoxysomes (gly-ox′ih-somz)   Membrane-bounded structures in cells of certain plant seeds; contain a large array of enzymes that convert stored fat to sugar.

gnetophyte (nee′toe-fite)   One of a small phylum of unusual gymnosperms that possess many features similar to flowering plants.

goblet cells   Unicellular glands that secrete mucus.

goiter (goy′ter)   An enlargement of the thyroid gland.

golden alga   A member of a phylum of algae, most of which are biflagellated, unicellular, and contain pigments, including chlorophyll *a* and *c* and carotenoids, including fucoxanthin.

Golgi complex (goal′jee)   Organelle composed of stacks of flattened, membranous sacs. Mainly responsible for modifying, packaging, and sorting proteins that will be secreted or targeted to other organelles of the internal membrane system or to the plasma membrane; also called Golgi body or Golgi apparatus.

gonad (goh′nad)   A gamete-producing gland; an ovary or a testis.

gonadotropin-releasing hormone (GnRH)   A hormone secreted by the hypothalamus that stimulates the anterior pituitary to secrete follicle-stimulating hormone or luteinizing hormone.

gonadotropins (go-nad″oh-troh′pins)   Hormones produced by the anterior pituitary gland and the embryo that stimulate the function of the testes and ovaries; include follicle-stimulating hormone, luteinizing hormone, and human chorionic gonadotropin.

gradualism   The idea that evolutionary change of a species is due to a slow, steady accumulation of changes over time. Compare with *punctuated equilibrium.*

graft rejection   An immune response directed against a transplanted tissue or organ.

grain   A simple, dry, one-seeded fruit in which the fruit wall is fused to the seed coat, making it impossible to separate the fruit from the seed, e.g., corn and wheat kernels.

granum (pl. *grana*)   A stack of thylakoids within a chloroplast.

gravitropism (grav″ih-troh′pizm)   Growth of a plant in response to gravity.

gray crescent   The grayish area of cytoplasm that marks the region where gastrulation begins in an amphibian embryo.

gray matter   Nervous tissue in the brain and spinal cord that contains cell bodies, dendrites, and unmyelinated axons. Compare with *white matter.*

green alga   A member of a diverse phylum of algae that contain the same pigments as plants (chlorophylls *a* and *b* and carotenoids).

greenhouse effect   The global warming of Earth's atmosphere produced by the buildup of carbon dioxide and other greenhouse gases, which trap the sun's radiation in much the same way that glass does in a greenhouse.

greenhouse gases   Trace gases in the atmosphere that allow the sun's energy to penetrate to the Earth's surface but do not allow as much of it to escape as heat.

gross primary productivity   The rate at which energy accumulates in an ecosystem (as biomass) during photosynthesis. Compare with *net primary productivity.*

ground state   The lowest energy state of an atom.

ground tissue system   All tissues in the plant body other than the dermal tissue system and vascular tissue system; consists of parenchyma, collenchyma, and sclerenchyma.

growth factor   An extracellular protein that signals certain cells to grow and divide.

growth hormone (GH)   A hormone secreted by the anterior lobe of the pituitary gland; stimulates growth of body tissues; also called somatotropin.

growth rate   The rate of change of a population; calculated by subtracting the death rate from the birth rate (in populations with little or no migration).

guanine (gwan′een) A nitrogenous purine base that is a component of nucleic acids and GTP.

guard cell One of a pair of epidermal cells that adjust their shape to form a stomatal pore for gas exchange.

guttation (gut-tay′shun) The appearance of water droplets on leaves, forced out through leaf pores by root pressure.

gymnosperm (jim′noh-sperm) Any of a group of seed plants in which the seeds are not enclosed in an ovary; gymnosperms frequently bear their seeds in cones. Includes four phyla: conifers, cycads, ginkgoes, and gnetophytes.

habitat The natural environment or place where an organism, population, or species lives.

habituation (hab-it″yoo-ay′shun) A type of learning in which an animal becomes accustomed to a repeated, irrelevant stimulus and no longer responds to it.

half-life The period of time required for a radioisotope to change into a different material.

haploid (hap′loyd) The condition of having one set of chromosomes per nucleus. Compare with *diploid* and *polyploid*.

Hardy-Weinberg principle The theorem that allele frequencies do not change from generation to generation in a large population in the absence of microevolutionary processes (mutation, genetic drift, gene flow, natural selection).

Hatch-Slack pathway See *C₄ plant*.

haustorium (hah-stor′ee-um) (pl. *haustoria*) (1) In parasitic fungi, a specialized hypha that penetrates a host cell and obtains nourishment from the cytoplasm; (2) In parasitic plants, a specialized root that penetrates a host plant.

Haversian canals (ha-vur′zee-un) Channels extending through the matrix of bone; contain blood vessels and nerves.

heat The total amount of kinetic energy in a sample of a substance.

heat energy Energy that flows between two objects that differ in temperature; unit of measurement is the calorie or kilocalorie.

heat of vaporization The amount of heat energy that must be supplied to change one gram of a substance from the liquid phase to the vapor phase.

helicases Enzymes that unwind the two strands of a DNA double helix.

heliotropism The ability of leaves or flowers of certain plants to follow the sun by aligning themselves either perpendicular or parallel to the sun's rays; also called solar tracking.

helper T cell T lymphocyte that activates B lymphocytes and can stimulate cytotoxic T cell production.

hemichordates A phylum of sedentary, wormlike deuterostomes.

hemizygous (hem″ih-zy′gus) Possessing only one allele for a particular locus; a human male is hemizygous for all X-linked genes. Compare with *homozygous* and *heterozygous*.

hemocoel Blood cavity characteristic of animals with an open circulatory system.

hemoglobin (hee′moh-gloh″bin) The red, iron-containing protein pigment in blood that transports oxygen and carbon dioxide and aids in regulation of pH.

hemophilia (hee″-moh-feel′ee-ah) A hereditary disease in which blood does not clot properly; the form known as hemophilia A has an X-linked, recessive inheritance pattern.

Hensen's node See *primitive streak*.

hepatic (heh-pat′ik) Pertaining to the liver.

hepatic portal system The portion of the circulatory system that carries blood from the intestine through the liver.

herbivore (herb′uh-vore) An animal that feeds on plants or algae. Also called primary consumer.

hermaphrodite (her-maf′roh-dite) An organism that possesses both male and female sex organs.

heterochromatin (het″ur-oh-kroh′mah-tin) Highly coiled and compacted chromatin in an inactive state. Compare with *euchromatin*.

heterocyst (het′ur-oh-sist″) An oxygen-excluding cell of cyanobacteria that is the site of nitrogen fixation.

heterogametic Term describing an individual that produces two classes of gametes with respect to their sex chromosome constitutions. Human males (XY) are heterogametic, producing X and Y sperm. Compare with *homogametic*.

heterospory (het″ur-os′pur-ee) Production of two types of *n* spores, microspores (male) and megaspores (female). Compare with *homospory*.

heterothallic (het″ur-oh-thal′ik) Pertaining to certain algae and fungi that have two mating types; only by combining a plus strain and a minus strain can sexual reproduction occur. Compare with *homothallic*.

heterotroph (het′ur-oh-trof) Organism that cannot synthesize its own food from inorganic raw materials and therefore must obtain energy and body-building materials from other organisms. Also called consumer. Compare with *autotroph*. See *chemoheterotroph* and *photoheterotroph*.

heterozygote advantage A phenomenon in which the heterozygous condition confers some special advantage on an individual that either homozygous condition does not (i.e., *Aa* has a higher degree of fitness than does *AA* or *aa*).

heterozygous (het-ur′oh-zye′gus) Possessing a pair of unlike alleles for a particular locus. Compare with *homozygous*.

hexose A monosaccharide containing six carbon atoms.

histamine (his′tah-meen) Substance released from mast cells that is involved in allergic and inflammatory reactions.

histones (his′tones) Small, positively charged (basic) proteins in the cell nucleus that bind to the negatively charged DNA.

holdfast The basal structure for attachment to solid surfaces found in multicellular algae.

holoblastic cleavage Cleavage pattern in which the entire embryo cleaves; characteristic of eggs with little or moderate yolk (isolecithal or moderately telolecithal), e.g., the eggs of echinoderms, amphioxus, and mammals. Compare with *meroblastic cleavage*.

homeobox A DNA sequence of approximately 180 base pairs found in many homeotic genes and some other genes that are important in development; genes containing homeobox sequences code for certain transcription factors.

home range A geographical area that an animal seldom or never leaves.

homeostasis (home″ee-oh-stay′sis) The balanced internal environment of the body; the automatic tendency of an organism to maintain such a steady state.

homeotic gene (home″ee-ah′tik) A gene that controls the formation of specific structures during development. Such genes were originally identified through insect mutants in which one body part is substituted for another.

hominid (hah′min-id) Any of a group of extinct and living humans. Also called hominine.

hominine See *hominid*.

hominoid (hah′min-oid) The apes and hominids.

Homo erectus An extinct hominid that lived between 2 million and 200,000 years ago; made stone tools and may have worn clothing, built fires, and lived in caves.

Homo ergaster An extinct hominid that is recognized by some paleoanthropologists as separate from *H. erectus;* other paleoanthropologists do not recognize *H. ergaster* as a separate species and instead consider it to be the African line of *H. erectus.*

homogametic Term describing an individual that produces gametes with identical sex chromosome constitutions. Human females (XX) are homogametic, producing all X eggs. Compare with *heterogametic.*

Homo habilis An extinct hominid that lived between about 2.3 and 1.5 million years ago; lived in Africa and fashioned crude stone tools.

Homo heidelbergensis See *archaic Homo sapiens.*

homologous chromosomes (hom-ol′ah-gus) Chromosomes that are similar in morphology and genetic constitution. In humans there are 23 pairs of homologous chromosomes; one member of each pair is inherited from the mother, and the other from the father.

homologous features Features similar in underlying structure in different species as a result of their derivation from a common ancestor. Compare with *homoplastic features.*

Homo neanderthalensis Primitive humans who lived in Europe and western Asia from about 230,000 to 30,000 years ago; possessed sophisticated stone tools, cared for the sick and elderly, and buried their dead; also called Neandertals.

homoplastic features Structures similar in function or appearance in species without a recent common ancestor as a result of convergent evolution. Compare with *homologous features.*

homoplasy Refers to the evolution of similar, but independently evolved, characters that are not homologous.

Homo sapiens The modern human species.

homospory (hoh″mos′pur-ee) Production of one type of *n* spore that gives rise to a bisexual gametophyte. Compare with *heterospory.*

homothallic (hoh″moh-thal′ik) Pertaining to certain algae and fungi that are self-fertile. Compare with *heterothallic.*

homozygous (hoh″moh-zy′gous) Possessing a pair of identical alleles for a particular locus. Compare with *heterozygous.*

hormone An organic chemical messenger in multicellular organisms that is produced in one part of the body and transported to another part where it signals cells to alter some aspect of metabolism.

hornwort A type of spore-producing, nonvascular, thallose plant with a life cycle similar to mosses.

horsetail A phylum of seedless vascular plants with a life cycle similar to ferns.

host mothering The introduction of an embryo from one species into the uterus of another species, where it implants and develops; the host mother subsequently gives birth and may raise the offspring as her own; also called embryo transfer.

human chorionic gonadotropin (hCG) A hormone secreted by cells surrounding the early embryo; signals the mother's corpus luteum to continue to function.

human genetics The science of inherited variation in humans.

human immunodeficiency virus (HIV) The retrovirus that causes AIDS (acquired immune deficiency syndrome).

human leukocyte antigen (HLA) See *major histocompatibility complex.*

humus (hew′mus) Organic matter in various stages of decomposition in the soil; gives soil a dark brown or black color.

Huntington disease A genetic disease that has an autosomal dominant inheritance pattern and causes mental and physical deterioration.

hybrid Offspring of two genetically dissimilar parents.

hybrid breakdown A postzygotic reproductive isolating mechanism in which, although an interspecific hybrid is fertile and produces a second (F_2) generation, the F_2 has defects that prevent it from successfully reproducing.

hybrid inviability A reproductive isolating mechanism (postzygotic) in which the embryonic development of an interspecific hybrid is aborted.

hybrid sterility A reproductive isolating mechanism (postzygotic) in which an interspecific hybrid cannot reproduce successfully.

hybrid vigor Genetic superiority of an F_1 hybrid over either parent, due to the presence of heterozygosity for a number of different loci.

hybridization (1) Interbreeding between members of two different taxa; (2) Interbreeding between genetically dissimilar parents; (3) In molecular biology, complementary base pairing between nucleic acid (DNA or RNA) strands from different sources.

hybrid zone An area of overlap between two closely related populations, subspecies, or species, in which interbreeding occurs.

hydration Process of association of a substance with the partial positive and/or negative charges of water molecules.

hydrocarbon An organic compound composed solely of hydrogen and carbon atoms.

hydrogen bond A weak attractive force existing between a hydrogen atom with a partial positive charge and an electronegative atom (usually oxygen or nitrogen) with a partial negative charge. Compare with *covalent bond* and *ionic bond.*

hydrologic cycle The water cycle, which includes evaporation, precipitation, and flow to the ocean. The hydrologic cycle supplies terrestrial organisms with a continual supply of fresh water.

hydrolysis Reaction in which a covalent bond between two subunits is broken through the addition of the equivalent of a water molecule; a hydrogen atom is added to one subunit and a hydroxyl group to the other. Compare with *condensation synthesis.*

hydrophilic Attracted to water. Compare with *hydrophobic.*

hydrophobic Repelled by water. Compare with *hydrophilic.*

hydroponics (hy″dra-paun′iks) Growing plants in an aerated solution of dissolved inorganic minerals; i.e., without soil.

hydrostatic skeleton A type of skeleton found in some invertebrates in which contracting muscles push against a tube of fluid.

hydroxide ion Anion (negatively charged particle) consisting of oxygen and hydrogen; usually written OH^-.

hydroxyl group (hy-drok′sil) Polar functional group; abbreviated —OH.

hypertonic Term referring to a solution having an osmotic pressure (or solute concentration) greater than that of the solution with which it is compared. Compare with *hypotonic* and *isotonic.*

hypha (hy′fah) (pl. *hyphae*) One of the threadlike filaments composing the mycelium of a water mold or fungus.

hypocotyl (hy′poh-kah″tl) The part of the axis of a plant embryo or seedling below the point of attachment of the cotyledons.

hypoosmotic See *hypotonic.*

hypothalamus (hy-poh-thal′uh-mus) Part of the mammalian brain that regulates the pituitary gland, the autonomic system, emotional responses, body temperature, water balance, and appetite; located below the thalamus.

hypothesis A testable statement about the nature of an observation or relationship. Compare with *theory* and *principle*.

hypothetico-deductive approach Emphasizes the use of deductive reasoning to test hypotheses. Compare with *hypothetico-inductive approach*. See *deductive reasoning*.

hypothetico-inductive approach Emphasizes the use of inductive reasoning to discover new general principles. Compare with *hypothetico-deductive approach*. See *inductive reasoning*.

hypotonic Term referring to a solution having an osmotic pressure (or solute concentration) less than that of the solution with which it is compared. Compare with *hypertonic* and *isotonic*.

imaginal discs Paired structures in an insect larva that develop into specific adult structures during complete metamorphosis.

imago (ih-may′go) The adult form of an insect.

imbibition (im″bi-bish′en) The absorption of water by a seed prior to germination.

immigration A type of migration in which individuals enter a population and thus increase its size. Compare with *emigration*.

immune response See *specific* and *nonspecific defense mechanisms*.

immunoglobulin (im-yoon″oh-glob′yoo-lin) See *antibody*.

imperfect flower A flower that lacks either stamens or carpels. Compare with *perfect flower*.

imperfect fungi See *deuteromycetes*.

implantation The embedding of a developing embryo in the wall (endometrium) of the uterus.

imprinting A form of learning by which a young bird or mammal forms a strong social attachment to an individual (usually a parent) or object within a few hours after hatching or birth.

inbreeding Mating of genetically similar individuals. Homozygosity increases with each successive generation of inbreeding. Compare with *outbreeding*.

inbreeding depression The phenomenon in which inbred offspring of genetically similar individuals have lower fitness (e.g., decline in fertility and high juvenile mortality) than noninbred individuals.

inclusive fitness The total of an individual's direct and indirect fitness; includes the genes contributed directly to offspring and those contributed by way of a relative.

incomplete dominance Condition in which neither member of a pair of contrasting alleles is completely expressed when the other is present.

incomplete flower A flower that lacks one or more of the four parts: sepals, petals, stamens, and/or carpels. Compare with *complete flower*.

independent assortment, principle of Genetic principle, first noted by Gregor Mendel, that states that the alleles of unlinked loci are randomly distributed to gametes.

indeterminate growth Unrestricted growth, as for example, in stems and roots. Compare with *determinate growth*.

index fossils Fossils restricted to a narrow period of time and found in the same sedimentary layers in different geographical areas.

indoleacetic acid See *auxin*.

induced fit Conformational change in the active site of an enzyme that occurs when it binds to its substrate.

inducer molecule Substance that binds to a repressor protein, converting it to its inactive form, which is unable to prevent transcription.

inducible operon Operon that is normally inactive because a repressor molecule is attached to its operator; transcription is activated when a specific inducer molecule binds to the repressor, making it incapable of binding to the operator, e.g., the lactose operon of *Escherichia coli*. Compare with *repressible operon*.

induction Process by which the differentiation of a cell or group of cells is influenced by interactions with neighboring cells.

inductive reasoning Reasoning that uses specific examples to draw a general conclusion or discover a general principle. Compare with *deductive reasoning*. See *hypothetico-inductive approach*.

infant mortality rate The number of infant deaths per 1000 live births. (A child is an infant during its first two years of life.)

inflammatory response The response of body tissues to injury or infection, characterized clinically by heat, swelling, redness, and pain, and physiologically by increased dilation of blood vessels and increased phagocytosis.

inflorescence A cluster of flowers on a common floral stalk.

ingestion Process of taking food (or other material) into the body.

inhibitory postsynaptic potential (IPSP) A change in membrane potential that takes a neuron farther from the firing level. Compare with *excitatory postsynaptic potential (EPSP)*.

initiation (of protein synthesis) The first steps of protein synthesis, in which the large and small ribosomal subunits and other components of the translation machinery bind to the 5′ end of mRNA. See *elongation* and *termination*.

initiation codon The codon AUG, which serves as the signal to begin translation of messenger RNA; also called a start codon. Compare with *termination codon*.

innate behavior Behavior that is inherited and typical of the species; also called instinct.

innate immune responses See *nonspecific defense mechanisms*.

inner cell mass The cluster of cells in the early mammalian embryo that gives rise to the embryo proper.

inorganic compound A simple substance that does not contain a carbon backbone.

inositol triphosphate (IP$_3$) A second messenger that increases intracellular calcium concentration and activates enzymes.

insight learning Complex learning process in which an animal adapts past experience to solve a new problem that may involve different stimuli.

in situ conservation Conservation efforts that concentrate on preserving biological diversity in the wild. Compare with *ex situ conservation*.

instinct See *innate behavior*.

instrumental conditioning See *operant conditioning*.

insulin (in′suh-lin) A hormone secreted by the pancreas that lowers blood glucose concentration. Compare with *glucagon*.

insulin-like growth factors (IGF) Somatomedins; proteins that mediate responses to growth hormone.

integrin A transmembrane protein that connects the cell to the extracellular matrix.

integral membrane protein A protein that is tightly associated with the lipid bilayer of a biological membrane; a transmembrane integral protein spans the bilayer. Compare with *peripheral membrane protein*.

integration Process of summing (adding and subtracting) incoming neural signals.

integumentary system (in-teg″yoo-men′tar-ee) The body's covering, including the skin and its nails, glands, hair, and other associated structures.

intercellular substance In connective tissues, the combination of matrix and fibers in which the cells are embedded.

interferons (in″tur-feer′on) Cytokines produced by animal cells when challenged by a virus; prevent viral reproduction and enable cells to resist a variety of viruses.

interkinesis The stage between meiosis I and meiosis II. Interkinesis is usually brief; the chromosomes may decondense, reverting at least partially to an interphase-like state, but DNA synthesis and chromosome duplication do not occur.

interleukins A diverse group of cytokines produced mainly by macrophages and lymphocytes.

intermediate-day plant A plant that flowers when it is exposed to days and nights of intermediate length but does not flower when the daylength is too long or too short. Compare with *long-day*, *short-day*, and *day-neutral plants*.

intermediate filaments Cytoplasmic fibers that are part of the cytoskeletal network and are intermediate in size between microtubules and microfilaments.

internal membrane system Group of membranous structures in eukaryotic cells that interact extensively through direct connections or by means of vesicles; includes the endoplasmic reticulum, outer membrane of the nuclear envelope, Golgi complex, lysosomes, and the plasma membrane; also called endomembrane system.

interneuron (in"tur-noor'on) A nerve cell that carries impulses from one nerve cell to another and is not directly associated with either an effector or a sensory receptor. Also known as an association neuron.

internode The region on a stem between two successive nodes. Compare with *node*.

interoceptor (in'tur-oh-sep"tor) A sense organ within a body organ that transmits information regarding chemical composition, pH, osmotic pressure, or temperature. Compare with *exteroceptor*.

interphase Stage of the cell cycle between successive mitotic divisions; its subdivisions are the G_1 (first gap), S (DNA synthesis), and G_2 (second gap) phases.

interspecific competition The interaction between members of different species that vie for the same resource in an ecosystem (such as food, living space, or other resources). Compare with *intraspecific competition*.

interstitial cells (of testis) The cells between the seminiferous tubules that secrete testosterone.

interstitial fluid The fluid that bathes the tissues of the body; also called tissue fluid.

intertidal zone The marine shoreline area between the high tide mark and the low tide mark.

intraspecific competition The interaction between members of the same species that vie for the same resource in an ecosystem (such as food, living space, or other resources). Compare with *interspecific competition*.

intron A non–protein-coding region of a eukaryotic gene and also of the pre-mRNA transcribed from such a region. Introns do not appear in mRNA. Compare with *exon*.

invertebrate An animal without a backbone (vertebral column); invertebrates account for about 95% of animal species.

in vitro Occurring outside a living organism (literally "in glass"). Compare with *in vivo*.

in vivo Occurring in a living organism. Compare with *in vitro*.

ion An atom or group of atoms bearing one or more units of electrical charge, either positive (cation) or negative (anion).

ionic bond Chemical attraction between a cation and an anion. Compare with *covalent bond* and *hydrogen bond*.

ionic compound Substance consisting of cations and anions, which are attracted by their opposite charges; ionic compounds do not consist of molecules. Compare with *covalent compound*.

ionization The dissociation of a substance to yield ions, e.g., the ionization of water yields H^+ and OH^-.

iris The pigmented portion of the vertebrate eye.

irreversible inhibitor A substance that permanently inactivates an enzyme. Compare with *reversible inhibitor*.

islets of Langerhans (eye'lets of Lahng'er-hanz) The endocrine portion of the pancreas that secretes glucagon and insulin, hormones that regulate the concentration of glucose in the blood.

isogamy (eye-sog'uh-me) Sexual reproduction involving motile gametes of similar form and size. Compare with *anisogamy* and *oogamy*.

isogenic Term describing a strain of organisms in which all individuals are genetically identical and homozygous at all loci.

isolecithal egg Egg containing a relatively small amount of uniformly distributed yolk. Compare with *telolecithal egg*.

isomer (eye'soh-mer) One of two or more chemical compounds having the same chemical formula, but different structural formulas, e.g., structural and geometric isomers and enantiomers.

isoprene units Five-carbon hydrocarbon monomers that make up certain lipids such as carotenoids and steroids.

isotonic (eye"soh-ton'ik) Term applied to solutions that have identical concentrations of solute molecules and hence the same osmotic pressure. Compare with *hypertonic* and *hypotonic*.

isotope (eye'suh-tope) An alternate form of an element with a different number of neutrons, but the same number of protons and electrons. See also *radioisotopes*.

jelly coat One of the acellular coverings of the eggs of certain animals, such as echinoderms.

joint Junction between two or more bones of the skeleton.

joule A unit of energy, equivalent to 0.239 calories.

juvenile hormone (JH) An arthropod hormone that preserves juvenile structure during a molt. Without it, metamorphosis toward the adult form takes place.

juxtaglomerular apparatus (juks"tah-glo-mer'yoo-lar) A structure in the kidney that secretes renin in response to a decrease in blood pressure.

K selection A reproductive strategy in which a species (known as a *K*-selected species or *K* strategist) typically has a large body size, slow development, long lifespan, and does not devote a large proportion of its metabolic energy to the production of offspring. Compare with *r selection*.

karyotype (kare'ee-oh-type) The chromosomal constitution of an individual. Representations of the karyotype are generally prepared by photographing the chromosomes and arranging the homologous pairs according to size, centromere position, and pattern of bands.

keratin (kare'ah-tin) A horny, water-insoluble protein found in the epidermis of vertebrates and in nails, feathers, hair, and horns.

ketone An organic molecule containing a carbonyl group bonded to two carbon atoms. Compare with *aldehyde*.

keystone species A species whose presence in an ecosystem largely determines the species composition and functioning of that ecosystem.

kidney The paired vertebrate organ important in excretion of metabolic wastes and in osmoregulation.

killer T cell See *cytotoxic T cell*.

kilobase (kb) 1000 bases or base pairs of a nucleic acid.

kilocalorie The amount of heat required to raise the temperature of 1 kg of water 1°C; also called Calorie.

kilojoule 1000 joules. See *joule*.

kinases Enzymes that catalyze the transfer of phosphate groups from ATP to acceptor molecules. Protein kinases activate or inactivate proteins through the addition of phosphates at specific locations.

kinetic energy Energy of motion. Compare with *potential energy.*

kinetochore (kin-eh′toh-kore) Portion of the chromosome centromere to which the mitotic spindle fibers attach.

kingdom A broad taxonomic category made up of related phyla; many biologists currently recognize six kingdoms of living organisms.

kin selection A type of natural selection that favors altruistic behavior toward relatives (kin), thereby ensuring that, although the chances of an individual's survival are lessened, some of its genes will survive through successful reproduction of close relatives; increases inclusive fitness.

Klinefelter syndrome Inherited condition in which the affected individual is a sterile male with an XXY karyotype.

Krebs cycle See *citric acid cycle.*

labyrinth The system of interconnecting canals of the inner ear of vertebrates.

labyrinthodonts The first successful group of tetrapods.

lactation (lak-tay′shun) The production or release of milk from the breast.

lacteal (lak′tee-al) One of the many lymphatic vessels in the intestinal villi that absorb fat.

lactic acid A three-carbon organic acid; also known as lactate.

lagging strand Strand of DNA that is synthesized as a series of short segments, called Okazaki fragments, which are then covalently joined by DNA ligase. Compare with *leading strand.*

lamins Proteins attached to the inner surface of the nuclear envelope that provide a type of skeletal framework.

large intestine The portion of the digestive tract of humans (and other vertebrates) consisting of the cecum, colon, rectum, and anus.

larva (pl. *larvae*) An immature form in the life history of some animals; may be unlike the parent.

larynx (lare′inks) The organ at the upper end of the trachea that contains the vocal cords.

lateral meristems Areas of localized cell division on the side of a plant that give rise to secondary tissues. Lateral meristems, including the vascular cambium and the cork cambium, cause an increase in the girth of the plant body. Compare with *apical meristem.*

leaching The process by which dissolved materials are washed away or carried with water down through the various layers of the soil.

leader sequence Noncoding sequence of nucleotides in mRNA that is transcribed from the region that precedes (is upstream to) the coding region.

leading strand Strand of DNA that is synthesized continuously. Compare with *lagging strand.*

learning A change in the behavior of an animal that results from experience.

legume (leg′yoom) (1) A simple, dry fruit that splits open at maturity along two sutures to release seeds; (2) Any member of the pea family, e.g., pea, bean, peanut, alfalfa.

lek A small territory in which males compete for females.

lens The oval, transparent structure located behind the iris of the vertebrate eye; bends incoming light rays and brings them to a focus on the retina.

lenticels (len′tih-sels) Porous swellings of cork cells in the stems of woody plants; facilitate the exchange of gases.

leukocytes (loo′koh-sites) White blood cells; colorless amoeboid cells that defend the body against disease-causing organisms.

leukoplasts Colorless plastids; include amyloplasts, which are used for starch storage in cells of roots and tubers.

LH See *luteinizing hormone.*

lichen (ly′ken) A compound organism composed of a symbiotic alga or cyanobacterium and a fungus.

ligament (lig′uh-ment) A connective tissue cable or strap that connects bones to each other or holds other organs in place.

ligand A molecule that binds to a specific site in a receptor or other protein.

light-dependent reactions Reactions of photosynthesis in which light energy absorbed by chlorophyll is used to synthesize ATP and usually NADPH.

lignin (lig′nin) A substance found in many plant cell walls that confers rigidity and strength, particularly in woody tissues.

limbic system In vertebrates, an action system of the brain. In humans, plays a role in emotional responses, motivation, autonomic function, and sexual response.

limiting factor An environmental factor that restricts the growth, distribution, or abundance of a particular population.

limnetic zone (lim-net′ik) The open water away from the shore of a lake or pond extending down as far as sunlight penetrates. Compare with *littoral zone* and *profundal zone.*

linkage The tendency for a group of genes located on the same chromosome to be inherited together in successive generations.

lipase (lip′ase) Fat-digesting enzyme.

lipid Any of a group of organic compounds that are insoluble in water but soluble in nonpolar solvents; lipids serve as energy storage and are important components of cell membranes.

littoral zone (lit′or-ul) The region of shallow water along the shore of a lake or pond. Compare with *limnetic zone* and *profundal zone.*

liver A large, complex organ that secretes bile, helps maintain homeostasis by removing or adding nutrients to the blood, and performs many other metabolic functions.

liverworts A phylum of spore-producing, nonvascular, thallose or leafy plants with a life cycle similar to mosses.

locus The place on the chromosome at which the gene for a given trait occurs, i.e., a segment of the chromosomal DNA containing information that controls some feature of the organism; also called gene locus.

logistic population growth Population growth that initially occurs at a constant rate of increase over time (i.e., exponential) but then levels out as the carrying capacity of the environment is approached. When the increase in number versus time is plotted on a graph, logistic growth produces a characteristic S-shaped curve. Compare with *exponential population growth.*

long-day plant A plant that flowers in response to shortening nights; also called short-night plants. Compare with *short-day, intermediate-day,* and *day-neutral plants.*

long-night plant See *short-day plant.*

loop of Henle (Hen′lee) The U-shaped loop of a mammalian kidney tubule, which extends down into the renal medulla.

loose connective tissue A type of connective tissue that is widely distributed in the body; consists of fibers strewn through a semifluid matrix.

lophoporate phyla Three related invertebrate deuterostome phyla, characterized by a ciliated ring of tentacles that surrounds the mouth.

lumen (loo´men)   (1) The space enclosed by a membrane, such as the lumen of the endoplasmic reticulum; (2) The cavity or channel within a tube or tubular organ, such as a blood vessel or the digestive tract.

lung   An internal respiratory organ that functions in gas exchange; enables an animal to breathe air.

luteinizing hormone (LH) (loot´-eh-ny-zing)   Gonadotropic hormone secreted by the anterior pituitary; stimulates ovulation and maintains the corpus luteum in the ovaries of females; stimulates testosterone production by interstitial cells in the testes of males.

lymph (limf)   The colorless fluid within the lymphatic vessels that is derived from blood plasma and resembles it closely in composition; contains white blood cells; ultimately lymph is returned to the blood.

lymph node   A mass of lymph tissue surrounded by a connective tissue capsule; manufactures lymphocytes and filters lymph.

lymphatic system   A subsystem of the cardiovascular system; returns excess interstitial fluid (lymph) to the circulation; defends the body against disease organisms.

lymphocyte (lim´foh-site)   White blood cell with nongranular cytoplasm that is responsible for immune responses. See *B cell* and *T cell*.

lysis (ly´sis)   The process of disintegration of a cell or some other structure.

lysogenic conversion   The change in properties of bacteria that result from the presence of a prophage.

lysosomes (ly´soh-somes)   Intracellular organelles present in many animal cells; contain a variety of hydrolytic enzymes.

lysozyme   An enzyme found in many tissues and in tears and other body fluids; attacks the cell wall of many gram-positive bacteria.

macroevolution (mak´´roh-eh-voh-loo´shun)   Large-scale evolutionary change over long time spans. Compare with *microevolution*.

macromolecule   A very large organic molecule, such as a protein or nucleic acid.

macronucleus   A large nucleus found, along with one or several micronuclei, in ciliates. The macronucleus regulates metabolism and growth. Compare with *micronucleus*.

macronutrient   An essential element that is required in fairly large amounts for normal growth. Compare with *micronutrient*.

macrophage (mak´roh-faje)   A large phagocytic cell capable of ingesting and digesting bacteria and cellular debris. Macrophages are also antigen-presenting cells.

major histocompatibility complex (MHC)   A group of membrane proteins, present on the surface of most cells, that are slightly different in each individual. In humans, the MHC is called the HLA (human leukocyte antigen) group.

malignant   Term used to describe cancer cells, tumor cells that are able to invade tissue and metastasize.

Malpighian tubules (mal-pig´ee-an)   The excretory organs of many arthropods.

mammals   Class of vertebrates characterized by hair, mammary glands, a diaphragm, and differentiation of teeth.

mandible (man´dih-bl)   (1) The lower jaw of vertebrates; (2) Jawlike, external mouthparts of insects.

mangrove forest   A wetland dominated by mangrove trees in which the salinity fluctuates between that of sea water and fresh water; mangrove trees grow on protected shores in the intertidal zone.

mantle   In the mollusk, a fold of tissue that covers the visceral mass and that usually produces a shell.

marine snow   The organic debris (plankton, dead organisms, fecal material, etc.) that "rains" into the dark area of the oceanic province from the lighted region above; the primary food of most organisms that live in the ocean's depths.

marsupials (mar-soo´pee-uls)   A subclass of mammals, characterized by the presence of an abdominal pouch in which the young are carried for some time after they are born; the young are born in a very undeveloped condition.

mass extinction   The extinction of numerous species during a relatively short period of geological time. Compare with *background extinction*.

mast cell   A type of cell found in connective tissue; contains histamine and is important in allergic reactions.

maternal effect genes   Genes of the mother that are transcribed during oogenesis and subsequently affect the development of the embryo. Compare with *zygotic genes*.

matrix (may´triks)   (1) The interior of the compartment enclosed by the inner mitochondrial membrane; (2) Nonliving material secreted by and surrounding connective tissue cells; contains a network of microscopic fibers.

matter   Anything that has mass and takes up space.

maxillae   Appendages used for manipulating food; characteristic of crustaceans.

mechanical isolation   A reproductive isolating mechanism (prezygotic) in which fusion of the gametes of two species is prevented by morphological or anatomical differences.

mechanoreceptor (meh-kan´oh-ree-sep´´tor)   A sensory cell or organ that perceives mechanical stimuli, e.g., touch, pressure, hearing, and balance.

medulla (meh-dul´uh)   (1) The inner part of an organ, such as the medulla of the kidney; compare with *cortex*; (2) The most posterior part of the vertebrate brain, lying next to the spinal cord.

medusa   A jellyfish-like animal; a free-swimming, umbrella-shaped stage in the life cycle of certain cnidarians. Compare with *polyp*.

megaphyll (meg´uh-fil)   Type of leaf found in horsetails, ferns, gymnosperms, and angiosperms; contains multiple vascular strands (i.e., complex venation). Compare with *microphyll*.

megaspore (meg´uh-spor)   The *n* spore in heterosporous plants that gives rise to a female gametophyte. Compare with *microspore*.

meiosis (my-oh´sis)   Process in which a 2*n* cell undergoes two successive nuclear divisions (meiosis I and meiosis II), potentially producing four *n* nuclei; leads to the formation of gametes in animals and spores in plants.

memory cell   B or T lymphocyte that permits rapid mobilization of immune response on second or subsequent exposure to a particular antigen.

meninges (meh-nin´jeez) (sing. *meninx*)   The three membranes that envelop the brain and spinal cord: the dura mater, arachnoid, and pia mater.

menopause   The period (usually from 45 to 55 years of age) in women when the recurring menstrual cycle ceases.

menstruation (men-stroo-ay´shun)   The monthly discharge of blood and degenerated uterine lining in the human female; marks the beginning of each menstrual cycle.

meristem (mer´ih-stem)   A localized area of mitotic cell division in the plant body. See *apical meristem* and *lateral meristem*.

meroblastic cleavage   Cleavage pattern observed in the telolecithal eggs of reptiles and birds, in which cleavage is restricted to a small disc of cytoplasm at the animal pole. Compare with *holoblastic cleavage*.

mesenchyme (mes'en-kime)   A loose, often jelly-like connective tissue containing undifferentiated cells; found in the embryos of vertebrates and the adults of some invertebrates.

mesoderm (mez'oh-derm)   The middle germ layer of the early embryo; gives rise to connective tissue, muscle, bone, blood vessels, kidneys, and many other structures. Compare with *ectoderm* and *endoderm*.

mesophyll (mez'oh-fil)   Photosynthetic tissue in the interior of a leaf; sometimes differentiated into palisade mesophyll and spongy mesophyll.

Mesozoic era   That part of geological time extending from roughly 248 to 65 million years ago.

messenger RNA (mRNA)   RNA that specifies the amino acid sequence of a protein; transcribed from DNA.

metabolic pathway   A series of chemical reactions in which the product of one reaction becomes the substrate of the next reaction.

metabolic rate   Energy production of an organism per unit time. See *basal metabolic rate*.

metabolism   The sum of all the chemical processes that occur within a cell or organism: the transformations by which energy and matter are made available for use by the organism. See *anabolism* and *catabolism*.

metamorphosis (met"ah-mor'fuh-sis)   Transition from one developmental stage to another, such as from a larva to an adult.

metanephridia (sing. *metanephridium*)   The excretory organs of annelids and mollusks; each metanephridium consists of a tubule open at both ends; at one end a ciliated funnel opens into the coelom, and the other end of the tube opens to the outside of the body.

metaphase (met'ah-faze)   The stage of mitosis, and of meiosis I and II, in which the chromosomes line up on the equatorial plane of the cell. Occurs after prophase and before anaphase.

metastasis (met-tas'tuh-sis)   The spreading of cancer cells from one organ or part of the body to another.

methyl group   Nonpolar functional group; abbreviated R—CH$_3$.

microclimate   Local variations in climate produced by differences in elevation, in the steepness and direction of slopes, and in exposure to prevailing winds.

microevolution   Small-scale evolutionary changes due to changes in allele or genotype frequencies that occur within a population over successive generations. Compare with *macroevolution*.

microfilaments   Thin fibers composed of actin protein subunits; form part of the cytoskeleton.

microfossils   Ancient traces (fossils) of microscopic life.

micronucleus   One or more smaller nuclei found, along with the macronucleus, in ciliates. The micronucleus is involved in sexual reproduction. Compare with *macronucleus*.

micronutrient   An essential element that is required in trace amounts for normal growth. Compare with *macronutrient*.

microphyll (mi'kro-fil)   Type of leaf found in club mosses; contains one vascular strand (i.e., simple venation). Compare with *megaphyll*.

microsphere   A protobiont formed by adding water to abiotically formed polypeptides.

microspore (mi'kro-spor)   The *n* spore in heterosporous plants that gives rise to a male gametophyte. Compare with *megaspore*.

microtubule (my-kroh-too'bewl)   Hollow cylindrical fiber composed of tubulin protein subunits; microtubules are major components of the cytoskeleton and found in mitotic spindles, cilia, flagella, centrioles, and basal bodies.

microtubule-associated proteins (MAPs)   Include fibrous proteins and motors, such as kinesin and dynein.

microtubule-organizing center (MTOC)   Region of the cell from which microtubules extend and which appears to serve as the major site of assembly of microtubules from tubulin subunits. The MTOCs of many organisms (including animals, but not flowering plants nor most gymnosperms) contain a pair of centrioles.

microvilli (sing. *microvillus*)   Minute projections of the plasma membrane that increase the surface area of the cell; found mainly in cells concerned with absorption or secretion, such as those lining the intestine or the kidney tubules.

middle lamella   Layer composed of pectin polysaccharides that serves to cement together the primary cell walls of adjacent plant cells.

migration   Movement of an organism (individual or population) from one place to another. See *immigration* and *emigration*.

mineralocorticoids (min"ur-al-oh-kor'tih-koidz)   Hormones produced by the adrenal cortex that regulate mineral metabolism and, indirectly, fluid balance. The principal mineralocorticoid is aldosterone.

minerals   Inorganic nutrients ingested in the form of salts dissolved in food and water.

missense mutation   A type of base-substitution mutation that causes one amino acid to be substituted for another in the resulting protein product. Compare with *nonsense mutation*.

mitochondria (my"toh-kon'dree-ah) (sing. *mitochondrion*)   Spherical or elongate intracellular organelles that are the sites of oxidative phosphorylation in eukaryotes; include an outer membrane and an inner membrane.

mitosis (my-toh'sis)   Division of the cell nucleus resulting in two daughter nuclei, each with the same number of chromosomes as the parent nucleus; mitosis consists of four phases: prophase, metaphase, anaphase, and telophase. Cytokinesis usually overlaps the telophase stage.

mitosis promoting factor (MPF)   A cyclin-dependent protein kinase that controls the transition of a cell from the G$_2$ stage of interphase to mitosis; formerly known as maturation promoting factor.

mitotic spindle   Structure consisting mainly of microtubules that provides the framework for chromosome movement during cell division.

mitral valve   See *atrioventricalar valve*.

mobile genetic element   See *transposon*.

mole   The atomic mass of an element or the molecular mass of a compound, expressed in grams; one mole of any substance has 6.02×10^{23} units (Avogadro's number).

molecular chaperones   Proteins that help other proteins fold properly. While not dictating the folding pattern, chaperones make the process more efficient.

molecular clock analysis   A comparison of the DNA nucleotide sequences of related organisms in order to estimate when they diverged from one another during the course of evolution.

molecular formula   Type of chemical formula that gives the actual numbers of each type of atom in a molecule. Compare with *structural formula*.

molecular mass   The sum of the atomic masses of the atoms that make up a single molecule of a compound; expressed in atomic mass units (amu) or daltons.

molecule The smallest particle of a covalently bonded element or compound that has the composition and properties of a larger part of the substance.

mollusks A phylum of coelomate protostome animals characterized by a soft body, visceral mass, mantle, and foot.

molting The shedding and replacement of an outer covering such as an exoskeleton.

molting hormone A steroid hormone that stimulates growth and molting in insects.

monoacylglycerol (mon"o-as"-il-glis'er-ol) A neutral fat consisting of glycerol combined chemically with a single fatty acid. Also called monoglyceride.

monocot (mon'oh-kot) One of the two classes of flowering plants; monocot seeds contain a single cotyledon, or seed leaf. Compare with *dicot*.

monocyte (mon'oh-site) A type of white blood cell; a large phagocytic, nongranular leukocyte that enters the tissues and differentiates into a macrophage.

monoecious (mon-ee'shus) Having male and female reproductive parts in separate flowers or cones on the same plant; compare with *dioecious*.

monogamy A mating system in which a male animal mates with a single female during a breeding season; common among birds.

monoglyceride See *monoacylglycerol*.

monohybrid cross A genetic cross that takes into account the behavior of alleles of a single locus. Compare with *dihybrid cross*.

monokaryotic (mon"o-kare-ee-ot'ik) Condition of having a single *n* nucleus per cell, characteristic of certain fungal hyphae. Compare with *dikaryotic*.

monomer (mon'oh-mer) A molecule of a compound of relatively low molecular weight that can be linked with other similar molecules to form a polymer.

monophyletic group (mon"oh-fye-let'ik) A group made up of organisms that evolved from a common ancestor. Compare with *polyphyletic group* and *paraphyletic group*.

monosaccharide (mon-oh-sak'ah-ride) A simple sugar that cannot be degraded by hydrolysis to a simpler sugar.

monosomy Condition in which only one member of a chromosome pair is present. Compare with *trisomy*.

monotremes (mon'oh-treems) Egg-laying mammals such as the duck-billed platypus of Australia.

monounsaturated fatty acid See *fatty acid*.

monozygotic twins Genetically identical twins that arise from the division of a single fertilized egg; commonly known as identical twins. Compare with *dizygotic twins*.

morphogen Any chemical agent thought to be responsible for the processes of cell differentiation and pattern formation that lead to morphogenesis.

morphogenesis (mor-foh-jen'eh-sis) The development of the form and structures of an organism and its parts; proceeds by a series of steps known as pattern formation.

mortality The number of deaths per 1000 people per year.

morula (mor'yoo-lah) An early embryo consisting of a solid ball of cells.

mosaic development Highly invariant developmental pattern in which the fate of each blastomere becomes determined at a very early stage. Compare with *regulative development*.

mosses A phylum of spore-producing nonvascular plants with an alternation of generations in which the dominant *n* gametophyte alternates with a 2*n* sporophyte that remains attached to the gametophyte.

motor neuron Efferent neuron; a neuron that transmits impulses away from the central nervous system.

motor unit All the skeletal muscle fibers that are stimulated by a single motor neuron.

mRNA cap An unusual nucleotide, 7-methylguanylate, that is added to the 5′ end of a eukaryotic messenger RNA. Capping enables eukaryotic ribosomes to bind to the message.

mucosa (mew-koh'suh) A mucous membrane, especially in the lining of the digestive and respiratory tracts.

mucus (mew'cus) A sticky secretion composed of covalently linked protein and carbohydrate; serves to lubricate body parts and trap particles of dirt and other contaminants. (The adjectival form is spelled *mucous*.)

Müllerian mimicry (mul-ler'ee-un mim'ih-kree) The resemblance of dangerous, unpalatable, or poisonous species to one another so that they are more easily recognized by potential predators, which learn to avoid all of them after tasting one. Compare with *Batesian mimicry*.

multiple alleles (al-leels') Three or more alleles of a single locus (in a population), such as the alleles governing the ABO series of blood types.

multiple fruit A fruit that develops from many ovaries of many separate flowers, e.g., pineapple. Compare with *simple, aggregate,* and *accessory fruits*.

muscle (1) A tissue specialized for contraction; (2) An organ that produces movement by contraction.

mutagen (mew'tah-jen) Any agent capable of producing mutations.

mutation Any change in DNA; may include a change in the nucleotide base pairs of a gene, a rearrangement of genes within the chromosomes so that their interactions produce different effects, or a change in the chromosomes themselves.

mutualism A symbiotic relationship in which both partners benefit from the association. Compare with *parasitism* and *commensalism*.

mycelium (my-seel'ee-um) (pl. *mycelia*) The vegetative body of most fungi and certain protists (water molds); consists of a branched network of hyphae.

mycorrhizae (my"kor-rye'zee) Mutualistic associations of fungi and plant roots that aid in the plant's absorption of essential minerals from the soil.

mycotoxins Poisonous chemical compounds produced by fungi, e.g., aflatoxins that harm the liver and are known carcinogens.

myelin sheath (my'eh-lin) The white fatty material that forms a sheath around the axons of certain nerve cells, which are then called myelinated fibers.

myocardial infarction (MI) Heart attack; serious consequence occurring when the heart muscle receives insufficient oxygen.

myofibrils (my-oh-fy'brilz) Tiny threadlike structures in the cytoplasm of striated and cardiac muscle that are responsible for contractions of the cell; contain myofilaments.

myofilament (my-oh-fil'uh-ment) One of the filaments making up a myofibril; the structural unit of muscle proteins in a muscle cell. See *thick (myosin) filament* and *thin (actin) filament*.

myoglobin (my'oh-glob"bin) A hemoglobin-like, oxygen-transferring protein found in muscle.

myosin (my'oh-sin) A protein that, together with actin, is responsible for muscle contraction.

n The chromosome number of a gamete. The chromosome number of a zygote is 2*n*. If an organism is not polyploid, the *n* gametes are haploid and the 2*n* zygotes are diploid.

NAD⁺/NADH Oxidized and reduced forms, respectively, of nicotinamide adenine dinucleotide; coenzyme that transfers electrons (as hydrogen), particularly in catabolic pathways, including cellular respiration.

NADP⁺/NADPH Oxidized and reduced forms, respectively, of nicotinamide adenine dinucleotide phosphate; coenzyme that acts as an electron (hydrogen) transfer agent, particularly in anabolic pathways, including photosynthesis.

nanoplankton Extremely minute (less than 70 to 75 μm in diameter) algae that are major producers in the ocean because of their great abundance; part of phytoplankton.

nastic movement A temporary, reversible movement of a plant organ in response to external stimuli; movement is caused by changes in the turgor of certain cells.

natality The number of births per 1000 people per year.

natural selection The mechanism of evolution proposed by Charles Darwin; the tendency of organisms that possess favorable adaptations to their environment to survive and become the parents of the next generation. Evolution occurs when natural selection results in changes in allele frequencies in a population.

natural killer cell (NK cell) A large, granular lymphocyte that functions in both nonspecific and specific defense; releases cytokines and proteolytic enzymes that target tumor cells and cells infected with viruses and other pathogens.

Neandertal See *Homo neanderthalensis*.

nectary (nek'ter-ee) A gland or other structure that secretes nectar.

negative feedback system A homeostatic mechanism in which a change in some condition triggers a response that counteracts, or reverses, the changed condition, e.g., how mammals maintain body temperature. Compare with *positive feedback system*.

nekton (nek'ton) Free-swimming aquatic organisms such as fish and turtles. Compare with *plankton*.

nematocyst (nem-at'oh-sist) A stinging structure found within cnidocytes (stinging cells) in cnidarians; used for anchorage, defense, and capturing prey.

nematodes Phylum of pseudocoelomate animals commonly known as roundworms.

nemerteans Phylum of functionally acoelomate animals commonly known as ribbon worms; possess a complete digestive tract.

neo-Darwinism See *synthetic theory of evolution*.

neonate Newborn individual.

neoplasm See *tumor*.

nephridial organ (neh-frid'ee-al) The excretory organ of many invertebrates; consists of simple or branching tubes that usually open to the outside of the body through pores.

nephron (nef'ron) The functional, microscopic unit of the vertebrate kidney.

neritic province (ner-ih'tik) Ocean water that extends from the shoreline to where the bottom reaches a depth of 200 m. Compare with *oceanic province*.

nerve A bundle of axons (or dendrites) wrapped in connective tissue that conveys impulses between the central nervous system and some other part of the body.

nerve net A system of interconnecting nerve cells found in cnidarians and echinoderms.

nervous tissue A type of animal tissue specialized for conducting electrochemical impulses.

net primary productivity Energy that remains in an ecosystem (as biomass) after cellular respiration has occurred. Compare with *gross primary productivity*.

neural crest (noor'ul) Group of cells along the neural tube that migrate and form structures of the peripheral nervous system and certain other structures.

neural transmission See *transmission, neural*.

neural tube The hollow, longitudinal structure in the early vertebrate embryo that gives rise to the brain and spinal cord. The neural tube forms from the neural plate, a flattened, thickened region of the ectoderm that rolls up and sinks below the surface.

neuron (noor'on) A nerve cell; a conducting cell of the nervous system that typically consists of a cell body, dendrites, and an axon.

neurotransmitter A chemical messenger used by neurons to transmit impulses across a synapse.

neutral fat A lipid used for energy storage, consisting of a glycerol covalently bonded to one, two or three fatty acids. See *monoacylglycerol, diacylglycerol,* and *triacylglycerol*.

neutral solution A solution of pH 7; there are equal concentrations of hydrogen ions (H⁺) and hydroxide ions (OH⁻). Compare with *acidic solution* and *basic solution*.

neutral variation Variation that does not appear to confer any selective advantage or disadvantage to the organism.

neutron (noo'tron) An electrically neutral particle with a mass of one atomic mass unit (amu) found in the atomic nucleus. Compare with *proton* and *electron*.

neutrophil (new'truh-fil) A type of granular leukocyte.

niche (nich) The totality of an organism's adaptations, its use of resources, and the lifestyle to which it is fitted in its community. The niche describes how an organism uses materials in its environment as well as how it interacts with other organisms; also called ecological niche.

nicotinamide adenine dinucleotide See NAD⁺/NADH.

nicotinamide adenine dinucleotide phosphate See NADP⁺/NADPH.

nitrification (nie"tra-fuh-kay'shun) The conversion of ammonia to nitrate by certain bacteria (nitrifying bacteria) in the soil; part of the nitrogen cycle.

nitrogen cycle The worldwide circulation of nitrogen from the abiotic environment into living things and back into the abiotic environment.

nitrogen fixation The conversion of atmospheric nitrogen (N_2) to ammonia (NH_3) by certain bacteria; part of the nitrogen cycle.

nitrogenase (nie-traa'jen-ase) The enzyme responsible for nitrogen fixation under anaerobic conditions.

node The area on a stem where each leaf is attached. Compare with *internode*.

nodules Swellings on the roots of plants, such as legumes, in which symbiotic nitrogen-fixing bacteria (*Rhizobium*) live.

noncompetitive inhibitor A substance that lowers the rate at which an enzyme catalyzes a reaction but does not bind to the active site. Compare with *competitive inhibitor*.

noncyclic photophosphorylation In photosynthesis, a series of reactions involving Photosystems I and II; results in the formation of ATP (by chemiosmosis), NADPH, and O_2. Compare with *cyclic photophosphorylation*.

nondisjunction Abnormal separation of sister chromatids or of homologous chromosomes caused by their failure to disjoin (move apart) properly during mitosis or meiosis.

nonpolar covalent bond Chemical bond formed by the equal sharing of electrons between atoms of approximately equal electronegativity. Compare with *polar covalent bond*.

nonpolar molecule   Molecule that does not have a positively charged end and a negatively charged end; nonpolar molecules are generally insoluble in water. Compare with *polar molecule*.

nonsense mutation   A type of base substitution mutation that results in an amino acid–specifying codon being changed to a termination (stop) codon; when the abnormal mRNA is translated, the resulting protein is usually truncated and nonfunctional. Compare with *missense mutation*.

nonspecific defense mechanisms   Mechanisms such as physical barriers (e.g., the skin) and phagocytosis that provide general protection against pathogens. Also called innate immune responses. Compare with *specific defense mechanisms*.

norepinephrine (nor-ep-ih-nef′rin)   A neurotransmitter that is also a hormone secreted by the adrenal medulla.

Northern blot hybridization   A technique in which RNA fragments, previously separated by gel electrophoresis, are transferred to a nitrocellulose membrane. A specific radioactive genetic probe is then allowed to hybridize to complementary fragments, thus marking their locations. Compare with *Southern blot hybridization* and *Western blotting*.

notochord (no′toe-kord)   The flexible, longitudinal rod in the anteroposterior axis that serves as an internal skeleton in the embryos of all chordates and in the adults of some.

nuclear area   Region of a bacterial cell that contains DNA but is not enclosed by a membrane.

nuclear envelope   The double membrane system that encloses the cell nucleus of eukaryotes.

nuclear equivalence   The concept that, with few exceptions, the nuclei of all differentiated cells of an adult organism are genetically identical to each other and to the nucleus of the zygote from which they were derived.

nuclear pores   Structures in the nuclear envelope that allow passage of certain materials between the cell nucleus and the cytoplasm.

nucleolus (new-klee′oh-lus) (pl. *nucleoli*)   Specialized structure in the cell nucleus formed from regions of several chromosomes; site of assembly of the ribosomal subunits.

nucleoplasm   The contents of the cell nucleus.

nucleoside triphosphate   Molecule consisting of a nitrogenous base, a pentose sugar, and three phosphate groups, e.g., adenosine triphosphate (ATP).

nucleosomes (new′klee-oh-somz)   Repeating units of chromatin structure, each consisting of a length of DNA wound around a complex of eight histone molecules (two of each of four different types). Adjacent nucleosomes are connected by a DNA linker region associated with a fifth histone protein.

nucleotide (noo′klee-oh-tide)   A molecule composed of one or more phosphate groups, a five-carbon sugar (ribose or deoxyribose), and a nitrogenous base (purine or pyrimidine).

nucleus (new′klee-us) (pl. *nuclei*)   (1) The central region of an atom, containing the protons and neutrons; (2) A cellular organelle in eukaryotes that contains the DNA and serves as the control center of the cell; (3) A mass of nerve cell bodies in the central nervous system.

nut   A simple, dry fruit that contains a single seed and is surrounded by a hard fruit wall.

nutrients   The chemical substances in food that are used as components for synthesizing needed materials and/or as energy sources.

nutrition   The process of taking in and using food (nutrients).

obligate anaerobe   Organism that grows only in the absence of oxygen. Compare with *facultative anaerobe*.

occipital lobes   Posterior areas of the mammalian cerebrum; interpret visual stimuli from the retina of the eye.

oceanic province   That part of the open ocean that overlies an ocean bottom deeper than 200 meters. Compare with *neritic province*.

Okazaki fragment   One of many short segments of DNA, each 100 to 1000 nucleotides long, that must be joined by DNA ligase to form the lagging strand in DNA replication.

olfactory epithelium   Tissue containing the odor-sensing neurons.

ommatidium (om″ah-tid′ee-um) (pl. *ommatidia*)   One of the light-detecting units of a compound eye, consisting of a lens and a crystalline cone that focus light onto photoreceptors called retinular cells.

omnivore (om′nih-vore)   An animal that eats a variety of plant and animal materials.

oncogene (on′koh-jeen)   An abnormally functioning gene implicated in causing cancer. Compare with *proto-oncogene* and *anti-oncogene*.

onycophorans (on″ih-kof′or-anz)   Phylum of rare, tropical, caterpillar-like animals, structurally intermediate between annelids and arthropods, possessing an annelid-like excretory system, and claw-tipped short legs.

oocytes (oh′oh-sites)   Meiotic cells that give rise to egg cells (ova).

oogamy (oh-og′uh-me)   The fertilization of a large, nonmotile female gamete by a small, motile male gamete. Compare with *isogamy* and *anisogamy*.

oogenesis (oh″oh-jen′eh-sis)   Production of female gametes (eggs) by meiosis. Compare with *spermatogenesis*.

oospore   A thick-walled, resistant spore formed from a zygote during sexual reproduction in water molds and certain algae.

open circulatory system   A type of circulatory system in which the blood bathes the tissues directly; characteristic of arthropods and many mollusks. Compare with *closed circulatory system*.

open system   An entity that can exchange energy and matter with its surroundings. Compare with *closed system*.

operant conditioning   A type of learning in which an animal is rewarded or punished for performing a behavior it discovers by chance; also called instrumental conditioning.

operator site   One of the control regions of an operon; the DNA segment to which a repressor binds, thereby inhibiting the transcription of the adjacent structural genes of the operon.

operculum   In bony fishes, a protective flap of the body wall that covers the gills.

operon (op′er-on)   In prokaryotes, a group of structural genes that are coordinately controlled and transcribed as a single message, plus their adjacent regulatory elements.

optimal foraging   The process of obtaining food in a manner that maximizes benefits and/or minimizes costs.

orbital   Region in which electrons occur in an atom or molecule.

order   A taxonomic category made up of related families.

organ   A specialized structure, such as the heart or liver, made up of tissues and adapted to perform a specific function or group of functions.

organ of Corti (kor′tie)   The structure within the inner ear of vertebrates that contains receptor cells that sense sound vibrations.

organ system   Body system; an organized group of tissues and organs that work together to perform a specialized set of functions.

organelle   One of the specialized structures within the cell, such as the mitochondria, Golgi complex, ribosomes, or contractile vacuole; most organelles are membrane-bounded.

organic compound   A compound composed of a backbone made up of carbon atoms.

organism   Any living system composed of one or more cells.

organismic respiration   See *respiration*.

organogenesis   The process of organ formation.

orgasm (or'gazm)   The climax of sexual excitement.

origin of replication   A specific site on the DNA where replication can begin.

osmoregulation (oz"moh-reg-yoo-lay'shun)   The active regulation of the osmotic pressure of body fluids so that they do not become excessively dilute or excessively concentrated.

osmosis (oz-moh'sis)   Net movement of water (the principal solvent in biological systems) by diffusion through a selectively permeable membrane from a region of higher concentration of water (a hypotonic solution) to a region of lower concentration of water (a hypertonic solution).

osmotic pressure   The pressure that must be exerted on the hypertonic side of a selectively permeable membrane to prevent diffusion of water (by osmosis) from the side containing pure water.

osteichthyes (os"tee-ick'thees)   The vertebrate class of bony fishes.

osteocyte (os'tee-oh-site)   A mature bone cell; an osteoblast that has become embedded within the bone matrix and occupies a lacuna.

osteon (os'tee-on)   Spindle-shaped unit of bone composed of concentric layers of osteocytes; Haversian system of bone.

otoliths (oh'toe-liths)   Small calcium carbonate crystals in the saccule and utricle of the inner ear; sense gravity and are important in static equilibrium.

outbreeding   The mating of individuals of unrelated strains. Compare with *inbreeding*.

ovary (oh'var-ee)   (1) In animals, one of the paired female gonads responsible for producing eggs and sex hormones; (2) In flowering plants, the base of the carpel that contains ovules; ovaries develop into fruits after fertilization.

oviduct (oh'vih-dukt)   Tube that carries ova from the ovary to the uterus, cloaca, or body exterior. Also called fallopian tube or uterine tube.

oviparous (oh-vip'ur-us)   Bearing young in the egg stage of development; egg-laying.

ovoviviparous (oh'voh-vih-vip"ur-us)   A type of development in which the young hatch from eggs incubated inside the mother's body.

ovulation (ov-u-lay'shun)   The release of an ovum from the ovary.

ovule (ov'yool)   The structure (i.e., megasporangium) in the plant ovary that develops into the seed following fertilization.

ovum (pl. *ova*)   Female gamete of an animal.

oxaloacetate   Four-carbon compound; important intermediate in the citric acid cycle and in the C_4 and CAM pathways of carbon fixation in photosynthesis.

oxidation   The loss of one or more electrons (or hydrogen atoms) by an atom, ion, or molecule. Compare with *reduction*.

oxidative phosphorylation (fos"for-ih-lay'shun)   The production of ATP using energy derived from the transfer of electrons in the electron transport system of mitochondria; occurs by chemiosmosis.

oxygen debt   The oxygen necessary to metabolize the lactic acid produced during strenuous exercise.

oxygen dissociation curve   A curve depicting the percentage saturation of hemoglobin with oxygen, as a function of certain variables such as oxygen concentration, carbon dioxide concentration, or pH.

oxyhemoglobin   Hemoglobin that has combined with oxygen.

oxytocin (ok"see-tow'sin)   Hormone secreted by the hypothalamus and released by the posterior lobe of the pituitary gland; stimulates contraction of the pregnant uterus and the ducts of mammary glands.

ozone   A blue gas, O_3, with a distinctive odor that is a human-made pollutant near Earth's surface (in the troposphere) but a natural and essential component of the stratosphere.

ozone depletion   The thinning of the ozone layer in the stratosphere.

P generation (parental generation)   Members of two different true-breeding lines that are crossed to produce the F_1 generation.

P680   Chlorophyll *a* molecules that serve as the reaction center of Photosystem II, transferring photoexcited electrons to a primary acceptor; named by their absorption peak at 680 nm.

P700   Chlorophyll *a* molecules that serve as the reaction center of Photosystem I, transferring photoexcited electrons to a primary acceptor; named by their absorption peak at 700 nm.

pacemaker (of the heart)   See *sinoatrial (SA) node*.

Pacinian corpuscle (pah-sin'-ee-an kor'pus-el)   A receptor located in the dermis of the skin that responds to pressure.

paedomorphosis   Retention of juvenile or larval features in a sexually mature animal.

pair bond   A stable relationship between animals of opposite sex that ensures cooperative behavior in mating and rearing the young.

paleoanthropology (pay"lee-o-an-thro-pol'uh-gee)   The study of human evolution.

Paleozoic era   That part of geological time extending from 570 to 248 million years ago.

palindromic   Reading the same forward and backward; some DNA sequences are palindromic because the base sequence of one strand is the reverse of the base sequence in its complement; for example, the complement of 5'-GAATTC-3' is 3'-CTTAAG-5'.

palisade mesophyll (mez'oh-fil)   The vertically stacked, columnar mesophyll cells near the upper epidermis in certain leaves. Compare with *spongy mesophyll*.

pancreas (pan'kree-us)   Large gland located in the vertebrate abdominal cavity. The pancreas produces pancreatic juice containing digestive enzymes; also serves as an endocrine gland, secreting the hormones insulin and glucagon.

panspermia   The idea that life did not originate on Earth, but began elsewhere in the galaxy and drifted through space to Earth.

parabronchi (sing. *parabronchus*)   Thin-walled ducts in the lungs of birds; gases are exchanged across their walls.

paracrine regulation   A type of regulation in which a signal molecule (e.g., certain hormones) diffuses through interstitial fluid and acts on nearby target cells. Compare with *autocrine regulation*.

paraphyletic group   A group of organisms made up of a common ancestor and some, but not all, of its descendants. Compare with *monophyletic group* and *polyphyletic group*.

parapodia (par"uh-poh'dee-ah) (sing. *parapodium*)   Paired, thickly bristled paddle-like appendages extending laterally from each segment of polychaete worms.

parasite   A heterotrophic organism that obtains nourishment from the living tissue of another organism (the host).

parasitism (par'uh-si-tiz"m)   A symbiotic relationship in which one member (the parasite) benefits and the other (the host) is adversely affected. Compare with *commensalism* and *mutualism*.

parasympathetic nervous system A division of the autonomic nervous system concerned with the control of the internal organs; functions to conserve or restore energy. Compare with *sympathetic nervous system.*

parathyroid glands Small, pea-sized glands closely adjacent to the thyroid gland; their secretion regulates calcium and phosphate metabolism.

parathyroid hormone (PTH) Hormone secreted by the parathyroid glands; regulates calcium and phosphate metabolism.

parenchyma (par-en′kih-mah) Highly variable living plant cells that have thin primary walls; function in photosynthesis, the storage of nutrients, and/or secretion.

parthenogenesis (par″theh-noh-jen′eh-sis) The development of an unfertilized egg into an adult organism; common among honey bees, wasps, and certain other arthropods.

partial pressure (of a gas) The pressure exerted by gas in a mixture, which is the same pressure it would exert if alone. For example, the partial pressure of atmospheric oxygen (PO_2) is 160 mm Hg at sea level.

passive immunity Temporary immunity that depends on the presence of immunoglobulins produced by another organism.

patch clamp technique A method that allows researchers to study the ion channels of a tiny patch of membrane by tightly sealing a micropipette to the patch and measuring the flow of ions through the channels.

patchiness See *clumped dispersion.*

pathogen (path′oh-gen) An organism, usually a microorganism, capable of producing disease.

pattern formation See *morphogenesis.*

pedigree analysis Technique by which an inheritance pattern is traced through multiple generations.

peduncle The stalk of a flower or inflorescence.

pellicle A flexible outer covering of protein; characteristic of certain protists, e.g., ciliates and euglenoids.

penis The male sexual organ of copulation in reptiles and mammals.

pentose A sugar molecule containing five carbons.

people overpopulation A situation in which there are too many people in a given geographical area; results in pollution, environmental degradation, and resource depletion. Compare with *consumption overpopulation.*

pepsin (pep′sin) An enzyme produced in the stomach that initiates digestion of protein.

peptide (pep′tide) A compound consisting of a chain of amino acid groups. A dipeptide consists of two amino acids, a polypeptide of many.

peptide bond A distinctive covalent carbon-to-nitrogen bond that links amino acids in peptides and proteins.

peptidoglycan (pep″tid-oh-gly′kan) A modified protein or peptide possessing an attached carbohydrate; component of the eubacterial cell wall.

perennial plant (purr-en′ee-ul) A woody or herbaceous plant that grows year after year, i.e., lives more than two years. Compare with *annual* and *biennial.*

perfect flower A flower that has both stamens and carpels. Compare with *imperfect flower.*

pericentriolar material Rather amorphous, somewhat fibrillar matter surrounding the centrioles in the microtubule organizing centers in cells of animals and other organisms possessing centrioles; chemically similar to the material in the microtubule-organizing centers of organisms lacking centrioles.

pericycle (pehr′eh-sy″kl) A layer of meristematic cells typically found between the endodermis and phloem in roots.

periderm (pehr′ih-durm) The outer bark of woody stems and roots; composed of cork cells, cork cambium, and cork parenchyma, along with traces of primary tissues.

period An interval of geological time that is a subdivision of an era. Each period is divided into epochs.

peripheral membrane protein A protein associated with one of the surfaces of a biological membrane. Compare with *integral membrane protein.*

peripheral nervous system (PNS) In vertebrates, the nerves and receptors that lie outside the central nervous system. Compare with *central nervous system (CNS).*

peristalsis (pehr″ih-stal′sis) Rhythmic waves of muscular contraction and relaxation in the walls of hollow tubular organs, such as the ureter or parts of the digestive tract, that serve to move the contents through the tube.

permafrost Permanently frozen subsoil characteristic of frigid areas such as the tundra.

peroxisomes (pehr-ox′ih-somz) Membrane-bounded organelles in eukaryotic cells containing enzymes that produce or degrade hydrogen peroxide.

persistence A characteristic of certain chemicals that are extremely stable and may take many years to be broken down into simpler forms by natural processes.

petal One of the parts of the flower attached inside the whorl of sepals; petals are usually colored.

petiole (pet′ee-ohl) The part of a leaf that attaches to a stem.

pH The negative logarithm of the hydrogen ion concentration of a solution (expressed as moles per liter). Neutral pH is 7, values less than 7 are acidic, and those greater than 7 are basic.

phage See *bacteriophage.*

phagocytosis (fag″oh-sy-toh′sis) Literally, "cell eating"; a type of endocytosis by which certain cells engulf food particles, microorganisms, foreign matter, or other cells.

pharynx (far′inks) Part of the digestive tract. In complex vertebrates it is bounded anteriorly by the mouth and nasal cavities and posteriorly by the esophagus and larynx; the throat region in humans.

phenetics (feh-neh′tiks) An approach to classification based on measurable similarities in phenotypic characters, without consideration of homology or other evolutionary relationships. Compare with *cladistics* and *classical evolutionary taxonomy.*

phenotype (fee′noh-type) The physical or chemical expression of an organism's genes. Compare with *genotype.*

phenylketonuria (PKU) (fee″nl-kee″toh-noor′ee-ah) An inherited disease in which there is a deficiency of the enzyme that normally converts phenylalanine to tyrosine; results in mental retardation if untreated.

pheromone (fer′oh-mone) A substance secreted by an organism to the external environment that influences the development or behavior of other members of the same species.

phloem (flo′em) The vascular tissue that conducts dissolved sugar and other organic compounds in plants.

phosphate group A weakly acidic functional group that can release one or two hydrogen ions.

phosphodiester linkage Covalent linkage between two nucleotides in a strand of DNA or RNA; includes a phosphate group bonded to the sugars of two adjacent nucleotides.

phosphoenolpyruvate (PEP) Three-carbon phosphorylated compound that is an important intermediate in glycolysis and is a reactant in the initial carbon fixation step in C_4 and CAM photosynthesis.

phosphoglycerate (PGA) Phosphorylated three-carbon compound that is an important metabolic intermediate.

phospholipids (fos″foh-lip′idz) Fatlike substances in which there are two fatty acids and a phosphorus-containing group attached to glycerol; major components of cellular membranes.

phosphorus cycle The worldwide circulation of phosphorus from the abiotic environment into living things and back into the abiotic environment.

phosphorylation (fos″for-ih-lay′shun) The introduction of a phosphate group into an organic molecule. Kinases are enzymes that catalyze certain phosphorylation reactions.

photic region The surface layer of the ocean (the top 100 m or so), where light penetrates and photosynthesis occurs.

photoautotroph Organism that obtains energy from light and synthesizes organic compounds from inorganic raw materials; includes plants, algae, and some bacteria. Compare with *chemoautotroph.*

photoheterotroph Organism that is able to carry out photosynthesis to obtain energy but is unable to fix carbon dioxide and therefore requires organic compounds as a carbon source; includes some bacteria. Compare with *chemoheterotroph.*

photolysis (foh-tol′uh-sis) The photochemical splitting of water in the light-dependent reactions of photosynthesis, catalyzed by a specific enzyme.

photon (foh′ton) A particle of electromagnetic radiation; one quantum of radiant energy.

photoperiodism (foh″teh-peer′ee-o-dizm) The physiological response (such as flowering) of plants to variations in the length of daylight and darkness.

photophosphorylation (foh″toh-fos-for-ih-lay′shun) The production of ATP in photosynthesis.

photoreceptor (foh″toh-ree-sep′tor) (1) A sense organ specialized to detect light; (2) A pigment that absorbs light before triggering a physiological response.

photorespiration (foh″toh-res-pur-ay′shun) Process that reduces the efficiency of photosynthesis in C_3 plants during hot spells in summer; consumes oxygen and produces carbon dioxide through the degradation of Calvin cycle intermediates.

photosynthesis (foh″toh-sin′thuh-sis) The biological process that captures light energy and transforms it into the chemical energy of organic molecules (such as carbohydrates), which are manufactured from carbon dioxide and water; performed by plants, algae, and certain bacteria.

photosystem I One of two complexes, consisting of chlorophyll molecules, accessory pigments, proteins, and associated electron acceptors, responsible for capturing light energy and transferring excited electrons; photosystem I best absorbs and uses light of about 700 nm. Compare with *photosystem II.*

photosystem II One of two complexes, consisting of chlorophyll molecules, accessory pigments, proteins, and associated electron acceptors; responsible for capturing light energy and transferring excited electrons; photosystem II best absorbs and uses light of about 680 nm. Compare with *photosystem I.*

phototropism (foh″toh-troh′pizm) The growth of a plant in response to the direction of light.

phycocyanin (fy″koh-sy′ah-nin) A blue pigment found in cyanobacteria and red algae.

phycoerythrin (fy″koh-ee-rih′thrin) A red pigment found in cyanobacteria and red algae.

phylogenetic systematics See *cladistics.*

phylogenetic tree A branching diagram that shows lines of descent among a group of related species.

phylogeny (fy-loj′en-ee) The complete evolutionary history of a group of organisms.

phylum (fy′lum) A taxonomic grouping of related, similar classes; a category beneath the kingdom and above the class.

phytochemicals Compounds found in plants that are not essential nutrients for humans but appear to play important roles in preventing certain diseases.

phytochrome (fy′toh-krome) A blue-green, proteinaceous pigment involved in a wide variety of physiological responses to light; occurs in two interchangeable forms depending on the ratio of red to far-red light.

phytoplankton (fy″toh-plank′tun) Microscopic floating algae and cyanobacteria that are the base of most aquatic food webs. Compare with *zooplankton.* See *plankton* and *nanoplankton.*

pia mater (pee′a may′ter) The inner membrane covering the brain and spinal cord; the innermost of the meninges.

pigment A substance that selectively absorbs light of different wavelengths.

pili (pie′lie) (sing. *pilus*) Hairlike structures on the surface of many bacteria. Function in conjugation or attachment.

pineal gland (pie-nee′al) Endocrine gland located in the brain.

pinocytosis (pin″oh-sy-toh′sis) Cell drinking; a type of endocytosis by which cells engulf and absorb droplets of liquids.

pioneer The first organism to colonize an area and begin the first stage of succession.

pistil The female reproductive organ of a flower; consists of either a single carpel or two or more fused carpels. See *carpel.*

pith The innermost tissue in the stems and roots of many herbaceous plants; primarily a storage tissue.

pituitary gland (pi-too′ih-tehr″ee) Endocrine gland located below the hypothalamus; secretes several hormones influencing a wide range of physiological processes.

placenta (plah-sen′tah) The partly fetal and partly maternal organ whereby materials are exchanged between fetus and mother in the uterus of placental mammals.

placoderms (plak′oh-durms) A group of extinct jawed fishes.

plankton Free-floating, mainly microscopic aquatic organisms found in the upper layers of the water; composed of phytoplankton and zooplankton. Compare with *nekton.*

planula larva (plan′yoo-lah) A ciliated larval form found in cnidarians.

plasma membrane The selectively permeable surface membrane that encloses the cell contents and through which all materials entering or leaving the cell must pass.

plasma cell Cell that secretes antibodies; a differentiated B lymphocyte.

plasma proteins Proteins such as albumins, globulins, and fibrinogen that circulate in the blood plasma.

plasmids (plaz′midz) Small circular DNA molecules that carry genes separate from the main bacterial DNA.

plasmodesmata (sing. *plasmodesma*) Cytoplasmic channels connecting adjacent plant cells and allowing for the movement of molecules and ions between cells.

plasmodial slime mold (plaz-moh′dee-uhl) A fungus-like protist whose feeding stage consists of a plasmodium.

plasmodium (plaz-moh′dee-um) (1) A multinucleate, ameboid mass of living matter; a coenocyte. (2) The feeding phase of the life cycle of plasmodial slime molds.

plasmolysis (plaz-mol′ih-sis)   The shrinkage of cytoplasm and the pulling away of the plasma membrane from the cell wall when a plant cell (or other walled cell) loses water, usually in a hypertonic environment.

plastids (plas′tidz)   A family of membrane-bounded organelles occurring in photosynthetic eukaryotic cells; include chloroplasts, chromoplasts, and amyloplasts and other leukoplasts.

platelets (playt′lets)   Cell fragments in vertebrate blood that function in clotting; also called thrombocytes.

platyhelminthes   Phylum of acoelomate animals commonly known as flatworms.

pleisomorphic characters   See *shared ancestral characters*.

pleiotropic (ply″oh-troh′pik)   Term referring to an allele that affects a number of characteristics of an individual.

pleural membrane (ploor′ul)   The membrane that lines the thoracic cavity and envelops each lung.

plumule (ploom′yool)   The embryonic shoot apex, or terminal bud, located above the point of attachment of the cotyledon(s).

pneumatophore (noo-mat′uh-for″)   Roots that extend up out of the water in swampy areas and are thought to provide aeration between the atmosphere and submerged roots.

polar body   A small *n* cell produced during oogenesis in female animals that does not develop into a functional ovum.

polar covalent bond   Chemical bond formed by the sharing of electrons between atoms that differ in electronegativity; the end of the bond near the more electronegative atom has a partial negative charge, the other end has a partial positive charge. Compare with *nonpolar covalent bond*.

polar molecule   Molecule that has one end with a partial positive charge and the other with a partial negative charge; polar molecules are generally soluble in water. Compare with *nonpolar molecule*.

polar nucleus   In flowering plants, one of two *n* cells in the embryo sac that fuse with a sperm during double fertilization to form the 3*n* endosperm.

pollen grain   The immature male gametophyte of seed plants (i.e., all gymnosperms and angiosperms) that produces sperm capable of fertilization.

pollen tube   In gymnosperms and flowering plants, a tube or extension that forms after germination of the pollen grain and through which male gametes (sperm cells) pass into the ovule.

pollination (pol″uh-nay′shen)   In seed plants, the transfer of pollen from the male to the female part of the plant.

polyadenylation (pol″ee-a-den-uh-lay′shun)   That part of eukaryotic mRNA processing in which multiple adenine-containing nucleotides (a poly A tail) are added to the 3′ end of the molecule.

polyandry   A mating system in which a female mates with several males during a breeding season. Compare with *polygyny*.

polygyny   A mating system in which a male animal mates with many females during a breeding season. Compare with *polyandry*.

poly A tail   See *polyadenylation*.

polygenic inheritance (pol″ee-jen′ik)   Inheritance in which several independently assorting or loosely linked non-allelic genes modify the intensity of a trait or contribute to the phenotype in additive fashion.

polymer (pol′ih-mer)   A molecule built up from repeating subunits of the same general type (monomers), such as a protein, nucleic acid, or polysaccharide.

polymerase chain reaction (PCR)   A method by which a targeted DNA fragment can be amplified in vitro to produce millions of copies.

polymorphism (pol″ee-mor′fizm)   (1) The existence of two or more phenotypically different individuals within a population; (2) The presence of more than one allele for a given locus in a population.

polyp (pol′ip)   A hydra-like animal; the sessile stage of the life cycle of certain cnidarians. Compare with *medusa*.

polypeptide   A compound consisting of many amino acids linked by peptide bonds.

polyphyletic group (pol″ee-fye-let′ik)   A group made up of organisms that evolved from two or more different ancestors. Compare with *monophyletic group* and *paraphyletic group*.

polyploid (pol′ee-ployd)   The condition of having more than two sets of chromosomes per nucleus. Compare with *diploid* and *haploid*.

polyribosome   A complex consisting of a number of ribosomes attached to an mRNA during translation; also known as a polysome.

polysaccharide (pol-ee-sak′ah-ride)   A carbohydrate consisting of many monosaccharide subunits, e.g., starch, glycogen, and cellulose.

polysome   See *polyribosome*.

polytene   Term describing a giant chromosome consisting of many (usually more than 1000) parallel DNA double helices. Polytene chromosomes are typically found in cells of the salivary glands and some other tissues of certain insects, such as the fruit fly, *Drosophila*.

polyunsaturated fatty acid   See *fatty acid*.

pons (ponz)   The white bulge that is the part of the brainstem between the medulla and the midbrain; connects various parts of the brain.

population   A group of organisms of the same species that live in the same geographical area at the same time.

population bottleneck   See *bottleneck*.

population crash   An abrupt decline in the size of a population.

population density   The number of individuals of a species per unit of area or volume at a given time.

population ecology   That branch of biology that deals with the numbers of a particular species that are found in an area and how and why those numbers change (or remain fixed) over time.

poriferans   Animal phylum that comprises the sponges.

positive feedback system   A homeostatic mechanism in which a change in some condition triggers a response that intensifies the changing condition. Compare with *negative feedback system*.

posterior   Toward the tail end of a bilaterally symmetrical animal. Compare with *anterior*.

postsynaptic neuron   A neuron that transmits an impulse away from a synapse. Compare with *presynaptic neuron*.

postzygotic barrier   One of several mechanisms that prevent gene flow between species after fertilization has taken place; examples include hybrid inviability, hybrid sterility, and hybrid breakdown. Compare with *prezygotic barrier*.

potential energy   Stored energy; energy that can do work as a consequence of its position or state. Compare with *kinetic energy*.

preadaptation   A novel evolutionary change in a preexisting biological structure that enables it to have a different function; feathers, which evolved from reptilian scales, represent a preadaptation for flight.

Precambrian time   All of geological time before the Paleozoic era, encompassing approximately the first 4 billion years of Earth's history.

predation   Relationship in which one organism (the predator) kills and devours another organism (the prey).

pre-mRNA   RNA precursor to mRNA in eukaryotes; contains both introns and exons.

pressure-flow hypothesis   The mechanism by which dissolved sugar is thought to be transported in phloem; caused by a pressure gradient between the source (where sugar is loaded into the phloem) and the sink (where sugar is removed from phloem).

prenatal   Term referring to the time before birth.

presynaptic neuron   A neuron that transmits an impulse to a synapse. Compare with *postsynaptic neuron.*

prezygotic barrier   One of several mechanisms that interfere with fertilization between male and female gametes of different species; examples include temporal isolation, behavioral isolation, mechanical isolation, and gametic isolation. Compare with *postzygotic barrier.*

primary consumer   See *herbivore.*

primary growth   An increase in the length of a plant that occurs at the tips of the shoots and roots due to the activity of apical meristems. Compare with *secondary growth.*

primary mycelium   A mycelium in which the cells are monokaryotic and haploid; a mycelium that grows from either an ascospore or a basidiospore. Compare with *secondary mycelium.*

primary producer   See *autotroph.*

primary response   The response of the immune system to first exposure to an antigen. Compare with *secondary response.*

primary structure (of a protein)   The complete sequence of amino acids in a polypeptide chain, beginning at the amino end and ending at the carboxyl end. Compare with *secondary, tertiary,* and *quaternary protein structure.*

primary succession   An ecological succession that occurs on land that has not previously been inhabited by plants; no soil is present initially. See *succession.* Compare with *secondary succession.*

primer   See *RNA primer.*

primitive groove   See *primitive streak.*

primitive streak   Dynamic, constantly changing structure that forms at the midline of the blastodisc in birds, mammals, and some other vertebrates, and is active in gastrulation. Cells from the surface enter the interior through the furrow at its center (primitive groove), which is the functional equivalent of the blastopore. The anterior end of the primitive streak is Hensen's node.

primosome   A complex of proteins responsible for synthesizing the RNA primers required in DNA synthesis.

principle   A scientific theory that has withstood repeated testing and has the highest level of scientific confidence. Compare with *hypothesis* and *theory.*

prion   An infectious agent that consists only of protein.

producer   See *autotroph.*

product   Substance formed by a chemical reaction. Compare with *reactant.*

product rule   Rule for combining the probabilities of independent events by multiplying their individual probabilities. Compare with *sum rule.*

profundal zone (pro-fun′dl)   The deepest zone of a large lake, located below the level of penetration by sunlight. Compare with *littoral zone* and *limnetic zone.*

progesterone (pro-jes′ter-own)   A steroid hormone secreted by the ovary (mainly by the corpus luteum) and placenta; stimulates the uterus and breasts.

progymnosperm (pro-jim′noh-sperm)   An extinct group of plants that are thought to have been the ancestors of gymnosperms.

prokaryote (pro-kar′ee-ote)   Cell that lacks a nucleus and other membrane-bounded organelles; includes the bacteria, members of kingdoms Eubacteria and Archaebacteria. Compare with *eukaryote.*

promoter   Nucleotide sequence in DNA to which RNA polymerase attaches to begin transcription.

prop root   An adventitious root that arises from the stem and provides additional support for a plant such as corn.

prophage (pro′faj)   A latent state of a bacteriophage in which the viral genome is inserted into the bacterial host chromosome.

prophase   The first stage of mitosis and of meiosis I and meiosis II. During prophase the chromosomes become visible as distinct structures, the nuclear envelope breaks down, and a spindle forms. (Meiotic prophase I is more complex and includes synapsis of homologous chromosomes and crossing-over.)

proplastids   Organelles that are plastid precursors; may mature into various specialized plastids, including chloroplasts, chromoplasts, leukoplasts, or amyloplasts.

proprioceptor (pro″pree-oh-sep′tor)   Receptors in muscles, tendons, and joints that respond to changes in movement, tension, and position; enable an animal to perceive the position of its body.

prostaglandins (pros″tah-glan′dinz)   Derivatives of unsaturated fatty acids that produce a wide variety of hormone-like effects; synthesized by most cells of the body; sometimes called local hormones.

prostate gland   A gland in male animals that produces an alkaline secretion that is part of the semen.

protein   A large, complex organic compound composed of covalently linked amino acid subunits; contains carbon, hydrogen, oxygen, nitrogen, and sulfur.

prothallus (pro-thal′us) (pl. *prothalli*)   The free-living, *n* gametophyte in ferns and other seedless vascular plants.

protist (proh′tist)   One of a vast kingdom of eukaryotic organisms, primarily unicellular or simple multicellular; mostly aquatic.

protobionts (proh″toh-by′ontz)   Assemblages of organic polymers that spontaneously form under certain conditions. Protobionts may have been involved in chemical evolution.

proton   A particle present in the nuclei of all atoms that has one unit of positive charge and a mass of one atomic mass unit (amu). Compare with *electron* and *neutron.*

proto-oncogene   A gene that normally promotes cell division in response to the presence of certain growth factors; when mutated it may become an oncogene, possibly leading to the formation of a cancer cell. Compare with *oncogene.*

protonema (proh″toh-nee′mah) (pl. *protonemata*)   In mosses, a filament of *n* cells that grows from a spore and develops into leafy moss gametophytes.

protonephridia (proh″toh-nef-rid′ee-ah) (sing. *protonephridium*)   The flame-cell excretory organs of flatworms and some other simple invertebrates.

protostome (proh′toh-stome)   Major division of the animal kingdom in which the blastopore develops into the mouth, and the anus forms secondarily; includes the annelids, arthropods, and mollusks. Compare with *deuterostome.*

protozoa (proh″toh-zoh′a) (sing. *protozoon*)   An informal group of unicellular, animal-like protists, including amoebas, foraminiferans, actinopods, ciliates, flagellates, and apicomplexans. (The adjectival form is spelled protozoan.)

provirus (pro-vy'rus)   A part of a virus, consisting of nucleic acid only, that has been inserted into a host genome.

proximal   Closer to the point of reference. Compare with *distal*.

proximal convoluted tubule   The part of the renal tubule that extends from Bowman's capsule to the loop of Henle. Compare with *distal convoluted tubule*.

proximate causes (of behavior)   Immediate causes of behavior, such as genetic, developmental, and physiological processes that permit the animal to carry out a specific behavior. Compare with *ultimate causes of behavior*.

pseudocoelom (sue"doh-see'lom)   A body cavity between the mesoderm and endoderm; derived from the blastocoel. Compare with *coelom*.

pseudocoelomate (sue"doh-seel'oh-mate)   Animal possessing a pseudocoelom. Compare with *coelomate* and *acoelomate*.

pseudoplasmodium (soo"doe-plaz-moh'dee-um)   In cellular slime molds, an aggregation of amoeboid cells that forms a spore-producing fruiting body during reproduction.

pseudopodium (sou"doe-poe'dee-um) (pl. *pseudopodia*)   A temporary extension of an amoeboid cell that is used for feeding and locomotion.

puff   In a polytene chromosome, a decondensed region that is a site of intense RNA synthesis.

pulmonary circulation   The part of the circulatory system that delivers blood to and from the lungs for oxygenation. Compare with *systemic circulation*.

pulse, arterial   Alternate expansion and recoil of an artery.

pulvinus (pul-vy'nus)   A special structure, often located at the base of the petiole, that functions in leaf movement by changes in turgor.

punctuated equilibrium   The idea that evolution proceeds with periods of little of no change within a species, followed by very active phases, so that major adaptations or clusters of adaptations appear suddenly in the fossil record. Compare with *gradualism*.

Punnett square   Grid structure, first developed by Reginald Punnett, that allows direct calculation of the probabilities of occurrence of all possible offspring of a genetic cross.

pupa (pew'pah) (pl. *pupae*)   A stage in the development of an insect, between the larva and the imago (adult); a form that neither moves nor feeds, and may be in a cocoon.

pure line   See *true-breeding line*.

purines (pure'eenz)   Nitrogenous bases with carbon and nitrogen atoms in two attached rings, e.g., adenine and guanine; components of nucleic acids, ATP, GTP, NAD^+, and certain other biologically active substances. Compare with *pyrimidines*.

pyramid of biomass   An ecological pyramid that illustrates the total biomass, as, for example, the total dry weight, of all living organisms at each successive trophic level in an ecosystem.

pyramid of energy   An ecological pyramid that shows the energy flow through each trophic level of an ecosystem.

pyramid of numbers   An ecological pyramid that shows the number of organisms at each trophic level in an ecosystem.

pyrimidines (pyr-im'ih-deenz)   Nitrogenous bases, each composed of a single ring of carbon and nitrogen atoms, e.g., thymine, cytosine, and uracil; components of nucleic acids. Compare with *purines*.

pyruvate (pyruvic acid)   A three-carbon organic acid; the end product of glycolysis.

quadrupedal (kwad'roo-ped"ul)   Walking on all fours.

quantitative trait   A trait that shows continuous variation in a population (e.g., human height). Quantitative traits usually have polygenic inheritance patterns.

quaternary structure (of a protein)   The overall conformation of a protein produced by the interaction of two or more polypeptide chains. Compare with *primary, secondary,* and *tertiary protein structure*.

r selection   A reproductive strategy in which a species (known as an *r*-selected species or *r* strategist) typically has a small body size, rapid development, short lifespan, and devotes a large proportion of its metabolic energy to the production of offspring. Compare with *K selection*.

radial cleavage   Pattern of blastomere production in which the cells are located directly above or below one another; characteristic of early deuterostome embryos. Compare with *spiral cleavage*.

radial symmetry   A body plan in which any section through the mouth and down the length of the body divides the body into similar halves. Jellyfish and other cnidarians have radial symmetry. Compare with *bilateral symmetry*.

radicle (rad'ih-kl)   The embryonic root of a seed plant.

radioactive decay   The process in which a radioactive element emits radiation and, as a result, its nucleus changes into the nucleus of a different element.

radioisotopes   Unstable isotopes that spontaneously emit radiation; also called radioactive isotopes.

radula (rad'yoo-lah)   A rasplike structure in the digestive tract of chitons, snails, squids, and certain other mollusks.

rain shadow   An area on the downwind side of a mountain range with very little precipitation. Deserts often occur in rain shadows.

random dispersion   The spatial distribution pattern of a population in which the presence of one individual has no effect on the distribution of other individuals. Compare with *clumped dispersion* and *uniform dispersion*.

range   The area where a particular species occurs.

ray   A chain of parenchyma cells (one to many cells thick) that functions for lateral transport in stems and roots of woody plants.

ray-finned fishes   Group of bony fishes that gave rise to most modern osteichthyes.

reabsorption   The selective removal of certain substances from the glomerular filtrate by the renal tubules and collecting ducts of the kidney, and their return into the blood.

reactant   Substance that participates in a chemical reaction. Compare with *product*.

reaction center   Portion of an antenna complex within a photosystem that includes chlorophyll *a* molecules capable of transferring electrons to a primary acceptor; the reaction center of Photosystem I is P700 and of Photosystem II is P680.

realized niche   The lifestyle that an organism actually pursues, including the resources that it actually uses. An organism's realized niche is narrower than its fundamental niche because of interspecific competition. Compare with *fundamental niche*.

receptacle   The end of a flower stalk where the flower parts (sepals, petals, stamens, and carpels) are attached.

reception   Process of detecting a stimulus.

receptor down-regulation   The process by which some hormone receptors decrease in number, thereby suppressing the sensitivity of target cells to the hormone. Compare with *receptor up-regulation*.

receptor-mediated endocytosis   A type of endocytosis in which extracellular molecules become bound to specific receptors on the cell surface and then enter the cytoplasm enclosed in vesicles.

receptor up-regulation   The process by which some hormone receptors increase in number, thereby increasing the sensitivity of the target cells to the hormone. Compare with *receptor down-regulation*.

recessive allele (al-leel′)   An allele that is not expressed in the heterozygous state. Compare with *dominant allele*.

recombinant DNA   Any DNA molecule made by combining genes from different organisms.

recombination, genetic   The appearance of new gene combinations. Recombination in eukaryotes generally results from meiotic events, either crossing-over or shuffling of chromosomes.

red alga   Member of a diverse phylum of algae that contain the pigments chlorophyll *a*, carotenoids, phycocyanin, and phycoerythrin.

red blood cell (RBC)   See *erythrocyte*.

red tide   A red or brown coloration of ocean water caused by a population explosion, or bloom, of dinoflagellates.

redox reaction (ree′dox)   Chemical reaction in which one or more electrons are transferred from one substance (the substance that becomes oxidized) to another (the substance that becomes reduced). See *oxidation* and *reduction*.

reduction   The gain of one or more electrons (or hydrogen atoms) by an atom, ion, or molecule. Compare with *oxidation*.

reflex action   An automatic, involuntary response to a given stimulus that generally functions to restore homeostasis.

refractory period   The brief period of time that must elapse after the response of a neuron or muscle fiber, during which it cannot respond to another stimulus.

regulative development   Very plastic developmental pattern in which each individual blastomere retains totipotency. Compare with *mosaic development*.

regulon   A group of operons that are coordinately controlled.

renal (ree′nl)   Pertaining to the kidney.

renal pelvis   The funnel-shaped chamber of the kidney that receives urine from the collecting ducts; urine then moves into the ureters.

renin (reh′nin)   An enzyme released by the kidney in response to a decrease in blood pressure, which activates a pathway leading to production of angiotensin II, a hormone that increases aldosterone release.

replacement-level fertility   The number of children a couple must produce in order to "replace" themselves. The average number is greater than two because some children die before reaching reproductive age.

replication fork   Y-shaped structure produced during the semiconservative replication of DNA.

replication   See *DNA replication*.

repressible operon   Operon that is normally active but can be controlled by a repressor protein, which becomes active when it binds to a corepressor (usually the end product of a metabolic pathway); the active repressor binds to the operator, making the operon transcriptionally inactive, e.g., the tryptophan operons of *Escherichia coli* and *Salmonella*. Compare with *inducible operon*.

repressor protein   A transcription factor that acts as a negative regulator, inhibiting transcription when bound to DNA; some repressors require a corepressor to be active; some other repressors become inactive when bound to an inducer molecule. Compare with *activator protein*.

reproduction   Process by which new individuals are produced. See *asexual reproduction* and *sexual reproduction*.

reproductive isolating mechanisms   The reproductive barriers that prevent a species from interbreeding with another species; as a result, each species' gene pool is isolated from other species.

reptile   Class of vertebrate animals characterized by dry skin with horny scales and adaptations for terrestrial reproduction; includes turtles, snakes, and alligators.

residual capacity   The volume of air that remains in the lungs at the end of a normal exhalation.

resin   A viscous organic material that certain plants produce and secrete into specialized ducts; may play a role in deterring disease organisms or plant-eating insects.

resolution   See *resolving power*.

resolving power   The ability of a microscope to show fine detail, defined as the minimum distance between two points at which they can be seen as separate images; also called resolution.

resource partitioning   The reduction of competition for environmental resources such as food that occurs among coexisting species as a result of each species' niche differing from the others in one or more ways.

respiration   (1) Cellular respiration is the process by which cells generate ATP through a series of redox reactions in which the terminal electron acceptor is an inorganic compound. In aerobic respiration the terminal electron acceptor is molecular oxygen; in anaerobic respiration the terminal acceptor is an inorganic molecule other than oxygen. (2) Organismic respiration is the process of gas exchange in a complex animal, generally through a specialized respiratory surface, such as a lung or gill.

respiratory centers   Centers in the medulla and pons that regulate breathing.

resting potential   The membrane potential (difference in electrical charge between the two sides of the plasma membrane) of a neuron in which no action potential is occurring. The typical resting potential is about −70 millivolts.

restriction enzyme   One of a class of enzymes that cleave DNA at specific base sequences; produced by bacteria to degrade foreign DNA; used in recombinant DNA technology.

restriction fragment length polymorphism (RFLP) analysis   A technique that permits assessment of the degree of relatedness among individuals within a population; individuals are compared on the basis of the different patterns of DNA fragments generated when their DNA is cut with the same restriction enzyme.

restriction map   A physical map of DNA in which sites cut by specific restriction enzymes serve as landmarks.

reticular activating system (RAS) (reh-tik′yoo-lur)   A diffuse network of neurons in the brain stem responsible for maintaining consciousness.

retina (ret′ih-nah)   The innermost of the three layers of the eyeball, which is continuous with the optic nerve and contains the light-sensitive rod and cone cells.

retrovirus (ret′roh-vy″rus)   An RNA virus that uses reverse transcriptase to produce a DNA intermediate in the host cell.

reverse transcriptase   Enzyme produced by retroviruses to enable the transcription of DNA from the viral RNA in the host cell.

reversible inhibitor   A substance that forms weak bonds with an enzyme, temporarily interfering with its function; a reversible inhibitor can be competitive or noncompetitive. Compare with *irreversible inhibitor*.

Rh factors   Red blood cell antigens, known as D antigens, first identified in *Rhesus* monkeys. Persons possessing these antigens are Rh^+, those lacking them are Rh^-. See *erythroblastosis fetalis*.

rhizome (ry′zome)   A horizontal underground stem that bears leaves and buds and often serves as a storage organ and a means of asexual reproduction, e.g., iris.

rhodopsin (rho-dop′sin)   Visual purple; a light-sensitive pigment found in the rod cells of the vertebrate eye; a similar molecule is employed by certain bacteria in the capture of light energy to make ATP.

ribonucleic acid (RNA) A family of single-stranded nucleic acids that function mainly in protein synthesis.

ribosomal RNA (rRNA) See *ribosomes.*

ribosomes (ry′boh-sohms) Organelles that are part of the protein synthesis machinery of both prokaryotic and eukaryotic cells; consist of a larger and smaller subunit, each composed of ribosomal RNA (rRNA) and ribosomal proteins.

ribozyme (ry′boh-zime) A molecule of RNA that has catalytic properties.

ribulose bisphosphate (RuBP) A five-carbon phosphorylated compound with a high energy potential that reacts with carbon dioxide in the initial step of the Calvin cycle.

ribulose bisphosphate carboxylase See *Rubisco.*

RNA polymerase Enzyme that catalyzes the synthesis of RNA from a DNA template. Also called DNA-dependent RNA polymerase.

RNA primer Sequence of about five RNA nucleotides that are synthesized during DNA replication to provide a 3′ end to which DNA polymerase can add nucleotides. The RNA primer is later degraded and replaced with DNA.

RNA world A model that proposes that, during the evolution of cells, RNA was the first informational molecule to evolve, followed at a later time by proteins and DNA.

rod One of the rod-shaped, light-sensitive cells of the retina that are particularly sensitive to dim light and mediate black and white vision. Compare with *cone.*

root cap A covering of cells over the root tip that protects the delicate meristematic tissue directly behind it.

root graft The process of roots from two different plants growing together and becoming permanently attached to one another.

root hair An extension, or outgrowth, of a root epidermal cell. Root hairs increase the absorptive capacity of roots.

root pressure The pressure in root xylem that occurs as a result of water moving into roots from the soil.

root system The underground portion of a plant that anchors it in the soil and absorbs water and dissolved minerals.

rough ER See *endoplasmic reticulum.*

Rubisco Common name of ribulose bisphosphate carboxylase, the enzyme that catalyzes the reaction of carbon dioxide with ribulose bisphosphate in the Calvin cycle.

rugae (roo′jee) Folds, such as those in the lining of the stomach.

runner See *stolon.*

S phase Stage in interphase of the cell cycle during which DNA and other chromosomal constituents are synthesized. Compare with G_1 and G_2 phases.

saccule The structure within the vestibule of the inner vertebrate ear that along with the utricle houses the receptors of static equilibrium.

salinity The concentration of dissolved salts (such as sodium chloride) in a body of water.

salivary glands Accessory digestive glands found in vertebrates; in humans there are three pairs; also present in some invertebrates.

salt Ionic compound consisting of an anion other than a hydroxide ion and a cation other than a hydrogen ion. A salt can be formed by the reaction between an acid and a base.

salt marsh A wetland dominated by grasses in which the salinity fluctuates between that of sea water and fresh water; salt marshes are usually located in estuaries.

saltatory conduction Transmission of a neural impulse along a myelinated neuron; ion activity at one node depolarizes the next node along the axon.

saprobe See *decomposer.*

saprotroph (sap′roh-trof) See *decomposer.*

sarcolemma (sar″koh-lem′mah) The muscle cell plasma membrane.

sarcomere (sar′koh-meer) A segment of a striated muscle cell located between adjacent Z-lines that serves as a unit of contraction.

sarcoplasmic reticulum System of vesicles in a muscle cell that surrounds the myofibrils and releases calcium in muscle contraction; a modified endoplasmic reticulum.

saturated fatty acid See *fatty acid.*

savanna (suh-van′uh) A tropical grassland containing scattered trees; found in areas of low rainfall or seasonal rainfall with prolonged dry periods.

scaffolding proteins Nonhistone proteins that help maintain the structure of a chromosome.

schizocoely (skiz′oh-seely) Process of coelom formation in which the mesoderm splits into two layers, forming a cavity between them; characteristic of protostomes. Compare with *enterocoely.*

Schwann cells Supporting cells found in nervous tissue outside the central nervous system.

sclereid (skler′id) In plants, a sclerenchyma cell that is variable in shape but typically not long and tapered. Compare with *fiber.*

sclerenchyma (skler-en′kim-uh) Cells that provide strength and support in the plant body, are often dead at maturity, and have extremely thick walls; includes fibers and sclereids.

scramble competition Intraspecific competition in which all of the individuals in a population "share" the limited resource equally, so that at high population densities none of them obtains an adequate amount. Compare with *contest competition.*

scrotum (skroh′tum) The external sac of skin found in most male mammals that contains the testes and their accessory organs.

second law of thermodynamics The physical law that states that the total amount of entropy in the universe continually increases. Compare with *first law of thermodynamics.*

second messenger A substance within a cell that relays a message and (usually) triggers a response to a hormone combined with a receptor at the cell's surface, e.g., cyclic AMP and calcium ions.

secondary consumer See *carnivore.*

secondary growth An increase in the girth of a plant due to the activity of the vascular cambium and cork cambium; secondary growth results in the production of secondary tissues, i.e., wood and bark. Compare with *primary growth.*

secondary mycelium A dikaryotic mycelium formed by the fusion of two primary hyphae. Compare with *primary mycelium.*

secondary response Rapid production of antibodies induced by a second exposure to an antigen several days, weeks, or even months after the initial exposure. Compare with *primary response.*

secondary structure (of a protein) A regular geometric shape produced by hydrogen bonding between the atoms of the uniform polypeptide backbone; includes the alpha helix and the beta-pleated sheet. Compare with *primary, tertiary,* and *quaternary protein structure.*

secondary succession An ecological succession that takes place after some disturbance destroys the existing vegetation; soil is already present. See *succession.* Compare with *primary succession.*

secretory vesicles Small cytoplasmic vesicles that move substances from an internal membrane system to the plasma membrane.

seed A plant reproductive body composed of a young, multicellular plant and nutritive tissue (food reserves), enclosed by a seed coat.

seed coat The outer protective covering of a seed.

seed fern An extinct group of seed-bearing woody plants with fernlike leaves; seed ferns probably descended from progymnosperms and gave rise to cycads and possibly ginkgoes.

segregation, principle of Genetic principle, first noted by Gregor Mendel, that states that two alleles of a locus become separated into different gametes.

selectively permeable membrane A membrane that allows some substances to cross it more easily than others. Biological membranes are generally permeable to water, but restrict the passage of many solutes.

semen Fluid composed of sperm suspended in various glandular secretions that is ejaculated from the penis during orgasm.

semicircular canals The passages in the vertebrate inner ear containing structures that control the sense of equilibrium (balance).

semiconservative replication See *DNA replication.*

semilunar valves Valves between the ventricles of the heart and the arteries that carry blood away from the heart; aortic and pulmonary valves.

seminal vesicles (1) In mammals, glandular sacs that secrete a component of seminal fluid; (2) In some invertebrates, structures that store sperm.

seminiferous tubules (sem-ih-nif′er-ous) Coiled tubules in the testes in which spermatogenesis takes place in male vertebrates.

senescence (se-nes′cents) The aging process.

sensory neuron A neuron that transmits an impulse from a receptor to the central nervous system.

sepal (see′pul) One of the outermost parts of a flower, usually leaflike in appearance, that protect the flower as a bud.

septum (pl. *septa*) A cross wall or partition, e.g., the walls that divide a hypha into cells.

serotonin A neurotransmitter of the biogenic amine group.

Sertoli cells (sur-tole′ee) Supporting cells of the tubules of the testis.

sessile (ses′sile) Permanently attached to one location, e.g., coral animals.

setae (sing. *seta*) Bristle-like structures that aid in annelid locomotion.

sex-influenced trait Genetic trait that is expressed differently in males and females.

sex-linked gene Gene carried on a sex chromosome. In mammals almost all sex-linked genes are borne on the X chromosome (are X-linked).

sexual isolation See *behavioral isolation.*

sexual reproduction Type of reproduction in which two gametes (usually, but not necessarily, contributed by two different parents) fuse to form a zygote. Compare with *asexual reproduction.*

sexual selection A type of natural selection that occurs when individuals of a species vary in their ability to compete for mates. Individuals with reproductive advantages are selected over others of the same sex.

shade avoidance The tendency of plants that are adapted to high light intensities to grow taller when they are closely surrounded by other plants.

shared ancestral characters Traits that were present in an ancestral species that have remained essentially unchanged; suggest a distant common ancestor. Also called pleisomorphic characters. Compare with *shared derived characters.*

shared derived characters Homologous traits found in two or more taxa that are present in their most recent common ancestor but not in earlier common ancestors. Also called synapomorphic characters. Compare with *shared ancestral characters.*

shoot system The above-ground portion of a plant, such as the stem and leaves.

short-day plant A plant that flowers in response to lengthening nights; also called long-night plants. Compare with *long-day, intermediate-day,* and *day-neutral plants.*

short-night plant See *long-day plant.*

sickle cell anemia An inherited form of anemia in which there is abnormality in the hemoglobin beta chains; the inheritance pattern is autosomal recessive.

sieve tube members Cells that conduct dissolved sugar in the phloem of flowering plants.

sign stimulus Any stimulus that elicits a fixed action pattern in an animal.

signal transduction A process in which a cell converts an extracellular signal into an intracellular signal that affects some function in the cell. Also see *cell signaling.*

simple fruit A fruit that develops from a single ovary. Compare with *aggregate, accessory,* and *multiple fruits.*

sinoatrial (SA) node Mass of specialized cardiac muscle in which the impulse triggering the heartbeat originates; the pacemaker of the heart.

skeletal muscle Striated (voluntary) muscle of vertebrates, so-called because it usually is directly or indirectly attached to some part of the skeleton. Compare with *cardiac muscle* and *smooth muscle.*

slash-and-burn agriculture A type of agriculture in which tropical rain forest is cut down, allowed to dry, and burned. The crops that are planted immediately afterwards thrive because the ashes provide nutrients; in a few years, however, the soil is depleted and the land must be abandoned.

small intestine Portion of the vertebrate digestive tract that extends from the stomach to the large intestine.

small nuclear ribonucleoprotein complexes (snRNP) Aggregations of RNA and protein responsible for binding to pre-mRNA in eukaryotes and catalyzing the excision of introns and the splicing of exons.

smooth ER See *endoplasmic reticulum.*

smooth muscle Involuntary muscle tissue that lacks transverse striations; found mainly in sheets surrounding hollow organs, such as the intestine. Compare with *cardiac muscle* and *skeletal muscle.*

social behavior Interaction of two or more animals, usually of the same species.

sociobiology The branch of biology that focuses on the evolution of social behavior through natural selection.

sodium-potassium pump Active transport system that transports sodium ions out of, and potassium ions into, cells.

soil erosion The wearing away or removal of soil from the land. Although soil erosion occurs naturally from precipitation and runoff, human activities (such as clearing the land) accelerate it.

solar tracking See *heliotropism.*

solute A dissolved substance. Compare with *solvent.*

solvent Substance capable of dissolving other substances. Compare with *solute.*

somatic cell In animals, a cell of the body not involved in formation of gametes. Compare with *germ line cell.*

somatic nervous system That part of the vertebrate peripheral nervous system that keeps the body in adjustment with the external environment; includes the sensory receptors on the body surface and within the muscles, and the nerves that link them with the central nervous system. Compare with *autonomic nervous system*.

somatotropin See *growth hormone*.

sonogram See *ultrasound imaging*.

soredium (sor-id′e-um) (pl. *soredia*) In lichens, a type of asexual reproductive structure that consists of a cluster of algal cells surrounded by fungal hyphae.

sorus (soh′rus) (pl. *sori*) In ferns, a cluster of spore-producing sporangia.

Southern blot hybridization A technique in which DNA fragments, previously separated by gel electrophoresis, are transferred to a nitrocellulose membrane. A specific radioactive genetic probe is then allowed to hybridize to complementary fragments, thus marking their locations. Compare with *Northern blot hybridization* and *Western blotting*.

speciation Evolution of a new species.

species According to the biological species concept, groups of populations whose members are capable of interbreeding in nature to produce fertile offspring and do not interbreed with members of other species.

species richness The number of species in a community.

specific defense mechanisms The production of antibodies or T cells in response to foreign antigens; also called adaptive immune responses. Compare with *nonspecific defense mechanisms*.

specific epithet The second part of the name of a species; designates a specific species belonging to that genus.

specific heat The amount of heat energy that must be supplied to raise the temperature of 1 gram of a substance 1 degree Celsius.

sperm The motile, *n* male reproductive cell of animals and some plants and protists; also called spermatozoan.

spermatid (spur′ma-tid) An immature sperm cell.

spermatocyte (spur-mah′toh-site) A meiotic cell that gives rise to spermatids and ultimately to mature sperm cells.

spermatogenesis (spur″mah-toh-jen′eh-sis) The production of sperm by meiosis. Compare with *oogenesis*.

spermatozoan (spur-mah-toh-zoh′un) See *sperm*.

sphincter (sfink′tur) A group of circularly arranged muscle fibers, the contractions of which close an opening, such as the pyloric sphincter at the exit of the stomach.

spinal cord In vertebrates, the dorsal, tubular nerve cord.

spinal nerves In vertebrates, the nerves that emerge from the spinal cord.

spindle See *mitotic spindle*.

spine A leaf that is modified for protection, such as a cactus spine.

spiracle (speer′ih-kl) An opening for gas exchange, such as the opening on the body surface of a trachea in insects.

spiral cleavage Distinctive spiral pattern of blastomere production in an early protostome embryo. Compare with *radial cleavage*.

spirillum (pl. *spirilla*) A long, rigid, helical bacterium. Compare with *spirochete, vibrio, bacillus*, and *coccus*.

spirochete A long, flexible, helical bacterium. Compare with *spirillum, vibrio, bacillus*, and *coccus*.

spleen Abdominal organ located just below the diaphragm that removes worn-out blood cells and bacteria from the blood and plays a role in immunity.

spongy mesophyll (mez′oh-fil) The loosely arranged mesophyll cells near the lower epidermis in certain leaves. Compare with *palisade mesophyll*.

spontaneous reaction See *exergonic reaction*.

sporangium (spor-an′jee-um) (pl. *sporangia*) A spore case, found in plants, certain protists, and fungi.

spore A reproductive cell that gives rise to individual offspring in plants, fungi, and certain algae and protozoa.

sporophyll (spor′oh-fil) A leaflike structure that bears spores.

sporophyte generation (spor′oh-fite) The 2*n*, spore-producing stage in the life cycle of a plant. Compare with *gametophyte generation*.

stabilizing selection Natural selection that acts against extreme phenotypes and favors intermediate variants; associated with a population well adapted to its environment. Compare with *directional selection* and *disruptive selection*.

stamen (stay′men) The male part of a flower; consists of a filament and anther.

standing-water ecosystem A lake or pond ecosystem.

starch A polysaccharide composed of alpha glucose subunits; made by plants for energy storage.

start codon See *initiation codon*.

stasis Long periods in the fossil record in which evolutionary change is extremely slow.

statocyst (stat′oh-sist) An invertebrate sense organ containing one or more granules (statoliths); senses gravity and motion.

statoliths (stat′uh-liths) Granules of loose sand or calcium carbonate found in statocysts.

stele The cylinder in the center of roots and stems that contains the vascular tissue.

stem cell A relatively undifferentiated cell capable of repeated cell division. At each division at least one of the daughter cells usually remains a stem cell, while the other may differentiate as a specific cell type.

sterilization A procedure that renders an individual incapable of producing offspring; the most common surgical procedures are vasectomy in the male and tubal ligation in the female.

steroids (steer′oids) Complex molecules containing carbon atoms arranged in four attached rings, three of which contain six carbon atoms each and the fourth of which contains five; e.g., cholesterol and certain hormones, including the male and female sex hormones of vertebrates.

stigma Portion of the carpel where pollen grains land during pollination (and before fertilization).

stipe A short stalk or stemlike structure that is a part of the body of certain multicellular algae.

stipule (stip′yule) One of a pair of scalelike or leaflike structures found at the base of certain leaves.

stolon (stow′lon) An above-ground, horizontal stem with long internodes; stolons often form buds that develop into separate plants, e.g., strawberry; also called runner.

stomach Muscular region of the vertebrate digestive tract, extending from the esophagus to the small intestine.

stomata (sing. *stoma*) Small pores located in the epidermis of plants that provide for gas exchange for photosynthesis; each stoma is flanked by two guard cells, which are responsible for its opening and closing.

stop codon See *termination codon*.

stratosphere The layer of the atmosphere between the troposphere and the mesosphere. It contains a thin ozone layer that protects life by filtering out much of the sun's ultraviolet radiation.

stratum basale (strat′um bah-say′lee)   The deepest sublayer of the human epidermis, consisting of cells that continuously divide. Compare with *stratum corneum.*

stratum corneum   The most superficial sublayer of the human epidermis. Compare with *stratum basale.*

strobilus (stroh′bil-us) (pl. *strobili*)   In certain plants, a conelike structure that bears spore-producing sporangia.

stroke volume   The volume of blood pumped by one ventricle during one contraction.

stroma   A fluid space of the chloroplast, enclosed by the chloroplast inner membrane and surrounding the thylakoids; site of the reactions of the Calvin cycle.

stromatolite (stroh-mat′oh-lite)   A column-like rock that is composed of many minute layers of prokaryotic cells, usually cyanobacteria.

structural formula   Type of chemical formula that shows the spatial arrangement of the atoms in a molecule. Compare with *molecular formula.*

structural isomer   One of two or more chemical compounds having the same chemical formula, but differing in the covalent arrangement of their atoms, e.g., glucose and fructose.

style   The neck connecting the stigma to the ovary of a carpel.

subsidiary cell   A structurally distinct epidermal cell associated with a guard cell.

substrate   A substance on which an enzyme acts; a reactant in an enzymatically catalyzed reaction.

succession   The sequence of vegetation changes in a community over time; also called ecological succession. See *primary succession* and *secondary succession.*

sucker   A shoot that develops adventitiously from a root; a type of asexual reproduction.

sulcus (sul′kus) (pl. *sulci*)   A groove, trench, or depression, especially one occurring on the surface of the brain, separating the convolutions.

sulfhydryl group   Functional group abbreviated —SH; found in organic compounds called thiols.

sum rule   Rule for combining the probabilities of mutually exclusive events by adding their individual probabilities. Compare with *product rule.*

summation   The process of adding together excitatory postsynaptic potentials (EPSPs).

suppressor T cell   T lymphocyte that suppresses the immune response.

supraorbital ridge (soop″rah-or′bit-ul)   Prominent bony ridge above the eye socket. Ape skulls have prominent supraorbital ridges.

surface tension   The attraction that the molecules at the surface of a liquid may have for each other.

survivorship   The proportion of individuals in a population that survive to a particular age; usually presented as a survivorship curve.

suspensor (suh-spen′sur)   In plant embryo development, a multicellular structure that anchors the embryo and aids in nutrient absorption from the endosperm.

swim bladder   Hydrostatic organ in bony fishes that permits the fish to hover at a given depth.

symbiosis (sim-bee-oh′sis)   An intimate relationship between two or more organisms of different species. See *commensalism, mutualism,* and *parasitism.*

sympathetic nervous system   A division of the autonomic nervous system; its general effect is to mobilize energy, especially during stress situations; prepares the body for fight-or-flight response. Compare with *parasympathetic nervous system.*

sympatric speciation (sim-pa′trik)   The evolution of a new species within the same geographical region as the parental species. Compare with *allopatric speciation.*

symplast   A continuum consisting of the cytoplasm of many plant cells, connected by plasmodesmata. Compare with *apoplast.*

synapomorphic characters   See *shared derived characters.*

synapse (sin′aps)   The junction between two neurons or between a neuron and an effector (muscle or gland).

synapsis (sin-ap′sis)   The process of physical association of homologous chromosomes during prophase I of meiosis.

synaptonemal complex   Structure, visible with the electron microscope, that forms between synapsed homologous chromosomes.

syngamy (sin′gah-mee)   The union of the gametes in sexual reproduction.

synthetic theory of evolution   The synthesis of previous theories, especially of Mendelian genetics, with Darwin's theory of evolution by natural selection to formulate a comprehensive explanation of evolution; also called neo-Darwinism.

systematics   The scientific study of the diversity of organisms and their evolutionary relationships. Taxonomy, the science of naming, describing, and classifying organisms, is an aspect of systematics.

systemic circulation   The part of the circulatory system that delivers blood to and from the tissues and organs of the body. Compare with *pulmonary circulation.*

systole (sis′tuh-lee)   Phase of the cardiac cycle when the heart is contracting. Compare with *diastole.*

T cell (T lymphocyte)   Type of white blood cell responsible for a wide variety of immune functions, particularly cell-mediated immunity. T cells are processed in the thymus. Compare with *B cell.*

T tubules   Transverse tubules; system of inward extensions of the muscle fiber plasma membrane.

taiga (tie′gah)   Northern coniferous forest biome found primarily in Canada, northern Europe, and Siberia; also called boreal forest.

taproot system   A root system in plants that has one main root with smaller roots branching off it. Compare with *fibrous root system.*

target tissue or cell   A tissue or cell that is affected by a hormone.

TATA box   Component of a eukaryotic promoter region; consists of a sequence of bases located about 30 base pairs upstream from the transcription initiation site.

taxon   A formal taxonomic group at any level, e.g., phylum or genus.

taxonomy (tax-on′ah-mee)   The science of naming, describing, and classifying organisms; see *systematics.*

Tay-Sachs disease   A serious genetic disease in which abnormal lipid metabolism in the brain causes mental deterioration in affected infants and young children; inheritance pattern is autosomal recessive.

tectorial membrane (tek-tor′ee-ul)   The roof membrane of the organ of Corti in the cochlea of the ear.

teleosts   Modern bony fishes.

telolecithal egg   Egg with a large amount of yolk, concentrated at the vegetal pole. Compare with *isolecithal egg.*

telophase (teel′oh-faze or tel′oh-faze)   The last stage of mitosis and of meiosis I and II when, having reached the poles, chromosomes become decondensed, and a nuclear envelope forms around each group.

temperate deciduous forest A forest biome that occurs in temperate areas where annual precipitation ranges from about 75 cm to 125 cm.

temperate grassland A grassland characterized by hot summers, cold winters, and less rainfall than is found in a temperate deciduous forest biome.

temperate rain forest A coniferous biome characterized by cool weather, dense fog, and high precipitation, e.g., the north Pacific coast of North America.

temperate virus A virus that can become integrated into the host DNA as a prophage.

temperature The average kinetic energy of the particles in a sample of a substance.

temporal isolation A prezygotic reproductive isolating mechanism in which genetic exchange is prevented between similar species because they reproduce at different times of the day, season, or year.

tendon A connective tissue structure that joins a muscle to another muscle, or a muscle to a bone. Tendons transmit the force generated by a muscle.

tendril A leaf or stem that is modified for holding or attaching onto objects.

tension-cohesion model The mechanism by which water and dissolved minerals are thought to be transported in xylem; water is pulled upward under tension due to transpiration while maintaining an unbroken column in xylem due to cohesion; also called transpiration-cohesion model.

teratogen Any agent capable of interfering with normal morphogenesis in an embryo, thereby causing malformations; examples include radiation, certain chemicals, and certain infectious agents.

terminal bud A bud at the tip of a stem. Compare with *axillary bud*.

termination (of protein synthesis) The final stage of protein synthesis, which occurs when a termination (stop) codon is reached, causing the completed polypeptide chain to be released from the ribosome. See *initiation* and *elongation*.

termination codon Also known as a stop codon. Any of the three codons in mRNA that do not code for an amino acid (UAA, UAG, or UGA). This stops translation at that point. Compare with *initiation codon*.

territoriality Behavior pattern in which one organism (usually a male) stakes out a territory of its own and defends it against intrusion by other members of the same species and sex.

tertiary structure (of a protein) (tur′she-air″ee) The overall three-dimensional shape of a polypeptide that is determined by interactions involving the amino acid side chains. Compare with *primary, secondary,* and *quaternary protein structure*.

test A shell.

test cross Genetic cross in which either an F_1 individual, or an individual of unknown genotype, is mated to a homozygous recessive individual.

testis (tes′tis) (pl. *testes*) The male gonad that produces sperm and the male hormone testosterone; in humans and certain other mammals the testes are located in the scrotum.

testosterone (tes-tos′ter-own) The principal male sex hormone (androgen); a steroid hormone produced by the interstitial cells of the testes; stimulates spermatogenesis and is responsible for primary and secondary sex characteristics in the male.

tetrad Chromosome complex formed by the synapsis of a pair of homologous chromosomes (i.e., four chromatids) during meiotic prophase I; also known as a bivalent.

tetrapods (tet′rah-podz) Four-limbed vertebrates: the amphibians, reptiles, birds, and mammals.

thalamus (thal′uh-mus) The part of the vertebrate brain that serves as a main relay center, transmitting information between the spinal cord and the cerebrum.

thallus (thal′us) (pl. *thalli*) The simple body of an alga, fungus, or nonvascular plant that lacks root, stems, or leaves, e.g., a liverwort thallus or a lichen thallus.

theory A widely accepted explanation supported by a large body of observations and experiments. A good theory relates facts that appear to be unrelated; it predicts new facts and suggests new relationships. Compare with *hypothesis* and *principle*.

therapsids (ther-ap′sids) A group of mammal-like reptiles of the Permian period; gave rise to the mammals.

thermal stratification The marked layering (separation into warm and cold layers) of temperate lakes during the summer. See *thermocline*.

thermocline (thur′moh-kline) A marked and abrupt temperature transition in temperate lakes between warm surface water and cold deeper water. See *thermal stratification*.

thermodynamics (thurm″oh-dy-nam′iks) Principles governing energy transfer (often expressed in terms of heat transfer). See *first law of thermodynamics* and *second law of thermodynamics*.

thermoreceptor A sensory receptor that responds to heat.

thick filaments Filaments composed mainly of myosin; found in muscle fibers. Compare with *thin filaments*.

thigmomorphogenesis A plant developmental response to mechanical stresses such as wind, rain, hail, and contact with passing animals.

thigmotropism (thig′moh-troh′pizm) Plant growth in response to contact with a solid object, such as the twining of plant tendrils.

thin filaments Filaments composed mainly of actin; found in muscle fibers. Compare with *thick filaments*.

threatened species A species in which the population is small enough for it to be at risk of becoming extinct throughout all or part of its range, but not so small that it is in imminent danger of extinction. Compare with *endangered species*.

threshold level The potential that a neuron or other excitable cell must reach for an action potential to be initiated.

thrombocytes See *platelets*.

thylakoids (thy′lah-koidz) Interconnected system of flattened, saclike membranous structures inside the chloroplast; the thylakoid membranes contain chlorophyll and the electron transport chain and enclose a compartment, the thylakoid space.

thymine (thy′meen) A nitrogenous pyrimidine base found in DNA.

thymus gland (thy′mus) An endocrine gland that functions as part of the lymphatic system; important in the ability to make immune responses.

thyroid gland An endocrine gland that lies anterior to the trachea and releases hormones that regulate the rate of metabolism.

thyroid hormones Hormones, including thyroxin, secreted by the thyroid gland; stimulate rate of metabolism.

tidal volume The volume of air moved into and out of the lungs with each normal resting breath.

tight junctions Specialized structures that form between some animal cells, producing a tight seal that prevents materials from passing through the spaces between the cells.

tissue A group of closely associated, similar cells that work together to carry out specific functions.

tissue culture The growth of tissue or cells in a synthetic growth medium under sterile conditions.

tissue fluid See *interstitial fluid*.

tolerance Decreased response to a drug over time.

tonoplast The membrane surrounding a vacuole.

topoisomerases Enzymes that relieve twists and kinks in a DNA molecule by breaking and rejoining the strands.

torsion Twisting of the visceral mass characteristic of gastropod mollusks.

total fertility rate The average number of children born to a woman during her lifetime.

totipotency (toh-ti-poh'tun-cee) Ability of a cell (or nucleus) to provide information for the development of an entire organism.

trace element An element required by an organism in very small amounts.

trachea (tray'kee-uh)(pl. *tracheae*) (1) Principal thoracic air duct of terrestrial vertebrates; windpipe; (2) One of the microscopic air ducts (or tracheal tubes) branching throughout the body of most terrestrial arthopods and some terrestrial mollusks.

tracheal tubes See *trachea*.

tracheid (tray'kee-id) A type of water-conducting and supporting cell in the xylem of vascular plants.

tract A bundle of nerve fibers within the central nervous system.

transcription The synthesis of RNA from a DNA template.

transcription factors DNA-binding proteins that regulate transcription; include positively acting activators and negatively acting repressors.

transduction The transfer of a genetic fragment from one cell to another, e.g., from one bacterium to another, by a virus.

transfer RNA (tRNA) RNA molecules that bind to specific amino acids and serve as adapter molecules in protein synthesis. The tRNA anticodons bind to complementary mRNA codons.

transformation (1) The incorporation of genetic material into a cell, thereby changing its phenotype; (2) The conversion of a normal cell to a malignant cell.

transgenic organism A plant or animal that has incorporated foreign DNA into its genome.

translation Conversion of information provided by mRNA into a specific sequence of amino acids in a polypeptide chain; process also requires transfer RNA and ribosomes.

translocation (1) The movement of organic materials (dissolved food) in the phloem of a plant; (2) Chromosome abnormality in which part of one chromosome has become attached to another; (3) Part of the elongation cycle of protein synthesis in which a transfer RNA attached to the growing polypeptide chain is transferred from the A site to the P site.

transmembrane protein An integral membrane protein that spans the lipid bilayer.

transmission, neural Conduction of a neural impulse along a neuron or from one neuron to another.

transpiration Loss of water vapor from the aerial surfaces of a plant (i.e., leaves and stems).

transpiration-cohesion model See *tension-cohesion model*.

transport vesicles Small cytoplasmic vesicles that move substances from one membrane system to another.

transposable element See *transposon*.

transposon (tranz-poze'on) A DNA segment that is capable of moving from one chromosome to another or to different sites within the same chromosome; also called a transposable element or mobile genetic element.

transverse tubules See *T tubules*.

triacylglycerol (try-as"il-glis'er-ol) A neutral fat consisting of a glycerol combined chemically with three fatty acids; also called triglyceride.

tricarboxylic acid (TCA) cycle See *citric acid cycle*.

trichocyst (trik'oh-sist) A cellular organelle found in certain ciliates that can discharge a threadlike structure that may aid in trapping and holding prey.

trichome (try'kohm) A hair or other appendage growing out from the epidermis of a plant.

tricuspid valve See *atrioventricular valve*.

triglyceride See *triacylglycerol*.

triose A sugar molecule containing three carbons.

triplet A sequence of three nucleotides that serves as the basic unit of genetic information.

triplet code The sequences of three nucleotides that compose the codons, the units of genetic information in mRNA that specify the order of amino acids in a polypeptide chain.

trisomy (try'sohm-ee) Condition in which a chromosome is present in triplicate, instead of the normal pair. Compare with *monosomy*.

trochophore larva (troh'koh-for) A larval form found in mollusks and many polychaetes.

trophic level (trow'fik) Each sequential step in a food web (from producers to primary, secondary, or tertiary consumer). Each organism in an ecosystem is assigned to a trophic level based on its primary source of nourishment; all producers belong to the first trophic level, all herbivores belong to the second trophic level, and so on.

trophoblast (troh'foh-blast) The outer cell layer of a late blastocyst which, in placental mammals, gives rise to the chorion and to the fetal contribution to the placenta.

tropic hormone (trow'pic) A hormone, produced by one endocrine gland, that targets another endocrine gland.

tropical dry forest A tropical forest where enough precipitation falls to support trees but not enough to support the lush vegetation of a tropical rain forest; often occurs in areas with pronounced rainy and dry seasons.

tropical rain forest A lush, species-rich forest biome that occurs in tropical areas where the climate is very moist throughout the year. Tropical rain forests are also characterized by old, infertile soils.

tropism (troh'pizm) In plants, a directional growth response that is elicited by an environmental stimulus.

tropomyosin (troh-poh-my'oh-sin) A regulatory muscle protein involved in contraction.

true-breeding line A genetically pure strain of organism, i.e., one in which all individuals are homozygous for the traits under consideration; also called a pure line.

tube feet Structures characteristic of echinoderms; function in locomotion and feeding.

tuber A thickened end of a rhizome that is fleshy and enlarged for food storage, e.g., white potato.

tubular transport maximum (Tm) The maximum rate at which a substance can be reabsorbed from the renal tubules of the kidney.

tumor Mass of tissue that grows in an uncontrolled manner; a neoplasm.

tumor necrosis factors (TNF) Cytokines that can kill tumor cells and can stimulate immune cells to initiate an inflammatory response.

tumor suppressor gene See *anti-oncogene*.

tundra (tun'dra)   A treeless biome (between the taiga in the south and the polar ice cap in the north) that consists of boggy plains covered by lichens and small plants such as mosses. The tundra is characterized by harsh, very cold winters and extremely short summers. Also called arctic tundra. Compare with *alpine tundra*.

tunicates   Chordates belonging to subphylum Urochordata; sea squirts.

turgor pressure (tur'gor)   Hydrostatic pressure that develops within a walled cell, such as a plant cell, when the osmotic pressure of the cell's contents is greater than the osmotic pressure of the surrounding fluid.

Turner syndrome   An inherited condition in which only one sex chromosome (an X chromosome) is present in cells; karyotype is designated XO. Affected individuals are sterile females.

ultimate causes (of behavior)   Evolutionary explanations for why a certain behavior occurs. Compare with *proximate causes of behavior*.

ultrasound imaging   Technique in which high frequency sound waves (ultrasound) are used to provide an image (sonogram) of an internal structure.

ultrastructure   The fine detail of a cell, generally only observable by use of an extremely sophisticated microscope, such as an electron microscope.

umbilical cord   In placental mammals, the organ that connects the embryo to the placenta.

uniform dispersion   The spatial distribution pattern of a population in which individuals are regularly spaced. Compare with *random dispersion* and *clumped dispersion*.

unsaturated fatty acid   See *fatty acid*.

upstream promoter elements (UPE)   Components of a eukaryotic promoter, found upstream of the RNA polymerase-binding site; the strength of a promoter is affected by the number and type of UPEs present.

upwelling   An upward movement of water that brings nutrients from the ocean depths to the surface. Where upwelling occurs, the ocean is very productive.

uracil (yur'ah-sil)   A nitrogenous pyrimidine base found in RNA.

urea (yur-ee'ah)   The principal nitrogenous excretory product of mammals; one of the water-soluble end products of protein metabolism.

ureter (yoo-ree'tur)   One of the paired tubular structures that conducts urine from the kidney to the bladder.

urethra (yoo-ree'thruh)   The tube that conducts urine from the bladder to the outside of the body.

uric acid (yoor'ik)   The principal nitrogenous excretory product of insects, birds, and reptiles; a relatively insoluble end product of protein metabolism; also occurs in mammals as an end product of purine metabolism.

urinary bladder   An organ that receives urine from the ureters and temporarily stores it.

urinary system   Body system in vertebrates that consists of kidneys, urinary bladder, and associated ducts.

urochordates   Subphylum of chordates; includes the tunicates.

uterine tube (yoo'tur-in)   See *oviduct*.

uterus (yoo'tur-us)   The hollow, muscular organ of the female reproductive tract in which the fetus undergoes development.

utricle   The structure within the vestibule of the vertebrate inner ear that, along with the saccule, houses the receptors of static equilibrium.

vaccine (vak-seen')   A commercially produced, weakened or killed antigen associated with a particular disease that stimulates the body to make antibodies.

vacuole (vak'yoo-ole)   A fluid-filled, membrane-bounded sac found within the cytoplasm; may function in storage, digestion, or water elimination.

vagina   The elastic, muscular tube, extending from the cervix to its orifice, that receives the penis during sexual intercourse and serves as the birth canal.

valence electrons   The electrons in the outer electron shell, known as the valence shell, of an atom; in the formation of a chemical bond an atom can accept electrons into its valence shell, or donate or share valence electrons.

van der Waals forces   Weak attractive forces between atoms; caused by interactions among fluctuating charges.

vas deferens (vas def'ur-enz)   One of the paired sperm ducts that connects the epididymis of the testis to the ejaculatory duct.

vascular cambium   A lateral meristem that produces secondary xylem (wood) and secondary phloem (inner bark). Compare with *cork cambium*.

vascular tissue system   The tissues specialized for translocation of materials throughout the plant body, i.e., the xylem and phloem.

vasoconstriction   Narrowing of the diameter of blood vessels.

vasodilation   Expansion of the diameter of blood vessels.

vector   (1) Any carrier or means of transfer; (2) Agent, such as a plasmid or virus, that transfers genetic information; (3) Agent that transfers a parasite from one host to another.

vegetal pole   The yolky pole of a vertebrate or echinoderm egg. Compare with *animal pole*.

vein   (1) A blood vessel that carries blood from the tissues toward a chamber of the heart; compare with *artery*. (2) A strand of vascular tissue that is part of the network of conducting tissue in a leaf.

veliger larva   Larval stage of many marine gastropods (snails) and bivalves (e.g., clams); often is a second larval stage that develops after the trochophore larva.

ventilation   The process of actively moving air or water over a respiratory surface.

ventral   Toward the lowermost surface or belly of an animal. Compare with *dorsal*.

ventricle   (1) A cavity in an organ; (2) One of the several cavities of the brain; (3) One of the chambers of the heart that receives blood from an atrium.

vernalization (vur"nul-uh-zay'shun)   The induction of flowering by a low temperature treatment.

vertebrates   Subphylum of chordates; possess a bony vertebral column; include fishes, amphibians, reptiles, birds, and mammals.

vesicle (ves'ih-kl)   Any small sac, especially a small spherical membrane-bounded compartment, within the cytoplasm.

vessel element   A type of water-conducting cell in the xylem of vascular plants.

vestibular apparatus   Collectively, the saccule, utricle, and semicircular canals of the inner ear.

vestigial (ves-tij'ee-ul)   Rudimentary; an evolutionary remnant of a formerly functional structure.

vibrio   A spiral-shaped bacterium that has the form of a short helix. Compare with *spirillum*, *spirochete*, *bacillus*, and *coccus*.

villus (pl. *villi*)   A multicellular, minute, elongated projection, from the surface of an epithelial membrane, e.g., villi of the mucosa of the small intestine.

viroid (vy′roid)   Tiny, naked infectious particle consisting only of nucleic acid.

virulent   Able to cause disease in a host. Compare with *avirulent*.

virus   A tiny pathogen composed of a core of nucleic acid usually encased in protein and capable of infecting living cells. A virus is characterized by total dependence upon a living host.

viscera (vis′ur-uh)   The internal body organs, especially those located in the abdominal or thoracic cavities.

visceral mass   Concentration of body organs (viscera) located above the foot; characteristic of mollusks.

vital capacity   The maximum volume of air a person can expire after filling the lungs to the maximum extent.

vitamin   A complex organic molecule required in very small amounts for normal metabolic functioning.

vitelline envelope   An acellular covering of the eggs of certain animals (e.g., echinoderms), located just outside the plasma membrane.

viviparous (vih-vip′er-us)   Bearing living young that develop within the body of the mother.

vulva   The external genital structures of the female.

warning coloration   The conspicuous coloring of a poisonous or distasteful organism that enables potential predators to easily see and recognize it. Compare with *cryptic coloration.*

water mold   A fungus-like protist with a body consisting of a coenocytic mycelium that reproduces asexually by forming motile zoospores and sexually by forming oospores.

water potential   Free energy of water; the water potential of pure water is zero and that of solutions is a negative value. Differences in water potential are used to predict the direction of water movement (always from a region of less negative water potential to a region of more negative water potential).

water vascular system   Unique hydraulic system of echinoderms; functions in locomotion and feeding.

wavelength   The distance from one wave peak to the next; the energy of electromagnetic radiation is inversely proportional to its wavelength.

weathering   A chemical or physical process that helps form soil from rock; during weathering, the rock is gradually broken into smaller and smaller pieces.

Western blotting   A technique in which proteins, previously separated by gel electrophoresis, are transferred to paper. A specific labeled antibody is generally used to mark the location of a particular protein. Compare with *Southern blot hybridization* and *Northern blot hybridization.*

whisk ferns   One of a phylum of seedless vascular plants with a life cycle similar to ferns.

white matter   Nervous tissue in the brain and spinal cord that contains myelinated axons. Compare with *gray matter.*

wild type   The phenotypically normal (naturally occurring) form of a gene or organism.

wobble   The ability of some tRNA anticodons to associate with more than one mRNA codon; in these cases the 5′ base of the anticodon is capable of forming hydrogen bonds with more than one kind of base in the 3′ position of the codon.

work   Any change in the state or motion of matter.

X-linked gene   Gene carried on an X chromosome.

x-ray diffraction   A technique for determining the spatial arrangement of the components of a crystal.

xylem (zy′lem)   The vascular tissue that conducts water and dissolved minerals in plants.

XYY karyotype   Chromosome constitution that causes affected individuals (who are fertile males) to be unusually tall, with severe acne.

yeast   A unicellular fungus (ascomycete) that reproduces asexually by budding or fission and sexually by ascospores.

yolk sac   One of the extraembryonic membranes; a pouchlike outgrowth of the digestive tract of certain vertebrate embryos that grows around the yolk, digests it, and makes it available to the rest of the developing embryo. The yolk sac of mammals, which lacks yolk, serves as the location for the formation of embryonic blood cells.

zero population growth   Point at which the birth rate equals the death rate. A population with zero population growth does not change in size.

zona pellucida (pel-loo′sih-duh)   The thick, transparent covering that surrounds the plasma membrane of a mammalian ovum.

zooflagellate   A unicellular, nonphotosynthetic protozoon that possesses one or more long, whiplike flagella.

zooplankton (zoh″oh-plank′tun)   The nonphotosynthetic organisms present in plankton, e.g., protozoa, tiny crustaceans, and the larval stages of many animals. See *plankton.* Compare with *phytoplankton.*

zoospore (zoh′oh-spore)   A flagellated motile spore produced asexually by certain algae, water molds, and other protists.

zooxanthellae (zoh″oh-zan-thel′ee) (sing. *zooxanthella*)   Endosymbiotic, photosynthetic dinoflagellates found in certain marine invertebrates; their mutualistic relationship with corals enhances the corals' reef-building ability.

zygomycetes (zy″gah-my′seats)   Fungi characterized by the production of nonmotile asexual spores and sexual zygospores.

zygospore (zy′gah-spor)   A thick-walled sexual spore produced by a zygomycete.

zygote   The 2*n* cell that results from the union of *n* gametes in sexual reproduction. Species that are not polyploid have haploid gametes and diploid zygotes.

zygotic genes   Genes that are transcribed after fertilization, either in the zygote or in the embryo. Compare with *maternal effect genes.*

zygotic segmentation genes   In *Drosophila*, genes transcribed in the embryo that are responsible for controlling formation of the segmented body.

Boldface page numbers indicate pages on which the index term is defined; "il" following the page number indicates an illustration, "t" a table, "f" a focus box, and "n" a footnote.

Nucleotides, 67–69, 67il, 68il, **250**–251, 252il
Nucleus, **10**, 81, **82**–83, 82t
    totipotent, 344–347, 347il
Nudibranch, **615**, 615il, 861–863, 861il, 862il,
    863il, 864, 1045, 1045il
Numerical taxonomy, 475
Nüsslein-Volhard, Christiane, 366
Nut, **752**, 752il
Nutrients, 167–168, **975**, 988–995
Nutrition, **975**
    animal modes of, 976–977
    as career, 196
    children's needs, 995
    disease and, 988, 990f, 991, 994
    embryo and, 1088, 1089t
    metabolic diversity and, 189

Oak, 12t, 670, 670il, 710il, 722il
Obelia, 597, 597il
Obesity, 922f, 988, **994**–995, 996–997f
Obligate anaerobes, 495
Occipital lobes, **855**, 856il
Ocean, 1180–1181
    atmosphere and, 1180–1181
    Ordovician, 434–435
    oxygen depletion, 438
    primordial, 427
    thermal expansion, 1223
    tides, 1204
Ocean currents, **1180**, 1180il, 1204
Oceanic province, **1205**–1206
Ocean sunfish, 416, 416il, 649, 651
Ocelli, 616
Octopus, 617
Odonata, 629t
Oil body, 755
Okazaki fragments, **258**, 260il
Okazaki, Reijii, 258
Olfaction, 887, **888**–890
Olfactory epithelium, 434t, 442, **888**–890, 889il
Oligochaeta, 617t, **618**, 618il
Oligodendrocyte, **828**
Oligosaccharin, 781
Ommatidia, **621**, **891**, 894il
Omnivores, **976**, 1142
Oncogenes, 356–357f, 368, 797f
One gene, one enzyme hypothesis, 247
Onion, 699il
Onycophora, **622**–623, 623il
Oocyte donation, 1056f
Oogamy, **519**
Oogenesis, **209**, **1054**, 1055il, **1074**
Oogonia, **1054**
Oospore, **524**
Oparin, Alexander, 427
Open circulatory system, **614**, **904**–905, 905il
Open system, **136**, 137il
Operant conditioning, **1099**, 1100il
Operator, **289**, 289il
Operculum, **649**, **960**, 960il
Operon, **288**–290, 289il
Ophiuroidea, 637, 639il, 641
Opossum, 660
Opposite leaf arrangement, 687, 688il
Optic chiasm, **900**, 9001il
Optic nerve, 866t, 867, **897**, 898il

Optimal foraging, **1103**–1104, 1104il
Oral contraceptives, **1065**, 1066t
Oral surface, **637**
Orangutan, 451, 452il, 453
Orbitals, **28**, 29il
Orca, 1105, 1144
Orchid, 726il, 745, 1215
Orders, **12**, 468il, 469, 469il
    of insects, 629–631t
    of mammals, 662–664t
Ordovician period, 434–435, 434–435t
Organ, **673**, **786**, 796
Organelle, **10**, **74**, 81–97, **82**–83t
Organic compounds, **24**, 44–70
Organic fertilizers, 737
Organic matter, 731–732, 732il
Organic molecules, 45, 46il, 427–429, 427il
Organic polymer, 427–428
Organism, 3–7, **10**
Organismic model of communities, **1158**, 1158il
Organismic respiration, 154, **957**
Organ of Corti, **885**, 886il
Organogenesis, **1081**–1083, 1081t, 1083il
Organ system, **10**, **786**, 796–802, 798–800t
Organ transplant, 947, 1013f
Orgasm, **1062**
Oriental realm, 1159
Origins of replication, **258**, 259, 261il
Ornithischian dinosaur, 441il
Orthoptera, 629t
Ortiz, Henry, 462–464
Oscillatoria, 605il
Osculum, **591**
Osmoconformers, **1002**
Osmoregulation, 587, **1000**, 1002–1015
Osmoregulators, **1002**
Osmosis, **116**–118, 116il, 724
Osmotic pressure, **116**, 117il
Osteichthyes, **644**, 646t, 649–652, 650ils
Osteoarthritis, 815
Osteoblasts, **814**
Osteoclasts, **795**, 815
Osteocytes, **791**–795, 794il, **814**, 814il
Osteon, 794il, **795**, 814
Osteoporosis, 815, 1013f, 1058f
Ostia, 591, **621**
Ostracoderms, 435, 436il, **644**–646
Otoliths, **883**, 884il
Outbreeding, **239**
Outer mitochondrial membrane, 95, 95il
Outgroup analysis, **475**, 475il, 476
Out of Africa hypothesis, 457, 458
Oval window, **885**, 885il, 886il
Ovary, 580, 742, **1052**
    in flowers, 579il, 580, 742, 742il
    hormones and, 1029t, 1052–1054, 1053il,
        1054il, 1059t
Overfishing, 599, 1206–1207
Overgrazing, 1193
Overproduction, 374
Oviduct, 1054il, **1055**, 1084
Oviparity, **649**
Ovipositor, 632
Ovoviviparity, **649**
Ovulation, 1054il, **1055**
Ovule, **570**, 741, 742il
Ovum, 75, **209**, **1045**, 1062, 1063

Oxaloacetate, **161**, 168, **191**
Oxidation, **35**, **143**
Oxidative decarboxylation, 160
Oxidative phosphorylation, **163**
Oxidizing agent, 35
Oxygen, 14–15
    in Earth's atmosphere, 426, 430–431, 1178
    as essential element, 734–735, 735t
    in gas exchange, 957, 966–968
    in photosynthesis, 182il
Oxygen carrying capacity, **967**
Oxygen content, **967**
Oxygen debt, **821**
Oxygen-hemoglobin dissociation curves, **967**,
    967il
Oxyhemoglobin, **967**
Oxytocin, 1020, 1021il, **1030**, 1032il, **1058**,
    1063
Oyster, 616–617
Ozone, **431**, 431il, 693f, 1221, 1221n, **1225**
Ozone depletion, **1225**–1227
Ozone hole, 1225–1226, 1226il, 1227
Ozone layer, 1226–1228, 1225il

PABA, 150–151, 151il
Pacemaker, **915**, 915il
Pacinian corpuscle, **880**, 881il
Packer, Craig, 1104
Paedomorphosis, **416**–417, 417il, 653
Pain, 892–893f
Painted lady, 192il
Painted trillium, 740, 740il
Pair bond, **1110**, 1111il
Pair-rule genes, **353**, 353il, 354t
Palearctic realm, 1159, 1160il
Paleoanthropology, **447**
Paleobotany, 583
Paleocene epoch, 434t, 442
Paleozoic era, **433**–438, 434t
Palindromic base sequences, **305**
Palisade layer, 689il, **690**
Palmately netted venation, 687, 688il
Palolo worm, 618
Pancreas, **984**, 984il
Pancreatic amylase, **984**
Panda, 977il
Pandorina, 525
Pangaea, 420il, 421
Pangolin, 380
Panspermia, 429
Paper birch, 710il, 730
Pap test, 797f
Parabronchi, **962**
Paracrine regulation, **1019**, 1020il
Parakeet, 1214
Parallel venation, 687, 688il
Paramecium, 4il, 117, 207, 208il, 510, 511,
    511il, 1127, 1150, 1151il
Paramylon, 513
Paraphyletic group, **472**, 472il
Parapodia, **618**
Parasite(s), **541**, **1148**
    apicomplexans, 511–512
    Ascaris, 606
    defenses against, 931
    fish, 1205

Pseudopodia, **505**, 506, 507il, 933il
Pseudostratified epithelium, **787**, 788t
Psilotum nudum, 561, 562il
Psychological dependence, 869–871
PTC, 887
Pterophyta, 559
Pterosaur, 439il, 440
Puccinia monoica, 541
Puff, **350**
Puffball, 536, 537il
Pulmonary circulation, **653**, 911, 920–921, 920il
Pulmonary emphysema, 970–972
Pulmonary valve, **914**
Pulmonate mollusk, **616**
Pulp cavity, **980**, 980il
Pulvinus, **769**
Punctuated equilibrium, 414–415, 415il
Punnett, Reginald, 223
Punnett square, **223**, 224, 225, 240f
Pupation, **350**
Pupfish, 409, 409il
Pupil, **893**, 896il
Purines, **251**
Purkinje fibers, **915**, 915il
Pursuit strategies, 1143–1144, 1144il
Pusey, Anne, 1110f
P wave, 916–917, 917il
Pylorus, **982**
Pyramid of biomass, **1167**, 1167il
Pyramid of energy, **1168**, 1168il
Pyramid of numbers, **1167**, 1167il
Pyrimidines, **251**
Pyruvate, 156, 160–161, 165, 168

QRS complex, 916–917, 917il
Quadrat study, 464
Quadrupeds, **451**
Quaternary period, 434t, 441
Quaternary structure, **65**, 66il
Quetzalcoatlus, 439il, 440
Quiring, R., 352

Rabies, 487
Raccoon, 1141il
Radial cleavage, **590**, 591il, **1077**, 1007il
Radial nervous system, **847**
Radial symmetry, **588**, 588il
Radiation, 175il, 283–285, 426, 1177il
Radiation therapy, 797f, 1058f
Radicle, 721, **749**
Radioactive decay, **379**, 379il
Radioactive labeling, 312
Radioisotopes, **28**, **378**–379, 1165
Radius, 380, 380il, 814, 814il
Radula, **614**
Rain shadow, **1181**, 1181il
Random dispersion, **1124**
Random mating, 391
Range, **1159**
Ras proteins, 114
Ray-finned fish, 436, **651**
Rays (chondrichthyes), 646t, 648, 648il
Rays (of stem), 709il, **710**
Reabsorption, 1009il, **1010**–1011

Reactant, **30**, 140
Reaction center, **183**, 183il
Reading frame, **280**, 281
Realized niche, **1149**, 1150il
Receptacle, **578**, 741, 742il
Reception, **828**, 829il
Receptor(s), 112, 844, 871, 1022–1026
Receptor down-regulation, **1023**
Receptor-mediated endocytosis, **123**–126, 125il
Receptor potential, **878**–879
Receptor up-regulation, **1023**
Recessive allele, **220**, 224
Reciprocal altruism, **1113**
Reciprocal translocation, **327**
Recognition helix, **297**
Recombinant DNA, **306**
Recombinant DNA technology, 231, 282–283f, **304**
   antibodies, 940–941
   applications, 304, 312, 332, 544
   cellular aging studies, 263
   immunity and, 940f
   interferon production, 931
   methods, 305–315
   See also Genetic engineering
Recombinant gametes, **229**, 230il
Recombinant plasmids, 307, 309il
Recombination, genetic, **229**, 230il
Rectum, **987**
Red algae, 509t, **519**–520, 521il, 525
Red blood cells (RBCs), 795, **906**–907, 907il, 908t
   human, 117, 117il, 336–337
   mammalian, 107il
   sickled, 328–329, 329il
Red dart frog, 652il
Red-eared slider, 171il
Red fibers, **822**
Redox reactions, **35**, 143–144
Red spruce, 693f
Red tides, 514–515
Reducing agent, 35
Reduction, **35**, 143
Reductionism, **10**, 730
Referred pain, 892f
Reflex action, **853**
Regeneration, 88
Regular dense connective tissue, **791**
Regulative development, **362**, **1078**
Regulon, **293**–294, 294il
Reindeer moss, 539, 554
Relapse prevention therapy, 870
Relapsing fever, 626
Relative refractory period, **834**
Relaxation techniques
Release, viral, **483**, 484il
Release factors, 276
Releaser, **1097**, 1098il
Releasing hormone, **1030**, 1033il
REM sleep, **857**–859
Renal arteries, **921**, 921il
Renal cortex, **1005**, 1007il
Renal medulla, **1005**, 1007il
Renal papilla, **1005**, 1007il
Renal pelvis, **1005**, 1007il
Renal pyramids, **1005**, 1007il
Renal threshold, **1011**

Renal tubule, **1008**, 1008il, 1011–1012, 1011il
Renal veins, **921**, 921il
Renin, **920**, 1014
Renin-angiotensin-aldosterone pathway, **1014**
Replacement-level fertility, **1135**, 1136n
Replication, 91, **483**, 484il
Replication fork, **258**, 260il
Repolarization, **833**
Repressible enzyme, **290**
Repressible genes, 293t
Repressor protein, **289**, 289il
Reproduction
   asexual, **6**, 755–758, **1045**, 1046
   human, 1047–1061
   lytic, 483
   sexual, **1045**, 1046–1047
Reproductive isolating mechanisms, **405**, 405t, 408–414
Reproductive strategies, 1130–1133, 1132il
Reptile, **653**–656
   brain, 851il
   circulation and heart, 911, 912, 912il
   classification, 655–656
   development, 383il
   evolutionary adaptations, 653–655
   lungs, 961il
   mesozoic, 653
Reptilia, **644**, 646t, 653, 655il
Reserpine, 859
Reservoir species, 487f
Residual capacity, **965**, 965t
Resin, 571
Resolution, 76, **1062**
Resolving power, 76
Resource consumption, 1137–1138, 1137il
Resource partitioning, **1150**, 1151il
Respiration, **957**
Respiratory centers, 963il, **968**
Respiratory pigments, **962**
Respiratory system, **796**
   amphibian, 961
   annelid, 618
   human, 799il, 800t, 962–972, 963il
   polluted air and, 969–972
   reptilian, 961il
Resting potential, **830**–832
Restoration ecology, 1199, 1216–1217, 1216il
Restriction enzymes, **305**, 305n, 306il, 309il, 314
Restriction fragment length polymorphisms, 314–315, 314il
Reticular activating system (RAS), **857**
Reticular connective tissue, **790**, 792t
Reticular fibers, **790**
Retina, **893**, 896il, 897, 898il, 899il, 900
Retinal, 59, **897**–898
Retrograde amnesia, 860
Retrograde signal, 864–865, 868–869f
Retroviruses, 282, **317**–318, **486**, 950–954
Reverse transcriptase. *See* Reverse transcription
Reverse transcription, 282–283f, 309, 317, **486**, 486il
Reversible inhibition, 149–150
Reznick, David, 385–386
RFLPs, **314**–315, 314il
Rhabdome, **891**, 894il
Rheumatoid arthritis, 815, 950

Scientific Measurement

Some Common Prefixes

		Examples
kilo	1000	a kilogram is 1000 grams
centi	0.01	a centimeter is 0.01 meter
milli	0.001	a milliliter is 0.001 liter
micro (μ)	one-millionth	a micrometer is 0.000001 (one-millionth) of a meter
nano (n)	one-billionth	a nanogram is 10^{-9} (one-billionth) of a gram
pico (p)	one-trillionth	a picogram is 10^{-12} (one-trillionth) of a gram

The relationship between mass and volume of water (at 20°C)

$$1 \text{ g} = 1 \text{ cm}^3 = 1 \text{ mL}$$

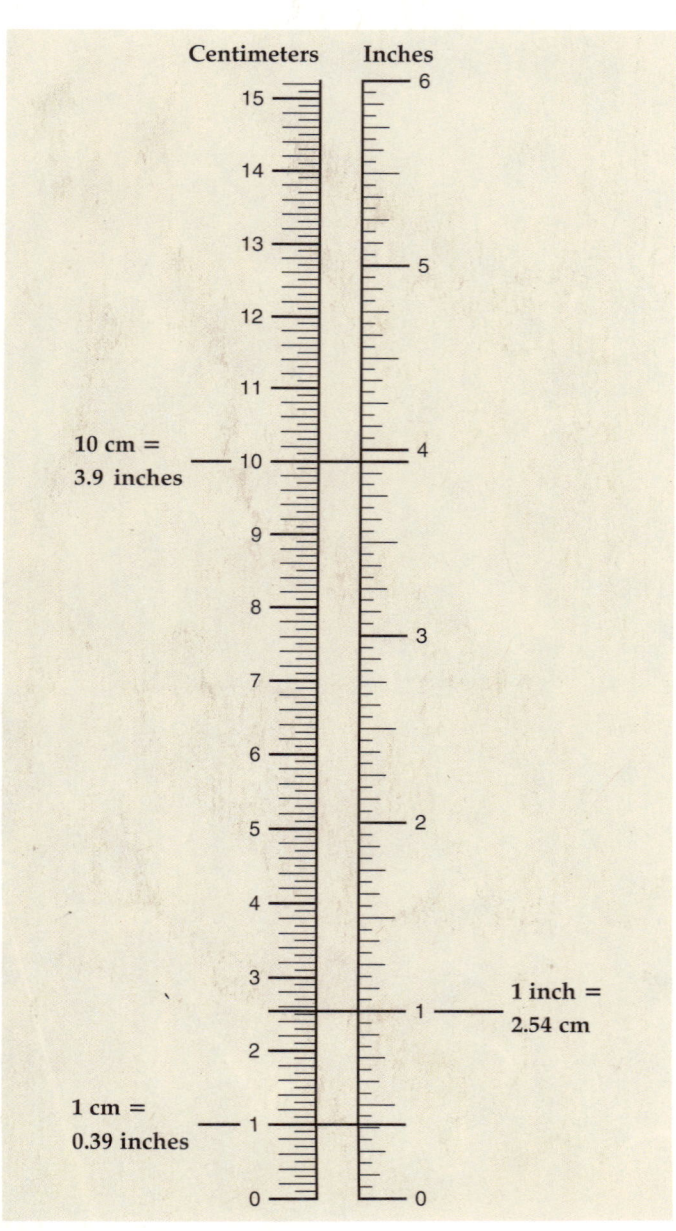

10 cm = 3.9 inches

1 inch = 2.54 cm

1 cm = 0.39 inches

Some Common Units of Length

Unit	Abbreviation	Equivalent
meter	m	approximately 39 in
centimeter	cm	10^{-2} m
millimeter	mm	10^{-3} m
micrometer	μm	10^{-6} m
nanometer	nm	10^{-9} m
angstrom	Å	10^{-10} m

Length Conversions

1 in = 2.5 cm	1 mm = 0.039 in
1 ft = 30 cm	1 cm = 0.39 in
1 yd = 0.9 m	1 m = 39 in
1 mi = 1.6 km	1 m = 1.094 yd
	1 km = 0.6 mi

To convert	Multiply by	To obtain
inches	2.54	centimeters
feet	30	centimeters
centimeters	0.39	inches
millimeters	0.039	inches

Standard Metric Units

		Abbreviation
Standard unit of mass	gram	g
Standard unit of length	meter	m
Standard unit of volume	liter	L